T0180334

Lecture Notes in Computer Science   **9134**

Commenced Publication in 1973
Founding and Former Series Editors:
Gerhard Goos, Juris Hartmanis, and Jan van Leeuwen

Advanced Research in Computing and Software Science

Subline of Lecture Notes in Computer Science

More information about this series at http://www.springer.com/series/7407

Magnús M. Halldórsson · Kazuo Iwama
Naoki Kobayashi · Bettina Speckmann (Eds.)

Automata, Languages, and Programming

42nd International Colloquium, ICALP 2015
Kyoto, Japan, July 6–10, 2015
Proceedings, Part I

 Springer

Editors
Magnús M. Halldórsson
Reykjavik University
Reykjavik
Iceland

Kazuo Iwama
Kyoto University
Kyoto
Japan

Naoki Kobayashi
The University of Tokyo
Tokyo
Japan

Bettina Speckmann
Technische Universiteit Eindhoven
Eindhoven
The Netherlands

ISSN 0302-9743   ISSN 1611-3349 (electronic)
Lecture Notes in Computer Science
ISBN 978-3-662-47671-0   ISBN 978-3-662-47672-7 (eBook)
DOI 10.1007/978-3-662-47672-7

Library of Congress Control Number: 2015941869

LNCS Sublibrary: SL1 – Theoretical Computer Science and General Issues

Springer Heidelberg New York Dordrecht London

Printed on acid-free paper

Springer-Verlag GmbH Berlin Heidelberg is part of Springer Science+Business Media
(www.springer.com)

Preface

ICALP 2015, the 42nd edition of the International Colloquium on Automata, Languages and Programming, was held in Kyoto, Japan during July 6–10, 2015. ICALP is a series of annual conferences of the European Association for Theoretical Computer Science (EATCS), which first took place in 1972. This year, the ICALP program consisted of the established track A (focusing on algorithms, automata, complexity, and games) and track B (focusing on logic, semantics, and theory of programming), and of the recently introduced track C (focusing on foundations of networking).

In response to the call for papers, the Program Committee received 507 submissions, the highest ever: 327 for track A, 115 for track B, and 65 for track C. Out of these, 143 papers were selected for inclusion in the scientific program: 89 papers for Track A, 34 for Track B, and 20 for Track C. The selection was made by the Program Committees based on originality, quality, and relevance to theoretical computer science. The quality of the manuscripts was very high indeed, and many deserving papers could not be selected.

The EATCS sponsored awards for both a best paper and a best student paper for each of the three tracks, selected by the Program Committees. The best paper awards were given to the following papers:

- Track A: Aaron Bernstein and Clifford Stein. "Fully Dynamic Matching in Bipartite Graphs"
- Track B: Jarkko Kari and Michal Szabados. "An Algebraic Geometric Approach to Nivat's Conjecture"
- Track C: Yiannis Giannakopoulos and Elias Koutsoupias. "Selling Two Goods optimally"

The best student paper awards, for papers that are solely authored by students, were given to the following papers:

- Track A: Huacheng Yu. "An Improved Combinatorial Algorithm for Boolean Matrix Multiplication"
- Track A: Radu Curticapean. "Block Interpolation: A Framework for Tight Exponential-Time Counting Complexity"
- Track B: Georg Zetzsche. "An Approach to Computing Downward Closures"

Track A gave out two student paper awards this year because of the very high quality of the two winning papers.

The conference was co-located with LICS 2015, the 30th ACM/IEEE Symposium on Logic in Computer Science.

Apart from the contributed talks, ICALP 2015 included invited presentations by Ken-ichi Kawarabayashi, Valerie King, Thomas Moscibroda, Anca Muscholl, Peter O'Hearn, of which the latter two were joint with LICS. Additionally, it contained tutorial sessions by Piotr Indyk, Andrew Pitts, and Geoffrey Smith, all joint with LICS, and a masterclass on games by Ryuhei Uehara. Abstracts of their talks are included in these proceedings as well. The program of ICALP 2015 also included presentation of the EATCS Award 2015 to Christos Papadimitriou.

This volume of the proceedings contains all contributed papers presented at the conference in Track A. A companion volume contains all contributed papers presented in Track B and Track C together with the papers and abstracts of the invited speakers. The following workshops were held as satellite events of ICALP/LICS 2015:

HOPA 2015 — Workshop on the Verification of Higher-Order Programs

LCC 2015 — 16th International Workshop on Logic and Computational Complexity

NLCS 2015 — Third Workshop on Natural Language and Computer Science

LOLA 2015 — Workshop on Syntax and Semantics for Low-Level Languages

QCC 2015 — Workshop on Quantum Computational Complexity

WRAWN 2015 — 6th Workshop on Realistic Models for Algorithms in Wireless Networks

YR-ICALP 2015 — Young Researchers Forum on Automata, Languages and Programming

We wish to thank all authors who submitted extended abstracts for consideration, the Program Committees for their scholarly effort, and all referees who assisted the Program Committees in the evaluation process.

We thank the sponsors (ERATO Kawarabayashi Large Graph Project; MEXT Grant-in-Aid for Scientific Research on Innovative Areas "Exploring the Limits of Computation"; Research Institute for Mathematical Sciences, Kyoto University; and Tateisi Science and Technology Foundation) for their support.

We are also grateful to all members of the Organizing Committee and to their support staff.

Thanks to Andrei Voronkov and Shai Halevi for writing the conference management systems EasyChair andWeb-Submission-and-Review software, which were used in handling the submissions and the electronic Program Committee meeting, as well as in assisting in the assembly of the proceedings.

Last but not least, we would like to thank Luca Aceto, the president of EATCS, for his generous advice on the organization of the conference.

May 2015

Magnús M. Halldórsson
Kazuo Iwama
Naoki Kobayashi
Bettina Speckmann

Organization

Program Committee

Track A

Peyman Afshani	Aarhus University, Denmark
Hee-Kap Ahn	POSTECH, South Korea
Hans Bodlaender	Utrecht University, The Netherlands
Karl Bringmann	Max-Planck Institut für Informatik, Germany
Sergio Cabello	University of Ljubljana, Slovenia
Ken Clarkson	IBM Almaden Research Center, USA
Éric Colin de Verdière	École Normale Supérieure Paris, France
Stefan Dziembowski	University of Warsaw, Poland
David Eppstein	University of California at Irvine, USA
Dimitris Fotakis	National Technical University of Athens, Greece
Paul Goldberg	University of Oxford, UK
MohammadTaghi Hajiaghayi	University of Maryland at College Park, USA
Jesper Jansson	Kyoto University, Japan
Andrei Krokhin	Durham University, UK
Asaf Levin	Technion, Israel
Inge Li Gørtz	Technical University of Denmark, Denmark
Pinyan Lu	Microsoft Research Asia, China
Frédéric Magniez	Université Paris Diderot, France
Kazuhisa Makino	Kyoto University, Japan
Elvira Mayordomo	Universidad de Zaragoza, Spain
Ulrich Meyer	Goethe University Frankfurt am Main, Germany
Wolfgang Mulzer	Free University Berlin, Germany
Viswanath Nagarajan	University of Michigan, USA
Vicky Papadopoulou	European University Cyprus, Cyprus
Michał Pilipczuk	University of Bergen, Norway
Liam Roditty	Bar-Ilan University, Israel
Ignaz Rutter	Karlsruhe Institute of Technology, Germany
Rocco Servedio	Columbia University, USA
Jens Schmidt	TU Ilmenau, Germany
Bettina Speckmann	TU Eindhoven, The Netherlands
Csaba D. Tóth	California State University Northridge, USA
Takeaki Uno	National Institute of Informatics, Japan
Erik Jan van Leeuwen	Max-Planck Institut für Informatik, Germany
Rob van Stee	University of Leicester, UK
Ivan Visconti	University of Salerno, Italy

Track B

Andreas Abel	Chalmers and Gothenburg University, Sweden
Albert Atserias	Universitat Politècnica de Catalunya, Spain
Christel Baier	TU Dresden, Germany
Lars Birkedal	Aarhus University, Denmark
Luís Caires	Universidade Nova de Lisboa, Portugal
James Cheney	University of Edinburgh, UK
Wei Ngan Chin	National University of Singapore, Singapore
Ugo Dal Lago	University of Bologna, Italy
Thomas Ehrhard	CNRS and Université Paris Diderot, France
Zoltán Ésik	University of Szeged, Hungary
Xinyu Feng	University of Science and Technology of China, China
Wan Fokkink	VU University Amsterdam, The Netherlands
Shin-ya Katsumata	Kyoto University, Japan
Naoki Kobayashi	The University of Tokyo, Japan
Eric Koskinen	New York University, USA
Antonín Kučera	Masaryk University, Czech Republic
Orna Kupferman	Hebrew University, Israel
Annabelle McIver	Macquarie University, Australia
Dale Miller	Inria Saclay, France
Markus Müller-Olm	University of Münster, Germany
Andrzej Murawski	University of Warwick, UK
Joel Ouaknine	University of Oxford, UK
Prakash Panangaden	McGill University, Canada
Pawel Parys	University of Warsaw, Poland
Reinhard Pichler	TU Vienna, Austria
Simona Ronchi Della Rocca	University of Turin, Italy
Jeremy Siek	Indiana University, USA

Track C

Ioannis Caragiannis	University of Patras, Greece
Katarina Cechlarova	Pavol Jozef Safarik University, Slovakia
Shiri Chechik	Tel Aviv University, Israel
Yuval Emek	Technion, Israel
Sándor Fekete	TU Braunschweig, Germany
Pierre Fraigniaud	CNRS, Université Paris Diderot, France
Leszek Gąsieniec	University of Liverpool, UK
Aristides Gionis	Aalto University, Finland
Magnús M. Halldórsson	Reykjavik University, Iceland
Monika Henzinger	Universität Wien, Austria
Bhaskar Krishnamachari	University of Southern California, USA
Fabian Kuhn	University of Freiburg, Germany
Michael Mitzenmacher	Harvard University, USA
Massimo Merro	University of Verona, Italy

Gopal Pandurangan   University of Houston, USA
Pino Persiano   University of Salerno, Italy
R. Ravi   Carnegie Mellon University, USA
Ymir Vigfusson   Emory University, USA
Roger Wattenhofer   ETH Zürich, Switzerland
Masafumi Yamashita   Kyushu University, Japan

Organizing Committee

Masahito Hasegawa   Kyoto University, Japan
Atushi Igarashi   Kyoto University, Japan
Kazuo Iwama   Kyoto University, Japan
Kazuhisa Makino   Kyoto University, Japan

Financial Sponsors

ERATO Kawarabayashi Large Graph Project
MEXT Grant-in-Aid for Scientific Research on Innovative Areas: "Exploring the Limits of Computation"
Research Institute for Mathematical Sciences, Kyoto University
Tateisi Science and Technology Foundation

Additional Reviewers

Abboud, Amir
Abdulla, Parosh
Abed, Fidaa
Abraham, Ittai
Ailon, Nir
Ajwani, Deepak
Albers, Susanne
Almeida, Jorge
Alt, Helmut
Alur, Rajeev
Alvarez, Victor
Alvarez-Jarreta, Jorge
Ambainis, Andris
Aminof, Benjamin
Anagnostopoulos, Aris
Andoni, Alexandr
Angelidakis, Haris
Anshelevich, Elliot
Antoniadis, Antonios

Arai, Hiromi
Aronov, Boris
Asada, Kazuyuki
Aspnes, James
Aubert, Clément
Augustine, John
Auletta, Vincenzo
Austrin, Per
Avin, Chen
Avni, Guy
Baelde, David
Baillot, Patrick
Bansal, Nikhil
Banyassady, Bahareh
Barnat, Jiri
Barth, Stephan
Barto, Libor
Basavaraju, Manu
Bassily, Raef

Baswana, Surender
Bateni, Mohammadhossein
Batu, Tugkan
Baum, Moritz
Béal, Marie-Pierre
Beigi, Salman
Beimel, Amos
Ben-Amran, Amr
Berenbrink, Petra
Bernáth, Attila
Berthé, Valérie
Bes, Alexis
Besser, Bert
Bevern, René Van
Bi, Jingguo
Bienstock, Daniel
Bille, Philip
Bilò, Vittorio
Bizjak, Ales
Björklund, Henrik
Blais, Eric
Bläsius, Thomas
Blömer, Johannes
Bogdanov, Andrej
Bojanczyk, Mikolaj
Bollig, Benedikt
Bonfante, Guillaume
Bonnet, Edouard
Bourhis, Pierre
Bousquet, Nicolas
Boyar, Joan
Bozzelli, Laura
Bradfield, Julian
Brandes, Philipp
Brandt, Sebastian
Braverman, Vladimir
Bresolin, Davide
Brzuska, Christina
Brânzei, Simina
Bucciarelli, Antonio
Buchbinder, Niv
Buchin, Kevin
Bulatov, Andrei
Cai, Jin-Yi
Cai, Zhuohong
Canonne, Clement

Cao, Yixin
Carayol, Arnaud
Carmi, Paz
Caron, Pascal
Caskurlu, Bugra
Cassez, Franck
Castagnos, Guilhem
Castellani, Ilaria
Castelli Aleardi, Luca
Cenzer, Douglas
Chakrabarty, Deeparnab
Chalermsook, Parinya
Chan, T.-H. Hubert
Chan, Timothy M.
Chattopadhyay, Arkadev
Chekuri, Chandra
Chen, Ho-Lin
Chen, Wei
Chen, Xi
Chen, Xujin
Chitnis, Rajesh
Chlamtac, Eden
Chlebikova, Janka
Cho, Dae-Hyeong
Chonev, Ventsislav
Christodoulou, George
Cicalese, Ferdinando
Cimini, Matteo
Clairambault, Pierre
Claude, Francisco
Clemente, Lorenzo
Cleve, Richard
Cloostermans, Bouke
Cohen-Addad, Vincent
Columbus, Tobias
Cording, Patrick Hagge
Coretti, Sandro
Cormode, Graham
Cornelsen, Sabine
Cosentino, Alessandro
Coudron, Matthew
Crouch, Michael
Cygan, Marek
Czerwiński, Wojciech
Czumaj, Artur
Dachman-Soled, Dana

Dahlgaard, Søren
Dalmau, Victor
Dantchev, Stefan
Daruki, Samira
Das, Anupam
Dasler, Philip
Datta, Samir
Daum, Sebastian
Dawar, Anuj
De Bonis, Annalisa
De Caro, Angelo
De, Anindya
Dehghani, Sina
Deligkas, Argyrios
Dell, Holger
Demangeon, Romain
Demri, Stéphane
Denzumi, Shuhei
Diakonikolas, Ilias
Dibbelt, Julian
Dietzfelbinger, Martin
Dinsdale-Young, Thomas
Dinur, Itai
Disser, Yann
Dobrev, Stefan
Doerr, Carola
Döttling, Nico
Dotu, Ivan
Doty, David
Dräger, Klaus
Drucker, Andrew
Duan, Ran
Dubslaff, Clemens
Duetting, Paul
van Duijn, Ingo
Duncan, Ross
Durand, Arnaud
Durand-Lose, Jérôme
Dürr, Christoph
Dvorák, Wolfgang
Dyer, Martin
Efthymiou, Charilaos
Eirinakis, Pavlos
Elbassioni, Khaled
Elmasry, Amr
Emanuele, Viola

Emmi, Michael
Emura, Keita
Englert, Matthias
Epelman, Marina
Epstein, Leah
Ergun, Funda
Erickson, Alejandro
Esfandiari, Hossein
Fahrenberg, Uli
Farinelli, Alessandro
Faust, Sebastian
Fawzi, Omar
Fefferman, Bill
Feldman, Moran
Feldmann, Andreas Emil
Feng, Yuan
Fernique, Thomas
Ferraioli, Diodato
Fijavz, Gasper
Filinski, Andrzej
Filmus, Yuval
Filos-Ratsikas, Aris
Find, Magnus Gausdal
Firsov, Denis
Fleiner, Tamas
Foerster, Klaus-Tycho
Fomin, Fedor
Fontes, Lila
Forbes, Michael A.
Forejt, Vojtech
Formenti, Enrico
François, Nathanaël
Fränzle, Martin
Frascaria, Dario
Friedrich, Tobias
Fu, Hongfei
Fuchs, Fabian
Fuchsbauer, Georg
Fukunaga, Takuro
Fuller, Benjamin
Funk, Daryl
Fürer, Martin
Gabizon, Ariel
Gaboardi, Marco
Gacs, Peter
Gaertner, Bernd

Galanis, Andreas
Galčík, František
Ganguly, Sumit
Ganor, Anat
Ganty, Pierre
Garg, Naveen
Gaspers, Serge
Gawrychowski, Pawel
Gazda, Maciej
Gehrke, Mai
Gemsa, Andreas
Georgiadis, Loukas
Gerhold, Marcus
van Glabbeek, Rob
Göller, Stefan
Goncharov, Sergey
Göös, Mika
Gopalan, Parikshit
Gorbunov, Sergey
Gouveia, João
Grandjean, Etienne
Grandoni, Fabrizio
Green Larsen, Kasper
Grigoriev, Alexander
Grohe, Martin
Groote, Jan Friso
Grossi, Roberto
Grunert, Romain
Guessarian, Irène
Guiraud, Yves
Guo, Heng
Gupta, Anupam
Hadfield, Stuart
Hague, Matthew
Hahn, Ernst Moritz
Haitner, Iftach
Halevi, Shai
Hamann, Michael
Hampkins, Joel
Hansen, Kristoffer Arnsfelt
Har-Peled, Sariel
Harrow, Aram
Hastad, Johan
Hatano, Kohei
Haverkort, Herman
He, Meng

Heindel, Tobias
Hendriks, Dimitri
Henze, Matthias
Hermelin, Danny
Herranz, Javier
Heunen, Chris
Heydrich, Sandy
Hlineny, Petr
Hoffmann, Frank
Hoffmann, Jan
Hofheinz, Dennis
Hofman, Piotr
Holm, Jacob
Holmgren, Justin
Hong, Seok-Hee
Houle, Michael E.
Høyer, Peter
Hsu, Justin
Huang, Shenwei
Huang, Zengfeng
Huang, Zhiyi
Hwang, Yoonho
van Iersel, Leo
Im, Sungjin
Immerman, Neil
Inaba, Kazuhiro
Iovino, Vincenzo
Ishii, Toshimasa
Italiano, Giuseppe F.
Ito, Takehiro
Ivan, Szabolcs
Iwata, Yoichi
Izumi, Taisuke
Jaberi, Raed
Jaiswal, Ragesh
Jancar, Petr
Janin, David
Jansen, Bart M.P.
Jansen, Klaus
Jayram, T.S.
Jeavons, Peter
Jeffery, Stacey
Jerrum, Mark
Jeż, Łukasz
Jhanwar, Mahabir Prasad
Johnson, Matthew

Johnson, Matthew P.
Jones, Mark
Jones, Neil
Jordan, Charles
Jørgensen, Allan Grønlund
Jovanovic, Aleksandra
Jukna, Stasys
Kakimura, Naonori
Kalaitzis, Christos
Kamiyama, Naoyuki
Kanade, Varun
Kanazawa, Makoto
Kane, Daniel
Kanellopoulos, Panagiotis
Kantor, Erez
Kanté, Mamadou Moustapha
Kaplan, Haim
Karhumaki, Juhani
Kari, Jarkko
Kärkkäinen, Juha
Kashefi, Elham
Katajainen, Jyrki
Katz, Matthew
Kawachi, Akinori
Kazana, Tomasz
Kelk, Steven
Keller, Barbara
Keller, Orgad
Kenter, Sebastian
Kerenidis, Iordanis
Khan, Maleq
Khani, Reza
Khoussainov, Bakhadyr
Kida, Takuya
Kiefer, Stefan
Kijima, Shuji
Kim, Eun Jung
Kim, Heuna
Kim, Min-Gyu
Kim, Ringi
Kim, Sang-Sub
Kishida, Kohei
Kiyomi, Masashi
Klauck, Hartmut
Klavík, Pavel
Klima, Ondrej

Klin, Bartek
Knauer, Christian
Kobayashi, Yusuke
Kollias, Konstantinos
Kolmogorov, Vladimir
Komusiewicz, Christian
König, Barbara
König, Michael
Konrad, Christian
Kontogiannis, Spyros
Kopczynski, Eryk
Kopelowitz, Tsvi
Kopparty, Swastik
Korman, Matias
Kortsarz, Guy
Korula, Nitish
Kostitsyna, Irina
Kotek, Tomer
Kothari, Robin
Kovacs, Annamaria
Kozen, Dexter
Kraehmann, Daniel
Kral, Daniel
Kralovic, Rastislav
Kratsch, Dieter
Kratsch, Stefan
Krcal, Jan
Krenn, Stephan
Kretinsky, Jan
Kreutzer, Stephan
van Kreveld, Marc
Kriegel, Klaus
Krinninger, Sebastian
Krishna, Shankara Narayanan
Krishnaswamy, Ravishankar
Krizanc, Danny
Krumke, Sven
Krysta, Piotr
Kulkarni, Raghav
Kumar, Amit
Kumar, Mrinal
Künnemann, Marvin
Kuperberg, Greg
Kuroda, Satoru
Kurz, Alexander
Kyropoulou, Maria

Mousset, Frank
Mucha, Marcin
Mueller, Tobias
Müller, David
Müller-Hannemann, Matthias
Murakami, Keisuke
Murano, Aniello
Musco, Christopher
Mustafa, Nabil
Nadathur, Gopalan
Nagano, Kiyohito
Nakazawa, Koji
Nanongkai, Danupon
Narayanan, Hariharan
Navarra, Alfredo
Navarro, Gonzalo
Nayyeri, Amir
Nederhof, Mark-Jan
Nederlof, Jesper
Newman, Alantha
Nguyen, Huy
Nguyen, Kim Thang
Nguyen, Viet Hung
Niazadeh, Rad
Nicholson, Patrick K.
Niedermann, Benjamin
Nielsen, Jesper Buus
Nielsen, Jesper Sindahl
Nies, André
Nikolov, Aleksandar
Nishimura, Harumichi
Nitaj, Abderrahmane
Nöllenburg, Martin
Nordhoff, Benedikt
Novotný, Petr
Obremski, Maciej
Ochremiak, Joanna
Oh, Eunjin
Okamoto, Yoshio
Oliveira, Igor
Onak, Krzysztof
Ordóñez Pereira, Alberto
Oren, Sigal
Orlandi, Claudio
Otachi, Yota
Ott, Sebastian

Otto, Martin
Oveis Gharan, Shayan
Ozeki, Kenta
Ozols, Maris
Padro, Carles
Pagani, Michele
Pagh, Rasmus
Paluch, Katarzyna
Panagiotou, Konstantinos
Panigrahi, Debmalya
Paolini, Luca
Parter, Merav
Pasquale, Francesco
Paul, Christophe
Pedersen, Christian Nørgaard Storm
Pelc, Andrzej
Penna, Paolo
Perdrix, Simon
Perelli, Giuseppe
Persiano, Giuseppe
Pettie, Seth
Peva, Blanchard
Philip, Geevarghese
Phillips, Jeff
Piccolo, Mauro
Pietrzak, Krzysztof
Pilaud, Vincent
Piliouras, Georgios
Pilipczuk, Marcin
Pinto, Joao Sousa
Piterman, Nir
Place, Thomas
Poelstra, Andrew
Pokutta, Sebastian
Polak, Libor
Polishchuk, Valentin
Pountourakis, Emmanouil
Prencipe, Giuseppe
Pruhs, Kirk
Prutkin, Roman
Qin, Shengchao
Quas, Anthony
Rabehaja, Tahiry
Räcke, Harald
Raghavendra, Prasad
Raghothaman, Mukund

Raman, Rajiv
Raskin, Jean-Francois
Razenshteyn, Ilya
Regev, Oded
Rehak, Vojtech
Reis, Giselle
van Renssen, André
Reshef, Yakir
Reyzin, Leonid
Reyzin, Lev
Riba, Colin
Richerby, David
Riely, James
Riveros, Cristian
Robere, Robert
Robinson, Peter
Roeloffzen, Marcel
Röglin, Heiko
Rote, Günter
Rotenberg, Eva
Roth, Aaron
Rothvoss, Thomas
de Rougemont, Michel
Rümmele, Stefan
Sabel, David
Sabok, Marcin
Sacchini, Jorge Luis
Sach, Benjamin
Saha, Ankan
Saha, Chandan
Saitoh, Toshiki
Sakavalas, Dimitris
Salvati, Sylvain
Sanchez Villaamil, Fernando
Sangnier, Arnaud
Sankowski, Piotr
Sankur, Ocan
Saptharishi, Ramprasad
Saraswat, Vijay
Satti, Srinivasa Rao
Saurabh, Saket
Sawant, Anshul
Scharf, Ludmila
Schieber, Baruch
Schlotter, Ildikó
Schneider, Stefan

Schnitger, Georg
Schoenebeck, Grant
Schrijvers, Okke
Schweitzer, Pascal
Schweller, Robert
Schwitter, Rolf
Schöpp, Ulrich
Scquizzato, Michele
Seddighin, Saeed
Segev, Danny
Seidel, Jochen
Seiferth, Paul
Sekar, Shreyas
Sen, Siddhartha
Senizergues, Geraud
Serre, Olivier
Seshadhri, C.
Seto, Kazuhisa
Seurin, Yannick
Shepherd, Bruce
Sherstov, Alexander
Shi, Yaoyun
Shinkar, Igor
Shioura, Akiyoshi
Siebertz, Sebastian
Singh, Mohit
Sitters, Rene
Sivignon, Isabelle
Skorski, Maciej
Skrzypczak, Michał
Skutella, Martin
Smith, Adam
Soares Barbosa, Rui
Sobocinski, Pawel
Solan, Eilon
Sommer, Christian
Son, Wanbin
Sorensen, Tyler
Sorge, Manuel
Sottile, Frank
Spalek, Robert
Spoerhase, Joachim
Srba, Jiri
Srivastava, Piyush
Staals, Frank
Stampoulis, Antonis

Staton, Sam
Stefankovic, Daniel
Stein, Clifford
Stein, Yannik
Stenman, Jari
Stephan, Frank
Stirling, Colin
Stokes, Klara
Stolz, David
Strasser, Ben
Streicher, Thomas
Sun, He
Sun, Xiaorui
Suomela, Jukka
Svendsen, Kasper
Sviridenko, Maxim
Swamy, Chaitanya
Takahashi, Yasuhiro
Takazawa, Kenjiro
Talebanfard, Navid
Tamaki, Suguru
Tan, Li-Yang
Tan, Tony
Tang, Bo
Tanigawa, Shin-Ichi
Tasson, Christine
Tavenas, Sébastien
Teillaud, Monique
Telelis, Orestis
Thaler, Justin
Thapper, Johan
Thomas, Rekha
Ting, Hingfung
Tiwary, Hans
Torán, Jacobo
Tov, Roei
Tovey, Craig
Treinen, Ralf
Triandopoulos, Nikos
Trung, Ta Quang
Tsukada, Takeshi
Tulsiani, Madhur
Tuosto, Emilio
Tzamos, Christos
Uchizawa, Kei
Ueno, Shuichi

Uitto, Jara
Ullman, Jon
Ullman, Jonathan
Umboh, Seeun
Unno, Hiroshi
Uno, Yushi
Uramoto, Takeo
Urrutia, Florent
Vagvolgyi, Sandor
Vahlis, Yevgeniy
Valiron, Benoît
Vanden Boom, Michael
Vdovina, Alina
Veith, David
Venkatasubramanian, Suresh
Venkitasubramaniam,
    Muthuramakrishnan
Ventre, Carmine
Vereshchagin, Nikolay
Vidick, Thomas
Vijayaraghavan, Aravindan
Vildhøj, Hjalte Wedel
Vinayagamurthy, Dhinakaran
Vishnoi, Nisheeth
Vitanyi, Paul
Vivek, Srinivas
Vondrak, Jan
Voudouris, Alexandros
Wahlström, Magnus
Walter, Tobias
Walukiewicz, Igor
Wasa, Kunihiro
Watanabe, Osamu
Wee, Hoeteck
Wegner, Franziska
Wei, Zhewei
Weichert, Volker
Weinberg, S. Matthew
Weinstein, Omri
Wenner, Alexander
Werneck, Renato
Wexler, Tom
White, Colin
Wichs, Daniel
Wiese, Andreas
Willard, Ross

Williams, Ryan
Williamson, David
Wilson, David
Wimmer, Karl
Winslow, Andrew
Woeginger, Gerhard J.
Wojtczak, Dominik
de Wolf, Ronald
Wolff, Alexander
Wong, Prudence W.H.
Woodruff, David
Wootters, Mary
Worrell, James
Wrochna, Marcin
Wu, Xiaodi
Wu, Zhilin
Xiao, Tao
Xie, Ning
Xu, Jinhui
Yamakami, Tomoyuki
Yamamoto, Masaki
Yamauchi, Yukiko
Yang, Kuan

Yaroslavtsev, Grigory
Yehudayoff, Amir
Yodpinyanee, Anak
Yogev, Eylon
Yoon, Sang-Duk
Yoshida, Yuichi
Yun, Aaram
Yuster, Raphael
Zampetakis, Emmanouil
Zanuttini, Bruno
Zemor, Gilles
Zhang, Chihao
Zhang, Jialin
Zhang, Qin
Zhang, Shengyu
Zhou, Gelin
Zhou, Yuan
Živný, Stanislav
Zois, Georgios
Zorzi, Margherita
van Zwam, Stefan
Zwick, Uri

Contents – Part I

Track A: Algorithms, Complexity and Games

Statistical Randomized Encodings: A Complexity Theoretic View 1
  Shweta Agrawal, Yuval Ishai, Dakshita Khurana,
  and Anat Paskin-Cherniavsky

Tighter Fourier Transform Lower Bounds 14
  Nir Ailon

Quantifying Competitiveness in Paging with Locality of Reference 26
  Susanne Albers and Dario Frascaria

Approximation Algorithms for Computing Maximin Share Allocations 39
  Georgios Amanatidis, Evangelos Markakis, Afshin Nikzad,
  and Amin Saberi

Envy-Free Pricing in Large Markets: Approximating Revenue
and Welfare 52
  Elliot Anshelevich, Koushik Kar, and Shreyas Sekar

Batched Point Location in SINR Diagrams via Algebraic Tools 65
  Boris Aronov and Matthew J. Katz

On the Randomized Competitive Ratio of Reordering Buffer Management
with Non-uniform Costs 78
  Noa Avigdor-Elgrabli, Sungjin Im, Benjamin Moseley, and Yuval Rabani

Serving in the Dark Should Be Done Non-uniformly 91
  Yossi Azar and Ilan Reuven Cohen

Finding the Median (Obliviously) with Bounded Space 103
  Paul Beame, Vincent Liew, and Mihai Pătrașcu

Approximation Algorithms for Min-Sum k-Clustering
and Balanced k-Median 116
  Babak Behsaz, Zachary Friggstad, Mohammad R. Salavatipour,
  and Rohit Sivakumar

Solving Linear Programming with Constraints Unknown 129
  Xiaohui Bei, Ning Chen, and Shengyu Zhang

Deterministic Randomness Extraction from Generalized and Distributed
Santha-Vazirani Sources   143
    Salman Beigi, Omid Etesami, and Amin Gohari

Limitations of Algebraic Approaches to Graph Isomorphism Testing   155
    Christoph Berkholz and Martin Grohe

Fully Dynamic Matching in Bipartite Graphs   167
    Aaron Bernstein and Cliff Stein

Feasible Interpolation for QBF Resolution Calculi   180
    Olaf Beyersdorff, Leroy Chew, Meena Mahajan, and Anil Shukla

Simultaneous Approximation of Constraint Satisfaction Problems   193
    Amey Bhangale, Swastik Kopparty, and Sushant Sachdeva

Design of Dynamic Algorithms via Primal-Dual Method   206
    Sayan Bhattacharya, Monika Henzinger, and Giuseppe F. Italiano

What Percentage of Programs Halt?   219
    Laurent Bienvenu, Damien Desfontaines, and Alexander Shen

The Parity of Set Systems Under Random Restrictions with Applications
to Exponential Time Problems   231
    Andreas Björklund, Holger Dell, and Thore Husfeldt

Spotting Trees with Few Leaves   243
    Andreas Björklund, Vikram Kamat, Łukasz Kowalik, and Meirav Zehavi

Constraint Satisfaction Problems over the Integers with Successor   256
    Manuel Bodirsky, Barnaby Martin, and Antoine Mottet

Hardness Amplification and the Approximate Degree
of Constant-Depth Circuits   268
    Mark Bun and Justin Thaler

Algorithms and Complexity for Turaev-Viro Invariants   281
    Benjamin A. Burton, Clément Maria, and Jonathan Spreer

Big Data on the Rise? – Testing Monotonicity of Distributions   294
    Clément L. Canonne

Unit Interval Editing Is Fixed-Parameter Tractable   306
    Yixin Cao

Streaming Algorithms for Submodular Function Maximization   318
    Chandra Chekuri, Shalmoli Gupta, and Kent Quanrud

Multilinear Pseudorandom Functions.   33
    Aloni Cohen and Justin Holmgren

Zero-Fixing Extractors for Sub-Logarithmic Entropy. 34:
    Gil Cohen and Igor Shinkar

Interactive Proofs with Approximately Commuting Provers . . . .   35!
    Matthew Coudron and Thomas Vidick

Popular Matchings with Two-Sided Preferences and One-Sided Ties . . . . . .   36"
    Ágnes Cseh, Chien-Chung Huang, and Telikepalli Kavitha

Block Interpolation: A Framework for Tight Exponential-Time
Counting Complexity . . . .   38(
    Radu Curticapean

On Convergence and Threshold Properties of Discrete Lotka-Volterra
Population Protocols 39:
    Jurek Czyzowicz, Leszek Gąsieniec, Adrian Kosowski,
    Evangelos Kranakis, Paul G. Spirakis, and Przemysław Uznański

Scheduling Bidirectional Traffic on a Path. . . . .   40(
    Yann Disser, Max Klimm, and Elisabeth Lübbecke

On the Problem of Approximating the Eigenvalues of Undirected Graphs
in Probabilistic Logspace. . . . .   41(
    Dean Doron and Amnon Ta-Shma

On Planar Boolean CSP. 43:
    Zdeněk Dvořák and Martin Kupec

On Temporal Graph Exploration.   44-
    Thomas Erlebach, Michael Hoffmann, and Frank Kammer

Mind Your Coins: Fully Leakage-Resilient Signatures
with Graceful Degradation. . . . .   45(
    Antonio Faonio, Jesper Buus Nielsen, and Daniele Venturi

A (1+ε)-Embedding of Low Highway Dimension Graphs into Bounded
Treewidth Graphs. . . . .   46(
    Andreas Emil Feldmann, Wai Shing Fung, Jochen Könemann,
    and Ian Post

Lower Bounds for the Graph Homomorphism Problem . . . .   48]
    Fedor V. Fomin, Alexander Golovnev, Alexander S. Kulikov,
    and Ivan Mihajlin

Parameterized Single-Exponential Time Polynomial Space Algorithm
for Steiner Tree 494
Fedor V. Fomin, Petteri Kaski, Daniel Lokshtanov, Fahad Panolan,
and Saket Saurabh

Relative Discrepancy Does not Separate Information
and Communication Complexity.................................. 506
Lila Fontes, Rahul Jain, Iordanis Kerenidis, Sophie Laplante,
Mathieu Laurière, and Jérémie Roland

A Galois Connection for Valued Constraint Languages of Infinite Size..... 517
Peter Fulla and Stanislav Živný

Approximately Counting H-Colourings Is #BIS-Hard 529
Andreas Galanis, Leslie Ann Goldberg, and Mark Jerrum

Taylor Polynomial Estimator for Estimating Frequency Moments......... 542
Sumit Ganguly

ETR-Completeness for Decision Versions of Multi-player (Symmetric)
Nash Equilibria.. 554
Jugal Garg, Ruta Mehta, Vijay V. Vazirani, and Sadra Yazdanbod

Separate, Measure and Conquer: Faster Polynomial-Space Algorithms
for Max 2-CSP and Counting Dominating Sets 567
Serge Gaspers and Gregory B. Sorkin

Submatrix Maximum Queries in Monge Matrices Are Equivalent
to Predecessor Search 580
Paweł Gawrychowski, Shay Mozes, and Oren Weimann

Optimal Encodings for Range Top-k, Selection, and Min-Max........... 593
Paweł Gawrychowski and Patrick K. Nicholson

2-Vertex Connectivity in Directed Graphs 605
Loukas Georgiadis, Giuseppe F. Italiano, Luigi Laura,
and Nikos Parotsidis

Ground State Connectivity of Local Hamiltonians 617
Sevag Gharibian and Jamie Sikora

Uniform Kernelization Complexity of Hitting Forbidden Minors 629
Archontia C. Giannopoulou, Bart M.P. Jansen, Daniel Lokshtanov,
and Saket Saurabh

Counting Homomorphisms to Square-Free Graphs, Modulo 2 642
Andreas Göbel, Leslie Ann Goldberg, and David Richerby

Approximately Counting Locally-Optimal Structures 654
 Leslie Ann Goldberg, Rob Gysel, and John Lapinskas

Proofs of Proximity for Context-Free Languages and Read-Once
Branching Programs (Extended Abstract)........................ 666
 Oded Goldreich, Tom Gur, and Ron D. Rothblum

Fast Algorithms for Diameter-Optimally Augmenting Paths............ 678
 Ulrike Große, Joachim Gudmundsson, Christian Knauer,
 Michiel Smid, and Fabian Stehn

Hollow Heaps 689
 Thomas Dueholm Hansen, Haim Kaplan, Robert E. Tarjan,
 and Uri Zwick

Linear-Time List Recovery of High-Rate Expander Codes.............. 701
 Brett Hemenway and Mary Wootters

Finding 2-Edge and 2-Vertex Strongly Connected Components
in Quadratic Time... 713
 Monika Henzinger, Sebastian Krinninger, and Veronika Loitzenbauer

Improved Algorithms for Decremental Single-Source Reachability
on Directed Graphs.. 725
 Monika Henzinger, Sebastian Krinninger, and Danupon Nanongkai

Weighted Reordering Buffer Improved via Variants of Knapsack
Covering Inequalities 737
 Sungjin Im and Benjamin Moseley

Local Reductions 749
 Hamid Jahanjou, Eric Miles, and Emanuele Viola

Query Complexity in Expectation.............................. 761
 Jedrzej Kaniewski, Troy Lee, and Ronald de Wolf

Near-Linear Query Complexity for Graph Inference 773
 Sampath Kannan, Claire Mathieu, and Hang Zhou

A QPTAS for the Base of the Number of Crossing-Free Structures
on a Planar Point Set...................................... 785
 Marek Karpinski, Andrzej Lingas, and Dzmitry Sledneu

Finding a Path in Group-Labeled Graphs with Two Labels Forbidden...... 797
 Yasushi Kawase, Yusuke Kobayashi, and Yutaro Yamaguchi

Lower Bounds for Sums of Powers of Low Degree Univariates 810
 Neeraj Kayal, Pascal Koiran, Timothée Pecatte, and Chandan Saha

Approximating CSPs Using LP Relaxation   822
  Subhash Khot and Rishi Saket

Comparator Circuits over Finite Bounded Posets   834
  Balagopal Komarath, Jayalal Sarma, and K.S. Sunil

Algebraic Properties of Valued Constraint Satisfaction Problem   846
  Marcin Kozik and Joanna Ochremiak

Towards Understanding the Smoothed Approximation Ratio
of the 2-Opt Heuristic   859
  Marvin Künnemann and Bodo Manthey

On the Hardest Problem Formulations for the 0/1 Lasserre Hierarchy   872
  Adam Kurpisz, Samuli Leppänen, and Monaldo Mastrolilli

Replacing Mark Bits with Randomness in Fibonacci Heaps   886
  Jerry Li and John Peebles

A PTAS for the Weighted Unit Disk Cover Problem   898
  Jian Li and Yifei Jin

Approximating the Expected Values for Combinatorial Optimization
Problems Over Stochastic Points   910
  Lingxiao Huang and Jian Li

Deterministic Truncation of Linear Matroids   922
  Daniel Lokshtanov, Pranabendu Misra, Fahad Panolan,
  and Saket Saurabh

Linear Time Parameterized Algorithms for Subset Feedback Vertex Set   935
  Daniel Lokshtanov, M.S. Ramanujan, and Saket Saurabh

An Optimal Algorithm for Minimum-Link Rectilinear Paths
in Triangulated Rectilinear Domains   947
  Joseph S.B. Mitchell, Valentin Polishchuk, Mikko Sysikaski,
  and Haitao Wang

Amplification of One-Way Information Complexity via Codes
and Noise Sensitivity   960
  Marco Molinaro, David P. Woodruff, and Grigory Yaroslavtsev

A (2+ε)-Approximation Algorithm for the Storage Allocation Problem   973
  Tobias Mömke and Andreas Wiese

Shortest Reconfiguration Paths in the Solution Space
of Boolean Formulas   985
  Amer E. Mouawad, Naomi Nishimura, Vinayak Pathak,
  and Venkatesh Raman

Computing the Fréchet Distance Between Polygons with Holes   997
  Amir Nayyeri and Anastasios Sidiropoulos

An Improved Private Mechanism for Small Databases   1010
  Aleksandar Nikolov

Binary Pattern Tile Set Synthesis Is NP-Hard.   1022
  Lila Kari, Steffen Kopecki, Pierre-Étienne Meunier, Matthew J. Patitz,
  and Shinnosuke Seki

Near-Optimal Upper Bound on Fourier Dimension of Boolean Functions
in Terms of Fourier Sparsity   1035
  Swagato Sanyal

Condensed Unpredictability   1046
  Maciej Skórski, Alexander Golovnev, and Krzysztof Pietrzak

Sherali-Adams Relaxations for Valued CSPs   1058
  Johan Thapper and Stanislav Živný

Two-Sided Online Bipartite Matching and Vertex Cover: Beating
the Greedy Algorithm   1070
  Yajun Wang and Sam Chiu-wai Wong

The Simultaneous Communication of Disjointness with Applications
to Data Streams   1082
  Omri Weinstein and David P. Woodruff

An Improved Combinatorial Algorithm for Boolean Matrix Multiplication. . . .   1094
  Huacheng Yu

Author Index   1107

Contents – Part II

Invited Talks

Towards the Graph Minor Theorems for Directed Graphs 3
Ken-Ichi Kawarabayashi and Stephan Kreutzer

Automated Synthesis of Distributed Controllers 11
Anca Muscholl

Track B: Logic, Semantics, Automata and Theory of Programming

Games for Dependent Types 31
Samson Abramsky, Radha Jagadeesan, and Matthijs Vákár

Short Proofs of the Kneser-Lovász Coloring Principle. 44
James Aisenberg, Maria Luisa Bonet, Sam Buss, Adrian Crăciun,
and Gabriel Istrate

Provenance Circuits for Trees and Treelike Instances 56
Antoine Amarilli, Pierre Bourhis, and Pierre Senellart

Language Emptiness of Continuous-Time Parametric Timed Automata. 69
Nikola Beneš, Peter Bezděk, Kim G. Larsen, and Jiří Srba

Analysis of Probabilistic Systems via Generating Functions and Padé
Approximation 82
Michele Boreale

On Reducing Linearizability to State Reachability 95
Ahmed Bouajjani, Michael Emmi, Constantin Enea, and Jad Hamza

The Complexity of Synthesis from Probabilistic Components. 108
Krishnendu Chatterjee, Laurent Doyen, and Moshe Y. Vardi

Edit Distance for Pushdown Automata 121
Krishnendu Chatterjee, Thomas A. Henzinger, Rasmus Ibsen-Jensen,
and Jan Otop

Solution Sets for Equations over Free Groups Are EDT0L Languages 134
Laura Ciobanu, Volker Diekert, and Murray Elder

Limited Set quantifiers over Countable Linear Orderings. 146
Thomas Colcombet and A.V. Sreejith

Reachability Is in DynFO 159
Samir Datta, Raghav Kulkarni, Anish Mukherjee, Thomas Schwentick,
and Thomas Zeume

Natural Homology 171
Jérémy Dubut, Éric Goubault, and Jean Goubault-Larrecq

Greatest Fixed Points of Probabilistic Min/Max Polynomial Equations,
and Reachability for Branching Markov Decision Processes 184
Kousha Etessami, Alistair Stewart, and Mihalis Yannakakis

Trading Bounds for Memory in Games with Counters................. 197
Nathanaël Fijalkow, Florian Horn, Denis Kuperberg,
and Michał Skrzypczak

Decision Problems of Tree Transducers with Origin 209
Emmanuel Filiot, Sebastian Maneth, Pierre-Alain Reynier,
and Jean-Marc Talbot

Incompleteness Theorems, Large Cardinals, and Automata
over Infinite Words.. 222
Olivier Finkel

The Odds of Staying on Budget 234
Christoph Haase and Stefan Kiefer

From Sequential Specifications to Eventual Consistency 247
Radha Jagadeesan and James Riely

Fixed-Dimensional Energy Games Are in Pseudo-Polynomial Time 260
Marcin Jurdziński, Ranko Lazić, and Sylvain Schmitz

An Algebraic Geometric Approach to Nivat's Conjecture 273
Jarkko Kari and Michal Szabados

Nominal Kleene Coalgebra 286
Dexter Kozen, Konstantinos Mamouras, Daniela Petrişan,
and Alexandra Silva

On Determinisation of Good-for-Games Automata 299
Denis Kuperberg and Michał Skrzypczak

Owicki-Gries Reasoning for Weak Memory Models 311
Ori Lahav and Viktor Vafeiadis

On the Coverability Problem for Pushdown Vector Addition Systems
in One Dimension.. 324
Jérôme Leroux, Grégoire Sutre, and Patrick Totzke

Compressed Tree Canonization. 337
  Markus Lohrey, Sebastian Maneth, and Fabian Peternek

Parsimonious Types and Non-uniform Computation 350
  Damiano Mazza and Kazushige Terui

Baire Category Quantifier in Monadic Second Order Logic 362
  Henryk Michalewski and Matteo Mio

Liveness of Parameterized Timed Networks. 375
  Benjamin Aminof, Sasha Rubin, Florian Zuleger, and Francesco Spegni

Symmetric Strategy Improvement. 388
  Sven Schewe, Ashutosh Trivedi, and Thomas Varghese

Effect Algebras, Presheaves, Non-locality and Contextuality 401
  Sam Staton and Sander Uijlen

On the Complexity of Intersecting Regular, Context-Free,
and Tree Languages 414
  Joseph Swernofsky and Michael Wehar

Containment of Monadic Datalog Programs via Bounded Clique-Width 427
  Mikołaj Bojańczyk, Filip Murlak, and Adam Witkowski

An Approach to Computing Downward Closures 440
  Georg Zetzsche

How Much Lookahead Is Needed to Win Infinite Games?. 452
  Felix Klein and Martin Zimmermann

**Track C: Foundations of Networked Computation: Models,
Algorithms and Information Management**

Symmetric Graph Properties Have Independent Edges 467
  Dimitris Achlioptas and Paris Siminelakis

Polylogarithmic-Time Leader Election in Population Protocols 479
  Dan Alistarh and Rati Gelashvili

Core Size and Densification in Preferential Attachment Networks 492
  Chen Avin, Zvi Lotker, Yinon Nahum, and David Peleg

Maintaining Near-Popular Matchings 504
  *Sayan Bhattacharya, Martin Hoefer, Chien-Chung Huang,
  Telikepalli Kavitha, and Lisa Wagner*

Ultra-Fast Load Balancing on Scale-Free Networks   516
Karl Bringmann, Tobias Friedrich, Martin Hoefer, Ralf Rothenberger,
and Thomas Sauerwald

Approximate Consensus in Highly Dynamic Networks: The Role
of Averaging Algorithms.   528
Bernadette Charron-Bost, Matthias Függer, and Thomas Nowak

The Range of Topological Effects on Communication.   540
Arkadev Chattopadhyay and Atri Rudra

Secretary Markets with Local Information   552
Ning Chen, Martin Hoefer, Marvin Künnemann, Chengyu Lin,
and Peihan Miao

A Simple and Optimal Ancestry Labeling Scheme for Trees   564
Søren Dahlgaard, Mathias Bæk Tejs Knudsen, and Noy Rotbart

Interactive Communication with Unknown Noise Rate   575
Varsha Dani, Mahnush Movahedi, Jared Saia, and Maxwell Young

Fixed Parameter Approximations for k-Center Problems in Low Highway
Dimension Graphs   588
Andreas Emil Feldmann

A Unified Framework for Strong Price of Anarchy in Clustering Games. . . .   601
Michal Feldman and Ophir Friedler

On the Diameter of Hyperbolic Random Graphs.   614
Tobias Friedrich and Anton Krohmer

Tight Bounds for Cost-Sharing in Weighted Congestion Games   626
Martin Gairing, Konstantinos Kollias, and Grammateia Kotsialou

Distributed Broadcast Revisited: Towards Universal Optimality   638
Mohsen Ghaffari

Selling Two Goods Optimally   650
Yiannis Giannakopoulos and Elias Koutsoupias

Adaptively Secure Coin-Flipping, Revisited.   663
Shafi Goldwasser, Yael Tauman Kalai, and Sunoo Park

Optimal Competitiveness for the Rectilinear Steiner Arborescence
Problem.   675
Erez Kantor and Shay Kutten

Normalization Phenomena in Asynchronous Networks   688
   Amin Karbasi, Johannes Lengler, and Angelika Steger

Broadcast from Minicast Secure Against General Adversaries   701
   Pavel Raykov

Author Index   713

Statistical Randomized Encodings:
A Complexity Theoretic View

Shweta Agrawal[1], Yuval Ishai[2], Dakshita Khurana[3]([✉]),
and Anat Paskin-Cherniavsky[4]

[1] IIT, Delhi, India
shweta@cse.iitd.ac.in
[2] Technion, Haifa, Israel
yuvali@cs.technion.ac.il
[3] UCLA and Center for Encrypted Functionalities, Los Angeles, USA
dakshita@cs.ucla.edu
[4] Ariel University and UCLA, Ariel, Israel
anatpc@ariel.ac.il

Abstract. A randomized encoding of a function $f(x)$ is a randomized function $\hat{f}(x,r)$, such that the "encoding" $\hat{f}(x,r)$ reveals $f(x)$ and essentially no additional information about x. Randomized encodings of functions have found many applications in different areas of cryptography, including secure multiparty computation, efficient parallel cryptography, and verifiable computation.

We initiate a complexity-theoretic study of the class SRE of languages (or boolean functions) that admit an efficient statistical randomized encoding. That is, $\hat{f}(x,r)$ can be computed in time poly($|x|$), and its output distribution on input x can be sampled in time poly($|x|$) given $f(x)$, up to a small statistical distance.

We obtain the following main results.

- **Separating SRE from efficient computation:** We give the first examples of promise problems and languages in SRE that are widely conjectured to lie outside P/poly. Our candidate promise problems and languages are based on the standard Learning with Errors (LWE) assumption, a non-standard variant of the Decisional Diffie Hellman (DDH) assumption and the "Abelian Subgroup Membership problem" (which generalizes Quadratic-Residuosity and a variant of DDH).

- **Separating SZK from SRE:** We explore the relationship of SRE with the class SZK of problems possessing statistical zero knowledge proofs. It is known that SRE ⊆ SZK. We present an oracle separation which demonstrates that a containment of SZK in SRE cannot be proved via relativizing techniques.

Y. Ishai–Research supported by the European Union's Tenth Framework Programme (FP10/2010-2016) under grant agreement no. 259426 ERC-CaC, ISF grants 1361/10 and 1709/14 and BSF grant 2012378.

M.M. Halldórsson et al. (Eds.): ICALP 2015, Part I, LNCS 9134, pp. 1–13, 2015.
DOI: 10.1007/978-3-662-47672-7_1

1   Introduction

A randomized encoding (RE) of a function [5,13] allows one to represent a complex function $f(x)$ by a "simpler" randomized function, $\hat{f}(x,r)$, such that the "encoding" $\hat{f}(x,r)$ reveals $f(x)$ but no other information about x[1]. More specifically, there should exist an (unbounded) *decoder* that computes $f(x)$ given $\hat{f}(x,r)$, and an efficient randomized *simulator* that simulates the output of the encoder $\hat{f}(x,r)$, only given $|x|$ and $f(x)$. We refer to the former decoding requirement as *correctness* and to the latter simulation requirement as *privacy*. Privacy can either be perfect, statistical, or computational, depending on the required notion of "closeness" between the simulated distribution and the output distribution of $\hat{f}$. The complexity class SRE (resp. PRE, CRE) is defined to be the class of boolean functions $f : \{0,1\}^* \rightarrow \{0,1\}$, or equivalently languages, admitting a randomized encoding $\hat{f}$ that can be computed in polynomial time and having statistical (resp. perfect, computational) privacy. In this paper, we initiate the study of the class SRE of functions admitting a statistical randomized encoding (SRE).

As a cryptographic primitive, randomized encodings were first studied explicitly by Ishai and Kushilevitz [13], although they were used implicitly in prior work in the context of secure multiparty computation [11,16,19]. They have found application in different areas of cryptography, such as parallel implementations of cryptographic primitives [5], verifiable computation and secure delegation of computations [6], secure multiparty computation [4,8,9,13,14], and even in algorithm design [15]. We refer the reader to [3] for a survey of such applications.

The *parallel complexity* of randomized encodings was studied by Applebaum et al. [5], who demonstrated that all functions in the complexity class NC^1 (and even certain functions that are conjectured not to be in NC [2]) admit an SRE in NC^0. This establishes a provable speedup in the context of parallel time complexity. It is natural to ask a similar question in the context of *sequential* time complexity. For which functions (if any) can an SRE enable a super-polynomial speedup? This question is the focus of our work.

Characterizing the Class SRE. Let us consider the power of the class SRE of all functions admitting a polynomial-time computable statistical randomized encoding. It is evident that $\mathsf{P} \subseteq \mathsf{SRE}$, where $\hat{f}(x,r)$ simply outputs $f(x)$. This satisfies both the correctness and privacy requirements. But is $\mathsf{SRE} \subseteq \mathsf{P}$?

- **SRE for trivial hard languages.** First, we consider unary languages, i.e., languages $L \subseteq \{0\}^*$. These languages admit the trivial SRE defined by $\hat{f}(x) = x$. Indeed, the decoder can be defined by $D(z) = f(z)$ and the simulator, on input $(1^n, b)$, can output 0^n. Privacy holds since there is only one input of every length. However, such unary languages may not even be

---

[1] It also reveals $|x|$. This is unavoidable, as otherwise the output of $\hat{f}$ is one of two disjoint distributions supported over a finite domain, which puts $f(x)$ in BPP.

decidable, as illustrated for example by the language U_{HP} - the unary encoding of the halting problem, which admits an SRE but is not decidable. This example also extends to "trivial" binary languages such that for a given input length, all inputs are either in the language or not. However, note that such trivial languages are always contained in the class P/poly, namely the class of functions admitting polynomial-size (but possibly non-uniform) circuits. This demonstrates that getting a candidate separation between SRE and P or even PSPACE is not enough; to demonstrate the power of randomized encodings over efficient computation in a meaningful way, we must separate the class SRE from P/poly.

- **Is SRE more powerful that P/poly?** Let us now examine the relationship of SRE and P/poly. To begin, observe that for functions with long outputs, it is easy to find candidate functions that are not known to be efficiently computable by non-uniform circuits, but admit an efficient SRE. For example, assume there exists a family of one way permutations $\{f_n\}_{n \in \mathbb{N}}$ secure against non-uniform adversaries. Then the seemingly hard function $f^{-1}(x)$ can be encoded by the identity $\hat{f}^{-1}(x) = x$. As f^{-1} is also a permutation, this encoding is both private and correct. However, for boolean functions, the question looks much more interesting. To the best of our knowledge, no previous candidates for languages or promise problems that are conjectured to lie outside P/poly but admit efficient SRE have been proposed. This is one of the questions we study in this work.

- **Is SZK more powerful that SRE?** Another natural question about randomized encodings is their relationship with the class SZK of languages admitting statistical zero knowledge proofs. It is not hard to show that SRE $\subseteq$ SZK [2].[2] This implies that SRE is unlikely to contain NP. Based on current examples for SZK languages it seems likely that the containment SRE $\subseteq$ SZK is strict, but no formal evidence was given in this direction. This motivates the question of finding an oracle relative to which SZK is not contained in SRE.

Why is the class SRE *interesting?* As has been pointed out already, for functions that are efficiently computable, the SRE can just compute the function itself. Therefore, the class SRE is interesting only when the functions themselves are *not* efficiently computable, in which case the complexity of the decoder must inherently be super-polynomial. While most known applications of randomized encodings of functions require the decoder to be efficient, there are some applications that do not (see [3]). Moreover, even in cases where the decoder is required to be efficient, SRE functions can be "scaled down" so that decoding takes a feasible time T whereas encoding time is sub-polynomial in T. For instance, the computation of an SRE function can be delegated from a weak client to a powerful but untrusted server by directly applying an SRE on instances of a small

---

[2] Here and in the following, when writing SRE $\subseteq$ SZK we restrict SRE to only contain languages L that are *non-trivial* in the sense that for every sufficiently large input length n there are inputs x_0, x_1 of length n such that $x_0 \in L$ and $x_1 \notin L$. This excludes languages such as the unary undecidable language mentioned earlier. The containment proof in [2] implicitly assumes non-triviality.

size n, such that the server may be allowed to run in time $\exp(n)$ while the client is only required to run in time $\mathsf{poly}(n)$. Indeed, many real-life problems require exponential time to solve using the best known algorithms.

1.1 Our Results

Our results can be summarized as follows.

1. **Separating SRE from P/poly:**
   We provide three candidates to separate SRE from efficient computation.
   - We give a candidate *language*, for which we conjecture hardness based on a *non-standard* variant of the DDH assumption. We give an efficient SRE for this language which builds on the random self reduction for DDH demonstrated by Naor and Reingold [17].
   - Next, we give a candidate (dense) *promise problem*, the hardness of which follows from the hardness of the *standard* Learning with Errors assumption. We devise an efficient SRE for this promise problem.
   - Last, we design a non-uniform SRE for the Abelian subgroup membership ASM family of promise problems. This problem generalizes quadratic residuosity and (an instance of an augmented) co-DDH problem. We also give a specific instance of this promise problem, which is a language, and conjecture that this language is outside of P/poly based on a variant of co-DDH, an assumption introduced in [12].
2. **Separating SZK from SRE:** We show the existence of an oracle, relative to which SZK $\not\subseteq$ SRE. This oracle separation implies that the containment SZK $\subseteq$ SRE (if true) cannot be proved via relativizing proof techniques.

1.2 Overview of Main Techniques

We now give an overview of the main techniques used for our separations.

Separating SRE from P/poly. We provide several SRE constructions for problems that are conjectured to lie outside P/poly. It may be helpful to point out here, that problems in SRE also admit an SZK proof, and the existence of hard problems in SZK implies the existence of one-way functions. Therefore, we cannot hope to get an unconditional result, or even one based on P $\neq$ NP. We have the following candidates based on various assumptions, which we later summarize in Table 1.

- **Candidate Language Related to DDH.**
  Our first candidate is a language, which we call DDH$'$, whose hardness is related to the Decisional Diffie Hellman (DDH) assumption. We consider inputs of the form $\langle g, g^a, g^b, g^c \rangle$ where g is any generator of a fixed DDH group per input length. Roughly, the input is in the language iff it corresponds to a DDH tuple, that is, if $g^c = g^{ab}$ in a fixed group generated by g. Our SRE for this problem builds on the random self-reduction given by Naor and Reingold [17]. However, not only do we randomize the

DDH exponents following [17], but also randomize the generator of the DDH group.

Finally, in order to devise a candidate language, we must fix the description of the group and its generator, given just the length of the input. We achieve this by suggesting an efficient, deterministic procedure to generate a DDH group and other parameters required by the encoding algorithm, given the input length. However, note that the hardness of DDH' cannot be reduced to the standard DDH. This is because DDH is an average case assumption where the public parameters are chosen randomly. In our case, we must fix the public parameters per input length, and DDH does not guarantee that this restriction preserves hardness. We conjecture however, that DDH' remains infeasible for fixed parameters.

- **Dense Promise Problem Based on LWE.**
  Our second example is a (dense) promise problem DLWE', whose hardness reduces to the hardness of the standard LWE problem. DLWE' approximately classifies noisy codewords $(\mathbf{A}, \mathbf{b} = \mathbf{A}\mathbf{s} + \mathbf{e})$ into Yes and No instances, depending upon on the size of the error vector $\mathbf{e}$. Roughly speaking, Yes instances correspond to small errors and No instances to large errors.

  Note that, an SRE encoding of input $(\mathbf{A}, \mathbf{b} = \mathbf{A}\mathbf{s} + \mathbf{e})$ must be oblivious of all information about $\mathbf{A}, \mathbf{s}, \mathbf{e}$ except the relative size of the error vector $\mathbf{e}$. We begin by using the additive homomorphism of the LWE secret to mask $\mathbf{s}$. Specifically, we choose a random vector $\mathbf{t}$ and compute $\mathbf{b}' = \mathbf{b} + \mathbf{A}\mathbf{t} = \mathbf{A}(\mathbf{s} + \mathbf{t}) + \mathbf{e}$. Now, $\mathbf{b}'$ no longer retains information about $\mathbf{s}$. To hide $\mathbf{A}$, we multiply $(\mathbf{A}, \mathbf{b})$ by a random low norm matrix $\mathbf{R}$ and invoke the leftover hash lemma to argue that $\mathbf{R}\mathbf{A}$ looks random even when $\mathbf{R}$'s entries are chosen from a relatively small range. For No instances, $\mathbf{e}$ is large enough that $\mathbf{R}\mathbf{e}$ also hides $\mathbf{e}$ via LHL, but to hide the smaller $\mathbf{e}$ of Yes instances, we must add additional noise $\mathbf{r}_0$. This extra noise is large enough to hide $\mathbf{e}$ but not large enough to affect correctness. For more details, please see Section 3.1.

- **Generalizing QR, and candidate language related to co-DDH.**
  Our final candidate is the Abelian Subgroup Membership (promise) problem ASM, which generalizes the quadratic residuosity problem QR_N for composite modulus N. ASM is specified by an abelian group G, and a subgroup H of G, such that $I(G/H) = \mathbb{Z}_q^t$ for prime q, integer t and some isomorphism I. We define Yes instances to be well-formed $x \in H$, and No instances to be well-formed $x \in G \setminus H$. We note that $QR_N \in \mathsf{P/poly}$, and therefore is not a candidate for separation. However, we present a different candidate language, which is an instance of ASM, and which we conjecture to lie outside $\mathsf{P/poly}$ based on a variant of the co-DDH assumption in [12].

  At a high level, our SRE for the generalized ASM promise problem is constructed as follows. Given input x,
  - Compute $y = x \cdot h$ for random $h \xleftarrow{\$} H$.

- Pick random elements $(x_1, x_2, \ldots x_{t-1}) \xleftarrow{\$} G$.
  Define $\mathbf{X} = [I(x_1), \ldots, I(x_{t-1}), I(y)]$.
- Pick $\mathbf{R} \xleftarrow{\$} \mathbb{Z}_q^{t \times t}$. Output $\mathbf{R} \cdot \mathbf{X}$.

The first step randomizes x *within* its coset[3], erasing all information except the coset of x. Next, observe that membership of x in the subgroup H is encoded by the rank of $\mathbf{X}$ – if $x \in H$ then $\mathbf{X}$ is singular, whereas if $x \notin H$, then $\mathbf{X}$ is non-singular with high probability. Thus, randomizing $\mathbf{X}$ via $\mathbf{RX}$ hides everything except the rank of $\mathbf{X}$, effectively erasing coset information about x. The decoder learns whether $x \in H$ by computing the rank of $\mathbf{RX}$. Finally, we amplify the privacy and correctness parameters by applying a generic masking technique, that may be of independent interest.

Table 1. Our Candidates. The SREs are uniform and private against non-uniform adversaries. If not a language, we exhibit a promise problem. The * denotes that a specific instance of ASM is a language, though ASM is in general a promise problem.

Candidate	Language	Hardness
DDH$'$	Language	Non-Std DDH
DLWE$'$	(Dense) Promise Problem	Std LWE
ASM(co-DDH)	Language*	Non-Std co-DDH

Separating SZK *from* SRE Applebaum [2] showed that any language that admits an SRE encoding also admits an SZK proof. This was done by reducing SRE to the statistical distance problem [18] which admits a two-round SZK protocol. The question of whether this containment is strict is still open.

We give an oracle separation between the classes SZK and SRE. We diagonalize over oracle SRE encoders to obtain a language that is not in oracle-SRE, but admits an oracle-SZK proof. Our technique involves generalizing the method of [1] that separates oracle-SZK machines from oracle-BPP machines, with the oracle being determined during diagonalization. This technique is reminiscent of the one in [7] showing that any proof for P=NP does not relativize. However, our setting diverges from that of [1] in two ways.

First, we diagonalize over SRE encoders such that decoders are unbounded. However, in the presence of unbounded machines, an oracle similar to [1] would be only as powerful as the plain model. To deal with this, we derive an alternate definition for SRE, where the output of PPT encoders falls into two distinct distributions over a polynomially large support (unlike binary output BPP machines). In order to derive an outlying language via diagonalization in this new setting, we must account for the size of the support. We stress here that our separation does not reduce to the SZK – BPP separation in [1], and can in fact, be viewed as a generalization of their result.

---

[3] This step is similar to the classic SRE for QR_p which encodes x by $x \cdot r^2$ for randomly chosen r. However, this is insufficient even for QR_N where N is composite (hence for ASM), as it leaks coset information of x.

1.3 Related Work

The classes PREN, SREN and CREN have been defined by Applebaum, Ishai and Kushilevitz [6] as the class of functions that admit perfect (resp. statistical computational) randomized encodings in NC^0 with a polynomial-time decoder. In contrast, in this work we do not restrict the complexity of decoding the output. Applebaum [2] observed that $QR_p \in$ SREN while not known to be in NC suggesting a separation between these classes.

Aiello and Håstad[1] gave a technique for the oracle separation of SZK from BPP, by diagonalizing over oracle-BPP machines. Our technique for the oracle separation of SZK from uniform SRE follows in their broad outline, but must be adapted to oracle-SRE machines whose outputs are over a large support. Also, note that SRE has been used in the past for reducing the complexity of complete problems for a subclass of SZK (more specifically, the class $SZKP_L$ of problems having statistical zero-knowledge proofs where the honest verifier and its simulator are computable in logarithmic space) [10].

2 Preliminaries

In this section, we define basic notation and recall some definitions which will be used in our paper. Given a vector x, $|x|$ denotes its size. We let $\mathsf{size}(C)$ denote the size of a circuit C and $\mathsf{size}(f)$ denote the size of the smallest circuit computing f. The statistical distance between two distributions $\mathcal{X}$ and $\mathcal{Y}$ over space Ω, is defined as $\Delta(\mathcal{X}, \mathcal{Y}) \equiv \frac{1}{2}\Sigma_{u \in \Omega}|\Pr_{X \sim \mathcal{X}}[X = u] - \Pr_{Y \sim \mathcal{Y}}[Y = u]|$.

The definition of a promise problem, the class P/poly (extended to also include promise problems) and the class SZK, are mostly standard in the literature.

We now formally define the notion of a statistical randomized encoding of a function, language or promise problem. Similarly to the previous definition from [5], our definition requires the encoding to be uniform by default.

Definition 1 (Statistical randomized encodings ((ϵ, δ)-SRE)). *[5] Let* $f : \{0,1\}^* \rightarrow \{0,1\}^*$ *be a function and* $l(n)$ *an output length function such that* $|f(x)| = l(|x|)$ *for every* $x \in \{0,1\}^*$. *We say that* $\hat{f} : \{0,1\}^* \times \{0,1\}^* \rightarrow \{0,1\}^*$ *is a* $\epsilon(n)$-private $\delta(n)$-correct *(uniform) statistical randomized encoding of* f *(abbreviated* (ϵ, δ)-SRE*), if the following holds:*

- **Length regularity.** *There exist polynomially-bounded and efficiently computable length functions* $m(n), s(n)$ *such that for every* $x \in \{0,1\}^n$ *and* $r \in \{0,1\}^{m(n)}$, *we have* $|\hat{f}(x,r)| = s(n)$.
- **Efficient encoding.** *There exists a polynomial-time encoding algorithm denoted by* $\mathsf{enc}(\cdot, \cdot)$ *that, given* $x \in \{0,1\}^*$ *and* $r \in \{0,1\}^{m(|x|)}$, *outputs* $\hat{f}(x,r)$.
- δ-**correctness.** *There exists an unbounded decoder* dec, *such that for every* $x \in \{0,1\}^n$ *we have* $\Pr[\mathsf{dec}(1^n, \hat{f}(x, U_{m(n)})) \neq f(x)] \leq \delta(n)$.

- **ϵ-privacy.** *There exists a probabilistic polynomial-time simulator S, such that for every $x \in \{0,1\}^n$ we have $\Delta(S(1^n, f(x)), \hat{f}(x, U_{m(n)})) \leq \epsilon(n)$.*

An (ϵ, δ)-SRE of a language $L \subset \{0,1\}^*$ is an (ϵ, δ)-SRE of the corresponding boolean function $f : \{0,1\}^* \to \{0,1\}$. When ϵ and δ are omitted, they are understood to be negligible functions.

Extensions. A *non-uniform* (ϵ, δ)-SRE of f is defined similarly, except that the encoding algorithm is implemented by a family of polynomial-size circuits. For a *partial function* f, defined over a subset $X \subseteq \{0,1\}^*$, the correctness and privacy requirements should only hold for every $x \in X$. An (ϵ, δ)-SRE of a promise problem (Yes, No) is an (ϵ, δ)-SRE of the corresponding partial boolean function.

Definition 2 (The class SRE[4]). *The class SRE is defined to be the set of all languages that admit an SRE (namely, an (ϵ, δ)-SRE for some negligible ϵ, δ). For concrete functions $\epsilon(n), \delta(n)$, we use (ϵ, δ)-SRE to denote the class of languages admitting an (ϵ, δ)-SRE.*

3   Separating SRE from Efficient Computation

We devise three candidates for separating SRE from efficient computation. In this section, we outline one candidate promise problem, that belongs to SRE and is unlikely to be in P/poly based on the standard LWE assumption.

We also devise a candidate language based on a non-standard, but plausible, hardness assumption related to DDH. The final candidate is based on the Abelian Subgroup Membership problem. Please refer to the full version for details on these candidates.

3.1   Learning with Errors (LWE)-based Promise Problem

In this section, we devise a candidate *promise problem* DLWE′ based on the hardness of the Learning with Errors (LWE) assumption.

Definition 3. DLWE′ = {Yes, No} *where* Yes *and* No *are defined as follows.*

$$\text{Yes} = \bigcup_n \text{Yes}_n, \quad \text{No} = \bigcup_n \text{No}_n$$

The parameters m, p, ϵ are set per input length n as $m = n^2, p = n^{40}, \delta = 0.05$.

$$\text{Yes}_n \triangleq \left\{ (\mathbf{A}, \mathbf{As} + \mathbf{e}) \mid \mathbf{A} \in \mathbb{Z}_p^{m \times n}, \mathbf{s} \in \mathbb{Z}_p^n, \mathbf{e} \in [-p^\delta, p^\delta]^m, \Delta(\mathcal{R}_\mathbf{A}, \mathcal{U}_{m \times n}) \leq p^{-0.16m} \right\}$$

$$\text{No}_n \triangleq \left\{ (\mathbf{A}, \mathbf{As} + \mathbf{e}) \mid \mathbf{A} \in \mathbb{Z}_p^{m \times n}, \mathbf{s} \in \mathbb{Z}_p^n, \mathbf{e} \in \mathbb{Z}_p^m \setminus [-p^{2/3}, p^{2/3}]^m, \right.$$

$$\left. \Delta((\mathcal{R}_\mathbf{A}, \mathcal{R}_\mathbf{e}), (\mathcal{U}_{m \times n}, \mathcal{U}_m)) \leq p^{-0.16m} \right\} \setminus \text{Yes}_n$$

---

[4] The difference between the class SRE and the class SREN defined in [5] is that SRE allows the encoding algorithm to run in polynomial time whereas SREN restricts the encoding algorithm to be in NC^0.

Here, $\mathcal{R}_\mathbf{A}$ denotes the distribution $\mathbf{RA} \pmod p$ induced by choosing $\mathbf{R}$ uniformly in $[-p^{2/3}, p^{2/3}]^{m \times m}$. Similarly, $\mathcal{R}_\mathbf{e}$ denotes the distribution $\mathbf{Re} \pmod p$ induced by choosing $\mathbf{R}$ (same as before) uniformly in $[-p^{2/3}, p^{2/3}]^{m \times m}$. $\mathcal{U}_{m \times n}$ and $\mathcal{U}_m$ denote the uniform distribution in $\mathbb{Z}_p^{m \times n}$ and $\mathbb{Z}_p^m$ respectively.

We must explicitly subtract Yes_n from No_n because there may exist $\mathbf{s}, \mathbf{e}$ and $\tilde{\mathbf{s}}, \tilde{\mathbf{e}}$ such that $\mathbf{As} + \mathbf{e} = \mathbf{A}\tilde{\mathbf{s}} + \tilde{\mathbf{e}}$ and $\tilde{\mathbf{e}} \in (\mathbb{Z}_p \setminus [-p^{2/3}, p^{2/3}])^m$ but $\mathbf{e} \in [-p^\delta, p^\delta]^m$ resulting in an overlap between the sets Yes_n and No_n. The condition involving the statistical distance is a technicality required for using the leftover hash lemma in the construction. The value $p^{-0.16m}$ in the definition is a representative inverse polynomial function in the input size n. We also define a new promise problem DLWE'' which is exactly the same as DLWE', except setting $p = 2^n$ for each input length n. The analysis of DLWE'' is the same except $p^{-0.16m}$ is $\mathsf{negl}(n)$.

It is easy to show that the hardness of DLWE' and DLWE'' against P/poly follows from the hardness of the standard decisional Learning with Errors problem DLWE for the same parameters.

Theorem 1. $\mathsf{DLWE}' \in (1/\mathrm{poly}, 1/\mathrm{poly})\text{-SRE}$ *and* $\mathsf{DLWE}'' \in (\mathsf{negl}, \mathsf{negl})\text{-SRE}$.

Proof. We construct an SRE for DLWE' here. On input an instance of size n, the encoder, decoder, simulator compute parameters m, ϵ, δ, p as functions of n.

Encoding. The algorithm $\mathsf{enc}_{\mathsf{SRE}}(1^n, \mathbf{A}, \mathbf{b})$ is defined as follows.

1. Pick $\mathbf{R} \xleftarrow{\$} [-p^{2/3}, p^{2/3}]^{m \times m}$, $\mathbf{r}_0 \xleftarrow{\$} [-p^{2/3+3\delta}, p^{2/3+3\delta}]^m$, $\mathbf{t} \xleftarrow{\$} \mathbb{Z}_p^n$.
2. Set $\mathbf{A}' = \mathbf{RA}$ and $\mathbf{b}' = \mathbf{r}_0 + \mathbf{Rb}$.
3. Output $(\mathbf{A}'', \mathbf{b}'') = (\mathbf{A}', \mathbf{A}'\mathbf{t} + \mathbf{b}')$.

Decoding. The algorithm $\mathsf{dec}_{\mathsf{SRE}}(1^n, \mathbf{A}'', \mathbf{b}'')$ accepts if and only if there exist $\mathbf{x} \in \mathbb{Z}_p^n$, $\mathbf{e} \in \mathbb{Z}_p^m$, such that $\mathbf{b}'' = \mathbf{A}''\mathbf{x} + \mathbf{e}''$, and $\mathbf{e}'' \in [-p^{2/3+4\delta}, p^{2/3+4\delta}]$.

Simulation. On input 1^n and a bit b where $b = 0/1$ represents membership in Yes/No respectively, the simulator does the following.

- If $b = 0$, pick $\mathbf{U} \xleftarrow{\$} \mathbb{Z}_p^{m \times n}$, $\mathbf{t} \xleftarrow{\$} \mathbb{Z}_p^n$, $\mathbf{e} \xleftarrow{\$} [-p^{2/3+3\delta}, p^{2/3+3\delta}]^m$. Output $(\mathbf{U}, \mathbf{Ut} + \mathbf{e})$.
- If $b = 1$, pick $\mathbf{U} \xleftarrow{\$} \mathbb{Z}_p^{m \times n}$ and $\mathbf{u} \xleftarrow{\$} \mathbb{Z}_p^m$. Output $(\mathbf{U}, \mathbf{u})$.

Analysis. We give a brief overview of the correctness and privacy arguments. Recall that,

$$\mathsf{enc}_{\mathsf{SRE}}(1^n, \mathbf{A}, \mathbf{As} + \mathbf{e}) = \Big(\mathbf{RA}, \; \mathbf{RA}(\mathbf{s} + \mathbf{t}) + (\mathbf{Re} + \mathbf{r}_0)\Big) \quad \text{where}$$

$$\mathbf{t} \xleftarrow{\$} \mathbb{Z}_p^n, \;\; \mathbf{R} \xleftarrow{\$} [-p^{2/3}, p^{2/3}]^{m \times m}, \;\; \mathbf{r}_0 \xleftarrow{\$} [-p^{2/3+3\delta}, p^{2/3+3\delta}]^m.$$

Thus, the secret in $\mathbf{b}''$, namely $\mathbf{s} + \mathbf{t}$, is distributed uniformly in $\mathbb{Z}_p^n$.

- **Case 1:** $(\mathbf{A}, \mathbf{As} + \mathbf{e}) \in \mathsf{Yes}_n$. In this case, $\mathbf{e} \in [-p^\delta, p^\delta]^m$.
  Then, for $\mathbf{R} \xleftarrow{\$} [-p^{2/3}, p^{2/3}]^m$, $\mathbf{Re} \in [-p^{2/3+2\delta}, p^{2/3+2\delta}]^m$.
  Moreover, by choice of $\mathbf{r}_0$, we have $\mathbf{Re} << \mathbf{r}_0$, thus $\Delta(\mathbf{Re} + \mathbf{r}_0, \mathbf{r}_0) \leq p^{-\delta m}$.
  By definition of the promise problem, we have that $\Delta(\mathcal{R}_\mathbf{A}, \mathbf{U}_{m \times n}) \leq p^{-0.16m}$.
  Then the following hold:

  - *Correctness.* $\mathbf{Re} + \mathbf{r}_0 \in [-p^{2/3+4\delta}, p^{2/3+4\delta}]$. Thus, correctness is perfect.
  - *Privacy.* By the above arguments on the distribution of $(\mathbf{RA})$, $(\mathbf{s} + \mathbf{t})$ and $(\mathbf{Re} + \mathbf{r}_0)$, and by the simulator's choice of $(\mathbf{U}, \mathbf{t}, \mathbf{e})$, we can argue that the output distribution is at most $p^{-0.16m}$-far from the distribution induced by SRE.enc on an instance of Yes_n.

- **Case 2:** $(\mathbf{A}, \mathbf{As} + \mathbf{e}) \in \mathsf{No}_n$. We have that $\mathbf{e} \in (\mathbb{Z}_p \setminus [-p^{2/3}, p^{2/3}])^m$ and $\Delta((\mathcal{R}_\mathbf{A}, \mathcal{R}_\mathbf{e}), (\mathbf{U}_{m \times n}, \mathbf{u}_m)) \leq p^{-0.16m}$. Then the following hold:

  - *Correctness.* Standard averaging arguments prove that all entries of $\mathbf{Re} + \mathbf{r}_0$ are larger than $p^{2/3+4\delta}$ with probability $\geq 1 - p^{-0.13m}$. Moreover, the probability that randomizing an instance in No_n results in an encoding that corresponds to some 'small' error vector[5], $\leq p^{-\delta m}$. Overall, we obtain $p^{-0.1m}$-correctness.
  - *Privacy.* We show that a random sample $(\mathbf{A}, \mathbf{b}) \xleftarrow{\$} \mathbb{Z}_p^{m \times n} \times \mathbb{Z}_p^m$ (simulator output) is $(1 - p^{-0.1m})$ close to the distribution induced by SRE.enc on a No_n instance. First, we show that randomly chosen $(\mathbf{A}, \mathbf{b})$ are such that, w.h.p. there exist no $(\mathbf{s}, \text{small}[5]\ \mathbf{e})$ such that $\mathbf{b} = \mathbf{As} + \mathbf{e}$. We also prove that w.h.p. $\mathbf{A}, \mathbf{e}$ corresponding to random $(\mathbf{A}, \mathbf{b})$ are such that the distributions $\mathcal{R}_\mathbf{A}$ and $\mathcal{R}_\mathbf{e}$ are close to uniform.

4   Oracle Separation Between SRE and SZK

In this section, we crucially use the following Lemma about the class (ϵ, δ)-SRE. This Lemma follows directly from the definition of (ϵ, δ)-SRE.

Lemma 1. *Let $\mathcal{E}_x$ denote the distribution $\mathsf{enc}(x, r)$ for the algorithm $\mathsf{enc}(\cdot, \cdot)$ of a language L admitting an (ϵ, δ)-SRE, induced for any input x by picking r uniformly at random in $\{0, 1\}^*$. Then, $\Delta(\mathcal{E}_x, \mathcal{E}_{x'}) \leq 2\epsilon$ iff $f(x) = f(x')$ (equivalently, both $x, x' \in L$ or both $x, x' \notin L$). Moreover, $\Delta(\mathcal{E}_x, \mathcal{E}_{x'}) \geq 1 - 2\delta$ iff $f(x) \neq f(x')$ (equivalently, either $x \in L, x' \notin L$ or $x \notin L, x' \in L$).*

In this section, we study the relation between the classes SRE and SZK. We recall the following theorem from [2].

Imported Theorem 1. *[2] Any non-trivial language that admits an (ϵ, δ)-SRE such that $(1 - 2\delta)^2 > 2\epsilon$, also admits an SZK proof.*

---

[5] Here, 'small' denotes error of magnitude less than $p^{2/3+4\delta}$, such that the instance wrongly decodes to Yes. However, in the rest of the paper, 'small' denotes error $\leq p^\delta$.

Here, we explore whether the containment $\mathsf{SRE} \subseteq \mathsf{SZK}/\mathsf{poly}$ is strict. We give an oracle separation between the classes SZK (more precisely, the class $\mathsf{SZK}[2]$ of languages that admit a 2-round SZK proof - note that this is the strongest separation) and SRE, but restricted to the uniform setting. For any oracle A, we denote by SRE^A the class SRE where encoders have oracle access to A. Similarly, we denote by SZK^A the class SZK where verifiers have oracle access to A.

Theorem 2. *There exists an oracle A, such that* $\mathsf{SZK}[2]^A \not\subset \mathsf{SRE}^A$.

Proof Overview. Broadly, we diagonalize over all oracle SRE-encoder machines to obtain a language which does not have any SRE encoding. We construct this language in rounds, one for each input length. Specifically, we will ensure that for every input length n, the output of the encoder on inputs 0^n and 1^n is either less than $(1 - 2\delta)$ or more than ϵ, violating the definition of SRE from Lemma 1[6].

This is done via classifying the characteristic vector of the language into unique and redundant sets, such that it is impossible for any encoder with polynomially many oracle queries to distinguish between unique versus redundant characteristic. Moreover, a contrived language is set such that 0^n is never in the language, and 1^n is in the language iff the characteristic vector is unique.

Intuitively, since encoders cannot distinguish between a unique versus redundant characteristic, one of the following cases will always occur. Either, there exists a redundant characteristic (implying that both 0^n and 1^n are not in the language) such that the encodings of 0^n and 1^n are more than ϵ-apart; or, there exists a unique characteristic (implying that 1^n is in the language while 0^n is not) such that the encodings of 0^n and 1^n are less than $(1 - 2\delta)$-apart. We set the language according to whichever of these cases is true. This ensures that the output of the encoders is not an SRE for this language.

However, proving either of the two cases is true is significantly more involved than in the BPP setting of [1] (refer to the full version for details). Finally, we can show that this language has an SZK proof, this follows in a similar manner as [1].

5   Conclusion and Open Problems

In this paper, we study the class SRE of languages and promise problems that admit efficient statistical randomized encodings. We present the first candidates for SRE problems that are not in P/poly. These include a candidate promise problem based on the hardness of standard LWE, as well as candidate languages based on variants of the DDH assumption and the co-DDH assumption of [12].

Then, we explore the relationship of the class SRE with the class SZK of languages admitting statistical zero knowledge proofs. While it is known that all non-trivial languages in SRE are also in SZK [2], whether the converse holds is open. However, we exhibit an oracle and a (non-trivial) language that has an

---

[6] It is interesting to note that unlike the BPP-SZK [1] separation, a unary language is not helpful for separation since such a language will always have an SRE. Thus, our contrived language will be non-trivial and binary.

oracle-based SZK proof but does not have an oracle-based SRE. This shows that a containment of SZK in SRE cannot be proved via relativizing techniques. Several natural questions remain open. The first is to identify a complete language in SRE, thereby obtaining a better characterization of this class. A second is to better understand the relation between statistical randomized encodings and random self-reductions (RSR). An RSR for a language or a promise problem can be viewed as a restricted form of SRE where the decoder just decides the problem itself. Our LWE-based language is a candidate for a problem in SRE which is not in RSR, thus supporting the conjecture that RSR ⊂ SRE. Is there an oracle separating these classes? Finally, it would be interesting to find additional (and preferably "useful") candidates for intractable problems in SRE, as well as natural polynomial-time solvable problems for which an SRE can provide polynomial speedup over the best known algorithms.

References

1. Aiello, W., Håstad, J.: Relativized perfect zero knowledge is not BPP. Inf. Comput. (1991)
2. Applebaum, B.: Cryptography in Constant Parallel Time. Ph.D. thesis, Technion (2007)
3. Applebaum, B.: Randomly encoding functions: a new cryptographic paradigm. In: Fehr, S. (ed.) ICITS 2011. LNCS, vol. 6673, pp. 25–31. Springer, Heidelberg (2011)
4. Applebaum, B., Ishai, Y., Kushilevitz, E.: Computationally private randomizing polynomials and their applications. In: IEEE Conference on Computational Complexity, pp. 260–274. IEEE Computer Society (2005)
5. Applebaum, B., Ishai, Y., Kushilevitz, E.: Cryptography in NC0. SIAM J. Comput. 36(4), 845–888 (2006)
6. Applebaum, B., Ishai, Y., Kushilevitz, E.: From secrecy to soundness: efficient verification via secure computation. In: Abramsky, S., Gavoille, C., Kirchner, C., Meyer auf der Heide, F., Spirakis, P.G. (eds.) ICALP 2010. LNCS, vol. 6198, pp. 152–163. Springer, Heidelberg (2010)
7. Baker, T.P., Gill, J., Solovay, R.: Relativizatons of the P =? NP question. SIAM J. Comput. 4(4), 431–442 (1975)
8. Ben-Or, M., Goldwasser, S., Wigderson, A.: Completeness theorems for non-cryptographic fault-tolerant distributed computation. In: STOC. ACM (1988)
9. Chaum, D., Crépeau, C., Damgard, I.: Multiparty unconditionally secure protocols. In: STOC, pp. 11–19. ACM, New York (1988)
10. Dvir, Z., Gutfreund, D., Rothblum, G.N., Vadhan, S.: On approximating the entropy of polynomial mappings. In: ICS, pp. 460–475 (2011)
11. Feige, U., Killian, J., Naor, M.: A minimal model for secure computation (extended abstract). In: STOC, pp. 554–563 (1994)
12. Galbraith, S.D., Rotger, V.: Easy decision-diffie-hellman groups (2004)
13. Ishai, Y., Kushilevitz, E.: Randomizing polynomials: a new representation with applications to round-efficient secure computation. In: FOCS, pp. 294–304. IEEE Computer Society (2000)
14. Ishai, Y., Kushilevitz, E.: Perfect constant-round secure computation via perfect randomizing polynomials. In: Widmayer, P., Triguero, F., Morales, R., Hennessy, M., Eidenbenz, S., Conejo, R. (eds.) ICALP 2002. LNCS, vol. 2380, pp. 244–256. Springer, Heidelberg (2002)

15. Ishai, Y., Kushilevitz, E., Paskin-Cherniavsky, A.: From randomizing polynomials to parallel algorithms. In: ITCS. ACM, New York (2012)
16. Kilian, J.: Founding crytpography on oblivious transfer. In: STOC, pp. 20–31 ACM, New York (1988)
17. Naor, M., Reingold, O.: Number-theoretic constructions of efficient pseudo-random functions. J. ACM **51**(2), Mar 2004
18. Sahai, A., Vadhan, S.: A complete problem for statistical zero knowledge. J. ACM **50**(2), 196–249 (2003). http://doi.acm.org/10.1145/636865.636868
19. Yao, A.C.C.: How to generate and exchange secrets (extended abstract). In: FOCS, pp. 162–167 (1986)

Tighter Fourier Transform Lower Bounds

Nir Ailon[(✉)]

Technion Israel Institute of Technology, Haifa, Israel
nailon@cs.technion.ac.il

Abstract. The Fourier Transform is one of the most important linear transformations used in science and engineering. Cooley and Tukey's Fast Fourier Transform (FFT) from 1964 is a method for computing this transformation in time $O(n \log n)$. Achieving a matching lower bound in a reasonable computational model is one of the most important open problems in theoretical computer science. In 2014, improving on his previous work, Ailon showed that if an algorithm speeds up the FFT by a factor of $b = b(n) \geq 1$, then it must rely on computing, as an intermediate "bottleneck" step, a linear mapping of the input with condition number $\Omega(b(n))$. Our main result shows that a factor b speedup implies existence of not just one but $\Omega(n)$ b-ill conditioned bottlenecks occurring at $\Omega(n)$ different steps, each causing information from independent (orthogonal) components of the input to either overflow or underflow. This provides further evidence that beating FFT is hard. Our result also gives the first quantitative tradeoff between computation speed and information loss in Fourier computation on fixed word size architectures. The main technical result is an entropy analysis of the Fourier transform under transformations of low trace, which is interesting in its own right.

1 Introduction

The (discrete) normalized Fourier transform (DFT) is a complex mapping sending input $x \in \mathbb{C}^n$ to $Fx \in \mathbb{C}^n$, where F is a unitary matrix defined by $F(k, \ell) = n^{-1/2} e^{-i2\pi k\ell/n}$. The Walsh-Hadamard transform is a real orthogonal mapping in $\mathbb{R}^n$ (for n an integer power of 2) sending an input x to Fx, where $F(k, \ell) = \frac{1}{\sqrt{n}}(-1)^{\langle [k-1], [\ell-1] \rangle}$, with $\langle \cdot, \cdot \rangle$ is dot-product, and $[p]$ denotes (here only) the bit representation of the integer $p \in \{0, \dots, n-1\}$ as a vector of $\log_2 n$ bits. Both transformations are special (and most important) cases of abstract Fourier transforms defined with respect to corresponding Abelian groups. The Fast Fourier Transform (FFT) of Cooley and Tukey [9] is a method for computing the DFT of $x \in \mathbb{C}^n$ in time $O(n \log n)$. The fast Walsh-Hadamard transform computes the Walsh-Hadamard transform in time $O(n \log n)$. Both fast transformations perform a sequence of rotations on pairs of coordinates, and are hence special cases of so-called linear algorithms, as defined in [17].

The DFT is instrumental as a subroutine in fast polynomial multiplication [10] (chapter 30), fast integer multiplication [11–13], cross-correlation and auto-correlation detection in images and time-series (via convolution) and, as a more

© Springer-Verlag Berlin Heidelberg 2015
M.M. Halldórsson et al. (Eds.): ICALP 2015, Part I, LNCS 9134, pp. 14–25, 2015.
DOI: 10.1007/978-3-662-47672-7_2

recent example, convolution networks for deep learning [16]. Both DFT and Walsh-Hadamard are useful for fast Johnson-Lindenstrauss transform for dimensionality reduction [4–6,15] and the related restricted isometry property (RIP) matrix construction [7,15,19]). It is beyond the scope of this work to survey all uses of Fourier transforms in both theory of algorithms and in complexity. For the sake of simplicity the reader is encouraged to assume that F is the Walsh-Hadamard transform, and that by the acronym "FFT" we refer to the fast Walsh-Hadamard transform. The modifications required for the DFT (rather, the real embedding thereof) require a slight modification to the potential function which we mention but do not elaborate on for simplicity. Our results nevertheless apply also to DFT.

It is not known whether $\Omega(n \log n)$ operations are necessary, and this problem is one of the most important open problems in theoretical computer science [8]. It is trivial that a linear number of steps is necessary, because every input coordinate must be probed. Papadimitriou derives in [18] an $\Omega(n \log n)$ lower bound for DFT over finite fields using a notion of an information flow network. It is not clear how to extend that result to the Complex field. There have also been attempts [20] to reduce the constants hiding in the upper bound of $O(n \log n)$, while also separately counting the number of additions versus the number of multiplications (by constants). In 1973, Morgenstern proved that if the moduli of the constants used in the computation are bounded by 1 then the number of steps required for computing the *unnormalized* Fourier transform, defined by $n^{1/2}F$ in the linear algorithm model is at least $\frac{1}{2}n \log_2 n$. He used a potential function related to matrix determinant, which makes the technique inapplicable for deriving lower bounds for the (normalized) F. Morgenstern's result also happens to imply that the transformation $\sqrt{n}\,\mathrm{Id}$ ($\sqrt{n}$ times the identity) has the same complexity as the Fourier transform, which is not a satisfying conclusion. Also note that stretching the input norm by a factor of $\sqrt{n}$ requires representing numbers of $\omega(\log n)$ bits, and it cannot be simply assumed that a multiplication or an addition over such numbers can be done in $O(1)$ time.

Ailon [1] studied the complexity of the (normalized) Fourier transform in a computational model allowing only orthogonal transformations acting on (and replacing in memory) two intermediates at each step. He showed that at least $\Omega(n \log n)$ steps were required. The proof was done by defining a potential function on the matrices $M^{(t)}$ defined by composing the first t gates. The potential function is simply the sum of Shannon entropy of the probability distributions defined by the squared modulus of elements in the matrix rows. (Due to orthogonality, each row, in fact, thus defines a probability distribution). That result had two shortcomings: (i) The algorithm was assumed not to be allowed to use extra memory in addition to the space used to hold the input. In other words, the computation was done *in place*. (ii) The result was sensitive to the normalization of F, and was not useful in deriving any lower bound for γF for $\gamma \notin \{\pm 1\}$.

In [2], Ailon took another step forward by showing a lower bound for computing *any scaling* of the Fourier transform in a stronger model of computation which we call *uniformly well conditioned*. At each step, the algorithm can perform a nonsingular linear transformation on at most two intermediates, as long as the matrix $M^{(t)}$ defining the composition of the first t steps must have condition number at most κ, for all i. We remind the reader that condition number of a matrix is defined as the ratio between its largest and smallest (nonzero) singular values. Otherwise stated, the result implies that if an algorithm computes the Fourier transform in time $(n \log n)/b$ for some $b > 1$, then some $M^{(t)}$ must have condition number at least $\Omega(b)$. This means that the computation output relies on an ill conditioned intermediate step. The result in [2] made a qualitative claim about compromise of numerical stability due to a ill condition.

1.1   Our Contribution

Here we establish (Theorem 51) that a b-factor speedup of FFT for $b = b(n) = \omega(1)$ either *overflows* at $\Omega(n)$ different time steps due to $\Omega(n)$ pairwise orthogonal input directions, or *underflows* at $\Omega(n)$ different time steps, losing accuracy of order $\Omega(b)$ at n orthogonal input directions. Note that achieving this could not be simply done by a more careful analysis of [2], but rather requires an intricate analysis of the entropy of Fourier transform under transformations of small trace. This analysis (Lemma 61) is interesting in its own right.

2   Computational Model

We remind the reader of the computational model discussed in [1,2], which is a special case of the linear computational model. The machine state represents a vector in $\mathbb{R}^\ell$ for some $\ell \geq n$, where it initially equals the input $x \in \mathbb{R}^n$ (with possible padding by zeroes, in case $\ell > n$). Each step (gate) is either a *rotation* or a *constant*. A rotation applies a 2-by-2 rotation mapping on a pair of machine state coordinates (rewriting the result of the mapping to the two coordinates). We remind the reader that a 2-by-2 rotation mapping is written in matrix form as $\begin{pmatrix} \cos\theta & \sin\theta \\ -\sin\theta & \cos\theta \end{pmatrix}$ for some real (angle) θ. A constant gate multiplies a single machine state coordinate (rewriting the result) by a nonzero constant. In case the constant equals -1, we call it a reflection gate.

In case $\ell = n$ we say that we are in the in-place model. Any nonsingular linear mapping over $\mathbb{R}^n$ can be decomposed into a sequence of rotation and constant gates in the in-place model, and hence our model is, in a sense, universal. FFT works in the in-place model, using rotations (and possibly reflections) only. A restricted method for dealing with $\ell > n$ was developed in [2], and can be applied here too in a certain sense (see Section 7 for a discussion). We focus in this work on the in-place model only.

Since both rotations and constants apply a linear transformation on the machine state, their composition is a linear transformation. If $\mathcal{A}_n$ is an in-place algorithm for computing a linear mapping over $\mathbb{R}^n$, it is convenient to write it as $\mathcal{A}_n = (M^{(0)} = \mathrm{Id}, M^{(1)}, \ldots, M^{(m)})$ where m is the number of steps (gates), $M^{(t)} \in \mathbb{R}^{n \times n}$ is the mapping that satisfies that for input $x \in \mathbb{R}^n$ (the initial machine state), $M^{(t)}x$ is the machine state after t steps. (Id is the identity matrix). The matrix $M^{(m)}$ is the target transformation, which will typically be F in our setting. In fact, due to the scale invariance of the potential function we use, we could take $M^{(m)}$ to be any nonzero scaling of F, but to reduce notation we simply assume a scaling of 1. For any $t \in [m]$, if the t'th gate is a rotation, then $M^{(t)}$ defers from $M^{(t-1)}$ in at most two rows, and if the t'th gate is a constant, then $M^{(t)}$ defers from $M^{(t-1)}$ in at most one row.

2.1 Numerical Architecture

The in-place model implicitly assumes representation of a vector in $\mathbb{R}^n$ in memory using n words. A typical computer word represents a coordinate (with respect to some fixed orthogonal basis) in the range $[-1, 1]$ to within some accuracy $\varepsilon = \Theta(1)$.[1] For sake of simplicity, ε should be thought of as 2^{-31} or 2^{-63} in modern computers of 32 or 64 bit words, respectively.

To explain the difficulties in speeding up FFT on computers of fixed precision in the in-place model, we need to understand whether (and in what sense) standard FFT is at all suitable on such machines. First, we must restrict the domain of inputs. Clearly this domain cannot be $\mathbb{R}^n$, because computer words can only represent coordinates in the range $[-1, 1]$, by our convention. We consider input from an n-ball of radius $\Theta(\sqrt{n})$, which we denote $\mathcal{B}(\Theta(\sqrt{n}))$. An n-ball is invariant under orthogonal transformations, and is hence a suitable domain. Encoding a single coordinate of such an input might require $\omega(1)$ bits (an overflow). However, using well known tools from high dimensional geometry, encoding a single coordinate of a *typical* input chosen randomly from $\mathcal{B}(\Theta(\sqrt{n}))$ requires $O(1)$ bits, fitting inside a machine word.[2] We hence take a statistical approach and define a state of overflow as trying to encode, in some fixed memory word (coordinate), a random number of $\omega(1)$ bits in expectation, at a fixed time step in the algorithm. This definition allows us to avoid dealing with accommodation of integers requiring super-constant bits and, in turn, with logical bit-operation complexity. Although the definition might seem impractical at first, it allows us to derive very interesting information vs computational speed tradeoffs. (In the future work Section 7 we shall discuss allowing varying word sizes and its implications on complexity.) By our definition, standard FFT for input drawn uniformly from $\mathcal{B}(\Theta(\sqrt{n}))$ does not overflow at all, because any coordinate of the machine state at any step is tightly concentrated (in absolute value) around $\Theta(1)$. It will be easier however to replace the uniform distribution from the ball with the multivariate Gaussian $\mathcal{N}(0, \Theta(n) \cdot \mathrm{Id})$, which is a good approximation of the former for large n. With this assumption, any coordinate of the standard FFT machine state at any step follows the law $\mathcal{N}(0, \Theta(1))$. By simple integration against the Gaussian measure, one can verify that the expected number of bits required to encode such a random variable (to within fixed accuracy ε) is

---

[1] The range $[-1, 1]$ is immaterial and can be replaced with any range of the form $[-a, a]$ for $a > 0$.

[2] By "encoding" here we simply mean the base-2 representation of the integer $\lfloor x(i)/\varepsilon \rfloor$.

$\Theta(1)$, hence no overflow occurs. This input assumption together with the no-overflow guarantee **will serve as our benchmark**. For further discussion on the numerical arhitecture and definition of *overflow* we refer the reader, due to lack of space, to the extended version [3].

3   The Matrix Quasi-Entropy Function

The set $\{1, \ldots, q\}$ is denoted by $[q]$. By $\mathbb{R}^{a \times b}$ we formally denote matrices of a rows and b columns. Matrix transpose is denoted by $(\cdot)^T$. We use $(\cdot)^{-T}$ as shorthand for $((\cdot)^{-1})^T = ((\cdot)^T)^{-1}$. If $A \in \mathbb{R}^{a \times b}$ is a matrix and I is a subset of $[b]$, then (borrowing from Matlab syntax) $A(:, I)$ is the submatrix obtained by stacking the columns corresponding to the indices in I side by side and $A(I, :)$ is the submatrix obtained by stacking the rows corresponding to the indices in I one on top of the other. We shall also write, for $i \in [b]$, $A(:, i)$ and $A(i, :)$ as shorthands for $A(:, \{i\})$ and $A(\{i\}, :)$, respectively. All logarithms are base 2.

We slightly abuse notation and extend the definition of the quasi-entropy function $\Phi(M)$ defined on nonsingular matrices M from [2], as follows. Given two matrix arguments $A, B \in \mathbb{R}^{a \times b}$ for some $a, b \geq 1$, $\Phi(A, B)$ is defined as $\sum_{i=1}^{a} \sum_{j=1}^{b} -A(i, j) B(i, j) \log |A(i, j) B(i, j)|$.

This naturally extends to vectors, namely for $u, v \in \mathbb{R}^a$, $\Phi(u, v)$ is as above by viewing $\mathbb{R}^a$ as $\mathbb{R}^{a \times 1}$. If $A, B \in \mathbb{R}^{a \times b}$ and a, b are even, then we define the *complex quasi-entropy* function $\Phi^{\mathbb{C}}(A, B)$ to be:

$$\sum_{i=1}^{a} \sum_{j=1}^{b/2} \Big(-(A(i, 2j-1)B(i, 2j-1) + A(i, 2j)B(i, 2j)) \times$$

$$\log |A(i, 2j-1)B(i, 2j-1) + A(i, 2j)B(i, 2j)| \Big) .$$

The function $\Phi^{\mathbb{C}}$ can be used for proving our results for the real representation of the complex DFT, which we omit from this manuscript for simplicity. The reason we need this modification to Φ for DFT is explained in the proof of Lemma 61, needed by Theorem 51 below. Elsewhere, we will work (for convenience and brevity) only with Φ. Abusing notation, and following [2], we define for any nonsingular matrix M: $\Phi(M) := \Phi(M, M^{-T}), \Phi^{\mathbb{C}}(M) := \Phi^{\mathbb{C}}(M, M^{-T})$. It is easy to see that $\Phi(F) = n \log n$ for the Walsh-Hadamard transform, because all matrix elements are $\pm 1/\sqrt{n}$. If F is a real representation of the $(n/2)$-DFT, then clearly $\Phi^{\mathbb{C}}(F) = n \log(n/2)$, because all matrix elements of the (complex representation of the) $(n/2)$-DFT are complex unit roots times $(n/2)^{-1/2}$.

It will be also useful to consider a generalization of the potential of a nonsingular matrix M, by allowing linear operators acting on the rows of M and M^{-T}, respectively. More precisely, we will let $\Phi_{P,Q}(M)$ be shorthand for $\Phi(MP, M^{-T}Q)$, where $P, Q \in \mathbb{R}^{n \times a}$ are some mappings. (We will only be working with projection matrices P, Q here). Similarly, $\Phi_{P,Q}^{\mathbb{C}}(M, M^{-T}) := \Phi^{\mathbb{C}}(MP, M^{-T}Q)$.

Further notation: For any matrix $A \in \mathbb{R}^{n \times n}$, let $\sigma_1(A), \ldots, \sigma_n(A)$ denote its singular values, where we use the convention $\sigma_1(A) \geq \cdots \geq \sigma_n(A)$. If A is nonsingular, then the condition number $\kappa(A)$ is defined by $\sigma_1(A)/\sigma_n(A)$. For any matrix A, we let $\|A\|$ denote its spectral norm and $\|A\|_F$ its Frobenius norm. If x is a vector, hence, $\|x\| = \|x\|_2 = \|x\|_F$. Let $\mathcal{B}$ denote the Euclidean unit ball in $\mathbb{R}^n$. For an integer a, let $[a]$ be shorthand for $\{1, \ldots, a\}$.

4   Generalized Ill Conditioned Bottleneck from Speedup

We show that if an in-place algorithm $\mathcal{A}_n = (M^{(0)} = \mathrm{Id}, \ldots, M^{(m)} = F)$ speeds up FFT by a factor of $b \geq 1$, then for some t the matrix $M^{(t)}$ is ill conditioned (in a generalized sense, to be explained). This is a generalization of the main result in [2], with a simpler proof that can be found in the extended version [3], for the sake of completeness.

Theorem 41. *Fix n, and let $\mathcal{A}_n = \{\mathrm{Id} = M^{(0)}, \ldots, M^{(m)}\}$ be an in-place algorithm computing some linear function in $\mathbb{R}^n$ and let $P, Q \in \mathbb{R}^{n \times n}$ be two matrices. For any $t \in [m]$, let $\{i_t, j_t\}$ denote the set of at most two indices that are affected by the t'th gate (if the t'th gate is a constant gate, then $i_t = j_t$, otherwise it's a rotation acting on indices i_t, j_t). Then for any $R \in [\lfloor n/2 \rfloor]$ there exists $t \in [m]$ such that*

$$\left\| (M^{(t)} P)(I_t, :) \right\|_F \left\| ((M^{(t)})^{-T} Q)(I_t, :) \right\|_F \geq \frac{R(\Phi_{P,Q}(M^{(m)}) - \Phi_{P,Q}(\mathrm{Id}))}{m \log 2R}, \tag{4.1}$$

where $I_t = \bigcup_{t'=t}^{t+R-1} \{i_{t'}, j_{t'}\}$. Additionally, if $R = 1$ then the t'th gate can be assumed to be a rotation.

In particular, if $M^{(m)} = F$ and $m = (n \log n)/b$ for some $b \geq 1$ ("$\mathcal{A}_n$ speeds up FFT by a factor of b") and $P = Q = \mathrm{Id}$, then

$$\left\| (M^{(t)})(I_t, :) \right\|_F \left\| ((M^{(t)})^{-T})(I_t, :) \right\|_F \geq \frac{Rb}{\log 2R}. \tag{4.2}$$

For the main result in this paper in the next section, we will only need the case $R = 1$ of the theorem. It is worthwhile, however, to state the case of general $R > 1$ because it gives rise to a stronger notion of ill-condition than is typically used. Since this is not the main focus of this work, we omit the details of this discussion. Henceforth, we will only use the theorem with $R = 1$.

We discuss the implication of the theorem, in case $R = 1, P = Q = \mathrm{Id}$. The theorem implies that an algorithm with $m = (n \log n)/b$ must exhibit an intermediate matrix $M^{(t)}$ and a pair of indices i_t, j_t such that the t'th gate is a rotation acting on i_t, j_t and additionally:

$$\sqrt{\left(\|M^{(t)}(i_t, :)\|^2 + \|M^{(t)}(j_t, :)\|^2 \right) \left(\|(M^{(t)})^{-T}(i_t, :)\|^2 + \|(M^{(t)})^{-T}(j_t, :)\|^2 \right)} \geq b.$$

Hence, either

$$(i) \quad \sqrt{\|M^{(t)}(i_t,:)\|^2 + \|M^{(t)}(j_t,:)\|^2} \geq \sqrt{b} \quad \text{-or-}$$

$$(ii) \quad \sqrt{\|(M^{(t)})^{-T}(i_t,:)\|^2 + \|(M^{(t)})^{-T}(j_t,:)\|^2} \geq \sqrt{b} \ .$$

Case (i). We can assume wlog that

$$\|M^{(t)}(i_t,:)\|^2 \geq b/2 \ . \tag{4.3}$$

Let $x_{\text{over}}^T := M^{(t)}(i_t,:)/\|M^{(t)}(i_t,:)\| \in \mathbb{R}^n$ (x_{over} is the normalized i_t'th row of $M^{(t)}$, transposed). Recall that the input x is distributed according to the law $\mathcal{N}(0, \Theta(1) \cdot \mathrm{Id})$. The i_t'th coordinate just before the t'th gate equals $\|M^{(t)}(i_t,:)\|x^T x_{\text{over}}$, and is hence distributed $\mathcal{N}(0, \Theta(\|M^{(t)}(i_t,:)\|^2))$. Using (4.3), this is $\mathcal{N}(0, \Omega(b))$. If $b = b(n) = \omega(1)$, then by our definition we reach overflow.

It is possible as a preprocessing step to replace x with $x-(x^T x_{\text{over}})x_{\text{over}}$ (eliminating the overflow component), and then to reintroduce the offending component by adding $(x^T x_{\text{over}})F x_{\text{over}}$ as a postprocessing step. In the next section, however, we shall show that, in fact, there must be $\Omega(n)$ pairwise orthonormal directions (in input space) that overflow at $\Omega(n)$ different time steps, so such a simple "hack" cannot work.

Case (ii). This scenario, as the reader guesses, should be called *underflow*. In case (ii), wlog

$$\|(M^{(t)})^{-T}(i_t,:)\|^2 \geq b/2 \ . \tag{4.4}$$

Now define $x_{\text{under}}^T = (M^{(t)})^{-T}(i_t,:)/\|(M^{(t)})^{-T}(i_t,:)\| \in \mathbb{R}^n$, and consider the orthonormal basis $u_1, \ldots u_n \in \mathbb{R}^n$ so that $u_1 = x_{\text{under}}$. For any $t' \in [m]$ (and in particular for $t' = t$):

$$g_1 := x_{\text{under}}^T x = (x_{\text{under}}^T (M^{(t')})^{-1}) \cdot (M^{(t')}x) \ .$$

Now notice that the i_t'th coordinate of $(x_{\text{under}}^T (M^{(t)})^{-1})$ has magnitude at least $\sqrt{b/2}$ by (4.4) and the construction of x_{under}. Also notice that for all $i \neq i_t$, the row $M^{(t)}(i,:)$ is orthogonal to x_{over}, by matrix inverse definition. This means that coordinate $i \neq i_t$ of $M^{(t)}x$ contains no information about g_1. All the information in g_1 is hence contained in $(M^{(t)}x)(i_t)$. More precisely, g_1 is given by $g_1 = ((M^{(t)})^{-T}x_{\text{under}})(i_t) \times (M^{(t)}x)(i_t) - e$, where e is a random variable independent of g_1. But $|((M^{(t)})^{-T}x_{\text{under}})(i_t)| \geq \sqrt{b/2}$, and $(M^{(t)}x)(i_t)$ is known only up to an additive error of ε, due to our assumptions on quantization in the numerical architecture. This means that g_1 can only be known up to an additive error of at least $\varepsilon\sqrt{b/2}$, for *any* value of e. It is important to note that this uncertainty cannot be "recovered" later by the algorithm, because at any step the machine state contains all the information about the input (aside from the input distribution prior). In other words, any information forgotten at any step cannot be later recalled. For an illustration, we refer the reader to the extended version [3].

Notice that at step 0, the input vector coordinates $x(1), \ldots, x(n)$ are repre-sented in individual words, each of which gives rise to an uncertainty interval of width ε. So merely storing the input in memory in the standard coordinate system implies knowing its location up to an uncertainty n-cube with side ε, and of diameter $\varepsilon\sqrt{n}$.[3] An uncertainty interval of size $\varepsilon\sqrt{b/2} = O(\varepsilon\sqrt{\log n})$ in a single direction is therefore relatively benign. The next section tells us, however, that the problem is amplified $\Omega(n)$-fold.

5 Many Independent Ill Conditioned Bottlenecks

Theorem 51. *Fix n, and let $\mathcal{A}_n = \{\mathrm{Id} = M^{(0)}, \ldots, M^{(m)} = F\}$ be an in-place algorithm computing F in time $m = (n \log n)/b$ for some $b \geq 1$. Then one of the following (i)-(ii) must hold:*

(i) *(Severe Overflow) There exists an orthonormal system $v_1, \ldots, v_{n'} \in \mathbb{R}^n$, integers $t_1, \ldots, t_{n'} \in [m]$ and $i_1, \ldots, i_{n'} \in [n]$ with $n' = \Omega(n)$ such that for all $j \in [n']$,*

$$M^{(t_j)}(i_j, :)P_j = \alpha_j v_j \quad \text{with } \alpha_j = \Omega(\sqrt{b}) , \qquad (5.1)$$

where P_j is projection onto the space orthogonal to $v_1, \ldots, v_{j-1}$.

(ii) *(Severe Underflow) There exists an orthonormal system $u_1, \ldots, u_{n'} \in \mathbb{R}^n$, integers $t_1, \ldots, t_{n'} \in [m]$ and $i_1, \ldots, i_{n'} \in [n]$ with $n' = \Omega(n)$ such that for all $j \in [n']$,*

$$(M^{(t_j)})^{-T}(i_j, :)Q_j = \gamma_j u_j \quad \text{with } \gamma_j = \Omega(\sqrt{b}) , \qquad (5.2)$$

where Q_j is projection onto the space orthogonal to $u_1, \ldots, u_{j-1}$.

In both cases (i) and (ii), the gates at time $t_1, \ldots t_{n'}$ are rotations, and for all $j \in [n']$ the index i_j is one of the two indices affected by the corresponding rotation. Additionally, the set $\{t_1, \ldots, t_{n'}\}$ is of cardinality at least $n'/2$.

The proof heavily relies on Lemma 61 (Section 6) and can be found in the extended version [3] due to lack of space. We discuss its numerical implications, continuing the discussion following Theorem 41. In the severe overflow case, Theorem 51 tells us that there exists an orthonormal collection $v_1, \ldots, v_{n'}$ (with $n' = \Omega(n)$) in input space, such that each v_i behaves like x_{over} from the previous section. This means that, if the speedup factor b is $\omega(1)$, we have overflow caused by a linear number of independent input components, occurring at $\Omega(n)$ different time steps (by the last sentence in the theorem). In the extreme case of speedup $b = \Theta(\log n)$ (linear number of gates), this means that in a constant fraction of time steps overflow occurs.

For the severe underflow case we offer a geometric interpretation. The theo-rem tells us that there exists an orthonormal collection $u_1, \ldots, u_{n'}$ in the input

---

[3] To be precise, we must acknowledge the prior distribution on x which also provides information about its whereabouts.

space that is bad in the following sense. For each $j \in [n']$, redefine $g_j = u_j^T x$ to be the input component in direction u_j. Again, the variables $g_1, \ldots, g_{n'}$ are iid $\mathcal{N}(0, \Theta(1))$. The first element in the series, u_1, can be analyzed as x_{under} (from the previous section) whereby it was argued that before the t_1'th step, the component $g_1 = u_1^T x$ can only be known to within an interval of width $\Omega(\gamma_1 \varepsilon)$, independently of information from components orthogonal to u_1. We remind the reader that by this we mean that the *width* of the interval is independent, but the location of the interval depends smoothly (in fact, linearly) on information from orthogonal components of x.

As for $u_2, \ldots, u_{n'}$: For each $j \in [n']$, let $z_j := (M^{(t_j)})^{-T}(i_j, :)$. Therefore $u_1 = z_1 / \|z_1\|$ and by (5.2), for $j > 1$ we can write $z_j = \gamma_j u_j + h_j$, where $h_j \in \text{span}\{u_1, \ldots, u_{j-1}\}$. Treating $z_j / \|z_j\|$ again as x_{under}, we conclude that the component $(z_j / \|z_j\|)^T x$ can only be known to within an interval of size $\Omega(\varepsilon \|z_j\|)$, given any value of the projection of input x onto the space orthogonal to z.

We extend the list of vectors $z_1, \ldots, z_{n'}$, orthonormal vectors $u_1, \ldots, u_{n'}$, numbers $\gamma_1, \ldots, \gamma_{n'}$ and projections $Q_1, \ldots, Q_{n'}$ to size n as follows. Having defined z_j, u_j, Q_j, γ_j for some $j \geq n'$, we inductively define Q_{j+1} as projection onto the space orthogonal to $\text{span}\{z_1, \ldots, z_j\} = \text{span}\{u_1, \ldots, u_j\}$ and z_{j+1} to be a standard basis vector such that $\|Q_{j+1} z_{j+1}\|^2 \geq 1 - j/n$. (Such a vector exists because there must exist an index $i_0 \in [n]$ such that $\sum_{j'=1}^{j} u_{j'}(i_0)^2 \leq j/n$, by orthonormality of the collection $u_1, \ldots, u_j$; Now set z_{j+1} to have a unique 1 at coordinate i_0 and 0 at all other coordinates.) We let u_{j+1} be $Q_{j+1} z_{j+1} / \|Q_{j+1} z_{j+1}\|$, that is, a normalized vector pointing to the component of z_{j+1} that is orthogonal to $\text{span}\{z_1, \ldots, z_j\} = \text{span}\{u_1, \ldots, u_j\}$. The number γ_{j+1} is defined as $\|Q_{j+1} z_{j+1}\|$. By construction, $\gamma_{j+1} \geq \sqrt{1 - j/n}$.

The above extends the partial construction arising from the severe underflow to a full basis, with the following property:

Proposition 52. *For any $j \in [n]$, even given exact knowledge of the exact projection $\tilde{x}$ of x onto the space orthogonal to z_j, the quantity $x^T(z_j / \|z_j\|)$ upon termination of the algorithm can only be known to within an interval of the form $[s, s + \varepsilon \|z_j\|]$ where s depends smoothly (in fact, linearly) on $\tilde{x}$.*

The proposition is simply a repetition of the analysis done for x_{under} in the previous section. For $j > n'$ it is a simple consequence of the fact that upon initialization of the algorithm with input x, each coordinate of x (and in particular $x^T z_j$) is stored in a single machine word, while all other machine words store information independent of $x^T z_j$. Hence the uncertainty of width $\varepsilon \|z_j\| = \varepsilon$.

What do we know about x upon termination of the algorithm? As stated earlier, any information that was lost during execution, cannot be later recovered. Let $\mathcal{I}$ denote the set of possible inputs, given the information we are left with upon termination. Consider the projection Q_2 onto the space orthogonal to $u_1 = z_1 / \|z_1\|$, as a function defined over $\mathcal{I}$. Let $\mathcal{I}_2 = Q_2 \mathcal{I}$ denote its image. The preimage of any point $w \in \mathcal{I}_2$ must contain a line segment of length at least $\varepsilon \gamma_1$ parallel to u_1, due to the uncertainty in $x^T u_1$. Hence the volume of $\mathcal{I}$ is at least $\varepsilon\gamma_1$ times the $(n-1)$-volume of $\mathcal{I}_2$.[4] Continuing inductively, we lower bound the $(n-j+1)$-volume of $\mathcal{I}_j := Q_j\mathcal{I} = Q_j\mathcal{I}_{j-1}$ for $j > 2$. Consider the projection Q_j as a function operating on $\mathcal{I}_{j-1}$, and any point w in the image $\mathcal{I}_j$. By definition of Q_j, there exists $\hat{w} \in \mathcal{I}$ such that $Q_j\hat{w} = w$. By proposition 52, the intersection of the line $\mathcal{L} = \{\hat{w} + \eta z_j : \eta \in \mathbb{R}\}$ with $\mathcal{I}$ must contain a segment Δ of size $\varepsilon\|z_j\|$. The projection $Q_j\Delta$ of this segment is contained in the line $Q_j\mathcal{L} = \{w + \eta u_j : \eta \in \mathbb{R}\}$. The size of the segment is $\varepsilon\|Q_j z_j\| = \varepsilon\gamma_j$. This means that the $(n-j+1)$-volume of $\mathcal{I}_{j+1}$ is at least $\varepsilon\gamma_j$ times the $(n-j)$-volume of $\mathcal{I}_{j+1} = Q_{j+1}\mathcal{I}_j$.

Concluding, we get that the volume of $\mathcal{I}$ is at least $\prod_{j=1}^{n}\gamma_j$. From the construction immediately preceding Proposition 52, we get (using the fact that $n' = \Omega(n)$): $\log\frac{\mathrm{vol}(\mathcal{I})}{\varepsilon^n} \geq n'\log\sqrt{b/2} + \sum_{j=n'+1}^{n}\log\sqrt{1 - \frac{j-1}{n}} = \Omega(n\log b)$. This tells us that the volume of uncertainty in the input (and hence, the output) of a b-speedup of FFT in the in-place model is at least $b^{\Omega(n)}$ times the volume of uncertainty incurred simply by storing the input in memory.

6 Main Technical Lemma

The following is the most important technical lemma in this work. Roughly speaking, it tells us that application of operators that are close to Id to the rows of F and F^{-T} does not reduce the corresponding potential by much. Similarly, assuming that P, Q are positive semidefinite with spectral norm at most 1, applying these transformations to the rows of Id does not increase the corresponding potential by much.

Lemma 61. *Let* $P, Q \in \mathbb{R}^{n\times n}$ *be two matrices. Let* $\hat{P} = \mathrm{Id} - P, \hat{Q} = \mathrm{Id} - Q$. *Then*

$$\Phi(FP, F^{-T}Q) \geq n\log n - (\mathrm{tr}\,\hat{P} + \mathrm{tr}\,\hat{Q})\log n - O\left((\|\hat{P}\|_F^2 + \|\hat{Q}\|_F^2)\log n\right) .$$

If, additionally, P *and* Q *are positive semi-definite contractions, then*

$$\Phi_{P,Q}(\mathrm{Id}) = \Phi(P, Q) \leq \mathrm{tr}\,\hat{P} + \mathrm{tr}\,\hat{Q} + O\left((\|\hat{P}\|_F^2 + \|\hat{Q}\|_F^2)\log n\right) .$$

The proof, available in the extended version [3], takes advantage of the smoothness of the matrices F and Id (that is, almost all matrix elements have exactly the same magnitude). This is the reason we needed to modify Φ and work with $\Phi^{\mathbb{C}}$ for the complex case: If F were the real representation of the $n/2$-DFT matrix, then it is not smooth in this sense. It does hold though that for any $i \in [n]$ and $j \in [n/2]$: $F(i, 2j-1)^2 + F(i, 2j)^2 = 2/n$, so the matrix is smooth only in the sense that all pairs of adjacent elements have the same norm (viewed as $\mathbb{R}^2$ vectors).

---

[4] We need to be precise about measurability, but this is a simple technical point from the fact that the interval endpoint depends smoothly on the projection, as claimed in Proposition 52.

7   Future Work

Taking into account bit operation complexity, and using state-of-the-art integer multiplication algorithms [11,12] it can be quite easily shown that both severe overflow and severe underflow could be resolved by allowing flexible word size, accommodating either large numbers (in the overflow case) or increased accuracy (in the underflow case). In fact, allowing $O(\log b)$-bit words at the time steps at which overflow (or underflow) occur, of which there are $\Omega(n)$ many by Theorem 51, suffice. Hence, this work does not rule out the possibility of (in the extreme case of $b = \Theta(\log n)$) a Fourier transform algorithm in the in-place model using a linear number of gates, in bit operation complexity of $\tilde{\Omega}(n \log \log n)$, where $\tilde{O}()$ here hides $\log \log \log n$ factors arising from fast integer multiplication algorithms. We conjecture that such an algorithm does not actually exist, and leave this as the main open problem.

Another problem that was left out in this work is going beyond the in-place model. In the more general model, the algorithm works in space $\mathbb{R}^\ell$ for $\ell > n$, where the $(\ell - n)$ extra coordinates can be assumed to be initialized with 0, and the first n are initialized with the input $x \in \mathbb{R}^n$. The final matrix $M^{(m)}$ of Fourier transform algorithm $\mathcal{A}_n = \{\text{Id} = M^{(0)}, \ldots, M^{(m)}\}$ contains F as a sub matrix, so that the output Fx can simply be extracted from a subset of n coordinates of $M^{(m)}x$, which can be assumed to be the first. The matrix $M^{(m)}$ (and its inverse-transpose) therefore contains $(\ell - n)$ extra rows. The submatrix defined by the extra rows (namely, the last $\ell - n$) and the first n columns were referred to in [2] as the "garbage" part of the computation. To obtain an $\Omega(n \log n)$ computational lower bound in the model assumed there,[5] it was necessary to show that $\Phi_{P,P}(M^{(m)})P = \Omega(n \log n)$, where $P \in \mathbb{R}^{\ell \times \ell}$ is projection onto the space spanned by the first n standard basis vectors.[6] To that end, it was shown that such a potential lower bound held as long as spectral norm of the "garbage" submatrices was properly upper bounded. That result, in fact, can be deduced as a simple outcome of Lemma 61 that was developed here. What's more interesting is how to generalize Theorem 51 to the non in-place model, and more importantly how to analyze the numerical accuracy implications of overflow and underflow to the non in-place model. Such a generalization is not trivial and is another immediate open problem following this work.

Another interesting possible avenue is to study the complexity of Fourier transform on input x for which some prior knowledge is known. The best example is when Fx is assumed sparse, for which much interesting work on the upper bound side has been recently done by Indyk et al. (see [14] and references therein).

Many algorithms use the Fourier transform as a subroutine. In certain cases (fast polynomial multiplication, fast integer multiplication [11,12], fast Johnson-Lindenstrauss transform for dimensionality reduction [4–6,15] and the related

---

[5] In [2], the model simply assumed that all matrices $M^{(t)}$ for $t - 1 \ldots m$ have bounded condition number. Quantifying the effect of ill condition on numerical stability, overflow and underflow, was not done there.

[6] The function $\Phi_{P,Q}(M)$ was not defined in [2], and was only implicitly used.

restricted isometry property (RIP) matrix construction [7,15,19]) the Fourier transform subroutine is the algorithm's bottleneck. Can we use the techniques developed here to derive lower bounds (or rather, time-accuracy tradeoffs) for those algorithms as well? Moreover, we can ask how the implications of speeding up the Fourier transform subroutine (as derived in this work) affect the numerical outcome of these algorithms, assuming they insist on using Fourier transform as a black box.

References

1. Ailon, N.: A lower bound for Fourier transform computation in a linear model over 2x2 unitary gates using matrix entropy. Chicago J. of Theo. Comp. Sci. (2013)
2. Ailon, N.: An $n \log n$ lower bound for Fourier transform computation in the well conditioned model (2014). arXiv:1403.1307
3. Ailon, N.: Tighter Fourier transform complexity tradeoffs. Technical report (2015). (arxiv:1404:1741)
4. Ailon, N., Chazelle, B.: The fast Johnson-Lindenstrauss transform and approximate nearest neighbors. SIAM J. Comput. 39(1), 302–322 (2009)
5. Ailon, N., Liberty, E.: Fast dimension reduction using rademacher series on dual BCH codes. Discrete & Computational Geometry 42(4), 615–630 (2009)
6. Ailon, N., Liberty, E.: An almost optimal unrestricted fast Johnson-Lindenstrauss transform. ACM Transactions on Algorithms 9(3), 21 (2013)
7. Ailon, N., Rauhut, H.: Fast and rip-optimal transforms. Discrete & Computational Geometry 52(4), 780–798 (2014)
8. Various Authors. List of unsolved problems in computer science. Wikipedia
9. Cooley, J.W., Tukey, J.W.: An algorithm for the machine computation of complex Fourier series. J. of American Math. Soc., 297–301 (1964)
10. Cormen,T.H., Leiserson, C.E., Rivest, R.L., Stein, C.: Introduction to Algorithms, 3rd edition. MIT Press (2009)
11. Anindya, De., Kurur, P.P., Saha, C., Saptharishi, R.: Fast integer multiplication using modular arithmetic. SIAM J. on Comp. 42 (2013)
12. Fürer, M.: Faster integer multiplication. SIAM J. Comp. 39(3), 979–1005 (2009)
13. Harvey, D., van der Hoeven, J., Lecerf, G.: Even faster integer multiplication. Technical report (2014). (arXiv:1407.3360)
14. Indyk, P., Kapralov, M., Price, E.: (Nearly) sample-optimal sparse fourier transform. In: Proceedings of the Twenty-Fifth Annual ACM-SIAM Symposium on Discrete Algorithms, SODA 2014, Portland, Oregon, USA, January 5–7, 2014, pp. 480–499 (2014)
15. Krahmer, F., Ward, R.: New and improved Johnson-Lindenstrauss embeddings via the restricted isometry property. SIAM J. Math. Analysis 43(3), 1269–1281 (2011)
16. Mathieu, M., Henaff, M., LeCun, Y.: Fast training of convolutional networks through FFTs. In: International Conference on Learning Representations (ICLR) (2014)
17. Morgenstern, J.: Note on a lower bound on the linear complexity of the fast Fourier transform. J. ACM 20(2), 305–306 (1973)
18. Papadimitriou, C.H.: Optimality of the fast Fourier transform. J. ACM 26(1), 95–102 (1979)
19. Rudelson, M., Vershynin, R.: Sampling from large matrices: An approach through geometric functional analysis. J. ACM, 54(4) (2007)
20. Winograd, S.: On computing the discrete Fourier transform. Proc. Nat. Assoc. Sci. 73(4), 1005–1006 (1976)

Quantifying Competitiveness in Paging with Locality of Reference

Susanne Albers[✉] and Dario Frascaria

Department of Computer Science, Technische Universität München, München, Germany
{albers,frascari}@informatik.tu-muenchen.de

Abstract. The classical paging problem is to maintain a two-level memory system so that a sequence of requests to memory pages can be served with a small number of faults. Standard competitive analysis gives overly pessimistic results as it ignores the fact that real-world input sequences exhibit locality of reference. In this paper we study the paging problem using an intuitive and simple locality model that records inter-request distances in the input. A characteristic vector $\mathcal{C}$ defines a class of request sequences that satisfy certain properties on these distances. The concept was introduced by Panagiotou and Souza [19].

As a main contribution we develop new and improved bounds on the performance of important paging algorithms. A strength and novelty of the results is that they express algorithm performance in terms of locality parameters. In a first step we develop a new lower bound on the number of page faults incurred by an optimal offline algorithm OPT. The bound is tight up to a small additive constant. Based on these expressions for OPT's cost, we obtain nearly tight upper and lower bounds on LRU's competitiveness, given any characteristic vector $\mathcal{C}$. The resulting ratios range between 1 and k, depending on $\mathcal{C}$. Furthermore, we compare LRU to FIFO and FWF. For the first time we show bounds that quantify the difference between LRU's performance and that of the other two strategies. The results imply that LRU is strictly superior on inputs with a high degree of locality of reference. In particular, there exist general input families for which LRU achieves constant competitive ratios whereas the guarantees of FIFO and FWF tend to k, the size of the fast memory. Finally, we report on an experimental study that demonstrates that our theoretical bounds are very close to the experimentally observed ones. Hence we believe that our contributions bring competitive paging again closer to practice.

1 Introduction

Paging is a fundamental resource management problem in computer science. In algorithms research it has been studied extensively ever since Sleator and Tarjan

Susanne Albers Work supported by the German Research Foundation, grant Al 464/7-1.

M.M. Halldórsson et al. (Eds.): ICALP 2015, Part I, LNCS 9134, pp. 26–38, 2015.
DOI: 10.1007/978-3-662-47672-7_3

published their seminal paper [20] on the competitive analysis of algorithms. In the *paging problem* we are given a two-level memory system consisting of a small fast memory and a large slow memory. At any time up to k pages, for some $k \in \mathbb{N}$, can reside in fast memory. A paging algorithm ALG is presented with a request sequence $\sigma = \sigma(1), \ldots, \sigma(m)$, where each request $\sigma(t)$ specifies a memory page. If the referenced page is in fast memory, a *memory hit* occurs. Otherwise $\sigma(t)$ is a *page fault* and the missing page must be loaded from slow memory into fast memory. If the fast memory is full, ALG must evict a page from fast memory; in the *online* setting this decision must be made without knowledge of any future requests. The goal is to serve σ so as to minimize the total number of faults.

For an online algorithm ALG and a request sequence σ, let ALG(σ) denote the number of page faults incurred. Let OPT(σ) be the number of faults generated by an optimal offline algorithm OPT. Strategy ALG is *c-competitive* if, for every σ, ALG(σ) is at most c times OPT(σ). The optimal competitive ratio achieved by deterministic online algorithms is equal to k [20]. Classical algorithms such as LRU (Least-Recently-Used), FIFO (First-In First-out) and FWF (Flush-When-Full) are all k-competitive.

It was soon observed that the competitiveness of k is overly pessimistic. In practice algorithms such as LRU and FIFO attain constant performance ratios in the range $[1.5, 4]$, see also [21]. Furthermore, LRU outperforms FIFO, which does not show in competitive analysis. The deficiency of the competitive measure is that it considers arbitrary request sequences whereas input sequences generated by real programs have a special structure. They exhibit *locality of reference*, i.e. whenever a page is requested it is likely to be referenced again in the near future. In a cornerstone paper Borodin et al. [9] initiated the investigation of paging with locality of reference. Over the years various frameworks modeling locality of reference have been proposed. Moreover, new and alternative performance measures have been introduced. In this paper we revisit paging with locality of reference, considering again the competitive performance measure. Compared to previous studies we present for the first time strong guarantees that quantify competitiveness in terms of locality parameters of the input. We analyze individual algorithms and relate pairs of strategies.

Input Model: We use a model for locality of reference introduced by Panagiotou and Souza [19]. The framework is simple, yet captures the essentials of locality of reference: Whenever a page is requested, it is likely to be re-accessed soon. Hence locality can appropriately be modeled by inter-request distances. Specifically, feasible input is defined by a *characteristic vector* $\mathcal{C} = (c_0, \ldots, c_{p-1})$, where p denotes the total number of distinct pages referenced. Again, let σ be a request sequence and $\sigma(t)$ be the request at time t. We refer to $\sigma(t)$ as a *distance-l request*, where $0 \leq l \leq p - 1$, if the following two conditions hold. (1) The page x referenced by $\sigma(t)$ has been requested before in σ and its most recent request was $\sigma(t')$. (2) The number of distinct pages requested between $\sigma(t')$ and $\sigma(t)$ is equal to l, i.e. $|\{\sigma(t'+1), \ldots, \sigma(t-1)\}| = l$. In a request sequence σ characterized by $\mathcal{C} = (c_0, \ldots, c_{p-1})$, there are exactly c_l distance-l requests, for $l = 0, \ldots, p-1$. The total number of requests in σ is $p + \sum_{l=0}^{p-1} c_l$. Given any $\mathcal{C}$, the competitive ratio of an algorithm ALG is defined as $R_{\mathrm{ALG}}(\mathcal{C}) = \max_\sigma \mathrm{ALG}(\sigma)/\mathrm{OPT}(\sigma)$, where the maximum ranges over all request sequences characterized by $\mathcal{C}$. As this set of sequences it finite, the maximum is well-defined.

Previous Work: There exists a considerable body of literature on paging with locality of reference. Due to the wealth of results we can only present a selection. A good survey article is [11]. In their initial paper [9] Borodin et al. introduced *access graphs* G, representing the execution of programs, to model locality of reference. The vertices of G correspond to the memory pages. Page x may be requested after y if they are adjacent in G. Borodin et al. showed that, for any G, the competitiveness $R_{\mathrm{LRU}}(G)$ of LRU depends on the number of articulation nodes whose removal separates G. They also developed an algorithm that achieves the best possible competitive ratio attainable for any given G, up to a constant factor [9,18]. Chrobak and Noga [12] proved that LRU is always at least as good as FIFO, i.e. for any G, $R_{\mathrm{LRU}}(G) \le R_{\mathrm{FIFO}}(G)$.

Articles [4,16,17] make probabilistic assumptions about the input. A diffuse adversary [17] generates a request sequence according to a probability distribution that belongs to a known family of distributions. In Markov paging [16] the input is generated by a Markov chain. Algorithms are evaluated in terms of the page fault rate. In [1] concave functions, modeling the working set sizes of programs, restrict the allowed input. Again page fault rates are evaluated.

Especially in recent years various alternative performance measures, in addition to the well-known page fault rate, have been proposed. These include (a) the max/max ratio [5], (b) bijective and average analysis [2,3], (c) the relative worst-order ratio [6,7], (e) relative interval analysis [8,13] and (e) parametrized analysis [10]. In a bijective analysis two algorithms ALG1 and ALG2 are compared on permutations of the same requests. Let $\mathcal{I}_n$ denote the request sequences of length n. ALG1 is *no worse* than ALG2, in signs ALG1 $\preceq$ ALG2, if for all $n \ge n_0$ there is a bijection $b : \mathcal{I}_n \to \mathcal{I}_n$ such that ALG1$(\sigma) \le$ ALG2$(b(\sigma))$ for all $\sigma \in \mathcal{I}_n$. In this setting LRU is no worse than any other online algorithm ALG assuming that locality is modeled by a concave function [3]. However, LRU $\preceq$ FIFO and FIFO $\preceq$ LRU, see [2], so that there is no strict separation between LRU and FIFO under bijective analysis.

The concept of characteristic vectors was defined by Panagiotou and Souza [19]. As a main result they lower bound the number of page faults incurred by OPT on a request sequence σ characterized by $\mathcal{C} = (c_0, \ldots, c_{p-1})$, i.e.

$$\mathrm{OPT}(\sigma) \ge \tfrac{1}{1 + \frac{k-1}{k} - \frac{k-1}{p-1}} \sum_{l=k}^{p-1} \tfrac{l-k+1}{l} c_l. \tag{1}$$

They define an (α, β)-adversary that chooses vectors $\mathcal{C}$ satisfying $\sum_{l=k}^{\alpha k - 1} c_l \le \beta \sum_{l=\alpha k}^{p-1} c_l$. Against this adversary LRU achieves a competitive ratio of $2(1 + \beta)\alpha/(\alpha - 1)$.

Our Contribution: We investigate paging using classical competitive analysis and adopt the concept of characteristic vectors $\mathcal{C} = (c_0, \ldots, c_{p-1})$ to model locality of reference. It is intuitive to represent input characteristics by a fingerprint of the inter-request distances: If a request sequence exhibits a high degree of locality, then a large majority of the requests are distance-l requests, for small l so that the corresponding vector entries c_l take large values. Given a real-world trace, the underlying $\mathcal{C}$ can be extracted easily by a single scan over the data.

We present new and significantly improved bounds on the performance of the most important paging strategies. A particular strength and novelty of the results is that they quantify algorithm performance in terms of locality parameters. Furthermore, the bounds very accurately predict the performance observed in practice. This finding results from an experimental study we conducted with traces from a benchmark library. These tests confirm the value of our theoretical bounds.

In Section 2, given any characteristic vector $\mathcal{C}$, we develop a new lower bound on the number of page faults incurred by OPT to serve any request sequence σ characterized by $\mathcal{C}$. Technically, the analysis relies on a new approach that relates the number of page faults to the number of memory hits and amortizes the values appropriately. Specifically, we show that

$$\text{OPT}(\sigma) \geq \max\left\{ p, k + \sum_{l=k}^{\lambda-1} c_l \frac{l-k+1}{k-1} + c_\lambda^* \frac{\lambda-k+1}{k-1} \right\}. \tag{2}$$

Here λ and c_λ^* are solutions of an equation that matches faults and hits, assuming that page faults preferably occur on long-distance requests. We prove that our lower bound is tight, up to an additive term of at most $2(\lambda - k + 1)$, which in turn is upper bounded by $2(p - k)$. More precisely, we construct an input sequence that OPT can serve with the stated number of faults. The construction and cost analysis of the sequence are involved. Additionally, we show that our lower bound (2) is always greater than that given in (1). In the experiments (2) significantly outperforms (1).

In Section 3 we evaluate the competitiveness of LRU. Given the analysis of $\text{OPT}(\sigma)$, we derive nearly tight upper and lower bounds on $R_{\text{LRU}}(\mathcal{C})$, for any $\mathcal{C}$. The resulting ratios range between 1 and k, depending on $\mathcal{C}$. In the experiments it shows that these refined ratios are very close to LRU's experimentally observed competitiveness. For all the traces and all values of k, the theoretical bounds are at most 2.5 times the experimentally observed performance. In most cases the gap is much smaller. To the best of our knowledge this is the first time that theoretical performance guarantees for paging match the experimental ones up to a constant factor, independently of k. We remark that our theoretical guarantees cannot exactly match the experimental ones because $R_{\text{LRU}}(\mathcal{C}) = \max_\sigma \text{LRU}(\sigma)/\text{OPT}(\sigma)$ is still a worst-case ratio. A real-world trace, in general, is not a worst-case input for the underlying $\mathcal{C}$.

In Section 4 we show that LRU is superior to other popular paging strategies. We focus on a comparison with FIFO and FWF, which have received considerable attention in the memory management literature. We first prove that LRU is always at least as good as the other two strategies, i.e. $R_{\text{LRU}}(\mathcal{C}) \leq R_{\text{FIFO}}(\mathcal{C})$ and $R_{\text{LRU}}(\mathcal{C}) \leq R_{\text{FWF}}(\mathcal{C})$ for any $\mathcal{C}$. This is not surprising; similar relations have been shown in other frameworks as well. In this paper we go one step further and quantify the performance difference between LRU and FIFO, respectively FWF. We make use of the fact that LRU's competitiveness can be expressed as

$R_{\mathrm{LRU}}(\mathcal{C}) = \mathrm{LRU}(\mathcal{C})/\mathrm{OPT}(\mathcal{C})$, where $\mathrm{OPT}(\mathcal{C})$ denotes the minimum number of page faults required to serve any request sequence defined by $\mathcal{C}$ and $\mathrm{LRU}(\mathcal{C})$ is LRU's fixed cost for every input specified by $\mathcal{C}$. We prove that

$$R_{\mathrm{FIFO}}(\mathcal{C}) \geq \frac{\mathrm{LRU}(\mathcal{C}) + c(k-1)}{\mathrm{OPT}(\mathcal{C}) + c(1 - 1/k) + 1},$$

where c depends on the vector entries c_l, $1 \leq l \leq p-1$. If the number of distance-l requests with $l \geq k$ is not too small, then FIFO's competitiveness tends to k as the locality in the input (captured by entries c_l, for small l) increases. In particular, there exist input classes $\mathcal{C}$ for which LRU's competitiveness is constant while that of FIFO is close to k. The same results hold for FWF, except that slightly "weaker" assumptions on the input are made.

Notation and Conventions: Throughout this paper we assume that the initial fast memory is empty. Furthermore we assume $p > k$ since otherwise a request sequence can be served without any faults. Moreover let $k \geq 2$. When constructing and analyzing a request sequence, a page is called *new* if it has not been referenced so far.

2   Analysis of OPT

Let $\mathcal{C} = (c_0, \ldots, c_{p-1})$ be an arbitrary characteristic vector. First we develop a lower bound on $\mathrm{OPT}(\sigma)$, for any σ defined by $\mathcal{C}$. Then we prove that our bound is nearly tight.

2.1   A lower Bound

Given any σ, let f_l denote the total number of page faults incurred by OPT on distance-l requests, $0 \leq l \leq p - 1$, and let $h_l = c_l - f_l$ be the number of hits on this type of requests. We relate the total number of faults to the number of hits.

Lemma 1. *Let σ be any request sequence characterized by $\mathcal{C}$. There holds a)* $\mathrm{OPT}(\sigma) = p + \sum_{l=k}^{p-1} f_l$ *and b)* $p + \sum_{l=k}^{p-1} f_l \geq k + \sum_{l=k}^{p-1} h_l \frac{l-k+1}{k-1}$.

Proof. We first prove part a). There holds $\mathrm{OPT}(\sigma) = p + \sum_{l=0}^{p-1} f_l$ because OPT incurs one page fault whenever any of the p distinct pages is requested for the first time. Moreover, by the definition of f_l, OPT has exactly f_l faults on the distance-l requests, for $l = 0, \ldots, p - 1$. It remains to argue that $f_l = 0$, for $l = 0, \ldots, k - 1$. Obviously, $f_0 = 0$. So assume $l \geq 1$. Consider a distance-l request $\sigma(t) = x$ and let $\sigma(t')$, where $t' < t$, be the most recent request when page x was referenced in σ. Immediately after OPT has served $\sigma(t')$, page x is in fast memory. Whenever OPT incurs a fault on a request $\sigma(s)$, $t' < s < t$, the set $\{\sigma(s), \ldots, \sigma(t)\}$ of pages referenced until and including $\sigma(t)$ contains at most $l + 1 \leq k$ pages. Hence the set contains at most $k - 1$ pages different from $y = \sigma(s)$. When serving $\sigma(s)$, OPT evicts a page whose next request is farthest in the future. Thus it drops a page not referenced by $\sigma(s+1), \ldots, \sigma(t)$.

We next prove part b). To this end we assign tokens to page faults whenever OPT has a hit on a distance-l request, where $l \geq k$, in σ. Let $\sigma(t) = x$ be such a request and let $\sigma(t')$ be the most recent request to x. A total of l distinct pages are referenced in the subsequence $\sigma(t' + 1), \ldots, \sigma(t - 1)$. Since $l \geq k$ OPT incurs at least $l - (k - 1)$ page faults in this subsequence because $\sigma(t)$ is a hit. Now we select the last $l - (k - 1)$ page faults occurring before $\sigma(t)$ and assign a token to each of these faults. By this process, exactly $\sum_{l=k}^{p-1} h_l(l - k + 1)$ tokens are placed.

In the following we upper bound the number of tokens a page fault may be assigned. A page fault on $\sigma(s)$ can receive a token whenever there is a hit on a request $\sigma(t)$ with $s < t$, page $x = \sigma(t)$ is not referenced in $\sigma(s+1), \ldots, \sigma(t-1)$ and x is in fast memory immediately after $\sigma(s)$ is served. By the latter property, there can be a most $k - 1$ such pages and hence $\sigma(t)$ is assigned not more than $k - 1$ tokens.

We next argue that the first k page faults in σ do not receive any token. Let $\sigma(t_1), \ldots, \sigma(t_k)$ be the requests where these first k page faults occur. Recall that the initial fast memory is empty. Hence $\sigma(t_1) = 1$, the k pages referenced by $\sigma(t_1), \ldots, \sigma(t_k)$ are pairwise distinct and the subsequence $\sigma(1), \ldots, \sigma(t_k)$ only contains requests to these pages. Furthermore, the first hit on a distance-l request with $l \geq k$ occurs after $\sigma(t_k)$. Let $\sigma(t)$, $t > t_k$, be such a hit and assume that the referenced page $x = \sigma(t)$ was requested most recently by $\sigma(t')$, where $t' < t_k$, so that any of the faults $\sigma(t_1), \ldots, \sigma(t_k)$ could potentially be assigned a token. The subsequence $\sigma(t' + 1), \ldots, \sigma(t - 1)$ contains l pages, at least $l - (k - 1)$ of which are different from those referenced by $\sigma(t_1), \ldots, \sigma(t_k)$. These pages different from $\sigma(t_1), \ldots, \sigma(t_k)$ are referenced after $\sigma(t_k)$ and the first request to each of these pages is a fault since, again, the initial fast memory is empty. Our token assignment scheme places $l - (k - 1)$ tokens on the last $l - (k - 1)$ page faults prior to $\sigma(t)$. Hence faults $\sigma(t_1), \ldots, \sigma(t_k)$ do not receive any token.

We conclude that the total number of tokens is upper bounded by $(p-k)(k-1) + \sum_{l=k}^{p-1} f_l(k - 1)$, i.e. $\sum_{l=k}^{p-1} h_l(l - k + 1) \leq (p - k)(k - 1) + \sum_{l=k}^{p-1} f_l(k - 1)$. Dividing the last inequality by $k - 1$ and adding k we obtain, as desired, $k + \sum_{l=k}^{p-1} h_l \frac{l-k+1}{k-1} \leq p + \sum_{l=k}^{p-1} f_l$. $\square$

For the further analysis, given a vector $\mathcal{C} = (c_0, \ldots, c_{p-1})$, we define two functions f and g as well as values λ and c_λ^*. For any integer j with $k \leq j \leq p-1$ and any real number γ with $0 \leq \gamma \leq c_j$, let

$$f(j, \gamma) = k + \sum_{l=k}^{j-1} c_l \frac{l-k+1}{k-1} + \gamma \frac{j-k+1}{k-1} \quad \text{and} \quad g(j, \gamma) = p + (c_j - \gamma) + \sum_{l=j+1}^{p-1} c_l.$$

Intuitively, $f(i, \gamma)$ is the number of page faults, as implied by Lemma 1, if memory hits occur on all the distance-l requests, for $l = k, \ldots, j - 1$, and γ distance-j requests. The corresponding $g(j, \gamma)$ is the number of requests where these faults can occur. If $f(p - 1, c_{p-1}) \leq g(p - 1, c_{p-1})$, then let $\lambda = p - 1$ and $c_\lambda^* = c_{p-1}$. Otherwise determine the largest λ and corresponding c_λ^* such that $f(\lambda, c_\lambda^*) = g(\lambda, c_\lambda^*)$. Note that in either case $f(\lambda, c_\lambda^*) \leq g(\lambda, c_\lambda^*)$. The following technical Lemma 2 is proven in the full paper. Part b) will be important in the sequel. Given any parameter pair j', γ' with $f(j',\gamma') \leq g(j',\gamma')$, it relates the function values to those of the pair λ, c_λ^*.

Lemma 2. *a) The values λ and c_λ^* are well-defined.*
b) Let j' and γ' be a pair such that $f(j',\gamma') \leq g(j',\gamma')$. Then $f(j',\gamma') \leq f(\lambda, c_\lambda^*) \leq g(\lambda, c_\lambda^*) \leq g(j',\gamma')$. Moreover, $j' \leq \lambda$. If $j' = \lambda$, then $\gamma' \leq c_\lambda^*$.

Theorem 1. *Let σ be any request sequence characterized by $\mathcal{C}$. There holds*

$$\mathrm{OPT}(\sigma) \geq \max\left\{p, k + \sum_{l=k}^{\lambda-1} c_l \frac{l-k+1}{k-1} + c_\lambda^* \frac{\lambda-k+1}{k-1}\right\}.$$

Proof. Obviously, $\mathrm{OPT}(\sigma) \geq p$. By Lemma 1

$$\mathrm{OPT}(\sigma) = p + \sum_{l=k}^{p-1} f_l \geq k + \sum_{l=k}^{p-1} h_l \frac{l-k+1}{k-1}. \tag{3}$$

Determine the largest j', where $k \leq j' \leq p-1$, and corresponding γ', where $0 \leq \gamma' \leq c_{j'}$ such that $\sum_{l=k}^{p-1} h_l = \sum_{l=k}^{j'-1} c_l + \gamma'$. Intuitively, we express the total number of hits in terms of a prefix of the c_l-values, for increasing $l \geq k$. Of course, the hits do not necessarily occur on all the distance-l requests, where $l \leq j'$. Taking into account that $f_l = c_l - h_l$, for any l, we obtain

$$\mathrm{OPT}(\sigma) = p + \sum_{l=k}^{p-1} c_l - \sum_{l=k}^{p-1} h_l = p + (c_{j'} - \gamma') + \sum_{l=j'+1}^{p-1} c_l = g(j',\gamma'). \tag{4}$$

In (3) expression $k + \sum_{l=k}^{p-1} h_l \frac{l-k+1}{k-1}$ is minimized if the hits occur on distance-l requests with smallest possible l subject to the constraint that at most c_l distance-l requests occur in σ. Hence

$$k + \sum_{l=k}^{p-1} h_l \frac{l-k+1}{k-1} \geq k + \sum_{l=k}^{j'-1} c_l \frac{l-k+1}{k-1} + \gamma' \frac{j'-k+1}{k-1} = f(j',\gamma').$$

Combining (3) and (4) together with the last inequality we obtain $\mathrm{OPT}(\sigma) = g(j',\gamma') \geq k + \sum_{l=k}^{p-1} h_l \frac{l-k+1}{k-1} \geq f(j',\gamma')$. Using Lemma 2, part b), we conclude $\mathrm{OPT}(\sigma) \geq f(\lambda, c_\lambda^*) = k + \sum_{l=k}^{\lambda-1} c_l \frac{l-k+1}{k-1} + c_\lambda^* \frac{\lambda-k+1}{k-1}$. $\qquad\square$

Proposition 1. *The lower bound on $\mathrm{OPT}(\sigma)$ stated in Theorem 1 is always greater than that in inequality (1).*

The proof is given in the full version of the paper.

2.2   Tightness of the Lower Bound

The lower bound of Theorem 1 is essentially best possible. We develop a strategy that, given an arbitrary $\mathcal{C} = (c_0, \ldots, c_{p-1})$, constructs a request sequence that can be served with the stated number of page faults, up to an additive constant of $2(\lambda - k + 1)$. The strategy is called *GenerateRequestSequence*, or *GRS* for short. It takes the original $\mathcal{C}$ and in a general step issues a distance-l request, $0 \leq l \leq p-1$, according to a specific protocol. The corresponding value c_l is reduced by 1. The process stops when all vector entries c_l, $0 \leq l \leq p-1$, are equal to 0 and all the p distinct pages have been requested. Due to space limitations a detailed presentation of GRS (along with pseudo-code) is given in the full version of the paper. In the following we just give a sketch of the strategy.

High-level description of GRS: First, starting with an empty fast memory, GRS requests k new pages. Then GRS generates a sequence of phases in which requests to new pages or distance-l requests with $k \leq l \leq p - 1$ are issued. The goal is to reduce the vector entries $c_k, \ldots, c_{p-1}$ to 0 while generating subsequences of requests that can be served with low cost. Each phase, except for possibly the last one, consists of exactly l^* requests, for some properly chosen l^* that depends of the state of C at the beginning of the phase. Such a phase is *complete*. The last phase may contain fewer requests.

The Phases: Each phase with l^* requests, for the calculated value l^*, contains $l^* - k + 1$ so called *long distance requests* followed by $k - 1$ *short distance requests*. When generating a long distance request, GRS either requests a new page or issues a distance-l request, for the largest index $l \geq k$ such that $c_l > 0$. In a short distance request GRS poses a distance-l request, for the smallest possible $l \geq k$ such that $c_l > 0$. We will prove that each phase can be served so that page faults occur only on the long distance requests.

Phase Lengths: An important component of GRS is the choice of l^*, for each phase. Loosely speaking, l^* is the smallest j such that (a) $\sum_{l=k}^{j} c_l \geq k - 1$ and (b) $\sum_{l=k}^{p-1} c_l \geq j$, provided that such a value exists. Condition (a) ensures that $k - 1$ short distance requests can be issued. Condition (b) guarantees that a complete phase can be generated.

The analysis of the request sequence constructed by GRS is involved. Again, a complete analysis is contained in the full paper. We state the main result.

Theorem 2. *Let σ be the request sequence generated by GRS. There holds*

$$\text{OPT}(\sigma) \leq k + \sum_{l=k}^{\lambda-1} c_l \frac{l-k+1}{k-1} + c_\lambda^* \frac{\lambda-k+1}{k-1} + 2(\lambda - k + 1).$$

3   The Competitiveness of LRU

We present upper and lower bounds on the competitive ratio $R_{\text{LRU}}(C)$, for any C. While the bounds involve a number of terms, we stress that they are nearly tight, up to an additive constant of $2(\lambda - k + 1)$ in the denominator of the ratios. Of course, one could simplify the expressions at the expense of weakening the bounds. In the experiments the gap between the upper and the lower bounds is extremely small. After stating the corollary we show that our expressions for $R_{\text{LRU}}(C)$ range between 1 and k.

Corollary 1. *Let $C = (c_0, \ldots, c_{p-1})$ be an arbitrary characteristic vector. Then*

$$R_{\text{LRU}}(C) \leq \frac{p + \sum_{l=k}^{p-1} c_l}{\max\left\{p, k + \sum_{l=k}^{\lambda-1} c_l \frac{l-k+1}{k-1} + c_\lambda^* \frac{\lambda-k+1}{k-1}\right\}} \tag{5}$$

and

$$R_{\text{LRU}}(\mathcal{C}) \geq \frac{p + \sum_{l=k}^{p-1} c_l}{k + \sum_{l=k}^{\lambda-1} c_l \frac{l-k+1}{k-1} + c_\lambda^* \frac{\lambda-k+1}{k-1} + 2(\lambda - k + 1)}.$$

Proof. For any σ, $\text{LRU}(\sigma) = p + \sum_{l=k}^{p-1} c_l$. This holds true because LRU never incurs a page fault on a distance-l request with $0 \leq l \leq k - 1$ as the referenced page is still in fast memory. Moreover, LRU has a fault on every distance-l request with $k \leq l \leq p - 1$ since the accessed page has been evicted from fast memory since its last reference. The corollary then follows from Theorems 1 and 2. □

We argue that the upper bound in (5) can be constant, and as low as 1, in particular when given vectors $\mathcal{C}$ modeling request sequences with a high degree of locality of reference. First consider the very simple case that $\mathcal{C} = (c_0, \ldots, c_{k-1}, 0, \ldots, 0)$. The ratio in (5) is equal to 1. A more interesting case is the scenario in which $\mathcal{C}$ has a small number of positive entries c_l with $l \geq k$. In the benchmark library we used there exist traces with this property. In order to keep the calculations simple we assume that there is a single positive entry c_l with $l \geq k$. W.l.o.g. $c_{p-1} > 0$, i.e. $\mathcal{C} = (c_0, \ldots, c_{k-1}, 0, \ldots, 0, c_{p-1})$. If $c_{p-1} \leq p$, then the ratio in (5) is upper bounded by 2. So assume $c_{p-1} > p$. In this case $f(p-1, c_{p-1}) > g(p-1, c_{p-1})$. Hence $\lambda = p - 1$ and c_λ^* satisfies $k + c_\lambda^*(p-k)/(k-1) = p + c_{p-1} - c_\lambda^*$. Solving the last equation for c_λ^*, we obtain that the ratio in (5) is upper bounded by $(p+c_{p-1})/(k+c_{p-1}(p-k)/(p-1))$. For increasing c_{p-1} the last ratio approaches $\frac{p-1}{p-k}$. If $p = k + 1$, then the latter expression is equal to k, which is consistent with the fact that LRU is k-competitive on sequences in which a total of $k + 1$ distinct pages are referenced. If $p = rk$, for some constant $r > 1$, then $\frac{p-1}{p-k}$ is smaller than $\frac{r}{r-1}$, i.e. we obtain constant competitive ratios if r is not too close to 1.

Finally, we check that the upper bound for $R_{\text{LRU}}(\mathcal{C})$ is at most k. First assume that, for the given $\mathcal{C}$, there holds $f(p-1, c_{p-1}) \leq g(p-1, c_{p-1})$. In this case $k + \sum_{l=k}^{p-1} c_l \frac{l-k+1}{k-1} \leq p$, which implies $\sum_{l=k}^{p-1} c_l \leq (k-1)p$. Thus the numerator in (5) is upper bounded by kp. On the other hand, if $f(p-1, c_{p-1}) > g(p-1, c_{p-1})$, then $f(\lambda, c_\lambda^*) = g(\lambda, c_\lambda^*)$. In this case the numerator $p + \sum_{l=k}^{p-1} c_l$ in (5) is

$$p + \sum_{l=k}^{\lambda-1} c_l + c_\lambda^* + (c_\lambda - c_\lambda^*) + \sum_{l=\lambda+1}^{p-1} c_l = \sum_{l=k}^{\lambda-1} c_l + c_\lambda^* + g(\lambda, c_{\lambda^*})$$

$$= \sum_{l=k}^{\lambda-1} c_l + c_\lambda^* + f(\lambda, c_\lambda^*) = k + \sum_{l=k}^{\lambda-1} c_l \frac{l}{k-1} + c_\lambda^* \frac{\lambda}{k-1},$$

which is at most k times the denominator in (5) as $\frac{l}{l-k+1} \leq k$, for any $l \geq k$.

4   Separating LRU from FIFO and FWF

We compare LRU to FIFO and FWF and start with a comparision to FIFO. The detailed proofs of all the results of this section are presented in the full paper.

Theorem 3. *For any $\mathcal{C}$, there holds $R_{\text{FIFO}}(\mathcal{C}) \geq R_{\text{LRU}}(\mathcal{C})$.*

The next Theorem 4 sharply separates LRU from FIFO. Observe that, for any $\mathcal{C}$, the competitiveness of LRU can be expressed as $R_{\text{LRU}}(\mathcal{C}) = \text{LRU}(\mathcal{C})/\text{OPT}(\mathcal{C})$, where $\text{LRU}(\mathcal{C}) = p + \sum_{l=k}^{p-1} c_l$ is the number of faults incurred by LRU on every input characterized by $\mathcal{C}$ and $\text{OPT}(\mathcal{C})$ denotes the minimum number of page faults required to serve any request sequence defined by $\mathcal{C}$. We use this notation in the following.

Theorem 4 presents a lower bound on $R_{\text{FIFO}}(\mathcal{C})$, for any $\mathcal{C}$, given $R_{\text{LRU}}(\mathcal{C}) = \text{LRU}(\mathcal{C})/\text{OPT}(\mathcal{C})$. In that lower bound c depends on the minimum c_l, where $1 \leq l \leq k-1$, and roughly $\sum_{l=k}^{p-1} c_l$. For increasing c, the competitiveness of FIFO can be made arbitrarily close to $(k-1)/(1-1/k) = k$. In Section 3 we analyzed vectors $\mathcal{C} = (c_0, \ldots, c_{k-1}, 0, \ldots, 0, c_{p-1})$ and showed that LRU's competitiveness is constant, for sufficiently large c_{p-1}, provided that p is not to close to k. Hence, for large $c_1, \ldots, c_{k-1}$ and c_{p-1}, the competitiveness of LRU is a small constant while that of FIFO is close to k. We remark that c cannot be larger than $\text{LRU}(\mathcal{C})$ but this is sufficient to establish a lower bound of at least $k/2$ on FIFO's competitiveness.

Theorem 4. *Let* $\mathcal{C} = (c_0, \ldots, c_{p-1})$ *be any vector. Let* $c_{\min} = \min_{1 \leq l \leq k-1} c_l$ *and* $c = \min\{\lfloor c_{\min}/2 \rfloor, p - k + \sum_{l=k}^{p-1} c_l\}$. *Then*

$$R_{\text{FIFO}}(\mathcal{C}) \geq \frac{\text{LRU}(\mathcal{C}) + c(k-1)}{\text{OPT}(\mathcal{C}) + c(1 - 1/k) + 1}.$$

Since Theorem 4 is a main result of this paper we give a proof sketch here. Again, a full analysis is provided in the full paper.

First, the proof relies on a characterization of request sequences that OPT can serve with low cost. For any $\mathcal{C}$, consider the sequences defined by $\mathcal{C}$ for which OPT incurs the smallest number $\text{OPT}(\mathcal{C})$ of page faults. We prove that, among these sequences, there exists one in which all distance-l requests with $0 \leq l \leq k-1$ occur at the end of the sequence. In fact the proofs of all the results presented in this section rely on this property.

The proof of Theorem 4 then starts out with a sequence σ^* just described: OPT incurs $\text{OPT}(\mathcal{C})$ faults; all distance-l requests with $0 \leq l \leq k-1$ occur at the end. Given σ^*, a nemesis sequence for FIFO is constructed in three steps. (1) We remove all distance-l requests with $0 \leq l \leq k-1$ from σ^*. From this truncated sequence we further remove the last c requests. (2) To the sequence obtained in step (1) we append c phases $P(1), \ldots, P(c)$. Any $P(i)$ consists of two parts. In the first part, for increasing $l = 1, \ldots, k-1$, a distance-l request is issued. The second part of the phase starts with a request to a new page or a distance-l request. Then again, for increasing $l = 1, \ldots, k-1$, a distance-l request is issued. (3) Finally, we append the missing distance-l requests, where $0 \leq l \leq k-1$. The resulting sequence σ is one characterized by $\mathcal{C}$. In the proof we show that FIFO has a page fault on every request in the prefix sequence of σ that was not modified by the construction described above. Furthermore, we show that FIFO incurs k page faults in every phase $P(i)$. This implies $\text{FIFO}(\sigma) \geq \text{LRU}(\mathcal{C}) + c(k-1)$. In a final step we prove that OPT's cost increases by at most $c(1 - 1/k) + 1$.

Next we address FWF and give results corresponding to those for FIFO. In the separation bound of Theorem 6 the vector entries $c_1, \ldots, c_{k-1}$ may be by a factor of 2 smaller compared to those in Theorem 4.

Theorem 5. *For any* $\mathcal{C}$*, there holds* $R_{\mathrm{FWF}}(\mathcal{C}) \geq R_{\mathrm{LRU}}(\mathcal{C})$.

Theorem 6. *Let* $\mathcal{C} = (c_0, \ldots, c_{p-1})$ *be any vector. Let* $c_{\min} = \min_{1 \leq l \leq k-1} c_l$ *and* $c = \min\{c_{\min}, p - k + \sum_{l=k}^{p-1} c_l\}$. *Then*

$$R_{\mathrm{FWF}}(\mathcal{C}) \geq \frac{\mathrm{LRU}(\mathcal{C}) + c(k-1)}{\mathrm{OPT}(\mathcal{C}) + c(1 - 1/k) + 1}.$$

5 Experiments

We briefly report on an experimental study we have performed with reference traces from a benchmark library. We summarize the main results; the full version of this paper contains a detailed report including many figures. In our experiments we used the benchmark library [14]. This test suite was specifically designed to evaluate the performance of memory systems. A detailed description can be found in the SIGMETRICS paper [15]. The trace library consists of 15 files that contain sequential logs of memory locations used by various programs. Standard applications from the Linux and the Windows NT operating systems were executed.

In a first step, for each trace, we have extracted the underlying characteristic vector, simply by counting the number of distance-l requests, for each $l \geq 0$, in it. Uniformly over all files, in each resulting vector, the entries basically form a non-increasing sequence, with a huge majority of the requests representing distance-l requests, for small values of l. Once again this confirms the fact that real-world sequences exhibit a high degree of locality.

In a second step we have compared, for each trace/request sequence σ, the optimum number of page faults $\mathrm{OPT}(\sigma)$ to our bounds given in Theorems 1 and 2. For all the traces the difference is small. Hence our lower bound on $\mathrm{OPT}(\sigma)$ in Theorem 1 quite accurately predicts the optimum cost. Furthermore, the additive expression of $2(\lambda - k + 1)$ in the bound of Theorem 2 is not critical. We point out that our lower bound on $\mathrm{OPT}(\sigma)$ cannot match the true service cost because the bound holds for *every* request sequence specified by a characteristic vector $\mathcal{C}$. The given trace σ, in general, is not a sequence that can be served with the minimum number of faults, among inputs characterized by the underlying $\mathcal{C}$. Additionally, we have evaluated the lower bound on $\mathrm{OPT}(\sigma)$ given by Panagiotou and Souza [19], cf. inequality (1). In the experiments our new lower bound developed in this paper is always significantly better. The gap increases as the fast memory size k increases. For larger values of k, the lower bound by Panagiotou and Souza is relatively weak. One could improve it by considering the maximum of p and (1). Even for small k, our new lower bound improves upon that of Panagiotou and Souza by at least 25% to 100%.

Finally we have compared, for each trace σ and underlying $\mathcal{C}$, the upper & lower bounds on LRU's competitiveness $R_{\mathrm{LRU}}(\mathcal{C})$ (see Corollary 1) to the experimentally observed competitiveness for σ. For all the files, our bounds give small constant competitive factors that are typically in the range $[1, 4]$. In a few exceptional cases the factor might be as high as 8. Our upper bound on $R_{LRU}(\mathcal{C})$ is always at most 2.5 times the experimentally observed competitiveness. Again, our bounds cannot exactly match the latter competitiveness since $R_{LRU}(\mathcal{C})$ is the maximum ratio of $LRU(\sigma)/OPT(\sigma)$, considering σ characterized by $\mathcal{C}$. A trace at hand, in general, is not such a worst-case sequence. Interestingly, for varying k, our bounds exhibit the same overall behavior as the experimentally observed competitiveness. Thus they correctly describe the general qualitative behavior of $R_{LRU}(\mathcal{C})$, depending on k. We finally remark that this is the first work on the paging problem in which theoretically proven and experimentally observed performance guarantees match up to small constant factors, independently of k.

References

1. Albers, S., Favrholdt, L.M., Giel, O.: On paging with locality of reference. J. Comput. Syst. Sci. **70**(2), 145–175 (2005)
2. Angelopoulos, S., Dorrigiv, R., López-Ortiz, A.: On the separation and equivalence of paging strategies. In: Proc. 18th ACM-SIAM SODA, pp. 229–237 (2007)
3. Angelopoulos, S., Schweitzer, P.: Paging and list update under bijective analysis. J. ACM **60**(2), 7 (2013)
4. Becchetti, L.: Modeling locality: a probabilistic analysis of LRU and FWF. In: Albers, S., Radzik, T. (eds.) ESA 2004. LNCS, vol. 3221, pp. 98–109. Springer, Heidelberg (2004)
5. Ben-David, S., Borodin, A.: A new measure for the study of on-line algorithms. Algorithmica **11**(1), 73–91 (1994)
6. Boyar, J., Favrholdt, L.M., Larsen, K.S.: The relative worst-order ratio applied to paging. J. Comput. Syst. Sci. **73**(5), 818–843 (2007)
7. Boyar, J., Gupta, S., Larsen, K.S.: Access graphs results for LRU versus FIFO under relative worst order analysis. In: Fomin, F.V., Kaski, P. (eds.) SWAT 2012. LNCS, vol. 7357, pp. 328–339. Springer, Heidelberg (2012)
8. Boyar, J., Gupta, S., Larsen, K.S.: Relative interval analysis of paging algorithms on access graphs. In: Dehne, F., Solis-Oba, R., Sack, J.-R. (eds.) WADS 2013. LNCS, vol. 8037, pp. 195–206. Springer, Heidelberg (2013)
9. Borodin, A., Irani, S., Raghavan, P., Schieber, B.: Competitive paging with locality of reference. J. Comput. Syst. Sci. **50**, 244–258 (1995)
10. Dorrigiv, R., Ehmsen, M.R., López-Ortiz, A.: Parameterized analysis of paging and list update algorithms. In: Bampis, E., Jansen, K. (eds.) WAOA 2009. LNCS, vol. 5893, pp. 104–115. Springer, Heidelberg (2010)
11. Dorrigiv, R., López-Ortiz, A.: On developing new models, with paging as a case study. SIGACT News **40**(4), 98–123 (2009)
12. Chrobak, M., Noga, J.: LRU is better than FIFO. Algorithmica **23**, 180–185 (1999)
13. Dorrigiv, R., López-Ortiz, A., Munro, J.I.: On the relative dominance of paging algorithms. Theor. Comput. Sci. **410**(38–40), 3694–3701 (2009)
14. S. Kaplan. Trace reduction for virtual memory simulation. Benchmark library at https://www3.amherst.edu/~sfkaplan/research/trace-reduction/
15. Kaplan, S.F., Smaragdakis, Y., Wilson, P.R.: Trace reduction for virtual memory simulations. In: Proc. International ACM SIGMETRICS Conference, pp. 47–58 (1999)

16. Karlin, A., Phillips, S., Raghavan, P.: Markov paging. SIAM J. Comput. **30**(3), 906–922 (2000)
17. Koutsoupias, E., Papadimitriou, C.H.: Beyond competitive analysis. SIAM J. Comput. **30**(1), 300–317 (2000)
18. Irani, S., Karlin, A.R., Phillips, S.: Strongly competitive algorithms for paging with locality of reference. SIAM J. Comput. **25**, 477–497 (1996)
19. Panagiotou, K., Souza, A.: On adequate performance measures for paging. In: Proc. 38th Annual ACM Symposium on Theory of Computing (STOC), pp. 487–496 (2006)
20. Sleator, D.D., Tarjan, R.E.: Amortized efficiency of list update and paging rules. Commun. ACM **28**, 202–208 (1985)
21. Young, N.E.: The k-server dual and loose competitiveness for paging. Algorithmica **11**, 525–541 (1994)

Approximation Algorithms for Computing Maximin Share Allocations

Georgios Amanatidis[1], Evangelos Markakis[1]([✉]), Afshin Nikzad[2], and Amin Saberi[2]

[1] Department of Informatics, Athens University of Economics and Business, Athens, Greece
{gamana,markakis}@aueb.gr
[2] Department of Management Science and Engineering, Stanford University, Stanford, USA
{nikzad,saberi}@stanford.edu

Abstract. We study the problem of computing maximin share guarantees, a recently introduced fairness notion. Given a set of n agents and a set of goods, the maximin share of a single agent is the best that she can guarantee to herself, if she would be allowed to partition the goods in any way she prefers, into n bundles, and then receive her least desirable bundle. The objective then in our problem is to find a partition, so that each agent is guaranteed her maximin share. In settings with indivisible goods, such allocations are not guaranteed to exist, hence, we resort to approximation algorithms. Our main result is a 2/3-approximation, that runs in polynomial time for any number of agents. This improves upon the algorithm of Procaccia and Wang [14], which also produces a 2/3-approximation but runs in polynomial time only for a constant number of agents. We then investigate the intriguing case of 3 agents, for which it is already known that exact maximin share allocations do not always exist. We provide a 6/7-approximation algorithm for this case, improving on the currently known ratio of 3/4. Finally, we undertake a probabilistic analysis. We prove that in randomly generated instances, with high probability there exists a maximin share allocation. This can be seen as a justification of the experimental evidence reported in [5,14], that maximin share allocations exist almost always.

1 Introduction

We study a fair division problem in the context of allocating indivisible goods. Fair division has attracted the attention of various scientific disciplines, including

---

A full version of this paper can be found at: http://arxiv.org/abs/1503.00941.
Georgios Amanatidis and Evangelos Markakis—Research co-financed by the European Union (European Social Fund - ESF) and Greek national funds through the Operational Program "Education and Lifelong Learning" of the National Strategic Reference Framework (NSRF) - Research Funding Program: "THALES - Investing in knowledge society through the European Social Fund".

M.M. Halldórsson et al. (Eds.): ICALP 2015, Part I, LNCS 9134, pp. 39–51, 2015.
DOI: 10.1007/978-3-662-47672-7_4

among others, mathematics, economics, and political science. Ever since the first attempt for a formal treatment by Steinhaus, Banach, and Knaster [17], many interesting and challenging questions have emerged. Over the past decades, a vast literature has developed, see e.g., [7,15], and several notions of fairness have been suggested. The area gradually gained popularity in computer science as well, as most of the questions are inherently algorithmic, see among others, [9,19] for earlier works, and the upcoming survey [13] on more recent results. The objective in fair division problems is to allocate a set of resources to a set of n agents in a way that leaves every agent satisfied. In the continuous case, the available resources are typically represented by the interval [0, 1], whereas in the discrete case, we have a set of distinct, indivisible goods. The preferences of each agent are represented by a valuation function, which is usually an additive function on the set of goods (a probability distribution in the continuous case). Given such a setup, many solution concepts have been proposed as to what constitutes a fair solution. Some of the standard ones include *proportionality, envy-freeness, equitability* and several variants of them.

All the above solutions can be attained in the case of divisible goods. In the presence of indivisible goods however, we cannot have any such guarantees; in fact in most cases we cannot even guarantee reasonable approximations. Instead, we focus on a concept recently introduced by Budish [8], that can be seen as a relaxation of proportionality. The rationale is as follows: suppose that an agent, say agent i, is asked to partition the goods into n bundles and then the rest of the agents make a choice before i. In the worst case, agent i will be left with her least valuable bundle. Hence, a risk-averse agent would choose a partition that maximizes the minimum value of a bundle in the partition. This value is called the maximin share of agent i. The objective then is to find an allocation where every person receives at least her maximin share. Even for this notion, existence is not guaranteed under indivisible goods (see [10,14]). But, it is possible to have constant factor approximations, as has been recently shown in [14].

Contribution: Our main result, in Section 4, is a $(2/3 - \varepsilon)$-approximation algorithm, for any constant $\varepsilon > 0$, that runs in polynomial time for any number of agents. That is, the algorithm produces an allocation where every agent receives a bundle worth at least $2/3 - \varepsilon$ of her maximin share. Our result improves upon the $2/3$-approximation of Procaccia and Wang [14], which runs in polynomial time only for a constant number of agents. To achieve this, we redesign certain parts of the algorithm in [14], arguing about the existence of appropriate, carefully constructed matchings in a bipartite graph representation of the problem. Before that, in Section 3, we provide a simpler, faster $1/2$-approximation algorithm. Despite the worse factor, this algorithm still has its own merit due to its simplicity. We then investigate the case of $n = 3$ agents. This case is an interesting turning point on the approximability of the problem, as we know that for $n = 2$, there always exist maximin share allocations. Adding a third agent makes the problem significantly more complex, and the best known ratio was $3/4$ by [14]. We provide an algorithm with an improved approximation guarantee of $6/7$, by examining more deeply the set of allowed matchings that we can

use to satisfy the agents. Finally, motivated by the apparent difficulty in find
ing impossibility results on the approximability of the problem, we undertake a
probabilistic analysis. Our analysis shows that in randomly generated instances
maximin share allocations exist with high probability. This may be seen as a
justification of the experimental evidence reported in [5,14], which show that
maximin share allocations exist in most cases even for a small number of agents

Related Work: For an overview of the classic fairness notions and related
results, we refer the reader to the books of [7] and [15]. The notion we study
here was introduced by Budish in [8] for ordinal utilities (where agents have
rankings over alternatives), building on concepts by Moulin [12]. The work of [5]
defined the notion for cardinal utilities, in the form that we study it here, and
also provided many important insights as well as experimental evidence. Further, in [14], a 2/3 approximation is provided along with constructions showing
instances where no maximin share allocation exists even for $n = 3$. The negative
results of [14], and more recently of [10], as well as the extensive experimentation by Bouveret and Lemaître [5], reveal that it has been challenging to produce
lower bounds, i.e., instances where no α-approximation of a maximin share allocation exists, even for α very close to 1. Finally, in [10], a probabilistic analysis,
similar in spirit but more general than ours, is provided, covering a wide range
of distributions on the valuation functions. However, their analysis, general as
it may be, needs very large values of n to guarantee relatively high probability,
hence it does not fully justify the experimental results discussed above.

A seemingly related problem is that of max-min fairness (or Santa Claus),
see e.g. [1–3]. In this problem we want to find an allocation where the value of
the least happy person is maximized. With identical agents, the two problems
coincide, but beyond this special case, they exhibit very different behavior.

2   Definitions and Notation

Let $N = \{1, ..., n\}$ be a set of n agents and $M = \{1, ..., m\}$ be a set of indivisible
goods. For any $k \in \mathbb{N}$, we will also be using $[k]$ to denote the set $\{1, ..., k\}$. We
assume each agent has an additive valuation function $v_i(\cdot)$, so that for every
$S \subseteq M$, $v_i(S) = \sum_{j \in S} v_i(\{j\})$. For $j \in M$, we will use v_{ij} for $v_i(\{j\})$.

Given any subset $S \subseteq M$, an allocation of S to the n agents is a partition
$T = (T_1, ..., T_n)$, where $T_i \cap T_j = \emptyset$ and $\bigcup T_i = S$. Let $\Pi_n(S)$ be the set of all
partitions of a set S into n bundles. The notions below were originally defined
by Budish [8] and later on by [5] in the same setting that we study here.

Definition 1. *Given a set of n agents, and any set $S \subseteq M$, the n-maximin
share of an agent i with respect to S, is:* $\boldsymbol{\mu}_i(n, S) = \max\limits_{T \in \Pi_n(S)} \min\limits_{T_j \in T} v_i(T_j)$.

We refer to $\boldsymbol{\mu}_i(n, M)$ simply as the maximin share of i. The solution concept
defined in [8] asks for a partition that gives each agent her maximin share.

Definition 2. *Given a set of agents N, and a set of goods M, a partition $T =
(T_1, ..., T_n) \in \Pi_n(M)$ is called a maximin share allocation if $v_i(T_i) \geq \boldsymbol{\mu}_i(n, M)$,
for every agent $i \in N$.*

As shown in [14], maximin share allocations do not always exist. Hence, our focus is on approximation algorithms, i.e. algorithms that produce a partition where each agent i receives a bundle worth at least $\rho \cdot \mu_i(n, M)$, for some $\rho \leq 1$.

3   Warmup: A Polynomial Time 1/2-approximation

We find it instructive to provide first a simpler and faster algorithm that achieves a worse approximation of $1/2$. In the course of obtaining this algorithm, we also identify some important properties and insights that we will use in the next sections.

We start with an upper bound on our solution for each agent. The maximin share guarantee is a relaxation of proportionality, so we trivially have:

Claim 1. *For every $i \in N$ and every $S \subseteq M$, $\mu_i(n, S) \leq \dfrac{v_i(S)}{n} = \dfrac{\sum_{j \in S} v_{ij}}{n}$.*

We now show how to get an additive approximation. Algorithm 1 below achieves an additive approximation of v_{max}, where $v_{max} = \max_{i,j} v_{ij}$. This simple algorithm, which we will refer to as the *Greedy Round-Robin Algorithm*, has also been discussed in [5], where it was shown that when all item values are in $\{0, 1\}$, it produces an exact maximin share allocation. Some variations of this algorithm have also been used in other allocation problems, see e.g., [6], or the protocol in [4]. We discuss further the properties of Greedy Round-Robin in Section 6. The set V_N in the statement of the algorithm is the set of valuation functions $V_N = \{v_i : i \in N\}$, which can be encoded as a valuation matrix since the functions are additive.

Algorithm 1. Greedy Round-Robin(N, M, V_N)

---

1 Set $S_i = \emptyset$ for each $i \in N$.
2 Fix an ordering of the agents arbitrarily.
3 **while** $\exists$ *unallocated items* **do**
4 $\quad$ $S_i = S_i \cup \{j\}$, where i is the next agent to be examined in the current round (proceeding in a round-robin fashion) and j is i's most desired item among the currently unallocated items.
5 **return** $(S_1, ..., S_n)$

Theorem 2. *If $(S_1, ..., S_n)$ is the output of Algorithm 1, then for every $i \in N$,*

$$v_i(S_i) \geq \frac{\sum_{j \in M} v_{ij}}{n} - v_{max} \geq \mu_i(n, M) - v_{max}.$$

The next important ingredient is the following monotonicity property, which says that we can allocate a single good to an agent without decreasing the maximin share of others.

Lemma 1 (Monotonicity property). *For any agent i and any good j, it holds that $\mu_i(n - 1, M \setminus \{j\}) \geq \mu_i(n, M)$.*

Algorithm 2. APX-MMS$_{1/2}(N, M, V_N)$

---

1 Set $S = M$
2 **for** $i = 1$ *to* $|N|$ **do**
3 $\quad$ Let $\alpha_i = \frac{\sum_{j \in S} v_{ij}}{|N|}$
4 **while** $\exists i, j$ *s.t.* $v_{ij} \geq \alpha_i/2$ **do**
5 $\quad$ Allocate j to i.
6 $\quad S = S \setminus \{j\}$
7 $\quad N = N \setminus \{i\}$
8 $\quad$ Recompute the α_is.
9 Run Greedy Round-Robin on the remaining instance.

We are now ready for the 1/2-approximation, obtained by Algorithm 2, which uses Greedy Round-Robin, but only after we allocate the most valuable goods.

Theorem 3. *Let N be a set of n agents with additive valuations, and let M be a set of goods. Algorithm 2 produces an allocation $(S_1, ..., S_n)$ such that*

$$v_i(S_i) \geq \frac{1}{2}\mu_i(n, M), \ \forall i \in N.$$

The proofs of this section are omitted due to space constraints.

4 A Polynomial Time $\left(\frac{2}{3} - \varepsilon\right)$-approximation

The main result of this section is Theorem 4, establishing a polynomial time algorithm for achieving a 2/3-approximation to the maximin share of each agent.

Theorem 4. *Let N be a set of n agents with additive valuation functions, and let M be a set of goods. For any constant $\varepsilon > 0$, there exists a polynomial time algorithm, producing an allocation $(S_1, ..., S_n)$, such that for all $i \in N$:*

$$v_i(S_i) \geq \left(\frac{2}{3} - \varepsilon\right)\mu_i(n, M).$$

Our result is based on the algorithm by Procaccia and Wang in [14], which also guarantees to each agent a 2/3-approximation; however, it runs in polynomial time only for a constant number of agents. Here, we identify the source of exponentiality and take a different approach regarding certain parts of their algorithm. For the sake of completeness, we first present the necessary related results of [14], before we discuss the steps that are needed to obtain our result.

First of all, note that for a single agent i, the problem of deciding whether $\mu_i(n, M) \geq k$ for a given k is NP-complete. However, a PTAS follows by Woeginger [18]. In the original paper, which is in the context of job scheduling, Woeginger gave a PTAS for maximizing the minimum completion time on identical machines. But this scheduling problem is identical to computing a maximin partition with respect to a given agent i. Indeed, it is enough to think of the machines as identical agents all having i's valuation function. Hence:

Theorem 5 (follows from [18]). *Suppose we have a set M of goods to be divided among n agents. Then, for each agent i, there exists a PTAS for approximating $\boldsymbol{\mu}_i(n, M)$.*

A central quantity in the algorithm of [14] is the *n-density balance parameter*, denoted by ρ_n and defined below. Before stating the definition, we give a high level idea for clarity. Assume that in the course of an algorithm, we have used a subset of the items to "satisfy" some of the agents, and that those items do not have "too much" value for the rest of the agents. Then we should expect to be able to "satisfy" the remaining agents using the remaining unallocated items. Essentially, the parameter ρ_n is the best guarantee one can hope to achieve for the remaining agents, based only on the fact that the complement of the set left to be shared is relatively small. After a quite technical analysis, Procaccia and Wang calculate the exact value of ρ_n in the following lemma.

Lemma 2 (Density Balance Lemma, Lemma 3.2 of [14]). *For any number of agents $n \geq 2$, let*

$$\rho_n = \max \left\{ \lambda \; \middle| \; \begin{array}{l} \forall M, \forall \text{ additive } v_i \in (\mathbb{R}^+)^{2^M}, \forall S \subseteq M, \forall k, \ell \text{ s.t. } k + \ell = n, \\ v_i(M \setminus S) \leq \ell \lambda \boldsymbol{\mu}_i(n, M) \Rightarrow \boldsymbol{\mu}_i(k, S) \geq \lambda \boldsymbol{\mu}_i(n, M) \end{array} \right\}.$$

Then, $\rho_n = \dfrac{2\lfloor n \rfloor_{odd}}{3\lfloor n \rfloor_{odd} - 1} > \dfrac{2}{3}$, where $\lfloor n \rfloor_{odd}$ denotes the largest odd integer less than or equal to n.

We are now ready to state the algorithm, referred to as APX-MMS (Algorithm 3 below). We elaborate on the crucial differences between Algorithm 3 and the result of [14] after Lemma 3. At first, the algorithm computes each agent's $(1 - \varepsilon')$-approximate maximin value using Woeginger's PTAS, where $\varepsilon' = \frac{3\varepsilon}{4}$. Let $\boldsymbol{\xi} = (\xi_1, \ldots, \xi_n)$ be the vector of these values. Hence, $\forall i$, $\boldsymbol{\mu}_i(n, M) \geq \xi_i \geq (1 - \varepsilon')\boldsymbol{\mu}_i(n, M)$. Then, APX-MMS makes a call to the recursive algorithm REC-MMS to compute a $\left(\frac{2}{3} - \varepsilon\right)$-approximate partition. REC-MMS takes the arguments $\varepsilon', n = |N|$, $\boldsymbol{\xi}$, S (the set of items that have not been allocated yet), K (the set of agents that have not received a share of items yet), and the valuation functions $V_K = \{v_i | i \in K\}$. The guarantee provided by REC-MMS is that as long as the already allocated goods are not worth too much for the currently active agents of K, we can satisfy them with the remaining goods. More formally, under the assumption that

$$\forall i \in K, \quad v_i(M \setminus S) \leq (n - |K|)\rho_n \boldsymbol{\mu}_i(n, M), \tag{1}$$

which we will show that it holds before each call, REC-MMS($\varepsilon', n, \boldsymbol{\xi}, S, K, V_K$) computes a $|K|$-partition of S, so that each agent receives items of value at least $(1 - \varepsilon')\rho_n \xi_i \geq (1 - \varepsilon')^2 \rho_n \boldsymbol{\mu}_i(n, M) > (1 - 2\varepsilon')\frac{2}{3}\boldsymbol{\mu}_i(n, M) = \left(\frac{2}{3} - \varepsilon\right)\boldsymbol{\mu}_i(n, M)$.

The initial call of the recursion is, of course, REC-MMS($\varepsilon', n, \boldsymbol{\xi}, M, N, V_N$). Before moving on to the next recursive call, REC-MMS appropriately allocates some of the items to some of the agents, so that they receive value at least

$(1 - \varepsilon')\rho_n \xi_i$ each. This is achieved by identifying an appropriate matching between some currently unsatisfied agents and certain bundles of items. In particular, the most important step in the algorithm is to first compute the set X^+ (line 6), which is the set of agents that will not be matched in the current call. The remaining active agents are then guaranteed to get matched in the current round, whereas X^+ will be satisfied in the next recursive calls. In order to ensure this for X^+, REC-MMS guarantees that for the rest of the items, and for the set X^+, inequality (1) holds. Note that (1) trivially holds for the initial call.

Algorithm 3. APX-MMS(ε, N, M, V_N)

---
1   $\varepsilon' = \frac{3\varepsilon}{4}$
2   **for** $i = 1$ *to* $|N|$ **do**
3 | Use Woeginger's PTAS to compute a $(1 - \varepsilon')$-approximation ξ_i of
        $\mu_i(|N|, M)$. Let $\boldsymbol{\xi} = (\xi_1, \ldots, \xi_n)$.
4   **return** REC-MMS($\varepsilon', |N|, \boldsymbol{\xi}, M, N, V_N$)

For simplicity, in the description of REC-MMS, we assume that $K = \{1, 2, \ldots, |K|\}$. Also, for the bipartite graph defined below in the algorithm, by $\Gamma(U)$ we denote the set of neighbors of the vertices in U.

Algorithm 4. REC-MMS($\varepsilon', n, \boldsymbol{\xi}, S, K, V_K$)

---
1   **if** $|K| = 1$ **then**
2 | Allocate all of S to agent 1.
3   **else**
4 | Use Woeginger's PTAS to compute a $(1 - \varepsilon')$-approximate $|K|$-maximin
        partition of S with respect to agent 1 from K, say $(S_1, \ldots, S_{|K|})$.
5 | Create a bipartite graph $G = (X \cup Y, E)$, where $X = Y = K$ and
        $E = \{(i, j) \mid i \in X, j \in Y, v_i(S_j) \geq (1 - \varepsilon')\rho_n \xi_i\}$.
6 | Find a set $X^+ \subset X$, as described in Lemma 3.
7 | Given a perfect matching A, between $X \setminus X^+$ and a subset of $Y \setminus \Gamma(X^+)$,
        allocate S_j to agent i iff $(i, j) \in A$ (the matching is a byproduct of line 6).
8 | **if** $X^+ = \emptyset$ **then**
9 | | Output the above allocation.
10 | **else**
11 | | Output the above allocation, together with REC-MMS($\varepsilon', n, \boldsymbol{\xi}, S^*$,
            X^+, V_{X^+}), where S^* is the subset of S not allocated in line 7.

To proceed with the analysis, and since the choice of X^+ plays an important role, we should clarify what properties of X^+ are needed for the algorithm to work. The following lemma is the most crucial part in the design of our algorithm.

Lemma 3. *Assume that for n, M, S, K, V_K, inequality (1) holds, and let $G = (X \cup Y, E)$ be the bipartite graph defined in line 5 of* REC-MMS. *Then there exists a subset X^+ of $X \setminus \{1\}$, such that:*

(i) X^+ can be found efficiently.

(ii)   There exists a perfect matching between $X \setminus X^+$ and a subset of $Y \setminus \Gamma(X^+)$.

(iii) If we allocate subsets to agents according to such a matching (as described in line 7) and $X^+ \neq \emptyset$, then inequality (1) holds for n, M, S^*, X^+, V_{X^+}, where S^* is the unallocated subset of S.

Before we prove Lemma 3, we elaborate on the main differences between our setup and the approach of Procaccia and Wang [14]:

Choice of X^+. In [14], X^+ is defined as $\arg\max_{Z \subseteq K \setminus \{1\}} \{|Z| \mid |Z| \geq |\Gamma(Z)|\}$. Clearly, when n is constant, so is $|K|$, and thus the computation of X^+ is trivial. However, it is not clear how to efficiently find such a set in general, when n is not constant. We propose an alternative definition of X^+, which is efficiently computable and has the desired properties. In short, our X^+ is any appropriately selected counterexample to Hall's Theorem for the graph G in line 5.

Choice of ε. The algorithm works for any $\varepsilon > 0$, but in [14] the choice of ε depends on n, and is chosen so that $(1 - \varepsilon)\rho_n \geq \frac{2}{3}$. This is possible since for any n, $\rho_n \geq \frac{2}{3}(1 + \frac{1}{3n-1})$. However, in this case, the running time of Woeginger's PTAS is not polynomial in n. Here, we consider any fixed ε, independent of n.

We present below the definition of X^+ and prove Lemma 3.

Proof of Lemma 3 :   We will show that either $X^+ = \emptyset$ (in the case where G has a perfect matching), or any set X^+ with

$$X^+ \in \{Z \subseteq X : |Z| > |\Gamma(Z)| \text{ and } \exists \text{ matching of size } |X \setminus Z| \text{ in } G \setminus Z\}$$

has the desired properties. We show how to find such a set efficiently. We first find a maximum matching B of G. If $|B| = |K|$, then we are done, since for $X^+ = \emptyset$, properties *(i)* and *(ii)* of Lemma 3 hold, while we need not check *(iii)*. If $|B| < |K|$, then there must be a subset of X violating the condition of Hall's Theorem. Let X_u, X_m be the partition of X in unmatched and matched vertices respectively, according to B. Similarly, we define Y_u, Y_m. Now, we construct a directed graph $G' = (X \cup Y, E')$, where we direct all edges of G from X to Y, and on top of that, we add one copy of each edge of the matching with direction from Y to X. In particular, $\forall i \in X, \forall j \in Y$ if $(i, j) \in E$ then $(i, j) \in E'$ and if moreover $(i, j) \in B$ then $(j, i) \in E'$. We claim that the following set satisfies the desired properties

$$X^+ := X_u \cup \{v \in X : v \text{ is reachable from } X_u\}.$$

Note that X^+ is easy to compute; after finding the maximum matching in G, and constructing G', we can run a depth-first search in each connected component of G', starting from the vertices of X_u.

Given the definition of X^+, we now show property *(ii)*. Back to the original graph G, we first claim that $|X^+| > |\Gamma(X^+)|$. To prove this, note that if $j \in$

$\Gamma(X^+)$, then $j \in Y_m$. If not, then it is not difficult to see that there is an augmenting path from a vertex in X_u to j, which contradicts the maximality of B. Indeed, since $j \in \Gamma(X^+)$, let i be a neighbor of j in X^+. If $i \in X_u$ then the edge (i, j) would enlarge the matching. Otherwise $i \in X_m$ and since also $i \in X^+$, there is a path in G' from some vertex of X_u to i. But this path by construction of the directed graph G' must consist of an alternation of unmatched and matched edges, hence together with (i, j) we have an augmenting path. Therefore, $\Gamma(X^+) \subseteq Y_m$, i.e., for any $j \in \Gamma(X^+)$, there is an edge (i, j) in the matching B. But then i has to belong to X^+, by the construction of G' (and since $j \in \Gamma(X^+)$). To sum up: for any $j \in \Gamma(X^+)$, there is exactly one distinct $i \in X^+ \cap X_m$, i.e., $|X^+ \cap X_m| \geq |\Gamma(X^+)|$. In fact, we have equality here, because it is also true that for any $i \in X^+ \cap X_m$, there is a distinct vertex $j \in Y_m$, which is trivially reachable from X^+. Hence, $|X^+ \cap X_m| = |\Gamma(X^+)|$. Since $X_u \neq \emptyset$, we have $|X^+| = |X_u| + |X^+ \cap X_m| \geq 1 + |\Gamma(X^+)|$. So, $|X^+| > |\Gamma(X^+)|$. Also, note that $X^+ \subseteq X \setminus \{1\}$, because for any $Z \subseteq X$ that contains vertex 1 we have $|\Gamma(Z)| = |K| \geq |Z|$, since $v_1(S_j) \geq (1 - \varepsilon')\boldsymbol{\mu}_1(k, S) \geq (1 - \varepsilon')\rho_n \boldsymbol{\mu}_1(n, M) \geq (1 - \varepsilon')\rho_n \xi_1(n, M)$, for all $1 \leq j \leq |K|$.

We now claim that if we remove X^+ and $\Gamma(X^+)$ from G, then the restriction of B on the remaining graph, still matches all vertices of $X \setminus X^+$, establishing property *(ii)*. Indeed, note first that for any $i \in X \setminus X^+$, $i \in X_m$, since X^+ contains X_u. Also, for any edge $(i, j) \in B$ with $i \in X$ and $j \in \Gamma(X^+)$, we have $i \in X^+$ by the construction of X^+. So, for any $i \in X \setminus X^+$, its pair in B belongs to $Y \setminus \Gamma(X^+)$. Equivalently, B induces a perfect matching between $X \setminus X^+$ and a subset of $Y \setminus \Gamma(X^+)$ (this is the matching A in line 7 of the algorithm).

What is left to prove is that property *(iii)* also holds for X^+. This can be done by the same arguments as in [14], thus we have the following lemma.

Lemma 4 ([14], end of Subsection 3.1). *Assume that inequality (1) holds for n, M, S, K, V_K, and let G be the graph defined in line 5. For any $Z \subseteq X$, if there exists a perfect matching between $X \setminus Z$ and a subset of $Y \setminus \Gamma(Z)$, say Y^*, and there are no edges between Z and Y^* in G, then property (iii) holds as well.*

Clearly, there can be no edges between X^+ and $Y \setminus \Gamma(X^+)$. Hence, Lemma 4 can be applied to X^+, completing the proof.   □

Given Lemma 3, we can now easily complete the proof for the correctness of APX-MMS and thus the proof of Theorem 4 (omitted due to space constraints).

5   The Case of $n = 3$ Agents

We now focus on the intriguing case of 3 agents. When $n = 2$, it is pointed out in [5] that maximin share allocations exist via an analog of the cut and choose protocol. Using the PTAS of [18], we can then have a $(1 - \varepsilon)$-approximation in polynomial time. In contrast, as soon as we move to $n = 3$, things become more interesting. It is proved in [14] that with 3 agents there exist instances where no maximin share allocation exists. The best known approximation guarantee is $\frac{3}{4}$, by observing that the quantity ρ_n, defined in Section 4, satisfies $\rho_3 \geq \frac{3}{4}$.

We provide a different algorithm, improving the approximation to $\frac{6}{7} - \varepsilon$. To do this, we combine ideas from both algorithms presented so far in Section 3 and Section 4. The main result of this subsection is as follows:

Theorem 6. *Let $N = \{1, 2, 3\}$ be a set of three agents with additive valuations, and let M be a set of goods. For any constant $\varepsilon > 0$, there exists a polynomial time algorithm that produces an allocation (S_1, S_2, S_3), such that for all $i \in N$:*

$$v_i(S_i) \geq \left(\frac{6}{7} - \varepsilon\right)\boldsymbol{\mu}_i(3, M).$$

The algorithm is shown below, while the proof of Theorem 6 is omitted. Here, we provide a brief outline of how the algorithm works.

Algorithm 5. APX-3-MMS$(\varepsilon, M, v_1, v_2, v_3)$

---

1 $\varepsilon' = \frac{7}{6}\varepsilon$
2 Compute a $(1 - \varepsilon)$-approximation ξ_i of $\boldsymbol{\mu}_i(3, M)$ for $i \in \{1, 2, 3\}$.
3 **if** $\exists i \in \{1, 2, 3\}, j \in M$ *such that* $v_{ij} \geq \frac{6}{7}\xi_i$ **then**
4   Give item j to agent i and divide $M \setminus \{j\}$ among the other two agents in a "cut-and-choose" fashion.

5 **else**
6   Agent 1 computes an $(1 - \varepsilon)$-approximate maximin partition of M into three sets, say (A_1, A_2, A_3).
7   **if** $\exists j_2, j_3 \in \{1, 2, 3\}$ *such that* $j_2 \neq j_3$, $v_2(A_{j_2}) \geq \frac{6}{7}\xi_2$ *and* $v_3(A_{j_3}) \geq \frac{6}{7}\xi_3$ **then**
8   Give set A_{j_2} to agent 2, set A_{j_3} to agent 3, and the last set to agent 1.
9   **else**
10   There are two sets that have value less than $\frac{6}{7}\xi_2$ w.r.t. agent 2, say for simplicity A_2 and A_3.
11   Agent 2 computes $(1 - \varepsilon')$-approximate 2-maximin partitions of $A_1 \cup A_2$ and $A_1 \cup A_3$, say (B_1, B_2) and (B_1', B_2') respectively, and discards the partition with the smallest maximin value, say (B_1', B_2').
12   Agent 3 takes the set she prefers from (B_1, B_2); agent 2 takes the other, and agent 1 takes A_3.

Algorithm Outline: First, approximate values for the $\boldsymbol{\mu}_i$s are calculated as before. Then, if there are items with large value to some agent, in analogy to Algorithm 2, we first allocate one of those, reducing this way the problem to the simple case of $n = 2$. If there are no items of large value, then the first agent partitions the items as in Algorithm 3. In the case where this partition does not satisfy all three agents, then the second agent repartitions two of the bundles of the first agent. Actually, she tries two different such repartitions, and we show that at least one of them works out. The definition of a bipartite preference graph and a corresponding matching (as in Algorithm 3) is never mentioned explicitly here. However, the main idea is that if there is more than one way to pick a perfect matching between $X \setminus X^+$ and a subset of $Y \setminus \Gamma(X^+)$, then we try them all and choose the best one.

6   A Probabilistic Analysis

Setting efficient computation aside, what is the best ρ for which a ρ-approximate allocation does exist? All we know so far is that $\rho \neq 1$ by the elaborate constructions in [10,14]. However, extensive experimentation in [5] (and also [14]) showed that in all generated instances, there always existed a maximin share allocation. Motivated by these experimental observations and by the lack of impossibility results, we present a probabilistic analysis, showing that indeed we expect that in most cases there exist allocations where every agent receives her maximin share. In particular, we analyze the Greedy Round-Robin algorithm from Section 3 when each v_{ij} is drawn from the uniform distribution over $[0, 1]$.

Recently, in [10], similar results are shown for a large set of distibutions over $[0, 1]$, including $U[0, 1]$. Although, asymptotically, their results yield a theorem that is more general than ours, we consider our analysis to be of independent interest, since we have much better bounds on the probabilities for the special case of $U[0, 1]$, even for relatively small values of n. We start with the following:

Theorem 7. *Let $N = [n]$ be a set of agents and $M = [m]$ be a set of goods, and assume that the $v_{ij}s$ are i.i.d. random variables that follow $U[0, 1]$. Then, for $m \geq 2n$ and large enough n, the Greedy Round-Robin algorithm allocates to each agent i a set of goods of total value at least $\frac{1}{n}\sum_{j=1}^{m} v_{ij}$ with probability $1 - o(1)$. The $o(1)$ term is $O(1/n)$ when $m > 2n$ and $O(\log n/n)$ when $m = 2n$.*

The proof is based on tools like Hoeffding's and Chebyshev's inequalities, and on a very careful estimation of the probabilities when $m < 2.5n$. Note that for $m \geq 2n$, this provides an even stronger guarantee than the maximin share.

We now state a similar result for any m. We use a modification of Greedy Round-Robin. While $m < 2n$, the algorithm picks any agent uniformly at random and gives her only her "best" item. When the number of available items becomes two times the number of active agents, the algorithm proceeds as usual.

Theorem 8. *Let $N = [n]$, $M = [m]$, and the $v_{ij}s$ be as in Theorem 7. Then, for any m and large enough n, the Modified Greedy Round-Robin algorithm allocates to each agent i a set of items of total value at least $\mu_i(n, M)$ with probability $1 - o(1)$. The $o(1)$ term is $O(1/n)$ when $m > 2n$ and $O(\log n/n)$ when $m \leq 2n$.*

Theorems 7 and 8 may leave the impression that n has to be large. Actually, there is no reason why we cannot consider n fixed and let m grow. Following very closely the proof of Theorem 7 for $m \geq 4n$, we get the next corollary.

Corollary 1. *Let $N = [n]$, $M = [m]$, and the $v_{ij}s$ be as in Theorem 7. Then, for fixed n and large enough m, the Greedy Round-Robin algorithm allocates to each agent i a set of goods of total value at least $\frac{1}{n}\sum_{j=1}^{m} v_{ij}$ with probability $1 - O(\log^2 m/m^2)$.*

7 Conclusions

The most interesting open question is undoubtedly whether one can improve on the 2/3-approximation. Going beyond 2/3 seems to require a drastically different approach. Even establishing better ratios for special cases could still provide new insights into the problem. It would be interesting, for example, to see if we can have an improved ratio for the special case studied in [2] for the Santa Claus problem. Obtaining negative results seems to be an even more challenging task, given our probabilistic analysis and the results of related works. The negative results in [10,14] require very elaborate constructions, which still do not yield an inapproximability factor far away from 1. Apart from improving the approximation quality, exploring practical aspects of our algorithms is another direction, see e.g. [16]. Finally, we have not addressed here the issues of truthfulness and mechanism design, a stimulating topic for future work. Given the impossibility results in [11] for a related problem, we expect similar negative results here too. When payments are allowed however, more interesting questions may arise.

References

1. Asadpour, A., Saberi, A.: An approximation algorithm for max-min fair allocation of indivisible goods. In: ACM Symposium on Theory of Computing (STOC), pp. 114–121 (2007)
2. Bansal, N., Sviridenko, M.: The santa claus problem. In: ACM Symposium on Theory of Computing (STOC), pp. 31–40 (2006)
3. Bezakova, I., Dani, V.: Allocating indivisible goods. ACM SIGecom Exchanges **5**, 11–18 (2005)
4. Bouveret, S., Lang, J.: A general elicitation-free protocol for allocating indivisible goods. In: Proceedings of the 22nd International Joint Conference on Artificial Intelligence, IJCAI 2011, pp. 73–78 (2011)
5. Bouveret, S., Lemaître, M.: Characterizing conflicts in fair division of indivisible goods using a scale of criteria. In: International conference on Autonomous Agents and Multi-Agent Systems, AAMAS 2014, pp. 1321–1328 (2014)
6. Brams, S.J., King, D.: Efficient fair division - help the worst off or avoid envy. Rationality and Society **17**(4), 387–421 (2005)
7. Brams, S.J., Taylor, A.D.: Fair Division: from Cake Cutting to Dispute Resolution. Cambrige University Press (1996)
8. Budish, E.: The combinatorial assignment problem: Approximate competitive equilibrium from equal incomes. Journal of Political Economy **119**(6), 1061–1103 (2011)
9. Edmonds, J., Pruhs, K.: Balanced allocations of cake. In: Symposium on Foundations of Computer Science (FOCS), pp. 623–634 (2006)
10. Kurokawa, D., Procaccia, A.D., Wang, J.: When can the maximin share guarantee be guaranteed? Manuscript (2015)
11. Markakis, E., Psomas, C.-A.: On worst-case allocations in the presence of indivisible goods. In: Chen, N., Elkind, E., Koutsoupias, E. (eds.) Internet and Network Economics. LNCS, vol. 7090, pp. 278–289. Springer, Heidelberg (2011)
12. Moulin, H.: Uniform externalities: Two axioms for fair allocation. Journal of Public Economics **43**(3), 305–326 (1990)

13. Procaccia, A.D.: Cake cutting algorithms. In: Brandt, F., Conitzer, V., Endriss, U Lang, J., Procaccia, A. (eds.) Handbook of Computational Social Choice, chap. 13 Cambridge University Press (2015)
14. Procaccia, A.D., Wang, J.: Fair enough: guaranteeing approximate maximin shares In: ACM Conference on Economics and Computation, EC 2014, pp. 675–692 (2014
15. Robertson, J.M., Webb, W.A.: Cake Cutting Algorithms: be fair if you can. AI Peters (1998)
16. Spliddit: Provably fair solutions (2015). http://www.spliddit.org/
17. Steinhaus, H.: The problem of fair division. Econometrica **16**, 101–104 (1948)
18. Woeginger, G.: A polynomial time approximation scheme for maximizing the min imum machine completion time. Operations Research Letters **20**, 149–154 (1997)
19. Woeginger, G., Sgall, J.: On the complexity of cake cutting. Discrete Optimizatior **4**(2), 213–220 (2007)

Envy-Free Pricing in Large Markets: Approximating Revenue and Welfare

Elliot Anshelevich, Koushik Kar, and Shreyas Sekar$^{(\boxtimes)}$

Rensselaer Polytechnic Institute, Troy, NY, USA
eanshel@cs.rpi.edu, koushik@ecse.rpi.edu, sekars@rpi.edu

Abstract. We study the classic setting of envy-free pricing, in which a single seller chooses prices for its many items, with the goal of maximizing revenue once the items are allocated. Despite the large body of work addressing such settings, most versions of this problem have resisted good approximation factors for maximizing revenue; this is true even for the classic unit-demand case. In this paper, we study envy-free pricing with unit-demand buyers, but unlike previous work we focus on *large* markets: ones in which the demand of each buyer is infinitesimally small compared to the size of the overall market. We assume that the buyer valuations for the items they desire have a nice (although reasonable) structure, i.e., that the aggregate buyer demand has a monotone hazard rate and that the values of every buyer type come from the same support.

For such large markets, our main contribution is a 1.88 approximation algorithm for maximizing revenue, showing that good pricing schemes can be computed when the number of buyers is large. We also give a $(e, 2)$-bicriteria algorithm that simultaneously approximates both maximum revenue and welfare, thus showing that it is possible to obtain both good revenue and welfare at the same time. We further generalize our results by relaxing some of our assumptions, and quantify the necessary tradeoffs between revenue and welfare in our setting. Our results are the first known approximations for large markets, and crucially rely on new lower bounds which we prove for the profit-maximizing solutions.

1 Introduction

How should a seller controlling multiple goods choose prices for these goods, so that the prices yield good revenue and yet are efficiently computable? This question is among the most fundamental of algorithmic challenges motivated by Economic paradigms. At a high level, this setting can be modeled as a two-stage game: the seller chooses prices, and the buyers respond by purchasing goods at these prices. A common constraint in this context is one of *envy-freeness*, i.e., every buyer receives items that maximize her utility, and thus would not want to "switch places" with any other buyer.

Despite the surge of papers studying envy-free pricing in recent years [3,14], even the simplest versions of this problem have resisted good approximation factors for maximizing revenue. This is true even for the common setting of

© Springer-Verlag Berlin Heidelberg 2015
M.M. Halldórsson et al. (Eds.): ICALP 2015, Part I, LNCS 9134, pp. 52–64, 2015.
DOI: 10.1007/978-3-662-47672-7_5

unit-demand buyers, where every buyer desires one unit of good from a demand set S_i (possibly different for each buyer i); she values all items in S_i equally and has no value for items outside of S_i. The problem of revenue-maximization with unit-demand buyers is among the most popular versions of the pricing problem. While the best known approximation algorithm for the item-pricing version of this problem has only a logarithmic factor [3,14], more sophisticated pricing mechanisms have yielded some beautiful, near-optimal mechanisms, but only by giving up envy-freeness [6,13].

In this paper, we study envy-free pricing with unit-demand buyers, and form good approximation algorithms for maximizing both revenue and welfare. Unlike most previous work on this subject, we focus on *large* markets: ones in which the demand of each buyer is infinitesimally small compared to the market size. For envy-free settings, studying large markets is much more reasonable than a market with only a few buyers. Indeed, in such a market, a seller may not be able to *price discriminate* (i.e., sell the same good to different buyers at different prices), and would instead simply post a price for each good, which would apply to all of the buyers. The fact that all buyers who receive a copy of the same good pay the same price, along with buyers always purchasing a unit of the *cheapest* good in their set S_i, would guarantee that the allocation is envy-free.

Our Model. We consider a single monopolist producing a set S of goods, which are near-substitutes. The seller can produce any desired quantity x_t of a good $t \in S$, for which he incurs a cost of $C_t(x_t)$. The seller's main objective is to set prices on the goods to maximize revenue; in addition to revenue, the seller may also be interested in welfare guarantees. The market consists of a set B of buyer types: for a given type $i \in B$, all the buyers having this type desire the same set $S_i \subseteq S$ of items. Every individual buyer's demand is infinitesimal compared to the market size. Therefore, we can represent every type $i \in B$ by a (inverse) demand function $\lambda_i(x)$ such that for any given v, we know how many buyers x have a valuation of $v = \lambda_i(x)$ or more for items in S_i.

Assumptions on Buyer Demand: In this work, we will assume that all the inverse demand functions considered have a *Monotone Hazard Rate* (See Section 2). This is a common assumption on the demand, encompassing several popular functions previously considered in the literature, including concave, power-law, and exponential demand [2]. Secondly, in large markets with similar items, it is typical to assume that the valuations across buyer types are correlated; for example when the valuations of all buyers are sampled (albeit differently) from some global distribution. Our main results hold under a *Uniform Peak* assumption that captures many such natural settings where the buyer valuations are correlated. Formally, an instance satisfies the Uniform Peak assumption if the peak of the supports of the demand functions are the same across buyer types. In Section 4, we generalize our results to settings without this assumption. Our model captures several scenarios of interest; we illustrate two of them below.

1. **PEV Charging**: As Plug-in Electric Vehicles become commonplace, it is expected that charging stations will be set up at many locations. Due to the variable cost of electricity generation, these stations may have different prices for charging during different time intervals. We can model each time slot as an item t; every buyer has a set of time slots during which she can charge, and the seller may be able to predict the demand using prior data.
2. **Display Advertising**: A publisher may have a set of items (e.g., advertising slots) being sold via simultaneous posted price auctions. The ad-slots are differentiated (in their position or location on the website) and a large number of buyers are interested in buying these items, each interested in some specific subset depending on their target audience.

Our model retains the combinatorial flavor of the general envy-free pricing problem: different buyer types have access to different subsets of items, and these subsets are not correlated in any way. It is this combinatorial aspect which contributes to the hardness of the problem. In fact, recent complexity results [3,4] indicate that the general unit-demand problem with uniform valuations may not admit approximation algorithms with factors better than $O(\log |B|)$. The starting point of our work is the fact that in large markets with many buyer types, $O(\log |B|)$ algorithms are not acceptable. A large body of work has circumvented this hardness by studying interesting instances in which the combinatorial aspect of the model is limited or rendered moot [5,7,11]. In contrast, we impose no such restriction on the model, instead making the assumption that the buyer valuations follow a nice (monotone hazard rate) structure, while the sets S_i can be arbitrary. For the large market settings we are interested in, our assumptions seem more reasonable than restricting demand sets.

Two aspects of large markets that we will feature in this paper warrant further discussion. First, while the majority of literature has focused on envy-free pricing to maximize revenue (see Related Work for exceptions), we focus on maximizing both revenue and social welfare, and the trade-offs therein. This is motivated by the fact that in large markets with repeated engagement, compromising on welfare may often lead to poor revenue in the long-run. Second, in our model sellers face convex production costs C_t for each item. This strictly generalizes models with limited or unlimited supply which are usually the norm. In large markets, assuming limited supply is too rigid as sellers may often be able to increase production, albeit at a higher cost. Costs, however, are a nontrivial addition to the envy-free model. Many of the standard techniques that previously yielded good algorithms, especially single-pricing for all items, fail to do so in our framework. The seller now faces the onerous task of balancing demand with production costs, which may be different for different items.

1.1   Our Results

Our main contribution in this work is an algorithm for envy-free pricing that extracts more than half the optimal revenue. Additionally, we show that by giving up a small amount of revenue, the seller can guarantee high social welfare.

(Main Theorems). For settings with Monotone Hazard Rate demand func tions that obey the Uniform Peak assumption, we present

1. A 1.88-Approximation algorithm for maximizing revenue
2. A $(e, 2)$-Bicriteria algorithm that simultaneously approximates both maximum revenue (factor e) and welfare (factor 2).

Although both of our results use a continuous ascending-price algorithm, we describe an efficient implementation for this algorithm. We also show that revenue maximization remains NP-Hard even with the Uniform Peak assumption.

We next generalize the uniform peak assumption and consider markets where every buyer type i has a (potentially different) support $[\lambda_i^{min}, \lambda_i^{max}]$. For this setting, our results are parameterized by a factor Δ that equals the ratio of the maximum λ_i^{max} to the minimum λ_i^{max} across buyer types. We show a $O(\log \Delta)$ approximation to the optimal revenue in this setting, and thus imply that as long as the valuations for different buyer types are not too different, we can still extract high revenue. Moreover, we show that this $O(\log \Delta)$ solution also guarantees one fourth of the optimum social welfare. Although the actual buyer demand may be quite asymmetric, our result depends only on the difference in the peak of the supports; it is reasonable to expect that this difference is not too large if the goods are similar.

We now summarize the two high-level contributions that enable our results.

1. We provide a general framework to derive good algorithms for large markets with *production costs*, extensively using techniques from the theory of mincost flows.
2. Our constant-approximation factors depend crucially on the insight that we gain on the prices in the revenue-maximizing solution. In contrast to previous work, where the approximation factor of the revenue of the computed solution is usually obtained by comparing it to the optimum social welfare [14] (which is an upper bound on optimum revenue), we are able to directly compare the revenue of our solution to the profit-maximizing solution.

1.2 Related Work

Our work is a part of a rather extensive body of literature studying envy-free or item-pricing; the field is too vast to survey here and we will only sample the most relevant results. The Unit-Demand Pricing (UDP) problem where buyers have different valuations for different items was first considered in [14], which gave a $O(\log |B|)$ approximation algorithm for maximizing revenue. The version that we study (each buyer has equal valuation for all items in S_i, and 0 otherwise) has been referred to as UDP-MIN or UDP with Uniform Valuations. Surprisingly, the addition of uniform values has not lead to any improved algorithms for the general UDP problem. Moreover, recent complexity results [3,4] indicate that a sub-logarithmic approximation factor may be unlikely for both problems.

Assuming more structure on the combinatorial aspect of UDP (i.e., sets S_i stating which buyers have access to which items) has yielded more tractable instances. For example, good approximation algorithms exist when each item is desired by at most k buyer types [7,15]. For settings with budgeted buyers who have access to all items but have a limit on the amount of money they can spend, [11] give a 0.5-approximation algorithm; we remark that budgeted buyers can be captured with an inverse demand $\lambda(x) = c/x$. In contrast, ours is among the few papers that makes no assumptions on the demand sets S_i but still obtains a constant approximation factor. Finally, another active line of work has looked at envy-free pricing when each buyer demands a single bundle of items. For more details, the reader is asked to refer to [3] and the references therein.

More broadly, our work bears certain similarities to algorithmic pricing mechanisms [5] in a Bayesian setting, especially posted price mechanisms. In fact, the aggregate demand that we consider can be interpreted as buyers deriving values from a known distribution. Although posted pricing provides excellent guarantees, even in multi-parameter settings [6,13], the mechanisms seldom result in envy-free allocations because it is assumed that buyers choose items in some order. At a high level, our work is a part of the literature exploring the space of multi-parameter settings with some structure. In addition to a valuation, buyers have a demand set (S_i) in our model, whereas researchers have looked at other models where the additional parameter is the quantity demanded [8] or a position in a metric space [9].

Finally, envy-free pricing to maximize welfare coincides with the notion of Walrasian Equilibrium minus the market clearing constraint. In large markets, Walrasian Equilibria are guaranteed to exist [1], although their revenue may be poor. In discrete markets, existence is not guaranteed and the focus has been on solutions that are *approximately envy-free* but still guarantee good welfare [10,12]. There has also been some work on approximating both revenue and welfare over a restricted space of solutions; for instance, the space of all equilibria in GSP [16], or all competitive equilibria for *sharp* multi-unit demand [8]. In contrast, bi-criteria approximations like ours, which compare both objectives for the same solution to the unrestricted global optima, have not been previously considered in the envy-free literature to the best of our knowledge.

2   Model and Preliminaries

We study the pricing problem faced by a central seller controlling a set S of goods with a large number of buyers, each belonging to one of the buyer types in B. All the buyers having a given type B_i have the same set of desired items $S_i \subseteq S$. We model the market structure as a bipartite graph $G = (B \cup S, E)$ where there is an edge between each buyer type B_i and every good in S_i. For every individual buyer $j \in B_i$, her valuation is v_j for items in her demand set S_i and 0 otherwise. Note that different buyers belonging to the same type B_i can have different valuations for the items in S_i.

Aggregate Demand and Production Cost: Every individual buyer's demand is infinitesimal compared to the market size. Therefore, we can model the aggregate demand of all buyers having type B_i using an inverse demand function $\lambda_i(x)$; $v = \lambda_i(x_i)$ means that x_i of these buyers have a value of v or more for the items in S_i. As an example, consider $\lambda_i(x) = 1 - x$ for $x \in [0,1]$. This means that the total population of buyers with type B_i is one; $\lambda_i(0.25) = 0.75$ implies that one-fourth of these buyers have a valuation of 0.75 or more. Finally, the seller incurs a production cost of $C_t(x)$ for producing x amount of good $t \in S$.

Best-Reponse and Envy-Freeness: A complete solution consists of prices and an allocation, and is specified by three vectors $(\boldsymbol{p}, \boldsymbol{x}, \boldsymbol{y})$. The seller's strategy is to select a price vector $\boldsymbol{p}$ where p_t is the price on item $t \in S$. We define $\boldsymbol{x}$ to be the buyer demand vector such that x_i is the amount of good allocated to buyers from type B_i. Finally, $\boldsymbol{y}$ is the allocation such that y_t is the total amount of good t allocated to buyers and $y_t(i)$, the amount to buyer type B_i. We only consider allocations $\boldsymbol{y}$ that are *feasible* with $\boldsymbol{x}$ and G: for all i, $\sum_t y_t(i)$ should equal x_i, and buyers in B_i must only receive allocations of items belonging to S_i. Then,

- Given $\boldsymbol{p}$, we let $\overline{p}_i$ denote the minimum price available to buyers from type B_i, i.e., $\overline{p}_i = \min_{t \in S_i} p_t$.
- The buyer demand $\boldsymbol{x}$ is said to be a **best-response** to the prices $\boldsymbol{p}$ iff $\forall B_i$, $\overline{p}_i = \lambda_i(x_i)$. That is, a population of x_i buyers from B_i have a value of $\overline{p}_i$ or larger, and thus are maximizing their utility by deciding to purchase items at a price of $\overline{p}_i$.
- Given $\boldsymbol{p}$ and $\boldsymbol{x}$, the allocation $\boldsymbol{y}$ is said to be **envy-free** if buyer demand is a best-response to the prices, and if for every buyer the items they are allocated are the lowest priced items available to them, i.e., $y_t(i) > 0 \Rightarrow p_t = \overline{p}_i$.

Our main objective is an envy-free solution that maximizes revenue. Given $(\boldsymbol{p}, \boldsymbol{x}, \boldsymbol{y})$, the *revenue* or *profit*[1] of the seller is the total payment minus costs incurred, i.e.,

$$\text{Revenue} = \sum_{t \in S} (p_t y_t - C_t(y_t)).$$

We also consider solutions with good social welfare, i.e., the total utility of all the buyers plus that of the seller. As long as the solution is envy-free, buyers are utility-maximizing, and so the aggregate utility of buyers belonging to type i is the sum of their values minus payments, which is $\int_0^{x_i} \lambda_i(x)dx - \overline{p}_i x_i$. Since the payments cancel out, the total social welfare of a solution is equal to

$$\text{Social Welfare} = \sum_{B_i \in B} \int_{x=0}^{x_i} \lambda_i(x)dx - \sum_{t \in S} C_t(y_t).$$

We make the following assumptions on the inverse demand and cost functions.

1. By definition, $\lambda_i(x)$ cannot increase with x. Additionally, we assume that $\lambda_i(x)$ is *continuously differentiable* on $(0, T_i)$ (here T_i is the population of buyers in B_i), and has a monotone hazard rate (see definition below).

---

[1] We use the terms revenue and profit interchangeably to be consistent with previous revenue-maximization literature.

2. For all $t \in S$, we take the production costs $C_t(y)$ to be convex, which is the norm in the literature. We also assume that $C_t(y)$ is continuously differentiable and define $c_t(y)$ to be its derivative. All our results hold if an item t has a limited supply of Y_t, and $C_t(y)$ is only differentiable until $y = Y_t$.

Definition 1. (MHR) *An inverse demand function $\lambda(x)$ is said to be log-concave or equivalently, have a monotone hazard rate if $\frac{\lambda'(x)}{\lambda(x)}$ is non-increasing with x.*

Many commonly used buyer demand functions belong to this class including uniform ($\lambda(x) = a$), linear ($\lambda(x) = a - x$) and exponential inverse demand ($\lambda(x) = e^{-x}$). Although the monotone hazard rate requirement gives the appearance of being somewhat restrictive, this assumption is actually rather weak. We show that even with only MHR demand, our framework encompasses the previously studied unit-demand pricing problem (UDP) in small markets.

Proposition 2. *Any UDP instance with uniform valuations in markets with a finite number of buyers can be reduced to an instance of our problem where all buyer types have MHR inverse demand.*

Therefore, our setting strictly generalizes previously studied UDP problems, and we show in Theorem 8 that our general problem does not admit approximation algorithms with any reasonable approximation factor. Our main contribution, however, is proving that the addition of a little bit of structure (via uniform peaks) to this general framework provides much greater insight into the nature of the revenue-maximizing solution, and leads to good algorithms.

Optimal Solutions. We use the notation $(\boldsymbol{p}^{opt}, \boldsymbol{x}^{opt}, \boldsymbol{y}^{opt})$ to denote an envy-free solution maximizing revenue, and $(\boldsymbol{x}^*, \boldsymbol{y}^*)$ to denote an allocation that maximizes welfare (since welfare does not depend on the prices). Given a graph G, functions λ_i and C_t, it is easy to see that the solution maximizing social welfare can be computed using a convex program. We also note that for the price vector $\boldsymbol{p}^*$ where $p_t^* = c_t(y_t^*)$, $(\boldsymbol{x}^*, \boldsymbol{y}^*)$ is actually an envy-free allocation. Finally, it is not difficult to show that $p_t^{opt} \geq p_t^*$; in Lemma 6 we show much stronger lower bounds on $\boldsymbol{p}^{opt}$ which enable us to prove our results.

Connection to Flows: We can view a feasible allocation $\boldsymbol{y}$ as a flow from the items S to the buyers with a demand of $\boldsymbol{x}$, assuming that G is fixed. Notice that there are several feasible flows for a given demand $\boldsymbol{x}$. We will be most interested in min-cost flows: the feasible allocation $\boldsymbol{y}$ that also minimizes the total production cost $\sum_t C_t(y_t)$. The min-cost flow is independent of the prices and given $\boldsymbol{x}$, can be computed efficiently using a convex program.

It is easy to see that $\boldsymbol{y}^*$ is a min-cost flow, but general envy-free solutions including $(\boldsymbol{p}^{opt}, \boldsymbol{x}^{opt}, \boldsymbol{y}^{opt})$, may not use min-cost flows, since envy-freeness constrains the buyers to use only the items with cheapest *price*, while min-cost flows form allocations to optimize production costs. We reiterate that given a price vector $\boldsymbol{p}$, the best-response buyer demand $\boldsymbol{x}$ can be computed using $\overline{p}_i = \lambda_i(x_i)$, and given $(\boldsymbol{p}, \boldsymbol{x})$, we can always determine an envy-free allocation $\boldsymbol{y}$. Interestingly, the solutions returned by our algorithms are not only envy-free, but also use min-cost flows for the corresponding buyer demand $\boldsymbol{x}$.

3 Large Markets with Uniform Peak Valuations

As argued in the Introduction, for markets with a large number of buyers it often makes sense to assume that the inverse demand functions λ_i have the same support $[\lambda^{min}, \lambda^{max}]$ for all $i \in B$. In fact, all our results in this section hold under a more general Uniform Peak assumption, which only requires that the peak support value $\lambda^{\max} = \lambda_i(0)$ is the same for all buyer types i. This would occur, for example, when a large population of buyers is assigned to different buyer types in a random way. However, this assumption does not affect the combinatorial nature of the optimization problem, i.e., choosing which items get a high price, and which receive a lower price. Therefore, the NP-Hardness proof for the unit-demand case in [14] can be adapted to our setting.

Claim 3. *The problem of envy-free revenue-maximization is NP-Hard even in large markets with Uniform Peak valuations and MHR Inverse Demand.*

In this section, we establish our main result: a 1.88 approximation algorithm for maximizing revenue when the inverse demand functions are MHR with uniform peaks. We begin with a general, parameter-dependent procedure for generating prices, which will be the building block of all our algorithms. Although the algorithm is described here as a more intuitive continuous-time procedure, it can be efficiently implemented using $O(|B| \log \lambda^{\max})$ min-cost flow computations. To simplify discussion, henceforth we will use "buyer" interchangeably with "buyer type" as long as the context is clear and use i instead of B_i for buyer types.

Algorithm 1 begins by pricing all the items at the price vector $\boldsymbol{p}^*$, which makes $(\boldsymbol{p}^*, \boldsymbol{x}^*, \boldsymbol{y}^*)$ an envy-free allocation. We gradually increase prices on the items belonging to an 'active set', initialized to the set of cheapest items in $\boldsymbol{p}^*$ and the buyers receiving these items. At each stage, every item in the active set has the same price (active price) allowing us to compute the min-cost flow for only the active buyers and items. As we increase the active price, if it equals p_t^* for some inactive t, we add t and buyers using t to the active set. This simple ascending-price algorithm continues until a stopping condition dependent on a parameter $k \geq 1$ is reached for some item t (Equation (1)); once this happens the price of item t becomes fixed, and item t is removed from the active set. We now discuss some properties of this algorithm that hold for all k. For a given parameter $k \geq 1$, we will use $(\boldsymbol{p}^k, \boldsymbol{x}^k, \boldsymbol{y}^k)$ to denote the solution returned by our algorithm. Recall that $(\boldsymbol{p}^{opt}, \boldsymbol{x}^{opt}, \boldsymbol{y}^{opt})$ is the revenue-maximizing solution.

Properties Satisfied by Algorithm 1 and Price Hierarchy. Every 'stage' of our algorithm corresponds to a unique value of the active price (i.e., price of all the active items), so we can refer to the allocation formed by the algorithm at some point as the allocation at active price p. Figure 1 describes the natural hierarchy between the Active, Inactive, and Finished sets at every value of the

---

Algorithm 1 Ascending-Price Procedure with Stop Parameter k

1: Set initial prices on the items, $p_t = p_t^*$.
2: ACTIVE ← All minimally priced items and all buyers using these items.
3: INACTIVE ← $B \cup S \setminus$ ACTIVE; FINISH ← $\emptyset$
4: **while** FINISH $\neq B \cup S$ **do**
5:     Increase the price of all ACTIVE items by an infinitesimal amount
        {All ACTIVE items have the same price, the active price.}
6:     Compute the min-cost flow for the sub-graph induced by ACTIVE {We prove
        later: At every stage active buyers only receive allocations of active items}
7:     **if** $t \in$ INACTIVE s.t p_t^* equals the active price **then**
8:         Remove t, buyers using t from INACTIVE and add to ACTIVE
9:     **end if**
10:    **if** $t \in$ ACTIVE meets the stopping criterion in the current solution **then**
11:        Remove t, buyers using t from ACTIVE and add to FINISH.
12:    **end if**
13: **end while**

$$\textbf{Stopping Criterion}(p_t, y_t, k): \quad p_t - c_t(y_t) = \frac{1}{k}(\lambda_{max} - c_t(y_t)) \qquad (1)$$

---

active price p. It is not difficult to show that the statement in Figure 1 always holds, starting with the initial envy-free solution $(\boldsymbol{p^*}, \boldsymbol{x^*}, \boldsymbol{y^*})$. Also notice that the following properties always hold.

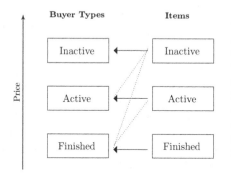

Fig. 1. At any stage of Algorithm 1, the order of prices for the different items is: Inactive > Active > Finished. Thick edges indicate that the buyers in a certain set (Active, Inactive or Finished) receive allocations only from the items in the same set. Dotted edges between a buyer set and an item set indicate although the buyers have access to the items in that set, they are not currently receiving any allocation of that item.

Lemma 4. *Suppose that at some stage of Algorithm 1 where the active price is* p, *the corresponding solution is* $\boldsymbol{p}, \boldsymbol{x}, \boldsymbol{y}$. *Then, the following invariants hold:*

1. $\boldsymbol{x}, \boldsymbol{y}$ *is an envy-free allocation to the corresponding prices* $\boldsymbol{p}$.
2. $\boldsymbol{y}$ *is a min-cost flow to the corresponding buyer demand* $\boldsymbol{x}$.

(Proof Sketch) A solution is a min-cost flow (respectively envy-free) if all buyers are using the items with the smallest marginal costs (price) available to them. We claim that analogous to the price hierarchy of Figure 1, there is a similar

hierarchy for the marginal costs of the items belonging to the three sets. Therefore, it suffices to show that buyers are using the items with the smallest prices or marginal costs within the same set, since all *cross-edges* go to items with a higher price or marginal cost. For inactive buyers and items, the prices and allocations are the same as in $(\boldsymbol{p}^*, \boldsymbol{x}^*, \boldsymbol{y}^*)$, so both envy-freeness and minimum cost follow immediately. For active buyers, this is trivially true since all active items have the same price and we compute a min-cost flow within the active set

Finally, consider a finished buyer i receiving some item t. Any t' with a lower price must have reached the stopping condition before t. So i was active when t' became finished which implies $(i, t') \notin E$. Similarly, a rearrangement of Equation 1 shows us that any t' with a smaller marginal cost than t must have finished strictly before t. By the same reasoning, i cannot have an edge to t'. So, i has no edges to items with smaller price or marginal cost than t.   ∎

Now we are ready to state the main result of this section.

Theorem 5. *Running Algorithm 1 for $k = e$ and $k = \sqrt{e}$ and returning the solution with higher profit results in an envy-free allocation which has a $(4\sqrt{e} - 2 - e) \approx 1.877$ approximation to the optimal profit.*

Since we already argued that Algorithm 1 returns envy-free solutions, we only need to establish the approximation bound. The crucial lemma that allows us to do this is the following lower bound which we prove on the prices in $\boldsymbol{p}^{opt}$. Specifically, this lower bound allows us to compare our solution directly to the revenue-maximizing solution, instead of using the welfare-maximizer as a proxy.

Lemma 6. *For every item t, its price in $\boldsymbol{p}^{opt}$ cannot be smaller than its price in $\boldsymbol{p}^e$, the prices returned by Algorithm 1 for $k = e$.*

(Proof Sketch.) The actual proof is rather technical and we provide only the basic ideas here. We proceed by contradiction. Suppose for some t, $p_t^{opt} < p_t^e$, w.l.o.g, let t be the item with the lowest price satisfying this inequality. We claim that the amount of item t allocated in $\boldsymbol{y}^{opt}$ cannot be smaller than the amount allocated in $\boldsymbol{y}^e$. Essentially, this is true because the decreased prices in $\boldsymbol{y}^{opt}$ lead to larger demand and larger allocations. Now in our solution, once an item becomes finished, we change neither its price nor its allocation. This means that every finished item must still obey the stopping condition. For t,

$$p_t^e - c_t(y_t^e) = \frac{1}{e}(\lambda^{max} - c_t(y_t^e)) \text{ (and) } p_t^{opt} - c_t(y_t^{opt}) < \frac{1}{e}(\lambda^{max} - c_t(y_t^{opt})) \quad (2)$$

The equality is from the stopping condition and the inequality from the fact that $y_t^{opt} \geq y_t^e$ and $p_t^{opt} < p_t^e$. We now claim that there exists a price p_2 such that if we take the prices $\boldsymbol{p}^{opt}$, and then increase the price of t and similarly-priced items to p_2, then the resulting envy-free allocation has higher revenue than the optimal one, giving us a contradiction. To prove this we make use of MHR properties, most crucially that for a non-increasing MHR function $f_i(x)$ with $f_i(0) > ef_i(x_1)$ for a given x_1, $f_i(x_1 - \epsilon)(x_1 - \epsilon) > f_i(x_1)x_1$ for some ϵ. Define $f_i(x) := \lambda_i(x) - c$, where c is the marginal cost of t. Recall that for any buyer i if at a price of p_1, the corresponding demand is x_1, then $p_1 = \lambda_i(x_1)$; this means that when we increase the price from p_1 to p_2, the change in profit is $f_i(x_2)x_2 - f_i(x_1)x_1$, where x_2 is the new flow. Now, take the optimal prices $\boldsymbol{p^{opt}}$ ($p_1 = p_t^{opt}$), and increase the price of all items with the same price as t ($p_2 = p_1 + \delta$); for some buyer type i, let the reduced flow be given by $x_2 = x_1 - \epsilon$. From Equation(2), we know $f_i(0) > ef_i(x_1)$, and so applying the MHR property and summing over all buyer types, we get that the new profit is higher. ∎

To complete the proof of Theorem 5, recall that we return the best of Algorithm 1 for $k = \{e, \sqrt{e}\}$. Note that $\boldsymbol{p^{\sqrt{e}}} \geq \boldsymbol{p^e}$ since decreasing k in Equation (1) only delays the stopping point. Define B^H to be the buyers whose payment in $\boldsymbol{p^{opt}}$ is larger than in $\boldsymbol{p^{\sqrt{e}}}$, and B^L to be the rest of the buyers. We can show that $\boldsymbol{p^e}$ extracts a large fraction of the optimum profit from the buyers in B^L and $\boldsymbol{p^{\sqrt{e}}}$ from B^H. A key lemma that completes the bound is that for MHR functions, for an increase in price from $\boldsymbol{p^e}$ to $\boldsymbol{p^{\sqrt{e}}}$, the profit loss is at most a factor two, and so $\boldsymbol{p^{\sqrt{e}}}$ extracts at least half the profit from the buyers in B^L.

The precise factor of 1.88 comes from carefully balancing these bounds; this leads to the choice of $\boldsymbol{p^{\sqrt{e}}}$ and $\boldsymbol{p^e}$. It is important to note that $\boldsymbol{p^{\sqrt{e}}}$ is not simply a scaled version of the prices in $\boldsymbol{p^e}$; its construction crucially depends on the stopping condition, which in turn depends on both the price and the production cost. The presence of production costs means that previous approaches (e.g., scale prices uniformly, choose a single price for all items) do not work well, as they can end up with solutions with high cost and thus low overall profit.

3.1 Approximating Revenue and Social Welfare Simultaneously

For sellers who care about both revenue and welfare, as is common in repeated mechanisms where you want the buyers to "leave happy", we also provide the following guarantees.

Theorem 7. *Algorithm 1 for $k = e$ provides an envy-free solution which is a e-approximation to the optimal profit with at least half the optimal welfare.*

This result actually provides an additional, stronger revenue-welfare trade-off. Suppose we run the algorithm in Theorem 7, and obtain welfare which is exactly $\frac{1}{\alpha}$ of optimum (we know that $\alpha \leq 2$). Then, our analysis guarantees that the profit of the resulting solution is actually at least $\max(\frac{1}{e}, \frac{\alpha-1}{\alpha})$ of the optimum; for instance if $\alpha = 2$, then we actually get half the optimal revenue.

4 Relaxing the Uniform Peak Valuation Assumption

In this section, we relax the assumption that for all demand functions, $\lambda_i(0)$ is the same. We capture the distortion in this quantity via a parameter Δ which is the ratio of the maximum value of $\lambda_i(0)$ over all i to the minimum value. Even though the λ_i's may not be the same, it is likely that they are closely distributed since all buyers are interested in a similar type of good. Unfortunately, the revenue-maximization problem (UDP) without the Uniform Peak assumption is so general (recall Proposition 2) that it does not admit approximation factors that are even polynomial in Δ. However, we show that we can still extract a good fraction of the optimum revenue and welfare as long as the production costs $C_t(x)$ are doubly convex, i.e., their derivatives are also convex with $c_t(0) = 0$.

Theorem 8. *1. There cannot be a $O(\Delta^k)$-approximation algorithm for any $k > 0$ for UDP in Large Markets with MHR inverse demand and convex costs unless $NP \subseteq DTIME(n^{(log^c n)})$ for some constant c.*

2. For any instance with MHR demand and Doubly Convex costs, we can compute an envy-free solution which is a $O(\log \Delta)$-approximation to the optimal revenue, and which also guarantees $\frac{1}{4}^{th}$ of the optimum welfare.

All the omitted proofs can be found in a full version of this paper available at http://arxiv.org/abs/1503.00340.

Acknowledgements. This work was partially supported by NSF awards CCF-1101495 and CNS-1218374.

References

1. Eduardo, M., Azevedo, E., Weyl, G., White, A.: Walrasian equilibrium in large, quasilinear markets. Theoretical Economics **8**(2), 281–290 (2013)
2. Bagnoli, M., Bergstrom, T.: Log-concave probability and its applications. Economic Theory **26**(2), 445–469 (2005)
3. Briest, P., Krysta, P.: Buying cheap is expensive: Approximability of combinatorial pricing problems. SIAM J. Comput. **40**(6), 1554–1586 (2011)
4. Chalermsook, P., Chuzhoy, J., Kannan, S., Khanna, S.: Improved hardness results for profit maximization pricing problems with unlimited supply. In: Gupta, A., Jansen, K., Rolim, J., Servedio, R. (eds.) APPROX 2012 and RANDOM 2012. LNCS, vol. 7408, pp. 73–84. Springer, Heidelberg (2012)
5. Chawla, S., Hartline, J.D., Kleinberg, R.D.: Algorithmic pricing via virtual valuations. In: Proceedings of EC (2007)
6. Chawla, S., Hartline, J.D., Malec, D.L., Sivan, B.: Multi-parameter mechanism design and sequential posted pricing. In: Proceedings of STOC (2010)
7. Chen, N., Deng, X.: Envy-free pricing in multi-item markets. ACM Transactions on Algorithms **10**(2), 7 (2014)
8. Chen, N., Deng, X., Goldberg, P.W., Zhang, J.: On revenue maximization with sharp multi-unit demands. Journal of Combinatorial Optimization 1–32 (2014)
9. Chen, N., Ghosh, A., Vassilvitskii, S.: Optimal envy-free pricing with metric substitutability. SIAM J. Comput. **40**(3), 623–645 (2011)
10. Chen, N., Rudra, A.: Walrasian equilibrium: Hardness, approximations and tractable instances. Algorithmica **52**(1), 44–64 (2008)
11. Feldman, M., Fiat, A., Leonardi, S., Sankowski, P.: Revenue maximizing envy-free multi-unit auctions with budgets. In: EC (2012)
12. Feldman, M., Gravin, N., Lucier, B.: Combinatorial walrasian equilibrium. In: STOC 2013 (2013)

13. Feldman, M., Gravin, N., Lucier, B.: Combinatorial auctions via posted prices. In: Proceedings of SODA (2015)
14. Guruswami, V., Hartline, J.D., Karlin, A.R., Kempe, D., Kenyon, C., McSherry, F.: On profit-maximizing envy-free pricing. In: Proceedings of SODA (2005)
15. Im, S., Pinyan, L., Wang, Y.: Envy-free pricing with general supply constraints for unit demand consumers. J. Comput. Sci. Technol. **27**(4), 702–709 (2012)
16. Lucier, B., Leme, R.P., Tardos, É.: On revenue in the generalized second price auction. In: Proceedings of WWW 2012 (2012)

Batched Point Location in SINR Diagrams via Algebraic Tools

Boris Aronov[1] and Matthew J. Katz[2]($\boxtimes$)

[1] Department of Computer Science and Engineering,
Polytechnic School of Engineering, New York University, Brooklyn, NY 11201, USA
boris.aronov@nyu.edu
[2] Department of Computer Science, Ben-Gurion University, Beer-Sheva, Israel
matya@cs.bgu.ac.il

Abstract. The *SINR model* for the quality of wireless connections has been the subject of extensive recent study. It attempts to predict whether a particular transmitter is heard at a specific location, in a setting consisting of n simultaneous transmitters and background noise. The SINR model gives rise to a natural geometric object, the *SINR diagram*, which partitions the space into n regions where each of the transmitters can be heard and the remaining space where no transmitter can be heard.

Efficient *point location* in the SINR diagram, i.e., being able to build a data structure that facilitates determining, for a query point, whether any transmitter is heard there, and if so, which one, has been recently investigated in several papers. These planar data structures are constructed in time at least quadratic in n and support logarithmic-time approximate queries. Moreover, the performance of some of the proposed structures depends strongly not only on the number n of transmitters and on the approximation parameter ε, but also on some geometric parameters that cannot be bounded *a priori* as a function of n or ε.

In this paper, we address the question of *batched* point location queries, i.e., answering many queries simultaneously. Specifically, in one dimension, we can answer n queries *exactly* in amortized polylogarithmic time per query, while in the plane we can do it approximately.

All these results can handle *arbitrary* power assignments to the transmitters. Moreover, the amortized query time in these results depends only on n and ε.

Finally, these results demonstrate the (so far underutilized) power of combining algebraic tools with those of computational geometry and other fields.

1 Introduction

The *SINR (Signal to Interference plus Noise Ratio) model* attempts to more realistically predict whether a wireless transmission is received successfully, in

Work on this paper by B.A. has been partially supported by NSF Grants CCF-11-17336 and CCF-12-18791. Work on this paper by M.K. has been partially supported by grant 1045/10 from the Israel Science Foundation. A more complete version of this paper is available on arXiv [3].

M.M. Halldórsson et al. (Eds.): ICALP 2015, Part I, LNCS 9134, pp. 65–77, 2015.
DOI: 10.1007/978-3-662-47672-7_6

a setting consisting of multiple simultaneous transmitters in the presence of background noise. In particular, it takes into account the attenuation of electromagnetic signals. The SINR model has been explored extensively in the literature [19].

Let $S = \{s_1, \ldots, s_n\}$ be a set of n points in the plane representing n transmitters. Let $p_i > 0$ be the transmission power of transmitter s_i, $i = 1, \ldots, n$. In the *SINR model*, a receiver located at point q is able to receive the signal transmitted by s_i if the following inequality holds:

$$\frac{\frac{p_i}{|q-s_i|^\alpha}}{\Sigma_{j \neq i} \frac{p_j}{|q-s_j|^\alpha} + N} \geq \beta,$$

where $|a - b|$ denotes the Euclidean distance between points a and b, and $\alpha > 0$, $\beta > 1$,[1] and $N > 0$ are given constants (N represents the background noise). This inequality is also called the *SINR inequality*, and when it holds, we say that q *receives* (or *hears*) s_i; we refer to the left hand side of the inequality as *SIN ratio* (for receiver q w.r.t. transmitter s_i).

Notice that, since $\beta > 1$, a necessary condition for q to receive s_i is that $p_i/|q - s_i| > p_j/|q - s_j|$, for any $j \neq i$. In particular, in the *uniform power setting* where $p_1 = p_2 = \cdots = p_n$, a necessary condition for q to receive s_i is that s_i is the closest to q among the transmitters in S. This simple observation implies that, for any point q in the plane, either exactly one of the transmitters is received by q or none of them is. Thus, one can partition the plane into n not necessarily connected reception regions R_i, one per transmitter in S, plus an additional region $R_\emptyset$ consisting of all points where none of the transmitters is received. This partition is called the *SINR diagram* of S. Consider the *multiplicatively-weighted Voronoi diagram* D of S in which the region V_i associated with s_i consists of all points q in the plane for which $\frac{1}{\sqrt[\alpha]{p_i}}|q - s_i| < \frac{1}{\sqrt[\alpha]{p_j}}|q - s_j|$, for any $j \neq i$ [4]. Then $R_i \subset V_i$.

In a seminal paper, Avin et al. [6] studied properties of SINR diagrams, focusing on the uniform power setting. Their main result is that in this setting the reception regions R_i are convex and fat. (Here, R_i is *fat* if the ratio between the radii of the smallest disk centered at s_i containing R_i and the largest disk centered at s_i contained in R_i is bounded by some constant.) In the non-uniform power setting, on the other hand, the reception regions are not necessarily connected, and their connected components are not necessarily convex or fat. In fact, they may contain holes [17].

A natural question that one may ask is: "Given a point q in the plane, does q receive one of the transmitters in S, and if yes which one?" Or equivalently: "Which region of the SINR diagram does q belong to?" The latter question is referred to as a *point-location query* in the SINR diagram of S. We can answer it in linear time by first finding the sole candidate, s_i, as the transmitter for

---

[1] In this paper, we assume $\beta > 1$. A variant of our techniques applies also when $\beta < 1$: up to $1/\beta$ receivers can be heard simultaneously, multiple nearest neighbors need to be identified as the candidates, and the algorithms slow down correspondingly.

which the ratio $\frac{1}{\sqrt[\alpha]{p}}|q - s|$ is minimum, and then evaluating the SIN ratio and comparing it to β. To facilitate multiple queries, one may want to build a data structure that can guarantee faster response. We can expedite the first step by constructing the appropriate Voronoi diagram $D = D(\mathcal{S})$ together with a point location structure, so that the sole candidate transmitter for a point q can be found in $O(\log n)$ time. However, the boundary of the region R_i is described by a degree-$\Theta(n)$ algebraic curve; it seems difficult (impossible, in general?) to build a data structure that can quickly determine the side of the curve a given point lies on. The answer is not even obvious in one dimension (where the transmitters and potential receivers all lie on a line), as there R_i is a collection of intervals delimited by roots of a polynomial of degree $\Theta(n)$.

The problem has been approached by constructing data structures for *approximate* point location in SINR diagrams. All approaches use essentially the same logic: first find the sole candidate s_i that the query point q may hear and then approximately locate q in R_i. This is done by constructing two sets R_i^+, R_i^- such that $R_i^+ \subset R_i \subset R_i^- \subset V_i$,[2] and preprocessing them for point location. In the region R_i^+ reception of s_i is guaranteed, so if $q \in R_i^+$, return "can hear s_i." Outside of R_i^- one cannot hear s_i, so if $q \notin R_i^-$, return "cannot hear anything." The set $R_i^- \setminus R_i^+$ is where the approximation occurs: s_i may or may not be heard there, so if $q \in R_i^-$ but $q \notin R_i^+$, return "may or may not hear s_i."

Two different notions of approximation have appeared in the literature. In the first [6,17], it is guaranteed that the uncertain answer is only given infrequently, namely that $area(R_i^- \setminus R_i^+) \le \varepsilon \cdot area(R_i)$, for a suitable parameter $\varepsilon > 0$. In the second [17], it is promised that the SIN ratio for every point in $R_i^- \setminus R_i^+$ lies within $[c_1\beta, c_2\beta]$ for suitable constants c_1, c_2 with $0 < c_1 < 1$, $c_2 > 1$.

We now briefly summarize previous work. Observing the difficulty of answering point-location queries exactly, Avin et al. [6] resorted to approximate query answers in the *uniform power* setting. Given an $\varepsilon > 0$ they build a data structure in total time $O(n^2/\varepsilon)$ and space $O(n/\varepsilon)$ that can be wrong only in a region of area $\varepsilon \cdot area(R_i)$ for each s_i (i.e., approximation of the first type described above). It supports logarithmic-time queries.

In a subsequent paper, Kantor et al. [17] studied properties of SINR diagrams in the *non-uniform power* setting. After revealing several interesting and useful properties, such as that the reception regions in the $(d + 1)$-dimensional SINR diagram of a d-dimensional scene are connected, they present several solutions to the problem of efficiently answering point-location queries. One of them uses the second type of approximation, with $c_1 = (1 - \varepsilon)^{2\alpha}$ and $c_2 = (1 + \varepsilon)^{2\alpha}$, for a prespecified $\varepsilon > 0$. Queries can be performed in time $O(\log(n \cdot \varphi/\varepsilon))$, where φ is an upper bound on the fatness parameters of the reception regions (which cannot be bounded as a function of n or ε). The size of this data structure is $O(n \cdot \varphi'/\varepsilon^2)$ and its construction time is $O(n^2 \cdot \varphi'/\varepsilon^2)$, where $\varphi' > \varphi^2$ is some function of the fatness parameters of the reception regions.

---

[2] Notice that we have not followed the original notation in the literature, for consistency with our notation below.

Although highly non-trivial, the known results for point location in the SINR model are unsatisfactory, in that they suffer from very large preprocessing times. Moreover, in the non-uniform setting, the bounds include geometric parameters such as φ and φ' above, which cannot be bounded as a function of n or ε. In this paper we focus on *batched* point location in the SINR model. That is, given a set $\mathcal{Q}$ of m query points, determine for each point $q \in \mathcal{Q}$ whether it receives one of the transmitters in $\mathcal{S}$, and if yes, which one. Often the set of query points is known in advance, for example, in the planning stage of a wireless network or when examining an existing network. In these cases, one would like to exploit the additional information to speed up query processing. We achieve this goal in the SINR model; that is, we devise efficient approximation and exact algorithms for batched point location in various settings. Our algorithms use a novel combination of sophisticated geometric data structures and tools from computer algebra for multipoint evaluation, interpolation, and fast multiplication of polynomials and rational functions. For example, consider 1-dimensional batched point location where $m = n$ and power is non-uniform. We can answer *exactly* a point-location query in amortized time $O(\log^2 n \log \log n)$. Considering the same problem in the plane, for any $\varepsilon > 0$, we can approximately answer a query in amortized time polylogarithmic in n and ε, as opposed to the result of Kantor et al. [17] mentioned above in which the bounds depend on additional geometric parameters which cannot be bounded as a function of n or ε.

1.1 Related Work

The papers most relevant to ours are those by Avin et al. [6] and Kantor et al. [17] discussed above. Avin et al. [5] also considered the problem of handling queries of the following form (in the uniform-power setting): Given a transmitter s_i and query point q, does q receive s_i by successively applying interference cancellation? (Interference cancellation is a technology that enables a point q to receive a transmitter s, even if s's signal is not the strongest one received at q; see [5] for further details.)

Gupta and Kumar [11] initiated an extensive study of the *maximum capacity* and *scheduling* problems in the SINR model. Given a set L of sender-receiver pairs (i.e., directional links), the *maximum capacity* problem is to find a *feasible* subset of L of maximum cardinality, where $L' \subseteq L$ is *feasible* if, when only the senders of the links in L' are active, each of the links in L' is feasible according to the SINR inequality. The *scheduling* problem is to partition L into a minimum number of feasible subsets (i.e., rounds). We mention several papers and results dealing with the maximum capacity and scheduling problems. Goussevskaia et al. [10] showed that both problems are NP-complete, even in the uniform power setting. Goussevskaia et al. [9], Halldórsson and Wattenhofer [14], and Wan et al. [24] gave constant-factor approximation algorithms for the maximum-capacity problem yielding an $O(\log n)$-approximation algorithm for the scheduling problem, assuming uniform power. In [9] they note that their $O(1)$-approximation algorithm also applies to the case where the ratio between the maximum and minimum power is bounded by a constant and for the case where the number of different power levels is constant. More recently, Halldórsson and Mitra [13] have considered the case of oblivious power. This is a special case of non-uniform power where the power of a link is a simple function of the link's length. They gave an $O(1)$-approximation algorithm for the maximum capacity problem, yielding an $O(\log n)$-approximation algorithm for scheduling. Finally the version where one assigns powers to the senders (i.e., with power control has also been studied, see, e.g., [2,12,13,18,22].

1.2   Our Tools and Goals

Besides making progress on the actual problems being considered here, we view this work as another demonstration of what we hope to be a developing trend of combining tools from the computer algebra world with those of computational geometry and other fields. Several relatively recent representatives of such synergy show examples of seemingly impossible speed-ups in geometric algorithms by expressing a subproblem in algebraic terms [1,20,21]. The algebraic tools themselves are mostly classical ones, such as Fast Fourier Transform, fast polynomial multiplication, multipoint evaluation, and interpolation [7,23]; see [3, Appendix A] for details. We combine them with only slightly newer tools from computational geometry, such as Voronoi diagrams, point location structures in the plane, fast exact and approximate nearest-neighbor query data structures, and range searching data structures [8]; refer to [3, Appendix B]. One very recent result we need is that of Har-Peled and Kumar [15] that, as a special case, allows one to build a compact data structure for approximating multiplicatively weighted nearest-neighbor queries in the plane; the exact version appears to require building the classical multiplicatively weighted Voronoi diagram, which is a quadratic-size object.

We hope that the current work will lead to further productive collaborations between computational geometry and computer algebra.

1.3   Our Results

We now summarize our main results. We use O^* notation to suppress logarithmic factors and O_ε to denote polynomial dependence on $1/\varepsilon$, where $\varepsilon > 0$ is the approximation parameter. In general, we present algorithms for both the uniform-power and non-uniform-power settings, where the algorithms of the former type are usually somewhat simpler.

- In one dimension, we can perform n queries among n transmitters exactly in $O^*(n)$ total time; see Section 2.
- In two dimensions, we can perform n queries among n transmitters approximately in $O_\varepsilon^*(n)$ total time; see Section 3.2.
- We can also facilitate exact batch queries when queries or transmitters form a grid; we omit the details in this version; see [3].

2   Batched Point Location on the Line

In this section $\mathcal{S}$ is a set of $n \geq 3$ point transmitters and $\mathcal{Q}$ is a set of m query points, both on the line. We first consider the *uniform-power version* of the problem, where each transmitter has transmission power 1 (i.e., $p_1 = \cdots = p_n = 1$), and then extend the approach to the arbitrary power version.

2.1   Uniform Power

A query point q receives s_i if and only if

$$\frac{\frac{1}{|q-s_i|^\alpha}}{\Sigma_{j\neq i}\frac{1}{|q-s_j|^\alpha} + N} \geq \beta.$$

Recall that, since $\beta > 1$, if q receives one of the transmitters, then it must be the transmitter that is closest to it; we call it the *candidate* transmitter for q and denote it by $s(q) = s(q, \mathcal{S})$.

Next, we define a univariate function f as

$$f(q) := \sum_{j=1}^{n} \frac{1}{|q - s_j|^\alpha}.$$

Then, q can hear its candidate transmitter $s(q)$ if and only if

$$E(q) := \frac{\frac{1}{|q-s(q)|^\alpha}}{f(q) - \frac{1}{|q-s(q)|^\alpha} + N} \geq \beta.$$

Theorem 1. *For any fixed positive even integer α, given a set $\mathcal{S}$ of transmitters (all of power 1) and a set $\mathcal{Q}$ of receivers, of sizes n and m respectively, we can determine which, if any, transmitter is received by each receiver in total time $O((n + m)\log^2 n \log\log n)$.*

Proof. As pointed out above, a receiver q can receive only the closest transmitter $s(q)$, if any, as the SINR inequality implies $\frac{1}{|q-s(q)|^\alpha} > \frac{1}{|q-s|^\alpha}$ for any $s \neq s(q)$, or equivalently, $|q - s(q)| < |q - s|$. So, as a first step, we identify the closest transmitter for each receiver, which can be done, for example, by sorting $\mathcal{S}$, and using binary search for each receiver, in total time $O((m + n)\log n)$. Moreover, we can compute the term $\frac{1}{|q-s(q)|^\alpha}$, for each $q \in Q$, in the same amount of time.

Observe that f is a sum of n low-degree fractional functions of a single real variable q, so according to [3, Corollary 1], we can now evaluate f on all points of $\mathcal{Q}$ simultaneously in time $O((n + m)\log^2 n \log\log n)$.

In $O(m)$ additional operations we can evaluate the expressions $E(q_1), \ldots, E(q_m)$ and determine for which receivers the SINR inequality holds, so that the signal is actually received.

Computing and evaluating the fraction dominates the computation cost, so the total running time is $O((n + m)\log^2 n \log\log n)$. $\qquad\square$

2.2 Arbitrary Power

We proceed in a similar manner, except the construction of the multiplicatively weighted Voronoi diagram on a line, which is more subtle; see [3].

Theorem 2. *For any fixed positive even integer α, given a set $\mathcal{S}$ of transmitters (not necessarily all of the same power) and a set $\mathcal{Q}$ of receivers, of sizes n and m respectively, we can determine which, if any, transmitter is received by each receiver in total time $O((n + m) \log^2 n \log \log n)$.*

3 Batched Point Location in the Plane

In this section $\mathcal{S} = \{s_i\}$ is a set of n point transmitters in the plane. We consider three versions of (batched) point location, where in the first two the answers we obtain are exactly correct, while in the third one the answer to a query q may be either "s" (meaning that q receives s), "no" (meaning that q does not receive any transmitter), or "maybe" (meaning that q may or may not be receiving some transmitter; the SIN ratio is too close to β and we are unable to decide quickly whether it is above or below β).

Specifically, we consider the following three versions of (batched) point location. In the first version, we assume that the *transmitters* form an $\sqrt{n} \times \sqrt{n}$ non-uniform grid and that each transmitter has power 1. We show how to solve a *single* point-location query in this setting in $O(\sqrt{n} \log^2 n \log \log n)$ (rather than linear) time. In the second version, we assume that the *receivers* form an $n \times n$ non-uniform grid, but the n transmitters, on the other hand, are located anywhere in the plane. Moreover, we allow arbitrary transmission powers. We show how to answer the n^2 queries in near-quadratic (rather than cubic) time. The details of these two versions are omitted due to space limitations; see [3].

Finally, in the third version (Section 3.2), we do not make any assumptions on the location of the devices (either transmitters or receivers). As a result of this, we might not be able to give a definite answer in borderline instances. Specifically, given n transmitters and m receivers, we compute (in total time near-linear in $n + m$), for each receiver q, its unique candidate transmitter s and a value $\tilde{E}(q)$, such that, if $\tilde{E}(q)$ is sufficiently greater than β, then q surely receives s, if $\tilde{E}(q)$ is sufficiently smaller than β, then q surely does not receive s, and otherwise, q may or may not receive some transmitter (i.e., $\tilde{E}(q)$ lies in the *uncertainty interval*). We first present a solution for which the uncertainty interval is $[2^{-\alpha/2}\beta, 2^{\alpha/2}\beta)$, i.e., a constant-factor approximation. We then generalize it so that the uncertainty region is $[(1 - \varepsilon)\beta, (1 + \varepsilon)\beta)$, for any $\varepsilon > 0$, i.e., a PTAS. We consider both the uniform- and arbitrary-power settings.

3.1 General Discussion

Once again, the SINR inequality determines which, if any, of the transmitters $s \in S$ can be heard by a receiver at point q and the only candidate transmitter $s(q)$ is the one that minimizes $|q - s|/p^{1/\alpha}$ among all transmitters s with corresponding power p. In the uniform-power case, this means the transmitter closest to q in Euclidean distance, and the matching space decomposition is the Euclidean Voronoi diagram which can be constructed in $O(n \log n)$ time (see [8]), where $n = |\mathcal{S}|$. In the non-uniform-power case, this corresponds to the multiplicatively weighted Voronoi diagram in the plane, which is a structure of worst-case complexity $\Theta(n^2)$ that can be constructed in time $O(n^2)$; see [4].

Once again we define the function $f(q)$, which represents the total signal strength at q from *all* transmitters, and express the decision of whether the transmitter $s(q)$ is received at q by computing $E(q)$ and comparing it with β. The difference from the one-dimensional case is that $f(q)$ is now a sum of low-degree *bivariate* fractions, with the two variables being the coordinates of q.

In all cases, the goal is to evaluate $f(q)$, for each receiver q, and to identify the suitable candidate transmitter $s(q)$, faster than by brute force. Given this information, the decision can be made in constant time per receiver.

Due to space constraints, we omit the discussion of transmitters on a grid and of receivers on a grid; see [3]. Therefore in the remainder of the section we focus on the last version of the problem.

3.2   Approximating the General Case

We now abandon the ambition to get exact answers and aim for an approximation algorithm, in the sense we will make precise below. Again, $\mathcal{S} = \{s_i\}$ is the set of n transmitters, with each s_i a point in the plane with power p_i; similarly $\mathcal{Q} = \{q_j\}$ is the set of m receivers, where a generic receiver is $q = (q_x, q_y)$.

For a query point q and a transmitter $s = (s_x, s_y)$ of power p, set $l(q, s) = \max\{|q_x - s_x|, |q_y - s_y|\}$; in other words, $l(q, s)$ is the L^∞ distance between points q and s. In complete analogy to our previous approach, put

$$\tilde{f}(q) := \sum_{i=1}^{n} \frac{p_i}{l(q, s_i)^\alpha} \quad \text{and} \quad \tilde{E}(q) := \frac{\frac{p}{l(q,s)^\alpha}}{\tilde{f}(q) - \frac{p}{l(q,s)^\alpha} + N}.$$

What is the significance of the quantity $\tilde{E}(q)$? Since for any two points s, q, $l(q, s) \le |q - s| \le \sqrt{2}\, l(q, s)$,

$$2^{-\alpha/2} \frac{p_j}{l(q, s_j)^\alpha} \le \frac{p_j}{|q - s_j|^\alpha} \le \frac{p_j}{l(q, s_j)^\alpha},$$

so $2^{-\alpha/2} \tilde{f}(q) \le f(q) \le \tilde{f}(q)$, and therefore $2^{-\alpha/2} \tilde{E}(q) \le E(q) \le 2^{\alpha/2} \tilde{E}(q)$. Informally, $\tilde{E}(q)$ is "pretty close" to $E(q)$.

This suggests an approximation strategy that begins by computing $\tilde{E}(q)$ instead of $E(q)$. If $\tilde{E}(q) \ge 2^{\alpha/2}\beta$, we know that $E(q) \ge \beta$ and the signal from the unique candidate transmitter $s(q)$ *is* received. If $\tilde{E}(q) < 2^{-\alpha/2}\beta$, then $E(q) < \beta$ and the signal from $s(q)$ is *not* received and therefore no signal is received by q. For intermediate values of $\tilde{E}(q)$, we cannot definitely determine whether $s(q)$'s signal is received at q.

Now we turn to the actual batch computation of $\tilde{E}(q)$ for all receivers in $\mathcal{Q}$ and point out a few additional caveats.

Computationally, $\tilde{E}(q)$ can be evaluated in constant time, given $\tilde{f}(q)$ and point $s(q) = s(q, \mathcal{S})$. So we focus on these two subproblems. For the uniform-power case, we can construct the Voronoi diagram of $\mathcal{S}$, preprocess it for point location, and query it with each receiver, for a total cost of $O((n + m) \log n)$ [8]. In the case of non-uniform power, if we are content with near-quadratic running time, we can determine $s(q)$ by computing the multiplicatively weighted Voronoi diagram of $\mathcal{S}$ as outlined above, and then querying it with each receiver in total time $O(n^2 + m \log n)$ (see [4,8], which is too much for $m \approx n$. We provide an alternative below.

We show how to compute the values $\tilde{f}(q_1), \ldots, \tilde{f}(q_m)$ in near-linear time, using a two-dimensional orthogonal range search tree. Indeed, observe that $l(s, q) = |q_x - s_x|$ provided $|q_x - s_x| \geq |q_y - s_y|$. For a fixed q, the region W_q containing the transmitters of $\mathcal{S}$ satisfying this inequality is a 90° double wedge. Using (a tilted version of) the orthogonal range search tree [8] (see, [Section B.1, Fact 14]), we can construct a pair decomposition $\{(\mathcal{S}_i, \mathcal{Q}_i)\}$ of small size, so that each pair (s, q) with $s \in W_q$ appears in exactly one product $\mathcal{S}_i \times \mathcal{Q}_i$.

We now denote by $\tilde{f}(q, Z)$ the sum analogous to $\tilde{f}(q)$, where the summation goes over the elements of the supplied set Z rather than those of $\mathcal{S}$. Clearly,

$$\tilde{f}(q, \mathcal{S} \cap W_q) = \sum_{i : q \in \mathcal{Q}_i} \tilde{f}(q, \mathcal{S}_i), \tag{1}$$

by the definition of the pair decomposition. The number of terms in the last sum is $O(\log^2 n)$. Notice that $\tilde{f}(q, \mathcal{S}_i)$, for a fixed i, is a sum of small fractional *univariate* functions, with $|\mathcal{S}_i|$ terms in it, since the expression for transmitters in W_q depends only on q_x and not on q_y. Now for each pair $(\mathcal{Q}_i, \mathcal{S}_i)$, we use [3, Corollary 1] to evaluate $\tilde{f}(q, \mathcal{S}_i)$ on each $q \in \mathcal{Q}_i$ in total time $O((|\mathcal{Q}_i| + |\mathcal{S}_i|) \log^2 |\mathcal{S}_i| \log \log |\mathcal{S}_i|) = O((|\mathcal{Q}_i| + |\mathcal{S}_i|) \log^2 n \log \log n)$. This gives us all the summands of (1) and therefore allows us to evaluate $\tilde{f}(q, \mathcal{S} \cap W_q)$ for all $q \in \mathcal{Q}$, in total time at most proportional to $\sum_i (|\mathcal{Q}_i| + |\mathcal{S}_i|) \log^2 n \log \log n = (\sum_i (|\mathcal{Q}_i| + |\mathcal{S}_i|)) \log^2 n \log \log n = O((m + n) \log^4 n \log \log n)$.

Of course, we have only treated those s that lie in W_q. But the calculation is repeated in the complementary double wedge, where now only the y-coordinates matter and $\tilde{f}(q)$ is the sum of the two values thus obtained.

Theorem 3. *For any fixed positive even integer α, given a set $\mathcal{S}$ of n transmitters (all of power 1) and a set $\mathcal{Q}$ of m receivers, we can do the following in total time $O((m + n) \log^4 n \log \log n)$. For each $q \in \mathcal{Q}$, we find its unique candidate transmitter $s(q)$ and compute a value $\tilde{E}(q)$, such that (i) if $\tilde{E}(q) \geq 2^{\alpha/2}\beta$, then q can definitely hear $s(q)$, (ii) if $\tilde{E}(q) < 2^{-\alpha/2}\beta$, then q definitely cannot hear $s(q)$, and (iii) if $2^{-\alpha/2}\beta \leq \tilde{E}(q) < 2^{\alpha/2}\beta$, then q may or may not hear $s(q)$.*

The algorithm for the non-uniform power case is hampered by the fact that the obvious way to identify the candidate transmitter each receiver might hear seems to involve constructing the multiplicatively weighted Voronoi diagram of quadratic complexity. However, we do not need the exact multiplicatively closest neighbor, but rather a reasonably-close approximation of the value $|q-s|/p(s)^{1/\alpha}$, over all $s \in \mathcal{S}$ (being off by a multiplicative factor of at most $2^{1/2}$ is sufficient; see the discussion below). Such an approximation is provided by an algorithm of Har-Peled and Kumar [15,16], by setting $\varepsilon = 2^{1/2} - 1$ (see [3]), yielding the following:

Theorem 4. *For any fixed positive even integer α and any $\beta > 2^{\alpha/2}$, given a set $\mathcal{S}$ of n transmitters of arbitrary powers and a set $\mathcal{Q}$ of m receivers, we can do the following in total time $O(n \log^7 n + m \log^4 n \log \log n)$ and $O(n \log^4 n + m \log^2 n)$ space: For each $q \in \mathcal{Q}$, we find a transmitter s_q and compute a value $\tilde{E}(q)$, such that (i) if $\tilde{E}(q) \geq 2^{\alpha/2}\beta$, then q can definitely hear s_q (implying that $s_q = s(q)$), (ii) if $\tilde{E}(q) < 2^{-\alpha/2}\beta$, then q definitely cannot hear any transmitter, and (iii) if $2^{-\alpha/2}\beta \leq \tilde{E}(q) < 2^{\alpha/2}\beta$, then q may or may not hear one of the transmitters.*

Note. The transmitter s_q in the theorem above is not necessarily the unique candidate transmitter $s(q)$. We would like to show that if $\tilde{E}(q) \geq 2^{\alpha/2}\beta$ (and therefore $E(q) \geq \beta$), then s_q is necessarily $s(q)$. Assume that they are different (i.e., that $s_q \neq s(q)$), and let e_q (resp., $e(q)$) be the strength of s_q's signal (resp., $s(q)$'s signal) at q. Then, we know that $e_q \leq e(q) \leq 2^{\alpha/2}e_q$. Notice that $E(q) \leq e(q)/e_q$, since $E(q)$ is maximized when there is no third transmitter and no noise, so $e(q)/e_q \geq \beta$ (since $E(q) \geq \beta$). Recall that we are assuming that $\beta > 2^{\alpha/2}$, so we get that $e(q)/e_q > 2^{\alpha/2}$, which is a contradiction.

We now turn the algorithm described above into a PTAS, in the sense that we will confine $\tilde{E}(q)$ to the range $((1-\varepsilon)E(q), (1+\varepsilon)E(q)]$, for a given $\varepsilon > 0$. We outline the approach below. Consider the regular k-gon K_k circumscribed around the Euclidean unit disk, for a large enough even $k \geq 4$ specified below. We modify the above algorithm, replacing the L^∞-norm whose "unit disk" is a square, with the norm $|\ldots|_k$ with K_k as the unit disk. Then $|v|_k \leq |v| \leq (1 + \Theta(k^{-2}))|v|_k$, for any vector v in the plane. In the range-searching data structure, wedges with opening angle $\pi/2 = 2\pi/4$ are replaced by wedges with opening angle $2\pi/k$, and we need $k/2$ copies of the structure.

In terms of the quality of approximation, the factor $2^{\alpha/2} = (\sqrt{2})^\alpha$ is replaced by $(1 + \Theta(k^{-2}))^\alpha \approx 1 + \alpha\Theta(k^{-2})$. Hence to obtain an approximation factor of $1 + \varepsilon$, we set $1 + \varepsilon = 1 + \alpha\Theta(k^{-2})$, or $k = c(\alpha/\varepsilon)^{1/2}$, for a suitable absolute constant c. In other words, it is sufficient to create $O(\varepsilon^{-1/2})$ copies of the data structure. To summarize, we have:

Theorem 5. *For a positive ε, any fixed positive even integer α, given a set $\mathcal{S}$ of n transmitters (all of power 1) and a set $\mathcal{Q}$ of m receivers, we can do the following in total time $O((m + n)\varepsilon^{-1/2} \log^4 n \log \log n)$. For each $q \in \mathcal{Q}$, we find its unique candidate transmitter $s(q)$ and compute a value $\tilde{E}(q)$, such that (i) if $\tilde{E}(q) \geq (1 + \varepsilon)\beta$, then q can definitely hear $s(q)$, (ii) if $\tilde{E}(q) < (1 - \varepsilon)\beta$, then q definitely cannot hear $s(q)$, and (iii) if $(1 - \varepsilon)\beta \leq \tilde{E}(q) < (1 + \varepsilon)\beta$, then q may or may not hear $s(q)$.*

Theorem 6. *For a positive ε, any fixed positive even integer α, and any $\beta > 1 - \varepsilon$,[3] given a set S of n transmitters of arbitrary powers and a set Q of m receivers we can do the following in total time $O(n\varepsilon^{-6}\log^7 n + m\varepsilon^{-1/2}\log^4 n \log\log n)$ and $O(n\varepsilon^{-6}\log^4 n + m\varepsilon^{-1/2}\log^2 n)$ space: For each $q \in Q$, we find a transmitter s_q and compute a value $\tilde{E}(q)$, such that (i) if $\tilde{E}(q) \geq (1+\varepsilon)\beta$, then q can definitely hear s_q (implying that $s_q = s(q)$), (ii) if $\tilde{E}(q) < (1-\varepsilon)\beta$, then q definitely cannot hear any transmitter, and (iii) if $(1-\varepsilon)\beta \leq \tilde{E}(q) < (1+\varepsilon)\beta$, then q may or may not hear one of the transmitters.*

4 Concluding Remarks

We described several algorithms that combine computational geometry techniques and methods of computer algebra to obtain very fast batched SINR diagram point-location queries.

We believe that Theorems 5 and 6 can be applied to speed up the preprocessing stage of existing point-location results. Consider, e.g., the data structure presented by Avin et al. [6] for a set of n uniform-power transmitters, whose construction time is $O(n^2/\delta)$. This data structure is actually a collection of n data structures, one per transmitter, where the data structure DS_i for transmitter s_i consists of an inner (R_i^+) and outer (R_i^-) approximation for reception region R_i, so that $area(R_i^- \setminus R_i^+) \leq \delta \cdot area(R_i)$, see the definitions in the introduction. The construction of DS_i is based on the convexity and fatness of region R_i and consists of two stages. In the first, explicit estimates for the radii of the largest disk centered at s_i and contained in R_i and the smallest such disk containing R_i are obtained, by applying a binary-search-like procedure (beginning with the distance between s_i to its nearest (other) transmitter in S), where each comparison is resolved by explicitly evaluating the SIN ratio at some point q and comparing it to β, i.e., by an *in/out* test. In the second stage, a $1/\delta \times 1/\delta$ grid scaled to exactly cover the outer disk is laid, and, by performing $O(1/\delta)$ additional in/out tests, the sets R_i^+ and R_i^- are obtained (as collections of grid cells). This algorithm thus performs $\Theta(\log n + 1/\delta)$ in/out tests per transmitter, at a cost of $\Theta(n)$ operations each; the high cost of each test is the bottleneck.

We believe that it is possible to speed up the algorithm by constructing the n individual data structures in parallel. During the construction, we will form $O(\log n + 1/\delta)$ batches of n queries each, and use Theorem 5 to deal with each of them in near-linear time. The only problem is that our query answers are not exact, but approximate; for some queries, instead of "in" or "out," we answer "maybe." We think that there is a way to overcome this problem, but we leave it for a full version.

Besides speeding up the construction time of known structures, we would like to find other applications of batched point location to other problems studied in the SINR model.

---

[3] This requirement is analogous to that in Theorem 4 to guarantee that the approximately highest-strength transmitter returned by the data structure is in fact the right one.

We note that our results are general, in the sense that analogous results can be obtained for diagrams that are induced by other inequalities similar to the SINR inequality.

Finally, on a larger scale, we are interested in further applications where algebraic and geometric tools can be combined to achieve significant improvements.

Acknowledgments. B.A. would like to acknowledge the help of Sariel Har-Peled in matters of approximation and of Guillaume Moroz in matters of algebra. He would also like to thank Pankaj K. Agarwal for general encouragement.

References

1. Ajwani, D., Ray, S., Seidel, R., Tiwary, H.R.: On computing the centroid of the vertices of an arrangement and related problems. In: Dehne, F., Sack, J.-R., Zeh, N. (eds.) WADS 2007. LNCS, vol. 4619, pp. 519–528. Springer, Heidelberg (2007)
2. Andrews, M., Dinitz, M.: Maximizing capacity in arbitrary wireless networks in the SINR model: Complexity and game theory. In: INFOCOM, pp. 1332–1340 (2009)
3. Aronov, B., Katz, M.J.: Batched point location in SINR diagrams via algebraic tools (2014). arXiv:1412.0962 [cs.CG]
4. Aurenhammer, F., Edelsbrunner, H.: An optimal algorithm for constructing the weighted Voronoi diagram in the plane. Pattern Recognition 251–257 (1984)
5. Avin, C., Cohen, A., Haddad, Y., Kantor, E., Lotker, Z., Parter, M., Peleg, D.: SINR diagram with interference cancellation. In: SODA, pp. 502–515 (2012)
6. Avin, C., Emek, Y., Kantor, E., Lotker, Z., Peleg, D., Roditty, L.: SINR diagrams: Convexity and its applications in wireless networks. J. ACM **59**(4), 18:1–318:4 (2012)
7. Bini, D., Pan, V.Y.: Polynomial and Matrix Computations: Fundamental Algorithms, vol. 1. Birkhauser Verlag, Basel (1994)
8. de Berg, M., Cheong, O., van Kreveld, M., Overmars, M.H.: Computational Geometry: Algorithms and Applications, 3rd edn. Springer-Verlag, Berlin (2008)
9. Goussevskaia, O., Halldórsson, M.M., Wattenhofer, R., Welzl, E.: Capacity of arbitrary wireless networks. In: INFOCOM, pp. 1872–1880 (2009)
10. Goussevskaia, O., Oswald, Y.A., Wattenhofer, R.: Complexity in geometric SINR. In: MobiHoc, pp. 100–109 (2007)
11. Gupta, P., Kumar, P.R.: The capacity of wireless networks. IEEE Trans. Information Theory **46**(2), 388–404 (2000)
12. Halldórsson, M.M.: Wireless scheduling with power control. ACM Transactions on Algorithms **9**(1) (2012)
13. Halldórsson, M.M., Mitra, P.: Wireless capacity with oblivious power in general metrics. In: SODA, pp. 1538–1548 (2011)
14. Halldórsson, M.M., Wattenhofer, R.: Wireless communication is in APX. In: Albers, S., Marchetti-Spaccamela, A., Matias, Y., Nikoletseas, S., Thomas, W. (eds.) ICALP 2009, Part I. LNCS, vol. 5555, pp. 525–536. Springer, Heidelberg (2009)
15. Har-Peled, S., Kumar, N.: Approximating minimization diagrams and generalized proximity search. SIAM J. Comput. Accepted for publication. http://sarielhp.org/p/12/wann/wann.pdf
16. Har-Peled, S., Kumar, N.: Approximating minimization diagrams and generalized proximity search. In: FOCS, pp. 717–726 (2013)

17. Kantor, E., Lotker, Z., Parter, M., Peleg, D.: The topology of wireless communication. In: STOC, pp. 383–392 (2011)
18. Kesselheim, T.: A constant-factor approximation for wireless capacity maximization with power control in the SINR model. In: SODA, pp. 1549–1559 (2011)
19. Lotker, Z., Peleg, D.: Structure and algorithms in the SINR wireless model. SIGACT News **41**(2), 74–84 (2010)
20. Moroz, G., Aronov, B.: Computing the distance between piecewise-linear bivariate functions. In: SODA, pp. 288–293 (2012)
21. Moroz, G., Aronov, B.: Computing the distance between piecewise-linear bivariate functions. ACM Transactions on Algorithms (2013). Accepted for publication arXiv:1107.2312 [cs.CG]
22. Moscibroda, T., Wattenhofer, R.: The complexity of connectivity in wireless networks. In: INFOCOM, pp. 23–29 (2006)
23. von zur Gathen, J.: Modern Computer Algebra. Cambridge University Press, Cambridge (1999)
24. Wan, P.-J., Jia, X., Yao, F.: Maximum independent set of links under physical interference model. In: Liu, B., Bestavros, A., Du, D.-Z., Wang, J. (eds.) WASA 2009. LNCS, vol. 5682, pp. 169–178. Springer, Heidelberg (2009)

On the Randomized Competitive Ratio of Reordering Buffer Management with Non-Uniform Costs

Noa Avigdor-Elgrabli[1], Sungjin Im[2]($\boxtimes$), Benjamin Moseley[3], and Yuval Rabani[4]

[1] Yahoo! Labs Haifa, MATAM, Haifa 31095, Israel
noaa@yahoo-inc.com
[2] University of California, Merced, CA 95344, USA
sim3@ucmerced.edu
[3] Washington University in St. Louis, St. Louis, MO 63130, USA
bmoseley@wustl.edu
[4] The Hebrew University of Jerusalem, Jerusalem 91904, Israel
yrabani@cs.huji.ac.il

Abstract. Reordering buffer management (RBM) is an elegant theoretical model that captures the tradeoff between buffer size and switching costs for a variety of reordering/sequencing problems. In this problem, colored items arrive over time, and are placed in a buffer of size k. When the buffer becomes full, an item must be removed from the buffer. A penalty cost is incurred each time the sequence of removed items switches colors. In the non-uniform cost model, there is a weight w_c associated with each color c, and the cost of switching to color c is w_c. The goal is to minimize the total cost of the output sequence, using the buffer to rearrange the input sequence.

Recently, a randomized $O(\log \log k)$-competitive online algorithm was given for the case that all colors have the same weight (FOCS 2013). This is an exponential improvement over the nearly tight bound of $O(\sqrt{\log k})$ on the deterministic competitive ratio of that version of the problem (Adamaszek et al., STOC 2011). In this paper, we give an $O((\log \log k\gamma)^2)$-competitive algorithm for the non-uniform case, where γ is the ratio of the maximum to minimum color weight. Our work demonstrates that randomness can achieve exponential improvement in the competitive ratio even for the non-uniform case.

1 Introduction

Motivation and background. In the reordering buffer management problem (RBM) a stream of colored items enters a buffer of limited capacity k, which is used to permute the input stream. Once the buffer is full, any item can be removed from the buffer to the permuted output stream to make room for the

Sungjin Im—Supported in part by NSF grant CCF-1409130.

M.M. Halldórsson et al. (Eds.): ICALP 2015, Part I, LNCS 9134, pp. 78–90, 2015.
DOI: 10.1007/978-3-662-47672-7_7

next input item. This is repeated until the buffer is empty. The goal is to minimize the context switching cost of the output stream due to color changes. The literature considers various cost models. The simplest version is the uniform cost model, where each color switch costs 1. In this paper, we are concerned with the so-called non-uniform cost model, where each color c has a weight w_c, and a switch in the output stream to color c costs w_c. In the online version of the problem, the decision on which item to remove from the buffer must be made on-the-fly without knowing the future input stream. In the offline version of the problem, the entire input stream is known in advance.

RBM models a wide range of applications in production engineering, logistics, computer systems, network optimization, and information retrieval (see, e.g., [6,12,14,15]). In essence, RBM, introduced in [15], gives a nice theoretical framework which allows us to study the tradeoff between buffer size and context switching costs. This tradeoff is evident in many applications. From the perspective of the theory of algorithms, this seemingly simple problem is NP-hard [8], and it presents significant algorithmic challenges both in the offline and the online settings. For instance, simple algorithms such as greedy or FIFO are known to have poor performance.

RBM was studied mostly in the online setting [1,3,5,9–11,15]. The performance guarantees for uniform RBM were essentially resolved in a sequence of papers. There is a deterministic $O(\sqrt{\log k})$-competitive online algorithm and a nearly matching $\Omega(\sqrt{\log k / \log \log k})$ lower bound [1]. The randomized competitive ratio is $\Theta(\log \log k)$. The lower bound is from [1] and the upper bound was recently proved in [5]. So, similar to some other online problems such as paging, randomness gives an exponential improvement in the competitive ratio. In the offline setting, there is an $O(1)$-approximation algorithm [4] (see alternative algorithms in [5,13]), but no hardness of approximation result beyond NP-hardness of the exact solution.

In contrast, there is a wide gap in our understanding of non-uniform RBM. The best known upper bound on the competitive ratio of non-uniform RBM is $\min\{\log k / \log \log k, \sqrt{\log k\gamma}\}$, where γ is the ratio of maximum to minimum color weight. This bound combines the results for two deterministic algorithms from [3] and [1]. For γ which is polynomial in k, the algorithm in [1] nearly matches the deterministic lower bound for the uniform case. The above-mentioned uniform case upper bounds of $O(\log \log k)$ on the randomized competitive ratio and of $O(1)$ on the approximation guarantee seem to use uniformity inherently. So the randomized competitive ratio and the approximability of non-uniform RBM were far from settled. Very recently, an offline approximation guarantee of $O(\log \log k\gamma)$ was shown [13]. Hence, the looming question concerning non-uniform RBM was if randomness can give an exponential improvement of the competitive ratio as it did for the uniform case. (We note that in the case of paging, for instance, the analogous question regarding the randomized competitive ratio of weighted caching remained open for a very long time.)

Our Results. In this paper, we answer the above question in the affirmative. Specifically, we prove the following theorem.

Theorem 1. *There is a randomized $O((\log \log k\gamma)^2)$-competitive online algorithm for the non-uniform RBM problem.*

Our algorithm is based on the online primal-dual schema (see [7] for a survey). The algorithm consists of two phases. In the first phase, the algorithm computes deterministically a feasible fractional solution to an LP relaxation for non-uniform RBM. The LP solution is computed online. In parallel, the algorithm examines the partial LP solution and rounds it online using randomness to get an integral RBM solution which is the output of the algorithm. We lose a factor of $O(\log \log k\gamma)$ in each of the two phases. Interestingly, both phases use a resource augmentation argument to bound the cost of the online solution they produce. In the first phase, the cost of the online generated LP solution is compared against the cost of a dual LP solution for a smaller buffer (see [2, 5, 10] for previous application of this idea in similar contexts). In the second phase, resource augmentation is used to give the integral solution a bit more buffer space than the LP solution that is rounded to generate it. Overall, the integral solution uses a buffer of size k (thus respecting the buffer capacity constraint), the LP primal solution uses a buffer of size $k - \frac{k}{\log k\gamma}$, and the LP dual that is used for bounding the LP cost is for a buffer of size roughly $k - \frac{2k}{\log k\gamma}$. We note that all the previous resource augmentation proofs for RBM either did not apply to the non-uniform case, or they did not prove sufficiently tight bounds. Our proof is new and different from previous proofs.

The first phase of computing the LP solution is generally framed after the algorithm for the uniform case in [5]. The algorithm combines two methods. One method uses the online version of the multiplicative weights update method (see [7]) and works well as long as the color blocks in the buffer do not exceed a size of $O(k/\log k\gamma)$. The other method uses an integral dual fitting-based algorithm that works well when all the color blocks in the buffer have size $\Omega(k/\log k\gamma)$ when they are removed. In [5], the main difficulty was to combine the two algorithms to work well when the buffer contains a mixture of the two types of color blocks. However, the way the two algorithms were combined in [5] inherently uses uniformity, because whenever there was a switch between the two types in one color, other completely arbitrary colors could be charged. In order to facilitate the combination in the non-uniform case, the algorithm and its analysis had to be modified. The result happens to be a simpler and cleaner algorithm and analysis.

The second phase of the algorithm is motivated by the recent offline approximation algorithm in [13]. There, a solution to a slightly different LP was rounded to give an $O(\log \log k\gamma)$ approximation guarantee (without using resource augmentation). However, the algorithm in [13] had several steps that rely crucially on offline information about the LP solution. In particular, that algorithm makes decisions based on when the LP removes certain items in the future. Here we show how to round an LP solution without using future information, exploiting resource augmentation instead. The algorithm is substantially different, simpler than the offline rounding algorithm, and even simpler than the rounding algorithms for the uniform case. (The uniform case rounding algorithms relied crucially on uniformity, and it does not seem that they could be modified to handle color weights.)

Due to lack of space, most of the proofs are deferred to the full version of the paper. We give some informal intuition on the analysis.

2 Preliminaries

In the reordering buffer management problem we consider, there is a sequence $\mathcal{I}$ of n items that arrive over time online. Each item i is associated with a specific color $c(i)$ which stands for the item's type. A single item arrives at every time step from 1 to n and we assume items are indexed in increasing order. Each color c has a positive weight w_c and we denote the ratio of maximum to minimum weight by γ. There is a buffer of size k, and we are allowed to hold items up to the buffer size. Once the buffer becomes full, we are forced to output an item. The goal is to reorder the items using the buffer to minimize the total cost of color switches in the output. Each color switch costs the weight of the color switched to.

Another useful view of the output is to view the sequence of items output as a partition of items into color blocks – a color block or simply block refers to a sequence of items of the same color. In this view, each block of color c contributes w_c to the objective. We assume without loss of generality that each block I is a contiguous sequence of items of the same color ordered in first-in-first-out manner starting with the first arriving item in I. Let $c(I)$ denote the color of the items in I. When a block I is associated with the time t that its first item is removed from the buffer, we call the pair (I, t) a *batch*. For a batch $b = (I, t)$ and an item $i \in I$, we denote by $M_b(i)$ the time that i is removed from the buffer. Note that the total number of all possible blocks is polynomial in n, and so is the total number of possible batches.

For a given input instance, we let OPT_k denote the optimal solution with a buffer of size k. Throughout the analysis, we will compare an algorithm with a buffer of size k to an optimal or linear program solution with a buffer of size smaller than k. This will be clearly indicated when we are making the comparison. We appeal to the following theorem when comparing against a solution with a smaller buffer size. A similar theorem was shown for the unweighted version of the problem and we extend this to the weighted version. In our analysis, we will set k' to be roughly $k - \frac{k}{\log k\gamma}$, which can increase the cost of the optimal solution by at most a constant factor.

Theorem 2. *For any input sequence and $k' < k$, respectively, $\mathrm{OPT}_{k'} \leq O(1) \cdot (\frac{k}{k'} + (k - k')\frac{\log k'\gamma}{k'})\mathrm{OPT}_k$, where OPT_s denotes the cost of the optimal solution using a buffer of size s.*

We use the following linear programming relaxation for the problem, which is defined over $x \geq 0$. It is similar to the relaxation introduced in [3].

$$\min \sum_{I,j} w_{c(I)} x_{I,j} \quad \text{s.t.} \quad \sum_{(I,j), i \in I} x_{I,j} \geq 1 \qquad \forall i = 1, 2, \ldots, n \qquad (1)$$

$$\sum_{(I,j'): j' \leq j < j' + |I|} x_{I,j'} \leq 1 \qquad \forall j = k+1, \ldots, k+n \qquad (2)$$

The quantity $x_{I,j}$, which we call the *height* of batch (I, j), refers to the amount by which the batch (I, j) is scheduled. It is an easy exercise to see this is a valid LP relaxation. The first constraint ensures that each item is processed by an amount of 1. The second constraint ensures that the total height of the intervals at a time step is at most 1. Put $\beta_{i,j} = \sum x_{I,j'}$, where the sum is taken over batches (I, j') such that $i \in I$ and $M_{I,j'}(i) \leq j$. So $\beta_{i,j}$ denotes the total amount item i is processed by time j. Also put $v_{i,j} = 1 - \beta_{i,j}$; this is the remaining "volume" of item i that still needs to be processed at time j. The dual of the linear program is over $y, z \geq 0$ and is given as follows.

$$\max \sum_{i=1}^{n} y_i - \sum_{j=k+1}^{k+n} z_j \quad \text{s.t.} \quad \sum_{i \in I} y_i - \sum_{j'=j}^{j+|I|-1} z_{j'} \leq w_{c(I)} \qquad \forall (I, j) \qquad (3)$$

We will denote the LP for a buffer of size k as LP_k and the dual for a buffer of size k as DP_k.

Our online algorithm will use this LP to guide its decisions. In particular, the algorithm approximately solves this LP in an online fashion. The algorithm simultaneously rounds this LP online to construct the solution. Formally the following is what we mean by solving the LP online. Consider any fixed time t. All batches considered so far end no later than time t – at the next time step $t + 1$, some of batches reaching this time moment t can be extended to time $t + 1$ by adding an extra element to the batch if there is an available element of the same color to be scheduled. In this case, the height of such a batch must remain the same. Note that formally, the batch changes to a larger batch. Also, new batches can start at the current time, and Constraints (2) must be satisfied at each time until time t. Finally, Constraints (1) must be eventually satisfied. Our algorithm and analysis are split into two parts. In Section 3 we show how to construct the LP solution online and in Section 4 we show how to round the LP solution in an online fashion.

3   Solving the LP Online

3.1   The Algorithm

We give an online algorithm that constructs a primal fractional LP solution x for a buffer of size k. We prove that the cost of x, which is $\sum_{(I,j)} w_{c(I)} \cdot x_{I,j}$ is at most $O(\log \log(k\gamma))$ times the optimal cost. In order to prove this bound, the algorithm also constructs a dual solution (y, z) for a buffer of a smaller size $k' = k - \frac{k}{2 \ln(k\gamma)}$. The bound is then obtained by comparing the costs of the primal and dual solutions. The construction of (y, z) is done by scaling an infeasible solution $(\hat{y} + \bar{y}, \hat{z})$, where $(\hat{y}, \hat{z})$ is generated through a version of the online primal-dual schema, and $\bar{y}$ is an extra penalty imposed via a dual fitting procedure. Informally, $(\hat{y}, \hat{z})$ pays for removing from the buffer "small" blocks, and $\bar{y}$ pays for removing "large" blocks. The meaning of "small" and "large" will be made precise in the discussion below. In addition to all of the above variables, we also maintain pseudo-primal variables $\tilde{x}$ that will help us construct the fractional solution x.

The algorithm proceeds as follows. Initially, all primal and dual variables are set to 0 (this includes x, y, z, $\tilde{x}$, $\hat{y}$, $\hat{z}$, $\bar{y}$). Our initial output slot is $t = k+1$, and the first k input items are fully in the buffer. We raise some of the dual variables at a uniform rate, so it is convenient to think about the solution as a function of $\mu \in [0, \infty)$, where $\mu = 0$ denotes the initial state. (Of course, the implementation is not a continuous process—there is a finite sequence of "interesting" values of μ where something happens, and the algorithm can compute those thresholds. However, it is convenient to describe the continuous process.)

The algorithm increases all the variables $\hat{y}_i$ for all input items i in that are in the buffer and have not been scheduled to be removed completely from the buffer (see below), and all the variables $\hat{z}_j$ for all output slots $j \geq t$ at the same rate $d\mu$. Notice that this affects future i-s and j-s. We don't need their values until we reach them, and at that point the value can be computed given the past. Raising some of the variables in $(\hat{y}, \hat{z})$ changes the primal solution x. In order to see how this is done, consider the buffer's contents. Of the total volume of k, there might be some volume that we already decided to remove, but its removal will happen past the current output slot t. We'll call it *phantom volume* and the rest *real volume*. Part of the real buffer volume is kept as *frozen volume* (it will consist only of integral items). We'll call the real volume that is not frozen *active volume*.

Consider a dual constraint indexed (I, j). Put $\sigma_{I,j} = \sum_{i \in I} \hat{y}_i - \sum_{j'=j}^{j+|I|-1} \hat{z}_{j'}$. This is the current *dual cost* of the batch (I, j). Notice that we know the current dual cost even if the batch is matched to output slots we haven't yet reached (and even if it includes items we haven't yet seen). As we raise $(\hat{y}, \hat{z})$, the dual cost of some of the batches may increase. We want to measure only part of this increase, the part that is due to items that contribute to the active volume in the buffer. We call this part the *pseudo-dual cost* and we denote it by $\tilde{\sigma}_{I,j}$. In order to explain this, notice that $\frac{d\sigma_{I,j}}{d\mu}$ is precisely the number of items of color $c(I)$ that contribute to the real volume and are scheduled by (I, j) before the current output slot t. Thus, we raise $\tilde{\sigma}_{I,j}$ at a rate $\frac{d\tilde{\sigma}_{I,j}}{d\mu}$ which is the number of items of color $c(I)$ that contribute to the active volume (i.e., excluding items that are frozen) and are scheduled by (I, j) before the current output slot t. This is what normally happens with $\tilde{\sigma}$. However, there are special "events" that trigger a reset of $\tilde{\sigma}_{I,j}$ to 0. After a reset, $\tilde{\sigma}_{I,j}$ grows again at the rate defined above.

The pseudo-dual costs determine the values of the pseudo-primal variables. We maintain at all times the equation

$$\tilde{x}_{I,j} = \begin{cases} \frac{1}{\ln(k\gamma)} \cdot \frac{\tilde{\sigma}_{I,j}}{w_{c(I)}} & \tilde{\sigma}_{I,j} < w_{c(I)}, \\ \frac{1}{\ln(k\gamma)} \cdot e^{\tilde{\sigma}_{I,j}/w_{c(I)}-1} & \tilde{\sigma}_{I,j} \geq w_{c(I)}. \end{cases}$$

(It should be noted that when we reset $\tilde{\sigma}_{I,j}$ this also resets $\tilde{x}_{I,j}$. However, the reader will soon notice that by Equation (4) this does not reset any actual primal variable—such a reset would violate our intention to construct the solution online.) Now, the items that contribute to the active volume are further classified as *fractional* or *integral*. For each color present in the active volume there are either fractional or integral items (contributing to the active volume), but not both. We say that the active items of a specific color constitute an *active block* in the buffer, which is either a fractional active block or an integral active block.

We are now ready to explain how the schedule up to time $t - 1$ is extended (i.e., how to update the primal solution x). It will be convenient to present the algorithm as choosing batches to schedule and then increasing their height continuously until some event stops the increase and sets the final height of the scheduled batch. Also, when we decide to schedule a batch, we may not know its full extent, because it may end with items that we haven't yet reached in the input stream. However, we will be able to extend the batch as we go along, so in describing the algorithm, we also specify the rule that determines the extent of the batch, and this rule is checked as we go along. Notice that the current output slot t might be already partially filled with previously scheduled batches that haven't reached their end (the partial schedule from t onward is precisely the phantom volume). So our goal is to fill up output slot t and then move on to the next output slot that is not completely filled up.

If an output slot gets filled up, or an item gets scheduled completely, this stops the increase of the height of the current batches, and we execute the following procedure, depending on the event.

Filling up an output slot: When we fill up the output slot t, we have to advance to a later output slot and start the extention process afresh. In this case, new items enter into the buffer, replacing the volume that is removed from the buffer in the filled up output slots (t and possibly later slots). When an item enters the buffer, it is usually frozen, unless the buffer contains an integral active block of this color. In the latter case, the item is sometimes appended to the integral block, according to the rule that specifies the end of the batches that will remove this block from the buffer. If the item is not appended to the integral block, it is frozen as usual.

Scheduling an item entirely: At some point, the initial items of some batch may get scheduled with total height 1. This means that they are either removed from the buffer, or (if they are scheduled in the future) they no longer contribute to the real volume (but they still contribute to the phantom volume). In this case the height of the relevant batch is fixed, and we may continue scheduling a new batch of this color that begins with the items that still contribute to the real volume.

We now describe how an output slot t gets filled up. There are a few cases to consider:

Evicting integral blocks: We first consider the integral active blocks. If there exists (I, j) for which $\tilde{x}_{I,j}$ reached 1 and the items of color $c(I)$ in the buffer are an integral active block B, we set $\bar{y}_i = \frac{w_{c(I)}}{2|B|}$ for all $i \in B$. We reset $\tilde{\sigma}_{I',j'}$ (and hence $\tilde{x}_{I',j'}$) for all (I', j') of color $c(I)$. Then, we schedule batches consisting of this block followed by all the items that can be appended to it assuming it is removed starting from output slot t. The total height of the batches we schedule is 1 (i.e., we remove the block and the appended items completely from the buffer), but we may have to split the height across several batches because some of the output volume beyond time $t - 1$ might be already taken by previously scheduled batches.

Releasing frozen items: We next consider the frozen items in the buffer. We release frozen items in two cases. Firstly, if there is a color c with more than $\frac{k}{100\ln(k\gamma)}$ frozen items in the buffer, we first schedule batches to remove all the volume of the fractional active block of color c from the buffer (they all end with the same last unfrozen item of color c; notice that while we schedule these batches, t might move forward). Then, we reset $\tilde{\sigma}_{I,j}$ to 0 for all batches (I, j) with $c(I) = c$. Finally, we move the frozen items (including additional items that may have been added while removing the preceding fractional volume) to form an integral active block. Secondly, if there is a fractional active block with fewer than $\frac{k}{10\ln(k\gamma)}$ items, we add all the frozen items of this color to the fractional active block. Notice that this event can happen while we are filling up output slot t (because some items get scheduled completely).

Scheduling fractional blocks: We finally consider fractional active blocks (assuming none of the above cases can now be applied). We schedule them in batches in parallel. Such a batch (J, t) consists of the sequence of items in the fractional active block, followed by the items of this color that are in the fractional active block at the time that they are needed to continue the batch. Thus, a fractional batch ends in one of three cases: (i) we haven't reached the next input item of this color; (ii) the next input item of this color is frozen (in this case we say that the batch is *interrupted*); (iii) the next input item of this color begins an integral block. (Notice, that when a batch is being scheduled, we may know only a prefix of the sequence of items in the batch. However, we can extend this sequence on-the-fly and transfer the fractional weight from the prefix to the extended batch as we go along. This does not change the packing of the items in the past time slots, only in future time slots.)

All these fractional batches are scheduled in parallel. Their height is increased as μ grows by the following rate.

$$\frac{dx_{J,t}}{d\mu} = \max_{(I,j)} \left\{ \frac{d\tilde{x}_{I,j}}{d\mu} \; : \; c(I) = c(J) \right\}. \tag{4}$$

We increase their height until, as explained above, some event triggers a change in the batch or in t. A batch (J, t) is said to be *relevant to* (the dual cost of)

(I, j) for every (I, j) that has at some point μ a positive value in the right-hand side of the above expression (i.e., $c(I) = c(J)$ and $\tilde{x}_{I,j}$ grows while $x_{J,t}$ grows).

Regular resets: Occasionally while scheduling fractional batches, we reset some $\tilde{\sigma}_{I,j}$ to 0. We will call this a *regular reset* (to distinguish it from other resets that happen while dealing with integral blocks). Suppose that a fractional batch (J, t) is interrupted at output slot $t' > t$. Let i be the interrupting item (i.e., i is frozen when we reach t'). We consider the set of batches that (J, t) is relevant to. For such a batch (I, j) we reset $\tilde{\sigma}_{I,j}$ if and when the following three conditions hold: (i) The block I contains i; (ii) item i is the first item of I that ever interrupted a batch that is relevant to (I, j); (iii) more than half of the items of color $c(i)$ that contribute to the real volume are frozen. Notice that for any (I, j), a regular reset happens at most once. We denote the value of μ at the time of this regular reset by $\mu_0(I, j)$ and the interrupting item i by $f(I, j)$. If (I, j) never experiences a regular reset, we put $\mu_0(I, j) = \infty$. Also recall that if $\tilde{\sigma}_{I,j}$ is reset to 0, automatically $\tilde{x}_{I,j}$ is reset to 0.

Occasional cleanup: We sometimes clean up the buffer of a color c. The condition for cleaning up color c is as follows: since the previous execution of this step, we just moved past the end of scheduled fractional batches of color c of total height at least $\frac{1}{10}$. (For this purpose we count only batches that are removed while μ increases and not batches that are removed during cleanup.) In this case, we append the frozen items of color c to the color c batches that occupy the current output slot. Then, if there are still items of color c that contribute to the real volume, we schedule additional batches to remove all color c items from the real volume. Obviously, all the frozen items of color c will now be part of the phantom volume.

3.2 Competitive Analysis

Clearly, the algorithm computes a feasible primal solution x. We show that the primal cost of x (which uses a buffer of size k) is proportional to the dual cost of the infeasible solution $(\hat{y} + \bar{y}, \hat{z})$ (which uses a smaller buffer size k'). Then we prove an upper bound on the factor that is needed to scale $(\hat{y} + \bar{y}, \hat{z})$ to a feasible solution (y, z).

Properties of the Primal Solution. We begin with a bound on the phantom volume. This justifies the choice of k'.

Lemma 1. *At any time during the execution of the algorithm, the phantom volume never exceeds* $\frac{12k}{100 \ln(k\gamma)}$.

Lemma 1 immediately implies the following corollary.

Corollary 1. *At any given time, the real volume in the buffer is more than* $k - \frac{12k}{100 \ln(k\gamma)} \geq k'$.

Next we show that the pseudo-primal variables are bounded.

Claim. For every batch (I, j), it holds that $\tilde{x}_{I,j} \leq \frac{11}{10}$ always.

The main idea behind the proof is that $\tilde{x}_{I,j}$ is bounded by the total height of color $c(I)$ batches that are removed since the last reset of $\tilde{x}_{I,j}$. The total height of batches that extend beyond the current output slot is at most 1, and the total height of batches that ended is less than $\frac{1}{10}$, otherwise we would have executed a cleanup step.

Bounding the primal cost. We show that the primal cost of x is proportional to the dual cost of $(\hat{y} + \bar{y}, \hat{z})$.

Lemma 2. *At the end,* $\sum_{(I,j)} x_{I,j} = O(1) \cdot \left(\sum_{i=1}^{n} \hat{y}_i + \sum_{i=1}^{n} \bar{y}_i - \sum_{j=k'+1}^{k'+n} \hat{z}_j \right)$.

The main idea of the proof is the following. We bound separately the cost of scheduling fractional blocks, the cost of evicting integral blocks, and the cost of cleanup. For fractional blocks, we relate the rate by which the primal cost is increased to the rate by which the dual cost is increased. We use the gap between the primal and dual buffer size and the fact that the real volume is most of the buffer (Corollary 1) to show that the dual cost increases sufficiently fast. For integral blocks, the increase in $\sum_{i=1}^{n} \bar{y}_i$ directly bounds the primal cost of evicting those blocks. The cleanup cost is charged against the primal cost of the fractional batches that caused the cleanup.

Dual Feasibility. Here we show that if we scale $(\hat{y} + \bar{y}, \hat{z})$ by a factor of $O(\log \log(k\gamma))$, then we get a feasible dual solution (x, y), namely, for every batch (I, j), $\sum_{i \in I} y_i - \sum_{j'=j}^{j+|I|-1} z_{j'} \leq w_{c(I)}$. So fix a batch (I, j). The main idea of the proof is to partition I into *segments*, according to what the algorithm does with these items. A segment is a maximal substring of items that were all scheduled as a fractional block or an integral block. So there are alternating fractional and integral segments. (Notice that a fractional segment includes also items that were removed during cleanup.) We then partition (I, j) into two sub-batches (I_1, j), (I_2, j') as follows. Let $i \in I$ be the first item that still contributes to the real volume when the algorithm reaches the output slot that (I, j) matches to i. Then, I_1 contains all the items in I that precede i, and I_2 contains the rest of I's items (so $j' = M_{I,j}(i)$). The cost of each sub-batch is bounded using a different argument. Roughly speaking, in (I_1, j) the fractional segments do not incur a positive cost, and at most $O(\log \log(k\gamma))$ integral segments incur a positive cost of $O(w_{c(I)})$. In (I_2, j') there are $O(1)$ segments, and each fractional segment incurs a cost of $O(w_{c(I)}) \cdot \log \log(k\gamma)$. This discussion leads to the following lemma.

Lemma 3. *The pair (y, z) is a feasible dual solution for a buffer of size k'.*

We conclude with the main result of this section.

Theorem 3. *The primal cost of the output x of the LP algorithm is within a factor of $O(\log \log(k\gamma))$ of the LP optimum.*

Proof. Notice that $\sum_{I,j} w_{c(I)} \cdot x_{I,j} \leq O(1) \cdot \left(\sum_{i=1}^{n} \hat{y}_i + \sum_{i=1}^{n} \bar{y}_i - \sum_{j=k'+1}^{k'+n} \hat{z}_j \right) = O(\log \log(k\gamma)) \cdot \left(\sum_{i=1}^{n} y_i - \sum_{j=k'+1}^{k'+n} z_j \right) \leq O(\log \log(k\gamma)) \cdot \mathrm{DP}_{k'} = O(\log \log(k\gamma)) \cdot$

$P_{k'} \leq O(\log \log(k\gamma)) \cdot \mathrm{LP}_k$. The first inequality uses Lemma 2. The second equality uses Lemma 3. The third inequality uses Theorem 2.

Rounding the LP Online

In this section we give an algorithm that rounds the linear program solution of the previous section in an online fashion. Our online rounding requires a sampling which we name α-sampling. The α-sampling is essentially a "boosted-up" independent rounding. Let $0 < \alpha \leq 1$ be a constant to be fixed later. We sample each batch b starting at time t independently with probability $\min\{\alpha, x_b\}/\alpha$, and add it to a pool Bag. Define an item i's α-ready time, t_i^α as the first time t such that there is a batch $b \in$ Bag that schedules i at time t – if no such batch b exists, then set $t_i^\alpha = \infty$. At any time $t \geq t_i^\alpha$, we say that i is α-ready at time t.

To see that we can do the sampling online, note that each batch in the LP keeps the same height from when it starts until it ends. Hence we can immediately decide whether to add a batch to Bag or not when the batch starts in the LP solution.

4.1 Online Rounding Algorithm

The online rounding algorihtms takes as input an online LP solution with a buffer of size $k' = k - \frac{k}{\log k\gamma}$ and returns an online algorithm using a buffer of size k. Recall that reducing the optimal solution's buffer by $\frac{k}{\log k\gamma}$ only increases its cost by a $O(1)$ factor, as we have shown in Lemma 2. In the previous section, we presented how to construct an LP solution online assuming the buffer size is k for notaitonal simplicity. The actual LP solution should has a buffer of size k' and the dual LP's buffer size should be scaled appropriately.

The algorithm at any time always outputs an item for the color that was previously output in the last time step if possible. Otherwise, the algorithm needs to decide which color to switch to. The algorithm has several rules on which color to switch to at time t and attempts to execute the rules in the following order. The first three rules are easy cases and the crux of the algorithm is the final two rules. The rules are similar to the algorithm in [13]. However, the algorithm in [13] required an additional rule and also the main rules in their algorithm used future offline information from the LP.

We require some notation to define formally the algorithm. Let ϵ be any constant between 0 and $1/100$ and α be a constant at most ϵ. We will later set $\epsilon = 1/100$ and $\alpha = \epsilon$. Let $\mathcal{B}(t)$ denote (the set of items in) the algorithm's buffer at time t. Let $n_c^A(t)$ denote the number of items for color c in $\mathcal{B}(t)$. Let $n_c^O(t)$ be the number of items in the LP at time t for color c that have been processed by at most $1/2 + \epsilon$. Let $C_s(t)$ contain all colors c where $0 < n_c^A(t) \leq \frac{k}{\log^3 k\gamma}$ and $C_b(t)$ contain all colors c where $n_c^A(t) > \frac{k}{\log^3 k\gamma}$. Let $E^O(t)$ be the set of items that have been processed by at most $1/2 + 2\epsilon$ in the LP at time t that are not in $\mathcal{B}(t)$, i.e. $E^O(t) := \{i \notin \mathcal{B}(t) \mid i \leq t, \beta_{i,t} \leq 1/2 + 2\epsilon\}$. Let $c^*(t)$ be the color such that batches in the LP for color $c^*(t)$ that intersect time t is greater than $1/2$ if it exists. Let $v^O_{c,t} = \sum_{i,c(i)=c} 1 - \beta_{i,t}$ denote the remaining volume of items for color c in the LP at time t.

Algorithm:

Rule (i) If there is an item in $i \in \mathcal{B}(t)$ processed by ϵ in the LP, switch to color $c(i)$.

Rule (ii) If there is an item $i \in \mathcal{B}(t)$ that is α ready at time t, switch to color $c(i)$.

Rule (iii) If there is a color c where $n^A_c(t) \geq k/10$, switch to color c.

Rule (iv) If the LP has processed items in $\mathcal{B}(t)$ corresponding to colors in $C_s(t)$ by a total of at least $\frac{|E^O(t)|}{8} + \frac{k}{2\log k\gamma}$ by time t then switch to the color of minimum weight that is not $c^*(t)$.

Rule (v) We perform this rule if none of the others apply. In this case, the algorithm switches to a color $c \in C_b(t)$ such that $n^A_c(t) \geq \frac{10}{11}v^O_{c,t}$. (We can show that such a color exists.)

References

1. Adamaszek, A., Czumaj, A., Englert, M., Räcke, H.: Almost tight bounds for reordering buffer management. In: STOC, pp. 607–616 (2011)
2. Adamaszek, A., Czumaj, A., Englert, M., Räcke, H.: Optimal online buffer scheduling for block devices. In: STOC, pp. 589–598 (2012)
3. Avigdor-Elgrabli, N., Rabani, Y.: An improved competitive algorithm for reordering buffer management. In: SODA, pp. 13–21 (2010)
4. Avigdor-Elgrabli, N., Rabani, Y. : An improved competitive algorithm for reordering buffer management. In: FOCS, pp. 1–10 (2013)
5. Avigdor-Elgrabli, N., Rabani, Y.: An optimal randomized online algorithm for reordering buffer management (2013). CoRR, 1303.3386
6. Blandford, D., Blelloch, G.: Index compression through document reordering. In: Proceedings of the Data Compression Conference, DCC 2002, pp. 342-. IEEE Computer Society, Washington, DC (2002)
7. Buchbinder, N., Naor, J.: The design of competitive online algorithms via a primal-dual approach. Foundations and Trends in Theoretical Computer Science **3**(2–3), 93–263 (2009)
8. Chan, H.-L., Megow, N., Sitters, R., van Stee, R.: A note on sorting buffers offline. Theor. Comput. Sci. **423**, 11–18 (2012)
9. Englert, M., Räcke, H., Westermann, M.: Reordering buffers for general metric spaces. Theory of Computing **6**(1), 27–46 (2010)
10. Englert, M., Westermann, M.: Reordering buffer management for non-uniform cost models. In: Caires, L., Italiano, G.F., Monteiro, L., Palamidessi, C., Yung, M. (eds.) ICALP 2005. LNCS, vol. 3580, pp. 627–638. Springer, Heidelberg (2005)
11. Gamzu, I., Segev, D.: Improved online algorithms for the sorting buffer problem on line metrics. ACM Transactions on Algorithms **6**(1) (2009)
12. Gutenschwager, K., Spiekermann, S., Vos, S.: A sequential ordering problem in automotive paint shops. Intl. J. of Production Research **42**(9), 1865–1878 (2004)

13. Im, S., Moseley, B.: New approximations for reordering buffer management. In: SODA, pp. 1093–1111 (2014)
14. Krokowski, Jens, Räcke, Harald, Sohler, Christian, Westermann, Matthias: Reducing state changes with a pipeline buffer. In: VMV, p. 217 (2004)
15. Räcke, H., Sohler, C., Westermann, M.: Online scheduling for sorting buffers. In: Möhring, R.H., Raman, R. (eds.) ESA 2002. LNCS, vol. 2461, pp. 820–832. Springer, Heidelberg (2002)

Serving in the Dark should be done Non-Uniformly

Yossi Azar and Ilan Reuven Cohen[✉]

Blavatnik School of Computer Science, Tel-Aviv University, Tel Aviv, Israel
azar@tau.ac.il, ilanrcohen@gmail.com

Abstract. We study the following balls and bins stochastic game between a player and an adversary: there are B bins and a sequence of ball arrival and extraction events. In an arrival event a ball is stored in an empty bin chosen by the adversary and discarded if no bin is empty. In an extraction event, an algorithm selects a bin, clears it, and gains its content. We are interested in analyzing the gain of an algorithm which serves in the dark without any feedback at all, i.e., does not see the sequence, the content of the bins, and even the content of the cleared bins (i.e. an oblivious algorithm). We compare that gain to the gain of an optimal, open eyes, strategy that gets the same online sequence. We name this gain ratio the "loss of serving in the dark".

The randomized algorithm that was previously analyzed is choosing a bin independently and uniformly at random, which resulted in a competitive ratio of about 1.69. We show that although no information is ever provided to the algorithm, using non-uniform probability distribution reduces the competitive ratio. Specifically, we design a 1.55-competitive algorithm and establish a lower bound of 1.5. We also prove a lower bound of 2 against any deterministic algorithm. This matches the performance of the round robin 2-competitive strategy. Finally, we present an application relating to a prompt mechanism for bounded capacity auctions.

1 Introduction

The behavior of an algorithm inherently depends on its input. In some cases the input is only partially known to the algorithm (e.g. online algorithms, distributed algorithms and incentive compatible algorithms) and it may still perform well. In extreme cases the input is virtually unknown to the algorithm. In these cases the algorithm needs to act (almost) independently of the input. Such algorithms are called *oblivious algorithms*. Typically, oblivious algorithms act uniformly at random over their choices. For example, consider a case where there are m weighted balls and n bins. The algorithm needs to assign the balls to the bins as to minimize the maximum load over all bins (where the load of a bin is the sum of weights of balls which are assigned to it). Consider a simple case where $m = n^2$, n of them are of weight 1 and the others are of weight 0. Clearly, the optimal solution is 1. An oblivious algorithm does not know the

---

Supported in part by the Israel Science Foundation and by the Israeli Centers of Research Excellence (I-CORE) program (Center No. 4/11).

M.M. Halldórsson et al. (Eds.): ICALP 2015, Part I, LNCS 9134, pp. 91–102, 2015.
DOI: 10.1007/978-3-662-47672-7_8

weights (it only knows n and m). Clearly any deterministic oblivious algorithm may encounter a maximum of load of n. Fortunately, using randomization an algorithm which assigns each ball uniform at random achieves an expected maximum load of $\log n / \log \log n$. In this paper, we consider a problem where the best previous known oblivious algorithm is to select uniformly at random. Interestingly, we show that using a non-uniform distribution improves the performance. This problem is called *serving in the dark* and has an application in prompt mechanism design for packet scheduling.

The Serving in the Dark Game. In this game, there is an arbitrary sequence of ball arrival events and ball extraction events. On the arrival of a new ball, the adversary assigns the ball to an unoccupied bin of its choice. The ball is discarded only if all bins are occupied. On an extraction event the algorithm chooses one of the bins, clears it, and gains its contents. Once the sequence ends, all the bins that contain balls are cleared and their content is added to the total gain. The goal of the algorithm is to maximize the number of cleared balls for the sequence. If the algorithm can see the content of the bins, at any extraction step it would choose a bin with a ball, if one exists, thereby maximizing the total gain. This gain is defined as the **optimal gain** (note that in such a case, the adversary's choices of which bin to assign the ball to are irrelevant). We consider an algorithm which *serves in the dark*. Specifically, the algorithm is not aware of the arrival events and of the content of the bins. Moreover, when the algorithm clears a bin, it does not see the bin content. Equivalently, the algorithm does not get any feedback during the sequence (as such, it can also be called an *oblivious algorithm*). We can describe any sequence which contains N extractions as a sequence of N time units $X = \langle X_1, \ldots, X_N \rangle$, where at time j, $X_j \geq 0$ balls arrive and then one extraction event takes place. In this paper we compare the gain achieved by an algorithm that serves in the dark to the **optimal gain** on the worst possible sequence.

The most natural algorithm is the round robin on the bins, which is $(2-1/B)$-competitive. We show in this paper that this is the best possible deterministic algorithm. Hence, in order to improve this bound one needs to use randomization. The most natural randomized algorithm is to choose a bin independently and uniformly at random. For such an algorithm, the choices the adversary makes for the assignment of the balls become irrelevant and the game becomes somewhat degenerate. For the uniform algorithm, the exact competitive ratio for the worst sequence has been determined in [4] to be approximately 1.69.

On one hand, it may seem that the best possible strategy for an algorithm is to choose a bin uniformly at random, since the algorithm does not get any feedback during the sequence. Hence, if some bin is chosen with a smaller probability, then the adversary is more likely to put the next ball in that bin. On the other hand, although no information is provided, the algorithm might want to choose a bin that has not been examined recently. Here we show that by using a **non-uniform** distribution we can substantially improve the competitive ratio to 1.55 and get relatively close to the lower bound of 1.5 that we establish.

Application: Prompt Mechanism Design for Packet Scheduling. Con sider the basic packet scheduling mechanism in which an online sequence c packets with arbitrary private values arrives to a network device that can accom modate up to B packets. The device can transmit one packet in each time step The goal is to maximize the overall value of the transmitted packets. A trivia greedy mechanism keeps the B packets with the highest values at any moment i time, and transmits the packet with the highest value when possible. This mech anism is optimal, truthful, but not prompt, i.e., the price cannot be determinec at the time of transmission (see [7]). A prompt mechanism can be designec by using a *value-oblivious* algorithm. Such algorithms have the property tha during transmission no preference is given to a packet with a higher value. We note that value-oblivious algorithms may inspect the values of packets on thei arrival. Therefore, one can assume, without loss of generality, that any value-oblivious algorithm keeps the B packets with the highest values at any moment in time. One example of a value-oblivious algorithm is the FIFO algorithm. which transmits the earliest packet in the buffer. This algorithm is known tc be $2 - 1/B$-competitive against the absolute optimum [10]. An algorithm which transmits a packet independently and uniformly at random is approximately 1.69-competitive [4]. A natural question is whether one can gain from using a non-uniform distribution. This question can be reduced, by the zero-one princi ple [5], to the the *Serving in the dark Game* described above.

1.1   Our Results

In this paper, we provide a *time-order based* algorithm that uses a non-uniform distribution over the bins. This algorithm is approximately 1.55-competitive for the *serving in the dark game*, which improves the previously known results. Recall that the competitive ratio of a randomized algorithm is the worst ratio over all sequences between the **optimal gain** and the *expected gain* of the algorithm.

Theorem 1. *There exists a randomized algorithm for serving in the dark which is* $(1.55 + o(1))$*-competitive, where* $o(1)$ *is a function of* B.

We also show a relatively close lower bound for **any** randomized algorithm for serving in the dark.

Theorem 2. *Any randomized algorithm for serving in the dark is at least* 1.5-*competitive.*

The lower bound for Theorem 2, and all other missing details and proofs are in the appendix. In order to prove Theorem 1, we actually prove a more general theorem, for any *time-order based* algorithm. A *time-order based* algorithm is described by a probability distribution on B ordered bins, where the order is determined by the last time a bin has been cleared. A probability distribution is called monotone non-decreasing, if for any two bins, the most recent bin in the order does not have a higher probability than the least recent bin. We analyze any monotone *time-order based* algorithm as follows:

Theorem 3. *For any p a monotone non-decreasing and bounded probability distribution on $[0, 1]$, let $H(x) = \int_x^1 p(y)dy$. Let f be the solution to the differential equation $f'(x) = -H(f(x))$, with $f(0) = 1$. The competitive ratio of a block time-order based algorithm which uses p is $\max_{x \geq 1} \left\{ \frac{x}{x - f(x) + f(x-1) - 1} \right\} (1 + o(1))$, where $o(1)$ is a function of B.*

In the above theorem $f(x)$ corresponds to the fraction of balls in the bins starting with B balls followed by xB extractions step with no arrivals.

Application: Prompt Mechanisms for Bounded Capacity Auctions. We can use the a serving in the dark algorithm to establish a truthful and prompt selection mechanism for bounded capacity auctions. A bounded capacity auction is a single-item periodic auction for bidders that arrive online, in which the number of participating bidders is bounded, e.g., when the auction room has a limited size. We can apply the serving in the dark algorithm for designing a mechanism for packet scheduling. Specifically, we design a truthful prompt mechanism that is approximately $(1.55 + o(1))$-competitive.

1.2   Our Approach and Techniques

An essential component in our approach is to utilize a deterministic fractional algorithm, which describes in vector form the 'expected' content of the bins, since we do not know how to analyze directly the randomized algorithm. The *deterministic* fractional algorithm will be used as a proxy for the analysis. We analyze the gain of this fractional algorithm compared with the gain of the optimal gain-maximizing strategy. This fractional algorithm is designed in a natural way to correspond to the randomized algorithm and depends on its probability density function.

It is important to note that our analysis is significantly more complicated than the analysis of the uniform distribution case. Specifically, when using the uniform distribution the state of all the bins can be described by a single number, the number of balls in the bins. For arbitrary distributions, presenting the state as a vector is crucial in analyzing the behavior of the algorithm, since different bins are chosen with different probabilities and the probability of choosing one specific bin changes over time. We carefully examine the arrival events and the extraction events for this vector. Our techniques enable us to consider any monotone probability density function for a time-order based algorithm, and characterize up to one parameter the worst input sequence for a fractional algorithm that uses this distribution.

Next, we compare the randomized algorithm with the fractional one. We observe that the gain of this algorithm is not the expected gain of the randomized algorithm, but rather dominates it. Nevertheless, we still establish that it is within a $1 + o(1)$ factor away from the expected gain of the randomized algorithm. For the uniform distribution this was previously done by defining a simple supermartingale on the Markov process of the difference between the number of balls in the fractional algorithm and that in the randomized algorithm. Here, we have to define a chain of separate Markov processes for groups of consecutive bins. In addition, each Markov process in the chain influences the next one. In order to deal with this complex process, we design sequence of hybrid algorithms, each has some randomized part followed by some fractional part. We compare the fractional algorithm to the randomized algorithm by performing a sequence of comparisons between a consecutive hybrid algorithms. Finally by combining the result comparing the fractional algorithm and the optimum algorithm with the result comparing randomized algorithm to the fractional one enables us to prove the upper bound.

1.3   Further Related Work

In the example that we previously considered there are m weighted balls and n bins. The algorithm needs to assign the balls to the bins to minimize the maximum load. Sanders [17] considered the case where the size of the balls are unknown to the algorithm. He analyzed an oblivious algorithm which assigns each ball uniformly at random to a bin and proved that the worst ratio of the maximum load to the optimal maximum load is achieved when a subset of balls have equal size and the rest are 0. Obliviously, this ratio is bounded by $\log n/\log\log n$. As mentioned before any deterministic oblivious algorithm will achieve a ratio of at least n. Another version of the balls and bins problem is assigning B balls to B bins, where the goal is to maximize the number of non empty bins, where the bins may be permuted by an adversary. A simple result states that if the balls are placed independently and uniformly at random in the bins, then the expected fraction of full bins is $1 - 1/e$. If this procedure could have been performed under light i.e. the permutation at each step was known, then one could deterministically place each ball in a different bin, and hence the fraction of full bins would have been 1.

There are other randomized balls and bins stochastic processes that have been analyzed using various techniques such as martingales and Azuma's inequality. We refer the reader to the papers [1,2,8,9,11–14] and to the references therein for a more comprehensive review of the literature.

Another example of an algorithm that behaves oblivious to the input is an algorithm for scheduling jobs with release times on identical machines in order to minimize the weighted completion time, introduced by Schulz and Skutella [18]. Their 2-competitive algorithm assigns each job uniformly to a random machine, independently of the assignment of other jobs, while the order of processing the jobs on each machine depends on the input. For the flow time Chekuri et al. [6] showed a constant competitive ratio using the same random oblivious dispatching with extra resources.

Another problem that can be viewed as serving requests independently of the input is the well studied oblivious routing [3,15,16]. Here, a graph is given together with a set of requests to connect pairs of vertices with arbitrary demands. At the preprocessing stage, the graph is given without the requested pairs. The oblivious routing algorithm determines a route (a flow) between any two vertices, independently of their demands and the existence of other pairs.

For each request pair, the service is performed using the predetermined flow for this pair scaled by their demand. This achieves logarithmic approximation with respect to the optimal solution for the specific pairs and demands.

2 The Model

Given B bins, consider an arbitrary sequence of arrival events and extraction events.

- Arrival event: a new ball is stored in an unoccupied bin determined by the adversary. If all the bins are occupied, then the ball is discarded.
- Extraction event: the algorithm chooses one of the bins, clears it, and gains its content.

The goal of the algorithm is to maximize the number of extracted balls for the sequence. We assume that all the balls that remain in the bins at the end of the sequence are extracted (this is not required if the optimal gain is large enough).

We consider an algorithm which serves in the dark, i.e., without any input during the whole process (except for the value of B). The algorithm can be viewed as a probability distribution over all infinite sequences of numbers in the set $\{1, \ldots, B\}$. Note that the input sequence is arbitrary and the algorithm does not know the sequence or does not know when the sequence ends. We assume that the adversary knows the algorithm, and that at any moment of time it sees the contents of all bins (even if the algorithm is randomized).

We can describe any sequence which contains N extractions as a sequence of N time steps $X = \langle X_1, \ldots, X_N \rangle$, where at time step j, $X_j \geq 0$ balls arrive and then one extraction event takes place. For a given algorithm ALG, we set $G_i = 1$ if a ball is extracted in extraction step i, and $G_i = 0$ otherwise. Let L_i be the number of balls in the bins before extraction step i: $L_i = \min\{L_{i-1} + X_i - G_{i-1}, B\}$. Denote by O_i the number of overflown (discarded) balls at arrival time i. We have $O_i = \max\{L_{i-1} + X_i - G_{i-1} - B, 0\}$. Using these notations, let $G(X)$ bet the total gain of ALG on a sequence X. By definition, $G(X) = \sum_{i=1}^{N} G_i + L_{N+1} = \sum_{i=1}^{N} X_i - \sum_{i=1}^{N} O_i$. The gain of a randomized algorithm is its expected gain over all algorithm's coins tosses.

The open eye optimal gain: We compare the gain achieved by an algorithm that serves in the dark with the optimal gain on the worst possible sequence. The optimal algorithm can see the contents of the bins at any extraction step and would always choose a bin with a ball if one exists. By that it would maximize the total gain (defined as **optimal gain**). Since the optimal algorithm (called OPT) sees the content of the bins, the choices of the adversary are irrelevant. The gain of OPT in each step i is $G_i^{\mathrm{OPT}} = \min\{L_i^{\mathrm{OPT}}, 1\}$. We use the standard measure to compare a general algorithm with the optimal one (denoted as ϱ): $\rho(X) = \frac{G^{\mathrm{OPT}}(X)}{G(X)}$ and $\varrho = \max_X \rho(X)$.

3   Deterministic Serving Algorithms

One simple deterministic serving algorithm is to perform a round robin over the bins, i.e., on an extraction event the algorithm chooses the least recent bin that it had cleared.

Theorem 4. *The round robin serving algorithm is* $(2 - 1/B)$*-competitive.*

The proof is a simpler version of the proof given for FIFO packet scheduling [10]. Next, we give a bound for the competitiveness of any deterministic serving algorithm.

Theorem 5. *Any deterministic serving algorithm is at least* $(2 - 1/B)$*-competitive.*

4   Randomized Algorithms and their Analysis

In this section, we design and analyze randomized serving in the dark algorithms.

4.1   Time-Order Based Randomized Algorithms

Let $p : [0, 1] \to \mathbb{R}$ be a monotone non-decreasing probability density function (i.e., $\int_0^1 p(x)dx = 1$). We define a *time-order based* algorithm, denoted by TOB_p, as follows:

On each extraction event:

- Order the bins according to their last extraction step (latest is first).
- Clear one bin, where the probability to clear the j'th ordered bin is $\displaystyle\int_{(j-1)/B}^{j/B} p(x)dx$.

Algorithm 1: Time-Order Based Algorithm TOB_p

The algorithm may be described as follows: the bins are ordered in a line of length B. At each step a position in the line is chosen with a fixed monotone non-decreasing probability distribution function p on the positions. The ball (if exists) is extracted from the corresponding bin and then the bin is moved to the beginning of the line.

4.2   Grouping Bins Together

In order to use a concentration result for the bins, we generalize the algorithm so that instead of B ordered bins, we keep $b_1, \ldots, b_{B/K}$ ordered blocks of volume K (i.e., with K bins each), for some constant $K \geq 1$. We impose no internal order inside a block of bins. The algorithm, denoted as TOB_p^K, uses a data structure of list of B/K blocks, where each block contains K bin indices. On an extraction

vent, it chooses a block r with probability $q^r = \int_{K(r-1)/B}^{Kr/B} p(x)dx$. Afterwards,
chooses one of the bins in the block uniformly at random and clear it. Finally,
or each block $r' < r$ a bin is chosen uniformly at random from it and associate
to the next ordered block, where the extracted bin is associated to the first
lock in the order.

Algorithm TOB_p^K on extraction event i:

- Choose block c_i with probability $Pr[c_i = r] = q^r = \int_{K(r-1)/B}^{Kr/B} p(x)dx$.
- Choose a bin $j_r \in b_r$ from each block $r \leq c_i$ uniformly at random.
- Clear bin j_{c_i} (from block c_i).
- Associate bin j_r with block b_{r+1} (for $r < c_i$), associate bin j_{c_i} with block b_1.

Algorithm 2: The Block Time-Order Based Algorithm - TOB_p^K

Note that the block time-order based algorithm with $K = 1$ is exactly the
time-order based algorithm introduced above. We introduce the following notation with respect to the K-block time-order based algorithm with monotone
distribution p ,called TOB_p^K, (we omit K and p if they are clear from the context). Let c_i be the block chosen in step i. Let E_i^r be the indicator of whether
in extraction step i a ball is extracted from block r, i.e., $E_i^r = 1$ if $r \leq c_i$ and j_r
contains a ball, 0 otherwise. The gain in step i is $G_i = E_i^{c_i}$. Clearly, the gain is
equal to 1 if we extracted a ball from the chosen block. Let L_i^r be the number
of balls in the r'th ordered block before extraction step i. By the definition of
the algorithm, the load of block r after the i'th extraction is $L_i^r + E_i^{r-1} - E_i^r$ if
$r \leq c_i$, otherwise it remains L_i^r.

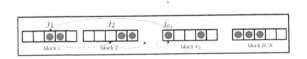

Fig. 1. The algorithm's selection in some step i: the selected block is c_i, j_{c_i} contains
a ball therefore $G_i = E_i^{c_i} = 1$. Note also that the algorithm choses a bin from each
block before c_i and associates this bin with the next block, and that the algorithm
associates the extracted bin j_{c_i} with the first block. Specifically, in the above example,
$E_i^1 = 1, E_i^2 = 0$, therefore, the load in the first block decreased by one and the load in
the second block increased by one.

Next, let us consider arrival events. Since the algorithm uses a fixed monotone
non-decreasing distribution over the ordered blocks, it is easy to determine the
optimal strategy of the adversary.

Observation 6. *For the block time-order based algorithm, on an arrival event the adversary assigns a ball in the block with the smallest index that has an empty bin.*

By the above observation, on an arrival event the number of balls in the minimum index block block whose load is smaller than K increases by one. Note that for a given sequence X, the load in each block is a random variable. Since a new ball is stored in the first vacant bin, the block index of this new ball is also a random variable, which makes the analysis of the algorithm complicated. In order to circumvent this difficulty, we next introduce a deterministic fractional algorithm that is close to the randomized one.

4.3   Fractional Deterministic Algorithms

We define a deterministic algorithm that 'behaves like' the expectation of the TOB_p^K algorithm. Given an input sequence X, the gain and the current loads of the blocks in TOB_p^K are (integer) random variables, since there is randomization in the extraction events. Alternatively, we define FRC_p^K algorithm as a deterministic fractional algorithm, where a fractional of a ball is the deterministically extracted. In each step, the fraction of the balls that is extracted in FRC_p^K corresponds to the the probability that a ball is extracted in TOB_p^K given the current state. Specifically, the load of a block after an extraction event is defined as (we omit FRC, K, p, X if those are clear from the context)

$$L_{i-1}^r + \left(E_i^{r-1} - E_i^r \right) \sum_{j=r}^{B/K} q^j, \tag{1}$$

where $E_i^r = L_i^r / K$. The gain in each step is $G_i = \sum_{r=1}^{B/K} q^r E_i^r$. The arrival of balls is defined as for the randomized algorithm. Note that since the load is fractional, a ball can be split into parts lying in several different blocks.

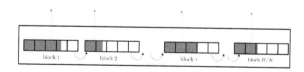

Fig. 2. The fractional block time-order based algorithm, FRC in which L_i^j, E_i^j, G_i are fractional numbers. In the example above $E_i^r = 3/5$.

4.4   Analyzing the Fractional Algorithm versus the Optimal Algorithm

The analysis consists of two parts. In the first part we characterize the worst sequence for any distribution p. In the second part we analyze the worst gain ratio of that sequence. By combining the two parts we bound the maximum gain ratio using p. The proof is in the appendix.

Theorem 7. *Given an arbitrary monotone non-decreasing and bounded probability density function p, let $H^p(x) = \int_x^1 p(y)dy$. Let f be a function that satisfies $f(0) = 1$ and $f'(x) = -H^p(f(x))$. The competitive ratio ϱ of the fractional algorithm that uses the function p:* $\varrho^{\mathrm{FRC}} \leq \max_{x \geq 1} \left\{ \frac{x}{x - f(x) + f(x-1) - 1} \right\} (1 + o(1))$.

4.5 Analysis of the Randomized Algorithm versus the Fractional One

In order to compare the fractional algorithm with the randomized one, it is sufficient to analyze input instances in which the fractional algorithm does not overflow. The reason is that removing balls which overflow in the fractional algorithm from the sequence, does not decrease the gain of the fractional algorithm and does not increase the gain of the randomized algorithm. We prove that with high probability a randomized algorithm with slightly larger volume does not overflow on such sequences. First, we compare a single fractional block to a single randomized block:

We define the **extraction probability** of a block with index i as the probability that a block with index at least i will be chosen. We define the **input sequence** of a block as the sum of: (A) the volume overflown from the previous block and (B) the volume extracted from the previous block that was not added to the gain. Additionally, we define the **output sequence** of a block is its extracted volume plus its overflown volume. Note that, the load of a block depends on the block's input sequence and on its extraction probability.

We prove that any input sequence that does not overflow a fractional block, does not overflow a 'slightly larger' randomized block with high probability. A slightly larger means that we increase the randomized block size as well as increase its extraction probability . We prove that this implies that their output sequences are close for any input sequence. Finally, we introduce a hybrid algorithm HYB_m, in HYB_m the first m blocks are randomized and the rest are fractional. Note that in the HYB_m algorithm the input sequence for the block $m+1$ is a random sequence. We compare a HYB_m algorithm to a HYB_{m+1} algorithm by replacing block $m + 1$ (a fractional block) with a randomized block. Specifically, using coupling on the randomized choices in the first m blocks we get that the input sequences for the block $m+1$ are the same. Next, we compare the output sequence of the block $m+1$ in HYB_{m+1} with the deterministic output (after the coupling) of block $m+1$ in HYB_m. Specifically, given a sequence and a coupling for which HYB_m does not overflow then HYB_{m+1} with a slightly larger fractional block does not overflow with high probability. By applying this iteratively for $m = 0$ to B/K, we prove that with high probability the randomized algorithm will not overflow and deduce the following theorem:

Theorem 8. *For any fractional block algorithm FRC there exists a time order base TOB algorithm such that $G^{\mathrm{TOB}}(X) \geq G^{\mathrm{FRC}}(X)(1 - o(1))$, for any input sequence X.*

Single Fractional Block versus Randomized Block. Recall that for a fractional or a randomized block, the load in each step depends only on its **input sequence**, and its **extraction probability** as defined above. First, we bound (with high probability) the difference in the load between a fractional block and a randomized block for sequences where the fractional block does not overflow. Let $\epsilon_{N,\Delta K} = N^3 \exp\left(-(\Delta K)^2/8N\right)$ (we omit $N, \Delta K$).

Lemma 1. *Let* B_{FRC} *be a fractional block of size* K *and extraction probability* $Q^{B_{\mathrm{FRC}}}$ *and* B_{TOB} *be a randomized block of size* $K+\Delta K$ *and extraction probability* $Q^{B_{\mathrm{TOB}}} = Q^{B_{\mathrm{FRC}}} \frac{K+\Delta K}{K}$. *Then for any input sequence* X *that* B_{FRC} *does not overflow,*

$$\Pr(\exists i \leq N : |L_i^{B_{\mathrm{FRC}}}(X) - L_i^{B_{\mathrm{TOB}}}(X)| \geq \Delta K) \leq N^3 \exp\left(-(\Delta K)^2/8N\right) = \epsilon.$$

Next, we examine the output sequence Y of the FRC block compared to output sequence of the TOB block for any input sequence X. The output sequence is defined as the extracted volume plus the overflow volume, i.e., $Y_i = E_i + O_i$.

Lemma 2. *Let* B_{FRC} *be a fractional block of size* K *and extraction probability* $Q^{B_{\mathrm{FRC}}}$, *and let* TOB *be a randomized block of size* $K + \Delta K$ *and extraction probability* $Q^{B_{\mathrm{TOB}}} = Q^{B_{\mathrm{FRC}}} \frac{K+\Delta K}{K}$. *For any input sequence* X *we have with probability of at least* $1-\epsilon$ *that* $-\Delta K \leq \sum_{j=1}^{i} \left(Y_j^{B_{\mathrm{TOB}}}(X) - Y_j^{B_{\mathrm{FRC}}}(X)\right) \leq 3\Delta K$.

The Hybrid Algorithm. We define the hybrid algorithm HYB_m in which the first m blocks are randomized and the rest are fractional. On extraction step i a block c_i is chosen. A randomized block r ($r \leq m$) will extract from one of its bins if $r \leq c_i$. The extraction from the fractional block is done as in the fractional algorithm independent of the choice c_i. Note that HYB_0 is a fractional algorithm and that $\mathrm{HYB}_{B/K}$ is a randomized algorithm. We design HYB_{m+1} such that all the blocks except block $m+1$ and block B/K are with the same size and extraction probability as in HYB_m. In HYB_{m+1} we set block $m+1$ to be of size $K + \Delta K$ and extraction probability $Q \cdot (K + \Delta K)/K$, where K and Q are the size and extraction probability of block $m+1$ in HYB_m. In addition, we set the last block of HYB_{m+1} to be of size $\tilde{K} + 4\Delta K$ and set its extraction probability to $\tilde{Q} \cdot (\tilde{K} + 4\Delta K)/\tilde{K}$, where $\tilde{K}$ and $\tilde{Q}$ are the size and the extraction probability of the last block in HYB_m. Denote X^m as the (random) input sequence to the block m. The following observation follows immediately from the above construction of HYB_{m+1}.

Observation 9. *For any input sequence* X^m *such that* $\mathrm{HYB}_{m-1}(X^m)$ *does not overflow then* $\mathrm{HYB}_m(X^m)$ *has at least* $4\Delta K$ *vacant volume in each step.*

Lemma 3. *If a sequence* X *does not overflow* HYB_{m-1} *with probability of at least* $(1-\epsilon)^{m-1}$ *then* X *does not overflow* HYB_m *with probability of at least* $(1-\epsilon)^m$.

By applying Lemma 3 B/K times we obtain the following

Corollary 1. *If a sequence* X *does not overflow* HYB_0 *then* X *does not overflow* $\mathrm{HYB}_{B/K}$ *with probability of at least* $(1-\epsilon)^{B/K}$.

Putting Everything Together. The summary of the proof of Theorem 8 is in the appendix. By combining Theorem 7 and Theorem 8, we conclude that $\rho^{\text{TOB}} \leq \max_{x \geq 1} \left\{ \frac{x}{x - f(x) - 1 + f(x-1)} \right\} (1 + o(1))$, which completes the proof for Theorem 3. The specific distribution function to prove Theorem 1 is in the appendix.

References

1. Alon, N., Spencer, J.H.: The Probabilistic Method, 2nd edn. Wiley, New York (2000)
2. Azar, Y., Broder, A.Z., Karlin, A.R., Upfal, E.: Balanced allocations. SIAM J. Comput **29**(1), 180–200 (1999)
3. Azar, Y., Cohen, E., Fiat, A., Kaplan, H., Räcke, H.: Optimal oblivious routing in polynomial time. J. Comput. Syst. Sci **69**(3), 383–394 (2004)
4. Azar, Y., Cohen, I.R., Gamzu, I.: The loss of serving in the dark. In: Proceedings 45th Annual ACM Symposium on Theory of Computing, pp. 951–960 (2013)
5. Azar, Y., Richter, Y,: The zero-one principle for switching networks. In: Proceedings 36th Annual ACM Symposium on Theory of Computing, pp. 64–71 (2004)
6. Chekuri, C., Goel, A., Khanna, S., Kumar, A.: Multi-processor scheduling to minimize flow time with epsilon resource augmentation. In: Proceedings of the 36th Annual ACM Symposium on Theory of Computing, Chicago, IL, USA, June 13–16, 2004, pp. 363–372 (2004)
7. Cole, R., Dobzinski, S., Fleischer, L.K.: Prompt mechanisms for online auctions. In: Monien, B., Schroeder, U.-P. (eds.) SAGT 2008. LNCS, vol. 4997, pp. 170–181. Springer, Heidelberg (2008)
8. Dubhashi, P.D., Panconesi, A.: Concentration of Measure for the Analysis of Randomized Algorithms. Cambridge University Press (2009)
9. Johnson, N.L., Kotz, S.: Urn Models and Their Applications. John Wiley & Sons (1977)
10. Kesselman, A., Lotker, Z., Mansour, Y., Patt-Shamir, B., Schieber, B., Sviridenko, M.: Buffer overflow management in qos switches. SIAM J. Comput **33**(3), 563–583 (2004)
11. Kolchin, V.F., Sevastyanov, B.A., Chistyakov, V.P.: Random Allocations. John Wiley & Sons (1978)
12. McDiarmid, C.: Concentration. In: Probabilistic Methods for Algorithmic Discrete Mathematics, Springer (1998)
13. Mitzenmacher, M., Richa, A.W., Sitaraman, R.: The power of two random choices: a survey of techniques and results. In: Handbook of Randomized Computing. Springer
14. Mitzenmacher, M., Upfal, E.: Probability and computing - randomized algorithms and probabilistic analysis. Cambridge University Press (2005)
15. Räcke, H.: Minimizing congestion in general networks. In: 43rd Symposium on Foundations of Computer Science, pp. 43–52. IEEE Computer Society (2002)
16. Räcke, H.: Optimal hierarchical decompositions for congestion minimization in networks. In: Proceedings 40th Annual ACM Symposium on Theory of Computing, pp. 255–264 (2008)
17. Sanders, P.: On the competitive analysis of randomized static load balancing. In: Proceedings of the first Workshop on Randomized Parallel Algorithms, RANDOM (1996)
18. Schulz, A.S., Skutella, M.: Scheduling unrelated machines by randomized rounding. SIAM J. Discrete Math. **15**(4), 450–469 (2002)

Finding the Median (Obliviously) with Bounded Space

Paul Beame[✉], Vincent Liew, and Mihai Pătraşcu

University of Washington, Seattle, WA, USA
beame@cs.washington.edu

Abstract. We prove that any oblivious algorithm using space S to find the median of a list of n integers from $\{1, \ldots, 2n\}$ requires time $\Omega(n \log \log_S n)$. This bound also applies to the problem of determining whether the median is odd or even. It is nearly optimal since Chan, following Munro and Raman, has shown that there is a (randomized) selection algorithm using only s registers, each of which can store an input value or $O(\log n)$-bit counter, that makes only $O(\log \log_s n)$ passes over the input. The bound also implies a size lower bound for read-once branching programs computing the low order bit of the median and implies the analog of $\mathsf{P} \neq \mathsf{NP} \cap \mathsf{coNP}$ for length $o(n \log \log n)$ oblivious branching programs.

1 Introduction

The problem of selection or, more specifically, finding the median of a list of values is one of the most basic computational problems. Indeed, the classic deterministic linear-time median-finding algorithm of [9], as well as the more practical expected linear-time randomized algorithm QuickSelect are among the most widely taught algorithms.

Though these algorithms are asymptotically optimal with respect to time, they require substantial manipulation and re-ordering of the input during their execution. Hence, they require the ability to write into a linear number of memory cells. (These algorithms can be implemented with only $O(1)$ memory locations in addition to the input if they are allowed to overwrite the input memory.) In many situations, however, the input is stored separately and cannot be overwritten unless it is brought into working memory. The number of bits S of working memory that an algorithm with read-only input uses is its *space*. This naturally leads to the question of the tradeoffs between the time T and space S required to find the median, or for selection more generally.

Munro and Paterson [18] gave multipass algorithms that yield deterministic time-space tradeoff upper bounds for selection for small space algorithms and

---

P. Beame—Research supported by NSF grants CCF-1217099 and CCF-0916400.

V. Liew—Research supported by NSF grant CCF-1217099.

M. Pătraşcu—Much of this work was done with Mihai in 2009 and 2010 when the lower bounds for oblivious algorithms were obtained. This paper is dedicated to his memory.

© Springer-Verlag Berlin Heidelberg 2015
M.M. Halldórsson et al. (Eds.): ICALP 2015, Part I, LNCS 9134, pp. 103–115, 2015.
DOI: 10.1007/978-3-662-47672-7_9

1owed that the number of passes p must be $\Omega(\log_s n)$ where $S = s \log_2 n$. Build-
1g on this work, Frederickson [14] extended the range of space bounds to nearly
near space, deriving a multipass algorithm achieving a time-space tradeoff of
he form $T = O(n \log^* n + n \log_s n)$. In the case of randomly ordered inputs,
1unro and Raman [19] showed that on average an even better upper bound of
$= O(\log \log_s n)$ passes and hence $T = O(n \log \log_s n)$ is possible. Chakrabarti,
ayram, and Pătraşcu [12] showed that this is asymptotically optimal for mul-
ipass computations on randomly ordered input streams. Their analysis also
pplied to algorithms that perform arbitrary operations during their execution.

Chan [13] showed how to extend the ideas of Munro and Raman [19] to yield
, randomized median-finding algorithm achieving the same time-space tradeoff
apper bound as in the average case that they analyze. The resulting algorithm,
ike all of those discussed so far, only accesses its input using comparisons. Chan
oupled this algorithm with a corresponding time-space tradeoff lower bound
of $T = \Omega(n \log \log_S n)$ for randomized comparison branching programs, which
mplies the same lower bound for the randomized comparison RAM model. This
s the first lower bound for selection allowing more than multipass access to the
nput; the input access can be input-dependent but the algorithm must base all
ts decisions on the input order. Though a small gap remains because $S \neq s$, the
nain question left open by [13] is that of finding time-space tradeoff lower bounds
or median-finding algorithms that are not restricted to the use of comparisons.

Comparison-based versus general algorithms. Though comparison-based algo-
·ithms for selection may be natural, when the input consists of an array of
$O(\log n)$-bit integers, as one often assumes, there are natural alternatives to comparisons such as hashing that might potentially yield more efficient algorithms.
Though comparison-based algorithms match the known time-space tradeoff lower
bounds in efficiency for sorting when time T is $\Omega(n \log n)$ [4,10,21], they are powerless in the regime when T is $o(n \log n)$. Moreover, if one considers the closely
related problem of element distinctness, determining whether or not the input
has duplicates, the known time-space tradeoff lower bound of $T = \Omega(n^{2-o(1)}/S)$
for (randomized) comparison branching programs [22] can be beaten for S up
to $n^{1-o(1)}$ by an algorithm using hashing [5] that achieves $T = \tilde{O}(n^{3/2}/S^{1/2})^1$.
Therefore, the restriction to comparison-based algorithms can be a significant
limitation on efficiency.

Our results. We prove a tight $T = \Omega(n \log \log_S n)$ lower bound for median-finding
using arbitrary oblivious algorithms. Oblivious algorithms are those that can access
the data in any order, not just in a fixed number of sweeps across the input, but
that order cannot be data dependent. Our lower bound applies even for the decision problem of computing MEDIANBIT, the low order bit of the median, when the
input consists of n integers chosen from $\{1, \ldots, 2n\}$. This bound substantially generalizes the lower bound of [12] for multipass median-finding algorithms. Though
our lower bound does not apply when there is input-dependent access to the input,

---

^{1} We use $\tilde{O}$ and $\tilde{\Omega}$ notations to hide logarithmic factors.

it allows one to hash the input data values into working storage, and to organize and manipulate working storage in arbitrary ways.

The median can be computed by a simple nondeterministic oblivious read-once branching program of polynomial size that guesses and verifies which input integer is the median. When expressed in terms of size for time-bounded oblivious branching programs our lower bound therefore shows that for every time bound T that is $o(n \log \log n)$, MEDIANBIT and its complement have nondeterministic oblivious branching programs of polynomial size but MEDIANBIT requires super-polynomial size deterministic oblivious branching programs, hence separating the analogs of P from NP ∩ coNP.

We derive our lower bound using a reduction from a new communication complexity lower bound for two players to find the low order bit of median of their joint set of input integers in a bounded number of rounds. The use of communication complexity lower bounds in the "best partition" model to derive lower bounds for oblivious algorithms is not new, but the necessity of bounded rounds is. We derive our bound via a round-preserving reduction from oblivious computation to best-partition communication complexity [2, 20]. This reduction is asymptotically less efficient than the reductions of [3, 11] but the latter do not preserve the number of rounds, which is essential here since there is a very efficient $O(\log n)$-bit communication protocol using an unbounded number of rounds [17]. Moreover, the loss in efficiency does not prevent us from achieving asymptotically optimal lower bounds.

We further show that the fact that the median function is symmetric in its inputs implies that our oblivious branching program lower bound also applies to the case of non-oblivious read-once branching programs. Ideally, we would like to extend our non-oblivious results to larger time bounds. However, we show that extending our lower bound even to read-twice branching programs in the non-oblivious case would require fundamentally new lower bound techniques. The hardness of the median problem is essentially that of a decision problem: Though the median problem has $\Theta(\log n)$ bits of output, the high order bits of the median are very easy to compute; it is really the low order bit, MEDIAN-BIT, that is the hardest to produce and encapsulates all of the difficulty of the problem. Moreover, all current methods for time-space tradeoff lower bounds for decision problems on general branching programs, and indeed for read-k branching programs for $k > 1$, also apply to nondeterministic algorithms computing either the function or its complement and hence cannot apply to the median because it is easy for such algorithms.

2   Preliminaries

Let D and R be finite sets. We first define branching programs that compute functions $f : D^n \to R$: A D-way branching program is a connected directed acyclic multigraph with special nodes: the source node and possibly many sink nodes, a sequence of n input values and one output. Each non-sink node is labeled with an input index and every edge is labeled with a symbol from D, which corresponds to the value of the input indexed at the originating node; there is precisely one out-edge from each non-sink node labeled by each element of D. We assume that each sink node is labeled by an element of R. The time T required by a branching program is the length of the longest path from the source to a sink and the space S is $\log_2$ of the number of nodes in the branching program. A branching program is *leveled* iff all the paths from the source to any given node in the program are of the same length; a branching program can be leveled by adding at most $\log_2 T$ to its space.

A branching program B computes a function $f_B : D^n \to R$ by starting at the source and then proceeding along the nodes of the graph by querying the input locations associated with each node and following the corresponding edges until it reaches a sink node; the label of the sink node is the output of the function.

A branching program is *oblivious* iff on every path from the source node to a sink node, the sequence of input indices is precisely the same. It is (syntactic) *read-k* iff no input index appears more than k times on any path from the source to a sink.

Branching programs can easily simulate any sequential model of computation using the same time and space bounds. In particular branching programs using time T and space S can simulate random-access machine (RAM) algorithms using time T measured in the number of input locations queried and space S measured in the number of bits of read/write storage required. The same applies to the simulation of randomized RAM algorithms by randomized branching programs.

We also find it useful to discuss nondeterministic branching programs for (non-Boolean) functions, which simulate nondeterministic RAM algorithms for function computation. These have the property that multiple outedges from a single node can have the same label and outedges for some labels may not be present. Every input must have at least one path that leads to a sink and all paths followed by an input vector that lead to a sink must lead to the same one, whose label is the output value of the program. This is different from the usual version for decision problems in which one only considers accepting paths and infers the output value for those that are not accepting. When we consider Boolean functions we will typically assume the usual version based on accepting paths only.

We consider bounded-round versions of deterministic and randomized two-party communication complexity in which two players Alice and Bob receive $x \in \mathcal{X}$ and $y \in \mathcal{Y}$ and cooperate to compute a function $f : \mathcal{X} \times \mathcal{Y} \to \mathcal{Z}$. A round in a protocol is a maximal segment of communication in which the player who speaks does not change. For a distribution $\mathcal{D}$ on $\mathcal{X} \times \mathcal{Y}$, we say that a 2-party deterministic communication protocol computes f with error at most $\varepsilon < 1/2$ under $\mathcal{D}$ iff the probability over $\mathcal{D}$ that the output of the protocol on input $(x, y) \sim \mathcal{D}$ is equal to $f(x, y)$ is at least $1 - \varepsilon$. As usual, via Yao's lemma, for any such distribution $\mathcal{D}$, the minimum number of bits communicated by any deterministic protocol that computes f with error at most ε is a lower bound on the number of bits communicated by any (public coin) randomized protocol that computes f with error at most ε.

We say that a 2-party deterministic communication protocol has parameters $[P, \varepsilon; m_1, m_2, \ldots]$ for f over a distribution $\mathcal{D}$ if:

- the first player to speak is $P \in \{A, B\}$;
- it has error $\varepsilon < \frac{1}{2}$ under input distribution $\mathcal{D}$;
- the players alternate turns, sending messages of $m_1, m_2, \ldots$ bits, respectively.

For probability distributions P and Q on a domain U, the statistical distance between P and Q, is $\|P - Q\| = \max_{A \subseteq U} |P(A) - Q(A)|$, which is $1/2$ of the L_1 distance between P and Q. Let log denote $\log_2$ unless otherwise specified. i Let $H(X)$ be the binary entropy of random variable X, $H(X|Y) = \mathbb{E}_{y \sim Y} H(X|Y=y)$, and let $I(X; Y|Z)$ be the mutual information between random variables X and Y conditioned on random variable Z. We have $I(X; Y|Z) \le H(X|Z) \le H(X)$.

3 Round Elimination

Let $f : \mathcal{X} \times \mathcal{Y} \to \{0, 1\}$ and consider a distribution $\mathcal{D}$ on $\mathcal{X} \times \mathcal{Y}$. We define the 2-player communication problem $f^{[k]}$ as follows: Alice receives $x \in \mathcal{X}^k$, while Bob receives $y \in \mathcal{Y}^k$ and $j \in [k]$; together they want to find $f(x_j, y_j)$. Also, given $\mathcal{D}$ we define an input distribution $\mathcal{D}^{[k]}$ for $f^{[k]}$ by choosing each (x_i, y_i) pair independently from $\mathcal{D}$, and independently choosing j uniformly from $[k]$.

The following lemma is a variant of standard techniques and was suggested to us by Anup Rao; its proof is in the full paper.

Lemma 1. *Assume that there exists a 2-party deterministic protocol for $f^{[k]}$ with parameters $[A, \varepsilon; m_1, m_2, m_3, \ldots]$ over $\mathcal{D}^{[k]}$ where $m_1 = \delta^2 k/(8 \ln 2)$. Then there exists a 2-party deterministic protocol for f with parameters $[B, \varepsilon + \delta; m_2, m_3, \ldots]$ over $\mathcal{D}$.*

The intuition for this lemma is that, since $f^{[k]}$ has k independent copies of the function f and Alice's first message has length at most m_1 which is only a small fraction of k, there must be some copy of f on which B learns very little information. This is so much less than one bit that B could forego this information in computing f and still only lose δ in his probability of correctness. The quadratic difference between the number of bits of information per copy, $\delta^2/(8 \ln 2)$, and the probability difference, δ, comes from Pinsker's inequality which relates information and statistical distance.

4 The Bounded-Round Communication Complexity of (the Least-Significant Bit of) the Median

We consider the complexity of the following communication game. Given a set A of n elements from $[2n]$ partitioned equally between Alice and Bob, determine the least significant bit of the median of A. (Since n must be even in order for A to be partitioned evenly, we take the median to be $n/2$-th largest element of A.) We consider the number of rounds of communication required when the length of each message is at most m for any $m \ge \log n$.

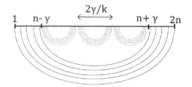

Fig. 1. Recursive construction of the pairing for the hard instances

A Hard Distribution on Median Instances. For our hard instances we first define a pairing of the elements of $[2n]$ that depends on the value of m. The set A will include precisely one element from each pair. For the input to the communication problem, we randomly partition the pairs equally between the two players which will therefore also automatically equally partition the set A. We then show how to randomly choose one element from each pair to include in A.

In the construction, we define the pairing of $[2n]$ recursively; the parameters of each recursive pairing will depend on the initial value n_0 of n. Let $k = k(m, n_0) = m \log^2 n_0$. If $\sqrt{n} < k \log^3 n_0$ then the elements of $[1, 2n]$ are simply paired consecutively. If $\sqrt{n} \geq k \log^3 n_0$ then the pairing of $[2n]$ consists of a "core" of $\gamma = \sqrt{n}/\log^2 n_0$ pairs, plus $n - \gamma$ "shell" pairs on $[1, n - \gamma] \cup [n + 1 + \gamma, 2n]$. In the shell, i and $2n + 1 - i$ are paired. The core pairs are obtained by embedding k recursive instances (using the same values of m and n_0) of $n' = \frac{\gamma}{k}$ pairs each on consecutive sets of $\frac{2\gamma}{k}$ elements, and placing them back-to-back in the value range $[n - \gamma + 1, n + \gamma]$, see Figure 1. The size of the problem at each level of recursion decreases from n to $n' = \gamma/k = \sqrt{n}/(m \log^4 n_0)$. In determining the median, the only relevant information about the shell elements is how many are below n; let this number be $\frac{n}{2} - x$. If $x \in [1, \gamma]$, the median of the entire array A will be the x-th order statistic of the core.

If furthermore, $x = \frac{\gamma}{k}(j - \frac{1}{2})$ for an integer j, the median of A will be exactly the median of the j-th embedded subproblem. In our distribution of hard instances, we will ensure that x has this nice form.

Formally, the distribution $\mathcal{D}^n_{m,n_0}$ of the hard instances A of size n on $[2n]$ is the following. Generate k recursive instances on $\mathcal{D}^{\gamma/k}_{m,n_0}$ and place shifted versions of them back-to-back inside the core. Choose $j \in [k]$ uniformly at random. Choose $\frac{n}{2} - \frac{\gamma}{k}(j - \frac{1}{2})$ uniformly random shell elements in $[1, n - \gamma]$ to include in A; for every $i \in [1, n - \gamma] \setminus A$, we have $2n + 1 - i \in A$. This will ensure that the median of A is precisely the median of the j-th recursive instance inside the core.

Initially we have $n = n_0$ and the recursion only continues when $\gamma = \sqrt{n}/\log^2 n_0 \geq k \log n_0$, so in the base case we have at least $\log n_0$ elements. In this case, the i-th element is chosen randomly and uniformly from the paired elements $2i - 1$ and $2i$ and so the least significant bit of the median is uniformly chosen in $\{0, 1\}$.

The size of the problem after t levels of recursion remains at least $n_0^{1/2^t}/(m \log^4 n_0)^{2-1/2^{t-1}}$ and our definition gives at least t levels provided that this size $n_0^{1/2^t}/(m \log^4 n_0)^{2-1/2^{t-1}} \geq \log n_0$; i.e., $n_0 \geq m^{2^{t+1}-2} \log^{9 \cdot 2^t - 2} n_0$. We will show that after one message for each level of recursion, the answer is still not determined.

The general idea of the lower bound is that each round of communication, which consists of at most m bits and is much smaller than the branching factor k, will give almost no information about a typical recursive subproblem in the core.

We use the round elimination lemma to make this precise, and with it derive the following theorem:

Theorem 1. *If, for A chosen according to $\mathcal{D}^n_{m,n}$ and partitioned randomly, Alice and Bob determine the least significant bit of the median of A with bounded error $\varepsilon < 1/2$ using t messages of at most $m \geq \log n$ bits each, then $m^{2^{t+1}-2} > n/\log^{9\cdot 2^t-2} n$, which implies that $t \geq \log\log_m n - c$ for some constant c.*

The Partition Between the Players. To ensure that neither player has enough information to skip a level of the recursion, we insist that the shell for each subproblem be nicely partitioned between the two players. For any given shell there is a set of $n' > m^2/2 \geq 0.5\log^2 n_0$ shell pairs. Since a player receives a random $1/2$ of all pairs, by Hoeffding's inequality, with probability $2^{-\Omega(n')}$, which is $n_0^{-\Omega(\log n_0)}$, at least $\frac{n'}{3}$ pairs go to each player. We can use this to say that with high probability at least $1/3$ of all shell elements at a level go to each player at every level of the recursion: This follows easily because over all levels of the recursive pairing, there are only a total of $o(\sqrt{n_0})$ different shells associated with subproblems and each one fails only with probability $n_0^{-\Omega(\log n_0)}$.

From now on, fix a partition satisfying the above requirement at all recursion nodes. We will prove a lower bound for any partition satisfying this property. Since we are discarding $o(1)$ of possible partitions, the error of the protocol may increase by $o(1)$, which is negligible.

The Induction. Our proof of Theorem 1 will work by induction, using the following message elimination lemma:

Lemma 2. *Assume that there is a protocol for the median on instances of size n, with error ε on $\mathcal{D}^n_{m,n_0}$ for $\sqrt{n} \geq k\log n_0 = m\log^3 n_0$, using t messages of size at most m starting with Alice. Then, there is a protocol for a subproblem of size γ/k, with error $\varepsilon + O(\frac{1}{\log n_0})$ on $\mathcal{D}^{\gamma/k}_{m,n_0}$, using $t-1$ messages of size at most m starting with Bob.*

We use Lemma 2 to prove Theorem 1 by inductively eliminating all messages. Let $n_0 = n$. At each application we remove one message to get an error increase of $O(\frac{1}{\log n_0})$. If the number of rounds is less than the number of levels of recursion, i.e., $m^{2^{t+1}-2} \leq n/\log^{9\cdot 2^t-2} n$, then the MEDIANBIT value of the subproblem will still be a uniformly random bit on the remaining input, but the protocol will have no communication and the error will have increased to at most $\epsilon + O(\frac{t}{\log n}) < 1/2$ since t is $O(\log\log_m n)$, which is a contradiction.

To prove Lemma 2 we want to apply Lemma 1 using the k subproblems in the core, but the assumption of Lemma 1 requires that (1) Alice does not know anything about which subproblem $j \in [k]$ is chosen by Bob, and (2) that subproblem j is chosen uniformly at random. The choice of subproblem j is determined by the shell elements at this level.

Denote Alice's shell elements by x^s, and Bob's shell elements by y^s. Let Alice's part of the core subproblems be $x_1, \ldots, x_k$, and Bob's part be $y_1, \ldots, y_k$. Note that the choice of the relevant subproblem j is some function of (x^s, y^s), and the median of the whole array is the median of $x_j \cup y_j$.

The proof of Lemma 2 proceeds in two stages:

Fixing x^s. We first fix the value of x^s so that the choice of subproblem does not depend on Alice's input and, moreover, so that the probabilities for different values of j over Bob's input y^s will not be very different from each other because they are still near the middle binomial coefficients.

By the niceness of the partition of the pairs, we know that the number of Alice's shell pairs is $|x^s| \in \left[\frac{1}{3}(n - \gamma), \frac{2}{3}(n - \gamma)\right]$. Let a be the number of elements in x^s that are below n. We want to fix x^s such that the error does not increase too much, and $|a - \frac{|x^s|}{2}| \leq \sqrt{n} \cdot \log n_0$:

No matter which value of $j \in [k]$ is chosen in the input distribution, the shell elements chosen to be below n consist of a random subset of $x^s \cup y^s$ of a fixed size that is between $n/2 - \gamma$ and $n/2 + \gamma$; i.e., of fractional size p_j between $\frac{1}{2} - \frac{\gamma}{n}$ and $\frac{1}{2} + \frac{\gamma}{n}$. By Hoeffding's inequality, the probability that the actual number a of these elements that land in x^s deviates from $|x^s|/2$ by more than $(t + \frac{\gamma}{n})|x^s|$ is at most $2e^{-2t^2|x^s|}$. Since $(n - \gamma)/3 \leq |x^s| \leq 2(n - \gamma)/3$, the probability that this deviates from $|x^s|/2$ by more than $\sqrt{n} \log n_0$ is at most $n_0^{-O(\log n_0)}$. We discard all values of x^s that lead to a outside this range. Now fix x^s to be the value that minimizes the conditional error.

Making j uniform. Once x^s is fixed, j is a function only of y^s. Thus, we are close to the setup of Lemma 1: Alice receives $x_1, \ldots, x_k$, Bob receives $y_1, \ldots, y_k$ and $j \in [k]$, and they want to compute a function $f(x_j, y_j)$. The only problem is that the lemma requires a uniform distribution of j, whereas our distribution is no longer uniform (having fixed x^s). However, we will argue that it is not far from uniform.

For each fixed $j_0 \in [k]$, if a shell elements from Alice's part are below n, then Bob must have $\frac{n}{2} - a - \frac{\gamma}{k}(j_0 - \frac{1}{2})$ shell elements below n. Therefore, $\Pr[j = j_0]$ is proportional to $\binom{|y^s|}{\frac{n}{2} - a - \frac{\gamma}{k}(j_0 - \frac{1}{2})}$. More precisely $\Pr[j = j_0]$ is this binomial coefficient divided by the sum of the coefficients for all j_0. Thus, to understand how close j is to uniform, we must understand the the dependence of these binomial coefficients on j_0.

Let $\Delta = a - |x^s|/2$. This satisfies $|\Delta| \leq \sqrt{n} \log n_0$. Since $|y^s| = n - |x^s| \geq \frac{n - \gamma}{3} > n/4$ we have $\binom{|y^s|}{\frac{n}{2} - a - \frac{\gamma}{k}(j_0 - \frac{1}{2})} = \binom{|y^s|}{|y^s|/2 - \Delta - \delta_{j_0}}$ where $0 < \delta_{j_0} < \gamma$. Assume wlog that $\Delta \geq 0$. The ratio between different binomial coefficients is at most the ratio

$$\binom{n/4}{n/8-\Delta} \Big/ \binom{n/4}{n/8-\Delta-\gamma} = \frac{(n/8+\Delta+\gamma)\cdots(n/8+\Delta+1)}{(n/8-\Delta)\cdots(n/8-\Delta-\gamma+1)}$$

$$\leq \left(1 + \frac{10(2\Delta+\gamma)}{n}\right)^{\gamma}$$

which is $1 + O(\frac{\Delta\gamma}{n}) = 1 + O(\frac{1}{\log n_0})$ given the values of Δ and γ.

Therefore we have shown that the statistical distance between the induced distribution on j and the uniform distribution is $O(\frac{1}{\log n_0})$. We can thus consider the following alternative distribution for the problem: pick j uniformly at random, and manufacture y^s conditioned on this j. The error on the new distribution increases by at most $O(\frac{1}{\log n_0})$. Now we can apply Lemma 1. As $k \geq m \log^2 n_0$, the round elimination will increase the error by $O(\frac{1}{\log n_0})$.

5   Oblivious Branching Programs and the Median

The following result is essentially due to Okol'nishnikova [20], who used it with slightly different parameters for read-k branching programs, and was independently derived by Ajtai [2] in the context of general branching programs.

Proposition 1. *Let s be a sequence of of kn elements from $[n]$. If s is divided into $r = 4k^2$ segments $s_1, \ldots, s_r$, each of length $n/(4k)$, then there is an assignment of $2k$ segments s_j to a set L_A and all remaining segments s_j to L_B so that the number n_A (n_B) of elements of $[n]$ whose only appearances are in segments in L_A (respectively, L_B) satisfy $n_A \geq n/(2\binom{4k^2}{2k})$ and $n_B \geq n/2$.*

Proof. There is a subset V of at least $n/2$ elements of $[n]$ that occur at most $2k$ times in s and hence appear in at most $2k$ segments of s. Choose the $2k$ sets s_j to include in L_A uniformly at random. For a given $i \in V$, i will contribute to n_A if and only if all of the the at most $2k$ segments that contain its occurrences are chosen for L_A. This occurs with probability at least $1/\binom{r}{2k}$; hence the expected number of elements in V that only occur in segments of L_A is at least $|V|/\binom{r}{2k}$. Therefore we can select a fixed assignment that contains has at least this number. Since the total length of segments in L_A is at most $2kn/(4k) \leq n/2$, at least $n/2$ elements of $[n]$ only occur in segments in L_B.

Lemma 3. *Suppose that there is a $2n$-way oblivious branching program of size 2^S running in time $T = kn$ that computes MEDIANBIT for n distinct inputs from $[2n]$. Then there is deterministic 2-party communication protocol using at most $4k$ messages of S bits each plus a final 1-bit message to compute MEDIANBIT for $N = \lceil n/\binom{4k^2}{2k}\rceil$ distinct inputs from $[2N]$ that are divided evenly between the two players.*

Proof. Let s be the length T sequence of indices of inputs queried by the oblivious branching program. Let $k = T/n$, $r = 4k^2$, and $N = \lceil n/\binom{r}{2k}\rceil$. Fix the assignment of segments to L_A and L_B given by Proposition 1. Arbitrarily select subset I_A of $N/2$ of the n_A indices that only appear in L_A and give those inputs to player A. Similarly, select a subset I_B of $N/2$ of the n_B indices that only appear in L_B and give those inputs to player B. Let Q be the remaining set of $n - N$ input indices.

Fix any input assignment to the indices in Q that assigns $(n - N)/2$ distinct values from $[n - N]$ to half the elements of Q and the same number of distinct values from $[n + N + 1, 2n]$ to the other half of the elements of Q. After fixing this partial assignment we restrict the remaining inputs to have values in the segment $[n - N + 1, n + N]$ of length $2N$.

The communication protocol is derived as follows: Alice (resp. Bob) interprets her $N/2$ inputs from $[2N]$ as assignments from $[2n]$ to the elements of I_A (resp. I_B) by adding $n - N$ to each value. Alice will simulate the branching program executing the segments in L_A and Bob will simulate the branching program executing the segments in L_B. A player will continue the simulation until the next segment is held by the other player, at which point that player communicates the name of the node in the branching program reached at the end of its layer. Since L_A has only $2k$ segments, there are at most $4k$ alternations between players as well as the final output bit which gives the total communication. By construction, the median of the whole problem is the median of the N elements and the final answer for MEDIANBIT on $[2N]$ is computed by XOR-ing the result with the low order bit of $n - N$.

Theorem 2. *Any oblivious branching program computing* MEDIANBIT *for n inputs from $[2n]$ in time $T \leq kn$ requires size at least $2^{\tilde{\Omega}(n^{1/2^{4k+2}})}$; in particular, if it uses space S, any oblivious branching program requires time $T \geq 0.25n \log \log_S n - c\,n$ for some constant c.*

Proof. Since $T/n \leq k$, applying Lemma 3 we derive a 2-party communication protocol sending $t = 4k + 1$ messages of at most $S \geq \log n$ bits each to compute MEDIANBIT on $N \geq n/\binom{4k^2}{2k} \geq n/(2ek)^{2k}$ inputs from $[2N]$. By Theorem 1, $S > N^{1/(2^{t+1}-2)}/\log^{(9 \cdot 2^t - 2)/(2^{t+1}-2)} N > N^{1/(2^{4k+2}-2)}/\log^{71/15} N$ since $t \geq 4$ and hence $S \geq n^{1/(2^{4k+2}-2)}/\log^5 n$. The size of the branching program is 2^S where S is its space. Moreover, taking logarithms base S and then base 2 we have $4k \geq \log \log_S n - c'$ for some constant c'. $\quad\blacksquare$

Analog of P $\neq$ NP $\cap$ coNP for time-bounded oblivious BPs

Corollary 1. *Any oblivious branching program of length $T \leq kn$ computing the low order bit of the median requires size at least $exp(\tilde{\Omega}(n^{1/2^{4k+2}}))$; in particular, this size is super-polynomial when T is $o(n \log \log n)$.*

On the other hand, the median can be computed by a nondeterministic oblivious read-once branching program using only $O(\log n)$ space.

Lemma 4. *There is a nondeterministic oblivious read-once branching program of size $O(n^4)$ that computes the median on n integers from $[2n]$.*

Proof. The branching program guesses the value of the median in $[2n]$ and keep track of the number of elements that it has seen both less than the median and equal to the median in order to check that the value is correct.

In particular, in contrast to Corollary 1, Lemma 4 implies that MEDIAN-BIT can be computed in polynomial size by length n nondeterministic and co-nondeterministic oblivious branching programs, hence we have shown the analog of P $\neq$ NP $\cap$ coNP for oblivious branching programs of length $o(n \log \log n)$.

6   Beyond Oblivious Branching Programs

We first observe that our lower bounds for the median problem extend to the case of read-once branching programs by using the fact that such programs for the median can also be assumed to be oblivious without loss of generality. (Oblivious read-once branching programs are also known as *ordered binary decision diagrams (OBDDs)*.)

Lemma 5. *If $f : D^n \to R$ is a symmetric function of its inputs then for every read-once branching B computing f there is an oblivious read-once branching program, of precisely the same size as B, that computes f.*

Proof. With each node v in a read-once branching program, we can associate a set $I_v \subseteq [n]$ of input indices that are read along paths from the source node to v. We make B into an oblivious branching program by replacing the index at node v by $|I_v| + 1$. This yields an oblivious read-once branching program (not necessarily leveled) that reads its inputs in the order $x_1, x_2, \ldots, x_n$ along every path (possibly skipping over some inputs on the path). Since f is a symmetric function, a path of length $t \leq n$ in B queries t different input locations and the value of the function on the partial inputs is the same because the function is symmetric and the values in those t input locations are the same.

Corollary 2. *For any $\varepsilon < 1/2$, any read-once branching program computing* MEDIANBIT *for n integers from $[2n]$ requires size $2^{n^{\Omega(1)}}$.*

In particular this means that MEDIANBIT is another example, after those in [16], of a problem showing the analogue of P $\neq$ NP $\cap$ coNP for read-once branching programs. However, proving the analogous property even for read-twice branching programs remains open and will require a fundamentally new technique for deriving branching program lower bounds.

The approach in all lower bounds for general branching programs (or even for read-k branching programs) computing decision problems [1,2,6–8,11,20] applies equally well to nondeterministic computation. (For example, the fact that the technique also works for nondeterministic computation is made explicit in [11].) Though this technique has been used to separate nondeterministic from deterministic computation [2] computing a Boolean function f, it is achieved by proving a nondeterministic lower bound for computing $\overline{f}$. Since the nondeterministic oblivious read-once branching program computing the median has

$T = n$ and $S = O(\log n)$, the core of the median's hardness, MEDIANBIT, and its complement do not have non-trivial lower bounds; hence current time-space tradeoff lower bound techniques are powerless for computing the median. We conjecture that the lower bound $T = \Omega(n \log \log_S n)$ also holds for finding the median using general non-oblivious algorithms.

References

1. Ajtai, M.: A non-linear time lower bound for boolean branching programs. In: Proceedings 40th IEEE FOCS Conference, pp. 60–70. New York, NY (1999)
2. Ajtai, M.: Determinism versus non-determinism for linear time RAMs with memory restrictions. Journal of Computer and System Sciences 65(1), 2–37 (2002)
3. Alon, N., Maass, W.: Meanders and their applications in lower bounds arguments. Journal of Computer and System Sciences 37, 118–129 (1988)
4. Beame, P.: A general sequential time-space tradeoff for finding unique elements. SIAM Journal on Computing 20(2), 270–277 (1991)
5. Beame, P., Clifford, R., Machmouchi, W.: Element distinctness, frequency moments, and sliding windows. In: Proceedings 54th IEEE FOCS Conference, pp. 290–299. Berkeley, CA (2013)
6. Beame, P., Saks, M., Sun, X., Vee, E.: Time-space trade-off lower bounds for randomized computation of decision problems. J. ACM 50(2), 154–195 (2003)
7. Beame, P., Jayram, T.S., Saks, M.: Time-space tradeoffs for branching programs. Journal of Computer and System Sciences 63(4), 542–572 (2001)
8. Beame, P., Vee, E.: Time-space tradeoffs, multiparty communication complexity, and nearest-neighbor problems. In: Proceedings 34th ACM STOC Conference, pp. 688–697. Montreal, Quebec, Canada (2002)
9. Blum, M., Floyd, R.W., Pratt, V.R., Rivest, R.L., Tarjan, R.E.: Time bounds for selection. Journal of Computer and System Sciences 7(4), 448–461 (1972)
10. Borodin, A., Cook, S.A.: A time-space tradeoff for sorting on a general sequential model of computation. SIAM Journal on Computing 11(2), 287–297 (1982)
11. Borodin, A., Razborov, A.A., Smolensky, R.: On lower bounds for read-k times branching programs. Computational Complexity 3, 1–18 (1993)
12. Chakrabarti, A., Jayram, T.S., Patrascu, M.: Tight lower bounds for selection in randomly ordered streams. In: Proceedings 19th ACM-SIAM SODA Conference, pp. 720–729. San Francisco, CA (2008)
13. Chan, T.M.: Comparison-based time-space lower bounds for selection. ACM Transactions on Algorithms 6(2), 26:1–26:16 (2010)
14. Frederickson, G.N.: Upper bounds for time-space trade-offs in sorting and selection. Journal of Computer and System Sciences 34(1), 19–26 (1987)
15. Holenstein, T.: Parallel repetition: Simplification and the no-signaling case. Theory of Computing 5(1), 141–172 (2009)
16. Jukna, S., Razborov, A.A., Savický, P., Wegener, I.: On P versus NP∩ co-NP for decision trees and read-once branching programs. Computational Complexity 8(4), 357–370 (1999)
17. Kushilevitz, E., Nisan, N.: Communication Complexity. Cambridge University Press, Cambridge, England; New York (1997)
18. Munro, J.I., Paterson, M.S.: Selection and sorting with limited storage. Theoretical Computer Science 12, 315–323 (1980)

19. Munro, J.I., Raman, V.: Selection from read-only memory and sorting with minimum data movement. Theoretical Computer Science **165**(2), 311–323 (1996)
20. Okol'nishnikova, E.: On lower bounds for branching programs. Siberian Advances in Mathematics **3**(1), 152–166 (1993)
21. Pagter, J., Rauhe, T.: Optimal time-space trade-offs for sorting. In: Proceedings 39th IEEE FOCS Conference, pp. 264–268. Palo Alto, CA (1998)
22. Yao, A.C.-C.: Near-optimal time-space tradeoff for element distinctness. SIAM Journal on Computing **23**(5), 966–975 (1994)

Approximation Algorithms for Min-Sum k-Clustering and Balanced k-Median

Babak Behsaz, Zachary Friggstad$^{(\boxtimes)}$, Mohammad R. Salavatipour, and Rohit Sivakumar

Department of Computing Science, University of Alberta, Edmonton, AB, Canada
{behsaz,zacharyf,mreza,rohit2}@ualberta.ca

Abstract. We consider two closely related fundamental clustering problems in this paper. In the *Min-Sum k-Clustering* problem, one is given a metric space and has to partition the points into k clusters while minimizing the total pairwise distances between the points assigned to the same cluster. In the *Balanced k-Median* problem, the instance is the same and one has to obtain a partitioning into k clusters $C_1, \ldots, C_k$, where each cluster C_i has a center c_i, while minimizing the total assignment costs for the points in the metric; here the cost of assigning a point j to a cluster C_i is equal to $|C_i|$ times the distance between j and c_i in the metric.

In this paper, we present an $O(\log n)$-approximation for both these problems where n is the number of points in the metric that are to be served. This is an improvement over the $O(\epsilon^{-1} \log^{1+\epsilon} n)$-approximation (for any constant $\epsilon > 0$) obtained by Bartal, Charikar, and Raz [STOC '01]. We also obtain a quasi-PTAS for Balanced k-Median in metrics with constant doubling dimension.

As in the work of Bartal et al., our approximation for general metrics uses embeddings into tree metrics. The main technical contribution in this paper is an $O(1)$-approximation for Balanced k-Median in hierarchically separated trees (HSTs). Our improvement comes from a more direct dynamic programming approach that heavily exploits properties of standard HSTs. In this way, we avoid the reduction to special types of HSTs that were considered by Bartal et al., thereby avoiding an additional $O(\epsilon^{-1} \log^{\epsilon} n)$ loss.

1 Introduction

One of the most ubiquitous problems encountered in computing science is clustering. At a high level, a clustering problem arises when we want to aggregate data points into groups of similar objects. Often, there are underlying metric distances $d(u, v)$ between data points u, v that quantify their similarities. Ideally, we want to cluster the objects into few clusters while ensuring that the distances within a cluster are small.

M.R. Salavatipour—Supported by NSERC.

© Springer-Verlag Berlin Heidelberg 2015
M.M. Halldórsson et al. (Eds.): ICALP 2015, Part I, LNCS 9134, pp. 116–128, 2015.
DOI: 10.1007/978-3-662-47672-7_10

In this paper, we focus on two closely related problems, which are referred to in the literature as *Min-Sum k-clustering* (MSkC) and *Balanced k-Median* (BkM). In both problems, we are given a metric space over a set of n points V, which we assume is given as a weighted graph $G = (V, E)$ with shortest-path distances $d(u, v)$ between any two vertices $u, v \in V$. In the MSkC problem the goal is to partition the points V into k clusters $C_1, \ldots, C_k$ to minimize the sum of pair-wise distances between points assigned to the same cluster: $\sum_{i=1}^{k} \sum_{\{j,j'\} \subseteq C_i} d(j, j')$.

This problem (MSkC) was first introduced by Sahni and Gonzalez [14] and is the complement of the Max k-Cut problem. Bartal et al. [4] gave an $O(\epsilon^{-1} \log^{\epsilon} n)$-approximation for any constant $\epsilon > 0$, for the case of Hierarchically Separated Trees (HSTs), which in turn (using the $O(\log n)$ bound for approximating metrics using HSTs [8]), gives an $O(\epsilon^{-1} \log^{1+\epsilon} n)$-approximation for general metrics. To do this, Bartal et al. consider BkM, where the input is the same as MSkC and the goal is to select k points $c_1, \ldots, c_k \in V$ as the centers of the clusters and partition the nodes V into clusters $C_1, \ldots, C_k$ to minimize $\sum_{i=1}^{k} |C_i| \sum_{v \in C_i} d(v, c_i)$. The multiplier $|C_i|$ on the contribution of $d(v, c_i)$ to the objective function penalizes clusters for being too large, hence the term *balanced*. As observed in [4], it is easy to show that an α-approximation for either MSkC or BkM implies a 2α-approximation for the other problem in metric graphs. The approximation of [4] for MSkC was obtained by presenting such an approximation for BkM.

1.1   Related Work

The facility location interpretation of the BkM leads to a natural generalization of the problem. In this generalization, we are given a set of clients $C \subseteq V$ and a set of facilities $F \subseteq V$. We need to choose k facilities from F to open and the clients in C must be served by these k facilities. In other words, the set of clients must be partitioned into k clusters and the center assigned to each partition must be chosen from F. Note that C and F can have common vertices. The special case that $C = F = V$ is the original problem we defined. We often use the term "facility" to refer to the center of a cluster in BkM and the points assigned to that center are the "clients" that get served by that facility.

The $O(\epsilon^{-1} \log^{1+\epsilon} n)$-approximation of [4] stands as the best approximation for both MSkC and BkM after fourteen years. They also describe a bicriteria $O(1)$-approximation (for BkM) that uses $O(k)$ clusters. Fernandez de la Vega et al. [9] gave a $(1+\epsilon)$-approximation for MSkC with running time of $O(n^{3k} 2^{\epsilon^{-k^2}})$.

BkM and MSkC have been further studied in more restricted settings. BkM can be solved in time $n^{O(k)}$ by "guessing" the center locations and their capacities, and then finding a minimum-cost assignment from the clients to these centers [10]. This yields a 2-approximation for MSkC when k is regarded as a constant. Furthermore, Indyk gives a PTAS [11] for MSkC when $k = 2$.

The factor-2 reduction between BkM and MSkC fails to hold when the distances are not in a metric space. Indeed, one can still solve non-metric instances of BkM in $n^{O(k)}$ time, however no $n^{2-\epsilon}$-approximation is possible for non-metric MSkC for any constant $\epsilon > 0$ and any $k \geq 3$ [12]. An $O(\sqrt{\log n})$-approximation for non-metric MSkC for $k = 2$ is known as this is just a reformulation of the Minimum Uncut problem [1].

These problems have been studied in geometric spaces as well. For point sets in $\mathbb{R}^d$ and a constant k Schulman [15] gave an algorithm for MSkC that either outputs a $(1 + \epsilon)$-approximation, or a solution that agrees with the optimum clustering on $(1 - \epsilon)$-fraction of the points but may have a much larger than optimum cost. Finally, Czumaj and Sohler [7] have developed a $(4 + \epsilon)$-approximation algorithm for MSkC for the case when $k = o(\log n / \log \log n)$ and constant ϵ.

Perhaps the most well studied related problem is the classical k-*Median* problem where one has to find a partition of the point set into k sets $C_1, \ldots, C_k$, each having a center c_i while minimizing the total sum of distances of the points to their respective center. Some of the most recent results, following a long line of research, are [5, 13, 17], which bring down the approximation ratio to $2.592 + \epsilon$. It is worth pointing out that both MSkC and BkM seem significantly more difficult than the classical k-median problem. For instance, for the case of k-median if one is given the set of k centers the clustering of the points is immediate as each point will be assigned to the nearest center point; this has been used in a simple local search algorithm that is proved to have approximation ratio $3 + \epsilon$ [2]. However, for the case of BkM, even if one is given the location of k centers it is not clear how to cluster the points optimally.

1.2   Results and Techniques

Our two primary results are $O(\log n)$-approximation algorithms for both BkM and MSkC, improving over their previous $O(\epsilon^{-1} \log^{1+\epsilon} n)$-approximations for any constant $\epsilon > 0$ [4], and a quasi-polynomial time approximation scheme (QPTAS) for BkM in metrics with constant doubling dimension (a.k.a. doubling metrics). Note that this includes Euclidean spaces of constant dimension. Before this work, there were no results known for Euclidean metrics apart from what was known about general metrics.

Similar to the approximation in [4], our improved $O(\log n)$-approximation for general metrics uses Hierarchically Separated Trees (HSTs), defined formally in Section 2. Specifically, we give a deterministic constant-factor approximation for BkM on HSTs. As is well-known, an arbitrary metric can be probabilistically embedded into an HST with the expected stretch of each edge being $O(\log n)$ [8], thus our algorithm leads immediately to a randomized, polynomial time algorithm that computes a solution with expected cost $O(\log n)$ times the optimum solution cost.

The approximation in [4] relied on slightly non-conventional HSTs where the diameters of the subtrees drop by an $O(\log^\epsilon n)$-factor instead of the usual $O(1)$ factor. One can obtain such HSTs with $O(\frac{1}{\epsilon} \log n / \log \log n)$ height which was necessary in order to ensure that their algorithm runs in polynomial time. Our dynamic programming approach is quite different and requires a few observations about the structure of optimal solutions in 2-HSTs. In this way, we avoid dependence on the height of the tree in the running time of our algorithm, thereby obtaining a polynomial-time, constant-factor approximation for 2-HSTs and ultimately, a $O(\log n)$-approximation in general metrics.

Our second result, which is a QPTAS for BkM, is essentially a dynamic programming algorithm which builds on the hierarchical decomposition of a metric space with constant doubling dimensions. We start this by presenting a QPTAS for BkM for the case of a tree metric and show how this can be extended to metrics with constant doubling dimensions. This result strongly suggests that the problem is not APX-hard and therefore should have a PTAS for these metrics.

For our algorithms we consider a special case of the BkM problem in which each cluster has a *type* based on rounding up the size of the cluster to the nearest power of $(1 + \epsilon)$ for some given constant $\epsilon > 0$; we call this the ϵ-*Restricted Balanced k-median* (RBkM) problem. Here each cluster has one of the types $0, 1, \ldots, \lceil \log_{1+\epsilon} n \rceil$, where n denotes the number of clients, i.e., $n = |C|$. A cluster that is of type i can serve at most $(1 + \epsilon)^i$ clients and the cost of serving each client j in a type i cluster with center (facility) c is $(1 + \epsilon)^i \cdot d(c, j)$ (regardless of how many clients are served by the facility). We sometimes refer to $(1 + \epsilon)^i$ as the capacity or the multiplier of the center (facility) of the cluster. We also say that the center of the cluster and all the clients of that cluster are of type i. It is not hard to see that an α-approximation algorithm for this version results in a $((1 + \epsilon)\alpha)$-approximation algorithm for the BkM problem.

Section 2 outlines our approach for the general $O(\log n)$-approximation, including specific definitions of the HSTs we use. The dynamic programming approach for HSTs appears in Section 3. We present the QPTAS for BkM in doubling metrics in Section 4.

2   An $O(\log n)$-Approximation for General BkM

As noted earlier, our $O(\log n)$-approximation uses embeddings into tree metrics. In particular, we use the fact that an arbitrary metric can be probabilistically approximated by Hierarchically Separated Trees with $O(\log n)$ distortion. We begin by listing some properties of μ-HSTs that we use in our algorithm.

Definition 1. *For $\mu > 1$, a μ-Hierarchical Well Separated Tree (μ-HST) is a metric space defined on the leaves of a rooted tree T. Let the level of an internal node in the tree be the number of edges on the path to the root. Let Δ denote the diameter of the resulting metric space. For a vertex $u \in T$, let $\Delta(u)$ denote the diameter of the subtree rooted at u. Then the tree has the following properties:*

- *All edges at a particular level have the same weight.*
- *All leaves are at the same level.*
- *For any internal node u at level i, $\Delta(u) = \Delta \cdot \mu^{-i}$.*

By this definition, any two leaf nodes u and v with a least common ancestor w are at distance exactly $\Delta(w)$ from each other. If T is a μ-HST then we let $d_T(u, v)$ denote the distance between u and v in T. It follows from [8] that for any

teger $\mu > 1$, any metric can be probabilistically embedded into μ-HSTs with
retch $O(\mu \cdot \log_\mu n)$. Furthermore, we can sample a μ-HST from this distribution
polynomial time.

In an instance of BkM on μ-HSTs T, only the leaf nodes of T correspond
clients and all the cluster centers must be leaf nodes of T. We use this in a
andard way to get a randomized $O(\log n)$-approximation for BkM and MSkC.

Note: Our techniques can be used to get a PTAS for μ-HSTs for any constant μ
y solving the ϵ-RBkM problem exactly for appropriately small values of ϵ, but it
enough to describe a 2-approximation for BkM in 2-HSTs to get an $O(\log n)$-
pproximation in general metrics. Thus, we focus on this case for simplicity.

Dynamic Programming for BkM in 2-HSTs

Recall that in ϵ-RBkM, the capacity of each facility (or the size of each cluster)
s rounded up to the nearest power of $1 + \epsilon$. For ease of exposition, we focus
n the 1-RBkM problem (i.e. where all cluster sizes are powers of two) and
resent an exact algorithm for this problem on 2-HSTs. Clearly, this implies a
-approximation for the BkM problem on such graphs. In this section we simply
se RBkM to refer to 1-RBkM. We prove the following:

Theorem 1. *RBkM instances in 2-HSTs can be solved in polynomial time.*

To solve RBkM exactly on 2-HSTs using Dynamic Programming, we start by
lemonstrating the existence of an optimal solution with certain helpful proper-
ies. Let $T = (V, E)$ denote the 2-HST rooted at a vertex $r \in V$. For any vertex
$v \in V$, let T_v denote the subtree of T rooted at v. It is obvious that T_v itself
s a 2-HST. A client (or facility) is said to be located in the subtree T_v if its
orresponding vertex in the tree belongs to T_v. In the same vein, a client (or
acility) is located outside T_v if it is located in the subtree $T \backslash T_v$.

We say that a facility at location v_f serves a client at location v_c if v_c is part
f the cluster with center v_f. We emphasize that only the leaf nodes of a 2-HST
re clients and we can only open facilities at leaf nodes. We say that a facility
it v_f is of type i if it is open with capacity 2^i. Thus, each client v being served
oy v_f is being served with cost $2^i \cdot d(v_f, v)$.

The following two lemmas are helpful in narrowing our search for the optimum solution. There proofs are omitted due to lack of space.

Lemma 1. *In an optimal solution, each open facility serves its collocated client.*

Lemma 2. *For every optimal solution and for each vertex v, there is at most
one type i of facility in T_v that serves clients located outside T_v. Also, any other
facility in T_v has type at least i.*

We record a few more simple observations before describing our recurrence.

Observation 1. *In an optimal solution to RBkM with two vertices $u, v \in V$ such that T_u and T_v are disjoint, there cannot exist two facilities f_u and f_v and clients c_u and c_v in the subtrees rooted at u and v, respectively, such that f_u serves c_v and f_v serves c_u.*

If this were not the case, we can reduce the cost by swapping the clients and having f_u serving to c_u and f_v serving to c_v to get a cheaper solution.

Observation 2. *For any feasible solution to RBkM and a vertex v in the tree, if $u, w \in T_v$ are two clients served by two facilites $f_u, f_w \notin T_v$ then the cost of pairing u with f_u and w with f_w is the same as the cost of pairing u with f_w and w with f_u.*

This is because for every vertex $v \in T$, all clients and facilities in T_v are equidistant from v by Definition 1. For the next observation, recall that all the leaves in T are located at the same level.

Observation 3. *For a facility with multiplier m_f located at v_f and a client located at v_c, let v_{lca} denote their least common ancestor. Then the cost of serving v_c at v_f is $2 \cdot m_f \cdot d(v_f, v_{lca})$.*

This will be helpful in our algorithm because, in some sense, it only keeps track of the distance between v_f and v_{lca} for a client v_c served by v_f. For an edge e between v_f and v_{lca}, we call $2 \cdot m_f \cdot d(e)$ the *actual cost* of the edge e for the (v_c, v_f) pair, where $d(e)$ is the weight of e in the metric. Note that the sum of the actual costs of edges between v_f and v_{lca} is precisely $m_f \cdot d(v_f, v_c)$.

Definition 2. *For a subtree T_v of T and any feasible solution to RBkM, we use $cost_{T_v}^{in}$ to refer to the sum of the actual costs of edges within T_v accrued due to all the facility-client pairs (v_f, v_c) where $v_f \in T_v$.*

Thus, for any feasible solution to RBkM, $cost_{T_r}^{in}$ is the cost of this solution.

Definition 3. *In a partial assignment of clients to facilities, the slack of a facility f is the difference between its capacity and the number of clients assigned to f. The slack of a subtree T_v rooted at a v is the total slack of facilities in T_v.*

We first present our dynamic programming algorithm under the assumption that the 2-HST is a full binary tree. This cannot be assumed in general, but we present this first because it is simpler than the general case and still introduces the key ideas behind our algorithm.

The general case is more technical and requires two levels of DP; the details will appear in the full version of this paper. Some intuition regarding this case is discussed at the end of this section.

.1   The Special Case of Full Binary Trees

'o define a subproblem for the DP, let us consider an arbitrary feasible solution nd focus on a subtree T_v, for $v \in T$. We start by defining a few parameters:

- k_v is the number of facilities opened in the subtree T_v.
- t_v denotes the type of the facility, if any, in T_v which serves clients located outside T_v (c.f. Lemma 2). We assign a value of -1 to t_v if no client in $T \backslash T_v$ is served by a facility in T_v.
- u_v is the number of clients in $T \backslash T_v$ that are served by facilities in T_v.
- d_v is the number of clients in T_v that are served by facilities in $T \backslash T_v$ (and)
- o is the slack of T_v.

£ach table entry is of the form $\mathcal{A}[v, k_v, t_v, u_v, d_v, o]$. For a vertex $v \in V$, the value tored in this table entry is the minimum of $cost_{T_v}^{in}$ over all feasible solutions with)arameters k_v, t_v, u_v, d_v, o if the cell is a non-pessimal state (defined below).

Observation 3 in the previous section provides insight on why it is sufficient .o keep track of the d_v values without caring about the type or the location of .he facilities outside of T_v for calculating the cost of the solution. Our algorithm 'or RBkM fills the table for all permissible values of parameters v, k_v, t_v, u_v, d_v and o for every vertex v in a bottom-up fashion (from leaf to root). For vertices .n the same level, ties are broken arbitrarily.

Pessimal States and Base Cases. An entry of the dynamic programming :able is said to be *trivially suboptimal* if it is forced to contain a facility that does not cover its collocated client and is said to be *infeasible* when either the number of clients to be covered or the number of facilities to be opened within a subtree is greater than the total number of nodes in the subtree. We call an entry of the table *pessimal* when it is either infeasible or trivially suboptimal. It is easy to determine the pessimal states in the DP table at the leaf level of the tree. For other subproblems, a cell in the table is pessimal if and only if all its subproblems are pessimal states. For the ease of execution of our DP, we assign a value of ∞ to these cells in our table.

Notice that, at the leaf level of a 2-HST, all the vertices are client nodes. But some of these nodes may also have a collocated facility opened. At this stage, the only non-pessimal subproblems are the following:

(a) Facility nodes that correspond to subproblems of the kind $\mathcal{A}[v, 1, t_v, u_v, 0, o]$ satisfying the capacity constraint that $u_v + o + 1 = 2^{t_v}$, where the number 1 indicates the facility's collocated client from Lemma 1 (and)
(b) Client nodes which have subproblems of the form $\mathcal{A}[v, 0, -1, 0, 1, 0]$.

The value stored in these entries are zero.

The Recurrence. If the vertex v has two children v_1 and v_2 and the values for the dynamic program are already computed for all subproblems of T_{v_1} and T_{v_2} then the recurrence we use is given as follows:

$$\mathcal{A}[v, k_v, t_v, u_v, d_v, o] = \min_{k', k'', t_1^*, t_2^*, u_1^*, u_2^*, d_1^*, d_2^*, o_1, o_2} (\mathcal{A}[v_1, k', t_1^*, u_1^*, d_1^*, o_1]$$

$$+ \mathcal{A}[v_2, k'', t_2^*, u_2^*, d_2^*, o_2] + 2 \sum_{i \in \{1,2\}, t_i^* \geq 0} 2^{t_i^*} \cdot u_i^* \cdot d(v, v_i)),$$

where the subproblems in the above equation satisfy the following "consistency constraints":

Type Consistency: We consider two cases for the type t_v assuming that $u_v > 0$. If $u_v = 0$, the problem boils down to the case where $t_v = -1$.

1. If $t_v = -1$, then no facility in T_v serves clients located in $T \backslash T_v$. Therefore, all the clients served by facilities in T_{v_1} are located within T_{v_1} or in T_{v_2}. Similarly, for the subtree T_{v_2}, every client served by a facility in T_{v_2} is either located in T_{v_1} or in T_{v_2}. But it is clear from Observation 1 that an optimal solution cannot simultaneously have a facility in T_{v_1} serving a client in T_{v_2} and a facility in T_{v_2} serving a client in T_{v_1}. Hence, $\min(t_{v_1}, t_{v_2}) = t_v = -1$.
2. If $t_v \geq 0$, then there exists at least one client in $T \backslash T_v$ that will be served by a facility in T_v. Without loss of generality, if one of the two subtrees, say T_{v_1} has a type $t_{v_1} = -1$, then the type of the other subtree t_{v_2} must be equal to the type of the facility leaving its parent, t_v. Otherwise, if both the values t_{v_1} and t_{v_2} are non-negative, Lemma 2 implies that $\min(t_{v_1}, t_{v_2}) = t_v$.

Slack Consistency: The slack of T_v comes from the combined slack of facilities in both its subtrees, T_{v_1} and T_{v_2}. Therefore, $o = o_1 + o_2$.

Consistency in the Number of Facilities : k_v is the number of facilities opened in T_v. Since these facilities belong to either of the two subtrees T_{v_1} and T_{v_2}, we have that $k_v = k' + k''$.

Flow Consistency: $u_1^* + u_2^* + d_v = d_1^* + d_2^* + u_v$. This constraint ensures that the subproblems we are looking at are consistent with the u_v and d_v values in hand. More specifically, note that u_1^* is the number of clients in $T \backslash T_{v_1}$ served by facilities in T_{v_1} and that these u_1^* clients can either be located in T_{v_2} or in the subtree $T \backslash T_v$. Let us denote by u_{1a}^*, the number of such clients in $T \backslash T_v$ and by u_{1b}^*, the number of clients in T_{v_2} served by facilities in T_{v_1}. Likewise, let u_{2a}^* be the number of clients in $T \backslash T_v$ and u_{2b}^*, the number of clients in T_{v_1} which are served by facilities in T_{v_2}. It is easy to see that $u_{1a}^* + u_{1b}^* = u_1^*$ and $u_{2a}^* + u_{2b}^* = u_2^*$. Also, by accounting for the clients in $T \backslash T_v$ served by facilities in T_v we see

$$u_v = u_{1a}^* + u_{2a}^* \tag{1}$$

Out of the d_1^* clients in T_{v_1} and d_2^* clients in T_{v_2} which are served by facilities located outside their respective subtrees, d_v of these clients are served by facilities in $T \backslash T_v$, while the remaining clients $d_1^* + d_2^* - d_v$ must either be served by the u_{1b}^* facilities situated in T_{v_1} and u_{2b}^* situated in T_{v_2}. Hence,

$$d_1^* + d_2^* = d_v + u_{1b}^* + u_{2b}^* \tag{2}$$

umming up the Equations (1) and (2) and from the observation that $u_{1a}^* + u_{1b}^* = u_1^*$ and $u_{2a}^* + u_{2b}^* = u_2^*$, we get the flow constraint stated above.

The last term in the recurrence gives the sum of actual costs of the edges etween v and its children for the client-facility pairs where the facility is inside ne of the two subtrees T_{v_1} or T_{v_2}. From Definition 2, this value is equal to the ifference, $cost_{T_v}^{in} - (cost_{T_{v_1}}^{in} + cost_{T_{v_2}}^{in})$.

The optimal RBkM solution is the minimum value from among the entries $A[r, k, -1, 0, 0, o]$ for all values of o. Note that the number of different values each arameter can take is bounded by the number of nodes in the tree (we assume k s at most the number of leaves, or else the problem is trivial) and the number of ecursive calls made to compute a single entry is also polynomially-bounded, so hese values can be computed in polynomial time using dynamic programming.

ntuition Behind General HSTs

n HSTs that are not necessarily binary, we still computes the values A as lescribed in the binary case. However, computing these values for subproblems ooted at a vertex v with multiple children $u_1, \ldots, u_\ell$ requires a more sophisicated approach. For this, we use an "inner" dynamic programming algorithm hat, for each $0 \le i \le k$, tracks the movement of clients between $\{T_{u_1}, \ldots, T_{u_i}\}$ and $\{T_{u_{i+1}}, \ldots, T_{u_\ell}\}$, as well as movement in and out of T_v. Using observations ike in the binary case, we only have to keep track of the number of clients from a constant number of types.

4   QPTAS for Doubling Metrics

In this section we consider the generalization of BkM where C and F are not necessarily equal and present a QPTAS for it when the input metric has constant doubling dimension. We also assume that $\epsilon > 0$ is a fixed constant (error parameter) and present an exact algorithm for ϵ-RBkM which clearly implies a $(1 + \epsilon)$-approximation for BkM.

For simplicity of explanation, we will describe the QPTAS only for tree metrics and defer the details for doubling metrics to the full version of this paper. At a high level, the extension to doubling metrics uses similar ideas as our QPTAS in trees, modified appropriately to work with the hierarchical decomposition of doubling metrics described by by Talwar [16].

4.1   A QPTAS for Tree Metrics

In this section, we present an exact quasi-polynomial time algorithm for the ϵ-RBkM problem on trees. Without loss of generality, we assume the tree is rooted at an arbitrary vertex r. We repeatedly remove leaves with no client or facility until there is no such leaf in the tree. We also repeatedly remove internal vertices of degree two with no client or facility by consolidating their incident edges into one edge of the total length. Also, it is not hard to see that by introducing dummy vertices and zero length edges, we can convert this modified rooted tree into an equivalent binary tree[1] in which the clients and facilities are only located on *distinct* leaves. In other words, each leaf has either a client or a facility. The number of vertices and edges in this binary tree remains linear in the size of the original instance.

Let $p = \lceil \log_{1+\epsilon} n \rceil$. In a solution for the ϵ-RBkM problem, we say a client or facility has type i if it belongs to a type i cluster for some $0 \leq i \leq p$. We first observe a structural property in an optimal solution of an instance of ϵ-RBkM. We think of the clients get connected to facilities (the center of the cluster) to get some service. Having said this, we prove that there is an optimal solution in which type i clients either enter or leave a subtree but not both. In other words, in this solution, there are no two clients of the same type such that one enters the subtree to get connected to a facility and one leaves the subtree to get connected to a facility. To see this, let T_v be the subtree rooted at an arbitrary vertex v, and assume clients j_1 and j_2 have the same type, j_1 is not in T_v but enters this subtree to be served by facility i_1, and client j_2 is in T_v but leaves this subtree to be served by a facility i_2. Then, it is not hard to see that because j_1 and j_2 have the same type, if we send j_1 to i_2 and j_2 to i_1, we get another feasible clustering with no more cost. Therefore, starting from an optimal solution, one can transform it to a new optimal solution satisfying the above property. We now present a dynamic programming to compute the optimal solution for the given instance of ϵ-RBkM in quasi-polynomial time.

The Table. The table in our dynamic programming algorithm captures "snapshots" of solutions in a particular subtree which includes the information of how many clients of each type either enter or leave this subtree. The subproblems have the form $(v, k', \mathbf{Q})$, where v is a vertex of the tree, $k' \leq k$, and $\mathbf{Q}$ is a vector of length $p + 1$ of integers; we describe these parameters below. We want to find the minimum cost solution to cover all the clients in T_v, the subtree rooted at v, such that:

1. There are at most $0 \leq k' \leq k$ open facilities in T_v. These facilities serve clients inside or outside T_v.
2. The clients in T_v are covered by the facilities inside T_v or outside T_v.
3. $\mathbf{Q}$ is a $p + 1$ dimensional vector. The ith component of this vector q_i determines the number of type i clients that enter or leave T_v. When $0 \leq q_i \leq n$, q_i is the number of type i clients that enter T_v and when $-n \leq q_i \leq 0$, $|q_i|$ is the number of type i clients that leave T_v.

In a partial solution for the subproblem, the types of clients in T_v and the at most k' facilities to be opened in T_v must be determined. Each client must be assigned to an open facility of the same type in T_v or sent to v to be serviced outside, and each client shipped from outside to v must be assigned to a facility of its type inside T_v. The cost of a partial solution accounts for the cost of sending a client in T_v to a facility inside or to v (i.e., distance to the facility of v times $(1 + \epsilon)^i$ where i is the type client) plus, for the clients shipped from outside of

---

[1] A tree in which every node other than the leaves has two children.

T_v to v, the cost of sending them from v to their designated facility in T_v (i.e. the distance from v to the facility times $(1 + \epsilon)^i$ where i is the type client). We keep the value of a minimum cost partial solution in table entry $A[v, k', \mathbf{Q}]$. After filling this table, the final answer will be in the entry $A[r, k, \mathbf{0}]$ where $\mathbf{0}$ is a vector with $p + 1$ zero components.

Base Case 1: There is a client on v. Then, for each $0 \leq j \leq p$, we do as follows. We form a vector $\mathbf{Q}$ with $p + 1$ components such that the ith component $q_i = 0$ for all $i \neq j$ and $q_i = -1$ for $i = j$. Then, we set $A[v, 0, \mathbf{Q}] = 0$. We set all other entries of the form $A[v, ., .]$ to infinity.

Base Case 2: There is a facility on v. Then, for each type $0 \leq j \leq p$ and for each integer $(1 + \epsilon)^{j-1} < t \leq (1 + \epsilon)^j$, we do as follows. We form a vector $\mathbf{Q}$ with $p + 1$ components such that the ith component $q_i = 0$ for all $i \neq j$ and $q_i = t$ for $i = j$. We set $A[v, 1, \mathbf{Q}] = 0$ and all other entries of the form $A[v, ., .]$ to infinity.

Recursive Case: Consider a subtree rooted at a vertex v with two children v_1 and v_2. We say the subproblem corresponding $(v, k', \mathbf{Q})$ is *consistent* with subproblems $(v_1, k_1', \mathbf{Q}_1)$ and $(v_2, k_2', \mathbf{Q}_2)$ if $k_1' + k_2' \leq k'$ and $\mathbf{Q}_1 + \mathbf{Q}_2 = \mathbf{Q}$.

To find the value of a subproblem $(v, k', \mathbf{Q})$, we initialize $A[v, k', \mathbf{Q}] = \infty$ and enumerate over all subproblems for its children v_1 and v_2. For each pair of consitent subproblems $(v_1, k_1', \mathbf{Q}_1)$ and $(v_2, k_2', \mathbf{Q}_2)$, we update the entry to the minimum of its current value and:

$$\sum_{i=1}^{2} (A[v_i, k_i', \mathbf{Q}_i] + \sum_{j=0}^{p} |q_j^{(i)}| \cdot (1 + \epsilon)^j \cdot d(v_i, v)),$$

where $q_j^{(i)}$ is the jth component of $\mathbf{Q}_i$.

Note that the size of the DP table is $O(n^{p+3})$ and we can compute each entry in time $n^{O(p)}$, therefore:

Theorem 2. *There is a QPTAS for the BkM problem on tree metrics.*

5 Conclusion

In this paper, we have given an $O(\log n)$-approximation for BkM and MSkC in general metrics and also a quasi-PTAS for BkM in doubling metrics. Of course, the most natural open problem is to determine if either of these problems admits a true constant-factor approximation in arbitrary metric spaces. A PTAS for BkM in doubling dimension metrics or even Euclidean metrics seems quite plausible but even obtaining a constant-factor approximation in such cases is an interesting open problem. Perhaps one direction of attack would be to consider LP relaxation for the problem. It can be shown that the most natural configuration based LP (where we would have a variable $x_{i,C}$ for every possible facility location i and a set C of clients assigned to it) is equivalent to the natural LP relaxation. One of the difficulties of using LP for BkM is that most of the standard rounding techniques that have been used successfully for facility location or the k-median problem (such as filtering, clustering, etc) do not seem to work for the BkM due to the multiplier of cluster sizes. For example, the bicriteria approximation of [4] relies on a correspondence between BkM and a variant of capacitated k-median on a semi-metric space. They then used a Lagrangian relaxation and a primal-dual method to solve the capacitated k-median; the end result though opens $O(k)$ centers. Chuzhoy and Rabani [6] presented a better approximation for capacitated k-median where there are at most k locations of centers while up to $O(1)$ centers may be open at each location. Adapting their algorithm to work for the semi-metric space resulting from the work of [4] breaks down at a technical point. In particular, where one has to combine two solutions obtained from the primal-dual method with $k_1 < k < k_2$ number of centers. If one could overcome this technical difficulty then it could lead to a $O(1)$-approximation for MSkC and BkM on general metrics. Overall, it would be interesting to see if the standard LP relaxation has a constant integrality gap.

References

1. Agarwal, A., Charikar, M., Makarychev, K., Makarychev, Y.: $O(\sqrt{\log n})$-approximation algorithms for Min UnCut, Min-2CNF deletion, and directed cut problems. In: Proc. of STOC (2005)
2. Arya, V., Garg, N., Khandekar, R., Meyerson, A., Munagala, K., Pandit, V.: Local Search Heuristics for k-Median and Facility Location Problem. SIAM Journal on Computing **33**, 544–562 (2004)
3. Bartal, Y.: Probabilistic approximation of metric spaces and its algorithmic application. In: Proc. of FOCS (1996)
4. Bartal, Y., Charikar, M., Raz, D.: Approximating min-sum k-Clustering in metric spaces. In: Proc. of STOC (2001)
5. Byrka, J., Pensyl, T., Rybicki, B., Srinivasan, A., Trinh, K.: An improved approximation for k-median, and positive correlation in budgeted optimization. In: Proc. of SODA (2015)
6. Chuzhoy, J., Rabani, Y.: Approximating k-median with non-uniform capacities. In: Proc. of SODA (2005)
7. Czumaj, A., Sohler, C.: Small space representations for metric min-sum k-clustering and their applications. In: Thomas, W., Weil, P. (eds.) STACS 2007. LNCS, vol. 4393, pp. 536–548. Springer, Heidelberg (2007)
8. Fakcharoenphol, J., Rao, S., Talwar, K.: A tight bound on approximating arbitrary metrics by tree metrics. In: Proc. of STOC (2003)
9. de la Vega, W.F., Karpinski, M., Kenyon, C., Rabani, Y.: Approximation schemes for clustering problems. In: Proc. STOC (2003)
10. Guttman-Beck, N., Hassin, R.: Approximation algorithms for min-sum p-clustering. Discrete Applied Mathematics **89**, 125–142 (1998)
11. Indyk, P.: A sublinear time approximation scheme for clustering in metric spaces. In: Proc. of FOCS (1999)
12. Kann, V., Khanna, S., Lagergren, J., Panconessi, A.: On the hardness of max k-cut and its dual. In: Israeli Symposium on Theoretical Computer Science (1996)
13. Li, S., Svensson, O.: Approximating k-median via pseudo-approximation. In: Proc. of STOC (2013)

4. Sahni, S., Gonzalez, T.: P-Complete Approximation Problems. J. of the ACM (JACM) **23**(3), 555–565 (1976)
5. Schulman, L.J.: Clustering for edge-cost minimization. In: Proc. of STOC (2000)
6. Talwar, K.: Bypassing the embedding: algorithms for low dimensional metrics. In: Proc. of STOC (2004)
7. Wu, C., Xu, D., Du, D., Wang, Y.: An improved approximation algorithm for k-median problem using a new factor-revealing LP. http://arxiv.org/abs/1410.4161

Solving Linear Programming with Constraints Unknown

Xiaohui Bei[1](✉), Ning Chen[2], and Shengyu Zhang[3](✉)

[1] Max Planck Institute for Informatics, Saarbrücken, Germany
xbei@mpi-inf.mpg.de
[2] Nanyang Technological University, Singapore, Singapore
[3] The Chinese University of Hong Kong, Shatin, Hong Kong

Abstract. What is the value of input information in solving linear programming? The celebrated ellipsoid algorithm tells us that the full information of input constraints is not necessary; the algorithm works as long as there exists an oracle that, on a proposed candidate solution, returns a violation in the form of a separating hyperplane. Can linear programming still be efficiently solved if the returned violation is in other formats?

Motivated by some real-world scenarios, we study this question in a trial-and-error framework: there is an oracle that, upon a proposed solution, returns the *index* of a violated constraint (with the content of the constraint still hidden). When more than one constraint is violated, two variants in the model are investigated. (1) The oracle returns the index of a "most violated" constraint, measured by the Euclidean distance of the proposed solution and the half-spaces defined by the constraints. In this case, the LP can be efficiently solved (under a mild condition of non-degeneracy). (2) The oracle returns the index of an arbitrary (i.e., worst-case) violated constraint. In this case, we give an algorithm with running time exponential in the number of variables. We then show that the exponential dependence on n is unfortunately necessary even for the query complexity. These results put together shed light on the amount of information that one needs in order to solve a linear program efficiently.

The proofs of the results employ a variety of geometric techniques, including the weighted spherical Voronoi diagram and the furthest Voronoi diagram.

1 Introduction

Solving linear programming (LP) is a central question studied in operations research and theoretical computer science. The existence of efficient algorithms for LP is one of the cornerstones of a broad class of designs in, for instance, approximation algorithms and combinatorial optimization. The feasibility problem of linear programming asks to find an $x \in \mathbb{R}^n$ to satisfy a number of linear constraints $Ax > b$. Some previous algorithms, such as the simplex and interior point algorithms, assume that the constraints are explicitly given. In contrast, the ellipsoid method is able to find a feasible solution even without full knowledge of the constraints. This remarkable property grants the ellipsoid method an important role in many theoretical applications.

© Springer-Verlag Berlin Heidelberg 2015
M.M. Halldórsson et al. (Eds.): ICALP 2015, Part I, LNCS 9134, pp. 129–142, 2015.
DOI: 10.1007/978-3-662-47672-7_11

A central ingredient in the ellipsoid method is an oracle that, for a proposed (infeasible) point $x \in \mathbb{R}^n$, provides a violation that separates x and the feasible region of the LP in the format of a hyperplane. Such a separation oracle captures situations in which the input constraints are unavailable or cannot be accessed affordably, and the available information is from separating hyperplanes for proposed solutions. A natural question is what if the feedback for a proposed solution is not a separating hyperplane. Aside from theoretical curiosity, the question relates to practical applications, where the acquired violation information is actually rather different and even more restricted and limited.

Transmit power control in cellular networks has been extensively studied in the past two decades, and the techniques developed have become foundations in the CDMA standards in today's 3G networks. In a typical scenario, there are a number of pairs of transmitters and receivers, and the transmission power of each transmitter needs to be determined to ensure that the signal is strong enough for the target receiver, yet not so strong that it interferes with other receivers. This requirement can be written as an LP of the form $Ax > b$, where each constraint i corresponds to the requirement that the Signal to Interference Ratio (SIR) is no less than a certain threshold. In general the power control is a well-known hard problem (except for very few cases, such as power minimization [9]); a major difficulty is that matrix A depends mainly on the "channel gains", which are largely unknown in many practical scenarios [4]. Thus the LP $Ax > b$ needs to be solved despite the unavailability of (A, b). What is available here is that the system can try some candidate solution x and observe violation information (namely whether the SIR exceeds the threshold). The system can then adjust and propose new solutions until finally finding an x to satisfy $Ax > b$.

There are more examples in other areas (e.g., normal form games and product design and experiments [20]) with input information hidden. In these examples, for any proposed solution that does not satisfy all the constraints, only certain salient phenomena of violation (such as signal interference) are exhibited, which give *indices* of violated constraints only. With so little information obtained from violations, is it still possible to solve LP efficiently? In general, what is the least amount of input information, in what format, that one needs to solve a linear program efficiently? This work attempts to address these questions on the value of input information in solving LP.

1.1   Model and Results

Our model is defined as follows. In an LP $Ax > b$, the constraints $a_i x > b_i$ are hidden from us. We can propose candidate solutions $x \in \mathbb{R}^n$ to a *verification oracle*[1]. If x satisfies $Ax > b$, then the oracle returns Yes and the job is done. If x is not a feasible solution, then the oracle returns the index of a violated constraint. The algorithm continues until it either finds a feasible solution or

---

[1] The verification oracle is simply a means of determining whether a solution is feasible. It arises from the nature of LP as shown from the foregoing examples. For infeasible solutions, the feedback is a signaled violation.

concludes that no feasible solution exists. The algorithm is adaptive in the sens that future queries may depend on the information returned during previou queries. We focus only on the feasibility problem, to which an optimization LF can be transformed by a standard binary search.

Note that when the proposed solution is not feasible, the oracle returns only the *index i* of a violation rather than the constraint $a_i x > b_i$ itself. We make this assumption for two reasons. First, consistent with the aforementioned examples we are often only able to observe unsatisfactory phenomena (such as a strong interference in the power control problem). However, the exact reasons (corre sponding to the content of violated constraints) for these problems may still be unknown. Second, as our major focus is on the value of information in solving linear programming, a weaker assumption on the information obtained implies stronger algorithmic complexity results. Indeed, as will be shown, in some settings efficient algorithms exist even with this seeming deficit of information.

For a proposed solution x, if there are multiple violated constraints, the oracle returns the index of one of them[2]. This raises the question of which violation the oracle returns, and two variants are studied in this paper. In the first one, the oracle gives more information by returning the index of a "most violated" constraint, where the extent of a violation is measured by $(b_i - \langle a_i, x \rangle)/\|a_i\|$, the Euclidean distance of the proposed solution x and the half-space defined by the constraint. This oracle, referred to as the *furthest oracle*, attempts to capture the situation in which the first violation that occurs or is observed is usually the most severe and dominant one. The second variant follows the tradition of worst-case analysis in theoretical computer science, and makes no assumption about the returned violation. This oracle is referred to as the *worst-case oracle*.

We will denote by UnknownLP the problem of solving LP with unknown constraints in the above model. In either oracle model, the time complexity is the minimum amount of time needed for any algorithm to solve the UnknownLP problem, where each query, as in the standard query complexity, costs a unit of time.

Our results are summarized below. In a nutshell, when given a furthest oracle, a polynomial-time algorithm exists to solve LP (under a mild condition of non-degeneracy). On the other hand, if only a worst-case oracle is given, the best time cost is polynomial in m, the number of constraints, but exponential in n, the number of variables. Note that it is efficient when n is small, a well-studied scenario called *fixed-dimensional LP*. The exponential dependence on n is unfortunately necessary even for the query complexity. This lower bound, when combined with the positive result for the furthest oracle case, yields an illustration of the boundary of tractable LP.

---

[2] It is also natural to consider the case where the oracle returns the indices of all violated constraints. That model turns out to be so strong as to make the linear program easily solvable. By moving the proposed points and observing the change of the set of violated constraints, one can quickly identify the value of each (a_i, b_i).

Theorem 1. *The* UnknownLP *problem can be solved in time polynomial in the input size[3] in the furthest oracle model, provided that the input is non-degenerate.*

The exact definition of non-degeneracy is given in Section 3. The condition is mild; actually a random perturbation on inputs yields non-degeneracy, thus the theorem implies that the smooth complexity is polynomial.

The main idea of the algorithm design is as follows. Instead of searching for a solution x directly, we consider the point $(A, b) \in \mathbb{R}^{m(n+1)}$ as a degenerate polyhedron, and use the ellipsoid method to find (A, b). In each iteration take the center (A', b') of the current ellipsoid in $\mathbb{R}^{m(n+1)}$, and aim to construct a separating hyperplane between (A, b) and (A', b') through queries to the furthest oracle. The main difficulty lies in the case when (A', b') is infeasible, in which a separating hyperplane cannot be constructed explicitly. It can be observed that upon a query x, with the help of the furthest oracle, the information returned from the oracle has a strong connection to the Voronoi diagram. Specifically, if x is not a feasible solution, then the returned index is always the furthest Voronoi cell that contains x. We can manage to compute the Voronoi diagram, but this does not uniquely determine the constraints that define the LP. To handle this difficulty, we give a sufficient and necessary characterization reducing the input LP to that of a new and homogeneous LP, for which the constraints can be identified using the structure of a corresponding weighted spherical closest Voronoi diagram.

For the worst-case oracle, we first establish the following upper bound which is exponential in the number of variables only.

Theorem 2. *The* UnknownLP *problem with m constraints, n variables, and input size L can be deterministically solved in time $(mnL)^{poly(n)}$. In particular, the algorithm is of polynomial time for constant dimensional LP (i.e. constant number of variables n).*

At the heart of the efficiency guarantee of our algorithm is a technical bound of $\sum_{i=0}^{n} \binom{m}{i}$ on the number of "holes" formed by the union of m convex bodies in $\mathbb{R}^n$.

The above theorem implies a polynomial time algorithm when the number n of variables is a constant. This is a well-studied scenario, called *fixed dimensional LP* in which n is much smaller than the number of constraints m; see [5,8,15, 18,19] and the survey [7].

On the other hand, a natural question is whether the exponential dependence is necessary; at the very least, can we improve the bound to subexponential, as Kalai [15] and Matoušek et al. [18] have done for simplex-like algorithms? Unfortunately, the next lower bound theorem indicates that this is impossible.

Theorem 3. *Any algorithm that solves the* UnknownLP *problem with m constraints and n variables needs $\Omega(m^{\lfloor n/2 \rfloor})$ queries to the oracle, regardless of its time cost.*

---

[3] The notion of input size in the unknown input setting is explained in Section 2.

We prove this by constructing a family of $2k = \Theta(m^{\lfloor n/2 \rfloor})$ LPs, such that the first k LPs P_i have disjoint feasible regions, and the last k LPs P_i' are the infeasible variants of the first k LPs. Unless the algorithm proposes a point in one of feasible regions, the oracle is designed to return a fixed (thus meaningless) constraint index. After $k-1$ queries, the algorithm still cannot distinguish LP P_i and P_i' for some i. Thus even the feasibility problem cannot be solved with less than k queries.

It is worth comparing the exponential hardness of UnknownLP with the complexities of Nash and CE, the problems of finding a Nash or correlated equilibrium in a normal-form game, in the trial-and-error model. In our previous work [2], we presented algorithms with *polynomial* numbers of queries for Nash and CE with unknown payoff matrices in the model with worst-case oracle[4]. Nash and CE can be written as quadratic and linear programs, respectively, but why is the general UnknownLP hard while the unknown-input Nash and CE are easy (especially when all are given unlimited computational power)? The most critical reason is that in normal-form games, there *always exists* a Nash and a correlated equilibrium, but a general linear program may not have feasible solutions. Indeed, if a feasible solution is guaranteed to exist (even for only a random instance), such as when the number of constraints is no more than that of variables, then an efficient algorithm for UnknownLP does exist: see the full version [3]. (In our algorithms for UnknownLP, the major effort is devoted to handling infeasible LP instances.) It is interesting to see that the *solution-existing* property plays a fundamental role in developing efficient algorithms.

Related Work. There were a few work studying LP with restricted input information [22,23,25], in settings different than the current paper; see the full version [3] for detailed comparisons. The trial-and-error model was proposed in [2], where a number of specific questions were studied. In [14], the model of [2] is extended to probabilistic queries and systematic studies about Constraint Satisfaction Problems (CSP) are conducted.

2   Preliminaries

Consider the following linear program (LP): $Ax > b$, where $A = (a_{ij})_{m \times n} \in \mathbb{R}^{m \times n}$ and $b = (b_1, \ldots, b_m)^T \in \mathbb{R}^m$. The *feasibility* problem asks to find a feasible solution $x \in \mathbb{R}^n$ that satisfies $Ax > b$ (or report that such a solution does not exist). Equivalently, this is to find a point $x \in \mathbb{R}^n$ that satisfies m linear constraints $\{a_i x > b_i : i \in [m]\}$, where each $a_i = (a_{i1}, \ldots, a_{in})$.

In the *unknown-constraint LP feasibility* problem, denoted by UnknownLP, the coefficient matrix A and the vector b are unknown to us, and we need to determine whether the LP has a feasible solution and find one if it does. We can propose candidate solutions $x \in \mathbb{R}^n$ to a *verification oracle*. If a query x is indeed a feasible solution, the oracle returns Yes and the problem is solved.

---

[4] An algorithm proposes a candidate equilibrium and a verification oracle returns the index of an arbitrary better response of some player as a violation.

therwise, the oracle returns an index i satisfying $a_i x \leq b_i$, i.e., the index of violated constraint. Note that from this, the algorithm knows only the *index* but not a_i and b_i. In addition, if multiple constraints are violated, only the ⸱dex of *one* of them is returned.

We will analyze the complexity for two types of oracles: the *furthest oracle* ⸱hich returns the index of a "most" violated constraint (Section 3), and the ⸱orst-case oracle which can return an arbitrary index among those violated ⸱nstraints (Section 4). In either variant, a query to the oracle takes unit time.

⸱put Size and Solution Precision. A clarification is needed for the size of ⸱he input. Since the input LP instance (A, b) is unknown, neither do we know its ⸱inary size. To handle this issue, we assume that we are given the information ⸱hat there are m constraints[5], n variables, and the binary size of the input ⸱stance (A, b) is at most L. Note that L is $O(mn \log(N))$, where N is the ⸱aximum entry (in absolute value) in A and b. We say that an algorithm solves ⸱nknownLP efficiently if its running time is $poly(m, n, L)$.

Given an LP with input size $L = O(mn \log(N))$, it is known [16] that if the LP ⸱as a feasible solution, then there is one whose numerators and denominators of ⸱ll components are bounded by $(nN)^n$. Hence, an alternative way to describe our ⸱ssumption is that, instead of knowing the input size bound L, there is a required ⸱recision for feasible solutions. That is, we only look for a feasible solution in ⸱hich the numerators and denominators of all components are bounded by the ⸱equired precision. These two assumptions, i.e., giving an input size bound and ⸱iving a solution precision requirement, are equivalent, and it is necessary to ⸱ave one of them in our algorithms.[6] In the rest of the paper, we will use the ⸱irst one, the input size bound, to analyze the running time of our algorithms.

The unit sphere in $\mathbb{R}^n$ is denoted by $S^{n-1} = \{x \in \mathbb{R}^n : \|x\| = 1\}$, where, ⸱hroughout this paper, $\|\cdot\|$ refers to the ℓ_2-norm. A set $C \subseteq \mathbb{R}^n$ is a *convex* ⸱one if for any $x, y \in C$ and any $\alpha, \beta > 0$, $\alpha x + \beta y$ is also in C. The *normalized* ⸱olume of a convex cone C is defined as the ratio $v(C) = \frac{\text{vol}_n(C \cap B^n)}{\frac{1}{2} \cdot \text{vol}_n(B^n)}$ where B^n ⸱s the closed unit ball in $\mathbb{R}^n$ and vol_n refers to the n-dimensional volume. For ⸱ny set $C \in \mathbb{R}^n$, its *polar cone* C^* is the set $C^* = \{y \in \mathbb{R}^n : \langle x, y \rangle \leq 0, \forall x \in C\}$.

Lemma 4 [24]. *Let $C_1, C_2, \ldots, C_k$ be k closed convex cones, then $(\bigcap_i C_i)^* = conv(\bigcup_i C_i^*)$.*

---

[5] Indeed, the number of constraints can be unknown to us as well: In an algorithm, we only need to track those violated constraints that have ever been returned by the oracle.

[6] Otherwise, we may not be able to distinguish between cases when there are no feasible solutions (e.g., $x > 0, x < 0$) and when there are feasible solutions but the feasible set is very small (e.g., $x > 0, x < \epsilon$). For any queried solution $y > 0$, the oracle always returns that the second constraint is violated. However, we cannot distinguish whether it is $x < 0$ in the first LP or $x < \epsilon$ in the second LP, as ϵ can be arbitrarily small and we have no information on how small it is.

It was shown in [21] (Lemma 8.14) that if an LP has a feasible solution, then the set of solutions within the ball $\{x \in \mathbb{R}^n : \|x\| \leq n2^L\}$ has volume at least $2^{-(n+2)L}$. Given this lemma, we can easily derive the following claim.

Lemma 5. *If a linear program $Ax > 0$ has a feasible solution, then the feasible region is a convex cone in $\mathbb{R}^n$ and has normalized volume no less than $2^{-(2n+3)L}$.*

3   Furthest Oracle

In this section, we will consider the UnknownLP problem $Ax > b$ with the *furthest oracle*, formally defined as follows. For a proposed candidate solution x, if x is not a feasible solution, instead of returning the index of an arbitrary (worse case) violated constraint, the oracle returns the index of a "most violated" constraint, measured by the Euclidean distance from the proposed solution x and the half-space defined by the constraint. More precisely, the oracle returns the index of a constraint which, among all i with $\langle a_i, x \rangle \leq b_i$, maximizes $\frac{b_i - \langle a_i, x \rangle}{\|a_i\|}$, the distance from x to the half-space $\{z \in \mathbb{R}^n : \langle a_i, z \rangle \geq b_i\}$. If there are more than one maximizer, the oracle returns an arbitrary one.

Compared to the worse-case oracle, the furthest oracle reveals more information about the unknown LP system, and indeed, it can help us to derive a more efficient algorithm. Our main theorem in this section is the following.

Theorem 6. *The UnknownLP problem $Ax > b$ with a non-degenerate matrix A in the furthest oracle model can be solved in time polynomial in the input size.*

We call a matrix $A = (a_1, \ldots, a_m)^T$ *non-degenerate* if for each point $p \in S^{n-1}$, at most n points in $\left\{ \frac{a_1}{\|a_1\|}, \ldots, \frac{a_m}{\|a_m\|} \right\}$ have the same spherical distance to p on S^{n-1}. This assumption is with little loss of generality; it holds for almost all real instances and can be derived easily by a small perturbation.

Next we describe our algorithm for the special case of $Ax > 0$.

3.1   Algorithm Solving $Ax > 0$

We assume without loss of generality that $\|a_i\| = 1$ for all i. Furthermore, we can also always propose points in S^{n-1} for the same reason.

Ellipsoid Method and Issues. The main approach of the algorithm is to use the ellipsoid method to find the unknown matrix $A = (a_{ij})_{m \times n}$, which can be viewed as a point in the dimension $\mathbb{R}^{mn}$, i.e., a degenerate polyhedron in $\mathbb{R}^{mn}$. Initially, for the given input size information m, n and L, we choose a sufficiently large ellipsoid that contains the candidate region of A, and pick the center $A' \in \mathbb{R}^{mn}$ of the ellipsoid. To further the ellipsoid method, we need a hyperplane separating A' from A.

Consider the linear system $A'x > 0$. If it has a feasible solution x, then $\{x : A'x > 0\}$ is a full-dimensional cone. We query an x in this cone to the oracle. If the oracle returns an affirmative answer, then x is a feasible solution

$Ax > 0$ as well, and the job is done. Otherwise, the oracle returns an index i, meaning that $\langle a_i, x \rangle \le 0$. Hence, we have $\langle a_i', x \rangle > 0 \ge \langle a_i, x \rangle$, which defines a separating hyperplane between A and A' with normal vector $(\underbrace{0, \ldots, 0}_{(i-1)n}, x, \underbrace{0, \ldots, 0}_{(m-i)n})$

note that a hyperplane in $\mathbb{R}^{mn}$ has a normal vector of dimension mn, and x is vector of dimension n, also that we know the information of A' and x). Thus, we can cut the candidate region of A by a constant fraction and continue with the ellipsoid method.

Note that there is a small issue: In our problem, the solution polyhedron degenerates to a point $A \in \mathbb{R}^{mn}$ and has volume 0. As the input A is unknown, we cannot use the standard approach in the ellipsoid method to introduce a positive volume for the polyhedron by adding a small perturbation. This issue can be handled by a more involved machinery developed by Grötschel, Lovász, and Schrijver [11,12], which solves the strong nonemptiness problem for well-described polyhedra given by a strong separation oracle, as long as a *strong separation oracle* exists. In the algorithms described below, we will construct such oracles, thereby circumventing the issue of perturbation of the unknown point A. The same idea has been used in [2] to find a Nash equilibrium when the payoff matrix is unknown and degenerates to a point in a high-dimensional space. More discussions refer to [2,11,12].

The main difficulty is when the LP $A'x > 0$ is infeasible. In the following part of this section we will discuss how to find a proper separating hyperplane in this case.

Spherical (Closest) Voronoi Diagram. Note that $Ax > 0$ is equivalent to $-Ax < 0$, and i minimizes $\langle a_i, x \rangle$ if and only if it maximizes $\langle -a_i, x \rangle$. In the rest of this subsection, for notational convenience, we use $x \in S^{n-1}$ to denote a proposed solution point, and let $y = -x$. Since the distance from a proposed solution x to a half-space $\{z \in \mathbb{R}^n : \langle a_i, z \rangle \ge 0\}$ is $-\langle a_i, x \rangle = \langle a_i, y \rangle$, the oracle returns us an index $i \in \arg\max_i \{\langle a_i, y \rangle : \langle a_i, y \rangle \ge 0\}$ if x is not feasible. Note that $\|z - a_i\| \le \|z - a_j\|$ if and only if $\langle a_i, z \rangle \ge \langle a_j, z \rangle$ for any $z \in S^{n-1}$; thus, $\langle a_i, y \rangle$ is closely related to the distance between a_i and y on S^{n-1}. That is, the oracle actually provides information about the closest Voronoi diagram of $a_1, \ldots, a_m$ on S^{n-1}.

The (closest) *Voronoi diagram* of a set of points $\{a_i\}_i$ in S^{n-1} is a partition of S^{n-1} into cells, such that each point a_i is associated with the cell $\{z \in S^{n-1} : d(z, a_i) \le d(z, a_j), \forall j\}$, where d in our case is the spherical distance on S^{n-1}. We denote by Vor the spherical (closest) Voronoi diagram of the points $a_1, \ldots, a_m$ on S^{n-1} and denote by $\mathsf{Vor}(i)$ the cell in the diagram associated with a_i, i.e.,

$$\mathsf{Vor}(i) = \left\{ z \in S^{n-1} : \langle a_i, z \rangle \ge \langle a_j, z \rangle, \ \forall j \in [m] \right\} \tag{1}$$
$$= \left\{ z \in S^{n-1} : \|z - a_i\| \le \|z - a_j\|, \ \forall j \in [m] \right\}.$$

If the oracle returns i upon a query $x = -y \in S^{n-1}$, then $y \in \mathsf{Vor}(i)$.

Representation. Note that for a general (spherical) Voronoi diagram formed by m points, it is possible that some of its cells contain exponential number of vertices, which is unaffordable for our algorithm. However, in the H representation of a convex polytope, every cell can be represented by at most m linear inequalities, as shown in Formula (1). In the following, we will see that the information of these linear inequalities is sufficient to implement our algorithm efficiently.

Weighted Spherical (Closest) Voronoi Diagram. For the presumed matrix A', note that it can be an arbitrary point in the space $\mathbb{R}^{mn}$ and each row in A may not necessarily fall into S^{n-1}. Our solution is to consider a *weighted spherical Voronoi diagram*, denoted by Vor', of points $\frac{a'_1}{\|a'_1\|}, \ldots, \frac{a'_m}{\|a'_m\|}$ on S^{n-1} as follows for each point $\frac{a'_i}{\|a'_i\|}$, its associated cell is defined as

$$\mathsf{Vor}'(i) = \left\{ z \in S^{n-1} : \langle a'_i, z \rangle \geq \langle a'_j, z \rangle, \forall j \in [m] \right\}.$$

Note that Vor' is a partition of S^{n-1}; and if we assign a weight $\|a'_i\|$ to each point $\frac{a'_i}{\|a'_i\|}$, then for each point $p \in \mathsf{Vor}'(i)$, the site among $\frac{a'_1}{\|a'_1\|}, \ldots, \frac{a'_m}{\|a'_m\|}$ that has the smallest *weighted* distance to p is $\frac{a'_i}{\|a'_i\|}$.[7] Note that each cell of Vor' is defined by a set of linear inequalities (other than the unit norm requirement) and each of them can be computed efficiently.

Now we have two diagrams: Vor, which is unknown, and Vor', which can be represented efficiently using the H-representation. If $\mathsf{Vor} \neq \mathsf{Vor}'$, then there exists a point $y \in S^{n-1}$ such that $y \in \mathsf{Vor}(i)$ and $y \notin \mathsf{Vor}'(i)$. Suppose that $y \in \mathsf{Vor}'(j)$ for some $j \neq i$. According to the definition, we have $\langle a_i, y \rangle \geq \langle a_j, y \rangle$ and $\langle a'_i, y \rangle < \langle a'_j, y \rangle$; this gives us a separating hyperplane between A and A'. The questions are then (1) how to find such a point y when $\mathsf{Vor} \neq \mathsf{Vor}'$, and (2) what if $\mathsf{Vor} = \mathsf{Vor}'$.

Consistency Check. In this part we will show how to check whether $\mathsf{Vor} = \mathsf{Vor}'$, and if not equal, how to find a y as above. Although we know neither the positions of points $a_1, \ldots, a_m$, nor the corresponding spherical Voronoi diagram Vor, we can still efficiently compare it with Vor', with the help of the oracle.

For each cell $\mathsf{Vor}'(i)$, assume that it has k facets (i.e., $(n-1)$-dimensional faces). Note that $k \leq m$ and that $\mathsf{Vor}'(i)$ is uniquely determined by these facets. Further, each facet is defined by a hyperplane $H'_{ij} = \{ z \in S^{n-1} : \langle a'_i, z \rangle = \langle a'_j, z \rangle \}$ for some $j \neq i$. To decide whether $\mathsf{Vor} = \mathsf{Vor}'$, for each i and j such that $\mathsf{Vor}'(i) \cap \mathsf{Vor}'(j) \neq \emptyset$, we find a sufficiently small ϵ_y and three points y, $y + \epsilon_y$, $y - \epsilon_y$, such that

$$y \in \mathsf{Vor}'(i) \cap \mathsf{Vor}'(j) \subset H'_{ij}, \quad y + \epsilon_y \in \mathsf{Vor}'(i) \setminus \mathsf{Vor}'(j), \quad y - \epsilon_y \in \mathsf{Vor}'(j) \setminus \mathsf{Vor}'(i).$$

Notice that such y and ϵ_y exist and can be found efficiently. We now query points $y + \epsilon_y$ and $y - \epsilon_y$ to the oracle. If the oracle does return us the expected answers,

---

[7] The reason of defining such a weighted spherical Voronoi diagram is that we want to have a separating hyperplane between A and $A' = (a'_1, \ldots, a'_m)^T$, rather than $\left(\frac{a'_1}{\|a'_1\|}, \ldots, \frac{a'_m}{\|a'_m\|} \right)^T$.

i.e., i and j, respectively, then, with $\|\epsilon_y\|$ sufficiently small (up to $2^{-poly(L)}$), we can conclude that y must also be in the facet of $\mathsf{Vor}(i)$ and $\mathsf{Vor}(j)$ of the hidden diagram Vor. That is, $y \in H_{ij} = \{z \in S^{n-1} : \langle a_i, z \rangle = \langle a_j, z \rangle\}$. We implement the above procedure $n - 1$ times to look for $n - 1$ linearly independent points $y_1, \ldots, y_{n-1} \in \mathsf{Vor}'(i) \cap \mathsf{Vor}'(j)$. If the oracle always returns the expected answers i and j, respectively, for all $k = 1, \ldots, n - 1$, then we know that $H_{ij} = H'_{ij}$.

The procedure described above can be implemented in polynomial time. Now we can use this approach to check all facets of all of the cells of Vor'. If none of them returns us an unexpected answer, we know that every facet of every cell $\mathsf{Vor}'(i)$ is also a facet of cell $\mathsf{Vor}(i)$, i.e., the set of linear constraints that defines $\mathsf{Vor}'(i)$ is a subset of those that define $\mathsf{Vor}(i)$. Thus, we have $\mathsf{Vor}(i) \subseteq \mathsf{Vor}'(i)$ for each i. Together with the fact that both Vor and Vor' are tessellations of S^{n-1}, we can conclude that $\mathsf{Vor} = \mathsf{Vor}'$.

Lemma 7. *For the hidden matrix $A \in \mathbb{R}^{mn}$ with spherical Voronoi diagram Vor and proposed matrix $A' \in \mathbb{R}^{mn}$ with weighted spherical Voronoi diagram Vor', we can in polynomial time*

- *either conclude that $\mathsf{Vor} = \mathsf{Vor}'$, or*
- *find a separating hyperplane between A and A'.*

A formal and detailed description of this consistency check procedure and its correctness proof can be found in the full version [3].

Voronoi Diagram Recognition. If the above process concludes that $\mathsf{Vor} = \mathsf{Vor}'$, we have successfully found the Voronoi diagram Vor (in its H-representation) for the hidden points $a_1, \ldots, a_m$. It was shown by Hartvigsen [13] that given a Voronoi diagram with its H-representation, a set of points that generates the diagram can be computed efficiently. Further, Ash and Bolker [1] showed that the set of points that generates a non-degenerate Voronoi diagram is unique. Therefore, by coupling these two results and the assumption that the input matrix A is non-degenerate, we are able to identify the positions of $a_1, \ldots, a_m$ given the computed Voronoi diagram Vor, and easily determine if the LP $Ax > 0$ has a feasible solution, and compute one if it exists.

The general case of $Ax > b$ New difficulties arise in the general case of $Ax > b$. A particular one is that, even for the non-degenerate input A, the Voronoi diagram may correspond to *multiple* sets of points a_i, which makes it hard to recover the a_i's. To handle this difficulty, we give a sufficient and necessary characterization reducing the input LP to that of a new and homogeneous LP, for which the constraints can be identified using the structure of a corresponding weighted spherical closest Voronoi diagram. We unfortunately have to leave this part to the full version [3] due to space limit.

4   Worst-Case Oracle

In this section, we consider the worst-case oracle. Recall that in this setting, the oracle plays as an adversary by giving the worst-case violation index to force an algorithm to use the maximum amount of time to solve the problem.

For any linear program $Ax > b$, we can introduce another variable y and transform the linear program into the following form:

$$Ax - by > 0, \quad y > 0$$

It is easy to check that $Ax > b$ is feasible if and only if the new LP is feasible, and the solutions of these two linear systems can be easily transformed to each other. Given the oracle for $Ax > b$, one can also get another oracle for the new LP easily. (On a query (x, y), if $y \leq 0$, return the index $m + 1$; otherwise, query x/y to the oracle for $Ax > b$.) This means that the UnknownLP problem of the homogeneous form $Ax > 0$ is no easier than the problem of the general form. In all the analysis of this section, we will therefore only consider the problem of form $Ax > 0$.

Geometric Explanations. Let us consider the problem from a geometric viewpoint. Any matrix $A = (a_{ij})_{m \times n}$ can be considered as m points $a_1, a_2, \ldots, a_m$ in the n-dimensional space $\mathbb{R}^n$, where each $a_i = (a_{i1}, a_{i2}, \ldots, a_{in})$. The positions of these points are unknown to us. Finding a feasible solution $x \in \mathbb{R}^n$ that satisfies $Ax > 0$ is equivalent to finding an open half-space

$$H_x = \left\{ y \in \mathbb{R}^n : \langle x, y \rangle \triangleq x_1 y_1 + x_2 y_2 + \cdots + x_n y_n > 0 \right\}$$

containing all points a_i.

In an algorithm, we propose a sequence of candidate solutions. When a query $x \in \mathbb{R}^n$ violates a constraint i, we know that $\langle a_i, x \rangle \leq 0$. Hence, a_i cannot be contained in the half-space H_x, and we are able to cut H_x off from the possible region of a_i. Based on this observation, we maintain a set region(i), the region of possible positions of point a_i consistent with the information obtained from the previous queries. Initially, no information is known about the position of any point; thus, region(i) $= \mathbb{R}^n$ for all $1 \leq i \leq m$.

Let us have a closer look at these regions. For each i, suppose that $x_1^i, x_2^i, \ldots, x_k^i$ are the queried points we have made so far for which the oracle returns index i. Then all information we know about a_i till this point is that the possible region is region(i) $= \bigcap_{j=1}^{k} \{ y \in \mathbb{R}^n : \langle x_j^i, y \rangle \leq 0 \}$. Since region($i$) is the intersection of k closed half-spaces, it is a convex set. Equivalently, this means that any feasible solution to the LP, if existing, cannot be in region(i)*, the polar cone of region(i). Since the polar cone of a half-space $\{ y \in \mathbb{R}^n \mid \langle x_j^i, y \rangle \leq 0 \}$ is the ray along its normal vector, i.e., $\{ \lambda x_j^i \mid \lambda \geq 0 \}$, we have by Lemma 4 that

$$\text{region}(i)^* = conv\left(\bigcup_j \{ y \mid \langle x_j^i, y \rangle \leq 0 \}^* \right) = conv\left(\{ \lambda x_j^i \mid 1 \leq j \leq k, \lambda \geq 0 \} \right).$$

Since region(i)*'s are the forbidden areas for any feasible solution, we can conclude that the LP has no feasible solution if $\bigcup_i \text{region}(i)^* = \mathbb{R}^n$.

Convex Hull Covering Algorithms. Based on above observations, we now sketch a framework of *convex hull covering algorithms* that solves the UnknownLP problem. The algorithm maintains a list of m convex cones

$$\text{region}(1)^*, \text{region}(2)^*, \ldots, \text{region}(m)^* \subseteq \mathbb{R}^n.$$

Initially, $\mathsf{region}(i)^* = \emptyset$ for all $1 \le i \le m$. On each query $x \in \mathbb{R}^n$, the oracle either returns Yes, indicating that the problem is solved, or returns us an index i, in which case we update $\mathsf{region}(i)^*$ to $conv\,(\mathsf{region}(i)^*, \{\lambda x \mid \lambda > 0\})$. The algorithm terminates when either the oracle returns Yes, or when $\mathbb{R}^n - \bigcup_i \mathsf{region}(i)^*$ does not contain a convex cone with normalized volume at least $2^{-(2n+3)L}$, which indicates that the given instance has no feasible solution. The above discussion can be formalized into the following theorem.

Theorem 8. *Any algorithm that falls into the convex hull covering algorithm framework solves the* UnknownLP *problem.*

Though the framework guarantees the correctness, it does not specify how to make queries to control complexity. Next we will show an algorithm with nearly optimal complexity. The basic idea is to use induction on dimension. That is, we pick an $(n-1)$-dimensional subspace and recursively solve the problem on the subspace. The subroutine either finds a point x in the subspace that satisfies $Ax > 0$ (in which case the algorithm ends), or finds out that there is no feasible solution in the entire subspace. In the latter case, the whole space of candidate solutions can be divided into two open half-spaces, and we will work on each of them separately. In general, we have a collection of connected regions that can still contain a valid solution. These regions are the "holes", formally called *chambers*, separated by $\bigcup_i \mathsf{region}(i)^*$ (recall that points in $\mathsf{region}(i)^*$ cannot be a feasible solution). We can then pick a chamber with the largest volume, and cut it into two balanced halves by calling the subroutine on the hyperplane slicing the chamber.

There are several issues for the above approach. The main one is that there may be too many chambers: *a priori*, the number can grow exponentially with m. There are also other technical issues to be handled, such as how to represent chambers (which are generally concave), how to compute (even approximately) the volume of chambers, how to find a hyperplane to cut a chamber into two balanced halves, etc.

For the first and main issue, it can be shown that the number of chambers cannot be too large. In general, Kovalev [17] showed that any m convex sets in $\mathbb{R}^n$ cannot form more than $\sum_{i=1}^n \binom{m}{i}$ chambers. For the rest of the technical issues, we deal with them in the following way. Instead of keeping track of all actual chambers, in our algorithm, we maintain a collection of disjoint *sector cylinders*, which can be shown to be supersets of chambers. Furthermore, we keep only cylinders that contain at least one chamber, thus, the bound for the number of chambers also bounds the number of cylinders from above.

Theorem 2 can be proved based on the ideas described above. The details of the algorithm and its analysis can be found in the full version [3].

5 Concluding Remarks

We consider solving LP when the input constraints are unknown, and show that different kinds of violation information yield different computational complexities. LP is a powerful tool employed in real applications dealing with objects that are largely unknown. For example, in the node localization of sensor network where the locations of targets are unknown [6], the computation of the location in some settings can be formulated as a linear program with constraints that measure partial information obtained from data [10]. However, the estimation usually has various levels of error, which may lead to violations of the presumed constraints. Interesting questions that deserve further explorations are what can be theoretically analyzed there, and in general, what other natural formats of violations there are in linear programming and what complexities they impose.

References

1. Ash, P.F., Bolker, E.D.: Recognizing dirichlet tessellations. Geometriae Dedicata **19**(2), 175–206 (1985)
2. Bei, X., Chen, N., Zhang, S.: On the complexity of trial and error. In: Proceedings of the 45th ACM Symposium on Theory of Computing, pp. 31–40 (2013)
3. Bei, X., Chen, N., Zhang, S.: Solving linear programming with constraints unknown (2013). arXiv:1304.1247
4. Chiang, M., Hande, P., Lan, T., Tan, C.-W.: Power control in wireless cellular networks. Foundations and Trends in Networking **2**(4), 381–533 (2007)
5. Clarkson, K.L.: Las vegas algorithms for linear and integer programming when the dimension is small. Journal of the ACM **42**(2), 488–499 (1995)
6. Doherty, L., Pister, K., Ghaoui, L.E.: Convex position estimation in wireless sensor networks. In: Proceedings of the Twentieth IEEE Annual Joint Conference of the IEEE Computer and Communications Societies (INFOCOM), pp. 1655–1663 (2001)
7. Dyer, M., Megiddo, N., Welzl, E.: Linear programming. In: Goodman, J.E., O'Rourke, J. (eds.) Handbook of Discrete and Computational Geometry. CRC Press (2004)
8. Dyer, M.E.: On a multidimensional search technique and its application to the euclidean one-centre problem. SIAM Journal on Computing **15**(3), 725–738 (1986)
9. Foschini, G.J., Miljanic, Z.: A simple distributed autonomous power control algorithm and its convergence. IEEE Transactions on Vehicular Technology **42**(3), 641–646 (1993)
10. Gentile, C.: Distributed sensor location through linear programming with triangle inequality constraints. In: Proceedings of IEEE Conference on Communications, pp. 3192–3196 (2005)
11. Grötschel, M., Lovász, L., Schrijver, A.: Geometric methods in combinatorial optimization. In: Progress in Combinatorial Optimization, pp. 167–183 (1984)
12. Grötschel, M., Lovász, L., Schrijver, A.: Geometric Algorithms and Combinatorial Optimization. Springer (1988)
13. Hartvigsen, D.: Recognizing voronoi diagrams with linear programming. INFORMS Journal on Computing **4**(4), 369–374 (1992)
14. Ivanyos, G., Kulkarni, R., Qiao, Y., Santha, M., Sundaram, A.: On the complexity of trial and error for constraint satisfaction problems. In: Esparza, J., Fraigniaud, P., Husfeldt, T., Koutsoupias, E. (eds.) ICALP 2014. LNCS, vol. 8572, pp. 663–675. Springer, Heidelberg (2014)
15. Kalai, G.: A subexponential randomized simplem algorithm. In: Proceedings of the ACM Symposium on Theory of Computing (STOC), pp. 475–482 (1992)

16. Khachiyan, L.: A polynomial algorithm in linear programming. Doklady Akademii Nauk SSSR **244**, 1093–1096 (1979)
17. Kovalev, M.: A property of convex sets and its application. Matematicheskie Zametki, pp. 89–99. English translation: Mathematical Notes, **44**, 537–543 (1988)
18. Matousek, J., Sharir, M., Welzl, E.: A subexponential bound for linear programming. Algorithmica **16**(4/5), 498–516 (1996)
19. Megiddo, N.: Linear programming in linear time when the dimension is fixed. Journal of the ACM **31**(1), 114–127 (1984)
20. Montgomery, D.: Design and Analysis of Experiments, 7 edn. Wiley (2008)
21. Papadimitriou, C.H., Steiglitz, K.: Combinatorial Optimization: Algorithms and Complexity. Dover Publications (1998)
22. Papadimitriou, C.H., Yannakakis, M.: Linear programming without the matrix. In: Proceedings of the ACM Symposium on Theory of Computing (STOC), pp. 121–129 (1993)
23. Ryzhov, I.O., Powell, W.B.: Information collection for linear programs with uncertain objective coefficients. SIAM Journal on Optimization **22**(4), 1344–1368 (2012)
24. Sandgren, L.: On convex cones. Mathematica Scandinavica **2**, 19–28 (1954)
25. Yudin, D.B., Nemirovskii, A.S.: Informational complexity and efficient methods for the solution of convex extremal problems. Ekonomika i Matematicheskie Metody, **12**, 357–369 (1976). English translation: Matekon **13**(3), 25–45 (1977)

Deterministic Randomness Extraction from Generalized and Distributed Santha-Vazirani Sources

Salman Beigi[1], Omid Etesami[1]([✉]), and Amin Gohari[1,2]

[1] School of Mathematics, Institute for Research in Fundamental Sciences (IPM),
Tehran, Iran
salman.beigi@gmail.com, etesami@ipm.ir
[2] Department of Electrical Engineering,
Sharif University of Technology, Tehran, Iran
aminzadeh@sharif.edu

Abstract. A Santha-Vazirani (SV) source is a sequence of random bits where the conditional distribution of each bit, given the previous bits, can be partially controlled by an adversary. Santha and Vazirani show that deterministic randomness extraction from these sources is impossible. In this paper, we study the generalization of SV sources for non-binary sequences. We show that unlike the binary case, deterministic randomness extraction in the generalized case is sometimes possible. We present a necessary condition and a sufficient condition for the possibility of deterministic randomness extraction. These two conditions coincide in "non-degenerate" cases.

Next, we turn to a distributed setting. In this setting the SV source consists of a random sequence of pairs $(a_1, b_1), (a_2, b_2), \ldots$ distributed between two parties, where the first party receives a_i's and the second one receives b_i's. The goal of the two parties is to extract common randomness without communication. Using the notion of *maximal correlation*, we prove a necessary condition and a sufficient condition for the possibility of common randomness extraction from these sources. Based on these two conditions, the problem of common randomness extraction essentially reduces to the problem of randomness extraction from (non-distributed) SV sources. This result generalizes results of Gács and Körner, and Witsenhausen about common randomness extraction from i.i.d. sources to adversarial sources.

1 Introduction

Randomized algorithms are simpler and more efficient than their deterministic counterparts in many applications. In some settings such as communication complexity and distributed computing, it is even possible to prove unconditionally that allowing randomness improves the efficiency of algorithms (see e.g., [14,19, 30]). However, access to sources of randomness (especially common randomness) may be limited, or the quality of randomness in the source may be far from perfect. Having such an imperfect source of randomness, one may be able to extract

© Springer-Verlag Berlin Heidelberg 2015
M.M. Halldórsson et al. (Eds.): ICALP 2015, Part I, LNCS 9134, pp. 143–154, 2015.
DOI: 10.1007/978-3-662-47672-7_12

lmost) unbiased and independent random bits using *randomness extractors*. randomness extractor is a function applied to an imperfect source of randomness whose outcome is an almost perfect source of randomness.

The problem of randomness extraction from imperfect sources of randomness as perhaps first considered by Von Neumann [28]. A later important work in his area is [23] where Santha and Vazirani introduced the imperfect sources f randomness now often called Santha-Vazirani (SV) sources. These sources an easily be defined in terms of an adversary with two coins. Consider an dversary who has two different coins, one of which is biased towards heads (e.g., 'r(heads) = 2/3) and the other one is biased towards tails (e.g., Pr(heads) = /3). The adversary, in each time step, chooses one of the two coins and tosses . Adversary's choice of coin may depend (probabilistically) on the previous utcomes of the tosses. The sequence of random outcomes of these coin tosses is alled a SV source.

Santha and Vazirani [23] show that randomness extraction from the above ources through a deterministic method is impossible. More precisely, they show hat for every deterministic way of extracting one random bit, there is a strategy or the adversary such that the extracted bit is biased, or more specifically, the xtracted bit is 0 with probability either $\geq 2/3$ or $\leq 1/3$. Subsequently, other roofs for this result have been found (see e.g., [1,21]). Fig 1 shows a more refined ersion of this result, which provides a more detailed picture of the limits of what he adversary can achieve.

Despite this negative result, such imperfect sources of randomness are enough or many applications. For example, as shown by Vazirani and Vazirani [25,26], andomized polynomial-time algorithms that use perfect random bits can be imulated using SV sources. This fact can also be verified using the fact that he min-entropy of SV sources is linear in the size of the source (where min-entropy, in the context of extractors, was first introduced by [9]). Indeed, by the ater theory of randomness extraction (e.g., see [31]), it is possible to efficiently xtract polynomially many almost random bits from such sources with high min-entropy if we are, in addition to the imperfect source, endowed with a perfectly random seed of logarithmic length. (In fact, for the special case of SV sources, a seed of constant length is enough [27, Problem6.6]). For the application of randomized polynomial-time algorithms, we can enumerate in polynomial time over all possible seeds.

Enumerating over all seeds may be inefficient for some applications, or does not work at all, e.g., in interactive proofs and one-shot scenarios such as cryptography. Therefore, it is natural to ask whether deterministic randomness extraction from imperfect sources of randomness is possible. For most applications, it is also necessary to require that the extractor be explicit, i.e., extraction can be done efficiently (in polynomial time). Previous to this work, explicit deterministic extractors had been constructed for many different classes of sources, including i.i.d. bits with unknown bias [28], Markov chains [5], affine sources [7,16], polynomial sources [11,12], and sources consisting of independent blocks [6].

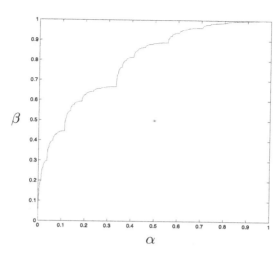

β

α

Fig. 1. Given any deterministic extractor, the pair (α, β) is above the curve specified in this figure, where α and β are the minimum and maximum value of probability of the output being zero that the adversary can achieve by choosing its strategy. The plot is for the binary SV source with two coins with probability of heads respectively equal to $1/3$ and $2/3$. The point $(1/2, 1/2)$ is specified by a red star in the figure. The curve has fractal-like self-similiarity: The curve can be split at point $(1/3, 2/3)$ into two curves each of which is a normalized version of the whole curve. To see how the curve is obtained, see Appendix A of the full version [2].

Deterministic Extractors for Generalized SV Sources. Although [23] proves the impossibility of deterministic randomness extraction from SV sources, this impossibility is shown only for binary sources. In this paper we show that if we consider a generalization of SV sources over *non-binary* alphabets, deterministic randomness extraction is indeed possible under certain conditions.

To generalize SV sources over non-binary alphabets, we assume that the adversary, instead of coins, has some multi-faceted (say 6-sided) dice. The numbers written on the faces of different dice are the same, but each die may have a different probability for a given face value. The adversary throws these dice n times, each time choosing a die to throw depending on the results of the previous throws. Again, the outcome is an imperfect source of randomness, for which we may ask whether deterministic randomness extraction is possible or not.

When the dice are non-degenerate, i.e., all faces of all dice have non-zero probability, we give a necessary and sufficient condition for the existence of a deterministic strategy for extracting one bit with arbitrarily small bias. For example, when the dice are 6-sided, the necessary and sufficient condition implies that we can deterministically extract an almost unbiased bit when the adversary has access to any arbitrary set of five non-degenerate dice, but randomness extraction is not possible in general when the adversary has access to six non-degenerate 6-sided dice. More precisely, a set of non-degenerate dice leads to extractable generalized SV sources if and only if the convex hull of the set of probability distributions associated with the set of dice does not have full dimension in the "probability simplex". We emphasize that when we prove the possibility of deterministic extraction, we also provide an explicit extractor.

Relation to Block-Sources. The generalized SV sources considered in this paper are also a generalization of "block-sources" defined by Chor and Goldreich [9], where the source is divided into several blocks such that each block has min-entropy at least k conditioned on the value of the previous blocks. Such a block-source can be thought as a generalized SV source where the adversary can generate each block (given previous blocks) using any "flat" distribution with support 2^k. Being a special case of generalized SV sources (defined here), block-sources have another difference as well: Since it is impossible to extract from a single block-source deterministically, the common results regarding extraction from block-sources are about either seeded extractors (e.g. [18]) or extraction from at least two independent block-sources (e.g. [20]).

Common Randomness Extractors. Common random bits, shared by distinct parties, constitute an important resource for distributed algorithms; common random bits can be used by the parties to synchronize the randomness of their local actions. We may ask the question of randomness extraction in this setting too. Assuming that the parties are provided with an imperfect source of common randomness, the question is whether perfect common randomness can be extracted from this source or not.

Gács and Körner [15] and Witsenhausen [29] have looked at the problem of extraction of common random bits from a very special class of imperfect sources, namely i.i.d. sources. In this case, the *bipartite* source available to the parties is generated as follows: In each time step, a pair (A, B) with some predetermined distribution (known by the two parties) and independent of the past is generated; A is revealed to the first party and B is revealed to the second party. After receiving arbitrarily many repetitions of random variables A and B, the two parties aim to extract a common random bit. It is known that in this case, the two parties (who are not allowed to communicate) can generate a common random bit if and only if A and B have a common data [29]. This means that common randomness generation is possible if A and B can be expressed as $A = (A', C)$ and $B = (B', C)$ for a nonconstant common part C, i.e., there are nonconstant functions f, g such that $C = f(A) = g(B)$. Observe that when a common part exists, common randomness can be extracted by the parties by applying the same extractor on the sequence of C's. That is, the problem of common randomness extraction in the i.i.d. case is reduced to the problem of ordinary randomness extraction. These results are obtained using a measure of correlation called *maximal correlation*. The key feature of this measure of correlation that helps proving the above result is the *tensorization property*, i.e., the maximal correlation between random variables A and B is equal to that of A^n and B^n for any n, where A^n and B^n denote n i.i.d. repetitions of A and B.

In this paper we consider the problem of common randomness extraction from *distributed SV sources* defined as follows. In a distributed SV source, the adversary again has some multi-faceted dice, but here, instead of a single number, a pair of numbers (A, B) is written on each face. As before, the set of values written on the faces of the dice is the same, but the probabilities of face values may differ in different dice. In each time step, the adversary depending on the results of the previous throws, picks a die and throws it. If (A, B) is the result of the throw, A is given to the first party and B to second party. Thus, the two parties will observe random variables A and B whose joint distribution depends on the choice of die by the adversary. An application of this distributed case would be a key-agreement scenario under tampering.

Again consider the non-degenerate case where all faces on all the dice of the adversary have positive probability. We show that in this case, we can extract a common random bit from the distributed SV source if and only if it is possible to extract randomness from the common part of A and B. That is, similar to the i.i.d. case, the problem of common randomness extraction from distributed SV sources is reduced to the problem of randomness extraction from non-binary generalized SV sources. Since by our results, we know when randomness extraction from generalized SV sources is possible, we obtain a complete answer to the problem in the distributed case too.

In cases more general than non-degenerate cases we have the following: If C is the common data of A and B, then if there does not exist a nonzero real function of C which has zero expectation under all the different dice of the adversary, then common randomness extraction is impossible. This shows that the relation between the problem of common randomness extraction and the problem of randomness extraction from the common part holds also in some settings other than non-degenerate cases. For example, it resolves the problem of common randomness extraction from the following interesting distributed SV source.

Example. A concrete example of a distributed SV source is as follows. Let us start with the original source considered by Santha and Vazirani with two coins. Assume that the adversary chooses coin $S \in \{1, 2\}$ (where coin 1 is biased towards heads and coin 2 is biased towards tails) and let the outcome of the throw of the coin be denoted by random variable C. The first party, Alice, is assumed to observe both the identity of the coin chosen by the adversary, i.e., S, and the outcome of the coin, which is C. The second party, Bob, observes the outcome of the coin C, but only gets to see the choice of the adversary with probability 0.99. That is, Bob gets $B = (C, \tilde{S})$ where $\tilde{S}$ is the result of passing S through a binary erasure channel with erasure probability 0.01. Here the common part of $A = (C, S)$ and $B = (C, \tilde{S})$ is just C. Our result (Theorem 3) then implies that Alice and Bob cannot benefit from their knowledge of the actions of adversary, and should only consider the C sequence. But then from the result of [23], we can conclude that common random bit extraction is impossible in this example.

Proof Techniques. We briefly explain the techniques used in the proof of the above results. For the full proofs, we refer the reader to the full version of the paper [2].

To show the possibility of deterministic extraction, we use a nonzero real function of the die face values that has zero expectation under all distributions induced by the different dice of the adversary. Then as we throw the dice several times, we consider the sum of the value of this function applied to the outcome of the dice throws. This sum forms a martingale. We stop the martingale once its absolute value exceeds a particular bound. Since the function used was nonzero, the martingale has large variance after a few throws, and therefore the martingale will be stopped with high probability. Also by the theorem of stopping times, the martingale has zero mean whenever we stop it. Then the extracted bit, determined by whether the stopped martingale is positive or is negative, would be unbiased.

To show the impossibility of deterministic extraction, we view a deterministic extractor that extracts one bit from a generalized SV source as labeling the leaves of a rooted tree with zeros and ones. Each sequence of dice throws corresponds to a path from the root to one of the leaves, and at each node, the adversary has some limited control of which branch to take while moving from the root towards the leaves. We need to show that either the minimum or the maximum of the probability of the output bit being zero, over all adversary's strategies, is far from $1/2$. Our idea is to track these maximum and minimum probabilities in a recursive way, i.e., to find these probabilities for any node of the tree in terms of these values for its children. We then by induction show that for each node of the tree either the minimum probability or the maximum probability is far from $1/2$.

To be more precise, given a deterministic extractor, let α be the minimum probability of output bit being zero (over all strategies of the adversary). Similarly, let β be the maximum probability of output bit being zero (over all strategies of the adversary). Then we show that under certain conditions, there exists a *continuous* function $g(\cdot)$ on the interval $[0, 1]$, such that $\beta \geq g(\alpha)$ and furthermore $g(1/2) > 1/2$. We prove $\beta \geq g(\alpha)$ inductively using the tree structure discussed above. This implies the desired impossibility result, as by the continuity of $g(\cdot)$, both α and β cannot be close to $1/2$. For instance, for the binary SV source with two coins having probability of heads respectively equal to $1/3$ and $2/3$, Figure 1 shows a curve where (α, β) always lies above it. This curve is clearly isolated from $(1/2, 1/2)$.

We follow similar ideas for proving our impossibility result for common randomness extraction from a distributed SV source; again we construct a continuous function, which somehow captures not only the minimum and maximum of the probability of the extracted common bit being zero, but also the probability that the two parties agree on their extracted bits. The construction of this function is more involved in the distributed case; it has two terms one of which is similar to the function in the non-distributed case, and the other is inspired by the definition of maximal correlation mentioned above.

Contributions to Information Theory. As mentioned above, the problem of common randomness extraction from i.i.d. sources has been studied in the information theory community. Then our work provides a generalization and an alternative proof of known results in the i.i.d. case. In particular, we give a new proof of Witsenhausen's result [29] on the impossibility of common randomness extraction from certain i.i.d. sources.

We also would like to point out that a generalized SV source as we define, is indeed an arbitrarily varying source (AVS) [10,13] with a causal adversary. These sources are studied in the information theory literature from the point of view of source coding [4].

Notations. In this paper we consider functions $X : \mathcal{C} \to \mathbb{R}$. Such a function can be thought of as a random variable $X = X(C)$. We sometimes for simplicity use the notation $X(c) = x_c$. The expected value and variance of X are denoted by $\mathbb{E}[X]$ and $\mathrm{Var}[X]$ respectively.

We sometimes have several distributions over the same set $\mathcal{C}$ which are indexed by elements $s \in \mathcal{S}$. In this case to avoid confusions, the expectation value and variance are specified by a subscript s.

For simplicity of notation a sequence $C_1, \ldots, C_n$ of (not necessarily i.i.d.) random variables is denoted by C^n. Similarly for $c_1, \ldots, c_n \in \mathcal{C}$ we use $c^n = (c_1, \ldots, c_n)$. We also use the notation $c_{[k:k+\ell]} = (c_k, c_{k+1}, \ldots, c_{k+\ell})$.

2   Randomness Extraction from Generalized SV Sources

Definition 1 (Generalized SV source). *Let $\mathcal{C}$ be a finite alphabet set. Consider a finite set of distributions over $\mathcal{C}$ indexed by a set $\mathcal{S}$. That is, assume that for any $s \in \mathcal{S}$ we have a distribution over $\mathcal{C}$ determined by numbers $p_s(c)$ for all $c \in \mathcal{C}$. A sequence $C_1, C_2, \cdots$ of random variables, each over alphabet set $\mathcal{C}$, is said to be a generalized SV source with respect to distributions $p_s(c)$, if the sequence is generated as follows: Assume that $C_1, \ldots, C_{i-1}$ are already generated. In order to determine C_i, an adversary chooses $S_i = s_i \in \mathcal{S}$, depending only on $C_1, \ldots, C_{i-1}$. Then C_i is sampled from the distribution $p_{s_i}(c)$.*

We can think of specifying s as choosing a particular multi-faceted die, and c as the facet that results from throwing the die. The joint probability distribution $p(c_1, c_2, \cdots, c_n, s_1, s_2, \cdots, s_n)$ of random variables $C_1, \ldots, C_n$ and $S_1, \ldots, S_n$ in a generalized SV source factorizes as follows:

$$q(s_1)p_{s_1}(c_1)q(s_2|c_1)p_{s_2}(c_2) \cdots q(s_n|c_1 \cdots c_{n-1})p_{s_n}(c_n),$$

where $q(s_i|c_1 \cdots c_{i-1})$ describes the action of the adversary at time i. Here, first the adversary chooses $S_1 = s_1$ with probability $q(s_1)$, and then $C_1 = c_1$ is generated with probability $p_{s_1}(c_1)$. Then the adversary chooses $S_2 = s_2$ with probability $q(s_2|c_1)$ and then $C_2 = c_2$ is generated with probability $p_{s_2}(c_2)$, and so on.

Generalized SV sources can be alternatively characterized as follows: Given i and $C_1 = c_1, \ldots, C_{i-1} = c_{i-1}$, the distribution of C_i should be a convex combination of the set of $|\mathcal{S}|$ distributions $\{p_s(\cdot) : s \in \mathcal{S}\}$. We emphasize that even after fixing distributions $p_s(c)$, the generalized SV source (similar to ordinary SV sources) is not a fixed source, but rather a class of sources. This is because in each step s_i is chosen arbitrarily by the adversary as a (probabilistic) function of $C_1, \ldots, C_{i-1}$. Nevertheless, once we fix adversary's strategy, the generalized SV source is fixed in that class of sources.

Definition 2 (Deterministic extraction). *We say that deterministic randomness extraction from the generalized SV source determined by distributions $p_s(c)$ is possible if for every $\epsilon > 0$ there exist n and $\Gamma_n : \mathcal{C}^n \to \{0,1\}$ such that for every strategy of the adversary, the distribution of $\Gamma_n(C^n)$ is ϵ-close, in total variation distance, to the uniform distribution. That is, independent of adversary's strategy, $\Gamma_n(C^n)$ is an almost uniform bit.*

In the following we present a necessary condition and separately a sufficient condition for the existence of deterministic extractors for generalized SV sources. In the non-degenerate case, i.e., when $p_s(c) > 0$ for all s, c, these two conditions coincide. Thus we fully characterize the possibility of deterministic randomness extraction from generalized SV sources in the non-degenerate case.

2.1   A Sufficient Condition for the Existence of Randomness Extractors

Theorem 1. *Consider a generalized SV source with alphabet $\mathcal{C}$, set of dice $\mathcal{S}$, and probability distributions $p_s(c)$. Suppose that there exists $\psi : \mathcal{C} \to \mathbb{R}$ such that for every $s \in \mathcal{S}$ we have $\mathbb{E}_{(s)}[\psi(C)] = 0$ and $\mathrm{Var}_{(s)}[\psi(C)] > 0$, where $\mathbb{E}_{(s)}$ and $\mathrm{Var}_{(s)}$ are expectation and variance with respect to the distribution $p_s(\cdot)$. Then randomness can be extracted from this SV source.*

Observe that if $p_s(c) > 0$ for all s, c, then this theorem can equivalently be stated as follows: Thinking of each distribution $p_s(\cdot)$ as a point in the probability simplex, if the convex hull of the set of points $\{p_s(\cdot) : s \in \mathcal{S}\}$ in the probability simplex does not have full dimension, then deterministic randomness extraction is possible. For instance if $|\mathcal{S}| < |\mathcal{C}|$ this condition is always satisfied and then we can deterministically extract randomness.

Remark 1. The analysis of the proof of Theorem 1 would show that the bias could be polynomially small, namely a bias of $\Theta(n^{-1/3})$.

2.2   A Necessary Condition for the Existence of Randomness Extractors

The main result of this subsection is the following theorem.

Theorem 2. *Consider a generalized SV source with alphabet $\mathcal{C}$, set of dice $\mathcal{S}$, and probabilities $p_s(c)$. Suppose that there is no non-zero function $\psi : \mathcal{C} \to \mathbb{R}$*

such that for all $s \in \mathcal{S}$ we have $\mathbb{E}_{(s)}[\psi(C)] = 0$. Then deterministic randomness extraction from this generalized SV source is impossible.

Again, let us consider the case where $p_s(c) > 0$ for all s, c. In this case ψ being non-zero is equivalent to $\mathrm{Var}_{(s)}[\psi] > 0$ for all s. Then comparing to Theorem 1 we find that the necessary and sufficient condition for the possibility of deterministic extraction is the existence of a non-zero ψ with $\mathbb{E}_{(s)}[\psi] = 0$.

Corollary 1. *Consider a generalized SV source with alphabet $\mathcal{C}$, set of dice $\mathcal{S}$, and probabilities $p_s(c)$. Let $\mathcal{S}'$ be a subset of $\mathcal{S}$ and let $\mathcal{C}'$ be the set of all c for which there exists some $s \in \mathcal{S}'$ such that $p_{s'}(c) > 0$. Suppose that there is no non-zero function $\psi : \mathcal{C} \to \mathbb{R}$ such that (i) ψ is zero on $\mathcal{C} - \mathcal{C}'$, and (ii) for all $s \in \mathcal{S}'$ we have $\mathbb{E}_{(s)}[\psi(C)] = 0$. Then deterministic randomness extraction from this generalized SV source is impossible.*

3   Distributed SV Sources

Distributed SV sources can be defined similarly to generalized SV sources except that in this case, the outcome in each time step is a pair that is distributed between two parties.

Definition 3. *Fix finite sets $\mathcal{A}, \mathcal{B}, \mathcal{S}$. Let $p_s(ab)$ define a probability distribution over $\mathcal{A} \times \mathcal{B}$ for any $s \in \mathcal{S}$. The distributed SV source with respect to distributions $p_s(ab)$ is defined as follows. The adversary in each time step i, depending on the previous outcomes $(A_1, B_1) = (a_1, b_1), \ldots, (A_{i-1}, B_{i-1}) = (a_{i-1}, b_{i-1})$ chooses some $S_i = s_i$. Then $(A_i, B_i) = (a_i, b_i)$ is sampled from the distribution $p_{s_i}(a_i b_i)$. The sequence of random variables $(A_1, B_1), (A_2, B_2), \ldots$, is called a distributed SV source.*

Here we assume that the outcomes of this SV source are distributed between two parties, say Alice and Bob. That is, in each time step i, A_i is revealed to Alice and B_i is revealed to Bob. So Alice receives the sequence $A_1, A_2, \ldots$, and Bob receive the sequence $B_1, B_2, \ldots$.

In this section we are interested in whether two parties can generate a common random bit from distributed SV sources. To be more precise, let us first define the problem more formally.

Definition 4. *We say that common randomness can be extracted from the distributed SV source $(A_1, B_1), (A_2, B_2), \ldots$ if for every $\epsilon > 0$ there is n and functions $\Gamma_n : \mathcal{A}^n \to \{0, 1\}$ and $\Lambda_n : \mathcal{B}^n \to \{0, 1\}$ such that for every strategy of adversary, the distributions of $K_1 = \Gamma_n(A^n)$ and $K_2 = \Lambda_n(B^n)$ are ϵ-close (in total variation distance) to uniform distribution, and that $\Pr[K_1 \neq K_2] < \epsilon$.*

In the above definition we considered only deterministic protocols for extracting a common random bit. We could also consider probabilistic protocols where Γ_n and Λ_n are random functions depending on *private* randomnesses of Alice and Bob respectively. More precisely, we could take $K_1 = \Gamma_n(A^n, R_1)$ and

$_2 = \Lambda_n(B^n, R_2)$ with the above conditions on K_1, K_2, where R_1 and R_2 are private randomnesses of Alice and Bob respectively, which are independent of the SV source and of each other. Nevertheless, if a common random bit can be extracted with probabilistic protocols, then common randomness extraction with deterministic protocols is also possible.

Lemma 1. *In the problem of common random bit extraction, with no loss of generality we may assume that the parties do not have private randomness.*

.1   Common Data

As discussed in the introduction, the notion of the common data of two random variables A, B first appeared in the problem of common randomness extraction from i.i.d. sources. Briefly speaking, common data of A and B is the finest random variable C that can be computed both as a function $C = C_1(A)$ of A, and as a function $C = C_2(B)$ of B. In the full version of this paper [2], we give a new proof of Witsenhausen's theorem that randomness extraction from i.i.d. repetitions of (A, B) is feasible if and only if common data exists, if and only if maximal correlation is equal to 1.

Here we are interested in common randomness extraction from distributed SV sources. So we need to define common data for such sources. The common data of a distributed SV source (given by distributions $p_s(ab)$ indexed by $s \in \mathcal{S}$) is the finest random variable C that can be computed both as a function $C = C_1(A)$ of A, and as a function $C = C_2(B)$ of B. Here we need $C_1(A) = C_2(B)$ to hold with probability 1 under all distributions $p_s(ab)$.

3.2   Common Random Bit Extraction from Distributed SV Sources

Theorem 3. *Consider a distributed SV source (as in Definition 3) with corresponding sets $\mathcal{S}$, $\mathcal{A}$, and $\mathcal{B}$ and corresponding distributions $p_s(ab)$. Let C be the common data of the distributed SV source. Let $p_s(abc)$ denote the induced joint distribution of A, B, and C. Suppose that there is no non-zero function $\psi : \mathcal{C} \to \mathbb{R}$ such that $\mathbb{E}_{(s)}[\psi(C)] = 0$ for all s. Then common randomness cannot be extracted from this distributed SV source.*

An algorithm to extract common random bits is to focus on the common part C that can be computed by both Alice and Bob. Indeed C itself can be thought of as a generalized SV source. If deterministic randomness extraction from C is possible, then Alice and Bob can obtain a common random bit by individually applying the randomness extraction protocol. Comparing with Theorems 1 and 2, and assuming $p_s(c) > 0$ for all s, c, the above theorem states that a common random bit can be extracted if and only if deterministic randomness extraction from C is possible.

4   Future Work

In this paper we completely characterized the randomness extraction problem for non-degenerate cases. A future work could be to solve this problem for the degenerate cases. In the degenerate cases, for generalized non-distributed sources Corollary 1 gives a mildly stronger necessary condition than Theorem 2, but there is still a gap between this necessary condition and the sufficient condition of Theorem 1.

We note that our randomness extractor in Theorem 1 extracts a bit whose bias is inverse polynomially small in the length of the source sequence. It is interesting to see if this extractor could be improved to yield a bit with an exponentially small bias. Furthermore, if we want to produce more than one bit of randomness, the tradeoff between the number of produced random bits and their quality is open.

Another interesting problem is to look at efficient adversaries, similar to the work of [1]. Our proofs only show existence of inefficient adversaries.

Another way to restrict the adversary is to put limitations on the number of times the adversary can choose a strategy $s \in \mathcal{S}$, i.e. there can be a cost associated to each strategy s.

A different type of limitation can be on the adversary's knowledge about the sequence generated so far. More specifically, the adversary might have *noisy or partial* access to the previous outcomes in the sequence (these sources are called "active sources" [22]). These sources model adversaries with limited memory. Space bounded sources have been studied in [17,24].

Finally, the problem of common randomness extraction can be studied for three or more parties instead of just two parties.

References

1. Austrin, P., Chung, K.-M., Mahmoody, M., Pass, R., Seth, K.: On the impossibility of cryptography with tamperable randomness. In: Garay, J.A., Gennaro, R. (eds.) CRYPTO 2014, Part I. LNCS, vol. 8616, pp. 462–479. Springer, Heidelberg (2014)
2. Beigi, S., Etesami, O., Gohari, A.: Deterministic Randomness Extraction from Generalized and Distributed Santha-Vazirani Sources (2014). arXiv:1412.6641
3. Beigi, S., Tse, D.: under preparation
4. Berger, T.: The source coding game. IEEE Trans. on Information Theory **IT−17**(1), 71–76 (1971)
5. Blum, M.: Independent unbiased coin flips from a correlated biased source - a finite state Markov chain. Combinatorica **6**(2), 97–108 (1986)
6. Bourgain, J.: More on the sum-product phenomenon in prime fields and its applications. International Journal of Number Theory (2005)
7. Bourgain, J.: On the construction of affine extractors. Geometric And Functional Analysis **17**(1), 33–57 (2007)
8. Chor, B., Goldreich, O., Håstad, J., Freidmann, J., Rudich, S., Smolensky, R.: The bit extraction problem of t-resilient functions. In: Proceedings of the 26th Annual Symposium on Foundations of Computer Science, pp. 396–407 (1985)

9. Chor, B., Goldreich, O.: Unbiased Bits from Sources of Weak Randomness and Probabilistic Communication Complexity. SIAM J. Comput. **17**(2), 230–261 (1988)
10. Dobrusin, R.L.: Individual methods for transmission of information for discrete channels without memory and messages with independent components. Sov. Math. **4**, 253–256 (1963)
11. Dvir, Z.: Extractors for varieties. Computational Complexity **21**(4), 515–572 (2012)
12. Dvir, Z., Gabizon, A., Wigderson, A.: Extractors and rank extractors for polynomial sources. In: FOCS 2007: Proceedings of the 48th Annual IEEE Symposium on Foundations of Computer Science, pp. 52–62 (2007)
13. Dobrusin, R.L.: Unified methods of optimal quantizing of messages. Sov. Math. **4**, 284–292 (1963)
14. Fischer, M.J., Lynch, N.A.: A lower bound for the time to assure interactive consistency. Information Processing Letters **14**, 183–186 (1982)
15. Gács, P., Körner, J.: Common information is far less than mutual information. Problems of Control and Information Theory **2**(2), 119–162 (1972)
16. Gabizon, A., Raz, R.: Deterministic extractors for affine sources over large fields. In: Proceedings of the 46th FOCS, pp. 407–418 (2005)
17. Kamp, J., Rao, A., Vadhan, S., Zuckerman, D.: Deterministic extractors for small-space sources. In: Proceedings of the thirty-eighth annual ACM symposium on Theory of computing, pp. 691–700 (2006)
18. Nisan, N., Zuckerman, D.: Randomness is Linear in Space. Journal of Computer and System Sciences **52**(1), 43–52 (1996)
19. Rabin, M.O.: Randomized byzantine generals. In: Proceedings of the 24th Annual Symposium on Foundations of Computer Science, pp. 403–409 (1983)
20. Rao, A.: Extractors for a constant number of polynomially small min-entropy independent sources. In: Proceedings of the 38th STOC, pp. 497–506 (2006)
21. Reingold, O., Vadhan, S., Wigderson, A.: A note on extracting randomness from Santha-Vazirani sources. Unpublished manuscript (2004)
22. Palaiyanur, H., Chang, C., Sahai, A.: Lossy compression of active sources. In: IEEE International Symposium on Information Theory, pp. 1977–1981 (2008)
23. Santha, M., Vazirani, U.: Generating quasi-random sequences from slightly-random sources. In: Proceedings of Symposium on the Foundations of Computer Science (1984). Jourdnal of Computer and System Sciences, **33**(1), 75–87 (1986)
24. Vazirani, U.V.: Efficiency considerations in using semi-random sources. In: Proceedings of the Nineteenth STOC, pp. 160–168 (1987)
25. Vazirani, U.V., Vazirani, V.V.: Random polynomial time is equal to slightly-random polynomial time. In: Proc. 26th Annual IEEE Symposium on the Foundations of Computer Science, pp. 417–428 (1985)
26. Vazirani, U.V., Vazirani, V.V.: Sampling a population with a single semi-random source. In: Proc. 6th FST & TCS Conf. (1986)
27. Vadhan, S.: Pseudorandomness. Now Publishers (2012)
28. von Neumann, J.: Various techniques used in connection with random digits. Applied Math Series **12**, 36–38 (1951)
29. Witsenhausen, H.S.: On sequences of pairs of dependent random variables. SIAM Journal on Applied Mathematics **28**(1), 100–113 (1975)
30. Yao, A.C.: Some Complexity Questions Related to Distributed Computing. In: Proc. of 11th STOC, vol. 14, pp. 209–213 (1979)
31. Zuckerman, D.: Randomness-optimal oblivious sampling. Random Structures and Algorithms **11**, 345–367 (1997)

Limitations of Algebraic Approaches to Graph Isomorphism Testing

Christoph Berkholz$^{(\boxtimes)}$ and Martin Grohe

RWTH Aachen University, Aachen, Germany
{berkholz,grohe}@informatik.rwth-aachen.de

Abstract. We investigate the power of graph isomorphism algorithms based on algebraic reasoning techniques like Gröbner basis computation. The idea of these algorithms is to encode two graphs into a system of equations that are satisfiable if and only if if the graphs are isomorphic, and then to (try to) decide satisfiability of the system using, for example, the Gröbner basis algorithm. In some cases this can be done in polynomial time, in particular, if the equations admit a bounded degree refutation in an algebraic proof systems such as Nullstellensatz or polynomial calculus. We prove linear lower bounds on the polynomial calculus degree over all fields of characteristic $\neq 2$ and also linear lower bounds for the degree of Positivstellensatz calculus derivations.

We compare this approach to recently studied linear and semidefinite programming approaches to isomorphism testing, which are known to be related to the combinatorial Weisfeiler-Lehman algorithm. We exactly characterise the power of the Weisfeiler-Lehman algorithm in terms of an algebraic proof system that lies between degree-k Nullstellensatz and degree-k polynomial calculus.

1 Introduction

The graph isomorphism problem (GI) is notorious for its unresolved complexity status. While there are good reasons to believe that GI is not NP-complete, it is wide open whether it is in polynomial time.

Complementing recent research on linear and semidefinite programming approaches to GI [1,7,10,13,14], we investigate the power of GI-algorithms based on algebraic reasoning techniques like Gröbner basis computation. The idea of all these approaches is to encode isomorphisms between two graphs as solutions to a system of equations and possibly inequalities and then try to solve this system or relaxations of it. Most previous work is based on the following encoding: let G, H be graphs with adjacency matrices A, B, respectively. Note that G and H are isomorphic if and only if there is a permutation matrix X such that $AX = XB$. If we view the entries x_{vw} of the matrix X as variables, we obtain a system of linear equations. We introduce equations forcing all row- and column

---

RWTH Aachen University—The first author is currently at KTH Stockholm, supported by a fellowship within the Postdoc-Program of the German Academic Exchange Service (DAAD).

M.M. Halldórsson et al. (Eds.): ICALP 2015, Part I, LNCS 9134, pp. 155–166, 2015.
DOI: 10.1007/978-3-662-47672-7_13

ums of X to be 1 and add the inequalities $x_{vw} \geq 0$. It follows that the integer solutions to this system are $0/1$-solutions that correspond to isomorphisms between G and H. Of course this does not help to solve GI, because we cannot find integer solutions to a system of linear inequalities in polynomial time. The first question to ask is what happens if we drop the integrality constraints. Almost thirty years ago, Tinhofer [17] proved that the system has a rational (or, equivalently, real) solution if and only if the so-called colour refinement algorithm does not distinguish the two graphs. *Colour refinement* is a simple combinatorial algorithm that iteratively colours the vertices of a graph according to their "iterated degree sequences", and, to distinguish two graphs, tries to detect a difference in their colour patterns. For every k, there is a natural generalisation of the colour refinement algorithm that colours k-tuples of vertices instead of single vertices; this generalisation is known as the k-*dimensional Weisfeiler-Lehman algorithm (k-WL)*. Atserias and Maneva [1] and independently Malkin [13] proved that the Weisfeiler-Lehman algorithm is closely tied to the *Sherali-Adams* hierarchy [16] of increasingly tighter LP-relaxations of the integer linear program for GI described above: the distinguishing power of k-WL is between that of the $(k-1)$st and kth level of the Sherali-Adams hierarchy. Otto and the second author of this paper [10] gave a precise correspondence between k-WL and the nonnegative solutions to a system of linear equations between the $(k-1)$st and kth level of the Sherali-Adams hierarchy. Already in 1992, Cai, Fürer, and Immerman [5] had proved that for every k there are non-isomorphic graphs G_k, H_k (called *CFI-graphs* in the following) of size $O(k)$ that are not distinguished by k-WL, and combined with the results of Atserias-Maneva and Malkin, this implies that no sublinear level of the Sherali-Adams hierarchy suffices to decide isomorphism. O'Donnell, Wright, Wu, and Zhou [14] (also see [7]) studied the Lasserre hierarchy [12] of semi-definite relaxations of the integer linear program for GI. They proved that the same CFI-graphs cannot even be distinguished by sublinear levels of the Lasserre hierarchy.

However, there is a different way of relaxing the integer linear program to obtain a system that can be solved in polynomial time: we can drop the nonnegativity constraints, which are the only inequalities in the system. Then we end up with a system of linear equalities, and we can ask whether it is solvable over some finite field or over the integers. As this can be decided in polynomial time, it gives us a new polynomial time algorithm for graph isomorphism: we solve the system of equations associated with the given graphs. If there is no solution, then the graphs are nonisomorphic. (We say that the system of equations *distinguishes* the graphs.) If there is a solution, though, we do not know if the graphs are isomorphic or not. Hence the algorithm is "sound", but not necessarily "complete". Actually, it is not obvious that the algorithm is not complete. If we interpret the linear equations over $\mathbb{F}_2$ or over the integers, the system does distinguish the CFI-graphs (which is not very surprising because these graphs encode systems of linear equations over $\mathbb{F}_2$). Thus the lower bound techniques applied in all previous results do not apply here. However, we construct nonisomorphic graphs that cannot be distinguished by this system (see Theorem 6.4).

In the same way, we can drop the nonnegativity constraints from the levels o the Sherali-Adams hierarchy and then study solvability over finite fields or over the integers, which gives us increasingly stronger systems. Even more powerfu algorithms can be obtained by applying algebraic techniques based on Gröbner basis computations. Proof complexity gives us a good framework for proving lower bounds for such algorithms. There are algebraic proof systems such as the polynomial calculus [6] and the weaker Nullstellensatz system [2] that characterise the power of these algorithms. The degree of refutations in the algebraic systems roughly corresponds to the levels of the Sherali-Adams and Lasserre hierarchies for linear and semi-definite programming, and to the dimension of the Weisfeiler-Lehman algorithm. We identify a fragment of the polynomial calculus, called the monomial polynomial calculus, such that degree-k refutations in this system precisely characterise distinguishability by k-WL (see Theorem 4.4).

As our main lower bounds, we prove that for every field $\mathbb{F}$ of characteristic $\neq 2$, there is a family of nonisomorphic graphs G_k, H_k of size $O(k)$ that cannot be distinguished by the polynomial calculus in degree k. Furthermore, we prove that there is a family of nonisomorphic graphs G_k, H_k of size $O(k)$ that cannot be distinguished by the Positivstellensatz calculus in degree k. The Positivstellensatz calculus [9] is an extension of the polynomial calculus over the reals and subsumes semi-definite programming hierarchies. Thus, our results slightly generalise the results of O'Donnell et al. [14] on the Lasserre hierarchy (described above). Technically, our contribution is a low-degree reduction from systems of equations describing so-called Tseitin tautologies to the systems for graph isomorphism. Then we apply known lower bounds [3,9] for Tseitin tautologies.

2   Algebraic Proof Systems

Polynomial calculus (PC) is a proof system to prove that a given system of (multivariate) polynomial equations P over a field $\mathbb{F}$ has no 0/1-solution. We always normalise polynomial equations to the form $p = 0$ and just write p to denote the equation $p = 0$. The derivation rules are the following (for polynomial equations $p \in \mathsf{P}$, polynomials f, g, variables x and field elements a, b):

$$\frac{}{p}, \quad \frac{}{x^2 - x}, \quad \frac{f}{xf}, \quad \frac{g \quad f}{ag + bf}.$$

The *axioms* of the systems are all $p \in \mathsf{P}$ and $x^2 - x$ for all variables x. A PC *refutation* of P is a derivation of 1. The polynomial calculus is sound and complete, that is, P has a PC refutation if and only if it is unsatisfiable. The *degree* of a PC derivation is the maximal degree of all polynomials in the derivation. Originally, Clegg et. al. [6] introduced the polynomial calculus to model Gröbner basis computation. Moreover, using the Gröbner basis algorithm, it can be decided in time $n^{O(d)}$ whether a given system of polynomial equations has a PC refutation of degree d (see [6]).

We introduce the following restricted variant of the polynomial calculus. A *monomial-PC* derivation is a PC-derivation where we require that the polynomial f in the multiplication rule $\frac{f}{xf}$ is either a monomial or the product of a monomial and an axiom.

If we restrict the application of the multiplication rule even further and require f to be the product of a monomial and an axiom, we obtain the Nullstellensatz proof system [2]. This proof system is usually stated in the following static form. A *Nullstellensatz* refutation of a system P of polynomial equations consists of polynomials f_p, for $p \in$ P, and g_x, for all variables x, such that

$$\sum_{p \in \mathsf{P}} f_p p + \sum_x g_x (x^2 - x) = 1.$$

The degree of a Nullstellensatz refutation is the maximum degree of all polynomials $f_p p$.

2.1   Low-Degree Reductions

To compare the power of the polynomial calculus for different systems of polynomial equations, we use *low degree reductions* [4]. Let P and R be two sets of polynomials in the variables $\mathcal{X}$ and $\mathcal{Y}$, respectively. A *degree-(d_1, d_2) reduction* from P to R consist of the following:

- for each variable $y \in \mathcal{Y}$ a polynomial $f_y(x_1, \ldots, x_k)$ of degree at most d_1 in variables $x_1, \ldots, x_k \in \mathcal{X}$;
- for each polynomial $r(y_1, \ldots, y_\ell) \in$ R a degree-d_2 PC derivation of

$$r\big(f_{y_1}(x_{11}, \ldots, x_{1k_1}), \ldots, f_{y_\ell}(x_{\ell 1}, \ldots, x_{\ell k_\ell})\big)$$

from P.
- for each variable $y \in \mathcal{Y}$ a degree-d_2 PC derivation of

$$f_y(x_1, \ldots, x_k)^2 - f_y(x_1, \ldots, x_k)$$

from P.

Lemma 2.1 ([4]). *If there is a degree-(d_1, d_2) reduction from P to R and R has a polynomial calculus refutation of degree k, then P has a polynomial calculus refutation of degree $\max(d_2, kd_1)$.*

2.2   Linearisation

For a system of polynomial equations P over variables x_i let P^r be the set of all polynomial equations of degree at most r obtained by multiplying a polynomial in P by a monomial over the variables x_i. Furthermore, for a system of polynomial equations P let MLIN(P) be the the multi-linearisation of P obtained by replacing every monomial $x_{i_1} \cdots x_{i_\ell}$ by a variable $X_{\{i_1, \ldots, i_\ell\}}$. Observe that if P has a 0/1-solution α, then so does MLIN(P) as we can set $\alpha(X_{\{i_1, \ldots, i_\ell\}}) := \alpha(x_{i_1}) \cdots \alpha(x_{i_\ell})$. The converse however does not hold since a solution α for MLIN(P) does not have to satisfy $\alpha(X_{\{ab\}}) = \alpha(X_{\{a\}})\alpha(X_{\{b\}})$. The next lemma states a well-known connection between Nullstellensatz and Linear Algebra.

Lemma 2.2 [3]. *Let* P *be a system of polynomial equations. The following state-ments are equivalent.*

- P *has a degree* r *Nullstellensatz refutation.*
- *The system of linear equations* MLIN(P^r) *has no solution.*

This characterisation of Nullstellensatz proofs in terms of a linear system of equations (also called *design* [3]) is a useful tool for proving lower bounds on the Nullstellensatz degree. Unfortunately, a similar characterisation for bounded degree PC is not in sight. However, for the newly introduced system monomial-PC, which lies between Nullstellensatz and PC, we have a similar criterion for the non-existence of refutations. The proof is deferred to the appendix.

Lemma 2.3. *If* MLIN(P^d) *has a solution* α *that additionally satisfies*

$$\alpha(X_\pi) = 0 \implies \alpha(X_\rho) = 0, \text{ for all } \pi \subseteq \rho,$$

then P *has no degree* d *monomial-PC derivation.*

2.3   Linear and Semidefinite-Programming Approaches

In the previous section we have seen that degree-d Nullstellensatz corresponds to solving a system of *linear equations* of size $n^{O(d)}$, which can be done in time $n^{O(d)}$. Over the reals, this approach can be strengthened by considering hierarchies of relaxations for linear and semi-definite programming.

In this setting one additionally adds linear inequalities, typically $0 \leq x \leq 1$. In the same way as for the Nullstellensatz, one lifts this problem to higher dimensions, by multiplying the inequalities and equations with all possible monomials of bounded degree. Afterwards, one linearises this system as above to obtain a system of *linear inequalities* of size $n^{O(d)}$, which can also be solved in polynomial time using linear programming techniques. This lift-and-project technique is called Sherali-Adams relaxation of level d [16].

Another even stronger relaxation is based on semidefinite programming techniques. This technique has different names: Positivstellensatz, Sum-of-Squares (SOS), or Lasserre Hierarchy. Here we take the view point as a proof system, which was introduced by Grigoriev and Vorobjov [9] and directly extends the Nullstellensatz over the reals. A degree-d *Positivstellensatz* refutation of a system P of polynomial equations consists of polynomials f_p, for $p \in$ P, and g_x, for all variables x, and in addition polynomials h_i such that

$$\sum_{p \in P} f_p p + \sum_x g_x(x^2 - x) = 1 + \sum_i h_i^2.$$

The degree of a Positivstellensatz refutation is the maximum degree of all poly-nomials $f_p p$ and h_i^2. It is important to note that Positivstellensatz refutations can be found in time $n^{O(d)}$ using semi-definite programming. This has been inde-pendently observed by Parrilo [15] in the context of algebraic geometry and by Lasserre [12] in the context of linear optimisation.

Grigoriev and Vorobjov [9] also introduced a proof system called Positivstellensatz calculus, which extends polynomial calculus in the same way as Positivstellensatz extends Nullstellensatz. A *Positivstellensatz calculus* refutation of
system of polynomials P is a polynomial calculus derivation over the reals of
$+\sum_i h_i^2$. Again, the degree of such a refutation is the maximum degree of every
polynomial in the derivation.

3 Equations for Graph Isomorphism

We find it convenient to encode isomorphism using different equations than
those from the system $AX = XB$ described in the introduction. However, the
equations $AX = XB$ can easily be derived in our system, and thus lower bounds
for our system imply lower bounds for the $AX = XB$-system.

Throughout this section, we fix graphs G and H, possibly with coloured
vertices and/or edges. Isomorphisms between coloured graphs are required to
preserve the colours. We assume that either $|V(G)| \geq 2$ or $|V(H)| \geq 2$. We
shall define a system $\mathsf{P}_{\mathrm{iso}}(G, H)$ of polynomial equations that has a solution if
and only if G and H are isomorphic. The equations are defined over variables
$x_{vw}, v \in V(G), w \in V(H)$. A solution to the system is intended to describe an
isomorphism ι from G to H, where $x_{vw} \mapsto 1$ if $\iota(v) = w$ and $x_{vw} \mapsto 0$ otherwise.
The system $\mathsf{P}_{\mathrm{iso}}(G, H)$ consists of the following linear and quadratic equations:

$$\sum_{v \in V(G)} x_{vw} - 1 = 0 \qquad \text{for all } w \in V(H) \tag{3.1}$$

$$\sum_{w \in V(H)} x_{vw} - 1 = 0 \qquad \text{for all } v \in V(G) \tag{3.2}$$

$$x_{vw} x_{v'w'} = 0 \qquad \begin{array}{l}\text{for all } v, v' \in V(G), w, w' \in V(H) \text{ such} \\ \text{that } \{(v, w), (v', w')\} \text{ is not a local iso-} \\ \text{morphism.}\end{array} \tag{3.3}$$

A *local isomorphism* from G to H is an injective mapping π with domain in $V(G)$
and range in $V(H)$ (often viewed as a subset of $V(G) \times V(H)$) that preserves
adjacencies, that is $vw \in E(G) \iff \pi(v)\pi(w) \in E(H)$. If G and H are coloured
graphs, local isomorphisms are also required to preserve colours.

To enforce 0/1-assignments we add the following set Q of quadratic equalities

$$x_{vw}^2 - x_{vw} = 0 \qquad \text{for all } v \in V(G), w \in V(H). \tag{3.4}$$

We treat these equations separately because they are axioms of the polynomial
calculus anyway. Observe that the equations (3.1) and (3.2) in combination with
(3.4) make sure that every solution to the system describes a bijective mapping
from $V(G)$ to $V(H)$. The equations (3.3) make sure that this bijection is an
isomorphism. Thus, for every field $\mathbb{F}$, the system $\mathsf{P}_{\mathrm{iso}}(G, H) \cup \mathsf{Q}$ has a solution
over $\mathbb{F}$ if and only G and H are isomorphic.

4   Weisfeiler-Lehman Is Located Between Nullstellensatz and Polynomial Calculus

To relate the Weisfeiler-Lehman algorithm to our proof systems, we use the following combinatorial game. The *bijective k-pebble game* [11] on graphs G and H is played by two players called *Spoiler* and *Duplicator*. Positions of the game are sets $\pi \subseteq V(G) \times V(H)$ of size $|\pi| \leq k$. The game starts in an initial position π_0. If $|V(G)| \neq |V(H)|$ or if π_0 is not a local isomorphism, then Spoiler wins the game immediately, that is, after 0 rounds, Otherwise, the game is played in a sequence of *rounds*. Suppose the position after the ith round is π_i. In the $(i+1)$st round, Spoiler chooses a subset $\pi \subseteq \pi_i$ of size $|\pi| < k$. Then Duplicator chooses a bijection $f : V(G) \rightarrow V(H)$. Then Spoiler chooses a vertex $v \in V(G)$, and the new position is $\pi_{i+1} := \pi \cup \{(v, f(v))\}$. If π_{i+1} is not a local isomorphism, then Spoiler wins the play after $(i+1)$ rounds. Otherwise, the game continues with the $(i+2)$nd round. Duplicator wins the play if it lasts forever, that is, if Spoiler does not win after finitely many rounds. *Winning strategies* for either player in the game are defined in the natural way.

Lemma 4.1 ([5,11]). *k-WL distinguishes G and H if and only if Spoiler has a winning strategy for the bijective k-pebble game on G, H with initial position $\emptyset$.*

Observe that each game position $\pi = \{(v_1, w_1), \ldots, (v_\ell, w_\ell)\}$ of size ℓ corresponds to a multilinear monomial $\boldsymbol{x}_\pi = x_{v_1 w_1} \ldots x_{v_\ell w_\ell}$ of degree ℓ; for the empty position we let $\boldsymbol{x}_\emptyset := 1$.

Lemma 4.2. *Let $\mathbb{F}$ be a field of characteristic 0. If Spoiler has a winning strategy for the r-round bijective k-pebble game on G, H with initial position π_0, then there is a degree k monomial-PC derivation of $\boldsymbol{x}_{\pi_0}$ from $\mathsf{P}_{\mathrm{iso}}(G, H)$ over $\mathbb{F}$.*

Proof. The proof is by induction over r. For the base case $r = 0$, suppose that Spoiler wins after round 0. If $|V(G)| \neq |V(H)|$, the system $\mathsf{P}_{\mathrm{iso}}(G, H)$ has the following degree-1 Nullstellensatz refutation:

$$\sum_{v \in V(G)} \frac{1}{a} \left(\sum_{w \in V(H)} x_{vw} - 1 \right) + \sum_{w \in V(H)} -\frac{1}{a} \left(\sum_{v \in V(G)} x_{vw} - 1 \right) = 1,$$

where $a = |V(G) - V(H)|$. It yields a degree-1 monomial PC refutation of $\mathsf{P}_{\mathrm{iso}}(G, H)$ and thus a derivation of $\boldsymbol{x}_{\pi_0}$ of degree $|\pi_0| \leq k$. Otherwise, π_0 is not a local isomorphism. Then there is a 2-element subset $\pi := \{(v, w), (v', w')\} \subseteq \pi_0$ that is not a local isomorphism. Multiplying the axiom $x_{vw} x_{v'w'} = \boldsymbol{x}_\pi$ with the monomial $\boldsymbol{x}_{\pi_0 \setminus \pi}$, we obtain a monomial-PC derivation of $\boldsymbol{x}_{\pi_0}$ of degree $|\pi_0| \leq k$.

For the inductive step, suppose that Spoiler has a winning strategy for the $(r+1)$-round game starting in position π_0. Let $\pi \subseteq \pi_0$ with $|\pi| < k$ be the set chosen by Spoiler in the first round of the game. We can derive $\boldsymbol{x}_{\pi_0}$ from $\boldsymbol{x}_\pi$ by multiplying with the monomial $\boldsymbol{x}_{\pi_0 \setminus \pi}$. Hence it suffices to show that we can derive $\boldsymbol{x}_\pi$ in degree k.

Consider the bipartite graph B on $V(G) \uplus V(H)$ which has an edge vw for all $v \in V(G), w \in V(H)$ such that Spoiler *cannot* win from position $\pi \cup \{(v, w)\}$

at most r rounds. As from position π, Spoiler wins in $r + 1$ rounds, there is a bijection $f : V(G) \to V(H)$ such that $(v, f(v)) \in E(B)$ for all $v \in V(G)$. By Hall's Theorem, it follows that there is a set $S \subseteq V(G)$ such that $|N^B(S)| < |S|$. Let S be a maximal set with this property and let $T := N^B(S)$.

We claim that $N^B(T) = S$. To see this, suppose for contradiction that there is a vertex $v \in N^B(T) \setminus S$. By the maximality of S, we have $N^B(v) \not\subseteq T$. Let $w \in N^B(v) \setminus T$. Moreover, let $w' \in N^B(v) \cap T$ (exists because $v \in N^B(T)$) and $v' \in N^B(w') \cap S$ (exists because $T = N^B(S)$). Then by the definition of B, Duplicator has a winning strategy for the r-round bijective k-pebble game with initial positions $\pi \cup \{(v', w')\}$, $\pi \cup \{(v, w')\}$, and $\pi \cup \{(v, w)\}$, which implies that he also has a winning strategy for the game with initial position $\pi \cup \{(v', w)\}$. Here we use the fact that the relation "duplicator has a winning strategy for the r-round bijective k-pebble game" defines an equivalence relation on the initial positions. Thus $(v', w) \in E(B)$, which contradicts $w \notin N^B(S)$. This proves the claim.

By the induction hypothesis and the claim we know that $(\star)$ $x_\pi x_{vw}$ has a degree-k monomial PC derivation if $v \in S, w \notin T$ or $v \notin S, w \in T$. Furthermore, we can derive

$$\sum_{v \in S} x_\pi \left(\sum_{w \in V(H)} x_{vw} - 1 \right) - \sum_{w \in T} x_\pi \left(\sum_{v \in V(G)} x_{vw} - 1 \right) \quad (4.1)$$

by multiplying the axioms (3.1), (3.2) with x_π and building a linear combination. By subtracting and adding monomials from $(\star)$, this polynomial simplifies to $(|T| - |S|)x_\pi$. After dividing by the coefficient $|T| - |S| \neq 0$, we get x_π. We can divide by $|T| - |S|$ because the characteristic of the field $\mathbb{F}$ is 0. $\square$

The following lemma is, at least implicitly, from [10]. As the formal framework is different there, we nevertheless give the proof in the appendix.

Lemma 4.3. *Let $\mathbb{F}$ be a field of characteristic 0 and $k \geq 2$. If Duplicator has a winning strategy for the bijective k-pebble game on G, H then there is a solution α of $\mathrm{MLIN}(\mathrm{P}_{\mathrm{iso}}(G, H)^k)$ over $\mathbb{F}$ that additionally satisfies $\alpha(X_\pi) = 0 \implies \alpha(X_\rho) = 0$ for all $\pi \subseteq \rho$.*

Theorem 4.4. *Let $\mathbb{F}$ be a field of characteristic 0. Then the following statements are equivalent for two graphs G and H.*

(1) The graphs are distinguishable by k-WL.
(2) There is a degree-k monomial-PC refutation of $\mathrm{P}_{\mathrm{iso}}(G, H)$ over $\mathbb{F}$.

Proof. Follows immediately from lemmas 2.3, 4.2 and 4.3. $\square$

We do not now the exact relation between Nullstellensatz and monomial-PC for the graph isomorphism polynomials. In particular, we do not know whether degree-k Nullstellensatz is as strong as the k-dimensional Weisfeiler-Lehman algorithm and leave this as open question. In the other direction, we remark that, at least for degree 2, full polynomial calculus is strictly stronger than degree-2 monomial-PC and hence the Colour Refinement Algorithm. However, we believe that the gap is not large. Our intuition is supported by Theorem 6.2, which implies that low-degree PC is not able to distinguish Cai-Fürer-Immerman graphs. Thus, polynomial calculus has similar limitations as the Weisfeiler-Lehman algorithm [5], Resolution [18], the Sherali-Adams hierarchy [1,10] and the Positivstellensatz [14].

5   Groups CSPs and Tseitin Polynomials

5.1   From Group CSPs to Graph Isomorphism

We start by defining a class of constraint satisfaction problems (CSPs) where the constraints are co-sets of certain groups. Throughout this section, we let Γ be a finite group. Recall that a *CSP-instance* has the form $(\mathcal{X}, D, \mathcal{C})$, where $\mathcal{X}$ is a finite set of *variables*, D is a finite set called the *domain* and $\mathcal{C}$ a finite set of *constraints* of the form $(\bar{x}, R)$, where $\bar{x} \in \mathcal{X}^k$ and $R \subseteq D^k$, for some $k \geq 1$. A *solution* to such an instance is an assignment $\alpha : X \to D$ such that $\alpha(\bar{x}) \in R$ for all constraints $(\bar{x}, R) \in \mathcal{C}$. An instance of a Γ-*CSP* has domain Γ and constraints of the form $(\bar{x}, \Delta\gamma)$, where $\Delta \leq \Gamma^k$ is a subgroup of a k-fold direct product Γ^k of Γ and $\gamma \in \Gamma^k$, so that $\Delta\gamma$ is a right coset of Δ. We specify instances as sets $\mathcal{C}$ of constraints; the variables are given implicitly. With each constraint $C = ((x_1, \ldots, x_k), \Delta\gamma)$, we associate the *homogeneous* constraint $\widetilde{C} = ((x_1, \ldots, x_k), \Delta)$. For an instance $\mathcal{C}$, we let $\widetilde{\mathcal{C}} = \{\widetilde{C} \mid C \in \mathcal{C}\}$.

Next, we reduce Γ-CSP to GI. Let $\mathcal{C}$ be a Γ-CSP in the variable set $\mathcal{X}$. We construct a coloured graph $G(\mathcal{C})$ as follows.

- For every variable $x \in \mathcal{X}$ we take vertices $\gamma^{(x)}$ for all $\gamma \in \Gamma$. We colour all these vertices with a fresh colour $L^{(x)}$.
- For every constraint $C = ((x_1, \ldots, x_k), \Delta\gamma) \in \mathcal{C}$ we add vertices $\beta^{(C)}$ for all $\beta \in \Delta\gamma$. We colour all these vertices with a fresh colour $L^{(C)}$. If $\beta = (\beta_1, \ldots, \beta_k)$, we add an edge $\{\beta^{(C)}, \beta_i^{(x_i)}\}$ for all $i \in [k]$. We colour this edge with colour $M^{(i)}$.

We let $\widetilde{G}(\mathcal{C})$ be the graph $G(\widetilde{\mathcal{C}})$ where for all constraints $C \in \mathcal{C}$ we identify the two colours $L^{(C)}$ and $L^{(\widetilde{C})}$.

Lemma 5.1. *A Γ-CSP instance $\mathcal{C}$ is satisfiable if and only if the graphs $G(\mathcal{C})$ and $\widetilde{G}(\mathcal{C})$ are isomorphic.*

Example 1 (The Tseitin Tautologies and the CFI-construction). For every graph H and set $T \subseteq V(H)$ we define the following $\mathbb{Z}_2$-CSP $\mathcal{TS} = \mathcal{TS}(H, T)$.

- For every edge $e \in E(H)$ we have a variable z_e.
- For every vertex $v \in V(H)$ we define a constraint C_v. Suppose that v is incident with the edges $e_1, \ldots, e_k$ (in an arbitrary order), and let $z_i := z_{e_i}$.

Let $\Delta := \{(i_1, \ldots, i_k) \in \mathbb{Z}_2^k \mid \sum_{i=1}^k i_j = 0\} \leq \mathbb{Z}_2^k$. We will also use the coset $\Delta + (1, 0, \ldots, 0) = \{(i_1, \ldots, i_k) \in \mathbb{Z}_2^n \mid \sum_{i=1}^k i_j = 1\}$ If $v \notin T$, we let $C_v := (z_1, \ldots, z_k, \Delta)$, and if $v \in T$ we let $C_v := (z_1, \ldots, z_k, \Delta + (1, 0, \ldots, 0))$.

Observe that $\mathcal{TS}$ is a set of Boolean constraints, all of them linear equations over the field $\mathbb{F}_2$; they are known as the *Tseitin tautologies* associated with H and T. We think of assigning a "charge" 1 to every vertex in T and charge 0 to all remaining vertices. Now we are looking for a set $F \subseteq E(H)$ of edges such that for every vertex v, the number of edges in F incident with v is congruent to the charge of v modulo 2. A simple double counting argument shows that $\mathcal{TS}$ is unsatisfiable if $|T|$ is odd. (The sum of degrees in the graph $(V(H), F)$ is even and, by construction, equal to the sum $|T|$ of the charges, which is odd.)

It turns out that the graphs $G(\mathcal{TS})$ and $\widetilde{G}(\mathcal{TS})$ are precisely the *CFI-graphs* defined from H with all vertices in T "twisted". These graphs have been introduced by Cai, Fürer, and Immerman [5] to prove lower bounds for the Weisfeiler-Lehman algorithm and have found various other applications in finite model theory since then.

5.2   Low-Degree Reduction From Tseitin to Isomorphism

For every graph H and set $T \subseteq V(H)$, we let $\mathsf{P}_{\mathrm{Ts}}(H, T)$ be the following system of polynomial equations:

$$z_e^2 - 1 = 0 \quad \text{for all } e \in E(H), \tag{5.1}$$

$$1 + z_{e_1} z_{e_2} \cdots z_{e_k} = 0 \quad \text{for all } v \in T \text{ with incident edges } e_1, \ldots, e_k, \tag{5.2}$$

$$1 - z_{e_1} z_{e_2} \cdots z_{e_k} = 0 \quad \text{for all } v \in V(H) \setminus T \text{ with incident edges } e_1, \ldots, e_k. \tag{5.3}$$

Observe that for every field $\mathbb{F}$ of characteristic $\neq 2$ there is a one-to-one correspondence between solutions to the system $\mathsf{P}_{\mathrm{Ts}}(H, T)$ over $\mathbb{F}$ and solutions for the CSP-instance $\mathcal{TS}(H, T)$ (see Example 1) via the "Fourier" correspondence $1 \mapsto 0, -1 \mapsto 1$.

Lemma 5.3. *Let $\mathbb{F}$ be a field of characteristic $\neq 2$. Let $k \geq 2$ be even and H a k-regular graph, and let $T \subseteq V(H)$. Let $G := G(\mathcal{TS}(H, T))$ and $\widetilde{G} := \widetilde{G}(\mathcal{TS}(H, T))$.*
 Then there is a degree-$(k, 2k)$ reduction from $\mathsf{P}_{\mathrm{Ts}}(H, T)$ to $\mathsf{P}_{\mathrm{iso}}(G, H)$.

6   Lower Bounds

We obtain our lower bounds combining the low-degree reduction of the previous section with known lower bounds for Tseitin polynomials due to Buss et al. [4] for polynomial calculus and Grigoriev [8] for the Positivstellensatz calculus.

Theorem 6.1 ([4,8]). *For every $n \in \mathbb{N}$ there is a 6-regular graph H_n of size $O(n)$ such that $\mathsf{P}_{\mathrm{Ts}}(H_n, V(H_n))$ is unsatisfiable, but:*

(1) *there is no degree-n polynomial calculus refutation of* $\mathsf{P}_{\mathrm{Ts}}(H_n, V(H_n))$ *over any field* $\mathbb{F}$ *of characteristic* $\neq 2$;

(2) *there is no degree-n Positivstellensatz calculus refutation of* $\mathsf{P}_{\mathrm{Ts}}(H_n, V(H_n))$ *over the reals.*

Now our main lower bound theorem reads as follows.

Theorem 6.2. *For every* $n \in \mathbb{N}$ *there are non-isomorphic graphs* G_n, $\widetilde{G}_n$ *of size* $O(n)$, *such that*

(1) *there is no degree-n polynomial calculus refutation of* $\mathsf{P}_{\mathrm{iso}}(G_n, \widetilde{G}_n)$ *over any field* $\mathbb{F}$ *of characteristic* $\neq 2$;

(2) *there is no degree-n Positivstellensatz calculus refutation of* $\mathsf{P}_{\mathrm{iso}}(G_n, \widetilde{G}_n)$ *over the reals.*

Proof. This follows from Lemmas 2.1 and 5.3 and Theorem 6.1.   □

It follows that over finite fields, polynomial calculus has similar shortcomings than over fields of characteristic 0. However, a remarkable exception is $\mathbb{F}_2$, where we are not able to prove linear lower bounds on the degree. Here the approach to reduce from Tseitin fails, as the Tseitin Tautologies are satisfiable over $\mathbb{F}_2$. As a matter of fact, the next theorem shows that CFI-graphs can be distinguished with Nullstellensatz of degree 2 over $\mathbb{F}_2$.

Theorem 6.3. *Let* H *be a graph* $T \subseteq V(H)$ *such that* $|T|$ *is odd. Then there is a degree-2 Nullstellensatz refutation over* $\mathbb{F}_2$ *of* $\mathsf{P}_{\mathrm{iso}}(G, \widetilde{G})$, *where* $G = G(\mathcal{TS}(H, T))$ *and* $\widetilde{G} = \widetilde{G}(\mathcal{TS}(H, T))$.

Thus, to prove lower bounds for algebraic proof systems over $\mathbb{F}_2$ we need new techniques. Our final theorem, which even derives lower bound over $\mathbb{Z}$, is a first step.

Theorem 6.4. *There are non-isomorphic graphs* G, H *such that the system of linear Diophantine equations* $\mathrm{MLIN}(\mathsf{P}_{\mathrm{iso}}(G, H)^2)$ *has a solution over* $\mathbb{Z}$.

Corollary 6.5. *There are non-isomorphic graphs* G, H *such that* $\mathsf{P}_{\mathrm{iso}}(G, H)$ *has no degree-2 Nullstellensatz refutation over* $\mathbb{F}_q$ *for any prime* q.

7   Concluding Remarks

Employing results and techniques from propositional proof complexity, we prove strong lower bounds for algebraic algorithms for graph isomorphism testing, which show that these algorithm are not much stronger than known algorithms such as the Weisfeiler-Lehman algorithm.

Our results hold over all fields except—surprisingly—fields of characteristic 2. For fields of characteristic 2, and also for the ring of integers, we only have very weak lower bounds. It remains an challenging open problem to improve these.

Acknowledgments. We thank Anuj Dawar and Erkal Selman for many inspiring discussions in the initial phase of this project.

References

1. Atserias, A., Maneva, E.: Sherali-Adams relaxations and indistinguishability in counting logics. SIAM J. Comput. **42**(1), 112–137 (2013)
2. Beame, P., Impagliazzo, R., Krajicek, J., Pitassi, T., Pudlak, P.: Lower bounds on Hilbert's nullstellensatz and propositional proofs. In: Proceedings of the 35th Annual Symposium on Foundations of Computer Science, pp. 794–806 (1994)
3. Buss, S.: Lower bounds on nullstellensatz proofs via designs. In: Proof Complexity and Feasible Arithmetics, pp. 59–71. American Mathematical Society (1998)
4. Buss, S., Grigoriev, D., Impagliazzo, R., Pitassi, T.: Linear gaps between degrees for the polynomial calculus modulo distinct primes. Journal of Computer and System Sciences **62**(2), 267–289 (2001)
5. Cai, J., Fürer, M., Immerman, N.: An optimal lower bound on the number of variables for graph identification. Combinatorica **12**, 389–410 (1992)
6. Clegg, M., Edmonds, J., Impagliazzo, R.: Using the Groebner basis algorithm to find proofs of unsatisfiability. In: Proceedings of the 28th Annual ACM Symposium on Theory of Computing, pp. 174–183 (1996)
7. Codenotti, P., Schoenbeck, G., Snook, A.: Graph isomorphism and the Lasserre hierarchy (2014). CoRR arXiv:1107.0632v2
8. Grigoriev, D.: Linear lower bound on degrees of positivstellensatz calculus proofs for the parity. Theoretical Computer Science **259**(1–2), 613–622 (2001)
9. Grigoriev, D., Vorobjov, N.: Complexity of null- and positivstellensatz proofs. Annals of Pure and Applied Logic **113**(1–3), 153–160 (2001)
10. Grohe, M., Otto, M.: Pebble games and linear equations. In: Cégielski, P., Durand, A. (eds.) Proceedings of the 26th International Workshop on Computer Science Logic. Leibniz International Proceedings in Informatics (LIPIcs), vol. 16, pp. 289–304 (2011)
11. Hella, L.: Logical hierarchies in PTIME. Information and Computation **129**, 1–19 (1996)
12. Lasserre, J.B.: Global optimization with polynomials and the problem of moments. SIAM Journal on Optimization **11**(3), 796–817 (2001)
13. Malkin, P.: Sherali-Adams relaxations of graph isomorphism polytopes. Discrete Optimization **12**, 73–97 (2014)
14. O'Donnell, R., Wright, J., Wu, C., Zhou, Y.: Hardness of robust graph isomorphism, Lasserre gaps, and asymmetry of random graphs. In: Proceedings of the 25th Annual ACM-SIAM Symposium on Discrete Algorithms, pp. 1659–1677 (2014)
15. Parrilo, P.: Structured Semidefinite Programs and Semialgebraic Geometry Methods in Robustness and Optimization. Ph.D. thesis, California Institute of Technology (2000)
16. Sherali, H.D., Adams, W.P.: A hierarchy of relaxations between the continuous and convex hull representations for zero-one programming problems. SIAM Journal on Discrete Mathematics **3**(3), 411–430 (1990)
17. Tinhofer, G.: Graph isomorphism and theorems of Birkhoff type. Computing **36**, 285–300 (1986)
18. Torán, J.: On the resolution complexity of graph non-isomorphism. In: Järvisalo, M., Van Gelder, A. (eds.) SAT 2013. LNCS, vol. 7962, pp. 52–66. Springer, Heidelberg (2013)

Fully Dynamic Matching in Bipartite Graphs

Aaron Bernstein[1] and Cliff Stein[2]([✉])

[1] Department of Computer Science, Columbia University, New York, NY, USA
bernstei@gmail.com
[2] Department of IEOR and Computer Science, Columbia University,
New York, NY, USA
cliff@ieor.columbia.edu

Abstract. We present two fully dynamic algorithms for maximum cardinality matching in bipartite graphs. Our main result is a *deterministic* algorithm that maintains a $(3/2 + \epsilon)$ approximation in *worst-case* update time $O(m^{1/4}\epsilon^{-2.5})$. This algorithm is polynomially faster than all previous *deterministic* algorithms for *any* constant approximation, and faster than all previous algorithms (randomized included) that achieve a better-than-2 approximation. We also give stronger results for bipartite graphs whose arboricity is at most α, achieving a $(1+\epsilon)$ approximation in worst-case update time $O(\alpha(\alpha+\log(n))+\epsilon^{-4}(\alpha+\log(n))+\epsilon^{-6})$, which is $O(\alpha(\alpha+\log n))$ for constant ϵ. Previous results for small arboricity graphs had similar update times but could only maintain a maximal matching (2-approximation). All these previous algorithms, however, were not limited to bipartite graphs.

1 Introduction

Finding a maximum cardinality matching in a bipartite graph is a classic problem in computer science and combinatorial optimization. There are efficient polynomial time algorithms (e.g. [11]), and well-known applications, ranging from early algorithms to minimize transportation costs (e.g. [10,13]) to recent applications in on-line advertising and social media (e.g. [7,15]). For matching, the restriction to bipartite graphs is natural and models many real-world applications. Furthermore, in many of these applications, the graph is actually changing over time. We study the *fully dynamic* variant of bipartite matching in which the goal is to maintain a near-maximum matching in a graph subject to a sequence of edge insertions and deletions. When an edge change occurs, the goal is to maintain the matching in time significantly faster than simply recomputing it from scratch.

One of our results is for bipartite *small-arboricity* graphs, which we define here. The *arboricity* of a graph, denoted by $\alpha(G)$ is $\max_J \frac{|E(J)|}{V(J)-1}$ where $J = (V(J), E(J))$ is any subgraph of G induced by at least two vertices. Many classes of graphs in practice have constant arboricity, including planar graphs, graphs with bounded genus and graphs with bounded tree width. Every graph has arboricity at most $O(\sqrt{m})$.

A. Bernstein—Supported in part by an NSF Graduate Fellowship and a Simons Foundation Graduate Fellowship.

C. Stein—Supported in part by NSF grants CCF-1349602 and CCF-1421161.

M.M. Halldórsson et al. (Eds.): ICALP 2015, Part I, LNCS 9134, pp. 167–179, 2015.
DOI: 10.1007/978-3-662-47672-7_14

.1 Previous Work

1 addition to exact algorithms on static graphs, there is previous work on pproximating matching and on finding online matchings. Duan and Pettie howed how to find a $(1+\epsilon)$-approximate weighted matching in nearly linear time 5]; their paper also contains an excellent summary of the history of matching lgorithms. Motivated partly by online advertising, there has also been signif-:ant work on "online matching" (e.g. [7,15]), both exact and approximate. In 1ost online matching work, the graph is dynamic, but with a restricted set of .pdates. Typically, one side of the bipartite graph is fixed at the beginning of the lgorithm. The vertices on the other side arrive, one at a time, and when a vertex .rrives, we learn about all of its incident edges. Deletions are not allowed, nor ypically are changes to the matching, although some work also studies models hat measure the number of changes needed to maintain a matching [4,5,8].

We now turn to fully dynamic matchings. Algorithms can be classified by 1pdate time, approximation ratio, whether they are randomized or deterministic .nd whether they have a worst-case or amortized update time. The distinction)etween deterministic and randomized is particularly important here as all of the ·xisting randomized algorithms require the assumption of an *oblivious* adversary hat does not see the algorithm's random bits; thus, in addition to working)nly with high probability, randomized dynamic algorithms must make an extra ssumption on the model which makes them inadequate in certain settings.

For maintaining an *exact* maximum matching, the best known update time s $O(n^{1.495})$ (Sankowski [19]), which in dense graphs is much faster than recon-structing the matching from scratch. If we restrict the model to bipartite graphs .nd to the incremental or decremental setting – where we allow only edge inser-:ions or only edge deletions (but not both) – Bosek *et al.*([4]) show that we can .chieve total update time (over all insertions or all deletions) $m\sqrt{n}$ for an exact natching and $m\epsilon^{-1}$ for a $(1+\epsilon)$-matching, which is optimal in that it matches :he best known bounds for the static case. For the special case of *convex* bipar-:ite graphs in the fully dynamic setting, Brodal *et al.* showed how to maintain ιn *implicit* (exact) matching with very fast update but slow query time.

Going back to the general problem of maintaining an explicit matching in a fully dynamic setting, we can achieve a much faster update time than $O(n^{1.495})$ .f we allow approximation. One can trivially maintain a *maximal* (and so 2-approximate) matching in $O(n)$ time per update. Ivkovic and Lloyd [12] showed how to improve the update time to $O((m+n)^{\sqrt{2}/2})$. Onak and Rubinfeld [18] were :o first to achieve truly fast update times, presenting a randomized algorithm that maintains a $O(1)$-approximate matching in amortized update time $O(\log^2 n)$ time (with high probability). Baswana *et al.*[2] improved upon this with a ran-domized algorithm that maintains a maximal matching (2-approximation) in .mortized update $O(\log n)$ time per update. These two algorithms are extremely fast, but suffer from being amortized and inherently randomized, and also from the fact their techniques focus on local changes, and so seem unable to break through the barrier of a 2-approximation.

The first result to achieve a better-than-2 approximation was by Neiman and Solomon [17], who presented a *deterministic, worst-case* algorithm for maintaining a 3/2-approximate matching. However, the price of this improvement was a huge increase in update time: from $O(\log n)$ to $O(\sqrt{m})$. Gupta and Peng [9] later improved upon the approximation, presenting a deterministic algorithm that maintains a $(1+\epsilon)$-approximate matching in worst-case update time $O(\sqrt{m}\epsilon^{-2})$.

The two deterministic algorithms are strongly tethered to the $\sqrt{m}$ bound and do not seem to contain any techniques for breaking past it. An important open question was thus: can we achieve $o(\sqrt{m})$ update with a deterministic algorithm? (In fact Onak and Rubinfeld [18] presented a deterministic algorithm with amortized update time $O(\log^2 n)$, but it only achieves a $\log(n)$-approximation.) Very recently, Bhattacharya, Henzinger, and Italiano [3] presented a deterministic algorithm with worst-case update time $O(m^{1/3}\epsilon^{-2})$ that maintains a $(4+\epsilon)$ approximation; this can be improved to $(3+\epsilon)$ at the cost of introducing amortization. The same paper presents a deterministic algorithm with amortized update time only $O(\epsilon^{-2}\log n)$ that maintains a $(2+\epsilon)$ *fractional* matching. Finally, Neiman and Solomon [17] showed that in graphs of constant arboricity we can maintain a maximal (so 2-approximate) matching in amortized time $O(\log(n)/\log\log(n))$; using a recent dynamic orientation algorithm of Kopelowitz *et al.*[14], this algorithm yields a $O(\log(n))$ *worst-case* update time.

Abboud and Williams [1] recently showed a conditional lower bound for dynamic matching in general graphs assuming that 3-sum cannot be solved in $o(n^2)$ time; they show that there exists a constant $k \in [2, 10]$ with the following property: any algorithm that maintains an approximate matching in which every augmenting path has length at least $2k - 1$ has amortized update time $\Omega(m^{1/3})$.

1.2   Results

If we disregard special cases such as small arboricity or fractional matchings, we see that existing algorithms for dynamic matching seem to fall into two groups: there are fast (mostly randomized) algorithms that do not break through the 2-approximation barrier, and there are slow algorithms with $O(\sqrt{m})$ update that achieve a better-than-2 approximation. Thus the obvious question is whether we can design an algorithm – deterministic or randomized – that achieves a tradeoff between these two: a $o(\sqrt{m})$ update and a better-than-2 approximation. We answer this question in the affirmative for bipartite graphs.

Theorem 1. *Let G be a bipartite graph subject to a series of edge insertions and deletions, and let ϵ be $< 2/3$. Then, we can maintain a $(3/2 + \epsilon)$-approximate matching in G in deterministic worst-case update time $O(m^{1/4}\epsilon^{-2.5})$.*

This theorem achieves a new trade-off even if one considers existing randomized algorithms. Focusing on only deterministic algorithms the improvement is even more drastic: our algorithm improves upon not just $\sqrt{m}$ but $m^{1/3}$, and so achieves the fastest known deterministic update time (excluding the $\log(n)$-approximation of [18]), while still maintaining a better-than-2 approximation.

lso, since $m^{1/4} = O(\sqrt{n})$, our algorithm is the first to achieve a better-than-2 pproximation in time strictly sublinear in the number of nodes. Of course, our lgorithm has the disadvantage of only working on bipartite graphs.

For small arboricity graphs we also show how to break through the maximal natching (2-approximation) barrier and achieve a $(1 + \epsilon)$-approximation.

Theorem 2. *Let G be a bipartite graph subject to a series of edge insertions nd deletions, and let ϵ be < 1. Say that at all times G has arboricity at most $\cdot$. Then, we can maintain a $(1 + \epsilon)$-approximate matching in G in deterministic vorst-case update time $O(\alpha(\alpha + \log(n)) + \epsilon^{-4}(\alpha + \log(n)) + \epsilon^{-6})$ For constant α nd ϵ the update time is $O(\log(n))$, and for α and ϵ polylogarithmic the update ime is polylogarithmic.*

Jote that a $(1 + \epsilon)$-approximation with polylog update time is pretty much the test we can hope for. The conditional lower bound of Abboud and Williams [1] rovides an indication that such a result might not be possible for general graphs, ut we have presented the first class of graphs (bipartite, polylog arboricity) for vhich it is achievable.

.3   Techniques

Ve can think of the dynamic matching problem as follows: We are given a lynamic graph G and want to maintain a large subgraph M of maximum legree 1. This task turns out to be quite hard because, as the graph evolves, M s unstable and has few appropriate structural properties.

Very recently, Bhattacharya *et al.*[3] presented the idea of using a transition uubgraph H, which they refer to as a *kernel* of G: the idea is to maintain H as G changes, and then maintain M in H. Maintaining an approximate matching M is significantly easier in a bounded degree graph, so we need a graph H that nas the following properties: it should have bounded degree, it should be easy to naintain in G, and most importantly, a large matching using edges in H should ve a good approximation to the maximum matching in G.

Our algorithm uses the same basic idea of transition subgraph with bounded legree, but the details are entirely different from those in [3] . Their subgraph H is just a maximal B-matching with B around $m^{1/3}$, that allows some slack on the maximality constraint. The use of a maximal matching is a natural choice n a dynamic setting because maximality is a purely *local* constraint, and so easier to maintain dynamically. The downside is that as long as one relies on maximality, one can never achieve a better-than-2 approximation; due to other lifficulties, their paper in fact only achieves a $(3 + \epsilon)$-approximation.

The main technical contribution of this paper is to present a new type of bounded-degree subgraph, which we call an *edge degree constrained subgraph* (EDCS). The problem with a simple B-matching is that the edges are not sufficiently "spread out" to all the vertices: imagine that G consists of 4 sets L_1, L_2, R_1, R_2, each of size $n/2$, where the edges form a complete graph except that there are no edges between L_2 and R_2. One possible maximal B-matching includes many edges between L_1 and R_1 while leaving L_2 and R_2 completely isolated. The resulting matching is only 2-approximate, which is what we are trying to overcome. Our EDCS circumvents this problem by trying to spread out edges. For each edge, instead of separately upper bounding the matching-degree of both endpoints (B-matching) it upper bounds the *sum* of the matching-degrees of the endpoints, and then captures the notion of maximality by also lower bounding this sum for edges not in the matching. Using an EDCS prevents the above scenario as the sum of the matching-degrees of edges from L_1 to R_2 will be illegally small unless the matching-degree of R_2 is raised by adding some of those edges to the graph, thus ensuring a larger matching in H.

Although the definition is somewhat similar, the structure of an edge degree constrained subgraph is entirely different from that of a maximal B-matching, and for this reason both our analysis of the approximation factor and our algorithm for maintaining this subgraph are entirely different from those in [3]. In particular, while the constraints in an EDCS seem purely local in that they concern only the degrees of the endpoints of an edge, they in fact have a global effect in a way that they do not in a maximal B-matching. In the latter, as long as an edge does not directly violate the degree constraints, it can *always* be added to the maximal B-matching, without concern for the edges elsewhere in the graph. But as seen from the above example, this is not true in an EDCS: although the edges from L_1 and R_1 do not themselves violate any constraints, they prevent the constraints between L_1 and R_2 or L_2 and R_1 from being satisfied. An analysis of this global structure is what allows us to go beyond the 2-approximation. On the other hand, the same global structure makes the EDCS more difficult to maintain dynamically; we end up showing that an EDCS contains something akin to augmenting paths, although more locally well behaved. We also develop a general new technique for maintaining a transition subgraph based on dynamic graph orientation, which allows us to reduce the update time from $O(m^{1/3})$ to $O(m^{1/4})$. That being said, the additional complications inherent in an EDCS have so far prevented us from extending our results to non-bipartite graphs.

We omit many details in this extended abstract and refer the reader to the full paper for details.

2   Preliminaries

Let $G = (L \bigcup R, E)$ be an undirected, unweighted bipartite graph where $|L| = |R| = n$ and $|E| = m$. Unless otherwise specified, "graph" will always refer to a bipartite graph. In general, we will often be dealing with graphs other than G, so all of our notation will be explicit about the graph in question. We define $d_G(v)$ to be the degree of a vertex v in G; if the graph in question is weighted, then $d_G(v)$ is the sum of the weights of all incident edges. We define *edge degree* as $\delta(u, v) = d(u) + d(v)$. If H is a subgraph of G, we say that an edge in G is *used* if it is also in H, and *unused* if it is not in H. Throughout this paper we will only be dealing with subgraphs H that contain the full vertex set of G, so we will use the notion of a subgraph and of a subset of edges of G interchangeably.

A matching in a graph G is a set of disjoint edges in G. We let $\mu(G)$ denote the size of the maximum matching in G. A vertex is called *matched* if it is incident to one of the sets in the matching, and *free* or *unmatched* otherwise. We now state a simple corollary of an existing result of [9].

Lemma 1 ([9]). *If a dynamic graph G has maximum degree B at all times, then we can maintain a $(1 + \epsilon)$-approximation matching under insertions and deletions in worst-case update time $O(B\epsilon^{-2})$ per update.*

Proof. This lemma immediately follows from a simple algorithm presented in sect. 3.2 of [9] which shows how to achieve update time $|E(G)|\epsilon^{-2}/\mu(G)$ (for the transition from amortized to worst-case see appendix A.3 of the same paper), as well as the fact that we always have $|E(G)|/\mu(G) \leq 2B$ because all edges must be incident to one of the $2\mu(G)$ matched vertices in the maximum matching, and each of those vertices have degree at most B.

Orientations An orientation of an undirected graph G is an assignment of a direction to each edge in E. Given an orientation of edge (u, v) from u to v, we say that u *owns* edge (u, v) and will define the *load* of a vertex v to be the number of edges owned by v. Orientations of small max load are closely linked to arboricity: every graph with arboricity α has an α-orientation [16]. Our algorithms will at all times maintain an orientation of the *dynamic* graph G. We rely on two results to do this: one by Kopelowitz *et al.*[14], and a second simple result new to this paper whose proof we leave for the full version.

Theorem 3. *[14] Given a dynamic graph G that at all times has arboricity $\leq \alpha$, there exists an algorithm that maintains an orientation with max load $O(\alpha \log(n))$ such that every insertion/deletion to G is processed in worst-case update time $O(\alpha(\alpha + \log(n)))$ and requires at most $O(\alpha + \log(n))$ edge reorientations.*

Theorem 4. *Given a dynamic graph G, we can maintain an orientation with max load $O(\sqrt{m})$ in worst-case update time $O(1)$ per insertion/deletion to G.*

3   The Framework

We now define the transition subgraph H mentioned in Sect. 1.3.

Definition 1. *An unweighted edge degree constrained subgraph(EDCS) (G, β, β^-) is a subset of the edges $H \subseteq E$ with the following properties:*

(P1) if (u, v) is used (in H) then $d_H(u) + d_H(v) \leq \beta$,
(P2) if (u, v) is unused (in $G - H$) then $d_H(u) + d_H(v) \geq \beta^-$.

We also define a similar subgraph where edges in H have weights, effectively allowing them to be used more than once. The properties change somewhat as now used edges can always take more weight, so it makes sense to lower bound the degrees of used edges as well. Recall that the degree of a vertex in a weighted graph is the sum of the weights of the incident edges. returnpoint

Definition 2. *A weighted edge degree constrained subgraph(EDCS)* (G, β, β^-) *is a subset of the edges* $H \subseteq E$ *with positive integer weights that has properties:*

(P1) if (u, v) *is used then* $d_H(u) + d_H(v) \leq \beta$
(P2) for all edges (u, v), *we have* $d_H(u) + d_H(v) \geq \beta^-$

Algorithm Outline: To process an edge insertion/deletion in G: First, we update the small-max-load edge orientation (Theorem 3 or 4. Second, we update the subgraph H so it remains a valid EDCS of the changed graph G (Sect. 5); this relies on the graph orientation for efficiency. Third, we update the $(1+\epsilon)$-approximate matching in H with respect to the changes to H from the previous step (See Lemma 1). The maintained $(1 + \epsilon)$-approximate matching of H is also our final matching in G; the central claim of this paper is that because H is an EDCS, $\mu(H)$ is not too far from $\mu(G)$, so a good approximation to $\mu(H)$ is also a decent approximation to $\mu(G)$ (see Sectionr̃efsec:matching).

There is a subtle difficulty that arises from using a transition graph in a dynamic algorithm. By Lemma 1, as long as H has degree bounded by Δ_H, we can maintain a $(1 + \epsilon)$-approximate matching in H in time $O(\Delta_H)$ *per update in H*. But a single change in G could in theory causes many changes in H, each of which would take $O(\Delta_H)$ time to process. This motivates the following definition: given an algorithm A that maintains a subgraph H in a dynamic graph G, we define the *update ratio* of A to be the maximum number of edge changes (insertions or deletions) that A could make to H given a single edge change in G.

We can now state the main theorems of the paper. We present general and small arboricity graphs separately, but the basic framework described above remains the same in both cases. In all the theorems below, the parameter ϵ corresponds to the desired approximation ratio (either $(1 + \epsilon)$ or $(3/2 + \epsilon)$).

3.1   General Bipartite Graphs

For the sake of intuition, think of β in the two theorems below as roughly $m^{1/4}$.

Theorem 5. *Let G be a bipartite graph, and let $\lambda = \epsilon/4$. Let H be an unweighted EDCS with $\beta^- = \beta(1 - \lambda)$, where β is a parameter we will choose later. Then $\mu(H) \geq (2/3 - \epsilon)\mu(G)$.*

Theorem 6. *Let G be a bipartite graph. Let H be an unweighted EDCS with $\beta^- = \beta(1 - \lambda)$, where λ is a positive constant less than 1. There is an algorithm that maintains H over updates in G (i.e. maintains H as a valid edge degree constrained subgraph) with the following properties:*

  – *The algorithm has worst case update time* $O((\frac{1}{\lambda})(\beta + \frac{\sqrt{m}}{\lambda\beta}))$.
  – *The update ratio of the algorithm is* $O(1/\lambda)$.

Proof of Theorem 1 We use the algorithm outline presented near the beginning of Sect. 3. We let be transition subgraph H be an unweighted $EDCS(G, \beta, \beta(1 - \lambda))$ with $\lambda = 4\epsilon^{-1} = O(\epsilon^{-1})$ and $\beta = m^{1/4}\epsilon^{1/2}$. By Theorem 6 we can maintain H in worst-case update time $O((\frac{1}{\lambda})(\beta + \frac{\sqrt{m}}{\lambda\beta})) = O(m^{1/4}\epsilon^{-2.5} + m^{1/4}\epsilon^{-.5}) = O(m^{1/4}\epsilon^{-2.5})$. The update ratio is $O(\lambda^{-1}) = O(\epsilon^{-1})$. Since degrees in H are clearly bounded by β, by Lemma 1 we can maintain a $(1 + \epsilon)$-approximate matching in H in time $O(\beta\epsilon^{-2})$; multiplying by the update ratio of maintaining H in G, we need $O(\beta\epsilon^{-3}) = O(m^{1/4}\epsilon^{-2.5})$ time to maintain the matching per change in G. By Theorem 7, $\mu(H)$ is a $(3/2 + \epsilon)$-approximation to $\mu(G)$, so our matching is a $(3/2 + \epsilon)(1 + \epsilon) = (3/2 + \epsilon)$-approximate matching in G.   □

3.2   Small Arboricity Graphs

Theorem 7. *Let G be a bipartite graph, and let $\beta > 4\epsilon^{-2}$. Let H be a weighted EDCS with $\beta^- = \beta - 1$. Then $\mu(H) \geq \mu(G)(1 - \epsilon)$.*

Theorem 8. *Let G be a bipartite graph with arboricity α. Let H be a weighted EDCS with $\beta^- = \beta - 1$. There is an algorithm that maintains H over updates in G with the following properties:*

- *The algorithm has worse-case update time $O(\beta^2(\alpha + \log n) + \alpha(\alpha + \log n))$.*
- *The update ratio of the algorithm is $O(\beta)$.*

The proof of Theorem 2 is analogous to that of Theorem 1 with β set to ϵ^{-2}.

4   An EDCS Contains an Approximate Matching

In this section we prove Theorems 5 and 7. Both proofs will be by contradiction; for example, for Theorem 5 to be false, there must be an unweighted $EDCS(G, \beta, \beta(1 - \lambda))$ H such that $\mu(H) < (2/3 - \epsilon)\mu(G)$. To exhibit the contradiction, we start by establishing a property that must hold of *any* subgraph H defined on the full vertex set of G for which $\mu(H)$ is smaller than $\mu(G)$; the smaller $\mu(H)$, the more constraining the property. Loosely speaking, the property is a generalization of the fact that the maximum matching on H establishes an (S, T) cut with no edges crossing in H, but at least $\mu(G) - \mu(H)$ edges crossing in G. We use the convention that the subscript L or R refer to the side of the bipartition in which the vertices lie. The proof of the following lemma involves a careful accounting of augmenting paths and is left for the full version.

Lemma 2. *Let $G = (V, E_G)$ be a bipartite graph, and let $H = (V, E_H)$ be a subgraph of G. Then, there exist vertex sets $S_L^*, S_L, S_R, T_R^*, T_R, T_L$ with the following properties:*

1. $|S_L| + |T_L| = |S_R| + |T_R| = \mu(H)$.
2. In E_H, all edges incident to $S_L \bigcup S_L^*$ go to S_R and all edges incident to $T_R \bigcup T_R^*$ go to T_L.

3. G contains a perfect matching between S_L and S_R and between T_L and T_R $(|S_L| = |S_R|, |T_L| = |T_R|)$.

4. $|S_L^*| = |T_R^*| = \mu(G) - \mu(H)$ and G contains a perfect matching between these sets.

Let us say, for contradiction, that $\mu(H)$ is much smaller than $\mu(G)$. Then according to Lemma 2, there is a perfect matching between S_L^* and T_R^* in G but not H. Thus, by property P2 of an EDCS, for every edge (v, w) on that matching $d_H(v) + d_H(w)$ must be almost β. This implies that the average degree in H of vertices in S_L^* and T_R^* must be at least around $\beta/2$. But all the edges in H incident to S_L^* and T_R^* can only go to S_R and T_L, which are relatively small if $\mu(H)$ is much smaller than $\mu(G)$. To close the contradiction we argue that because of property P1 of an EDCS, we simply won't be able to fit all those edges from S_L^* to S_R and T_R^* to T_L. We argue this by bounding how high degrees can get in an EDCS. Intuitively, if U and V have equal size and all edges are between U and V, we expect the average degree on each side to be no more than $\beta/2$, as if each vertex had degree $\beta/2$ then all edge degrees would be β – the maximum allowed by property P1. We now state a generalization of this intuition which shows that if one of the sets U, V is larger than the other, it will have average degree below $\beta/2$; the proof is left for the full version.

Lemma 3. *Let us say that in some graph we have disjoint sets (U, V) such that $|U| = c|V|$, and all edges incident to U go to V (but there may be edges incident to V which do not go to U). Let $d(v)$ be the degree of vertex v in this graph, and say that for every edge (u, v) in the graph $d(u) + d(v) \leq \beta$ for some parameter β. Then, the average degree of vertices in U is at most $\frac{\beta}{c+1}$.*

Proof of Theorem 5: Let us say, for the sake of contradiction, that we had $\mu(H) < (2/3 - \epsilon)\mu(G)$. Then, we have sets $S_L^*, S_L, S_R, T_R^*, T_R, T_L$ as in Lemma 2. By property 4 of this lemma, S_L^* and T_R^* have a perfect matching between them consisting of $\mu(G) - \mu(H)$ edges in $E_G - E_H$ – that is, a perfect matching of *unused* edges. Thus, by the property P2 of an EDCS, for each edge (u, v) in this matching we have $d_H(u) + d_H(v) \geq \beta(1 - \lambda)$, which implies that the total degree of vertices in $S_L^* \bigcup T_R^*$ is at least $\beta(1 - \lambda)(\mu(G) - \mu(H))$. Now, by property 4 of Lemma 2 we know that $|S_L^*| = |T_R^*| = \mu(G) - \mu(H)$, so $|S_L^* \bigcup T_R^*| = 2(\mu(G) - \mu(H))$, so we have:

$$\text{average degree of } S_L^* \bigcup T_R^* \geq \frac{\beta(1-\lambda)(\mu(G) - \mu(H))}{2(\mu(G) - \mu(H))} = \beta\frac{(1-\lambda)}{2}. \tag{1}$$

We argue such a high average degree is not possible. Since $\mu(H) < (2/3 - \epsilon)\mu(G)$:

$$|S_L^* \bigcup T_R^*| = 2(\mu(G) - \mu(H)) > \mu(H)(1 + \epsilon). \tag{2}$$

Observe that we are now in the situation described in Lemma 3: $S_L^* \bigcup T_R^*$ corresponds to U, and $S_R \bigcup T_L$ corresponds to V. Property 2 of Lemma 2 precisely tells us that all edges from U go to V, as needed in Lemma 3. We know from properties 3 and 1 of Lemma 2 that $|V| = |S_R \bigcup T_L| = |S_R| + |T_L| = |S_L| + |T_L| = (H)$ so by Eq. 2 we have $|U| = |S_L^* \bigcup T_R^*| = c|V|$ for some $c > (1 + \epsilon)$. Thus emma 3 tells us that the average degree of U is at most $\beta/(1 + c) \le \beta/(2 + \epsilon)$, hich some simple algebra shows is strictly less than $\beta(1 - \lambda)/2$ because we set $= \epsilon/4$. We have thus arrived at a contradiction with Eq. 1, so our original ssumption that $\mu(H) < (2/3 - \epsilon)\mu(G)$ must be false. □

mall Arboricity Graphs: We now turn to Theorem 7. The full proof is left for he full version, but we give some intuition here. The statement is very similar to 'heorem 5, but with two crucial differences: we are now dealing with a *weighted* DCS H, and the approximation we need to guarantee is $1 - \epsilon$ instead of $2/3 - \epsilon$. Note that Theorem 7 is true of general graphs as well; we only use it for small rboricity graphs, however, because a weighted EDCS is difficult to maintain in eneral graphs.) It may seem unintuitive that a weighted EDCS contains a better natching than an unweighted one since it will in fact have fewer total edges to vork with. To show why a weighted EDCS is better, see for a simple example vhere an unweighted EDCS only contains a $(3/2)$-approximate matching, but a veighted one does not suffer the same issues.

In the proof of Theorem 5 we constructed the sets S_L^*, S_L, S_R, T_R^*, T_R, T_L rom Lemma 2 and then argued that S_L^* (and analogously T_R^*) must have low verage degree because all of its edges go to S_R, so we simply cannot fit that nany edges before violating property P1 of an EDCS. Now, we could upper ound the average degree of S_L^* even better if we could argue that there also ad to be other edges coming into S_R, taking up space. The natural candidate vould be the edges on the matching from S_L to S_R guaranteed by property 3 f Lemma 2. In Theorem 5 we were unable to take advantage of these edges ecause we were dealing with an *unweighted* EDCS, so a single matching worth f edges did not count for much. The properties of a weighted EDCS, however, an force this single matching to be used multiple times, thus leaving even less pace for edges leaving S_L^*. The proof of Theorem 7 is thus analogous to that of Theorem 5 but requires a stronger version of Lemma 3.

5   Maintaining an Edge Degree Constrained Subgraph

n this section, we outline the proofs of Theorems 6 and 8, leaving the details or the full version of the paper.

Recall that $\delta(u, v)$ denotes the edge degree of (u, v), $d_H(u) + d_H(v)$. We define an edge to be *full* if it is in H and has edge degree β. We define it to be *deficient* f it is not in H and has the minimum allowable edge degree β^-: this is $\beta - 1$ or the weighted EDCS in Theorem 8 and $\beta(1 - \lambda)$ for the unweighted EDCS f Theorem 6. We define a vertex to be *increase-safe* if it has no incident full dges and *decrease-safe* if it has no incident deficient edges; it is easy to see that ncreasing (decreasing) the degree of an increase-safe (decrease-safe) vertex by ne does not lead to a violation of any EDCS constraints.

Now, let us say that we delete some edge (u, v) from G. If (u, v) was not in the EDCS H then all constraints remain satisfied. Otherwise, deleting (u, v) causes the degree of u and v to decrease by one. Let us focus on fixing up vertex v; vertex u can then be handled analogously. If v was decrease-safe, then all constraints relating to v remain satisfied and we are done. Otherwise, it must have had some incident deficient edge (v, v_2). Adding this edge to H rebalances the degree of v to what it was before the deletion, but now the degree of v_2 has increased by one. If v_2 was increase-safe, the degree increase does not violate any constraints, and we are done. Otherwise, v_2 must have an incident full edge (v_2, v_3) which we delete from the graph; this rebalances v_2 but decreases the degree of v_3, so we look for an incident deficient edge. We continue in this fashion until we end on an increase/decrease-safe vertex.

We can thus fix up an edge deletion by finding an alternating path of full and deficient edges that ends in an increase/decrease-safe vertex. Insertions are handled analogously. This is similar to finding an augmenting path in a matching, except that this latter case is much harder because we might hit a dead end and have to back track; but we can fix up an EDCS by following *any* sequence of full/deficient edges. Moreover, the resulting alternating path is always simple and contains few edges: for the small arboricity case (Theorem 8) where $\beta^- = \beta - 1$, it is not hard to see that in any such alternating path the vertex degrees $d_H(v)$ on either side of the bipartition are either increasing or decreasing by 1, so since $d_H(v)$ is always between 0 and β, the path has length $O(\beta)$; in the small arboricity case, $O(\beta)$ is small because we set $\beta = O(1/\epsilon^2)$. In the general case (Theorem 6), β is large but the gap between β and β^- is $\beta\lambda$, so degrees on either side change by $\beta\lambda$ and the path has length only $O(1/\lambda)$.

To find such an alternating path of full and deficient edges we maintain a data structure that for any vertex v can return an incident full or deficient edge (whichever is asked for), or indicate that none exists. Since the alternating path will always be short, this data structure will only be queried a small number of times per insertion/deletion in G. We maintain this data structure using a dynamic orientation, in which each edge is owned by one of its endpoints (see end of Sect. 2). Let us focus on the small arboricity case, where the dynamic orientation maintains a small max load. Each vertex will maintain fullness/deficiency information about the edges it does *not* own, storing each category of edge (full/deficient) in its own list. To find a full/deficient edge incident to some vertex v, the data structure simply picks an edge from the corresponding list in $O(1)$ time; if the list is empty, the data structure then manually checks all the edges that v *does* own: since the max load is small, this can be done efficiently. When the status of a vertex v changes, to maintain itself the data structure must transfer this information along all edges (v, u) that are *not* owned by u, but since these are precisely the edges owned by v, there can only be a small number of them.

The basic idea is the same for general bipartite graphs (Theorem 6), except that now the max load is $O(\sqrt{m})$, and we cannot afford to spend $O(\sqrt{m})$ per update. Note that in this case, however, there is a gap of $\beta\lambda$ between full and deficient edges, so intuitively, the degree of a vertex has to change $\beta\lambda$ time before it must be updated in the data structure. This leads to an update time f around $\sqrt{m}/(\beta\lambda)$, as needed in Theorem 6. The details, however, are quite involved, especially since we need a *worst-case* update time.

Conclusion

We have presented the first fully dynamic matching algorithm to achieve a $o(\sqrt{m})$ update time while maintaining a better-than-2-approximate bipartite matching. It is also the fastest known deterministic algorithm for achieving *any* constant approximation, and certainly any better-than-2 approximation. The main open questions are in how far we can push this tradeoff. Can we achieve a *randomized* better-than-2 approximation with update time polylog(n)? For *deterministic* algorithms, can we achieve a constant approximation with update time polylog(n), or a $(1 + \epsilon)$-approximation with update time $o(\sqrt{m})$?

The other natural question is whether our results can be extended to general (non-bipartite) graphs and non-bipartite graphs of small arboricity. The definition of an edge degree constrained subgraph does not inherently rely on bipartiteness, and neither do many of the techniques in this paper. The main obstruction to the generalization seems to lie in the structural property exhibited in Lemma 2. Is there an analogue for non-bipartite graphs?

Acknowledgments. We thank Tsvi Kopelowitz for several helpful discussions and for pointing us towards useful information about orientations.

References

1. Abboud, A., Williams, V.V.: Popular conjectures imply strong lower bounds for dynamic problems. In: Proceedings of FOCS 2014, pp. 434–443 (2014)
2. Baswana, S., Gupta, M., Sen, S.: Fully dynamic maximal matching in O (log n) update time. In: Proceedings of FOCS 2011, pp. 383–392 (2011)
3. Bhattacharya, S., Henzinger, M., Italiano, G.F.: Deterministic fully dynamic data structures for vertex cover and matching. In: SODA, pp. 785–804 (2015)
4. Bosek, B., Leniowski, D., Sankowski, P., Zych, A.: Online bipartite matching in offline time. In: Proceedings of FOCS 2014, pp. 384–393 (2014)
5. Chaudhuri, K., Daskalakis, C., Kleinberg, R.D., Lin, H.: Online bipartite perfect matching with augmentations. In: INFOCOM, pp. 1044–1052 (2009)
6. Duan, R., Pettie, S.: Linear-time approximation for maximum weight matching. J. ACM **61**(1), 1 (2014)
7. Feldman, J., Henzinger, M., Korula, N., Mirrokni, V.S., Stein, C.: Online stochastic packing applied to display ad allocation. In: de Berg, M., Meyer, U. (eds.) ESA 2010, Part I. LNCS, vol. 6346, pp. 182–194. Springer, Heidelberg (2010)
8. Gupta, A., Kumar, A., Stein, C.: Maintaining assignments online: matching, scheduling, and flows. In: SODA, pp. 468–479 (2014)
9. Gupta, M., Peng, R.: Fully dynamic (1+ e)-approximate matchings. In: Proceedings of FOCS 2013, pp. 548–557 (2013)
10. Hitchcock, F.: The distribution of a product from several sources to numberous localities. J. Math Phys. **20**, 224–230 (1941)

11. Hopcroft, J.E., Karp, R.M.: An $n^{5/2}$ algorithm for maximum matching in bipartite graphs. SIAM Journal on Computing **2**, 225–231 (1973)
12. Ivković, Z., Lloyd, E.L.: Fully dynamic maintenance of vertex cover. In: van Leeuwen, Jan (ed.) WG 1993. LNCS, vol. 790, pp. 99–111. Springer, Heidelberg (1994)
13. Kantorovitch, L.: On the translocation of masses. Doklady Akad. Nauk SSSR **37**, 199–201 (1942)
14. Kopelowitz, T., Krauthgamer, R., Porat, E., Solomon, S.: Orienting fully dynamic graphs with worst-case time bounds. In: Esparza, J., Fraigniaud, P., Husfeldt, T., Koutsoupias, E. (eds.) ICALP 2014, Part II. LNCS, vol. 8573, pp. 532–543. Springer, Heidelberg (2014)
15. Mehta, A., Saberi, A., Vazirani, U., Vazirani, V.: Adwords and generalized on-line matching. In: Proceedings of FOCS 2005, pp. 264–273 (2005)
16. Nash-Williams, C.S.J.A.: Edge disjoint spanning trees of finite graphs. Journal of the London Mathematical Society **36**, 445–450 (1961)
17. Neiman, O., Solomon, S.: Simple deterministic algorithms for fully dynamic maximal matching. In: Proceedings of STOC 2013, pp. 745–754 (2013)
18. Onak, K., Rubinfeld, R.: Maintaining a large matching and a small vertex cover. In: Proceedings of STOC 2010, pp. 457–464 (2010)
19. Sankowski, P.: Faster dynamic matchings and vertex connectivity. In: Proceedings of SODA 2007, pp. 118–126 (2007)

Feasible Interpolation for QBF
Resolution Calculi

Olaf Beyersdorff[1]([✉]), Leroy Chew[1], Meena Mahajan[2], and Anil Shukla[2]

[1] School of Computing, University of Leedds, Leedds, UK
{o.beyersdorff,mm12lnc}@leeds.ac.uk
[2] The Institute of Mathematical Sciences, Chennai, India
{meena,anilsh}@imsc.res.in

Abstract. In sharp contrast to classical proof complexity we are currently short of lower bound techniques for QBF proof systems. We establish the feasible interpolation technique for all resolution-based QBF systems, whether modelling CDCL or expansion-based solving. This both provides the first general lower bound method for QBF calculi as well as largely extends the scope of classical feasible interpolation. We apply our technique to obtain new exponential lower bounds to all resolution-based QBF systems for a new class of QBF formulas based on the clique problem. Finally, we show how feasible interpolation relates to the recently established lower bound method based on strategy extraction [7].

1 Introduction

The main aim in proof complexity is to understand the complexity of theorem proving. Arguably, what is even more important is to establish techniques for lower bounds, and the recent history of computational complexity speaks volumes on how difficult it is to develop general lower bound techniques. Understanding the size of proofs is important for at least two reasons. The first is its tight relation to the separation of complexity classes: NP vs. coNP for propositional proofs, and NP vs. PSPACE in the case of proof systems for quantified boolean formulas (QBF). New superpolynomial lower bounds for specific proof systems rule out specific classes of non-deterministic poly-time algorithms for problems in co-NP or PSPACE, thereby providing an orthogonal approach to the predominantly machine-oriented view of computational complexity.

The second reason to study lower bounds for proofs is the analysis of SAT and QBF solvers: powerful algorithms that efficiently solve the classically hard problems of SAT and QBF for large classes of practically relevant formulas. Modern SAT solvers routinely solve industrial instances in millions of variables for various applications. Although QBF solving is at a much earlier state, due to its greater expressivity, QBF even applies to further fields such as formal verification

Feasible Interpolation for QBF Resolution Calculi—This work was supported by the EU Marie Curie IRSES grant CORCON, grant no. 48138 from the John Templeton Foundation, EPSRC grant EP/L024233/1, and a Doctoral Training Grant from EPSRC (2nd author).

M.M. Halldórsson et al. (Eds.): ICALP 2015, Part I, LNCS 9134, pp. 180–192, 2015.
DOI: 10.1007/978-3-662-47672-7_15

or planning [5, 13, 24]. Each successful run of a solver on an unsatisfiable instance can be interpreted as a proof of unsatisfiability; and modern SAT solvers based on conflict-driven clause learning (CDCL) are known to implicitly generate resolution proofs. Thus, understanding the complexity of resolution proofs directly translates into sharp bounds for the performance of CDCL-based SAT solvers.

The picture is more complex for QBF solving, as there exist two main, yet conceptually very different paradigms: CDCL-based and expansion-based solving. A variety of QBF resolution systems have been designed to capture the power of QBF solvers based on these paradigms. The core system is Q-Resolution (Q-Res), introduced in [17]. This has been augmented to capture ideas from CDCL solving, leading to long-distance resolution (LD-Q-Res) [2], universal resolution (QU-Res) [25], or its combinations like LQU+-Res [3]. Powerful proof systems for expansion-based solving were recently developed in the form of ∀Exp+Res [16], and the stronger IR-calc and IRM-calc [6]. Latest findings show that CDCL and expansion are indeed orthogonal paradigms as the underlying proof systems from the two categories are incomparable with respect to simulations [7].

Understanding which general techniques can be used to show lower bounds for proof systems is of paramount importance in proof complexity. For propositional proof systems we have a number of effective techniques, most notably the size-width technique [4], deriving size from width bounds, game characterisations (e.g. [9, 23]), the approach via proof-complexity generators (cf. [19]), and feasible interpolation. Feasible interpolation, first introduced by Krajíček [18], is a particularly successful paradigm that transfers circuit lower bounds to proof size lower bounds. The technique has been shown to be effective for resolution [18], cutting planes [22] and even Frege systems for modal and intuitionistic logics [15]. However, feasible interpolation fails for strong propositional systems as Frege systems under plausible cryptographic and number-theoretic assumptions [10, 11, 20].

The situation is drastically different for QBF proof systems, where we currently possess a very limited bag of techniques. At present we only have the very recent strategy extraction technique [7], which works only for Q-Res, a game characterisation of the very weak tree-like Q-Res [8], and ad-hoc lower bound arguments for various systems [7, 17]. In addition, the recent paper [3] develops methods to lift some previous lower bounds from Q-Res to stronger systems.

Our Contributions

1. A General Lower Bound Technique. We show that the feasible interpolation technique applies to all resolution-type QBF proof systems, whether expansion or CDCL based. This provides the first truly general lower bound technique for QBF proof systems, and—at the same time—hugely extends the scope of the feasible interpolation method.

In a nutshell, feasible interpolation works for true implications $A(\boldsymbol{p}, \boldsymbol{q}) \rightarrow B(\boldsymbol{p}, \boldsymbol{r})$ (or, equivalently, false conjunctions $A(\boldsymbol{p}, \boldsymbol{q}) \wedge \neg B(\boldsymbol{p}, \boldsymbol{r})$), which by Craig's interpolation theorem [12] possess interpolants $C(\boldsymbol{p})$ in the common variables $\boldsymbol{p}$. Such interpolants, even though they exist, may not be of polynomial size [21].

however, it may be the case that we can always efficiently extract such interpolants from a proof of the implication in a particular proof system P, and in his case, the system P is said to admit feasible interpolation. If we know that particular class of formulas does not admit small interpolants (either unconditional or under suitable assumptions), then there cannot exist small proofs of the formulas in the system P. Here we show that this feasible interpolation theorem holds for arbitrarily quantified formulas $A(p, q)$ and $B(p, r)$ above, when the common variables p are existentially quantified before all other variables.

2. New Lower Bounds for QBF Systems. As our second contribution we exhibit new hard formulas for QBF resolution systems. It is fair to say that we are currently quite short of hard examples: research so far has mainly considered formulas of Kleine Büning et al. [17] and their modifications [3,7], a principle from [16], and parity formulas recently introduced in [7]. This again is in sharp contrast with classical proof complexity where a wealth of different combinatorial principles as well as random formulas are known to be hard for resolution.

Our new hard formulas are QBF contradictions formalising the easy and appealing fact that a graph cannot both have and not have a k-clique. The trick is that in our formulation, each interpolant for these formulas has to solve the k-clique problem. Using our interpolation theorem together with the exponential lower bound for the monotone circuit complexity of clique [1], we obtain exponential lower bounds for the clique-co-clique formulas in all CDCL and expansion-based QBF resolution systems.

We remark that conceptually our clique-co-clique formulas are different from and indeed simpler than the clique-colour formulas used for the interpolation technique in classical proof systems. This is due to the greater expressibility of QBF. Indeed it is not clear how the clique-co-clique principle could even be formulated succinctly in propositional logic.

3. Comparison to Strategy Extraction. On a conceptual level, we uncover a tight relationship between feasible interpolation and strategy extraction. Strategy extraction is a very desirable property of QBF proof systems and is known to hold for the main resolution-based systems, cf. eg. [6]. From a refutation of a false QBF, a winning strategy for the universal player can be efficiently extracted.

Like feasible interpolation, the lower bound technique based on strategy extraction from [7] also transfers circuit lower bounds to proof size bounds. However, instead of monotone circuit bounds as in the case of feasible interpolation, the strategy extraction technique imports AC^0 circuit lower bounds. Here we show that each feasible interpolation problem can be transformed into a strategy extraction problem, where the interpolant corresponds to the winning strategy of the universal player on the first universal variable. This clarifies that indeed feasible interpolation can be viewed as a special case of strategy extraction.

Organisation of the Paper. In Sect. 2 we review definitions and relations of relevant QBF calculi. In Sect. 3 we start by recalling the overall idea for feasible interpolation and show interpolation theorems for the strongest CDCL-based system $\mathsf{LQU^+\text{-}Res}$ as well as the strongest expansion-based proof system $\mathsf{IRM\text{-}calc}$. This implies feasible interpolation for all QBF resolution-based systems. Further we show that all these systems even admit monotone feasible interpolation. In Sect. 4 we obtain the new lower bounds for the clique-co-clique formulas. Section 5 reformulates interpolation as a strategy extraction problem.

2   Preliminaries

A literal is a boolean variable or its negation. We say that literals x and $\neg x$ are complementary. A *clause* is a disjunction of literals and a *term* is a conjunction of literals. The empty clause is denoted by $\square$, and is semantically equivalent to false. A formula in *conjunctive normal form* (CNF) is a conjunction of clauses. For a literal $l = x$ or $l = \neg x$, we write $\text{var}(l)$ for x and extend this notation to $\text{var}(C)$ for a clause C. For any partial assignment α and clause C, we write $C|_\alpha$ for the clause obtained after applying the partial assignment α to C.

Quantified Boolean Formulas (QBFs) extend propositional logic with boolean quantifiers with the standard semantics: $\forall x.F$ is satisfied by the same truth assignments as $F|_{x=0} \wedge F|_{x=1}$ and $\exists x.F$ as $F|_{x=0} \vee F|_{x=1}$. We assume that QBFs are in *closed prenex form* with a CNF matrix, i.e., we consider the form $\mathcal{Q}_1 X_1 \ldots \mathcal{Q}_k X_k.\phi$, where $\mathcal{Q}_i \in \{\exists, \forall\}$, $\mathcal{Q}_i \neq \mathcal{Q}_{i+1}$, and X_i are pairwise disjoint sets of variables. The formula ϕ is in CNF and is defined only on variables $X_1 \cup \ldots \cup X_k$. The propositional part ϕ is called the *matrix* and the rest the *prefix*. If $x \in X_i$, we say that x is at *level* i and write $\text{lv}(x) = i$; we write $\text{lv}(l)$ for $\text{lv}(\text{var}(l))$. The *index* $\text{ind}(x)$ provides more detailed information on the position of x in the prefix, i.e. all variables are indexed by $1, \ldots, n$ from left to right.

$$\frac{}{C}\ (\text{Ax}) \qquad \frac{D \cup \{u\}}{D}\ (\forall\text{-Red}) \qquad \frac{D \cup \{u^*\}}{D}\ (\forall\text{-Red}^*)$$

C is a clause in the matrix. Literal u is universal and $\text{lv}(u) \geq \text{lv}(l)$ for all $l \in D$.

$$\frac{C_1 \cup U_1 \cup \{x\} \qquad C_2 \cup U_2 \cup \{\neg x\}}{C_1 \cup C_2 \cup U}\ (\text{Res})$$

We consider four instantiations of the Res-rule:

S∃R: x is existential.
If $z \in C_1$, then $\neg z \notin C_2$. $U_1 = U_2 = U = \emptyset$.
S∀R: x is universal. Other conditions same as S∃R.
L∃R: x is existential.
If $l_1 \in C_1, l_2 \in C_2$, $\text{var}(l_1) = \text{var}(l_2) = z$ then $l_1 = l_2 \neq z^*$. U_1, U_2 contain only universal literals with $\text{var}(U_1) = \text{var}(U_2)$. $\text{ind}(x) < \text{ind}(u)$ for each $u \in \text{var}(U_1)$.
If $w_1 \in U_1, w_2 \in U_2, \text{var}(w_1) = \text{var}(w_2) = u$ then $w_1 = \neg w_2$, $w_1 = u^*$ or $w_2 = u^*$. $U = \{u^* \mid u \in \text{var}(U_1)\}$.
L∀R: x is universal. Other conditions same as L∃R.

Fig. 1. The rules of CDCL-based proof systems

A QBF $\mathcal{Q}_1 X_1 \ldots \mathcal{Q}_k X_k.\phi$ can be thought of as a *game* between the *universal* and the *existential player*. In the i-th step of the game, the player $\mathcal{Q}_i$ assigns values to all the variables X_i. The existential player wins the game iff the matrix ϕ

valuates to 1 under the assignment constructed in the game; otherwise the universal player wins. Given a universal variable u with index i, a *strategy for u* a function from all variables of index $< i$ to $\{0,1\}$. A QBF is false iff there xists a *winning strategy* for the universal player, i.e. if the universal player has strategy for all universal variables that wins any possible game [14].

Resolution-Based Calculi for QBF. We now give a brief overview of the nain existing resolution-based calculi for QBF. We start by describing the proof ystems modelling *CDCL-based QBF solving*; their rules are summarized in ig. 1. The most basic and important system is *Q-resolution (Q-Res)* by Kleine Büning et al. [17]. It is a resolution-like calculus that operates on QBFs in prenex orm with CNF matrix. In addition to the axioms, Q-Res comprises the resolution ule S∃R and universal reduction ∀-Red (cf. Fig. 1).

$$\overline{\left\{ x^{[\tau]} \mid x \in C, x \text{ is exist.} \right\}} \text{ (Ax)}$$

C is a non-tautological clause from the matrix.
$\tau = \{0/u \mid u \text{ is universal in } C\}$, where the notation $0/u$ for literals u is shorthand for $0/x$ if $u = x$ and $1/x$ if $u = \neg x$.

$$\frac{C}{\text{inst}(\tau, C)} \text{ (Instantiation)}$$

τ is an assignment to universal variables with $\text{rng}(\tau) \subseteq \{0,1\}$.

$$\frac{x^{\tau \cup \xi} \vee C_1 \qquad \neg x^{\tau \cup \sigma} \vee C_2}{\text{inst}(\sigma, C_1) \cup \text{inst}(\xi, C_2)} \text{ (Res)}$$

$\text{dom}(\tau)$, $\text{dom}(\xi)$ and $\text{dom}(\sigma)$ are mutually disjoint. $\text{rng}(\tau) = \{0,1\}$

$$\frac{C \vee b^\mu \vee b^\sigma}{C \vee b^\xi} \text{ (Merging)}$$

$\text{dom}(\mu) = \text{dom}(\sigma)$.
$\xi = \{c/u \mid c/u \in \mu, c/u \in \sigma\} \cup \{*/u \mid c/u \in \mu, d/u \in \sigma, c \neq d\}$

Fig. 2. The rules of IRM-calc [6]

Long-distance resolution (LD-Q-Res) appears originally in the work of Zhang and Malik [26] and was formalized into a calculus by Balabanov and Jiang [2]. It merges complementary literals of a universal variable u into the special literal u^*. LD-Q-Res uses the rules L∃R, ∀-Red and ∀-Red* (cf. Fig. 1).

QU-resolution (QU-Res) [25] removes the restriction from Q-Res that the resolved variable must be an existential variable and allows resolution of universal variables. The rules of QU-Res are S∃R, S∀R and ∀-Red.

LQU+-Res [3] extends LD-Q-Res and QU-Res by allowing short and long distance resolution over arbitrary pivots; however, the pivot is never a merged literal z^*. LQU+-Res uses the rules L∃R, L∀R, ∀-Red and ∀-Red*.

The second type of calculi models *expansion-based QBF solving*. These calculi are based on *instantiation* of universal variables: ∀Exp+Res [16], IR-calc, and IRM-calc [6]. All these calculi operate on clauses that comprise only existential variables from the original QBF, which are additionally *annotated* by a substitution to some universal variables,

e.g. $\neg x^{0/u_1 1/u_2}$. For any annotated literal l^σ, the substitution σ must not make assignments to variables at a higher quantification level than l, i.e. if $u \in \text{dom}(\sigma)$, then u is universal and $\text{lv}(u) < \text{lv}(l)$. To preserve this invariant, we use the *auxiliary notation* $l^{[\sigma]}$, which for an existential literal l and an assignment σ to

the universal variables filters out all assignments that are not permitted, i.e
$l^{[\sigma]} = l^{\{c/u \in \sigma \mid \mathrm{lv}(u) < \mathrm{lv}(l)\}}$.

On partial assignments we use auxiliary operations of *completion* and *instantiation*. For assignments τ and μ, we write $\tau \overset{\vee}{} \mu$ for the assignment σ defined as $\sigma(x) = \tau(x)$ if $x \in \mathrm{dom}(\tau)$, otherwise $\sigma(x) = \mu(x)$ if $x \in \mathrm{dom}(\mu)$. The operation $\tau \overset{\vee}{} \mu$ is called *completion* as μ provides values for variables not defined in τ. The operation is associative and therefore we can omit parentheses.

For an assignment τ and an annotated clause C, the function $\mathrm{inst}(\tau, C)$ returns the annotated clause $\{l^{[\sigma \overset{\vee}{} \tau]} \mid l^\sigma \in C\}$. The system IR-calc uses annotations and instantiation [6]. The calculus IRM-calc further extends IR-calc by enabling annotations containing $*$. The rules of IRM-calc are shown in Fig. 2. The symbol $*$ may be introduced by the merge rule, e.g. by collapsing $x^{0/u} \vee x^{1/u}$ into $x^{*/u}$.

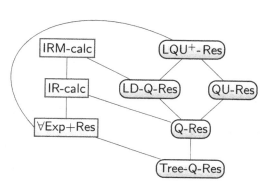

Fig. 3. The simulation order of QBF resolution systems. Systems on the left correspond to expansion-based solving, whereas the systems on the right are CDCL based.

The simulation order of QBF resolution systems is shown in Fig. 3. All proof systems have been exponentially separated (cf. [7] and references therein).

Definition 1. *For clauses C, D we write $C \preceq D$ if for any literal $l \in C$ we have $l \in D$ or $l^* \in D$ and for any $l^* \in C$ we have $l^* \in D$.*

For annotations τ and σ we say that $\tau \preceq \sigma$ if $\mathrm{dom}(\tau) = \mathrm{dom}(\sigma)$ and for any $c/u \in \tau$ we have $c/u \in \sigma$ or $*/u \in \sigma$ and for any $*/u \in \tau$ we have $*/u \in \sigma$.

If C, D are annotated clauses, we write $C \preceq D$ if there is an injective function $f : C \hookrightarrow D$ such that for all $l^\tau \in C$ we have $f(l^\tau) = l^\sigma$ with $\tau \preceq \sigma$.

3 Feasible (Monotone) Interpolation

In this section we show that feasible interpolation and feasible monotone interpolation hold for LQU⁺-Res and IRM-calc. We adapt the technique first used by Pudlák [22] to re-prove and generalise the result of Krajíček [18].

3.1 The Setting

Consider a false QBF sentence $\mathcal{F}$ of the form $\exists \boldsymbol{p} \mathcal{Q} \boldsymbol{q} \mathcal{Q} \boldsymbol{r}. [A(\boldsymbol{p}, \boldsymbol{q}) \wedge B(\boldsymbol{p}, \boldsymbol{r})]$, where $\boldsymbol{p}$, $\boldsymbol{q}$, and $\boldsymbol{r}$ are mutually disjoint sets of propositional variables, $A(\boldsymbol{p}, \boldsymbol{q})$ is a CNF formula on variables $\boldsymbol{p}$ and $\boldsymbol{q}$, and $B(\boldsymbol{p}, \boldsymbol{r})$ is a CNF formula on variables $\boldsymbol{p}$ and $\boldsymbol{r}$. Thus $\boldsymbol{p}$ are the common variables between them. The $\boldsymbol{q}$ and $\boldsymbol{r}$ variables can be quantified arbitrarily, with any number of quantification levels. The sentence is equivalent to the following, not in prenex form $\exists \boldsymbol{p} [\mathcal{Q} \boldsymbol{q}. A(\boldsymbol{p}, \boldsymbol{q}) \wedge \mathcal{Q} \boldsymbol{r}. B(\boldsymbol{p}, \boldsymbol{r})]$.

Definition 2. *Let $\mathcal{F}$ be a false QBF of the form $\exists p Q q Q r. [A(p, q) \wedge B(p, r)]$. An interpolation circuit for $\mathcal{F}$ is a boolean circuit G such that on every $0, 1$ assignment a for p we have*

$$G(a) = 0 \implies Q q. A(a, q) \text{ is false.}$$
$$G(a) = 1 \implies Q r. B(a, r) \text{ is false.}$$

We say that a QBF proof system S has feasible interpolation *if for any S-proof π of a QBF $\mathcal{F}$ of the form above, we can extract from π an interpolation circuit for $\mathcal{F}$ of size polynomial in the size of π.*

We say that S has monotone feasible interpolation *if the following holds: in the same setting as above, if p appears only positively in $A(p, q)$, then we can extract from π a monotone interpolation circuit for $\mathcal{F}$.*

As our main results, we show that both LQU$^+$-Res and IRM-calc have monotone feasible interpolation. We first outline the general idea.

Proof Idea. Fix a proof system $S \in \{$LQU$^+$-Res, IRM-calc$\}$ and an S-proof π of $\mathcal{F}$. Consider the following definition of a q-clause and an r-clause.

Definition 3. *We call a clause C in π a q-clause (resp. r-clause), if C contains only variables p, q (resp. p, r). We also call C a q-clause (resp. r-clause), if C contains only p variables, but all its descendant clauses in the proof π (all clauses with a directed path to C in π) are q (resp. r)-clauses. In the case of IRM-calc the annotations are not considered and can be from either set.*

From π we construct a circuit C_π with the p-variables as inputs: For each node u with clause C_u in the proof π, associate a gate g_u (or a constant-size circuit) in the circuit C_π. Next, we inductively construct, for any assignment a to the p variables, another proof-like structure $\pi'(a)$. For each node u with clause C_u in the proof π, associate a clause $C'_{u,a}$ in the structure $\pi'(a)$. Finally, we obtain $\pi''(a)$ from the structure $\pi'(a)$ by instantiating p variables to the assignment a and doing some pruning, and show that $\pi''(a)$ is a valid proof in S. We then find that if $C_\pi(a) = 0$, then $\pi''(a)$ uses only q-clauses and thus is a refutation of $Q q. A(a, q)$, and if $C_\pi(a) = 1$, then $\pi''(a)$ uses only r-clauses and thus is a refutation of $Q r. B(a, r)$. Thus C_π is the desired interpolant circuit.

More precisely, we show by induction on the height of u in π (that is, the length of the longest path to u from a source node in π) that:

1. $C'_{u,a} \preceq C_u$.
2. $g_u(a) = 0 \implies C''_{u,a}$ is a q-clause and can be obtained from the clauses of $A(a, q)$ alone using the rules of S.
3. $g_u(a) = 1 \implies C''_{u,a}$ is an r-clause and can be obtained from the clauses of $B(a, r)$ alone using the rules of S.

From the above, we have the following conclusion. Let r be the root of π. Then on any assignment a to the p variables we have:

(1) $C'_{r,a} \preceq C_r = \square$, so $C'_{r,a} = \square$. Therefore, $C''_{r,a} = C'_{r,a}|a = \square$.
(2) $g_r(a) = 0 \implies \square$ is a q-clause and can be obtained from the clauses of $A(a, q)$ alone using the rules of system S. Hence by soundness of S, $Q q. A(a, q)$ is false.

(3) $g_r(a) = 1 \implies \square$ is an r-clause and can be obtained from the clauses of $B(a, r)$ alone using the rules of system S. Hence by soundness of S $Qq.B(a, r)$ is false.

Thus g_r, the output gate of the circuit, computes an interpolant.

When $\mathcal{F}$ has only existential quantification, π is a classical resolution proof, and this is exactly the interpolant computed by Pudlák's method in [22]. The challenge here is to construct π' and π'' appropriately when the stronger proof systems are used for general QBF, while maintaining the inductive invariants.

3.2   Interpolants from LQU⁺-Res Proofs

We now implement the idea described above for LQU⁺-Res.

Theorem 1. *LQU⁺-Res has feasible interpolation.*

Proof Sketch. As mentioned in the proof outline, for an LQU⁺-Res proof π of $\mathcal{F}$ we describe the circuit C_π with input p, and the proof-like structure $\pi'(a)$. All the claims given in the outline can be established using induction on the height of u, completing the proof.

The DAG underlying the circuit C_π and the structure $\pi'(a)$ is exactly the same as the DAG underlying the proof π. For each node u with clause C_u in π we associate a gate g_u and clause $C'_{u,a}$ as follows:

1. u is a leaf node. Then g_u is a constant gate, valued 0 if $C_u \in A(p, q)$ and valued 1 if $C_u \in B(p, r)$, and $C'_{u,a} = C_u$.
2. u is a universal reduction node with child v, where a literal l with $\mathrm{var}(l) = x$ is reduced. Then g_u is a no-operation gate, and $C'_{u,a} = C'_{v,a} \setminus \{x, \neg x, x^*\}$.
3. u corresponds to a resolution step with an existential variable $x \in p$ as pivot. Let v and w be the children of u, and $C_v = C_1 \vee x$, $C_w = C_2 \vee \neg x$, $C_u = C$. In this case, g_u is the selector "gate" $\mathrm{sel}(x, g_v, g_w)$, where $\mathrm{sel}(x, a, b) = (\neg x \wedge a) \vee (x \wedge b)$. Note that all the variables in p are existential variables. We define $C'_{u,a}$ as $C'_{v,a} \setminus \{x\}$ or $C'_{w,a} \setminus \{\neg x\}$ to correspond with the selector gate.
4. u corresponds to a resolution step with a variable $x \in q$ as pivot. Then g_u is an OR gate. To define $C'_{u,a}$, we consider several cases. Let v and w be the children of u, with $C_v = C_1 \vee U_1 \vee x$, $C_w = C_2 \vee U_2 \vee \neg x$, $C_u = C_1 \vee C_2 \vee U$.
   (a) If $g_v(a) = 1$, then $C'_{u,a} = C'_{v,a}$.
   (b) Else if $g_w(a) = 1$, then $C'_{u,a} = C'_{w,a}$.
   (c) Else if $x \notin C'_{v,a}$, then $C'_{u,a} = C'_{v,a}$.
   (d) Else if $\neg x \notin C'_{w,a}$, then $C'_{u,a} = C'_{w,a}$.
   (e) Else $C'_{u,a}$ is obtained by resolving $C'_{v,a}$ and $C'_{w,a}$ on x.
5. u corresponds to a resolution step with a variable $x \in r$ as pivot. Then g_u is an AND gate. The definition of $C'_{u,a}$ is dual to the above case. $\square$

3.3   Interpolants from IRM-calc Proofs

We now establish the interpolation theorem for the expansion-based calculi, following the same overall idea described in Sect. 3.1.

Theorem 2. *IRM-calc has feasible interpolation.*

Proof Sketch. We construct the circuit in the same way as in Theorem 1 for the resolution rules and unary rules. The structure π' is constructed similarly; we copy clauses or use applicable resolution in almost the same situations. However, we now have to respect the definition of $\preceq$ for annotated clauses. We use the instantiation rule to ensure that we have the same domains as in the original proof as required for $C'_{u,a} \preceq C_u$. However, even if these domains are identical the values may differ slightly (as we cannot use the instantiation rule for $*$) and this may cause more literals to appear ($l^{0/u}$ and $l^{*/u}$ may appear where originally there was just $l^{*/u}$). We adjust for this by using the merging rule, which will give us the injection as required for $\preceq$.   □

3.4   Monotone Interpolation

To transfer known circuit lower bounds into size of proof bounds, we need a monotone version of the previous interpolation theorems, which we prove next.

Theorem 3. *LQU⁺-Res and IRM-calc have monotone feasible interpolation.*

Proof. In previous subsections, we have shown that the circuit $C_\pi(p)$ is a correct interpolant for the QBF sentence $\mathcal{F}$. That is, if $C_\pi(p) = 0$ then $\mathcal{Q}q.A(a,q)$ is false, and if $C_\pi(p) = 1$ then $\mathcal{Q}r.B(a,r)$ is false.

However, if p occurs only positively in $A(p,q)$ then we construct a monotone circuit $C_\pi^{mon}(p)$ such that, on every $0,1$ assignment a to p we have

$$C_\pi^{mon}(a) = 0 \implies \mathcal{Q}q.A(a,q) \text{ is false, and}$$
$$C_\pi^{mon}(a) = 1 \implies \mathcal{Q}r.B(a,r) \text{ is false.}$$

We obtain $C_\pi^{mon}(p)$ from $C_\pi(p)$ by replacing all selector gates $g_u = \mathsf{sel}(x, g_v, g_w)$ by the following monotone ternary connective: $g_u = (x \vee g_v) \wedge g_w$ where nodes v and w are the children of u in π. We also change the proof-like structure $\pi'(a)$; the construction is the same as before except that at p-resolution nodes, the rule for fixing $C'_{u,a}$ is also changed to reflect the monotone function used instead.

More precisely, the functions $\mathsf{sel}(x, g_v, g_w)$ and $g_u = (x \vee g_v) \wedge g_w$ differ only when $x = 0$, $g_v(a) = 1$, and $g_w(a) = 0$. We set $C'_{u,a}$ to $C'_{w,a} \setminus \{\neg x\}$ if $x = 1$ or if $x = 0$, $g_v(a) = 1$ and $g_w(a) = 0$, and to $C'_{v,a} \setminus \{x\}$ otherwise.

We need to show that at the differing setting, the inductive statements relating the modified $C'_{u,a}$, $g_u(a)$ and $C''_{u,a}$ continue to hold. The relation $C''_{u,a} \preceq C_u$ holds by induction. Now consider the gate values.

We know by induction that $g_v(a) = 1$ means that $C''_{v,a}$ is an r-clause and can be derived from $B(a,r)$ alone. When $x = 0$, $C'_{u,a} = C'_{v,a}$ and the selector gate will output the value of $g_v(\boldsymbol{a})$ which is a 1. Hence $C''_{u,\boldsymbol{a}}$ is an $\boldsymbol{r}$-clause. However observe that at this setting, $g_w(\boldsymbol{a}) = 0$, which means by induction that $C''_{w,\boldsymbol{a}}$ is a $\boldsymbol{q}$-clause and can be derived using $A(\boldsymbol{a}, \boldsymbol{q})$ clauses alone via the appropriate proof system. Thus by our assumption about $\boldsymbol{p}$ variables appearing only positively in A, the clause $C'_{w,\boldsymbol{a}}$ does not contain $\neg x$. Thus we can safely assign $C'_{u,\boldsymbol{a}} = C'_{w,\boldsymbol{a}}$. This completes the proof. □

4 New Lower Bounds for IRM-calc and LQU⁺-Res

We now apply our interpolation theorems to obtain new exponential lower bounds for a new class of QBFs. The lower bound will be directly transferred from the following monotone circuit lower bound for the problem $\text{CLIQUE}(n, k)$, asking whether a given graph with n nodes has a clique of size k.

Theorem 4 (Alon, Boppana 87 [1]). *All monotone circuits that compute* $\text{CLIQUE}(n, n/2)$ *are of exponential size.*

We now build the QBF. Fix an integer n (indicating the number of vertices of the graph) and let $\boldsymbol{p}$ be the set of variables $\{p_{uv} \mid 1 \le u < v \le n\}$. An assignment to $\boldsymbol{p}$ picks a set of edges, and thus an n-vertex graph. Let $\boldsymbol{q}$ be the set of variables $\{q_{iu} \mid i \in [\frac{n}{2}], u \in [n]\}$. We use the following clauses.

$$
\begin{aligned}
C_i &= q_{i1} \vee \cdots \vee q_{in} && \text{for } i \in [\tfrac{n}{2}]\\
D_{i,j,u} &= \neg q_{iu} \vee \neg q_{ju} && \text{for } i,j \in [\tfrac{n}{2}], i < j \text{ and } u \in [n]\\
E_{i,u,v} &= \neg q_{iu} \vee \neg q_{iv} && \text{for } i \in [\tfrac{n}{2}] \text{ and } u,v \in [n], u < v\\
F_{i,j,u,v} &= \neg q_{iu} \vee \neg q_{jv} \vee p_{uv} && \text{for } i,j \in [\tfrac{n}{2}], i < j \text{ and } u \ne v \in [n].
\end{aligned}
$$

We can now express $\text{CLIQUE}(n, n/2)$ as a polynomial-size QBF $\exists \boldsymbol{q}.A_n(\boldsymbol{p}, \boldsymbol{q})$:

$$
A_n(\boldsymbol{p}, \boldsymbol{q}) = \bigwedge_{i \in [\frac{n}{2}]} C_i \wedge \bigwedge_{i<j, u \in [n]} D_{i,j,u} \wedge \bigwedge_{i \in [\frac{n}{2}], u < v} E_{i,u,v} \wedge \bigwedge_{i<j, u \ne v} F_{i,j,u,v}.
$$

The F-part verifies that the size-$n/2$ subset picked out by the C, D, E-parts is a clique. Note that the edge variables $\boldsymbol{p}$ appear monotone in $A_n(\boldsymbol{p}, \boldsymbol{q})$.

Likewise co-$\text{CLIQUE}(n, n/2)$ can be written as a QBF $\forall \boldsymbol{r_1} \exists \boldsymbol{r_2}.B_n(\boldsymbol{p}, \boldsymbol{r_1}, \boldsymbol{r_2})$ of polynomial size. To construct this we use a polynomial-size circuit that checks whether the nodes specified by $\boldsymbol{r_1}$ fail to form a clique in the graph given by $\boldsymbol{p}$. We then use existential variables $\boldsymbol{r_2}$ for the gates of the circuit and can then form a CNF $B_n(\boldsymbol{p}, \boldsymbol{r_1}, \boldsymbol{r_2})$ that represents the circuit computation.

Now we can form a sequence of false QBFs, stating that the graph encoded in $\boldsymbol{p}$ both has a clique of size $n/2$ (as witnessed by $\boldsymbol{q}$) and likewise does not have such a clique as expressed in the B part:

$$
\Phi_n(\boldsymbol{p}, \boldsymbol{q}, \boldsymbol{r}) = \exists \boldsymbol{p} \exists \boldsymbol{q} \forall \boldsymbol{r_1} \exists \boldsymbol{r_2}.A_n(\boldsymbol{p}, \boldsymbol{q}) \wedge B_n(\boldsymbol{p}, \boldsymbol{r_1}, \boldsymbol{r_2}).
$$

This formula has the unique interpolant $\text{CLIQUE}(n, n/2)(\boldsymbol{p})$. But since all monotone circuits for this are of exponential size by Theorem 4 and monotone circuits of size polynomial in IRM-calc and LQU⁺-Res proofs can be extracted by Theorem 3, all such proofs must be of exponential size, yielding:

Theorem 5. *The QBFs $\Phi_n(p, q, r)$ require exponential-size proofs in IRM-calc and LQU$^+$-Res.*

Feasible Interpolation Vs Strategy Extraction

Recall the two player game semantics of a QBF: every false QBF has a winning strategy for the universal player, where the strategy for each variable depends only on the variables played before. We now explain the relation between strategy extraction — one of the main paradigms for QBF systems — and feasible interpolation. In Sect. 3 we studied QBFs of the form $\mathcal{F} = \exists p \mathcal{Q} q \mathcal{Q} r. [A(p, q) \wedge B(p, r)]$. If we add a common universal variable b we can change it to an equivalent QBF

$$\mathcal{F}^b = \exists p \, \forall b \, \mathcal{Q} q \, \mathcal{Q} r. [(A(p, q) \vee b) \wedge (B(p, r) \vee \neg b)].$$

If $\mathcal{F}$ is false, then also $\mathcal{F}^b$ is false and thus the universal player has a winning strategy, including a strategy for $b = \sigma(p)$ for the common universal variable b.

Remark 1. Every winning strategy $\sigma(p)$ for b is an interpolant for $\mathcal{F}$, i.e., for every $0, 1$ assignment a of p, $\sigma(a) = 0 \implies \mathcal{Q} q. A(a, q)$ is false, and $\sigma(a) = 1 \implies \mathcal{Q} r. B(a, r)$ is false.

Theorem 6. *Fix a proof system $S \in \{LQU^+\text{-}Res, IRM\text{-}calc\}$.*

1. *From each S-refutation π of $\mathcal{F}^b$ we can extract in polynomial time a boolean circuit for $\sigma(p)$, i.e., the part of the winning strategy for variable b.*
2. *If in the same setting as above for $\mathcal{F}^b$, the variables p appear only positively in $A(p, q)$, then we can extract a monotone boolean circuit for $\sigma(p)$ from an S-refutation π of $\mathcal{F}^b$ in polynomial time (in the size of π).*

Proof Sketch. As we can compute the (monotone) interpolant when b is absent, we use the same proof with a few modifications for the new formula.

We first change the definition of q and r-clauses to allow for b and $\neg b$ literals: We call any clause in the proof a q-clause (resp. r-clause) if it contains only variables p, q or literal b (resp. p, r or literal $\neg b$). We retain the inheritance property for clauses only containing p variables. We also slightly modify the invariants to include the new definitions. Additionally we make a small change to the first invariant: we now claim that $C'_{u,a} \backslash \{b, \neg b\} \preceq C_u$.

While constructing the circuit C_π, we also need to consider a resolution step on the common universal variable b. Here we arbitrarily pick one of v or w. For example here we pick v and let $g_u = g_v$, disregarding the input from g_w.

Similarly while constructing π', we need to consider a resolution step on b. Let C_u be obtained by resolving C_v and C_w on b, with $b \in C_v$ and $\neg b \in C_w$. To find $C'_{u,a}$ we look at our choice of wiring in the circuit construction. If g_u is wired to g_v ($g_u = g_v$) then we take $C'_{u,a}$ to equal $C'_{v,a}$.

The rest of the proof, showing that all the (modified) invariants hold, goes through by induction as before, and allows us to conclude that g_r, the output gate of the circuit, computes $\sigma(p)$. □

As a corollary, the versions $\Phi_n^b(p, q, r)$ of the formulas from Sect. 4 also require exponential-size proofs in IRM-calc and LQU$^+$-Res.

Acknowledgements. We thank Pavel Pudlák and Mikoláš Janota for helpfu discussions on the relation between feasible interpolation and strategy extractior during the recent Dagstuhl Seminar 'Optimal Algorithms and Proofs' (14421).

References

1. Alon, N., Boppana, R.B.: The monotone circuit complexity of boolean functions Combinatorica **7**(1), 1–22 (1987)
2. Balabanov, V., Jiang, J.-H.R.: Unified QBF certification and its applications. Formal Methods in System Design **41**(1), 45–65 (2012)
3. Balabanov, V., Widl, M., Jiang, J.-H.R.: QBF resolution systems and their proof complexities. In: Sinz, C., Egly, U. (eds.) SAT 2014. LNCS, vol. 8561, pp. 154–169. Springer, Heidelberg (2014)
4. Ben-Sasson, E., Wigderson, A.: Short proofs are narrow - resolution made simple. Journal of the ACM **48**(2), 149–169 (2001)
5. Benedetti, M., Mangassarian, H.: QBF-based formal verification: Experience and perspectives. JSAT **5**(1–4), 133–191 (2008)
6. Beyersdorff, O., Chew, L., Janota, M.: On unification of QBF resolution-based calculi. In: Csuhaj-Varjú, E., Dietzfelbinger, M., Ésik, Z. (eds.) MFCS 2014, Part II. LNCS, vol. 8635, pp. 81–93. Springer, Heidelberg (2014)
7. Beyersdorff, O., Chew, L., Janota, M.: Proof complexity of resolution-based QBF calculi. In : STACS, pp. 76–89 (2015)
8. Beyersdorff, O., Chew, L., Sreenivasaiah, K.: A game characterisation of tree-like Q-resolution size. In: Dediu, A.-H., Formenti, E., Martín-Vide, C., Truthe, B. (eds.) LATA 2015. LNCS, vol. 8977, pp. 486–498. Springer, Heidelberg (2015)
9. Beyersdorff, O., Kullmann, O.: Unified characterisations of resolution hardness measures. In: Sinz, C., Egly, U. (eds.) SAT 2014. LNCS, vol. 8561, pp. 170–187. Springer, Heidelberg (2014)
10. Bonet, M.L., Domingo, C., Gavaldà, R., Maciel, A., Pitassi, T.: Non-automatizability of bounded-depth Frege proofs. Computational Complexity **13**(1–2), 47–68 (2004)
11. Maria Luisa Bonet: Toniann Pitassi, and Ran Raz. On interpolation and automatization for Frege systems. SIAM Journal on Computing **29**(6), 1939–1967 (2000)
12. Craig, W.: Three uses of the Herbrand-Gentzen theorem in relating model theory and proof theory. The Journal of Symbolic Logic **22**(3), 269–285 (1957)
13. Egly, U., Kronegger, M., Lonsing, F., Pfandler, A.: Conformant planning as a case study of incremental QBF solving (2014). CoRR, abs/1405.7253
14. Goultiaeva, A., Van Gelder, A., Bacchus, F.: A uniform approach for generating proofs and strategies for both true and false QBF formulas. In: IJCAI, pp. 546–553 (2011)
15. Hrubeš, P.: On lengths of proofs in non-classical logics. Annals of Pure and Applied Logic **157**(2–3), 194–205 (2009)
16. Janota, M., Marques-Silva, J.: Expansion-based QBF solving versus Q-resolution. Theor. Comput. Sci. **577**, 25–42 (2015)
17. Hans Kleine Büning: Marek Karpinski, and Andreas Flögel. Resolution for quantified Boolean formulas. Inf. Comput. **117**(1), 12–18 (1995)
18. Krajíček, J.: Interpolation theorems, lower bounds for proof systems and independence results for bounded arithmetic. J. Symb. Log. **62**(2), 457–486 (1997)

19. Krajíček, J.: Forcing with random variables and proof complexity, vol. 382, Lecture Note Series. London Mathematical Society (2011)
20. Krajíček, J., Pudlák, P.: Some consequences of cryptographical conjectures for S_2^1 and EF. Information and Computation **140**(1), 82–94 (1998)
21. Mundici, D.: Tautologies with a unique Craig interpolant, uniform vs. nonuniform complexity. Annals of Pure and Applied Logic **27**, 265–273 (1984)
22. Pudlák, P.: Lower bounds for resolution and cutting planes proofs and monotone computations. The Journal of Symbolic Logic **62**(3), 981–998 (1997)
23. Pudlák, P.: Proofs as games. American Math. Monthly, pp. 541–550 (2000)
24. Rintanen, J.: Asymptotically optimal encodings of conformant planning in QBF. In: AAAI, pp. 1045–1050. AAAI Press (2007)
25. Van Gelder, A.: Contributions to the theory of practical quantified boolean formula solving. In: Milano, M. (ed.) CP 2012. LNCS, vol. 7514, pp. 647–663. Springer, Heidelberg (2012)
26. Zhang, L., Malik, S.: Conflict driven learning in a quantified Boolean satisfiability solver. In: ICCAD, pp. 442–449 (2002)

Simultaneous Approximation of Constraint Satisfaction Problems

Amey Bhangale[1]([✉]), Swastik Kopparty[2], and Sushant Sachdeva[3]

[1] Department of Computer Science, Rutgers University, New Brunswick, USA
amey.bhangale@rutgers.edu
[2] Department of Mathematics and Department of Computer Science,
Rutgers University, New Brunswick, USA
swastik.kopparty@rutgers.edu
[3] Department of Computer Science, Yale University, New Haven, USA
sachdeva@cs.yale.edu

Abstract. Given k collections of 2SAT clauses on the same set of variables V, can we find one assignment that satisfies a large fraction of clauses from *each* collection? We consider such *simultaneous* constraint satisfaction problems, and design the first nontrivial approximation algorithms in this context.

Our main result is that for every CSP $\mathcal{F}$, for $k < \tilde{O}(\log^{1/4} n)$, there is a polynomial time constant factor *Pareto* approximation algorithm for k simultaneous MAX-$\mathcal{F}$-CSP instances. Our methods are quite general, and we also use them to give an improved approximation factor for simultaneous MAX-w-SAT (for $k < \tilde{O}(\log^{1/3} n)$). In contrast, for $k = \omega(\log n)$, no nonzero approximation factor for k simultaneous MAX-$\mathcal{F}$-CSP instances can be achieved in polynomial time (assuming the Exponential Time Hypothesis).

These problems are a natural meeting point for the theory of constraint satisfaction problems and multiobjective optimization. We also suggest a number of interesting directions for future research.

1 Introduction

The theory of approximation algorithms for constraint satisfaction problems (CSPs) is a very central and well developed part of modern theoretical computer science. Its study has involved fundamental theorems, ideas, and problems such as the PCP theorem, linear and semidefinite programming, randomized rounding, the Unique Games Conjecture, and deep connections between them [3, 4, 15, 21, 29, 30].

Amey Bhangale—Research supported in part by NSF grant CCF-1253886.
Swastik Kopparty—Research supported in part by a Sloan Fellowship and NSF grant CCF-1253886.
Sushant Sachdeva—Research supported by the NSF grants CCF-0832797, CCF-1117309, and Daniel Spielman's & Sanjeev Arora's Simons Investigator Grants. Part of this work was done when this author was at the Simons Institute for the Theory of Computing, UC Berkeley, and at the Department of Computer Science, Princeton University.

M.M. Halldórsson et al. (Eds.): ICALP 2015, Part I, LNCS 9134, pp. 193–205, 2015.
DOI: 10.1007/978-3-662-47672-7_16

In this paper, we initiate the study of *simultaneous approximation algorithms* for constraint satisfaction problems. A typical such problem is the simultaneous MAX-CUT problem: Given a collection of k graphs $G_i = (V, E_i)$ on the same vertex set V, the problem is to find a single cut (i.e., a partition of V) so that in *every* G_i, a large fraction of the edges go across the cut.

More generally, let $\mathcal{F}$ be a set of bounded-arity predicates on $[q]$-valued variables. Let V be a set of n $[q]$-valued variables. An $\mathcal{F}$-CSP is a weighted collection $\mathcal{W}$ of constraints on V, where each constraint is an application of a predicate from $\mathcal{F}$ to some variables from V. For an assignment $f : V \to [q]$ and a $\mathcal{F}$-CSP instance $\mathcal{W}$, we let $\mathsf{val}(f, \mathcal{W})$ denote the total weight of the constraints from $\mathcal{W}$ satisfied by f. The MAX-$\mathcal{F}$-CSP problem is to find f which maximizes $\mathsf{val}(f, \mathcal{W})$. If $\mathcal{F}$ is the set of all predicates on $[q]$ of arity w, then MAX-$\mathcal{F}$-CSP is also called MAX-w-CSP$_q$.

We now describe the setting for the problem we consider: k-*fold simultaneous* MAX-$\mathcal{F}$-CSP. Let $\mathcal{W}_1, \ldots, \mathcal{W}_k$ be $\mathcal{F}$-CSPs on V, each with total weight 1. Our high level goal is to find an assignment $f : V \to [q]$ for which $\mathsf{val}(f, \mathcal{W}_\ell)$ is large for all $\ell \in [k]$.

These problems fall naturally into the domain of multi-objective optimization: there is a common search space, and multiple objective functions on that space. Since even optimizing one of these objective functions could be NP-hard, it is natural to resort to approximation algorithms. Below, we formulate some of the approximation criteria that we will consider, in decreasing order of difficulty:

1. **Pareto approximation:** Suppose $(c_1, \ldots, c_k) \in [0, 1]^k$ is such that there is an assignment f^* with $\mathsf{val}(f^*, \mathcal{W}_\ell) \geq c_\ell$ for each $\ell \in [k]$.
   An α-Pareto approximation algorithm in this context is an algorithm, which when given $(c_1, \ldots, c_k)$ as input, finds an assignment f such that $\mathsf{val}(f, \mathcal{W}_\ell) \geq \alpha \cdot c_\ell$, for each $\ell \in [k]$.
2. **Minimum approximation:** This is basically the Pareto approximation problem when $c_1 = c_2 = \ldots = c_k$. Define OPT to be the maximum, over all assignments f^*, of $\min_{\ell \in [k]} \mathsf{val}(f^*, \mathcal{W}_\ell)$.
   An α-minimum approximation algorithm in this context is an algorithm which finds an assignment f such that $\min_{\ell \in [k]} \mathsf{val}(f, \mathcal{W}_\ell) \geq \alpha \cdot \text{OPT}$.
3. **Detecting Positivity:** This is a very special case of the above, where the goal is simply to determine whether there is an assignment f which makes $\mathsf{val}(f, \mathcal{W}_\ell) > 0$ for all $\ell \in [k]$.

When $k = 1$, minimum approximation and Pareto approximation correspond to the classical MAX-CSP approximation problems (which have received much attention). Our focus in this paper is on general k. It is useful to think of k as $O(1)$, or a slowly growing function of n, say $\log \log n$. As we will see in the discussions below, the nature of the problem changes quite a bit for $k > 1$. In particular, direct applications of classical techniques like random assignments and convex programming relaxations fail to give even a constant factor approximation for values of k greater than certain threshold.

The theory of *exact* multiobjective optimization has been very well studied, (see eg. [10, 27] and the references therein). A common theme in this area is to consider a classical optimization problem, optimizing a given objective function over a given search space, and to then consider the problem of simultaneously optimizing many such objective functions on a common search space.

This viewpoint has motivated the study of multiobjective versions of many exactly optimizable problems, including shortest paths, minimum spanning trees matchings, etc. [12, 27].

Here we undergo a systematic study of multiobjective optimization in the context of MAX-CSPs. This gives rise to a number of interesting problems and phenomena that seem ripe for study given the current technology in the study of CSPs, and also worthy of study in their own right. This is the main motivation for our paper. We discuss the existing works on multiobjective approximation of certain CSPs in Section 1.6.

We have two further motivations for studying simultaneous approximations for constraint satisfaction problems. Firstly, these are very natural algorithmic questions that capture optimization constraints in a way which more naïve formulations (such as taking linear combinations of the given CSPs) cannot. Secondly, the study of simultaneous approximation algorithms for CSPs sheds new light on various aspects of standard approximation algorithms for CSPs. For example, as we will see later, the trivial random assignment based algorithms turn out to be more useful in the simultaneous setting for us than the SDP-based algorithms, even though SDP-based algorithms in general give better approximation factors for many CSPs.

1.1 Observations About Simultaneous Approximation

We now discuss why a direct application of the classical CSP algorithms fails in this setting, and limitations on the approximation ratios that can be achieved.

We begin with a trivial remark. Finding an α-minimum (or Pareto) approximation to the k-fold MAX-$\mathcal{F}$-CSP is at least as hard as finding an α-approximation the classical MAX-$\mathcal{F}$-CSP problem (i.e., $k = 1$). Thus the known limits on polynomial-time approximability extend naturally to our setting.

Max-1-SAT. The simplest simultaneous CSP is MAX-1-SAT. The problem of getting a 1-Pareto or 1-minimum approximation to k-fold simultaneous MAX-1-SAT is essentially the **NP**-hard SUBSET-SUM problem. There is a simple $2^{\mathrm{poly}(k/\varepsilon)} \cdot \mathrm{poly}(n)$-time $(1-\varepsilon)$-Pareto approximation algorithm based on dynamic programming.

It is easy to see that detecting positivity of a k-fold simultaneous MAX-1-SAT is exactly the same problem as detecting satisfiability of a SAT formula with k clauses (a problem studied in the fixed parameter tractability community). Thus, this problem can be solved in time $2^{O(k)} \cdot \mathrm{poly}(n)$ (see [26]), and under the Exponential Time Hypothesis, one does not expect a polynomial time algorithm when $k = \omega(\log n)$.

Random Assignments. Let us consider algorithms based on random assignments. A typical example is MAX-CUT. A uniformly random partition in a weighted graph cuts $1/2$ the total weight in expectation. This gives a $1/2$ - approximation to the classical MAX-CUT problem.

In the simultaneous setting, if we happened to know that the cut value of all the instances is concentrated around $1/2$ with high probability, then using union bound we would obtain a cut that is simultaneously good for all instances. However, in general, such concentration does not hold. For $k \geq 3$, the probability that a random assignment cuts a constant fraction of the edges in each of the k instances, can be zero. In particular, there is no "trivial" random-assignment-based constant factor approximation algorithm for simultaneous CSPs.

SDP Algorithms. How do algorithms based on semi-definite programming (SDP) generalize to the simultaneous setting?

For the usual MAX-CUT problem ($k = 1$), the celebrated Goemans-Williamson SDP algorithm [15] gives a 0.878-approximation. The SDP relaxation generalizes naturally to the simultaneous setting; it allows us to find a vector solution which is a simultaneously good cut for $G_1, \ldots, G_k$. Perhaps we can apply hyperplane rounding to the SDP solution to obtain a simultaneously good cut for all G_i? We know that each G_i gets a good cut in expectation, but we need each G_i to get a good cut *with high probability* to guarantee a simultaneously good cut.

However, there are cases where the hyperplane rounding fails completely. For weighted instances, the SDP does not have *any* constant integrality gap. For unweighted instances, for every fixed k, we find an instance of k-fold simultaneous MAX-CUT (with arbitrarily many vertices and edges) where the SDP relaxation has value $1 - \Omega\left(\frac{1}{k^2}\right)$, while the optimal simultaneous cut has value only $1/2$.

Furthermore, applying the hyperplane rounding algorithm to this vector solution gives (with probability 1) a simultaneous cut value of *0*. These integrality gaps are described in the full version of this paper. [6]

Thus the natural extension of SDP based techniques for simultaneous approximation fail quite spectacularly. A-priori, this failure is quite surprising, since SDPs (and LPs) generalize to the multiobjective setting seamlessly.

Matching the Random Assignment Threshold? Given the ease and simplicity of algorithms based on random assignments for $k = 1$, giving algorithms in the simultaneous setting that match their approximation guarantees is a natural benchmark. Perhaps, it is always possible to do as well in the simultaneous setting as a random assignment for one instance.

Somewhat surprisingly, this is incorrect. For simultaneous MAX-Ew-SAT (CNF-SAT where every clause has exactly w distinct literals), a simple reduction from MAX-E3-SAT (with $k = 1$) shows that it is **NP**-hard to give a $(7/8+\varepsilon)$-minimum approximation for k-fold simultaneous MAX-Ew-SAT for large enough constants k (see full version [6] for the proof).

Proposition 1. *For every integer $w \geq 4$ and a real number $\varepsilon > 0$, given 2^{w-3} instances of* MAX-Ew-SAT *that are simultaneously satisfiable, it is* **NP***-hard to find a $(7/8 + \varepsilon)$-minimum (or Pareto) approximation.*

On the other hand, a random assignment to a single MAX-Ew-SAT instance satisfies a $1 - 2^{-w}$ fraction of constraints in expectation. In particular, it shows that simultaneous CSPs can have worse approximation factors than their classical ($k = 1$) counterparts.

1.2   Results

Our results address the approximability of k-fold simultaneous MAX-$\mathcal{F}$-CSP for large k. Our main algorithmic result shows that for every $\mathcal{F}$, and k not too large, k-fold simultaneous MAX-$\mathcal{F}$-CSP has a constant factor Pareto approximation algorithm.

Theorem 1. *Let q, w be constants. For every $\varepsilon > 0$, there is a $2^{O(k^4/\varepsilon^2 \log(k/\varepsilon))}$. $\mathrm{poly}(n)$-time $\left(\frac{1}{q^{w-1}} - \varepsilon\right)$-Pareto approximation algorithm for k-fold simultaneous MAX-w-CSP$_q$.*

The dependence on k implies that the algorithm runs in polynomial time up to $k = \tilde{O}((\log n)^{1/4})$ simultaneous instances.[1]

For particular CSPs, our methods allow us to do significantly better, as demonstrated by our following result for MAX-w-SAT.

Theorem 2. *Let w be a constant. For every $\varepsilon > 0$, there is a $2^{O(k^3/\varepsilon^2 \log(k/\varepsilon))}$. $\mathrm{poly}(n)$-time $(3/4 - \varepsilon)$-Pareto approximation algorithm for k-fold MAX-w-SAT.*

Given a single MAX-Ew-SAT instance, a random assignment satisfies a $1 - 2^{-w}$ fraction of the constraints in expectation. The approximation ratio achieved by the above theorem seems modest in comparison (even though it is for general MAX-w-SAT). However, Proposition 1 demonstrates it is **NP**-hard to do much better. The proofs of the above theorems appear in the full version [6].

Remarks:

1. As demonstrated by Proposition 1, it is sometimes impossible to match the approximation ratio achieved by a random assignment for $k = 1$. By comparison, the approximation ratio given by Theorem 1 is slightly better than that achieved by a random assignment ($1/q^w$). This is comparable to the best possible approximation ratio for $k = 1$, which is w/q^{w-1} up to constants [8, 25]. Our methods also prove that picking the best assignment out of $2^{O(k^4/\varepsilon^2 \log(k/\varepsilon))}$ independent and uniformly random assignments achieves a $(1/q^w - \varepsilon)$-Pareto approximation with high probability.
2. Our method is quite general. For any CSP with a convex relaxation and an associated rounding algorithm that assigns each variable independently from a distribution with certain *smoothness property*, it can be combined with our techniques to achieve essentially the same approximation ratio for k simultaneous instances. The distribution associated with rounding algorithm is called *smooth* if every variable has at least some constant probability, bounded away from 0, of getting any given element in the domain (see full version [6] for the formal definition).

---

[1] The $\tilde{O}(\cdot)$ hides $\mathrm{poly}(\log \log n)$ factors.

3. We reiterate that Pareto approximation algorithms achieve a multiplicative approximation for each instance. One could also consider the problem of achieving simultaneous approximations with an α-multiplicative and ε-additive error. This problem can be solved by a significantly simpler algorithm and analysis (but note that this variation does not even imply an algorithm for detecting positivity).

4. Our analysis, in fact, proves that a uniformly random assignment achieves a $(1/q^k - \varepsilon)$ Pareto-approximation with tiny but noticeable probability (about $q^{-\text{poly}(k/\varepsilon)}$). Note that it is not true that a random assignment, with non-trivial probability, satisfies a constant fraction of constraints in every instance, as the simultaneous objective value may be arbitrarily small. Our analysis is involved, and does not simply follow by analyzing the behavior of a random assignment on each of the k instances individually.

1.3   Complementary Results

Refined Hardness Results. As we saw earlier, assuming ETH, there is no algorithm for even detecting positivity of k-fold simultaneous MAX-1-SAT for $k = \omega(\log n)$. A predicate $P : \{0, 1\}^w \to \{\text{TRUE}, \text{FALSE}\}$ is said to be 0-*valid*/1-*valid* if the all-0-assignment/all-1-assignment satisfies P. We call a collection $\mathcal{F}$ of predicates *trivial* if either all predicates in $\mathcal{F}$ are 0-valid or all of them are 1-valid. Clearly, if $\mathcal{F}$ is trivial, then the simultaneous MAX-$\mathcal{F}$-CSP instances can be solved exactly (by considering the all-0-assignment/all-1-assignment). Here we prove that for any "nontrivial" collection of Boolean predicates $\mathcal{F}$, assuming ETH, there is no polynomial time algorithm for detecting positivity for k-fold simultaneous MAX-$\mathcal{F}$-CSP instances for $k = \omega(\log n)$. In particular, it is hard to obtain any poly-time constant factor approximation for $k = \omega(\log n)$. This implies a complete *dichotomy theorem* for constant factor approximations of k-fold simultaneous Boolean CSPs.

Theorem 3. *Assume the Exponential Time Hypothesis [18, 19]. Let $\mathcal{F}$ be a fixed finite set of Boolean predicates. If $\mathcal{F}$ is non trivial in the above sense, then for $k = \omega(\log n)$, detecting positivity of k-fold simultaneous MAX-$\mathcal{F}$-CSP on n variables requires time super-polynomial in n.*

Simultaneous Approximations via SDPs. It is a tantalizing possibility that one could use SDPs to improve the LP-based approximation algorithms that we develop. Especially for constant k, it is not unreasonable to expect that one could obtain a constant factor Pareto or minimum approximation, for k-fold simultaneous CSPs, better than what can be achieved by linear programming methods.

In this direction, we show how to use simultaneous SDP relaxations to obtain a polynomial time $(1/2 + \Omega(1/k^2))$-minimum approximation for k-fold simultaneous MAX-CUT on *unweighted graphs*.

Theorem 4. *For large enough n, there is an algorithm that, given k-fold simultaneous unweighted* MAX-CUT *instances on n vertices, runs in time $2^{2^{2^{O(k)}}} \cdot \mathrm{poly}(n)$, and computes a $\left(\frac{1}{2} + \Omega\left(\frac{1}{k^2}\right)\right)$-minimum approximation.*

See the full version [6] for proofs of Theorems 3 and 4.

1.4 Techniques

For the initial part of this discussion, we focus on the $q = w = 2$ case, and only achieve a $(1/4 - \varepsilon)$-Pareto approximation. Explicitly, we have a set of Boolean variables V, and k instances $\mathcal{W}_1, \dots, \mathcal{W}_k$ of width-2 CSPs on this set of variables. We know that there is an assignment $f^* : V \to \{0, 1\}$ s.t. for each instance $\ell \in [k]$, $\mathsf{val}(f^*, \mathcal{W}_\ell) \geq c_\ell$. Our goal is to find an assignment $f : V \to \{0, 1\}$ s.t. for each instance $\ell \in [k]$, $\mathsf{val}(f, \mathcal{W}_\ell) \geq (\frac{1}{4} - \varepsilon) \cdot c_\ell$.

Preliminary Observations. Consider a uniformly random assignment $g : V \to \{0, 1\}$. It is easy to see that for each instance $\ell \in [k]$, $\mathbf{E}[\mathsf{val}(g, \mathcal{W}_\ell)]$ (i.e., the expected satisfied weight in instance ℓ) is at least $1/4$ the total weight of all constraints in instance ℓ. If for some reason we knew that for each $\ell \in [k]$, the random variable $\mathsf{val}(g, \mathcal{W}_\ell)$ is concentrated around its expected value *with high probability*, then we could take a union bound over all the instances and conclude that with high probability, $\mathsf{val}(g, \mathcal{W}_\ell)$ is large for *every* $\ell \in [k]$.

It turns out that for any instance where the desired concentration does not occur, there is some variable $x \in V$ which has high degree in that instance (i.e., the weight of all constraints involving that variable is a constant fraction of the total weight of all constraints). This high degree variable seems very useful for our goal of finding a good assignment, since we can potentially influence the satisfaction of the instance quite a bit by just by changing this one variable.

This motivates a high level plan: either all the instances are well-concentrated, in which case a random assignment works, or else the some instance has a high degree variable, in which case we can try to set that variable and repeat.

A Recursive Algorithm. The simplest possible implementation of the above high-level plan leads to the following algorithm.

This algorithm is initially called with ρ being the trivial partial assignment.

Algorithm1(ρ)

Given variables V, instances $\mathcal{W}_1, \mathcal{W}_2, \dots, \mathcal{W}_k$, target values $c_1, \dots, c_k$.
$\rho : S_\rho \to \{0, 1\}$, with $S_\rho \subseteq V$, is a partial assignment.

1. For a uniformly random assignment $g : V \setminus S_\rho \to \{0, 1\}$, compute, for each $\ell \in [k]$, $\mu_\ell = \mathbf{E}[\mathsf{val}(\rho \cup g, \mathcal{W}_\ell)]$ and $\sigma_\ell^2 = \mathbf{Var}[\mathsf{val}(\rho \cup g, \mathcal{W}_\ell)]$.
2. If for each $\ell \in [k]$, we have $\sigma_\ell^2 \ll \mu_\ell^2$ (i.e., concentration), then:
   (a) pick a uniformly random assignment $g : V \setminus S_\rho \to \{0, 1\}$, and consider the total assignment $\rho \cup g$.
   (b) return $\min_{\ell \in [k]} \frac{\mathsf{val}(\rho \cup g, cal\mathcal{W}_\ell)}{c_\ell}$

3. Otherwise, pick some ℓ with $\sigma_\ell^2 = \Omega(\mu_\ell^2)$.
   (a) Find a variable $x \in V \setminus S_\rho$ with high degree in the residual instance $\mathcal{W}_\ell|_\rho$.
   (b) Consider the two partial assignments ρ_0, ρ_1 obtained by extending the domain of ρ to include x, with $\rho_0(x) = 0$ and $\rho_1(x) = 1$.
   (c) return $\max(\mathsf{Algorithm1}(\rho_0), \mathsf{Algorithm1}(\rho_1))$

Based on the discussion earlier, one can easily show that the above algorithm is a $(1/4 - \varepsilon)$-Pareto approximation algorithm. Indeed, these recursive calls partition the assignment space $\{0, 1\}^V$ into subcubes (based on the partial assignments), such that within each subcube, we know that a uniformly random assignment to the unset variables makes the satisfied weight of any instance concentrate around its expected value (which is at least $(1/4 - \varepsilon)$ times the maximum possible satisfied weight within that subcube).

On the other hand, this algorithm **need not terminate in polynomial time**; the recursive calls might lead the algorithm to take exponential time. In what follows, we force (a small variation of) the above algorithm to terminate any recursion that has reached depth $\mathrm{poly}(\frac{k}{\varepsilon})$. The main challenge which we need to overcome, and this occupies most of this paper, is to show that this new algorithm is still a $(1/4 - \varepsilon)$-Pareto approximation algorithm.

The Truncated Recursion. We now give the outline of a better version of the previous algorithm in which the recursion is truncated to run in polynomial time (for ease of exposition, some details have been swept under the rug).

This algorithm is initially called with ρ being the trivial partial assignment, and count being an array of all 0's.

$\mathsf{Algorithm2}(\rho, \mathsf{count})$
Given variables V, instances $\mathcal{W}_1, \mathcal{W}_2, \ldots, \mathcal{W}_k$, target values $c_1, \ldots, c_k$.
Parameter t is set to $\mathrm{poly}(k/\varepsilon)$.
$\rho : S_\rho \to \{0, 1\}$, with $S_\rho \subseteq V$, is a partial assignment.
count $: [k] \to \mathbb{N}$ maintains a count of how many times the recursion was caused by each instance.

1. For a uniformly random assignment $g : V \setminus S_\rho \to \{0, 1\}$, compute, for each $\ell \in [k]$, $\mu_\ell = \mathbf{E}[\mathsf{val}(\rho \cup g, \mathcal{W}_\ell)]$ and $\sigma_\ell^2 = \mathbf{Var}[\mathsf{val}(\rho \cup g, \mathcal{W}_\ell)]$.
2. If for each $\ell \in [k]$, we have $((\sigma_\ell^2 \ll \mu_\ell^2)$ OR $(\mathsf{count}(\ell) \geq t))$, then:
   (a) Pick a uniformly random partial assignment $g : V \setminus S_\rho \to \{0, 1\}$.
   (b) For each $h : S_\rho \to \{0, 1\}$, consider the total assignment $h \cup g : V \to \{0, 1\}$.
   (c) return $\max_h \min_{\ell \in [k]} \frac{\mathsf{val}(h \cup g, calW_\ell)}{c_\ell}$
3. Otherwise, pick some ℓ with $(\sigma_\ell^2 = \Omega(\mu_\ell^2))$ AND $(\mathsf{count}(\ell) < t)$.
   (a) Find a variable $x \in V \setminus S_\rho$ with high degree in the residual instance $\mathcal{W}_\ell|_\rho$.
   (b) Consider the two partial assignments ρ_0, ρ_1 obtained by extending the domain of ρ to include x, where $\rho_0(x) = 0$ and $\rho_1(x) = 1$.
   (c) Let $\mathsf{count}' = \mathsf{count}$. Let $\mathsf{count}'(\ell) = \mathsf{count}'(\ell) + 1$.
   (d) return $\max(\mathsf{Algorithm2}(\rho_0, \mathsf{count}'), \mathsf{Algorithm2}(\rho_1, \mathsf{count}'))$

The two most notable differences between this algorithm and the previous one are (1) the recursion is truncated so that no instance can cause more than $t = \text{poly}(k/\varepsilon)$ recursive calls, and (2) at the end of the recursion, the algorithm considers every possible assignment to S_ρ along with a uniformly random assignment to $V \setminus S_\rho$.

Analyzing the Truncated Recursion. The analysis starts by considering the assignment $f^* : V \to \{0,1\}$ satisfying $\text{val}(f^*, \mathcal{W}_\ell) \geq c_\ell$ for each $\ell \in [k]$. There is a unique branch of the recursion in which the argument ρ is consistent with f^* (i.e., satisfies $\rho|_{S_\rho} = f^*|_{S_\rho}$. Let us follow this branch of the recursion until the recursion stops, i.e., the condition in step 2 is satisfied. At this point we know that every instance ℓ is either 'low-variance" (namely the satsfied weight of a uniformly random assignment is close to its expectation with high probability), or else it is "high-variance", in which case we know know that $\text{count}(\ell) = t$. It is easy to show that for every low variance instance $\ell \in [k]$, the output f of the algorithm will have $\text{val}(f, \mathcal{W}_\ell) \geq (1/4 - \varepsilon) \cdot c_\ell$.

To argue about a high variance instance $\ell \in [k]$, we need to make several observations. Since $\text{count}(\ell) = t$, we know that instance ℓ caused t of the recursive calls on this branch. Every time a recursive call is made because of instance ℓ, it means that we found some variable $x \in V$ whose degree in the residual instance is large, and we brought x into the domain of ρ (thus reducing the total weight of all constraints, counted with multiplicity of the number of unset variables in it, in the new residual instance). Thus for high variance instances, where t recursive calls have been made, we can conclude that the total weight of all constraints (counted with multiplicity) remaining in the residual instance is exponentially small in t. If we choose t to be sufficiently large in k and ε, then this total weight can be made small.

These ingredients are already enough to show that the algorithm outputs an assignment f with the following additive-multiplicative approximation guarantee: for each $\ell \in [k]$, $\text{val}(f, \mathcal{W}_\ell) \geq (1/4) \cdot c_\ell - \varepsilon \cdot \text{totalwt}(\mathcal{W}_\ell)$, where $\text{totalwt}(\mathcal{W}_\ell)$ is the total weight of all constraints in the instance $\mathcal{W}_\ell$. This is still far from a pure multiplicative approximation guarantee, which seems to require a significantly more delicate analysis.

To achieve the pure multiplicative approximation guarantee, we show that there is a perturbation h of $f^*|_{S_\rho}$ which simultaneously (1) satisfies a lot of weight from every high variance instance, and (2) preserves the property that for a uniformly random assignment $g : V \setminus S_\rho \to \{0,1\}$, for every low variance instance $\ell \in [k]$ we have that $\text{val}(h \cup g, \mathcal{W}_\ell)$ is well-concentrated around its expectation. The procedure that constructs this perturbation itself is quite involved (but this happens only in the analysis). It is based on (1) the fact that there are many variables which were brought into the domain of ρ which have high weight in instance ℓ, and (2) understanding how much changing the value of these variables affects the other instances.

Improved Approximation, and Generalization. To get the claimed $(\frac{1}{2} - \varepsilon)$-Pareto approximation for the $q = w = 2$ case, we replace the uniformly random choice of $g : V \setminus S \to \{0, 1\}$ by a suitable LP relaxation + randomized rounding strategy. Concretely, we write an LP relaxation of the residual CSP, and round it to obtain $g : V \setminus S_\rho \to \{0, 1\}$. The rounding is via a certain rounding algorithm of Trevisan (which has some desirable smoothness properties). The analysis is nearly identical (but crucially uses the smoothness of the rounding), and the improved approximation comes from the improved approximation factor of the classical LP relaxation for MAX-2-CSP.

The generalization of this algorithm to general q, w is notationally technical, but conceptually there is only one new ingredient. Instead of bringing a high-degree variable x into the domain of the assignment ρ, we bring in a set of variables X, such that the total weight of all constraints which involve *all* the variables in X is large. The size of this X may vary from 1 to w. This turns out to be the appropriate generalization of the previous algorithm, and the analysis goes along the lines of the previous analysis, except for an additional appearance of a concentration bound for Lipschitz functions.

The algorithm for MAX-w-SAT uses the fact that there is an LP-based $3/4$ approximation for MAX-w-SAT *with a smooth rounding algorithm*. The original LP-based $3/4$-approximation algorithms [14, 35] did not have smooth rounding algorithms, but it turns out that Trevisan's rounding algorithm (which is smooth) also gives a $3/4$-factor approximation for MAX-w-SAT, and this suffices for our purposes. We also use the fact that a MAX-w-SAT constraint can be satisfied by perturbing any one variable. This leads to some significant simplification in the actual algorithm.

1.5   Discussion

We have only made initial progress on what we believe is a large number of interesting problems in the realm of simultaneous approximation of CSPs. We list here a few of the interesting directions for further research:

1. When designing SDP-based algorithms for the classical MAX-CSP problems, we are usually only interested in the expected value of the rounded solution. For k-fold simultaneous MAX-$\mathcal{F}$-CSP with $k > 1$, we are naturally led to the question of how concentrated the value of the solution output by the rounding is around its mean.

   Decorrelation of SDP rounding arises in recent algorithms [5, 16, 31] based on SDP hierarchies. It would be interesting to see if such ideas could be useful in this context.

2. When $k = O(1)$, for each $\mathcal{F}$, one can ask the question: what is the best Pareto approximation factor achievable for k-fold MAX-$\mathcal{F}$-CSP in polynomial time? While in Theorem 1 we do not focus on giving improved approximation factors for special $\mathcal{F}$, our methods will give better approximation factors for any $\mathcal{F}$ which has a good LP relaxation that comes equipped with a sufficiently smooth independent-rounding algorithm. It would be very interesting if one could employ SDPs for approximating simultaneous MAX-$\mathcal{F}$-CSP. A particularly nice question here: *Is there a polynomial time 0.878-Pareto approximation algorithm for $O(1)$-fold simultaneous* MAX-CUT? We do not even

know a $(1/2 + \varepsilon)$-Pareto approximation algorithm (but note that Theorem 4 does give this for $O(1)$-fold simultaneous *unweighted* MAX-CUT).

3. As demonstrated by hardness result for MAX-w-SAT given in Proposition 1, even for constant k, the achievable approximation factor can be strictly smaller than its classical counterpart. It would be very interesting to have a systematic theory of hardness reductions for simultaneous CSPs for $k = O(1)$. The usual paradigm for proving hardness of approximation based on label cover and long codes seems to break down completely for simultaneous CSPs.

1.6 Related Work

The theory of *exact* multiobjective optimization has been very well studied, (see eg. [10,27] and the references therein).

The only directly comparable work for simultaneous approximation algorithms for CSPs we are aware of is the work of Glaßer *et al.* [13]. They give a $1/2$-Pareto approximation for MAX-SAT with a running time of $n^{O(k^2)}$. For bounded width clauses, our algorithm does better in both approximation guarantee and running time.

For MAX-CUT, there are a few results of a similar flavor. For two graphs, the results of Angel *et al.* [2] imply a 0.439-Pareto approximation algorithm (though their actual results are incomparable to ours). Bollobás and Scott [7] asked what is the largest simultaneous cut in two unweighted graphs with m edges each. Kuhn and Osthus [22], using the second moment method, proved that for k simultaneous unweighted instances, there is a simultaneous cut that cuts at least $m/2 - O(\sqrt{km})$ edges in each instance, and give a deterministic algorithm to find it (this leads to a $(\frac{1}{2} - o(1))$-Pareto approximation for unweighted instances with sufficiently many edges). Our main theorem implies the same Pareto approximation factor for simultaneous MAX-CUT on general weighted instances, while for k-fold simultaneous MAX-CUT on unweighted instances, our Theorem 4 gives a $\left(\frac{1}{2} + \Omega(\frac{1}{k^2})\right)$-minimum approximation algorithm.

References

1. Alon, N., Gutin, G., Kim, E.J., Szeider, S., Yeo, A.: Solving MAX- r-SAT above a tight lower bound. Algorithmica **61**(3), 638–655 (2011)
2. Angel, E., Bampis, E., Gourvs, L.: Approximation algorithms for the bi-criteria weighted MAX-CUT problem. Discrete Applied Mathematics **154**(12), 1685–1692 (2006)
3. Arora, S., Lund, C., Motwani, R., Sudan, M., Szegedy, M.: Proof verification and the hardness of approximation problems. J. ACM **45**(3), 501–555 (1998)
4. Arora, S., Safra, S.: Probabilistic checking of proofs: A new characterization of NP. J. ACM **45**(1), 70–122 (1998)
5. Barak, B., Raghavendra, P., Steurer, D.: Rounding semidefinite programming hierarchies via global correlation. In: FOCS, pp. 472–481 (2011)

6. Bhangale, A., Kopparty, S., Sachdeva, S.: Simultaneous approximation of constraint satisfaction problems. CoRR abs/1407.7759 (2014). http://arxiv.org/abs/1407.7759
7. Bollobás, B., Scott, A.D.: Judicious partitions of bounded-degree graphs. Journal of Graph Theory 46(2), 131–143 (2004)
8. Chan, S.O.: Approximation resistance from pairwise independent subgroups. In: STOC 2013, pp. 447–456. ACM (2013)
9. Charikar, M., Makarychev, K., Makarychev, Y.: Note on MAX-2SAT. Electronic Colloquium on Computational Complexity (ECCC) 13(064) (2006)
10. Diakonikolas, I.: Approximation of Multiobjective Optimization Problems. Ph.D. thesis, Columbia University (2011)
11. Dinur, I., Regev, O., Smyth, C.D.: The hardness of 3 - uniform hypergraph coloring. In: FOCS 2002, p. 33. IEEE Computer Society, Washington (2002)
12. Ehrgott, M., Gandibleux, X.: A survey and annotated bibliography of multiobjective combinatorial optimization. OR-Spektrum 22(4), 425–460 (2000)
13. Glaßer, C., Reitwießner, C., Witek, M.: Applications of discrepancy theory in multiobjective approximation. In: FSTTCS 2011, pp. 55–65 (2011)
14. Goemans, M.X., Williamson, D.P.: A new $\frac{3}{4}$-approximation algorithm for MAX SAT. In: IPCO, pp. 313–321 (1993)
15. Goemans, M.X., Williamson, D.P.: Improved approximation algorithms for maximum cut and satisfiability problems using semidefinite programming. J. ACM 42(6), 1115–1145 (1995)
16. Guruswami, V., Sinop, A.K.: Lasserre hierarchy, higher eigenvalues, and approximation schemes for graph partitioning and quadratic integer programming with psd objectives. In: FOCS, pp. 482–491 (2011)
17. Håstad, J.: Some optimal inapproximability results. J. ACM 48(4), 798–859 (2001)
18. Impagliazzo, R., Paturi, R.: On the complexity of k-SAT. Journal of Computer and System Sciences 62(2), 367–375 (2001)
19. Impagliazzo, R., Paturi, R., Zane, F.: Which problems have strongly exponential complexity? J. Comput. Syst. Sci. 63(4), 512–530 (2001)
20. Khanna, S., Sudan, M., Trevisan, L., Williamson, D.P.: The approximability of constraint satisfaction problems. SIAM J. Comput. 30(6), 1863–1920 (2001)
21. Khot, S.: On the power of unique 2-prover 1-round games, pp. 767–775 (2002)
22. Kühn, D., Osthus, D.: Maximizing several cuts simultaneously. Comb. Probab. Comput. 16(2), 277–283 (2007)
23. Mahajan, M., Raman, V.: Parameterizing above guaranteed values: Maxsat and maxcut. J. Algorithms 31(2), 335–354 (1999)
24. Mahajan, M., Raman, V., Sikdar, S.: Parameterizing above or below guaranteed values. J. Comput. Syst. Sci. 75(2), 137–153 (2009)
25. Makarychev, Konstantin, Makarychev, Yury: Approximation algorithm for non-boolean MAX k-CSP. In: Gupta, Anupam, Jansen, Klaus, Rolim, José, Servedio, Rocco (eds.) APPROX 2012 and RANDOM 2012. LNCS, vol. 7408, pp. 254–265. Springer, Heidelberg (2012)
26. Marx, D.: Slides: CSPs and fixed-parameter tractability (2013). http://www.cs.bme.hu/dmarx/papers/marx-bergen-2013-csp.pdf
27. Papadimitriou, C.H., Yannakakis, M.: On the approximability of trade-offs and optimal access of web sources. In: Proceedings, 41st Annual Symposium on Foundations of Computer Science, 2000, pp. 86–92 (2000)
28. Patel, V.: Cutting two graphs simultaneously. J. Graph Theory 57(1), 19–32 (2008)
29. Raghavendra, P.: Optimal algorithms and inapproximability results for every CSP? In: STOC 2008, pp. 245–254. ACM, New York (2008)

30. Raghavendra, P., Steurer, D.: How to round any CSP. In: In Proc. 50th IEEE Symp. on Foundations of Comp. Sci. (2009)
31. Raghavendra, P., Tan, N.: Approximating csps with global cardinality constraints using sdp hierarchies. In: SODA, pp. 373–387 (2012)
32. Rautenbach, D., Szigeti, Z.: Simultaneous large cuts. Forschungsinstitut für Diskrete Mathematik, Rheinische Friedrich-Wilhelms-Universität (2004)
33. Schaefer, T.J.: The complexity of satisfiability problems. In: STOC 1978, pp. 216–226. ACM, New York (1978)
34. Trevisan, L.: Parallel approximation algorithms by positive linear programming. Algorithmica 21(1), 72–88 (1998)
35. Yannakakis, M.: On the approximation of maximum satisfiability. In: SODA 1992, pp. 1–9 (1992)

Design of Dynamic Algorithms via Primal-Dual Method

Sayan Bhattacharya[1]([⊠]), Monika Henzinger[2], and Giuseppe F. Italiano[3]

[1] Institute of Mathematical Sciences, Chennai, India
bsayan@imsc.res.in
[2] University of Vienna, Vienna, Austria
monika.henzinger@univie.ac.at
[3] Università di Roma "Tor Vergata", Rome, Italy
giuseppe.italiano@uniroma2.it

Abstract. In this paper, we develop a dynamic version of the primal-dual method for optimization problems, and apply it to obtain the following results. (1) For the dynamic set-cover problem, we maintain an $O(f^2)$-approximately optimal solution in $O(f \cdot \log(m + n))$ amortized update time, where f is the maximum "frequency" of an element, n is the number of sets, and m is the maximum number of elements in the universe at any point in time. (2) For the dynamic b-matching problem, we maintain an $O(1)$-approximately optimal solution in $O(\log^3 n)$ amortized update time, where n is the number of nodes in the graph.

1 Introduction

The primal-dual method lies at the heart of the design of algorithms for combinatorial optimization problems. The basic idea, contained in the "Hungarian Method" [8], was extended and formalized by Dantzig et al. [5] as a general framework for linear programming, and thus it became applicable to a large variety of problems. Few decades later, Bar-Yehuda et al. [1] were the first to apply the primal-dual method to the design of approximation algorithms. Subsequently, this paradigm was applied to obtain approximation algorithms for a wide collection of NP-hard problems [11]. When the primal-dual method is applied to approximation algorithms, an approximate solution to the problem and a feasible solution to the dual of an LP relaxation are constructed simultaneously, and the performance guarantee is proved by comparing the values of both solutions. The primal-dual method was also extended to online problems [4]. Here, the input is revealed only in parts, and an online algorithm is required to respond to each new input upon its arrival (without being able to see the future). The algorithm's performance is compared against the benchmark of an optimal omniscient algorithm that can view the entire input sequence in advance.

Monika Henzinger—Supported by the European Research Council under the European Union's Seventh Framework Programme (FP/2007-2013) / ERC Grant Agreement no. 340506.
Giuseppe F. Italiano—Partially supported by MIUR under Project AMANDA.

M.M. Halldórsson et al. (Eds.): ICALP 2015, Part I, LNCS 9134, pp. 206–218, 2015.
DOI: 10.1007/978-3-662-47672-7_17

In this paper, we focus on dynamic algorithms for optimization problems. In the dynamic setting, the input of a problem is being changed via a sequence of updates, and after each update one is interested in maintaining the solution to the problem much faster than recomputing it from scratch. We remark that the dynamic and the online setting are completely different: in the dynamic scenario one is concerned more with guaranteeing fast (worst-case or amortized) update times rather than comparing the algorithms' performance against optimal offline algorithms. As a main contribution of this paper, we develop a dynamic version of the primal-dual method, thus opening up a completely new area of application of the primal-dual paradigm to the design of dynamic algorithms. With some careful insights, our recent algorithms for dynamic matching and dynamic vertex cover [3] can be reinterpreted in this new framework. In this paper, we apply the new dynamic primal-dual framework to the design of two other optimization problems: the dynamic set-cover problem and the dynamic b-matching problem.

Definition 1 (Set-Cover). *We are given a universe $\mathcal{U}$ of at most m elements, and a collection $\mathcal{S}$ of n sets $S \subseteq \mathcal{U}$. Each set $S \in \mathcal{S}$ has a (polynomially bounded by n) "cost" $c_S > 0$. The goal is to select a subset $\mathcal{S}' \subseteq \mathcal{S}$ such that each element in $\mathcal{U}$ is covered by some set $S \in \mathcal{S}'$ and the total cost $\sum_{S \in \mathcal{S}'} c(S)$ is minimized.*

Definition 2 (Dynamic Set-Cover). *Consider a dynamic version of the problem specified in Definition 1, where the collection $\mathcal{S}$, the costs $\{c_S\}, S \in \mathcal{S}$, the upper bound f on the maximum frequency $\max_{u \in \mathcal{U}} |\{S \in \mathcal{S} : u \in S\}|$, and the upper bound m on the maximum size of the universe $\mathcal{U}$ remain fixed. The universe $\mathcal{U}$, on the other hand, keeps changing dynamically. In the beginning, we have $\mathcal{U} = \emptyset$. At each time-step, either an element u is inserted into the universe $\mathcal{U}$ and we get to know which sets in $\mathcal{S}$ contain u, or some element is deleted from the universe. The goal is to maintain an approximately optimal solution to the set-cover problem in this dynamic setting.*

Definition 3 (b-Matching). *We are given an input graph $G = (V, E)$ with $|V| = n$ nodes, where each node $v \in V$ has a capacity $c_v \in \{1, \ldots, n\}$. A b-matching is a subset $E' \subseteq E$ of edges such that each node v has at most c_v edges incident to it in E'. The goal is to select the b-matching of maximum cardinality.*

Definition 4 (Dynamic b-Matching). *Consider a dynamic version of the problem specified in Definition 3, where the node set V and the capacities $\{c_v\}, v \in V$ remain fixed. The edge set E, on the other hand, keeps changing dynamically. In the beginning, we have $E = \emptyset$. At each time-step, either a new edge is inserted into the graph or some existing edge is deleted from the graph. The goal is to maintain an approximately optimal solution to the b-matching problem in this dynamic setting.*

As stated in [4,11], the set-cover problem has played a pivotal role both for approximation and for online algorithms, and thus it seems a natural problem to consider in our dynamic setting. Our definition of dynamic set-cover is inspired by the standard formulation of the online set-cover problem [4], where the elements arrive online.

Our Techniques. Roughly speaking, our dynamic version of the primal-dual method works as follows. We start with a feasible primal solution and an infeasible dual solution for the problem at hand. Next, we consider the following process: gradually increase all the primal variables at the same rate, and whenever a primal constraint becomes tight, stop the growth of all the primal variables involved in that constraint, and update accordingly the corresponding dual variable. This primal growth process is used to define a suitable data structure based on a hierarchical partition. A level in this partition is a set of the dual variables whose corresponding primal constraints became (approximately) tight at the same time-instant. To solve the dynamic problem, we maintain the data structure, the hierarchical partition and the corresponding primal-dual solution dynamically using a simple greedy procedure. This is sufficient for solving the dynamic set-cover problem. For the dynamic b-matching problem, we need some additional ideas. We first get a fractional solution to the problem using the previous technique. To obtain an integral solution, we perform randomized rounding on the fractional solution in a dynamic setting. This is done by sampling the edges with probabilities that are determined by the fractional solution.

Our Results. Our new dynamic primal-dual framework yields efficient dynamic algorithms for both the dynamic set-cover problem and the dynamic b-matching problem. In particular, for the dynamic set-cover problem we maintain a $O(f^2)$-approximately optimal solution in $O(f \cdot \log(m + n))$ amortized update time (see Corollary 1 in Section 2). On the other hand, for the dynamic b-matching problem, we maintain a $O(1)$-approximation in $O(\log^3 n)$ amortized time per update (see Theorem 7 in Section 3). Further, we can show that an edge insertion/deletion in the input graph, on average, leads to $O(\log^2 n)$ changes in the set of matched edges maintained by our algorithm.

Related Work. The design of dynamic algorithms is one of the classic areas in theoretical computer science with a countless number of applications. Dynamic graph algorithms have received special attention, and there have been many efficient algorithms for several dynamic graph problems, including dynamic connectivity, minimum spanning trees, transitive closure, shortest paths and matching problems (see, e.g., the survey in [6]). The b-matching problem contains as a special case matching problems, for which many dynamic algorithms are known [2,3,7,9,10]. Unfortunately, none of the results on dynamic matching extends to the dynamic b-matching problem. To the best of our knowledge, no previous result was known for dynamic set-cover problem.

2　Maintaining a Set-Cover in a Dynamic Setting

We define a problem called "fractional hypergraph b-matching" (Definition 5). Later, we show that this generalizes the well-known set-cover problem (Lemma 1). Our main result is Theorem 2, which, along with Lemma 1, implies Corollary 1.

Definition 5 (Fractional Hypergraph b-Matching). *We are given an input hypergraph $G = (V, E)$ with $|V| = n$ nodes and at most $m \geq |E|$ edges. Let $\mathcal{E}_v \subseteq E$ denote the set of edges incident upon a node $v \in V$, and let $\mathcal{V}_e = \{v \in V : e \in \mathcal{E}_v\}$ denote the set of nodes an edge $e \in E$ is incident upon. Let $c_v > 0$ denote the "capacity" of a node $v \in V$, and let $\mu \geq 1$ denote the "multiplicity" of an edge. We assume that the μ and the c_v values are polynomially bounded by n. Our goal is to assign a "weight" $x(e) \in [0, \mu]$ to each edge $e \in E$ in such a way that (a) $\sum_{e \in \mathcal{E}_v} x(e) \leq c_v$ for all nodes $v \in V$, and (b) the sum of the weights of all the edges is maximized.*

Below, we write a linear program for the above problem and its dual.

Primal LP: Maximize $\displaystyle\sum_{e \in E} x(e)$ (1)

subject to: $\displaystyle\sum_{e \in \mathcal{E}_v} x(e) \leq c_v$ $\forall v \in V.$ (2)

$0 \leq x(e) \leq \mu$ $\forall e \in E.$ (3)

Dual LP: Minimize $\displaystyle\sum_{v \in V} c_v \cdot y(v) + \sum_{e \in E} \mu \cdot z(e)$ (4)

subject to: $\displaystyle z(e) + \sum_{v \in \mathcal{V}_e} y(v) \geq 1$ $\forall e \in E.$ (5)

$y(v), z(e) \geq 0$ $\forall v \in V, e \in E.$ (6)

Definition 6. *A feasible solution to LP (1) is λ-maximal, $\lambda \geq 1$, iff for every edge $e \in E$ with $x(e) < \mu$, there is some node $v \in \mathcal{V}_e$ with $\sum_{e' \in \mathcal{E}_v} x(e') \geq c_v/\lambda$.*

Theorem 1. *Let $f \geq \max_{e \in E} |\mathcal{V}_e|$ be an upper bound on the maximum possible "frequency" of an edge. Let OPT be the optimal objective value of LP (1). Any λ-maximal solution to LP (1) has an objective value that is at least $OPT/(\lambda f + 1)$.*

Proof (Sketch). Follows from LP duality.   □

Definition 7 (Dynamic Fractional Hypergraph b-Matching). *Consider a dynamic version of the problem specified in Definition 5, where the node-set V, the capacities $\{c_v\}, v \in V$, the upper bound f on the maximum frequency $\max_{e \in E} |\mathcal{V}_e|$, and the upper bound m on the maximum number of edges remain fixed. The edge-set E, on the other hand, keeps changing dynamically. In the beginning, we have $E = \emptyset$. At each time-step, either an edge is inserted into the graph or an edge is deleted from the graph. The goal is to maintain an approximately optimal solution to the problem in this dynamic setting.*

Theorem 2. *We can maintain a $(f + 1 + \epsilon f)$-maximal solution to dynamic fractional hypergraph b-matching in $O(f \cdot \log(m + n)/\epsilon^2)$ amortized update time.*

We now compare fractional hypergraph b-matching with set-cover.

Lemma 1. *The dual LP (4) is an LP-relaxation of the set-cover problem.*

Proof (Sketch). Given an instance of the set-cover problem, we create an instance of the hypergraph b-matching problem as follows. For each element $u \in \mathcal{U}$ create an edge $e(u) \in E$, and for each set $S \in \mathcal{S}$, create a node $v(S) \in V$ with capacity $c_{v(S)} = c_S$. Ensure that an element u belongs to a set S iff $e(u) \in \mathcal{E}_{v(S)}$. Set $\mu = \max_{v \in V} c_v + 1$. Since $\mu > \max_{v \in V} c_v$, it can be shown that an optimal solution to the dual LP (4) will set $z(e) = 0$ for every edge $e \in E$. Thus, we can remove the variables $\{z(e)\}$ from the constraints and the objective function of LP (4) to get a new LP with the same optimal objective value. This new LP is an LP-relaxation for the set-cover problem. □

Corollary 1. *We can maintain an $(f^2 + f + \epsilon f^2)$-approximately optimal solution to the dynamic set cover problem in $O(f \cdot \log(m+n)/\epsilon^2)$ amortized update time.*

Proof (Sketch). We map the set cover instance to a fractional hypergraph b-matching instance as in the proof of Lemma 1. By Theorem 2, in $O(f \log(m + n)/\epsilon^2)$ amortized update time, we can maintain a feasible solution $\{x^*(e)\}$ to LP (1) that is λ-maximal, where $\lambda = f + 1 + \epsilon f$. Consider a collection of sets $\mathcal{S}^* = \{S \in \mathcal{S} : \sum_{e \in \mathcal{E}_{v(S)}} x(e) \geq c_{v(S)}/\lambda\}$. Since we can maintain the fractional solution $\{x^*(e)\}$ in $O(f \log(m + n)/\epsilon^2)$ amortized update time, we can also maintain $\mathcal{S}^*$ without incurring any additional overhead in the update time. Now, using complementary slackness conditions, we can show that each element $e \in \mathcal{U}$ is covered by some $S \in \mathcal{S}^*$, and the sum $\sum_{S \in \mathcal{S}^*} c_S$ is at most (λf)-times the size of the primal solution $\{x^*(e)\}$. The corollary follows from LP duality. □

For the rest of this section, we focus on proving Theorem 2. First, in the static setting, inspired by the primal-dual method for set-cover we consider the following algorithm for the fractional hypergraph b-matching problem.

- Consider a primal solution with $x(e) \leftarrow 0$ for all $e \in E$, and let $F \leftarrow E$.
- WHILE there is some primal constraint that is not tight:
  - Keep increasing the primal variables $\{x(e)\}, e \in F$, uniformly at the same rate till some primal constraint becomes tight. At that instant, "freeze" all the primal variables involved in that constraint and delete them from the set F, and set the corresponding dual variable to one.

Figure 1 defines a variant of the above procedure that happens to be easier to maintain in a dynamic setting. The main idea is to discretize the continuous primal growth process. Define $c_{\min} = \min_{v \in V} c_v$, and without any loss of generality, assume that $c_{\min} > 0$. Fix $\alpha, \beta > 1$, and define $L = \lceil \log_\beta(m\mu\alpha/c_{\min}) \rceil$.

Claim 3. *If $x(e) = \mu \cdot \beta^{-L}$ for all $e \in E$, then we have a feasible primal solution.*

Proof. Clearly, $x(e) \leq \mu$ for all $e \in E$. Now, consider any node $v \in V$. We have $\sum_{e \in \mathcal{E}_v} x(e) = |\mathcal{E}_v| \cdot \mu \cdot \beta^{-L} \leq |E| \cdot \mu \cdot \beta^{-L} \leq m \cdot \mu \cdot \beta^{-L} \leq m \cdot \mu \cdot (c_{\min}/(m\mu\alpha)) = c_{\min}/\alpha < c_v$. Hence, all the primal constraints are satisfied. □

01.	Set $x(e) \leftarrow \mu \cdot \beta^{-L}$ for all $e \in E$, and define $c_v^* = c_v/(f\alpha\beta)$ for all $v \in V$.
02.	Set $V_L \leftarrow \{v \in V : \sum_{e \in \mathcal{E}_v} x(e) \geq c_v^*\}$, and $E_L \leftarrow \bigcup_{v \in V_L} \mathcal{E}_v$.
03.	FOR $i = L - 1$ to 1:
04.	Set $x(e) \leftarrow x(e) \cdot \beta$ for all $e \in E \setminus \bigcup_{k=i+1}^{L} E_i$.
05.	Set $V_i \leftarrow \{v \in V \setminus \bigcup_{k=i+1}^{L} V_k : \sum_{e \in \mathcal{E}_v} x(e) \geq c_v^*\}$.
06.	Set $E_i \leftarrow \bigcup_{v \in V_i} \mathcal{E}_v$.
07.	Set $V_0 \leftarrow V \setminus \bigcup_{k=1}^{L} V_i$, and $E_0 \leftarrow \bigcup_{v \in V_0} \mathcal{E}_v$.
08.	Set $x(e) \leftarrow x(e) \cdot \beta$ for all $e \in E_0$.

Fig. 1. DISCRETE-PRIMAL-DUAL()

Our new algorithm is described in Figure 1. We initialize our primal solution by setting $x(e) \leftarrow \mu\beta^{-L}$ for every edge $e \in E$, as per Claim 3. Say that a node v is *nearly-tight* if its corresponding primal constraint is tight within a factor of $f\alpha\beta$, and *slack* otherwise. Say that an edge is *nearly-tight* if it is incident upon some *nearly-tight* node, and *slack* otherwise. Let $V_L \subseteq V$ and $E_L \subseteq E$ respectively denote the sets of *nearly-tight* nodes and edges, immediately after the initialization step. The algorithm then performs $L - 1$ iterations.

At iteration $i \in \{L - 1, \ldots, 1\}$, the algorithm increases the weight $x(e)$ of every *slack* edge e by a factor of β. Since the total weight received by every *slack* node v (from its incident edges) never exceeds $c_v/(f\alpha\beta)$, this weight-increase step does not violate any primal constraint. The algorithm then defines V_i (resp. E_i) to be the set of new nodes (resp. edges) that become *nearly-tight* due to this weight-increase step.

Finally, the algorithm defines V_0 (resp. E_0) to be the set of nodes (resp. edges) that are *slack* at the end of iteration $i = 1$. It terminates after increasing the weight of every edge in E_0 by a factor of β.

When the algorithm terminates, it is easy to check that $x(e) = \mu \cdot \beta^{-i}$ for every edge $e \in E_i, i \in \{0, \ldots, L\}$. We also have $c_v^* \leq \sum_{e \in \mathcal{E}_v} x(e) \leq \beta \cdot c_v^*$ for every node $v \in \bigcup_{k=1}^{L} V_k$, and $\sum_{e \in \mathcal{E}_v} x(e) \leq c_v^*$ for every node $v \in V_0$. Furthermore, at the end of the algorithm, every edge $e \in E$ is either *nearly-tight*, or it has weight $x(e) = \mu$. We, therefore, reach the following conclusion.

Claim 4. *The algorithm described in Figure 1 returns an $(f\alpha\beta)$-maximal solution to the fractional hypergraph b-matching problem.*

Our goal is to make a variant of the procedure in Figure 1 work in a dynamic setting. Towards this end, we introduce the concept of an (α, β)-partition (see Definition 8) satisfying a certain invariant (see Invariant 5). The reader is encouraged to notice the similarities between this construct and Figure 1.

Definition 8. *Fix any two parameters $\alpha, \beta > 1$ and let $c_{\min} = \min_{v \in V} c_v > 0$. An (α, β)-partition of the hypergraph G partitions its node-set V into subsets $V_0 \ldots V_L$, where $L = \lceil \log_\beta(m\mu\alpha/c_{\min}) \rceil$. For $i \in \{0, \ldots, L\}$, we identify the subset V_i as the ith "level" of this partition, and denote the level of a node v by*

v). *Thus, we have* $v \in V_{\ell(v)}$ *for all* $v \in V$. *We also define the level of each edge* $\in E$ *as* $\ell(e) = \max_{v \in V_e}\{\ell(v)\}$, *and assign a "weight"* $w(e) = \mu \cdot \beta^{-\ell(e)}$ *to* e.

Given an (α, β)-partition, let $\mathcal{E}_v(i) = \{e \in \mathcal{E}_v : \ell(e) = i\}$ denote the set of dges incident to v that are in the ith level, and let $\mathcal{E}_v(i,j) = \bigcup_{k=i}^{j} \mathcal{E}_v(k)$ denote ne set of edges incident to v whose levels are in the range $[i,j]$. Similarly, we efine the notations $\mathcal{D}_v = |\mathcal{E}_v|$ and $\mathcal{D}_v(i,j) = |\mathcal{E}_v(i,j)|$. Let $W_v = \sum_{e \in \mathcal{E}_v} w(e)$ enote the total weight a node $v \in V$ receives from the edges incident to it. We lso define the notation $W_v(i) = \sum_{e \in \mathcal{E}_v} \mu \cdot \beta^{-\max(\ell(e),i)}$. It gives the total weight ne node v would receive from the edges incident to it, *if the node* v *itself were o go to the* i^{th} *level*. It is easy to check that an (α, β)-partition satisfies the ollowing conditions for all nodes $v \in V$.

$$W_v(L) \leq c_{\min}/\alpha \tag{7}$$

$$W_v(L) \leq \cdots \leq W_v(i) \leq \cdots \leq W_v(0) \tag{8}$$

$$W_v(i) \leq \beta \cdot W_v(i+1) \quad \forall i \in \{0, \ldots, L-1\}. \tag{9}$$

nvariant 5. *Define* $c_v^* = c_v/(f\alpha\beta)$. *For every node* $v \in V$, *if* $\ell(v) = 0$, *then* $V_v \leq f\alpha\beta \cdot c_v^*$. *Else if* $\ell(v) \geq 1$, *then* $c_v^* \leq W_v \leq f\alpha\beta \cdot c_v^*$.

Lemma 2. *Consider an* (α, β)-*partition that satisfies Invariant 5. The edge-weights* $\{w(e)\}, e \in E$, *give an* $(f\alpha\beta)$-*maximal solution to LP (1)*.

Proof (Sketch). Similar to the proof of Claim 4. □

Handling an Edge Insertion/Deletion. Consider an (α, β)-partition in the graph G. A node is called *dirty* if it violates Invariant 5, and *clean* otherwise. Since the edge-set E is initially empty, every node is clean and at level zero before the first update. Now consider the time instant just prior to the tth update. By induction hypothesis, at this instant every node is clean. Then the tth update akes place, which inserts (resp. deletes) an edge e in E with weight $w(e) = \mu\beta^{-\ell(e)}$. This increases (resp. decreases) the weights $\{W_x\}, x \in V_e$. Due to this change, the nodes $x \in V_e$ might become dirty. To recover from this, we call the subroutine in Figure 2.

Consider any node $v \in V$ and suppose that $W_v > f\alpha\beta c_v^* = c_v \geq c_{\min}$. In this event, since $\alpha > 1$, equation 7 implies that $W_v(L) < W_v(\ell(v))$ and hence we have $L > \ell(v)$. In other words, when the procedure described in Figure 2 decides o increment the level of a dirty node v (Step 02), we know for sure that the current level of v is strictly less than L (the highest level in the (α, β)-partition).

Next, consider an edge $e \in \mathcal{E}_v$. If we change $\ell(v)$, then this may change the weight $w(e)$, and this in turn may change the weights $\{W_z\}, z \in V_e$. Thus, a single iteration of the WHILE loop in Figure 2 may lead to some clean nodes becoming dirty, and some other dirty nodes becoming clean. If and when the WHILE loop terminates, however, we are guaranteed that every node is clean and that Invariant 5 holds.

Bounding the Amortized Update Time. For each node $v \in V$ and each $i \in \{0, \ldots, L\}$, we store the set of edges $\{e \in \mathcal{E}_v : \ell(e) = i\}$ in a doubly linked

---

01. WHILE there exists a dirty node v
02.   IF $W_v > f\alpha\beta c_v^*$, THEN
            // If true, then by equation 7, we have $\ell(v) < L$.
03.   Increment the level of v by setting $\ell(v) \leftarrow \ell(v) + 1$.
04.   ELSE IF ($W_v < c_v^*$ and $\ell(v) > 0$), THEN
05.   Decrement the level of v by setting $\ell(v) \leftarrow \ell(v) - 1$.

Fig. 2. RECOVER()

list NEIGHBORS$[v, i]$. The update time of our algorithm is dominated by the time taken to update these lists. Next, note that each time the level of an edge changes, we have to update at most f lists (one corresponding to each node $v \in \mathcal{V}_e$). Hence, the time taken to update the lists is given by $f \cdot \delta_l$, where δ_l is the number of times the procedure in Figure 2 changes the level of an edge. Using a potential function, in the full version of the paper we show that $\delta_l \leq t \cdot O(L/\epsilon)$ after t edge insertions/deletions in G starting from an empty graph, for $\alpha = 1 + 1/f + 3\epsilon$, $\beta = 1 + \epsilon$. Since $L = \lceil \log_\beta(m\mu\alpha/c_{\min}) \rceil$ and $\mu, c_{\min}$ are polynomially bounded by n, the amortized update time is $O(f\delta_l/t) = O(f\log(m + n)/\epsilon^2)$.

3 Maintaining a b-Matching in a Dynamic Setting

In this section, we will present a dynamic algorithm for the problem specified in Definitions 3, 4 (see Theorem 7). Given any subset of edges $E' \subseteq E$ and any node $v \in V$, let $\mathcal{N}(v, E') = \{u \in V : (u, v) \in E'\}$ denote the set of neighbors of v with respect to the edge-set E', and let $\deg(v, E') = |\mathcal{N}(v, E')|$. Next, consider any "weight" function $w : E' \to \mathbf{R}^+$. For every node $v \in V$, we define $W_v = \sum_{u \in \mathcal{N}(v,E)} w(u, v)$. Finally, for every subset of edges $E' \subseteq E$, we define $w(E') = \sum_{e \in E'} w(e)$. Next, we show how to maintain a "fractional" b-matching.

Theorem 6. *Fix a constant $\epsilon \in (0, 1/4)$, and let $\lambda = 4$. In $O(\log n)$ amortized update time, we can maintain a fractional b-matching $w : E \to [0, 1]$ in $G = (V, E)$ such that:*

$$W_v \leq c_v/(1 + \epsilon) \text{ for all nodes } v \in V. \qquad (10)$$
$$w(u, v) = 1 \text{ for each edge } (u, v) \in E \text{ with } W_u, W_v < c_v/\lambda. \qquad (11)$$

Further, the size of the optimal b-matching in G is $O(1)$ times the sum $\sum_{e \in E} w(e)$.

Proof (Sketch). Note that the fractional b-matching problem is a special case of fractional hypergraph b-matching (Definitions 5, 7) where $\mu = 1$, $m = n^2$, and $f = 2$.

We scale down the capacity of each node $v \in V$ by a factor of $(1 + \epsilon)$, by defining $\tilde{c}_v = c_v/(1 + \epsilon)$ for all $v \in V$. Next, we apply Theorem 2 on the input graph $G = (V, E)$ with $\mu = 1$, $m = n^2$, $f = 2$, and the reduced capacities

$\{\tilde{c}_v\}, v \in V$. Let $\{w(e)\}, e \in E$, be the resulting $(f + 1 + \epsilon f)$-maximal matching (see Definition 6). Since $\epsilon < 1/3$ and $f = 2$, we have $\lambda \geq f + 1 + \epsilon f$. Since ϵ is a constant, the amortized update time for maintaining the fractional b-matching becomes $O(f \cdot \log(m+n)/\epsilon^2) = O(\log n)$. Finally, by Theorem 1, the fractional b-matching $\{w(e)\}$ is an $(\lambda f + 1) = 9$-approximate optimal b-matching in G in the presence of the reduced capacities $\{\tilde{c}_v\}$. But scaling down the capacities reduces the objective of LP (1) by at most a factor of $(1 + \epsilon)$. Hence, the size of the optimal b-matching in G is at most $9(1 + \epsilon) = O(1)$ times the sum $\sum_{e \in E} w(e)$. This concludes the proof.   □

Set $\lambda = 4$ for the rest of this section. We will show how to dynamically convert the fractional b-matching $\{w(e)\}$ from Theorem 6 into an integral b-matching, by losing a constant factor in the approximation ratio. The main idea is to randomly sample the edges $e \in E$ based on their $w(e)$ values. But, first we introduce the following notations.

Say that a node $v \in V$ is "nearly-tight" if $W_v \geq c_v/\lambda$ and "slack" otherwise. Let T denote the set of all nearly-tight nodes. We also partition the node-set V into two subsets: $B \subseteq V$ and $S = V \setminus B$. Each node $v \in B$ is called "big" and has $\deg(v, E) \geq c \log n$, for some large constant $c > 1$. Each node $v \in S$ is called "small" and has $\deg(v, E) < c \log n$. Define $E_B = \{(u, v) \in E : \text{either } u \in B \text{ or } v \in B\}$ to be the subset of edges with at least one endpoint in B, and let $E_S = \{(u, v) \in E : \text{either } u \in S \text{ or } v \in S\}$ be the subset of edges with at least one endpoint in S. We define the subgraphs $G_B = (V, E_B)$ and $G_S = (V, E_S)$.

Overview of Our Approach. We maintain the following quantitates. (1) A random subset $H_B \subseteq E_B$, and a weight function $w^B : H_B \to [0, 1]$ in the subgraph $G_B(H) = (V, H_B)$, as per Definition 9. (2) A random subset $H_S \subseteq E_S$, and a weight function $w^S : H_S \to [0, 1]$ in the subgraph $G_S(H) = (V, H_S)$, as per Definition 10. (3) A maximal b-matching $M_S \subseteq H_S$ in the subgraph $G_S(H)$, that is, for every edge $(u, v) \in H_S \setminus M_S$, there is a node $q \in \{u, v\}$ such that $\deg(q, M_S) = c_q$. (4) The set of edges $E^* = \{e \in E : w(e) = 1\}$.

We will show that with high probability, one of the edge-sets H_B, M_S, E^* is an $O(1)$-approximation to the optimal b-matching in G. The rest of this section is organized as follows. In Lemma 3 (resp. Lemma 4), we prove some properties of the random set H_B (resp. H_S) and the weight function w^B (resp. w^S). In Lemma 5, we show that the edge-sets H_B, H_S, M_S and E^* can be maintained in a dynamic setting in $O(\log^3 n)$ amortized update time. We prove our main result in Theorem 7.

Definition 9. *Let $Z_B(e) \in \{0, 1\}$ be a random variable such that (a) it is set to one if $e \in H_B$ and zero otherwise, and (b) the following properties are satisfied.*

$$\text{With probability one, } \deg(v, H_B) \leq c_v \text{ for every small node } v \in S. \quad (12)$$

$$\Pr[e \in H_B] = \mathbf{E}[Z_B(e)] = w(e) \text{ for every edge } e \in E_B. \quad (13)$$

$$\forall v \in B, \text{ variables } \{Z_B(u, v)\}, u \in \mathcal{N}(v, E_B), \text{ are mutually independent.} \quad (14)$$

$$\text{For each edge } e \in H_B, \text{ we have } w^B(e) = 1 \quad (15)$$

Definition 10. *Let $Z_S(e) \in \{0,1\}$ be a random variable such that (a) it is set to one if $e \in H_S$ and zero otherwise, and (b) the following properties hold.*

$$\Pr[e \in H_S] = \mathbf{E}[Z_S(e)] = p_e = \min(1, w(e) \cdot (c\lambda \log n/\epsilon)) \quad \forall e \in E_S. \quad (16)$$

$$\text{The variables } \{Z_S(e)\}, e \in E_S, \text{ are mutually independent.} \quad (17)$$

$$\text{For each edge } e \in H_S, \text{ we have } w^S(e) = \begin{cases} w(e) & \text{if } p_e \geq 1; \\ \epsilon/(c\lambda \log n) & \text{if } p_e < 1. \end{cases} \quad (18)$$

Lemma 3. *For every node $v \in V$, define $W_v^B = \sum_{u \in \mathcal{N}(v, H_B)} w^B(u, v)$. The following conditions hold with high probability. (a) For every node $v \in V$, we have $W_v^B \leq c_v$. (b) For every node $v \in B \cap T$, we have $W_v^B \geq (1 - \epsilon) \cdot (c_v/\lambda)$.*

Proof (Sketch).

Consider any small node $v \in S$. By equations 12, 15, we have $W_v^B = \deg(v, H_B) \leq c_v$ with high probability. Now, consider any big node $v \in B$. By equations 13, 15 and linearity of expectation, we have $\mathbf{E}[W_v^B] = W_v \leq c_v/(1 + \epsilon)$. Furthermore, if $v \in B \cap T$, then we have $\mathbf{E}[W_v^B] = W_v \geq c_v/\lambda$. Since $c_v \geq c \log n$, the Lemma now follows from equation 14 and Chernoff bound. □

Lemma 4. *For every node $v \in V$, define $W_v^S = \sum_{u \in \mathcal{N}(v, H_S)} w^S(u, v)$. The following conditions hold with high probability. (a) For each node $v \in V$, we have $W_v^S \leq c_v$. (b) For each node $v \in S$, we have $\deg(v, H_S) = O(\log^2 n)$. (c) For each node $v \in S \cap T$, we have $W_v^S \geq (1 - \epsilon) \cdot (c_v/\lambda)$.*

Proof (Sketch). In order to highlight the main idea subject to space constraints, we assume that $p_e < 1$ for every edge $e \in E_S$. First, consider any small node $v \in S$. Since $\mathcal{N}(v, E_S) = \mathcal{N}(v, E)$, from equations 10, 16, 18 and linearity of expectation, we infer that $\mathbf{E}[\deg(v, H_S)] = (c\lambda \log n/\epsilon) \cdot W_v \leq (c\lambda \log n/\epsilon) \cdot (c_v/(1 + \epsilon))$. Since $c_v \in [1, c \log n]$, from equation 17 and Chernoff bound we infer that $\deg(v, H_S) \leq (c\lambda \log n/\epsilon) \cdot c_v = O(\log^2 n)$ with high probability. Next, note that $W_v^S = \deg(v, H_S) \cdot (\epsilon/(c\lambda \log n))$. Hence, we also get $W_v^S \leq c_v$ with high probability. Next, suppose that $v \in S \cap T$. In this case, we have $\mathbf{E}[\deg(v, H_S)] = (c\lambda \log n/\epsilon) \cdot W_v \geq (c\lambda \log n/\epsilon) \cdot (c_v/\lambda)$. Again, since this expectation is sufficiently large, applying Chernoff bound we get $\deg(v, H_S) \geq (c\lambda \log n/\epsilon) \cdot (1 - \epsilon) \cdot (c_v/\lambda)$ with high probability. It follows that $W_v^S = (\epsilon/(c\lambda \log n)) \cdot \deg(v, H_S) \geq (1 - \epsilon) \cdot (c_v/\lambda)$ with high probability.

Finally, applying a similar argument we can show that for every big node $v \in B$, we have $W_v^S \leq c_v$ with high probability. □

Lemma 5. *With high probability, we can maintain the edge-sets H_B, E^*, H_S, and a maximal b-matching M_S in $G_S(H) = (V, H_S)$ in $O(\log^3 n)$-amortized update time.*

Proof (Sketch). We maintain the fractional b-matching $\{w(e)\}$ as per Theorem 6. This requires $O(\log n)$ amortized update time, and starting from an empty graph, edge insertions/deletions in G lead to $O(t \log n)$ many changes in the edge-weights $\{w(e)\}$ (see Section 2). Thus, we can easily maintain the edge-set $E^* = \{e \in E : w(e) = 1\}$ in $O(\log n)$ amortized update time.

Next, we show to maintain the edge-set H_S. We do this by independently sampling each edge $e \in E_S$ with probability p_e. This probability is completely determined by the weight $w(e)$. So we need to resample the edge each time its weight changes. Thus, the amortized update time for maintaining H_S is $O(\log n)$.

Next, we show how to maintain the maximal b-matching M_S in H_S. Every edge $e \in H_S$ has at least one endpoint in S, and each node $v \in S$ has $\deg(v, H_S) = O(\log^2 n)$ with high probability (see Lemma 4). Due to this fact, for each node $v \in B$, we can maintain the set of its free neighbors $F_v(S) = \{u \in \mathcal{N}(v, H_S) : u$ is unmatched in $M_S\}$ in $O(\log^2 n)$ worst case time per update in H_S, with high probability (w.h.p.). Using the sets $\{F_v(S)\}, v \in B$, after each edge insertion/deletion in H_S, we can update the maximal b-matching M_S in $O(\log^2 n)$ worst case time w.h.p. [9]. Since each edge insertion/deletion in G, on average, leads to $O(\log n)$ edge insertions/deletions in H_S, we spend $O(\log^3 n)$ amortized update time for maintaining M_S, w.h.p.

Finally, we show how to maintain the set H_B. The edges $(x, y) \in E_B$ with both endpoints $x, y \in B$ are sampled independently with probability $w(x, y)$. This requires $O(\log n)$ amortized update time. Next, each small node $v \in S$ randomly selects some neighbors $u \in \mathcal{N}(v, E_B)$ and adds the corresponding edges (u, v) to the set H_B, ensuring that $\Pr[(u, v) \in H_B] = w(u, v)$ for all $u \in \mathcal{N}(v, E_B)$ and that $\deg(v, H_B) \leq c_v$. The random choices made by the different small nodes are mutually independent, which implies equation 14. But, for a given node $v \in S$, the random variables $\{Z_B(u, v)\}, u \in \mathcal{N}(v, E_B)$, are completely correlated. They are determined as follows.

In the beginning, we pick a number η_v uniformly at random from the interval $[0, 1)$, and, in a predefined manner, label the set of big nodes as $B = \{v_1, \ldots, v_{|B|}\}$. For each $i \in \{1, \ldots, |B|\}$, we define $a_i(v) = w(v, v_i)$ if $v_i \in \mathcal{N}(v, E_B)$ and zero otherwise. We also define $A_i(v) = \sum_{j=1}^{i} a_j(v)$ for each $i \in \{1, \ldots, |B|\}$ and set $A_0(v) = 0$. At any given point in time, we define $\mathcal{N}(v, H_B) = \{v_i \in B : A_{i-1}(v) \leq k + \eta_v < A_i(v)$ for some nonnegative integer $k < c_v\}$. Under this scheme, for every node $v_i \in B$, we have $\Pr[v_i \in \mathcal{N}(v, H_B)] = A_i(v) - A_{i-1}(v) = a_i(v)$. Thus, we get $\Pr[v_i \in \mathcal{N}(v, H_B)] = w(v, v_i)$ for all $v_i \in \mathcal{N}(v, E_B)$, and $\Pr[v_i \in \mathcal{N}(v, H_B)] = 0$ for all $v_i \neq \mathcal{N}(v, E_B)$. Also note that $\deg(v, H_B) \leq \lceil \sum_{v_i \in \mathcal{N}(v, E_B)} w(v, v_i) \rceil \leq \lceil W_v \rceil \leq \lceil c_v/(1 + \epsilon) \rceil \leq c_v$. Hence, equations 12, 13 are satisfied.

In the full paper, we show that the sums $\{A_i(v)\}, v \in S, i$, and the sets $\{\mathcal{N}(v, H_B)\}, v \in S$, can be maintained using a balanced binary tree data structure in $O(\log^3 n)$ amortized update time. This means that the set H_B can also be maintained in $O(\log^3 n)$ amortized update time.   □

Theorem 7. *With high probability, we can maintain an $O(1)$-approximately optimal b-matching in the input graph G in $O(\log^3 n)$ amortized update time.*

Proof (Sketch). We maintain the random sets of edges H_B and H_S, a maximal b-matching M_S in the subgraph $G_S(H) = (V, H_S)$, and the set of edges $E^* = \{e \in E : w(e) = 1\}$ as per Lemma 5. This requires $O(\log^3 n)$ amortized update time with high probability (w.h.p.). We will show that w.h.p., one of the edge-sets E^*, H_B and M_S is an $O(1)$-approximately optimal b-matching in G. But, first, we claim that:

$$w(E^*) + \sum_{v \in B \cap T} W_v + \sum_{v \in S \cap T} W_v \geq w(E) \tag{19}$$

equation 19 holds since each edge $e \in E \setminus E^*$ has at least one endpoint in T (by equation 11), and hence each edge $e \in E$ contributes at least $w(e)$ to the left hand side and exactly $w(e)$ to the right hand side. Next, note that by Lemmas 3, 4, the weight functions w^B, w^S are fractional b-matchings in G with high probability. For the rest of proof, we condition on this event, and consider three possible cases based on equation 19.

Case 1. $w(E^*) \geq (1/3) \cdot w(E)$. In this case, since $w(e) = 1$ for all $e \in E^*$, Theorem 6 imply that E^* is an $O(1)$-approximately optimal b-matching in G.

Case 2. $\sum_{v \in B \cap T} W_v \geq (1/3) \cdot w(E)$. Here, Lemma 3 and equation 15 imply that: $\sum_{v \in B \cap T} W_v \leq \sum_{v \in B \cap T} c_v \leq \sum_{v \in B \cap T} O(1) \cdot W_v^B \leq O(1) \cdot 2 \cdot w^B(H^B) = O(1) \cdot |H_B|$. Since $O(1) \cdot |H_B| \geq \sum_{v \in B \cap T} W_v \geq (1/3) \cdot w(E)$, Theorem 6 implies that the edge-set H_B is an $O(1)$-approximately optimal b-matching in G.

Case 3. $\sum_{v \in S \cap T} W_v \geq (1/3) \cdot w(E)$. Here, Lemma 4 implies that: $\sum_{v \in S \cap T} W_v \leq \sum_{v \in S \cap T} c_v \leq O(1) \cdot \sum_{v \in S \cap T} W_v^S \leq O(1) \cdot 2 \cdot w^S(H_S) \leq O(1) \cdot |M_S|$. The last inequality holds since M_S is a 1-maximal b-matching in $G_S(H) = (V, H_S)$, and hence we have $w^S(H_S) \leq 3 \cdot |M_S|$ (see Theorem 1). Finally, since $O(1) \cdot |M_S| \geq \sum_{v \in S \cap T} W_v \geq (1/3) \cdot w(E)$, Theorem 6 implies that the edge-set M_S is an $O(1)$-approximately optimal b-matching in G. $\qquad\square$

References

1. Bar-Yehuda, R., Even, S.: A linear time approximation algorithm for the weighted vertex cover problem. Journal of Algorithms **2**, 198–203 (1981)
2. Baswana, S., Gupta, M., Sen, S.: Fully dynamic maximal matching in $O(\log n)$ update time. In: FOCS, pp. 383–392 (2011)
3. Bhattacharya, S., Henzinger, M., Italiano, G.F.: Deterministic fully dynamic data structures for vertex cover and matching. In: Procs. 26th Annual ACM-SIAM Symposium on Discrete Algorithms (SODA 2015), pp. 785–804 (2015)
4. Buchbinder, N., Naor, J.: The design of competitive online algorithms via a primal-dual approach. Foundations and Trends in Theoretical Computer Science **3**(2–3), 93–263 (2009)
5. Dantzig, G.B., Ford, L.R., Fulkerson, D.R.: A primal-dual algorithm for linear programs. In: Kuhn, H.W., Tucker, A.W. (eds.) Linear Inequalities and Related Systems, pp. 171–181. Princeton University Press (1956)

5. Eppstein, D., Galil, Z., Italiano, G.F.: Dynamic graph algorithms. In: Atallah, M.J., Blanton, M. (eds.) Algorithms and Theory of Computation Handbook, 2nd edn., vol. 1, pp. 9.1–9.28. CRC Press (2009)

7. Gupta, M., Peng, R.: Fully dynamic $(1 + \epsilon)$-approximate matchings. In: FOCS, pp. 548–557 (2013)

8. Kuhn, H.W.: The Hungarian method for the assignment problem. Naval Research Logistics Quarterly 2, 83–97 (1955)

9. Neiman, O., Solomon, S.: Simple deterministic algorithms for fully dynamic maximal matching. In: STOC, pp. 745–754 (2013)

10. Onak, K., Rubinfeld, R.: Maintaining a large matching and a small vertex cover. In: STOC, pp. 457–464 (2010)

11. Vazirani, V.: Approximation Algorithms. Springer-Verlag, NY (2001)

What Percentage of Programs Halt?

Laurent Bienvenu[1]([✉]), Damien Desfontaines[2], and Alexander Shen[3,4]

[1] LIAFA - CNRS and Université Paris 7, Paris, France
`laurent.bienvenu@liafa.univ-paris-diderot.fr`
[2] Google Inc., Zurich, Switzerland
`damien@desfontain.es`
[3] LIRMM - CNRS and Université Montpellier, Montpellier, France
[4] On leave from IITP RAS, Moscow, Russia
`alexander.shen@lirmm.fr`

Abstract. Fix an optimal Turing machine U and for each n consider the ratio ρ_n^U of the number of halting programs of length at most n by the total number of such programs. Does this quantity have a limit value? In this paper, we show that it is not the case, and further characterise the reals which can be the limsup of such a sequence ρ_n^U. We also study, for a given optimal machine U, how hard it is to approximate the domain of U from the point of view of coarse and generic computability.

1 Introduction

1.1 Motivation

The title of this paper, 'What percentage of programs halt?' is intentionally provocative; obviously, the answer depends on the programming language. To make this question reasonable, we need to put some restrictions on the programming language (=interpreter). Following the theory of algorithmic information, we consider "optimal programming languages". That is, we consider an optimal Turing machine U (see below for the exact definition) and look, for each n, at the fraction ρ_n^U of inputs of length at most n on which U halts (among all inputs of those lengths). It is well known that the sequence ρ_n^U is not computable (knowing the exact values of ρ_n^U, one can solve the halting problem). What else can be said about it? For example, can ρ_n^U converge to some limit? As we will see, this cannot happen (Theorem 4). What can then be said about the limit points of ρ_n^U? They are Martin-Löf random numbers, even relative to $\mathbf{0}'$ (Theorem 5). What are the possible values of $\limsup \rho_n^U$? All $\mathbf{0}'$-lower semicomputable $\mathbf{0}'$-random numbers (Theorem 6; for liminf similar question remains open).

In the second part of the paper we build on these results to study a related question: can we somehow approximate the domain of U? That is, can we find an algorithm that tells us whether $U(p)$ terminates or not, giving the correct answer for most inputs p? This question may be formalized in different ways. For most of them, the answer will not depend on the particular choice of optimal machine, with the notable exception of Theorem 15.

© Springer-Verlag Berlin Heidelberg 2015
M.M. Halldórsson et al. (Eds.): ICALP 2015, Part I, LNCS 9134, pp. 219–230, 2015.
DOI: 10.1007/978-3-662-47672-7_18

All these questions are quite natural and similar results appeared in different settings. In 1974 Nancy Lynch [11] considered similar questions for a more restricted class of machines that are optimal in some effective sense, as defined by Schnorr [15]. Later the question for some specific universal machine was studied by Hamkins and Miasnikov who showed [7] that the halting problem can be approximated in this case. They considered Turing machines with one-sided tape. Their result implies that corresponding universal machine is not optimal and thus not effectively optimal.) The criterion for the domains of optimal machines (a set is a domain for some optimal machine if and only if it is a computably enumerable set such that the complexity of the number of strings of length at most n in this set is $n - O(1)$) was obtained by Calude, Nies, Staiger, and Stephan [4]. The recursion-theoretic properties of different versions of approximate computability have been studied by Downey, Jockusch, and Schupp [6]. See also Antti Valmari [16] who provides a survey of some other results, including the ones from [14] and [8]. Our goal in this paper is to provide a unified approach that allows us to give simple proofs of known results (sometimes in a more general form) and establish some new ones.

1.2 Definitions and Notation

For a set E, by χ_E we denote the characteristic function of this set. If E is finite, $|E|$ denotes its cardinality. We write 'log' for base 2 logarithms.

We denote by $\{0,1\}^*$ the set of all (finite) binary strings, by $\{0,1\}^n$ the set of strings of length n and by $\{0,1\}^{\leqslant n}$ the set of strings of length at most n. The length of a string x is denoted by $|x|$. We denote by $\{0,1\}^\omega$ the set of infinite binary sequences. They are also identified with real numbers in $[0,1]$ in binary notation; we mention the cases when the non-uniqueness (the same number has two representations) creates problems.

For a partial computable function f, the domain of f, denoted by $\mathrm{dom}(f)$, is the set of inputs on which f halts. A *machine* is a partial computable function from $\{0,1\}^*$ to $\{0,1\}^*$. An input p of a machine M is sometimes referred to as a *program*, and if $M(p) = x$, we say that p is a *description of x* (relative to M), or that x is the *output* of program p.

By C and K we respectively denote the plain and prefix-free versions of Kolmogorov complexity. We assume that the reader has some background in computability theory, Kolmogorov complexity and algorithmic randomness (see, e.g., [5,10,13,17]).

Definition 1. *A machine U is said to be* optimal *if for every machine M there is a constant c_M such that whenever $M(p) = x$, there is a q such that $|q| \leqslant |p| + c_M$ and $U(q) = x$.*

This definition is used to define plain Kolmogorov complexity: if U is optimal, then $C_U(x) = \min\{|q|\colon U(q) = x\}$ is the plain Kolmogorov complexity function (defined up to $O(1)$ additive term).

In the rest of this paper, we assume that U is a fixed optimal Turing machine. Let H_n be the number of programs of length at most n on which U halts, and let ρ_n^U be the *fraction* of programs of length at most n on which U halts among all programs of lengths at most n. For simplicity, we define $\rho_n^U = H_n/2^{n+1}$, even though technically there are only $2^{n+1} - 1$ programs of length at most n. Since we are only concerned in the asymptotic behaviour, this does not matter (to make things completely formal we could also add an extra program of length 0).

2 Counting How Many Programs Halt

2.1 The Complexity of H_n

The following easy lemma is well-known (see for example [17]).

Lemma 2. *For all n, $C(H_n|n) = C(H_n) = n$ with $O(1)$-precision.*

Proof. Indeed, $C(H_n|n) \leqslant C(H_n) \leqslant n$ with $O(1)$-precision since $H_n \leqslant 2^{n+1}$. Conversely, if we have a program q that maps n to H_n and is d bits shorter than n, we may take $O(\log d)$-bit self-delimiting description of d and append q; the resulting string allows us to reconstruct d, then q, then $n = |q| + d$, then $H_n = q(n)$. Then we find all strings of length at most n where U is defined, and a string z that has no description of size at most n. This gives $C(z) > n$ and $C(z) \leqslant O(\log d) + n - d + O(1)$ at the same time, so $d = O(1)$.   □

The next lemma extends this result to *approximations* for H_n.

Lemma 3. *Let N be an integer such that $|N - H_n| \leqslant 2^k$. Then $C(N|n) \geqslant n - k - K(k|n) - O(1)$ and $K(N|n) \geqslant n - k - O(1)$.*

Proof. Let N be an approximation of H_n with error at most 2^k, and let t be the program of length $C(N|n)$ that maps n to N. We can reconstruct H_n given n, t and the difference $N - H_n$ (first we reconstruct N and then H_n). So $C(H_n|n) \leqslant C(t, N-H_n|n)$. The pair $(t, N-H_n)$ can be described by appending t to the self-delimiting description of $N-H_n$ (the latter requires $k+K(k|n)$ bits), or we could use self-delimiting program t and append plain description of $N - H_n$ (the latter requires $k+O(1)$ bits).[1] This gives respectively $n \leqslant C(N|n)+k+K(k|n)+O(1)$ and $n \leqslant K(N|n) + k + O(1)$.   □

2.2 Limit Points of ρ_n^U

Lemma 3 can be used to get some information about ρ_n^U. Assume that r_n is some computable sequence of rational numbers. How close can it approximate ρ_n^U? The complexity $K(r_n|n)$ is $O(1)$, so Lemma 3 gives a constant upper bound for $n - k$, so $k = n - O(1)$, which means an absolute error for H_n of size at least $\Omega(2^n)$, so $|\rho_n - r_n|$ is separated from 0 for sufficiently large n.

In particular, taking $r_n = 0$ or $r_n = 1$, we see that $\varepsilon \leqslant \rho_n^U \leqslant 1 - \varepsilon$ for some $\varepsilon > 0$ and for all sufficiently large n.

---

[1] In other words, we use the inequality $C(u, v) \leqslant K(u) + C(v)$ in two different ways.

We will use this lemma to show that ρ_n^U has no limit. Note first that it is very easy to construct a particular optimal machine V such that ρ_n^V does not converge. (For example, it is easy to construct an optimal machine V defined only on inputs of even length, then ρ_{2n}^V and ρ_{2n+1}^V differ by factor 2: the numerator is the same and the denominators differ by factor 2.) The next theorem shows that ρ_n^U *never* converges, no matter which optimal machine we choose.

Theorem 4. *The sequence* $(\rho_n^U)_{n \in \mathbb{N}}$ *does not converge.*

Proof. Consider a computable sequence r_n that is everywhere dense in $[0, 1]$ (say, enumerates all rational numbers in $[0, 1]$). If ρ_n^U has some limit ρ, then ρ_n^U is close to ρ for *all sufficiently large* n while r_n is close to ρ for *infinitely many* n, so the difference $r_n - \rho_n^U$ cannot be separated from 0. □

Now that we know that the sequence ρ_n^U does not converge, one can study its limit points. The next theorem shows that any limit point of the sequence must be quite complex, indeed, Martin-Löf random relative to $\mathbf{0}'$.

Theorem 5. *All limit points of* $(\rho_n^U)_{n \in \mathbb{N}}$ *are Martin-Löf random relative to* $\mathbf{0}'$.

Proof. For this proof we need to use a theorem by Miller [12] (see also [1] for a simple proof): a real number (a bit sequence) $x \in [0, 1]$ is Martin-Löf random relative to $\mathbf{0}'$ if and only if there is a constant c such that for every prefix σ of x, there is a finite string τ extending σ such that $C(\tau) \geqslant |\tau| - c$.

Suppose x is a limit point of ρ_n^U. First note that x cannot be a rational number (otherwise the constant sequence $r_n = r$ approximates ρ_n^U), so x has a unique binary representation. Let σ be a prefix of x and let k be the length of σ. Split $[0, 1]$ into 2^k equal intervals of size 2^{-k}. Then x is strictly inside one of these intervals (this interval consists of all binary extensions of σ). Since x is a limit point, some ρ_n^U also belongs to this interval. Recall that ρ_n^U is a binary fraction $H_n/2^{n+1}$ (here it is important that we use this denominator, not $2^{n+1} - 1$; of course, this does not change the limit points). Therefore, H_n (considered as a string of length $n + 1$ with leading zeros) is an extension of σ, and $C(H_n) \geqslant |H_n| - O(1)$ due to Lemma 2, so it remains to use Miller's result. □

We do not know whether the converse holds, i.e., whether any real that is Martin-Löf random relative to $\mathbf{0}'$ is a limit point of some sequence ρ_n^U for some optimal U. However, we can give a full characterisation of the reals that are lim sup's of those sequences.

Theorem 6. *The* lim sup *of* $(\rho_n^U)_{n \in \mathbb{N}}$ *is upper semicomputable relative to* $\mathbf{0}'$ *(and Martin-Löf random relative to* $\mathbf{0}'$ *by the previous theorem). Moreover, the converse holds: every real in* $[0, 1]$ *that is upper semicomputable relatively to* $\mathbf{0}'$ *and Martin-Löf random relative to* $\mathbf{0}'$ *is the* lim sup *of* ρ_n^V *for some optimal machine* V.

Proof. Let us consider first a simpler question. Assume that X is an arbitrary computably enumerable set, i.e., the domain of some machine, not necessarily an optimal one; x_n is the number of strings of length n in X, and X_n is the number of strings of length at most n in X (so $X_n = x_0 + \ldots + x_n$). Consider the upper density of X, i.e., $\limsup X_n/2^{n+1}$. Which reals can appear as upper densities of computably enumerable sets?

Lemma 7. *A real number x in $[0,1]$ is the upper density of some computably enumerable set X if and only if x is upper semicomputable relative to $\mathbf{0}'$.*

Proof. In one direction: $X_n/2^{n+1}$ is a uniformly lower semicomputable sequence of reals, and one can show (see, e.g., [6]) that $\limsup$ of such a sequence is upper semicomputable relative to $\mathbf{0}'$.

Reverse direction: assume that x is upper semicomputable relative to $\mathbf{0}'$. It is known that x can be represented as $\limsup k_n$ for some computable sequence k_n of rational numbers (see [6] or [17]). Then $x = \lim_n K_n$, where $K_n = \sup(k_n, k_{n+1}, \ldots)$ form a uniformly lower semicomputable sequence. We may assume without loss of generality that $K_n \in [0,1]$ (since the limit is in $[0,1]$) and that K_n are rational numbers with denominator 2^n (by rounding; note that the resulting sequence K_n may not be computable, only lower semicomputable). Then we consider a computably enumerable set X that contains exactly K_n strings of length n (here we use that K_n are lower semicomputable). It is easy to see that the upper density of X is x; in fact, the density (the limit, not only $\limsup$) exists and is equal to x, since the fraction of n-bit strings in X converges to x as $n \to \infty$.   □

It remains to show that for x that are not only upper semicomputable relative to $\mathbf{0}'$ but also Martin-Löf random relative to $\mathbf{0}'$, the set X can be made a domain of an optimal machine. Our next step is the following simple observation.

Lemma 8. *If some real $x \in [0,1]$ is the upper density of the domain of some optimal machine, the same is true for $x/2$ and $(1+x)/2$.*

(In terms of binary representation $x/2$ is $0x$, and $(1+x)/2$ is $1x$.)

Proof. For $x/2$ we just "shift" the domain of the optimal machine by adding leading 0 to all the arguments. For $(1+x)/2$ we do the same and also add all strings starting with 1 to the domain (with arbitrary values, e.g., they all can be mapped to an empty string). In both cases the machine remains optimal, the complexity increases only by 1.   □

Deleting the first bit preserves randomness, so we may assume without loss of generality that x (that is random and upper semicomputable relative to $\mathbf{0}'$) is smaller than $1/2$ (starts with 0), and then apply Lemma 8 to add leading ones.

Now we are ready to use another known result: *every random upper semicomputable x is Solovay complete among upper semicomputable reals* (all properties are considered relative to $\mathbf{0}'$); according to one of the equivalent definitions of Solovay completeness, this means that for every other upper semicomputable relative to $\mathbf{0}'$) y and for large enough N there exists another upper semicomputable (relative to $\mathbf{0}'$) z such that $x = y/N + z$. This result combines the work of Calude et al. [3] and Kučera-Slaman [9] (see [2] for a simplified proof). Technically these papers consider lower semicomputable reals instead of upper semicomputable ones. However, an upper semicomputable real is just the opposite of a lower semicomputable real, randomness is stable under sign change, and Solovay reducibility, although often restricted to numbers in $[0, 1]$, extends naturally to all real numbers (again, see [2]), so the result also holds for upper semicomputable reals. Also, we need a relativized version of their result to $\mathbf{0}'$; as usual, relativization is straightforward.

So let us assume that $x \in (0, 1/2)$ and $x = y/2^d + z$ where y is the upper density for some optimal machine U and z is upper semicomputable relative to $\mathbf{0}'$. (The large denominator N is chosen to be a power of 2.) Now we combine two tricks used for Lemmas 7 and 8. Namely, we apply Lemma 7 to $2z$ (note that $z < 1/2$), and then add leading 1's to all the strings in the corresponding set. This gives us density z while using only right half of the binary tree (strings that start with 1). Then we add d zeros to all strings in the domain of U as we did when proving Lemma 8; this gives us density $y/2^d$ using only left half of the binary tree (actually, a small part of it, if d is large). Then we combine both parts and get an optimal machine (since the left part is optimal) with upper density $y/2^d + z$ as required. (Note that in general lim sup is not additive, but in our case we have not only lim sup, but limit in one of the parts, so additivity holds.)   □

3   Approximating the Halting Problem

3.1   Generic and Coarse Computability

Instead of just counting the terminating programs of bounded length, one can also look at a related question: *is there an algorithm which, given p, predicts whether or not $p \in \mathrm{dom}(U)$, and is right "most of the time"?* This is a rather informal question; to make it formal we have to specify what we mean by 'predict', and 'most of the time'. There are several ways to do this, and two paradigms in particular have received a lot of attention in the recent literature, the so-called *coarse computability* and *generic computability*. For both of them, "being right most of the time" is understood as "being right on a set of density 1". (Recall that the upper density $\bar{\rho}(A)$ of a set $A \subseteq \{0, 1\}^*$ is $\limsup_n |A \cap \{0, 1\}^{\leqslant n}|/2^{n+1}$, the lower density $\underline{\rho}(A)$ is $\liminf_n |A \cap \{0, 1\}^{\leqslant n}|/2^{n+1}$, and when the two are equal, their common value is called the density of A. Sometimes the density is defined for sets of natural numbers and all the initial segments are considered, not only powers of 2, but for density 1 this does not matter.)

The difference between coarse computability and generic computability lies in the prediction model. In coarse computability, the predictor is a *total* computable function which given an input $p \in \{0, 1\}^*$ should always return 0 or 1 (meaning "$p \notin \mathrm{dom}(U)$" and "$p \in \mathrm{dom}(U)$" respectively), but is allowed to be incorrect sometimes, as long as the set of errors has density zero. In the generic computability model, the predictor function is still $0/1$-valued, but is allowed to be partial as long as its domain has density 1, *and whenever a $0/1$-prediction is made, it must be correct.* Formally, we have the following definitions.

Definition 9. *A set $A \subseteq \{0,1\}^*$ is* coarsely computable *if there exists a total computable function $f : \{0,1\}^* \to \{0,1\}$ such that the set $\{p \mid f(p) = \chi_A(p)\}$ has density 1. A set $A \subseteq \{0,1\}^*$ is* generically computable *if there exists a partial computable function $f : \{0,1\}^* \to \{0,1\}$ such that $\mathrm{dom}(f)$ has density 1 and $f(p) = \chi_A(p)$ for all $p \in \mathrm{dom}\, f$.*

These two notions are incomparable: a computably enumerable set can be coarsely computable but not generically computable and vice-versa (see [6]). The initial informal question we started with can now be precisely formulated: if U is an optimal machine, can $\mathrm{dom}(U)$ be coarsely computable? generically computable? The answer is no, even if we allow the approximating function f to be both non-total and sometimes wrong — still requiring that it is correct for most inputs. Moreover, f has $\Omega(1)$ fraction of errors among strings of length at most n, *for all sufficiently large n*, not only for infinitely many n (as it is needed to show that f is not coarsely/generically computable). Similar results were obtained in a slightly different setting in [14]; we provide a simple argument that requires only optimality and covers both generic and coarse computability.

Theorem 10. *For every partial computable function $f : \{0,1\}^* \to \{0,1\}$ there exists some $\varepsilon > 0$ so that the fraction of strings x of size at most n where f is undefined or gives a wrong answer $(f(x) \neq \chi_{\mathrm{dom}\, U}(x))$ exceeds ε for all sufficiently large n.*

Proof. We repeat the proof of Lemma 3. Knowing n and some bound 2^{n-d} for the number of errors (of both types: f is undefined or the value is wrong) that f makes for strings of length at most n, we wait until f becomes defined on all strings of those lengths except for 2^{n-d} many. Then we count the number of positive answers; it differs from H_n by at most $O(2^{n-d})$. The difference can be specified by $n-d+O(1)$ bits, so the complexity $C(H_n|n)$ is bounded by $K(d|n) + (n-d) + O(1)$, where $O(1)$-constant depends on f. The bound $C(H_n|n) \geqslant n - O(1)$ then implies that $d - K(d|n) \leqslant O(1)$, so $d = O(1)$. This provides the required bound $2^{-O(1)}$ for the fraction of errors for all large enough n. □

3.2 Allowing a Small Density of Errors and 'Infinitely Often'-Success

The constant ε in Theorem 10 may depend on the predictor f. Can we prove a stronger result where the same ε is used for all predictors? It is indeed possible if we only want the predictor to have a lot of errors for *infinitely many* lengths, not for all sufficiently large ones. The result of this type was obtained in [8]; we provide a simple proof of its version for arbitrary optimal machines. More precisely, let us consider the following definition (which makes sense when α is close to 1).

Definition 11. *Let* $\alpha \in [0,1]$*. A set* $A \subseteq \{0,1\}^*$ *is* α-*coarsely computable if there exists a total computable function* $f : \{0,1\}^* \to \{0,1\}$ *such that the set* $\{p \mid f(p) = \chi_A(p)\}$ *has lower density at least* α*. A set* $A \subseteq \{0,1\}^*$ *is* α-*generically computable if there exists a partial computable function* $f : \{0,1\}^* \to \{0,1\}$ *such that* $\mathrm{dom}(f)$ *has lower density at least* α *and* $f(p) = \chi_A(p)$ *for all* $p \in \mathrm{dom}(f)$*.*

Although we saw that there was no implication between being generically computable and coarsely computable, there *is* such a link in the quantified setting. Namely, if a set is α-generically computable, it is β-coarsely computable for any $\beta < \alpha$. Indeed, consider some rational threshold r between β and α. We now that for infinitely many lengths the fraction of answers provided by generic predictor f, exceeds r. These lengths can be ultimately discovered (by waiting until the fraction exceeds r). Let us consider a fast growing computable sequence that contains only these lengths (not necessarily all of them). For lengths in this sequence we know r-fraction of correct answers and give arbitrary answers or the rest. (There is a small technical problem since these answers could be incompatible with the answers chosen previously, for smaller lengths. But if the lengths in the sequence grow fast enough, this small change is compensated by the difference between β and r.)

Theorem 12. *There exists* $\alpha < 1$ *such that* $\mathrm{dom}(U)$ *is neither* α-*coarsely computable nor* α-*generically computable.*

Proof. This result, like Theorem 10, remains true even if we allow errors of both types (as before), and the proof is similar. Proving Theorem 10, we noted that $C(H_n|n) \leqslant K(d|n) + (n-d) + O(1)$, if some (fixed) algorithm f has fraction of errors at most 2^{-d} on strings of length at most n. Here the constant in $O(1)$ depends on f. We also can treat f as a parameter; the same argument gives then $C(H_n|n) \leqslant K(d|n) + (n-d) + K(f|n) + O(1)$, where $K(f|n)$ is the prefix conditional complexity of an algorithm computing f, given n. Now $O(1)$ is the same for all f. It remains to note that for every computable f there are infinitely many n such that $K(f|n) = O(1)$, where the constant does not depend on f (or n). For example, we may consider n whose binary representation starts with self-delimited encoding of the program for f. The rest of proof remains the same, and we get the same ε for all f (but only for n that make f simple). $\square$

The next natural question is whether we can combine both results and beat each predictor for all sufficiently large lengths, still using the same ε for all predictors. The following definition formalizes this question; we use dual notions where lower density is replaced by upper density (and "i.o." stands for "infinitely often").

Definition 13. *Let* $\alpha \in [0,1]$*. A set* $A \subseteq \{0,1\}^*$ *is* α-*i.o.-coarsely computable if there exists a total computable function* $f : \{0,1\}^* \to \{0,1\}$ *such that the set* $\{p \mid f(p) = \chi_A(p)\}$ *has an upper density of at least* α*. A set* $A \subseteq \{0,1\}^*$ *is* α-*i.o.-generically computable if there exists a partial computable function* $f : \{0,1\}^* \to \{0,1\}$ *such that* $\mathrm{dom}(f)$ *has an upper density of at least* α *and* $f(p) = \chi_A(p)$ *for all* $p \in \mathrm{dom}(f)$*.*

Now the situation changes.

Theorem 14. *For any $\alpha < 1$,* $\mathrm{dom}(U)$ *is α-i.o.-coarsely computable.*

Note however that $\mathrm{dom}(U)$ is never 1-i.o.-computable due to Theorem 10.

Proof sketch. The proof is similar to the argument above (that relates generic and coarse computations). Consider the value $\rho = \limsup \rho_n^U$, and consider some rational number r that is smaller than ρ but very close to it (the difference is less that $1 - \alpha$). There are infinitely many lengths for which the fraction of terminating computations exceeds r, and these lengths can be discovered ultimately, so we can consider a computable fast increasing sequence containing only those "good" lengths (not necessarily all of them). For each length in this sequence, we run U until we get r-fraction of terminating programs, and use the results for coarse prediction. The positive answers are guaranteed to be correct, while the negative answers may be incorrect. But the fraction of incorrect answers ultimately becomes less than $1 - \alpha$, since for large n the values ρ_n^U can only slightly exceed ρ (and therefore r). Again we should be careful enough to consider a fast growing sequence of lengths, so that small lengths do not interfere with large ones. □

This argument provides i.o.-coarse computability but not i.o.-generic computability. In fact, the latter may depend on the choice of the optimal machine (the rare situation we mentioned in the introduction).

Theorem 15. *There exists an optimal machine U_1 such that for any $\alpha < 1$ the set $\mathrm{dom}(U_1)$ is α-i.o.-generically computable. But there also exists an optimal machine U_2 such that $\mathrm{dom}(U_2)$ is not α-i.o.-generically computable for some $\alpha < 1$.*

The second statement appeared (in a bit different setting) in [11].
Proof sketch. In fact, we can use any "left-total" optimal machine as U_1. A machine is called *left-total* if for each n it is defined on some initial segment of $\{0, 1\}^n$ in lexicographical order. (Any other computable ordering on $\{0, 1\}^n$ will work.) In other words, if such a machine is defined on some string, it is also defined on all preceding strings of the same length.

It is easy to construct a left-total optimal machine U_1 by transforming a given optimal machine U into a total one: when a new description of length n is discovered for U, we add to U_1 a description of the same object using the lexicographically first string not used earlier.

Now we need to show that for a left-total machine U_1 its domain $\mathrm{dom}(U_1)$ is α-i.o.-generically computable. The idea is simple: for the left-total machine knowing the number of n-bits strings in its domain determines what are these strings. And if we know this number with some precision, we can guarantee both the positive and negative answers except for some interval in the middle (its length is the difference between the upper and lower bounds). So we use the same trick as before, but for strings of the same length. Let us see how this can be done.

Let ρ'_n be the number of strings of length n in the domain of U_1, and let $= \limsup \rho'_n$. Fix some rational threshold that is smaller than ρ' but very close to it. If it is given to us as an advice, together with the position after which ρ'_n exceed ρ' only by a very small margin, we can effectively find lengths where we can generically compute U_1 with a small fraction of omissions. Again we can form a computable increasing sequence of lengths with this property, and construct a generic predictor that is quite precise for these lengths and undefined in all other lengths (to avoid false answers for the cases where we do not have enough information).

However, this is not enough for us, since in our definition the fraction of prediction failures is calculated in the set of all strings of length *at most* n, and even if we know everything for n-bit strings, this covers only half of the strings in question. So in this way we cannot make the error less than $1/2$.

But we can repeat the trick: consider the lengths that just precede the lengths in the subsequence. For them we have no information yet, but we may consider corresponding ρ'_n and guess the limsup *for this subsequence*. Then, using some rational threshold close to this limsup, and the position after which ρ'_n exceed this limsup only by a small margin, we can get a (computable increasing) subsequence of lengths where we have guaranteed good approximations *for two subsequent lengths*, thus reducing the error from $1/2$ to $1/4$ (approximately).

Now we can repeat the trick finitely many times and get arbitrary small error. Note that for this we need only finitely many bits of advice, so this still gives a partial computable predictor. The first statement is proven.

For the second part we use the argument provided by Lynch [11]. We can take the standard universal machine as U_2: let $U_2(0^e 1p) = M_e(p)$ where M_e is eth machine in a standard enumeration. It remains to show that the domain of U_2 is not α-i.o.-generically computable for some $\alpha < 1$. It is because some special computably enumerable set is embedded into this domain with fixed density. Here are the details.

Post has shown that there exist *simple sets*, i.e., computably enumerable sets whose complement is infinite but does not contain an infinite enumerable subset. It is easy to construct a very sparse simple set S (either by adapting the original Post's construction or taking the set of strings whose Kolmogorov complexity is very small compared to their length). Such a set can be α-i.o.-generically computable only for very small α. Indeed, our predictor gives only finitely many negative answers (otherwise we get an enumerable infinite subset of the complement). Also it can give positive answers only for a very sparse set (in all lengths), since the entire set S is sparse (and positive answers form a subset).

It remains to take machine M_e whose domain is S, and note that the strings of the form $0^e 1p$ form a fixed-density subset in the set of all strings; let δ be this density. A generic predictor for the domain of U_2 that gives error less than $\delta/2$ for infinitely many n, will provide i.o.-generic prediction for S with threshold approximately $\delta/2$, which is not possible. □

3.3 The Probabilistic Case

Most of the results proven in this section are negative, i.e., they show that $\mathrm{dom}(U)$ is hard to approximate *deterministically*. Does the situation change if try to get such approximations *probabilistically*? This can be understood in several ways; in the sequel we use the approach motivated by mass problems in Medvedev's sense. Let us start by giving the corresponding definitions. We consider machines with random oracle (a sequence of independent fair coin tosses).

Definition 16. *A set $A \subseteq \{0,1\}^*$ is* coarsely probabilistically computable *if there exists an oracle machine Γ^X with random oracle X such that the event "Γ^X computes a total function such that the set $\{p \mid \Gamma^X(p) = \chi_A(p)\}$ has density 1" has positive probability. A set $A \subseteq \{0,1\}^*$ is* generically probabilistically computable *if there exists an oracle machine Γ^X such that the event "$\mathrm{dom}(\Gamma^X)$ has density 1 and $\Gamma^X(p) = \chi_A(p)$ for all $p \in \mathrm{dom}(\Gamma^X)$" has positive probability. The notions of α-coarsely probabilistically computable, α-generically probabilistically computable, α-i.o.-coarsely probabilistically computable, and α-i.o.-generically probabilistically computable are defined in a similar way.*

One could think that allowing probabilistic computations does not change the situation. Admittedly, it does not change it much. All the results above remain the same in the probabilistic case (with more complicated proofs), with the exception of 1-i.o.-coarse computability.

Theorem 17. *Just like in the deterministic case, for α sufficiently close to 1:*

 (i) $\mathrm{dom}(U)$ *is neither α-coarsely probabilistically computable nor α-generically probabilistically computable;*
 (ii) *whether $\mathrm{dom}(U)$ is α-i.o.-generically probabilistically computable or not depends on the particular choice of machine U;*
 (iii) $\mathrm{dom}(U)$ *is not 1-i.o.-generically probabilistically computable.*

However, unlike in the deterministic case:

 (iv) $\mathrm{dom}(U)$ *is always 1-i.o.-coarsely probabilistically computable.*

The full proof of these statements is quite long, and is omitted due to space restrictions.

Acknowledgments. This paper is based on the work done while D.D. was visiting LIRMM (Montpellier) and Poncelet laboratory (Moscow). We thank our colleagues from both laboratories (in particular the ESCAPE team, and Kolmogorov seminar group) for hostpitality. A.S. thanks Antti Valmari for interesting discussion (during RuFiDiM seminar in Turku) that was the starting point for some of the arguments in this paper. Thanks also go to three anonymous referees for helpful feedback. The authors also acknowledge the support of the Templeton Foundation.

References

1. Bienvenu, L., Muchnik, A., Shen, A., Vereshchagin, N.: Limit complexities revisited [once more]. Technical report (2012). arxiv:1204.0201
2. Bienvenu, L., Shen, A.: Random semicomputable reals revisited. In: Dinneen, M.J., Khoussainov, B., Nies, A. (eds.) Computation, Physics and Beyond. LNCS, vol. 7160, pp. 31–45. Springer, Heidelberg (2012)
3. Calude, C.S., Hertling, P.H., Khoussainov, B., Wang, Y.: Recursively enumerable reals and chaitin omega numbers. In: Meinel, C., Morvan, M. (eds.) STACS 1998. LNCS, vol. 1373, pp. 596–606. Springer, Heidelberg (1998)
4. Calude, C., Nies, A., Staiger, L., Stephan, F.: Universal recursively enumerable sets of strings. Theoretical Computer Science **412**(22), 2253–2261 (2011)
5. Downey, R., Hirschfeldt, D.: Algorithmic randomness and complexity. Theory and Applications of Computability. Springer, New York (2010)
6. Downey, R.G., Jockusch Jr., C.G., Schupp, P.E.: Asymptotic density and computably enumerable sets. Journal of Mathematical Logic **13**(02) (2013)
7. Hamkins, J.D., Miasnikov, A.: The halting problem is decidable on a set of asymptotic probability one. Notre Dame Journal of Formal Logic **47**(4) (2006)
8. Köhler, S., Schindelhauer, C., Ziegler, M.: On approximating real-world halting problems. In: Liśkiewicz, M., Reischuk, R. (eds.) FCT 2005. LNCS, vol. 3623, pp. 454–466. Springer, Heidelberg (2005)
9. Kučera, A., Slaman, T.: Randomness and recursive enumerability. SIAM Journal on Computing **31**, 199–211 (2001)
10. Li, M., Vitányi, P.: An Introduction to Kolmogorov Complexity and Its Applications, 3rd edn. Springer, New York (2007)
11. Lynch, N.: Approximations to the halting problem. Journal of Computer and System Sciences, 9–143 (1974)
12. Miller, J.S.: Every 2-random real is Kolmogorov random. Journal of Symbolic Logic **69**(3), 907–913 (2004)
13. Nies, A.: Computability and randomness. Oxford University Press, Oxford Logic Guides (2009)
14. Schindelhauer, C., Jakoby, A.: The non-recursive power of erroneous computation. In: Pandu Rangan, C., Raman, V., Sarukkai, S. (eds.) FST TCS 1999. LNCS, vol. 1738, p. 394. Springer, Heidelberg (1999)
15. Claus Peter Schnorr: Optimal enumerations and optimal Gödel numberings. Mathematical Systems Theory **8**(2), 181–191 (1974)
16. Valmari, A.: The asymptotic proportion of hard instances of the halting problem. Technical report, November 2014. arxiv:1307.7066v2
17. Vereshchagin, N., Uspensky, V., Shen, A.: Kolmogorov complexity and algorithmic randomness (In Russian. See www.lirmm.fr/~ashen for the draft translation.). MCCME (2013)

The Parity of Set Systems Under Random Restrictions with Applications to Exponential Time Problems

Andreas Björklund[1], Holger Dell[2]([✉]), and Thore Husfeldt[1,3]

[1] Lund University, Lund, Sweden
[2] Saarland University and Cluster of Excellence (MMCI), Saarbrucken, Germany
hdell@mmci.uni-saarland.de
[3] IT University of Copenhagen, Copenhagen, Denmark

Abstract. We reduce the problem of detecting the existence of an object to the problem of computing the parity of the number of objects in question. In particular, when given any non-empty set system, we prove that randomly restricting elements of its ground set makes the size of the restricted set system an odd number with significant probability. When compared to previously known reductions of this type, ours excel in their simplicity: For graph problems, restricting elements of the ground set usually corresponds to simple deletion and contraction operations, which can be encoded efficiently in most problems. We find three applications of our reductions:

1. An exponential-time algorithm: We show how to decide Hamiltonicity in directed n-vertex graphs with running time 1.9999^n provided that the graph has at most 1.0385^n Hamiltonian cycles. We do so by reducing to the algorithm of Björklund and Husfeldt (FOCS 2013) that computes the parity of the number of Hamiltonian cycles in time 1.619^n.
2. A new result in the framework of Cygan et al. (CCC 2012) for analyzing the complexity of NP-hard problems under the Strong Exponential Time Hypothesis: If the parity of the number of Set Covers can be determined in time 1.9999^n, then Set Cover can be decided in the same time.
3. A structural result in parameterized complexity: We define the parameterized complexity class $\oplus W[1]$ and prove that it is at least as hard as $W[1]$ under randomized fpt-reductions with bounded one-sided error; this is analogous to the classical result $NP \subseteq RP^{\oplus P}$ by Toda (SICOMP 1991).

1 Introduction

A set family $\mathscr{F}$ with an odd number of elements is of course nonempty. In the present paper we study randomized reductions where the opposite holds with significant probability: We reduce the *decision problem* of determining if $|\mathscr{F}|$ is non-zero to the *parity problem* of determining if $|\mathscr{F}|$ is odd. Originally such decision-to-parity reductions were obtained as mere corollaries to various

© Springer-Verlag Berlin Heidelberg 2015
M.M. Halldórsson et al. (Eds.): ICALP 2015, Part I, LNCS 9134, pp. 231–242, 2015.
DOI: 10.1007/978-3-662-47672-7_19

solation lemmas," such as the one of Valiant and Vazirani [16], where the
·duction is to the *unambiguous problem* of distinguishing between $|\mathscr{F}| = 0$ and
$\mathscr{F}| = 1$. More recently, Gupta [11] designed one that is not isolating but achieves
significantly better success probability.

Our decision-to-parity reductions are not isolating, but they have a *much*
mpler structure than the existing options in that they compute random restric-
ons of the universe. In other words, they randomly delete or contract elements
f the universe. Surprisingly, the success probabilities that this approach yields
o not depend on the size of the universe – they do depend on the cardinality
f the sets in $\mathscr{F}$ or the cardinality of $\mathscr{F}$ itself, and they lie between the success
robabilities of Valiant and Vazirani [16] and Gupta [11].

Organization. In §1.1, we state the main lemma of this paper and discuss its
elationship with and consequences for probabilistic polynomial identity tests as
vell as various isolation lemmas. Before we prove the Main Lemma in §2, we
.rst state its applications for Hamiltonicity in §1.2, for Set Cover in §1.3, and
or W[1] in §1.4. The formal proofs of these results appear in the full version of
his paper.

.1 Set Systems Under Random Reductions

.et $\mathscr{F}$ denote a family of sets. We present our reductions in a general combina-
orial setting, but for the sake of concreteness we invite the reader to think of
$\mathscr{F}$ as the family of all vertex subsets that form a k-clique, or the family of all
·dge subsets that form a Hamiltonian cycle. For instance, in the house graph in
·ig. 1, the family $\{\{1,2,3,4,7\},\{1,3,4,6,8\}\}$ corresponds to the Hamiltonian
·ycles.

Fig. 1

Let U be the ground set of $\mathscr{F}$, that is, $\mathscr{F} \subseteq 2^U$. A *restriction* is a function
$\rho: U \to \{0,1,*\}$. The *restricted family* $\mathscr{F}\!\restriction_\rho$ consists of all sets $F \in \mathscr{F}$ that
·satisfy $i \in F$ for all i with $\rho(i) = 1$ and $i \notin F$ for all i with $\rho(i) = 0$. A *random*
restriction is a distribution over restrictions ρ where $\rho(i)$ is randomly sampled for
·each i independently subject to $\Pr_\rho(\rho(i) = 0) = p_0$ and $\Pr_\rho(\rho(i) = 1) = p_1$. We
·are interested in the event that the number of sets in the restricted family $\mathscr{F}\!\restriction_\rho$
·s odd, which we write as $\oplus \mathscr{F}\!\restriction_\rho$.

Lemma 1 (Main Lemma). *Let $\mathscr{F}$ be a nonempty family of sets, each of size
·at most k. Let ρ denote a random restriction with $p_1 = 0$.*

(i) If $p_0 \geq \frac{1}{2}$, then

$$\Pr_{\rho}(\oplus\mathscr{F}\restriction_{\rho}) \geq (1 - p_0)^k . \tag{1}$$

(ii) If $p_0 < \frac{1}{2}$, then

$$\Pr_{\rho}(\oplus\mathscr{F}\restriction_{\rho}) \geq (1 - p_0)^k \left(\frac{p_0}{1 - p_0}\right)^{\min\{\log|\mathscr{F}|, k\}} . \tag{2}$$

All of our applications are based on random restrictions with $p_1 = 0$; in this case, the success probabilities do not depend on the size of the underlying ground set – they do when $p_1 > 0$: As we will see in Lemma 9, the bound we get for the success probability has a factor of $(1 - p_1)^{n-k}$. Since in our setting n is big compared to k, the case $p_1 > 0$ leads to success probabilities that are exponentially small in n, which is no good.

Examples. Consider the graph of Fig. 1, where $|U| = 8$, $k = 5$, and $|\mathscr{F}| = 2$, and assume $p_1 = 0$. The restriction ρ results in an odd number of Hamiltonian cycles exactly if $\rho(1) = \rho(3) = \rho(4) = *$ and either $\rho(2) = \rho(7) = *$ or $\rho(6) = \rho(8) = *$ (but not both). For $p_0 = \frac{1}{2}$ this happens with probability $\frac{12}{256} = \frac{3}{64}$, slightly better than the bound $\frac{1}{32}$ promised by (1). If we set $p_0 = \frac{1}{5}$ then (2) promises the better bound $\Pr_{\rho}(\oplus\mathscr{F}\restriction_{\rho}) = (\frac{4}{5})^5 \cdot \frac{1}{4} = \frac{256}{3125} \geq 0.081$. For completeness, direct calculation shows that $\Pr_{\rho}(\oplus\mathscr{F}\restriction_{\rho}) = 4 \cdot (\frac{4}{5})^6 \cdot \frac{1}{5} + 2 \cdot (\frac{4}{5})^5 \cdot (\frac{1}{5})^2 = \frac{18432}{78125} \geq 0.235$, so the bound is far from tight in this example.

A simple example that attains (1) with equality is the singleton family $\mathscr{F}$ consisting only of the set $\{1, \ldots, k\}$. Then one easily computes $\Pr_{\rho}(\oplus\mathscr{F}\restriction_{\rho}) = \Pr_{\rho}(\rho(1) = \cdots = \rho(k) = *) = (1 - p_0)^k$. For an example attaining (2) with equality, consider the family $\mathscr{F}$ of sets F satisfying $\{1, \ldots, (1 - \epsilon)k\} \subseteq F \subseteq \{1, \ldots, k\}$, where $0 < \epsilon \leq \frac{1}{2}$ holds and ϵk is an integer. Then $|\mathscr{F}| = 2^{\epsilon k}$. There is but one restriction ρ for which the event $\oplus\mathscr{F}\restriction_{\rho}$ happens, namely when $\rho(i) \neq 0$ for all $i \leq (1 - \epsilon)k$ and $\rho(i) = 0$ for all $i > (1 - \epsilon)k$. Thus, with $p_0 = \epsilon$ we have

$$\Pr_{\rho}(\oplus\mathscr{F}\restriction_{\rho}) = (1 - p_0)^{(1-\epsilon)k} p_0^{\epsilon k} = (1 - p_0)^k \left(\frac{p_0}{1 - p_0}\right)^{\log|\mathscr{F}|} .$$

Connection with Probabilistic Polynomial Identity Tests. The Main Lemma can be expressed in terms of polynomials over finite fields instead of restricted set systems by considering the nonempty set system $\mathscr{F}$ as the nonzero polynomial

$$p(x_1, \ldots, x_n) = \sum_{F \in \mathscr{F}} \prod_{i \in F} x_i$$

in the polynomial ring $\mathrm{GF}(2)[x]$. The Main Lemma then says that if $a \in \mathrm{GF}(2)^n$ is chosen uniformly at random, we have

$$\Pr_{a}\Big(p(a_1, \ldots, a_n) = 0\Big) \leq 1 - 2^{-k} ,$$

here k is the total degree of p; since p is multilinear, k corresponds to the maximum number of variables occurring in a monomial.

Thus, our Main Lemma can be understood as a variant of the well-known probabilistic polynomial identity test of DeMillo and Lipton (1978), Schwartz (1980), and Zippel (1979) (cf. [1, Lemma 7.5]). In its standard form, this lemma bounds the probability by $k/2$, where the 2 stems from the size of the finite field GF(2); we usually have $k/2 \geq 1$ and so the bound is vacuous. Nevertheless, variants of the lemma for small finite fields have been studied. In particular, the basic form of the Main Lemma where $p_0 = \frac{1}{2}$ and $p_1 = 0$ appears in Cohen and Tal [7, Lemma 2.2] and Vassilevska Williams et al. [17, Lemma 2.2.]. For smaller values of p_0, the Main Lemma yields as a corollary the following probabilistic polynomial identity test, which may be new and of independent interest; it applies to *sparse* polynomials over GF(2).

Corollary 2. *Let $p(x_1, \ldots, x_n)$ be a non-zero polynomial over $GF(2)$ in n variables, with total degree k and at most $2^{\epsilon k}$ monomials. Let $a \in \{0,1\}^n$ be sampled from the distribution where $\Pr(a_i = 1) = \epsilon$ holds for each i independently. Then*

$$\Pr_a \Big(p(a_1, \ldots, a_n) = 0 \Big) \leq 1 - 2^{-H(\epsilon)k} .$$

Comparison to isolation lemmas based on linear equations. In their seminal paper, Valiant and Vazirani [16] prove an isolation lemma that can be described for non-empty set systems $\mathscr{F}$ over a ground set U of size n as follows: Suppose we know $s = \log |\mathscr{F}|$ for some s. Then we sample a function $h : \{0,1\}^U \to \{0,1\}^s$ at random from a family of pairwise uniform hash functions. We interpret h as mapping subsets of U to vectors in $\{0,1\}^s$. We define the restricted family $\mathscr{F}_{h=0}$ as

$$\mathscr{F}_{h=0} = \Big\{ F \in \mathscr{F} \ : \ h(F) = (0, \ldots, 0) \Big\} .$$

Valiant and Vazirani [16] prove that $\mathscr{F}_{h=0}$ has exactly one element with probability at least some positive constant. Since the cardinality of $\mathscr{F}$ is not known, the value of s must be guessed at random from $\{1, \ldots, n\}$, and the success probability for the whole construction becomes $\Omega(1/n)$. In particular,

$$\Pr_h(\oplus \mathscr{F}_{h=0}) \geq \Omega(\tfrac{1}{n}) .$$

The procedure we just described is useful for problems that are sufficiently rich to express the condition $h(F) = 0$. In particular, the set of all affine linear functions $h : GF(2)^n \to GF(2)^s$ is often used as the family of hash functions; these functions have the form $h(x) = Ax + b$ for some suitable matrix A and vector b over GF(2). Thus the condition $h(F) = 0$ becomes a set of linear equations over GF(2), which can be expressed as a polynomial-size Boolean formula – in fact most natural NP-complete problems are able to express linear constraints with only a polynomial overhead in instance size.

In the exponential time setting, we cannot afford such polynomial blow-up and many problems, including the satisfiability of k-CNF formulas, are not known to be able to efficiently express arbitrary linear constraints. Nevertheless, Calabro *et al.* [6] are able to design an isolation lemma for k-CNF satisfiability, essentially by considering *sparse* linear equation systems, that is, systems where each equation depends only on k variables. Things seem to get even worse for problems such as Set Cover, where we are unable to efficiently express sparse linear equations. This is where our random restrictions come into play since they are much simpler than linear equations; in terms of CNF formulas, they correspond to adding singleton clauses like (x_i) or $(\neg x_i)$.

Neglecting, for a moment, the fact that we may be unable to express the necessary constraints, let us compare the guarantees of Valiant and Vazirani [16] and the Main Lemma: we only achieve oddness instead of isolation, but we do so with probability 2^{-k} instead of $\Omega(\frac{1}{n})$ — our probability is better if $n \geq 2^k$.

Comparison to isolation lemmas based on minimizing weight. Another isolation lemma for k-CNF satisfiability suitable for the exponential-time setting is due to Traxler [15] and is based on the isolation lemma of Mulmuley, Vazirani, and Vazirani [13]. Their construction associates random weights $w(x) \in \{1, \ldots, 2|U|\}$ with each element in the ground set. One then considers for each $r \in \{0, \ldots, 2k|U|\}$ the subfamily of sets of weight exactly r, formally defined as

$$\mathscr{F}_{w,r} = \left\{ F \in \mathscr{F} : \ \sum_{x \in F} w(x) = r \right\}.$$

The isolation lemma of Mulmuley, Vazirani, and Vazirani [13] says that there is a unique set $F \in \mathscr{F}$ of minimum weight r with probability at least $\frac{1}{2}$. In particular, for this $r = r(\mathscr{F}, w)$ we have $\Pr_w\big(\oplus \mathscr{F}_{w,r} \mid r = r(\mathscr{F}, w)\big) \geq \frac{1}{2}$. Since r is not known, we sample it uniformly at random, which yields the overall success probability

$$\Pr_{w,r}\big(\oplus \mathscr{F}_{w,r}\big) \geq \Omega(\tfrac{1}{kn}).$$

The difficulty with this approach is that, when the weighted instance of, say, Set Cover is translated back to an unweighted instance, the parameters are not preserved because the weights are taken from a set of nonconstant size. On the other hand, the weights 0 and 1 can be expressed in many problems as simple deletions or contractions.

We can view the Main Lemma in the weight-minimization framework as follows: sample random weights $w(x) \in \{0, 1\}$ independently for each x such that $w(x) = 0$ holds with probability p_0, and define the weight of $F \in \mathscr{F}$ as $\prod_{x \in F} w(x)$; note by taking the logarithm that minimizing the product is identical to minimizing the sum. The Main Lemma yields a lower bound on the probability that the number of sets with nonzero weight is odd. For comparison with Traxler [15], note that we only achieve oddness instead of isolation, but we do so with probability 2^{-k} instead of $\Omega(\frac{1}{kn})$, which is much better when k is small.

Other parity lemmas and optimality. Not all decision-to-parity reductions are based on an isolation procedure: Gupta [11] uses a construction of small-bias ample spaces to design a randomized polynomial-time procedure that maps ny Boolean formula F, whose set of satisfying assignments corresponds to a et family $\mathscr{F}$, to a formula F', whose family of satisfying assignments $\mathscr{F}'$ is a ıbfamily of $\mathscr{F}$; the guarantee is that, if $\mathscr{F}$ is not empty, then $\Pr(\oplus\mathscr{F}') \geq \frac{1}{2}$.

The constraints in the construction of Gupta [11] are arbitrary linear equa-ıons, which we do not know how to encode into less expressive problems such s Set Cover. On the other hand, restrictions of families often correspond to ontractions or deletions, which are typically easy to express. Nevertheless, the ıccess probability of Gupta [11] is much better than the one guaranteed by the lain Lemma, and one may wonder whether this is an artifact of our proof. Alas, ve prove in the full version of this paper that this is not the case: no decision-to-•arity reduction that is based on random restrictions can have a better success •robability than what is achieved by the Main Lemma.

.2   Consequences for Directed Hamiltonicity

The most straightforward algorithmic application of our reductions is to trans-ate a decision problem to its corresponding parity problem. This is useful in case , faster variant is known for the parity version. In the regime of exponential time >roblems, we currently know a single candidate for this approach: Björklund and lusfeldt [2] recently found an algorithm that computes the parity of the number •f Hamiltonian cycles in a directed n-vertex graph in $O(1.619^n)$ time, but we do ıot know how to decide Hamiltonicity in directed graphs in time $(2 - \Omega(1))^n$. Ve devise such an algorithm in the special case that the number of Hamiltonian •ycles is guaranteed to be small. Let $H \colon [0,1] \to \mathbf{R}$ denote the binary entropy unction given by $H(\epsilon) = -(1 - \epsilon)\log_2(1 - \epsilon) - \epsilon\log_2\epsilon$.

Theorem 3. *For all $\epsilon > 0$, there is a randomized $O(2^{(0.6942+H(\epsilon))n})$ time algo-•ithm to detect a Hamiltonian cycle in a given directed n-vertex graph G with at ʻnost $2^{\epsilon n}$ Hamiltonian cycles.*

In particular, if the number of Hamiltonian cycles is known to be bounded ›y 1.0385^n, we decide Hamiltonicity in time $O(1.9999^n)$.

Discussion and related work. The best time bound currently known for directed Hamiltonicity is $2^n/\exp(\Omega(\sqrt{n/\log n}))$ due to Björklund [3]. In particular, no 1.9999^n algorithm is known. There are no insightful hardness arguments to ıccount for this situation; for instance, there is no lower bound under the Strong Exponential Time Hypothesis. We do know an $O(1.657^n)$ time algorithm for Hamiltonicity detection in undirected graphs [4] and an $O(1.888^n)$ time algo-rithm for bipartite directed graphs [8]. The existence of a $(2 - \Omega(1))^n$ algorithm for the general case is currently an open question.

Is Theorem 3 further evidence for a $(2 - \Omega(1))^n$ time algorithm for directed Hamiltonicity? We are undecided about this. For a counterargument, consider another problem where a restriction of the solution set leads to a $(2-\Omega(1))^n$ time algorithm, without making the general case seem easier: Counting the number of perfect matchings in a bipartite $2n$-vertex graph. It is not known how to solve the general problem faster than $2^n / \exp(\Omega(\sqrt{n/\log n}))$, but when there are not too many matchings, they can be counted in time $(2 - \Omega(1))^n$ [5].

We remark that when the input graph is bipartite, we could reduce to the faster parity algorithm of Björklund and Husfeldt [2], which runs in time $1.5^n \operatorname{poly}(n)$. For this class of graphs, our constructions imply that there is a randomized algorithm to detect a Hamiltonian cycle in time $O(2^{(0.5848+H(\epsilon))n})$ if the input graph has at most $2^{\epsilon n}$ Hamiltonian cycles. In particular, if the number of Hamiltonian cycles is at most $O(1.0431^n)$, the resulting bound is better than the bound $O(1.888^n)$ of Cygan, Kratsch, and Nederlof [8]. Similarly, for the undirected (non-bipartite) case, we can beat the $O(1.657^n)$ bound of Björklund [4] for the undirected case for instances with at most $O(1.0024^n)$ cycles.

In summary, detecting a Hamiltonian cycle seems to become easier when we know that there are few of them. Currently, this result appears to be the most interesting application of the Main Lemma. However, it is unclear if future work on Hamiltonicity will prove it to be a central linchpin in our final understanding, or render it completely useless—it could still turn out that the decision problem in the general case is *easier* than the parity problem.

1.3   Consequences for Set Cover and Hitting Set

For Set Cover and Hitting Set, we establish a strong connection between the parity and decision versions, namely that computing the parity of the number of solutions cannot be much easier than finding one.

Consider as input a family $\mathscr{F}$ of m subsets of some universe U with n elements. A subfamily $\mathscr{C} \subseteq \mathscr{F}$ is *covering* if the union of all $C \in \mathscr{C}$ equals U. The Set Cover problem is given a set family $\mathscr{F}$ and a positive integer t to decide if there is a covering subfamily with at most t sets. The problem's parity analogue $\oplus$ Set Covers is to determine the parity of the number covering subfamilies with at most t sets.

Dually, a set $H \subseteq U$ is a *hitting set* if H intersects F for every $F \in \mathscr{F}$. The Hitting Set problem is given a set family $\mathscr{F}$ and a positive integer t to decide if there exists a hitting set of size at most t. The parity analogue $\oplus$ Hitting Sets is to determine the parity of the number of hitting sets of size at most t.

Theorem 4. *Let $c \geq 1$.*

(i) If $\oplus$ Set Covers can be solved in time $d^n \cdot \operatorname{poly}(n+m)$ for all $d > c$, then the same is true for Set Cover.

(ii) If $\oplus$ Hittings Sets can be solved in time $d^m \cdot \operatorname{poly}(n+m)$ for all $d > c$, then the same is true for Hitting Set.

Discussion and related work. Theorem 4 should be understood in the framework of Cygan *et al.* [9], where it establishes a new reduction in their network of reductions. Our results are complementary to the alternative parameterization, with n and m exchanged in Theorem 4, which is already known: The isolation lemma of Calabro *et al.* [6] in combination with Cygan *et al.* [9] implies that if $\oplus$ Hitting Sets can be solved in time $d^n \cdot \text{poly}(n + m)$ for all $d > c$, then the same is true for Hitting Set.

1.4 Consequences for W[1]

We define the parameterized complexity class $\oplus$W[1] in terms of its complete problem $\oplus$ Multicolored Cliques: This problem is given a graph G and a coloring $c: V(G) \rightarrow [k]$ to decide if there is an odd number of *multicolored cliques*, that is, cliques of size exactly k where each color is used exactly once. Formally we treat $\oplus$ Multicolored Cliques as an ordinary decision problem. We let $\oplus$W[1] be the class of all parameterized problems that have an *fpt-reduction* to $\oplus$ Multicolored Clique. We recall from Flum and Grohe [10, Def. 2.1] that fpt-reductions are deterministic many-to-one reductions that run in fixed-parameter tractable time and that map an instance with parameter k to an instance with parameter at most $f(k)$. The following connection between W[1] and $\oplus$W[1] is a consequence of the Main Lemma.

Theorem 5. *There is a randomized fpt-reduction from* Multicolored Clique *to* $\oplus$ Multicolored Cliques *with one-sided error at most $\frac{1}{2}$; errors may only occur on yes-instances.*

Discussion and related work. Our motivation for Theorem 5 stems from structural complexity: Toda's theorem [14] states that PH $\subseteq$ P$^{\#P}$, that is, every problem in the polynomial-time hierarchy reduces to counting satisfying assignments of Boolean formulas. Theorem 5 aspires to be a step towards an interesting analogue of Toda's theorem in parameterized complexity. In particular, the first step of Toda's proof is

$$NP \subseteq RP^{\oplus P}, \tag{3}$$

or in words: there is a randomized polynomial-time oracle reduction from Sat to $\oplus$ Sat with bounded error and which can only err on positive instances; the existence of such a reduction follows from the isolation lemma. Using a trick that we also rely on in the proof of Theorem 5, Toda [14] is able to turn this reduction into a many-to-one reduction. In terms of structural complexity, the existence of such a many-to-one reduction from Sat to $\oplus$ Sat then implies

$$NP \subseteq RP^{\oplus P[1]}, \tag{4}$$

where the notation [1] indicates that the number of queries to the $\oplus$P-oracle is at most one. Theorem 5 is a natural and direct parameterized complexity analogue of (4), but for obvious reasons we decided not to state it as W[1] $\subseteq$ RFPT$^{\oplus W[1][1]}$.

Montoya and Müller [12, Theorem 8.6] prove a parameterized complexity analogue of the isolation lemma. Implicit in their work is a W[1]-analogue of (3); more precisely, they obtain a reduction with similar specifications as the one in Theorem 5, but with two main differences: While their reduction guarantees uniqueness rather than just oddness, it is only a many-to-many and not a many-to-one reduction. Moving from (3) to (4) is almost automatic in the polynomial-time setting, however we do not see how Theorem 5 could be obtained directly from its weaker many-to-many version.

We remark that Theorem 5 reveals a body of algorithmic open problems, the most intriguing of which, perhaps, is the question whether $\oplus$ k-Paths is fixed-parameter tractable or $\oplus$W[1]-hard. Note that $\oplus$ k-Matchings is polynomial-time solvable by a reduction to the determinant, which is established using a standard interpolation argument in the matching polynomial.

2 Proof of the Main Lemma

Let $U = [n] = \{1, \ldots, n\}$. We define the distribution $\mathcal{D}(p_0, p_1, n)$ over the set of all restrictions $\rho : [n] \to \{0, 1, *\}$ as follows: For each $i \in [n]$ independently, we sample $\rho(i)$ at random so that $\rho(i) = b$ holds with probability exactly p_b for $b \in \{0, 1, *\}$ where p_* is defined as $1 - (p_0 + p_1)$. The following cancellation trick lies at the heart of the Main Lemma.

Lemma 6 (Cancellation Lemma). *Let $\mathscr{F}$ be a family of subsets of $[n]$ and let $i \in [n]$. We define the family of sets for which i is relevant as*

$$\mathscr{F}_{\triangle i} \doteq \left\{ f \in \mathscr{F} \ : \ (f \triangle \{i\}) \notin \mathscr{F} \right\}.$$

Then, for all $\rho : [n] \to \{0, 1, *\}$ with $\rho(i) = *$, we have $\oplus \mathscr{F} \!\restriction_\rho = \oplus \mathscr{F}_{\triangle i} \!\restriction_\rho$.

Proof. We prove that $\mathscr{F}' \doteq (\mathscr{F} \!\restriction_\rho) \setminus (\mathscr{F}_{\triangle i} \!\restriction_\rho)$ has an even number of elements by defining a fixed-point free involution $\pi : \mathscr{F}' \to \mathscr{F}'$. For each $f \in \mathscr{F}'$, we define $\pi(f) = f \triangle \{i\}$. Note that $\pi(f)$ is indeed a member of $\mathscr{F}'$ because $\rho(i) = *$ and $\pi(f) \in (\mathscr{F} \!\restriction_\rho) \setminus \mathscr{F}_{\triangle i}$. It is clear that $\pi(f) \neq f$ and $\pi(\pi(f)) = f$, and so π is a fixed-point free involution. ∎

The proof of the Main Lemma works by an induction. The base case for the induction is a set family that is *extremal* in the following sense.

Definition 7. Let $\mathscr{F}$ be a family over $[n]$. Let $I = \{ i \in [n] \ : \ \mathscr{F}_{\triangle i} = \emptyset \}$ be the set of all *irrelevant* vertices of $\mathscr{F}$ and let $F_b = \{ i \in [n] \ : \ \mathscr{F} \!\restriction_{[i \mapsto \bar{b}]} = \emptyset \}$ for $b \in \{0, 1\}$ be the set of all vertices of $\mathscr{F}$ that are *forced* to b. The family $\mathscr{F}$ is *extremal* if it is non-empty and satisfies $[n] = F_0 \cup F_1 \cup I$, that is, each variable is either forced or irrelevant.

We collect a few basic observations about these sets in the following lemma.

Lemma 8. *Let $\mathscr{F}$ be an extremal family over $[n] = F_0 \cup F_1 \cup I$, let $k_+ \doteq \max_{f \in \mathscr{F}} |f|$ and $k_- \doteq \min_{f \in \mathscr{F}} |f|$. Then*

(i) the sets I, F_1, and F_0 are pairwise disjoint,
(ii) the number of irrelevant vertices is $|I| \leq k_+ - k_-$,

(ii) the number of vertices forced to one is $|F_1| \le k_-$,

(iv) the number of vertices forced to zero is $|F_0| \le n - k_+$, and

(v) $\mathscr{F}$ is extremal if and only if $\mathscr{F} = \{ F_1 \cup g \; : \; g \subseteq I \}$.

Proof. (i) follows immediately from the definitions of the sets. Let $f \in \mathscr{F}$. Then $f \setminus I \in \mathscr{F}$ and $f \cup I \in \mathscr{F}$. Clearly $k_- \le |f \setminus I| = |f \cup I| - |I| \le k_+ - |I|$, which proves (ii). By definition, F_1 is contained in all sets $f \in \mathscr{F}$, which implies (iii). Symmetrically, any set $f \in \mathscr{F}$ satisfies $f \cap F_0 = \emptyset$, which implies (iv). Finally, for (v), let $f \in \mathscr{F}$. By definition, $F_1 \subseteq f \subseteq [n] \setminus F_0$. Since $[n] = F_0 \,\dot\cup\, F_1 \,\dot\cup\, I$, this implies $f \subseteq F_1 \cup I$. For the reverse inclusion, let $f = F_1 \cup g$ for some $g \subseteq I$. Since $\mathscr{F}$ is not empty, it must contain a set, which by the first inclusion has the form $F_1 \cup g'$ for some $g' \subseteq I$. Since $g \subseteq I$ we have $f = F_1 \cup g = (F_1 \cup g') \triangle (g \triangle g') \in \mathscr{F}$. ∎

Lemma 9. *Let $\mathscr{F}$ be an extremal family over $[n] = F_0 \cup F_1 \cup I$. Let $\mathcal{D} \doteq \mathcal{D}(p_0, p_1, n)$. Then*

$$\Pr_{\rho \sim \mathcal{D}} \left(\oplus \mathscr{F} \restriction_\rho \right) = (1 - p_1)^{|F_0|} (1 - p_0)^{|F_1|} (p_0 + p_1)^{|I|}.$$

Proof. Note that $F_0 \cap F_1 = \emptyset$ and $(F_0 \cup F_1) \cap I = \emptyset$ holds for all $\mathscr{F} \ne \emptyset$. The assumption is that every vertex is either forced to 0, forced to 1, or irrelevant. By Lemma 6, for all $i \in I$, the event $\rho(i) = *$ implies that $\mathscr{F} \restriction_\rho$ is of even size. Hence we need to condition on the event $\rho(i) \in \{0, 1\}$ for all $i \in I$. This event occurs with probability $(p_0 + p_1)^{|I|}$. Furthermore, since the vertices in I are all irrelevant, if we set them to 0 or 1 arbitrarily, we end up with a family $\mathscr{F}'$ that has the same cardinality regardless of the assignment on the irrelevant vertices. In particular, since all vertices in $[n] \setminus I$ are forced, each set $\mathscr{F}'$ has exactly one element. The probability that the unique element of $\mathscr{F}'$ survives after further restricting the vertices of $F_0 \cup F_1$ randomly is equal to $(1 - p_1)^{|F_0|} (1 - p_0)^{|F_1|}$. ∎

To analyze the probability of the event $\oplus \mathscr{F} \restriction_\rho$ for non-extremal families $\mathscr{F} \ne \emptyset$, we consider the following type of branching process:

○ If $\mathscr{F}$ is extremal, return $\mathscr{F}$ as a leaf.

○ If $\mathscr{F} \ne \emptyset$ is not extremal, let $i \in [n] \setminus F_0 \cup F_1 \cup I$ and add the following two children to the branching tree:

$$\mathscr{F}_0 \doteq \left\{ f \subseteq [n] \setminus \{i\} \; : \; f \in \mathscr{F} \right\}$$

$$\mathscr{F}_* \doteq \left\{ f \subseteq [n] \setminus \{i\} \; : \; f \in \mathscr{F} \text{ xor } f \triangle \{i\} \in \mathscr{F} \right\}$$

Note that $\mathscr{F}_0 = \mathscr{F} \restriction_{[i \mapsto 0]}$ and $\mathscr{F}_* = \mathscr{F}_{\triangle i} - i$ where

$$\mathscr{F} - i \doteq \left\{ f \subseteq [n] \setminus \{i\} \; : \; f \in \mathscr{F} \text{ or } f \cup \{i\} \in \mathscr{F} \right\}.$$

Therefore, $i \notin F_1$ implies $\mathscr{F}_0 \ne \emptyset$ and $i \notin I$ implies $\mathscr{F}_* \ne \emptyset$.

The above process defines a finite branching tree T of $\mathscr{F} \neq \emptyset$, which is generally not unique since we can choose which i should be branched on next. We let the *cost* of a branching tree T of $\mathscr{F}$ be the maximum $|I(\mathscr{F}')|$ over all leaves $\mathscr{F}'$ of T, that is, it is the maximum number of irrelevant vertices that any leaf has. The *cost* of $\mathscr{F}$ is the minimum cost over all branching trees of $\mathscr{F}$. We provide the following simple upper bound on the cost of any branching tree.

Lemma 10. *Let* $\mathscr{F} \neq \emptyset$. *Any branching tree* T *of* $\mathscr{F}$ *has cost* $\leq$ $\min\{k_+, \log|\mathscr{F}|\}$.

Proof. Note that $\mathscr{F}_0$ and $\mathscr{F}_*$ have size at most $|\mathscr{F}|$ and contain sets of size at most k_+. Thus, by induction, any leaf $\mathscr{F}'$ in T has size at most $|\mathscr{F}|$ and contains sets of size at most k_+. We apply Lemma 8 to the extremal family $\mathscr{F}'$. We get $|I(\mathscr{F}')| \leq k_+$ as well as $2^{|I(\mathscr{F}')|} = |\mathscr{F}'| \leq |\mathscr{F}|$, which proves the claim. ∎

Lemma 11 (Main Lemma Based on Family Cost). *Let* $\mathscr{F}$ *be a non-empty family of cost at most* c *and with sets of size at most* k. *Let* $\mathcal{D} \doteq \mathcal{D}(p, 0, n)$.

If $p \geq \frac{1}{2}$, *then* $\Pr_{\rho \sim \mathcal{D}}\left(\oplus \mathscr{F} \restriction_\rho\right) \geq (1 - p)^k$.

If $p < \frac{1}{2}$, *then* $\Pr_{\rho \sim \mathcal{D}}\left(\oplus \mathscr{F} \restriction_\rho\right) \geq (1 - p)^{k-c} p^c$.

Clearly our bound for $p \geq \frac{1}{2}$ is maximized at $p = \frac{1}{2}$. Moreover, note that $(1 - p)^{k-c} p^c \geq 2^{-H(p)k}$ holds if $c/k \leq p \leq \frac{1}{2}$ and is maximized at $p = c/k$. Thus once c and k are fixed, Lemma 11 gives its best guarantee for $p = \min\{\frac{1}{2}, \frac{c}{k}\}$.

Proof. Let T be a branching tree for $\mathscr{F}$ that has cost at most c. We prove the claim by induction on the structure of T. If $\mathscr{F}$ is a leaf of T, then $\mathscr{F}$ is extremal and we are in the situation of Lemma 9 with $p_0 = p = 1 - p_*$ and $p_1 = 0$. This yields $\Pr_\rho(\oplus \mathscr{F} \restriction_\rho) = (1 - p)^{|F_1|} \cdot p^{|I|} \geq (1 - p)^{k-|I|} \cdot p^{|I|}$, where the inequality follows from Lemma 8. If $p < \frac{1}{2}$, the function $x \mapsto h_p(x) \doteq (1 - p)^{k-x} p^x$ is strictly decreasing and the assumption $|I| \leq c$ implies $h_p(|I|) \geq h_p(c)$, which proves the claimed inequality. If $p \geq \frac{1}{2}$, then $h_p(x)$ is non-decreasing and we use the trivial lower bound $h_p(|I|) \geq h_p(0)$ to obtain the claimed inequality.

Now let $\mathscr{F}$ be an inner vertex of T, where T selects some $i \in [n] \setminus F_0 \cup F_1 \cup I$ and produces the children $\mathscr{F}_0$ and $\mathscr{F}_*$. We estimate the probability of the event $\oplus \mathscr{F} \restriction_\rho$ by conditioning on the i-th coordinate:

$$\Pr_\rho\left(\oplus \mathscr{F} \restriction_\rho\right) = p_0 \cdot \Pr_\rho\left(\oplus \mathscr{F} \restriction_\rho \;\middle|\; \rho(i) = 0\right) + p_* \cdot \Pr_\rho\left(\oplus \mathscr{F} \restriction_\rho \;\middle|\; \rho(i) = *\right) .$$

Recall that $p_1 = 0$ and $p_0 + p_* = 1$. Thus it remains to prove that $(1 - p)^k$ or $(1 - p)^{k-c} p^c$ are lower bounds for the two remaining conditional probabilities. The event $\rho(i) = 0$ implies $\mathscr{F} \restriction_\rho = \mathscr{F}_0 \restriction_{\rho'}$ where ρ' is identical to ρ except that it is undefined on i. Furthermore, if $\rho(i) = *$, then the Cancellation Lemma 6 implies $\oplus \mathscr{F} \restriction_\rho = \oplus \mathscr{F}_* \restriction_{\rho'}$. Since i is neither in F_1 nor in I, the families $\mathscr{F}_0$ and $\mathscr{F}_*$ are not empty. Moreover, their maximum set sizes are bounded by k. Also note that, by definition of the cost, the cost of $\mathscr{F}_0$ and the cost of $\mathscr{F}_*$ are each at most c. Thus we can apply the induction hypothesis on the families $\mathscr{F}_0$ and $\mathscr{F}_*$. This finishes the proof of the lemma. ∎

Acknowledgments. We would like to thank Radu Curticapean for reminding some of us of the matching polynomial, Moritz Müller for helping us understand the relationship between our decision-to-parity reduction and their variant of the isolation lemma, and Ryan Williams for pointing us to [17]. AB and TH are supported by the Swedish Research Council, grant VR 2012-4730: Exact Exponential-time Algorithms.

References

1. Arora, S., Barak, B.: Computational Complexity: A Modern Approach. Cambridge University Press (2009)
2. Björklund, A., Husfeldt, T.: The parity of directed hamiltonian cycles. In: Proc. 54th Annual IEEE Symposium on Foundations of Computer Science, FOCS, Berkeley, CA, USA, October 26–29, pp. 727–735 (2013)
3. Björklund, A.: Below all subsets for permutational counting problems, (2012) arXiv:1211.0391 [cs:DS]
4. Björklund, A.: Determinant sums for undirected Hamiltonicity. SIAM J. Comput. **43**(1), 280–299 (2014)
5. Björklund, A., Husfeldt, T., Lyckberg, I.: Computing the permanent modulo a prime power. In: preparation (2015)
6. Calabro, C., Impagliazzo, R., Kabanets, V., Paturi, R.: The complexity of unique k-SAT: An isolation lemma for k-CNFs. In: Proc. 18th IEEE Conference on Computational Complexity, CCC, Aarhus, Denmark, July 7–10 (2003)
7. Cohen, G., Tal, A.: Two structural results for low degree polynomials and applications. Electronic Colloquium on Computational Complexity (ECCC). Tech report TR13-145 (2013)
8. Cygan, M., Kratsch, S., Nederlof, J.: Fast Hamiltonicity checking via bases of perfect matchings. In: Proc. 45th Symposium on Theory of Computing, STOC, Palo Alto, CA, USA, June 1–4, pp. 301–310 (2013)
9. Cygan, M., Dell, H., Lokshtanov, D., Marx, D., Nederlof, J., Okamoto, Y., Paturi, R., Saurabh, S., Wahlström, M.: On problems as hard as CNFSAT. In: Proc. 27th IEEE Conference on Computational Complexity, CCC, Porto, Portugal, June 26–84 (2012)
10. Flum, J., Grohe, M.: Parameterized Complexity Theory. Springer (2006)
11. Gupta, S.: Isolating an odd number of elements and applications in complexity theory. Theor. Comput. Syst. **31**(1), 27–40 (1998)
12. Montoya, J.A., Müller, M.: Parameterized random complexity. Theor. Comput. Syst. **52**(2), 221–270 (2013)
13. Mulmuley, K., Vazirani, U.V., Vazirani, V.V.: Matching is as easy as matrix inversion. Combinatorica **7**(1), 105–113 (1987)
14. Toda, S.: PP is as hard as the polynomial-time hierarchy. SIAM J. Comput. **20**(5), 865–877 (1991)
15. Traxler, P.: The time complexity of constraint satisfaction. In: Grohe, M., Niedermeier, R. (eds.) IWPEC 2008. LNCS, vol. 5018, pp. 190–201. Springer, Heidelberg (2008)
16. Valiant, L.G., Vazirani, V.V.: NP is as easy as detecting unique solutions. Theor. Comput. Sci. **47**, 85–93 (1986)
17. Williams, V.V., Wang, J., Williams, R., Yu, H.: Finding four-node subgraphs in triangle time. In: Proc. 26th Annual ACM-SIAM Symposium on Discrete Algorithms, SODA, San Diego, CA, USA, January 4–6, 2015, pp. 1671–1680 (2015)

Spotting Trees with Few Leaves

Andreas Björklund[1], Vikram Kamat[2], Łukasz Kowalik[2(✉)], and Meirav Zehavi[3]

[1] Department of Computer Science, Lund University, Lund, Sweden
[2] Faculty of Mathematics, Informatics and Mechanics, University of Warsaw,
Warsaw, Poland
kowalik@mimuw.edu.pl
[3] Department of Computer Science, Technion – Israel Institute of Technology,
Haifa, Israel

Abstract. We show two results related to the HAMILTONICITY and
k-PATH algorithms in undirected graphs by Björklund [FOCS'10], and
Björklund et al., [arXiv'10]. First, we demonstrate that the technique
used can be generalized to finding some k-vertex tree with l leaves in
an n-vertex undirected graph in $O^*(1.657^k 2^{l/2})$ time. It can be applied
as a subroutine to solve the k-INTERNAL SPANNING TREE (k-IST) problem in $O^*(\min(3.455^k, 1.946^n))$ time using polynomial space, improving
upon previous algorithms for this problem. In particular, for the first
time, we break the natural barrier of $O^*(2^n)$. Second, we show that the
iterated random bipartition employed by the algorithm can be improved
whenever the host graph admits a vertex coloring with few colors; it can
be an ordinary proper vertex coloring, a fractional vertex coloring, or
a vector coloring. In effect, we show improved bounds for k-PATH and
HAMILTONICITY in any graph of maximum degree $\Delta = 4, \ldots, 12$ or with
vector chromatic number at most 8.

1 Introduction

Given an undirected host graph G on n vertices, the (k, l)-TREE problem asks
if G contains a tree T on k vertices, such that the number of leaves in T is exactly
l. This problem is a natural generalization of the classic k-PATH problem: for
$l = 2$, the definitions of (k, l)-TREE and k-PATH coincide. For $k = n$ and $l = 2$, we
get the classic HAMILTONIAN PATH. Furthermore, (k, l)-TREE is tightly linked
to the well-studied k-INTERNAL SPANNING TREE (k-IST) problem, which asks if
a given graph G on n vertices contains a spanning tree T with at least k internal
vertices. Indeed, it is well-known that a yes-instance of k-IST is a yes-instance
of $(k + l, l)$-TREE for some $l \leq k$, and vice versa [9]. Because of the connections
to HAMILTONIAN PATH, the (k, l)-TREE, k-IST and k-PATH problems, even in
bipartite graphs or in graphs of bounded degree 3, are NP-hard.

In this paper, we study parameterized algorithms, which attempt to solve
NP-hard problems by confining the combinatorial explosion to a parameter k.

Work partially supported by the National Science Centre of Poland, grant number
2013/09/B/ST6/03136 and ERC StG project PAAl no. 259515 (ŁK). The paper was
prepared while the second author held a post-doctoral position at Warsaw Center of
Mathematics and Computer Science.

M.M. Halldórsson et al. (Eds.): ICALP 2015, Part I, LNCS 9134, pp. 243–255, 2015.
DOI: 10.1007/978-3-662-47672-7_20

More precisely, a problem is *fixed-parameter tractable (FPT)* with respect to a parameter k if it can be solved in time $O^*(f(k))$ for some function f, where O^* hides factors polynomial in the input size. Our results have also consequences in the field of moderately exponential-time algorithms, where one aims at providing an $O(c^n)$-time algorithm, with the constant $c > 1$ being as small as possible.

We develop an FPT algorithm for (k, l)-TREE in general graphs that relies upon a non-trivial generalization of the technique underlying the HAMILTONIC-ITY and k-PATH algorithms in [2,4]. We thus break the natural barrier of $O^*(2^n)$ in the running time bound for k-IST. Then, we conduct a thorough examination of (k, l)-TREE in special classes of graphs that admit a vertex coloring with few colors. This in turn implies faster algoritms for HAMILTONICITY, k-PATH and k-IST in bounded degree graphs. Apart from the classic vertex coloring, we consider fractional coloring and vector coloring, thus showing that the latter tool, famous in approximation algorithms, is also helpful in parameterized complexity.

1.1   Related Work

The k-IST problem and its directed version, k-INTERNAL OUT-BRANCHING (k-IOB),[1] are of interest in database systems [12] and water supply network design [30]. Note that any k-IOB algorithm also solves k-IST (after replacing every edge of the input graph by two oppositely oriented arcs). Over the last decade k-IST and k-IOB were heavily researched, resulting in a number of algorithms using a variety of approaches. The first FPT algorithm, running in time $O^*(2^{O(k \log k)})$ was due to Prieto et al. [29]. Cohen et al [9] obtained an algorithm running in time $O^*(49.4^k)$, which was the first $O^*(2^{O(k)})$-time bound. Currently the fastest algorithms are due to Zehavi [36] and run in $O^*(3.617^k)$ randomized time or $O^*(5.139^k)$ deterministic time. However, both these algorithms use exponential space. Prior to this work the best time bound of a polynomial space algorithm was $O^*(4^k)$: first, Daligault [11] and Zehavi [35] obtained a randomized algorithm and very recently, Li et al. [24] showed a $2k$-kernel implying a deterministic algorithm. For the special case of graphs of bounded degree Δ, Zehavi [35] shows a k-IOB algorithm running in time $O^*(4^{(1 - \frac{\Delta+1}{2\Delta(\Delta-1)})k})$. Another specialized algorithm, due to Raible et al. [30], solves k-IST in graphs of maximum degree 3 in time $O^*(2.137^k)$. There also has been quite some interest in moderately exponential time algorithms for k-IST. Unlike in many other graph problems, even an $O^*(2^n)$ algorithm is not completely trivial. Algorithms achieving this bound were shown by Raible et al. [30] (with exponential space) and Nederlof [28] (polynomial space). Note that $O^*(2^n)$ is a natural barrier, as it corresponds to the number of all subsets of vertices. During recent years, researchers managed to surpass this barrier for a number of problems, like DOMINATING SET by Grandoni [19], FEEDBACK VERTEX SET by Razgon [31] and HAMILTONICITY by Björklund [2]. Raible et al. [30] explicitly posed an open question asking whether the approach of Björklund [2] can be extended to k-IST either for general graphs or for graphs of

---

[1] In the k-IOB problem, we need to decide if a directed graph G contains a spanning tree T with exactly one vertex of in-degree 0 and at least k internal vertices.

large vertex cover. Raible et al. [30] were able to cross the $O^*(2^n)$ barrier for graphs of bounded degree — they get an algorithm running in time $O^*((2^{\Delta+1}-1)^{\frac{n}{\Delta-1}}$ and present a further improvement to $O^*(1.862^n)$ when $\Delta = 3$.

The (k, l)-TREE problem was implicitly introduced by Cohen et al. [9] as a tool for solving k-IOB. They obtain an $O^*(6.14^k)$-time algorithm. Currently, the fastest randomized algorithms, due to Daligault [11] and Zehavi [35], run in time $O^*(2^k)$ and have a polynomial space complexity. Note that we meet the $O^*(2^k)$ barrier again. For the deterministic case, Zehavi [36] (relying on [33]) shows an $O^*(2.597^k)$-time algorithm for (k, l)-TREE with an exponential space complexity.

The fundamental k-PATH problem is well-studied in the field of Parameterized Complexity. In the past three decades, it enjoyed a race towards obtaining the fastest FPT algorithm (see [1,4,8,15,16,23,26,34,36]). Currently, the best randomized algorithm, due to Björklund et al. [4], runs in time $O^*(1.657^k)$ and has a polynomial space complexity, and the best deterministic algorithm, due to Zehavi [36], runs in time $O^*(2.597^k)$ and has an exponential space complexity.

The result in [4] extends Björklund's $O^*(1.657^n)$ time algorithm for HAMILTONICITY in [2]. The same paper contains also a better bound of $O^*(2^{n/2})$ for bipartite graphs. HAMILTONICITY in graphs of bounded maximum degree Δ received a considerable attention beginning with the paper of Eppstein [14], followed by works of Iwama and Nakashima [22], Gebauer [20] and Björklund et al. [3]. Currently the best algorithm for $\Delta = 3$, due to Cygan et al. [10], runs in time $O^*(1.201^n)$ [10], while prior to this work, for $\Delta \geq 4$ the best bound was that of the general Björklund's $O^*(1.657^n)$-time algorithm.

1.2 Our Contribution

We obtain improved algorithms for (k, l)-TREE, k-IST, k-PATH and HAMILTONICITY. Our algorithms are randomized Monte-Carlo with one sided error (they never report a false positive, and they report false negatives with constant probability), having polynomial space complexities. Our contribution is twofold. While we focus on *decision* problems, the corresponding search versions can be solved with an additional $O(k \log n)$ running time overhead, see [7].

From Paths to Trees. First, we develop an algorithm that solves (k, l)-TREE in general graphs in time $O^*(1.657^k 2^{\ell/2})$. This can be seen as a generalization of the technique of Björklund et al. [2,4] from detecting paths to detecting trees. In the original technique one enumerates walks of length k (rather than paths, which are walks without repeating vertices) and then uses an algebraic tool to sieve-out the walks which are not paths. The algebraic tool is to design a polynomial which is a sum of monomials, each corresponding to a walk. The trick is that thanks to the use of bijective labelings, the non-path walks can be paired-up so that both corresponding monomials are the same and thus cancel over a field of characteristic 2. It is quite clear that in the tree case walks should be replaced by branching walks (see [27]). However, the main difficulty lies in the pairing argument, which requires a new labeling scheme and becomes much more delicate. Indeed, this extension is non-trivial as it is exactly the topic of the open problem posed by Raible et al. [30] mentioned in the previous section.

Table 1. Running times of our k-PATH algorithm in bounded degree graphs (left), and bounded vector chromatic number (right). For bounds for HAMILTONICITY, set $k = n$.

Δ	Running Time
3	$O^*(1.5705^k)$
4	$O^*(2^{2k/3}) = O^*(1.5874^k)$
5, 6	$O^*(2^{7k/10}) = O^*(1.6245^k)$
7, 8	$O^*(2^{5k/7}) = O^*(1.6406^k)$
9, 10	$O^*(2^{13k/18}) = O^*(1.6497^k)$
11, 12	$O^*(2^{8k/11}) = O^*(1.6555^k)$

$\chi_v(G)$	Running Time
4	$O^*(1.6199^k)$
5	$O^*(1.6356^k)$
6	$O^*(1.6448^k)$
7	$O^*(1.6510^k)$
8	$O^*(1.6554^k)$

Our algorithm, similarly as in [4], uses a random bipartition of the vertices. The running time depends in a crucial way on a random variable (called the number of needed labels) in the resulting probability space. The second difficulty was to determine the distribution of this variable (Lemma 7), and again this turned out to be much more demanding than in the path case.

The (k, l)-TREE algorithm described above already implies an improved result for k-IST in general graphs. However, it works much faster if the hidden spannning tree has few leaves. We design a different strategy, based on finding a maximum matching, to accelerate the algorithm when it looks for solutions with many leaves. By merging the two strategies, we obtain an $O^*(3.455^k)$ time algorithm for k-IST. This is the first $O^*((4 - \Omega(1))^k)$-time algorithm that uses polynomial space. It immediately implies a moderately exponential time algorithm running in time $O^*(1.946^n)$, which breaks the $O^*(2^n)$ barrier.

Paths and Trees in Colored Graphs. Next, we study (k, l)-TREE in graphs that admit a vertex coloring with d colors. This can be seen as an extension of Björklund's $O^*(2^{n/2})$-time algorithm for HAMILTONICITY in bipartite graphs [2]. However, the insight in [2] was to find a small vertex cover of the hidden solution. Here, we use a different insight: by choosing roughly half of the color classes, we get a small subset of the hidden solution vertices which covers many (but not all) of its edges. The resulting algorithm runs in $O^*(2^{(1 - \frac{\lfloor d/2 \rfloor \lceil d/2 \rceil}{d(d-1)})k})$ time when $= O(1)$ (see Section 3 for a more complicated bound in the general case). For a graph of bounded degree Δ that is neither complete nor an odd cycle, one can construct a proper Δ-coloring in linear time (e.g., by Lovász's proof of Brooks' theorem [25]). This immediately results in a fast algorithm for (k, l)-TREE in such graphs (see Table 1 for the special cases of k-PATH and HAMILTONICITY; the case $\Delta = 3$ is solved by a special algorithm, see below), along with an improved algorithm for k-IST in such graphs (for details see the full version [5]).

Fractional coloring is a well studied generalization of the classical vertex coloring. Our algorithm for graphs of low chromatic number generalizes quite easily to the case of low fractional chromatic number. This has consequences in improved algorithms for some special graph classes, e.g. an $O^*(1.571^k 1.274^l)$-time algorithm for (k, l)-TREE in subcubic triangle-free graphs, or $O^*(1.571^k)$-time algorithm for k-PATH in general subcubic graphs. For subcubic graphs of even larger girth, we get further improved bounds.

Another relaxation of the classical coloring is vector coloring, known for its importance in approximation algorithms (e.g. [18]). Its important advantage is that, contrary to the classical or fractional coloring, a $(1 + \epsilon)$-approximation can be found in polynomial time [17]. We provide an algorithm for (k, l)-TREE that applies vector coloring. It results in improved running time when the vector chromatic number is at most 8.

Organization. Section 2 developes an algorithm for (k, l)-TREE in general graphs. Then, our contributions for colored graphs are given in Section 3 (proper coloring, along with a consequence for (k, l)-TREE in bounded degree graphs). Due to space limitations, we omitted some details in this extended abstract. We also had to skip the description of the remaining constrictions, i.e., our algorithm for k-IST in general graphs, algorithms for graphs of low fractional coloring number or vector coloring number, consequences for k-IST in bounded degree graphs and for (k, l)-TREE in graphs with small chromatic number, and specialized algorithms for (k, l)-TREE and k-PATH in subcubic graphs. All this missing material can be found in the technical report [5].

Notation. Throughout the paper we consider undirected graphs. For an integer k, by $[k]$ we denote the set $\{1, 2, \ldots, k\}$. For a set S and an integer k, by $\binom{S}{k}$ we denote the family of all subsets of S of size k. Let us write $V = V(G) = [n]$ for the vertex set and $E = E(G)$ for the edge set of the host graph G.

2   Finding Trees on k Vertices with l Leaves

In this section we generalize the k-PATH algorithm by Björklund et al. [2,4] to finding subtrees on k nodes including l leaves. Throughout this section, assume we have a fixed partition $V = V_1 \cup V_2$ of the vertices of the input graph. For the promised generalization, we will use a *random* bipartition, similarly as in [2,4].

Branching Walks. The notion of branching walk was introduced by Nederlof [27]. A mapping $h : V(T) \to V(G)$ is a *homomorphism* from a graph T to the host G if $\{h(a), h(b)\} \in E(G)$ for all $\{a, b\} \in E(T)$. We adopt the convention of calling the elements of $V(T)$ *nodes* and the elements of $V(G)$ *vertices*.

A *branching walk* in G is a pair $B = (T, h)$ where T is an unordered rooted tree and $h : V(T) \to V(G)$ is a homomorphism from T to G. The walk *starts* from the vertex $h(1)$ in G, and its *size* is $|V(T)|$. We say that the walk is *simple* if h is injective, and *weakly simple* if for any node $x \in V(T)$, the homomorphism h is injective on children of x. The walk is *U-turn-free* if for any node a of T, every child b of a maps to a different vertex than the parent c of a, i.e., $h(b) \neq h(c)$.

A *proper order* of B is any permutation $\pi : V(T) \to \{1, \ldots, |V(T)|\}$ such that for every two nodes $a, b \in V(T)$,

(i) if $\mathrm{depth}(a) < \mathrm{depth}(b)$ then $\pi(a) < \pi(b)$,
(ii) if $a, b \neq \mathrm{root}(T)$ and $\pi(\mathrm{parent}(a)) < \pi(\mathrm{parent}(b))$ then $\pi(a) < \pi(b)$,
(iii) if a and b are siblings and $h(a) < h(b)$ then $\pi(a) < \pi(b)$.

The following proposition is immediate.

Proposition 1. *Any weakly simple branching walk has exactly one proper order.*

We say that a weakly simple branching walk B is *properly ordered* if $V(T) = \{1, \ldots, |V(T)|\}$ and the proper order from Proposition 1 is the identity function.

Labeling. For a tree T, by $L(T)$ we note the set of leaves of T and by $I(T)$ we denote the set of internal vertices of T, i.e., $I(T) = V(T) \setminus L(T)$.

Like in [4] or [6] our crucial tool are *labelled* branching walks. In [6], every node in the tree T of a branching walk (T, h) was assigned a label. Here, similarly as in [4], we do not assign labels to some nodes, but we assign labels to some edges of T. We define the set of *labellable elements* of a branching walk $B = (T, h)$ as $\mathrm{la}(B) = L(T) \cup (h(I(T)) \cap V_1) \cup \{uv \in E(T) \; : \; h(u), h(v) \in V_2\}$. Similarly, for a subtree T of graph G, let $\mathrm{la}(T) = L(T) \cup (I(T) \cap V_1) \cup \{uv \in E(T) \; : \; u, v \in V_2\}$.

We say that a branching walk $B = (T, h)$ is *admissible* when B is weakly simple, U-turn-free, and properly ordered. For nonnegative integers k, l, r, we also say that B is (k, l, r)-*fixed* if T has k nodes and l leaves, and $|\mathrm{la}(B)| = r$.

The Polynomial. Let r be an integer. We use three kinds of variables in our polynomial. First, for any edge $uv \in E(G)$, where $u < v$, we have a variable x_{uv}. For simplicity we will denote $x_{vu} = x_{uv}$. Second, for each $q \in V \cup E$ and for each $l \in [r]$ we have a variable $y_{q,l}$. Third, for each $v \in V(G)$, we have a variable z_v. By $\mathbf{x}$ we denote the sequence of all x_{uv}-type variables, while by $\mathbf{y}$ we denote the sequence of all $y_{q,l}$-type variables.

For a branching walk $B = (T, h)$ and a labeling $\ell : \mathrm{la}(B) \to [|\mathrm{la}(B)|]$, we define the monomial

$$\mathrm{mon}(B, \ell) = z_{h(1)} \prod_{\substack{\{u,v\} \in E(T) \\ u < v}} x_{h(u),h(v)} \prod_{q \in \mathrm{la}(B)} y_{h(q),\ell(q)} \,,$$

where for $uv \in E(T)$, $h(uv)$ denotes the edge $h(u)h(v) \in E(G)$. We define a multivariate polynomial P_i with coefficients in a field of characteristic 2 by

$$P_i = \sum_{\substack{B=(T,h) \\ B \text{ is admissible} \\ B \text{ is } (k,l,i)\text{-fixed}}} \sum_{\substack{\ell : \mathrm{la}(B) \to [|\mathrm{la}(B)|] \\ \ell \text{ bijective}}} \mathrm{mon}(B, \ell) \,.$$

Finally, let $P_{r\downarrow} = \sum_{i=2}^{r} P_i$.

Lemma 1. *The set of pairs* (B, ℓ)*, where B is a non-simple, admissible and (k, l, i)-fixed branching walk and $\ell : \mathrm{la}(B) \to [|\mathrm{la}(B)|]$ is a bijection, can be partitioned into pairs, and the two monomials corresponding to each pair are identical.*

The proof consists of three cases (see [5]). If a non-simple branching walk $B = (T, h)$ contains two elements $e_1, e_2 \in \mathrm{la}(B)$ (vertices or edges) which h maps to the same element of G, then we pair up (B, ℓ) with (B, ℓ'), where ℓ' is obtained from ℓ by swapping the labels of e_1 and e_2. In the second case we assume there is a pair of vertices u, v such that $h(u) = h(v)$ and u is an ancestor of v. Then, we modify the tree in B by reversing the order of vertices on the path between

u and v. Finally, in the third case we assume there is a pair of vertices u, v such that $h(u) = h(v)$ and neither u is an ancestor of v nor v is an ancestor of u. Then, we modify the tree in B by swapping the subtrees rooted at u and v. In all of the three cases we show that the constructed new labelled branching walk is non-simple, admissible, (k, l, i)-fixed, and corresponds to an identical monomial as B. The most delicate issue, however, is to guarantee that if we start from the new labelled branching walk, and we follow the same way of assignment, then we get B back. This is obtained by a very careful way of choosing the case to apply (if more than one applies), and the pair of elements that map to the same place in G (if there are several such pairs).

The following lemma follows quite easily from Lemma 1 (see [5]).

Lemma 2. *The polynomial $P_{\downarrow r}$ is non-zero iff the input graph contains a subtree T_G with k nodes and l leaves, such that $\mathrm{la}(T_G) \leq r$.*

Evaluating the Polynomial. Due to lack of space, we only sketch an algorithm that evaluates the polynomial $P_{\downarrow r}$ in a given point $(\mathbf{x}, \mathbf{y}, \mathbf{z})$ in time $O^*(2^r)$. For a detailed description, see [5]. Clearly, it suffices to show this bound for every polynomial P_i. To this end, we rewrite P_i as a sum of 2^r polynomials such that each of them can be evaluated in polynomial time. For each $X \subseteq [r]$, let

$$P_i^X = \sum_{\substack{B=(T,h) \\ B \text{ is admissible} \\ B \text{ is } (k,l,i)\text{-fixed}}} \sum_{\ell:\mathrm{la}(B)\to X} \mathrm{mon}(B, \ell) \,.$$

Note that the labelings in the second summation may not be bijective. By the Principle of Inclusion and Exclusion, and since the coefficients of P_i are from a field of characteristic 2, it can be shown that $P_i = \sum_{X \subseteq \{1,2,\dots,k\}} P_i^X$. Therefore, it suffices to evaluate each of the polynomials P_i^X in polynomial time, which can be done by a complicated, but standard, dynamic programming. We get that

Lemma 3. *$P_{\downarrow r}$ can be evaluated in time $O^*(2^r)$ and polynomial space.*

A Single Evaluation Algorithm. Assume that if there is a (k, l)-tree T_G in G, then parameter r is at least as large as $|\mathrm{la}(T_G)|$. Then, by Lemma 2, we can test the existence of a (k, l)-tree by testing whether the polynomial $P_{\downarrow r}$ is non-zero. The latter task can be performed efficiently using a single evaluation of the polynomial $P_{\downarrow r}$. For this purpose, we need the Schwartz-Zippel Lemma, shown independently by DeMillo and Lipton [13], Schwartz [32] and Zippel [37].

Lemma 4. *Let $p(x_1, x_2, \dots, x_n) \in F[x_1, \dots, x_n]$ be a polynomial of degree at most d over a field F, and assume p is not identically zero. Let S be a finite subset of F. Sample values $a_1, a_2, \dots, a_n$ from S uniformly at random. Then,*

$$\Pr(p(a_1, a_2, \dots, a_n) = 0) \leq d/|S|.$$

Lemma 5. *Let $V(G) = V_1 \cup V_2$ be a fixed bipartition of the vertex set of G. There is an algorithm running in $O^*(2^r)$ time and polynomial space such that*
- *If G does not contain a (k,l)-tree, then the algorithm always answers NO,*
- *If G contains a (k,l)-tree T_G such that $|\mathrm{la}(T_G)| \le r$, then the algorithm answers YES with probability at least $\frac{1}{2}$.*

Proof. The algorithm is as follows: using the algorithm from Lemma 3, evaluate the polynomial $P_{\downarrow r}$ over the field $\mathrm{GF}(2^{\lceil \log_2(k+r)\rceil + 1})$, substituting the variables by independently chosen random field elements. The time bound follows from Lemma 3. If there is no (k,l)-tree in the input graph, by Lemma 2 the evaluation returns 0, so we report the correct answer.

Now assume there is a (k,l)-tree T_G such that $\mathrm{la}(T_G) \le r$. Then, by Lemma 2, P is a non-zero polynomial. Note that $\deg(P_i) = k + i$, hence $\deg(P_{\downarrow r}) \le k + r$. Hence, by the Schwartz-Zippel Lemma, P evaluates to the zero field element with probability at most $\frac{1}{2}$. This finishes the proof.   □

The Random Bipartition Algorithm. Now we assume that $V = V_1 \cup V_2$ is a random bipartition, i.e., every vertex goes to V_1 independently with probability $1/2$. We aim to choose the value of parameter r large enough so that if there is a (k,l)-tree T_G in G, then with high probability $|\mathrm{la}(T_G)| \le r$. Then, by Lemma 5, we are done. Of course, putting $r = k$ would perfectly achieve the above goal, but then we only get the running time of $O^*(2^k)$, matching that of Zehavi [35].

A natural choice is to set the value of r close to the expectation of $\mathrm{la}(T_G)|$. The next lemma follows from the definition of $\mathrm{la}(T_G)$, by the linearity of expectation.

Lemma 6. *For every (k,l)-tree T_G in G, we have $\mathrm{E}(|\mathrm{la}(T_G)|) = \frac{3}{4}k + \frac{1}{2}l - \frac{1}{4}$.*

By the lemma above and Markov's inequality, if we put $r = \frac{3}{4}k + \frac{1}{2}l$, then the probability that $|\mathrm{la}(T_G)| \le r$ is $\Omega(\frac{1}{k+l})$. Hence it suffices to repeat the algorithm from Lemma 5 (i.e., evaluate the polynomial $P_{\downarrow r}$) $O(k + l)$ times, answering true iff at least one evaluation was non-zero, to get a Monte-Carlo algorithm for testing the existence of a (k,l)-tree. The complexity of this algorithm is $O^*(2^{(3k+2l)/4})$. However, similarly as in [2,4], we can do better. The idea is to use a value of r smaller than that appearing in the expectation by an $\Omega(k)$ term; then the probability that a (k,l)-subtree is admissible is inverse-exponential. Hence, we need to repeat the algorithm from Lemma 5 exponentially many times, every time for a different random bipartition. It turns out that for carefully selected values of r, this pays off. To find this value, the following lemma is crucial.

Lemma 7. *Fix an arbitrary (k,l)-tree T_G in G. For any integer t such that $0 \le t \le (k-1)/2$, we have $\Pr\left(|\mathrm{la}(T_G)| \le k + \frac{l}{2} - t\right) \ge \dfrac{1}{2^{k+1}}\dbinom{k-1}{2t}$.*

Proof. Root T_G at an arbitrary vertex r. Let the random variable X_{22} denote the number of edges $uv \in E(T_G)$ such that $u, v \in V_2$. Also, let $X_{1,i}$ denote the number of internal vertices in V_1. Then, by the definition of $\mathrm{la}(T_G)$, we have $|\mathrm{la}(G_T)| = l + X_{1,i} + X_{22}$.

Fix a subset of edges $S \in \binom{E(T_G)}{2t}$. For $a = 0, 1$, let $c_a : V(T_G) \rightarrow \{1, 2\}$ be the assignment of vertices of T_G to sets V_1, V_2 such that for every $v \in V(T_G)$, we have $c_a(v) = 1$ if and only if on the path from r to v in T_G the number of edges from S is congruent to a modulo 2. Since every vertex is colored 1 in exactly one of the colorings c_0, c_1, we infer that $X_{1,i}(c_0) + X_{1,i}(c_1) = k - l$. Similarly every edge in $E(T_G) \setminus S$ is colored 22 in exactly one of the colorings c_0, c_1; hence $X_{22}(c_0) + X_{22}(c_1) = k - 1 - 2t$. It follows that

$$\min\{X_{1,i}(c_0) + X_{22}(c_0), X_{1,i}(c_1) + X_{22}(c_1)\} \leq (k - l + k - 1 - 2t)/2 < k - \tfrac{l}{2} - t.$$

Hence, for at least one of the colorings c_0, c_1, we have $|\,\mathrm{la}(T_G)| < k + \tfrac{l}{2} - t$. For all choices of S there are at least $\tfrac{1}{2}\binom{k-1}{2t}$ such colorings, so the claim follows.   □

The following lemma follows immediately from Stirling's approximation.

Lemma 8. *For any fixed α, $0 < \alpha < 1$,* $\binom{n}{\alpha n} = O^*\left(\left(\dfrac{1}{\alpha^\alpha(1-\alpha)^{1-\alpha}}\right)^n\right).$

Theorem 1. *There is a randomized $O^*(1.66^k 2^{l/2})$-time polynomial space algorithm for (k, l)-Tree.*

Proof. Fix $\epsilon \geq 0$. Let $t = \lfloor(\tfrac{1}{4} + \epsilon)k\rfloor$ and $r = k - t + \lceil\tfrac{l}{2}\rceil = \lceil(\tfrac{3}{4} - \epsilon)k\rceil + \lceil\tfrac{l}{2}\rceil$. We choose a random bipartition $V = V_1 \cup V_2$, and apply the algorithm from Lemma 5. We repeat this $\lceil 2^{k+1}/\binom{k-1}{2t}\rceil$ times, returning YES iff at least one of the executions of the algorithm from Lemma 5 returned YES. If there is no (k, l)-tree in the input graph, by Lemma 5 we report the correct answer. Now assume there is a (k, l)-tree T_G. Call a bipartition *nice* if $|\,\mathrm{la}(T_G)| \leq r$. By Lemma 7, a random bipartition is nice with probability at least $p = \tfrac{1}{2^{k+1}}\binom{k-1}{2t}$. Hence, at least one of the tried bipartitions is nice with probability at least $1 - (1 - p)^{1/p} \geq 1 - 1/e$. For such a bipartition, the algorithm from Lemma 5 answers YES with probability at least $\tfrac{1}{2}$. Hence we report a false-negative with probability at most $1/e + \tfrac{1}{2} < 1$.

By Lemma 5, the running time is

$$O^*\left(2^{r+k} / \binom{k-1}{2t}\right) = O^*\left(2^{(7/4-\epsilon)k+l/2} / \binom{k}{(1/2 + 2\epsilon)k}\right).$$

By Lemma 8, we can express this by $O^*((f(\epsilon))^k 2^{l/2})$, for $f(\epsilon) = 2^{7/4-\epsilon}(\tfrac{1}{2} + 2\epsilon)^{\frac{1}{2}+2\epsilon}(\tfrac{1}{2} - 2\epsilon)^{\frac{1}{2}-2\epsilon}$. The function f attains a minimum smaller than 1.65685 for $\epsilon = 0.042894$. Hence the claim.   □

3   Colored Graphs

In this section, to improve the running times of algorithms from Section 2 in restricted settings, we adjust the partition $V = V_1 \cup V_2$ to particular graph classes where vertex colorings guide us in making the partition. We will consider three ways of coloring the vertices. The first is ordinary proper vertex coloring of the graph, i.e., color the vertices so that no edge has both its endpoints colored by the same color. The least number of colors needed is denoted by χ_G. The second way is fractional vertex coloring, that assigns a subset of b colors to each vertex from a palette of a colors so that the endpoints of each edge receive disjoint subsets of colors. The smallest possible ratio a/b is denoted by $\chi_f(G)$. The third way is vector coloring, that assigns unit length vectors to the vertices so the minimum angle α between every edge's endpoints' vectors is as large as possible. The smallest possible value of $1 + cos^{-1}(\alpha)$ is denoted by $\chi_v(G)$.

The following chain of inequalities holds (see e.g. [21]), where $\omega(G)$ is the clique number, i.e., the size of the largest clique in G.

Theorem 2. *For any graph G, $\omega(G) \leq \chi_v(G) \leq \chi_f(G) \leq \chi(G)$.*

Consider a proper d-coloring $c : V \to \{1, \ldots, d\}$ of the host graph G. Fix a number $t \in \{0, \ldots, d\}$. Our idea is to define V_1 as the union of t color classes, and V_2 as the remaining vertices. Clearly, for some choices of the colors the set $\mathrm{la}(T_G)$ for a solution T_G can be large, and then by Lemma 2 we need to set the parameter r high (which makes the running time slow). However, if we try *all* the possible choices of t colors, in at least one of them the set $\mathrm{la}(T_G)$ will be small enough. This is stated in the following lemma.

Lemma 9. *Let c be a given d-coloring of G. Let T_G be a fixed (k, l)-tree in G. There is a choice of t color classes $c_1, \ldots, c_t$ such that for $V_1 = \bigcup_{i=1}^{t} c^{-1}(i)$,*

$$\mathrm{la}(T_G) \leq \left(1 - \frac{x(d - x)}{d(d - 1)}\right) k + \left(1 - \frac{x}{d}\right) l,$$

where $x = \lfloor \frac{d + (l/k)(d-1)}{2} \rceil$. In particular, $|\,\mathrm{la}(T_G)| \leq \left(1 - \frac{\lfloor \frac{d}{2} \rfloor \lceil \frac{d}{2} \rceil}{d(d-1)}\right) k + \frac{l}{2}$.

Proof. For $i = 1, \ldots, d$, let k_i denote the number of nodes of T_G colored by i. Similarly, let l_i denote the number of leaves of T_G colored by i. Finally, for $i, j = 1, \ldots, d$, let $k_{i,j}$ denote the number of edges in T_G with one endpoint colored with i and the other colored with j. Note that $\sum_{i=1}^{d} k_i = k$, $\sum_{i=1}^{d} l_i = l$, and $\sum_{1 \leq i < j \leq d} k_{i,j} = k - 1$. Fix $t = 0, \ldots, d$. It follows that the average size of $|\,\mathrm{la}(T_G)|$, over all possible choices of the set S of t colors, equals

$$\frac{1}{\binom{d}{t}} \sum_{S \in \binom{[d]}{t}} \left(\sum_{i \in S} k_i + \sum_{i \notin S} l_i + \sum_{\{i,j\} \cap S = \emptyset} k_{i,j}\right) = \frac{\binom{d-1}{t-1}k + \binom{d-1}{t}l + \binom{d-2}{t}(k - 1)}{\binom{d}{t}} \leq$$

$$\left(1 - \frac{t(d - t)}{d(d - 1)}\right) k + \frac{d - t}{d}l.$$

To minimize the above expression, we choose $t = x$. The choice $t = \lceil d/2 \rceil$ is optimal when l is small (as e.g. in the application for k-PATH). $\square$

By combining Lemma 5 with Lemma 9, we get the following theorem.

Theorem 3. *Assume we are given a proper d-coloring of the host graph, for some fixed d. Then, there is a randomized polynomial space algorithm for finding a (k, l)-tree running in time $O^*(2^{(1- \frac{x(d-x)}{d(d-1)})k+(1-\frac{x}{d})l})$, where $x = \lfloor \frac{d+(l/k)(d-1)}{2} \rceil$. In particular, the running time can be bounded by $O^*(2^{(1- \frac{\lfloor \frac{d}{2} \rfloor \lceil \frac{d}{2} \rceil}{d(d-1)})k+\frac{l}{2}})$.*

If the input graph is a clique or an odd cycle, (k, l)-TREE can clearly be solved in polynomial time. Otherwise, one can find a proper Δ-coloring in linear time by Lovász's proof of Brooks' theorem [25], so Theorem 3 implies the following.

Corollary 1. *There is a randomized polynomial space algorithm for (k, l)-TREE in graphs of bounded degree Δ which runs in time $O^*(2^{(1- \frac{x(\Delta-x)}{\Delta(\Delta-1)})k+(1-\frac{x}{\Delta})l})$, where $x = \lfloor \frac{\Delta+(l/k)(\Delta-1)}{2} \rceil$. In particular, the running time can be bounded by $O^*(2^{(1- \frac{\lfloor \frac{\Delta}{2} \rfloor \lceil \frac{\Delta}{2} \rceil}{\Delta(\Delta-1)})k+\frac{l}{2}})$. With $l = 2$, the same result holds for k-PATH, and with $k = n$, for HAMILTONICITY.*

References

1. Alon, N., Yuster, R., Zwick, U.: Color coding. J. ACM **42**(4), 844–856 (1995)
2. Björklund, A.: Determinant sums for undirected Hamiltonicity. SIAM J. on Computing **43**(1), 280–299 (2014)
3. Björklund, A., Husfeldt, T., Kaski, P., Koivisto, M.: The travelling salesman problem in bounded degree graphs. In: Aceto, L., Damgård, I., Goldberg, L.A., Halldórsson, M.M., Ingólfsdóttir, A., Walukiewicz, I. (eds.) ICALP 2008, Part I. LNCS, vol. 5125, pp. 198–209. Springer, Heidelberg (2008)
4. Björklund, A., Husfeldt, T., Kaski, P., Koivisto, M.: Narrow sieves for parameterized paths and packings (2010). CoRR, abs/1007.1161
5. Björklund, A., Kamat, V., Kowalik, L., Zehavi, M.: Spotting trees with few leaves (2015). CoRR, abs/1501.00563
6. Björklund, A., Kaski, P., Kowalik, L.: Probably optimal graph motifs. In: Proc. STACS 2013. LIPIcs, vol. 20, pp. 20–31 (2013)
7. Björklund, A., Kaski, P., Kowalik, Ł.: Fast witness extraction using a decision oracle. In: Schulz, A.S., Wagner, D. (eds.) ESA 2014. LNCS, vol. 8737, pp. 149–160. Springer, Heidelberg (2014)
8. Chen, J., Kneis, J., Lu, S., Molle, D., Richter, S., Rossmanith, P., Sze, S.H., Zhang, F.: Randomized divide-and-conquer: Improved path, matching, and packing algorithms. SIAM J. on Computing **38**(6), 2526–2547 (2009)
9. Cohen, N., Fomin, F.V., Gutin, G., Kim, E.J., Saurabh, S., Yeo, A.: Algorithm for finding k-vertex out-trees and its application to k-internal out-branching problem. J. Comput. Syst. Sci. **76**(7), 650–662 (2010)
10. Cygan, M., Nederlof, J., Pilipczuk, M., Pilipczuk, M., van Rooij, J.M.M., Wojtaszczyk, J.O.: Solving connectivity problems parameterized by treewidth in single exponential time. In: Proc. FOCS 2011, pp. 150–159 (2011)
11. Daligault, J.: Combinatorial techniques for parameterized algorithms and kernels, with applications to multicut. PhD thesis, Universite Montpellier II (2011)
12. Demers, A., Downing, A.: Minimum leaf spanning tree. US Patent no. 6,105,018, August 2013

3. DeMillo, R.A., Lipton, R.J.: A probabilistic remark on algebraic program testing. Inf. Process. Lett. **7**, 193–195 (1978)
4. Eppstein, D.: The traveling salesman problem for cubic graphs. J. Graph Algorithms Appl. **11**(1), 61–81 (2007)
5. Fomin, F., Lokshtanov, D., Saurabh, S.: Efficient computation of representative sets with applications in parameterized and exact agorithms. In: SODA, pp. 142–151 (2014)
6. Fomin, F.V., Lokshtanov, D., Panolan, F., Saurabh, S.: Representative sets of product families. In: Schulz, A.S., Wagner, D. (eds.) ESA 2014. LNCS, vol. 8737, pp. 443–454. Springer, Heidelberg (2014)
7. Gärtner, B., Matoušek, J.: Approximation algorithms and semidefinite programming. Springer, Heidelberg (2012)
8. Goemans, M.X., Williamson, D.P.: Improved approximation algorithms for maximum cut and satisfiability problems using semidefinite programming. Journal of the ACM **42**(6), 1115–1145 (1995)
9. Grandoni, F.: A note on the complexity of minimum dominating set. J. Discrete Algorithms **4**(2), 209–214 (2006)
10. Gebauer, H.: On the number of hamilton cycles in bounded degree graphs. In: Proc. ANALCO 2008, pp. 241–248 (2008)
11. Gvozdenovic, N., Laurent, M.: The operator psi for the chromatic number of a graph. SIAM Journal on Optimization **19**(2), 572–591 (2008)
12. Iwama, K., Nakashima, T.: An improved exact algorithm for cubic graph TSP. In: Lin, G. (ed.) COCOON 2007. LNCS, vol. 4598, pp. 108–117. Springer, Heidelberg (2007)
13. Koutis, I.: Faster algebraic algorithms for path and packing problems. In: Aceto, L., Damgård, I., Goldberg, L.A., Halldórsson, M.M., Ingólfsdóttir, A., Walukiewicz, I. (eds.) ICALP 2008, Part I. LNCS, vol. 5125, pp. 575–586. Springer, Heidelberg (2008)
14. Li, W., Wang, J., Chen, J., Cao, Y.: A 2k-vertex kernel for maximum internal spanning tree (2014). CoRR abs/1412.8296
15. Lovász, L.: Three short proofs in graph theory. J. Combin. Theory Ser. **19**, 269–271 (1975)
16. Monien, B.: How to find long paths efficiently. Annals of Discrete Mathematics **25**, 239–254 (1985)
17. Nederlof, J.: Fast polynomial-space algorithms using möbius inversion: improving on steiner tree and related problems. In: Albers, S., Marchetti-Spaccamela, A., Matias, Y., Nikoletseas, S., Thomas, W. (eds.) ICALP 2009, Part I. LNCS, vol. 5555, pp. 713–725. Springer, Heidelberg (2009)
18. Nederlof, J.: Fast polynomial-space algorithms using inclusion-exclusion. Algorithmica **65**(4), 868–884 (2013)
19. Prieto, E., Sloper, C.: Reducing to independent set structure - the case of k-internal spanning tree. Nord. J. Comput. **12**(3), 308–318 (2005)
20. Raible, D., Fernau, H., Gaspers, D., Liedloff, M.: Exact and parameterized algorithms for max internal spanning tree. Algorithmica **65**(1), 95–128 (2013)
21. Razgon, I.: Exact computation of maximum induced forest. In: Arge, L., Freivalds, R. (eds.) SWAT 2006. LNCS, vol. 4059, pp. 160–171. Springer, Heidelberg (2006)
22. Schwartz, J.T.: Fast probabilistic algorithms for verification of polynomial identities. J. ACM **27**(4), 701–717 (1980)

33. Shachnai, H., Zehavi, M.: Representative families: a unified tradeoff-based approach. In: Schulz, A.S., Wagner, D. (eds.) ESA 2014. LNCS, vol. 8737, pp. 786–797 Springer, Heidelberg (2014)

34. Williams, R.: Finding paths of length k in $O^*(2^k)$ time. Inf. Process. Lett. **109**(6) 315–318 (2009)

35. Zehavi, M.: Algorithms for k-internal out-branching. In: Gutin, G., Szeider, S (eds.) IPEC 2013. LNCS, vol. 8246, pp. 361–373. Springer, Heidelberg (2013)

36. Zehavi, M.: Mixing color coding-related techniques (2014). CoRR, abs/1410.5062

37. Zippel, R.: Probabilistic algorithms for sparse polynomials. In: Ng, K.W. (ed.) EUROSAM 1979 and ISSAC 1979. LNCS, vol. 72, pp. 216–226. Springer, Heidelberg (1979)

Constraint Satisfaction Problems over the Integers with Successor

Manuel Bodirsky[1]([⊠]), Barnaby Martin[2], and Antoine Mottet[3]

[1] Institut für Algebra, TU Dresden, Dresden, Germany
Manuel.Bodirsky@tu-dresden.de
[2] School of Science and Technology, Middlesex University, London, UK
[3] École Normale Supérieure de Cachan, Cachan, France

Abstract. A *distance constraint satisfaction problem* is a constraint satisfaction problem (CSP) whose constraint language consists of relations that are first-order definable over $(\mathbb{Z}; \mathrm{succ})$, i.e., over the integers with the successor function. Our main result says that every distance CSP is in P or NP-complete, unless it can be formulated as a finite domain CSP in which case the computational complexity is not known in general.

1 Introduction

> *"Die ganzen Zahlen hat der liebe Gott gemacht, alles andere ist Menschenwerk."*[1] Leopold Kronecker

A *constraint satisfaction problem* is a computational problem where the input consists of a finite set of variables and a finite set of constraints, and where the question is whether there exists a mapping from the variables to some fixed domain such that all the constraints are satisfied. When the domain is finite, and arbitrary constraints are permitted in the input, the CSP is NP-complete. However, when only constraints for a restricted set of relations are allowed in the input, it might be possible to solve the CSP in polynomial time. The set of relations that is allowed to formulate the constraints in the input is often called the *constraint language*. The question which constraint languages give rise to polynomial-time solvable CSPs has been the topic of intensive research over the past years. It has been conjectured by Feder and Vardi [8] that CSPs for constraint languages over finite domains have a complexity dichotomy: they are in P or NP-complete.

A famous CSP over an infinite domain is *feasibility of linear inequalities over the integers*. It is of great importance in practice and theory of computing,

---

M. Bodirsky—The first author has received funding from the European Research Council under the European Community's Seventh Framework Programme (FP7/2007-2013 Grant Agreement no. 257039).
B. Martin—The second author was supported by EPSRC grant EP/L005654/1.
[1] *"God made the integers, all the rest is the work of man."* Quoted in *Philosophies of Mathematics*, page 13, by Alexander George, Daniel J. Velleman, Philosophy, 2002.

© Springer-Verlag Berlin Heidelberg 2015
M.M. Halldórsson et al. (Eds.): ICALP 2015, Part I, LNCS 9134, pp. 256–267, 2015.
DOI: 10.1007/978-3-662-47672-7_21

and NP-complete. In order to obtain a systematic understanding of polynomial time solvable restrictions and variations of this problem, Jonsson and Lööw [13] proposed to study the class of CSPs where the constraint language Γ is definable in *Presburger arithmetic*; that is, it consists of relations that have a first-order definition over $(\mathbb{Z}; \leq, +)$. Equivalently, each relation $R(x_1, \ldots, x_n)$ in Γ can be defined by a disjunction of conjunctions of the atomic formulas of the form $p \leq 0$ where p is a linear polynomial with integer coefficients and variables from $\{x_1, \ldots, x_n\}$. The constraint satisfaction problem for Γ, denoted by $\mathrm{CSP}(\Gamma)$, is the problem of deciding whether a given conjunction of formulas of the form $R(y_1, \ldots, y_n)$, for some n-ary R from Γ, is satisfiable in Γ. By appropriately choosing such a constraint language Γ, a great variety of problems over the integers can be formulated as $\mathrm{CSP}(\Gamma)$. Several constraint languages Γ over the integers are known where the CSP can be solved in polynomial time. However, a complete complexity classification for the CSPs of Jonsson-Lööw languages appears to be a very ambitious goal.

In this paper, we study one of the most basic classes of constraint languages that falls into the framework of Jonsson and Lööw, namely the class of *distance constraint satisfaction problems* [1]. A distance constraint satisfaction problem is a CSP for a constraint language over the integers whose relations have a first-order definition over $(\mathbb{Z}; \mathrm{succ})$ where succ is the successor function. The structure $(\mathbb{Z}; \mathrm{succ})$ has quantifier-elimination, and it is easy to see that a relation is first-order definable over $(\mathbb{Z}; \mathrm{succ})$ if and only if it can be defined by a disjunction of conjunctions of literals of the form $x = \mathrm{succ}^c(y)$ or $x \neq \mathrm{succ}^c(y)$ for $c \in \mathbb{N}$.

It has been shown previously that distance CSPs for constraint languages whose relations have *bounded Gaifman degree* are either NP-complete, or in P, or can also be formulated with a constraint language over a finite domain [1]. The finite Gaifman degree assumption is quite strong; however, here we prove that the same is true even if we drop this assumption. In other words, we show that if the Feder-Vardi dichotomy conjecture for finite domain CSPs is true, then also the class of all distance CSPs exhibits a complexity dichotomy.

Our proof relies on the so-called universal-algebraic approach; this is the first time that this approach has been used for constraint languages that are not finite or countably infinite ω-*categorical*. The central insight of the universal-algebraic approach to constraint satisfaction is that the computational complexity of a CSP is captured by the set of *polymorphisms* of the constraint language. One of the ideas of the present paper is that in order to use polymorphisms when the constraint language is not ω-categorical, we have to pass to the countably saturated model of the integers with successor. The relevance of saturated models for the universal-algebraic approach has already been pointed out in joint work of the authors with Martin Hils [2], but this is the first time that this perspective has been used to perform complexity classification for a large class of concrete computational problems.

The formal definitions of CSPs and distance CSPs can be found in Section 2. The border between distance CSPs in P and NP-complete distance CSPs can be most elegantly stated using the terminology of the mentioned universal-algebraic pproach to constraint satisfaction. This is why we first give a brief introduction) this approach in Section 3, and only then give the technical description of our ‖ain result in Section 4. Section 5 gives a classification of distance constraint ιnguages that might be of independent interest; this classification is the basis f our classification of the complexity of distance CSPs. Our algorithmic results ιn be found in Section 6. Finally, we put all the results together to prove our ‖ain result in Section 7. We discuss our result and promising future research uestions in Section 8.

Distance CSPs

ιet Γ be a structure with a finite relational signature τ. When R is a relation ymbol from τ, we write R^Γ for the relation it denotes in the structure Γ.

A τ-*formula* is a first-order formula built from the relations from τ, and quality. A τ-formula is *primitive positive (pp)* if it is of the form $\exists x_1, \ldots, x_k (\psi_1 \wedge \cdots \wedge \psi_m)$ where each ψ_i is an atomic τ-formula. *Sentences* are formulas without ree variables.

Definition 1 (CSP(Γ)). *The* constraint satisfaction problem *for Γ is the following computational problem.*

nput: *A primitive positive τ-sentence Φ.*
Question: $\Gamma \models \Phi$?

The structure Γ will also be called the *constraint language* of CSP(Γ). A elational structure Γ is a *reduct* of a structure Δ if it has the same domain as Δ and every relation R^Γ of arity k is *first-order definable* over Δ, that is, there xists a first-order formula φ in the signature of Δ with k free variables such that or all elements $u_1, \ldots, u_k$ of Γ we have $R^\Gamma(u_1, \ldots, u_k) \Leftrightarrow \Delta \models \varphi(u_1, \ldots, u_k)$.

We write $(\mathbb{Z}; \mathrm{succ})$ for the structure of the integers with the successor function.

Definition 2 (Distance CSP). *A* distance CSP *is a constraint satisfaction problem where the constraint language is finite and a reduct of $(\mathbb{Z}; \mathrm{succ})$.*

It is well-known that $(\mathbb{Z}; \mathrm{succ})$ admits quantifier elimination (this is easy to rove, and can be found explicitly in [9]). Moreover, it is easy to see that every quantifier-free formula is over $(\mathbb{Z}; \mathrm{succ})$ equivalent to a quantifier-free formula n conjunctive normal form (CNF) where every atomic formula is of the form $y = \mathrm{succ}^n(x)$ for $n \in \mathbb{N}$, where $\mathrm{succ}^n(x)$ is defined inductively by $\mathrm{succ}^0(x) = x$, and $\mathrm{succ}^{n+1}(x) = \mathrm{succ}(\mathrm{succ}^n(x))$. We will call formulas of this form *standardized*.

Example 1. We give examples of reducts of $(\mathbb{Z}; \mathrm{succ})$; the relations from those xamples will re-appear in later sections.

1. $(\mathbb{Z}; \mathrm{Diff}_S)$, where $\mathrm{Diff}_S := \{(x, y) : x, y \in \mathbb{Z}, y - x \in S\}$ for a finite set $S \subset \mathbb{Z}$.
2. $(\mathbb{Z}; \mathrm{Diff}_{\{2\}}, \{(x, y) : |x - y| \leq 2\})$.

3. $(\mathbb{Z}; F)$ where F is the 4-ary relation $\{(x, y, u, v) : x = \text{succ}(y) \Leftrightarrow u = \text{succ}(v)\}$.
4. $(\mathbb{Z}; \neq, \text{Dist}_i)$ where $\text{Dist}_i := \{(x, y) : |x - y| = i\}$.

The last two examples have unbounded Gaifman degree (see Section 5.1), so they do not fall into the scope of [1]. The following is easy to see.

Proposition 1. *All distance CSPs are in NP.*

3   The Algebraic Approach

The starting point of the universal algebraic approach to analyze the complexity of CSPs is the observation that when a relation R can be defined by a primitive positive formula over Γ, then $\text{CSP}(\Gamma)$ allows to simulate the 'richer' problem $\text{CSP}(\Delta)$ where $\Delta = (\Gamma, R)$ has been obtained from Γ by adding R as another relation. The proof of this fact given by Jeavons, Cohen, and Gyssens [12] works for all structures Γ over finite or over infinite domains. Since we will use this fact very frequently, we will not explicitly refer back to it from now on.

Polymorphisms are an important tool to study the question of which relations are primitive positive definable in Γ. We say that a function $f \colon D^n \to D$ *preserves* a relation $R \subseteq D^m$ if for all $t_1, \ldots, t_n \in R$ the tuple $f(t_1, \ldots, t_n)$ obtained by applying f componentwise to the tuples $t_1, \ldots, t_n$ is also in R; otherwise, f *violates* R. A *polymorphism* of a relational structure Γ with domain D is a function from D^n to D, for some finite n, which preserves *all* relations of Γ. We write $\text{Pol}(\Gamma)$ for the set of all polymorphisms of Γ. It is clear that a polymorphism of a structure Γ also preserves all relations that are primitive positive definable in Γ; this holds for arbitrary finite and infinite structures Γ. If Γ is finite or ω-categorical [5], then a relation is preserved by all polymorphisms *if and only if* it is primitive positive definable in Γ.

The structures that we consider in this paper will not be ω-categorical; however, following the philosophy in [2], one can refine these universal-algebraic methods to apply them also in our situation. The *(first-order) theory* of a structure Γ, denoted by $\text{Th}(\Gamma)$, is the set of all first-order sentences that are true in Γ. We define some notation to conveniently work with models of $\text{Th}(\Gamma)$ and their reducts.

Definition 3 ($\kappa.\mathbb{Z}$). *Let κ be a cardinal. We write $\kappa.\mathbb{Z}$ for κ copies of $\mathbb{Z}$ indexed by the elements of κ; formally, $\kappa.\mathbb{Z}$ is the set $\{(a, z) : a \in \kappa, z \in \mathbb{Z}\}$. Then $(\kappa.\mathbb{Z}; \text{succ})$ is the structure where succ denotes the function that maps (a, z) to $(a, z + 1)$.*

It is well-known and easy to see that the models of $\text{Th}(\mathbb{Z}; \text{succ})$ are precisely the structures isomorphic to $(\kappa.\mathbb{Z}; \text{succ})$, for some cardinal κ. When $k \in \mathbb{Z}$ and $u = (a, z) \in \kappa.\mathbb{Z}$, we write $u + k$ for $(a, z + k)$.

Definition 4 ($\kappa.\Gamma$). *Let Γ be a reduct of $(\mathbb{Z}; \text{succ})$ with signature τ. Then $\kappa.\Gamma$ denotes the 'corresponding' reduct of $(\kappa.\mathbb{Z}; \text{succ})$ with signature τ. Formally, when $R \in \tau$ and φ_R is a formula that defines R^Γ, then $R^{\kappa.\Gamma}$ is the relation defined by φ_R over $(\kappa.\mathbb{Z}; \text{succ})$.*

We use ω to denote the smallest infinite cardinal throughout the article. Note that $(\omega.\mathbb{Z}; \text{succ})$ is isomorphic to the structure $(\mathbb{Q}; x \mapsto x + 1)$. In the following, we identify $(\mathbb{Z}; \text{succ})$ with the copy of $(\mathbb{Z}; \text{succ})$ induced by $0.\mathbb{Z}$ in $(\omega.\mathbb{Z}; \text{succ})$. That is, we view $(\mathbb{Z}; \text{succ})$ as a substructure of $(\omega.\mathbb{Z}; \text{succ})$, and consequently Γ is a substructure of $\omega.\Gamma$ for each reduct Γ of $(\mathbb{Z}; \text{succ})$.

A *type* of a structure Δ is a set p of formulas with one free variable x such that $\cup \text{Th}(\Delta)$ is satisfiable (that is, $\{\varphi(c) : \varphi \in p\} \cup \text{Th}(\Delta)$, for a new constant symbol c, has a model). A τ-structure Γ is ω-*saturated* if for all choices of finitely many constants $c_1, \ldots, c_n$ for elements of Γ, and every type p of $(\Gamma, c_1, \ldots, c_n)$, there exists an element d of Γ such that $(\Gamma, c_1, \ldots, c_n) \models \varphi(d)$ for all $\varphi \in p$. When Γ and Δ are two countable ω-saturated structures with the same first-order theory, then Γ and Δ are isomorphic [11]. Note that $(\omega.\mathbb{Z}; \text{succ})$ is ω-saturated. More generally, $\omega.\Gamma$ is ω-saturated for every reduct Γ of $(\mathbb{Z}; \text{succ})$.

We define the function $- \colon (\kappa.\mathbb{Z})^2 \to (\mathbb{Z} \cup \{\omega\})$ for $x, y \in \kappa.\mathbb{Z}$ by

$$x - y := z \in \mathbb{Z} \quad \text{if } x = \text{succ}^z(y) \text{ for } z \geq 0,$$
$$\text{or } y = \text{succ}^{-z}(x) \text{ for } z < 0;$$
$$x - y := \omega \quad \text{otherwise.}$$

When Γ and Δ are two structures with the same relational signature τ, then a *homomorphism* from Γ to Δ is a function from the domain of Γ to the domain of Δ such that for every $R \in \tau$ of arity k we have $R^\Gamma(u_1, \ldots, u_k) \Rightarrow R^\Delta(f(u_1), \ldots, f(u_k))$. It is straightforward to see that if there is a homomorphism from Γ to Δ, and vice versa, then $\text{CSP}(\Gamma)$ and $\text{CSP}(\Delta)$ are the same computational problem.

Lemma 1 (See Lemma 2.1 in [2]). *Let Γ be ω-saturated, let Δ be countable, let $d_1, \ldots, d_k$ be elements of Δ, and let $c_1, \ldots, c_k$ be elements of Γ. Suppose that for all primitive positive formulas φ such that $\Delta \models \varphi(d_1, \ldots, d_k)$ we have $\Gamma \models \varphi(c_1, \ldots, c_k)$. Then there exists a homomorphism from Δ to Γ that maps d_i to c_i for all $i \leq k$.*

An *endomorphism* is a unary polymorphism. To classify the computational complexity of the CSP for all reducts of a structure Γ, it often turns out to be important to study the possible endomorphisms of those reducts first, before studying the polymorphisms, e.g. for the reducts of $(\mathbb{Q}; <)$ in [4] and the reducts of the countably infinite random graph in [6].

We are now in the position to state a general result, Theorem 1, that might explain the importance of ω-saturated models for the universal-algebraic approach. When Γ is a structure, then the *orbit* of a k-tuple $(a_1, \ldots, a_k)$ of elements of Γ is the set $\{(\alpha(a_1), \ldots, \alpha(a_k)) \mid \alpha \in \text{Aut}(\Gamma)\}$.

Theorem 1. *Let Γ be a countable ω-saturated structure, let Δ be a reduct of Γ, and R a relation with a first-order definition in Γ. Then*

- *R has a first-order definition in Δ if and only if R is preserved by the auto morphisms of Δ;*
- *R has an existential positive definition in Δ if and only if R is preserved by the endomorphisms of Δ;*
- *if R consists of n orbits of k-tuples in Γ, then R has a primitive positive definition in Δ if and only if R is preserved by all polymorphisms of Δ of arity n.*

4   Statement of Results

The border between NP-complete successor CSPs and successor CSPs in P can be described as follows, modulo the Feder-Vardi dichotomy conjecture. A reduct Γ of $(\mathbb{Z}; \mathrm{succ})$ is *positive* if all relations of Γ have a *positive* first-order definition in $(\mathbb{Z}; \mathrm{succ})$, this is, by a first-order formula without negation. We write $\mathbb{N}$ for the natural numbers including 0, and $\mathbb{N}^+$ for the set of positive natural numbers.

Definition 5. *For $d \in \mathbb{N}^+$, the d-modular maximum, $\max_d \colon \mathbb{Z}^2 \to \mathbb{Z}$, is defined by $\max_d(x, y) := \max(x, y)$ if $x = y \mod d$ and $\max_d(x, y) := x$ otherwise. The d-modular minimum is defined analogously.*

Note that these two operations are not commutative when $d > 1$.

Theorem 2. *Let Γ be a reduct of $(\mathbb{Z}; \mathrm{succ})$ with finite signature. Then there exists a structure Δ such that $\mathrm{CSP}(\Delta)$ equals $\mathrm{CSP}(\Gamma)$ and one of the following cases applies.*

1. *Δ has a finite domain, and the CSP for Γ is conjectured to be in P or NP-complete [8].*
2. *Δ is a reduct of $(\mathbb{Z}; \mathrm{succ})$ and preserved by a modular max or modular min. In this case, $\mathrm{CSP}(\Gamma)$ is in P.*
3. *Δ is a reduct of $(\mathbb{Z}; \mathrm{succ})$ such that $\omega.\Delta$ is preserved by an (equivalently, all) isomorphisms between $(\omega.\mathbb{Z}; \mathrm{succ})^2$ and $(\omega.\mathbb{Z}; \mathrm{succ})$. In this case, $\mathrm{CSP}(\Gamma)$ is in P.*
4. *$\mathrm{CSP}(\Gamma)$ is NP-complete.*

5   Definability of Successor

The goal of this section is a proof that the CSPs for reducts of $(\mathbb{Z}; \mathrm{succ})$ fall into four classes. This will allow us to focus in later sections on reducts of $(\mathbb{Z}; \mathrm{succ})$ where succ is pp-definable, where succ is now used to denote the graph of the successor function, that is, $\mathrm{succ} = \{(x, y) \in \mathbb{Z}^2 \mid y = x + 1\}$.

Theorem 3. *Let Γ be a reduct of $(\mathbb{Z}; \mathrm{succ})$ with finite signature. Then $\mathrm{CSP}(\Gamma)$ equals $\mathrm{CSP}(\Delta)$ where Δ is one of the following:*

1. *a finite structure;*
2. *a reduct of $(\mathbb{Z}; =)$;*
3. *a reduct of $(\mathbb{Z}; F)$ where Dist_k is pp-definable for all $k \geq 1$ (see Example 1);*
4. *a reduct of $(\mathbb{Z}; \mathrm{succ})$ where succ is pp-definable.*

The proof of this result requires some effort and spreads over the following subsections. Before we go into this, we explain the significance of the four classes for the CSP.

It is easy to see that there exists a structure Δ with a finite domain such that $\mathrm{CSP}(\Gamma)$ equals $\mathrm{CSP}(\Delta)$ if and only if Γ has an endomorphism with finite range. So we will assume in the following that this is not the case.

The CSPs for reducts of $(\mathbb{Z}; =)$ have been studied in [3]; they are either in P or NP-complete. Hence, we are also done if there exists a reduct Δ of $(\mathbb{Z}; =)$ such that $\mathrm{CSP}(\Delta) = \mathrm{CSP}(\Gamma)$. Several equivalent characterizations of those reducts Γ will be given in Section 5.2. This is essential for proving Theorem 3.

When Γ is a reduct of $(\mathbb{Z}; \mathrm{succ})$ where for all $k \geq 1$ the relation Dist_k is pp-definable, then $\mathrm{CSP}(\Gamma)$ is NP-complete; this is a consequence of the following proposition from [1].

Proposition 2 (Proposition 26 in [1]). *Suppose that the relations Dist_1 and Dist_5 are pp-definable in Γ. Then $\mathrm{CSP}(\Gamma)$ is NP-hard.*

The previous paragraphs explain why Theorem 3 indeed reduces the complexity classification of CSPs for finite-signature reducts Γ of $(\mathbb{Z}; \mathrm{succ})$ to the case where succ is pp-definable in Γ.

5.1   Degrees

We consider three notions of *degree* for relations R that are first-order definable in $(\mathbb{Z}; \mathrm{succ})$:

- For $x \in \mathbb{Z}$, we consider the number of $y \in \mathbb{Z}$ that appear together with x in a tuple from R; this number is the same for all $x \in \mathbb{Z}$, and called the *Gaifman-degree* of R (it is the degree of the Gaifman graph of $(\mathbb{Z}; R)$).
- The *distance degree* of R is the supremum of d such that there are $x, y \in \mathbb{Z}$ that occur together in a tuple of R and $|x - y| = d$.
- The *quantifier-elimination-degree (qe-degree)* of R is the minimal q so that there is a quantifier-free definition of R containing no nesting of succ that is greater than q.

The degree of a reduct of $(\mathbb{Z}; \mathrm{succ})$ is the supremum of the degrees of its relations, for any of the three notions of degree. The paper [1] considered reducts of $(\mathbb{Z}; \mathrm{succ})$ with finite Gaifman-degree. Note that the Gaifman-degree is finite if and only if the distance degree is finite. In this paper, qe-degree will play the central role, as any reduct of $(\mathbb{Z}; \mathrm{succ})$ with finite relational signature clearly has finite qe-degree. We call a binary relation *trivial* if it is pp-definable over $(\mathbb{Z}; \mathrm{succ})$, and *non-trivial* otherwise.

5.2 Petrus

The following theorem is the rock upon which we build our church.

Theorem 4 (Petrus). *Let Γ be a reduct of $(\mathbb{Z}; \mathrm{succ})$ with finite relational signature and without an endomorphism of finite range. Then the following are equivalent:*

1. *there exists a reduct Δ of $(\mathbb{Z}; =)$ such that $\mathrm{CSP}(\Delta)$ equals $\mathrm{CSP}(\Gamma)$;*
2. *$\omega.\Gamma$ has an endomorphism whose range induces a structure isomorphic to a reduct of $(\mathbb{Z}; =)$;*
3. *for all ℓ greater than the qe-degree of Γ, there exists $e \in \mathrm{End}(\Gamma)$ so that the range of e is included in $\{\ell z \mid z \in \mathbb{Z}\}$;*
4. *for all $t \geq 1$, there is an $e \in \mathrm{End}(\Gamma)$, $z \in \mathbb{Z}$, such that $|e(z+t) - e(z)| > t$;*
5. *for all $t \geq 1$, there is an $e \in \mathrm{End}(\omega.\Gamma)$, $z \in \omega.\mathbb{Z}$, such that $|e(z+t) - e(z)| > t$;*
6. *all binary relations with a primitive positive definition in Γ are either the equality relation or have unbounded distance degree;*
7. *for all distinct $z_1, z_2 \in \mathbb{Z}$ there is a homomorphism $h \colon \Gamma \to \omega.\Gamma$ such that $h(z_1) - h(z_2) = \omega$;*
8. *for all distinct $z_1, z_2 \in \mathbb{Z}$ there is an $e \in \mathrm{End}(\omega.\Gamma)$ such that $e(z_1) - e(z_2) = \omega$, and for all $x, y \in \omega.\mathbb{Z}$ with $x - y = \omega$ we have $e(x) - e(y) = \omega$;*
9. *there exists an $e \in \mathrm{End}(\omega.\Gamma)$ with infinite range such that $e(x) - e(y) = \omega$ or $e(x) = e(y)$ for any two distinct $x, y \in \omega.\Gamma$.*

We would like to mention that the finite-signature assumption in the statement of Theorem 4 is necessary.

Example 2. Consider the reduct $\Gamma := (\mathbb{Z}; I_1, I_2, \dots)$ of $(\mathbb{Z}; \mathrm{succ})$ where $I_i := \{(x, y) : x \neq \mathrm{succ}^i(y)\}$. Then the endomorphisms of Γ are precisely the automorphisms of $(\mathbb{Z}; \mathrm{succ})$, and hence Γ does not satisfy items (3) and (4), but it does satisfy the remaining items.

5.3 Boundedness and Rank

Let Γ be a reduct of $(\mathbb{Z}; \mathrm{succ})$ without a finite-range endomorphism. Theorem 4 (Petrus) characterized the "degenerate case" when $\mathrm{CSP}(\Gamma)$ is the CSP for a reduct of $(\mathbb{Z}; =)$. For such Γ, as we have mentioned before, the complexity of the CSP has already been classified. In the following we will therefore assume that the equivalent items of Theorem 4, and in particular, item (5), do *not* apply. To make the best use of those findings, we introduce the following terminology.

Definition 6. *Let $k \in \mathbb{N}^+, c \in \mathbb{N}$. A function $e \colon \kappa_1.\mathbb{Z} \to \kappa_2.\mathbb{Z}$ is (k, c)-bounded if for all $u \in \kappa_1.\mathbb{Z}$ we have $|e(u + k) - e(u)| \leq c$.*

We say that e is *tightly-k-bounded* if it is (k, k)-bounded, and *k-bounded* if it is (k, c)-bounded for some $c \in \mathbb{N}$. We say that $\kappa.\Gamma$ is (k, c)-*bounded* if all its endomorphisms are; similarly, $\kappa.\Gamma$ is *tightly-k-bounded* if all its endomorphisms are. We call the smallest $t \in \mathbb{N}^+$ such that $\kappa.\Gamma$ is tightly-t-bounded the *tight rank* of $\kappa.\Gamma$. Similarly, we call the smallest $r \in \mathbb{N}^+$ such that $\kappa.\Gamma$ is r-bounded the *rank* of $\kappa.\Gamma$. The negation of item (5) in Theorem 4 says that there exists $t \in \mathbb{N}^+$ such that $\omega.\Gamma$ is tightly-t-bounded. Clearly, being tightly-t-bounded implies being t-bounded. Hence, the negation of item (5) in Theorem 4 also implies that $\omega.\Gamma$ has finite rank $r \leq t$.

Example 3. There are rank one reducts of $(\mathbb{Z}; \mathrm{succ})$ which do have non-injective endomorphisms, but no finite-range endomorphisms. Consider the second structure in the Example 1:

$$\Gamma := (\mathbb{Z}; \mathrm{Diff}_{\{2\}}, \{(x, y) : |x - y| \leq 2\}) .$$

Note that Γ has rank one: as e preserves the relation $\{(x, y) : |x - y| \leq 2\})$ we have $|e(x+1) - e(x)| \leq 2$. Also note that Γ has the non-injective endomorphism defined by $e(x) = x$ for even x, and $e(x) = x + 1$ for odd x.

These two notions of rank are the key to generalize the results from [1] about reducts of $(\mathbb{Z}; \mathrm{succ})$ with finite distance degree to general finite-signature reducts.

Remark. All reducts of $(\mathbb{Z}; \mathrm{succ})$ are *strongly minimal* (see [11][14]), another important concept from model theory. Our notion of rank resembles the notion of *dimension* in this context. However, the two notions are different. Consider for instance the structure

$$(\mathbb{Z}; \mathrm{succ}^2, \neq, \{(x, y) : x \neq \mathrm{succ}^3(y)\}) .$$

This structure has dimension one, since the algebraic closure of any of its elements is all of $\mathbb{Z}$. However, the rank of this structure is two and not one.

In order to understand the relations pp-definable in a reduct of $(\omega.\mathbb{Z}, \mathrm{succ})$ with finite rank, we start with the structures which have rank 1, and then show how to factor structures with higher rank to structures of rank 1.

Theorem 5. *Let Γ be a finite-signature reduct of $(\mathbb{Z}; \mathrm{succ})$ so that $\omega.\Gamma$ has rank one. Then $\mathrm{CSP}(\Gamma)$ equals $\mathrm{CSP}(\Delta)$ where Δ is one of the following:*

1. *a finite structure;*
2. *a reduct of $(\mathbb{Z}; F)$ where Dist_k is pp-definable for all $k \geq 1$ (see Example 1);*
3. *a reduct of $(\mathbb{Z}; \mathrm{succ})$ where succ is pp-definable.*

Definition 7. *Let Γ be a reduct of $(\mathbb{Z}; \mathrm{succ})$ and $k \in \mathbb{N}^+$. Then we write Γ/k for the substructure of Γ induced by the set $\{z \in \mathbb{Z} : z = 0 \bmod k\}$.*

For instance, in Example 3 the structure $\Gamma/2$ is isomorphic to

$$(\mathbb{Z}; \mathrm{succ}, \{(x, y) : |x - y| \leq 1\}) .$$

Proposition 3. *Let Γ be a reduct of $(\mathbb{Z}; \text{succ})$ such that $\omega.\Gamma$ has rank $r \in \mathbb{N}$ Then Γ/r has the same CSP as Γ, and is isomorphic to a reduct Δ of $(\mathbb{Z}; \text{succ})$ such that $\omega.\Delta$ has rank one.*

Theorem 3 can now be proved using a combination of Proposition 3 Theorem 5, and Theorem 4.

6   Algorithms

We treat items 2 and 3 in Theorem 2. Let si be any isomorphism between $(\omega.\mathbb{Z}, \text{succ})^2$ and $(\omega.\mathbb{Z}, \text{succ})$. A standardized formula is *Horn* if all its clauses have at most one *positive literal*, i.e., a literal of the form $x = \text{succ}^p(y)$.

Proposition 4. *Let Γ be a reduct of $(\mathbb{Z}; \text{succ})$. If $\omega.\Gamma$ is preserved by si then every relation of Γ has a quantifier-free Horn definition over $(\mathbb{Z}; \text{succ})$. In this case, $\text{CSP}(\Gamma)$ is in P.*

The key algorithmic result here is that satisfiability of Horn formulas can be decided as follows: when the positive unit clauses imply that a literal in the input is false (this can be checked in polynomial time), remove this literal. Repeat this step. If we derive an empty clause in this way, there is no satisfying assignment. Otherwise, we are finally in a situation in which every literal is satisfied by a solution to the positive clauses. Using the assumption that si is a polymorphism of $\omega.\Gamma$, we obtain a satisfying assignment for all clauses in the input.

Theorem 6. *Let Γ be a finite-signature reduct of $(\mathbb{Z}; \text{succ})$ preserved by $\max_d$ or $\min_d$ for some $d \in \mathbb{N}$. Then $\text{CSP}(\Gamma)$ is in P.*

We describe two ideas for the proof of Theorem 6. The first is to reduce $\text{CSP}(\Gamma)$ to $\text{CSP}(\Gamma/d)$. We prove that Γ/d is preserved by max or min. The second idea is to solve $\text{CSP}(\Gamma/d)$ using the (still polynomial-time) *uniform version* of the *arc-consistency procedure*, where both the instance and the (finite) template are given in the input. It suffices to work with templates that are finite substructure of Γ/d whose size is linear in the size of the instance of $\text{CSP}(\Gamma/d)$.

7   The Classification

In this section we prove Theorem 2. By Theorem 3, we are essentially left with the task to classify the CSP for finite-signature *expansions* of $(\mathbb{Z}; \text{succ})$, i.e., reducts of $(\mathbb{Z}; \text{succ})$ which have succ among their relations.

Theorem 7. *Let Γ be a first-order expansion of $(\mathbb{Z}; \text{succ})$. Then at least one of the following is true:*

1. *Γ is positive and preserved by $\max_d$ or $\min_d$ for some $d \in \mathbb{N}$,*
2. *Γ is non-positive and $\omega.\Gamma$ is preserved by si,*
3. *$\text{CSP}(\Gamma)$ is NP-hard.*

To show this theorem, we first prove the following lemma. A standardized formula over the signature of $(\mathbb{Z}; \text{succ})$ in DNF is called *reduced* when every formula obtained by removing literals or clauses is not equivalent over $(\mathbb{Z}; \text{succ})$. It is clear that every quantifier-free formula is equivalent to a reduced formula.

Lemma 2. *For a first-order expansion Γ of $(\omega.\mathbb{Z}; \mathrm{succ})$, are equivalent:*

1. *every reduced DNF that defines a relation of Γ is positive,*
2. *Γ has an endomorphism that violates the binary relation given by $|x-y| = \omega$,*
3. *Γ does not pp-define a non-trivial binary relation of infinite distance degree.*

Using Lemma 2, we treat positive and non-positive expansions Γ of $(\mathbb{Z}; \mathrm{succ})$ separately. In the non-positive case, we first show that when $\omega.\Gamma$ omits si as a polymorphism, then there exists a non-trivial binary relation with finite distance degree with a pp-definition in Γ. Together with the non-trivial binary relation of infinite distance degree from Lemma 2, one can then prove hardness of $\mathrm{CSP}(\Gamma)$ by a reduction from CSPs for finite undirected graphs G, using the classic result that $\mathrm{CSP}(G)$ is hard if G contains an odd cycle [10].

To treat the positive case, we make essential use of results and techniques that have been developed for reducts with finite distance degree in [1], based on the following lemma.

Lemma 3. *Let Γ be a positive first-order expansion of $(\mathbb{Z}; \mathrm{succ})$ that does not admit a modular max or modular min polymorphism. Then there is a non-trivial finite binary relation pp-definable in Γ.*

One of the concepts needed in the proof of Lemma 3 above and Proposition 5 below is the notion of *decomposability*. A relation R of arity n is r-*decomposable* if $R(x_1, \ldots, x_n)$ is equivalent to $\bigwedge_J \exists_{j \notin J} x_j . R(x_1, \ldots, x_n)$ where J ranges over all the r-element subsets of $\{1, \ldots, n\}$.

Definition 8. *A d-progression is a set of the form $[a, b \mid d] := \{a, a + d, a + 2d, \ldots, b\}$, for $a \leq b$ with $b - a$ divisible by d.*

One can show that if there is a non-trivial finite binary relation R pp-definable in Γ, and $\{b - a \in \mathbb{Z} \mid (a, b) \in R\}$ is not a d-progression for any $d \geq 1$, then $\mathrm{CSP}(\Gamma)$ is NP-hard. By considering Γ/d instead of Γ, we can reduce to the case $d = 1$. In order to prove Theorem 7, it thus suffices to show the following.

Proposition 5. *Let Γ be a positive first-order expansion of $(\mathbb{Z}; \mathrm{succ})$, and $S \subset \mathbb{Z}$ a 1-progression, $|S| > 1$, such that Diff_S is pp-definable in Γ. Then Γ is preserved by max or min; or $\mathrm{CSP}(\Gamma)$ is NP-hard.*

In the proof of this proposition we use known results about finite domain CSPs. More specifically, we apply these results to substructures Δ of $(\Gamma, 0)$ induced by $\{-n, \ldots, n\}$. All singleton unary relations are pp definable in Δ. Then it is known that $\mathrm{CSP}(\Delta)$ is NP-hard, or Δ has a so-called *weak near unanimity* polymorphism of arity $k \geq 2$ (combining a result from [7] with a result from [15]). We show that in our situation, such polymorphisms must generate min or max on $\{-n, \ldots, n\}$, which then implies that also Γ is preserved by min or max.

8 Discussion

The structure $(\mathbb{Z}; \text{succ})$ is among the simplest structures that is not ω-categorical. Note that $(\mathbb{Z}; \text{succ})$ and its reducts are uncountably categorical and ω-stable. They are also *automatic* in the sense of algorithmic model theory.

We want to stress that the difficulties we had to overcome when classifying reducts of $(\mathbb{Z}; \text{succ})$ will be present in classifications of reducts of richer structures, such as $(\mathbb{Z}; \text{succ}, \leq)$ (which has the same reducts as $(\mathbb{Z}; <)$), $(\mathbb{Z}; +)$, or even $(\mathbb{Z}; +, \leq)$, i.e., Presburger arithmetic, and we view it as an interesting question which of our techniques might generalise to such more general contexts.

References

1. Bodirsky, M., Dalmau, V., Martin, B., Pinsker, M.: Distance constraint satisfaction problems. In: Hliněný, P., Kučera, A. (eds.) MFCS 2010. LNCS, vol. 6281, pp. 162–173. Springer, Heidelberg (2010)
2. Bodirsky, M., Hils, M., Martin, B.: On the scope of the universal-algebraic approach to constraint satisfaction. Logical Methods in Computer Science (LMCS) 8(3), 13 (2012). An extended abstract that announced some of the results appeared in the proceedings of Logic in Computer Science (LICS 2010)
3. Bodirsky, M., Kára, J.: The complexity of equality constraint languages. Theory of Computing Systems 3(2), 136–158 (2008). A conference version appeared in the proceedings of Computer Science Russia (CSR 2006)
4. Bodirsky, M., Kára, J.: The complexity of temporal constraint satisfaction problems. Journal of the ACM 57(2), 1–41 (2009). An extended abstract appeared in the Proceedings of the Symposium on Theory of Computing (STOC 2008)
5. Bodirsky, M., Nešetřil, J.: Constraint satisfaction with countable homogeneous templates. Journal of Logic and Computation 16(3), 359–373 (2006)
6. Bodirsky, M., Pinsker, M.: Schaefer's theorem for graphs. In: Proceedings of the Annual Symposium on Theory of Computing (STOC), pp. 655–664 (2011). Preprint of the long version available at arxiv.org/abs/1011.2894
7. Bulatov, A.A., Jeavons, P., and Krokhin, A.A.: The complexity of constraint satisfaction: an algebraic approach (a survey paper). In: Structural Theory of Automata, Semigroups and Universal Algebra (Montreal, 2003), NATO Science Series II: Mathematics, Physics, Chemistry 207, 181–213 (2005)
8. Feder, T., Vardi, M.Y.: The computational structure of monotone monadic SNP and constraint satisfaction: a study through Datalog and group theory. SIAM Journal on Computing 28, 57–104 (1999)
9. Hedman, S.: A First Course in Logic: An Introduction to Model Theory, Proof Theory, Computability, and Complexity (Oxford Texts in Logic). Oxford University Press Inc, New York (2004)
10. Hell, P., Nešetřil, J.: On the complexity of H-coloring. Journal of Combinatorial Theory, Series B 48, 92–110 (1990)
11. Hodges, W.: A shorter model theory. Cambridge University Press, Cambridge (1997)
12. Jeavons, P., Cohen, D., Gyssens, M.: Closure properties of constraints. Journal of the ACM 44(4), 527–548 (1997)
13. Jonsson, P., Lööw, T.: Computation complexity of linear constraints over the integers. Artificial Intelligence 195, 44–62 (2013)
14. Marker, D.: Model Theory: An Introduction. Springer, New York (2002)
15. Maróti, M., McKenzie, R.: Existence theorems for weakly symmetric operations. Algebra Universalis 59, 3 (2008)

Hardness Amplification and the Approximate Degree of Constant-Depth Circuits

Mark Bun[1][(✉)] and Justin Thaler[2]

[1] Harvard University, Cambridge, Massachusetts
mbun@seas.harvard.edu
[2] Yahoo! Labs, New York, USA
jthaler@fas.harvard.edu

Abstract. We establish a generic form of hardness amplification for the approximability of constant-depth Boolean circuits by polynomials. Specifically, we show that if a Boolean circuit cannot be pointwise approximated by low-degree polynomials to within constant error in a certain one-sided sense, then an OR of disjoint copies of that circuit cannot be pointwise approximated even with very high error. As our main application, we show that for every sequence of degrees $d(n)$, there is an explicit depth-three circuit $F : \{-1,1\}^n \to \{-1,1\}$ of polynomial-size such that any degree-d polynomial cannot pointwise approximate F to error better than $1 - \exp(-\tilde{\Omega}(nd^{-3/2}))$. As a consequence of our main result, we obtain an $\exp(-\tilde{\Omega}(n^{2/5}))$ upper bound on the the discrepancy of a function in AC^0, and an $\exp(\tilde{\Omega}(n^{2/5}))$ lower bound on the threshold weight of AC^0, improving over the previous best results of $\exp(-\Omega(n^{1/3}))$ and $\exp(\Omega(n^{1/3}))$ respectively.

Our techniques also yield a new lower bound of $\Omega(n^{1/2}/\log^{(d-2)/2}(n))$ on the approximate degree of the AND-OR tree of depth d, which is tight up to polylogarithmic factors for any constant d, as well as new bounds for read-once DNF formulas. In turn, these results imply new lower bounds on the communication and circuit complexity of these classes, and demonstrate strong limitations on existing PAC learning algorithms.

1 Introduction

The ε-approximate degree of a Boolean function $f : \{-1,1\}^n \to \{-1,1\}$, denoted $\widetilde{\deg}_\varepsilon(f)$, is the minimum degree of a real polynomial that approximates f to error ε in the ℓ_∞ norm. Approximate degree has pervasive applications in theoretical

---

The full version of this paper is available at http://arxiv.org/abs/1311.1616.
Supported by an NDSEG Fellowship and NSF grant CNS-1237235.
Parts of this work were done while the author was a graduate student at Harvard University, and a Research Fellow at the Simons Institute for the Theory of Computing. This work was supported by an NSF Graduate Research Fellowship, NSF grants CNS-1011840 and CCF-0915922, and a Research Fellowship from the Simons Institute for the Theory of Computing.

M.M. Halldórsson et al. (Eds.): ICALP 2015, Part I, LNCS 9134, pp. 268–280, 2015.
DOI: 10.1007/978-3-662-47672-7_22

computer science. For example, lower bounds on approximate degree underly many tight lower bounds on quantum query complexity (e.g., [2,3,5,31]), and have been used to resolve several long-standing open questions in communication complexity [27]. Meanwhile, upper bounds on approximate degree underly many of the fastest known learning algorithms, including PAC learning DNF and read-once formulas [4,14], agnostically learning disjunctions [12], and PAC learning in the presence of irrelevant information [15,24].

Despite the range and importance of these applications, large gaps remain in our understanding of approximate degree. The approximate degree of any *symmetric* Boolean function has been understood since Paturi's 1992 paper [22], but once we move beyond symmetric functions, few general results are known.

In this paper, we perform a careful study of the approximate degree of constant-depth Boolean circuits. In particular, we establish a generic form of hardness amplification for the pointwise approximation of small depth circuits by low-degree polynomials: we show that if a Boolean circuit f cannot be pointwise approximated to within constant error in a certain one-sided sense by polynomials of a given degree, then the circuit F obtained by taking an OR of disjoint copies of f cannot be approximated even with error exponentially close to 1. Notice that if f is computed by a circuit of polynomial size and constant depth, then so is F.

Our proof extends a recent line of work [8,18,25,33] that seeks to prove approximate degree lower bounds by constructing explicit *dual polynomials*, which are dual solutions to a linear program that captures the approximate degree of any function. Specifically, we show that given a dual polynomial demonstrating that f cannot be approximated to within constant error, we can construct a dual polynomial demonstrating that F cannot be approximated even with error exponentially close to 1.

As the main application of our hardness amplification technique, for any $d > 0$ we exhibit an explicit function $F : \{-1,1\}^n \to \{-1,1\}$ computed by a polynomial size circuit of depth three for which any degree-d polynomial cannot pointwise approximate F to error $1 - \exp(-\tilde{\Omega}(nd^{-3/2}))$. We then use this result to obtain new bounds on two quantities that play central roles in learning theory, communication complexity, and circuit complexity: *discrepancy* and *threshold weight*. Specifically, we prove a new upper bound of $\exp(-\tilde{\Omega}(n^{2/5}))$ for the discrepancy of a function in AC^0, and a new lower bound of $\exp(\tilde{\Omega}(n^{2/5}))$ for the threshold weight of AC^0. As a second application, our hardness amplification result allows us to resolve, up to polylogarithmic factors, the approximate degree of AND-OR trees of arbitrary constant depth. Finally, our techniques also yield new lower bounds for read-once DNF formulas.

2 Hardness Amplification

Recall that the ε-approximate degree of a Boolean function f is the minimum degree of a real polynomial that pointwise approximates f to error ε. Another fundamental measure of the complexity of f is its *threshold degree*, denoted

$\deg_{\pm}(f)$. The threshold degree of f is the least degree of a real polynomial that agrees in sign with f at all Boolean inputs.

Central to our results is a measure of the complexity of a Boolean function that we call *one-sided approximate degree*. This quantity, which we denote by $\widetilde{\deg}_{\varepsilon}(f)$, is an intermediate complexity measure that lies between ε-approximate degree and threshold degree. Unlike approximate degree and threshold degree, one-sided approximate degree treats inputs in $f^{-1}(1)$ and inputs in $f^{-1}(-1)$ asymmetrically.

More specifically, $\widetilde{\mathrm{odeg}}_{\varepsilon}(f)$ captures the least degree of a *one-sided approximation* for f. Here, a one-sided approximation p for f is a polynomial that approximates f to error at most ε at all points $x \in f^{-1}(1)$, and satisfies the threshold condition $p(x) \le -1 + \varepsilon$ at all points $x \in f^{-1}(-1)$. Notice that $\widetilde{\deg}_{\varepsilon}(f)$ is always *at most* $\widetilde{\deg}_{\varepsilon}(f)$, but can be smaller. Similarly, $\widetilde{\mathrm{odeg}}_{\varepsilon}(f)$ is always *at least* $\deg_{\pm}(f)$, but can be larger.

One-sided approximate degree is the complexity measure that we amplify for constant-depth circuits: given a depth k circuit f on m variables that has one-sided approximate degree greater than d, we show how to generically transform f into a depth $k + 1$ circuit F on $t \cdot m$ variables such that F cannot be pointwise approximated by degree d polynomials even to error $1 - 2^{-t}$.[1]

Theorem 1. *Suppose $f : \{-1, 1\}^m \to \{-1, 1\}$ has one-sided approximate degree* $\widetilde{\mathrm{odeg}}_{1/2}(f) > d$. *Denote by $F : \{-1, 1\}^{m \cdot t} \to \{-1, 1\}$ the block-wise composition* $\mathrm{OR}_t(f, \ldots, f)$, *where OR_t denotes the OR function on t variables. Then F cannot be pointwise approximated by degree-d polynomials to within error $1 - 2^{-t}$ by degree-d polynomials. That is, the $(1 - 2^{-t})$-approximate degree of F is greater than d.*

Remark: Theorem 1 demonstrates that one-sided approximate degree admits a form of hardness amplification within AC^0, which does not generally hold for the ordinary approximate degree. Indeed, Theorem 1 fails badly if the condition $\widetilde{\mathrm{odeg}}_{1/2}(f) > d$ is replaced with the weaker condition $\widetilde{\deg}_{1/2}(f) > d$ (in fact, $f = \mathrm{OR}_m$ is a counter-example).

A *dual formulation* of one-sided approximate degree was previously exploited by Gavinsky and Sherstov to separate the multi-party communication versions of NP and co-NP [9], as well as by the current authors [8] and independently by Sherstov [25] to resolve the approximate degree of the two-level AND-OR tree. In this paper, we introduce the primal formulation of one-sided approximate degree, which allows us to express Theorem 1 as a hardness amplification result. We also argue for the importance of one-sided approximate degree as a complexity measure in its own right.

Prior Work on Hardness Amplification for Approximate Degree. For the purposes of this discussion, we informally consider a hardness amplification result for approximate degree to be any statement of the following form:

---

[1] Follow-up work by Sherstov [26] has established a lower bound on the *threshold degree* of F. Specifically, he has shown that there is some constant c such that $\deg_{\pm}(F) > \min\{ct, d\}$. See Section 6 for further discussion of this result.

Fix two functions $f : \{-1,1\}^m \to \{-1,1\}$ and $g : \{-1,1\}^t \to \{-1,1\}$. Then the composed function $g(f,\ldots,f) : \{-1,1\}^{m \cdot t} \to \{-1,1\}$ is strictly harder to approximate in the ℓ_∞ norm by low-degree polynomials than is the function f.

We think of such a result as establishing that application of the outer function g to t disjoint copies of f amplifies the hardness of f. Here we consider polynomial degree to be a resource, and "harder to approximate" can refer either to the amount of resources required for the approximation, to the error of the approximation, or to a combination of the two.

Two particular kinds of hardness amplification results for approximate degree have received particular attention. *Direct-sum* theorems focus on amplifying the degree required to obtain an approximation, but do not focus on amplifying the error. For example, a typical direct-sum theorem identifies conditions on f and g that guarantee that $\widetilde{\deg}_\varepsilon(g(f,\ldots,f)) \geq \widetilde{\deg}_\varepsilon(g) \cdot \widetilde{\deg}_\varepsilon(f)$. In contrast, a *direct-product* theorem focuses on amplifying both the error and the minimum degree required to achieve this error. An *XOR lemma* is a special case of either type of theorem where the combining function g is the XOR function. Ideally, an XOR lemma of the direct-product form establishes that there exists a sufficiently small constant $\delta > 0$ such that $\widetilde{\deg}_{1-2^{-\delta t}}(\mathrm{XOR}_t(f,\ldots,f)) \geq t \cdot \widetilde{\deg}_{1/3}(f)$. That is, an XOR lemma establishes that approximating the XOR of t disjoint copies of f requires a t-fold blowup in degree relative to f, even if one allows error exponentially close to 1.

O'Donnell and Servedio [21] proved an XOR lemma for *threshold degree*, establishing that $\mathrm{XOR}_t(f,\ldots,f)$ has threshold degree t times the threshold degree of f. In later work, Sherstov [33] proved a direct sum result for approximate degree that holds whenever the combining function g has low block-sensitivity. His techniques also capture O'Donnell and Servedio's XOR lemma for threshold degree as a special case. In [31], Sherstov proved a number of hardness amplification results for approximate degree. Most notably, he proved an optimal XOR lemma, as well as a direct-sum theorem that holds whenever the combining function has close to maximal approximate degree (i.e., approximate degree $\Omega(t)$). Sherstov used his XOR lemma to prove direct product theorems for quantum query complexity, and in subsequent work [32], to show direct product theorems for the multiparty communication of set disjointness.

Comparison to Prior Work. In this paper, we are interested in establishing approximate degree lower bounds for constant-depth circuits over the basis $\{\mathrm{AND},\mathrm{OR},\mathrm{NOT}\}$. For this purpose, it is essential to consider combining functions (such as OR, see Theorem 1) that are themselves in AC^0, ruling out the use of XOR as a combining function. Our hardness amplification result (Theorem 1) is orthogonal to direct-sum theorems: direct-sum theorems focus on amplifying degree but not error, while Theorem 1 focuses on amplifying error but not degree. Curiously, Theorem 1 is nonetheless a critical ingredient in our proof of a direct-sum type theorem for AND-OR trees of constant depth (Theorem 3).

Proof Idea. As discussed in the introduction, our proof of Theorem 1 relies on a dual characterization of one-sided approximate degree (see the full version of this ork). Specifically, for any m-variate Boolean function f satisfying $\widetilde{\mathrm{odeg}}_{1/2}(f) >$, there exists a dual object $\psi : \{-1,1\}^m \to \mathbb{R}$ that witnesses this fact — we fer to ψ as a "dual polynomial" for f. The dual polynomial ψ satisfies three nportant properties: (1) ψ has high correlation with f, (2) ψ has zero correlation ith all polynomials of degree at most d, and (3) $\psi(x)$ agrees in sign with $f(x)$ or all $x \in f^{-1}(-1)$. We refer to the second property by saying ψ has *pure high egree d*, and we refer to the third property by saying that ψ has *one-sided error*.

Our proof proceeds by taking a dual witness ψ to the high one-sided approxi-nate degree of f, and a certain dual witness Ψ for the function OR_t, and combin-ıg them to obtain a dual witness ζ for the fact that $\widetilde{\mathrm{deg}}_{1-2^{-t}}(\mathrm{OR}_t(f,\ldots,f)) >$. Our analysis of the combined dual witness crucially exploits two properties: rst, that ψ has one-sided error and second, that the vector whose entries are ll equal to -1 has very large (in fact, maximal) Hamming distance from the nique input in $\mathrm{OR}_t^{-1}(1)$.

Our method of combining the two dual witnesses was first introduced by herstov [33, Theorem 3.3] and independently by Lee [18]. This method was lso used by the present authors in [8] to resolve the approximate degree of the wo-level AND-OR tree, and by Sherstov [31] to prove direct sum and direct ›roduct theorems for polynomial approximation. However, as discussed above, ›rior work used this method of combining dual witnesses exclusively to amplify he *degree* in the resulting lower bound; in contrast, we use the combining method ın the proof of Theorem 1 to amplify the *error* in the resulting lower bound.

From a technical perspective, the primary novelty in the proof of Theorem lies in our choice of an appropriate (and simple) dual witness Ψ for OR_t, ınd the subsequent analysis of the correlation of the combined witness ζ with $\mathrm{OR}_t(f,\ldots,f)$. By our choice of Ψ, we are able to show that ζ has correlation vith $\mathrm{OR}_t(f,\ldots,f)$ that is *exponentially* close to 1, yielding a lower bound even ›n the degree of approximations with very high error.

3 Lower Bounds For AC^0

3.1 A New One-Sided Approximate Degree Lower Bound for AC^0

Ɔur ultimate goal is to use Theorem 1 to construct a function F in AC^0 that s hard to approximate by low-degree polynomials even with error exponentially :lose to 1. However, in order to apply Theorem 1, we must first identify an AC^0 `unction f such that $\widetilde{\mathrm{odeg}}_{1/2}(f)$ is large.

To this end, we identify fairly general conditions guaranteeing that the one-sided approximate degree of a function is *equal* to its approximate degree, up :o a logarithmic factor. To express our result, let $[N] = \{1,\ldots,N\}$, and let m, N, R be a triple of positive integers such that $R \geq N$, and $m = N \cdot \log_2 R$. In most cases, we will take $R = N$. We specifically consider Boolean functions f ɔn $\{-1,1\}^m$ that interpret their input x as the values of a function g_x mapping $[N] \to [R]$. That is, we break x up into N blocks each of length $\log_2 R$, and regard each block x_i as the binary representation of $g_x(i)$. Hence, we think of

f as computing some *property* ϕ_f of functions $g_x : [N] \rightarrow [R]$. We say that a property ϕ is *symmetric* if for all $g : [N] \rightarrow [R]$, all permutations σ on $[R]$, and all permutations π on $[N]$, it holds that $\phi(g) = \phi(\sigma \circ g \circ \pi)$.

Theorem 2. *Let* $f : \{-1, 1\}^m \rightarrow \{-1, 1\}$ *be a Boolean function corresponding to a symmetric property* ϕ_f *of functions* $g_x : [N] \rightarrow [R]$. *Suppose that for every pair* $x, y \in f^{-1}(-1)$, *there is a pair of permutations* σ *on* $[R]$ *and* π *on* $[N]$ *such that* $g_x = \sigma \circ g_y \circ \pi$. *Then* $\widetilde{\mathrm{odeg}}_\varepsilon(f) \geq \frac{1}{\log_2 R} \cdot \widetilde{\mathrm{deg}}_\varepsilon(f)$ *for all* $\varepsilon > 0$.

Proof Idea. It is enough to show that any one-sided ε-approximation p to f can be transformed into an actual ε-approximation r to f in a manner that does not increase the degree by too much (i.e., in a manner guaranteeing that $\deg(r) \leq (\log_2 R) \deg(p)$).

Our transformation from p to r consists of two steps. In the first step, we turn p into a "symmetric" polynomial $p^{\mathrm{sym}}(x) := \mathbb{E}_{y \sim x}[p(y)]$ where $y \sim x$ if $g_y = \sigma \circ g_x \circ \pi$ for some permutations σ on $[R]$ and π on $[N]$. It follows from work of Ambainis [3] that the map $p \mapsto p^{\mathrm{sym}}$ increases the degree of p by a factor of at most $\log_2 R$. In the second step, we argue that there is an affine transformation r of p^{sym} that is an actual ε-approximation to f, completing the construction.

The existence of the affine transformation r of p^{sym} follows from two observations: (1) if p is a one-sided approximation for f, then so is p^{sym} (this holds because ϕ_f is symmetric), and (2) p^{sym} takes on a constant value v on $f^{-1}(-1)$, i.e., $p^{\mathrm{sym}}(x) = v$ for all $x \in f^{-1}(-1)$ (this holds because $x \sim y$ for every pair of inputs $x, y \in f^{-1}(-1)$). Thus even if p^{sym} poorly approximates f on $f^{-1}(-1)$, we can still obtain a good approximation r by applying an affine transformation to the range of p^{sym} that maps v to -1 and moves all values closer to 1.

In our primary application of Theorem 2, we let $f : \{-1, 1\}^m \rightarrow \{-1, 1\}$ be the ELEMENT DISTINCTNESS function. Aaronson and Shi [2] showed that the approximate degree of ELEMENT DISTINCTNESS is $\Omega((m/\log m)^{2/3})$. ELEMENT DISTINCTNESS is computed by a CNF of polynomial size, and Aaronson and Shi's result remains essentially the best-known lower bound for the approximate degree of a function in AC^0. Theorem 2 applies to ELEMENT DISTINCTNESS, yielding the following corollary.

Corollary 1. *Let* $f : \{-1, 1\}^m \rightarrow \{-1, 1\}$ *denote the* ELEMENT DISTINCTNESS *function. Then* $\widetilde{\mathrm{odeg}}(f) = \tilde{\Omega}(m^{2/3})$.

The best known lower bound on the one-sided approximate degree of an AC^0 function that followed from prior work was $\Omega(m^{1/2})$ (which holds for the AND function [9,20]). Section 6 describes some further implications of Theorem 2.

3.2   Accuracy-Degree Tradeoff Lower Bounds for AC^0

By Corollary 1, we can apply Theorem 1 to ELEMENT DISTINCTNESS to obtain a depth-three Boolean circuit F with $t \cdot m$ inputs such that $\widetilde{\mathrm{deg}}_\varepsilon(F) = \tilde{\Omega}(m^{2/3})$, or $\varepsilon = 1 - 2^{-t}$. By choosing t and m appropriately, we obtain a depth-three circuit on $n = t \cdot m$ variables of size poly(n) such that any degree-d polynomial cannot pointwise approximate F to error better than $1 - \exp(-\tilde{\Omega}(nd^{-3/2}))$.

Corollary 2. *For every $d > 0$, there is a depth-3 Boolean circuit $F : \{-1,1\}^n \rightarrow \{-1,1\}$ of size poly(n) such that any degree-d polynomial cannot pointwise approximate F to error better than $1 - \exp(-\tilde{\Omega}(nd^{-3/2}))$. In particular, there is a depth-3 circuit F such that any polynomial of degree at most $n^{2/5}$ cannot pointwise approximate F to error better than $1 - \exp(-\tilde{\Omega}(n^{2/5}))$.*

3.3 Discrepancy Upper Bound

Discrepancy is a central quantity in communication complexity and circuit complexity. For instance, upper bounds on the discrepancy of a function f immediately yield lower bounds on the cost of small-bias communication protocols for computing f (The full version of this work has details). The first exponentially small discrepancy upper bounds for AC^0 were proved by Burhman et al. [7] and Sherstov [29,30], who exhibited constant-depth circuits with discrepancy $\exp(-\Omega(n^{1/3}))$. We improve the best-known upper bound to $\exp(-\tilde{\Omega}(n^{2/5}))$.

Table 1. Comparison of our new discrepancy bound for AC^0 to prior work. The circuit depth column lists the depth of the circuit used to exhibit the bound.

Reference	Discrepancy Bound	Circuit Depth
Sherstov [30]	$\exp(-\Omega(n^{1/5}))$	3
Buhrman et al. [7]	$\exp(-\Omega(n^{1/3}))$	3
Sherstov [29]	$\exp(-\Omega(n^{1/3}))$	3
This work	$\exp(-\tilde{\Omega}(n^{2/5}))$	4

Our result relies on a powerful technique developed by Sherstov [29], known as the pattern-matrix method. This technique allows one to automatically translate lower bounds on the ε-approximate degree of a Boolean function F into upper bounds on the *discrepancy* of a related function F' as long as ε is exponentially close to one. By applying the pattern-matrix method to Corollary 2, we obtain the following result.

Corollary 3. *There is a depth-4 Boolean circuit $F' : \{-1,1\}^n \rightarrow \{-1,1\}$ with discrepancy $\exp(-\tilde{\Omega}(n^{2/5}))$.*

3.4 Threshold Weight Lower Bound

A *polynomial threshold function* (PTF) for a Boolean function f is a multilinear polynomial p with integer coefficients that agrees in sign with f on all Boolean inputs. The *weight* of an n-variate polynomial p is the sum of the absolute value of its coefficients. The *degree-d threshold weight* of a Boolean function $f : \{-1,1\}^n \rightarrow \{-1,1\}$, denoted $W(f,d)$, refers to the least weight of a degree-d

PTF for f. We let $W(f)$ denote the quantity $W(f, n)$, i.e., the least weight of any threshold function for f regardless of its degree. As discussed in the full version of this work, threshold weight has important applications in learning theory.

Threshold weight is closely related to ε-approximate degree when ε is very close to 1. This allows us to translate Corollary 2 into a lower bound on the degree-d threshold weight of AC^0.

Corollary 4. *For every $d > 0$, there is a depth-3 Boolean circuit $F : \{-1, 1\}^n \to \{-1, 1\}$ of size* $\text{poly}(n)$ *such that $W(F, d) \geq \exp(\tilde{\Omega}(nd^{-3/2}))$. In particular, $W(F, n^{2/5}) = \exp(\tilde{\Omega}(n^{2/5}))$.*

A result of Krause [16] allows us to extend our new degree-d threshold weight lower bound for F into a *degree independent* threshold weight lower bound for a related function F'. The previous best lower bound on the threshold weight of AC^0 was $\exp(\Omega(n^{1/3}))$, due to Krause and Pudlák [17].

Corollary 5. *There is a depth-4 Boolean circuit $F' : \{-1, 1\}^n \to \{-1, 1\}$ satisfying $W(F') = \exp(\tilde{\Omega}(n^{2/5}))$.*

Moreover, while the threshold weight bound of Corollary 5 is stated for polynomial threshold functions over $\{-1, 1\}^n$, we show that the same threshold weight lower bound also holds for polynomials over $\{0, 1\}^n$.

4   Approximate Degree Lower Bounds for AND-OR Trees

The d-level AND-OR tree on n variables is a function described by a read-once circuit of depth d consisting of alternating layers of AND gates and OR gates. We assume for simplicity that all gates have fan-in $n^{1/d}$. For example, the two-level AND-OR tree is a read-once CNF in which all gates have fan-in $n^{1/2}$.

Until recently, the approximate degree of AND-OR trees of depth two or greater had resisted characterization, despite 19 years of attention [3,8,10,20, 25,33,34]. The case of of depth two was reposed as a challenge problem by Aaronson in 2008 [1], as it captured the limitations of existing lower bound techniques. This case was resolved last year by the current authors [8], and independently by Sherstov [25], who proved a lower bound of $\Omega(\sqrt{n})$, matching an upper bound of Høyer, Mosca, and de Wolf [10]. However, the case of depth three or greater remained open. To our knowledge, the best known lower bound for $d \geq 3$ was $\Omega(n^{1/4+1/2d})$, which follows by combining the depth-two lower bound [8,25] with an earlier direct-sum theorem of Sherstov [33, Theorem 3.1].

By combining the techniques of our earlier work [8] with our hardness amplification result (Theorem 1), we improve this lower bound to $\Omega(n^{1/2}/\log^{(d-2)/2}(n))$ for any constant $d \geq 2$. A line of work on quantum query algorithms [4,10,23] established an upper bound of $O(n^{1/2})$ for AND-OR trees of any depth, demonstrating that our result is optimal up to polylogarithmic factors.

Theorem 3. *Let $AND\text{-}OR_{d,n}$ denote the d-level AND-OR tree on n variables. Then $\widetilde{\deg}(AND\text{-}OR_{d,n}) = \Omega(n^{1/2}/\log^{(d-2)/2} n)$ for any constant $d \geq 2$.*

Proof Idea. To introduce our proof technique, we first describe the method used in [8] to construct an optimal dual polynomial in the case $d = 2$, and we identify why this method breaks down when trying to extend to the case $d = 3$. We then explain how to use our hardness amplification result (Theorem 1) to construct a different dual polynomial that does extend to the case $d = 3$.

Let M denote the fan-in of all gates in OR-AND$_{2,M^2}$. In our earlier work [8], we constructed a dual polynomial for OR-AND$_{2,M^2}$ as follows. It is known that there is a dual polynomial γ_1 witnessing the fact that $\widetilde{\text{odeg}}(\text{AND}_M) = \Omega(M^{1/2})$, and a dual polynomial γ_2 witnessing the fact that $\widetilde{\deg}(\text{OR}_M) = \Omega(M^{1/2})$. We then combined the dual witnesses γ_1 and γ_2, using the same "combining" technique as in the proof of Theorem 1, to obtain a dual witness $\gamma_3 : \{-1,1\}^{M^2} \to \mathbb{R}$ for the high approximate degree of OR-AND$_{2,M^2}$.

Recall that we say a dual witness has *pure high degree* d if it has zero correlation with every polynomial of degree at most d. It followed from earlier work [33] that γ_3 has pure high degree equal to the product of the pure high degrees of γ_1 and γ_2, yielding an $\Omega(M)$ lower bound on the pure high degree of γ_3. The new ingredient of the analysis in [8] was to use the one-sided error of the "inner" dual witness γ_1 to argue that γ_3 also had good correlation with OR-AND$_{2,M^2}$.

Extending to Depth Three. Let $M = n^{1/3}$ denote the fan-in of all gates in AND-OR$_{3,n}$. To construct a dual witness for AND-OR$_{3,n}$ = AND$_M$(OR-AND$_{2,M^2}, \dots,$ OR-AND$_{2,M^2}$), it is natural to try the following approach. Let γ_4 be a dual polynomial witnessing the fact that the approximate degree of AND$_M = \Omega(\sqrt{M})$. Then we can combine γ_3 and γ_4 as above to obtain a dual function γ_5.

The difficulty in establishing that γ_5 is a dual witness to the high approximate degree of AND-OR$_{3,n}$ is in showing that γ_5 has good correlation with AND-OR$_3$. In our earlier work, we showed γ_3 has large correlation with OR-AND$_{2,n}$ by exploiting the fact that the inner dual witness γ_1 had one-sided error, i.e., $\gamma_1(y)$ agrees in sign with AND$_M$ whenever $y \in \text{AND}_M^{-1}(-1)$. However, γ_3 itself does not satisfy an analogous property: there are inputs $x_i \in \text{OR-AND}_{2,M^2}^{-1}(-1)$ such that $\gamma_3(x_i) > 0$, *and* there are inputs $x_i \in \text{OR-AND}_{2,M^2}^{-1}(1)$ such that $\gamma_3(x_i) < 0$.

To circumvent this issue, we use a different inner dual witness γ_3' in place of γ_3. Our construction of γ_3' utilizes our hardness amplification analysis to achieve the following: while γ_3' has error "on both sides", the error from the "wrong side" is very small. The hardness amplification step causes γ_3' to have pure high degree that is lower than that of the dual witness γ_3 constructed in [8] by a $\sqrt{\log n}$ factor. However, the hardness amplification step permits us to prove the desired lower bound on the correlation of γ_5 with AND-OR$_{3,n}$. The proof for the general case, which is quite technical, appears in the full version of this work.

5   Lower Bounds for Read-Once DNFs and CNFs

Our techniques also yield new lower bounds on the approximate degree and degree-d threshold weight of read-once DNF and CNF formulas. Before stating our results, we discuss relevant prior work.

In their seminal work on perceptrons, Minsky and Papert exhibited a read-once DNF $f : \{-1,1\}^n \to \{-1,1\}$ with *threshold degree* $\Omega(n^{1/3})$ [19]. That is, a real polynomial requires degree $\Omega(n^{1/3})$ just to agree with f in sign. However, to our knowledge no non-trivial lower bound on the degree-d threshold *weight* of read-once DNFs was known for any $d = \omega(n^{1/3})$.

In an influential result, Beigel [6] exhibited a polynomial-size (read-many) DNF called ODD-MAX-BIT satisfying the following: there is some constant $\delta > 0$ such that $\widetilde{\deg}_{1-2^{-\delta n/d^2}}(\text{ODD-MAX-BIT}) > d$, and hence also $W(\text{ODD-MAX-BIT}, d) = \exp(\Omega(n/d^2))$. Motivated by applications in computational learning theory, Klivans and Servedio showed that Beigel's lower bound is essentially tight for $d < n^{1/3}$ [15]. Very recently, Servedio, Tan, and Thaler showed an alternative lower bound on the degree-d threshold weight of ODD-MAX-BIT. Specifically, they showed that $W(\text{ODD-MAX-BIT}, d) = \exp(\Omega(\sqrt{n/d}))$ [24]. The lower bound of Servedio et al. improves over Beigel's for any $d > n^{1/3}$, and is essentially tight in this regime (i.e., when $d > n^{1/3}$).

While ODD-MAX-BIT is a relatively simple DNF (in fact, it is a *decision list*), it is not a read-once DNF. Our results extend the lower bounds of Servedio et al. and Beigel from decision lists to read-once DNFs and CNFs. In the statement of the results below, we restrict ourselves to DNFs, as the case of CNFs is entirely analogous.

5.1   Extending Servedio et al.'s Lower Bound to Read-Once DNFs

In order to extend the lower bound of Servedio et al. to read-once DNFs and CNFs, we extend our hardness amplification techniques from one-sided approximate degree to a new quantity we call *degree-d one-sided non-constant approximate weight*. This quantity captures the least L_1 *weight* (excluding the constant term) of a polynomial of degree at most d that is a one-sided approximation of f. We denote the degree-d one-sided approximate weight of a Boolean function f by $W^*_\varepsilon(f, d)$, where ε is an error parameter. We prove the following analog of Theorem 1.

Theorem 4. *Fix $d > 0$. Let $f : \{-1,1\}^m \to \{-1,1\}$, and suppose $W^*_{3/4}(f, d) > w$. Let $F : \{-1,1\}^{m \cdot t} \to \{-1,1\}$ denote the function $\text{OR}_t(f, \ldots, f)$. Then any degree-$d$ polynomial that approximates F to error $1 - 2^{-t}$ requires weight $2^{-5t}w$.*

Adapting a proof of Servedio et al., we can show that $W^*_{3/4}(\text{AND}_m, d) \geq 2^{\Omega(m/d)}$. By applying Theorem 4 with $f = \text{AND}_m$, along with standard manipulations, we are able to extend the lower bound of Servedio et al. to read-once CNFs and DNFs.

Corollary 6. *For each $d = o(n/\log^4 n)$, there is a read-once DNF F satisfying $W(F, d) = \exp(\Omega(\sqrt{n/d}))$.*

In particular, there is a read-once DNF that cannot be computed by any PTF of $\text{poly}(n)$ weight, unless the degree is $\tilde{\Omega}(n)$.

.2   Extending Beigel's Lower Bound to Read-Once DNFs

t is known that $\widetilde{\mathrm{odeg}}(\mathrm{AND}_m) = \Omega(m^{1/2})$. By applying Theorem 1 with $f = $ AND_m, we obtain the following result.

Corollary 7. *There is an (explicit) read-once DNF* $F : \{-1, 1\}^n \to \{-1, 1\}$ *with* $\widetilde{\mathrm{deg}}_{1-2^{-n/d^2}}(F) = \Omega(d)$.

We remark that for $d < n^{1/3}$, Corollary 7 is subsumed by Minsky and Papert's seminal result that exhibited a read-once DNF F with threshold degree $\Omega(n^{1/3})$ [19]. However, for $d > n^{1/3}$, it is not subsumed by Minsky and Papert's esult, nor by Corollary 6. Indeed, Corollary 6 yields a lower bound on the degree-d threshold weight of read-once DNFs, but not a lower bound on the *approximate-degree* of read-once DNFs.

3   Discussion

Subsequent Work by Sherstov. In 1969, Minsky and Papert gave a lower bound of $\Omega(n^{1/3})$ on the threshold degree of an explicit read-once DNF formula. Klivans and Servedio [14] proved their lower bound to be tight within a logarithmic factor for DNFs of polynomial size, but it remained a well-known open question to give a threshold degree lower bound of $\Omega(n^{1/3+\delta})$ for a function in AC^0; the only progress prior to our work was due to O'Donnell and Servedio [21], who established an $\Omega(n^{1/3} \log^k n)$ lower bound for any constant $k > 0$.

Let f denote the ELEMENT DISTINCTNESS function on $n^{3/5}$ variables. In an earlier version of this work, we conjectured that the function $F = \mathrm{OR}_{n^{2/5}}(f, \ldots, f)$ appearing in Corollary 2 in fact satisfies $\mathrm{deg}_\pm(f) = \tilde{\Omega}(n^{2/5})$, and observed that this would yield the first polynomial improvement on Minsky and Papert's lower bound. Sherstov [26, Theorem 7.1] has recently proved our conjecture. His proof, short and elegant, extends our dual witness construction in the proof of Theorem 1 to establish a different form of hardness amplification, from one-sided approximate degree to threshold degree. Specifically, he shows that if a Boolean function f has one-sided approximate degree d, then the block-wise composition $\mathrm{OR}_t(f, \ldots, f)$ has threshold degree at least $\min\{ct, d\}$ for some constant c. This result is incomparable to our Theorem 1 when $t \leq d$, but when $t \gg d$, Sherstov's result is a substantial strengthening of Theorem 1.

In the same work, Sherstov has also proven a much stronger and more difficult result: for any $k > 2$, he gives a read-once formula of depth k with threshold degree $\Omega(n^{(k-1)/(2k-1)})$. Notice that for any constant $\delta > 0$, this yields an AC^0 function with threshold degree $\Omega(n^{1/2-\delta})$. This in turn yields an improvement of our discrepancy upper bound (Corollary 3) for AC^0 to $\exp(-\Omega(n^{1/2-\delta}))$, and of our threshold weight lower bound (Corollary 5) to $\exp(\Omega(n^{1/2-\delta}))$.

Subsequent Work by Kanade and Thaler. Existing applications of onesided approximate degree [8,9,25,26] have all been of a negative nature (proving communication or circuit lower bounds, establishing limitations on PAC learning algorithms, etc.). Kanade and Thaler [13] have identified a positive (algorithmic) application of one-sided approximate degree. Specifically, they show that one-sided approximate degree upper bounds imply fast algorithms in the reliable agnostic learning framework of Kalai et al. [11]. This framework captures learning tasks in which one type of error (such as false negative errors) is costlier than other types. Kanade and Thaler use this result to give the first sub-exponential time algorithms for distribution-independent reliable learning of several fundamental concept classes.

In light of these developments, we are optimistic that the notion of one-sided approximate degree will continue to enable progress on questions within the analysis of Boolean functions and computational complexity theory.

Acknowledgments. We are grateful to Sasha Sherstov, Robert Špalek, Li-Yang Tan, and the anonymous reviewers for valuable feedback on earlier versions of this manuscript.

References

1. Aaronson, S.: The polynomial method in quantum and classical computing. In: FOCS (2008). (Slides available at www.scottaaronson.com/talks/polymeth.ppt)
2. Aaronson, S., Shi, Y.: Quantum lower bounds for the collision and the element distinctness problems. Journal of the ACM **51**(4), 595–605 (2004)
3. Ambainis, A.: Polynomial degree and lower bounds in quantum complexity: Collision and element distinctness with small range. Theory Comput. **1**(1), 37–46 (2005)
4. Ambainis, A., Childs, A.M., Reichardt, B., Špalek, R., Zhang, S.: Any AND-OR formula of size N can be evaluated in time $N^{1/2+o(1)}$ on a quantum computer. SIAM J. Comput. **39**(6), 2513–2530 (2010)
5. Beals, R., Buhrman, H., Cleve, R., Mosca, M., de Wolf, R.: Quantum lower bound by polynomials. J. ACM **48**(4), 778–797 (2001)
6. Beigel, R.: Perceptrons, PP, and the polynomial hierarchy. Computational Complexity **4**, 339–349 (1994)
7. Buhrman, H., Vereshchagin, N.K., de Wolf, R.: On computation and communication with small bias. CCC, pp. 24–32 (2007)
8. Bun, M., Thaler, J.: Dual lower bounds for approximate degree and markov-bernstein inequalities. In: Fomin, F.V., Freivalds, R., Kwiatkowska, M., Peleg, D. (eds.) ICALP 2013, Part I. LNCS, vol. 7965, pp. 303–314. Springer, Heidelberg (2013)
9. Gavinsky, D., Sherstov, A.A.: A separation of NP and coNP in multiparty communication complexity. Theory of Computing **6**(1), 227–245 (2010)
10. Høyer, P., Mosca, M., de Wolf, R.: Quantum search on bounded-error inputs. In: ICALP, pp. 291–299 (2003)
11. Kalai, A., Kanade, V., Mansour, Y.: Reliable agnostic learning. J. Comput. Syst. Sci. **78**(5), 1481–1495 (2012)
12. Kalai, A., Klivans, A., Mansour, Y., Servedio, R.: Agnostically learning halfspaces. SIAM Journal on Computing **37**(6), 1777–1805 (2008)
13. Kanade, V., Thaler, J.: Distribution-independent reliable learning. In: COLT (2014)

4. Klivans, A.R., Servedio, R.A.: Learning DNF in time $2^{\tilde{O}(n^{1/3})}$. J. of Comput. and System Sci. **68**(2), 303–318 (2004)
5. Klivans, A.R., Servedio, R.A.: Toward attribute efficient learning of decision lists and parities. Journal of Machine Learning Research **7**, 587–602 (2006)
6. Krause, M.: On the computational power of Boolean decision lists. Computational Complexity **14**(4), 362–375 (2005)
7. Krause, M., Pudlák, P.: On the computational power of depth-2 circuits with threshold and modulo gates. Theor. Comput. Sci. **174**(1–2), 137–156 (1997)
8. Lee, T.: A note on the sign degree of formulas (2009). CoRR abs/0909.4607
9. Minsky, M.L., Papert, S.A.: Perceptions: An Introduction to Computational Geometry. MIT Press, Cambridge (1969)
10. Nisan, N., Szegedy, M.: On the degree of boolean functions as real polynomials. Computational Complexity **4**, 301–313 (1994)
11. O'Donnell, R., Servedio, R.: New degree bounds for polynomial threshold functions. Combinatorica **30**(3), 327–358 (2010)
12. Paturi, R.: On the degree of polynomials that approximate symmetric Boolean functions (Preliminary Version). STOC, pp. 468–474 (1992)
13. Reichardt, B.: Reflections for quantum query algorithms. In: SODA (2011)
14. Servedio, R.A., Tan, L.-Y., Thaler, J.: Attribute-Efficient learning and weight-degree tradeoffs for polynomial threshold functions. COLT **23**, 14.1–14.19 (2012)
15. Sherstov, A.A.: Approximating the AND-OR Tree. Theory of Computing (2013)
16. Sherstov, A.A.: Breaking the Minsky-Papert Barrier for constant-depth circuits. STOC (2014)
17. Sherstov, A.A.: Communication lower bounds using dual polynomials. Bulletin of the EATCS **95**, 59–93 (2008)
18. Sherstov, A.A.: Optimal bounds for sign-representing the intersection of two half-spaces by polynomials. STOC, pp. 523–532 (2010)
19. Sherstov, A.A.: The pattern matrix method. SIAM J. Comput. **40**(6), 1969–2000 (2011)
20. Sherstov, A.A.: Separating AC^0 from depth-2 majority circuits. SIAM J. Comput. **28**(6), 2113–2129 (2009)
21. Sherstov, A.A.: Strong direct product theorems for quantum communication and query complexity. SIAM J. Comput. **41**(5), 1122–1165 (2012)
22. Sherstov, A.A.: The multiparty communication complexity of set disjointness. STOC, pp. 525–524 (2012)
23. Sherstov, A.A.: The intersection of two halfspaces has high threshold degree. FOCS, pp. 343–362 (2009). (To appear in SIAM J. Comput. (special issue for FOCS 2009))
24. Shi, Y.: Approximating linear restrictions of Boolean functions. Manuscript (2002) web.eecs.umich.edu/shiyy/mypapers/linear02-j.ps

Algorithms and Complexity
for Turaev-Viro Invariants

Benjamin A. Burton$^{(\boxtimes)}$, Clément Maria, and Jonathan Spreer

The University of Queensland, Brisbane, QLD 4072, Australia
bab@maths.uq.edu.au, {c.maria,j.spreer}@uq.edu.au

Abstract. The Turaev-Viro invariants are a powerful family of topological invariants for distinguishing between different 3-manifolds. They are invaluable for mathematical software, but current algorithms to compute them require exponential time.

The invariants are parameterised by an integer $r \geq 3$. We resolve the question of complexity for $r = 3$ and $r = 4$, giving simple proofs that computing Turaev-Viro invariants for $r = 3$ is polynomial time, but for $r = 4$ is #P-hard. Moreover, we give an explicit fixed-parameter tractable algorithm for arbitrary r, and show through concrete implementation and experimentation that this algorithm is practical—and indeed preferable—to the prior state of the art for real computation.

Keywords: Computational Topology · 3-Manifolds · Invariants · #P-hardness · Parameterised complexity

1 Introduction

In geometric topology, testing homeomorphism (topological equivalence) is a fundamental algorithmic problem. However, beyond dimension two it is remarkably difficult. In dimension three—the focus of this paper—an algorithm follows from Perelman's proof of the geometrisation conjecture [12], but it is extremely intricate, its complexity is unknown and it has never been implemented.

As a result, practitioners in computational topology rely on simpler *invariants*—properties of a topological space that can tell different spaces apart. One of the best known invariants is homology, but for 3-manifolds (the 3-dimensional generalisation of surfaces) this is weak: there are many topologically different 3-manifolds with same homology. Therefore major software packages in 3-manifold topology rely on invariants that are stronger but more difficult to compute.

In the discrete setting, among the most useful invariants for 3-manifolds are the *Turaev-Viro invariants* [19]. These are analogous to the Jones polynomial for knots: they derive from quantum field theory, but offer a much simpler combinatorial interpretation that lends itself well to algorithms and exact computation.

A full version of this article is available at arXiv:1503.04099.

J. Spreer — Supported by the Australian Research Council (projects DP1094516, DP140104246).

© Springer-Verlag Berlin Heidelberg 2015
M.M. Halldórsson et al. (Eds.): ICALP 2015, Part I, LNCS 9134, pp. 281–293, 2015.
DOI: 10.1007/978-3-662-47672-7_23

hey are implemented in the major software packages *Regina* [5] and the *Mani-ld Recogniser* [14,15], and they play a key role in developing census databases, hich are analogous to the well-known dictionaries of knots [3,14]. Their main ifficulty is that they are slow to compute: current implementations [5,15] are ased on backtracking searches, and require exponential time.

The aims of this paper are to (i) introduce the Turaev-Viro invariants to ne wider computational topology community; (ii) understand the complexity of omputing them; and (iii) develop new algorithms suitable for practical software.

The Turaev-Viro invariants are parameterised by two integers r and q, with ≥ 3; we denote these invariants by $TV_{r,q}$. A typical algorithm for computing $V_{r,q}$ will take as input a triangulated 3-manifold, composed of n tetrahedra ttached along their triangular faces; we use n to indicate the input size. For ll known algorithms, the difficulty of computing $TV_{r,q}$ grows significantly as r ncreases (but in contrast, the difficulty is essentially independent of q).

Our main results are as follows.

– Kauffman and Lins [9] state that for $r = 3, 4$ one can compute $TV_{r,q}$ via "simple and efficient methods of linear algebra", but they give no details on either the algorithms or the complexity. We show here that in fact the situations for $r = 3$ and $r = 4$ are markedly different: computing $TV_{r,q}$ for orientable manifolds and $r = 3$ is polynomial time, but for $r = 4$ is #P-hard.
– We give an explicit algorithm for computing $TV_{r,q}$ for general r that is fixed-parameter tractable (FPT). Specifically, for any fixed r and any class of input triangulations whose dual graphs have bounded treewidth, the algorithm has running time linear in n. Furthermore, we show through comprehensive experimentation that this algorithm is *practical*—we implement it in the open-source software package *Regina* [5], run it through exhaustive census databases, and find that this new FPT algorithm is comparable to—and often significantly faster than—the prior backtracking algorithm.
– We give a new geometric interpretation of the formula for $TV_{r,q}$, based on systems of "normal arcs" in triangles. This generalises earlier observations of Kauffman and Lins for $r = 3$ based on embedded surfaces [9], and offers an interesting potential for future algorithms based on Hilbert bases.

The #P-hardness result for $r = 4$ is the first classical hardness result for the Turaev-Viro invariants.[1] However, the proofs for this and the polynomial-time · = 3 result are simple: the algorithm for $r = 3$ derives from a known homological ormulation [14], and the result for $r = 4$ adapts Kirby and Melvin's NP-hardness proof for the more complex Witten-Reshetikhin-Turaev invariants [10].

The FPT algorithm for general r is significant in that it is not just theoretical, out also practical—and indeed *preferable*—for real software. It was previously known that computing $TV_{r,q}$ is FPT [6], but that prior result was purely existen-cial (based on Courcelle's theorem), and would lead to infeasibly large constants n the running time if translated to a concrete algorithm. More generally, FPT algorithms do not always translate well into practical software tools, and this

---

[1] For *quantum* computation, approximating Turaev-Viro invariants is universal [1].

paper is significant in giving the first demonstrably practical FPT algorithm in 3-manifold topology.

2 Preliminaries

Let M be a closed 3-manifold. A *generalised triangulation* of M is a collection of n abstract tetrahedra $\Delta_1, \ldots, \Delta_n$ equipped with affine maps that identify (or "glue together") their $4n$ triangular faces in pairs, so that the underlying topological space is homeomorphic to M. See the full paper for details.

Generalised triangulations are widely used across major 3-manifold software packages. They are (as the name suggests) more general than simplicial complexes, which allows them to express a rich variety of different 3-manifolds using very few tetrahedra. For instance, with just $n \leq 11$ tetrahedra one can create $13\,400$ distinct prime orientable 3-manifolds [4,14].

2.1 The Turaev-Viro Invariants

Let $\mathfrak{T}$ be a generalised triangulation of a closed 3-manifold M, and let r and q be integers with $r \geq 3$, $0 < q < 2r$, and $\gcd(r, q) = 1$. We define the Turaev-Viro invariant $\mathrm{TV}_{r,q}(\mathfrak{T})$ as follows.

Let V, E, F and T denote the set of vertices, edges, triangles and tetrahedra respectively of the triangulation $\mathfrak{T}$. Let $I = \{0, 1/2, 1, 3/2, \ldots, (r-2)/2\}$; note that $|I| = r - 1$. We define a *colouring* of $\mathfrak{T}$ to be a map $\theta \colon E \to I$; that is, θ "colours" each edge of $\mathfrak{T}$ with an element of I. A colouring θ is *admissible* if, for each triangle of $\mathfrak{T}$, the three edges e_1, e_2, and e_3 bounding the triangle satisfy:

- the *parity condition* $\theta(e_1) + \theta(e_2) + \theta(e_3) \in \mathbb{Z}$;
- the *triangle inequalities* $\theta(e_1) \leq \theta(e_2) + \theta(e_3)$, $\theta(e_2) \leq \theta(e_1) + \theta(e_3)$, and $\theta(e_3) \leq \theta(e_1) + \theta(e_2)$; and
- the *upper bound constraint* $\theta(e_1) + \theta(e_2) + \theta(e_3) \leq r - 2$.

More generally, we refer to any triple $(i, j, k) \in I \times I \times I$ satisfying these three conditions as an *admissible triple* of colours.

For each admissible colouring θ and for each vertex $v \in V$, edge $e \in E$, triangle $f \in F$ or tetrahedron $t \in T$, we define *weights* $|v|_\theta, |e|_\theta, |f|_\theta, |t|_\theta \in \mathbb{C}$. Their precise values are unimportant, but depend only on the colours of the incident edges; see the full paper for details. What is important though is that the weights are all polynomials on ζ with rational coefficients, where $\zeta = e^{i\pi q/r}$. Using these weights, we define the *weight of the colouring* to be

$$|\mathfrak{T}|_\theta = \prod_{v \in V} |v|_\theta \times \prod_{e \in E} |e|_\theta \times \prod_{f \in F} |f|_\theta \times \prod_{t \in T} |t|_\theta, \qquad (1)$$

and the Turaev-Viro invariant to be the sum over all admissible colourings

$$\mathrm{TV}_{r,q}(\mathfrak{T}) = \sum_{\theta \text{ admissible}} |\mathfrak{T}|_\theta.$$

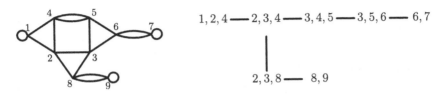

Fig. 1. The dual graph and a tree decomposition of a 3-manifold triangulation

In [19], Turaev and Viro show that $TV_{r,q}(\mathfrak{T})$ is indeed an invariant of the manifold; that is, if $\mathfrak{T}$ and $\mathfrak{T}'$ are generalised triangulations of the same closed -manifold M, then $TV_{r,q}(\mathfrak{T}) = TV_{r,q}(\mathfrak{T}')$ for all r, q. Although $TV_{r,q}(\mathfrak{T})$ is efined on the complex numbers $\mathbb{C}$, it always takes a real value (more precisely, : is the square of the modulus of a Witten-Reshetikhin-Turaev invariant) [21].

.2   Treewidth and Parameterised Complexity

Throughout this paper we always refer to *nodes* and *arcs* of graphs, to clearly distinguish these from the *vertices* and *edges* of triangulations.

Robertson and Seymour introduced the concept of the *treewidth* of a graph 17], which now plays a major role in parameterised complexity. Here, we adapt his concept to triangulations in a straightforward way.

Definition 1. *Let $\mathfrak{T}$ be a generalised triangulation of a 3-manifold, and let T e the set of tetrahedra in $\mathfrak{T}$. A* tree decomposition *$(X, \{B_\tau\})$ of $\mathfrak{T}$ consists of a ree X and bags $B_\tau \subseteq T$ for each node τ of X, for which:*

- *each tetrahedron $t \in T$ belongs to some bag B_τ;*
- *if a face of some tetrahedron $t_1 \in T$ is identified with a face of some other tetrahedron $t_2 \in T$, then there exists a bag B_τ with $t_1, t_2 \in B_\tau$;*
- *for each tetrahedron $t \in T$, the bags containing t correspond to a connected subtree of X.*

The width *of this tree decomposition is defined as $\max |B_\tau| - 1$. The* treewidth *of $\mathfrak{T}$, denoted $\mathrm{tw}(\mathfrak{T})$, is the smallest width of any tree decomposition of $\mathfrak{T}$.*

The relationship between this definition and the classical graph-theoretical notion of treewidth is simple: $\mathrm{tw}(\mathfrak{T})$ is the treewidth of the *dual graph* of $\mathfrak{T}$, the 1-valent multigraph whose nodes correspond to tetrahedra of $\mathfrak{T}$ and whose arcs represent pairs of tetrahedron faces that are identified together.

Figure 1 shows the dual graph of a 9-tetrahedra triangulation of a 3-manifold, along with a possible tree decomposition. The largest bags have size three, and so the width of this tree decomposition is $3 - 1 = 2$.

Definition 2. *A* nice tree decomposition *of a generalised triangulation* $\mathfrak{T}$ *is* a *tree decomposition* $(X, \{B_\tau\})$ *of* $\mathfrak{T}$ *whose underlying tree* X *is rooted, and where*

- *The bag* B_ρ *at the root of the tree is empty (*B_ρ *is called the* root bag*);*
- *If a bag* B_τ *has no children, then* $|B_\tau| = 1$ *(such a* B_τ *is called a* leaf bag*);*
- *If a bag* B_τ *has two children* B_σ *and* B_μ, *then* $B_\tau = B_\sigma = B_\mu$ *(such a* B_τ *is called a* join bag*);*
- *Every other bag* B_τ *has precisely one child* B_σ, *and either:*
  - $|B_\tau| = |B_\sigma| + 1$ *and* $B_\tau \supset B_\sigma$ *(such a* B_τ *is called an* introduce bag*), or*
  - $|B_\tau| = |B_\sigma| - 1$ *and* $B_\tau \subset B_\sigma$ *(such a* B_τ *is called a* forget bag*).*

Given a tree decomposition of a triangulation $\mathfrak{T}$ of width k and $O(n)$ bags, we can convert this in $O(n)$ time into a *nice* tree decomposition of $\mathfrak{T}$ that also has width k and $O(n)$ bags [13].

3   Algorithms for Computing Turaev-Viro Invariants

All of the algorithms in this paper use exact arithmetic. This is crucial if we wish to avoid floating-point numerical instability, since computing $\mathrm{TV}_{r,q}$ may involve exponentially many arithmetic operations.

We briefly describe how this exact arithmetic works. Since all weights in the definition of $\mathrm{TV}_{r,q}$ are rational polynomials in $\zeta = e^{i\pi q/r}$, all arithmetic operations remain within the rational field extension $\mathbb{Q}(\zeta)$. If ζ is a primitive nth root of unity then this field extension is called the nth *cyclotomic field*. This in turn is isomorphic to the polynomial field $\mathbb{Q}[X]/\Phi_n(X)$, where $\Phi_n(X)$ is the nth cyclotomic polynomial with degree $\varphi(n)$ (Euler's totient function). Therefore we can implement exact arithmetic using degree $\varphi(n)$ polynomials over $\mathbb{Q}$.

If r is odd and q is even, then ζ is a primitive rth root of unity, and $\mathbb{Q}(\zeta) \cong \mathbb{Q}[X]/\Phi_r(X)$. Otherwise ζ is a primitive $(2r)$th root of unity, and $\mathbb{Q}(\zeta) \cong \mathbb{Q}[X]/\Phi_{2r}(X)$. In this paper we give our complexity results in terms of arithmetic operations in $\mathbb{Q}(\zeta)$; see the full paper for details on the underlying complexity of arithmetic in this field.

3.1   The Backtracking Algorithm for Computing $\mathrm{TV}_{r,q}$

There is a straightforward but slow algorithm to compute $\mathrm{TV}_{r,q}$ for arbitrary r, q. The core idea is to use a backtracking algorithm to enumerate all admissible colourings of edges, and compute and sum their weights. Both major software packages that compute Turaev-Viro invariants—the *Manifold Recogniser* [15] and *Regina* [5]—currently employ optimised variants of this.

Let $\mathfrak{T}$ be a 3-manifold triangulation, with ℓ edges $e_1, \ldots, e_\ell$. A simple Euler characteristic argument gives $\ell = n + v$ where n is the number of tetrahedra and v is the number of vertices in $\mathfrak{T}$. Therefore $\ell \in \Theta(n)$.

To enumerate colourings, since each edge admits $r - 1$ possible colours, the backtracking algorithm traverses a search tree of $O((r-1)^\ell)$ nodes: a node at depth i corresponds to a partial colouring of the edges $e_1, \ldots, e_i$, and each node as degree $r-1$ (one edge per colour). Each leaf on the tree represents a (possibly not admissible) colouring of all the edges. At each node we maintain a "weight" of the current partial colouring, and update this weight as we traverse the tree. we reach a leaf whose colouring is admissible, we add this weight to our total.

Lemma 3. *If we sort the edges $e_1, \ldots, e_\ell$ by decreasing degree, the backtracking algorithm terminates in $O((r-1)^\ell)$ arithmetic operations in $\mathbb{Q}(\zeta)$.*

The proof is simple. The main complication is to ensure that updating the weight of the current partial colouring takes amortised constant time. For this we use Chebyshev's inequality, plus the observation that the average edge degree is ≤ 6. See the full version of this paper for the detailed proof.

To obtain a bound in the number of tetrahedra n, we note that a closed and connected 3-manifold triangulation with $n > 2$ tetrahedra must have $v \leq n + 1$ vertices. Combined with $n = \ell - v$ above, we have a worst-case running time of $O((r-1)^{2n+1})$ arithmetic operations in $\mathbb{Q}(\zeta)$.

3.2 A Polynomial-Time Algorithm for $r = 3$

In this section, we assume some basic knowledge on homology theory. We refer to the full version of this article or [16] for an overview. Throughout this section, $\mathfrak{T}$ will denote an n-tetrahedra triangulation of an orientable 3-manifold M.

The value of $\mathrm{TV}_{3,q}(\mathfrak{T})$, $q \in \{1, 2\}$, is closely related to $\mathbf{H}_2(M, \mathbb{Z}_2)$, the 2-dimensional homology group of M with $\mathbb{Z}_2$ coefficients. $\mathbf{H}_2(M, \mathbb{Z}_2)$ is a $\mathbb{Z}_2$-vector space whose dimension is the second Betti number $\beta_2(M, \mathbb{Z}_2)$. Its elements are (for our purposes) equivalence classes of 2-cycles, called *homology classes*, which can be represented by 2-dimensional triangulated surfaces S embedded in $\mathfrak{T}$.[2]

The *Euler characteristic* of a triangulated surface S, denoted by $\chi(S)$, is $\chi(S) = v - e + f$, where v, e and f denote the number of vertices, edges and triangles of S respectively. We define the Euler characteristic $\chi(c)$ of a 2-cycle c to be the Euler characteristic of the embedded surface it represents. Given $\mathfrak{T}$, the dimension $\beta_2(M, \mathbb{Z}_2)$ of $\mathbf{H}_2(M, \mathbb{Z}_2)$ may be computed in $O(\mathrm{poly}(n))$ operations. The following result is well known [14]:

Proposition 4. *Let M be a closed orientable 3-manifold. Then $\mathrm{TV}_{3,2}(M) =$ the order of $\mathbf{H}_2(M, \mathbb{Z}_2)$. Moreover, if M contains a 2-cycle with odd Euler characteristic then $\mathrm{TV}_{3,1}(M) = 0$, and otherwise $\mathrm{TV}_{3,1}(M) = \mathrm{TV}_{3,2}(M)$.*

Consequently $\mathrm{TV}_{3,2}(M) = 2^{\beta_2(M, \mathbb{Z}_2)}$, and one can compute $\mathrm{TV}_{3,2}(M)$ in polynomial time. The parity of the Euler characteristic of 2-cycles does not change within a homology class; moreover, given two 2-cycles c and c', $\chi(c + c') \equiv \chi(c) + \chi(c') \bmod 2$. Consequently, one can check whether $\mathrm{TV}_{3,1}(M) = 0$ or $\mathrm{TV}_{3,1}(M) = \mathrm{TV}_{3,2}(M)$ by computing the Euler characteristic of a cycle in each of the $\beta_2(M, \mathbb{Z}_2)$ homology classes that generate $\mathbf{H}_2(M, \mathbb{Z}_2)$. Because $\beta_2(M, \mathbb{Z}_2) = O(n)$, this leads to a polynomial time algorithm also.

---

[2] We use *discrete normal surfaces*, which are transversal to the 1-skeleton of $\mathfrak{T}$.

3.3   #P-Hardness of TV$_{4,1}$

The complexity class #P is a function class that counts accepting paths of a non-deterministic Turing machine [20]. Informally, given an NP decision problem C asking for the existence of a solution, its #P analogue #C is a counting problem asking for the number of such solutions. A problem is #P-*hard* if every problem in #P polynomially reduces to it. For example, the problem #$3SAT$, which asks for the number of satisfying assignments of a $3CNF$ formula, is #P-hard.

Naturally, counting problems are "harder" than their decision counterpart, and so #P-hard problems are at least as hard as NP-complete problems—specifically, #P complete problems are as hard as any problem in the polynomial hierarchy [18]. Hence proving #P hardness is a strong complexity statement.

Kirby and Melvin [11] prove that computing the Witten-Reshetikhin-Turaev invariant τ_r is #P hard for $r = 4$. This is a more complex 3-manifold invariant which is closely linked to the Turaev-Viro invariant TV$_{r,1}$ by the formula TV$_{r,1}(M) = |\tau_r(M)|^2$. Although computing TV$_{r,1}$ is "easier" than computing τ_r, the we can adapt the Kirby-Melvin hardness proof to fit our purposes.

To prove their result, Kirby and Melvin reduce the problem of counting the zeros of a cubic form to the computation of τ_4. Given a cubic form

$$c(x_1,\ldots,x_n) = \sum_i c_i\, x_i + \sum_{i,j} c_{ij}\, x_i x_j + \sum_{i,j,k} c_{ijk}\, x_i x_j x_k$$

in n variables over $\mathbb{Z}/2\mathbb{Z}$ and with #c zeros, they define a triangulation of a 3-manifold M_c with $O(\mathrm{poly}(n))$ tetrahedra satisfying $\tau_4(M_c) = 2\#c - 2^n$ and hence TV$_{r,1} = (2\#c - 2^n)^2$.

Consequently, counting the zeros of $c(x_1,\ldots,x_n)$ reduces to computing $\tau_4(M_c)$, and so computing TV$_{4,1}$ determines #c up to a $\pm$ sign ambiguity (depending on whether or not c admits more than half of the input as zeros).

Establishing the existence of a zero for a cubic form is an NP-complete problem, which implies that counting the number of zeros is #P complete. Consequently, computing τ_4 is #P hard. Kirby and Melvin prove this claim explicitly by reducing #$3SAT$ to the problem of counting the zeros of a cubic form; moreover, we observe that their construction ensures that this cubic form admits *more than half of its inputs as zeros*. See the full paper for details of the argument.

Thus the same reduction process as for τ_4 applies for TV$_{4,1}$, and so:

Corollary 5. *Computing* TV$_{4,1}$ *is* #P *hard.*

4   A Fixed-Parameter Tractable Algorithm

We present an explicit FPT algorithm for computing TV$_{r,q}$ for fixed r. As is common for treewidth-based methods, the algorithm involves dynamic programming over a tree decomposition $(X, \{B_\tau\})$. We first describe the data that we compute and store at each bag B_τ, and then give the algorithm itself.

Our first step is to reorganise the formula for TV$_{r,q}(\mathfrak{T})$ to be a product over tetrahedra only. This makes it easier to work with "partial colourings" corresponding to triangulation edges.

Definition 6. *Let $\mathfrak{T}$ be a generalised triangulation of a 3-manifold, and let V, E, F and T denote the vertices, edges, triangles and tetrahedra of $\mathfrak{T}$ respectively. For each vertex $x \in V$, each edge $x \in E$ and each triangle $x \in F$, we arbitrarily choose some tetrahedron $\Delta(x)$ that contains x.*

Now consider the definition of $\mathrm{TV}_{r,q}(\mathfrak{T})$. For each admissible colouring $\theta : E \to I$ and each tetrahedron $t \in T$, we define the adjusted tetrahedron weight $|t|'_\theta$:

$$|t|'_\theta = |t|_\theta \times \prod_{\substack{v \in V \\ \Delta(v)=t}} |v|_\theta \times \prod_{\substack{e \in E \\ \Delta(e)=t}} |e|_\theta \times \prod_{\substack{f \in F \\ \Delta(f)=t}} |f|_\theta.$$

It follows from equation (1) that the full weight of the colouring θ is just

$$|\mathfrak{T}|_\theta = \prod_{t \in T} |t|'_\theta.$$

Notation 7. *Let X be a rooted tree. For any non-root node τ of X, we denote the parent node of τ by $\hat{\tau}$. For any two nodes σ, τ of X, we write $\sigma \prec \tau$ if σ is a descendant node of τ.*

Definition 8. *Let $\mathfrak{T}$ be a generalised triangulation of a 3-manifold, and let V, E, F and T denote the vertices, edges, triangles and tetrahedra of $\mathfrak{T}$ respectively. Let $(X, \{B_\tau\})$ be a nice tree decomposition of $\mathfrak{T}$. For each node τ of the rooted tree X, we define the following sets:*

- *$T_\tau \subseteq T$ is the set of all tetrahedra that appear in bags beneath τ but not in the bag B_τ itself. More formally: $T_\tau = (\bigcup_{\sigma \prec \tau} B_\sigma) \backslash B_\tau$.*
- *$F_\tau \subseteq F$ is the set of all triangles that appear in some tetrahedron $t \in T_\tau$.*
- *$E_\tau \subseteq E$ is the set of all edges that appear in some tetrahedron $t \in T_\tau$.*
- *$E^*_\tau \subseteq E_\tau$ is the set of all edges that appear in some tetrahedron $t \in T_\tau$ and also some other tetrahedron $t' \notin T_\tau$; we refer to these as the current edges at node τ.*

We can make the following immediate observations:

Lemma 9. *If τ is a leaf of the tree X, then we have $T_\tau = F_\tau = E_\tau = E^*_\tau = \emptyset$. If τ is the root of the tree X, then we have $T_\tau = T$, $F_\tau = F$, $E_\tau = E$, and $E^*_\tau = \emptyset$.*

The key idea is, at each node τ of the tree, to store explicit colours on the "current" edges $e \in E^*_\tau$ and to aggregate over all colours on the "finished" edges $e \in E_\tau \backslash E^*_\tau$. For this we need some further definitions and notation.

Definition 10. *Again let $\mathfrak{T}$ be a generalised triangulation of a 3-manifold, and let $(X, \{B_\tau\})$ be a nice tree decomposition of $\mathfrak{T}$. Fix some integer $r \geq 3$, and consider the set of colours $I = \{0, 1/2, 1, 3/2, \ldots, (r-2)/2\}$ as used in defining the Turaev-Viro invariants $\mathrm{TV}_{r,q}$.*

Let τ be any node of X. We examine "partial colourings" that only assign colours to the edges in E_τ:

- *Consider any colouring $\theta\colon E_\tau \to I$. We call θ admissible if, for each triangle in F_τ, the three edges e, f, g bounding the triangle yield an admissible triple $(\theta(e), \theta(f), \theta(g))$.*
- *Define Ψ_τ to be the set of all colourings $\psi\colon E_\tau^* \to I$ that can be extended to some admissible colouring $\theta\colon E_\tau \to I$.*
- *Consider any colouring $\psi \in \Psi_\tau$ (so $\psi\colon E_\tau^* \to I$). We define the "partial invariant"*

$$\mathrm{TV}_{r,q}(\mathfrak{T}, \tau, \psi) = \sum_{\substack{\theta \text{ admissible} \\ \theta = \psi \text{ on } E_\tau^*}} \prod_{t \in T_\tau} |t|_\theta'.$$

Essentially, the partial invariant $\mathrm{TV}_{r,q}(\mathfrak{T}, \tau, \psi)$ considers all admissible ways θ of extending the colouring ψ from the current edges E_τ^* to also include the "finished" edges in E_τ, and then sums the weights of the partial colourings $\prod |t|_\theta'$ for all such extensions θ using only the tetrahedra in T_τ.

We can now give our full fixed-parameter tractable algorithm for $\mathrm{TV}_{r,q}$.

Algorithm 11. *Let $\mathfrak{T}$ be a generalised triangulation of a 3-manifold. We compute $\mathrm{TV}_{r,q}(\mathfrak{T})$ for given r, q as follows.*

Build a nice tree decomposition $(X, \{B_\tau\})$ of $\mathfrak{T}$. Then work through each node τ of X from the leaves of X to the root, and compute Ψ_τ and $\mathrm{TV}_{r,q}(\mathfrak{T}, \tau, \psi)$ for each $\psi \in \Psi_\tau$ as follows.

1. *If τ is a leaf bag, then $E_\tau^* = E_\tau = \emptyset$, Ψ_τ contains just the trivial colouring ψ on $\emptyset$, and $\mathrm{TV}_{r,q}(\mathfrak{T}, \tau, \psi) = 1$.*
2. *If τ is some other introduce bag with child node σ, then $T_\tau = T_\sigma$. This means that $\Psi_\tau = \Psi_\sigma$, and for each $\psi \in \Psi_\tau$ we have $\mathrm{TV}_{r,q}(\mathfrak{T}, \tau, \psi) = \mathrm{TV}_{r,q}(\mathfrak{T}, \sigma, \psi)$.*
3. *If τ is a forget bag with child node σ, then $T_\tau = T_\sigma \cup \{t\}$ for the unique "forgotten" tetrahedron $t \in B_\tau \backslash B_\sigma$. Moreover, E_τ^* extends E_σ^* by including the six edges of t (if they were not already present).*

   For each colouring $\psi \in \Psi_\sigma$, enumerate all possible ways of colouring the six edges of t that are consistent with ψ on any edges of t that already appear in E_σ^*, and are admissible on the four triangular faces of t. Each such colouring on t yields an extension $\psi'\colon E_\tau^* \to I$ of $\psi\colon E_\sigma^* \to I$. We include ψ' in Ψ_τ, and record the partial invariant $\mathrm{TV}_{r,q}(\mathfrak{T}, \tau, \psi') = \mathrm{TV}_{r,q}(\mathfrak{T}, \sigma, \psi)$.
4. *If τ is a join bag with child nodes σ_1, σ_2, then T_τ is the disjoint union $T_{\sigma_1} \dot\cup T_{\sigma_2}$. Here E_τ^* is a subset of $E_{\sigma_1}^* \cup E_{\sigma_2}^*$.*
   For each pair of colourings $\psi_1 \in \Psi_{\sigma_1}$ and $\psi_2 \in \Psi_{\sigma_2}$, if ψ_1 and ψ_2 agree on the common edges in $E_{\sigma_1}^* \cap E_{\sigma_2}^*$ then record the pair (ψ_1, ψ_2).
   Each such pair yields a "combined colouring" in Ψ_τ, which we denote by $\psi_1 \cdot \psi_2\colon E_\tau^* \to I$; note that different pairs (ψ_1, ψ_2) might yield the same colouring $\psi_1 \cdot \psi_2$ since some edges from $E_{\sigma_1}^* \cup E_{\sigma_2}^*$ might not appear in E_τ^*. Then Ψ_τ consists of all such combined colourings $\psi_1 \cdot \psi_2$ from recorded pairs (ψ_1, ψ_2). Moreover, for each combined colouring $\psi \in \Psi_\tau$ we compute the partial invariant $\mathrm{TV}_{r,q}(\mathfrak{T}, \tau, \psi)$ by aggregating over all duplicates:

$$\mathrm{TV}_{r,q}(\mathfrak{T}, \tau, \psi) = \sum_{\substack{(\psi_1, \psi_2) \text{ recorded} \\ \psi_1 \cdot \psi_2 = \psi}} \mathrm{TV}_{r,q}(\mathfrak{T}, \sigma_1, \psi_1) \cdot \mathrm{TV}_{r,q}(\mathfrak{T}, \sigma_2, \psi_2).$$

Once we have processed the entire tree, the root node ρ of X will have $E_\rho^* = \emptyset$, ρ will contain just the trivial colouring ψ on $\emptyset$, and $\mathrm{TV}_{r,q}(\mathfrak{T}, \rho, \psi)$ for this trivial colouring will be equal to the Turaev-Viro invariant $\mathrm{TV}_{r,q}(\mathfrak{T})$.

The time complexity of this algorithm is simple to analyse. Each leaf bag or introduce bag can be processed in $O(1)$ time (of course for the introduce bag we must avoid a deep copy of the data at the child node). Each forget bag produces $\Psi_\tau| \leq (r-1)^{|E_\tau^*|}$ colourings, each of which takes $O(|E_\tau^*|)$ time to analyse.

Naïvely, each join bag requires us to process $|\Psi_{\sigma_1}| \cdot |\Psi_{\sigma_2}| \leq (r-1)^{|E_{\sigma_1}^*|+|E_{\sigma_2}^*|}$ pairs of colourings (ψ_1, ψ_2). However, we can optimise this. Since we are only interested in colourings that agree on $E_{\sigma_1}^* \cap E_{\sigma_2}^*$, we can first partition Ψ_{σ_1} and Ψ_{σ_2} into buckets according to the colours on $E_{\sigma_1}^* \cap E_{\sigma_2}^*$, and then combine pairs from each bucket individually. This reduces our work to processing at most $(r-1)^{|E_{\sigma_1}^* \cup E_{\sigma_2}^*|}$ pairs overall. Each pair takes $O(|E_\tau^*|)$ time to process, and the preprocessing cost for partitioning Ψ_{σ_i} is $O\left(|\Psi_{\sigma_i}| \cdot \log |\Psi_{\sigma_i}| \cdot |E_{\sigma_i}^*|\right) = O\left((r-1)^{|E_{\sigma_i}^*|} \cdot |E_{\sigma_i}^*|^2 \log r\right)$.

Suppose that our tree decomposition has width k. At each tree node τ, every edge in E_τ^* must belong to some tetrahedron in the bag B_τ, and so $|E_\tau^*| \leq 6(k+1)$. Likewise, at each join bag described above, every edge in $E_{\sigma_1}^*$ or $E_{\sigma_2}^*$ must belong to some tetrahedron in the bag B_{σ_i} and therefore also the parent bag B_τ, and so $|E_{\sigma_1}^* \cup E_{\sigma_2}^*| \leq 6(k+1)$. From the discussion above, it follows that every bag can be processed in time $O\left((r-1)^{6(k+1)} \cdot k^2 \log r\right)$, and so:

Theorem 12. *Given a generalised triangulation $\mathfrak{T}$ of a 3-manifold with n tetrahedra, and a nice tree decomposition of $\mathfrak{T}$ with width k and $O(n)$ bags, Algorithm 11 computes $\mathrm{TV}_{r,q}(\mathfrak{T})$ in $O\left(n \cdot (r-1)^{6(k+1)} \cdot k^2 \log r\right)$ arithmetic operations in $\mathbb{Q}(\zeta)$.*

Theorem 12 shows that, for fixed r, if we can keep the treewidth small then computing $\mathrm{TV}_{r,q}$ becomes linear time, even for large inputs. This of course is the main benefit of fixed-parameter tractability. In our setting, however, we have an added advantage: $\mathrm{TV}_{r,q}$ is a topological invariant, and does not depend on our particular choice of triangulation.

Therefore, if we are faced with a large treewidth triangulation, we can *retriangulate* the manifold (for instance, using bistellar flips and related local moves), in an attempt to make the treewidth smaller. This is extremely effective in practice, as seen in Section 5.

Even if the treewidth is large, every tree node has $|E_\tau^*| \leq \ell$, where ℓ is the number of edges in the triangulation. Therefore the time complexity of Algorithm 11 reduces to $O\left(n \cdot (r-1)^\ell \cdot \ell^2 \log r\right)$, which is only a little slower than the backtracking algorithm (Lemma 3). This is in sharp contrast to many FPT algorithms from the literature, which—although fast for small parameters—suffer from extremely poor performance when the parameter becomes large.

5   Implementation and Experimentation

Here we implement Algorithm 11 (the fixed-parameter tractable algorithm), and subject both it and the backtracking algorithm to exhaustive experimentation.

The FPT algorithm is implemented in the open-source software package *Regina* [5]: the source code is available from *Regina*'s public git repository, and will be included in the next release. For consistency we compare it to *Regina*'s long-standing implementation of the backtracking algorithm.[3]

In our implementation, we do not compute treewidths precisely (an NP-complete problem)—instead, we implement the quadratic-time `GreedyFillIn` heuristic [2], which is reported to produce small widths in practice [7]. This way, costs of building tree decompositions are insignificant (but included in the running times). For both algorithms, we use relatively naïve implementations of arithmetic in cyclotomic fields—these are asymptotically slower than described in Section 3, but have very small constants.

We use two data sets for our experiments, both taken from large "census databases" of 3-manifolds to ensure that the experiments are comprehensive and not cherry-picked.

The first census contains all 13 400 closed prime orientable manifolds that can be formed from $n \leq 11$ tetrahedra [4, 14]. This simulates "real-world" computation—the Turaev-Viro invariants were used to build this census. Since the census includes all minimal triangulations of these manifolds, we choose the representative whose heuristic tree decomposition has smallest width (since we are allowed to retriangulate).

The second data set contains the first 500 triangulations from the (much larger) Hodgson-Weeks census of closed hyperbolic manifolds [8]. This shows performance on larger triangulations, with n ranging from 9 to 20.

We compare the performance of both algorithms for each data set, measuring running times for $TV_{7,1}$ (the largest r for which the experiments were feasible); see the full paper for details and plots. The results are striking: the FPT algorithm runs faster in over 99% of cases, including most of the cases with largest treewidth. In the worst example the FPT algorithm runs $3.7\times$ slower than the backtracking, but both data sets have examples that run $> 440\times$ faster. It is also pleasing to see a clear impact of the treewidth on the performance of the FPT algorithm, as one would expect.

6   An Alternate Geometric Interpretation

In this section, we give a geometric interpretation of admissible colourings on a triangulation of a 3-manifold $\mathfrak{T}$ in terms of *normal arcs*, i.e., line segments passing through triangles of $\mathfrak{T}$ that each connect two distinct edges, do not meet any vertices of $\mathfrak{T}$, and are pairwise disjoint. More precisely, we have the following:

---

[3] The *Manifold Recogniser* [15] also implements a backtracking algorithm, but it is not open-source and so comparisons are more difficult.

Theorem 13. *Given a 3-manifold triangulation $\mathfrak{T}$, and $r \geq 3$, an admissible colouring of the edges of $\mathfrak{T}$ with $r - 1$ colours corresponds to a system of normal arcs in the 2-skeleton with $\leq r - 2$ arcs per triangle forming a collection of cycles in the boundary of each tetrahedron of $\mathfrak{T}$.*

See the full version of this paper for a proof of this statement, as well as more details on the system of normal arcs.

Now, let $\mathfrak{T}$ be a closed n-tetrahedron 3-manifold triangulation, t a tetrahedron of $\mathfrak{T}$, f_1 and f_2 two triangles of t with common edge e of colour $\phi(e)$, and a_i and b_i the respective non-negative numbers of the two normal arc types in f_i meeting e, $i \in \{1, 2\}$. Since the system of normal arcs on t forms a collection of cycles on the boundary of t, we must have $a_1 + b_1 = a_2 + b_2 \leq r - 2$, giving rise to a total of $6n$ linear equations and $12n$ linear inequalities on $6n$ variables which all admissible colourings on $\mathfrak{T}$ must satisfy. Thus, finding admissible colourings on $\mathfrak{T}$ translates to the enumeration of integer lattice points within the polytope defined by the above equalities and inequalities.

Now, if we drop the upper bound constraint above, we get a cone. Computing the Hilbert basis of integer lattice points of this cone yields a finite description of all admissible colourings for any $r \geq 3$ and, thus, the essential information to compute $\mathrm{TV}_{r,q}(\mathfrak{T})$ for arbitrary r. Transforming this approach into a practical algorithm is work in progress.

References

1. Alagic, G., Jordan, S.P., König, R., Reichardt, B.W.: Estimating Turaev-Viro three-manifold invariants is universal for quantum computation. Physical Review A **82**(4), 040302(R) (2010)
2. Bodlaender, H.L., Koster, A.M.C.A.: Treewidth computations. I. Upper bounds. Inform. and Comput. **208**(3), 259–275 (2010)
3. Burton, B.A.: Structures of small closed non-orientable 3-manifold triangulations. J. Knot Theory Ramifications **16**(5), 545–574 (2007)
4. Burton, B.A.: Detecting genus in vertex links for the fast enumeration of 3-manifold triangulations. In: ISSAC 2011: Proceedings of the 36th International Symposium on Symbolic and Algebraic Computation, pp. 59–66. ACM (2011)
5. Burton, B.A., Budney, R., Pettersson, W., et al.: Regina: Software for 3-manifold topology and normal surface theory. http://regina.sourceforge.net/
6. Burton, B.A., Downey, R.G.: Courcelle's theorem for triangulations, March 2014. arXiv:1403.2926 (Preprint)
7. van Dijk, T., van den Heuvel, J.P., Slob, W.: Computing treewidth with LibTW (2006). http://www.treewidth.com
8. Hodgson, C.D., Weeks, J.R.: Symmetries, isometries and length spectra of closed hyperbolic three-manifolds. Experiment. Math. **3**(4), 261–274 (1994)
9. Kauffman, L.H., Lins, S.: Computing Turaev-Viro invariants for 3-manifolds. Manuscripta Math. **72**(1), 81–94 (1991)
10. Kirby, R., Melvin, P.: The 3-manifold invariants of Witten and Reshetikhin-Turaev for sl(2, **C**). Invent. Math. **105**(3), 473–545 (1991)
11. Kirby, R., Melvin, P.: Local surgery formulas for quantum invariants and the Arf invariant. Geom. Topol. Monogr. 7, 213–233 (2004) (Geom. Topol. Publ.)

12. Kleiner, B., Lott, J.: Notes on Perelman's papers. Geom. Topol. **12**(5), 2587–285? (2008)
13. Kloks, T.: Treewidth: Computations and Approximations, vol. 842. Springer, Heidelberg (1994)
14. Matveev, S.: Algorithmic Topology and Classification of 3-Manifolds. No. 9 in Algorithms and Computation in Mathematics. Springer, Heidelberg (2003)
15. Matveev, S., et al.: Manifold recognizer. http://www.matlas.math.csu.ru/
16. Munkres, J.R.: Elements of algebraic topology. Addison-Wesley (1984)
17. Robertson, N., Seymour, P.D.: Graph minors. II. Algorithmic aspects of tree-width. J. Algorithms **7**(3), 309–322 (1986)
18. Toda, S.: PP is as hard as the polynomial-time hierarchy. SIAM J. Comput. **20**(5), 865–877 (1991)
19. Turaev, V.G., Viro, O.Y.: State sum invariants of 3-manifolds and quantum $6j$-symbols. Topology **31**(4), 865–902 (1992)
20. Valiant, L.G.: The complexity of computing the permanent. Theor. Comput. Sci. **8**, 189–201 (1979)
21. Walker, K.: On Witten's 3-manifold invariants (1991). http://canyon23.net/math/

Big Data on the Rise?
Testing Monotonicity of Distributions

Clément L. Canonne[(✉)]

Columbia University, New York, USA
ccanonne@cs.columbia.edu

Abstract. The field of property testing of probability distributions, or distribution testing, aims to provide fast and (most likely) correct answers to questions pertaining to specific aspects of very large datasets. In this work, we consider a property of particular interest, *monotonicity of distributions*. We focus on the complexity of monotonicity testing across different models of access to the distributions [5,7,8,20]; and obtain results in these new settings that differ significantly (and somewhat surprisingly) from the known bounds in the standard sampling model [1].

1 Introduction

Before even the advent of data, information, records and insane amounts thereof to treat and analyze, probability distributions have been everywhere, and understanding their properties has been a fundamental problem in Statistics.[1] Whether it be about the chances of winning a (possibly rigged) game in a casino, or about predicting the outcome of the next election; or for social studies or experiments, or even for the detection of suspicious activity in networks, hypothesis testing and density estimation have had a role to play. And among these distributions, *monotone* ones have often been of paramount importance: is the probability of getting a cancer decreasing with the distance from, say, one's microwave? Are aging voters more likely to vote for a specific party? Is the success rate in national exams correlated with the amount of money spent by the parents in tutoring?

All these examples, however disparate they may seem, share one unifying aspect: *data* may be viewed as the probability distributions it defines and originates from; and understanding the properties of this data calls for testing these distributions. In particular, our focus here will be on testing whether the data – its underlying distribution – happens to be *monotone*,[2] or on the contrary far from being so.

The full version of this work is available as [4].

[1] As well as – crucially – in crab population analysis [16].

[2] Recall that a distribution D on $\{1, \dots, n\}$ is said to be *monotone* (non-increasing) if $D(1) \geq \cdots \geq D(n)$, i.e. if its probability mass function is non-increasing. We hereafter denote by $\mathcal{M}$ the class of monotone distributions.

© Springer-Verlag Berlin Heidelberg 2015
M.M. Halldórsson et al. (Eds.): ICALP 2015, Part I, LNCS 9134, pp. 294–305, 2015.
DOI: 10.1007/978-3-662-47672-7_24

Since the seminal work of Batu, Kumar, and Rubinfeld [1], this fundamental property has been well-understood in the usual model of access to the data, which only assumes independent samples. However, a recent trend in distribution testing has been concerned with introducing and studying new models which provide additional flexibility in observing the data. In these new settings, our understanding of what is possible and what remains difficult is still in its infancy; and this is in particular true for monotonicity, for which very little is known. This work intends to mitigate this state of affairs.

We hereafter assume the reader's familiarity with the broad field of property testing, and the more specific setting of distribution testing. For detailed surveys of the former, she or he is referred to, for instance, [11,12,17,18]; an overview of the latter can be found e.g. in [19], or [3]. Details of the models we consider (besides the usual sampling oracle setting, denoted by SAMP) are described in [5,6,8] (for the conditional sampling oracle COND, and its variants INTCOND and PAIRCOND restricted respectively to interval and pairwise queries); [1,7, 14] for the Dual and Cumulative Dual models; and [20] for the evaluation-only oracle, EVAL. The reader confused by the myriad of notations featured in the previous sentence may find the relevant definitions in Section 2 (as well as in the aforementioned papers).

Results. In this paper, we provide both upper and lower bounds for the problem of testing monotonicity, across various types of access to the unknown distribution. A summary of results, including the best currently known bounds on monotonicity testing of distributions, can be found in Table 1 below. As noted in Section 3, many of the lower bounds are implied by the corresponding lower bound on testing uniformity.

Table 1. Summary of results for monotonicity testing. The highlighted ones are new; bounds with an asterisk* hold for non-adaptive testers.

MODEL	UPPER BOUND	LOWER BOUND
SAMP	$\tilde{O}\left(\frac{\sqrt{n}}{\varepsilon^6}\right)$	$\Omega\left(\frac{\sqrt{n}}{\varepsilon^2}\right)$
COND	$\tilde{O}\left(\frac{1}{\varepsilon^{22}}\right), \tilde{O}\left(\frac{\log^2 n}{\varepsilon^3} + \frac{\log^4 n}{\varepsilon^2}\right)$	$\Omega\left(\frac{1}{\varepsilon^2}\right)$
INTCOND	$\tilde{O}\left(\frac{\log^5 n}{\varepsilon^4}\right)$	$\Omega\left(\sqrt{\frac{\log n}{\log\log n}}\right)$
EVAL	$O\left(\max\left(\frac{\log n}{\varepsilon}, \frac{1}{\varepsilon^2}\right)\right)^*$	$\Omega\left(\frac{\log n}{\varepsilon}\right)^*, \Omega\left(\frac{\log n}{\log\log n}\right)$
Cumulative Dual	$\tilde{O}\left(\frac{1}{\varepsilon^4}\right)$	$\Omega\left(\frac{1}{\varepsilon}\right)$

Techniques. Two main ideas are followed in obtaining our upper bounds: the first one, illustrated in Section 3 and Section 4.1, is the approach of Batu et al. [1], which reduces monotonicity testing to uniformity testing on polylogarithmically any intervals. This relies on a structural result for monotone distributions which asserts that they admit a succinct partition in intervals, such that on each interval the distribution is either close to uniform (in ℓ_2 distance), or puts very little weight.

The second approach, on which Section 4.2 is based (as well as the results for the EVAL and Dual models) also leverages a structural result, due this time to Birgé [2]. As before, this theorem states that each monotone distribution admits a succinct "flat approximation," but in this case the partition *does not depend on the distribution itself* (see Section 2 for a more rigorous exposition). From there, the high-level idea is to perform two different checks: first, that the distribution D is close to its "flattening" $\bar{D}$; and then that this flattening itself is close to monotone – where to be efficient the latter exploits the fact that the effective support of $\bar{D}$ is very small, as there are only polylogarithmically many intervals in the partition. If both tests succeed, then it must be the case that D is close to monotone.

Organization. In this extended abstract, we focus on the upper bounds for the conditional models, Theorem 5 and Theorem 6, which can be found in Section 4. Indeed, these two results illustrate both of our key approaches, and many of the ideas that are used to obtain our bounds in the other access models are already developed in the proofs of these two theorems. Due to space constraints, the pseudocode of our algorithms, as well as the full proofs of the theorems covered in this extended abstract, are deferred to the full version [4]. (This full version also contains the statements and details of the results pertaining to the EVAL and Cumulative Dual models, as well as additional results on tolerant testing and learning in some of the models considered.)

2   Preliminaries

All throughout this paper, we denote by $[n]$ the set $\{1, \ldots, n\}$, and by log the logarithm in base 2. A *probability distribution* over a (finite) domain Ω is a non-negative function $D: \Omega \to [0,1]$ such that $\sum_{x \in \Omega} D(x) = 1$. We denote by $\mathcal{U}(\Omega)$ the uniform distribution on Ω. Given a distribution D over Ω and a set $S \subseteq \Omega$, we write $D(S)$ for the total probability weight $\sum_{x \in S} D(x)$ assigned to S by D. Finally, for $S \subseteq \Omega$ such that $D(S) > 0$, we denote by D_S the *conditional* distribution of D restricted to S, that is $D_S(x) = \frac{D(x)}{D(S)}$ for $x \in S$ and $D_S(x) = 0$ otherwise. As is usual in distribution testing, in this work the distance between two distributions D_1, D_2 on Ω will be the *total variation distance* $d_{\mathrm{TV}}(D_1, D_2) \stackrel{\text{def}}{=} \frac{1}{2}\|D_1 - D_2\|_1 = \max_{S \subseteq \Omega}(D_1(S) - D_2(S))$ which takes value in $[0,1]$.

Models and access to the distributions. We shall work in the framework of *property testing*, where a testing algorithm for some fixed property is a randomized algorithm which, on input ε, must (with high probability) accept any input that has the property; and reject any input that is at a distance ε from any object satisfying the property. In our case, the inputs are probability distributions over a (known) domain $[n]$, and a property is a subset of all distributions over $[n]$. We now describe (informally) the settings we shall work in, which define the *type of access* the testing algorithms are granted to the input distribution.[3] In the first and most common setting (SAMP), the testers access the unknown distribution by getting independent and identically distributed samples from it.

A natural extension, COND, allows the algorithm to provide a query set $S \subseteq [n]$, and get a sample from the conditional distribution induced by D on S: that is, the distribution D_S on S defined by $D_S(i) = D(i)/D(S)$. By restricting the type of allowed query sets to the class of intervals $\{a, \ldots, b\} \subseteq [n]$, one gets a weaker version of this model, INTCOND (for "interval-cond").

Of a different flavor, providing (only) *evaluation* queries to the probability mass function (pmf) (resp. to the cumulative distribution function (cdf)) of the distribution an EVAL (resp. CEVAL) oracle access. When the algorithm is provided with both SAMP and EVAL (resp. SAMP and CEVAL) oracles to the distribution, we say it has *Dual (resp. Cumulative Dual) access* to it.

Monotone distributions. We now state here a few crucial facts about monotone distributions, namely that they admit a succinct approximation, itself monotone:

Definition 1 (Oblivious decomposition). *Given a parameter $\varepsilon > 0$, the corresponding oblivious decomposition of $[n]$ is the partition $\mathcal{I}_\varepsilon = (I_1, \ldots, I_\ell)$, where $\ell = \Theta\left(\frac{\log n}{\varepsilon}\right)$ and $|I_k| = \lfloor (1+\varepsilon)^k \rfloor$, $1 \leq k \leq \ell$.*

For a distribution D and parameter ε, define $\Phi_\varepsilon(D)$ to be the *flattened distribution* with relation to the decomposition $\mathcal{I}_\varepsilon$: $\Phi_\varepsilon(D)(i) = \frac{D(I_k)}{|I_k|}$ for $k \in [\ell], \forall i \in I_k$. Note that while $\Phi_\varepsilon(D)$ (obviously) depends on D, the partition $\mathcal{I}_\varepsilon$ itself does *not*; i.e., it can be computed prior to getting any sample from D.

Theorem 1 ([2]). *If D is monotone non-increasing, $\mathrm{d_{TV}}(D, \Phi_\varepsilon(D)) \leq \varepsilon$.*

Remark 1. The first use of this result in this discrete learning setting is due to Daskalakis et al. [9]. For a proof for discrete distributions (whereas the original paper by Birgé is intended for continuous ones), the reader is referred to [10] (Section 3.1, Theorem 5).

Corollary 1 (Robustness). *Suppose D is ε-close to monotone non-increasing. Then $\mathrm{d_{TV}}(D, \Phi_\alpha(D)) \leq 2\varepsilon + \alpha$; furthermore, $\Phi_\alpha(D)$ is also ε-close to monotone non-increasing.*

Other tools. Finally, we will use as subroutines the following results of Canonne, Ron, and Servedio. The first one, restated below, provides a way to "compare" the probability weight of disjoint subsets of elements in the COND model:

---

[3] For a formal definition of these models, the reader is referred to the full version [4].

Lemma 1 ([5, Lemma2]). *Given as input two disjoint subsets of points* $X, Y \subseteq \Omega$ *together with parameters* $\eta \in (0,1]$, $K \geq 1$, *and* $\delta \in (0,1/2]$, *as well as* COND *query access to a distribution* D *on* Ω, *there exists a procedure* COMPARE *that either outputs a value* $\rho > 0$ *or outputs* High *or* Low, *and satisfies the following:*

(i) *If* $D(X)/K \leq D(Y) \leq K \cdot D(X)$ *then with probability at least* $1 - \delta$ *the procedure outputs a value* $\rho \in [1 - \eta, 1 + \eta]D(Y)/D(X);$

(ii) *If* $D(Y) > K \cdot D(X)$ *then with probability at least* $1 - \delta$ *the procedure outputs either* High *or a value* $\rho \in [1 - \eta, 1 + \eta]D(Y)/D(X);$

(iii) *If* $D(Y) < D(X)/K$ *then with probability at least* $1 - \delta$ *the procedure outputs either* Low *or a value* $\rho \in [1 - \eta, 1 + \eta]D(Y)/D(X).$

The procedure performs $O\left(\frac{K \log(1/\delta)}{\eta^2}\right)$ COND *queries on the set* $X \cup Y$.

The second estimates the distance between the uniform distribution and an unknown distribution, given a conditional oracle for the latter:

Theorem 2 ([5, Theorem14]). *Given as input* $\varepsilon \in (0,1]$ *and* $\delta \in (0,1]$, *as well as* PAIRCOND *query access to a distribution* D *on* Ω, *there exists an algorithm that outputs a value* $\hat{d}$ *and has the following guarantee. The algorithm performs* $\tilde{O}\left(1/\varepsilon^{20} \log(1/\delta)\right)$ *queries and, with probability at least* $1 - \delta$, *the value it outputs satisfies* $\left|\hat{d} - \mathrm{d}_{\mathrm{TV}}(D, \mathcal{U})\right| \leq \varepsilon$.

3   Previous Work: Standard Model

In this section, we describe the currently known results for monotonicity testing in the standard (sampling) oracle model. These bounds on the sample complexity, tight up to logarithmic factors, are due to Batu et al. [1];[4] while not directly applicable to the other access models we will consider, we note that some of the techniques they use will be of interest to us in Section 4.1.

Theorem 3 ([1, Theorem10]). *There exists an* $O\left(\frac{\sqrt{n}}{\varepsilon^6} \operatorname{polylog} n\right)$*-query tester for monotonicity in the* SAMP *model.*

Proof (sketch). Their algorithm works by taking this many samples from D, and then using them to recursively split the domain $[n]$ in half, as long as the conditional distribution on the current interval is not close enough to uniform (or not enough samples fall into it). If the binary tree created during this recursive process exceeds $O(\log^2 n/\varepsilon)$ nodes, the tester rejects. Batu et al. then show that this succeeds with high probability, the leaves of the recursion yielding a partition of $[n]$ in $\ell = O(\log^2 n/\varepsilon)$ intervals $I_1, \ldots, I_\ell$, such that either (a) the conditional distribution D_{I_j} is $O(\varepsilon)$-close to uniform on this interval; or (b) I_j is "light," i.e. has weight at most $O(\varepsilon/\ell)$ under D. This implies this partition

---

[4] [1] originally states an $\tilde{O}(\sqrt{n}/\varepsilon^4)$ sample complexity, but their argument seems to only result in an $\tilde{O}(\sqrt{n}/\varepsilon^6)$ bound.

defines an ℓ-flat distribution $\bar{D}$ which is $\varepsilon/2$-close to D, and can be easily learnt from another batch of samples; once this is done, it only remains to test (e.g. via linear programming, which can be done efficiently) whether this $\bar{D}$ is itself $\varepsilon/2$-close to monotone, and accept if and only this is the case.

Theorem 4 ([1, Theorem11]). *Any tester for monotonicity in the* SAMP *model must perform* $\Omega\left(\frac{\sqrt{n}}{\varepsilon^2}\right)$ *queries.*

To prove this lower bound, they reduce the problem of uniformity testing to monotonicity testing: the result then follows from the $\Omega(\sqrt{n}/\varepsilon^2)$ lower bound of [15] for testing uniformity.[5] We note that the argument above extends to all models: that is, any lower bound for testing uniformity directly implies a corresponding lower bound for monotonicity in the same access model (giving the bounds in Table 1).

4 With Conditional Samples

In this section, we focus on testing monotonicity with a stronger type of access to the underlying distribution, that is given the ability to ask conditional queries. More precisely, we prove the following theorem:

Theorem 5. *There exists an* $\tilde{O}\left(\frac{1}{\varepsilon^{22}}\right)$*-query tester for monotonicity in the* COND *model.*

Furthermore, assuming only a (restricted) type of conditional queries are allowed, one can still get an exponential improvement from the standard sampling model:

Theorem 6. *There exists an* $\tilde{O}\left(\frac{\log^5 n}{\varepsilon^4}\right)$*-query tester for monotonicity in the* INTCOND *model.*

We now prove these two theorems, starting with Theorem 6. In doing so, we will also derive a weaker, poly($\log n, 1/\varepsilon$)-query tester for COND; before turning in Section 4.2 to the constant-query tester of Theorem 5.

4.1 A poly($\log n, 1/\varepsilon$)-Query Tester for INTCOND

Our algorithm (Algorithm 1) follows the same overall idea as the one from [1], which a major difference. As in theirs, the first step will be to partition $[n]$ into a small number of intervals, such that the conditional distribution D_I on each interval I is close to uniform; that is,

$$d_{\mathrm{TV}}(D_I, \mathcal{U}_I) = \sum_{i \in I} \left| \frac{D(i)}{D(I)} - \frac{1}{|I|} \right| \le \frac{\varepsilon}{4} . \tag{1}$$

---

[5] While [1] only shows a $\Omega(\sqrt{n})$ lower bound, as they invoke the (previously best known) lower bound of [13] for uniformity testing, their argument straightforwardly extends to the result of Paninski.

he original approach (in the sampling model) of Batu et al. was based on
timating the ℓ_2 norm of the conditional distribution *via* the number of collisions
om a sufficiently large sample; this yielded a $\tilde{O}(\sqrt{n})$ sample complexity.

However, using directly as a subroutine (in the COND model) an algorithm for
olerantly) testing uniformity, one can perform this first step with $\ell_{\max} \log \frac{1}{\delta} =$
$_{\max} \log \ell_{\max}$ calls[6] to this subroutine, each with approximation parameter $\frac{\varepsilon}{4}$
he proof of correctness of [1] does not depend on how the test of uniformity
actually performed, in the partitioning step). A first idea would be to use for
iis the following result:

act 1 ([6]). *One can test ε-uniformity of a distribution D_r over $[r]$ in the
onditional sampling model:*

- *with $\tilde{O}(1/\varepsilon^2)$ samples, given access to a COND_{D_r} oracle;*
- *with $\tilde{O}(\log^3 r/\varepsilon^3)$ samples, given access to a $\mathsf{INTCOND}_{D_r}$ oracle.*

Iowever, this does *not* suffice for our purpose: indeed, Algorithm 1 needs in
tep 6 not only to reject distributions that are too far from uniform, *but also
o accept those that are close enough*. A standard uniformity tester as the one
bove does not ensure the latter condition: for this, one would *a priori* need
olerant tester for uniformity. While [6] does describe such a tolerant tester
see Theorem 2), it only applies to COND – and we aim at getting an INTCOND
ester.

To resolve this issue, we observe that what the algorithm requires is slightly
veaker: namely, to distinguish distributions on an interval I that (a) are $\Omega(\varepsilon)$-far
rom uniform from those that are (b) $O(\varepsilon/|I|)$-close to uniform *in ℓ_∞ distance*.
t is not hard to see that the two testers of Fact 1 can be adapted in a straight-
orward fashion to meet this guarantee, with the same query complexity. Indeed,
b) is equivalent to asking that the ratio $D(x)/D(y)$ of any two points in I be
1 $[1-\varepsilon, 1+\varepsilon]$, which is exactly what both testers check.
As a corollary, we get:

Corollary 2. *Given access to a conditional oracle $\mathcal{O}$ for a distribution D over
n], the algorithm* $\textsc{TestMonCond}^{\mathcal{O}}$ *outputs* yes *when D is monotone and* no
vhen it is ε-far from monotone, with probability at least 2/3. The algorithm uses

- $\tilde{O}\left(\frac{\ell_{\max}}{\varepsilon} + \frac{\ell_{\max}}{\varepsilon^2} + \frac{\log^4 n}{\varepsilon^2}\right) = \tilde{O}\left(\frac{\log^2 n}{\varepsilon^3} + \frac{\log^4 n}{\varepsilon^2}\right)$ *samples, when $\mathcal{O} = \mathsf{COND}_D$;*
- $\tilde{O}\left(\frac{\ell_{\max}}{\varepsilon} + \ell_{\max}\frac{\log^3 n}{\varepsilon^3} + \frac{\log^4 n}{\varepsilon^2}\right) = \tilde{O}\left(\frac{\log^5 n}{\varepsilon^4}\right)$ *samples, when $\mathcal{O} = \mathsf{INTCOND}_D$.*

This is turn implies Theorem 6. Note that we make sure in Step 9 that each of
he intervals we recurse on contains at least one of the "reference samples" h_i:
his is in order to guarantee all conditional queries made on a set with non-zero
orobability. Discarding the "light intervals" can be done without compromising
he correctness, as with high probability each of them has probability weight at

---

[6] Where the logarithmic dependence on δ aims at boosting the (constant) success
probability of the uniformity testing algorithm, in order to apply a union bound
over the $O(\ell_{\max})$ calls.

Algorithm 1 General algorithm $\textsc{TestMonCond}^{\mathcal{O}}$

Require: $\mathcal{O} \in \{\textsf{COND}, \textsf{INTCOND}\}$ access to D

1: Define $\ell_{\max} \stackrel{\text{def}}{=} O\big(\log^2 n/\varepsilon\big)$, $\delta \stackrel{\text{def}}{=} O(1/\ell_{\max})$.

2: Draw $m \stackrel{\text{def}}{=} O\big(\frac{\varepsilon}{\ell_{\max}} \log \frac{1}{\delta}\big)$ samples $h_1, \ldots, h_m$.

3: **PartitionStart**

4:     Start with interval $I \leftarrow [n]$

5:     **repeat**

6:         Test (with probability $\geq 1 - \delta$) if D_I is $\varepsilon/4$-close to the $\mathcal{U}(I)$

7:         **if** $d_{\mathrm{TV}}(D_I, \mathcal{U}_I) > \frac{\varepsilon}{4}$ **then**

8:             bisect I in half

9:             recursively test each half that contains some h_i, mark the others as
    "light"

10:         **else if** $\ell_{\max}$ splits have been made **then**

11:             **return** no

12:         **end if**

13:     **until** all intervals are close to uniform or have been marked "light"

14: **PartitionEnd**

15: Let $\mathcal{I}_\ell = \langle I_1, \ldots, I_\ell \rangle$ denote the partition of $[n]$ into intervals induced by the leaves
    of the recursion from the previous step.

16: Obtain an additional sample T of size $O\big(\frac{\log^4 n}{\varepsilon^2}\big)$.

17: Let $\hat{D}$ denote the ℓ-flat distribution described by $(\mathbf{w}, \mathcal{I}_\ell)$ where ω_j is the fraction
    of samples from T falling in I_j.

18: **if** $\hat{D}$ is $(\varepsilon/2)$-close to monotone **then** $\triangleright$ Can be checked in poly(ℓ)-time

19:     **return** yes

20: **end if**

21: **return** no

---

most $\frac{\varepsilon}{4\ell_{\max}}$, and therefore in total the light intervals can amount to at most $\varepsilon/4$
of the probability weight of D – as in the original argument of Batu et al., we
can still conclude that with high probability $\hat{D}$ is $\varepsilon/2$-close to D.

4.2 A poly($1/\varepsilon$)-Query Tester for COND

The idea in proving Theorem 5 is to reduce the task of testing monotonicity to
another property, but on a (related) distribution *over a much smaller domain*.
We begin by some relevant notations and definitions:

Reduction from Testing Properties Over $[\ell]$. For fixed α and D, let D_α^{red}
be the *reduced* distribution on $[\ell]$ with respect to the oblivious decomposition $\mathcal{I}_\alpha$,
where all throughout $\ell = \ell(\alpha, n)$ as per Definition 1; i.e, $\forall k \in [\ell]$, $D_\alpha^{\text{red}}(k) = D(I_k) = \Phi_\alpha(D)(I_k)$. (Note that given oracle access $\textsf{SAMP}_D$, it is easy to simulate
$\textsf{SAMP}_{D_\alpha^{\text{red}}}$.)

Definition 2 (Exponential Property). *Fix n, α, and the corresponding $\ell = \ell(n, \alpha)$. For distributions over $[\ell]$, let the property $\mathcal{P}_\alpha$ be defined as "$Q \in \mathcal{P}_\alpha$ if and only if there exists $D \in \mathcal{M}$ over $[n]$ such that $Q = D_\alpha^{\text{red}}$."*

Fact 2. *Given a distribution Q over $[\ell]$, let* $\text{expand}_\alpha(Q)$ *denote the distribution over $[n]$ obtained by "spreading" uniformly $Q(k)$ over I_k (again, considering the oblivious decomposition of $[n]$ for α). Then,*

$$Q \in \mathcal{P}_\alpha \Leftrightarrow \text{expand}_\alpha(Q) \in \mathcal{M} \tag{2}$$

Fact 3. *Given a distribution Q over $[\ell]$, the following also holds:[7] $Q \in \mathcal{P}_\alpha$ if and only if $\forall k < \ell$, $Q(k+1) \le (1+\alpha)Q(k)$. Moreover, by Fact 2, we have that for D over $[n]$, $\Phi_\alpha(D) \in \mathcal{M}$ if and only if $D_\alpha^{\text{red}} \in \mathcal{P}_\alpha$.*

We shall also use the following result on flat distributions (adapted from [1, Lemma7]) and whose proof is deferred to the full version.

Fact 4. *$\Phi_\alpha(D)$ is ε-close to monotone if and only if it is ε-close to a $\mathcal{I}_\alpha$-flat monotone distribution (that is, a monotone distribution piecewise constant, according to the same partition $\mathcal{I}_\alpha$).*

Observe that Facts 2, 3 and 4 altogether imply that, for $\mathcal{I}_\alpha$-flat distributions, distance to monotonicity and distance to $\mathcal{P}_\alpha$ of the reduced distribution are equal.

Efficient Approximation of Distance to $\Phi(D)$.

Lemma 2. *Given* COND *access to a distribution D over $[n]$, there is an algorithm that, on input α and $\varepsilon, \delta \in (0,1]$, makes $\tilde{O}(\frac{1}{\varepsilon^{22}} \log \frac{1}{\delta})$ queries (independent of α) and outputs $\hat{d}$ such that, with probability at least $1 - \delta$, $|\hat{d} - d_{\text{TV}}(D, \Phi_\alpha(D))| \le \varepsilon$.*

Proof. We describe such algorithm for a constant probability of success; boosting the success probability to $1 - \delta$ at the price of a multiplicative $\log \frac{1}{\delta}$ factor can then be achieved by standard techniques (repetition, and taking the median value). Let D, ε and $\mathcal{I}_\alpha$ be defined as before; define Z to be a random variable taking values in $[0,1]$, such that, for $k \in [\ell]$, Z is equal to $d_{\text{TV}}(D_{I_k}, \mathcal{U}_{I_k})$ with probability $\omega_k = D(I_k)$. An easy computation then shows that $\mathbb{E}Z = \sum_{k=1}^\ell \omega_k d_{\text{TV}}(D_{I_k}, \mathcal{U}_{I_k}) = d_{\text{TV}}(D, \Phi_\alpha(D))$. Putting aside for now the fact that we only have (using as a subroutine the COND algorithm from Theorem 2 to estimate the distance to uniformity) access to *additive approximations* of the $d_{\text{TV}}(D_{I_k}, \mathcal{U}_{I_k})$'s, one can simulate independent draws from Z by taking each time a fresh sample $i \sim D$, looking up the k for which $i \in I_k$, and calling the COND subroutine to get the corresponding value. Applying a Chernoff bound, only $O(1/\varepsilon^2)$ such draws are needed, each of them costing $\tilde{O}(1/\varepsilon^{20})$ COND queries.

---

[7] We point out that the equivalence stated here once again ignores, for the sake of conceptual clarity, technical details arising from the discrete setting. Taking these into account would yield a slightly weaker characterization, with a twofold implication instead of an equivalence; which would still be good enough for our purpose.

Dealing with approximation. It suffices to estimate $\mathbb{E}Z$ within an additive $\varepsilon/2$, which can be done with probability $9/10$ by simulating $m = O(1/\varepsilon^2)$ samples from Z. To get each sample, for the index k drawn we can call the COND subroutine with parameters $\varepsilon/2$ and $\delta = 1/(10m)$ to obtain an estimate of $d_{\mathrm{TV}}(D_{I_k}, \mathcal{U}_{I_k})$. By a union bound we get that, with probability at least $9/10$, all estimates are within an additive $\varepsilon/2$ of the true value, incurring only a $O(\log 1/\varepsilon)$ additional factor in the overall sample complexity $\tilde{O}(1/\varepsilon^{20})$. Conditioned on this, we get that the approximate value we compute instead of $\mathbb{E}Z$ is off by at most $\varepsilon/2 + \varepsilon/2 = \varepsilon$ (where the first term corresponds to the approximation of the value of Z for each draw, and the second comes from the additive approximation of $\mathbb{E}Z$ by sampling).

The Algorithm. The tester is described in Algorithm 2. The second step, as argued in Lemma 2, uses $\tilde{O}(1/\varepsilon^{22})$ samples; we will show shortly after that *efficiently testing ε-farness to $\mathcal{P}_\gamma$ is also achievable with $\tilde{O}(1/\varepsilon^6)$ COND queries* – concluding the proof of Theorem 5.

Algorithm 2 Algorithm TestMonCond

Require: COND access to D
1: Simulating $\mathrm{COND}_{D_\alpha^{\mathrm{red}}}$, check if $\Phi_\alpha(D)$ is $(\varepsilon/4)$-close to monotone by testing $(\varepsilon/4)$-farness (of D_α^{red}) to $\mathcal{P}_\alpha$; **return** no if not.
2: Test whether $\Phi_\alpha(D)$ is $(\varepsilon/4)$-close to D using the sampling approach discussed above; **return** no if not.
3: **return** yes

Correctness of Algorithm 2. Assume we can efficiently perform the two steps, and condition on their execution being correct (as each of them is run with for instance parameter $\delta = 1/10$, this happens with probability at least $3/4$).

- If D is monotone non-increasing, so is $\Phi_\alpha(D)$; by Fact 3, this means that $\mathcal{P}_\alpha(D_\alpha^{\mathrm{red}})$ holds, and the first step passes. Theorem 1 then ensures that D and $\Phi_\alpha(D)$ are α-close, and the algorithm outputs yes;
- If D is ε-far from monotone, then either (a) $\Phi_\alpha(D)$ is $\frac{\varepsilon}{2}$-far from monotone or (b) $d_{\mathrm{TV}}(D, \Phi_\alpha(D)) > \frac{\varepsilon}{2}$; if (b) holds, no matter how the algorithm behaves in first step, the algorithm not go further that the second step, and output no. Assume now that (b) does not hold, i.e. only (a) is satisfied. By putting together Facts 2 to 4, we conclude that (a) implies that D_α^{red} is $\frac{\varepsilon}{2}$-far from $\mathcal{P}_\alpha$, and the algorithm outputs no in the first step.

Testing ε-farness to $\mathcal{P}_\gamma$. To achieve this objective, we begin with the following lemmas, which relate the distance between a distribution Q and $\mathcal{P}_\alpha$ to the total weight of points that violate the property.

Algorithm 3 TESTINGEXPONENTIALPROPERTY

Require: PAIRCOND access to Q, $\alpha \in [0,1)$ ▷ Useful for $\alpha = \Theta(\varepsilon) < 1$
Ensure: with probability at least $3/4$ returns **no** if Q is $O(\varepsilon)$-close to $\mathcal{P}_\alpha$, and **yes** if it satisfies $\mathcal{P}_\alpha$.

Set $\tau \overset{\text{def}}{=} \varepsilon\alpha^2$
Draw $m \overset{\text{def}}{=} \Theta\big(\frac{1}{\varepsilon\alpha}\big)$ samples $s_1, \ldots, s_m$ from Q ▷ Contains an element from W_τ
w.h.p.
for $i = 1$ **to** m **do**
  if $s_i \geq 2$ **then**
    Call COMPARE (from Lemma 1) on $\{s_i - 1\}$, $\{s_i\}$ with $\eta = \frac{\tau}{2}$, $K = 2$ and
$\delta = \frac{1}{10m}$.
    if the procedure outputs High **then return no**
    else if it outputs a value ρ **then** ▷ $\frac{1-\eta}{\rho} \cdot Q(s_i) \leq Q(s_i - 1) \leq \frac{1+\eta}{\rho} \cdot Q(s_i)$
      if $\rho < \frac{1+\eta}{1+\alpha+\tau}$ **then return no**
      end if
    end if
  end if
end for
return yes

Lemma 3. *For a distribution Q over $[\ell]$, let $W = \{\, i : \; Q(i) > (1+\alpha)Q(i-1)\,\}$ be the set of* witnesses *(points which violate the property). Then, the distance from Q to the property $\mathcal{P}_\alpha$ is $O(1/\alpha)Q(W)$.*

This implies that when Q is ε-far from having the property, it suffices to get $O(1/(\alpha\varepsilon))$ samples from Q and compare them to their neighbors to detect a violation with high probability. While this last step would be easy with an *exact* EVAL oracle, for the purpose of this section we can only use an approximate one. The lemma below addresses this issue, by ensuring that there will be many points *"patently"* violating the property.

Lemma 4. *For Q as above and $\tau > 0$, let $W_\tau = \{\, i : \; Q(i) > (1+\alpha+\tau)\, Q(i-1)\,$ be the set of τ-witnesses (so that $W = \bigcup_{\tau>0} W_\tau$). Then, the distance from Q to the property $\mathcal{P}_\alpha$ is at most $O(1/(\alpha+\tau))Q(W_\tau) + O\big(\tau/\alpha^2\big)$.*

Corollary 3. *Taking $\alpha = \Theta(\varepsilon)$ and $\tau = \varepsilon\alpha^2$, we get that if $Q(W_\tau) \leq \varepsilon^2$, then Q is $O(\varepsilon)$-close to $\mathcal{P}_\alpha$.*

By leveraging Corollary 3, we are able to obtain efficient approximation of the distance of a distribution to the "exponential property":

Theorem 7. *There exists a constant $0 < c < 1$ such that, for any $\varepsilon > 0$: if Q satisfies $\mathcal{P}_\alpha$ (where $\alpha = c\varepsilon$), then with probability at least $2/3$ Algorithm TEST-EXPPROPERTY returns* yes, *and if Q is $\Omega(\varepsilon)$-far from $\mathcal{P}_\alpha$, then with probability at least $2/3$ Algorithm* TESTEXPPROPERTY *returns* no. *The number of* PAIR-COND *queries performed by the algorithm is $\tilde{O}\big(1/\varepsilon^8\big)$.*

The algorithm can be found in Algorithm 3. (The proofs of Lemmas 3, 4, and Theorem 7 are deferred to the full version.)

References

1. Batu, T., Kumar, R., Rubinfeld, R.: Sublinear algorithms for testing monotone and unimodal distributions. In: Proceedings of STOC, pp. 381–390. ACM New York (2004)
2. Birgé, L.: On the risk of histograms for estimating decreasing densities. The Annals of Statistics **15**(3), 1013–1022 (1987)
3. Canonne, C.L.: A Survey on Distribution Testing: Your Data is Big. But is it Blue? Electronic Colloquium on Computational Complexity (ECCC) 22, 63 (2015)
4. Canonne, C.L.: Big Data on the Rise: Testing monotonicity of distributions. ArXiV:abs/1501.06783 (2015)
5. Canonne, C.L., Ron, D., Servedio, R.A.: Testing probability distributions using conditional samples. ArXiV:abs/1211.2664, November 2012
6. Canonne, C.L., Ron, D., Servedio, R.A.: Testing equivalence between distributions using conditional samples. In: Proceedings of SODA, pp. 1174–1192. SIAM (2014), see also [5] (full version)
7. Canonne, C.L., Rubinfeld, R.: Testing probability distributions underlying aggregated data. In: Proceedings of ICALP, pp. 283–295 (2014)
8. Chakraborty, S., Fischer, E., Goldhirsh, Y., Matsliah, A.: On the power of conditional samples in distribution testing. In: Proceedings of ITCS, pp. 561–580. ACM, New York (2013)
9. Daskalakis, C., Diakonikolas, I., Servedio, R.A.: Learning k-modal distributions via testing. In: Proceedings of SODA, pp. 1371–1385. SIAM (2012)
10. Daskalakis, C., Diakonikolas, I., Servedio, R.A., Valiant, G., Valiant, P.: Testing k-modal distributions: Optimal algorithms via reductions. In: Proceedings of SODA, pp. 1833–1852. SIAM (2013)
11. Fischer, E.: The art of uninformed decisions: A primer to property testing. BEATCS **75**, 97–126 (2001)
12. Goldreich, O. (ed.): Property Testing: Current Research and Surveys. LNCS, vol. 6390. Springer, Heidelberg (2010)
13. Goldreich, O., Ron, D.: On testing expansion in bounded-degree graphs. Technical report, TR00-020, Electronic Colloquium on Computational Complexity (ECCC) (2000)
14. Guha, S., McGregor, A., Venkatasubramanian, S.: Streaming and sublinear approximation of entropy and information distances. In: Proceedings of SODA, pp. 733–742. SIAM, Philadelphia (2006)
15. Paninski, L.: A coincidence-based test for uniformity given very sparsely sampled discrete data. IEEE Transactions on Information Theory **54**(10), 4750–4755 (2008)
16. Pearson, K.: Contributions to the Mathematical Theory of Evolution. Philosophical Transactions of the Royal Society of London. (A.) 185, 71–110 (1894)
17. Ron, D.: Property Testing: A Learning Theory Perspective. Foundations and Trends in Machine Learning **1**(3), 307–402 (2008)
18. Ron, D.: Algorithmic and analysis techniques in property testing. Foundations and Trends in Theoretical Computer Science **5**, 73–205 (2010)
19. Rubinfeld, R.: Taming Big Probability Distributions. XRDS **19**(1), 24–28 (2012)
20. Rubinfeld, R., Servedio, R.A.: Testing monotone high-dimensional distributions. Random Structures and Algorithms **34**(1), 24–44 (2009)

Unit Interval Editing
Is Fixed-Parameter Tractable

Yixin Cao[✉]

Department of Computing, Hong Kong Polytechnic University, Hong Kong, China
yixin.cao@polyu.edu.hk

Abstract. Given a graph G and integers k_1, k_2, and k_3, the unit interval editing problem asks whether G can be transformed into a unit interval graph by at most k_1 vertex deletions, k_2 edge deletions, and k_3 edge additions. We give an algorithm solving the problem in $2^{O(k \log k)} \cdot (n+m)$ time, where $k := k_1 + k_2 + k_3$, and n, m denote respectively the numbers of vertices and edges of G. Therefore, it is fixed-parameter tractable parameterized by the total number of allowed operations.

This implies the fixed-parameter tractability of the unit interval edge deletion problem, for which we also present a more efficient algorithm running in time $O(4^k \cdot (n + m))$. Another result is an $O(6^k \cdot (n + m))$-time algorithm for the unit interval vertex deletion problem, significantly improving the best-known algorithm running in time $O(6^k \cdot n^6)$.

1 Introduction

A graph is a *unit interval graph* if its vertices can be assigned to unit-length intervals on the real line such that there is an edge between two vertices if and only if their corresponding intervals intersect. Most important applications of unit interval graphs were found in computational biology [7,9], where data are mainly obtained by unreliable experimental methods. Thus, the graph representing the raw data is very unlikely to be a unit interval graph, and an important step toward understanding the data is to find out and fix the hidden errors. Various modification problems to unit interval graphs have been formulated: Given a graph G on n vertices and m edges, can we make G a unit interval graph by at most k modifications [7,9]. In particular, edge additions (completion) and edge deletions are used to fix false negatives and false positives respectively, while vertex deletions can be viewed as the elimination of outliers. We have thus three variants, all known to be NP-complete [7,10,16].

The problems unit interval completion and unit interval vertex deletion were known to be fixed-parameter tractable (FPT) [1,9]. We show that the edge deletion variant is FPT as well.

Theorem 1. *The problems unit interval vertex deletion and unit interval edge deletion can be solved in time $O(6^k \cdot (n + m))$ and $O(4^k \cdot (n + m))$ respectively.*

Work partially done at Institute for Computer Science and Control, Hungarian Academy of Sciences, supported by ERC 280152 and OTKA NK105645.

M.M. Halldórsson et al. (Eds.): ICALP 2015, Part I, LNCS 9134, pp. 306–317, 2015.
DOI: 10.1007/978-3-662-47672-7_25

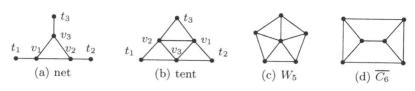

Fig. 1. Small forbidden induced graphs

The algorithm for unit interval vertex deletion significantly improves the currently best-known result, which takes $O(6^k \cdot n^6)$ time [8]. We also improve the running time of the $O(n^7)$-time approximation algorithm [8] to $O(nm)$, while preserving the approximation ratio 6.

We leave it as an open problem the existence of polynomial kernels of the unit interval edge deletion problem. Recall that polynomial kernels were known for unit interval completion [9] and unit interval vertex deletion [6]. However, the kernel size $O(k^{53})$ for unit interval vertex deletion is way too large, so further efforts are needed to make it reasonably small.

We further consider the unit interval editing problem, which, given a graph G, asks whether there are a set V_- of at most k_1 vertices, a set E_- of at most k_2 edges, and a set E_+ of at most k_3 non-edges, such that the deletion of V_- and E_- and the addition of E_+ make G a unit interval graph. This formulation generalizes all the three single-type modifications, and is also natural from the aspect of the aforementioned applications for de-noising data, where different types of errors are commonly found coexisting. Indeed, the assumption that the input data contains only a single type of errors is somewhat counterintuitive. We show that this general editing problem is also FPT, parameterized by the total number of allowed operations.

Theorem 2. *The unit interval editing problem can be solved in* $2^{O(k \log k)} \cdot (n+m)$ *time, where* $k := k_1 + k_2 + k_3$.

The study of general editing problems was initiated by Cai [3], who observed that the problem is FPT if the objective graph class has a finite number of minimal forbidden induced subgraphs. More challenging is to devise parameterized algorithms for those graph classes whose minimal forbidden induced subgraphs are not finite. Prior to this paper, the only known nontrivial graph class on which the general editing problem is FPT is the chordal graphs [5]. We extend this territory by including another well-studied graph class. As a corollary, Thm. 2 implies the fixed-parameter tractability of the unit interval edge editing problem [2], which allows both edge operations but not vertex deletions.

It is known that a graph is a unit interval graph if and only if it contains no claw, net, tent, (see Fig. 1,) or any hole (i.e., a cycle induced by at least four vertices) [15]. Unit interval graphs are thus a subclass of chordal graphs, which are those graphs containing no holes. Modification problems to chordal graphs and unit interval graphs are among the earliest studied problems in parameterized computation, and their study had been closely related. For example, the

gorithm for unit interval completion in [9] is a natural spin-off of their algo-
thm for chordal completion. A better analysis was shortly done by Cai [3], who
lso made explicit the use of bounded-search tree in disposing of finite forbidden
iduced subgraphs. This observation and the parameterized algorithm of [13] for
ie chordal vertex deletion problem immediately imply that unit interval vertex
eletion is FPT. However, neither approach can be adapted to the edge deletion
ersion in a simple way. Compared to completion that needs to add $\Omega(\ell)$ edges
ɔ fill a C_ℓ (i.e., a hole of length ℓ) in, an arbitrarily large hole can be fixed
y a single edge deletion. On the other hand, the deletion of vertices leaves an
iduced subgraph, which allows us to focus on holes once all claws, nets, and
ents have been eliminated; however, the deletion of edges to fix holes of a {claw,
.et, tent}-free graph may introduce new claw(s), net(s), and/or tent(s). There-
ɔre, it is not obvious how to use the parameterized algorithm for chordal edge
.eletion [13] to solve the unit interval edge deletion problem.

Direct algorithms for unit interval vertex deletion were later discovered [1,8],
oth of which use a two-phase approach. The first phase of their algorithms
reaks all forbidden induced subgraphs on at most six vertices. Although this
hase is conceptually intuitive, how to efficiently carry it out is rather nontrivial,
.g., the simple brute-force way used by [1,8] introduces an n^6 factor to the
unning time. Their approaches diverse completely in the second phase. A high-
omplexity procedure was used in [1], while [8] showed that a graph after the
irst phase is a proper circular-arc graph, on which the problem is linear-time
olvable.

Although the algorithm of [8] is nice and simple, its self-contained proof is
xcruciatingly complex. We revisit the relation between unit interval graphs and
ome subclasses of proper circular-arc graphs, and study it in a structured way.
n particular, we observe that unit interval graphs are precisely the intersection
ɔf chordal graphs and proper Helly circular-arc graphs. They inspire us to show
.hat a connected {claw, net, tent, C_4, C_5}-free graph is a proper Helly circular-
irc graph. Hence, the unit interval vertex deletion problem can be solved in
inear time on them [8]. Likewise, using the structural properties of proper Helly
:ircular-arc graphs, we can derive a linear-time algorithm for unit interval edge
leletion on them. By further characterizing connected {claw, net, tent, C_4}-free
graphs that are not proper Helly circular-arc graphs, we are able to show a
stronger result.

Theorem 3. *The problems unit interval vertex deletion and unit interval edge
deletion can be solved in $O(n + m)$ time on {claw, net, tent, C_4}-free graphs.*

[t is then quite simple to use bounded-search tree to develop the parameterized
algorithms stated in Thm. 1, though some nontrivial analysis is required to
ɔbtain the time bound for unit interval edge deletion.

Our algorithm for unit interval editing also uses the two-phase approach.
However, we are not able to show that it can be solved in polynomial time on
proper Helly circular-arc graphs. Therefore, in the first phase, we do away with
not only claws, nets, nets, and C_4's, also all holes of length at most $k_3 + 3$, in all

possible ways. The high exponential factor in the running time is purely due to this phase. After that, every hole has length at least $k_3 + 4$, and has to be fixed by vertex or edge deletions. We show that an inclusion-wise minimal solution of this reduced graph does not add edges, and the problem can then be solved in linear time.

2 {Claw, net, tent, C_4}-free Graphs

All graphs discussed in this paper are undirected and simple. All input graphs are assumed to be connected, hence $n = O(m)$. If we add a new vertex to a C_ℓ and make it adjacent to no or all vertices in the hole, then we end with a C_ℓ^* or W_ℓ, respectively. The complement graph of a graph G is denoted by $\overline{G}$.

An interval graph is the intersection graph of a set of intervals on the real line. A natural way to extend interval graphs is to use arcs and a circle in the place of intervals and the real line, and the intersection graph of arcs on a circle is a *circular-arc graph*. The set of intervals or arcs is called an *interval model* or *arc model* respectively, and it can be specified by their $2n$ endpoints. In a *unit interval* or a *unit arc model*, every interval or arc has length 1. An interval or arc model is *proper* if no interval or arc in it properly contains another interval or arc. A graph is a *unit/proper interval/circular-arc graph* if has a unit/proper interval/arc model respectively.

Clearly, any (unit/proper) interval model can be viewed as a (unit/proper) arc model with some point uncovered, and hence all (unit/proper) interval graphs are always (unit/proper) circular-arc graphs. A unit interval/arc model is necessarily proper, but the other way does not hold true in general. A well-known result states that a proper interval model can always be made unit, and thus these two graph classes coincide [14,15]. This fact will be heavily used in the present paper; e.g., most of our proofs consist in modifying a proper arc model into a proper interval model, which represents the desired unit interval graph. On the other hand, the class of unit circular-arc graphs is only a proper subclass of proper circular-arc graphs, evidenced by, say, the tent. An arc model is *Helly* if every set of pairwise intersecting arcs has a common intersection. A circular-arc graph is *proper Helly* if it has an arc model that is both proper and Helly. The set of minimal forbidden induced subgraphs of proper Helly circular-arc graphs includes claw, net, tent, W_4, W_5, $\overline{C_6}$, and C_ℓ^* for all $\ell \geq 4$ [11]. As a consequence, if a proper Helly circular-arc graph is chordal, then it is a unit interval graph.

Let $\mathcal{F}$ denote the set {claw, net, tent, C_4}. We use *fat W_5* to denote a graph obtained from a W_5 by replacing each of its six vertices by a distinct clique, where five cliques make its *fat hole*, and the other clique is its *hub*.

Theorem 4. *Let G be a connected graph.*

(1) If G is $\mathcal{F}$-free, then it is either a fat W_5 or a proper Helly circular-arc graph.

(2) In $O(m)$ time we can detect an induced subgraph in $\mathcal{F}$, partition $V(G)$ into 6 cliques constituting a fat W_5, or build a proper and Helly arc model for G.

he reader is referred to the full version of this paper [4] for more discussions as
ell as all proofs omitted here.

A unit interval model is always a proper and Helly arc model, but a unit
nterval graph might have an arc model that is neither proper nor Helly. On
he other hand, if a proper Helly circular-arc graph G is not chordal, then the
et of arcs for vertices in a hole necessarily covers the circle, and it is minimal.
nterestingly, the converse holds true as well.

Proposition 1. *[11] In a proper and Helly arc model for a non-chordal graph,
minimal set of arcs whose union covers the circle corresponds to a hole.*

Proposition 1 forbids a proper and Helly arc model to have two or three arcs
hat cover the entire circle (i.e., the model must be normal and Helly, though not
ecessarily proper). These observations enable us to find a shortest hole from
proper Helly circular-arc graph by finding a minimal set of arcs covering the
ircle. This is another important step of our algorithm for unit interval editing.

Lemma 1. *There is an $O(m)$-time algorithm for finding a shortest hole of a
proper Helly circular-arc graph.*

In this paper, all intervals and arcs are closed, and no distinct intervals or
rcs are allowed to share an endpoint. In an interval model, the interval I_v for
ertex v is given by $[\text{lp}(v), \text{rp}(v)]$, where $\text{lp}(v) < \text{rp}(v)$ are its *left and right
endpoints* respectively. In an arc model, the arc A_v for vertex v is given by
$[\text{ccp}(v), \text{cp}(v)]$, where $\text{ccp}(v)$ and $\text{cp}(v)$ are its *counterclockwise and clockwise
endpoints* respectively. All points in an arc model are assumed to be nonnegative.
We point out that possibly $\text{ccp}(v) > \text{cp}(v)$; such an arc A_v necessarily passes
hrough the point 0. We say that an arc model is *canonical* if the perimeter of
he circle is $2n$, and every endpoint is a different integer in $\{0, 1, \ldots, 2n - 1\}$.

Each point α in an interval model $\mathcal{A}$ or arc model $\mathcal{I}$ defines a clique, denoted
by $K_{\mathcal{A}}(\alpha)$ or $K_{\mathcal{I}}(\alpha)$ respectively, which is the set of vertices whose intervals
or arcs contain α. There are at most $2n$ distinct cliques defined as such. For
any point ρ, we can find a positive value ϵ such that the only possible endpoint
n $[\rho - \epsilon, \rho + \epsilon]$ is ρ. Here the value of ϵ should be understood as a function—
depending on the model as well as the point ρ—instead of a constant.

In a proper and Helly arc model $\mathcal{A}$ for graph G, if $uv \in E(G)$, then either
ccp(v) or $\text{cp}(v)$ (but not both) is contained in A_u. Thus, we can define for them
a left-right relation, which can be understood from the viewpoint of an observer
placed at the center of the model. We say that arc A_v intersects arc A_u from
the left when $\text{cp}(v) \in A_u$, denoted by $v \to u$.

3   Vertex Deletion

A set V_- of vertices is a *hole cover* of G if $G - V_-$ is chordal. The hole covers of
proper Helly circular-arc graphs are characterized by the following lemma.

Lemma 2. *Let $\mathcal{A}$ be a proper and Helly arc model for a non-chordal graph G.
A set V_- is a hole cover of G iff it contains $K_{\mathcal{A}}(\alpha)$ for some point α in $\mathcal{A}$.*

It is easy to verify that in a fat W_5, it suffices to delete the clique from the fat hole with the minimum size. Therefore, Thm. 4 and Lem. 2 imply the following.

Corollary 1. *The unit interval vertex deletion problem can be solved in $O(m)$ time (1) on proper Helly circular-arc graphs and (2) on $\mathcal{F}$-free graphs.*

Our parameterized algorithm calls Thm. 4(2), and then based on the outcome it solves the problem by making recursive calls to itself, or calling the algorithm of Cor. 1. Note that a subgraph in $\mathcal{F}$ has at most 6 vertices, and at least one of them needs to be deleted. For each subgraph, at most 6 recursive calls are made. For the approximation algorithm, instead of branching, we delete all vertices of a subgraph in $\mathcal{F}$.

Lemma 3. *There are an $O(6^k \cdot m)$-time algorithm for unit interval vertex deletion and an $O(nm)$-time 6-approximation algorithm for its minimization version.*

4   Edge Deletion

Our algorithm for unit interval edge deletion goes similarly as Lem. 3. Therefore, the focus of this section is on the disposal of $\mathcal{F}$-free graphs, i.e., the proof of the second part of Thm. 3. Let G be a proper Helly circular-arc graph. For each point α in a proper and Helly arc model $\mathcal{A}$ for G, we can define the following set of edges:

$$\overrightarrow{E}_{\mathcal{A}}(\alpha) = \{vu : v \in K_{\mathcal{A}}(\alpha), u \notin K_{\mathcal{A}}(\alpha), v \to u\}.$$

It is easy to verify that the following gives a proper interval model for $G - \overrightarrow{E}_{\mathcal{A}}(0)$:

$$I_v := \begin{cases} [\mathsf{ccp}(v), \mathsf{cp}(v) + \ell] & \text{if } v \in K_{\mathcal{A}}(0), \\ [\mathsf{ccp}(v), \mathsf{cp}(v)] & \text{otherwise,} \end{cases}$$

where ℓ is the perimeter of the circle in $\mathcal{A}$. For an arbitrary point α, the model $G - \overrightarrow{E}_{\mathcal{A}}(\alpha)$ can be given analogously, e.g., we may rotate the model first to make $\alpha = 0$. Note that rotating all arcs in the model does not change the intersections among them.

Proposition 2. *Let $\mathcal{A}$ be a proper and Helly arc model for a non-chordal graph G. For any point α in $\mathcal{A}$, the subgraph $G - \overrightarrow{E}_{\mathcal{A}}(\alpha)$ is a unit interval graph.*

The other direction is more involved and more challenging. For two disjoint sets X, Y of vertices, let $E_G(X, Y)$ denote the set of edges of G that has one end in X and the other end in Y, i.e., $(X \times Y) \cap E(G)$. A unit interval graph $\underline{G}$ is called a *(spanning) unit interval subgraph* of G if $V(\underline{G}) = V(G)$ and $E(\underline{G}) \subseteq E(G)$; it is called *maximum* if it has the largest number of edges among all unit interval subgraphs of G. To prove all maximum unit interval subgraphs have a certain property, we use the following argument by contradiction. Given a unit interval subgraph $\underline{G}$ not having the property, we locally modify a unit interval model $\mathcal{I}$ for $\underline{G}$ to a *proper* interval model $\mathcal{I}'$ such that the represented graph $\underline{G}'$ satisfies $E(\underline{G}') \subseteq E(G)$ and $|E(\underline{G}')| > |E(\underline{G})|$.

Lemma 4. *Let $\mathcal{A}$ be a proper and Helly arc model for a non-chordal graph G. For any maximum unit interval subgraph $\underline{G}$ of G, the deleted edges $E_- := E(G) \setminus E(\underline{G})$ is $\overrightarrow{E}_{\mathcal{A}}(\rho)$ for some point ρ in $\mathcal{A}$.*

Proof. We fix a unit interval model $\mathcal{I}$ for $\underline{G}$. Let v be the vertex with the leftmost interval $[0,1]$ in $\mathcal{I}$. All arcs in the proof are referred to the model $\mathcal{A}$ for G. Denote by u and w the vertices of $N_{\underline{G}}[v]$ that have the leftmost and the rightmost arcs respectively; possibly $u = v$ and/or $w = v$. Note that $N_{\underline{G}}[v] = K_{\mathcal{I}}(1)$, which is a clique; hence, if $u \neq w$, then $uw \in E(\underline{G}) \subset E(G)$, and in particular, $u \to w$. Let $\alpha := \mathrm{ccp}(u) - \epsilon$ and $\beta := \mathrm{cp}(w) + \epsilon$; for notational convenience, we may assume that $[\alpha, \beta]$ does not cover the point 0 (the union of A_u, A_v, and A_w does not cover the circle). Note that an arc covering α or β has to intersect A_u or A_w respectively. Thus, since the model is proper and by Prop. 1, no arc contains both α and β. On the other hand, the arc A_x for any $x \in N_{\underline{G}}[v]$ is in (α, β). Therefore, $K_{\mathcal{A}}(\alpha)$, $K_{\mathcal{A}}(\beta)$, and $N_{\underline{G}}[v]$ are pairwise disjoint.

We argue first that $K_{\mathcal{A}}(\alpha)$ and $K_{\mathcal{A}}(\beta)$ cannot be both adjacent to $N_{\underline{G}}[v]$ in $\underline{G}$. This holds vacuously if $N_{\underline{G}}[v]$ is a single component of $\underline{G}$. Hence we assume otherwise: let I_x be the first interval with $\mathrm{lp}(x) > \mathrm{rp}(v)$, then $\mathrm{lp}(x) < 2$ and $N_{\underline{G}}(x)$ intersects $N_{\underline{G}}[v]$. Note that $x \notin N_{\underline{G}}[v]$, as otherwise setting I_v to $[\mathrm{lp}(x) + -1, \mathrm{lp}(x) + \epsilon]$ gives a spanning unit interval subgraph of G with edges $E(\underline{G}) \cup \{vx\}$, a contradiction. Hence, x is in either $K_{\mathcal{A}}(\alpha)$ or $K_{\mathcal{A}}(\beta)$. Assume first that $x \in K_{\mathcal{A}}(\alpha) \setminus N_G[v]$ and we show that $K_{\mathcal{A}}(\beta)$ is nonadjacent to $N_{\underline{G}}[v]$ in $\underline{G}$. Suppose for contradiction, that there is some vertex $y \in K_{\mathcal{A}}(\beta)$ such that I_y intersects I_z for $z \in N_{\underline{G}}[v]$. Then $\mathrm{lp}(y) < \mathrm{rp}(z) < 2 < \mathrm{rp}(x)$, which means that I_y intersects I_x as well (noting $\mathrm{lp}(x) < \mathrm{lp}(y)$). As a result, $xy \in E(\underline{G}) \subseteq E(G)$, and A_x intersects A_y; by the selection of α and β, we must have $y \to x$, but then A_x, A_y, and A_z do not satisfy the Helly property, a contradiction. A symmetric argument implies that $K_{\mathcal{A}}(\alpha)$ is nonadjacent to $N_{\underline{G}}[v]$ in $\underline{G}$ when $z \in K_{\mathcal{A}}(\beta) \setminus N_G[v]$.

Assume without loss of generality that $K_{\mathcal{A}}(\alpha)$ is not adjacent to $N_{\underline{G}}[v]$ in $\underline{G}$. Let $\mu := \mathrm{cp}(u)$; note that $\mu \in A_v$ and $K_{\mathcal{A}}(\mu) \subseteq N_{\underline{G}}[v]$. Since the model $\mathcal{A}$ is proper and Helly, no arc in $\mathcal{A}$ can contain both α and μ. Therefore, $K_{\mathcal{A}}(\alpha)$ and $K_{\mathcal{A}}(\mu)$ are disjoint, and from the definition of α, we can conclude that

$$\overrightarrow{E}_{\mathcal{A}}(\alpha) = E_G\big(K_{\mathcal{A}}(\alpha), K_{\mathcal{A}}(\mu)\big).$$

Let $E_- := E(G) \setminus E(\underline{G})$; by Prop. 2, $|E_-| \leq |\overrightarrow{E}_{\mathcal{A}}(\alpha)|$. We argue that they have to be equal. Suppose for contradiction, $E_- \neq \overrightarrow{E}_{\mathcal{A}}(\alpha)$, then $\overrightarrow{E}_{\mathcal{A}}(\alpha) \nsubseteq E_-$. There must be some vertices in $K_{\mathcal{A}}(\mu)$ that are adjacent to $K_{\mathcal{A}}(\alpha)$ in $\underline{G}$; by assumption (that $K_{\mathcal{A}}(\alpha)$ is not adjacent to $N_{\underline{G}}[v]$ in $\underline{G}$), these vertices are not in $N_{\underline{G}}[v]$. Let $X := K_{\mathcal{A}}(\mu) \setminus N_{\underline{G}}[v]$. We take a vertex $x \in X$ such that $N_{\underline{G}}(x) \cap K_{\mathcal{A}}(\alpha)$ has the largest cardinality, which is positive. Recall that $u \in N_{\underline{G}}[v]$; hence $x \neq u$ and $u \to x$. We may assume that $\mathrm{lp}(x)$ is contained in some interval for a vertex in $K_{\mathcal{A}}(\alpha)$, and the other case follows by symmetry. Note that $N_G(x) \setminus N_G(u)$ is disjoint from $K_{\mathcal{A}}(\alpha)$, and by the Helly property, it cannot be adjacent to $N_G(x) \cap K_{\mathcal{A}}(\alpha)$. Therefore, for any $y \in N_{\underline{G}}(x) \setminus N_G(u) \subseteq N_G(x) \setminus N_G(u)$, the interval I_y has to contain $\mathtt{rp}(x)$; in other words, $\mathtt{lp}(y) \in I_x$. Let

$$\gamma := \begin{cases} \mathtt{rp}(x) & \text{if } N_{\underline{G}}(x) \setminus N_G(u) = \emptyset, \\ \min_{y \in N_{\underline{G}}(x) \setminus N_G(u)} \mathtt{lp}(y) & \text{otherwise.} \end{cases}$$

Setting $I'_u = [\min_{y:\gamma \in I_y} \mathtt{lp}(y) - \epsilon, \gamma - \epsilon]$ gives also a proper interval model $\mathcal{I}'$ To see that $\mathcal{I}'$ represents a subgraph of G, note that $N_G[X] \cap K_{\mathcal{A}}(\alpha) \subseteq N_G(u)$ Since $\underline{G}$ is a maximum spanning unit interval subgraph of G, it follows that $|N_{\underline{G}}(x) \cap K_{\mathcal{A}}(\alpha)| \le |N_{\underline{G}}(u)|$. Likewise, since $N_{\underline{G}}(u) \subseteq N_G[v]$, it follows that $N_{\underline{G}}[u] = N_{\underline{G}}[v]$ (otherwise we can set $I'_v = [\mathtt{lp}(u) - \epsilon, \mathtt{rp}(u) - \epsilon]$ to get a larger spanning unit interval subgraph of G). Therefore, for every $x' \in X$, it holds that

$$|N_{\underline{G}}(x') \cap K_{\mathcal{A}}(\alpha)| \le |N_{\underline{G}}(x) \cap K_{\mathcal{A}}(\alpha)| \le |N_{\underline{G}}(u)| < |N_{\underline{G}}[u]| = |N_{\underline{G}}[v]|.$$

The first inequality is ensured by the selection of x. However, noting that $N_{\underline{G}}[v] \subseteq N_G(x')$ for every $x' \in X$, it can be inferred

$$\begin{aligned}
|E_-| &\ge |E_G(K_{\mathcal{A}}(\alpha), N_{\underline{G}}[v])| + |E_G(X, N_{\underline{G}}[v])| + |E_G(K_{\mathcal{A}}(\alpha), X) \setminus E_{\underline{G}}(K_{\mathcal{A}}(\alpha), X)| \\
&= |E_G(K_{\mathcal{A}}(\alpha), N_{\underline{G}}[v])| + |E_G(K_{\mathcal{A}}(\alpha), X)| + |E_G(X, N_{\underline{G}}[v])| - |E_{\underline{G}}(K_{\mathcal{A}}(\alpha), X)| \\
&= |E_G(K_{\mathcal{A}}(\alpha), N_{\underline{G}}[v])| + |E_G(K_{\mathcal{A}}(\alpha), X)| + \sum_{x' \in X} (|N_{\underline{G}}[v]| - |N_{\underline{G}}(x') \cap K_{\mathcal{A}}(\alpha)|) \\
&> |E_G(K_{\mathcal{A}}(\alpha), N_{\underline{G}}[v])| + |E_G(K_{\mathcal{A}}(\alpha), X)| \\
&= |\overrightarrow{E}_{\mathcal{A}}(\alpha)|,
\end{aligned}$$

which contradicts Prop. 2. Thus, $E_- = \overrightarrow{E}_{\mathcal{A}}(\alpha)$, and this concludes the proof.
$\square$

There is a linear number of different places to check, and thus the edge deletion problem can also be solved in linear time on proper Helly circular-arc graphs. The problem is also simple on fat W_5's.

Theorem 5. *The unit interval edge deletion problem can be solved in $O(m)$ time (1) on proper Helly circular-arc graphs and (2) on $\mathcal{F}$-free graphs.*

Proof. For (1), we may assume that the input graph G is not an unit interval graph; it is then connected and not chordal. We build a proper and Helly arc model $\mathcal{A}$ for G; without loss of generality, assume that it is canonical. According to Lem. 4, the problem reduces to finding a point α in $\mathcal{A}$ such that $\overrightarrow{E}_{\mathcal{A}}(\alpha)$ is minimized. It suffices to consider the $2n$ points $i + 0.5$ for $i \in \{0, \ldots, 2n - 1\}$. We calculate first $\overrightarrow{E}_{\mathcal{A}}(0.5)$, and then for $i = 1, \ldots, 2n - 1$, we deduce $\overrightarrow{E}_{\mathcal{A}}(i + 0.5)$ from $\overrightarrow{E}_{\mathcal{A}}(i - 0.5)$ as follows. If i is a clockwise endpoint of some arc, then $\overrightarrow{E}_{\mathcal{A}}(i + 0.5) = \overrightarrow{E}_{\mathcal{A}}(i - 0.5)$. Otherwise, $i = \mathtt{ccp}(v)$ for some vertex v, then the difference between $\overrightarrow{E}_{\mathcal{A}}(i + 0.5)$ and $\overrightarrow{E}_{\mathcal{A}}(i - 0.5)$ is the set of edges incident to v. In particular, $\{uv : u \to v\} = \overrightarrow{E}_{\mathcal{A}}(i - 0.5) \setminus \overrightarrow{E}_{\mathcal{A}}(i + 0.5)$, while $\{uv : v \to u\} = \overrightarrow{E}_{\mathcal{A}}(i + 0.5) \setminus \overrightarrow{E}_{\mathcal{A}}(i - 0.5)$. Note that the initial value $\overrightarrow{E}_{\mathcal{A}}(0.5)$ can be alculated in $O(m)$ time, and then each vertex and its adjacency list is scanned xactly once. It follows that the total running time is $O(m)$.

For (2), we may assume that the input graph G is connected, as otherwise e work on its components one by one. According to Thm. 4(1), G is either a roper Helly circular-arc graph or a fat W_5. The former case has been considered bove, and now assume G is a fat W_5. Let $K_0, \ldots, K_4$ be the five cliques in the it hole, and let K_5 be the hub. Consider a pair of vertices u, v in K_i, where $\in \{0, \ldots, 5\}$. By definition, $N_G[u] = N_G[v]$. We argue that $N_{\underline{G}}[u] = N_{\underline{G}}[v]$ for ny maximum spanning unit interval subgraph $\underline{G}$ of G. Suppose the contrary, nen setting I_u to I_v or I_v to I_u will end with a spanning unit interval subgraph f G with strictly more edges than $\underline{G}$. Therefore, we need to delete $E_G(K_i, K_{i+1})$ s well as $E_G(K_5, K_i)$ or $E_G(K_5, K_{i+1})$ for some $i \in \{0, \ldots, 4\}$ (all subscripts re modulo 5). Once the sizes of all six cliques have been calculated, which can e done in $O(m)$ time, the minimum set of edges can be decided in constant ime. Therefore, the total running time is $O(m)$. The proof is now complete. □

Theorems 4 and 5 already imply a bound-search tree algorithm for the unit nterval edge deletion problem running in time $O(9^k \cdot m)$. Here the constant is decided by the tent, which has 9 edges. However, a closer look at it tells s that deleting one edge from a tent introduces a claw or C_4, which forces us o delete some other edge(s). The disposal of a net is similar. This observation nd a refined analysis will yield the running time claimed in Theorem 1. What lominates the branching step is the disposal of C_4's. With the technique the uthor developed in [12], one may (slightly) improve the runtime to $O(c^k \cdot m)$ or some constant $c < 4$.

5   General Editing

et $V_- \subseteq V(G)$, and let E_- and E_+ be a set of edges and a set of non-edges f $G - V_-$ respectively. We say that (V_-, E_-, E_+) is an *editing set* of G if the leletion of E_- from and the addition of E_+ to $G - V_-$ create a unit interval graph. Its *size* is defined to be the 3-tuple $(|V_-|, |E_-|, |E_+|)$, and we say that it s *smaller* than (k_1, k_2, k_3) if all of $|V_-| \leq k_1$ and $|E_-| \leq k_2$ and $|E_+| \leq k_3$ hold rue and at least one inequality is strict. The unit interval editing problem is ormally defined as follows.

> *Input:* A graph G and three nonnegative integers k_1, k_2, and k_3.
> *Task:* Either construct an editing set (V_-, E_-, E_+) of G that has size
>   at most (k_1, k_2, k_3), or report that no such set exists.

By and large, our algorithm for the unit interval editing also uses the same wo-phase approach as the previous algorithms. The main discrepancy lies in the irst phase, when we are not satisfied with an $\mathcal{F}$-free graph; in particular, we also vant to do away with all holes of length at most $k_3 + 3$, which are precisely those ioles fixable by merely adding edges. In the very special cases where $k_3 = 0$ or 1, a fat W_5 satisfies these conditions. It is not hard to solve the problem on fat W_5's, but to make the rest of the section more focused and to simplify the presentation, we exclude these cases by disposing of all C_5's in the first phase.

A graph is called *reduced* if it contains no claw, net, tent, C_4, C_5, or C_ℓ with $\ell \leq k_3 + 3$. A reduced graph G is a proper Helly circular-arc graph, and hence if it happens to be chordal, then it must be a unit interval graph, and we terminate the algorithm. Otherwise, our algorithm enters the second phase. Now that G is reduced, every minimal forbidden induced subgraph is a hole C_ℓ with $\ell > k_3 + 3$, which can only be fixed by deleting vertices and/or edges. Here we again exploit a proper and Helly arc model $\mathcal{A}$ for G. According to Lem. 2, if there exists some point ρ in the model such that $|K_{\mathcal{A}}(\rho)| \leq k_1$, then it suffices to return $(K_{\mathcal{A}}(\rho), \emptyset, \emptyset)$ as the solution. Therefore, we may assume hereafter that no such point exists, then G remains reduced and non-chordal after at most k_1 vertex deletions. As a result, we have to delete edges as well.

Consider an (inclusion-wise minimal) editing set (V_-, E_-, E_+) to a reduced graph G. It is easy to verify that $(\emptyset, E_-, E_+)$ is an (inclusion-wise minimal) editing set of the reduced graph $G - V_-$. In particular, E_- intersects all holes of $G - V_-$. We use $\mathcal{A} - V_-$ as a shorthand for $\{A_v \in \mathcal{A} : v \notin V_-\}$. One may want to use Lem. 4 to find a minimum set E_- of edges (i.e., $\overrightarrow{E}_{\mathcal{A}-V_-}(\alpha)$ for some point α) to finish the task. However, Lem. 4 has not ruled out the possibility that we delete less edges to break all long holes, and subsequently add edges to fix the incurred subgraphs in $\{$claw, net, tent, C_4, C_5, $C_\ell\}$ with $\ell \leq k_3 + 3$. So we need the following lemma.

Lemma 5. *Let (V_-, E_-, E_+) be an inclusion-wise minimal editing set of a reduced graph G. If $|E_+| \leq k_3$, then $E_+ = \emptyset$.*

Proof. We may assume without loss of generality $V_- = \emptyset$, as otherwise it suffices to consider the inclusion-wise minimal editing set $(\emptyset, E_-, E_+)$ to the still reduced graph $G - V_-$. Let $\mathcal{A}$ be a proper and Helly arc model for G. Let E'_- be an inclusion-wise minimal subset of E_- such that for every hole in $G - E'_-$, the union of arcs for its vertices does not cover the circle of $\mathcal{A}$. Note that E'_- exists because E_- itself satisfies this condition: suppose that there exists in $G - E_-$ a hole whose arcs cover the circle, then it has at least $k_3 + 4$ vertices (Prop. 1) and cannot be fixed by the addition of E_+. We argue that $\underline{G} := G - E'_-$ is already a unit interval graph. It follows that $E_- = E'_-$ and $E_+ = \emptyset$.

Suppose for contradiction, there is $X \subseteq V(G)$ inducing a claw, net, tent, or a hole in $\underline{G}[X]$. We find three vertices $u, v, w \in X$ such that $uw \in E'_-$ and $uv, vw \in E(\underline{G})$ as follows. Note that $\bigcup_{v \in X} A_v$ cannot cover the whole circle: by assumption, this is true when $\underline{G}[X]$ is a hole; on the other hand, by Lem. 1, and noting that G is $\{C_4, C_5\}$-free, at least 6 arcs are needed to cover the circle (Prop. 1), but a claw, net, or tent has at most 6 vertices, and cannot be a subgraph of a C_6. Thus, $G[X]$ is a unit interval graph. So we can find two vertices x, z from X having $xz \in E'_-$. We find a shortest x-z path in $\underline{G}[X]$. If the path has more than one inner vertex, then it makes a hole together with xz, which means that there exists an inner vertex y of this path such that $xy \in E'_-$ or $yz \in E'_-$. We consider then the new pair x, y or y, z accordingly. Note that

eir distance in $\underline{G}[X]$ is smaller than xz, and hence repeating this argument
t most $|X| - 3$ times) will end with two vertices with distance precisely 2 in
$[X]$. They are the desired u and w, while any common neighbor of them in
$[X]$ can be v. By the minimality of E'_-, in $\underline{G} + uw$ there exists a hole H such
1at arcs for its vertices cover the circle in $\mathcal{A}$. This hole H necessarily passes
w, and we denote it by $x_1 x_2 \cdots x_{\ell-1} x_\ell$, where $x_1 = u$ and $x_\ell = w$. Note that
$_u$ intersects A_w, and since $\mathcal{A}$ is proper and Helly, A_u, A_v, A_w cannot cover the
ircle. From $x_1 x_2 \cdots x_{\ell-1} x_\ell$ we can find p and q such that $1 \le p < q \le \ell$ and
$x_p, vx_q \in E(\underline{G})$ but $vx_i \notin E(\underline{G})$ for every $p < i < q$. Here possibly $p = 1$ and/or
$= \ell$. Then $vx_p \cdots x_q$ makes a hole of $\underline{G}$, and the union of its arcs covers the
ircle, contradicting the definition of E'_-.   □

Therefore, a yes-instance on a reduced graph always has a solution adding
o edges. We present here a stronger algorithmic result on deleting vertices and
dges to make a graph a unit interval graph.

Lemma 6. *Given a proper Helly circular-arc graph G and a nonnegative integer*
, we can calculate in $O(m)$ time the minimum number q such that G has an
diting set of size $(p, q, 0)$. In the same time we can find such an editing set.

Proof. Let us fix a proper and Helly arc model $\mathcal{A}$ for G. We may assume that G
s not chordal and $K_{\mathcal{A}}(\rho) > p$ for any point ρ. Hence, $q > 0$: for any subset V_-
f at most p vertices, $G - V_-$ remains reduced and non-chordal. For each point ρ
n $\mathcal{A}$, we can define an editing set $(V_-, E_-, \emptyset)$ by taking the p vertices in $K_{\mathcal{A}}(\rho)$
vith the rightmost arcs as V_- and $\overrightarrow{E}_{\mathcal{A}-V_-}(\rho)$ as E_-. We argue first that the
ninimum cardinality of this edge set, taken among all points in $\mathcal{A}$ is the desired
number q.

Let $(V^*_-, E^*_-, \emptyset)$ be an editing set of G with size $(p, q, 0)$. According to Lem. 4,
here is a point α such that the deletion of $E'_- := \overrightarrow{E}_{\mathcal{A}-V^*_-}(\alpha)$ from $G - V^*_-$ makes
t a unit interval graph and $|E'_-| \le |E^*_-|$. We now consider the original model $\mathcal{A}$.
Note that a vertex in V^*_- is in either $K_{\mathcal{A}}(\alpha)$ or $\{v \notin K_{\mathcal{A}}(\alpha) : u \to v, u \in K_{\mathcal{A}}(\alpha)\}$;
otherwise replacing this vertex by any end of an edge in E^*_-, and removing
his edge from E^*_- gives an editing set of size $(p, q - 1, 0)$. Let V_- comprise
he $|V^*_- \cap K_{\mathcal{A}}(\alpha)|$ vertices of $K_{\mathcal{A}}(\alpha)$ whose arcs are the rightmost in them, as
vell as the first $|V^*_- \setminus K_{\mathcal{A}}(\alpha)|$ vertices whose arcs are to the right of α. And let
$E_- := \overrightarrow{E}_{\mathcal{A}-V_-}(\alpha)$. It is easy to verify that $|E_-| \le |E^*_-| = q$ and $(V_-, E_-, \emptyset)$ is
also an editing set of G (Lem. 2). Note that arcs for V_- are consecutive in $\mathcal{A}$.
Let v be the vertex in V_- with the rightmost arc, and then $\mathrm{ccp}(v) - \epsilon$ is the
desired point ρ.

We give now the $O(m)$-time algorithm for finding the desired point, for which
ve assume that $\mathcal{A}$ is canonical. It suffices to consider the $2n$ points $i + 0.5$ for
$\in \{0, \ldots, 2n - 1\}$. We calculate first the V_- and E_- for 0.5, and maintain a
queue of p elements, which are the vertices corresponding to the p rightmost
arcs containing 0.5. For $i = 1, \ldots, 2n - 1$, we deduce the new sets for $i + 0.5$
from the previous point as follows. If i is a clockwise endpoint of some arc,
chen both of them do not change. Otherwise, $i = \mathrm{ccp}(v)$ for some vertex v,
chen we enqueue v, and dequeue u, and the difference between $\overrightarrow{E}_{\mathcal{A}}(i + 0.5)$ and

$\overrightarrow{E}_\mathcal{A}(i-0.5)$ is the number of edges incident to u. In particular, $\{xu : x \to u\} = \overrightarrow{E}_\mathcal{A}(i-0.5) \setminus \overrightarrow{E}_\mathcal{A}(i+0.5)$, while $\{xu : x \to u\} = \overrightarrow{E}_\mathcal{A}(i+0.5) \setminus \overrightarrow{E}_\mathcal{A}(i-0.5)$. Note that the initial value $\overrightarrow{E}_\mathcal{A}(0.5)$ can be found in $O(m)$ time, and then each vertex and its adjacency is scanned exactly once. The total running time is $O(m)$.

The combinatorial characterization on mixed hole covers consisting of both vertices and edges, thereby extending Lems. 2 and 4. Lemmas 5 and 6 have the following consequence: it suffices to call the algorithm with $p = k_1$, and returns the found editing set if $q \le k_2$, or "NO" otherwise.

Corollary 2. *The unit interval editing problem can be solved in $O(m)$ time on reduced graphs.*

Putting together these steps, Thm. 2 follows.

References

1. van Bevern, R., Komusiewicz, C., Moser, H., Niedermeier, R.: Measuring indifference: unit interval vertex deletion. In: Thilikos, D.M. (ed.) WG 2010. LNCS, vol. 6410, pp. 232–243. Springer, Heidelberg (2010)
2. Burzyn, P., Bonomo, F., Durán, G.: NP-completeness results for edge modification problems. Discrete Appl. Math. **154**(13), 1824–1844 (2006)
3. Cai, L.: Fixed-parameter tractability of graph modification problems for hereditary properties. Inf. Proc. Letters **58**(4), 171–176 (1996)
4. Cao, Y.: Unit interval editing is fixed-parameter tractable. arXiv:1504.04470 (2015)
5. Cao, Y., Marx, D.: Chordal editing is fixed-parameter tractable. In: Mayr, E.W., Portier, N. (eds.) STACS. LIPIcs, vol. 25, pp. 214–225. Schloss Dagstuhl (2014)
6. Fomin, F.V., Saurabh, S., Villanger, Y.: A polynomial kernel for proper interval vertex deletion. SIAM J. Discr. Math. **27**(4), 1964–1976 (2013)
7. Goldberg, P.W., Golumbic, M.C., Kaplan, H., Shamir, R.: Four strikes against physical mapping of DNA. J. Comput. Biol. **2**(1), 139–152 (1995)
8. van't Hof, P., Villanger, Y.: Proper interval vertex deletion. Algorithmica **65**(4), 845–867 (2013)
9. Kaplan, H., Shamir, R., Tarjan, R.E.: Tractability of parameterized completion problems on chordal, strongly chordal, and proper interval graphs. SIAM J. Comput. **28**(5), 1906–1922 (1999)
10. Lewis, J.M., Yannakakis, M.: The node-deletion problem for hereditary properties is NP-complete. J. Comput. System Sci. **20**(2), 219–230 (1980)
11. Lin, M.C., Soulignac, F.J., Szwarcfiter, J.L.: Normal Helly circular-arc graphs and its subclasses. Discrete Appl. Math. **161**(7–8), 1037–1059 (2013)
12. Liu, Y., Wang, J., You, J., Chen, J., Cao, Y.: Edge deletion problems: Branching facilitated by modular decomposition. Theor. Comp. Sci. **573**, 63–70 (2015)
13. Marx, D.: Chordal deletion is fixed-parameter tractable. Algorithmica **57**(4),break 747–768 (2010)
14. Roberts, F.S.: Indifference graphs. In: Harary, F. (ed.) Proof Techniques in Graph Theory, pp. 139–146. Academic Press, New York (1969)
15. Wegner, G.: Eigenschaften der Nerven homologisch-einfacher Familien im R^n. Ph.D. thesis, Universität Göttingen (1967)
16. Yannakakis, M.: Computing the minimum fill-in is NP-complete. SIAM J. Alg. Discrete Methods **2**(1), 77–79 (1981)

Streaming Algorithms for Submodular Function Maximization

Chandra Chekuri, Shalmoli Gupta, and Kent Quanrud$^{(\boxtimes)}$

Department of Computer Science, University of Illinois, Urbana, IL 61801, USA
{chekuri,sgupta49,quanrud2}@illinois.edu

Abstract. We consider the problem of maximizing a nonnegative submodular set function $f : 2^{\mathcal{N}} \to \mathbb{R}^+$ subject to a p-matchoid constraint in the single-pass streaming setting. Previous work in this context has considered streaming algorithms for modular functions and monotone submodular functions. The main result is for submodular functions that are *non-monotone*. We describe deterministic and randomized algorithms that obtain a $\Omega(\frac{1}{p})$-approximation using $O(k \log k)$-space, where k is an upper bound on the cardinality of the desired set. The model assumes value oracle access to f and membership oracles for the matroids defining the p-matchoid constraint.

1 Introduction

Let $f : 2^{\mathcal{N}} \to \mathbb{R}$ be a set function defined over a ground set $\mathcal{N}$. f is *submodular* if it exhibits decreasing marginal values in the following sense: if $e \in \mathcal{N}$ is any element, and $A, B \subseteq \mathcal{N}$ with $A \subseteq B$ are any two nested sets, then $f(A + e) - f(A) \geq f(B + e) - f(B)$. The gap $f(A + e) - f(A)$ is called the *marginal value* of e with respect to f and A, and denoted $f_A(e)$. An equivalent characterization for submodular functions is that for any two sets $A, B \subseteq \mathcal{N}$, $f(A \cup B) + f(A \cap B) \leq f(A) + f(B)$.

Submodular functions play a fundamental role in classical combinatorial optimization where rank functions of matroids, edge cuts, coverage, and others are instances of submodular functions (see [24,37]). More recently, there is a large interest in constrained submodular function optimization driven both by theoretical progress and a variety of applications in computer science. The needs of the applications, and in particular the sheer bulk of large data sets, have brought into focus the development of fast algorithms for submodular optimization. Recent work on the theoretical side include the development of faster worst-case approximation algorithms in the traditional sequential model of computation [3,13,27], algorithms in the streaming model [2,12] as well as in the map-reduce model of computation [30].

C. Chekuri—Work on this paper supported in part by NSF grant CCF-1319376.
S. Gupta—Work on this paper supported in part by NSF grant CCF-1319376.
K. Quanrud—Work on this paper supported in part by NSF grants CCF-1319376, CCF-1421231, and CCF-1217462.

© Springer-Verlag Berlin Heidelberg 2015
M.M. Halldórsson et al. (Eds.): ICALP 2015, Part I, LNCS 9134, pp. 318–330, 2015.
DOI: 10.1007/978-3-662-47672-7_26

In this paper we consider constrained submodular function *maximization*. The goal is to find $\max_{S\in\mathcal{I}} f(S)$ where $\mathcal{I} \subseteq 2^{\mathcal{N}}$ is a *downward-closed* family of sets; i.e. $A \in \mathcal{I}$ and $B \subseteq A$ implies $B \in \mathcal{I}$. $\mathcal{I}$ is also called an *independence family* and any set $A \in \mathcal{I}$ is called an *independent set*. Submodular maximization under various independence constraints has been extensively studied in the literature. The problem can be easily seen to be NP-hard even for a simple cardinality constraint as it encompasses standard NP-hard problems like the Max-k-cover problem. Constrained submodular maximization has found several new applications in recent years. Some of these include data summarization [17,34,39], influence maximization in social networks [15,16,25,28,38], generalized assignment[10], mechanism design [1], and network monitoring [33].

In some of these applications, the amount of data involved is much larger than the main memory capacity of individual computers. This motivates the design of space-efficient algorithms which can process the data in *streaming* fashion, where only a small fraction of the data is kept in memory at any point. There has been some recent work on submodular function maximization in the streaming model focused on *monotone* functions (i.e. $f(A) \leq f(B)$, whenever $A \subseteq B$). This assumption is restrictive from both a theoretical and practical point of view.

In this paper we present streaming algorithms for non-monotone submodular function maximization subject to various combinatorial constraints, the most general being a p-matchoid. p-matchoid's generalize many basic combinatorial constraints such as the cardinality constraint, the intersection of p matroids, and matchings in graphs and hyper-graphs. A formal definition of a p-matchoid is given in Section 2. We consider the abstract p-matchoid constraint for theoretical reasons, and most constraints in practice should be simpler. We explicitly consider the cardinality constraint and obtain an improved bound.

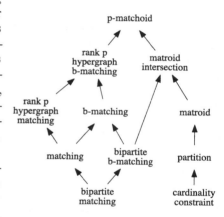

Fig. 1. Hierarchy of set systems

We now describe the problem formally. We are presented a groundset of elements $\mathcal{N} = \{e_1, e_2, \ldots e_n\}$, with no assumption made on the order or the size of the datastream. The goal is to select an independent set $S \subseteq \mathcal{N}$ (where independence is defined by the p-matchoid), which maximizes a nonnegative submodular function f while using as little space as possible. We make the following assumptions: (i) the function f is available via a value oracle, that takes as input a set $S \subseteq \mathcal{N}$ and returns the value $f(S)$; (ii) the independence family $\mathcal{I}$ is available via a membership oracle with some additional information needed in the p-matchoid setting; and (iii) the constraints specify explicitly, and a priori, an upper bound k on the number of elements to be chosen. We discuss these turn. The availability of a value oracle for f is a reasonable and standard assumption in the sequential model of computation, but needs some justification in restrictive models of computation such as streaming where the goal is to store at any point of time only a small subset of the elements of $\mathcal{N}$. Can $f(S)$ be evaluated without having access to all of $\mathcal{N}$? This of course depends on f. [2] gives several examples of interesting and useful functions where this is indeed possible. The second assumption is also reasonable if, as we remarked, the p-matchoid constraint is in practice going to be a simple one that combines basic matroids such as cardinality, partition and laminar matroid constraints that can be specified compactly and implicitly. Finally, the third assumption is guided by the fact that an abstract model of constraints can in principle lead to every element being chosen. In many applications the goal is to select a small and important subset of elements from a much larger set; and it is therefore reasonable to expect knowledge of an upper bound on how many can be chosen. Submodular set functions are ubiquitous and arise explicitly and implicitly in a variety of settings. The model we consider in this paper may not be directly useful in some important scenarios of interest. Nevertheless, the ideas underlying the analysis in the streaming model that we consider here may still be useful in speeding up existing algorithms and/or reduce their space usage.

As is typical for streaming algorithms, we measure performance in four basic dimensions: (i) the approximation ratio $f(S)/$OPT, where S is the output of the algorithm and OPT is the value of an optimal solution; (ii) the space usage of the algorithm; (iii) the update time or the time required to process each stream element; and (iv) the number of passes the algorithm makes over the data stream.

Our Results. We develop randomized and deterministic algorithms that yield an $\Omega(1/p)$-approximation for maximizing a non-negative submodular function under a p-matchoid constraint in the one-pass streaming setting. The space usage is $O(k \log k)$, essentially matching recent algorithms for the simpler setting of maximizing a monotone submodular function subject to a cardinality constraint [2]. The randomized algorithm achieves better constants than the deterministic algorithm. As far as we are aware, we present the first streaming algorithms for non-monotone submodular function maximization under constraints beyond cardinality. We give an improved bound of $\frac{1-\epsilon}{2+e}$ for the cardinality constraint. For the monotone case our bounds match those of Chakrabarti and Kale [12] for a single pass; we give a self-contained algorithm and analysis. Table 1 summarizes our results for a variety of constraints.

A brief overview of techniques. Streaming algorithms for constrained modular and submodular function optimization are usually clever variations of the greedy algorithm, which picks elements in iterations to maximize the gain in each iteration locally while maintaining feasibility. For monotone functions, in the offline setting, greedy gives a $1/(p + 1)$-approximation for the p-matchoid constraint and a $(1 - 1/e)$-approximation for the cardinality constraint [23]. The offline greedy algorithm cannot be directly implemented in streams, but we outline two different strategies that are still greedy in spirit. For the cardinality constraint,

Table 1. Best known approximation bounds for submodular maximization. Bounds for randomized algorithms that hold in expectation are marked (R). For hypergraph b-matchings and matroid intersection, p is fixed. In the results for p-matchoids, $o(1)$ goes to zero as p increases. New bounds attained in this paper are marked (⋆). All new bounds except for the cardinality constraint are the first bounds for their class. The best previous bound for the cardinality constraint is about .0893, by [9].

	offline		streaming	
constraint	monotone	nonnegative	monotone	nonnegative
cardinality	$1-1/e$ [36]	$1/e + .004$ [8]	$\frac{1-\epsilon}{2}$ [2]	$\frac{1-\epsilon}{2+e}$ (R,⋆)
matroid	$1-1/e$ (R) [11]	$\frac{1-\epsilon}{e}$ (R) [20]	$1/4$ [12]	$\frac{1-\epsilon}{4+e}$ (R,⋆)
matchings	$\frac{1}{2+\epsilon}$ [21]	$\frac{1}{4+\epsilon}$ [21]	$4/31$ [12]	$\frac{1-\epsilon}{12+e}$ (R,⋆)
b-matchings	$\frac{1}{2+\epsilon}$ [21]	$\frac{1}{4+\epsilon}$ [21]	$1/8$ (⋆)	$\frac{1-\epsilon}{12+e}$ (R,⋆)
rank p hypergraph b-matching	$\frac{1}{p+\epsilon}$ (R) [21]	$\frac{p-1}{p^2+\epsilon}$ [21]	$1/4p$ (⋆)	$\frac{(1-\epsilon)(p-1)}{5p^2-4p+\epsilon}$ (R,⋆)
intersection of p matroids	$\frac{1}{p+\epsilon}$ [32]	$\frac{p-1}{p^2+(p-1)\epsilon}$ [32]	$1/4p$ [12]	$\frac{(1-\epsilon)(p-1)}{5p^2-4p}$ (R,⋆)
p-matchoids	$\frac{1}{p+1}$ [11,23]	$\frac{(1-\epsilon)(2-o(1))}{ep}$ (R) [14,20]	$1/4p$ (⋆)	$\frac{(1-\epsilon)(2-o(1))}{(8+e)p}$ (R,⋆)

Badanidiyuru *et al.* [2] designed an algorithm that adds an element to its running solution S only if the marginal gain is at least a threshold of about $\mathrm{OPT}/2k$. Although the quantity $\mathrm{OPT}/2k$ is not known a priori, they show that it lies in a small and identifiable range, and can be approximated with $O(\log k)$ well-spaced guesses. The algorithm then maintains $O(\log k)$ solutions in parallel, one for each guess. Another strategy from Chakrabarti and Kale [12], based on previous work for matchings [19,35] and matroid constraints [4] with modular weights, will consider deleting elements from S when adding a new element to S is infeasible. More specifically, when a new element e is encountered, the algorithm finds a subset $C \subseteq S$ such that $(S \setminus C) + e$ is feasible, and compare the gain $f((S \setminus C) + e) - f(S)$ to a quantity representing the value that C adds to S. In the modular case, this may be the sum of weights of elements in C; for monotone submodular functions, Chakrabarti and Kale used marginal values, fixed for each element when the element is added to S, as proxy weights instead.

The non-monotone case is harder because marginal values can be negative even when f is non-negative. The natural greedy algorithm fails for even the simple cardinality constraint, and the best offline algorithms for nonnegative submodular maximization are uniformly weaker (see Table 1). To this end, we adapt techniques from the recent work of Buchbinder *et al.* [8] in our randomized algorithm, and techniques from Gupta *et al.* [26] for the deterministic version. Buchbinder *et al.* randomized the standard greedy algorithm (for cardinality) y repeatedly gathering the top (say) k remaining elements, and then randomly icking only one of them. We adapt this to the greedy setting by adding the top ements to a buffer B as they appear in the stream, and randomly adding an ement from B to S only when B fills up. What remains of B at the end of the ream is post-processed by an offline algorithm. Gupta *et al.* gave a framework or adapting any monotone submodular maximization algorithm to nonnegative ibmodular functions, by first running the algorithm once to generate one inde-endent set S_1, then running the algorithm again on the complement of S_1 to enerate a second set S_2, and running an unconstrained maximization algorithm n S_1 to produce a third set S_3, finally returning the best of S_1, S_2, and S_3. Our eterministic streaming algorithm is a natural adaptation, piping the rejected lements of one instance of a streaming algorithm directly into a second instance f the same algorithm, and post-processing all the elements taken by the first treaming instance. Both of our algorithms require that we limit the number of lements ever added to S, which then limits the size of the input for the post-rocessor. This limit is enforced by the idea of additive thresholds from [2] and simple but subtle notion of value that ensures the properties we desire.

Related Work. There is substantial literature on constrained submodular unction optimization, and we only give a quick overview. Many of the basic iroblems are NP-Hard, so we will mainly focus on the development of approx-mation algorithms. The (offline) problem $\max_{S \in \mathcal{I}} f(S)$ for various constraints ias been extensively explored starting with the early work of Fisher, Nemhauser, Volsey on greedy and local search algorithms [23,36]. Recent work has obtained nany new and powerful results based on a variety of methods including vari-nts of greedy [7,8,26], local search [22,31,32], and the multilinear relaxation 5,11,14,29]. Monotone submodular functions admit better bounds than non-nonotone functions (see Table 1). For a p-matchoid constraint, which is our pri-nary consideration, an $\Omega(1/p)$-approximation can be obtained for non-negative unctions. Recent work has also obtained new lower bounds on the approximation atio achievable in the oracle model via the so-called symmetery gap technique 40]; this also yields lower bounds in the standard computational models [18].

Streaming algorithms for submodular functions are a very recent phe-iomenon with algorithms developed recently for monotone submodular functions 2,12]. [2] gives a $1/2 - \epsilon$ approximation for monotone functions under cardinality :onstraint using $O(k \log k/\epsilon)$ space. [12] focuses on more general constraints like nteresctions of p-matroids and rank p hypergraphs, giving an approximation of ./4p using a single pass. Their algorithm extends to multiple passes, with an ipproximation bound of $1/(p + 1 + \epsilon)$ with $O(\epsilon^{-3} \log p)$ passes. The main focus of [30] is on the map-reduce model although they claim some streaming results is well.

Related to the streaming models are two *online* models where elements arrive n an online fashion and the algorithm is required to maintain a feasible solution S at all times; each element on arrival has to be processed and any element which s discarded from S at any time cannot be added back later. Strong lower bounds :an be shown in this model and two relaxations have been considered. In the *secretary model*, the elements arrive according to a random permutation of the ground set and an element added to S cannot be discarded later. In the secretary model, constant factor algorithms are known for the cardinality constraint and some special cases of a single matroid constraint [6,26]. These algorithms assume the stream is randomly ordered and their performance degrades badly against adversarial streams; the best competitive ratio for a single general matroid is $O(\log k)$ (where k is the rank of the matroid). Recently, Buchbinder *et al.* [9] considered a different relaxation of the online model where *preemptions* are allowed: elements added to S can be discarded later. Algorithms in the preemptive model are usually streaming algorithms, but the converse is not true (although the one-pass algorithms in [12] are preemptive). For instance, the algorithm in [2] maintains multiple feasible solutions and our algorithms maintain a buffer of elements neither accepted nor rejected. The space requirement of an algorithm in the online model is not necessarily constrained since in principle an algorithm is allowed to keep track of all the past elements seen so far. The main result in [9], as it pertains to this work, is a randomized 0.0893-competitive algorithm for cardinality constraints using $O(k)$-space. As Table 1 shows, we obtain a $(1-\epsilon)/(2+e)$-competitive algorithm for this case using $O(k \log k/\epsilon^2)$-space.

Paper Organization. Section 2 reviews combinatorial definitions and introduces the notion of incremental values. Section 3 analyzes an algorithm that works for monotone submodular functions, and Section 4 adapts this algorithm to the non-monotone case. Due to space constraints, we defer all proofs as well as a deterministic streaming algorithm to the full version[1].

2 Preliminaries

Matroids. A *matroid* is a finite set system $\mathcal{M} = (\mathcal{N}, \mathcal{I})$, where $\mathcal{N}$ is a set and $\mathcal{I} \subseteq 2^{\mathcal{N}}$ is a family of subsets such that: (i) $\emptyset \in \mathcal{I}$, (ii) If $A \subseteq B \subseteq \mathcal{N}$, and $B \in \mathcal{I}$, then $A \in \mathcal{I}$, (iii) If $A, B \in \mathcal{I}$ and $|A| < |B|$, then there is an element $b \in B \setminus A$ such that $A + b \in \mathcal{I}$. In a matroid $\mathcal{M} = (\mathcal{N}, \mathcal{I})$, $\mathcal{N}$ is called the *ground set* and the members of $\mathcal{I}$ are called *independent sets* of the matroid. The bases of $\mathcal{M}$ share a common cardinality, called the *rank* of $\mathcal{M}$.

Matchoids. Let $\mathcal{M}_1 = (\mathcal{N}_1, \mathcal{I}_1), \ldots, \mathcal{M}_q = (\mathcal{N}_q, \mathcal{I}_q)$ be q matroids over overlapping groundsets. Let $\mathcal{N} = \mathcal{N}_1 \cup \cdots \cup \mathcal{N}_q$ and $\mathcal{I} = \{S \subseteq \mathcal{N} : S \cap \mathcal{N}_\ell \in \mathcal{I}_\ell \text{ for all } \ell\}$. The finite set system $\mathcal{M}^p = (\mathcal{N}, \mathcal{I})$ is a *p-matchoid* if for every element $e \in \mathcal{N}$, e is a member of $\mathcal{N}_\ell$ for at most p indices $\ell \in [q]$. *p*-matchoids generalizes matchings and intersections of matroids, among others (see Figure 1).

Maximizing submodular functions under a p-matchoid constraint. Let $\mathcal{N}$ be a set of elements, $f : 2^{\mathcal{N}} \to \mathbb{R}_{\geq 0}$ a nonnegative submodular function on $\mathcal{N}$, and $\mathcal{M}^p = (\mathcal{N}, \mathcal{I})$ a *p*-matchoid for some integer p. We want to approximate

---

[1] available on arXiv

PT $= \max_{S \in \mathcal{I}} f(S)$. There are several polynomial-time approximation algorithms that give an $\Omega(1/p)$-approximation for this problem, with better bounds or simpler constraints (see Table 1). These algorithms are used as a black box called `Offline`, with approximation ratio denoted by γ_p: if `Offline` returns $\in \mathcal{I}$, then $\mathbf{E}[f(S)] \geq \gamma_p \text{OPT}$ (possibly without expectation, if `Offline` is deterministic).

Incremental Value. Let $\mathcal{N}$ be a ground set, and let $f : 2^{\mathcal{N}} \to \mathbb{R}$ be a submodular function. For a set $S \subseteq \mathcal{N}$ and an element $e \in S$, what is the value that e adds to S? One idea is to take the margin $f_{S-e}(e) = f(S) - f(S - e)$ of adding e to $S - e$. However, because f is not necessarily modular, we can only say that $\sum_{e \in S} f_{S-e}(e) \leq f(S)$ without equality. It is natural to ask for a different notion of value where the values of the parts sum to the value of the whole.

Let $\mathcal{N}$ be an *ordered* set and $f : 2^{\mathcal{N}} \to \mathbb{R}$ be a set function. For a set $S \subseteq \mathcal{N}$ and element $e \in \mathcal{N}$, the *incremental value* of e in S, denoted $\nu(f, S, e)$, is defined as

$$\nu(f, S, e) = f_{S'}(e), \text{ where } S' = \{s \in S : s < e\}.$$

The key point of incremental values is that they capture the entire value of a set. The following holds for *any* set function.

Lemma 1. *Let $\mathcal{N}$ be an ordered set, $f : 2^{\mathcal{N}} \to \mathbb{R}$ a set function, and $S \subseteq \mathcal{N}$ a set. Then $f(S) = \sum_{e \in S} \nu(f, S, e)$.*

When f is submodular, we have decreasing incremental values analogous (and closely related) to decreasing marginal returns of submodular function.

Lemma 2. *Let $S \subseteq T \subseteq \mathcal{N}$ be two nested subsets of an ordered set $\mathcal{N}$, let $f : 2^{\mathcal{N}} \to \mathbb{R}$ be submodular, and let $e \in \mathcal{N}$. Then $\nu(f, T, e) \leq \nu(f, S, e)$.*

The following is also an easy consequence of submodularity.

Lemma 3. *Let $\mathcal{N}$ be an ordered set of elements, let $f : 2^{\mathcal{N}} \to \mathbb{R}$ be a submodular function, $S, Z \subseteq \mathcal{N}$ two sets, and $e \in S$. Then $\nu(f_Z, S, e) \leq \nu(f, Z \cup S, e)$.*

3 Streaming Greedy

Let $\mathcal{M}^p = (\mathcal{N}, \mathcal{I})$ be a p-matchoid and f a submodular function. The elements of $\mathcal{N}$ are presented in a stream, and we order $\mathcal{N}$ by order of appearance. We assume value oracle access to f, that given $S \subseteq \mathcal{N}$, returns the value $f(S)$. We also assume membership oracles for each of the q matroids defining $\mathcal{M}^p$: given $S \subseteq \mathcal{N}_\ell$, there is an oracle for $\mathcal{M}_\ell$ that returns whether or not $S \in \mathcal{I}_\ell$.

We first present a deterministic streaming algorithm `Streaming-Greedy` that yields an $\Omega(1/p)$-approximation for monotone submodular functions, but performs poorly for non-monotone functions. The primary motivation in presenting `Streaming-Greedy` is as a building block for a randomized algorithm `Randomized-Streaming-Greedy` presented in Section 4, and a deterministic

```
            Streaming-Greedy (α, β)

   S ← ∅
   while (stream is not empty)
     e ← next element in the stream
     C ← Exchange-Candidates(S,e)
     // C satisfies S − C + e ∈ 𝓘
     if f_S(e) ≥ α + (1 + β) Σ_{c∈C} ν(f, S, c)
       S ← S \ C + e
   end while
   return S
```

```
            Exchange-Candidates(S,e)
   C ← ∅
   for ℓ = 1, …, q
     if e ∈ 𝒩_ℓ and (S + e) ∩ 𝒩_ℓ ∉ 𝓘_ℓ
       S_ℓ = S ∩ 𝒩_ℓ
       X ← {s ∈ S_ℓ : (S_ℓ − s + e) ∈ 𝓘_ℓ}
       // X + e is a circuit
       c_ℓ ← arg min_{x∈X} ν(f, S, x)
       C ← C + c_ℓ
     end if
   end for
   return C
```

algorithm deferred to the full version. The analysis for these algorithms relies crucially on properties of Streaming-Greedy.

Streaming-Greedy maintains an independent set $S \in \mathcal{I}$; as an element arrives in the stream, it is either discarded or added to S in exchange for a well-chosen subset of S. The threshold for exchanging is tuned by two nonnegative parameters α and β. At the end of the stream, Streaming-Greedy outputs S.

The overall strategy is similar to previous algorithms developed for matchings [19,35] and intersections of matroids [4] when f is modular, and generalized by [12] to monotone submodular functions. There are two main differences. One is the use of the additive threshold α. The second is the use of the incremental value ν. By using incremental value, the value of an element $e \in S$ is not fixed statically when e is first added to S, and increases over time as other elements are dropped from S. These two seemingly minor modifications are crucial to the eventual algorithms for non-monotone functions.

We remark that Streaming-Greedy also fits the online preemptive model.

Outline of the Analysis: Let $T \in \mathcal{I}$ be some fixed feasible set (we can think of T as an optimum set). In the offline analysis of the standard greedy algorithm one can show that $f(S \cup T) \leq (p+1)f(S)$, where S is the output of greedy; for the monotone case this implies that $f(S) \geq f(T)/(p+1)$. The analysis here hinges on the fact that each element of $T \setminus S$ is available to greedy when it chooses each element. In the streaming setting, this is no longer feasible and hence the need to remove elements in favor of new high-value elements. To relate $\tilde{S}$, the final output, to T, we consider U, the set of all elements ever added to S. The analysis proceeds in two steps.

First, we upper bound $f(U)$ by $f(\tilde{S})$ as $f(U) \leq (1 + \frac{1}{\beta}) \cdot f(\tilde{S}) - \frac{\alpha}{\beta}|U|$. Second, we upper bound $f(T \cup U)$ as $f(T \cup U) \leq k\alpha + \frac{(1+\beta)^2}{\beta} \cdot p \cdot f(\tilde{S})$. For $\alpha = 0$, we obtain $f(T \cup U) \leq \frac{(1+\beta)^2}{\beta} \cdot p \cdot f(\tilde{S})$, which yields $f(T) \leq 4p f(\tilde{S})$ when f is monotone (for $\beta = 1$); this gives the same bound as [12]. The crucial difference is that we are able to prove an upper bound on the size of U, namely, $|U| \leq \text{OPT}/\alpha$; hence, if we choose the threshold α to be $c\text{OPT}/k$ for some parameter c we have $|U| \leq k/c$. This plays a critical role in analyzing the non-monotone case in the subsequent sections that use Streaming-Greedy as a black box. The upper bound on $|U|$

achieved by the definition of ν and the threshold α; we stress that this is not
so obvious as it may seem because the function f can be non-monotone and the
marginal values can be negative.

Randomized Streaming Greedy

Randomized-Streaming-Greedy adapts Streaming-Greedy to nonnegative submodular functions by employing a randomized buffer B to limit the probability
that any element is added to the running solution S. Like Streaming-Greedy,
Randomized-Streaming-Greedy
maintains the invariant $S \in \mathcal{I}$.
However, when a "good" element
would have been added to S by
Streaming-Greedy, it is instead
placed in B. Once the number of
elements in B hits a limit K, we
pick one element in B uniformly
at random and add it to S just as
Streaming-Greedy would. Modifying S may break the invariant
that the buffer only contains good
elements. Since f is submodular,
the incremental value $\nu(f, S, e)$ of
each $e \in S$ may increase if a preceding element is deleted. Furthermore, the marginal value $f_S(b)$ of
each buffered element $b \in B$ may
decrease as elements are added to
S. Thus, after modifying S, we
reevaluate each $b \in B$ and discard
elements that are no longer good.

Let $\tilde{B}$ be the set of elements
remaining in the buffer B when the
stream ends. We process $\tilde{B}$ with an
offline algorithm to produce a second solution S', and finally return the set $\hat{S}$
which is the better of S and S'.

```
Randomized-Streaming-Greedy(α, β)

S ← ∅, B ← ∅
while (stream is not empty)
    e ← next element in the stream
    if Is-Good(S,e) then B ← B + e
    if |B| = K then
        e ← uniformly random from B
        C ← Exchange-Candidates(S,e)
        B ← B - e,  S ← (S \ C) + e
        for all e' ∈ B
            unless  Is-Good(S,e')
                B ← B - e'
    end if
end while
S' ← Offline(B)
return  arg max_{Z∈{S,S'}} f(Z)

Is-Good(S,e)

C ← Exchange-Candidates(S,e)
if f_S(e) ≥ α + (1 + β) Σ_{e'∈C} ν(f,S,e')
    return TRUE
else return FALSE
```

Outline of the Analysis. Let $T \in \mathcal{I}$ be an arbitrary independent set. Let $T' = T \setminus \tilde{B}$ be the portion fully processed by the online portion and $T'' = T \cap \tilde{B}$ the remainder left over in the buffer and processed offline.

In the full version, we first show that the analysis for T' largely reduces to that of Section 3. In particular, this gives us a bound on $f(U \cup T')$. We combine this with a bound on $f(T'')$, guaranteed by the offline algorithm, to obtain an overall bound on $f(U \cup T)$ by $f(\hat{S})$. We then bound $f(T)$ with respect to $f(U)$, leveraging the fact that the buffer limits the probability of elements being added to S. We tie together the analysis to bound $f(T)$ by $f(\hat{S})$ for fixed α and β.

The analysis reveals that the optimal choice for β is 1, and that α should be chosen in proportion to OPT/k, where k is the rank of the $\mathcal{M}^p$. Since OPT is not known a prioriwe leverage a technique by Badanidiyuru *et al.* [2] that efficiently guesses the α to within a constant factor of the target value. The final algorithm is then $\log k$ copies of `Randomized-Streaming-Greedy` run in parallel, each instance corresponding to a "guess" for α. One of these guesses is approximately correct, and attains the bounded asserted in Theorem 1.

Theorem 1. *Let $\mathcal{M}^p = (\mathcal{N}, \mathcal{I})$ be a p-matchoid of rank k, let $f : 2^{\mathcal{N}} \to \mathbb{R}_{\geq 0}$ a nonnegative submodular function over $\mathcal{N}$, and let $\epsilon > 0$ be fixed. Suppose there exists an algorithm for the offline instance of the problem with approximation ratio γ_p. Then there exists a streaming algorithm using total space $O\left(\frac{k \log k}{\epsilon^2}\right)$ that, given a stream over $\mathcal{N}$, returns a set $\hat{S} \in \mathcal{I}$ such that*

$$(1 - \epsilon)\mathrm{OPT} \leq \left(4p + \frac{1}{\gamma_p}\right) \boldsymbol{E}\left[f(\hat{S})\right].$$

4.1 Simpler Algorithm and Better Bound for Cardinality Constraint

When the p-matchoid is simply a cardinality constraint with rank k, we can do better. If we set $\beta = \infty$ in `Randomized-Streaming-Greedy` α, β, then the algorithm will only try to add to S without exchanging while $|S| < k$, effectively halting once we meet the cardinality constraint $|S| = k$. On the right, we rewrite `Randomized-Streaming-Greedy` α, ∞ with the unnecessary logic removed.

```
Randomized-Streaming-Greedy(α,∞)
B ← ∅,  S ← ∅
while (stream is not empty)
    e ← next element in the stream
    if |S| ≤ k and f_S(e) > α then
        B ← B + e
    if |B| = K then
        e ← uniformly random from B
        B ← B - e,  S ← S + e
        for all e' ∈ B s.t. f_S(e') ≤ α
            B ← B - e'
    end if
end while
S' ← Offline(B)
return arg max_{Z∈{S,S'}} f(Z)
```

The analysis, provided in the full version, reveals that the appropriate choice for α is $\mathrm{OPT}/(2 + e)k$, where $\mathrm{OPT} = \max\{f(T) : |T| \leq k\}$ is the maximum value attainable by a set of k elements, and that a sufficiently large choice for K is k/ϵ. As before, we can efficiently approximate α by guessing α in increasing powers of $(1+\epsilon)$, maintaining at most $O(\log_{1+\epsilon} k) = O(\epsilon^{-1} \log k)$ instances of `Randomized-Streaming-Greedy`(α, ∞) at any instant. The resulting bound is stronger than derived above for a 1-matchoid.

Theorem 2. *Let $f : 2^{\mathcal{N}} \to \mathbb{R}_{\geq 0}$ be a nonnegative submodular function over a ground set $\mathcal{N}$, and let $\epsilon > 0$ be fixed. Then there exists a streaming algorithm* *sing total space* $O\left(\frac{k \log k}{\epsilon^2}\right)$ *that, given a stream over* $\mathcal{N}$, *returns a set* $\hat{S}$ *such that* $|\hat{S}| \leq k$ *and* $f(\hat{S}) \geq \frac{1-\epsilon}{2+\epsilon} \cdot OPT$, *where* $OPT = \max\{f(T) : |T| \leq k\}$ *is the maximum value attainable by a set of* k *elements.*

References

1. Babaioff, M., Immorlica, N., Kleinberg, R.: Matroids, secretary problems, and online mechanisms. In: Proc. 18th ACM-SIAM Sympos. Discrete Algs. (SODA), pp. 434–443. Philadelphia, PA, USA (2007)
2. Badanidiyuru, A., Mirzasoleiman, B., Karbasi, A., Krause, A.: Streaming submodular optimization: massive data summarization on the fly. In: Proc. 20th ACM Conf. Knowl. Disc. and Data Mining (KDD), pp. 671–680 (August 2014)
3. Badanidiyuru, A., Vondrák, J.: Fast algorithms for maximizing submodular functions. In: Proc. 25th ACM-SIAM Sympos. Discrete Algs. (SODA), pp. 1497–1514 (2014)
4. Badanidiyuru Varadaraja, A.: Buyback problem - approximate matroid intersection with cancellation costs. In: Aceto, L., Henzinger, M., Sgall, J. (eds.) ICALP 2011, Part I. LNCS, vol. 6755, pp. 379–390. Springer, Heidelberg (2011)
5. Bansal, N., Korula, N., Nagarajan, V., Srinivasan, A.: Solving packing integer programs via randomized rounding with alterations. Theo. Comput. 8(1), 533–565 (2012)
6. Bateni, M., Hajiaghayi, M., Zadimoghaddam, M.: Submodular secretary problem and extensions. ACM Trans. Algs. 9(4), 32:1–32:23 (2013)
7. Buchbinder, N., Feldman, M., Naor, J., Schwartz, R.: A tight linear time (1/2)-approximation for unconstrained submodular maximization. In: Proc. 53rd Annu. IEEE Sympos. Found. Comput. Sci. (FOCS), pp. 649–658 (2012)
8. Buchbinder, N., Feldman, M., Naor, J., Schwartz, R.: Submodular maximization with cardinality constraints. In: Proc. 25th ACM-SIAM Sympos. Discrete Algs. (SODA), pp. 1433–1452 (2014)
9. Buchbinder, N., Feldman, M., Schwartz, R.: Online submodular maximization with preemption. In: Proc. 26th ACM-SIAM Sympos. Discrete Algs. (SODA), pp. 1202–1216 (2015)
10. Calinescu, G., Chekuri, C., Pál, M., Vondrák, J.: Maximizing a submodular set function subject to a matroid constraint (extended abstract). In: Fischetti, M., Williamson, D.P. (eds.) IPCO 2007. LNCS, vol. 4513, pp. 182–196. Springer, Heidelberg (2007)
11. Calinescu, G., Chekuri, C., Pál, M., Vondrák, J.: Maximizing a monotone submodular function subject to a matroid constraint. SIAM J. Comput. 40(6), 1740–1766 (2011)
12. Chakrabarti, A., Kale, S.: Submodular maximization meets streaming: matchings, matroids, and more. In: Lee, J., Vygen, J. (eds.) IPCO 2014. LNCS, vol. 8494, pp. 210–221. Springer, Heidelberg (2014)
13. Chekuri, C., Jayram, T.S., Vondrák, J.: On multiplicative weight updates for concave and submodular function maximization. In: Proceedings of ITCS (2015)
14. Chekuri, C., Vondrák, J., Zenklusen, R.: Submodular function maximization via the multilinear relaxation and contention resolution schemes. In: Proc. 43th Annu. ACM Sympos. Theory Comput. (STOC), pp. 783–792 (2011)

15. Chen, W., Wang, C., Wang, Y.: Scalable influence maximization for prevalent viral marketing in large-scale social networks. In: Proc. 16th ACM Conf. Knowl. Disc and Data Mining (KDD), pp. 1029–1038 (2010)

16. Chen, W., Wang, Y., Yang, S.: Efficient influence maximization in social networks. In: Proc. 15th ACM Conf. Knowl. Disc. and Data Mining (KDD), pp. 199–208 New York, NY, USA (2009)

17. Dasgupta, A., Kumar, R., Ravi, S.: Summarization through submodularity and dispersion. In: Proc. 51st Ann. Meet. Assoc. for Comp. Ling. (ACL), vol. 1 pp. 1014–1022 (2013)

18. Dobzinski, S., Vondrak, J.: From query complexity to computational complexity. In Proc. 44th Annu. ACM Sympos. Theory Comput. (STOC), pp. 1107–1116 (2012)

19. Feigenbaum, J., Kannan, S., McGregor, A., Suri, S., Zhang, J.: On graph problems in a semi-streaming model. Theo. Comp. Sci. 348(2–3), 207–216 (2005)

20. Feldman, M., Naor, J., Schwartz, R.: A unified continuous greedy algorithm for submodular maximization. In: Proc. 52nd Annu. IEEE Sympos. Found. Comput. Sci. (FOCS), pp. 570–579 (2011)

21. Feldman, M., Naor, J.S., Schwartz, R., Ward, J.: Improved approximations for k-exchange systems. In: Demetrescu, C., Halldórsson, M.M. (eds.) ESA 2011 LNCS, vol. 6942, pp. 784–798. Springer, Heidelberg (2011)

22. Filmus, Y., Ward, J.: Monotone submodular maximization over a matroid via non-oblivious local search. SIAM J. Comput. 43(2), 514–542 (2014)

23. Fisher, M.L., Nemhauser, G.L., Wolsey, L.A.: An analysis of approximations for maximizing submodular set functions - II. Math. Prog. Studies 8, 73–87 (1978)

24. Fujishige, S.: Submodular functions and optimization, vol. 58. Elsevier (2005)

25. Goyal, A., Bonchi, F., Lakshmanan, L.V.S.: A data-based approach to social influence maximization. Proc. VLDB Endow. 5(1), 73–84 (2011)

26. Gupta, A., Roth, A., Schoenebeck, G., Talwar, K.: Constrained non-monotone submodular maximization: offline and secretary algorithms. In: Saberi, A. (ed.) WINE 2010. LNCS, vol. 6484, pp. 246–257. Springer, Heidelberg (2010)

27. Iyer, R., Jegelka, S., Bilmes, J.: Fast semidifferential-based submodular function optimization. In: Proc. 30th Int. Conf. Mach. Learning (ICML), vol. 28, pp. 855–863 (2013)

28. Kempe, D., Kleinberg, J., Tardos, É.: Maximizing the spread of influence through a social network. In: Proc. 9th ACM Conf. Knowl. Disc. and Data Mining (KDD), pp. 137–146. New York, NY, USA (2003)

29. Kulik, A., Shachnai, H., Tamir, T.: Approximations for monotone and nonmonotone submodular maximization with knapsack constraints. Math. Oper. Res. 38(4), 729–739 (2013)

30. Kumar, R., Moseley, B., Vassilvitskii, S., Vattani, A.: Fast greedy algorithms in mapreduce and streaming. In: Proc. 25th Ann. ACM Sympos. Parallelism Alg. Arch. (SPAA), pp. 1–10 (2013)

31. Lee, J., Mirrokni, V.S., Nagarajan, V., Sviridenko, M.: Maximizing nonmonotone submodular functions under matroid or knapsack constraints. SIAM J. Discrete Math. 23(4), 2053–2078 (2010)

32. Lee, J., Sviridenko, M., Vondrák, J.: Submodular maximization over multiple matroids via generalized exchange properties. Math. Oper. Res. 35, 795–806 (2010)

33. Leskovec, J., Krause, A., Guestrin, C., Faloutsos, C., VanBriesen, J., Glance, N.: Cost-effective outbreak detection in networks. In: Proc. 13th ACM Conf. Knowl. Disc. and Data Mining (KDD), pp. 420–429. New York, NY, USA (2007)

4. Lin, H., Bilmes, J.: A class of submodular functions for document summarization. In: Proc. 49th Ann. Meet. Assoc. Comput. Ling.: Human Lang. Tech. (HLT), vol. 1, pp. 510–520 (2011)
5. McGregor, A.: Finding graph matchings in data streams. In: 8th Intl. Work. Approx. Algs. Combin. Opt. Problems, pp. 170–181 (2005)
6. Nemhauser, G.L., Wolsey, L.A., Fisher, M.L.: An analysis of approximations for maximizing submodular set functions - I. Math. Prog. 14(1), 265–294 (1978)
7. Schrijver, A.: Combinatorial optimization: polyhedra and efficiency, vol. 24. Springer Verlag (2003)
8. Seeman, L., Singer, Y.: Adaptive seeding in social networks. In: Proc. 54th Annu. IEEE Sympos. Found. Comput. Sci. (FOCS), pp. 459–468 (2013)
9. Sipos, R., Swaminathan, A., Shivaswamy, P., Joachims, T.: Temporal corpus summarization using submodular word coverage. In: Proc. 21st ACM Int. Conf. Inf. and Know. Management (CIKM), pp. 754–763 (2012)
10. Vondrák, J.: Symmetry and approximability of submodular maximization problems. SIAM J. Comput. 42(1), 265–304 (2013)

Multilinear Pseudorandom Functions

Aloni Cohen$^{(\boxtimes)}$ and Justin Holmgren$^{(\boxtimes)}$

MIT, Cambridge, MA, USA
{aloni,holmgren}@mit.edu

Abstract. We define the new notion of a *multilinear* pseudorandom function (PRF), and give a construction with a proof of security assuming the hardness of the decisional Diffie-Hellman problem. A direct application of our construction yields (non-multilinear) PRFs with aggregate security from the same assumption, resolving an open question in [CGV15]. Additionally, multilinear PRFs give a new way of viewing existing algebraic PRF constructions: our main theorem implies they too satisfy aggregate security.

1 Introduction

Pseudorandom functions (PRFs) are of fundamental importance in modern cryptography. A PRF is efficiently computable and succinctly described, but is indistinguishable from a random function. But a random functions are too unstructured for many applications, and as a result many specialized pseudorandom functions with more structure have emerged. Goldreich, Goldwasser, and Nussboim [GGN03] define a general notion of pseudo-implementing huge random objects with extra structure. In this paper, we define and construct *multilinear* pseudorandom functions assuming the decisional Diffie-Hellman assumption (DDH), and we show applications to prior work. Before presenting an informal definition, we discuss the motivating applications.

The recent work of Cohen, Goldwasser, and Vaikuntanathan [CGV15] introduced the notion of an aggregate pseudorandom function family $\mathcal{F}$ with extra efficiency and security properties. First, the key K for a PRF f enables efficient computation of $\mathsf{Agg}_f(S) = \sum_{x \in S} f(x)$ for some class of succinctly described, but possibly exponentially large, sets S. Second, no efficient algorithm can distinguish oracle access to $f(\cdot)$ and $\mathsf{Agg}_f(\cdot)$ from oracle access to $g(\cdot)$ and $\mathsf{Agg}_g(\cdot)$, where g is a truly random function. The main constructions of [CGV15] were proven secure assuming the subexponential hardness of DDH.

A different line of work studies a notion of algebraic pseudorandom functions by Benabbas, Gennaro, and Vahlis [BGV11]. This notion is incomparable to aggregate PRFs: algebraic PRFs generalize efficient aggregation, but provide no security guarantees when an adversary has an aggregation oracle. Algebraic PRFs have a number of applications in verifiable computation and multiparty computation [BGV11,Haz15]. Because of their restricted security, algebraic PRFs have thus far only been considered for polynomially-sized domains – security over large domains would require subexponential hardness of DDH.

© Springer-Verlag Berlin Heidelberg 2015
M.M. Halldórsson et al. (Eds.): ICALP 2015, Part I, LNCS 9134, pp. 331–342, 2015.
DOI: 10.1007/978-3-662-47672-7_27

In both works, the reliance on subexponential hardness of DDH (or the small-domain restriction) is unsatisfying. Subexponential hardness is a significantly stronger assumption, and is, for example, false in $\mathbb{Z}_p^*$. Additionally, security reductions from subexponential hardness assumptions necessitate larger security parameters and thus lose efficiency.

We define a multilinear PRF family as a family of functions $\{\mathcal{F} : V_1 \times \cdots \times V_n \to Y\}$ mapping a product of vector spaces V_i to another vector space Y, in which a random function from the family is indistinguishable from a random *multilinear* function with the same domain and codomain. One case of particular interest is when each V_i is $\mathbb{F}_p^2$. Then any multilinear function from $V_1 \times \cdots \times V_n \to W$ is defined by 2^n values, which we think of as inducing a PRF mapping $\{0,1\}^n \to Y$. Multilinearity allows us to efficiently compute specific weighted sums of exponentially many PRF values from the above works. This encompasses "hypercube" aggregation from [CGV15] and the closed-form efficiency requirements of [BGV11].

2   Preliminaries

Notation For a set S, we use $x \leftarrow S$ to mean that x is sampled uniformly at random from S. We denote the finite field of order p by $\mathbf{F}_p$. All vectors $\mathbf{v}$ are column vectors, and $\mathbf{v}^t$ denotes the transpose.

2.1   Linear Maps

Given a vector spaces V and W over a field $\mathbb{F}$, we say that a map $T : V \to W$ is *linear* if $T(c_1\mathbf{v}_1 + c_2\mathbf{v}_2) = c_1 T(\mathbf{v}_1) + c_2 T(\mathbf{v}_2)$ is a valid identity for all vectors $\mathbf{v}_1$, $\mathbf{v}_2$ in V and scalars c_1, c_2 in $\mathbb{F}$. We say that a map $T : V_1 \times \cdots \times V_n \to W$ is *multilinear* if it is linear in each component.

When V and W have finite dimensions d_V and d_W (which is the only case we consider in this paper), a linear map from V to W can be represented by a matrix in $\mathbb{F}^{d_W \times d_V}$. This representation depends on the choice of bases for V and W. When $\mathbb{F}$ is a finite field (also the only case we consider), a *random* linear map from $V \to W$ can be sampled by picking an arbitrary basis for V and W and sampling a uniformly random matrix from $\mathbb{F}^{d_W \times d_V}$. The set of all linear maps from V to W will be denoted as W^V.

2.2   Tensor Products of Vector Spaces

Given vector spaces V and W of dimensions d_V and d_W over $\mathbb{F}$, the tensor product $V \otimes W$ is defined as a $d_V d_W$-dimensional vector space over $\mathbb{F}$. For any $v \in V$ and $w \in W$, their tensor product $v \otimes w \in V \otimes W$ can be defined by the following laws:

1. $(v_1 + v_2) \otimes w = v_1 \otimes w + v_2 \otimes w$
2. $v \otimes (w_1 + w_2) = v \otimes w_1 + v \otimes w_2$
3. For any $c \in \mathbb{F}$, $(cv) \otimes w = v \otimes (cw) = c(v \otimes w)$.

If V has a basis $v_1, \ldots, v_{d_V}$, and W has a basis $w_1, \ldots, w_{d_W}$, there is a natural basis for $V \otimes W$, namely $\{v_i \otimes w_j\}_{i \in [d_V], j \in [d_W]}$. Expanding $v = \sum_{i \in [d_V]} a_i v_i$ and $w = \sum_{j \in [d_W]} b_j w_j$ in the respective bases of V and W, and applying the above laws yields $v \otimes w = \sum_{i \in [d_V], j \in [d_W]} a_i b_j (v_i \otimes w_j)$. A *simple* tensor is defined as one which can be written as $v \otimes w$.

One can repeat the tensor product operation to obtain a space $V_1 \otimes \cdots \otimes V_n$ for any n vector spaces, with simple tensors of the form $v_1 \otimes \cdots \otimes v_n$. Vectors in such a space are sometimes called tensors, and the vector spaces in which they reside are called tensor spaces.

It is easy to observe that the mapping $\phi : V_1 \times \cdots \times V_n \to V_1 \otimes \cdots \otimes V_n$ given by $\phi(v_1, \ldots, v_n) = v_1 \otimes \cdots \otimes v_n$ is multilinear. In fact, this map is in some sense the most general multilinear map on $V_1 \times \cdots V_n$. Given any other vector space Z and a multilinear map $h : V_1 \times \cdots V_n \to Z$, there exists a unique linear map $f_h : V_1 \otimes \cdots \otimes V_n \to Z$ such that $h = f_h \circ \phi$. As a result, multilinear functions mapping $V_1 \times \cdots \times V_n \to Z$ naturally correspond to linear functions mapping $V_1 \otimes \cdots \otimes V_n \to Z$, and vice versa. This is known as the universal property of a tensor product space. This correspondence will be essential to proving correctness of Algorithm 2 and thereby the main theorem of this work.

Abusing our notation for the set of linear maps, we will write $Y^{V_1 \otimes \cdots \otimes V_n}$ to denote the set of all multilinear functions mapping $V_1 \times \cdots \times V_n$ into Y.

SpanSearch Solver for Simple Tensors. We will also need an algorithm which can solve the following problem: Suppose we are given $m + 1$ simple tensors $\mathbf{u}^0, \ldots, \mathbf{u}^m$ in $V_1 \otimes \cdots \otimes V_n$, with each $\mathbf{u}^i$ given in the form $\mathbf{v}_1^i \otimes \cdots \otimes \mathbf{v}_n^i$. Can we find coefficients $c_1, \ldots, c_m \in \mathbb{F}$ such that $\mathbf{u}^0 = \sum_{i=1}^m c_i \mathbf{u}^i$? Or is $\mathbf{u}^0$ independent of $\mathbf{u}^1, \ldots, \mathbf{u}^m$? The standard linear algebra algorithm of Gaussian elimination takes time which is $\mathsf{poly}(\prod_i \dim(V_i))$, which is exponential in the problem description length n due to the simple tensors' succinct representation.

[BW04] gives a deterministic polynomial-time algorithm solving this problem, which we will use in our security proof. For completeness, we reproduce a version of their simpler randomized algorithm in the full version of this paper.

2.3   Decisional Diffie-Hellman Assumption

We define an adversary's *advantage* in distinguishing distributions:

Definition 1. *We say that a probabilistic algorithm $\mathcal{A}$ has advantage $|\epsilon|$ in distinguishing distributions $\mathcal{D}_0$ and $\mathcal{D}_1$ if*

$$\Pr\left[\mathcal{A}(x_b) = b | x_0 \leftarrow \mathcal{D}_0, x_1 \leftarrow \mathcal{D}_1, b \leftarrow \{0,1\}\right] = \frac{1}{2} + \epsilon.$$

We now recall the standard DDH assumption. For self-consistency of notation we will denote the group operation additively, even though the DDH assumption is more commonly presented with a multiplicative group operation[1]. Suppose a

---

[1] This is probably because the first groups suspected to satisfy the Diffie-Hellman assumption were subgroups of $\mathbb{Z}_p^*$.

group G with generator g and prime order p are fixed, and denote by T_G the time to perform the group operation.

Definition 2. *Define $DDH_\mathbf{R}$ as the distribution of*

$$(ag, bg, cg)$$

where a, b, and c are chosen independently and uniformly at random from $\mathbf{F}_p$.

Definition 3. *Define $DDH_\mathbf{PR}$ as the distribution of*

$$(ag, bg, abg)$$

where a and b are chosen independently and uniformly at random from $\mathbf{F}_p$.

Assumption 1 $((\tau, \epsilon)$-**DDH**) *All probabilistic algorithms $\mathcal{A}$ running in time at most τ have advantage at most ϵ in distinguishing $DDH_\mathbf{R}$ from $DDH_\mathbf{PR}$.*

The standard DDH assumption postulates an ensemble of groups $\{\mathcal{G}_\lambda\}_{\lambda \in \mathbb{N}}$ such that when $G \leftarrow \mathcal{G}_\lambda$, G satisfies $(\mathsf{poly}(\lambda), \mathsf{negl}(\lambda))$-DDH.

$(d \times T)$-Matrix DDH. Our proof of security will use the Matrix DDH assumption of Boneh et al. [BHHO08], which is known to follow from the standard DDH assumption.

Definition 4. *Define $I_\mathbf{R}^{d \times T}$ as the distribution of $\mathbf{C}g$ when $\mathbf{C}$ is chosen uniformly at random from $\mathbf{F}_p^{d \times T}$.*

Definition 5. *Define $I_\mathbf{PR}^{d \times T}$ as the distribution of $\mathbf{ab}^t g$ where $\mathbf{a}$ and $\mathbf{b}$ are chosen uniformly at random from $\mathbf{F}_p^d$ and $\mathbf{F}_p^T$ respectively.*

Boneh et al. prove the following (which in their paper is also Lemma 1).

Lemma 1 ([BHHO08]). *For every (d, T), if there is an adversary $\mathcal{A}$ distinguishing $I_\mathbf{PR}^{d \times T}$ from $I_\mathbf{R}^{d \times T}$ with advantage ϵ in time τ, there is a distinguisher $\mathcal{D}$ which distinguishes $DDH_\mathbf{PR}$ from $DDH_\mathbf{R}$ with advantage ϵ/d in time $\tau + O(T_G \cdot d \cdot T \log p)$, where T_G is the time to perform the group operation in G.*

3   Definition

The security definition for a multilinear pseudorandom function family parallels the usual definition of a pseudorandom function family. That is, oracle access to a multilinear pseudorandom function must be indistinguishable from oracle access to a random multilinear function.

Definition 6. *Syntactically, a multilinear pseudorandom function family consists of a probabilistic polynomial-time algorithm KeyGen and a deterministic polynomial-time algorithm Eval.*

- KeyGen(1^λ): KeyGen *takes a security parameter in unary.* KeyGen *outputs* *secret key* K, *and also outputs as public parameters a field* $\mathbb{F}$, *input vecto spaces* $V_1, \ldots, V_n$, *and a codomain vector space* Y.
- Eval($K, \mathbf{v}_1, \ldots, \mathbf{v}_n$): Eval *takes as input a secret key* K *and vectors* $(\mathbf{v}_1, \ldots \mathbf{v}_n)$, *and outputs a vector* $y = F_K(\mathbf{v}_1, \ldots, \mathbf{v}_n)$ *in the codomain.*

KeyGen *and* Eval *must satisfy* security*: for all probabilistic polynomial-time algorithms* $\mathcal{A}$,

$$\Pr\left[\mathcal{A}^{F_b}(PP, 1^\lambda) = b \,\middle|\, \begin{array}{l} (K, PP) \leftarrow \mathsf{KeyGen}(1^\lambda), b \leftarrow \{0, 1\} \\ F_0 = \mathsf{Eval}(K, \cdot), F_1 \leftarrow Y^{V_1 \otimes \cdots \otimes V_n} \end{array}\right] \leq \frac{1}{2} + \mathsf{negl}(\lambda)$$

Truthfulness. While our definition only requires *indistinguishability* from a random multilinear function, Construction 1 is actually multilinear itself. This fact allows our construction to satisfy the definition of an aggregate PRF, as discussed in Section 5. We adopt the terminology of [GGN03], calling this property "truthfulness".

Remark 1. One can imagine variants on Definition 6. Specifically, we imagine specifying the domain and codomain arbitrarily rather than receiving them as outputs of KeyGen. Our construction achieves this in a limited sense; we can specify n and $\dim(V_1), \ldots, \dim(V_n)$, but Y must always be a DDH-hard group of large prime order, and $\mathbb{F}$ must be $\mathbf{F}_p$. Constructing a multilinear pseudorandom function family over arbitrary finite fields or rings is an intriguing open question. A special case of this question was posed by [GGN03].Paraphrased in our terminology, they asked whether there is a multilinear pseudorandom function family mapping $\mathbf{F}_2^2 \times \cdots \times \mathbf{F}_2^2 \to \mathbf{F}_2$.

One might naively attempt to solve this by composing our construction with a homomorphism from Y to $\mathbf{F}_2$. Unfortunately, in our construction Y must be a DDH-hard group, so no such homomorphism can be efficiently computable.

4   Construction

We now construct a multilinear pseudorandom function family based on DDH-hard groups. Given as public parameters a DDH-hard group G of order p with generator g, and arbitrary dimensions $d_1, \ldots, d_n$, we construct a multilinear pseudorandom function family $\mathcal{F}_{d_1, \ldots, d_n}$ mapping $\mathbf{F}_p^{d_1} \times \cdots \times \mathbf{F}_p^{d_n} \to G$. The security of our construction is determined only by the choice of G, and so we have no explicit security parameter, in contrast to Definition 6. To match the definition, one can easily let KeyGen generate a group in which the (assumed) hardness of the DDH problem corresponds to the given security parameter.

Construction 1. $\mathcal{F}_{d_1, \ldots, d_n}$ is defined by

- KeyGen(): KeyGen samples vectors $\mathbf{w}_1, \ldots, \mathbf{w}_n$, where $\mathbf{w}_i \leftarrow \mathbf{F}_p^{d_i}$ is sampled uniformly at random. It returns the secret key $K = (\mathbf{w}_1, \ldots, \mathbf{w}_n)$.
- Eval($K, (\mathbf{v}_1, \ldots, \mathbf{v}_n)$): Eval returns $\left(\prod_{i=1}^n \langle \mathbf{w}_i, \mathbf{v}_i \rangle\right) g$, where $\langle \cdot, \cdot \rangle$ denotes the inner product.

Remark 2. This construction generalizes the Naor-Reingold PRF[NR04], but we allow richer queries. Specifically, to recover the Naor-Reingold construction, set each $d_i = 2$, and restrict each $\mathbf{v}_i$ to be a basis vector.

Remark 3 (Truthfulness). Every function in $\mathcal{F}_{d_1,\ldots,d_n}$ is truly multilinear (not just indistinguishable from multilinear). This follows from the bilinearity of the inner product and the multilinearity of multiplication (e.g. $(x, y, z) \mapsto xyz$ is multilinear).

4.1 Proof of Security

Our main security proof is the following theorem.

Theorem 1. *When instantiated with a DDH-hard group G, Construction 1 satisfies Definition 6.*

Specifically, if there is an algorithm $\mathcal{A}$ running in time T such that

$$\Pr\left[\mathcal{A}^{F_b}(1^\lambda) = b \,\middle|\, \begin{array}{l} K \leftarrow \mathsf{KeyGen}(), F_0 = \mathsf{Eval}(K, \cdot), \\ F_1 \leftarrow G^{\mathbf{F}_p^{d_1} \otimes \cdots \otimes \mathbf{F}_p^{d_n}}, b \leftarrow \{0,1\} \end{array}\right] = \frac{1}{2} + \epsilon$$

then there is a distinguisher $\mathcal{D}$ running in time $\mathsf{poly}(T, T_G, \sum_i d_i)$ which distinguishes $DDH_{\mathbf{PR}}$ from $DDH_{\mathbf{R}}$ with advantage at least $\frac{|\epsilon|}{n \cdot \max_i d_i}$.

Proof Overview. In this overview, we outline a proof by induction on n. In our actual proof we "unroll" the induction and prove the theorem directly.

When $n = 1$, our construction is a truly random linear function mapping $V_1 \to G$, given by $\mathbf{v} \mapsto \langle \mathbf{w}, \mathbf{v} \rangle g$ for randomly chosen $\mathbf{w}$ and generator g.

We now show that an oracle implementing our construction is pseudorandom for $n > 1$. By definition, $F_0(\mathbf{v}_1, \ldots, \mathbf{v}_n)$ is equal to $\langle \mathbf{w}_n, \mathbf{v}_n \rangle \prod_{i=1}^{n-1} \langle \mathbf{w}_i, \mathbf{v}_i, \rangle g$. By the inductive hypothesis, oracle access to $\langle \mathbf{w}_n, \mathbf{v}_n \rangle R_{n-1}(\mathbf{v}_1, \ldots, \mathbf{v}_n)$ is indistinguishable, where R_{n-1} is a truly random multilinear function in $G^{V_1 \otimes \cdots \otimes V_{n-1}}$. Although $\dim(V_1 \otimes \cdots \otimes V_{n-1})$ is exponential in n, we are able to efficiently implement an oracle to R_{n-1} in a stateful manner.

It remains to show that oracle access to $\langle \mathbf{w}_n, \mathbf{v}_n \rangle R_{n-1}$ is indistinguishable from a random multilinear function $F_1 = R_n \leftarrow G^{V_1 \otimes \cdots \otimes V_n}$. We show that a distinguisher $\mathcal{A}$ of oracle access to R_n from oracle access to $\langle \mathbf{w}_n, \mathbf{v}_n \rangle R_{n-1}$ violates the Matrix DDH assumption. This indistinguishability relies on two different ways of statefully implementing any R_n, given in Algorithm 1 and Algorithm 2.

While we described the proof as an induction, directly applying these ideas in our main proof does not yield an efficient reduction. Below, we use the standard hybrid argument technique to avoid this pitfall.

Our proof relies on two different algorithms for statefully and efficiently implementing oracle access to a random multilinear function, R_n from $V_1 \times \cdots \times V_n$ to G. We can instead consider R_n as a random linear function from $V = V_1 \otimes \cdots \otimes V_n$ to G, using the correspondence described in the preliminaries. Because we consider linear functions on $V_1 \otimes \cdots \otimes V_n$ only as a tool to describe multilinear functions on $V_1 \times \cdots \times V_n$, we are able to restrict our attention to *simple* tensors in the analysis.

Algorithm 1. We maintain a map M which stores a subset of a mapping $V \to Y$. That is M stores a collection of pairs $(\mathbf{u} \mapsto \mathbf{y})$; we say that $M(\mathbf{u}) = \mathbf{y}$ if such an entry for $\mathbf{u}$ exists in M, and that $M(\mathbf{u}) = \bot$ otherwise. Initially M is the empty set. A query $\mathbf{v}$ is answered by executing the following steps:

1. Check whether $\{\mathbf{v}\} \cup \{\mathbf{u} : M(\mathbf{u}) \neq \bot\}$ is linearly independent. If it is, sample a random vector $\mathbf{y} \leftarrow Y$ and add the mapping $(\mathbf{v} \mapsto \mathbf{y})$ to M.
2. Compute $\mathbf{v} = \sum_j c_j \mathbf{u}_j$ where for each j, $M(\mathbf{u}_j) \neq \bot$.
3. Return $\sum_j c_j M(\mathbf{u}_j)$.

The efficiency of Steps 1 and 2 relies on the SpanSearch algorithm, which works for simple tensors.

Proposition 1. *Algorithm 1 implements a random linear function mapping* $V \to Y$.

Proof. Suppose the queries up to time t are given by $\mathbf{v}_1, \ldots, \mathbf{v}_t \in V$. Let $(\mathbf{v}_{i_1} \mapsto \mathbf{y}_{i_1}), \ldots, (\mathbf{v}_{i_j} \mapsto \mathbf{y}_{i_j})$ be the first j entries in M. The vectors $\mathbf{v}_{i_1}, \ldots, \mathbf{v}_{i_j}$ are a basis for $\mathsf{span}(\mathbf{v}_1, \ldots, \mathbf{v}_t)$. It is easy to see that Algorithm 1 implements a linear map on $\mathsf{span}(\mathbf{v}_1, \ldots, \mathbf{v}_t)$ which is given by a random matrix. In particular, this matrix has columns $\mathbf{y}_{i_1}, \ldots, \mathbf{y}_{i_j}$.

We now give an alternate algorithm implementing a random linear function mapping $U \otimes W \to Y$ for any vector spaces U, W, and Y. In particular, we will take $U = V_1 \otimes \cdots \otimes V_{j-1}$ and $W = V_j$.

Algorithm 2. Queries are of the form $\mathbf{u} \otimes \mathbf{w} \in U \otimes W$. We maintain a map M which stores a subset of a mapping $U \to Y^W$. That is M stores a collection of pairs $(\mathbf{z} \mapsto f)$, where each f is a linear map from W to Y. We say that $M(\mathbf{z}) = f$ if such an entry for $\mathbf{z}$ exists in M, and that $M(\mathbf{z}) = \bot$ otherwise. Initially M is the empty set. A query $\mathbf{u} \otimes \mathbf{w}$ is answered by executing the following steps:

1. Check whether $\{\mathbf{u}\} \cup \{\mathbf{z} : M(\mathbf{z}) \neq \bot\}$ is linearly independent. If it is, sample a random linear map $f : W \to Y$ and add the mapping $(\mathbf{z} \mapsto f)$ to M.
2. Write $\mathbf{u} = \sum_j c_j \mathbf{z}_j$ where for each j, $M(\mathbf{z}_j) = f_j$.
3. Return $\sum_j c_j f_j(\mathbf{w})$.

Proposition 2. *Algorithm 2 implements a random linear function mapping* $U \otimes W \to Y$.

Proof. A linear function mapping U to the space Y^W of linear functions from W to Y can be equivalently viewed as a bilinear function mapping $U \times W \to Y$. As discussed in the preliminaries, there is a bijective correspondence between such bilinear functions and linear functions mapping $U \otimes W \to Y$. Then Proposition 2 is just a special case of Proposition 1.

The main lemma used in the proof of Theorem 1 is that the following two distributions on linear functions are indistinguishable.

Definition 7. *For* $j > 0$*, let* $\mathcal{RF}_j$ *denote* $G^{\mathbf{F}_p^{d_1} \otimes \cdots \otimes \mathbf{F}_p^{d_j}}$. *Let* $\mathcal{RF}_0 = G$.

Definition 8. *For $j > 0$, let $\mathcal{PRF}_j$ denote the distribution of multilinear functions defined by*

$$(\mathbf{v}_1 \otimes \cdots \otimes \mathbf{v}_j) \mapsto \langle \mathbf{w}, \mathbf{v}_j \rangle R(\mathbf{v}_1 \otimes \cdots \otimes \mathbf{v}_{j-1})$$

where $\mathbf{w}$ is sampled from $\mathbf{F}_p^{d_j}$ and R is sampled from $\mathcal{RF}_{j-1}$.

Lemma 2. *If there is an oracle algorithm $\mathcal{A}$ running in time T such that*

$$\Pr\left[\mathcal{A}^{F_b}() = b \big| F_0 \leftarrow \mathcal{PRF}_j, F_1 \leftarrow \mathcal{RF}_j, b \leftarrow \{0,1\}\right] = \frac{1}{2} + \epsilon$$

then there is a distinguisher $\mathcal{D}$ running in time $\mathsf{poly}(T, \sum_{i \leq j} d_i)$ such that $\mathcal{D}$ breaks Matrix DDH with the same advantage. That is,

$$\Pr\left[\mathcal{D}(M_b) = b \big| M_0 \leftarrow I_{\mathbf{PR}}^{d_j \times T}, M_1 \leftarrow I_{\mathbf{PR}}^{d_j \times T}, b \leftarrow \{0,1\}\right] = \frac{1}{2} + \epsilon.$$

The distinguisher $\mathcal{D}$ is defined to execute the following steps:

1. Take $\tilde{\mathbf{C}}g$ as input. Here $\tilde{\mathbf{C}}$ is either equal to $\mathbf{ab}^t$ for random $\mathbf{a} \in \mathbf{F}_p^{d_j}$ and $\mathbf{b} \in \mathbf{F}_p^T$ or is sampled uniformly at random $C \leftarrow \mathbf{F}_p^{d_j \times T}$. Denote the kth column of $\tilde{\mathbf{C}}g$ by $\boldsymbol{\gamma}_k$
2. Create an (initially empty) map M to store a subset of $\mathbf{F}_p^{d_1} \times \cdots \times \mathbf{F}_p^{d_{j-1}} \to G^{d_j}$. That is M stores a collection of pairs $(\mathbf{v}_1 \otimes \cdots \otimes \mathbf{v}_{j-1} \mapsto \mathbf{g})$, where each $\mathbf{g} \in G^{d_j}$. We will preserve the invariant that $\{\mathbf{u} : M(\mathbf{u}) \neq \bot\}$ is linearly independent.
3. Run the adversary $\mathcal{A}()$, answering queries as follows:
   On the ith query $\mathbf{v}_1^i \otimes \cdots \otimes \mathbf{v}_{j-1}^i \otimes \mathbf{v}_j^i$, first define $\mathbf{v}_{-j}^i = \mathbf{v}_1^i \otimes \cdots \otimes \mathbf{v}_{j-1}^i$. Use our SpanSearch solver to check whether $\{\mathbf{v}_{-j}^i\} \cup \{\mathbf{u} : M(\mathbf{u}) \neq \bot\}$ is linearly independent. If it is, add the mapping $(\mathbf{v}_{-j}^i \mapsto \boldsymbol{\gamma}_i)$ to M.
   Otherwise, our SpanSearch solver tells us how to write $\mathbf{v}_{-j}^i$ as $\sum_k \alpha_k \mathbf{u}_k$, where each $M(\mathbf{u}_k)$ is not $\bot$. $\mathcal{D}$ then answers $\mathcal{A}$'s query with $\sum_k \alpha_k \langle M(\mathbf{u}_k), \mathbf{v}_j^i \rangle$.
4. Finally, $\mathcal{D}$ outputs the same answer that $\mathcal{A}$ outputs.

Lemma 2 follows from the following two claims.

Claim. When $\tilde{\mathbf{C}}$ is uniformly random, then $\mathcal{D}$ answers queries according to the same distribution as $\mathcal{RF}_j$.

Proof. This follows from Proposition 2. Namely, when $\tilde{C}$ is uniformly random, the columns $\boldsymbol{\gamma}_i$ define independent and uniformly random linear maps from $\mathbf{F}_p^{d_j} \to G$. $\mathcal{A}$'s queries are therefore answered according to a random multilinear function, which is the same as $\mathcal{RF}_j$.

Claim. When $\tilde{\mathbf{C}}$ is generated as $\mathbf{a}\mathbf{b}^t$, then $\mathcal{D}$ answers queries according to the same distribution as $\mathcal{PRF}_j$.

Proof. Suppose that $\tilde{\mathbf{C}}$ is $\mathbf{a}\mathbf{b}^t$. Then each $\boldsymbol{\gamma}_i$ is $\mathbf{a}b_i$, where each b_i is sampled independently and uniformly at random from $\mathbf{F}_p$. So $\mathcal{D}$ can equivalently change M to only store $(\mathbf{v}_{-j}^i \mapsto b_i g)$ and now answers queries with $\langle \mathbf{a}, \mathbf{v}_j^i \rangle (\sum_k \alpha_k M(\mathbf{u}_k))$. By Proposition 1, this is the same as $\langle \mathbf{a}, \mathbf{v}_j^i \rangle R(\phi(\mathbf{v}_{-j}^i))$ with R sampled from $\mathcal{RF}_j$. By definition, this is the same as answering queries with a randomly sampled R from $\mathcal{PRF}_j$.

We can now prove Theorem 1.

Proof (of Theorem 1). We define distinguishers $\mathcal{D}_J$ for each $J \in \{0, \ldots, n\}$ that execute the following steps:

1. Prepare a stateful implementation $R \leftarrow \mathcal{RF}_J$ using Algorithm 1 backed by our SpanSearch solver.
2. Sample $\mathbf{w}_i$ uniformly at random from $\mathbf{F}_p^{d_i}$ for each $i \in \{J+1, \ldots, n\}$.
3. Run $\mathcal{A}$, answering its queries $(\mathbf{v}_1, \ldots, \mathbf{v}_n)$ with $\left(\prod_{i=J+1}^n \langle \mathbf{w}_i, \mathbf{v}_i \rangle\right) R(\mathbf{v}_1, \ldots, \mathbf{v}_J)$
4. Output whatever $\mathcal{A}$ outputs.

First, it is clear that the output of $\mathcal{D}_0$ is the same as the output of $\mathcal{A}^{\mathsf{Eval}(K, \cdot)}()$ where $K \leftarrow \mathsf{KeyGen}()$, and the output of $\mathcal{D}_n$ is the same as the output of $\mathcal{A}^{F_1}()$ where $F_1 \leftarrow \mathcal{RF}_n$. By a standard hybrid argument, there must exist some $j \in \{0, \ldots, n-1\}$ such that $\mathcal{D}_j$ and $\mathcal{D}_{j+1}$ output 1 with probabilities differing by at least $|\epsilon|/n$.

But if we replace $\mathcal{D}_{j+1}$'s (black-box) usage of $F \leftarrow \mathcal{RF}_{j+1}$ by $F \leftarrow \mathcal{PRF}_{j+1}$, then $\mathcal{D}_{j+1}$ is functionally equivalent to $\mathcal{D}_j$. So $\mathcal{D}_{j+1}$ can be used to distinguish oracle access to $\mathcal{RF}_{j+1}$ from oracle access to $\mathcal{PRF}_{j+1}$. Lemma 2 implies that $\mathcal{D}_{j+1}$ can be used to distinguish $I_{\mathbf{R}}^{d_{j+1} \times Q}$ from $I_{\mathbf{PR}}^{d_{j+1} \times Q}$ with advantage at least $|\epsilon|/n$, where Q is any bound on the number of linearly independent queries made by $\mathcal{A}$. In particular $Q \leq T$. Lemma 1 implies that $\mathcal{D}_{j+1}$ can be used to distinguish $DDH_{\mathbf{R}}$ from $DDH_{\mathbf{PR}}$ with advantage at least $\frac{|\epsilon|}{n \cdot d_{j+1}}$. $\square$

5   Applications

In this section, we show how our multilinear PRF simplifies and improves PRF constructions in [BGV11] and [CGV15]. We instantiate the vector spaces $F_p^{d_i}$ of Construction 1 appropriately, and show that oracle access to a multilinear PRF suffices to perfectly simulate oracle access to the functions from those works.

Aggregate PRFs [CGV15] are PRF families with extra efficiency and security properties. First, the key K for a PRF f enables efficient computation of $\mathsf{Agg}_f(S) = \sum_{x \in S} f(x)$ for some class of succinctly described, but possibly exponentially large, sets S. Second, no efficient algorithm can distinguish oracle access to $f(\cdot)$ and $\mathsf{Agg}_f(\cdot)$ from oracle access to $g(\cdot)$ and $\mathsf{Agg}_g(\cdot)$, where g is a truly random function.

One specific setting that [CGV15] addresses is when S can be any "hypercube". A hypercube $H_p \subset \{0,1\}^n$ is described by a pattern $p \in \{\{0\}, \{1\}, \{0,1\}\}^n$. H_p is defined as $\{x \in \{0,1\}^n : x_i \in p_i\}$. Informally, H_p is the set obtained by fixing the bits of x at particular indices, and allowing all other bits to vary freely. [CGV15] showed a construction with efficent evaluation, but security relied on the subexponential hardness of the DDH problem.

We show that the hypercube construction in [CGV15] is a special case of Construction 1. The correctness of the aggregate queries is implied by the truthfulness of our construction. Thus we prove aggregate security of their construction relying only on the standard DDH assumption.

Corollary 1. *Assuming the (polynomial) hardness of DDH over the group G, [CGV15]'s PRFs for hypercubes[2] and decision trees [3] are secure aggregate PRFs.*

Proof. As shown in [CGV15], it suffices to prove the case of hypercubes.

Let B denote the 2-dimensional vector space whose basis vectors are $|0\rangle$ and $|1\rangle$. Our construction gives a pseudorandom multilinear function F mapping B^n to G. This function F induces a pseudorandom function $f : \{0,1\}^n \to G$ given by $f(b_1 \ldots b_n) = F(|b_1\rangle, \ldots, |b_n\rangle)$.

First observe we can compute the sum of $f(x)$ for all x as

$$F(|0\rangle + |1\rangle, \ldots, |0\rangle + |1\rangle).$$

To fix a bit x_i to b – thus aggregating over a smaller hypercube – we replace the ith argument of F above with $|b\rangle$. That is, to compute $\mathsf{Agg}_f(H_p)$ for some hypercube H_p, we evaluate

$$F\left(\sum_{b \in p_1} |b\rangle, \ldots, \sum_{b \in p_n} |b\rangle\right)$$

This yields the correct aggregate value by the truthfulness (multilinearity) of our construction. Therefore oracle access to Agg_f can be simulated with even a restricted oracle to F. Namely, we only make queries where each argument is either $|0\rangle$, $|1\rangle$, or $|0\rangle + |1\rangle$. Theorem 1 then implies aggregate PRF security of f. □

We can actually achieve the more generalized aggregation, as required in the work of Benabbas, Gennaro, and Vahlis [BGV11] on algebraic PRFs, while maintaining aggregate security. For example, efficiently evaluating

$$p_f(z) = \sum_{x \in \{0,1\}^n} f(x) z^n$$

has applications in verifiable and multiparty computation [BGV11, Haz15]. We can achieve this functionality with oracle access to our multilinear PRF F, thus

---

[2] Section 3.2.

[3] Section 3.3.

keeping aggregate security. Specifically, instantiate Construction 1 as above. One can then compute

$$p_f(z) = F\big(|0\rangle + z^{2^{n-1}}|1\rangle \ , \ \ldots \ , \ |0\rangle + z\,|1\rangle\big).$$

Correctness and security follow directly because F is a pseudorandom multilinear function. This can easily be extended to cover the more general multivariable algebraic PRF considered in [BGV11] (Section 4.2), along with a number of other immediate generalizations.

Each of the above applications uses only the simplest of vector spaces. There are many other ways in which a multilinear PRF can be invoked, but we highlight these two examples as applications which have already appeared in the literature.

6   Extensions

Two other classes of functions which are fundamental in mathematics are *symmetric* and *skew-symmetric* multilinear functions. Informally, a function from $V^n \to Y$ is symmetric if swapping any arguments x_i and x_j does not affect the value, and is skew-symmetric if such a swap negates the value. Pseudorandom implementations of these classes of functions are interesting open problems.

Definition 9. *A function $F : V^n \to Y$ is said to be* symmetric *if for all $i \neq j$,*

$$F(x_1, \ldots, x_n) = F(x_1', \ldots, x_n'),$$

where

$$x_k' = \begin{cases} x_i & \text{if } k = j \\ x_j & \text{if } k = i \\ x_k & \text{otherwise.} \end{cases}$$

F is said to be skew-symmetric *if $F(x_1, \ldots, x_n) = -F(x_1', \ldots, x_n')$.*

Given a group G of order p with generator g, we present a candidate construction of a symmetric multilinear pseudorandom function family,

Construction 2. $\mathcal{F}_{d,n}$ is defined by

– KeyGen(): KeyGen samples a vector $\mathbf{w}$ uniformly at random from $\mathbf{F}_p^d$.
– Eval($\mathbf{w}, \mathbf{v}_1, \ldots, \mathbf{v}_n$)): Eval returns $\big(\prod_{i=1}^n \langle \mathbf{w}, \mathbf{v}_i \rangle\big) g$, where $\langle \cdot, \cdot \rangle$ denotes the inner product.

This is a modification to Construction 1 in which $\mathbf{w}_1 = \cdots = \mathbf{w}_n$, which clearly yields symmetric multilinear functions, but security is less clear.

In case $d = 2$, security reduces to the n-Strong DDH assumption. This assumption states that $(h, xh, \ldots, x^n h)$ is indistinguishable from $n + 1$ random elements of G, when h is a randomly chosen generator of G, and x is a random element of $\mathbf{F}_p$. This is because a symmetric multilinear function on $(\mathbf{F}_p^2)^n$ is defined by $n + 1$ "basis" values, which in the above construction correspond to this tuple.

Conjecture 1. With a suitably chosen group G, Construction 2 defines a symmetric multilinear family of pseudorandom functions.

Acknowledgements. The authors would like to thank Shafi Goldwasser and Vinod Vaikuntanathan for their helpful discussions and mentorship. Aloni Cohen's research was supported in part by the NSF Graduate Student Fellowship. Justin Holmgren's research was supported in part by the NSF MACS project.

References

[BGV11] Benabbas, S., Gennaro, R., Vahlis, Y.: Verifiable delegation of computation over large datasets. In: Rogaway, P. (ed.) CRYPTO 2011. LNCS, vol. 6841, pp. 111–131. Springer, Heidelberg (2011)

[BHHO08] Boneh, D., Halevi, S., Hamburg, M., Ostrovsky, R.: Circular-secure encryption from decision diffie-hellman. In: Wagner, D. (ed.) CRYPTO 2008. LNCS, vol. 5157, pp. 108–125. Springer, Heidelberg (2008)

[BW04] Bogdanov, A., Wee, H.M.: A stateful implementation of a random function supporting parity queries over hypercubes. In: Jansen, K., Khanna, S., Rolim, J.D.P., Ron, D. (eds.) RANDOM 2004 and APPROX 2004. LNCS, vol. 3122, pp. 298–309. Springer, Heidelberg (2004)

[CGV15] Cohen, A., Goldwasser, S., Vaikuntanathan, V.: Aggregate pseudorandom functions and connections to learning. In: Dodis, Y., Nielsen, J.B. (eds.) TCC 2015, Part II. LNCS, vol. 9015, pp. 61–89. Springer, Heidelberg (2015)

[GGN03] Goldreich, O., Goldwasser, S., Nussboim, A.: On the implementation of huge random objects. In: Proceedings of the 44th Annual Symposium on Foundations of Computer Science, pp. 68–79 (2003)

[Haz15] Hazay, C.: Oblivious polynomial evaluation and secure set-intersection from algebraic PRFs. In: Dodis, Y., Nielsen, J.B. (eds.) TCC 2015, Part II. LNCS, vol. 9015, pp. 90–120. Springer, Heidelberg (2015)

[NR04] Naor, M., Reingold, O.: Number-theoretic constructions of efficient pseudorandom functions. Journal of the ACM (JACM) **51**(2), 231–262 (2004)

Zero-Fixing Extractors
for Sub-Logarithmic Entropy

Gil Cohen[1] and Igor Shinkar[2]([✉])

[1] Department of Computer Science and Applied Mathematics,
Weizmann Institute of Science, Rehovot 76100, Israel
[2] Courant Institute of Mathematical Sciences, New York University, New York, USA
ishinkar@cims.nyu.edu

Abstract. An (n, k)-bit-fixing source is a distribution on n bit strings, that is fixed on $n - k$ of the coordinates, and jointly uniform on the remaining k bits. Explicit constructions of bit-fixing extractors by Gabizon, Raz and Shaltiel [SICOMP 2006] and Rao [CCC 2009], extract $(1 - o(1)) \cdot k$ bits for $k = \operatorname{poly} \log n$, almost matching the probabilistic argument. Intriguingly, unlike other well-studied sources of randomness, a result of Kamp and Zuckerman [SICOMP 2006] shows that, for *any* k, some small portion of the entropy in an (n, k)-bit-fixing source can be extracted. Although the extractor does not extract all the entropy, it does extract $\log(k)/2$ bits.

In this paper we prove that when the entropy k is small enough compared to n, this exponential entropy-loss is unavoidable. More precisely, we show that for $n > \operatorname{Tower}(k^2)$ one cannot extract more than $\log(k)/2 + O(1)$ bits from (n, k)-bit-fixing sources. The remaining entropy is inaccessible, information theoretically. By the Kamp-Zuckerman construction, this negative result is tight. For small enough k, this strengthens a result by Reshef and Vadhan [RSA 2013], who proved a similar bound for extractors computable by space-bounded streaming algorithms.

Our impossibility result also holds for what we call *zero-fixing* sources. These are bit-fixing sources where the fixed bits are set to 0. We complement our negative result, by giving an explicit construction of an (n, k)-zero-fixing extractor that outputs $\Omega(k)$ bits for $k \geq \operatorname{poly} \log \log n$. Finally, we give a construction of an (n, k)-bit-fixing extractor, that outputs $k - O(1)$ bits, for entropy $k = (1 + o(1)) \cdot \log \log n$, with running-time $n^{O((\log \log n)^2)}$. This answers an open problem by Reshef and Vadhan [RSA 2013].

1 Introduction

Randomness is an invaluable resource in many areas of theoretical computer science, such as algorithm design, data structures and cryptography. For many

---

G. Cohen—Supported by an ISF grant and by the I-CORE Program of the Planning and Budgeting Committee.

I. Shinkar—Research supported by NSF grants CCF 1422159, 1061938, 0832795 and Simons Collaboration on Algorithms and Geometry grant.

© Springer-Verlag Berlin Heidelberg 2015
M.M. Halldórsson et al. (Eds.): ICALP 2015, Part I, LNCS 9134, pp. 343–354, 2015.
DOI: 10.1007/978-3-662-47672-7_28

omputational tasks, the best known algorithms assume that random bits are
o their disposal. In cryptography and in distributed computing, randomness is,
rovably, a necessity. Nevertheless, truly random bits are not always available. A
ource of randomness might be defective, producing random bits that are biased
nd correlated. Even a sample from an ideal source of randomness can suffer
uch defects due to information leakage. Motivated by this problem, the notion
f randomness extractors was introduced.

Broadly speaking, a *randomness extractor* is a function that extracts
lmost truly random bits given a sample from a defective source of random-
ess. Well-known instantiations are seeded extractors [NZ96, GUV09, DKSS09],
wo-source extractors [CG88, Raz05, Bou05], and more generally multi-source
xtractors [Raz05, BKS+05, BIW06, Rao09a, Li11a, Li13], as well as affine extrac-
ors [Bou07, Yeh11, Li11b]. Randomness extractors are central objects in pseudo-
andomness, with many applications beyond their original motivation. Over the
ast 30 years, a significant research effort was directed towards the construction
f randomness extractors in different settings. We refer the reader to Shaltiel's
ntroductory survey on randomness extractors [Sha11] for more information.

3it-fixing Extractors. A well-studied defective source of randomness is a
oit-fixing source. An (n, k)-*bit-fixing source* is a distribution X over $\{0, 1\}^n$,
vhere some $n - k$ of the bits of X are fixed, and the joint distribution of
he remaining k bits is uniform. The problem of extracting randomness from
oit-fixing sources was initiated in the works of [Vaz85, BBR85, CGH+85], moti-
rated by applications to fault-tolerance, cryptography and communication com-
olexity. More recently, bit-fixing extractors have found applications to formulae
ower bounds [KRT13], and for compression algorithms for "easy" Boolean func-
ions [CKK+13].

The early works on bit-fixing extractors were concentrated on positive and
negative results for extracting a truly uniform string. In [CGH+85], it was
observed that one can efficiently extract a uniform bit even from $(n, 1)$-bit-
fixing sources, simply by XOR-ing all the input bits. In a sharp contrast, it
was shown that extracting two jointly uniform bits cannot be done even from
$(n, n/3 - 1)$-bit-fixing sources. Given this state of affairs, early works dealt
with what we call "the high-entropy regime". Using a relation to error correcting codes, Chor et al. [CGH+85] showed how to efficiently extract roughly
$n - t \cdot \log_2(n/t)$ truly uniform output bits from $(n, n - t)$-bit-fixing sources, with
$t = o(n)$. The authors complemented this result by an almost matching upper
bound of $n - (t/2) \cdot \log_2(n/t)$ on the number of truly uniform output bits one can
extract. In the same paper, some results were obtained also for (n, k)-bit-fixing
sources, where k is slightly below $n/2$. Further lower bounds for this regime of
parameters were obtained by Friedman [Fri92].

These negative results naturally led to study the relaxation, where the output of the extractor is only required to be close to uniform, in statistical distance.[1] A simple probabilistic argument can be used to show that, computational aspects aside, one can extract $m = k - 2\log(1/\varepsilon) - O(1)$ bits that are ε-close to uniform from any (n, k)-bit-fixing source, as long as $k \geq \log(n) + 2\log(1/\varepsilon) + O(1)$. For simplicity, in the rest of this section we think of ε as a small constant. Thus, in particular, by allowing for some small constant error $\varepsilon > 0$, one can extract almost all the entropy k from any (n, k)-bit-fixing source, even for k as low as $\log(n) + O(1)$. We call the range $\log n \leq k \leq o(n)$, "the low-entropy regime".

The probabilistic argument mentioned above only yields an existential proof, whereas efficiently computable extractors are far more desired. Kamp and Zuckerman [KZ06] gave the first explicit construction of an (n, k)-bit-fixing extractor with $k = o(n)$. More precisely, for any constant $\gamma > 0$, an explicit $(n, n^{1/2 + \gamma})$-bit-fixing extractor was given, with $\Omega(n^{2\gamma})$ output bits. In a subsequent work Gabizon, Raz and Shaltiel [GRS06] obtained an explicit $(n, \log^c n)$-bit-fixing extractor, where $c > 1$ is some universal constant. Moreover, the latter extractor outputs $(1 - o(1))$-fraction of the entropy, thus getting very close to the parameters of the non-explicit construction obtained by the probabilistic method. Using different techniques, Rao [Rao09b] obtained a bit-fixing extractor with improved dependency on the error ε.

For a vast majority of randomness extraction problems, such as the problem of constructing two-source extractors and affine extractors, a naïve probabilistic argument yields (non-explicit) extractors with essentially optimal parameters. Interestingly, this is not the case for bit-fixing extractors. The first evidence for that comes from the observation mentioned above. Namely, the XOR function is an extractor for $(n, 1)$-bit-fixing sources. A result of Kamp and Zuckerman [KZ06] shows that this is not an isolated incident, and in fact, for any $k \geq 1$ there is an (explicit and simple) extractor for (n, k)-bit-fixing sources, that outputs $\Omega(\log_2(k))$ random bits that are ε-close to uniform for $\varepsilon = \exp(-k^{\Omega(1)})$. This result was later improved and simplified by Reshef and Vadhan [RV13], who showed how to extract $0.5 \cdot (\log k - \log\log(1/\varepsilon))$ bits. On the other hand, one can show that, with high probability, a random function with a single output bit is constant on some bit-fixing source with entropy, say, $\log(n)/10$. Thus, in this setting, *structured* functions outperform *random* functions, in the sense that the former can extract a logarithmic amount of the entropy from bit-fixing sources with arbitrarily low entropy, whereas the latter are constant, with high probability, on some $(n, \log(n)/10)$-bit-fixing source.

Reshef and Vadhan [RV13] considered k that is sub-logarithmic in n – a regime we call the "very low entropy regime". In [RV13] it is shown that any

---

[1] Friedman [Fri92] studied other notions of closeness. Although different measures are of interest, when analyzing extractors, the gold standard measure of closeness between distributions is statistical distance. In this paper we follow the convention, and measure the error of an extractor by the statistical distance of its output to the uniform distribution.

xtractor that is computable by a space-bounded streaming algorithm can output nly $O(\log k)$ bits in this regime.

.1 Our Contribution

)ur first result states that when the entropy k is small enough compared to n, ne cannot extract more than $0.5 \cdot \log_2(k) + O(1)$ bits from an (n, k)-bit-fixing ource, information theoretically. That is, for small enough k, the computational ssumption on the extractor imposed in [RV13] can be removed. Note that this egative result is tight as implied by the constructions of [KZ06, RV13].

In fact, the following impossibility result holds also for what we call *zero-fixing* ources. A random variable X is an (n, k)-zero-fixing source if it is an (n, k)-bit-xing source, where all the fixed bits are set to zero. Zero-fixing sources are atural as they model bit-fixing sources in which the fixed bits are set to some efault value rather than to an arbitrary value.

To state the result, we introduce the function $\mathsf{Tower} \colon \mathbb{N} \to \mathbb{N}$ that is defined s follows: $\mathsf{Tower}(0) = 1$, and for an integer $n \geq 1$, $\mathsf{Tower}(n) = 2^{\mathsf{Tower}(n-1)}$.

Theorem 1. *For any integers* n, k *such that* $\mathsf{Tower}(k^{3/2}) < n$, *the following olds. Let* $\mathsf{Ext} \colon \{0,1\}^n \to \{0,1\}^m$ *be an* (n, k)-*zero-fixing extractor with error* ε. *f* $m > 0.5 \cdot \log_2(k) + O(1)$, *then* $\varepsilon \geq 0.99$.

Since the impossibility result stated in Theorem 1 holds for the more estricted type of sources, namely for zero-fixing sources, it is natural to try nd complement it with feasibility results. Using a naïve probabilistic argu-nent, one can prove the existence of an (n, k)-zero-fixing extractor, for any $\geq \log \log n + \log \log \log n + O(1)$, with $m = k - O(1)$ output bits, where ve treat the error ε as constant, for simplicity. Our second result is an almost natching explicit construction.

Theorem 2. *For any constant* $\mu > 0$, $n, k \in \mathbb{N}$, *such that* $k \geq (\log \log n)^{2+\mu}$, *here exists an efficiently computable function*

$$\mathsf{ZeroBFExt} \colon \{0,1\}^n \to \{0,1\}^m,$$

where $m = \Omega(k)$, *with the following property. For any* (n, k)-*zero-fixing source* X, *it holds that* $\mathsf{ZeroBFExt}(X)$ *is* $(2^{-k^{\Omega(1)}} + (k \log n)^{-\Omega(1)})$-*close to uniform.*

We remark that the techniques used in [GRS06, Rao09b] for the constructions of bit-fixing extractors seem to work only for $k \geq \operatorname{poly} \log n$, even for zero-fixing sources, and new ideas are required in order to exploit the extra structure of zero-fixing sources in order to extract $\Omega(k)$ bits from such sources with sub-logarithmic entropy.

As mentioned, Reshef and Vadhan [RV13] proved that for $k = o(\log n)$, any space-bounded streaming algorithm can extract at most $O(\log k)$ bits. The authors left open the problem of whether or not one can extract $\Omega(k)$ bits for $k = o(\log n)$. Theorem 1 shows that this is impossible for k which is very small compared to n. Nevertheless, in the following theorem we answer the open problem of [RV13] positively and show that one can extract $k - O(1)$ bits even when $k = O(\log \log n)$. For simplicity, we state here the theorem for a constant error ε.

Theorem 3. *For any integers n, k, and constant $\varepsilon > 0$, such that $k > \log \log n + 2 \log \log \log n + O_\varepsilon(1)$, there exists a function*

$$\mathsf{QuasiBFExt} \colon \{0,1\}^n \to \{0,1\}^m,$$

where $m = k - O_\varepsilon(1)$, with the following property. Let X be an (n, k)-bit-fixing source. Then, $\mathsf{QuasiBFExt}(X)$ is ε-close to uniform. The running-time of evaluating $\mathsf{QuasiBFExt}$ is $n^{O_\varepsilon((\log \log n)^2)}$.

Due to space limitations the full proofs of Theorem 1, Theorem 2 and Theorem 3 are omitted from this extended abstract. In Section 3 we give an overview for the proofs of Theorem 1 and Theorem 2. For the sake of clarity, in this section we allow ourselves to be informal and somewhat imprecise.

2 Preliminaries

Throughout the paper we denote by log the logarithm to the base 2. For $n \in \mathbb{N}$, we denote the set $\{1, 2, \ldots, n\}$ by $[n]$. For $n, r \in \mathbb{N}$, we let $\log^{(r)}(n)$ be the composition of the log function with itself r times, applied to n. Formally, $\log^{(0)}(n) = n$, and for $r \geq 1$, we define $\log^{(r)}(n) = \log(\log^{(r-1)}(n))$. For an integer $h \in \mathbb{N}$, we let $\mathsf{Tower}(h)$ be a height h tower of exponents of 2. More formally, $\mathsf{Tower}(0) = 1$, and for $h \geq 1$, $\mathsf{Tower}(h) = 2^{\mathsf{Tower}(h-1)}$.

Sources of Randomness. In this paper we use the following sources of randomness.

Definition 1 (Bit-fixing sources). *Let n, k be integers such that $n \geq k$. A random variable X on n bits is called an (n, k)-bit-fixing source, if there exists $S \subseteq [n]$ with size $|S| = k$, such that $X|_S$ is uniformly distributed, and each X_i with $i \notin S$ is fixed.*

Definition 2 (Affine sources). *Let n, k be integers, with $n \geq k$. A random variable X on n bits is called an (n, k)-affine source, if X is uniformly distributed on some affine subspace $U \subseteq \mathbb{F}_2^n$ of dimension k.*

Definition 3 (Weak sources). *Let n, k be integers such that $n \geq k$. A random variable X on n bits is called an (n, k)-weak source, if for any $x \in \mathsf{supp}(X)$, it holds that $\mathbf{Pr}[X = x] \geq 2^{-k}$.*

Note that any (n, k)-bit-fixing source is an (n, k)-affine source, and any (n, k)-affine source is an (n, k)-weak source. We introduce the following two sources of randomness.

Definition 4 (Zero-fixing sources). *Let n, k be integers such that $n \geq k$. A random variable X on n bits is called an (n, k)-zero-fixing source, if there exists $S \subseteq [n]$ with size $|S| = k$, such that $X|_S$ is uniformly distributed, and each X_i with $i \notin S$ is fixed to zero.*

Definition 5 (Fixed-weight sources). *Let n, k, w be integers, with $n \geq k \geq w$. A random variable $X \subseteq \{0, 1\}^n$ is called an (n, k, w)-fixed-weight source, if there exists $S \subseteq [n]$, with size $|S| = k$, such that a sample from $x \sim X$ is obtained as follows. First, one samples a string $x' \in \{0, 1\}^k$ of weight w, uniformly at random from all $\binom{k}{w}$ such strings. Then, $x|_S = x'$, and $x_i = 0$ for all $i \notin S$.*

We will need the following known constructions of an extractor and a condenser.

Theorem 4 ([Li13]). *For every constant $\mu > 0$ and all integers n, k with $k \geq \log^{2+\mu} n$, there exists an explicit function $\mathsf{Li}: (\{0, 1\}^n)^c \to \{0, 1\}^m$, with $m = \Omega(k)$ and $c = O(1/\mu)$, such that the following holds. If $X_1, \ldots, X_c$ are independent (n, k)-weak sources, then*

$$\mathsf{Li}(X_1, \ldots, X_c) \approx_\varepsilon U_m,$$

where $\varepsilon = n^{-\Omega(1)} + 2^{-k^{\Omega(1)}}$.

Theorem 5 ([Rao09b]). *For all integers n, k, there exists an efficiently computable linear transformation $\mathsf{Cond}: \{0, 1\}^n \to \{0, 1\}^{k \log n}$, such that for any (n, k)-bit-fixing source X it holds that Cond restricted to X is one-to-one.*

We further use of the following well-known fact.

Fact 1. *For any integer n, and $0 < \alpha < 1/2$, it holds that*

$$\sum_{k=0}^{\lfloor \alpha n \rfloor} \binom{n}{k} \leq 2^{H(\alpha) \cdot n},$$

where $H(p) = -p \log_2(p) - (1 - p) \log_2(1 - p)$ is the binary entropy function.

3 Proof Overviews

In this section we give an overview for the proofs of Theorem 1 and Theorem 2. For the sake of clarity, in this section we allow ourselves to be informal and somewhat imprecise.

3.1 Proof Overview for Theorem 1

To give an overview for the proof of Theorem 1, we start by considering a related problem. Instead of proving an upper bound on the number of output bits of an (n, k)-zero-fixing extractor, we prove an upper bound for zero-error dispersers. Generally speaking, a *zero-error disperser* for a class of sources is a function that obtains all outputs, even when restricted to any source in the class. More concretely, an (n, k)-*zero-fixing zero-error disperser* is a function ZeroErrDisp: $\{0,1\}^n \rightarrow \{0,1\}^m$, such that for any (n, k)-zero-fixing source X, it holds that supp(ZeroErrDisp(X)) $= \{0,1\}^m$. We show that for any such zero-error disperser, if k is small enough compared to n, then $m \leq \log_2(k + 1)$. More specifically, we prove that for any integers n, k such that Tower$(k^2) < n$ and $m = \lfloor\log_2(k + 1)\rfloor + 1$, for any function $f: \{0,1\}^n \rightarrow \{0,1\}^m$, there exists an (n, k)-zero-fixing source, restricted to which f is a symmetric function, i.e., f depends only on the input's weight. In particular, f does not obtain all possible outputs.[2] This implies that if $f: \{0,1\}^n \rightarrow \{0,1\}^m$ is a (n, k)-zero-fixing zero-error dispersers and Tower$(k^2) < n$, then $m \leq \log_2(k + 1)$.

Given $f: \{0,1\}^n \rightarrow \{0,1\}^m$, we construct the required source X in a level-by-level fashion, as follows. Trivially, f is symmetric on any $(n, 1)$-zero-fixing source, regardless of the value of m. Next, we find an $(n, 2)$-zero-fixing source on which f is symmetric. By the pigeonhole principle, there exists a set of indices $I_1 \subseteq [n]$, with size $|I_1| \geq n/2^m$, such that $f(e_i) = f(e_j)$ for all $i, j \in I_1$. Here, for an index $i \in [n]$, we denote by e_i the unit vector with 1 at the ith coordinate. If $n > 2^m$, then $|I_1| \geq 2$, and so there exist two distinct $i, j \in I_1$. Thus, f restricted to the $(n, 2)$-zero-fixing source $\{0, e_i, e_j, e_i + e_j\}$ is symmetric.

We take a further step, and find an $(n, 3)$-zero-fixing source on which f is symmetric. We restrict ourselves to the index set I_1 above, and consider the complete graph with vertex set I_1, where for every two distinct vertices $i, j \in I_1$, the edge connecting them is colored by the color $f(e_i + e_j)$, where we think of $\{0,1\}^m$ as representing 2^m colors. By the multi-color variant of Ramsey theorem, there exists a set $I_2 \subseteq I_1$, of size

$$|I_2| \geq \log(|I_1|)/\text{poly}(2^m),$$

such that the complete graph induced by I_2 is monochromatic. Therefore, if $n > 2^{2^{O(m)}} = 2^{\text{poly}(k)}$, then $|I_2| \geq 3$, and so there exist distinct $i_1, i_2, i_3 \in I_2$ such that

$$f(e_{i_1}) = f(e_{i_2}) = f(e_{i_3}),$$
$$f(e_{i_1}+e_{i_2}) = f(e_{i_1} + e_{i_3}) = f(e_{i_2} + e_{i_3}).$$

Thus, f is symmetric on the $(n, 3)$-zero-fixing source spanned by $\{e_{i_1}, e_{i_2}, e_{i_3}\}$.

To construct an $(n, 4)$-zero-fixing source on which f is symmetric, we consider the complete 3-uniform hypergraph on vertex set I_2 as above, where an edge $\{i_1, i_2, i_3\}$ is colored by $f(e_{i_1} + e_{i_2} + e_{i_3})$. Applying the multi-color Ramsey theorem for hypergraphs, we obtain a subset of the vertices $I_3 \subseteq I_2$, with size

$$|I_3| \geq \log\log(|I_2|)/\text{poly}(2^m),$$

---

[2] If $m > \lfloor\log_2(k + 1)\rfloor + 1$, then the same result can be obtained by restricting the output to the first $\lfloor\log_2(k + 1)\rfloor + 1$ output bits.

ıch that the induced complete hypergraph by the vertex set I_3 is monochro-
ıatic. Therefore, if $\log \log \log n \geq \text{poly}(k)$, then $|I_3| \geq 4$, and thus there are
ıstinct coordinates $i_1, i_2, i_3, i_4 \in I_3$ such that f is symmetric on the $(n, 4)$-zero-
xing source spanned by $\{e_{i_1}, e_{i_2}, e_{i_3}, e_{i_4}\}$.

We continue this way, and find an (n, k)-zero-fixing source on which f is
ymmetric, by applying similar Ramsey-type arguments on r-uniform complete
ypergraphs, with 2^m colors, for $r = 4, 5, \ldots, k - 1$. A calculation shows that as
ıng as $\text{Tower}(k^2) < n$, such a source can be found.

To obtain the negative result for (n, k)-bit-fixing extractors, we follow a sim-
ar argument. The only difference is that in this case, it is enough to find an
$n, k)$-bit-fixing source X, such that f is symmetric restricted only to the $O(\sqrt{k})$
ıiddle levels of X. Since most of the weight of X sits in these levels, an (n, k)-bit-
xing extractor cannot be symmetric restricted to these middle levels, regardless
f the values obtained by the extractor in the remaining points of X.

.2   Proof Overview for Theorem 2

nformally speaking, the advantage one should exploit when given a sample from
ın (n, k)-zero-fixing source X, as apposed to a sample from a more general bit-
ıxing source, is that "1 hits randomness". More formally, if $X_i = 1$, then we
an be certain that $i \in S$, where $S \subset [n]$ is the set of indices for which $X|_S$ is
ıniform. How should we exploit this advantage?

A natural attempt would be the following. Consider all (random) indices
$\leq i_1 < i_2 < \cdots < i_W \leq n$, such that $X_{i_1} = \cdots = X_{i_W} = 1$. Note that W, the
Iamming weight of the sample, is a random variable concentrated around $k/2$.
.et $M = i_{W/2}$ be the median of these random indices. One can show that, with
ıigh probability with respect to the value of M, both the prefix $(X_1, X_2, \ldots, X_M)$
ınd the suffix $(X_{M+1}, X_{M+2}, \ldots, X_n)$ have entropy roughly $k/2$. Intuitively, this
s because the "hidden" random bits, namely bits in coordinates $i \in S$ such that
$X_i = 0$, must be somewhat intertwined with the "observed" random bits – bits
n coordinates $i \in S$ for which $X_i = 1$. In particular, except with probability
$2^{-\Omega(k)}$ over the value of M, both the prefix and the suffix have entropy at least
).49k. Thus, by appending these prefix and suffix with zeros, one can get two n
ıit sources $X_{\text{left}}, X_{\text{right}}$, each having entropy at least $0.49k$.

We observe that conditioned on the value of the median M, the random vari-
ıbles X_{left} and X_{right} preserve the zero-fixing structure. Unfortunately, however,
$X_{\text{left}}, X_{\text{right}}$ are *dependent*. In this proof overview, we rather continue with the
lescription of the zero-fixing extractor as if $X_{\text{left}}, X_{\text{right}}$ were independent, and
leal with the dependencies later on.

After obtaining X_{left} and X_{right}, we apply the lossless-condenser of Rao from
Theorem 5 on each of these random variables. This is an efficiently computable
function $\text{Cond} \colon \{0, 1\}^n \to \{0, 1\}^{k \log n}$, that is one-to-one when restricted to any
(n, k)-bit-fixing source. We compute $Y_{\text{left}} = \text{Cond}(X_{\text{left}})$ and $Y_{\text{right}} = \text{Cond}(X_{\text{right}})$
to obtain two $(k \log n, 0.49k)$-weak sources. Note that the one-to-one guarantee
implies that no entropy is lost during the condensing, and so the entropy of
$Y_{\text{left}}, Y_{\text{right}}$ equals the entropy of $X_{\text{left}}, X_{\text{right}}$, respectively.

At this point, for simplicity, assume we have an explicit optimal two-source extractor

$$\mathsf{TwoSourceExt} \colon \{0,1\}^{k \log n} \times \{0,1\}^{k \log n} \to \{0,1\}^m$$

to our disposal Given this extractor, the output of our zero-fixing extractor is $\mathsf{TwoSourceExt}(Y_{\mathsf{left}}, Y_{\mathsf{right}})$. Working out the parameters, one can see that an optimal two-source extractor would yield an (n, k)-zero-fixing extractor for $k > \log \log n + O(\log \log \log n)$, error $2^{-\Omega(k)}$ and output length, say, $0.9k$.

Constructing two-source extractors for even sub-linear entropy, let alone for logarithmic entropy, as used in the last step, is a major open problem in pseudorandomness. Even for our short input length $k \log n = \tilde{O}(\log n)$, no poly$(n)$-time construction is known. In this proof overview however, we choose to rely on such an assumption for the sake of clarity. In the real construction, we apply the split-in-the-median process above, recursively, to obtain c weak-sources, for any desired constant c. In a recent breakthrough, Li [Li13] gave an explicit construction of a multi-source extractor, that extracts a constant fraction of the entropy, from a constant number of weak-sources with poly-logarithmic entropy. In the actual construction, instead of using a two-source extractor, we use the extractor of Li with the appropriate constant c, as stated in Theorem 4.

Working Around the Dependencies. So far we ignored the dependencies between X_{left} and X_{right}, even though their condensed images are given as inputs to a two-source extractor, and the latter expects its inputs to be independent. As we now explain, the dependencies between X_{left} and X_{right} can be worked around.

The crucial observation is the following: conditioned on the fixing of the Hamming weight W of the sample X, and conditioned on any fixing of the median M, the random variables $X_{\mathsf{left}}, X_{\mathsf{right}}$ are independent! To see this, fix $W = w$. Then, conditioned on the event $M = m$, the value of the prefix $X_1, \ldots, X_m$ gives no information whatsoever about the suffix. More precisely, conditioned on any fixing of the prefix $X_1, \ldots, X_m$, the suffix is distributed uniformly at random over all $n - m$ bit strings, with zeros outside $S \cap \{m + 1, \ldots, n\}$, and exactly $w/2$ ones in $S \cap \{m + 1, \ldots, n\}$.

This observation motivates the following definition. We say that a random variable X is an (n, k, w)-*fixed-weight source*, if there exists $S \subseteq [n]$, with size $|S| = k$, such that a sample $x \sim X$ is obtained as follows. First, one samples a string $x' \in \{0,1\}^k$ of weight w, uniformly at random from all $\binom{k}{w}$ such strings, and then sets $X|_S = x'$, and $X_i = 0$ for all $i \notin S$. It is easy to see that any (n, k)-zero-fixing source is $2^{-\Omega(k)}$-close to a convex combination of (n, k, w)-fixed-weight sources, with w ranges over $k/3, \ldots, 2k/3$. Therefore, any extractor for (n, k, w)-fixed-weight sources, for all such values of w, is also an extractor for (n, k)-zero-fixing sources.

We now reanalyze the algorithm described above. Since an (n, k)-zero-fixing source is $2^{-\Omega(k)}$-close to a convex combination of (n, k, w)-fixed-weight sources, with $k/3 \le w \le 2k/3$, we may assume, for the analysis sake, that the input is sampled from an (n, k, w)-fixed-weight source for some *fixed* $k/3 \le w \le 2k/3$. Fix also the median M to some value $m \in [n]$. Note that X_{left} is an

$(n, k_{\text{left}}(m), w/2)$-fixed-weight source[3], and X_{right} is an $(n, k_{\text{right}}(m), w/2)$-fixed-weight source, with $k_{\text{left}}(m)$ and $k_{\text{right}}(m)$ being deterministic functions of m, satisfying $k_{\text{left}}(m) + k_{\text{right}}(m) = k$. Moreover, by the discussion above, we have that conditioned on the fixing $M = m$, the two random variables $X_{\text{left}}, X_{\text{right}}$ are independent.

To summarize, conditioned on any fixing $M = m$, the two random variables $X_{\text{left}}, X_{\text{right}}$ are independent and preserve their fixed-weight structure. We further note that, with probability $1 - 2^{-\Omega(k)}$ over the value of M, it holds that $k_{\text{left}}, k_{\text{right}} \geq 0.49k$.

Recall that at this point we apply Rao's lossless-condenser on both X_{left} and X_{right}, to obtain shorter random variables $Y_{\text{left}}, Y_{\text{right}}$. Rao's condenser is one-to-one when restricted to bit-fixing sources. Since X_{left} and X_{right} are fixed-weight sources, they are in particular contained in some (n, k)-bit-fixing sources, and so the random variables $Y_{\text{left}}, Y_{\text{right}}$ have the same entropy as $X_{\text{left}}, X_{\text{right}}$, respectively.

It is worth mentioning that Rao's condenser Cond is linear, and as a result, if X_{left} were a bit-fixing source, then the resulting $Y_{\text{left}} = \text{Cond}(X_{\text{left}})$ would have been an affine source. This property was crucial for Rao's construction of bit-fixing extractors. Since we wanted to maintain independence between $X_{\text{left}}, X_{\text{right}}$, in our case these random variables are no longer bit-fixing sources, but rather fixed-weight sources. Thus, the resulting $Y_{\text{left}}, Y_{\text{right}}$ are not affine sources, but only weak sources, with min-entropy $\log_2\left(\binom{0.49k}{w/2}\right) = \Omega(k)$. This is good enough for our needs, as in the next step we use a two-source extractor, and do not rely on the affine-ness.

Lastly, we apply a two-source extractor on the condensed random variables $Y_{\text{left}}, Y_{\text{right}}$, which is a valid application, as these sources are independent, and with probability $1 - 2^{-\Omega(k)}$, both have entropy $\Omega(k)$.

4   Conclusion and Open Problems

The Number of Extractable Bits in Terms of the Dependency of k in n.
In this paper we study the intriguing behavior of the number of output bits one can extract from zero-fixing sources (and bit-fixing sources) in terms of the dependency of k in n. Theorem 2 and Theorem 3 imply that when $k > (1 + o(1)) \cdot \log \log n$, one can extract essentially all the entropy of the source, whereas when $\text{Tower}(k^{3/2}) < n$, one cannot extract more than a logarithmic amount of the entropy. The remaining entropy is inaccessible, information theoretically.

Is there a threshold phenomena behind this problem? Namely, is there some function $\tau \colon \mathbb{N} \to \mathbb{N}$, such that when $k > \tau(n)$, one can extract $\Omega(k)$ bits, whereas when $k < o(\tau(n))$, one can extract only $O(\log k)$ bits? Or perhaps the number of extractable bits in terms of the dependency of k in n is more gradual? Are there different behaviors for zero-fixing and bit-fixing sources? Theorem 3 shows that if there is such a threshold $\tau(n)$, then the function $\tau(n)$ is asymptotically not larger than $\log \log n$.

---

[3] To be more precise, X_{left} is not an $(n, k_{\text{left}}(m), w/2)$-fixed-weight source per se, as its mth bit is constantly 1. Ignoring this bit would make X_{left} a fixed-weight source.

Explicit Bit-fixing Extractors for Sub-logarithmic Entropy. Theorem ? gives a bit-fixing extractor QuasiBFExt that outputs essentially all the entropy of the source, even when the entropy is double-logarithmic in the input length Although the running-time of evaluating QuasiBFExt is not polynomial in n, it is not very high, and we feel that constructing a polynomial-time bit-fixing extractor for sub-logarithmic, or even double-logarithmic entropy, should be attainable We suspect that such a construction would require new ideas, as the ideas used in [GRS06, Rao09b] inherently require the entropy to be at least logarithmic in the input length. Furthermore, the split-in-the-median idea used in the proof of Theorem 2, is based on the "1 hits randomness" property that is unique to zero-fixing sources, and does not seem to be helpful for general bit-fixing sources.

Acknowledgement. We are thankful to Ran Raz and Avishay Tal for many fruitful discussions regarding this work. We thank the anonymous referees for their valuable comment.

References

[BBR85] Bennett, C.H., Brassard,G., Robert, J.M.: How to reduce your enemys information. In: Advances in Cryptology (CRYPTO), vol. 218, pp. 468–476. Springer (1985)

[BIW06] Barak, B., Impagliazzo, R., Wigderson, A.: Extracting randomness using few independent sources. SIAM Journal on Computing **36**(4), 1095–1118 (2006)

[BKS+05] Barak, B., Kindler, G., Shaltiel, R., Sudakov, B., Wigderson, A.: Simulating independence: new constructions of condensers, ramsey graphs, dispersers, and extractors. In: Proceedings of the Thirty-Seventh Annual ACM Symposium on Theory of Computing, pp. 1–10. ACM (2005)

[Bou05] Bourgain, J.: More on the sum-product phenomenon in prime fields and its applications. International Journal of Number Theory **1**(1), 1–32 (2005)

[Bou07] Bourgain, J.: On the construction of affine extractors. GAFA Geometric And Functional Analysis **17**(1), 33–57 (2007)

[CG88] Chor, B., Goldreich, O.: Unbiased bits from sources of weak randomness and probabilistic communication complexity. SIAM Journal on Computing **17**(2), 230–261 (1988)

[CGH+85] Chor, B., Goldreich, O., Håstad, J., Freidmann, J., Rudich, S., Smolensky, R.: The bit extraction problem or t-resilient functions. In: Proceedings of the 26th Annual Symposium on Foundations of Computer Science, pp. 396–407. IEEE (1985)

[CKK+13] Chen, R., Kabanets, V., Kolokolova, A., Shaltiel, R., Zuckerman, D.: Mining circuit lower bound proofs for meta-algorithms. In: Electronic Colloquium on Computational Complexity (ECCC), vol. 20, pp. 57 (2013)

[DKSS09] Dvir, Z., Kopparty, S., Saraf, S., Sudan, M.: Extensions to the method of multiplicities, with applications to Kakeya sets and mergers. In: Proceedings of the 50th Annual IEEE Symposium on Foundations of Computer Science, pp. 181–190. IEEE (2009)

[Fri92] Friedman, J.: On the bit extraction problem. In: Proceedings of the 33rd Annual Symposium on Foundations of Computer Science, pp. 314–319. IEEE (1992)

[GRS06] Gabizon, A., Raz, R., Shaltiel, R.: Deterministic extractors for bit-fixing sources by obtaining an independent seed. SIAM Journal on Computing 36(4), 1072–1094 (2006)

[GUV09] Guruswami, V., Umans, C., Vadhan, S.: Unbalanced expanders and randomness extractors from Parvaresh-Vardy codes. Journal of the ACM 56(4), 20 (2009)

[KRT13] Komargodski, I., Raz, R., Tal, A.: Improved average-case lower bounds for DeMorgan formula size. In: Proceedings of the 54th Annual IEEE Symposium on Foundations of Computer Science (FOCS), pp. 588–597. IEEE (2013)

[KZ06] Kamp, J., Zuckerman, D.: Deterministic extractors for bit-fixing sources and exposure-resilient cryptography. SIAM Journal on Computing 36(5), 1231–1247 (2006)

[Li11a] Li, X.: Improved constructions of three source extractors. In: Proceedings of the 26th IEEE Annual Conference on Computational Complexity (CCC), pp. 126–136. IEEE (2011)

[Li11b] Li, X.: A new approach to affine extractors and dispersers. In: Proceedings of the 26th IEEE Annual Conference on Computational Complexity (CCC), pp. 137–147. IEEE (2011)

[Li13] Li, X.: Extractors for a constant number of independent sources with polylogarithmic min-entropy. In: Proceedings of the 54th IEEE Annual Symposium on Foundations of Computer Science (FOCS), pp. 100–109. IEEE (2013)

[NZ96] Nisan, N., Zuckerman, D.: Randomness is linear in space. Journal of Computer and System Sciences 52(1), 43–52 (1996)

[Rao09a] Rao, A.: Extractors for a constant number of polynomially small min-entropy independent sources. SIAM Journal on Computing 39(1), 168–194 (2009)

[Rao09b] Rao, A.: Extractors for low-weight affine sources. In: Proceedings of 24th Annual IEEE Conference on Computational Complexity, (CCC 2009), pp. 95–101. IEEE (2009)

[Raz05] Raz, R.: Extractors with weak random seeds. In: Proceedings of the Thirty-Seventh Annual ACM Symposium on Theory of Computing, pp. 11–20. ACM (2005)

[RV13] Reshef, Y., Vadhan, S.: On extractors and exposure-resilient functions for sublogarithmic entropy. Random Structures & Algorithms 42(3), 386–401 (2013)

[Sha11] Shaltiel, R.: An introduction to randomness extractors. In: Aceto, L., Henzinger, M., Sgall, J. (eds.) ICALP 2011, Part II. LNCS, vol. 6756, pp. 21–41. Springer, Heidelberg (2011)

[Vaz85] Vazirani, V.U.: Towards a strong communication complexity theory or generating quasi-random sequences from two communicating slightly-random sources. In: Proceedings of the Seventeenth Annual ACM Symposium on Theory of Computing, pp. 366–378. ACM (1985)

[Yeh11] Yehudayoff, A.: Affine extractors over prime fields. Combinatorica 31(2), 245–256 (2011)

Interactive Proofs with Approximately Commuting Provers

Matthew Coudron[1]([✉]) and Thomas Vidick[2]

[1] Massachusetts Institute of Technology, Cambridge, MA, USA
mcoudron@mit.edu
[2] California Institute of Technology, Pasadena, CA, USA
vidick@cms.caltech.edu

Abstract. The class MIP* of promise problems that can be decided through an interactive proof system with multiple entangled provers provides a complexity-theoretic framework for the exploration of the nonlocal properties of entanglement. Very little is known in terms of the power of this class. The only proposed approach for establishing upper bounds is based on a hierarchy of semidefinite programs introduced independently by Pironio et al. and Doherty et al. in 2006. This hierarchy converges to a value, the field-theoretic value, that is only known to coincide with the provers' maximum success probability in a given proof system under a plausible but difficult mathematical conjecture, Connes' embedding conjecture. No bounds on the rate of convergence are known.

We introduce a rounding scheme for the hierarchy, establishing that any solution to its N-th level can be mapped to a strategy for the provers in which measurement operators associated with distinct provers have pairwise commutator bounded by $O(\ell^2/\sqrt{N})$ in operator norm, where ℓ is the number of possible answers per prover.

Our rounding scheme motivates the introduction of a variant of quantum multiprover interactive proof systems, called MIP^*_δ, in which the soundness property is required to hold against provers allowed to operate on the same Hilbert space as long as the commutator of operations performed by distinct provers has norm at most δ. Our rounding scheme implies the upper bound $\text{MIP}^*_\delta \subseteq \text{DTIME}(\exp(\exp(\text{poly})/\delta^2))$. In terms of lower bounds we establish that $\text{MIP}^*_{2^{-\text{poly}}}$ contains NEXP with completeness 1 and soundness $1 - 2^{-\text{poly}}$. We discuss connections with the mathematical literature on approximate commutation and applications to device-independent cryptography.

1 Introduction

In a multiprover interactive proof system, a *verifier* with bounded resources (a polynomial-time Turing machine) interacts with multiple all-powerful but noncommunicating *provers* in an attempt to verify the truth of a mathematical statement — the membership of some input x, a string of bits, in a language L, such as 3-SAT. The provers always collaborate to maximize their chances of making the verifier accept the statement, and their maximum probability of success in

© Springer-Verlag Berlin Heidelberg 2015
M.M. Halldórsson et al. (Eds.): ICALP 2015, Part I, LNCS 9134, pp. 355–366, 2015.
DOI: 10.1007/978-3-662-47672-7_29

doing so is called the *value* $\omega = \omega(x)$ of the protocol. We will sometimes refer to a given protocol as an "interactive game" and call the provers "players". A proof system's *completeness* c is the smallest value of $\omega(x)$ over all $x \in L$, while its soundness s is the largest value of $\omega(x)$ over $x \notin L$; a protocol is sound if $s < c$.

The class of all languages that have multiprover interactive proof systems with $c \geq 2/3$ and $s \leq 1/3$, denoted MIP, is a significant broadening of its non-interactive, single-prover analogue MA, as is witnessed by the characterization MIP = NEXP [BFL91]. This result is one of the cornerstones on which the PCP theorem [AS98, ALM+98] was built, with consequences ranging from cryptography [BOGKW88] to hardness of approximation [FGL+96].

Quantum information suggests a natural extension of the class MIP. The laws of quantum mechanics assert that, in the physical world, a set of non-communicating provers may share an arbitrary entangled quantum state, a physical resource which strictly extends their set of strategies but provably does not allow them to communicate. The corresponding extension of MIP is the class MIP* of all languages that have multiprover interactive proof systems with entangled provers [KM03].

Physical intuition for the significance of the prover's new resource, entanglement, dates back to Einstein, Podolsky and Rosen's paradoxical account [EPR35] of the consequences of quantum entanglement, later clarified through Bell's pioneering work [Bel64]. To state the relevance of Bell's results more precisely in our context we first introduce the mathematical formalism used by Bell to model locality. With each prover's private space is associated a separate Hilbert space. The joint quantum state of the provers is specified by a unit vector $|\Psi\rangle$ in the tensor product of their respective Hilbert spaces. Upon receiving its query from the verifier, each prover applies a local measurement (a positive operator supported on its own Hilbert space) the outcome of which is sent back to the verifier as its answer. The supremum of the provers' probability of being accepted by the verifier, taken over all Hilbert spaces, states in their joint tensor product, and local measurements, is called the entangled value ω^* of the game. The analogue quantity for "classical" provers (corresponding to shared states which are product states) is denoted ω.

Bell's work and the extensive literature on Bell inequalities [CHSH69, Ara02] and quantum games [CHTW04] establishes that there are protocols, or interactive games, for which $\omega^* > \omega$. This simple fact has important consequences for interactive proof systems. First, a proof system sound with classical provers may no longer be so in the presence of entanglement. Cleve et al. [CHTW04] exhibit a class of restricted interactive proof systems, XOR proof systems, such that the class with classical provers equals NEXP while the same proof systems with entangled provers cannot decide any language beyond EXP. Second, the completeness property of a proof system may also increase through the provers' use of entanglement. As a result optimal strategies may require the use of arbitrarily large Hilbert spaces for the provers — no explicit bound on the dimension of these spaces is known as a function of the size of the game.

In fact no better upper bound on the class MIP* is known other than its lan guages being recursively enumerable: they may not even be decidable! Thi unfortunate state of affairs stems from the fact that, while the value ω^* may b approached from below through exhaustive search in increasing dimensions there is no verifiable criterion for the termination of such a procedure.

Bounding entangled-Prover Strategies. The question of deriving algorithmic methods for placing upper bounds on the entangled value ω^* of a given proto col has long frustrated researchers' efforts. Major progress came in 2006 through the introduction of a hierarchy of relaxations based on semidefinite program ming [DLTW08, NPA07] that we will refer to as the QCSDP hierarchy. These relaxations follow a similar spirit as e.g. the Lasserre hierarchy in combinato rial optimization [Lau03], and can be formulated using the language of sums of squares of *non-commutative* polynomials. In contrast with the commutative set ting, this leads to a hierarchy that is in general infinite and need not converge at any finite level.

The limited convergence results that are known for the QCSDP hierarchy involve a formalization of locality for quantum provers which originates in the study of infinite-dimensional systems such as those that arise in quantum field theory. Here the idea is that observations made at different space-time locations should be represented by operators which, although they may act on the same Hilbert space, should nevertheless commute — a minimal requirement ensur ing that the joint outcome of any two measurements made by distinct parties should be well-defined and independent of the order in which the measure ments were performed.

For the case of finite-dimensional systems this seemingly weaker condition is equivalent to the existence of a tensor product representation [DLTW08]. In contrast, for the case of infinite-dimensional systems the two formulations are not known to be equivalent. This question, known as Tsirelson's problem in quantum information, was recently shown to be equivalent to a host of deep mathematical conjectures [SW08, JNP+11], in particular Connes' embed ding conjecture [Con76] and Kirchberg's QWEP conjecture [Kir93]. The valid ity of these conjectures has a direct bearing on our understanding of MIP*. The QCSDP hierarchy is known to converge to a value called the *field-theoretic value* ω^f of the game, which is the maximum success probability achievable by com muting strategies of the type described above. A positive answer to Tsirelson's conjecture thus implies that $\omega^* = \omega^f$ and both quantities are computable. However, even assuming the conjecture and in spite of strong interest (the use of the first few levels of the hierarchy has proven extremely helpful to study a range of questions in device independence [BSS14, YVB+14] and the study of nonlocality [PV10]) absolutely no bounds have been obtained on the conver gence rate of the hierarchy. It is only known that if a certain technical condition, called a rank loop, holds, then convergence is achieved [NPA08]; unfortunately the condition is computationally expensive to verify (even for low levels of the hierarchy) and, in general, may not be satisfied at any finite level.

Beyond the obvious limitations for practical applications, these severe computational difficulties are representative of the intrinsic difficulty of working with the model of entangled provers. Our work is motivated by this state of affairs: we establish the first quantitative convergence results for the quantum SDP hierarchy. Our main observation is that successive levels of the hierarchy place bounds on the value achievable by provers employing a relaxed notion of strategy in which measurements applied by distinct provers are allowed to *approximately commute*: their commutator is bounded, in operator norm, by a quantity that goes to zero with the level in the hierarchy.

In this abstract we describe our quantitative results, use them to motivate the introduction of a sub-class MIP^*_{ac} of MIP^* and prove non-trivial lower and upper bounds on that class. We discuss the relevance of the study of MIP^*_{ac} for that of MIP^* and closely related results from the mathematical literature. We refer to the full version for precise definitions as well as complete proofs of the results announced here.

2   A Rounding Scheme for the QCSDP Hierarchy

Our main technical result is a rounding procedure for the QCSDP hierarchy of semidefinite programs [NPA07, DLTW08]. The procedure maps any feasible solution to the N-th level of the hierarchy to a set of measurement operators for the provers that approximately commute. For simplicity we state and prove our results for the case of a single round of interaction with two provers and classical messages only. Extension to multiple provers is straightforward; we expect generalizations to multiple rounds and quantum messages to be possible but leave them for future work.

Definition 1. *An (m, ℓ) strategy for the provers specified by two sets of m POVMS $\{A_x^a\}_{1 \leq a \leq \ell}$ and $\{B_y^b\}_{1 \leq b \leq \ell}$ with ℓ outcomes each, where $x, y \in \{1, \dots, m\}$.*

A strategy is said to be δ-AC if for every x, y, a and b, $\| A_x^a B_y^b - B_y^b A_x^a \| \leq \delta$, where $\| \cdot \|$ denotes the operator norm.

Our results apply to the QCSDP hierarchy of semidefinite programs as defined in [NPA07].

Theorem 1. *Let G be a 2-prover one-round game with classical messages in which each player has ℓ possible answers, and $\omega^N_{QCSDP}(G)$ the optimum of the N-th level of the QCSDP hierarchy. Then there exists a $\delta = O(\ell^2 / \sqrt{N})$ and a δ-AC strategy for the provers with success probability $\omega^N_{QCSDP}(G)$ in G.[1]*

---

[1] Due to the approximate commutation of the provers' strategies the success probability of δ-AC strategies may a priori depend on the order in which the measurement operators are applied. In our context the parameter δ will always be small enough that we can neglect this effect. Moreover, for the particular kind of strategies constructed in our rounding scheme the value will not be affected by the order.

Our result is the first to derive the condition that the *operator norm* of commutators is small. In contrast it is not hard to show that a feasible solution to the first level of the hierarchy already gives rise to measurement operators that exactly satisfy a commutation relation *when evaluated on the state* (corresponding to the zeroth-order vector provided by the hierarchy). While the latter condition can be successfully exploited to give an exact rounding procedure from the first level for the class of XOR games [CHTW04], and an approximate rounding for the more general class of unique games [KRT10], we do not expect it to be sufficient in general. In particular, even approximate tightness of the first level of the hierarchy for three-player games would imply EXP = NEXP [Vid13]. We will furtherore show that the problem of optimizing over strategies which approximately commute, to within sufficiently small error and in *operator norm*, is NEXP-hard (see Section 3 for details).

The proof of Theorem 1 is constructive: starting from any feasible solution to the N-th level of the QCSDP hierarchy we construct measurement operators for the provers with pairwise commutators bounded by δ in operator norm, and which achieve a value in the game that equals the objective value of the N-th level SDP. Recall that this SDP has $O(m\ell)^N$ vector variables indexed by strings of length at most N over the formal alphabet $\{P_x^a, Q_y^b\}$ containing a symbol for each possible (question,answer) pair to any of the provers. Our main idea is to introduce a "graded" variant of the construction in [NPA08] (which was used to show convergence under the rank loop constraint). Rather informally, the rounded measurement operators, $\{\tilde{P}_x^a\}$ for the first prover and $\{\tilde{Q}_y^b\}$ for the second, can be defined as follows:

$$\tilde{P}_x^a \equiv \frac{1}{N-1} \sum_{i=1}^{N-1} \Pi_{\leq i} \Pi_{P_x^a} \Pi_{\leq i} \quad \text{and} \quad \tilde{Q}_y^b \equiv \frac{1}{N-1} \sum_{j=1}^{N-1} \Pi_{\leq j} \Pi_{Q_y^b} \Pi_{\leq j}.$$

Here $\Pi_{P_x^a}$ and $\Pi_{Q_y^b}$ are projectors as defined in [NPA08], i.e. as the projection onto vectors associated with strings ending in the formal label P_x^a, Q_y^b of the corresponding operator. The novelty is the introduction of the $\Pi_{\leq i}$, which project onto the subspace spanned by all vectors associated with strings of length at most i. Thus $\tilde{P}_x^a$ itself is not a projector, and it gives more weight to vectors indexed by shorter strings.

The intuition behind this rounding scheme is as follows. The winning probability is unchanged because it is determined by the action of the measurement operators on the subspace $\text{Im}(\Pi_{\leq 1})$. On the other hand, the rounded operators approximately commute in the operator norm because the original operators commuted exactly on the subspace $\text{Im}(\Pi_{\leq N-1})$, and we have now shifted the weight of the operators so that they are supported on that subspace. Furthermore, while truncating the operators abruptly at level $N-1$ (by conjugating by $\Pi_{\leq N-1}$ for example) could result in a large commutator, we perform a "smooth" truncation across vectors indexed by strings of increasing length.

3   Interactive Proofs with Approximately Commuting Provers

Motivated by the rounding procedure ascertained in Theorem 1 we propose a modification of the class MIP* in which the assumption that isolated provers must perform perfectly commuting measurements is relaxed to a weaker condition of *approximately commuting* measurements.

Definition 2. *Let* $\mathrm{MIP}^*_\delta(k, c, s)$ *be the class of promise problems* (L_{yes}, L_{no}) *that can be decided by an interactive proof system in which the verifier exchanges a single round of classical messages with k quantum provers* $P_1, \ldots, P_k$ *and such that:*

- *If the input* $x \in L_{yes}$ *then there exists a perfectly commuting strategy for the provers that is accepted with probability at least c,*
- *If* $x \in L_{no}$ *then any* δ-*AC strategy is accepted with probability at most s.*

Note that the definition of MIP^*_δ requires the completeness property to be satisfied with perfectly commuting provers; indeed we would find it artificial to seek protocols for which optimal strategies in the "honest" case would be required to depart from the commutation condition. Instead, only the soundness condition is relaxed by giving *more* power to the provers, who are now allowed to apply any "approximately commuting" strategy. The "approximately" is quantified by the parameter δ,[2] and for any $\delta' \leq \delta$ the inclusions $\mathrm{MIP}^*_\delta \subseteq \mathrm{MIP}^*_{\delta'} \subseteq \mathrm{MIP}^*$ trivially hold. It is important to keep in mind that while δ can be a function of the size of the protocol it must be independent of the dimension of the provers' operators, which is unrestricted.

δ-AC strategies were previously considered by Ozawa [Oza13] in connection with Tsirelson's problem. Ozawa proposes a conjecture, the "Strong Kirchberg Conjecture (I)", which if true implies the equality $\mathrm{MIP}^* = \cup_{\delta>0} \mathrm{MIP}^*_\delta$. We state and discuss the conjecture further as Conjecture 1 below. Unfortunately the conjecture seems well beyond the reach of current techniques (Ozawa himself formulates doubts as to its validity). However, in our context less stringent formulations of the conjecture would still imply conclusive results relating MIP^*_δ to MIP^*; we discuss such variants in Section 4.

Further motivation for the definition of MIP^*_δ may be found by thinking operationally — with e.g. cryptographic applications in mind, how does one ascertain that "isolated" provers indeed apply commuting measurements? The usual line of reasoning applies the laws of quantum mechanics and special relativity to derive the tensor product structure from space-time separation. However, not only is strict isolation virtually impossible to enforce in all but the simplest experimental scenarios, but the implication "separation $\Longrightarrow$ tensor product" may itself be subject to questioning — in particular it may not be a testable prediction, at least not to precision that exceeds the number of measurements, or observations, performed. Relaxations of the tensor product condition

---

[2] As a first approximation the reader may think of δ as a parameter that is inverse exponential in the input length $|x|$. In terms of games, this corresponds to δ being inverse polynomial in the number of questions in the game, which is arguably the most natural setting of parameters.

have been previously considered in the context of device-independent cryptography; for instance Silman et al. [SPM13] require that the joint measurement performed by two isolated devices be close, in operator norm, to a tensor product measurement. Our approximate commutation condition imposes a weaker requirement, and thus our convergence results on the hierarchy also apply to their setting; we discuss this in more detail in Section 4.2.

A computationally Tractable Class? Theorem 1 can be interpreted as evidence that the hierarchy converges at a polynomial rate to the maximum success probability for MIP^*_{ac} provers. More formally, it implies the inclusion $\text{MIP}^*_\delta \subseteq \text{TIME}(\exp(\exp(\text{poly})/\delta^2))$ for any $\delta > 0$, thereby justifying our claim that the class MIP^*_δ is tractable. This stands in stark contrast with $\text{MIP}^* = \text{MIP}^*_0$, for which no upper bound is known.

Having shown that the new class has "reasonable" complexity, it is natural to ask whether the additional power granted to the provers might actually make the class trivial — could provers that are δ-AC be no more useful than a single quantum prover, even for very small δ? We show this is not the case by establishing the inclusion $\text{NEXP} \subseteq \text{MIP}^*_{2-\text{poly}}(2,1,1-2^{-\text{poly}})$. This is a direct analogue of the same lower bound for MIP^* [IKM09], and is proven using the same technique. We conjecture that the inclusion $\text{NEXP} \subseteq \text{MIP}^*_{2-\text{poly}}(3,1,2/3)$ also holds, and that this can be derived by a careful extension of the results in [IV12, Vid13].

4   Discussion

Our introduction of MIP^*_{ac} is motivated by a desire to develop a framework for the study of quantum multiprover interactive proof systems that is both computationally tractable and relevant for typical applications of such proof systems. Our main technical result, Theorem 1, demonstrates the first aspect. In this section we discuss the relevance of the new model, its connection with the standard definition of MIP^*, and possible applications to quantum information.

4.1   Commuting Approximants: Some Results, Limits, and Possibilities

While we believe MIP^*_{ac} is of interest in itself, we do not claim that approximately commuting provers are more natural than commuting provers, or provers in tensor product form; the main goal in introducing the new class is to shed light on its thus-far-intractable parent MIP^*. In light of the results from Section 2 the relationship between the two classes seems to hinge on the general mathematical problem of finding exactly commuting approximants to approximately commuting matrices.

Limits for Commuting Approximants. The main objection to the existence of a positive answer for the "commuting approximants" question is revealed by a beautiful construction of Voiculescu who exhibits a surprisingly simple scenario in which commuting approximants provably do not exist [Voi83]. The following is a direct consequence of Voiculescu's result.

Theorem 2 (Voiculescu). *For every $d \in \mathbb{N}$ there exists a pair of unitary matrices $U_1, U_2 \in \mathbb{C}^{d \times d}$ with $\|[U_1, U_2]\| \leq O(\frac{1}{d})$, such that for any pair of complex matrices $A, B \in \mathbb{C}^{d \times d}$ satisfying $[A, B] = 0, \max(\|U_1 - A\|, \|U_2 - B\|) = \Omega(1)$.*

In Voiculescu's example U_1 is a d-dimensional cyclic permutation matrix, and U_2 is a diagonal matrix whose eigenvalues are the dth roots of unity. The proof draws on a connection to homology, in particular using a homotopy invariant to establish the lower bound on distance to commuting approximants. A succinct and elementary proof of the result is given by Exel and Loring [EL89].

In the context of non-local entangled strategies one is most concerned with Hermitian matrices representing measurements, rather than unitaries. However, as a consequence of Theorem 2 we see that if one considers the Hermitian operators $M_k^j = \frac{(-i)^j}{2}(U_k + (-1)^j U_k^\dagger)$ ($j \in \{0, 1\}$) we have that $\|[M_1^j, M_2^{j'}]\| \leq O(\frac{1}{d})$, and yet any exactly commuting set of matrices must be a constant distance away in the operator norm. Thus Theorem 2 rules out the strongest form of a "commuting approximants" statement, which would ask for approximants in the same space as the original matrices, and with a commutator bound that does not depend on the dimension of the matrices.

Theorem 2 invites us to refine the "commuting approximants" question and distinguish the ways in which it may avoid the counter-example.

Ozawa's Conjecture. Motivated by the study of Tsirelson's problem and the relationship with Tsirelson's conjecture, Ozawa [Oza13] introduces two equivalent conjectures, the "Strong Kirchberg Conjecture (I)" and "Strong Kirchberg Conjecture (II)" respectively, which conjecture the existence of commuting approximants to approximately commuting sets of POVM measurements and unitaries respectively. The novelty of these conjectures, which allows them to avoid the immediate pitfall given by Voiculescu's example, is that Ozawa considers approximants in a larger Hilbert space than the original approximately commuting operators. Precisely, his Strong Kirchberg Conjecture (I) states the following:

Conjecture 1 (Ozawa). Let $m, \ell \geq 2$ be such that $(m, \ell) \neq (2, 2)$ [3]. For every $\kappa > 0$ there exists $\varepsilon > 0$ such that, if $\dim \mathcal{H} < \infty$ and (P_i^k), (Q_j^l) is a pair of m projective ℓ-outcome POVMs on $\mathcal{H}$ satisfying $\|[P_i^k, Q_j^l]\| \leq \varepsilon$, then there is a finite-dimensional Hilbert space $\tilde{H}$ containing $\mathcal{H}$ and projective POVMs $\tilde{P}_i^k, \tilde{Q}_j^l$

---

[3] The case $(m, \ell) = (2, 2)$ is the only nontrivial setting for which we have some understanding. In particular nonlocal games with two inputs and two outputs per party can be analyzed via an application of Jordan's lemma [Mas05].

on $\tilde{\mathcal{H}}$ such that $\|[\tilde{P}_i^k, \tilde{Q}_j^l]\| = 0$ and $\|\Phi_{\mathcal{H}}(\tilde{P}_i^k) - P_i^k\| \leq \kappa$ and $\|\Phi_{\mathcal{H}}(\tilde{Q}_j^l) - Q_j^l\| \leq \kappa$. Here $\Phi_{\mathcal{H}}$ denotes the compression to $\mathcal{H}$, defined by $\Phi_{\mathcal{H}}(M) \equiv P_{\mathcal{H}} M P_{\mathcal{H}}$, where $P_{\mathcal{H}}$ is the projection onto $\mathcal{H}$.

Ozawa gives an elegant proof of a variant of the conjecture that applies to just two approximately commuting unitaries, thereby establishing that extending the Hilbert space can allow one to avoid the complications in Voiculescu's example. He also establishes that the conjecture is *stronger* than Kirchberg's conjecture (itself equivalent to Tsirelson's problem and Connes' embedding conjecture), casting doubt, if not on its validity, at least on its approachability.

Nevertheless, we can mention the following facts. First, Conjecture 1 implies the equality $\mathrm{MIP}_{ac}^* = \mathrm{MIP}^*$; in fact it implies that $\mathrm{MIP}_\delta^* = \mathrm{MIP}^*$ for small enough δ, depending on how the parameter ε in Conjecture 1 depends on κ, m and d. For this it suffices to verify that a state ρ optimal for a strategy based on POVMs P_i^k and Q_j^l in a given protocol can be lifted to a state $\tilde{\rho}$ on $\tilde{\mathcal{H}}$ such that the correlations exhibited by performing the POVMs $\tilde{P}_i^k$, $\tilde{Q}_j^l$ on $\tilde{\rho}$ approximately reproduce those generated by P_i^k, Q_j^l on ρ; this is easily seen to be the case provided κ is small enough.

Second, Conjecture 1 can be weakened in several ways without losing the implication that $\mathrm{MIP}_{ac}^* = \mathrm{MIP}^*$. For instance, it is not necessary for the exactly commuting $\tilde{P}_i^k$, $\tilde{Q}_j^l$ to approximate the P_i^k, Q_j^l in operator norm — in the context of interactive games, only the correlations obtained by measuring a particular state need to be preserved, and this does not in general imply an approximation as strong as that promised in Conjecture 1.

Dimension Dependent Bounds. An alternative relaxation for the "commuting approximants" question is to allow the approximation error to depend explicitly on the dimension of the matrices. A careful analysis of the rounding scheme from Theorem 1 shows that it produces d-dimensional POVM elements with an $O(1/\sqrt{\log(d)})$ bound on the commutators (this is because the dimension of the subspace $\mathrm{Im}(\Pi_{\leq N-1})$ is exponential in N). Unfortunately, Voiculescu's result (Theorem 2) shows that one can only hope for good approximants in the operator norm if the commutator bound is $o(1/d)$. It remains instructive to find *any* explicit existence result for commuting approximants in the general case, regardless of dimension dependence. Concretely, we conjecture that Conjecture 1 may be true with a parameter κ that scales with the dimension d of the operators $\{P_i^k, Q_j^l\}$ as $\kappa = \varepsilon^c \operatorname{poly}(d)^{(ml)^2}$ for some constant $0 < c \leq 1$.

An Alternative Norm. Another relaxation of the "commuting approximants" question, which would be sufficient to imply $\mathrm{MIP}_{ac}^* = \mathrm{MIP}^*$, is to allow for any set of commuting approximants which approximately preserves the winning probability of the game. For concreteness we include a precise version of a possible statement along these lines:

Conjecture 2. There exists a function $f(\varepsilon, k) : \mathbb{R}^+ \times \mathbb{N} \to \mathbb{R}^+$ satisfying $\lim_{\varepsilon \to 0} f(\varepsilon, k) = 0$ for all $k \in \mathbb{N}$, such that for every game G and (m, ℓ) strategy (A_x^a, B_y^b, ρ) which is δ-AC, there exists a 0-AC strategy $(\tilde{A}_x^a, \tilde{B}_y^b, \rho)$ for G satisfying

$$\left| \omega^* \left(((A_x^a, B_y^b, \rho); G) - \omega^* ((\tilde{A}_x^a, \tilde{B}_y^b, \rho); G) \right) \right| \leq f(\delta, m\ell).$$

4.2 Device-Independent Randomness Expansion and Weak Cross-Talk

A device-independent randomness expansion (DIRE) protocol is a protocol which may be used by a classical verifier to certify that a pair of untrusted devices are producing true randomness. Under the sole assumptions that the devices do not communicate with each other, and that the verifier has access to a small initial seed of uniform randomness, the protocol allows for the generation of much larger quantities of certifiably uniform random bits; hence the term "randomness expansion". This conclusion relies only on the assumption that the two devices do not communicate, and in particular does not require any limit on the computational power of the devices, as is typically the case in the study of pseudorandomness. The precise formalization of DIRE protocols is rather involved, and we direct the interested reader to the flourishing collection of works on the topic [CK11, PAM10, MS14].

Our definition of MIP_{ac}^* is directly relevant to the notion of devices with *weak cross-talk* introduced in [SPM13] as a model which relaxes the assumption that the devices must not communicate, leading to protocols that are more robust to leakage than the traditional model of device-independence. [SPM13] proposes the use of the QCSDP hierarchy in order to optimize over the set of "weakly interacting" quantum strategies that they introduce, but no bounds are shown on the rate of convergence. This is where MIP_{ac}^* becomes relevant. Our notion of δ-AC strategies is easily seen to be a relaxation of weak cross-talk, and thus the analogue of the approach in [SPM13] when performed with a δ-AC constraint is at least as robust as the weak cross-talk approach. Our rounding scheme for the QCSDP hierarchy thus provides a specific algorithm and complexity bound that applies to both δ-AC strategies and strategies with weak cross-talk.

References

[ALM+98] Arora, S., Lund, C., Motwani, R., Sudan, M., Szegedy, M.: Proof verification and the hardness of approximation problems. J. ACM **45**(3), 501–555 (1998)

[Ara02] Aravind, P.K.: The magic squares and Bell's theorem. Technical report (2002 arXiv:quant-ph/0206070

[AS98] Arora, S., Safra, S.: Probabilistic checking of proofs: A new characterization of NP. J. ACM **45**(1), 70–122 (1998)

[Bel64] John, S.: Bell. On the Einstein-Podolsky-Rosen paradox. Physics **1**, 195–200 (1964)

[BFL91] Babai, L., Fortnow, L., Lund, C.: Non-deterministic exponential time has two-prover interactive protocols. Comput. Complexity **1**, 3–40 (1991)

[BOGKW88] Ben-Or, M., Goldwasser, S., Kilian, J., Wigderson, A.: Multi-prover interac
tive proofs: How to remove intractability assumptions. In: Proceedings o
the 20th Annual ACM Symposium on Theory of Computing (STOC), pp
113–131 (1988)

[BSS14] Bancal, J.-D., Sheridan, L., Scarani, V.: More randomness from the same
data. New Journal of Physics 16(3), 033011 (2014)

[CHSH69] Clauser, J.F., Horne, M.A., Shimony, A., Holt, R.A.: Proposed experimen
to test local hidden-variable theories. Phys. Rev. Lett. 23, 880–884 (1969)

[CHTW04] Cleve, R., Høyer, P., Toner, B., Watrous, J.: Consequences and limits of non
local strategies. In: Proc. 19th IEEE Conf. on Computational Complexity
(CCC 2004), pp. 236–249. IEEE Computer Society (2004)

[CK11] Colbeck, R., Kent, A.: Private randomness expansion with untrusted
devices. Journal of Physics A: Mathematical and ..., 1–11 (2011)

[Con76] Connes, A.: Classification of injective factors cases ii_1, ii_∞, iii_λ, $\lambda \neq 1$
Annals of Mathematics 104(1), 73–115 (1976)

[DLTW08] Doherty, A.C., Liang, Y-C., Toner, B., Wehner, S.: The quantum moment
problem and bounds on entangled multi-prover games. In: Proc. 23rd
IEEE Conf. on Computational Complexity (CCC 2008), pp. 199–210 (2008)

[EL89] Exel, R., Loring, T.: Almost commuting unitary matrices. In: Proceedings
of the American Mathematical Society 106(4), 913–915 (1989)

[EPR35] Einstein, A., Podolsky, B., Rosen, N.: Can quantum-mechanical description of physical reality be considered complete? Physical Review 47, 777–
780 (1935)

[FGL+96] Feige, U., Goldwasser, S., Lovász, L., Safra, S., Szegedy, M.: Interactive
proofs and the hardness of approximating cliques. J. ACM 43(2), 268–292
(1996)

[IKM09] Ito, T., Kobayashi, H., Matsumoto, K.: Oracularization and two-prover
one-round interactive proofs against nonlocal strategies. In: Proc. 24th
IEEE Conf. on Computational Complexity (CCC 2009), pp. 217–228. IEEE
Computer Society (2009)

[IV12] Ito, T., Vidick, T., A multi-prover interactive proof for NEXP sound against
entangled provers. In: Proc. 53rd FOCS, pp. 243–252 (2012)

[JNP+11] Junge, M., Navascues, M., Palazuelos, C., Perez-Garcia, D., Scholz, V.B.,
Werner, R.F.: Connes' embedding problem and tsirelson's problem. J.
Math. Physics 52(1) (2011)

[Kir93] Kirchberg, E.: On non-semisplit extensions, tensor products and exactness
of group C^*-algebras. Inventiones mathematicae 112(1), 449–489 (1993)

[KM03] Kobayashi, H., Matsumoto, K.: Quantum multi-prover interactive proof
systems with limited prior entanglement. Journal of Computer and System Sciences 66(3), 429–450 (2003)

[KRT10] Kempe, J., Regev, O., Toner, B.: Unique games with entangled provers are
easy. SIAM J. Comput. 39(7), 3207–3229 (2010)

[Lau03] Laurent, M.: A comparison of the Sherali-Adams, Lovász-Schrijver, and
Lasserre relaxations for 0–1 Programming. Mathematics of Operations
Research 28(3), 470–496 (2003)

[Mas05] Ll. Masanes. Extremal quantum correlations for n parties with
two dichotomic observables per site. Technical report (2005).
arXiv:quant-ph/0512100

[MS14] Miller, C.A., Shi, Y.: Robust protocols for securely expanding randomness
and distributing keys using untrusted quantum devices. In: Proc. 46th
STOC. ACM New York (2014)

[NPA07] Navascués, M., Pironio, S., Acín, A.: Bounding the set of quantum correlations. Phys. Rev. Lett. **98**, 010401 (2007)

[NPA08] Navascués, M., Pironio, S., Acín, A.: A convergent hierarchy of semidefinite programs characterizing the set of quantum correlations. New Journal of Physics, 10(073013) (2008)

[Oza13] Ozawa, N.: Tsirelson's problem and asymptotically commuting unitary matrices. Journal of Mathematical Physics 54(3) (2013)

[PAM10] Pironio, S., Acín, A., Massar, S.: Random numbers certified by Bell's theorem. Nature, 1–26 (2010)

[PV10] Pál, K.F., Vértesi, T.: Maximal violation of a bipartite three-setting, two-outcome Bell inequality using infinite-dimensional quantum systems. Phys. Rev. A **82**, 022116 (2010)

[SPM13] Silman, J., Pironio, S., Massar, S.: Device-independent randomness generation in the presence of weak cross-talk. Phys. Rev. Lett. **110**, 100504 (2013)

[SW08] Scholz, V.B., Werner, R.F.: Tsirelson's problem. Technical report (2008). arXiv:0812.4305v1 [math-ph]

[Vid13] Vidick, T.: Three-player entangled XOR games are NP-hard to approximate. In: Proc. 54th FOCS (2013)

[Voi83] Voiculescu, D.: Asymptotically commuting finite rank unitary operators without commuting approximants. Acta Sci. Math. (Szeged) **45**, 429–431 (1983)

[YVB+14] Yang, T.H., Vertesi, T., Bancal, J-D., Scarani, V., Navascues, M.: Robust and Versatile Black-Box Certification of Quantum Devices. Phys. Rev. Lett. 113(4), (July 22, 2014)

Popular Matchings with Two-Sided Preferences and One-Sided Ties

Ágnes Cseh[1](✉), Chien-Chung Huang[2](✉), and Telikepalli Kavitha[3](✉)

[1] TU Berlin, Berlin, Germany
cseh@math.tu-berlin.de
[2] Chalmers University, Göteborg, Sweden
huangch@chalmers.se
[3] Tata Institute of Fundamental Research, Mumbai, India
kavitha@tcs.tifr.res.in

Abstract. We are given a bipartite graph $G = (A \cup B, E)$ where each vertex has a preference list ranking its neighbors: in particular, every $a \in A$ ranks its neighbors in a strict order of preference, whereas the preference lists of $b \in B$ may contain ties. A matching M is *popular* if there is no matching M' such that the number of vertices that prefer M' to M exceeds the number that prefer M to M'. We show that the problem of deciding whether G admits a popular matching or not is NP-hard. This is the case even when every $b \in B$ either has a strict preference list or puts all its neighbors into a single tie. In contrast, we show that the problem becomes polynomially solvable in the case when each $b \in B$ puts all its neighbors into a single tie. That is, all neighbors of b are tied in b's list and b desires to be matched to any of them. Our main result is an $O(n^2)$ algorithm (where $n = |A \cup B|$) for the popular matching problem in this model. Note that this model is quite different from the model where vertices in B have no preferences and do *not* care whether they are matched or not.

1 Introduction

We are given a bipartite graph $G = (A \cup B, E)$ where the vertices in A are called applicants and the vertices in B are called posts, and each vertex has a preference list ranking its neighbors in an order of preference. Here we assume that vertices in A have strict preferences while vertices in B are allowed to have ties in their preference lists. Thus each applicant ranks all posts that she finds interesting in a strict order of preference, while each post need not come up with a total order on all interested applicants – here applicants may get grouped together in terms of their suitability, thus equally competent applicants are tied together at the same rank.

Our goal is to compute a *popular* matching in G. The definition of popularity uses the notion of each vertex casting a "vote" for one matching versus another.

Á. Cseh—Work done while visiting TIFR, supported by the Deutsche Telekom Stiftung.

© Springer-Verlag Berlin Heidelberg 2015
M.M. Halldórsson et al. (Eds.): ICALP 2015, Part I, LNCS 9134, pp. 367–379, 2015.
DOI: 10.1007/978-3-662-47672-7_30

A vertex v *prefers* matching M to matching M' if either v is unmatched in M' and matched in M or v is matched in both matchings and $M(v)$ (v's partner in M) is ranked better than $M'(v)$ in v's preference list. In an election between matchings M and M', each vertex v votes for the matching that it prefers or it abstains from voting if M and M' are equally preferable to v. Let $\phi(M, M')$ be the number of vertices that vote for M in an election between M and M'.

Definition 1. *A matching M is* popular *if $\phi(M, M') \geq \phi(M', M)$ for every matching M'.*

If $\phi(M', M) > \phi(M, M')$, then we say M' is *more popular* than M and denote it by $M' \succ M$; else $M \succeq M'$. Observe that popular matchings need not always exist. Consider an instance where $A = \{a_1, a_2, a_3\}$ and $B = \{b_1, b_2, b_3\}$ and for $i = 1, 2, 3$, each a_i has the same preference list which is b_1 followed by b_2 followed by b_3 while each b_i ranks a_1, a_2, a_3 the same, i.e. a_1, a_2, a_3 are tied together in b_i's preference list. It is easy to see that for any matching M here, there is another matching M' such that $M' \succ M$, thus this instance admits no popular matching.

The popular matching problem is to determine if a given instance $G = (A \cup B, E)$ admits a popular matching or not, and if so, to compute one. This problem has been studied in the following two models.

- *1-sided model:* here it is only vertices in A that have preferences and cast votes; vertices in B are objects with no preferences or votes.
- *2-sided model:* vertices on both sides have preferences and cast votes.

Popular matchings need not always exist in the 1-sided model and the problem of whether a given instance admits one or not can be solved efficiently using the characterization and algorithm from [1]. In the 2-sided model when all preference lists are strict, it can be shown that any stable matching is popular [3]; thus a popular matching can be found in linear time using the Gale-Shapley algorithm. However when ties are allowed in preference lists on both sides, Biró, Irving, and Manlove [3] showed that the popular matching problem is NP-complete. In this paper we focus on the following variant:

∗ it is only vertices in A that have preference lists ranking their neighbors, however vertices on *both* sides cast votes.

That is, vertices in B have no preference lists ranking their neighbors – however b desires to be matched to any of its neighbors. Thus in an election between two matchings, b abstains from voting if it is matched in both or unmatched in both, else it votes for the matching where it is matched. An intuitive understanding of such an instance is that A is a set of applicants and B is a set of tasks – while each applicant has a preference list over the tasks that she is interested in, each task just cares to be assigned to anyone who is interested in performing it. We will see in Section 2 that the above problem is significantly different from the popular matching problem in the 1-sided model where vertices in B do not cast votes. We show the following results here, complementing our polynomial time algorithm in Theorem 1 with our hardness result in Theorem 2.

Theorem 1. *Given a bipartite graph $G = (A \cup B, E)$ where each $a \in A$ has strict preference list over its neighbors while each $b \in B$ puts all its neighbor into a single tie, the popular matching problem in G can be solved in $O(n^2)$ time where $|A \cup B| = n$.*

Theorem 2. *The popular matching problem is NP-complete in $G = (A \cup B, E$ where each $a \in A$ has a strict preference list while each $b \in B$ either has a stric preference list or puts all its neighbors into a single tie.*

Note that our NP-hardness reduction needs B to have $\Omega(|B|)$ vertices with strict preference lists and $\Omega(|B|)$ vertices with single ties as their preference lists. Theorem 2 follows from a simple reduction from the (2,2)-E3-SAT problem which is NP-complete [2]. Our reduction shows that the 2-sided popular matching problem in $G = (A \cup B, E)$ where every vertex in A has a strict preference list of length 2 or 4 and every vertex in B has either a strict preference list of length 2 or a single tie of length 2 or 3 as a preference list is NP-complete.

We show Theorem 1 by partitioning the set B into three sets: the first set X is a subset of top posts and, roughly speaking, the second set Y consists of *mid-level* posts, and the third set Z consists of *unwanted* posts (see Figure 1) Applicants get divided into two sets: the set of those with one or more neighbors in the set Z (call this set $\mathsf{nbr}(Z)$) and the rest (this set is $A \setminus \mathsf{nbr}(Z)$).

Our algorithm performs the partition of B into X, Y, and Z over several iterations. Initially $X = F$, where F is the set of top posts, $Y = B \setminus F$, and $Z = \emptyset$. In each iteration, certain top posts get *demoted* from X to Y and certain non-top posts get demoted from Y to Z. With new posts entering Z, we also have applicants moving from $A \setminus \mathsf{nbr}(Z)$ to $\mathsf{nbr}(Z)$. Using the partition $\langle X, Y, Z \rangle$ of B, we will build a graph H where each applicant keeps at most two edges: either to its most preferred post in X and also in Y or to its most preferred post in Z and also in Y. Some dummy posts may be included in Y.

We prove that G admits a popular matching if and only if H admits an A-complete matching, i.e., one that matches all vertices in A. We show that corresponding to any popular matching in G, there is a partition $\langle L_1, L_2, L_3 \rangle$ of B into *top posts*, *mid-level posts*, and *unwanted posts* such that $X \supseteq L_1$ and $Z \subseteq L_3$, where $\langle X, Y, Z \rangle$ is the partition computed by our algorithm. This allows us to show that if H does not admit an A-complete matching, then G has no popular matching. In fact, not every popular matching in G becomes an A-complete matching in H. However it will be the case that if G admits popular matchings, then at least one of them becomes an A-complete matching in H.

Background. Popular matchings have been well-studied in the 1-sided model [1,9–15] where only vertices of A have preferences and cast votes. Abraham et al. [1] gave polynomial time algorithms to determine if a given instance admits a popular matching or not – their algorithm also works when preference lists of vertices in A admit ties. Gärdenfors [5], who introduced the notion of popular matchings, considered this problem in the domain of 2-sided preference lists. In any instance $G = (A \cup B, E)$ with 2-sided strict preference lists, a stable matching is actually a minimum size popular matching and polynomial algorithms for computing a maximum size popular matching were given in [7,8]. **Organization of the Paper.** Section 2 has preliminaries, Section 3 has our algorithm and our proof of correctness. Due to the space constraints, certain proofs (incl. the proof of Theorem 2) have been omitted from this version of the paper. These proofs will be included in the full version of the paper.

2 Preliminaries

For any $a \in A$, let $f(a)$ denote a's most desired post. Let $F = \{f(a) : a \in A\}$ be the set of top posts. We will refer to posts in F as f-posts and to those in $B \setminus F$ as non-f-posts. For any $a \in A$, let r_a be the rank of a's most preferred non-f-post in a's preference list; when all of a's neighbors are in F, we set $r_a = \infty$. The following theorem characterizes popular matchings in the 1-sided voting model.

Theorem 3 (from [1]). *Let $G = (A \cup B, E)$ be an instance of the 1-sided popular matching problem, where each $a \in A$ has a strict preference list. Let M be any matching in G. M is popular if and only if the following two properties are satisfied:*

(i) M matches every $b \in F$ to some applicant a such that $b = f(a)$;
(ii) M matches each applicant a to either $f(a)$ or its neighbor of rank r_a.

Thus the only applicants that may be left unmatched in a popular matching here are those $a \in A$ that satisfy $r_a = \infty$.

Let us consider the following example where $A = \{a_1, a_2, a_3\}$ and $B = \{b_1, b_2, b_3\}$: both a_1 and a_2 have the same preference list which is $b_1 > b_2$ (b_1 followed by b_2) while a_3's preference list is $b_1 > b_2 > b_3$. Assume first that only applicants cast votes. The only posts that any of a_1, a_2, a_3 can be matched to in a popular matching here are b_1 and b_2. As there are three applicants and only two possible partners in a popular matching, there is no popular matching here. However in our 2-sided voting model, where posts also care about being matched and all neighbors are in a single tie, we have a popular matching $\{(a_1, b_1), (a_2, b_2), (a_3, b_3)\}$. Note that b_3 is ranked third in a_3's preference list, which is worse than $r_{a_3} = 2$, however such edges are permitted in popular matchings in our 2-sided model.

Consider the following example: $A = \{a_0, a_1, a_2, a_3\}$ and $B = \{b_0, b_1, b_2, b_3\}$; both a_1 and a_2 have the same preference list which is $b_1 > b_2$ while a_3's preference list is $b_1 > b_0 > b_2$ and a_0's preference list is $b_0 > b_3$. There is again no popular matching here in the 1-sided model, however in our 2-sided voting model, we have a popular matching $\{(a_0, b_3), (a_1, b_1), (a_2, b_2), (a_3, b_0)\}$. Note that $b_0 \in F$ and here it is matched to a_3 and $f(a_3) \neq b_0$; also a_3 is matched to its second ranked post: this is neither its top post nor its r_{a_3}-th ranked post ($r_{a_3} = 3$ here).

Thus popular matchings in our 2-sided voting model are quite different from the characterization given in Theorem 3 for popular matchings in the 1-sided model. Our algorithm (presented in Section 3) uses the following decomposition.

Dulmage-Mendelsohn Decomposition [4]. Let M be a maximum matching in a bipartite graph $G = (A \cup B, E)$. Using M, we can partition $A \cup B$ into three disjoint sets: a vertex v is *even* (similarly, *odd*) if there is an even (resp., odd) length alternating path (with respect to M) from an unmatched vertex to v. Similarly, a vertex v is *unreachable* if there is no alternating path from an unmatched vertex to v. Denote by $\mathcal{E}$, $\mathcal{O}$, and $\mathcal{U}$ the sets of even, odd, and unreachable vertices, respectively. The following properties (proved in [6]) will be used in our algorithm and analysis.

- $\mathcal{E}$, $\mathcal{O}$, and $\mathcal{U}$ are pairwise disjoint. Let M' be any maximum matching in G and let $\mathcal{E}'$, $\mathcal{O}'$, and $\mathcal{U}'$ be the sets of even, odd, and unreachable vertices with respect to M', respectively. Then $\mathcal{E} = \mathcal{E}'$, $\mathcal{O} = \mathcal{O}'$, and $\mathcal{U} = \mathcal{U}'$.
- Every maximum matching M matches all vertices in $\mathcal{O} \cup \mathcal{U}$ and has size $|\mathcal{O}| + |\mathcal{U}|/2$. In M, every vertex in $\mathcal{O}$ is matched with some vertex in $\mathcal{E}$, and every vertex in $\mathcal{U}$ is matched with another vertex in $\mathcal{U}$.
- The graph G has no edge in $\mathcal{E} \times (\mathcal{E} \cup \mathcal{U})$.

3   Finding Popular Matchings in a 2-sided Voting Model

The input is $G = (A \cup B, E)$ where each applicant $a \in A$ has a strict preference list while each post $b \in B$ has a single tie as its preference list. Our algorithm below builds a graph H using a partition $\langle X, Y, Z \rangle$ of B that is constructed in an iterative manner. Initialize $X = F$, $Y = B \setminus F$, and $Z = \emptyset$.

For any $a \in A$, recall that r_a is the rank of a's most preferred non-f-post. For any $U \subseteq B$, let $\mathsf{nbr}(U)$ (similarly, $\mathsf{nbr}_H(U)$) denote the set of neighbors in G (resp., in H) of the vertices in U. Note that our algorithm will maintain $\mathsf{nbr}_H(X) \cap \mathsf{nbr}(Z) = \emptyset$ by ensuring that $\mathsf{nbr}_H(X) \subseteq A \setminus \mathsf{nbr}(Z)$.

(I) While true do
    0. H is the empty graph on $A \cup B$.
    1. For each $a \in A \setminus \mathsf{nbr}(Z)$ do:
       – if $f(a) \in X$ then add the edge $(a, f(a))$ to H.
    2. For every $b \in X$ that is isolated in H do:
       – delete b from X and add b to Y.
    3. For each $a \in A$ do:
       – let b be a's most preferred post in the set Y; if the rank of b in a's preference list is $\leq r_a$ (i.e., r_a or better), then add (a, b) to H.
    4. Consider the graph H constructed in steps 1-3. Compute a maximum matching in H. *[This is to identify "even" posts in H.]*
       – If there exist even posts in Y then delete all even posts from Y and add them to Z.
       – Else quit the While-loop.
(II) Every $a \in \mathsf{nbr}(Z)$ adds the edge (a, b) to H where b is a's most preferred post in the set Z.
(III) Add all posts in $D = \{\ell(a) : a \in A \text{ and } r_a = \infty\}$ to Y, where $\ell(a)$ is the *dummy* last resort post of applicant a. For every applicant a such that $\mathsf{nbr}(\{a\}) \subseteq X$, add the edge $(a, \ell(a))$ to H.

Note that if a matching M includes the edge $(a, \ell(a))$, it means a is unmatched in M. The condition for exiting the While-loop ensures that all posts in Y, and hence all in $X \cup Y$, are odd/unreachable in the subgraph of H with the set of posts restricted to *real* posts in $X \cup Y$ (i.e., the non-dummy ones). So starting with a maximum matching in this subgraph and augmenting it after adding the edges on posts in Z in Step (II) and the edges on dummy posts in Step (III), we get a maximum matching in H that matches all real posts in $X \cup Y$. After the construction of H, our algorithm for the popular matching problem in G is given below.

- If H admits an A-complete matching, then return one that matches all real posts in $X \cup Y$; else output "G has no popular matching".

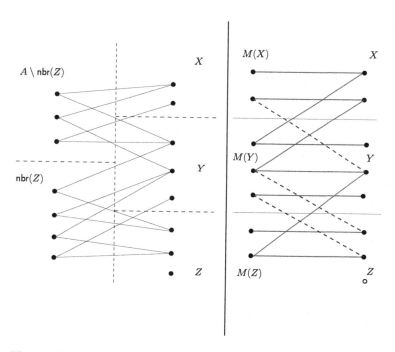

Fig. 1. The set B gets partitioned into X, Y, and Z. We have $\mathsf{nbr}_H(X) \cap \mathsf{nbr}(Z) = \emptyset$. In the figure on the right, the horizontal edges belong to M. Only the edges of $(M(Y) \times X) \cup (M(Z) \times (X \cup Y))$ can be labeled $+1$.

In the rest of this section, we prove the following theorem.

Theorem 4. *G admits a popular matching if and only if H admits an A-complete matching, i.e., one that matches all vertices in A.*

The Sufficient Part. We first show that if H admits an A-complete matching, then G admits a popular matching. We have already observed that if H admits an A-complete matching, then H has an A-complete matching that matches all real posts in $X \cup Y$. Call this matching M; other than the dummy last resort posts, all posts that are unmatched in M have to be in Z.

A useful observation is that $Z \subseteq B \setminus F$. This is because in Step 4 of the While-loop in our algorithm, all f-posts in Y are odd/unreachable in H as they are the only neighbors in H of applicants who regard them as f-posts.

We now assign edge labels in $\{\pm 1\}$ to all edges in $G \setminus M$: for an edge (a, b) in $G \setminus M$, if a prefers b to $M(a)$, then we label this edge $+1$, else we label this -1. The label of (a, b) is basically a's vote for b vs $M(a)$. Figure 1 is helpful here.

For any $U \in \{X, Y, Z\}$, let $M(U) \subseteq A$ be the set of applicants matched in M to posts in U. The following lemma is important.

Lemma 1. *Every edge of G in $M(X) \times Y$ is labeled -1; similarly, every edge in $M(Y) \times Z$ is labeled -1. Any edge labeled $+1$ has to be either in $M(Y) \times X$ or in $M(Z) \times (X \cup Y)$.*

Proof. Every edge of $\mathsf{nbr}(X) \times X$ that is present in H is a top ranked edge. Since M belongs to H, the edges of M from $\mathsf{nbr}(X) \times X$ are top ranked edges. Thus it is clear that every edge of G in $M(X) \times Y$ is labeled -1. Regarding $M(Y) \times Z$, every edge of $\mathsf{nbr}(Y) \times Y$ that is present in the graph H is an edge (a, b) where the rank of b in a's preference list is $\leq r_a$ (i.e., r_a or better); on the other hand, every edge of $\mathsf{nbr}(Z) \times Z$ that is present in the graph H is an edge (a, b') where the rank of b' in a's preference list is $\geq r_a$ (because $b' \in B \setminus F$). Since M belongs to H, the edges of M from $\mathsf{nbr}(Y) \times Y$ are ranked better than edges of $\mathsf{nbr}(Z) \times Z$. Thus every edge of G in $M(Y) \times Z$ is labeled -1.

We now show that any edge labeled $+1$ has to be in either $M(Y) \times X$ or $M(Z) \times (X \cup Y)$ (see Figure 1). Consider any edge $(a, b) \notin M$ such that $b \in U$ and $a \in M(U)$, where $U \in \{X, Y, Z\}$. It follows from the construction of the graph H that a vertex in $\mathsf{nbr}(U)$ can be adjacent in H to only its most preferred post in U. Thus any edge $(a, b) \notin M$ where $b \in U$ and $a \in M(U)$ is ranked -1. We have already seen that all edges in $M(X) \times Y$ and in $M(Y) \times Z$ are labeled -1. There are no edges in $M(X) \times Z$ since $M(X) \subseteq A \setminus \mathsf{nbr}(Z)$. Thus any edge labeled $+1$ has to be in either $M(Y) \times X$ or $M(Z) \times (X \cup Y)$. $\qquad \square$

Let M' be any matching in G. The symmetric difference of M' and M is denoted by $M' \oplus M$: this consists of alternating paths and alternating cycles – note that edges here alternate between M and M'. It will be convenient to assume that last resort posts are used only in M and not in M'. The claim that $M \succeq M'$ follows easily from Lemma 2. This proves the popularity of M.

Lemma 2. *Consider $M' \oplus M$. The following three properties hold:*

(i) *in any alternating cycle in $M' \oplus M$, the number of edges that are labeled -1 is at least the number of edges that are labeled $+1$.*

(ii) *in any alternating path in $M' \oplus M$, the number of edges that are labeled $+1$ is at most two plus the number of edges that are labeled -1; in case one of the endpoints of this path is a last resort post, then the number of edges labeled $+1$ is at most one plus the number of edges labeled -1.*

(iii) *in any even length alternating path in* $M' \oplus M$, *the number of edges that are labeled* -1 *is at least the number of edges that are labeled* $+1$; *in case one of the endpoints of this path is a last resort post, then the number of edges labeled* -1 *is at least one plus the number of edges labeled* $+1$.

The Necessary Part. We now show the other side of Theorem 4. That is, if G admits a popular matching, then H admits an A-complete matching. Let M^* be a popular matching in G. Lemma 3 will be useful to us.

Lemma 3. *If* $(a,b) \in M^*$ *and* $b \in F$, *then* b *has rank better than* r_a *in* a's *preference list.*

Label the edges of $G \setminus M^*$ by $+1$ or -1: the label of an edge (a,b) in $G \setminus M^*$ is the vote of a for b vs $M^*(a)$. In case a is not matched in M^*, then $\mathsf{vote}(a,b) = +1$ for any neighbor b of a. Due to the popularity of M^*, the following two properties hold on these edge labels (otherwise $M^* \oplus \rho \succ M^*$).

$(*)$ there is no alternating path ρ such that the edge labels in $\rho \setminus M^*$ are $\langle +1, +1, +1, \cdots \rangle$, i.e., no three consecutive non-matching edges labeled $+1$.

$(**)$ there is no alternating path ρ where the edge labels in $\rho \setminus M^*$ are $\langle +1, +1, -1, +1, +1, \cdots \rangle$.

From the matching M^* and the edge labels on $G \setminus M^*$, we partition B into $L_1 \cup L_2 \cup L_3$ as follows. This partition uses property $(*)$ in a crucial way.

0. Initialize $L_1 = L_2 = \emptyset$ and $L_3 = \{b \in B : b$ is unmatched in $M^*\}$. We now add more posts to the sets L_1, L_2, L_3 as described below.
1. Any alternating path with respect to M^* can have at most two consecutive non-matching edges that are labeled $+1$. For each length-5 alternating path $\rho = a_0\text{-}b_0\text{-}a_1\text{-}b_1\text{-}a_2\text{-}b_2$ where $(a_0, b_0), (a_1, b_1), (a_2, b_2) \in M^*$ and both (a_1, b_0) and (a_2, b_1) are marked $+1$, add b_{i-1} to L_i, for $i = 1, 2, 3$.
2. Now consider those $b \in B$ that are matched in M^* but b is not a part of any length-5 alternating path where both the non-matching edges are labeled $+1$. We repeat the following two steps till there are no more posts to be added to either L_2 or L_3 via these rules:
   - suppose $M^*(b)$ has no $+1$ edge incident on it: if $M^*(b) \in \mathsf{nbr}(L_3)$, then add b to L_2.
   - if $M^*(b)$ has a $+1$ edge to a vertex in L_2, then add b to L_3.
3. For each b such that $M^*(b)$ has no $+1$ edge incident on it:
   - if $M^*(b) \notin \mathsf{nbr}(L_3)$, then add b to L_1.
4. For each b not yet in $L_2 \cup L_3$ and $M^*(b)$ has a $+1$ edge to a vertex in L_1:
   - add b to L_2.

Lemma 4. *The above partition* $\langle L_1, L_2, L_3 \rangle$ *satisfies the following properties:*

(1) $F \subseteq L_1 \cup L_2$, *where* F *is the set of top posts.*
(2) $M^*(L_1) \cap \mathsf{nbr}(L_3) = \emptyset$.

We will use the partition $\langle L_1, L_2, L_3 \rangle$ of B to build the following subgraph $G' = (A \cup B, E')$ of G. For each $a \in A$, include the following edges in E':

(i) if $a \notin \mathsf{nbr}(L_3)$, then add the edge $(a, f(a))$ to E'.
(ii) if a has a neighbor of rank $\leq r_a$ in L_2, then add the edge (a, b) to E', where b is a's most preferred neighbor in L_2.
(iii) if $a \in \mathsf{nbr}(L_3)$, then add the edge (a, b) to E', where b is a's most preferred neighbor in L_3.

Lemma 5. *Every edge of the matching M^* belongs to the graph G'.*

Proof. The set B has been partitioned into $L_1 \cup L_2 \cup L_3$. We will now show that for each post b_0 that is matched in M^*, the edge $(M^*(b_0), b_0)$ belongs to G'.

– *Case 1.* The post $b_0 \in L_1$. Hence there is no $+1$ edge incident on $a_0 = M^*(b_0)$, in other words, $b_0 = f(a_0)$. Lemma 4.2 tells us that $M^*(L_1) \cap \mathsf{nbr}(L_3) = \emptyset$; hence a_0 has no neighbor in L_3 and by rule (i) above, the edge $(a_0, f(a_0)) = (a_0, b_0)$ belongs to the edge set of G'.

– *Case 2.* Next we consider the case when $b_0 \in L_2$. It is easy to see that b_0 has to be a_0's most preferred post in L_2, where $a_0 = M^*(b_0)$. Otherwise there would have been an edge (a_0, b_1) labeled $+1$ with $b_1 \in L_2$, where b_1 is a_0's most preferred post in L_2. Then either $b_1 \in L_1$ or $b_0 \in L_3$ (from how we construct the sets L_1, L_2, L_3), a contradiction. We now have to show that the rank of b_0 in a_0's preference list is $\leq r_a$, otherwise the edge (a_0, b_0) does not belong to G'.

Suppose $b_0 \in F$. Since the edge $(a_0, b_0) \in M^*$, which is a popular matching, it follows from Lemma 3 that b_0 is ranked better than r_{a_0} in a_0's preference list; thus the edge (a_0, b_0) would belong to G'. So the case left is when $b_0 \notin F$. If b_0 is not a_0's most preferred post outside F, then there is the length-5 alternating path $\rho = b_0\text{-}a_0\text{-}b_1\text{-}a_1\text{-}f(a_1)\text{-}M^*(f(a_1))$, where b_1 is the most preferred post of a_0 outside F and $a_1 = M^*(b_1)$. The alternating path ρ has two consecutive non-matching edges (a_0, b_1) and $(a_1, f(a_1))$ that are labeled $+1$. This contradicts the presence of b_0 in L_2 as such a post would have to be in L_3. Thus if $b_0 \notin F$, then b_0 has to be a_0's most preferred post outside F, i.e. b_0 has rank r_{a_0} in a_0's preference list.

– *Case 3.* We finally consider the case when the post $b_0 \in L_3$. We need to show that b_0 is the most preferred post of $a_0 = M^*(b_0)$ in L_3. Suppose not. Let b_1 be a_0's most preferred post in L_3. Since $b_1 \in L_3$ while $F \cap L_3 = \emptyset$ (by Lemma 4.1), we know that there is an edge labeled $+1$ incident on $a_1 = M^*(b_1)$. Let this edge be (a_1, b_2) and let a_2 be $M^*(b_2)$. So there is a length-5 alternating path $p = b_0\text{-}a_0\text{-}b_1\text{-}a_1\text{-}b_2\text{-}a_2$ where both the non-matching edges (a_0, b_1) and (a_1, b_2) are labeled $+1$. This contradicts the presence of b_1 in L_3 as such a post would have to be in L_2. Thus b_0 is a_0's most preferred post in L_3. □

The following lemma shows the relationship between the partition $\langle L_1, L_2, L_3 \rangle$ and the partition $\langle X, Y, Z \rangle$ constructed by our algorithm earlier.

Lemma 6. *The set $X \supseteq L_1$ and the set $Z \subseteq L_3$, where X and Z are the sets in the partition $\langle X, Y, Z \rangle$ constructed by our algorithm that builds the graph H.*

The matching M^* need not be A-complete. However it would help us to assume that M^* is A-complete, so we augment M^* by adding $(a, \ell(a))$ edges for every $a \in A$ that is unmatched in M^*. Recall that $\ell(a)$ is the dummy last resort post of a. However the augmented matching M^* need not belong to the graph G' any longer – hence we augment G' also by adding some dummy vertices and some edges as described below.

The augmentation of G' is analogous to Step (III) of our algorithm – we augment G' as follows: let $L_2 = L_2 \cup D$, where $D = \{\ell(a) : a \in A$ and $r_a = \infty\}$; if $\mathsf{nbr}(\{a\}) \subseteq L_1$, then add $(a, \ell(a))$ to G'. Thus when compared to G', the augmented G' has some new vertices (all these are dummy last resort posts) and some new edges – each new edge is of the form $(a, \ell(a))$ where $\ell(a)$ is a's only neighbor in $L_2 \cup L_3$. These new edges are enough to show the following lemma.

Lemma 7. *The augmented matching M^* belongs to the augmented graph G'.*

Since the augmented M^* is an A-complete matching, it follows from Lemma 7 that the augmented graph G' admits an A-complete matching. Theorem 5 uses Lemma 6 to show that if the augmented graph G' admits an A-complete matching, then so does the graph H constructed by our algorithm.

Theorem 5. *If H does not admit an A-complete matching, then the augmented graph G' cannot admit an A-complete matching.*

Proof. We will use G' to refer to the *augmented* graph G' in this proof. The rules for adding edges in H and in G' are exactly the same – the only difference is in the partition $\langle X, Y, Z \rangle$ on which H is based vs the partition $\langle L_1, L_2, L_3 \rangle$ on which G' is based. If $\langle X, Y, Z \rangle = \langle L_1, L_2, L_3 \rangle$, then the graphs H and G' are exactly the same.

Refer to Figure 2. This denotes how the partition $\langle X, Y, Z \rangle$ can be modified to the partition $\langle L_1, L_2, L_3 \rangle$. We know from Lemma 6 that $X \supseteq L_1$ and $Z \subseteq L_3$. Consider the subgraph G'_0 of G' induced on the vertex set $A' = (A \setminus \mathsf{nbr}(Z)) \cup (\mathsf{nbr}(Z) \cap \mathsf{nbr}_H(L_3 \setminus Z))$ and $B' = X \cup Y$. This is the part bounded by the box in Figure 2. In our analysis, we can essentially separate G' into G'_0 and the part outside G'_0 due to the following claim that says G' has no edges between A' and Z.

Claim 1. *G' has no edge (a, b) where $a \in A'$ and $b \in Z$.*

Proof. Any applicant $a \in A'$ has to belong to either $A \setminus \mathsf{nbr}(Z)$ or to $\mathsf{nbr}(Z) \cap \mathsf{nbr}_H(L_3 \setminus Z)$ (see Figure 2). There is obviously no edge in G between a vertex in $A \setminus \mathsf{nbr}(Z)$ and any vertex in Z. So suppose $a \in \mathsf{nbr}(Z) \cap \mathsf{nbr}_H(L_3 \setminus Z)$. For $b \in L_3$, if the edge (a, b) is in G', then b has to be a's most preferred post in L_3. We will now show that $b \in L_3 \setminus Z$, equivalently $b \notin Z$. Thus G' has no edge (a, b) where $a \in A'$ and $b \in Z$.

Since $a \in \mathsf{nbr}_H(L_3 \setminus Z)$, the graph H contains an edge between a and some $b' \in L_3 \setminus Z$. Recall that an element of $L_3 \setminus Z$ is a real post in Y. By the rules of including edges in H, it follows that the rank of b' in a's preference list is $\leq r_a$.

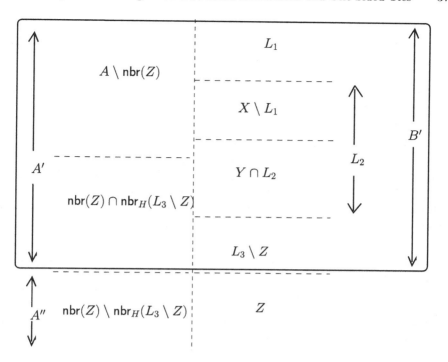

Fig. 2. The part of G' inside the box will be called G'_0. The graph G' has no edge between any applicant in A' and any post in Z.

The entire set L_3 cannot contain any post of rank better than r_a for any $a \in A$ since any post of rank better than r_a in a's list belongs to F while $L_3 \cap F = \emptyset$ (by Lemma 4.1). So b' has rank r_a in a's list. Thus a's most preferred neighbor in L_3 belongs to $L_3 \setminus Z$. $\qquad \square$

Let G_0 be the subgraph of G'_0 obtained by deleting from G'_0 the edges that are absent in H. Thus G_0 is a subgraph of both G' and H. The following claim will be useful to us.

Claim 2. *All posts in $(X \setminus L_1) \cup (L_3 \setminus Z)$ are odd/unreachable in G_0. Moreover, every edge (a,b) in G' that is missing in H satisfies $b \in (X \setminus L_1) \cup (L_3 \setminus Z)$.*

Consider the graph G_1 whose edge set is the intersection of the edge sets of G' and H. Equivalently, G_1 can be constructed by adding to the edge set of G_0 the edges incident on $A'' = \mathsf{nbr}(Z) \setminus \mathsf{nbr}_H(L_3 \setminus Z)$ that are present in both G' and H (see Fig. 2). This is due to the fact that G' has no edge in $A' \times Z$.

We claim that all posts in $(X \setminus L_1) \cup (L_3 \setminus Z)$ are odd/unreachable in G_1. This is because Claim 2 tells us that each post in this set is odd/unreachable in G_0 and due to the absence of $A' \times Z$ edges in G', the graph G_1 has no *new* edge (new when compared to G_0) incident on the set A' of applicants in G_0. Hence all posts in $(X \setminus L_1) \cup (L_3 \setminus Z)$ remain odd/unreachable in G_1.

Claim 2 also tells us that all edges in G' that are missing in H are incident on posts in $(X \setminus L_1) \cup (L_3 \setminus Z)$. We know that all these posts are odd/unreachable in G_1, hence G' has no *new* edge (new when compared to G_1) on posts that are *even* in G_1. Thus the size of a maximum matching in G' equals the size of a maximum matching in G_1. This is at most the size of a maximum matching in H, since G_1 is a subgraph of H. Hence if H has no A-complete matching, then neither does G'. □

Theorem 5, along with Lemma 7, finishes the proof of the necessary part of Theorem 4 and this completes the proof of correctness of our algorithm.

It is easy to see that each iteration of our algorithm takes $O(n)$ time (where $|A \cup B| = n$) since it involves finding a maximum matching in a subgraph where each vertex in A has degree at most 2. Thus the running time of our algorithm is $O(n^2)$ and Theorem 1 stated in Section 1 follows.

Conclusions and an Open Problem. We gave an $O(n^2)$ algorithm for the popular matching problem in $G = (A \cup B, E)$ where vertices in A have strict preference lists while each vertex in B puts all its neighbors into a single tie and $n = |A \cup B|$. Our algorithm needs the preference lists of vertices in A to be strict and the complexity of this problem when ties are allowed in the preference lists of vertices in A is currently unknown.

References

1. Abraham, D.J., Irving, R.W., Kavitha, T., Mehlhorn, K.: Popular matchings. SIAM Journal on Computing **37**(4), 1030–1045 (2007)
2. Berman, P., Karpinski, M., Scott, A.D.: Approximation hardness of short symmetric instances of MAX-3SAT. Electronic Colloquium on Computational Complexity Report, number 49 (2003)
3. Biró, P., Irving, R.W., Manlove, D.F.: Popular Matchings in the Marriage and Roommates Problems. In: Calamoneri, T., Diaz, J. (eds.) CIAC 2010. LNCS, vol. 6078, pp. 97–108. Springer, Heidelberg (2010)
4. Dulmage, A., Mendelsohn, N.: Coverings of bipartite graphs. Canadian Journal of Mathematics **10**, 517–534 (1958)
5. Gärdenfors, P.: Match making: assignments based on bilateral preferences. Behavioural Science **20**, 166–173 (1975)
6. Graham, R.L., Grötschel, M., Lovasz, L., (eds.) The Handbook of Combinatorics, chapter 3, Matchings and Extensions, by W. R. Pulleyblank, pp. 179–232. North Holland (1995)
7. Huang, C.-C., Kavitha, T.: Popular Matchings in the Stable Marriage Problem. In: Aceto, L., Henzinger, M., Sgall, J. (eds.) ICALP 2011, Part I. LNCS, vol. 6755, pp. 666–677. Springer, Heidelberg (2011)
8. Kavitha, T.: Popularity vs Maximum cardinality in the stable marriage setting. In: Proceedings of the 23rd SODA, pp. 123–134 (2012)
9. Kavitha, T., Mestre, J., Nasre, M.: Popular Mixed Matchings. In: Albers, S., Marchetti-Spaccamela, A., Matias, Y., Nikoletseas, S., Thomas, W. (eds.) ICALP 2009, Part I. LNCS, vol. 5555, pp. 574–584. Springer, Heidelberg (2009)

10. Kavitha, T., Nasre, M.: Note: Optimal popular matchings. Discrete Applied Mathematics **157**(14), 3181–3186 (2009)
11. Mahdian, M.: Random popular matchings. In: Proceedings of the 7th EC, pp. 238–242 (2006)
12. Manlove, D.F., Sng, C.T.S.: Popular Matchings in the Capacitated House Allocation Problem. In: Azar, Y., Erlebach, T. (eds.) ESA 2006. LNCS, vol. 4168, pp. 492–503. Springer, Heidelberg (2006)
13. McCutchen, R.M.: The Least-Unpopularity-Factor and Least-Unpopularity-Margin Criteria for Matching Problems with One-Sided Preferences. In: Laber, E.S., Bornstein, C., Nogueira, L.T., Faria, L. (eds.) LATIN 2008. LNCS, vol. 4957, pp. 593–604. Springer, Heidelberg (2008)
14. McDermid, E., Irving, R.W.: Popular Matchings: Structure and Algorithms. In: Ngo, H.Q. (ed.) COCOON 2009. LNCS, vol. 5609, pp. 506–515. Springer, Heidelberg (2009)
15. Mestre, J.: Weighted Popular Matchings. In: Bugliesi, M., Preneel, B., Sassone, V., Wegener, I. (eds.) ICALP 2006. LNCS, vol. 4051, pp. 715–726. Springer, Heidelberg (2006)

Block Interpolation: A Framework for Tight Exponential-Time Counting Complexity

Radu Curticapean[1,2](✉)

[1] Department of Computer Science, Saarland University, Saarbrücken, Germany
curticapean@cs.uni-sb.de
[2] Institute for Computer Science and Control,
Hungarian Academy of Sciences (MTA SZTAKI), Budapest, Hungary

Abstract. We devise a framework for proving tight lower bounds under the counting exponential-time hypothesis #ETH introduced by Dell et al. Our framework allows to convert many known #P-hardness results for counting problems into results of the following type: If the given problem admits an algorithm with running time $2^{o(n)}$ on graphs with n vertices and $\mathcal{O}(n)$ edges, then #ETH fails. As exemplary applications of this framework, we obtain such tight lower bounds for the evaluation of the zero-one permanent, the matching polynomial, and the Tutte polynomial on all non-easy points except for two lines.

1 Introduction

Counting complexity is a classical sub-field of complexity theory, launched by the seminal paper [17] that introduced the class #P and proved #P-hardness of the permanent. Since then, various counting problems were systematically proven to be #P-hard, including the evaluation of graph polynomials such as the Tutte polynomial [12] and the cover polynomial [1], and counting solutions to constraint-satisfaction problems [2,3] as well as so-called Holant problems [5].

We depart from the classical setting of #P-hardness and follow the route taken by [7], who proved conditional lower bounds on the running times required to solve counting problems. Our results assume the exponential-time hypothesis #ETH, introduced in [7], which postulates that the satisfying assignments to a 3-CNF formula φ on n variables and $\mathcal{O}(n)$ clauses cannot be *counted* in time $2^{o(n)}$. This hypothesis is trivially implied by its decision version ETH, introduced in [11], which assumes the same lower bound for *deciding* satisfiability of φ.

We obtain our lower bounds by a play on polynomial interpolation, which is arguably the most important technique for non-parsimonious reductions between counting problems, as used in [4,7,9,10,12,14,16]. As a first example of this technique, let us reduce counting perfect matchings to counting matchings that are not necessarily perfect, using a simplification of an argument in [16].

Step 1: Given a graph G, let m_k for $k \in \mathbb{N}$ denote its number of matchings with exactly k unmatched vertices. Define the polynomial $\mu(x) = \sum_{k=0}^{n} m_k x^k$ and

R. Curticapean—Supported by ERC Starting Grant PARAMTIGHT (No. 280152).

M.M. Halldórsson et al. (Eds.): ICALP 2015, Part I, LNCS 9134, pp. 380–392, 2015.
DOI: 10.1007/978-3-662-47672-7_31

observe that its coefficient m_0 is equal to the number of perfect matchings in G. Since μ has maximum degree n, we can use Lagrange interpolation to recover all of its coefficients (and m_0 in particular) from evaluations of μ at $n + 1$ distinct input points.

Step 2: We show how to evaluate $\mu(t)$ for $t \in \mathbb{N}$ by a reduction to counting matchings: Let G_t be obtained from G by adding, for each $v \in V(G)$, a *gadget* consisting of $t - 1$ fresh vertices adjacent to v. Then it can be checked that $\mu(t)$ is equal to the number of matchings in G_t.

By evaluating $\mu(t)$ for all $t \in [n + 1]$ via Step 2 and an oracle for counting matchings, we can use Lagrange interpolation as in Step 1 to obtain m_0. This also provides a lower bound for counting matchings, which is however far from tight: If counting *perfect* matchings on n-vertex graphs has a lower bound of $2^{\Omega(n)}$, then only a $2^{\Omega(\sqrt{n})}$ lower bound for counting matchings follows from the above argument. This is because G_{n+1} has a gadget of size n at each vertex, and thus n^2 vertices in total. Using more sophisticated gadgets with $\mathcal{O}(\log^c n)$ vertices, similar reductions (for other problems) were obtained in [7,9,10], implying $2^{\Omega(n/\log^c n)}$ lower bounds for these problems, which are however still not tight. In particular, the tight lower bound for the source problem of computing a hard coefficient was "shrinked" in the reduction.

Let us call a reduction gadget-interpolation-based if it proceeds along the two steps above: First encode a hard problem into the coefficients of a polynomial p, then find "local" gadgets that can be placed at vertices or edges, and which allow to evaluate $p(\xi)$ at sufficiently many points ξ by reduction to the target problem. Finally use Lagrange interpolation to recover p from these evaluations. This is a well-trodden route for obtaining #P-hardness proofs.

When carried out on n-vertex graphs G, such reductions typically yield polynomials p of degree n, hence require $n+1$ evaluations of p at *distinct* points, and thus in turn require $n + 1$ *distinct* gadgets to be placed at vertices of G. Since there are only finitely many simple graphs on $\mathcal{O}(1)$ vertices, the size of such gadgets must necessarily grow as some unbounded function $\alpha(n)$, and we can hence only obtain $2^{\Omega(n/\alpha(n))}$ lower bounds for $\alpha \in \omega(1)$. Additionally, such reductions run in polynomial time, which is required for the setting of #P-hardness, but nonessential in exponential-time complexity: To obtain a lower bound of $2^{\Omega(n)}$, we might as well use a reduction that requires $2^{o(n)}$ time and issues $2^{o(n)}$ queries to the target problem. The limitations we observed in this paragraph are immanent to every known lower bound under #ETH.

In this paper, we circumvent these barriers by introducing a framework that allows to apply the full power of subexponential reductions to counting problems. To this end, we use a simple trick based on *multivariate* polynomial interpolation: In this setting, we are not given an unknown univariate polynomial p of degree n, which we have to interpolate from $n + 1$ oracle calls, but rather a multivariate polynomial $\boldsymbol{p}$ with total degree n, but maximum degree $c = \mathcal{O}(1)$ in each indeterminate, which we can interpolate from $2^{o(n)}$ evaluations. Each evaluation $\boldsymbol{p}(\xi)$ is performed at a tuple ξ whose entries are contained in a set of size $c + 1$, and this will enable us to compute $\boldsymbol{p}(\xi)$ by attaching only $c + 1$

distinct gadgets to G. The catch here is that different vertices may obtain different gadgets, which was not feasible in the univariate setting.

Our technique is phrased as a general framework that allows to convert a large body of gadget-interpolation-based #P-hardness proofs into tight lower bounds under #ETH. The growth of the gadgets used in such proofs is *irrelevant* to the framework, as only a constant number of gadgets will be used. This allows us to use luxuriously large gadgets, and in particular, we do not need to invoke involved gadget constructions, e.g., for simulating weights in the Tutte polynomial [7, 10] or in the independent set polynomial [9]. To showcase our framework, we show that #ETH implies $2^{\Omega(n)}$ lower bounds for the following problems on unweighted simple graphs G with n vertices and $\mathcal{O}(n)$ edges[1], all of which admit trivial $2^{\mathcal{O}(n)}$ algorithms on such graphs.

- Counting perfect matchings, even for bipartite unweighted graphs G. In [7], only a lower bound of $2^{\Omega(n/\log n)}$ was shown under #ETH. A tight lower bound of $2^{\Omega(n)}$ was obtained only (a) under rETH, which implies ETH, which in turn implies #ETH, but no converse direction is known, or (b) under #ETH, but by introducing negative edge weights. Negative edge weights are generally worrying, e.g., because perfect matchings in bipartite graphs can be approximately counted on graphs with non-negative edge weights, but are inapproximable when negative edge weights are present [13].
- Evaluating the matching polynomial $\mu(G; \xi)$ at fixed $\xi \in \mathbb{Q}$. No lower bounds for this problem are stated in the literature.
- Evaluating the independent set polynomial $I(G; \xi)$ at fixed $\xi \in \mathbb{Q} \setminus \{0\}$. In [9], a lower bound of $2^{\Omega(n/\log^3 n)}$ was shown at general $\xi \in \mathbb{Q} \setminus \{0\}$, and $2^{\Omega(n)}$ at $\xi = 1$, but neither of these bounds assume sparse graphs.
- Evaluating the Tutte polynomial at all points except for two lines. In [7], only lower bounds of $2^{\Omega(n/\log^c n)}$ could be shown on sparse simple graphs.

2 Preliminaries

The graphs in this paper are finite, undirected and simple. They may feature edge- or vertex-weights within intermediate steps of arguments, but all such weights will ultimately be removed to obtain hardness results on *unweighted* graphs. For simplicity, we phrase our results using only rational numbers, but they could be easily adapted to $\mathbb{R}$ and, with some care, also to $\mathbb{C}$.

Our arguments and results use *graph polynomials*, which are functions that map graphs G to polynomials $p(G) \in \mathbb{Q}[\boldsymbol{x}]$, where $\boldsymbol{x}$ is some set of indeterminates. We abbreviate $p(G; \xi) := (p(G))(\xi)$. The arguably most famous graph polynomial is the Tutte polynomial, which we define in the following, along with the matching polynomial and the independent set polynomial.

---

[1] As is common to exponential-time complexity, it is crucial to obtain hardness results for *sparse* graphs, which feature $\mathcal{O}(n)$ edges on n vertices: Many reductions proceed by placing gadgets at edges, and this would map graphs with $\omega(n)$ edges to target graphs on $\omega(n)$ vertices, thus ruling out tight lower bounds of the type $2^{\Omega(n)}$.

Definition 1. *Let G be a graph. Let $\mathcal{M}[G]$ denote the set of (not necessaril*[y] *perfect) matchings in G and let $\mathrm{usat}(G, M)$ for $M \in \mathcal{M}[G]$ denote the set o*[f] *unmatched vertices of G in M. Then we define the* matching polynomial μ *(als*[o] *called* matching defect polynomial*) as $\mu(G; x) = \sum_{M \in \mathcal{M}[G]} x^{|\mathrm{usat}(G,M)|}$.*

Let $\mathcal{I}[G]$ denote the set of independent sets of G. Then we define the inde[-]*pendent set polynomial I as $I(G; x) = \sum_{S \in \mathcal{I}[G]} x^{|S|}$.*

Let $k(G, A)$ denote the number of connected components in the edge-induce[d] *subgraph $G[A]$. Then define the classical parameterization of the* Tutte polyno[-]*mial as $T(G; x, y) = \sum_{A \subseteq E(G)} (x-1)^{k(G,A) - k(G,E)} (y-1)^{k(G,A)+|A|-|V|}$. We wil*[l] *also work with the following* random-cluster formulation *of the Tutte polynomial which is defined by $Z(G; q, w) = \sum_{A \subseteq E(G)} q^{k(G,A)} w^{|A|}$.*

The polynomials Z and T are essentially the same polynomial up to repa[-]rameterization. As in [7], with $q = (x-1)(y-1)$ and $w = y - 1$, we have

$$T(G; x, y) = (x-1)^{-k(G,E)} (y-1)^{-|V(G)|} Z(G; q, w). \tag{1}$$

For the following definition, let $\mathcal{E}$ denote the set of all edges of all graphs.

Definition 2. *Write $\mathcal{PM}[G]$ for the set of perfect matchings of G, and let $\boldsymbol{x} = (x_e)_{e \in \mathcal{E}}$ be a set of indeterminates. Then the* perfect matching polynomial *is defined as $\mathrm{PerfMatch}(G) = \sum_{M \in \mathcal{PM}[G]} \prod_{e \in M} x_e$. Note that only finitely many*[y] *indeterminates are present in $\mathrm{PerfMatch}(G)$. If G is bipartite, we also denote $\mathrm{PerfMatch}(G)$ by the* permanent $\mathrm{perm}(G)$.

For any graph polynomial p, we define two problems $\mathsf{Coeff}(p)$ and $\mathsf{Eval}(p)$, and a family of problems $\mathsf{Eval}_S(p)$ for fixed subsets $S \subseteq \mathbb{Q}$.

$\mathsf{Coeff}(p)$:   On input G, compute all coefficients of $p(G)$.

$\mathsf{Eval}(p)$:   On input G and a tuple ξ, evaluate $p(G; \xi)$. We often see ξ as vertex-
              or edge-weights that are substituted into indeterminates of $p(G)$.

$\mathsf{Eval}_S(p)$:   On input G and a tuple ξ whose entries are from S, evaluate $p(G; \xi)$.
              If p is univariate and $S = \{a\}$, this simply asks asks to compute
              $p(G; a)$, and we write $\mathsf{Eval}_a(p)$ in this case.

Rather than evaluating a multivariate graph polynomial $\boldsymbol{p}$ like PerfMatch on an unweighted graph G and a tuple ξ, we often annotate edges/vertices of G with the entries of ξ, assuming $V(G)$ and $E(G)$ to be ordered. We then speak of evaluating $\boldsymbol{p}(G')$ on the weighted graph G' derived from G, ξ this way.

Given a univariate polynomial $p \in \mathbb{Q}[x]$ of degree n, we can use Lagrange interpolation to compute the coefficients of p when provided with the set $\{(\xi, p(\xi)) \mid \xi \in \Xi\}$ for any $\Xi \subseteq \mathbb{Q}$ of size $n + 1$. This can be generalized to multivariate polynomials $p \in \mathbb{Q}[\boldsymbol{x}]$, for instance, if Ξ is a sufficiently large grid.

Lemma 1. *Let $p \in \mathbb{Q}[x_1, \ldots, x_n]$, and for $i \in [n]$, let the degree of x_i in p be bounded by $d_i \in \mathbb{N}$. Let $\Xi = \Xi_1 \times \ldots \times \Xi_n$ where $\Xi_i \subseteq \mathbb{Q}$ and $|\Xi_i| = d_i + 1$ for all $i \in [n]$. Then we can compute the coefficients of p in time $\mathcal{O}(|\Xi|^3)$ when given as input the set $\{(\xi, p(\xi)) \mid \xi \in \Xi\}$.* $\square$

We also adapt subexponential-time Turing reduction families [11] for our use.

Definition 3. *A* subexponential reduction family *from problem* A *to* B *is an algorithm* $\mathbb{T}$ *with oracle access for* B. *Its inputs are pairs* (G, ϵ) *where* G *is an input graph for* A, *and* ϵ *with* $0 < \epsilon \leq 1$ *is a runtime parameter, such that*

1. $\mathbb{T}$ *computes* $A(G)$, *and it does so in time* $f(\epsilon) \cdot 2^{\epsilon |V(G)|} \cdot |V(G)|^{\mathcal{O}(1)}$, *and*
2. $\mathbb{T}$ *only invokes the oracle for* B *on graphs* G' *with at most* $g(\epsilon) \cdot (|V(G)| + |E(G)|)$ *vertices and edges.*

In these statements, f *and* g *are computable functions that depend only on* ϵ. *We write* A $\leq_{serf}$ B *if such a reduction exists.*

That is, the runtime of $\mathbb{T}$ (and hence, the number of oracle queries) can be chosen as $2^{\epsilon n}$ for arbitrarily small ϵ, in particular for $\epsilon = 1/\omega(1)$. We can hence ensure that the runtime of $\mathbb{T}$ is $2^{o(n)}$, but this comes at the cost of incurring a "blowup factor" of $g(\epsilon)$ in the reduction images. It can be verified that subexponential reductions preserve lower bounds as expected, see [11]:

Lemma 2. *If* A *admits a subexponential reduction family to* B *and* B *can be solved in time* $2^{o(n)} n^{\mathcal{O}(1)}$ *on graphs with* n *vertices and* $\mathcal{O}(n)$ *edges, then* A *can be solved in time* $2^{o(n)} n^{\mathcal{O}(1)}$ *on graphs with* n *vertices and* $\mathcal{O}(n)$ *edges.* □

The paper is organized as follows: In Section 3, we introduce our interpolation framework, and in Section 4, we present examples for graph polynomials that fit into this framework, and for which we consequently obtain tight lower bounds.

3   The Block Interpolation Framework

For a general class of univariate graph polynomials p, we show that $\mathsf{Coeff}(p) \leq_{serf} \mathsf{Eval}_\xi(p)$ at fixed $\xi \in \mathbb{Q}$. In the introduction, we have seen an example for such a reduction involving the polynomial μ, which was however not tight. In this section, we generalize this argument and ensure that it yields tight lower bounds.

To this end, we first describe, in Section 3.1, the "format" required from p for our framework to apply. Then we show in Section 3.2 how to reduce $\mathsf{Coeff}(p) \leq_{serf} \mathsf{Eval}_S(\boldsymbol{p})$, where $\boldsymbol{p}$ is a multivariate version of p and $S \subseteq \mathbb{Q}$ has size $\mathcal{O}(1)$. In Section 3.3, we then show how to reduce to $\mathsf{Eval}_\xi(p)$.

3.1   Admissible Graph Polynomials

Our framework applies to all univariate graph polynomials that admit "obvious" multivariate generalizations. More specifically, we call p *subset-admissible* if p is induced by a *sieving function* χ which filters the structures counted by p, and a *weight selector* ω which assigns a weight to each of these structures.

Definition 4. *Let $\mathcal{G}$ denote the set of all graphs, and let $\mathcal{F}$ denote the set of all vertices and edges of graphs. For $\chi : \mathcal{G} \times 2^{\mathcal{F}} \to \mathbb{C}$ and $\omega : \mathcal{G} \times 2^{\mathcal{F}} \to 2^{\mathcal{F}}$, we say that (χ, ω) induce the graph polynomial*

$$p_{\chi,\omega}(G; x) = \sum_{A \subseteq V(G) \cup E(G)} \chi(G, A) \cdot x^{|\omega(G,A)|}. \tag{2}$$

The polynomial p is subset-admissible if $p = p_{\chi,\omega}$ for some (χ, ω) as above.

Note that χ and ω may be partial functions, since, e.g., the value of $\chi(G, A)$ is irrelevant if $A \not\subseteq V(G) \cup E(G)$.

In the following, we observe that μ and I from Definition 1 are subset-admissible. It would be nice to show the same for T and Z, but this fails for syntactic reasons, since admissible polynomials are univariate by definition. Instead, we work with restrictions of Z to $Z_q := Z(q, \cdot)$ for fixed $q \in \mathbb{Q}$.

Example 1. Given a sentence ϕ, let $[\phi] = 1$ if ϕ is true, and $[\phi] = 0$ otherwise. The polynomial μ is induced by $\chi : (G, A) \mapsto [A \in \mathcal{M}[G]]$ and $\omega : (G, A) \mapsto \mathrm{usat}(G, A)$, and I is induced by $\chi : (G, A) \mapsto [A \in \mathcal{I}[G]]$ and $\omega : (G, A) \mapsto A$.

For $q \in \mathbb{Q} \setminus \{0\}$, the polynomial $Z_q = Z(q, \cdot)$ is induced by $\chi : (G, A) \mapsto q^{\kappa(G,A)}$ and $\omega : (G, A) \mapsto A$. We stress again that $Z_q \in \mathbb{Q}[x]$ is a univariate restriction of Z for fixed $q \in \mathbb{Q}$.

Every graph polynomial of the form $p_{\chi,\omega}$ admits a canonical *multivariate generalization* $\boldsymbol{p}_{\chi,\omega}$ on indeterminates $\mathbf{x} = \{x_a \mid a \in \mathcal{F}\}$, which is given by

$$\boldsymbol{p}_{\chi,\omega}(G; \mathbf{x}) = \sum_{A \subseteq V(G) \cup E(G)} \chi(G, A) \prod_{a \in \omega(G,A)} x_a. \tag{3}$$

Compare this to (2). The polynomial $\boldsymbol{p}_{\chi,\omega}$ coincides with $p_{\chi,\omega}$ when substituting $x_a \leftarrow x$ for all $b \in \mathcal{F}$. Note also that $\boldsymbol{p}$ is multilinear by definition. Similar multivariate generalizations were known, e.g., for the Tutte polynomial [15].

Example 2. Consider $p = p_{\chi,\omega}$ with $\chi(G, A) = [A \in \mathcal{PM}[G]]$ and $\omega(G, A) = A$. Then $p = m_G \cdot x^{|V(G)|/2}$, where $m_G = |\mathcal{PM}[G]|$. We also have $\boldsymbol{p} = \mathrm{PerfMatch}$.

We remark also that the coefficients of p can be recovered from those of $\boldsymbol{p}$:

Lemma 3. *For any monomial θ, let c_θ denote the coefficient of θ in $\boldsymbol{p}$. For $k \in \mathbb{N}$, let C_k denote the set of monomials in $\boldsymbol{p}$ with total power k. For $k \in \mathbb{N}$, the coefficient of x^k in p is equal to $\sum_{\theta \in C_k} c_\theta$.* $\square$

3.2   First Reduction Step: Multivariate Interpolation

Let $p = p_{\chi,\omega}$ be subset-admissible. For ease of presentation, we assume for now that $\omega : \mathcal{G} \times 2^{\mathcal{F}} \to 2^{\mathcal{E}}$, that is, ω maps only into edge-subsets rather than subsets of edges and vertices. The general case is shown identically, with more notation.

We reduce $\mathsf{Coeff}(p) \leq_{serf} \mathsf{Eval}(\boldsymbol{p})$ by means of interpolation, where $\boldsymbol{p}$ denotes the multivariate generalization of p. Recall that, in the univariate case, to obtain

$p(G)$ for an m-edge graph G, we require the evaluations of $p(G; \xi)$ at $m+1$ distinct points ξ. For the multivariate generalization $\boldsymbol{p}$, we can interpolate via Lemma 1: Since $\boldsymbol{p}$ is multilinear, this requires the evaluations of $\boldsymbol{p}(G; \xi)$ on a grid with two distinct values per coordinate, say $\Xi = [2]^m$. By Lemma 3, the coefficients of p can be obtained from those of $\boldsymbol{p}$, so we could interpolate $\boldsymbol{p}$ to recover p.

While this detour seems extremely wasteful due to its 2^m incurred evaluations, it yields the following reward: For each variable x_e in $\boldsymbol{p}$, the setting of Lemma 1 only requires us to substitute *two* distinct values (or *weights*) into x_e, whereas interpolation on p requires $m + 1$ distinct substitutions to its only variable x. A small number of distinct weights will be very useful, since each such weight will be simulated by a certain gadget at e. If there are only two weights to simulate, then we require only two fixed gadgets, whose sizes are trivially bounded by $\mathcal{O}(1)$.

However, to interpolate $\boldsymbol{p}$, we still need the prohibitively large number of 2^m evaluations. To overcome this, we trade off the number of evaluations with the numbers of distinct values that need to be simulated at each edge, and thus, with the size of the gadgets ultimately required.

Lemma 4. *Let p be subset-admissible, with multivariate generalization $\boldsymbol{p}$, and let $W = (w_0, w_1, \ldots)$ be an infinite recursively enumerable sequence of pairwise distinct numbers in $\mathbb{Q}$. Then $\mathsf{Coeff}(p) \leq_{serf} \mathsf{Eval}(\boldsymbol{p})$ holds by a reduction that, on input (G, ϵ), only asks queries $\boldsymbol{p}(G')$ on graphs G' obtained from G by introducing edge-weights from $W_d = \{w_0, \ldots, w_d\}$ for $d = f(\epsilon)$.*

When invoking Lemma 4, the list W contains the weights that can be simulated by gadgets. Note that *any* such list W can be used if W is infinite and recursively enumerable. Furthermore, note that $\boldsymbol{p}$ is evaluated only on edge-weighted versions of G itself; properties such as bipartiteness are hence trivially preserved. In the following, we write $x \leftarrow y$ for substituting y into x.

Proof (of Lemma 4). Let $d \in \mathbb{N}$ be a parameter, to be chosen later depending on ϵ, and let $G = (V, E)$ be an m-edge graph for which we want to determine the coefficients of $p = p(G)$. Let $\boldsymbol{x} = \{x_e \mid e \in E\}$ denote the indeterminates of $\boldsymbol{p}$ and note that both p and $\boldsymbol{p}$ have maximum degree m.

In the first step, partition E into $t = \lceil m/d \rceil$ blocks $E_1, \ldots, E_t$ of size at most d each, using an arbitrary equitable assignment of edges to blocks. Define new indeterminates $\boldsymbol{y} = \{y_1, \ldots, y_t\}$ and a new multivariate polynomial $\boldsymbol{q} \in \mathbb{Q}[\boldsymbol{y}]$ by substituting $x_e \leftarrow y_i$ for all $i \in [t]$ and $e \in E_i$. We are working with three polynomials, namely p, $\boldsymbol{p}$ and $\boldsymbol{q}$. While the total degree of $\boldsymbol{q}$ is bounded by m, the degree of each indeterminate y_i in $\boldsymbol{q}$ is bounded by d, since each block contains at most d edges. Hence, the number of monomials in $\boldsymbol{q}$ is at most $(d+1)^t = 2^{d'm}$ with $d' = \mathcal{O}(\log(d)/d)$. Note that $d' \to 0$ as $d \to \infty$.

We will obtain the coefficients of $\boldsymbol{q}$ via interpolation, and we observe that the coefficients of $\boldsymbol{q}$ allow to determine those of the univariate version p. Write c_k^p for the coefficient of x^k in p and c_θ^q for the coefficient of the monomial θ in $\boldsymbol{q}$. Analogously to Lemma 3, we have $c_k^p = \sum_{\theta \in C_k} c_\theta^q$ where C_k for $k \in \mathbb{N}$ is the set of all monomials with total power k in $\boldsymbol{q}$. This allows us to compute the coefficients of p from those of $\boldsymbol{q}$.

It remains to describe how to obtain the coefficients of q. For this, recall the definition of W_d from the statement. We evaluate q on the grid $\Xi = (W_d)$ using the oracle for $\mathsf{Eval}(p)$: For each $\xi \in \Xi$, substitute $y_i \leftarrow \xi_i$ for all $i \in [t]$ to obtain an edge-weighted graph G_ξ that contains only weights from W_d, and for which we can thus compute $p(G_\xi)$. Using $|\Xi| = (d+1)^t = 2^{d'm}$ oracle calls and grid interpolation via Lemma 1, we obtain all coefficients of q in time $\mathcal{O}(2^{3d'm})$. Since $d' \to 0$ as $d \to \infty$, we can pick d large enough such that $3d' \leq \epsilon$ and thus achieve running time $\mathcal{O}(2^{\epsilon m})$. No vertices or edges are added to G.   □

3.3   Second Reduction Step: Weight Simulation by Gadgets

With Lemma 4, we can reduce $\mathsf{Coeff}(p)$ on a graph G to $\mathsf{Eval}(p)$ on versions G_ξ obtained from G by introducing $\mathcal{O}(1)$ distinct edge-weights. For the full reduction, this latter problem must be reduced to $\mathsf{Eval}_\xi(p)$ for fixed $\xi \in \mathbb{Q}$. This may not work for all $\xi \in \mathbb{Q}$: For instance, $\mathsf{Eval}_0(I)$ for I at 0 is trivial. We must hence impose conditions on ξ to enable this reduction.

Definition 5. *Let p be subset-admissible, let $\xi \in \mathbb{Q}$ and*

- *let $W = (w_0, w_1, \ldots)$ be a sequence of pairwise distinct values in $\mathbb{Q}$,*
- *let $\mathcal{H} = (H_0, H_1, \ldots)$ be a sequence of edge-gadgets, which are triples (H, u, v) with a graph H and attachment vertices $u, v \in V(H)$, and*
- *let $F : \mathcal{G} \times \mathbb{Q} \to \mathbb{Q} \setminus \{0\}$ be a polynomial-time computable factor function.*

If G is edge-weighted with weights from W, let $T(G)$ be obtained by replacing, for $i \in \mathbb{N}$, each $uv \in E(G)$ of weight w_i with a fresh copy of H_i by identifying u, v across G and H_i. We say that $(\mathcal{H}, F)$ allows to reduce $\mathsf{Eval}_W(p)$ to $\mathsf{Eval}_\xi(p)$ if the following holds: Whenever G is a graph with edge-weights from W, then

$$p(G) = \frac{p(T(G); \xi)}{F(G, \xi)}. \tag{4}$$

The same definition applies to vertex-weighted graphs; here we use vertex-gadgets, *which are pairs (H, v) with an attachment vertex $v \in V(H)$. Vertex-gadgets are inserted at a vertex $v \in V(G)$ by identifying v in H and G.*

As a first example, we consider (well-known) edge-gadgets for PerfMatch.

Example 3. Let p denote the polynomial from Example 2 with $p = \mathrm{PerfMatch}$. Let $\mathcal{H} = (H_1, H_2, \ldots)$ be such that H_k for $k \in \mathbb{N}$ is the graph obtained by placing k parallel edges between u and v and then subdividing each edge twice. Let $\mathbb{N} = (1, 2, 3, \ldots)$ and let F denote the function that maps all inputs to 1. Then $(\mathcal{H}, F)$ allows to reduce $\mathsf{Eval}_\mathbb{N}(p)$ to $\mathsf{Eval}_1(p)$.

We then easily observe the following lemma.

Lemma 5. *Let $W = (w_0, w_1, \ldots)$ and let $(\mathcal{H}, F)$ allow to reduce $\mathsf{Eval}_W(p)$ to $\mathsf{Eval}_\xi(p)$. Let G feature only edge-weights from W. Then we can use (4) to compute $p(G)$ from $p(T(G); \xi)$. If G has n vertices and m edges, and only contains edge-weights w_i with $i \leq t$ for some $t \in \mathbb{N}$, then $T(G)$ has $\mathcal{O}(n + sm)$ vertices and edges, where $s = \max_{i \in [t]} |V(H_i)| + |E(H_i)|$ depends only on $\mathcal{H}$ and t.*

By combining Lemmas 4 and 5, we obtain the wanted reduction from $\mathsf{Coeff}(p)$ to $\mathsf{Eval}_\xi(p)$ at fixed points $\xi \in \mathbb{Q}$ and finish the set-up of our framework.

Theorem 1. *Let p be subset-admissible and let $\xi \in \mathbb{Q}$. Assuming #ETH, the problem $\mathsf{Eval}_\xi(p)$ admits no $2^{o(n)}$ time algorithm on unweighted graphs with $\mathcal{O}(n)$ vertices and edges, provided that the following holds:*

(C1) *Assuming #ETH, the problem $\mathsf{Coeff}(p)$ admits no $2^{o(n)}$ algorithm on graphs with n vertices and $\mathcal{O}(n)$ edges.*

(C2) *There is a recursively enumerable sequence $W = (w_0, w_1, \ldots)$ of pairwise distinct weights, a sequence of gadgets $\mathcal{H} = (H_0, H_1, \ldots)$ and a factor function F such that $(\mathcal{H}, F)$ allows to reduce $\mathsf{Eval}_W(\boldsymbol{p})$ to $\mathsf{Eval}_\xi(p)$.*

Proof. We present a subexponential reduction family from $\mathsf{Coeff}(p)$ to $\mathsf{Eval}_\xi(p)$. Given $\epsilon > 0$ and a graph G with n vertices and $\mathcal{O}(n)$ edges, apply Lemma 4 to reduce $\mathsf{Coeff}(p)$ to multiple instances of $\mathsf{Eval}(\boldsymbol{p})$ in time $2^{\epsilon n}$ such that each instance uses only weights $w_0, \ldots, w_s$ with $s = f(\epsilon)$.

Since $(\mathcal{H}, F)$ allows to reduce $\mathsf{Eval}_W(\boldsymbol{p})$ to $\mathsf{Eval}_\xi(p)$, we can invoke Lemma 5 and reduce each instance G' for $\mathsf{Eval}(\boldsymbol{p})$ to an instance of $\mathsf{Eval}_\xi(p)$ on the graph $T(G)$, which features $\mathcal{O}(sn)$ vertices and edges. □

Remark 1. If the source instance G for $\mathsf{Coeff}(p)$ has maximum degree Δ, then the reduction images $T(G)$ feature maximum degree $\Delta + \mathcal{O}(1)$. By suitable choice of $\mathcal{H}$, we can also ensure other properties on $T(G)$: For instance, if G is bipartite and all edge-gadgets $(H, u, v) \in \mathcal{H}$ can be 2-colored such that u and v receive different colors (as can be verified for Example 3), then $T(G)$ is bipartite as well.

4 Applications of the Framework

In the following subsections, we apply Theorem 1 to obtain tight lower bounds for counting problems, including the unweighted permanent in Section 4.1, the matching polynomial in Section 4.2 and the Tutte polynomial in Section 4.3.

4.1 The Unweighted Permanent

As stated in the introduction, it was shown in [7] that, unless #ETH fails, the problem $\mathsf{Eval}_{\{-1,1\}}(\mathrm{perm})$ on graphs with n vertices and $\mathcal{O}(n)$ edges admits no algorithm with runtime $2^{o(n)}$. It was also shown that an algorithm for the *unweighted* permanent on such graphs would falsify rETH, the randomized version of ETH. We improve upon this by showing that it is sufficient to assume #ETH, which is a priori weaker than ETH and is a more natural assumption for lower bounds on counting problems.

Theorem 2. *Assuming #ETH, the problem $\mathsf{Eval}_1(\mathrm{perm})$ of counting unweighted perfect matchings in bipartite graphs cannot be solved in time $2^{o(n)}$ on graphs with n vertices and $\mathcal{O}(n)$ edges.*

Proof. We invoke Theorem 1 to show $\mathsf{Eval}_{\{-1,1\}}(\mathrm{perm}) \leq_{serf} \mathsf{Eval}_1(\mathrm{perm})$. Let G be a graph with edge-weights from $\{-1,1\}$ on n vertices and $\mathcal{O}(n)$ edges and let $E_{-1}(G)$ denote the set of edges with weight -1 in G. Define a sieve $\chi(G, A) = [A \in \mathcal{PM}[G]]$ and weight selector $\omega(G, A) = A \cap E_{-1}(G)$ and observe that these induce a univariate graph polynomial $p = p_{\chi,\omega}$ with $p(G; -1) = \mathrm{perm}(G)$. Since knowledge of the coefficients of $p(G)$ allows to evaluate $p(G; -1)$, we obtain from [7, Thm. 1.3] that $\mathsf{Coeff}(p)$ admits no $2^{o(n)}$ time algorithm on sparse graphs under #ETH. Hence (C1) of Theorem 1 is satisfied.

To check (C2), recall the pair $(\mathcal{H}, F)$ from Example 3 that allows to reduce $\mathsf{Eval}_{\mathbb{N}}(p)$ to $\mathsf{Eval}_1(p)$. By Remark 1, the reduction images $T(G)$ constructed by Theorem 1 are bipartite as well.   □

We collect a series of corollaries for other counting problems from this theorem. Let $L(G)$ denote the *line graph* of a graph $G = (V, E)$: This graph has vertex set E, and $e, e' \in E(G)$ are adjacent in $L(G)$ iff $e \cap e' \neq \emptyset$. A graph is *line* if it is the line graph of some graph.

Corollary 1. *Assuming #ETH, the following cannot be solved in time $2^{o(n)}$:*

1. *$\mathsf{Eval}_1(\mathrm{perm})$ on graphs with n vertices and maximum degree 3.*
2. *Counting maximum independent sets (or minimum vertex covers), even in line graphs with n vertices and maximum degree 4.*
3. *Counting minimum-weight satisfying assignments to monotone 2-CNF formulas on n variables, even if every variable appears in at most four clauses.*

Proof. For the first statement, we use a reduction from the permanent on general graphs to graphs of maximum degree 3, shown in [6], which maps graphs with n vertices and m edges to graphs with $\mathcal{O}(n + m)$ vertices and edges.

For the second statement, if G has m edges and maximum degree $\Delta = \Delta(G)$, then $L(G)$ has m vertices and maximum degree $2(\Delta - 1)$. The set $\mathcal{PM}[G]$ corresponds bijectively to the independent sets of size $n/2$ in $L(G)$, which are the maximum independent sets in $L(G)$, unless G has no perfect matching, which we can test efficiently. The maximum independent sets in turn stand in bijection with the minimum vertex covers of $L(G)$ via complementation. We thus obtain the statement by reduction from $\mathsf{Eval}_1(\mathrm{perm})$ on graphs of maximum degree 3.

For the third statement, observe that the minimum vertex covers of a graph $H = (V, E)$ correspond bijectively to the minimum-weight satisfying assignments of the following monotone 2-CNF formula φ: Create a variable x_v for each $v \in V$ and a clause $(x_u \vee x_v)$ for each $uv \in E$. This is standard, noted also in [16].   □

4.2   The Matching and Independent Set Polynomials

We prove a tight lower bound for $\mathsf{Eval}_\xi(\mu)$ at fixed $\xi \in \mathbb{Q}$ by invoking Theorem 1. The perfect matchings of G are counted by the coefficient of x^0 in $\mu(G)$, so $\mathsf{Coeff}(\mu)$ and $\mathsf{Eval}_0(\mu)$ have the same lower bound as $\mathrm{perm}^{0,1}$ on graphs of maximum degree 3, settling (C1). In the full version, we show (C2) and obtain:

Theorem 3. *If there is some $\xi \in \mathbb{Q}$ such that $\mathsf{Eval}_\xi(\mu)$ can be solved in time $2^{o(n)}$ on graphs with n vertices and maximum degree $\mathcal{O}(1)$, then #ETH fails. This holds especially for $\mathsf{Eval}_1(\mu)$, which amounts to counting matchings.*

As in Corollary 1, we can easily obtain corollaries for the independent set polynomial and for monotone 2-SAT, improving upon [7,9].

Corollary 2. *Assuming #ETH, the following cannot be solved in time $2^{o(n)}$:*

1. *$\mathsf{Eval}_\xi(I)$ on line graphs of maximum degree $\mathcal{O}(1)$, for $\xi \in \mathbb{Q} \setminus \{0\}$, especially at $\xi = 1$, which amounts to counting independent sets (or vertex covers).*
2. *Counting satisfying assignments to monotone 2-CNF formulas, even if every variable appears in at most $\mathcal{O}(1)$ clauses.*

To prove the first statement, recall that the matchings with k edges in G correspond bijectively to the independent k-sets in $L(G)$. In $\mu(G; \xi)$, a matching with k edges is weighted by ξ^{n-2k}. We can therefore obtain the claim by reduction from $\mathsf{Eval}_{\xi'}(\mu)$ at fixed ξ'. For the second statement, see Corollary 1.

4.3 The Tutte Polynomial

We use univariate restrictions of Z, as discussed in Example 1. Writing $Z_q = Z(q, \cdot)$ for fixed $q \in \mathbb{Q} \setminus \{0\}$ and $Z_0(G; q, w) = \sum_{A \subseteq E(G)} q^{k(G,A)-k(G,E)} w^{|A|}$ for $q = 0$ as in [7], we use Theorem 1 to prove lower bounds for $\mathsf{Eval}_w(Z_q)$ at fixed $w \in \mathbb{Q}$. As in the previous examples, we require a lower bound for $\mathsf{Coeff}(Z_q)$, which we adapt from [7], and weight simulation.

Lemma 6. *[7, Propositions 4.1 and 4.3] Assuming #ETH, the problem $\mathsf{Coeff}(Z_q)$ for $q \in \mathbb{Q} \setminus \{1\}$ cannot be solved in time $2^{o(n)}$ on n-vertex graphs with $\mathcal{O}(n)$ edges.*

In [7], the problem $\mathsf{Coeff}(Z_q)$ is reduced to unweighted evaluation via *Theta graphs* and *wumps*, families of edge-gadgets that incur only $\mathcal{O}(\log^c n)$ blowup. This economical (but still not constant) factor however requires a quite involved analysis. Using block interpolation, we can instead use mere paths, and hence perform *"stretching"*, a classical weight simulation technique for the Tutte polynomial [7,8,12]. Recall $\boldsymbol{Z_q}$, as defined by Example 1 and (3).

Lemma 7. *For $k \in \mathbb{N}$, let P_k denote the path on k edges with distinguished start/end vertices $u, v \in V(P_k)$ and let $\mathcal{P} = (P_1, P_2, \ldots)$. Let $w, q \in \mathbb{Q}$ be fixed with $w \neq 0$ and $q \notin \{1, -w, -2w\}$. Then there is an infinite sequence of pairwise distinct weights W and a factor function F such that $(\mathcal{P}, F)$ allows to reduce $\mathsf{Eval}_W(\boldsymbol{Z_q})$ to $\mathsf{Eval}_w(\boldsymbol{Z_q})$.*

The proof is given in the full version. By combining Lemma 6 for (C1) and Lemma 7 for (C2), we can then invoke Theorem 1 and obtain:

Theorem 4. *Let $w \neq 0$ and $q \notin \{1, -w, -2w\}$. Assuming #ETH, the problem $\mathsf{Eval}_w(Z_q)$ admits no $2^{o(n)}$ algorithm on graphs with n vertices and $\mathcal{O}(n)$ edges.*

Using the substitution (1) that maps $Z(\cdot, \cdot)$ to the classical parameterization $T(\cdot, \cdot)$ of the Tutte polynomial, we obtain the following corollary.

Corollary 3. *Assuming #ETH, the Tutte polynomial $T(x, y)$ cannot be evaluated in time $2^{o(n)}$ on graphs with n vertices and $\mathcal{O}(n)$ edges, provided that $y \notin \{0, 1\}$ and $(x, y) \notin \{(1, 1), (-1, -1), (0, -1), (-1, 0)\}$ and $(x - 1)(y - 1) \neq 1.*

If (x, y) satisfies either of the last two conditions of the corollary, then $T(x, y)$ admits a polynomial-time algorithm. The remaining points with $y \in \{0, 1\}$ are however not covered by Corollary 3, and we obtain no lower bounds.

Acknowledgments. The author thanks Holger Dell and the anonymous reviewers for proofreading earlier versions of this paper.

References

1. Bläser, M., Dell, H.: Complexity of the Cover Polynomial. In: Arge, L., Cachin, C., Jurdziński, T., Tarlecki, A. (eds.) ICALP 2007. LNCS, vol. 4596, pp. 801–812. Springer, Heidelberg (2007)
2. Bulatov, A.A.: The Complexity of the Counting Constraint Satisfaction Problem. In: Aceto, L., Damgård, I., Goldberg, L.A., Halldórsson, M.M., Ingólfsdóttir, A., Walukiewicz, I. (eds.) ICALP 2008, Part I. LNCS, vol. 5125, pp. 646–661. Springer, Heidelberg (2008)
3. Cai, J.-Y., Chen, X.: Complexity of counting CSP with complex weights. STOC **2012**, 909–920 (2012)
4. Cai, J.-Y., Lu, P., Xia, M.: A Computational Proof of Complexity of Some Restricted Counting Problems. In: Chen, J., Cooper, S.B. (eds.) TAMC 2009. LNCS, vol. 5532, pp. 138–149. Springer, Heidelberg (2009)
5. Cai, J., Pinyan, L., Xia, M.: Dichotomy for Holant* problems of boolean domain. SODA **2011**, 1714–1728 (2011)
6. Dagum, P., Luby, M.: Approximating the permanent of graphs with large factors. Theor. Comput. Sci. **102**(2), 283–305 (1992)
7. Dell, H., Husfeldt, T., Marx, D., Taslaman, N., Wahlen, M.: Exponential time complexity of the permanent and the tutte polynomial. ACM Transactions on Algorithms **10**(4), 21 (2014)
8. Leslie Ann Goldberg and Mark Jerrum: The complexity of computing the sign of the tutte polynomial. SIAM J. Comput. **43**(6), 1921–1952 (2014)
9. Hoffmann, C.: Exponential Time Complexity of Weighted Counting of Independent Sets. In: Raman, V., Saurabh, S. (eds.) IPEC 2010. LNCS, vol. 6478, pp. 180–191. Springer, Heidelberg (2010)
10. Husfeldt, T., Taslaman, N.: The Exponential Time Complexity of Computing the Probability That a Graph Is Connected. In: Raman, V., Saurabh, S. (eds.) IPEC 2010. LNCS, vol. 6478, pp. 192–203. Springer, Heidelberg (2010)
11. Impagliazzo, R., Paturi, R., Zane, F.: Which problems have strongly exponential complexity? J. Computer and Sys. Sci. **63**(4), 512–530 (2001)
12. Jaeger, F., Vertigan, D.L., Welsh, D J.A.: On the computational complexity of the Jones and Tutte polynomials. Mathematical Proceedings of the Cambridge Philosophical Society 108(1), 35–53 (1990)

13. Jerrum, M., Sinclair, A., Vigoda, E.: A polynomial-time approximation algorithm for the permanent of a matrix with nonnegative entries. J. ACM **51**(4), 671–697 (2004)
14. Linial, N.: Hard enumeration problems in geometry and combinatorics. SIAM Journal on Algebraic and Discrete Methods **7**(2), 331–335 (1986)
15. Sokal, A.D.: The multivariate Tutte polynomial (alias Potts model) for graphs and matroids. Surveys in Combinatorics **327**, 173–226 (2005)
16. Vadhan, S.P.: The complexity of counting in sparse, regular, and planar graphs. SIAM J. Comput. **31**(2), 398–427 (2001)
17. Valiant, L.G.: The complexity of computing the permanent. Theoretical Computer Science **8**(2), 189–201 (1979)

On Convergence and Threshold Properties of Discrete Lotka-Volterra Population Protocols

Jurek Czyzowicz[1], Leszek Gąsieniec[2], Adrian Kosowski[3](✉)
Evangelos Kranakis[4], Paul G. Spirakis[2,5], and Przemysław Uznański[6]

[1] Department d'Informatique, Université du Québec en Outaouais,
Gatineau, QC, Canada
jurek.czyzowicz@uqo.ca
[2] Department of Computer Science, University of Liverpool, Liverpool, UK
L.A.Gasieniec@liverpool.ac.uk
[3] Inria Paris and LIAFA, Université Paris Diderot, Paris, France
adrian.kosowski@inria.fr
[4] Carleton University, School of Computer Science, Ottawa, ON, Canada
kranakis@scs.carleton.ca
[5] CTI, Patras, Greece
P.Spirakis@liverpool.ac.uk
[6] Helsinki Institute for Information Technology HIIT,
Aalto University, Espoo, Finland
przemyslaw.uznanski@aalto.fi

Abstract. In this work we focus on a natural class of population protocols whose dynamics are modeled by the discrete version of Lotka-Volterra equations with no linear term. In such protocols, when an agent a of type (species) i interacts with an agent b of type (species) j with a as the initiator, then b's type becomes i with probability P_{ij}. In such an interaction, we think of a as the predator, b as the prey, and the type of the prey is either converted to that of the predator or stays as is. Such protocols capture the dynamics of some opinion spreading models and generalize the well-known Rock-Paper-Scissors discrete dynamics. We consider the pairwise interactions among agents that are scheduled uniformly at random.

We start by considering the convergence time and show that any Lotka-Volterra-type protocol on an n-agent population converges to some absorbing state in time polynomial in n, w.h.p., when any pair of agents is allowed to interact. By contrast, when the interaction graph is a star, there exist protocols of the considered type, such as Rock-Paper-Scissors, which require exponential time to converge. We then study threshold effects exhibited by Lotka-Volterra-type protocols with 3 and more species under interactions between any pair of agents. We present a simple 4-type protocol in which the probability difference of reaching

This research was partially funded by the EU IP FET Proactive project MULTI-PLEX, by ERC Project ALGAME, by ANR project DISPLEXITY, by NCN grant DEC-2011/02/A/ST6/00201, by University of Liverpool EEE/CS NeST initiative, and by NSERC. A full version of the paper is available at http://arxiv.org/abs/1503.09168.

© Springer-Verlag Berlin Heidelberg 2015
M.M. Halldórsson et al. (Eds.): ICALP 2015, Part I, LNCS 9134, pp. 393–405, 2015.
DOI: 10.1007/978-3-662-47672-7_32

the two possible absorbing states is strongly amplified by the ratio of the initial populations of the two other types, which are transient, but "control" convergence. We then prove that the Rock-Paper-Scissors protocol reaches each of its three possible absorbing states with almost equal probability, starting from any configuration satisfying some sub-linear lower bound on the initial size of each species. That is, Rock-Paper-Scissors is a realization of a "coin-flip consensus" in a distributed system. Some of our techniques may be of independent value.

1 Introduction

Population protocols are a recent model of computation that captures the way in which the complex behavior of systems (biological, sensor nets, etc.) emerges from the underlying local interactions of agents. Agents are modeled as anonymous automata with a finite number of states, and interactions (changes of state) occur between randomly chosen pairs of agents under some fixed set of local rules. The interaction follows from the mobility of agents in the population, as in the case of birds flying past each other in a flock in the setting originally described by Angluin et al. [2,4]. More generally, we can model agents as nodes of an interaction graph G, and assume interactions take place along the edges of this graph.

Population protocols provide a way of describing dynamical effects which may occur in a population. For example, one can imagine that members of a population can be either healthy or infected, and whenever two individuals meet, if one is infected, then the other one also becomes infected. Thus the interesting question becomes: how fast can the infection spread? Quite naturally, population protocols are also used to model opinion spread in populations under interactions. An interaction between a pair of agents, one holding opinion A and the other opinion B, results in a possible change of opinion by one of the interacting agents. Eventually, the population protocol may lead the system to converge to a state in which one type, A or B, becomes dominant in the population. The probability of convergence to a given dominant type may potentially depend on the initial state of the population in different ways, e.g., exhibiting linear behavior, or transitions at one or more thresholds (cf. Fig. 1).

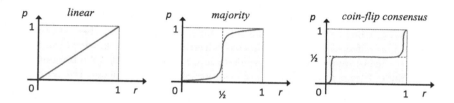

Fig. 1. Examples of objective functions for opinion spreading: probability p that a given type becomes dominant in the population as a function of the fraction r of its supporters in the initial population

In this work, we focus on a natural scenario of interactions modeled by the discrete version of Lotka-Volterra equations, with the goal of better understanding their applicability in the computational framework of opinion spreading and voting protocols. In their original form, the (continuous) Lotka-Volterra differential equation were initially applied in the modeling of periodic chemical reactions and also in the predator-prey dynamics of fish in the Adriatic Sea [12, p.11], and are perhaps best known for their connection to replicator dynamics and to evolving strategies in game theory [12,19]. In discrete Lotka-Volterra-type (LV-type) population protocols, during an interaction, the initiating agent (holding some state A) tries to impose its state on the other agent (holding some other state B) and succeeds with some probability P_{AB}. LV-type interactions are natural both in the context of predator-prey protocols, in that they correspond to a possible expansion of the predating (initiating) agent into the ecological niche of its prey, and in opinion propagation, in that they do not allow a new derived state C to be created as a result of an interaction.

Our Results. We start by proving in Section 2 a general convergence result: any LV-type protocol on a n-agent population converges to some absorbing state in time $\mathcal{O}(\text{poly}(n))$, w.h.p., under the model of uniformly random interactions between agents (i.e., when the interaction graph is the complete graph $G = K_n$). By contrast, we also show in Section 2 that introducing an interaction constraint can severely impact the convergence time for LV-type protocol. We consider a specific LV-type protocol known as *rock-paper-scissors* (RPS), in which each of the three types overcomes exactly one other type in cyclic manner, and show that RPS requires exponential time to converge to an absorbing state when the interaction graph is a star ($G = K_{1,n}$).

Next, we look at the applicability of LV-type protocols in the context of their threshold behavior in voting problems which require a consensus of opinion. For the case of 2 types, the only unbiased LV-type protocol encompasses the so-called "game of life and death" between the 2 types, converging to a given absorbing state with probability proportional to its initial representation in the population (regardless of the interaction graph G). This captures the linear behavior shown in Fig. 1. We show, however, that for 3 and more types, threshold effects become apparent even under uniform interactions ($G = K_n$). We start by proposing in Section 3 a simple 4-type majority-type protocol, in which the probability difference of reaching the two absorbing types is amplified with respect to the ratio of the initial populations of the two other states. We close the paper by exhibiting in Section 4, for the before-mentioned RPS protocol a completely different type of threshold effect in the small population region. We prove that RPS reaches each of its three absorbing types with almost equal probability ($1/3 \pm o(1)$), starting from any configuration satisfying some sub-linear lower bound on the sizes of the three types. Our proof proceeds by a Martingale-type analysis and takes into account the symmetries of the state space of the protocol. We can thus view the RPS protocol as an embodiment of the "coin-flip consensus" illustrated in Fig. 1: any opinion with non-negligible representation in the population, even a minority one, has an equal chance of success in the opinion-spreading process. To the best of our knowledge, this is the first population protocol with polynomial-time convergence for which such a property has been identified.

Related Work. The population protocol model of Angluin et al. [2,3] captures random interactions between finite-state agents, motivated by applications in sensor mobility. Despite the limited computational capabilities of individual sensors, such protocols permit at least (depending on available extensions to the model) the computation of two important classes of functions: threshold predicates, which decide if the weighted average of types appearing in the population exceeds a certain value, and modulo remainders of similar weighted averages. The majority function, which belongs to the class of threshold functions, was shown to be stably computable for the complete interaction graph [2]. Another majority protocol for the complete interaction graph, converging to a population of a single type, was proposed in [3]. This protocol relies on 3 types, two of which represent the original types present in the population, while the third is a transient type representing a blank opinion. The type reached by the protocol is the initial majority type, w.h.p., provided that the initial difference between the majority and minority type is $\omega(\sqrt{n}\log n)$ for a n-agent population. A 4-state protocol for finding a majority is presented in [5], based on a different principle of "leader" and "follower" agents, and achieves similar performance guarantees. Finally, [17] presented a 4-state protocol which converges in expected polynomial time to the initial majority type with probability 1, even when the difference between types in the original population is constant and when interactions are not spread uniformly over the population, but restricted to a connected subgraph of agent pairs. Other applications and models of population protocols are surveyed in [5,18].

Spreading of Opinion and Voting. The spread of trust and opinion in a social network was one of the original motivations for the study of population protocols [9]. Problems in which a set of nodes has to converge to a consensus decision chosen from a candidate set of values proposed by the participating nodes, are also of fundamental importance in distributed computing, in tasks such as serialization of database operations or leader election [11]. Models of voting processes, which solve such questions, involve the propagation of opinion through multiple push- or pull-operations between pairs of agents, usually performed in parallel throughout the system. From the perspective of security and simplicity of design, a desirable property of the protocols is that at any time during the execution, the state of the node should describe its current opinion, belonging to the set of opinions initially represented in the population. Under this constraint, given a set of only 2 initial opinions, it is impossible to obtain convergence to the majority opinion w.h.p. of correctness in the standard model of voting (cf. e.g. [7]). However, majority voting can be achieved in many graph classes by extensions of the population protocol framework, allowing simultaneous interactions between more than 2 nodes. Specifically, protocols in which a node polls a constant number k of randomly chosen neighbors in the interaction graph and changes its

opinion as the majority opinion in the chosen neighborhood set, have been con
sidered in the literature [1,6–8].

Discrete LV Dynamics and Cyclic Games. The continuous Lotka-Volterra
dynamics, first defined in [16], gave rise to several discrete variants of predator
prey models of interaction in a population, which differ essentially in the way
the population size is maintained after the prey is attacked by the predator
The LV-type model is particularly worthy of study due to its transient stability in a setting in which several species are in a cyclic predator-prey relation
useful for maintaining biodiversity, e.g., in bacterial colonies [13,14]. Cycles o:
length 3, in which type 1 attacks type 2, type 2 attacks type 3, and type 3
attacks type 1, form the basis of the best-known such protocol, called rock-
paper-scissors (RPS). The transient properties of RPS and related protocols
describing in particular the time until the system collapses to an absorbing state.
have been studied in the statistical physics literature using a variety of experimental and analytical techniques, under various scheduler models. The original
analytical estimation method applied to RPS was based on approximation with
the Fokker-Planck equation [21]. A subsequent analysis of cyclic 3- and 4-species
models using Khasminskii stochastic averaging can be found in [10]. A mean field
approximation-based analysis of RPS was performed in [20]. All of these results
provide a qualitative understanding of cyclic protocols, and at a quantitative
level, provide evidence that the RPS protocol reaches an absorbing state after
roughly $\mathcal{O}(n^2)$ interactions.

Model and Preliminaries. We consider population protocols in the following
setting. The population V with k types (species) is a set of n agents, with
each agent $v \in V$ assigned a state variable s, whose value at time t is denoted
$s_t(v) \in \{1, \dots, k\} \equiv [k]$, describing its current type. The elements of V are
connected into an (undirected, connected) *interaction graph* $G = (V, E)$. Agents
assigned to type i at time t, $1 \le i \le k$, are called the *population* of type i at
time t.

The population protocol P is a probability distribution over $[k]^2$, taking values in $[k]^2$. In an execution of protocol P, at each time step $t = 1, 2, 3, \dots$,
a scheduler daemon picks a pair of interacting agents $u, v \in V$ such that
$(u, v) \in E$ u.a.r., and updates the state variables of these agents, sampling the
pair $(s_{t+1}(u), s_{t+1}(v))$ according to the distribution $P(s_t(u), s_t(v))$. We will say
that the population protocol is of the *Lotka-Volterra type* (LV-type for short) if
the state of the initiating agent (the predator) never changes during an interaction, and the state of the other agent (the prey) either remains unchanged
or changes to that of the initiator, i.e., for any transition which occurs with
non-zero probability, we have $s_{t+1}(u) = s_t(u)$ and $s_{t+1}(v) \in \{s_t(u), s_t(v)\}$.

For $i \in [k]$ and a fixed execution of protocol P, we will denote the size of
the i-th population as $n_i(t) = |\{v \in V : s_t(v) = i\}|$, and its relative size as
$x_i(t) = n_i(t)/n$. The set of states of all n agents at time t is referred to as
the *state* or *configuration* of the system. When the interaction graph G is the

complete graph K_n, then we identify the state of the system with the vector $x(t)$. For $G = K_n$, the protocol P defines a Markov chain on the set X of possible states $x(t)$. We note that in this case, the size of the state space can be trivially bounded as $\mathcal{O}(n^k)$, i.e., is polynomial in n for any fixed protocol.

Throughout the paper, any LV-type protocol P will be identified with its $k \times k$ probability matrix P, such that for an interaction (u, v), we have $s_{t+1}(v) = s_t(u)$ with probability $P_{s_t(u), s_t(v)}$, and $s_{t+1}(v) = s_t(v)$ with probability $1 - P_{s_t(u), s_t(v)}$. (Informally, we may write: "$ij \rightarrow ii$ with probability P_{ij}".) In general, matrix P need not be symmetric nor stochastic. We only assume that $P_{ii} = 0$, for $1 \le i \le k$, and that every type interacts in some way with at least one other type (for every i, $1 \le i \le k$, there exists j, $1 \le i \le k$, such that $P_{ij} > 0$ or $P_{ji} > 0$). We will denote the value of the minimal non-zero entry of matrix P as $P_{\min}$. For every LV-type protocol, we construct the corresponding digraph $D(P)$, whose vertex set is the set of types $[k]$, and an arc (i, j) exists if $P_{i,j} > 0$. We call the dynamics *irreducible* if the digraph $D(P)$ has no sources (i.e., there are no types without a predator, so each column of matrix P has at least one non-zero entry) and is connected.

We remark that as $n \rightarrow \infty$, our random process converges to its (deterministic) limit continuous dynamics, given by the following set of first-order differential equations (a special case of the continuous Lotka-Volterra equations, with no linear term):

$$\frac{dx_i(t)}{dt} = x_i(t) \sum_{j=1}^{k} \left[(P_{ij} - P_{ji}) x_j(t) \right], \quad \text{for } 1 \le i \le k. \tag{1}$$

Our discrete population case with finite n can be informally seen as a special form of "noise" introduced into the Lotka-Volterra equation (1).

In this paper, we also give our attention to two specific LV-type protocols:

- *Rock-Paper-Scissors (RPS)* is the LV-type protocol with $k = 3$ types (denoted $1, 2, 3$), whose probability matrix P has the following non-zero entries: $P_{12} = P_{23} = P_{31} = 1$.
- *Wolves-and-Sheep (WS)* is the LV-type protocol with $k = 4$ types (denoted X, Y, x, y), whose probability matrix P has the following non-zero entries: $P_{XY} = P_{Xx} = P_{YX} = P_{Yy} = 1$, $P_{Xy} = P_{Yx} = 1/2$.

We use the term "with very high probability" (w.v.h.p.) to denote events occurring with probability at least $1 - e^{-\Omega(\log^2 n)}$ and the term "with high probability" (w.h.p.) for events occurring with probability at least $1 - n^{-\Omega(1)}$. In time and distance analysis, we will use the notation $\widetilde{\mathcal{O}}$ to conceal poly-logarithmic factors ($\widetilde{\mathcal{O}}(f) = \mathcal{O}(f \text{polylog}(n))$).

2   Convergence of Discrete LV-type Protocols

We start by showing that any LV-type protocol on a population of size n converges to an absorbing state in time $\mathcal{O}(\text{poly}(n))$, when there are no population constraints (the interaction graph is K_n).

Theorem 1 (LV-type convergence for complete interactions). *For any probability matrix* P*, there exists a constant* c *such that the LV-type protocol defined by* P *converges for the complete interaction graph to an absorbing state in* $\mathcal{O}(n^c)$ *steps, w.v.h.p.*

Before proceeding with the proof, we introduce some auxiliary notation. For a fixed matrix P, we define the skew-symmetric *net interaction matrix* A as $A = P - P^T$. Observe that $A_{ij} = P_{ij} - P_{ji}$ and equation (1) describing the continuous dynamics now takes the simpler form:

$$\frac{dx_i}{dt} = x_i(A_i x), \qquad (2)$$

where we treat x as a column vector, and A_i is the i-th row vector of matrix A (cf. [12, Chapter 7] for a more detailed exposition of the properties of this continuous dynamics).

For a fixed real vector $b \in \mathbf{R}^k$, which we will appropriately choose later, we define the potential U of a system state x as (compare [12, equation (5.3)]):

$$U(x) = \sum_{i=1}^{k} b_i \ln x_i,$$

and by $U(t)$ we will mean $U(x(t))$. Observe that under evolution of the system given by the continuous dynamics (2), we have:

$$\frac{dU}{dt} = \sum_{i=1}^{k} b_i \frac{1}{x_i} \frac{dx_i}{dt} = \sum_{i=1}^{k} b_i(A_i x) = \sum_{j=1}^{k} \left(\left(\sum_{i=1}^{k} b_i A_{ij} \right) x_j \right) = b^T A x.$$

We define vector b as follows:
(i) if there exists a non-zero vector $b \geq 0$, such that $b^T A = 0$, choose b as any such vector with $\|b\|_\infty = 1$.
(ii) otherwise, choose b as any vector satisfying $b^T A > 0$, with $\|b\|_\infty = 1$.
The completeness of the above definition follows from a basic theorem of linear optimization, known as the "no arbitrage theorem" in financial mathematics (cf. also [12, proof of Thm. 5.2.1]).

Proof (of Theorem 1, sketch). Observe that by the definition of the LV-type process, if a certain type i has been eliminated by time t ($x_i(t) = 0$), then it will never reappear ($x_i(\tau) = 0$, for all $\tau \geq t$). We will now show that the number of non-zero values of x_i, $1 \leq i \leq k$, is reduced by at least one within a polynomial number of steps, w.h.p. Note that this is sufficient to obtain the claim of the theorem, since we may iterate the argument, each time restricting the definition of the dynamics and the matrix A to those of the k types which are non-empty. In the rest of the proof, we will be assuming w.l.o.g. that $x_i > 0$, for all i. We will also be assuming that the dynamics is irreducible; otherwise, if digraph $D(P)$ is disconnected, we can consider each of the weakly connected components separately, and if any of the weakly connected components has a source, then one can easily show that all the prey of this source is eliminated in a polynomial number of steps.

The main part of the proof is contained in the following claim.

Claim. For any irreducible LV-type protocol, there exists a constant n_{min}, such that for any $n > 0$, for any initialization of the protocol with n agents, w.v.h.p. there exists a time step $T \in \mathcal{O}(\text{poly}(n))$ in which $n_i(T) < n_{min}$, for some type i, $1 \leq i \leq k$.

The proof of the claim proceeds by a careful analysis of the change of potential U in time. Depending on case (i) and (ii) in the definition of vector b, we provide an absolute lower bound on the expectation $\mathbf{E}(\delta(t)|x(1), \ldots, x(t-1))$ for the stochastic process $\delta(t) = U(t+1) - U(t)$, given that $n_i(T) \geq n_{min}$ for all types i. We then perform a super/sub-martingale analysis for the deviation of $\delta(t)$ from this expected change of potential for the two cases (i) and (ii), applying Azuma's inequality to bound the number of steps of the process until we reach $n_i(T) < n_{min}$, for some type i. The details are provided in the full version of the paper.

To obtain the claim of the theorem, we now need to notice that whenever the population of some type drops below a constant threshold n_{min}, the probability that the population is eliminated completely within the next $\mathcal{O}(1)$ steps of the irreducible protocol is polynomially large in n. Overall, after at most $\mathcal{O}(n^{2n_{min}+1})$ occurrences of the event "there exists $1 \leq i \leq k$ such that $n_i < n_{min}$", each of which takes place every polynomial number of steps w.v.h.p. by the Claim, one of the species will have been eliminated completely w.v.h.p., which gives the claim of the theorem.   □

It turns out that for LV-type protocols, the convergence time may become exponential when the interaction graph is not complete. Whereas all LV-type protocols with 2 species (e.g., the game of life-and-death [12]) converge in polynomial time to an absorbing state for any interaction graph, this is no longer true when the number of species is at least 3. We observe this for the rock-paper-scissors (RPS) protocol on the star.

Theorem 2 (RPS convergence on the star). *The RPS protocol with a $K_{1,n}$ interaction graph, initialized so that initially each type has at least $n_{min} \geq n/3 - n/200$ agents, reaches the absorbing state in expected time $T_{abs} \geq e^{n^{\Omega(1)}}$.*

3   The Wolves-and-Sheep (WS) Protocol

In this section, we investigate the dynamics of the Wolves-and-Sheep LV-type protocol, aiming at replicating dynamics of infection spreading for two different infections and two types of partial immunity to infections. In the considered setting, initially almost all the population consists of types x and y (susceptible agents known as "sheep"). A constant number of infected agents of types X and Y (the "wolves") are introduced into the population. Following the definition of the protocol, a wolf acting as a predator infects a sheep of a type denoted by the same lower-case with probability 1, and a sheep of the opposite lower-case type with smaller probability $(1/2)$. Thus, in the protocol, population x of sheep has affinity towards X (or resistance for Y), and population y has affinity towards Y (or resistance for X). We note that in the definition of the WS protocol, we also add some random drift between the species X and Y, which does not affect the nature of the process, but allows us to achieve an absorbing state in which eventually only the dominant type is represented.

Theorem 3 (majority amplification by WS). *Let $n_X(0) = 1$, $n_Y(0) = 1$, $n_x(0) = \Theta(n)$ and $n_y(0) = \Theta(n)$, such that $\frac{n_x(0)}{n_y(0)} = \frac{1+\varepsilon}{1-\varepsilon}$ for some absolute constant $\varepsilon > 0$. Then the system reaches the absorbing state with only population X, w.h.p.*

4 The Rock-Paper-Scissors (RPS) Protocol

In this section, our goal is to show that the RPS protocol reaches each of absorbing states with almost equal probability, given that the initial population of each species is linear (or slightly sub-linear) in n. The RPS protocol admits a cyclic symmetry of behavior with respect to its species. For each of the species $a \in \{1, 2, 3\}$, the relative change Δx_a in population of this species in the given step can be expressed as:

$$\Delta x_a = \frac{1}{n} \cdot \Delta n_a = \begin{cases} +1/n, & \text{with probability } x_a x_{a+1}, \\ -1/n, & \text{with probability } x_{a+2} x_a, \\ 0, & \text{otherwise}, \end{cases} \quad (3)$$

where the population of at most one species changes in every step. The indices of populations are always $1, 2$, or 3, and other values should be treated as $\mod 3$, in the given range. We also introduce the continuous dynamics $\bar{x}(t)$ corresponding to the RPS process, given for each species by the differential equation:

$$\frac{d\bar{x}_a}{dt} = \frac{\bar{x}_a}{n}(\bar{x}_{a+1} - \bar{x}_{a+2}), \quad (4)$$

which corresponds precisely to the continuous dynamics (1), up to an additional time-scaling factor of n introduced for easier comparison with the discrete process. In all further considerations, we set the potential U used in the analysis as: $U(x) = \ln x_1 + \ln x_2 + \ln x_3$. Lines $U = \text{const}$ correspond to orbits in the continuous setting (4).

Theorem 4 (coin-flip consensus property of RPS). *For any state x such that $x_a > n^{-0.002}$ for all $a \in \{1, 2, 3\}$, the probability of the system reaching any one of its three possible absorbing states is $\frac{1}{3} \pm \tilde{\mathcal{O}}(n^{-0.05})$.*

The proof of the theorem relies on the observation that the discrete RPS protocol approximately follows the limit cycle (orbit) of its continuous version.

More precisely, we will observe that for an appropriately chosen starting state $x(0) = (x_1, x_2, x_3)$ of the system, there is a time moment t (corresponding to an approximate traversal of $1/3$ of the limit cycle) for which the state is given as $x(t) = (x_3 + \Delta x_3, x_1 + \Delta x_1, x_2 + \Delta x_2)$, with Δx_i sufficiently small. We will then use this to observe that if the probability of reaching any fixed absorbing state i from state (x_1, x_2, x_3) is p and of reaching absorbing state i from state $(x_1 + \Delta x_1, x_2 + \Delta x_2, x_3 + \Delta x_3)$ is p_Δ, then by cyclic symmetry of populations, the probability of reaching state $(i + 1) \bmod 3$ from state $x(t)$ is also p_Δ. If $p \approx p_\Delta$, then state $x(0)$ leads to absorbing states i and $(i + 1) \bmod 3$ with almost the same probability.

At an intuitive level, the main arguments of the proof are the following. To show that the probability of reaching an absorbing states are almost the same for points $x(0) = (x_1, x_2, x_3)$ and $y(0) = (x_3, x_1, x_2)$, we perform a coupling of walks starting from $x(0)$ and $y(0)$. Here, coupling of Markovian processes is understood in the usual sense (cf. e.g.[15]), though it is worth noting that since we are interested only in reaching an absorbing state, we can in some steps of the coupling decide to delay one of the walks, allowing the other to run, provided that each of the processes remains unbiased. For simplicity, suppose that a walk x is located at a point at which all populations are of linear size in N and the difference in size between the largest and smallest population is also linear in n (e.g., $U(x) = -20$). The behavior of the (undelayed) walk x under our evolution in the next t steps (for t sufficiently small with respect to n) can be seen as a superposition of three types of motion:

1. Propagation along the trajectory $U(x) = \text{const}$ at a speed approximately given by the evolution of the continuous process (4). The Euclidean distance traversed in a single step is $\Theta(1/n)$, or $\Theta(t/n)$ over t steps.

2. Random drift along the trajectory $U(x) = \text{const}$ (slowing or accelerating with respect to the average speed). Over a short interval time of length t, this drift shifts the point by $\pm\widetilde{\mathcal{O}}(\sqrt{t}/n)$ along its trajectory.

3. Random drift orthogonal to the trajectory $U(x) = \text{const}$. Over a short interval time of length t, this drift shifts the potential U of the point by $\pm\widetilde{\mathcal{O}}(\sqrt{t}/n)$.

The analysis of the process is somewhat technical, since the two types of random drift have slightly biased averages, the probabilities of different moves are changing over time, and the motion in different directions is not independent. The drift and the propagation speed also depend on the relation between the maximum and minimum of the sizes of the three populations, which change in time. We recall the following simple property of a simple random walk on a line: a walk starting from point 0 and proceeding for T steps is confined to an interval of the form $[-\widetilde{\mathcal{O}}(\sqrt{T}), \widetilde{\mathcal{O}}(\sqrt{T})]$ w.v.h.p, but is likely to hit all points at a distance of $o(\sqrt{T})$ from 0. The random drifts of our process behaves closely enough to a combination of independent random walks that by a Doob martingale analysis, we can apply a generalization of this property to our process (see the full version of the paper for details). We first introduce two measures of distance of a pair of points $x^{(a)}$, $x^{(b)}$ in our state space:

- $d_U(x^{(a)}, x^{(b)}) = |U(x^{(a)}) - U(x^{(b)})|$,
- $d_\infty(x^{(a)}, x^{(b)}) = \|x^{(a)} - x^{(b)}\|_\infty = \max\{|x_1^{(a)} - x_1^{(b)}|, |x_2^{(a)} - x_2^{(b)}|, |x_3^{(a)} - x_3^{(b)}|\}$

We next show a sequence of claims bounding the distances d_U and d_∞ of a process progressing under the discrete RPS protocol from that under the continuous dynamics (4). For compactness, these are given here in the form of the following summary lemma.

Lemma 1. *Let $x(0)$ be a point in the state space of RPS with $U(x(0)) > -\gamma \ln n$ for some absolute constant $0 < \gamma < 1/6$, let $x(t)$ be the random variable representing the point reached after following the population protocol for t steps starting from point $x(0)$. Next, consider the process $\bar{x}(t)$ governed by the continuous RPS dynamics (4) with any starting point $\bar{x}(0)$ such that $d_\infty(\bar{x}(0), x(0)) \leq \Delta$, for some $\Delta > 0$. Then, for sufficiently large n and any integer $T > 0$, the following claim holds:*

- *If $T \leq n^{5/3}$, then: $\forall_{t \in \{1,\ldots,T\}} \; d_U(x(t), x(0)) = \widetilde{\mathcal{O}}(T^{0.5}/n^{1-\gamma})$, w.v.h.p.*
- *If $T \leq n^{2/3}$, then: $\forall_{t \in \{1,\ldots,T\}} \; d_\infty(x(t), \bar{x}(t)) = \widetilde{\mathcal{O}}(\Delta + T^{0.5}/n)$, w.v.h.p.*
- *If $T \leq n^{5/3}$ and $-10 > U(x(0)) > -\gamma \ln n$ for some absolute constant $0 < \gamma < 1/6$, then: $\forall_{t \in \{1,\ldots,T\}} \; d_\infty(x(t), \bar{x}(t)) = \widetilde{\mathcal{O}}((Tn^{\gamma-2/3}+1)\cdot(\Delta+T^{0.5}n^{\gamma-1}))$, w.v.h.p.*
- *If $n^{6\gamma} \leq T \leq n^{4/3-8\gamma}$, $\Delta \leq T^{0.5}n^{\gamma-1}$, and $-10 > U(x(0)) > -\gamma \ln n$ for some absolute constant $0 < \gamma < 1/6$, then there exists an integer time step $T' = (1 + o(1))T$, such that $d_\infty(x(T'), \bar{x}(T)) = \widetilde{\mathcal{O}}(T^{0.5}n^{\gamma-1})$, w.v.h.p.*

We are now ready to apply the coupling technique to obtain the main technical result of this section.

Lemma 2. *Fix $\gamma = 0.005$ and $\varepsilon = 0.05$. Let $x(0) = (x_1(0), x_2(0), x_3(0))$ be arbitrarily fixed with $-12 > U(x(0)) > -\gamma \ln n$, and let $y(0) = (x_3(0), x_1(0), x_2(0))$. Then, there exist a coupling of x and y which leads to the same absorbing state with probability $1 - \widetilde{\mathcal{O}}(n^{-\varepsilon})$.*

Proof (sketch). The proof proceeds by a coupling of walks originating from x and y. By a slight abuse of notation, we will denote by $x(t)$ and $y(t)$ the position of each of the two walks in the state space after t steps, which may include steps in which a given walk is delayed. Our goal is to make points $x(t)$ and $y(t)$ coalesce within a small number of steps T, i.e., to obtain $x(T) = y(T)$ with probability $1 - \mathcal{O}(n^{-\varepsilon})$, where $T \ll n^{1.33}$.

The coupling proceeds in five phases. We limit this proof sketch to a high-level overview, thinking for now of the potential $U(x(0)) = \Theta(1)$ to simplify calculations. In Phase 1, point x approaches point y, which is stopped. In this way, the infinity norm distance between x and y is reduced, at the cost of increasing the d_U distance to slightly over $n^{-0.5}$. Next, in Phase 2 we run both walks independently, so that the distance d_U in time follows a random evolution resembling a random walk, and after slightly more than n steps, the value $d_U = 0$ is hit with sufficiently high probability. Whereas the walks are now orthogonally aligned, we also need to align them along the orbit, since we may at this point have them at a distance of $d_\infty > n^{-1/3}$ apart. By allowing the slower walk to atch up, we reduce d_∞ to slightly more than $n^{-2/3}$, at the cost of increasing v to a similar value. In this way, we have decreased the norm in both distances from about $n^{-0.5}$ to about $n^{-2/3}$). We iterate Phase 2, reducing each time the istance between the two walks in both norms, up to an iteration in which the ize of the populations in x and y differ by an arbitrarily small polynomial in n. t this point, only a very small number of time steps remains until coalescence. Ve first align the two states by evolving one of them until the size of one of the hree populations is identical for x and y at the end of Phase 3, and then perform , standard coupling by correlating the evolution of x and y in Phase 4, so as to nake the sizes of the other two populations meet for x and y, while maintaining quality on the size of the population coalesced in Phase 3. After the coupling is ,chieved, we evolve the coalesced state into an absorbing state in Phase 5.   □

The assumptions of Lemma 2 hold for any point x satisfying the assumptions of Theorem 4, either at time 0, or after a certain number of steps, once potential J has been sufficiently reduced. This completes the proof of Theorem 4.

References

1. Abdullah, M.A., Draief, M.: Global majority consensus by local majority polling on graphs of a given degree sequence. Discrete Applied Mathematics **180**, 1–10 (2015)
2. Angluin, D., Aspnes, J., Diamadi, Z., Fischer, M.J., Peralta, R.: Computation in networks of passively mobile finite-state sensors. Distributed Computing **18**(4), 235–253 (2006)
3. Angluin, D., Aspnes, J., Eisenstat, D.: A simple population protocol for fast robust approximate majority. Distributed Computing **21**(2), 87–102 (2008)
4. Angluin, D., Aspnes, J., Eisenstat, D., Ruppert, E.: The computational power of population protocols. Distributed Computing **20**(4), 279–304 (2007)
5. Aspnes, J., Ruppert, E.: An introduction to population protocols. In: Middleware for Network Eccentric and Mobile Applications, pp. 97–120. Springer Verlag (2009)
6. Becchetti, L., Clementi, A.E.F., Natale, E., Pasquale, F., Silvestri, R., Trevisan, L.: Simple Dynamics for Majority Consensus. In: Proc. SPAA, pp. 247–256 (2014)
7. Cooper, C., Elsässer, R., Radzik, T.: The power of two choices in distributed voting. In: Esparza, J., Fraigniaud, P., Husfeldt, T., Koutsoupias, E. (eds.) ICALP 2014, Part II. LNCS, vol. 8573, pp. 435–446. Springer, Heidelberg (2014)
8. Cruise, J., Ganesh, A.: Probabilistic consensus via polling and majority rules. Queueing Systems: Theory and Applications **78**(2), 99–120 (2014)
9. Diamadi, Z., Fischer, M.J.: A simple game for the study of trust in distributed systems. Wuhan University Journal of Natural Sciences **6**(1–2), 72–82 (2001)
10. Dobrinevski, A., Frey, E.: Extinction in neutrally stable stochastic Lotka-Volterra models. Phys. Rev. E **85**, 051903 (2012)
11. Hassin, Y., Peleg, D.: Distributed probabilistic polling and applications to proportionate agreement. Information & Computation **171**(2), 248–268 (2001)
12. Hofbauer, J., Sigmund, K.: Evolutionary Games and Population Dynamics. Cambridge University Press (1998)
13. Kerr, B., Riley, M.A., Feldman, M.W., Bohannan, B.J.M.: Local dispersal promotes biodiversity in a real-life game of rock-paper-scissors. Nature **418**(6894), 171–174 (2002)

14. Kirkup, B.C., Riley, M.A.: Antibiotic-mediated antagonism leads to a bacteria game of rock-paper-scissors in vivo. Nature **428**(6981), 412–414 (2004)
15. Levin, D.A., Peres, Y., Wilmer, E.L.: Markov chains and mixing times. American Mathematical Society (2006)
16. Lotka, A.J.: Contribution to the Theory of Periodic Reactions. J. Phys. Chem **14**(3), 271–274 (1910)
17. Mertzios, G.B., Nikoletseas, S.E., Raptopoulos, C.L., Spirakis, P.G.: Determining majority in networks with local interactions and very small local memory. In: Esparza, J., Fraigniaud, P., Husfeldt, T., Koutsoupias, E. (eds.) ICALP 2014 LNCS, vol. 8572, pp. 871–882. Springer, Heidelberg (2014)
18. Michail, O., Chatzigiannakis, I., Spirakis, P.G.: New Models for Population Protocols. Morgan & Claypool Synthesis Lectures on Distributed Computing Theory (2011)
19. Szolnoki, A., Mobilia, M., Jiang, L.-L., Szczesny, B., Rucklidge, A.M., Perc, M. Cyclic dominance in evolutionary games: a review. J. R. Soc. Interface **11**, 2014073 (2014)
20. Parker, M., Kamenev, A.: Extinction in the Lotka-Volterra model. Phys. Rev. E **80**, 021129 (2009)
21. Reichenbach, T., Mobilia, M.: M, and E. Frey. Coexistence versus extinction in the stochastic cyclic Lotka-Volterra model. Phys. Rev. E **74**, 051907 (2006)

Scheduling Bidirectional Traffic on a Path

Yann Disser, Max Klimm, and Elisabeth Lübbecke[(⊠)]

Department of Mathematics, Technische Universität Berlin, Berlin, Germany
{disser,klimm,eluebbecke}@math.tu-berlin.de

Abstract. We study the fundamental problem of scheduling bidirectional traffic along a path composed of multiple segments. The main feature of the problem is that jobs traveling in the same direction can be scheduled in quick succession on a segment, while jobs in opposing directions cannot cross a segment at the same time. We show that this tradeoff makes the problem significantly harder than the related flow shop problem, by proving that it is NP-hard even for identical jobs. We complement this result with a PTAS for a single segment and non-identical jobs. If we allow some pairs of jobs traveling in different directions to cross a segment concurrently, the problem becomes APX-hard even on a single segment and with identical jobs. We give polynomial algorithms for the setting with restricted compatibilities between jobs on a single and any constant number of segments, respectively.

Keywords: Bidirectional traffic · Scheduling · Packet routing · Computational complexity · PTAS · APX-hardness

1 Introduction

The scheduling of bidirectional traffic on a path is essential when operating single-track infrastructures such as single-track railway lines, canals, or communication channels. Roughly speaking, the schedule governs when to move jobs from one node of the path to another along the segments of the path. The goal is to schedule all jobs such that the sum of their arrival times at their respective destinations is minimized. A central feature of real-world single-track infrastructures is that after one job enters a segment of the path, further jobs moving in the *same* direction can do so with relatively little headway, while traffic in the *opposite* direction usually has to wait until the whole segment is empty again (cf. Fig. 1a for a schematic illustration).

Formally, in the bidirectional scheduling problem we are given a path of consecutive segments connected at nodes, and a set of jobs, each with a release date and a designated start and destination node. The time job j needs to traverse segment i is governed by two quantities: its *processing time* p_{ij} and its *transit time* τ_{ij}. While the former prevents the segment from being used by any

---

M. Klimm and E. Lübbecke—This research was carried out in the framework of MATHEON supported by Einstein Foundation Berlin.

M.M. Halldórsson et al. (Eds.): ICALP 2015, Part I, LNCS 9134, pp. 406–418, 2015.
DOI: 10.1007/978-3-662-47672-7_33

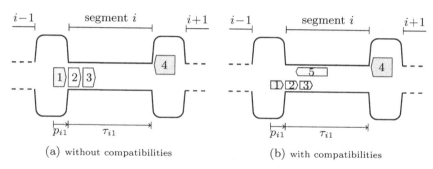

$$
\begin{array}{cc}
\text{(a) without compatibilities} & \text{(b) with compatibilities}
\end{array}
$$

Fig. 1. Bidirectional scheduling of ship traffic through a canal, with and without compatibilities. The processing time p_{ij} of job j is the time needed to enter segment i with sufficient security headway, i.e., the delay before other jobs in the same direction may enter the segment. The travel time τ_{ij} is the time needed to traverse the entire segment once entered. In both (a) and (b), jobs $1, 2, 3$ can enter the segment in quick succession, while job 4 has to wait until they left the segment. In (b), job 5 is compatible with jobs $1, 2, 3$ so that they may cross concurrently. The time to cross turnouts is assumed to be negligible.

other job (running in *either* direction), the latter only blocks the segment from being used by jobs running in *opposite* direction. For example, this allows us to model settings with bidirectional train traffic on a railway line split into single-track segments that are connected by turnouts (cf. Lusby et al. [17, Section 2]). In this setting, jobs correspond to trains, the processing time of a job is the time needed for the train to fully enter the next segment, and the transit time is the time to traverse the segment (and entirely move into the next turnout). While a train is entering a single-track segment of the line, no other train may do so. The next train in the same direction can enter immediately afterwards, whereas trains in opposite direction have to wait until the segment is clear again in order to prevent a collision.

Fig. 2 shows the path-time-diagram of a feasible schedule for two segments and four jobs. Jobs are represented by parallelograms of the same color. The processing time of a job on a segment is reflected by the height of the corresponding parallelogram, while the transit time is the remaining time (y-distance) to the lowest point of the parallelogram. In a feasible schedule, jobs may not intersect, and, in particular, a job can only begin being processed at a segment once it has fully exited the previous segment. Note that in the example it makes sense for the two rightbound jobs to switch order while waiting at the central node.

We also study a generalization of the model to situations where some of the jobs are allowed to pass each other when traveling in different directions (cf. Fig. 1b). This is a natural assumption, e.g., when scheduling the ship traffic on a canal, where smaller ships are allowed to pass each other while larger ships are not (cf. Lübbecke et al. [16]). In practice, the rules that decide which ships are allowed to pass each other are quite complex and depend on multiple parameters of the ships such as length, width, and draught (e.g., cf. [5]). We

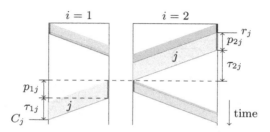

Fig. 2. Representation of a schedule on two segments ($i = 1, 2$) and four jobs as a path-time-diagram. In this example, all jobs are processed immediately at their release date. Job j is released at time r_j at the right end of segment 2 and needs to reach the left end of segment 1. Since it never has to wait, its completion time is smallest possible: $C_j = r_j + p_{2j} + \tau_{2j} + p_{1j} + \tau_{1j}$.

model these complex rules in the most general way by a bipartite compatibility graph for each segment, where vertices correspond to jobs and two jobs running in different directions are connected by an edge if they can cross the segment concurrently.

Our Results. Table 1 gives a summary of our results. We first show that scheduling bidirectional traffic is hard, even without processing times and with identical transit times (Section 3). The proof is via a non-standard reduction from MAXCUT. The key challenge is to use the local interaction of the jobs on the path to model global interaction between the vertices in the MAXCUT. We overcome this issue by introducing polynomially many vertex gadgets encoding the partition of each vertex and synchronizing these copies along the instance. We complement this result with a polynomial time approximation scheme (PTAS) for a single segment and arbitrary processing times (Section 4) using the $(1 + \epsilon)$-rounding technique of Afrati et al. [1].

We then show that bidirectional scheduling with arbitrary compatibility graphs is APX-hard already on a single segment and with identical processing times (Section 5). The proof is via a reduction from a variant of MAX-3-SAT which is NP-hard to approximate within a factor smaller than 1016/1015, as shown by Berman et al. [3]. As a byproduct, we obtain that also minimizing the makespan is APX-hard in this setting. We again complement our hardness result by polynomial algorithms for identical jobs on constant numbers of segments and with a constant number of compatibility types (Section 6).

Significance. With this paper we initiate the mathematical study of optimized dispatching of traffic in networks with bidirectional edges, e.g. train networks, ship canals, communication channels, etc. In all of these settings, traffic in one direction limits the possible throughput in the other direction. While in the past decades a wealth of results has been established for the unidirectional case (i.e., classical scheduling, and, in particular, flow shop models), surprisingly, and

Table 1. Overview of our results for bidirectional scheduling.
[1] even if $p = 0$, $\tau_i = 1$, [2] only if $p = 1, \tau_i \leq$ const, [3] even if $\tau_i = p = 1$.

compatibilities	Number m of segments		
	$m = 1$	m const.	m arbitrary
Different jobs $p_{ij} = p_j$, $\tau_{ij} = \tau_i$			
none/all compatible	PTAS [Thm. 2]		NP-hard[1] [Thm. 1]
		NP-hard [15]	
Identical jobs $p_{ij} = p$, $\tau_{ij} = \tau_i$ none compatible			
const. # types	polynomial [Thm. 5]	polynomial[2] [Thm. 6]	NP-hard[1] [Thm. 1]
arbitrary		APX-hard[3] [Thm. 4]	

despite their practical importance, bidirectional infrastructures have not received a similar attention so far.

The bidirectional scheduling model that we propose captures the essence of bidirectional traffic by distinguishing processing and transit times. This simple framework already allows to exhibit the computational key challenges of this setting. In particular, we show that bidirectional scheduling is already hard for identical jobs on a path, which is in contrast to the unidirectional case. We observe another increase in complexity when allowing specific types of traffic to use an edge concurrently in both directions. In practice, this is reasonable e.g. for ship traffic in a canal, where small vessels may pass each other. In that sense, we show that scheduling ship traffic is already hard on a single edge and, thus, considerably harder than scheduling train traffic.

While bidirectional scheduling is hard in general, we show that certain features of real-world scenarios can make the problem tractable, e.g., a small number of turnouts along a single path and/or a small number of different vessels. In this work we restrict ourselves to simple paths, but we hope that our results are a first step towards understanding traffic in general bidirectional networks.

Related Work. Scheduling problems are a fundamental class of optimization problems with a multitude of known hardness and approximation results (cf. Lawler et al. [13] for a survey). To the best of our knowledge, the bidirectional scheduling model that we propose and study in this paper has not been considered in the past nor is it contained as a special case in any other scheduling model. We give an overview of known results for related models.

For a single segment and jobs traveling from left to right, bidirectional scheduling reduces to the classical single machine scheduling problem, which Lenstra et al. [15] showed to be hard when minimizing total completion time. Afrati et al. [1] gave a PTAS with generalizations to multiple identical or a constant number of unrelated machines. Chekuri and Khanna [6] further generalized the result to related machines. We give a different generalization for

idirectional scheduling. For unrelated machines Hoogeveen et al. [11] showed
hat the completion time cannot be approximated efficiently within arbitrary
recision, unless P = NP.

Bidirectional scheduling also has similarities to scheduling of two job families
with a setup time that is required between jobs of different families. The general
omments in Potts and Kovalyov [19] on dynamic programs for such kinds of
roblems apply in part to our technique for Theorem 5.

When all jobs need to be processed on all segments in the same order and all
ransit times are zero, bidirectional scheduling reduces to flow shop scheduling.
Garey et al. [10] showed that it is NP-hard to minimize the sum of completion
imes in flow shop scheduling, even when there are only two machines and no
elease dates. They showed the same result for minimizing the makespan on
hree machines. Hoogeveen et al. [11] showed that there is no PTAS for flow
hop scheduling without release dates, unless P = NP. In contrast, Brucker
t al. [4] showed that flow shop problems with unit processing times can be
olved efficiently, even when all jobs require a setup on the machines that can
e performed by a single server only.

Job shop scheduling is a generalization of flow shop scheduling that allows
obs to require processing by the machines in any (not necessarily linear) order,
f. Lawler et al. [13, Section 14] for a survey. In this setting, the minimization of
he sum of completion times was proven to even be MAX-SNP-hard by Hoogeveen
t al. [11]. Queyranne and Sviridenko [20] gave a $\mathcal{O}((\log(m\mu)/\log\log(m\mu))^2)$-
pproximation for the weighted case with release dates, where μ denotes the
naximum number of operations per job. Fishkin et al. [8] gave a PTAS for a
onstant number of machines and operations per job. It is worth noting that job
hop scheduling does not contain bidirectional scheduling as a special case, since
t does not incorporate the distinction between processing and transit times for
obs passing a machine in different directions.

Job shop scheduling problems with unit jobs are strongly related to packet
outing problems where general graphs are considered, see the discussion in
eminal paper by Leighton et al. [14]. They proved that the makespan of any
oacket routing problem is linear in two trivial lower bounds, called the congestion
ind the dilation. For more recent progress in this direction, see, e.g., Scheideler
21] and Peis and Wiese [18]. All these works, however, consider minimizing the
nakespan and assume that the orientation of the graph is fixed. Antoniadis et al.
2] also consider average flow time on a directed line. They give lower bounds for
ompetitive ratios in the online setting and $\mathcal{O}(1)$ competitive algorithms with
esource augmentation for the maximum flow time.

2   Preliminaries

In the bidirectional scheduling problem, we are given a set $M = \{1, \ldots, m\}$
of segments which we imagine to be ordered from left to right. Further, we are
given two disjoint sets of J^r and J^l of *rightbound* and *leftbound* jobs, respectively,
with $J = J^r \cup J^l$ and $n = |J|$. Each job is associated with a *release date* $r_j \in$

$\mathbb{N}$, a *start segment* s_j and a *target segment* t_j, where $s_j \leq t_j$ for rightbound jobs and $s_j \geq t_j$ for leftbound jobs. A rightbound job j needs to cross the segments $s_j, s_j+1, \ldots, t_j-1, t_j$, and a leftbound job needs to cross the segments $s_j, s_j-1, \ldots, t_j+1, t_j$. We denote by M_j the set of segments that job j needs to cross. Each job j is associated with a processing time $p_j \in \mathbb{N}$ and each segment i is associated with a transit time $\tau_i \in \mathbb{N}$. Note that we restrict ourselves to identical processing times for a single job and identical transit times for a single segment. We call $p_j + \tau_i$ the *running time* of job j on segment i.

A *schedule* is defined by fixing the start times S_{ij} for each job j on each segment $i \in M_j$. The *completion time* of job j on segment i is then defined as $C_{ij} = S_{ij} + p_j + \tau_i$. The overall completion time of job j is $C_j = C_{t_j j}$. A schedule is feasible if it has the following properties.

1. Release dates are respected, i.e., $r_j \leq S_{s_j j}$ for each $j \in J$.
2. Jobs travel towards their destination, i.e., $C_{ij} \leq S_{i+1,j}$ (resp. $C_{ij} \leq S_{i-1,j}$) for rightbound (resp. leftbound) jobs j and $i \in M_j \setminus \{t_j\}$.
3. Jobs j, j' traveling in the same direction are not processed on segment $i \in M_j \cap M_{j'}$ concurrently, i.e., $[S_{ij}, S_{ij} + p_j) \cap [S_{ij'}, S_{ij'} + p_{j'}) = \emptyset$.
4. Jobs j, j' traveling in different directions are neither processed nor in transit on segment $i \in M_j \cap M_{j'}$ concurrently, i.e., $[S_{ij}, C_{ij}) \cap [S_{ij'}, C_{ij'}) = \emptyset$.

Our objective is to minimize the *total completion time* $\sum C_j = \sum_{j \in J} C_j$.

Other natural objectives are the minimization of the *makespan* $C_{\max} = \max\{C_j \mid j \in J\}$ or the *total waiting time* $\sum W_j = \sum_{j \in J} W_j$ where the individual waiting time of a job j is $W_j = C_j - \sum_{i \in M_j}(p_j + \tau_i) - r_j$. Note that minimizing the total waiting time is equivalent to minimizing the total completion time.

We also consider a generalization of the model, where some of the jobs traveling in different directions are allowed to pass each other. Formally, for each segment i, we are given a bipartite *compatibility graph* $G_i = (J^r \dot\cup J^l, E_i)$ with $E_i \subseteq J^r \times J^l$. Two jobs j, j' that are connected by an edge in G_i are allowed to run on segment i concurrently, i.e., condition 4 above need not be satisfied. Specifically, jobs j, j' may be processed or be in transit simultaneously.

In the following sections we give an intuitive overview of our constructions and refer for the details of all omitted proofs to the full version [7].

3   Hardness of Bidirectional Scheduling

First, we show that scheduling bidirectional traffic is hard, even when all processing times are zero and all transit times coincide. In other words, we eliminate all interaction between jobs in the same direction and show that hardness is merely due to the decision when to switch between left- and rightbound operation of each segment. This is in contrast to one-directional (flow shop) scheduling with identical processing times, which is trivial. Formally, we show the following result.

Theorem 1. *The bidirectional scheduling problem is* NP-*hard even if* $p_j = 0$ *and* $\tau_i = 1$ *for each* $j \in J$ *and* $i \in M$.

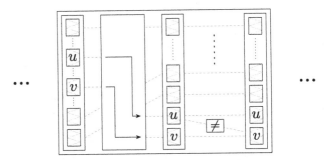

Fig. 3. Illustration of the vertex gadget in the leftbound (left) and the rightbound (right) state. At each time $t = 0, \ldots, 11$ multiple right- and leftbound jobs are released. Since all jobs have processing time 0, jobs in the same direction can be processed simultaneously. The only two sensible schedules differ in whether leftbound jobs are processed at even or odd times.

We reduce from the MaxCut problem which is contained in Karp's list of 21 NP-complete problems [12]. Given an undirected graph $G = (V, E)$ and some $k \in \mathbb{N}$ we ask for a partition $V = V_1 \cup V_2$ with $|E \cap (V_1 \times V_2)| \geq k$.

For a considered instance $\mathcal{I}$ of MaxCut we construct an instance of the bidirectional scheduling problem which can be scheduled without exceeding some specific waiting time if and only if $\mathcal{I}$ admits a solution. The translation to sum of completion times is then straightforward.

A cornerstone of our construction is the *vertex gadget* that occupies a fixed time interval on a single segment and can only be (sensibly) scheduled in two ways (cf. Fig. 4), which we interpret as the choice whether to put the corresponding vertex in the first or second part of the partition, respectively. We introduce multiple *vertex segments* that each have exactly one vertex gadget for each vertex in $\mathcal{I}$ and add further gadgets that ensure that the state of all vertex gadgets for the same vertex is the same across all segments. These gadgets allow us to synchronize vertex gadgets on consecutive vertex segments in two ways. We can either simply synchronize vertex gadgets that occupy the same time interval on the two vertex segments (*copy gadget*), or we can synchronize pairs of vertex gadgets occupying the same consecutive time intervals on the two vertex segments by linking the first gadget on the first segment with the second one on the second segment and vice-versa, i.e., we can transpose the order of two consecutive gadgets from one vertex segment to the next (*transposition gadget*).

We construct an edge gadget for each edge in $\mathcal{I}$ that incurs a small waiting time if two vertex gadgets in consecutive time intervals and segments are in different states and a slightly higher waiting time if they are in the same state. By tuning the multiplicity of each job, we can ensure that only schedules make sense where vertex gadgets are scheduled consistently. Minimizing the waiting time then corresponds to maximizing the number of edge gadgets that link vertex gadgets in different states, i.e., maximizing the size of a cut.

In order to fully encode the given MaxCut instance $\mathcal{I}$, we need to introduce an edge gadget for each edge in $\mathcal{I}$. However, edge gadgets can only link vertex

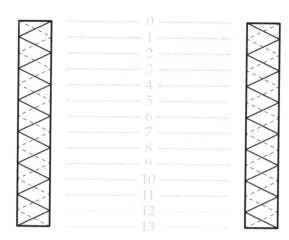

Fig. 4. Illustration of our hardness construction for a single edge $e = \{u, v\}$. First, a sequence of segments is used to change the order of vertex gadgets, such that the vertex gadgets corresponding to u and v occupy consecutive time intervals. Then, an edge gadget is added that incurs an increased waiting time if the vertex gadgets for u and v are in the same state.

gadgets in consecutive time intervals. We can overcome this limitation by adding a sequence of vertex segments and transposing the order of two vertex gadgets from one segment to the next as described before. With a linear number of vertex segments we can reach an order where the two vertex gadgets we would like to connect with an edge gadget are adjacent. At that point, we can add the edge gadget, and then repeat the process for all other edges in $\mathcal{I}$ (cf. Fig. 4).

We can reformulate Theorem 1 for nonzero processing times, simply by making the transit time large enough that the processing time does not matter.

Corollary 1. *The bidirectional scheduling problem is* NP-*hard even if $p_j = 1$ and $\tau_i = \tau$ for each $j \in J$ and $i \in M$.*

4   A PTAS for Bidirectional Scheduling

We give a polynomial time approximation scheme (PTAS), i.e., a polynomial $(1 + \varepsilon)$-approximation algorithm for each $\varepsilon > 0$, for bidirectional scheduling on a single segment with general processing times. This problem is hard even if all jobs have the same direction [15]. We extend the machine scheduling PTAS of Afrati et al. [1] to the bidirectional case, provided that the jobs are either all pairwise in conflict or pairwise compatible. The main issue when trying to adopt the technique of [1] is to account for the different roles of processing and transit times for the interaction of jobs in the same and different directions.

Theorem 2. *The bidirectional scheduling problem on a single segment and with compatibility graph $G_1 \in \{K_{n_r,n_1}, \emptyset\}$ admits a PTAS.*

The first part of the proof in [1] is to restrict to processing times and release ates of the form $(1+\varepsilon)^x$ for some $x \in \mathbb{N}$ and $r_j \geq \varepsilon(p_j+\tau_1)$. Allowing fractional rocessing and release times we can show that any instance can be adapted to ave these properties, without making the resulting schedule worse by a factor f more than $(1 + \varepsilon)$. We may thus partition the time horizon into intervals $x = [(1 + \varepsilon)^x, (1 + \varepsilon)^{x+1}]$, such that every job is released at the beginning f an interval. Since jobs are not released too early, we may conclude that he maximum number of intervals σ covered by the running time of a sin-le job is constant. This allows us to group intervals together in blocks $B_t = I_{t\sigma}, I_{t\sigma+1}, \ldots, I_{(t+1)\sigma-1}\}$ of σ intervals each, such that every job scheduled to tart in block B_t will terminate before the end of the next block B_{t+1}.

To use the fact that each block only interacts with the next block in our lynamic program, we need to specify an interface for this interaction. For that urpose we introduce the notion of a *frontier*. A block *respects an incoming frontier* $F = (f_1, f_r)$ if no leftbound (rightbound) job scheduled to start in the lock starts earlier than f_1 (f_r). Similarly, a block *respects an outgoing fron-ier* $F = (f_1, f_r)$ if no leftbound or rightbound job scheduled to start in the lock would interfere with a leftbound (rightbound) job starting at time f_1 (f_r). The symmetrical structure of the compatibility graph (K_{n_r,n_l} or $\emptyset$) allows us o use this simple interface. We introduce a dynamic programming table with ntries $T[t, F, U]$ that are designed to hold the minimum total completion time of scheduling all jobs in $U \subseteq J$ to start in block B_t or earlier, such that B_t respects he outgoing frontier F. We define $C(t, F_1, F_2, V)$ to be the minimum total com-pletion time of scheduling all jobs in V to start in B_t with B_t respecting the ncoming frontier F_1 and the outgoing frontier F_2 (and ∞ if this is impossible). We have the following recursive formula for the dynamic programming table:

$$T[t, F, U] = \min_{F', V \subseteq U} \{T[t - 1, F', U \setminus V] + C(t, F', F, V)\}.$$

To turn this into an efficient dynamic program, we need to limit the depen-dencies of each entry and show that $C(\cdot)$ can be computed efficiently. The number of blocks to be considered can be polynomially bounded by $\log D$, where $D = \max_j r_j + n \cdot (\max_j p_j + \tau_1)$ is an upper bound on the makespan. The following lemma shows that we only need to consider polynomially many other entries to compute $T[t, F, U]$ and we only need to evaluate $C(\cdot)$ for job sets of constant size, which we can do in polynomial time by simple enumeration.

Lemma 1. *There is a schedule with a sum of completion times within a factor of $(1 + \varepsilon)$ of the optimum and with the following properties:*
1. The number of jobs scheduled in each block is bounded by a constant.
2. Every two consecutive blocks respect one of constantly many frontiers.

Proof (sketch). Partitioning the released jobs of each interval direction-wise by processing time into *small* and *large* jobs and bundling small jobs into packages of roughly the same size allows us to bound the number of released jobs per interval by a constant, similarly as in [1]. Furthermore, we establish that we may assume jobs to remain unscheduled only for constantly many blocks.

For the second property, we stretch all time intervals by a factor of $(1 + \varepsilon)$ which gives enough room to decrease the start times of those jobs interfering with two blocks such that an $1/\varepsilon^2$-fraction of an interval separates jobs starting in two consecutive blocks. Thus, we only need to consider $\frac{\sigma}{\varepsilon^2}$ possible frontier values per direction, or a total of $\left(\frac{\sigma}{\varepsilon^2}\right)^2$ possible frontiers.   ⬜

5   Hardness of Custom Compatibilities

In Section 3, we showed that bidirectional scheduling is hard on an unbounded number of machines, even for identical jobs. As the main result of this section we show that for arbitrary compatibility graphs the problem is APX-hard already on a single segment and with unit processing and transit times. For ease of exposition, we first show that the minimization of the makespan is NP-hard. Later we extend this result towards minimum completion time and APX-hardness.

Theorem 3. *The bidirectional scheduling problem on a single segment and with an arbitrary compatibility graph is* NP-*hard even if $p_j = \tau_1 = 1$ for each $j \in J$.*

We give a reduction from an NP-hard variant of SAT (cf. [9]). $(\leq 3, 3)$-SAT considers a formula with a set of clauses C of size three over a set of variables X, where each variable appears in at most three clauses and asks if there is a truth assignment of X satisfying C. Note the difference to the polynomially solvable $(3, 3)$-SAT, where each variable appears in *exactly* three clauses [22].

For a given $(\leq 3, 3)$-SAT formula we construct a bidirectional scheduling instance that can be scheduled within some specific makespan T if and only if the given formula is satisfiable. Our construction is best explained by partitioning the time horizon $[0, T]$ into four parts (cf. Fig. 5 along with the following).

We use a frame of blocking jobs that need to be scheduled at their release date. We can enforce this by making sure that at least one blocking job is released at (almost) each unit time step and that blocking jobs that are not supposed to run concurrently are incompatible. We release variable jobs that have to be scheduled into gaps between the blocking jobs. More precisely, in the first part of the construction we release 6 jobs within a separate time interval for each variable. Two of these jobs are leftbound and need to be scheduled within the first two parts of the construction, which implies that one of the two remaining pairs of rightbound jobs must be scheduled after the second part. If the first pair is delayed we interpret this as an assignment of *true* to the variable and otherwise as *false*.

The third part of the construction has a gap for each clause, with compatibilities ensuring that only variable jobs can be scheduled into the gap which satisfy the clause. Since each literal can only appear in at most two clauses, there are enough variable jobs to satisfy all clauses if the formula is satisfied. Finally, the last part has $2|X| - |C|$ gaps that fit any variable job. In order to schedule all variable jobs before the end of the last part, we thus need to schedule a variable job into each gap of a clause. This is possible if and only if the given $(\leq 3, 3)$-SAT formula is satisfiable. We can easily extend our result to completion or waiting

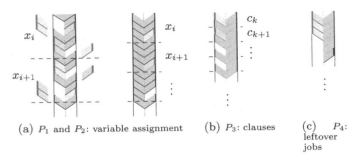

(a) P_1 and P_2: variable assignment   (b) P_3: clauses   (c) P_4: leftover jobs

Fig. 5. Illustration (colored) of the four parts of our construction. Time is directed downwards, rightbound (leftbound) jobs are depicted on the left (right) of each figure.

times by adding many blocking jobs after the last part, such that violating the makespan also ruins the the total completion time.

With a slight adaption of the construction and more involved arguments, we can even show APX-hardness of the problem. We reduce from a specific variant of MAX-3-SAT, where each literal occurs exactly twice, and which is NP-hard to approximate within a factor of 1016/1015, see Berman et al. [3].

Theorem 4. *The bidirectional scheduling problem on a single segment and with an arbitrary compatibility graph is APX-hard even if $p_j = \tau_1 = 1$ for each $j \in J$.*

5 Dynamic Programs for Restricted Compatibilities

After establishing the hardness of bidirectional scheduling with a general compatibility graph in the last section, in this section we turn to the case of a constant number of different compatibility types. Due to the identical processing times, the jobs in each direction can be scheduled in the order of their release dates. The only decision left is when to switch between left- and rightbound operation of the segments. This decision is hard in the general case (Theorem 1), but we are able to formulate a dynamic program for any constant number of segments.

Our result generalizes to the case when some jobs of different directions are compatible as long as the number of *compatibility types* is constant, where two jobs j_1, j_2 in the same direction are defined to have the same compatibility type if the set of jobs compatible with j_1 is equal to the set of jobs compatible with j_2 on each segment. Formally, j_1 and j_2 have the same compatibility type if $\{j : \{j_1, j\} \in E_i\} = \{j : \{j_2, j\} \in E_i\}$ for the compatibility graphs $G_i = (J^l \cup J^r, E_i)$ of each segment i.

For a single segment we partition J into κ subsets of jobs $J^1, \ldots, J^\kappa$ where all jobs of J^c, $c \in 1, \ldots, \kappa$, have the same compatibility type c, and let $n_c = |J^c|$. Since the jobs of each subset only differ in their release dates, they can again be scheduled in the order of their release dates. This observation allows us to define a dynamic program that decides how to merge the job sets $J^1, \ldots, J^\kappa$ such that the resulting schedule has minimum total completion time.

Theorem 5. *The bidirectional scheduling problem can be solved in polynomial time if $m = 1$, κ is constant and $p_j = p$ for each $j \in J$.*

We now consider a constant number of segments $m > 1$. The main complication in this setting is that decisions on one segment can influence decisions on other segments, and, in general, every job can influence every other job in this way. In particular, we need to keep track of how many jobs of each type are in transit at each segment, and we can thus not easily adapt the dynamic program for a single segment. We propose a different dynamic program that relies on all transit times being bounded by a constant and can be adapted for assumptions complementary to Theorem 1.

Theorem 6. *The bidirectional scheduling problem can be solved in polynomial time if m and κ are constant and either $p_j = 1$ for each $j \in J$ and τ_i is constant for each $i \in M$ or $p_j = 0$ for each $j \in J$ and $\tau_i = 1$ for each $i \in M$.*

References

1. Afrati, F., Bampis, E., Chekuri, C., Karger, D., Kenyon, C., Khanna, S., Milis, I., Queyranne, M., Skutella, M., Stein, C., Sviridenko, M.: Approximation schemes for minimizing average weighted completion time with release dates. In: Proc. 40th Symposium on Foundations of Computer Science (FOCS), pp. 32–43 (1999)
2. Antoniadis, A., Barcelo, N., Cole, D., Fox, K., Moseley, B., Nugent, M., Pruhs, K.: Packet forwarding algorithms in a line network. In: Pardo, A., Viola, A. (eds.) LATIN 2014. LNCS, vol. 8392, pp. 610–621. Springer, Heidelberg (2014)
3. Berman, P., Karpinski, M., Scott, A.D.: Approximation hardness of short symmetric instances of MAX-3SAT. Electronic Colloquium on Computational Complexity (ECCC) **10**(49) (2003)
4. Brucker, P., Knust, S., Wang, G.: Complexity results for flow-shop problems with a single server. European J. Oper. Res. **165**, 398–407 (2005)
5. Bundesamt für Seeschifffahrt und Hydrographie (BSH). German Traffic Regulations for Navigable Maritime Waterways. Hamburg and Rostock, Germany (2013)
6. Chekuri, C., Khanna, S.: A PTAS for minimizing weighted completion time on uniformly related machines. In: Orejas, F., Spirakis, P.G., van Leeuwen, J. (eds.) ICALP 2001. LNCS, vol. 2076, pp. 848–861. Springer, Heidelberg (2001)
7. Disser, Y., Klimm, M., Lübbecke, E.: Scheduling bidirectional traffic on a path. arXiv:1504.07129 (2015)
8. Fishkin, A.V., Jansen, K., Mastrolilli, M.: On minimizing average weighted completion time: A PTAS for the job shop problem with release dates. In: Ibaraki, T., Katoh, N., Ono, H. (eds.) ISAAC 2003. LNCS, vol. 2906, pp. 319–328. Springer, Heidelberg (2003)
9. Garey, M.R., Johnson, D.S.: Computers and intractability: A Guide to the Theory of NP-Completeness. W.H. Freeman & Co., New York (1979)
10. Garey, M.R., Johnson, D.S., Sethi, R.: The complexity of flowshop and jobshop scheduling. Math. Oper. Res. **1**(2), 117–129 (1976)
11. Hoogeveen, H., Schuurman, P., Woeginger, G.J.: Non-approximability results for scheduling problems with minsum criteria. In: Bixby, R.E., Boyd, E.A., Ríos-Mercado, R.Z. (eds.) IPCO 1998. LNCS, vol. 1412, pp. 353–366. Springer, Heidelberg (1998)

2. Karp, R.M.: Reducibility among combinatorial problems. In: Miller, R.E., Thatcher, J.W., Bohlinger, J.D. (eds.) Complexity of Computer Computations. The IBM Research Symposia Series, pp. 85–103 (1972)
3. Lawler, E.L., Lenstra, J.K., Rinnooy Kan, A.H.G., Shmoys, D.B.: Sequencing and scheduling: algorithms and complexity. In: Handbooks in Operations Research and Management Science, vol. 4, pp. 445–522 (1993)
4. Leighton, F.T., Maggs, B.M., Rao, S.B.: Packet routing and job-shop scheduling in o(congestion+dilation) steps. Combinatorica $\mathbf{14}$(2), 167–186 (1994)
5. Lenstra, J.K., Rinnooy Kan, A.H.G., Brucker, P.: Complexity of machine scheduling problems. Ann. Discrete Math. $\mathbf{1}$, 343–362 (1977)
6. Lübbecke, E., Lübbecke, M.E., Möhring, R.H.: Ship traffic optimization for the Kiel Canal. Technical Report 4681, Optimization. Online 12 (2014)
7. Lusby, R.M., Larsen, J., Ehrgott, M., Ryan, D.: Railway track allocation: models and methods. OR Spectrum $\mathbf{33}$(4), 843–883 (2011)
8. Peis, B., Wiese, A.: Universal packet routing with arbitrary bandwidths and transit times. In: Günlük, O., Woeginger, G.J. (eds.) IPCO 2011. LNCS, vol. 6655, pp. 362–375. Springer, Heidelberg (2011)
19. Potts, C.N., Kovalyov, M.Y.: Scheduling with batching: A review. European J. Oper. Res. $\mathbf{120}$(2), 228–249 (2000)
20. Queyranne, M., Sviridenko, M.: New and improved algorithms for minsum shop scheduling. In: Proc. 11th Symposium on Discrete Algorithms (SODA), pp. 871–878 (2000)
21. Scheideler, C.: Offline routing protocols. In: Universal Routing Strategies for Interconnection Networks, pp. 57–71 (1998)
22. Tovey, C.A.: A simplified NP-complete satisfiability problem. Discrete Appl. Math. $\mathbf{8}$(1), 85–89 (1984)

On the Problem of Approximating the Eigenvalues of Undirected Graphs in Probabilistic Logspace

Dean Doron[(✉)] and Amnon Ta-Shma

The Blavatnik School of Computer Science,
Tel-Aviv University, 69978 Tel Aviv, Israel
deandoron@mail.tau.ac.il, amnon@tau.ac.il

Abstract. We introduce the problem of *approximating* the eigenvalues of a given stochastic/symmetric matrix in the context of classical space-bounded computation.

The problem can be *exactly* solved in $\mathsf{DET} \subseteq \mathsf{NC}^2$. Recently, it has been shown that the approximation problem can be solved by a *quantum* logspace algorithm. We show a BPL algorithm that approximates any eigenvalue with a *constant* accuracy. The result we obtain falls short of achieving the polynomially-small accuracy that the quantum algorithm achieves. Thus, at our current state of knowledge, we can achieve polynomially-small accuracy with quantum logspace algorithms, constant accuracy with probabilistic logspace algorithms, and no nontrivial result is known for deterministic logspace algorithms. The quantum algorithm also has the advantage of working over arbitrary, possibly non-stochastic Hermitian operators.

Our work raises several challenges. First, a derandomization challenge, trying to achieve a deterministic algorithm approximating eigenvalues with some non-trivial accuracy. Second, a de-quantumization challenge, trying to decide whether the quantum logspace model is strictly stronger than the classical probabilistic one or not. It also casts the deterministic, probabilistic and quantum space-bounded models as problems in linear algebra with differences between symmetric, stochastic and arbitrary operators. We therefore believe the problem of approximating the eigenvalues of a graph is not only natural and important by itself, but also important for understanding the relative power of deterministic, probabilistic and quantum logspace computation.

1 Introduction

A graph G can be associated with a linear operator A that describes a random walk on G. The operator A takes an especially simple form when G is undirected:

D. Doron—Supported by the Israel science Foundation grant no. 994/14, by the United States – Israel Binational Science Foundation grant no. 2010120 and by the Blavatnik Fund.

A. Ta-Shma—Supported by the Israel science Foundation grant no. 994/14 and by the United States – Israel Binational Science Foundation grant no. 2010120.

M.M. Halldórsson et al. (Eds.): ICALP 2015, Part I, LNCS 9134, pp. 419–431, 2015.
DOI: 10.1007/978-3-662-47672-7_34

f G **is regular and undirected** then A is symmetric and has a complete basis of orthonormal eigenvectors with real eigenvalues. In other words, there exists a unitary basis under which A is diagonal with real eigenvalues λ_i on the diagonal.

f G **is undirected** but not necessarily regular, then A is s diagonalizable with real eigenvalues, i.e., the picture is the same as before except that the basis is not necessarily unitary.[1]

f G **is directed** then A does not necessarily have a full basis of eigenvectors. In this case A (like any other linear operator) can be brought to its canonical Jordan Normal Form, where there exists a basis under which A is block-diagonal and each block has an eigenvalue λ on the main diagonal and 1 on the diagonal above it.

In this paper we raise the following natural questions:

– How difficult is it to approximate the largest eigenvalues of a general (not necessarily stochastic or non-negative) operator?
– How difficult is it to approximate all the spectrum of an operator?
– Does the problem become easy and belong to L when the graph is undirected?
– How about approximating the singular values of a graph?

1.1   The Bigger Picture

Derandomization is a major challenge of theoretical computer science. In the space-bounded model, Nisan [1] constructed a pseudo-random generator (PRG) against logarithmic space-bounded non-uniform algorithms that uses seed length $O(\log^2 n)$. Using that he showed BPL is contained in the class having simultaneously polynomial time and $O(\log^2 n)$ space. Saks and Zhou [2] showed BPL is contained in $\mathsf{DSPACE}(\log^{1.5} n)$. Reingold [3] showed that undirected st-connectivity (which was shown to be in RL by [4]) already belongs to L. These results seem to indicate that randomness does not add additional power to the model and many conjecture that in fact BPL = L. Yet, we currently do not know a PRG with seed length $o(\log^2 n)$, nor a general derandomization result that simultaneously uses $o(\log^2 n)$ space and polynomial time.

One can look up and ask which upper bounds we know on BPL. We then know the following:

$$\mathsf{NC}^1 \subseteq \mathsf{L} \subseteq \mathsf{RL} \subseteq \mathsf{NL} \subseteq \mathsf{DET} \subseteq \mathsf{NC}^2 \subseteq \mathsf{DSPACE}(O(\log^2 n)),$$

where DET is the class of languages that are NC^1 Turing-reducible to the problem *intdet* of computing the determinant of an integer matrix (see [5] for a definition

---

[1] If G is undirected and irregular, then the adjacency matrix $\tilde{A}$ is symmetric but the transition matrix $A = D^{-1}\tilde{A}$, where D is the diagonal degrees matrix, is not symmetric. Yet, consider the matrix $L = D^{-1/2}\tilde{A}D^{-1/2}$. L is symmetric and thus has an eigenvector basis with real eigenvalues. $A = D^{-1/2}LD^{1/2}$ is conjugate to L and thus is diagonalizable and has the same eigenvalues. As A is stochastic its eigenvalues are in the range $[-1, 1]$.

of DET). As it turns out, many important problems in linear algebra, such as inverting a matrix, or equivalently, solving a set of linear equations are in DET, and often complete for it (see, e.g., [5]). The fact that NL ⊆ DET is due to [5] who showed that the directed connectivity problem, STCON is reducible to *intdet*. DET ⊆ NC^2 follows from Csansky's algorithm [6] for the parallel computation of the determinant. In addition to the above we also know that BPL ⊆ DET (e.g., using the fact that matrix powering is DET complete).

While matrix powering is complete for DET, *approximating* matrix powering of *stochastic* matrices is in BPL. To see that, assume A represents a stochastic matrix. Then one can approximate $A^k[s,t]$ by estimating the probability a random walk over A starting at s reaches t after k steps.[2] Conversely, it is possible to convert a BPL machine to a stochastic operator A such that the probability the machine moves from s to t in k steps is $A^k[s,t]$.[3] Thus, in a sense, approximating matrix-powering of stochastic operators is complete for BPL.

We now deviate from the classical picture we had so far and consider a quantum space-bounded model. In 1999, Watrous [7] defined the model of quantum logspace computation, and proved several facts on it. The definition was modified several times, see, [8]. Roughly speaking, a language is in BQL if there exists an L–uniform family of quantum circuits solving the language with only $O(\log n)$ qubits. The quantum circuits are over some universal basis of gates (e.g., CNOT, HAD, T) plus intermediate measurements (that in particular may simulate a stream of random coins). For details we refer the reader to [8,9]. The works of Watrous, van Melkebeek and Watson showed that BQL is also contained in NC^2.

Recently, it was shown in [9], building on an earlier work by [10], that it is possible to *approximate* the singular value decomposition (SVD) of a given linear operator in BQL. This also implies that it is possible to approximately invert a matrix in BQL. A natural question left open by this work is:

Open Problem: Is it possible to approximate the SVD of an arbitrary linear operator in BPL? The problem is also open for Hermitian operators, where singular values and eigenvalues coincide (up to their sign).

In fact, this question is open also when the operator is the transition matrix arising from a walk on a regular, undirected graph.

Thus, somewhat surprisingly, we see that the deterministic / probabilistic / quantum space-bounded classes and the class DET are capable of doing some sort of linear algebra on corresponding operators. Namely,

– In DET we can compute exactly the determinant which is the product of all eigenvalues as well as the product of all singular values. We can also solve matrix powering. Both problems are *complete* in DET. With that we can approximately invert an operator or perform the SVD decomposition.

---

[2] For completeness we include a proof of this in Appendix A. We also extend the class for which this works to matrices with non-negative or complex entries as long as their infinity norm is at most 1.

[3] This reduction is standard and appears in many papers, e.g., already in [1].

In a sense, DET is an *exact* computation of the spectrum (e.g., in terms of the characteristic polynomial) of an *arbitrary* linear operator.

- BQL is capable of approximating the whole singular value decomposition of any operator. This is somewhat equivalent to saying that BQL is capable of *approximating* the eigenvalues of *Hermitian* operators.
- BPL is capable of *approximating* matrix powering. In this paper we will show BPL can approximate any eigenvalue of an *undirected* graph with constant accuracy. We do not know yet whether we can do the same for *directed* graphs or whether we can approximate the *whole spectrum* of undirected graphs.
- In L we do not know how to do any of the above, but Reingold showed L is capable of solving USTCON, i.e., connectivity on undirected graphs. Notice that undirected graphs roughly correspond to the intersection of stochastic and Hermitian operators.

1.2   On the Problem of Approximating Arbitrary Eigenvalues of Undirected Graphs in BPL

We define the following promise problem:

Definition 1. *($EV_{\alpha,\beta}$) The input is a stochastic, Hermitian matrix A, $\lambda \in [-1, 1]$ and $\alpha < \beta$.*

Yes instances : *There is an eigenvalue λ_i of A such that $|\lambda_i - \lambda| \leq \alpha$.*
No instances : *All eigenvalues of A are β–far from λ.*

One way to design a BPL algorithm for the problem is by "de-quantumizing" the quantum algorithm.[4] The BQL algorithm solves the above problem for any Hermitian operator A whose eigenvalues are τ–separated, for, say, $\tau = n^{-c}$, $\alpha = \frac{\tau}{4}$ and $\beta = 2\alpha$. That is, the quantum algorithm can handle any polynomially small accuracy. With such accuracy one can turn the solution of the promise problem to a procedure approximating the whole spectrum.

We develop a BPL algorithm that follows the main idea of the quantum algorithm, and in that sense we de-quantumize the quantum algorithm, but we achieve much worse parameters. Specifically, we prove that the promise problem $EV_{\alpha,\beta}$ belongs to BPL, for *constant* parameters $\alpha < \beta$. On the one hand the result is disappointing because the quantum algorithm does so much better and can handle polynomially small gaps. On the other hand, we remark that we do not know how to achieve even constant approximation with a deterministic logspace algorithm. We are not aware of many natural promise problems in BPL that are not known to be in L. This paper shows $EV_{\alpha,\beta}$ is such a promise problem.

---

[4] We remark that Ben-Or and Eldar [11] recently de-quantumized the SVD quantum algorithm and obtained a classical probabilistic algorithm for inverting matrices that achieves the state of the art running time, using a completely new approach that is derived from the quantum algorithm. We would like to do the same in the space-bounded model.

1.3   Our Technique

The usual way of describing the quantum algorithm is that it applies quantum phase estimation on the completely mixed state. The completely mixed state is a uniform mixture of the pure states that are formed from the eigenvectors of A, and on each such eigenvector, the quantum phase estimation estimates the corresponding eigenvalue. Thus, if the procedure can be run in (quantum) logarithmic space, we essentially sample a random eigenvector/eigenvalue pair and from that we can approximately get the SVD decomposition of A.

Another (less standard) way of viewing the quantum algorithm is that it manipulates the eigenvalues of an input matrix A without knowing the decomposition of A to eigenvectors and eigenvalues. This can be done using the simple fact that if $\lambda_1, \ldots, \lambda_n$ are the roots of the characteristic polynomial of A, and if p is an arbitrary univariate polynomial, then $p(\lambda_1), \ldots, p(\lambda_n)$ are the roots of the characteristic polynomial of the matrix $p(A)$. The probability the algorithm measures λ is proportional to Tr $(p(A))$, where p is a shift of the Fejér kernel by λ (see, e.g., [12, Chapter2]). Applying p on A amplifies the eigenvalues that are close to λ to a value close to 1, and damps eigenvalues far from λ close to 0. Thus, Tr $(p(A))$ approximately counts the number of eigenvalues close to λ.

We would like to follow the same approach but with a probabilistic algorithm rather than a quantum one. We say a matrix A is *simulatable* if a probabilistic logspace algorithm can approximate $A^k[s, t]$ for any k polynomial in n and with polynomially-small accuracy (see Definition 2 for the exact details). From the discussion above it is clear that if A is the transition matrix of a (directed or undirected) graph then A is simulatable (see Lemma 1). We remark that in the appendix we show that even non-stochastic matrices A with negative or complex entries are simulatable as long as A has infinity norm at most 1, namely, those matrices A for which all rows $i \in [n]$ have ℓ_1 norm at most 1, $\sum_j |A[i, j]| \leq 1$.

If A is simulatable and the coefficients of $p(x) = \sum_i c_i x^i$ are not too large (i.e., only polynomially large in n), then we can approximate in BPL the matrix $p(A) = \sum_i c_i A^i$. In particular, we can also approximate Tr $(p(A))$. By taking p to be a threshold polynomial with degree logarithmic in n (that guarantees the size of the coefficients c_i is polynomial in n) and a threshold around λ, we can solve $EV_{\alpha,\beta}(A)$ for constants $\alpha < \beta$ (see Section 3).

There are many other possible candidate functions for a threshold polynomial p. However, we prove in Theorem 2 that no polynomial can do significantly better than a threshold polynomial. The reason the quantum algorithm works better is because it is able to take p up to some polynomial degree (rather than logarithmic degree) not worrying about the (quite large) size of the coefficients, thus leading to much better accuracy. The quantum algorithm also has the advantage that it works for any normal operator A, not necessarily stochastic or simulatable.

Thus, the algorithm we give for $EV_{\alpha,\beta}$ is simple: Approximate Tr $(p(A))$ to a simple logarithmic degree polynomial p. Nevertheless, we believe it features a new component that has not been used before by probabilistic space-bounded algorithms. An algorithm that takes a random walk on a graph and takes a decision based on the walk length and connectivity properties of the graph (as,

,g., [4]) works with some power of the input matrix A. More generally, such an lgorithm can work with a convex combination of powers of the input matrix by probabilistically choosing which power to take). The algorithm we present tilizes *arbitrary* (positive or *negative*) combinations of matrix powers and we elieve it is a crucial feature of the solution. We are not aware of previous BPL lgorithms using such a feature.

The approach above does not work for approximating the eigenvalues of a *directed* graph G. It is still true that the resulting operator A is stochastic and herefore simulatable. Also, it remains true that if λ is an eigenvalue of A (i.e., a oot of the characteristic polynomial) then $p(\lambda)$ is a root of $p(A)$. However, since A is not Hermitian, the eigenvalues λ of A may be complex and we do not know how to control $p(\lambda)$ when p may have both negative and positive coefficients. We believe it should be possible to approximate in BPL an arbitrary eigenvalue of any stochastic operator (not necessarily Hermitian) to within constant accuracy, but we have not been able to show it so far.

1.4   A Short Discussion

We believe the problem of approximating the eigenvalues of an undirected graph is natural and important. Also, at our current state of knowledge, it simultaneously separates deterministic, probabilistic and quantum complexity: In BQL we can solve it with polynomially-small accuracy, in BPL with constant accuracy and in L we do not know how to solve it at all. Thus it poses several challenges:

- First, there is the natural question of whether one can approximate eigenvalues in BPL with better accuracy. A positive answer would imply BPL approximations to many important linear algebra problems that are currently only known to be in NC^2. A negative answer would imply a separation between BQL and BPL.
- Second, it raises the natural question of derandomization. Can one design a *deterministic* algorithm approximating eigenvalues to constant accuracy?

We believe the solution of this problem is not only important by itself, but may also shed new light on the strengths and weaknesses of the space-bounded model, and the relative strengths of the deterministic, probabilistic and quantum models of space-bounded computation.

2   Preliminaries

Often we are interested in approximating a *value* (e.g., an entry in a matrix with integer values or the whole matrix) with a probabilistic machine. More precisely, assume there exists some value $u = u(x) \in \mathbb{R}$ that is determined by the input $x \in \{0,1\}^n$. We say a probabilistic TM $M(x, y)$ (ε, δ)–approximates $u(x)$ if:

$$\forall_{x \in \{0,1\}^n} \quad \Pr_y \left[|M(x, y) - u(x)| \geq \varepsilon \right] \leq \delta. \tag{1}$$

A random walk on a graph G (or its transition matrix A) can be simulated by a probabilistic logspace machine. As a consequence, a probabilistic logspace machine can approximate powers of A well. Here we try to extend this notion to arbitrary linear operators A, not necessarily stochastic. We say a matrix A is *simulatable* if any power of it can be approximated by a probabilistic algorithm running in small space. Formally:

Definition 2. *We say that a family of matrices $\mathcal{A}$ is* simulatable *if there exists a probabilistic algorithm that on input $A \in \mathcal{A}$ of dimension n with $\|A\| \leq poly(n)$, $k \in \mathbb{N}$, $s, t \in [n]$, runs in space $O(\log \frac{nk}{\varepsilon\delta})$ and (ε, δ)-approximates $A^k[s, t]$.*

In Appendix A we give for completeness a proof that:

Lemma 1. *The family of transition matrices of (directed or undirected) graphs is simulatable.*

We say $\|A\|_\infty \leq c$ if for every $i \in [n]$, $\sum_j |A[i,j]| \leq c$. In the same Appendix we also show:

Lemma 2. *The family of real matrices with infinity norm at most 1 is simulatable.*

3   Approximating Eigenvalues with Constant Accuracy

In this section we prove:

Theorem 1. *There exists a probabilistic algorithm that on input a stochastic matrix B with real eigenvalues in $[0, 1]$, constants $\beta > \alpha > 0$ and $\lambda \in [0, 1]$ such that:*

- *There are d eigenvalues λ_i satisfying $|\lambda - \lambda_i| \leq \alpha$,*
- *All other eigenvalues λ_i satisfy $|\lambda - \lambda_i| \geq \beta$,*

outputs d with probability at least $2/3$. Furthermore the algorithm runs in probabilistic space $O(\log n)$.

We remark that Theorem 1 covers the case of transition matrices of undirected graphs. As mentioned earlier, a transition matrix A of an undirected graph has an eigenvector basis with real eigenvalues in the range $[-1, 1]$. Taking $B = \frac{1}{2}A + \frac{1}{2}I_{n \times n}$ we get a stochastic matrix with eigenvalues in the range $[0, 1]$, and whose eigenvectors are in a natural one-to-one correspondence with A's eigenvalues.

Proof. (Of Theorem 1) The input to the algorithm is $n, B, \lambda, \alpha, \beta$. We assume a univariate polynomial $p(x) = \sum_{i=0}^{M} c_i x^i$ with the following properties:

- p has a sharp peak around λ, i.e., $p(x) \geq 1 - \eta$ for $x \in [\lambda - \alpha, \lambda + \alpha]$ and $p(x) \leq \eta$ for $x \in [0, 1] \setminus (\lambda - \beta, \lambda + \beta)$, where $\eta = \eta(n) = n^{-2}$.

– p can be computed in L. Formally, $M = \deg(p)$ and $|c_i|$ are at most $\text{poly}(n)$ and for every i, c_i can be computed (exactly) by a deterministic Turing machine that uses $O(\log n)$ space.

In the next subsection we show how to obtain such a polynomial p with $M = 2(\beta - \alpha)^{-2} \log n$ and $|c_i| \leq 2^{O(M)}$.

Choose $\varepsilon = \frac{1}{n}$ and $\delta = \frac{1}{3}$. Set $\varepsilon' = \varepsilon \cdot 2^{-2M}$ and $\delta' = \delta \cdot 2^{-M}$. The output of the algorithm is the integer closest to

$$R = \sum_{i=0}^{M} c_i \cdot \text{TP}(B, n, i, \varepsilon', \delta')$$

where TP is the probabilistic algorithm guaranteed by Lemma 2 that (ε', δ')–approximates $\text{Tr}\,(B^i)$.

It is easy to check that:

Claim. $\Pr[|R - \text{Tr}\,(p(B))| \geq \varepsilon] \leq \delta$.

As $\text{Tr}\,(p(B)) = \sum_{i=1}^{n} p(\lambda_i)$, $\Pr[|R - \sum_{i=1}^{n} p(\lambda_i)| \geq \varepsilon] \leq \delta$. However, $p(\lambda_i)$ is large when λ_i is α–close to λ and small when it is β–far from λ, and we are promised that *all* eigenvalues λ_i are either α–close or β–far from λ. Thus,

$$|\text{Tr}\,(p(B)) - d| \leq n\eta.$$

Altogether, except for probability δ, $|R - d| \leq \varepsilon + n\eta \leq \frac{1}{3}$, and the nearest integer closest to R is d. The correctness follows. It is also straightforward to check that the space complexity is $O(\log(n\varepsilon^{-1}\delta^{-1})) = O(\log n)$.

The constant accuracy we achieve is far from being satisfying. The matrix B has n eigenvalues in the range $[0, 1]$, so the average distance between two neighboring eigenvalues is $1/n$. Thus, the assumption that there is an interval of length $\beta - \alpha$ with no eigenvalue is often not true. The desired accuracy we would like to get is $o(1/n)$. Having such accuracy would enable outputting an approximation of the whole spectrum of B, using methods similar to those in [9], thus getting a true classical analogue to the quantum algorithm in [9]. However, we do not know how to achieve subconstant accuracy. The question whether better accuracy is possible in BPL is one of the main questions raised by this work.

3.1   Using the Symmetric Threshold Functions

There are several natural candidates for the function p above. In this subsection we use the threshold function to obtain such a function p. For $\lambda = \frac{k}{M}$ for some integers k and M, define:

$$p_\lambda(x) = \sum_{i=k}^{M} \binom{M}{i} x^i (1-x)^{M-i}.$$

p_λ approximates well the threshold function $\mathbf{Th}_\lambda(x) : [0,1] \to \{0,1\}$ that is one for $x \geq \lambda$ and zero otherwise. Specifically, using the Chernoff bound, we obtain:

Lemma 3. *Let* $x \in [0,1]$. $p_\lambda(x)$ *approximates* $\mathbf{Th}_\lambda(x)$ *over* $[0,1]$ *with accuracy* $(\xi(\varepsilon))^{Mx}$, *where* $\varepsilon = \frac{\lambda-x}{x}$ *and* $\xi(\varepsilon) = \frac{e^\varepsilon}{(1+\varepsilon)^{1+\varepsilon}}$.

As a polynomial in x, $p_\lambda(x) = \sum_{i=0}^M c_i x^i$ with $c_i = (-1)^i \sum_{j=\lambda M}^i \binom{M}{j}\binom{M-j}{i-j}(-1)^j$ and therefore $|c_i| \leq \sum_{j=\lambda M}^i \binom{M}{j}\binom{M-j}{i-j} \leq M\binom{M}{M/2}^2 = 2^{O(M)}$. Furthermore, c_i can be computed (exactly) by a deterministic Turing machine that uses $O(M)$ space by simply running through the loop over j, each time updating the current result by $(-1)^j\binom{M}{j}\binom{M-j}{i-j}$.

To obtain our polynomial p, define p as the difference between the threshold polynomial around $\lambda + \Delta$ and the threshold polynomial around $\lambda - \Delta$,

$$p(x) = p_{\lambda-\Delta}(x) - p_{\lambda+\Delta}(x)$$

where $M = 32(\beta - \alpha)^{-2}\log n$ and $\Delta = (\alpha + \beta)/2$. It is easy to check that

Lemma 4. $p(x) \geq 1 - n^{-2}$ *for every* x *that is* α*–close to* λ *(i.e.,* $|x - \lambda| < \alpha$*) and* $p(x) \leq n^{-2}$ *for every* x *that is* β*–far from* λ *(i.e.,* $|x - \lambda| \geq \beta$*).*

3.2 The Limitation of the Technique

In this subsection, we prove the accuracy of the above technique cannot be enhanced merely by choosing a different polynomial p. Approximating threshold functions by a polynomial is well-studied and well understood (see, for example, [13–15] and references therein). However, we need to adapt this work to our needs because we have an additional requirement that the magnitude of the polynomial's coefficients is small.

We start by formalizing the properties of p that were useful to us. We say that $\mathcal{P} = \{p_{\lambda,n}\}_{\lambda \in [0,1], n \in \mathbb{N}}$ is a family of polynomials if for every $\lambda \in [0,1]$ and $n \in \mathbb{N}$, $p_{\lambda,n}$ is a univariate polynomial with coefficients in $\mathbb{R}$.

Definition 3. *(Small family) Let* $\mathcal{P}$ *be a family of polynomials and fix* $\lambda \in [0,1]$. *For every* $n \in \mathbb{N}$, *write* $p_{\lambda,n}(x) = \sum_{i=0}^{\deg(p_{\lambda,n})} c_{\lambda,n,i} x^i$. *We say the family is* $s(n)$*–small if,*

- $\deg(p_{\lambda,n}) \leq 2^{s(n)}$,
- *For every* $0 \leq i \leq \deg(p_{\lambda,n})$, $|c_{\lambda,n,i}| \leq 2^{s(n)}$, *and*
- *There exists a deterministic Turing machine running in space* $s(n)$ *that outputs* $c_{\lambda,n,0}, \ldots, c_{\lambda,n,\deg(p_{\lambda,n})}$.

Definition 4. *(Distinguisher family) Let* $\mathcal{P}$ *be a family of polynomials and fix* $n \in \mathbb{N}$. *Given* $\alpha < \beta$ *in* $(0,1)$ *and* $\eta < 1/2$, *we say the family is* (α, β, η)*–distinguisher for* $\lambda \in [0,1]$ *if,*

– For every $x \in [0,1]$ that is α–close to λ, $p_{\lambda,n}(x) \in [1-\eta, 1]$, and
– For every $x \in [0,1]$ that is β–far from λ, $p_{\lambda,n}(x) \in [0,\eta]$.

Theorem 2. Let $\alpha, \beta, \lambda, \eta$ be such that $\alpha \leq \beta$, $\beta = o(1)$, $\eta = o(n^{-1})$ and $\lambda + \beta \leq \frac{1}{2}$. Then there is no (α, β, η)–distinguisher family for λ that is $O(\log n)$–small.

Proof. Assume there exists such a family $\{p_{\lambda,n}\}_{\lambda \in [0,1], n \in \mathbb{N}}$ with $s(n) = c' \log n$. We first show that without loss of generality p has logarithmic degree. Let $r_{\lambda,n}(x)$ be the residual error of truncating $p_{\lambda,n}(x)$ after $c \log n$ terms, for c that will soon be determined. Also, w.l.o.g., assume $x \in [0,1)$ is bounded away from 1. Then:

$$r_{\lambda,n}(x) \leq \sum_{i=c\log n+1}^{\deg(p_{\lambda,n})} |c_{\lambda,n,i}| \cdot x^i \leq n^{c'} \cdot \frac{x^{c\log n}}{1-x} \leq \frac{1}{1-x} n^{c'-c\log(1/x)}.$$

So, by taking $c = \lceil \frac{c'+2-\log(1-x)}{\log(1/x)} \rceil$ we obtain $r_{\lambda,n}(x) \leq n^{-2}$.

We now show that $O(\log n)$–degree polynomials cannot decay around λ fast enough. Assume to the contrary that there exists such a distinguisher family, so $|p_{\lambda,n}(x)| < n^{-1}$ for $x \in [\lambda+\beta, 1]$. The following lemma states that if a function has a small value on an interval, than it cannot be too large outside it. Namely,

Lemma 5. *[16, Theorem 2.9.11]* Let $T_n(x)$ be the Chebyshev polynomial (of the first kind) of degree n. Then, if the polynomial $P_n(x) = \sum_{i=0}^{n} c_i x^i$ satisfies the inequality $|P_n(x)| \leq L$ on the segment $[a,b]$ then at any point outside the segment we have

$$|P_n(x)| \leq L \cdot \left| T_n\left(\frac{2x-a-b}{b-a}\right) \right|.$$

For properties of the Chebyshev polynomials see [17, Chapter 1.1]. We mention a few properties that we use. An explicit representation of $T_n(x)$ is given by $T_n(x) = \frac{(x-\sqrt{x^2-1})^n + (x+\sqrt{x^2-1})^n}{2}$. $|T_n(-x)| = |T_n(x)|$ and T_n is monotonically increasing for $x > 1$. Also,

$$|T_n(1+\delta)| \leq \left(1+\delta+\sqrt{(1+\delta)^2-1}\right)^n \leq \left(1+4\sqrt{\delta}\right)^n \leq e^{4n\sqrt{\delta}} \leq 2^{8n\sqrt{\delta}} \quad (2)$$

for $0 \leq \delta \leq 1$. Then:

$$
\begin{aligned}
|p_{\lambda,n}(\lambda)| &\leq n^{-1} \cdot \left| T_{c\cdot\log n}\left(\frac{\lambda-\beta-1}{-\lambda-\beta+1}\right) \right| \\
&= n^{-1} \cdot \left| T_{c\cdot\log n}\left(1+\frac{2\beta}{1-\lambda-\beta}\right) \right| & \text{By } |T_n(x)| = |T_n(-x)| \\
&\leq n^{-1} \cdot |T_{c\cdot\log n}(1+4\beta)| & \text{By the monotonicity of } T_n(x) \text{ for } x > 1 \text{ and } \lambda+\beta \leq \tfrac{1}{2}
\end{aligned}
$$

By Equation (2) $|p_{\lambda,n}(\lambda)| \leq n^{-1} 2^{32c\sqrt{\beta}\log n} \leq n^{-1+32c\sqrt{\beta}}$. As $\beta = o(1)$ for n large enough we have $|p_{\lambda,n}(\lambda)| \leq n^{-1/2}$, contradicting the fact that $|p_{\lambda,n}(\lambda)| \geq 1 - n^{-1}$.

We note that for values very close to 1, polynomials of higher degrees are useful, and indeed better approximations are possible. In particular, one can separate a 1 eigenvalue from $1 - \frac{1}{n}$ by using the polynomial x^{n^2}.

References

1. Nisan, N.: Pseudorandom generators for space-bounded computation. Combinatorica **12**, 449–461 (1992)
2. Saks, M.E., Zhou, S.: $BP_H SPACE(S) \subseteq DSPACE(S^{3/2})$. J. Comput. Syst. Sci. **58** 376–403 (1999)
3. Reingold, O.: Undirected connectivity in log-space. J. ACM **55** (2008)
4. Aleliunas, R., Karp, R.M., Lipton, R., Lovasz, L., Rackoff, C.: Random walks, universal traversal sequences, and the complexity of maze problems. In: 20th Annual Symposium on Foundations of Computer Science, pp. 218–223 (1979)
5. Cook, S.A.: A taxonomy of problems with fast parallel algorithms. Information and Control **64** (1985); International Conference on Foundations of Computation Theory
6. Csansky, L.: Fast parallel matrix inversion algorithms. SIAM Journal of Computing **5**, 618–623 (1976)
7. Watrous, J.: Space-bounded quantum complexity. Journal of Computer and System Sciences **59**, 281–326 (1999)
8. van Melkebeek, D., Watson, T.: Time-space efficient simulations of quantum computations. Electronic Colloquium on Computational Complexity (ECCC) **17**, 147 (2010)
9. Ta-Shma, A.: Inverting well conditioned matrices in quantum logspace. In: Proceedings of the 45th Annual ACM Symposium on Symposium on Theory of Computing, STOC 2013, pp. 881–890. ACM, New York (2013)
10. Harrow, A.W., Hassidim, A., Lloyd, S.: Quantum algorithm for linear systems of equations. Phys. Rev. Lett. **103**, 150502 (2009)
11. Ben-Or, M., Eldar, L.: Optimal algorithms for linear algebra by quantum inspiration. CoRR abs/1312.3717 (2013)
12. Hoffman, K.: Banach Spaces of Analytic Functions. Dover Books on Mathematics Series. Dover Publications, Incorporated (2007)
13. Saff, E.B., Totik, V.: Polynomial approximation of piecewise analytic functions. Journal of the London Mathematical Society **s2-39**, 487–498 (1989)
14. Eremenko, A., Yuditskii, P.: Uniform approximation of sgn(x) by polynomials and entire functions. Journal d'Analyse Mathématique **101**, 313–324 (2007)
15. Diakonikolas, I., Gopalan, P., Jaiswal, R., Servedio, R.A., Viola, E.: Bounded independence fools halfspaces. SIAM Journal on Computing **39**, 3441–3462 (2010)
16. Timan, A.: Theory of Approximation of Functions of a Real Variable. Dover books on advanced mathematics. Pergamon Press (1963)
17. Rivlin, T.: The Chebyshev polynomials. Pure and applied mathematics. Wiley (1974)

A   Simulatable Matrices

Lemma 1. *The family of transition matrices of (directed or undirected) graphs is simulatable.*

Proof. Let $G = (V, E)$ be a graph with n vertices and let A be its transition matrix. Let $k \in \mathbb{N}$, $s, t \in [n]$ and $\delta, \varepsilon > 0$. Consider the algorithm that on input k, s, t, takes T independent random walks of length k over G starting at vertex

. The algorithm outputs the ratio of walks that reach vertex t. Let Y_i be the random value that is 1 if the i-th trial reached t and 0 otherwise. Then, for every , $\mathbb{E}[Y_i] = A^k[s,t]$. Also, $Y_1, \ldots, Y_T$ are independent. By Chernoff,

$$\Pr[|\frac{1}{T}\sum_{i=1}^{T} Y_i - A^k[s,t]| \geq \varepsilon] \leq 2e^{-2\varepsilon^2 T}$$

Taking $T = \mathrm{poly}(\varepsilon^{-1}, \log \delta^{-1})$, the error probability (i.e., getting an estimate that is ε far from the correct value) is at most δ. Altogether, the algorithm runs in pace $O(\log(Tnk|E|)) = O(\log(nk\varepsilon^{-1}) + \log\log \delta^{-1})$, assuming $|E| = \mathrm{poly}(n,k)$.

We say $\|A\|_\infty \leq c$ if for every $i \in [n]$, $\sum_j |A[i,j]| \leq c$. We show:

Lemma 2. *The family of real matrices with infinity norm at most* 1 *is simulatable.*

Proof. We prove the result to real matrices, with positive or negative entries, as long as they have bounded infinity norm. By generalizing the sign of an entry to its *phase*, the result easily applies to complex matrices as well.

Let A be a real matrix of dimension n such that $\|A\|_\infty \leq 1$. Let $d_i(A) = \sum_j |A[i,j]|$. Let $k \in \mathbb{N}$, $s,t \in [n]$ and $\delta, \varepsilon > 0$. Note that:

$$A^k[s,t] = \sum_{i_1=1}^{n}\sum_{i_2=1}^{n}\cdots\sum_{i_{k-1}=1}^{n} A[s,i_1] \cdot A[i_1,i_2]\cdot \ldots \cdot A[i_{k-1},t]$$

$$= \sum_{i_1=1}^{n}\sum_{i_2=1}^{n}\cdots\sum_{i_{k-1}=1}^{n} \frac{|A[s,i_1]|}{d_s(A)} \cdot \frac{|A[i_1,i_2]|}{d_{i_1}(A)} \cdot \ldots \cdot$$

$$\times \frac{|A[i_{k-1},t]|}{d_{i_{k-1}}(A)} \cdot p\left(A, \langle s, i_1, i_2, \ldots, i_{k-1}, t\rangle\right),$$

where

$$p\left(A, \langle s, i_1, i_2, \ldots, i_{k-1}, t\rangle\right) = \frac{d_s(A) \cdot d_{i_1}(A) \cdot \ldots \cdot d_{i_{k-1}}(A)}{\mathrm{sgn}\left(A[s,i_1] \cdot A[i_1,i_2] \cdot \ldots \cdot A[i_{k-1},i_t]\right)}.$$

Consider the algorithm that on input k, s, t, takes T independent random walks of length k over G starting from vertex s. Iterating over all random walks, the algorithm approximates $\frac{1}{T}\sum_i y(i)$, where $y(i) = p(A,i)$ if the walk i reached t, and 0 otherwise. Correspondingly, let Y_i be the random value that is $p(A,i)$ if the i'th walk reached t and 0 if it did not. Then,

$$\mathbb{E}[Y_i] = \sum_{i_1=1}^{n} \sum_{i_2=1}^{n} \cdots \sum_{i_{k-1}=1}^{n} A[s, i_1] \cdot A[i_1, i_2] \cdot \ldots \cdot A[i_{k-1}, t] \cdot p(A, \langle s, i_1, \ldots, i_{k-1}, t \rangle) = A^k[s, t]$$

Denote the algorithm's outcome by $M(k, s, t)$. As in Lemma 1, and using the fact that $|p(A, i)| \leq 1$, the algorithm can (ε, δ)–approximates $\mathbb{E}[Y_i]$ by choosing T which is $\text{poly}(\varepsilon^{-1}, \log \delta^{-1})$. Following the same analysis as of Lemma 1, the algorithm runs in $O(\log nk\varepsilon^{-1} + \log \log \delta^{-1})$ space. We conclude that A is simulatable.

On Planar Boolean CSP

Zdeněk Dvořák[✉] and Martin Kupec

Computer Science Institute, Charles University in Prague, Prague, Czech Republic
{rakdver,magon}@iuuk.mff.cuni.cz

Abstract. We give a partial classification of the complexity of Planar Boolean CSP, including a complete dichotomy for templates containing only relations of arity at most 5.

1 Introduction

The Constraint Satisfaction Problem (CSP) is a far-reaching generalization of many natural satisfaction and coloring problems. It is usually parameterized by its domain and template. A *domain* D is an arbitrary finite set. In this paper, we almost exclusively consider the *boolean* CSP, where the domain is the set $\{0, 1\}$. A *template* $\mathcal{T}$ is a finite set of *constraint types*; each constraint type $R \in \mathcal{T}$ is a relation over D of finite arity $a(R)$, i.e., R is a subset of $D^{a(R)}$.

Let $\mathcal{T}$ be a template. An *instance* A over $\mathcal{T}$ consists of a finite set $V(A)$ of *variables*, and a finite set $C(A)$ of *constraints*; each constraint $c \in C(A)$ is a tuple $(R, v_1, \ldots, v_{a(R)})$, where R is a constraint type belonging to $\mathcal{T}$ and $v_1, \ldots, v_{a(R)}$ are variables belonging to $V(A)$. The instance A is *satisfiable* if there exists a function $f : V(A) \to D$ such that for every constraint $(R, v_1, \ldots, v_{a(R)}) \in C(A)$, the tuple $(f(v_1), \ldots, f(v_{a(R)}))$ satisfies the relation R. The function f is called a *satisfying assignment*. For a template $\mathcal{T}$, the $\mathcal{T}$-*CSP* is the algorithmic problem of deciding whether an input instance over $\mathcal{T}$ is satisfiable.

One of the best known open questions in computer science is the CSP Dichotomy Conjecture of Feder and Vardi [9], stating that for every finite domain D and for every template $\mathcal{T}$ over D, the $\mathcal{T}$-CSP problem is either polynomial-time solvable, or NP-complete. Although there has been much progress recently [1–4], the dichotomy conjecture is still open. However, the complexity of $\mathcal{T}$-CSP has been characterized for many natural classes of templates $\mathcal{T}$. Most relevant to the topic of this paper is the celebrated result of Schaefer [16] which proved the dichotomy conjecture over the boolean domain. In a more modern language, we can state his result in the terms of polymorphisms.

For integers $a, k \geq 0$, a domain D, a function $f : D^k \to D$ and a-tuples $T_1 = (x_1^1, x_2^1, \ldots, x_a^1)$, ..., $T_k = (x_1^k, x_2^k, \ldots, x_a^k)$ of elements of D, let $f(T_1, \ldots, T_k)$ denote the a-tuple

$$(f(x_1^1, x_1^2, \ldots, x_1^k), f(x_2^1, x_2^2, \ldots, x_2^k), \ldots, f(x_a^1, x_a^2, \ldots, x_a^k)).$$

Z. Dvořák—Supported by project GA14-19503S (Graph coloring and structure) of Czech Science Foundation.

© Springer-Verlag Berlin Heidelberg 2015
M.M. Halldórsson et al. (Eds.): ICALP 2015, Part I, LNCS 9134, pp. 432–443, 2015.
DOI: 10.1007/978-3-662-47672-7_35

A function $f : D^k \to D$ is a *polymorphism* of a relation $R \subseteq D^a$ if all $T_1, \ldots, T_k \in$ R satisfy $f(T_1, \ldots, T_k) \in R$. We say that f is a *polymorphism* of a template $\mathcal{T}$ i. it is a polymorphism of R for every $R \in \mathcal{T}$. Let us define several special functions that often appear as polymorphisms:

- ZERO : $\{0,1\} \to \{0,1\}$ given by $\text{ZERO}(x) = 0$ for every $x \in \{0,1\}$,
- ONE : $\{0,1\} \to \{0,1\}$ given by $\text{ONE}(x) = 1$ for every $x \in \{0,1\}$,
- AND : $\{0,1\}^2 \to \{0,1\}$ given by $\text{AND}(x,y) = xy$ for every $x,y \in \{0,1\}$,
- OR : $\{0,1\}^2 \to \{0,1\}$ given by $\text{OR}(x,y) = x + y - xy$ for every $x,y \in \{0,1\}$
- XOR_3 : $\{0,1\}^3 \to \{0,1\}$ given by $\text{XOR}_3(x,y,z) = (x + y + z) \bmod 2$ for every $x,y,z \in \{0,1\}$,
- MAJ_3 : $\{0,1\}^3 \to \{0,1\}$ given by $\text{MAJ}_3(x,x,y) = \text{MAJ}_3(x,y,x) = \text{MAJ}_3(y,x,x) = x$ for every $x,y \in \{0,1\}$, and
- NOT : $\{0,1\} \to \{0,1\}$ given by $\text{NOT}(x) = 1 - x$ for every $x \in \{0,1\}$.

We say that a template $\mathcal{T}$ over boolean domain is *Schaefer-easy* if ZERO, ONE, AND, OR, XOR_3, or MAJ_3 is a polymorphism of $\mathcal{T}$.

Theorem 1 (Schaefer [16]). *Let $\mathcal{T}$ be a template over boolean domain. If $\mathcal{T}$ is Schaefer-easy, then $\mathcal{T}$-CSP is polynomial-time solvable, otherwise $\mathcal{T}$-CSP is NP-complete.*

Suppose that $\mathcal{T}$ is not Schaefer-easy. We are interested in restrictions that can be imposed on the input instances of $\mathcal{T}$-CSP that make the problem polynomial-time solvable. A natural way is to restrict the *incidence graph* of the instance A, that is, the bipartite multigraph with vertex set $V(A) \cup C(A)$ and with edges cv_i for every $c = (R, v_1, \ldots, v_{a(R)}) \in C(A)$ and $i = 1, \ldots, a(R)$. We study the planar variant of the boolean constraint satisfaction problem, where the incidence graph of the input instance is required to be planar. It is well-known that there can be a difference in the complexities; for instance, letting NAE = $\{0,1\}^3 \setminus \{(0,0,0), (1,1,1)\}$, note that {NAE}-CSP is NP-complete by Theorem 1, but it can be solved in polynomial time when restricted to instances whose incidence graph is planar [14].

Additionally, in this paper we require the order of the edges around the constraint vertices in the plane drawing of the incidence graph to respect the order of the arguments of the corresponding constraint. Note that the variant without this additional restriction can be modelled: simply replace each constraint type R by all constraint types obtained from R by permuting the order of the inputs. Hence, this choice leads to a finer classification.

Let us remark that dealing only with the boolean domain is natural, as it avoids a number of difficulties not encountered in the non-planar case; for example, planar $\{\neq\}$-CSP over 4-element domain is polynomial-time solvable, but for highly non-trivial reasons (Four Color Theorem). Let us also point out somewhat related works of Cai et al. [5] and Guo and Williams [11], who considered the counting version of the problem (restricted to symmetric constraints).

Let us now describe the considered problem formally. Rather than working with the incidence graph, we equivalently define the problem in the terms of a elated plane graph where constraints correspond to faces (this simplifies some of the transformations that we describe later). Throughout the paper, graphs re allowed to have loops, possibly several at a single vertex, and parallel edges. For a connected plane graph G, let $F(G)$ denote the set of faces of G, let $f_o(G)$ enote the outer face of G, and let $F'(G) = F(G)\setminus\{f_o(G)\}$. For a face $f \in F(G)$, et $b(f)$ denote the closed walk bounding f, enumerated in the clockwise order round f. A *plane instance* over a template T is an instance A over T together ith a connected plane graph G with vertex set $V(A)$, and an injective (but not ecessarily surjective) function $\varphi : C(A) \to F(G)$, such that every constraint $= (R, v_1, \ldots, v_{a(R)}) \in C(A)$ satisfies $b(\varphi(c)) = v_1 v_2 \ldots v_{a(R)} v_1$. The *planar T-CSP* is the algorithmic problem of deciding whether for an input plane instance $A, G, \varphi)$, the instance A is satisfiable. Let us remark that given an instance A, it s possible to decide in linear time whether there exists a plane instance (A, G, φ), y a straightforward dynamic programming algorithm over the SPQR-tree [13] f the incidence graph of A; hence, including the plane representation of A in he input does not affect the complexity of the problem.

In order to partially classify the complexity of planar T-CSP, we need several more definitions. A relation or a template over the boolean domain is *self-complementary* if NOT is its polymorphism. Let $\oplus$ denote addition (or equivalently, subtraction) modulo 2. Consider an a-tuple $T = (x_1, \ldots, x_a) \in \{0,1\}^a$. By dT, denote the a-tuple $(x_1 \oplus x_2, x_2 \oplus x_3, \ldots, x_{a-1} \oplus x_a, x_a \oplus x_1)$. Let $R \subseteq \{0,1\}^a$ be a self-complementary relation. Let $dR = \{dT : T \in R\}$. Note that since R is self-complementary, $|dR| = |R|/2$ and dR uniquely determines R. Furthermore, every element of dR has even number of entries equal to 1. For a self-complementary emplate T, let $dT = \{dR : R \in T\}$.

For two a-tuples $T = (x_1, \ldots, x_a) \in \{0,1\}^a$ and $T' = (x'_1, \ldots, x'_a) \in \{0,1\}^a$, et $T \oplus T' = (x_1 \oplus x'_1, x_2 \oplus x'_2, \ldots, x_a \oplus x'_a)$. Let $I(T) = \{i \in \{1, \ldots, a\} : x_i = 1\}$. Let $e_{a,i}$ denote the a-tuple with $I(e_{a,i}) = \{i\}$. A set $S \subseteq \{0,1\}^a$ is an *even Δ-matroid* if

- $|I(T)|$ has the same parity for every $T \in S$, and
- for every $T_1, T_2 \in S$ and for every $i \in I(T_1 \oplus T_2)$, there exists $j \in I(T_1 \oplus T_2) \setminus \{i\}$ such that $T_1 \oplus e_{a,i} \oplus e_{a,j} \in S$.

An instance A over a template T is *binary* if each variable appears exactly twice in the constraints. A *plane-binary instance* consists of a binary instance A, a connected plane graph G and a bijective function $\varphi : C(A) \to F(G)$, such that $V(A) = E(G)$ and every $c = (R, e_1, \ldots, e_{a(R)}) \in C(A)$ satisfies $b(\varphi(c)) = e_1 e_2 \ldots e_{a(R)}$. Note the key distinction between plane instances and plane-binary instances: in the former, the variables correspond to the vertices of the graph, while in the latter, they correspond to the edges. For a template T, we define the binary T-CSP and planar-binary T-CSP algorithmic problems in the natural way.

For any integer $a \geq 1$, let $\text{EVEN}_a = \{(x_1, \ldots, x_a) \in \{0,1\}^a : x_1 + \ldots + x_a \text{ is even}\}$. Let $\text{EVENS} = \{\text{EVEN}_1, \text{EVEN}_2, \text{EVEN}_3\}$.

Let T be a template over the boolean domain. As the main result of this paper, we give the following partial classification of planar T-CSP:

- If T is Schaefer-easy, then (planar) T-CSP is polynomial-time solvable by Theorem 1.
- If T is neither Schaefer-easy nor self-complementary, then planar T-CSP is NP-complete (see Section 2)
- If T is self-complementary, then planar T-CSP is polynomially equivalent to planar-binary $(dT \cup \text{EVENS})$-CSP (see Section 3).
  - If additionally there exists $R \in T$ such that dR is not an even Δ-matroid and T is not Schaefer-easy, then planar T-CSP is NP-complete (see Section 4).
  - Otherwise, the complexity of planar T-CSP is open, except for the polynomial-time solvable cases discussed below.

Note that we know no template T' over the boolean domain such that every $R' \in T'$ is an even Δ-matroid and binary T'-CSP is NP-complete (even without the assumption of planarity). Hence, it is plausible that in the last case, planar T-CSP is always polynomial-time solvable.

The motivation for the even Δ-matroid restriction comes from the study of the complexity of binary T-CSP by Feder [8], who showed that the complexities of T-CSP and binary T-CSP coincide unless R is a Δ-matroid for every $R \in T$ (the Δ-matroids do not have to be even—we omit the definition of a general Δ-matroid, since it is not relevant to our study). If R is a Δ-matroid for every $R \in T$, then binary T-CSP becomes a special case of the parity problem in Δ-matroids. A number of additional restrictions ensuring a polynomial-time algorithm were identified before—this is the case if all the Δ-matroids $R \in T$ are compact [12], or all are co-independent [8], or all are local [6], or all are linear [10] (see the respective papers for the definitions of these classes of Δ-matroids). Ultimately, though, the full classification of the complexity of binary boolean CSP is still an open problem.

Let us point out another case solvable in polynomial time. Let G be a graph and let $v_1, \ldots, v_a \in V(G)$ be pairwise distinct vertices of G. For an a-tuple $T = (x_1, \ldots, x_a) \in \{0,1\}^a$, let $G_T = G - \{v_i : 1 \le i \le a, x_i = 1\}$, and let

$$M(G, v_1, \ldots, v_a) = \{T \in \{0,1\}^a : G_T \text{ has a perfect matching}\}.$$

We say that a relation $R \subseteq \{0,1\}^a$ is *matching-realizable* if $R = M(G, v_1, \ldots, v_a)$ for some (not necessarily planar) graph G and some pairwise distinct vertices $v_1, \ldots, v_a \in V(G)$. By considering alternating paths in pairs of matchings, it is easy to see that every matching-realizable set is an even Δ-matroid. We say that a template T over the boolean domain is *matching-realizable* if R is matching-realizable for every $R \in T$.

For a matching-realizable template T, the (not necessarily planar) binary T-CSP reduces to testing whether a graph has a perfect matching (for each constraint in the instance, add a copy of the graph showing that its constraint type is matching-realizable, and join by edges the vertices corresponding to the same ariable). This can be decided in polynomial time [7]. Since EVEN_a is matching-realizable for every $a \geq 1$ (by a clique on a or $a + 1$ vertices, whichever is even), we have the following consequence of the last case of the partial classification.

Corollary 1. *If $\mathcal{T}$ is a self-complementary template and $d\mathcal{T}$ is matching-realizable, then planar $\mathcal{T}$-CSP is polynomial-time solvable.*

For example, let G_{NAE} be the graph with vertex set $\{w, v_1, v_2, v_3\}$ and edges $v_i w$ for $i = 1, 2, 3$, then $d\text{NAE} = \{(0, 1, 1), (1, 0, 1), (1, 1, 0)\} = M(G_{\text{NAE}}, v_1, v_2, v_3)$. Hence, the corollary implies the well-known result that planar $\{\text{NAE}\}$-CSP can be solved in polynomial time. More importantly, in Section 5, we show that all even Δ-matroids of arity at most 5 are matching-realizable. This implies the following corollary, which completes the classification of the complexity of planar $\mathcal{T}$-CSP over boolean domain under the assumption that all relations in $\mathcal{T}$ have arity at most 5.

Corollary 2. *Suppose that $\mathcal{T}$ is a self-complementary template such that dR is an even Δ-matroid for every $R \in \mathcal{T}$. If $a(R) \leq 5$ for every $R \in \mathcal{T}$, then planar $\mathcal{T}$-CSP is polynomial-time solvable.*

The rest of paper is structured as follows. In Section 2, we study the templates that are neither Schaefer-easy nor self-complementary, and show the NP-hardness of the planar CSP for such templates. In Section 3, we consider the templates $\mathcal{T}$ that are self-complementary, and establish the polynomial-time equivalence with planar-binary CSP for the template $(d\mathcal{T} \cup \text{EVENS})$. In Section 4, we show that for a self-complementary, non-Schaefer easy template $\mathcal{T}$ such that not all relations in $d\mathcal{T}$ are even Δ-matroids, the planar $\mathcal{T}$-CSP is NP-complete. Section 5 is devoted to the study of matching-realizable sets, establishing Corollary 2.

2   Non-Self-Complementary Templates

Let $\mathcal{T}$ be a template over the binary domain, let (A, G, φ) be a plane instance over $\mathcal{T}$ such that $\varphi^{-1}(f_o(G)) = \emptyset$, and let $b(f_o(G)) = v_a v_{a-1} \ldots v_1 v_a$ (where the vertices $v_1, \ldots, v_a$ need not be pairwise distinct). Let $R(A, G, \varphi) \subseteq \{0, 1\}^a$ consist of the a-tuples $(x_1, \ldots, x_a) \in \{0, 1\}^a$ such that there exists a satisfying assignment $f : V(A) \rightarrow \{0, 1\}$ with $f(v_i) = x_i$ for $i = 1, \ldots, a$. If a relation $R \subseteq \{0, 1\}^a$ is equal to $R(A, G, \varphi)$ for some plane instance (A, G, φ) over $\mathcal{T}$, then we say that R is *planarly $\mathcal{T}$-expressible*.

The following observations are standard.

Lemma 1. *Let $\mathcal{T}$ be a template over the boolean domain and let $R \subseteq \{0, 1\}^a$ and $R' \subseteq \{0, 1\}^b$ be arbitrary relations. Suppose that R is planarly $\mathcal{T}$-expressible. Then*

— *every polymorphism of $\mathcal{T}$ is also a polymorphism of R,*

— *planar $\mathcal{T}$-CSP is polynomially equivalent to planar $(\mathcal{T} \cup \{R\})$-CSP, and*

— *if R' is planarly $\mathcal{T} \cup \{R\}$-expressible, then R' is also planarly $\mathcal{T}$-expressible.*

Let us remark that without the planarity restriction, the expressibility is exactly characterized by polymorphisms. However, planar expressibility and especially non-expressibility seem much harder to demonstrate.

Let C_0, C_1 and N be the relations defined as $C_0 = \{(0)\}$, $C_1 = \{(1)\}$, and $N = \{(0,1),(1,0)\}$. Let $R \subseteq \{0,1\}^a$ be a relation and let I be a subset of $\{1,\ldots,a\}$, $I = \{i_1,\ldots,i_b\}$ with $i_1 < i_2 < \ldots < i_b$. The *projection* of R to coordinates I is the set $R[I] \subseteq \{0,1\}^b$ such that $(x'_1,\ldots,x'_b) \in R[I]$ if and only if there exists some $(x_1,\ldots,x_a) \in R$ satisfying $x'_1 = x_{i_1}, \ldots, x'_b = x_{i_b}$. For $v = 0,1$, the *fixation* $R[I \to v]$ denotes the relation such that $T = (x_1,\ldots,x_a) \in R[I \to v]$ if and only if $T \in R$ and $x_i = v$ for every $i \in I$. Let $R[I \xrightarrow{\cdot} v]$ denote the relation $R[I \to v][\{1,\ldots,a\} \setminus I]$. The *twist* $R \odot I$ of R is the relation $\{T \oplus \bigoplus_{i \in I} e_{a,i} : T \in R\}$. For $i \in \{1,\ldots,a\}$, the *=-restriction* of R at i is the relation $R[i = i+1] \subseteq R$ such that $T = (x_1,\ldots,x_a) \in R[i = i+1]$ if and only if $T \in R$ and $x_i = x_{i+1}$ (where $x_{a+1} = x_1$), and the *≠-restriction* of R at i is the set $R[i \neq i+1] \subseteq R$ such that $T = (x_1,\ldots,x_a) \in R[i \neq i+1]$ if and only if $T \in R$ and $x_i \neq x_{i+1}$.

Lemma 2. *Let $\mathcal{T}$ be a template over the boolean domain. Suppose that a relation $R \subseteq \{0,1\}^a$ is planarly $\mathcal{T}$-expressible.*

1. *All projections and =-restrictions of R are planarly $\mathcal{T}$-expressible.*
2. *If N is planarly $\mathcal{T}$-expressible, then all ≠-restrictions and twists of R are planarly $\mathcal{T}$-expressible.*
3. *If C_0 and C_1 are planarly $\mathcal{T}$-expressible, then all fixations of R are planarly $\mathcal{T}$-expressible.*

For a relation $R \subseteq \{0,1\}^a$, let $\|R\| = (a,|R|)$. Let us start with a key expressibility result.

Lemma 3. *Let $\mathcal{T}$ be a template over the boolean domain that is not Schaefer-easy. Then N is planarly $\mathcal{T}$-expressible, and if $\mathcal{T}$ is not self-complementary, then C_0 and C_1 are planarly $\mathcal{T}$-expressible.*

Proof. Let $R \subseteq \{0,1\}^a$ be a non-empty planarly $\mathcal{T}$-expressible relation such that ZERO is not a polymorphism of R and subject to these conditions, $\|R\|$ is lexicographically minimal. Note that ZERO is a polymorphism of a non-empty set if and only if the set contains the zero tuple $(0,\ldots,0)$. By the minimality, ZERO is a polymorphism of every projection of R to $a-1$ coordinates. Hence, $e_{a,1}, e_{a,2}, \ldots, e_{a,a} \in R$. If $a \geq 3$, then let $R' = R[1 = 2]$. Note that $e_{a,3} \in R'$ and $e_{a,1} \notin R'$, and since ZERO is not a polymorphism of R', we obtain a contradiction with the minimality of R. Hence, $a \leq 2$. If $a = 2$, then the same argument shows that $(1,1) \notin R$, and thus $R = N$. If $a = 1$, then $R = C_1$. Hence,

either N or C_1 is planarly $\mathcal{T}$-expressible, and symmetrically either N or C_0 is planarly $\mathcal{T}$-expressible.

$$\tag{1}$$

Suppose now that C_0 and C_1 are planarly $\mathcal{T}$-expressible. Let $R \subseteq \{0,1\}^a$ be a non-empty planarly $\mathcal{T}$-expressible relation such that AND is not a polymorphism of R and subject to these conditions, $\|R\|$ is lexicographically minimal. Let $T = x_1, \ldots, x_a) \in R$ and $T' = (x_1', \ldots, x_a') \in R$ be a-tuples such that $\text{AND}(T, T') \notin R$. If there exists $i \in \{1, \ldots, a\}$ such that $x_i = x_i'$, then let $R' = R[\{i\} \xrightarrow{} x_i]$. Then AND is not a polymorphism of R' and we obtain a contradiction with the minimality of R. Hence, $x_i \neq x_i'$ for $i = 1, \ldots, a$ for all $T, T' \in R$ such that $\text{AND}(T, T') \notin R$. Since AND is a polymorphism of every projection of R to $a - 1$ coordinates, any such projection contains a projection of $\text{AND}(T, T') = 0, \ldots, 0)$. Hence $e_{a,1}, e_{a,2}, \ldots, e_{a,a} \in R$. Since AND is not a polymorphism of R, we have $a \geq 2$. Note that $\text{AND}(e_{a,1}, e_{a,a}) = (0, \ldots, 0) \notin R$, and as we observed before, $e_{a,1}$ and $e_{a,a}$ differ in all coordinates. Therefore, $a = 2$. We conclude that $N \subseteq R \subseteq N \cup \{(1,1)\}$.

Similarly, since OR is not a polymorphism of $\mathcal{T}$, it follows that some set R' such that $N \subseteq R' \subseteq N \cup \{(0,0)\}$ is planarly $\mathcal{T}$-expressible. Let A be the instance containing constraints (R, v_1, v_2) and (R', v_1, v_2), let G be the graph consisting of three edges between vertices v_1 and v_2, and let φ be the function mapping the elements of $C(A)$ to the two non-outer faces of G. Then (A, G, φ) is a plane instance showing that N is planarly $\{R, R'\}$-expressible, and by Lemma 1, N is planarly $\mathcal{T}$-expressible. Hence,

if C_0 and C_1 are planarly $\mathcal{T}$-expressible, then N is planarly $\mathcal{T}$-expressible.

$$(2)$$

Let us now consider the case that N is planarly $\mathcal{T}$-expressible and that $\mathcal{T}$ is not self-complementary. Let $R \subseteq \{0,1\}^a$ be a non-empty planarly $\mathcal{T}$-expressible relation such that NOT is not a polymorphism of R and subject to these conditions, $\|R\|$ is lexicographically minimal. Let $T = (x_1, \ldots, x_a) \in R$ be such that $\text{NOT}(T) \notin R$. If $a \geq 2$, then either $x_1 = x_2$ or $x_1 \neq x_2$. In the former case, let $R' = R[1 = 2][\{2, \ldots, a\}]$. In the latter case, let $R' = R[1 \neq 2][\{2, \ldots, a\}]$. In both cases, NOT is not a polymorphism of R', which contradicts the minimality of R. Hence, $a = 1$, and thus $R = C_0$ or $R = C_1$. By Lemma 2, $R \odot \{1\}$ is also planarly $\mathcal{T}$-expressible. Hence,

if N is planarly $\mathcal{T}$-expressible and $\mathcal{T}$ is not self-complementary, then C_0 and C_1 are planarly $\mathcal{T}$-expressible.

$$(3)$$

If $\mathcal{T}$ is self-complementary, then every planarly $\mathcal{T}$-expressible relation is self-complementary by Lemma 1, and thus neither C_0 nor C_1 is planarly $\mathcal{T}$-expressible. By (1), N is planarly $\mathcal{T}$-expressible.

If $\mathcal{T}$ is not self-complementary, then (1) implies that either N or both C_0 and C_1 are planarly $\mathcal{T}$-expressible. In the latter case, (2) implies that N also is planarly $\mathcal{T}$-expressible. Hence, C_0 and C_1 are planarly $\mathcal{T}$-expressible by (3). □

Using Lemmas 2 and 3 as well as ideas similar to the proof of Lemma 3, we prove the following. Let 1-IN-3 $= \{(1,0,0), (0,1,0), (0,0,1)\}$.

Lemma 4. *Let $\mathcal{T}$ be a template over the boolean domain that is not Schaefer-easy. If $\mathcal{T}$ is not self-complementary, then 1-IN-3 is planarly $\mathcal{T}$-expressible.*

Note that planar {1-IN-3}-CSP is known to be NP-complete [15]. Hence, we obtain the main result of this section.

Corollary 3. *If T is a template over the boolean domain that is neither Schaefer-easy nor self-complementary, then planar T-CSP is NP-complete.*

3   Self-Complementary Templates and Binary CSP

We now aim to prove the equivalence between planar T-CSP and planar-binary $(dT \cup \text{EVENS})$-CSP for any self-complementary template T.

Let T be a template over the binary domain. A *partial plane-binary instance* over T consists of an instance B over T, a connected plane graph H such that the outer face of H is not incident with a bridge, and a bijective function $\theta : C(B) \to F'(H)$, such that $V(B) = E(H)$ and every $c = (R, e_1, \ldots, e_{a(R)}) \in C(B)$ satisfies $b(\theta(c)) = e_1 e_2 \ldots e_{a(R)}$. Let $b(f_o(H)) = e_a e_{a-1} \ldots e_1$. Note that the variables e_1, $\ldots$, e_a appear exactly once in B, while all other variables appear exactly twice. Let $R_b(B, H, \theta) \subseteq \{0,1\}^a$ consist of all a-tuples $(x_1, \ldots, x_a) \in \{0,1\}^a$ such that there exists a satisfying assignment $f : V(B) \to \{0,1\}$ with $f(e_i) = x_i$ for $i = 1, \ldots, a$. If a relation $R \subseteq \{0,1\}^a$ is equal to $R_b(B, H, \theta)$ for some partial plane-binary instance (B, H, θ) over T, then we say that R is *plane-binary T-expressible*.

We say that a plane instance (A, G, φ) is *near-total* when $\varphi^{-1}(f) = \emptyset$ if and only if f is the outer face of G, and the outer face of G is not incident with a bridge. Let T be a self-complementary template. Let (A, G, φ) be a near-total plane instance over T. Let (B, G, θ) be a partial plane-binary instance over dT defined as follows. We set $V(B) = E(G)$. For each constraint $c = (R, v_1, \ldots, v_{a(R)}) \in C(A)$, letting $e_1 e_2 \ldots e_{a(R)}$ be the walk bounding the face $\varphi(c)$, where $e_1 = v_1 v_2$, $e_2 = v_2 v_3$, $\ldots$, $e_{a(R)} = v_{a(R)} v_1$, we add the constraint $c' = (dR, e_1, \ldots, e_{a(R)})$ to $C(B)$ and set $\theta(c') = \varphi(c)$. We say that the partial plane-binary instance (B, G, θ) is *derived from* (A, G, φ).

Lemma 5. *Let T be a self-complementary template. Let (A, G, φ) be a near-total plane instance over T, and let (B, G, θ) be the partial plane-binary instance over dT derived from it. Then $R_b(B, G, \theta) = dR(A, G, \varphi)$.*

For any $i \geq 1$, let $\text{ALL}_i = \{0,1\}^i$, and note that $d\text{ALL}_i = \text{EVEN}_i$. Let $\text{ALLS} = \{\text{ALL}_1, \text{ALL}_2, \text{ALL}_3\}$.

Corollary 4. *Let T be a self-complementary template. A relation $R \subseteq \{0,1\}^a$ is planarly T-expressible if and only if the relation dR is plane-binary $(dT \cup EVENS)$-expressible.*

This gives the sought correspondence with planar-binary CSP.

Theorem 2. *For any self-complementary template T, planar T-CSP and planar-binary $(dT \cup EVENS)$-CSP are polynomially equivalent.*

Proof. Let (A, G, φ) be any planar instance over $\mathcal{T}$. Replace an arbitrary edge v of G by two parallel edges, add a loop e at v to the resulting 2-face, and make the face bounded by the loop e the outer face of G. Let $R = R(A, G, \varphi)$. Let $B, H, \theta)$ be the partial plane-binary instance over $d\mathcal{T} \cup \text{EVENS}$ such that $dR = R_b(B, H, \theta)$, which we constructed in Corollary 4 (where $|B| + |V(H)| + |E(H)|$ is bounded by a polynomial in $|A| + |V(G)| + |E(G)|$). Note that A is satisfiable if and only if $R \neq \emptyset$, which is in turn equivalent to $dR \neq \emptyset$ and B being satisfiable.

Similarly, we can transform any plane-binary instance (B, H, θ) over $(d\mathcal{T} \cup \text{EVENS})$ to an equivalent plane instance over $\mathcal{T}$: replace any edge of H by a double edge and make the resulting 2-face into the outer one, thus obtaining a partial plane-binary instance (B', H', θ') over $d\mathcal{T} \cup \text{EVENS}$ such that $R_b(B', H', \theta') \neq \emptyset$ if and only if B is satisfiable, and apply Corollary 4 in the opposite direction. □

4   NP-Hardness of the Non-Δ-Matroid Case

Let $\text{CROSS} = \{(0,0,0,0), (0,1,0,1), (1,0,1,0), (1,1,1,1)\}$. Let $\mathcal{T}$ be a template over the boolean domain. Note that given any instance over $\mathcal{T}$ with non-planar incidence graph, CROSS can be used to replace the crossings and to obtain an equivalent plane instance over $\mathcal{T} \cup \{\text{CROSS}\}$. In conjunction with Lemma 1, we have the following.

Observation 1. *Let $\mathcal{T}$ be a template over the boolean domain. If CROSS is planarly $\mathcal{T}$-expressible, then $\mathcal{T}$-CSP and planar $\mathcal{T}$-CSP are polynomially equivalent.*

We say that a self-complementary relation $R' \subseteq \{0,1\}^4$ is a *standardized inconsistent relation* if $\text{CROSS} \subseteq R'$ and every $(s_1, s_2, s_3, s_4) \in R'$ satisfies $s_1 = s_2$ or $s_1 = s_3 \neq s_2 = s_4$.

Lemma 6. *Let $\mathcal{T}$ be a self-complementary template such that N is planarly $\mathcal{T}$-expressible. Suppose that there exists a relation $R' \subseteq \{0,1\}^a$ such that dR' is not an even Δ-matroid and R' is planarly $\mathcal{T}$-expressible. Then there exists a standardized inconsistent relation R that is planarly $\mathcal{T}$-expressible.*

Next, we consider the sets that can be expressed from a standardized inconsistent relation. Let $\text{NEAR-CROSS} = \text{CROSS} \cup \{(0,0,0,1), (1,1,1,0)\}$. Given a standardized inconsistent relation R and the relation N, the plane instance depicted in Figure 1(a) shows that either CROSS or NEAR-CROSS is planarly $\{R, N\}$-expressible. Furthermore, the plane instance depicted in Figure 1(b) shows that CROSS is planarly $\{\text{NEAR-CROSS}\}$-expressible. Using these facts, we can now describe exactly when CROSS is planarly $\mathcal{T}$-expressible for a self-complementary template $\mathcal{T}$ that is not Schaefer-easy.

Lemma 7. *Let $\mathcal{T}$ be a self-complementary template such that N is planarly $\mathcal{T}$-expressible. Then CROSS is planarly $\mathcal{T}$-expressible if and only if dR is not an even Δ-matroid for some $R \in \mathcal{T}$.*

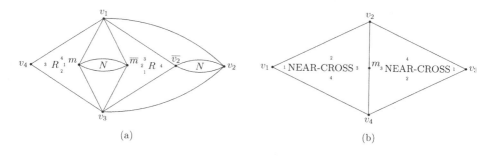

Fig. 1. Expressing CROSS from a standardized inconsistent relation

We can now easily prove the rest of the NP-hardness part of our partial characterization.

Theorem 3. *Let* T *be a self-complementary template that is not Schaefer-easy, and suppose that there exists* $R \in T$ *such that* dR *is not an even* Δ-*matroid. Then planar* T-*CSP is NP-complete.*

Proof. By Lemma 3, N is planarly T-expressible. By Lemma 7, CROSS is planarly T-expressible. Therefore, Observation 1 and Theorem 1 imply that planar T-CSP is NP-complete. ☐

5 Matching-Realizability of Even Δ-Matroids of Arity at Most 5

In this section, we prove that all even Δ-matroids of arity at most 5 are matching-realizable, which finishes the classification of the complexity of plane boolean CSP restricted to arity at most 5.

Let π be a permutation of $\{1, \ldots, a\}$. For $T = (t_1, \ldots, t_a) \in D^a$, let $\pi \circ T = (t_{\pi(1)}, t_{\pi(2)}, \ldots, t_{\pi(a)})$, and for a relation $R \subseteq \{0,1\}^a$, let $\pi \circ R$ denote the set $\{\pi \circ T : T \in R\}$.

Lemma 8. *Let* $R \subseteq \{0,1\}^a$ *be an even* Δ-*matroid.*

(a) *The twist* R' *of* R *at any subset* I *of* $\{1, \ldots, a\}$ *is an even* Δ-*matroid. Furthermore, if* R *is matching-realizable, then* R' *is matching-realizable.*

(b) *For any permutation* π *of* $\{1, \ldots, a\}$, *the set* $\pi \circ R$ *is an even* Δ-*matroid. Furthermore, if* R *is matching-realizable, then* $\pi \circ R$ *is matching-realizable.*

We say that sets $R, R' \subseteq \{0,1\}^a$ are *similar* if $R' = \pi \circ R''$ for a twist R'' of R and some permutation π. Note that R is an even Δ-matroid if and only if R' is an even Δ-matroid.

Theorem 4. *Let* $a \leq 5$ *be a positive integer and let* $R \subseteq \{0,1\}^a$ *be a relation. Then* R *is an even* Δ-*matroid if and only if* R *is matching-realizable.*

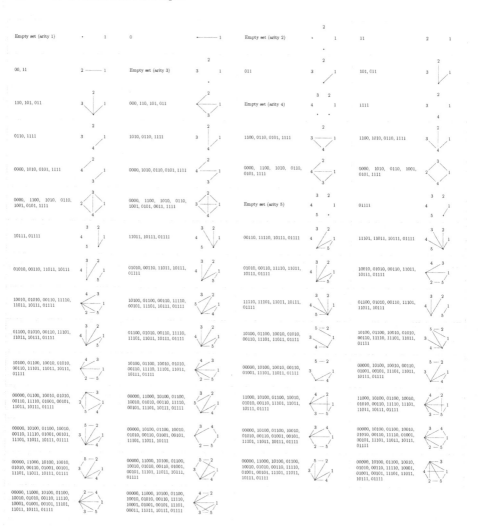

Fig. 2. Even Δ-matroids and graphs by that they are matching-realized

Proof. For every even Δ-matroid R, there exists a similar relation R' such that all elements of R' have even number of entries equal to 1. There are $2^{2^{a-1}} \leq 65536$ subsets of $\{0,1\}^a$ whose elements have even number of entries equal to 1. Using the program which can be found at http://atrey.karlin.mff.cuni.cz/~rakdver/ consistent.c, we enumerated all such subsets, and checked that every subset of $\{0,1\}^a$ that is an even Δ-matroid is similar to one of the sets listed in Figure 2. For each set R' listed in the figure, we also show a graph G and its vertices $1, \ldots, a$ such that $R' = M(G, 1, \ldots, a)$. By Lemma 8, we conclude that R is matching-realizable. $\qquad\square$

Acknowledgments. We would like to thank Victor Dalmau for useful discussions leading to the connection with binary CSP and with Δ-matroids.

References

1. Barto, L., Kozik, M.: Constraint satisfaction problems of bounded width. In: Proceedings of the 2009 50th Annual IEEE Symposium on Foundations of Computer Science, pp. 595–603. IEEE Computer Society, Washington, DC (2009)
2. Barto, L., Kozik, M., Niven, T.: The CSP dichotomy holds for digraphs with no sources and no sinks (a positive answer to a conjecture of Bang-Jensen and Hell). SIAM Journal on Computing **38**(5), 1782–1802 (2009)
3. Bulatov, A.: A dichotomy theorem for constraint satisfaction problems on a 3-element set. Journal of the ACM (JACM) **53**(1), 66–120 (2006)
4. Bulatov, A., Jeavons, P., Krokhin, A.: Classifying the complexity of constraints using finite algebras. SIAM Journal on Computing **34**(3), 720–742 (2005)
5. Cai, J.Y., Lu, P., Xia, M.: Holographic algorithms with matchgates capture precisely tractable planar #CSP. In: Proceedings of the 2010 51st Annual IEEE Symposium on Foundations of Computer Science, pp. 427–436. IEEE Computer Society, Washington, DC (2010)
6. Dalmau, V., Ford, D.K.: Generalized satisfiability with limited occurrences per variable: a study through delta-matroid parity. In: Rovan, B., Vojtáš, P. (eds.) MFCS 2003. LNCS, vol. 2747, pp. 358–367. Springer, Heidelberg (2003)
7. Edmonds, J.: Paths, trees, and flowers. Canad. J. Math. **17**, 449–467 (1965)
8. Feder, T.: Fanout limitations on constraint systems. Theor. Comput. Sci. **255**, 281–293 (2001)
9. Feder, T., Vardi, M.Y.: The computational structure of monotone monadic SNP and constraint satisfaction: A study through Datalog and group theory. SIAM Journal on Computing **28**(1), 57–104 (1998)
10. Geelen, J.F., Iwata, S., Murota, K.: The linear Delta-matroid parity problem. Journal of Combinatorial Theory, Series B **88**(2), 377–398 (2003)
11. Guo, H., Williams, T.: The complexity of planar boolean #CSP with complex weights. In: Fomin, F.V., Freivalds, R., Kwiatkowska, M., Peleg, D. (eds.) ICALP 2013, Part I. LNCS, vol. 7965, pp. 516–527. Springer, Heidelberg (2013)
12. Istrate, G.: Looking for a version of Schaefer's dichotomy theorem when each variable occurs at most twice, technical Report TR652, The University of Rochester (1997)
13. Mac Lane, S.: A structural characterization of planar combinatorial graphs. Duke Mathematical Journal **3**, 460–472 (1937)
14. Moret, B.M.E.: Planar NAE3SAT is in P. SIGACT News **19**(2), 51–54 (1988)
15. Mulzer, W., Rote, G.: Minimum-weight triangulation is NP-hard. J. ACM **55**(2) (2008)
16. Schaefer, T.: The complexity of satisfiability problems. In: Proceedings of the Tenth Annual ACM Symposium on Theory of Computing, pp. 216–226. ACM (1978)

On Temporal Graph Exploration

Thomas Erlebach[1]($\boxtimes$), Michael Hoffmann[1], and Frank Kammer[2]

[1] Department of Computer Science, University of Leicester, Leicester, England
{te17,mh55}@leicester.ac.uk
[2] Institut für Informatik, Universität Augsburg, Augsburg, Germany
kammer@informatik.uni-augsburg.de

Abstract. A temporal graph is a graph in which the edge set can change from step to step. The temporal graph exploration problem TEXP is the problem of computing a foremost exploration schedule for a temporal graph, i.e., a temporal walk that starts at a given start node, visits all nodes of the graph, and has the smallest arrival time. We consider only temporal graphs that are connected at each step. For such temporal graphs with n nodes, we show that it is **NP**-hard to approximate TEXP with ratio $O(n^{1-\varepsilon})$ for any $\varepsilon > 0$. We also provide an explicit construction of temporal graphs that require $\Theta(n^2)$ steps to be explored. We then consider TEXP under the assumption that the underlying graph (i.e. the graph that contains all edges that are present in the temporal graph in at least one step) belongs to a specific class of graphs. Among other results, we show that temporal graphs can be explored in $O(n^{1.5}k^2 \log n)$ steps if the underlying graph has treewidth k and in $O(n \log^3 n)$ steps if the underlying graph is a $2 \times n$ grid. We also show that sparse temporal graphs with regularly present edges can always be explored in $O(n)$ steps.

Keywords: Inapproximability · Planar graphs · Bounded treewidth · Regularly present edges · Irregularly present edges

1 Introduction

Many networks are not static and change over time. For example, connections in a transport network may only operate at certain times. Connections in social networks are created and removed over time. Links in wired or wireless networks may change dynamically. Dynamic networks have been studied in the context of faulty networks, scheduled networks, time-varying networks, etc. For an overview, see [5,15,18]. We consider a model of time-varying networks called *temporal graphs*. A temporal graph $\mathcal{G}$ is given by a sequence of graphs $G_0 = (V, E_0)$, $G_1 = (V, E_1)$, $G_2 = (V, E_2)$, ..., $G_L = (V, E_L)$ that all share the same vertex set V, but whose edge sets may differ. The number L is called the *lifetime* of $\mathcal{G}$. We assume that the whole temporal graph is presented to the algorithm.

Standard algorithms for well known problems such as connected components, diameter, reachability, shortest paths, graph exploration, etc. cannot be used directly in temporal graphs. In particular, Berman [2] observes that the vertex version of Menger's theorem does not hold for temporal graphs. Kempe et al. [10]

© Springer-Verlag Berlin Heidelberg 2015
M.M. Halldórsson et al. (Eds.): ICALP 2015, Part I, LNCS 9134, pp. 444–455, 2015.
DOI: 10.1007/978-3-662-47672-7_36

characterize the temporal graphs in which Menger's theorem holds and show that it is **NP**-complete to decide whether there are two node-disjoint time-respecting paths between a given source and sink. Mertzios et al. [14] show that there is a natural variation of Menger's theorem that holds for temporal graphs. Moreover, the standard algorithms usually optimize only one parameter, but problems in temporal graphs usually have more than one parameter to optimize, e.g., one can search for a *shortest*, a *foremost*, or a *fastest s-t-path* [3], i.e., a path from s to t with a minimal number of edges, earliest arrival time, and a shortest duration, respectively.

We consider the temporal graph exploration problem, introduced in [16] and denoted TEXP, whose goal is to compute a schedule (or temporal walk) with the earliest arrival time such that an agent can visit all vertices in V. The agent is initially located at a start node $s \in V$. In step i $(i \geq 0)$ the agent can either remain at its current node or move to an adjacent node via an edge that is present in E_i. We remark that static undirected graphs can easily be explored in less than $2|V|$ steps using depth-first search, while there are static directed graphs for which exploration requires $\Theta(|V|^2)$ steps. The problem to explore a graph (as part of an exploration of a maze) was already formulated by Shannon [19] in 1951.

Flocchini et al. [7] consider the graph exploration problem on temporal graphs with periodicity defined by the periodic movements of carriers. Much of the research is based on models where edges appear with a certain probability [1, 9, 11] or with some kind of periodicity [4, 13]. Except in Sect. 5, we do not assume that edges appear with some periodicity or certain probabilistic properties. Instead, unless stated otherwise, we only assume that the given temporal graph is always connected. Michail and Spirakis [16] observe that without the assumption that the given temporal graph is connected at all times, it is even **NP**-complete to decide if the graph can be explored at all. They also show that, under this assumption, any temporal graph can be explored with an arrival time n^2. They also prove that there is no $(2 - \varepsilon)$-approximation for TEXP for any $\varepsilon > 0$ unless **P** = **NP**. They define the dynamic diameter of a temporal graph to be the minimum integer d such that for any time i and any vertex v, any other vertex w can be reached in d steps on a temporal walk that starts at v at time i. They provide a d-approximation algorithm for TEXP, where d is the dynamic diameter of the temporal graph. We note that d can be as large as $n - 1$, and hence the approximation ratio of their algorithm in terms of n is only $n - 1$. Thus, there is a significant gap between the lower bound of $2 - \varepsilon$ and the upper bound of $n - 1$ on the best possible approximation ratio, which we address in this paper.

Our contributions. We close the gap between the upper and lower bound on the approximation ratio of TEXP by proving that it is **NP**-hard to approximate TEXP with ratio $O(n^{1-\varepsilon})$ for any $\varepsilon > 0$. Furthermore, we provide an explicit construction of undirected temporal graphs that require $\Theta(n^2)$ steps to be explored. We then consider TEXP under the assumption that the underlying graph (i.e. the graph that contains all edges that are present in the temporal graph in at least one step) belongs to a specific class of graphs. We show that temporal graphs can be explored in $O(n^{1.5}k^2 \log n)$ steps if the underlying graph as treewidth k, in $O(n \log^3 n)$ steps if the underlying graph is a $2 \times n$ grid, and in $O(n)$ steps if the underlying graph is a cycle or a cycle with a chord. Several of these results use a technique by which we specify an exploration schedule for multiple agents and then apply a general reduction from the multi-agent case to the single-agent case. We also show that there exist temporal graphs where the underlying graph is a bounded-degree planar graph and each G_i is a path such that the optimal arrival time of the exploration walk is $\Omega(n \log n)$. Finally, we consider a setting where the underlying graph is sparse and edges are present with a certain regularity and show that temporal graphs can always be explored with an arrival time $O(n)$. A full version of our paper can be found in [6].

The remainder of the paper is structured as follows. In Sect. 2, we give some definitions and preliminary results. Section 3 presents our inapproximability result for general temporal graphs. The results for temporal graphs with restricted underlying graphs are given in Sect. 4. Temporal graphs with regularly present edges are considered in Sect. 5, and Sect. 6 concludes the paper.

2   Preliminaries

Definitions. A *temporal graph* $\mathcal{G}$ with vertex set V and lifetime L is given by a sequence of graphs $(G_i)_{0 \le i \le L}$ with $G_i = (V, E_i)$. Throughout the paper, we only consider temporal graphs for which each G_i is connected and undirected. We refer to i, $0 \le i \le L$, as *time* i or *step* i. The graph $G = (V, E)$ with $E = \bigcup_{0 \le i \le L} E_i$ is called the *underlying graph* of $\mathcal{G}$. If the underlying graph is an X, we call the temporal graph a *temporal* X or a *temporal realization of* X. For example, a temporal cycle is a temporal graph whose underlying graph is a cycle, and a temporal graph of bounded treewidth is a temporal graph whose underlying graph has bounded treewidth.

If an edge e is in E_i, we use the edge-time pair (e, i) to denote the existence of e at time i. A *temporal* (or *time-respecting*) *walk* from $v_0 \in V$ starting at time t to $v_k \in V$ is an alternating sequence of vertices and edge-time pairs $v_0, (e_0, i_0), v_1, \ldots, (e_{k-1}, i_{k-1}), v_k$ such that $e_j = \{v_j, v_{j+1}\} \in E_{i_j}$ for $0 \le j \le k - 1$ and $t \le i_0 < i_1 < \cdots < i_{k-1}$. The walk reaches v_k at time $i_{k-1} + 1$. We often explain the construction of a temporal walk by describing the actions of an agent that is initially located at v and can in every step i either stay at its current node or move to a node that is adjacent to v in E_i.

For a given temporal graph $\mathcal{G}$ with source node s, an *exploration schedule* $\mathcal{S}$ is a temporal walk that starts at s at time 0 and visits all vertices. The *arrival time* of $\mathcal{S}$ is the time step in which the walk reaches the last unvisited vertex. An exploration schedule with smallest arrival time is called *foremost*. The temporal exploration problem TEXP is defined as follows: Given a temporal graph $\mathcal{G}$ with source node s and lifetime at least $|V|^2$, compute a foremost exploration schedule. To ensure the existence of a feasible solution, we assume that the lifetime of the given temporal graph $\mathcal{G}$ is at least $|V|^2$. We also consider a multi-agent variant k-TEXP of TEXP in which there are k agents initially located at s. An exploration schedule $\mathcal{S}$ comprises temporal walks for all k agents such that each node of $\mathcal{G}$ is visited by at least one agent. The arrival time of $\mathcal{S}$ is ther the time when the last unvisited node is reached by an agent.

A ρ-approximation algorithm for TEXP or k-TEXP is an algorithm that runs in polynomial time and outputs an exploration schedule whose arrival time is at most ρ times the arrival time of the optimal exploration schedule.

Preliminary Results. We establish some preliminary results that will be useful for the proofs of our main results. The following lemma allows us to bound the steps of a temporal walk from one vertex to another vertex in a temporal graph

Lemma 1 (Reachability). *Let $\mathcal{G}$ be a temporal graph with vertex set V. Assume that an agent is at vertex u. Let v be another vertex and H a subset of the vertices that includes u and v and has size k. If in each of $k-1$ steps the subgraph induced by H contains a path from u to v (which can be a different path in each step), then the agent can move from u to v in these $k-1$ steps.*

Proof. For $i \geq 0$, let S_i be the set of vertices that the agent could have reached after i steps. We have $S_0 = \{u\}$. We claim that as long as $v \notin S_i$, at least one vertex of H is added to S_i to form S_{i+1}. To see this, consider the graph in step $i+1$. By the assumption, the graph induced by H contains a path from u to v. The first vertex on this path that is not in S_i is added to S_{i+1}. As H contains only k vertices, there can be at most $k-1$ steps until v is reached. □

We now show that a solution to k-TEXP yields a solution to TEXP.

Lemma 2 (Multi-agent to Single-agent). *Let G be a graph with n vertices. If any temporal realization of G can be explored in t steps with k agents, any temporal realization of G can be explored in $O((t+n)k \log n)$ steps with one agent.*

Proof. Let $\mathcal{G}$ be a temporal realization of G. Consider the exploration schedule constructed as follows: In the first t steps, the k agents explore $\mathcal{G}$ in t steps. Then all k agents move back to the start vertex in n steps. Refer to these $t+n$ steps as a *phase*. Note that the phase can be repeated as often as we like. We construct a schedule for a single agent x by copying one of the k agents in each phase. In each phase, the k agents together visit all n vertices, so the agent that visits the largest number of vertices that have not yet been explored by x must visit at least a $1/k$ fraction of these unexplored vertices. We let x copy that agent in this phase. This is repeated until x has visited all vertices.

The number of unexplored vertices is n initially. Each iteration takes $t+n$ steps and reduces the number of unexplored vertices by a factor of $1-1/k$. Then after $\lceil k \ln n \rceil + 1$ iterations, the number of unexplored vertices is less than $n \cdot (1-1/k)^{k \ln n} \leq ne^{-\ln n} = 1$ and therefore all vertices are explored. □

The next lemma shows that edge contractions do not increase the arrival time of an exploration in the worst case.

Lemma 3 (Edge Contraction). *Let G be a graph such that any temporal realization of G can be explored in t steps. Let G' be a graph that is obtained from G by contracting edges. Then any temporal realization of G' can also be explored in t steps.*

Proof. Consider a temporal realization of G'. Consider the corresponding temporal realization of G in which all the contracted edges are always present. Let S be a schedule with an arrival time t that explores the temporal realization of G. S can be executed in t steps in the temporal realization of G' simply by ignoring moves along edges that were contracted.   □

Corollary 1. *Let $c < 1$ be a constant and $t(n)$ a function that is monotone increasing and satisfies $t(kn) = O(t(n))$ for any constant $k > 0$, e.g., a polynomial. Let C be a class of graphs such that any temporal realization of a graph G in the class can be explored in $t(n)$ steps, where n is the number of nodes of G. Let $\mathcal{D}$ be the class of graphs that contains all graphs that can be obtained from a graph G in C with n vertices by at most cn edge contractions. Then any temporal realization of a graph in $\mathcal{D}$ with n' vertices can be explored in $O(t(n'))$ steps.*

Proof. Let G be a graph in the class C, and let H be obtained from G by at most cn edge contractions. Furthermore, let n and n' be the number of vertices of G and H, respectively. Thus, $n' \geq (1 - c)n$. Since any temporal realization of G can be explored in $t(n)$ steps, by Lemma 3, any realization of H can also be explored in $t(n) \leq t(n'/(1 - c)) = O(t(n'))$ steps.   □

3   Lower Bounds for General Temporal Graphs

While static undirected graphs with n nodes can always be explored in less than $2n$ steps, the following lemma shows that there are temporal graphs that require $\Omega(n^2)$ steps.

Lemma 4. *There is an infinite family of temporal graphs that, for every $n \geq 1$, contains a $2n$-vertex temporal graph $\mathcal{G}$ that requires $\Omega(n^2)$ steps to be explored.*

Proof. Let $V = \{c_j, \ell_j \mid 0 \leq j \leq n - 1\}$ be the vertex set of $\mathcal{G}$. For any step $i \geq 0$, the graph G_i is a star with center $c_{i \bmod n}$. The start vertex is c_0. If an agent is at a vertex that is not the current center, the agent can only wait or travel to the current center. As in the next step the center will have changed, the agent is again at a vertex that is not the current center. Hence, to get from one vertex ℓ_j to another vertex ℓ_k for $k \neq j$, n steps are needed: The fastest way is to move from ℓ_j to the center of the current star, and then to wait for $n - 1$ steps until that vertex is again the center of a star, and then to move to ℓ_k. The total number of steps is $\Omega(n^2)$.   □

Lemmas 2 and 4 also imply the following.

Corollary 2. *For any constant number of agents, there is an infinite family of temporal graphs such that each n-vertex temporal graph in the family cannot be explored in $o(n^2/\log n)$ steps.*

The underlying graph of the temporal graph in the proof of Lemma 4 has maximum degree $|V| - 1$. For graphs with maximum degree bounded by d, we can show a lower bound of $\Omega(dn)$ in the following lemma.

Lemma 5. *For every even $d \geq 2$, there is an infinite family of temporal graph with underlying graphs of maximum degree d that require $\Omega(dn)$ steps to be explored, where n is the number of vertices of the graph.*

Proof. Without loss of generality, n is a multiple of d. We construct $\mathcal{G}$ in two steps. First, we construct n/d copies of a temporal graph $\mathcal{G}'$, which we connect in the end. $\mathcal{G}'$ is the graph with d vertices constructed as in the proof of Lemma 4 (by setting the n to $d/2$). Note that moving from a vertex ℓ_j in a copy of $\mathcal{G}'$ to a vertex ℓ_k for $k \neq j$ in the same copy of $\mathcal{G}'$ requires $\Omega(d)$ steps.

Let $\mathcal{G}_1, \ldots, \mathcal{G}_{n/d}$ be the n/d copies of $\mathcal{G}'$. For all $i = 1, \ldots, n/d-1$, connect $\mathcal{G}_i$ and $\mathcal{G}_{i+1}$ by merging vertex ℓ_1 of $\mathcal{G}_i$ with ℓ_0 of $\mathcal{G}_{i+1}$. Let $\mathcal{G}$ be the graph obtained. Note that the underlying graph of $\mathcal{G}$ has maximum degree d (the vertices that have been merged have degree d, all other vertices ℓ_j have degree $d/2$, and all vertices c_j have degree $d-1$). Note that, by our way of merging, $\mathcal{G}$ is connected at all times as this is true for all copies of $\mathcal{G}'$.

Let us consider an exploration schedule of $\mathcal{G}$. Similar to the arguments used in the proof of Lemma 4, we can now observe that getting from any ℓ_i in one copy of $\mathcal{G}'$ to a different vertex ℓ_j in the same or another copy of $\mathcal{G}'$ takes at least $d/2$ steps (in most of these, the agent may not move). As there are at least $n/d \cdot (d/2 - 2) = \Omega(n)$ such pairs in every exploration schedule of $\mathcal{G}$, we need $\Omega(dn)$ steps in total. $\square$

Theorem 1. *Approximating temporal graph exploration with ratio $O(n^{1-\varepsilon})$ is **NP**-hard.*

Proof. We give a reduction from the Hamiltonian s-t path problem, which is **NP**-hard [8]. Assume we are given an instance I' of the Hamiltonian s-t path problem consisting of an undirected n'-vertex graph G', a start vertex s, and an end vertex t. We now construct an instance I of the temporal graph exploration problem as follows: Take the temporal graph as constructed in the proof of Lemma 4 with $n = (n')^c$ for some constant c. In addition, replace each ℓ_i by a copy of G'. Call it the *ith copy* of G'. The edges in each copy of G' are present in every step. The edge $\{c_j, \ell_i\}$ is replaced by an edge connecting c_j and vertex s in the ith copy. We also call the vertices c_i the *center vertices*. In addition, we have so-called *quick links*. Each quick link is an edge that connects the vertex t of the i-th copy with the vertex s of the $(i+1)$-th one only in step $i \cdot n'$ for every $1 \leq i < n-1$. Denote by $\mathcal{G}$ the resulting temporal graph. Note that $\mathcal{G}$ has $n^* = n(1 + n')$ vertices and that $n = \Theta((n^*)^{c/(c+1)})$.

Clearly, if G' has a Hamiltonian path from s to t, then $\mathcal{G}$ can be explored in $O(n^*)$ steps: The agent starts at c_0 and then explores the first copy of G' in n' steps by following the Hamiltonian s-t-path. The agent arrives at t in the first copy of G' at step n', and we can use a quick link in step n' to move to s in the second copy of G', etc. After exploring all copies of G', we can explore all remaining center vertices c_i in $O(n^*)$ steps, i.e., $\mathcal{G}$ can be explored in $O(n^*)$ steps.

Now assume that G' does not have a Hamiltonian s-t-path. This means that a copy of G' cannot be explored in one visit while using both available quick link connections. Hence in the exploration, every copy must either be visited or left via a center vertex. As moving from one copy to another via a center vertex takes n steps, exploring the n copies takes at least $\frac{1}{2}n(n-1)$ steps. So a total of at least $\Omega(n^2) = \Omega((n^*)^{2c/(c+1)}) = \Omega((n^*)^{2-\varepsilon})$ steps are needed, where ε can be made arbitrarily small by choosing c large enough.

Distinguishing whether $\mathcal{G}$ can be explored in $O(n^*)$ steps or whether it requires $\Omega((n^*)^{2-\varepsilon})$ steps therefore solves the Hamiltonian s-t-path problem, and the theorem follows.   □

4   Restricted Underlying Graphs

In Sect. 3, we showed that arbitrary temporal graphs may require $\Omega(n^2)$ steps to be explored and that it is **NP**-hard to approximate the optimal arrival time of an exploration schedule within $O(n^{1-\varepsilon})$ for any $\varepsilon > 0$. This motivates us to consider the case where the underlying graph is from a restricted class of graphs. In particular, the underlying graph of the construction from Lemma 4 is dense (it contains $\Omega(n^2)$ edges) and has large maximum degree. For the case of underlying graphs with degree bound d, we could only show that there are graphs that require $\Omega(dn)$ steps. It is therefore interesting to consider cases of underlying graphs that are sparse, or have bounded degree, or are planar. We consider several such cases in this section.

4.1   Lower Bound for Planar Bounded-Degree Graphs

First, we show that even the restriction to underlying graphs that are planar and have bounded degree is not sufficient to ensure the existence of an exploration schedule with a linear number of steps.

Theorem 2. *Even if the underlying graph $G = (V, E)$ of a temporal graph $\mathcal{G}$ is planar with maximum degree 4 and the graph G_i in every step $i \geq 0$ is a simple path, an optimal exploration can take $\Omega(n \log n)$ steps, where $n = |V|$.*

Proof (sketch). Without loss of generality, we assume that $n = 2^k$ for some $k \geq 3$. Consider the following underlying graph G: It contains vertices $V_0 = \{t_i, b_i \mid 0 \leq i \leq n/4 - 1\}$, the edges $\{t_i, t_{i+1}\}$, $\{b_i, b_{i+1}\}$, $\{t_i, b_{i+1}\}$ and $\{b_i, t_{i+1}\}$ for $0 \leq i < n/4 - 1$, and a path P of $n/2$ additional vertices that connects t_0 and b_0. It is not hard to see that G is planar: Arrange the vertices as in Figure 1. For each $0 \leq i < n/4 - 1$, draw the edge $\{b_i, t_{i+1}\}$ as shown in the figure and the edge $\{t_i, b_{i+1}\}$ around the outside. We refer to the edges $\{t_{i-1}, t_i\}$ and $\{b_{i-1}, b_i\}$ as *horizontal* edges of column i, and the edges $\{t_{i-1}, b_i\}$ and $\{b_{i-1}, t_i\}$ as *cross* edges of column i. Consider the following temporal realization of G:

The path P is always present. We divide the time into *rounds*, the first round consists of the first $n/2$ steps, etc. For the first round, the graph additionally contains the horizontal edges of all columns. For the next round, the horizontal edges of column $n/8$ are replaced by the cross edges. For the next round, the

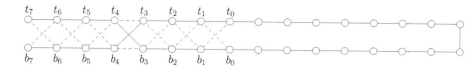

Fig. 1. The underlying graph constructed in the proof of Theorem 2 for $n = 32$. Edges present at the second round are drawn solid, the remaining edges are drawn dashed.

horizontal edges of columns $n/16$ and $3n/16$ are replaced by the cross edges. Following the same pattern of replacements (each time the horizontal edges of the middle column in each stretch of horizontal edges are replaced by the cross edges), this is repeated for $O(\log n)$ rounds.

Observe that with $n/2$ steps, any agent can explore either the vertices in V_0 connected to t_0 or those connected to b_0. Furthermore, no matter which of the two sets of vertices the algorithm visits, in the next $n/2$ steps half of the unvisited vertices will be connected to t_0 and half to b_0. Thus, for all start positions of an agent, it requires $\Omega(\log n)$ rounds until all vertices are visited. □

4.2   Underlying Graphs with Bounded Treewidth

Theorem 3. *Any temporal graph whose underlying graph has treewidth at most k can be explored in $O(n^{1.5}k^2 \log n)$ steps.*

Proof. Consider a nice tree decomposition [12,17] of the underlying graph, i.e., the tree is a binary tree and all nodes are so-called join nodes, introduce nodes, or forget nodes. Select bags as separators via the following procedure: Visit the bags in a post-order traversal of the tree. Select a bag B as a *separator* if the number of unmarked vertices below the bag exceeds $\sqrt{n}$, or if the number of selected bags that are below B and are not descendants of another selected bag is at least 2. If a bag B is selected, mark all vertices in B and below B. Vertices in B are called *separator vertices*. The number of bags selected as separators is $O(\sqrt{n})$. This can be shown as follows. At any point of the procedure, call a selected bag a topmost bag if it is not a descendant of another selected bag. If a bag is selected because there are more than $\sqrt{n}$ unmarked vertices below, the number of topmost bags increases by at most one and $\sqrt{n}$ unmarked vertices become marked. This can happen at most $\sqrt{n}$ times. If a bag is selected because there are two topmost bags below it, the number of topmost bags decreases by one. As the number of topmost bags increases by one at most $\sqrt{n}$ times, it can also decrease at most $\sqrt{n}$ times, and hence at most $\sqrt{n}$ bags are selected because there are two topmost selected bags immediately below them.

The selected separators split the graph into $O(\sqrt{n})$ components (that are not necessarily connected) such that each component contains at most $2\sqrt{n}$ vertices (not counting separators) and is connected to a constant number of separators, i.e., to at most ck separator vertices for some constant c. The algorithm now explores the components one by one. Each component H is explored with ck agents as follows: First, in n steps, move one virtual agent to each of the ck

vertices in the separators that separate the component from the rest of the graph. Then repeat the following operation: Let v be an arbitrary unvisited vertex in H. In each of the next $4ck\sqrt{n}$ steps, v is connected to at least one of the ck separator vertices, so there exists one separator vertex s to which v is connected in at least $4\sqrt{n}$ steps. The agent from s can visit v and return to s in these steps. Therefore, all of the up to $2\sqrt{n}$ vertices in H can be visited in $2\sqrt{n} \cdot 4ck\sqrt{n} = O(kn)$ steps by ck agents. Using the idea in the proof of Lemma 2, this implies that one agent can explore H in $O(k^2 n \log n)$ steps. As there are $O(\sqrt{n})$ components, the whole graph can be explored in $O(n^{1.5} k^2 \log n)$ steps.   □

4.3 Cycles and Cycles with Chords

Theorem 4. *Any temporal cycle C of length n can be explored in $3n$ steps and the optimal number of steps can be computed in polynomial time.*

Proof. Consider two virtual agents, one moving clockwise and one counterclockwise. Since C is connected, at most one edge of C is missing at all times. Thus, in each step, one of the two agents can move, except when the agents are in adjacent places and the edge between them is absent. If the edge stays absent for the next n steps, one of the agents can visit the whole cycle by turning around and traversing the cycle. If the edge is present in one of the next n steps, the agents can use the edge to pass each other and continue the traversal of the cycle. One of the virtual agents will have completed the traversal of the whole cycle in at most $3n$ steps. Pick that agent and use it as the solution.

By shortcutting backward and forward moves of the agents such that no vertices are skipped completely, the optimal schedule is of one of a constant number of types: move clockwise around the cycle; move counter-clockwise around the cycle; move clockwise to some vertex v, then counter-clockwise until the cycle is explored; move counter-clockwise to some vertex w, then clockwise until the cycle is explored. The types can be enumerated in polynomial time, and the optimal schedule for each can be calculated in a greedy way.   □

Observation 1. *There is a temporal cycle graph in which the optimal exploration requires at least $2n - 3$ steps.*

Proof (sketch). Assume that u, v, w is a subpath of the cycle and the agent is initially at u. Let the edge $\{u, v\}$ be absent for the first $n - 2$ steps, and let the edge $\{v, w\}$ be absent in all steps after that.   □

Theorem 5. *A temporal cycle with one chord can be explored in $O(n)$ time.*

Proof. Let the left and right cycle be the two cycles that contain the chord. Check how often the chord is present in the first $10n$ steps. If the chord is present in more than $7n$ steps, use $3n$ of these to explore the (left or right) cycle in which the start node is contained, n to move to the other cycle, and $3n$ to explore that cycle. Otherwise, there are $3n$ steps in which the chord is absent and the remaining graph is a cycle instance. The cycle can be explored in these steps.   □

We conjecture that Theorem 5 can be extended to $O(1)$ chords.

4.4   The $2 \times n$ Grid

Theorem 6. *Any temporal $2 \times n$ grid can be explored in $O(n \log n)$ steps with 4 log n agents.*

Proof. We show a slightly more general statement. We show that, if we are given an underlying graph G' being a grid of size $2 \times n'$ and a subgrid G'' of size $2 \times n''$ of G' such that each pair of vertices in G'' is connected in G', then $4 \log n'$ agents initially on some vertices of G'' can explore G'' in $T(n') = O(n'(\log n'))$ time. The theorem follows by taking $G' = G'' = G$.

We start with exploring the left half H' of G''. The idea is to move 4 agents to the corners of H', one to each corner, and all remaining $4(\log n') - 4$ agents to a suitable *middle location* of H'—specified below—using the first $2n'$ steps. This is possible by Lemma 1. For the next $T(n'/2) + n'/2$ steps, in each step where it is possible, we move the 2 agents ℓ_1 and ℓ_2 on the left corners of H' in parallel to the right using only horizontal edges. Similarly, we move the 2 agents r_1 and r_2 on the right corners to the left in parallel. Let i and j be the number of steps of ℓ_1 and r_1, respectively. The middle location is any position between the final position of ℓ_1 and ℓ_2 on the left and the final position of r_1 and r_2 on the right. If the agents on the left and on the right meet, they stop moving and H' is explored. In particular, if H' is a 2×1 grid, ℓ_1 and r_1 (as well as ℓ_2 and r_2) are at the same vertex, i.e., we can stop immediately and $T(1) = O(1)$. Otherwise, in the same $T(n'/2) + n'/2$ steps where the 4 agents move, we explore recursively the subgrid H'' of H' consisting of the columns that are not visited by the 4 corner agents. More precisely, whenever neither the 2 agents ℓ_1 and ℓ_2 nor the 2 agents r_1 and r_2 move, each pair of vertices of H'' is connected in H' and the agents starting in the middle location can explore H'' in $T(n'/2)$ steps. Consequently, after the first $2n'$ steps to place the agents, the next $T(n'/2) + i + j \leq T(n'/2) + n'/2$ steps are enough to explore H'.

We subsequently explore the right half in the same way. The total time to explore G'' is $T(n') \leq 2(2n' + T(n'/2) + n'/2) = O(n' \log n')$.   □

Using Lemma 2, we can reduce the number of agents to one.

Corollary 3. *A temporal $2 \times n$ grid can be explored in $O(n \log^3 n)$ steps by one agent.*

5   Temporal Graphs with Regularly Present Edges

We say that a temporal graph has regularly present edges if for every edge e there is a constant integer I_e such that the number of consecutive steps in which e is absent from the temporal graph is at most I_e and at least I_e/c for some constant $c > 1$.

Theorem 7. *A temporal graph $\mathcal{G}$ with regularly present edges that has n vertices and $O(n)$ edges can be explored in $O(n)$ steps.*

Proof (sketch). Round all I_e down to the nearest power of 2; denote the result by J_e. Calculate a minimum spanning tree T with respect to edge weights J_e. Explore the graph by following an Euler tour of T. Moving over an edge e takes at most $I_e \leq 2J_e$ steps, so the total exploration takes at most $2\sum_{e \in T} J_e$ steps.

We next show that $\sum_{e \in T} J_e = O(n)$. Consider any $k \geq 0$ such that T contains at least one edge e with $J_e = 2^k$. Consider the connected components $C_1, \ldots, C_{r_k}$ of $T \setminus \{e \in T \mid J_e = 2^k\}$. Observe that every edge *leaving* a component C_i (i.e., with one endpoint in C_i) must have weight at least 2^k. Let E_i be the set of edges of the underlying graph of $\mathcal{G}$ that leave C_i. Since in each step the graph is connected and hence in each step at least one of the edges of E_i must be present, $\sum_{e \in E_i} 1/(I_e/c) \geq 1$. Thus, $\sum_{e \in E_i} \frac{c}{J_e} \geq 1$. Assign a charge of $c2^k/J_e$ to each $e \in E_i$. The total charge that C_i assigns to E_i is $\sum_{e \in E_i} c2^k/J_e = 2^k \sum_{e \in E_i} c/J_e \geq 2^k$. As an edge receives charge $c2^k/J_e$ from at most two components C_i, no edge receives more than $2c2^k/J_e$ of charge for every fixed k.

The total weight of edges of weight 2^k in T is $2^k(r_k - 1)$. Each of the r_k components assigns a charge of 2^k to edges, so the total charge of the r_k components is greater than the total cost of edges of weight 2^k in T. To bound the total charge that an edge e of G can receive, let the weight of e be $J_e = 2^j$. For $k > j$, e does not receive any charge. For each $k \leq j$, e receives charge at most $2c2^k/2^j$. The total charge received by e is then at most $\sum_{k \leq j} \frac{2c2^k}{2^j} \leq \frac{2c2^{j+1}}{2^j} = 4c$.

So we have that all the weight of T is charged to edges of G, and no edge of G receives more than $8c$ of charge. As G has $O(n)$ edges, the total charge is at most $O(4cn) = O(n)$, and hence the weight of T is $O(n)$.   □

6   Conclusion

The study of temporal graphs is still in its infancy, and we do not yet have intuition and a range of techniques comparable to what has been developed over many years for static graphs. Even seemingly simple tasks such as constructing temporal graphs (possibly with an underlying graph from a given family) that cannot be explored quickly is surprisingly difficult. We hope that the methods used in this paper to prove results for temporal graphs, e.g., the general conversion of multi-agent solutions to single-agent solutions, contribute to the formation of a growing toolbox for dealing with temporal graphs.

Our results directly suggest a number of questions for future work. In particular, deriving tight bounds on the largest number of steps required to explore a temporal graph whose underlying graph is an $m \times n$ grid, a bounded degree graph, or a planar graph would be interesting. It would also be interesting to study the approximability of TEXP for restricted underlying graphs, and to identify further cases of underlying graphs, where the temporal exploration problem can be solved optimally in polynomial time.

An interesting variation of TEXP is to allow the agent to make two moves (instead of one) in every time step. The temporal graph constructed in the proof of Lemma 4 can be explored with an arrival time $O(n)$ in the modified model. It would be interesting to determine tight bounds for the modified model.

References

1. Ajtai, M., Komlós, J., Szemerédi, E.: Largest random component of a k-cube Combinatorica **2**(1), 1–7 (1982)
2. Berman, K.A.: Vulnerability of scheduled networks and a generalization of Menger's theorem. Networks **28**(3), 125–134 (1996)
3. Bui-Xuan, B., Ferreira, A., Jarry, A.: Computing shortest, fastest, and foremost journeys in dynamic networks. Int. J. Found. Comput. Sci. **14**(2), 267–285 (2003)
4. Casteigts, A., Flocchini, P., Mans, B., Santoro, N.: Measuring temporal lags in delay-tolerant networks. In: Proc. 25th Conference on IEEE International Symposium on Parallel and Distributed Processing (IPDPS 2011), pp. 209–218. IEEE (2011)
5. Casteigts, A., Flocchini, P., Quattrociocchi, W., Santoro, N.: Time-varying graphs and dynamic networks. IJPEDS **27**(5), 387–408 (2012)
6. Erlebach, T., Hoffmann, M., Kammer, F.: On temporal graph exploration. CoRR abs/1504.07976 (2015). arXiv:1504.07976
7. Flocchini, P., Mans, B., Santoro, N.: Exploration of periodically varying graphs. In: Dong, Y., Du, D.-Z., Ibarra, O. (eds.) ISAAC 2009. LNCS, vol. 5878, pp. 534–543. Springer, Heidelberg (2009)
8. Garey, M.R., Johnson, D.S.: Computers and Intractability, A Guide to the Theory of NP-Completeness. W.H. Freeman and Co., San Francisco (1979)
9. Karlin, A.R., Nelson, G., Tamaki, H.: On the fault tolerance of the butterfly. In: Proc. 26th Annual ACM Symposium on Theory of Computing, STOC 1994, pp. 125–133. ACM (1994)
10. Kempe, D., Kleinberg, J.M., Kumar, A.: Connectivity and inference problems for temporal networks. J. Comput. Syst. Sci. **64**(4), 820–842 (2002)
11. Kesten, H.: The critical probability of bond percolation on the square lattice equals $\frac{1}{2}$. Comm. Math. Phys. **74**(1), 41–59 (1980)
12. Kloks, T. (ed.): Treewidth, Computations and Approximations. LNCS, vol. 842. Springer, Heidelberg (1994)
13. Liu, C., Wu, J.: Scalable routing in cyclic mobile networks. IEEE Trans. Parallel Distrib. Syst. **20**(9), 1325–1338 (2009)
14. Mertzios, G.B., Michail, O., Chatzigiannakis, I., Spirakis, P.G.: Temporal network optimization subject to connectivity constraints. In: Fomin, F.V., Freivalds, R., Kwiatkowska, M., Peleg, D. (eds.) ICALP 2013, Part II. LNCS, vol. 7966, pp. 657–668. Springer, Heidelberg (2013)
15. Michail, O.: An introduction to temporal graphs: An algorithmic perspective. CoRR abs/1503.00278 (2015). arXiv: 1503.00278
16. Michail, O., Spirakis, P.G.: Traveling salesman problems in temporal graphs. In: Csuhaj-Varjú, E., Dietzfelbinger, M., Ésik, Z. (eds.) MFCS 2014, Part II. LNCS, vol. 8635, pp. 553–564. Springer, Heidelberg (2014)
17. Scheffler, P.: A practical linear time algorithm for disjoint paths in graphs with bounded tree-width. Technical Report 396, Department of Mathematics, Technische Universität Berlin (1994)
18. Scheideler, C.: Models and techniques for communication in dynamic networks. In: Alt, H., Ferreira, A. (eds.) STACS 2002. LNCS, vol. 2285, pp. 27–49. Springer, Heidelberg (2002)
19. Shannon, C.: Presentation of a maze-solving machine. In: Proc. 8th Conference of the Josiah Macy Jr. Found (Cybernetics), pp. 173–180 (1951)

Mind Your Coins: Fully Leakage-Resilient Signatures with Graceful Degradation

Antonio Faonio[1]($\boxtimes$), Jesper Buus Nielsen[1], and Daniele Venturi[2]

[1] Aarhus University, Aarhus, Denmark
{antfa,jbn}@cs.au.dk
[2] Sapienza University of Rome, Rome, Italy
venturi@di.uniromal.it

Abstract. We construct a new leakage-resilient signature scheme. Our scheme remains unforgeable in the noisy leakage model, where the only restriction on the leakage is that it does not decrease the min-entropy of the secret key by too much. The leakage information can depend on the entire state of the signer; this property is sometimes known as *fully* leakage resilience.

An additional feature of our construction, is that it offers a graceful degradation of security in situations where standard existential unforgeability is impossible. This property was recently put forward by Nielsen *et al.* (PKC 2014) in the bounded leakage model, to deal with settings in which the secret key is much larger than the size of a signature.

For security parameter κ, our scheme tolerates leakage on the entire state of the signer until $\omega(\log \kappa)$ bits of min-entropy are left in the secret key, and is proven secure in the standard model. While we describe our scheme in terms of generic building blocks, we also explain how to instantiate it efficiently under fairly standard number-theoretic assumptions.

Keywords: Cryptography · Leakage-resilience · Signature schemes

1 Introduction

Cryptography relies on secret information and random sources to accomplish its tasks. In order for a given cryptographic primitive to be secure, it is typically required that its secrets and randomness are well-protected, and cannot be influenced by an attacker. In practice, however, it is not always possible to fulfil this requirement, and partial information about the secret state of a cryptosystem can leak to an external adversary, e.g., via so-called side-channel attacks exploiting physical characteristics of a crypto-device, such as power consumption [1], electromagnetic radiation [2], and running times [3].

A. Faonio and J.B. Nielsen—Supported by European Research Council Starting Grant 279447.
D. Venturi—Partially supported by the European Commission (Directorate-General Home Affairs) under the GAINS project HOME/2013/CIPS/AG/4000005057, and by the European Union's Horizon 2020 research and innovation programme under grant agreement No 644666.

M.M. Halldórsson et al. (Eds.): ICALP 2015, Part I, LNCS 9134, pp. 456–468, 2015.
DOI: 10.1007/978-3-662-47672-7_37

Recently a lot of effort has been put into constructing cryptographic primitives that come along with some form of leakage resilience, meaning that the scheme should remain secure even in case the adversary obtains some type of leakage on the secrets used within the system. A common way to model leakage attacks, is to empower the adversary with access to a leakage oracle, taking as input (adaptively chosen) functions f_i and returning $f_i(st)$ where st is the current secret state of the cryptosystem under attack. Clearly some restriction on the functions f_i has to be put, as otherwise there is no hope for security. By now, a plethora of leakage models (corresponding to different ways how to restrict the functions f_i) have been proposed. The most relevant to our work is the so-called *noisy leakage* model, which assumes the total amount of leakage does not reduce the entropy of the secret key by too much. This setting is a natural strengthening of the *bounded leakage* setting (see, among others, [4–16]).), where the leakage has to obey to the stricter restriction that the total length of the leakage information is bounded by some a-priori fixed value.[1] Known leakage-resilient primitives in the noisy leakage model include one-way relations, public-key encryption, and signature schemes [4,9,17].

For a signature scheme to remain existentially unforgeable in the presence of leakage, it must necessarily be the case that the signature algorithm is not in the set of the allowed leakage functions, as otherwise an adversary could simply leak a forgery. A first consequence of this is that signatures must be very long, as the goal is to enlarge the secret key to tolerate more and more leakage, which is impractical. A second consequence is that we cannot make any meaningful security statement (even for the case of bounded leakage) for schemes where the size of the secret key is much larger than the size of a single signature. One remarkable such case is the setting of the Bounded Retrieval Model [18–20] (BRM), where one intentionally inflates the size of the secret key while keeping constant the size of a signature and the verification key, as well as the computational complexity of the scheme (w.r.t. signature computation/verification). Still, we would like to not consider a scheme completely insecure if the adversary cannot do better than leaking a few signatures.

A first step towards addressing the above issues was recently taken by Nielsen *et al.* [15] (for the bounded leakage model) who introduced a "graceful degradation" property requiring that an adversary should not be able to produce more forgeries than what he could have leaked. More precisely, in order to break unforgeability, an adversary has to produce n forgeries where $n \approx \lambda/(\gamma \cdot s) + 1$ for signatures of size s, a total of λ bits of leakage, and a "slack parameter" $\gamma \in (0,1]$ measuring how close to optimal security a scheme is. The main advantage is that one can design schemes where the size of the secret key is independent of the signature size, leading to shorter signatures. This flavour of leakage resilience still allows for interesting applications, e.g., leaky identification [15].

Our Contribution. We generalize the graceful degradation property to the setting of *fully* leakage resilience in the *noisy* leakage model. Our main notion, dubbed fully-leakage one-more unforgeability, is essentially the same as the one of [15],

---

[1] Physical leakage rarely obeys to this restriction, e.g., a power trace could be much longer than the secret key, making schemes in the noisy leakage model more desirable.

with the twist that leakage functions can be applied to the entire state of the signer and are not bounded in their length.

We construct a *fully* leakage-resilient signature scheme in the noisy leakage model; our signature scheme is based on generic cryptographic building blocks and improves over previous works. The scheme tolerates leakage on the entire state of the signer until $\omega(\log \kappa)$ bits of min-entropy are left in the secret key, and offers graceful degradation with slack parameter $O(1/q_s)$, where q_s is the number of adversarial signature queries, allowing to have short signatures of size independent of the size of the secret key.

We refer the reader to Section 4 for a description of our scheme along with an outline of its proof of security.

Extensions. In the full version of this paper [21], we analyse several extensions of the above result. First off, we propose a "middle-ground" notion which models a setting where secure erasures of the state are available. In particular, we consider that the random coins sampled by the signer are completely erased after each invocation of the signing algorithm. In this model we construct two schemes that obtain optimal leakage resilience and graceful degradation with slack parameter $O(1/\kappa)$ (independent from the number of signature queries made by the adversary). While requiring perfect erasure is a strong assumption (see, e.g., [22]), we believe our notion might still make sense for some applications, as it in particular allows to design simpler and more efficient schemes.

Second, we construct a practical scheme secure in the BRM with optimal slack parameter and leakage resilience, but which requires a random oracle.

Related work. On a high level, our scheme follows the pattern of [5]; our techniques are mostly related to the ones in [11,12]. We stress that these schemes are known to be secure only in the bounded leakage setting (and, in fact, some of them can be shown to be insecure for noisy leakage).

In the full version [21] we review the schemes of [5,11,12], and provide a more detailed comparison with our scheme. We also explain how to instantiate our scheme efficiently, under fairly standard number-theoretic assumptions.

Signature schemes with bounded leakage resilience are also constructed in [16,23,24]. The setting of noisy leakage is also studied in the context of leakage-resilient circuit compilers; see, e.g., [25,26].

2   Preliminaries

If x is a string, we denote its length by $|x|$; if $\mathcal{X}$ is a set, $|\mathcal{X}|$ represents the number of elements in $\mathcal{X}$. Vectors and matrices are typeset in boldface. For a vector $\mathbf{v} = (v_1, \ldots, v_n)$ we sometimes write $\mathbf{v}[i]$ for the i-th element of $\mathbf{v}$. When x is chosen randomly in $\mathcal{X}$, we write $x \leftarrow_{\$} \mathcal{X}$. When A is an algorithm, we write $y \leftarrow \mathsf{A}(x)$ to denote a run of A on input x and output y; if A is randomized, then y is a random variable and $\mathsf{A}(x; r)$ denotes a run of A on input x and randomness r. An algorithm A is *probabilistic polynomial-time* (ppt) if A is randomized and for any input $x, r \in \{0, 1\}^*$ the computation of $\mathsf{A}(x; r)$ terminates in at most $\mathsf{poly}(|x|)$ steps. Throughout the paper we let κ denote the security parameter. We say that a function $\nu : \mathbb{N} \to \mathbb{R}$ is negligible in the security parameter κ if $\nu(\kappa) = \kappa^{-\omega(1)}$. A positive function f is noticeable if there exist a positive polynomial $p(\cdot)$ and a number κ_0 such that $f(\kappa) \geqslant 1/p(\kappa)$ for all $\kappa \geqslant \kappa_0$. For two ensembles $\mathcal{X} = \{X_\kappa\}_{\kappa \in \mathbb{N}}$ and $\mathcal{Y} = \{Y_\kappa\}_{\kappa \in \mathbb{N}}$, we write $\mathcal{X} \equiv \mathcal{Y}$ if they are identically distributed and $\mathcal{X} \approx \mathcal{Y}$ to denote that the two distributions are statistically or computationally close.

Random variables and min-entropy. The min-entropy of a random variable X over a set $\mathcal{X}$ is defined as $\mathbb{H}_\infty(X) := -\log \max_x \mathbb{P}[X = x]$ and represents the best chance of guessing X by an unbounded adversary. Conditional average min-entropy captures how hard it is to guess X on average, given some side information Z, and it is denoted as $\widetilde{\mathbb{H}}_\infty(X|Z) := -\log \mathbb{E}_z [\max_x \Pr[X = x | Z = z]]$.

Commitment schemes. A (non-interactive) commitment scheme (CS) is a tuple of algorithms (Setup, Com), defined as follows: (1) Algorithm Setup takes as input the security parameter and outputs a verification key ϑ; (2) Algorithm Com takes as input a message $m \in \mathcal{M}$, randomness $r \in \mathcal{R}$, the verification key ϑ and outputs a value $com \in \mathcal{C}$. To open a commitment com we output (m, r); an opening is valid if and only if $com = \mathsf{Com}(\vartheta, m; r)$. A commitment scheme has two standard properties, known as binding and hiding. Whenever $\mathcal{M}$ and $\mathcal{R}$ are a finite field $\mathbb{F}$, we say that a commitment is *linearly homomorphic* if given commitments $com = \mathsf{Com}(\vartheta, m; r)$ and $com' = \mathsf{Com}(\vartheta, m'; r')$ and a field element $c \in \mathbb{F}$, one can efficiently compute the commitments $com^* := \mathsf{Com}(\vartheta, m + m'; r + r')$ and $com'' := \mathsf{Com}(\vartheta, c \cdot m; c \cdot r)$. We write $com \cdot com'$ and com^c for the mappings $(com, com') \mapsto com^*$ and $(c, com) \mapsto com''$.

Trapdoor commitments. A trapdoor CS is a tuple of algorithms (ESetup, Com, ECom, Equiv) specified as follows: (1) ESetup takes as input the security parameter and outputs a pair $(\vartheta, \tau) \leftarrow \mathsf{ESetup}(1^\kappa)$; (2) The tuple $(\mathsf{ESetup}_1, \mathsf{Com})$ is a computationally binding commitment scheme with message space $\mathcal{M}$, randomness space $\mathcal{R}$, and commitment space $\mathcal{C};$[2] (3) ECom takes as input a pair (ϑ, τ) and outputs a pair $(com, r') \leftarrow \mathsf{ECom}(\vartheta, \tau)$; (4) Equiv takes as input (τ, m, r') and outputs $r \leftarrow \mathsf{Equiv}(\tau, m, r')$.

For a trapdoor linearly homomorphic CS we require the following additional property. Let $(\vartheta, \tau) \leftarrow \mathsf{ESetup}(1^\kappa)$, $(com_1, r_1') \leftarrow \mathsf{ECom}(\vartheta, \tau)$ and $(com_2, r_2') \leftarrow \mathsf{ECom}(\vartheta, \tau)$. Then we can use randomness $r_1' + r_2'$ to equivocate $com_1 \cdot com_2$, and randomness $c \cdot r_1'$ to equivocate com_1^c (for any c).

Hybrid commitments. A *hybrid* [27] CS is either perfectly binding or trapdoor hiding, depending on the setup of the verification key.

Definition 1. *We say that* $\mathcal{COM} = (\mathsf{Setup}, \mathsf{Com}, \mathsf{ESetup}, \mathsf{ECom}, \mathsf{Equiv})$ *is a hybrid commitment scheme if the following holds.*

---

[2] Algorithm ESetup_1 outputs the first output of ESetup.

Perfectly Binding: (Setup, Com) *forms a perfectly binding* CS;

Trapdoor Hiding: (ESetup, Com, ECom, Equiv) *forms a trapdoor* CS;

Hybridness: *The distribution ensemble* $\{\vartheta : (\vartheta, \tau) \leftarrow_\$ \mathsf{ESetup}(1^\kappa)\}_{\kappa \in \mathbb{N}}$ *is computationally indistinghuishable from* $\{\vartheta : \vartheta \leftarrow_\$ \mathsf{Setup}(1^\kappa)\}_{\kappa \in \mathbb{N}}$.

We call a verification key *equivocable* (resp. *binding*) if it is generated by algorithm ESetup (resp. Setup). Similarly, a commitment is *equivocable* (resp. *binding*) if it is generated using an equivocable (resp. binding) verification key.

NIWI arguments. For a NP-relation $\mathfrak{R} \subseteq \{0,1\}^* \times \{0,1\}^*$, the language associated with $\mathfrak{R}$ is $\mathcal{L}_\mathfrak{R} = \{x : \exists w \text{ s.t. } (x,w) \in \mathfrak{R}\}$. A non-interactive argument system (Init, Prove, Ver) is a tuple of algorithms specified as follows: (1) Init takes as input the security parameter and outputs a common reference string $\mathsf{crs} \leftarrow \mathsf{Init}(1^\kappa)$; (2) Prove takes as input a pair $(x,w) \in \mathfrak{R}$, and outputs a proof $\pi \leftarrow \mathsf{Prove}(\mathsf{crs}, x, w)$; (3) Ver takes as input a statement x and a proof π, and outputs a bit $b \leftarrow \mathsf{Ver}(\mathsf{crs}, x, \pi)$.

A non-interactive witness indistinguishable argument system satisfies a property known as completeness, and two additional properties known as adaptive soundness and statistical witness indistinguishability.

3 Fully-Leakage One-More Unforgeability

We explain how to extend the notion of one-more unforgeability from [15] in two directions: (i) all intermediate values generated within the lifetime of the system (and not just the secret key) are subject to leakage; (ii) the amount of leakage is constraint only by the min-entropy left in the secret key (and not by its total length). Following Dodis et al. [9], we need a notion of a function being ℓ-leaky.

Definition 2. *A (possibly randomized) function* $f : \{0,1\}^* \to \{0,1\}^*$ *is* ℓ-*leaky, if for all* $\kappa \in \mathbb{N}$ *we have that* $\widetilde{\mathbb{H}}_\infty(U_\kappa | f(U_\kappa)) \geq \kappa - \ell$, *where* U_κ *is the uniform distribution over* $\{0,1\}^\kappa$.

A signature scheme is a triple of ppt algorithms $\mathcal{SS} = (\mathsf{KGen}, \mathsf{Sign}, \mathsf{Verify})$ defined as follows: (1) KGen takes as input the security parameter κ and outputs a verification key/signing key pair (vk, sk); (2) Sign takes as input a message $m \in \mathcal{M}$ and the signing key sk and outputs a signature σ; (3) Verify takes as input the verification key vk and a pair (m, σ) and outputs a bit $\mathsf{Verify}(vk, (m, \sigma)) \in \{0,1\}$. We denote by $\tilde{s} := |\sigma|$ the size of a signature output via $\mathsf{Sign}(sk, \cdot)$ and with $\mathsf{s} := \widetilde{\mathbb{H}}_\infty(sk | pk) - \max_{m \in \mathcal{M}} \widetilde{\mathbb{H}}_\infty(sk | pk, \mathsf{Sign}(sk, m))$ the amount of information of the secret key that one signature carries. We say that $\mathcal{SS}$ satisfies correctness if for all messages $m \in \mathcal{M}$ and for all pairs of keys (vk, sk) generated via KGen, we have that $\mathsf{Verify}(vk, (m, \mathsf{Sign}(sk, m)))$ returns 1 with overwhelming probability over the randomness of the signing algorithm.

Given a signature scheme $\mathcal{SS}$, consider the experiment $\mathsf{Exp}_{\mathcal{SS}, \mathsf{A}}^{\mathsf{one-more}}(\kappa, \ell, q_s, \gamma)$ running with a ppt adversary A and parametrized by the security parameter

$\kappa \in \mathbb{N}$, the leakage parameter $\ell \in \mathbb{N}$, and the slack parameter $\gamma := \gamma(\kappa)$ defined as follow:

1. Sample $r_0 \in \{0,1\}^*$, run the key generation algorithm to obtain a pair $(vk, sk) := \mathsf{KGen}(1^\kappa; r_0)$, and return vk to A; let $st = \{r_0\}$.
2. The adversary A can adaptively issue signing queries. Upon input a message $m \in \mathcal{M}$, the adversary is given a signature $\sigma := \mathsf{Sign}(sk, m; r)$ computed using fresh coins r. The state is updated to $st := st \cup \{r\}$. Let $\mathcal{Q}$ be the set of signing queries issued by A.
3. The adversary A can adaptively issue leakage queries. Upon input an arbitrary efficiently computable function f described as a circuit, the adversary is given $f(st)$ where st is the current state.
4. The adversary A outputs n pairs $(m_1^*, \sigma_1^*), \ldots, (m_n^*, \sigma_n^*)$.
5. The experiment outputs 1 if and only if the following conditions are satisfied:
   (a) $\mathsf{Verify}(vk, (m_i^*, \sigma_i^*)) = 1$ and $m_i^* \notin \mathcal{Q}$, for all $i \in [n]$.
   (b) The messages $m_1^*, \ldots, m_n^*$ are pairwise distinct.
   (c) Each leakage function is ℓ_i-leaky, $\sum_i \ell_i \leq \ell$, and $|\mathcal{Q}| \leq q_s$.
   (d) If $s = 0$ then $n \geq \lfloor \tilde{\ell}/(\gamma \cdot \tilde{s}) \rfloor + 1$ otherwise $n \geq \lfloor \ell/(\gamma \cdot s) \rfloor + 1$ where $\tilde{\ell}$ is the total length of the leakage.

Definition 3. *We say that $\mathcal{SS}$ is $(\ell, q_s, \gamma, \varepsilon)$-fully-leakage one-more unforgeable w.r.t. noisy leakage if for every ppt adversary A asking q_s signature queries we have that $\mathbb{P}[\mathsf{Exp}_{\mathcal{SS},A}^{one-more}(\kappa, \ell, q_s, \gamma) = 1] \leq \varepsilon$.*

The above definition requires that an adversary should produce a number of forgeries strictly larger than the ones he could have leaked (up-to the slack factor γ), however the number of leaked signatures now might depend on the amount of information that a signature actually carries. We distinguish two cases, depending on whether the signature algorithm (statistically) reveals partial information on the secret key or not. In the first case, the parameter n in the winning condition is related to the leakage parameter ℓ, since a forgery is de-facto a leaky function of the secret; in the second case we need to be more pessimistic, and let the parameter n be related to the actual leakage $\tilde{\ell}$ performed by the adversary. Note that in the bounded leakage setting $\tilde{\ell} = \ell$; in particular Definition 3 implies [15, Definition1].

As pointed out in [15], the slack parameter γ specifies how close to optimal security $\mathcal{SS}$ is. In particular, in case $\gamma = 1$ one-more unforgeability requires that A cannot forge even a single signature more than what it could have leaked via leakage queries. As γ decreases, so does the strength of the signature scheme (the extreme case being $\gamma = |\mathcal{M}|^{-1}$, where we have no security).

4 The Signature Scheme

We start by abstracting away a special type of hybrid commitment scheme (whose properties are used in a modular way in the security proof), called a *secret sharing* hybrid commitment (SSHCS). We describe it in Section 4.1. The signature scheme and an outline of the security proof are given in Section 4.2.

.1 *Secret Sharing* Hybrid Commitment

Let $\mathcal{COM} = (\mathsf{Setup}, \mathsf{Com}, \mathsf{ESetup}, \mathsf{ECom}, \mathsf{Equiv})$ be a hybrid linearly homomorphic CS, with $\mathcal{M} = \mathbb{F}^\mu$ and $\mathcal{R} = \mathbb{F}^\nu$ for a finite field $\mathbb{F}$ and parameters $\mu, \nu \in \mathbb{N}$. For parameters $p, t \in \mathbb{N}$ such that $0 \le p \le t$, consider the following algorithms:

– The key generation algorithm $\mathsf{Setup}_{ss}(1^\kappa)$ picks t different and independent verification keys using $\mathsf{Setup}(1^\kappa)$, and lets the output be $\bar{\vartheta} = (\vartheta_1, \ldots, \vartheta_t)$.
– The commitment algorithm $\mathsf{Com}_{ss}(\bar{\vartheta}, m)$ first picks uniformly random values $s_1, \ldots, s_p \leftarrow\!\!{}^\$ (\mathbb{F}^\mu)^p$ such that $m = \sum_i s_i$ (i.e., it defines a secret sharing of the message m); then for any $i \in [p]$ it commits to the share s_i picking a uniformly chosen verification key ϑ_{j_i}, i.e., $com_i := \mathsf{Com}(\vartheta_{j_i}, s_i; r_i)$ where $r_i \leftarrow\!\!{}^\$ \mathbb{F}^\nu$. The output is $(\mathbf{j}, \mathbf{com}) = ((J_1, \ldots, J_p), (com_1, \ldots, com_p))$; the randomness is $\mathbf{r} = (r_1, \ldots, r_p) \in (\mathbb{F}^\nu)^p := \mathcal{R}_{ss}$.

The size of a commitment is $O(\mu p \kappa)$. We extend $(\mathsf{Setup}_{ss}, \mathsf{Com}_{ss})$ with a special key generation algorithm and a corresponding equivocal commitment algorithm:

– The *probable binding* key generation algorithm $\overline{\mathsf{Setup}}(1^\kappa, c)$ takes as input an auxiliary parameter $1 \le c \le \log \kappa$. The algorithm follows the same procedure of Setup_{ss} but for a random subset I^* of $[t]$ with cardinality $2^{-c}t$ picks the verification keys using the equivocal key generation algorithm $(\vartheta, \tau) \leftarrow\!\!{}^\$ \mathsf{ESetup}(1^\kappa)$. It outputs $\bar{\vartheta}$ and the trapdoor $\bar{\tau} := \left(I^*, \{\tau_i\}_{i \in [t] \setminus I^*} \right)$.
– The equivocal commitment algorithm $\overline{\mathsf{ECom}}$ takes as input the trapdoor $\bar{\tau}$. It picks $SSS\mathbf{j} \leftarrow\!\!{}^\$ [t]^p$, and if all the sampled indexes are not in I^* then it outputs a special symbol $\perp$, otherwise it commits to $p - 1$ uniformly random shares $S_1, \ldots, S_{p-1}$ and generates an equivocal commitment using as verification key the one with the smallest index lying in $I^* \cap \mathbf{j}$. To equivocate to the message m the algorithm $\overline{\mathsf{Equiv}}$ sets the equivocal commitment to $\left(m - \sum_{I=1}^{p-1} s_i \right)$.

It follows by hybridness of $\mathcal{COM}$ that verification keys and commitments generated with $\overline{\mathsf{Setup}}$, $\overline{\mathsf{ECom}}$ and $\overline{\mathsf{Equiv}}$ are computationally indistinguishable from the ones generated with Setup_{ss} and Com_{ss}. We say that a commitment $(\mathbf{j}, \mathbf{com})$ is *Binding* if the intersection of $\mathbf{j}$ and I^* is empty. We state the following important properties for $\mathcal{COM}_{ss} := (\mathsf{Setup}_{ss}, \overline{\mathsf{Setup}}, \mathsf{Com}_{ss}, \overline{\mathsf{ECom}}, \overline{\mathsf{Equiv}})$.

Lemma 1 (Probable Binding Property, informal). *For any $c \in \mathbb{N}$ such that $0 < c \le \log \kappa$ the following holds. If $p \ge \log \kappa$ and $t \ge \kappa$:*

(a) Any adversarial commitment is binding with probability at least $\frac{1}{\kappa^c} - \mathsf{negl}(\kappa)$.
(b) For any verification key $\bar{\vartheta}$ produced by $\overline{\mathsf{Setup}}(1^\kappa, c)$ the probability that algorithm $\overline{\mathsf{ECom}}(\bar{\vartheta}, \bar{\tau})$ outputs $\perp$ is $1/\kappa^c$.

The lemma is proven in the full version [21].

4.2   Scheme Description

Let $\mathcal{COM}$ be a trapdoor hiding, linearly homomorphic CS. Let $\mathcal{COM}_{ss}$ be the SSHCS described in Section 4.1. Our scheme $\mathcal{SS} = (\mathsf{KGen}, \mathsf{Sign}, \mathsf{Verify})$ has message space[3] equal to $\mathbb{F}$ and is described below:

Key Generation. Let $t, d, \mu \in \mathbb{N}$ be parameters. Run $\mathsf{crs} \leftarrow \mathsf{Init}(1^\kappa)$, sample $\vartheta \leftarrow \mathsf{Setup}(1^\kappa)$, and $\bar{\vartheta} := (\vartheta_1, \ldots, \vartheta_t) \leftarrow \mathsf{Setup}_{ss}(1^\kappa)$. Sample $\boldsymbol{\Delta} \leftarrow_\$ (\mathbb{F}^\mu)^{d+1}$ and $\mathbf{r} = (r_0, \ldots, r_d) \leftarrow_\$ \mathbb{F}^{d+1}$, and compute commitments $com_i = \mathsf{Com}(\vartheta, \boldsymbol{\delta}_i; r_i)$ for $i \in [0, d]$, where $\boldsymbol{\delta}_i \in \mathbb{F}^\mu$ is the i-th column of $\boldsymbol{\Delta}$. Output

$$sk = (\boldsymbol{\Delta}, \mathbf{r}) \qquad vk = (\mathsf{crs}, \vartheta, \{\vartheta_i\}_{i=1}^t, \{com_i\}_{i=0}^d).$$

Signature. For $j \in [\mu]$, let $\delta_j(X)$ be the degree d polynomial having as coefficients the elements in the j-th row of $\boldsymbol{\Delta}$, similarly let $r(X)$ be the degree d polynomial having as coefficients the elements of $\mathbf{r}$;[4] moreover, let $com(X) := \prod_i com_i^{X^i}$. Define $\boldsymbol{\Delta}(X)$ to be the vector of polynomials $(\delta_1(X), \ldots, \delta_\mu(X))$. Consider the following polynomial-time relation:

$$\mathfrak{R} := \left\{ (\vartheta, \bar{\vartheta}, c\tilde{o}m, c\bar{o}m); (\tilde{m}, \tilde{r}, \bar{r}) \left| \begin{array}{l} c\tilde{o}m = \mathsf{Com}(\vartheta, \tilde{m}; \tilde{r}) \\ c\bar{o}m = \mathsf{Com}_{ss}(\bar{\vartheta}, \tilde{m}; \bar{r}) \end{array} \right. \right\}.$$

To sign a message $m \in \mathbb{F}$ compute $\tilde{m} = \boldsymbol{\Delta}(m)$ and $\tilde{r} = r(m)$, and let $c\tilde{o}m := \mathsf{Com}(\vartheta, \tilde{m}; \tilde{r})$ and $c\bar{o}m := \mathsf{Com}_{ss}(\bar{\vartheta}, \tilde{m}; \bar{r})$ where $\bar{r} \leftarrow_\$ \mathcal{R}_{ss}$. Using crs as common reference string, generate a NIWI argument π for $(\vartheta, \bar{\vartheta}, c\tilde{o}m,) \in \mathcal{L}_\mathfrak{R}$, the language generated by the above relation $\mathfrak{R}$. Output $\sigma = (\pi, c\bar{o}m)$.

Verification. Given a pair (m, σ), parse σ as $\sigma = (\pi, c\bar{o}m)$. Output the same as $\mathsf{Ver}(\mathsf{crs}, \pi, (\vartheta, \bar{\vartheta}, com(m), c\bar{o}m))$.

Theorem 1. *Let $\mu \in \mathbb{N}$, and let $\mathbb{F}$ be a finite field of size $\log |\mathbb{F}| = \kappa$ for security parameter $\kappa \in \mathbb{N}$. For any constant $0 \leqslant \xi < \mu/(\mu+1)$, let $\theta \in \mathbb{N}$ be such that*

$$\theta \geqslant \log(2e\mu) - \log\left(\frac{\mu}{\mu+1} - \xi\right).$$

Whenever $d = \kappa^\theta$, $t = \kappa$ and $p = \log\kappa$ the above signature scheme is $(\xi|sk|, q_s, O(\frac{1}{q_s}), \mathsf{negl}(\kappa))$-fully-leakage one-more unforgeable.

The slack parameter $\gamma(\kappa, q_s)$ is linear in the number of signature queries, but still polynomial in the security parameter as $q_s = \mathsf{poly}(\kappa)$. As shown in [15, Section5] this is good enough for some applications.

We outline the proof of Theorem 1 and refer the reader to the full version for the details. We start by defining a series of hybrid experiments computationally close to each other by the cryptographic security of the primitives we used. Then we prove two lemmas on the conditional min-entropy of the secret key in the

---

[3] To obtain a signature scheme with message space $\{0,1\}^*$ it is sufficient to first apply a collision resistant hash function from $\{0,1\}^*$ to $\mathbb{F}$ to the message, and then sign.

[4] Namely, we set $\delta_j(X) := \sum_{i=0}^d \delta_{j,i} \cdot X^i$ and $r(X) := \sum_{i=0}^d r_i \cdot X^i$.

nal hybrid experiment, and show that if an adversary wins the security game with noticeable probability then the two lemmas are in contradiction.

Let A be an adversary asking $q_s = \kappa^c$ (for some constant c) signature queries, such that

$$\mathbb{P}[\mathsf{Exp}_{\mathcal{SS},\mathsf{A}}^{\mathsf{one-more}}(\kappa, \ell, q_s, \gamma) = 1] = \varepsilon(\kappa).$$

The goal of the sequence of hybrids is to reach a final experiment in which signature queries *on average* do not reveal information about the secret polynomial (X). For space reasons, we omit the description of the intermediate hybrids here and present directly the final experiment $\mathcal{H}_3^{\mathsf{Leak}_\Delta(\cdot)}$.

Hybrid $\mathcal{H}_3^{\mathsf{Leak}_\Delta(\cdot)}$ does not sample Δ as part of the signing key, but can instead access it via $\mathsf{Leak}_\Delta(\cdot)$. The commitments $\{com_i\}_{i=0}^d$ in the verification key of the signature scheme are equivocal (let $\{r_i'\}_{i=0}^d$ be the second output of the equivocal commitment algorithm) and the verification key $\bar{\vartheta}$ of the SSHCS is generated using the probable binding key generation algorithm $\overline{\mathsf{Setup}}(1^\kappa, c)$. Moreover, signature queries are answered using the equivocal commitment algorithm $\overline{\mathsf{ECom}}(\bar{\vartheta}, \bar{\tau})$. Specifically, upon input a message m the hybrid replies as follow: i) it produces an equivocal commitment $c\bar{o}m \leftarrow_\$ \overline{\mathsf{ECom}}(\bar{\vartheta}, \bar{\tau})$; (ii) if $\overline{\mathsf{ECom}}(\bar{\vartheta}, \bar{\tau})$ does not return $\perp$, it produces an argument π that both the commitment $c\bar{o}m$ and the commitment $com(m) = \prod_i com_i^{m^i}$ open to $(0)^\mu$.

The equivocation algorithm guarantees a valid witness and maintains the hybrid indistinguishable to the real experiment; in fact, it allows to simulate the *"real"* randomness for the commitment. Similarly, the statistical property of the NIWI allows to simulate the *"real"* randomness for the argument π. In particular, we write $st(\Delta)$ to stress that the real randomness can be expressed as a randomized function of Δ. Therefore, any leakage query $f(sk, st)$ can be re-defined as a query $f''(sk) := f(sk, st(sk))$, as Δ is part of sk. The hybrid makes two kind of oracle queries to $\mathsf{Leak}_\Delta(\cdot)$:

T1 For each signature query $m \in \mathbb{F}$ where $\overline{\mathsf{ECom}}(\bar{\vartheta}, \bar{\tau})$ returns $\perp$, the experiment defines the leakage function $f_m' := \Delta(m)$ and queries $\mathsf{Leak}_\Delta(\cdot)$ on f_m'. We call such signature query a *bad* query (otherwise it is called *good*). Given the value $\Delta(m)$, the signature $\sigma = (c\bar{o}m, \pi)$ is computed by committing to $\Delta(m)$, namely $c\bar{o}m = \mathsf{Com}_{ss}(\bar{\vartheta}, m; \bar{r})$; then the hybrid equivocates $com(m)$ to $\Delta(m)$ and, given the witness $(\Delta(m), \bar{r}, r)$ where $r \leftarrow \mathsf{Equiv}(vk, \Delta(m), r'(m))$, it produces the argument π.

T2 For each leakage query f the experiment defines a function f'' and queries $\mathsf{Leak}_\Delta(\cdot)$ on f''. The function $f''(\Delta) := f'(\Delta, st(\Delta))$ hard-wires all values necessary to reconstruct the current state $st = st(\Delta)$; this includes the trapdoor information needed in order to reconstruct $\mathbf{r}$, and all random coins used to simulate previous signature queries.

One can show that $\varepsilon_3 \geqslant \varepsilon - \mathsf{negl}(\kappa)$, where ε_3 is the advantage of A in $\mathcal{H}_3$. We prove two lemmas on the conditional min-entropy of the secret vector Δ given the view View_3 in the hybrid experiment $\mathcal{H}_3$.

Lemma 2 (Lower Bound). *The following inequality holds*

$$\mathbb{P}\left[\mathbb{H}_\infty\left(\boldsymbol{\Delta}\mid \mathsf{View}_3 = v\right) \geqslant |\boldsymbol{\Delta}| - 2e\mu \log|\mathbb{F}| - \log 4 - \ell\right] > \frac{2}{3}, \qquad (1)$$

where the probability is taken over the randomness of the experiment.

Intuitively, the only relevant information that the view View_3 reveals on $\boldsymbol{\Delta}$ comes from the bad signature queries (leakage of type T1) and from the leakage queries made by the adversary (leakage of type T2); in fact, both the verification key and the answers to good signatures queries are completely independent of $\boldsymbol{\Delta}$. Let Z be the number of bad signature queries. By a Chernoff bound, Z does not diverge significantly from the average:

$$\mathbb{E}\left[Z\right] = \sum_{i=1}^{q_s} \Pr[\mathsf{ECom}_{ss}(\bar{\vartheta}) = \bot] = \sum_{i=1}^{q_s} \frac{1}{q_s} = 1,$$

where the last equality holds because of point (a) in Lemma 1. Therefore the total size of leakage is w.h.p. $2e\mu \log|\mathbb{F}|$ from type T1 queries and ℓ from type T2 queries. The lemma follows by the chain rule for leaky functions, and by a Markov-like argument on the average conditional min-entropy.

Lemma 3 (Upper Bound). *For any adversary* A *such that* $\mathbb{P}[\mathsf{Exp}_{\mathcal{SS},\mathsf{A}}^{\mathsf{one-more}}(\kappa, \ell, q_s, \gamma) = 1] = \varepsilon$, *there exists a constant $c' > 0$ such that:*

$$\mathbb{E}\left[\mathbb{H}_\infty\left(\boldsymbol{\Delta}\mid \mathsf{View}_3 = v\right)\right] \leqslant \left(d + 1 - \frac{n}{q_s}\right)\mu \log|\mathbb{F}| - c' \log\varepsilon(\kappa), \qquad (2)$$

where the expectation is taken over the randomness of the experiment.

We define a predictor strategy that outputs $\boldsymbol{\Delta}$ with "high probability" and uses as subroutine the adversary A. The predictor runs the hybrid experiment $\mathcal{H}_3$ with A. Eventually, the adversary A outputs a set of forgeries $(m_1^*, \sigma_1^*), \ldots, (m_n^*, \sigma_n^*)$. We say that a signature $\sigma^* = (\pi^*, c\bar{o}m^*)$ is *binding* if the commitment $c\bar{o}m^*$ is binding. Let K the number of binding forged signatures; w.l.o.g. let us assume that the first K forged signatures are binding. The predictor finds, for all $i \in [K]$, the unique vector $\boldsymbol{\Delta}(m_i^*)$ such that $,_i^* = \mathsf{Com}_{ss}(\bar{\vartheta}, \boldsymbol{\Delta}(m_i^*); \bar{r}_i^*)$ (for some randomness $\bar{r}_i^* \in \mathcal{R}_{ss}$). For any $i \in [K]$, if the argument π_i^* is for a valid statement then the found vector must be exactly $\boldsymbol{\Delta}(m_i^*)$. The predictor can succeed only if the adversary A outputs n valid forgeries and does not break the soundness of the NIWI argument system in hybrid experiment $\mathcal{H}_3$, which happens with probability greater or equal to ε_3. This explains the term $-c' \log\varepsilon(\kappa)$ in Eq. (2).

For any $j \in [\mu]$ the predictor knows K points on the polynomial[5] $\delta_j(X)$; moreover any type T1 query reveals other points. Therefore, if Z is the number of bad signatures, the predictor obtains $K + Z$ points of the polynomial $\delta_j(X)$. The predictor simply guesses $d+1-K-Z$ points and uses polynomial interpolation to

---

[5] Recall that $\delta_j(X)$ is the polynomial defined by the coefficients in the j-th row of $\boldsymbol{\Delta}$.

etrieve the coefficients of δ_j. We compute the average number of points known y the predictor for any polynomial δ_j. By linearity of expectation:

$$\mathbb{E}[X + Z] \geq n(1/q_s - \mathsf{negl}(\kappa)) + 1 > n/q_s,$$

where we used that $\mathbb{E}[Z] = 1$ and $\mathbb{E}[K] \geqslant n(1/q_s - \mathsf{negl}(\kappa))$ by point (b) of Lemma 1. Therefore, if A wins the one-more unforgeability game the predictor needs to guess on average $\mu(d + 1 - n/q_s)$ elements in $\mathbb{F}$.

We are now ready to prove Theorem 1. Recall that $\log |\mathbb{F}| = \kappa$; by Markov's inequality on Eq. (2):

$$\mathbb{P}\left[\mathbb{E}\left[\mathbb{H}_\infty\left(\Delta|\mathsf{View}_3 = v\right)\right] \leq 3\left(\left(d + 1 - \frac{n}{q_s}\right)\mu\kappa - c'\log\epsilon(\kappa)\right)\right] > \frac{2}{3}. \quad (3)$$

f the inequality below (see Eq. (4)) holds then the event in Eq. (3) is the negation of the event in Eq. (1). However the sum of the probabilities is bigger than 1, yielding a contradiction.

$$3\left(\left(d + 1 - \frac{n}{q_s}\right)\mu\kappa - c'\log\epsilon(\kappa)\right) \leqslant (d+1)\mu\kappa - 2e\mu\kappa - \log 4 - \ell. \quad (4)$$

First we note that the left hand side of the equation above is a positive value, therefore it must be that $\ell \leqslant (d+1)\mu\kappa - 2e\mu\kappa - \log 4$. For any $0 \leqslant \xi \leqslant \frac{\mu}{\mu+1}$, the last equation holds if we set θ (recall that $d = \kappa^\theta$) as in the statement of the theorem. The inequality holds if

$$\frac{n\mu\kappa}{q_s} \geqslant \left(\frac{2e}{3} + d + 1\right)\mu\kappa + \frac{\ell}{3} - c'\log\varepsilon + \frac{\log 4}{3}.$$

Therefore:

$$\frac{n\kappa}{q_s} = \Omega(\ell - \log\varepsilon(\kappa)). \quad (5)$$

Since $n \geqslant \lfloor \frac{\ell}{\gamma \cdot s} \rfloor + 1$, $s = \Theta(\kappa)$ and ε is noticeable, Eq. (5) holds if we set $\gamma = O(\frac{1}{q_s})$. $\square$

References

1. Kocher, P.C., Jaffe, J., Jun, B.: Differential power analysis. In: Wiener, M. (ed.) CRYPTO 1999. LNCS, vol. 1666, pp. 388–397. Springer, Heidelberg (1999)
2. Quisquater, J.-J., Samyde, D.: ElectroMagnetic analysis (EMA): measures and counter-measures for smart cards. In: Attali, S., Jensen, T. (eds.) E-smart 2001. LNCS, vol. 2140, pp. 200–210. Springer, Heidelberg (2001)
3. Kocher, P.C.: Timing attacks on implementations of Diffie-Hellman, RSA, DSS, and other systems. In: Koblitz, N. (ed.) CRYPTO 1996. LNCS, vol. 1109, pp. 104–113. Springer, Heidelberg (1996)
4. Naor, M., Segev, G.: Public-Key cryptosystems resilient to key leakage. In: Halevi, S. (ed.) CRYPTO 2009. LNCS, vol. 5677, pp. 18–35. Springer, Heidelberg (2009)

5. Katz, J., Vaikuntanathan, V.: Signature schemes with bounded leakage resilience. In: Matsui, M. (ed.) ASIACRYPT 2009. LNCS, vol. 5912, pp. 703–720. Springer, Heidelberg (2009)

6. Alwen, J., Dodis, Y., Wichs, D.: Leakage-Resilient public-key cryptography in the bounded-retrieval model. In: Halevi, S. (ed.) CRYPTO 2009. LNCS, vol. 5677, pp. 36–54. Springer, Heidelberg (2009)

7. Davì, F., Dziembowski, S., Venturi, D.: Leakage-Resilient storage. In: Garay, J.A., De Prisco, R. (eds.) SCN 2010. LNCS, vol. 6280, pp. 121–137. Springer, Heidelberg (2010)

8. Brakerski, Z., Kalai, Y.T., Katz, J., Vaikuntanathan, V.: Overcoming the hole in the bucket: Public-key cryptography resilient to continual memory leakage. In FOCS, pp. 501–510 (2010)

9. Dodis, Y., Haralambiev, K., López-Alt, A., Wichs, D.: Cryptography against continuous memory attacks. In: FOCS, pp. 511–520 (2010)

10. Brakerski, Z., Goldwasser, S.: Circular and leakage resilient public-key encryption under subgroup indistinguishability. In: Rabin, T. (ed.) CRYPTO 2010. LNCS, vol. 6223, pp. 1–20. Springer, Heidelberg (2010)

11. Boyle, E., Segev, G., Wichs, D.: Fully leakage-resilient signatures. In: Paterson, K.G. (ed.) EUROCRYPT 2011. LNCS, vol. 6632, pp. 89–108. Springer, Heidelberg (2011)

12. Malkin, T., Teranishi, I., Vahlis, Y., Yung, M.: Signatures resilient to continual leakage on memory and computation. In: Ishai, Y. (ed.) TCC 2011. LNCS, vol. 6597, pp. 89–106. Springer, Heidelberg (2011)

13. Bitansky, N., Canetti, R., Halevi, S.: Leakage-Tolerant interactive protocols. In: Cramer, R. (ed.) TCC 2012. LNCS, vol. 7194, pp. 266–284. Springer, Heidelberg (2012)

14. Nielsen, J.B., Venturi, D., Zottarel, A.: On the connection between leakage tolerance and adaptive security. In: Kurosawa, K., Hanaoka, G. (eds.) PKC 2013. LNCS, vol. 7778, pp. 497–515. Springer, Heidelberg (2013)

15. Nielsen, J.B., Venturi, D., Zottarel, A.: Leakage-Resilient signatures with graceful degradation. In: Krawczyk, H. (ed.) PKC 2014. LNCS, vol. 8383, pp. 362–379. Springer, Heidelberg (2014)

16. Dagdelen, Ö., Venturi, D.: A second look at fischlin's transformation. In: Pointcheval, D., Vergnaud, D. (eds.) AFRICACRYPT. LNCS, vol. 8469, pp. 356–376. Springer, Heidelberg (2014)

17. Garg, S., Jain, A., Sahai, A.: Leakage-Resilient zero knowledge. In: Rogaway, P. (ed.) CRYPTO 2011. LNCS, vol. 6841, pp. 297–315. Springer, Heidelberg (2011)

18. Di Crescenzo, G., Lipton, R.J., Walfish, S.: Perfectly secure password protocols in the bounded retrieval model. In: Halevi, S., Rabin, T. (eds.) TCC 2006. LNCS, vol. 3876, pp. 225–244. Springer, Heidelberg (2006)

19. Dziembowski, S.: Intrusion-Resilience via the bounded-storage model. In: Halevi, S., Rabin, T. (eds.) TCC 2006. LNCS, vol. 3876, pp. 207–224. Springer, Heidelberg (2006)

20. Dziembowski, S.: On forward-secure storage. In: Dwork, C. (ed.) CRYPTO 2006. LNCS, vol. 4117, pp. 251–270. Springer, Heidelberg (2006)

21. Faonio, A., Nielsen, J.B., Venturi, D.: Mind your coins: Fully leakage-resilient signatures with graceful degradation. IACR Cryptology ePrint Archive 2014, 913 (2014)

2. Canetti, R., Eiger, D., Goldwasser, S., Lim, D.-Y.: How to protect yourself without perfect shredding. In: Aceto, L., Damgård, I., Goldberg, L.A., Halldórsson, M.M., Ingólfsdóttir, A., Walukiewicz, I. (eds.) ICALP 2008, Part II. LNCS, vol. 5126, pp. 511–523. Springer, Heidelberg (2008)

3. Faust, S., Kiltz, E., Pietrzak, K., Rothblum, G.N.: Leakage-Resilient signatures. In: Micciancio, D. (ed.) TCC 2010. LNCS, vol. 5978, pp. 343–360. Springer, Heidelberg (2010)

4. Faust, S., Kohlweiss, M., Marson, G.A., Venturi, D.: On the non-malleability of the fiat-shamir transform. In: Galbraith, S., Nandi, M. (eds.) INDOCRYPT 2012. LNCS, vol. 7668, pp. 60–79. Springer, Heidelberg (2012)

5. Faust, S., Rabin, T., Reyzin, L., Tromer, E., Vaikuntanathan, V.: Protecting circuits from leakage: the computationally-bounded and noisy cases. In: Gilbert, H. (ed.) EUROCRYPT 2010. LNCS, vol. 6110, pp. 135–156. Springer, Heidelberg (2010)

6. Duc, A., Dziembowski, S., Faust, S.: Unifying leakage models: from probing attacks to noisy leakage. In: Nguyen, P.Q., Oswald, E. (eds.) EUROCRYPT 2014. LNCS, vol. 8441, pp. 423–440. Springer, Heidelberg (2014)

7. Catalano, D., Visconti, I.: Hybrid trapdoor commitments and their applications. In: Caires, L., Italiano, G.F., Monteiro, L., Palamidessi, C., Yung, M. (eds.) ICALP 2005. LNCS, vol. 3580, pp. 298–310. Springer, Heidelberg (2005)

A $(1 + \varepsilon)$-Embedding of Low Highway Dimension Graphs into Bounded Treewidth Graphs

Andreas Emil Feldmann$^{(\boxtimes)}$, Wai Shing Fung, Jochen Könemann, and Ian Post

Department of Combinatorics and Optimization, University of Waterloo,
Waterloo, Canada
{andreas.feldmann,wsfung,jochen}@uwaterloo.ca, ian@ianpost.org

Abstract. Graphs with bounded *highway dimension* were introduced in [Abraham et al., SODA 2010] as a model of transportation networks. We show that any such graph can be embedded into a distribution over bounded treewidth graphs with arbitrarily small distortion. More concretely, if the highway dimension of G is constant we show how to randomly compute a subgraph of the shortest path metric of the input graph G with the following two properties: it distorts the distances of G by a factor of $1 + \varepsilon$ in expectation and has a treewidth that is polylogarithmic in the aspect ratio of G. In particular, this result implies quasi-polynomial time approximation schemes for a number of optimization problems that naturally arise in transportation networks, including Travelling Salesman, Steiner Tree, and Facility Location.

To construct our embedding for low highway dimension graphs we extend Talwar's [STOC 2004] embedding of low doubling dimension metrics into bounded treewidth graphs, which generalizes known results for Euclidean metrics. We add several non-trivial ingredients to Talwar's techniques, and in particular thoroughly analyze the structure of low highway dimension graphs. Thus we demonstrate that the geometric toolkit used for Euclidean metrics extends beyond the class of low doubling metrics.

1 Introduction

In [12,13], Bast et al. studied shortest-path computations in road networks and observed that such networks are highly structured: there is a small number of *transit* or *access* nodes such that when travelling from any point A to a distant location B along a shortest path, one will visit at least one of these nodes. The authors presented a shortest-path algorithm (called *transit node routing*) that capitalizes on this structure in road networks and demonstrated experimentally that it improves over previously best algorithms by several orders of magnitude. Motivated by Bast et al.'s work (among others), Abraham et al. [1–3] introduce a formal model for transportation networks and define the notion of *highway dimension*. Informally speaking, an edge-weighted graph $G = (V, E)$ has small *highway dimension* if, for any *scale* $r \geq 0$ and for all vertices $v \in V$, shortest

---

A.E. Feldmann—Supported by ERC Starting Grant PARAMTIGHT (No. 280152).

M.M. Halldórsson et al. (Eds.): ICALP 2015, Part I, LNCS 9134, pp. 469–480, 2015.
DOI: 10.1007/978-3-662-47672-7_38

aths of length at least r that are close (in terms of r) to v are *hit* by a small set of *ub* vertices. In the following formal definition, if dist(u,v) denotes the shortest-ath distance between vertices u and v, let $B_r(v) = \{u \in V \mid \text{dist}(u,v) \leq r\}$ be ne *ball* of radius r centered at v.

Definition 1. The *highway dimension* of a graph G is the smallest integer k uch that, for some universal constant $c \geq 4$, for every $r \in \mathbb{R}^+$, and every ball $B_{cr}(v)$ of radius cr, there are at most k vertices in $B_{cr}(v)$ hitting all shortest aths in $B_{cr}(v)$ of length more than r.

Rather than working with the above definition directly, we often consider the losely related notion of *shortest path covers* (also introduced in [1]).

Definition 2. For a graph G and $r \in \mathbb{R}^+$, a *shortest path cover* SPC$(r) \subseteq V$ s a set of *hubs* that cover all shortest paths of length in $(r, cr/2]$ of G. Such a over is called *locally s-sparse* for scale r, if no ball of radius $cr/2$ contains more han s vertices from SPC(r).

In particular, a graph with highway dimension k can be seen to have a *locally* $:$-sparse shortest path cover for any scale r [1]. In both definitions above Abra-am et al. [1] specifically choose $c = 4$ but also note that this choice is, to some xtent, arbitrary. In the present paper, the flexibility of being able to choose a lightly larger value of c is crucial as we will explain shortly. In the following, we vill let $\lambda = c - 4$ and call it the *violation* of Abraham et al.'s original definition. While we believe that a small positive violation does not stray from the intended neaning of highway dimension, we also point out that there are graphs whose uighway dimension is highly sensitive to the value of c. Hence this is not an entirely innocuous change.

Abraham et al. [1–3] focus on the shortest-path problem and formally investi-gate the performance of various prominent heuristics as a function of the highway limension of the underlying metric. They also point out that, "conceivably, bet-er algorithms for other [optimization] problems can be developed and analyzed under the small highway dimension assumption". The latter statement is the starting point of this paper.

We study three prominent NP-hard optimization problems that arise natu-rally in transportation networks: *Travelling Salesman*, *Steiner Tree* and *Facil-ity Location* (cf. [25] for formal definitions). Each of these was first studied in the context of transportation networks, and as we will show they admit quasi-polynomial time approximation schemes (QPTASs) on graphs with bounded uighway dimension. Our work thereby provides a complexity-theoretic separa-tion between the class of low highway dimension and general graphs, in which the aforementioned problems are APX-hard [16,17,19].

Technically, we achieve the above results by employing the powerful machin-ery of metric space embeddings [10,18]. Specifically, we compute a distribution over metrics induced by weighted low-treewidth graphs, each of which dominates the original metric, and whose expected *distortion* is (arbitrarily) small. The fol-lowing is the main result of this paper, where the *aspect ratio* is the maximum distance divided by the minimum distance between any vertices.

Theorem 3. *Let G be a graph with highway dimension k of violation $\lambda > 0$, and aspect ratio α. For any $\varepsilon > 0$, there is a polynomial-time computable probabilistic embedding H of G with treewidth $(\log \alpha)^{O\left(\log^2\left(\frac{k}{\varepsilon\lambda}\right)/\lambda\right)}$ and expected distortion $1 + \varepsilon$.*

Low highway dimension graphs do not exclude fixed-size minors and therefore do not have low treewidth [23]: the complete graph on vertices $\{1, \ldots, n\}$ where each edge $\{i, j\}$ with $i < j$ has length c^i, has highway dimension 1. The example also shows that the aspect ratio of a low-highway dimension graph can be exponential. We can show that the aspect ratio may be assumed to be polynomial for our considered problems when aiming for $1 + \varepsilon$ approximations. Existing algorithms for bounded treewidth graphs [5,14] then imply QPTASs on graphs with constant highway dimension.[1]

While Travelling Salesman, Facility Location, and Steiner Tree are APX-hard in general graphs, improved algorithms are known in special cases. For example, polynomial time approximation schemes (PTASs) for all three of these problems are known if the input metric is low-dimensional Euclidean or planar [4,6,8,14, 15,21,22]. Talwar [24] also showed that the work in [6,8,22] extends (albeit with quasi-polynomial running time) to low *doubling dimension* metrics. Bartal et al. [11] later presented a PTAS for Travelling Salesman instances in this class.

The concept of doubling dimension was first studied by Gupta et al. [20], and captures metrics that have restricted *volume growth*. Formally, a metric has doubling dimension at most d if every ball of radius $2r$ can be covered by 2^d balls of radius r, for any r. The class of constant doubling dimension metrics strictly generalizes that of Euclidean metrics in constant dimensions. Doubling dimension and highway dimension (as defined here) are incomparable metric parameters, however: Abraham et al. [1] noted that grids have doubling dimension 2 but highway dimension $\Theta(\sqrt{n})$, while stars have doubling dimension $\Theta(\log n)$ and highway dimension 1.

We briefly note here that there are alternative definitions of highway dimension. In particular, the more restrictive definition in [3] *implies* low doubling-dimension and hence Talwar [24] readily yields a QPTAS for the optimization problems we study. Our choice of definition is deliberate, however, and motivated by the fact that Definition 1 captures natural transportation networks that the more restrictive definition does not. For instance, typical *hub-and-spoke* networks used in airtraffic models are non-planar and have high doubling dimension, since they feature high-degree stars. This immediately renders them incompatible with the highway dimension definition in [3]. Nevertheless they have low highway dimension by Definition 1, since the airports act as hubs, which become sparser with growing scales as longer routes tend to be serviced by bigger airports. We can also prove that our definition is a strict generalization of the one in [3]: any graph with highway dimension k according to [3] has highway dimension $O(k^2)$ according to Definition 1, while a corresponding lower bound is not possible in general.

---

[1] All missing details and proofs of this extended abstract are deferred to the full version of the paper.

Our results not only provide further evidence that the highway dimension parameter is useful in characterizing the complexity of graph theoretic problems in combinatorial optimization. Importantly, they also show that the geometric toolkit of [6,8,22] extends beyond the class of low doubling dimension metrics, since the proof of Theorem 3 heavily relies on the embedding techniques proposed in [24].

1.1   Our Techniques

The embedding constructed in the proof of Theorem 3 heavily relies on previous work by Talwar [24] but needs many non-trivial new ideas, a few of which we sketch here. First, we give a quick overview of Talwar's embedding. The rough idea is to recursively decompose balls of points called *clusters* into child clusters of half the radius. This results in a hierarchy of clusters at different scales, which gives rise to a so called *split-tree*. In addition, each cluster is associated with a set of *net points*, which is a small set of well-spaced points covering the cluster. For each cluster, only the edges between the net points of its child clusters are kept. The shortest path between two points can then be approximated by a path that exits each cluster only via the net points. The error introduced due to the shifting of points to net points, as well as the total distortion, can be bounded as the sum of errors over all the scales. In the tree decomposition of the resulting embedding, each bag corresponds to a cluster and consists of the net points of its child clusters. Using the bounded doubling dimension assumption, the number of child clusters and number of net points per cluster can be bounded by constants depending on the doubling dimension and the desired stretch, which bounds the embedding's treewidth.

We want to construct a similar recursive decomposition for metrics with low highway dimension, which, however, turns out to be a non-trivial task. In order to obtain a decomposition we observe that the hubs in the shortest path cover induce a natural clustering of the vertices in G for any r (see Figure 1). Each vertex $v \in V$ whose distance from any hub is larger than $2r$ is said to belong to a *town* that is contained in the ball of radius r centered at v. All vertices that are not part of a town (and hence at distance no more than $2r$ from some hub) are said to be part of the *sprawl*. We will show that towns are nicely separated from other towns and the sprawl, and

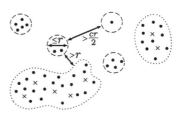

Fig. 1. The sprawl (enclosed by dotted lines) contains vertices close to hubs (crosses). Each town (dashed circles) has small diameter and is far from other vertices.

that the degree of separation is highly sensitive to the choice of c in Definition 1. It turns out that choosing $c = 4$ yields a separation that is just barely too small.

Based on this clustering, we compute a hierarchical decomposition of the graph that we call the *towns decomposition*. It is a laminar family of towns and recursively separates the graph into towns of decreasing scales, and our embedding is computed recursively on this decomposition. The towns decomposition is analogous to the quadtree decomposition in PTASs for Euclidean graphs [6–9] or the split-tree decomposition for low doubling dimension metrics [24], though the particulars differ greatly. At a high level, a town is similar to a cluster in Talwar's split-tree decomposition, though a town can contain many child towns. However a town belongs to the sprawl at a higher scale, which at that scale can be covered by a constant number of balls centered at the hubs. Roughly speaking, we can apply Talwar's decomposition technique to the sprawl and recursively construct a low treewidth embedding for each child town as long as we can somehow attach these embeddings to the embedding of the sprawl.

We prove that to preserve all distances within a town T it suffices to connect tree decompositions of T's subtowns in the towns decomposition via a carefully chosen set of so-called *core hubs* within T. It is noteworthy that unlike the nets in Talwar's split-tree decomposition, the hubs do not form a hierarchy, i.e., a hub at some scale may not be a hub at a lower scale. Nevertheless, we show that core hubs at different scales can be *aligned*: they can be shifted slightly in order to obtain a nested structure that is similar to a hierarchy. We are able to show that this alignment process does not affect the target stretch of our embedding and, most importantly, ensures that the resulting set of *approximate core hubs* within T has small doubling dimension. We first apply Talwar's [24] embedding of low doubling dimension metrics into bounded treewidth graphs to the approximate core hubs and then connect the recursively computed embeddings of the subtowns of smaller scales with each other through the embedding of these hubs. The details are described in section 4.

The most intricate part of our result is to prove low doubling dimension of the approximate core hubs. The general idea is to rely on the local sparsity of the shortest path covers: by definition, the core hubs lie in the sprawls of various scales, and for scale r the sprawl can be covered by balls of radius $2r$ around the hubs of the shortest path cover. In a low highway dimension graph, any ball B of radius $cr/2$ contains only a small number of hubs. Hence, to bound the doubling dimension, we attempt to use these hubs as centers of balls of smaller radius to cover the core hubs. Since these balls have radius $2r < cr/2$, this scheme can be applied recursively in order to cover the core hubs in B with balls of half the radius. Several issues arise with this approach though. To give only one example, part of the sprawl for scale r in B might be covered by balls centered at hubs outside of B. However a key insight of our work is that in fact the number of hubs in the *vicinity* of a ball is also bounded when using Definition 1 for the highway dimension.

2   Embeddings for Low Doubling Dimension Metrics

Next we formally define the treewidth and summarize the properties of Talwar's [24] embedding for low doubling dimension metrics that we require. Let $G = (V, E)$ be a graph. For $u, v \in V$ we denote the length of the shortest path between u and v by $\mathrm{dist}(u, v)$ and the distance between two sets $S, T \subset V$ by $\mathrm{dist}(S, T) = \min_{u \in S, v \in T} \mathrm{dist}(u, v)$. If the metric used for distances is ambiguous we specify the graph in the subscript, such as $\mathrm{dist}_G(u,v)$ or $\mathrm{dist}_H(u,v)$. The diameter $\mathrm{diam}(\cdot)$ of a graph or set of vertices is the maximum distance between any two vertices.

Definition 4. A *tree decomposition* D of a graph G is a (rooted) tree with vertices $b_1,\ldots,b_t$, where each b_i, for $i \in \{1,\ldots,t\}$, is called a *bag* and is a subset of V. Additionally it satisfies the following properties: (a) $\bigcup_{i=1}^{t} b_i = V$, (b) for every edge $\{u,v\} \in E$ there is a bag b with $u,v \in b$, and (c) for every $v \in V$ the bags containing v form a connected subtree of D. The *width* of the tree decomposition is $\max\{|b_i - 1| \mid i \in \{1,\ldots,t\}\}$. The *treewidth* of a graph G is the minimum width of any tree decomposition for G.

To construct our embedding we will mainly focus on the shortest path metric of the graph G. We let the distance function of every considered metric be the function $\mathrm{dist}(\cdot,\cdot)$ of the underlying graph. Though the treewidth is a property of a graph's edge set, whereas doubling dimension is a property of the metric it defines, Talwar [24] shows that low doubling dimension graphs can be approximated to within $1 + \varepsilon$ by bounded treewidth graphs. Formally this means the following.

Definition 5. Let (X,dist) be a metric, and $\mathcal{D}$ be a distribution over metrics (X,dist'). If for all $x,y \in X$, $\mathrm{dist}(x,y) \leq \mathrm{dist}'(x,y)$ for each $\mathrm{dist}' \in \mathcal{D}$, and $\mathbf{E}_{\mathrm{dist}'\in\mathcal{D}}[\mathrm{dist}'(x,y)] \leq a \cdot \mathrm{dist}(x,y)$, then $\mathcal{D}$ is an *embedding* with (expected) *stretch* or *distortion* a. If every $\mathrm{dist}' \in \mathcal{D}$ is the shortest path metric of some graph class $\mathcal{G}$, then $\mathcal{D}$ is a *(probabilistic) embedding into $\mathcal{G}$*.

The main result of Talwar [24] that we use for our embedding of low highway dimension graphs into bounded treewidth graphs, is the following.

Theorem 6 ([24]). *Let (X,dist) be a metric with doubling dimension d and aspect ratio α. For any $\varepsilon > 0$, there is a polynomial-time computable probabilistic embedding H of (X,dist) with treewidth $(d\log(\alpha)/\varepsilon)^{O(d)}$ and expected distortion $1 + \varepsilon$.*

As described in the introduction, Talwar's embedding employs a randomized *split-tree* decomposition, which is a hierarchical decomposition of the vertices X of a metric into *clusters* of smaller and smaller diameter. A cluster is a subset of X, where the highest cluster is X itself and the lowest ones are individual vertices. Each level of this hierarchy is associated with an index. Our construction of the embedding for low highway dimension graphs also has levels associated with indices, but these have different growth rates. To avoid confusion we will denote the levels of Talwar's split-tree decomposition with indices $\bar{i},\bar{j}$, etc., and ours with indices i,j etc.

The tree decomposition constructed from the split-tree has a bag for each cluster. The tree on the bags exactly corresponds to the split-tree. Each bag contains a coarse set of points of the cluster. More concretely, for a metric (X,dist), a subset $Y \subseteq X$ is called a δ-*cover* if for every $u \in X$ there is a $v \in Y$ such that $\mathrm{dist}(u,v) \leq \delta$. A δ-*net* is a δ-cover with the additional property that dist$(u, v) > \delta$ for all vertices $u, v \in Y$. For a cluster C on level $\bar{i}$ the corresponding bag contains a $\Theta(\varepsilon 2^{\bar{i}}/(d \log \alpha))$-net of C. For every bag b the graph embedding contains a complete graph on the nodes in b. The net in each bag thus serves as a set of *portals*, through which connections leaving the cluster are routed, analogous to those in [7].

3 Properties of Low Highway Dimension Graphs

We assume w.l.o.g. that every shortest path is unique by slightly perturbing edge lengths. Thus it is possible to compute locally $O(k \log k)$-sparse shortest path covers in polynomial time [2] (or locally k-sparse covers in time $n^{O(k)}$). We can show that computing the highway dimension is NP-hard even for graphs with unit edge lengths, so in general approximations are needed.

An important observation is that the vertices of low highway dimension graphs are grouped together in all regions that are far from the hubs. This gives rise to our main observation on the structure of low highway dimension graphs, as summarized in the following definition: for any scale the vertices are partitioned into one *sprawl* and several *towns* with large separations in between.

Definition 7. Given a shortest path cover SPC(r) for scale r, and a vertex $v \in V$ such that dist$(v, \text{SPC}(r)) > 2r$, we call the set $T = \{u \in V \mid \text{dist}(u, v) \leq r\}$ a *town* for scale r. The *sprawl* for scale r is the set of all vertices that are not in towns.

Note that the vertices of the sprawl are at most $2r$ away from a hub, but there can be vertices in towns that are closer than $2r$ to some hub, as long as the town has some other vertex that is farther away. Note also that the towns are defined with respect to a shortest path cover SPC(r), and using two distinct shortest path covers will not always result in the same set of towns. We will fix a minimal shortest path cover SPC(r) for any scale r and only consider towns with respect to this cover. We summarize the basic properties of towns below.

Lemma 8. *Let T be a town of scale r. Then* diam$(T) \leq r$ *and* dist$(T, V \setminus T) > r$. *For any vertex v of the sprawl of scale r,* dist$(v, \text{SPC}(r)) \leq 2r$.

We will exploit this structure for growing scales to construct our embedding. More concretely, we will consider scales $r_i = (c/4)^i$ for values $i \in \mathbb{N}_0$ and call i the *level* of the sprawl, towns, and shortest path cover of scale r_i. We choose our scales in this way since $2r_i = cr_{i-1}/2$. As a consequence, a ball of radius $2r_i$ around a hub of level i that covers part of the sprawl contains at most s hubs of the next lower level $i - 1$ if the shortest path covers are locally s-sparse. We will exploit this in our analysis in order to bound the treewidth of our embedding.

Note that the scales do not grow if $c = 4$, i.e., if the highway dimension definition is not violated. In the introduction we claimed that we need the violation in order to obtain large separations between towns and other vertices of the graph. It turns out that for violation $\lambda = 0$ it is technically possible to have growing scales with similar properties that can be used recursively. However the growth of the scales and the separation between a town and the rest of the graph, as given by Lemma 8, are inevitably connected. In particular, if $\lambda = 0$ the largest separation obtainable is at most r. Our reason for introducing non-zero violations in Definition 1 is that we need separations greater than r for our construction.

By scaling we can assume that the shortest distance between any two vertices is slightly more than $c/2$. Hence $\text{SPC}(r_0) = \emptyset$ since there are no paths of length in $(r_0, cr_0/2]$. Throughout this paper we will assume that the shortest path covers are minimal. In particular this means that on level 0 there is no sprawl, and each vertex forms a singleton town. The highest level we consider is $m = \lceil \log_{c/4} \text{diam}(G) \rceil$. At this level $\text{SPC}(r_m) = \emptyset$ and hence the whole vertex set V of the graph is a town.

We show next that towns of different levels form a laminar family $\mathcal{T}$. Due to this laminar structure of towns we will use tree terminology such as *parents, children, siblings, ancestors,* and *descendants* of towns in $\mathcal{T}$. The *root* of the laminar family is the highest level town V.

Lemma 9. *Given a graph G, the set $\mathcal{T} := \{T \subseteq V \mid T$ is a town on level $i \in \mathbb{N}_0\}$ forms a laminar family. Furthermore, any town $T \in \mathcal{T}$ on level i either has 0 or at least 2 child towns, and in the latter case these are towns on levels below i.*

We refer to the laminar family $\mathcal{T}$ as the *towns decomposition* of G. Note that although a town $T \in \mathcal{T}$ appears once in $\mathcal{T}$, T can be a town on multiple levels of the shortest path covers, if it is a town with respect to both $\text{SPC}(r_i)$ and $\text{SPC}(r_{i+1})$. From now on we will consider the graph metric (V, dist_G) induced by G instead of G itself. All properties of towns and sprawl, such as given by Lemma 8 and 9, are still valid in the metric.

4 Constructing the Embedding

We now describe our algorithm in more detail. All missing parts leading to the proof of Theorem 3 are deferred to the full version of the paper. PTASs for Euclidean and low doubling graphs [7,24] use graph decomposition coupled with a small number of "portal" nodes: paths leaving a cluster in the decomposition must do so via an appropriate portal, resulting in a small "interface" between distinct clusters in the decomposition. Intuitively, the hubs are natural choices for portals, since long paths through some ball must pass through a hub. However problems crop up almost immediately because hubs are not guaranteed to be well-spaced or consistent between levels, and although all long paths through a ball may be hit by portals, there may be many short paths that go nowhere near one.

We overcome these difficulties using the towns decomposition. Lemma 8 guarantees that towns are isolated from both each other and the sprawl. Consequently, any approximate shortest paths between nodes in a town must remain within that town. The embedding is constructed recursively on the metric using the structure of the towns decomposition $\mathcal{T}$. That is, for a town $T \in \mathcal{T}$ we assume that we have already computed an embedding (and accompanying tree decomposition) with expected stretch $1 + \varepsilon$ for each child town of T. We then connect these embeddings so that distances between them are preserved within a $1 + \varepsilon$ factor in expectation. This gives an embedding for T, and since V itself is the root of the towns decomposition, eventually yields an embedding for G.

The key insight that lets us connect the child towns of T is that there exists a set of so-called *approximate core hubs* X_T in T with low doubling dimension that can serve as the crossroads through which child towns connect. We will compute a low-treewidth embedding of the set X_T based on Theorem 6 and connect the embeddings of the child towns to it. In particular, for every child town T' we will identify a bag b of the tree decomposition of X_T containing hubs that are close to T'. We call b the *connecting bag* of T'. The embedding of T is constructed by connecting every vertex in each child town to every hub in the corresponding connecting bag. This means that short connections between child towns can be routed directly through hubs in the connecting bags. Long connections on the other hand can be routed through the embedding of the core hubs X_T at only a small overhead.

The tree decomposition for T is constructed by connecting each tree decomposition $D_{T'}$ for a child town T' to the corresponding connecting bag b of the tree decomposition D_X for the hubs in X_T. Even though this yields a tree of bags containing all vertices of the town T, properties (b) and (c) of Definition 4 might be violated by this initial attempt. We need to make two modifications to the bags: first we need to add all vertices of b to each bag of $D_{T'}$. Since the treewidth of D_X is bounded by Theorem 6, this does not let the bags grow by too much. Then we also need to add all hubs of X_T in the child town T' to each bag of $D_{T'}$, and to b and all descendants of b in D_X. To bound the growth of the bags in this step, we will bound the number of hubs in X_T in a child town T'.

The set X_T is an approximate hub set of T. To define the set properly we need some additional insights on the structure of hubs of different levels in T. The *core* of T is the intersection of sprawls formed by removing all subtowns of T above a given level:

Definition 10. Let $T \in \mathcal{T}$ be a town on level j, and let S_i be the sprawl of V on level $i \leq j$. The *core* C_i *of* T *on level* i is inductively defined as follows: $C_j = T$, and $C_i = S_i \cap C_{i+1}$ for $i \leq j - 1$. The *core hubs of* T are given by the set $\bigcup_{i=1}^{j-1} C_i \cap \mathrm{SPC}(r_i)$.

By this definition a town T on level j can be partitioned into its core on level i and its child towns on levels i and higher. Observe also that the set system $\{C_i\}_{i=0}^{j}$ forms a chain, i.e., $C_{i-1} \subseteq C_i$. Intuitively, the core hubs should have low doubling dimension: if the shortest path covers are locally s-sparse, then in a ball around a hub at level i there will be at most s hubs in that ball on level $i - 1$ that cover the core on that level. In fact one can show that the doubling dimension of the core hubs is fairly small but unfortunately not small enough for our purposes. In particular, we need the doubling dimension to be ndependent of the aspect ratio α of the metric. To circumvent this issue, roughly speaking, we shift each core hub so that it overlaps with lower level core hubs if possible, making the hubs nested to some degree. However, in order to preserve distances we will only shift them by at most an ε fraction. This shifting produces the set X_T of *approximate core hubs* of T, which we use to construct our core embedding. Note that we do not use the approximate hubs X_T to define our towns decomposition, only to produce a low-treewidth core embedding. We rely on the following non-trivial properties, which require an intricate proof.

Theorem 11. *Let $\mathcal{T}$ be a towns decomposition of a graph of highway dimension k, given by locally s-sparse shortest path covers on all levels with violation $\lambda > 0$. For any town $T \in \mathcal{T}$ of a level j there exists a polynomially computable set of approximate core hubs $X_T \subseteq T$ such that for any core hub $h \in C_i \cap \mathrm{SPC}(r_i)$ of T on level $i \in \{1, \ldots, j-1\}$, there is a vertex $h' \in X_T$ with $\mathrm{dist}_G(h, h') \le \varepsilon r_i$, and the doubling dimension of X_T is $d = O(\log(\frac{ks \log(1/\varepsilon)}{\lambda})/\lambda)$.*

From now on, we use d to denote the above doubling dimension bound. Unfortunately we cannot apply the embedding of Theorem 6 to the set X_T directly because we must show the existence of a valid, low-width tree decomposition of the resulting embedding of T, after connecting the embeddings of T's child towns to the embedding H_X of X_T. For this to work we need to make sure that the approximate core hubs contained in the same child town T' do not end up in different bags in the tree decomposition D_T of H_T. Our solution is to pick a representative core hub for each child town T'. Specifically, let $Y_T \subseteq X_T$ contain one arbitrary approximate core hub for each child town T' of T for which $T' \cap X_T \ne \emptyset$. We say that a vertex $v \in Y_T$ of a child town T' *represents* the nodes in $X_T \cap T'$ (including v itself). Since Y_T is a sub-metric of X_T it inherits the doubling dimension bound of Theorem 11. Therefore we can compute an embedding for the metric (Y_T, dist_G) with bounded treewidth by Theorem 6.

Given the embedding of Y_T, we convert it into an embedding H_X of X_T by replacing a vertex $v \in Y_T$ with the clique on all approximate core hubs that v represents in the embedding. We obtain the tree decomposition D_X of H_X from the decomposition of the embedding for Y_T by also replacing v with all the hubs it represents in each bag containing v. It is easy to see that D_X is a valid tree decomposition, i.e., it satisfies all properties of Definition 4. We can show that the number of approximate core hubs in each child town is bounded, and therefore the growth of the treewidth caused by replacing a vertex by its represented hubs is also bounded. We also need to bound the extra distortion incurred by going from Y_T to X_T and show that a $1+\varepsilon$ distortion on Y_T translates into a $1 + O(\varepsilon)$ distortion on X_T, which entails reproving the relevant parts of Theorem 6.

After computing the embedding H_X for X_T, we connect each recursively computed embedding for the child towns of T to H_X to form the final embedding H_T. We need to argue that H_X exists every time there are child towns to connect. From Lemma 9 we know that T has at least two child towns if it has any. We can show that there is a core hub h in T on any shortest path between a pair of child towns. By Theorem 11, there is an approximate core hub in X_T close to h. Since X_T is non-empty, H_X exists. Once we compute H_X we connect every vertex of a child town T' to all hubs in a bag b of the tree decomposition D_X of H_X. This bag b is $\log_2(1/\varepsilon) + \log d$ levels higher in the split-tree decomposition than the level corresponding to the shortest distance that needs to be bridged from T' to any other vertex in T. At the same time we will make sure that the net defining b is fine enough so that lengths of connections passing through b are preserved to a sufficient degree. This way, short connections from T' to core hubs with length up to $O(1/\varepsilon)$ times the separation of T' are preserved in expectation by routing through the hubs in b. Connections to more distant hubs can be rerouted from a hub close to T' through the embedding H_X with only an ε overhead.

Recall that levels of the split-tree decomposition are denoted by $\bar{i}, \bar{j}$ etc. To determine the level of the bag b, note that due to our growth rate of $c/4 = 1 + \lambda/4$ of the levels (and the assumption that the violation λ is at most 4) the intervals $(r_i, 2r_i]$ of the shortest path covers might overlap. Let i be the level for which the distance between T' and its closest sibling town lies in the interval $(r_i, r_{i+1}]$, and let $\bar{i} = \lceil \log_2 r_i \rceil$ be the corresponding level of the split tree decomposition of D_X. Now let $h \in X_T$ be the closest approximate core hub to T' (which might lie inside of T'). If $\bar{j}$ is the highest level of D_X, i.e. it is the level of the cluster containing all of X_T, then the bag b of the tree decomposition D_X is the one on level $\bar{l} = \min\{\bar{j}, \bar{i} + \log_2(1/\varepsilon) + \log_2 d\}$ for which the corresponding cluster C contains h. All edges between vertices of T' and b are added to the embedding for T, and we call the bag b the *connecting bag* for T'.

Note that there are several parameters ε we can adjust independently: the target distortion of Talwar's algorithm, the level in the split-tree decomposition at which a child town is attached, and the amount of adjustment permitted in defining X_T. The latter two parameters we set to ε, but the distortion in Theorem 6 needs to be smaller. We use ε' for the target distortion of this embedding and set $\varepsilon' = \varepsilon^2$.

References

1. Abraham, I., Fiat, A., Goldberg, A.V., Werneck, R.F.: Highway dimension, shortest paths, and provably efficient algorithms. In: Proceedings, ACM-SIAM Symposium on Discrete Algorithms, pp. 782–793 (2010)
2. Abraham, I., Delling, D., Fiat, A., Goldberg, A.V., Werneck, R.F.: VC-dimension and shortest path algorithms. In: Aceto, L., Henzinger, M., Sgall, J. (eds.) ICALP 2011, Part I. LNCS, vol. 6755, pp. 690–699. Springer, Heidelberg (2011)
3. Abraham, I., Delling, D., Fiat, A., Goldberg, A.V., Werneck, R.F.: Highway dimension, shortest paths, and provably efficient shortest path algorithms. Technical Report (2013)
4. Ageev, A.A.: An approximation scheme for the uncapacitated facility location problem on planar graphs
5. Ageev, A.A.: A criterion of polynomial-time solvability for the network location problem. In: Proceedings, MPS Conference on Integer Programming and Combinatorial Optimization, pp. 237–245 (1992)

6. Arora, S.: Polynomial time approximation schemes for euclidean traveling salesman and other geometric problems. J. ACM **45**(5), 753–782 (1998)
7. Arora, S.: Approximation schemes for np-hard geometric optimization problems: A survey. Math. Programming **97**(1–2), 43–69 (2003)
8. Arora, S., Raghavan, P., Rao, S.: Approximation schemes for euclidean k-medians and related problems. In: Proceedings, ACM Symp. on Theory of Computing, pp. 106–113 (1998)
9. Arora, S.: Polynomial time approximation schemes for euclidean traveling salesman and other geometric problems. In: Proceedings, ACM Symp. on Theory of Computing, pp. 2–11 (1996)
10. Bartal, Y.: On approximating arbitrary metrices by tree metrics. In: Proceedings, ACM Symp. on Theory of Computing, pp. 161–168 (1998)
11. Bartal, Y., Gottlieb, L.-A., Krauthgamer, R.: The traveling salesman problem: low-dimensionality implies a polynomial time approximation scheme. In: Proceedings, ACM Symp. on Theory of Computing, pp. 663–672 (2012)
12. Bast, H., Funke, S., Matijevic, D., Sanders, P., Schultes, D.: In transit to constant time shortest-path queries in road networks. In: Algorithm Engineering & Experiments. SIAM (2007)
13. Bast, H., Funke, S., Matijevic, D.: Ultrafast shortest-path queries via transit nodes. The Shortest Path Problem: Ninth DIMACS Implementation Challenge **74**, 175–192 (2009)
14. Bateni, M., Chekuri, C., Ene, A., Hajiaghayi, M.T., Korula, N., Marx, D.: Prize-collecting steiner problems on planar graphs. In: Proceedings, ACM-SIAM Symposium on Discrete Algorithms, pp. 1028–1049 (2011)
15. Borradaile, G., Kenyon-Mathieu, C., Klein, P.: A polynomial-time approximation scheme for steiner tree in planar graphs. In: Proceedings, ACM-SIAM Symposium on Discrete Algorithms, pp. 1285–1294 (2007)
16. Chlebík, M., Chlebíková, J.: Approximation hardness of the steiner tree problem on graphs. In: Penttonen, M., Schmidt, E.M. (eds.) SWAT 2002. LNCS, vol. 2368, pp. 170–179. Springer, Heidelberg (2002)
17. Engebretsen, L., Karpinski, M.: Approximation hardness of TSP with bounded metrics. In: Orejas, F., Spirakis, P.G., van Leeuwen, J. (eds.) ICALP 2001. LNCS, vol. 2076, pp. 201–212. Springer, Heidelberg (2001)
18. Fakcharoenphol, J., Rao, S., Talwar, K.: A tight bound on approximating arbitrary metrics by tree metrics. In: Proceedings, ACM Symp. on Theory of Computing, pp. 448–455. ACM (2003)
19. Guha, S., Khuller, S.: Greedy strikes back: Improved facility location algorithms. J. Algorithms **31**(1), 228–248 (1999)
20. Gupta, A., Krauthgamer, R., Lee, J.R.: Bounded geometries, fractals, and low-distortion embeddings. In: Proceedings, IEEE Symposium on Foundations of Computer Science, pp. 534–543 (2003)
21. Klein, P.: A linear-time approximation scheme for tsp in undirected planar graphs with edge-weights. SIAM Journal on Computing **37**(6), 1926–1952 (2008)
22. Mitchell, J.S.B.: Guillotine subdivisions approximate polygonal subdivisions: A simple polynomial-time approximation scheme for geometric tsp, k-mst, and related problems. SIAM Journal on Computing **28**(4), 1298–1309 (1999)
23. Robertson, N., Seymour, P.D.: Graph minors. ii. algorithmic aspects of tree-width. J. Algorithms **7**(3), 309–322 (1986)
24. Talwar, K.: Bypassing the embedding: algorithms for low dimensional metrics. In: Proceedings, ACM Symp. on Theory of Computing, pp. 281–290 (2004)
25. Vazirani, V.V.: Approximation Algorithms. Springer-Verlag, New York (2001). Inc

Lower Bounds for the Graph Homomorphism Problem

Fedor V. Fomin[1,3], Alexander Golovnev[2,3]([✉]), Alexander S. Kulikov[3],
and Ivan Mihajlin[3,4]

[1] University of Bergen, Bergen, Norway
[2] New York University, New York, USA
alexgolovnev@gmail.com
[3] St. Petersburg Department of Steklov Institute of Mathematics,
Saint Petersburg, Russia
[4] UC San Diego, San Diego, USA

Abstract. The graph homomorphism problem (HOM) asks whether the vertices of a given n-vertex graph G can be mapped to the vertices of a given h-vertex graph H such that each edge of G is mapped to an edge of H. The problem generalizes the graph coloring problem and at the same time can be viewed as a special case of the 2-CSP problem. In this paper, we prove several lower bounds for HOM under the Exponential Time Hypothesis (ETH) assumption. The main result is a lower bound $2^{\Omega\left(\frac{n \log h}{\log \log h}\right)}$. This rules out the existence of a single-exponential algorithm and shows that the trivial upper bound $2^{\mathcal{O}(n \log h)}$ is almost asymptotically tight.

We also investigate what properties of graphs G and H make it difficult to solve HOM(G, H). An easy observation is that an $\mathcal{O}(h^n)$ upper bound can be improved to $\mathcal{O}(h^{\mathrm{vc}(G)})$ where $\mathrm{vc}(G)$ is the minimum size of a vertex cover of G. The second lower bound $h^{\Omega(\mathrm{vc}(G))}$ shows that the upper bound is asymptotically tight. As to the properties of the "right-hand side" graph H, it is known that HOM(G, H) can be solved in time $(f(\Delta(H)))^n$ and $(f(\mathrm{tw}(H)))^n$ where $\Delta(H)$ is the maximum degree of H and $\mathrm{tw}(H)$ is the treewidth of H. This gives single-exponential algorithms for graphs of bounded maximum degree or bounded treewidth. Since the chromatic number $\chi(H)$ does not exceed $\mathrm{tw}(H)$ and $\Delta(H)+1$, it is natural to ask whether similar upper bounds with respect to $\chi(H)$ can be obtained. We provide a negative answer by establishing a lower bound $(f(\chi(H)))^n$ for every function f. We also observe that similar lower bounds can be obtained for locally injective homomorphisms.

1 Introduction

A *homomorphism* $G \to H$ from an undirected graph G to an undirected graph H is a mapping from the vertex set G to that of H such that the image of every edge of G is an edge of H. Then the GRAPH HOMOMORPHISM problem HOM(G, H)

The full version of the paper is available at http://arxiv.org/abs/1502.05447

© Springer-Verlag Berlin Heidelberg 2015
M.M. Halldórsson et al. (Eds.): ICALP 2015, Part I, LNCS 9134, pp. 481–493, 2015.
DOI: 10.1007/978-3-662-47672-7_39

is the problem to decide for given graphs G and H, whether $G \to H$. Many combinatorial structures in G, for example independent sets and proper vertex colorings, may be viewed as graph homomorphisms to a particular graph H, see the book of Hell and Nešetřil [18] for a thorough introduction to the topic. It was shown by Feder and Vardi in [8] that the CONSTRAINT SATISFACTION PROBLEM (CSP) can be interpreted as a homomorphism problem on relational structures, and thus GRAPH HOMOMORPHISM encompasses a large family of problems generalizing COLORING but less general than CSP.

Hell and Nešetřil showed that for any fixed simple graph H, the problem whether there exists a homomorphism from G to H is solvable in polynomial time if H is bipartite, and NP-complete if H is not bipartite [17]. Since then, algorithms for and the complexity of graph homomorphisms (and homomorphisms between other discrete structures) have been studied studied intensively [1, 2, 15, 26, 27].

There are two different ways graph homomorphisms are used to extract useful information about graphs. Let us consider two homomorphisms, from a "small" graph F into a "large" graph G and from a "large" graph G into a "small" graph H, which can be represented by the following formula (here we borrow the intuitive description from the book of Lovász [25]): $F \to G \to H$. Then "left-homomorphisms" from various small graphs F into G are useful to study the local structure of G. For example, if F is a triangle, then the number of "left-homomorphisms" from F into G is the number of triangles in graph G. This type of information is closely related to sampling, and we refer to the book of Lovász [25] which provides many applications of homomorphisms. "Right-homomorphisms" into "small" different graphs H are related to global properties of graph G.

The trivial brute-force algorithm solving "left-homomorphism" from an f-vertex graph F into an n-vertex graph G runs in time $2^{\mathcal{O}(f \log n)}$: we try all possible vertex subsets of G of size at most f, which is $n^{\mathcal{O}(f)}$ and then for each subset try all possible f^f mappings into it from F. Interestingly, this naïve algorithm is asymptotically optimal. Indeed, as it was shown by Chen et al. [4], assuming Exponential Time Hypothesis (ETH), there is no $g(k)n^{o(k)}$ time algorithm deciding if an input n-vertex graph G contains a clique of size at least k, for any computable function g. Since this is a very special case of GRAPH HOMOMORPHISM HOM(F, G) with F being a clique of size k, the result of Chen et al. rules out algorithms for GRAPH HOMOMORPHISM of running time $g(f)2^{o(f \log n)}$, from F to G, when the number of vertices f in F is significantly smaller than the number of vertices n in G.

Brute-force for "right-homomorphism" HOM(G, H), checking all possible mappings from G into H, also runs in time $2^{\mathcal{O}(n \log h)}$, where h is the number of vertices in H. However, prior to our work there were no results indicating that asymptotically better algorithms, say of running time $2^{\mathcal{O}(n)}$, are highly unlikely.

Our interest in "right-homomorphisms" is due to the recent developments in the area of exact exponential algorithms for COLORING and 2-CSP (CSP where all constraints have arity at most 2) problems. The area of exact exponential algorithms is about solving intractable problems significantly faster than the trivial exhaustive search, though still in exponential time [12]. For example, as for GRAPH HOMOMORPHISM, a naïve brute-force algorithm for coloring an n-vertex graph G in h colors is to try for every vertex a possible color, resulting in the running time $\mathcal{O}^*(h^n) = 2^{\mathcal{O}(n \log h)}$.[1] Since h can be of order $\Omega(n)$, the brute-force algorithm computing the chromatic number runs in time $2^{\mathcal{O}(n \log n)}$. It was already observed in 1970s by Lawler [21] that the brute-force for the COLORING problem can be beaten by making use of dynamic programming over maximal independent sets resulting in single-exponential running time $\mathcal{O}^*((1 + \sqrt[3]{3})^n) = \mathcal{O}(2.45^n)$. Almost 30 years later Björklund, Husfeldt, and Koivisto [3] succeeded to reduce the running time to $\mathcal{O}^*(2^n)$. It is well-known that COLORING is a special case of graph homomorphism. More precisely, graph G is colored in at most h colors if and only if $G \to K_h$, where K_h is a complete graph on h vertices. Due to this, very often in the literature $\mathrm{HOM}(G, H)$, when $h = |V(H)| \leq n$, is referred as H-coloring of G. And as we observed already, for H-coloring, the brute-force algorithm solving H-coloring runs in time $2^{\mathcal{O}(n \log h)}$. In spite of all the similarities between graph coloring and homomorphism, no substantially faster algorithm was known and it was an open question in the area of exact algorithms if there is a single-exponential algorithm solving H-coloring in time $2^{\mathcal{O}(n+h)}$ [11,28,31,32], see also [12, Chapter12].

On the other hand, GRAPH HOMOMORPHISM is a special case of 2-CSP with n variables and domain of size h. It was shown by Traxler [30] that unless the Exponential Time Hypothesis (ETH) fails, there is no algorithm solving 2-CSP with n variables and domain of size h in time $h^{o(n)} = 2^{o(n \log h)}$. This excludes (up to ETH) the existence of a single-exponential c^n time algorithm for some constant $c > 1$ for 2-CSP.

Our Results. In this paper we show that from the algorithmic perspective, the behavior of "right-homomorphism" is, unfortunately, much closer to 2-CSP than to COLORING. The main result of this paper is the following theorem, which excludes (up to ETH) resolvability of $\mathrm{HOM}(G, H)$ in time $2^{o\left(\frac{n \log h}{\log \log h}\right)}$.

Theorem 1. *Unless ETH fails, for any constant $d > 0$ there exists a constant $c = c(d) > 0$ such that for any function $3 \leq h(n) \leq n^d$, there is no algorithm solving $\mathrm{HOM}(G, H)$ for an n-vertex graph G and $h(n)$-vertex graph H in time*

$$\mathcal{O}^* \left(2^{\frac{cn \log h(n)}{\log \log h(n)}} \right). \tag{1}$$

Remark 1. In order to obtain more general results, in all lower bounds proven in this paper we assume implicitly that the number h of vertices of the graph H is a function of the number n of the vertices of the graph G. At the same

---

[1] $\mathcal{O}^*(\cdot)$ hides polynomial factors in the input length. Most of the algorithms considered in this paper take graphs G and H as an input. By saying that such an algorithm has a running time $\mathcal{O}^*(f(G, H))$ we mean that the running time is upper bounded by $p(|V(G)| + |E(G)| + |V(H)| + |E(H)|) \cdot f(G, H)$ for a fixed polynomial p.

time, to exclude some pathological cases we assume that the function $h(n)$ is "reasonable" meaning that it is non-decreasing and time-constructible.

While Theorem 1 rules out the existence of a single-exponential algorithm for GRAPH HOMOMORPHISM, single-exponential algorithms can be found in the literature for a number of restricted conditions on the "right hand" graph H. For example, when the treewidth of H is at most t, or more generally, when the clique-width of the core of H does not exceed t, the problem is solvable in time $f(t)^n$ for some function f [32]. Another example is when the maximum vertex degree $\Delta(H)$ of H is bounded by a constant. In this case, it is easy to see that a simple branching algorithm also resolves HOM(G, H) in single-exponential time. The chromatic number $\chi(H)$ of H does not exceed the treewidth of H, nor does it exceed $\Delta(H)$ (plus one). Therefore, it is natural to ask if a single-exponential algorithm exists when the chromatic number of H is bounded. Unfortunately, this is unlikely to happen.[2]

Theorem 2 (∗). *Unless ETH fails, for any function $f\colon \mathbb{N} \to \mathbb{N}$ there is no algorithm solving* HOM(G, H) *for an n-vertex graph G and a graph H in time* $\mathcal{O}^*\left((f(\chi(H)))^n\right)$.

Another interesting question about homomorphisms concerns the complexity of the problem when graph G poses a specific structure. In particular, when the treewidth of G does not exceed t, then HOM(G, H) is solvable in time $\mathcal{O}^*(h^t)$ [7]. Let vc(G) be the minimum size of a vertex cover in graph G. We prove that

Theorem 3 (∗). *Unless ETH fails, for any constant d there exists a constant $c = c(d) > 0$ such that for any function $3 \le h(n) \le n^d$, there is no algorithm solving* HOM(G, H) *for an n-vertex graph G and $h(n)$-vertex graph H in time* $\mathcal{O}^*\left(h(n)^{c \cdot \mathrm{vc}(G)}\right)$.

Since vc(G) is always at most the treewidth of G, Theorem 3 shows that the known bounds $\mathcal{O}^*(h^t) = \mathcal{O}^*(h^{\mathrm{vc}(G)})$ on the complexity of homomorphisms from graphs of bounded treewidth and vertex cover are asymptotically optimal (Note that the minimum vertex cover of G can be found in time $1.28^{\mathrm{vc}(G)} \cdot n^{\mathcal{O}(1)}$ [5]). It is interesting to compare Theorem 3 with existing results on variants of graph homomorphism parameterized by the vertex cover and the treewidth of an input graph. The techniques of obtaining lower bounds developed by Lokshtanov, Marx, and Saurabh in [23], can be used to show that COLORING cannot be computed in time $2^{o(\mathrm{vc}(G) \log \mathrm{vc}(G))}$, unless ETH fails [22]. However, the question if coloring in h colors of a given graph G can be done in time $h^{o(\mathrm{vc}(G))}$ remains open. Another work related to Theorem 3 is the paper of Marx [26] providing lower bounds on the running time of algorithms for "left-homomorphisms" on classes of structures of bounded treewidth.

As a byproduct of our proof of Theorem 1, we obtain similar lower bounds for locally injective graph homomorphisms. A homomorphism $f\colon G \to H$ is

---

[2] Proofs of the statements marked with (∗) are omitted due to space restrictions.

called *locally injective* if for every vertex $u \in V(G)$, its neighborhood is mapped injectively into the neighborhood of $f(u)$ in H, i.e., if every two vertices with a common neighbor in G are mapped onto distinct vertices in H. As graph homomorphism generalizes graph coloring, locally injective graph homomrohism can be seen as a generalization of graph distance constrained labelings. An $L(2,1)$-labeling of a graph G is a mapping from $V(G)$ into the nonnegative integers such that the labels assigned to vertices at distance 2 are different while labels assigned to adjacent vertices differ by at least 2. This problem was studied intensively in combinatorics and algorithms, see e.g. Griggs and Yeh [14] or Fiala et al. [9]. Fiala and Kratochvíl suggested the following generalization of $L(2,1)$-labeling, we refer [10] for the survey. For graphs G and H, an $H(2,1)$-labeling is a mapping $f : V(G) \to V(H)$ such that for every pair of distinct adjacent vertices $u, v \in V(G)$, images $f(u)\, f(v)$ are distinct and nonadjacent in H. Moreover, if the distance between u and v in G is two, then $f(u) \neq f(v)$. It is easy to see that a graph G has an $L(2,1)$-labeling with maximum label at most k if and only if there is an $H(2,1)$-labeling for H being a k-vertex path. Then the following is known, see for example [10], there is an $H(2,1)$-labeling of a graph G if and only if there is a locally injective homomorphism from G to the complement of H.

Several single-exponential algorithms for $L(2,1)$-labeling can be found in the literature, the most recent algorithm is due to Junosza-Szaniawski et al. [20] which runs in time $\mathcal{O}(2.6488^n)$. For $H(2,1)$-labeling, or equivalently for locally injective homomorphisms, single-exponential algorithms were known only for special cases when the maximum degree of H is bounded [16] or when the bandwidth of the complement of H is bounded [28]. The following theorem explains why no such algorithms were found for arbitrary graph H.

Theorem 4 (∗). *Unless ETH fails, for any constant $d > 0$ there exists a constant $c = c(d) > 0$ such that for any function $3 \leq h(n) \leq n^d$, there is no algorithm deciding if there is a locally injective homomorphism from an n-vertex graph G and $h(n)$-vertex graph H in time $\mathcal{O}^* \left(2^{\frac{cn \log h(n)}{\log \log h(n)}} \right)$.*

To establish lower bounds for graph homomorhisms, we proceed in two steps. First we obtain lower bounds for LIST GRAPH HOMOMORPHISM by reducing it to the 3-coloring problem on graphs of bounded degree. More precisely, for a given graph G with vertices of small degrees, we construct an instance (G', H') of LIST GRAPH HOMOMORPHISM, such that G is 3-colorable if and only if there exists a list homomorphism from G' to H'. Moreover, our construction guarantees that a "fast" algorithm for list homomorphism parameterized by the number of vertices, size of a vertex cover or the chromatic number, implies an algorithm for 3-coloring violating ETH. The reduction is based on a "grouping" technique, however, to do the required grouping we need a trick exploiting the condition that G has a bounded maximum vertex degree and thus can be colored in a bounded number of colors in polynomial time. In the second step of reductions we proceed from list homomorphisms to normal homomorphisms. Here we need

pecific gadgets with a property that any homomorphism from such a graph to
:self preserves an order of its specific structures.

The remaining part of the paper is organized as follows. In Section 2 we
ive all the necessary definitions. Section 3 contains all the necessary reductions
/hich are used to prove lower bounds for the GRAPH HOMOMORPHISM problem
n Section 4.

2 Preliminaries

Graphs. We consider simple undirected graphs, where $V(G)$ denotes the set of
vertices and $E(G)$ denotes the set of edges of a graph G. For a given subset S
of $V(G)$, $G[S]$ denotes the subgraph of G induced by S, and $G - S$ denotes the
graph $G[V(G) \setminus S]$. A vertex set S of G is an *independent set* if $G[S]$ is a graph
with no edges, and S is a *clique* if $G[S]$ is a complete graph. The set of neighbors
of a vertex v in G is denoted by $N_G(v)$, and the set of neighbors of a vertex set
S is $N_G(S) = \bigcup_{v \in S} N_G(v) \setminus S$. By $N_G[S]$ we denote the closed neighborhood of
the set S, i.e., the set S together with all its neighbors: $N_G[S] = S \cup N_G(S)$. For
an integer n, we use $[n]$ to denote the set of integers $\{1, \ldots, n\}$.

The complete graph on k vertices is denoted by K_k. A *coloring* of a graph
G is a function assigning a color to each vertex of G such that adjacent vertices
have different colors. A k-coloring of a graph uses at most k colors, and the
chromatic number $\chi(G)$ is the smallest number of colors in a coloring of G. By
Brook's theorem, for any connected graph G with maximum degree $\Delta > 2$, the
chromatic number of G is at most Δ unless G is a complete graph, in which case
the chromatic number is $\Delta + 1$. Moreover, a $(\Delta + 1)$-coloring of a graph can be
found in polynomial time by a straightforward greedy algorithm.

Throughout the paper we implicitly assume that there is a total order on the
set of vertices of a given graph. This allows us to treat a k-coloring of a n-vertex
graph simply as a vector in $[k]^n$.

A set $S \subseteq V(G)$ is a vertex cover of G, if for every edge of G at least one of
its endpoints belongs to S.

Let G be an n-vertex graph, $1 \leq r \leq n$ be an integer, and $V(G) = B_1 \sqcup B_2 \sqcup
\ldots \sqcup B_{\lceil \frac{n}{r} \rceil}$ be a partition of the set of vertices of G into sets of size r with the
last set possibly having less than r vertices. Then *the edge preserving r-grouping*
of G with respect to the partition $V(G) = B_1 \sqcup B_2 \sqcup \ldots \sqcup B_{\lceil \frac{n}{r} \rceil}$ is a graph G_r
with vertices $B_1, \ldots, B_{\lceil \frac{n}{r} \rceil}$ such that B_i and B_j are adjacent if and only if there
exist $u \in B_i$ and $v \in B_j$ such that $\{u, v\} \in E(G)$. To distinguish vertices of the
graphs G and G_r, the vertices of G_r will be called *buckets*.

For a graph G, its *square* G^2 has the same set of vertices as G and $\{u, v\} \in
E(G^2)$ if and only if there is a path of length at most 2 between u and v in G
(thus, $E(G) \subseteq E(G^2)$). It is easy to see that if the degree of G is less than Δ
then the degree of G^2 is less than Δ^2 and hence a Δ^2-coloring of G^2 can be
easily found.

Homomorphisms and list homomorphisms. Let G and H be graphs. A mapping $\varphi : V(G) \to V(H)$ is a *homomorphism* if for every edge $\{u, v\} \in E(G)$ its image $\{\varphi(u), \varphi(v)\} \in E(H)$. If there exists a homomorphism from G to H, we often write $G \to H$. The GRAPH HOMOMORPHISM problem HOM(G, H) asks whether or not $G \to H$.

Assume that for each vertex v of G we are given a list $\mathcal{L}(v) \subseteq V(H)$. A *list homomorphism* of G to H, also known as a *list H-colouring* of G, with respect to the lists $\mathcal{L}$, is a homomorphism $\varphi : V(G) \to V(H)$, such that $\varphi(v) \in \mathcal{L}(v)$ for all $v \in V(G)$. The LIST GRAPH HOMOMORPHISM problem LIST-HOM(G, H) asks whether or not graph G with lists $\mathcal{L}$ admits a list homomorphism to H with respect to $\mathcal{L}$.

Exponential Time Hypothesis. Our lower bounds are based on a well-known complexity hypothesis formulated by Impagliazzo, Paturi, and Zane [19].

> **Exponential Time Hypothesis (ETH)**: There is a constant $s > 0$ such that 3-CNF-SAT with n variables and m clauses cannot be solved in time $2^{sn}(n + m)^{\mathcal{O}(1)}$.

This hypothesis is widely applied in the theory of exact exponential algorithms, we refer to [6, 24] for an overview of ETH and its implications.

In our paper we are using the following application of ETH with respect to 3-COLORING. The 3-COLORING problem is the problem to decide whether the given graph can be properly colored in 3 colors.

Proposition 1 (Theorem 3.2 in [24], and Exercise 7.27 in [29]). *Unless ETH fails, there exists a constant $\alpha > 0$ such that 3-COLORING on n-vertex graphs of average degree four cannot be solved in time $\mathcal{O}^*(2^{\alpha n})$.*

It is well known that 3-COLORING remains NP-complete on graphs of maximum vertex degree four. Moreover, the classical reduction, see e.g. [13], allows for a given n-vertex graph G to construct a graph G' with maximum vertex degree at most four and $|V(G')| = \mathcal{O}(|E(G)|)$ such that G is 3-colorable if and only if G' is. Thus Proposition 1 implies the following (folklore) lemma which will be used in our proofs.

Lemma 1. *Unless ETH fails, there exists a constant $\beta > 0$ such that there is no algorithm solving 3-COLORING on n-vertex graphs of maximum degree four in time $\mathcal{O}^*(2^{\beta n})$.*

3 Reductions

This section constitutes the main technical part of the paper and contains all the necessary reductions used in the lower bounds proofs. Using these reductions as building blocks the lower bounds follow from careful calculations. The general pipeline is as follows. To prove a lower bound with respect to a given graph

omplexity measure we take a graph G of maximum degree four that needs to
e 3-colored and construct an equisatisfiable instance (G', H') of LIST-HOM
using Lemma 2 or Lemma 3). We then use Lemma 5 to transform (G', H')
into an equisatisfiable instance (G'', H'') of HOM. Thus, an algorithm checking
whether there exists a homomorphism from G'' to H'' can be used to check
whether the initial graph G can be 3-colored. At the same time we know a lower
bound for 3-COLORING under ETH (Lemma 1). This gives us a lower bound for
HOM. We emphasize that our reductions provide almost tight lower bounds for
HOM under ETH.

Lemma 2 (3-Coloring$(G) \rightarrow$ LIST-HOM(G', H') with small $|V(G')|$).
*There exists an algorithm that given an n-vertex graph G of maximum degree
four and an integer $2 \leq r \leq n$ constructs an instance (G', H') of LIST-HOM
such that $|V(G')| = \lceil n/r \rceil$ and $|V(H')| \leq r^{50r}$ which is satisfiable if and only if
the initial graph G is 3-colorable. The running time of the algorithm is polynomial in n and the size of the output graphs.*

Proof. Constructing G'. Partition the vertices of G into sets of size r (this is
possible since $r \leq n$) arbitrarily and let $G' = G_r$ be the edge preserving r-
grouping of G with respect to this partition. The maximum vertex degree in
graph G' does not exceed $4r$, hence its square can be properly colored with at
most $L = 16r^2 + 1$ colors (and such a coloring can be computed efficiently).
Fix any such coloring and denote by $\ell(B)$ the color of a bucket $B \in V(G')$.
To distinguish this coloring from a 3-coloring of G that we are looking for, in
the following we call $\ell(B)$ a *label* of B. An important property of this labelling
is that all the neighbors of any bucket have different labels. Thus to specify a
neighbor of a given bucket B it is sufficient to specify the label of this neighbor.
This will be crucial for the construction of the graph H' given below.

Constructing H'. The graph H' is constructed as follows. Roughly, it contains
all possible "configurations" of buckets from G', where a configuration of $B \in
V(G')$ contains its label $\ell(B)$, a 3-coloring of all r vertices of the bucket $B \subseteq
V(G)$, and a 3-coloring of all the neighbors of these r vertices in G. We will use
lists to allow mapping of a bucket $B \in V(G')$ to only those configurations that
are consistent with a 3-coloring of the closed neighborhood $N_G[B]$.

A configuration is a tuple $C = (\ell, c, (p_1, \ell_1, q_1, c_1), \ldots, (p_{4r}, \ell_{4r}, q_{4r}, c_{4r}))$ from
$[L] \times [3]^r \times ([r] \times [L] \times [r] \times [3])^{4r}$ and the set of vertices of H' is the set of all
configurations. Thus, the number of vertices in H' is equal to (recall that $r \geq 2$)

$$L \cdot 3^r \cdot (r^2 \cdot L \cdot 3)^{4r} \leq r^{4r} \cdot r^{2r} \cdot (r^2 \cdot r^7 \cdot r^2)^{4r} \leq r^{50r} . \tag{2}$$

For a given bucket $B \in V(G')$ such a configuration C sets the following.
Integer $\ell \in [L]$ is a label of B, $c \in [3]^r$ is a 3-coloring of $B \subseteq V(G)$ (we assume
a fixed order on the vertices of the graph G so that the vector $c \in [3]^r$ can be
uniquely decoded to a 3-coloring of B). The rest of C defines a 3-coloring of
all the vertices adjacent to B in G as follows. Let $\{u_1, v_1\}, \ldots, \{u_k, v_k\} \in E(G)$
be all the edges in the lexicographic order such that $u_i \in B$ and $v_i \notin B$ for all

$i \in [k]$. Note that $k \leq 4r$ since the degree of G is at most 4 and $|B| \leq r$. Then $(p_i, \ell_i, q_i, c_i) \in [r] \times [L] \times [r] \times [3]$ defines an edge $\{v_i, w_i\}$ and a color of v_i as follows: $p_i \in [r]$ is the number of u_i in B, $\ell_i \in [L]$ is the label of the unique possible neighbor $B' \ni v_i$ of B in G', $q_i \in [r]$ is the number of v_i in B', and $c_i \in [3]$ is the color of v_i.

Two configurations $C_1 = (\ell^1, c^1, \{(p_i^1, \ell_i^1, q_i^1, c_i^1)\}_{i=1}^{4r})$ and $C_2 = (\ell^2, c^2, \{(p_i^2, \ell_i^2, q_i^2, c_i^2)\}_{i=1}^{4r})$ are adjacent if their colorings do not contradict each other. I.e., C_1 contains colors of vertices from a bucket labeled by ℓ^2. We require them to be the same as the ones from the coloring c^2 (and similarly for the second configuration). More formally, C_1 and C_2 are adjacent if for every $i \in [4r]$, if $\ell_i^1 = \ell_2$ then c_i^1 is equal to the color of q_i^1-th vertex in the vector c_2, and if $\ell_i^2 = \ell_1$ then c_i^2 is equal to the color of q_i^2-th vertex in c_1.

Defining lists of allowed vertices. We allow to map a bucket $B \in V(G')$ to a configuration $C = (\ell, \ldots) \in V(H')$ if and only if $\ell(B) = \ell$ and C defines a valid 3-coloring of $N_G[B]$ (that is, any two adjacent vertices from $N_G[B]$ are given different colors).

Correctness. We now show that G is 3-colorable if and only if there is a list-homomorphism from G' to H'. The forward direction is clear: given a 3-coloring of G, one can map each bucket B to the configuration containing the label of this bucket and the coloring of $N_G[B]$. For the reverse direction, we take a homomorphism $\phi \colon G' \to H'$ and for each bucket B we decode from $\phi(B)$ the 3-coloring of all the vertices of $N_G[B]$. Note that if $N_G[B] \cap N_G[B'] \neq \emptyset$ for buckets $B, B' \in V(G')$, then $\{B, B'\} \in E(G')$. In this case, the edges of H' guarantee that $\phi(B)$ and $\phi(B')$ assign the same color to each vertex in $N_G[B] \cup N_G[B']$. Hence such a decoding of a 3-coloring from the homomorphism ϕ is well defined. The list constraints of the LIST-HOM instance further guarantee that the resulting 3-coloring is valid.

Running time of the reduction. Clearly, the algorithm takes time polynomial in n and the size of the graphs G' and H'. □

Lemma 3 ($(*)$ **3-Coloring**$(G) \to$ **LIST-HOM**(G', H') **with small** $vc(G')$**).** *There exists an algorithm that given an n-vertex graph G of maximum degree 4 and an integer $2 \leq r \leq n$ constructs an instance (G', H') of LIST-HOM such that $vc(G') = \lceil n/r \rceil$ and $|V(H')| \leq 300^r$ which is satisfiable if and only if the initial graph G is 3-colorable. The running time of the algorithm is polynomial in n and the size of the output graphs.*

Lemma 4 ($(*)$ **LIST-HOM** $\to$ **LIST-HOM with small** $\chi(H')$**).** *Given an instance (G, H) of LIST-HOM and a k-coloring of G one can construct in polynomial time a graph H' such that $\chi(H') \leq k$, $|V(H')| = k|V(H)|$, and (G, H) is equisatisfiable to (G, H').*

Lemma 5 ((∗) LIST-HOM → HOM). *There is a polynomial-time algorithm that from an instance (G, H) of LIST-HOM where $|V(G)| = n$, $|V(H)| = h \geq 3$, $\chi(H) \leq t$ constructs an equisatisfiable instance (G', H') of HOM where $|V(G')| \leq n + \Delta$, $\mathrm{vc}(G') \leq \mathrm{vc}(G) + \Delta$, $|V(H')| \leq \Delta$ for $\Delta = (h+1)(t+11)$, $\chi(H') \leq t+10$.*

The main technical contribution of the paper is Lemma 5. Due to space restrictions we only sketch the proof here. Our goal is to reduce an instance (G, H) of LIST-HOM to an instance (G, H) of HOM. Intuitively, we want to incorporate the list constraints into a new instance of HOM. The first approach is to take $G' = G \cup H$, $H = H$. For every list constraint that forbids $u \in V(G)$ to be mapped to $v \in V(H)$, we add an edge $\{u, v\}$ to $E(G)$. If the graph H from G' is mapped to the graph H' identically, then this edge represents the constraint. Unfortunately, the left-hand side copy of H might not be mapped into the right-hand side one identically(for example, if H has non-trivial automorphisms). In order to get around this, we introduce $h = |H|$ vertices in both G and H. These new vertices play the role of the graph H in the previous example, namely, using simple gadgets they implement the list constraints. The main challenge now is to construct two gadgets L and R such that they have h selected vertices $L \subseteq L$, $R \subseteq R$, and L can be mapped to R only. We construct one gadget $T_{t,k}$ and take $L = R = T_{t,k}$ such that $T_{t,k}$ can be mapped to itself only. First we show a gadget D on 6 vertices that has a fixed point $z \in V(D)$, i.e. any homomorphism $D \to D$ maps z to z. We then combine k such gadgets in a row to get the $T_{t,k}$ gadget that has t fixed points. Then it remains to note that we can add t large cliques to $T_{t,k}$ so that no part of this gadget can be mapped into the graph H.

4 Lower Bounds for the Graph Homomorphism Problem

Proof (of Theorem 1).

Let $\gamma > 4$ be a large enough constant such that $\frac{\log x}{100 \log \log x} \geq 2$ for $x \geq \gamma$. If $h(n) < \gamma$ for all values of n, then an algorithm with running time (1) would solve 3-COLORING in time $\mathcal{O}^* \left(2^{\frac{cn \log h(n)}{\log \log h(n)}} \right) = \mathcal{O}^* \left(2^{cn \log \gamma} \right)$ (recall that $h(n) \geq 3$). Therefore, by choosing a small enough constant c such that $c \log \gamma < \beta$, we arrive to a contradiction with Lemma 1.

From now on we assume that $h(n) \geq \gamma$ for large enough values of n. Let G be an n-vertex graph of maximum degree 4 that needs to be 3-colored. We first use Lemma 2, for a parameter $2 \leq r \leq n$ to be defined later, to get an equisatisfiable instance (G', H') of LIST-HOM with $|V(G')| = n/r$ and $|V(H')| \leq r^{50r}$. Note that $\chi(H') \leq |V(H')| \leq r^{50r}$. Hence Lemma 5 provides us with an equisatisfiable instance (G'', H'') of HOM with $|V(G'')| \leq n/r + (r^{50r} + 1)(r^{50r} + 11) \leq n/r + r^{102r}$ and $|V(H'')| \leq (r^{50r} + 1)(r^{50r} + 11) \leq r^{102r}$. Let

$$r' = \frac{\log n}{204 \log \log n}, \quad r = \min \left(r', \frac{\log h(\frac{2n}{r'})}{102 \log \log h(\frac{2n}{r'})} \right).$$

Note that $r \leq r' < n$. Also, $h(n) \geq \gamma$ implies that $r \geq 2$ for sufficiently large values of n. Let us show that

$$r \geq \frac{\log h(\frac{2n}{r'})}{d \cdot 204 \log \log h(\frac{2n}{r'})}. \tag{3}$$

This clearly holds if $r < r'$, so consider the case $r = r'$. The function $\log x / \log \log x$ increases for $x > 4$. Recall that $h(n) \geq \gamma$ for large enough values of n, hence $h(2n/r') \geq \gamma > 4$ for large enough values of n. Hence

$$\frac{\log h\left(\frac{2n}{r'}\right)}{\log \log h\left(\frac{2n}{r'}\right)} \leq \frac{d \log n}{\log \log n + \log d} \leq \frac{d \log n}{\log \log n} = 204 dr' = 204 dr$$

which implies (3). Then $|V(G'')| \leq \frac{n}{r} + r^{102r} \leq \frac{n}{r} + (\log n)^{\frac{\log n}{2 \log \log n}} \leq \frac{2n}{r}$,

$$|V(H'')| \leq r^{102r} \leq \left(\log h\left(\frac{2n}{r'}\right)\right)^{\frac{\log h(\frac{2n}{r'})}{\log \log h(\frac{2n}{r'})}} = h\left(\frac{2n}{r'}\right) \leq h\left(\frac{2n}{r}\right) \leq h(|V(G'')|).$$

Hence one can add isolated vertices to both G'' and H'' (clearly this does not change the problem) such that $|V(G'')| = 2n/r$ and $|V(H'')| = h(2n/r)$ and run an algorithm from the theorem statement on the instance (G'', H'').

Note that the running time of the reduction is polynomial in $|G|, |G'|, |G''|$, $|H|, |H'|, |H''|$, i.e. $\text{poly}(n, h(2n/r)) = \mathcal{O}^*(1)$. Thus, an algorithm with running time (1) for HOM implies an algorithm for 3-COLORING with running time

$$\mathcal{O}^*\left(2^{c \cdot \frac{2n}{r} \cdot \frac{\log h(\frac{2n}{r})}{\log \log h(\frac{2n}{r})}}\right) = \mathcal{O}^*\left(2^{408cdn}\right)$$

(recall the inequality (3)). Therefore, by choosing a small enough constant $c > 0$ such that $408cd < \beta$, we arrive to a contradiction with Lemma 1.

$\square$

Acknowledgments. We are grateful to Daniel Lokshtanov and Saket Saurabh for helpful discussions as well as for anonymous referees for many useful comments. The results from Section 3 are obtained with a partial support by the Government of the Russian Federation (grant 14.Z50.31.0030). Results presented in Section 4 are supported by the Grant of the President of the Russian Federation (MK-6550.2015.1).

References

1. Austrin, P.: Towards sharp inapproximability for any 2-CSP. SIAM J. Comput. **39**(6), 2430–2463 (2010)
2. Barto, L., Kozik, M., Niven, T.: Graphs, polymorphisms and the complexity of homomorphism problems. In: Proceedings of the 40th Annual ACM Symposium on Theory of Computing (STOC), pp. 789–796 (2008)

3. Björklund, A., Husfeldt, T., Koivisto, M.: Set partitioning via inclusion-exclusion. SIAM J. Computing **39**(2), 546–563 (2009)
4. Chen, J., Huang, X., Kanj, I.A., Xia, G.: Strong computational lower bounds via parameterized complexity. J. Computer and System Sciences **72**(8), 1346–1367 (2006)
5. Chen, J., Kanj, I.A., Xia, G.: Improved upper bounds for vertex cover. Theoretical Computer Science **411**(40–42), 3736–3756 (2010)
6. Cygan, M., Fomin, F.V., Kowalik, L., Lokshtanov, D., Marx, D., Pilipczuk, M., Pilipczuk, M., Saurabh, S.: Parameterized Algorithms. Springer (2015)
7. Diaz, J., Serna, M., Thilikos, D.M.: Counting H-colorings of partial k-trees. Theoretical Computer Science **281**, 291–309 (2002)
8. Feder, T., Vardi, M.Y.: The computational structure of monotone monadic SNP and constraint satisfaction: A study through datalog and group theory. SIAM J. Comput. **28**(1), 57–104 (1998)
9. Fiala, J., Golovach, P.A., Kratochvíl, J.: Computational complexity of the distance constrained labeling problem for trees (extended abstract). In: Aceto, L., Damgård, I., Goldberg, L.A., Halldórsson, M.M., Ingólfsdóttir, A., Walukiewicz, I. (eds.) ICALP 2008, Part I. LNCS, vol. 5125, pp. 294–305. Springer, Heidelberg (2008)
10. Fiala, J., Kratochvíl, J.: Locally constrained graph homomorphisms - structure, complexity, and applications. Computer Science Review **2**(2), 97–111 (2008)
11. Fomin, F.V., Heggernes, P., Kratsch, D.: Exact algorithms for graph homomorphisms. Theory of Computing Systems **41**(2), 381–393 (2007)
12. Fomin, F.V., Kratsch, D.: Exact Exponential Algorithms. Springer (2010)
13. Garey, M.R., Johnson, D.S.: Computers and Intractability: A Guide to the Theory of NP-Completeness. W. H. Freeman (1979)
14. Griggs, J.R., Yeh, R.K.: Labelling graphs with a condition at distance 2. SIAM J. Discrete Math. **5**(4), 586–595 (1992)
15. Grohe, M.: The complexity of homomorphism and constraint satisfaction problems seen from the other side. J. ACM **54**(1) (2007)
16. Havet, F., Klazar, M., Kratochvíl, J., Kratsch, D., Liedloff, M.: Exact algorithms for L(2, 1)-labeling of graphs. Algorithmica **59**(2), 169–194 (2011)
17. Hell, P., Nešetřil, J.: On the complexity of H-coloring. J. Combinatorial Theory Ser. B **48**(1), 92–110 (1990)
18. Hell, P., Nešetřil, J.: Graphs and homomorphisms. Oxford Lecture Series in Mathematics and its Applications, vol. 28. Oxford University Press, Oxford (2004)
19. Impagliazzo, R., Paturi, R., Zane, F.: Which problems have strongly exponential complexity. J. Computer and System Sciences **63**(4), 512–530 (2001)
20. Junosza-Szaniawski, K., Kratochvíl, J., Liedloff, M., Rossmanith, P., Rzazewski, P.: Fast exact algorithm for l(2, 1)-labeling of graphs. Theor. Comput. Sci. **505**, 42–54 (2013)
21. Lawler, E.L.: A note on the complexity of the chromatic number problem. Inf. Process. Lett. **5**(3), 66–67 (1976)
22. Lokshtanov, D.: Private communication (2014)
23. Lokshtanov, D., Marx, D., Saurabh, S.: Slightly superexponential parameterized problems. In: Proceedings of the 21st Annual ACM-SIAM Symposium on Discrete Algorithms (SODA), pp. 760–776. SIAM (2011)
24. Lokshtanov, D., Marx, D., Saurabh, S.: Lower bounds based on the exponential time hypothesis. Bulletin of EATCS **3**(105) (2013)
25. Lovász, L.: Large networks and graph limits, vol. 60. American Mathematical Soc. (2012)

26. Marx, D.: Can you beat treewidth? Theory of Computing **6**(1), 85–112 (2010)
27. Raghavendra, P.: Optimal algorithms and inapproximability results for every CSP♪ In: Proceedings of the 40th Annual ACM Symposium on Theory of Computing (STOC), pp. 245–254 (2008)
28. Rzażewski, P.: Exact algorithm for graph homomorphism and locally injective graph homomorphism. Inf. Process. Lett. **114**(7), 387–391 (2014)
29. Sipser, M.: Introduction to the Theory of Computation. Cengage Learning (2005)
30. Traxler, P.: The time complexity of constraint satisfaction. In: Grohe, M., Niedermeier, R. (eds.) IWPEC 2008. LNCS, vol. 5018, pp. 190–201. Springer, Heidelberg (2008)
31. Wahlström, M.: Problem 5.21. time complexity of graph homomorphism. In: Thore Husfeldt, Dieter Kratsch, R.P., Sorkin, G. (eds.) Exact Complexity of NP-Hard Problems. Dagstuhl Seminar 10441 Final Report. Dagstuhl (2010)
32. Wahlström, M.: New plain-exponential time classes for graph homomorphism. Theory of Computing Systems **49**(2), 273–282 (2011)

Parameterized Single-Exponential Time Polynomial Space Algorithm for Steiner Tree

Fedor V. Fomin[1], Petteri Kaski[2], Daniel Lokshtanov[1], Fahad Panolan[1,3](✉),
and Saket Saurabh[1,3]

[1] University of Bergen, Bergen, Norway
{fomin,daniello}@uib.no
[2] Aalto University, Espoo, Finland
petteri.kaski@aalto.fi
[3] Institute of Mathematical Sciences, Chennai, India
{fahad,saket}@imsc.res.in

Abstract. In the Steiner tree problem, we are given as input a connected n-vertex graph with edge weights in $\{1, 2, \ldots, W\}$, and a subset of k terminal vertices. Our task is to compute a minimum-weight tree that contains all the terminals. We give an algorithm for this problem with running time $\mathcal{O}(7.97^k \cdot n^4 \cdot \log W)$ using $\mathcal{O}(n^3 \cdot \log nW \cdot \log k)$ space. This is the first single-exponential time, polynomial-space FPT algorithm for the weighted STEINER TREE problem.

1 Introduction

In the STEINER TREE problem, we are given as input a connected n-vertex graph, a non-negative weight function $w : E(G) \to \{1, 2, \ldots, W\}$, and a set of terminal vertices $T \subseteq V(G)$. The task is to find a minimum-weight connected subgraph ST of G containing all terminal nodes T. In this paper we use the parameter $k = |T|$.

STEINER TREE is one of the central and best-studied problems in Computer Science with various applications. We refer to the book of Prömel and Steger [16] for an overview of the results and applications of the Steiner tree problem. STEINER TREE is known to be APX-complete, even when the graph is complete and all edge costs are either 1 or 2 [2]. On the other hand the problem admits a constant factor approximation algorithm, the currently best such algorithm (after a long chain of improvements) is due to Byrka et al. and has approximation ratio $\ln 4 + \varepsilon < 1.39$ [6].

STEINER TREE is a fundamental problem in parameterized algorithms [7]. The classic algorithm for STEINER TREE of Dreyfus and Wagner [8] from 1971 might well be the first parameterized algorithm for *any* problem. The study of parameterized algorithms for STEINER TREE has led to the design of important techniques,

---

The research leading to these results has received funding from the European Research Council under the European Union's Seventh Framework Programme (FP/2007-2013) / ERC Grant Agreements 267959, 338077 and 306992

M.M. Halldórsson et al. (Eds.): ICALP 2015, Part I, LNCS 9134, pp. 494–505, 2015.
DOI: 10.1007/978-3-662-47672-7_40

such as Fast Subset Convolution [3] and the use of branching walks [13]. Research
on the parameterized complexity of STEINER TREE is still on-going, with very
recent significant advances for the planar version of the problem [14,15].

Algorithms for STEINER TREE are frequently used as a subroutine in fixed-
parameter tractable (FPT) algorithms for other problems; examples include vertex cover problems [11], near-perfect phylogenetic tree reconstruction [4], and
connectivity augmentation problems [1].

Motivation and Earlier Work. For more than 30 years, the fastest FPT algorithm for STEINER TREE was the $3^k \cdot \log W \cdot n^{\mathcal{O}(1)}$-time dynamic programming
algorithm by Dreyfus and Wagner [8]. Fuchs et al. [10] gave an improved algo-
rithm with running time $\mathcal{O}((2 + \varepsilon)^k n^{f(1/\varepsilon)} \log W)$. For the unweighted version
of the problem, Björklund et al. [3] gave a $2^k n^{\mathcal{O}(1)}$ time algorithm. All of these
algorithms are based on dynamic programming and use exponential space.

Algorithms with high space complexity are in practice more constrained
because the amount of memory is not easily scaled beyond hardware constraints
whereas time complexity can be alleviated by allowing for more time for the
algorithm to finish. Furthermore, algorithms with low space complexity are typically easier to parallelize and more cache-friendly. These considerations motivate
a quest for algorithms whose memory requirements scale polynomially in the
size of the input, even if such algorithms may be slower than their exponential-
space counterparts. The first polynomial space $2^{\mathcal{O}(k)} n^{\mathcal{O}(1)}$-time algorithm for the
unweighted STEINER TREE problem is due to Nederlof [13]. This algorithm runs
in time $2^k n^{\mathcal{O}(1)}$, matching the running time of the best known exponential space
algorithm. Nederlof's algorithm can be extended to the weighted case, unfortunately this comes at the cost of a $\mathcal{O}(W)$ factor both in the time and the space
complexity. Lokshtanov and Nederlof [12] showed that the $\mathcal{O}(W)$ factor can be
removed from the space bound, but with a factor $\mathcal{O}(W)$ in the running time.
The algorithm of Lokshtanov and Nederlof [12] runs in $2^k \cdot n^{\mathcal{O}(1)} \cdot W$ time and
uses $n^{\mathcal{O}(1)} \log W$ space. Note that both the algorithm of Nederlof [13] and the
algorithm of Lokstanov and Nederlof [12] have a $\mathcal{O}(W)$ factor in their running
time. Thus the running time of these algorithms depends exponentially on the
input size, and therefore these algorithms are not FPT algorithms for weighted
STEINER TREE.

For weighted STEINER TREE, the only known polynomial space FPT algo-
rithm has a $2^{\mathcal{O}(k \log k)}$ running time dependence on the parameter k. This algorithm follows from combining a $(27/4)^k \cdot n^{\mathcal{O}(\log k)} \cdot \log W$ time, polynomial space
algorithm by Fomin et al. [9] with the Dreyfus–Wagner algorithm. Indeed, one
runs the algorithm of Fomin et al. [9] if $n \leq 2^k$, and the Dreyfus–Wagner algorithm if $n > 2^k$. If $n \leq 2^k$, the running time of the algorithm of Fomin et al. is
bounded from above by $2^{\mathcal{O}(k \log k)}$. When $n > 2^k$, the Dreyfus–Wagner algorithm
becomes a polynomial time (and space) algorithm.

Prior to this work the existence of a polynomial space algorithm with running
time $2^{\mathcal{O}(k)} \cdot n^{\mathcal{O}(1)} \cdot \log W$, i.e a single exponential time polynomial space FPT
algorithm, was an open problem asked explicitly in [9,12].

Contributions and Methodology. The starting point of our present algorithm is the $(27/4)^k \cdot n^{\mathcal{O}(\log k)} \cdot \log W$-time, polynomial-space algorithm by Fomin et al. [9]. This algorithm crucially exploits the possibility for *balanced separation* (cf. Lemma 1 below). Specifically, an optimal Steiner tree ST can be partitioned into two trees ST_1 and ST_2 containing the terminal sets T_1 and T_2 respectively, so that the following three properties are satisfied: (a) The two trees share exactly one vertex v and no edges. (b) Neither of the two trees ST_1 or ST_2 contain more than a $2/3$ fraction of the terminal set T. (c) The tree ST_1 is an optimal Steiner tree for the terminal set $T_1 \cup \{v\}$, and ST_2 is an optimal Steiner tree for the terminal set $T_2 \cup \{v\}$.

Dually, to find the optimal tree ST for the terminal set T it suffices to (a) guess the vertex v, (b) partition T into T_1 and T_2, and (c) recursively find optimal trees for the terminal sets $T_1 \cup \{v\}$ and $T_2 \cup \{v\}$. Since there are n choices for v, and $\binom{k}{k/3}$ ways to partition T into two sets T_1 and T_2 such that $|T_1| = |T|/3$, the running time of the algorithm is essentially governed by the recurrence

$$T(n,k) \leq n \cdot \binom{k}{k/3} \cdot (T(n, k/3) + T(n, 2k/3)). \tag{1}$$

Unraveling (1) gives the $(27/4)^k \cdot n^{\mathcal{O}(\log k)} \cdot \log W$ upper bound for the running time, and it is easy to see that the algorithm runs in polynomial space. However, this algorithm is not an FPT algorithm because of the $n^{\mathcal{O}(\log k)}$ factor in the running time.

The factor $n^{\mathcal{O}(\log k)}$ is incurred by the factor n in (1), which in turn originates from the need to iterate over all possible choices for the vertex v in each recursive call. In effect the recursion tracks an $\mathcal{O}(\log k)$-sized set S of *split vertices* (together with a subset T' of the terminal vertices T) when it traverses the recursion tree from the root to a leaf.

The key idea in our new algorithm is to redesign the recurrence for optimal Steiner trees so that we obtain control over the size of S using an alternation between

1. balanced separation steps (as described above), and
2. novel *resplitting* steps that maintain the size of S at no more than 3 vertices throughout the recurrence.

In essence, a resplit takes a set S of size 3 and splits that set into three sets of size 2 by combining each element in S with an arbitrary vertex v, while at the same time splitting the terminal set T' into three parts in all possible (not only balanced) ways. While the combinatorial intuition for resplitting is elementary (cf. Lemma 2 below), the implementation and analysis requires a somewhat careful combination of ingredients.

Namely, to run in polynomial space, it is not possible to use extensive amounts of memory to store intermediate results to avoid recomputation. Yet, if no memoization is used, the novel recurrence does not lead to an FPT algorithm, let alone to a single-exponential FPT algorithm. Thus neither a purely dynamic programming nor a purely recursive implementation will lead to the desired algorithm. *A combination of the two will, however, give a single-exponential time algorithm that uses polynomial space.*

Roughly, our approach is to employ recursive evaluation over subsets T' of the terminal set T, but each recursive call with T' will compute and return the optimal solutions for every possible set S of split vertices. Since by resplitting we have arranged that S always has size at most 3, this hybrid evaluation approach will use polynomial space. Since each recursive call on T' yields the optimum weights for every possible S, we can use dynamic programming to efficiently combine these weights so that single-exponential running time results.

In precise terms, our main result is as follows:

Theorem 1. STEINER TREE *can be solved in time* $\mathcal{O}(7.97^k n^4 \log nW)$ *time using* $\mathcal{O}(n^3 \log nW \log k)$ *space.*

Whereas our main result seeks to optimize the polynomial dependency in n for both the running time and space usage, it is possible to trade between polynomial dependency in n and the single-exponential dependency in k to obtain faster running time as a function k, but at the cost of increased running time and space usage as a function of n. In particular, we can use larger (but still constant-size) sets S to avoid recomputation and to arrive at a somewhat faster algorithm:

Theorem 2. *There exists a polynomial-space algorithm for* STEINER TREE *running in* $\mathcal{O}(6.751^k n^{O(1)} \log W)$ *time.*

2 Preliminaries

Given a graph G, we write $V(G)$ and $E(G)$ for the set of vertices and edges of G, respectively. For subgraphs G_1, G_2 of G, we write $G_1 + G_2$ for the subgraph of G with vertex set $V(G_1) \cup V(G_2)$ and edge set $E(G_1) \cup E(G_2)$. For a graph G, $S \subseteq V(G)$ and $v \in V(G)$, we use $G - S$ and $G - v$ to denote the induced subgraphs $G[V(G) \backslash S]$ and $G[V(G) \backslash \{v\}]$ respectively. For a path $P = u_1 u_2 \cdots u_\ell$ in a graph G, we use $\overleftarrow{P}$ to denote the reverse path $u_\ell u_{\ell-1} \cdots u_1$. The minimum weight of a Steiner tree of G on terminals T is denoted by $st_G(T)$. When graph G is clear from the context, we will simply write $st(T)$. For a set U and a non negative integer i, we use $\binom{U}{i}$ and $\binom{U}{\leq i}$ to denote the set of all subsets of U, of size exactly i and the set of all subsets of U, of size at most i respectively. For a set U, we write $U_1 \uplus U_2 \uplus \cdots \uplus U_\ell = U$ if $U_1, U_2, \ldots, U_\ell$ is a partition of U.

Separation and Resplitting. A set of nodes S is called an α-*separator* of a graph G, $0 < \alpha \leq 1$, if the vertex set $V(G) \setminus S$ can be partitioned into sets V_L and V_R of size at most αn each, such that no vertex of V_L is adjacent to any vertex of V_R. We next define a similar notion, which turns out to be useful for Steiner trees. Given a Steiner tree ST on terminals T, an α-Steiner separator S of ST is a subset of nodes which partitions $ST - S$ in two forests $\mathcal{R}_1$ and $\mathcal{R}_2$, each one containing at most αk terminals from T.

Lemma 1 (Separation). [5,9] *Every Steiner tree ST on terminal set T, $|T| \geq$* *, has a 2/3-Steiner separator $S = \{s\}$ of size one.*

The following easy lemma enables us to control the size of the split S set at no more than 3 vertices.

Lemma 2 (Resplitting $(\star)$[1]). *Let F be a tree and $S \in \binom{V(F)}{3}$. Then there is a vertex $v \in V(F)$ such that each connected component in $F - v$ contains at most one vertex of S.*

3 Algorithm

In this section we design an algorithm for STEINER TREE which runs in time $\mathcal{O}(7.97^k n^4 \log nW)$ time using $\mathcal{O}(n^3 \log nW \log k)$ space. Most algorithms for STEINER TREE, including ours, are based on recurrence relations that reduce finding the optimal Steiner tree to finding optimal Steiner trees in the same graph, but with a smaller terminal set. We will define four functions f_i for $\in \{0, 1, 2, 3\}$. Each function f_i takes as input a vertex set S of size at most and a subset T' of T. The function $f_i(S, T')$ returns a real number. We will define the functions using recurrence relations, and then prove that $f_i(S, T')$ is exactly $st_G(T' \cup S)$.

In the recurrences we will work with the following partitioning schemes for the current set of terminals T'. Let $\mathcal{P}(T')$ be the set of all possible partitions (T_1, T_2, T_3) of T' into three parts and let $\mathcal{B}(T')$ be the set of all possible partitions (T_1, T_2) of T' into two parts such that $|T_1|, |T_2| \leq 2k/3$.

For $T' \subseteq T$, $i \in \{0, 1, 2, 3\}$, and $S \in \binom{V(G)}{\leq i}$, we define $f_i(S, T')$ as follows. When $|T'| \leq 2$, $f_i(S, T') = st_G(T' \cup S)$. For $|T'| \geq 3$, we define $f_i(S, T')$ using the following recurrences.

Separation. For $i \in \{0, 1, 2\}$, let us define

$$f_i(S, T') = \min_{\substack{(T_1, T_2) \in \mathcal{B}(T')}} \min_{\substack{v \in V(G) \\ S_1 \uplus S_2 = S}} f_{i+1}\big(S_1 \cup \{v\}, T_1\big) + f_{i+1}\big(S_2 \cup \{v\}, T_2\big) \quad (2)$$

Resplitting. For $i = 3$, let us define

$$f_i(S, T') = \min_{\substack{(T_1, T_2, T_3) \in \mathcal{P}(T')}} \min_{\substack{S_1 \uplus S_2 \uplus S_3 = S \\ |S_1|, |S_2|, |S_3| \leq i-2 \\ v \in V(G)}} \sum_{r=1}^{3} f_{i-1}\big(S_r \cup \{v\}, T_r\big) \quad (3)$$

The recurrences (2) and (3) are recurrence relations for STEINER TREE:

Lemma 3 $(\star)$. *For all $T' \subseteq T$, $0 \leq i \leq 3$, and $S \in \binom{V(G)}{\leq i}$ it holds that $f_i(S, T') = st_G(T' \cup S)$.*

---

[1] Proofs of results marked with a $\star$ are deferred to the full version of the paper

---

Algorithm 1. Implementation of procedure F_i for $i \in \{0, 1, 2\}$

Input: $T' \subseteq T$
Output: $st_G(T' \cup S)$ for all $S \in \binom{V(G)}{\leq i}$

1   **if** $|T'| \leq 2$ **then**
2   **for** $S \in \binom{V(G)}{\leq 3}$ **do**
3       $A[S] \leftarrow st_G(T' \cup S)$ (compute using the Dreyfus–Wagner algorithm)
4   **return** A

5   **for** $S \in \binom{V(G)}{\leq i}$ **do**
6     $A[S] \leftarrow \infty$
7   **for** $T_1, T_2 \in \mathcal{B}(T')$ **do**
8     $A_1 \leftarrow F_{i+1}(T_1)$
9     $A_2 \leftarrow F_{i+1}(T_2)$
10   **for** $S_1 \uplus S_2 \in \binom{V(G)}{\leq i}$ *such that* $|S_2| \leq |S_1|$ *and* $v \in V(G)$ **do**
11   **if** $A[S_1 \uplus S_2] > A_1[S_1 \cup \{v\}] + A_2[S_2 \cup \{v\}]$ **then**
12         $A[S_1 \uplus S_2] \leftarrow A_1[S_1 \cup \{v\}] + A_2[S_2 \cup \{v\}]$

13 **return** A.

Our algorithm uses (2) and (3) to compute $f_0(\emptyset, T)$, which is exactly the cost of an optimum Steiner tree. A naïve way of turning the recurrences into an algorithm would be to simply make one recursive procedure for each f_i, and apply (2) and (3) directly. However, this would result in a factor $n^{O(\log k)}$ in the running time, which we seek to avoid. As the naïve approach, our algorithm has one recursive procedure F_i for each function f_i. The procedure F_i takes as input a subset T' of the terminal set, and returns an array that, for every $S \in \binom{V(G)}{\leq i}$, contains $f_i(S, T')$.

The key observation is that if we seek to compute $f_i(S, T')$ for a fixed T' and *all* choices of $S \in \binom{V(G)}{\leq i}$ using recurrence (2) or (3), we should not just iterate over every choice of S and then apply the recurrence to compute $f_i(S, T')$ because it is much faster to compute all the entries of the return array of F_i simultaneosly, by iterating over every eligible partition of T, making the required calls to F_{i+1} (or F_{i-1} if we are using recurrence (3)), and updating the appropriate array entries to yield the return array of F_i. Next we give pseudocode for the procedures F_0, F_1, F_2, F_3.

The procedure F_i for $0 \leq i \leq 2$ operates as follows. (See Algorithm 1.) Let $T' \subseteq T$ be the input to the procedure F_i. If $|T'| \leq 2$, then F_i computes $st_G(T' \cup S)$ for all $S \in \binom{V(G)}{\leq i}$ using the Dreyfus–Wagner algorithm and returns these values. The procedure F_i has an array A indexed by $S \in \binom{V(G)}{\leq i}$. At the end of the procedure F_i, $A[S]$ will contain the value $st_G(T' \cup S)$ for all $S \in \binom{V(G)}{\leq i}$. For each $(T_1, T_2) \in \mathcal{B}(T')$ (line 7), F_i calls $F_{i+1}(T_1)$ and $F_{i+1}(T_2)$ and it returns two sets of values $\{f_{i+1}(S, T_1) \mid S \in \binom{V(G)}{\leq i+1}\}$ and $\{f_i(S, T_2) \mid S \in$

Algorithm 2. Implementation of procedure F_3

Input: $T' \subseteq T$
Output: $st_G(T' \cup S)$ for all $S \in \binom{V(G)}{\leq 3}$

1 **if** $|T'| \leq 2$ **then**
2   **for** $S \in \binom{V(G)}{\leq 3}$ **do**
3          $A[S] \leftarrow st_G(T' \cup S)$ (compute using the Dreyfus–Wagner algorithm)
4   **return** A

5 **for** $S \in \binom{V(G)}{\leq 3}$ **do**
6      $A[S] \leftarrow \infty$
7 **for** $T_1, T_2, T_3 \in \mathcal{P}(T')$ **do**
8      $A_1 \leftarrow F_2(T_1)$
9      $A_2 \leftarrow F_2(T_2)$
10      $A_3 \leftarrow F_2(T_3)$
11   **for** $S_1, S_2, S_3 \in \binom{V(G)}{\leq 3}$ **and** $v \in V(G)$ **do**
12   **if** $A[S_1 \cup S_2 \cup S_3] > A_1[S_1 \cup \{v\}] + A_2[S_2 \cup \{v\}] + A_3[S_3 \cup \{v\}]$ **then**
13            $A[S_1 \cup S_2 \cup S_3] \leftarrow A_1[S_1 \cup \{v\}] + A_2[S_2 \cup \{v\}] + A_3[S_3 \cup \{v\}]$
14 **return** A.

$\binom{V(G)}{\leq i}$)}, respectively. Let A_1 and A_2 be two arrays used to store the return values of $F_{i+1}(T_1)$ and $F_{i+1}(T_2)$ respectively. That is, $A_1[S] = f_{i+1}(S, T_1)$ for all $S \in \binom{V(G)}{\leq i+1}$ and $A_2[S'] = f_i(S', T_2)$ for all $S' \in \binom{V(G)}{\leq i+1}$. Now we update A as follows. For each $S_1 \uplus S_2 \in \binom{V(G)}{\leq i}$ and $v \in V(G)$ (line 10), if $A[S_1 \uplus S_2] > A_1[S_1 \cup \{v\}] + A_2[S_2 \cup \{v\}]$, then we update the entry $A[S_1 \uplus S_2]$, with the value $A_1[S_1 \cup \{v\}] + A_2[S_2 \cup \{v\}]$. So at the end the inner **for** loop, $A[S]$ contains the value

$$\min_{\substack{v \in V(G) \\ S_1 \uplus S_2 = S}} f_{i+1}(S_1 \cup \{v\}, T_1) + f_i(S_2 \cup \{v\}, T_2).$$

Since we do have a outer **for** loop which runs over $(T_1, T_2) \in \mathcal{B}(T')$, we have updated $A[S]$ with

$$\min_{(T_1, T_2) \in \mathcal{B}(T')} \min_{\substack{v \in V(G) \\ S_1 \uplus S_2 = S}} f_{i+1}(S_1 \cup \{v\}, T_1) + f_i(S_2 \cup \{v\}, T_2).$$

at the end of the procedure. Then F_i will return A.

The procedure F_3 works as follows. (See Algorithm 2.) Let $T' \subseteq T$ be the input to the procedure F_3. If $|T'| \leq 2$, then F_3 computes $st_G(T' \cup S)$ for all $S \in \binom{V(G)}{\leq 3}$ using the Dreyfus–Wagner algorithm and returns these values. The procedure F_3 has an array A indexed by $S \in \binom{V(G)}{\leq 3}$. At the end of the procedure F_3, $A[S]$ will contain the value $st_G(T' \cup S)$ for all $S \in \binom{V(G)}{\leq 3}$. For each $(T_1, T_2, T_3) \in \mathcal{P}(T')$ (line 7), F_3 calls $F_2(T_1)$, $F_2(T_2)$ and $F_2(T_3)$, and it returns three sets of values $\{f_2(S, T_1) \mid S \in \binom{V(G)}{\leq 2}\}$, $\{f_2(S, T_2) \mid S \in \binom{V(G)}{\leq 2}\}$

and $\{f_2(S, T_3) \mid S \in \binom{V(G)}{\leq 2}\}$, respectively. Let A_1, A_2 and A_3 be three arrays used to store the outputs of $F_2(T_1)$, $F_2(T_2)$ and $F_2(T_3)$ respectively. That is, $A_r[S] = f_2(S, T_r)$ for $r \in \{1, 2, 3\}$. Now we update A as follows. For each $S_1, S_2, S_3 \in \binom{V(G)}{\leq 1}$ and $v \in V(G)$ (line 11), if $A[S_1 \cup S_2 \cup S_3] > A_1[S_1 \cup \{v\}] + A_2[S_2 \cup \{v\}] + A_3[S_3 \cup \{v\}]$, then we update the entry $A[S_1 \cup S_2 \cup S_3]$, with the value $A_1[S_1 \cup \{v\}] + A_2[S_2 \cup \{v\}] + A_3[S_3 \cup \{v\}]$. So at the end the inner **for** loop, $A[S]$ contains the value

$$\min_{\substack{S_1 \cup S_2 \cup S_3 = S \\ |S_1|, |S_2|, |S_3| \leq 1 \\ v \in V(G)}} \sum_{r=1}^{3} f_2(S_r \cup \{v\}, T_r).$$

Since we do have a outer **for** loop which runs over $(T_1, T_2, T_3) \in \mathcal{P}(T')$, we have updated $A[S]$ with

$$\min_{(T_1, T_2, T_3) \in \mathcal{P}(T')} \min_{\substack{S_1 \cup S_2 \cup S_3 = S \\ |S_1|, |S_2|, |S_3| \leq 1 \\ v \in V(G)}} \sum_{r=1}^{3} f_2(S_r \cup \{v\}, T_r).$$

at the end of the procedure. Then F_3 will return A as the output.

In what follows we prove the correctness and analyze the running time and memory usage of the call to the procedure $F_0(T)$.

Lemma 4. *For every $i \leq 3$, $T' \subseteq T$ the procedure $F_i(T')$ outputs an array that for every $S \in \binom{V(G)}{\leq i}$, contains $f_i(S, T')$.*

Proof. Correctness of Lemma 4 follows directly by an induction on $|T|$. Indeed, assuming that the lemma statement holds for the recursive calls made by the procedure F_i, it is easy to see that each entry of the output table is exactly equal to the right hand side of recurrence (2) (recurrence (3) in the case of F_3).   □

Observation 1. *The recursion tree of the procedure $F_0(T)$ has depth $\mathcal{O}(\log k)$.*

Proof. For every $i \leq 2$ the procedure $F_i(T')$ only makes recursive calls to $F_{i+1}(T'')$ where $|T''| \leq 2|T'|/3$. The procedure $F_3(T')$ makes recursive calls to $F_2(T'')$ where $|T''| \leq |T'|$. Therefore, on any root-leaf path in the recursion tree, the size of the considered terminal set T' drops by a constant factor every second step. When the terminal set reaches size at most 2, no further recursive calls are made. Thus any root-leaf path has length at most $\mathcal{O}(\log k)$.   □

Lemma 5. *The procedure $F_0(T)$ uses $\mathcal{O}(n^3 \log nW \log k)$ space.*

Proof. To upper bound the space used by the procedure $F_0(T)$ it is sufficient to upper bound the memory usage of every individual recursive call, not taking into account the memory used by its recursive calls, and then multiply this upper bound by the depth of the recursion tree.

Each individual recursive call will at any point of time keep a constant number of tables, each containing at most $\mathcal{O}(n^3)$ entries. Each entry is a number less than or equal to nW, therefore each entry can be represented using at most $\mathcal{O}(\log nW)$ bits. Thus each individual recurisve call uses at most $\mathcal{O}(n^3 \log nW)$ bits. Combining this with Observation 1 proves the lemma. $\square$

Next we analyze the running time of the algorithm. Let $\tau_i(k)$ be the total number of arithmetic operations of the procedure $F_i(T')$ for all $i \leq 3$, where $c = |T'|$ on an n-vertex graph. It follows directly from the structure of the procedures F_i for $i \leq 2$, that there exits a constant C such that the following recurrences hold for τ_i, $i \leq 2$:

$$\tau_i(k) \leq \sum_{\frac{k}{3} \leq j \leq \frac{2k}{3}} \binom{k}{j}\left(\tau_{i+1}(j) + \tau_{i+1}(k-j) + Cn^3\right)$$

$$\leq 2 \sum_{\frac{k}{3} \leq j \leq \frac{2k}{3}} \binom{k}{j}\left(\tau_{i+1}(j) + Cn^3\right) \leq 2k \max_{\frac{k}{3} \leq j \leq \frac{2k}{3}} \binom{k}{j}\left(\tau_{i+1}(j) + Cn^3\right) \quad (4)$$

Let $\binom{k}{i_1,i_2,i_3}$ be the number of partitions of k distinct elements into sets of sizes i_1, i_2, and i_3. It follows directly from the structure of the procedure F_3, that there exists a constant C such that the following recurrence holds for τ_3:

$$\tau_3(k) = \sum_{i_1+i_2+i_3=k} \binom{k}{i_1,i_2,i_3}\left(\tau_2(i_1) + \tau_2(i_2) + \tau_2(i_3) + Cn^4\right)$$

$$\leq \sum_{i_1 \geq i_2,i_3} \binom{k}{i_1,i_2,i_3} 3 \cdot \left(\tau_2(i_1) + Cn^4\right) \leq 3 \sum_{i_1 \geq \frac{k}{3}} \binom{k}{i_1} 2^{k-i_1} \cdot \left(\tau_2(i_1) + Cn^4\right)$$

$$\leq 3k \max_{i_1 \geq \frac{k}{3}} \binom{k}{i_1} 2^{k-i_1} \cdot \left(\tau_2(i_1) + Cn^4\right) \quad (5)$$

Now we will bound $\tau_3(k)$ from above using (4) and (5). The following facts are required for the proof.

Fact 1. *By Stirling's approximation, $\binom{k}{\alpha k} \leq \left(\alpha^{-\alpha}(1-\alpha)^{(\alpha-1)}\right)^k$ [17].*

Fact 2. *For every fixed $x \geq 4$, function $f(y) = \frac{x^y}{y^y(1-y)^{1-y}}$ is increasing on interval $(0, 2/3]$.*

Lemma 6. *There exists a constant C such that $\tau_3(k) \leq C \cdot 11.7899^k n^4$*

Proof. We prove by induction on k, that $\tau_2(k) \leq \hat{C}k^{(c \log k)}9.78977^k n^4$ and $\tau_3(k) \leq \hat{C}k^{(c \log k)}11.7898^k n^4$. We will pick $\hat{C}$ to be a constant larger than the constants of (4) and (5), and sufficiently large so that the base case of the induction holds. We prove the inductive step. By the induction hypothesis and (4), we have that

$$\tau_2(k) \le 2k \max_{\frac{1}{3} \le \alpha \le \frac{2}{3}} \binom{k}{\alpha k} \left(\hat{C}(\alpha k)^{(c \log \alpha k)} 11.7898^{\alpha k} n^4 + \hat{C} n^3 \right)$$

$$\le 2k \left(\frac{11.7898^{2/3}}{(2/3)^{2/3} (1/3)^{1/3}} \right)^k \cdot \left(\hat{C} \left(\frac{2k}{3} \right)^{(c \log 2k/3)} n^4 + \hat{C} n^3 \right) \quad \text{(Fact 1, 2)}$$

$$\le (9.78977)^k \cdot 2k \cdot \left(\hat{C} \left(\frac{2k}{3} \right)^{(c \log 2k/3)} n^4 + \hat{C} n^3 \right)$$

$$\le 9.78977^k \cdot \hat{C} k^{(c \log k)} n^4$$

The last inequality holds if c is a sufficiently large constant (independent of k). By the induction hypothesis and (5), we have that

$$\tau_3(k) \le 3k \max_{1 \ge \alpha \ge \frac{1}{3}} \binom{k}{\alpha k} 2^{(1-\alpha)k} \cdot \left(9.78977^{\alpha k} \cdot \hat{C}(\alpha k)^{(c \log \alpha k)} n^4 + \hat{C} n^4 \right)$$

$$\le 3k \max_{1 \ge \alpha \ge \frac{1}{3}} \left(\alpha^{-\alpha} (1-\alpha)^{(\alpha-1)} 2^{(1-\alpha)} 9.78977^{\alpha} \right)^k \cdot \left(\hat{C}(\alpha k)^{(c \log \alpha k)} + \hat{C} n^4 \right)$$

$$\le 11.7898^k \cdot \hat{C} k^{(c \log k)} n^4$$

The last inequality holds for sufficiently large constants $\hat{C}$ and c. For a sufficiently large constant C it holds that

$$C \cdot 11.7899^k n^4 \ge 11.7898^k \cdot \hat{C} k^{(c \log k)} n^4,$$

completing the proof.   □

Lemma 7. *For every* $i \le 2$ *and constants* C_{i+1} *and* $\beta_{i+1} \ge 4$ *such that for every* $k \ge 1$ *we have* $\tau_{i+1}(k) \le C_{i+1} \beta_{i+1}^k n^4$, *there exists a constant* C_i *such that* $\tau_i(k) \le C_i \cdot 1.8899^k \cdot \beta_{i+1}^{2k/3} \cdot n^4$.

Proof. By (4) we have that

$$\tau_i(k) \le 2k \max_{\frac{k}{3} \le i \le \frac{2k}{3}} \binom{k}{j} (\tau_{i+1}(j) + C n^3)$$

$$\le (2k + C) \max_{\frac{k}{3} \le i \le \frac{2k}{3}} \binom{k}{j} (C_{i+1} \beta_{i+1}^j n^4)$$

$$\le C_{i+1} \cdot (2k + C) \cdot (\frac{3}{2^{2/3}})^k \cdot \beta_{i+1}^{2k/3} \cdot n^4$$

$$\le C_i \cdot 1.8899^k \cdot \beta_{i+1}^{2k/3} \cdot n^4$$

The last inequality holds for a sufficiently large C_i depending on C_{i+1} and β_{i+1} but not on k.   □

Lemma 8. *The procedure* $F_0(T)$ *uses* $\mathcal{O}(7.97^k n^4 \log nW)$ *time.*

Proof. We show that $\tau_0(k) = \mathcal{O}(7.9631^k n^4)$. Since each arithmetic operation takes at most $\mathcal{O}(\log nW)$ time the lemma follows. Applying Lemma 7 on the upper bound for $\tau_3(k)$ from Lemma 6 proves that

$$\tau_2(k) = \mathcal{O}(1.8899^k \cdot 11.7899^{2k/3} n^4) = \mathcal{O}(9.790^k n^4).$$

Re-applying Lemma 7 on the above upper bound for $\tau_2(k)$ yields

$$\tau_1(k) = \mathcal{O}(1.8899^k \cdot 9.790^{2k/3} n^4) = \mathcal{O}(8.6489^k n^4).$$

Re-applying Lemma 7 on the above upper bound for $\tau_1(k)$ yields

$$\tau_0(k) = \mathcal{O}(1.8899^k \cdot 8.6489^{2k/3} n^4) = \mathcal{O}(7.9631^k n^4).$$

This completes the proof. □

We are now in position to prove our main theorem.

Proof (of Theorem 1). The algorithm calls the procedure $F_0(T)$ and returns the value stored for $f_0(\emptyset, T)$. By Lemma 4 the procedure $F_0(T)$ correctly computes $f_0(\emptyset, T)$, and by Lemma 3 this is exactly equal to the cost of the optimal Steiner tree. By Lemma 5 the space used by the algorithm is at most $\mathcal{O}(n^3 \log nW \log k)$, and by Lemma 8 the time used is $\mathcal{O}(7.97^k n^4 \log nW)$. □

Obtaining Better Parameter Dependence. The algorithm from Theorem 1 is based on defining and computing the functions f_i, $0 \leq i \leq 3$. The functions f_i, $i \leq 2$ are defined using recurrence (2), while the function f_3 is defined using recurrence (3). For every constant $t \geq 4$ we could obtain an algorithm for STEINER TREE by defining functions f_i, $0 \leq i \leq t - 1$ using (2) and f_t using (3). A proof identical to that of Lemma 3 shows that $f_i(S, T') = ST_G(S \cup T')$ for every $i \leq t$.

We can now compute $f_0(\emptyset, T)$ using an algorithm almost identical to the algorithm of Theorem 1, except that now we have $t + 1$ procedures, namely a procedure F_i for each $i \leq t$. For each i and terminal set $T' \subseteq T$ a call to the procedure $F_i(T')$ computes an array containing $f_i(S, T')$ for every set S of size at most i.

For $i < t$, the procedure F_i is based on (2) and is essentially the same as Algorithm 1. Further, the procedure F_t is based on (3) and is essentially the same as Algorithm 2. The correctness of the algorithm and an $\mathcal{O}(n^t \log(nW))$ upper bound on the space usage follows from arguments identical to Lemma 4 and Lemma 5 respectively.

For the running time bound, an argument identical to Lemma 6 shows that $\tau_t(k) = \mathcal{O}(11.7899^k n^{t+1})$. Furthermore, Lemma 7 now holds for $i \leq t - 1$. In the proof of Lemma 8 the bound for $\tau_0(k)$ is obtained by starting with the $\mathcal{O}(11.7899^k n^4)$ bound for τ_3 and applying Lemma 7 three times. Here we can upper bound $\tau_0(k)$ by starting with the $\mathcal{O}(11.7899^k n^{t+1})$ bound for τ_t and applying Lemma 7 t times. This yields a $C_0 \cdot \beta_0^k$ upper bound for $\tau_0(k)$, where

$$\beta_0 = (11.7899^{(2/3)^t}) 1.8899^{\sum_{i=0}^{t-1}(2/3)^i}$$

It is easy to see that as t tends to infinity, the upper bound for β_0 tends to a number between 6.75 and 6.751. This proves Theorem 2.

References

1. Basavaraju, M., Fomin, F.V., Golovach, P., Misra, P., Ramanujan, M.S., Saurabh S.: Parameterized algorithms to preserve connectivity. In: Esparza, J., Fraigniaud P., Husfeldt, T., Koutsoupias, E. (eds.) ICALP 2014. LNCS, vol. 8572, pp. 800–811 Springer, Heidelberg (2014)
2. Bern, M.W., Plassmann, P.E.: The Steiner problem with edge lengths 1 and 2. Inf. Process. Lett. **32**(4), 171–176 (1989)
3. Björklund, A., Husfeldt, T., Kaski, P., Koivisto, M.: Fourier meets Möbius: fast subset convolution. In: Proceedings of the 39th Annual ACM Symposium on Theory of Computing (STOC), pp. 67–74. ACM, New York (2007)
4. Blelloch, G.E., Dhamdhere, K., Halperin, E., Ravi, R., Schwartz, R., Sridhar, S.: Fixed parameter tractability of binary near-perfect phylogenetic tree reconstruction. In: Bugliesi, M., Preneel, B., Sassone, V., Wegener, I. (eds.) ICALP 2006. LNCS, vol. 4051, pp. 667–678. Springer, Heidelberg (2006)
5. Bodlaender, H.L.: A partial k-arboretum of graphs with bounded treewidth. Theoretical Computer Science **209**(1–2), 1–45 (1998)
6. Byrka, J., Grandoni, F., Rothvoß, T., Sanità, L.: Steiner tree approximation via iterative randomized rounding. J. ACM **60**(1), 6 (2013)
7. Downey, R.G., Fellows, M.R.: Fundamentals of Parameterized Complexity. Texts in Computer Science. Springer (2013)
8. Dreyfus, S.E., Wagner, R.A.: The Steiner problem in graphs. Networks **1**(3), 195–207 (1971)
9. Fomin, F.V., Grandoni, F., Kratsch, D., Lokshtanov, D., Saurabh, S.: Computing optimal Steiner trees in polynomial space. Algorithmica **65**(3), 584–604 (2013)
10. Fuchs, B., Kern, W., Mölle, D., Richter, S., Rossmanith, P., Wang, X.: Dynamic programming for minimum Steiner trees. Theory of Computing Systems **41**(3), 493–500 (2007)
11. Guo, J., Niedermeier, R., Wernicke, S.: Parameterized complexity of generalized vertex cover problems. In: Dehne, F., López-Ortiz, A., Sack, J.-R. (eds.) WADS 2005. LNCS, vol. 3608, pp. 36–48. Springer, Heidelberg (2005)
12. Lokshtanov, D., Nederlof, J.: Saving space by algebraization. In: Proceedings of the 42nd Annual ACM Symposium on Theory of Computing (STOC), pp. 321–330. ACM (2010)
13. Nederlof, J.: Fast polynomial-space algorithms using inclusion-exclusion. Algorithmica **65**(4), 868–884 (2013)
14. Pilipczuk, M., Pilipczuk, M., Sankowski, P., van Leeuwen, E.J.: Subexponential-time parameterized algorithm for Steiner tree on planar graphs. In: Proceedings of the 30th International Symposium on Theoretical Aspects of Computer Science (STACS), Leibniz International Proceedings in Informatics (LIPIcs). Schloss Dagstuhl-Leibniz-Zentrum fuer Informatik, vol. 20, pp. 353–364. Dagstuhl, Germany (2013)
15. Pilipczuk, M., Pilipczuk, M., Sankowski, P., van Leeuwen, E.J.: Network sparsification for Steiner problems on planar and bounded-genus graphs. In: Proceedings of the 55th Annual Symposium on Foundations of Computer Science (FOCS), pp. 276–285. IEEE (2014)
16. Prömel, H.J., Steger, A.: The Steiner Tree Problem. Advanced Lectures in Mathematics. Friedr. Vieweg & Sohn, Braunschweig (2002)
17. Robbins, H.: A remark on Stirling's formula. Amer. Math. Monthly **62**, 26–29 (1955)

Relative Discrepancy Does not Separate Information and Communication Complexity

Lila Fontes[1]([⊠]), Rahul Jain[2], Iordanis Kerenidis[3], Sophie Laplante[1], Mathieu Laurière[1]([⊠]), and Jérémie Roland[4]

[1] LIAFA, Université Paris-Diderot, Paris, France
{fontes,laplante,mathieu.lauriere}@liafa.univ-paris-diderot.fr
[2] CQT, National University of Singapore, Singapore, Singapore
rahul@comp.nus.edu.sg
[3] LIAFA, CNRS, Université Paris-Diderot, Paris, France
jkeren@liafa.univ-paris-diderot.fr
[4] ULB, QuIC, Ecole Polytechnique de Bruxelles, Brussel, Belgium
jroland@ulb.ac.be

Abstract. Does the information complexity of a function equal its communication complexity? We examine whether any currently known techniques might be used to show a separation between the two notions. Ganor *et al.* recently provided such a separation in the distributional case for a specific input distribution. We show that in the non-distributional setting, the relative discrepancy bound is smaller than the information complexity, hence it cannot separate information and communication complexity. In addition, in the distributional case, we provide a linear program formulation for relative discrepancy and relate it to variants of the partition bound, resolving also an open question regarding the relation of the partition bound and information complexity. Last, we prove the equivalence between the adaptive relative discrepancy and the public-coin partition, implying that the logarithm of the adaptive relative discrepancy bound is quadratically tight with respect to communication.

1 Introduction

The question of whether information complexity equals communication complexity is one of the most important outstanding questions in communication complexity. Communication complexity measures the amount of bits Alice and Bob need to communicate to each other in order to compute a function whose input is shared between them. On the other hand, information complexity measures the amount of information Alice and Bob must reveal about their inputs in order to compute the function. Equality between information and communication complexity is equivalent to a compression theorem in the interactive setting. It is known that a single message can be compressed to its information content [1–4] and here the question is whether such a compression is possible for an interactive conversation.

An important application of information complexity is to prove direct sum theorems for communication complexity, namely show that computing k

© Springer-Verlag Berlin Heidelberg 2015
M.M. Halldórsson et al. (Eds.): ICALP 2015, Part I, LNCS 9134, pp. 506–516, 2015.
DOI: 10.1007/978-3-662-47672-7_41

instances of a function costs k times the communication of computing a single instance. This has been shown to be true in the simultaneous and one-way models [5,6], for bounded-round two-way protocols under product distributions [3,7] or non-product distributions [1], and also for specific functions like Disjointness [8]; non-trivial direct sum theorems have also been shown for general two-way randomized communication complexity [9]. Since the information complexity is equal to amortized communication complexity [1], the question of whether information and communication complexity are equal is equivalent to whether communication complexity has a direct sum property [1,10]. Note that in the case of deterministic, zero-error protocols, a separation between information and communication complexity is known for Equality [10].

Since information complexity deals with the information Alice and Bob transmit about their inputs, it is necessary to define a distribution on these inputs. For each fixed distribution μ, we define the distributional information complexity of a function f (also known as the information cost) as the information Alice and Bob transmit about their inputs in any protocol that solves f with small error according to μ [1,5]. The (non-distributional) information complexity of the function f is defined as its distributional information complexity for the worst distribution μ [10]. In this paper we consider the internal information complexity.

Similarly, for communication complexity, one may also consider a model with a distribution μ over the inputs, and the error probability of the protocol is taken over this distribution. This is called a distributional model, and Yao's minmax principle [11] states that the randomized communication complexity of f is equal to its distributional communication complexity for the worst distribution μ, where the randomized communication complexity of a function f is defined as the minimum number of bits exchanged, in the worst case over the inputs, for a randomized protocol to compute the function with small error [12].

One can therefore ask whether the following stronger relation holds: is the distributional communication complexity equal to the distributional information complexity for all input distributions μ? A positive answer to this question would also imply a positive answer to the initial question, proving the equality of information and communication complexity.

In a recent breakthrough, Ganor et al. [13,14] defined a function f and a distribution μ, for which there is an exponential separation between the distributional information and communication complexity. Does this settle the question of communication versus information? First, let us note that the gap, although exponential, is very small compared to the input size: a $\log\log(n)$ communication lower bound and a $\log\log\log(n)$ information upper bound, for inputs of size n. More importantly, Ganor et al.'s results prove that the *distributional* information and communication complexities are not equal for all distributions μ.

How could we settle the question in the non-distributional setting? To prove a separation it is necessary to show that the communication complexity of a specific function is large, while its information complexity is small. In other words, we

eed a lower bound technique which provides a lower bound for communication
ut *not* for information.

In previous work, Kerenidis *et al.* [15] showed that almost all known lower
ound techniques for communication also provide lower bounds for information.
More precisely, they studied the relaxed partition bound and proved that it sub-
umes all known lower bound techniques (except the partition bound [16]). In
ddition, they proved that for any distribution μ, the distributional informa-
ion complexity can be lower bounded by the relaxed partition bound. This also
olds in the non-distributional setting. An open question was whether the parti-
ion bound remained a candidate for separating information and communication
omplexity.

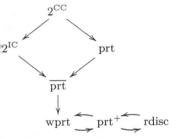

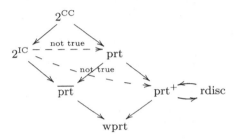

a) The non-distributional case. The
act that rdisc is upper bounded by the
rt is given in Theorem 3.

(b) The distributional case. The equivalence between prt$^+$ and rdisc is given
in Theorem 5. The separation given by
Ganor *et al.* is between IC and rdisc.

Fig. 1. Definitions follow in Sections 3, 4, and 5. An arrow from one bound to another
ndicates that the former is at least as large as the latter.

The main question we ask is whether the techniques developed by Ganor *et*
al. can help in proving, or disproving, the equality of information and communication complexity of a function f in the non-distributional setting. For their
separation, Ganor *et al.* introduced a new communication lower bound called
relative discrepancy. They showed that for a specific function f and a specific
distribution μ, this quantity is high, while the distributional information complexity is low. We study how large this new bound is compared to the other
known lower bound techniques, and whether it can be used to separate information and communication complexity in the non-distributional setting. Our main
results are:

Result 1: In the non-distributional case, we show that relative discrepancy is
bounded above by the relaxed partition bound (**Theorem 3**). By the results of
[15], this means that relative discrepancy cannot be used to separate information
and communication complexity.

Result 2: In the distributional case, we provide a clear relation between relative discrepancy, relaxed partition and partition bound. We give an equivalent linear program formulation for relative discrepancy (**Theorem 5**) and show how relative discrepancy and relaxed partition can be derived from the partition bound by imposing some simple extra constraints. This also answers negatively to the open question in [15] regarding whether the partition bound is a lower bound on information.

Recently, lower bound techniques that use partitions instead of considering just rectangles have been proposed. Jain *et al.* defined the public coin partition bound, and showed that its logarithm is quadratically related to communication complexity [17]. In addition, Ganor *et al.* introduced the adaptive relative discrepancy [14]. We study the relation between them and show the following:

Result 3: For any μ, adaptive relative discrepancy and public-coin partition bound are equivalent (**Theorem 6**). Hence the logarithm of the adaptive relative discrepancy is quadratically tight to communication.

In addition to providing a linear program for relative and adaptive relative discrepancies, the different variants of the partition bound have several other advantages. They can be defined for a wider range of problems, including non-boolean functions; they have natural interpretations in terms of zero-communication protocols, a fact used for relating information complexity to these bounds [15] and for recent advances in the log rank conjecture [18].

In Section 2 we provide the necessary background and definitions. In Section 3 we prove that relative discrepancy is less than relaxed partition (in the non-distributional setting). In Section 4 we consider the setting with a fixed μ, and compare the partition bound and its variants to the relative discrepancy bound. In Section 5, we consider the adaptive relative discrepancy and compare it to the public coin partition bound. The full version of the paper appears in [19].

2   Preliminaries

Let $\mathbf{X}$ and $\mathbf{Y}$ be the sets of inputs to the two players, and $\mathbf{Z}$ be the set of possible outputs. Since the discrepancy-based bounds studied in this paper apply naturally only to boolean functions, f will usually denote a (possibly partial) function over $\mathbf{X} \times \mathbf{Y}$ taking values in $\mathbf{Z} = \{0, 1\}$, while μ denotes a probability distribution over $\mathbf{X} \times \mathbf{Y}$. [1]

2.1   Information and Communication Complexity

For any (possibly partial) function f over inputs $\mathbf{X} \times \mathbf{Y}$, and any $\epsilon \in (0, 1/2)$, the communication cost of a protocol that computes f with error probability at most ϵ is the number of bits sent for the worst case input.

---

[1] The partition-based definitions apply to non-boolean functions, relations, and bipartite distributions as well, but we do not give the full definitions in this paper for those settings.

Definition 1. *The (public-coin) communication complexity of f, denoted $R_\epsilon(f)$, is the best communication cost for any protocol that computes f using public coins with error at most ϵ for any input (x, y). For any distribution μ over the inputs, the distributional (public-coin) communication complexity of f, denoted $R_\epsilon(f, \mu)$, is the cost of the best protocol that computes f with error at most ϵ, where the error probability is taken over the input distribution.*

For information complexity, we are interested not in the number of bits exchanged, but the amount of information revealed about the inputs. We consider the internal information complexity in this paper. Here $I(X; Y)$ denotes the mutual information between random variables X and Y, and $I(X; Y|Z)$ is the mutual information conditioned on Z.

Definition 2 (Information complexity). *Fix f, μ, ε. Let (X, Y, Π) be the tuple distributed according to (X, Y) sampled from μ and then Π being the transcript of the protocol π applied to X, Y. Then define:*

1. $\mathrm{IC}_\mu(\pi) = I(X; \Pi \mid Y) + I(Y; \Pi \mid X)$
2. $\mathrm{IC}_\mu(f, \varepsilon) = \inf_\pi \mathrm{IC}_\mu(\pi)$, *where π computes f with error at most ϵ*
3. $\mathrm{IC}(f, \varepsilon) = \max_\mu \mathrm{IC}_\mu(f, \varepsilon)$

2.2 Lower Bound Techniques

For any family of variables $\{\beta_{x,y}\}_{(x,y) \in \mathbf{X} \times \mathbf{Y}}$ and any subset $E \subseteq \mathbf{X} \times \mathbf{Y}$, we will denote $\beta(E) = \sum_{(x,y) \in E} \beta_{x,y}$, and $\beta = \beta(\mathbf{X} \times \mathbf{Y})$. Unless otherwise specified "$\forall x, y$" means "$\forall x, y \in \mathbf{X} \times \mathbf{Y}$", "$\forall z$" means "$\forall z \in \mathbf{Z}$", "$\forall R$" means "for all rectangles R in $\mathbf{X} \times \mathbf{Y}$", and "$\forall P$" means "for all partitions P of $\mathbf{X} \times \mathbf{Y}$ into labeled rectangles (R, z)". We also denote by $|P|$ the size of the partition, that is, the number of rectangles (R, z) it contains.

Following Ganor et al. (with small changes that do not affect the value of the bound), we define the relative discrepancy bound $\mathrm{rdisc}_\varepsilon(f, \mu)$, as follows. Without loss of generality, we assume $\mathrm{supp}(\mu) = \mathrm{supp}(f)$.

Definition 3 (Relative discrepancy bound [14]). *Let μ be a distribution over $\mathbf{X} \times \mathbf{Y}$ and let $f : \mathrm{supp}(\mu) \to \{0, 1\}$ be a function.*

$$\mathrm{rdisc}_\varepsilon(f, \mu) = \sup_{\kappa, \delta, \rho_{xy}} \frac{1}{\delta}\left(\tfrac{1}{2} - \kappa - \varepsilon\right)$$

$$\text{subject to} \quad \left(\tfrac{1}{2} - \kappa\right) \cdot \rho(R) \le \mu(R \cap f^{-1}(z)) \quad \forall R, z \ s.t. \ \rho(R) \ge \delta$$

$$\sum_{xy} \rho_{xy} = 1, \ 0 \le \kappa < \frac{1}{2}, \ 0 < \delta < 1, \ \rho_{xy} \ge 0 \quad \forall (x, y).$$

For the non-distributional case, we define $\mathrm{rdisc}_\varepsilon(f) = \max_\mu \mathrm{rdisc}_\varepsilon(f, \mu)$, where the maximum is over distributions μ over $\mathbf{X} \times \mathbf{Y}$ (which implicitly adds nonnegativity and normalization constraints on μ).

Note that neither the constraints nor the objective function are linear in the variables. Intuitively, the distribution ρ rebalances the weight of the 0-region and the 1-region of any rectangle R by putting weights on all (x, y) and not just the ones in the support of μ. If this rebalancing is possible even for rectangles with very small weight (i.e. δ is small), then the relative discrepancy increases.

Using this formulation, Ganor *et al.* show:

Theorem 1 ([14]). *Let f : supp$(\mu) \to \{0, 1\}$ be a (possibly partial) function. Then $\log(\mathrm{rdisc}_\varepsilon(f, \mu)) \leq R_\varepsilon(f, \mu)$.*

The relaxed partition bound was introduced by Kerenidis *et al.* [15] who proved that for any function, it is bounded above by its information complexity. Their result holds also relative to any input distribution.[2]

Definition 4 (Relaxed partition bound [15]). *Let μ be a distribution over $\mathbf{X} \times \mathbf{Y}$ and let f : supp$(\mu) \to \{0, 1\}$ be a function.*

$$\overline{\mathrm{prt}}_\varepsilon(f, \mu) = \max_{\alpha, \beta_{xy}} \qquad \beta - \alpha\epsilon$$

$$subject\ to\ : \qquad \beta(R) - \alpha\mu(R \cap f^{-1}(z)) \leq 1 \qquad \forall R, z$$

$$\alpha \geq 0, \quad \alpha\mu_{xy} - \beta_{xy} \geq 0 \qquad \forall(x, y),$$

where R ranges over all rectangles, $(x, y) \in \mathbf{X} \times \mathbf{Y}$ and $z \in \{0, 1\}$. The non-distributional relaxed partition bound is $\overline{\mathrm{prt}}_\varepsilon(f) = \max_\mu \overline{\mathrm{prt}}_\varepsilon(f, \mu)$. For the non-distributional case, we use $\alpha_{x,y}$ instead of $\alpha\mu_{x,y}$ (which is not linear if μ is no longer fixed), with $\alpha_{x,y}$ positive but not normalized.

Kerenidis *et al.* [15] provided both a primal and dual formulation of the relaxed partition bound. The above is the dual formulation. The corresponding primal formulation can be interpreted in terms of the highest non-abort probability of a zero-communication protocol for f.

Theorem 2 ([15]). *For all μ, boolean functions f over the support of μ and all $\varepsilon \in (0, \frac{1}{4}]$, $\Omega\left(\varepsilon^2 \log \overline{\mathrm{prt}}_{2\varepsilon}(f, \mu)\right) = \mathrm{IC}_\mu(f, \varepsilon) \leq R_\varepsilon(f, \mu)$.*

3   Relative Discrepancy Is Bounded by Relaxed Partition

We show that the non-distributional relative discrepancy is bounded above by the relaxed partition, which implies that a stronger technique is necessary in order to separate information and communication complexity. (See Figure 1a).

Theorem 3. *For any boolean f, and $\epsilon \in (0, 1/3)$, $\mathrm{rdisc}_{\frac{3}{2}\epsilon}(f) \leq \overline{\mathrm{prt}}_\epsilon(f)$.*

---

[2] Compared with the original formulation [15], there is an implicit change of variables: we use $\beta_{x,y}$ here to denote what was $\alpha_{x,y} - \beta_{x,y}$ in the original notation.

Proof. It suffices to show that for any feasible solution of rdisc, there exists feasible solution for $\overline{\mathrm{prt}}$ whose objective value is at least as large. Let $\kappa, \delta, \{\rho_{x,y}\}_{x,y}, \{\mu_{x,y}\}_{x,y})$ be a feasible solution of relative discrepancy for f. Define for any $(x,y) \in \mathbf{X} \times \mathbf{Y}$, $\alpha_{x,y} = \frac{1}{\delta}(\frac{1}{2}-\kappa)\rho_{x,y}+\frac{1}{\delta}\mu_{x,y}$ and $\beta_{x,y} = \frac{1}{\delta}(\frac{1}{2}-\kappa)\rho_{x,y}$. We show that the relaxed partition constraints are satisfied. First, the sign constraints are satisfied. Moreover, for any R, z,

$$\beta(R) - \alpha(R \cap f^{-1}(z))$$
$$= \frac{1}{\delta}(\tfrac{1}{2} - \kappa)\rho(R) - \frac{1}{\delta}\mu(R \cap f^{-1}(z)) - \frac{1}{\delta}(\tfrac{1}{2} - \kappa)\rho(R \cap f^{-1}(z))$$
$$\leq \frac{1}{\delta}(\tfrac{1}{2} - \kappa)\rho(R) - \frac{1}{\delta}\mu(R \cap f^{-1}(z)) \qquad \text{(since } \rho_{xy} \geq 0 \text{ for any } (x,y))$$

There are two cases: if $\rho(R) \geq \delta$, then $\frac{1}{\delta}(\frac{1}{2} - \kappa)\rho(R) - \frac{1}{\delta}\mu(R \cap f^{-1}(z)) \leq 0 \leq 1$ by the relative discrepancy constraint; otherwise $\rho(R) < \delta$ and $\frac{1}{\delta}(\frac{1}{2} - \kappa)\rho(R) - \frac{1}{\delta}\mu(R \cap f^{-1}(z)) < (\frac{1}{2} - \kappa) - \frac{1}{\delta}\mu(R \cap f^{-1}(z)) \leq \frac{1}{2} \leq 1$.

Finally we compare the objective values. Since ρ and μ are distributions, $\alpha = \frac{1}{\delta}(\frac{3}{2} - \kappa)$ and $\beta = \frac{1}{\delta}(\frac{1}{2} - \kappa)$, so $\beta - \epsilon\alpha = \frac{1}{\delta}[\frac{1}{2} - \kappa - (\frac{3}{2} - \kappa)\epsilon] \geq \frac{1}{\delta}(\frac{1}{2} - \kappa - \epsilon) = \mathrm{rdisc}_{\frac{3}{2}\epsilon}(f)$. □

Combining Theorem 2 and Theorem 3 gives us that relative discrepancy is a lower bound on information complexity.

Corollary 1. *For all functions* $f : \mathbf{X} \times \mathbf{Y} \to \{0,1\}$ *and all* $\varepsilon \in (0, \frac{1}{6}]$, $\Omega(\varepsilon^2 \log(\mathrm{rdisc}_{3\varepsilon}(f))) = \mathrm{IC}(f, \varepsilon) \leq R_\varepsilon(f).$

Remark 1. Our change of variables satisfies an additional constraint :

$$\beta_{x,y} \geq 0 \text{ for any } (x,y) \in \mathbf{X} \times \mathbf{Y}. \tag{1}$$

since $\rho_{x,y} \geq 0$. We will examine the role of this constraint in Section 4. It turns out to be a key point in understanding how relative discrepancy relates to the partition bound and its variants. Also notice that $\alpha_{x,y}$ is not proportional to $\mu_{x,y}$, so this change of variable does not carry over to the distributional case, since $\alpha_{x,y}$ cannot be written as $\alpha\mu_{xy}$.

4   The Distributional Case

In this section we study how the various bounds relate, relative to a fixed distribution μ, and uncover an elegant relationship between the bounds by adding simple positivity constraints to the partition bound.

We start with a fixed-distribution version of the partition bound [16], which we define below. It follows easily from the original proof that this is a lower bound on distributional communication complexity and that it equals the partition bound in the worst case distribution.

Definition 5 (Partition bound).

$$\text{prt}_\epsilon(f,\mu) = \max_{\alpha,\beta_{xy}} \qquad \beta - \epsilon\alpha$$

$$\text{subject to}: \qquad \beta(R) - \alpha\mu(R \cap f^{-1}(z)) \leq 1 \qquad \forall R, z$$

$$\alpha \geq 0.$$

The non-distributional bound is $\text{prt}_\epsilon(f) = \max_\mu \text{prt}_\epsilon(f,\mu)$. *Going from the non-distributional setting to a fixed distribution* μ, $\alpha_{x,y}$ *is replaced by* $\alpha \cdot \mu_{x,y}$, *that is,* $\{\alpha_{x,y}\}$ *is* $\{\mu_{x,y}\}$ *scaled by a factor* α.

Theorem 4 ([16]). *Let* $f : \text{supp}(\mu) \to \{0,1\}$ *be a (possibly partial) function. Then* $\log(\text{prt}_\varepsilon(f,\mu)) \leq R_\varepsilon(f,\mu)$.

Note that the relaxed partition bound (Definition 4) is obtained from the partition bound by adding the constraint $\alpha\mu_{x,y} - \beta_{x,y} \geq 0$ for all (x,y).

As suggested in the proof of Theorem 3, we now consider the constraint $\beta_{x,y} \geq 0$ for all x,y. Adding this constraint to the partition bound results in a new bound which we call the positive partition bound.

Definition 6 (Positive partition bound).

$$\text{prt}^+_\epsilon(f,\mu) = \max_{\alpha,\beta_{xy}} \qquad \beta - \epsilon\alpha$$

$$\text{subject to}: \qquad \beta(R) - \alpha\mu(R \cap f^{-1}(z)) \leq 1 \qquad \forall R, z$$

$$\alpha \geq 0, \quad \beta_{xy} \geq 0 \qquad \forall(x,y).$$

We also define $\text{prt}^+_\epsilon(f) = \max_\mu \text{prt}^+_\epsilon(f,\mu)$, *and use* $\alpha_{x,y}$ *instead of* $\alpha\mu_{x,y}$.

The weak partition bound is obtained by adding both constraints.

Definition 7 (Weak partition bound).

$$\text{wprt}_\epsilon(f,\mu) = \max_{\alpha,\beta_{xy}} \qquad \beta - \epsilon\alpha$$

$$\text{subject to}: \quad \beta(R) - \alpha\mu(R \cap f^{-1}(z)) \leq 1 \qquad \forall R, z,$$

$$\alpha \geq 0, \quad \beta_{xy} \geq 0, \quad \alpha\mu_{xy} - \beta_{xy} \geq 0 \qquad \forall(x,y).$$

We also define $\text{wprt}_\epsilon(f) = \max_\mu \text{wprt}_\epsilon(f,\mu)$.

Because we have added a constraint to a maximization problem, it is easy to see that the following holds (see Figure 1b).

Proposition 1. *For all* f,μ,ϵ,

$$\text{wprt}_\epsilon(f,\mu) \leq \text{prt}^+_\epsilon(f,\mu) \leq \text{prt}_\epsilon(f,\mu) \text{ and } \text{wprt}_\epsilon(f,\mu) \leq \overline{\text{prt}}_\epsilon(f,\mu) \leq \text{prt}_\epsilon(f,\mu).$$

In [19], we show the following equivalence:

Theorem 5. *Let* μ *be a distribution on* $\mathbf{X} \times \mathbf{Y}$ *and* f *be a boolean function on the support of* μ *such that either* $\text{rdisc}_\epsilon(f,\mu) \geq 1$ *or* $\text{prt}^+_{4\epsilon}(f,\mu) > 2$. *Then for any* $\epsilon \in (0,1/4)$, $\frac{\epsilon}{2}\text{prt}^+_{4\epsilon}(f,\mu) \leq \text{rdisc}_\epsilon(f,\mu) \leq \text{prt}^+_\epsilon(f,\mu)$.

Each inequality is proven by a different change of variables. At a high level, $\rho_{x,y}$ is proportional to $\beta_{x,y}$ and δ is a scaling factor.

Revisiting the non-distributional case For the change of variables in the proof of Theorem 3, we have noted that the constraint $\beta_{xy} \geq 0$ holds $\forall (x, y)$ (see Inequality 1). This shows that, in the non-distributional case, relative discrepancy is, in fact, no larger than the weak partition bound, i.e. $\mathrm{rdisc}_\epsilon(f) \leq \mathrm{wprt}_{\frac{2}{3}\epsilon}(f)$.

Lemma 1. *For any boolean* f, *and* $\epsilon \in (0, 1/2)$, $\mathrm{prt}^+_\epsilon(f) \leq \mathrm{wprt}_{\frac{\epsilon}{2}}(f) + \frac{\epsilon}{2}$.

Proof. Let $\alpha_{x,y}, \beta_{x,y}$ be a feasible solution for prt^+, and consider the following assignment for wprt: $\alpha'_{x,y} = \alpha_{x,y} + \beta_{x,y}$, $\beta'_{x,y} = \beta_{x,y}$. The constraint on rectangles is still satisfied, and the added positivity constraint $\alpha'_{x,y} - \beta'_{x,y} = \alpha_{x,y} \geq 0$ is also satisfied. Finally, the objective function for wprt with error $\frac{\epsilon}{2}$ is $\beta' - \frac{\epsilon}{2}\alpha' = \beta - \frac{\epsilon}{2}\beta - \frac{\epsilon}{2}\alpha \geq \beta - \epsilon\alpha - \frac{\epsilon}{2}$ (where we have used the constraint on $R = \mathbf{X} \times \mathbf{Y}$), as claimed. □

The change of variables in the proof of Theorem 3 is just the composition of the two changes of variables in Theorem 5 and Lemma 1. It is also now clearer how the distributional and the non-distributional settings differ. It cannot be the case that $\mathrm{prt}^+_\epsilon(f, \mu) \leq \mathrm{wprt}_\epsilon(f, \mu)$ for fixed distribution, since Ganor *et al.* provide a counterexample. We can also see that for this specific change of variable, by setting $\alpha'_{x,y} = \alpha_{x,y} + \beta_{x,y}$, $\alpha'_{x,y}$ cannot be written as $\alpha_{x,y} = \alpha\mu_{x,y}$, as we would need in the distributional case, since it is a combination of α and β.

5   Adaptive Relative Discrepancy Is Equivalent to the Public Coin Partition

In this section, we compare two lower bound techniques for communication complexity introduced recently. We give below a distributional version of the public-coin partition[3].

Definition 8 (Public coin partition bound [17]).

$$\mathrm{pprt}_\epsilon(f, \mu) = \max_{\alpha,\beta} \quad \beta - \epsilon\alpha$$

$$\text{subject to :} \quad \beta - \sum_{(R,z)\in P} \alpha\mu(R \cap f^{-1}(z)) \leq |P| \qquad \forall P$$

$$\alpha \geq 0, \beta \geq 0.$$

Ganor *et al.* introduced the following notion, which is not a linear program:

---

[3] Note that this is a simplified definition with respect to the original one by means of removing redundant variables and constraints in the primal formulation, taking the dual of the resulting expression, and replacing $\alpha_{x,y}$ by $\alpha\mu_{x,y}$, where the distribution μ is fixed.

Definition 9 (Adaptive relative discrepancy [14]).

$$\text{ardisc}_\varepsilon(f,\mu) = \sup_{\kappa,\delta,\rho^P_{x,y}} \tfrac{1}{\delta}\left(\tfrac{1}{2} - \kappa - \varepsilon\right) \qquad \text{subject to :}$$

$$\left(\tfrac{1}{2} - \kappa\right)\rho^P(R) \leq \mu(R \cap f^{-1}(z)), \; \forall P, \, \forall (z,R) \in P : \rho^P(R) \geq \delta$$

$$0 \leq \kappa < \tfrac{1}{2}, \quad 0 < \delta < 1, \quad \rho^P = 1, \rho^P_{x,y} \geq 0, \quad \forall P, \forall(x,y).$$

Then $\text{ardisc}_\varepsilon(f) = \max_\mu \text{ardisc}_\varepsilon(f,\mu)$.

In [19], we prove the following result :

Theorem 6. *For any distribution* μ, *any function* $f : \text{supp}(\mu) \to \{0,1\}$ *and* $\epsilon \in (0, \tfrac{1}{4})$ *such that either* $\text{ardisc}_\epsilon(f,\mu) \geq 1$ *or* $\text{pprt}_{4\epsilon}(f,\mu) > 2$,

$$\frac{\epsilon}{2}\text{pprt}_{4\epsilon}(f,\mu) \leq \text{ardisc}_\epsilon(f,\mu) \leq \text{pprt}_\epsilon(f,\mu).$$

Since the logarithm of the public coin partition bound is polynomially related to randomized communication complexity [17], this tells us that the logarithm of the adaptive relative discrepancy is also polynomially related to communication complexity.

Corollary 2. *For any* μ, $f : \text{supp}(\mu) \to \{0,1\}$ *and* $\epsilon \in (0, \tfrac{1}{8})$,

$$\log(\text{ardisc}_\epsilon(f,\mu)) \leq R_\epsilon(f,\mu) \leq \left(\log \text{ardisc}_{\epsilon/8}(f,\mu) + 2\log\frac{1}{\epsilon} + 6\right)^2.$$

Acknowledgments. We are grateful to Nikos Leonardos for useful discussions and we also thank Virginie Lerays for many fruitful discussions and for the simplification of the public coin partition bound. R.J. would like to thank Anurag Anshu, Prahladh Harsha, Priyanka Mukhopadhyay and Vankatesh Srinivasan for helpful discussions. L.F., I.K., S.L. and M.L. acknowledge support from the French ANR Blanc project RDAM ANR-12-BS02-005, European Union CHIST-ERA grant DIQIP, the European Union Seventh Framework Programme (FP7/2007-2013) under grant agreement n. 600700 (QALGO) and the ERC project QCC. J.R. acknowledges support from the Belgian ARC project COPHYMA. The work of R.J. is supported by the Singapore Ministry of Education Tier 3 Grant, the National University of Singapore Young Researcher Award 2012 and the Core Grants of the Center for Quantum Technologies, Singapore. Part of the work done while visiting the Banff International Research Station for Mathematical Innovation and Discovery, Banff, Canada and while visiting the Simon's Institute, U.C. Berkeley, USA.

References

1. Braverman, M., Rao, A.: Information equals amortized communication. IEEE Transactions on Information Theory **60**(10), 6058–6069 (2014)
2. Fano, R.M.: The transmission of information, Technical Report 65, Research Laboratory for Electronics. MIT, Cambridge (1949)

3. Jain, R., Radhakrishnan, J., Sen, P.: A direct sum theorem in communication-complexity via message compression. In: Baeten, J.C.M., Lenstra, J.K., Parrow, J., Woeginger, G.J. (eds.) ICALP 2003. LNCS, vol. 2719, pp. 300–315. Springer, Heidelberg (2003)
4. Shannon, C.E.: A Mathematical Theory of Computation. The Bell System Technical Journal **27**, 379–423, 623–656 (1948)
5. Chakrabarti, A., Wirth, A., Yao, A., Shi, Y.: Informational complexity and the direct sum problem for simultaneous message complexity. In: FOCS, pp. 270–278 (2001)
6. Jain, R., Radhakrishnan, J., Sen, P.: Optimal direct sum and privacy trade-off results for quantum and classical communication complexity. In: CoRR, vol. abs/0807.1267, pp. 285–296 (2008)
7. Harsha, P., Jain, R., McAllester, D., Radhakrishnan, J.: The communication complexity of correlation. In: CCC, pp. 10–23 (2007)
8. Bar-Yossef, Z., Jayram, T.S., Kumar, R., Sivakumar, D.: An information statistics approach to data stream and communication complexity. Journal of Computer and System Sciences **68**(4), 702–732 (2004)
9. Barak, B., Braverman, M., Chen, X., Rao, A.: How to compress interactive communication. In: STOC, pp. 67–76 (2010)
10. Braverman, M.: Interactive information complexity. In: STOC, pp. 505–524 (2012)
11. Yao, A.C.-C.: Lower bounds by probabilistic arguments. In: FOCS, pp. 420–428 (1983)
12. Yao, A.C.C.: Some complexity questions related to distributive computing (preliminary report). In: STOC, pp. 209–213 (1979)
13. Ganor, A., Kol, G., Raz, R.: Exponential separation of information and communication. In: ECCC, vol. 21, p. 49 (2014)
14. Ganor, A., Kol, G., Raz, R.: Exponential separation of information and communication for boolean functions. In: ECCC, vol. 113 (2014)
15. Kerenidis, I., Laplante, S., Lerays, V., Roland, J., Xiao, D.: Lower bounds on information complexity via zero-communication protocols and applications. In: FOCS, pp. 500–509 (2012)
16. Jain, R., Klauck, H.: The partition bound for classical communication complexity and query complexity. In: CCC, pp. 1–28 (2010)
17. Jain, R., Lee, T., Vishnoi, N.: A quadratically tight partition bound for classical communication complexity and query complexity. In: CoRR, vol. abs/1401.4512 (2014)
18. Gavinsky, D., Lovett, S.: En route to the log-rank conjecture: new reductions and equivalent formulations. In: Esparza, J., Fraigniaud, P., Husfeldt, T., Koutsoupias, E. (eds.) ICALP 2014. LNCS, vol. 8572, pp. 514–524. Springer, Heidelberg (2014)
19. Fontes, L., Jain, R., Kerenidis, I., Laplante, S., Lauriere, M., Roland, J.: Relative Discrepancy does not separate Information and Communication Complexity (2015). http://eccc.hpi-web.de/report/2015/028/

A Galois Connection for Valued Constraint Languages of Infinite Size

Peter Fulla and Stanislav Živný[(✉)]

Department of Computer Science, University of Oxford, Oxford, UK
{peter.fulla,standa.zivny}@cs.ox.ac.uk

Abstract. A Galois connection between clones and relational clones on a fixed finite domain is one of the cornerstones of the so-called algebraic approach to the computational complexity of non-uniform Constraint Satisfaction Problems (CSPs). Cohen et al. established a Galois connection between *finitely-generated* weighted clones and *finitely-generated* weighted relational clones [SICOMP'13], and asked whether this connection holds in general. We answer this question in the affirmative for weighted (relational) clones with *real* weights and show that the complexity of the corresponding Valued CSPs is preserved.

1 Introduction

The constraint satisfaction problem (CSP) is a general framework capturing decision problems arising in many contexts of computer science [13]. The CSP is NP-hard in general but there has been much success in finding tractable fragments of the CSP by restricting the types of relations allowed in the constraints. A set of allowed relations has been called a *constraint language* [11]. For some constraint languages, the associated constraint satisfaction problems with constraints chosen from that language are solvable in polynomial-time, whilst for other constraint languages this class of problems is NP-hard [11]; these are referred to as *tractable languages* and *NP-hard languages*, respectively. Dichotomy theorems, which classify each possible constraint language as either tractable or NP-hard, have been established for constraint languages over two-element domains [19], three-element domains [5], for conservative (containing all unary relations) constraint languages [7], for maximal constraint languages [4,8], for graphs (corresponding to languages containing a single binary symmetric relation) [12], and for digraphs without sources and sinks (corresponding to languages containing a single binary relations without sources and sinks) [2]. The most successful approach to classifying the complexity of constraint languages has been the algebraic approach [1,6,15].

The *valued* constraint satisfaction problem (VCSP) is a general framework that captures not only feasibility problems but also optimisation problems [10,14]. A VCSP instance represents each constraint by a *weighted relation*,

The authors were supported by a Royal Society Research Grant. Stanislav Živný was supported by a Royal Society University Research Fellowship.

© Springer-Verlag Berlin Heidelberg 2015
M.M. Halldórsson et al. (Eds.): ICALP 2015, Part I, LNCS 9134, pp. 517–528, 2015.
DOI: 10.1007/978-3-662-47672-7_42

which is a $\overline{\mathbb{Q}}$-valued function where $\overline{\mathbb{Q}} = \mathbb{Q} \cup \{\infty\}$, and the goal is to find a labelling of variables minimising the sum of the values assigned by the constraints to that labelling. Tractable fragments of the VCSP have been identified by restricting the types of allowed weighted relations that can be used to define the valued constraints. A set of allowed weighted relations has been called a *valued constraint language* [10]. Classifying the complexity of *all* valued constraint languages is a challenging task as it includes as a special case the classification of $\{0, \infty\}$-valued languages (i.e. constraint languages); this would answer the conjecture of Feder and Vardi [11], which asserts that every constraint language is either tractable or NP-hard, and its algebraic refinement, which specifies the precise boundary between tractable and NP-hard languages [6]. However, several nontrivial results are known, see [14] for a recent survey. Dichotomy theorems, which classify each possible valued constraint language as either tractable or NP-hard, have been established for valued constraint languages over two-element domains [10], for conservative (containing all $\{0, 1\}$-valued unary cost functions) valued constraint languages [17], and also for finite-valued (all weighted relations are $\mathbb{Q}$-valued) constraint languages [22]. Moreover, the power of the basic linear programming relaxation for valued constraint languages has been characterised [16, 21].

Cohen et al. have introduced an algebraic theory of weighted clones [9], further extended in [18], for classifying the computational complexity of valued constraint languages. This theory establishes a one-to-one correspondence between valued constraint languages closed under expressibility (which does not change the complexity of the associated class of optimisation problems), called weighted relational clones, and weighted clones [9]. This is an extension of (a part of) the algebraic approach to CSPs which relies on a one-to-one correspondence between constraint languages closed under pp-definability (which does not change the complexity of the associated class of decision problems), called relational clones, and clones [6], thus making it possible to use deep results from universal algebra. In fact, the recent progress on the power of the basic linear programming relaxation [16] and the classification of finite-valued constraint languages [22], as well as results on special cases of Valued CSPs such as Min-Sol-Hom [23], rely on the work of Cohen et al [9].

Contributions

The Galois connection between weighted clones and weighted relational clones established in [9] was proved only for weighted (relational) clones generated by a set of a *finite* size. The authors asked whether such a correspondence holds also for weighted (relational) clones in general. In this paper we answer this question in the affirmative.

Firstly, we show that the Galois connection from [9] (using only rational weights) does *not* work for general weighted (relational) clones. Secondly, we alter the definition of weighted (relational) clones and establish a new Galois connection that holds even when the generating set has an infinite size. We allow weighted relations and weightings to assign real weights instead of rational, require weighted relational clones to be closed under operator Opt, and prove that these changes preserve tractability of a constraint language.

Including the Opt operator in the definition of weighted relational clones simplifies the structure of the space of all weighted clones, and guarantees that every non-projection polymorphism of a weighted relational clone Γ is assigned a positive weight by some weighted polymorphism of Γ.

The proof of the Galois connection in [9] relies on results on linear programming duality; we used their generalisation from the theory of convex optimisation in order to establish the connection even for infinite sets.

2 Background

2.1 Valued CSPs

Throughout the paper, let D be a fixed finite set of size at least two.

Definition 1. An m-ary relation[1] over D is any mapping $\phi : D^m \to \{c, \infty\}$ for some $c \in \mathbb{Q}$. We denote by $\mathbf{R}_D^{(m)}$ the set of all m-ary relations and let $\mathbf{R}_D = \bigcup_{m \geq 1} \mathbf{R}_D^{(m)}$.

Given an m-tuple $\mathbf{x} \in D^m$, we denote its ith entry by $\mathbf{x}[i]$ for $1 \leq i \leq m$.
Let $\overline{\mathbb{Q}} = \mathbb{Q} \cup \{\infty\}$ denote the set of rational numbers with (positive) infinity.

Definition 2. An m-ary weighted relation over D is any mapping $\gamma : D^m \to \overline{\mathbb{Q}}$. We denote by $\mathbf{\Phi}_D^{(m)}$ the set of all m-ary weighted relations and let $\mathbf{\Phi}_D = \bigcup_{m \geq 1} \mathbf{\Phi}_D^{(m)}$.

From Definition 2 we have that relations are a special type of weighted relations.

Example 1. An important example of a (weighted) relation is the binary equality $\phi_=$ on D defined by $\phi_=(x, y) = 0$ if $x = y$ and $\phi_=(x, y) = \infty$ if $x \neq y$.
Another example of a relation is the unary empty relation $\phi_\emptyset$ defined on D by $\phi_\emptyset(x) = \infty$ for all $x \in D$.

For any m-ary weighted relation $\gamma \in \mathbf{\Phi}_D^{(m)}$, we denote by $\mathrm{Feas}(\gamma) = \{\mathbf{x} \in D^m \mid \gamma(\mathbf{x}) < \infty\} \in \mathbf{R}_D^{(m)}$ the underlying *feasibility relation,* and by $\mathrm{Opt}(\gamma) = \{\mathbf{x} \in \mathrm{Feas}(\gamma) \mid \gamma(\mathbf{x}) \leq \gamma(\mathbf{y})$ for every $\mathbf{y} \in D^m\} \in \mathbf{R}_D^{(m)}$ the relation of minimal-value tuples.

---

[1] An m-ary relation over D is commonly defined as a subset of D^m. Note that Definition 1 is equivalent to the standard definition as any mapping ϕ can be seen as set $R = \{\mathbf{x} \in D^m \mid \phi(\mathbf{x}) < \infty\}$, and any set $R \subseteq D^m$ can be represented by mapping ϕ such that $\phi(\mathbf{x}) = 0$ when $\mathbf{x} \in R$ and $\phi(\mathbf{x}) = \infty$ otherwise. Consequently, we shall use both definitions interchangeably.

Definition 3. *Let* $V = \{x_1, \ldots, x_n\}$ *be a set of variables. A* valued constraint *over* V *is an expression of the form* $\gamma(\mathbf{x})$ *where* $\gamma \in \Phi_D^{(m)}$ *and* $\mathbf{x} \in V^m$. *The number* m *is called the* arity *of the constraint, the weighted relation* γ *is called the* constraint weighted relation, *and the tuple* $\mathbf{x}$ *the* scope *of the constraint.*

We call D the *domain*, the elements of D *labels* (for variables), and say that the weighted relation in Φ_D take *values* or *weights*.

Definition 4. *An instance of the* valued constraint satisfaction problem, *VCSP, is specified by a finite set* $V = \{x_1, \ldots, x_n\}$ *of variables, a finite set* D *of labels, and an* objective function *I expressed as follows:*

$$I(x_1, \ldots, x_n) = \sum_{i=1}^{q} \gamma_i(\mathbf{x}_i), \tag{1}$$

where each $\gamma_i(\mathbf{x}_i)$, $1 \leq i \leq q$, *is a valued constraint over* V. *Each constraint can appear multiple times in* I.

The goal is to find an assignment *(or a* labelling*) of labels to the variables that minimises* I.

CSPs are a special case of VCSPs using only (unweighted) relations with the goal to determine the existence of a feasible assignment.

Definition 5. *Any set* $\Gamma \subseteq \Phi_D$ *is called a* (valued) constraint language *over* D, *or simply a* language. *We will denote by* VCSP(Γ) *the class of all VCSP instances in which the constraint weighted relations are all contained in* Γ.

Definition 6. *A constraint language* Γ *is called* tractable *if* VCSP(Γ') *can be solved (to optimality) in polynomial time for every finite subset* $\Gamma' \subseteq \Gamma$, *and* Γ *is called* intractable *if* VCSP(Γ') *is NP-hard for some finite* $\Gamma' \subseteq \Gamma$.

We are interested in the computational complexity of various constraint languages, see [14] for a recent survey on this topic.

2.2   Weighted Relational Clones

Definition 7. *A weighted relation* γ *of arity* r *can be obtained by* addition *from the weighted relation* γ_1 *of arity* s *and the weighted relation* γ_2 *of arity* t *if* γ *satisfies the identity*

$$\gamma(x_1, \ldots, x_r) = \gamma_1(y_1, \ldots, y_s) + \gamma_2(z_1, \ldots, z_t) \tag{2}$$

for some (fixed) choice of $y_1, \ldots, y_s$ *and* $z_1, \ldots, z_t$ *from amongst the* $x_1, \ldots, x_r$.

Definition 8. *A weighted relation* γ *of arity* r *can be obtained by* minimisation *from the weighted relation* γ' *of arity* $r + s$ *if* γ *satisfies the identity*

$$\gamma(x_1, \ldots, x_r) = \min_{y_1 \in D, \ldots, y_s \in D} \gamma'(x_1, \ldots, x_r, y_1, \ldots, y_s). \tag{3}$$

Definition 9. *A constraint language $\Gamma \subseteq \Phi_D$ is called a* weighted relational *clone if it contains the binary equality relation $\phi_=$ and the unary empty relation $\phi_\emptyset$,[2] and is closed under addition, minimisation, scaling by non-negative rational constants, and addition of rational constants.*

For any Γ, we define wRelClone(Γ) *to be the smallest weighted relational clone containing Γ.*

Note that for any weighted relational clone Γ, if $\gamma \in \Gamma$ then Feas(γ) $\in \Gamma$ as Feas(γ) = 0γ (we define $0 \cdot \infty = \infty$).

Definition 10. *Let $\Gamma \subseteq \Phi_D$ be a constraint language, $I \in$ VCSP(Γ) an instance with variables V, and $L = (v_1, \ldots, v_r)$ a list of variables from V. The projection of I onto L, denoted $\pi_L(I)$, is the r-ary weighted relation on D defined as*

$$\pi_L(I)(x_1, \ldots, x_r) = \min_{\{s:V \to D \;\mid\; (s(v_1),\ldots,s(v_r))=(x_1,\ldots,x_r)\}} I(s). \tag{4}$$

We say that a weighted relation γ is expressible *over a constraint language Γ if $\gamma = \pi_L(I)$ for some $I \in$ VCSP(Γ) and list of variables L. We call the pair (I, L) a* gadget *for expressing γ over Γ.*

The list of variables L in a gadget may contain repeated entries. The minimum over an empty set is ∞.

Example 2. For any $\Gamma \subseteq \Phi_D$, we can express the binary equality relation $\phi_=$ on D over language Γ using the following gadget. Let $I \in$ VCSP(Γ) be the instance with a single variable v and no constraints, and let $L = (v, v)$. Then, by Definition 10, $\pi_L(I) = \phi_=$.

We may equivalently define a weighted relational clone as a set $\Gamma \subseteq \Phi_D$ that contains the unary empty relation $\phi_\emptyset$ and is closed under expressibility, scaling by non-negative rational constants, and addition of rational constants [9, Proposition 4.5].

The following result has been shown in [9].

Theorem 1. *A constraint language Γ is tractable if and only if* wRelClone(Γ) *is tractable, and Γ is intractable if and only if* wRelClone(Γ) *is intractable.*

Consequently, when trying to identify tractable constraint languages, it is sufficient to consider only weighted relational clones.

2.3   Weighted Clones

Any mapping $f : D^k \to D$ is called a k-ary *operation*. We will apply a k-ary operation f to k m-tuples $\mathbf{x}_1, \ldots, \mathbf{x}_k \in D^m$ coordinatewise, that is,

$$f(\mathbf{x}_1, \ldots, \mathbf{x}_k) = (f(\mathbf{x}_1[1], \ldots, \mathbf{x}_k[1]), \ldots, f(\mathbf{x}_1[m], \ldots, \mathbf{x}_k[m])) \in D^m. \tag{5}$$

---

[2] Although the definition in [9] does not require inclusion of $\phi_\emptyset$, the proofs there implicitly assume its presence in any weighted relational clone.

Definition 11. *Let γ be an m-ary weighted relation on D and let f be a k-ary operation on D. Then f is a* polymorphism *of γ if, for any $X = (\mathbf{x}_1, \ldots, \mathbf{x}_k) \in (\mathrm{Feas}(\gamma))^k$, we have that $f(X) = f(\mathbf{x}_1, \ldots, \mathbf{x}_k) \in \mathrm{Feas}(\gamma)$.*

For any constraint language Γ over a set D, we denote by $\mathrm{Pol}(\Gamma)$ the set of all operations on D which are polymorphisms of all $\gamma \in \Gamma$. We write $\mathrm{Pol}(\gamma)$ for $\mathrm{Pol}(\{\gamma\})$.

A k-ary *projection* is an operation of the form $e_i^{(k)}(x_1, \ldots, x_k) = x_i$ for some $1 \le i \le k$. Projections are (trivial) polymorphisms of all constraint languages.

Definition 12. *The* superposition *of a k-ary operation $f : D^k \to D$ with k ℓ-ary operations $g_i : D^\ell \to D$ for $1 \le i \le k$ is the ℓ-ary function $f[g_1, \ldots, g_k] : D^\ell \to D$ defined by*

$$f[g_1, \ldots, g_k](x_1, \ldots, x_\ell) = f(g_1(x_1, \ldots, x_\ell), \ldots, g_k(x_1, \ldots, x_\ell)). \qquad (6)$$

Definition 13. *A* clone *of operations, C, is a set of operations on D that contains all projections and is closed under superposition. The k-ary operations in a clone C will be denoted by $C^{(k)}$.*

Example 3. For any D, let $\mathbf{J}_D$ be the set of all projections on D. By Definition 13, $\mathbf{J}_D$ is a clone.

It is well known that $\mathrm{Pol}(\Gamma)$ is a clone for all constraint languages Γ.

Definition 14. *A k-ary* weighting *of a clone C is a function $\omega : C^{(k)} \to \mathbb{Q}$ such that $\omega(f) < 0$ only if f is a projection and*

$$\sum_{f \in C^{(k)}} \omega(f) = 0. \qquad (7)$$

We will call a function $\omega : C^{(k)} \to \mathbb{Q}$ that satisfies Equation (7) but assigns a negative weight to some operation $f \notin \mathbf{J}_D^{(k)}$ an improper weighting. *In order to emphasise the distinction we may also call a weighting a* proper weighting.

Definition 15. *For any clone C, a k-ary weighting ω of C, and $g_1, \ldots, g_k \in C^{(\ell)}$, the* superposition *of ω and $g_1, \ldots, g_k$, is the function $\omega[g_1, \ldots, g_k] : C^{(\ell)} \to \mathbb{Q}$ defined by*

$$\omega[g_1, \ldots, g_k](f') = \sum_{\{f \in C^{(k)} \mid f[g_1, \ldots, g_k] = f'\}} \omega(f). \qquad (8)$$

If the result of a superposition is a proper weighting (that is, negative weights are only assigned to projections), then that superposition will be called a proper superposition.

Definition 16. *A* weighted clone, *Ω, is a non-empty set of weightings of some fixed clone C, called the* support clone *of Ω, which is closed under scaling by non-negative rational constants, addition of weightings of equal arity, and proper superposition with operations from C.*

We now link weightings and weighted relations by the concept of weighted polymorphism, which will allow us to establish a useful correspondence between weighted clones and weighted relational clones.

Definition 17. *Let γ be an m-ary weighted relation on D and let ω be a k-ary weighting of a clone C of operations on D. We call ω a* weighted polymorphism *of γ if $C \subseteq \mathrm{Pol}(\gamma)$ and for any $X = (\mathbf{x}_1, \mathbf{x}_2, \ldots, \mathbf{x}_k) \in (\mathrm{Feas}(\gamma))^k$, we have*

$$\sum_{f \in C^{(k)}} \omega(f) \cdot \gamma(f(X)) = \sum_{f \in C^{(k)}} \omega(f) \cdot \gamma(f(\mathbf{x}_1, \mathbf{x}_2, \ldots, \mathbf{x}_k)) \leq 0. \qquad (9)$$

If ω is a weighted polymorphism of γ, we say that γ is improved by ω.

Example 4. Consider the class of submodular functions. These are precisely the functions γ defined on $D = \{0, 1\}$ satisfying $\gamma(\min(\mathbf{x}_1, \mathbf{x}_2)) + \gamma(\max(\mathbf{x}_1, \mathbf{x}_2)) - \gamma(\mathbf{x}_1) - \gamma(\mathbf{x}_2) \leq 0$, where min and max are the two binary operations that return the smaller and larger of their two arguments respectively (with respect to the usual order $0 < 1$). In other words, the set of submodular functions is the set of weighted relations improved by the binary weighting ω_{sub} defined by: $\omega_{sub}(f) = -1$ if $f \in \{e_1^{(2)}, e_2^{(2)}\}$, $\omega_{sub}(f) = +1$ if $f \in \{\min, \max\}$, and $\omega_{sub}(f) = 0$ for all other binary operations on D.

Definition 18. *For any $\Gamma \subseteq \mathbf{\Phi}_D$, we define $\mathrm{wPol}(\Gamma)$ to be the set of all weightings of $\mathrm{Pol}(\Gamma)$ which are weighted polymorphisms of all weighted relations $\gamma \in \Gamma$. We write $\mathrm{wPol}(\gamma)$ for $\mathrm{wPol}(\{\gamma\})$.*

Definition 19. *We denote by $\mathbf{W}_C$ the set of all possible weightings of clone C, and define $\mathbf{W}_D$ to be the union of the sets $\mathbf{W}_C$ over all clones C on D.*

Any $\Omega \subseteq \mathbf{W}_D$ may contain weightings of *different* clones over D. We can then extend each of these weightings with zeros, as necessary, so that they are weightings of the same clone C, where C is the smallest clone containing all the clones associated with weightings in Ω.

Definition 20. *We define $\mathrm{wClone}(\Omega)$ to be the smallest weighted clone containing this set of extended weightings obtained from Ω.*

For any $\Omega \subseteq \mathbf{W}_D$, we denote by $\mathrm{Imp}(\Omega)$ the set of all weighted relations in $\mathbf{\Phi}_D$ which are improved by all weightings $\omega \in \Omega$.

The main result in [9] establishes a 1-to-1 correspondence between weighted relational clones and weighted clones.

Theorem 2 ([9]).

1. *For any finite D and any finite $\Gamma \subseteq \mathbf{\Phi}_D$, $\mathrm{Imp}(\mathrm{wPol}(\Gamma)) = \mathrm{wRelClone}(\Gamma)$.*
2. *For any finite D and any finite $\Omega \subseteq \mathbf{W}_D$, $\mathrm{wPol}(\mathrm{Imp}(\Omega)) = \mathrm{wClone}(\Omega)$.*

Thus, when trying to identify tractable constraint languages, it is sufficient to consider only languages of the form $\mathrm{Imp}(\Omega)$ for some weighted clone Ω.

Results

First we show that Theorem 2 can be slightly extended to certain constraint languages and sets of weightings of infinite size.

Theorem 3.

1. Let $\Gamma \subseteq \Phi_D$. Then $\mathrm{Imp}(\mathrm{wPol}(\Gamma)) = \mathrm{wRelClone}(\Gamma)$ *if and only if* $\mathrm{wRelClone}(\Gamma) = \mathrm{Imp}(\Omega)$ *for some* $\Omega \subseteq \mathbf{W}_D$.
2. Let $\Omega \subseteq \mathbf{W}_D$. Then $\mathrm{wPol}(\mathrm{Imp}(\Omega)) = \mathrm{wClone}(\Omega)$ *if and only if* $\mathrm{wClone}(\Omega) = \mathrm{wPol}(\Gamma)$ *for some* $\Gamma \subseteq \Phi_D$.

Proof. We will only prove the first case as the second one is analogous. Suppose that $\mathrm{wRelClone}(\Gamma) = \mathrm{Imp}(\Omega)$ for some $\Omega \subseteq \mathbf{W}_D$. As $\Gamma \subseteq \mathrm{wRelClone}(\Gamma)$, every weighting in Ω improves Γ, hence $\Omega \subseteq \mathrm{wPol}(\Gamma)$ and $\mathrm{Imp}(\mathrm{wPol}(\Gamma)) \subseteq \mathrm{Imp}(\Omega) = \mathrm{wRelClone}(\Gamma)$. The inclusion $\mathrm{wRelClone}(\Gamma) \subseteq \mathrm{Imp}(\mathrm{wPol}(\Gamma))$ follows from the fact that $\mathrm{Imp}(\mathrm{wPol}(\Gamma))$ is a weighted relational clone [9, Proposition 6.2] that contains Γ.

The converse implication holds trivially for $\Omega = \mathrm{wPol}(\Gamma)$.

We remark that any *finitely generated* weighted relational clone on a finite domain satisfies, by Theorem 2 (1), the condition of Theorem 3 (1). Similarly, any finitely generated weighted clone on a finite domain, by Theorem 2 (2), satisfies the condition of Theorem 3 (2).

However, our next result shows that Theorem 2 does *not* hold for all infinite constraint languages and infinite sets of weightings.

Theorem 4. *There is a finite D and an infinite $\Gamma \subseteq \Phi_D$ with $\mathrm{Imp}(\mathrm{wPol}(\Gamma)) \neq \mathrm{wRelClone}(\Gamma)$. Moreover, there is a finite D and an infinite $\Omega \subseteq \mathbf{W}_D$ with $\mathrm{wPol}(\mathrm{Imp}(\Omega)) \neq \mathrm{wClone}(\Omega)$.*

Our aim is to establish a Galois connection even for infinite sets of weighted relations and weightings. As we demonstrate in the proof of Theorem 4, this cannot be done when restricted to rational weights; hence we allow weighted relations and weightings to assign *real-valued* weights. To distinguish them from their formerly defined rational-valued counterparts, we will use a subscript/superscript $\mathbb{R}$.

We will show that $\mathrm{wPol}_{\mathbb{R}}(\Gamma)$ is a *closed* weighted clone for any set of weighted relations Γ; analogously, we will show that $\mathrm{Imp}_{\mathbb{R}}(\Omega)$ is a *closed* weighted relational clone for any set of weightings Ω. Therefore, the one-to-one correspondence between weighted relational clones and weighted clones which we want to establish cannot possibly hold for sets that are not closed. As there exist (infinite) sets $\Gamma \subseteq \Phi_D^{\mathbb{R}}$, $\Omega \subseteq \mathbf{W}_D^{\mathbb{R}}$ such that $\mathrm{wRelClone}_{\mathbb{R}}(\Gamma)$, $\mathrm{wClone}_{\mathbb{R}}(\Omega)$ are not closed, we need to include the closure operator in the statement defining the Galois connection.

Inspired by weighted pp-definitions [20], we extend the notion of weighted relational clones: we require them to be closed under the Opt operator. This change is justified by a result in which we prove that the inclusion of Opt preserves tractability. In order to retain the one-to-one correspondence with weighted clones, we need to alter their definition too: weightings now assign weights to all operations and hence are independent of the support clone (which becomes meaningless and we discard it).

Including the Opt operator brings two advantages to the study of weighted clones. Firstly, it slightly simplifies the structure of the space of all weighted clones. According to the original definition, a weighted clone is determined by its support clone and the set of weightings it consists of; by our definition a weighted clone equals the set of its weightings. Secondly, any non-projection polymorphism of a weighted relational clone Γ is assigned a positive weight by some weighted polymorphism of Γ.

Our main result is the following theorem, which holds for our new definition of real-valued weightings and weighted relations.

Theorem 5 (Main).

1. *For any finite D and any $\Gamma \subseteq \Phi_D^{\mathbb{R}}$, $\mathrm{Imp}_{\mathbb{R}}(\mathrm{wPol}_{\mathbb{R}}(\Gamma)) = \overline{\mathrm{wRelClone}_{\mathbb{R}}(\Gamma)}$. Moreover, if Γ is finite, then $\mathrm{Imp}_{\mathbb{R}}(\mathrm{wPol}_{\mathbb{R}}(\Gamma)) = \mathrm{wRelClone}_{\mathbb{R}}(\Gamma)$.*
2. *For any finite D and any $\Omega \subseteq \mathbf{W}_D^{\mathbb{R}}$, $\mathrm{wPol}_{\mathbb{R}}(\mathrm{Imp}_{\mathbb{R}}(\Omega)) = \overline{\mathrm{wClone}_{\mathbb{R}}(\Omega)}$. Moreover, if Ω is finite, then $\mathrm{wPol}_{\mathbb{R}}(\mathrm{Imp}_{\mathbb{R}}(\Omega)) = \mathrm{wClone}_{\mathbb{R}}(\Omega)$.*

Finally, we show that taking the weighted relational clone of a constraint language preserves solvability with an absolute error bounded by ϵ (for any $\epsilon > 0$).

4   New Galois Connection

Let $\overline{\mathbb{R}} = \mathbb{R} \cup \{\infty\}$ denote the set of real numbers with (positive) infinity. We will allow weights in relations and weighted relations, as defined in Definition 1 and 2 respectively, to be real numbers. In other words, an m-ary weighted relation γ on D is a mapping $\gamma : D^m \to \overline{\mathbb{R}}$. We will add a subscript/superscript $\mathbb{R}$ to the notation introduced in Section 2 in order to emphasise the use of real weights.

For any fixed arity m and any $F \subseteq D^m$, consider the set of all m-ary weighted relations $\gamma \in \Phi_D^{\mathbb{R}}$ with $\mathrm{Feas}(\gamma) = F$. Let us denote this set by H and equip it with the inner product defined as

$$\langle \alpha, \beta \rangle = \sum_{\mathbf{x} \in F} \alpha(\mathbf{x}) \cdot \beta(\mathbf{x}) \tag{10}$$

for any $\alpha, \beta \in H$; H is then a real Hilbert space. Set $\Phi_D^{\mathbb{R}}$ is a disjoint union of such Hilbert spaces for all m and F, and therefore a topological space with the disjoint union topology induced by inner products on the underlying Hilbert spaces. When we say a set of weighted relations is open/closed, we will be referring to this topology.

Definition 21. *A constraint language $\Gamma \subseteq \Phi_D^{\mathbb{R}}$ is called a* weighted relational clone *if it contains the binary equality relation $\phi_=$ and the unary empty relation $\phi_\emptyset$, and is closed under addition, minimisation, scaling by non-negative real constants, addition of real constants, and under the* Opt *operator.*
For any Γ, we define $\text{wRelClone}_{\mathbb{R}}(\Gamma)$ *to be the smallest weighted relational clone containing Γ.*

For a weighted relational clone Γ, its topological closure $\overline{\Gamma}$ is also a weighted relational clone, as all the operations that we require weighted relational clones to be closed under are continuous mappings.

As opposed to Definition 9, our new definition requires weighted relational clones to be closed under operator Opt. In order to establish a Galois connection now, we need to make an adjustment to the definition of weighted clone too. We will discard the explicit underlying support clone; instead, (k-ary) weightings will assign weights to all (k-ary) operations. The role of the support clone of a weighted clone Ω is then taken over by $\text{supp}(\Omega)$ (see Lemma 1).

We denote by $\mathcal{O}_D^{(k)}$ the set of all k-ary operations on D and let $\mathcal{O}_D = \bigcup_{k\geq 0} \mathcal{O}_D^{(k)}$.

Definition 22. *A k-ary* weighting *is a function $\omega : \mathcal{O}_D^{(k)} \to \mathbb{R}$ such that $\omega(f) < 0$ only if f is a projection and*

$$\sum_{f \in \mathcal{O}_D^{(k)}} \omega(f) = 0. \tag{11}$$

We define $\text{supp}(\omega) = \{f \in \mathcal{O}_D^{(k)} \mid \omega(f) > 0 \lor f \in \mathbf{J}_D^{(k)}\}$.

We will call a function $\omega : \mathcal{O}_D^{(k)} \to \mathbb{R}$ that satisfies Equation (11) but assigns a negative weight to some operation $f \notin \mathbf{J}_D^{(k)}$ an improper *weighting. In order to emphasise the distinction we may also call a weighting a* proper *weighting.*

We denote by $\mathbf{W}_D^{\mathbb{R}}$ the set of all weightings on domain D. For any fixed arity k, consider the set H of all functions $\mathcal{O}_D^{(k)} \to \mathbb{R}$ equipped with the inner product defined as

$$\langle \alpha, \beta \rangle = \sum_{f \in \mathcal{O}_D^{(k)}} \alpha(f) \cdot \beta(f) \tag{12}$$

for any $\alpha, \beta \in H$; H is then a real Hilbert space. Set $\mathbf{W}_D^{\mathbb{R}}$ lies in the disjoint union of such Hilbert spaces for all k, which is a topological space with the disjoint union topology induced by inner products on the underlying Hilbert spaces. When we say a set of weightings is open/closed, we will be referring to this topology. Clearly, any closure point of a set of weightings is itself a weighting.

Definition 23. *Let Ω be a non-empty set of weightings on a fixed domain D. We define* $\text{supp}(\Omega) = \mathbf{J}_D \cup \bigcup_{\omega \in \Omega} \text{supp}(\omega)$.
We call Ω a weighted clone *if it is closed under scaling by non-negative real constants, addition of weightings of equal arity, and proper superposition with operations from $\text{supp}(\Omega)$.*

For any weighted clone Ω, its topological closure $\overline{\Omega}$ is also a weighted clone as all the operations that we require weighted clones to be closed under are continuous mappings.

Again, we link weightings and weighted relations by the concept of weighted polymorphism.

Definition 24. *Let γ be an m-ary weighted relation on D and let ω be a k-ary weighting on D. We call ω a weighted polymorphism of γ if $\mathrm{supp}(\omega) \subseteq \mathrm{Pol}(\gamma)$ and for any $X = (\mathbf{x}_1, \mathbf{x}_2, \ldots, \mathbf{x}_k) \in (\mathrm{Feas}(\gamma))^k$, we have*

$$\sum_{f \in \mathrm{supp}(\omega)} \omega(f) \cdot \gamma(f(X)) = \sum_{f \in \mathrm{supp}(\omega)} \omega(f) \cdot \gamma(f(\mathbf{x}_1, \mathbf{x}_2, \ldots, \mathbf{x}_k)) \leq 0. \quad (13)$$

If ω is a weighted polymorphism of γ we say that γ is improved by ω.

In the proof of Theorem 5, we will often use the following characterisation of weighted polymorphisms. Let $\gamma \in \boldsymbol{\Phi}_D^{\mathbb{R}}$ be a weighted relation and $\omega \in \mathbf{W}_D^{\mathbb{R}}$ a k-ary weighting such that $\mathrm{supp}(\omega) \subseteq \mathrm{Pol}(\gamma)$. Let us denote by H the Hilbert space of functions $\mathrm{Pol}^{(k)}(\gamma) \to \mathbb{R}$ with the inner product analogous to (12). As weighting ω assigns non-zero weights only to operations from $\mathrm{supp}(\omega) \subseteq \mathrm{Pol}^{(k)}(\gamma)$, we can identify ω with its restriction to $\mathrm{Pol}^{(k)}(\gamma)$. For any $X \in (\mathrm{Feas}(\gamma))^k$, we define $\gamma[X] \in H$ as $\gamma[X](f) = \gamma(f(X))$. Inequality (13) is then equivalent to $\langle \omega, \gamma[X] \rangle \leq 0$.

The (internal) polar cone K° of a set $K \subseteq H$ is defined as

$$K^\circ = \{ \alpha \in H \mid \langle \alpha, \beta \rangle \leq 0 \text{ for all } \beta \in H \}. \quad (14)$$

It is well known ([3]) that K° is a convex cone, i.e. K° is closed under addition of vectors and scaling by non-negative constants. Moreover, K° is a closed set, and $K^{\circ\circ} = (K^\circ)^\circ$ is the closure of the smallest convex cone containing K. If K is a finite set, then the smallest convex cone containing K is closed. Let $K = \{ \gamma[X] \mid X \in (\mathrm{Feas}(\gamma))^k \}$; weighting ω is then a weighted polymorphism of γ if and only if $\omega \in K^\circ$.

The following lemma (and its corollary) shows that $\mathrm{supp}(\Omega)$ consists of all polymorphisms of $\mathrm{Imp}_{\mathbb{R}}(\Omega)$ and hence fulfills the same role as the support clone in Definition 16.

Lemma 1. *Let $\Omega \subseteq \mathbf{W}_D^{\mathbb{R}}$ be a weighted clone. Then $\mathrm{supp}(\Omega) = \mathrm{Pol}(\mathrm{Imp}_{\mathbb{R}}(\Omega))$.*

Corollary 1. *Let $\Gamma \subseteq \boldsymbol{\Phi}_D^{\mathbb{R}}$ be a weighted relational clone. Then we have that $\mathrm{supp}(\mathrm{wPol}_{\mathbb{R}}(\Gamma)) = \mathrm{Pol}(\Gamma)$.*

Theorem 6. *Let $\Gamma, \Gamma' \subseteq \boldsymbol{\Phi}_D^{\mathbb{R}}$ be finite constraint languages such that Γ contains only weighted relations of the form $c \cdot \gamma'$ for $c \geq 0, \gamma' \in \Gamma'$. For any $\epsilon > 0$ there is a polynomial-time reduction that for any instance $I \in \mathrm{VCSP}(\Gamma)$ outputs an instance $I' \in \mathrm{VCSP}(\Gamma')$ such that for any optimal assignment s' of I' it holds $I(s') \in [v, v + \epsilon]$, where v is the value of an optimal assignment of I.*

References

1. Barto, L., Kozik, M.: Constraint Satisfaction Problems Solvable by Local Consistency Methods. Journal of the ACM **61**(1), article No. 3
2. Barto, L., Kozik, M., Niven, T.: The CSP dichotomy holds for digraphs with no sources and no sinks. SIAM Journal on Computing **38**(5), 1782–1802 (2009)
3. Boyd, S.P., Vandenberghe, L.: Convex Optimization, CUP (2004)
4. Bulatov, A.: A graph of a relational structure and constraint satisfaction problems. In: Proc. LICS 2004. IEEE Computer Society, pp. 448–457 (2004)
5. Bulatov, A.: A dichotomy theorem for constraint satisfaction problems on a 3-element set. Journal of the ACM **53**(1), 66–120 (2006)
6. Bulatov, A., Krokhin, A., Jeavons, P.: Classifying the Complexity of Constraints using Finite Algebras. SIAM Journal on Computing **34**(3), 720–742 (2005)
7. Bulatov, A.A.: Complexity of conservative constraint satisfaction problems. ACM Transactions on Computational Logic **12**(4), article 24
8. Bulatov, A.A., Krokhin, A.A., Jeavons, P.G.: The complexity of maximal constraint languages. In: Proc. STOC 2001, pp. 667–674 (2001)
9. Cohen, D.A., Cooper, M.C., Creed, P., Jeavons, P., Živný, S.: An algebraic theory of complexity for discrete optimisation. SIAM Journal on Computing **42**(5), 915–1939 (2013)
10. Cohen, D.A., Cooper, M.C., Jeavons, P.G., Krokhin, A.A.: The Complexity of Soft Constraint Satisfaction. Artificial Intelligence **170**(11), 983–1016 (2006)
11. Feder, T., Vardi, M.Y.: The Computational Structure of Monotone Monadic SNP and Constraint Satisfaction: A Study through Datalog and Group Theory. SIAM Journal on Computing **28**(1), 57–104 (1998)
12. Hell, P., Nešetřil, J.: On the Complexity of H-coloring. Journal of Combinatorial Theory, Series B **48**(1), 92–110 (1990)
13. Hell, P., Nešetřil, J.: Colouring, constraint satisfaction, and complexity. Computer Science Review **2**(3), 143–163 (2008)
14. Jeavons, P., Krokhin, A., Živný, S.: The complexity of valued constraint satisfaction. Bulletin of the European Association for Theoretical Computer Science (EATCS) **113**, 21–55 (2014)
15. Jeavons, P.G., Cohen, D.A., Gyssens, M.: Closure Properties of Constraints. Journal of the ACM **44**(4), 527–548 (1997)
16. Kolmogorov, V., Thapper, J., Živný, S.: The power of linear programming for general-valued CSPs. SIAM Journal on Computing **44**(1), 1–36 (2015)
17. Kolmogorov, V., Živný, S.: The complexity of conservative valued CSPs. Journal of the ACM **60**(2), article No. 10
18. Kozik, M., Ochremiak, J.: Algebraic properties of valued constraintsatisfaction problem. In: Proc. ICALP 2015. Springer (2015)
19. Schaefer, T.J.: The complexity of satisfiability problems. In: Proc. STOC 1978, pp. 216–226. ACM (1978)
20. Thapper, J.: Aspects of a constraint optimisation problem, Ph.D. thesis, Department of Computer Science and Information Science, Linköping University (2010)
21. Thapper, J., Živný, S.: The power of linear programming for valued CSPs. In: Proc. FOCS 2012, pp. 669–678. IEEE (2012)
22. Thapper, J., Živný, S.: The complexity of finite-valued CSPs. In: Proc. STOC 2013, pp. 695–704. ACM (2013)
23. Uppman, H.: The complexity of three-element min-sol and conservative min-cost-hom. In: Fomin, F.V., Freivalds, R., Kwiatkowska, M., Peleg, D. (eds.) ICALP 2013, Part I. LNCS, vol. 7965, pp. 804–815. Springer, Heidelberg (2013)

Approximately Counting H-Colourings is #BIS-Hard

Andreas Galanis[1][(✉)], Leslie Ann Goldberg[1], and Mark Jerrum[2]

[1] Department of Computer Science, University of Oxford, Oxford, UK
agalanis@cs.ox.ac.uk
[2] School of Mathematical Sciences, Queen Mary University of London, London, UK

Abstract. We consider counting H-colourings from an input graph G to a target graph H. We show that for any fixed graph H without trivial components, this is as hard as the well-known problem #BIS, the problem of (approximately) counting independent sets in a bipartite graph. #BIS is a complete problem in an important complexity class for approximate counting, and is believed not to have an FPRAS. If this is so, then our result shows that for every graph H without trivial components, the H-colouring counting problem has no FPRAS. This problem was studied a decade ago by Goldberg, Kelk and Paterson. They were able to show that approximately sampling H-colourings is #BIS-hard, but it was not known how to get the result for approximate counting. Our solution builds on non-constructive ideas using the work of Lovász. The full version is available at `arxiv.org/abs/1502.01335`. The theorem numbering here matches the full version.

1 Introduction

The independent set and k-colouring models are well-known statistical physics models which have also been studied in computer science. A particularly interesting question is the complexity of counting and approximate counting in these models. Given an input graph G, the problem is to approximate the number of independent sets (or proper k-colourings) of G. Both of these problems can be viewed as special cases of the more general problem of approximately counting H-colourings. This paper studies the complexity of the more general problem.

We begin with few definitions. Let $H = (V(H), E(H))$ be a fixed graph which is allowed to have self-loops, but not parallel edges. An H-colouring of a graph $G = (V(G), E(G))$ is a homomorphism from G to H, i.e., an assignment $h : V(G) \to V(H)$ that maps every edge (u, v) of G to an edge of H. Given an input graph G, we are interested in computing the number of H-colourings of G. We

The research leading to these results has received funding from the European Research Council under the European Union's Seventh Framework Programme (FP7/2007-2013) ERC grant agreement no. 334828. The paper reflects only the authors' views and not the views of the ERC or the European Commission. The European Union is not liable for any use that may be made of the information contained therein.

© Springer-Verlag Berlin Heidelberg 2015
M.M. Halldórsson et al. (Eds.): ICALP 2015, Part I, LNCS 9134, pp. 529–541, 2015.
DOI: 10.1007/978-3-662-47672-7_43

refer to this problem as the #H-Col problem. Also, we denote by #H-Col(G) the number of H-colourings of G. The examples mentioned earlier correspond to H-colourings as follows. Proper k-colourings of G correspond to H-colourings of G when H is a k-clique. Independent sets of G correspond to H-colourings of G when H is the connected 2-vertex graph with exactly one self-loop.

Our goal in this work is to quantify the computational complexity of approximately counting H-colourings. In particular, we seek to determine for which graphs H the problem #H-Col admits a fully polynomial randomised approximation scheme (FPRAS). Dyer and Greenhill [3] have completely classified the computational complexity of *exactly* counting H-colourings in terms of the parameter graph H. We say that a connected graph is *trivial* if it is either a clique with self-loops on every vertex or a complete bipartite graphs with no self-loops. Dyer and Greenhill showed that the problem #H-Col is polynomial-time solvable when each connected component of H is *trivial*; otherwise it is #P-complete. The complexity of the corresponding decision problem has also been characterised. Hell and Nešetřil [7] showed that deciding whether an input graph G admits an H-colouring is NP-complete unless H contains a self-loop or H is bipartite (in which case it admits a trivial polynomial-time algorithm).

The problem of approximately *sampling* H-colourings has been shown to be #BIS-hard [6] provided that H contains no trivial components (the existence of trivial components may lead to artificial approximation schemes, see [6, Section 7] for an explicit example). More precisely, for any such H, a fully polynomial approximate sampler (FPAS) for H-colourings would imply that there is an FPRAS for #BIS, which is the problem of counting the independent sets of a bipartite graph. #BIS plays an important role in approximation complexity. Despite many attempts, nobody has found an FPRAS for #BIS and it is conjectured that none exists (even though it is unlikely that approximating #BIS is NP-hard). Various natural algorithms have been ruled out as candidate FPRASes for #BIS [4,5,11]. Moreover, Dyer et al. [1] showed that #BIS is complete under approximation-preserving (AP) reductions in a logically defined class of problems, called #RHΠ_1, to which an increasing variety of problems have been shown to belong.

Perhaps surprisingly, the hardness result of [6] for sampling H-colourings does not imply hardness for approximately counting H-colourings. This might be puzzling at first since, for the independent set and k-colouring models, approximate counting is well-known to be equivalent to approximate sampling (this equivalence has been proved in [8] for the so-called class of self-reducible problems in #P). However, for general graphs H it is only known [2] that an FPAS for sampling H-colourings implies an FPRAS for counting H-colourings (but not the reverse direction). For a thorough discussion of this point we refer the reader to [2] (where also an example of a problem in #P is given which, under usual complexity theory assumptions, admits an FPRAS but not an FPAS).

In this paper, we address the following questions: "Is there a graph H for which approximately counting H-colourings is substantially easier than approximately sampling H-colourings?" "Is there a graph H such that #H-Col lies between P and the class of #BIS-hard problems?" We present the analogue of the hardness result of [6] in the counting setting, therefore providing evidence that the answers to the previous questions are negative. To formally state the result, recall the notion of an approximation preserving reduction $\leq_{AP}$ (introduced in [1]). For counting problems #A and #B, $\#A \leq_{AP} \#B$ implies that an FPRAS for #B yields an FPRAS for #A. Our main result is the following.

Theorem 1. *Let H be a graph with no trivial components. (H has no parallel edges but can have self loops.) Then* $\#BIS \leq_{AP} \#H\text{-Col}$.

Interestingly, in the proof of Theorem 1 we use a non-constructive approach, partly inspired by tools from graph-homomorphism theory introduced by Lovász [10].

2   Reductions for Sampling versus Reductions for Counting

We start by considering the closely related work [6]. The assumptions on the graph H are the same as in Theorem 1 — namely that H does not have trivial components. The proof in [6] shows that approximately *sampling* H-colourings is at least as hard as #BIS. We first overview the approach of this paper since we will use several ingredients of their proof. We will also describe the new ingredients which will allow us to leap from the sampling setting to the counting setting.

Let H be a graph for which we wish to show that $\#BIS \leq_{AP} \#H\text{-Col}$. To do this, it clearly suffices to find a subgraph H' of H such that $\#BIS \leq_{AP} \#H'\text{-Col}$ and $\#H'\text{-Col} \leq_{AP} \#H\text{-Col}$. For a subset S of $V(H)$, let $H[S]$ be the subgraph of H induced by the set S. Further, let $N_\cup(S)$ denote the neighbourhood of S in H, i.e., the set of vertices in H which are adjacent to a vertex in S.

Restricting H-colourings to Induced-subgraph Colourings. To motivate how to find such a subgraph H', we give a high-level reduction scheme. This will reveal that the subgraphs induced by the neighbourhoods of maximum-degree vertices of H are natural choices for H'.[1] To see this, let's temporarily suppose that we already have a subgraph H' of H in mind, and let G' be an input for $\#H'\text{-Col}$ such that $|V(G')| = n'$. We next construct an instance G of $\#H\text{-Col}$ by adding to G' a special vertex w and a large independent set I with $|I| \gg n'$. We also add all edges between the special vertex w and the vertices of G' and I. See the first graph G in the figure (on page 12). Let h be a homomorphism from G to H with $h(w) = v_i$. Observe that the edges between w and the rest of G enforce that the restriction of h to G' is an $N_\cup(v_i)$-colouring of G'. Similarly, the restriction of h to I is a $N_\cup(v_i)$-colouring of I. There are $(\deg_H(v_i))^{|I|}$ choices for the restriction of h on the independent set I. From this, it is not hard to show that the effect of the large independent set I is to enforce that in all but a negligible fraction of H-colourings of G, the special vertex w is assigned a colour among the maximum-degree vertices of H and thus G' is coloured using the subgraph of H induced by the neighbourhood of such a vertex.

---

[1] We will later modify this reduction scheme in a non-trivial way, but still the aspects we highlight will carry over to the modified reduction scheme.

In an ideal scenario, there is a unique vertex v in H with maximum degree nd, further, the subgraph $H' = H[N_\cup(v)]$ of H is non-trivial and different from I. If both of these hypotheses hold, the reduction above can be used to show hat #H'-Col $\leq_{AP}$ #H-Col (since the uniqueness of v implies that $h(w) = v$ in almost all" H-colorings) and (say, by induction) we have #BIS $\leq_{AP}$ #H'-Col. 3ut what happens when these hypotheses do not hold?

A significant problem which arises at this point is the existence of multiple elevant neighbourhoods. That is, there may be several maximum-degree vertices n H, and the subgraphs induced by their neighbourhoods may not be isomor-•hic. It is much easier to deal with this problem in the sampling setting than n the counting setting. We now describe the difference between these settings, .nd our approach to this (initial) hurdle.

Counting Subgraph-induced Colourings — the Case of Multiple Sub-graphs. To illustrate concretely the part of the sampling argument in [6] which oreaks down in the counting setting, we consider the toy example H (2nd graph n the figure) and overview how the argument in [6] works. Since H is regular, the elevant (induced) neighbourhoods of the vertices in H are given by the graphs H_1 and H_2 (depicted in the figure immediately after H). These correspond to he neighbourhoods of the vertices v_1 and v_2, respectively (the remaining neigh-oourhoods are isomorphic to H_2). Note that #BIS $\leq_{AP}$ #H_1-Col and #BIS $\leq_{AP}$ #H_2-Col (see [9] where all graphs H with up to four vertices are classified).

So we already have sampling reductions from #BIS to #H_1-Col and #H_2-Col .nd we want a sampling reduction from #BIS to #H-Col, e.g., an algorithm for :ampling bipartite independent sets using an oracle for sampling H-colourings. Here's how it works. Let G' be an input to #BIS. Using the sampling reduc-ions to #H_1-Col and #H_2-Col, construct from G' two graphs G_1 and G_2 such hat an (approximately) uniform H_1-colouring of G_1 allows us to construct an approximately) uniform independent set of G' and similarly an (approximately) ıniform H_2-colouring of G_2 also allows this. Then let G be the graph obtained oy taking the disjoint union of G_1 and G_2, adding a special vertex w, and adding ıll edges between w and G_1 and G_2 (note, there is no need for the independent :et I used previously since H is regular; see the graph G in the bottom row of he figure).

Given a random H-colouring h of G, revealing the colour of the special vertex v allows us to generate either a random H_1-colouring of G_1 or a random H_2-:olouring of G_2. In particular, if $h(w)$ is v_1, by considering the restriction of h on G_1 we obtain a random H_1-colouring of G_1. Similarly, if $h(w)$ is any other vertex, we obtain a random H_2-colouring of G_2. By our assumptions for G_1 and G_2, in each case we can then obtain a random independent set of G'.

In contrast, the aforementioned reduction scheme fails in the counting setting. Namely, considering cases for the colour $h(w)$, we obtain the following equality

$$\#H\text{-Col}(G) = \#H_1\text{-Col}(G_1)\,\#H_1\text{-Col}(G_2) + 4\#H_2\text{-Col}(G_1)\,\#H_2\text{-Col}(G_2). \quad (1)$$

Given an approximation of #H-Col(G), say Z, observe that (1) yields little ınformation about whether Z is a good approximation for #H_1-Col(G_1) or

#H_2-Col(G_2). This issue goes away in the sampling setting precisely because we can distinguish between the two cases by just looking at the colour of the special vertex w in the random H-colouring of G.

Thus, to proceed with the reduction in the counting setting we have to focus our attention on one of H_1 or H_2, say H_1, and somehow prove that #H_1-Col $\leq_{AP}$ #H-Col. The question which arises is how to choose between H_1 and H_2. This becomes more complicated for general graphs H since it is not hard to imagine that instead of just two graphs H_1, H_2 we will typically have a collection of graphs $H_1, \ldots, H_t$ corresponding to the induced neighbourhoods of vertices $v_1, \ldots, v_t$ of H, for some t which can be arbitrarily large (depending on the graph H). To make matters worse, apart from very basic information on the H_i's (such as connectedness or number of vertices/edges), we will not be able to control significantly their graph structure.

At this point, we employ a non-constructive approach using a tool from [10]: for arbitrary non-isomorphic graphs H_1, H_2 there exists a (fixed) graph J depending only on H_1 and H_2 such that #H_1-Col(J) $\neq$ #H_2-Col(J). In **Lemma 7 of the full version**, we extend this to an arbitrary collection of pairwise non-isomorphic graphs $H_1, \ldots, H_t$ as follows: we prove the existence of a graph J so that for some $i^* \in [t]$ it holds that #H_{i^*}-Col(J) > #H_i-Col(J) for all $i \neq i^*$. Intuitively, the graph J will be used to "select" the subgraph H_{i^*}. Note, we will not require any further knowledge about what H_{i^*} or J is, freeing us from the cumbersome (and perhaps difficult) task of looking into the finer details of the graph structure of $H_1, \ldots, H_t$. With the graph J in hand, we then take sufficiently many disjoint copies of J and connect them to the special vertex w. This ensures that in most H-colourings the vertex w gets coloured with the vertex v_i.

To utilise the above, we will further need to ensure that #BIS $\leq_{AP}$ #H_i-Col for every i. If we could ensure that the H_i's are proper subgraphs of H and non-trivial, then using the arguments above, we could complete the proof using induction. However, this is clearly not possible in general since for example, as we noted earlier, there may exist a vertex v in H such that $N_\cup(v) = V(H)$. Dealing with such cases is the bulk of the work in the sampling setting of [6] and these cases cause even more problems for us. To deal with them, we need a further non-constructive argument — one that turns out to be more technical than, and substantially different from the ideas in Lovász [10].

3   Proof Outline

Since we are interested in instances of #BIS, which are bipartite, we will need to consider H-colourings of bipartite graphs. We will assume that every bipartite graph G comes with a (fixed) proper 2-colouring of its parts with colours $\{L, R\}$. We will use $L(G)$ and $R(G)$ to denote the vertices of G coloured with L and R respectively. We will refer to a bipartite graph as a 2-coloured graph to emphasise the proper 2-colouring of its vertices. Colour-preserving homomorphisms are those that map $L(G)$ to $L(H)$ and $R(G)$ to $R(H)$. Given input G, #Fixed$H$-Col

the problem of computing the colour-preserving homomorphisms from G to
I, which we denote $\#\mathsf{Fixed}H\text{-}\mathsf{Col}(G)$. This problem is key to our analysis.

A bipartite graph H will be called *full* if there exist vertices $u \in L(H)$ and
$\in R(H)$ such that u is adjacent to every vertex in $R(H)$ and v is adjacent to
very vertex in $L(H)$. In this case, vertices u and v are also called full. If H is
ull, then it is also connected. The full version of our paper proves the following
emma, which is an analogue of Lemma 7 in [6] and allows us to restrict our
ttention to the $\#\mathsf{Fixed}H\text{-}\mathsf{Col}$ problem. The proof is along the lines described in
ection 1, albeit with some modifications to account for technical details.

Lemma 3. *Let H be a graph without trivial components. There exists a full and
on-trivial 2-coloured graph H' such that $\#\mathsf{Fixed}H'\text{-}\mathsf{Col} \leq_{\mathsf{AP}} \#H\text{-}\mathsf{Col}$.*

Theorem 1 follows easily from Lemma 3 and the following central lemma.

Lemma 4. *Let H be a 2-coloured graph which is full and not trivial. Then
$\#\mathsf{BIS} \leq_{\mathsf{AP}} \#\mathsf{Fixed}H\text{-}\mathsf{Col}$.*

1 Overview of Proof of Lemma 4

Let (V_L, V_R) denote the vertex partition of H and let F_L, F_R be the subsets of
ull vertices in V_L, V_R, respectively (i.e., every vertex in F_L is connected to every
ertex in V_R and every vertex in F_R is connected to every vertex in V_L). Since H
s full, we have $F_L, F_R \neq \emptyset$. For a subset S of $V(H)$, let $H[S]$ be the subgraph of
H induced by the set S. We will use $N_\cap(S)$ to denote the *joint* neighbourhood
of S in H, i.e., the set of vertices in H which are adjacent to *every* vertex in S.

An Inductive Approach Using Maximal Bicliques of H. The proof of
Lemma 4 will be by induction on the number of vertices of H. Our goal will
be to find a subgraph H' of H (which will also be 2-coloured, full and not
rivial) with $|V(H')| < |V(H)|$ such that $\#\mathsf{Fixed}H'\text{-}\mathsf{Col} \leq_{\mathsf{AP}} \#\mathsf{Fixed}H\text{-}\mathsf{Col}$. If we
ind such an H', we will finish using the inductive hypothesis that $\#\mathsf{BIS} \leq_{\mathsf{AP}}$
$\#\mathsf{Fixed}H'\text{-}\mathsf{Col}$. When we are not able to find such a subgraph H', we will use an
alternative method to show that $\#\mathsf{BIS} \leq_{\mathsf{AP}} \#\mathsf{Fixed}H\text{-}\mathsf{Col}$.

To select H', we consider the set $\mathcal{C}$ of bicliques in H. Bicliques will be
denoted as (S_L, S_R), where S_L, S_R are the parts of the biclique belonging to
V_L, V_R, respectively. Formally, $\mathcal{C} = \{(S_L, S_R) : S_L \subseteq V_L, S_R \subseteq V_R, S_L \times S_R \subseteq E(H)\}$. In fact, we will be more interested in (inclusion) *maximal
bicliques* of H, i.e., bicliques which are not contained in another biclique. Note,
$(F_L, V_R), (V_L, F_R)$ are maximal bicliques in H. For lack of better terminology,
we refer to these two special bicliques as the *extremal* bicliques. Our interest in
maximal bicliques is justified by the following simple claim on which we base
our inductive step.

Lemma 8. *Let (S_L, S_R) be a maximal biclique which is not extremal, i.e, $S_L \neq F_L$, $S_R \neq F_R$. We have that $S_L \neq V_L$, $S_R \neq V_R$, $N_\cup(S_L) = V_R$ and $N_\cup(S_R) = V_L$. Let $H_1 = H[S_L \cup V_R]$ and $H_2 = H[V_L \cup S_R]$. Then, for $i \in \{1, 2\}$, we have
that H_i is full and not trivial and further satisfies $|V(H_i)| < |V(H)|$.*

The Basic Gadget. We now discuss a gadget that is used in [6]. While it will not work for us, it will nevertheless help to motivate our later selection of a more elaborate gadget for our needs. Consider a complete bipartite graph $K_{a,b}$ with a vertices on the left and b vertices on the right. The integers a and b should be thought of as sufficiently large numbers which may depend on the size of the input to #Fixed$H$-Col. Roughly speaking, we will be interested in the colours appearing on the left and right of $K_{a,b}$ in a typical colour-preserving homomorphism from $K_{a,b}$ to H.

To make this precise, for a colour-preserving homomorphism $h : K_{a,b} \to H$, the *phase* of h is the pair $\big(h(L(K_{a,b})), h(R(K_{a,b}))\big)$, i.e., the subsets of V_L and V_R appearing on the left and right of $K_{a,b}$ under the homomorphism h, respectively. Since $K_{a,b}$ is a complete bipartite graph, we have that a phase is a biclique of H, i.e., an element of $\mathcal{C}$. Let $(S_L, S_R) \in \mathcal{C}$. For convenience, we will refer to the total number of colour-preserving homomorphisms whose phase equals (S_L, S_R) as the contribution of the phase/biclique (S_L, S_R) to the gadget. Our induction step crucially depends on analysing the *dominant phases* of the gadget, i.e., the phases with the largest contribution.

It is not hard to see that the contribution of a phase/biclique (S_L, S_R) to the gadget $K_{a,b}$ is roughly equal to $|S_L|^a |S_R|^b$. Thus, the dominant phases are determined by the ratio a/b. Rather than restricting ourselves to integers a and b it will be convenient to consider positive *real* numbers $\alpha, \beta > 0$ and the corresponding phases with dominant contribution.

Definition 9 (The set of dominating bicliques $\mathcal{C}_{\alpha,\beta}$). *Let α and β be positive real numbers. Define $\mathcal{C}_{\alpha,\beta}$ to be the set of bicliques (S_L, S_R) which maximize $|S_L|^\alpha |S_R|^\beta$. Note that for positive α, β the bicliques in $\mathcal{C}_{\alpha,\beta}$ are in fact maximal.*

For the purpose of the following discussion and to avoid delving into (at this point) unnecessary technical details, we will assume for now that α and β are rationals so that, for an integer Q we have that $a = Q\alpha$ and $b = Q\beta$ are integers and $\alpha/\beta = a/b$. In the full version, we use Dirichlet's approximation.

A Reduction Scheme. The structure of our reduction scheme expands on the work of [6]. The following are implicit in [6]:

1. if the set of the dominating bicliques $\mathcal{C}_{\alpha,\beta}$ consists *only* of the extremal bicliques, and both of these are in $\mathcal{C}_{\alpha,\beta}$, then #BIS reduces to #Fixed$H$-Col.
2. if $|\mathcal{C}_{\alpha,\beta}| = 1$ and the unique dominating biclique in $\mathcal{C}_{\alpha,\beta}$ is *not* extremal, then #FixedH'-Col reduces to #Fixed$H$-Col for some subgraph H' of H.

Unfortunately, there are graphs H (such as the graph H in the bottom row of the figure) such that, for every choice of α and β, we do not fall into case 1 or case 2. This graph is analysed in the full version. Despite such bad examples, It will be useful to see how the gadget $K_{a,b}$ is used, so we give a quick overview of the reductions which yield Items 1 and 2 (since these are only implicit in [6]).

For Item 1, let G' be a (2-coloured) bipartite graph which is an input to #BIS. To construct an instance of #Fixed$H$-Col, replace each vertex of G' with a distinct copy of $K_{a,b}$. Further, for each edge (u, v) in G' with $u \in L(G')$ and

$\in R(G')$ add all edges between the right part of u's copy of $K_{a,b}$ and the left part of v's copy of $K_{a,b}$. In the final graph, say G, by scaling a, b to be much larger than the size of G' (while keeping fixed the ratio $a/b = \alpha/\beta$), the phases of the gadgets $K_{a,b}$ in "almost all" colour-preserving homomorphisms from G to H are elements of $\mathcal{C}_{\alpha,\beta}$ and in particular are extremal bicliques. It then remains to observe that independent sets of G' are encoded by those homomorphisms where the phase of a gadget corresponding to a vertex in $L(G')$ is (F_L, V_R) if the vertex is in the independent set and (V_L, F_R) otherwise. Similarly, the phase of a gadget corresponding to a vertex in $R(G')$ is (V_L, F_R) if the vertex is in the independent set and (F_L, V_R) otherwise.

For Item 2, the use of the gadget $K_{a,b}$ is depicted at the beginning of the bottom row in the figure. Namely, for a 2-coloured connected bipartite graph G', consider the graph obtained by adding all edges between $L(G')$ and $R(K_{a,b})$. We will typically denote the graph obtained by this construction as $K_{a,b}(G')$. For the following discussion, we set $G := K_{a,b}(G')$. In the setting of Item 2, we have that $\mathcal{C}_{\alpha,\beta}$ consists of a unique maximal biclique (S_L, S_R) which is not extremal. Once again, by making a, b large relative to the size of G' (while maintaining the ratio $a/b = \alpha/\beta$), the phase of the gadget $K_{a,b}$ in "almost all" homomorphisms h of the graph G will be the dominating biclique (S_L, S_R). Let us consider such a homomorphism h whose restriction on $K_{a,b}$ has as a phase the maximal biclique (S_L, S_R). The edges between $R(K_{a,b})$ and $L(G')$ enforce that $h(L(G')) \subseteq N_\cap(S_R) = S_L$, where in the latter equality we used that (S_L, S_R) is a maximal biclique. It follows that $h(R(G')) \subseteq N_\cup(S_L) = V_R$ (see Lemma 8 for the latter equality). Thus, the restriction of h on G' is an H_1-colouring of the graph G', where $H_1 = H[S_L \cup V_R]$ is the same graph as in Lemma 8. Viewing G' as an instance of $\#H_1$-Col and G as an instance of $\#H$-Col, one obtains $\#H_1$-Col $\leq_{AP} \#H$-Col. Since H_1 is full, not trivial and has fewer vertices than H, one can use the inductive hypothesis to conclude $\#BIS \leq_{AP} \#H_1$-Col. We remark here that using the non-constructive approach of Lemma 7 and along the lines we described in Section 1, we will be able to remove the restriction that $|\mathcal{C}_{\alpha,\beta}| = 1$ as long as $\mathcal{C}_{\alpha,\beta}$ does not include an extremal biclique of H.

In view of Items 1 and 2, the scheme pursued in [6] (and which we will also follow to a certain extent) is to fix $1 > \alpha, \beta > 0$ such that $|F_L|^\alpha |V_R|^\beta = |V_L|^\alpha |F_R|^\beta$, so that the contribution of the extremal bicliques (F_L, V_R) and (V_L, F_R) to the gadget $K_{a,b}$ is equal. This has the beneficial effect that $\mathcal{C}_{\alpha,\beta}$ includes either none or both of the extremal bicliques. The only very bad scenario remaining is when $\mathcal{C}_{\alpha,\beta}$ includes both the extremal bicliques as well as (at least) one non-extremal biclique, since then not only $|\mathcal{C}_{\alpha,\beta}| > 1$ (which is already a problem for the approach implicit in [6]) but also the coexistence of extremal and non-extremal bicliques in $\mathcal{C}_{\alpha,\beta}$ impedes the non-constructive approach of Lemma 7. While for the sampling problem studied in [6] the coexistence of extremal and non-extremal bicliques was recoverable by "gluing" the reductions together (as we explained in a simplified setting in Section 1), this is no longer the case in the counting setting. More precisely, for the counting problem we will have to understand for which graphs H the coexistence of extremal and non-extremal bicliques occurs and consider more elaborate gadgets in the reduction to overcome thi coexistence.

A Non-constructive Gadget. The key idea is to introduce another non constructive argument (in addition to the approach suggested by Lemma 7 by viewing the construction of $K_{a,b}(G')$ (see again the bottom row of the figure as a gadget parameterised by the graph G'. To emphasize that G' is no longe an input graph, let us switch notation from G' to Γ, i.e., Γ is a 2-coloured graph and $K_{a,b}(\Gamma)$ is the graph in the figure where G' is replaced by the graph Γ We will choose a, b sufficiently large so that the graph Γ is "small" relative to the graph $K_{a,b}$, so its effect on the dominant phases will be of second order We stress here that we will never try to specify Γ explicitly; all we need is the existence of a helpful Γ. In the following, we expand on this point and set up some relevant quantities for the proof.

As for the basic gadget, we define the phase of a colour-preserving homo-morphism $h : K_{a,b}(\Gamma) \to H$ as the pair $\big(h(L(K_{a,b})), h(R(K_{a,b}))\big)$. Note that the phase of h is determined by its restriction on $K_{a,b}$ (but not on Γ) and, thus, as before the phases are supported on bicliques of H. We once again set $a = Q\alpha, b = Q\beta$ and let Q be a large integer relative to the size of Γ. With this setup, the phases with the dominant contribution in $K_{a,b}(\Gamma)$ are related to those in $K_{a,b}$ and in particular we will make a and b sufficiently large to ensure that they are a subset of $\mathcal{C}_{\alpha,\beta}$. Note however that the graph Γ has the effect of reweighting each phase contribution in $K_{a,b}(\Gamma)$ relative to the one in $K_{a,b}$.

To understand the reweighted contribution, consider a homomorphism $h :$ $K_{a,b}(\Gamma) \to H$ whose phase is a biclique $(S_L, S_R) \in \mathcal{C}$. The edges between $R(K_{a,b})$ and $L(\Gamma)$ enforce that $h(L(\Gamma)) \subseteq N_\cap(S_R)$ and thus (since we will ensure that Γ has no isolated vertices) we obtain that $h(R(\Gamma)) \subseteq N_\cup(N_\cap(S_R))$; it follows that the restriction of h to Γ is supported by vertices in $H[N_\cap(S_R) \cup N_\cup(N_\cap(S_R))]$. It is useful to see what happens when the phase (S_L, S_R) of the homomorphism is a maximal biclique (say in $\mathcal{C}_{\alpha,\beta}$): then, $N_\cap(S_R) = S_L$ and $N_\cup(S_L) = V_R$ (from Lemma 8). Thus, in the case where the phase of h corresponds to a maximal biclique, the restriction of h to Γ is supported by vertices in $H[S_L \cup V_R]$. It will be useful to distill the following definitions from the above remarks.

Definition 10 (The graph H_{S_L,S_R}). *Let (S_L, S_R) be a biclique in H, i.e., $(S_L, S_R) \in \mathcal{C}$. Define H_{S_L,S_R} to be the (bipartite) graph $H[N_\cap(S_R) \cup N_\cup(N_\cap(S_R))]$, whose 2-colouring is naturally induced by the 2-colouring of H. Note that when (S_L, S_R) is a maximal biclique, we have that $H_{S_L,S_R} = H[S_L \cup V_R]$.*

Definition 11 (The parameter $\zeta(S_L, S_R, \Gamma)$). *Let Γ be a 2-coloured graph and let (S_L, S_R) be a biclique in H, i.e., $(S_L, S_R) \in \mathcal{C}$. We will use $\zeta(S_L, S_R, \Gamma)$ to denote $\#\mathsf{Fixed}H_{S_L,S_R}\text{-Col}(\Gamma)$, where H_{S_L,S_R} is as in Definition 10.*

Utilising the above definitions and the remarks earlier, we obtain that the con-tribution of the biclique (S_L, S_R) to the gadget $K_{a,b}(\Gamma)$ is roughly equal to $\zeta(S_L, S_R, \Gamma)|S_L|^a|S_R|^b$. The guiding principle will be to choose a, b, and Γ

ppropriately so that the dominant phases are supported either on (both of) he extremal bicliques or on the non-extremal bicliques (but not a combination f both). Roughly, the choice of a, b will restrict the dominant phases in $K_{a,b}(\Gamma)$ o be a subset of $\mathcal{C}_{\alpha,\beta}$, while the graph Γ will pick out either the extremal icliques or a set of non-extremal bicliques. (In the latter case, we will further eed to ensure that exactly one non-extremal biclique makes a significant contribution to the gadget. To do this, we will utilise Lemma 7.) When there is no uch graph Γ, we will use an alternative method to find a (2-coloured) subgraph H' of H which is also full and not trivial such that $\#\mathsf{Fixed}H'\text{-Col}$ reduces to $\#\mathsf{Fixed}H\text{-Col}$. In other words, the non-existence of a "helpful" gadget Γ will stablish a useful property for $\#\mathsf{Fixed}H\text{-Col}$ on an arbitrary input.

We will equalise the contribution of the extremal bicliques in the gadget $K_{a,b}(\Gamma)$, so we we will use the following special case of Definition 11.

Definition 12 (The parameters $\zeta_1^{\mathrm{ex}}(\Gamma)$, $\zeta_2^{\mathrm{ex}}(\Gamma)$). *Let Γ be a 2-coloured graph. Let $\zeta_1^{\mathrm{ex}}(\Gamma)$, $\zeta_2^{\mathrm{ex}}(\Gamma)$ be the values of $\zeta(S_L, S_R, \Gamma)$ when (S_L, S_R) is the extremal biclique $(F_L, V_R), (V_L, F_R)$ respectively. By definition, $\zeta_1^{\mathrm{ex}}(\Gamma)$ equals $\zeta(F_L, V_R, \Gamma)$ which equals $|F_L|^{|L(\Gamma)|}|V_R|^{|R(\Gamma)|}$ and $\zeta_2^{\mathrm{ex}}(\Gamma) := \zeta(V_L, F_R, \Gamma) = \#\mathsf{Fixed}H\text{-Col}(\Gamma)$.*

To see that $\zeta(F_L, V_R, \Gamma) = |F_L|^{|L(\Gamma)|}|V_R|^{|R(\Gamma)|}$, note that $H[F_L \cup V_R]$ is a complete bipartite graph; to see the second equality in the def'n of $\zeta_2^{\mathrm{ex}}(\Gamma)$, note that $H[V_L \cup V_R] = H$. The "asymmetry" in the definitions of $\zeta_1^{\mathrm{ex}}(\Gamma)$ and $\zeta_2^{\mathrm{ex}}(\Gamma)$ is caused by the choice of connecting the right part of $K_{a,b}$ to the left part of Γ.)

To equalise the contribution of the extremal bicliques in the final gadget we will need to slightly perturb our selection of a, b. Instead of setting $a = Q\alpha$ and $b = Q\beta$, we will choose $\hat{a} = Q\alpha$ and $\hat{b} = Q\beta + \gamma$ for some appropriate γ (note that we only perturb the size of b). Now, for a phase (S_L, S_R) the multiplicative correction to its contribution in $K_{\hat{a},\hat{b}}(\Gamma)$ relative to the one in $K_{a,b}$ is given by $\zeta(S_L, S_R, \Gamma)|S_R|^{\gamma}$. Thus, to equalise the contribution of the extremal bicliques in $K_{\hat{a},\hat{b}}(\Gamma)$, we will need the following parameter $\gamma = \gamma(\Gamma)$.

Definition 13 (The parameter $\gamma(\Gamma)$). *Let Γ be a 2-coloured graph. Define $\gamma(\Gamma)$ to be the unique (real) solution to the following equation: $\zeta_1^{\mathrm{ex}}(\Gamma)|V_R|^{\gamma(\Gamma)} = \zeta_2^{\mathrm{ex}}(\Gamma)|F_R|^{\gamma(\Gamma)}$. Note that $F_R \subset V_R$ so $\gamma(\Gamma)$ is well-defined for all Γ.*

With these definitions, for a 2-coloured Γ, we define the following subset of $\mathcal{C}_{\alpha,\beta}$:

Definition 15 (The set of dominating bicliques $\mathcal{C}_{\alpha,\beta}^{\Gamma}$). *Let $1 > \alpha, \beta > 0$ and Γ be a 2-coloured graph. Define $\mathcal{C}_{\alpha,\beta}^{\Gamma}$ to be set of (maximal) bicliques $(S_L, S_R) \in \mathcal{C}_{\alpha,\beta}$ which further maximize $\zeta(S_L, S_R, \Gamma)|S_R|^{\gamma(\Gamma)}$.*

The Cases in the Proof of Lemma 4 (Overview with Examples). Consider the set of maximal bicliques $\mathcal{C}_{\alpha,\beta}$ where $1 > \alpha, \beta > 0$ satisfy $|F_L|^{\alpha}|V_R|^{\beta} = |V_L|^{\alpha}|F_R|^{\beta}$. For the discussion in this section we may assume that $\mathcal{C}_{\alpha,\beta}$ includes both extremal bicliques and at least one non-extremal biclique. Let $\big(S_L^{(1)}, S_R^{(1)}\big)$,

$\ldots, (S_L^{(t)}, S_R^{(t)})$ be an enumeration of the non-extremal bicliques in $\mathcal{C}_{\alpha,\beta}$. Recall that all elements of $\mathcal{C}_{\alpha,\beta}$ are maximal bicliques of H. For convenience, in this section let H_i denote the subgraph $H[S_L^{(i)} \cup V_R]$ (this corresponds to the graph $H_{S_L^{(i)}, S_R^{(i)}}$ in Definition 10) and set $\zeta_i(\Gamma) = \zeta(S_L^{(i)}, S_R^{(i)}, \Gamma) = \#\mathsf{Fixed}H_i\text{-Col}(\Gamma)$ (cf. Definition 11). For the extremal bicliques we will instead use the notation $H_1^{\mathrm{ex}}, H_2^{\mathrm{ex}}$ to denote the graphs $H_{F_L, V_R}, H_{V_L, F_R}$ respectively. Note that H_1^{ex} is a complete bipartite graph with bipartition $\{F_L, V_R\}$ while H_2^{ex} is H itself.

There are three complementary cases to consider for the proof of Lemma 4. Recall by construction that for every 2-coloured graph Γ, the extremal bicliques have equal contribution in the graph $K_{\hat{a}, \hat{b}}(\Gamma)$ (where $\hat{a} = Qa$, $\hat{b} = Qb + \gamma(\Gamma)$ for some large Q). The three cases are as follows.

1. There exists $i \in [t]$ and a 2-coloured graph Γ such that the biclique $(S_L^{(i)}, S_R^{(i)})$ dominates over the extremal bicliques in the gadget $K_{\hat{a}, \hat{b}}(\Gamma)$.
2. There exists $i \in [t]$ so that for every 2-coloured graph Γ the biclique $(S_L^{(i)}, S_R^{(i)})$ has the same contribution as the extremal bicliques in the gadget $K_{\hat{a}, \hat{b}}(\Gamma)$.
3. For all $i \in [t]$ and every 2-coloured graph Γ, the contribution of the biclique $(S_L^{(i)}, S_R^{(i)})$ is at most the contribution of the extremal bicliques in $K_{\hat{a}, \hat{b}}(\Gamma)$. Further, for all $i \in [t]$ there exists a 2-coloured graph Γ_i such that the biclique $(S_L^{(i)}, S_R^{(i)})$ is dominated by the extremal bicliques in the gadget $K_{\hat{a}, \hat{b}}(\Gamma_i)$.

Case 1. An example of Case 1 is the graph H at the right of the top row of the figure. Let us first see why the example is in Case 1. The full vertices of H are vertices 1 and $1'$ and the extremal bicliques of H are $(\{1\}, [9'])$ and $([9], \{1'\})$. Thus, the α, β pairs which equalise the contribution of the extremal bicliques in $K_{a,b}$ satisfy $\alpha = \beta$. The dominating bicliques $\mathcal{C}_{\alpha,\beta}$ for $\alpha = \beta$ are the extremal bicliques and the two bicliques $(\{1, 2, 3\}, \{1', 2', 3'\})$ and $(\{1, 8, 9\}, \{1', 8', 9'\})$. In the full version of the paper, we show that when the graph Γ is an edge, the only dominating biclique in $\mathcal{C}_{\alpha,\beta}^\Gamma$ for the gadget $K_{\hat{a}, \hat{b}}(\Gamma)$ is $(\{1, 2, 3\}, \{1', 2', 3'\})$.

Now, in the general setting of Case 1, we have that, in the gadget $K_{\hat{a}, \hat{b}}(\Gamma)$, the extremal bicliques are not dominating (since they are dominated by the biclique $(S_L^{(i)}, S_R^{(i)})$). Note however that there may still be more than one element in $\mathcal{C}_{\alpha,\beta}^\Gamma$, unlike the example in the figure. To pick out only one biclique from $\mathcal{C}_{\alpha,\beta}^\Gamma$ we further apply Lemma 7 on the graphs H_i corresponding to bicliques in $\mathcal{C}_{\alpha,\beta}^\Gamma$. This yields a graph J which "prefers" a particular graph, say, H_j. Then, by an argument analogous to the one in Section 1 (i.e., paste sufficiently many disjoint copies of J in $K_{\hat{a}, \hat{b}}(\Gamma)$), one can show that $\#\mathsf{Fixed}H_j\text{-Col} \leq_{\mathsf{AP}} \#\mathsf{Fixed}H\text{-Col}$. By induction, we have $\#\mathsf{BIS} \leq_{\mathsf{AP}} \#\mathsf{Fixed}H_j\text{-Col}$ and hence $\#\mathsf{BIS} \leq_{\mathsf{AP}} \#\mathsf{Fixed}H\text{-Col}$.

Case 2. An example of Case 2 is the graph H in the final row of the figure. In the full version of the paper, we show that H falls into Case 2 of our analysis based on the following equality which holds for the graphs H_1 and H_1^{ex} in the bottom row of the figure and any 2-coloured graph Γ (note that $H_2^{\mathrm{ex}} = H$):

$$(\#\mathsf{Fixed}H_1\text{-}\mathsf{Col}(\varGamma))^2 = \#\mathsf{Fixed}H_1^{\mathrm{ex}}\text{-}\mathsf{Col}(\varGamma)\ \#\mathsf{Fixed}H_2^{\mathrm{ex}}\text{-}\mathsf{Col}(\varGamma). \qquad (2)$$

A "quick" way to derive (2) is to observe that the tensor product of H_1 with self is the same graph as the tensor product of the graphs H_1^{ex} and H_2^{ex}.

Now, in the general setting of Case 2, the non-existence of a gadget $K_{\hat{a},\hat{b}}(\varGamma)$ that distinguishes between $(S_L^{(i)}, S_R^{(i)})$ and the extremal bicliques for any choice of $\varGamma$ allows us to obtain an equality (analogous to (2)) relating $\#\mathsf{Fixed}H\text{-}\mathsf{Col}(\varGamma)$ and $\#\mathsf{Fixed}H_i\text{-}\mathsf{Col}(\varGamma)$ on every "input" $\varGamma$. It follows that $\#\mathsf{Fixed}H_i\text{-}\mathsf{Col} \leq_{\mathsf{AP}}$ $\#\mathsf{Fixed}H\text{-}\mathsf{Col}$ and we thus obtain that $\#\mathrm{BIS} \leq_{\mathsf{AP}} \#\mathsf{Fixed}H\text{-}\mathsf{Col}(\varGamma)$ as in Case 1.

Case 3. To obtain an example of Case 3, modify the example graph H from Case 1 adding edges $(5, 5')$ and $(6, 6')$. The dominating bicliques $\mathcal{C}_{\alpha,\beta}$ for $\alpha = \beta$ are once again the extremal bicliques and the two bicliques $(\{1, 2, 3\}, \{1', 2', 3'\})$, $\{1, 8, 9\}, \{1', 8', 9'\})$. In the full version, we show that when $\varGamma$ is an edge, the dominating bicliques in $\mathcal{C}_{\alpha,\beta}^{\varGamma}$ for the gadget $K_{\hat{a},\hat{b}}(\varGamma)$ are the extremal bicliques.

Now, in the general setting of Case 3, we have that, for every i, in the gadget $K_{\hat{a},\hat{b}}(\varGamma_i)$, the extremal bicliques are dominating over the biclique $(S_L^{(i)}, S_R^{(i)})$. Consider the graph $\varGamma$ which is the disjoint union of the $\varGamma_i$'s. It is not hard then to show that in the gadget $K_{\hat{a},\hat{b}}(\varGamma)$ the extremal bicliques are dominating over all non-extremal bicliques. One can then use the reduction for Item 1 in Section 4 to show that $\#\mathrm{BIS} \leq_{\mathsf{AP}} \#\mathsf{Fixed}H\text{-}\mathsf{Col}$ (instead of using the gadget $K_{a,b}$ as discussed there, we instead use the gadget $K_{\hat{a},\hat{b}}(\varGamma)$).

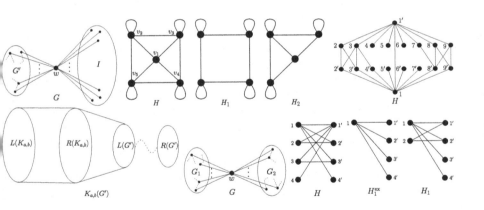

References

1. Dyer, M.E., Goldberg, L.A., Greenhill, C.S., Jerrum, M.: The relative complexity of approximate counting problems. Algorithmica **38**(3), 471–500 (2003)
2. Dyer, M.E., Goldberg, L.A., Jerrum, M.: Counting and sampling H-colourings. Information and Computation **189**(1), 1–16 (2004)
3. Dyer, M.E., Greenhill, C.: The complexity of counting graph homomorphisms. Random Structures and Algorithms **17**(3–4), 260–289 (2000)

4. Ge, Q., Štefankovič, D.: A graph polynomial for independent sets of bipartite graphs. In: FSTTCS, pp. 240–250 (2010)
5. Goldberg, L., Jerrum, M.: A counterexample to rapid mixing of the Ge-Štefankovič process. Electron. Commun. Probab. **17**(5), 1–6 (2012)
6. Goldberg, L.A., Kelk, S., Paterson, M.: The complexity of choosing an H-coloring (Nearly) uniformly at random. SIAM J. Comput. **33**(2), 416–432 (2004)
7. Hell, P., Nešetřil, J.: On the complexity of H-coloring. Journal of Combinatorial Theory, Series B **48**(1), 92–110 (1990)
8. Jerrum, M.R., Valiant, L.G., Vazirani, V.V.: Random generation of combinatorial structures from a uniform distribution. TCS **43**, 169–188 (1986)
9. Kelk, S.: On the relative complexity of approximately counting H-colourings. Ph.D. Thesis, University of Warwick (2004)
10. Lovász, L.: Operations with structures. Acta Math. Acad. Sci. Hungar. **18**, 321–328 (1967)
11. Mossel, E., Weitz, D., Wormald, N.: On the hardness of sampling independent sets beyond the tree threshold. Prob. Theory Related. Fields **143**, 401–439 (2009)

Taylor Polynomial Estimator for Estimating Frequency Moments

Sumit Ganguly$^{(\boxtimes)}$

Indian Institute of Technology, Kanpur, India
sganguly@cse.iitk.ac.in

Abstract. We present a randomized algorithm for estimating the pth moment F_p of the frequency vector of a data stream in the general update (turnstile) model to within a multiplicative factor of $1 \pm \epsilon$, for $p > 2$, with high constant confidence. For $0 < \epsilon \le 1$, the algorithm uses space $O(n^{1-2/p}\epsilon^{-2} + n^{1-2/p}\epsilon^{-4/p}\log(n))$ words. This improves over the current bound of $O(n^{1-2/p}\epsilon^{-2-4/p}\log(n))$ words by Andoni et al. in [2]. Our space upper bound matches the lower bound of Li and Woodruff [17] for $\epsilon = (\log(n))^{-\Omega(1)}$ and the lower bound of Andoni et al. [3] for $\epsilon = \Omega(1)$.

1 Introduction

The data stream model is relevant for online applications over massive data, where an algorithm may use only sub-linear memory and a single pass over the data to summarize a large data-set that appears as a sequence of incremental updates. Queries may be answered using only the data summary. A data stream is viewed as a sequence of m records of the form (i, v), where, $i \in [n] = \{1, 2, \ldots, n\}$ and $v \in \{-M, -M + 1, \ldots, M - 1, M\}$. The record (i, v) changes the ith coordinate f_i of the n-dimensional *frequency vector* f to $f_i + v$. The pth moment of the frequency vector f is defined as $F_p = \sum_{i \in [n]} |f_i|^p$, for $p \ge 0$. The (randomized) F_p estimation problem is: Given p and $\epsilon \in (0, 1]$, design an algorithm that makes one pass over the input stream and returns $\hat{F}_p$ such that $\Pr[|\hat{F}_p - F_p| \le \epsilon F_p] \ge 0.6$ (where, the constant 0.6 can be replaced by any other constant $> 1/2$.) In this paper, we consider estimating F_p for the regime $p > 2$, called the *high moments* problem. The problem was posed and studied in the seminal work of Alon, Matias and Szegedy in [1].

Space lower bounds. Since a deterministic estimation algorithm for F_p requires $\Omega(n)$ bits [1], research has focussed on randomized algorithms [3,5,10, 13,16,17,22,23]. Andoni et al. in [3] present a bound of $\Omega(n^{1-2/p}\log(n))$ *words* assuming that the algorithm is a *linear sketch*. Li and Woodruff in [17] show a lower bound of $\Omega(n^{1-2/p}\epsilon^{-2}\log(n))$ bits in the turnstile streaming model. For *linear* sketch algorithms, the lower bound is the sum of the above two lower bounds, namely, $\Omega(n^{1-2/p}(\epsilon^{-2} + \log(n)))$ words.

Space upper bounds. The table in Figure 1 chronologically lists algorithms and their properties for estimating F_p for $p > 2$ of data streams in the *turnstile mode*. Algorithms for *insertion-only* streams are not directly comparable to algorithms

© Springer-Verlag Berlin Heidelberg 2015
M.M. Halldórsson et al. (Eds.): ICALP 2015, Part I, LNCS 9134, pp. 542–553, 2015.
DOI: 10.1007/978-3-662-47672-7_44

for update streams—however, we note that the best algorithm for insertion-only streams is by Braverman et al. in [7] that uses $O(n^{1-2/p})$ bits, for $p \geq 3$ and $\epsilon = \Omega(1)$.

Contribution. We show that for each fixed $p > 2$ and $0 < \epsilon \leq 1$, there is an algorithm for estimating F_p in the general update streaming model that uses space $O(n^{1-2/p}(\epsilon^{-2} + \epsilon^{-4/p}\log(n)))$ words, with word size $O(\log(nmM))$ bits. It is the most space economical algorithm as a function of n and $1/\epsilon$. The space bound of our algorithm matches the lower bound of $\Omega(n^{1-2/p}\epsilon^{-2})$ of Li and Woodruff in [17] for $\epsilon \leq (\log n)^{-p/(2(p-2))}$ and the lower bound $\Omega(n^{1-2/p}\log(n))$ words of Andoni et al. in [3] for linear sketches and $\epsilon = \Omega(1)$.

Algorithm	Space in $O(\cdot)$ words	Update time $O(\cdot)$
IW[15]	$n^{1-2/p}\left(\epsilon^{-1}\log(n)\right)^{O(1)}$	$(\log^{O(1)} n)(\log(mM))$
Hss[6]	$n^{1-2/p}\epsilon^{-2-4/p}\log(n)\log^2(nmM)$	$\log(n)\log(nmM)$
MW [18]	$n^{1-2/p}(\epsilon^{-1}\log(n))^{O(1)}$	$n^{1-2/p}(\epsilon^{-1}\log n)^{O(1)}$
AKO[2]	$n^{1-2/p}\epsilon^{-2-4/p}\log(n)$	$\log n$
BO-I [8]	$n^{1-2/p}\epsilon^{-2-4/p}\log(n)\log^{(c)}(n)$	$\log n$
this paper	$n^{1-2/p}\epsilon^{-2} + n^{1-2/p}\epsilon^{-4/p}\log(n)$	$\log^2(n)$

Fig. 1. Space requirement of published algorithms for estimating F_p, $p > 2$. Word-size is $O(\log(nmM))$ bits for algorithms for update streams. $\log^{(c)}(n)$ denotes c times iterated logarithm for $c = O(1)$.

Techniques and Overview. We design the Geometric-Hss algorithm for estimating F_p that builds upon the Hss technique presented in [6,12]. It uses a layered data structure with $L + 1 = O(\log n)$ levels numbered from 0 to L and uses an ℓ_2-heavy-hitter structure based on CountSketch [11] at each level to identify and estimate $|f_i|^p$ for each heavy-hitter. The heavy-hitters structure at each level has the same number of $s = O(\log n)$ hash tables with each hash table having the number of buckets (height of table). The main new ideas are as follows. The height of any CountSketch table at level l is α^l times the height of any of the tables of the level 0 structure, where, $0 < \alpha < 1$ is a constant. The geometric decrease ensures that the total space required is a constant times the space used by the lowest level and avoids increasing space by a factor of $O(\log n)$ as in the Hss algorithm.

In all previous works, an estimate for $|f_i|^p$ for a *sampled item* i was obtained by retrieving an estimate $\hat{f}_i$ of f_i from the heavy-hitter structure of an appropriately chosen level, and then computing $|\hat{f}_i|^p$. In order for $|\hat{f}_i|^p$ to lie within $(1 \pm \epsilon)|f_i|^p$, $|\hat{f}_i - f_i|$ had to be constrained to be at most $O(\epsilon|f_i|/p)$. By the lower bound results of [20], the estimation error for CountSketch is in general optimal and cannot be improved. We circumvent this problem by designing a more accurate estimator $\bar{\vartheta}(\lambda, k)$ for $|f_i|^p$ directly. If λ is an estimate for $|f_i|$ that is accurate to within a constant relative error, that is, $\lambda \in (1 \pm O(1/p))|f_i|$ and there are independent, identically distributed and unbiased estimates $X_1, X_2, \ldots, X_{\Theta(k)}$

of $|f_i|$ with standard deviation $\sigma[X_j] \leq O(|f_i|/p)$, then, it is shown that (i) $\mathbb{E}\left[\bar{\vartheta}(\lambda, k)\right] \in (1 \pm O(1/p)^k)|f_i|^p$, and (ii) $\mathsf{Var}\left[\bar{\vartheta}(\lambda, k)\right] \leq O(|f_i|^{2p-2}\sigma^2[X_j])$. The estimator $\bar{\vartheta}$ is designed using a *Taylor polynomial estimator*. Given an estimate $\lambda = |\hat{f}_i|$ for $|f_i|$ such that $\lambda \in (1 \pm O(1/p))|f_i|$, the $k+1$ term *Taylor polynomial estimator* denotes $\vartheta(\lambda, k) = \sum_{j=0}^{k} \binom{p}{j}\lambda^{p-j}(X_1 - \lambda)(X_2 - \lambda)\ldots(X_j - \lambda)$, where, $X_1, \ldots, X_k$ are independent and identically distributed estimators of $|f_i|$. Note that replacing the X_j's by $|f_i|$ gives the expression $\sum_{j=0}^{k-1} \binom{p}{j}\lambda^{p-j}(|f_i| - \lambda)^j$, which is the degree-$k$ term Taylor polynomial expansion of $|f_i|^p$ around λ (i.e., $(\lambda + (|f_i| - \lambda))^p$. A new estimator $\bar{\vartheta}(\lambda, k, r)$ is defined as the average of r *dependent* Taylor polynomial estimators ϑ's, where, each of these r ϑ-estimators is obtained from a certain k-subset of random variables $X_1, \ldots, X_s$, with $s = O(k)$, and each k-subset is drawn from an appropriate code and has a controlled overlap with another k-subset from the code. Note that now, only a constant factor (i.e., within a factor of $1 \pm O(1/p)$) accuracy for the estimate λ of $|f_i|$ is needed, rather than an $O(\epsilon)$-accuracy needed earlier.

Finally, we note that Hss algorithm [12] used full independence of hash functions and then invoked Indyk's method [14] of using Nisan's pseudo-random generator to fool space-bounded computations [19]. In our algorithm, we show that it suffices to use only limited $d = O(\log n)$-wise independence of hash families, by changing the way the hash functions are composed.

2   Taylor Polynomial Estimator

Let X be a random variable with $\mathbb{E}[X] = \mu$ and $\mathsf{Var}[X] = \sigma^2$. Singh in [21] considered the following problem: given a function $\psi : \mathbb{R} \to \mathbb{R}$, design an unbiased estimator θ for $\psi(\mathbb{E}[X])$ (i.e., $\mathbb{E}[\theta] = \psi(\mathbb{E}[X])$). Singh proposed the following solution for an analytic function $\psi(t) = \sum_{k \geq 0} \gamma_k(0)t^k$. Choose any distribution ν over $\mathbb{N}$ (naturals) with probability function $p_\nu(n)$, choose $n \sim \nu$ and define the estimator $\theta = (p_\nu(n))^{-1}\gamma_n(0) \cdot X_1 \cdot X_2 \ldots \cdot X_n$ where the X_i's are independent copies of X. The estimator satisfies $\mathbb{E}[\theta] = \sum_{n \geq 0}(p_\nu(n))^{-1} \cdot p_\nu(n) \cdot \gamma_n(0)\mathbb{E}[X_1]\mathbb{E}[X_2]\ldots\mathbb{E}[X_n] = \sum_{n \geq 0}\gamma_n(0)\mu^n = \psi(\mu)$. However, the variance can be large; for the geometric distribution ν with $p_\nu(n) = q(1-q)^n$, for $n \geq 0$ and $0 < q \leq 1$, it is shown in [9] that $\mathbb{E}[\theta^2] = (1/q)\sum_{n \geq 0}\gamma_n^2(0)((\mu^2 + \sigma^2)/(1-q))^n$.

The Taylor polynomial estimator (abbreviated as TP estimator) is derived from the Taylor's series of $\psi(\mu) = \psi(\lambda + (\mu - \lambda))$ by expanding it around λ, an estimate of μ, and then truncating it after the first $k+1$ terms. Let $X_1, \ldots, X_k$ be independent and identically distributed as X. Define

$$\vartheta(\psi, \lambda, k, \{X_l\}_{l=1}^k) = \sum_{j=0}^{k} \gamma_j(\lambda)(X_1 - \lambda)(X_2 - \lambda)\ldots(X_j - \lambda) \ .$$

where, $\gamma_j(t)$ is the function $\psi^{(j)}(t)/j!$, for $j = 0, 1, \ldots$. Its expectation and variance properties are given below. Denote by $\sigma^2 = \mathsf{Var}[X_j]$ and $\eta^2 = \mathbb{E}\left[(X_j - \lambda)^2\right] = \sigma^2 + (\mu - \lambda)^2$, for $j = 1, \ldots, k$.

Lemma 1. *Let $\{X_l\}_{l=1}^k$ be independent random variables with expectation μ and standard deviation σ. Let $\eta = (\sigma^2 + (\mu - \lambda)^2)^{1/2}$ and let ψ be analytic in the region $[\lambda, \mu]$. Then the following hold.*

1. For some $\lambda' \in (\mu, \lambda)$, $\left| \mathbb{E}\left[\vartheta(\psi, \lambda, k, \{X_l\}_{l=1}^k) \right] - \psi(\mu) \right| \leq |\gamma_{k+1}(\lambda')| \cdot |\mu - \lambda|^{k+1}$.

2. $\mathsf{Var}\left[\vartheta(\psi, \lambda, k, \{X_l\}_{l=1}^k) \right] \leq \left(\sum_{j=1}^k |\gamma_j(\lambda)| \eta^j \right)^2$.

Corollaries 1 and 2 apply the Taylor polynomial estimator to $\psi(t) = t^p$.

Corollary 1. *Assume the premises of Lemma 1. Further, let $\psi(t) = t^p$, $p \geq 1$, $\mu > 0$, $|\lambda - \mu| \leq \alpha\mu$, for some $0 \leq \alpha < 1/2$ and $k + 1 > p$. Then, $\left| \mathbb{E}\left[\vartheta(x^p, \lambda, k, \{X_l\}_{l=1}^k) \right] - \mu^p \right| \leq \left(\frac{\alpha}{1-\alpha} \right)^{(k+1)} \cdot \mu^p \cdot \left(\frac{p}{k+1} \right)^{\lfloor p \rfloor + 1}$. In particular, for p integral, $\mathbb{E}\left[\vartheta(x^p, \lambda, k, \{X_l\}_{l=1}^k) \right] = \mu^p$.*

Corollary 2. *Assume the premises of Lemma 1. Further let $\psi(t) = t^p$, $p \geq 2$, $\mu > 0$, $|\lambda - \mu| \leq \min(\mu, \lambda)/(25p)$ and $\sigma \leq \lambda/(25p)$. Then $\mathsf{Var}[\vartheta(x^p, \lambda, \{X_l\}_{l=1}^k] \leq (1.08)p^2\mu^{2p-2}\eta^2$.*

Averaged Taylor polynomial estimator. We use a version of the Gilbert-Varshamov theorem from [4].

Theorem 1 (Gilbert-Varshamov). *For positive integers $q \geq 2$ and $k > 1$, and real value $0 < \epsilon < 1 - 1/q$, there exists a set $\mathcal{C} \subset \{0, 1\}^{qk}$ of binary vectors with exactly k ones such that $\mathcal{C}$ has minimum Hamming distance $2\epsilon k$ and $\log |\mathcal{C}| > (1 - H_q(\epsilon))k \log q$, where, H_q is the q-ary entropy function $H_q(x) = -x \log_q \frac{x}{q-1} - (1 - x) \log_q (1 - x)$.*

Corollary 3. *For $k \geq 150$, there exists a code $Y \subset \{0, 1\}^{8k}$ such that $|Y| \geq 16k$, each $y \in Y$ has exactly k 1's, and the minimum Hamming distance among distinct codewords in Y is $3k/2$.*

Let Y be a code as given by Corollary 3. Each $y \in Y$ is a boolean vector $y = (y(1), y(2), \ldots, y(s))$ of dimension $s = 8k$ with exactly k 1's. It can be equivalently viewed as a k-dimensional ordered sequence $y \equiv (y_1, y_2, \ldots, y_k)$ where $1 \leq y_1 < y_2 < \ldots < y_k \leq s$, and y_j is the index of the jth occurrence of 1 in y.

We first define the TP estimator for $\psi(\mu)$ given an estimate λ for $\mu = \mathbb{E}[X_j]$ corresponding to a codeword $y \in Y$ and ordered according to a given permutation $\pi : [k] \rightarrow [k]$. Let $\pi : [k] \rightarrow [k]$ be a permutation and $y = (y_1, \ldots, y_k)$ be an ordered sequence of size k. Then, $\pi(y)$ denotes the sequence of indices $(y_{\pi(1)}, \ldots, y_{\pi(k)})$. The TP estimator corresponding to $y \in Y$ and permutation π is defined as

$$\vartheta(\psi, \lambda, k, s, y, \pi, \{X_t\}_{t=1}^s) = \sum_{v=0}^k \gamma_v(\lambda) \prod_{l=1}^v \left(X_{y_{\pi(l)}} - \lambda \right) .$$

Let $\{\pi_y\}_{y \in Y}$ denote a set of $|Y|$ randomly and independently chosen permutations that map $[k] \rightarrow [k]$ that is placed in (arbitrary) 1-1 correspondence with Y.

The averaged Taylor polynomial estimator AVGTP averages the $|Y|$ TP estimators corresponding to each codeword in Y, ordered by the permutations $\{\pi_y\}_{y \in Y}$ respectively, as follows.

$$\bar{\vartheta}(\psi, \lambda, k, s, Y, \{\pi_y\}_{y \in Y}, \{X_l\}_{l=1}^s) = \frac{1}{|Y|} \sum_{y \in Y} \vartheta(\psi, \lambda, k, s, y, \pi_y, \{X_l\}_{l=1}^s) \quad (1)$$

The Taylor polynomial estimator in RHS of Eqn. (1) corresponding to each $y \in Y$ is referred to simply as ϑ_y, when the other parameters are clearly understood from context. Note that for any $y \in Y$ and permutation π_y, $\mathbb{E}[\vartheta_y]$ is the same. Therefore, due to averaging, the AVGTP estimator has the same expectation as the ϑ_y's.

Lemma 2. *Let* $p \geq 2, q = 8$, $k \geq \max(150, 40(\lfloor p \rfloor + 2))$ *and* $s = qk$. *Let* $Y \subseteq \{0,1\}^s$ *such that, (a)* $|Y| \geq 16k$, *(b) each* $y \in Y$ *has exactly* k *ones, and (c) the minimum Hamming distance among distinct codewords in* Y *is* $3k/2$. *Let* $\{X_1, \ldots, X_s\}$ *be a family of independent and identically distributed random variables with expectation* $\mu > 0$ *and variance* σ^2. *Let* λ *be an estimate for* μ *satisfying* $|\lambda - \mu| \leq \min(\mu, \lambda)/(25p)$ *and let* $\sigma < \min(\mu, \lambda)/(25p)$. *Let* $\eta = ((\lambda - \mu)^2 + \sigma^2)^{1/2} > 0$. *Let* $\bar{\vartheta}$ *denote* $\bar{\vartheta}(t^p, \lambda, k, s, Y, \{\pi_y\}_{y \in Y}, \{X_l\}_{l=1}^s)$. *Then* $\text{Var}[\bar{\vartheta}] \leq \left(\frac{5p^2}{12k}\right)\mu^{2p-2}\eta^2$.

3   Algorithm

The Geometric-Hss algorithm uses a level-wise structure corresponding to levels $= 0, 1, \ldots, L$, where, the values of L and the other parameters are given in Figure 2. The original stream $\mathcal{S}$ is sub-sampled hierarchically to produce random sub-streams for each of the levels $\mathcal{S}_0 = \mathcal{S} \supset \mathcal{S}_1 \supset \mathcal{S}_2 \supset \cdots \mathcal{S}_L$, where, $\mathcal{S}_l$ is the sub-stream that maps to level l. The stream $\mathcal{S}_0$ is the entire input stream. $\mathcal{S}_1$ is obtained by sampling each item i appearing in $\mathcal{S}_0$ with probability $1/2$; if i is sampled, then all its records (i, v) are included in $\mathcal{S}_1$, otherwise none of its records are included. In general, $\mathcal{S}_{l+1}$ is obtained by sampling items from $\mathcal{S}_l$ with probability $1/2$, so that $\Pr[i \in \mathcal{S}_{l+1} \mid i \in \mathcal{S}_l] = 1/2$. This is done by a sequence of independently chosen random hash functions $g_1, g_2, \ldots, g_L$ each mapping $[n] \to \{0, 1\}$. Then i is included in $\mathcal{S}_l$ iff $g_1(i), g_2(i), \ldots, g_l(i)$ are each equal to 1. The g_l's are chosen from a $d = O(\log n)$-wise independent hash family. Corresponding to each level $l = 0, 1, \ldots, L - 1$, a pair of structures $(\text{HH}_l, \text{TPEst}_l)$ are kept, where, HH_l is a CountSketch$(16C_l, s)$ structure with $s = O(\log n)$ hash tables each consisting of $16C_l$ buckets. The TPEst$_l$ structure is used by the Taylor polynomial estimator at level l and is a standard CountSketch$(16C_l, 2s)$ structure except as follows. (a) The hash functions h_{lr}'s used for the hash tables T_{lr}'s are 5-wise independent. (b) The Rademacher family $\{\xi_{lr}(i)\}_{i \in [n]}$ is 4-wise independent for each table index r, and is independent across the r's, $r \in [2s]$. The hash tables $\{T_{lr}\}_{r \in [2s]}$ have $16C_l$ buckets each and use the hash function h_{lr}, for $r \in [2s]$. Corresponding to the final

Reduction factor	$\alpha = 1 - (1 - 2/p)\nu,\ \nu = 0.01$
Basic space parameters	$B = 425(2\alpha)^{p/2}n^{1-2/p}\epsilon^{-2}/\min(\epsilon^{4/p-2}, \log(n))$ $C = (27p)^2 B$
Number of levels L	$L = \lceil \log_{2\alpha} \frac{n}{C} \rceil$
Degree of independence of $g_1, \ldots, g_L$	$d = 50\lceil \log n \rceil$
Level-wise space parameters	$B_l = 4\alpha^l B,\quad l = 0, 1, \ldots, L-1$ $C_l = 4\alpha^l C,\quad l = 0, 1, \ldots, L-1$ $C_L = 16(4\alpha^L C),$
Taylor Polynomial Estimator Parameters	$k = 100\lceil \log n \rceil, r = 16k, s = 8k$
Degree of independence of table hash functions	$t = 6$

Fig. 2. Parameters used by the Geometric-Hss algorithm

level L, there is only an HH_L structure which is a $\mathsf{CountSketch}(C_L^*, s)$ structure, where $C_L^* = 16C_L$. The structure at level L uses $O(1)$ times larger space for HH_L to facilitate the discovery of all items and their frequencies mapping to this level (with very high probability). By using random bits independent of the ones used in the above structures, we assume that there exists an estimate $\hat{F}_2$ satisfying $F_2 \leq \hat{F}_2 \leq (1 + 0.01/(2p))F_2$ with probability $1 - n^{-25}$. Let $\bar{\epsilon} = (B/C)^{1/2} = 1/(27p)$ and define the level-wise thresholds as follows.

$$T_0 = \left(\frac{\hat{F}_2}{B}\right)^{1/2},\ T_l = \left(\frac{1}{2\alpha}\right)^{l/2} T_0, l \in [L-1],\ \text{and}$$

$$Q_l = T_l - \bar{\epsilon}T_l,\ l \in \{0\} \cup [L-1],\quad Q_L = 1/2\ . \tag{2}$$

Sampled Groups $\bar{G}_l$. Let $\hat{f}_{il}$ be the estimate for f_i obtained from level l using HH_l. For $l \in \{0\} \cup [L-1]$, we say that i is "discovered" at level l, or that $l_d(i) = l$ if l is the smallest level such that $|\hat{f}_{il}| \geq Q_l$. Define $\hat{f}_i = \hat{f}_{i,l_d(i)}$. $l_d(i)$ is set to L iff $i \in S_L$ and i has not been discovered at any earlier level. Items are placed into sample groups, denoted by $\bar{G}_l$, for $l \in \{0\} \cup [L]$, as follows. If i is discovered at level l and $|\hat{f}_{il}| \geq T_l$, then, i is included in $\bar{G}_l$. If i is discovered at level l but $|\hat{f}_{il}| < T_l$, then, i is placed in $\bar{G}_{l+1}$ with probability $1/2$ if the flip of an unbiased coin K_i turns up heads. That is,

$$\bar{G}_0 = \{i : |\hat{f}_i| \geq T_0\},$$

$$\bar{G}_l = \{i : (l_d(i) = l \text{ and } |\hat{f}_i| \geq T_l) \text{ or } (l_d(i) = l-1 \text{ and } |\hat{f}_i| < T_{l-1} \text{ and } K_i = 1)\}$$

$$\bar{G}_L = \{i : l_d(i) = L \text{ or } (l_d(i) = L-1 \text{ and } |\hat{f}_i| < T_{L-1} \text{ and } K_i = 1)\}$$

where, the definition for $\bar{G}_l$ in the second equation above applies to $1 \leq l \leq L-1$. *The event* NoColl. Let $\widehat{\mathrm{TopK}}_l(C_l)$ be the set of the top-C_l elements in terms the estimates $|\hat{f}_{il}|$ at level l. For $l \in [L]$, NoColl$_l$ is said to hold if for each $i \in \widehat{\mathrm{TopK}}_l(C_l)$, there exists a set $R_l(i) \subset [2s]$ of indices of hash tables of the tructure TPEST$_l$ such that $|R_l(i)| \geq s$ and that i does not collide with any other em of $\widehat{\text{TOPK}}_l(C_l)$ in the buckets $h_{lq}(i)$, for $q \in R_l(i)$. More precisely,

$$\text{NOCOLL}_l \equiv \forall i \in \widehat{\text{TOPK}}_l(C_l), \exists R_l(i) \subset [2s]\, (|R_l(i)| \geq s \text{ and}$$
$$\forall q \in R_l(i), \forall j \in \widehat{\text{TOPK}}_l(C_l) \setminus \{i\}\ \ h_{lq}(i) \neq h_{lq}(j)\Big)\ . \quad (3)$$

he event NOCOLL is defined as $\text{NOCOLL} \equiv \wedge_{l=0}^{L}\text{NOCOLL}_l$. The analysis shows 4OCOLL to be a high probability event, however, if NOCOLL fails, then, the stimate for F_p returned is 0.

For each sampled item i whose discovery level $l_d(i) < L$, the averaged Taylor)olynomial estimator is used to obtain an estimate of $|f_i|^p$ using the struc- ure TPEST at level $l_d(i)$. If $l_d(i) = L$, then with high probability $\hat{f}_i = f_i$ ind one can use the simpler estimator $|\hat{f}_i|^p$ instead. For $i \in [n]$ such that $_d(i) < L$, the parameter λ_i used by the Taylor polynomial estimator for stimating $|f_i|^p$ is set to $|\hat{f}_i| = |\hat{f}_{i,l_d(i)}|$. Let $l = l_d(i)$. By NOCOLL, let $?_l(i) = \{t_1, t_2, \ldots, t_s\} \subset [2s]$. Let X_{ijl} be the (standard) estimate for $|f_i|$)btained from table $T_{l,j}$, that is, $X_{ijl} = T_{lj}[h_{lj}(i)] \cdot \xi_{lj}(i) \cdot \text{sgn}(\hat{f}_i)$, for $j \in R_l(i)$. [he estimator $\bar{\vartheta}_i = \bar{\vartheta}(t^p, |\hat{f}_i|, k, s, Y, \{\pi_j\}_{j=1}^{s}, \{X_{ijl}\}_{j \in R_l(i)})$ where, Y is a code iatisfying Corollary 3 and is $s = qk$-dimensional and of size at least $16k$. The)arameters k and s are given in Figure 2. The estimator $\hat{F}_p$ for F_p is defined)elow.

$$\hat{F}_p = \sum_{l=0}^{L} \sum_{i \in \bar{G}_l, l_d(i) < L} 2^l \cdot \bar{\vartheta}_i + \sum_{i \in \bar{G}_L, l_d(i) = L} 2^L \cdot |\hat{f}_i|^p\ . \quad (4)$$

4 Analysis

_n this section, we analyze the Geometric-Hss algorithm.

Let the permutation rank($\cdot$) place items in non-decreasing order by their absolute frequencies. The k-residual second moment of f is defined as $F_2^{\text{res}}(k) = \sum_{i \in [n], \text{rank}(i) > k} f_i^2$. Let $F_2^{\text{res}}(k, l)$ denote the (random) k-residual second moment of the frequency vector corresponding to S_l. The analysis is conditioned on the conjunction of the following set of events, collectively denoted as $\mathcal{G}$.

(1) GOODF$_2 \equiv F_2 \leq \hat{F}_2 \leq (1 + 0.001/(2p))F_2$,

(2) NOCOLL   defined in (3),

(3) GOODEST $\equiv \forall l : 0 \leq l \leq L,\ \forall i \in [n],\ |\hat{f}_{il} - f_i| \leq \left(F_2^{\text{res}}(C_l, l)/C_l\right)^{1/2}$,

(4) SMALLRES $\equiv \forall l : 0 \leq l \leq L,\ F_2^{\text{res}}(2C_l, l) \leq 1.5 F_2^{\text{res}}\left(\lceil (2\alpha)^l C \rceil\right)/2^{l-1}$.

(5) ACCUEST $\equiv \forall l : 0 \leq l \leq L,\ \forall i \in [n],\ |\hat{f}_{il} - f_i| \leq \left(F_2^{\text{res}}\left(\lceil (2\alpha)^l C \rceil\right)/(2(2\alpha)^l C)\right)^{1/2}$,

(6) GOODL $\equiv \forall i \in S_L, \hat{f}_{iL} = f_i$.

(7) SMALLHH $\equiv \forall l : 0 \leq l \leq L, \{i : |\hat{f}_{il}| \geq Q_l\} \subset \overline{\text{TOPK}}(C_l)$.

The events comprising $\mathcal{G}$ may be explained as follows. The event that $\hat{F}_2$ is an $1 + O(1/p)$-factor approximation of F_2 is given by GOODF$_2$. The event GOODEST states that for all items and all levels, the frequency estimation errors incurred by the HH structure at any level and for any item remains within the high-probability error bound as given by the CountSketch algorithm [11]. However, the bounds in GOODEST have to be expressed in terms of $F_2^{\text{res}}(C_l, l)$, which are themselves random variables. The event SMALLRES gives some control on this random variable by giving an upper bound on $F_2^{\text{res}}(C_l, l)$ as $1.5 F_2^{\text{res}}\left((2\alpha)^l C)/2^{l-1}\right)$. This is then used by the event ACCUEST to assert that the frequency estimation for an item i at a certain level l has an additive accuracy of $F_2^{\text{res}}\left((2\alpha)^l C)/(2\alpha)^l C)\right)$, which is a non-random function of l. An item i is classified as a heavy-hitter at level l if $\hat{f}_{il} \geq Q_l$, that is, its estimate obtained from the HH structure at level l exceeds the threshold Q_l. The event SMALLHH is said to hold if at each level, the set of heavy hitters are a subset of the set of the items with the top-C_l estimated frequencies. The NOCOLLISION event is used only by the TPEST family of structures at each level, and ensures that each heavy-hitter remains isolated from all the other heavy-hitters of that level in at least s of the tables of the TPEST structure at that level. Lemma 3 shows that $\mathcal{G}$ holds except with inverse polynomially low probability.

Lemma 3. *For the choice of parameters in Figure 2, $\mathcal{G}$ holds with probability $1 - O(n^{-24})$.*

Items are divided into groups according to their frequencies, as follows.

$$G_0 = \{i : |f_i| \geq T_0\}, G_l = \{i : T_l \leq |f_i| < T_{l-1}\}, \quad l \in [L-1], G_L = \{i : 1 \leq |f_i| < T_{L-1}\} .$$

The groups are partitioned into subsets $\text{lmargin}(G_l)$, $\text{mid}(G_l)$ and $\text{rmargin}(G_l)$.

$$\text{lmargin}(G_l) = \{i : T_l \leq |f_i| < T_l(1 + \bar{\epsilon})\}, \quad l = 0, \ldots, L-1,$$
$$\text{rmargin}(G_l) = \{i : T_{l-1}(1 - 2\bar{\epsilon}) \leq |f_i| < T_{l-1}\}, \quad l \in [L]$$
$$\text{mid}(G_l) = \{i : T_l + T_l\bar{\epsilon} \leq |f_i| < T_{l-1} - 2T_{l-1}\bar{\epsilon}\}, \quad l \in [L-1],$$
$$\text{mid}(G_0) = \{i : |f_i| \geq T_0(1 + \bar{\epsilon})\}$$
$$\text{mid}(G_L) = \{1 \leq |f_i| < T_{L-1}(1 - 2\bar{\epsilon})\} .$$

G_0 and G_L have no $\text{rmargin}(G_0)$ and $\text{lmargin}(G_L)$ defined, respectively. These definitions are similar (though not identical) to the Hss algorithm [12]. The group G_l consists of all items in the frequency range (T_{l-1}, T_l). The ratio $T_{l-1}/T_l = (2\alpha)^{1/2}$, except for the last group G_L, whose frequency range is $[1, T_{L-1})$ and the frequency ratio $T_{L-1}/1 \geq (\frac{2\alpha}{\bar{\epsilon}})^{1/2}(\frac{F_2}{n})^{1/2}$ and can be large.

Properties of the Sampling Scheme. Lemma 4 presents basic properties of the sampling scheme. In the remainder of this paper, we assume that $c > 23$ is a constant satisfying $\Pr[\neg\mathcal{G}]/\Pr[\mathcal{G}] \leq n^{-c}$.

Lemma 4. *Let $i \in G_l$.*

1. If $i \in mid(G_l)$, then, $\left| 2^l \Pr\left[i \in \bar{G}_l \mid \mathcal{G}\right] - 1 \right| \leq 2^l n^{-c}$. Further, conditional on $\mathcal{G}$, (i) $i \in \bar{G}_l$ iff $i \in \mathcal{S}_l$, and, (ii) i may not belong to any $\bar{G}_{l'}$, for $l' \neq l$.

2. If $i \in lmargin(G_l)$, then $\left| 2^{l+1} \Pr\left[i \in \bar{G}_{l+1} \mid \mathcal{G}\right] + 2^l \Pr\left[i \in \bar{G}_l \mid \mathcal{G}\right] - 1 \right| \leq 2^l n^{-c}$. Further, conditional on $\mathcal{G}$, i may belong to either $\bar{G}_l$ or $\bar{G}_{l+1}$, but not to any other sampled group.

3. If $i \in rmargin(G_l)$, then $\left| 2^l \Pr\left[i \in \bar{G}_l \mid \mathcal{G}\right] + 2^{l-1} \Pr\left[i \in \bar{G}_{l-1} \mid \mathcal{G}\right] - 1 \right| \leq O(2^l n^{-c})$. Further, conditional on $\mathcal{G}$, i can belong to either $\bar{G}_{l-1}$ or $\bar{G}_l$ and not to any other sampled group.

Lemma 4 is essentially true (with minor changes) for the original Hss method [6, 2], although the Hss analysis used full-independence of hash functions whereas here we work with limited independence.

Lemma 5 essentially repeats the results of Lemma 4, conditional upon the event that another item maps to some sampled group. This property is useful in variance calculations later.

Lemma 5. Let $i, j \in [n]$, $i \neq j$ and $j \in G_r$. Then, $\sum_{r'=0}^{L} 2^{r'} \Pr\left[j \in \bar{G}_{r'} \mid i \in \mathcal{S}_l, \mathcal{G}\right] = 1 \pm O(2^r n^{-c})$. In particular, the following hold.

1. If $j \in mid(G_r)$, then $2^r \Pr\left[j \in \bar{G}_r \mid i \in \mathcal{S}_l, \mathcal{G}\right] = 1 \pm 2^r n^{-c}$ and for any $r \neq r'$, $\Pr\left[j \in \bar{G}_{r'} \mid i \in \mathcal{S}_l, \mathcal{G}\right] = 0$.

2. If $j \in lmargin(G_r)$, then, $2^{r+1} \Pr\left[j \in \bar{G}_{r+1} \mid i \in \mathcal{S}_l, \mathcal{G}\right] + 2^r \Pr\left[j \in \bar{G}_r \mid i \in \mathcal{S}_l, \mathcal{G}\right] = 1 \pm 2^{r+1} n^{-c}$. Further, for any $r' \notin \{r, r+1\}, \Pr\left[j \in \bar{G}_{r'} \mid i \in \mathcal{S}_l, \mathcal{G}\right] = 0$.

3. If $j \in rmargin(G_r)$, then $2^r \Pr\left[j \in \bar{G}_r \mid i \in \mathcal{S}_l, \mathcal{G}\right] + 2^{r-1} \Pr\left[j \in \bar{G}_{r-1} \mid i \in \mathcal{S}_l, \mathcal{G}\right] = 1 \pm 2^{r+1} n^{-c}$. Further, for any $r' \notin \{r-1, r\}, \Pr\left[j \in \bar{G}_{r'} \mid i \in \mathcal{S}_l, \mathcal{G}\right] = 0$.

For variance calculations, we need good control on the joint probability distribution $\Pr\left[i \in \bar{G}_r, j \in \bar{G}_{r'} \mid \mathcal{G}\right]$, which is shown in Lemma 6.

Lemma 6. For $i \in G_l$, $j \in G_m$ and i, j distinct, $\sum_{r,r'} 2^{r+r'} \Pr\left[i \in \bar{G}_r, j \in \bar{G}_{r'} \mid \mathcal{G}\right] - 1 \leq O((2^l + 2^m) n^{-c})$.

Application of Taylor Polynomial Estimator. Let $i \in \bar{G}_{l'}$ for some $l' \in \{0\} \cup [L-1]$. Then, i has been discovered at a level $l_d(i) = l$ (say). The algorithm estimates $|f_i|^p$ from the TPEST structure at the discovery level l using the estimator $\bar{\vartheta}_i = \bar{\vartheta}(\psi(t) = t^p, |\hat{f}_i|, k, s, Y, \{\pi_j\}_{j=1}^s, \{X_{ijl}\}_{j \in R_l(i)})$. If $l_d(i) = l$ and $i \in \bar{G}_{l'}$ for some l', then, $\hat{f}_i$ is defined as $\hat{f}_{il}$ and for any $j \in R_l(i)$, $\sigma_{il} = (\text{Var}\left[X_{ijl}\right])^{1/2}$ and $\eta_{il} = (\sigma_{il}^2 + (|f_i| - |\hat{f}_{il}|)^2$. We first show that the premises of Corollary 1 and Lemma 2 are satisfied so that we can use their implications.

Lemma 7. Assume the parameter values listed in Figure 2 and assume that $\mathcal{G}$ holds. Then, if $l_d(i) = l$ for some $l \in \{0\} \cup [L-1]$, then, the following properties hold. (i) $|\hat{f}_{il} - f_i| \leq |f_i|/(26p)$, (ii) $\mathbb{E}\left[X_{ijl} \mid l_d(i) = l, |\hat{f}_{il}| > Q_l, j \in R_l(i), \mathcal{G}\right] =$

$|f_i|$ *(iii)* $|f_i| \geq 15pn_{il}$, *(iv)* $\eta_{il}^2 \leq 2.7(\bar{\epsilon}T_l)^2$, *(v)* $|\hat{f}_{il} - f_i| \leq |\hat{f}_i|/(26p)$ *and (vi* $|\hat{f}_i|/\eta_{il} \geq 16p$. Further, *(vii)* if $l_d(i) = L$, then, $\hat{f}_i = f_i$ and $\eta_{iL} = 0$.

For $i, k \in S_l$, $j \in [2s]$, let $u_{ikjl} = 1$ iff $h_{lj}(i) = h_{lj}(k)$ and 0 otherwise.

Lemma 8. *Assume the parameters in Figure 2 and $p \geq 2$. Suppose $i \in \bar{G}_l$, for some $l \in \{0\} \cup [L-1]$ Then, $\left| \mathbb{E}\left[\bar{\vartheta}_i \mid \mathcal{G} \right] - |f_i|^p \right| \leq n^{-2500p}|f_i|^p$. Further if p is integral, then, $\mathbb{E}\left[\bar{\vartheta}_i \mid \mathcal{G} \right] = |f_i|^p$.*

We denote by $\bar{\xi}$ the set of random bits defining the family of Rademacher random variables used by the TPEST structures, that is, the set of random bits that defines the family $\{\xi_{lj}(i) \mid i \in [n], j \in [2s], l \in \{0\} \cup [L]\}$. Lemma 9 shows that the event NOCOLL gives an *uncorrelated* property for product of Taylor polynomial estimators.

Lemma 9. *Suppose $i \in \bar{G}_r$ and $i' \in \bar{G}_{r'}$. Then,*
$$\mathbb{E}_{\bar{\xi}}[\bar{\vartheta}_i \bar{\vartheta}_{i'} \mid \hat{f}_i, \hat{f}_{i'}, \mathcal{G}] = \mathbb{E}_{\bar{\xi}}[\bar{\vartheta}_i \mid \hat{f}_i, \mathcal{G}] \mathbb{E}_{\bar{\xi}}[\bar{\vartheta}_{i'} \mid \hat{f}_{i'}, \mathcal{G}].$$

Expectation and Variance of $\hat{F}_p$ Estimator. For uniformity of notation, let $\bar{\vartheta}_i$ denote $|\hat{f}_i|$ when $l_d(i) = L$ and otherwise, let its meaning be unchanged. Let z_{il} be an indicator variable that is 1 if $i \in \bar{G}_l$ and 0 otherwise. Since an item may be sampled into at most one group, $\sum_{l \in [L]} z_{il} \in \{0, 1\}$. Using the extended definition of $\bar{\vartheta}_i$ mentioned above, we can write $\hat{F}_p$ as,

$$\hat{F}_p = \sum_{l=0}^{L} \sum_{i \in \bar{G}_l} 2^l \bar{\vartheta}_i = \sum_{i \in [n]} \sum_{l=0}^{L} z_{il} \cdot 2^l \cdot \bar{\vartheta}_i = \sum_{i \in [n]} Y_i, \text{ where, } Y_i = \sum_{l'=0}^{L-1} 2^{l'} z_{il'} \bar{\vartheta}_i. \quad (5)$$

Lemma 10 shows that $\hat{F}_p$ is almost an unbiased estimator for F_p. This easily follows from Lemma 8.

Lemma 10. $\mathbb{E}\left[\hat{F}_p \mid \mathcal{G} \right] = F_p(1 \pm O(n^{-c+1}))$.

We will use the following facts.

$$F_2 \leq n^{1-2/p} F_p^{2/p} \text{ and } F_{2p-2} \leq F_p^{2-2/p}, \quad \text{for } p \geq 2. \quad (6)$$

Lemma 11. *Let $B = Kn^{1-2/p}\epsilon^{-2}/\log(n)$ and $C = (27p)^2 B$. Then,*

$$\mathsf{Var}\left[Y_i \mid \mathcal{G} \right] \leq \begin{cases} \dfrac{\epsilon^2 |f_i|^{2p-2} F_p^{2/p}}{(5)(10)^4 K} & \text{if } i \in mid(G_0) \\ 2^{l+1}(1.002)|f_i|^{2p} & (i \in G_l \text{ for some } l \geq 1) \text{ or } (i \in lmargin(G_0)). \end{cases}$$

Lemma 12 builds on the approximate pair-wise independence of the sampling scheme (Lemmas 5 and 6) and of the $\bar{\vartheta}_i$ estimators (Lemma 9) to show that the contribution of the cross terms of the form $|\mathbb{E}[Y_i Y_j \mid \mathcal{G}] - \mathbb{E}[Y_i \mid \mathcal{G}]\mathbb{E}[Y_j \mid \mathcal{G}]|$, for $i \neq j$ is very small.

52 S. Ganguly

Lemma 12. *Let* $i \neq j$. *Then,* $\left| \mathbb{E}\left[Y_i Y_j \mid \mathcal{G}\right] - \mathbb{E}\left[Y_i \mid \mathcal{G}\right] \mathbb{E}\left[Y_j \mid \mathcal{G}\right]\right| \leq O(n^{-c+1})|f_i|^p|f_j|^p$.

Lemma 13. $\mathrm{Var}\left[\hat{F}_p \mid \mathcal{G}\right] \leq \epsilon^2 F_p^2/50$.

Theorem 2. *For each fixed* $p > 2$ *and* $0 < \epsilon \leq 1$, *there exists an algorithm in the general update data stream model that returns* $\hat{F}_p$ *satisfying* $\left|\hat{F}_p - F_p\right| < \epsilon F_p$ *with probability* $3/4$. *The algorithm uses space* $O(n^{1-2/p}\epsilon^{-2}+n^{1-2/p}\epsilon^{-4/p}\log(n))$ *words of size* $O(\log(nmM))$ *bits. The time taken to process each stream update is* $O(\log^2 n)$.

Acknowledgments. The author thanks Venugopal G. Reddy for correcting an error in the analysis.

References

1. Alon, N., Matias, Y., Szegedy, M.: The space complexity of approximating frequency moments. Journal of Computer Systems and Sciences **58**(1), 137–147 (1998). Preliminary version appeared in Proceedings of ACM Symposium on Theory of Computing (STOC) 1996, pp. 1–10
2. Andoni, A., Krauthgamer, R., Onak, K.: Streaming algorithms via precision sampling. In: Proceedings of IEEE Foundations of Computer Science (FOCS) (2011). A version appears in arXiv:1011.1263v1 [cs.DS] November 2010
3. Andoni, A., Nguyen, H.L., Polyanskiy, Y., Wu, Y.: Tight lower bound for linear sketches of moments. In: Fomin, F.V., Freivalds, R., Kwiatkowska, M., Peleg, D. (eds.) ICALP 2013, Part I. LNCS, vol. 7965, pp. 25–32. Springer, Heidelberg (2013)
4. Ba, K.D., Indyk, P., Price, E., Woodruff, D.: Lower bounds for sparse recovery. In: Proceedings of ACM Symposium on Discrete Algorithms (SODA) (2008)
5. Bar-Yossef, Z., Jayram, T.S., Kumar, R., Sivakumar, D.: An information statistics approach to data stream and communication complexity. In: Proceedings of ACM Symposium on Theory of Computing STOC, pp. 209–218 (2002)
6. Bhuvanagiri, L., Ganguly, S., Kesh, D., Saha, C.: Simpler algorithm for estimating frequency moments of data streams. In: Proceedings of ACM Symposium on Discrete Algorithms (SODA), pp. 708–713 (2006)
7. Braverman, V., Katzman, J., Seidell, C., Vorsanger, G.: Approximating large frequency moments with $O(n^{1-2/k})$ bits. In: Proceedings of International Workshop on Randomization and Computation (RANDOM) (2014). Published earlier as arXiv:1401.1763, January 2014
8. Braverman, V., Ostrovsky, R.: Recursive Sketching For Frequency Moments (November 2010). arXiv:1011.2571v1 [cs.DS]
9. Cesa-Bianchi, N., Shwartz, S.S., Shamir, O.: Online learning of noisy data with kernels. In: Proceedings of ACM International Conference on Learning Theory (COLT) (2010)
10. Chakrabarti, A., Khot, S., Sun, X.: Near-optimal lower bounds on the multi-party communication complexity of set disjointness. In: Proceedings of International Conference on Computational Complexity (CCC) (2003)
11. Charikar, M., Chen, K., Farach-Colton, M.: Finding frequent items in data streams. Theoretical Computer Science **312**(1), 3–15 (2004). Preliminary version appeared inProceedings of ICALP 2002, pp. 693–703

12. Ganguly, S., Bhuvanagiri, L.: Hierarchical Sampling from Sketches: Estimating Functions over Data Streams. Algorithmica **53**, 549–582 (2009)
13. Ganguly, S.: A Lower Bound for Estimating High Moments of a Data Stream (December 2011). arXiv:1201.0253
14. Indyk, P.: Stable distributions, pseudorandom generators, embeddings, and data stream computation. J. ACM **53**(3), 307–323 (2006). Preliminary Version appeared in Proceedings of IEEE FOCS 2000, pp. 189–197
15. Indyk, P., Woodruff, D.: Optimal approximations of the frequency moments. In: Proceedings of ACM Symposium on Theory of Computing STOC, pp. 202–298. Baltimore, Maryland, USA (June 2005)
16. Jayram, T.S., Woodruff, D.: Optimal bounds for johnson-lindenstrauss transforms and streaming problems with low error. In: Proceedings of ACM Symposium on Discrete Algorithms (SODA) (2011)
17. Li, Y., Woodruff, D.P.: A tight lower bound for high frequency moment estimation with small error. In: Raghavendra, P., Raskhodnikova, S., Jansen, K., Rolim, J.D.P. (eds.) RANDOM 2013 and APPROX 2013. LNCS, vol. 8096, pp. 623–638. Springer, Heidelberg (2013)
18. Monemizadeh, M., Woodruff, D.: 1-pass relative-error l_p-sampling with applications. In: Proceedings of ACM Symposium on Discrete Algorithms (SODA) (2010)
19. Nisan, N.: Pseudo-random generators for space bounded computation. In: Proceedings of ACM Symposium on Theory of Computing STOC, pp. 204–212 (May 1990)
20. Price, E., Woodruff, D.: $(1 + \epsilon)$-approximate sparse recovery. In: Proceedings of IEEE Foundations of Computer Science (FOCS) (2011)
21. Singh, R.: Existence of unbiased estimates. Sankhya: The Indian Journal of Statistics **26**(1), 93–96 (1964)
22. Woodruff, D.P.: Optimal space lower bounds for all frequency moments. In: Proceedings of ACM Symposium on Discrete Algorithms (SODA), pp. 167–175 (2004)
23. Woodruff, D.P., Zhang, Q.: Tight bounds for distributed functional monitoring. In: Proceedings of ACM Symposium on Theory of Computing STOC (2012)

ETR-Completeness for Decision Versions of Multi-player (Symmetric) Nash Equilibria

Jugal Garg[1], Ruta Mehta [2(✉)], Vijay V. Vazirani[2], and Sadra Yazdanbod[2]

[1] Max-Planck-Institut für Informatik, Saarbrücken, Germany
jgarg@mpi-inf.mpg.de
[2] College of Computing, Georgia Institute of Technology, Atlanta, GA, USA
{rmehta,vazirani,syazdanb}@cc.gatech.edu

Abstract. As a result of some important works [4–6,10,15], the complexity of 2-player Nash equilibrium is by now well understood, even when equilibria with special properties are desired and when the game is symmetric. However, for multi-player games, when equilibria with special properties are desired, the only result known is due to Schaefer and Štefankovič [18]: that checking whether a 3-player NE (3-Nash) instance has an equilibrium in a ball of radius half in l_∞-norm is ETR-complete, where ETR is the class Existential Theory of Reals.

Building on their work, we show that the following decision versions of 3-Nash are also ETR-complete: checking whether (i) there are two or more equilibria, (ii) there exists an equilibrium in which each player gets at least h payoff, where h is a rational number, (iii) a given set of strategies are played with non-zero probability, and (iv) all the played strategies belong to a given set.

Next, we give a reduction from 3-Nash to symmetric 3-Nash, hence resolving an open problem of Papadimitriou [14]. This yields ETR-completeness for symmetric 3-Nash for the last two problems stated above as well as completeness for the class FIXP$_a$, a variant of FIXP for strong approximation. All our results extend to k-Nash, for any constant $k \geq 3$.

1 Introduction

Nash equilibrium (NE) is arguably the most important and well-studied solution concept within game theory and understanding its complexity has led to an impressive theory which was discovered largely over the last decade. We denote by k-Nash the problem of computing a NE in a k-player game for a constant k. For the case of 2-Nash, the seminal results of Daskalakis, Goldberg and Papadimitriou [6], and Chen and Deng [4] exactly characterize the complexity of this problem, namely it is PPAD-complete. This leads us to another basic question: of finding a k-Nash solution that satisfies special properties, e.g., has a payoff of at least h for each player. These questions were first studied by Gilboa and Zemel [10]: they considered 2-Nash under numerous special properties and showed them all to be NP-complete [5]. Thus the complexity of the 2-player case is very well understood.

Supported by NSF Grants CCF-0914732 and CCF-1216019.

M.M. Halldórsson et al. (Eds.): ICALP 2015, Part I, LNCS 9134, pp. 554–566, 2015.
DOI: 10.1007/978-3-662-47672-7_45

Although the 2-player case is the most classical and well studied case, it is also important to study the complexity of the multi-player, especially in the context of new applications arising on the Internet and other large networks where multiple players are locked in strategic situations. Indeed there has been much activity on this front, e.g., see [1,11,16], but the picture is not as clear as the 2-player case. A fundamental difference between 2-Nash and k-Nash, for $k \geq 3$, is that whereas the former always admits an equilibrium that can be written using rational numbers [12], the latter require irrational numbers in general, as shown by Nash himself [13] (we will assume that all numbers in the given instance are rational). It is easy to see that in the latter case, equilibria are algebraic numbers. This difference makes the multi-player case much harder.

Daskalakis, Goldberg and Papadimitriou [6], showed that for k-player games, $k \geq 3$, finding an ϵ-approximate Nash equilibrium is PPAD-complete. The complexity of exact equilibrium was resolved by Etessami and Yannakakis [7], who showed this case to be complete for their class FIXP. How about the complexity of finding a k-Nash solution that satisfies special properties? Due to the inherent difficulty of dealing with irrational numbers, this problem remained open until 2009, when Schaefer and Štefankovič [18] formally defined class *Existential Theory of Reals* (ETR), and showed that checking if a 3-player game has a NE in which every strategy is played with probability at most 0.5 (**InBox**) is ETR-complete. ETR is the class of "yes" instances of existentially quantified formulas with bases $\{+, -, *, \wedge, \vee, =, <, >\}$ on real numbers; we note that this class was informally known and used earlier than [18], e.g., see [2].

Our first set of results extend ETR-completeness to NE computation with a number of special properties in ≥ 3 player games: (*i*) checking if a game has more than one NE (**NonUnique**). NE where, (*ii*) each player gets at least h payoff (**MaxPayoff**), (*iii*) a given set of strategies are played with +ve probability (**Subset**), or (*iv*) all the played strategies belong to a given set (**Superset**).

Our second set of results deal with symmetric games. Symmetry arises naturally in numerous strategic situations and with the growth of the Internet, on which typically users are indistinguishable, such situations are only becoming more ubiquitous. In a *symmetric game* all players participate under identical circumstances, i.e., strategy sets and payoffs. Thus the payoff of player i depends only on the strategy, s, played by her and the multiset of strategies, S, played by the others, without reference to their identities. Furthermore, if any other player j were to play s and the remaining players S, the payoff to j would be identical to that of i. A *symmetric Nash equilibrium* (SNE) is a NE in which all players play the same strategy. Nash [13], while providing game theory with its central solution concept, also defined the notion of a symmetric game and proved, in a separate theorem, that such games always admit a symmetric equilibrium.

A simple reduction is known from 2-Nash to symmetric 2-Nash, and it shows that the latter is also PPAD-complete. The questions studied by Gilboa and Zemel [10] for 2-player games were studied by Conitzer and Sandholm [5] for symmetric games and were shown to be NP-complete. On the other hand, no reduction is known from 3-Nash to symmetric 3-Nash. Indeed, after giving the duction from 2-Nash to symmetric 2-Nash, Papadimitriou [14] states, "Amazingly, it is not clear how to generalize this proof for three player games!"

Our second set of results deals with symmetric k-player games, for $k \geq 3$. We rst give a reduction from 3-Nash to symmetric 3-Nash, hence settling the open roblem of [14]. This also enables us to show that symmetric 3-Nash is complete or the class FIXP_a, Strong Approximation FIXP, which is a variant of FIXP hat is meant for the Turing machine model. It also yields ETR-completeness or **Superset** and **Subset** in such games. Once the 3-player case is settled, we rove analogous results for symmetric k-player games, for $k > 3$.

[8] gave a dichotomy for NE, showing a qualitative difference between 2-Nash and k-Nash along three different criteria, see Table 1. The results of this paper add a fourth criterion to this dichotomy, namely complexity of decision roblems. Additionally, we get an analogous dichotomy for symmetric NE, see Table 2. Results of current paper are indicated by $\mathcal{CP}$ in the tables.

Table 1. Dichotomy for Nash equilibrium

	2-Nash	k-Nash, $k \geq 3$
Nature of solution	Rational [12]	Algebraic; irrational example [13]
Complexity	PPAD-complete [3,6,15]	FIXP-complete [7]
Practical algorithms	Lemke-Howson [12]	?
Decision problems	NP-complete [5,10]	ETR-complete: [18] $\mathcal{CP}$ (Theorems 13, 14)

Table 2. Dichotomy for symmetric Nash equilibrium

	Symmetric 2-Nash	Symmetric k-Nash, $k \geq 3$
Nature of solution	Rational [12]	Algebraic; irrational example $\mathcal{CP}$ together with [13]
Complexity	PPAD-complete [3,6,15]	FIXP_a-complete: $\mathcal{CP}$ (Theorem 24)
Practical algorithms	Lemke-Howson [12,17]	?
Decision problems	NP-complete [5]	ETR-complete: $\mathcal{CP}$ (Theorems 23)

1.1 Technical Overview

We first give the main idea behind our reduction from 3-Nash to symmetric 3-Nash (Theorem 19). We will reduce the given game (A, B, C), where each tensor is $m \times n \times p$, to a symmetric game, D, of dimension $l \times l \times l$, where $l = m + n + p$ (see Section 2.1 for the description of (symmetric) games). In this game, under each symmetric NE, the strategy of each player can be decomposed into three vectors, say x, y, z, of dimension m, n, p, respectively. An essential condition for recovering a Nash equilibrium for the original game (A, B, C) is that each of these three vectors be non-zero; this is also the most difficult part of the reduction.

To achieve this we construct a $3 \times 3 \times 3$ symmetric game G all of whose symmetric NE are of full support, even though it is only partially specified (see (4)). We "blow up" G to derive D, which is $l \times l \times l$, and the unspecified entries of G create room where tensors A, B, C are "inserted". Now, if (x, y, z) is a symmetric NE of D then so is $(\sum_i x_i, \sum_j y_j, \sum_k z_k)$ of G. As a result, each vector,

$\boldsymbol{x}, \boldsymbol{y}, \boldsymbol{z} \neq 0$. Next we show that if these vectors are scaled to probability vectors they form a NE for (A, B, C). Additional arguments yield ETR-completeness for **Subset** and **Superset** for symmetric k-Nash (Theorems 20 and 21).

Next we give idea for showing that symmetric 3-Nash is complete for the class FIXP$_a$ (Theorem 22), Strong Approximation FIXP, which is a variant of FIXP that is meant for the Turing machine model. Note that we are unable to show that symmetric 3-Nash is complete for the class FIXP itself, since we don't see how to express the solution to the given instance as a rational linear projection of the solution of the reduced instance.

Under FIXP$_a$, given an instance I and a rational $\epsilon > 0$, we need to compute a vector $\boldsymbol{x}$ that is within (additive) ϵ distance from some solution, i.e., $\exists \boldsymbol{x}^* \in Sol(I)$ such that $|\boldsymbol{x}^* - \boldsymbol{x}|_\infty \leq \epsilon$, in time polynomial in $size[I]$ and $\log(1/\epsilon)$. In the above reduction, obtaining a solution of (A, B, C) involves e.g., dividing $\boldsymbol{x}$ by $\sum_i x_i$. If the latter is very small, this may give us a vector that is very far away from a solution of (A, B, C), even though x may be close to a solution of D.

We get around this problem by a small change in the above reduction, namely, we need to multiply the tensors A, B, C by a small constant ϵ' before they are "inserted" at the appropriate places in G' to get symmetric game D. This ensures that $(\sum_i x_i, \sum_j y_j, \sum_k z_k)$ is approximately $(1/3, 1/3, 1/3)$. As a result, given a point close to a solution of D, we can get a point "close" to a solution of (A, B, C).

Next, we describe how we show ETR-completeness for the four decision problems, mentioned in the previous section, for k-Nash. To show hardness in case of 3-players, we reduce **InBox**, which is known to be ETR-complete for 3-Nash [18], to each of **MaxPayoff**, **Subset** and **Superset**, and then from **MaxPayoff** to **NonUnique**. Hardness for the k-Nash, $k > 3$, follows since 3-Nash reduces to k-Nash trivially by introducing dummy players. To show containment in ETR we give a Non-linear complementarity problem (NCP) formulation that exactly captures NE of a given game.

Next, we briefly explain the reduction from **InBox** to **MaxPayoff** for the 2-player case (see Section 3.1 for details); 3-player case is an extension of it. Let the given game be represented by two payoff matrices (A, B) of dimension $m \times n$, one for each player. The **InBox** problem is to check if it has a NE in which all strategies are played with at most 0.5 probability. We reduce it to checking if another game (C, D) has a NE in which every player gets payoff at least $h > 0$ (**MaxPayoff**). Wlog we can assume that $A, B > 0$.

We construct $m(n + 1) \times n(m + 1)$ matrices C and D, where the top-left block is set to $A + h$ and $B + h$ respectively. This ensures that if each player gets payoff h at a NE, then strategies from this block are played with non-zero probability, and normalizing them gives a NE of (A, B). The latter follows since NE set remains invariant under additive scaling of payoffs. In order to retrieve a NE in 0.5 ball, we ensure that if any of these strategies is (relatively) played with more than 0.5 probability then a sequence of deviations leads to both players playing only among their last mn strategies where payoff is zero ($< h$).

In particular suppose the second player plays $\boldsymbol{y}$ in the top-left block. The last mn strategies of the row player are divided in to n blocks of size m, one for each y_j, $j \leq n$ such that if $y_j > 0.5$ then best response of the first player is to deviate to j^{th} block. The payoff of the second player is set to -1 in these blocks, so then y_j fetches -1 and second player is forced to deviate to her last mn strategies where both get zero. Similarly for the first player.

Due to space constraints, next we present overview of our two results (i) ETR-hardness for **MaxPayoff**, **Subset** and **Superset**, through reduction from **InBox**, (ii) reduction from 3-Nash to symmetric 3-Nash, and ETR-hardness results for the latter. For missing proofs and details refer to full-version [9].

2   Preliminaries

In this section we formally define the (symmetric) k-Nash problem, and their decision problems. Further, we discuss the complexity classes ETR and FIXP.

Notations: All vectors are in bold-face letters, and i^{th} coordinate of vector $\boldsymbol{x}$ is denoted by x_i, and $\boldsymbol{x}^{-i}$ denotes the vector $\boldsymbol{x}$ with i^{th} coordinate removed. $\mathbf{1}$ and $\mathbf{0}$ represent all ones and all zeros vector respectively of appropriate dimension. For integers $k < l$, $\boldsymbol{x}(k : l) = (x_k, x_{k+1}, \ldots, x_l)$. We use $[n]$ to denote set $\{1, \ldots, n\}$ and $[k : l]$ to denote $\{k, k+1, \ldots, l\}$. If $\boldsymbol{x}$ is of m dimension, then by $\sigma(\boldsymbol{x})$ we mean $\sum_{i=1}^{m} x_i$, and $\eta(\boldsymbol{x}) = \boldsymbol{x}/\sigma(\boldsymbol{x})$. Concatenation of vectors $\boldsymbol{x}$ and $\boldsymbol{y}$ is denoted by $(\boldsymbol{x}|\boldsymbol{y})$. Given a matrix A and $h \in \mathbb{R}$, $A + h$ denotes the matrix A with h added to each of its entries. Further, $A(i, :)$ is its ith row and $A(:, j)$ is its jth column.

2.1   (Symmetric) k-Nash

For a given k-player game let $S_i, i \in [k]$ be the set of pure strategies of player i, and let $\boldsymbol{S} = \times_i S_i$. The payoffs of player i can be represented by a k-dimensional tensor A_i, such that $A_i(\boldsymbol{s})$ denotes the payoff she gets when $\boldsymbol{s} \in \boldsymbol{S}$ is played. Players may randomize among their strategies. Let Δ_i denote the set of mixed strategy profiles of player i, and let $\boldsymbol{\Delta} = \times_i \Delta_i$. Expected payoff of player i from $\boldsymbol{x} = (\boldsymbol{x}^1, \ldots, \boldsymbol{x}^k) \in \boldsymbol{\Delta}$ is $\pi_i(\boldsymbol{x}) = \sum_{\boldsymbol{s} \in \boldsymbol{S}} (\Pi_{i \in [k]} \boldsymbol{x}_{s_i}^i) A_i(\boldsymbol{s})$.

Definition 1. *(Nash Equilibrium (NE) [13]) $\boldsymbol{x} \in \boldsymbol{\Delta}$ is said to be a NE if no player gains by unilateral deviation. Formally, $\forall i$, $\forall \boldsymbol{x}' \in \Delta_i$, $\pi_i(\boldsymbol{x}) \geq \pi_i(\boldsymbol{x}', \boldsymbol{x}^{-i})$.*

Let $\pi_i(s, \boldsymbol{x}^{-i})$ denote the payoff i receives when she plays $s \in S_i$ and others play as per $\boldsymbol{x}^{-i}$. It is easy to see that, $\boldsymbol{x}$ is a NE iff [13],

$$\forall i \in [k], \quad \forall s \in S_i, \quad x_s^i > 0 \quad \Rightarrow \quad \pi_i(s, \boldsymbol{x}^{-i}) = \max_{t \in S_i} \pi_i(t, \boldsymbol{x}^{-i}) \tag{1}$$

Symmetric k-Nash: In a symmetric game the players are indistinguishable. Their strategy sets are identical (S) and payoffs are symmetric represented by one tensor A. For a player, the payoff she gets by playing $s' \in S$, when others are playing $\boldsymbol{s} \in S^{k-1}$, is $A(s', \boldsymbol{s})$. Further, who is playing what in $\boldsymbol{s}$ does not matter. Formally, A satisfies $A(s', \boldsymbol{s}) = A(s', \boldsymbol{s}_\tau)$ for all permutations τ of $(1, \ldots, k-1)$, where $\boldsymbol{s}_\tau$ is the corresponding permuted vector.

A profile $\boldsymbol{x} \in \Delta$ is called *symmetric* if $\boldsymbol{x}^i = \boldsymbol{x}^j$, $\forall i, j$, thus one vector $\boldsymbol{x} \in \Delta$ is enough to denote a symmetric profile. At a symmetric strategy profile all the players get the same payoff, and we denote it by $\pi(\boldsymbol{x})$. The problem of computing symmetric NE (SNE) of a symmetric game is called *symmetric k-Nash*.

Note that description of a (symmetric) k-player game takes $O(km^k)$ space, where $m = \max_i |S_i|$, which is exponential in m and k. To keep it polynomial, we consider k as a constant. Further, wlog $(A_1, \ldots, A_k) > 0$ because adding a constant to the tensors does not change the set of NE.

2-Nash: The payoff tensors in case of 2-player game are matrices, say (A, B), A for player one and B for player two. If the first player plays i and second plays j, then their respective payoff are A_{ij} and B_{ij}. Game is said to be symmetric if $B = A^T$. A mixed strategy is $(\boldsymbol{x}, \boldsymbol{y}) \in \Delta_1 \times \Delta_2$, and respective payoffs at such a strategy are $\boldsymbol{x}^T A \boldsymbol{y}$ and $\boldsymbol{x}^T B \boldsymbol{y}$. The NE characterization of (1) reduces to:

$$\forall i \in S_1, \ x_i > 0 \Rightarrow (A\boldsymbol{y})_i = \max_{k \in S_1}(A\boldsymbol{y})_k; \ \forall j \in S_2, \ y_j > 0 \Rightarrow (\boldsymbol{x}^T B)_j = \max_{k \in S_2}(\boldsymbol{x}^T B)_k (2)$$

3-Nash: It is the k-Nash problem with 3 players. Such a game can be represented by 3-dimensional tensors (A, B, C); A for player one, B for player two, and C for player three. If player one plays i, two plays j and three plays k, then their respective payoffs are A_{ijk}, B_{ijk}, and C_{ijk}. If the game is symmetric then we have $A_{ijk} = A_{ikj} = B_{jik} = B_{kij} = C_{jki} = C_{kji}$. A mixed strategy is denoted by $(\boldsymbol{x}, \boldsymbol{y}, \boldsymbol{z}) \in \Delta_1 \times \Delta_2 \times \Delta_3$. Thus NE characterization of (1) reduces to:

$$\begin{aligned}
\forall i \in S_1, \ x_i > 0 &\Rightarrow \textstyle\sum_{j,k} A_{ijk} y_j z_k = \max_{l \in S_1} \sum_{j,k} A_{ljk} y_j z_k \\
\forall j \in S_2, \ y_j > 0 &\Rightarrow \textstyle\sum_{i,k} B_{ijk} x_i z_k = \max_{l \in S_2} \sum_{i,k} B_{ilk} x_i z_k \qquad (3) \\
\forall k \in S_3, \ z_k > 0 &\Rightarrow \textstyle\sum_{i,j} C_{ijk} x_i y_j = \max_{l \in S_3} \sum_{i,j} C_{ijl} x_i y_j
\end{aligned}$$

Decision Problems: Computational complexity of numerous decision problems have been studied for 2-Nash and 3-Nash [5,10]. Here are some interesting ones:

- **NonUnique:** If there exists more than one NE.
- **MaxPayoff:** Given a rational number h, if there exists a NE where every player gets payoff at least h.
- **Subset:** Given sets $T_i \subset \mathcal{S}_i$, $\forall i \in [1:k]$, if there exists a NE where every strategy in T_i is played with positive probability by player i.
- **Superset:** Given sets $T_i \subset \mathcal{S}_i$, $\forall i \in [1:k]$, if there exists a NE where all the strategies outside T_i are played with zero probability by player i.
- **InBox:** If there is a NE where every strategy is played with ≤ 0.5 probability.

All but last have been shown to be NP-complete in case of 2-Nash [5,10], and the last one is shown to be ETR-complete in case of 3-Nash [18]. In this paper, we show ETR-completeness for the first four decision problems for k-Nash, and for third and fourth for symmetric k-Nash.

2.2   Existential Theory of Reals (ETR)

In order to capture decision problems arising in *existential theory of reals* (ETR), Schaefer and Štefankovič [18] defined complexity class ETR as follows: An instance I of class ETR consists of a sentence of the form,

$$(\exists x_1, \ldots, x_n) \quad \phi(x_1, \ldots, x_n),$$

where ϕ is a quantifier-free $(\wedge, \vee, \neg)$-Boolean formula over the predicates (sentences) defined by signature $\{0, 1, -1, +, *, <, \leq, =\}$ over variables that take real values. The question is if the sentence is true. The size of the problem is $n + size(\phi)$, where n is the number of variables and $size(\phi)$ is the minimum number of signatures needed to represent ϕ (we refer the reader to [18] for more details on ETR, and its relation with other classes like PSPACE). Schaefer and Štefankovič showed that for 3-Nash, problem **InBox** is ETR-complete.

2.3 The Class FIXP and Its Variant FIXP$_a$

Etessami and Yannakakis [7] defined the class FIXP to capture complexity of the exact fixed point problems with algebraic solutions. A FIXP problem is to find a fixed-point of a function $F : D \to D$ over a convex, compact domain D, *i.e.*, find $\boldsymbol{x} \in D$ s.t. $F(\boldsymbol{x}) = \boldsymbol{x}$. The function is given by an arithmetic circuit C with $\{\min, \max, +, -, *, /\}$ gates, rational constants, and n input/output; $size[C] = n + \#$ gates + bit-length(constants). Given $\ll \in D$ to C as an input, all its gates are well defined.

Fixed-points of F may be irrational. To remain faithful to Turing machine computation, Etessami and Yannakakis [7] defined a discrete class FIXP$_a$.

(Strong) Approximation FIXP$_a$: Given circuit C defining function F, and a rational $\epsilon > 0$, compute a vector $\boldsymbol{x}$ that is within (additive) ϵ distance from $\boldsymbol{x}^*$ where $F(\boldsymbol{x}^*) = \boldsymbol{x}^*$ (a fixed-point), in time polynomial in $size[C]$ and $\log(1/\epsilon)$.

Theorem 2. *[7] Given a 3-player game (A_1, A_2, A_3), computing its NE is FIXP-complete. The corresponding (Strong) Approximation is complete for* FIXP$_a$.

3 k-Nash: ETR-completeness for Decision Problems

In this section we show that **MaxPayoff**, **Subset**, **Superset** and **NonUnique** are ETR-hard in k-player games, for any constant $k \geq 3$; refer to full-version [9] for containment in ETR. It suffices to show the results for 3-Nash, as a 3-player game can be reduced to a k-player game trivially by adding $k - 3$ dummy players, with one strategy each. To show hardness for **MaxPayoff**, **Subset** and **Superset** we reduce from **InBox**, and for **NonUnique** we reduce from **MaxPayoff**.

3.1   InBox to MaxPayoff, Subset and Superset

To convey the main ideas, we first describe the reduction in 2-player games and later generalize it to the 3-player case (see [9]). We show the reduction from

InBox to **MaxPayoff**, and from the intermediate lemmas, reduction to **Subset** and **Superset** will follow. Let the given two player game be represented by $m \times n$ dimensional payoff matrices $(A, B) > 0$.

For $a \geq 0$, let $\mathcal{B}_a = [0, \ a]^{m+n}$ be a ball of radius a at origin in l_∞ norm. We will construct another game (C, D), with $m(n + 1) \times n(m + 1)$-dimensional matrices, and show that it has a NE where each player gets at least $h > 0$ payoff (**MaxPayoff**) if and only if the game (A, B) has a NE in ball $\mathcal{B}_{0.5}$ (**InBox**). First we define a couple of notations required for the construction.

Definition 3. *Let i and j be integers where $i \in [m]$ and $j \in [n]$, and h be a real number. We define the following operators:*

$A_{(i,:)+h}$: *matrix A with h added to the entries in its ith row.*

$A_{(:,j)+h}$: *matrix A with h added to the entries in its jth column.*

Definition 4. *Given a matrix M of dimension $a \times b$ and integers r, s such that $a + r - 1 \leq m(n + 1)$ and $b + s - 1 \leq n(m + 1)$, define $[M]_{r,s}$ to be an $m(n + 1) \times n(m + 1)$-dimensional matrix where M is copied starting at position (r, s), and all other coordinates are set to zero.*

Using the above notations we construct matrices C, D as follows, where $h > 0$.

$$C = [A + h]_{1,1} + [(-1)_{m \times mn}]_{1,n+1} + \sum_{j \in [n]} [A_{(:,j)+2h}]_{jm+1,1}$$
$$D = [B + h]_{1,1} + [(-1)_{mn \times n}]_{m+1,1} + \sum_{i \in [m]} [B_{(i,:)+2h}]_{1,in+1}$$

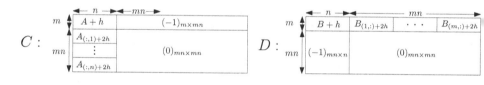

The next lemma follows from the construction of C, D. Recall $\sigma(\boldsymbol{x}) = \sum_i x_i$.

Lemma 5. *Given a strategy $(\boldsymbol{x}', \boldsymbol{y}')$ of game (C, D), let $\boldsymbol{x} = \boldsymbol{x}'(1 : m)$, $\boldsymbol{y} = \boldsymbol{y}'(1 : n)$, $\alpha = h * \sigma(\boldsymbol{y}) - \sigma(\boldsymbol{y}'(n+1 : (m+1)n))$, and $\beta = h * \sigma(\boldsymbol{x}) - \sigma(\boldsymbol{x}'(m+1 : (n+1)m))$. Then,*

$$(C\boldsymbol{y}')_i = \begin{cases} \alpha + (A\boldsymbol{y})_i & \text{if } i \in [m] \\ 2hy_{\lfloor(i-1)/m\rfloor} + (A\boldsymbol{y})_r & \text{if } i \in [m+1, m(n+1)], \ r = ((i-1) \bmod m) + 1. \end{cases}$$

$$(\boldsymbol{x}'^T D)_j = \begin{cases} \beta + (\boldsymbol{x}^T B)_j & \text{if } j \in [n] \\ 2hx_{\lfloor(j-1)/n\rfloor} + (\boldsymbol{x}^T B)_r & \text{if } j \in [n+1, n(m+1)], r = ((j-1) \bmod n) + 1. \end{cases}$$

Before the formal reduction, here is a brief intuition. Note that in (C, D) we have copied $(A + h, B + h)$ in the top-left $m \times n$ block, we call it *first block* now on. Since adding a constant does not change NE of a game, if strategies from the first block are played with non-zero probability at a NE of (C, D), then it may ive a NE of (A, B). This is ensured if payoffs achieved at the NE are positive or at least $h > 0$; a solution of **MaxPayoff**), using Lemma 5.

To guarantee a in $\mathcal{B}_{0.5}$ for game (A, B) (solution of **InBox**), we make use f the blocks added after the first block in both the directions. In particular, a Lemma 5, if $\exists j \in [n]$, $y_j > 0.5 * \sigma(\boldsymbol{y})$, then for the first player her first m strategies are worse than those from block $[mj + 1 : mj + m]$, forcing her to play nly from her last mn strategies. This will force the second player to move away rom the first block too (or else he gets $-ve$ payoff), and thereby leading to a NE where both play from last mn strategies and both get zero payoff (not a solution f **MaxPayoff**). We will use these observations crucially in the reduction.

For game (A, B) only those NE $(\boldsymbol{x}, \boldsymbol{y})$ are interesting which satisfy $\boldsymbol{x}, \boldsymbol{y} \leq 0.5$ solutions of **InBox**). We show that such NE are retained as NE of (C, D). The roof uses the fact that in C and D, top-left block encodes A and B respectively.

Lemma 6. (A, B) has a NE $(\boldsymbol{x}, \boldsymbol{y}) \in \mathcal{B}_{0.5}$ iff $((\boldsymbol{x}, 0_{mn}), (\boldsymbol{y}, 0_{mn}))$ is a NE of $C, D)$.

Lemma 6 maps a solution of **InBox** in game (A, B) to a NE of (C, D) where layers play only among their first m, n strategies respectively. Next we show a reverse mapping: a NE of (C, D) where both players play some of first m, n strategies, gives a NE of game (A, B). Recall that for vector $\boldsymbol{x}$, $\eta(\boldsymbol{x}) = \boldsymbol{x}/\sigma(\boldsymbol{x})$.

Lemma 7. If $(\boldsymbol{x}', \boldsymbol{y}')$ is a NE of game (C, D) s.t. $\boldsymbol{x} = \boldsymbol{x}'[1 : m]$ and $\boldsymbol{y} = \boldsymbol{y}'[1 : n]$ are non zero, then $(\eta(\boldsymbol{x}), \eta(\boldsymbol{y}))$ is a NE for game (A, B), and $(\eta(\boldsymbol{x}), \eta(\boldsymbol{y})) \in \mathcal{B}_{0.5}$.

Lemmas 6 and 7 implies that game (A, B) has a NE in $\mathcal{B}_{0.5}$ if and only if game (C, D) has a NE where both the players play some of first m, n strategies respectively. If we show that to get payoff of at least h in the latter game, players have to play some of first m, n strategies, then clearly the reduction will follow.

Lemma 8. Given a strategy profile $(\boldsymbol{x}', \boldsymbol{y}')$, if $\boldsymbol{x}'^T C \boldsymbol{y}' \geq h$ and $\boldsymbol{x}'^T D \boldsymbol{y}' \geq h$ then $\boldsymbol{x} = \boldsymbol{x}'(1 : m)$ and $\boldsymbol{y} = \boldsymbol{y}'(1 : n)$ are non-zero.

The next theorem follows using Lemmas 6, 7, and 8.

Theorem 9. Game (A, B) has a NE in ball $\mathcal{B}_{0.5}$ if and only if game (C, D) has a NE where every player gets payoff at least h.

Next theorem shows reduction from **InBox** to **Superset** using Lemma 6.

Theorem 10. Game (A, B) has a NE in $\mathcal{B}_{0.5}$ if and only if game (C, D) has a NE where all the strategies played with non-zero probability by first and second player are from $T_1 = [1 : m]$ and $T_2 = [1 : n]$.

Lemmas 6 and 7 imply that, one of first m, n strategies are played with non-zero probability by respective players in game (C, D) if and only if game (A, B) has a NE in ball $\mathcal{B}_{0.5}$. Thus next theorem gives a Turing (and not a many-one) reduction from **InBox** to **Subset**.

Theorem 11.   *Game (A, B) has a NE in ball $\mathcal{B}_{0.5}$ if and only if $\exists i \in [m], \exists j \in [n]$ such that for $T_1 = \{i\}$ and $T_2 = \{j\}$, game (C, D) has a NE where all strategies of T_1 and T_2 are played with non-zero probability.*

We can extend Theorems 9, 10 and 11 to 3-player games; see full-version [9] for details. These together with ETR-hardness of **InBox** in 3-Nash [18], and containment in ETR gives the next result.

Theorem 12.   *Problems* **MaxPayoff**, **Subset** *and* **Superset** *are ETR-complete in 3-player games.*

A 3-player game can be reduced to a k-player game trivially, without changing its set of NE, by adding $k - 3$ dummy players with one strategy each (payoff tensor $A_i = [h], i > 3$ for **MaxPayoff**). And therefore, the next theorem follows.

Theorem 13.   *Given a k-player game $(A_1, \ldots, A_k)$, for a constant $k \geq 3$, problems of* **NonUnique**, **MaxPayoff**, **Subset** *and* **Superset** *are ETR-complete.*

Finally, to show ETR-completeness for **NonUnique**, we reduce **MaxPayoff** to **NonUnique** in 3-player games (see [9]), and thereby obtain,

Theorem 14.   *Given a k-player game $(A_1, \ldots, A_k)$, for a constant $k \geq 3$, problem of* **NonUnique** *is ETR-complete.*

4   Symmetric 3-Nash: ETR and FIXP_a Completeness

In this section, we give a reduction from 3-Nash to symmetric 3-Nash, and thereby obtain ETR-hardness for **Subset** and **Superset**, and FIXP_a-hardness; for containment in ETR and FIXP_a see full-version [9].

Let the given game be (A, B, C), where each tensor is $m \times n \times p$. Let D denote the reduced symmetric game, which will be of dimension $l \times l \times l$, where $l = m + n + p$. Let $(\boldsymbol{x}, \boldsymbol{y}, \boldsymbol{z})$ be a NE of (A, B, C). We will show that there are positive numbers α, β, γ such that $(\boldsymbol{d}, \boldsymbol{d}, \boldsymbol{d})$ is a NE of the reduced game, where $\boldsymbol{d}$ is a l-dimensional vector $(\alpha\boldsymbol{x}|\beta\boldsymbol{y}|\gamma\boldsymbol{z})$. Furthermore, let $(\boldsymbol{d}, \boldsymbol{d}, \boldsymbol{d})$ be a NE of the reduced game, where $\boldsymbol{d}$ decomposes into vectors $\boldsymbol{x}', \boldsymbol{y}', \boldsymbol{z}'$ of dimension m, n, p respectively. Scaling these vectors gives a NE $(\boldsymbol{x}, \boldsymbol{y}, \boldsymbol{z})$ of game (A, B, C). This will yield mapping in both directions.

Essential to this reduction is the $3 \times 3 \times 3$ symmetric game $G(a, b, c)$ given below. We represent the payoff tensor of the first player by three 3×3 matrices, one for each of her pure strategy. Here a, b, c are any non-negative reals.

$$
\begin{bmatrix} 0 & 0 & 0 \\ 0 & 1 & a \\ 0 & a & 0 \end{bmatrix}, \quad
\begin{bmatrix} 0 & 0 & b \\ 0 & 0 & 0 \\ b & 0 & 1 \end{bmatrix}, \quad
\begin{bmatrix} 1 & c & 0 \\ c & 0 & 0 \\ 0 & 0 & 0 \end{bmatrix} \tag{4}
$$

Lemma 15.   *If (α, β, γ) is a symmetric NE of game G, then $\alpha, \beta, \gamma > 0$.*

From G, we derive symmetric game D, which is $l \times l \times l$, by blowing up each of the three strategies of G to m, n, p number of strategies respectively. Copy 0s and 1s to their respective blocks, and replace blocks corresponding to a, b, c by A, B, C respectively. For a formal description of D see full-version [9].

In the above game, suppose two players are playing mixed-strategy $\boldsymbol{d} = (\boldsymbol{x}|\boldsymbol{y}|\boldsymbol{z})$, where $\boldsymbol{x}, \boldsymbol{y}, \boldsymbol{z}$ are of dimensions m, n, p respectively. Then from strategy s the third player receives payoff:

$$
\pi^D(s, \boldsymbol{d}) = \begin{cases} (\sigma(\boldsymbol{y}))^2 + 2\sum_{j\in[n],k\in[p]} A_{sjk}y_j z_k, & \text{if } s \leq m, \\ (\sigma(\boldsymbol{z}))^2 + 2\sum_{i\in[m],k\in[p]} B_{isk}x_i z_k & \text{if } m < s \leq m+n \\ (\sigma(\boldsymbol{x}))^2 + 2\sum_{i\in[m],j\in[n]} C_{ijs}x_i y_j & \text{if } m+n < s \leq l \end{cases} \tag{5}
$$

Wlog we assume that $A, B, C \geq 0$ and hence $D \geq 0$. We consider $\frac{0}{0}$ as 0.

Lemma 16. *If* $\boldsymbol{d} = (\boldsymbol{x}|\boldsymbol{y}|\boldsymbol{z})$ *is a SNE of game* D *then* $(\sigma(\boldsymbol{x}), \sigma(\boldsymbol{y}), \sigma(\boldsymbol{z}))$ *is a NE of* $G(a, b, c)$ *where* $a = \frac{\max_{s\leq m}\sum_{jk}A_{sjk}y_j z_k}{\sigma(\boldsymbol{y})\sigma(\boldsymbol{z})}$, $b = \frac{\max_{s\leq n}\sum_{i,k}B_{isk}x_i z_k}{\sigma(\boldsymbol{x})\sigma(\boldsymbol{z})}$, $c = \frac{\max_{s\leq p}\sum_{i,j}C_{ijs}x_i y_j}{\sigma(\boldsymbol{x})\sigma(\boldsymbol{y})}$.

Lemmas 15 and 16 imply that at any SNE $\boldsymbol{d} = (\boldsymbol{x}|\boldsymbol{y}|\boldsymbol{z})$, all three components $\boldsymbol{x}, \boldsymbol{y}, \boldsymbol{z}$ of the strategy profile are non-zero. Next we show that normalizing each gives a NE of the original game (A, B, C).

Lemma 17. *If* $\boldsymbol{d} = (\boldsymbol{x}|\boldsymbol{y}|\boldsymbol{z})$ *is a SNE of game* D, *then* $(\eta(\boldsymbol{x}), \eta(\boldsymbol{y}), \eta(\boldsymbol{z}))$ *is a NE of game* (A, B, C).

The mapping from SNE of game D to NE of game (A, B, C) established in Lemma 17 implies that computing SNE in symmetric games is no easier than computing a NE in normal games. We can extend this reduction to k-Nash (see [9]). Next, we show a mapping in reverse direction, i.e., from NE of (A, B, C) to a SNE of D, to obtain ETR-hardness results for a number of decision problems in symmetric 3-Nash.

Lemma 18. *Let* $(\boldsymbol{x}, \boldsymbol{y}, \boldsymbol{z})$ *be a NE of* (A, B, C), *and let* (α, β, γ) *be a NE of game* $G(a, b, c)$ *where* a, b, c *are set to payoffs of the first, second and third players respectively at the NE of game* (A, B, C). *Then* $\boldsymbol{d} = (\alpha\boldsymbol{x}|\beta\boldsymbol{y}|\gamma\boldsymbol{z})$ *is a SNE of game* D.

The next theorem summaries the relation between NE of game (A, B, C) and SNE of game D, and follows using Lemmas 17 and 18.

Theorem 19. *Profile* $\boldsymbol{d} = (\boldsymbol{x}|\boldsymbol{y}|\boldsymbol{z})$ *is a SNE of game* D *iff* $(\eta(\boldsymbol{x}), \eta(\boldsymbol{y}), \eta(\boldsymbol{z}))$ *is a NE of game* (A, B, C).

We showed a number of ETR-completeness results for 3-Nash in Section 3. Since, support of NE remains intact in the reduction from 3-Nash to symmetric 3-Nash as shown in Theorem 19, next we show ETR-completeness of **Subset** and **Superset** problems for symmetric 3-Nash.

Theorem 20. *Given a symmetric game D and a subset $T \subset S$, it is ETR complete to check if there exists a SNE x s.t. $x_s > 0$, $\forall s \in T$ (**Subset**).*

The next theorem follows similarly using Theorems 12 and 19.

Theorem 21. *Given a symmetric game D and a subset $T \subset S$, it is ETR-complete to check if there exists a SNE x s.t. $x_s = 0$, $\forall s \in S \setminus T$ (**Superset**).*

Even though Theorem 19 reduces 3-Nash, which is known to be FIXP-complete [7], to symmetric 3-Nash, we do not get FIXP-harness for the latter. This is because to obtain a solution, say x, of former requires *division* among the coordinates of a solution, say d, of latter. While FIXP reduction requires that every x_i is a linear function of some d_j, with rational coefficients (because of irrational solutions). Instead, we show FIXP_a-completeness for symmetric 3-Nash which always has a rational solution (see [9]), and obtain the following.

Theorem 22. *Symmetric 3-Nash is FIXP_a-complete.*

Since there is no trivial extension of symmetric 3-player game to symmetric k-player game, we can extend Theorems 20, 21 and 22 to symmetric k-Nash, and get the following.

Theorem 23. *For symmetric k-Nash, problems **Subset** and **Superset** are ETR-complete, where $k \geq 3$ is a constant.*

Theorem 24. *For a constant $k \geq 3$, symmetric k-Nash is FIXP_a-complete.*

We refer the reader to full-version [9] for a discussion on the significance of our results and open questions.

References

1. Babichenko, Y.: Query complexity of approximate Nash equilibria. In: STOC, pp. 535–544 (2014)
2. Canny, J.: Some algebraic and geometric computations in PSPACE. In: STOC, pp. 460–467 (1988)
3. Chen, X., Deng, X., Teng, S.H.: Settling the complexity of computing two-player Nash equilibria. Journal of the ACM **56**(3) (2009)
4. Chen, X., Deng, X.: Settling the complexity of two-player Nash equilibrium. In: FOCS, pp. 261–272 (2006)
5. Conitzer, V., Sandholm, T.: New complexity results about Nash equilibria. Games and Economic Behavior **63**(2), 621–641 (2008)
6. Daskalakis, C., Goldberg, P.W., Papadimitriou, C.H.: The complexity of computing a Nash equilibrium. SIAM Journal on Computing **39**(1), 195–259 (2009)
7. Etessami, K., Yannakakis, M.: On the complexity of Nash equilibria and other fixed points. SIAM Journal on Computing **39**(6), 2531–2597 (2010)
8. Garg, J., Mehta, R., Vazirani, V.V.: Dichotomies in equilibrium computation, and complementary pivot algorithms for a new class of non-separable utility functions. In: ACM Symposium on the Theory of Computing, pp. 525–534 (2014)

9. Garg, J., Mehta, R., Vazirani, V.V., Yazdanbod, S.: ETR-completeness for decision versions of multi-player (symmetric) nash equilibria (2015). http://www.cc.gatech.edu/~vazirani/3NASH.pdf
10. Gilboa, I., Zemel, E.: Nash and correlated equilibria: Some complexity considerations. Games Econ. Behav. 1, 80–93 (1989)
11. Jiang, A.X., Leyton-Brown, K.: Polynomial-time computation of exact correlated equilibrium in compact games. In: ACM EC, pp. 119–126 (2011)
12. Lemke, C.E., Howson Jr., J.T.: Equilibrium points of bimatrix games. SIAM J. on Applied Mathematics 12(2), 413–423 (1964)
13. Nash, J.F.: Non-cooperatie games. Annals of Mathematics 54(2), 286–295 (1951)
14. Papadimitriou, C.H.: http://www.cs.berkeley.edu/~christos/agt11/notes/lect3.pdf
15. Papadimitriou, C.H.: On the complexity of the parity argument and other inefficient proofs of existence. JCSS 48(3), 498–532 (1994)
16. Papadimitriou, C.H., Roughgarden, T.: Computing equilibria in multi-player games. In: SODA, pp. 82–91 (2005)
17. Savani, R., von Stengel, B.: Hard-to-solve bimatrix games. Econometrica 74(2), 397–429 (2006)
18. Schaefer, M., Štefankovič, D.: Fixed points, Nash equilibria, and the existential theory of the reals. manuscript (2011)

Separate, Measure and Conquer: Faster Polynomial-Space Algorithms for Max 2-CSP and Counting Dominating Sets

Serge Gaspers[1]($\boxtimes$) and Gregory B. Sorkin[2]

[1] UNSW Australia and NICTA, Sydney, Australia
sergeg@cse.unsw.edu.au
[2] London School of Economics, London, UK
g.b.sorkin@lse.ac.uk

Abstract. We show a method resulting in the improvement of several polynomial-space, exponential-time algorithms. The method capitalizes on the existence of small balanced separators for sparse graphs, which can be exploited for branching to disconnect an instance into independent components. For this algorithm design paradigm, the challenge to date has been to obtain improvements in worst-case analyses of algorithms, compared with algorithms that are analyzed with advanced methods, such as Measure and Conquer. Our contribution is the design of a general method to integrate the advantage from the separator-branching into Measure and Conquer, for an improved running time analysis.

We illustrate the method with improved algorithms for MAX $(r, 2)$-CSP and #DOMINATING SET. For MAX $(r, 2)$-CSP instances with domain size r and m constraints, the running time improves from $r^{m/6}$ to $r^{m/7.5}$ for cubic instances and from $r^{0.19 \cdot m}$ to $r^{0.18 \cdot m}$ for general instances, omitting subexponential factors. For #DOMINATING SET instances with n vertices, the running time improves from 1.4143^n to 1.2458^n for cubic instances and from 1.5673^n to 1.5183^n for general instances. It is likely that other algorithms relying on local transformations can be improved using our method, which exploits a non-local property of graphs.

1 Introduction

Graph separators have been used for divide-and-conquer algorithms since the 70s [21]. For classes of instances with sublinear separators, e.g., planar graphs, this often gives subexponential- or polynomial-time algorithms. It is natural to design a branching strategy that strives to disconnect an instance into components, even when no sublinear separators are known. While this has successfully been done experimentally [3,4,11,17,20], we are not aware of worst-case analyses of branching algorithms that are based on linear separators. Our algorithms exploit small separators, specifically, balanced separators of size about $n/6$ for cubic graphs of order n. Their existence is known since 2001, they have been used

© Springer-Verlag Berlin Heidelberg 2015
M.M. Halldórsson et al. (Eds.): ICALP 2015, Part I, LNCS 9134, pp. 567–579, 2015.
DOI: 10.1007/978-3-662-47672-7_46

n pathwidth-based algorithms using exponential time and exponential space, and it is natural to try to exploit them for polynomial-space algorithms.

We now introduce the main separation results. We use standard graph notation from [5]. In a graph $G = (V, E)$, the *contraction* of an edge $uv \in E$ is an operation replacing u and v by one new vertex c_{uv} that is adjacent to $N_G(\{u, v\})$. The graph G is *cubic* or *3-regular* if each vertex has degree 3 and *subcubic* if each vertex has degree at most 3.

Let (L, S, R) be a partition of the vertex set of a graph G such that there is no edge in G with one endpoint in L and the other endpoint in R. We say that (L, S, R) is a *separation* of G, and that S is a *separator* of G, *separating* L and R. The following lemma follows from results in [9,22].

Lemma 1. *For any subcubic graph G with n vertices, a separation (L, S, R) with $|S| \leq \frac{n}{6} + o(n)$ and $|L|, |R| \leq \frac{n - |S| + 1}{2}$ can be computed in polynomial time.*

A direct application of branching on the vertices in such a separator yields algorithms inferior to existing Measure and Conquer ones. Our improvements have their origin in a simple observation: if an algorithm can always branch on vertices in the separator, then the usual measure of improvement is achieved at each step, and the splitting of the graph into two parts when the separator is emptied is a bonus. We get the best of both. The technical challenges are to amortize this bonus over the previous branches to prove a better running time, and to control the balance of the separation as the algorithm proceeds so that the bonus is significant.

We illustrate for cubic MAX 2-CSP. We will be optimistic in this sketch, doing the analysis rigorously in Section 2. The problem class will also be defined there, but for now one may think of MAX CUT, with domain size $r = 2$. Let us "pivot" on a vertex $v \in S$, i.e., sequentially assign it each possible value, eliminate it and its incident edges (see rule R3 below), and solve each case recursively. It is possible that v has neighbors within S, but this is a favorable case, reducing the number of subsequent branches needed. So, suppose that v has neighbors only in L and R. If all neighbors were in one part, the separator could be made smaller, so let us skip over this case as well. The cases of interest, then, are when v has two neighbors in L and one in R, or vice-versa. Suppose that these cases occur equally often; this is the bit of optimism that will require more care to get right. In that case, after all $|S|$ branchings, the sizes of L and R are each reduced by $\frac{3}{2}|S|$, since degree-2 vertices get contracted away. This would lead to a running time bound $t(n)$ satisfying the recurrence

$$t(n) = r^{n/6} \cdot 2t(\tfrac{5}{12}n - \tfrac{3}{2} \cdot \tfrac{1}{6}n),$$

leading to a solution with $t(n) = O^\star(r^{n/5})$. This conjectured bound would improve on the best previous time bound of $O^\star(r^{n/4})$, and Section 2 establishes that the bound is true, modulo a subexponential factor in the running time.

Our algorithms exploit a *global* graph structure, the separator, while executing an algorithm based on *local* simplification and branching rules. The use of global structure may also make it possible to circumvent lower bounds for classes of algorithms restricted to local information [1,2].

Results. Section 2 gives a first analysis of a separator-based algorithm. It solves cubic instances of MAX 2-CSP in time $r^{(1/5+o(1))\,n}$, where n is the number of vertices, improving on the previously fastest $O^\star(r^{n/4})$ time polynomial-space algorithm [25]. By [25, Theorem22], this cubic result allows solution of general instances of MAX 2-CSP in time $r^{(9/50+o(1))\,m}$, improving on the previously fastest $O^\star(r^{(19/100)\,m})$ polynomial-space algorithm [25]. The latter improvement holds also for MAX CUT, an important special case of MAX 2-CSP, and for Polynomial and Ring CSP, generalizations encompassing graph bisection, the Ising model, and counting problems (see [26]).

While MAX 2-CSP is a central problem in exponential-time algorithms, the analysis of its branching algorithms is typically easier than for other problems, largely because the branching creates isomorphic subinstances. In Section 3, we develop the Separate, Measure and Conquer method in full generality, and use this in Section 4 to design faster polynomial-space algorithms for counting dominating sets. For graphs with maximum degree 3, we obtain an algorithm with a time bound of $3^{(1/5+o(1))\,n} = O(1.2458^n)$, improving on the previous best $O^\star(2^{(1/2)\,n}) = O(1.4143^n)$ [19]. For general graphs, we obtain a different algorithm, with time bound $O(1.5183^n)$, improving on the previous best $O(1.5673^n)$ [27]. For details and proofs omitted from this conference version, see [14].

2   Max 2-CSP

Using the notation from [25], an instance (G, S) of MAX 2-CSP (also called MAX $(r, 2)$-CSP) is given by a *constraint graph* $G = (V, E)$ and a set S of *score functions*. Writing $[r] = \{1, ..., r\}$ for the set of available vertex colors, we have a *dyadic* score function $s_e : [r]^2 \to \mathbb{R}$ for each edge $e \in E$, a *monadic* score function $s_v : [r] \to \mathbb{R}$ for each vertex $v \in V$, and a single *niladic* score "function" $s_\emptyset : [r]^0 \to \mathbb{R}$ which is just a constant convenient for bookkeeping. A *candidate solution* is a function $\phi : V \to [r]$ assigning colors to the vertices (ϕ is an *assignment* or *coloring*), and its score is

$$s(\phi) := s_\emptyset + \sum_{v \in V} s_v(\phi(v)) + \sum_{uv \in E} s_{uv}(\phi(u), \phi(v)).$$

An *optimal solution* ϕ is one which maximizes $s(\phi)$.

Let us recall the reductions from [25]. R0–R2 are simplification rules, creating one subinstance, and R3 is a branching rule, creating r subinstances. An optimal solution for (G, S) can be found in polynomial time from optimal solutions of the subinstances.

R0   If $d(y) = 0$, then set $s_\emptyset = s_\emptyset + \max_{C \in [r]} s_y(C)$ and delete y from G.
R1   If $N(y) = \{x\}$, then replace the instance with (G', S') where $G' = (V', E') = G - y$ and S' is the restriction of S to V' and E' except that for all $C \in [r]$ we set

$$s'_x(C) = s_x(C) + \max_{D \in [r]} \{s_{xy}(C, D) + s_y(D)\}.$$

2 If $N(y) = \{x, z\}$, then replace the instance with (G', S') where $G' = (V', E') = (V - y, (E \setminus \{xy, yz\}) \cup \{xz\})$ and S' is the restriction of S to V' and E', except that for $C, D \in [r]$ we set

$$s'_{xz}(C, D) = s_{xz}(C, D) + \max_{F \in [r]} \{s_{xy}(C, F) + s_{yz}(F, D) + s_y(F)\}$$

if there was already an edge xz, discarding the first term $s_{xz}(C, D)$ otherwise.

3 Let y be a vertex of degree at least 3. There is one subinstance (G', s^C) for each color $C \in [r]$, where $G' = (V', E') = G - y$ and s^C is the restriction of s to V' and E', except that we set

$$(s^C)_\emptyset = s_\emptyset + s_y(C), \quad \text{and} \quad (s^C)_x(D) = s_x(D) + s_{xy}(D, C)$$

for every neighbor x of y and every $D \in [r]$.

We will now describe a new separator-based algorithm for cubic MAX 2-CSP, outperforming the algorithm from [25]. Using it as a subroutine in the algorithm for general instances [25] also gives a faster running time for MAX 2-CSP.

2.1   Background

For a cubic instance of MAX 2-CSP, an instance whose constraint graph G is 3-regular, the fastest known polynomial-space algorithm makes simple use of the reductions above. The algorithm branches on a vertex v of degree 3, giving r instances with a common constraint graph G', where v has been deleted. In G', the three G-neighbors of v each have degree 2. Simplification rules are applied to rid G' of degree-2 vertices, and further vertices of degree 0, 1, or 2 that may result, until the constraint graph becomes another cubic graph G''. This results in r instances with the common constraint graph G'', to which the same algorithm is applied recursively. The running time of the algorithm is exponential in the number of branchings, and since each branching destroys 4 degree-3 vertices (the pivot vertex v and its three neighbors), the running time is bounded by $O^\star(r^{n/4})$; details may be found in [24].

Here, we break this $r^{n/4}$ barrier by selecting pivot vertices using global properties of the graph. Our algorithm pivots only on vertices in a separator; when the separator is exhausted, G has been split into components L and R which can be solved independently. The efficiency gain comes from the component splitting: if the time to solve an instance with n vertices is $O^\star(r^{cn})$, the time to solve an instance consisting of components L and R is $O^\star(r^{c|L|}) + O^\star(r^{c|R|})$, which (for L and R of comparable sizes) is hugely less than the time bound $O^\star(r^{c(|L|+|R|)})$ for a single component of the same total order. This efficiency gain comes at no cost: until the separator is exhausted, branching on vertices in the separator is just as efficient as branching on any other vertex.

2.2   Analysis

To analyze the algorithm, we use the Measure and Conquer method. Our measure associates a non-negative real to each instance. As in [13], we use penalty terms in the measure to treat tricky cases. We also take from [25] and [13] the treatment of vertices of degrees 1 and 2 within the Measure and Conquer framework.

Recall [8,12,13] that the Measure and Conquer analysis applies to an algorithm which polynomially transforms an instance I to one or more instances $I_1, \ldots, I_k$, solves those instances recursively, and obtains a solution to I in polynomial time from the solutions of $I_1, \ldots, I_k$. The measure $\mu(I)$ of an instance I should satisfy that for any instance,

$$\mu(I) \geq 0, \tag{1}$$

and for any transformation of I into $I_1, \ldots, I_k$,

$$r^{\mu(I_1)} + \cdots + r^{\mu(I_k)} \leq r^{\mu(I)}. \tag{2}$$

Given these hypotheses, the algorithm solves any instance I in time $O^\star(r^{\mu(I)})$ if the number of recursive calls from the root to a leaf of the search tree is polynomial.

Here, we present an instance of MAX 2-CSP in terms of a separation (L, S, R) of its constraint graph $G = (V, E)$. We write L_3, S_3, and R_3 for the subsets of degree-3 vertices of L, S, and R, respectively, and we will always assume that $|L_3| \leq |R_3|$, if necessary swapping the roles of L and R to make it so. We write $|S_2|$ for the number of degree-2 vertices in S. We define the measure of an instance as

$$\mu(L, S, R) = w_s|S_3| + w_{s,2}|S_2| + w_r|R_3| + w_b\mathbb{1}(|R_3| = |L_3|)$$
$$+ w_c\mathbb{1}(|R_3| = |L_3| + 1) + w_d\log_{3/2}(|R_3| + |S_3|), \tag{3}$$

where the values w_s, $w_{s,2}$, w_r, w_b, w_c, and w_d are constants to be determined and the indicator function $\mathbb{1}$(event) takes the value 1 if the event is true and 0 otherwise. For the constraint (1) that $\mu \geq 0$, it suffices to constrain each of the constants to be nonnegative:

$$w_s, w_{s,2}, w_r, w_b, w_c, w_d \geq 0. \tag{4}$$

Intuitively, the terms w_b and w_c are the only representations of the size of L in μ, and account for the greater time needed when the left side is as large (or nearly as large) as the right. The logarithmic term offsets increases in penalty terms that may result when a new separator is computed, where the instance may go from imbalanced to balanced.

Concretely, from (2), each reduction imposes a constraint on the measure. We treat the reductions in their order of priority: when presenting one reduction, we assume that no previous reduction can be applied. Denote by μ the value of the measure before the reduction is applied and by μ' its value after the reduction.

Degree 0. If the instance contains a vertex v of degree 0, then perform R0 on v. Removing v has no effect on the measure and Condition (2) is satisfied.

Half-edge deletion. A half-edge deletion occurs when the degree of a vertex v decreases. We will require that a degree decrease does not increase the measure, which will validate R1, the collapse of parallel edges, and R2 for vertices in $L \cup R$. If $d(v) \leq 2$, Condition (2) is satisfied since $w_{s,2} \geq 0$ by (4), and v only affects the measure if $v \in S$ and $d(v) = 2$. Now, assume $d(v) = 3$. Taking into account changes in imbalance, and separately analyzing the cases where v is in L, S, and R, we obtain the constraints:

$$-w_b + w_c \leq 0 \qquad (5) \qquad\qquad -w_r + w_c \leq 0 \qquad (7)$$

$$-w_s + w_{s,2} \leq 0 \qquad (6) \qquad \text{and } -w_r + w_b - w_c \leq 0. \qquad (8)$$

Separation. This reduction is the only one special to separation, and its constraint looks quite different from those in previous works. The reduction applies when $S = \emptyset$, which arises in two cases. One is at the beginning of the algorithm, when the instance has not been separated, and may be represented by the trivial separation $(\emptyset, \emptyset, V)$. The second is when reductions on separated instances have exhausted the separator, so that S is empty but L and R are nonempty, and the instance is solved by solving the instances on L and R independently, via a new separation (L', S', R') for R and another such separation (L'', S'', R'') for L. The reduction is applied to a graph $G = (V, E)$ that is cubic and can be assumed to be of at least some constant order, $|V| \geq k$, since a smaller instance can be solved in constant time. By Lemma 1 we know that, for any constant $\epsilon > 0$, there is a size $k = k(\epsilon)$ such that any cubic graph G of order at least k has a separation (L, S, R) with $|S| \leq (\frac{1}{6} + \epsilon)|V|$, $|S|, |R| \leq \frac{5}{12}|V|$. From (2), making worst-case assumptions about balance, it suffices to constrain that

$$r^{w_s|S_3'|+w_r|R_3'|+w_b+w_d \log(|R_3'|+|S_3'|)} + r^{w_s|S_3''|+w_r|R_3''|+w_b+w_d \log(|R_3''|+|S_3''|)}$$

$$\leq r^{w_r|R_3|+w_d \log(|R_3|)}.$$

From the separator properties, this in turn is implied by

$$2 \cdot r^{w_s(1/6+\epsilon)|R_3|+w_r(5/12 \cdot |R_3|)+w_b+w_d \log(8/12 \cdot |R_3|)} \leq r^{w_r|R_3|+w_d \log(|R_3|)},$$

where we have estimated $|L_3'|, |R_3'| \leq \frac{5}{12}|R_3|$ and $|S_3'| \leq (\frac{1}{6} + \epsilon)|R_3| \leq \frac{3}{12}|R_3|$ in the log term on the left hand side. Since $r \geq 2$, it suffices to constrain that

$$1 + w_s(\tfrac{1}{6} + \epsilon)|R_3| + w_r(\tfrac{5}{12}|R_3|) + w_b + w_d \log(\tfrac{8}{12}|R_3|) \leq w_r|R_3| + w_d \log(|R_3|).$$

Taking $\frac{3}{2} = \frac{12}{8}$ to be the logarithm's base and setting $w_d = w_b + 1$, the left term $w_d \log(\frac{8}{12}|R_3|)$ is equal to $-(w_b + 1) + (w_b + 1) \log(|R_3|)$, and it suffices to have

$$(\tfrac{1}{6} + \epsilon)w_s + \tfrac{5}{12}w_r \leq w_r. \qquad (9)$$

Degree 2 in S. If the instance has a vertex $s \in S$ of degree 2, then perform R2 on s. Let $N(s) = \{u_1, u_2\}$. The vertex s is removed and the edge u_1u_2 is added if it was not present already. If L or R contain no neighbor of s, Condition (2) is implied by the constraints of the half-edge deletions. If $u_1 \in L$ and $u_2 \in R$ (or the symmetric case), then S is not a separator any more. The algorithm removes u_2 from R and adds it to S. If $d(u_2) = 2$, we have that $\mu' - \mu \leq 0$. Otherwise $d(u_2) = 3$ and $\mu' - \mu \leq -w_{s,2} + w_s - w_r + \max(0, w_c, w_b - w_c)$. Since $w_c \geq 0$ by (4) it suffices to constrain

$$-w_{s,2} + w_s - w_r + w_c \leq 0 \quad (10) \quad \text{and} \quad -w_{s,2} + w_s - w_r + w_b - w_c \leq 0. \quad (11)$$

No neighbor in L. If the separation (L, S, R) has a vertex $v \in S$ with no neighbor in L, "drag" v into R, i.e., transform the instance by changing the separation to $(L', S', R') := (L, S \setminus \{v\}, R \cup \{v\})$. It is easily checked that this is a valid separation, and with $|L_3'| \leq |R_3'|$ implied by $|L_3| \leq |R_3|$. Indeed the new instance is no more balanced than the old, so that the difference between the new and old measures is $\mu' - \mu \leq -w_s + w_r$, and to satisfy condition (2) it suffices that

$$-w_s + w_r \leq 0, \quad (12)$$

since by (4) and (5) an increase in imbalance does not increase the measure.

No neighbor in R. A vertex $v \in S$ with no neighbor in R is dragged into L. The case where $|R_3| = |L_3|$ is covered by the previous case, reversing the roles of L and R. Otherwise, $|R_3| \geq |L_3| + 1$, and $\mu' - \mu \leq -w_s + \max(0, w_c, w_b - w_c)$. We constrain that

$$-w_s + w_c \leq 0 \quad (13) \quad \text{and} \quad -w_s + w_b - w_c \leq 0. \quad (14)$$

With the above cases covered, we may assume that the pivot vertex $s \in S$ has degree 3 and at least one neighbor in each of L and R.

One neighbor in each of L, S, and R. To branch on a vertex $s \in S$ with one neighbor in each of L, S, and R, perform R3 on s, deleting it from the constraint graph. Since both L and R lose a degree-3 vertex, there is no change in balance and the constraint is

$$1 - 2w_s + w_{s,2} - w_r \leq 0. \quad (15)$$

The form and the initial 1 come from the reduction's generating r instances with common measure μ', so the constraint is $r \cdot r^{\mu'} \leq r^\mu$, or equivalently $1 + \mu' - \mu \leq 0$. The value of $\mu' - \mu$ comes from S losing two degree-3 vertices but gaining a degree-2 vertex, and R losing a degree-3 vertex.

Two neighbors in L. If $s \in S$ has two neighbors in L and one neighbor in R, applying R3 removes s, reduces the degree of a degree-3 vertex in R, and increases the imbalance by one. The algorithm performs R3 if $|R_3| \leq |L_3| + 1$, where $\mu' - \mu \leq -w_s - w_r + \max(-w_b + w_c, -w_c)$. Thus, we constrain

$$1 - w_s - w_r - w_b + w_c \leq 0 \quad (16) \quad \text{and} \quad 1 - w_s - w_r - w_c \leq 0. \quad (17)$$

If, instead, $|R_3| \geq |L_3| + 2$, then the algorithm drags s into L and its neighbor $r \in R$ into S, replacing (L, S, R) by $(L \cup \{s\}, (S \setminus \{s\}) \cup \{r\}, R \setminus \{r\})$. We need to ensure that $-w_r + \max(w_b, w_c) \leq 0$, which, since $w_c \leq w_b$ by (5), is satisfied if we constrain

$$-w_r + w_b \leq 0. \tag{18}$$

Two neighbors in R. If $s \in S$ has two neighbors in R and one neighbor in L, the algorithm performs R3, which removes s, reduces the degree of two degree-3 vertices in R, and decreases the imbalance by one. For the case where $R_3| = |L_3|$, we refer to (16) since L and R are swapped after the reduction. For the other cases, we constrain

$$1 - w_s - 2w_r - w_c + w_b \leq 0 \quad (19) \qquad \text{and } 1 - w_s - 2w_r + w_c \leq 0. \tag{20}$$

This describes all the constraints on the measure. To minimize the running time proven by the analysis, we minimize w_r, obtaining the following optimal, feasible weights:

$$w_r = 0.2 + \varepsilon \qquad w_s = 0.7 \qquad w_{s,2} = 0.6 \qquad w_b = 0.2 \qquad w_c = 0.1.$$

All constraints are satisfied and $\mu \leq (0.2 + \varepsilon)n = (1/5 + o(1))n$.

It only remains to verify that the depth of the search trees is polynomial. Since not every reduction removes a vertex (some only modify the separation (L, S, R)), it is crucial to guarantee some kind of progress for each reduction. Since each reduction decreases another polynomially-bounded measure $\eta(L, S, R, E) := 3|S_3| + 2|R_3| + |L_3| + 2|E|$, by at least one, the depth of the search trees is indeed polynomial.

Theorem 1. *On input of a MAX 2-CSP instance on a constraint graph G with n vertices and m edges, the described algorithm solves G in time $r^{n/5+o(n)} = r^{2m/15+o(m)}$ if G is cubic, time $r^{7m/40+o(m)}$ if G has maximum degree 4, and time $r^{9m/50+o(m)}$ in general, using polynomial space.*

This improves on the previous best running times [25] of $O^\star(r^{m/6})$, $O^\star(r^{3m/16})$, and $O^\star(r^{19m/100})$. The same improvements also hold for Max Cut, an important special case of MAX $(2, 2)$-CSP. Theorem 1 extends instantly to Polynomial CSP and Ring CSP, where the scores are multivariate formal polynomials, or take values in an arbitrary ring. The setting is precisely defined in [26], and the extensions follow immediately from the fact that the algorithm here depends only on R0–R3. Plugging our algorithm into the analysis of [16] also improves that running time from $O^\star(r^{n \cdot \left(1 - \frac{3}{d+1}\right)})$ to $r^{n \cdot (1 - \frac{3}{d+1}) + o(n)}$ for any Max 2-CSP instance with n vertices and average degree $d \geq 5$.

3   The Separate, Measure and Conquer Technique

Our MAX 2-CSP algorithm illustrates that one can exploit separator-based branching to design a more efficient exponential-time algorithm. However, MAX 2-CSP algorithms have certain features that make the analysis simpler than for other problems. First, R3 produces instances with the same constraint graph, and therefore the same measure. Second, the measure of R depends only on the number of degree-3 vertices in R. This implies a discretized change in the measure for R whenever L and R are swapped. In general the change in measure when swapping L and R could take values dense within a continuous domain. Our measure for the MAX 2-CSP algorithm also implies that the initial separator only needs to balance the *number* of vertices in L and R instead of the *measure* of L and R, which is what is needed more generally. Finally, a general method is needed to combine the separator-based branching, which would typically be done for sparse instances, with the general case, where vertex degrees are arbitrary.

Our general method of analysis resolves all the complications mentioned. It applies to recursive algorithms that label vertices of a graph, and where an instance can be decomposed into two independent subinstances when all the vertices of a separator have been labeled in a certain way. Let $G = (V, E)$ be a graph and $\ell : V \to L$ be a labeling of its vertices by labels in the finite set L. For a subset of vertices $W \subseteq V$, denote by $\mu_r(W)$ and $\mu_s(W)$ two measures for the vertices in W in the graph G labeled by ℓ. The measure μ_r is used for the vertices on the right hand side of the separator and μ_s for the vertices in the separator. Let (L, S, R) be a separation of G. Initially, we use the separation $(L, S, R) = (\emptyset, \emptyset, V)$. We define the measure

$$\mu(L, S, R) = \mu_s(S) + \mu_r(R) + \max\left(0, B - \frac{\mu_r(R) - \mu_r(L)}{2}\right)$$
$$+ (1 + B) \cdot \log_{1+\epsilon}(\mu_r(R) + \mu_s(S)), \tag{21}$$

where $\epsilon > 0$ is a constant that will be chosen small enough to satisfy constraint (24) below, and B is an arbitrary constant greater than the maximum change in imbalance in each transformation in the analysis, except the Separation transformation. The *imbalance* of an instance is $\mu_r(R) - \mu_r(L)$, and we assume, as previously, that

$$\mu_r(R) \geq \mu_r(L). \tag{22}$$

To make sure a balanced separator can be computed efficiently, we will assume that adding a vertex to R changes $\mu_r(R)$ by at most B (adjusting B if necessary):

$$|\mu_r(R \cup \{v\}) - \mu_r(R)| \leq B \qquad \text{for each } R \subseteq V \text{ and } v \in V. \tag{23}$$

We also assume that $\mu_r(R)$ can be computed in time polynomial in $|V|$ for each $R \subseteq V$.

Let us now look more closely at the measure (21). The terms $\mu_s(S)$ and $\mu_r(R)$ naturally define measures for the vertices in S and R. No term of the measure directly accounts for the vertices in L; we merely enforce that $\mu_r(R) \geq \mu_r(L)$. The term $\max\left(0, B - \frac{\mu_r(R) - \mu_r(L)}{2}\right)$ is a penalty term based on how balanced the instance is: the more balanced the instance, the larger the penalty term. The penalty term has become continuous, varying from 0 to B. The final logarithmic term amortizes the increase in measure of at most B due to the balance terms each time the instance is separated.

Let us now formulate some generic constraints that the measure should obey.

Separation. We assume that an instance with a separation (L, S, R) can be separated into two independent subinstances $(L, S, \emptyset)$ and $(\emptyset, S, R)$ when the labeling of S allows it; specifically, when all vertices in S have been labeled by a subset $L_s \subseteq L$. This arises in two cases. The first is at the beginning of the algorithm when the graph has not been separated, which is represented by the trivial separation $(\emptyset, \emptyset, V)$. The second is when our reductions have produced a separable instance.

Let (L, S, R) be such that $\ell(s) \in L_s$ for each $s \in S$. The algorithm recursively solves the subinstances $(L, S, \emptyset)$ and $(\emptyset, S, R)$. Let us focus on the instance $(\emptyset, S, R)$; the treatment of the other instance is symmetric. After a cleanup phase, where simplification rules are applied, the next step is to compute a new separator of $S \cup R$. This can be done in various ways, depending on the graph class. For example, polynomial-time computable balanced separators can be derived from upper bounds on the pathwidth of graphs with bounded maximum or average degree [6,7,12]. After a balanced separator (L', S', R') has been computed for $S \cup R$, the instance is solved recursively, and so is the instance $L \cup S$, separated into (L'', S'', R''). Both solutions are then combined into a solution for the instance $L \cup S \cup R$. Without loss of generality, assume $\mu(L', S', R') \geq \mu(L'', S'', R'')$. Assuming that the separation and combination are done in polynomial time, the imposed constraint on the measure is

$$2 \cdot 2^{\mu_r(R') + \mu_s(S') + B + (1+B) \cdot \log_{1+\epsilon}(\mu_r(R') + \mu_s(S'))}$$
$$\leq 2^{\mu_r(R) + \mu_s(S) + (1+B) \cdot \log_{1+\epsilon}(\mu_r(R) + \mu_s(S))}.$$

To satisfy the constraint, it suffices to constrain that

$$\mu_r(R) + \mu_s(S) \geq (1 + \epsilon)(\mu_r(R') + \mu_s(S')). \tag{24}$$

This is the only constraint involving the size of a separation. It constrains that separating $(\emptyset, S, R)$ to (L', S', R') should reduce $\mu_r(R) + \mu_s(S)$ by a constant factor, namely $1 + \epsilon$.

Branching. Suppose a transformation taking (L, S, R, ℓ) to (L', S', R', ℓ') decreases $\mu_r(R) + \mu_r(L)$ by d. Since the measure includes roughly (and at least) half of $\mu_r(R) + \mu_r(L)$, ideally $\mu_r(R) + \max\left(0, B - \frac{\mu_r(R) - \mu_r(L)}{2}\right)$ decreases by $d/2$. One can show that this is indeed the case for our measure if the the following condition holds:

If $\mu_r(R) - \mu_r(L) > B$, then $\mu_r(R) - \mu_r(R') \geq \mu_r(L) - \mu_r(L')$. (25)

Condition (25) is very natural, expressing that, if the instance is imbalanced or risks becoming imbalanced we would like to make more progress on the large side. Thus, if Condition (25) holds, then the analysis is at least as good as a non-separator based analysis, but with the additional improvement due to the separator branching.

Integration into a standard Measure and Conquer analysis. The Separate, Measure and Conquer analysis will typically be used when the instance has become sufficiently sparse that one can guarantee that a small separator exists. We can view the part played by the Separate, Measure and Conquer analysis as a subroutine with measure μ', and integrate it into any other Measure and Conquer analysis with different measure μ. We only need guarantee that the measure of an instance does not increase when transitioning to the subroutine, by constraining that $\mu'(I) \leq \mu(I)$ for all instances I [12].

4   Counting Dominating Sets

The #DS problem is to compute, for a given graph G, the function d such that $d(k)$ is the number of dominating sets of G of size k. Its current fastest polynomial-space algorithm runs in time $O(1.5673^n)$ [27]. While many algorithms for domination problems rely on a transformation to SET COVER, the current fastest polynomial-space algorithm for subcubic graphs works directly on the input graph and runs in time $O^\star(2^{n/2})$ [19]. We can apply the Separate, Measure and Conquer method to design and analyze faster algorithms for #DS for subcubic graphs and, separately, for general graphs.

Theorem 2. *#DS can be solved in time $3^{n/5+o(n)}$ on subcubic graphs and in time $O(1.5183^n)$ on general graphs, using only polynomial space.*

Our algorithm for subcubic graphs uses a new 3-way branching inspired by the inclusion/exclusion branching of [28]. The algorithm of [19] had running time $O^\star(4^{n/4})$. Our 3-way branching improves its running time bound to $O^\star(3^{n/4}) = O(1.3161^n)$. Using separation improves it further to $3^{n/5+o(n)} = O(1.2458^n)$. Our algorithm for general graphs essentially just adds separation to [27].

5   Conclusions

We have presented a new method to analyze separator-based branching algorithms within the Measure and Conquer framework. It uses a novel kind of measure that amortizes the sudden large gain when an instance decomposes into independent subinstances. The key feature needed to apply the method is that an algorithm eventually reaches instances where small balanced separators can be computed efficiently. This is so for algorithms that reach sparse graphs in their final stages, but could also include cases where the treewidth of the graph is bounded, or where a graph with small treewidth can be reached by branching on a few vertices [10,15,18].

There are problems for which traditional algorithms are already so fast that branching on separators does not seem to offer an advantage. For example, the current fastest algorithm for MAXIMUM INDEPENDENT SET on subcubic graphs runs in $O(1.0836^n)$ time [29], and merely branching on the vertices of the separator would take $2^{n/6+o(n)} = \Omega(1.1225^n)$ time. A second limitation is that Separate, Measure and Conquer subroutines can often be replaced by treewidth-based ynamic programming subroutines [7], leading to the same or smaller running imes; for example, #DS can be solved in time $O(1.5002^n)$ [23]. However, such lgorithms use exponential space.

We believe that the Separate, Measure and Conquer method is widely applicable, but poses fresh challenges, as it presents more choices in the design of lgorithms and more complications in the analysis. It also provides impetus to ooking for other global properties that may be exploited to derive efficient algoithms.

Acknowledgments. The research was supported in part by the DIMACS 2006–010 Special Focus on Discrete Random Systems, NSF grant DMS-0602942. Serge Jaspers is the recipient of an Australian Research Council Discovery Early Career Researcher Award (project number DE120101761) and a Future Fellowship (project number FT140100048). NICTA is funded by the Australian Government through the Department of Communications and the Australian Research Council through the ICT Centre of Excellence Program. The research was done in part at Dagstuhl Seminars 0441 (Exact Complexity of NP-hard problems, 2010) and 13331 (Exponential Algoithms: Algorithms and Complexity Beyond Polynomial Time, 2013), and was prented at the latter.

References

1. Achlioptas, D., Sorkin, G.B.: Optimal myopic algorithms for random 3-SAT. In: Proc. FOCS 2000, pp. 590–600 (2000)
2. Alekhnovich, M., Hirsch, E.A., Itsykson, D.: Exponential lower bounds for the running time of DPLL algorithms on satisfiable formulas. J. Autom. Reasoning **35**(1–3), 51–72 (2005)
3. Biere, A., Sinz, C.: Decomposing SAT problems into connected components. JSAT **2**(1–4), 201–208 (2006)
4. Dechter, R., Mateescu, R.: And/or search spaces for graphical models. Artif. Intell. **171**(2–3), 73–106 (2007)
5. Diestel, R.: Graph Theory. Springer (2010)
6. Edwards, K., McDermid, E.: A general reduction theorem with applications to pathwidth and the complexity of MAX 2-CSP. Algorithmica. (to appear)
7. Fomin, F.V., Gaspers, S., Saurabh, S., Stepanov, A.A.: On two techniques of combining branching and treewidth. Algorithmica **54**(2), 181–207 (2009)
8. Fomin, F.V., Grandoni, F., Kratsch, D.: A measure & conquer approach for the analysis of exact algorithms. J. ACM **56**(5) (2009)
9. Fomin, F.V.: Høie, K.: Pathwidth of cubic graphs and exact algorithms. Inform. Process. Lett. **97**(5), 191–196 (2006)
10. Fomin, F.V., Lokshtanov, D., Misra, N., Saurabh, S.: Planar F-deletion: approximation, kernelization and optimal FPT algorithms. In: Proc. FOCS 2012, pp. 470–479 (2012)
11. Freuder, E.C., Quinn, M.J.: Taking advantage of stable sets of variables in constraint satisfaction problems. In: Proc. IJCAI 1985, pp. 1076–1078 (1985)
12. Gaspers, S.: Exponential Time Algorithms - Structures, Measures, and Bounds. VDM (2010)
13. Gaspers, S., Sorkin, G.B.: A universally fastest algorithm for Max 2-Sat, Max 2-CSP, and everything in between. J. Comput. System Sci. **78**(1), 305–335 (2012)

14. Gaspers, S., Sorkin, G.B.: Separate, Measure and Conquer: Faster algorithms for Max 2-CSP and counting dominating sets (2014). arXiv:1404.0753 [cs.DS]
15. Gaspers, S., Szeider, S.: Strong backdoors to bounded treewidth SAT. In: Proc FOCS 2013, pp. 489–498 (2013)
16. Golovnev, A., Kutzkov, K.: New exact algorithms for the 2-constraint satisfaction problem. Theor. Comput. Sci. **526**, 18–27 (2014)
17. Gottlob, G., Leone, N., Scarcello, F.: A comparison of structural CSP decomposition methods. Artif. Intell. **124**(2), 243–282 (2000)
18. Kim, E.J., Langer, A., Paul, C., Reidl, F., Rossmanith, P., Sau, I., Sikdar, S.: Linear kernels and single-exponential algorithms via protrusion decompositions. In: Fomin, F.V., Freivalds, R., Kwiatkowska, M., Peleg, D. (eds.) ICALP 2013, Part I. LNCS, vol. 7965, pp. 613–624. Springer, Heidelberg (2013)
19. Kneis, Joachim, Mölle, Daniel, Richter, Stefan, Rossmanith, Peter: Algorithms based on the treewidth of sparse graphs. In: Kratsch, Dieter (ed.) WG 2005. LNCS, vol. 3787, pp. 385–396. Springer, Heidelberg (2005)
20. Li, W., van Beek, P.: Guiding real-world SAT solving with dynamic hypergraph separator decomposition. In: Proc. ICTAI 2004, pp. 542–548 (2004)
21. Lipton, R.J., Tarjan, R.E.: Application of a planar separator theorem. In: Proc. FOCS 1977, pp. 162–170 (1977)
22. Monien, B., Preis, R.: Upper bounds on the bisection width of 3- and 4-regular graphs. J. Discrete Algorithms **4**(3), 475–498 (2006)
23. Nederlof, J., van Rooij, J.M.M., van Dijk, T.C.: Inclusion/exclusion meets measure and conquer. Algorithmica **69**(3), 685–740 (2014)
24. Scott, A.D., Sorkin, G.B.: Solving sparse random instances of Max Cut and Max 2-CSP in linear expected time. Comb. Probab. Comput. **15**(1–2), 281–315 (2006)
25. Scott, A.D., Sorkin, G.B.: Linear-programming design and analysis of fast algorithms for Max 2-CSP. Discrete Optim. **4**(3–4), 260–287 (2007)
26. Scott, A.D., Sorkin, G.B.: Polynomial constraint satisfaction problems, graph bisection, and the Ising partition function. ACM Trans. Algorithms **5**(4), Art. 45, 27 (2009)
27. van Rooij, J.M.M.: Polynomial space algorithms for counting dominating sets and the domatic number. In: Calamoneri, T., Diaz, J. (eds.) CIAC 2010. LNCS, vol. 6078, pp. 73–84. Springer, Heidelberg (2010)
28. van Rooij, J.M.M., Nederlof, J., van Dijk, T.C.: Inclusion/exclusion meets measure and conquer. In: Fiat, A., Sanders, P. (eds.) ESA 2009. LNCS, vol. 5757, pp. 554–565. Springer, Heidelberg (2009)
29. Xiao, M., Nagamochi, H.: Confining sets and avoiding bottleneck cases: A simple maximum independent set algorithm in degree-3 graphs. Theor. Comput. Sci. **469**, 92–104 (2013)

Submatrix Maximum Queries in Monge Matrices Are Equivalent to Predecessor Search

Paweł Gawrychowski[1]([✉]), Shay Mozes[2], and Oren Weimann[3]

[1] University of Warsaw, Warsaw, Poland
gawry@mimuw.edu.pl
[2] IDC Herzliya, Herzliya, Israel
smozes@idc.ac.il
[3] University of Haifa, Haifa, Israel
oren@cs.haifa.ac.il

Abstract. We present an optimal data structure for submatrix maximum queries in $n \times n$ Monge matrices. Our result is a two-way reduction showing that the problem is equivalent to the classical predecessor problem in a universe of polynomial size. This gives a data structure of $O(n)$ space that answers submatrix maximum queries in $O(\log \log n)$ time, as well as a matching lower bound, showing that $O(\log \log n)$ query-time is optimal for any data structure of size $O(n \operatorname{polylog}(n))$. Our result settles the problem, improving on the $O(\log^2 n)$ query-time in SODA'12, and on the $O(\log n)$ query-time in ICALP'14.

In addition, we show that partial Monge matrices can be handled in the same bounds as full Monge matrices. In both previous results, partial Monge matrices incurred additional inverse-Ackerman factors.

1 Introduction

Data structures for range queries and for predecessor queries are among the most studied data structures in computer science. Given an $n \times n$ matrix M, a *range maximum* (also called submatrix maximum) data structure can report the maximum entry in any query submatrix (a set of consecutive rows and a set of consecutive columns) of M. Given a set $S \subseteq [0, U)$ of n integers from a polynomial universe U, a *predecessor* data structure can report the predecessor (and successor) in S of any query integer $x \in [0, U)$. In this paper, we prove that these two seemingly unrelated problems are in fact equivalent when the matrix M is a *Monge* matrix.

A full version of this paper can be found as Arxiv preprint arXiv:1502.07663.
P. Gawrychowski, S. Mozes, and O. Weimann—PG is currently holding a post-doctoral position at Warsaw Center of Mathematics and Computer Science. SM and OW partially supported by Israel Science Foundation grant 794/13. SM partially supported by the Israeli ministry of absorption.

© Springer-Verlag Berlin Heidelberg 2015
M.M. Halldórsson et al. (Eds.): ICALP 2015, Part I, LNCS 9134, pp. 580–592, 2015.
DOI: 10.1007/978-3-662-47672-7_47

Range Maximum Queries. A long line of research over the last three decades including [3,9,10,13,20] achieved range maximum data structures of $\tilde{O}(n^2)$ space and $\tilde{O}(1)$ query time[1], culminating with the $O(n^2)$-space $O(1)$-query data structure of Yuan and Atallah [20]. In general matrices, this is optimal since representing the input matrix already requires $\Theta(n^2)$ space. In fact, reducing the additional space to $O(n^2/c)$ is known to incur an $\Omega(c)$ query-time [5] and such tradeoffs can indeed be achieved for any value of c [4,5].

However, in many applications, the matrix M is not stored explicitly but any entry of M can be computed when needed in $O(1)$ time. One such case is when M is a sparse matrix with $N = o(n^2)$ nonzero entries. In this case the problem is known in computational geometry as the *orthogonal range searching* problem on the $n \times n$ grid. Various data structures with $\tilde{O}(N)$-space and $\tilde{O}(1)$-query appear in a long history of results including [2,7,8,11,13]. For a survey on orthogonal range searching see [18]. Another case where the additional space can be made $o(n^2)$ (and in fact even $O(n)$) is when the matrix is a Monge matrix.

Range Maximum Queries in Monge Matrices. A matrix M is Monge if for any pair of rows $i < j$ and columns $k < \ell$ we have that $M[i,k] + M[j,\ell] \geq M[i,\ell] + M[j,k]$. Submatrix maximum queries on Monge matrices have various important applications in combinatorial optimization and computational geometry such as problems involving distances in the plane, and in problems on convex n-gons. See [6] for a survey on Monge matrices and their uses in combinatorial optimization. Submatrix maximum queries on Monge matrices are used in algorithms that efficiently find the largest empty rectangle containing a query point, in dynamic distance oracles for planar graphs, and in algorithms for maximum flow in planar graphs. See [15] for more details.

Given an $n \times n$ Monge matrix M it is possible to obtain compact data structures of only $\tilde{O}(n)$ space that can answer submatrix maximum queries in $\tilde{O}(1)$ time. The first such data structure was given by Kaplan, Mozes, Nussbaum and Sharir [15]. They presented an $O(n \log n)$-space data structure with $O(\log^2 n)$ query time. This was improved in [14] to $O(n)$ space and $O(\log n)$ query time.

Breakpoints and Partial Monge Matrices. Given an $m \times n$ Monge matrix M, let $r(c)$ be the row containing the maximum element in the c-th column of M. It is easy to verify that the $r(\cdot)$ values are monotone, i.e., $r(1) \leq r(2) \leq \ldots \leq r(n)$. Columns c such that $r(c-1) < r(c)$ are called the *breakpoints* of M. A Monge matrix consisting of $m < n$ rows has $O(m)$ breakpoints, which can be found in $O(n)$ time using the SMAWK algorithm [1].

Some applications involve *partial* Monge matrices rather than full Monge matrices. A partial Monge matrix is a Monge matrix where some of the entries are undefined, but the defined entries in each row and in each column are contiguous. The total number of breakpoints in a partial Monge matrix is still $O(m)$ [14], and they can be found in $O(n \cdot \alpha(n))$ time[2] using an algorithm of Klawe and

---

[1] The $\tilde{O}(\cdot)$ notation hides polylogarithmic factors in n.
[2] Here $\alpha(n)$ is the inverse-Ackerman function.

Kleitman [16]. This was used in [14,15] to extend their solutions to partial Monge matrices at the cost of an additional $\alpha(n)$ factor to the query time.[3]

Our Results. In this paper, we fully resolve the submatrix maximum query problem in $n \times n$ Monge matrices by presenting a data structure of $O(n)$ space and $O(\log \log n)$ query time. Consequently, we obtain an improved query time for other applications such as finding the largest empty rectangle containing a query point. We compliment our upper bound with a matching lower bound, showing that $O(\log \log n)$ query-time is optimal for any data structure of size $O(n \operatorname{polylog}(n))$. In fact, implicit in our upper and lower bound is an equivalence between the predecessor problem in a universe of polynomial size and the range maximum query problem in Monge matrices. The upper bound essentially reduces a submatrix query to a predecessor problem, and vice versa, the lower bound reduces the predecessor problem to a submatrix query problem.

Finally, we extend our result to partial Monge matrices with the exact same bounds (i.e., $O(n)$ space and $O(\log \log n)$ query time). Our result is the first to achieve such extension with no overhead.

Techniques. Let M be an $n \times n$ Monge matrix[4]. Consider a full binary tree $\mathcal{T}$ whose leaves are the rows of M. Let M_u be the submatrix of M composed of all rows (i.e., leaves) in the subtree of a node u in $\mathcal{T}$. Both existing data structures for submatrix maximum queries [14,15] store, for each node u in $\mathcal{T}$ a data structure D_u. The goal of D_u is to answer submatrix maximum queries for queries that include an arbitrary interval of columns and *exactly all rows* of M_u. This way, an arbitrary query is covered in [14,15] by querying the D_u structures of $O(\log n)$ canonical nodes of $\mathcal{T}$. An $\Omega(\log n)$ bound is thus inherent for any solution that examines the canonical nodes. We overcome this obstacle by designing a stronger data structure D_u. Namely, one that supports queries that include an arbitrary interval of columns and *a prefix of rows* or *a suffix of rows* of M_u. This way, an arbitrary query can be covered by just two D_us. The idea behind the new design is to efficiently encode the changes in column maxima as we add rows to M_u one by one. Retrieving this information is done using weighted ancestor search and range maximum queries on trees. This is a novel use of these techniques.

For our lower bound, we show that for any set of n integers $S \subseteq [0, n^2)$ there exists an $n \times n$ Monge matrix M such that the predecessor of x in S can be found with submatrix minimum queries on M. The predecessor lower bound of Pătraşcu and Thorup [19] then implies that $O(n \operatorname{polylog}(n))$ space requires $\Omega(\log \log n)$ query time. We overcome two technical difficulties here: First, M should be Monge. Second, there must be an $O(n \operatorname{polylog}(n))$-size representation of M which can retrieve any entry $M[i, j]$ in $O(1)$ time.

Finally, for handling partial Monge matrices, and unlike previous solutions for this case, we do not directly adapt the solution for the full Monge case to partial Monge matrices. Instead we decompose the partial Monge matrix into many full Monge matrices, that can be preprocessed to be queried cumulatively

---

[3] In [15], there was also an additional $\log n$ factor to the space.

[4] We consider $m \times n$ matrices, but for simplicity we sometimes state the results for $n \times n$ matrices.

in an efficient way. This requires significant technical work and careful use of the structure of the decomposition.

Roadmap. In Sect. 2 we present an $O(n \log n)$-space data structure for Monge matrices that answers submatrix maximum queries in $O(\log \log n)$ time. In Sect. 3 we reduce the space to $O(n)$. Our lower bound is given in Sect. 4. The extension to partial Monge matrices, that we believe is a significant contribution of our paper, is deferred to the full version due to lack of space.

2 Data Structure for Monge Matrices

Our goal in this section is to construct, for a given $m \times n$ Monge matrix M, a data structure of size $O(m \log n)$ that answers submatrix maximum queries in $O(\log \log n)$ time. In Sect. 3 we show how to reduce the space from $O(n \log n)$ to $O(n)$ when $m = n$. We will actually show a stronger result, namely the structure allows us to reduce in $O(1)$ time a submatrix maximum query into $O(1)$ predecessor queries on a set consisting of n integers from a polynomial universe.

We denote by $pred(m, n)$ the complexity of a predecessor query on a set of m integers from a universe $\{0, \ldots, n-1\}$. It is well known that there are $O(m)$ data structures achieving $pred(m, n) = \min\{O(\log m), O(\log \log n)\}$.

Recall that a submatrix maximum query returns the maximum $M[i, j]$ over all $i \in [i_0, i_1]$ and $j \in [j_0, j_1]$ for a given $i_0 \leq i_1$ and $j_0 \leq j_1$. We start by answering the easier *subcolumn maximum queries* within these space and time bounds. That is, finding the maximum $M[i, j]$ over all $i \in [i_0, i_1]$ for a given $i_0 \leq i_1$ and j.

We construct a full binary tree $\mathcal{T}$ over the rows of M. Every leaf of the tree corresponds to a single row of M, and every inner node corresponds to the range of rows in its subtree. To find the maximum $M[i, j]$ over all $i \in [i_0, i_1]$ for a given $i_0 \leq i_1$ and j, we first locate the lowest common ancestor (lca) u of the leaves corresponding to i_0 and i_1 in the tree. Then we decompose the query into two parts: one fully within the range of rows M_ℓ of the left child of u, and one fully within the range of rows M_r of the right child of u. The former ends at the last row of M_ℓ and the latter starts at the first row of M_r. We equip every node with two data structures allowing us to answer such simpler subcolumn maximum queries. Because of symmetry (if M is Monge, so is M', where $M'[i, j] = M[n + 1 - i, n + 1 - j]$) it is enough to show how to answer subcolumn maximum queries starting at the first row.

Lemma 1. *Given an $m \times n$ Monge matrix M, a data structure of size $O(m)$ can be constructed in $O(m \log n)$ time to answer in $O(pred(m, n))$ time subcolumn maximum queries starting at the first row of M.*

Proof. Consider queries spanning an *entire* column c of M. To answer such a query, we only need to find the corresponding $r(c)$. If we store the breakpoints of M in a predecessor structure, where every breakpoint c links to its corresponding value of $r(c)$, a query can be answered with a single predecessor search. More precisely, to determine the maximum in the c-th column of M, we locate the largest reakpoint $c' \le c$, and set $r(c) = r(c')$. Hence we can construct a data structure f size $O(m)$ to answer *entire column* maximum queries in $O(pred(m,n))$ time. Let M_i be a Monge matrix consisting of the first i rows of M. By applying the bove reasoning to every M_i separately, we immediately get a structure of size $)(m^2)$ answering subcolumn maximum queries starting at the first row of M in $)(pred(m,n))$ time. We want to improve on this by utilizing the dependency of he structures constructed for different i's. Namely it can be observed that the ist of breakpoints of M_{i+1} is a prefix of the list of breakpoints of M_i to which we .ppend at most one new element. In other words, if the breakpoints of M_i are tored on a stack, we need to pop zero or more elements and push at most one new lement to represent the breakpoints of M_{i+1}. Consequently, instead of storing . separate list for every M_i, we can succinctly describe the content of all stacks vith a single tree T on at most $m + 1$ nodes. For every i, we store a pointer to a iode $s(i) \in T$, such that the ancestors of $s(i)$ (except for the root) are exactly the breakpoints of M_i. Whenever we pop an element from the current stack, we move o the parent of the current node, and whenever we push an element, we create i new node and make it a child of the current node. Initially, the tree consists of ust the root. Every node is labelled with a column number and by construction hese numbers are strictly increasing on any path starting at the root (the root is abelled with $-\infty$). Therefore, a predecessor search for j among the breakpoints of M_i reduces to finding the leftmost ancestor of $s(i)$ whose label is at most j. This is known as the *weighted ancestor* problem. Weighted ancestor queries on i tree of size $O(m)$ are equivalent to predecessor searching on a number of sets of $O(m)$ total size [17][5], achieving the claimed space and query time bounds.

   To finish the proof, we need to bound the construction time. The bottleneck s constructing the tree T. Let $c_1 < c_2 < \ldots < c_k$ for some $k \le i$ be the break-points of M_i. As long as $M[i + 1, c_k] \ge M[r(c_k), c_k]$ we decrease k by one, i.e., remove the last breakpoint. This process is repeated $O(m)$ times in total. If $k = 0$ we create a new breakpoint $c_1 = 1$. If $k \ge 1$ and $M[i + 1, c_k] < M[r(c_k), c_k]$, we check if $M[i + 1, n] \ge M[r(c_k), n]$. If so, we need to create a new breakpoint. To his end, we need to find the smallest j such that $M[i + 1, j] \ge M[r(c_k), j]$. This can be done in $O(\log n)$ using binary search. Consequently, T can be constructed n $O(m \log n)$ time. Then augmenting it with a weighted ancestor structure takes $)(m)$ time.   □

   We apply Lemma 1 twice to every node of the full version tree $\mathcal{T}$. Once for subcolumn maximum queries starting at the first row and once for queries end-ng at the last row. Since the total size of all structures at the same level of the ree is $O(m)$, the total size of our subcolumn maximum data structure becomes $)(m \log m)$, and it can be constructed in $O(m \log m \log n)$ time to answer queries in $O(pred(m,n))$ time. Hence we have proved the following.

Theorem 1. *Given an $m \times n$ Monge matrix M, a data structure of size $)(m \log m)$ can be constructed in $O(m \log m \log n)$ time to answer subcolumn maximum queries in $O(pred(m,n))$ time.*

---

[5] Technically, the reduction adds $O(\log^* m)$ to the query time, but this can be avoided.

By symmetry (a transpose of a Monge matrix is Monge) we can answer sub-row maximum queries (where the query is a single row and a range of columns) in $O(pred(n, m))$ time. We are now ready to tackle general submatrix maximum queries.

At a high level, the idea is identical to the one used for subcolumn maximum queries: we construct a full binary tree $\mathcal{T}$ over the rows of M, where every node corresponds to a range of rows. To find maximum $M[i, j]$ over all $i \in [i_0, i_1]$ and $j \in [j_0, j_1]$ for a given $i_0 \leq i_1$ and $j_0 \leq j_1$, we locate the lowest common ancestor of the leaves corresponding to i_0 and i_1 and decompose the query into two parts, the former ending at the last row of M_ℓ and the latter starting at the first row of M_r. Every node is equipped with two data structures allowing us to answer submatrix maximum queries starting at the first row or ending at the last row. As before, it is enough to show how to answer submatrix maximum queries starting at the first row.

Lemma 2. *Given an $m \times n$ Monge matrix M, and a data structure that answers subrow maximum queries on M in $O(pred(n, m))$ time, one can construct in $O(m \log m)$ time a data structure consuming $O(m)$ additional space, that answers submatrix maximum queries starting at the first row of M in $O(pred(m, n) + pred(n, m))$ time.*

Proof. We extend the proof of Lemma 1. Let $c_1 < c_2 < \ldots < c_k$ be the breakpoints of M stored in a predecessor structure. For every $i \geq 2$ we precompute and store the value $m_i = \max_{j \in [c_{i-1}, c_i)} M[r(c_{i-1}), j]$. These values are augmented with a (one dimensional) range maximum query data structure. To begin with, consider a submatrix maximum query starting at the first row of M and ending at the last row of M, i.e., we need to calculate the maximum $M[i, j]$ over all $i \in [1, m]$ and $j \in [j_0, j_1]$. We find in $O(pred(m, n))$ the successor of j_0, denoted c_i, and the predecessor of j_1, denoted $c_{i'}$. There are three possibilities:

1. The maximum is reached for $j \in [j_0, c_i)$,
2. The maximum is reached for $j \in [c_i, c_{i'})$,
3. The maximum is reached for $j \in [c_{i'}, j_1]$.

The first and the third possibilities can be calculated with subrow maximum queries in $O(pred(n, m))$, because both ranges span an interval of columns and a single row. The second possibility can be calculated with a range maximum query on the range $(i, i']$. Consequently, we can construct a data structure of size $O(m)$ to answer such submatrix maximum queries in $O(pred(m, n) + pred(n, m))$ time.

The above solution can be generalized to queries that start at the first row of M but do not necessarily end at the last row of M. This is done by considering the Monge matrices M_i consisting of the first i rows of M. For every such matrix, we need a predecessor structure storing all of its breakpoints, and additionally a range maximum structure over their associated values. Hence now we need to construct a similar tree T as in Lemma 1 on $O(m)$ nodes, but now every node has both a weight and a value. The weight of a node is the column number of the corresponding breakpoint c_k, and the value is its m_k (or undefined if $k = 1$). As in Lemma 1, the breakpoints of M_i are exactly the ancestors of the node $s(i)$. Note that every m_k is defined in terms of c_{k-1} and c_k, but this is not a problem because the predecessor of a breakpoint does not change during the whole construction. We maintain a weighted ancestor structure using the weights (in order to find i and $c_{i'}$ in $O(pred(m, n))$ time), and a *generalized range maximum structure* using the values. A generalized range maximum structure of a tree T, given two query nodes u and v, returns the maximum value on the unique u-to-v path in T. It can be implemented in $O(m)$ space and $O(1)$ query time after $O(m \log m)$ preprocessing [10] once we have the values. The values can be computed with subrow maximum queries in $O(m \cdot pred(n, m)) = O(m \log m)$ total time.   □

By applying Lemma 2 twice to every node of the full binary tree T, we construct in $O(m \log^2 m)$ time a data structure of size $O(m \log m)$ to answer submatrix maximum queries in $O(pred(m, n) + pred(n, m))$ time. In order to apply Lemma 2 to a node of T we need a subrow maximum query data structure for the corresponding rows of the matrix M. Note, however, that a single subrow maximum query data structure for M can be used for all nodes of T.

Theorem 2. *Given an $m \times n$ Monge matrix M, and a data structure answering subrow maximum queries on M in $O(pred(n, m))$ time, one can construct in $O(m \log^2 m)$ time a data structure taking $O(m \log m)$ additional space, that answers submatrix maximum queries on M in $O(pred(m, n) + pred(n, m))$ time.*

By combining Theorem 1 with Theorem 2, given an $n \times n$ Monge matrix M, a data structure of size $O(n \log n)$ can be constructed in $O(n \log^2 n)$ time to answer submatrix maximum queries in $O(pred(n, n))$ time.

3   Obtaining Linear Space

In this section we show how to decrease the space of the data structure presented in Sect. 2 to be linear. We extend the idea developed in our previous paper [14]. The previous linear space solution was based on partitioning the matrix M into n/x matrices $M_1, M_2, \ldots, M_{n/x}$, where each M_i is a *slice* of M consisting of $x = \log n$ consecutive rows. Then, instead of working with the matrix M, we worked with the $(n/x) \times n$ matrix M', where $M'[i, j]$ is the maximum entry in the j-th column of M_i.

Subcolumn Queries. Consider a subcolumn query. Suppose the query is entirely contained in some M_i. This means it spans less than $x = \log n$ rows. In [14], since the desired query time was $O(\log n)$, a query simply inspected all elements of the subcolumn. In our case however, since the desired query time is only $O(\log \log n)$, we apply the above partitioning scheme twice. We explain this now.

We start with the following lemma, that provides an efficient data structure for queries consisting of a single column and *all* rows in rectangular matrices. The statement of the lemma was taken almost verbatim from the previous solution [14]. Its query time was originally stated in terms of query to a predecessor structure, but here we prefer to directly plug in the bounds implied by atomic heaps [12] (which support predecessor searches in constant time provided x is $O(\log n)$). This requires only an additional $O(n)$ time and space preprocessing.

Lemma 3 ([14]). *Given an $x \times n$ Monge matrix, a data structure of size $O(x)$ can be constructed in $O(x \log n)$ time to answer entire-column maximum queries in $O(1)$ time, if $x = O(\log n)$.*

Our new subcolumn data structure is summarized in the following theorem. It uses the above lemma and two applications of the partitioning scheme.

Theorem 3. *Given an $m \times n$ Monge matrix M, a data structure of size $O(m)$ can be constructed in $O(m \log n)$ time to answer subcolumn maximum queries in $O(\log \log(n + m))$ time.*

Proof. We first partition M into n/x matrices $M_1, M_2, \ldots, M_{n/x}$, where $x = \log m$. Every M_i is a slice of M consisting of x consecutive rows. Next, we partition every M_i into x/x' matrices $M_{i,1}, M_{i,2}, \ldots, M_{i,x'}$, where $x' = \log \log m$. Every $M_{i,j}$ is a slice of M_i consisting of x' consecutive rows (without loss of generality, assume that x divides m and x' divides x). Now we define a new $(m/x) \times n$ matrix M', where $M'[i, j]$ is the maximum entry in the j-th column of M_i. Similarly, for every M_i we define a new $(x/x') \times n$ matrix M_i', where $M_i'[j, k]$ is the maximum entry in the k-th column of $M_{i,j}$.

We apply Lemma 3 on every M_i and $M_{i,j}$ in $O(m \log n)$ total time and $O(m)$ total space, so that any $M'[i, j]$ or $M_i'[j, k]$ can be retrieved $O(1)$ time. Furthermore, it can be easily verified that M' and all M_i's are also Monge. Therefore, we can apply Theorem 1 on M' and every M_i'. The total construction time is $O((m/x) \log(m/x) \log n + (m/x)(x/x') \log(x/x') \log n) = O(m \log n)$, and the total size of all structures constructed so far is $O((m/x) \log(m/x) + (m/x)(x/x') \log(x/x')) = O(m)$.

Now consider a subcolumn maximum query. If the range of rows is fully within a single $M_{i,j}$, the query can be answered naively in $O(x') = O(\log \log m)$ time. Otherwise, if the range of rows is fully within a single M_i, the query can be decomposed into a prefix fully within some $M_{i,j}$, an infix corresponding to a range of rows in M_i', and a suffix fully within some $M_{i,j'}$. The maximum in the prefix and the suffix can be computed naively in $O(x') = O(\log \log m)$ time, and the maximum in the infix can be computed in $O(\log \log n)$ time using the structure constructed for M_i'. Finally, if the range of rows starts inside some M_i and ends inside another $M_{i'}$, the query can be decomposed into two queries fully within M_i and $M_{i'}$, respectively, which can be processed in $O(\log \log n)$ time as explained before, and an infix corresponding to a range of rows of M'. The maximum in the infix can be computed in $O(\log \log n)$ time using the structure constructed for M'. □

Submatrix Queries. We are ready to present the final version of our data structure. It is based on two applications of the partitioning scheme, and an additional trick of transposing the matrix.

Theorem 4. *Given an $n \times n$ Monge matrix M, a data structure of size $O(n)$ can be constructed in $O(n \log n)$ time to answer submatrix maximum queries in $O(\log \log n)$ time.*

Proof. We partition M as described in the proof of Theorem 3, i.e., M is partitioned into n/x matrices $M_1, M_2, \ldots, M_{n/x}$, where $x = \log n$, and every M_i is then partitioned into x/x' matrices $M_{i,1}, M_{i,2}, \ldots, M_{i,x'}$, where $x' = \log \log n$. Then we define smaller Monge matrices M' and M_i', and provide $O(1)$ time access to their entries with Lemma 3. We apply Theorem 3 to the transpose of M' to get a subrow maximum query data structure for M'. This takes $O(n)$ space and $O(n \log n)$ time. With this data structure we can apply Theorem 2 on M', which takes an additional $O(\frac{n}{\log n} \log \frac{n}{\log n}) = O(n)$ space and $O(n \log n)$ time. We would have liked to apply Theorem 3 to the transpose of all M_i' as well, but this would require $O(n)$ space for each matrix, which we cannot afford. Since we do not have subrow maximum query data structure for the M_i's, we cannot apply Theorem 2 to them directly. However, note that the subrow maximum query data structure is used in Theorem 2 in two ways (see the proof of Lemma 2). The first use is in directly finding the subrow maximum in cases 1 and 3 in the proof of Lemma 2. In the absence of the subrow structure, we can still report the two rows containing the candidate maximum, although not the maximum itself. The second use is in computing the values for the generalized range maximum structure required to handle case 2 in that proof. In this case, we do not really need the fast query of the data structure of Theorem 3, and can use instead the slower linear space data structure from [14, Lemma 2] to compute the values in $O(n \log n)$ time. Thus, we can apply Theorem 2 to each M_i', and get at most two candidate rows of M_i' (from cases 1 and 3), and one candidate entry of M_i' (from case 2), with the guarantee that the submatrix maximum is among these candidates.

We repeat the above preprocessing on the transpose of M. Now consider a submatrix maximum query. If the range of rows starts inside some M_i and ends inside another $M_{i'}$, the query can be decomposed into two queries fully within M_i and $M_{i'}$, respectively, and an infix corresponding to a range of rows of M'. The maximum in the infix can be computed in $O(\log \log n)$ time using the structure constructed for M'. Consequently, it is enough to show how to answer a query in $O(\log \log n)$ time when the range of rows is fully within a single M_i. In such case, if the range of rows starts inside some $M_{i,j}$ and ends inside another $M_{i,j'}$, the query can be decomposed into a prefix fully within $M_{i,j}$, an infix corresponding to a range of rows in M_i' and a suffix fully within some $M_{i,j'}$. As we explained above, even though we cannot locate the maximum in the infix exactly, we can isolate at most 2 rows (plus a single entry) of M_i', such that the maximum lies in one of these rows. Each row of M_i' corresponds to a range of rows fully inside some $M_{i,j}$. Consequently, we reduced the query in $O(\log \log n)$ time to a constant number of queries such that the range of rows in each query is fully within a single $M_{i,j}$. Since each $M_{i,j}$ consists of $O(\log \log n)$ rows of M, we have identified, in $O(\log \log n)$ time, a set of $O(\log \log n)$ rows of M that contain the desired submatrix maximum.

Now we repeat the same procedure on the transpose of M to identify a set of $O(\log \log n)$ columns of M that contain the desired submatrix maximum. Since a submatrix of a Monge matrix is also Monge, the submatrix of M corresponding to these sets of candidate rows and columns is an $O(\log \log n) \times O(\log \log n)$

Monge matrix. By running the SMAWK algorithm [1] in $O(\log\log n)$ time or this small Monge matrix, we can finally determine the answer.   ⌐

4   Lower Bound

A predecessor structure stores a set of n integers $S \subseteq [0, U)$, so that given x we can determine the largest $y \in S$ such that $y \leq x$. As shown by Pǎtraşcu and Thorup [19], for $U = n^2$ any predecessor structure consisting of $O(n\,\text{polylog}(n))$ words needs $\Omega(\log\log n)$ time to answer queries, assuming that the word size is $\Theta(\log n)$. We will use their result to prove that our structure is in fact optimal.

Given a set of n integers $S \subseteq [0, n^2)$ we want to construct $n \times n$ Monge matrix M such that the predecessor of any x in S can be found using one submatrix minimum query on M and $O(1)$ additional time (to decide which query to ask and then return the final answer). Then, assuming that for any $n \times n$ Monge matrix there exists a data structure of size $O(n\,\text{polylog}(n))$ answering submatrix minimum queries in $o(\log\log n)$ time, we can construct a predecessor structure of size $O(n\,\text{polylog}(n))$ answering queries in $o(\log\log n)$ time, which is not possible. The technical difficulty here is twofolds. First, M should be Monge. Second, we are working in the indexing model, i.e., the data structure for submatrix minimum queries can access the matrix. Therefore, for the lower bound to carry over, M should have the following property: there is a data structure of size $O(n\,\text{polylog}(n))$ which retrieves any $M[i, j]$ in $O(1)$ time. Guaranteeing that both properties hold simultaneously is not trivial.

Before we proceed, let us comment on the condition $S \subseteq [0, n^2)$. While quadratic universe is enough to invoke the $\Omega(\log\log n)$ lower bound for structures of size $O(n\,\text{polylog}(n))$, our reduction actually implies that even for larger polynomially bounded universes, i.e., $S \subseteq [0, n^c)$, for any fixed c, it is possible to construct $n \times n$ Monge matrix M such that the predecessor of x in S can be found with $O(1)$ submatrix minimum queries on M and $O(1)$ additional time (and, as previously, any $M[i, j]$ can be retrieved in $O(1)$ time with a structure of size $O(n)$). This is because any predecessor queries on a set of n integers $S \subseteq [0, n^c)$ can be reduced in $O(1)$ time to $O(1)$ predecessor queries on a set of n integers $S' \subseteq [0, n^2)$ with a structure of size $O(n)$. See full version of this paper.

The following propositions are easy to verify:

Proposition 1. *A matrix M is Monge iff $M[i, j] + M[i + 1, j + 1] \leq M[i + 1, j] + M[i, j + 1]$ for all i, j such that all these entries are defined.*

Proposition 2. *If a matrix M is Monge, then for any vector H the matrix M', where $M'[i, j] = M[i, j] + H[j]$ for all i, j, is also Monge.*

Theorem 5. *For any set of n integers $S \subseteq [0, n^2)$, there exists a data structure of size $O(n)$ returning any $M[i, j]$ in $O(1)$ time, where M is a Monge matrix such that the predecessor of x can be found using $O(1)$ time and one submatrix minimum query on M.*

Proof. We partition the universe $[0, n^2)$ into n parts $[0, n), [n, 2n), \ldots$ The i-th part $[i \cdot n, (i+1) \cdot n)$ defines a Monge matrix M_i consisting of $|S \cap [i \cdot n, (i+1) \cdot n)|$ rows and n columns. The idea is to encode the predecessor of $x \in [0, n^2)$ by the minimum element in the $(x \bmod n + 1)$-th column of $M_{\lfloor x/n \rfloor}$. We first describe how these matrices are defined, and then show how to stack them together.

Consider any $0 \le i < n$. Every element in $S \cap [i \cdot n, (i+1) \cdot n) = \{a_1, a_2, \ldots, a_k\}$ has a unique corresponding row in M_i. Let $a_j = i \cdot n + a'_j$, so that $a'_1 < a'_2 < \ldots < a'_k$ and $a'_j \in [0, n)$ for all j, and also define $a'_{k+1} = n$. We describe an incremental construction of M_i. For technical reasons, we start with an artificial top row containing $1, 2, 3, \ldots, n$. Then we add the rows corresponding to $a'_1, a'_2, \ldots, a'_k$. The row corresponding to a'_j consists of three parts. The middle part starts at the $(a'_j + 1)$-th column, ends at the a'_{j+1}-th column, and contains only 1's. The elements in the left part decrease by 1 and end with 2 at the a'_j-th column, similarly the elements in the right part (if any) start with 2 at the $(a'_{j+1} + 1)$-th column and increase by 1. Formally, the k-th element of the $(j + 1)$-th row, denoted $M_i[j + 1, k]$, is defined as follows.

$$
M_i[j + 1, k] = \begin{cases} a'_j - k + 2 & \text{if } k \in [1, a'_j] \\ 1 & \text{if } k \in [a'_j + 1, a'_{j+1}] \\ k - a'_{j+1} + 1 & \text{if } k \in [a'_{j+1} + 1, n] \end{cases} \tag{1}
$$

Finally, we end with an artificial bottom row containing $n, n - 1, \ldots, 1$. We need to argue that every M_i is Monge. By Proposition 1, it is enough to consider every pair of adjacent rows r_1, r_2 there. Define $r'_1[j] = r_1[j] - r_1[j - 1]$ and similarly $r'_2[j] = r_2[j] - r_2[j - 1]$. To prove that M_i is Monge, it is enough to argue that $r'_2[j] \ge r'_1[j]$ for all $j \ge 2$. By construction, both r'_1 and r'_2 are of the form $-1, -1, \ldots, -1, 0, 0, \ldots, 0, 1, 1, \ldots, 1$, and all 0's in r'_2 are on the right of all 0's in r'_1. Therefore, M_i is Monge.

Now one can observe that the predecessor of $x \in [0, n^2)$ can be found by looking at the $(x \bmod n+1)$-th column of $M_{\lfloor x/n \rfloor}$. We check if $x < a_1$, and if so return the predecessor of a_1 in the whole S. This can be done in $O(1)$ time and $O(n)$ additional space by explicitly storing a_1 and its predecessor for every i. Otherwise we know that the predecessor of x is a_j such that $x \bmod n \in [a'_j, a'_{j+1})$, and, by construction, we only need to find $j \in [1, k]$ such that the $(x \bmod n + 1)$-th element of row $j + 1$ in M_i is 1. This is exactly a subcolumn minimum query.

We cannot simply concatenate all M_i's to form a larger Monge matrix. We use Proposition 2 instead. Initially, we set $M = M_0$. Then we consider every other M_i one-by-one maintaining invariant that the current M is Monge and its last row is $n, n - 1, \ldots, 1$. In every step we add the vector $H = [-n + 1, -n + 3, \ldots, n - 1]$ to the current matrix M, obtaining a matrix M' whose last row is $1, 2, \ldots, n$. By Proposition 2, M' is Monge. Then we can construct the new M by appending M_i without its first row to M'. Because the first row of M_i is also $1, 2, \ldots, n$, the new M is also Monge. Furthermore, because we add the same value to all elements in the same column of M_i, answering subcolumn minimum queries on M_i can be done with subcolumn minimum queries on the final M.

We need to argue that elements of M can be accessed in $O(1)$ using a data structure of size $O(1)$. To retrieve $M[j, k]$, first we lookup in $O(1)$ time the appropriate M_i from which it originates. This can be preprocessed and stored for every j in $O(n)$ total space and allows us to reduce the question to retrieving $M_i[j', k]$. Because Proposition 2 is applied exactly $n - 1 - i$ times after appending M_i to the current M, then we can return $M_i[j', k] + (n - 1 - i)H[k]$. To find $M_i[j', k]$, we just directly use Eq. 1, which requires only storing $a'_1, a'_2, \ldots, a'_n$ in $O(n)$ total space.   □

References

1. Aggarwal, A., Klawe, M.M., Moran, S., Shor, P., Wilber, R.: Geometric applications of a matrix-searching algorithm. Algorithmica **2**(1), 195–208 (1987)
2. Alstrup, S., Brodal, G.S., Rauhe, T.: New data structures for orthogonal range searching. In: 41st FOCS, pp. 198–207 (2000)
3. Amir, A., Fischer, J., Lewenstein, M.: Two-dimensional range minimum queries. In: Ma, B., Zhang, K. (eds.) CPM 2007. LNCS, vol. 4580, pp. 286–294. Springer, Heidelberg (2007)
4. Brodal, G.S., Davoodi, P., Lewenstein, M., Raman, R., Srinivasa Rao, S.: Two dimensional range minimum queries and Fibonacci lattices. In: Epstein, L., Ferragina, P. (eds.) ESA 2012. LNCS, vol. 7501, pp. 217–228. Springer, Heidelberg (2012)
5. Brodal, G.S., Davoodi, P., Rao, S.S.: On space efficient two dimensional range minimum data structures. In: de Berg, M., Meyer, U. (eds.) ESA 2010, Part II. LNCS, vol. 6347, pp. 171–182. Springer, Heidelberg (2010)
6. Burkard, R.E., Klinz, B., Rudolf, R.: Perspectives of Monge properties in optimization. Discrete Appl. Math. **70**, 95–161 (1996)
7. Chan, T.M., Larsen, K.G., Pătraşcu, M.: Orthogonal range searching on the RAM, revisited. In: 27th SOCG, pp. 354–363 (2011)
8. Chazelle, B.: A functional approach to data structures and its use in multidimensional searching. SIAM Journal on Computing **17**, 427–462 (1988)
9. Chazelle, B., Rosenberg, B.: Computing partial sums in multidimensional arrays. In: 5th SOCG, pp. 131–139 (1989)
10. Demaine, E.D., Landau, G.M., Weimann, O.: On Cartesian trees and range minimum queries. Algorithmica **68**(3), 610–625 (2014)
11. Farzan, A., Munro, J.I., Raman, R.: Succinct indices for range queries with applications to orthogonal range maxima. In: Czumaj, A., Mehlhorn, K., Pitts, A., Wattenhofer, R. (eds.) ICALP 2012, Part I. LNCS, vol. 7391, pp. 327–338. Springer, Heidelberg (2012)
12. Fredman, M.L., Willard, D.E.: Trans-dichotomous algorithms for minimum spanning trees and shortest paths. J. Comput. Syst. Sci. **48**(3), 533–551 (1994)
13. Gabow, H., Bentley, J.L., Tarjan, R.E.: Scaling and related techniques for geometry problems. In: 16th STOC, pp. 135–143 (1984)
14. Gawrychowski, P., Mozes, S., Weimann, O.: Improved submatrix maximum queries in monge matrices. In: Esparza, J., Fraigniaud, P., Husfeldt, T., Koutsoupias, E. (eds.) ICALP 2014. LNCS, vol. 8572, pp. 525–537. Springer, Heidelberg (2014)

15. Kaplan, H., Mozes, S., Nussbaum, Y., Sharir, M.: Submatrix maximum queries in Monge matrices and Monge partial matrices, and their applications. In: 23rd SODA, pp. 338–355 (2012)
16. Klawe, M.M., Kleitman, D.J.: An almost linear time algorithm for generalized matrix searching. SIAM Journal Discret. Math. **3**(1), 81–97 (1990)
17. Kopelowitz, T., Lewenstein, M.: Dynamic weighted ancestors. In: 18th SODA, pp. 565–574 (2007)
18. Nekrich, Y.: Orthogonal range searching in linear and almost-linear space. Comput. Geom. **42**(4), 342–351 (2009)
19. Pătraşcu, M., Thorup, M.: Time-space trade-offs for predecessor search. In: 38th STOC, pp. 232–240 (2006)
20. Yuan, H., Atallah, M.J.: Data structures for range minimum queries in multidimensional arrays. In: 21st SODA, pp. 150–160 (2010)

Optimal Encodings for Range Top-k, Selection, and Min-Max

Paweł Gawrychowski[1] and Patrick K. Nicholson[2]([✉])

[1] Institute of Informatics, University of Warsaw, Warsaw, Poland
[2] Max-Planck-Institut für Informatik, Saarbrücken, Germany
pnichols@mpi-inf.mpg.de

Abstract. We consider encoding problems for range queries on arrays. In these problems the goal is to store a structure capable of recovering the answer to all queries that occupies the information theoretic minimum space possible, to within lower order terms. As input, we are given an array $A[1..n]$, and a fixed parameter $k \in [1, n]$. A *range top-k* query on an arbitrary range $[i, j] \subseteq [1, n]$ asks us to return the ordered set of indices $\{\ell_1, ..., \ell_k\}$ such that $A[\ell_m]$ is the m-th largest element in $A[i..j]$, for $1 \le m \le k$. A *range selection* query for an arbitrary range $[i, j] \subseteq [1, n]$ and query parameter $k' \in [1, k]$ asks us to return the index of the k'-th largest element in $A[i..j]$. We completely resolve the space complexity of both of these heavily studied problems—to within lower order terms—for all $k = o(n)$. Previously, the constant factor in the space complexity was known only for $k = 1$. We also resolve the space complexity of another problem, that we call *range min-max*, in which the goal is to return the indices of both the minimum and maximum elements in a range.

1 Introduction

Many important algorithms make use of range queries over arrays of values as subroutines [14,17]. As a prime example, text indexes that support pattern matching queries often maintain an array storing the lengths of the longest common prefixes between consecutive suffixes of the text. During a search for a pattern this array is queried in order to find the position of the minimum value in a given range. That is, a subroutine is needed that can preprocess an array A in order to answer *range minimum queries*. Formally, as input to such a query we are given a range $[i, j] \subseteq [1, n]$, and wish to return the index $k = \arg\min_{i \le \ell \le j} A[\ell]$. In text indexing applications memory is often the constraining factor, so the question of how many bits are needed to answer range minimum queries has been heavily studied. After a long line of research (see [2,16]), it has been determined that such queries can be answered in constant time, by storing a data structure of size $2n + o(n)$ bits [7]. Furthermore, this space bound is optimal to within lower order terms (see [7, Sec. 1.1.2]). The interesting thing is that the space does not depend on the number of bits required to store individual

P. Gawrychowski—Currently holding a post-doctoral position at Warsaw Center of Mathematics and Computer Science.

© Springer-Verlag Berlin Heidelberg 2015
M.M. Halldórsson et al. (Eds.): ICALP 2015, Part I, LNCS 9134, pp. 593–604, 2015.
DOI: 10.1007/978-3-662-47672-7_48

elements of the array A. After constructing the data structure we can discard the array A, while still retaining the ability to answer range minimum queries.

Results of this kind, where it is shown that the solutions to all queries can be stored using less space than is required to store the original array, fall into the category of *encodings*, and, more generally, *succinct* data structures [11]. Specifically, given a set of combinatorial objects χ we wish to represent an arbitrary member of χ using $\lg |\chi| + o(\lg |\chi|)$ bits[1], while still supporting queries, if possible. If queries can be supported by the representation then we refer to it as a data structure, but if not, then we refer to it as an encoding. For the case of range minimum queries or range maximum queries, the set χ turns out to be *Cartesian trees*, which were introduced by Vuillemin [18]. For a given array A, the Cartesian tree encodes the solution to all range minimum queries, and similarly, if two arrays have the same solutions to all range minimum queries, then their Cartesian trees are identical [7].

Recently, there has been a lot of interest the following two problems, that generalize range maximum queries in two different ways. The input to each of the following problems is an array $A[1..n]$, that we wish to preprocess into an encoding occupying as few bits as possible, such that the answers to all queries are still recoverable. We assume a value $k \geq 1$ is fixed at preprocessing time.

– **Range top-k:** Given an arbitrary query range $[i, j] \subseteq [1, n]$ and $k' \in [1, k]$, return the indices of the k' largest values in $[i, j]$. This problem is the natural generalization of range maximum queries and has been the focus of a several papers, leading to asymptotically optimal lower and upper space bounds of $\Omega(n \lg k)$ and $\mathcal{O}(n \lg k)$ bits, proved by Grossi et al. [10] and Navarro, Raman, and Rao [15], respectively. The latter upper bound is a data structure that can answer range top-k' queries in optimal $\mathcal{O}(k')$ time.

– **Range k-selection:** Given an arbitrary query range $[i, j] \subseteq [1, n]$ and $k' \leq k$, return the index of the k'-th largest value in $[i, j]$. This problem was studied in a series of recent papers (see [8] and [3] for further references), culminating in data structures that occupy a linear number of words, and can answer queries in $\mathcal{O}(\lg k' / \lg \lg n + 1)$ time [4]. This query time matches a cell-probe lower bound for near-linear space data structures [12]. It is straightforward to see that any encoding of range top-k queries is also an encoding for range k-selection queries, though the question of how much time is required during a query remains unclear [15]. Very recently, Navarro, Raman, and Rao [15] described a data structure that can be used to answer range k-selection queries in optimal $\mathcal{O}(\lg k' / \lg \lg n + 1)$ time [15], and, like the range top-k data structure, occupies $\mathcal{O}(n \lg k)$ bits of space.

Our Results. We present the first space-optimal encodings to range top-k— and therefore range selection also—as well as a new problem that we call *range min-max*, in which the goal is to return the indices of both the minimum and maximum element in the array. We emphasize that, on their own, the encodings

---

[1] We use $\lg x$ to denote $\log_2 x$.

Table 1. Old and new results. Both upper and lower bounds are expressed in bits. Our bounds make use of the binary entropy function $H(x) = x \lg(\frac{1}{x}) + (1 - x) \lg(\frac{1}{1-x})$. For the entry marked with a † the claimed bound holds when $k = o(n)$.

Ref.	Query	Lower Bound	Upper Bound	Query Time
[7]	max	$2n - \Theta(\lg n)$	$2n + o(n)$	$\mathcal{O}(1)$
[10,15]	top-k	$\Omega(n \lg k)$	$\mathcal{O}(n \lg k)$	$\mathcal{O}(k')$
[5]	top-2	$2.656n - \Theta(\lg n)$	$3.272n + o(n)$	$\mathcal{O}(1)$
Thm. 1, 2	min-max	$3n - \Theta(\lg(n))$	$3n + o(n)$	$\mathcal{O}(1)$
Thm. 3, 4	top-2	$3nH(\frac{1}{3}) - \Theta(\text{polylog}(n))$	$3nH(\frac{1}{3}) + o(n)$	—
Thm. 3, 4	top-k	$(k+1)nH(\frac{1}{k+1})(1 - o(1))$†	$(k+1)nH(\frac{1}{k+1}) + o(n)$	—

for range top-k and selection do not support queries efficiently: they merely store the solutions to all queries in a compressed form. However, our encoding for range min-max can be augmented with $o(n)$ additional bits of data to create a data structure that supports queries in $\mathcal{O}(1)$ time. Furthermore, even without query support, our encodings for range top-k and selection address a problem posed in the papers of Grossi et al. [10] and Navarro et al. [15].

In Table 1 we present a summary of previous and new results. Prior to this work, the only value for which the exact coefficient of n was known was the case in which $k = 1$ (i.e., range maximum queries). For even $k = 2$ the best previous estimate was that the coefficient of n is between 2.656 and 3.272 [5]. The lower bound of 2.656 was derived using generating functions and an extensive computational search [5]. In contrast, our method is purely combinatorial and gives the exact coefficient for all $k = o(n)$. For $k = 2, 3, 4$ the coefficients are (rounding up) 2.755, 3.245, and 3.610, respectively.

As mentioned above, a negative aspect of our encodings is that they appear to be somewhat difficult to use as the basis for a data structure. However, in the full version [9], we present a data structure based on our encoding that *nearly* matches the optimal space bound. Explicitly, we can achieve a space bound of $(k + 1.5)nH(\frac{1.5}{k+1.5}) + o(n \lg k)$ bits with query time $\mathcal{O}(\text{poly}(k \lg n))$. Thus, our data structure achieves space much closer to the optimal bound than the previous best result [15], but the query time is worse. We leave the following data structure problem open: how can range top-k and selection queries be supported with optimal query time using space matching our encodings (to within lower order terms)?

Finally, we wish to point out that although our formulation of the range top-k problem returns the indices in sorted order, the constant factor in our lower bound also holds for the *unsorted* version, in which we return the indices in an arbitrary order, provided $k = o(n)$. This follows since any encoding strategy for unsorted range top-k can be used to construct a sorted top-k encoding, by padding the end of the input array with $k - 1$ values larger than any other. The unsorted encoding of this padded array can be used to infer the solution to an arbitrary sorted top-k query $[i, j]$ by examining the solutions to queries $[i, j], [i, j + 1], ..., [i, n + k - 1]$: see the full version for details [9].

Discussion of Techniques and Road Map. Prior work for top-k, for $k \geq 2$, focused on encoding a decomposition of the array, called a shallow cutting [10, 15]. Since shallow cuttings are a general technique used to solve many other range searching problems [12, 13], these previous works [10, 15] required additional information beyond storing the shallow cutting in order to recover the answers to top-k queries. Furthermore, in these works the exact constant factor is not disclosed, though we estimate it to be at least twice as large as the bounds we present. For the specific case of range top-2 queries a different encoding has been proposed based on *extended Cartesian trees* [5]. In contrast to both of the previous approaches, our encoding is based the approach of Fischer and Heun [7], who describe what is called a 2D min-heap (resp. max-heap) in order to encode range minimum queries (resp. range maximum queries). We begin in Section 2 by showing how to generalize their technique to simultaneously answer both range minimum and range maximum queries. Our encoding provides the answer to both using $3n + o(n)$ bits in total, compared to $4n + o(n)$ bits using the trivial approach of constructing both encodings separately. We then show this bound is optimal by proving that any encoding for range min-max queries can be used to distinguish a certain class of permutations. We move on in Section 3 to generalize Fischer and Heun's technique in a clean and natural way to larger values of k. Indeed, the encoding we present—like that of Fischer and Heun—is simple enough to implement. The main difficulty is proving that the bound achieved by our technique is optimal. For this we enumerate a particular class of walks, via an application of the so-called cycle lemma of Dvoretzky and Motzkin [6].

Due to lack of space we focus primarily on space lower bounds for encodings. However, in the full version of this paper [9] we show our encoding can be used as the basis for a range top-k data structure. Though the resultant space bound and query time are suboptimal, we note that interesting challenges had to be overcome to design a data structure based on our encoding. Concisely, we required the ability to decompose the encoding into smaller blocks in order to support queries efficiently. To do this we, in some sense, generalized the pioneers approach of Jacobson [11] via a non-trivial decomposition theorem. Since balanced parentheses representations appear in many succinct data structures, we believe this will likely be of independent interest.

2 Optimal Encodings of Range Min-Max Queries

In this section we describe our encoding for range min-max queries. We use $\text{RMINMAX}(A[i..j])$ to denote a range min-max query on a subarray $A[i..j]$. The solution to the query is the ordered set of indices $\{\ell_1, \ell_2\}$ such that $\ell_1 = \arg\max_{\ell \in [i,j]} A[\ell]$ and $\ell_2 = \arg\min_{\ell \in [i,j]} A[\ell]$.

2.1 Review of Fischer and Heun's Technique

We review the algorithm of Fischer and Heun [7] for constructing the encoding of range minimum (resp. maximum) queries.

Consider an array $A[1..n]$ storing n numbers. Without loss of generality we can alter the values of the numbers so that they are a permutation, breaking ties in favour of the leftmost element. To construct the encoding for range minimum queries we sweep the array from left to right[2], while maintaining a stack. A string of bits $T_{\min}$ (resp. $T_{\max}$) will be emitted in reverse order as we scan the array. Whenever we push an element onto the stack, we emit a one bit, and whenever we pop we emit a zero bit. Initially the stack is empty, so we push the position of the first element we encounter on the stack, in this case, 1. Each time we increment the current position, i, we compare the value of $A[i]$ to that of the element in the position t, that is stored on the top of the stack. While $A[t]$ is not less than (resp. not greater than) $A[i]$, we pop the stack. Once $A[t]$ is less than (resp. greater than) the current element or the stack becomes empty, we push i onto the stack. When we reach the end of the array, we pop all the elements on the stack, emitting a zero bit for each element popped, followed by a one bit.

Fischer and Heun showed that the string of bits output by this process can be used to encode a rooted ordinal tree in terms of its *depth first unary degree sequence* or DFUDS [7]. To extract the tree from a sequence, suppose we read d zero bits until we hit the first one bit. Based on this, we create a node v of degree d, and continue building first child of v recursively. Since there are at most $2n$ stack operations, the tree is therefore represented using $2n$ bits. We omit the technical details of how a query is answered, but the basic idea is to augment this tree representation with succinct data structures supporting navigation operations.

2.2 Upper Bound for Range Min-Max Queries

We propose the following encoding for a simultaneous representation of $T_{\min}$ and $T_{\max}$. Scan the array from left to right and maintain two stacks: a min-stack for range minimum queries, and a max-stack for range maximum queries. Notice that in each step except for the first and last, we are popping an element from exactly one of the two stacks. This crucial observation allows us to save space. We describe our encoding in terms of the min-stack and the max-stack maintained as above. Unlike before however, we maintain two separate bit strings, T and U. If the new element causes $\delta \geq 1$ elements on the min-stack to be popped, then we prepend $0^{\delta-1}1$ to the string T, and prepend 0 to the string U. Otherwise, if the new element causes δ elements on the max-stack to be popped, we prepend $0^{\delta-1}1$ to the string T, and 1 to the string U. Since exactly $2n$ elements are popped during n push operations, the bit string T has length $2n$, and the bit string U has length n, for a total of $3n$ bits.

In the full version [9] we show that by using techniques from succinct data structures it is possible to also support queries on this encoding in $\mathcal{O}(1)$ time.

Theorem 1. *There is a data structure that occupies $3n + o(n)$ bits of space, such that any query* RMINMAX$(A[i..j])$ *can be answered in* $\mathcal{O}(1)$ *time.*

---

[2] In the original paper the sweeping process moves from right to left, but either direction yields a correct algorithm by symmetry.

2.3   Lower Bound for Range Min-Max Queries

Given a permutation $\pi = (p_1, ..., p_n)$, we say π contains the permutation pattern $s_1\text{-}s_2\text{-}...\text{-}s_m$ if there exists a subsequence of π whose elements have the same relative ordering as the elements in the pattern. That is, there exist some $x_1 < x_2 < ... < x_m \in [1, n]$ such that for all $i, j \in [1, m]$ we have that $\pi(x_i) < \pi(x_j)$ if and only if $s_i < s_j$. For example, if $\pi = (1, 4, 2, 5, 3)$ then π contains the permutation pattern 1-3-4-2: we use this hyphen notation to emphasize that the indices need not be consecutive. In this case, the series of indices in π matching the pattern are $x_1 = 1$, $x_2 = 2$, $x_3 = 4$ and $x_4 = 5$. If no hyphen is present between elements s_i and s_{i+1} in the permutation pattern, then the indices x_i and x_{i+1} must be consecutive: i.e., $x_{i+1} = x_i + 1$. In terms of the example, π does not contain the permutation pattern 1-34-2.

A permutation $\pi = (p_1, ..., p_n)$ is a *Baxter permutation* if there exist no indices $1 \leq i < j < k \leq n$ such that $\pi(j + 1) < \pi(i) < \pi(k) < \pi(j)$ or $\pi(j) < \pi(k) < \pi(i) < \pi(j+1)$. Thus, Baxter permutations are those that do not contain 2-41-3 and 3-14-2. Permutations with less than 4 elements are trivially Baxter permutations, and for permutations on 4 elements the non-Baxter permutations are exactly $(2, 4, 1, 3)$ and $(3, 1, 4, 2)$. Baxter permutations are well studied, and their asymptotic behaviour is known (see, e.g., OEIS A001181 [1]).

We have the following lemma:

Lemma 1. *Suppose π is a Baxter permutation, stored in an array $A[1..n]$ such that $A[i] = \pi(i)$. If an encoding that can recover all range minimum and maximum queries is constructed on A, then π can be recovered from the encoding.*

Proof. In order to recover the permutation, it suffices to show that we can perform pairwise comparisons on any two elements in A using range minimum and range maximum queries. The proof follows by induction on n.

For the base case, for $n = 1$ there is exactly one permutation, so there is nothing to recover. Thus, let us assume that the lemma holds for all permutations on less than $n \geq 2$ elements. For a permutation on n elements, consider the subpermutation induced by the array prefix $A[1..(n-1)]$ and suffix $A[2..n]$. These subpermutations must be Baxter permutations, since deleting elements from the prefix or suffix of a Baxter permutation cannot create a 2-41-3 or a 3-14-2. Thus, it suffices to show that we can compare $A[1]$ and $A[n]$, as all the remaining pairwise comparisons can be performed by the induction hypothesis.

Let $x = \text{RMin}(A[1..n])$ and $y = \text{RMax}(A[1..n])$ be the indices of the minimum and maximum elements in the array, respectively. If $x \in \{1, n\}$ or $y \in \{1, n\}$ we can compare $A[1]$ and $A[n]$, so assume $x, y \in [2, n-1]$. Without loss of generality we consider the case where $x < y$: the opposite case is symmetric (i.e., replacing 3-14-2 with 2-41-3), and $x \neq y$ because $n \geq 2$. Consider an arbitrary index $i \in [x, ..., y]$, and the result of comparing $A[1]$ to $A[i]$ and $A[i]$ to $A[n]$ (that can be done by the induction hypothesis, as $i \in [2, n-1]$). The result is a partial order on three elements, and is either:

1. One of the two chains $A[1] < A[i] < A[n]$ or $A[n] < A[i] < A[1]$, in which case we are done since $A[1]$ and $A[n]$ can be compared; or

2. A partial order in which $A[i]$ is the minimum or maximum element, and $A[1]$ is incomparable with $A[n]$.

If we are in the latter case for all $i \in [x,y]$, then let $f(i) = 0$ if $A[i]$ is the minimum element in this partial order, and $f(i) = 1$ otherwise. Because of how x and y were chosen, $f(x) = 0$ and $f(y) = 1$. If we consider the values of $f(i)$ for all $i \in [x,y]$, there must exist two indices $i, i+1 \in [x,y]$ such that $f(i) = 0$ and $f(i+1) = 1$. Therefore, the indices $1, i, i+1, n$ form the forbidden pattern 3-14-2, unless $A[1] < A[n]$. $\qquad\square$

Theorem 2. *Any data structure encoding range minimum and maximum queries simultaneously must occupy $3n - \Theta(\log n)$ bits, for sufficiently large values of n.*

Proof. Let $L(n)$ be the number of Baxter permutations on n elements. It is known (cf. [1]) that $\lim_{n\to\infty} \frac{L(n)\pi\sqrt{3}n^4}{2^{3n+5}} = 1$. Since we can encode and recover each one by the procedure discussed in Lemma 1, our encoding data structure must occupy at least $\lg L(n) = 3n - \Theta(\log n)$ bits, if n is sufficiently large. $\qquad\square$

3 Optimal Encodings for Top-k Queries

In this section we use $\text{RTOPK}(A[i..j])$ to denote a range top-k query on the sub-array $A[i..j]$. The solution to such a query is an ordered list of indices $\{\ell_1, ..., \ell_k\}$ such that $A[\ell_m]$ is the m-th largest element in $A[i..j]$.

3.1 Upper Bound for Encoding Top-k Queries

Like the encoding for range min-max queries, our encoding for range top-k queries is based on representing the changes to a certain structure as we scan through the array A. Each prefix in the array will correspond to a different structure. We denote the structure, that we will soon describe, for prefix $A[1..j]$ as $S_k(j)$, for all $1 \le j \le n$. The structure $S_k(j)$ will allow us to answer $\text{RTOPK}(A[i..j])$ for any $i \in [1,j]$. Our encoding will store the differences between $S_k(j)$ and $S_k(j+1)$ for all $j \in [1, n-1]$. Let us begin by defining a single instance for an arbitrary j.

We first define the directed graph $G_j = (V, E)$ with vertices labelled $\{1, ..., j\}$, and where an edge $(i', j') \in E$ iff both $i' < j'$ and $A[i'] < A[j']$ for all $1 \le i' < j' \le j$. We call G_j the *dominance graph* of $A[1..j]$, and say j' *dominates* i', or i' is *dominated by* j', if $(i', j') \in E$. Next consider the out-degree $d_j(\ell)$ of the vertex labelled $\ell \in [1,j]$ in G_j. We define an array $S[1..j]$, where $S[\ell] = d_j(\ell)$ for $1 \le \ell \le j$. The structure $S_k(j)$ is defined as follows: take the array $S[1..j]$, and for each entry $\ell \in [1,j]$ such that $S[\ell] > k$, replace $S[\ell]$ with k. We use the notation $S_k(j, \ell)$ to refer to the ℓ-th array entry in the structure $S_k(j)$. We refer to an index ℓ to be *active* iff $S_k(j, \ell) < k$, and as *inactive* otherwise. We note that $S_k(n)$ is reminiscent of the one-sided top-k structure of Grossi et al. [10].

Lemma 2. *The total ordering of elements $A[i_1], ..., A[i_{j'}]$, where $\{i_1, ..., i_{j'}\}$ are the active indices in $S_k(j)$, can be recovered by examining only $S_k(j)$.*

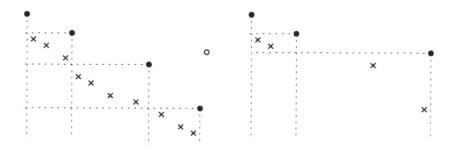

Fig. 1. Geometric interpretation of how the structure $S_k(j)$ is updated to $S_k(j+1)$. In the example $k = 2$, and the value of each active element in the array is represented by its height. Black circles denote 0 values in the array $S_2(j)$, whereas crosses represent 1 values, and 2 values (inactive elements) are not depicted. When the new point (empty circle) is inserted to the structure on the left, it increments the counters of the smallest 10 active elements, resulting in the picture on the right representing $S_2(j+1)$.

Proof. We scan the structure $S_k(j)$ from index j down to 1, maintaining a total ordering on the active elements seen so far. Initially, we have an empty total ordering. At each active location ℓ the value $S_k(j, \ell)$ indicates how many active elements in locations $[\ell+1, j]$ are larger than $A[\ell]$. This follows since an inactive element cannot dominate an active element in the graph G_j. Thus, we can insert $A[\ell]$ into the current total ordering of active elements.   □

We define the *size* of $S_k(j)$ as follows: $|S_k(j)| = \sum_{\ell=1}^{j}(k - S_k(j, \ell))$. The key observation is that the structure $S_k(j+1)$ can be constructed from $S_k(j)$ using the following procedure:

1. Compute the value $\delta_j = |S_k(j)| - |S_k(j+1)| + k$. This quantity is always non-negative, as we add one new element to the large staircase, which increases the size by at most k.
2. Find the δ_j indices among the active elements in $S_k(j)$ such that their values in A are the smallest via Lemma 2. Denote this set of indices as $\mathcal{I}$.
3. For each $\ell \in [1, j]$, set $S_k(j+1, \ell) = S_k(j, \ell) + 1$ iff $\ell \in \mathcal{I}$, and $S_k(j+1, \ell) = S_k(j, \ell)$ otherwise.
4. Add the new element at the end of the array, setting $S_k(j+1, j+1) = 0$.

Thus, to construct $S_k(j+1)$ all that is needed is $S_k(j)$ and the value δ_j: see Figure 1. This implies that by storing δ_j for $j \in [1, n-1]$ we can build any $S_k(j)$.

Theorem 3. *Solutions to all queries* $\mathrm{RTOPK}(A[i..j])$ *can be encoded in at most* $(k+1)nH(\frac{1}{k+1})$ *bits of space.*

Proof. Suppose we store the bitvector $0^{\delta_1}10^{\delta_2}1\ldots0^{\delta_{n-1}}1$. This bitvector contains no more than kn zero bits. This follows since each active counter can be incremented k times before it becomes inactive. Thus, storing the bitvector requires no more than $\lg\binom{(k+1)n}{n} \le (k+1)nH(\frac{1}{k+1})$ bits.

Next we prove that this is all we need to answer a query $\text{RTOPK}(A[i..j])$. We use the encoding to construct $S_k(j)$. We know that for every element at inactive index ℓ in $S_k(j)$ there are at least k elements with larger value in $A[\ell+1..j]$. Consequently, these elements need not be returned in the solution, and it is enough to recover the indices of the top-k values among the elements at active indices at least i. We apply Lemma 2 on $S_k(j)$ to recover these indices and return them as the solution.   □

3.2   Lower Bound for Encoding Top-k Queries

The goal of this section is to show that the encoding from Section 3.1 is, in fact, optimal. The first observation is that all structures $S_k(j)$ for $j \in [1, n]$ can be reconstructed with RTOPK queries.

Lemma 3. *Any $S_k(j)$ can be reconstructed with* RTOPK *queries.*

Proof. To reconstruct $S_k(j)$, we execute the query $\text{RTOPK}(A[\ell..j])$ for each $\ell \in [1, j]$. If index ℓ is returned as the k'-th largest element in $[\ell, j]$, then by definition there are exactly $k'-1$ elements in locations $A[\ell+1..j]$ with value larger than $A[\ell]$. Thus, ℓ is an active location and $S_k(j, \ell) = k'-1$. If ℓ is not returned by the query, then it is inactive and we set $S_k(j, \ell) = k$.   □

Recall that we encode all structures by specifying $\delta_1, \delta_2, \ldots, \delta_{n-1}$. We call an $(n-1)$-tuple of nonnegative integers $(\delta_1, \delta_2, \ldots, \delta_{n-1})$ *valid* if it encodes some $S_k(1), S_k(2), \ldots, S_k(n)$, i.e., if there exists at least one array $A[1..n]$ consisting of distinct integers such that the structure constructed for $A[1..j]$ is exactly the encoded $S_k(j)$, for every $j = 1, 2, \ldots, n$. Then the number of bits required by the encoding is at least the logarithm of the number of valid $(n-1)$-tuples $(\delta_1, \delta_2, \ldots, \delta_{n-1})$. Our encoding from Section 3.1 shows this number is at most $\binom{(k+1)n}{n}$, but we need to argue in the other direction, which is far more involved.

Recall that the size of a particular $S_k(j)$ is $|S_k(j)| = \sum_{i=1}^{j}(k - S_k(j, i))$. We would like to argue that there are many valid $(n-1)$-tuples $(\delta_1, \delta_2, \ldots, \delta_{n-1})$. This will be proven in a series of transformations.

Lemma 4. *If $(\delta_1, \delta_2, \ldots, \delta_{n-1})$ is valid, then for any $\delta_n \in \{0, 1, \ldots, \lceil \frac{M}{k} \rceil\}$ where $M = \sum_{i=1}^{n-1}(k - \delta_i)$, the tuple $(\delta_1, \delta_2, \ldots, \delta_{n-1}, \delta_n)$ is also valid.*

Proof. Let $A[1..n]$ be an array such that the structure constructed for $A[1..j]$ is exactly $S_k(j)$, for every $j = 1, 2, \ldots, n$. By definition of δ_j, we have that $M = \sum_{i=1}^{n-1}(k - \delta_i) < |S_k(n)|$. Denote the number of active elements in $S_k(j)$ with the corresponding entry set to α as m_α for $\alpha \in [0, k-1]$. For any $s \in \{0, 1, \ldots, \sum_{\alpha=0}^{k-1} m_\alpha\}$, we can adjust $A[n+1]$ so that it is larger than exactly the s smallest active elements in $S_k(n)$. Thus, choosing any $\delta_n \in \{0, 1, \ldots, \sum_{\alpha=1}^{k} m_\alpha\}$ results in a valid $(\delta_1, \delta_2, \ldots, \delta_n)$. Since $|S_k(n)| = \sum_{\alpha=0}^{k-1}(k - \alpha)m_\alpha \le k \sum_{\alpha=0}^{k-1} m_\alpha$, we have $\sum_{\alpha=0}^{k-1} m_\alpha \ge \lceil \frac{|S_k(n)|}{k} \rceil$, proving the claim.   □

Every valid $(n-1)$-tuple $(a_1, a_2, \ldots, a_{n-1})$ corresponds in a natural way to walk of length $n-1$ in a plane, where we start at $(0,0)$ and perform steps of the form $(1, a_i)$, for $i = 1, 2, \ldots, n-1$. We consider a subset of all such walks. Denoting the current position by (x_i, y_i), we require that a_i is an integer from $k - \lceil \frac{y_i}{k} \rceil, k]$. Under such conditions, any walk corresponds to a valid $(n-1)$-tuple $(\delta_1, \delta_2, \ldots, \delta_{n-1})$, because we can choose $\delta_i = k - a_i$ and apply Lemma 4. Therefore, we can focus on counting such walks.

The condition $[k - \lceil \frac{y_i}{k} \rceil, k]$ is not easy to work with, though. We will count more restricted walks instead. A Y-restricted nonnegative walk of length n starts at $(0,0)$ and consists of n steps of the form $(1, a_i)$, where $a_i \in Y$ for $= 1, 2, \ldots, n$, such that the current y-coordinate is always nonnegative. Y is an arbitrary set of integers.

Lemma 5. *The number of valid $(n-1)$-tuples is at least as large as the number of $[k - \Delta, k]$-restricted nonnegative walks of length $n - 1 - \Delta$.*

Proof. We have already observed that the number of valid $(n-1)$-tuples is at least as large as the number of walks consisting of $n-1$ steps of the form $(1, a_i)$, where $a_i \in [k - \lceil \frac{y_i}{k} \rceil, k]$ for $i = 1, 2, \ldots, n-1$. We distinguish a subset of such walks, where the first Δ steps are of the form $(1, k)$, and then we always stay above (or on) the line $y = k\Delta$. Under such restrictions, $a_i \in [k - \Delta, k]$ implies $u_i \in [k - \lceil \frac{y_i}{k} \rceil, k]$, so counting $[k - \Delta, k]$-restricted nonnegative walks gives us a lower bound on the number of valid $(n-1)$-tuples. □

We move to counting Y-restricted nonnegative walks of length n. Again, counting them directly is non-trivial, so we introduce a notion of Y-restricted returning walk of length n, where we ignore the condition that the current y-coordinate should be always nonnegative, but require the walk ends at $(n, 0)$.

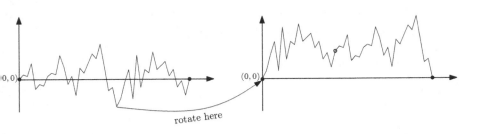

Fig. 2. Left: a Y-restricted walk ending at $(n, 0)$. Right: a cyclic rotation of the walk on the left such that the walk is always nonnegative.

Lemma 6. *The number of Y-restricted nonnegative walks of length n is at least as large as the number of Y-restricted returning walks of length n divided by n.*

Proof. This follows from the so-called cycle lemma [6], but we prefer to provide a simple direct proof. We consider only Y-restricted nonnegative walks of length

n ending at $(n, 0)$, and denote their set by W_1. The set of Y-restricted returning walks of length n is denoted by W_2. The crucial observation is that a cyclic rotation of any walk in W_2 is also a walk in W_2. Moreover, there is always at least one such cyclic rotation which results in the walk becoming nonnegative (see Figure 2). Therefore, we can define a total function $f : W_2 \to W_1$, that takes a walk w and rotates it cyclically as to make it nonnegative. Because there are just n cyclic rotations of a walk of length n, any element of W_1 is the image of at most n elements of W_2 through f. Therefore, $|W_1| \geq \frac{|W_2|}{n}$ as claimed. $\square$

The only remaining step is to count $[k - \Delta, k]$-restricted returning walks of length $n - 1 - \Delta$. This is equivalent to counting ordered partitions of $k(n - 1 - \Delta)$ into parts $a_1, a_2, \ldots, a_{n-1-\Delta}$, where $a_i \in [0, \Delta]$ for every $i = 1, 2, \ldots, n - 1 - \Delta$. This follows since a partition of size ℓ corresponds to a step of size $k - \ell$.

Lemma 7. *The number of ordered partitions of N into g parts, where every part is from $[0, B]$, is at least $\binom{N - 2g' + g - 1}{g - g' - 1}$, where $g' = \lfloor \frac{N}{B} \rfloor$.*

Proof. The number of ordered partitions of N into g parts, where there are no restrictions on the sizes of the parts, is simply $\binom{N + g - 1}{g - 1}$. To take the restrictions into the account, we first split N into blocks of length B (except for the last block, which might be shorter). This creates $g' + 1$ blocks. Then, we additionally split the blocks into smaller parts, which ensures that all parts are from $[0, B]$. We restrict the smaller parts, so that the first and the last smaller part in every block is strictly positive. This ensures that given the resulting partition into parts, we can uniquely reconstruct the blocks. Therefore, we only need to count the number of ways we can split the blocks into such smaller parts, and by standard reasoning this is at least $\binom{N - 2g' + g - 1}{g - g' - 1}$. This follows by conceptually merging the last element in block i with the first element in block $i + 1$, so that no further partitioning can happen between them, and then partitioning the remaining set into $g - g'$ pieces. Every such partition corresponds to a distinct restricted partition obtained by splitting between the merged elements, which creates g' additional blocks. $\square$

We are ready to combine all the ingredients. Setting $N = k(n - 1 - \Delta)$, $g = n - 1 - \Delta$, $g' = \lfloor \frac{k(n-1-\Delta)}{\Delta} \rfloor = \lfloor \frac{k(n-1)}{\Delta} \rfloor - k$ and substituting, the number of bits required by the encoding is:

$$\lg \binom{N - 2g' + g - 1}{g - g' - 1} > \lg \binom{(k+1)(n - 2 - \Delta - g')}{n - 2 - \Delta - g'}.$$

Using the entropy function as a lower bound, this is at least $(k + 1)n'H(\frac{1}{k+1}) - \Theta(\log n')$, where $n' = n - 2 - \Delta - g' \geq n(1 - \frac{k}{\Delta}) + \frac{k}{\Delta} + k - 2 - \Delta$. Thus, we have the following theorem:

Theorem 4. *For sufficiently large values of n, any data structure that encodes range top-k queries must occupy $(k+1)n'H(\frac{1}{k+1}) - \Theta(\log n')$ bits of space, where $n' \geq n(1 - \frac{k}{\Delta}) + \frac{k}{\Delta} + k - 2 - \Delta$, and $\Delta \geq 1$ can be selected to be any positive integer. If $k = o(n)$, then Δ can be chosen such that $\Delta = \omega(k)$ and $\Delta = o(n)$, yielding that the lower bound is $(k + 1)nH(\frac{1}{k+1})(1 - o(1))$ bits.*

References

1. OEIS Foundation Inc., The On-Line Encyclopedia of Integer Sequences, Number of Baxter permutations of length n (2011). http://oeis.org/A001181 (Accessed 24 September 2014)
2. Bender, M.A., Farach-Colton, M., Pemmasani, G., Skiena, S., Sumazin, P.: Lowest common ancestors in trees and directed acyclic graphs. Journal of Algorithms **57**(2), 75–94 (2005)
3. Brodal, G.S., Gfeller, B., Jørgensen, A.G., Sanders, P.: Towards optimal range medians. Theoretical Computer Science **412**(24), 2588–2601 (2011)
4. Chan, T.M., Wilkinson, B.T.: Adaptive and approximate orthogonal range counting. In: Proc. of the 24th Annual ACM-SIAM Symposium on Discrete Algorithms (SODA), pp. 241–251. SIAM (2013)
5. Davoodi, P., Navarro, G., Raman, R., Rao, S.: Encoding Range Minima and Range Top-2 Queries. Phil. Trans. R. Soc. A **372**(2016), 1471–2962 (2014)
6. Dvoretzky, A., Motzkin, T.: A problem of arrangements. Duke Mathematical Journal **14**(2), 305–313 (1947)
7. Fischer, J., Heun, V.: Space-efficient preprocessing schemes for range minimum queries on static arrays. SIAM J. Comput. **40**(2), 465–492 (2011)
8. Gagie, T., Puglisi, S.J., Turpin, A.: Range quantile queries: another virtue of wavelet trees. In: Karlgren, J., Tarhio, J., Hyyrö, H. (eds.) SPIRE 2009. LNCS, vol. 5721, pp. 1–6. Springer, Heidelberg (2009)
9. Gawrychowski, P., Nicholson, P.K.: Optimal Encodings for Range Min-Max and Top-k. CoRR abs/1411.6581 (2014). http://arxiv.org/abs/1411.6581
10. Grossi, R., Iacono, J., Navarro, G., Raman, R., Rao, S.S.: Encodings for range selection and top-k queries. In: Bodlaender, H.L., Italiano, G.F. (eds.) ESA 2013. LNCS, vol. 8125, pp. 553–564. Springer, Heidelberg (2013)
11. Jacobson, G.: Space-efficient static trees and graphs. In: Proc. of the 30th Annual Symposium on Foundations of Computer Science, pp. 549–554. IEEE (1989)
12. Jørgensen, A.G., Larsen, K.G.: Range selection and median: tight cell probe lower bounds and adaptive data structures. In: Proc. of the Twenty-Second Annual ACM-SIAM Symposium on Discrete Algorithms (SODA), pp. 805–813. SIAM (2011)
13. Matoušek, J.: Reporting points in halfspaces. Computational Geometry **2**(3), 169–186 (1992)
14. Navarro, G.: Spaces, trees, and colors: The algorithmic landscape of document retrieval on sequences. ACM Comput. Surv. **46**(4), 52 (2013)
15. Navarro, G., Raman, R., Satti, S.R.: asymptotically optimal encodings for range selection. In: Proc. 34th International Conference on Foundation of Software Technology and Theoretical Computer Science (FSTTCS). LIPIcs, vol. 29, pp. 291–301. Schloss Dagstuhl - Leibniz-Zentrum fuer Informatik (2014)
16. Sadakane, K.: Succinct data structures for flexible text retrieval systems. Journal of Discrete Algorithms **5**(1), 12–22 (2007)
17. Skala, M.: Array range queries. In: Brodnik, A., López-Ortiz, A., Raman, V., Viola, A. (eds.) Ianfest-66. LNCS, vol. 8066, pp. 333–350. Springer, Heidelberg (2013)
18. Vuillemin, J.: A unifying look at data structures. Communications of the ACM **23**(4), 229–239 (1980)

2-Vertex Connectivity in Directed Graphs

Loukas Georgiadis[1], Giuseppe F. Italiano[2], Luigi Laura[3(✉)],
and Nikos Parotsidis[1]

[1] University of Ioannina, Ioannina, Greece
{loukas,nparotsi}@cs.uoi.gr
[2] Università di Roma "Tor Vergata", Rome, Italy
giuseppe.italiano@uniroma2.it
[3] "Sapienza" Università di Roma, Roma, Italy
laura@dis.uniroma1.it

Abstract. Given a directed graph, two vertices v and w are *2-vertex-connected* if there are two internally vertex-disjoint paths from v to w and two internally vertex-disjoint paths from w to v. In this paper, we show how to compute this relation in $O(m + n)$ time, where n is the number of vertices and m is the number of edges of the graph. As a side result, we show how to build in linear time an $O(n)$-space data structure, which can answer in constant time queries on whether any two vertices are 2-vertex-connected. Additionally, when two query vertices v and w are not 2-vertex-connected, our data structure can produce in constant time a "witness" of this property, by exhibiting a vertex or an edge that is contained in all paths from v to w or in all paths from w to v. We are also able to compute in linear time a sparse certificate for 2-vertex connectivity, i.e., a subgraph of the input graph that has $O(n)$ edges and maintains the same 2-vertex connectivity properties as the input graph.

1 Introduction

Let $G = (V, E)$ be a directed graph (digraph), with m edges and n vertices. G is *strongly connected* if there is a directed path from each vertex to every other vertex. The *strongly connected components* of G are its maximal strongly connected subgraphs. Two vertices $u, v \in V$ are *strongly connected* if they belong to the same strongly connected component of G. A vertex (resp., an edge) of G is a *strong articulation point* (resp., a *strong bridge*) if its removal increases the number of strongly connected components. A digraph G is *2-vertex-connected* if it has at least three vertices and no strong articulation points; G is *2-edge-connected* if it has no strong bridges. The *2-vertex-* (resp., *2-edge-*) *connected components* of G are its maximal 2-vertex- (resp., 2-edge-) connected subgraphs.

Differently from undirected graphs, in digraphs 2-vertex and 2-edge connectivity have a much richer and more complicated structure. To see an example

Giuseppe F. Italiano—Partially supported by the Italian Ministry of Education, University and Research (MIUR) under Project AMANDA (Algorithmics for MAssive and Networked DAta).

© Springer-Verlag Berlin Heidelberg 2015
M.M. Halldórsson et al. (Eds.): ICALP 2015, Part I, LNCS 9134, pp. 605–616, 2015.
DOI: 10.1007/978-3-662-47672-7_49

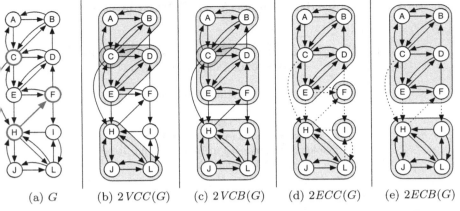

(a) G (b) $2VCC(G)$ (c) $2VCB(G)$ (d) $2ECC(G)$ (e) $2ECB(G)$

Fig. 1. (a) A strongly connected digraph G, with strong articulation points and strong bridges shown in red (better viewed in color). (b) The 2-vertex-connected components of G. (c) The 2-vertex-connected blocks of G. (d) The 2-edge-connected components of G. (e) The 2-edge-connected blocks of G.

of this, let v and w be two distinct vertices and consider the following natural 2-vertex and 2-edge connectivity relations, defined in [3,7,11]. Vertices v and w are said to be *2-vertex-connected* (resp., *2-edge-connected*), and we denote this relation by $v \leftrightarrow_{2v} w$ (resp., $v \leftrightarrow_{2e} w$), if there are two internally vertex-disjoint (resp., two edge-disjoint) directed paths from v to w and two internally vertex-disjoint (resp., two edge-disjoint) directed paths from w to v (note that a path from v to w and a path from w to v need not be edge- or vertex-disjoint). A *2-vertex-connected block* (resp., *2-edge-connected block*) of a digraph $G = (V, E)$ is defined as a maximal subset $B \subseteq V$ such that $u \leftrightarrow_{2v} v$ (resp., $u \leftrightarrow_{2e} v$) for all $u, v \in B$. In undirected graphs, the 2-vertex- (resp., 2-edge-) connected blocks are identical to the 2-vertex- (resp., 2-edge-) connected components. As shown in Figure 1, this is not the case for digraphs. Put in other words, differently from the undirected case, in digraphs 2-vertex- (resp., 2-edge-) connected components do not encompass the notion of pairwise 2-vertex (resp., 2-edge) connectivity among its vertices. We note that pairwise 2-connectivity may be relevant in several applications, where one is interested in local properties, e.g., checking whether two vertices are 2-connected, rather than in global properties.

It is thus not surprising that 2-connectivity problems on directed graphs appear to be more difficult than on undirected graphs. For undirected graphs it has been known for over 40 years how to compute all bridges, articulation points, 2-edge- and 2-vertex-connected components in linear time, by simply using depth first search [12]. In the case of digraphs, however, the very same problems have been much more challenging. Indeed, it has been shown only few years ago that all strong bridges and strong articulation points of a digraph can be computed in linear time [6]. Furthermore, the best current bound for computing the 2-edge- and the 2-vertex-connected components in digraphs is not even linear, but it is $O(n^2)$, and it was achieved only very recently by Henzinger et al. [5], improving previous $O(mn)$ time bounds [8, 10]. Finally, it was shown also very recently how to compute the 2-edge-connected blocks of digraphs in linear time [3].

In this paper, we complete the picture on 2-connectivity for digraphs by presenting the first algorithm for computing the 2-vertex-connected blocks in $O(m + n)$ time. Our bound is asymptotically optimal and it improves sharply over a previous $O(mn)$ time bound by Jaberi [7]. As a side result, our algorithm constructs an $O(n)$-space data structure that reports in constant time if two vertices are 2-vertex-connected. Additionally, when two query vertices v and w are not 2-vertex-connected, our data structure can produce, in constant time, a "witness" by exhibiting a vertex (i.e., a strong articulation point) or an edge (i.e., a strong bridge) that separates them. We are also able to compute in linear time a sparse certificate for 2-vertex connectivity, i.e., a subgraph of the input graph that has $O(n)$ edges and maintains the same 2-vertex connectivity properties. Our algorithm follows the high-level approach of [3] for computing the 2-edge-connected blocks. However, the algorithm for computing the 2-vertex-connected blocks is much more involved and requires several novel ideas and non-trivial techniques to achieve the claimed bounds. In particular, the main technical difficulties that need to be tackled when following the approach of [3] are the following.

First, the algorithm in [3] maintains a partition of the vertices into approximate blocks, and refines this partition as the algorithm progresses. Unlike 2-edge-connected blocks, however, 2-vertex-connected blocks do not partition the vertices of a digraph, and therefore it is harder to maintain approximate blocks throughout the algorithm's execution. To cope with this problem, we show that these blocks can be maintained using a more complicated forest representation, and we define a set of suitable operations on this representation in order to refine and split blocks. We believe that our forest representation of the 2-vertex-connected blocks of a digraph can be of independent interest.

Second, in [3] we used a properly defined *canonical decomposition* of the input digraph G, in order to obtain smaller *auxiliary* digraphs (not necessarily subgraphs of G) that maintain the original 2-edge-connected blocks of G. A key property of this decomposition was the fact that any vertex in an auxiliary graph G_r is reachable from a vertex outside G_r only through a single strong bridge. In the computation of the 2-vertex-connected blocks, we have to decompose the graph according to strong articulation points, and so the above crucial property is completely lost. To overcome this problematic issue, we need to design and to implement efficiently a different and more sophisticated decomposition.

Third, differently from 2-edge connectivity, 2-vertex connectivity in digraphs is plagued with several degenerate special cases, which are not only more tedious but also more cumbersome to deal with. For instance, the algorithm in [3] exploits implicitly the property that two vertices v and w are 2-edge-connected if and only if the removal of any edge leaves v and w in the same strongly connected component. Unfortunately, this property no longer holds for 2-vertex connectivity, as for instance two mutually adjacent vertices are always left in the same strongly connected component by the removal of any other vertex, but they are ot necessarily 2-vertex-connected. To handle this more complicated situation, re introduce the notion of *vertex-resilient blocks* and prove some useful proper- ies about the vertex-resilient and 2-vertex-connected blocks of a digraph.

Another difference with [3] is that now we are able to provide a witness or two vertices not being 2-vertex-connected. This approach can be applied to rovide a witness for two vertices not being 2-edge-connected, thus extending he result in [3]. For lack of space, proofs and some details are omitted from this xtended abstract.

2 Flow Graphs, Dominators, and Bridges

n this section we introduce some terminology that will be useful throughout he paper. A *flow graph* is a digraph such that every vertex is reachable from a distinguished start vertex. Let $G = (V, E)$ be the input digraph, which we assume to be strongly connected. (If not, we simply treat each strongly connected omponent separately.) For any vertex $s \in V$, we denote by $G(s) = (V, E, s)$ the orresponding flow graph with start vertex s; all vertices in V are reachable rom s since G is strongly connected. The *dominator relation* in $G(s)$ is defined as follows: A vertex u is a *dominator* of a vertex w (u *dominates* w) if every path from s to w contains u; u is a *proper dominator* of w if u dominates w and $u \neq w$. The dominator relation is reflexive and transitive. Its transitive reduction s a rooted tree, the *dominator tree* $D(s)$: u dominates w if and only if u is an ancestor of w in $D(s)$. If $w \neq s$, $d(w)$, the parent of w in $D(s)$, is the *immediate dominator* of w: it is the unique proper dominator of w that is dominated by all proper dominators of w. The dominator tree $D(s)$ has the following *parent property* [4]: For all $(v, w) \in E$, v is a descendant of $d(w)$ in $D(s)$. An edge (u, w) s a *bridge* in $G(s)$ if all paths from s to w include (u, w). The dominator tree of a flow graph can be computed in linear time, see, e.g., [1,2]. Italiano et al. 6] showed that the strong articulation points of G can be computed from the dominator trees of $G(s)$ and $G^R(s)$, where s is an arbitrary start vertex and G^R s the digraph that results from G after reversing edge directions; similarly, the strong bridges of G correspond to the bridges of $G(s)$ and $G^R(s)$.

3 Vertex-resilient Blocks and 2-vertex-connected Blocks

Let v and w be two distinct vertices in a digraph. By Menger's Theorem [9], $v \leftrightarrow_{2e} w$ if and only if the removal of any edge leaves v and w in the same strongly connected component, i.e., two vertices are 2-edge-connected if and only if they are resilient to the deletion of a single edge. The situation for 2-vertex connectivity is more complicated. Indeed, two mutually adjacent vertices are left in the same strongly connected component by the removal of any other vertex, although they are not necessarily 2-vertex-connected. To handle this situation, we use the following notation, which was also considered in [7]. Vertices v and w are said to be *vertex-resilient*, denoted by $v \leftrightarrow_{vr} w$ if the removal of any vertex different from v and w leaves v and w in the same strongly connected component.

We define a *vertex-resilient block* of a digraph $G = (V, E)$ as a maximal subset $B \subseteq V$ such that $u \leftrightarrow_{vr} v$ for all $u, v \in B$. Note that, as a (degenerate) special case, a vertex-resilient block might consist of a singleton vertex only: we denote this as a *trivial vertex-resilient block*. In the following, we will consider only non-trivial vertex-resilient blocks. Since there is no danger of ambiguity, we will call them simply vertex-resilient blocks. We remark that two vertices v and w that are vertex-resilient are not necessarily 2-vertex-connected: this is indeed the case for vertices H and F in the digraph of Figure 1(a). If, however, v and w are not adjacent then $v \leftrightarrow_{2v} w$ if and only if $v \leftrightarrow_{vr} w$.

We next provide some basic properties of vertex-resilient and 2-vertex-connected blocks. Denote by $VRB(u)$ the vertex-resilient blocks that contain u. Define the *block graph* $F = (V_F, E_F)$ of G as follows. The vertex set V_F consists of the vertices in V and also contains one *block node* for each vertex-resilient block of G. The edge set E_F consists of the edges $\{u, B\}$ where $B \in VRB(u)$.

Lemma 1. *Graph F is acyclic.*

Lemma 2. *The number of vertex-resilient blocks in a digraph G is at most $n-1$.*

Lemma 3. *The total number of vertices in all vertex-resilient blocks is at most $2n - 2$.*

Lemma 4. *Let u and v be any vertices that are not vertex-resilient but are connected by a path P in F. Then, for any vertex $w \in V \setminus \{u, v\}$ on P, u and v are not strongly connected in digraph $G \setminus w$.*

We consider F as a forest of rooted trees by choosing an arbitrary vertex as the root of each tree. Then $u \leftrightarrow_{vr} w$ if and only if u and w are siblings or one is the grandparent of the other. We can perform both tests in constant time simply by storing the parent of each vertex in F. Thus, we can test in constant time if two vertices are vertex-resilient. Note that we cannot always apply Lemma 4 to find a strong articulation point that separates two vertices u and w that are not vertex-resilient. Indeed, two vertices that are strongly connected but not vertex-resilient may not even be connected by a path in the forest F. So if we wish to return a witness that u and w are not vertex-resilient, we cannot rely on F. We deal with this problem in Section 4.

Now we turn to 2-vertex-connected blocks. Menger's Theorem [9] implies that if v and w are not adjacent then $v \leftrightarrow_{2v} w$ if and only if $v \leftrightarrow_{vr} w$. If, on the other hand, $v \leftrightarrow_{vr} w$ but v and w are not 2-vertex-connected, then at least one of the edges (v, w) and (w, v) exists in G and it must be a strong bridge.

Lemma 5. *Let v and w be two distinct vertices of G such that $v \leftrightarrow_{vr} w$. Then, v and w are not 2-vertex connected if and only if at least one of the edges (v, w) and (w, v) is a strong bridge in G.*

The following corollary, which relates 2-vertex-connected, 2-edge-connected and vertex-resilient blocks, is an immediate consequence of Lemma 5.

Corollary 6. *For any two distinct vertices v and w, $v \leftrightarrow_{2v} w$ if and only if $v \leftrightarrow_{vr} w$ and $v \leftrightarrow_{2e} w$.*

By Corollary 6 we have that the 2-vertex-connected blocks are refinements of the vertex-resilient blocks, formed by the intersections of the vertex-resilient blocks and the 2-edge-connected blocks of the digraph G. Since the 2-edge-connected blocks are a partition of the vertices of G, these intersections partition each vertex-resilient block. From this property we conclude that Lemmas 1, 2, and 3 also hold for the 2-vertex-connected blocks.

4   Computing the Vertex-resilient Blocks

In this section we present new algorithms for computing the vertex-resilient blocks of a digraph G. We can assume that G is strongly connected, so $m \geq n$. If not, then we process each strongly connected component separately; if $u \leftrightarrow_{vr} v$ then u and v are in the same strongly connected component S of G, and moreover, any vertex on a path from u to v or from v to u also belongs in S. We begin with a simple algorithm that removes a single strong articulation point at a time. In order to get a more efficient solution, we need to consider simultaneously how different strong articulation points divide the vertices into blocks, which we do with the help of dominator trees. We achieve linear running time by combining the simple algorithm with the dominator-tree-based division, and by applying suitable operations on the block forest structure.

A Simple Algorithm. A simple way to compute the vertex-resilient blocks is by removing the strong articulation points of G one at a time. Let u and v be two distinct vertices. We say that a strong articulation point x *separates u from v* if all paths from u to v contain x. In this case u and v belong to different strongly connected components of $G \setminus x$. This observation implies that we can compute the vertex-resilient blocks by computing the strongly connected components of $G \setminus x$ for every strong articulation point x. To do this efficiently we define an operation that refines the currently computed blocks. Let $\mathcal{B}$ be a set of blocks, let $\mathcal{S}$ be a partition of a set $U \subseteq V$, and let x be a vertex not in U.

refine$(\mathcal{B}, \mathcal{S}, x)$: For each block $B \in \mathcal{B}$, substitute B by the sets $B \cap (S \cup \{x\})$ of size at least two, for all $S \in \mathcal{S}$.

In Section 5, where we will compute the 2-vertex-connected blocks from the vertex-resilient blocks and the 2-edge-connected blocks, we will use the notation *refine*$(\mathcal{B}, \mathcal{S})$ as a shorthand for *refine*$(\mathcal{B}, \mathcal{S}, x)$ with $x = null$.

Lemma 7. *Let N be the total number of elements in all sets of $\mathcal{B}$ ($N = \sum_{B \in \mathcal{B}} |B|$), and let K be the number of elements in U. Then, the operation refine$(\mathcal{B}, \mathcal{S}, x)$ can be executed in $O(N + K)$ time.*

In our simple algorithm, that we refer to as SimpleVRB, we initialize the current set of blocks as $\mathcal{B} = \{V\}$, i.e., we begin from the trivial set containing only one block. Then we compute the strong articulation points of G, and perform the following computations for each strong articulation point x. We compute the strongly connected components $S_1, \ldots, S_k$ of $G \setminus x$, and let $\mathcal{S}$ be the partition of $V \setminus x$ defined by the strongly connected components S_i. Then, we execute $refine(\mathcal{B}, \mathcal{S}, x)$.

Lemma 8. *Algorithm SimpleVRB runs in $O(mp^*)$ time, where p^* is the number of strong articulation points of G. This is $O(mn)$ in the worst case.*

Auxiliary Graphs. We will show how to obtain a faster algorithm by applying the framework developed in [3] for the computation of the 2-edge-connected blocks, namely by using dominator trees and auxiliary graphs. As already mentioned, auxiliary graphs need to be defined in a substantially different way, which complicates several technical details.

As a warm up, first consider the computation of $VRB(v)$, i.e., the vertex-resilient blocks that contain a specific vertex v. Consider the flow graph $G(v)$ with start vertex v and its reverse $G^R(v)$, obtained after reversing edge directions. Let w be a vertex other than v. Clearly, v and w are vertex-resilient if and only if v is the only proper dominator of w in both $G(v)$ and $G^R(v)$, i.e., $d(w) = v$ and $d^R(w) = v$. Now let u be a sibling of w in both $D(v)$ and $D^R(v)$. The fact that $d^R(w) = v$ and $d(u) = v$ implies that for any vertex $x \in V \setminus \{v, w, u\}$ there is a path from w to u through v that avoids x. So w and u are in a common vertex-resilient block that contains v if and only if they lie in the same strongly connected component of $G \setminus v$. This observation implies the following linear-time algorithm to compute the vertex-resilient blocks that contain v. Compute the dominator trees $D(v)$ and $D^R(v)$ of $G(v)$ and $G^R(v)$ respectively. Let $C(v)$ (resp., $C^R(v)$) be the set of children of v in $D(v)$ (resp., $D^R(v)$). Set $U = C(v) \cap C^R(v)$ and initialize the set of blocks $\mathcal{B} = \{U\}$. Compute the strongly connected blocks $S_1, S_2, \ldots, S_k$ of $G \setminus v$. Let $\mathcal{S}$ be the set that contains the nonempty restrictions of the S_i sets to U, i.e., $\mathcal{S}$ contains the nonempty sets $S_i \cap U$. Finally, execute $refine(\mathcal{B}, \mathcal{S}, v)$.

Note that all the vertex-resilient blocks can be computed in $O(mn)$ time by applying the above algorithm to all vertices v. To avoid the repeated applications of this algorithm we develop a new concept of *auxiliary graphs* for 2-vertex connectivity. Before doing that, we state two properties regarding information that a dominator tree can provide about vertex-resilient blocks and paths. The proof of Lemma 9 is immediate.

Lemma 9. *Let $G = (V, E)$ be a strongly connected graph, and let $s \in V$ be an arbitrary start vertex. Any two vertices x and y are vertex-resilient only if they are siblings in $D(s)$ or one is the immediate dominator of the other in $G(s)$.*

Lemma 10. *Let r be a vertex, and let v be any vertex that is not a descendant of r in $D(s)$. Then there is a path from v to r that does not contain any proper descendants of r in $D(s)$. Moreover, all simple paths from v to any descendant of r in $D(s)$ contain r.*

As in [3], *auxiliary graphs* are a key concept in our algorithm that provides a decomposition of the input digraph G into smaller digraphs (not necessarily subgraphs of G) that maintain the original vertex-resilient blocks. In [3] we sed a *canonical decomposition* of the input digraph, in order to obtain auxiliary graphs that maintain the 2-edge-connected blocks. A key property of this ecomposition was the fact that any vertex in an auxiliary graph G_r is reachable om a vertex outside G_r only though a single strong bridge. In the computation of the vertex-resilient blocks, however, we have to decompose the input igraph according to strong articulation points, and thus the above property is ompletely lost. To overcome this critical issue, we apply a different and more involved decomposition.

Let s be an arbitrarily chosen start vertex in G. Recall that we denote by $G(s)$ the flow graph with start vertex s, by $G^R(s)$ the flow graph obtained from $G(s)$ after reversing edge directions, by $D(s)$ and $D^R(s)$ the dominator trees of $G(s)$ and $G^R(s)$ respectively, and by $C(v)$ and $C^R(v)$ the set of children of v in $D(s)$ and $D^R(s)$ respectively. For each vertex r, let $C^k(r)$ denote the level k descendants of r, i.e., $C^0(r) = \{r\}$, $C^1(r) = C(r)$, etc. For each vertex $r \neq s$ that s not a leaf in $D(s)$ we build the *auxiliary graph* $G_r = (V_r, E_r)$ *of* r as follows. The vertex set of G_r is $V_r = \cup_{k=0}^{3} C^k(r)$ and it is partitioned into a set of *ordinary* vertices $V_r^o = C^1(r) \cup C^2(r)$ and a set of *auxiliary* vertices $V_r^a = C^0(r) \cup C^3(r)$. The auxiliary graph G_r results from G by contracting the vertices in $V \setminus V_r$ as ollows. All vertices that are not descendants of r in $D(s)$ are contracted into r. For each vertex $w \in C^3(r)$, we contract all descendants of w in $D(s)$ into w. See Figure 2. We use the same definition for the auxiliary graph G_s of s, with he only difference that we let s be an ordinary vertex. Also note that when we orm G_s from G, no vertex is contracted into s. In order to bound the size of all auxiliary graphs, we eliminate parallel edges during those contractions.

Lemma 11. *The auxiliary graphs G_r have at most $4n$ vertices and $4m+n$ edges in total.*

The following lemmas show that the auxiliary graphs are strongly connected and maintain the vertex-resilient relation of the original digraph.

Lemma 12. *Each auxiliary graph G_r is strongly connected.*

Lemma 13. *Let v and w be any two distinct vertices of G. Then v and w are vertex-resilient in G if and only if they are both ordinary vertices in an auxiliary graph G_r and they are vertex-resilient in G_r.*

Now we specify how to compute all the auxiliary graphs $G_r = (V_r, E_r)$ in $O(m + n)$ time. Observe that the edge set E_r contains all edges in $G = (V, E)$ induced by the vertices in V_r (i.e., edges $(u, v) \in E$ such that $u \in V_r$ and $v \in V_r$). We also add in E_r the following types of *shortcut* edges that correspond to paths in G. (a) If G contains an edge (u, v) such that $u \notin V_r$ is a descendant of r in $D(s)$ and $v \in V_r$ then we add the shortcut edge (z, v) where z is the ancestor of u in $D(s)$ such that $z \in C^3(r)$. (b) If G contains an edge (u, v) such that u but not v is a descendant of r in $D(s)$ then we add the shortcut edge (z, r) where z is the nearest ancestor of u in $D(s)$ such that $z \in V_r$ ($z = u$ if $u \in V_r$). We note that we do not keep multiple (parallel) shortcut edges. See Figure 2. We also note that G_s does not contain type-(b) shortcut edges.

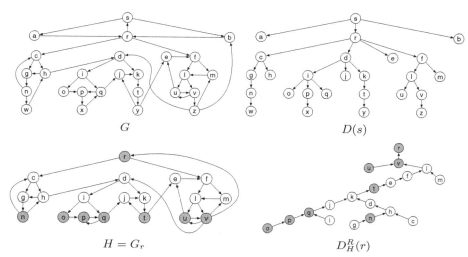

Fig. 2. A strongly connected graph G, the dominator tree $D(s)$ of flow graph $G(s)$, the auxiliary graph $H = G_r$ and the dominator tree $D_H^R(r)$ of the flow graph $H^R(r)$. (The edges of the dominator tree $D_H^R(r)$ are shown directed from child to parent.) The auxiliary vertices of H are shown gray.

To construct the auxiliary graphs $G_r = (V_r, E_r)$ we need to specify how to compute the shortcut edges of type (a) and (b). To do this efficiently we need to test ancestor-descendant relations in $D(s)$, which can be done in $O(1)$-time after $O(n)$-time preprocessing [13], e.g., we number the vertices of $D(s)$ in preorder and compute the number of descendants of each vertex. Suppose (u, v) is an edge of type (a). We need to find the ancestor z of u in $D(s)$ such that $z \in C^3(r)$. We process all such arcs of G_r as follows. We create a list B_r that contains the edges (u, v) of type (a), and sort B_r in increasing preorder of u. We create a second list B_r' that contains the vertices in $C^3(r)$, and sort B_r' in increasing preorder. Then, the shortcut edge of (u, v) is (z, v), where z is the last vertex in the sorted list B_r' such that $pre(z) \le pre(u)$, where $pre(v)$ is the preorder number of v in $D(s)$. Thus the shortcut edges of type (a) can be computed in linear time by bucket sorting and merging. Now we consider the edges of type (b). For each vertex $w \in C^3(r)$ we need to test if there is an edge (u, v) in G such that u is a proper descendant of w and v is not a descendant of r in $D(s)$. In this case, we add in G_r the edge (w, r). To do this test efficiently, we assign to each edge (u, v) a tag $t(u, v)$ which we set equal to the preorder number of the nearest common ancestor of u and v in $D(s)$. We can do this easily by using the parent property of $D(s)$ and the $O(1)$-time test of the ancestor-descendant relation as follows: $t(u, v) = pre(u)$ if u is an ancestor of v in $D(s)$, $t(u, v) = pre(v)$ if v is an ancestor of u in $D(s)$, and $t(u, v) = pre(d(v))$ otherwise. At each node $w \ne s$ in $D(s)$ we store a label $\ell(w)$ which is the minimum tag of among the edges (w, v). Using these labels we compute for each $w \ne s$ in $D(s)$ the values $low(w) = \min\{\ell(v) \mid v \text{ is a descendant of } w \text{ in } D(s)\}$. These computations can be done in $O(m)$ time by processing the tree $D(s)$ in a bottom-up order. Now

onsider the auxiliary graph G_r. We process the vertices in $C^3(r)$. For each such
ertex w we add the shortcut edge (w, r) if $low(w) < pre(r)$.

Lemma 14. *We can compute all auxiliary graphs G_r in $O(m + n)$ time.*

The Linear-time Algorithm. Our linear-time algorithm FastVRB is illustrated
n Figure 3. It uses two levels of auxiliary graphs and applies one iteration of
Algorithm SimpleVRB for each auxiliary graph of the second level. The algorithm
uses different dominator trees, and applies Lemma 9 in order to identify the
ertex-resilient blocks. Since different dominator trees may define different blocks
which by Lemma 9 are supersets of the vertex-resilient blocks), we will use an
operation that we call *split* to combine the different blocks.

  We begin by computing the dominator tree $D(s)$ for an arbitrary start vertex
s. For any vertex v, we let $\hat{C}(v)$ denote the set containing v and the children
of v in $D(s)$, i.e., $\hat{C}(v) = C(v) \cup \{v\}$. Lemma 9 gives an initial division of the
ertices into blocks that are supersets of the vertex-resilient blocks. Specifically,
he vertex-resilient blocks that contain v are subsets of $\hat{C}(v)$ or $\hat{C}(d(v))$ (for
$v \neq s$). During the course of the algorithm, each vertex v becomes associated
with a set of blocks $\mathcal{B}(v)$ that contain v, which are subsets of $\hat{C}(v)$ and $\hat{C}(d(v))$
f $v \neq s$. The blocks are refined by applying the operation *refine*, defined above,
and operation *split* that we define next, and at the end of the algorithm each set
of blocks $\mathcal{B}(v)$ will be equal to $VRB(v)$. Let B be a block and T be a tree with
ertex set $V(T) \supseteq B$. For any vertex $v \in V(T)$, let $\hat{C}_T(v)$ be the set containing
v and the children of v in T.

$split(B, T)$: Return the set that consists of the blocks $B \cap \hat{C}_T(v)$ of size at least
    two, for all $v \in V(T)$.

Lemma 15. *Let N be the number of vertices in $V(T)$. Then, the operation
$split(B, T)$ can be executed in $O(N)$ time.*

  At a high level, the algorithm begins with a "coarse" block tree, induced by
the $\hat{C}(v)$ sets of $D(s)$, which is then refined by the blocks defined from the dominator trees of the auxiliary graphs. The final vertex-resilient block forest is then
computed by considering the strongly connected components of the second level
auxiliary graphs, after removing their designated start vertex. The algorithm
needs to keep track of the blocks that contain a specific vertex, and, conversely,
of the vertices that are contained in a specific block. To facilitate this search we
explicitly store the adjacency lists of the current block forest F. Recall that F
is bipartite, so the adjacency list of a vertex v stores the blocks that contain
v, and the adjacency list of a block node B stores the vertices in B. Initially F
contains one block for each set $\hat{C}(v)$, for all vertices v that are not leaves in $D(s)$.
These blocks are later refined by executing the *split* and *refine* operations, which
maintain the invariant that F is a forest, and that any two distinct blocks have
at most two vertices in common. When we execute a *split* or a *refine* operation
we can update the adjacency lists of F, while maintaining the bounds given in
Lemmas 7 and 15. During the execution of the algorithm the number of blocks
in F remains at most $n - 1$. blocks at any given time. This fact implies that
Lemma 3 holds, so the total number of vertices and edges in F is $O(n)$.

Algorithm FastVRB: Linear-time computation of the vertex-resilient blocks of a strongly connected digraph $G = (V, E)$

Step 1: Choose an arbitrary vertex $s \in V$ as a start vertex. Compute the dominator tree $D(s)$. For any vertex v, let $\hat{C}(v)$ be the set containing v and the children of v in $D(s)$. For every vertex v that is not a leaf in $D(s)$, associate block $\hat{C}(v)$ with every vertex $w \in \hat{C}(v)$.

Step 2: Compute the auxiliary graphs G_r for all vertices r that are not leaves in $D(s)$.

Step 3: Process the vertices of $D(s)$ in bottom-up order. For each auxiliary graph $H = G_r$ with r not a leaf in $D(s)$ do:

    Step 3.1: Compute the dominator tree $T = D_H^R(r)$.

    Step 3.2: Compute the set $\mathcal{B}$ of blocks that contain vertices in $C(r)$.

    Step 3.3: For each block $B \in \mathcal{B}$ execute $split(B, T)$.

    Step 3.4: Compute the auxiliary graphs H_q^R for all vertices q that are not leaves in T.

    Step 3.5: For each auxiliary graph H_q^R with q not a leaf do:

        Step 3.5.1: Compute the set $\mathcal{B}_q$ of blocks that contain at least two ordinary vertices in H_q^R.

        Step 3.5.2: Compute the set $\mathcal{S}$ of the strongly connected components of $H_q^R \setminus q$.

        Step 3.5.3: Refine the blocks in $\mathcal{B}_q$ by executing $refine(\mathcal{B}_q, \mathcal{S}, q)$.

Fig. 3. Algorithm FastVRB

Theorem 16. *Algorithm FastVRB is correct and runs in $O(m + n)$ time.*

Queries. Algorithm FastVRB computes the vertex-resilient blocks of the input digraph G and stores them in the block forest F of Section 3, which makes it straightforward to test in constant time if two query vertices v and w are vertex-resilient. Here we show that if v and w are not vertex-resilient, then we can report a witness of this fact, that is, a strong articulation point x such that v and w are not in the same strongly connected component of $G \setminus x$. Using this witness, it is straightforward to verify in $O(m)$ time that v and w are not vertex-resilient; it suffices to check that v is not reachable from w in $G \setminus x$ or vice versa.

To obtain our witness, we would like to apply Lemma 4, but this requires v and w to be in the same tree of the block forest. Fortunately, we can find the witness fast by applying Lemmas 9 and 10. To that end, it suffices to store the dominator tree $D(s)$ of $G(s)$, and the dominator trees $D_H^R(r)$ of all auxiliary graphs $H^R = G_r^R$. The space required for these data structures is $O(n)$ by Lemma 11. The details are provided in the full version.

Theorem 17. *Let G be a digraph with n vertices and m edges. We can compute the vertex-resilient blocks of G in $O(m + n)$ time and store them in a data structure of $O(n)$ space. Given this data structure, we can test in $O(1)$ time if any two vertices are vertex-resilient. Moreover, if the two vertices are not vertex-resilient, then we can report in $O(1)$ time a strong articulation point that separates them.*

5 Computing the 2-vertex-connected Blocks

We can compute the 2-vertex-connected blocks of the input digraph $G = (V, E)$ by applying Corollary 6 as follows. Given the vertex-resilient blocks $\mathcal{B}$ and the 2-edge-connected blocks $\mathcal{S}$ of G, we simply execute $refine(\mathcal{B}, \mathcal{S})$. This takes $O(n)$ time by Lemma 7. Also, since the 2-vertex-connected blocks have a block forest representation, we can test if two given vertices are 2-vertex-connected in $O(1)$ time as described in Section 3. If we only wish to answer queries of whether two vertices v and w are 2-vertex-connected, without computing explicitly the 2-vertex and the 2-edge-connected blocks, then we can use a simpler alternative, as suggested by Lemma 5. This way we can also obtain a strong bridge that separates a pair of vertices that are vertex-resilient but not 2-vertex-connected.

Theorem 18. *Let G be a digraph with n vertices and m edges. We can compute the 2-vertex-connected blocks of G in $O(m + n)$ time and store them in a data structure of $O(n)$ space. Given this data structure, we can test in $O(1)$ time if any two vertices are 2-vertex-connected. Moreover, if the two vertices are not 2-vertex-connected, then we can report in $O(1)$ time a strong articulation point or a strong bridge that separates them.*

References

1. Alstrup, S., Harel, D., Lauridsen, P.W., Thorup, M.: Dominators in linear time. SIAM Journal on Computing **28**(6), 2117–32 (1999)
2. Buchsbaum, A.L., Georgiadis, L., Kaplan, H., Rogers, A., Tarjan, R.E., Westbrook, J.R.: Linear-time algorithms for dominators and other path-evaluation problems. SIAM Journal on Computing **38**(4), 1533–1573 (2008)
3. Georgiadis, L., Italiano, G.F., Laura, L., Parotsidis, N.: 2-edge connectivity in directed graphs. In: Proc. 26th ACM-SIAM Symp. on Discrete Algorithms, pp. 1988–2005 (2015)
4. Georgiadis, L., Tarjan, R.E.: Dominator tree certification and independent spanning trees (2012). CoRR, abs/1210.8303
5. Henzinger, M., Krinninger, S., Loitzenbauer, V.: Finding 2-edge and 2-vertex strongly connected components in quadratic time. In: Proc. 42nd International Colloquium on Automata, Languages, and Programming (ICALP 2015) (2015)
6. Italiano, G.F., Laura, L., Santaroni, F.: Finding strong bridges and strong articulation points in linear time. Theoretical Computer Science **447**, 74–84 (2012)
7. Jaberi, R.: Computing the 2-blocks of directed graphs (2014). CoRR, abs/1407.6178
8. Jaberi, R.: On computing the 2-vertex-connected components of directed graphs (2014) CoRR, abs/1401.6000
9. Menger, K.: Zur allgemeinen kurventheorie. Fund. Math. **10**, 96–115 (1927)
10. Nagamochi, H., Watanabe, T.: Computing k-edge-connected components of a multigraph. IEICE Transactions on Fundamentals of Electronics, Communications and Computer Sciences **E76A**(4), 513–517 (1993)
11. Reif, J.H., Spirakis, P.G.: Strong k-connectivity in digraphs and random digraphs. Technical Report TR-25-81, Harvard University (1981)
12. Tarjan, R.E.: Depth-first search and linear graph algorithms. SIAM Journal on Computing **1**(2), 146–160 (1972)
13. Tarjan, R.E.: Finding dominators in directed graphs. SIAM Journal on Computing **3**(1), 62–89 (1974)

Ground State Connectivity of Local Hamiltonians

Sevag Gharibian[1] and Jamie Sikora[2]([⊠])

[1] Department of Computer Science,
Virginia Commonwealth University, Richmond, USA
[2] Centre for Quantum Technologies and MajuLab, CNRS-UNS-NUS-NTU
International Joint Research Unit, UMI 3654, National University of Singapore,
Singapore, Singapore
cqtjwjs@nus.edu.sg

Abstract. The study of ground state energies of local Hamiltonians has played a fundamental role in quantum complexity theory. In this paper, we take a new direction by introducing the physically motivated notion of "ground state connectivity" of local Hamiltonians, which captures problems in areas ranging from quantum stabilizer codes to quantum memories. We show that determining how "connected" the ground space of a local Hamiltonian is can range from QCMA-complete to PSPACE-complete, as well as NEXP-complete for an appropriately defined "succinct" version of the problem. As a result, we obtain a natural QCMA-complete problem, a goal which has generally proven difficult since the conception of QCMA over a decade ago. Our proofs rely on a new technical tool, the Traversal Lemma, which analyzes the Hilbert space a local unitary evolution must traverse under certain conditions. We show that this lemma is essentially tight with respect to the length of the unitary evolution in question.

1 Introduction

Over the last fifteen years, the merging of condensed matter physics and computational complexity theory has given rise to a new field of study known as *quantum Hamiltonian complexity*. The cornerstone of this field is arguably Kitaev's [1] quantum version of the Cook-Levin theorem [2,3], which says that the problem of estimating the ground state energy of a local Hamiltonian is complete for the class Quantum Merlin Arthur (QMA), where QMA is a natural generalization of NP. Here, a k-local Hamiltonian is an operator $H = \sum_i H_i$ acting on n qubits, such that each local Hermitian constraint H_i acts non-trivially on k qubits. The *ground state energy* of H is simply the smallest eigenvalue of H, and the corresponding eigenspace is known as the *ground space* of H.

Kitaev's result spurred a long line of subsequent works on variants of the ground energy estimation problem, known as the k-local Hamiltonian problem (k-LH). For example, Oliveira and Terhal showed that LH remains QMA-complete in the physically motivated case of qubits arranged on a 2D lattice [4].

© Springer-Verlag Berlin Heidelberg 2015
M.M. Halldórsson et al. (Eds.): ICALP 2015, Part I, LNCS 9134, pp. 617–628, 2015.
DOI: 10.1007/978-3-662-47672-7_50

Bravyi and Vyalyi proved [5] that the *commuting* variant of 2-LH is in NP. More recently, the complexity of the version of 2-LH in which large positive and negative weights on local terms are allowed[1] was characterized by Cubitt and Montanaro [7] in a manner analogous to Schaeffer's dichotomy theorem for Boolean satisfiability [8]. Thus, k-LH has served as an excellent "benchmark" problem for delving into the complexity of problems encountered in the study of local Hamiltonians. Yet, one can also ask about the *properties of the ground space* itself. For example, is it topologically ordered? Can we evaluate local observables against it (e.g. for non-degenerate ground state $|\psi\rangle$ and 2-local observable O, can one estimate $\langle\psi|\, I \otimes O\, |\psi\rangle$)? It is this direction which we pursue in this paper.

Specifically, in this paper we define a notion of *connectivity* of the ground space of H, which roughly asks: Given ground states $|\psi\rangle$ and $|\phi\rangle$ of H as input, are they "connected" through the ground space of H? Somewhat more formally, we have (see Section 2 for a formal definition):

Definition 1 (Ground State Connectivity (GSCON) (informal)). *Given as input a local Hamiltonian H and ground states $|\psi\rangle$ and $|\phi\rangle$ of H (specified via quantum circuits), as well as parameters m and l, does there exist a sequence of l-qubit unitaries $(U_i)_{i=1}^m$ such that:*

1. *($|\psi\rangle$ mapped to $|\phi\rangle$) $U_m \cdots U_1 |\psi\rangle \approx |\phi\rangle$, and*
2. *(intermediate states in ground space) $\forall\, i \in [m]$, $U_i \cdots U_1 |\psi\rangle$ is in the ground space of H?*

In other words, GSCON asks whether there exists a sequence of m unitaries, each acting on (at most) l qubits, mapping the initial state $|\psi\rangle$ to the final state $|\phi\rangle$ *through* the ground space of H. We stress that the parameters m (i.e. number of unitaries) and l (i.e. the locality of each unitary) are key; as we discuss shortly, depending on their setting, the complexity of GSCON can vary greatly.

Physics Motivation. The original inspiration for this work came from the classical study of *reconfiguration* problems (see *Previous work* below for details). For example, the reconfiguration problem for 3SAT asks: Given a 3SAT formula ϕ and satisfying assignments x and y for ϕ, does there exist a sequence of bit flips mapping x to y, such that each intermediate assignment encountered is also a satisfying assignment for ϕ? Although classically, reconfiguration problems are arguably mostly interesting from a theoretical perspective, their quantum variant (i.e. GSCON) turns out to be physically relevant. To illustrate, we now discuss connections to *quantum memories* and *stabilizer codes*.

Quantum Memories. A key challenge in building quantum computers is the implementation of long-lived qubit systems. In low-temperature systems, one approach is to encode a qubit in the ground state of a gapped Hamiltonian with

---

[1] Note that certain physically motivated local Hamiltonian models, such as the Heisenberg anti-ferromagnet (see, e.g., [6] for a definition), require unit weights on all constraints, and are thus not captured by the dichotomy theorem of [7].

a degenerate ground space. Here, the degeneracy ensures the ground space has at least two basis states, logical $|\widetilde{0}\rangle$ and $|\widetilde{1}\rangle$, and the gap ensures that external noise does not (easily) take a ground state out of the ground space. However this is not sufficient — although environmental noise may not take the state out of the ground space, it can still alter the state *within* the ground space (e.g. inadvertently map $|\widetilde{0}\rangle$ to $|\widetilde{1}\rangle$). Thus, making the typical assumption that errors act locally, it should ideally not be possible for $|\widetilde{0}\rangle$ to be mapped to $|\widetilde{1}\rangle$ through the ground space via a sequence of local operations. This is precisely the principle behind Kitaev's toy chain model [9], and the motivation behind the toric code [10] (see also [11]). This notion of how "robust" a quantum memory is can thus be phrased as an instance of GSCON: Given a gapped Hamiltonian H, a ground state $|\psi\rangle$ to which the quantum memory is initialized, and an undesired ground state $|\phi\rangle$, is there a sequence of local errors mapping the state of our quantum memory through the ground space from $|\psi\rangle$ to $|\phi\rangle$?

Stabilizer Codes. Roughly, a stabilizer code is a quantum error-correcting code defined by a set of commuting Hermitian operators, $S = \{G_1, \ldots, G_k\}$, such that $G_i \neq -I$ and $\|G_i\|_\infty \leq 1$ for all $G_i \in S$. The *codespace* for S is the set of all $|\psi\rangle$ satisfying $G_i |\psi\rangle = |\psi\rangle$ for all $i \in [k]$. In other words, defining G_i^+ as the projection onto the $+1$ eigenspace of G_i, the codespace is the ground space of the positive semidefinite Hamiltonian $H := \sum_{i=1}^{k}(I - G_i^+)$. Typically, errors are assumed to occur on a small number of qubits at a time; with this assumption in place, the following is a special case of GSCON: Given H and codewords $|\psi\rangle$ and $|\phi\rangle$, does there exist a sequence of at most m local errors mapping $|\psi\rangle$ to $|\phi\rangle$, such that the entire error process is undetectable, i.e. each intermediate state remains in the codespace?

Results. Having motivated GSCON, we now informally state our results.

Theorem 1 (Informal, see Theorem 5 for a formal statement). GSCON *for polynomially large m (i.e. for polynomially many local unitaries U) and $l = 2$ (i.e. 2-qubit unitaries) is* QCMA-*complete.*

Here, QCMA is QMA except with a classical prover [12]. See Section 2 for a formal definition. Theorem 1 says that determining whether there exists a polynomial-size quantum circuit mapping $|\psi\rangle$ to $|\phi\rangle$ through the ground space of H is QCMA-complete.

Theorem 2 (Informal, see full version [13] for a formal statement). GSCON *for exponentially large m (i.e. for exponentially many local unitaries U) and $l = 1$ (i.e. 1-qubit unitaries) is* PSPACE-*complete.*

Theorem 2 says that determining whether there exists an exponential length sequence of 1-qubit unitaries mapping $|\psi\rangle$ to $|\phi\rangle$ through the ground space of H is PSPACE-complete.

Finally, in the full version we define a succinct variant of GSCON, called SUCCINCT GSCON, in which the Hamiltonian H has a succinct circuit description, and the initial and final states $|\psi\rangle$ and $|\phi\rangle$ are product states. We show:

Theorem 3 (Informal, see full version [13] for a formal statement). *The problem* SUCCINCT GSCON *for exponentially large m (i.e. for exponentially many local unitaries U) and l = 1 (i.e. 1-qubit unitaries) is* NEXP-*complete.*

We remark that the choices of m and l above are key to our results. For example, Theorem 1 holds for any constant $l \geq 2$ (see remarks after its proof); however, for $l \in \omega(\log N)$ (for N the input size) the problem is likely no longer in QCMA, as the prover cannot send a classical description of each local unitary. Similarly, attempting to extend Theorem 2 by setting $l = 2$ appears problematic, as then any intermediate state in the unitary evolution seems to require exponential space to represent. This latter problem is, however, in NEXP. We thus conjecture that it is actually NEXP-complete.

Proof Techniques. Our results rely on a new technical lemma called the Traversal Lemma, as well as the use of ϵ-nets and ϵ-pseudo-nets (also known as *improper covering sets*). We now outline the proof techniques behind Theorem 5 (QCMA-completeness) in more detail; using similar ideas, we have that Theorems 2 (PSPACE-completeness) and 3 (NEXP-completeness) follow analogously.

Specifically, we outline both QCMA-hardness and containment in QCMA. Beginning with the former, the central idea behind the construction is as follows. Let V be an arbitrary QCMA verification circuit, and let H' be the local Hamiltonian obtained from V via Kitaev's circuit-to-Hamiltonian construction [1]. Then, we design the input Hamiltonian H to GSCON so that "traversing its ground space" is equivalent to simulating the following protocol: Starting from the all-zeroes state, prepare the ground state of H' (which can be done efficiently since V is a QCMA circuit), and subsequently flip a set of special qubits called GO qubits. This latter step "activates" the check Hamiltonian H, which now "verifies" that the ground state prepared is indeed correct. Finally, uncompute the ground state to arrive at a target state of all-zeroes except in the GO register, which is now set to all ones.

To prove correctness of this construction, our main technical tool is a new lemma we call the *Traversal Lemma*, which analyzes the Hilbert space a local unitary evolution must traverse in certain settings. Specifically, define two states $|\psi\rangle$ and $|\phi\rangle$ as k-*orthogonal* if for any k-local unitary U, we have $\langle\phi|U|\psi\rangle = 0$. In other words, any application of a k-local unitary leaves $|\psi\rangle$ and $|\phi\rangle$ orthogonal. Then, the Traversal Lemma roughly says that for k-orthogonal states $|\psi\rangle$ and $|\phi\rangle$, if we wish to map $|\psi\rangle$ to $|\phi\rangle$ via a sequence of k-local unitaries, then at some step in this evolution we must leave the space spanned by $|\psi\rangle$ and $|\phi\rangle$, i.e. we must have "large" inner product with $I - |\psi\rangle\langle\psi| - |\phi\rangle\langle\phi|$. (Here, "large" means the inner product scales at least as $\Omega(1/m^2)$, for m the number of k-local unitaries applied.) To prove the Traversal Lemma, we use a combination of the Gentle Measurement Lemma of Winter [14] and an idea inspired by the quantum Zeno effect.

As the Traversal Lemma is a key technical contribution of this paper, we also study its properties further (i.e. independently of its application to our complexity theoretic results). For example, we show the lemma is tight up to a polynomial factor in the number of unitaries, m. To do so, we give a pair of 2-orthogonal states $|\psi\rangle$, $|\phi\rangle$ with the following property: For any $0 < \Delta < 1/2$ we construct a carefully selected sequence of $O(1/\Delta^2)$ 2-local unitaries mapping $|\psi\rangle$ to $|\phi\rangle$, such that at any point in this mapping, the inner product with $I - |\psi\rangle\langle\psi| - |\phi\rangle\langle\phi|$ is at most Δ. We also delve further into the study of k-orthogonality, including giving an intuitive characterization of the notion.

Finally, containment of GSCON in QCMA is shown via a simple and natural verification procedure, wherein the prover sends a classical description of the local unitaries $\{U_i\}$, and the verifier prepares many copies of the starting, final, and all intermediate states and checks that all required properties hold. To make this rigorous, we construct an ϵ-pseudo-net, which allows us to easily discretize the space of d-dimensional unitary operators for any $d \geq 2$. Such pseudo-nets come with a tradeoff: On the negative side, they contain non-unitary operators. On the positive side, they are not only straightforward to construct, but more importantly, they have the following property: Given any element A in the pseudo-net, there are efficient *explicit* protocols for checking if A is close to unitary, and if so, for "rounding" it to such a unitary.

Previous Work. To the best of our knowledge, our work is the first to study reconfiguration in the quantum setting. In the classical setting, the inspiration for our work came from the paper of Gopalan, Kolaitis, Maneva, and Papadimitriou [15], which shows that determining whether two solutions x and y of a Boolean formula are connected through the solution space is either in P or is PSPACE-complete, depending on the constraint types allowed in the formula. Recently, Mouawad, Nishimura, Pathak and Raman [16] studied the variant of this problem in which one seeks the *shortest* possible Boolean reconfiguration path; they show this problem is either in P, NP-complete, or PSPACE-complete. In this sense, our definition of GSCON can be thought of as a quantum generalization of the problem studied in [16]. More generally, since the work of [15], a flurry of papers have appeared studying reconfiguration for problems ranging from Boolean satisfiability to vertex cover to graph coloring (e.g. [17–20]).

Significance to Complexity Theory. We now discuss the significance of our results from a complexity theoretic perspective. We begin by focusing on QCMA, which is a natural class satisfying MA $\subseteq$ QCMA $\subseteq$ QMA. Although QCMA was introduced over a decade ago by Aharonov and Naveh [12], we still have an unfortunately small number of complete problems for it. In particular, to the best of our knowledge, the following is an exhaustive list at the time of writing:

- Does a given local Hamiltonian have an efficiently preparable ground state [21]?
- Does a given quantum circuit act almost as the identity on computational basis states [21]?

- Given a braid, can it be conjugated by another braid from a given class such that the Jones polynomial of its plat closure is nearly maximal [22]?
- Given a continuous-time classical random walk on a restricted class of graphs, and time T, do there exist vertices i and j such that the difference of the probabilities of being at i and j is at least $c \cdot \exp(-\mu T)$ [23]?
- Given a quantum circuit C accepting a non-empty monotone set, what is the smallest Hamming weight string accepted by C [24]?

In this regard, the pursuit of natural complete problems for QCMA has arguably proven rather difficult. Our results add a new, physically-motivated problem (i.e. GSCON) to the short list of QCMA-complete problems.

Second, a common focus in quantum complexity theory has been the problem of estimating the ground state energy of a given local Hamiltonian. However, less attention has been given to the complexity of determining other properties of local Hamiltonians. For example, Brown, Flammia, and Schuch showed [25] that computing the ground state degeneracy and density of states for a local Hamiltonian is #BQP-complete. Gharibian and Kempe showed [24] that determining the smallest subset of interaction terms of a given local Hamiltonian which yields a high energy ground space is cq-Σ_2-complete. Ambainis has shown [26] (among other results) that evaluating local observables against a local Hamiltonian is $P^{QMA[\log n]}$-complete, and that determining the spectral gap of a local Hamiltonian is in $P^{QMA[\log n]}$. Continuing in this vein, our work initiates a new direction of study regarding properties of local Hamiltonians beyond estimating the ground state energy, namely the study of ground state connectivity.

Finally, regarding the use of our proof techniques in the study of quantum algorithms and verification procedures, we hope the Traversal Lemma may prove useful in its own right. For example, in quantum adiabatic algorithms, it is often notoriously difficult to understand how a quantum state evolves in time from an easy-to-prepare initial state to some desired final state. The Traversal Lemma gives us a tool for studying the behaviour of such evolutions, playing a crucial role in our analysis here. We remark, however, that in quantum adiabatic evolution, the Hamiltonian itself changes with time, whereas here our Hamiltonian is fixed and we apply local unitary gates to our quantum state.

Organization. This paper is organized as follows. In Section 2, we state notation, definitions, and known results. Section 3 introduces the notion of k-orthogonality and the Traversal Lemma, and states our result regarding the latter's tightness. QCMA-hardness of GSCON via the Traversal Lemma is sketched in Section 4. The definition of the succinct version of GSCON, as well as all omitted technical details (e.g. full proofs of Theorems 5, 2, and 3 and our further study of k-orthogonality) can be found in the full version [13].

2   Preliminaries

We now state definitions and known tools useful to this work; an expanded such section, as required for our technical proofs, is given in the full version [13].

Notation. For $x \in \{0,1\}^n$, $|x\rangle \in (\mathbb{C}^2)^{\otimes n}$ denotes the computational basis state labeled by x. For complex Euclidean space $\mathcal{X}$, let $\mathrm{L}(\mathcal{X})$, $\mathrm{Herm}(\mathcal{X})$ and $\mathrm{U}(\mathcal{X})$ denote the sets of linear, Hermitian and unitary operators acting on $\mathcal{X}$, respectively. Define matrix norms: $\|A\|_{\max} := \max_{ij} |A(i,j)|$, the spectral norm $\|A\|_\infty := \max\{\|A|v\rangle\|_2 : \||v\rangle\|_2 = 1\}$, and trace norm $\|A\|_{\mathrm{tr}} := \mathrm{Tr}\sqrt{A^\dagger A}$. We treat the local dimension d of quantum systems as a constant.

Definitions and Tools. We now formally define the problem studied in this paper. (To ease parsing, the input parameters are highlighted in maroon online.)

Definition 2 (Ground State Connectivity (GSCON)).

- *Input parameters:*
  1. *k-local Hamiltonian $H = \sum_i H_i$ acting on n qubits with Hermitian H_i satisfying $\|H_i\|_\infty \leq 1$.*
  2. *$\eta_1, \eta_2, \eta_3, \eta_4, \Delta \in \mathbb{R}$, and integer $m \geq 0$, such that $\eta_2 - \eta_1 \geq \Delta$ and $\eta_4 - \eta_3 \geq \Delta$.*
  3. *Polynomial size quantum circuits U_ψ and U_ϕ generating "starting" and "target" states $|\psi\rangle$ and $|\phi\rangle$ (starting from $|0\rangle^{\otimes n}$), respectively, satisfying $\langle\psi| H |\psi\rangle \leq \eta_1$ and $\langle\phi| H |\phi\rangle \leq \eta_1$.*

- *Output:*
  1. *If there exists a sequence of l-local unitaries $(U_i)_{i=1}^m \in \mathrm{U}\left(\mathbb{C}^2\right)^{\times m}$ s.t.:*
     (a) (Intermediate states remain in low energy space) For all $i \in [m]$ and intermediate states $|\psi_i\rangle := U_i \cdots U_2 U_1 |\psi\rangle$, one has $\langle\psi_i| H |\psi_i\rangle \leq \eta_1$, and
     (b) (Final state close to target state) $\|U_m \cdots U_1 |\psi\rangle - |\phi\rangle\|_2 \leq \eta_3$, then output YES.
  2. *If for all l-local sequences of unitaries $(U_i)_{i=1}^m \in \mathrm{U}\left(\mathbb{C}^2\right)^{\times m}$, either:*
     (a) (Intermediate state obtains high energy) There exists $i \in [m]$ and intermediate state $|\psi_i\rangle := U_i \cdots U_2 U_1 |\psi\rangle$, s.t. $\langle\psi_i| H |\psi_i\rangle \geq \eta_2$, or
     (b) (Final state far from target state) $\|U_m \cdots U_1 |\psi\rangle - |\phi\rangle\|_2 \geq \eta_4$, then output NO.

A few remarks are in order. First, in the Hamiltonian complexity literature the gap size Δ for energy levels of local Hamiltonians is often taken to be inverse polynomial. Some of our results require this gap to be exponentially small. Allowing Δ to be specified as input thus allows us to precisely formulate such results. Second, the circuits U_ψ and U_ϕ are assumed to be given in terms of 1 and 2-qubit unitary gates. Third, all input parameters are specified with rational entries, each using $O(\mathrm{poly}(n))$ bits of precision.

Next, let us recall the definition of the complexity class QCMA [12].

Definition 3 (QCMA). *A promise problem $A = (A_{\mathrm{yes}}, A_{\mathrm{no}})$ is in QCMA if and only if there exist polynomials p, q and a polynomial-time uniform family of quantum circuits $\{Q_n\}$, where Q_n takes as input a string $x \in \Sigma^*$ with $|x| = n$, a classical proof $y \in \{0,1\}^{\otimes p(n)}$, and $q(n)$ ancilla qubits in state $|0\rangle^{\otimes q(n)}$, s.t.:*

– *(Completeness) If $x \in A_{\text{yes}}$, then there exists a proof $y \in \{0,1\}^{\otimes p(n)}$ such that Q_n accepts (x,y) with probability at least $2/3$.*

– *(Soundness) If $x \in A_{\text{no}}$, then for all proofs $y \in \{0,1\}^{\otimes p(n)}$, Q_n accepts (x,y) with probability at most $1/3$.*

We next state a lemma regarding the 3-local circuit-to-Hamiltonian construction of Kempe and Regev [27].

Lemma 1 (Kempe and Regev [27]). *Kempe and Regev's construction maps a quantum circuit V to a 3-local Hamiltonian H with parameters α and β s.t.:*

- *If there exists a proof $|\psi\rangle$ accepted by V with probability at least $1 - \epsilon$, then there exists a state $|\psi_{\text{hist}}\rangle$ achieving $\mathrm{Tr}(H |\psi_{\text{hist}}\rangle\langle\psi_{\text{hist}}|) \leq \alpha := \epsilon/(L+1)$.*
- *If V rejects all proofs $|\psi\rangle$ with probability at least $1 - \epsilon$, then the smallest eigenvalue of H is at least $\beta \in \Omega\left(\frac{1}{L^3}\right)$.*

3 k-Orthogonality and the Traversal Lemma

The key technical tool for proving our hardness results is the Traversal Lemma (Lemma 2). In this section, we state this lemma and study its tightness. All technical proofs, as well as a further study on the notion of k-orthogonality, are deferred to the full version [13]. We begin by introducing the notions of k-orthogonal states and k-orthogonal subspaces.

Definition 4 (k-orthogonal states and subspaces). *For $k \geq 1$, a pair of states $|v\rangle, |w\rangle \in (\mathbb{C}^d)^{\otimes n}$ is k-orthogonal if for all k-qudit unitaries U, we have $\langle w| U |v\rangle = 0$. We call subspaces $S, T \subseteq (\mathbb{C}^d)^{\otimes n}$ k-orthogonal if any pair of vectors $|v\rangle \in S$ and $|w\rangle \in T$ are k-orthogonal.*

Remarks on k-orthogonality: First, k-orthogonality implies orthogonality, but not vice versa. For example, $|000\rangle$ and $|111\rangle$ are 2-orthogonal and hence orthogonal. In contrast, $|000\rangle$ and $|100\rangle$ are orthogonal but not k-orthogonal for any $k \geq 1$ (i.e. simply apply Pauli X to qubit 1 to map $|000\rangle$ to $|100\rangle$). Similarly, letting S and T denote the $+1$ eigenspaces of $I \otimes |000\rangle\langle000|$ and $I \otimes |111\rangle\langle111|$, respectively, we have that S and T are 2-orthogonal subspaces.

We now state the Traversal Lemma, which says: For any two k-orthogonal subspaces S and T with $|v\rangle \in S$ and $|w\rangle \in T$, any sequence of m k-qudit unitaries mapping $|v\rangle$ to $|w\rangle$ must induce an evolution which has "large" overlap with the orthogonal complement of both S and T at some time step $i \in [m]$.

Lemma 2 (Traversal Lemma). *Let $S, T \subseteq (\mathbb{C}^d)^{\otimes n}$ be k-orthogonal subspaces. Fix arbitrary states $|v\rangle \in S$ and $|w\rangle \in T$, and consider a sequence of k-qudit unitaries $(U_i)_{i=1}^m$ such that $\| |w\rangle - U_m \cdots U_1 |v\rangle \|_2 \leq \epsilon$ for some $0 \leq \epsilon < 1/2$. Define $|v_i\rangle := U_i \cdots U_1 |v\rangle$ and $P := I - \Pi_S - \Pi_T$. Then, there exists an $i \in [m]$ such that $\langle v_i| P |v_i\rangle \geq \left(\frac{1-2\epsilon}{2m}\right)^2$.*

Proof. We sketch the proof here. We proceed by contradiction. Suppose that for all $i \in [m]$, we have that $\langle v_i | P | v_i \rangle$ is "small". Then, imagine that after each unitary U_i is applied, we measure our state $|v_i\rangle$ with operators $(P, I - P)$ and postselect onto outcome $I - P$. Since our overlap with P is "always small", the Gentle Measurement Lemma implies that each post-selected state must be "close" to the corresponding pre-measurement state. Based on this idea, one can show that even if we measure after each step i, the resulting final state we get is "close" to the target state $|w\rangle$.

However, since $|v\rangle \in S$ and since S and T are k-orthogonal subspaces, we have that each time we postselect onto $I - P$ after a $(k - 1)$-local unitary is applied, we are "snapped" back to subspace S with certainty. In other words, measuring after each unitary is applied results in a final state which lies in S. But any state in S is orthogonal to the target state $|w\rangle$, yielding the contradiction.

We next ask whether the Traversal Lemma is tight in the following sense: In Lemma 2, the lower bound on $\langle v_i | P | v_i \rangle$ scales as $\Theta(1/m^2)$ (for m the number of unitaries and for fixed ϵ). This intuitively suggests that one can better "avoid" the subspace P projects onto if one uses a longer sequence of local unitaries. Is such behavior possible? Or can the lower bound in Lemma 2 be improved to a constant independent of m? Our next result shows that a dependence on m in Lemma 2 is indeed necessary.

Theorem 4. *Assume the notation of Lemma 2. Fix any $0 < \Delta < 1/2$, and consider 2-orthogonal states $|v\rangle = |000\rangle$ and $|w\rangle = |111\rangle$, with $P := I - |v\rangle\langle v| - |w\rangle\langle w|$. Then, there exists a sequence of m 2-local unitary operations mapping $|v\rangle$ to $|w\rangle$ through intermediate states $|v_i\rangle$, each of which satisfy $\langle v_i | P | v_i \rangle \leq \Delta$, and where $m \in O(1/\Delta^2)$.*

The idea behind the proof is based on the following rough analogy: Suppose one wishes to map the point $(1, 1)$ (corresponding to $|000\rangle$) in the 2D Euclidean plane to $(-1, -1)$ (corresponding to $|111\rangle$) via a sequence of moves with the following two restrictions: (1) For each current point (x, y), the next move must leave precisely one of x or y invariant (analogous to 2-local unitaries acting on a 3-qubit state), and (2) the Euclidean distance between (x, y) and the line through $(1, 1)$ and $(-1, -1)$ never exceeds Δ (analogous to the overlap with P not exceeding Δ). In other words, we wish to stay close to a diagonal line while making only horizontal and vertical moves. This can be achieved by making a sequence of "small" moves resembling a "staircase". The smaller the size of each "step" in the staircase, the better we approximate the line, at the expense of requiring more moves (analogous to increasing the number of unitaries, m). Although the idea in this analogy is appealing in its simplicity, applying it to the setting of the Traversal Lemma is non-trivial, requiring a careful selection of 2-local unitary operations. A full proof is given in the full version [13].

4   QCMA-completeness

We now prove one of the main results of this paper.

Theorem 5. *There exists a polynomial p such that GSCON is QCMA-complete for $m \in O(p(n))$, $\Delta \in \Theta(1/m^5)$, $l = 2$, and $k \geq 5$, where n denotes the number of qubits H acts on.*

Intuitively, this says that GSCON is QCMA-complete when the unitaries U_i are at most 2-local, the number of unitaries scales polynomially, and the gap Δ scales inverse polynomially. We now show QCMA-hardness. Containment in QCMA goes via an argument utilizing pseudo-ϵ-nets, and is given in the full version [13].

Lemma 3. *There exists a polynomial p such that GSCON is QCMA-hard for $n \in O(p(n))$, $\Delta \in O(1/m^5)$, $l = 2$, and $k \geq 5$, where n denotes the number of qubits H acts on.*

Proof. We sketch the proof here. At a high level, our approach is as follows. Given a QCMA verification circuit V, let H' be the 3-local Hamiltonian output by Kempe and Regev's circuit-to-Hamiltonian construction. Then, our aim is to construct another Hamiltonian H such that "traversing the ground space of H" forces one to simulate the following protocol — starting with an initial state of all zeroes:

1. Apply a sequence of 2-qubit gates to prepare a ground state $|\psi_{H'}\rangle$ of H'.
2. Flip a first "GO" qubit to initiate a "check" that $|\psi_{H'}\rangle$ is indeed a ground state of H'.
3. Flip a second and third "GO" qubit to end the "check".
4. Uncompute $|\psi_{H'}\rangle$ to obtain a target state which is all zeroes, except for the "GO" qubits, which are set to all ones.

More formally, let Π' be an instance of a QCMA problem with verification circuit V' acting on a classical proof register p and ancilla register a. Define V as a new circuit which first measures the proof register in the computational basis, and then runs V'. We then define our Hamiltonian H based on V as follows. Let H' denote the 3-local Hamiltonian obtained from V using Kempe and Regev's circuit-to-Hamiltonian construction [27]. Then, we define H to act on a *Hamiltonian* register denoted h and GO register denoted G, such that $H := H'_h \otimes P_G$ for $P := I - |000\rangle\langle000| - |111\rangle\langle111|$. Since it is possible to re-write P 2-locally, we have that H is 5-local. We define our initial and final states as $|\psi\rangle := |0\cdots0\rangle_h \otimes |000\rangle_G$ and $|\psi\rangle := |0\cdots0\rangle_h \otimes |111\rangle_G$.

We now sketch correctness of this construction. First, suppose there exists a proof $x \in \{0,1\}^{n_p}$ accepted by V. The following unitary evolution maps $|\psi\rangle$ to $|\phi\rangle$ while remaining in the low-energy space of H (where recall in Kempe and Regev's construction that the Hamiltonian register h is itself composed of three sub-registers h_1, h_2, and h_3, corresponding to the *proof*, *ancilla*, and *clock* registers for H, respectively):

1. Prepare classical proof x in register h_1 using Pauli X gates.
2. In register h, prepare the history state $|\text{hist}_x\rangle$ of H'.
3. Apply $(X \otimes X \otimes I)_G$ to "initiate" checking of $|\text{hist}_x\rangle$.

4. Apply $(I \otimes I \otimes X)_G$ to "complete" checking of $|\text{hist}_x\rangle$.
5. In register h, uncompute the history state $|\text{hist}_x\rangle$ of H'.
6. Uncompute the classical proof x in register h_1 using Pauli X gates.

To see that this works, note first that Steps 2 and 5 can be carried out efficiently since V' is a QCMA verifier. Second, after Step 6, we have successfully mapped to state $|\phi\rangle$. Third, every intermediate state encountered is in the null space of H except for possibly after Step 3. As for after Step 3, let $|a_3\rangle$ denote our state at this point. Then, since a valid history state $|\text{hist}_x\rangle$ obtains low energy against H', we have that $\langle a_3| H |a_3\rangle$ will also be "small", as desired.

Conversely, suppose Π' is a NO instance, i.e., for all $x \in \{0, 1\}^{n_p}$, V rejects with high probability. Then, by Lemma 1, H' has no "small" eigenvalues. Now, let S and T denote the $+1$ eigenspaces of projections $I_h \otimes |000\rangle\langle 000|_G$ and $I_h \otimes |111\rangle\langle 111|_G$, respectively. Observe that S and T are 2-orthogonal subspaces, and that $|\psi\rangle \in S$ and $|\phi\rangle \in T$. Thus, for any sequence of two-qubit unitaries mapping start state $|\psi\rangle$ to $|\psi_m\rangle$, either $\| |\psi_m\rangle - |\phi\rangle \|_2$ is "large" (in which case we have a NO instance of GSCON and we are done), or we can apply the Traversal Lemma to conclude there exists an $i \in [m]$ such that $\langle \psi_i| P' |\psi_i\rangle$ is large, where we define $|\psi_i\rangle := U_i \cdots U_1 |\psi\rangle$ and $P' = I - \Pi_S - \Pi_T$. Since $P' = I_h \otimes P$, it now follows that $\langle \psi_i| H |\psi_i\rangle = \langle \psi_i| H' \otimes P |\psi_i\rangle \geq \beta \langle \psi_i| I_h \otimes P |\psi_i\rangle$ is also large, where we have used the fact that β is "large" given by Lemma 1, and where the inequality follows since $H' \succeq \beta I$.

Acknowledgments. We thank Joel Klassen, Barbara Terhal, Roberto Oliveira, Sarvagya Upadhyay, Damian Markham, Eleni Diamanti, Attila Pereszlenyi, Amer Mouawad, Vinayak Pathak, and David Gosset for insightful discussions, and anonymous referees for helpful feedback. SG acknowledges support from an NSERC Banting Postdoctoral Fellowship and the Simons Institute for the Theory of Computing at UC Berkeley. JS acknowledges support from an NSERC Postdoctoral Fellowship, the French National Research Agency (ANR-09-JCJC-0067-01), the European Union (ERC project QCC 306537), the Singapore Ministry of Education and National Research Foundation (MOE2012-T3-1-009).

References

1. Kitaev, A., Shen, A., Vyalyi, M.: Classical and Quantum Computation. American Mathematical Society (2002)
2. Cook, S.: The complexity of theorem proving procedures. In: Proceedings of the 3rd ACM Symposium on Theory of Computing (STOC 1972), pp. 151–158 (1972)
3. Levin, L.: Universal search problems. Problems of Information Transmission **9**(3), 265–266 (1973)
4. Oliveira, R., Terhal, B.M.: The complexity of quantum spin systems on a two-dimensional square lattice. Quantum Information & Computation **8**(10), 0900–0924 (2008)
5. Bravyi, S., Vyalyi, M.: Commutative version of the local Hamiltonian problem and common eigenspace problem. Quantum Information & Computation **5**(3), 187–215 (2005)
6. Gharibian, S., Huang, Y., Landau, Z., Shin, S.W.: Quantum Hamiltonian complexity (2014). arXiv.org e-Print quant-ph/1401.3916v1

7. Cubitt, T., Montanaro, A.: Complexity classification of local hamiltonian problems (2013). arXiv.org e-Print quant-ph/1311.3161
8. Schaefer, T.J.: The complexity of satisfiability problems. In: Proceedings of the 10th Symposium on Theory of computing, pp. 216–226 (1978)
9. Kitaev, A.: Unpaired majorana fermions in quantum wires. Physics-Uspekhi **44**, 131 (2001)
10. Kitaev, A.: Fault-tolerant quantum computation by anyons. Annals of Physics **303**(1), 2–30 (2003)
11. Kitaev, A., Laumann, C.: Topological phases and quantum computation (2009). arXiv.org e-Print quant-ph/0904.2771
12. Aharonov, D., Naveh, T.: Quantum NP - A survey (2002). arXiv.org e-Print quant-ph/0210077v1
13. Gharibian, S., Sikora, J.: Ground state connectivity of local hamiltonians (2014). arXiv.org e-Print quant-ph/1409.3182
14. Winter, A.: Coding theorem and strong converse for quantum channels, **45**(7), 2481–2485 (1999)
15. Gopalan, P., Kolaitis, P.G., Maneva, E.N., Papadimitriou, C.: The connectivity of boolean satisfiability: computational and structural dichotomies. In: Bugliesi, M., Preneel, B., Sassone, V., Wegener, I. (eds.) ICALP 2006. LNCS, vol. 4051, pp. 346–357. Springer, Heidelberg (2006)
16. Mouawad, A., Nishimura, N., Pathak, V., Raman, V.: Shortest reconfiguration paths in the solution space of Boolean formulas (2014). arXiv.org e-Print cs.CC/1404.3801v2
17. Cereceda, L., van den Heuvel, J., Johnson, M.: Connectedness of the graph of vertex-colorings. Discrete Mathematics **308**(56), 913–919 (2008)
18. Bonsma, P., Cereceda, L.: Finding paths between graph colourings: PSPACE-completeness and superpolynomial distances. Theoretical Computer Science **410**(50), 5215–5226 (2009)
19. Cereceda, L., van den Heuvel, J., Johnson, M.: Finding paths between 3-colorings. Journal of Graph Theory **67**(1), 69–82 (2011)
20. Ito, T., Kamiński, M., Demaine, E.D.: Reconfiguration of list edge-colorings in a graph. Discrete Applied Mathematics **160**(15), 2199–2207 (2012)
21. Wocjan, P., Janzing, D., Beth, T.: Two QCMA-complete problems. Quantum Information & Computation **3**(6), 635–643 (2003)
22. Wocjan, P., Yard, J.: The Jones polynomial: quantum algorithms and applications in quantum complexity theory. Quantum Information & Computation **8**(1), 147–180 (2008)
23. Janzing, D., Wocjan, P.: BQP-complete problems concerning mixing properties of classical random walks on sparse graphs (2006). arXiv.org e-Print quant-ph/0610235v2
24. Gharibian, S., Kempe, J.: Hardness of approximation for quantum problems. In: Czumaj, A., Mehlhorn, K., Pitts, A., Wattenhofer, R. (eds.) ICALP 2012, Part I. LNCS, vol. 7391, pp. 387–398. Springer, Heidelberg (2012)
25. Brown, B., Flammia, S., Schuch, N.: Computational difficulty of computing the density of states. Physical Review Letters **104**, 040501 (2011)
26. Ambainis, A.: On physical problems that are slightly more difficult than QMA. In: Proceedings of 29th IEEE Conference on Computational Complexity (CCC 2014), pp. 32–43 (2014)
27. Kempe, J., Regev, O.: 3-local Hamiltonian is QMA-complete. Quantum Information & Computation **3**(3), 258–264 (2003)

Uniform Kernelization Complexity of Hitting Forbidden Minors

Archontia C. Giannopoulou[1], Bart M.P. Jansen[2]([✉]), Daniel Lokshtanov[3], and Saket Saurabh[4]

[1] University of Warsaw, Warsaw, Poland
archontia.giannopoulou@gmail.com
[2] Eindhoven University of Technology, Eindhoven, The Netherlands
b.m.p.jansen@tue.nl
[3] University of Bergen, Bergen, Norway
daniello@ii.uib.no
[4] Institute of Mathematical Sciences, Chennai, India
saket@imsc.res.in

Abstract. The $\mathcal{F}$-MINOR-FREE DELETION problem asks, for a fixed set $\mathcal{F}$ and an input consisting of a graph G and integer k, whether k vertices can be removed from G such that the resulting graph does not contain any member of $\mathcal{F}$ as a minor. Fomin et al. (FOCS 2012) showed that the special case when $\mathcal{F}$ contains at least one planar graph has a kernel of size $f(\mathcal{F}) \cdot k^{g(\mathcal{F})}$ for some functions f and g. They left open whether this PLANAR $\mathcal{F}$-MINOR-FREE DELETION problem has kernels whose size is uniformly polynomial, of the form $f(\mathcal{F}) \cdot k^c$ for some universal constant c. We prove that some PLANAR $\mathcal{F}$-MINOR-FREE DELETION problems do not have uniformly polynomial kernels (unless NP $\subseteq$ coNP/poly), not even when parameterized by the vertex cover number. On the positive side, we consider the problem of determining whether k vertices can be removed to obtain a graph of treedepth at most η. We prove that this problem admits uniformly polynomial kernels with $\mathcal{O}(k^6)$ vertices for every fixed η.

Keywords: Kernelization · Treedepth · Minor-free deletion

1 Introduction

Kernelization is the subfield of parameterized and multivariate algorithmics that investigates the power of provably effective preprocessing procedures for hard combinatorial problems. In kernelization we study *parameterized problems*: decision problems where every instance x is associated with a parameter k that

Supported by ERC Grant 267959 and the Warsaw Center of Mathematics and Computer Science (A.G.), NWO Veni grant "Frontiers in Parameterized Preprocessing" and NWO Gravity grant "Networks" (B.M.P.J.), Bergen Research Foundation grant BeHard (D.L.), and ERC Starting Grant "Parameterized Approximation" (S.S.).

© Springer-Verlag Berlin Heidelberg 2015
M.M. Halldórsson et al. (Eds.): ICALP 2015, Part I, LNCS 9134, pp. 629–641, 2015.
DOI: 10.1007/978-3-662-47672-7_51

measures some aspect of its structure. A parameterized problem is said to admit a kernel of size $f : \mathbb{N} \to \mathbb{N}$ if every instance (x, k) can be reduced in polynomial time to an equivalent instance with both size and parameter value bounded by $f(k)$. For practical and theoretical reasons we are primarily interested in kernels whose size is polynomial, so-called *polynomial kernels*.

One of the fundamental challenges in the area is the possibility of characterizing general classes of parameterized problems possessing a kernel of polynomial size. In other words, to obtain "kernelization meta-theorems". In general, algorithmic meta-theorems have the following form: problems definable in a certain logic admit a certain kind of algorithms on certain inputs. A typical example of a meta-theorem is Courcelle's celebrated theorem which states that all graph properties definable in monadic second order logic can be decided in linear time on graphs of bounded treewidth. It seems very difficult to find a fragment of logic for which every problem expressible in this logic admits a polynomial kernel on all undirected graphs. The main obstacle in obtaining such results stems from the fact that even a simplest form of logic can formalize problems that are not even fixed parameter tractable (FPT). In graph theory, one can define a general family of problems as follows. Let $\mathcal{F}$ be a family of graphs. Given an undirected graph G and a positive integer k, is it possible to do at most k edits of G such that the resulting graph does not contain a graph from $\mathcal{F}$? Here one can define edits as either vertex/edge deletions, edge additions, or edge contraction. Similarly, one may consider containment as a subgraph, induced subgraph, or a minor. The topic of this paper is one such generic problem, namely, the $\mathcal{F}$-MINOR-FREE DELETION problem. It asks, for a fixed set of graphs $\mathcal{F}$ and an input consisting of a graph G and integer k, whether k vertices can be removed from G such that the resulting graph does not contain any member of $\mathcal{F}$ as a minor. The problem can also be viewed as finding a set of k vertices that hit all the minor models of $H \in \mathcal{F}$ in G, which explains the title. The parameterized complexity of this general problem is well understood: for every k there is an algorithm solving the problem in time $f(k) \cdot n^3$ [1,20]. Thus, the $\mathcal{F}$-MINOR-FREE DELETION problem is an interesting subject from the kernelization perspective: for which sets $\mathcal{F}$ does $\mathcal{F}$-MINOR-FREE DELETION admit a polynomial kernel?

Fomin et al. [11] studied the special case where $\mathcal{F}$ contains at least one planar graph, known as PLANAR $\mathcal{F}$-MINOR-FREE DELETION. It is much more restricted than $\mathcal{F}$-MINOR-FREE DELETION, but still generalizes problems such as VERTEX COVER and FEEDBACK VERTEX SET. These problems are essentially about deleting k vertices to get a graph of constant treewidth: graphs that exclude a planar graph H as a minor have treewidth at most $|V(H)|^{\mathcal{O}(1)}$ [4]. Fomin et al. [11] exploited the properties of graphs of bounded treewidth and obtained a constant factor approximation algorithm, a $2^{\mathcal{O}(k \log k)} \cdot n$ time parameterized algorithm, and—most importantly, from our perspective—a polynomial sized kernel for every PLANAR $\mathcal{F}$-MINOR-FREE DELETION problem. More precisely, they showed that PLANAR $\mathcal{F}$-MINOR-FREE DELETION admits a kernel of size $f(\mathcal{F}) \cdot k^{g(\mathcal{F})}$ for some functions f and g. The degree g of the polynomial in the kernel size grows very quickly; it is not even known to be computable. This result is the starting point of our research.

> Does PLANAR $\mathcal{F}$-MINOR-FREE DELETION have kernels whose size is *uniformly polynomial*, of the form $f(\mathcal{F}) \cdot k^c$ for a universal constant c that does not depend on $\mathcal{F}$?

We prove that some families of PLANAR $\mathcal{F}$-MINOR-FREE DELETION problems *do not* have uniformly polynomial kernels (unless NP $\subseteq$ coNP/poly). Since a graph class has bounded treewidth if and only if it excludes a planar graph as a minor, a canonical PLANAR $\mathcal{F}$-MINOR-FREE DELETION problem is TREEWIDTH-η DELETION: can k vertices be removed to obtain a graph of treewidth at most η? We denote by K_d and P_d a clique and path on d vertices, respectively. Our first theorem is the following lower bound result.

Theorem 1. *Let $d \geq 3$ be a fixed integer and $\epsilon > 0$. If the parameterization by solution size k of one of the problems*

1. $\{K_{d+1}\}$-MINOR-FREE DELETION,
2. $\{K_{d+1}, P_{4d}\}$-MINOR-FREE DELETION, and
3. TREEWIDTH-$(d-1)$ DELETION

admits a compression of bitsize $\mathcal{O}(k^{\frac{d}{2}-\epsilon})$, or a kernel with $\mathcal{O}(k^{\frac{d}{4}-\epsilon})$ vertices, then NP $\subseteq$ coNP/poly. In fact, even if the parameterization by the size x of a vertex cover of the input graph admits a compression of bitsize $\mathcal{O}(x^{\frac{d}{2}-\epsilon})$ or a kernel with $\mathcal{O}(x^{\frac{d}{4}-\epsilon})$ vertices, then NP $\subseteq$ coNP/poly.

Theorem 1 shows that the kernelization result of Fomin et al. [11] is tight in the following sense: the degree g of the polynomial in the kernel sizes for PLANAR $\mathcal{F}$-MINOR-FREE DELETION must depend on the family $\mathcal{F}$. In fact, the theorem gives the stronger result that even parameterized by the *vertex cover number* of the graph (a larger parameter), the TREEWIDTH-η DELETION problem does not admit uniformly polynomial kernels unless NP $\subseteq$ coNP/poly. This resolves an open problem of Cygan et al. [5].

A graph class has bounded treewidth if and only if it excludes a planar graph as a minor. Thus, by restricting the $\mathcal{F}$-MINOR-FREE DELETION problem to those $\mathcal{F}$ that contain a planar graph, one exploits the properties of graphs of bounded treewidth to design polynomial kernels for PLANAR $\mathcal{F}$-MINOR-FREE DELETION. It is a natural question whether further restrictions on $\mathcal{F}$ lead to uniformly polynomial kernels. However, the second item of Theorem 1 shows that even when $\mathcal{F}$ contains a *path*, the degree of the polynomial must, in general, depend on the set $\mathcal{F}$. This raises the question whether there are any general families of $\mathcal{F}$-MINOR-FREE DELETION problems that admit uniformly polynomial kernels.

Excluding planar minors results in graphs of bounded treewidth [19]; excluding forest minors results in graphs of bounded pathwidth [18]; and excluding path minors results in graphs of bounded treedepth [16]. A canonical $\mathcal{F}$-MINOR-FREE DELETION problem when $\mathcal{F}$ contains a path is therefore:

TREEDEPTH-η DELETION   **Parameter:** k
Input: An undirected graph G and a positive integer k.
Question: Does there exist a subset $Z \subseteq V(G)$ of size at most k such that $\mathrm{td}(G - Z) \leq \eta$?

Here $\mathbf{td}(G)$ denotes the treedepth of a graph G. The set Z is called a *treedepth-η modulator* of G. Surprisingly, we show that TREEDEPTH-η DELETION admits uniformly polynomial kernels. More precisely, we obtain the following theorem.

Theorem 2. TREEDEPTH-η DELETION *admits a kernel with* $2^{\mathcal{O}(\eta^2)}k^6$ *vertices.*

We prove several new results about the structure of optimal treedepth decompositions and exploit this to obtain the desired kernel for TREEDEPTH-η DELETION. Unlike the kernelization algorithm of Fomin et al. [11], our kernel is completely explicit. It does not use the machinery of protrusion replacement, which was introduced to the context of kernelization by Bodlaender et al. [2] and has subsequently been applied in various scenarios [8,10,12,15]. Using protrusion replacement one can prove that kernelization algorithms exist, but the technique generally does not explicitly give the algorithm nor a concrete size bound for the resulting kernel.

Techniques. The kernelization lower bound of Theorem 1 is obtained by reduction from EXACT d-UNIFORM SET COVER, parameterized by the number of sets in the solution. Existing lower bounds exist for these problems due to Dell and Marx [6] and Hermelin and Wu [14], showing that the degree of the kernel size must grow linearly with the cardinality d of the sets in the input. While the construction that proves Theorem 1 is relatively simple in hindsight, the fact that the construction applies to all three mentioned problems, and also applies to the parameterization by vertex cover number, makes it interesting.

Our main technical contribution lies in the kernelization algorithm for TREE-DEPTH-η DELETION. Our algorithm starts by enriching the graph G by adding edges between vertices that are connected by many internally vertex-disjoint paths. Like in prior work on TREEWIDTH-η DELETION [5], adding such edges does not change the answer to the problem. We then apply an algorithm by Reidl et al. [17] to compute an approximate treedepth-η modulator S of the resulting graph. The remainder of the algorithm strongly exploits the structure of the bounded-treedepth graph $G - S$. By combining separators for vertices that are not linked through many disjoint paths, we compute a small set Y such that all the bounded-treedepth connected components of $G - (S \cup Y)$ have a special structure: their neighborhood in S forms a clique, while they have less than η neighbors in Y. For such components C we can prove that optimal treedepth-η modulators contain at most 2η vertices from C. This important fact allows us to infer that optimal solutions cannot disturb the structure of the graph $G[C]$ too much. While it is relatively easy to bound the number of connected components of $G-(S \cup Y)$, the main work consists of reducing the size of each such component.

We formulate three lemmata that analyze under which circumstances the structure of optimal treedepth-η modulators is preserved when adding edges, removing edges, and removing vertices of the graph. By exploiting the fact that the solution size within a particular part C of the graph is constant, these lemmata ensure that even after deleting an optimal modulator from C, the remainder of C forces a structure of treedepth decompositions of the remaining graph that is compatible with the graph modifications. Of particular interest is the lemma showing that if v dominates the neighborhood of component C, then edges of v into the component may be safely discarded if certain other technical conditions are met.

The three described lemmata are the main tool in the reduction algorithm. To shrink components of $G - (S \cup Y)$ we have to add some edges, while removing other edges, to create settings where vertices can be removed from the instance without changing its answer. The fact that we have to combine edge additions and removals makes our reduction algorithm quite delicate: we cannot simply formulate reduction rules for adding and removing edges and apply them exhaustively, as they would work against each other. We therefore present a recursive algorithm that processes a treedepth-η decomposition of $G - S$ from top to bottom, making suitable transformations that bound the degree of the modulator S into the remainder of the component C. Using a careful measure expressed in terms of this degree, we can then prove that our algorithm achieves the desired size reduction.

Related Results. PLANAR $\mathcal{F}$-MINOR-FREE DELETION has received considerable attention [11,15] resulting in approximation, kernelization, and FPT algorithms. Cygan et al. [5] studied TREEWIDTH-η DELETION parameterized by the vertex cover number of a graph and obtained a kernel of size $k^{\mathcal{O}(\eta)}$. In a later paper, Fomin et al. [9] studied $\mathcal{F}$-MINOR-FREE DELETION parameterized by the vertex cover number of the graph. They obtained kernels of size $k^{\mathcal{O}(\Delta(\mathcal{F}))}$, where $\Delta(\mathcal{F})$ is an upper bound on the maximum degree of any graph in $\mathcal{F}$. Notable work involving the parameter treedepth includes the $2^{\mathcal{O}(t^2)} \cdot n$-time algorithm for testing treedepth by Reidl et al. [17] and the kernelization meta-theorems for problems parameterized by a treedepth-η modulator by Gajarský et al. [12].

2 Preliminaries

Notation not defined here is standard. All graphs we consider are finite, undirected, and simple. We write $H \subseteq G$ if H is a subgraph of G. Given two distinct vertices u and v we define $\lambda_G(u, v)$ as the maximum cardinality of a set of pairwise internally vertex-disjoint uv-paths in the graph G.

Treedepth. A *rooted tree* T is a tree with one distinguished vertex $r \in V(T)$, called the *root* of T. A *rooted forest* is a disjoint union of rooted trees. The roots introduce natural parent-child and ancestor-descendant relations between vertices in forest. A vertex x is a proper ancestor (proper descendant) of a vertex y if x is an ancestor (descendant) of y and $x \neq y$. We denote by $\mathbf{anc}_F(x)$

he proper ancestors of x; this set is empty if x is a root. We denote by $\pi(x)$ the arent of x in F. The parent of the root of the tree is $\perp$. For a rooted forest F nd a vertex $v \in V(F)$, we denote by F_v the subtree rooted at v that contains ll v's descendants, including v itself. The *depth* of a vertex x in a rooted forest F s the number of vertices on the unique simple path from x to the root of the tree o which x belongs; it is denoted **depth**(x, F). The *height* of v is the maximum umber of vertices on a simple path from v to a leaf in F_v. The height of F is he maximum height of a vertex of F and is denoted **height**(F). Two vertices x nd y are in *ancestor-descendant* relation if x is an ancestor of y or vice versa.

Definition 1 (Treedepth) *A treedepth decomposition of a graph G is a rooted forest F on the vertex set $V(G)$ (i.e., $V(G) = V(F)$) such that for every edge $\{u, v\}$ of G, the endpoints u and v are in ancestor-descendant relation. The treedepth of G, denoted* **td**(G)*, is the least $d \in \mathbb{N}$ such that there exists a treedepth decomposition F of G with* **height**$(F) = d$.

The following properties follow from this definition. The treedepth of a disconnected graph is the maximum treedepth of its connected components. If F is a treedepth decomposition of G and $S \subseteq V(G)$ induces a clique in G, then there is one root-to-leaf path in F containing all vertices of S. If H is a connected subgraph of G, then all vertices of H belong to the same tree in any treedepth decomposition. If $u, v \in V(H)$ are not in ancestor-descendant relation in T, then some vertex of H is a common ancestor of u and v.

We will work with the notion of a *nice treedepth decomposition*. A treedepth decomposition F of a graph G is a nice treedepth decomposition if, for every $v \in V(F)$, the subgraph of G induced by the vertices in F_v is connected. The following lemma shows that any graph has a minimum-height treedepth decomposition that is also nice.

Lemma 1 ([17]). *For every fixed η there is a polynomial-time algorithm that, given a graph G, either determines that* **td**$(G) > \eta$ *or computes a nice treedepth decomposition F of G of depth* **td**(G).

Lemma 2 ([12, Lemma 2]). *Fix $\eta \in \mathbb{N}$. Given a graph G, one can in polynomial time compute a subset $S \subseteq V(G)$ such that* **td**$(G - S) \leq \eta$ *and $|S|$ is at most 2^η times the size of a minimum treedepth-η modulator of G.*

3   Kernelization Lower Bounds

We turn our attention to kernelization and compression lower bounds. To prove that $\mathcal{F}$-MINOR-FREE DELETION does not have uniformly polynomial kernels for suitable families $\mathcal{F}$, we give a polynomial-parameter transformation from a problem for which a compression lower bound is known. The following problem is the starting point for our transformation.

EXACT d-UNIFORM SET COVER   **Parameter:** The universe size n.
Input: A finite set U of size n, an integer k, and a set family $\mathcal{F} \subseteq 2^U$ of size-d subsets of U.
Question: Is there a subfamily $\mathcal{F}' \subseteq \mathcal{F}$ consisting of at most k sets such that every element of U is contained in exactly one subset of $\mathcal{F}'$?

Observe that since all subsets in $\mathcal{F}$ have size exactly d, the requirement that each universe element is contained in exactly one subset in $\mathcal{F}'$ implies that a set $\mathcal{F}'$ can only be a solution if it consists of n/d subsets. This implies that $k = n/d$ for all nontrivial instances of the problem. Hermelin and Wu [14] obtained a compression lower bound for EXACT d-UNIFORM SET COVER. The same problem was also studied by Dell and Marx [6] under the name PERFECT d-SET MATCHING. They obtained a slightly stronger compression lower bound, which forms the starting point for our reduction.

Theorem 3 ([6, Theorem 1.2]). *For every fixed $d \geq 3$ and $\epsilon > 0$, there is no compression of size $\mathcal{O}(k^{d-\epsilon})$ for EXACT d-UNIFORM SET COVER unless* $\mathsf{NP} \subseteq \mathsf{coNP/poly}$.

We remark that, while Dell and Marx stated their main theorem in terms of kernelizations, the same lower bounds indeed hold for compressions. We present the construction that will be used to prove Theorem 1.

Lemma 3. *For every fixed d there is a polynomial-time algorithm that, given a set U of size n, an integer k, and a d-uniform set family $\mathcal{F} \subseteq \binom{U}{d}$, computes a graph G' with vertex cover number $\mathcal{O}(k^2)$ and an integer $k' \in \mathcal{O}(k^2)$, such that:*

1. *If there is a set $S' \subseteq V(G')$ of size at most k' such that $G' - S'$ is K_{d+1}-minor-free, then there is an exact set cover of U consisting of k sets from $\mathcal{F}$.*
2. *If there is an exact set cover of U consisting of k sets from $\mathcal{F}$, then there is a set $S' \subseteq V(G')$ of size at most k' such that $G' - S'$ is K_{d+1}-minor-free, P_{4d}-minor-free, and has treewidth at most $d - 1$.*

Proof. Given U of size n, the integer k, and the d-uniform set family $\mathcal{F}$, the algorithm proceeds as follows. If $k \neq n/d$ then no exact set cover with k sets exists; we output $G' := K_{d+1}$ and $k' := 0$. We focus on the case that $k = n/d$. The main idea behind the construction is to create an $n \times k$ matrix with one vertex per cell. Each one of the k columns contains n vertices that correspond to the n universe elements. By turning columns into cliques and adding small gadgets, we will ensure that solutions to the vertex deletion problem must take the following form: they delete all vertices of the matrix except for exactly d per column. By enforcing that from each row, all vertices but one are deleted, and that the d surviving vertices in a column form a subset in $\mathcal{F}$, we relate the minor-free deletion sets to solutions of the exact covering problem. The formal construction proceeds as follows. Without loss of generality we can assume that the universe U consists of $[n] = \{1, 2, \dots, n\}$, which simplifies the exposition.

1. Initialize G' as the graph consisting of $n \times k$ vertices $v_{i,j}$ for $i \in [n]$ and $j \in [k]$. For each column index $j \in [k]$ turn the vertex set $\{v_{i,j} \mid i \in [n]\}$ into a clique. We refer to $M := \{v_{i,j} \mid i \in [n], j \in [k]\}$ as the *matrix vertices*.

2. For every row index $i \in [n]$ add a dummy clique D_i consisting of $d-1$ vertices to G'. Make all vertices in D_i adjacent to vertices $\{v_{i,j} \mid j \in [k]\}$ of the i-th row.

3. As the last step we encode the set family $\mathcal{F}$ into the graph. For every set $X \in \binom{U}{d} \setminus \mathcal{F}$, which is a size-$d$ subset of $[n]$ that is not in the set family $\mathcal{F}$, we do the following. For each column index $j \in [k]$, we create an *enforcer vertex* $f_{j,X}$ for the set X into column j. The neighborhood of $f_{j,X}$ consists of the d vertices $\{v_{i,j} \mid i \in X\}$, i.e., the vertices in column j corresponding to set X.

Observation 3.1. $M \cup (\bigcup_{i \in n} D_i)$ *is a vertex cover of* G' *of size* $n(k + d) \in \mathcal{O}(k^2)$.

This concludes the construction of G'. It is easy to see that it can be performed in polynomial time for fixed d, since G' has $\mathcal{O}(n^{d+1})$ vertices. Define $k' := k(n - d)$. Since d is fixed we may absorb it into the $\mathcal{O}$-notation. As $n = kd$ this implies $k' \in \mathcal{O}(k^2)$. The proof that the construction satisfies the desired properties is deferred to the full version. □

The proof of Theorem 1 follows by combining Lemma 3 with standard kernelization lower bound tools and Theorem 3. It can be found in the full version [13].

4   Uniformly Polynomial Kernelization for Treedepth-η Deletion

In this section we discuss the kernelization procedure for TREEDEPTH-η DELETION. As this material spans twenty pages, space limitations prohibit us from giving full details here. For this extended abstract we have therefore chosen to give an intuitive high-level overview of the exploited structure and the preprocessing algorithm; details can be found in the full version [13]. As described in the introduction, the two main ingredients are a decomposition algorithm and a reduction algorithm, to be applied to each piece of the decomposition. Throughout this section, the reader should be aware of the two uses for the word decomposition employed here: on the one hand we are decomposing the input instance (G, k) of the deletion problem into several subgraphs that have a certain structure, while on the other hand the deletion problem we are solving asks for a set $S \subseteq V(G)$ whose removal ensures that $G - S$ has a bounded-height treedepth decomposition.

4.1   Structural Decomposition of the Input Graph

The first step of the decomposition phase enriches the input instance (G, k) with extra edges. In an analogue of previous work on TREEWIDTH-η DELETION [5], we show that when there are non-adjacent vertices u and v in G such that $\lambda_G(u, v) \geq$

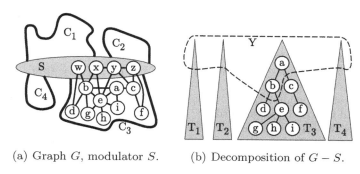

(a) Graph G, modulator S.   (b) Decomposition of $G - S$.

Fig. 1. Schematic illustration of an instance that has been decomposed. 1(a) The resulting graph G and the suboptimal treedepth-4 modulator S in G used when decomposing. Graph $G - S$ has four connected components, of which the third is drawn in detail. 1(b) Illustration of the treedepth-4 decomposition F of $G - S$. The forest F contains four decomposition trees $T_1, \ldots, T_4$, one for each component of $G - S$. By the properties of a treedepth decomposition, for any vertex $v \in V(G) \setminus S$, each neighbor $u \in N_G(v)$ is an ancestor of v in F, descendant of v in F, or contained in S. The decomposition ensures that for each connected component C of $G - (S \cup Y)$, the set $N_G(C) \cap S$ is a clique. This is illustrated for the connected component consisting of $\{e, g, h, i\}$, whose neighbors among S are $\{x, y\}$, a 2-clique. As the set Y is closed under taking ancestors, it consists of the top parts of decomposition trees in F.

$k + \eta$, then adding the edge $\{u, v\}$ does not change the answer to the instance. After exhaustively adding connecting pairs for which this condition is satisfied, we are guaranteed that for any remaining non-adjacent pair of vertices $\{u, v\}$ in G we have $\lambda_G(u, v) < k + \eta$. By Menger's theorem, this implies that there is a uv-vertex separator of size less than $k + \eta$.

After enriching the graph, we use the polynomial-time approximation algorithm for TREEDEPTH-η DELETION of Lemma 2 to find a suboptimal treedepth-η modulator S in the input instance (G, k) of size $\mathcal{O}(k)$. We use the structure that this modulator reveals in the bounded-treedepth subgraph $G - S$ to guide further processing, and compute a treedepth-η decomposition F of $G - S$ using Lemma 1. For every pair $\{u, v\}$ of remaining non-adjacent vertices in S, we compute a minimum uv-separator Y_{uv} and add $Y_{uv} \setminus S$ to a set Y. Since there are $\mathcal{O}(k^2)$ pairs of vertices among S, by the earlier bound this yields a set Y of size $\mathcal{O}(k^2(k + \eta))$. We then add all F-ancestors of vertices in Y to the set Y. Since each vertex has less than η ancestors in a treedepth-η decomposition, the size of Y increases by at most a factor η and remains $\mathcal{O}(k^3)$.

The resulting sets S and Y decompose the graph in a useful way. For every connected component C of $G - (S \cup Y)$, we know that $N_G(C) \cap S$ is a clique, since Y contains separators for all pairs of non-adjacent vertices in S. In addition, for every such component C we have $|N_G(C) \cap Y| < \eta$ since all such neighbors are contained on one root-to-leaf path of the height-η decomposition F. All such components C are therefore what we call η-*nearly clique separated*: there is a

---

Algorithm 1 Reduce(Graph G, treedepth-η modulator S, treedepth-η decomposition F of $G - S$, node v of F, $k \in \mathbb{N}$)

1: Let T be the tree in F containing v
2: **while** $\exists p, q \in N_G(T_v) \cup \{v\}$ with $\{p, q\} \notin E(G)$ and $\lambda_{G[\{p,q\} \cup T_v]}(p, q) \geq 3\eta$ **do**
3:     Add the edge $\{p, q\}$ to G
4: **while** $\exists$ distinct children $c_0, c_1, \ldots, c_{3\eta}$ of v s.t. c_0 has a neighbor $s \in S$, $N_G(T_{c_0}) \subseteq$
        $N_G[s]$, and for $i \in [3\eta]$ we have $\mathbf{td}(G[T_{c_i}]) \geq \mathbf{td}(G[T_{c_0}])$ and $s \in N_G(T_{c_i})$ **do**
5:     Remove the edges between s and members of T_{c_0} from graph G
6: **while** $\exists$ a child c^* of v such that $N_G(T_{c^*})$ is a clique, and for every $w \in N_G(T_{c^*})$
        there are 3η distinct children $c_1^w, \ldots, c_{3\eta}^w \neq c^*$ of v such that for all $i \in [3\eta]$ we
        have $\mathbf{td}(G[T_{c_i^w}]) \geq \mathbf{td}(G[T_{c^*}])$ and $w \in N_G(T_{c_i^w})$ **do**
7:     Remove the vertices in T_{c^*} from F and from G
8: **for each** remaining child c of v in T **do**
9:     Reduce(G, S, F, c, k)

---

clique in G containing all but η vertices of $N_G(C)$. We prove that minimum treedepth-η modulators contain at most 2η vertices of such components.[1]

The fact that minimum solutions delete at most 2η vertices (a constant independent of k) from components C of $G - (S \cup Y)$ will be extremely useful later on. The last part of the decomposition phase bounds the number of connected components of $G - (S \cup Y)$. The number of *non-simplicial components* (components whose S-neighborhood is not a clique) is already $\mathcal{O}(k^2(k + \eta))$, since each component provides a path between non-adjacent vertices $\{u, v\} \in \binom{S}{2}$ for which $\lambda_G(u, v) < \eta + k$. To bound the *simplicial components* (those with $N_G(C) \cap S$ a clique) requires more work. We give a structural lemma showing how to find a simplicial component whose deletion does not change the answer to the problem, in the case that there are many of such components. This step is inspired by earlier work [3, Rule 6] on PATHWIDTH. The resulting reduced graph is given as the output of the decomposition phase, together with the suboptimal modulator S and the treedepth-η decomposition F of $G - S$. See Fig. 1 for a schematic illustration.

4.2   Reduction Algorithm

After the decomposition phase, the goal of the reduction phase is to shrink the size of the connected components of $G - (S \cup Y)$; since S and Y have size $\mathcal{O}(k)$ and $\mathcal{O}(k^3)$, respectively, and the number of components of $G - (S \cup Y)$ is also bounded uniformly polynomially in k, bounding the size of each such component suffices to bound the size of G. Using the notion of a *nice* treedepth decomposition, we can ensure that the connected components of $G - (S \cup Y)$ correspond to

---

[1] If a solution S contains more than 2η vertices from C, then one would get a smaller solution by leaving C untouched and instead deleting the at most η vertices of $N_G(C)$ that are not part of the clique, and the vertices of the clique in $N_G(C)$ that are not deleted by S; there are at most η of the latter since treedepth-η graphs contain no $\eta + 1$-cliques.

the vertex sets of subtrees of the decomposition forest F rooted at vertices that are not in Y, but whose parent is in Y. Observe that if we could ensure that the maximum degree in the decomposition forest F is bounded by some function of η (but independent of k), then we would immediately get a size bound as desired: any subtree of maximum degree $f(\eta)$ has at most $f(\eta)^\eta$ vertices, since its height is at most η. Such a degree reduction is therefore our goal. However, we are not able to bound the degree by a function that is independent of k. Instead, by a top-down reduction algorithm on the decomposition forest F we can guarantee that the degree of a node v in the decomposition forest F, is bounded linearly in $|N_G(F_v) \cap S|$, which is the number of vertices of S that are adjacent to a node in the subtree of F rooted at v. This fact alone is not sufficient to bound the sizes of components C of $G - (S \cup Y)$ by a polynomial of degree independent of η, for it does not rule out the possibility of a complete degree-$|S|$ tree of height η, containing $\Omega(k^\eta)$ nodes.

The main challenge in obtaining uniformly polynomial kernels is to overcome this obstacle. To do so, we go through the decomposition trees from top to bottom, at every stage reducing the degree of the current node v using three new structural insights on treedepth. We ensure that every vertex $s \in S$ that has neighbors in any subtree rooted at a child of v, has a neighbor in at most $2^\eta \cdot 3\eta$ subtrees rooted at children of v. If this is violated, then we can first introduce new edges from s to ancestors of v and other members of S using an edge addition lemma, and afterward discard edges from s to descendants of v using an edge deletion lemma. Then we reduce the number of children whose subtrees contain no neighbors of S to constant, using a vertex deletion lemma. The procedure achieving this is given as the Reduce algorithm; its initial call is for the node v for which F_v contains the nodes of the component C we are shrinking. A careful induction reveals that this process is successful in reducing the total number of nodes in a connected component C of $G - (S \cup Y)$ to $f(\eta) \cdot k$. This achieves the desired total size reduction and yields a proof of Theorem 2.

5   Conclusion

In this paper we (re-)studied the PLANAR $\mathcal{F}$-MINOR-FREE DELETION problem from the perspective of (uniform) kernelization. We answered the question whether all PLANAR $\mathcal{F}$-MINOR-FREE DELETION problems have uniformly polynomial kernels negatively, but showed that the special case TREEDEPTH-η DELETION (which is a PLANAR $\mathcal{F}$-MINOR-FREE DELETION problem for every η, where every $\mathcal{F}$ contains a path) has uniformly polynomial kernels.

The distinction between uniformly versus non-uniformly polynomial kernels is similar to the distinction between algorithms whose parameter dependence is fixed-parameter tractable (FPT) versus slicewise-polynomial (XP), and opens up a similarly broad area of investigation. The kernelization complexity of $\mathcal{F}$-MINOR-FREE DELETION is still wide open. Some notable open problems in this direction are: (1) Does $\mathcal{F}$-MINOR-FREE DELETION admit a polynomial kernel for any fixed set $\mathcal{F}$, even when $\mathcal{F}$ contains no planar graphs? Even for the

pecial case of deleting k vertices to get a planar graph (VERTEX PLANARIZA-
ION), we do not know the answer. (2) Is it possible to obtain a dichotomy the-
rem, characterizing the families $\mathcal{F}$ for which PLANAR $\mathcal{F}$-MINOR-FREE DELE-
ION admits uniformly polynomial kernels? These questions are part of a large
esearch program into the complexity of $\mathcal{F}$-MINOR-FREE DELETION problems,
vhose importance was recognized by its listing in the *Research Horizons* section
f the recent textbook by Downey and Fellows [7, Chapter 33.2].

References

1. Adler, I., Grohe, M., Kreutzer, S.: Computing excluded minors. In: Proceedings of the 19th Annual ACM-SIAM Symposium on Discrete Algorithms (SODA 2008), pp. 641–650. ACM-SIAM (2008)
2. Bodlaender, H., Fomin, F.V., Lokshtanov, D., Penninkx, E., Saurabh, S., Thilikos, D.M.: (Meta) Kernelization. In: Proc. 50th FOCS, pp. 629–638. IEEE (2009)
3. Bodlaender, H.L., Jansen, B.M.P., Kratsch, S.: Kernel bounds for structural parameterizations of pathwidth. In: Fomin, F.V., Kaski, P. (eds.) SWAT 2012. LNCS, vol. 7357, pp. 352–363. Springer, Heidelberg (2012)
4. Chekuri, C., Chuzhoy, J.: Polynomial bounds for the grid-minor theorem. In: Proc. 46th STOC, pp. 60–69 (2014)
5. Cygan, M., Lokshtanov, D., Pilipczuk, M., Pilipczuk, M., Saurabh, S.: On the hardness of losing width. In: Marx, D., Rossmanith, P. (eds.) IPEC 2011. LNCS, vol. 7112, pp. 159–168. Springer, Heidelberg (2012)
6. Dell, H., Marx, D.: Kernelization of packing problems. In: Proc. 23rd SODA, pp. 68–81 (2012)
7. Downey, R.G., Fellows, M.R.: Fundamentals of Parameterized Complexity. Texts in Computer Science. Springer (2013)
8. Fomin, F., Lokshtanov, D., Saurabh, S., Thilikos, D.M.: Bidimensionality and kernels. In: Proc. 21st SODA, pp. 503–510 (2010)
9. Fomin, F.V., Jansen, B.M.P., Pilipczuk, M.: Preprocessing subgraph and minor problems: When does a small vertex cover help? J. Comput. System Sci. **80**(2), 468–495 (2014)
10. Fomin, F.V., Lokshtanov, D., Misra, N., Philip, G., Saurabh, S.: Hitting forbidden minors: approximation and kernelization. In: Proc. 28th STACS, pp. 189–200 (2011)
11. Fomin, F.V., Lokshtanov, D., Misra, N., Saurabh, S.: Planar $\mathcal{F}$-Deletion: approximation, kernelization and optimal FPT algorithms. In: Proc. 53rd FOCS, pp. 470–479 (2012)
12. Gajarský, J., Hliněný, P., Obdržálek, J., Ordyniak, S., Reidl, F., Rossmanith, P., Sánchez Villaamil, F., Sikdar, S.: Kernelization using structural parameters on sparse graph classes. In: Bodlaender, H.L., Italiano, G.F. (eds.) ESA 2013. LNCS, vol. 8125, pp. 529–540. Springer, Heidelberg (2013)
13. Giannopoulou, A.C., Jansen, B.M.P., Lokshtanov, D., Saurabh, S.: Uniform kernelization complexity of hitting forbidden minors (2015). CoRR, abs/1502.03965
14. Hermelin, D., Wu, X.: Weak compositions and their applications to polynomial lower bounds for kernelization. In: Proc. 23rd SODA, pp. 104–113 (2012)
15. Kim, E.J., Langer, A., Paul, C., Reidl, F., Rossmanith, P., Sau, I., Sikdar, S.: Linear kernels and single-exponential algorithms via protrusion decompositions. In: Fomin, F.V., Freivalds, R., Kwiatkowska, M., Peleg, D. (eds.) ICALP 2013, Part I. LNCS, vol. 7965, pp. 613–624. Springer, Heidelberg (2013)

16. Nesetril, J., Ossona de Mendez, P.: Tree-depth, subgraph coloring and homomorphism bounds. Eur. J. Comb. **27**(6), 1022–1041 (2006)
17. Reidl, F., Rossmanith, P., Villaamil, F.S., Sikdar, S.: A faster parameterized algorithm for treedepth. In: Esparza, J., Fraigniaud, P., Husfeldt, T. Koutsoupias, E. (eds.) ICALP 2014. LNCS, vol. 8572, pp. 931–942. Springer, Heidelberg (2014)
18. Robertson, N., Seymour, P.D.: Graph minors. I. Excluding a forest. J. Comb. Theory, Ser. B **35**(1), 39–61 (1983)
19. Robertson, N., Seymour, P.D.: Graph minors. V. Excluding a planar graph. J. Combin. Theory Ser. B **41**(1), 92–114 (1986)
20. Robertson, N., Seymour, P.D.: Graph minors. XIII. The disjoint paths problem. J. Combin. Theory Ser. B **63**(1), 65–110 (1995)

Counting Homomorphisms to Square-Free Graphs, Modulo 2

Andreas Göbel, Leslie Ann Goldberg, and David Richerby[(✉)]

Department of Computer Science, University of Oxford, Oxford, UK
davidr@cs.ox.ac.uk

Abstract. We study the problem ⊕HomsToH of counting, modulo 2, the homomorphisms from an input graph to a fixed undirected graph H. A characteristic feature of modular counting is that cancellations make wider classes of instances tractable than is the case for exact (non-modular) counting, so subtle dichotomy theorems can arise. We show the following dichotomy: for any H that contains no 4-cycles, ⊕HomsToH is either in polynomial time or is ⊕P-complete. This partially confirms a conjecture of Faben and Jerrum that was previously only known to hold for trees and for a restricted class of tree-width-2 graphs called cactus graphs. We confirm the conjecture for a rich class of graphs including graphs of unbounded tree-width. In particular, we focus on square-free graphs, which are graphs without 4-cycles. These graphs arise frequently in combinatorics, for example in connection with the strong perfect graph theorem and in certain graph algorithms. Previous dichotomy theorems required the graph to be tree-like so that tree-like decompositions could be exploited in the proof. We prove the conjecture for a much richer class of graphs by adopting a much more general approach.

1 Introduction

A homomorphism from a graph G to a graph H is a function from $V(G)$ to $V(H)$ that preserves edges, in the sense of mapping every edge of G to an edge of H; non-edges of G may be mapped to edges or non-edges of H. Many structures arising in graph theory can be represented naturally as homomorphisms. For example, the proper q-colourings of a graph G correspond to the homomorphisms from G to a q-clique. For this reason, homomorphisms from G to a graph H are often called "H-colourings" of G. Independent sets of G correspond to the homomorphisms from G to the connected graph with two vertices and one self-loop (vertices of G which are mapped to the self-loop are out of the

Theorem numbering in this extended abstract matches matches the full version: ArXiv, CoRR, abs/1501.07539. The research leading to these results has received funding from the European Research Council under the European Union's Seventh Framework Programme (FP7/2007–2013) ERC grant agreement no. 334828. The paper reflects only the authors' views and not the views of the ERC or the European Commission. The European Union is not liable for any use that may be made of the information contained therein.

M.M. Halldórsson et al. (Eds.): ICALP 2015, Part I, LNCS 9134, pp. 642–653, 2015.
DOI: 10.1007/978-3-662-47672-7_52

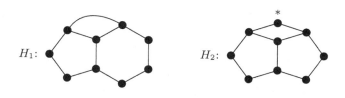

Fig. 1. Theorem 1.2 shows that $\oplus\mathrm{HomsToH}_1$ is $\oplus$P-complete, whereas $\oplus\mathrm{HomsToH}_2$ is in P. The role of the starred vertex is explained later in this section.

corresponding independent set; vertices which are mapped to the other vertex are in it). Homomorphism problems can also be seen as constraint satisfaction problems (CSPs) in which the constraint language consists of a single symmetric binary relation. Partition functions in statistical physics such as the Ising, Potts and hard-core models arise naturally as weighted sums of homomorphisms [2,8].

In this paper, we study the complexity of counting homomorphisms modulo 2. For graphs G and H, $\mathrm{Hom}(G \to H)$ denotes the set of homomorphisms from G to H. For each fixed H, we study the computational problem $\oplus\mathrm{HomsToH}$, which is the problem of computing $|\mathrm{Hom}(G \to H)|$ mod 2, for an input graph G.

The structure of H strongly influences the complexity of $\oplus\mathrm{HomsToH}$. For example, consider the graphs H_1 and H_2 in Figure 1. Our result (Theorem 1.2) shows that $\oplus\mathrm{HomsToH}_1$ is $\oplus$P-complete, whereas $\oplus\mathrm{HomsToH}_2$ is in P.

The aim of research in this area is to understand for which graphs H the problem $\oplus\mathrm{HomsToH}$ is in P, for which graphs H the problem is $\oplus$P-complete, and to prove that, for all graphs H, one or the other is true. Note that it isn't obvious, a priori, that there are no graphs H for which $\oplus\mathrm{HomsToH}$ has intermediate complexity – proving that there are no such graphs H is the main work of a so-called *dichotomy theorem*.

This line of work was introduced by Faben and Jerrum [6]. They made the following important conjecture (which requires a few definitions to state). An *involution* of a graph is an automorphism of order 2, i.e., an automorphism ρ that is not the identity but for which ρ^2 is the identity. Given a graph H and an involution ρ, H^ρ denotes the subgraph of H induced by the fixed points of ρ. We write $H \Rightarrow H'$ if there is an involution ρ of H such that $H^\rho = H'$ and we write $H \Rightarrow^* H'$ if either H is isomorphic to H' (written $H \cong H'$) or, for some positive integer k, there are graphs $H_1, \ldots, H_k$ such that $H \cong H_1$, $H_1 \Rightarrow \cdots \Rightarrow H_k$, and $H_k \cong H'$. Faben and Jerrum showed [6, Theorem 3.7] that for every graph H there is (up to isomorphism) exactly one involution-free graph H^* such that $H \Rightarrow^* H^*$. This graph H^* is called the *involution-free reduction* of H.

Conjecture 1.1. (Faben and Jerrum [6]) Let H be a graph. If its involution-free reduction H^* has at most one vertex, then $\oplus\mathrm{HomsToH}$ is in P; otherwise, $\oplus\mathrm{HomsToH}$ is $\oplus$P-complete.

Note that our claim in Figure 1 is consistent with Conjecture 1.1. H_1 is involution-free, so it is its own involution-free reduction, but the involution-free reduction of H_2 is the single vertex marked * in the figure.

Faben and Jerrum [6, Theorem 3.8] proved Conjecture 1.1 for the case in which H is a tree. Subsequently, the present authors [7, Theorem 1.6] proved the conjecture for a well-studied class of tree-width-2 graphs, namely *cactus graphs*, which are graphs in which each edge belongs to at most one cycle.

The main result of this paper is to prove the conjecture for a much richer class of graphs. In particular, we prove the conjecture for every graph H whose involution-free reduction has no 4-cycle. Graphs without 4-cycles are called "square-free" graphs. These graphs arise frequently in combinatorics, for example in connection with the strong perfect graph theorem [4] and certain graph algorithms [1]. Our main theorem is the following.

Theorem 1.2. *Let H be a graph whose involution-free reduction H^* is square-free. $\oplus\textsc{HomsTo}H$ is in P if H^* has at most one vertex; otherwise, $\oplus\textsc{HomsTo}H$ is $\oplus$P-complete.*

If H is square-free, then so is every induced subgraph, including its involution-free reduction H^*. Thus, we have the following corollary.

Corollary 1.3. *Let H be a square-free graph. If its involution-free reduction H^* has at most one vertex, then $\oplus\textsc{HomsTo}H$ is in P; otherwise, $\oplus\textsc{HomsTo}H$ is $\oplus$P-complete.*

In Section 1.3 we will discuss the reasons that we require H^* to be square-free in the proof of Theorem 1.2. First, in Section 1.1, we will describe the background to counting modulo 2. In Section 1.2, we will explain why Conjecture 1.1 is so much more difficult to prove for graphs with unbounded tree-width. Very briefly, in order to prove that $\oplus\textsc{HomsTo}H$ is $\oplus$P-hard without having a bound on the tree-width of H, it is necessary to take a much more abstract approach. Since it is not possible to decompose H using a tree-like decomposition as we did in [7, Theorem 1.6], we have instead come up with an abstract characterisation of graph-theoretic structures in H which lead to $\oplus$P-hardness. As we shall see, the proof that such structures always exist in square-free graphs involves interesting non-constructive elements, leading to a more abstract, and less technical (graph-theoretic) proof than [7], while applying to a substantially richer set of graphs H, including graphs with unbounded tree width.

1.1   Counting Modulo 2

Although counting modulo 2 produces a one-bit answer, the complexity of such problems has a rather different flavour from the complexity of decision problems. The complexity class $\oplus$P was first studied by Papadimitriou and Zachos [13] and by Goldschlager and Parberry [10]. $\oplus$P consists of all problems of the form "compute $f(x)$ mod 2" where computing $f(x)$ is a problem in #P. Toda [15] has shown that there is a randomised polynomial-time reduction from every problem in the polynomial hierarchy to some problem in $\oplus$P. As such, $\oplus$P is a large complexity class and $\oplus$P-completeness seems to represent a high degree of intractability.

The unique flavour of modular counting is exhibited by Valiant's famous restricted version of 3-SAT [16] for which counting solutions is #P-complete [17], counting solutions modulo 7 is in polynomial-time but counting solutions modulo 2 is $\oplus$P-complete [16]. The seemingly mysterious number 7 was subsequently explained by Cai and Lu [3], who showed that the k-SAT version of Valiant's problem is tractable modulo any prime factor of $2^k - 1$.

Counting modulo 2 closely resembles ordinary, non-modular counting, but is still very different. Clearly, if a counting problem can be solved in polynomial time, the corresponding decision and parity problems are also tractable, but the converse does not necessarily hold. A characteristic feature of modular counting is cancellations, which can make the modular versions of hard counting problems tractable. For example, consider not-all-equal SAT, the problem of assigning values to Boolean variables such that each of a given set of clauses contains both true and false literals. The number of solutions is always even, since solutions can be paired up by negating every variable in one solution to obtain a second solution. This makes counting modulo 2 trivial, while determining the exact number of solutions is #P-complete [9] and even deciding whether a solution exists is NP-complete [14].

We use cancellations extensively in this paper. For example, if we wish to compute the size of a set S modulo 2 then, for any even-cardinality subset $X \subseteq S$, we have $|S| \equiv |S \setminus X|$ mod 2. This means that we can ignore the elements of X. It is also helpful to partition the set S into disjoint subsets $S_1, \ldots, S_\ell$ exploiting the fact that $|S|$ is congruent modulo 2 to the number of odd-cardinality S_i. We use this idea frequently.

1.2   Going Beyond Bounded Tree-Width

Trees. All known hardness results for counting homomorphisms modulo 2 start with the following basic "pinning" approach. Let p be a function from $V(G)$ to $2^{V(H)}$. A homomorphism $f \in \text{Hom}(G \to H)$ *respects* the pinning function p if, for every $v \in V(G)$, $f(v)$ is in the set $p(v)$. Let $\text{PinHom}(G, H, p)$ be the set of homomorphisms from G to H that respect the pinning function p and let $\oplus$PINNEDHOMSTOH be the problem of counting, modulo 2, the number of homomorphisms in $\text{PinHom}(G, H, p)$, given an input graph G and a pinning function p.

Faben and Jerrum [6, Corollary 4.18] give a polynomial-time Turing reduction from the problem $\oplus$PINNEDHOMSTOH to the problem $\oplus$HOMSTOH for the special case in which the pinning function pins only two vertices of G, and these are both pinned to entire orbits of the automorphism group of H. The reduction relies on a result of Lovász [12].

In order to use the reduction, it is necessary to show that the special case of the problem $\oplus$PINNEDHOMSTOH is itself $\oplus$P-hard. Faben and Jerrum restrict their attention to the case in which H is a tree, and this is helpful.

Every involution-free tree is asymmetric (so the orbit of every vertex is trivial), so the pinning function p is actually able to pin two vertices of G to any two *particular* vertices of H. The reduction that they used to prove hardness of $\oplus$PINNEDHOMSTOH is from $\oplus$IS, the problem of counting independent sets modulo 2, which was shown to be $\oplus$P-complete by Valiant [16].

We first give an informal description of a general reduction from $\oplus$IS to the problem $\oplus$PINNEDHOMSTOH. (The general description is actually based on our current approach in this paper, but we can also present past approaches in this context.) The vertices and edges of an input G of $\oplus$IS are replaced by gadgets to give a graph J. In J, the gadget corresponding to the vertex v of G has a vertex y^v. We also choose an appropriate vertex i in H. Any homomorphism σ from J to the target graph H defines a set $I(\sigma) = \{v \in V(G) \mid \sigma(y^v) = i\}$ (mnemonic: "i" means "in" because $\sigma(y^v)$ is i exactly when v is in $I(\sigma)$). The configuration of the gadgets ensures that a set $I \subseteq V(G)$ has an odd number of homomorphisms σ with $I(\sigma) = I$ if and only if I is an independent set of G. Next, the homomorphisms $\sigma \in \text{Hom}(J \to H)$ can be partitioned according to the value of $I(\sigma)$. By the partitioning argument mentioned at the end of Section 1.1, the number of independent sets in G is equivalent to $|\text{Hom}(J \to H)|$, modulo 2.

The gadgets are chosen according to the structure and properties of H. Since Faben and Jerrum were working with trees, they were able to use gadgets with very simple structure: their gadgets are essentially paths and they exploit the fact that any non-trivial involution-free tree has at least two even-degree vertices and, of course, these have a unique path between them (which turns out to be useful).

Cactus Graphs. The situation for cactus graphs is much more complicated. Non-trivial involution-free cactus graphs still contain even-degree vertices but the presence of cycles means that paths, even shortest paths, are no longer guaranteed to be unique. Our solution in [7] was to use more complicated gadgets. They are still (loosely) based on paths, since they are defined in terms of numbers of walks between vertices of H. However, rather than requiring appropriate even-degree vertices (which might not exist), we used a second, and more complicated, gadget to "select" an even-cardinality subset of a vertex's neighbours. To find such gadgets in H, we used tree-like decompositions. Given a decomposition that breaks H into independent fragments, we inductively found gadgets (or, sometimes, partial gadgets) in the fragments, carefully putting them together across the join of the decomposition. All of this led to a very technical, very graph-theoretic solution, and also to a solution that does not generalise to graphs without tree-like decompositions.

The proof is complicated by the fact that there are involution-free graphs (even involution-free cactus graphs!) that have non-trivial automorphisms, unlike the situation for trees. Thus, the fact that the pinning function pins vertices to entire orbits (rather than to particular vertices) causes complications. The solution in [7, Section 8] relies on special properties of cactus graphs, and it is not clear how it could be generalised.

Unbounded Tree-Width. Since they are based around a tree-like decomposition, the techniques of [7] are not suitable for graphs with unbounded tree-width. To prove Conjecture 1.1 for a richer class of graphs, we adopt a much more abstract approach. Since we do not have tree-like decompositions, we instead mostly use structural properties of the whole graph to find gadgets. The structural properties do not always require technical detail – as we will see below, re-examining a result of Lovász [12] even allows us to demonstrate non-constructively the existence of some of the gadgets that we use.

In order to support our more general approach, we first have to generalise the pinning problem ⊕PINNEDHOMSTOH. We use the following **important definitions, which will be used later.** For any graph H, a *partially H-labelled graph* $J = (G, \tau)$ consists of an *underlying graph* G and a *pinning function* τ, which in this paper is a partial function from $V(G)$ to $V(H)$. Thus, every vertex v in the domain of τ is pinned to a *particular* vertex of H and *not* to a subset such as an orbit. A homomorphism from a partially labelled graph $J = (G, \tau)$ to H is a homomorphism $\sigma \colon G \to H$ such that, for all vertices $v \in \mathrm{dom}(\tau)$, $\sigma(v) = \tau(v)$. The intermediate problem that we study then is ⊕PARTLABHOMSTOH, the problem of computing $|\mathrm{Hom}(J \to H)| \bmod 2$, given a partially H-labelled graph J. In Section 3, we generalise the application of Lovász's theorem to show (Theorem 3.1) that ⊕PARTLABHOMSTOH ≤ ⊕HOMSTOH.

Armed with a stronger pinning technique, we then abstract away most of the complications that arose for graphs with small tree-width by instead using more general gadgets, defined in Section 4. Because they are not based on paths, they do not rely on uniqueness of any path in H. Instead, the gadgets have three main parts. Our new reduction from ⊕IS to ⊕HOMSTOH can be seen informally as assigning colours to both the vertices and the edges of G, where each "colour" is a vertex of H. One part of the gadget controls which colours can be assigned to each vertex, one controls which colours can be assigned to each edge and a third part determines how many homomorphisms there are from G to H, given the choice of colours for the vertices and edges. In addition to all of this, we identify two special vertices of H, one of which is the vertex i mentioned above.

The much more general nature of our gadgets compared to those used previously makes them much easier to find and, in some cases, allows us to find the parts of them non-constructively. We no longer need to find unique shortest paths in H or, indeed, any paths at all. In fact, all the gadgets that we construct in this paper use a "caterpillar gadget" (Definition 4.3) which allows us to use *any* specified path in the graph H instead of relying on a unique shortest path. Rather than finding hardness gadgets in components in some decomposition of H, we mostly find gadgets "in situ".

When a graph has two even-degree vertices, we can directly use those vertices and a caterpillar gadget to produce a hardness gadget (see Lemma 5.3). This already provides a self-contained proof of Faben and Jerrum's dichotomy for trees. Next, for graphs with only one even-degree vertex, we show (Corollary 5.5) that deleting an appropriate set of vertices leaves a component with two even-degree vertices and show (Lemma 5.7) how to simulate that vertex deletion with gadgets. This leaves only graphs in which every vertex has odd degree. In such a graph, we are able to use any shortest odd-length cycle to construct a gadget (Lemma 5.13). If there are no odd cycles, the graph is bipartite. In this interesting case (Lemma 5.15) we use our version of Lovász's result to find a gadget non-constructively.

2.3 Squares and Related Work

It is natural to ask why the involution-free reduction H^* in Theorem 1.2 is required to be square-free. We do not believe that the restriction to square-free graphs is fundamental, since our results on pinning apply to all involution-free graphs (Section 3) and neither our definition of hardness gadgets (Definition 4.1) nor our proof that the existence of a hardness gadget for H implies that $\oplus\textsc{HomsTo}H$ is $\oplus$P-complete (Theorem 4.2) requires H to be square-free. However, all the actual hardness gadgets that we find for graphs do rely on the absence of 4-cycles, as discussed in the full version, and removing this restriction seems technically challenging. We note that dealing with 4-cycles also caused significant difficulties in cactus graphs [7].

We have already mentioned earlier work on counting graph homomorphisms modulo 2. The problem of counting graph homomorphisms (exactly, rather than modulo a fixed constant) was previously studied by Dyer and Greenhill [5]. They showed the problem of counting homomorphisms to a fixed graph H is solvable in polynomial time if every connected component of H is a complete graph with a self-loop on every vertex or a complete bipartite graph with no self-loops, and is #P-complete, otherwise. Their work builds on an earlier dichotomy by Hell and Nešetřil [11] for the complexity of the graph homomorphism decision problem (the problem of distinguishing between the case where there are no homomorphisms and the case where there is at least one).

Note that much of the notation that we use below has been defined in the introduction. In addition, we write $[n] = \{1, \ldots, n\}$ and, for a set S and an element x, we often write $S - x$ for $S \setminus \{x\}$.

3   Partially Labelled Graphs and Pinning

It is often convenient to regard a graph as having some distinguished vertices $x_1, \ldots, x_r$ and we denote such a graph by $(G, x_1, \ldots, x_r)$. The distinguished vertices need not be distinct. A homomorphism from a graph $(G, x_1, \ldots, x_r)$ to $(H, y_1, \ldots, y_r)$ is a homomorphism σ from G to H with the property that $\sigma(x_i) = y_i$ for each $i \in [r]$. Isomorphisms of these graphs are defined similarly. In the full version, we generalise a result of Lovász [12] to prove the following.

Lemma 3.6. Let $(H, \bar{y})$ and $(H', \bar{y}')$ be involution-free graphs, each with r distinguished vertices. $(H, \bar{y}) \cong (H', \bar{y}')$ if and only if, for all (not necessarily connected) graphs $(G, \bar{x})$ with r distinguished vertices, $|\mathrm{Hom}((G, \bar{x}) \to (H, \bar{y}))| \equiv |\mathrm{Hom}((G, \bar{x}) \to (H', \bar{y}'))| \pmod{2}$.

Recall that $\oplus$PARTLABHOMSTOH is the problem of computing |Hom($J \to H$)| mod 2, given a partially H-labelled graph J. Using Lemma 3.6, and the implementation technique of Faben and Jerrum [6], we prove the following.

Theorem 3.1. $\oplus$PARTLABHOMSTOH $\leq$ $\oplus$HOMSTOH *for any involution-free graph H.*

The difference between Lemma 3.6 and similar previous lemmas is the inclusion of the distinguished vertices. This is necessary both for our more general pinning technique (Theorem 3.1) and because we will use Lemma 3.6 to non-constructively find hardness gadgets in Section 4.

4   Hardness Gadgets

In this section, we define the gadgets that we will use to prove $\oplus$P-completeness of $\oplus$HOMSTOH problems, by reduction from the parity independent set problem $\oplus$IS, i.e., the problem of computing the number of independent sets in an input graph, modulo 2. $\oplus$IS was shown to be $\oplus$P-complete by Valiant [16].

The gadgets that we use are considerably more general than the ones we defined for cactus graphs in [7]. This allows us to quickly prove hardness for large classes of square-free graphs and even to find gadgets non-constructively.

In the discussion that follows, we will choose a set $\Omega_y \subseteq V(H)$ and a vertex $i \in \Omega_y$. Given a graph G whose independent sets we wish to count modulo 2, we will construct a partially H-labelled graph $J = (G(J), \tau(J))$ and consider homomorphisms from J to H. $G(J)$ will contain a copy of $V(G)$ and we will be interested in homomorphisms that map every vertex in this copy to Ω_y. Vertices mapped to i will be in the independent set under consideration; vertices mapped to $\Omega_y - i$ will not be in the independent set.

Given a partially labelled graph $J = (G(J), \tau(J))$ and vertices $x_1, \ldots, x_r$ of $G(J)$ that are not in dom($\tau(J)$) and given vertices $y_1, \ldots, y_r$ of H, a homomorphism from $(J, x_1, \ldots, x_r)$ to $(H, y_1, \ldots, y_r)$ is a homomorphism from J to H which maps each x_i to y_i (for $i \in \{1, \ldots, r\}$).

Definition 4.1. *A hardness gadget $(i, s, (J_1, y), (J_2, z), (J_3, y, z))$ for a graph H consists of vertices i and s of H together with three connected, partially H-labelled graphs with distinguished vertices that satisfy the following properties. Let*

$$\Omega_y = \{a \in V(H) \mid |\text{Hom}((J_1, y) \to (H, a))| \text{ is odd}\}$$
$$\Omega_z = \{b \in V(H) \mid |\text{Hom}((J_2, z) \to (H, b))| \text{ is odd}\}$$
$$\Sigma_{a,b} = \text{Hom}((J_3, y, z) \to (H, a, b)).$$

The properties that we require are that, for each $o \in \Omega_y - i$ and each $x \in \Omega_z - s$, (1) $|\Omega_y|$ is even and $i \in \Omega_y$, (2) $|\Omega_z|$ is even and $s \in \Omega_z$, (3) $|\Sigma_{o,x}|$ is even, and (4) $|\Sigma_{o,s}|$, $|\Sigma_{i,x}|$ and $|\Sigma_{i,s}|$ are odd.

The following theorem shows that the presence of a hardness gadget implies that $\oplus$HOMSTOH is $\oplus$P-complete.

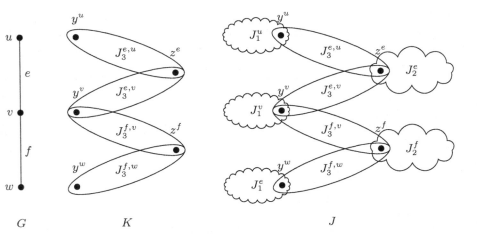

G K J

Fig. 2. The construction of the partially labelled graphs K and J from an example graph G, as in the proof of Theorem 4.2

Theorem 4.2. $\oplus$HomsToH *is* $\oplus$P-*complete for any involution-free graph* H *that has a hardness gadget.*

The proof of Theorem 4.2 consists of a reduction from the $\oplus$P-complete problem $\oplus$IS to $\oplus$PartLabHomsToH together with Theorem 3.1. The reduction from $\oplus$IS to $\oplus$PartLabHomsToH is illustrated in Figure 2.

Given an input graph G to $\oplus$IS, we first construct the partially H-labelled graph K from G by replacing every edge of G with two disjoint copies of J_3, as shown in the figure. To construct the partially H-labelled graph J, we then take K and add a disjoint copy of J_1 for every vertex $v \in G$ and a disjoint copy of J_2 for every edge $e \in G$ as shown in the figure. In the full version, we calculate the number of homomorphisms from J to H and show that $|\mathrm{Hom}(J \to H)|$ is equivalent modulo 2 to the number of independent sets in G. Intuitively, the role of J_1^u is to cancel all homomorphisms, apart from those in which the vertex y^u is mapped to a vertex in Ω_y. Similarly, J_2^e cancels all homomorphisms, apart from those in which the vertex z^e is mapped to Ω_z. Then the four properties in the definition of hardness gadget and the connections using J_3 cancel all homomorphisms apart from those in which the set of vertices y^u that are mapped to the special vertex "i" form an independent set of G.

In the paper, we use a particular gadget called a "caterpillar gadget" as the partially H-labelled graph J_3.

Definition 4.3. *Given a path* $P = v_0 \ldots v_k$ *in* H *of length at least 1, define the* caterpillar gadget $J_P = (G, \tau)$ *with distinguished vertices* y *and* z *as follows.* $V(G) = \{u_1, \ldots, u_{k-1}, w_1, \ldots, w_{k-1}, y, z\}$ *and* G *is the path* $y u_1 \ldots u_{k-1} z$ *together with edges* (u_j, w_j) *for* $1 \le j \le k-1$. $\tau = \{w_1 \mapsto v_1, \ldots, w_{k-1} \mapsto v_{k-1}\}$. *(See Figure 3).*

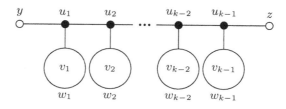

Fig. 3. The caterpillar gadget corresponding to a path $v_0 \ldots v_k$. The vertices $w_1, \ldots, w_{k-1}$ in the gadget are pinned to vertices $v_1, \ldots, v_{k-1}$ in H, respectively. A label next to a vertex indicates its identity; a label inside a white circle indicates what that vertex is pinned to.

The following lemma explains why we use caterpillar gadgets as the J_3 gadgets that appear in hardness gadgets. The point is that the properties guaranteed here coincide with the ones required in the definition of hardness gadgets (Definition 4.2). We write $\Gamma_H(v)$ for the neighbourhood of a vertex v in a graph H.

Lemma 4.5. *Let H be a square-free graph. Let $k > 0$ and let $P = v_0 \ldots v_k$ be a path in H with $\deg_H(v_j)$ odd for all $j \in \{1, \ldots, k-1\}$. Let $\Omega_y \subseteq \Gamma_H(v_0)$ and $\Omega_z \subseteq \Gamma_H(v_k)$, with $i = v_1 \in \Omega_y$ and $s = v_{k-1} \in \Omega_z$. For each $o \in \Omega_y - i$ and each $x \in \Omega_z - s$ the following properties hold. (1) $|\mathrm{Hom}((J_P, y, z) \to (H, o, x))| = 0$, (2) $|\mathrm{Hom}((J_P, y, z) \to (H, o, s))| = 1$, (3) $|\mathrm{Hom}((J_P, y, z) \to (H, i, x))| = 1$, and (4) $|\mathrm{Hom}((J_P, y, z) \to (H, i, s))|$ is odd.*

The proof of Lemma 4.5 relies on the fact that H is square-free. It can be found in the full version. The point is that, even if the proof is a little bit technical — it is sufficiently general that it applies to every square-free graph H. As long as H is square-free, *any* caterpillar gadget has the desired properties, so J_3 can always be taken to be a caterpillar, without requiring a detailed structural analysis of H.

5   Finding Hardness Gadgets

In this section we show how to identify hardness gadgets in different graphs. If H has two or more even-degree vertices, we can directly use them to construct a hardness gadget. In this case, the partially H-labelled graphs J_1 and J_2 will just be edges. In each of these, exactly one vertex is pinned, and it is pinned to an even-degree vertex of H. This is captured in the following lemma, which already provides a self-contained proof of Faben and Jerrum's dichotomy for trees.

Lemma 5.3. *Let H be a connected, square-free graph with at least two even-degree vertices. Then H has a hardness gadget.*

If H has exactly one even-degree vertex, we first show that deleting an appropriate set of vertices leaves a component with two even-degree vertices.

Corollary 5.5. *Let H be an involution-free graph that has exactly one vertex v of positive, even degree. For some r, the graph formed from H by deleting the ball at distance r around v has an involution-free component H^* that does not contain v but does contain at least two even-degree vertices.*

Corollary 5.5 allows us to construct a hardness gadget for H by attaching a path to the gadget already constructed in Lemma 5.3. We prove in the full version that the additional path essentially simulates the vertex deletion from Corollary 5.5. After calculations we are able to prove the following.

Lemma 5.7. *Any involution-free, square-free graph H that has exactly one vertex v of positive, even degree has a hardness gadget.*

This leaves only graphs H in which every vertex has odd degree. If such a graph has an odd-length cycle then we can use it to construct an appropriate hardness gadget. In the full version, we prove the following lemma.

Lemma 5.13. *Let H be a square-free graph in which every vertex has odd degree. If H contains an odd cycle, then it has a hardness gadget.*

The most interesting case, and the only one left, is the case in which H is a bipartite graph in which every vertex has odd degree. We use the following definition.

Definition 5.14. *An even gadget for a bipartite graph H is a connected bipartite graph G with a distinguished edge (w, x) such that $|\mathrm{Hom}((G, w, x) \to (H, a, b))|$ is even, for some edge (a, b) in H.*

Using our extended version of Lovász's result (Lemma 3.6) we are able to prove the following. The key point is that every bipartite (G, w, x) has exactly one homomorphism to the single edge (a, b). Since H is not a single edge, Lemma 3.6 says there is a (G, w, x) with an even number of homomorphisms to (H, a, b). This is not necessarily an even gadget but it allows us to construct one.

Lemma 5.15. *Every connected, bipartite graph except K_2 has an even gadget.*

An even gadget turns out to be useful for the following reason. If G and H are bipartite, then there is always at least one homomorphism from (G, w, x) to (H, a, b), since the whole of G can be mapped to the edge (a, b). Thus, the definition of even gadget implies that $|\mathrm{Hom}((G, w, x) \to (H, a, b))|$ is even and positive. Using this fact, and the additional fact that H is square-free, we are able to apply some additional pinning to the even gadget that is guaranteed to exist, in order to obtain a hardness gadget, so we obtain the following.

Lemma 5.16. *Let H be a connected, bipartite, square-free graph in which every vertex has odd degree. H has a hardness gadget.*

6   Main Theorem

Theorem 1.2 follows rather directly from Lemma 5.3, Lemma 5.7, Lemma 5.13 and Lemma 5.16. A technical issue arises concerning the connectivity of the involution-free reduction. This is dealt with in the full version.

References

1. Arends, F., Ouaknine, J., Wampler, C.W.: On searching for small Kochen-Specker vector systems. In: Kolman, P., Kratochvíl, J. (eds.) WG 2011. LNCS, vol. 6986, pp. 23–34. Springer, Heidelberg (2011)
2. Bulatov, A.A., Grohe, M.: The complexity of partition functions. Theor. Comput. Sci. **348**(2–3), 148–186 (2005)
3. Cai, J.-Y., Lu, P.: Holographic algorithms: From art to science. J. Comput. Syst. Sci. **77**(1), 41–61 (2011)
4. Conforti, M., Cornuéjols, G., Vušković, K.: Square-free perfect graphs. J. Combin. Theory Ser. B **90**(2), 257–307 (2004)
5. Dyer, M.E., Greenhill, C.S.: The complexity of counting graph homomorphisms. Random Struct. Algorithms **17**(3–4), 260–289 (2000)
6. Faben, J., Jerrum, M.: The complexity of parity graph homomorphism: an initial investigation. Theor. Comput. **11**, 35–57 (2015)
7. Göbel, A., Goldberg, L.A., Richerby, D.: The complexity of counting homomorphisms to cactus graphs modulo 2. ACM T. Comput. Theory, **6**(4), article 17 (2014)
8. Goldberg, L.A., Grohe, M., Jerrum, M., Thurley, M.: A complexity dichotomy for partition functions with mixed signs. SIAM J. Comput. **39**(7), 3336–3402 (2010)
9. Goldberg, L.A., Gysel, R., Lapinskas, J.: Approximately counting locally-optimal structures. CoRR, abs/1411.6829 (2014)
10. Goldschlager, L.M., Parberry, I.: On the construction of parallel computers from various bases of Boolean functions. Theor. Comput. Sci. **43**, 43–58 (1986)
11. Hell, P., Nešetřil, J.: On the complexity of H-coloring. J. Comb. Theory, Ser. B **48**(1), 92–110 (1990)
12. Lovász, L.: Operations with structures. Acta Math. Acad. Sci. Hungar. **18**(3–4), 321–328 (1967)
13. Papadimitriou, C.H., Zachos, S.: Two remarks on the power of counting. In: Cremers, A.B., Kriegel, H.-P. (eds.) Theoretical Computer Science. LNCS, vol. 145, pp. 269–275. Springer, Heidelberg (1982)
14. Schaefer, T.J.: The complexity of satisfiability problems. In: Proc. 10th Annual ACM Symposium on Theory of Computing (STOC 1978), pp. 216–226. ACM Press (1978)
15. Toda, S.: PP is as hard as the polynomial-time hierarchy. SIAM J. Comput. **20**(5), 865–877 (1991)
16. Valiant, L.G.: Accidental algorithms. In: Proc. 47th Annual IEEE Symposium on Foundations of Computer Science (FOCS 2006), pp. 509–517. IEEE (2006)
17. Xia, M., Zhang, P., Zhao, W.: Computational complexity of counting problems on 3-regular planar graphs. Theoret. Comput. Sci. **384**(1), 111–125 (2007)

Approximately Counting Locally-Optimal Structures

Leslie Ann Goldberg[1], Rob Gysel[2], and John Lapinskas[1]([⊠])

[1] Department of Computer Science, University of Oxford, Oxford, UK
lapinskas@cs.ox.ac.uk
[2] Department of Computer Science, University of California, Davis, USA

Abstract. A *locally-optimal* structure is a combinatorial structure that cannot be improved by certain (greedy) local moves, even though it may not be globally optimal. An example is a maximal independent set in a graph. It is trivial to construct an independent set in a graph. It is easy to (greedily) construct a maximal independent set. However, it is NP-hard to construct a globally-optimal (maximum) independent set.This situation is typical. Constructing a locally-optimal structure is somewhat more difficult than constructing an arbitrary structure, and constructing a globally-optimal structure is more difficult than constructing a locally-optimal structure. The same situation arises with listing. The differences between the problems become obscured when we move from listing to counting because nearly everything is #P-complete. However, we highlight an interesting phenomenon that arises in approximate counting, where approximately counting locally-optimal structures is apparently more difficult than approximately counting globally-optimal structures. Specifically, we show that counting maximal independent sets is complete for #P with respect to approximation-preserving reductions, whereas counting all independent sets, or counting maximum independent sets is complete for an apparently smaller class, #RHΠ_1 which has a prominent role in the complexity of approximate counting. Motivated by the difficulty of approximately counting maximal independent sets in bipartite graphs, we also study counting problems involving minimal separators and minimal edge separators (which are also locally-optimal structures). Minimal separators have applications via fixed-parameter-tractable algorithms for constructing triangulations and phylogenetic trees. Although exact (exponential-time) algorithms exist for listing these structures, we show that the counting problems are as hard as they could possibly be. All of the exact counting problems are #P-complete, and all of the approximation problems are complete for #P with respect to approximation-preserving reductions. A full version [14] containing detailed proofs is available at http://arxiv.org/abs/1411.6829. Theorem-numbering here matches the full version.

L.A. Goldberg and J. Lapinskas—The research leading to these results has received funding from the European Research Council under the European Union's Seventh Framework Programme (FP7/2007–2013) ERC grant agreement no. 334828. The paper reflects only the authors' views and not the views of the ERC or the European Commission. The European Union is not liable for any use that may be made of the information contained therein.

M.M. Halldórsson et al. (Eds.): ICALP 2015, Part I, LNCS 9134, pp. 654–665, 2015.
DOI: 10.1007/978-3-662-47672-7_53

1   Introduction

A *locally-optimal* structure is a combinatorial structure that cannot be improved by certain (greedy) local moves, even though it may not be globally optimal. An example is a maximal independent set in a graph. It is trivial to construct an independent set in a graph (for example, the singleton set containing any vertex is an independent set). It is easy to construct a maximal independent set (the greedy algorithm can do this). However, it is NP-hard to construct a globally-optimal independent set, which in this case means a maximum independent set. In the setting in which we work, this situation is typical. Constructing a locally-optimal structure is somewhat more difficult than constructing an arbitrary structure, and constructing a globally-optimal structure is more difficult than constructing a locally-optimal structure. For example, in bipartite graphs, it is trivial to construct an independent set, easy to (greedily) construct a maximal independent set, and more difficult to construct a maximum independent set (even though this can be done in polynomial time). This general phenomenon has been well-studied. In 1987, Johnson, Papadimitriou and Yannakakis [19] defined the complexity class PLS (for "polynomial-time local search") that captures local optimisation problems where one iteration of the local search algorithm takes polynomial time. As the authors point out, practically all empirical evidence leads to the conclusion that finding locally-optimal solutions is much easier than solving NP-hard problems, and this is supported by complexity-theoretic evidence, since a problem in PLS cannot be NP-hard unless NP=co-NP. An example that illustrates this point is the graph partitioning problem. For this problem it is trivial to find a valid partition, and it is NP-hard to find a globally-optimal (minimum weight) partition but Schäffer and Yannakakis [23] showed that finding a locally-optimal solution (with respect to a particular swapping-dynamics) is PLS-complete, so is presumably of intermediate complexity.

For listing combinatorial structures, a similar pattern emerges. Self-reducibility gives a nearly-trivial polynomial-space polynomial-delay algorithm for listing the independent sets of a graph [13]. A polynomial-space polynomial-delay algorithm for listing the *maximal* independent sets exists, due to Tsukiyama et al. [26], but it is more complicated. On the other hand, there is no polynomial-space polynomial-delay algorithm for listing the *maximum* independent sets unless P=NP. There is a polynomial-space polynomial-delay algorithm for listing the maximum independent sets of a bipartite graph [20], but this is substantially more complicated than any of the previous algorithms.

When we move from constructing and listing to counting, these differences become obscured because nearly everything is #P-complete. For example, counting independent sets, maximal independent sets, and maximum independent sets of a graph are all #P-complete problems, even if the graph is bipartite [27]. Furthermore, even *approximately* counting independent sets, maximal independent sets, and maximum independent sets of a graph are all #P-complete with respect to approximation-preserving reductions [8].

The purpose of this paper is to highlight an interesting situation that arises in approximate counting where, contrary to the situations that we have just discussed, approximately counting locally-optimal structures is apparently more difficult than counting globally-optimal structures.

In order to explain the result, we first briefly summarise what is known about the complexity of approximate counting within #P. This will be explained in more detail in Sect. 2. There are three relevant complexity classes — the class containing problems which admit a fully-polynomial randomised approximation scheme (FPRAS), the class #RHΠ_1, and #P itself. Dyer et al. [8] showed that #BIS, the problem of counting independent sets in a bipartite graph, is complete for #RHΠ_1 with respect to approximation-preserving (AP) reductions and that #IS, the problem of counting independent sets in a (general) graph is #P-complete with respect to AP-reductions. It is generally believed that the #RHΠ_1-complete problems are not FPRASable, but that they are of intermediate complexity, and are not as difficult to approximate as the problems which are #P-complete with respect to AP-reductions. Many problems have subsequently been shown to be #RHΠ_1-complete and #P-complete with respect to AP-reductions. More examples will be given in Sect. 2.

We can now describe the interesting situation which emerges with respect to independent sets in bipartite graphs. Dyer et al. [8] showed that approximately counting independent sets and approximately counting *maximum* independent sets are both #RHΠ_1-complete with respect to AP-reductions. Thus, the pattern outlined above would suggest that approximately counting *maximal* independent sets in bipartite graphs ought to also be #RHΠ_1-complete. However, we show (Theorem 1, below) that approximately counting *maximal* independent sets in bipartite graphs is actually #P-complete with respect to AP-reductions. Thus, either #RHΠ_1 and #P are equivalent in approximation complexity (contrary to the picture that has been emerging in earlier papers), or this is a scenario where approximately counting locally-optimal structures is actually more difficult than approximately counting globally-optimal ones.

Motivated by the difficulty of approximately counting maximal independent sets in bipartite graphs, we also study the problem of approximately counting other locally-optimal structures that arise in algorithmic applications. The problem of counting the *minimal separators* of a graph arises in diverse applications from triangulation theory to phylogeny construction in computational biology. A minimal separator is a particular type of vertex separator. Definitions are given in Sect. 1.1. Algorithmic applications arise because fixed-parameter-tractable algorithms are known whose running time is polynomial in the number of minimal separators of a graph. These algorithms were originally developed by Bouchitté and Todinca [5,6] (and improved in [9]) to exactly solve the so-called *treewidth* and *minimum-fill* problems; the former is widely studied due to its applicability to a number of other NP-complete problems [4]. The technique has recently been generalized [12] to cover problems including *treecost* [2] and *treelength* [22]. The algorithm can also be used to find a minimum-width *tree-decomposition* of a graph, a key data structure that is used to solve a variety of NP-complete problems in polynomial time when the width of the tree-decomposition is fixed [4]. In recent years, much research has been dedicated to exact-exponential algorithms for treewidth [3], the fastest of which [10] has running time closely connected to the number of minimal separators in the graph. Indeed, there exist polynomials p_L and p_U such that if the graph has n vertices and M minimal separators, then the running time is at least $p_L(n)M$ and at most $p_U(n)M^2$.

Bouchitté and Todinca's approach has also recently been applied to solve the *perfect phylogeny problem* and two of its variants [18]. In this problem, the input is a set of phylogenetic characters, each of which may be viewed as a partition of a subset of *species*. The goal is to find a phylogenetic tree such that every character is *convex* on that tree — that is, the parts of each partition form connected subtrees that do not overlap. Such a tree is called a *perfect phylogeny*.

In all of these applications, it would be useful to count the minimal separators of a graph, since this would give an a priori bound on the algorithms' running times. Thus, we consider the difficulty of this problem, whose complexity was previously unresolved, even in terms of exact computation. Theorem 2 shows that counting minimal separators is #P-complete, both with respect to Turing reductions (for exact computation) and with respect to AP-reductions. Thus, this problem is as difficult to approximate as any problem in #P.

Motivated by applications to treewidth [9] and phylogeny [17,18], we also consider various heuristic approximations to the minimal separator problem. The number of inclusion-minimal separators is a natural choice for a lower bound on the number of minimal separators. Conversely, the number of (s,t)-minimal separators, taken over all vertices s and t, is a natural choice for an upper bound on the number of minimal separators. Theorem 2 shows that both of these bounds are difficult to compute, either exactly or approximately. Finally, the number and structure of 2-component minimal separators is important in computational biology. These separators arise naturally in the problem of determining whether a subset of "quartet phylogenies" can be assembled uniquely [17]. Thus, we study the problem of counting such minimal separators. Theorem 2 shows that they are complete for #P with respect to exact and approximate computation.

Our new results about counting minimal vertex separators are obtained by first considering the problem of counting minimal edge separators. These locally-optimal structures are also known as *bonds* or *minimal cuts*, and are well-studied in other contexts — see e.g. Diestel [7]. Theorem 3 gives the first hardness result for counting these structures, either exactly or approximately.

1.1 Detailed Results

We now give formal definitions of the problems that we study, and state our results precisely. Our first result is that counting maximal independent sets in a bipartite graph is #P-complete with respect to Approximation-Preserving (AP) reductions (even though counting maximum independent sets in bipartite graphs is only #RHΠ_1-complete with respect to these reductions). (AP-reductions are discussed in Sect. 2.)

Definition 1. *Let G be a graph. We say that an independent set $X \subseteq V(G)$ of G is* maximal *if no proper superset of X is an independent set of G.*

Problem 1. #MaximalBIS.
Input: A bipartite graph G.
Output: The number of maximal independent sets of G.

The following theorem is proved in Sect. 3.

Theorem 1. #MaximalBIS $\equiv_{AP}$ #SAT.

Next we state our results relating to counting minimal separators.

Definition 2. *Let $G = (V, E)$ be a graph, and let $X \subseteq V$. For distinct $s, t \in V$, we say X is an (s,t)-separator of G if s and t lie in different components of $G - X$. If, in addition, no proper subset of X is an (s,t)-separator of G, then we say that X is a* minimal (s,t)-separator *of G. We say X is a* minimal separator *of G if X is a minimal (a,b)-separator of G for some $a, b \in V$.*

For example, let $V = \{1, 2, 3, 4, 5\}$, let $E = \{\{1,2\}, \{2,3\}, \{3,4\}, \{4,1\}, \{1,5\}\}$, and let G be the graph (V, E). G is a four-edge cycle with a pendant vertex. Then $\{1, 3\}$ is a minimal separator of G since it is a minimal $(2, 4)$-separator.

We have already seen that algorithms for counting and approximately counting minimal separators are useful in algorithmic applications. There is also lots of existing work on listing minimal separators. Given a graph G, let n be the number of vertices and let m be the number of edges. Kloks and Kratsch, and independently, Sheng and Liang, showed how to compute all (s, t)-minimal separators in $O(n^3)$ time per (s, t)-minimal separator [21,24]. Computing all minimal separators by computing (s, t)-minimal separators for each possible vertex pair in this way leads to an $O(n^5)$ time per minimal separator listing algorithm. Berry, Bordat, and Cogis [1] improved this approach, computing all minimal separators in $O(n^3)$ time per minimal separator. Each of these algorithms require storing minimal separators in an adequate data structure. Takata's algorithm [25] generates the set of minimal separators in $O(n^3 m)$ time per minimal separator but linear space. A graph has at most $O(1.6181^n)$ minimal separators [11]. We study the following computational problems, based on our desire to count and to approximately count minimal separators.

Problem 2. #(s,t)-BiMinimalSeps.
Input: A bipartite graph G and two vertices $s, t \in V(G)$.
Output: The number of minimal (s,t)-separators of G, denoted by $\mathrm{MS}(G, s, t)$.

Problem 3. #BiMinimalSeps.
Input: A bipartite graph G.
Output: The number of minimal separators of G, denoted by $\mathrm{MS}(G)$.

Theorem 2 below shows that both problems are #P-complete to solve exactly and are complete for #P with respect to approximation-preserving reductions.

Motivated by considerations in phylogeny [17] we also consider various heuristic approximations to the minimal separator problem. We start by defining the notion of an inclusion-minimal separator, since the number of these is a natural lower bound for the number of minimal separators.

Definition 4. *Let G be a graph. A minimal separator X of G is said to be an inclusion-minimal separator if no proper subset of X is a minimal separator.*

In the five-vertex example above, the minimal separator $\{1, 3\}$ is not an inclusion-minimal separator since $\{1\} \subset \{1, 3\}$ is a minimal $(5, 4)$-separator. However $\{1\}$ is an inclusion-minimal separator. We consider the following computational problem.

Problem 4. #BiInclusionMinimalSeps.
Input: A bipartite graph G.
Output: The number of inclusion-minimal separators of G, denoted by $\mathrm{IMS}(G)$.

We also consider the problem of counting 2-component minimal separators since these arise in phylogenetic assembly.

Problem 5. #(s, t)-BiConnMinimalSeps.
Input: A bipartite graph G and two vertices $s, t \in V(G)$.
Output: The number of minimal (s, t)-separators X of G such that $G - X$ has exactly two connected components.

Problem 6. #BiConnMinimalSeps.
Input: A bipartite graph G.
Output: The number of minimal separators X of G such that $G - X$ has exactly two connected components.

Our main theorem about minimal separators shows that all of these problems are #P-complete and are also complete for #P with respect to AP-reductions.

Theorem 2. *The problems* #(s, t)-BiMinimalSeps, #BiMinimalSeps, #(s, t)-BiConnMinimalSeps, #BiConnMinimalSeps *and* #BiInclusionMinimalSeps *are* #P-*complete and are equivalent to* #SAT *under AP-reduction.*

In order to prove Theorem 2, we first study algorithmic problems related to other natural locally-optimal structures, namely minimal edge-separators. These problems are also interesting for their own sake.

Definition 5. *Let $G = (V, E)$ be a graph, and let $F \subseteq E$. For distinct $s, t \in V$, we say F is an (s, t)-edge separator of G if s and t lie in different components of $G - F$. If in addition no proper subset of F is an (s, t)-edge separator of G then we say that F is a* minimal (s, t)-edge separator *of G. We say F is a* minimal edge separator *of G if it is a minimal (a, b)-edge separator for some $a, b \in V$.*

There is no need to define inclusion-minimal edge separators, since these turn out to be the same as minimal edge separators (unlike the situation for vertex separators). We show that both of the following problems are #P-complete with respect to AP-reductions, and that both are #P-complete to compute exactly.

Problem 7. #(s, t)-BiMinimalEdgeSeps.
Input: A bipartite graph G and two vertices $s, t \in V(G)$.
Output: The number of minimal (s, t)-edge separators of G, denoted by $\mathrm{MES}(G, s, t)$.

Problem 8. #BiMinimalEdgeSeps.
Input: A bipartite graph G.
Output: The number of minimal edge separators of G, denoted MES(G).

Theorem 3. *The   problems   #BiMinimalEdgeSeps   and   #(s,t)-BiMinimalEdgeSeps are #P-complete and are equivalent to #SAT under AP-reduction.*

In [14] we also study two other locally-optimal structures related to maximal independent sets in bipartite graphs. Theorem 4 shows that counting *dominating sets* in bipartite graphs is also #P-hard with respect to AP-reductions. Also, maximal independent sets in bipartite graphs can be represented as unions of sets, so (Theorems 5 and 6) a set union problem is also #P-hard with respect to AP-reductions, and so is its inverse.

2   Preliminaries

Most of our notation is standard, and we therefore defer it to Sect. 2 of [14]. The notions of a *fully polynomial randomised approximation scheme* (or *FPRAS*) and an *approximation-preserving reduction* (or *AP-reduction*) are standard in the field. If there is an AP-reduction from f to g, we write $f \leq_{AP} g$.

Dyer et al. [8] studied counting problems in #P and identified three classes of counting problems that are interreducible under AP-reductions. The first class, containing the problems that have an FPRAS, are trivially equivalent under AP-reduction since all the work can be embedded into the reduction (which declines to use the oracle). The second class is the equivalence class of #SAT, the problem of counting satisfying assignments to a Boolean formula in CNF, under AP-reduction. These problems are complete for #P with respect to AP-reductions. Zuckerman [29] has shown that #SAT cannot have an FPRAS unless RP = NP, so the same is true of any problem to which #SAT is AP-reducible.

The third class appears to be of intermediate complexity. It contains all of the counting problems expressible in a certain logically-defined complexity class, #RHΠ_1. Typical complete problems include counting the downsets in a partially ordered set [8], computing the partition function of the ferromagnetic Ising model with local external magnetic fields [15], and counting the independent sets in a bipartite graph, which is formally defined as follows.

Problem 12. #BIS.
Input: A bipartite graph G.
Output: The number of independent sets in G, denoted by IS(G).

In [8] it was shown that #BIS is complete for the logically-defined complexity class #RHΠ_1 with respect to AP-reductions. Goldberg and Jerrum [16] have conjectured that there is no FPRAS for #BIS. Early indications point to the fact that it may be of intermediate complexity, between the FPRASable problems and those that are complete for #P with respect to AP-reductions.

3   Hardness of #MaximalBIS

We first prove that #MaximalBIS is complete for #P with respect to AP-reductions. We reduce from the well-known problem of counting independent sets in an arbitrary graph.

Problem 13. #IS.
Input: A graph G.
Output: The number of independent sets in G.

Dyer et al. [8, Theorem3] shows that #IS is complete for #P with respect to AP-reductions. Using this we can now prove Theorem 1.

Proof. Since #MaximalBIS is in #P, #MaximalBIS $\leq_{AP}$ #SAT follows from [8]. To go the other direction, we will show #IS $\leq_{AP}$ #MaximalBIS. Let $MIS(G)$ denote the number of maximal independent sets in a graph G. Let $G = (V, E)$ be an instance of #IS. Without loss of generality let $V = [n]$ for some $n \in \mathbb{N}$, let $m = |E|$, and let $t = n + 2$. We shall construct an instance G' of #MaximalBIS with the property that $IS(G) \leq MIS(G')/2^{tm} \leq IS(G) + \frac{1}{4}$, which will be sufficient for the reduction. See Fig. 1 for an example.

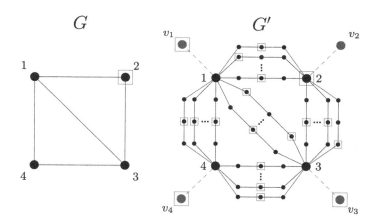

Fig. 1. An example of the reduction from an instance G of #IS to an instance G' of #MaximalBIS used in the proof of Theorem 1. The boxes around vertices indicate a non-maximal independent set in G and one of its maximal counterparts in G'. Note that the presence of v_4 ensures that vertex 4 has an occupied neighbour in G'.

Informally, we obtain a bipartite graph G' (an instance of #MaximalBIS) from G by first t-thickening and then 4-stretching each of G's edges and by also adding a bristle to each of G's vertices. Formally, we define G' as follows. For each $e \in E$ let X_e, Y_e and Z_e be sets of t vertices. We require all of these sets to be disjoint from each other and from $[n]$. Write $X_e = \{x_e^k \mid k \in [t]\}$,

$'_e = \{y_e^k \mid k \in [t]\}$, and $Z_e = \{z_e^k \mid k \in [t]\}$. Also, let $W = \bigcup_{e \in E} X_e \cup Y_e \cup Z_e$. Let $V' = \{v_1, \ldots, v_n\}$ be a set of distinct vertices which is disjoint from $[n] \cup W$. Then we define $V(G') = [n] \cup V^* \cup W$ and

$$E(G') = \{\{i, v_i\} \mid i \in [n]\} \cup \bigcup_{\substack{e=\{i,j\}\in E \\ i<j \\ k\in[t]}} \{\{i, x_e^k\}, \{x_e^k, y_e^k\}, \{y_e^k, z_e^k\}, \{z_e^k, j\}\} \ .$$

Let $S \subseteq [n]$ be arbitrary. We shall determine the number $\mathrm{MIS}_S(G')$ of maximal independent sets $T \subseteq V(G')$ with $T \cap [n] = S$, and thereby bound $\mathrm{MIS}(G')$. First, note that for every $S \subseteq [n]$, the set $S \cup \{v_i \in V^* \mid i \notin S\} \cup \bigcup_e Y_e$ is a maximal independent set of G', so $\mathrm{MIS}_S(G')$ is non-zero. Also, if T is a maximal independent set of G' and $T \cap [n] = S$ then $T \cap V^* = \{v_i \in V^* \mid i \notin S\}$. In particular, this implies that every unoccupied vertex in $[n]$ has an occupied neighbour in V^*.

Consider an edge $e = \{i, j\} \in E$, where $i < j$, and a value $k \in [t]$. If T is a maximal independent set of G' containing both i and j then $T \cap \{x_e^k, y_e^k, z_e^k\} = \{y_e^k\}$. However, if T is a maximal independent set of G' containing i but not j then $T \cap \{x_e^k, y_e^k, z_e^k\}$ can either be $\{y_e^k\}$ or $\{z_e^k\}$. This choice can be made independently for each $k \in [t]$. Similarly, if T is a maximal independent set of G' containing neither i nor j then $T \cap \{x_e^k, y_e^k, z_e^k\}$ can either be $\{x_e^k, z_e^k\}$, or $\{y_e^k\}$.

Given $S \subseteq [n]$, let $\mu(S)$ be the number of edges of G with both endpoints in S. We conclude from the previous observations that $\mathrm{MIS}_S(G') = 2^{(m-\mu(S))t}$ so $\mathrm{MIS}(G') = \sum_{S \subseteq [n]} 2^{(m-\mu(S))t}$. Since each independent set S of G has $\mu(S) = 0$, $\mathrm{MIS}(G') \geq \mathrm{IS}(G)2^{mt}$. Furthermore, since there are at most 2^n sets $S \subseteq [n]$ that are not independent sets of G, and each of these has $\mu(S) \geq 1$, we have

$$\mathrm{IS}(G) \leq \frac{\mathrm{MIS}(G')}{2^{tm}} \leq \mathrm{IS}(G) + 2^n 2^{-t} = \mathrm{IS}(G) + \frac{1}{4} \ . \tag{1}$$

Equation (1) implies that there is an AP-reduction from #IS to #MaximalBIS. The details of the reduction showing how to tune the accuracy parameter in the oracle call for approximating $\mathrm{MIS}(G')$ in order to get a sufficiently good approximation to $\mathrm{IS}(G)$ are exactly as in the proof of Theorem 3 of [8]. □

4   Minimal Separator Problems

The definition of minimal edge separator generalises naturally to multigraphs (see [14]). In order to prove Theorems 2 and 3 we consider two intermediate problems related to counting *maximum* minimal edge separators.

Problem 14. #LargeMinimalEdgeSeps.
Input: *A multigraph G and the maximum cardinality x of any minimal edge separator in G.*
Output: *The number of minimal edge separators of G with maximum cardinality, denoted by $\mathrm{LMES}(G)$.*

Problem 15. #(s, t)-LargeMinimalEdgeSeps.
Input: *A multigraph G, two distinct vertices $s, t \in V$, and the maximum cardinality y of any minimal (s, t)-edge separator in G.*
Output: *The number of minimal (s, t)-edge separators of G with maximum cardinality, which we denote by $\text{LMES}(G, s, t)$.*

The following proposition, due to Whitney [28], implies that minimal edge separators can be expressed in terms of vertex cuts.

Proposition 12. *Let $G = (V, E)$ be a connected multigraph. Then a multiset $F \subseteq E$ is a minimal edge separator of G if and only if $G - F$ has exactly two non-empty components, and F is the multiset of edges between them.* □

Since MAX-CUT is an intractable optimisation problem, we can show (see Lemma 16 of [14]) that the problems #LargeMinimalEdgeSeps and #(s, t)-LargeMinimalEdgeSeps are #SAT-hard to approximate and are #P-complete. In order to prove Theorem 3 it is then necessary to relate these problems to #BiMinimalEdgeSeps and #(s, t)-BiMinimalEdgeSeps. This is achieved by the following technical lemma, which is the heart of the proof.

Lemma 17. *Let $G = (V, E)$ be a connected multigraph, writing $n = |V|$ and $m = |E|$. Suppose (G, x) is an instance of #LargeMinimalEdgeSeps, and (G, s, t, y) is an instance of #(s, t)-LargeMinimalEdgeSeps. Let $k = \lceil m + \log_2(m) + 10 \rceil$. Then there exists a graph G' such that the following properties hold.*

(i) G' *is bipartite,* $V \subseteq V(G')$, *and* $|V(G')| \leq |E|k + |V|$.
(ii) $\text{LMES}(G) \leq \text{MES}(G')/2^{kx} \leq \text{LMES}(G) + \frac{1}{4}$.
(iii) $\text{LMES}(G, s, t) \leq \text{MES}(G', s, t)/2^{ky} \leq \text{LMES}(G, s, t) + \frac{1}{4}$.

The bipartite graph G' is constructed from G by first k-thickening and then 2-stretching each edge of G. The construction works because almost every minimal edge separator of G' has the following properties: (a) it contains at most one of the edges of a 2-path corresponding to an edge of G, and (b) if there is an intersection, then it intersects every such 2-path (so can be viewed as cutting the edge of G). It turns out that any minimal edge separator F of G corresponds to precisely $2^{k|F|}$ such minimal edge separators of G', and there aren't too many other minimal edge separators of G', so the construction goes through.

In order to prove Theorem 2, which is about vertex separators and not about edge separators, we need a similar, but more difficult, lemma.

Lemma 18. *Let $G = (V, E)$ be a connected multigraph, writing $n = |V|$ and $m = |E|$. Suppose (G, x) is an instance of #LargeMinimalEdgeSeps, and (G, s, t, y) is an instance of #(s, t)-LargeMinimalEdgeSeps. Let $k = \lceil m + n + \log_3(n^2) + 16 \rceil$. Then there exists a graph G' such that the following properties hold.*

(i) G' is bipartite, $V \subseteq V(G')$, and $|V(G')| \leq 3|E|k + |V|$.
(ii) $\text{LMES}(G) \leq \text{MS}(G')/3^{kx} \leq \text{LMES}(G) + \frac{1}{4}$.
(iii) $\text{LMES}(G, s, t) \leq \text{MS}(G', s, t)/3^{ky} \leq \text{LMES}(G, s, t) + \frac{1}{4}$.
(iv) $\text{LMES}(G) \leq \text{IMS}(G')/3^{kx} \leq \text{LMES}(G) + \frac{1}{4}$.

The construction of the bipartite graph G' is similar to the earlier one — G' is constructed from G by first k-thickening and then 4-stretching each edge of G. We are able to associate minimal (vertex) separators of G' with minimal edge separators of G in a similar way to the proof of Lemma 17, but the correspondence is significantly messier since a minimal separator of G' may contain vertices of V. Indeed, there may be exponentially many such separators (as a function of k)!

We define our correspondence as follows. If X is a minimal (vertex) separator of G' we define $\pi(X)$, the corresponding edge-separator of G, to be the set of edges e of G such that X contains some new vertex in the stretched thickening of e. The point is to identify a set of "good" minimal separators of G' so that every minimal edge separator F of G corresponds to exactly $3^{k|F|}$ good minimal separators of G', and not too many minimal separators of G' are not good. The details are somewhat complicated, and the notion of "good" needs to be refined. For $z \in \mathbb{N}$, we say that X is *z-good*, if it satisfies the following conditions: (a) it intersects at most one of the edges of a 4-path corresponding to an edge of G, (b) if there is an intersection, then it intersects every such 4-path (so it cuts the edge of G), (c) X contains no vertices of G, (d) $|\pi(X)| = z$. The key to the proof is showing that all but at most $3^{kx}/4$ minimal separators of G' are x-good, and all but at most $3^{ky}/4$ minimal (s, t)-separators of G' are y-good. The argument involves several steps. First, we show that there are at most $2^5 mk$ minimal separators in G' which are not minimal (b, c)-separators for some vertices b, c of G. Then we consider $a \in \mathbb{N}$ and distinct vertices b and c of G. We show that there are at most $2^{m+n}3^{k(a-1)}$ minimal (b, c)-separators X of G' with $|\pi(X)| < a$. Finally, we consider distinct vertices b and c of G and let z be the maximum cardinality of any minimal (b, c)-edge separator G. We show that, if X is a minimal (b, c)-separator of G' with $|\pi(X)| \geq z$, then X is z-good. This is the most difficult step. The details are included in [14].

Acknowledgements. We thank Luca Manzoni and Yuri Pirola.

References

1. Berry, A., Bordat, J.P., Cogis, O.: Generating all the minimal separators of a graph. International Journal of Foundations of Computer Science **11**(3), 397–403 (2000)
2. Bodlaender, H.L., Fomin, F.V.: Tree decompositions with small cost. Discrete Applied Mathematics **145**(2), 143–154 (2005)
3. Bodlaender, H.L., Fomin, F.V., Koster, A.M., Kratsch, D., Thilikos, D.M.: On exact algorithms for treewidth. ACM Trans. Algorithms **9**(1), 12:1–12:23 (2012)
4. Bodlaender, H.L., Koster, A.M.: Combinatorial optimization on graphs of bounded treewidth. The Computer Journal **51**(3), 255–269 (2008)

5. Bouchitté, V., Todinca, I.: Treewidth and minimum fill-in: grouping the minimal separators. SIAM Journal on Computing **31**(1), 212–232 (2001)
6. Bouchitté, V., Todinca, I.: Listing all potential maximal cliques of a graph. Theoretical Computer Science **276**(1–2), 17–32 (2002)
7. Diestel, R.: Graph Theory, 4th Edition. Graduate texts in mathematics, vol. 173. Springer (2012)
8. Dyer, M., Goldberg, L.A., Greenhill, C., Jerrum, M.: The relative complexity of approximate counting problems. Algorithmica **38**(3), 471–500 (2004)
9. Fomin, F.V., Kratsch, D., Todinca, I., Villanger, Y.: Exact algorithms for treewidth and minimum fill-in. SIAM Journal on Computing **38**(3), 1058–1079 (2008)
10. Fomin, F.V., Villanger, Y.: Finding induced subgraphs via minimal triangulations. In: 27th STACS, pp. 383–394 (2010)
11. Fomin, F.V., Villanger, Y.: Treewidth computation and extremal combinatorics. Combinatorica **32**(3), 289–308 (2012)
12. Furuse, M., Yamazaki, K.: A revisit of the scheme for computing treewidth and minimum fill-in. Theoretical Computer Science **531**(0), 66–76 (2014)
13. Goldberg, L.A.: Efficient Algorithms for Listing Combinatorial Structures. Cambridge University Press (1993). Cambridge Books Online
14. Goldberg, L.A., Gysel, R., Lapinskas, J.: Approximately counting locally-optimal structures. CoRR abs/1411.6829 (2014)
15. Goldberg, L.A., Jerrum, M.: The complexity of ferromagnetic Ising with local fields. Combin. Probab. Comput. **16**(1), 43–61 (2007)
16. Goldberg, L.A., Jerrum, M.: Approximating the partition function of the ferromagnetic Potts model. J. ACM **59**(5), 31 (2012). Art. 25
17. Gysel, R.: Unique perfect phylogeny characterizations via uniquely representable chordal graphs. CoRR abs/1305.1375 (2013)
18. Gysel, R.: Minimal triangulation algorithms for perfect phylogeny problems. In: Dediu, A.-H., Martín-Vide, C., Sierra-Rodríguez, J.-L., Truthe, B. (eds.) LATA 2014. LNCS, vol. 8370, pp. 421–432. Springer, Heidelberg (2014)
19. Johnson, D.S., Papadimitriou, C.H., Yannakakis, M.: How easy is local search? J. Comput. Syst. Sci. **37**(1), 79–100 (1988)
20. Kashiwabara, T., Masuda, S., Nakajima, K., Fujisawa, T.: Generation of maximum independent sets of a bipartite graph and maximum cliques of a circular-arc graph. J. Algorithms **13**(1), 161–174 (1992)
21. Kloks, T., Kratsch, D.: Listing all minimal separators of a graph. SIAM Journal on Computing **27**, 605–613 (1998)
22. Lokshtanov, D.: On the complexity of computing treelength. Discrete Applied Mathematics **158**(7), 820–827 (2010)
23. Schäffer, A.A., Yannakakis, M.: Simple local search problems that are hard to solve. SIAM J. Comput. **20**(1), 56–87 (1991)
24. Shen, H., Liang, W.: Efficient enumeration of all minimal separators in a graph. Theoretical Computer Science **180**(1–2), 169–180 (1997)
25. Takata, K.: Space-optimal, backtracking algorithms to list the minimal vertex separators of a graph. Discrete Appl. Math. **158**(15), 1660–1667 (2010)
26. Tsukiyama, S., Ide, M., Ariyoshi, H., Shirakawa, I.: A new algorithm for generating all the maximal independent sets. SIAM J. Comput. **6**(3), 505–517 (1977)
27. Vadhan, S.P.: The complexity of counting in sparse, regular, and planar graphs. SIAM J. Comput. **31**(2), 398–427 (2001)
28. Whitney, H.: Planar graphs. Fundamenta Mathematicae **21**(1), 73–84 (1933)
29. Zuckerman, D.: On unapproximable versions of NP-complete problems. SIAM J. Comput. **25**(6), 1293–1304 (1996)

Proofs of Proximity for Context-Free Languages and Read-Once Branching Programs
(Extended Abstract)

Oded Goldreich, Tom Gur[⊠], and Ron D. Rothblum

Department of Computer Science and Applied Mathematics,
Weizmann Institute of Science, 76100 Rehovot, Israel
{oded.goldreich,tom.gur,ron.rothblum}@weizmann.ac.il

Abstract. Proofs of proximity are probabilistic proof systems in which the verifier only queries a *sub-linear* number of input bits, and soundness only means that, with high probability, the input is close to an accepting input. In their minimal form, called *Merlin-Arthur proofs of proximity* ($\mathcal{MAP}$), the verifier receives, in addition to query access to the input, also free access to an explicitly given short (sub-linear) proof. A more general notion is that of an *interactive proof of proximity* ($\mathcal{IPP}$), in which the verifier is allowed to interact with an all-powerful, yet untrusted, prover. $\mathcal{MAP}$s and $\mathcal{IPP}$s may be thought of as the $\mathcal{NP}$ and $\mathcal{IP}$ analogues of property testing, respectively.

In this work we construct proofs of proximity for two natural classes of properties: (1) context-free languages, and (2) languages accepted by small read-once branching programs. Our main results are:

1. $\mathcal{MAP}$s for these two classes, in which, for inputs of length n, both the verifier's query complexity and the length of the $\mathcal{MAP}$ proof are $\widetilde{O}(\sqrt{n})$.

2. $\mathcal{IPP}$s for the same two classes with constant query complexity, polylogarithmic communication complexity, and logarithmically many rounds of interaction.

1 Introduction

The field of property testing, initiated by Rubinfeld and Sudan [RS96] and Goldreich, Goldwasser and Ron [GGR98], studies a computational model that consists of probabilistic algorithms, called *testers*, that need to decide whether a given object has a certain global property or is far (say, in Hamming distance) from all objects that have the property, based only on a local view of the object.

A line of work [EKR04,BSGH+06,DR06,RVW13,GR15,FGL14,KR14] has considered the question of designing *proof systems* within the property testing model. The minimal type of such a proof system, which was recently studied

---

Full version can be found in ECCC TR15-024 [GGR15].

This research was partially supported by the Israel Science Foundation (grant No. 671/13).

M.M. Halldórsson et al. (Eds.): ICALP 2015, Part I, LNCS 9134, pp. 666–677, 2015.
DOI: 10.1007/978-3-662-47672-7_54

by Gur and Rothblum [GR15], augments the property testing framework by replacing the tester with a *verifier* that receives, in addition to oracle access to the input, also free access to an explicitly given short (i.e., sub-linear length) proof. The guarantee is that for inputs that have the property there exists a proof that makes the verifier accept with high probability, whereas, for inputs that are far from the property, the verifier will reject *every* alleged proof with high probability. These proof systems can be thought of as the $\mathcal{NP}$ (or more accurately $\mathcal{MA}$) analogue of *property testing*, and are called *Merlin-Arthur proofs of proximity (MAP)*.[1]

A more general notion was considered by Rothblum, Vadhan and Wigderson [RVW13] (prior to [GR15]). Their proof system, which can be thought of as the $\mathcal{IP}$ analogue of property testing, consists of an all powerful (but untrusted) prover who interacts with a verifier that only has oracle access to the input x. The prover tries to convince the verifier that x has a particular property Π. Here, the guarantee is that for inputs in Π, there exists a prover strategy that will make the verifier accept with high probability, whereas for inputs that are far from Π, the verifier will reject with high probability no matter what prover strategy is employed. The latter proof systems are known as *interactive proofs of proximity (IPPs)*.[2]

The focus of this paper is identifying natural classes of properties that are known to be hard to test, but become easy to *verify* using the power of a proof ($\mathcal{MAP}$) or interaction with a prover ($\mathcal{IPP}$).

1.1   Our Results

One well-known class of properties that is hard to test is the class of *context-free languages*. Alon *et al.* [AKNS00] showed that there exists a context-free language that requires $\Omega\left(\sqrt{n}\right)$ queries to test (where here and throughout this work, n denotes the size of the input) and a context-free language that requires $\Omega(n)$ queries to test with *one-sided error*. Furthermore, there are no known (non-trivial) testers for general context-free languages.

Another interesting class is the class of languages that are accepted by small *read-once branching programs (ROBPs)*. Newman [New02] showed that the set of strings accepted by any small width ROBP can be efficiently tested.[3] More specifically, Newman showed that width w ROBPs can be tested using $(2^w/\varepsilon)^{O(w)}$ queries, where ε is the proximity parameter. Bollig [Bol05] showed that Newman's result cannot be extended to polynomial-sized ROBPs, by exhibiting an $O(n^2)$-sized ROBP that requires $\Omega(\sqrt{n})$ queries to test. No (non-trivial) testers for general ROBPs are known for width $\Omega(\sqrt{\log n})$.

---

[1] A related notion is that of a *probabilistically checkable proof of proximity (PCPP)* [BSGH+06,DR06]. $\mathcal{PCPP}$s differ from $\mathcal{MAP}$s in that the verifier is only given *query* (i.e., oracle) access to the proof, whereas in $\mathcal{MAP}$s, the verifier has free (*explicit*) access to the proof. Hence, $\mathcal{PCPP}$s are a $\mathcal{PCP}$ analogue of property testing.

[2] Indeed, $\mathcal{MAP}$s can be thought of as a restricted case of $\mathcal{IPP}$s, in which the interaction is limited to a single message sent from the prover to the verifier.

[3] The result in [New02] is stated only for *oblivious* ROBPs but in [Bol05, Section 1.3] it is stated that Newman's result holds also for general *non-oblivious* ROBPs.

In this work we consider the question of constructing *efficient* $\mathcal{MAP}$s and $\mathcal{IPP}$s for these two classes.[4] Here, by "efficient", we mean that *both* the *query complexity* (i.e., the number of queries performed by the verifier to the input) and the *proof complexity* (i.e., the length of the $\mathcal{MAP}$ proof) or *communication complexity* (i.e., the amount of communication with the $\mathcal{IPP}$ prover) are small and, in particular, sub-linear[5].

Our first pair of results are efficient $\mathcal{MAP}$s for context-free languages and for ROBPs. These $\mathcal{MAP}$s offer a multiplicative trade-off between the query and proof complexities. Here and throughout this work, $n \in \mathbb{N}$ specifies the length of the main input and $\varepsilon \in (0,1)$ denotes the proximity parameter.

Theorem 1. *For every context-free language $\mathcal{L}$ and every $k = k(n)$ such that $2 \leq k \leq n$, there exists an $\mathcal{MAP}$ for $\mathcal{L}$ that uses a proof of length $O(k \cdot \log n)$ and has query complexity $O\left(\frac{n}{k} \cdot \varepsilon^{-1}\right)$. Furthermore, the $\mathcal{MAP}$ has one-sided error.*

Theorem 2. *If a language $\mathcal{L}$ is recognized by a size $s = s(n)$ ROBP, then for every $k = k(n)$ such that $2 \leq k \leq n$, there exists an $\mathcal{MAP}$ for $\mathcal{L}$ that uses a proof of length $O(k \cdot \log s)$ and has query complexity $O\left(\frac{n}{k} \cdot \varepsilon^{-1}\right)$. Furthermore, the $\mathcal{MAP}$ has one-sided error.*

Hence, by setting $k = \sqrt{n}$, every context-free language and every language accepted by an ROBP of size at most $2^{\mathsf{polylog}(n)}$, has an $\mathcal{MAP}$ in which both the proof and query complexity are $\widetilde{O}(\sqrt{n})$ (w.r.t. constant proximity parameter).

Next, we ask whether the query and proof complexity in Theorems 1 and 2 can be significantly reduced by allowing more extensive *interaction* between the verifier and the prover (i.e., arbitrary interactive communication rather than just a fixed non-interactive proof). Very relevant to this question is a recent result of [RVW13] by which, loosely speaking, every language in $\mathcal{NC}$ (which contains all context-free languages [Ruz81] and languages accepted by small ROBPs[6]) has an $\mathcal{IPP}$ with $\widetilde{O}(\sqrt{n})$ query and communication complexities. While the [RVW13] result is more general, for context-free languages and ROBPs it achieves roughly the same query and communication complexities as the $\mathcal{MAP}$s in Theorems 1 and 2, but uses much more interaction (i.e., at least logarithmically many rounds of interaction compared to just a single message in our $\mathcal{MAP}$s).

---

[4] To see that these two classes do not contain each other, observe that the language $\{0^i 1^j 2^i 3^j : i, j \geq 1\}$, which is *not* a context-free language [HMU06, Example 7.20], has a $\mathsf{poly}(n)$-width ROBP (which simply counts the number of repeated occurrences of 0, 1, 2 and 3). On the other hand, Kriegal and Waack [KW88] showed that every ROBP for the Dyck_2 language, which is a context-free language, has size $2^{\Omega(n)}$.

[5] As pointed out in [GR15], if we do not restrict the length of the proof, then *every* property Π can be verified trivially using only a constant amount of queries, by considering an $\mathcal{MAP}$ proof that contains a full description of the input.

[6] See the full version [GGR15] for a discussion on why languages accepted by ROBPs can be computed in small depth.

Using cryptographic assumptions[7], Kalai and Rothblum [KR14] recently showed that there exists a language in $\mathcal{NC}_1$ for which every $\mathcal{IPP}$ requires that either the query or communication complexity be $\Omega(\sqrt{n})$. Hence, we cannot hope to improve the [RVW13] result in general. Still, for the special case of context-free languages and ROBPs, we show that we can actually extend the $\mathcal{MAP}$ protocols in Theorems 1 and 2 into highly efficient $\mathcal{IPP}$s with only *poly-logarithmic* complexity (using a sub-logarithmic number of rounds). More generally, our $\mathcal{IPP}$s offer a trade-off between the number of rounds of interaction and the query and communication complexities.

Theorem 3. *For every context-free language $\mathcal{L}$, every $k = k(n) \geq 2$ and $r = r(n) \geq 1$ such that $k^r \leq n$, there exists an r-round $\mathcal{IPP}$ for $\mathcal{L}$ with communication complexity $O\big((rk\log n) \cdot \varepsilon^{-1}\big)$ and query complexity $O\big(\frac{n}{k^r} \cdot \varepsilon^{-1}\big)$. Furthermore, the $\mathcal{IPP}$ is public-coin and has one-sided error.*

Theorem 4. *If a language $\mathcal{L}$ is recognized by a size $s = s(n)$ ROBP, then for every $k = k(n) \geq 2$ and $r = r(n) \geq 1$ such that $k^r \leq n$, there exists an r-round $\mathcal{IPP}$ for $\mathcal{L}$ with communication complexity $O\big((rk\log s)\cdot\varepsilon^{-1}\big)$ and query complexity $O\big(\frac{n}{k^r} \cdot \varepsilon^{-1}\big)$. Furthermore, the $\mathcal{IPP}$ is public-coin and has one-sided error.*

(Interestingly, and in contrast to Theorems 1 and 2, here the communication complexity also depends on the proximity parameter ε.) In particular, by setting $k = \log n$ and $r = \frac{\log n}{\log \log n}$, we obtain $\mathcal{IPP}$s for context-free languages and size $2^{\mathsf{polylog}(n)}$ ROBPs, with a sub-logarithmic number of rounds, constant query complexity, and poly-logarithmic communication complexity (w.r.t. constant proximity parameter).

A Remark on Computational Complexity. Following the property testing literature, we view the query complexity and the proof complexity (resp., communication complexity) as the primary resources of an $\mathcal{MAP}$ (resp., $\mathcal{IPP}$). Still, the running time of the verifier and of the prover are also important resources. The proofs/provers in our $\mathcal{MAP}$s and $\mathcal{IPP}$s are indeed efficient; that is, polynomial in the main input x (and in the case of ROBPs also in the size of the ROBP).

As for our verifiers, those in Theorems 1 and 3 run in polynomial time (i.e., $\mathsf{poly}(|x|)$ time) rather than in *sub-linear* time as one might hope. However, by increasing the round complexity in Theorem 3 by a poly-logarithmic factor, we can obtain an $\mathcal{IPP}$ with sub-linear time verification. Constructing an $\mathcal{MAP}$ for context-free languages with sub-linear time verification remains an interesting open question. The verifiers in Theorems 2 and 4 run in *sub-linear time* if they

---

[7] A sufficient assumption for [KR14] is the existence of (length-doubling) PRGs that can be computed in $\mathcal{NC}_1$ and whose output cannot be distinguished from random by circuits of size $2^{o(n)}$.

are given a *suitable* (natural) representation of the ROBP.[8] See the full version [GGR15] for further details.

Improved Results for Specific Languages. The paradigm used for the general results in Theorems 1-4 can be extended to yield better results for specific languages. A notable class of languages for which we obtain such an improvement is the class of languages of balanced parentheses expressions (a.k.a the Dyck languages), which are context-free languages, for which Parnas *et al.* [PRR01] showed a lower bound of $\widetilde{\Omega}(n^{1/11})$ for ordinary testers. Using special properties of the Dyck languages, we can improve on the general result in Theorem 1 in this special case and obtain a somewhat more efficient $\mathcal{MAP}$ for the Dyck languages. See details in the full version [GGR15].

A Remark on $\mathcal{LOGCFL}$. The well studied complexity class $\mathcal{LOGCFL}$ consists of all languages that are logspace reducible to a context-free language (see [Coo71]).[9] We stress that, while Theorems 1 and 3 hold for every *context-free language*, they do not necessarily extend to all languages in $\mathcal{LOGCFL}$, since the reductions may not preserve the classes $\mathcal{MAP}$ and $\mathcal{IPP}$. In fact, by the aforementioned lower bound of Kalai and Rothblum [KR14], assuming sufficiently strong cryptographic PRGs, there exists a language in $\mathcal{NC}_1 \subseteq \mathcal{LOGCFL}$ for which every $\mathcal{IPP}$ must have complexity $\Omega(\sqrt{n})$. Hence, an extension of Theorem 3 to $\mathcal{LOGCFL}$ is not likely to hold.

1.2   Proof Overview

The proofs of Theorems 1 and 2 (i.e., the $\mathcal{MAP}$ results) will follow (roughly) as special cases of the proofs of Theorems 3 and 4 (i.e., the $\mathcal{IPP}$ results), respectively. Hence, in this overview we focus on the proofs of Theorems 3 and 4, while explaining how to derive Theorems 1 and 2 as special cases.

The proofs of Theorems 3 and 4 share a common theme: For $\mathcal{L}$ that is either a context-free language or is accepted by a ROBP, we show that every input $x \in \mathcal{L}$ can be broken-down into k sub-problems (related to $\mathcal{L}$) such that the following holds:

1. On the one hand, if $x \in \mathcal{L}$, then there exists (1) a partition of $[n]$ into sets $S_1, \ldots, S_k$ (each of size roughly n/k); and (2) languages $\mathcal{L}_1, \ldots, \mathcal{L}_k$ such that both (1) and (2) have a concise representation, and, for every $i \in [k]$, the projection of x on S_i, denoted $x[S_i]$, is in the language $\mathcal{L}_i$. Furthermore, if $\mathcal{L}$ is a context-free language (resp., accepted by an ROBP), then the languages

---

[8] Indeed, the running time of the verifier crucially relies on the specific representation of the ROBP. We remark that there are other natural representations of ROBPs than the one we use, and for some of these representations obtaining sub-linear running time may not be feasible.

[9] Note that $\mathcal{LOGCFL}$ contains languages that are *not* context-free (e.g., the language $\{a^n b^n c^n : n \in \mathbb{N}\}$ is not context-free [HMU06, Example 7.19] but is computable in logspace (and hence also in $\mathcal{LOGCFL}$)).

$\mathcal{L}_1, \ldots, \mathcal{L}_k$ are all "variants" of context-free languages[10] (resp., accepted by ROBPs).

2. On the other hand, if x is "far" from $\mathcal{L}$, then for every concise representation of a partition $S_1, \ldots, S_k$ of $[n]$ and languages $\mathcal{L}_1, \ldots, \mathcal{L}_k$ (of the type used in 1), for an average $i \in [k]$, it holds that $x[S_i]$ is proportionally "far" from $\mathcal{L}_i$.

By design, the partition $S_1, \ldots, S_k$ as well as the corresponding languages $\mathcal{L}_1, \ldots, \mathcal{L}_k$ depend on the entire input x, and so the verifier (who only has query access to x) cannot generate them by itself. Instead, the concise representation of $S_1, \ldots, S_k$ and $\mathcal{L}_1, \ldots, \mathcal{L}_k$ will be specified by the prover (as a single message in the case of an $\mathcal{IPP}$, or as the entire proof string in the case of an $\mathcal{MAP}$).

Given the latter, we construct an $\mathcal{MAP}$ as follows. The $\mathcal{MAP}$ verifier selects at random a small subset $I \subseteq [k]$ and, for every $i \in I$, reads *all* of $x[S_i]$ (which is of length roughly n/k) and checks that $x[S_i] \in \mathcal{L}_i$. Indeed, by the two foregoing conditions, if $x \in \mathcal{L}$, then $x[S_i] \in \mathcal{L}_i$ for every $i \in [k]$, whereas if x is "far" from $\mathcal{L}$, then, by an averaging argument, for many $i \in [k]$, it holds that $x[S_i]$ is proportionally "far" from $\mathcal{L}_i$ (and in particular $x[S_i] \notin \mathcal{L}_i$), and the verifier will reject.

A natural approach for extending the foregoing $\mathcal{MAP}$ to an $\mathcal{IPP}$ is to have the verifier send the set I (where I is chosen at random as in the $\mathcal{MAP}$) to the prover, and then *recursively* run $|I|$ $\mathcal{IPP}$ protocols to check that $x[S_i]$ is close to $\mathcal{L}_i$, for every $i \in I$. In each recursive call the input shrinks by (roughly) a factor of k. After the recursion reaches depth r, where r is a predetermined bound on the number of rounds, the verifier can simply read its entire current input (of length $O(n/k^r)$) and decide whether to accept or reject.

The foregoing approach indeed works, but because there is more than one recursive call in each round, the complexity of the resulting $\mathcal{IPP}$ depends *exponentially* on the number of rounds r. Instead, we use a more economical approach, which avoids the exponential dependence on r, based on the notion of a *proximity oblivious tester* [GR11]. Recall that a proximity oblivious tester for a property Π is a tester that does not receive the proximity parameter ε as input and is only required to reject inputs that are ε-far from Π with probability proportional to ε (rather than probability $2/3$). To present a more economical recursion, the $\mathcal{IPP}$ that we design is similarly "proximity oblivious". The idea is to have the verifier select at random only a single index $i \in [k]$, send i to the prover, and then have the two parties recursively run an $\mathcal{IPP}$ protocol for verifying that $x[S_i]$ is close to $\mathcal{L}_i$. Indeed, if $x \in \mathcal{L}$ then $x[S_i] \in \mathcal{L}_i$, whereas if x is ε-far from $\mathcal{L}$, then, since i was chosen at random, on the average $x[S_i]$ is ε-far from $\mathcal{L}_i$, and therefore, by inductive reasoning, the verifier will reject with probability ε. To obtain constant soundness we can just repeat[11] the entire proximity oblivious protocol $O(1/\varepsilon)$ times in parallel.

---

[10] If $\mathcal{L}$ is a context-free language, then the languages $\mathcal{L}_1, \ldots, \mathcal{L}_k$ will be variants of context-free languages, which we call "partial derivation languages". However, if $\mathcal{L}$ is accepted by an ROBP, then the languages $\mathcal{L}_1, \ldots, \mathcal{L}_k$ are also accepted by (different) ROBPs.

[11] As expected, parallel repetition reduces the soundness error of $\mathcal{IPP}$s at an exponential rate. See the full version [GGR15] for details.

This concludes the high-level description of our $\mathcal{MAP}$s and $\mathcal{IPP}$s. Of course, the way in which the partition is generated is quite different in the case of context-free languages and in the case of ROBP, and different technical problems arise in each case. In the following subsections we discuss the specific details. In Section 1.2 we give an overview of how to partition read-once branching programs. Partitioning context-free languages is more involved, and so, in Section 1.2, as a warm-up, we first consider partitioning into *two parts* (i.e., $k = 2$). Then, in Section 1.2 we show how to extend the technique to *multiple parts* (i.e., general $k \geq 2$).

Partitioning ROBPs. Recall that a branching program on n variables is a directed acyclic graph with a unique source vertex with in-degree 0 and (possibly) multiple sink vertices with out-degree 0. Each sink vertex is labeled with either 0 (i.e., *reject*) or 1 (i.e., *accept*). Each non-sink vertex is labeled by an index $i \in [n]$ and has exactly 2 outgoing edges, which are labeled by 0 and 1. The output of the branching program B on input $x \in \{0,1\}^n$, denoted $B(x)$, is computed in a natural way by starting at the source vertex and taking a walk such that at a vertex labeled by $i \in [n]$, we traverse the outgoing edge labeled by x_i. Once a sink is reached, we output its label. The branching program is *read-once* (ROBP for short) if along every path from source to sink, every index ($i \in [n]$) appears at most once. The *size* of a branching program B, denoted $|B|$, is the number of vertices in it.

For any fixed ROBP B, we construct an $\mathcal{IPP}$ (and an $\mathcal{MAP}$, which is a special case of the $\mathcal{IPP}$) for the language accepted by B, denoted $\mathcal{L}_B \overset{\text{def}}{=} \{x \in \{0,1\}^n : B(x) = 1\}$. In this overview, we make a simplifying assumption that B is both *layered* and *ordered* (a.k.a., an *ordered binary decision diagram* or OBDD). That is, we assume that the vertices of B are partitioned into $n + 1$ layers such that, for every $i \in [n]$, edges only go from layer i to layer $i + 1$; and vertices in layer i are labeled by the index i (i.e., the ROBP reads its input "in order").

The key idea, which enables the $\mathcal{IPP}$ verifier to generate the aforementioned partition $S_1, \ldots, S_k$ (together with the corresponding languages), is to have the prover specify k evenly-spaced vertices along the accepting path corresponding to the input $x \in \mathcal{L}_B$. More specifically, observe that x induces a path $\varphi_0 \to \varphi_1 \to \cdots \to \varphi_n$ from the start vertex φ_0 to some accepting sink φ_n. The prover sends to the verifier a subsequence of this walk, specifically the subsequence $\varphi_{n/k}, \ldots, \varphi_{i \cdot n/k}, \ldots, \varphi_n$.

Given the subsequence, we can reduce the problem of verifying that there exists a path of length n from φ_0 to φ_n to verifying that there exists a path of length n/k between each pair of consecutive vertices in the sequence $\varphi_0, \varphi_{n/k}, \ldots, \varphi_{i \cdot n/k}, \ldots, \varphi_n$. In other words, for every $i \in [k]$ we consider the ROBP B_i that consists only of layers $(i - 1) \cdot n/k$ up to $i \cdot n/k$ of B, with the starting state $\varphi_{(i-1) \cdot n/k}$ and the (only) accepting state $\varphi_{i \cdot n/k}$. Verifying that $x \in \mathcal{L}_B$ can be reduced to verifying that $x[S_i] \in \mathcal{L}_{B_i}$, for every $i \in [k]$, where $S_i \subseteq [n]$ is the set of coordinates of x that are read by B_i and

$\mathcal{L}_{B_i} \overset{\text{def}}{=} \{z \in \{0,1\}^{n/k} : B_i(z) = 1\}$. Moreover, since $S_1, \ldots, S_k$ is a partition of $[n]$, if x is ε-far from $\mathcal{L}_B$, then $x[S_i]$ is ε-far from $\mathcal{L}_{B_i}$, for an average $i \in [k]$. Hence, we can follow the high-level outline that was suggested in Section 1.2; that is, the $\mathcal{IPP}$ verifier selects $i \in [k]$ at random, sends i to the prover, and then the two parties recursively run an $\mathcal{IPP}$ protocol to verify that $x[S_i]$ is close to the $\mathcal{L}_{B_i}$.

The foregoing intuition almost works but there is a subtle problem: What if the message sent by a *cheating* prover is such that $\mathcal{L}_{B_{i^*}}$ is empty, for some $i^* \in [k]$. This corresponds to a situation in which the branching program B contains no path from $\varphi_{(i^*-1) \cdot n/k}$ to $\varphi_{i^* \cdot n/k}$. In such case, with high probability (i.e., if the verifier chooses i such that $i \neq i^*$) the verifier, as described so far, will not notice this fact and may accept inputs that are far from $\mathcal{L}_B$.

We overcome this difficulty by observing that when the verifier interacts with the honest prover, it holds that $x[S_i] \in \mathcal{L}_{B_i}$ for every $i \in [k]$, and therefore $\mathcal{L}_{B_i} \neq \emptyset$. Hence, we can have the verifier explicitly check that $\mathcal{L}_{B_i} \neq \emptyset$ for *every* $i \in [k]$ (i.e., that there exists *some* input that leads from $\varphi_{(i-1) \cdot n/k}$ to $\varphi_{i \cdot n/k}$ in B). This check requires direct and full access to the branching program B (which is fixed) but does *not* require any queries to the input x, and so we can perform it for *every*[12] $i \in [k]$.

Given this additional check, we can show that the foregoing $\mathcal{IPP}$ works. To do so, we argue by induction on the number of rounds that if the input x is ε-far from $\mathcal{L}$ then the verifier rejects with probability at least ε. Indeed, if x is ε-far from $\mathcal{L}_B$, then in the first round we have that:

$$
\begin{aligned}
\Pr\left[\text{Verifier for } \mathcal{L}_B \text{ rejects } x\right] &= \mathop{\mathbf{E}}_i\left[\Pr\left[\text{Verifier for } \mathcal{L}_{B_i} \text{ rejects } x[S_i]\right]\right] \\
&\geq \mathop{\mathbf{E}}_i\left[\varepsilon_i\right] \\
&\geq \varepsilon,
\end{aligned}
$$

where ε_i denotes the relative distance of $x[S_i]$ from $\mathcal{L}_{B_i}$, for every $i \in [k]$, and the first inequality follows from the induction hypothesis.

We remark that when dealing with general ROBPs, rather than OBDDs, there are several additional technical difficulties. In particular, since B is not layered, we have to modify our definition of B_i (which previously consisted of layers $(i-1) \cdot n/k$ to $i \cdot n/k$ of B). A natural approach is to define B_i to consist of all paths (in B) of length n/k starting at $\varphi_{(i-1) \cdot n/k}$.[13] The difficulty is that B_i may depend on many, possibly even all, of the bits of x (since different paths may look at different bits), rather than just n/k bits (as was the case for OBDDs). Hence, the input does not necessarily shrink in the recursive step. Nevertheless,

---

[12] However, this check does increase the running time of the verifier (which we view as a secondary resource) to $\mathsf{poly}(|B|)$. This computation can be minimized by using a pre-processing step in which we compute a $|B| \times |B|$-sized table whose $(v,u)^{\text{th}}$ entry says whether the vertices v and u are connected in B.

[13] The actual definition of B_i that we use is different; see the full version [GGR15] for details.

we resolve this issue by showing that the *effective length* of the input, which is the number of bits that need to be read in order to determine whether the ROBP accepts, does shrink, and this suffices to make progress in the recursion. For further details, see the full version [GGR15].

Partitioning Context-Free Languages into Two Parts. Recall that a *context-free grammar* is a tuple $G = (V, \Sigma, R, A_{\mathsf{start}})$, where $V = \{A_1, A_2, \dots\}$ denotes a (finite) set of variables, $\Sigma = \{\sigma_1, \sigma_2, \dots\}$ denotes a (finite) set of terminal symbols (i.e., the alphabet), R is a set of production rules (e.g., rules of the form $A_7 \rightarrow \sigma_5 A_3 A_9 \sigma_8 A_2$) and $A_{\mathsf{start}} \in V$ denotes a special "start" variable. We say that a string $\alpha \in (\Sigma \cup V)^*$ is *derived* from a variable A_j, denoted by $A_j \stackrel{*}{\Rightarrow} \alpha$, if α can be obtained from A_j by iteratively applying production rules in R. Each such derivation can be described by a *derivation tree*, which is a rooted, directed, ordered, and labeled tree (with edges oriented away from the root), where the root is labeled by A_j, the leaves are labeled by the symbols of α (in order), and the children of each vertex in the tree correspond to an application of a production rule in G. The language $\mathcal{L} \subseteq \Sigma^*$ *generated* by G consists of all strings that can be derived from A_{start} using the production rules in R.

Let $\mathcal{L}$ be a context-free language and let $G = (V, \Sigma, R, A_{\mathsf{start}})$ be the context-free grammar that generates $\mathcal{L}$. In this section we show how to partition $x \in \mathcal{L}$ into two parts. Next, in Section 1.2, we show how to extend this technique to multiple parts.

For $x \in \mathcal{L}$ (i.e., $A_{\mathsf{start}} \stackrel{*}{\Rightarrow} x$), there exists a derivation tree T corresponding to the derivation $A_{\mathsf{start}} \stackrel{*}{\Rightarrow} x$. For simplicity, let us assume that T is a *binary* tree. The root of T is labeled by A_{start} and the leaves are labeled, in order, by $x_1, \dots, x_n$, where $n \stackrel{\text{def}}{=} |x|$. Recall that the Lewis-Stearns-Hartmanis Lemma [LSH65] states that every binary tree on n leaves has a subtree[14] with a number of leaves between $n/3$ and $2n/3$. Applying this lemma to T, we can find such a subtree T' of T. Observe that T' induces a partition of $[n]$ into two parts $S_1, S_2 \subseteq [n]$, where S_1 (which is actually an interval) contains all the leaves of T that belong to T' and $S_2 \stackrel{\text{def}}{=} [n] \backslash S_1$ contains all other leaves. The $\mathcal{IPP}$ prover finds T' and sends S_1 and A_1 to the verifier, where A_1 is the label of the root of T'. Since S_1 is an interval, the latter requires only $O(\log n)$ communication.

Given (S_1, A_1), the verifier can construct the partition and the corresponding languages, where the partition is simply (S_1, S_2) and the languages are

$$\mathcal{L}_1 \stackrel{\text{def}}{=} \left\{ w \in \Sigma^{|S_1|} \, : \, A_1 \stackrel{*}{\Rightarrow} w \right\}$$

and

$$\mathcal{L}_2 \stackrel{\text{def}}{=} \left\{ w \in \Sigma^{|S_2|} \, : \, A_2 \stackrel{*}{\Rightarrow} w[1, \dots, s-1] \circ A_1 \circ w[s, \dots, |S_2|] \right\},$$

where $A_2 \stackrel{\text{def}}{=} A_{\mathsf{start}}$ and $s \in [n]$ is the starting position of the interval S_1 in $[n]$.

---

[14] Here and throughout this work, by a subtree, we mean a node of the tree together with *all* of its descendants.

Note that $\mathcal{L}_2$ is not quite a context-free language (although $\mathcal{L}_1$ is). Rather, $\mathcal{L}_2$ consists of strings that correspond to *partial derivations* (i.e., derivation processes that end before all symbols are terminals) starting from A_{start} that produce strings that have the variable A_1 in their s^{th} coordinate. We refer to such languages, which we view as generalization of context-free languages, as *partial derivation languages*, and for the recursion to go through, we actually design the original protocol to handle not only context-free languages but also partial derivation languages.

Observe that if $x \in \mathcal{L}$, then clearly $x[S_1] \in \mathcal{L}_1$ and $x[S_2] \in \mathcal{L}_2$. On the other hand, suppose that $x[S_1]$ is ε_1-close to a string $z_1 \in \mathcal{L}_1$ and $x[S_2]$ is ε_2-close to a string $z_2 \in \mathcal{L}_2$. If we choose $i \in \{1, 2\}$ at random, such that $\Pr[i = 1] = |S_1|/n$ and $\Pr[i = 2] = |S_2|/n$, then x is $\mathbf{E}_i[\varepsilon_i]$-close to the string $z = z_2[1, \dots, s-1] \circ z_1 \circ z_2[s, |S_2|]$. Since $A_1 \overset{*}{\Rightarrow} z_1$ and $A_{\text{start}} \overset{*}{\Rightarrow} z_2[1, \dots, s-1] \circ A_1 \circ z_2[s, \dots, |S_2|]$ (because $z_1 \in \mathcal{L}_1$ and $z_2 \in \mathcal{L}_2$), we deduce that $A_{\text{start}} \overset{*}{\Rightarrow} z$, and therefore $z \in \mathcal{L}$. Hence, x is $\mathbf{E}_i[\varepsilon_i]$-close to $\mathcal{L}$.

Given the above, we can design an $\mathcal{IPP}$ for $\mathcal{L}$ similarly to the $\mathcal{IPP}$ for ROBP that was described in Section 1.2. Specifically, given (S_1, A_1), the verifier chooses at random $i \in \{1, 2\}$ according to the distribution above, sends i to the prover, and both parties run the protocol recursively, with respect to the language $\mathcal{L}_i$ and the input $x[S_i]$.

Partitioning Context-Free Languages into Multiple Parts. The first step in partitioning context-free languages into *multiple* parts is a generalization of the Lewis-Stearns-Hartmanis lemma that shows that, for every desired parameter $t \in [n]$, every (constant degree) tree T with n leaves has a subtree with roughly t leaves. The precise statement of the lemma and its proof are given in the full version [GGR15].

Using the generalized Lewis-Stearns-Hartmanis lemma, we can partition an input $x \in \mathcal{L}$ into k parts of (roughly) the same size in the following way. As before, we construct a derivation tree T corresponding to the derivation $A_{\text{start}} \overset{*}{\Rightarrow} x$. However, this time we use the generalized Lewis-Stearns-Hartmanis lemma to find a subtree T_1 with roughly n/k leaves. The coordinates of the leaves of T_1 constitute the first part of the partition (denoted by S_1). To find the second subtree, we remove the entire subtree T_1 from T, *except for its root*. We obtain a new tree T' with (roughly) $n - \frac{n}{k}$ leaves, where one of the leaves of T' is labeled by a variable rather than a terminal. By applying the generalized Lewis-Stearns-Hartmanis lemma again on the new tree T', we can find a subtree T_2 of T' with roughly n/k leaves. The second part (denoted by S_2) of our partition will consist of the coordinates of all the leaves of T_2 that are labeled by terminals (i.e., are also leaves of the original tree T). We stress that S_2 may not be an interval (but rather two intervals separated by S_1).

We proceed similarly, where in each iteration we remove the subtree that was found in the previous iteration (except for its root) and find a new subtree T_i of T with roughly n/k leaves. The subtrees $T_1, T_2, \dots, T_k$ induce a partition of $[n]$ where the ith part, denoted S_i (of size roughly n/k), consists of all leaves of T_i

hat are labeled by terminals (i.e., are leaves of the original tree T) but do not
belong to $S_1 \cup \cdots \cup S_{i-1}$.

While the representation of a general partition of $[n]$ into k parts requires
$\iota \cdot \log_2(k)$ bits, we show that the partition $S_1, \ldots, S_\ell$ actually has a concise
representation. Indeed, each subtree T_i induces an interval $I_i \subseteq [n]$, which contains all of its leaves (but potentially also coordinates of other parts in the
partition). Given $I_1, \ldots, I_\ell$, the partition $S_1, \ldots, S_\ell$ is uniquely determined (by
setting $S_i = I_i \backslash (I_1 \cup \cdots \cup I_{i-1})$). We remark that each pair of *intervals* can be
either disjoint or nested (i.e., either $I_i \cap I_j = \emptyset$ or $I_i \subsetneq I_j$).

In light of the foregoing discussion, the prover can send to the verifier the
intervals $I_1, \ldots, I_k$ and the variables $A_1, \ldots, A_\ell$ of the roots of the subtrees
$T_1, \ldots, T_k$ (respectively). Note that the root of the last subtree T_k is in fact
the root of the original derivation tree T (and thus $A_k = A_{\text{start}}$) and that its
corresponding interval I_k is $[n]$.

Let $I_{i_1}, \ldots, I_{i_k}$ be the ordered (from left to right) maximal intervals of $I_k = [n]$. That is, the (disjoint) intervals that are contained in I_k but are not contained
in any of the other intervals. Observe that if the intervals were generated as
prescribed, then A_{start} yields a string x' (composed of terminals and variables)
that results from x by replacing the substring $x[I_{i_j}]$ with the variable A_{i_j}, for
every $j \in [k]$. Denote the language that contains all such strings by $\mathcal{L}_k$. Similarly,
for any interval $I_{i_j} \in \{I_{i_1}, \ldots, I_{i_k}\}$, observe that A_{i_j} yields the string that results
from $x[I_{i_j}]$ by replacing coordinates in the maximal intervals that I_{i_j} contains
with the corresponding variables. Denote the language of all such strings by $\mathcal{L}_{i_j}$.
We show that by applying this idea iteratively we obtain languages $\mathcal{L}_1, \ldots, \mathcal{L}_k$
such that (1) if $x \in \mathcal{L}$, then $x[S_i] \in \mathcal{L}_i$ for every $i \in [k]$; and (2) if x is ε-far
from $\mathcal{L}$, then $x[S_i]$ is ε-far from $\mathcal{L}_i$, for an average $i \in [k]$, where the average is
weighted proportionally to the sizes of $S_1, \ldots, S_k$.

Given the partition above, verifying that $x \in \mathcal{L}$ is reduced to testing that
the sub-input $x[S_i]$ is close to $\mathcal{L}_i$, for $i \in [k]$ distributed as above. Hence, as
before, the verifier chooses i at random, sends i to the prover and the two parties
recursively run an $\mathcal{IPP}$ for verifying that $x[S_i]$ is ε-close to $\mathcal{L}_i$.

We emphasize that, as was the case for $k = 2$, the languages $\mathcal{L}_1, \ldots, \mathcal{L}_k$
are not necessarily *context-free languages* but are rather "partial derivation languages". Indeed, for the recursion to go through, we design the $\mathcal{IPP}$ to work for
such languages (rather than just context-free languages).

Acknowledgments. We thank Moni Naor and Avi Wigderson for pointing out the
connection to $\mathcal{LOGCFL}$.

References

[AKNS00] Alon, N., Krivelevich, M., Newman, I., Szegedy, M.: Regular languages
are testable with a constant number of queries. SIAM J. Comput. **30**(6),
1842–1862 (2000)

[Bol05] Bollig, B.: Property testing and the branching program size of boolean functions. In: Liśkiewicz, M., Reischuk, R. (eds.) FCT 2005. LNCS, vol 3623, pp. 258–269. Springer, Heidelberg (2005)

[BSGH+06] Ben-Sasson, E., Goldreich, O., Harsha, P., Sudan, M., Vadhan, S.P.: Robust PCPs of proximity, shorter PCPs, and applications to coding. SIAM J. Comput. **36**(4), 889–974 (2006)

[Coo71] Cook, S.A.: The complexity of theorem-proving procedures. In: Proceedings of the Third Annual ACM Symposium on Theory of Computing, pp. 151–158. ACM (1971)

[DR06] Dinur, I., Reingold, O.: Assignment testers: Towards a combinatorial proof of the PCP theorem. SIAM J. Comput. **36**(4), 975–1024 (2006)

[EKR04] Ergün, F., Kumar, R., Rubinfeld, R.: Fast approximate probabilistically checkable proofs. Inf. Comput. **189**(2), 135–159 (2004)

[FGL14] Fischer, E., Goldhirsh, Y., Lachish, O.: Partial tests, universal tests and decomposability. In: Innovations in Theoretical Computer Science, ITCS 2014, Princeton, NJ, USA, January 12–14, pp. 483–500 (2014)

[GGR98] Goldreich, O., Goldwasser, S., Ron, D.: Property testing and its connection to learning and approximation. Journal of the ACM (JACM) **45**(4), 653–750 (1998)

[GGR15] Goldreich, O., Gur, T., Rothblum, R.D.: Proofs of proximity for context-free languages and read-once branching programs. Electronic Colloquium on Computational Complexity (ECCC) 22, 24 (2015)

[GR11] Goldreich, O., Ron, D.: On proximity-oblivious testing. SIAM Journal on Computing **40**(2), 534–566 (2011)

[GR15] Gur, T., Rothblum, R.D.: Non-interactive proofs of proximity. In: Proceedings of the 2015 Conference on Innovations in Theoretical Computer Science, ITCS 2015, Rehovot, Israel, January 11-13, pp. 133–142. ACM (2015)

[HMU06] Hopcroft, J.E., Motwani, R., Ullman, J.D.: Introduction to Automata Theory, Languages, and Computation, 3rd edn. Addison-Wesley Longman Publishing Co. Inc., Boston (2006)

[KR14] Tauman Kalai, Y., Rothblum, R.D.: Arguments of proximity (2014) (manuscript)

[KW88] Kriegel, K., Waack, S.: Lower bounds on the complexity of real-time branching programs. ITA **22**(4), 447–459 (1988)

[LSH65] Lewis, P.M., Stearns, R.E., Hartmanis, J.: Memory bounds for recognition of context-free and context-sensitive languages. In: SWCT (FOCS), pp. 191–202 (1965)

[New02] Newman, I.: Testing membership in languages that have small width branching programs. SIAM Journal on Computing **31**(5), 1557–1570 (2002)

[PRR01] Parnas, M., Ron, D., Rubinfeld, R.: Testing parenthesis languages. In: Goemans, M.X., Jansen, K., Rolim, J.D.P., Trevisan, L. (eds.) RANDOM-APPROX 2001. LNCS, vol. 2129, pp. 261–272. Springer, Heidelberg (2001)

[RS96] Rubinfeld, R., Sudan, M.: Robust characterizations of polynomials with applications to program testing. SIAM J. Comput. **25**(2), 252–271 (1996)

[Ruz81] Ruzzo, W.L.: On uniform circuit complexity. J. Comput. Syst. Sci. **22**(3), 365–383 (1981)

[RVW13] Rothblum, G.N., Vadhan, S., Wigderson, A.: Interactive proofs of proximity: Delegating computation in sublinear time. In: Proceedings of the 45th Annual ACM Symposium on Theory of Computing (STOC) (2013)

Fast Algorithms for Diameter-Optimally Augmenting Paths

Ulrike Große[1], Joachim Gudmundsson[2], Christian Knauer[1],
Michiel Smid[3], and Fabian Stehn[1(✉)]

[1] Institut für Angewandte Informatik, Universität Bayreuth, Bayreuth, Germany
fabian.stehn@uni-bayreuth.de
[2] School of Information Technology, University of Sydney, Sydney, Australia
[3] School of Computer Science, Carleton University, Ottawa, Canada

Abstract. We consider the problem of augmenting a graph with n vertices embedded in a metric space, by inserting one additional edge in order to minimize the diameter of the resulting graph. We present an exact algorithm for the cases when the input graph is a path that runs in $O(n \log^3 n)$ time. We also present an algorithm that computes a $(1+\varepsilon)$-approximation in $O(n + 1/\varepsilon^3)$ time for paths in $\mathbb{R}^d$, where d is a constant.

1 Introduction

Let $G = (V, E)$ be a graph in which each edge has a positive weight. The weight (or length) of a path is the sum of the weights of the edges on this path. For any two vertices x and y in V, we denote by $\delta_G(x, y)$ their shortest-path distance, i.e., the minimum weight of any path in G between x and y. The diameter of G is defined as $\max\{\delta_G(x, y) : x, y \in V\}$.

Assume that we are also given weights for the non-edges of the graph G. In the *Diameter-Optimal k-Augmentation Problem*, DOAP(k), we have to compute a set F of k edges in $(V \times V) \setminus E$ for which the diameter of the graph $(V, E \cup F)$ is minimum.

In this paper, we assume that the given graph is a path embedded in a metric space, and the weight of any edge and non-edge is equal to the distance between its vertices. We consider the case when $k = 1$; thus, we want to compute one non-edge which, when added to the graph, results in an augmented graph of minimum diameter. Surprisingly, no non-trivial results were known even for this restricted case.

Throughout the rest of the paper, we assume that $(V, |\cdot|)$ is a metric space, consisting of a set V of n elements (called points). The distance between any two points x and y is denoted by $|xy|$. We assume that an oracle is available that returns the distance between any pair of points in $O(1)$ time. Our contribution is as follows:

The research on this topic has been initiated during the *Korean Workshop on Computational Geometry 2014* (KW2014).

M.M. Halldórsson et al. (Eds.): ICALP 2015, Part I, LNCS 9134, pp. 678–688, 2015.
DOI: 10.1007/978-3-662-47672-7_55

1. If G is a path, we solve problem DOAP(1) in $O(n \log^3 n)$ time.
2. If G is a path and the metric space is $\mathbb{R}^d$, where d is a constant, we compute a $(1 + \varepsilon)$-approximation for DOAP(1) in $O(n + 1/\varepsilon^3)$ time.

1.1 Related Work

The Diameter-Optimal k-Augmentation Problem for edge-weighted graphs, and many of its variants, have been shown to be NP-hard [16], or even W [2]-hard [9, 10]. Because of this, several special classes of graphs have been considered. Chung and Gary [5] and Alon et al. [1] considered paths and cycles with unit edge weights and gave upper and lower bounds on the diameter that can be achieved. Ishii [11] gave a constant factor approximation algorithm (approximating both k and the diameter) for the case when the input graph is outerplanar. Erdős et al. [7] investigated upper and lower bounds for the case when the augmented graph must be triangle-free.

The general problem: The Diameter-Optimal Augmentation Problem can be seen as a bicriteria optimization problem: In addition to the weight, each edge and non-edge has a cost associated with it. Then the two optimization criteria are (1) the total cost of the edges added to the graph and (2) the diameter of the augmented graph. We say that an algorithm is an (α, β)-approximation algorithm for the DOAP problem, with $\alpha, \beta \geq 1$, if it computes a set F of non-edges of total cost at most $\alpha \cdot B$ such that the diameter of $G' = (V, E \cup F)$ is at most $\beta \cdot D_{\mathrm{opt}}^B$, where D_{opt}^B is the diameter of an optimal solution that augments the graph with edges of total cost at most B.

For the restricted version when all costs and all weights are identical [2, 4,6,12,13], Bilò et al. [2] showed that, unless P=NP, there does not exist a $(c \log n, \delta < 1 + 1/D_{\mathrm{opt}}^B)$-approximation algorithm for DOAP if $D_{\mathrm{opt}}^B \geq 2$. For the case in which $D_{\mathrm{opt}}^B \geq 6$, they proved that, again unless P=NP, there does not exist a $(c \log n, \delta < \frac{5}{3} - \frac{7 - (D_{\mathrm{opt}}^B + 1) \bmod 3}{3 D_{\mathrm{opt}}^B})$-approximation algorithm.

Li et al. [13] showed a $(1, 4 + 2/D_{\mathrm{opt}}^B)$-approximation algorithm. The analysis of the algorithm was later improved by Bilò et al. [2], who showed that it gives a $(1, 2 + 2/D_{\mathrm{opt}}^B)$-approximation. In the same paper they also gave an $(O(\log n), 1)$-approximation algorithm.

For general costs and weights, Dodis and Khanna [6] gave an $O(n \log D_{\mathrm{opt}}^B, 1)$-approximation algorithm. Their result is based on a multi-commodity flow formulation of the problem. Frati et al. [9] recently considered the DOAP problem with arbitrary integer costs and weights. Their main result is a $(1, 4)$-approximation algorithm with running time $O((3^B B^3 + n + \log(Bn))Bn^2)$.

Geometric graphs: In the geometric setting, when the input is a geometric graph embedded in the Euclidean plane, there are very few results on graph augmentation in general. Rutter and Wolff [15] proved that the k-connectivity and k-edge-connectivity augmentation problems are NP-hard on plane geometric graphs, for $k = 2, 3, 4,$ and 5; the problem is infeasible for $k \geq 6$ because every planar graph as a vertex of degree at most 5. Currently, there are no known approximation algorithms for this problem. Farshi et al. [8] gave approximation algorithms for the problem of adding one edge to a geometric graph while minimizing the dilation. There were several follow-up papers [14,17], but there is still no non-trivial result known for the case when $k > 1$.

2 Augmenting a Path with One Edge

We are given a path $P = (p_1, \ldots, p_n)$ on n vertices in a metric space and assume that it is stored in an array $P[1, \ldots, n]$. To simplify notation, we associate a vertex with its index, that is $p_k = P[k]$ is also referred to as k for $1 \le k \le n$. This allows us to extend the total order of the indices to the vertex set of P. We denote the start vertex of P by s and the end vertex of P by e.

For $1 \le k < l \le n$, we denote the subpath $(p_k, \ldots, p_l)$ of P by $P[k, l]$, the cycle we get by adding the edge $\overline{p_k p_l}$ to $P[k, l]$ by $C[k, l]$, and the (unicyclic) graph we get by adding the edge $\overline{p_k p_l}$ as a *shortcut* to P by $\overline{P}[k, l]$; the length of $X \in \{P, P[k, l], C[k, l]\}$ is denoted by $|X|$. We will consider the functions $\delta_{k,l} := \delta_{\overline{P}[k,l]}$ and $c_{k,l} := \delta_{C[k,l]}$, where δ_G is the length of the shortest path between two vertices in G. For $1 \le k < l \le n$, we let

$$M(k, l) := \max_{1 \le x < y \le n} \overline{P}_{k,l}(x, y)$$

denote the *diameter* of the graph $\overline{P}[k, l]$.

Our goal is to compute a shortcut $\overline{p_k p_l}$ for P that minimizes the diameter of the resulting unicyclic graph, i.e., we want to compute

$$m(P) := \min_{1 \le k < l \le n} M(k, l).$$

We will prove the following result:

Theorem 1. *Given a path P on n vertices in a metric space, we can compute $m(P)$, and a shortcut realizing that diameter, in $O(n \log^3 n)$ time.*

The algorithm consists of two parts. We first describe a sequential algorithm for the *decision problem*. Given P and a threshold parameter $\lambda > 0$, decide if $m(P) \le \lambda$ (see Lemma 1 a) below). In a second step, we argue that the sequential algorithm can be implemented in a parallel fashion (see Lemma 1 b) below), thus enabling us to use the parametric search paradigm of Megiddo.

Lemma 1. *Given a path P on n vertices in a metric space and a real parameter $\lambda > 0$, we can decide in*

a) *$O(n \log n)$ time, or in*
b) *$O(\log n)$ parallel time using n processors*

whether $m(P) \le \lambda$; the algorithms also produce a feasible shortcut if it exists.

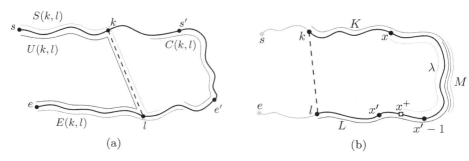

Fig. 1. (a) Illustration of the four distances that define the diameter of a shortcut $\overline{p_k p_l}$: $U(k, l)$ is the length of the shortest path connecting s and e; $O(k, l)$ is the length of the longest shortest path between any two points in $C[k, l]$; $S(k, l)$ ($E(k, l)$) is the length of the longest shortest path from s (e) to any vertex in $C(k, l)$. (b) Illustration of the computation of $O(k, l)$.

To prove this lemma, observe that

$$m(P) \leq \lambda \text{ iff } \bigvee_{1 \leq k < l \leq n} M(k, l) \leq \lambda.$$

The algorithm checks, for each $1 \leq k < n$, whether there is some $k < l \leq n$ such that $M(k, l) \leq \lambda$. If one such index k is found, we know that $m(P) \leq \lambda$; otherwise $m(P) > \lambda$. Clearly this approach also produces a feasible shortcut if it exists.

We decompose the function $M(k, l)$ into four monotone parts. This will facilitate our search for a feasible shortcut and enable us to do (essentially) binary search: For $1 \leq k < l \leq n$, we let

$$S(k, l) := \max_{k \leq x \leq l} \overline{p}_{k,l}(s, x), \qquad E(k, l) := \max_{k \leq x \leq l} \overline{p}_{k,l}(x, e),$$

$$U(k, l) := \overline{p}_{k,l}(s, e), \qquad O(k, l) := \max_{k \leq x < y \leq l} c_{k,l}(x, y).$$

Then we have $M(k, l) = \max\{S(k, l), E(k, l), U(k, l), O(k, l)\}$. The triangle inequality implies that

$$S(k, l) \leq S(k, l + 1), \qquad E(k, l) \geq E(k, l + 1),$$
$$U(k, l) \geq U(k, l + 1), \qquad O(k, l) \leq O(k, l + 1).$$

The function U is easy to evaluate once we have the array $D[1, \ldots, n]$ of the prefix-sums of the edge lengths: $D[i] := \sum_{1 \leq j < i} |p_j p_{j+1}|$. These sums can be computed in $O(n)$ time sequentially or in $O(\log n)$ time using n processors. If in addition to D, the vertices $s' = \max\{v \mid \delta_P(s, v) \leq \lambda\}$ and $e' = \min\{v \mid \delta_P(v, e) \leq \lambda\}$ are computed for a fixed λ in $O(\log n)$ time (via binary search on D), the following decision problems can be answered in constant time:

$$S(k, l) \leq \lambda, \quad E(k, l) \leq \lambda, \quad U(k, l) \leq \lambda.$$

We denote the maximum of these three functions by

$$N(k,l) = \max(S(k,l), E(k,l), U(k,l)).$$

Now clearly

$$M(k,l) = \max(N(k,l), O(k,l))$$

and, consequently

$$M(k,l) \leq \lambda \text{ iff } N(k,l) \leq \lambda \text{ and } O(k,l) \leq \lambda.$$

For fixed $1 \leq k < n$, the algorithm will first check whether there is some $k < l \leq n$ with $N(k,l) \leq \lambda$. If no such l exists, we can conclude that $M(k,l) > \lambda$ for all $k < l \leq n$. The monotonicity of S, E, and U implies that, for fixed $1 \leq k < n$, the set

$$N_k := \{k < l \leq n \mid N(k,l) \leq \lambda\}$$

is an *interval*. This interval can be computed (using binary search in P and in D as described above) in $O(\log n)$ time. If $N_k = \emptyset$ we can conclude that for the $1 \leq k < n$ under consideration and for all $k < l \leq n$, we have that $M(k,l) > \lambda$.

If N_k is non-empty, the monotonicity of O implies that it is sufficient to check for $l_k = \min N_k$ (i.e. the starting point of the interval) whether $O(k,l_k) \leq \lambda$:

$$\exists k < l \leq n : O(k,l) \leq \lambda \text{ iff } O(k,l_k) \leq \lambda.$$

Note that in this case we know that $N(k,l_k) \leq \lambda$.

Deciding the diameter of small cycles: We now describe how to decide for a given shortcut $1 \leq k < l \leq n$ if $O(k,l) \leq \lambda$, given that *we already know that* $N(k,l) \leq \lambda$. To this end, consider the following sets of vertices from $C[k,l]$: $K := \{k \leq x \leq l \mid \delta_P(k,x) \leq \lambda\}$, $L := \{k \leq x \leq l \mid \delta_P(x,l) \leq \lambda\}$, $M := K \cap L$, $K' := K \setminus L$, $L' := L \setminus K$.

These sets are intervals and can be computed in $O(\log n)$ time by binary search. Since $N(k,l) \leq \lambda$, we can conclude the following:

- the set of vertices of $C[k,l]$ is $K \cup L$
- $c_{k,l}(x,y) \leq \lambda$ for all $x,y \in K$
- $c_{k,l}(x,y) \leq \lambda$ for all $x,y \in L$
- $c_{k,l}(x,y) \leq \lambda$ for all $x \in M$, $y \in C[k,l]$

Consequently, if $c_{k,l}(x,y) > \lambda$ for $x,y \in C[k,l]$, we can conclude that $x \in K'$ and $y \in L'$. In order to establish that $O(k,l) \leq \lambda$, it therefore suffices to verify that

$$\bigwedge_{x \in K', y \in L'} c_{k,l}(x,y) \leq \lambda.$$

Note that on P any vertex x of K' is at least λ away from the vertex l, i.e., $\delta_P(x,l) > \lambda$. Let x^+ be point on (a vertex or an edge of) P that is closer (along P) by a distance of λ to l than to x, i.e., x^+ is the unique point on P such that

$$\delta_P(x^+,l) < \delta_P(x,l) \text{ and } \delta_P(x,x^+) = \lambda.$$

The next (in the direction of l) *vertex* of P will be denoted by x', i.e., $x < x' \leq l$ is the unique vertex of P such that

$$\delta_P(x, x' - 1) \leq \lambda \text{ and } \delta_P(x, x') > \lambda.$$

Since x is a vertex of K', x' is a vertex of L'. For the following discussion we denote the distance achieved in $C[k, l]$ by using the shortcut by $c^+_{k,l}$ and the distance achieved by travelling along P only by $c^-_{k,l}$, i.e.,

$$c^-_{k,l}(x, y) := \delta_P(x, y) \text{ and } c^+_{k,l}(x, y) := \delta_P(x, k) + |\overline{p_k p_l}| + \delta_P(l, y).$$

Clearly

$$c_{k,l}(x, y) = \min(c^+_{k,l}(x, y), c^-_{k,l}(x, y)), \text{ and } |C[k, l]| = c^+_{k,l}(x, y) + c^-_{k,l}(x, y).$$

For every vertex $y < x'$ on L' we have that $c_{k,l}(x, y) \leq c^-_{k,l}(x, y) \leq \lambda$, so if there is some vertex $x' \neq y \in L'$ such that $c_{k,l}(x, y) > \lambda$, we know that $x' < y \leq l$; in that case we have that $c^+_{k,l}(x, y) \leq c^+_{k,l}(x, x')$. Since we assume that $c_{k,l}(x, y) > \lambda$, we also know that $c^+_{k,l}(x, y) > \lambda$ and we can conclude that $c^+_{k,l}(x, x') > \lambda$, and consequently that $c_{k,l}(x, x') > \lambda$, i.e., for all $x \in K'$ we have that

$$\bigwedge_{y \in L'} c_{k,l}(x, y) \leq \lambda \text{ iff } c_{k,l}(x, x') \leq \lambda.$$

The distance between (the point) x^+ and (the vertex) x' on P is called the *defect* of x and is denoted by $\Delta(x)$, i.e., $\Delta(x) = \delta_P(x^+, x')$.

Lemma 2.

$$c_{k,l}(x, x') \leq \lambda \text{ iff } |C[k, l]| \leq \Delta(x) + 2\lambda$$

Proof. Observe that

$$\begin{aligned} |C[k, l]| &= \delta_P(x, k) + |\overline{p_k p_l}| + \delta_P(l, x') + \delta_P(x', x^+) + \delta_P(x^+, x) \\ &= \delta_P(x, k) + |\overline{p_k p_l}| + \delta_P(l, x') + \Delta(x) + \lambda \\ &= c^+_{k,l}(x, x') + \Delta(x) + \lambda. \end{aligned}$$

Since $c^-_{k,l}(x, x') > \lambda$, we have that $c_{k,l}(x, x') \leq \lambda$ iff $c^+_{k,l}(x, x') \leq \lambda$; the claim follows. □

To summarize the above discussion we have the following chain of equivalences (here $\Delta_{k,l} := |C[k, l]| - 2\lambda$):

$$O(k, l) \leq \lambda \Leftrightarrow \bigwedge_{x \in K'} c_{k,l}(x, x') \leq \lambda \Leftrightarrow \bigwedge_{x \in K'} \Delta_{k,l} \leq \Delta(x) \Leftrightarrow \min_{x \in K'} \Delta(x) \geq \Delta_{k,l}.$$

Since K' is an interval, the last condition can be tested easily after some preprocessing: To this end we compute a $1d$-range tree on D and associate with each vertex in the tree the minimum Δ-value of the corresponding canonical subset. For every *vertex* x of P that is at least λ away from the end vertex of P we can compute $\Delta(x)$ in $O(\log n)$ time by binary search in D. With these values the range tree can be built in $O(n)$ time. A query for an interval K' then gives us $\mu := \min_{x \in K'} \Delta(x)$ in $O(\log n)$ time and we can check the above condition in $O(1)$ time.

We describe the algorithm in pseudocode; see Algorithm 1.

Algorithm 1. Algorithm for deciding if $m(P) \le \lambda$

DECISIONALGORITHM(P, λ) ;   // Decide if $m(P) \le \lambda$
1 **begin**
    global $D \leftarrow$ COMPUTEPREFIXSUMS(P);
    global $s' \leftarrow \max\{v \mid \delta_P(s, v) \le \lambda\}$;
    global $e' \leftarrow \min\{v \mid \delta_P(v, e) \le \lambda\}$;
    global $T \leftarrow$ COMPUTERANGETREE(P, λ);
    for $1 \le k < n$ **do**
        $N_k \leftarrow$ COMPUTEFEASIBLEINTERVALFORN(k, λ);
        if $N_k \neq \emptyset$ **and** CHECKOFORSHORTCUT$(k, \min(N_k), \lambda)$ **then**
             return TRUE
    return FALSE
end

CHECKOFORSHORTCUT(k, l, λ) ;   // Decide if $O(k, l) \le \lambda$
2 **begin**
    $K' \leftarrow \{k \le x \le l \mid \delta_P(k, x) \le \lambda \wedge \delta_P(x, l) > \lambda\}$;   // Compute the interval
    by binary search
    $\mu \leftarrow \min_{x \in K'} \Delta(x)$;   // Query the range tree T
    return $(\mu \ge |C[k, l]| - 2\lambda)$
end

The correctness of the algorithm follows from the previous discussion. COMPUTEPREFIXSUMS runs in $O(n)$ time, COMPUTERANGETREE runs in $O(n \log n)$ time, COMPUTEFEASIBLEINTERVALFORN runs in $O(\log n)$ time, a call to CHECKOFORSHORTCUT requires $O(\log n)$ time. The total runtime is therefore $O(n \log n)$. It is easy to see that with n processors, the steps COMPUTEPREFIXSUMS and COMPUTERANGETREE can be realized in $O(\log n)$ parallel time and that with this number of processors, all calls to CHECKOFORSHORTCUT can be handled in parallel. Therefore, the entire algorithm can be parallelized and has a parallel runtime of $O(\log n)$, as stated in Lemma 1 b). This concludes the proof of Lemma 1.

When we plug this result into the parametric search technique of Megiddo, we get the algorithm for the optimization problem as claimed in Theorem 1.

From the above discussion, we note that, since there are only four possible distances to compute to determine the diameter of a path augmented with one shortcut edge, the following corollary follows immediately.

Corollary 1. *Given a path P on n vertices in a metric space and a shortcut (u, v), the diameter of $P \cup (u, v)$ can be computed in $O(n)$ time.*

3 An Approximation Algorithm in Euclidean Space

In Section 2, we presented an $O(n \log^3 n)$-time algorithm for the problem when the input graph is a path in a metric space. Here we show a simple $(1 + \varepsilon)$-approximation algorithm with running time $O(n + 1/\varepsilon^3)$ for the case when the input graph is a path in $\mathbb{R}^d$, where d is a constant. The algorithm will use two ideas: clustering and the well-separated pair decomposition (WSPD) as introduced by Callahan and Kosaraju [3].

Definition 1 ([3]). *Let $s > 0$ be a real number, and let A and B be two finite sets of points in $\mathbb{R}^d$. We say that A and B are* well-separated *with respect to s, if there are two disjoint d-dimensional balls C_A and C_B, having the same radius, such that (i) C_A contains A, (i) C_B contains B, and (ii) the minimum distance between C_A and C_B is at least s times the radius of C_A.*

The parameter s will be referred to as the *separation constant*. The next lemma follows easily from Definition 1.

Lemma 3 ([3]). *Let A and B be two finite sets of points that are well-separated w.r.t. s, let x and p be points of A, and let y and q be points of B. Then (i) $|xy| \leq (1 + 4/s) \cdot |pq|$, and (ii) $|px| \leq (2/s) \cdot |pq|$.*

Definition 2 ([3]). *Let S be a set of n points in $\mathbb{R}^d$, and let $s > 0$ be a real number. A* well-separated pair decomposition *(WSPD) for S with respect to s is a sequence of pairs of non-empty subsets of S, $(A_1, B_1), \ldots, (A_m, B_m)$, such that*

1. *$A_i \cap B_i = \emptyset$, for all $i = 1, \ldots, m$,*
2. *for any two distinct points p and q of S, there is exactly one pair (A_i, B_i) in the sequence, such that (i) $p \in A_i$ and $q \in B_i$, or (ii) $q \in A_i$ and $p \in B_i$,*
3. *A_i and B_i are well-separated w.r.t. s, for $1 \leq i \leq m$.*

The integer m is called the size *of the WSPD.*

Callahan and Kosaraju showed that a WSPD of size $m = \mathcal{O}(s^d n)$ can be computed in $\mathcal{O}(s^d n + n \log n)$ time.

Algorithm. We are given a polygonal path P on n vertices in $\mathbb{R}^d$. We assume without loss of generality that the total length of P is 1. Partition P into $m = 1/\varepsilon_1$ subpaths $P_1, \ldots, P_m$, each of length ε_1, for some constant $0 < \varepsilon_1 < 1$ to be defined later. Note that a subpath may have one (or both) endpoint in the interior of an edge. For each subpath P_i, $1 \leq i \leq m$, select an arbitrary vertex r_i along P_i as a representative vertex, if it exists. The set of representative vertices is denoted R_P; note that the size of this set is at most $m = 1/\varepsilon_1$. Let $P(R)$

be the path consisting of the vertices of R_P, in the order in which they appear along the path P. We give each edge (u, v) of $P(R)$ a weight equal to $\delta_P(u, v)$. he interior of an edge of P, then $\delta_P(u, v)$ is defined in the natural way.)

Imagine that we "straighten" the path $P(R)$, so that it is contained on a line. n this way, the vertices of this path form a point set in $\mathbb{R}^1$; we compute a well-separated pair decomposition $\mathcal{W}$ for the one-dimensional set R_P, with separation constant $1/\varepsilon_2$, with $0 < \varepsilon_2 < 1/4$ to be defined later. Then, we go through all pairs $\{A, B\}$ in $\mathcal{W}$ and compute the diameter of $P(R) \cup \{(rep(A), rep(B))\}$, where $rep(A)$ and $rep(B)$ are representative points of A and B, respectively, which are arbitrarily chosen from their sets. Note that the number of pairs in $\mathcal{W}$ is $O(1/\varepsilon_1\varepsilon_2)$. Finally the algorithm outputs the best shortcut.

Analysis. We first discuss the running time and then turn our attention to the approximation factor of the algorithm.

The clustering takes $O(n)$ time, and constructing the WSPD of R_P takes $O(\frac{1}{\varepsilon_1\varepsilon_2} + \frac{1}{\varepsilon_1} \log \frac{1}{\varepsilon_1})$ time. For each of the $O(1/\varepsilon_1\varepsilon_2)$ well-separated pairs in $\mathcal{W}$, computing the diameter takes, by Corollary 1, time linear in the size of the uni-cyclic graph, that is, $O(\frac{1}{\varepsilon_1^2\varepsilon_2})$ time in total.

Lemma 4. *The running time of the algorithm is* $O(n + \frac{1}{\varepsilon_1^2\varepsilon_2})$.

Before we consider the approximation bound, we need to define some notation. Consider any vertex p in P. Let $r(p)$ denote the representative vertex of the subpath of P containing p. For any two vertices p and q in P, let $\{A, B\}$ be the well-separated pair such that $r(p) \in A$ and $r(q) \in B$. The representative points of A and B will be denoted $w(p)$ and $w(q)$, respectively.

Lemma 5. *For any shortcut* $e = (p, q)$ *and for any two vertices* $x, y \in P$, *we have*

$$(1 - 4\varepsilon_2) \cdot \delta_G(x, y) - 6\varepsilon_1 \leq \delta_H(w(x), w(y)) \leq (\frac{1}{1 - 4\varepsilon_2}) \cdot \delta_G(x, y) + 6\varepsilon_1,$$

where $G = P \cup (p, q)$ *and* $H = P(R) \cup (w(p), w(q))$.

Proof. We only prove the second inequality, because the proof of the first inequality is almost identical.

Consider two arbitrary vertices x, y in P, and consider a shortest path in G between x and y. We have two cases:

Case 1: If $\delta_G(x, y) = \delta_P(x, y)$, then $\delta_H(r(x), r(y)) \leq \delta_P(x, y) + 2\varepsilon_1$.

Case 2: If $\delta_G(x, y) < \delta_P(x, y)$, then the shortest path in G between x and y must traverse (p, q). Assume that the path is $x \rightsquigarrow p \rightarrow q \rightsquigarrow t$, thus $\delta_G(x, y) = \delta_P(x, p) + |pq| + \delta_P(q, y)$. Consider the following three observations:

(1) $|pq| \geq |r(p)r(q)| - 2\varepsilon_1$ and $|w(p)w(q)| \leq (1 + 4\varepsilon_2) \cdot |r(p)r(q)|$. Consequently, $|w(p)w(q)| \leq (1 + 4\varepsilon_2) \cdot (|pq| + 2\varepsilon_1)$.

(2) We have

$$
\begin{aligned}
\delta_P(x,p) &\geq \delta_P(w(x),w(p)) - \delta_P(w(x),x) - \delta_P(w(p),p) \\
&\geq \delta_P(w(x),w(p)) - (\varepsilon_1 + \delta_P(w(x),r(x))) - (\varepsilon_1 + \delta_P(w(y),r(y))) \\
&\geq \delta_P(w(x),w(p)) - (\varepsilon_1 + 2\varepsilon_2\delta_P(w(x),w(p))) - (\varepsilon_1 + 2\varepsilon_2\delta_P(w(x),w(p)) \\
&= (1 - 4\varepsilon_2) \cdot \delta_P(w(x),w(p)) - 2\varepsilon_1 \\
&\geq (1 - 4\varepsilon_2) \cdot \delta_H(w(x),w(p)) - 2\varepsilon_1
\end{aligned}
$$

That is, $\delta_H(w(x),w(p)) \leq \frac{1}{1-4\varepsilon_2} \cdot \delta_P(x,p) + 2\varepsilon_1$.

(3) We have, $\delta_H(w(y),w(q)) \leq \frac{1}{1-4\varepsilon_2} \cdot \delta_P(y,q) + 2\varepsilon_1$, following the same arguments as in (2).

Putting together the three observations we get:

$$
\begin{aligned}
\delta_H(w(x),w(y)) &\leq \delta_H(w(x),w(p)) + |w(p)w(q)| + \delta_H(w(q),w(y)) \\
&\leq (\frac{1}{1-4\varepsilon_2}) \cdot \delta_P(x,p) + 2\varepsilon_1) + ((1 + 4\varepsilon_2) \cdot (|pq| + 2\varepsilon_1)) \\
&\quad + (\frac{1}{1-4\varepsilon_2}) \cdot \delta_P(y,q) + 2\varepsilon_1) \\
&< (\frac{1}{1-4\varepsilon_2}) \cdot \delta_G(x,y) + 6\varepsilon_1,
\end{aligned}
$$

where the last inequality follows from the fact that $0 < \varepsilon_2 < 1/4$. This concludes the proof of the lemma. □

By setting $\varepsilon_1 = \varepsilon/60$ and $\varepsilon_2 = \varepsilon/32$ and using the fact that the diameter of H is at least $1/2$, we obtain the following theorem that summarizes this section.

Theorem 2. *Given a path P with n vertices in $\mathbb{R}^d$ and a real number $\varepsilon > 0$, we can compute a shortcut to P in $O(n + 1/\varepsilon^3)$ time such that the resulting uni-cyclic graph has diameter at most $(1 + \varepsilon) \cdot d_{\text{opt}}$, where d_{opt} is the diameter of an optimal solution.*

References

1. Alon, N., Gyárfás, A., Ruszinkó, M.: Decreasing the diameter of bounded degree graphs. Journal of Graph Theory **35**, 161–172 (1999)
2. Bilò, D., Gualà, L., Proietti, G.: Improved approximability and non-approximability results for graph diameter decreasing problems. Theoretical Computer Science **417**, 12–22 (2012)
3. Callahan, P.B., Kosaraju, S.R.: A decomposition of multidimensional point sets with applications to k-nearest-neighbors and n-body potential fields. Journal of the ACM **42**, 67–90 (1995)

4. Chepoi, V., Vaxès, Y.: Augmenting trees to meet biconnectivity and diameter constraints. Algorithmica **33**(2), 243–262 (2002)

5. Chung, F.R.K., Garey, M.R.: Diameter bounds for altered graphs. Journal of Graph Theory **8**(4), 511–534 (1984)

6. Dodis, Y., Khanna, S.: Designing networks with bounded pairwise distance. In: Proceedings of the 31st Annual ACM Symposium on Theory of Computing (STOC), pp. 750–759 (1999)

7. Erdős, P., Gyárfás, A., Ruszinkó, M.: How to decrease the diameter of triangle-free graphs. Combinatorica **18**(4), 493–501 (1998)

8. Farshi, M., Giannopoulos, P., Gudmundsson, J.: Improving the stretch factor of a geometric network by edge augmentation. SIAM Journal on Computing **38**(1), 226–240 (2005)

9. Frati, F., Gaspers, S., Gudmundsson, J., Mathieson, L.: Augmenting graphs to minimize the diameter. Algorithmica, 1–16 (2014)

10. Gao, Y., Hare, D.R., Nastos, J.: The parametric complexity of graph diameter augmentation. Discrete Applied Mathematics **161**(10–11), 1626–1631 (2013)

11. Ishii, T.: Augmenting outerplanar graphs to meet diameter requirements. Journal of Graph Theory **74**, 392–416 (2013)

12. Kapoor, S., Sarwat, M.: Bounded-diameter minimum-cost graph problems. Theory of Computing Systems **41**(4), 779–794 (2007)

13. Li, C.-L., McCormick, S.T., Simchi-Levi, D.: On the minimum-cardinality-bounded-diameter and the bounded-cardinality-minimum-diameter edge addition problems. Operations Research Letters **11**(5), 303–308 (1992)

14. Luo, J., Wulff-Nilsen, C.: Computing best and worst shortcuts of graphs embedded in metric spaces. In: Hong, S.-H., Nagamochi, H., Fukunaga, T. (eds.) ISAAC 2008. LNCS, vol. 5369, pp. 764–775. Springer, Heidelberg (2008)

15. Rutter, I., Wolff, A.: Augmenting the connectivity of planar and geometric graphs. Journal of Graph Algorithms and Applications **16**(2), 599–628 (2012)

16. Schoone, A.A., Bodlaender, H.L., van Leeuwen, J.: Diameter increase caused by edge deletion. Journal of Graph Theory **11**, 409–427 (1997)

17. Wulff-Nilsen, C.: Computing the dilation of edge-augmented graphs in metric spaces. Computational Geometry - Theory and Applications **43**(2), 68–72 (2010)

Hollow Heaps

Thomas Dueholm Hansen[1], Haim Kaplan[2(✉)],
Robert E. Tarjan[3,4], and Uri Zwick[2]

[1] Department of Computer Science, Aarhus University, Aarhus, Denmark
tdh@cs.au.dk
[2] Blavatnik School of Computer Science, Tel Aviv University, Tel Aviv-Yafo, Israel
zwick@tau.ac.il, haimk@post.tau.ac.il
[3] Department of Computer Science, Princeton University, Princeton, NJ 08540, USA
[4] Intertrust Technologies, Sunnyvale, CA 94085, USA
ret@CS.Princeton.EDU

Abstract. We introduce the *hollow heap*, a very simple data structure
with the same amortized efficiency as the classical Fibonacci heap. All
heap operations except *delete* and *delete-min* take $O(1)$ time, worst case
as well as amortized; *delete* and *delete-min* take $O(\log n)$ amortized time.
Hollow heaps are by far the simplest structure to achieve this. Hollow
heaps combine two novel ideas: the use of lazy deletion and re-insertion to
do *decrease-key* operations, and the use of a dag (directed acyclic graph)
instead of a tree or set of trees to represent a heap. Lazy deletion produces
hollow nodes (nodes without items), giving the data structure its name.

1 Introduction

A *heap* is a data structure consisting of a set of *items*, each with a *key* selected
from a totally ordered universe. Heaps support the following operations:

make-heap(): Return a new, empty heap.
find-min(h) : Return an item of minimum key in heap h, or *null* if h is empty.
insert(e, k, h): Return a heap formed from heap h by inserting item e, with key k.
Item e must be in no heap.
delete-min(h): Return a heap formed from non-empty heap h by deleting the
item returned by *find-min*(h).
meld(h_1, h_2): Return a heap containing all items in item-disjoint heaps h_1 and h_2.
decrease-key(e, k, h): Given that e is an item in heap h with key greater than k,
return a heap formed from h by changing the key of e to k.
delete(e, h) : Return a heap formed by deleting e, assumed to be in h, from h.

The original heap h passed to *insert*, *delete-min*, *decrease-key*, and *delete*,
and the heaps h_1 and h_2 passed to *meld*, are destroyed by the operations. Heaps
do *not* support search by key; operations *decrease-key* and *delete* are given the
location of item e in heap h. The parameter h can be omitted from *decrease-key*
and *delete*, but then to make *decrease-key* operations efficient if there are inter-
mixed *meld* operations, a separate disjoint set data structure is needed to keep
track of the partition of items into heaps. (See the discussion in [12].)

© Springer-Verlag Berlin Heidelberg 2015
M.M. Halldórsson et al. (Eds.): ICALP 2015, Part I, LNCS 9134, pp. 689–700, 2015.
DOI: 10.1007/978-3-662-47672-7_56

Fredman and Tarjan [8] invented the *Fibonacci heap*, an implementation of heaps that supports *delete-min* and *delete* on an n-item heap in $O(\log n)$ amortized time and each of the other operations in $O(1)$ amortized time. Applications of Fibonacci heaps include a fast implementation of Dijkstra's shortest path algorithm [4,8] and fast algorithms for undirected and directed minimum spanning trees [6,9]. Since the invention of Fibonacci heaps, a number of other heap implementations with the same amortized time bounds have been proposed [1–3,7,10,11,13,16,18]. Notably, Brodal [1] invented a very complicated heap implementation that achieves the time bounds of Fibonacci heaps in the worst case. Brodal et al. [2] later simplified this data structure, but it is still significantly more complicated than any of the amortized-efficient structures. For further discussion of these and related results, see [10]. We focus here on the *amortized* efficiency of heaps.

In spite of its many competitors, Fibonacci heaps remain one of the simplest heap implementations to describe and code, and are taught in numerous undergraduate and graduate data structures courses. We present *hollow heaps*, a data structure that we believe surpasses Fibonacci heaps in its simplicity. Our data structure has two novelties: it uses lazy deletion to do *decrease-key* operations in a simple and natural way, avoiding the *cascading cut* process used by Fibonacci heaps, and it represents a heap by a dag (directed acyclic graph) instead of a tree or a set of trees. The amortized analysis of hollow heaps is simple, yet nontrivial. We believe that simplifying fundamental data structures, while retaining their performance, is an important endeavor.

In a Fibonacci heap, a *decrease-key* produces a heap-order violation if the new key is less than that of the parent node. This causes a *cut* of the violating node and its subtree from its parent. Such cuts can eventually destroy the "balance" of the data structure. To maintain balance, each such cut may trigger a cascade of cuts at ancestors of the originally cut node. The cutting process results in loss of information about the outcomes of previous comparisons. It also makes the worst-case time of a *decrease-key* operation $\Theta(n)$ (although modifying the data structure reduces this to $\Theta(\log n)$; see e.g., [14]). In a hollow heap, the item whose key decreases is merely moved to a new node, preserving the existing structure. Doing such lazy deletions carefully is what makes hollow heaps simple but efficient.

The remainder of this paper consists of six sections. Section 2 describes hollow heaps at a high level. Section 3 analyzes them. Section 4 presents an alternative version of hollow heaps that uses a tree representation instead of a dag representation. Section 5 describes a rebuilding process that can be used to improve the time and space efficiency of hollow heaps. Section 6 gives implementation details for the data structure in Section 2. The full version of this paper also contains implementation details of the data structure in Section 4 and further explores the design space of the data structures, identifying variants that are efficient and variants that are not.

2 Hollow Heaps

Our data structure extends and refines a well-known generic representation of heaps. The structure is *exogenous* rather than *endogenous* [19]: nodes *hold* items rather than *being* items. Moving items among nodes precludes the possibility of making the data structure endogenous.

Many previous heap implementations, including Fibonacci heaps, represent a heap by a set of heap-ordered trees: each node holds an item, with each child holding an item having key no less than that of the item in its parent. We extend this idea from trees to dags, and to dags whose nodes may or may not hold items. Since the data structure is an extension of a tree, we extend standard tree terminology to describe it. If (u, v) is a dag arc, we say u is a *parent* of v and v is a *child* of u. A node that is not a child of any other node is a *root*.

We represent a non-empty heap by a dag whose nodes hold the heap items, at most one per node. If e is an item, $e.node$ is the node holding e. We call a node *full* if it holds an item and *hollow* if not. If u is a full node, $u.item$ is the item u holds. Thus if e is an item, $e.node.item = e$. A node is full when created but can later become hollow, by having its item moved to a newly created node or deleted. A hollow node remains hollow until it is destroyed. Each node, full or hollow, has a key. The key of a full node is the key of the item it holds. The key of a hollow node is the key of the item it once held, just before that item was moved to another node or deleted. A full node is a child of at most one other node; a hollow node is a child of at most two other nodes.

The dag is topologically ordered by key: if u is a parent of v, then $u.key \le v.key$. Henceforth we call this *heap order*. Except in the middle of a *delete* operation, the dag has one full root and no hollow roots. Heap order guarantees that the root holds an item of minimum key. We access the dag via its root. We call the item in the root the *root item*.

We do the heap operations with the help of the *link* primitive. Given two full roots v and w, $link(v, w)$ compares the keys of v and w and makes the root of larger key a child of the other; if the keys are equal, it makes v a child of w. The new child is the *loser* of the link, its new parent is the *winner*. Linking eliminates one full root, preserves heap order, and gives the loser a parent, its *first parent*.

To make a heap, return an empty dag. To do *find-min*, return the item in the root. To meld two heaps, if one is empty return the other; if both are non-empty, link the roots of their dags and return the winner. To insert an item into a heap, create a new node, store the item in it (making the node full), and meld the resulting one-node heap with the existing heap.

We do *decrease-key* and *delete* operations using lazy deletion. To decrease the key of item e in heap h to k, let $u = e.node$. If $u = h$ (u is the root of the dag), merely set $u.key = k$. Otherwise (u is a child), proceed as follows. Create a new node v; move e from u to v, making u hollow; set $v.key = k$; do $link(h, v)$; and, if v is the loser of this link, make u a child of v. If u becomes a child of v, then v is the *second parent* of u, in contrast to its first parent, previously acquired via a link with a full node. A node only becomes hollow once, so it acquires a second parent at most once.

Remark. The arc (v, u) added to the dag by decrease-key represents the inequality $v.key < u.key$. If such arcs are not added, the resulting algorithm does not have the desired efficiency, as we show in the full version of this paper.

To do a *delete-min*, do a *find-min* followed by a deletion of the returned item. To delete an item e, remove e from the node holding it, say u, making u hollow. A node u made hollow in this way never acquires a second parent. If u is not the root of the dag, the deletion is complete. Otherwise, repeatedly destroy hollow roots and link full roots until there are no hollow roots and at most one full root. The proof of the following theorem is immediate.

Theorem 1. *The hollow heap operations perform the heap operations correctly and maintain the invariants that the graph representing a heap is a heap-ordered dag; each full node has at most one parent; each hollow node has at most two parents; and, except in the middle of a delete operation, the dag representing a heap has no hollow roots and at most one full root.*

The only flexibility in this implementation is the choice of which links to do in deletions of root items. To keep the number of links small, we give each node u a non-negative integer rank $u.rank$. We use ranks in a special kind of link called a *ranked link*. A ranked link of two roots is allowed only if they have the same rank; it links them and increases the rank of the winner (the remaining root) by 1. In contrast to a ranked link, an *unranked link* links any two roots and changes no ranks. We call a child *ranked* or *unranked* if it most recently acquired a first parent via a ranked or unranked link, respectively.

When linking two roots of equal rank, we can do either a ranked or an unranked link. We do ranked links only when needed to guarantee efficiency. Specifically, links in *meld* and *decrease-key* are unranked. Each *delete-min* operation destroys hollow roots and does ranked links until none are possible (there are no hollow roots and all full roots have different ranks); then it does unranked links until there is at most one root.

The last design choice is the initial node ranks. We give a node created by an *insert* a rank of 0. In a *decrease-key* that moves an item from a node u to a new node v, we give v a rank of $\max\{0, u.rank - 2\}$. The latter choice is what makes hollow heaps efficient.

We conclude this section by mentioning some benefits of using hollow nodes and a dag representation. Hollow nodes allow us to treat *decrease-key* as a special kind of insertion, allowing us to avoid cutting subtrees as in Fibonacci heaps. As a consequence, *decrease-key* takes $O(1)$ time worst case: there are no cascading cuts as in [8], no cascading rank changes as in [10,14], and no restructuring steps to eliminate heap-order violations as in [2,5,13]. The dag representation explicitly maintains all key comparisons between undeleted items, allowing us to avoid restructuring altogether: links are cut only when hollow roots are destroyed.

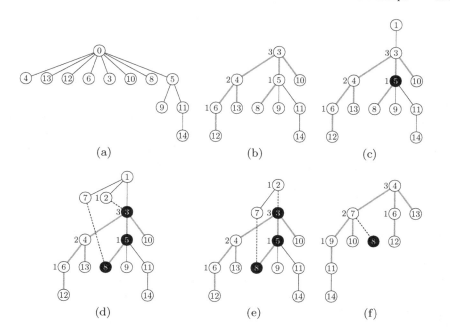

Fig. 1. Operations on a hollow heap. Numbers in nodes are keys; black nodes are hollow. Bold gray, solid, and dashed lines denote ranked links, unranked links, and second parents, respectively. Numbers next to nodes are non-zero ranks. (a) Successive insertions of items with keys 14, 11, 5, 9, 0, 8, 10, 3, 6, 12, 13, 4 into an initially empty heap. (b) After a *delete-min* operation. All links during the *delete-min* are ranked. (c) After a decrease of key 5 to 1. (d) After a decrease of key 3 to 2 followed by a decrease of key 8 to 7. The two new hollow nodes both have two parents. (e) After a second *delete-min*. The only hollow node that becomes a root is the original root. One unranked link, between the nodes holding keys 2 and 7 occurs. (f) After a third *delete-min*. Two hollow nodes become roots; the other loses one parent. All links are ranked.

3 Analysis

The most mysterious detail of hollow heaps is the way ranks are updated in *decrease-key* operations. Our analysis reveals the reason for this choice. We need to show that the rank of a heap node is at most logarithmic in the number of nodes in the dag representing the heap, and that the amortized number of ranked children per node is also at most logarithmic.

To do both, we assign *virtual parents* to certain nodes. We use virtual parents in the analysis only; they are not part of the data structure in Section 2. (Section 4 presents a version of hollow heaps that *does* use them.)

A node may acquire a virtual parent, have its virtual parent changed, or lose its virtual parent. As we shall see, virtual parents define a *virtual forest*. In particular, each node has at most one virtual parent at a time. If v is the virtual parent of u, we say that u is a *virtual child* of v. A node u is a *virtual descendant* of a node v if there is a path from v to u via virtual children.

When a node is created, it has no virtual parent. When a root u loses a link to a node v, v becomes the virtual parent of u (as well as its first parent). If u already has a virtual parent, v replaces it. (By Lemma 1 below, a root cannot have a virtual parent, so such a replacement never happens.) When a *decrease-key* moves an item from a node u to a new node v, if u has more than two ranked virtual children, two of its ranked virtual children of highest ranks remain virtual children of u, and the rest of its virtual children become virtual children of v. (By Lemma 2 below, the ranked virtual children of a node have distinct ranks, so the two that remain virtual children of u are uniquely defined.) If the virtual parent of a node u is destroyed, u loses its virtual parent. If u is null it can subsequently acquire a new virtual parent by losing a link.

Lemma 1. *If w is a virtual child of u, there is a path in the dag from u to w.*

Proof. We prove the lemma for a given node w by induction on time. When w is created it has no virtual parent. It may acquire a virtual parent only by losing a link to a node u, which then becomes both its parent and its virtual parent, so the lemma holds after the link. Suppose that u is currently the virtual parent of w. By the induction hypothesis, there is a path from u to w in the dag, so w is not a root and cannot participate in link operations. The virtual parent of w can change only as a result of a *decrease-key* operation on the item $e = u.item$. If $u \neq h$, such a *decrease-key* operation creates a new node v, moves e to v, and then links v and h. The operation may also make v the new virtual parent of w. If v wins the link, it becomes the unique root, so there is a path from v to w in the dag. If v loses the link, the arc (v, u) is added to the dag, making v the second parent of u. Since there was a path in the dag from u to w, there is now also a path from v to w. Finally, note that dag arcs are only destroyed when hollow roots are destroyed. Thus a path to w from its virtual parent u in the dag, present when u becomes the virtual parent of w, cannot be destroyed unless u is destroyed, in which case w loses its virtual parent, so the lemma holds vacuously.   □

Corollary 1. *Virtual parents define a forest. If w is a root of the dag, it has no virtual parent. If w is a virtual child of u, then w stops being a virtual child of u only when u is destroyed or when a decrease-key operation is applied to the item residing in u.*

Lemma 2. *Let u be a node of rank r. If u is full, or u is a node made hollow by a delete, u has exactly one ranked virtual child of each rank from 0 to $r - 1$ inclusive, and none of rank r or greater. If u was made hollow by a decrease-key and $r > 1$, u has exactly two ranked virtual children, of ranks $r - 1$ and $r - 2$. If u was made hollow by a decrease-key and $r = 1$, u has exactly one ranked virtual child, of rank 0. If u was made hollow by a decrease-key and $r = 0$, u has no ranked virtual children.*

Proof. The proof is by induction on the number of operations. The lemma is immediate for nodes created by insertions. Both ranked and unranked links preserve the truth of the lemma, as does the removal of an item from a node by a *delete*. By Corollary 1, a node loses virtual children only as a result of a *decrease-key* operation. Suppose the lemma is true before a *decrease-key* on the item in a node u of rank r. By the induction hypothesis, u has exactly one ranked virtual child of rank i for $0 \leq i < r$, and none of rank r or greater. If the *decrease-key* makes u hollow, the new node v created by the *decrease-key* has rank $\max\{0, u.rank - 2\}$, and v acquires all the virtual children of u except the two ranked virtual children of ranks $r - 1$ and $r - 2$ if $r > 1$, or the one ranked virtual child of rank 0 if $r = 1$. Thus the lemma holds after the *decrease-key*. □

Recall the definition of the Fibonacci numbers: $F_0 = 0$, $F_1 = 1$, $F_i = F_{i-1} + F_{i-2}$ for $i \geq 2$. These numbers satisfy $F_{i+2} \geq \phi^i$, where $\phi = (1 + \sqrt{5})/2$ is the golden ratio [15].

Corollary 2. *A node of rank r has at least $F_{r+3} - 1$ virtual descendants.*

Proof. The proof is by induction on r using Lemma 2. The corollary is immediate for $r = 0$ and $r = 1$. If $r > 1$, the virtual descendants of a node u of rank r include itself and all virtual descendants of its virtual children v and w of ranks $r - 1$ and $r - 2$, which it has by Lemma 2. By Corollary 1, virtual parents define a forest, so the sets of virtual descendants of v and w are disjoint. By the induction hypothesis, u has at least $1 + F_{r+2} - 1 + F_{r+1} - 1 = F_{r+3} - 1$ virtual descendants. □

Theorem 2. *The maximum rank of a node in a hollow heap of N nodes is at most $\log_\phi N$.*

Proof. Immediate from Corollary 2 since $F_{r+3} - 1 \geq F_{r+2} \geq \phi^r$ for $r \geq 0$. □

To complete our analysis, we need to bound the time of an arbitrary sequence of heap operations that starts with no heaps. It is straightforward to implement the operations so that the worst-case time per operation other than *delete-min* and *delete* is $O(1)$, and that of a *delete* on a heap of N nodes is $O(1)$ plus $O(1)$ per hollow node that loses a parent plus $O(1)$ per link plus $O(\log N)$. In Section 6 we give an implementation that satisfies these bounds and is space-efficient. We shall show that the amortized time for a *delete* on a heap of N nodes is $O(\log N)$ by charging the parent losses of hollow nodes and some of the links to other operations, $O(1)$ per operation.

Suppose a hollow node u loses a parent in a *delete*. This either makes u a root, in which case u is destroyed by the same *delete*, or it reduces the number of parents of u from two to one. We charge the former case to the *insert* or *decrease-key* that created u, and the latter case to the *decrease-key* that gave u its second parent. Since an *insert* or *decrease-key* can create at most one node, and a *decrease-key* can give at most one node a second parent, the total charge, and hence the total number of parent losses of hollow nodes, is at most 1 per *insert* and 2 per *decrease-key*.

A *delete* does unranked links only once there is at most one root per rank. Thus the number of unranked links is at most the maximum node rank, which is at most $\log_\phi N$ by Theorem 2. To bound the number of ranked links, we use a potential argument. We give each root and each unranked child a potential of 1. We give a ranked child a potential of 0 if it has a full virtual parent, 1 otherwise (its virtual parent is hollow or has been deleted). We define the potential of a set of dags to be the sum of the potentials of their nodes. With this definition the initial potential is 0 (there are no nodes), and the potential is always non-negative. Each ranked link reduces the potential by 1: a root becomes a ranked child of a full node. It follows that the total number of ranked links over a sequence of operations is at most the sum of the increases in potential produced by the operations.

An unranked link does not change the potential: a root becomes an unranked child. An *insert* increases the potential by 1: it creates a new root $(+1)$ and does an unranked link $(+0)$. A *decrease-key* increases the potential by at most 3: it creates a new root $(+1)$, it creates a hollow node that has at most two ranked virtual children by Lemma 2 $(+2)$, and it does an unranked link $(+0)$. Removing the item in a node u during a *delete* increases the potential by $u.rank$, also by Lemma 2: each of the $u.rank$ ranked virtual children of u gains 1 in potential. By Theorem 2, $u.rank = O(\log N)$. We conclude that the total number of ranked links is at most 1 per *insert* plus 3 per *decrease-key* plus $O(\log N)$ per *delete* on a heap with N nodes. Combining our bounds gives the following theorem:

Theorem 3. *The amortized time per hollow heap operation is $O(1)$ for each operation other than a delete, and $O(\log N)$ per delete on a heap of N nodes.*

4 Eager Hollow Heaps

It is natural to ask whether there is a way to represent a hollow heap by a tree instead of a dag. The answer is yes: we maintain the structure defined by the virtual parents instead of that defined by the parents. We call this the *eager version* of hollow heaps: it moves children among nodes, which the *lazy version* in Section 2 does not do. As a result it can do different links than the lazy version, but it has the same amortized efficiency.

To obtain eager hollow heaps, we modify *decrease-key* as follows: When a new node v is created to hold the item previously in a node u, if $u.rank > 2$, make v the parent of all but the two ranked children of u of highest ranks; optionally, make v the parent of some or all of the unranked children of u. Do not make u a child of v.

In an eager hollow heap, each node has at most one parent. Thus each heap is represented by a tree, accessed via its root. The analysis of eager hollow heaps differs from that of lazy hollow heaps only in using parents instead of virtual parents. Only the parents of ranked children matter in the analysis.

The proofs of the following results are essentially identical to the proofs of the results in Section 2, with the word "virtual" deleted.

Lemma 3. *Let u be a node of rank r in an eager hollow heap. If u is full, or u is a node made hollow by a delete, u has exactly one ranked child of each rank from 0 to $r-1$ inclusive, and none of rank r or greater. If u was made hollow* *by a decrease-key and $r > 1$, u has exactly two ranked children, of ranks $r - 1$ and $r - 2$. If u was made hollow by a decrease-key and $r = 1$, u has exactly one ranked child, of rank 0. If u was made hollow by a decrease-key and $r = 0$, u has no ranked children.*

Corollary 3. *A node of rank r in an eager hollow heap has at least $F_{r+3} - 1$ descendants.*

Theorem 4. *The maximum rank of a node in an eager hollow heap of N nodes is at most $\log_\phi N$.*

Theorem 5. *The amortized time per eager hollow heap operation is $O(1)$ for each operation other than a delete, and $O(\log N)$ per delete on an N-node heap.*

An alternative way to think about eager hollow heaps is as a variant of Fibonacci heaps. In a Fibonacci heap, the cascading cuts that occur during a *decrease-key* prune the tree in a way that guarantees that ranks remain logarithmic in subtree sizes. Eager hollow heaps guarantee logarithmic ranks by leaving (at least) two children and a hollow node behind at the site of the cut. This avoids the need for cascading cuts or rank changes, and makes the *decrease-key* operation $O(1)$ time in the worst case.

5 Rebuilding

The number of nodes N in a heap is at most the number of items n plus the number of *decrease-key* operations on items that were ever in the heap or in heaps melded into it. If the number of *decrease-key* operations is polynomial in the number of insertions, $\log N = O(\log n)$, so the amortized time per *delete* is $O(\log n)$, the same as for Fibonacci heaps. In applications in which the storage required for the problem input is at least linear in the number of heap operations, the extra space needed for hollow nodes is linear in the problem size. Both of these conditions hold for the heaps used in many graph algorithms, including Dijkstra's shortest path algorithm [4,8], various minimum spanning tree algorithms [4,8,9,17], and Edmonds' optimum branching algorithm [6,9]. In these applications there is at most one *insert* per vertex and one or two *decrease-key* operations per edge or arc, and the number of edges or arcs is at most quadratic in the number of vertices. In such applications hollow heaps are asymptotically as efficient as Fibonacci heaps.

For applications in which the number of *decrease-key* operations is huge compared to the heap sizes, we can use periodic rebuilding to guarantee that $N = O(n)$ for every heap. To do this, keep track of N and n for every heap. When $N > cn$ for a suitable constant $c > 1$, rebuild. We offer two ways to do the rebuilding. The first is to completely disassemble the dag and reinsert all its items into a new, initially empty heap. A second method that does no key comparisons is to convert the dag into a tree containing only full nodes, as follows: For each node that has two parents, eliminate the second parent, making the dag a tree. Give each full child a rank of 0 and a parent equal to its nearest full proper ancestor. Delete all the hollow nodes. To extend the analysis in Sections 3 and 4 to cover the second rebuilding method, we define every child to be unranked after rebuilding. Either way of rebuilding can be done in a single traversal of the dag, taking $O(N)$ time. Since $N > cn$ and $c > 1$, $O(N) = O(N - n)$. That is, the rebuilding time is $O(1)$ per hollow node. By charging the rebuilding time to the *decrease-key* and *delete* operations that created the hollow nodes, $O(1)$ per operation, we obtain the following theorem:

Theorem 6. *With rebuilding, the amortized time per hollow heap operation is $O(1)$ for each operation other than a delete-min or delete, and $O(\log n)$ per delete-min or delete on a heap of n items. These bounds hold for both lazy and eager hollow heaps.*

By making c sufficiently large, we can arbitrarily reduce the rebuilding overhead, at a constant factor cost in space and an additive constant cost in the amortized time of *delete*. Whether rebuilding is actually a good idea in any particular application is a question to be answered by experiments.

6   Implementation of Hollow Heaps

In this section we develop an implementation of the data structure in Section 2 that satisfies the time bounds in Section 3 and that is tuned to save space. We store each set of children in a list. Each new child of a node v is added to the front of the list of children of v. Since hollow nodes can be in two lists of children, it might seem that we need to make the lists of children exogenous. But we can make them endogenous by observing that only hollow nodes can have two parents, and a hollow node with two parents is last on the list of children of its second parent (since it is the earliest child, and later children are added to the front of the list). This allows us to use two pointers per node u to represent lists of children: $u.child$ is the first child of u, *null* if u has no children; $u.next$ is the next sibling of u on the list of children of its first parent.

With this representation, given a child u of a node v, we need ways to answer three questions: (i) Is u last on the list of children of v? (ii) Does u have two parents? (iii) Is v the first or the second parent of u? If u has only one parent, the first question is easy to answer: u is the last child of v if and only if $u.next = null$. There are several ways to answer the second two questions in $O(1)$ time. We develop a detailed implementation using one method, and we discuss alternatives in the full version of the paper.

Each node u stores a pointer $u.item$ to the item it holds if it is full; if u is hollow, $u.item = null$. Each hollow node u stores a pointer to its second parent $u.sp$; if u is hollow but has at most one parent, $u.sp = null$. A *decrease-key* operation makes a newly hollow node u a child of a new node v by setting $v.child = u$ but not changing $u.next$: $u.next$ is the next sibling of u on the list of children of the first parent of u. We answer the three questions as follows: (i) A child u of v is last on the list of children of v if and only if $u.next = null$ (u is last on any list of children containing it) or $u.sp = v$ (u is hollow with two parents and v is its second parent); (ii) u has two parents if and only if $u.sp \neq null$; (iii) v is the second parent of u if and only if $u.sp = v$.

Each node u also stores its key and rank, and each item e stores the node $e.node$ holding it. The total space needed is four pointers, a key and a rank per node, and one pointer per item. Ranks are small integers, requiring $\lg \lg N + O(1)$ bits each.

Implementation of *delete* requires keeping track of roots as they are deleted and linked. To do this, we maintain a list L of hollow roots, singly linked by *next* pointers. We also maintain an array A of full roots, indexed by rank, at most one per rank. When a *delete* makes a root hollow, do the following. First, initialize L to contain the hollow root and A to be empty. Second, repeat the following until L is empty: Delete a node x from L, apply the appropriate one of the following cases to each child u of x, and then destroy x:

(i) u is hollow and v is its only parent: Add u to L: deletion of x makes u a root.

(ii) u has two parents and v is the second: Set $u.sp = null$ and stop processing children of x: u is the last child of x. Since u still has its first parent, it does not become a root.

(iii) u has two parents and v is the first: Set $u.sp = null$ and $u.next = null$.

(iv) u is full: Add u to A unless A contains a root of the same rank. If it does, link u with this root via a ranked link and repeat this with the winner until A does not contain a root of the same rank; then add the final winner to A.

Third and finally (once L is empty), empty A and link full roots via unranked links until there is at most one.

With this implementation, the worst-case time per operation is $O(1)$ except for *delete* operations that remove root items. A *delete* that removes a root item takes $O(1)$ time plus $O(1)$ time per hollow node that loses a parent plus $O(1)$ time per link plus $O(\log_\phi N)$ time, where N is the number of nodes in the tree just before the *delete*, since $max\text{-}rank = O(\log_\phi N)$ by Theorem 2. These are the bounds claimed in Section 3.

Acknowledgement. Thomas Dueholm Hansen is supported by The Danish Council for Independent Research | Natural Sciences (grant no. 12-126512); and the Sino-Danish Center for the Theory of Interactive Computation, funded by the Danish National Research Foundation and the National Science Foundation of China (under the grant 61061130540). Haim Kaplan is supported by the Israel Science Foundation grants no. 822-10 and 1841/14, the German-Israeli Foundation for Scientific Research and Development (GIF) grant no. 1161/2011, and the Israeli Centers of Research Excellence (I-CORE) program (Center No. 4/11). Uri Zwick is supported by BSF grant no. 2012338 and by The Israeli Centers of Research Excellence (I-CORE) program (Center No. 4/11).

References

1. Brodal, G.S.: Worst-case efficient priority queues. In: Proceedings of the 7th ACM-SIAM Symposium on Discrete Algorithms (SODA), pp. 52–58 (1996)

2. Brodal, G.S., Lagogiannis, G., Tarjan, R.E.: Strict Fibonacci heaps. In: Proc. of the 44th ACM STOC, pp. 1177–1184 (2012)
3. Chan, T.M.: Quake heaps: a simple alternative to fibonacci heaps. In: Brodnik, A., López-Ortiz, A., Raman, V., Viola, A. (eds.) Space-Efficient Data Structures, Streams, and Algorithms. LNCS, vol. 8066, pp. 27–32. Springer, Heidelberg (2013)
4. Dijkstra, E.W.: A note on two problems in connexion with graphs. Numerische Mathematik 1, 269–271 (1959)
5. Driscoll, J.R., Gabow, H.N., Shrairman, R., Tarjan, R.E.: Relaxed heaps: an alternative to Fibonacci heaps with applications to parallel computation. Communications of the ACM 31(11), 1343–1354 (1988)
6. Edmonds, J.: Optimum branchings. J. Res. Nat. Bur. Standards 71B, 233–240 (1967)
7. Elmasry, A.: The violation heap: a relaxed Fibonacci-like heap. Discrete Math., Alg. and Appl., 2(4), 493–504 (2010)
8. Fredman, M.L., Tarjan, R.E.: Fibonacci heaps and their uses in improved network optimization algorithms. Journal of the ACM 34(3), 596–615 (1987)
9. Gabow, H.N., Galil, Z., Spencer, T.H., Tarjan, R.E.: Efficient algorithms for finding minimum spanning trees in undirected and directed graphs. Combinatorica 6, 109–122 (1986)
10. Haeupler, B., Sen, S., Tarjan, R.E.: Rank-pairing heaps. SIAM Journal on Computing 40(6), 1463–1485 (2011)
11. Høyer, P.: A general technique for implementation of efficient priority queues. In: Proceedings of the 3rd Israeli Symposium on the Theory of Computing and Systems (ISTCS), pp. 57–66 (1995)
12. Kaplan, H., Shafrir, N., Tarjan, R.E.: Meldable heaps and boolean union-find. In: Proc. of the 34th ACM STOC, pp. 573–582 (2002)
13. Kaplan, H., Tarjan, R.E.: Thin heaps, thick heaps. ACM Transactions on Algorithms 4(1), 1–14 (2008)
14. Kaplan, H., Tarjan, R.E., Zwick, U.: Fibonacci heaps revisited. CoRR, abs/1407.5750 (2014)
15. Knuth, D.E.: Sorting and searching. The art of computer programming, vol. 3, 2nd edn. Addison-Wesley (1998)
16. Peterson, G.L.: A balanced tree scheme for meldable heaps with updates. Technical Report GIT-ICS-87-23, School of Informatics and Computer Science, Georgia Institute of Technology, Atlanta, GA (1987)
17. Prim, R.C.: Shortest connection networks and some generalizations. Bell System Technical Journal 36, 1389–1401 (1957)
18. Takaoka, T.: Theory of 2–3 heaps. Discrete Appl. Math. 126(1), 115–128 (2003)
19. Tarjan, R.E.: Data structures and network algorithms. SIAM (1983)

Linear-Time List Recovery of High-Rate Expander Codes

Brett Hemenway[1]([✉]) and Mary Wootters[2]

[1] University of Pennsylvania, Philadelphia, USA
fbrett@cis.upenn.edu
[2] Carnegie Mellon University, Pittsburgh, USA
marykw@cs.cmu.edu

Abstract. We show that expander codes, when properly instantiated, are high-rate list recoverable codes with linear-time list recovery algorithms. List recoverable codes have been useful recently in constructing efficiently list-decodable codes, as well as explicit constructions of matrices for compressive sensing and group testing. Previous list recoverable codes with linear-time decoding algorithms have all had rate at most $1/2$; in contrast, our codes can have rate $1 - \varepsilon$ for any $\varepsilon > 0$. We can plug our high-rate codes into a framework of Alon and Luby (1996) and Meir (2014) to obtain linear-time list recoverable codes of arbitrary rates R, which approach the optimal trade-off between the number of non-trivial lists provided and the rate of the code.

While list-recovery is interesting on its own, our primary motivation is applications to list-decoding. A slight strengthening of our result would imply linear-time and optimally list-decodable codes for all rates. Thus, our result is a step in the direction of solving this important problem.

1 Introduction

In the theory of error correcting codes, one seeks a code $\mathcal{C} \subset \mathbb{F}^n$ so that it is possible to recover any *codeword* $c \in \mathcal{C}$ given a corrupted version of that codeword. The most standard model of corruption is from errors: some constant fraction of the symbols of a codeword might be adversarially changed. Another model of corruption is that there is some uncertainty: in each position $i \in [n]$, there is some small list $S_i \subset \mathbb{F}$ of possible symbols. In this model of corruption, we cannot hope to recover c exactly; indeed, suppose that $S_i = \{c_i, c_i'\}$ for some codewords $c, c' \in \mathcal{C}$. However, we can hope to recover a short list of codewords that contains c. Such a guarantee is called *list recoverability*.

While this model is interesting on its own—there are several settings in which this sort of uncertainty may arise—one of our main motivations for studying list-recovery is *list-decoding*. We elaborate on this more in Section 1.1 below.

We study the list recoverability of *expander codes*. These codes—introduced by Sipser and Spielman in [29]—are formed from an expander graph and an

---

M. Wootters–Research funded by NSF MSPRF grant DMS-1400558.

M.M. Halldórsson et al. (Eds.): ICALP 2015, Part I, LNCS 9134, pp. 701–712, 2015.
DOI: 10.1007/978-3-662-47672-7_57

inner code $\mathcal{C}_0$. One way to think about expander codes is that they preserve some property of $\mathcal{C}_0$, but have some additional useful structure. For example, [29] showed that if $\mathcal{C}_0$ has good distance, then so does the the expander code; the additional structure of the expander allows for a linear-time decoding algorithm. In [20], it was shown that if $\mathcal{C}_0$ has some good (but not great) locality properties, then the larger expander code is a good locally correctable code. In this work, we extend this list of useful properties to include list recoverability. We show that if $\mathcal{C}_0$ is a list recoverable code, then the resulting expander code is again list recoverable, but with a linear-time list recovery algorithm.

1.1 List Recovery

List recoverable codes were first studied in the context of list-decoding and soft-decoding: a list recovery algorithm is at the heart of the celebrated Guruswami-Sudan list-decoder for Reed-Solomon codes [17] and for related codes [16]. Guruswami and Indyk showed how to use list recoverable codes to obtain good list- and uniquely-decodable codes [12–14]. More recently, list recoverable codes have been studied as interesting objects in their own right, and have found several algorithmic applications, in areas such as compressed sensing and group testing [7,23,28].

We consider list recovery from erasures, which was also studied in [8,14]. That is, some fraction of symbols may have no information; equivalently, $S_i = \mathbb{F}$ for a constant fraction of $i \in [n]$. Another, stronger guarantee is list recovery from errors. That is, $c_i \notin S_i$ for a constant fraction of $i \in [n]$. We do not consider this stronger guarantee here, and it is an interesting question to extend our results for erasures to errors. It should be noted that the problem of list recovery is interesting even when there are neither errors nor erasures. In that case, the problem is: given $S_i \subset \mathbb{F}$, find all the codewords $c \in \mathcal{C}$ so that $c_i \in S_i$ for all i.

There are two parameters of interest. First, the rate $R := \log_q(|\mathcal{C}|)/n$ of the code: ideally, we would like the rate to be close to 1. Second, the efficiency of the recovery algorithm: ideally, we would be able to perform list-recovery in time linear in n. We survey the relevant results on list recoverable codes in Figure 1. While there are several known constructions of list recoverable codes with high rate, and there are several known constructions of list recoverable codes with linear-time decoders, there are no known prior constructions of codes which achieve both at once.

In this work, we obtain the best of both worlds, and give constructions of high-rate, linear-time list recoverable codes. Additionally, our codes have constant (independent of n) list size and alphabet size. As mentioned above, our codes are actually expander codes—in particular, they retain the many nice properties of expander codes: they are explicit linear codes which are efficiently (uniquely) decodable from a constant fraction of errors.

We can use these codes, along with a construction of Alon and Luby [1], recently highlighted by Meir [26], to obtain linear-time list recoverable codes of any rate R, which obtain the optimal trade-off between the fraction $1 - \alpha$ of erasures and the rate R. More precisely, for any $R \in [?\ ?\ ]$, $\ell \in \mathbb{N}$, and $\eta > 0$, there is some $L = L(\eta, \ell)$ so that we can construct rate R codes which are $(R + \eta, \ell, L)$-list recoverable in linear time. The fact that our codes from the previous paragraph have rate approaching 1 is necessary for this construction. To the best of our knowledge, linear-time list-decodable codes obtaining this trade-off were also not known.

It is worth noting that if our construction worked for list recovery from *errors*, rather than erasures, then the reduction above would obtain linear-time list decodable codes, of rate R and tolerating $1 - R - \eta$ errors. (In fact, it would yield codes that are list-recoverable from errors, which is a strictly stronger notion). So far, all efficiently list-decodable codes in this regime have polynomial-time decoding algorithms. In this sense, our work is a step in the direction of linear-time optimal list decoding, which is an important open problem in coding theory.[1]

1.2   Expander Codes

Our list recoverable codes are actually properly instantiated *expander codes.* Expander codes are formed from a d-regular expander graph, and an *inner code* C_0 of length d, and are notable for their extremely fast decoding algorithms. We give the details of the construction below in Section 2. The idea of using a graph to create an error correcting code was first used by Gallager [6], and the addition of an inner code was suggested by Tanner [30]. Sipser and Spielman introduced the use of an expander graph in [29]. There have been several improvements over the years by Barg and Zemor [2–4,31].

Recently, Hemenway, Ostrovsky and Wootters [20] showed that expander codes can also be *locally corrected,* matching the best-known constructions in the high-rate, high-query regime for locally-correctable codes. That work showed that as long as the inner code exhibits suitable locality, then the overall expander code does as well. This raised a question: what other properties of the inner code does an expander code preserve? In this work, we show that as long as the inner code is list recoverable (even without an efficient algorithm), then the expander code itself is list recoverable, but with an extremely fast decoding algorithm.

It should be noted that the works of Guruswami and Indyk cited above on linear-time list recovery are also based on expander graphs. However, that construction is different from the expander codes of Sipser and Spielman. In particular, it does not seem that the Guruswami-Indyk construction can achieve a high rate while maintaining list recoverability.

---

[1] In fact, adapting our construction to handle errors, even if we allow polynomial-time decoding, is interesting. First, it would give a new family of efficiently-decodable, optimally list-decodable codes, very different from the existing algebraic constructions. Secondly, there are no known uniformly constructive explicit codes (that is, constructible in time $\text{poly}(n) \cdot C_\eta$) with both constant list-size and constant alphabet size—adapting our construction to handle errors, even with polynomial-time recovery, could resolve this.

1.3 Our Contributions

We summarize our contributions below:

1. **The first construction of linear-time list-recoverable codes with rate approaching 1.** As shown in Figure 1, existing constructions have either low rate or substantially super-linear recovery time. The fact that our codes have rate approaching 1 allows us to plug them into a construction of [26], to achieve the next bullet point:
2. **The first construction of linear-time list-recoverable codes with optimal rate/erasure trade-off.** We will show in Section 3.2 that our high-rate codes can be used to construct list-recoverable codes of arbitrary rates R, where we are given information about only an $R + \varepsilon$ fraction of the symbols. As shown in Figure 1, existing constructions which achieve this trade-off have substantially super-linear recovery time.
3. **A step towards linear-time, optimally list decodable codes.** Our results above are for list-recovery from *erasures*. While this has been studied before [14], it is a weaker model than a standard model which considers *errors*. As mentioned above, a solution in this more difficult model would lead to algorithmic improvements in list decoding (as well as potentially in compressed sensing, group testing, and related areas). It is our hope that understanding the erasure model will lead to a better understanding of the error model, and that our results will lead to improved list decodable codes.
4. **New tricks for expander codes.** One take-away of our work is that expander codes are extremely flexible. This gives a third example (after unique- and local- decoding) of the expander-code construction taking an inner code with some property and making that property efficiently exploitable. We think that this take-away is an important observation, worthy of its own bullet point. It is a very interesting question what other properties this may work for.

2 Definitions and Notation

An error correcting code is (α, ℓ, L) *list recoverable* (from errors) if given lists of ℓ possible symbols at every index, there are at most L codewords whose symbols lie in a α fraction of the lists. We will use a slightly different definition of list recoverability, matching the definition of [14]; to distinguish it from list recovery from errors, we will call it list recoverability from *erasures*.

Definition 1 (List recoverability from erasures). *An error correcting code* $\mathcal{C} \subset \mathbb{F}_q^n$ *is* (α, ℓ, L)-*list recoverable from erasures if the following holds. Fix any sets* $S_1, \ldots, S_n$ *with* $S_i \subset \mathbb{F}_q$, *so that* $|S_i| \leq \ell$ *for at least* αn *of the* i's *and* $S_i = \mathbb{F}_q$ *for all remaining* i. *Then there are most* L *codewords* $c \in \mathcal{C}$ *so that* $c \in S_1 \times S_2 \times \cdots \times S_n$.

Source	Rate	List size L	Alphabet size	Agreement α	Recovery time	Explicit Linear
Random code	$1-\gamma$	$O(\ell/\gamma)$	$\ell^{O(1/\gamma)}$	$1-O(\gamma)$		
Random pseudolinear code [11]	$1-\gamma$	$O\left(\frac{\ell\log(\ell)}{\gamma^2}\right)$	$\ell^{O(1/\gamma)}$	$1-O(\gamma)$		
Random linear code [9]	$1-\gamma$	$\ell^{O(\ell/\gamma^2)}$	$\ell^{O(1/\gamma)}$	$1-O(\gamma)$		L
Folded Reed-Solomon codes [16]	$1-\gamma$	$n^{O(\log(\ell)/\gamma)}$	$n^{O(\log(\ell)/\gamma^2)}$	$1-O(\gamma)$	$n^{O(\log(\ell)/\gamma^2)}$	EL
Folded RS subcodes: evaluation points in an explicit subspace-evasive set [5]	$1-\gamma$	$(1/\gamma)^{O(\ell/\gamma)}$	$n^{O(\ell/\gamma^2)}$	$1-O(\gamma)$	$n^{O(\ell/\gamma^2)}$	E
Folded RS subcodes: evaluation points in a non-explicit subspace-evasive set [10]	$1-\gamma$	$O\left(\frac{\ell}{\gamma^2}\right)$	$n^{O(\ell/\gamma^2)}$	$1-O(\gamma)$	$n^{O(\ell/\gamma^2)}$	
(Folded) AG subcode [18,19]	$1-\gamma$	$O(\ell/\gamma)$	$\exp(\tilde{O}(\ell/\gamma^2))$	$1-O(\gamma)$	$C_{\ell,\gamma}n^{O(1)}$	
[13]	$2^{-2^{O(\ell)}}$	ℓ	$2^{2^{2^{O(\ell)}}}$	$1-2^{-2^{\ell^{O(1)}}}$	$O(n)$	E
[14]	$\ell^{-O(1)}$	ℓ	$2^{\ell^{O(1)}}$	$.999\ (\star)$	$O(n)$	E
This work	$1-\gamma$	$\ell^{\gamma^{-4}\ell^{C\ell/\gamma^2}}$	$\ell^{O(1/\gamma)}$	$1-O(\gamma^3)\ (\star)$	$O(n)$	EL

Fig. 1. Results on high-rate list recoverable codes and on linear-time decodable list recoverable codes. Above, n is the block length of the (α,ℓ,L)-list recoverable code, and $\gamma > 0$ is sufficiently small and independent of n. Agreement rates marked $(\star)$ are for erasures, and all others are from errors. An empty "recovery time" field means that there are no known efficient algorithms. We remark that [19], along with the explicit subspace designs of [15], also give explicit constructions of high-rate AG subcodes with polynomial time list-recovery and somewhat complicated parameters; the list-size L becomes super-constant.

The results listed above of [5,10,16,18,19] also apply for any rate R and agreement $R+\gamma$. In Section 3.2, we show how to achieve the same trade-off (for erasures) in linear time using our codes.

Our construction will be based on *expander graphs*. We say a d-regular graph H is a *spectral expander* with parameter λ, if λ is the second-largest eigenvalue of the normalized adjacency matrix of H. Intuitively, the smaller λ is, the better connected H is—see [22] for a survey of expanders and their applications. We will take H to be a *Ramanujan graph*, that is, so that $\lambda \leq \frac{2\sqrt{d-1}}{d}$; explicit constructions of Ramanujan graphs are known [24,25,27] for arbitrarily large values of d. For a graph H with vertices $V(H)$ and edges $E(H)$, we use the following notation. For a set $S \subset V(H)$, we use $\Gamma(S)$ to denote the neighborhood

$$\Gamma(S) = \{v \,:\, \exists u \in S, (u,v) \in E(H)\}.$$

For a set of edges $F \subset E(H)$, we use $\Gamma_F(S)$ to denote the neighborhood restricted to F:

$$\Gamma_F(S) = \{v \,:\, \exists u \in S, (u,v) \in F\}.$$

Given a d-regular H and an inner code $\mathcal{C}_0$, we define the *Tanner code $\mathcal{C}(H, \mathcal{C}_0)$* as follows.

Definition 2 (Tanner code [30]). *If H is a d-regular graph on n vertices and $\mathcal{C}_0$ is a linear code of block length d, then the* Tanner code *created from $\mathcal{C}_0$ and H is the linear code $\mathcal{C} \subset \mathbb{F}_q^{E(H)}$, where each edge H is assigned a symbol in $\mathbb{F}_q$ and the edges adjacent to each vertex form a codeword in $\mathcal{C}_0$.*

$$\mathcal{C} = \{c \in \mathbb{F}_q^{E(H)} \,:\, \forall v \in V(H), c|_{\Gamma(v)} \in \mathcal{C}_0\}$$

Because codewords in $\mathcal{C}_0$ are *ordered* collections of symbols whereas edges adjacent to a vertex in H may be unordered, creating a Tanner code requires choosing an ordering of the edges at each vertex of the graph. Although different orderings lead to different codes, our results (like all previous results on Tanner codes) work for all orderings. As our constructions work with any ordering of the edges adjacent to each vertex, we assume that some arbitrary ordering has been assigned, and do not discuss it further.

When the underlying graph H is an expander graph,[2] we call the resulting Tanner code an *expander code*. Sipser and Spielman showed that expander codes are efficiently uniquely decodable from about a δ_0^2 fraction of errors. We will only need unique decoding from erasures; the same bound of δ_0^2 obviously holds for erasures as well, but for completeness we state the following lemma, which we prove in the full version [21].

Lemma 1. *If $\mathcal{C}_0$ is a linear code of block length d that can recover from an $\delta_0 d$ number of erasures, and H is a d-regular expander with normalized second eigenvalue λ, then the expander code $\mathcal{C}$ can be recovered from a $\frac{\delta_0}{k}$ fraction of erasures in linear time whenever $\lambda < \delta_0 - \frac{2}{k}$.*

Throughout this work, $\mathcal{C}_0 \subset \mathbb{F}_q^d$ will be (α_0, ℓ, L)-list recoverable from erasures, and the distance of $\mathcal{C}_0$ is δ_0. We choose H to be a Ramanujan graph, and $\mathcal{C} = \mathcal{C}(H, \mathcal{C}_0)$ will be the expander code formed from H and $\mathcal{C}_0$.

---

[2] Although many expander codes rely on bipartite expander graphs (e.g. [31]), we find it notationally simpler to use the non-bipartite version.

3 Results and Constructions

In this section, we give an overview of our constructions and state our results. Our main result (Theorem 1) is that list recoverable inner codes imply list recoverable expander codes. We then instantiate this construction to obtain the high-rate list recoverable codes claimed in Figure 1. Next, in Theorem 3 we show how to combine our codes with a construction of Meir [26] to obtain linear-time list recoverable codes which approach the optimal trade-off between α and R.

3.1 High-Rate Linear-Time List Recoverable Codes

Our main theorem is that list recoverable codes imply list recoverable expander codes:

Theorem 1. *Suppose that C_0 is (α_0, ℓ, L)-list recoverable from erasures, of rate R_0, length d, and distance δ_0, and suppose that H is a d-regular expander graph with normalized second eigenvalue λ, if*

$$\lambda < \frac{\delta_0^2}{12\ell^L}$$

Then the expander code C formed from C_0 and H has rate at least $2R_0 - 1$ and is (α, ℓ, L')-list recoverable from erasures, where

$$L' \leq \exp_\ell \left(\frac{72\, \ell^{2L}}{\delta_0^2(\delta_0 - \lambda)^2} \right)$$

and α satisfies

$$1 - \alpha \geq (1 - \alpha_0) \left(\frac{\delta_0(\delta_0 - \lambda)}{6} \right).$$

Further, the running time of the list recovery algorithm is $O_{L,\ell,\delta_0,d}(n)$.

Above, the notation $\exp_\ell(\cdot)$ means $\ell^{(\cdot)}$. Before we discuss the proof of Theorem 1 and the recovery algorithm, we show how to instantiate these codes to give the parameters claimed in Figure 1.

We will use a random linear code as the inner code. A probabilistic argument shows that there exist inner codes with $R_0 = 1 - \gamma$, distance $\delta_0 = \gamma(1 + \mathcal{O}(\gamma))$ that are (α_0, ℓ, L)-list recoverable, over an alphabet of size $q^{O(1/\gamma)}$. (See the full version of this paper [21]). Plugging all this into Theorem 1, we get explicit codes of rate $1 - 2\gamma$ which are (α, ℓ, L')-list recoverable in linear time, for

$$L' = \exp_\ell \left(\gamma^{-4} \exp_\ell \left(\exp_\ell \left(C\ell/\gamma^2 \right) \right) \right)$$

and for $\alpha = 1 - C'\gamma^3$ for some constants C, C'. This recovers the parameters claimed in Figure 1. Above, we can choose $d = O\left(\frac{\ell^{2L}}{\gamma^4} \right)$ so that the Ramanujan graph would have parameter λ obeying the conditions of Theorem 1. Thus, when ℓ, γ are constant, so is the degree d, and the running time of the recovery algorithm is linear in n, and thus in the block length nd of the expander code. By Lemma 1, because the distance of the inner code is $\delta_0 = \gamma(1 + O(\gamma))$, the distance of our construction is $\delta = \Omega(\gamma^2)$.

Remark 1. Both the alphabet size and the list size L' are constant, if ℓ and γ are constant. However, L' depends rather badly on ℓ, even compared to the other high-rate constructions in Figure 1. This is because the bound on random linear codes that we use for our inner code is likely not tight; it would be interesting to either improve this bound or to give an inner code with better list size L. The key restrictions for such an inner code are that (a) the rate of the code must be close to 1; (b) the list size L must be constant, and (c) the code must be linear. Notice that (b) and (c) prevent the use of either Folded Reed-Solomon codes or their restriction to a subspace evasive set, respectively.

3.2 List Recoverable Codes Approaching Capacity

We can use our list recoverable codes, along with a construction of Alon and Luby [1] (which has also been used for similar purposes by Guruswami and Indyk [12], and was recently used and highlighted by Meir [26]), to construct codes which approach the optimal trade-off between the rate R and the agreement α. To quantify this, we state the following analog of the list-decoding capacity theorem.

Theorem 2 (List recovery capacity theorem). *For every $R > 0$, and $L \geq \ell$, there is some code $\mathcal{C}$ of rate R over $\mathbb{F}_q$ which is $(R + \eta(\ell, L), \ell, L)$-list recoverable from erasures, for any*

$$\eta(\ell, L) \geq \frac{4\ell}{L} \qquad and \qquad q \geq \ell^{2/\eta}.$$

Further, for any constant $R > 0$, any integer ℓ, and any sufficiently small $\eta > 0$, any code of rate R which is $(R - \eta, \ell, L)$-list recoverable from erasures must have $L = q^{\Omega(n)}$.

The proof is a straightforward probabilistic argument and is given in the full version [21]. Although Theorem 2 ensures the existence of certain list-recoverable codes, the proof of Theorem 2 is probabilistic, and does not provide a means of efficiently identifying (or list recovering) these codes. Using the approach of [26] we can turn our construction of linear-time list recoverable codes into linear-time list recoverable codes approaching capacity.

Theorem 3. *For any $R > 0$, $\ell > 0$, and for all sufficiently small $\eta > 0$, there is some L, depending only on ℓ and η, and some constant d, depending only on η, so that whenever $q \geq \ell^{6/\eta}$ there is a family of (α, ℓ, L)-list recoverable codes $\mathcal{C} \subset \mathbb{F}_{q^d}^n$ with rate at least R, for $\alpha = R + \eta$. Further, these codes can be list-recovered in linear time.*

We follow the approach of [26], which adapts a construction of [1] to take advantage of high-rate codes with a desirable property. Informally, the takeaway of [26] is that, given a family of codes with any nice property and rate approaching 1, one can make a family of codes with the same nice property that acheives the Singleton bound. The proof of Theorem 3, as well as the construction, can be found in the full version [21].

4   Recovery Procedure and Proof of Theorem 1

In this section, we outline at a high level the ideas and techniques in our list recovery algorithm. A detailed description of the recovery algorithm and the proof of correctness can be found in the full version [21]. The complete list recovery algorithm is presented in the full version. The algorithm proceeds in three steps, which we describe below. Due to space constraints, we omit the details of these steps, which can be found in the full version of the paper [21].

1. First, we list recover locally at each vertex, using the list recoverability of the inner code.
   This step yields a list of L codewords at each vertex.
2. Next, we choose an edge, and one of the ℓ possible symbols on that edge. The crux of the decoding algorithm is identifying how this choice propagates through the graph.
   This propagation will cover a constant fraction of the edges in the graph. We repeat this propagation for each of the ℓ choices of symbol for the chosen edge. This yields a collection of ℓ possible partial codewords.
3. Step 2 yields partial assignments (that assign values to a constant fraction of the symbols in the expander code). To turn these partial assignments into full assignments, we repeat Step 2 a constant number of times until we have partial assignments that cover essentially the entire graph. We stop once we have covered enough edges, and we use the minimum distance of the expander code to uniquely fill in the unknown edges. Since each iteration of Step 2 yields ℓ possible partial assignments (all to the same set of edges), if we repeat Step 2 t times, we can stitch them together to obtain ℓ^t possible assignments.

The difficulty in analyzing this algorithm comes from determining how a choice of a symbol in Step 2 propagates through the graph. We sketch the intuition below; the formal discussion can be found in [21].

For simplicity, suppose there are no erasures—our final algorithm can recover from a constant fraction of erasures, but the intuition is cleaner if there are no erasures—and suppose that each edge of H holds a list of ℓ possible symbols. Suppose (v, u) is the edge chosen at Step 2. We might hope that a choice of a symbol on this edge (or even the choice of a codeword at vertex u) would determine the codeword at v. This is unfortunately not likely to be true because $L > \ell$: a choice of one of ℓ symbols on (v, u) is not sufficient to uniquely determine one of the L codewords on vertex v. Instead of analyzing propagation at a vertex level, we focus on propagation at the edge level.

To do this, we introduce the notion of equivalence classes of *edges*. Suppose that the neighbors of v are $u_1, \ldots, u_d$. There are L possible codewords at v, and there are ℓ possible choices of symbol at (v, u_i); thus there are at most ℓ^L possible maps from codeword at v to symbol at (v, u_i). If $d \gg \ell^L$, then by the pigeonhole principle some of these maps must be identical. We call edges (v, u_i) and (v, u_j) equivalent with respect to v if their maps are identical. In particular, this means that a choice of symbol of (v, u_i) defines the choice of symbol on (v, u_j). Thus a choice of symbol on (v, u_1) (say), will determine symbols of (v, u_i) for all (v, u_i) in the same equivalence class as (v, u_1).

We can then repeat this logic at each of these vertices u_i: the choice of symbol on (u_i, v) will determine symbols on edges (u_i, w) that are equivalent to (u_i, v) with respect to u_i. In this way, the choice of a single symbol propagates through a large portion of the graph. We use the expansion of the graph to show that this propagation ends up covering a constant fraction of the graph. Thus, after making a constant number of choices (and using the distance of the expander code to take care of the small fraction of untouched edges), we will have recovered every assignment of symbols which is consistent with the given lists.

There are several details omitted from the sketch above. For example, we argued above that *some* equivalence classes are large. Of course, some may also be small. What if (v, u) belongs to a small equivalence class and our choice does not propagate? We show in the appendix that there is a large subgraph H' of H so that every equivalence class in H' is large. The full details, and a complete description of the recovery algorithm, can be found in the full version of the paper [21].

5   Conclusion and Open Questions

We have shown that expander codes, properly instantiated, are high-rate list recoverable codes with constant list size and constant alphabet size, which can be list recovered in linear time. To the best of our knowledge, no such construction was known. Our work leaves several open questions. Most notably, our algorithm can handle *erasures,* but it seems much more difficult to handle errors. As mentioned above, handling list recovery from errors would open the door for many of the applications of list recoverable codes, to list-decoding and other areas. Extending our results to errors with linear-time recovery would be most interesting, as it would immediately lead to optimal linear-time list-decodable codes. However, even polynomial-time recovery would be interesting: in addition to given a new, very different family of efficient locally-decodable codes, this could lead to explicit (uniformly constructive), efficiently list-decodable codes with constant list size and constant alphabet size, which is (to the best of our knowledge) currently an open problem. Second, the parameters of our construction could be improved: our choice of inner code (a random linear code), and its analysis, is clearly suboptimal. Our construction would have better performance with a better inner code. As mentioned in Remark 1, we would need a high-rate linear code which is list recoverable with constant list-size (the reason that this is not begging the question is that this inner code need not have a fast recovery algorithm). We are not aware of any such constructions.

Acknowledgments. We thank Venkat Guruswami for raising the question of obtaining high-rate linear-time list-recoverable codes, and for very helpful conversations. We also thank Or Meir for pointing out [26].

References

1. Alon, N., Luby, M.: A linear time erasure-resilient code with nearly optimal recovery. IEEE Transactions on Information Theory **42**(6), 1732–1736 (1996)
2. Barg, A., Zemor, G.: Error exponents of expander codes. IEEE Transactions on Information Theory **48**(6), 1725–1729 (2002)
3. Barg, A., Zemor, G.: Concatenated codes: serial and parallel. IEEE Transactions on Information Theory **51**(5), 1625–1634 (2005)
4. Barg, A., Zemor, G.: Distance properties of expander codes. IEEE Transactions on Information Theory **52**(1), 78–90 (2006)
5. Dvir, Z., Lovett, S.: Subspace evasive sets. In: Proceedings of the 44th Annual ACM Symposium on Theory of Computing (STOC), pp. 351–358. ACM (2012)
6. Gallager, R.G.: Low Density Parity-Check Codes. Technical report. MIT (1963)
7. Gilbert, A.C., Ngo, H.Q., Porat, E., Rudra, A., Strauss, M.J.: ℓ_2/ℓ_2-foreach sparse recovery with low risk. In: Fomin, F.V., Freivalds, R., Kwiatkowska, M., Peleg, D. (eds.) ICALP 2013, Part I. LNCS, vol. 7965, pp. 461–472. Springer, Heidelberg (2013)
8. Guruswami, V.: List decoding from erasures: Bounds and code constructions. IEEE Transactions on Information Theory **49**(11), 2826–2833 (2003)
9. Guruswami, V.: List decoding of error-correcting codes. LNCS, vol. 3282. Springer, Heidelberg (2004)
10. Guruswami, V.: Linear-algebraic list decoding of folded reed-solomon codes. In: Proceedings of the 26th Annual Conference on Computational Complexity (CCC), pp. 77–85. IEEE (2011)
11. Guruswami, V., Indyk, P:. Expander-based constructions of efficiently decodable codes. In: Proceedings of the 42nd Annual IEEE Symposium on Foundations of Computer Science (FOCS), pp. 658–667. IEEE (October 2001)
12. Guruswami, V., Indyk, P.: Near-optimal linear-time codes for unique decoding and new list-decodable codes over smaller alphabets. In: Proceedings of the 34th Annual ACM Aymposium on Theory of computing (STOC), pp. 812–821. ACM (2002)
13. Guruswami, V., Indyk, P.: Linear time encodable and list decodable codes. In: Proceedings of the 35th Annual ACM Symposium on Theory of Computing (STOC), pp. 126–135. ACM, New York (2003)
14. Guruswami, V., Indyk, P.: Linear-time list decoding in error-free settings. In: Díaz, J., Karhumäki, J., Lepistö, A., Sannella, D. (eds.) ICALP 2004. LNCS, vol. 3142, pp. 695–707. Springer, Heidelberg (2004)
15. Guruswami, V., Kopparty, S.: Explicit subspace designs. In: Proceedings of the 54th Annual IEEE Symposium on Foundations of Computing (FOCS), pp. 608–617. IEEE (2013)
16. Guruswami, V., Rudra, A.: Explicit codes achieving list decoding capacity: Error-correction with optimal redundancy. IEEE Transactions on Information Theory **54**(1), 135–150 (2008)
17. Guruswami, V., Sudan, M.: Improved decoding of Reed-Solomon and algebraic-geometry codes. IEEE Transactions on Information Theory 45(6) (1999)
18. Guruswami, V., Xing, C.: Folded codes from function field towers and improved optimal rate list decoding. In: Proceedings of the 44th Annual ACM Symposium on Theory of Computing (STOC), pp. 339–350. ACM (2012)
19. Guruswami, V., Xing, C.: List decoding reed-solomon, algebraic-geometric, and gabidulin subcodes up to the singleton bound. In: Proceedings of the 45th Annual ACM Symposium on Theory of Computing (STOC), pp. 843–852. ACM (2013)

20. Hemenway, B., Ostrovsky, R., Wootters, M.: Local correctability of expander codes. Information and Computation (2014)
21. Hemenway, B., Wootters, M.: Linear-time list recovery of high-rate expander codes. ArXiv preprint 1503.01955 (2015)
22. Hoory, S., Linial, N., Wigderson, A.: Expander graphs and their applications. Bulletin of the American Mathematical Society **43**(4), 439–561 (2006)
23. Indyk, P., Ngo, H.Q., Rudra, A.: Efficiently decodable non-adaptive group testing. In: Proceedings of the 21st Annual ACM-SIAM Symposium on Discrete Algorithms (SODA), pp. 1126–1142. Society for Industrial and Applied Mathematics (2010)
24. Lubotzky, A., Phillips, R., Sarnak, P.: Ramanujan graphs. Combinatorica **8**(3), 261–277 (1988)
25. Margulis, G.A.: Explicit Group-Theoretical Constructions of Combinatorial Schemes and Their Application to the Design of Expanders and Concentrators. Probl. Peredachi Inf. **24**(1), 51–60 (1988)
26. Meir, O.: Locally correctable and testable codes approaching the singleton bound, ECCC Report TR14-107 (2014)
27. Morgenstern, M.: Existence and Explicit Constructions of q + 1 Regular Ramanujan Graphs for Every Prime Power q. Journal of Combinatorial Theory, Series B **62**(1), 44–62 (1994)
28. Ngo, H.Q., Porat, E., Rudra, A.: Efficiently decodable compressed sensing by list-recoverable codes and recursion. In: Proceedings of the Symposium on Theoretical Aspects of Computer Science (STACS), vol. 14, pp. 230–241 (2012)
29. Sipser, M., Spielman, D.A.: Expander codes. IEEE Transactions in Information Theory 42(6) (1996)
30. Tanner, R.: A recursive approach to low complexity codes. IEEE Transactions on Information Theory **27**(5), 533–547 (1981)
31. Zemor, G.: On expander codes. IEEE Transactions on Information Theory **47**(2), 835–837 (2001)

Finding 2-Edge and 2-Vertex Strongly Connected Components in Quadratic Time

Monika Henzinger, Sebastian Krinninger, and Veronika Loitzenbauer$^{(\boxtimes)}$

Faculty of Computer Science, University of Vienna, Vienna, Austria
veronika.loitzenbauer@univie.ac.at

Abstract. We present faster algorithms for computing the 2-edge and 2-vertex strongly connected components of a directed graph. While in *undirected* graphs the 2-edge and 2-vertex connected components can be found in linear time, in *directed* graphs with m edges and n vertices only rather simple $O(mn)$-time algorithms were known. We use a hierarchical sparsification technique to obtain algorithms that run in time $O(n^2)$. For 2-edge strongly connected components our algorithm gives the first running time improvement in 20 years. Additionally we present an $O(m^2/\log n)$-time algorithm for 2-edge strongly connected components, and thus improve over the $O(mn)$ running time also when $m = O(n)$. Our approach extends to k-edge and k-vertex strongly connected components for any constant k with a running time of $O(n^2 \log n)$ for k-edge-connectivity and $O(n^3)$ for k-vertex-connectivity.

1 Introduction

Problem Description. In a directed graph G two vertices u and v are *2-edge strongly connected* if from u to v and from v to u, respectively, there are two paths that have no common edge. A *2-edge strongly connected component* (2eSCC) of G is a maximal subgraph of G such that in the subgraph every pair of distinct vertices is 2-edge strongly connected. Two vertices u and v are *2-vertex strongly connected* in G if they remain strongly connected after the removal of any single vertex except u and v from G. A *2-vertex strongly connected component* (2vSCC) of G is a maximal subgraph of G such that in the subgraph every pair of distinct vertices is 2-vertex strongly connected. Edge and vertex connectivity are central properties of graphs and have many applications [1,24], for example in the construction of reliable communication networks [2] and in the analysis of the structure of networks [26].

Our Results. In this work we present algorithms that compute the 2eSCCs and the 2vSCCs of a directed graph in $O(n^2)$ time. For 2eSCCs we additionally provide an algorithm that runs in $O(m^2/\log n)$ time, which is faster than $O(n^2)$ if $m = O(n)$. Thus we significantly improve upon the previous $O(mn)$-time algorithms for both 2eSCCs [14,25] and 2vSCCs [21]. For 2eSCCs the previous upper

---

Full version available at http://arxiv.org/abs/1412.6466

M.M. Halldórsson et al. (Eds.): ICALP 2015, Part I, LNCS 9134, pp. 713–724, 2015.
DOI: 10.1007/978-3-662-47672-7_58

bound stood for 20 years. Our approach immediately generalizes to computing the *k-edge strongly connected components* (keSCCs) and the *k-vertex strongly connected components* (kvSCCs). We give algorithms that, for any integral constant $k > 2$, compute (1) the keSCCs in time $O(n^2 \log n)$ (improving upon the previous upper bound of $O(mn)$ [25]) and (2) the kvSCCs in time $O(n^3)$ (improving upon the previous upper bound of $O(mn^2)$ [22]).

Related Work. The 2-edge and 2-vertex connected components of an *undirected* graph can be determined in linear time [19,27]. In *directed* graphs several related problems can be solved in linear time: Testing whether a graph is 2-edge or 2-vertex strongly connected [10,13,28], finding all *strong bridges* and *strong articulation points* [20], and determining the *2-edge* and *2-vertex strongly connected blocks* [14,15]. An edge is a strong bridge and a vertex is a strong articulation point, respectively, if its removal from the graph increases the number of strongly connected components (SCCs) of the graph. Note the difference between 'blocks' and 'components' in directed graphs: In a 2-edge strongly connected block every pair of distinct vertices is 2-edge strongly connected; however, as opposed to a 2eSCC, the paths to connect the vertices in a block might use vertices that are *not* in the same block. Each 2eSCC is completely contained in one 2-edge strongly connected block, i.e., the 2eSCCs refine the 2-edge strongly connected blocks. In the full version of this paper we provide a construction that shows that knowing the 2-edge strongly connected blocks of a graph does not help in finding its 2eSCCs. The relation between blocks and components for vertex connectivity is analogous.

Georgiadis et al. [14] and Jaberi [21] described simple algorithms to compute the 2eSCCs and 2vSCCs in $O(mn)$-time, respectively, and posed as an open problem whether this can be improved to linear time as well. An $O(mn)$ running time for computing the 2eSCCs was already achieved by Nagamochi and Watanabe in 1993 [25], which in fact solved the more general problem of computing the keSCCs. To the best of our knowledge, the fastest known algorithm for computing the kvSCCs is by Makino [22] and has a running time of $O(mn^2)$ (when combined with an $O(mn)$-time algorithm for finding minimum vertex-separators [7,9,11,18]; combined with [13] and [16] it also gives an $O(mn)$-time algorithm for 2vSCCs). In *undirected* graphs there are linear-time algorithms for computing both the 3-edge [12] and the 3-vertex [19] connected components. The k-edge connected components of an undirected graph can be computed in time $O(n^2)$ [25]. The runtime of Makino's algorithm can for k-vertex connected components in undirected graphs be reduced to $O(n^3)$ by a preprocessing step [23]. Thus, to the best of our knowledge, our algorithms for keSCCs and kvSCCs match the runtimes for undirected graphs for $k > 3$ (up to a logarithmic factor).

Techniques. We use a *hierarchical graph sparsification* that was introduced by Henzinger et al. [17] for undirected graphs and extended to directed graphs and game graphs in [4,5]. Roughly speaking, this sparsification technique allows us to replace the 'm' in the $O(mn)$ running time by an 'n', yielding $O(n^2)$. Our main technical contribution is to find structural properties of connectivity in directed graphs that allow us to apply this technique. Note that while various ways o. sparsification are used in algorithms for *undirected* graphs, such approaches are rarely found for *directed* graphs.

We briefly present the main ideas behind our algorithm for 2-vertex connectiv ity. The approach for edge connectivity is similar. The fastest known asymptotic running time of $O(mn)$ for computing 2vSCCs can be achieved with the follow ing approach: Assume that the graph is strongly connected. First find a strong articulation point of the graph, i.e., a vertex whose removal increases the number of SCCs. Then remove the strong articulation point and compute the SCCs. For each SCC, recurse on the subgraph it induces together with the strong artic ulation point. The recursion stops when no strong articulation point is found anymore. The SCCs remaining in the end are the 2vSCCs. We now explain in which way our algorithm deviates from this scheme.

Let for 2-vertex connectivity a 2-*isolated set* S be a set of vertices that (a) cannot be reached by the vertices of $V \setminus S$ or (b) that can be reached from $V \setminus S$ only through one vertex v. We show that every 2vSCC of G contains either only vertices of $S \cup \{v\}$ or only vertices of $V \setminus S$. Thus the algorithm can recurse on the subgraphs induced by $S \cup \{v\}$ and $V \setminus S$, respectively. The difference to the straightforward approach is thus the following: Instead of repeatedly identifying strong articulation points, we focus on separating 2-isolated sets of vertices. To see why this is useful, note that the incoming edges of the vertices of a 2-isolated set S consist of the incoming vertices from other vertices of S and edges from at most one vertex of $V \setminus S$ to S. Thus the number of incoming edges of each vertex in S is bounded by the number of vertices in S. We use this insight as follows: When searching for a 2-isolated set, we start the search in a subgraph of G that includes all vertices but only the first incoming edge of each vertex. If no 2-isolated set is found, we repeatedly double the number of incoming edges per vertex in the subgraph until the search is successful. In this way the search will take time $O(n)$ per vertex in the 2-isolated set. This will allow us to bound the total runtime by $O(n^2)$. Note that to achieve this running time we cannot afford to compute all SCCs in each recursive call because the recursion depth might be $\Theta(n)$; we therefore do not assume that the input graph is strongly connected.

To correctly identify 2-isolated sets by a search in a proper subgraph of G, the algorithm finds *vertex-dominators* in slightly modified *flow graphs*. A flow graph is a directed graph with a designated root where all vertices are reachable from the root. A vertex is a vertex-dominator in a flow graph if some other vertex can be reached from the root only through this vertex. Our algorithms use the linear-time algorithms for finding dominators [3,10,16,28] and SCCs [27] as subroutines.

In the $O(m^2/\log n)$-algorithm for 2eSCCs we search for 2-(edge-)isolated sets in subgraphs that are obtained by local breadth-first searches from vertices that lost edges in the previous iteration of the algorithm. Such local breadth-first searches were first used for Büchi games by Chatterjee et al. [6].

Outline. In Section 2 the main definitions and the notation are introduced. In Section 3 we show when and how we can identify a 2-isolated set in a proper ubgraph of G. In Section 4 we present the $O(n^2)$-algorithm for 2vSCCs. In Section 5 we outline how the results from Sections 3 and 4 extend to keSCCs and kvSCCs. The proofs are given in the full version of this paper.

2 Preliminaries

Let $G = (V, E)$ be a directed graph with $m = |E|$ edges and $n = |V|$ vertices. Except when mentioned explicitly, we only consider simple graphs, i.e., graphs without parallel edges. The reverse graph $Rev(G)$ of G is equal to (V, E^R) where E^R is the set containing for each edge $(u, v) \in E$ its reverse (v, u). We use $S \subseteq V$ to denote a subset S of V and $S \subsetneq V$ to denote a proper subset S of V. For any set $S \subseteq V$ we denote by $G[S]$ the subgraph of G induced by the vertices in S, i.e., the graph $(S, E \cap (S \times S))$. We call edges from some $u \in V \setminus S$ to some $v \in S$ the *incoming edges* of S. The incoming edges of a vertex v in G are denoted by $\text{In}_G(v)$, the number of incoming edges by $\text{Indeg}_G(v)$; analogously we use $\text{Out}_G(v)$ and $\text{Outdeg}_G(v)$ for outgoing edges. We denote by $G \setminus V'$ the graph $G[V \setminus V']$ and by $G \setminus E'$ the graph $(V, E \setminus E')$ for an arbitrary set of vertices $V' \subseteq V$ and an arbitrary set of edges $E' \subseteq E$.

Strong Connectivity. A subgraph $G[S]$ induced by some set of vertices S is *strongly connected* if for every pair of distinct vertices u and v in S there exists a path from u to v and a path from v to u in $G[S]$. A single vertex is considered strongly connected. The *strongly connected components* (SCCs) of G are its maximal strongly connected subgraphs and form a partition of V. A strongly connected subgraph with no outgoing edges is a *bottom SCC* (bSCC), a strongly connected subgraph with no incoming edges is a *top SCC* (tSCC). By definition, bSCCs and tSCCs are maximal. Every graph G contains at least one bSCC and at least one tSCC. If G is not strongly connected, then there exist both a bSCC and a tSCC that are disjoint and thus one of them contains at most half of the vertices of G. Note that a bSCC in G is a tSCC in $Rev(G)$ and vice versa. We further use that when a set of vertices S cannot be reached by any vertex of $V \setminus S$ in G, then $G[S]$ contains a tSCC of G.

Strong 2-Vertex Connectivity. A vertex $v \in V$ is a *strong articulation point* if the removal of v from G increases the number of SCCs in G. Two (simple) paths are *internally vertex-disjoint* if they do not share a vertex except possibly their endpoints. Two distinct vertices u and v are *2-vertex strongly connected* in G if they are strongly connected and remain strongly connected after the removal of any vertex except u and v from G. If there is no edge between u and v, then it holds that u and v are 2-vertex strongly connected if and only if there exists two internally vertex-disjoint paths from u to v and two internally vertex-disjoint paths from v to u [14]. A subgraph $G[S]$ induced by some set of vertices S is *2-vertex strongly connected* if every pair of distinct vertices u and v in S is 2-vertex strongly connected in $G[S]$. The *2-vertex strongly connected components*[1]

---

[1] Our definitions follow [14], while [15,21] use slightly different definitions. The 2vSCCs of [15,21] can be determined in $O(n)$ time from the 2vSCCs defined here.

(2vSCCs) of a graph are its maximal 2-vertex strongly connected subgraphs. Equivalently, the 2vSCCs are the maximal strongly connected subgraphs such that none of the subgraphs contains a strong articulation point. This definition of 2vSCCs allows for *degenerate* 2vSCCs with less than three vertices. While the 2eSCCs form a partition of the vertices of the graph, the 2vSCCs form a partition of a subset of the edges. In the remainder of the paper we omit "strong(ly)" from the above definitions whenever it is clear from the context.

Flow Graphs. We define the *flow graph* $G(r)$ to be the graph G with a vertex $r \in V$ designated as the root and with all vertices not reachable from r removed. A *vertex-dominator* in $G(r)$ is a vertex $v \in V \setminus \{r\}$ for which there exists a vertex $u \in V \setminus \{r, v\}$ such that u is reachable from r and every path from r to u contains v. We say that v *dominates* u in $G(r)$. Note that in contrast to articulation points the removal of a vertex-dominator from G might not increase the number of SCCs but instead might remove edges between SCCs.

3 New Top SCCs and Dominators in Subgraphs

Let an *isolated set* S *w.r.t. 2-vertex-connectivity* (*2-isolated set*) be a set of vertices with (1) incoming edges from at most one vertex and for which (2) there exist vertices without edges to S in G. 2-isolated sets can be used to design a divide-and-conquer based algorithm for the following reason: Let T be the vertex set of a 2vSCC. The 2vSCC $G[T]$ is (1) strongly connected and (2) for any proper subset S of T such that there exists a set of vertices U in T that has no edge to any vertex of S, there are at least two vertices in $T \setminus (S \cup U)$ that connect U with the vertices in S. Thus if we detect a set of vertices S that (a)

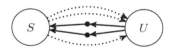

cannot be reached by the vertices of $V \setminus S$ or (b) that can be reached from $V \setminus S$ only through one vertex v, then we know that each 2vSCC of G contains either only vertices of $S \cup \{v\}$ or only vertices of $V \setminus S$. A 2-isolated set satisfies (a) or (b). Our algorithm repeatedly identifies specific 2-isolated sets S and recurses on the subgraphs induced by $S \cup \{v\}$ and $V \setminus S$, respectively. As the recursion depth can be $\Theta(n)$, to achieve an $o(mn)$ running time, we cannot afford to look at all edges in each level of recursion. Thus our algorithms are based on the following question: *Can we identify 2-isolated sets by searching in a proper subgraph of G?* Note that whenever an articulation point v is removed from a strongly connected graph G, then there exist both a tSCC and a bSCC in $G \setminus \{v\}$ that were adjacent to v in G and are disjoint. Let T be the vertices in the tSCC in $G \setminus \{v\}$. Observe that T is a 2-isolated set in G. Further, if T contains only a few vertices, then each vertex in T has a low in-degree in G because all incoming edges to vertices in T in G come from v or other vertices of T. In our algorithm we search for such "almost tSCCs" $G[T]$ in the subgraph of G induced by vertices with low

n-degree, which only takes time linear in the number of edges in this subgraph.
We do the same on $Rev(G)$ to detect small almost bSCCs.

Definition 1. *A set of vertices T induces an* almost tSCC *in G with respect to a vertex v if $G[T]$ is a tSCC in $G \setminus \{v\}$ but has incoming edges from v in G.*

Given a vertex v such that an almost tSCC induced by T w.r.t. v exists, the top SCC $G[T]$ can be identified in a subgraph of $G \setminus \{v\}$ in time linear in the number of edges in the subgraph as long as it is contained in the subgraph. But how can we identify the vertex v without looking at the whole graph? Assume there exists a vertex $r \neq v$ that is not in T but can reach v. Since $G[T]$ is a tSCC in $G \setminus \{v\}$, it follows that v dominates every vertex of T in the flow graph $G(r)$. This still holds in any subgraph of G as long as r can reach T in the subgraph. If additionally all incoming edges of the vertices in T are present in the subgraph, we can identify v and T in time linear in the number of edges in the subgraph by finding the vertex-dominator v in the flow graph with root r and the tSCC $G[T]$ in the subgraph with v removed. Thus, instead of finding the right v, we only have to find the right r. As edges are missing in the subgraph, it is not a-priori clear how to choose r, but, as shown below, we can use an artificial vertex as root r. Hence our approach is to first search for vertex-dominators v in a subgraph with an additional artificial root and then for a tSCC in the subgraph with v removed. When the search is successful, we recurse separately on the almost tSCC and the remaining graph.

In our algorithm we cannot afford to identify all SCCs in the current graph G as we only want to spend time proportional to the edges in a proper subgraph of G; thus we cannot assume that the graph we are considering is strongly connected. This means that, in contrast to strongly connected graphs [20], when we identify a vertex-dominator v in $G(r)$, the vertex v might not necessarily be an articulation point in G. However, for an almost tSCC w.r.t. v we still know that the set of vertices T in the almost tSCC is a 2-isolated set, i.e., all vertices of $V \setminus (T \cup \{v\})$ that can reach T in G can reach T only through v. Thus there cannot be two internally vertex-disjoint paths from any vertex of $G \setminus (T \cup \{v\})$ to any vertex of T. This intuition about almost tSCCs is summarized in the following lemma, which we use to show the correctness of our approach.

Lemma 2. *Let v be a vertex such that some set of vertices T induces an almost tSCC with respect to v in G. Let $W = V \setminus (T \cup \{v\})$. If $W \neq \emptyset$, then there do not exist two internally vertex-disjoint paths from any vertex of W to any vertex of T in G, i.e., no vertex of W is 2-vertex-connected to any vertex of T. Additionally, the vertex v is a vertex-dominator in $G(r)$ for every $r \in W$ that can reach v in G.*

Let $G_h = (V_h, E_h)$ be a subgraph of a directed graph $G = (V, E)$, i.e., $V_h \subseteq V$ and $E_h \subseteq G[V_h]$. We use the index h to identify specific subgraphs. In the remainder of this section we want to characterize which almost tSCCs in G we can identify in G_h. Let v be a vertex such that an almost tSCC w.r.t. v exists in G. To identify v as a vertex-dominator in a flow graph, we define below a graph created from G_h with an auxiliary root. Let the *white* vertices $A_{G,h} \subseteq V_G$

be the set of vertices for which we have the guarantee that for each vertex in $A_{G,h}$ its incoming edges in G_h are the same as in G. Let $B_{G,h} = V_h \setminus A_{G,h}$ be the *blue* vertices, which might miss incoming edges in G_h compared to G. We show that as long as the vertices in the almost tSCC are white, i.e., are not missing incoming edges in G_h, an almost tSCC w.r.t. a vertex v in G_h is an almost tSCC w.r.t. v in G and vice versa. In contrast, no conclusions can be drawn from an almost tSCC in G_h that includes blue vertices.

Definition 3. *For a given subgraph $G_h = (V_h, E_h)$ of a directed graph $G = (V, E)$ and a set of blue vertices $B_{G,h}$ that contains all vertices that have fewer incoming edges in G_h than in G, we define the flow graph $F_{G,h}(r_{G,h})$ as follows. If $|B_{G,h}| \geq 2$, let $F_{G,h}$ be the graph G_h with an additional vertex $r_{G,h}$ and an additional edge from $r_{G,h}$ to each vertex in $B_{G,h}$. If $B_{G,h}$ contains a single vertex, we name it $r_{G,h}$ and let $F_{G,h} = G_h$.*

In the following consider a subgraph G_h and a set of vertices V_h partitioned into $B_{G,h}$ and $A_{G,h}$ as above; the statements for $F_{G,h}$ hold whenever $F_{G,h}$ is defined.

Lemma 4. *A set of white vertices $T \subseteq A_{G,h}$ induces a tSCC in G_h and $F_{G,h}$, respectively, if and only if it induces a tSCC in G.*

If white vertices T induce an almost tSCC $G[T]$ with respect to v, all incoming edges, and thus v, are present in G_h. This implies the following corollary.

Corollary 5. *A set of white vertices $T \subseteq A_{G,h}$ induces an almost tSCC with respect to a vertex $v \in V$ in G_h and $F_{G,h}$, respectively, if and only if it induces an almost tSCC with respect to v in G.*

The following lemma specifies which almost tSCCs w.r.t. a vertex v in G we can identify by searching for vertex-dominators in $F_{G,h}(r_{G,h})$ based on the reachability of v from the vertices in $B_{G,h}$.

Lemma 6. *Assume $B_{G,h} \neq \emptyset$, let $T \subseteq A_{G,h}$ be a set of white vertices, and let $v \in V$ be such that there exists an almost tSCC $G[T]$ with respect to v in G. If v is either not in $B_{G,h}$ and can be reached from a vertex of $B_{G,h}$ or v is in $B_{G,h}$ and $|B_{G,h}| \geq 2$, then v is a dominator in $F_{G,h}(r_{G,h})$.*

In the following section we define specific subgraphs G_h that allow us to identify an almost tSCC in G that has at most a certain size by searching for vertex-dominators v in $F_{G,h}(r_{G,h})$ and tSCCs in $G_h \setminus \{v\}$. We additionally have to consider one special case, namely if v is the only vertex in $B_{G,h}$ and an almost tSCC w.r.t. v exists. In this case we have $r_{G,h} = v$. We explicitly identify almost tSCCs with respect to this vertex.

4   2vSCCs in $O(n^2)$ time

In this section we provide some intuition for the algorithm and outline its analysis. To find vertex-dominators, articulation points, and SCCs the known linear time algorithms are used (see Section 1).

Let $G = (V, E)$ be a simple directed graph. We consider for $i \in \mathbb{N}$ the subgraphs $G_i = (V, E_i)$ of G where E_i contains for each vertex of V its first 2^i incoming edges in E (for some arbitrary but fixed ordering of the incoming edges of each vertex). Note that when $i \geq \log(\max_{v \in V} \mathrm{Indeg}_G(v))$, then $G_i = G$. Let γ be the minimum of $\max_{v \in V} \mathrm{Indeg}_G(v)$ and $\max_{v \in V} \mathrm{Outdeg}_G(v)$. Following Definition 3, the set $B_{G,i}$ contains all vertices with in-degree more than 2^i in G.

Procedure 2vSCC(G)

```
1  for i ← 1 to ⌈log γ⌉ − 1 do
2  │   (S, Z) ← 2IsolatedSetLevel(G, i)
   │   /* Z contains v if G[S] is almost top or bottom SCC w.r.t. v  */
3  │   if S ≠ ∅ then
4  │   └   return 2vSCC(G[S ∪ Z]) ∪ 2vSCC(G[V \ S])

5  (S, Z) ← 2IsolatedSet(G)
   /* Z contains v if G[S] is almost top SCC w.r.t. v              */
6  if S ≠ ∅ then
7  │   return 2vSCC(G[S ∪ Z]) ∪ 2vSCC(G[V \ S])
8  else
9  └   return {G}
```

Let S be a set of at most 2^i vertices that induces a strongly connected subgraph $G[S]$ of G such that $G[S]$ is a top SCC or an almost top SCC with respect to some vertex v. Since the only edges from vertices of $V \setminus S$ to S are from v, the in-degree of each vertex in S can be at most 2^i. By applying the results from the previous section, we show that we can detect such a set S by searching for SCCs and vertex-dominators in the graphs $F_{G,i}$ constructed from G_i with the artificial root $r_{G,i}$ as in Definition 3.

Lemma 7. *If a set of vertices S with $|S| \leq 2^i$ induces a tSCC or an almost tSCC in G with respect to some vertex v, then $S \subseteq V \setminus B_{G,i}$.*

To find bSCCs and almost bSCCs we also search for top SCCs in $Rev(G)$. The search for both top and bottom SCCs ensures that whenever an (almost) tSCC and a disjoint (almost) bSCC exist in G, we only spend time proportional to the smaller one. This search is performed in Procedure 2IsolatedSetLevel, which fulfills the following guarantee.

Lemma 8. *If for some integer $1 \leq i < \log \gamma$ and $\mathcal{G} \in \{G, Rev(G)\}$ there exists a set of vertices $T \subseteq V \setminus B_{\mathcal{G},i}$ that induces in $\mathcal{G}$ a tSCC or an almost tSCC with respect to some vertex v with $T \subsetneq V \setminus \{v\}$, then 2IsolatedSetLevel(G, i) returns a non-empty set S.*

In Procedure 2vSCC we start the search for (almost) top SCCs at $i = 1$. Whenever the search is not successful, we increase i by one, until we have $G_i = G$ or $Rev(G)_i = Rev(G)$. For the search the Procedure 2IsolatedSetLevel is used as long as $2^i < \gamma$, i.e., both $B_{G,i}$ and $B_{Rev(G),i}$ are non-empty, and the Procedure 2IsolatedSet afterwards. Procedure 2IsolatedSet identifies an (almost) top SCC in G if one exists by using the known procedures for finding SCCs and articulation points. In this way we can show that whenever we had to go up to i^* or had to use Procedure 2IsolatedSet to identify an (almost) top or bottom SCC in G, the identified subgraph contains $\Omega(2^{i^*})$ vertices, where $i^* = \lceil \log \gamma \rceil$ for Procedure 2IsolatedSet. This will imply that the search in G_i and $Rev(G)_i$ for i up to i^* takes time $O(n \cdot 2^{i^*})$ which is $O(n \cdot \min\{|S|, |V \setminus S|\})$. This will allow us to bound the total running time by $O(n^2)$.

Procedure `2IsolatedSetLevel`(G, i)

1 **foreach** $\mathcal{G} \in \{G, Rev(G)\}$ **do**
/* $2^i < \max_{v \in V} \mathrm{Indeg}_{\mathcal{G}}(v) \Longrightarrow B_{\mathcal{G},i} \neq \emptyset$ */
2   construct $\mathcal{G}_i = (V, E_i)$ with $E_i = \cup_{v \in V}\{\text{first } 2^i \text{ edges in } \mathrm{In}_{\mathcal{G}}(v)\}$
3     $B_{\mathcal{G},i} = \{v \mid \mathrm{Indeg}_{\mathcal{G}}(v) > 2^i\}$
4     $S \leftarrow$ `TopSCCWithout`$(\mathcal{G}_i, B_{\mathcal{G},i})$
5   **if** $S \neq \emptyset$ **then**
6   **return** $(S, \emptyset)$

7   construct flow graph $F_{\mathcal{G},i}(r_{\mathcal{G},i})$ /* see Definition 3 */
8   **if** *exists vertex-dominator* v *in* $F_{\mathcal{G},i}(r_{\mathcal{G},i})$ **then**
9       $S \leftarrow$ `TopSCCWithout`$(\mathcal{G}_i \setminus \{v\}, B_{\mathcal{G},i})$
10   **return** $(S, \{v\})$

11   **else if** $|B_{\mathcal{G},i}| = 1$ *and* $\exists\, tSCC \subsetneq V \setminus \{r_{\mathcal{G},i}\}$ *in* $\mathcal{G}_i \setminus \{r_{\mathcal{G},i}\}$ **then**
12      $S \leftarrow$ `TopSCC`$(\mathcal{G}_i \setminus \{r_{\mathcal{G},i}\})$
13   **return** $(S, \{r_{\mathcal{G},i}\})$

14 **return** $(\emptyset, \emptyset)$

Let $\mathcal{G}_i \in \{G_i, Rev(G)_i\}$. The Procedure 2IsolatedSetLevel first searches for a tSCC in $\mathcal{G}_i$ that does not contain a vertex of $B_{\mathcal{G},i}$. If no such tSCC is found, the flow graph $F_{\mathcal{G},i}(r_{\mathcal{G},i})$ is constructed and searched for vertex-dominators. If a vertex-dominator v is found, a tSCC in $\mathcal{G}_i \setminus \{v\}$ that does not contain a vertex of $B_{\mathcal{G},i}$ is found; one can show that such a tSCC always exists. We additionally have to consider the special case when $|B_{\mathcal{G},i}| = 1$. In this case we have $B_{\mathcal{G},i} = \{r_{\mathcal{G},i}\}$ and we want to detect when there exists an almost tSCC $\mathcal{G}[T]$ induced by some set of vertices T with respect to $r_{\mathcal{G},i}$ in $\mathcal{G}_i$ such that $V \setminus (T \cup \{r_{\mathcal{G},i}\})$ is not empty. We use Procedure `TopSCCWithout`(H, B) to denote the search for a tSCC induced by vertices S in a graph H such that S does not contain a vertex of B. Such a tSCC can simply be found by marking tSCCs in a standard SCC algorithm. We let all procedures that search for an SCC return the set of vertices S in the SCC instead of the subgraph $G[S]$.

If no call to Procedure 2IsolatedSetLevel could identify an (almost) top or bottom SCC, we check in Procedure 2IsolatedSet whether the graph is strongly connected and either make progress by separating strongly connected components from each other or by finding an articulation point in the strongly connected graph. If an articulation point v is found, disjoint top and bottom SCCs exist after the removal of the articulation point v. Procedure 2IsolatedSet returns a top SCC in $G \setminus \{v\}$ in this case. If the graph G is strongly connected and does not contain

---

Procedure 2IsolatedSet(G)

---

1 $S \leftarrow \text{TopSCC}(G)$
2 **if** $S \subsetneq V$ **then**
3 $\quad \lfloor$ **return** $(S, \emptyset)$
4 **if** *exists articulation point v in G* **then**
5 $\quad \mid \quad S \leftarrow \text{TopSCC}(G \setminus \{v\})$
6 $\quad \lfloor$ **return** $(S, \{v\})$
7 **return** $(\emptyset, \emptyset)$

---

an articulation point, then G is a 2vSCC. In this case the Procedure 2IsolatedSet returns the empty set, the recursion stops, and 2vSCC(G) returns G.

Whenever the algorithm identifies an (almost) top or bottom SCC induced by a set of vertices S, it recursively calls itself on $G[S \cup Z]$ and $G[V \setminus S]$ for $Z = \emptyset$ or $Z = \{v\}$, respectively. We use Lemma 2 to show that in this case every 2vSCC of G is completely contained in either $G[S \cup Z]$ or $G[V \setminus S]$, which will imply the correctness of the algorithm.

Theorem 9 (Correctness). *Let G be a simple directed graph.* 2vSCC(G) *computes the 2vSCCs of G.*

By stopping the recursion when the number of vertices is a small constant and distinguishing between the number of vertices n' at the current level of the recursion and the total number of vertices n, we can show that the runtime of $O(n' \cdot \min\{|S|, |V \setminus S|\})$ without recursion leads to a total runtime of $O(n^2)$.

Theorem 10 (Runtime). *Procedure* 2vSCC *can be implemented in time $O(n^2)$.*

5   Extension to kSCCs

For any integral constant $k > 2$ the presented algorithm extends to computing the k-edge and the k-vertex strongly connected components. In this section we outline the necessary changes, see the full version for more details.

Let an *element* of a graph G denote an edge when keSCCs are searched for and a vertex when kvSCCs are searched for. We first extend the concepts of bridges, articulation points, and dominators from a single element to sets of elements with size less than k. A *separator w.r.t. k-connectivity (k-separator)* is a minimal set of elements such that the set contains less than k elements and its removal from the graph increases the number of SCCs in the graph. Two distinct vertices u and v are *k-(strongly-)connected* if they are strongly connected and they remain strongly connected after the removal of any less than k elements different from u and v from G. The *k-strongly connected components* (kSCCs) of a graph G are its maximal subgraphs $G[S]$ such that every pair of distinct vertices u and v in S is k-connected in $G[S]$.

In a flow graph $G(r)$ a *dominator Z w.r.t. k-connectivity (k-dominator)* is a *minimal* set of less than k elements in $G(r) \setminus \{r\}$ such that there exists a vertex $u \in G(r) \setminus (\{r\} \cup Z)$ such that u is reachable from r and every path from r to u

contains an element of Z. A k-dominator in a flow graph $G(r)$ and a k-separator in a graph G can for edge-connectivity be found in time $O(m \log n)$ [8] and for vertex-connectivity in time $O(mn)$ [7,9,11,18].

A set of vertices T induces an *almost tSCC w.r.t. k-connectivity* (*k-almost tSCC*) in G with respect to a set of elements Z with $|Z| < k$ if $G[T]$ is a tSCC in $G \setminus Z$ but has, for vertex-connectivity, incoming edges from *each of* the vertices in Z, or, for edge-connectivity, *all* the edges in Z as incoming edges in G.

We adapt our algorithm as follows. For edge-connectivity we use different flow graphs: (1) We contract all vertices in $B_{G,i}$ to a single vertex, while keeping all edges between the vertices in $B_{G,i}$ and the remaining vertices as parallel edges. (2) We take the new contracted vertex as the root of the flow graph. With these definitions it is rather straightforward to extend the algorithm to keSCCs.

The extension to $k > 2$ is more complicated for vertex-connectivity. In particular, we have to deal with the case $0 < |B_{G,i}| < k$. Note that in this case we cannot use an additional vertex that we connect to the vertices of $B_{G,i}$ as the root in the flow graph because the vertices of $B_{G,i}$ would be a k-dominator in this flow graph independent of the underlying graph G. To be able to identify a set $Z \cap B_{G,i} \neq \emptyset$ with $|Z| < k$ for which a k-almost tSCC exists in G, we use $|B_{G,i}| < k$ different flow graphs. If the search in the $|B_{G,i}|$ flow graphs is not successful, we additionally search for a $(k - |B_{G,i}|)$-separator in $G_i \setminus B_{G,i}$. These changes give the following result.

Theorem 11. *For any integral constant $k > 2$ keSCCs can be computed in time $O(n^2 \log n)$ and kvSCCs in time $O(n^3)$. 2eSCCs can be computed in time $O(n^2)$.*

Acknowledgements. We would like to thank Giuseppe Italiano for suggesting the problem and Slobodan Mitrović for helpful discussions. V. L. would like to thank Christian Tschabuschnig for his help in improving the readability of the algorithms. This work was supported by the Austrian Science Fund (FWF): P23499-N23. Additionally, the research leading to these results has received funding from the European Research Council under the European Union's Seventh Framework Programme (FP/2007-2013) / ERC Grant Agreement no. 340506.

References

1. Bang-Jensen, J., Gutin, G.: Digraphs: Theory, algorithms and applications. Springer Monographs in Mathematics, 2nd edn. Springer, London (2009)
2. Bondy, J.A., Murty, U.S.R.: Graph Theory with Applications. Macmillan, London (1976)
3. Buchsbaum, A.L., Georgiadis, L., Kaplan, H., Rogers, A., Tarjan, R.E., Westbrook, J.R.: Linear-time algorithms for dominators and other path-evaluation problems. SIAM J. Comput. **38**(4), 1533–1573 (2008)
4. Chatterjee, K., Henzinger, M.: Efficient and Dynamic Algorithms for Alternating Büchi Games and Maximal End-component Decomposition. J. ACM **61**(3), 15:1–15:40 (2014). Announced at SODA 2011 and SODA 2012
5. Chatterjee, K., Henzinger, M., Loitzenbauer, V.: Improved algorithms for one-pair and k-pair Streett objectives. In: LICS (2015, to appear)

6. Chatterjee, K., Jurdziński, M., Henzinger, T.A.: Simple stochastic parity games. In: Baaz, M., Makowsky, J.A. (eds.) CSL 2003. LNCS, vol. 2803, pp. 100–113. Springer, Heidelberg (2003)
7. Even, S.: An Algorithm for Determining Whether the Connectivity of a Graph is at Least k. SIAM J. Comput. **4**(3), 393–396 (1975)
8. Gabow, H.N.: A matroid approach to finding edge connectivity and packing arborescences. J. Comput. Syst. Sci. **50**(2), 259–273 (1995). Announced at STOC 1991
9. Gabow, H.N.: Using expander graphs to find vertex connectivity. J. ACM **53**(5), 800–844 (2006). Announced at FOCS 2000
10. Gabow, H.N., Tarjan, R.E.: A linear-time algorithm for a special case of disjoint set union. In: STOC, pp. 246–251 (1983)
11. Galil, Z.: Finding the vertex connectivity of graphs. SIAM J. Comput. **9**(1), 197–199 (1980)
12. Galil, Z., Italiano, G.F.: Reducing edge connectivity to vertex connectivity. ACM SIGACT News **22**(1), 57–61 (1991)
13. Georgiadis, L.: Testing 2-vertex connectivity and computing pairs of vertex-disjoint s-t paths in digraphs. In: Abramsky, S., Gavoille, C., Kirchner, C., Meyer auf der Heide, F., Spirakis, P.G. (eds.) ICALP 2010. LNCS, vol. 6198, pp. 738–749. Springer, Heidelberg (2010)
14. Georgiadis, L., Italiano, G.F., Laura, L., Parotsidis, N.: 2-Edge connectivity in directed graphs. In: SODA, pp. 1988–2005 (2015)
15. Georgiadis, L., Italiano, G.F., Laura, L., Parotsidis, N.: 2-Vertex connectivity in directed graphs. In: ICALP (2015). arXiv:1409.6277 (to appear)
16. Georgiadis, L., Tarjan, R.E.: Finding dominators revisited. In: SODA, pp. 862–871 (2004)
17. Henzinger, M., King, V., Warnow, T.: Constructing a Tree from Homeomorphic Subtrees, with Applications to Computational Evolutionary Biology. Algorithmica **24**(1), 1–13 (1999). Announced at SODA 1996
18. Henzinger, M.R., Rao, S., Gabow, H.N.: Computing Vertex Connectivity: New Bounds from Old Techniques. Journal of Algorithms **34**(2), 222–250 (2000). Announced at FOCS 1996
19. Hopcroft, J.E., Tarjan, R.E.: Dividing a graph into triconnected components. SIAM J. Comput. **2**(3), 135–158 (1973)
20. Italiano, G.F., Laura, L., Santaroni, F.: Finding strong bridges and strong articulation points in linear time. Theor. Comput. Sci. **447**, 74–84 (2012)
21. Jaberi, R.: On computing the 2-vertex-connected components of directed graphs, January 2014. arXiv:1401.6000v1
22. Makino, S.: An algorithm for finding all the k-components of a digraph. International Journal of Computer Mathematics **24**(3–4), 213–221 (1988)
23. Nagamochi, H., Ibaraki, T.: A linear-time algorithm for finding a sparse k-connected spanning subgraph of a k-connected graph. Algorithmica, 583–596 (1992)
24. Nagamochi, H., Ibaraki, T.: Algorithmic aspects of graph connectivity. Cambridge University Press, New York (2008)
25. Nagamochi, H., Watanabe, T.: Computing k-edge-connected components of a multigraph. IEICE TRANSACTIONS on Fundamentals of Electronics, Communications and Computer Sciences **E76−A**(4), 513–517 (1993)
26. Newman, M.E.J.: Networks: An Introduction. Oxford University Press (2010)
27. Tarjan, R.E.: Depth-first search and linear graph algorithms. SIAM J. Comput. **1**(2), 146–160 (1972)
28. Tarjan, R.E.: Edge-disjoint spanning trees and depth-first search. Acta Inf. **6**(2), 171–185 (1976)

Improved Algorithms for Decremental Single-Source Reachability on Directed Graphs

Monika Henzinger[1], Sebastian Krinninger[1]([✉]), and Danupon Nanongkai[2]

[1] Faculty of Computer Science, University of Vienna, Vienna, Austria
sebastian.krinninger@univie.ac.at
[2] KTH Royal Institute of Technology, Stockholm, Sweden

Abstract. Recently we presented the first algorithm for maintaining the set of nodes reachable from a source node in a directed graph that is modified by edge deletions with $o(mn)$ total update time, where m is the number of edges and n is the number of nodes in the graph [Henzinger et al. STOC 2014]. The algorithm is a combination of several different algorithms, each for a different m vs. n trade-off. For the case of $m = \Theta(n^{1.5})$ the running time is $O(n^{2.47})$, just barely below $mn = \Theta(n^{2.5})$. In this paper we simplify the previous algorithm using new algorithmic ideas and achieve an improved running time of $\tilde{O}(\min(m^{7/6}n^{2/3}, m^{3/4}n^{5/4+o(1)}, m^{2/3}n^{4/3+o(1)} + m^{3/7}n^{12/7+o(1)}))$. This gives, e.g., $O(n^{2.36})$ for the notorious case $m = \Theta(n^{1.5})$. We obtain the same upper bounds for the problem of maintaining the strongly connected components of a directed graph undergoing edge deletions. Our algorithms are correct with high probabililty against an oblivious adversary.

Keywords: Dynamic graph algorithms · Reachability

1 Introduction

In this paper we study the decremental reachability problem. Given a directed graph G with n nodes and m edges and a source node s in G a *decremental single-source reachability algorithm* maintains the set of nodes reachable from s (i.e., all nodes v for which there is a path from s to v in the current version of G) during a sequence of edge deletions. The goal is to minimize the *total update time*, i.e., the total time needed to process *all* deletions such that reachability queries can be answered in constant time. A *decremental s-t reachability algorithm* is given a graph G undergoing edge deletions, a source node s, and a sink node t and it determines after every deletion in G whether s can still reach t.

A full version combining the findings of this paper and its predecessor [4] is available at http://arxiv.org/abs/1504.07959.

M. Henzinger and S. Krinninger—Supported by the Austrian Science Fund (FWF): P23499-N23 and the University of Vienna (IK I049-N). The research leading to these results has received funding from the European Research Council under the European Union's Seventh Framework Programme (FP/2007-2013) / ERC Grant Agreement no. 340506.

D. Nanongkai—Work partially done while at University of Vienna, Faculty of Computer Science, Austria.

© Springer-Verlag Berlin Heidelberg 2015
M.M. Halldórsson et al. (Eds.): ICALP 2015, Part I, LNCS 9134, pp. 725–736, 2015.
DOI: 10.1007/978-3-662-47672-7_59

Related Work. The incremental version of the single-source reachability problem, in which edges are *inserted* into the graph, can be solved with a total update time of $O(m)$ by performing an incremental graph search, where m is the final number of edges. Italiano [6] showed that in directed acyclic graphs the decremental problem can be solved in time $O(m)$ as well. In general directed graphs however, the problem could for a long time only be solved in time $O(mn)$ using the more general decremental single-source shortest paths algorithm of Even and Shiloach [2,3,7], which maintains a breadth-first search tree rooted at s, called *ES-tree*. This upper bound of $O(mn)$ is also achieved for the seemingly more complex decremental *all-pairs* reachability problem (also known as transitive closure) [9,13]. In the fully dynamic version of single-source reachability both insertions and deletions of edges are possible. The matrix-multiplication based transitive closure algorithms of Sankowski [10] give fully dynamic algorithms for single-source reachability and *s-t* reachability with worst-case running times of $O(n^{1.575})$ and $O(n^{1.495})$ *per update*, respectively.

These upper bounds have recently been complemented by Abboud and Vassilevska Williams [1] as follows. For the decremental *s-t* reachability problem, a combinatorial algorithm with a *worst-case* running time of $O(n^{2-\delta})$ (for some $\delta > 0$) per update or query implies a faster combinatorial algorithm for Boolean matrix multiplication and, as has been shown by Vassilevska Williams and Williams [12], for other problems as well. (For non-combinatorial algorithms, Henzinger et al. [5] showed that there is no algorithm with worst-case $O(n^{1-\delta})$ update and $O(n^{2-\delta})$ query time, assuming the so-called Online Matrix-Vector Multiplication conjecture.) Furthermore, for the problem of maintaining the number of nodes reachable from a source under deletions (which our algorithms can do) a worst-case running time of $O(m^{1-\delta})$ (for some $\delta > 0$) per update or query falsifies the strong exponential time hypothesis. Thus, amortization is indeed necessary to bypass these bounds.

In [4] we recently improved upon the long-standing upper bound of $O(mn)$ for decremental single-source reachability in directed graphs. In particular, we developed several algorithms whose combined expected running time is polynomially faster than $O(mn)$ for all values of m (i.e., for all possible densities of the initial graph). By a reduction from single-source reachability, our results in [4] immediately give an $o(mn)$ algorithm for maintaining strongly connected components under edge deletions. Previously, the fastest decremental algorithms for this problem had a total update time of $O(mn)$ as well [8,9,13].

Our Results. In this paper we improve upon the upper bounds provided in [4]. Furthermore, the running times achieved in this paper are arguably more natural than those in [4]. Although we previously broke the $O(mn)$ barrier for all values of m, we barely did so, giving a bound of $O(n^{2.47})$, when $m = \Theta(n^{1.5})$. In this paper we also get a better improvement, namely $O(n^{2.36})$ in this notorious case. In general, we can combine the algorithms of this paper to obtain a running time of $O(mn^{0.9+o(1)})$, whereas in [4] we obtained $\tilde{O}(mn^{0.984})$.

In [4] the starting point was to solve the decremental *s-t* reachability problem, which is also the case here. For this problem we obtain two algorithms with total update times of $\tilde{O}(\min(m^{5/4}n^{1/2}, m^{2/3}n^{4/3+o(1)}))$ and $O(m^{2/3}n^{4/3+o(1)} + m^{3/7}n^{12/7+o(1)})$, respectively. Just as in [4], extensions of these algorithms solve the decremental single-source reachability problem with total update times of $\tilde{O}(\min(m^{7/6}n^{2/3}, m^{3/4}n^{5/4+o(1)}))$ and $O(m^{2/3}n^{4/3+o(1)}+m^{3/7}n^{12/7+o(1)})$, respectively. Furthermore, it follows from a reduction [4,9] that there are algorithms for the decremental strongly connected components problem whose running times are the same up to a logarithmic factor. We compare these new results to the ones of [4] in Figure 1. All our algorithms are correct with high probability who fixes its sequence of updates and queries before the algorithm is initialized and their running time bounds hold in expectation. Due to space constraints this paper only contains an overview of the algorithm that has a total update time of $O(m^{2/3}n^{4/3+o(1)} + m^{3/7}n^{12/7+o(1)})$ and is thus the current fastest for dense graphs. The other algorithm and all omitted proofs can be found in the full version of this paper.

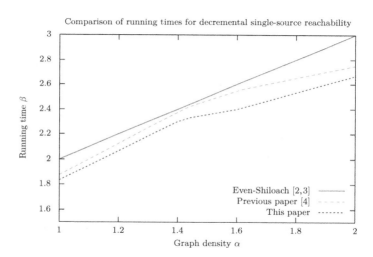

Fig. 1. Running times of decremental single-source reachability algorithms dependent on the density of the initial graph. A point (α, β) in this diagram means that for a graph with $m = \Theta(n^{\alpha})$ the algorithm has a running time of $O(n^{\beta+o(1)})$.

Techniques. There are two novel technical contributions: (1) The algorithm of [4] uses two kinds of randomly selected nodes, called *hubs* and *centers*, each fulfilling a different purpose. Maintaining an ES-tree for each hub up to depth h, it quickly tests for every pair of centers (x, y) whether there is a path of length at most $2h$ from x to y going through a hub. If there is no such path, we build a special graph, called *path union graph*, for the pair (x, y) that contains all paths of length $O(h)$ from x to y. Since there no longer is a path from x to y through a hub of length at most h, we know that their path union graph s "smaller" than the original graph. In this paper we show how to extend this approach multiple layers of path unions graphs. Hubs and centers of the previous algorithms become level k, resp. $k-1$ centers in the new approach. Level $k-1$ centers serve as hubs for the level $k-2$ centers, and more generally level i centers serve as hubs for level $i-1$ centers. To do this efficiently we build the ES-tree for a level i center x *inside* the path-union graph of x and another, potentially higher-level center. The fact that we use the smaller path-union graph instead of the original graph for these ES-trees (together with an improved data structure for computing path-union graphs, see (2) below) gives the improvement in the running time.

(2) In [4] we maintain for each center x an approximate path union data structure that computes a superset of the path union of x and any other center y. This superset is an approximation of the path union graph for (x, y) as it might contain paths between the two centers of length $O(h \log n)$ (and *not* as desired $O(h)$), but no longer. The total time spent in this data structure for x is (a) the size of the constructed path union graph and (b) a one-time "global charge" for using this data structure of $O(n^2)$. It is based on a hierarchical graph decomposition technique. Here we present a much simpler data structure that also constructs an approximate path union graph, but that does not require any hierarchical graph decomposition. This reduces the global charge per center from $O(n^2)$ to $O(m)$. We believe that this data structure is of independent interest.

Outline. In Section 2 we give the preliminaries. In Section 3 we present our new path union data structure. Finally, in Section 4 we show how to combine this idea with the multi-layer path union approach to obtain a faster decremental single-source reachability algorithm for dense graphs.

2   Preliminaries

In this section we review some notions and basic facts that we will use in the rest of this paper. We use the following notation: We consider a directed graph $G = (V, E)$ undergoing edge deletions, where V is the set of nodes of G and E is the set of edges of G. We denote by n the number of nodes of G and by m the number of edges of G *before the first edge deletion*. For every pair of nodes u and v we denote the distance from u to v in G by $d_G(u, v)$. For every subset of nodes $U \subseteq V$, we define $E(U) = E \cap U^2$ and denote by $G[U] = (U, E[U])$ the *subgraph of G induced by U*. For sets of nodes $U \subseteq V$ and $U' \subseteq V$ we define $E(U, U') = E \cap (U \times U')$, i.e., $E(U, U')$ is the set of edges $(u, v) \in E$ such that $u \in U$ and $v \in U'$. We write $\hat{O}(T(m, n))$ as an abbreviation for $O(T(m, n) \cdot n^{o(1)})$.

Like many decremental shortest paths and reachability algorithms, our algorithms internally use a data structure for maintaining a shortest paths tree up to a relatively small depth.

Theorem 1 (Even-Shiloach tree [2,3,7]). *There is a decremental algorithm, called* Even-Shiloach tree *(short: ES-tree), that, given a directed graph G undergoing edge deletions, a source node s, and a parameter $h \geq 1$, maintains a* *shortest paths tree from s and the corresponding distances up to depth h with total update time $O(mh)$, i.e., the algorithm maintains $d_G(s, v)$ and the parent of v in the shortest paths tree for every node v such that $d_G(s, v) \leq h$. By reversing the edges of G it can also maintain the distance from v to s for every node v in the same time.*

The central concept in the algorithmic framework introduced in [4] is the notion of the path union of a pair of nodes.

Definition 1. *For every directed graph G, every $h \geq 1$, and all pairs of nodes x and y of G, the path union $\mathcal{P}(x, y, h, G) \subseteq V$ is the set containing all nodes that lie on some path π from x to y in G of weight at most h.*

The path union has a simple characterization and can be computed efficiently.

Lemma 1 ([4]). *For every directed graph G, every $h \geq 1$ and all pairs of nodes x and y of G, we have $\mathcal{P}(x, y, h, G) = \{v \in V \mid d_G(x, v) + d_G(v, y) \leq h\}$. We can compute this set in time $O(m)$.*

Our algorithms use randomization in the following way: by sampling a set of nodes with a sufficiently large probability we can guarantee that certain sets of nodes contain at least one of the sampled nodes with high probability. To the best of our knowledge, the first use of this technique in graph algorithms goes back to Ullman and Yannakakis [11].

Lemma 2. *Let T be a set of size t and let $S_1, S_2, \ldots, S_k$ be subsets of T of size at least q. Let U be a subset of T that was obtained by choosing each element of T independently with probability $p = (a \ln(kt))/q$, for some parameter a. Then, for every $1 \leq i \leq k$, the set S_i contains a node of U with high probability (whp), i.e., probability at least $1 - 1/t^a$, and the size of U is $O((t \log(kt))/q)$ in expectation.*

3   Approximate Path Union Data Structure

In this section we present a data structure for a graph G undergoing edge deletions, a fixed node x, and a parameter h. Given a node y, it computes an "approximation" of the path union $\mathcal{P}(x, y, h, G)$. Using a simple static algorithm the path union can be computed in time $O(m)$ for each pair (x, y). We give an (almost) output-sensitive data structure for this problem, i.e., using our data structure the time will be proportional to the size of the approximate path union which might be $o(m)$. Additionally, we have to pay a global cost of $O(m)$ that is amortized over *all* approximate path union computations for the node x and *all* nodes y. This will be useful because in our reachability algorithm we can use probabilistic arguments to bound the size of the approximate path unions.

Proposition 1. *There is a data structure that, given a graph G undergoing edge deletions, a fixed node x, and a parameter h, provides a procedure APPROX-IMATEPATHUNION such that, given sequence of nodes $y_1, \ldots, y_k$, this procedure computes sets $F_1, \ldots F_k$ guaranteeing $\mathcal{P}(x, y, h, G) \subseteq F_i \subseteq \mathcal{P}(x, y, (\log m + 3)h, G)$ for all $1 \leq i \leq k$. The total running time is $O(\sum_{1 \leq i \leq k} |F_i| + m)$.*

3.1   Algorithm Description

Internally, the data structure maintains a set $R(x)$ of nodes, initialized with $R(x) = V$, such that the following invariant is fulfilled at any time: all nodes that can be reached from x by a path of length at most h are contained in $R(x)$ but $R(x)$ might contain other nodes as well). Observe that thus $R(x)$ contains the path union $\mathcal{P}(x, y, h, G)$ for every node y.

To gain some intuition for our approach consider the following way of computing an approximation of the path union $\mathcal{P}(x, y, h, G)$ for some node y. First, compute $B_1 = \{v \in R(x) \mid d_{G[R(x)]}(v, y) \le h\}$ using a backward breadth-first search (BFS) to y in $G[R(x)]$, the subgraph of G induced by $R(x)$. Second, compute $F = \{v \in R(x) \mid d_{G[B_1]}(x, v) \le h\}$ using a forward BFS from x in $G[B_1]$. It can be shown that $\mathcal{P}(x, y, h, G) \subseteq F \subseteq \mathcal{P}(x, y, 2h, G)$.[1] Given B_1, we could charge the time for computing F to the set F itself, but we do not know how to pay for computing B_1 as $B_1 \setminus F$ might be much larger than F.

Our idea is to additionally identify a set of nodes $X \subseteq \{v \in V \mid d_G(x, v) > h\}$ and remove it from $R(x)$. Consider a second approach where we first compute B_1 as above and then compute $B_2 = \{v \in R(x) \mid d_{G[R(x)]}(v, y) \le 2h\}$ and $F = \{v \in R(x) \mid d_{G[B_2]}(x, v) \le h\}$. It can be shown that $\mathcal{P}(x, y, h, G) \subseteq F \subseteq \mathcal{P}(x, y, 3h, G)$. Additionally, all nodes in $X = B_1 \setminus F$ are at distance more than h from x and therefore we can remove X from $R(x)$. Thus, we can charge the work for computing B_1 and F to X and F, respectively.[2] However, we now have a similar problem as before as we do not know whom to charge for computing B_2.

We resolve this issue by simply computing $B_i = \{v \in R(x) \mid d_{G[R(x)]}(v, y) \le ih\}$ for increasing values of i until we arrive at some i^* such that the size of B_{i^*} is at most double the size of B_{i^*-1}. We then return $F = \{v \in R(x) \mid d_{G[B_i]}(x, v) \le h\}$ and charge the time for computing B_i to $X = B_{i-1} \setminus F$ and F, respectively. As the size of B_i can double at most $O(\log n)$ times we have $\mathcal{P}(x, y, h, G) \subseteq F \subseteq \mathcal{P}(x, y, O(h \log n), G)$, as we show below. Procedure 1 shows the pseudocode of this algorithm. Note that in the special case that x cannot reach y the algorithm returns the empty set. In the analysis below, let i^* denotes the final value of i before Procedure 1 terminates.

3.2   Correctness

We first prove Invariant (I): the set $R(x)$ always contains all nodes that are at distance at most h from x in G. This is true initially as we initialize $R(x)$ to be V and we now show that it continues to hold because we only remove nodes at distance more than h from x.

Lemma 3. *If $R(x) \subseteq \{v \in V \mid d_G(x, v) \le h\}$, then for every node $v \in X$ removed from $R(x)$, we have $d_G(x, v) > h$.*

---

[1] Indeed, F might contain some node v with $d_G(x, v) = h$ and $d_G(v, y) = h$, but it will not contain any node w with either $d_G(x, w) > h$ or $d_G(w, y) > h$.

[2] Note that in our first approach removing $B_1 \setminus F$ would not have been correct as F was computed w.r.t to $G[B_1]$ and not w.r.t. $G[B_2]$.

Procedure 1: APPROXIMATEPATHUNION(y)

// All calls of APPROXIMATEPATHUNION(y) use fixed x and h.

1 Compute $B_1 = \{v \in R(x) \mid d_{G[R(x)]}(v, y) \leq h\}$ // backward BFS to y in subgraph induced by $R(x)$

2 **for** $i = 2$ **to** $\lceil \log m \rceil + 1$ **do**

3   Compute $B_i = \{v \in R(x) \mid d_{G[R(x)]}(v, y) \leq ih\}$ // backward BFS to y in subgraph induced by $R(x)$

4   **if** $|E(B_i)| \leq 2|E(B_{i-1})|$ **then**

5   Compute $F = \{v \in B_i \mid d_{G[B_i]}(x, v) \leq h\}$ // forward BFS from x in subgraph induced by B_i

6          $X \leftarrow B_{i-1} \setminus F$, $R(x) \leftarrow R(x) \setminus X$

7   **return** F

Proof. Let $v \in X = B_{i^*-1} \setminus F$ and assume by contradiction that $d_G(x, v) \leq h$. Since $v \in B_{i^*-1}$ we have $d_{G[R(x)]}(v, y) \leq (i^* - 1)h$. Now consider the shortest path π from x to v in G, which has length at most h. By the assumption, every node on π is contained in $G[R(x)]$. Therefore, for every node v' on π, we have $d_{G[R(x)]}(v', v) \leq h$ and thus

$$d_{G[R(x)]}(v', y) \leq d_{G[R(x)]}(v', v) + d_{G[R(x)]}(v, y) \leq h + (i^* - 1)h \leq i^* h$$

which implies that $v' \in B_{i^*}$. Thus, every node on π is contained in B_{i^*}. As π is a path from x to v of length at most h it follows that $d_{G[B_{i^*}]}(x, v) \leq h$. Therefore $v \in F$, which contradicts the assumption $v \in X$. □

We now complete the correctness proof by showing that the set of nodes returned by the algorithm approximates the path union.

Lemma 4. *Procedure 1 returns a set of nodes F such that $\mathcal{P}(x, y, h, G) \subseteq F \subseteq \mathcal{P}(x, y, (\log m + 3)h, G)$.*

Proof. We first argue that the algorithm actually returns some set of nodes F. Note that in Line 1 of the algorithm we always have $|E(B_i)| \geq |E(B_{i-1})|$ as $B_{i-1} \subseteq B_i$. As $E(B_i)$ is a set of edges and the total number of edges is at most m, the condition $|E(B_i)| \leq |E(B_{i-1})|$ therefore must eventually be fulfilled for some $2 \leq i \leq \lceil \log m \rceil + 1$.

We now show that $\mathcal{P}(x, y, h, G) \subseteq F$. Let $v \in \mathcal{P}(x, y, h, G)$, which implies that v lies on a path π from x to y of length at most h. For every node v' on π we have $d_G(x, v') \leq h$, which by Invariant (I) implies $v' \in R(x)$. Thus, the whole path π is contained in $G[R(x)]$. Therefore $d_{G[R(x)]}(v', y) \leq h$ for every node v' on π which implies that π is contained in $G[B_{i^*}]$. Then clearly we also have $d_{G[B_{i^*}]}(x, v) \leq h$ which implies $v \in F$.

Finally we show that $F \subseteq \mathcal{P}(x, y, (\log m + 3)h, G)$ by proving that $d_G(x, v) + d_G(v, y) \leq (\log m + 3)h$ for every node $v \in F$. As $G[B_{i^*}]$ is a subgraph of G, we have $d_G(x, v) \leq d_{G[B_{i^*}]}(x, v)$ and $d_G(v, y) \leq d_{G[B_{i^*}]}(v, y)$. By the definition of F we have $d_{G[B_{i^*}]}(x, v) \leq h$. As $F \subseteq B_{i^*}$ we also have $d_{G[B_{i^*}]}(v, y) \leq$

$^*h \leq (\lceil \log m \rceil + 1)h \leq (\log m + 2)h$. It follows that $d_G(x,v) + d_G(v,y) \leq h + (\log m + 2)h = (\log m + 3)h$. $\square$

3.3 Running Time Analysis

To bound the total running time we prove that each call of Procedure 1 takes time proportional to the number of edges in the returned approximation of the path union plus the number of edges incident to the nodes removed from $R(x)$. As each node is removed from $R(x)$ at most once, the time spent on all calls of Procedure 1 is then $O(m)$ plus the sizes of the subgraphs induced by the approximate path unions returned in each call.

Lemma 5. *The running time of Procedure 1 is $O(|E(F)|+|E(X,R(x))|+|E(R(x), X)|)$ where F is the set of nodes returned by the algorithm, and X is the set of nodes the algorithm removes from $R(x)$.*

Proof. The running time in iteration $2 \leq j \leq i^* - 1$ is $O(|E(B_j)|)$ as this is the cost of the breadth-first-search performed to compute B_j. In the last iteration i^*, the algorithm additionally has to compute F and X and remove X from $R(x)$. As F is computed by a BFS in $G[B_{i^*}]$ and $X \subseteq B_{i^*-1} \subseteq B_{i^*}$, these steps take time $O(|E(B_{i^*})|)$. Thus the total running time is $O(\sum_{1 \leq j \leq i^*} |E(B_j)|)$.

By checking the size bound in Line 4 of Procedure 1 we have $|E(B_j)| > 2|E(B_{j-1})|$ for all $1 \leq j \leq i^* - 1$ and $|E(B_{i^*})| \leq 2|E(B_{i^*-1})|$. By repeatedly applying the first inequality it follows that $\sum_{1 \leq j \leq i^*-1} |E(B_j)| \leq 2|E(B_{i^*-1})|$. Therefore we get

$$\sum_{1 \leq j \leq i^*} |E(B_j)| = \sum_{1 \leq j \leq i^*-1} |E(B_j)| + |E(B_{i^*})|$$
$$\leq 2|E(B_{i^*-1})| + 2|E(B_{i^*-1})| = 4|E(B_{i^*-1})|$$

and thus the running time is $O(|E(B_{i^*-1})|)$. Now observe that by $X = B_{i^*-1} \setminus F$ we have $B_{i^*-1} \subseteq X \cup F$ and thus

$$E(B_{i^*-1}) \subseteq E(F) \cup E(X) \cup E(X,F) \cup E(F,X)$$
$$\subseteq E(F) \cup E(X, R(x)) \cup E(R(x), X).$$

Therefore the running time is $O(|E(F)| + |E(X, R(x))| + |E(R(x), X)|)$. $\square$

4 Reachability via Center Graph

We now show how to combine the approximate path union data structure with a hierarchical approach to get an improved decremental reachability algorithm for dense graphs. The algorithm has a parameter $1 \leq k \leq \log n$ and for each $1 \leq i \leq k$ a parameter $c_i \leq n$. We determine suitable choices of these parameters in Section 4.2. For each $1 \leq i \leq k - 1$, our choice will satisfy $c_i \geq c_{i+1}$ and $c_i = \hat{O}(c_{i+1})$. Furthermore, we set $h_i = (3 + \log m)^{i-1} n/c_1$ for $1 \leq i \leq k$. At the initialization, the algorithm determines sets of nodes $C_1 \supseteq C_2 \supseteq \cdots \supseteq C_k$ such that $s, t \in C_1$ as follows. For each $1 \leq i \leq k$, we sample each node of the graph with probability $a c_i \ln n / n$ (for a large enough constant a), where the value of c_i will be determined later. The set C_i then consists of the sampled nodes, and if $i \leq k - 1$, it additionally contains the nodes in C_{i+1}. For every $1 \leq i \leq k$ we call the nodes in C_i i-centers. In the following we describe an algorithm for maintaining pairwise reachability between all 1-centers.

4.1   Algorithm Description

Data Structures. The algorithm uses the following data structures:

- For every i-center x and every $i \leq j \leq k$ an approximate path union data structure (see Proposition 1) with parameter h_j.
- For every k-center x an incoming and an outgoing ES-tree of depth h_k in G.
- For every pair of an i-center x and a j-center y such that $l := \max(i, j) \leq k - 1$, a set of nodes $Q(x, y, l) \subseteq V$. Initially, $Q(x, y, l)$ is empty and at some point the algorithm might compute $Q(x, y, l)$ using the approximate path union data structure of x.
- For every pair of an i-center x and a j-center y such that $l := \max(i, j) \leq k-1$ an ES-tree of depth h_l from x in $Q(x, y, l)$.
- For every pair of an i-center x and a j-center y such that $l := \max(i, j) \leq k-1$ a set of $(l + 1)$-centers certifying that x can reach y.

Certified Reachability Between Centers (Links). The algorithm maintains the following limited path information between centers, called *links*, in a top-down fashion. Let x be a k-center and let y be an i-center for some $1 \leq i \leq k - 1$. The algorithm links x to y if and only if y is contained in the outgoing ES-tree of depth h_k of x. Similarly the algorithm links y to x if and only if y is contained in the incoming ES-tree of depth h_k of x. Let x be an i-center and let y be a j-center such that $l := \max(i, j) \leq k - 1$. If there is an $(l + 1)$-center z such that x is linked to z and z is linked to y, the algorithm links x to y (we also say that z links x to y). Otherwise, the algorithm computes $Q(x, y, l)$ using the approximate path union data structure of x and starts to maintain an ES-tree from x up to depth h_l in $G[Q(x, y, l)]$. It links x to y if and only if y is contained in the ES-tree of x. Using a list of centers z certifying that x can reach y, maintaining the links between centers is straightforward.

Center Graph. The algorithm maintains a graph called *center graph*. Its nodes are the 1-centers and it contains the edge (x, y) if and only if x is linked to y. The algorithm maintains the transitive closure of the center graph. A query asking whether a center y is reachable from a center x in G is answered by checking the reachability in center graph. As s and t are 1-centers this answers s-t reachability queries.

Correctness. For the algorithm to be correct we have to show that there is a path from s to t in the center graph if and only if there is a path from s to t in G. We can in fact show more generally that this is the case for any pair of 1-centers.

Lemma 6. *For every pair of 1-centers x and y, there is a path from x to y in the center graph if and only if there is a path from x to y in G.*

4.2   Running Time Analysis

The key to the efficiency of the algorithm is to bound the size of the graphs $Q(x, y, l)$.

Lemma 7. *Let x be an i-center and let y be a j-center such that $l := \max(i, j) \leq k - 1$. If x is not linked to y by an $(l+1)$-center, then $Q(x, y, l)$ contains at most n/c_{l+1} nodes with high probability.*

With the help of this lemma we first analyze the running time of each part of the algorithm and argue that our choice of parameters gives the desired total update time.

Parameter Choice. We carry out the running time analysis with regard to two parameters $1 \leq b \leq c \leq n$ which we will set at the end of the analysis. We set $k = \lceil (\log (c/b))/(\sqrt{\log n \cdot \log \log n}) \rceil + 1$, $c_k = b$ and $c_i = 2^{\sqrt{\log n \cdot \log \log n}} c_{i+1} = \hat{O}(c_{i+1})$ for $1 \leq i \leq k - 1$. Note that the number of i-centers is $\tilde{O}(c_i)$ in expectation. Observe that

$$(3 + \log m)^{k-1} = O((\log n)^k) \leq O((\log n)^{\sqrt{\log n/ \log \log n}})$$
$$= O(2^{\sqrt{\log n \cdot \log \log n}}) = O(n^{\sqrt{\log \log n/ \log n}}) = O(n^{o(1)}).$$

Furthermore we have

$$c_1 = \left(2^{\sqrt{\log n \cdot \log \log n}}\right)^{k-1} c_k \geq 2^{\log (c/b)} b = \frac{c}{b} \cdot b = c$$

and by setting $k' = (\log (c/b))/(\sqrt{\log n \cdot \log \log n})$ we have $k \leq k' + 2$ and thus

$$c_1 = \left(2^{\sqrt{\log n \cdot \log \log n}}\right)^{k-1} c_k \leq \left(2^{\sqrt{\log n \cdot \log \log n}}\right)^{k'+1} c_k = 2^{\sqrt{\log n \cdot \log \log n}} c = \hat{O}(c).$$

Remember that $h_i = (3 + \log m)^{i-1} n/c_1$ for $1 \leq i \leq k$. Therefore we have $h_i = \hat{O}(n/c_1) = \hat{O}(n/c)$.

Maintaining ES-Trees. For every k-center we maintain an incoming and an outgoing ES-tree of depth h_k, which takes time $O(mh_k)$. As there are $\tilde{O}(c_k)$ k-centers, maintaining all these trees takes time $\tilde{O}(c_k m h_k) = \hat{O}(bmn/c)$.

For every i-center x and every j-center y such that $l := \max(i, j) \leq k - 1$, we maintain an ES-tree up to depth h_l in $G[Q(x, y, l)]$. By Lemma 7 $Q(x, y, l)$ has at most n/c_{l+1} nodes and thus $G[Q(x,y,l)]$ has at most n^2/c_{l+1}^2 edges. Maintaining this ES-tree therefore takes time $O((n^2/c_{l+1}^2) \cdot h_l) = \hat{O}(n^2/c_{l+1}^2(n/c_1)) = \hat{O}(n^3/(c_1 c_{l+1}^2))$. In total, maintaining all these trees takes time

$$\hat{O}\left(\sum_{1 \le i \le k-1} \sum_{1 \le j \le i} c_i c_j \frac{n^3}{c_1 c_{i+1}^2}\right) = \hat{O}\left(\sum_{1 \le i \le k-1} \sum_{1 \le j \le i} \frac{c_i c_1 n^3}{c_{i+1} c_1 c_k}\right)$$

$$= \hat{O}\left(\sum_{1 \le i \le k-1} \sum_{1 \le j \le i} \frac{n^3}{c_k}\right) = \hat{O}\left(k^2 \frac{n^3}{c_k}\right) = \hat{O}\left(\frac{n^3}{b}\right).$$

Computing Approximate Path Unions. For every i-center x and every $i \le j \le k$ we maintain an approximate path union data structure with parameter h_j. By Proposition 1 this data structures has a total running time of $O(m)$ and an additional cost of $O(|E(Q(x,y,j))|)$ each time the approximate path union $Q(x,y,j)$ is computed for some j-center y. By Lemma 7 the number of nodes of $Q(x,y,j)$ is n/c_{j+1} with high probability and thus its number of edges is n^2/c_{j+1}^2. Therefore, computing all approximate path unions takes time

$$\tilde{O}\left(\sum_{1 \le i \le k-1} \sum_{i \le j \le k} \left(c_i m + c_i c_j \frac{n^2}{c_{j+1}^2}\right)\right) = \tilde{O}\left(\sum_{1 \le i \le k-1} \sum_{i \le j \le k} \left(c_1 m + \frac{c_1 c_j n^2}{c_{j+1} c_k}\right)\right)$$

$$= \hat{O}\left(\sum_{1 \le i \le k-1} \sum_{i \le j \le k} \left(c_1 m + \frac{c_1 n^2}{c_k}\right)\right) = \hat{O}(k^2 c_1 m + k^2 c_1 n^2/c_k) = \hat{O}(cm + cn^2/b)$$

Maintaining Links Between Centers. For each pair of an i-center x and a j-center y there are at most $\tilde{O}(c_{l+1})$ $(l+1)$-centers that can possibly link x to y. Each such $(l+1)$-center is added to and removed from the list of $(l+1)$-centers linking x to y at most once. Thus, the total time needed for maintaining all these links is $\tilde{O}(\sum_{1 \le i \le k-1} \sum_{1 \le j \le i} c_i c_j c_{i+1}) = \tilde{O}(k^2 c_1^3) = \tilde{O}(c^3)$.

Maintaining Transitive Closure in Center Graph. The center graph has $\tilde{O}(c_1)$ nodes and thus $\tilde{O}(c_1^2)$ edges. During the algorithm edges are only deleted from the center graph and never inserted. Thus we can use known $O(mn)$-time decremental algorithms for maintaining the transitive closure [9,13] in the center graph in time $\tilde{O}(c_1^3) = \tilde{O}(c^3)$.

Total Running Time. Since the term cn^2/b is dominated by the term n^3/b, we obtain a total running time of $\hat{O}\left(bmn/c + n^3/b + cm + c^3\right)$. By setting $b = n^{5/3}/m^{2/3}$ and $c = n^{4/3}/m^{1/3}$ the running time is $\hat{O}(m^{2/3}n^{4/3} + n^4/m)$ and by setting $b = n^{9/7}/m^{3/7}$ and $c = m^{1/7}n^{4/7}$ the running time is $\hat{O}(m^{3/7}n^{12/7} + m^{8/7}n^{4/7})$.

4.3   Decremental Single-Source Reachability

The algorithm above works for a set of randomly chosen centers. Note that the algorithm stays correct if we add any number of nodes to C_1, thus increasing the number of 1-centers for which the algorithm maintains pairwise reachability. If the number of additional centers does not exceed the expected number of randomly chosen centers, then the same running time bounds still apply. Using the reductions of [4] this immediately implies decremental algorithms for maintaining single-source reachability and strongly connected components.

Theorem 2. *There are decremental algorithms for maintaining single-source reachability and strongly connected components with constant query time and expected total update time* $\hat{O}(m^{2/3}n^{4/3} + m^{3/7}n^{12/7})$ *that are correct with high probability against an oblivious adversary.*

References

1. Abboud, A., Vassilevska Williams, V.: Popular conjectures imply strong lower bounds for dynamic problems. In: Symposium on Foundations of Computer Science (FOCS), pp. 434–443 (2014)
2. Even, S., Shiloach, Y.: An on-line edge-deletion problem. Journal of the ACM **28**(1), 1–4 (1981)
3. Henzinger, M., King, V.: Fully dynamic biconnectivity and transitive closure. In: Symposium on Foundations of Computer Science (FOCS), pp. 664–672 (1995)
4. Henzinger, M., Krinninger, S., Nanongkai, D.: Sublinear-time decremental algorithms for single-source reachability and shortest paths on directed graphs. In: Symposium on Theory of Computing (STOC), pp. 674–683 (2014)
5. Henzinger, M., Krinninger, S., Nanongkai, D., Saranurak, T.: Unifying and strengthening hardness for dynamic problems via the online matrix-vector multiplication conjecture. In: Symposium on Theory of Computing (STOC) (2015)
6. Italiano, G.F.: Finding paths and deleting edges in directed acyclic graphs. Information Processing Letters **28**(1), 5–11 (1988)
7. King, V.: Fully dynamic algorithms for maintaining all-pairs shortest paths and transitive closure in digraphs. In: Symposium on Foundations of Computer Science (FOCS), pp. 81–91 (1999)
8. Roditty, L.: Decremental maintenance of strongly connected components. In: Symposium on Discrete Algorithms (SODA), pp. 1143–1150 (2013)
9. Roditty, L., Zwick, U.: Improved dynamic reachability algorithms for directed graphs. SIAM Journal on Computing **37**(5), 1455–1471 (2008). announced at FOCS 2002
10. Sankowski, P.: Dynamic transitive closure via dynamic matrix inverse. In: Symposium on Foundations of Computer Science (FOCS), pp. 509–517 (2004)
11. Ullman, J.D., Yannakakis, M.: High-probability parallel transitive-closure algorithms. SIAM Journal on Computing **20**(1), 100–125 (1991). announced at SPAA 1990
12. Vassilevska Williams, V., Williams, R.: Subcubic equivalences between path, matrix and triangle problems. In: Symposium on Foundations of Computer Science (FOCS), pp. 645–654 (2010)
13. Łącki, J.: Improved deterministic algorithms for decremental reachability and strongly connected components. ACM Transactions on Algorithms **9**(3), 27 (2013). announced at SODA 2011

Weighted Reordering Buffer Improved via Variants of Knapsack Covering Inequalities

Sungjin Im[1] and Benjamin Moseley[2]([✉])

[1] Department of Electrical Engineering and Computer Science,
University of California, Merced, CA 95344, USA
sim3@ucmerced.edu
[2] Washington University in St. Louis, St. Louis, MO 63130, USA
bmoseley@wustl.edu

Abstract. We consider the weighted Reordering Buffer Management problem. In this problem a set of n elements arrive over time one at a time and the elements can be stored in a buffer of size k. When the buffer becomes full, an element must be output. Elements are colored and if two elements are output consecutively and they have different colors then a switching cost is incurred. If the new color output is c, the cost is w_c. The objective is to reorder the elements to minimize the total switching cost in the output sequence.

In this paper, we give an improved randomized $O(\log \log \log k\gamma)$-approximation for this problem where γ is the ratio of the maximum to minimum weight of a color, improving upon the previous best $O(\log \log k\gamma)$-approximation. Our improvement builds on strengthening the standard linear program for the problem with non-standard knapsack covering inequalities. In particular, by leveraging the structure of these inequalities, our algorithm manages to render several random procedures more powerful and combine them effectively, thereby giving an exponential improvement upon the previous work.

1 Introduction

Buffer management theory focuses on studying how a buffer, typically of limited size, can be used to support an application. Due to the numerous applications of buffers, such as in networking and memory management, a rich and diverse theory has been developed. One well studied problem is the Reordering Buffer Management problem. In this problem, there is a set of n elements that arrive over time and the elements are colored. It is assumed that one element arrives at each time from 1 to n. There is a buffer of size k where the arriving elements can be stored in, and when the buffer becomes full an element must be output. If an element is output that has the same color as the previous element output, then there is no cost for outputting the element. Otherwise when the color changes, if the element output has color c then a cost of w_c is incurred. The goal is to reorder elements in the buffer to minimize the total cost incurred. Note that if

S. Im—Supported in part by NSF grant CCF-1409130.

M.M. Halldórsson et al. (Eds.): ICALP 2015, Part I, LNCS 9134, pp. 737–748, 2015.
DOI: 10.1007/978-3-662-47672-7_60

$v_c = 1$ for all colors c, then the goal is just to minimize the number of times the color changes in the output sequence.

The Reordering Buffer Management problem, which was elegantly formulated in [20], seeks to understand the fundamental tradeoff between the limited buffer size and the context switching cost. This simple, yet powerful, model captures several practical problems seen in paint shops, graphics rendering, as well as network buffering. For example, consider a server that forwards messages to clients. When switching from sending messages from one client to another, a cost is paid representing the overhead of the context switch. One may desire to limit the number of times the server switches between clients by buffering messages and reordering them to minimize the context switches. See [3,10,19,20] for more applications of this model. Besides the practical importance of the model, the model has been well studied theoretically. The main theoretical interest comes from the simplicity of the model and the fact that, yet being simple, the model is algorithmically challenging. Indeed, the model has been extensively studied both online and offline [1,2,4,6–8,12,13,20]. However, even though this problem has been rigorously studied for over a decade, the complexity of the problem is not well understood.

The challenges of the model emerge even when the weights of the colors are uniform. The uniform weight problem is known to be NP-Hard [4,12]. Further, algorithms that initially would seem to be ideal candidates for the problem fail to have a small approximation ratio. For example, simple algorithms such as Largest Color First, which outputs color that has the largest number of elements in the buffer, First-In First-Out and Least Recently Used all have strong lower bounds on their approximation ratios [20]. Due to this, previous work has focused on developing more sophisticated algorithms for the problem.

Initially an $O(\log^2 k)$-approximation was shown for the problem when the weights are all uniform [20]. This been improved through a sequence of woks [2,6,8,13]. Recently, the complexity of the unweighted case has been resolved up to constant factors and $O(1)$-approximation algorithms are known [7,16]. For the weighted version of the problem, the currently best known approximation is a randomized $O(\log \log k\gamma)$-approximation where here γ is the ratio of the maximum to minimum weight of a color [16]. Several algorithms are known which use resource augmentation where the algorithm is given a larger buffer than the optimal solution [12,20]. A key open question in the area is determining the right approximation ratio for the non-uniform weight version of the problem.

Results: In this work we improve upon the best known approximation ratio for the non-uniform weighted Reordering Buffer Management problem. We develop an algorithm that exponentially improves upon the best previously known algorithm's approximation guarantee of $O(\log \log k\gamma)$. Our main result is the following.

Theorem 1. *There exists a randomized $O(\log \log \log k\gamma)$-approximation algorithm for the weighted Reordering Buffer Management problem where γ is the ratio of the maximum to minimum weight of a color.*

For the online version of the problem, an $\Omega(\log \log k)$ lower bound is known on randomized algorithms as well as an $O((\log \log k\gamma)^2)$ upper bound [2,5]. This is the first case where the offline problem has been shown to have an approximation ratio better than the best possible competitive ratio. To show the main result, we introduce new linear program rounding techniques. In particular, we add knapsack covering inequalities to the standard linear program for the problem. See [11] for details on knapsack covering inequalities. We extend the definition of traditional knapsack covering inequalities by adding additional parameters to the inequalities, which prove to be very useful. By leveraging the structural properties given to us by these inequalities, we can circumvent barriers faced in previous rounding techniques. The inequalities were used in the context of unweighted reordering buffer in the authors' previous work [16]. However, they were only able to use the inequalities to give a small constant factor improvement for the unweighted case, but did not know how to use them for the weighted case. In this work, we demonstrate the power of our variants of knapsack covering inequalities by giving an exponential improvement for the weighted case. The inequalities will be further discussed in Section 2 and 3. We will give an overview of our algorithm and analysis in Section 3 together with the discussion on how this work is differentiated from the previous work.

Related Work: Besides the mentioned work on the offline buffer reordering problem, the problem has also been considered online. It is known that in the online setting that there is lower bound of $\Omega(\sqrt{\frac{\log k}{\log \log k}})$ on the competitive ratio of deterministic schedulers and $\Omega(\log \log k)$ on randomized schedulers [2]. The work of [2] gave the first $O(\sqrt{\log k})$-competitive deterministic online scheduler, essentially resolving the deterministic case when colors are unweighted. The recent work of [8] has resolved the randomized case when colors are unweighted by giving an $O(\log \log k)$-competitive online algorithm. For the weighted version of the problem, previous the best known online algorithm is a deterministic $O(\sqrt{\log k\gamma})$-competitive algorithm, which has recently been improved to $O((\log \log k\gamma)^2)$ randomized algorithm [5]. The problem has also been considered in the stochastic setting [14].

The Reordering Buffer Management problem has been generalized and been studied in several other settings. Most generalizations consider extending the definition of the cost function when switching colors. The work of [15,17] considers when the cost of switching between two colors forms a line metric and [9,18] considers when the costs form a general metric.

Organization: The paper is organized as follows. In Section 2 we start by introducing the linear programming relaxation we will consider throughout the paper as well as some useful lemmas and a simple randomized sampling procedure. In Section 3 we give a high-level sketch of our algorithm and analysis to show the intuition guiding our work. In Section 4 we formally introduce our algorithm and finally in Section 5 we give the formal proof of the algorithm's guarantees.

2 Preliminaries

In this section, we introduce our linear program, a few useful lemmas as well as a simple sampling scheme that our algorithm will utilize. We begin by introducing our linear program. We call a continuous sequence of elements of the same color in the output sequence a *color block*. We require elements for a color are output in first-in first-out order without loss of generality. Each color block (or simply block) b is a triple (i, t, ℓ) specifying the first element in the color block e_i, the time the block is scheduled t and the length of the sequence ℓ. Note that one can deduce all ℓ elements that are output in the block from the triple, and we let $(i', t') \in b$ if element $e_{i'}$ is output at time t' in the color block. Let B be the set of all possible color blocks in the output sequence. Note that B is polynomial in n. Let E denote the set of all elements.

Below is an integer programming formulation for the problem. The variable x_b specifies if the color block b is in the output sequence. The variable $y_{i,t}$ specifies if element e_i is output at time t and $\beta_{i,t}$ specifies if the element e_i was output at or before time t. We use the notation $E_{b, \leq t}$ to denote all the elements in the color block b which were output in b at or before time t. For a color block b let $c(b)$ be the color of the elements in b and, likewise, let $c(e_i)$ denote the color of the element e_i. Let $p(i)$ to denote the element for color $c(e_i)$ which is the previous element of this color that arrives before e_i – that is, the latest arriving element for color $c(e_i)$ that arrives before e_i.

$$\min \quad \sum_{b \in B} w_{c(b)} x_b \qquad\qquad\qquad\qquad\qquad \text{(IP)}$$

$$\text{s.t.} \quad y_{i,t} = \sum_{(i,t) \in b} x_b \qquad\qquad \forall i, t \qquad\qquad (1)$$

$$\sum_{i \in [n]} y_{i,t} = 1 \qquad\qquad \forall t \geq k+1 \qquad\qquad (2)$$

$$\sum_{t \in [k+1, k+n]} y_{i,t} = 1 \qquad\qquad \forall i \in [n] \qquad\qquad (3)$$

$$\beta_{i,t} = \sum_{i, t' \leq t} y_{i,t'} \qquad\qquad \forall i \in [n], t \in [k+1, k+n] \qquad (4)$$

$$\beta_{p(i), t-1} \geq \beta_{i,t} \qquad\qquad \forall i \in [n], t \geq k+1 \qquad\qquad (5)$$

$$\sum_{b \in B \setminus B'} (|E_{b, \leq t} \setminus E'|) x_b \geq (t - k - |E'|)(1 - \sum_{b \in B'} x_b) \qquad \forall t \in [k+1, k+n], B' \subseteq B, E' \subseteq E$$
$$\qquad\qquad\qquad\qquad\qquad\qquad\qquad\qquad\qquad\qquad\qquad (6)$$

$$x_b \in \{0, 1\} \qquad\qquad \forall b \in B \qquad\qquad (7)$$

Constraint (2) ensures that at most one element is output at each time. Constraint (3) ensures that each element is output at some time. Constraints (1) and (4) set the y and β variables according to the x variables. Constraint (5) ensures that elements are output in first-in-first-out order. Finally, the knapsack covering inequality is given in Constraint (6). We obtain an LP relaxation by replacing (7) with $x_b \in [0, 1]$.

Variants of Knapsack Covering Inequalities: The key constraints (6) deserve special attentions. The constraints are over all $B' \subseteq B$ and $E' \subseteq E$; therefore, there are exponentially many such constraints. To get a feel of the constraints, consider the simplest case that $B' = \emptyset$ and $E' = \emptyset$ with a fixed time t. Then the left-hand-side is simply the total number of elements output by time t, where each color block counts the number of elements it outputs by time t, and adds it to the summation. The right-hand-side is $t - k$, hence (6) states that at least $t - k$ elements must be output by time t due to the space limit of k for the buffer. Now consider an arbitrary E' with $B' = \emptyset$. Then, (6) lower bounds the total number of elements that has to be output in individual blocks by time t with the elements in E' excluded. In fact, (6) is a standard knapsack covering inequality if $B' = \emptyset$. Intuitively, this prevents the LP from cheating with elements.

In contrast, the power of having B' does not seem immediate – if there is a $b \in B'$ where $x_b = 1$, then the inequality is trivially satisfied, otherwise it becomes a standard knapsack inequality. However, having B' turns out to be very useful in randomized rounding. Recall that in the Reordering Buffer Problem the costs are determined by color blocks output, hence the complexity cannot be understood well without having a good control over blocks. In the fractional LP solution, we will be able to exclude some fractional color blocks and focus on "good" fractional blocks to derive nice probabilistic properties. The overall analysis is done with carefully chosen E' and B'.

To see why having $B' \neq \emptyset$ is useful, consider adding a color block b which has k^2 elements in it output by time t, but $x_b = \frac{1}{k}$ in an LP solution. From this color block, a total 'volume' of elements output is k. However, the color block is chosen by very little in the LP. By adding b to B', the right hand side decreases by a multiplicative factor of $1 - \frac{1}{k}$ while the left hand side decreases by an additive factor of k. This strengthens the LP. For instance, in the case that E' is chosen such that, $t - k - |E'| \leq k$, then the right hand side only decreases by an additive $k \cdot \frac{1}{k} = 1$, while the left hand side decreases by k. The added power is that the LP cannot output a large volume of elements using color blocks with many elements, but only choosing those color blocks themselves by a small amount.

Finally, we discuss the separation oracle regarding the constraints (6). Unfortunately, we do not know if there is a polynomial-time separation oracle when the constraint is defined over all $B' \subseteq B, E' \subseteq E$. However, there is a very easy separation oracle if either B' or E' is fixed. It turns out that we only need to consider polynomially many different E' for our analysis. That is, even though such a collection of E' is determined by $\{x_b\}$, we only need to look at polynomially many E', and this will allow us to solve the LP in polynomial time to the extent of our need. We defer the proof of solving the LP in polynomial time to a full version of this paper.

Useful Lemmas and Observations: Now we show some lemmas that will be useful throughout the paper. We will refer to x_b as the *height* of the color block

b in the LP solution. The following lemma will allow our algorithm to output a color at time t if the elements in the algorithm's buffer for the color at time t have been processed by a set of color blocks of substantial height in the LP by time t. This is similar to lemmas used in [2, 16] and is standard for the problem. The proof is omitted.

Lemma 1. *Consider any color block b output by our algorithm A which starts at time t and ends at time t''. Let $t' \geq k + 1$ be the earliest time before t such that A scheduled no element of color $c(b)$ during $[t', t)$. Suppose that the LP has a set of color blocks S of total height at least ϵ (i.e. $\sum_{b' \in S} x_{b'} \geq \epsilon$) that each have processed at least one element in b by time t – in particular, such a set S exists if the first element e_i in b is processed by at least ϵ by time t in the LP. Then there is a set of color blocks of total height at least ϵ for color $c(e_i)$ in the LP's solution that end during $(t', t'']$.*

The following proposition follows from constraint (1) in the LP.

Proposition 1. *Suppose that the LP has a set $\mathcal{I}$ of color c color blocks of color c and total height at least h, all starting no later than some time t. Further, suppose that each of blocks scheduled after time step t at least ℓ (possibly different) elements that entered the buffer no later than time step t. Then it is the case that LP has at least a total volume of $h\ell$ of elements of color c in its buffer at time t.*

Next we state a lemma that will allow us to compare against an LP with a slightly smaller buffer size. In particular, we will solve the LP with a buffer of size $k' = k - \frac{k}{\log k\gamma}$. This can be done by losing only an $O(1)$ factor in the approximation ratio as the lemma shows. The following lemma was shown in [5] and similar lemmas are known for the unweighted version of the problem. The proof of the lemma is omitted.

Lemma 2. *For any input sequence and $k' < k$, respectively, $\mathrm{OPT}_{k'} \leq O(1) \cdot (\frac{k}{k'} + (k - k')\frac{\log k'\gamma}{k'})\mathrm{OPT}_k$, where OPT_s denotes the cost of the optimal solution using a buffer of size s.*

Finally we introduce a sampling scheme which was originally used in [5] which is independent rounding coupled with a threshold rounding. We refer to a color block as *maximal* in the algorithm's output sequence if when the color block ends there are no more elements of the same color in the buffer at that time. In the sampling we will sample a color block b in the LP solution with probability $\frac{1}{\alpha}x_b$ if $\frac{1}{\alpha}x_b < 1$ and with probability 1 if $\frac{1}{\alpha}x_b \geq 1$. We call this the α-sampling. Let Bag denote the pool of color blocks sampled. Let t_i^α denote the earliest time that a color block in Bag schedules the element e_i and if no such color block exists set $t_i^\alpha = \infty$. We say that element e_i is α-ready at time t_i^α or at any time later. The proof of the following lemma is an extension of a proof found in [5]. The proof is deferred to a full version of this paper. For any set of color blocks A, let $x_{i,A}(t)$ denote the amount by which the element e_i is processed by color blocks in A by time t.

Lemma 3. *For any constant $0 < \alpha < 1$, the α-sampling satisfies the following properties :*

- *For any set of blocks A, the element e_i is α ready by time t with probability at least $(1 - 1/e)\min\{x_{i,A}(t)/\alpha, 1\}$. For any distinct elements e_i and e_j that are not processed by the same blocks in A by time t, the events that they become α ready by sampling color blocks from A are independent.*
- *The previous property implies that that for any element e_i and time step t such that $\beta_{i,t} \le \alpha$, $\Pr[t_i^\alpha \le t] \ge (1 - 1/e)\beta_{i,t}/\alpha$. This probability occurs independently for two elements if they are not processed by the same color blocks ever by time t. In particular, this is always true for elements of different colors.*
- *Consider any collection B' of disjoint maximal color blocks where each block $b' \in B'$ schedules at least one element i at time $t \ge t_i^\alpha$. The expected total cost of the blocks in B' is at most $(1/\alpha)\mathsf{Cost}_{\mathsf{LP}}$.*

The properties of the sampling scheme will be very useful for our analysis. The first property ensures that an element e_i can be scheduled by time t with probability proportional to amount it has been processed by the LP at time t, $\beta_{i,t}$. The second property ensures that the sampling is independent for elements of different colors or for elements where we can identify a set of color blocks that do not process both of them. The third property shows that the cost of outputting elements after their α-ready time can be charged to the LP.

3   Algorithm and Analysis Overview

In this section we give an outline of our algorithm and the analysis. Due to space constraints, the main analysis is deferred to a full version of this paper. The actual analysis is more involved but our goal here is to give the underlying intuition while ignoring lower level details.

Our algorithm begins by solving the linear program for the problem where the buffer size is set to be $k' = k - \frac{k}{\log k\gamma}$. The solution to the linear program is used to guide the algorithm on how elements should be output. The algorithm itself, works like an online algorithm that outputs elements sequentially from time $k + 1$ to time $n + k$. At any time t where there is an element in the algorithm's buffer $\mathcal{B}(t)$ that has the same color as the previous element output, the algorithm will output such an element. Otherwise, the algorithm needs to choose a color to switch to. At these points in time, the algorithm will use a set of rules to decide which color to switch to. These rules on the color to switch to are guided by the LP solution.

We now discuss the rules that the algorithm uses to decide the color to switch to. These rules are inspired by the previous work of [16] on the Buffer Reordering Management problem. The first set of rules are simple and similar to previous work. The algorithm is free to switch to any color c where (1) the elements in $\mathcal{B}(t)$ for color c have been processed by color blocks in the LP of total height at least ϵ (2) there is a an element for color c in $\mathcal{B}(t)$ that is α-ready or (3) there are more than $k/10$ elements in $\mathcal{B}(t)$ for color c. The cost of execution rule (1) is easily charged to the LP using Lemma 1 and the same is holds for rule (2) using Lemma 3. The cost of rule (3) can by charged to the LP using observations used in [16]. Intuitively, a color cannot be output many times if it occupies $\Theta(k)$ space in the buffer without the LP also needing to output the color. This is because the LP would need to store all of these elements, contradicting its buffer size and, therefore, we can charge to the LP.

The first three simple rules are used to give structural properties on the algorithm and LP's status when these rules cannot be applied. The interesting rules are the final two rules to be mentioned soon. Recall that the LP solution has buffer of smaller size than the algorithm. This implies that at any time t, the LP must have processed the elements in $\mathcal{B}(t)$ by a $\frac{k}{\log k\gamma}$ aggregate amount. The final two rules are based on whether this aggregate amount of work is focused mostly on elements for colors which occupy a large portion of the algorithm's buffer or a smaller portion of the buffer.

Let $C_s(t)$ be the set of colors where the algorithm has less than $\frac{k}{\log^3 k\gamma}$ elements for each of these colors in its buffer and let $C_b(t)$ be the remaining colors where the algorithm has more than $\frac{k}{\log^3 k\gamma}$ elements for these colors. In [16] it was shown that if a constant fraction of the work the LP has done on elements in $\mathcal{B}(t)$ are for colors in $C_s(t)$ then we should have sampled an element in $\mathcal{B}(t)$ for a colors in $C_s(t)$ with probability at least $1 - \frac{1}{k^2}$. Intuitively, a large volume of work was focused on these colors. Further, knowing that color blocks for colors in $C_s(t)$ can only include $\frac{k}{\log^3 k\gamma}$ elements from $\mathcal{B}(t)$, one can use concentration inequalities to show that we should have sampled such an element. Since we fail to sample an element with low probability, it can be shown that there is some color we can switch to such that the expected cost of switch to this color is small compared to the LP's cost. This will be rule (4).

The final rule and analysis of this rule is where our work differs from [16] and is where the knapsack covering inequalities proves to very useful. The algorithm will only perform this rule so long as the previous rules do not apply. In particular, since we do not execute rule (4), we know that a constant fraction of the work the LP has done by time t on the elements in $\mathcal{B}(t)$ are on elements that have colors in $C_b(t)$. Let $n_c^A(t)$ denote the number of elements for color c in $\mathcal{B}(t)$. Let $n_c^O(t)$ be the number of elements in the LP at time t for color c that have been processed by at most $1/2 + 2\epsilon$. The first step is showing that there is a color $c \in C_b(t)$ where $n_c^A(t) \geq \frac{3}{5} n_c^O(t)$. This will follow from the fact that if it were not true, then the LP has many elements the algorithm does not have for colors in $C_b(t)$. But then, we also know that no element in $\mathcal{B}(t)$ is processed by ϵ, since we did not use rule (1). Thus, the LP must have all elements in $\mathcal{B}(t)$ and these extra elements in its buffer, but this will cause a contradiction to the LP's buffer size. Rule (5) will allow the algorithm to switch to a color c where $n_c^A(t) \geq \frac{3}{5} n_c^O(t)$. Then we will show that we can execute this rule at most $O(\log \log k\gamma)$ times for a fixed color before the LP must output this color, allowing us to charge to this point in time in the LP. The argument follows by observing that if we output a color with at least $\frac{k}{\log^3 k\gamma}$ elements and $n_c^A(t) \geq \frac{3}{5} n_c^O(t)$ more than

$O(\log\log k\gamma)$ times and the LP does not do this color, then the LP must have $(\frac{k}{\log^3 k\gamma})(1 + \frac{3}{5})^{O(\log\log k\gamma)} > \Omega(k)$ elements for this color in its buffer at some time. This will draw a contradiction and therefore we can only output a color $O(\log\log k\gamma)$ times using this rule before we can find a time to charge to in the LP solution.

Naively, rule (5) will show our algorithm is a $O(\log\log k\gamma)$-approximation. However, we can improve this by showing that, in fact, we only perform rule (5) with low probability. Say with probability at most $\frac{1}{\log\log k\gamma}$. This will allow us to show that in expectation we only need to charge $O(1)$ to the LP. Showing this event happens with low probability will follow from the knapsack covering inequalities and by bosting the probability a block is randomly sampled in the LP by a $\Theta(\log\log\log k\gamma)$ factor. We note that these knapsack inequalities were not used in [16] to show a $O(\log\log k\gamma)$-approximation and this is how we circumvent hurdles faced in the analysis of [16]. In particular, it seems perfectly plausible using the standard LP that we could output a color $O(\log\log k\gamma)$ in this step with good probability. To see why this event happens with low probability, consider the knapsack covering inequality for time t.

$$\sum_{b\in B\setminus B'} (|E_{b,\leq t}\setminus E'|)x_b \geq (t - k' - |E'|)(1 - \sum_{b\in B'} x_b) \quad \forall B' \subseteq B, E' \subseteq E$$

Our goal is to show that there is $\Omega(1)$ height of color blocks the LP has scheduled on elements in $\mathcal{B}(t)$ if we execute rule (5). We will use this coupled with setting $\alpha < \frac{1}{\Theta(1)\log\log\log k\gamma}$ for the sample. If we can find such a height on color blocks in the LP, then the probability no element in $\mathcal{B}(t)$ is α-ready is at most $2^{-\Theta(\log\log\log k\gamma)} = \frac{1}{(\log\log k\gamma)^{\Theta(1)}}$ by Lemma 3. Further, if we did sample such an interval then an element in $\mathcal{B}(t)$ would be α ready at time t. Thus, we will have the desired probability and here one can see why we required that we boosted the probabilities in the sampling by a factor of $\Theta(\log\log\log k\gamma)$. To see why such a such a height exists, consider setting E' to be all elements that arrived by time t except those $\mathcal{B}(t)$ and B' to be the height of color block including that process at least one element in $\mathcal{B}(t)$ before time t. The left hand side must be 0, but $(t - k' - |E'|) = \frac{k}{\log k\gamma}$ since $|E'| = t - k$. Thus, it must be the case that $\sum_{b\in B'} x_b = 1$.

This is the intuition on how we can show that the cost accumulated by the algorithm by rule (5) is at most $O(1)$ multiplied by the cost of the LP in expectation. Unfortunately, the actual proof is much more involved. In particular, there is a dependency at different times on whether or not elements are α-ready. The proof needs to deal with these dependencies delicately. We handle this by showing that, in fact, a very large number of elements will become α-ready with good probability. Then using this we can group time steps together in such a way that if we succeed at a particular time, we will succeed at the later times where there are significant dependencies. This will then allow us to bound the cost of rule (5).

4   Algorithm

We require some notation to define formally the algorithm. Let ϵ be $\Theta(\frac{1}{\log\log\log k\gamma})$ and α at most ϵ. We will later set $\epsilon = \frac{1}{2^{20}\log\log\log k\gamma}$ and $\alpha = \epsilon$. Let $\mathcal{B}(t)$ denote (the set of elements in) the algorithm's buffer at time t. Let $n_c^A(t)$ denote the number of elements for color c in $\mathcal{B}(t)$. Let $n_c^O(t)$ be the number of elements in the LP at time t for color c that have been processed by at most $1/2 + 2\epsilon$. Intuitively, one should think of these elements as the ones not done by the LP. Let $C_s(t)$ contain all colors c where $0 < n_c^A(t) \leq \frac{k}{\log^3 k\gamma}$ and $C_b(t)$ contain all colors c where $n_c^A(t) > \frac{k}{\log^3 k\gamma}$. Let $E^O(t)$ be the set of elements that have been processed by at most $1/2 + 2\epsilon$ in the LP at time t that are not in $\mathcal{B}(t)$, i.e. $E^O(t) := \{e_i \,|e_i \notin \mathcal{B}(t), i \leq t, \beta_{i,t} \leq 1/2 + 2\epsilon\}$. Let $c^*(t)$ be the color such that color blocks in the LP for color $c^*(t)$ that intersect time t have height greater than $1/2$, if it exists. Note that there can only be one such color. Let $v_{c,t}^O = \sum_{i,c(e_i)=c} 1 - \beta_{i,t}$ denote the remaining volume of elements for color c in the LP at time t.

Let $t_{c,1}$ be the first time the LP accumulates cost ϵw_c for color c. That is, there exists a set of color blocks for color c of height at least ϵ which *start* at time $t_{c,1}$ or earlier. Assuming $t_{c,i-1}$ is defined, let $t_{c,i}$ be the earliest time that the LP accumulates cost ϵw_c for color c since time $t_{c,i-1}$. That is, during $(t_{c,i-1}, t_{c,i}]$ there exists a set of color blocks for color c of total height at least ϵ that start during $(t_{c,i-1}, t_{c,i}]$. Let $\mathcal{T}_c$ be the set of such times for color c. With these definitions in place, the algorithm can be defined as follows. The algorithm attempts to execute the rules in the order presented.

Algorithm:

RULE (I) If there is a set of color blocks S in the LP of total aggregate height ϵ (i.e. $\sum_{b\in S} x_b \geq \epsilon$) that each processes at least one element in $\mathcal{B}(t)$ for color c by time t then output color c. In particular, in a special case, if there is an element in $e_i \in \mathcal{B}(t)$ processed by ϵ in the LP, then switch to color $c(e_i)$.

RULE (II) If there is an element $e_i \in \mathcal{B}(t)$ that is α ready at time t then switch to color $c(e_i)$.

RULE (III) If there is a color c where $n_c^A(t) \geq k/10$, switch to color c.

RULE (IV) If the LP has processed elements in $\mathcal{B}(t)$ corresponding to colors in $C_s(t)$ by a total of at least $(|E^O(t)| + \frac{k}{\log k\gamma})/10$ by time t then switch to a color $c \in C_s(t)$ such that earliest time $t' \in \mathcal{T}_c$ after t is also the earliest time in $\cup_{c'\in C_s(t)}\mathcal{T}_{c'}$ after t.

RULE (V) We perform this rule if none of the others apply. Let L be the set of colors $c \in C_b(t)$ such that $n_c^A(t) \geq \frac{3}{5}n_c^O(t)$. The algorithm switches to a color $c \in L$ such that the earliest time t' in $\mathcal{T}_c$ after t is also the earliest time in $\cup_{c'\in L}\mathcal{T}_{c'}$ after t. We will show that $L \neq \emptyset$ if the earlier rules cannot be used.

---

5   Analysis

In this section our goal is to prove Theorem 1 by analyzing the algorithm given in the previous section. Recall that we solve the LP with a buffer size $k' := k - \frac{k}{\log k}$ and our algorithm has a buffer of size k. Throughout the proof, we will let LP denote the cost of the LP solution. To prove the approximation ratio of our algorithm, we bound the cost of each of the rules in the algorithm separately. First consider RULE (I). The following lemma is immediately implied by Lemma 1.

Lemma 4. *The total cost accumulated by the algorithm due to executing* RULE (I) *is at most* $O(\frac{1}{\epsilon})$LP.

Next consider the cost accumulated by RULE (II). By applying Lemma 3, we have the following lemma.

Lemma 5. *The total expected cost incurred when the algorithm executes* RULE (II) *is at most* $O(\frac{1}{\alpha})$LP.

Next we consider the cost accumulated by RULE (III) and RULE (IV). In this case, we appeal to the proofs shown in [16]. We note that this proof relies on structural properties in the elements in $\mathcal{B}(t)$ have since the algorithm did not use RULE (I) or RULE (II). These structural properties are sufficient for the proofs shown in [16].

Lemma 6 ([16]). *The total cost incurred when the algorithm executes* RULE (III) *is at most* $O(\frac{1}{\epsilon})$LP.

Lemma 7 ([16]). *The total expected cost incurred when the algorithm executes* RULE (IV) *is at most* $O(\frac{1}{\epsilon})$LP.

We now focus on bounding the expected number of times an element can be output due to RULE (V). The main analysis focuses on proving the following lemma.

Lemma 8. *Consider any time* $t_1 \in \mathcal{T}$ *and let* t_2 *be the next time in* $\mathcal{T}$ *after* t_1. *The expected number of times we execute* RULE (V) *is at most* $O(1)$ *during* $[t_1, t_2)$.

Once we have this lemma, combining it and the previous four lemmas proves Theorem 1. This is because between any two times t_1 and t_2 in $\mathcal{T}_c$ the LP accumulates a cost of at least ϵw_c and we can charge to this cost to bound the expected cost of executing RULE (V) by $O(1)$LP. Showing this lemma will complete the analysis. Due to space constraints, the proof is deferred.

References

1. Aboud, A.: Correlation clustering with penalties and approximating the reordering buffer management problem. Masters thesis, Computer Science Department, The Technion - Israel Institute of Technology (2008)

2. Adamaszek, A., Czumaj, A., Englert, M., Räcke, H.: Almost tight bounds for reordering buffer management. In: STOC, pp. 607–616 (2011)
3. Alborzi, H., Torng, E., Uthaisombut, P., Wagner, S.: The k-client problem. J. Algorithms 41(2), 115–173 (2001)
4. Asahiro, Y., Kawahara, K., Miyano, E.: Np-hardness of the sorting buffer problem on the uniform metric. Discrete Applied Mathematics 160(10–11), 1453–1464 (2012)
5. Avigdor-Elgrabli, N., Im, S., Moseley, B., Rabani, Y.: On the randomized competitive ratio of reordering buffer management with non-uniform costs. Manuscript (2014)
6. Avigdor-Elgrabli, N., Rabani, Y.: An improved competitive algorithm for reordering buffer management. In: SODA, pp. 13–21 (2010)
7. Avigdor-Elgrabli, N., Rabani, Y.: A constant factor approximation algorithm for reordering buffer management. In: SODA (2013)
8. Avigdor-Elgrabli, N., Rabani, Y.: An improved competitive algorithm for reordering buffer management. In: 54th Annual IEEE Symposium on Foundations of Computer Science, FOCS, October 26–29, 2013, Berkeley, CA, USA, pp. 1–10 (2013)
9. Bar-Yehuda, R., Laserson, J.: Exploiting locality: approximating sorting buffers. J. Discrete Algorithms 5(4), 729–738 (2007)
10. Blandford, D.K., Blelloch, G.E.: Index compression through document reordering. In: DCC, pp. 342–351 (2002)
11. Carr, R.D., Fleischer, L.K., Leung, V.J., Phillips, C.A.: Strengthening integrality gaps for capacitated network design and covering problems. In: Proceedings of the eleventh annual ACM-SIAM symposium on Discrete algorithms, SODA 2000, Philadelphia, PA, USA, pp. 106–115. Society for Industrial and Applied Mathematics (2000)
12. Chan, H.-L., Megow, N., Sitters, R., van Stee, R.: A note on sorting buffers offline. Theor. Comput. Sci. 423, 11–18 (2012)
13. Englert, M., Westermann, M.: Reordering buffer management for non-uniform cost models. In: Caires, L., Italiano, G.F., Monteiro, L., Palamidessi, C., Yung, M. (eds.) ICALP 2005. LNCS, vol. 3580, pp. 627–638. Springer, Heidelberg (2005)
14. Esfandiari, H., Hajiaghayi, M.T., Khani, M.R., Liaghat, V., Mahini, H., Räcke, H.: Online stochastic reordering buffer scheduling. In: Esparza, J., Fraigniaud, P., Husfeldt, T., Koutsoupias, E. (eds.) ICALP 2014. LNCS, vol. 8572, pp. 465–476. Springer, Heidelberg (2014)
15. Gamzu, I., Segev, D.: Improved online algorithms for the sorting buffer problem on line metrics. ACM Transactions on Algorithms, 6(1) (2009)
16. Im, S., Moseley, B.: New approximations for reordering buffer management. In: SODA, pp. 1093–1111 (2014)
17. Khandekar, R., Pandit, V.: Online sorting buffers on line. In: Durand, B., Thomas, W. (eds.) STACS 2006. LNCS, vol. 3884, pp. 584–595. Springer, Heidelberg (2006)
18. Kohrt, J.S., Pruhs, K.R.: A constant approximation algorithm for sorting buffers. In: Farach-Colton, M. (ed.) LATIN 2004. LNCS, vol. 2976, pp. 193–202. Springer, Heidelberg (2004)
19. Krokowski, J., Räcke, H., Sohler, C., Westermann, M.: Reducing state changes with a pipeline buffer. In: VMV, pp. 217 (2004)
20. Räcke, H., Sohler, C., Westermann, M.: Online Scheduling for Sorting Buffers. In: Möhring, R.H., Raman, R. (eds.) ESA 2002. LNCS, vol. 2461, pp. 820–832. Springer, Heidelberg (2002)

Local Reductions

Hamid Jahanjou[1], Eric Miles[2]([✉]), and Emanuele Viola[1]

[1] Northeastern University, Boston, MA, USA
{hamid,viola}@ccs.neu.edu
[2] UCLA, Los Angeles, CA, USA
enmiles@cs.ucla.edu

Abstract. We reduce non-deterministic time $T \geq 2^n$ to a 3SAT instance ϕ of quasilinear size $|\phi| = T \cdot \log^{O(1)} T$ such that there is an explicit circuit C that on input an index i of $\log |\phi|$ bits outputs the ith clause, and each output bit of C depends on $O(1)$ input bits. The previous best result was C in NC^1. Even in the simpler setting of polynomial size $|\phi| = \text{poly}(T)$ the previous best result was C in AC^0.

More generally, for any time $T \geq n$ and parameter $r \leq n$ we obtain $\log_2 |\phi| = \max(\log T, n/r) + O(\log n) + O(\log \log T)$ and each output bit of C is a decision tree of depth $O(\log r)$.

As an application, we tighten Williams' connection between satisfiability algorithms and circuit lower bounds (STOC 2010; SIAM J. Comput. 2013).

1 Introduction

The efficient reduction of arbitrary non-deterministic computation to 3SAT is a fundamental result with widespread applications. For many of these, two aspects of the efficiency of the reduction are at a premium. The first is the length of the 3SAT instance. A sequence of works shows how to reduce non-deterministic time-T computation to a 3SAT instance ϕ of quasilinear size $|\phi| = \tilde{O}(T) := T \log^{O(1)} T$ [HS66, Sch78, PF79, Coo88, GS89, Rob91]. This has been extended to PCP reductions [BGH+05, Mie09, BCGT13, BCGT12].

The second aspect is the computational complexity of producing the 3SAT instance ϕ given a machine M, an input $x \in \{0,1\}^n$, and a time bound $T = T(n) \geq n$. It is well-known and easy to verify that a ϕ of size $\text{poly}(T)$ is computable even by circuits from the restricted class NC^0. More generally, Agrawal, Allender, Impagliazzo, Pitassi, and Rudich show [AAI+01] that such NC^0 reductions exist whenever AC^0 reductions do.

A stronger requirement on the complexity of producing ϕ is critical for many applications. The requirement may be called *clause-explicitness*. It demands that the ith clause of ϕ be computable, given $i \leq |\phi|$ and $x \in \{0,1\}^n$, with resources $\text{poly}(|i|) = \text{poly}\log|\phi| = \text{poly}\log T$. In the case $|\phi| = \text{poly}(T)$, this is known to be possible by an unrestricted circuit D of size $\text{poly}(|i|)$. (The circuit has either

Supported by NSF grants CCF-0845003, CCF-1319206.

M.M. Halldórsson et al. (Eds.): ICALP 2015, Part I, LNCS 9134, pp. 749–760, 2015.
DOI: 10.1007/978-3-662-47672-7_61

random access to x, or, if $T \geq 2^n$, it may have x hardwired.) As a corollary, so-called succinct versions of NP-complete problems are complete for NEXP. Arora, Steurer, and Wigderson [ASW09] note that the circuit D may be taken from the restricted class AC^0. They use this to argue that, unless EXP = NEXP, standard NP-complete graph problems cannot be solved in time $\text{poly}(2^n)$ on graphs of size 2^n that are described by AC^0 circuits of size $\text{poly}(n)$.

Interestingly, applications to unconditional complexity lower bounds rely on reductions that are clause-explicit and simultaneously optimize the length of the 3SAT instance ϕ and the complexity of the circuit D computing clauses. For example, the time-space tradeoffs for SAT need to reduce non-deterministic time T to a 3SAT instance ϕ of quasilinear size $\tilde{O}(T)$ such that the ith clause is computable in time $\text{poly}(|i|) = \text{poly} \log |\phi|$ and space $O(\log |\phi|)$, see e.g. [FLvMV05] or Van Melkebeek's survey [vM06]. More recently, the importance of optimizing both aspects of the reduction is brought to the forefront by Williams' approach to obtain lower bounds by satisfiability algorithms that improve over brute-force search by a super-polynomial factor [Wil13a, Wil11b, Wil11a, SW12, Wil13b]. To obtain lower bounds against a circuit class C using this technique, one needs a reduction of non-deterministic time $T = 2^n$ to a 3SAT instance of size $\tilde{O}(T)$ whose clauses are computable by a circuit D of size $\text{poly}(n)$ that belongs to the class C. For example, for the ACC^0 lower bounds [Wil11b, Wil13b] one needs to compute them in ACC^0. However it has seemed "hard (perhaps impossible)" [Wil11b] to compute the clauses with such restricted resources.

Two workarounds have been devised [Wil11b, SW12]. Both exploit the fact that, under an assumption such as $P \subseteq ACC^0$, non-constructively there does exist such an efficient circuit computing clauses; the only problem is constructing it. They accomplish the latter using either nondeterminism [Wil11b] or brute-force [SW12] (cf. [AK10]). The overhead in these arguments limits the consequences of satisfiability algorithms: before this work, for a number of well-studied circuit classes C (discussed later) a lower bound against C did not follow from a satisfiability algorithm for circuits in C.

2   Our Results

We show that, in fact, it is possible to reduce non-deterministic computation of time $T \geq 2^n$ to a 3SAT formula ϕ of quasilinear size $|\phi| = \tilde{O}(T)$ such that given an index of $\ell = \log |\phi|$ bits to a clause, one can compute (each bit of) the clause by looking at a constant number of bits of the index. Such maps are also known as local, NC^0, or junta. More generally our results give a trade-off between decision-tree depth and $|\phi|$. The results apply to any time bound T, paying an inevitable loss in $|x| = n$ for T close to n.

Theorem 1 (Local reductions). *Let M be an algorithm running in time $T = T(n) \geq n$ on inputs of the form (x, y) where $|x| = n$. Given $x \in \{0, 1\}^n$ one can output a circuit $D : \{0, 1\}^\ell \to \{0, 1\}^{3v+3}$ in time $\text{poly}(n, \log T)$ mapping an index to a clause of a 3CNF ϕ in v-bit variables, for $v = \Theta(\ell)$, such that*

1. ϕ is satisfiable iff there is $y \in \{0,1\}^T$ such that $M(x,y) = 1$, and
2. for any $r \leq n$ we can have $\ell = \max(\log T, n/r) + O(\log n) + O(\log \log T)$ and each output bit of D is a decision tree of depth $O(\log r)$.

Note that for $T = 2^{\Omega(n)}$ we get that D is in NC^0 and ϕ has size $2^\ell = T \cdot \log^{O(1)} T$, by setting $r := n/\log T$. We also point out that the only place where locality $O(\log r)$ (as opposed to $O(1)$) is needed in D is to index bits of the string x.

The previous best result was D in NC^1 [BGH+05]. Even in the simpler setting of $|\phi| = \mathrm{poly}(T)$ the previous best result was D in AC^0 [ASW09].

Tighter connections between satisfiability and lower bounds. The quest for non-trivial satisfiability algorithms has seen significant progress recently, see e.g. [Wil11b, Her11, IMP12, BIS12, IPS13, CKS13]. Our results lower the bar for obtaining new circuit lower bounds from such algorithms. Previously, a lower bound for circuits of depth d and size s was implied by a satisfiability algorithm for depth $c \cdot d$ and size s^c for a constant $c > 1$ (for typical settings of s and d). With our proof it suffices to have a satisfiability algorithm for depth $d+c$ and size $c \cdot s$ for a constant c. This can be extended and optimized for several well-studied circuit classes. In particular we obtain the following new connections.

Corollary 1. *For each of the following classes C, if the satisfiability of circuits in C can be solved in time $2^n/n^{\omega(1)}$ then there is a problem $f \in \mathrm{E}^{\mathrm{NP}}$ that is not solvable by circuits in C:*

(1) linear-size circuits,

(2) linear-size series-parallel circuits,

(3) linear-size log-depth circuits,

(4) quasi-polynomial-size SYM-AND circuits.

Recall that available size lower bounds for unrestricted circuits are between $3n - o(n)$ and $5n - o(n)$, depending on the basis [Blu84, LR01, IM02]. Although Corollary 1 and Corollary 2 below are stated in terms of linear-size circuits, the proofs provide a close correspondence between the running time for satisfiability and the parameters of the circuit class. In particular, the constant hidden by the circuit size in class (1) can be optimized, as discussed in the paragraph "Subsequent work" below. At the moment this approach does not match known lower bounds, due to the (in)efficiency of known satisfiability algorithms.

In 1977 Valiant [Val77] focused attention on classes (2) and (3). (Some missing details about series-parallel graphs are provided in [Cal08].) The class (4) contains ACC [Yao90, BT94], and can be simulated by number-on-forehead protocols with a polylogarithmic number of players and communication [HG91]. Williams [Wil11b] gives a quasilinear-time algorithm to evaluate a SYM-AND circuit on all inputs.

For class (4) one can in fact obtain $f \in \mathrm{NE}$ using the seminal work by Impagliazzo, Kabanets, and Wigderson [IKW01] and its extension by Williams [Wil13a, Wil11b]. But to do so for classes (1)-(3), one would need a strengthening of [IKW01] to linear-size circuits, which we raise as an open problem.

It has long been known that the satisfiability of classes (1)-(3) in Corollary 1 can be linked to kSAT. Using Corollary 1, we can link kSAT to circuit lower bounds. (In the following, a kSAT instance has n variables and $O(n)^k$ clauses.)

Corollary 2.

(1) Assume that the exponential time hypothesis (ETH) is false [IP01]; i.e., for every $\epsilon > 0$, 3SAT is in time $2^{\epsilon n}$. Then there is a problem $f \in \mathrm{E}^{\mathrm{NP}}$ that is not solvable by linear-size circuits.

(2) Assume that the strong exponential time hypothesis (SETH) is false [IP01]; i.e., there is $\epsilon < 1$ such that for every k, kSAT is in time $2^{\epsilon n}$. Then there is a problem $f \in \mathrm{E}^{\mathrm{NP}}$ that is not solvable by linear-size series-parallel circuits.

(3) Assume that there is $\alpha > 0$ such that n^α-SAT is in time $2^{n-\omega(n/\log\log n)}$. Then there is a problem $f \in \mathrm{E}^{\mathrm{NP}}$ that is not solvable by linear-size log-depth circuits.

In Corollary 2, only (1) was known [Wil13a, Theorem 6.1]. Our proof is different: we obtain it immediately from (1) in Corollary 1 by the Cook-Levin theorem.

For context, the best algorithms for kSAT run in time $2^{n(1-O(1/k))}$ [DGH+02, PPSZ05].

Finally, we consider the class of polynomial-size depth-d circuits of threshold gates, which may have unbounded or bounded weights. (The latter case corresponds to Majority.) Recall that anything computed by a poly-size depth-d circuit with unbounded weights can be computed by a depth $d + 1$ circuit with bounded weights [HMP+93, GHR92], and that it is not known if EXP^{NP} has poly-size unbounded-weight circuits of depth $d = 2$. For these classes (and others) we show that a lower bound for depth d follows from a satisfiability algorithm for depth $d + 2$.

Corollary 3. *Consider unbounded fan-in circuits consisting of threshold gates (either bounded- or unbounded-weight). Let d be an integer.*

Suppose that for every c, given a circuit of depth $d+2$ and size n^c on n input bits one can decide its satisfiability in time $2^n/n^{\omega(1)}$.

Then NE does not have circuits of polynomial size and depth d.

A diagram of some of the classes mentioned above, and their relative power, can be found in [Vio13].

Our results have a few other consequences. For example they imply that the so-called succinct version of various NP-complete problems remain NEXP-complete even if described by an NC^0 circuit. In particular we obtain this for 3SAT and 3Coloring. Our techniques are also relevant to the notion of circuit uniformity. A standard notion of uniformity is log-space uniformity, requiring that the circuit is computable in logarithmic space or, equivalently, that given an index to a gate in the circuit one can compute its type and its children in linear space. Equivalences with various other uniformity conditions are given by Ruzzo [Ruz81], see also [Vol99]. We consider another uniformity condition which is stronger than previously considered ones in some respects. Specifically, we describe the circuit by showing how to compute children by an NC^0 circuit. i.e. a function with constant locality.

Theorem 2 (L-uniform ⇔ local-uniform). *Let $f : \{0,1\}^* \to \{0,1\}$ be a function computable by a family of log-space uniform polynomial-size circuits. Then f is computable by a family of polynomial-size circuits $C = \{C_n : \{0,1\}^n \to \{0,1\}\}_n$ such that there is a Turing machine that on input n (in binary) runs in time $O(\text{poly} \log n)$ and outputs a circuit $D : \{0,1\}^{O(\log n)} \to \{0,1\}^{O(\log n)}$ such that*
(i) D has constant locality: every output bit depends on $O(1)$ input bits, and
(ii) on input a label g of a gate in C_n, D outputs the type of g and labels for each child.

Does this paper simplify the proof that NEXP is not in ACC?. Recall that the proof [Wil11b] that NEXP is not in ACC uses as a black-box a result like Theorem 1 but with the requirement on the efficiency of D relaxed to polynomial-size circuits. If one instead uses as a black-box Theorem 1, one obtains a simpler proof, reported for completeness in the full version of this paper.

In fact, to obtain the separation of NEXP from ACC it suffices to prove a weaker version of Theorem 1 where D is, say, in AC^0. This weaker version has a simpler proof, as explained in §3. Independently of our work, Kowalski and Van Melkebeek proved this AC^0 result (personal communication).

Subsequent work. The announcement of our results as (ECCC Technical Report 13-099, July 2013) contained the same results as above except it did not mention Corollary 2 and items (2) and (4) in Corollary 1. After that announcement several related works have appeared. Oliveira's survey [Oli13] contains an alternative connection between satisfiability and circuit lower bounds, which yields a different proof of our Corollary 3 establishing a depth-2 overhead in that connection. Williams [Wil14] shows that the ability to count the number of satisfying assignments to circuits faster than brute-force search yields lower bounds against related circuits. His connection preserves the type of the gates in the input layer, a feature which is used to obtain some new lower bounds.

The work [BV14] builds on our results and is concurrent with [Wil14]. It gives a connection between derandomization and lower bounds that also preserves the type of the gates in the input layer. Thus, derandomization (or satisfiability), as opposed to counting, is sufficient for the lower bounds in [Wil14]. [BV14] also improves the depth loss of 2 in Corollary 3 to 1. Finally, they make a step in the direction we suggested of optimizing the constants in Item (1) of Corollary 1. In combination with the standard Cook-Levin reduction to 3SAT, they obtain that if 3SAT is in deterministic time c^n for any $c < 2^{1/10} = 1.07\ldots$ then E^{NP} does not have circuits of size $3n$ over the standard, full basis. Note that such a lower bound does not easily follow from diagonalization because the description length of a circuit of size $3n$ is superlinear. (Also recall the available lower bounds have the form $3n - o(n)$). The current record for solving 3SAT deterministically has $c = 1.33\ldots$ [MTY11], cf. [Her11].

As a corollary to [BV14], in this revision we show that even a somewhat more modest improvement to 3SAT algorithms would imply new lower bounds for non-boolean functions with range $m = 2$ bits. Such lower bounds do not seem known for any $m = o(n)$, cf. [KMM12].

Corollary 4 (Corollary to [BV14]). *If 3SAT is in time c^n for any $c < 2^{1/7} = 1.10\ldots$, then there exists a (non-Boolean) function $f : \{0,1\}^n \to \{0,1\}^2$ in E^{NP} such that any circuit over the full basis computing it requires at least $3n$ (non-input) gates.*

3 Techniques

Proofs of the theorems and corollaries above are omitted due to space constraints, but they can be found in the full version of this paper at the authors' websites. We now give an overview of the techniques used.

Background: Reducing non-deterministic time T to size-$\tilde{O}(T)$ 3SAT. Our starting point is the reduction of non-deterministic time-T computation to 3SAT instances of quasilinear size $T' = \tilde{O}(T)$. The classical proof of this result [HS66, Sch78, PF79, Coo88, GS89, Rob91] hinges on the oblivious Turing machine simulation by Pippenger and Fischer [PF79]. However computing connections in the circuit induced by the oblivious TM is a somewhat complicated recursive procedure, and we have not been able to use this construction for our results.

Instead, we use a proof by Van Melkebeek [vM06, §2.3.1] which replaces this simulation by coupling an argument due to Gurevich and Shelah [GS89] with sorting circuits. We note that the idea of using sorting is already in [GS89], but if one follows their paper one ends up using again the oblivious simulation. Van Melkebeek's observation is that essentially all that needs to be done obliviously is sorting, and so one can use a sorting network, a more familiar construction than the oblivious simulation. Specifically, Van Melkebeek uses Batcher's odd-even mergesort networks [Bat68]. This proof was rediscovered by a superset of the authors as a class project [VN12]. We now recall it in more detail.

Consider any general model of (non-deterministic) computation, such as RAM or random-access Turing machines. (One nice feature of this proof is that it directly handles models with random-access, aka direct-access, capabilities.) The proof reduces computation to the satisfiability of a circuit C. The latter is then reduced to 3SAT via the textbook reduction. Only the first reduction to circuit satisfiability is problematic and we will focus on that one here. Consider a non-deterministic time-T computation. The proof constructs a circuit of size $\tilde{O}(T)$ whose inputs are (non-deterministic guesses of) T configurations of the machine. Each configuration has size $O(\log T)$ and contains the state of the machine, all registers, and the content of the memory locations indexed by the registers. This computation is then verified in two steps. First, one verifies that every configuration C_i yields configuration C_{i+1} assuming that all bits read from memory are correct. This is a simple check of adjacent configurations. Then to verify correctness of read/write operations in memory, one sorts the configurations by memory indices, and within memory indices by timestamp. Now verification is again a simple check of adjacent configurations. The resulting circuit is outlined in Figure 1 (for a $2k$-tape random-access Turing machine). Using a sorting network of quasilinear size $\tilde{O}(T)$ results in a circuit of size $\tilde{O}(T)$.

Making low-space computation local. We employ a general technique that we call *spreading computation.* This shows that any circuit C whose connections can be computed in space linear in the description of a gate (i.e., space $\log |C|$) has an equivalent circuit C' of size $|C'| = \text{poly}|C|$ whose connections can be computed with constant locality.

The main idea in the proof is simply to let the gates of C' represent configurations of the low-space algorithm computing children in C. Then computing a child amounts to performing one step of the low-space algorithm, (each bit of) which can be done with constant locality in a standard Turing machine model.

We note that the technique of labeling gates by configurations goes back at least to the work of Ruzzo [Ruz81] who uses it to show the equivalence of some uniformity conditions involving alternating Turing machines that are simultaneously time and space restricted. However, [Ruz81] does not show how to compute gate connections with small locality, which is our aim here. We note that this task is non-trivial. For example, with constant locality one cannot even check the validity of a configuration. This means that the circuit C' has many invalid gates, i.e., gates that do not correspond to the computation of the low-space algorithm on a label of C. These gates could induce loops that do not correspond to computation, and make the final 3SAT instance always unsatisfiable. We avoid cycles by augmenting the low-space algorithm with a preliminary check for the validity of the configuration, and by including a clock in the configurations. These allow us to ensure that each invalid gate leads to a sink.

We apply spreading computation to the various sub-circuits checking consistency of configurations, corresponding to the triangles in Figure 1. These sub-circuits operate on configurations of size $O(\log T)$ and have size poly $\log T$. Hence, we can tolerate the polynomial increase in their complexity given by the spreading computation technique.

There remain however tasks for which we cannot use spreading computation. One is the sorting sub-circuit. Since it has size $> T$ we cannot afford a polynomial increase. Another task is indexing adjacent configurations. We now discuss these two in turn.

Sorting. We first mention a natural approach that gets us close but not quite to our main theorem. The approach is to define an appropriate labeling of the sorting network so that its connections can be computed very efficiently. We are able to define a labeling of bit-length $t + O(\log t) = \log \tilde{O}(T)$ for comparators in the odd-even mergesort network of size $\tilde{O}(2^t)$ (and depth t^2) that sorts $T = 2^t$ elements such that given a label one can compute the labels of its children by a decision tree of depth logarithmic in the length of the label, i.e. depth $\log \log \tilde{O}(T)$. With a similar labeling we can get linear size circuits. Or we can get constant locality at the price of making the 3SAT instance of size $T^{1+\epsilon}$. The details appear in the separate work [JMV14].

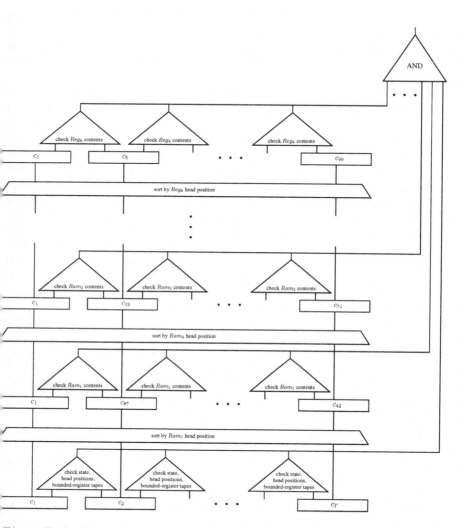

Fig. 1. Each of the T configurations has size $O(\log T)$. The checking circuits have size poly $\log T$. The sorting circuits have size $\tilde{O}(T)$. k is a constant. Hence overall circuit has size $\tilde{O}(T)$.

To obtain constant locality we use a variant by Ben-Sasson, Chiesa, Genkin and Tromer [BCGT13]. They replace sorting networks with routing networks based on De Bruijn graphs. We note that routing networks have been used extensively in the PCP literature starting, to our knowledge, with the work of Polishchuk and Spielman [PS94]. They have been used mostly for their algebraic properties, whereas we exploit the small locality of these networks. Specifically, the connections of these networks involve computing bit-shift, bit-xor, and addition by 1. The first two operations can easily be computed with constant locality, but the latter cannot in the standard binary representation. However, this addition by 1 is only on $O(\log \log T)$ bits. Hence we can afford an alternative, redundant representation which gives us an equivalent network where all the operations can be computed with constant locality. This representation again introduces invalid labels; those are handled in a manner similar to our spreading computation technique.

Plus one. Regardless of whether we are using sorting or routing networks, another issue that comes up in all previous proofs is addition by 1 on strings of $> \log T$ bits. This is needed to index adjacent configurations C_i and C_{i+1} for the pairwise checks in Figure 1. As mentioned before, this operation cannot be performed with constant locality in the standard representation. Also, we cannot afford a redundant representation (since strings of length $c \log T$ would correspond to an overall circuit of size $> T^c$).

For context, we point out an alternative approach to compute addition by 1 with constant locality which however cannot be used because it requires an inefficient pre-processing. The approach is to use primitive polynomials over $GF(2)^{\log T}$. These are polynomials modulo which x has order $2^{\log T} - 1$. Addition by 1 can then be replaced by multiplication by x, which can be shown to be local. This is similar to *linear feedback registers*. However, it is not known how to construct such polynomials efficiently w.r.t. their degrees, see [Sho92].

To solve this problem we use routing networks in a different way from previous works. Instead of letting the network output an array $C_1, C_2, \ldots$ representing the sorted configurations, we use the network to represent the "next configuration" map $C_i \to C_{i+1}$. Viewing the network as a matrix whose first column is the input and the last column is the output, we then perform the pairwise checks on every pair of input and output configurations that are in the same row. The bits of these configurations will be in the same positions in the final label, thus circumventing addition by one.

As we mentioned earlier, for a result such as NEXP not in ACC [Wil11b] it suffices to prove a weaker version of our Theorem 1 where the reduction is computed by, say, an AC^0 circuit. For the latter, it essentially suffices to show that either the sorting or the routing network's connections are in that class.

Acknowledgments. We are very grateful to Eli Ben-Sasson for a discussion on routing networks which led us to improving our main result, cf. §3. We also thank Ryan Williams for feedback on the write-up.

References

[AAI+01] Agrawal, M., Allender, E., Impagliazzo, R., Pitassi, T., Rudich, S.: Reducing the complexity of reductions. Computational Complexity **10**(2), 117–138 (2001)

[AK10] Allender, E., Koucký, M.: Amplifying lower bounds by means of self-reducibility. J. of the ACM, **57**(3) (2010)

[ASW09] Arora, S., Steurer, D., Wigderson, A.: Towards a study of low-complexity graphs. In: Albers, S., Marchetti-Spaccamela, A., Matias, Y., Nikoletseas, S., Thomas, W. (eds.) ICALP 2009, Part I. LNCS, vol. 5555, pp. 119–131. Springer, Heidelberg (2009)

[Bat68] Batcher, K.E.: Sorting networks and their applications. AFIPS Spring Joint Computing Conference **32**, 307–314 (1968)

[BCGT12] Ben-Sasson, E., Chiesa, A., Genkin, D., Tromer, E.: On the concrete-efficiency threshold of probabilistically-checkable proofs. Electronic Colloquium on Computational Complexity (ECCC) **19**, 45 (2012)

[BCGT13] Ben-Sasson, E., Chiesa, A., Genkin, D., Tromer, E.: Fast reductions from RAMs to delegatable succinct constraint satisfaction problems. In: ACM Innovations in Theoretical Computer Science Conf. (ITCS), pp. 401–414 (2013)

[BGH+05] Ben-Sasson, E., Goldreich, O., Harsha, P., Sudan, M., Vadhan, S.P.: Short PCPs verifiable in polylogarithmic time. In: IEEE Conf. on Computational Complexity (CCC), pp. 120–134 (2005)

[BIS12] Beame, P., Impagliazzo, R., Srinivasan, S.: Approximating AC^0 by small height decision trees and a deterministic algorithm for $\#AC^0$sat. In: IEEE Conf. on Computational Complexity (CCC), pp. 117–125 (2012)

[Blu84] Blum, N.: A boolean function requiring 3n network size. Theoretical Computer Science **28**, 337–345 (1984)

[BT94] Beigel, R., Tarui, J.: On ACC. Computational Complexity **4**(4), 350–366 (1994)

[BV14] Ben-Sasson, E., Viola, E.: Short PCPs with projection queries (2014). http://www.ccs.neu.edu/home/viola/

[Cal08] Calabro, C.: A lower bound on the size of series-parallel graphs dense in long paths. Electronic Colloquium on Computational Complexity (ECCC), **15**(110) (2008)

[CKS13] Chen, R., Kabanets, V., Saurabh, N.: An improved deterministic #SAT algorithm for small De Morgan formulas. Technical Report TR13-150, Electronic Colloquium on Computational Complexity (2013). http://www.eccc.uni-trier.de/

[Coo88] Cook, S.A.: Short propositional formulas represent nondeterministic computations. Information Processing Letters **26**(5), 269–270 (1988)

[DGH+02] Dantsin, E., Goerdt, A., Hirsch, E.A., Kannan, R., Kleinberg, J., Papadimitriou, C., Raghavan, P., Schöning, U.: A deterministic $(2 - 2/(k + 1))n$ algorithm for k-SAT based on local search. Theoretical Computer Science **289**(1), 69–83 (2002)

[FLvMV05] Fortnow, L., Lipton, R., van Melkebeek, D., Viglas, A.: Time-space lower bounds for satisfiability. J. of the ACM **52**(6), 835–865 (2005)

[GHR92] Goldmann, M., Håstad, J., Razborov, A.A.: Majority gates vs. general weighted threshold gates. Computational Complexity **2**, 277–300 (1992)

[GS89] Gurevich, Y., Shelah, S.: Nearly linear time. In: Logic at Botik, Symposium on Logical Foundations of Computer Science, pp. 108–118 (1989)

[Her11] Hertli, T.: 3-SAT faster and simpler - unique-SAT bounds for PPSZ hold in general. In: IEEE Symp. on Foundations of Computer Science (FOCS), pp. 277–284 (2011)

[HG91] Håstad, J., Goldmann, M.: On the power of small-depth threshold circuits. Comput. Complexity 1(2), 113–129 (1991)

[HMP$^+$93] Hajnal, A., Maass, W., Pudlák, P., Szegedy, M., Turán, G.: Threshold circuits of bounded depth. J. of Computer and System Sciences 46(2), 129–154 (1993)

[HS66] Hennie, F., Stearns, R.: Two-tape simulation of multitape turing machines. J. of the ACM 13, 533–546 (1966)

[IKW01] Impagliazzo, R., Kabanets, V., Wigderson, A.: In search of an easy witness: Exponential time vs. probabilistic polynomial time. In: IEEE Conf. on Computational Complexity (CCC) (2001)

[IM02] Iwama, K., Morizumi, H.: An explicit lower bound of $5n - o(n)$ for boolean circuits. In: Symp. on Math. Foundations of Computer Science (MFCS), pp. 353–364 (2002)

[IMP12] Impagliazzo, R., Matthews, W., Paturi, R.: A satisfiability algorithm for AC^0. In: ACM-SIAM Symp. on Discrete Algorithms (SODA), pp. 961–972 (2012)

[IP01] Impagliazzo, R., Paturi, R.: On the complexity of k-SAT. J. of Computer and System Sciences 62(2), 367–375 (2001)

[IPS13] Impagliazzo, R., Paturi, R., Schneider, S.: A satisfiability algorithm for sparse depth-2 threshold circuits. IEEE Symp. on Foundations of Computer Science (FOCS) (2013)

[JMV14] Jahanjou, H., Miles, E., Viola, E.: Succinct and explicit circuits for sorting and connectivity (2014). http://www.ccs.neu.edu/home/viola/

[KMM12] Kulikov, A.S., Melanich, O., Mihajlin, I.: A $5n - o(n)$ lower bound on the circuit size over U_2 of a linear boolean function. In: Cooper, S.B., Dawar, A., Löwe, B. (eds.) CiE 2012. LNCS, vol. 7318, pp. 432–439. Springer, Heidelberg (2012)

[LR01] Lachish, O., Raz, R.: Explicit lower bound of 4.5n - o(n) for boolena circuits. In: ACM Symp. on the Theory of Computing (STOC), pp. 399–408 (2001)

[Mie09] Mie, T.: Short pcpps verifiable in polylogarithmic time with o(1) queries. Ann. Math. Artif. Intell. 56(3–4), 313–338 (2009)

[MTY11] Makino, K., Tamaki, S., Yamamoto, M.: Derandomizing HSSW algorithm for 3-SAT (2011). CoRR, abs/1102.3766

[Oli13] Oliveira, I.C.: Algorithms versus circuit lower bounds (2013). CoRR, abs/1309.0249

[PF79] Pippenger, N., Fischer, M.J.: Relations among complexity measures. J. of the ACM 26(2), 361–381 (1979)

[PPSZ05] Paturi, R., Pudlák, P., Saks, M.E., Zane, F.: An improved exponential-time algorithm for k-sat. J. of the ACM 52(3), 337–364 (2005)

[PS94] Polishchuk, A., Spielman, D.A.: Nearly-linear size holographic proofs. In: ACM Symp. on the Theory of Computing (STOC), pp. 194–203 (1994)

[Rob91] Robson, J.M.: An O(T log T) reduction from RAM computations to satisfiability. Theoretical Computer Science 82(1), 141–149 (1991)

[Ruz81] Ruzzo, W.L.: On uniform circuit complexity. J. of Computer and System Sciences 22(3), 365–383 (1981)

[Sch78] Schnorr, C.-P.: Satisfiability is quasilinear complete in NQL. J. of the ACM **25**(1), 136–145 (1978)

[Sho92] Shoup, V.: Searching for primitive roots in finite fields. Math. Comp. **58**, 369–380 (1992)

[SW12] Santhanam, R., Williams, R.: Uniform circuits, lower bounds, and qbf algorithms. Electronic Colloquium on Computational Complexity (ECCC) **19**, 59 (2012)

[Val77] Valiant, L.G.: Graph-theoretic arguments in low-level complexity. In: Gruska, J. (ed.) MFCS 1977. LNCS, vol. 53, pp. 162–176. Springer, Heidelberg (1977)

[Vio13] Viola, E.: Challenges in computational lower bounds (2013). http://www.ccs.neu.edu/home/viola/

[vM06] van Melkebeek, D.: A survey of lower bounds for satisfiability and related problems. Foundations and Trends in Theoretical Computer Science **2**(3), 197–303 (2006)

[VN12] Viola, E., NEU. From RAM to SAT (2012). http://www.ccs.neu.edu/home/viola/

[Vol99] Vollmer, H.: Introduction to circuit complexity. Springer-Verlag, Berlin (1999)

[Wil11a] Williams, R.: Guest column: a casual tour around a circuit complexity bound. SIGACT News **42**(3), 54–76 (2011)

[Wil11b] Williams, R.: Non-uniform ACC circuit lower bounds. In: IEEE Conf. on Computational Complexity (CCC), pp. 115–125 (2011)

[Wil13a] Williams, R.: Improving exhaustive search implies superpolynomial lower bounds. SIAM J. on Computing **42**(3), 1218–1244 (2013)

[Wil13b] Williams, R.: Natural proofs versus derandomization. In: ACM Symp. on the Theory of Computing (STOC) (2013)

[Wil14] Williams, R.: New algorithms and lower bounds for circuits with linear threshold gates (2014)

[Yao90] Yao, A.C.-C.: On ACC and threshold circuits. In: IEEE Symp. on Foundations of Computer Science (FOCS), pp. 619–627 (1990)

Query Complexity in Expectation

Jedrzej Kaniewski[1,2], Troy Lee[1,3], and Ronald de Wolf[4,5]($\boxtimes$)

[1] Centre for Quantum Technologies,
National University of Singapore, Singapore, Singapore
[2] QuTech, Delft University of Technology, Delft, The Netherlands
[3] School of Physical and Mathematical Sciences, NTU, Singapore, Singapore
[4] Centrum Wiskunde en Informatica, Amsterdam, The Netherlands
[5] University of Amsterdam, Amsterdam, The Netherlands
j.kaniewski@nus.edu.sg, troyjlee@gmail.com, rdewolf@cwi.nl

Abstract. We study the query complexity of computing a function $f : \{0,1\}^n \rightarrow \mathbb{R}_+$ *in expectation*. This requires the algorithm on input x to output a nonnegative random variable whose expectation equals $f(x)$, using as few queries to the input x as possible. We exactly characterize both the randomized and the quantum query complexity by two polynomial degrees, the nonnegative literal degree and the sum-of-squares degree, respectively. We observe that the quantum complexity can be unboundedly smaller than the classical complexity for some functions, but can be at most polynomially smaller for Boolean functions. These query complexities relate to (and are motivated by) the extension complexity of polytopes. The *linear* extension complexity of a polytope is characterized by the randomized *communication* complexity of computing its slack matrix in expectation, and the *semidefinite* (psd) extension complexity is characterized by the analogous quantum model. Since query complexity can be used to upper bound communication complexity of related functions, we can derive some upper bounds on psd extension complexity by constructing efficient quantum query algorithms. As an example we give an exponentially-close entrywise approximation of the slack matrix of the perfect matching polytope with psd-rank only $2^{n^{1/2+\varepsilon}}$. Finally, we show randomized and quantum query complexity in expectation corresponds to the Sherali-Adams and Lasserre hierarchies, respectively.

1 Introduction

We study the complexity of computing a function $f : \{0,1\}^n \rightarrow \mathbb{R}_+$ *in expectation*, where our algorithm on input x should output a nonnegative real number whose expectation (over the algorithm's internal randomness) exactly equals $f(x)$. Getting the expectation right is easier than computing the function value $f(x)$ itself, and suffices in some applications. Suppose we want to approximate $F(x) = \sum_{i=1}^{m} f_i(x)$ that depends on $x \in \{0,1\}^n$. Then we can just compute each $f_i(x)$ *in expectation* and output the sum of the results. By linearity of expectation, the output will have expectation $F(x)$, and it will be tightly concentrated around its expectation if the random variables are not too wild (so the

© Springer-Verlag Berlin Heidelberg 2015
M.M. Halldórsson et al. (Eds.): ICALP 2015, Part I, LNCS 9134, pp. 761–772, 2015.
DOI: 10.1007/978-3-662-47672-7_62

Central Limit Theorem applies). It is not necessary to compute or even approximate any of the values $f_i(x)$ themselves for this. This illustrates that computing functions in expectation is an interesting model in its own right. Additionally, it is motivated by connections with the *extension complexity* of polytopes that are used in combinatorial optimization (roughly: the minimal size of linear or semidefinite programs for optimizing over such a polytope), as described below.

The complexity of computing f can be measured in different ways, and here we will focus on *query* complexity. We measure the complexity of computing a function in expectation by the (worst-case) number of queries to the input $x \in \{0,1\}^n$ that the best algorithm uses. We study both *randomized* and *quantum* versions of this model and show that both of these query complexities can be exactly characterized by natural notions of polynomial degree. In Section 3 we show that the randomized query complexity of computing f in expectation equals the "nonnegative literal degree" of f, which is the minimal d such that f can be written as a nonnegative linear combination of products of up to d variables or negations of variables. In Section 4 we show that the quantum complexity equals the "sum-of-squares degree", which is the minimal d such that there exist polynomials p_i of degree at most d satisfying $f(x) = \sum_i p_i(x)^2$ for all $x \in \{0,1\}^n$.

In Section 5 we observe that quantum and classical query complexities (equivalently: the above two types of polynomial degree) can be arbitrarily far apart. For example, the function $f(x) = (\sum_{i=1}^n x_i - 1)^2$ is the square of a degree-1 polynomial and hence computable in expectation with only 1 quantum query, while randomized algorithms need n queries to get this expectation right. In contrast, we show that for functions with range $\{0,1\}$ the gap can be at most cubic.

Lower bounds on the quantum query complexity can be obtained from lower bounding the sum-of-squares degree of the function at hand, which is often non-trivial. Using techniques from approximation theory, we prove that $f(x) = (\sum_{i=1}^n x_i - 1)(\sum_{i=1}^n x_i - 2)$ has sum-of-squares degree $\Omega(\sqrt{n})$. Hence quantum algorithms require $\Omega(\sqrt{n})$ queries to compute this function in expectation.

Our main motivation for studying query complexity in expectation comes from combinatorial optimization, in particular from linear and semidefinite programs. Many optimization problems can be formulated as maximizing or minimizing a linear function over a polytope. For example, in the Traveling Salesman Problem on n-vertex undirected graphs, one wants to minimize a linear function (the length of the tour) over the polytope $P \subseteq \mathbb{R}^{\binom{n}{2}}$ that is the convex hull of all Hamiltonian cycles in the complete n-vertex graph K_n. Representing this polytope as the feasible region of a small linear or semidefinite program would allow us to efficiently solve the problem using the ellipsoid or interior-point methods.

Informally, the **linear extension complexity** of a polytope $P \subseteq \mathbb{R}^d$ is the minimum number of linear inequalities (over the d variables of P and possibly auxiliary variables) whose feasible region projects down to P. Small linear extension complexity means there is a small linear program to optimize over P.

Motivated by erroneous claims [33] that the TSP polytope had polynomial linear extension complexity (implying P = NP), Yannakakis [36] showed that "symmetric" linear extensions of the Traveling Salesman Polytope need $2^{\Omega(n)}$

linear inequalities. He showed the same for the perfect matching polytope (which is spanned by all perfect matchings in K_n), despite the fact that finding a maximum matching can be done efficiently! For a long time, generalizing these lower bounds to arbitrary (possibly non-symmetric) linear extensions was an open question. However, recently Fiorini et al. [15] proved a $2^{\Omega(n^{1/2})}$ lower bound on the linear extension complexity of the TSP polytope. Subsequently Rothvoß [30] proved a $2^{\Omega(n)}$ lower bound for the perfect matching polytope, which via a reduction implies the same bound for TSP. Chan et al. [10] obtained lower bounds on linear extension complexity for constraint satisfaction problems via a different route: roughly put, they showed that arbitrary linear extensions are not much more powerful than the specific linear extensions produced by the "Sherali-Adams Hierarchy"; hence they could obtain lower bounds on linear extension complexity from known bounds on the Sherali-Adams hierarchy.

The **positive semidefinite (psd) extension complexity** of polytope P, which replaces the linear programs by potentially more powerful semidefinite programs, is the minimal dimension of a semidefinite program whose feasible region projects down to P. In contrast to the case of linear extension complexity, very few lower bounds on psd extension complexity are known. Until recently, there were only a few lower bounds for "symmetric" psd extensions [14,24]. However, in a *very* recent breakthrough, Lee et al. [23] generalized the approach of [10] to show that arbitrary psd extensions are not much more powerful than the specific psd extensions produced by the "Lasserre Hierarchy". In particular they showed that the TSP polytope has psd extension complexity $2^{\Omega(n^{1/13})}$.

Surprisingly, there is a very close connection between these extension complexities and the model of computing functions in expectation, albeit for the *communication complexity* of computing a 2-input function. More precisely, suppose Alice receives input x, Bob receives input y, and they want to compute some function $g(x, y)$ (which may also be viewed as a matrix). In the usual setting of communication complexity [20], one of the parties (let's say Bob) has to output this value $g(x, y)$ exactly, either with probability 1 or with high probability. However, we may also consider how much communication they need to compute $g(x, y)$ *in expectation*, i.e., now Bob needs to output a nonnegative random variable whose expected value equals $g(x, y)$. Faenza et al. [13] showed that the logarithm of the linear extension complexity of a polytope P equals the randomized communication complexity of computing (in expectation) a matrix associated with P, known as the *slack matrix*. Lifting this result to the quantum/psd case, Fiorini et al. [15] showed that the logarithm of the *psd* extension complexity equals the one-way *quantum* communication complexity of computing the slack matrix of P in expectation; in this model Alice sends a single quantum message to Bob. These connections show that studying (linear and psd) extension complexity of a polytope P is *equivalent* to studying (randomized and one-way quantum) communication complexity in expectation, of the slack matrix of P.

How is the *query* complexity of computing a function in expectation related to this *communication* complexity? Many functions of interest in communication complexity are of the form $g(x, y) = f(x \wedge y)$ for some Boolean function

$f : \{0,1\}^n \to \{0,1\}$, where the AND-connective is applied bitwise. Functions of this form also arise as (submatrices of) slack matrices of interesting polytopes, e.g. the correlation polytope. Quite generally across the usual models of worst-case complexity (deterministic, randomized or quantum) upper bounds on the *query complexity* of f imply upper bounds on the *communication complexity* of g. In Section 7 we show that this also holds for the randomized and quantum models of computing a function in expectation. As this leads to multi-round communication protocols, it implies that the one-way and two-way quantum communication complexity of computing a function in expectation are equal.

In Section 7.1 we give an application of the connection between query algorithms and communication complexity (equivalently, *psd rank*), by deriving an exponentially-close entrywise approximation of the slack matrix S of the perfect matching polytope with psd rank $2^{n^{1/2+\varepsilon}}$. This psd rank is surprisingly low in view of the fact that Rothvoß [30] showed that the nonnegative rank of S is $2^{\Omega(n)}$, and Braun and Pokutta [5] showed that any $\tilde{S}$ that is $O(1/n)$-close to S still needs nonnegative rank $2^{\Omega(n)}$. This result about approximating the slack matrix for matching in low psd rank, fits in a recent line of non-quantum results derived using tools and techniques from quantum information theory (see [11]).

Communication protocols derived from query algorithms have a specific structure. In spirit, this is somewhat similar to looking at linear/psd extensions derived from hierarchies of specific linear or semidefinite programs like the Sherali-Adams and Lasserre hierarchies. In Section 2.3 we show these two relaxations actually correspond in a precise sense: just as the linear and psd extension complexities are characterized by models of communication complexity in expectation, the Sherali-Adams and Lasserre hierarchies are characterized by randomized and quantum models of query complexity in expectation, respectively. This follows from known characterizations of these hierarchies in terms of polynomial degrees that exactly correspond to the ones considered here.

Remark: Due to space limitations, many of the proofs have been omitted from this version. These can be found in the longer version at `arXiv:1411.7280`.

2   Preliminaries

2.1   Polytopes and Extension Complexity

While most of this paper is about *query* complexity in expectation, much of it is motivated by (the hope to port our results to) *communication* complexity in expectation and its consequences for linear and semidefinite extension complexity of polytopes. Hence we start with the latter. A polytope $P \subseteq \mathbb{R}^d$ has both an *inner description* as the convex hull of a set $V \subseteq \mathbb{R}^d$ of points, $P = \mathrm{conv}(V)$; and an *outer description* as the intersection of halfspaces, $P = \{x \in \mathbb{R}^d : Ax \leq b\}$. A *slack matrix* integrates information from these two descriptions:

Definition 1. *Let $P = \mathrm{conv}(V) = \{x : Ax \leq b\}$ be a polytope. The slack matrix M of P has columns labeled by $v \in V$ and rows labeled by constraints $A_i x \leq b_i$, with entries $M(i,v) = b_i - A_i v$.*

Definition 2. *Let M be a nonnegative matrix. A nonnegative factorization of M of size d consists of two sets of d-dimensional nonnegative vectors $\{a_x\}, \{b_y\}$ such that $M(x,y) = a_x^T b_y$ for all x, y. The nonnegative rank of M, denoted $\mathrm{rk}_+(M)$, is the minimal size among all nonnegative factorizations of M. Equivalently, it is the minimum number of nonnegative rank-one matrices whose sum is M.*

Definition 3. *Let M be a nonnegative matrix. A psd factorization of M of size d consists of two sets of d-by-d psd matrices $\{A_x\}, \{B_y\}$ such that $M(x,y) = \mathrm{Tr}(A_x B_y)$ for all x, y. The psd rank of M, denoted $\mathrm{rk}_{\mathrm{psd}}(M)$, is the minimal size among all psd factorizations of M.*

A nonnegative factorization is a psd factorization by diagonal matrices.

The *linear extension complexity* of a polytope P is the minimum number of facets of a (higher-dimensional) polytope which projects to P. The *semidefinite (psd) extension complexity* of P is the minimum d such that an affine slice of the cone of d-by-d positive semidefinite matrices projects to P. These complexity measures can be captured in terms of the above notions of rank of a slack matrix:

Theorem 1 ([16,36]). *The linear extension complexity of a polytope P is the nonnegative rank of a slack matrix of P. The semidefinite (psd) extension complexity of P is the psd rank of a slack matrix of P.*

A polytope may have different slack matrices associated with it, depending on which inner and outer description are used. By Theorem 1 these slack matrices all have the same nonnegative and psd rank.

One of our targets is the correlation polytope: $\mathrm{COR}_n = \{xx^T : x \in \{0,1\}^n\}$. Fiorini et al. [15] showed that lower bounds on the linear/semidefinite extension complexity of the correlation polytope imply lower bounds on several other polytopes of interest, including the Traveling Salesman Polytope. The next lemma from [28] gives a family of submatrices of the slack matrix of COR_n.

Lemma 1. *Let $p(z) = a + bz + cz^2$ be a single-variate degree-2 polynomial nonnegative on $\{0,1,\ldots,n\}$. The matrix $M(x,y) = p(|x \wedge y|)$ for $(x,y) \in \{0,1\}^n$ is a submatrix of a slack matrix for the correlation polytope COR_n.*

In Section 6 we consider the matrix $M(x,y) = (|x \wedge y| - 1)(|x \wedge y| - 2)$ and its associated query problem $f(x) = (|x| - 1)(|x| - 2)$, where $|x|$ is Hamming weight.

2.2 Polynomials

We will study two types of polynomials that are obviously nonnegative on the Boolean cube: nonnegative literal polynomials and sum-of-squares polynomials.

Definition 4 (nonnegative literal degree). *A nonnegative literal polynomial is a nonnegative linear combination of products of variables and negations of variables, i.e., it can be written as*

$$p(x) = \sum_{S \subseteq [n]} \sum_{b \in \{0,1\}^{|S|}} \alpha_{S,b} \prod_{i \in S} ((-1)^{b_i} x_i + b_i) \tag{1}$$

where each $\alpha_{S,b} \geq 0$. *Its degree is* $\max\{|S| : \alpha_{S,b} \neq 0\}$. *The nonnegative literal degree of* $f : \{0,1\}^n \to \mathbb{R}_+$, *denoted* $\mathrm{ldeg}_+(f)$, *is the minimum degree of a nonnegative literal polynomial* p *that equals* f *on* $\{0,1\}^n$.

This measure has also been called the *nonnegative junta certificate degree* [23].

Definition 5 (sum-of-squares degree). *Let* d *be a natural number. A sum-of-squares polynomial of degree* d *is a polynomial* p *that can be written in the form* $p(x) = \sum_{i \in \mathcal{P}} p_i(x)^2$, *where* $\mathcal{P}$ *is a finite index set and the* p_i *are polynomials of degree* $\leq d$. *The sum-of-squares (sos) degree of* $f : \{0,1\}^n \to \mathbb{R}_+$, *denoted* $\deg_{sos}(f)$, *is the minimum* d *for which such a* p *equals* f *on* $\{0,1\}^n$.

Note that a sum-of-squares polynomial of degree d is actually a polynomial of degree $2d$; we allow this slight abuse of notation in order to give a clean characterization in Theorem 3 below.

2.3   The Sherali-Adams and Lasserre Hierarchies

Consider the optimization problem

$$\alpha(f) = \max_{x \in \{0,1\}^n} f(x) \tag{2}$$

where f is given by a multilinear polynomial. Many important optimization problems can be cast in this framework, including NP-hard ones. For example finding the maximum cut in a graph $G = (V, E)$ with n vertices corresponds to the quadratic function $f(x) = \sum_{\{i,j\} \in E} x_i(1 - x_j) + x_j(1 - x_i)$.

If $c \geq \alpha(f)$, then $c - f$ is nonnegative on $\{0,1\}^n$. One way we can witness this is by expressing $c - f$ as a polynomial which is obviously nonnegative for all $x \in \{0,1\}^n$. The *Sherali-Adams hierarchy* [31] looks for a witness in the form of a nonnegative literal polynomial. The sum-of-squares or *Lasserre hierarchy* looks for a witness in the form of a sum-of-squares polynomial [21,29,32].

If we can find a nonnegative literal polynomial p of degree d such that $c - f(x) = p(x)$, then this witnesses that the optimal value is upper bounded as $\alpha(f) \leq c$. Moreover, determining if the nonnegative literal polynomial degree of $c - f(x)$ is at most d can be formulated as a linear program of size $n^{O(d)}$. The value of the d-round Sherali-Adams relaxation for (2) is the smallest value of c such that $c - f(x)$ is a degree-d nonnegative literal polynomial. Thus the smallest d for which a Sherali-Adams relaxation certifies an *optimal* upper bound, is exactly the nonnegative literal degree $\mathrm{ldeg}_+(\alpha(f) - f)$ of the function $\alpha(f) - f$.

Similarly, if we can find $p_i : \{0,1\}^n \to \mathbb{R}$ of degree at most d, such that $c - f(x) = \sum_i p_i(x)^2$, then this witnesses that $\alpha(f) \leq c$. Searching for such polynomials p_i can be expressed as a semidefinite program of size $n^{O(d)}$. The smallest value of c such that $c - f$ is degree-d sum-of-squares is known to be equivalent to the relaxation of (2) given by the dth level of the Lasserre hierarchy. The level of the Lasserre hierarchy required to exactly capture (2) is thus $\deg_{sos}(\alpha(f) - f)$.

3 Randomized Query Complexity in Expectation

In this section we define and characterize classical randomized query complexity in expectation, characterize it by the nonnegative literal degree, and relate it to the Sherali-Adams hierarchy. A *randomized decision tree* is a probability distribution μ over deterministic decision trees. We consider deterministic decision trees with leaves labeled by nonnegative real numbers. A randomized decision tree computes a function $f : \{0,1\}^n \to \mathbb{R}_+$ *in expectation* if for every $x \in \{0,1\}^n$ the expected output of the tree on input x is $f(x)$. The *cost* of such a tree is, as usual, the maximum cost, that is the length of a longest path from the root to a leaf, of a deterministic decision tree that has nonzero μ-probability.

Definition 6. *The randomized query complexity of computing f in expectation, denoted* $\mathrm{RE}(f)$, *is the minimum cost among all randomized decision trees that compute f in expectation.*

Theorem 2. *Let $f : \{0,1\}^n \to \mathbb{R}_+$. Then* $\mathrm{RE}(f) = \mathrm{ldeg}_+(f)$.

Referring back to Section 2.3, this gives a connection between randomized query complexity in expectation and the Sherali-Adams hierarchy: the smallest d such that the d-round Sherali-Adams relaxation certifies the optimal upper bound $\alpha(f)$ on the maximization problem (2), is exactly $\mathrm{RE}(\alpha(f) - f)$.

4 Quantum Query Complexity in Expectation

Here we study *quantum* query complexity in expectation, characterize it by sum-of-squares degree, and relate it to the Lasserre hierarchy. We assume familiarity with quantum computing [27] and query complexity [9].

We define the quantum query complexity of computing a function $f : \{0,1\}^n \to \mathbb{R}_+$ in expectation. A T-query algorithm is described by unitaries $U_0, \ldots, U_T$ and a final POVM measurement $\{E_\theta\}_{\theta \in \Theta}$, where each E_θ is a psd matrix labeled by nonnegative real θ, and $\sum_{\theta \in \Theta} E_\theta = I$. As usual, on input x the query algorithm proceeds from the initial state $|\bar{0}\rangle$ by alternately applying a unitary and the query oracle O_x (which maps $|i, b\rangle \mapsto |i, b \oplus x_i\rangle$), so that the final state of the algorithm after T queries is $|\psi_x^T\rangle = U_T O_x \ldots O_x U_1 O_x U_0 |\bar{0}\rangle$. Let $E = \sum_{\theta \in \Theta} \theta E_\theta$. As the probability of output θ upon measuring $|\psi_x^T\rangle$ is $\mathrm{Tr}(E_\theta |\psi_x^T\rangle\langle\psi_x^T|)$, the expected value of the output is $\mathrm{Tr}(E|\psi_x^T\rangle\langle\psi_x^T|)$. The algorithm *computes f in expectation* if $f(x) = \mathrm{Tr}(E|\psi_x^T\rangle\langle\psi_x^T|)$ for every $x \in \{0,1\}^n$.

Definition 7. *The quantum query complexity of computing f in expectation, denoted* $\mathrm{QE}(f)$, *is the minimum T for which there is a T-query quantum algorithm computing f in expectation.*

Theorem 3. *Let $f : \{0,1\}^n \to \mathbb{R}_+$. Then* $\mathrm{QE}(f) = \deg_{sos}(f)$.

Proof. $\text{QE}(f) \geq \deg_{sos}(f)$. Say there is a T-query algorithm to compute f in expectation. Let $|\psi_x^T\rangle$ denote its state on input x after T queries. By the polynomial method [2], the amplitude of each basis state in $|\psi_x^T\rangle$ is an n-variate multilinear polynomial in x of degree $\leq T$. We have $f(x) = \sum_\theta \theta \langle \psi_x^T | E_\theta | \psi_x^T \rangle$. Let $E_\theta = \sum_i \lambda_i |e_\theta^i\rangle\langle e_\theta^i|$ be the eigenvalue decomposition of E_θ, where each $\lambda_i \geq 0$. Then $\langle \psi_x^T | E_\theta | \psi_x^T \rangle = \sum_i \lambda_i |\langle \psi_x^T | e_\theta^i \rangle|^2$. Since $\langle \psi_x^T | e_\theta^i \rangle$ is a linear combination of amplitudes of $|\psi_x^T\rangle$, it is a degree $\leq T$ polynomial in x. Since the coefficients θ and λ_i are nonnegative, this gives a representation of $\langle \psi_x^T | E_\theta | \psi_x^T \rangle$ as a sum-of-squares polynomial of degree $\leq T$.

$\underline{\text{QE}(f) \leq \deg_{sos}(f)}$. Let $d = \deg_{sos}(f)$. We first exhibit a quantum algorithm for the special case where $f = p^2$ for some degree-d polynomial p. This is inspired by the proof of [35, Theorem 2.3]. Let $p = \sum_s \widehat{p}(s)(-1)^{x \cdot s}$ be the Fourier representation of p, where s ranges over $\{0,1\}^n$. Because p has degree d, we have $\widehat{p}(s) \neq 0$ only if $|s| \leq d$. The algorithm is as follows:

1. Prepare n-qubit state $c \sum_s \widehat{p}(s)|s\rangle$, where $c = 1/\sqrt{\sum_s \widehat{p}(s)^2}$ is a constant.
2. Apply a unitary that maps $|s\rangle \mapsto (-1)^{x \cdot s}|s\rangle$ for all s of weight $|s| \leq d$; one can show that this can be implemented using d queries.
3. Apply the n-qubit Hadamard transform to the state.
4. Measure the state and output $2^n/c^2$ if the result was 0^n, otherwise output 0.

Note that the amplitude of the basis state $|0^n\rangle$ after step 3 is $\frac{c}{\sqrt{2^n}} \sum_s \widehat{p}(s)(-1)^{x \cdot s} = \frac{c}{\sqrt{2^n}} p(x)$. Hence the probability that the final measurement results in outcome 0^n is $(\frac{c}{\sqrt{2^n}} p(x))^2$, and the expected value of the output is $(\frac{c}{\sqrt{2^n}} p(x))^2 \cdot 2^n/c^2 = p(x)^2 = f(x)$, as desired. Now consider the general case where $f = \sum_{i \in \mathcal{P}} p_i^2$. The algorithm chooses one $i \in \mathcal{P}$ uniformly at random and runs the above algorithm to produce an output with expected value $p_i(x)^2$. It finally outputs that output multiplied by $|\mathcal{P}|$. Clearly, this uses at most d queries to x, and the expected value of its final output is $\frac{1}{|\mathcal{P}|} \sum_i p_i(x)^2 |\mathcal{P}| = \sum_i p_i(x)^2 = f(x)$. □

This connects quantum query complexity in expectation and the Lasserre hierarchy: the smallest level d of the Lasserre hierarchy that certifies the optimal upper bound $\alpha(f)$ on the maximization problem (2), is exactly $\text{QE}(\alpha(f) - f)$.

5   Gaps and Relations between $\text{RE}(f)$ and $\text{QE}(f)$

For some $f : \{0,1\}^n \to \mathbb{R}_+$, the quantum query complexity in expectation $\text{QE}(f)$ can be *much* smaller than its classical counterpart $\text{RE}(f)$. An extreme example is the n-bit function $f(x) = (|x| - 1)^2$, where $\text{QE}(f) = 1$ by Theorem 3, but $\text{RE}(f) = n$. The latter holds because on the all-0 input the algorithm needs to produce a nonzero output with positive probability, but on weight-1 inputs it can never output anything nonzero, hence a classical algorithm needs n queries on the all-0 input. In contrast, if the range of f is Boolean, then we can show that $\text{QE}(f)$ is at most polynomially smaller than $\text{RE}(f)$:

Theorem 4. *For every $f : \{0,1\}^n \to \{0,1\}$ we have $\text{RE}(f) \leq 16\text{QE}(f)^3$.*

The main reason this query complexity result is interesting is that the analogous statement for *communication* complexity is equivalent to the longstanding log-rank conjecture! The communication version of Theorem 4 would say that for all *Boolean* matrices M, the quantum and classical communication complexity of computing M in expectation are at most polynomially far apart. As noted by Fiorini et al. [15], this is equivalent to $\log \mathrm{rk}_+(M) \leq \mathrm{polylog}(\mathrm{rk}_{\mathrm{psd}}(M))$, which in turn is equivalent to the log-rank conjecture. Presumably such a communication version will be substantially harder to prove than the above query version. However, in many cases results in query complexity "mirror" (often much harder) results in communication complexity, so our Theorem 4 may be viewed as (weak) evidence for the log-rank conjecture.

6   A Quantum Query Complexity Lower Bound

Here we show that the n-bit function $f(x) = (|x|-1)(|x|-2)$ has $\mathrm{QE}(f) = \Omega(\sqrt{n})$. This result is motivated by the fact that a strong lower bound on the psd rank of the closely related matrix $M(x,y) = (|x \wedge y|-1)(|x \wedge y|-2)$ would have important consequences for the correlation polytope (M is a submatrix of the slack matrix for the correlation polytope, see Lemma 1). We hope that the methods of this section may in the future help lower bound this psd rank as well.

We prove our query complexity lower bound by showing the corresponding lower bound on the sum-of-squares degree of f. As is common in query complexity lower bounds by the polynomial method [2], we will use a symmetrization argument to define a single-variate polynomial $Q : \mathbb{R} \to \mathbb{R}$ that behaves well on $[n]$, and then use Markov's lemma from approximation theory to bound the degree of Q. A new complication in our setting is the following. If $f(x) = \sum_i p_i(x)^2$ then we would like to define a "symmetrized" polynomial $g : [n] \to \mathbb{R}$ where $g(k) = \mathbb{E}_{x:|x|=k}\left[\sum_i p_i(x)^2\right]$. However, we do not know how to prove that g remains a nonnegative polynomial. To get around this, we define symmetrized polynomials $q_i(k) = \mathbb{E}_{x:|x|=k}[p_i(x)]$ for each p_i individually, then recombine the symmetrized polynomials as $Q(k) = \sum_i q_i(k)^2$. We are then able to bound the sum-of-squares degree of Q.

Theorem 5. *If* $f(x) = (|x| - 1)(|x| - 2)$ *for* $x \in \{0,1\}^n$, $\deg_{sos}(f) \geq \sqrt{n/48}$.

7   Psd Rank and Query Complexity in Expectation

Fiorini et al. [15] defined a *one-way* model of quantum communication to compute a matrix in expectation, and showed that this complexity is characterized by the logarithm of the psd rank. We show below that this characterization still holds for the more general *two-way* communication model, which allows multiple rounds of communication. Hence one-way and two-way quantum communication complexity are the same for computation in expectation.

We will not formally define the model of two-way quantum communication complexity (see [34] for more technical details), instead just highlighting the differences of the model of computing a function in expectation to the normal model. As usual, Alice and Bob each start with their own input, x and y respectively, and then the protocol specifies whose turn it is to speak and what message they send to the other party. At the end of the protocol Bob must output a *nonnegative* number, which is a random variable z that depends on the inputs x and y as well as on the internal randomness of the protocol.

The major difference with the usual model is the notion of when a protocol is correct. Let M be a matrix with nonnegative real entries whose rows are indexed by Alice's possible inputs, and whose columns are indexed by Bob's inputs. We say a protocol *computes the matrix M in expectation* if, for every (x, y), $M(x, y)$ equals the expected value of the output z on input (x, y). As usual, the *cost* of the protocol is the worst-case number of qubits communicated (over all rounds).

Definition 8. *The quantum communication complexity of computing a matrix M in expectation, denoted* QCE(M), *is the minimum q such that there exists a quantum protocol of cost q that computes M in expectation. The minimum q when we restrict to one-way protocols is denoted* QCE$^1(M)$.

It turns out that two-way quantum communication complexity is not more powerful than its one-way cousin: both correspond to the psd rank.

Theorem 6. $\log \mathrm{rk}_{\mathrm{psd}}(M) \leq \mathrm{QCE}(M) \leq \mathrm{QCE}^1(f) \leq \lceil \log(\mathrm{rk}_{\mathrm{psd}}(M) + 1) \rceil$.

7.1 Upper Bounds on psd Rank from Quantum Algorithms

We can show that efficient quantum query algorithms for computing functions $f : \{0, 1\}^n \to \mathbb{R}_+$ in expectation give rise to an efficient quantum communication protocol to compute the matrix $M_f(x, y) = f(x \wedge y)$ in expectation, and hence to a low-rank psd factorization of M_f. We state it more generally:

Theorem 7. *Let Y be a finite set. For every $y \in Y$, let $f_y : \{0, 1\}^n \to \mathbb{R}_+$ satisfy* QE(f_y) $\leq T$. *Define a $2^n \times |Y|$ matrix M by $M(x, y) = f_y(x)$. Then* QCE(M) $\leq 2T(\log(n) + 1)$, *and hence* $\mathrm{rk}_{\mathrm{psd}}(M) \leq (2n)^{2T}$.

Lee et al. [23] independently proved a similar upper bound on psd rank in terms of the sos-degree of f_y rather than quantum query complexity.

As an application we will derive an exponentially-close entrywise approximation of the slack matrix S of the perfect matching polytope, by a matrix with psd rank not much bigger than $2^{\sqrt{n}}$. This shows a big difference to the case of nonnegative rank: Braun and Pokutta [5] show that any $\tilde{S}$ that is $O(1/n)$-close to S needs nonnegative rank $2^{\Omega(n)}$.

Edmonds gave a complete description of the facets of the perfect matching polytope for the complete n-vertex graph K_n [12]. The key are the *odd-set* inequalities: for a perfect matching M, viewed as a vector $M \in \{0, 1\}^{\binom{n}{2}}$ of weight $m = n/2$, and an odd-sized set $U \subseteq [n]$, the associated inequality says $|\delta(U) \cap M| \geq 1$, where $\delta(U) \in \{0, 1\}^{\binom{n}{2}}$ denotes the cut induced by U. In addition, there are $O(n^2)$ degree and nonnegativity constraints. Thus the corresponding slack matrix S has columns indexed by all perfect matchings M in K_n and rows indexed by odd-sized sets U with entries $S_{UM} = |\delta(U) \cap M| - 1$. There are $O(n^2)$ additional rows for the degree and nonnegativity constraints.

In the full version of this paper we show that the m-bit function $g(z) = |z| - 1$ can be approximated (in expectation) up to exponentially small error with quantum query complexity $O(m^{1/2+\varepsilon} \log m)$. Define $f_M(x) = g(x_M)$, where x_M denotes the restriction of n-bit string x to the m positions in the support of M. Applying Theorem 7 and adding $O(n^2)$ rows for the other constraints gives:

Theorem 8. $\forall \varepsilon > 0$ there is a matrix $\tilde{S}$ of psd rank $2^{O(n^{1/2+\varepsilon}(\log n)^2)}$ s.t.

1. $S_{UM} - 2^{-(n/2)^{2\varepsilon}} \leq \tilde{S}_{UM} \leq S_{UM}$ for the entries where $|\delta(U) \cap M| > (n/2)^{2\varepsilon}$;
2. $\tilde{S}_{xy} = S_{xy}$ for all other entries.

Acknowledgments. We thank Srinivasan Arunachalam, David Steurer, Mario Szegedy and Henry Yuen for useful discussions, Sebastian Pokutta for useful discussions and for pointing us to [5], and James Lee for sending us a version of [23]. Troy Lee is supported in part by the Singapore National Research Foundation under NRF RF Award No. NRF-NRFF2013-13. Ronald de Wolf is partially supported by a Vidi grant from the Netherlands Organization for Scientific Research (NWO) which ended in 2013, ERC Consolidator Grant QPROGRESS, and by the European Commission IST STREP project QALGO 600700.

References

1. Arunachalam, S., Yuen, H., de Wolf, R.: Unpublished manuscript, August 2014
2. Beals, R., Buhrman, H., Cleve, R., Mosca, M., de Wolf, R.: Quantum lower bounds by polynomials. Journal of the ACM **48**(4), 778–797 (2001)
3. Blekherman, G., Gouveia, J., Pfeiffer, J.: Sums of squares on the hypercube, February 18, 2014. arXiv/1402.4199
4. Brassard, G., Høyer, P., Mosca, M., Tapp, A.: Quantum amplitude amplification and estimation. In: Quantum Computation and Quantum Information: A Millennium Volume, AMS Contemporary Mathematics Series, vol. 305, pp. 53–74 (2002)
5. Braun, G., Pokutta, S.: The matching polytope does not admit fully-polynomial size relaxation schemes. In: Proc. of 26th SODA, pp. 837–846 (2015)
6. Buhrman, H., Cleve, R., Wigderson, A.: Quantum vs. classical communication and computation. In: Proc. of 30th ACM STOC, pp. 63–68 (1998)
7. Buhrman, H., Cleve, R., de Wolf, R., Zalka, C.: Bounds for small-error and zero-error quantum algorithms. In: Proc. of 40th IEEE FOCS, pp. 358–368 (1999)
8. Buhrman, H., de Wolf, R.: Communication complexity lower bounds by polynomials. In: Proc. of 16th IEEE Complexity (CCC), pp. 120–130 (2001)
9. Buhrman, H., de Wolf, R.: Complexity measures and decision tree complexity: A survey. Theoretical Computer Science **288**(1), 21–43 (2002)
10. Chan, S.O., Lee, J.R., Raghavendra, P., Steurer, D.: Approximate constraint satisfaction requires large LP relaxations. In: Proc. of 54th IEEE FOCS, pp. 350–359 (2013)
11. Drucker, A., de Wolf, R.: Quantum proofs for classical theorems. Theory of Computing (2011). ToC Library, Graduate Surveys 2

12. Edmonds, J.: Maximum matching and a polyhedron with 0,1-vertices. Journal of research of the National Bureau of Standards-B **69B**(1,2), 125–130 (1965)
13. Faenza, Y., Fiorini, S., Grappe, R., Tiwary, H.R.: Extended formulations, non-negative factorizations, and randomized communication protocols. In: Mahjoub, A.R., Markakis, V., Milis, I., Paschos, V.T. (eds.) ISCO 2012. LNCS, vol. 7422, pp. 129–140. Springer, Heidelberg (2012)
14. Fawzi, H., Saunderson, J., Parrilo, P.: Equivariant semidefinite lifts and sum-of-squares hierarchies, December 23, 2013. arXiv:1312.6662
15. Fiorini, S., Massar, S., Pokutta, S., Tiwary, H.R., de Wolf, R.: Linear vs. semidefinite extended formulations: exponential separation and strong lower bounds. In: Proc. of 44th ACM STOC, pp. 95–106 (2012)
16. Gouveia, J., Parrilo, P., Thomas, R.: Lifts of convex sets and cone factorizations. Mathematics of Operations Research **38**(2), 248–264 (2013). arXiv:1111.3164
17. Grigoriev, D.: Complexity of Positivstellensatz proofs for the knapsack. Computational Complexity **10**, 139–154 (2001)
18. Grover, L.K.: A fast quantum mechanical algorithm for database search. In: Proc. of 28th ACM STOC, pp. 212–219 (1996). quant-ph/9605043
19. Kremer, I.: Quantum Communication. MSc thesis, Hebrew University (1995)
20. Kushilevitz, E., Nisan, N.: Communication complexity. Cambridge UP (1997)
21. Lasserre, J.B.: Global optimization with polynomials and the problem of moments. SIAM Journal on Optimization **11**(3), 796–817 (2001)
22. Laurent, M.: Lower bound for the number of iterations in semidefinite hierarchies for the cut polytope. Mathematics of operations research **28**(4), 871–883 (2003)
23. Lee, J.R., Raghavendra, P., Steurer, D.: Lower bounds on the size of semidefinite programming relaxations, November 24, 2014. To appear in STOC 2015. arXiv:1411.6317
24. Lee, J.R., Raghavendra, P., Steurer, D., Tan, N.: On the power of symmetric LP and SDP relaxations. In: Proc. of 29th IEEE Complexity (CCC), pp. 13–21 (2014)
25. Midrijanis, G.: Exact quantum query complexity for total Boolean functions, March 23, 2004. quant-ph/0403168
26. Minsky, M., Papert, S.: Perceptrons. MIT Press (1987)
27. Nielsen, M.A., Chuang, I.L.: Quantum Computation and Quantum Information. Cambridge University Press (2000)
28. Padberg, M.: The boolean quadric polytope. Math. prog. **45**, 139–172 (1989)
29. Parrilo, P.: Structured semidefinite programs and semialgebraic geometry methods in robustness and optimization. Ph.D. thesis, Caltech (2000)
30. Rothvoß, T.: The matching polytope has exponential extension complexity. In: Proc. of 46th ACM STOC, pp. 263–272 (2014)
31. Sherali, H.D., Adams, W.P.: A hierarchy of relaxations between the continuous and convex hull representations for zero-one programming. SIAM Journal on Discrete Mathematics **3**, 411–430 (1990)
32. Shor, N.Z.: An approach to obtaining global extremums in polynomial mathematical programming problems. Cybernetics **23**, 695–700 (1987)
33. Swart, T.: P = NP. Tech. rep., University of Guelph (1986), revision 1987
34. de Wolf, R.: Quantum communication and complexity. Theoretical Computer Science **287**(1), 337–353 (2002)
35. de Wolf, R.: Nondeterministic quantum query and quantum communication complexities. SIAM Journal on Computing **32**(3), 681–699 (2003)
36. Yannakakis, M.: Expressing combinatorial optimization problems by linear programs. Journal of Computer and System Sciences **43**(3), 441–466 (1991)
37. Yao, A.C.C.: Quantum circuit complexity. In: Proc. of 34th IEEE FOCS, pp. 352–360 (1993)

Near-Linear Query Complexity for Graph Inference

Sampath Kannan[1], Claire Mathieu[2], and Hang Zhou[2(⊠)]

[1] Department of Computer and Information Science,
University of Pennsylvania, Philadelphia, PA, USA
kannan@cis.upenn.edu
[2] Département d'Informatique UMR CNRS 8548,
École Normale Supérieure, Paris, France
{cmathieu,hangzhou}@di.ens.fr

Abstract. How efficiently can we find an unknown graph using distance or shortest path queries between its vertices? Let $G = (V, E)$ be a connected, undirected, and unweighted graph of bounded degree. The edge set E is initially unknown, and the graph can be accessed using a *distance oracle*, which receives a pair of vertices (u, v) and returns the distance between u and v. In the *verification* problem, we are given a hypothetical graph $\hat{G} = (V, \hat{E})$ and want to check whether G is equal to $\hat{G}$. We analyze a natural greedy algorithm and prove that it uses $n^{1+o(1)}$ distance queries. In the more difficult *reconstruction* problem, $\hat{G}$ is not given, and the goal is to find the graph G. If the graph can be accessed using a *shortest path oracle*, which returns not just the distance but an actual shortest path between u and v, we show that extending the idea of greedy gives a reconstruction algorithm that uses $n^{1+o(1)}$ shortest path queries. When the graph has bounded treewidth, we further bound the query complexity of the greedy algorithms for both problems by $\tilde{O}(n)$. When the graph is chordal, we provide a randomized algorithm for reconstruction using $\tilde{O}(n)$ distance queries.

1 Introduction

How efficiently can we find an unknown graph using distance or shortest path queries between its vertices? This is a natural theoretical question from the standpoint of recovery of hidden information. This question is related to the *reconstruction* of Internet networks. Discovering the topology of the Internet is a crucial step for building accurate network models and designing efficient algorithms for Internet applications. Yet, this topology can be extremely difficult to find, due to the dynamic structure of the network and to the lack of centralized control. The network reconstruction problem has been studied extensively [1,2, 5,6,10,12]. Sometimes we have some idea of what the network should be like, based perhaps on its state at some past time, and we want to check whether our image of the network is correct. This is network *verification* and has received

---

The full version of the paper is available on the authors' websites.

© Springer-Verlag Berlin Heidelberg 2015
M.M. Halldórsson et al. (Eds.): ICALP 2015, Part I, LNCS 9134, pp. 773–784, 2015.
DOI: 10.1007/978-3-662-47672-7_63

attention recently [2,3,6]. This is an important task for routing, error detection, or ensuring service-level agreement (SLA) compliance, etc. For example, Internet service providers (ISPs) offer their customers services that require quality of service (QoS) guarantees, such as voice over IP services, and thus need to check regularly whether the networks are correct.

The topology of Internet networks can be investigated at the router and autonomous system (AS) level, where the set of routers (ASs) and their physical connections (peering relations) are the vertices and edges of a graph, respectively. Traditionally, we use tools such as traceroute and mtrace to infer the network topology. These tools generate path information between a pair of vertices. It is a common and reasonably accurate assumption that the generated path is the shortest one, i.e., minimizes the hop distance between that pair. In our first theoretical model, we assume that we have access to any pair of vertices and get in return their shortest path in the graph. Sometimes routers block traceroute and mtrace requests (e.g., due to privacy and security concerns), thus the inference of topology can only rely on delay information. In our second theoretical model, we assume that we get in return the hop distance between a pair of vertices. The second model was introduced in [10].

Graph inference using queries that reveal partial information has been studied extensively in different contexts, independently stemming from a number of applications. Beerliova et al. [2] studied network verification and reconstruction using an oracle, which, upon receiving a node q, returns all shortest paths from q to all other nodes, instead of one shortest path between a pair of nodes as in our first model. Erlebach et al. [6] studied network verification and reconstruction using an oracle which, upon receiving a node q, returns the distances from q to all other nodes in the graph, instead of the distance between a pair of nodes as in our second model. They showed that minimizing the number of queries for verification is NP-hard and admits an $O(\log n)$-approximation algorithm. In the *network realization* problem, we are given the distances between certain pairs of vertices and asked to determine the sparsest graph (in the unweighted case) or the graph of least total weight that realizes these distances. This problem was shown to be NP-hard [4]. In evolutionary biology, a well-studied problem is reconstructing evolutionary trees, thus the hidden graph has a tree structure. See for example [7,9,11]. One may query a pair of species and get in return the distance between them in the (unknown) tree. In our reconstruction problem, we allow the hidden graph to have an arbitrarily connected topology, not necessarily a tree structure.

1.1 The Problem

Let $G = (V, E)$ be a hidden graph that is connected, undirected, and unweighted, where $|V| = n$. We consider two query oracles. A *shortest path oracle* receives a pair $(u, v) \in V^2$ and returns a shortest path between u and v.[1] A *distance oracle*

---

[1] If there are several shortest paths between u and u, the oracle returns an arbitrary one.

receives a pair $(u, v) \in V^2$ and returns the number of edges on a shortest path between u and v.

In the *graph reconstruction* problem, we are given the vertex set V and have access to either a distance oracle or a shortest path oracle. The goal is to find every edge in E.

In the *graph verification* problem, again we are given V and have access to either oracle. In addition, we are given a connected, undirected, and unweighted graph $\hat{G} = (V, \hat{E})$. The goal is to check whether $\hat{G}$ is correct, that is, whether $\hat{G} = G$.

The efficiency of an algorithm is measured by its *query complexity*[2], i.e., the number of queries to an oracle. We focus on query complexity, while all our algorithms are of polynomial time and space. We note that $O(n^2)$ queries are enough for both reconstruction and verification via a distance oracle or a shortest path oracle: we only need to query every pair of vertices.

Let Δ denote the maximum degree of any vertex in the graph G. Unless otherwise stated, we assume that Δ is bounded, which is reasonable for real networks that we want to reconstruct or verify. Indeed, when Δ is $\Omega(n)$, both reconstruction and verification require $\Omega(n^2)$ distance or shortest path queries.

Let us focus on bounded degree graphs. It is not hard to see that $\Omega(n)$ distance or shortest path queries are required. The central question in this line of work is therefore: **Is the query complexity linear, quadratic, or somewhere in between?** In [10], Mathieu and Zhou provide a first answer: the query complexity for reconstruction via a distance oracle is subquadratic: $\tilde{O}(n^{3/2})$. In this paper, we show that the query complexity for reconstruction via a shortest path oracle or verification via either oracle is near-linear: $n^{1+o(1)}$. It is open whether there is an algorithm for reconstruction using a near-linear number of distance queries.

1.2   Our Results

Verification

Theorem 1. *For graph verification using a distance oracle, there is a deterministic algorithm (Algorithm 1) with query complexity $n^{1+O\left(\sqrt{(\log\log n + \log \Delta)/\log n}\right)}$, which is $n^{1+o(1)}$ when the maximum degree $\Delta = n^{o(1)}$. If the graph has treewidth w, the query complexity can be further bounded by $O(\Delta(\Delta + w \log n)n \log^2 n)$, which is $\tilde{O}(n)$ when Δ and w are $O(\text{polylog } n)$.*

The main task for verification is to confirm the *non-edges* of the graph. Algorithm 1 is greedy: every time it makes a query that confirms the largest number of non-edges that are not yet confirmed. To analyze the algorithm, first, we show that its query complexity is roughly $\ln n$ times the optimal number of queries OPT for verification. This is based on a reduction to the SET-COVER problem, see Section 3.1. It only remains to bound OPT.

---

[2] Expected query complexity in the case of randomized algorithms.

Table 1. Results (for bounded degree graphs). New results are in bold.

Objective	Query complexity
verification via either oracle	$n^{1+o(1)}$
reconstruction via a shortest path oracle	bounded treewidth: $\tilde{O}(n)$ **(Thm 1, Cor 2, and Thm 3)**
reconstruction via a distance oracle	$\tilde{O}(n^{3/2})$ [10] $\Omega(n \log n / \log \log n)$ **(Thm 5)** outerplanar: $\tilde{O}(n)$ [10] chordal: $\tilde{O}(n)$ **(Thm 4)**

To bound *OPT* and get the first statement in Theorem 1, it is enough to prove the desired bound for a different verification algorithm. This algorithm is a more sophisticated recursive version of the algorithm in [10]. Recursion is a challenge because, when we query a pair (u, v) in a recursive subgraph, the oracle returns the distance between u and v in the entire graph, not just within the subgraph. Thus new ideas are introduced for the algorithmic design. See Section 3.3.

To show the second statement in Theorem 1, similarly, we design another recursive verification algorithm with query complexity $\tilde{O}(n)$ for graphs of bounded treewidth. The algorithm uses some bag of a tree decomposition to separate the graph into balanced subgraphs, and then recursively verifies each subgraph. The same obstacle to recursion occurs. Our approach here is to add a few weighted edges to each subgraph in order to preserve the distance metric. The complete proof is in the full version of the paper.

We note that each query to a distance oracle can be simulated by the same query to a shortest path oracle. So from Theorem 1, we have:

Corollary 2. *For graph verification using a shortest path oracle, Algorithm 1 achieves the same query complexity as in Theorem 1.*

Reconstruction

Theorem 3. *For graph reconstruction using a shortest path oracle, there is a deterministic algorithm (Algorithm 4) that achieves the same query complexity as in Theorem 1.*

The key is to formulate this problem as a problem of verification using a distance oracle, so that we get the same query complexity as in Theorem 1. We extend the idea of greedy in Algorithm 1, and we show that each query to a shortest path oracle makes as much progress for reconstruction as the corresponding query to a distance oracle would have made for verifying a given graph. The main realization here is that reconstruction can be viewed as the verification of a dynamically changing graph. See Section 4.

Theorem 4. *For reconstruction of chordal graphs using a distance oracle, there is a randomized algorithm with query complexity* $O(\Delta^3 2^\Delta \cdot n(2^\Delta + \log^2 n) \log n)$, *which is* $\tilde{O}(n)$ *when the maximum degree* Δ *is* $O(\log \log n)$.

The algorithm in Theorem 4 first finds a separator using random sampling and statistical estimates, as in [10]. Then it partitions the graph into subgraphs with respect to this separator and recurses on each subgraph. However, the separator here is a clique instead of an edge in [10] for outerplanar graphs. Thus the main difficulty is to design and analyze a more general tool for partitioning the graph. The proof of the theorem is in the full version of the paper.

Lower Bounds. For graphs of bounded degree, both reconstruction and verification require $\Omega(n)$ distance or shortest path queries. In addition, there is a slightly better lower bound for reconstruction using a distance oracle, as in the following theorem.

Theorem 5. *For graph reconstruction using a distance oracle, assuming the maximum degree* $\Delta \geq 3$ *is such that* $\Delta = o\left(n^{1/2}\right)$, *any algorithm has query complexity* $\Omega(\Delta n \log n / \log \log n)$.

The proof of Theorem 5 is in the full version of the paper.

2   Notation

Let δ be the distance metric of G. For a subset of vertices $S \subseteq V$ and a vertex $v \in V$, define $\delta(S, v)$ to be $\min_{s \in S} \delta(s, v)$. For $v \in V$, let $N(v) = \{u \in V : \delta(u, v) \leq 1\}$ and let $N_2(v) = \{u \in V : \delta(u, v) \leq 2\}$. We define $\hat{\delta}$, $\hat{N}$, and $\hat{N}_2$ similarly with respect to the graph $\hat{G}$.

For a graph $G = (V, E)$, a distinct pair of vertices $uv \in V^2$ is an *edge* of G if $uv \in E$, and is a *non-edge* of G if $uv \notin E$.

For a subset of vertices $S \subseteq V$, let $G[S]$ be the subgraph induced by S. For a subset of edges $H \subseteq E$, we identify H with the subgraph induced by the edges of H. Let δ_H denote the distance metric of the subgraph H.

For a vertex $s \in V$ and a subset $T \subseteq V$, define $\text{QUERY}(s, T)$ as $\text{QUERY}(s, t)$ for every $t \in T$. For subsets $S, T \subseteq V$, define $\text{QUERY}(S, T)$ as $\text{QUERY}(s, t)$ for every $(s, t) \in S \times T$.

In the verification problem, an algorithm performs a set of queries, and its output is *no* if some query gives the wrong distance (or shortest path), and is *yes* if all queries give the right distances (or shortest paths).

3   Proof of Theorem 1

3.1   Greedy Algorithm

The task of verification comprises verifying that every edge of $\hat{G}$ is an edge of G, and verifying that every non-edge of $\hat{G}$ is a non-edge of G. The second part is called *non-edge verification*. In the second part, we assume that the first part is already done, which guarantees that $\hat{E} \subseteq E$. For graphs of bounded degree, the first part requires only $O(\Delta n)$ queries, thus the focus is on non-edge verification.

Theorem 6. *For graph verification using a distance oracle, there is a deterministic greedy algorithm (Algorithm 1) that uses at most $\Delta n + (\ln n + 1) \cdot OPT$ queries, where OPT is the optimal number of queries for non-edge verification.*

Now we prove Theorem 6. Let $\widehat{NE}$ be the set of the non-edges of $\hat{G}$. For each pair of vertices $(u, v) \in V^2$, we define $S_{u,v} \subseteq \widehat{NE}$ as follows:

$$S_{u,v} = \left\{ ab \in \widehat{NE} : \hat{\delta}(u, a) + \hat{\delta}(b, v) + 1 < \hat{\delta}(u, v) \right\}. \tag{1}$$

The following two lemmas relate the sets $S_{u,v}$ with non-edge verification.

Lemma 7. *Assume that $\hat{E} \subseteq E$. For every $(u, v) \in V^2$, if $\delta(u, v) = \hat{\delta}(u, v)$, then every pair $ab \in S_{u,v}$ is a non-edge of G.*

Proof. Consider any pair $ab \in S_{u,v}$. By the triangle inequality, $\delta(u, a) + \delta(a, b) + \delta(b, v) \geq \delta(u, v) = \hat{\delta}(u, v)$. By the definition of $S_{u,v}$ and using $\hat{E} \subseteq E$, we have $\hat{\delta}(u, v) > \hat{\delta}(u, a) + \hat{\delta}(b, v) + 1 \geq \delta(u, a) + \delta(b, v) + 1$. Thus $\delta(a, b) > 1$, i.e., ab is a non-edge of G. □

Lemma 8. *If a set of queries T verifies that every non-edge of $\hat{G}$ is a non-edge of G, then $\bigcup_{(u,v) \in T} S_{u,v} = \widehat{NE}$.*

Proof. Assume, for a contradiction, that some $ab \in \widehat{NE}$ does not belong to any $S_{u,v}$ for $(u, v) \in T$. Consider adding ab to the set of edges of $\hat{E}$: this will not create a shorter path between u and v, for any $(u, v) \in T$. Thus including ab in $\hat{E}$ is consistent with the answers of all queries in T. This contradicts the assumption that T verifies that ab is a non-edge of G. □

From Lemmas 7 and 8, the non-edge verification is equivalent to the SET-COVER problem with the universe $\widehat{NE}$ and the sets $\{S_{u,v} : (u, v) \in V^2\}$. The SET-COVER instance can be solved using the well-known greedy algorithm [8], which gives a $(\ln n + 1)$-approximation. Hence our greedy algorithm for verification (Algorithm 1). For the query complexity, first, verifying that $\hat{E} \subseteq E$ takes at most Δn queries, since the graph has maximum degree Δ. The part of non-edge verification uses a number of queries that is at most $(\ln n + 1)$ times the optimal number of queries. This proves Theorem 6.

3.2 Bounding *OPT* to Prove Theorem 1

From Theorems 6, in order to prove Theorem 1, we only need to bound *OPT*, as in the following two theorems.

Theorem 9. *For graph verification using a distance oracle, the optimal number of queries OPT for non-edge verification is $n^{1+O\left(\sqrt{(\log \log n + \log \Delta)/\log n}\right)}$.*

Algorithm 1. Greedy Verification

1: **procedure** VERIFY($\hat{G}$)
2:    **for** $uv \in \hat{E}$ **do** QUERY(u, v)
3:    $Y \leftarrow \emptyset$
4:    **while** $\hat{E} \cup Y$ does not cover all vertex pairs **do**
5:        choose (u, v) that maximizes $|S_{u,v} \setminus Y|$ $\triangleright$ $S_{u,v}$ defined in Equation (1)
6:        QUERY(u, v)
7:        $Y \leftarrow Y \cup S_{u,v}$

Theorem 10. *For graph verification using a distance oracle, if the graph has treewidth w, then the optimal number of queries OPT for non-edge verification is $O(\Delta(\Delta + w \log n)n \log n)$.*

Theorem 1 follows trivially from Theorems 6, 9, and 10, by noting that both Δ and $\log n$ are smaller than $n^{\sqrt{(\log \log n + \log \Delta)/\log n}}$. The proof of Theorem 9 is in Section 3.3, and the proof of Theorem 10 is in the full version of the paper.

3.3   Proof of Theorem 9

To show Theorem 9, we provide a recursive algorithm for non-edge verification with the query complexity in the theorem statement. As in [10], the algorithm selects a set of *centers* partitioning V into Voronoi cells and expands them slightly so as to cover all edges of G. But unlike [10], instead of using exhaustive search inside each cell, the algorithm verifies each cell recursively. The recursion is a challenge because the distance oracle returns the distance in the entire graph, not in the cell. Straightforward attempts to use recursion lead either to subcells that do not cover all edges of the cell, or to excessively large subcells. Our approach is to allow selection of centers *outside* the cell, while still limiting the subcells to being contained *inside* the cell (Figure 1). This simple but subtle setup is one novelty of the algorithmic design.

The verification algorithm uses the function SUBSET-CENTERS (Algorithm 2), which takes as input a graph $\hat{G} = (V, \hat{E})$, a subset of vertices $U \subseteq V$, and an integer $s \in [1, n]$, and outputs a set of *centers* $A \subseteq V$ such that in the graph $\hat{G}$, the vertices of the subset U are roughly equipartitioned into the Voronoi cells centered at vertices in A. This algorithm is a generalization of the CENTER algorithm by Thorup and Zwick [13]: when the subset U equals V, the SUBSET-CENTERS algorithm becomes their CENTER algorithm. For every $w \in V$, we define w's *cluster* in the graph G as $C_A(w) = \{v \in V : \delta(w, v) < \delta(A, v)\}$. We note that if $w \in A$, then $C_A(w) = \emptyset$, since $\delta(w, v) \geq \delta(A, v)$, for every $v \in V$. Similarly, we define w's cluster in the graph $\hat{G}$ as $\hat{C}_A(w) = \{v \in V : \hat{\delta}(w, v) < \hat{\delta}(A, v)\}$. The subscript A is omitted when clear from the context.

The following lemma is a straightforward extension of Theorem 3.1 in [13].

Algorithm 2. Finding Centers for a Subset

1: **function** SUBSET-CENTERS($\hat{G}, U, s$)
2:     $A \leftarrow \emptyset$
3:     **while** there exists $w \in V$ such that $|\hat{C}(w) \cap U| > 4|U|/s$ **do**
4:         $W \leftarrow \{w \in V : |\hat{C}(w) \cap U| > 4|U|/s\}$
5:         Add each element of W to A with probability $\min(s/|W|, 1)$
6:     **return** A

Lemma 11. *The function* SUBSET-CENTERS *(Algorithm 2) outputs a set $A \subseteq V$, such that, with probability at least $1/2$, we have $|A| \leq 4s \log n$ and $|\hat{C}(w) \cap U| \leq 4|U|/s$ for every $w \in V$. It uses no queries and its running time is polynomial.*

Next, we design a recursive algorithm for non-edge verification. Let $U \subseteq V$ represent the set of vertices for which we are currently verifying the induced subgraph. Verifying that every non-edge of $\hat{G}[U]$ is a non-edge of $G[U]$ is equivalent to verifying that every edge of $G[U]$ is an edge of $\hat{G}[U]$.

Let A be a set of centers computed by SUBSET-CENTERS. We define, for each $a \in A$, its *extended Voronoi cell* D_a as

$$D_a = \left(\bigcup \{C(b) : b \in N_2(a)\} \cup N_2(a) \right) \cap U. \tag{2}$$

Similarly, with respect to the graph $\hat{G}$, we define

$$\hat{D}_a = \left(\bigcup \{\hat{C}(b) : b \in \hat{N}_2(a)\} \cup \hat{N}_2(a) \right) \cap U. \tag{3}$$

The following lemma is a trivial extension of Lemma 3 in [10].

Lemma 12. $\bigcup_{a \in A} G[D_a]$ *covers every edge of $G[U]$.*

From Lemma 12, in order to verify that every edge of $G[U]$ is an edge of $\hat{G}[U]$, we only need to verify that every edge of $G[D_a]$ is an edge of $\hat{G}[D_a]$, for every $a \in A$. So we can apply recursion on each D_a.

The main difficulty is: **How to obtain D_a efficiently?** If we compute D_a from its definition, we first need to compute $N_2(a)$, which requires $\Omega(n)$ queries since $N_2(a)$ may contain nodes outside U. Instead, a careful analysis shows that we can check whether $D_a = \hat{D}_a$ without even knowing $N_2(a)$, whereas $\hat{D}_a$ can be inferred from the graph $\hat{G}$ with no queries. This is shown in Lemma 13, which is the main novelty of the algorithmic design.

Lemma 13. *Assume that $\hat{E} \subseteq E$. If $\delta(u, v) = \hat{\delta}(u, v)$ for every pair (u, v) from $\bigcup_{a \in A} \hat{N}_2(a) \times U$, then $D_a = \hat{D}_a$ for all $a \in A$.*

Proof. The proof is delicate but elementary. For every $b \in \bigcup_{a \in A} \hat{N}_2(a)$, we have $\hat{C}(b) \cap U = C(b) \cap U$, because $\hat{\delta}(b, u) = \delta(b, u)$ and $\hat{\delta}(A, u) = \delta(A, u)$ for every $u \in U$. Therefore, $\hat{D}_a$ can be rewritten as

$$\hat{D}_a = \left(\bigcup \{C(b) : b \in \hat{N}_2(a)\} \cup \hat{N}_2(a) \right) \cap U.$$

Algorithm 3. Recursive Verification

1: **procedure** VERIFY-SUBGRAPH($\hat{G}, U$)
2:　　**if** $|U| > n_0$ **then**
3:　　　　**repeat**
4:　　　　　　$A \leftarrow$ SUBSET-CENTERS($\hat{G}, U, s$)
5:　　　　**until** $|A| \leq 4s \log n$ **and** $|\hat{C}(w) \cap U| \leq 4|U|/s$ for every $w \in V$
6:　　　　**for** $a \in A$ **do**
7:　　　　　　QUERY($\hat{N}_2(a), U$)
8:　　　　　　VERIFY-SUBGRAPH($\hat{G}, \hat{D}_a$)　　　　　　　$\triangleright \hat{D}_a$ defined in Equation (3)
9:　　**else**
10:　　　　QUERY(U, U)

Since $\hat{E} \subseteq E$, we have $\hat{N}_2(a) \subseteq N_2(a)$. Therefore $\hat{D}_a \subseteq D_a$.

On the other hand, we have $N_2(a) \cap U \subseteq \hat{N}_2(a) \cap U$, because $\hat{\delta}(a, u) = \delta(a, u)$ for every $u \in N_2(a) \cap U$. To prove $D_a \subseteq \hat{D}_a$, it only remains to show that, for any vertex $u \notin N_2(a)$ such that $u \in C(b) \cap U$ for some $b \in N_2(a)$, we have $u \in C(x) \cap U$ for some $x \in \hat{N}_2(a)$. We choose x to be the vertex at distance 2 from a on a shortest a-to-u path in $\hat{G}$. By the assumption and the definition of x, we have:

$$\delta(x, u) = \hat{\delta}(x, u) = \hat{\delta}(a, u) - 2 = \delta(a, u) - 2.$$

By the triangle inequality, and using $b \in N_2(a)$ and $u \in C(b)$, we have:

$$\delta(a, u) \leq \delta(a, b) + \delta(b, u) \leq 2 + \delta(b, u) < 2 + \delta(A, u).$$

Therefore $\delta(x, u) < \delta(A, u)$. Thus $u \in C(x) \cap U$.　　　　　　　□

The recursive algorithm for non-edge verification is in Algorithm 3. It queries every $(u, v) \in \bigcup_{a \in A} \hat{N}_2(a) \times U$ and then recurses on each extended Voronoi cell $\hat{D}_a$. See Figure 1. The parameters n_0 and s are defined later. Correctness of the algorithm follows from Lemmas 12 and 13.

Now we bound the query complexity of VERIFY-SUBGRAPH($\hat{G}, V$). To provide intuition, we analyze an algorithm of 4 recursive levels, and show that its query complexity is $\tilde{O}(n^{4/3})$. The complete proof of the complexity stated in Theorem 9 is in the full version of the paper.

To simplify the presentation, we assume $\Delta = O(1)$. Let $s = n^{1/3}$ and let n_0 be some well-chosen constant. Consider any recursive call VERIFY-SUBGRAPH($\hat{G}, U$) where $|U| > n_0$. Let $A \subseteq V$ be the centers at the end of the **repeat** loop. By Lemma 11, the expected number of **repeat** loops is constant. For every $a \in A$, $\hat{N}_2(a)$ has constant size, since the graph has bounded degree. Every $\hat{C}(w) \cap U$ has size $O(|U|/n^{1/3})$, so every $\hat{D}_a$ has size $O(|U|/n^{1/3})$. Since $|A| = \tilde{O}(n^{1/3})$, the number of recursive calls on the next level is $\tilde{O}(n^{1/3})$. Therefore during the recursion, on the second level, there are $\tilde{O}(n^{1/3})$ recursive calls, where every subset has size $O(n^{2/3})$; on the third level, there are $\tilde{O}(n^{2/3})$ recursive calls, where every subset has size $O(n^{1/3})$; and on the fourth level, there are $\tilde{O}(n)$ recursive calls, where every subset has size $O(1)$. Every recursive call

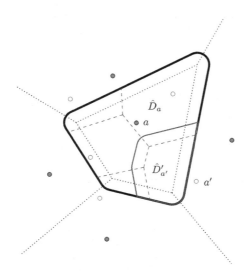

Fig. 1. Two levels of recursive calls of VERIFY-SUBGRAPH($\hat{G}, V$): The solid points are top-level centers returned by SUBSET-CENTERS($\hat{G}, V, s$). The dotted lines indicate the partition of V into Voronoi cells by those centers. The region inside the outer curve represents the extended Voronoi cell $\hat{D}_a$ of a center a. On the second level of the recursive call for $\hat{D}_a$, the hollow points are the centers returned by SUBSET-CENTERS($\hat{G}, \hat{D}_a, s$). Observe that some of those centers lie outside $\hat{D}_a$. The dashed lines indicate the partition of $\hat{D}_a$ into Voronoi cells by those centers. The region inside the inner curve represents the extended Voronoi cell $\hat{D}'_{a'}$ of a second-level center a'.

with subset U uses $\tilde{O}(n^{1/3} \cdot |U|)$ queries. Therefore, the overall query complexity is $\tilde{O}(n^{4/3})$.

Remark. *The recursive algorithm (Algorithm 3) can be used for verification by itself. However, we only use its query complexity to provide guarantee for the greedy algorithm (Algorithm 1), because the greedy algorithm is much simpler.*

4   Proof of Theorems 3

The algorithm (Algorithm 4) constructs an increasing set X of edges so that in the end $X = E$. At any time, the candidate graph is X.[3] Initially, X is the union of the shortest paths given as answers by $n - 1$ queries, so that X is a connected subgraph spanning V. At each subsequent step, the algorithm makes a query that leads either to the confirmation of many non-edges of G, or to the discovery of an edge of G.

Formally, we define, for every pair $(u, v) \in V^2$,

$$S_{u,v}^X = \{ab \in \text{non-edges of } X : \delta_X(u, a) + \delta_X(b, v) + 1 < \delta_X(u, v)\}. \quad (4)$$

---

[3] We identify X with the subgraph induced by the edges of X.

Algorithm 4. Greedy Reconstruction

1: **procedure** RECONSTRUCT(V)
2:     $u_0 \leftarrow$ an arbitrary vertex
3:     **for** $u \in V \setminus \{u_0\}$ **do** QUERY(u, u_0) to get a shortest u-to-u_0 path
4:     $X \leftarrow$ the union of the above paths, $Y \leftarrow \emptyset$
5:     **while** $X \cup Y$ does not cover all vertex pairs **do**
6:         choose (u, v) that maximizes $|S_{u,v}^X \setminus Y|$ ▷ $S_{u,v}^X$ defined in Equation (4)
7:         QUERY(u, v) to get a shortest u-to-v path
8:         **if** $\delta_G(u, v) = \delta_X(u, v)$ **then**
9:             $Y \leftarrow Y \cup S_{u,v}^X$
10:         **else**
11:             $e \leftarrow$ some edge of the above u-to-v path that is not in X
12:             $X \leftarrow X \cup \{e\}$
13:     **return** X

This is similar to $S_{u,v}$ defined in Equation (1). From Lemma 7, the pairs in $S_{u,v}^X$ can be confirmed as non-edges of G if $\delta_G(u, v) = \delta_X(u, v)$. At each step, the algorithm queries a pair (u, v) that maximizes the size of the set $S_{u,v}^X \setminus Y$. As a consequence, either all pairs in $S_{u,v}^X \setminus Y$ are confirmed as non-edges of G, or $\delta_G(u, v) \neq \delta_X(u, v)$, and in that case, the query reveals an edge along a shortest u-to-v path in G that is not in X; we then add this edge to X.

To see the correctness, we note that the algorithm maintains the invariant that the pairs in X are confirmed edges of G, and that the pairs in Y are confirmed non-edges of G. Thus when $X \cup Y$ covers all vertex pairs, we have $X = E$.

For the query complexity, first, consider the queries that lead to $\delta_G(u, v) \neq \delta_X(u, v)$. For each such query, an edge is added to X. This can happen at most $|E| \leq \Delta n$ times, because the graph has maximum degree Δ.

Next, consider the queries that lead to $\delta_G(u, v) = \delta_X(u, v)$. Define R to be the set of vertex pairs that are not in $X \cup Y$. We analyze the size of R during the algorithm. For each such query, the size of R decreases by $|S_{u,v}^X \setminus Y|$. To lower bound $|S_{u,v}^X \setminus Y|$, we consider the problem of non-edge verification using a distance oracle on the input graph X, and let T be an (unknown) optimal set of queries. By Theorem 9, $|T|$ is at most $f(n, \Delta) = n^{1+O\left(\sqrt{(\log \log n + \log \Delta)/\log n}\right)}$. By Lemma 8, the sets $S_{u,v}^X$ for all pairs $(u, v) \in T$ together cover $R \cup Y$, hence R. Therefore, at least one of these pairs satisfies

$$|S_{u,v}^X \setminus Y| \geq |R|/|T| \geq |R|/f(n, \Delta).$$

Initially, $|R| \leq n(n-1)/2$, and right before the last query, $|R| \geq 1$, thus the number of queries with $\delta_G(u, v) = \delta_X(u, v)$ is $O(\log n) \cdot f(n, \Delta)$.

Therefore, the overall query complexity is $O(\Delta n + \log n \cdot f(n, \Delta))$. Thus we obtained the same query bound as in the first statement of Theorem 1. To prove the query bound for graphs of treewidth w as in the second statement, the analysis is identical as above, except that $f(n, \Delta) = O(\Delta(\Delta + w \log n)n \log n)$, which comes from Theorem 10.

Remark. *Note that the above proof depends crucially on the fact that $f(n, \Delta)$ is a uniform bound on the number of distance queries for the non-edge verification of any n-vertex graph of maximum degree Δ. Thus, even though the graph X changes during the course of the algorithm because of queries (u, v) such that $\delta_G(u, v) \neq \delta_X(u, v)$, each query for which the distance in G and the current X are equal confirms $1/f(n, \Delta)$ fraction of non-edges.*

Acknowledgments. We thank Uri Zwick for Theorem 5. We thank Fabrice Benhamouda, Mathias Bæk Tejs Knudsen, Mikkel Thorup, and Jacob Holm for discussions. The first author was partially supported by NSF Grant NRI 1317788. The last two authors were partially supported by the French *Agence Nationale de la Recherche* under reference ANR-12-BS02-005 (RDAM project).

References

1. Achlioptas, D., Clauset, A., Kempe, D., Moore, C.: On the bias of traceroute sampling: or, power-law degree distributions in regular graphs. Journal of the ACM (JACM) **56**(4), 21 (2009)
2. Beerliova, Z., Eberhard, F., Erlebach, T., Hall, A., Hoffmann, M., Mihaľák, M., Shankar Ram, L.: Network discovery and verification. In: Kratsch, D. (ed.) WG 2005. LNCS, vol. 3787, pp. 127–138. Springer, Heidelberg (2005)
3. Castro, R., Coates, M., Liang, G., Nowak, R., Yu, B.: Network tomography: recent developments. Statistical Science **19**, 499–517 (2004)
4. Chung, F., Garrett, M., Graham, R., Shallcross, D.: Distance realization problems with applications to internet tomography. Journal of Computer and System Sciences **63**, 432–448 (2001)
5. Dall'Asta, L., Alvarez-Hamelin, I., Barrat, A., Vázquez, A., Vespignani, A.: Exploring networks with traceroute-like probes: Theory and simulations. Theoretical Computer Science **355**(1), 6–24 (2006)
6. Erlebach, T., Hall, A., Hoffmann, M., Mihaľák, M.: Network discovery and verification with distance queries. In: Calamoneri, T., Finocchi, I., Italiano, G.F. (eds.) CIAC 2006. LNCS, vol. 3998, pp. 69–80. Springer, Heidelberg (2006)
7. Hein, J.J.: An optimal algorithm to reconstruct trees from additive distance data. Bulletin of Mathematical Biology **51**(5), 597–603 (1989)
8. Johnson, D.S.: Approximation algorithms for combinatorial problems. Journal of computer and system sciences **9**(3), 256–278 (1974)
9. King, V., Zhang, L., Zhou, Y.: On the complexity of distance-based evolutionary tree reconstruction. In: SODA, pp. 444–453. SIAM (2003)
10. Mathieu, C., Zhou, H.: Graph reconstruction via distance oracles. In: Fomin, F.V., Freivalds, R., Kwiatkowska, M., Peleg, D. (eds.) ICALP 2013, Part I. LNCS, vol. 7965, pp. 733–744. Springer, Heidelberg (2013)
11. Reyzin, L., Srivastava, N.: On the longest path algorithm for reconstructing trees from distance matrices. Information processing letters **101**(3), 98–100 (2007)
12. Tarissan, F., Latapy, M., Prieur, C.: Efficient measurement of complex networks using link queries. In: INFOCOM Workshops, pp. 254–259. IEEE (2009)
13. Thorup, M., Zwick, U.: Compact routing schemes. In: Symposium on Parallel Algorithms and Architectures, pp. 1–10. ACM (2001)

A QPTAS for the Base of the Number of Crossing-Free Structures on a Planar Point Set

Marek Karpinski[1], Andrzej Lingas [2(✉)], and Dzmitry Sledneu[3]

[1] Department of Computer Science, University of Bonn, Bonn, Germany
marek@cs.uni-bonn.de
[2] Department of Computer Science, Lund University, Lund, Sweden
andrzej.lingas@cs.lth.se
[3] Centre for Mathematical Sciences, Lund University, Lund, Sweden
dzmitry@maths.lth.se

Abstract. The number of triangulations of a planar n point set S is known to be c^n, where the base c lies between 2.43 and 30. Similarly, the number of spanning trees on S is known to be d^n, where the base d lies between 6.75 and 141.07. The fastest known algorithm for counting triangulations of S runs in $O^*(2^n)$ time while that for counting spanning trees runs in $O^*(7.125^n)$ time. The fastest known arbitrarily close approximation algorithms for the base of the number of triangulations of S and the base of the number of spanning trees of S, respectively, run in time subexponential in n. We present the first quasi-polynomial approximation schemes for the base of the number of triangulations of S and the base of the number of spanning trees on S, respectively.

1 Introduction

By a crossing-free structure in the Euclidean plane, we mean a *planar straight-line graph* (PSLG), i.e., a plane graph whose edges $\{v, u\}$ are represented by properly non-intersecting straight-line segments with endpoints v, u, respectively. Triangulations and spanning trees on finite planar point sets are the two most basic examples of crossing-free structures in the plane, i.e., PSLGs. The problems of counting the number of such structures for a given planar n-point set belong to the most intriguing in Computational Geometry [2,4–6,8,10,11].

Counting Triangulations. A *triangulation* of a set S of n points in the Euclidean plane is a PSLG on S with a maximum number of edges. Let $F_t(S)$ stand for the set of all triangulations of S.

The problem of computing the number of triangulations of S, i.e., $|F_t(S)|$, is easy when S is convex. Simply, by a straightforward recurrence, $|F_t(S)| = C_{n-2}$, where C_k is the k-th Catalan number, in this special case. However, in the general case, the problem of computing the number of triangulations of S is neither known to be $\#P$-hard nor known to admit a polynomial-time counting algorithm.

Marek Karpinski—Research partially supported by DFG grants and the Hausdorff Center grant.
Andrzej Lingas—Research supported in part by VR grant 621-2011-6179.

© Springer-Verlag Berlin Heidelberg 2015
M.M. Halldórsson et al. (Eds.): ICALP 2015, Part I, LNCS 9134, pp. 785–796, 2015.
DOI: 10.1007/978-3-662-47672-7_64

It is known that $|F_t(S)|$ lies between $\Omega(2.43^n)$ [11] and $O(30^n)$ [10]. Since the so called flip graph whose nodes are triangulations of S is connected [12], all triangulations of S can be listed in exponential time by a standard traversal of this graph. When the number of the so called onion layers of the input point set is constant, the number of triangulations and other crossing-free structures can be determined in polynomial time [3]. Only recently, Alvarez and Seidel have presented an elegant algorithm for the number of triangulations of S running in $O^*(2^n)$ time [4] which is substantially below the aforementioned lower bound on $|F(S)|$ (the O^* notation suppresses polynomial in n factors).

Also recently, Alvarez, Bringmann, Ray, and Seidel [2] have presented an approximation algorithm for the number of triangulations of S based on a recursive application of the planar simple cycle separator [9]. Their algorithm runs in subexponential $2^{O(\sqrt{n}\log n)}$ time and over-counts the number of triangulations by at most a subexponential $2^{O(n^{\frac{3}{4}}\sqrt{\log n})}$ factor. It also yields a subexponential-time approximation scheme for the base of the number of triangulations of S, i.e., for $|F_t(S)|^{\frac{1}{n}}$. The authors of [2] observe also that just the inequalities $\Omega(8.65^n) \le |F_t(S)| \le O(30^n)$ yield the large exponential approximation factor $O(\sqrt{30/8.65}^n)$ for $|F_t(S)|$ trivially computable in polynomial time.

Counting Spanning Trees. A *spanning tree* U on a set S of n points in the Euclidean plane is a connected PSLG on S that is cycle-free, equivalently, that has $n-1$ edges. Let $F_s(S)$ stand for the set of all spanning trees on S.

It is known that $|F_s(S)|$ lies between $\Omega(6.75^n)$ [6] and $O(141.07^n)$ [8]. The fastest known algorithms for computing $|F_s(S)|$ runs in $O^*(7.125^n)$ time [13].

The aforementioned approximation algorithm for $|F_t(S)|$ due to Alvarez, Bringmann, Ray, and Seidel can be adapted to compute $|F_s(S)|$ in the same asymptotic subexponential $2^{O(\sqrt{n}\log n)}$ time within the same asymptotic subexponential $2^{O(n^{\frac{3}{4}}\sqrt{\log n})}$ approximation factor [2]. The adaption also yields a subexponential-time approximation scheme for the base of the number of spanning trees on S, i.e., for $|F_s(S)|^{\frac{1}{n}}$.

Our Contributions. We take a similar approximation approach to the problems of counting triangulations of S and counting spanning trees on S as Alvarez, Bringmann, Ray, and Seidel in [2]. However, importantly, instead of using recursively the planar simple cycle separator [9], we shall apply recursively the so called balanced α-cheap l-cuts of maximum independent sets of triangles within a dynamic programming framework developed by Adamaszek and Wiese in [1]. By using the aforementioned techniques, the authors of [1] designed the first quasi-polynomial time approximation scheme for the maximum weight independent set of polygons belonging to the input set of polygons with poly-logarithmically many edges.

Observe that a triangulation of S can be viewed as a maximum independent set of triangles drawn from the set of all triangles with vertices in S that are free from other points in S (triangles, or in general polygons, are identified with their open interiors). Also, a spanning tree on S can be easily complemented to a full triangulation on S. These simple observations enable us to use the aforementioned balanced α-cheap l-cuts recursively in order to bound an approximation factor of our approximation algorithm. The parameter α specifies the maximum fraction of an independent set of triangles that can be destroyed by the l-cut, which is a polygon with at most l vertices in a specially constructed set of points of polynomial size.

Similarly as the approximation algorithm from [2], our algorithm may over-count the true number of triangulations or spanning trees because the same triangulation or spanning tree, respectively, can be partitioned recursively in many different ways. In contrast with the approximation algorithm in [2], our algorithm may also under-count the number of triangulations of S or spanning trees on S, since our partitions generally destroy a fraction of edges in a triangulation or a spanning tree on S.

Our approximation algorithm for the number of triangulations of (or, the number of spanning trees on, respectively) a set S of n points with integer coordinates in the plane runs in $n^{(\log(n)/\epsilon)^{O(1)}}$ time. For $\epsilon > 0$, it returns a number at most $2^{\epsilon n}$ times smaller and at most $2^{\epsilon n}$ times larger than the number of triangulations of S (or, the number of spanning trees on S, respectively). Note that even for $\epsilon = (\log n)^{-O(1)}$, the running time is still quasi-polynomial.

As a corollary, we obtain quasi-polynomial approximation schemes for the base of the number of triangulations of S, i.e., for $|F_t(S)|^{\frac{1}{n}}$, and the base of the number of spanning trees on S, i.e., for $|F_s(S)|^{\frac{1}{n}}$, respectively. This implies that the problems of approximating $|F_t(S)|^{\frac{1}{n}}$ and $|F_s(S)|^{\frac{1}{n}}$ cannot be APX-hard (under standard complexity theoretical assumptions).

Organization of the Paper. In Preliminaries, we introduce basic concepts of the dynamic programming framework from [1]. In the following section, we present five properties of an abstract family of (crossing-free) structures on which the analysis of our approximation algorithm relies. Section 4 presents our approximation counting algorithm for the number of such structures on S and its time-complexity analysis. In Sections 5, upper bounds on the under-counting and the over-counting of the algorithm are derived, respectively. In Section 6, we obtain our main results by showing that planar triangulations and spanning trees satisfy these five properties.

2 Preliminaries

The Maximum Weight Independent Set of Polygons Problem (MWISP) is defined as follows [1]. We are given a set Q of n polygons in the Euclidean plane. Each polygon has at most k vertices, each of the vertices has integer coordinates. Next, each polygon P in Q is considered as an open set, i.e., it is identified with the set of points forming its interior. Also, each polygon $P \in Q$ has weight $w(P) > 0$ associated with it. The task is to find a maximum weight independent set of polygons in Q, i.e., a maximum weight set $Q' \subseteq Q$ such that for all pairs P_i, P_j of polygons in Q', if $P_i \neq P_j$ then it holds $P_i \cap P_j = \emptyset$.

The *bounding box* of Q is the smallest axis aligned rectangle containing all polygons in Q.

Note that in particular if Q consists of all triangles with vertices in a finite planar point set S such that no other point in S lies inside them or on their perimeter, each having weight 1, then the set of all maximum independent sets of polygons in Q is just the set of all triangulations of S. Recall that the latter set is denoted by $F_t(S)$ while the set of all spanning trees on S is denoted by $F_s(S)$.

Adamaszek and Wiese have shown that if $k = poly(\log n)$ then MWISP admits a QPTAS [1].

Fact 1 ([1]). *Let k be a positive integer. There exists a $(1 + \epsilon)$-approximation algorithm with a running time of $(nk)^{(\frac{k}{\epsilon} \log n)^{O(1)}}$ for the Maximum Weight Independent Set of Polygons Problem provided that each polygon has at most k vertices.*

Recently, Har-Peled generalized Fact 1 to include arbitrary polygons [7].

We need the following tool from [1].

Definition 1. *Let $l \in \mathbb{N}$ and $\alpha \in \mathbb{R}$ where $0 < \alpha < 1$. Let T be a set of pairwise non-touching triangles. A polygon Γ is a balanced α-cheap l-cut of T if*

- *Γ has at most l edges,*
- *the total weight of all triangles in T that intersect Γ does not exceed an α fraction of the total weight of triangles in T,*
- *the total weight of the triangles in T contained in Γ does not exceed two thirds of the total weight of triangles in T,*
- *the total weight of the triangles in T outside Γ does not exceed two thirds of the total weight of triangles in T.*

For a set of triangles T in the plane, a *DP-point* is a *basic DP-point* or an *additional DP-point*. The set of basic DP-points contains the four vertices of the bounding box of T and each intersection of a vertical line passing through a corner of a triangle in T with any edge of a triangle in T or a horizontal edge of the bounding box. The set of additional DP-points consists of all intersections of pairs of straight-line segments whose endpoints are basic DP-points. The authors of [1] observe that the total number of DP-points is $O(n^4)$.

Fact 2 (Lemma 3.6 in [1]). *Let $\delta > 0$ and let T be a set of pairwise non-touching triangles in the plane such that the weight of no triangle in T exceeds one third of the weight of T. Then there exists a balanced $O(\delta)$-cheap $(\frac{1}{\delta})^{O(1)}$-cut with vertices at basic DP-points.*

3   An Abstract Crossing-Free Structure

Triangulations and spanning trees are special cases of planar straight-line graphs (PSLGs). We shall consider an abstract family F_a of finite PSLGs having five properties (satisfied by triangulations and spanning trees as shown in Section 6) presented later in this page.

We shall use the following conventions in order to specify these properties and design an approximation algorithm for counting the number of PSLGs in F_a whose vertex set is an n-point planar point set S. We shall denote the latter set by $F_a(S)$.

We shall call a member in F_a a (crossing-free) *structure*, and a member in $F_a(S)$ a structure on S. Next, we shall call any subgraph of a structure a *substructure*.

Let P be a polygon with holes. The restriction of a structure G to P is the substructure consisting of all edges and vertices of G within P. (E.g., if G is a triangulation then the restriction is a partial triangulation, and if G is a spanning tree then the restriction is a forest, in general).

We say that a substructure is within P if all its vertices and all its edges are within P. Next, we shall call a substructure $H = (V_H, E_H)$ within P *maximal* if there is no other substructure $H' = (V_{H'}, E_{H'})$ within P, where $V_H = V_{H'}$, and $E_H \subsetneqq E_{H'}$. (E.g., if H is a partial triangulation within P then it cannot be extended to any larger partial triangulation by adding more edges, similarly, if H is a forest within P then it cannot be extended to any larger forest within P by adding more edges.)

We shall assume that the family F_a has the following properties.

1. One can decide if a PSLG with at most n vertices is a structure, i.e., belongs to F_a, in at most $2^{O(n \log n)}$ time.
2. If a structure has n vertices then it has $\Omega(n)$ edges. Two structures with the same set of vertices have the same number of edges.
3. Any substructure is in particular a substructure of a structure on the vertex set of the substructure.
4. Any extension of the restriction of a structure G to a simple polygon P with holes to a maximal substructure on the vertices of G within P uses at most $O(l)$ additional edges, where l is the number of edges of G with endpoints in P crossed by the boundaries of P.
5. Suppose that polygons P_1, P_2 with holes form a partition of a polygon P with holes. The union of a substructure within P_1 with a substructure within P_2 is a substructure.

By the definitions, F_a has also the following properties.

Lemma 1. *(Property 6) A maximal substructure H within the bounding box of the structure that H is a subgraph is a structure.*

Lemma 2. *(Property 7). Suppose that for $j = 1, \ldots, k'$, R_j is a maximal substructure within the polygon P_j with holes, and the polygons P_1 through $P_{k'}$ are pairwise non-overlapping and their union forms a polygon P with holes. Let $R'_1, \ldots, R'_{k'}$ be another sequence of maximal substructures within $P_1, \ldots, P_{k'}$, respectively, where R_j and R'_j have the same vertex set for $j = 1, \ldots, k'$. If $R_i \neq R'_i$ for some $i \in \{1, \ldots, k'\}$, each edge extension of $\bigcup_{j=1}^{k'} R_j$ to a maximal substructure within P is different from any edge extension of $\bigcup_{j=1}^{k'} R'_j$ to a maximal substructure within P.*

Proof. The proof is by contradiction. The joint edge extension of both sequences would contain $R_j \cup R_{j'}$ within P_j which would contradict the maximality of both R_j and $R_{j'}$ within P_j. □

4   Dynamic Programming

Our dynamic programming approximation algorithm for $|F_a(S)|$ is termed Algorithm 1 and it is depicted in Fig. 1.

Input: A set S of n points with integer coordinates in the Euclidean plane and natural number parameters k and Δ.

Output: An approximate number of structures on the vertex set S, i.e., an approximate $|F_a(S)|$.

1: $T \leftarrow$ the set of all triangles with vertices in S that do not contain any other point in S;

2: $\mathbb{P} \leftarrow$ a list of polygons (possibly with holes) with at most k vertices in total at DP points induced by T, topologically sorted with respect to geometric containment;

3: **for each** polygon $Q \in \mathbb{P}$ containing at most Δ points in S **do**

4:     $as(Q) \leftarrow$ exact number of maximal substructures on the vertex set $S \cap Q$ within Q;

5: **end for**

6: **for each** polygon set $Q \in \mathbb{P}$ containing more than Δ points in S **do**

7:     $as(Q) \leftarrow 0$;

8:     **for each** partition of Q into polygons $Q_1, \ldots, Q_{k'} \in \mathbb{P}$, where $k' \leq k$, no Q_j contains more than two thirds of points in $S \cap Q$, and $as(Q_1)$ through $as(Q_{k'})$ are defined **do**

9:         $as(Q) \leftarrow as(Q) + \prod_{j=1}^{k'} as(Q_j)$;

10:     **end for**

11: **end for**

12: Output $as(B)$, where B is the bounding box of T.

Fig. 1. Algorithm 1 for approximately counting structures on a finite planar point set

Time Complexity. The cardinality of T does not exceed n^3. Then, by the analogy with the dynamic programming algorithm of Adamaszek and Wiese for nearly maximum independent set of triangles [1], we call a polygon in the list $\mathbb{P}$ in Algorithm 1 a *DP cell* and observe that the number of DP cells is $(3n^3)^{O(k)} = n^{O(k)}$ (see Proposition 2.1 in [1]). Consequently, the number of possible partitions of a DP cell into at most k DP cells is $O(\binom{n^{O(k)}}{k})$, i.e., $n^{O(k^2)}$.

It follows that if we neglect the cost of computing the exact number of maximal substructures contained within a DP cell including at most Δ input points, then Algorithm 1 runs in $n^{O(k^2)}$ time.

We can compute the exact number of maximal substructures contained within a DP cell with at most Δ input points in $2^{O(\Delta \log \Delta)}$ time as follows. By enumerating all PSLGs on the subset of S contained in the DP cell, and using Property 1 and the fact that the number of PSLGs on at most Δ vertices is $2^{O(\Delta \log \Delta)}$, we can list all structures on this subset in $2^{O(\Delta \log \Delta)} \times 2^{O(\Delta \log \Delta)}$ $= 2^{O(\Delta \log \Delta)}$ time. Hence, by Property 3, we can exactly count all maximal substructures (on this subset) within the cell by pruning the aforementioned structures and checking maximality also in $2^{O(\Delta \log \Delta)}$ time. We conclude with the following lemma.

Lemma 3. *Algorithm 1 runs in* $n^{O(k^2)}2^{O(\Delta \log \Delta)}$ *time.*

5   Approximation Factor

Under-Counting. The potential under-counting stems from the fact that when a DP cell is partitioned into at most k smaller DP cells then the possible combinations of structure edges crossing the boundaries of the cells are not counted. Furthermore, in the leaf DP cells, i.e., those including at most Δ points from S, we count only maximal substructures while the restriction of a structure on S to a DP cell does not have to be a maximal substructure within the cell. See Q_5 in Fig. 2.

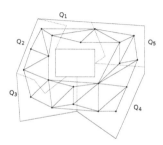

Fig. 2. An example of a maximal partial triangulation within a DP cell and a partition of the DP cell into smaller DP cells $Q_1, \ldots, Q_5$ crossing some triangles in the triangulation

Intuitively, the general idea of the proof of our upper bound on under-counting is as follows. For each structure $W \in F_a(S)$, there is a substructure counted by Algorithm 1 that can be obtained by removing $O(\epsilon n)$ edges from W and augmenting the resulting substructure with $O(\epsilon n)$ other edges. The final substructure is a union of maximal substructures contained in leaf DP cells.

Lemma 4. *Let S be a set of n points in the plane and let $\epsilon > 0$. For each $W \in F_a(S)$, there is a substructure $W^* \subseteq W$ on S containing at least a $1 - O(\epsilon)$ fraction of the edges of W and a substructure $M(W^*)$ on S which is an extension of W^* by $O(\epsilon n)$ edges such that the estimation returned by Algorithm 1 with k set to $\log^{O(1)}(n)/\epsilon^{O(1)}$ is not less than $|\bigcup_{W \in F(S)}\{M(W^*)\}|$.*

Proof. Let $W \in F_a(S)$ and let $T(W)$ be any triangulation of S that is an extension of W. By adapting the idea of the proof of the approximation ratio of the

QPTAS in [1], consider the following tree U of DP cells obtained by recursive applications of balanced α-cheap l-cuts.

At the root of U, there is the bounding box. By Fact 2, there is a balanced α-cheap l-cut, where $l = \alpha^{-O(1)}$, that splits the box into at most k children DP cells such that only α fraction of the triangular faces of $T(W)$ is crossed by the cut. The construction of U proceeds recursively in children DP cells and stops in DP cells that contain at most Δ points in S.

Note that the height of U is not greater than $\log_{3/2} n$.

For a node u of U, let W_u be the substructure that is the restriction of W to the vertices and edges of W contained in the DP cell Q_u associated with u. Analogously, let $T(W)_u$ be the partial triangulation of the points in $S \cap Q_u$ that is the restriction of $T(W)$ to (the vertices and edges of $T(W)$ contained in) Q_u. Clearly, W_u is a subgraph of $T(W)_u$. Next, let W_u^* be the substructure that is the union of W_t over the the leaves t of the subtree of U rooted at u. Note that W_u^* is a subgraph of W_u. Analogously, let $T(W)_u^*$ denote the restriction of $T(W)_u$ to the union of $T(W)_t$ over the the leaves t of the subtree of U rooted at u. Clearly, W_u^* is a subgraph of $T(W)_u^*$.

By induction on the height $h(u)$ of u in U, we obtain that the partial triangulation $T(W)_u^* \subseteq T(W)_u$ contains a $(1-\alpha)^{h(u)}$ fraction of triangular faces of $T(W)_u$. Set α to $\frac{O(\epsilon)}{\log(n/\epsilon)}$. It follows in particular that for the root r of U, $T(W)_r^* \subseteq T(W)$ contains at least a $(1-\alpha)^{\log_{3/2} n/\epsilon} \geq 1 - O(\epsilon)$ fraction of triangular faces in $T(W)$. Set $T(W)^*$ to $T(W)_r^*$ and W^* to W_r^*. By Property 2 ensuring that W has $\Omega(n)$ edges and the fact that each triangular face has three edges, we conclude that analogously W^* contains a $1 - O(\epsilon)$ fraction of the edges of W. Thus, the number of edges in W missing in W^* is $O(\epsilon n)$.

For a leaf t of U, let $M(W_t)$ be an (edge) extension of W_t to a maximal substructure within the leaf cell Q_t. By Property 4, the number of edges extending W_t to $M(W_t)$ is bounded by a constant times the number of edges in W crossing the boundary of Q_t and having an endpoint within Q_t.

For a node u of U, let $M(W_u^*)$ be a substructure within Q_u that is the union of $M(W_t)$ over the leaves t of the subtree of U rooted at u. We have also $M(W^*) = M(W_r^*)$ by $W^* = W_r^*$. It follows that the number of edges extending W^* to $M(W^*)$ is bounded by a constant times the number of edges of W missing in W^*, i.e., $O(\epsilon n)$.

We shall show by induction on $h(u)$ that Algorithm 1 counts at least the number of $M(W_u^*)$ while computing an estimation for Q_u.

If $h(u) = 0$, i.e., u is a leaf in U then $W_u^* = W_u$ and consequently in particular $M(W_u^*) = M(W_u)$ is counted by Algorithm 1.

Suppose in turn that u is an internal node in U with k' children $u_1, \ldots, u_{k'}$. When the estimation for Q_u is computed by Algorithm 1, the sum of products of estimations yielded by different partitions of Q_u into at most k DP cells is computed. In particular, the partition into $Q_{u_1}, \ldots, Q_{u_{k'}}$ is considered. By the induction hypothesis, the estimation for Q_{u_j} includes $M(W_{u_j}^*)$ for $j = 1, \ldots, k'$. Hence, the product of these estimations counts also $M(W_u^*) = \bigcup_{j=1}^{k'} M(W_{u_j}^*)$.

By $M(W^*) = M(W_r^*)$, to obtain the lemma it remains to show that the bound $\log^{O(1)}(n/\epsilon)/\epsilon^{O(1)}$ on k is sufficiently large. Following the proof of Lemma 2.1 in [1], observe that each DP cell Q_u at each level of U is an intersection of at most $O(\log(n/\epsilon))$ polygons, each with at most l edges and vertices at basic DP points. Hence, by $\alpha = \frac{O(\epsilon)}{\log(n/\epsilon)}$ and $l = \alpha^{-O(1)}$, the resulting polygons have at most $O(l^2 \log^2(n/\epsilon)) = \log^{O(1)}(n/\epsilon)/\epsilon^{O(1)}$ edges and vertices at basic and additional DP points.   □

Theorem 1. *The under-counting factor of Algorithm 1 with k set to $\log^{O(1)}(n/\epsilon)/\epsilon^{O(1)}$ is at most $2^{O(\epsilon n \log n)}$.*

Proof. Consider any structure $W \in F_a(S)$. By Lemma 4, the number of edges of W that are missing in the substructure $W^* \subseteq W$ is $O(\epsilon n)$. Since all structures in $F_a(S)$ have the same number of edges by Property 2, the number of edges completing W^* to any structure is $O(\epsilon n)$. It follows that the number of ways of completing W^* to a structure in $F_a(S)$ is not greater than the number of subsets of at most $O(\epsilon n)$ edges of the complete Euclidean graph on S, which is $2^{O(\epsilon n \log n)}$.

By Lemma 4, the estimation returned by Algorithm 1 with k set to $\log^{O(1)}(n/\epsilon)/\epsilon^{O(1)}$ is not less than $|\bigcup_{W \in F(S)}\{M(W^*)\}|$.

Now it remains to show that the maximum number of substructures $(W')^*$, $W' \in F_a(S)$, for which $M((W')^*) = M(W^*)$ is at most $2^{O(\epsilon n \log n)}$. By Lemma 4, the number of edges extending $(W')^*$ to $M((W')^*)$ is at most $O(\epsilon n)$. Consequently, the maximum number of such substructures $(W')^*$ is upper bounded by the number of subsets of at most $O(\epsilon n)$ edges of $M(W^*)$ (whose removal may form a substructure $(W')^*$ satisfying $M((W')^*) = M(W^*)$). The latter number is $2^{O(\epsilon n \log n)}$.

We conclude that for $W \in F(S)$, the number of other structures $W' \in F(S)$ for which $M((W')^*) = M(W^*)$ is at most $2^{O(\epsilon n \log n)}2^{O(\epsilon n \log n)} = 2^{O(\epsilon n \log n)}$. Now, the theorem follows from Lemma 4.   □

Over-Counting. The reason for over-counting in the estimation returned by our algorithm is as follows. The same structure or more generally substructure within a DP cell may be cut in the number of ways proportional to the number of considered partitions of the DP cell into at most k smaller DP cells. This reason is similar to that for over-counting of the approximation triangulation counting algorithm of Alvarez, Bringmann, Ray, and Seidel [2] based on the planar simple cycle separator theorem. Therefore, our initial recurrences and calculations are similar to those derived in the analysis of the over-counting from [2].

Lemma 5. *Let Q be an arbitrary DP cell processed by Algorithm 1 which contains more than Δ input points. Recall the calculation of the estimation for Q by summing the products of estimations for smaller DP cells $Q_1, \ldots, Q_l$ over $n^{O(k^2)}$ partitions of Q into $Q_1, \ldots, Q_l$, $l \leq k$. Substitute the true value of the number of maximal substructures (on input points) within each such smaller cell Q_i for the estimated one in the calculation. Let r be the resulting value. The number of maximal substructures (on input points) within Q is at least $r/n^{O(k^2)}$.*

Proof. Note that r is the sum of the number of different combinations of maximal structures within smaller DP cells $Q_1, \ldots, Q_l$ over $n^{O(k^2)}$ partitions of Q into smaller cells $Q_1, \ldots, Q_l$, $l \leq k$. Importantly, each such combination can be completed to some maximal substructure within Q (Property 5) but no two different combinations coming from the same partition $Q_1, \ldots, Q_l$ can be extended to the same maximal substructure within Q by Property 7 (Lemma 2).

Let M be the set of maximal substructures W within Q for which there is a partition into smaller DP cells $Q_1, \ldots, Q_l$, $l \leq k$, such that for $i = 1, \ldots, l$, W constrained to Q_i is a maximal substructure within Q_i. Note that for each $W \in M$, the number of the combinations that can be completed to W cannot exceed that of the considered partitions, i.e., $n^{O(k^2)}$, as each of the combinations has to come from a distinct partition $Q_1, \ldots, Q_l$.

Thus, there is a binary relationship between maximal substructures within Q that belong to M and the aforementioned combinations. It is defined on all the maximal substructures in M and on all the combinations, and a maximal substructures in M is in relation with at most $n^{O(k^2)}$ combinations. This yields the lemma. □

By Lemma 5, we can express the over-counting factor $L(Q, \Delta)$ of Algorithm 1 for a DP cell Q by the following recurrence:

$$L(Q, \Delta) = \sum_{(Q_1, \ldots, Q_{k'})} \prod_{j=1}^{k'} L(Q_j, \Delta) \leq n^{O(k^2)} \prod_{j=1}^{k^*} L(Q_j^*, \Delta)$$

where the summation is over all partitions of Q into DP cells $Q_1, \ldots, Q_{k'}$, such that $k' \leq k$, and $Q_1^*, \ldots, Q_{k^*}^*$ is a partition that maximizes the term $\prod_{j=1}^{k'} L(Q_j, \Delta)$. When Q contains at most Δ input points, Algorithm 1 computes the exact number of maximal substructures on these points within Q. Thus, we have $L(Q, \Delta) = 1$ in this case.

Following [2], it will be more convenient to transform our recurrence by taking logarithm of both sides. For any DP cell P, let $L'(P, \Delta) = \log L(P, \Delta)$. We obtain now:

$$L'(Q, \Delta) \leq O(k^2 \log n) + \sum_{j=1}^{k^*} L'(Q_j^*, \Delta)$$

Lemma 6. *Let B be a bounding box for a set S of n points in the plane. The equality $L'(B, \Delta) = O(k^2 \Delta^{-1} n \log^2 n)$ holds.*

Proof. Let U be the recurrence tree and let D be the set of non-leaf nodes whose all children are leaves in U. For each node $d \in D$, the corresponding DP cell includes at least $\Delta + 1$ points in S. It follows that $|D| \leq n/\Delta$. Any node in D has depth $O(\log n)$ in U. Hence, more generally, non-leaf nodes of U are placed on $O(\log n)$ height levels of U, where each level includes at most n/Δ nodes. Each subproblem corresponding to a non-leaf node of U contributes at most $O(k^2 \log n)$ to $L'(B, \Delta)$. Consequently, the total contribution of non-leaf nodes of U to $L'(B, \Delta)$ is $O(k^2 \log n \times (n/\Delta) \log n)$. Finally, recall that the subproblems corresponding to leaves of U do not contribute to the estimation.   □

Lemma 6 and Property 6 (Lemma 1) immediately yield the following corollary.

Theorem 2. *Let B be a bounding box for a set of n points in the plane. Set the parameter k in Algorithm 1 as in Theorem 1. If for $\epsilon > 0$ the parameter Δ in Algorithm 1 is set to $\frac{c}{\epsilon} k^2 \log^2 n$ for sufficiently large constant c then the over-counting factor is at most $2^{\epsilon n}$.*

6   Main Results

Lemma 7. *Triangulations and spanning trees on finite planar point sets satisfy the five properties of F_a.*

Proof. Properties 1, 2, 3 and 5 are clearly satisfied by triangulations and spanning trees.

To show that Property 4 holds for triangulations, consider an extension of the restriction of a triangulation G to a simple polygon P with holes to a maximal partial triangulation on the vertices of G within P. All the edges within P added by the extension have to be incident to vertices of triangular faces of G with at least one edge crossed by the boundary of P. Observe, that such a triangular face has to have at least one vertex within P that is an endpoint of an edge of G crossed by the boundaries of P. Let l be the number of edges of G with an endpoint within P crossed by the boundaries of P. It follows that the number of aforementioned triangles is at most $2l$ and consequently the number of the endpoints of the edges within P added by the extension does not exceed $3 \times 2l = O(l)$. Hence, the total number of the added edges is also $O(l)$.

To show in turn that Property 4 holds for spanning trees, consider the forest which is the restriction of a spanning tree G to a simple polygon P with holes. Let t be the number of connected components of the forest. It follows that the number l of edges of the spanning tree G with at least one endpoint within P crossed by the boundaries of P is at least $t - 1$. On the other hand, any edge extension of the forest to a maximal forest within P may add at most $t - 1 \leq l$ edges to the forest.   □

By combining Lemmata 7 and 3 with Theorems 1, 2 with ϵ set to $\epsilon/\log n$, we obtain our main result.

Theorem 3. *There exists an approximation algorithm for the number of triangulations of (or, the number of spanning trees on) a set S of n points with integer coordinates in the plane with a running time of at most $n^{(\log(n)/\epsilon)^{O(1)}}$ that returns a number at most $2^{\epsilon n}$ times smaller and at most $2^{\epsilon n}$ times larger than the number of triangulations of (or, spanning trees on, respectively) S.*

Corollary 1. *There exists a $(1 + \epsilon)$-approximation algorithm with a running time of at most $n^{(\log(n)/\epsilon)^{O(1)}}$ for the base of the number of triangulations of (or, spanning trees on) a set of n points with integer coordinates in the plane.*

Proof. Let c^n be the number of triangulations of (or, the number of spanning trees on) the input n point set, and let Λ be the number returned by the algorithm from Theorem 3. We have $\max\{\frac{c^n}{\Lambda}, \frac{\Lambda}{c^n}\} \leq 2^{\epsilon n}$ by Theorem 3. By taking the n-th root on both sides, we obtain $\max\{\frac{c}{\Lambda^{\frac{1}{n}}}, \frac{\Lambda^{\frac{1}{n}}}{c}\} \leq 2^\epsilon$. Now it is sufficient to observe that $2^\epsilon < 1 + \epsilon$ for $\epsilon < \frac{1}{2}$. $\qquad\square$

Final Remark. The other popular crossing-free structures like perfect matchings and cycle covers do not satisfy all the five properties of F_a. It is an intriguing open problem if they admit similar quasi-polynomial time approximation algorithms.

Acknowledgments. We are very grateful to unknown referees for many valuable comments.

References

1. Adamaszek, A., Wiese, A.: A QPTAS for maximum weight independent set of polygons with polylogarithmically many vertices. In: SODA 2014
2. Alvarez, V., Bringmann, K., Ray, S., Seidel, R.: Counting triangulations and other crossing-free structures approximately. Comput. Geom. **48**(5), 386–397 (2015)
3. Alvarez, V., Bringmann, K., Curticapean, R., Ray, S.: Counting crossing-free structures. In: SoCG 2012
4. Alvarez, V., Seidel, R.: A simple aggregative algorithm for counting triangulations of planar point sets and related problems. In: SoCG 2013
5. Dumitrescu, A., Schulz, A., Sheffer, A., Tóth, C.D.: Bounds on the maximum multiplicity of some common geometric graphs. SIAM J. Discrete Math. **27**(2), 802–826 (2013)
6. Flajolet, P., Noy, M.: Analytic combinatorics of non-crossing configurations. Discrete Mathematics **204**(1–3), 203–229 (1999)
7. Har-Peled, S.: Quasi-polynomial time approximation scheme for sparse subsets of polygons. In: SoCG 2014
8. Hoffmann, M., Sharir, M., Sheffer, A., Tóth, C.D., Welzl, E.: Counting plane graphs: flippability and its applications. In: Dehne, F., Iacono, J., Sack, J.-R. (eds.) WADS 2011. LNCS, vol. 6844, pp. 524–535. Springer, Heidelberg (2011)
9. Miller, G.L.: Finding small simple cycle separators for 2-connected planar graphs. J. Comput. Syst. Sci. **32**(3), 265–279 (1986)
10. Sharir, M., Sheffer, A.: Counting triangulations of planar point sets. Electr. J. Comb. **18**(1) (2011)
11. Sharir, M., Sheffer, A., Welzl, E.: On degrees in random triangulations of point sets. J. Comb. Theory, Ser. A **118**(7), 1979–1999 (2011)
12. Sibson, R.: Locally equiangular triangulations. Comput. J. **21**(3), 243–245 (1978)
13. Wettstein, M.: Counting and enumerating crossing-free geometric graphs. In: SoCG 2014

Finding a Path in Group-Labeled Graphs with Two Labels Forbidden

Yasushi Kawase[1], Yusuke Kobayashi[2], and Yutaro Yamaguchi[3]([✉])

[1] Tokyo Institute of Technology, Tokyo, Japan
kawase.y.ab@m.titech.ac.jp
[2] University of Tsukuba, Tsukuba, Japan
kobayashi@sk.tsukuba.ac.jp
[3] University of Tokyo, Tokyo, Japan
yutaro_yamaguchi@mist.i.u-tokyo.ac.jp

Abstract. The parity of the length of paths and cycles is a classical and well-studied topic in graph theory and theoretical computer science. The parity constraints can be extended to the label constraints in a group-labeled graph, which is a directed graph with a group label on each arc. Recently, paths and cycles in group-labeled graphs have been investigated, such as finding non-zero disjoint paths and cycles.

In this paper, we present a solution to finding an s–t path in a group-labeled graph with two labels forbidden. This also leads to an elementary solution to finding a zero path in a $\mathbb{Z}_3$-labeled graph, which is the first nontrivial case of finding a zero path. This situation in fact generalizes the 2-disjoint paths problem in undirected graphs, which also motivates us to consider that setting. More precisely, we provide a polynomial-time algorithm for testing whether there are at most two possible labels of s–t paths in a group-labeled graph or not, and finding s–t paths attaining at least three distinct labels if exist. We also give a necessary and sufficient condition for a group-labeled graph to have exactly two possible labels of s–t paths, and our algorithm is based on this characterization.

1 Introduction

1.1 Background

The parity of the length of paths and cycles in a graph is a classical and well-studied topic in graph theory and theoretical computer science. As the simplest example, one can easily check the bipartiteness of a given undirected graph, i.e., we can determine whether it contains a cycle of odd length or not. This can be done in polynomial time also in the directed case by using the ear decomposition. It is also an important problem to test whether a given directed graph contains

---

Y. Kawase—Supported by JSPS KAKENHI Grant Number 26887014.
Y. Kobayashi—Supported by JST, ERATO, Kawarabayashi Large Graph Project, and by JSPS KAKENHI Grant Number 24106002, 24700004.
Y. Yamaguchi—Supported by JSPS Fellowship for Young Scientists.

M.M. Halldórsson et al. (Eds.): ICALP 2015, Part I, LNCS 9134, pp. 797–809, 2015.
DOI: 10.1007/978-3-662-47672-7_65

a directed cycle of even length or not, which is known to be equivalent to Pólya's permanent problem [13] (see, e.g., [12]). A polynomial-time algorithm for this problem was devised by Robertson, Seymour, and Thomas [14].

In this paper, we focus on paths connecting two specified vertices s and t. It is easy to test whether a given undirected graph contains an s–t path of odd (or even) length or not, whereas the same problem is NP-complete in the directed case [11] (follows from [5]). A natural generalization of this problem is to consider paths of length p modulo q. One can easily see that, when $q = 2$, both of the following problems generalize the problem of finding an odd (or even) s–t path in an undirected graph:

- finding an s–t path of length p modulo q in an undirected graph, and
- finding an s–t path whose length is NOT p modulo q in an undirected graph, which is equivalent to determining whether all s–t paths are of length p modulo q or not.

Although these two generalizations are similar to each other, they are essentially different in the case of $q \geq 3$. In fact, a linear-time algorithm for the second generalization was given by Arkin, Papadimitriou, and Yannakakis [1] for any q, whereas not so much was known about the first generalization.

Recently, as another generalization of the parity constraints, paths and cycles in a group-labeled graph have been investigated, where a group-labeled graph is a directed graph with each arc labeled by a group element. In a group-labeled graph, the label of a walk is defined as the sum (or the ordered product when the underlying group is non-abelian) of the labels of the traversed arcs, where each arc can be traversed in the converse direction and then the label is inversed (see Section 2.1 for the precise definition). Analogously to paths of length p modulo q, it is natural to consider the following two problems: for a given element α,

(I) finding an s–t path of label α in a group-labeled graph, and

(II) finding an s–t path whose label is NOT α in a group-labeled graph, which is equivalent to determining whether all s–t paths are of label α or not.

Note that, when we consider Problem (I) or (II), by changing uniformly the labels of the arcs incident to s if necessary, we may assume that α is the identity of the underlying group. Hence, each problem is equivalent to finding a zero path or a non-zero path in a group-labeled graph. In what follows, we assume the black-box access to the underlying group, i.e., we can perform elementary operations for it in constant time (see Section 2.1 for the precise assumption).

If the underlying group is $\mathbb{Z}_2 = \mathbb{Z}/2\mathbb{Z} = (\{0,1\},+)$ and the label of each arc is 1, then the label of a path corresponds to the parity of its length because $-1 = 1$ in $\mathbb{Z}_2$. This shows that both of these two problems generalize the problem of finding an odd (or even) s–t path in an undirected graph. We note that, in a $\mathbb{Z}_2$-labeled graph, finding an s–t path of label $\alpha \in \mathbb{Z}_2$ is equivalent to finding an s–t path whose label is not $\alpha + 1 \in \mathbb{Z}_2$, but such equivalence cannot hold for any other nontrivial group.

As shown in Section 2.2, Problem (II) can be reduced to testing whether a group-labeled graph contains a non-zero cycle, whose label is not the identity. With this observation, Problem (II) can be easily solved in polynomial time for any underlying group. We mention that there are several results for packing non-zero paths [2,3,18,20] and non-zero cycles [9,19] with some conditions.

On the other hand, the difficulty of Problem (I) is heavily dependent on the underlying group Γ. When $\Gamma \simeq \mathbb{Z}_2$, since Problems (I) and (II) are equivalent as discussed above, it can be easily solved in polynomial time. When $\Gamma = \mathbb{Z}$, Problem (I) is NP-complete since the directed s–t Hamiltonian path problem reduces to this problem by labeling each arc with $1 \in \mathbb{Z}$ and letting $\alpha := n-1 \in \mathbb{Z}$, where n denotes the number of vertices. Huynh [8] showed the polynomial-time solvability of Problem (I) for any fixed finite abelian group, which is deeply dependent on the graph minor theory.

To investigate the gap between Problems (I) and (II), we make a new approach to these problems by generalizing Problem (II) so that multiple labels are forbidden. In this paper, we provide a solution to the case when two labels are forbidden. Our result also leads to an elementary solution to the first nontrivial case of Problem (I), i.e., when $\Gamma \simeq \mathbb{Z}_3 = \mathbb{Z}/3\mathbb{Z} = (\{0, \pm 1\}, +)$.

1.2   2-disjoint Paths Problem

Problem (I) in a $\mathbb{Z}_3$-labeled graph in fact generalizes the 2-disjoint paths problem, which also motivates us to consider the situation when two labels are forbidden. The 2-disjoint paths problem is to determine whether there exist two vertex-disjoint paths such that one is from s_1 to t_1 and the other from s_2 to t_2 for distinct vertices s_1, s_2, t_1, t_2 in a given undirected graph. We can reduce the 2-disjoint paths problem to Problem (I) in a $\mathbb{Z}_3$-labeled graph as follows: let $s := s_1$ and $t := t_2$, replace every edge in the given graph with an arc with label 0, add one arc from t_1 to s_2 with label 1, and ask whether the constructed $\mathbb{Z}_3$-labeled graph contains an s–t path of label 1 or not. If the answer is YES, then there exist desired two disjoint paths, and otherwise there do not.

The 2-disjoint paths problem can be solved in polynomial time [15–17], and the following theorem characterizes the existence of two disjoint paths.

Theorem 1 (Seymour [16]). *Let* $G = (V, E)$ *be an undirected graph and* $s_1, t_1, s_2, t_2 \in V$ *distinct vertices. Then, there exist two vertex-disjoint paths* P_i *connecting* s_i *and* t_i *($i = 1, 2$) if and only if there is no family of disjoint vertex sets* $X_1, X_2, \ldots, X_k \subseteq V \setminus \{s_1, t_1, s_2, t_2\}$ *such that*

1. $N_G(X_i) \cap X_j = \emptyset$ *for distinct* $i, j \in \{1, 2, \ldots, k\}$,
2. $|N_G(X_i)| \leq 3$ *for* $i = 1, 2, \ldots, k$, *and*
3. *if* G' *is the graph obtained from* G *by deleting* X_i *and adding a new edge joining each pair of distinct vertices in* $N_G(X_i)$ *for each* $i \in \{1, 2, \ldots, k\}$, *then* G' *can be embedded on a plane so that* s_1, s_2, t_1, t_2 *are on the outer boundary in this order.*

Our characterization (Theorem 12) of group-labeled graphs with exactly two possible labels of s–t paths is inspired by Theorem 1, which is used in the proof.

1.3 Our Contribution

Let Γ be an arbitrary group. For a Γ-labeled graph G and two distinct vertices s and t, let $l(G; s, t)$ be the set of all possible labels of s–t paths in G. Our first contribution is to give a characterization of Γ-labeled graphs G with two specified vertices s, t such that $l(G; s, t) = \{\alpha, \beta\}$, where α and β are distinct elements in Γ. Roughly speaking, we show that $l(G; s, t) = \{\alpha, \beta\}$ if and only if G is obtained from "nice" planar graphs (and some trivial graphs) by "gluing" them together (see Section 3.3). It is interesting that the planarity, which is a topological condition, appears in the characterization.

There exists an easy characterization of triplets (G, s, t) with $|l(G; s, t)| = 1$, which is used to solve Problem (II) (see Section 2.2 for details). Our characterization leads to the first nontrivial classification of Γ-labeled graphs in terms of the possible labels of s–t paths, and the classification is complete when $\Gamma \simeq \mathbb{Z}_3$.

We also show an algorithmic result, which is our second contribution. Based on the fact that our characterization can be tested in polynomial time, we present a polynomial-time algorithm for testing whether $|l(G; s, t)| \leq 2$ or not and finding at least three s–t paths whose labels are distinct if exist (see Theorem 9). In particular, our algorithm leads to an elementary solution to Problem (I) when $\Gamma \simeq \mathbb{Z}_3$, i.e., for each $\alpha \in \mathbb{Z}_3$, we can test whether $\alpha \in l(G; s, t)$ or not, and find an s–t path of label α if exists.

Note again that our results are not dependent on Γ, which can be non-abelian or infinite (as long as we can efficiently perform elementary operations for Γ).

The rest of this paper is organized as follows. In Section 2, we define several terms, notations, and operations, and describe well-known results. Section 3 is devoted to presenting our results: the efficient solvability of the problem to find an s–t path with two labels forbidden, and a characterization of Γ-labeled graphs with exactly two possible labels of s–t paths. Their verifications are sketched in Sections 4 and 5, and the complete proofs are left to the full version [10].

2 Preliminaries

2.1 Terms and Notations

Throughout this paper, let Γ be a group (which can be non-abelian or infinite), for which we usually use multiplicative notation with denoting the identity by 1_Γ (we sometimes use additive notation with denoting the identity by 0, e.g., when $\Gamma \simeq \mathbb{Z}_3$). We assume that elementary operations for Γ can be performed, i.e., the following procedures can be done in constant time for any $\alpha, \beta \in \Gamma$: getting the inverse element $\alpha^{-1} \in \Gamma$, computing the product $\alpha\beta \in \Gamma$, and testing the identification $\alpha = \beta$. A directed graph $G = (V, E)$ with a mapping $\psi_G \colon E \to \Gamma$ (called a *label function*) is called a Γ-*labeled graph*.

Graphs. Let $G = (V, E)$ be a directed graph. For vertices $v_0, v_1, \ldots, v_l \in V$ and arcs $e_1, e_2, \ldots, e_l \in E$ with $e_i = v_{i-1}v_i$ or $e_i = v_iv_{i-1}$ $(i = 1, 2, \ldots, l)$, a sequence $W = (v_0, e_1, v_1, e_2, v_2, \ldots, e_l, v_l)$ is called a *walk* in G. A walk W is called a *path* (in particular, a v_0–v_l *path*) if $v_0, v_1, \ldots, v_l$ are distinct, and a *cycle* if $v_0, v_1, \ldots, v_{l-1}$ are distinct and $v_0 = v_l$. We call v_0 and v_l (which may coincide) the *end vertices of* W, and each v_i $(1 \le i \le l-1)$ an *inner vertex on* W. Let $\bar{W}$ denote the reversed walk of W, i.e., $\bar{W} = (v_l, e_l, \ldots, v_1, e_1, v_0)$.

Let $X \subseteq V$ be a vertex set. We denote by $\delta_G(X)$ the set of arcs between X and $V \setminus X$ in G and by $N_G(X)$ the set of vertices adjacent to X in G, i.e., $\delta_G(X) := \{ e = xy \in E \mid |\{x,y\} \cap X| = 1 \}$ and $N_G(X) := \{ y \in V \setminus X \mid \delta_G(X) \cap \delta_G(\{y\}) \ne \emptyset \}$. We denote a singleton $\{x\}$ by its element x when it makes no confusion.

Let $G[X] := (X, E(X))$ denote the subgraph of G *induced by* X, where $E(X) := \{ e = xy \in E \mid \{x,y\} \subseteq X \}$. We denote by $G - X$ the subgraph of G obtained by removing all vertices in X, i.e., $G - X = G[V \setminus X]$. For an arc set $F \subseteq E$, we also denote by $G - F$ the subgraph of G obtained by removing all arcs in F, i.e., $G - F = (V, E \setminus F)$. Define $G[[X]] := G[X \cup N_G(X)] - E(N_G(X))$.

For an integer $k \ge 0$ and a vertex set $X \subsetneq V$ with $|X| = k$, we call X a *k-cut* in G if $G - X$ is not connected. A directed graph is called *k-connected* if it contains more than k vertices and no k'-cut for every $k' < k$. A *k-connected component* of G is a maximal k-connected induced subgraph $G[X]$ ($X \subseteq V$ with $|X| \ge k$).

Suppose that G is embedded on a plane. We call a unique unbounded face of G the *outer face* of G, and any other face an *inner face*. For a face F of G, let $\mathrm{bd}(F)$ denote the closed walk (whose end vertices coincide with each other) obtained by walking the boundary of F in an arbitrary direction from an arbitrary vertex on it.

Labels. Let $G = (V, E)$ be a Γ-labeled graph with a label function ψ_G, and $W = (v_0, e_1, v_1, \ldots, e_l, v_l)$ a walk in G. The *label* $\psi_G(W)$ of W is defined as the ordered product $\psi_G(e_l, v_l) \cdots \psi_G(e_2, v_2) \cdot \psi_G(e_1, v_1)$, where $\psi_G(e_i, v_i) := \psi_G(e_i)$ if $e_i = v_{i-1}v_i$ and $\psi_G(e_i, v_i) := \psi_G(e_i)^{-1}$ if $e_i = v_i v_{i-1}$. Note that, for the reversed walk $\bar{W}$ of W, we have $\psi_G(\bar{W}) = \psi_G(W)^{-1}$. In particular, since an arc uv with label α and an arc vu with label α^{-1} are equivalent, we identify such two arcs. We say that W is *balanced* (or a *zero walk*) if $\psi_G(W) = 1_\Gamma$ and *unbalanced* (or a *non-zero walk*) otherwise, and also that G is *balanced* if G contains no unbalanced cycle. Note that whether a cycle is balanced or not does not depend on the choices of the direction and the end vertex, since $\psi_G(\bar{C}) = \psi_G(C)^{-1}$ and $\psi_G(C') = \psi_G(e_1) \cdot \psi_G(C) \cdot \psi_G(e_1)^{-1}$, where $C = (v_0, e_1, v_1, \ldots, e_l, v_l = v_0)$ and $C' = (v_1, e_2, v_2, \ldots, e_l, v_l = v_0, e_1, v_1)$. Hence, when we consider whether a cycle is balanced or not, we can choose the direction and the end vertex arbitrarily.

For distinct vertices $s, t \in V$, let $l(G; s, t)$ be the set of all possible labels of s–t paths in G. When $l(G; s, t) = \{\alpha\}$ for some $\alpha \in \Gamma$, we also denote the element α itself by $l(G; s, t)$. Without loss of generality, we may assume that there is no vertex $v \in V$ that is not contained in any s–t path, since such a vertex does not make any effect on $l(G; s, t)$. To consider only such cases, let $\mathcal{D}$ be the set of all triplets (G', s, t) such that G' is a Γ-labeled graph with two specified vertices $s, t \in V(G')$ in which every vertex is contained in some s–t path. The following lemma guarantees that one can efficiently obtain a maximal induced subgraph G' of G such that $(G', s, t) \in \mathcal{D}$ and $l(G'; s, t) = l(G; s, t)$ by computing a 2-connected component of a graph (e.g., by [6]).

Lemma 2. *For a Γ-labeled graph $G = (V, E)$ and distinct vertices $s, t \in V$, $(G, s, t) \in \mathcal{D}$ if and only if the graph obtained from G by adding a new node $r \notin V$ and two arcs from r to s and from r to t is 2-connected.*

2.2   Finding a Non-zero Path

In this section, we show that a non-zero s–t path can be found (i.e., Problem (II) can be solved) efficiently by using well-known properties of Γ-labeled graphs. The following techniques are often utilized in dealing with Γ-labeled graphs (see, e.g., [2,3,18]).

Definition 3 (Shifting). Let $G = (V, E)$ be a Γ-labeled graph. For a vertex $v \in V$ and an element $\alpha \in \Gamma$, *shifting* (*a label function* ψ_G) *by* α *at* v means the following operation: update ψ_G to ψ'_G defined as, for each $e \in E$,

$$\psi'_G(e) := \begin{cases} \psi_G(e) \cdot \alpha^{-1} & (e \in \delta_G(v) \text{ leaves } v), \\ \alpha \cdot \psi_G(e) & (e \in \delta_G(v) \text{ enters } v), \\ \psi_G(e) & (\text{otherwise}). \end{cases}$$

Shifting at $v \in V$ does not change the label of any walk whose end vertices are not v, and neither that of any cycle C whose end vertex is v up to conjugate, i.e., $\psi'_G(C) = \alpha \cdot \psi_G(C) \cdot \alpha^{-1}$. Furthermore, when we apply shifting multiple times, the order of applications does not make any effect on the resulting label function, since each arc is affected only by shifting at its head or tail, which does not interfere with each other. We say that two Γ-labeled graphs G_1 and G_2 are (s, t)-*equivalent* if G_2 is obtained from G_1 by shifting by some $\alpha_v \in \Gamma$ at each $v \in V \setminus \{s, t\}$ (and then G_1 is obtained from G_2 by shifting by α_v^{-1} at each v). Note that $l(G_1; s, t) = l(G_2; s, t)$ if G_1 and G_2 are (s, t)-equivalent.

Lemma 4. *For a connected and balanced Γ-labeled graph $G = (V, E)$ and distinct vertices $s, t \in V$, one can find in polynomial time a Γ-labeled graph G' which is (s, t)-equivalent to G such that*

$$\psi_{G'}(e) = \begin{cases} \alpha & (e \in \delta_G(s) \text{ leaves } s), \\ \alpha^{-1} & (e \in \delta_G(s) \text{ enters } s), \\ 1_\Gamma & (\text{otherwise}), \end{cases}$$

for every arc $e \in E(G') = E$ and for some $\alpha \in \Gamma$ (in fact, $\alpha = l(G; s, t)$).

Lemma 5. *For any $(G, s, t) \in \mathcal{D}$, $|l(G; s, t)| = 1$ if and only if G is balanced.*

Lemmas 2, 4, and 5 lead to the following proposition.

Proposition 6. *Let $G = (V, E)$ be a Γ-labeled graph with a label function ψ_G and two specified vertices $s, t \in V$. Then, for any $\alpha \in \Gamma$, one can test whether $l(G; s, t) \subseteq \{\alpha\}$ or not in polynomial time. Furthermore, if $l(G; s, t) \not\subseteq \{\alpha\}$, then one can find an s–t path P with $\psi_G(P) \neq \alpha$ in polynomial time.*

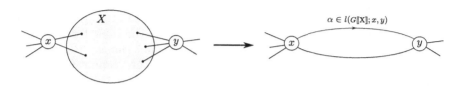

Fig. 1. 2-contraction

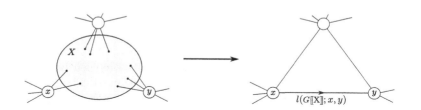

Fig. 2. 3-contraction

2.3   New Operations

For our characterization of triplets $(G, s, t) \in \mathcal{D}$ with $|l(G; s, t)| = 2$, we introduce a few new operations which do not change $l(G; s, t)$. Let $(G = (V, E), s, t) \in \mathcal{D}$, and recall that $G[\![X]\!] := G[X \cup N_G(X)] - E(N_G(X))$ for a vertex set $X \subseteq V$.

Definition 7 (2-contraction). For a vertex set $X \subseteq V \setminus \{s, t\}$ such that $N_G(X) = \{x, y\}$ for some distinct $x, y \in V$ and $G[\![X]\!]$ is connected, the 2-*contraction* of X is the following operation (see Fig. 1):

- remove all vertices in X, and
- add a new arc from x to y with label α for each $\alpha \in l(G[\![X]\!]; x, y)$ if there is no such arc.

The resulting graph is denoted by $G/_2 X$. A vertex set $X \subseteq V \setminus \{s, t\}$ is said to be 2-*contractible in* G if the 2-contraction of X can be performed in G and in particular $G[\![X]\!] \neq G$.

Definition 8 (3-contraction). For a vertex set $X \subseteq V \setminus \{s, t\}$ such that $|N_G(X)| = 3$, $G[X]$ is connected, and $G[\![X]\!]$ is balanced, the 3-*contraction* of X is the following operation (see Fig. 2):

- remove all vertices in X, and
- add a new arc from x to y with label $l(G[\![X]\!]; x, y)$ (which consists of a single element by Lemma 5) for each pair of $x, y \in N_G(X)$ if there is no such arc.

The resulting graph is denoted by $G/_3 X$. A vertex set $X \subseteq V \setminus \{s, t\}$ is said to be 3-*contractible in* G if the 3-contraction of X can be performed in G.

The 2-contraction and the 3-contraction are analogous to the operation which is performed in Condition 3 in Theorem 1, and we use the same term "contraction" to refer to each of them. Any contraction does not change $l(G; s, t)$, since each s–t path cannot enter $G[\![X]\!]$ after leaving it once (i.e., cannot traverse arcs in $G[\![X]\!]$ intermittently). Moreover, we also have $(G', s, t) \in \mathcal{D}$ for the resulting graph G' after any contraction.

3 Main Results

3.1 Algorithmic Results

As described in Section 2.2, Problem (II) can be solved efficiently, i.e., one can find a non-zero s–t path in polynomial time (Proposition 6). The following theorem, one of our main results, is the first nontrivial extension of this property, which claims that not only one label but also another can be forbidden simultaneously.

Theorem 9. *Let $G = (V, E)$ be a Γ-labeled graph with a label function ψ_G and two specified vertices $s, t \in V$. Then, for any distinct $\alpha, \beta \in \Gamma$, one can test whether $l(G; s, t) \subseteq \{\alpha, \beta\}$ or not in polynomial time. Furthermore, if $l(G; s, t) \not\subseteq \{\alpha, \beta\}$, then one can find an s–t path P with $\psi_G(P) \notin \{\alpha, \beta\}$ in polynomial time.*

Such an algorithm is constructed based on characterizations of Γ-labeled graphs with exactly two possible labels of s–t paths, which are shown in Section 3.2. We refer the readers to the full version [10] for our algorithm and a proof of this theorem, whose outline is shown in Section 4. It should be mentioned that this theorem leads to a solution to Problem (I) for $\Gamma \simeq \mathbb{Z}_3$.

Corollary 10. *Let $G = (V, E)$ be a $\mathbb{Z}_3$-labeled graph with a label function ψ_G and two specified vertices $s, t \in V$. Then one can compute $l(G; s, t)$ in polynomial time. Furthermore, for each $\alpha \in l(G; s, t)$, one can find an s–t path P with $\psi_G(P) = \alpha$ in polynomial time.*

3.2 Characterizations

Recall that $\mathcal{D}$ denotes the set of all triplets (G, s, t) such that G is a Γ-labeled graph with $s, t \in V(G)$ in which every vertex is contained in some s–t path. In this section, we provide a complete characterization of triplets $(G, s, t) \in \mathcal{D}$ with $l(G; s, t) = \{\alpha, \beta\}$ for some distinct $\alpha, \beta \in \Gamma$. We consider two cases separately: when $\alpha\beta^{-1} = \beta\alpha^{-1}$ and when $\alpha\beta^{-1} \neq \beta\alpha^{-1}$.

First, we give a characterization in the easier case: when $\alpha\beta^{-1} = \beta\alpha^{-1}$. Note that this case does not appear when $\Gamma \simeq \mathbb{Z}_3$. The following proposition holds analogously to Lemmas 4 and 5 in Section 2.2, which characterize triplets $(G, s, t) \in \mathcal{D}$ with $|l(G; s, t)| = 1$. A proof is left to the full version [10].

Proposition 11. *Let α and β be distinct elements in Γ with $\alpha\beta^{-1} = \beta\alpha^{-1}$. For any $(G, s, t) \in \mathcal{D}$, $l(G; s, t) = \{\alpha, \beta\}$ if and only if G is not balanced and there exists a Γ-labeled graph G' which is (s, t)-equivalent to G such that*

$$
\psi_{G'}(e) = \begin{cases} \alpha \text{ or } \beta & (e \in \delta_G(s) \text{ leaves } s), \\ \alpha^{-1} \text{ or } \beta^{-1} & (e \in \delta_G(s) \text{ enters } s), \\ 1_\Gamma \text{ or } \alpha\beta^{-1} & (\text{otherwise}), \end{cases}
$$

for every arc $e \in E(G') = E(G)$. Moreover, one can find such G' in polynomial time if exists.

We next discuss the main case, which is much more difficult: when $\alpha\beta^{-1} \neq \beta\alpha^{-1}$. The following theorem, one of our main results, completes a characterization of triplets $(G, s, t) \in \mathcal{D}$ with $l(G; s, t) = \{\alpha, \beta\}$ for some distinct $\alpha, \beta \in \Gamma$. The definition of the set $\mathcal{D}_{\alpha,\beta} \subseteq \mathcal{D}$, which appears in the theorem, is shown later through Definitions 13–15 in Section 3.3. In short, $(G, s, t) \in \mathcal{D}_{\alpha,\beta}$ if G is constructed by "gluing" together "nice" planar Γ-labeled graphs (and some trivial Γ-labeled graphs) and their derivations.

Theorem 12. *Let α and β be distinct elements in Γ with $\alpha\beta^{-1} \neq \beta\alpha^{-1}$. For any $(G, s, t) \in \mathcal{D}$, $l(G; s, t) = \{\alpha, \beta\}$ if and only if $(G, s, t) \in \mathcal{D}_{\alpha,\beta}$.*

Recall that $|l(G; s, t)| = 1$ if and only if G is balanced by Lemma 5, which can be easily tested by Lemma 4. Hence, these characterizations lead to the first nontrivial classification of Γ-labeled graphs in terms of the number of possible labels of s–t paths, and the classification is also complete when $\Gamma \simeq \mathbb{Z}_3$.

3.3   Definition of $\mathcal{D}_{\alpha,\beta}$

Fix distinct elements $\alpha, \beta \in \Gamma$ with $\alpha\beta^{-1} \neq \beta\alpha^{-1}$. To characterize triplets $(G, s, t) \in \mathcal{D}$ with $l(G; s, t) = \{\alpha, \beta\}$, let us define several sets of triplets $(G, s, t) \in \mathcal{D}$ for which it is easy to see that $l(G; s, t) = \{\alpha, \beta\}$. Theorem 12 claims that any triplet $(G, s, t) \in \mathcal{D}$ with $l(G; s, t) = \{\alpha, \beta\}$ is in fact contained in one of them.

Definition 13. *For distinct $\alpha, \beta \in \Gamma$ with $\alpha\beta^{-1} \neq \beta\alpha^{-1}$, let $\mathcal{D}^0_{\alpha,\beta}$ be the set of all triplets $(G, s, t) \in \mathcal{D}$ satisfying one of the following conditions.*

(A) *There exists a Γ-labeled graph G' which is not balanced and is (s, t)-equivalent to G such that either*

- *the label of every arc in $G' - s$ is 1_Γ and in $\delta_{G'}(s)$ is α or β, where all arcs in $\delta_{G'}(s)$ are assumed to leave s (see Fig. 3), or*
- *the label of every arc in $G' - t$ is 1_Γ and in $\delta_{G'}(t)$ is α or β, where all arcs in $\delta_{G'}(t)$ are assumed to enter t (see Fig. 4).*

(B) G is (s,t)-equivalent to the Γ-labeled graph which consists of six vertices s, v_1, v_2, v_3, v_4, t, six arcs $sv_1, sv_2, v_1v_2, v_3v_4, v_3t, v_4t$ with label 1_Γ, and two pairs of two parallel arcs from v_i to v_{i+2} ($i = 1, 2$) whose labels are both α and β (see Fig. 5).

(C) G can be embedded on a plane with the face set $\mathcal{F}$ satisfying the following conditions (see Fig. 6):

- both s and t are on the boundary of the outer face $F_0 \in \mathcal{F}$,
- one s–t path along $\mathrm{bd}(F_0)$ is of label α and the other is of β, and
- there exists a unique inner face F_1 whose boundary is unbalanced, i.e., $\psi_G(\mathrm{bd}(F_1)) \neq 1_\Gamma$ and $\psi_G(\mathrm{bd}(F)) = 1_\Gamma$ for any $F \in \mathcal{F} \setminus \{F_0, F_1\}$.

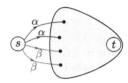

Fig. 3. The former of Case (A)

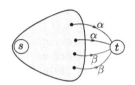

Fig. 4. The latter of Case (A)

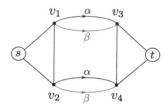

Fig. 5. Case (B)

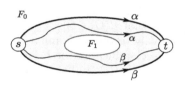

Fig. 6. Case (C)

It is not difficult to see that $l(G; s, t) = \{\alpha, \beta\}$ for any triplet $(G, s, t) \in \mathcal{D}^0_{\alpha,\beta}$.

For the following definitions, recall the operations called the "contractions," which are defined in Section 2.3 (see Definitions 7 and 8).

Definition 14. For distinct $\alpha, \beta \in \Gamma$ with $\alpha\beta^{-1} \neq \beta\alpha^{-1}$, we define $\mathcal{D}^1_{\alpha,\beta}$ as the minimal set of triplets $(G, s, t) \in \mathcal{D}$ with the following conditions:

- $\mathcal{D}^0_{\alpha,\beta} \subseteq \mathcal{D}^1_{\alpha,\beta}$, and
- if $(G/_3 X, s, t) \in \mathcal{D}^1_{\alpha,\beta}$ for some 3-contractible $X \subseteq V \setminus \{s, t\}$, then $(G, s, t) \in \mathcal{D}^1_{\alpha,\beta}$.

We are now ready to define $\mathcal{D}_{\alpha,\beta}$.

Definition 15. For distinct $\alpha, \beta \in \Gamma$ with $\alpha\beta^{-1} \neq \beta\alpha^{-1}$, we define $\mathcal{D}_{\alpha,\beta}$ as the minimal set of triplets $(G, s, t) \in \mathcal{D}$ with the following conditions:

- $\mathcal{D}^1_{\alpha,\beta} \subseteq \mathcal{D}_{\alpha,\beta}$, and

- if $(G/_2X, s, t) \in \mathcal{D}_{\alpha, \beta}$ for some $X \subseteq V \setminus \{s, t\}$ such that either $G[\![X]\!]$ is balanced or $(G[\![X]\!], x, y) \in \mathcal{D}^1_{\alpha', \beta'}$, where $N_G(X) = \{x, y\}$, and $\alpha', \beta' \in \Gamma$ satisfy $\alpha'\beta'^{-1} \neq \beta'\alpha'^{-1}$, then $(G, s, t) \in \mathcal{D}_{\alpha, \beta}$.

Note that the first condition can be replaced with $(G_0, s, t) \in \mathcal{D}_{\alpha, \beta}$, where G_0 consists of two parallel arcs from s to t whose labels are α and β.

It is easy to see that $l(G; s, t) = \{\alpha, \beta\}$ for any triplet $(G, s, t) \in \mathcal{D}_{\alpha, \beta}$ since any contraction does not change $l(G; s, t)$. A proof of the nontrivial direction ("only if" part of Theorem 12) is left to the full version [10] and its sketch is shown in Section 5.

4   Outline of Algorithm

In this section, we present an outline of an algorithm to test whether $|l(G; s, t)| \leq 2$ or not for a given Γ-labeled graph $G = (V, E)$ with $s, t \in V$. The complete description and the correctness are left to the full version [10], which also contains those of an algorithm to find three s–t paths whose labels are distinct when it has turned out that $|l(G; s, t)| \geq 3$. Note that, when $\Gamma \simeq \mathbb{Z}_3$, these algorithms can compute $l(G; s, t)$ itself and find s–t paths which attain all labels in $l(G; s, t)$.

- We may assume that $(G, s, t) \in \mathcal{D}$ by Lemma 2.
- Test whether $|l(G; s, t)| = 1$ or not by Lemmas 4 and 5.
- If $|l(G; s, t)| \geq 2$, then G contains an unbalanced cycle, which makes it possible to construct two s–t paths whose labels are distinct, say $\alpha, \beta \in \Gamma$.
- If $\alpha\beta^{-1} = \beta\alpha^{-1}$, test whether $l(G; s, t) = \{\alpha, \beta\}$ or not by Proposition 11. Otherwise, add an arc from s to t with label α if there is no such arc.
- Reduce G repeatedly as long as possible by 2-contraction (check all possible 2-cuts in G and solve subproblems recursively) or 3-contraction (check all possible 3-cuts in G). Note that if there exists a 2-contractible vertex set $X \subseteq V \setminus \{s, t\}$ with $N_G(X) = \{x, y\}$ in G such that $|l(G[\![X]\!]; x, y)| \geq 3$, then $|l(G; s, t)| \geq 3$ since $(G, s, t) \in \mathcal{D}$.
- Check whether $(G, s, t) \in \mathcal{D}^0_{\alpha, \beta}$ or not. This can be done easily for Cases (A) and (B) in Definition 13, and, for Case (C), by testing the planarity and computing an embedding (e.g., by [7]), which is almost unique due to the 3-connectivity of G (see, e.g., [4]).

5   Proof Sketch of Necessity Part of Theorem 12

In this section, we give a sketch of a proof of the necessity part of Theorem 12, whose proof is completed in the full version [10].

To derive a contradiction, assume that there exist distinct $\alpha, \beta \in \Gamma$ and a triplet $(G, s, t) \in \mathcal{D}$ such that $\alpha\beta^{-1} \neq \beta\alpha^{-1}$, $l(G; s, t) = \{\alpha, \beta\}$, and $(G, s, t) \notin \mathcal{D}_{\alpha, \beta}$. We choose such $\alpha, \beta \in \Gamma$ and $(G, s, t) \in \mathcal{D}$ so that G is as small as possible.

Fix an arbitrary arc e_0 in G leaving s, and consider the graph $G' := G - e_0$. By using the minimality of G, we can show that $(G', s, t) \in \mathcal{D}_{\alpha,\beta}$. We consider the following two cases separately: when $(G', s, t) \in \mathcal{D}^1_{\alpha,\beta}$ and when not.

In both cases, we can embed a graph $\tilde{G}$ obtained from G' (or $G - s$) by at most one 3-contraction on a plane so that the conditions of Case (C) in Definition 13 are satisfied (or derive a contradiction). By expanding a vertex set and adding e_0 (or s and $\delta_G(s)$), we try to extend the planar embedding of $\tilde{G}$ to G. Then, we have one of the following cases.

- Such an extension is possible, i.e., G can be embedded on a plane with the conditions of Case (C) in Definition 13. This contradicts that $(G, s, t) \notin \mathcal{D}_{\alpha,\beta}$.
- G contains a contractible vertex set, which contradicts that G is a minimal counterexample.
- We can construct an s–t path of label $\gamma \in \Gamma \setminus \{\alpha, \beta\}$ in G by using e_0 and some arcs in G', which contradicts that $l(G; s, t) = \{\alpha, \beta\}$.

In each case, we have a contradiction, which completes the proof. We note that Theorem 1 plays an important role in this case analysis.

References

1. Arkin, E.M., Papadimitriou, C.H., Yannakakis, M.: Modularity of cycles and paths in graphs. Journal of the ACM **38**, 255–274 (1991)
2. Chudnovsky, M., Geelen, J., Gerards, B., Goddyn, L., Lohman, M., Seymour, P.D.: Packing non-zero A-paths in group-labelled graphs. Combinatorica **26**, 521–532 (2006)
3. Chudnovsky, M., Cunningham, W., Geelen, J.: An algorithm for packing non-zero A-paths in group-labelled graphs. Combinatorica **28**, 145–161 (2008)
4. Diestel, R.: Graph Theory, 4th edn. Springer-Verlag, Heidelberg (2010)
5. Fortune, S., Hopcroft, J., Wyllie, J.: The directed subgraph homeomorphism problem. Theoretical Computer Science **10**, 111–121 (1980)
6. Hopcroft, J., Tarjan, R.: Efficient algorithm for graph manipulation. Communications of the ACM **16**, 372–378 (1973)
7. Hopcroft, J., Tarjan, R.: Efficient planarity testing. Journal of the ACM **21**, 549–568 (1974)
8. Huynh, T.: The Linkage Problem for Group-Labelled Graphs, Ph.D. Thesis, Department of Combinatorics and Optimization, University of Waterloo, Ontario (2009)
9. Kawarabayashi, K., Wollan, P.: Non-zero disjoint cycles in highly connected group labelled graphs. Journal of Combinatorial Theory, Ser. B **96**, 296–301 (2006)
10. Kawase, Y., Kobayashi, Y., Yamaguchi, Y.: Finding a path in group-labeled graphs with two labels forbidden (the full version is in preparation)
11. LaPaugh, A.S., Papadimitriou, C.H.: The even-path problem for graphs and digraphs. Networks **14**, 507–513 (1984)
12. McCuaig, W.: Pólya's permanent problem. The Electronic Journal of Combinatorics **11**, R79 (2004)
13. Pólya, G.: Aufgabe 424. Arch. Math. Phys. **20**, 271 (1913)
14. Robertson, N., Seymour, P.D., Thomas, R.: Permanents, Pfaffian orientations, and even directed circuits. Annals of Mathematics **150**, 929–975 (1999)

15. Shiloach, Y.: A polynomial solution to the undirected two paths problem. Journal of the ACM **27**, 445–456 (1980)
16. Seymour, P.D.: Disjoint paths in graphs. Discrete Mathematics **29**, 293–309 (1980)
17. Thomassen, C.: 2-linked graphs. European Journal of Combinatorics **1**, 371–378 (1980)
18. Tanigawa, S., Yamaguchi, Y.: Packing non-zero *A*-paths via matroid matching, Mathematical Engineering Technical Reports, METR 2013-08, University of Tokyo (2013)
19. Wollan, P.: Packing cycles with modularity constraint. Combinatorica **31**, 95–126 (2011)
20. Yamaguchi, Y.: Packing *A*-paths in group-labelled graphs via linear matroid parity. In: Proceedings of the 25th ACM-SIAM Symposium on Discrete Algorithms (SODA 2014), pp. 562–569 (2014)

Lower Bounds for Sums of Powers
of Low Degree Univariates

Neeraj Kayal[1], Pascal Koiran[2], Timothée Pecatte[2], and Chandan Saha[3](✉)

[1] Microsoft Research India, Bengaluru, India
[2] Ecole Normale Supérieure de Lyon, Lyon, France
[3] Indian Institute of Science, Bengaluru, India
ch.saha@gmail.com

Abstract. We consider the problem of representing a univariate polynomial $f(x)$ as a sum of powers of low degree polynomials. We prove a lower bound of $\Omega\left(\sqrt{\frac{d}{t}}\right)$ for writing an explicit univariate degree-d polynomial $f(x)$ as a sum of powers of degree-t polynomials.

Keywords: Arithmetic circuits · Lower bounds · Sums of powers · Wronskian · Shifted derivatives

1 Introduction

Valiant [22], defined the classes VP and VNP as the algebraic analogs of the classes P and NP. Informally, VP consists of (families of) efficiently computable (low-degree, multivariate) polynomials while VNP consists of (families of) explicit (low-degree, multivariate) polynomials. The problem of separating VNP from VP has since been one of the most important open problems in arithmetic complexity. Another basic question in complexity in general is whether computation can be efficiently parallelized. A seminal work by [23] showed that computation of low degree polynomials can indeed be efficiently parallelized - any *small* arithmetic circuit C computing a *low degree* multivariate polynomial $f(\mathbf{x})$ can be transformed to obtain another circuit C' of *low* depth and whose size is *not too large* computing the same polynomial $f(\mathbf{x})$. Subsequent refinements and improvements were obtained in a series of works [1,2,8,15,21]. This line of work in particular yields the following depth reduction result which shows that if a polynomial can be efficiently computed then it has a *not too large* representation as a sum of powers of low degree polynomials. Specifically:

Proposition 1 (Implicit in [21] and [8]). *Let $\{f_n(\mathbf{x}) : n \geq 1\}$ be a family of n-variate polynomials of degree $d = d(n)$ over an underlying field $\mathbb{F}$ which is algebraically closed and has characteristic zero. If this family is in VP then $f_n(\mathbf{x})$ admits a representation of the form*

$$f_n(\mathbf{x}) = \sum_{i=1}^{s} Q_i(\mathbf{x})^{e_i} \quad \text{where } \deg(Q_i) \leq \sqrt{d} \tag{1}$$

and where the number of summands s is at most $n^{O(\sqrt{d})}$.

© Springer-Verlag Berlin Heidelberg 2015
M.M. Halldórsson et al. (Eds.): ICALP 2015, Part I, LNCS 9134, pp. 810–821, 2015.
DOI: 10.1007/978-3-662-47672-7_66

Strong Enough Lower Bounds for Sums of Powers Imply General Circuit Lower Bounds. These depth reduction results also provide a potential approach towards the VP versus VNP problem – via proving strong enough lower bounds for low depth circuits. In particular, the contrapositive version of Proposition 1 means that a strong enough (at least $n^{\omega(\sqrt{d})}$) lower bound for representing an explicit family of polynomials $\{f_n(\mathbf{x}) : n \geq 1\}$ in the form (1) above will imply that this family is not in VP, thereby separating VP and VNP. Promising progress along this direction has recently been obtained. [9] considered representations of the form (1) above and introduced a complexity measure called *dimension of shifted partials* and obtained a $2^{\Omega(\sqrt{d})}$ lower bound for representations of the form (1) above. Follow-up work [7,12] obtained an $n^{\Omega(\sqrt{d})}$ lower bound for such representations, thereby coming tantalizingly close to the threshold required for obtaining superpolynomial lower bounds for general circuits. Since then, these techniques have been intensely investigated and followup work by [5,11,17] have used these techniques to obtain optimality of the known depth reduction results in many interesting cases. Some of these works also suggest that the dimension of shifted partials in itself might not be strong enough to separate VP from VNP. Further work [10,11,18] has suitably adapted and generalized the complexity measure to obtain lower bounds for more subclasses of arithmetic circuits.

Univariate Sums of Powers. Motivated by proposition 1, we introduce and study the problem of representing a *univariate* polynomial as a sum of powers of low-degree polynomials.

Definition 1. *Let $t \geq 1$ be an integer. For a polynomial $f(x) \in \mathbb{F}[x]$, define the sum of degree-t-powers complexity of f, denoted $s_t(f)$, as the smallest integer s such that f can be written as*

$$f(x) = \sum_{i=1}^{s} \alpha_i \cdot Q_i(x)^{e_i}, \quad \text{where } \forall i : \alpha_i \in \mathbb{F}, \deg(Q_i) \leq t.$$

We remark here that if the underlying field $\mathbb{F}$ is algebraically closed, we can assume without loss of generality that each scalar $\alpha_i = 1$. We seek to exhibit explicit polynomials $f(x)$ for which $s_t(f)$ is as large as possible. The motivation for this study is that univariate polynomials being much more well-known and easier to study than multivariate polynomials one can first try to develop proof techniques that yield improved lower bounds for the univariate case. In particular, the invariant theory of binary forms (aka univariate polynomials) is much better understood as compared to multivariate polynomials. One could also hope to apply some of the proof ideas from real/complex analysis or from the vast literature on Waring's problem[1] to obtain improved lower bounds on

---

[1] Waring's problem asks whether each natural number k has an associated positive integer $s(k)$ such that every natural number is the sum of at most s k-th powers of natural numbers. For example, every natural number is the sum of at most 4 squares, 9 cubes. Many variants of Waring's problem for algebraic integers and polynomials have also been studied.

$s_t(f)$. Our underlying hope is that some such improved proof technique or proof idea might admit a suitable generalization to the multivariate case as well. This could be one potential way to attack the VP versus VNP problem. We also note that there are formal results essentially following from the work of Koiran [14] which imply that seemingly mild lower bounds for a slight variant of the model being considered here directly implies a separation of VP from VNP.

Proposition 2. [Implicit in [14]]. *If there is an explicit family of* univariate *polynomials* $\{f_d(x) : d \geq 1\}$ *over an underlying algebraically closed field* $\mathbb{F}$ *of characteristic zero such that any representation of the form* $f_d(x) = \sum_{i=1}^{s} Q_i(x)^{e_i}$, *where* $\mathrm{Sparsity}(Q_i) \leq t$, *requires the number of summands* s *to be at least* $\left(\frac{d}{t}\right)^{\Omega(1)}$ *then* VP $\neq$ VNP.

This means that proving relatively mild lower bounds on a similar model (but with the degree bound replaced by the corresponding sparsity bound) already implies that VP is different from VNP.

Our results. In describing our results, we avoid floor/ceil notations for ease of presentation. Throughout this paper, the underlying field $\mathbb{F}$ will be of characteristic zero. We first note that a standard dimension counting argument implies that for a random polynomial $f(x)$ of degree d it is almost surely the case that $s_t(f) \geq \frac{d+1}{t+1}$. In comparison to this benchmark, we prove a lower bound of $s_t(f) \geq \Omega\left(\sqrt{\frac{d}{t}}\right)$ for an explicit family of polynomials of degree d.

Theorem 1. *Let* $d, t \geq 2$ *be integers. Let* $a_1, \ldots, a_{2t}$ *be any* $2t$ *distinct elements of the underlying field* $\mathbb{F}$. *Assume* $\mathbb{F}$ *is of characteristic zero. Let* $g \stackrel{def}{=} \prod_{k=1}^{2t} (x - a_k)$. *Define the univariate polynomial,*

$$f(x) \stackrel{def}{=} g(x)^{\frac{d}{2t}}. \tag{2}$$

Then $s_t(f) \geq \Omega\left(\sqrt{\frac{d}{t}}\right)$.

Our proof here employs the Wronskian[2] and is therefore quite different from the proof technique used in the recent works on homogeneous depth four circuits [7, 9, 12]. These works employ a complexity measure called the dimension of shifted partials to obtain lower bounds for a similar multivariate model. We also show that a suitable variant of shifted partials does yield a similar lower bound albeit for a different target polynomial. Specifically, we have:

---

[2] The Wronskian has been employed in arithmetic complexity previously in [16] to obtain nontrivial (but rather weak) lower bounds for writing a polynomial as a sum of powers of sparse polynomials. Indeed, [16] manage to prove something stronger - they obtain weak (but still nontrivial and interesting) bounds on the number of real roots of sums of powers of sparse polynomials.

Theorem 2. *Let $d, t \geq 2$ be integers such that $t < \frac{d}{4}$. Let the polynomial $f(x) = \sum_{i=1}^{m}(x - a_i)^d$, with distinct a_i's and let $m = \left\lfloor \sqrt{\frac{d}{t}} \right\rfloor$. Then $s_t(f) \geq \Omega\left(\sqrt{\frac{d}{t}}\right)$.*

Remark 1. 1. **Optimality of the lower bound.** The polynomial $f(x)$ in theorem 2 has the nice feature that it can also be expressed as a sum of $O(\sqrt{d/t})$ summands, each of which is a power of a polynomial of degree at most t. So, in this sense theorem 2 gives an optimal lower bound. The *target polynomial* in theorem 1 does not seem to have this property.

2. **Methods.** In the proof of theorem 2, we show that the dimension of shifted derivatives of the polynomial $f(x)$ is the maximum possible (for the appropriate choice of parameters). Since the polynomial $f(x)$ of theorem 2 also satisfies $s_t(f) \leq O\left(\sqrt{\frac{d}{t}}\right)$, it indicates that a lower bound better than $\Omega\left(\sqrt{\frac{d}{t}}\right)$ probably cannot be obtained via shifted derivatives. It is currently conceivable that the Wronskian-based proof could yield better lower bounds. A more detailed discussion on this may be found in Pecatte's internship report [19].

3. **On replacing the degree bound by the corresponding sparsity bound.** We also note that for multivariate polynomials, recent work by [10] successfully replaced the bound on the degrees of the Q_i's by the corresponding bound on the sparsity of the Q_i's. We note in passing that by proposition 2, if we could prove an analogous result as the one above but with the degree bound on the Q_i's replaced by a bound on their sparsities, then we would obtain a separation of VP from VNP. In this sparse setting, the best lower bound that is currently known is $\Omega\left(\sqrt{\frac{\log d}{\log t}}\right)$. It applies to any polynomial of degree d that has d distinct real roots [16].

4. **Upper bounds.** While the focus of this paper is on lower bounds, it is also natural to ask about upper bounds on $s_t(f)$. As mentioned above, the lower bound $s_t(f) \geq \frac{d+1}{t+1}$ follows from a simple dimension counting argument. Recent work on the Waring problem for polynomials [6] shows that this bound is tight for a generic polynomial of degree d when $t + 1$ divides $d + 1$. Moreover, a general result on "maximum rank versus generic rank" (Theorem 1 in [3]) shows that moving from a generic polynomial to a worst-case polynomial at most doubles $s_t(f)$. We conclude that the upper bound $s_t(f) \leq 2 \cdot \frac{d+1}{t+1}$ applies to *any* polynomial of degree d when $t+1$ divides $d+1$. Note that this upper boud is nonconstructive. A simple explicit construction shows that $s_t(f) = O((d/t)^2)$ for all f.

2 Preliminaries

2.1 The Wronskian

The *Wronskian* is a mathematical tool mainly used in the study of differential equations to show that a set of solutions is linearly independent.

Definition 2. [Wronskian]. *For n real functions $f_1, \ldots, f_n$, which are $n-1$ times differentiable, the Wronskian $W(f_1, \ldots, f_n)$ is defined by*

$$W(f_1, \ldots, f_n)(x) = \begin{vmatrix} f_1(x) & f_2(x) & \cdots & f_n(x) \\ f_1'(x) & f_2'(x) & \cdots & f_n'(x) \\ \vdots & \vdots & \ddots & \vdots \\ f_1^{(n-1)} & f_2^{(n-1)} & \cdots & f_n^{(n-1)} \end{vmatrix}.$$

We will use the following fact about the Wronskian whose proofs can be found in [20] (and which are known since the 19th century).

Proposition 3. *For any $f_1, \ldots, f_k$ which are $k-1$ times differentiable, if f_1 is a perfect power say if $f_1 = Q^e$ where $e \geq k$ then Q^{e-k+1} divides $W(Q^e, f_2, \ldots, f_k)$.*

Also, the Wronskian captures linear dependence of polynomials in $\mathbb{F}[x]$.

Proposition 4. *[4] Let $\mathbb{F}$ be a field of characteristic zero. For univariate polynomials $f_1, \ldots, f_n \in \mathbb{F}[x]$, they are linearly dependent if and only if the Wronskian $W(f_1, \ldots, f_n)$ vanishes everywhere.*

We will also use another result from [24] which gives a bound on the multiplicity of a root depending on the Wronskian. For a field element $\alpha \in \mathbb{F}$, and a polynomial $g(x) \in \mathbb{F}[x]$, let $N_\alpha(g)$ denote the multiplicity of g at α, i.e. the highest power of $(x - \alpha)$ which divides $g(x)$.

Lemma 1. *Let $\mathbb{F}$ be a field of characteristic zero. Let $Q_1, \ldots, Q_m$ be some linearly independent polynomial and $\alpha \in \mathbb{F}$, and let $F(x) = \sum_{i=1}^m Q_i(x)$. Then: $N_\alpha(F) \leq m - 1 + N_\alpha(W(Q_1, \ldots, Q_m))$, where $N_\alpha(W(Q_1, \ldots, Q_m))$ is finite since $W(Q_1, \ldots, Q_m) \not\equiv 0$.*

2.2   The Space of Shifted Derivatives

In Section 4 we give an alternate lower bound proof via a slight variant of a complexity measure first defined in [9]: the space of shifted partial derivatives. Using this complexity measure, [9] obtained exponential lower bounds on a similar multivariate model. The key intuition follows from the following simple observation: derivatives of Q^e of order $\leq k$ all share a large common factor, namely Q^{e-k}. We try to capture this property with the following complexity measure:

Definition 3 (Shifted derivatives space). *Let $f(x) \in \mathbb{F}[x]$ be a polynomial. The span of the l-shifted k-th order derivatives of f, denoted by $\left\langle x^{\leq i+l} \cdot f^{(i)} \right\rangle_{i \leq k}$, is defined as:*

$$\left\langle x^{\leq i+l} \cdot f^{(i)} \right\rangle_{i \leq k} \overset{def}{=} \mathbb{F}\text{-span}\left\{ x^j \cdot f^{(i)}(x) \; : \; i \leq k, \; j \leq i + l \right\}.$$

$\left\langle x^{\leq i+l} \cdot f^{(i)} \right\rangle_{i \leq k}$ *forms an $\mathbb{F}$-vector space and we denote by $\dim \left\langle x^{\leq i+l} \cdot f^{(i)} \right\rangle_{i \leq k}$ the dimension of this space.*

Remark 2. We have two trivial upper bounds on the dimension of the shifted derivatives space. First, for any polynomial f of degree d, the degree of any polynomial in $\langle x^{\leq i+l} \cdot f^{(i)} \rangle_{i \leq k}$ is less than $d + l$, hence $\dim \langle x^{\leq i+l} \cdot f^{(i)} \rangle_{i \leq k} \leq d+l+1$. Second, the dimension is less or equal than the cardinality of a generating family, thus $\dim \langle x^{\leq i+l} \cdot f^{(i)} \rangle_{i \leq k} \leq \sum_{i=0}^{k}(l+i+1)$. Thus, we have:

$$\dim \left\langle x^{\leq i+l} \cdot f^{(i)} \right\rangle_{i \leq k} \leq \min \left(d+l+1, \ (k+1)l + \binom{k+2}{2} \right).$$

We will see later some polynomials that achieve the above bounds and have full shifted derivative space. Since $\langle x^{\leq i+l} \cdot (f+g)^{(i)} \rangle_{i \leq k} \subseteq \langle x^{\leq i+l} \cdot f^{(i)} \rangle_{i \leq k} + \langle x^{\leq i+l} \cdot g^{(i)} \rangle_{i \leq k}$, the measure we defined is sub-additive.

3   Proof of Theorem 1

Suppose $f = \sum_{i=1}^{s} \alpha_i \cdot Q_i^{e_i}$. Since degree of every Q_i is bounded by t and $\deg(f) = d$, $e_i \geq \frac{d}{t}$ for some $i \in [s]$. Without loss of generality, let $e_1 \geq \frac{d}{t}$. Also, we can assume that $Q_1^{e_1}, \ldots, Q_s^{e_s}$ are $\mathbb{F}$-linearly independent - if not, we work with a basis and a smaller value for s. By taking derivatives of both sides of the equation $f = \sum_{i=1}^{s} \alpha_i \cdot Q_i^{e_i}$ with respect to x for j times we have,

$$\sum_{i=1}^{s} \alpha_i \cdot [Q_i^{e_i}]^{(j)} = f^{(j)}, \qquad \text{for every } j \in \{0, \ldots, s-1\} \,,$$

where $[Q_i^{e_i}]^{(j)}$ and $f^{(j)}$ are the j-th derivatives of $Q_i^{e_i}$ and f, respectively, with respect to x. The above equation defines a system of linear equations in $\alpha_1, \ldots, \alpha_s$. By applying Cramer's rule,

$$\alpha_1 = \frac{\mathrm{W}\left(f, Q_2^{e_2}, \ldots, Q_s^{e_s}\right)}{\mathrm{W}\left(Q_1^{e_1}, Q_2^{e_2}, \ldots, Q_s^{e_s}\right)}, \tag{3}$$

where $\mathrm{W}(g_1, \ldots, g_s)$ is the Wronskian determinant of the polynomials $g_1, \ldots, g_s$. Since $Q_1^{e_1}, Q_2^{e_2}, \ldots, Q_s^{e_s}$ are $\mathbb{F}$-linearly independent, $\mathrm{W}(Q_1^{e_1}, Q_2^{e_2}, \ldots, Q_s^{e_s}) \neq 0$. Observe that unless $s = \Omega\left(\frac{d}{t}\right)$, $Q_1^{e_1-(s-1)}$ divides $\mathrm{W}(Q_1^{e_1}, Q_2^{e_2}, \ldots, Q_s^{e_s})$ and $g^{\frac{d}{2t}-(s-1)}$ divides $\mathrm{W}(f, Q_2^{e_2}, \ldots, Q_s^{e_s})$. Let $\Delta \overset{\text{def}}{=} \{i \mid e_i \geq s \text{ and } 2 \leq i \leq s\}$. Then, $\prod_{i \in \Delta} Q_i^{e_i-(s-1)}$ divides both $\mathrm{W}(Q_1^{e_1}, Q_2^{e_2}, \ldots, Q_s^{e_s})$ and $\mathrm{W}(f, Q_2^{e_2}, \ldots, Q_s^{e_s})$. Thus, by analyzing the factors coming out common from the Wronskian determinants, we can express α_1 as

$$\alpha_1 = \frac{g^{\frac{d}{2t}-(s-1)} \cdot \prod_{i \in \Delta} Q_i^{e_i-(s-1)} \cdot W_1}{Q_1^{e_1-(s-1)} \cdot \prod_{i \in \Delta} Q_i^{e_i-(s-1)} \cdot W_2} = \frac{g^{\frac{d}{2t}-(s-1)} \cdot W_1}{Q_1^{e_1-(s-1)} \cdot W_2}. \tag{4}$$

Now observe that after taking $Q_1^{e_1-(s-1)}$ and $\prod_{i \in \Delta} Q_i^{e_i-(s-1)}$ common from $\mathrm{W}(Q_1^{e_1}, Q_2^{e_2}, \ldots, Q_s^{e_s})$, every polynomial in the r-th row of the Wronskian matrix of $Q_1^{e_1}, Q_2^{e_2}, \ldots, Q_s^{e_s}$ has degree upper bounded by $(s-1)t - (r-1)$. Hence, $\deg(W_2) \leq s(s-1)t - \sum_{r=1}^s (r-1) \leq s^2 t$. Since α_1 is a field element, $g^{\frac{d}{2t}-(s-1)}$ must divide $Q_1^{e_1-(s-1)} \cdot W_2$ (by Eq. 4). Polynomial g has $2t$ distinct roots, whereas polynomial Q_1 has at most t roots. Therefore, there are t distinct roots of g such that each of these roots divide W_2 with multiplicity $\frac{d}{2t} - (s-1)$. Since $\deg(W_2) \leq s^2 t$,

$$ s^2 t \;\geq\; t \cdot \left[\frac{d}{2t} - (s-1) \right] \quad \Rightarrow \quad s \;\geq\; \frac{1}{\sqrt{2}} \cdot \sqrt{\frac{d}{t}} - \frac{1}{2}. $$

The t=1 case. The argument can be strengthened to show the following when $t=1$: if $x^d + x^{d-1}$ is expressed as a sum of s-many d-th powers of linear polynomials then $s \geq d+1$. This optimum bound also follows from a work on representing homogeneous (multivariate) polynomials as sums of linear forms by Kleppe [13].

4 An Alternative Proof Using Shifted Partials

In this section, we will give a proof of theorem 2 using shifted derivatives. The proof will consist in first giving an upper bound on the dimension of shifted partials of a sum of powers of low degree polynomials. Thereafter, we give a lower bound on the dimension of shifted derivatives space of the polynomials of the form $f(x) = \sum_{i=1}^m (x - a_i)^d$. To do so, we will show that f does not satisfy a particular kind of differential equations, under some conditions.

4.1 Upper Bounding the Dimension of Shifted Partial Derivatives

We first show that in our model, polynomials have a small complexity according to the shifted partial dimension measure defined in Section 2:

Proposition 5. *For any polynomial f of degree d of the form $f = \sum_{i=1}^s \alpha_i Q_i^{e_i}$, with $\deg(Q_i) \leq t$ we have:* $\dim \langle x^{\leq i+l} \cdot f^{(i)} \rangle_{i \leq k} \leq s \cdot (l + kt + 1).$

Proof. Since the measure is sub-additive, we only have to show that for a simple building block f of the form Q^e, with $\deg Q \leq t$, we have $\dim \langle x^{\leq i+l} \cdot Q^{e(i)} \rangle_{i \leq k} \leq l + kt + 1$. Now note that any $g \in \langle x^{\leq i+l} \cdot Q^{e(i)} \rangle_{i \leq k}$ is of the form $g = Q^{e-k} \cdot R$. Moreover $\deg(R) \leq l + kt$ (since $\deg g \leq e \cdot t + l$). This directly gives the bound on the dimension. $\qquad\square$

4.2 Lower Bounding the Dimension of Shifted Derivatives for an Explicit Polynomial

Definition 4. Shifted Differential Equations *(SDE) are a kind of differential equations of the form* $\sum_{i=0}^k P_i(x) f^{(i)}(x) = 0$, *for some polynomials $P_i \in \mathbb{F}[x]$, not all zero, with $\deg(P_i) \leq i + l$. Here, k is called the* order *and l the* shift.

This kind of differential equations is linked with the notion of shifted derivatives.

Proposition 6. *For any $h(x) \in \mathbb{F}[x]$, if h does not satisfy any SDE of order k and of shift l, then $\left\langle x^{\leq i+l} \cdot h^{(i)} \right\rangle_{i \leq k}$ is full, i.e. :*

$$\dim \left\langle x^{\leq i+l} \cdot h^{(i)} \right\rangle_{i \leq k} = \sum_{i=0}^{k} (l+i+1) = (k+1)l + \binom{k+2}{2}.$$

In order to prove some conditions on the SDE satisfied by our target explicit polynomial $f(x)$, we first need to prove that the polynomials $(x - a_1)^d, \ldots, (x - a_m)^d$ cannot satisfy simultaneously a SDE if the order is not big enough:

Lemma 2. *For any $d, m \leq d$, for any distinct $(a_1, a_2, \ldots, a_m) \in \mathbb{F}^m$, the following property holds for the family $S = \{(x - a_1)^d, \ldots, (x - a_m)^d\}$: if a SDE is satisfied by every polynomial $h \in S$, then the order of the SDE must be greater than or equal to m.*

Proof. Assume that each polynomial in $S = \{(x - a_1)^d, \ldots, (x - a_m)^d\}$ satisfies the following SDE, with $k < m$:

$$\sum_{i=0}^{k} P_i(x) h^{(i)}(x) = 0 \quad \forall h \in S. \tag{5}$$

For all $j \in [m]$, we can factor out $(x - a_j)^{d-k}$ from the above equation to obtain a new SDE satisfied by the family $S' = \{(x - a_1)^k, \ldots, (x - a_m)^k\}$. i.e.:

$$\sum_{i=0}^{k} R_i(x) h^{(i)}(x) = 0 \quad \forall h \in S', \tag{6}$$

with $R_i(x) \stackrel{\text{def}}{=} \frac{d!}{k!} \frac{(k-i)!}{(d-i)!} P_i(x)$.

Since $k < m$, the family S' generate $\mathbb{F}_k[x]$ (the vector space of polynomials of degree at most k), and thus this implies that every polynomial of degree $\leq k$ should satisfy the SDE (6). We obtain the contradiction by plugging in $h(x) = x^{i_0}$ in SDE (6), where i_0 is the smallest integer such that $R_{i_0}(x) \neq 0$. $\qquad \square$

We can now prove the lower bound on the parameters of a SDE that f could satisfy, which will directly give the result.

Lemma 3. *For any $d, m \leq d$, for any m distinct elements $a_1, a_2, \ldots, a_m \in \mathbb{F}$, if the polynomial $f(x) = \sum_{i=1}^{m} (x - a_i)^d$ satisfies a SDE of parameters k, l then at least one of the two following conditions holds:*

  i) $k \geq m$, or,
  ii) $l > \frac{d}{m} - \frac{3}{2} \cdot m$.

Proof. We will prove the result by showing that if f satisfies a SDE and i) doesn't hold, then ii) must hold. Assume that f satisfies a differential equation of the following form:

$$\sum_{i=0}^{k} P_i(x) f^{(i)}(x) = 0, \tag{7}$$

with $k < m$ and $\deg(P_i) \leq i + l$.
For every $j \in [m]$, we denote by R_j the unique polynomial such that:

$$\sum_{i=0}^{k} P_i(x) \left((x - a_j)^d \right)^{(i)} (x) = R_j(x)(x - a_j)^{d-k}.$$

Notice that R_j is of degree at most $k + l$. By lemma 2, since $k < m$, not all R_j's can be 0, without loss of generality we have $R_1 \not\equiv 0$. For $j \in [m]$, we set $f_j(x) = R_j(x)(x - a_j)^{d-k}$ and, using linearity of differentiation, we rewrite differential equation (7) as: $-f_1(x) = \sum_{j=2}^{m} f_j(x)$.

Using Lemma 1, for a certain subset $J = \{j_1, \ldots, j_p\} \subseteq [2..m]$, we obtain

$$d - k \leq N_{a_1}(f_1) \leq p - 1 + N_{a_1}\left(W\left((f_j)_{j \in J} \right) \right). \tag{8}$$

We can factorize the Wronskian by $(x - a_j)^{d-k-(p-1)}$ for any $j \in J$:

$$N_{a_1}\left(W\left((f_j)_{j \in J} \right) \right) = N_{a_1} \begin{vmatrix} R_{1,1} & \cdots & R_{1,p} \\ \vdots & \ddots & \vdots \\ R_{p,1} & \cdots & R_{p,p} \end{vmatrix},$$

with $\deg(R_{i,j}) \leq l + k + p - i$. The determinant has degree $\leq p(l + k) + \binom{p}{2}$. Hence, inequality (8) becomes: $d - k \leq p - 1 + p(l + k) + \binom{p}{2}$. Using the fact that $p \leq m - 1$, we obtain: $d \leq (m-1) \cdot l + m \cdot k + \frac{(m-2)(m+1)}{2}$. Divide by m and drop negative terms to obtain: $\frac{d}{m} \leq l + k + \frac{m}{2}$. Using the hypothesis that $k < m$, we finally have: $l > \frac{d}{m} - \frac{3}{2}m$. □

4.3   Putting Things Together

We are now ready to give a proof of theorem 2.

Proof. We take k and l small enough to ensure that f does not satisfy any SDE of parameters k and l. Using lemma 3, it is enough to take:

- $k = m - 1 = \left\lfloor \sqrt{\frac{d}{t}} \right\rfloor - 1$ so that $k < m$,

- $l = \left\lfloor \sqrt{dt} - \frac{3}{2}\sqrt{\frac{d}{t}} \right\rfloor$ so that $l \leq \frac{d}{m} - \frac{3}{2}m$.

Using proposition 6, we thus establish a lower bound on the dimension of the shifted derivatives space:

$$\dim \left\langle x^{\leq i+l} \cdot f^{(i)} \right\rangle_{i \leq k} = (k+1)l + \binom{k+2}{2}$$

$$\geq \left(\sqrt{\frac{d}{t}} - 1 \right) \left(\sqrt{dt} - \frac{3}{2} \sqrt{\frac{d}{t}} - 1 \right) + \frac{1}{2} \left(\sqrt{\frac{d}{t}} \right)^2$$

$$= d \left(1 - \frac{1}{t} - \sqrt{\frac{t}{d}} + \frac{1}{2\sqrt{dt}} + \frac{1}{d} \right)$$

$$\geq d \left(1 - \frac{1}{t} - \sqrt{\frac{t}{d}} \right).$$

Now, assume that $f = \sum_{i=1}^{s} \alpha_i Q_i^{e_i}$, for some Q_i's with $\deg Q_i \leq t$. Proposition 5 gives the following upper bound on the dimension:

$$\dim \left\langle x^{\leq i+l} \cdot f^{(i)} \right\rangle_{i \leq k} \leq s \cdot (l + kt + 1) \leq s \cdot 2\sqrt{dt}.$$

Hence:

$$s \geq \frac{1 - \frac{1}{t} - \sqrt{\frac{t}{d}}}{2} \cdot \frac{d}{\sqrt{dt}}.$$

Now, since $t < \frac{d}{4}$, we have $\sqrt{\frac{t}{d}} < \frac{1}{2}$ and thus: $s = \Omega \left(\sqrt{\frac{d}{t}} \right).$

$\square$

5   Discussion

In this work, we introduce the model of sums of powers of univariates and gave a new proof technique (via the Wronskian) to prove a lower bound in this model. Even though the existing technique of shifted partials also yields a similar lower bound in this model, our proof (via the Wronskian) could nevertheless be interesting for it is different and perhaps some suitable generalization of it might yield improved lower bounds for some classes of multivariate circuits. In any case, we feel that the sum of powers of univariates model is easier to analyze and may serve as a testbed for other candidate techniques or complexity measures aiming to obtain improved circuit lower bounds. We conclude by mentioning a few open problems that are implicit in remark 1.

- Obtain a lower bound for sums of powers of t-sparse polynomials which is better than $\Omega(\sqrt{\frac{\log d}{\log t}})$.
- Obtain a $d^{O(1)}$-time algorithm for expressing a given degree d polynomial as a sum of $O(\frac{d}{t})$-many powers of degree-t polynomials.
- Improve the $\Omega(\sqrt{\frac{d}{t}})$ lower bound shown in this work.

References

1. Agrawal, M., Vinay, V.: Arithmetic circuits: a chasm at depth four. In: Foundations of Computer Science FOCS, pp. 67–75 (2008)
2. Allender, E., Jiao, J., Mahajan, M., Vinay, V.: Non-Commutative Arithmetic Circuits: Depth Reduction and Size Lower Bounds. Theor. Comput. Sci. **209**(1–2), 47–86 (1998)
3. Blekherman, G., Teitler, Z.: On maximum, typical and generic ranks. Mathematische Annalen, pp. 1–11 (2014)
4. Bocher, M.: The theory of linear dependence. Annals of Mathematics **2**(1/4), 81–96 (1900–1901)
5. Fournier, H., Limaye, N., Malod, G., Srinivasan, S.: Lower bounds for depth 4 formulas computing iterated matrix multiplication. In: Symposium on Theory of Computing, STOC 2014, pp. 128–135 (2014)
6. Fröberg, R., Ottaviani, G., Shapiro, B.: On the Waring problem for polynomial rings. Proceedings of the National Academy of Sciences **109**(15), 5600–5602 (2012)
7. Gupta, A., Kamath, P., Kayal, N., Saptharishi, R.: Approaching the chasm at depth four. In: Conference on Computational Complexity (CCC), pp. 65–73 (2013)
8. Gupta, A., Kamath, P., Kayal, N., Saptharishi, R.: Arithmetic circuits: a chasm at depth three. In: Foundations of Computer Science (FOCS), pp. 578–587 (2013)
9. Kayal, N.: An exponential lower bound for the sum of powers of bounded degree polynomials. Electronic Colloquium on Computational Complexity (ECCC) **19**, 81 (2012)
10. Kayal, N., Limaye, N., Saha, C., Srinivasan, S.: An exponential lower bound for homogeneous depth four arithmetic formulas. In: 55th IEEE Annual Symposium on Foundations of Computer Science, FOCS, pp. 61–70 (2014)
11. Kayal, N., Saha, C.: Lower bounds for depth three arithmetic circuits with small bottom fanin. Electronic Colloquium on Computational Complexity (ECCC) **21**, 89 (2014)
12. Kayal, N., Saha, C., Saptharishi, R.: A super-polynomial lower bound for regular arithmetic formulas. In: Symposium on Theory of Computing, STOC 2014, pp. 146–153 (2014)
13. Kleppe, J.: Representing a Homogenous Polynomial as a Sum of Powers of Linear Forms. Thesis for the degree of Candidatus Scientiarum (University of Oslo) (1999). http://folk.uio.no/johannkl/kleppe-master.pdf
14. Koiran, P.: Shallow circuits with high-powered inputs. In: Proceedings of the Innovations in Computer Science - ICS 2010, pp. 309–320. Tsinghua University, Beijing, 7–9 January 2011
15. Koiran, P.: Arithmetic circuits: The chasm at depth four gets wider. Theoretical Computer Science **448**, 56–65 (2012)
16. Koiran, P., Portier, N., Tavenas, S.: A Wronskian approach to the real τ-conjecture. J. Symb. Comput. **68**, 195–214 (2015)
17. Kumar, M., Saraf, S.: The limits of depth reduction for arithmetic formulas: it's all about the top fan-in. In: Symposium on Theory of Computing, STOC, pp. 136–145 (2014)
18. Kumar, M., Saraf, S.: On the power of homogeneous depth 4 arithmetic circuits. In: 55th IEEE Annual Symposium on Foundations of Computer Science, FOCS, pp. 364–373 (2014)

19. Pecatte, T.: Lower bounds for univariate polynomials: a Wronskian approach. M2 Internship Report (Ecole Normale Supérieure de Lyon) (2014). http://perso ens-lyon.fr/pascal.koiran/timothee_pecatte_master2report.pdf
20. Polya, G., Szego, G.: Problems and Theorems in Analysis, vol. II. Springer (1976)
21. Tavenas, S.: Improved bounds for reduction to depth 4 and depth 3. In: Chatterjee, K., Sgall, J. (eds.) MFCS 2013. LNCS, vol. 8087, pp. 813–824. Springer, Heidelberg (2013)
22. Valiant, L.G.: Completeness classes in algebra. In: Symposium on Theory of Computing STOC, pp. 249–261 (1979)
23. Valiant, L.G., Skyum, S., Berkowitz, S., Rackoff, C.: Fast parallel computation of polynomials using few processors. SIAM Journal on Computing **12**(4), 641–644 (1983)
24. Voorhoeve, M., Van Der Pooerten, A.J.: Wronskian determinants and the zeros of certain functions. Indagationes Mathematicae **78**(5), 417–424 (1975)

Approximating CSPs Using LP Relaxation

Subhash Khot[1] and Rishi Saket[2]($\boxtimes$)

[1] Computer Science Department, New York University, New York, USA
khot@cims.nyu.edu
[2] IBM Research, Bangalore, Karnataka, India
rissaket@in.ibm.com

Abstract. This paper studies how well the standard LP relaxation approximates a k-ary constraint satisfaction problem (CSP) on label set $[L]$. We show that, assuming the Unique Games Conjecture, it achieves an approximation within $O(k^3 \cdot \log L)$ of the optimal approximation factor. In particular we prove the following hardness result: let $\mathcal{I}$ be a k-ary CSP on label set $[L]$ with constraints from a *constraint class* $\mathcal{C}$, such that it is a (c, s)-integrality gap for the standard LP relaxation. Then, given an instance $\mathcal{H}$ with constraints from $\mathcal{C}$, it is NP-hard to decide whether,

$$\mathsf{opt}(\mathcal{H}) \geq \Omega\left(\frac{c}{k^3 \log L}\right), \quad \text{or} \quad \mathsf{opt}(\mathcal{H}) \leq 4 \cdot s,$$

assuming the Unique Games Conjecture. We also show the existence of an efficient LP rounding algorithm Round such that given an instance $\mathcal{H}$ from a *permutation invariant* constraint class $\mathcal{C}$ which is a (c, s)-*rounding gap* for Round, it is NP-hard to decide whether,

$$\mathsf{opt}(\mathcal{H}) \geq \Omega\left(\frac{c}{k^3 \log L}\right), \quad \text{or} \quad \mathsf{opt}(\mathcal{H}) \leq O\left((\log L)^k\right) \cdot s,$$

assuming the Unique Games Conjecture.

1 Introduction

A k-ary constraint satisfaction problem (CSP) over label set $[L]$ consists of a set of vertices and a set of k-uniform ordered hyperedges. For each hyperedge there is a constraint specifying the k-tuples of labels to the vertices in it that satisfy the hyperedge. The goal is to efficiently compute an assignment that satisfies the maximum number of hyperedges. This general definition includes many problems studied in computer science and combinatorial optimization such as MAXIMUM CUT, MAX-k-SAT and MAX-k-LIN[q]. Investigating the approximability of these problems has motivated a significant body of research.

One of the well studied methods of approximating a CSP is via the Linear Programming (LP) relaxation of the corresponding integer program[1]. For example, in its most basic formulation the LP relaxation gives a 2-approximation for

---

S. Khot—Research supported by NSF grants CCF 1422159, 1061938, 0832795 and Simons Collaboration on Algorithms and Geometry grant.

[1] We conveniently think of the problem as computing the *value* of the optimal labeling.

M.M. Halldórsson et al. (Eds.): ICALP 2015, Part I, LNCS 9134, pp. 822–833, 2015.
DOI: 10.1007/978-3-662-47672-7_67

MAXIMUM CUT and can do no better. On the other hand the seminal work of Goemans and Williamson [6] gave a 1.13823-approximation for MAXIMUM CUT using a semi-definite programming (SDP) relaxation. A matching *integrality gap* for this relaxation and its strengthening was shown by Feige and Schechtman [5], and Khot and Vishnoi [10] respectively. Moreover, this approximation factor was shown to be tight by Khot, Kindler, Mossel, and O'Donnell [8][2], assuming Khot's Unique Games Conjecture (UGC) [7]. A similar UGC-tight approximation via an SDP relaxation for the Unique Games problem itself was given by Charikar, Makarychev and Makarychev [2]. Greatly generalizing these results, Raghavendra [17] proved that a certain SDP relaxation achieves an approximation factor arbitrarily close to the optimal for *any* CSP, assuming the UGC. Raghavendra [17] formalized the connection between an integrality gap of the SDP relaxation and the corresponding UGC based hardness factor for a given CSP. For a general k-ary CSP over label set $[L]$, SDP relaxation yields a $O\left(L^k/Lk\right)$-approximation [14], and a corresponding hardness of approximation was recently shown by Chan [1].

While the above line of research underscores the theoretical importance of SDP relaxations, linear programs are usually more efficient in practice and are far more widely used as optimization tools. Thus, it is worthwhile to study how well LP relaxations perform for general classes of problems. In the first such result, Kumar, Manokaran, Tulsiani, and Vishnoi [12] showed a certain LP relaxation to be optimal for a large class of covering and packing problems, assuming the UGC. Dalmau and Krokhini [4] and Kun, O'Donnell, Tamaki, Yoshida, and Zhou [13] independently showed that *width*-1 (see for e.g. [13] for a formal definition) CSPs are *robustly decided* by LP relaxation, i.e. it satisfies almost all hyperedges on an almost satisfiable instance. In recent work, Dalmau, Krokhin, and Manokaran [3] have, assuming the UGC, classified CSPs for which the minimization version[3] admits a constant factor approximation via the LP relaxation.

In this work we study the linear programming analogue of the problem studied by Raghavendra [17], i.e. how well the standard LP relaxation approximates a CSP. We prove the following results.

1.1   Our Results

Let $\mathcal{C}$ be a *class* of constraints and let CSP-$[\mathcal{C}, k, L]$ be the k-ary constraint satisfaction problems over label set $[L]$ where each constraint is from the class $\mathcal{C}$. An instance $\mathcal{I}$ of CSP-$[\mathcal{C}, k, L]$ is a (c, s)-*integrality gap* instance if there is a solution to the LP relaxation $\mathsf{LP}(\mathcal{I})$ given in Fig. 1 with objective value at least c, and the optimum of $\mathcal{I}$ is at most s. The main result of this paper is as follows.

---

[2] [8] also assumed the *Majority is Stablest* conjecture which was later proved by Mossel, O'Donnell, and Oleszkiewicz [16].

[3] The goal in the minimization version of a CSP is to compute a labeling with the minimum number of unsatisfied constraints.

Theorem 1. *If $\mathcal{I}$ is a (c, s)-integrality gap instance of CSP-$[\mathcal{C}, k, L]$, then, assuming the Unique Games Conjecture it is NP-hard to distinguish whether a given instance $\mathcal{H}$ of CSP-$[\mathcal{C}, k, L]$ has*

$$\mathsf{opt}(\mathcal{H}) \geq \Omega\left(\frac{c}{k^3 \log L}\right), \quad or \quad \mathsf{opt}(\mathcal{H}) \leq 4 \cdot s.$$

The LP relaxation in Fig. 1 is given by a straightforward relaxation of the integer program for the CSP. The above theorem implies that this basic LP relaxation achieves an approximation factor within a multiplicative $O\left(k^3 \cdot \log L\right)$ of the optimal for any CSP-$[\mathcal{C}, k, L]$, assuming UGC. Note that Raghavendra [17] proved a stronger result: a transformation from a (c, s)-integrality gap for a certain SDP relaxation into a $(c - \varepsilon, s + \varepsilon)$-UGC hardness gap, which implies that the SDP relaxation essentially achieves the optimal approximation. We show that the LP relaxation is nearly as good, i.e. up to a multiplicative loss of $O\left(k^3 \cdot \log L\right)$ in the approximation. Before this work, the best known bound of L^{k-1} was implied by the results of Serna, Trevisan, and Xhafa [18]. In particular, [18] showed an L^{k-1}-approximation for any CSP-$[\mathcal{C}, k, L]$ obtained by the basic LP relaxation, generalizing a previous 2^{k-1}-approximation by Trevisan [19] for the boolean case.

Theorem 1 has tight dependence on L: for the Unique Games problem (which is a 2-CSP) on label set $[L]$, the standard LP relaxation has $\Omega(L)$ integrality gap (see Appendix I of [9]), whereas a very recent result of Kindler, Kolla, and Trevisan [11] gives an $O(L/\log L)$-approximate SDP rounding algorithm for any 2-CSP over label set $[L]$. The latter improves on a previous $O(L \log \log L/\log L)$-approximate SDP rounding algorithm for Unique Games given in [2].

Our second result pertains to CSPs with a *permutation invariant* set of constraints. Roughly speaking, a set of constraints is permutation invariant if it is closed under the permutation of labels on any of the vertices in the hyperedge. Most of the boolean CSPs such as Max-k-SAT, Max-k-AND, Max-k-XOR etc. are permutation invariant by definition. On larger label sets, Unique Games and Label Cover are well known examples of permutation invariant CSPs. We show that there is a simple randomized LP rounding algorithm such that a weaker version of Theorem 1 holds for a corresponding (c, s)-*rounding gap*, which is an instance of a permutation invariant CSP with an LP solution of value c on which the rounding algorithm has an expected payoff at most s. Our rounding algorithm independently rounds each vertex based only on the LP values associated with it. Thus, a *single* constraint suffices to capture its rounding gap. In particular, we prove the following theorem.

Theorem 2. *Let $\tilde{\mathcal{I}}$ be a single k-ary hyperedge $\tilde{e}$ with a constraint $C_{\tilde{e}}$ as an instance of a permutation invariant CSP-$[\mathcal{C}, k, L]$, which is a (c, s)-rounding gap for the algorithm* Round *given in Fig. 2. Then, assuming the Unique Games Conjecture it is NP-hard to distinguish whether a given instance $\mathcal{H}$ of CSP-$[\mathcal{C}, k, L]$ has*

$$\mathsf{opt}(\mathcal{H}) \geq \Omega\left(\frac{c}{k^3 \log L}\right), \quad or \quad \mathsf{opt}(\mathcal{H}) \leq O\left((\log L)^k\right) \cdot s.$$

1.2   Our Techniques

For proving Theorem 1, we follow the approach used in earlier works ([17], [12]) of converting an integrality gap instance for the LP relaxation into a UGC-hardness result, which translates the integrality gap into the hardness factor. This reduction essentially involves the construction of a *dictatorship gadget*, which is a toy instance of the CSP-$[\mathcal{C}, k, L]$ distinguishing between "dictator" labelings and "far from dictator" labelings. The construction is illustrated with the following simple example.

Consider an integrality gap instance consisting of just one edge $e = (u, v)$ over label set $[L]$, with the constraint given by the set $C_e \subseteq [L] \times [L]$ of satisfying assignments to (u, v). Let $(\overline{x}, \overline{y})$ be a solution to the corresponding LP relaxation given in Fig. 1. It is easy to see that the $\overline{x}$ variables corresponding to u (v) describe a distribution μ_u (μ_v) on $[L]$, and $\overline{y}$ describes a distribution ν_e on $[L] \times [L]$. Furthermore, the marginals of ν_e are μ_u and μ_v. Let $\tilde{\nu}_e = \rho\nu_e + (1 - \rho)(\mu_u \times \mu_v)$, for some parameter ρ. Clearly, the marginals of $\tilde{\nu}_e$ are also μ_u and μ_v.

The vertices of the dictatorship gadget are $\{u, v\} \times [L]^R$ where R is some large enough parameter. The weighted edges are formed as follows. Add an edge between $(u, \overline{r})$ and $(v, \overline{s})$ with weight $\tilde{\nu}_e^R(\overline{r}, \overline{s})$ with the constraint C_e. Here $\tilde{\nu}_e^R$ is the R-wise product distribution of $\tilde{\nu}_e$, i.e. the measure defined by choosing $\overline{r} = (r_1, \ldots, r_R)$ and $\overline{s} = (s_1, \ldots, s_R)$ such that (r_i, s_i) is sampled independently from $\tilde{\nu}_e$, for $i = 1, \ldots, R$.

It is easy to see that for any $i^* = 1, \ldots, R$, over the choice of $\overline{r}$ and $\overline{s}$ above, $(r_{i^*}, s_{i^*}) \in C_e$ with probability at least,

$$\rho \sum_{\overline{\ell} \in C_e} y_{e\overline{\ell}}. \tag{1}$$

Therefore, the above is the fraction of edges in the dictatorship gadget satisfied by labeling each $(u, (r_1, \ldots, r_R))$ with r_{i^*} and each $(v, (s_1, \ldots, s_R))$ with s_{i^*}. More formally, the expression in (1) is the *completeness* of the dictatorship gadget. Note that this is simply ρ times the objective value of the solution $(\overline{x}, \overline{y})$ to LP($\mathcal{I}$).

On the other hand, consider a labeling σ to the vertices of the dictatorship gadget. Define functions,

$$f_j(\overline{r}) := \mathbb{1}\{\sigma((u, \overline{r})) = j\}, \qquad g_j(\overline{s}) := \mathbb{1}\{\sigma((v, \overline{s})) = j\}, \tag{2}$$

for $j = 1, \ldots, L$, where $\mathbb{1}\{A\}$ denotes the indicator of the event A. We assume that the labeling σ is "far from dictator", i.e. each of the functions f_j and g_j are far from dictators. Estimating the weighted fraction of edges of the dictatorship gadget satisfied by σ entails analyzing expectations of the form,

$$\mathbb{E}_{\tilde{\nu}_e^R}\left[f_j(\overline{r})g_{j'}(\overline{s})\right], \tag{3}$$

for $1 \leq j, j' \leq L$. In the reduction of Raghavendra [17], such expressions essentially correspond to the payoff yielded by a randomized Gaussian rounding of the SDP solution, under the assumption that σ is far from a dictator. This is obtained by an application of the Invariance Principle developed by Mossel [15]. The parameter ρ is required to be set to only slightly less than 1 in [17] for the application of the Invariance Principle.

In our case the expectation in (3) does not *a priori* correspond to the payoff of any rounding of $(\overline{x}, \overline{y})$. However, we show that setting $\rho \approx (1/\log L)$ is sufficient to ensure,

$$\mathbb{E}_{\tilde{\nu}_e}[f_j g_{j'}] \approx \mathbb{E}[f_j]\mathbb{E}[g_{j'}], \tag{4}$$

when both $\mathbb{E}[f_j]$ and $\mathbb{E}[g_{j'}]$ are non-negligible. The RHS of the above corresponds to the payoff obtained by assigning u the label j with probability $\mathbb{E}[f_j]$, and independently assigning v label j with probability $\mathbb{E}[g_j]$, $j = 1, \ldots, L$. Thus, the fraction of edges of the dictatorship gadget satisfied by σ, i.e its *soundness*, is essentially bounded by the optimum of the integrality gap instance. There is a $O(\log L)$ loss in the hardness factor, as the completeness decreases due to the setting of ρ.

The proof of Theorem 2 proceeds by using a (c, s)-rounding gap $\tilde{\mathcal{I}}$ for the algorithm Round given in Fig. 2 to construct a CSP instance, with constraints being permutations of $\tilde{\mathcal{I}}$, which is a $(c/4, O((\log L)^k) \cdot s)$-integrality gap for the corresponding LP relaxation. A subsequent application of Theorem 1 with this integrality gap instance proves Theorem 2.

Organization of the Paper. Theorem 1 is restated in Sect. 3 as Theorem 3 which states a hardness reduction from Unique Games. Theorem 4 gives the transformation from a rounding gap to an integrality gap instance, and along with Theorem 3 proves Theorem 2. Due to lack of space, the proofs of Theorems 3 and 4 are omitted. The authors refer the reader to the full version of this paper [9] for all the missing proofs.

In the next section we define the constraint satisfaction problem and describe their LP relaxation that we study. The notion of correlated spaces and Gaussian stability bounds used in our reduction and analysis are also described.

2   Preliminaries

We begin by formally defining a constraint satisfaction problem and then describe the LP relaxation that we consider.

2.1 k-ary CSP over Label Set $[L]$

Let $k \geq 2$ and $L \geq 2$ be positive integers. We say that $C \subseteq [L]^k, C \neq \emptyset$, is a constraint. A collection of such constraints $\mathcal{C}$ is a (k, L)-*constraint class*, i.e.

$$\mathcal{C} \subseteq \left(2^{[L]^k} \setminus \{\emptyset\}\right).$$

We denote by CSP-$[\mathcal{C}, k, L]$ as the class of k-ary constraint satisfaction problems over label set $[L]$, where each constraint is from the class $\mathcal{C}$. Formally, an instance of $\mathcal{I}$ of CSP-$[\mathcal{C}, k, L]$ consists of a finite set of vertices $V_{\mathcal{I}}$, a set of k-uniform ordered hyperedges $E_{\mathcal{I}} \subseteq V_{\mathcal{I}}^k$ and constraints $\{C_e \in \mathcal{C} \mid e \in E\}$. In addition, the hyperedges have normalized weights $\{w_e \geq 0\}_{e \in E_{\mathcal{I}}}$ satisfying $\sum_{e \in E_{\mathcal{I}}} w_e = 1$. A labeling $\sigma : V_{\mathcal{I}} \mapsto [L]$ satisfies the hyperedge $e = (v_1, \ldots, v_k)$ if $(\sigma(v_1), \ldots, \sigma(v_k)) \in C_e$.

As an example, 3-SAT is a constraint satisfaction problem with $k = 3$ over the boolean domain, i.e. $L = 2$. The SAT predicate is over 3 variables. Allowing for negations of the boolean variables yields a constraint class $\mathcal{C}_{3-\text{SAT}}$ consisting of 8 constraints. Each constraint, being an OR over 3 literals, has 7 satisfying assignments (labelings).

Let us denote the weighted fraction of constraints satisfied by any labeling σ by $\mathsf{val}(\mathcal{I}, \sigma)$. The optimum value of the instance is given by,

$$\mathsf{opt}(\mathcal{I}) := \max_{\sigma : V \mapsto [L]} \mathsf{val}(\mathcal{I}, \sigma).$$

Permutation Invariant Constraints. Let $\pi_j : [L] \mapsto [L]$, $j = 1, \ldots, k$, be k permutations. For a constraint $C \subseteq [L]^k$, define the $[\pi_1, \ldots, \pi_k]$-permuted constraint as:

$$[\pi_1, \ldots, \pi_k]C := \{(\pi_1(j_1), \ldots, \pi_k(j_k)) \mid (j_1, \ldots, j_k) \in C\}. \tag{5}$$

A (k, L)-constraint class $\mathcal{C}$ is said to be *permutation invariant* if for every k permutations $\pi_j : [L] \mapsto [L]$ ($1 \leq j \leq k$), $C \in \mathcal{C}$ implies $[\pi_1, \ldots, \pi_k]C \in \mathcal{C}$. As mentioned earlier, boolean constraint classes such as k-SAT, k-AND and k-XOR are permutation invariant by definition since they are closed under negation of variables. For general L, Unique Games and Label Cover are well studied permutation invariant constraint classes.

2.2   LP Relaxation for CSP-$[\mathcal{C}, k, L]$

The standard linear programming relaxation for an instance $\mathcal{I}$ (as defined above) of CSP-$[\mathcal{C}, k, L]$ is obtained as follows. There is a variable $x_{v\ell}$ for each vertex $v \in V_{\mathcal{I}}$ and label $\ell \in [L]$. For each constraint C_e corresponding to hyperedge $e = (v_1, \ldots, v_k)$, and tuple $\bar{\ell} = (\ell_1, \ldots, \ell_k) \in [L]^k$ of labels, there is a variable $y_{e\bar{\ell}}$. In the integral solution these variables are $\{0, 1\}$-valued denoting the selection the particular label or tuple of labels for the corresponding vertex or hyperedge respectively. To ensure consistency they are appropriately constrained. Allowing the variables to take values in $[0, 1]$, we obtain the LP relaxation denoted by $\mathsf{LP}(\mathcal{I})$ and given in Fig. 1.

For a given instance $\mathcal{I}$, let

$$(\bar{x}, \bar{y}) = (\{x_{v\ell}\}_{v \in V_{\mathcal{I}}, \ell \in [L]}, \{y_{e\bar{\ell}}\}_{e \in E_{\mathcal{I}}, \bar{\ell} \in [L]^k}),$$

be a valid solution to $\mathsf{LP}(\mathcal{I})$. On this solution, the objective value of the LP is denoted by $\mathsf{lpval}(\mathcal{I}, (\bar{x}, \bar{y}))$. The *integrality gap*, i.e. how well the LP relaxation

$$\max \sum_{e \in E_{\mathcal{I}}} w_e \cdot \sum_{\overline{\ell} \in C_e} y_{e\overline{\ell}} \qquad (6)$$

subject to,

$$\forall v \in V_{\mathcal{I}}, \qquad \sum_{\ell \in [L]} x_{v\ell} = 1 \qquad (7)$$

$$\forall v \in V_{\mathcal{I}} \text{ and,}$$
$$e = (v_1, \ldots, v_{i-1}, v, v_{i+1}, \ldots, v_k) \in E_{\mathcal{I}} \text{ and,}$$
$$\ell^* \in [L], \qquad \sum_{\overline{\ell} \in [L]^{i-1} \times \{\ell^*\} \times [L]^{k-i}} y_{e\overline{\ell}} = x_{v\ell^*} \qquad (8)$$

$$\forall v \in V_{\mathcal{I}}, \ell \in [L], \qquad x_{v\ell} \geq 0. \qquad (9)$$
$$\forall e \in E_{\mathcal{I}}, \overline{\ell} \in [L]^k, \qquad y_{e\overline{\ell}} \geq 0. \qquad (10)$$

Fig. 1. LP Relaxation $\mathsf{LP}(\mathcal{I})$ for instance $\mathcal{I}$ of CSP-$[\mathcal{C}, k, L]$

approximates the integral optimum on $\mathcal{I}$, is given by,

$$\mathsf{intgap}(\mathcal{I}) := \frac{\mathsf{lpsup}(\mathcal{I})}{\mathsf{opt}(\mathcal{I})}, \qquad (11)$$

where,

$$\mathsf{lpsup}(\mathcal{I}) := \sup_{(\overline{x}, \overline{y})} \mathsf{lpval}(\mathcal{I}, (\overline{x}, \overline{y})). \qquad (12)$$

A smaller integrality gap – which is always at least 1 – indicates tightness of the LP relaxation. We say that $\mathcal{I}$ is a (c, s)-*integrality gap* instance if,

$$\mathsf{lpsup}(\mathcal{I}) \geq c, \quad \text{and} \quad \mathsf{opt}(\mathcal{I}) \leq s. \qquad (13)$$

Smooth LP Solutions. The following shows that the integrality gap is nearly attained by a solution to the LP relaxation which is discrete in the following sense.

Definition 1. *Given an instance $\mathcal{I}$ of CSP-$[\mathcal{C}, k, L]$, a solution $(\overline{x}, \overline{y})$ to $\mathsf{LP}(\mathcal{I})$ is δ-smooth if each variable $x_{v\ell}$ is at least δL^{-1} and each variable $y_{e\overline{\ell}}$ is at least δL^{-k}, for any $\delta > 0$.*

Due to lack of space we omit the proof of the following.

Lemma 1. *Given an instance $\mathcal{I}$ of CSP-$[\mathcal{C}, k, L]$, for any $\delta > 0$ and solution $(\overline{x}^*, \overline{y}^*)$ to $\mathsf{LP}(\mathcal{I})$, there is an (efficiently computable) δ-smooth solution $(\overline{x}, \overline{y})$ to $\mathsf{LP}(\mathcal{I})$ such that,*

$$\mathsf{lpval}(\mathcal{I}, (\overline{x}, \overline{y})) \geq (1 - \delta)\mathsf{lpval}(\mathcal{I}, (\overline{x}^*, \overline{y}^*)). \qquad (14)$$

In particular, there is a δ-smooth solution $(\overline{x}, \overline{y})$ to $\mathrm{LP}(\mathcal{I})$ such that,

$$\frac{\mathsf{lpval}(\mathcal{I}, (\overline{x}, \overline{y}))}{\mathsf{opt}(\mathcal{I})} \geq (1 - \delta)\mathsf{intgap}(\mathcal{I}). \tag{15}$$

2.3 A Rounding Algorithm for LP

Given an instance $\mathcal{I}$ of CSP-$[\mathcal{C}, k, L]$ and a solution $(\overline{x}^*, \overline{y}^*)$ to $\mathrm{LP}(\mathcal{I})$, the rounding algorithm Round is described in Fig. 2. The performance of the algorithm is

---

Round$(\mathcal{I}, (\overline{x}^*, \overline{y}^*))$:

1. Using Lemma 1 compute a 0.1-smooth solution $(\widehat{x}, \widehat{y})$ corresponding to $(\overline{x}^*, \overline{y}^*)$ satisfying Equation (14).
2. For each vertex $v \in V_{\mathcal{I}}$:
   a. Partition $[L]$ into subsets $\{S_t^v\}_{t=1}^T$, where $S_t^v = \{\ell \in [L] \mid (1/2^t) < \widehat{x}_{v\ell} \leq (1/2^{t-1})\}$. Note: $T = O(\log L)$, by 0.1-smoothness of $(\widehat{x}, \widehat{y})$.
   b. Choose u.a.r t_v^* from $\{t \mid S_t^v \neq \emptyset\}$.
   c. Label v with ℓ^* chosen u.a.r from $S_{t_v^*}^v$.

Fig. 2. Rounding Algorithm for $\mathrm{LP}(\mathcal{I})$ on instance $\mathcal{I}$ of CSP-$[\mathcal{C}, k, L]$

the expected (weighted) fraction of constraints satisfied by this labeling, and is denoted by $\mathsf{Roundval}(\mathcal{I}, (\overline{x}^*, \overline{y}^*))$. The rounding gap for $\mathcal{I}$ and $(\overline{x}^*, \overline{y}^*)$ is given by the following ratio.

$$\mathsf{RoundGap}(\mathcal{I}, (\overline{x}^*, \overline{y}^*)) := \frac{\mathsf{lpval}(\mathcal{I}, (\overline{x}^*, \overline{y}^*))}{\mathsf{Roundval}(\mathcal{I}, (\overline{x}^*, \overline{y}^*))}. \tag{16}$$

2.4 Gaussian Stability

We require the following notion of Gaussian stability in our analysis.

Definition 2. *Let $\Phi : \mathbb{R} \mapsto [0,1]$ be the cumulative distribution function of the standard Gaussian. For a parameter ρ, define,*

$$\Gamma_\rho(\mu, \nu) = \Pr[X \leq \Phi^{-1}(\mu), Y \leq \Phi^{-1}(\nu)], \tag{17}$$

where X and Y are two standard Gaussian random variables with covariance matrix $\begin{pmatrix} 1 & \rho \\ \rho & 1 \end{pmatrix}$. For $k \geq 3$, $(\rho_1, \ldots, \rho_{k-1}) \in [0,1]^{k-1}$, and $(\mu_1, \ldots, \mu_k) \in [0,1]^k$, inductively define,

$$\Gamma_{\rho_1, \ldots, \rho_{k-1}}(\mu_1, \ldots, \mu_k) = \Gamma_{\rho_1}(\mu_1, \Gamma_{\rho_2, \ldots, \rho_{k-1}}(\mu_2, \ldots, \mu_k)). \tag{18}$$

Due the lack of space we omit the proof of the following key lemma.

Lemma 2. *Let $k \geq 2$ be an integer and $T \geq 2$ such that $1 \geq \mu_i \geq (1/T)$ for $i = 1, \ldots, k$. Then, there exists a universal constant $C > 0$ such that for any $\varepsilon \in (0, 1/2]$,*

$$\rho = \frac{\varepsilon}{C(k-1)(\log T + \log(1/\varepsilon))}, \tag{19}$$

implies,

$$\Gamma_{\overline{\rho}_{k-1}}(\mu_1, \ldots, \mu_k) \leq (1+\varepsilon)^{k-1} \prod_{i=1}^{k} \mu_i,$$

where $\overline{\rho}_{k-1} = (\rho, \ldots, \rho)$, is a $(k-1)$-tuple with each entry ρ.

2.5 Correlated Spaces

The correlation between two correlated probability spaces is defined as follows.

Definition 3. *Suppose $(\Omega^{(1)} \times \Omega^{(2)}, \mu)$ is a finite correlated probability space with the marginal probability spaces $(\Omega^{(1)}, \mu)$ and $(\Omega^{(2)}, \mu)$. The correlation between these spaces is,*

$$\rho(\Omega^{(1)}, \Omega^{(2)}; \mu) = \sup\Big\{ |\mathbb{E}_\mu[fg]| \mid f \in L^2(\Omega^{(1)}, \mu), g \in L^2(\Omega^{(2)}, \mu),$$

$$\mathbb{E}[f] = \mathbb{E}[g] = 0; \mathbb{E}[f^2], \mathbb{E}[g^2] \leq 1 \Big\}.$$

Let $(\Omega_i^{(1)} \times \Omega_i^{(2)}, \mu_i)_{i=1}^{n}$ be a sequence of correlated spaces. Then,

$$\rho(\prod_{i=1}^{n} \Omega_i^{(1)}, \prod_{i=1}^{n} \Omega_i^{(2)}; \prod_{i=1}^{n} \mu_i) \leq \max_i \rho(\Omega_i^{(1)}, \Omega_i^{(2)}; \mu_i).$$

Further, the correlation of k correlated spaces $(\prod_{j=1}^{k} \Omega^{(j)}, \mu)$ is defined as follows:

$$\rho(\Omega^{(1)}, \Omega^{(2)}, \ldots, \Omega^{(k)}; \mu) := \max_{1 \leq i \leq k} \rho\left(\prod_{j=1}^{i-1} \Omega^{(j)} \times \prod_{j=i+1}^{k} \Omega^{(j)}, \Omega^{(i)}; \mu\right).$$

The Bonami-Beckner operator is defined as follows.

Definition 4. *Given a probability space (Ω, μ) and $\rho \geq 0$, consider the space $(\Omega \times \Omega, \mu')$ where $\mu'(x, y) = (1 - \rho)\mu(x)\mu(y) + \rho\mathbb{1}\{x = y\}\mu(x)$, where $\mathbb{1}\{x = y\} = 1$ if $x = y$ and 0 otherwise. The Bonami-Beckner operator T_ρ is defined by,*

$$(T_\rho f)(x) = \mathbb{E}_{(X,Y) \leftarrow \mu'} [f(Y) \mid X = x].$$

For product spaces $(\prod_{i=1}^{n} \Omega_i, \prod_{i=1}^{n} \mu_i)$, the Bonami-Beckner operator $T_\rho = \otimes_{i=1}^{n} T_\rho^i$, where T_ρ^i is the operator for the ith space (Ω_i, μ_i).

The influence of a function on a product space is defined as follows.

Definition 5. *Let f be a function on $(\prod_{i=1}^{n} \Omega_i, \prod_{i=1}^{n} \mu_i)$. The influence of the ith coordinate on f is:*

$$\mathsf{Inf}_i(f) = \mathbb{E}_{\{x_j | j \neq i\}} \left[Var_{x_i} \left[f(x_1, x_2, \ldots, x_i, \ldots, x_n) \right] \right].$$

The following is a folklore upper bound on the sum of influences of smoothed functions, and is proved as Lemma 1.13 in [20].

Lemma 3. *Let f be a function on $(\prod_{i=1}^{n} \Omega_i, \prod_{i=1}^{n} \mu_i)$ which takes values in $[-1, 1]$. Then,*

$$\sum_{i=1}^{n} \mathsf{Inf}_i(T_{1-\gamma}f) \leq \gamma^{-1}, \tag{20}$$

for any $\gamma \in (0, 1]$.

The analysis used in our results also requires invariance theorems along with bounds on the correlation of functions based on Mossel's work [15]. Due to lack of space we omit their statements.

2.6   Unique Games Conjecture

UNIQUEGAMES is the following constraint satisfaction problem.

Definition 6. *A UNIQUEGAMES instance $\mathcal{U}$ consists of a graph $G_{\mathcal{U}} = (V_{\mathcal{U}}, E_{\mathcal{U}})$, a label set $[R]$ and a set of bijections $\{\pi_e : [R] \mapsto [R] \mid e \in E_{\mathcal{U}}\}$. A labeling $\sigma : V_{\mathcal{U}} \mapsto [R]$ satisfies an edge $e = (u, v)$ if $\pi_e(\sigma(v)) = \sigma(u)$. The instance is called d-regular if $G_{\mathcal{U}}$ is d-regular.*

The UNIQUEGAMES problem is: given an instance of UNIQUEGAMES, find an assignment which satisfies the maximum fraction of edges. It is easy to see that if there exists an assignment that satisfies all edges, such an assignment can be efficiently obtained. In other words, the UNIQUEGAMES is easy on satisfiable instances. This is not known to be true for *almost* satisfiable instances, and the following conjecture on the hardness of UNIQUEGAMES on such instances was proposed by Khot [7].

Conjecture 1. *For any constant $\zeta > 0$, there is an integer $R > 0$, such that it is NP-hard, given a regular instance $\mathcal{U}$ of UNIQUEGAMES on label set $[R]$, to decide whether,*

YES Case. There is a labeling to the vertices of $\mathcal{U}$ which satisfies $(1 - \zeta)$ fraction of its edges.

NO Case. Any labeling satisfies at most ζ fraction of the edges.

3   Our Results Restated

The following is a restatement of Theorem 1 as a hardness reduction from UNIQUEGAMES.

Theorem 3. *Let $k \geq 2$ and $L \geq 2$ be positive integers. Let $\mathcal{I}$ be a (c, s)-integrality gap instance of CSP-$[\mathcal{C}, k, L]$. Then, there is a reduction from an instance $\mathcal{U}$ of* UNIQUEGAMES *given by Conjecture 1 with a small enough parameter ζ, to an instance $\mathcal{H}$ of CSP-$[\mathcal{C}, k, L]$ such that,*

YES Case. If $\mathcal{U}$ is a YES instance, then

$$\mathsf{opt}(\mathcal{H}) \geq \Omega\left(\frac{c}{k^3 \log L}\right).$$

NO Case. If $\mathcal{U}$ is a NO instance, then,

$$\mathsf{opt}(\mathcal{H}) \leq 4 \cdot s.$$

Theorem 3 is obtained by combining a *dictatorship gadget* with the hard instance of UNIQUEGAMES. As the name suggests, this gadget distinguishes between labelings defined by a dictator and those which are not. The dictatorship gadget illustrates the main ideas of the hardness reduction and is derived from the integrality gap instance $\mathcal{I}$ of CSP-$[\mathcal{C}, k, L]$, and is also a CSP-$[\mathcal{C}, k, L]$ instance. This notion is the same as defined by Raghavendra [17] and can be converted into a hardness reduction from UNIQUEGAMES using techniques from Sect. 6 of [17]. However, to avoid describing the framework of [17] in detail, the proof of Theorem 3 is via a direct hardness reduction from UNIQUEGAMES.

Our second result Theorem 2 is implied by the following theorem and an application of Theorem 3.

Theorem 4. *Let $k \geq 2$ and $L \geq 2$ be positive integers. Let $\tilde{\mathcal{I}}$ be an instance of CSP-$[\mathcal{C}, k, L]$ consisting of one hyperedge $\tilde{e}$ and its constraint $C_{\tilde{e}}$, and $(\overline{x^*}, \overline{y^*})$ be a solution to LP$(\tilde{\mathcal{I}})$ such that,*

$$\mathsf{lpval}(\tilde{\mathcal{I}}, (\overline{x^*}, \overline{y^*})) \geq \mathsf{Roundval}(\tilde{\mathcal{I}}, (\overline{x^*}, \overline{y^*})). \tag{21}$$

Then, there exists an instance $\mathcal{I}$ whose size depends only on L and k with constraints which are permutations of $C_{\tilde{e}}$, and a solution $(\overline{x}, \overline{y})$ to LP$(\mathcal{I})$ such that,

$$\mathsf{lpval}(\mathcal{I}, (\overline{x}, \overline{y})) \geq \frac{\mathsf{lpval}(\tilde{\mathcal{I}}, (\overline{x^*}, \overline{y^*}))}{4}, \tag{22}$$

and,

$$\mathsf{opt}(\mathcal{I}) \leq O\left((\log L)^k\right) \mathsf{Roundval}(\tilde{\mathcal{I}}, (\overline{x^*}, \overline{y^*})). \tag{23}$$

Acknowledgments. The authors thank Elchanan Mossel for helpful discussion on Gaussian stability bounds.

References

1. Chan, S.O.: Approximation resistance from pairwise independent subgroups. In Proc. STOC, pp. 447–456 (2013)
2. Charikar, M., Makarychev, K., Makarychev, Y.: Near-optimal algorithms for unique games. In: Proc. STOC, pp. 205–214 (2006)
3. Dalmau, V., Krokhin, A.A., Manokaran, R.: Towards a characterization of constant-factor approximable min CSPs. In: Proc. SODA, pp. 847–857 (2015)
4. Dinur, I., Kol, G.: Covering CSPs. In: Proc. CCC, pp. 207–218 (2013)
5. Feige, U., Schechtman, G.: On the optimality of the random hyperplane rounding technique for MAX CUT. Random Struct. Algorithms 20(3), 403–440 (2002)
6. Goemans, M.X., Williamson, D.P.: Improved approximation algorithms for maximum cut and satisfiability problems using semidefinite programming. Journal of the ACM 42(6), 1115–1145 (1995)
7. Khot, S.: On the power of unique 2-prover 1-round games. In: Proc. STOC, pp. 767–775 (2002)
8. Khot, S., Kindler, G., Mossel, E., O'Donnell, R.: Optimal inapproximability results for MAX-CUT and other 2-variable CSPs? SIAM Journal of Computing 37(1), 319–357 (2007)
9. Khot, S., Saket, R.: Approximating CSPs using LP relaxation (2015). http://researcher.ibm.com/researcher/files/in-rissaket/KS-icalp-full.pdf
10. Khot, S., Vishnoi, N.K.: The unique games conjecture, integrality gap for cut problems and embeddability of negative type metrics into ℓ_1. In: Proc. FOCS, pp. 53–62 (2005)
11. Kindler, G., Kolla, A., Trevisan, L.: Approximation of non-boolean 2CSP (2015). CoRR, abs/1504.00681. http://arxiv.org/pdf/1504.00681.pdf
12. Kumar, A., Manokaran, R., Tulsiani, M., Vishnoi, N.K.: On LP-based approximability for strict CSPs. In: Proc. SODA, pp. 1560–1573 (2011)
13. Kun, G., O'Donnell, R., Tamaki, S., Yoshida, Y., Zhou, Y.: Linear programming, width-1 CSPs, and robust satisfaction. In: Proc. ITCS, pp. 484–495 (2012)
14. Makarychev, K., Makarychev, Y.: Approximation algorithm for non-boolean Max-k-CSP. Theory of Computing 10, 341–358 (2014)
15. Mossel, E.: Gaussian bounds for noise correlation of functions. GAFA 19, 1713–1756 (2010)
16. Mossel, E., O'Donnell, R., Oleszkiewicz, K.: Noise stability of functions with low influences: invariance and optimality. Annals of Mathematics 171(1), 295–341 (2010)
17. Raghavendra, P.: Optimal algorithms and inapproximability results for every CSP? In: Proc. STOC, pp. 245–254 (2008)
18. Serna, M.J., Trevisan, L., Xhafa, F.: The (parallel) approximability of non-boolean satisfiability problems and restricted integer programming. In: Meinel, C., Morvan, M. (eds.) STACS 1998. LNCS, vol. 1373, pp. 488–498. Springer, Heidelberg (1998)
19. Trevisan, L.: Parallel approximation algorithms by positive linear programming. Algorithmica 21(1), 72–88 (1998)
20. Wenner, C.: Circumventing d-to-1 for approximation resistance of satisfiable predicates strictly containing parity of width at least four. Theory of Computing 9, 703–757 (2013)

Comparator Circuits over Finite Bounded Posets

Balagopal Komarath, Jayalal Sarma$^{(\boxtimes)}$, and K.S. Sunil

Department of Computer Science and Engineering,
Indian Institute of Technology Madras, Chennai, India
{baluks,jayalal,sunil}@cse.iitm.ac.in

Abstract. Comparator circuit model was originally introduced in [4] (and further studied in [2]) to capture problems which are not known to be P-complete but still not known to admit efficient parallel algorithms. The class CC is the complexity class of problems many-one logspace reducible to the Comparator Circuit Value Problem and we know that NLOG $\subseteq$ CC $\subseteq$ P. Cook *et al* [2] showed that CC is also the class of languages decided by polynomial size comparator circuits.

We study generalizations of the comparator circuit model that work over fixed finite bounded posets. We observe that there are universal comparator circuits even over arbitrary fixed finite bounded posets. Building on this, we show that general (resp. skew) comparator circuits of polynomial size over fixed finite *distributive* lattices characterizes CC (resp. LOG). Complementing this, we show that general comparator circuits of polynomial size over arbitrary fixed finite lattices exactly characterizes P and that when the comparator circuit is skew they characterize NLOG. In addition, we show a characterization of the class NP by a family of polynomial sized comparator circuits over fixed *finite bounded posets*. These results generalize the results in [2] regarding the power of comparator circuits. As an aside, we consider generalizations of Boolean formulae over arbitrary lattices. We show that Spira's theorem[5] can be extended to this setting as well and show that polynomial sized Boolean formulae over finite fixed lattices capture exactly NC^1.

Our techniques involve design of comparator circuits and finite posets. We then use known results from lattice theory to show that the posets that we obtain can be embedded into appropriate lattices. Our results gives new methods to establish CC upper bound for problems also indicate potential new approaches towards the problems P vs CC and NLOG vs LOG using lattice theoretic methods.

1 Introduction

Completeness for the class P for a problem, is usually considered to be an evidence that it is hard to design an efficient parallel algorithm for the problem. However, there are many computational problems in the class P, which are not known to be P-complete, yet designing efficient parallel algorithms for them

---

B. Komarath—Supported by TCS PhD Fellowship.

M.M. Halldórsson et al. (Eds.): ICALP 2015, Part I, LNCS 9134, pp. 834–845, 2015.
DOI: 10.1007/978-3-662-47672-7_68

has remained elusive. Some of the classical examples of such problems include
lex-least maximal matching problem and stable marriage problem [4].

Attempting to capture the exact bottleneck of computation in these problems
using a variant of Boolean circuit model, Mayr and Subramanian [4] (see also
[2]) studied the comparator circuit model.

A comparator circuit is a sorting network
working over the values 0 and 1. A comparator gate has 2 inputs and 2 outputs. The first
output is the AND of the two inputs and the
second output is the OR of the two inputs. A

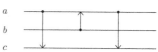

Fig. 1. A Comparator Circuit

comparator circuit is a circuit that has only comparator gates. In particular
fan-out gates are not allowed. We can assume wlog that NOT gates are used
only at the input level. A graphical representation of a comparator circuit is
shown in Figure 1. In this representation, we draw a set of parallel *lines*. Each
line carries a logical value which is updated by gates incident on that line. Each
gate is represented by a directed arrow from one line (Say i) to another (Say
j) and the gate updates the values of lines as follows. The value of line i (j) is
set to the OR (resp. AND) of values previously on lines i and j. The gates are
evaluated from left to right. The output of the circuit is the final value of a line
designated as the output line. We define the model formally in section 2.

In order to study the complexity theoretic significance of comparator circuits,
the corresponding circuit value problem was explored in [4]. That is, given a
comparator circuit and an input, test if the output wire carries a 1 or not.
The class CC is defined in [4] as the class of languages that are logspace many-
one reducible to the comparator circuit value problem. They also observed that
the class CC is contained in P. Feder's algorithm (described in [7]) for directed
reachability proves that the class CC contains NLOG as a subclass. These are the
best containments currently known about the complexity class CC.

There has been a recent spurt of activity in the characterization of CC. Cook
et al. [2] showed that the class CC is robust even if the complexity of the many-
one reduction to the comparator circuit value problem is varied from AC^0 to
NLOG. They also gave a characterization of the class CC in terms of a computational model (comparator circuit families). Their main contribution in this
regard is the introduction of a universal comparator circuit that can simulate
the computation of a comparator circuit given as input (to the universal circuit). Comparison of CC with the class NC has interesting implications to the
corresponding computational restrictions. For example, hardness for the class
CC is conjectured to be evidence that the problem is not efficiently paralleliz-
able. This intuition was further strengthened by Cook et al. [2] by showing that
there are oracle sets relative to which CC and NC are incomparable (NC is the
class of all languages efficiently solvable by parallel algorithms). In addition, it
is conjectured in [2] that the classes NC, SC and CC are pairwise incomparable.

Our Results & Techniques: In this paper, we study the computational power
of comparator circuits working over arbitrary fixed finite bounded posets (and
sub-families of the family of all finite bounded posets). Informally, instead of 0⁞

and 1, the values used while computation could be any element from the poset and the AND and OR gates compute (non-deterministically) maximal lower bounds and minimal upper bounds over the poset respectively. We define this model formally in section 3. We obtain the following results:

- There exist Universal Comparator Circuits for comparator circuits irrespective of the underlying bounded poset. (Proposition 2, Section 3)
- Comparator circuits of polynomial size over fixed finite distributive lattices capture the class CC. (Theorem 4, Section 4). This leads to a new way to show that a problem is in the class CC. That is, by designing a comparator circuit over a fixed finite lattice and then showing that the lattice is distributive (An application of this method to design CC algorithms for Stable Matching can be found in [4], See also Section 6.2 in [2]). There are lattice theoretic techniques known (cf. M_3-N_5 Theorem [3]) for showing that a lattice is distributive, this generalization might be independent interest.
- Going beyond distributivity, we show that comparator circuits of polynomial size over fixed finite lattices characterize the class P. (Theorem 5, Section 4). In particular, we design a fixed finite poset P over which, for any language $L \in$ P, there is a polynomial size comparator circuit family over P computing L. During computation, we only use lubs and glbs that exist in the poset P. This enables us to use Dedekind-MacNeille completion (DM completion) theorem to construct a fixed finite lattice completing the poset P while preserving the lubs and glbs of all pairs of elements and that lattice can be used to perform all computations in P. A potential drawback of the lattice thus obtained is that the complexity class captured by comparator circuits over it may vary depending on the element in the lattice used as the accepting element. By using standard tools from lattice theory, we derive that there is a fixed constant $i \geq 3$, such that comparator circuit over Π_i (where Π_i is the ith partition lattice - see section 2 for a definition) with polynomial size can compute all functions in P. Moreover, we show that comparator circuits over the lattice Π_i captures P irrespective of the accepting element used.

However, both partition lattice for $i \geq 3$ and the lattice given by DM completion are non-distributive. Exploring the possibility of another completion of the poset P into a distributive lattice that preserves existing lubs and glbs (which will show P = CC), we arrive at the following negative result : the poset P cannot be completed into any distributive lattice while preserving all existing lubs and glbs. (Theorem 6).

It is conceivable that the class P could be captured by a family of distributive lattices, while no finite fixed lattice capturing P can be distributive. Motivated by this, we also present an alternative proof of the main theorem using growing posets of much simpler structure (See Appendix A in the full paper[8]). However, we argue that this poset family also cannot be completed into a family of distributive lattices while preserving all existing lubs and glbs.

- Going beyond lattice structure, we show that comparator circuits over fixed finite bounded *posets* capture the class NP. (Theorem 7, Section 5). Here, we crucially use the fact that posets that are not lattices could have elements that does not have unique minimal upper bounds. Hence, any completion of this poset into a lattice will fail to capture NP, unless P = NP.

- Restricting the structure of the comparator circuit, we obtain an exact characterization of the class LOG using skew comparator circuits (Theorem 8). Noting that the polynomial sized skew Boolean circuits characterize exactly the class NLOG, this leads to a comparison between CC vs P and LOG vs NLOG problem : *both problems address the power of polynomial size Boolean circuits vs comparator circuits in general and skew circuits respectively.*

- We further study generalizations of skew comparator circuits to arbitrary lattices. When the lattice is distributive, it follows that the circuits capture exactly LOG. Complementing this, we show that are fixed finite fixed lattices P over which, the skew comparator circuits characterize exactly NLOG.(Theorem 9). This brings in a second comparison between CC vs P and NLOG vs LOG problems - *both problems address the power of polynomial size comparator circuits over arbitrary lattices vs distributive lattices in the general and skew comparator circuits case respectively.*

- We study generalizations of Boolean formulas to arbitrary lattices where the AND and OR gates compute the $\wedge$ and $\vee$ of the lattices. We generalize Spira's theorem[5] to this setting and show that polynomial sized Boolean formulae over finite fixed lattices capture exactly NC^1 (Theorem 10).

Thus, we observe that as the comparator circuit is allowed to compute over progressively general structures (From distributive lattices to arbitrary lattices to posets), the model captures classes of problems that are progressively harder to parallelize (From CC to P to NP).

The main technical contribution in our proofs is the design of posets and the corresponding comparator circuits for capturing complexity classes. We then use known ideas from lattice and order theory in order to derive lattices to which the constructed posets can be embedded.

2 Preliminaries

The standard definitions in complexity theory used in this paper can be found in standard textbooks [1]. All reductions in this paper are computable in logspace. In this section, we define comparator circuits, certain restrictions on comparator circuits and complexity classes based on those restrictions.

A *comparator circuit* has a set of n *lines* $\{w_1, \ldots, w_n\}$ and an ordered list of gates (w_i, w_j). Each line can be fed as input a value that is either (Boolean) 0 or 1. We define $val(w_i)$ to be the value of the line w_i. Each gate (w_i, w_j) updates the $val(w_i)$ to $val(w_i) \wedge val(w_j)$ and $val(w_j)$ to $val(w_i) \vee val(w_j)$ in order. After all gates have updated the values, the value of the line w_1 is the output of the circuit.

The COMPARATOR CIRCUIT VALUE PROBLEM is given (C, x) as input find the output of the comparator circuit C when fed x as input. We can think of

$\mathcal{C}$ being encoded according to the above definition of comparator circuits. We call this the ordered list representation as the gates are presented as an ordered list. Mayr and Subramanian [4] defined the complexity class CC as the set of all languages logspace reducible to the Comparator Circuit Value problem. Cook et al. [2] characterized the class CC as languages computed by AC^0-uniform families of annotated comparator circuits. In an annotated comparator circuit the initial value of a line could be an input variable x_i or its complement $\overline{x_i}$. In a family of annotated comparator circuits for a language L, the nth comparator circuit in the family has exactly n input variables $(x_1, \ldots, x_n)$ and the circuit computes $L \cap \{0,1\}^n$.

Skew Comparator Circuits: We now define skewness in comparator circuits. To begin with, we present an alternate definition of comparator circuits that is closer to the definition of standard Boolean circuits. A comparator gate is a 2-input, 2-output gate that takes a and b as inputs and outputs $a \wedge b$ and $a \vee b$. Then the comparator circuit is simply a circuit (in the usual sense) that consists of only comparator gates (In particular, fan-out gates are not allowed). Using this definition, we can encode comparator circuits by using DAGs as we encode standard Boolean circuits. It is easy to see that given a comparator circuit encoded as an ordered list of gates, we can obtain the DAG encoding the comparator circuit in logspace. Using this definition, we can talk about *wires* in the comparator circuit.

We say that an AND gate in a comparator gate is *used* if the AND output wire of that comparator gate is used in the circuit. An AND gate in the circuit is called *skew* if and only if at least one input to the gate is the constant 0 or the constant 1 or (in the case of annotated circuits) an input bit x_i or $\overline{x_i}$ for some i.

A comparator circuit is called a *skew comparator circuit* if and only if all used AND gates in the circuit are *skew*. The complexity class SkewCC consists of all languages that can be decided by poly-size skew comparator circuits. We define SkewCCVP to be the circuit evaluation problem for skew comparator circuits. Note that given the ordered list representation of a comparator circuit, it is easy to check whether an AND gate is used or not. For ex., if the ith gate is (w_1, w_2), then the AND output of this gate is unused iff there is no element in the list of gates with w_1 as a member at a position greater that i in the list.

The circuit family is LOG-uniform if and only if there exists a TM M that outputs the nth circuit in the family in $O(\log(n))$ space given 1^n as input. All circuits in this paper are LOG-uniform.

Preliminaries from Lattice and Order Theory: Basic definitions and terminology from standard lattice and order theory that are required later in the paper can be found in the full paper (See Section 2, [8]). A more detailed treatment can be found in standard textbooks [3]. We will now state some technical theorems from the theory which we crucially use. The following theorem shows that given a poset one can find a lattice that contains the poset.

Theorem 1 (Dedekind-Macneille Completion[3]). *For any poset P, there always exist a smallest lattice L that order embeds P. This lattice L is called the* Dedekind-MacNeille completion *of P.*

One crucial property of Dedekind-MacNeille completion is that it preserves all meets and joins that exist in the poset. i.e., if a and b are two elements in the poset and $a \vee b = x$ in the poset, then we have $f(a) \vee f(b) = f(x)$ in the Dedekind-MacNeille completion of the poset, where f is the embedding function that maps elements in P to elements in L.

Theorem 2 (Birkhoff's Representation Theorem[3]). *The elements of any finite distributive lattice can be represented as finite sets, in such a way that the join and meet operations over the finite distributive lattice correspond to unions and intersections of the finite sets used to represent those elements.*

The nth *partition lattice* for $n \geq 2$, denoted Π_n, is the lattice where elements are partitions of the set $\{1, \ldots, n\}$ ordered by refinement. Equivalently, the elements are equivalence relations on the set $\{1, \ldots, n\}$ where the glb is the intersection and lub is the transitive closure of the union.

Theorem 3 (Pudlák, Tůma[6]). *For any finite lattice L, there exists an i such that L can be embedded as a sublattice in Π_i.*

We can describe elements of the partition lattice Π_n by using undirected graphs on the vertex set $\{1, \ldots, n\}$. Given an undirected graph $G = (\{1, \ldots, n\}, E)$, the corresponding element $A_G \in \Pi_n$ is the equivalence relation $A_G = \{(i, j) : j \text{ is reachable from } i \text{ in } G\}$. The following proposition holds for partition lattices. A proof can be found in the full paper (See Proposition 2, [8]).

Proposition 1. *For any $A \in \Pi_i$, there exists a formula (over join, meet and arbitrary constants from the lattice) $\mathrm{GE'}_A(x)$ that evaluates to 1 if $x \geq A$ and $\mathrm{GE'}_A(x)$ evaluates to 0 otherwise.*

3   Generalization to Finite Bounded Posets and Universal Circuits

In this section, we consider comparator circuit models over arbitrary fixed finite bounded posets instead of the Boolean lattice on 2 elements. We then prove the existence of universal circuits for these models. The existence of these generalized universal comparator circuits imply that the classes characterized by comparator circuit families over fixed finite bounded posets also have canonical complete problems – The comparator circuit evaluation problem over the same fixed finite bounded poset.

Definition 1 (Comparator Circuits over Fixed Finite Bounded Posets). *A comparator circuit family over a finite bounded poset P with an accepting element $a \in P$ is a family of circuits $C = \{C_n\}_{n \geq 0}$ where $C_n =$*

(W, G, f) where $f : W \mapsto (P \cup \{(i, g) : 1 \leq i \leq n$ and $g : \Sigma \mapsto P\})$ is a comparator circuit. Here $W = \{w_1, \ldots, w_m\}$ is a set of lines and G is an ordered list of gates (w_i, w_j).

On input $x \in \Sigma^n$, we define the output of the comparator circuit C_n as follows. Each line is initially assigned a value according to f as follows. We denote the value of the line w_i by $val(w_i)$. If $f(w) \in P$, then the value is the element $f(w)$. Otherwise $f(w) = (i, g)$ and the initial value is given by $g(x_i)$. A gate (w_i, w_j) (non-deterministically) updates the value of the line w_i into $val(w_i) \wedge val(w_j)$ and the value of the line w_j into $val(w_i) \vee val(w_j)$. The values of lines are updated by each gate in G in order and the circuit accepts x if and only if $val(w_1) = a$ at the end of the computation for some sequence of non-deterministic choices.

Let Σ be any finite alphabet. A comparator circuit family C over a bounded poset P with an accepting element $a \in P$ decides $L \subseteq \Sigma^*$ if $\forall x \in \Sigma^*, C_{|x|}(x) = a \iff x \in L$.

Note that we can generalize any circuit model that uses only AND and OR gates to work over arbitrary bounded posets. We first prove that a universal comparator circuit exists even for comparator circuit model working over arbitrary finite fixed posets.

Proposition 2. *For any bounded poset P, there exists a universal comparator circuit $U_{n,m}$ over P that when given (C, x) as input, where C is a comparator circuit over P with n lines and m gates, simulates the computation of C. That is, $U_{n,m}$ has a sequence of non-deterministic choices that outputs $a \in P$ if and only if C has such a path, for any $a \in P$. Moreover, the size of $U_{n,m}$ is $\mathsf{poly}(n, m)$.*

Proof. We simply observe that the construction for a universal circuit for the class CC in [2] generalizes to arbitrary bounded posets. The gadget shown in Figure 2 enables/disables the gate $g = (y, x)$ depending on the "enable" input e. Now to simulate a single gate in the circuit C, the universal circuit uses $n(n-1)$ such gadgets where n is the number of lines in C. The inputs e and $\bar{e}$ for each gadget is set according to C. The circuit C can be simulated using $n(n-1)m$ gates where m is the number of gates in C.   □

Definition 2. *We define the complexity class (P, a)–CC as the set of all languages accepted by comparator circuit families over the finite bounded poset P with accepting element $a \in P$ where $|C_n| \leq \mathsf{poly}(n)$. If the complexity class does not change with the accepting element, we simply write P–CC.*

We note that for any bounded poset P with at least 2 elements, we can simulate a Boolean lattice by using 0 (least element) and some $a > 0$ in P. Therefore, we have $\mathsf{CC} \subseteq (\mathsf{P}, \mathsf{a})$–$\mathsf{CC}$.

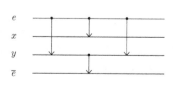

Fig. 2. Conditional Gadget

Definition 3. *For any finite bounded poset P and any $a \in P$, the comparator circuit evaluation problem (P, a)–CCVP is defined as the set of all tuples (C, x) such that C*

on input x *has a sequence of non-deterministic choices where it outputs* $a \in F$ *where* C *is a comparator circuit over* P.

The following proposition is a generalization of the corresponding theorem for Boolean comparator circuits in [2]. A proof can be found in the full version of this paper (See Prop. 4, [8]).

Proposition 3. *The language* (P, a)–CCVP *is complete for* (P, a)–CC *for all finite bounded posets* P *and any* $a \in P$.

4   Comparator Circuits over Lattices

First, we show that comparator circuits over distributive lattices is exactly the class CC.

Theorem 4. *Let* L *be any finite distributive lattice and* $a \in L$ *be an arbitrary element. Then* CC $= (L, a)$–CC.

Proof. By Birkhoff's representation theorem, every finite distributive lattice of k elements is isomorphic to a lattice where each element is some subset of $[k]$ (ordered by inclusion) and the join and meet operations in the original finite distributive lattice correspond to set union and set intersection operations in the new lattice. We will use this to simulate a circuit over an arbitrary finite distributive lattice L of size k using a circuit over the 0–1 lattice. Each line w in the original circuit is replaced by k lines $w_1, \ldots, w_k$. The invariant maintained is that whenever a line in the original circuit carries $a \in L$, these k lines carry the characteristic vector of the set corresponding to the element a. Now a gate (w, x) in the original circuit is replaced by k gates $(w_1, x_1), \ldots, (w_k, x_k)$ in the new circuit. The correctness follows from the fact that meet and join operations in the original circuit correspond to set union and set intersection which in turn correspond to AND and OR operations of the characteristic vectors.   □

The following lemma describes a fixed finite lattice over which comparator circuits capture P. In Theorem 5, we use this lemma to show that there exists a lattice that captures P irrespective of the accepting element. Figures 5 and 6 referred to in the following lemma can be found in the full paper.

Lemma 1. *Let* L *be the lattice in Figure 5. Then* P $= (L, 1)$–CC *(Note that* 1 *is not the maximum element in the lattice).*

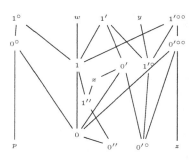

Fig. 3. Poset for simulating P

Proof. We will reduce the problem MCVP (Monotone Circuit Value Problem: Given (C, x) where C is a Boolean circuit where only AND and OR gates are allowed with inputs $x_1, \bar{x}_1, \cdots x_n, \overline{x_n}$, decide if $C(x) = 1$) which is complete for P to the comparator circuit value problem over the finite lattice given in Figure 5. Let (C, x) be the input to MCVP. For each wire in C, we add a line to our comparator circuit. The initial value of the lines that correspond to the input wires of C are set to 0 or 1 of the poset shown in Figure 3 according to whether they are 0 or 1 in x. The comparator circuit simulates C in a level by level fashion maintaining the invariant that the lines carry 0 or 1 depending on whether they carry 0 or 1 in C. We will show how our comparator circuit simulates a level 1 OR gate of fan-out 2. The proof then follows by an easy induction.

Since $0 \leq_P 1$ an AND (OR) gate in C can be simulated by a meet (join) operation in P. The gadget shown in Figure 6 is used to implement the fan-out operation. The idea is that the first gate in the gadget implements the AND/OR operation and the rest of the gates in this gadget "copies" the result of this operation into the lines o_1 and o_2 that correspond to the two output wires of the gate. The reader can verify that the elements of P satisfy the following meet and join identities. Figure 6 shows how one could use the following identities to copy the output of $a \vee b$ into two lines (labelled o_1 and o_2).

The identity $0 \vee 1 = 1$ is used to implement the Boolean AND/OR operation. This is used by the first gate in Figure 6. We then add a gate between the line carrying the result of the AND/OR operation and a line with value x. As the following identities show, this makes two "copies" of the result of the Boolean operation. $0 \vee x = 0'$, $1 \vee x = 1'$, $0 \wedge x = 0''$, $1 \wedge x = 1''$

Now, the following identities can be used to convert the first copy ($0'$ or $1'$) into the original value (0 or 1). $0' \wedge y = 0'^\circ$, $1' \wedge y = 1'^\circ$, $0'^\circ \vee z = 0'^{\circ\circ}$, $1'^\circ \vee z = 1'^{\circ\circ}$, $0'^{\circ\circ} \wedge w = 0$, $1'^{\circ\circ} \wedge w = 1$

Similarly, the following identities can be used to convert the second copy ($0''$ or $1''$) into the original value (0 or 1). $0'' \vee p = 0^\circ$, $1'' \vee p = 1^\circ$, $0^\circ \wedge w = 0$, $1^\circ \wedge w = 1$

The lattice in Figure 5 in the full paper is simply the Dedekind-MacNeille completion of P. Since the Dedekind-MacNeille completion preserves all existing meets and joins, the same computation can also be performed by this lattice.

To see that for any lattice L and any $a \in L$, (L, a)–CC is in P, observe that in poly-time we can evaluate the nth comparator circuit from the comparator circuit family for the language in (L, a)–CC.   □

Lemma 1 shows that the complexity class captured by the comparator circuit could change (Assuming CC ≠ P) depending on the underlying lattice *and* the accepting element. In the following theorem, we show that if we consider any partition lattice, say Π_i embedding the lattice L in Lemma 1, then the complexity class captured by comparator circuits over Π_i is also P *irrespective of the accepting element* from the lattice. Note that the universality of partition lattices only implies that *there exists* an element in Π_i such that comparator circuits with Π_i with that accepting element captures P. We crucially use the fact that comparator circuits in Lemma 1 outputs only the elements 0 or 1 in L to show the following theorem.

Theorem 5. *There exists a constant i such that* Π_i–CC = P.

Proof. We know that there exists a finite lattice L and an $a, b \in L$ such that for any language $M \in P$ there exists a comparator circuit family over L that decides M by using a to accept and b to reject. Also $b < a$. By Pudlak's theorem [6], we know that there exists a constant i such that L can be embedded in Π_i. It remains to show that the accepting element used does not change the complexity. In fact, we will show that for any $X, Y \in \Pi_i$ where $X \neq Y$, we can design a comparator circuit family over Π_i that accepts M using X and rejects using Y. Let A and B be the elements in Π_i that a and b gets mapped to by this embedding $(B < A)$. Then there exists a circuit family C over Π_i ,deciding M, that accepts using A and rejects using B. We will construct a circuit family C' over Π_i from C such that C' uses 1 to accept and 0 to reject. Here 1 and 0 are the maximum and minimum elements in Π_i. Now if we let x be the output of a circuit in the circuit family C, we can construct C' by computing $GE'_A(x)$. Similarly, we can construct a circuit family C'' that accepts using 0 and rejects using 1 by reducing the language M to $\overline{MCVP}$ and then applying the construction in Lemma 1 and then computing $GE'_A(x)$ on the output of this circuit. The required circuit family is then the one computing $(X \wedge C') \vee (Y \wedge C'')$. ☐

If we can show that there exists a finite distributive lattice such that the poset in Figure 3 can be embedded in that lattice while preserving all existing meets and joins, then $P = CC$. In the following theorem, we show that such an embedding is not possible. For a proof see full paper [8](Theorem 6).

Theorem 6. *The poset in Figure 3 cannot be embedded into any distributive lattice while preserving all meets and joins.*

5   Comparator Circuits over Bounded Posets

In this section, we consider the most general form of comparator circuits. i.e., we consider comparator circuits over fixed finite bounded posets. We show that the resulting complexity class is exactly the class NP.

Theorem 7. *Let P be any poset and let $a \in P$ be an arbitrary element in P, then $(P, a)-CC \subseteq NP$. Also, there exists a finite poset P and an $a \in P$ such that $NP = (P, a)-CC$.*

Proof (Sketch). The first part follows from observing that an NTM can evaluate a comparator circuit over any bounded poset by non-deterministically guessing one of the minimal upper bounds and maximal lower bounds at each gate.

For the second part, we reduce SAT (A well-known NP-complete problem) to the comparator circuit value problem over P with a as acceptor. The key idea behind the reduction is the same as that in the proof of Theorem 5. The only difference is we could have an input wire labelled with an x_i or an $\overline{x_i}$ for some i. In this case, we cannot initialize the corresponding line with an element from the lattice used in Theorem 5. Instead, we non-deterministically generate a 0 or a 1 as the initial value of that line by using the fact that in a poset two elements could have multiple minimal upper bounds. A proof can be found in the full paper (See Theorem 7, [8]). ☐

6   Skew Comparator Circuits

In this section, we study the skew comparator circuits defined in the preliminaries. We show that SkewCC is exactly the class LOG. Recall that the class NLOG can be characterized as the set of all languages computed by logspace-uniform Boolean circuits with skewed AND gates. So the result in this section draws a parallel between the P vs CC problem and the NLOG vs LOG problem. It follows that SkewCC over distributive lattices also characterizes exactly LOG.

We begin by considering a canonical complete problem for the class LOG. The language DGAP1 consists of all tuples (G, s, t) where $G = (V, E)$ is a directed graph where each vertex has out-degree at most one and $s, t \in V$ and there is a directed path from s to t. We use a variant of DGAP1 problem in our setting. The variant (called DGAP1') is that the out-degree constraint is not applied to s. It is easy to see that DGAP1' is also in LOG. Indeed, for each neighbour u of s, run the DGAP1 algorithm to check whether t is reachable from u.

Theorem 8. SkewCC = LOG

Proof. ($\subseteq$) Let $L \in$ SkewCC. We will prove that $L \in$ LOG by reducing L to DGAP1'. The reduction is as follows. Observe that we can reduce the language L to SkewCCVP by a logspace reduction (Using the uniformity algorithm). Then we reduce SkewCCVP to DGAP1'. The details of the proof can be found in the full paper (See Theorem 8, [8]).

($\supseteq$) Let $L \in$ LOG and let B be a poly-sized layered branching program deciding L. We will design a skew comparator circuit C to simulate B. Let s be a state in B reading x_i and let the edge labelled 1 be directed towards a state t and let the edge labelled 0 be directed towards a state u. We design a gadget that simulates this part of the BP B and this reduction can be implemented in NC^1. A detailed proof can be found in the full paper (See Theorem 8, [8]).   □

Since the construction in Theorem 4 preserves skewness of the circuit, we have the following corollary.

Corollary 1. *Let L be any distributive lattice and let a be any element in L, then $(L, a)^-$SkewCC = LOG.*

We now turn to skew comparator circuits working over arbitrary fixed finite lattices. We show that, over the partition lattice, there is a comparator circuit family of polynomial size that captures NLOG.

Theorem 9. *There exists[1] a constant i such that Π_i-SkewCC = NLOG*

Proof. NLOG $\subseteq \Pi_i$-SkewCC follows directly from the reduction in Theorem 5 and from NLOG=co-NLOG. We also use the fact that he reduction in Theorem 5 preserves the skewness of the circuit. To show that Π_i-SkewCC $\subseteq$ NLOG, we describe an NLOG evaluation algorithm for the circuit value problem Π_i-SkewCCVP.

---

[1] Here the constant i is the same as in Theorem 5

This algorithm is a generalization of the NLOG evaluation algorithm for skew (Boolean) circuits. The crucial idea in case of skew Boolean circuits is that the evaluation algorithm can use a depth first search without storing any backtracking information. In order to verify that the output of an OR gate is 1, the NLOG algorithm can non-deterministically guess one of the input gates and verify whether it is 1. This does not require storing any backtracking information as the output of the OR gate is 1 even if one of the inputs is 1. This is not true for computation over arbitrary lattices. For example, the output of an OR gate could be 1 (Maximum element) even if both its inputs are less than 1 (This could happen if inputs have values a and b and the lub of a and b is 1). So verifying that the output of an OR gate is 1 requires verifying values of both its inputs and the straightforward evaluation algorithm cannot do this in logspace. We overcome this difficulty by using certain properties of partition lattices. A proof can be found in the full paper (See Theorem 9, [8]).   □

7   Formulae over Lattices

It is well known that poly-size formulae capture the class NC^1. We can modify Definition 1 to define formulae over finite bounded posets. We denote by (L, a)-Formulae, where L is a lattice and $a \in L$, the class of all languages decided by a poly-size formula family over L using a as the accepting element. In this section, we show that poly-sized formulae over any fixed finite lattice is the class NC^1. The proof for the Boolean case is by [5] and it works by depth reducing an arbitrary formula of poly size to a Boolean formula of poly size and log depth. We show that a similar argument can be extended to the case of finite lattices as well. A proof can be found in the full paper (See Theorem 10, [8]).

Theorem 10. *Let L be any finite lattice and let a be an arbitrary element in L. We have (L, a)-Formulae $= \mathsf{NC}^1$.*

References

1. Arora, S., Barak, B.: Computational Complexity: A Modern Approach. Cambridge University Press (2009)
2. Cook, S.A., Filmus, Y., Lê, D.T.M.: The complexity of the comparator circuit value problem. ACM Trans. Comput. Theory **6**(4), 15:1–15:44 (2014)
3. Davey, B.A., Priestley, H.A.: Introduction to lattices and order. Cambridge University Press, Cambridge (1990)
4. Mayr, E.W., Subramanian, A.: The complexity of circuit value and network stability. J. Comput. Syst. Sci. **44**(2), 302–323 (1992)
5. Spira, P.M.: On time-hardware complexity tradeoffs for boolean functions. In: Proceedings of 4th Hawaii Symp. on System Sciences, pp. 525–527 (1971)
6. Pudlák, P., Tůma, J.: Every finite lattice can be embedded in a finite partition lattice. algebra universalis **10**(1), 74–95 (1980)
7. Subramanian, A.: The Computational Complexity of the Circuit Value and Network Stability Problems. PhD thesis, Stanford, CA, USA (1990). AAI9102356
8. Sunil, K.S., Komarath, B., Sarma, J.: Comparator circuits over finite bounded posets. Electronic Colloquium on Computational Complexity (ECCC) **22**, 35 (2015)

Algebraic Properties of Valued Constraint Satisfaction Problem

Marcin Kozik[1] and Joanna Ochremiak[2][(✉)]

[1] Jagiellonian University, Kraków, Poland
marcin.kozik@tcs.uj.edu.pl
[2] University of Warsaw, Warsaw, Poland
ochremiak@mimuw.edu.pl

Abstract. The paper presents an algebraic framework for optimization problems expressible as Valued Constraint Satisfaction Problems. Our results generalize the algebraic framework for the decision version (CSPs) provided by Bulatov et al. [SICOMP 2005].

We introduce the notions of weighted algebras and varieties, and use the Galois connection due to Cohen et al. [SICOMP 2013] to link VCSP languages to weighted algebras. We show that the difficulty of VCSP depends only on the weighted variety generated by the associated weighted algebra.

Paralleling the results for CSPs we exhibit a reduction to cores and rigid cores which allows us to focus on idempotent weighted varieties. Further, we propose an analogue of the Algebraic CSP Dichotomy Conjecture; prove the hardness direction and verify that it agrees with known results for VCSPs on two-element sets [Cohen et al. 2006], finite-valued VCSPs [Thapper and Živný 2013], and conservative VCSPs [Kolmogorov and Živný 2013].

1 Introduction

An instance of the Constraint Satisfaction Problem (CSP) consists of variables (to be evaluated in a domain) and constraints restricting the evaluations. The aim is to find an evaluation satisfying all the constraints or satisfying the maximal possible number of constraints or approximating the maximal possible number of satisfied constraints etc. depending on the version of the problem. Further one can divide constraint satisfaction problems with respect to the size of the domain, the allowed constraints or the shape of the instances.

A particularly interesting version of the CSP was proposed in a seminal paper of Feder and Vardi [11]. In this version the CSP is defined by a *language* which consists of relations over a finite set. An instance of such a CSP is allowed if all the constraint relations are from this set. The goal is to determine whether an instance has a solution satisfying all the constraints.

The first author was supported by the Polish National Science Centre (NCN) grant 2011/01/B/ST6/01006; the second author was supported by the Polish National Science Centre (NCN) grant 2012/07/B/ST6/01497.

© Springer-Verlag Berlin Heidelberg 2015
M.M. Halldórsson et al. (Eds.): ICALP 2015, Part I, LNCS 9134, pp. 846–858, 2015.
DOI: 10.1007/978-3-662-47672-7_69

Each language clearly defines a problem in NP; the whole family of problems is interesting for another reason: it is robust enough to include some well studied computational problems, e.g. 2-colorability, 3-SAT, solving systems of linear equations over $\mathbb{Z}_p$, and still is conjectured [11] not to contain problems of intermediate complexity. This conjecture (which holds for languages on two-element sets by the result of Schaefer [18]) is known as the Constraint Satisfaction Dichotomy Conjecture of Feder and Vardi. Confirming this conjecture would establish CSPs as one of the largest natural subclasses of NP without problems of intermediate complexity.

The conjecture always attracted a lot of attention, but the first results, even very interesting ones, were usually very specialized (e.g. [12]). A major breakthrough appeared with a series of papers establishing *the algebraic approach to CSP* [3,7,14]. This deep connection with an independently developed branch of mathematics introduced a new viewpoint and provided tools necessary to tackle wide classes of CSP languages at once. At the heart of this approach lies a Galois connection between languages and clones of operations called *polymorphisms* (which completely determine the complexity of the language).

Results obtained using these new methods include a full complexity classifications for CSPs on three-element sets [5] and those containing all unary relations [4,6]. Moreover, the algebraic approach to CSP allowed to propose a boundary between the tractable and NP-complete problems: this conjecture is known as the Algebraic Dichotomy Conjecture. Unfortunately, despite many efforts (e.g. [5]), both conjectures remain open.

The Valued Constraint Satisfaction Problem (VCSP) further extends the approach proposed by Feder and Vardi. The role of constraints is played by *cost functions* describing the price of choosing particular values for variables as a part of the solution. This generalization allows to construct languages modeling standard optimization problems, for example MAX-CUT. Moreover, by allowing ∞ as a cost of a tuple, a VCSP language can additionally model every problem that CSP can model, as well as hybrid problems like MIN-VERTEX-COVER. This makes the extended framework even more general (compare the survey [15]).

A number of classes of VCSPs have been thoroughly investigated. The underlying structure suggested capturing the properties of languages of cost functions using an amalgamation of algebraic and numerical techniques [10,20]. The first approach which provides a Galois correspondence (mirroring the Galois correspondence for CSPs) was proposed by Cohen et al. [9]. A weighted clone defined in this paper fully captures the complexity of a VCSP language.

The present paper builds on that correspondence imitating the line of research for CSPs [7]. It is organized in the following way: Section 2 contains preliminaries and basic definitions. In Section 3 we present a reduction to cores and rigid cores. Section 4 introduces a concept of a weighted algebra and a weighted variety, and shows that those notions are well behaved in the context of the Galois connection for VCSP. Reductions developed in Section 3 together with definitions from Section 4 allow us to focus on idempotent varieties. Section 5 states a conjecture postulating (for idempotent varieties) the division between the tractable and NP-hard cases of VCSP. The conjecture is clearly a strengthening of the Algebraic Dichotomy Conjecture [7]. Section 5 contains additionally the proof of the hardness direction of the conjecture as well as the reasoning showing that the conjecture agrees with complexity classifications for VCSPs on two-element sets [10], with finite-valued cost functions [20], and with conservative cost functions [16].

2   Preliminaries

2.1   The Valued Constraint Satisfaction Problem

Throughout the paper, let $\overline{\mathbb{Q}} = \mathbb{Q} \cup \{\infty\}$. We assume that $x + \infty = \infty$ and $y \cdot \infty = \infty$ for $y \geq 0$. An r-ary *relation* on a set D is a subset of D^r, a *cost function* on D of arity r is a function from D^r to $\overline{\mathbb{Q}}$. We denote by Φ_D the set of all cost functions on D. A cost function which takes only finite values is called *finite-valued*. A $\{0, \infty\}$-valued cost function is called *crisp* and can be viewed as a relation.

Definition 1. *An* instance of the valued constraint satisfaction problem (VCSP) *is a triple* $\mathcal{I} = (V, D, \mathcal{C})$ *with* V *a finite set of* variables, D *a finite* domain *and* $\mathcal{C}$ *a finite multi-set of* constraints. *Each constraint is a pair* $C = (\sigma, \varrho)$ *with* σ *a tuple of variables of length* r *and* ϱ *a cost function on* D *of arity* r.

An assignment *for* $\mathcal{I}$ *is a mapping* $s \colon V \to D$. *The* cost of an assignment s *is given by* $Cost_{\mathcal{I}}(s) = \sum_{(\sigma, \varrho) \in \mathcal{C}} \varrho(s(\sigma))$ *(where* s *is applied component-wise). To solve* $\mathcal{I}$ *is to find an assignment with a minimal cost, called an* optimal assignment.

Any set $\Gamma \subseteq \Phi_D$ is called a *valued constraint language* over D, or simply a *language*. If all cost functions from Γ are $\{0, \infty\}$-valued or finite-valued, we call it a *crisp* or *finite-valued* language, respectively.

By $\mathrm{VCSP}(\Gamma)$ we denote the class of all VSCP instances in which all cost functions in all constraints belong to Γ. $\mathrm{VCSP}(\Gamma_{crisp})$, where Γ_{crisp} is the language consisting of all crisp cost functions on some fixed set D, is equivalent to the classical CSP. For an instance $\mathcal{I} \in \mathrm{VCSP}(\Gamma)$ we denote by $\mathrm{Opt}_{\Gamma}(\mathcal{I})$ the cost of an optimal assignment. We say that a language Γ is *tractable* if, for every finite subset $\Gamma' \subseteq \Gamma$, there exists an algorithm solving any instance $\mathcal{I} \in \mathrm{VCSP}(\Gamma')$ in polynomial time, and we say that Γ is *NP-hard* if $\mathrm{VCSP}(\Gamma')$ is NP-hard for some finite $\Gamma' \subseteq \Gamma$.

Weighted Relational Clones. We follow the exposition of [9] and define a closure operator on valued constraint languages that preserves tractability.

Definition 2. *A cost function* ϱ *is* expressible *over a valued constraint language* $\Gamma \subseteq \Phi_D$ *if there exists an instance* $\mathcal{I}_\varrho \in \mathrm{VCSP}(\Gamma)$ *and a list* $(v_1, \ldots, v_r)$ *of variables of* $\mathcal{I}_\varrho$, *such that*

$$\varrho(x_1, \ldots, x_r) = \min_{\{s \colon V \to D \mid s(v_i) = x_i\}} Cost_{\mathcal{I}_\varrho}(s).$$

Note that the list of variables $(v_1, \ldots, v_r)$ in the definition above might contain repeated entries. Hence, it is possible that there are no assignments s such that $s(v_i) = x_i$ for all i. We define the minimum over the empty set to be ∞.

Definition 3. *A set $\Gamma \subseteq \Phi_D$ is a* weighted relational clone *if it is closed under expressibility, scaling by non-negative rational constants, and addition of rational constants. We define* wRelClo(Γ) *to be the smallest weighted relational clone containing Γ.*

If $\varrho(x_1, \ldots, x_r) = \varrho_1(y_1, \ldots, y_s) + \varrho_2(z_1, \ldots, z_t)$ for some fixed choice of arguments $y_1, \ldots, y_s, z_1, \ldots, z_t$ from amongst $x_1, \ldots, x_r$ then the cost function ϱ is said to be obtained by *addition* from the cost functions ϱ_1 and ϱ_2. It is easy to see that a weighted relational clone is closed under addition, and minimisation over arbitrary arguments.

The following result shows that we can restrict our attention to languages which are weighted relational clones.

Theorem 4 (Cohen et al. [9]). *A valued constraint language Γ is tractable if and only if* wRelClo(Γ) *is tractable, and it is NP-hard if and only if* wRelClo(Γ) *is NP-hard.*

Weighted Polymorphisms. A k-ary *operation* on D is a function $f \colon D^k \to D$. We denote by $\mathcal{O}_D$ the set of all finitary operations on D and by $\mathcal{O}_D^{(k)}$ the set of all k-ary operations on D. The k-ary *projections*, defined for all $i \in \{1, \ldots, k\}$, are the operations $\pi_i^{(k)}$ such that $\pi_i^{(k)}(x_1, \ldots, x_k) = x_i$. Let $f \in \mathcal{O}_D^{(k)}$ and $g_1, \ldots, g_k \in \mathcal{O}_D^{(l)}$. The l-ary operation $f[g_1, \ldots, g_k]$ defined by $f[g_1, \ldots, g_k](x_1, \ldots, x_l) = f(g_1(x_1, \ldots, x_l), \ldots, g_k(x_1, \ldots, x_l))$ is called the *superposition* of f and $g_1, \ldots, g_k$.

A set $C \subseteq \mathcal{O}_D$ is a *clone of operations* (or simply a *clone*) if it contains all projections on D and is closed under superposition. The set of k-ary operations in a clone C is denoted $C^{(k)}$. The smallest possible clone of operations over a fixed set D is the set of all projections on D, which we denote Π_D.

Following [9] we define a k-ary *weighting* of a clone C to be a function $\omega \colon C^{(k)} \to \mathbb{Q}$ such that $\sum_{f \in C^{(k)}} \omega(f) = 0$, and if $\omega(f) < 0$ then f is a projection. The set of operations to which a weighting ω assigns positive weights is called the *support* of ω and denoted supp(ω).

A new weighting of the same clone can be obtained by scaling a weighting by a non-negative rational, adding two weightings of the same arity and by the following operation called *superposition*.

Definition 5. *Let ω be a k-ary weighting of a clone C and let $g_1, \ldots, g_k \in C^{(l)}$. A* superposition *of ω and $g_1, \ldots, g_k$ is a function $\omega[g_1, \ldots, g_k] \colon C^{(l)} \to \mathbb{Q}$ defined by*

$$\omega[g_1, \ldots, g_k](f') = \sum_{\{f \in C^{(k)} \ | \ f[g_1, \ldots, g_k] = f'\}} \omega(f).$$

The sum of weights that any superposition $\omega[g_1, \ldots, g_k]$ assigns to the operations in $C^{(l)}$ is equal to zero, however, it may happen that a superposition assigns a negative value to an operation that is not a projection. A superposition is said to be *proper* if the result is a valid weighting.

A non-empty set of weightings over a fixed clone C is called a *weighted clone* if it is closed under non-negative scaling, addition of weightings of equal arity and proper superposition with operations from C. For any clone of operations C, the set of all weightings over C and the set of all zero-valued weightings of C are weighted clones.

We say that an r-ary relation R on D is *compatible* with an operation $f : D^k \rightarrow D$ if, for any list of r-tuples $\mathbf{x_1}, \ldots, \mathbf{x_k} \in R$ we have $f(\mathbf{x_1}, \ldots, \mathbf{x_k}) \in R$ (where f is applied coordinate-wise). Let $\varrho : D^r \rightarrow \overline{\mathbb{Q}}$ be a cost function. We define $\mathrm{Feas}(\varrho) = \{\mathbf{x} \in D^r \mid \varrho(\mathbf{x}) \text{ is finite}\}$ to be the *feasibility relation* of ϱ. We call an operation $f : D^k \rightarrow D$ a *polymorphism* of ϱ if the relation $\mathrm{Feas}(\varrho)$ is compatible with it. For a valued constraint language Γ we denote by $\mathrm{Pol}(\Gamma)$ the set of operations which are polymorphisms of all cost functions $\varrho \in \Gamma$. It is easy to verify that $\mathrm{Pol}(\Gamma)$ is a clone. The set of m-ary operations in $\mathrm{Pol}(\Gamma)$ is denoted $\mathrm{Pol}_m(\Gamma)$.

For crisp cost functions (relations) this notion of polymorphism corresponds precisely to the standard notion of polymorphism which has played a crucial role in the complexity analysis for the CSP [3,14].

Definition 6. *Take ϱ to be a cost function of arity r on D, and let $C \subseteq \mathrm{Pol}(\{\varrho\})$ be a clone of operations. A weighting $\omega : C^{(k)} \rightarrow \mathbb{Q}$ is called a* weighted poly-morphism *of ϱ if, for any list of r-tuples $\mathbf{x_1}, \ldots, \mathbf{x_k} \in \mathrm{Feas}(\varrho)$, we have*

$$\sum_{f \in C^{(k)}} \omega(f) \cdot \varrho(f(\mathbf{x_1}, \ldots, \mathbf{x_k})) \leq 0.$$

For a valued constraint language Γ we denote by $\mathrm{wPol}(\Gamma)$ the set of those weightings of the clone $\mathrm{Pol}(\Gamma)$ that are weighted polymorphisms of all cost functions $\varrho \in \Gamma$. The set of weightings $\mathrm{wPol}(\Gamma)$ is a weighted clone [9].

An operation f is *idempotent* if $f(x, ..., x) = x$. A weighted polymorphism is called *idempotent* if all operations in its support are idempotent. An operation $f \in \mathcal{O}_D^{(k)}$ is *cyclic* if for every $x_1, \ldots, x_k \in D$ we have that $f(x_1, x_2, \ldots, x_k) = f(x_2, \ldots, x_k, x_1)$. A weighted polymorphism is called *cyclic* if its support is non-empty and contains cyclic operations only.

A cost function ϱ is said to be *improved* by a weighting ω if ω is a weighted polymorphism of ϱ. For any set W of weightings over a fixed clone $C \subseteq \mathcal{O}_D$ we denote by $\mathrm{Imp}(W)$ the set of cost functions on D which are improved by all weightings $\omega \in W$.

By the result of Cohen et al. [9] for any finite valued constraint language Γ, we have $\mathrm{Imp}(\mathrm{wPol}(\Gamma)) = \mathrm{wRelClo}(\Gamma)$. This fact, together with Theorem 4, implies that tractable valued constraint languages can be characterized by their weighted polymorphisms.

2.2   Algebras and Varieties

An *algebraic signature* is a set of function symbols together with (finite) arities. An *algebra* **A** over a fixed signature Σ, has a *universe* A, and a set of *basic operations* that correspond to the symbols in the signature, i.e., if the signature contains a k-ary symbol f then the algebra has a basic operation $f^{\mathbf{A}}$, which is a function $f^{\mathbf{A}}\colon A^k \to A$.

A subset B of the universe of an algebra **A** is a *subuniverse* of **A** if it is closed under all operations of **A**. An algebra **B** is a *subalgebra* of **A** if B is a subuniverse of **A** and the operations of **B** are restrictions of all the operations of **A** to B. Let $(\mathbf{A}_i)_{i \in I}$ be a family of algebras (over the same signature). Their *product* $\Pi_{i \in I}\mathbf{A}_i$ is an algebra with the universe equal to the cartesian product of the A_i's and operations computed coordinate-wise. For two algebras **A** and **B** (over the same signature), a *homomorphism* from **A** to **B** is a function $h\colon A \to B$ that preserves all operations. It is easy to see, that an image of an algebra under a homomorphism $h\colon A \to B$ is a subalgebra of **B**.

Let $\mathcal{K}$ be a class of algebras over a fixed signature Σ. We denote by $\mathrm{S}(\mathcal{K})$ the class of all subalgebras of algebras in $\mathcal{K}$, by $\mathrm{P}(\mathcal{K})$ the class of all products of algebras in $\mathcal{K}$, by $\mathrm{P}_{fin}(\mathcal{K})$ the class of all finite products, and by $\mathrm{H}(\mathcal{K})$ the class of all homomorphic images of algebras in $\mathcal{K}$. If $\mathcal{K} = \{\mathbf{A}\}$ we write $\mathrm{S}(\mathbf{A})$, $\mathrm{P}(\mathbf{A})$, and $\mathrm{H}(\mathbf{A})$ instead of $\mathrm{S}(\{\mathbf{A}\})$, $\mathrm{P}(\{\mathbf{A}\})$, and $\mathrm{H}(\{\mathbf{A}\})$, respectively.

A *variety* $\mathcal{V}(\mathcal{K})$ is the smallest class of algebras closed under all three operations. For an algebra **A** the variety $\mathcal{V}(\{\mathbf{A}\})$ (denoted $\mathcal{V}(\mathbf{A})$) is the variety *generated* by **A**, and $\mathcal{V}_{fin}(\mathbf{A})$ is the class of finite algebras in $\mathcal{V}(\mathbf{A})$. Due to a result of Tarski [19] we know that for any finite algebra **A**, we have

$$\mathcal{V}(\mathbf{A}) = \mathrm{HSP}(\mathbf{A}) \quad \text{and} \quad \mathcal{V}_{fin}(\mathbf{A}) = \mathrm{HSP}_{fin}(\mathbf{A}).$$

We say that an equivalence relation $\sim$ on A is a *congruence* of **A** if it is a subalgebra of $\mathbf{A}^2$. Every congruence $\sim$ of **A** determines a *quotient* algebra $\mathbf{A}/\sim$.

A *term* t in a signature Σ is a formal expression built from variables and symbols in Σ that syntactically describes the composition of basic operations. For an algebra **A** over Σ a *term operation* $t^{\mathbf{A}}$ is an operation obtained by composing the basic operations of **A** according to t. Let s and t be a pair of terms in a signature Σ. We say that **A** satisfies the *identity* $s \approx t$ if the term operations $s^{\mathbf{A}}$ and $t^{\mathbf{A}}$ are equal. We say that a class of algebras $\mathcal{V}$ over Σ satisfies the identity $s \approx t$ if every algebra in $\mathcal{V}$ does.

It follows from Birkhoff's theorem [2] that the variety $\mathcal{V}(\mathbf{A})$ is the class of algebras that satisfy all the identities satisfied by **A**. An algebra **A** is *finitely generated* if there exists a finite subset F of its domain such that the only subalgebra of **A** containing F is **A**. If **A** is finite then $\mathcal{V}(\mathbf{A})$ is *locally finite*, i.e., every finitely generated algebra in $\mathcal{V}(\mathbf{A})$ is finite.

3   Core Valued Constraint Languages

For each valued constraint language Γ there is an associated algebra. It has universe D and the set of operations $\mathrm{Pol}(\Gamma)$. If all operations of any given algebra satisfy the identity $f(x, \ldots, x) \approx x$ (i.e. are *idempotent*) then we call the algebra *idempotent*. In this section we prove that every valued constraint language which is finite has a computationally equivalent valued constraint language whose associated algebra is idempotent.

Positive Clone. Those polymorphisms of a given language Γ which are assigned a positive weight by some weighted polymorphisms $\omega \in \mathrm{wPol}(\Gamma)$ are of special interest in the rest of the paper. We begin this section by proving that they form a clone.

Let $\mathcal{C}$ be a weighted clone over a set D. The following proposition shows that the set $\bigcup_{\omega \in \mathcal{C}} \mathrm{supp}(\omega)$, together with the set of projections Π_D, is a clone. We call it the *positive clone* of $\mathcal{C}$ and denote by C^+ (if $\mathcal{C}$ is $\mathrm{wPol}(\Gamma)$ then C^+ is denoted by $\mathrm{Pol}^+(\Gamma)$).

Proposition 7. *If $\mathcal{C}$ is a weighted clone then C^+ is a clone.*

Cores. Let Γ be a valued constraint language with a domain D. For $S \subseteq D$ we denote by $\Gamma[S]$ the valued constraint language defined on a domain S and containing the restriction of every cost function $\varrho \in \Gamma$ to S.

By generalizing the arguments for finite-valued languages given in [13,20], we show that Γ has a computationally equivalent valued constraint language Γ' such that $\mathrm{Pol}_1^+(\Gamma')$ contains only bijective operations. Such a language is called a *core*. Moreover, Γ' can be chosen to be equal to $\Gamma[S]$ for some $S \subseteq D$.

Proposition 8. *For every valued constraint language Γ there exists a core language Γ', such that the valued constraint language Γ is tractable if and only if Γ' is tractable, and it is NP-hard if and only if Γ' is NP-hard.*

For core languages we characterize the set of unary weighted polymorphisms as consisting of all weightings that assign positive weights only to bijective operations preserving all cost functions.

The proposition below witnesses the importance of the positive clone and is used to prove further results in the subsequent sections. Let Γ be a valued constraint language over a domain D which is finite and a core. For each arity m we fix an enumeration of all the elements of D^m. This allows us to treat every m-ary operation $f \in \mathcal{O}_D^{(m)}$ as a $|D^m|$-tuple. We define a $|D^m|$-ary cost function in $\mathrm{wRelClo}(\Gamma)$ that precisely distinguishes the m-ary operations in the positive clone from all the other m-ary polymorphisms.

Proposition 9. *Let Γ be a valued constraint language over a domain D which is finite and a core. For every m there exists a cost function $\varrho \colon \mathcal{O}_D^{(m)} \to \overline{Q}$ in $\mathrm{wRelClo}(\Gamma)$, and a rational number P, such that for every $f \in \mathcal{O}_D^{(m)}$ the following conditions are satisfied:*

1. *$\varrho(f) \geq P$,*
2. *$\varrho(f) < \infty$ if and only if $f \in \mathrm{Pol}(\Gamma)$,*
3. *$\varrho(f) = P$ if and only if $f \in \mathrm{Pol}^+(\Gamma)$.*

Rigid Cores. We further reduce the class of languages that we need to consider. Let Γ be a valued constraint language over an n-element domain $D = \{d_1, \ldots, d_n\}$ which is finite and a core. For each $i \in \{1, \ldots, n\}$, let

$$N_i(x) = \begin{cases} 0 & \text{if } x = d_i, \\ \infty & \text{otherwise,} \end{cases}$$

and let Γ_c denote the valued constraint language obtained from Γ by adding all cost functions N_i. Observe that $\mathrm{Pol}(\Gamma_c) = \mathrm{IdPol}(\Gamma)$, where by $\mathrm{IdPol}(\Gamma)$ we denote the set of idempotent polymorphisms of the language Γ. Hence, the only unary polymorphism of Γ_c is the identity, which also means that there is only one unary weighted polymorphism of Γ_c – the zero-valued weighted polymorphism.

A valued constraint language Γ is a *rigid core* if there is exactly one unary polymorphism of Γ, which is the identity. This notion corresponds to the classical notion of a rigid core considered in CSP [7]. The following proposition, together with Proposition 8, implies that for each finite language Γ, there is a computationally equivalent language that is a rigid core.

Proposition 10. *Let Γ be a valued constraint language which is finite and a core. The valued constraint language Γ_c is a rigid core. Moreover, Γ is tractable if and only if Γ_c is tractable, and Γ is NP-hard if and only if Γ_c is NP-hard.*

If Γ is a core language then the positive clone of Γ_c contains precisely the idempotent operations from the positive clone of Γ.

4   Weighted Varieties

One of the fundamental results of the algebraic approach to CSP [3,7,17] says that the complexity of a crisp language Γ depends only on the variety generated by the algebra $(D, \mathrm{Pol}(\Gamma))$. We generalize this fact to VCSP.

A k-ary *weighting* ω of an algebra $\mathbf{A}$ is a function that assigns rational weights to all k-ary term operations of $\mathbf{A}$ in such a way, that the sum of all weights is 0, and if $\omega(f) < 0$ then f is a projection. A (*proper*) *superposition* $\omega[g_1, \ldots, g_k]$ of a weighting ω with a list of l-ary term operations $g_1, \ldots, g_k$ from $\mathbf{A}$ is defined the same way as for clones (see Definition 5). An algebra $\mathbf{A}$ together with a set of weightings closed under non-negative scaling, addition of weightings of equal arity and proper superposition with term operations from $\mathbf{A}$ is called a *weighted algebra*.

For a variety $\mathcal{V}$ over a signature Σ and a term t we denote by $[t]_{\mathcal{V}}$ the equivalence class of t under the relation $\approx_{\mathcal{V}}$ such that $t \approx_{\mathcal{V}} s$ if and only if the variety $\mathcal{V}$ satisfies the identity $t \approx s$ (we skip the subscript, writing $[t]$ instead of $[t]_{\mathcal{V}}$, whenever the variety is clear from the context). Observe that if the variety is locally finite then there are finitely many equivalence classes of terms of a fixed arity [8].

Definition 11. *Let $\mathcal{V}$ be a locally finite variety over a signature Σ. A k-ary weighting ω of $\mathcal{V}$ is a function that assigns rational weights to all equivalence classes of k-ary terms over Σ in such a way, that the sum of all weights is 0, and if $\omega([t]) < 0$ then $\mathcal{V}$ satisfies the identity $t(x_1, \ldots, x_k) \approx x_i$ for some $i \in \{1, \ldots, k\}$. The variety $\mathcal{V}$ together with a nonempty set of weightings is called a weighted variety.*

Take any finite algebra $\mathbf{B} \in \mathcal{V}$. A k-ary weighting ω of $\mathcal{V}$ *induces* a weighting $\omega^{\mathbf{B}}$ of $\mathbf{B}$ in a natural way:

$$\omega^{\mathbf{B}}(f) = \sum_{\{[t] \mid t^{\mathbf{B}} = f\}} \omega([t]).$$

If $\omega([t]) < 0$ then the term operation $t^{\mathbf{B}}$ is a projection, and hence the weighting $\omega^{\mathbf{B}}$ is proper. For a weighted variety $\mathcal{V}$, by $\mathbf{B} \in \mathcal{V}$ we mean the algebra $\mathbf{B}$ together with the set of weightings induced by $\mathcal{V}$.

For every weighting ω of a finite weighted algebra $\mathbf{A}$ there is a corresponding weighting ω of the variety $\mathcal{V}(\mathbf{A})$ defined by $\omega([t]) = \omega(t^{\mathbf{A}})$. It follows from Birkhoff's theorem that it is well defined. A weighted variety $\mathcal{V}(\mathbf{A})$ *generated* by a weighted algebra $\mathbf{A}$ is the variety $\mathcal{V}(\mathbf{A})$ together with the set of weightings corresponding to the weightings of $\mathbf{A}$.

We prove that every finite algebra $\mathbf{B} \in \mathcal{V}(\mathbf{A})$ together with the set of weightings induced by $\mathcal{V}(\mathbf{A})$ is a weighted algebra. The only non-trivial part is to show that $\mathbf{B}$ is closed under proper superpositions.

Proposition 12. *For a finite weighted algebra $\mathbf{A}$ over a fixed signature Σ and a finite algebra $\mathbf{B} \in \mathcal{V}(\mathbf{A})$ let $\omega^{\mathbf{B}}$ be a k-ary weighting of $\mathbf{B}$ induced by the weighted variety $\mathcal{V}(\mathbf{A})$. If for some list $f_1^{\mathbf{B}}, \ldots, f_k^{\mathbf{B}}$ of l-ary term operations from $\mathbf{B}$ the composition $\omega^{\mathbf{B}}[f_1^{\mathbf{B}}, \ldots, f_k^{\mathbf{B}}]$ is proper then it is induced by some valid weighting of $\mathcal{V}(\mathbf{A})$.*

For a finite weighted algebra $\mathbf{A}$ let $\mathrm{Imp}(\mathbf{A})$ denote the set of those cost functions on A that are improved by all weightings of $\mathbf{A}$. We prove that for each finite weighted algebra $\mathbf{B} \in \mathcal{V}(\mathbf{A})$ the valued constraint language $\mathrm{Imp}(\mathbf{B})$ is not harder then $\mathrm{Imp}(\mathbf{A})$ i.e.:

Lemma 13. *Let $\mathbf{A}$ be a finite weighted algebra and let*

$$\mathbf{B} \in P_{fin}(\mathbf{A}) \ or \ \mathbf{B} \in S(\mathbf{A}) \ or \ \mathbf{B} \in H(\mathbf{A}) \ or \ finally \ \mathbf{B} \in \mathcal{V}(\mathbf{A})$$

then a VCSP defined by any finite subset of $\mathrm{Imp}(\mathbf{B})$ reduces in polynomial-time to a VCSP for some finite subset of $\mathrm{Imp}(\mathbf{A})$.

Therefore the complexity of Γ depends only on the weighted variety generated by the weighted algebra $(D, \mathrm{wPol}(\Gamma))$.

5 Dichotomy Conjecture

An operation t of arity k is called a *Taylor operation* of an algebra (or a variety), if t is idempotent and for every $j \leq k$ it satisfies an identity of the form

$$t(\Box_1, \Box_2, \ldots, \Box_k) \approx t(\triangle_1, \triangle_2, \ldots, \triangle_k),$$

where all $\Box_i$s and $\triangle_i$s are substituted with either x or y, but $\Box_j$ is x whenever $\triangle_j$ is y. In this section we prove the following theorem:

Theorem 14. *Let Γ be a finite core valued constraint language. If $\mathrm{Pol}^+(\Gamma)$ does not have a Taylor operation, then Γ is NP-hard.*

We conjecture[1] that these are the only cases of finite core languages which give rise to NP-hard VCSPs.

Conjecture. *Let Γ be a finite core valued constraint language. If $\mathrm{Pol}^+(\Gamma)$ does not have a Taylor operation, then Γ is NP-hard. Otherwise it is tractable.*

For crisp languages $\mathrm{Pol}^+(\Gamma) = \mathrm{Pol}(\Gamma)$. Therefore Theorem 14 generalizes the well-known result of Bulatov, Jeavons and Krokhin [3,7] concerning crisp core languages. Similarly the above conjecture is a generalization of The Algebraic Dichotomy Conjecture for CSP. Later on we show that it is supported by all known partial results on the complexity of VCSPs.

To prove Theorem 14 we use Proposition 9 and argue that any relation compatible with $\mathrm{Pol}^+(\Gamma)$ can be found as a set of tuples with minimal costs for some cost function improved by $\mathrm{wPol}(\Gamma)$. It is easy to notice that if $\mathrm{Pol}^+(\Gamma)$ does not have a Taylor operation, then such a relation with NP-complete CSP can be constructed.

As the Taylor operation is difficult to work with, in the reminder of the section we use a characterization of Taylor algebras as the algebras possessing a cyclic term. If Γ is a finite core constraint language then $(D, \mathrm{IdPol}^+(\Gamma))$ is a finite idempotent algebra. It follows that $\mathrm{IdPol}^+(\Gamma)$, and hence also $\mathrm{Pol}^+(\Gamma)$, has a Taylor operation if and only if it has an idempotent cyclic operation [1].

5.1 Two-Element Domain

A complete complexity classification for valued constraint languages over a two-element domain was established in [10]. All tractable languages have been defined via multimorphisms, which are a more restricted form of weighted polymorphisms. A k-ary *multimorphism* of a language Γ, specified as a k-tuple $\langle f_1, \ldots, f_k \rangle$ of k-ary operations on D, is a k-ary weighted polymorphism ω of Γ such that for each $i \in \{1, \ldots, k\}$, we have that $\omega(\pi_i) = -\frac{1}{k}$, and $\omega(f_i) = \frac{l}{k}$, where l is the number of times the operation f_i appears in the tuple.

---

[1] The conjecture was suggested in a conversation by Libor Barto, however it might have appeared independently earlier.

An operation $f \in \mathcal{O}_D^{(3)}$ is called a *majority* operation if for every $x, y \in D$ we have that $f(x, x, y) = f(x, y, x) = f(y, x, x) = x$. Similarly, an operation $f \in \mathcal{O}_D^{(3)}$ is called a *minority* operation if for every $x, y \in D$ it satisfies $f(x, x, y) = f(x, y, x) = f(y, x, x) = y$. We show the following proposition:

Proposition 15. *Let Γ be a finite core valued constraint language on $D = \{0, 1\}$. Then $\mathrm{Pol}^+(\Gamma)$ has an idempotent cyclic operation if and only if Γ admits at least one of the following six multimorphisms:* $\langle \min, \min \rangle$, $\langle \max, \max \rangle$, $\langle \min, \max \rangle$, $\langle \mathrm{Mjrty}, \mathrm{Mjrty}, \mathrm{Mjrty} \rangle$, $\langle \mathrm{Mnrty}, \mathrm{Mnrty}, \mathrm{Mnrty} \rangle$, $\langle \mathrm{Mjrty}, \mathrm{Mjrty}, \mathrm{Mnrty} \rangle$.

The proposition fully agrees with the classification of VCSP languages on two-element domain in [10].

5.2   Finite-Valued Languages

Theorem 16 (Thapper and Živný [20]). *Let Γ be a finite-valued constraint language which is a core. If Γ admits an idempotent cyclic weighted polymorphism of some arity $m > 1$, then Γ is tractable. Otherwise it is NP-hard.*

To show that our conjecture agrees with the above complexity classification we prove the following result (which holds for general-valued languages):

Proposition 17. *Let Γ be a core valued constraint language. Then Γ admits an idempotent cyclic weighted polymorphism of some arity $m > 1$ if and only if $\mathrm{Pol}^+(\Gamma)$ contains an idempotent cyclic operation of the same arity.*

5.3   Conservative Languages

A valued constraint language Γ over a domain D is called *conservative* if it contains all $\{0, 1\}$-valued unary cost functions on D. An operation $f \in \mathcal{O}_D^{(k)}$ is *conservative* if for every $x_1, \ldots, x_k \in D$ we have that $f(x_1, \ldots, x_k) \in \{x_1, \ldots, x_k\}$, and a weighted polymorphism is *conservative* if its support contains conservative operations only.

A *Symmetric Tournament Pair (STP)* is a conservative binary multimorphism $\langle \sqcap, \sqcup \rangle$, where both operations are commutative, i.e., $\sqcap(x, y) = \sqcap(y, x)$ and $\sqcap(x, y) = \sqcap(y, x)$ for all $x, y \in D$, and moreover $\sqcap(x, y) \neq \sqcup(x, y)$ for all $x \neq y$. A *MJN* is a ternary conservative multimorphism $\langle \mathrm{Mj}_1, \mathrm{Mj}_2, \mathrm{Mn}_3 \rangle$, such that $\mathrm{Mj}_1, \mathrm{Mj}_2$ are majority operations, and Mn_3 is a minority operation.

Theorem 18 (Kolmogorov and Živný [16]). *Let Γ be a conservative constraint language over a domain D. If Γ admits a conservative binary multimorphism $\langle \sqcap, \sqcup \rangle$ and a conservative ternary multimorphism $\langle \mathrm{Mj}_1, \mathrm{Mj}_2, \mathrm{Mn}_3 \rangle$, and there is a family M of two-element subsets of D, such that:*

- *for every $\{x, y\} \in M$, $\langle \sqcap, \sqcup \rangle$ restricted to $\{x, y\}$ is an STP,*
- *for every $\{x, y\} \notin M$, $\langle \mathrm{Mj}_1, \mathrm{Mj}_2, \mathrm{Mn}_3 \rangle$ restricted to $\{x, y\}$ is an MJN,*

then Γ is tractable. Otherwise it is NP-hard.

In this case, as well as in the others, it can be shown that the existence of an idempotent cyclic polymorphism in $\text{Pol}^+(\Gamma)$ is equivalent (for conservative Γ) to the tractability conditions from the theorem above.

Acknowledgments. We are grateful to Libor Barto and Jakub Bulin for inspiring discussions on VCSP.

References

1. Barto, L., Kozik, M.: Absorbing subalgebras, cyclic terms, and the constraint satisfaction problem. Logical Methods in Computer Science **8**(1) (2012)
2. Birkhoff, G.: On the structure of abstract algebras. Proceedings of the Cambridge Philosophical Society **31**, 433–454 (1935)
3. Bulatov, A., Jeavons, P., Krokhin, A.: Classifying the complexity of constraints using finite algebras. SIAM Journal on Computing **34**, 720–742 (2005)
4. Bulatov, A.A.: Tractable conservative constraint satisfaction problems. In: Proc. of the 18th Symposium on Logic in Computer Science, p. 321 (2003)
5. Bulatov, A.A.: A dichotomy theorem for constraint satisfaction problems on a 3-element set. J. ACM **53**(1), 66–120 (2006)
6. Bulatov, A.A.: Complexity of conservative constraint satisfaction problems. ACM Trans. Comput. Logic **12**(4), 24:1–24:66 (2011)
7. Bulatov, A.A., Krokhin, A.A., Jeavons, P.G.: Constraint satisfaction problems and finite algebras. In: Welzl, E., Montanari, U., Rolim, J.D.P. (eds.) ICALP 2000. LNCS, vol. 1853, pp. 272–282. Springer, Heidelberg (2000)
8. Burris, S., Sankappanavar, H.P.: A course in universal algebra. Graduate texts in mathematics. Springer (1981)
9. Cohen, D.A., Cooper, M.C., Creed, P., Jeavons, P.G., Živný, S.: An algebraic theory of complexity for discrete optimization. SIAM J. Comput. **42**(5), 1915–1939 (2013)
10. Cohen, D.A., Cooper, M.C., Jeavons, P.G., Krokhin, A.A.: The complexity of soft constraint satisfaction. Artif. Intell. **170**(11), 983–1016 (2006)
11. Feder, T., Vardi, M.Y.: The computational structure of monotone monadic SNP and constraint satisfaction: A study through datalog and group theory. SIAM J. Comput. **28**(1), 57–104 (1999)
12. Hell, P., Nešetřil, J.: On the complexity of h-coloring. Journal of Combinatorial Theory, Series B **48**(1), 92–110 (1990)
13. Huber, A., Krokhin, A., Powell, R.: Skew bisubmodularity and valued CSPs. In: Proc. SODA 2013, pp. 1296–1305. SIAM (2013)
14. Jeavons, P., Cohen, D., Gyssens, M.: Closure properties of constraints. J. ACM **44**(4), 527–548 (1997)
15. Jeavons, P., Krokhin, A., Živný, S.: The complexity of valued constraint satisfaction. Bulletin of the EATCS **113**, 21–55 (2014)
16. Kolmogorov, V., Živný, S.: The complexity of conservative valued CSPs. J. ACM **60**(2), 10:1–10:38 (2013)
17. Larose, B., Tesson, P.: Universal algebra and hardness results for constraint satisfaction problems. Theor. Comput. Sci. **410**, 1629–1647 (2009)

18. Schaefer, T.J.: The complexity of satisfiability problems. In: Proc. of the 10th ACM Symp. on Theory of Computing, STOC 1978, pp. 216–226 (1978)
19. Tarski, A.: A remark on functionally free algebras. Annals of Mathematics **47**(1), 163–166 (1946)
20. Thapper, J., Živný, S.: The complexity of finite-valued CSPs. In: Proc. of the 45th ACM Symp. on Theory of Computing, STOC 2013, pp. 695–704 (2013)

Towards Understanding the Smoothed Approximation Ratio of the 2-Opt Heuristic

Marvin Künnemann[1]([⊠]) and Bodo Manthey[2]

[1] Saarbrücken Graduate School of Computer Science,
Max Planck Institute for Informatics, Saarbrücken, Germany
marvin@mpi-inf.mpg.de
[2] University of Twente, Enschede, The Netherlands
b.manthey@utwente.nl

Abstract. The 2-Opt heuristic is a very simple, easy-to-implement local search heuristic for the traveling salesman problem. While it usually provides good approximations to the optimal tour in experiments, its worst-case performance is poor.

In an attempt to explain the approximation performance of 2-Opt, we analyze the smoothed approximation ratio of 2-Opt. We obtain a bound of $O(\log(1/\sigma))$ for the smoothed approximation ratio of 2-Opt. As a lower bound, we prove that the worst-case lower bound of $\Omega(\frac{\log n}{\log \log n})$ for the approximation ratio holds for $\sigma = O(1/\sqrt{n})$.

Our main technical novelty is that, different from existing smoothed analyses, we do not separately analyze objective values of the global and the local optimum on all inputs, but simultaneously bound them on the same input.

1 2-Opt and Smoothed Analysis

The traveling salesman problem (TSP) is one of the best-studied combinatorial optimization problems. Euclidean TSP is the following variant: given points $X \subseteq [0,1]^d$, find the shortest Hamiltonian cycle that visits all points in X (also called a *tour*). Even this restricted variant is NP-hard for $d \geq 2$ [17].

While Euclidean TSP admits a polynomial-time approximation scheme [1, 16], heuristics that are simpler and easier to implement are often used in practice. A very simple and popular heuristic for finding near-optimal tours quickly is the 2-Opt heuristic: starting from an initial tour, we iteratively replace two edges by two other edges to obtain a shorter tour until we have found a local optimum. Experiments indicate that 2-Opt converges to near-optimal solutions quickly and produces solutions that are within a few percent of the optimal solution [10,11]. In contrast to its success on practical instances, 2-Opt performs poorly in the worst case: the worst-case running-time is exponential even for $d = 2$ [8] and its worst-case approximation ratio of $O(\log n)$ has an almost matching lower bound of $\Omega(\log n / \log \log n)$ for Euclidean instances [6].

In order to explain the performance of algorithms whose worst-case performance guarantee does not reflect the observed performance, smoothed analysis

© Springer-Verlag Berlin Heidelberg 2015
M.M. Halldórsson et al. (Eds.): ICALP 2015, Part I, LNCS 9134, pp. 859–871, 2015.
DOI: 10.1007/978-3-662-47672-7_70

has been introduced [19], which is a hybrid of worst-case analysis (which is often too pessimistic) and average-case analysis (which is often dominated by completely random instances that have special properties not shared by typical instances). In smoothed analysis, an adversary specifies an instance, and then this instance is slightly randomly perturbed. The smoothed performance is the expected performance, where the expected value is taken over the random perturbation. The motivating assumption of smoothed analysis is that practical instances are often subjected to a small amount of random noise that can, e.g., come from measurement errors or numerical imprecision. Smoothed analysis often allows more realistic conclusions about the performance of an algorithm than mere worst-case or average-case analysis.

Smoothed analysis has been applied successfully to explain the running time of the 2-Opt heuristic [8,15] as well as other local search algorithms [2,3,14]. We refer to two surveys for an overview of smoothed analysis [13,20].

Much less is known about the smoothed approximation performance of algorithms. Karger and Onak have shown that multi-dimensional bin packing can be approximated arbitrarily well for smoothed instances [12] and there are frameworks to approximate Euclidean optimization problems such as TSP for smoothed instances [4,7]. However, these approaches mostly consider algorithms tailored to solving smoothed instances.

With respect to concrete algorithms, we are only aware of analyses of the jump and lex-jump heuristics for scheduling [5,9] and an upper bound of $O(\phi^{1/d})$ for the smoothed approximation ratio of 2-Opt in the so-called one-step model [8]. Here, ϕ is an upper bound on the density functions according to which the points are drawn. Translated to Gaussian perturbation, we would obtain an upper bound of $O(1/\sigma)$ if we truncate the Gaussian distribution such that all points lie in a hypercube of constant sidelength.

In order to explain the practical approximation performance of 2-Opt, we provide an improved smoothed analysis of its approximation ratio. More precisely, we provide bounds on the quality of the worst local optimum, when the n data points from $[0,1]^d$ are perturbed by Gaussian distributions of standard deviation σ. Our bound of $O(\log(1/\sigma))$ improves significantly upon the direct translation of the bound of Englert et al. [8] to Gaussian perturbations (see Section 3 for how to translate the bound to Gaussian perturbations). It smoothly interpolates between the average-case constant approximation ratio and the worst-case bound of $O(\log n)$.

In order to obtain our improved bound for the smoothed approximation ratio, we take into account the origins of the points, i.e., their unperturbed positions. Although this information is not available to the algorithm, it can be exploited in the analysis. The smoothed analyses of approximation ratios so far [4,5,7–9,12] essentially ignored this information. While this simplifies the analysis, being oblivious to the unperturbed positions seems to be too pessimistic. In fact, we see that the bound of Englert et al. [8] cannot be improved beyond $O(1/\sigma)$ by ignoring the positions of the points (Section 3). The reason for this limitation is that the lower bound for the global optimum is obtained if all points have the same origin, which corresponds to an average-case rather than a smoothed analysis. On the other hand, the upper bound for the local optimum has to hold for all choices of the unperturbed points, most of which yield higher costs for the global optimum than the average-case analysis. Taking this into account carefully yields our bound of $O(\log(1/\sigma))$ (Section 4).

To complement our upper bound, we show that the lower bound by Chandra et al. [6] remains true for $\sigma = O(1/\sqrt{n})$ (Section 5). We conclude our paper by discussing our results and pointing out open questions (Section 6). Due to lack of space some proofs had to be omitted, which we defer to a full version of this article.

2   Preliminaries

Throughout the paper, we consider input in the Euclidean space $[0,1]^d$ and assume the dimension d to be a fixed constant. Given a sequence of points $X = (X_1, \ldots, X_n)$ in $\mathbb{R}^d$, we call a collection $T \subseteq [n] \times [n]$ of edges a *tour*, if T is connected and every $i \in [n] = \{1, \ldots, n\}$ has in- and outdegree exactly one in T. (Note that we consider directed tours, which is useful in the analysis, but our distances are always symmetric.) Given any collection of edges S, its length is denoted by $L(S) = \sum_{(u,v) \in S} d(u, v)$, where $d(u, v)$ denotes the Euclidean distance between points X_u and X_v. We call a tour T *2-optimal*, if $d(u, v) + d(w, z) \leq d(u, w) + d(v, z)$ for all edge pairs $(u, v), (w, z) \in T$. Equivalently, it is not possible to obtain a shorter tour by replacing (u, v) and (w, z) in a 2-optimal tour T by two new edges. The 2-Opt heuristic replaces a pair of edges (u, v) and (w, z) by (u, w) and (v, z) if this decreases the tour length while this is possible. Thus, it terminates with a 2-optimal tour.

We call a collection $T \subseteq [n]^2$ a *partial 2-optimal tour* if T is a subset of a tour and $d(u, v) + d(w, z) \leq d(u, w) + d(v, z)$ for all edges $(u, v), (w, z) \in T$. Our main interests are the traveling salesman functional $\mathsf{TSP}(X) := \min_{\text{tour } T} L(T)$ and the following functional mapping the point set X to the length of the longest 2-optimal tour through X: $\mathsf{2OPT}(X) := \max_{\text{2-optimal tour } T} L(T)$.

We note that the results in Section 3 hold for metrics induced by arbitrary norms in $\mathbb{R}^d$ (Lemma 2 and 3) or typical ℓ_p norms (Lemma 4 and 5), not only for the Euclidean metric. We conjecture that also the upper bound in Section 4 holds for more general metrics, while the lower bound in Section 5 is probably specific for the Euclidean metric. Still, we think that the construction can be adapted to work for most natural metrics.

Perturbation models. In the Gaussian perturbation model (also called *two-step model*) for smoothed analysis, an adversary specifies points $x_1, \ldots, x_n$ in $[0,1]^d$ that serve as unperturbed *origins*. Each such point x_i is perturbed independently by adding a normally distributed random variable of mean 0 and standard deviation σ independently to each coordinate. Equivalently, we draw n random noise vectors $Z_i \sim \mathcal{N}(0, \sigma^2)$, where by abuse of notation $\mathcal{N}(0, \sigma^2)$ refers to the multivariate normal distribution with covariance matrix $\mathrm{diag}(\sigma^2)$, to obtain the *perturbed input* $X_1 = x_1 + Z_1, \ldots, X_n = x_n + Z_n$. For compactness, we denote the set of unperturbed points by $\overline{X} = \{x_1, \ldots, x_n\}$ and the set of perturbed points by $X = \{X_1, \ldots, X_n\}$. We write $X \leftarrow \mathrm{pert}_\sigma(\overline{X})$ to make explicit from which point set $\overline{X}$ the points in X are obtained.

Note that we may assume $\sigma \leq 1$ without loss of generality. If $\sigma > 1$, we can rescale the instance to be contained in $[0, 1/\sigma]^d$ and perturb the points by Gaussians with standard deviation 1 instead, which gives an equivalent instance. Thus, every upper bound for $\sigma = 1$ carries over to larger values of σ.

The ϕ-bounded perturbation model (also called *one-step model*) lets the adversary directly specify (not necessarily identical) distributions by choosing probability density functions $f_1, \ldots, f_n : [0, 1]^d \to [0, \phi]$. The perturbed input is then generated by independently sampling $X_1 \sim f_1, \ldots, X_n \sim f_n$. Note that the resulting input is always contained in $[0, 1]^d$ and with higher ϕ, the adversary can concentrate points to smaller regions of the input space. Roughly speaking, when translating Gaussian perturbations to the one-step model, ϕ is proportional to σ^{-d} for fixed d.

The following technical lemma provides a convenient way to bound the deviation of a perturbed point from its mean in the two-step model.

Lemma 1 (Chi-square bound [19, Cor. 2.19]). *Let x be a Gaussian random vector in $\mathbb{R}^d$ of standard deviation σ centered at the origin. Then, for $t \geq 3$, we have* $\mathrm{Pr}\left[\|x\| \geq \sigma 3\sqrt{d \ln t}\right] \leq t^{-2.9d}$.

3   Length of 2-optimal Tours Under Perturbations

In this section, we provide an upper bound for the length of any 2-optimal tour and a lower bound for the length of any global optimum. These two results yield an upper bound of $O(1/\sigma)$ for the approximation ratio.

Chandra et al. [6] proved a bound on the worst-case length of 2-optimal tours that, in fact, already holds for the more general notion of *partial* 2-optimal tours. For an intuition why this is true, let us point out that their proof strategy is to argue that not too many long arcs in a tour may have similar directions due to the 2-optimality of the edges, while short edges do not contribute much to the length. The claim then follows from a packing argument. It can be verified that it is never required that the collection of edges is closed or connected.

Lemma 2. *Let $d \geq 2$. There exists a constant c_d such that for every sequence X of n points in $[0, 1]^d$, any partial 2-optimal tour has length less than $c_d \cdot n^{1-1/d}$.*

While this bound directly applies to any perturbed instance under the one-step model, Gaussian perturbations fail to satisfy the premise of bounded support in $[0, 1]^d$. However, Gaussian tails are sufficiently light to enable us to translate the result to the two-step model by carefully taking care of outliers.

Lemma 3. *Let $d \geq 2$. There exists a constant b_d such that for any $\sigma \leq 1$ the following statement holds. The probability that any partial 2-optimal tour on X*

has length greater than $b_d \cdot n^{1-1/d}$, i.e., $2\mathsf{OPT}(X) \geq b_d \cdot n^{1-1/d}$, is bounded by $\exp(-\Omega(\sqrt{n}))$. *Furthermore,*

$$\mathrm{E}_{X \leftarrow \mathrm{pert}_\sigma(\overline{X})}\left[2\mathsf{OPT}(X)\right] \leq b_d \cdot n^{1-1/d}.$$

We complement the bound above by a lower bound on tour lengths of perturbed inputs, making use of the following result by Englert et al. [8] for the one-step model.

Lemma 4. *Let $X_1, \ldots, X_n$ be a ϕ-perturbed instance. Then with probability $1 - \exp(-\Omega(n))$, any tour on $X_1, \ldots, X_n$ has length at least $\Omega(n^{1-1/d}/\sqrt[d]{\phi})$.*

It also follows from their results that this bound translates to the two-step model consistently with the intuitive correspondence of $\phi \sim \sigma^{-d}$ between the one-step and the two-step model.

Lemma 5. *Let $X_1, \ldots, X_n$ be an instance of points in the unit cube perturbed by Gaussians of standard deviation $\sigma \leq 1$. Then with probability $1 - \exp(-\Omega(n))$ any tour on $X_1, \ldots, X_n$ has length at least $\Omega(\sigma n^{1-1/d})$.*

Note that Lemmas 3 and 5 almost immediately yield the following bound on the approximation performance for the two-step model.[1]

Observation 1. *Let $X_1, \ldots, X_n$ be an instance of points in the unit cube perturbed by Gaussians of standard deviation $\sigma \leq 1$. Then the approximation performance of 2-Opt is bounded by $O(1/\sigma)$ in expectation and with probability $1 - \exp(-\Omega(\sqrt{n}))$.*

We remark that this bound is best possible for an analysis of perturbed instances that separately bounds the lengths of any 2-optimal tour from above and gives a lower bound on any optimal tour.

4   Upper Bound on the Approximation Performance

In this section, we establish an upper bound on the approximation performance of 2-Opt under Gaussian perturbations. We achieve a bound of $O(\log 1/\sigma)$. Due to the lower bound presented in Section 5, we cannot expect an approximation ratio of $o(\log(1/\sigma)/\log\log(1/\sigma))$. Thus, our bound is almost tight.

As noted in the previous section, to beat $O(1/\sigma)$ it is essential to exploit the structure of the unperturbed input. This will be achieved by classifying edges of a tour into *long* and *short* edges and bounding the length of long edges by a (worst-case) global argument and short edges locally against the partial optimal tour on subinstances (by a reduction to an (almost-)average case). The local arguments for short edges will exploit how many unperturbed origins lie in the vicinity of a given region.

The global argument bounding long edges follows from the worst-case $O(\log n)$ bound on the worst-case approximation performance [6] that we rephrase here for our purposes.

---

[1] To show the expected approximation ratio, we additionally make use of Lemma 6.

Lemma 6. *Let T be a 2-optimal tour and* OPT *denote the length of the optimal traveling salesman tour T_{OPT}. Let T_i contain the set of all edges in T whose length is in $[\mathrm{OPT}/2^i, \mathrm{OPT}/2^{i-1}]$. Then $L(T_i) = O(\mathrm{OPT})$. In particular, it follows that $L(T) = O(\log n) \cdot \mathrm{OPT}$.*

In the proof of our bound of $O(\log 1/\sigma)$, the above lemma accounts for all edges of length $[\Omega(\sigma), O(1)]$. A central idea to bound all shorter edges is to apply the one-step model result to small parts of the input space. In particular, we will condition sets of points to be perturbed into cubes of side length σ. The following technical lemma helps to capture what values of ϕ suffice to express the conditional density function of these points depending on the distance of their unperturbed origins to the cube. This allows for appealing to the one-step model result of Lemma 4.

Lemma 7. *Let $c \in [0, \sigma]^d$ and $k = (k_1, \ldots, k_d) \in \mathbb{N}_0^d$. Let Y be the random variable $X \sim \mathcal{N}(c, \sigma^2)$ conditioned on $X \in Q := [k_1\sigma, (k_1+1)\sigma] \times \cdots \times [k_d\sigma, (k_d+1)\sigma]$ and f_Y be the corresponding probability density function. Then f_Y is bounded from above by $\exp(\|k\|_1 + (3/2)d)\sigma^{-d}$.*

The main result of this section is the following theorem.

Theorem 2. *Let $X = (X_1, \ldots, X_n)$ be an instance of points in $[0, 1]^d$ perturbed by Gaussians of standard deviation $\sigma \leq 1$. With probability $1 - \exp(-\Omega(n^{1/2-\varepsilon}))$ for any constant $\varepsilon > 0$, we have $2\mathrm{OPT}(X) \leq O(\log(1/\sigma)) \cdot \mathrm{TSP}(X)$. Furthermore, $\mathrm{E}\left[\frac{2\mathrm{OPT}(X)}{\mathrm{TSP}(X)}\right] = O(\log(1/\sigma))$.*

Since the approximation performance of 2-Opt is bounded by $O(\log n)$ in the worst-case, we may assume that $1/\sigma = O(n^\varepsilon)$ for all $\varepsilon > 0$, since otherwise our smoothed result is superseded by Lemma 6. In what follows, let T_{OPT} and T be any optimal and 2-optimal, respectively, traveling salesman tour on $X_1, \ldots, X_n$.

4.1 Outliers and Long Edges

We will first show that the contribution of almost all points outside $[0, 1]^d$ is bounded by $O(\sigma n^{1-1/d})$ with high probability and in expectation, similar to Lemma 3. For this, we subdivide C into growing cubes $A_i := [-a_i, 1 + a_i]^d$. Here, we set $a_i := 3\sigma\sqrt{di\ln(3/\sigma)}$ for $i \geq 1$ and $A_0 = [0, 1]^d$. Let n_i be the number of points not contained in A_{i-1}. For every point X_j, Lemma 1 with $t := (3/\sigma)^i$ bounds $\Pr[X_j \notin A_i] \leq (\sigma/3)^{2.9d(i-1)}$ (note that we have chosen the a_i such that $t \geq 3$). Thus, $\mathrm{E}[n_i] \leq n(\sigma/3)^{2.9d(i-1)}$. For any tour T, we define E_i as the set of edges of T contained in A_i with at least one endpoint in $A_i \setminus A_{i-1}$. We first bound the contribution of the E_i with $i \geq 2$.

Lemma 8. *With probability $1 - \exp(-\Omega(n^{1/2-\varepsilon}))$ for any constant $\varepsilon > 0$, we have $\sum_{i=2}^{\infty} L(E_i) = O(\sigma n^{1-1/d})$. Additionally, $\mathrm{E}[\sum_{i=2}^{\infty} L(E_i)] = O(\sigma n^{1-1/d})$.*

In the remainder of the proof, we bound the total length of edges inside A_1. Define $C := A_1$ and note that all edges in C have bounded length $\sqrt{d}(1 + a_1) = O(1)$. Recall that for any 2-optimal tour T, T_i contains the set of all edges in T whose length is in $[\text{OPT}/2^i, \text{OPT}/2^{i-1}]$. Let k_1 be such that $\sqrt{d}(1 + a_1) \in [\text{OPT}/2^{k_1}, \text{OPT}/2^{k_1-1}]$. Then $L(T_k) = 0$ for all $k < k_1$, since no longer edges exist. Let k_2 be such that $\sigma \in [\text{OPT}/2^{k_2}, \text{OPT}/2^{k_2-1}]$. Then $\sum_{k=k_1}^{k_2} L(T_k) = O((k_2 - k_1) \cdot \text{OPT}) = O(\log(1/\sigma)\text{OPT})$ by Lemma 6. This argument bounds the contribution of *long edges*, i.e., edges longer than σ, in the worst case, after observing the perturbation of the input points.

4.2   Short Edges

To account for the length of the remaining edges, we take a different route: Call an edge that is shorter than σ a *short edge* and partition the bounding box C into a grid of $(\sigma \times \cdots \times \sigma)$-cubes $C_1, \ldots, C_M$ with $M = \Theta((\sigma/(1+a_1))^{-d}) = \Theta(\sigma^{-d})$, which we call *cells*. All edges in T_k for $k \geq k_2$, i.e., short edges, are completely contained in a single cell or run from some cell C_i to one of its $3^d - 1$ neighboring cells. For a given tour T, let $E_{C_i}(T)$ denote the short edges of T for which at least one of the endpoints lies in C_i.

We aim to relate the length of the edges $E_{C_i}(T)$ for any 2-optimal tour T to the length of the edges $E_{C_i}(T_{\text{OPT}})$ of the optimal tour T_{OPT}. This local approach is justified by the following property.

Lemma 9. *For any tour T, the contribution $L(E_{C_i}(T))$ of cell C_i is lower bounded by $\text{TSP}(X \cap C_i) - O(\sigma |X \cap C_i|^{\frac{d-2}{d-1}})$.*

Intuitively, a cell C_i is of one of two kinds: either few points are expected to be perturbed into it and hence it cannot contribute much to the length of any 2-optimal tour (a *sparse cell*), or many unperturbed origins are close to the cell (a *heavy cell*). In the latter case, either the conditional densities of points perturbed into C_i are small, hence any optimal tour inside C_i has a large value by Lemma 4, or we find another cell close to C_i that has a very large contribution to the length of any tour.

To formalize this intuition, fix a cell C_i and let n_i be the expected number of points X_j with $X_j \in C_i$. Assume for convenience that a_1/σ and $(1 + a_1)/\sigma$ are integer. We describe the position of a cube C_i canonically by indices $\text{pos}(C_i) \in \{-\frac{a_i}{\sigma}, \ldots, \frac{1+a_i}{\sigma}\}^d$. For two cubes C_i and C_j, we define their distance as $\text{dist}(C_i, C_j) = \|\text{pos}(C_i) - \text{pos}(C_j)\|_1$. For $k \geq 0$, let D_k denote all cells of distance k to C_i and let $n(D_k)$ denote the cardinality of unperturbed origins located in a cell in D_k. We call a perturbed point $X_\ell \in C_i$ with unperturbed origin $x_\ell \in C_j$, for some $C_j \in D_k$, a *k-successful point*. Let S_k denote the set of all k-successful points. Then $n_i = \sum_{k=0}^{\infty} \text{E}[|S_k|]$.

Lemma 10. *Let $K \geq 0$ and define $S_{\leq K} := S_0 \cup \cdots \cup S_K$ as the set of k-successful points for $k \leq K$. Let $\mu := \text{E}[|S_{\leq K}|]$. If $K = o(\log \mu)$, then with probability $1 - \exp(\mu)$, we have*

$$L(E_{C_i}(T_{\text{OPT}})) \geq \frac{\sigma \mu^{1-1/d}}{\exp(O(K+1))}.$$

Proof (Sketch). The claim follows from Lemma 9 and by regarding $S_{\leq K}$ as a ϕ-perturbed instance. For this, Lemma 7 bounds the maximum density of the distributions and Lemma 4 bounds the optimal tour length from below. $\qquad\square$

Lemma 11. *Let* $\alpha := M^{\frac{d}{d-1}}$, $k_1 := \gamma \log \log(1/\sigma)$ *and* $k_2 := (1/\gamma')\sqrt{\log 1/\sigma}$ *for sufficiently small constants* γ, γ'. *Then we can classify each cell* C_i *with* $n_i \geq \frac{n}{\alpha}$ *into one of the following two types.*

(T1) With probability $1 - \exp(-\Omega(n^{1/2-\varepsilon}))$ *for any constant* $\varepsilon > 0$, *we have*

$$L(E_{C_i}(T)) = O(\log 1/\sigma)L(E_{C_i}(T_{\text{OPT}})).$$

(T2) There is some $C_j \in D_{k_1} \cup \cdots \cup D_{k_2}$ *such that for any* $f(1/\sigma) = \text{polylog}(1/\sigma)$, *we have*

$$L(E_{C_i}(T)) = \frac{L(E_{C_j}(T_{\text{OPT}}))}{f(1/\sigma)},$$

with probability $1 - \exp(-\Omega(n^{1/2-\varepsilon}))$ *for any constant* $\varepsilon > 0$.

Proof (Sketch). By Lemma 2, we can bound $L(E_{C_i}(T)) = O(\sigma n_i^{1-1/d})$. If we have $\mathbb{E}[|S_{\leq k_1}|] = \Omega(n_i)$, then Lemma 10 already proves C_i to have type T1. Otherwise, by tail bounds for the Gaussian distribution, we argue that some cell C_j in a cell of distance at most k_2 contains at least $n_i \exp(\Omega((\log \log n)^2))$ unperturbed origins. These are sufficiently many to let C_j contribute $f(1/\sigma)\sigma n_i^{1-1/d}$, for any $f(1/\sigma) = \text{polylog}(1/\sigma)$, to the optimal tour length. $\qquad\square$

4.3   The Total Length of 2-optimal Tours

To bound the total length of short edges, consider first sparse cells C_i, i.e., $n_i \leq n/\alpha$. For each such cell, Chernoff bounds yield that with probability $1 - \exp(-\Omega(n/\alpha))$, at most $2n/\alpha$ points are contained in C_i, since each point is perturbed independently. By union bound, no sparse cell contains more than $2n/\alpha$ points with probability at least $1 - M \exp(-\Omega(n/\alpha))$. In this event, Lemma 2 allows for bounding the contribution of sparse cells by

$$\sum_{i:n_i \leq n/\alpha} L(E_{C_i}(T)) \leq M(3\sigma)c_d \left(\frac{6n}{\alpha}\right)^{1-\frac{1}{d}} = O\left(\frac{M\sigma n^{1-\frac{1}{d}}}{\alpha^{1-\frac{1}{d}}}\right) = O(\sigma n^{1-\frac{1}{d}}). \quad (1)$$

For bounding the length in the remaining cells, the heavy cells, let $\mathcal{T}_1 := \{i \mid C_i \text{ has type T1}\}$ and $\mathcal{T}_2 := \{i \mid C_i \text{ has type T2}\}$. We observe that with probability at least $1 - M \exp(-\Omega(n^{1-\varepsilon})) = 1 - \exp(-\Omega(n^{1-\varepsilon}))$, all type-T1 cells C_i satisfy $L(E_{C_i}(T)) = O(\log 1/\sigma)L(E_{C_i}(T_{\text{OPT}}))$. Thus,

$$\sum_{i \in \mathcal{T}_1} L(E_{C_i}(T)) \leq \sum_{i \in \mathcal{T}_1} O(\log 1/\sigma)L(E_{C_i}(T_{\text{OPT}})) \leq O(\log 1/\sigma)\text{OPT}, \quad (2)$$

where the last inequality follows from $\sum_{i=1}^{M} L_{C_i}(T_{\text{OPT}}) \leq 2 \cdot \text{OPT}$, which holds since every edge in OPT (inside C) is counted twice on the left-hand side.

Let $A : \mathcal{T}_2 \to \{1, \ldots, M\}$ be any function that assigns to each type-T2 cell C_i a corresponding cell $C_{A(i)} \in D_{k_1} \cup \cdots \cup D_{k_2}$ satisfying the condition in (T2). We say that C_i *charges* $C_{A(i)}$. We can choose any $f(1/\sigma) = \text{polylog}(1/\sigma)$ and have with probability at least $1 - M \exp(-\Omega(n^{1-\varepsilon})) = 1 - \exp(-\Omega(n^{1-\varepsilon}))$ that $L(E_{C_i}(T)) = \frac{L(E_{C_{A(i)}}(T_{\text{OPT}}))}{f(1/\sigma)}$ for all $i \in \mathcal{T}_2$. Assume that this event occurs. Since every cell C_i can only be charged by cells in distance $k_1 \leq k \leq k_2$, each cell can only be charged $\sum_{k=k_1}^{k_2} |D_k| = O(k_2^d)$ times. Hence,

$$\sum_{i \in \mathcal{T}_2} L(E_{C_{A(i)}}(T_{\text{OPT}})) \leq O(k_2^d) \sum_{i=1}^{M} L(E_{C_i}(T_{\text{OPT}})) = O(k_2^d)\text{OPT}.$$

Since $k_2^d = \text{polylog}(1/\sigma)$, choosing $f(1/\sigma) = \text{polylog}(1/\sigma)$ sufficiently large yields

$$\sum_{i \in \mathcal{T}_2} L(E_{C_i}(T)) \leq \sum_{i \in \mathcal{T}_2} \frac{L(E_{C_{A(i)}}(T_{\text{OPT}}))}{f(1/\sigma)} \leq \frac{O(k_2^d)\text{OPT}}{f(1/\sigma)} = O(\text{OPT}). \qquad (3)$$

Proof (of Theorem 2). By a union bound, we can bound by $1-\exp(-\Omega(n^{1/2-\varepsilon}))$, for any constant $\varepsilon > 0$, the probability that (i) OPT $= \Omega(\sigma n^{1-1/d})$ (by Lemma 5), (ii) all edges outside C contribute $O(\sigma n^{1-1/d}) = O(\text{OPT})$ (by Lemma 8), (iii) all sparse cells contribute $O(\sigma n^{1-1/d}) = O(\text{OPT})$ (by (1)), (iv) the type-T1 cells C_i induce a cost of $O(\log 1/\sigma)\text{OPT}$ (by (2)), and (v) the type-2 cells induce a cost of $O(\text{OPT})$ (by (3)). Since the remaining edges are long edges and contribute only $O(\log(1/\sigma) \cdot \text{OPT})$, we obtain that every 2-optimal tour has a length of at most $O(\log 1/\sigma)\text{OPT}$ with probability $1 - \exp(-\Omega(n^{1/2-\varepsilon}))$.

Since a 2-optimal tour always constitutes a $O(\log n)$-approximation to the optimal tour length by Lemma 6, we also obtain that the expected cost of the worst 2-optimal tour is bounded by

$$O(\log 1/\sigma) \cdot \text{OPT} + \exp(-\Omega(n^{1/2-\varepsilon})) \cdot O(\log n) \cdot \text{OPT} = O(\log 1/\sigma) \cdot \text{OPT}.$$

$\square$

5   Lower Bound on the Approximation Ratio

We complement our upper bound on the approximation performance by the following lower bound: for $\sigma = O(1/\sqrt{n})$, the worst-case lower bound is robust against perturbations. For this, we face the technical difficulty that in general, a single outlier might destroy the 2-optimality of a desired long tour, potentially cascading into a series of 2-Opt iterations that result in a substantially different or even optimal tour.

Theorem 3. *Let $\sigma = O(1/\sqrt{n})$. For infinitely many n, there is an instance X of points in $\mathbb{R}^2$ perturbed by normally distributed noise of standard deviation σ such*

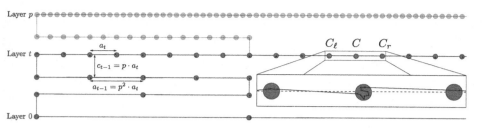

Fig. 1. Parts V_1 and V_3 of the lower bound instance. Each point is contained in a corresponding small container (depicted as brown circle) with high probability. The black lines indicate the constructed 2-optimal tour, which on V_2 runs analogously.

that with probability $1 - O(n^{-s})$ *for any constant* $s > 0$, *we have* $2\mathsf{OPT}(X) = \Omega(\log n / \log \log n) \cdot \mathsf{TSP}(X)$. *This also yields*

$$\mathrm{E}\left[\frac{2\mathsf{OPT}(X)}{\mathsf{TSP}(X)}\right] = \Omega\left(\frac{\log n}{\log \log n}\right).$$

We remark that our result transfers naturally to the one-step model with $\phi = \Omega(n)$ and interestingly, holds *with probability 1* over the random perturbations. Furthermore, even when we initialize the tour using the *nearest neighbor heuristic*, 2-Opt might, with probability $O(1)$, return a 2-optimal tour of length $\Omega(\log n / \log \log n) \cdot \mathsf{TSP}(X)$ on perturbed inputs. For space reasons, the necessary changes to the construction below are deferred to a full version of this article.

Proof of Theorem 3. We alter the construction of Chandra et al. [6] to strengthen it against Gaussian perturbations with standard deviation $\sigma = O(1/\sqrt{n})$ (see Figure 1). Let $p \geq 3$ be an odd integer and $P := 3p^{2p}$. The original instance of [6] is a subset of the $(P \times P)$-grid, which we embed into $[0,1]^2$ by scaling by $1/P$, and consists of three parts V_1, V_2 and V_3. The vertices in V_1 are partitioned into the layers $L_0, \ldots, L_p$. Layer i consists of $p^{2i} + 1$ equidistant vertices, each of which has a vertical distance of $c_i = p^{2p-2i-1}/P$ to the point above it in Layer $i + 1$ and a horizontal distance of $a_i = p^{2p-2i}/P$ to the nearest neighbor(s) in the same layer. The set V_2 is a copy of V_1 shifted to the right by a distance of $2/3$. The remaining part V_3 consists of a copy of Layer p of V_1 shifted to the right by $1/3$ to connect V_1 and V_2 by a path of points. We regard L_i as the set of Layer-i points in $V_1 \cup V_2 \cup V_3$.

As in the original construction, we will construct an instance of $n = \Theta(p^{2p})$ points, which implies $p = \Theta(\log n / \log \log n)$. Let $0 \leq t \leq p$ be the largest odd integer such that $p^{2t+1} \leq (3\sigma)^{-1}$. In our construction, we drop all Layers $t + 1, \ldots, p$ in both V_1 and V_2, as well as Layer p in V_3. Instead, we connect V_1 and V_2 already in Layer t by an altered copy of Layer t of V_1 shifted to the right by $1/3$. Let C be an arbitrary point of our construction, for convenience we will use the central point of Layer t in V_3. We introduce $p^{2p} - 1$ additional copies of this point C. These surplus points serve as a "padding" of the instance to ensure $n = \Theta(p^{2p})$. Note that the resulting instance has $t + 1$ layers $L_0, \ldots, L_t$.

We chose t such that the magnitude of perturbation is negligible compared to the pairwise distances of all non-padding points. Furthermore, the restriction on σ ensures that incorporating the padding points increases the optimal tour length only by a constant.

Lemma 12. *With probability $1 - O(n^{-s})$ for any constant $s > 0$, the optimal tour has length $O(1)$.*

We find a long 2-optimal tour on all non-padding points analogously to the original construction by taking a shortcut of the original 2-optimal tour, which connects V_1 and V_2 already in Layer t (see Figure 1).

Consider the padding points, which are yet to be connected. Let C_ℓ denote the nearest point in Layer t of V_3 that is to the left of C. Symmetrically, C_r is the nearest point to the right of C. Let T^p be any 2-optimal path from C_ℓ to C_r that passes through all the padding points (including C). We replace the edges (C_ℓ, C) and (C, C_r) by the path T^p, completing the construction of our tour T.

Lemma 13. *Let $s > 0$ be arbitrary. With probability $1 - O(n^{-s})$, T is 2-optimal and has a length of $\Omega(\log n / \log \log n)$.*

By Lemmas 12 and 13, Theorem 3 follows.

6   Discussions and Open Problems

We have proved an upper bound of $O(\log 1/\sigma)$ for the smoothed approximation ratio of 2-Opt. Furthermore, we have proved that the lower bound of Chandra et al. [6] remains robust even for $\sigma = O(1/\sqrt{n})$ and even if it is initialized with the nearest-neighbor heuristic. We leave as an open problem to generalize our upper bounds to the one-step model to improve the current bound of $O(\sqrt[d]{\phi})$ [8], but conjecture that this might be difficult.

While our bound significantly improves the previously known bound for the smoothed approximation ratio of 2-Opt, we readily admit that it still does not explain the performance observed in practice. A possible explanation is that when the initial tour is not picked by an adversary or the nearest neighbor heuristic, but using a construction heuristic such as the spanning tree heuristic or an insertion heuristic, an approximation factor of 2 is guaranteed even before 2-OPT has begun to improve the tour [18]. However, a smoothed analysis of the approximation ratio of 2-Opt initialized with a good heuristic might be difficult: even in the average-case, it is only known that the length of an optimal TSP is concentrated around $\gamma_d \cdot n^{\frac{d-1}{d}}$ for some constant $\gamma_d > 0$. But the precise value of γ_d is unknown [21]. Since experiments suggest that 2-Opt even with good initialization does not achieve an approximation ratio of $1 + o(1)$ [10,11], one has to deal with the precise constants, which seems challenging.

Finally, we conjecture that many examples for showing lower bounds for the approximation ratio of concrete algorithms for Euclidean optimization such as the TSP remain stable under perturbation for $\sigma = O(1/\sqrt{n})$. The question remains whether such small values of σ, although they often suffice to prove polynomial smoothed running-time, are essential to explain practical approximation ratios or if already slower decreasing σ provide a sufficient explanation.

References

1. Arora, S.: Polynomial time approximation schemes for Euclidean traveling salesman and other geometric problems. Journal of the ACM **45**(5), 753–782 (1998)
2. Arthur, D., Manthey, B., Röglin, H.: Smoothed analysis of the k-means method. Journal of the ACM **58**(5) (2011)
3. Arthur, D., Vassilvitskii, S.: Worst-case and smoothed analysis of the ICP algorithm, with an application to the k-means method. SIAM J. Comp. **39**(2), 766–782 (2009)
4. Bläser, M., Manthey, B., Rao, B.V.R.: Smoothed analysis of partitioning algorithms for Euclidean functionals. Algorithmica **66**(2), 397–418 (2013)
5. Brunsch, T., Röglin, H., Rutten, C., Vredeveld, T.: Smoothed performance guarantees for local search. Mathematical Programming **146**(1–2), 185–218 (2014)
6. Chandra, B., Karloff, H., Tovey, C.: New results on the old k-opt algorithm for the traveling salesman problem. SIAM J. Comp. **28**(6), 1998–2029 (1999)
7. Curticapean, R., Künnemann, M.: A quantization framework for smoothed analysis of euclidean optimization problems. In: Bodlaender, H.L., Italiano, G.F. (eds.) ESA 2013. LNCS, vol. 8125, pp. 349–360. Springer, Heidelberg (2013)
8. Englert, M., Röglin, H., Vöcking, B.: Worst case and probabilistic analysis of the 2-Opt algorithm for the TSP. Algorithmica **68**(1), 190–264 (2014)
9. Etscheid, M.: Performance guarantees for scheduling algorithms under perturbed machine speeds. Discrete Applied Mathematics (to appear)
10. Johnson, D.S., McGeoch, L.A.: The traveling salesman problem: A case study. In: Aarts, E., Lenstra, J.K. (eds.) Local Search in Combinatorial Optimization, chap. 8. John Wiley & Sons (1997)
11. Johnson, D.S., McGeoch, L.A.: Experimental analysis of heuristics for the STSP. In: Gutin, G., Punnen, A.P. (eds.) The Traveling Salesman Problem and its Variations, chap. 9. Kluwer Academic Publishers (2002)
12. Karger, D., Onak, K.: Polynomial approximation schemes for smoothed and random instances of multidimensional packing problems. In: Proc. of the 18th Ann. ACM-SIAM Symp. on Discrete Algorithms (SODA), pp. 1207–1216. SIAM (2007)
13. Manthey, B., Röglin, H.: Smoothed analysis: Analysis of algorithms beyond worst case. It - Information Technology **53**(6), 280–286 (2011)
14. Manthey, B., Röglin, H.: Worst-case and smoothed analysis of k-means clustering with Bregman divergences. J. of Comp. Geom. **4**(1), 94–132 (2013)
15. Manthey, B., Veenstra, R.: Smoothed analysis of the 2-Opt heuristic for the TSP: Polynomial bounds for Gaussian noise. In: Cai, L., Cheng, S.-W., Lam, T.-W. (eds.) ISAAC 2013. LNCS, vol. 8283, pp. 579–589. Springer, Heidelberg (2013)
16. Mitchell, J.S.B.: Guillotine subdivisions approximate polygonal subdivisions: A simple polynomial-time approximation scheme for Geometric TSP, k-MST, and related problems. SIAM J. Comp. **28**(4), 1298–1309 (1999)

17. Papadimitriou, C.H.: The Euclidean traveling salesman problem is NP-complete. Theoretical Computer Science **4**(3), 237–244 (1977)
18. Rosenkrantz, D.J., Stearns, R.E., Lewis II, P.M.: An analysis of several heuristics for the traveling salesman problem. SIAM J. Comp. **6**(3), 563–581 (1977)
19. Spielman, D.A., Teng, S.H.: Smoothed analysis of algorithms: Why the simplex algorithm usually takes polynomial time. Journal of the ACM **51**(3), 385–463 (2004)
20. Spielman, D.A., Teng, S.H.: Smoothed analysis: An attempt to explain the behavior of algorithms in practice. Communications of the ACM **52**(10), 76–84 (2009)
21. Yukich, J.E.: Probability Theory of Classical Euclidean Optimization Problems. Lecture Notes in Mathematics, vol. 1675. Springer (1998)

On the Hardest Problem Formulations for the 0/1 Lasserre Hierarchy

Adam Kurpisz[(⊠)], Samuli Leppänen, and Monaldo Mastrolilli

IDSIA, 6928 Manno, Switzerland
{adam,samuli,monaldo}@idsia.ch

Abstract. The Lasserre/Sum-of-Squares (SoS) hierarchy is a systematic procedure for constructing a sequence of increasingly tight semidefinite relaxations. It is known that the hierarchy converges to the 0/1 polytope in n levels and captures the convex relaxations used in the best available approximation algorithms for a wide variety of optimization problems.

In this paper we characterize the set of 0/1 integer linear problems and unconstrained 0/1 polynomial optimization problems that can still have an integrality gap at level $n - 1$. These problems are the hardest for the Lasserre hierarchy in this sense.

1 Introduction

The *Sum of Squares* (SoS) proof system introduced by Grigoriev and Vorobjov [20] is a proof system based on the *Positivstellensatz*. Shor [37], Nesterov [30], Parrilo [33] and Lasserre [24] show that it can be efficiently automatized using semidefinite programming (SDP) such that any n-variable degree-d proof can be found in time $n^{O(d)}$. The SDP, often called the Lasserre/SoS[1] hierarchy, is the dual of the SoS proof system, meaning that the Lasserre hierarchy value at "level d/2" of an optimization problem is equal to the best provable bound using a degree-d SoS proof (see the monograph by Laurent [26]). For a brief history of the different formulations from [20], [24], [33] and the relations between them and results in real algebraic geometry we refer the reader to [32].

The Lasserre hierarchy can be seen as a systematic procedure to strengthen a relaxation of an optimization problem by constructing a sequence of increasingly tight SDP relaxations. The tightness of the relaxation is parametrized by its *level* or *round*, which corresponds to the degree of the proof in the proof system. Moreover, it captures the convex relaxations used in the best available approximation algorithms for a wide variety of optimization problems. For example, the first round of the hierarchy for the INDEPENDENT SET problem implies the Lovász θ-function [28] and for the MAX CUT problem it gives the Goemans-Williamson

---

This work replaces and improves an early version of the paper titled "The Lasserre hierarchy in almost diagonal form" appeared in arXiv.

[1] For brevity, we will interchange Lasserre hierarchy with SoS hierarchy since they are essentially the same in our context.

© Springer-Verlag Berlin Heidelberg 2015
M.M. Halldórsson et al. (Eds.): ICALP 2015, Part I, LNCS 9134, pp. 872–885, 2015.
DOI: 10.1007/978-3-662-47672-7_71

relaxation [15]. The ARV relaxation of the SPARSEST CUT [2] problem is no stronger than the relaxation given in the third round of the Lasserre hierarchy, and the subexponential time algorithm for UNIQUE GAMES [1] is implied by a sublinear number of rounds [5,21]. More recently, it has been shown that $O(1)$ levels of the Lasserre hierarchy is equivalent in power to any polynomial size SDP extended formulation in approximating maximum constraint satisfaction problems [27]. Other approximation guarantees that arise from the first $O(1)$ levels of the Lasserre (or weaker) hierarchy can be found in [5,6,9,10,12,13,21,29,34]. For a more detailed overview on the use of hierarchies in approximation algorithms, see the surveys [11,25,26].

The limitations of the Lasserre hierarchy have also been studied. Most of the known lower bounds for the hierarchy originated in the works of Grigoriev [17,18] (also independently rediscovered later by Schoenebeck [36]). In [18] it is shown that random 3XOR or 3SAT instances cannot be solved by even $\Omega(n)$ rounds of SoS hierarchy. Lower bounds, such as those of [7,38] rely on [18,36] plus gadget reductions. For a different technique to obtain lower bounds, see the recent paper [4].

A particular weakness of the hierarchy revolves around the fact that it has hard time reasoning about terms of the form $x_1 + ... + x_n$ using the fact that all x_i's are 0/1. Grigoriev [17] showed that $\lfloor n/2 \rfloor$ levels of Lasserre are needed to prove that the polytope $\{x \in [0,1]^n \mid \sum_{i=1}^n x_i = \lfloor n/2 \rfloor + 1/2\}$ contains no integer point. A simplified proof can be found in [19].

In [8] Cheung considered a simple instance of the MIN KNAPSACK problem, i.e. the minimization of $\sum_{i=1}^n x_i$ for 0/1 variables such that $\sum_{i=1}^n x_i \geq \delta(n)$, for some $\delta(n) < 1$ that depends on n. Cheung proved that the Lasserre hierarchy requires n levels to converge to the integral polytope. This is shown by providing a feasible solution at level $n - 1$ of value $\frac{n}{n+1}$, whereas the smallest integral solution has value 1. This gives an integrality gap[2] of $1 + \frac{1}{n}$ that vanishes with n.

We emphasize that the main interest in the work of Cheung revolves around understanding how fast the Lasserre hierarchy converges to the integral polytope and not how fast the integrality gap reduces, therefore not ruling out the possibility that the integrality gap might decrease slowly with the number of levels. This is conceptually an important difference. For the MAX KNAPSACK (or MIN KNAPSACK) problem the presence of an integrality gap at some "large" level $t(n)$, that depends on n, is promptly implied by $P \neq NP$, whereas the existence of a "large" integrality gap at some "large" level $t(n)$ is not immediately clear (since both MAX KNAPSACK and MIN KNAPSACK problems admit an FPTAS). With this regard, note that Cheung's result also implies that for the MAX KNAPSACK the Lasserre hierarchy requires n levels to converge to the integral polytope. However, in [23] it is shown that only $O(1/\varepsilon)$ levels are needed to obtain an integrality gap of $1 - \varepsilon$, for any arbitrarily small constant $\varepsilon > 0$. It is also worth pointing out that currently the Cheung knapsack result [8] is the

---

[2] The *integrality gap* is defined to be the measure of the quality of the relaxation described by the ratio between the optimal integral value and the relaxed optimal value. If this ratio is different from 1 we will say that "there is an integrality gap".

only known integrality gap result for Lasserre/Sum-of-Squares hierarchy at level $n - 1$.

Our results. With n variables, the n-th level of the Lasserre hierarchy is sufficient to obtain the $0/1$ polytope, where the only feasible solutions are convex combinations of feasible integral solutions [24]. This can be proved by using the *canonical lifting lemma* (see Laurent [25]), where the feasibility of a solution to the Lasserre relaxation at level n reduces to showing that a certain diagonal matrix is positive semidefinite (PSD).

The main challenge in analyzing integrality gap instances at level smaller than n is showing that a candidate solution satisfies the positive semidefinite constraints. In this paper, we first show that the feasibility of a solution to the Lasserre relaxation at level $n - 1$ reduces to showing that a matrix differing from a diagonal matrix by a rank one matrix (almost diagonal form) is PSD. We analyze the eigenvalues of the almost diagonal matrices and obtain compact necessary and sufficient conditions for the existence of an integrality gap of the Lasserre relaxation at level $n - 1$. This result can be seen as the opposite of [16] where they consider the case when the first order Lasserre relaxation is exact.

Interestingly, for $0/1$ integer linear programs the existence of a gap at level $n - 1$ implies that the problem formulation contains only constraints of the form we call *Single Vertex Cutting* (SVC). An SVC constraint only excludes one vertex of the $\{0, 1\}^n$ hypercube. It can thus be seen as the most generic non-trivial form of constraint, since the feasible set of any integer linear program can be modeled using only constraints of this form.

This characterization allows us to show that n levels of Lasserre are needed to prove that a polytope defined by (exponentially many) SVC constraints contains no integer point. No other example of this kind was known at level n (the previously known example in [17] requires $\lfloor n/2 \rfloor$ levels).

One problem where SVC constraints can arise naturally is the KNAPSACK problem. By applying the computed conditions, we improve the Cheung [8] MIN KNAPSACK integrality gap of the Lasserre relaxation at level $n - 1$ from $1 + 1/n$ to any arbitrary large number. This shows a substantial difference between the MIN KNAPSACK and the MAX KNAPSACK when we take into consideration the integrality gap size of the Lasserre relaxation.

Furthermore, we show that a similar result holds beyond the class of integer linear programs. More precisely, we show that any unconstrained $0/1$ polynomial optimization problem exhibiting an integrality gap at level $n - 1$ of the Lasserre relaxation has necessarily an objective function given by a polynomial of degree n. This rules out the existence of any integrality gap at level $n - 1$ for any k-ary boolean constraint satisfaction problem with $k < n$. Finally, we provide an example of an unconstrained $0/1$ polynomial optimization problem with an integrality gap at level $n - 1$ of the Lasserre hierarchy, and discuss why the problem can be seen as a constraint satisfaction version of an SVC constraint. Our result complements the recent paper [14] where it is shown that the Lasserre relaxation does not have any gap at level $\lceil \frac{n}{2} \rceil$ when optimizing n-variate $0/1$ polynomials of degree 2.

2   The Lasserre Hierarchy

In this section we provide a definition of the Lasserre hierarchy [24]. For the applications that we have in mind, we restrict our discussion to optimization problems with 0/1-variables and linear constraints. More precisely, we consider the following general optimization problem $\mathbb{P}$: Given a multilinear polynomial $f : \{0,1\}^n \rightarrow \mathbb{R}$

$$\mathbb{P}: \quad \min\{f(x)|x \in \{0,1\}^n, g_\ell(x) \geq 0 \text{ for } \ell \in [m]\} \qquad (1)$$

where $\{g_\ell(x) : \ell \in [m]\}$ are linear functions of x.

Many basic optimization problems are special cases of $\mathbb{P}$. For example, any k-ary boolean constraint satisfaction problem, such as MAX CUT, is captured by (1) where a degree k function $f(x)$ counts the number of satisfied constraints, and no linear constraints $g_\ell(x) \geq 0$ are present. Also any 0/1 integer linear program is a special case of (1), where $f(x)$ is a linear function.

Lasserre [24] proposed a hierarchy of SDP relaxations for increasing δ,

$$\min\{L(f)|L : \mathbb{R}[X]_{2\delta} \rightarrow \mathbb{R}, L(1) = 1, \text{ and } L(u^2), L(u^2 g_\ell) \geq 0, \forall \text{ polynomial } u\} \quad (2)$$

where $L : \mathbb{R}[X]_{2\delta} \rightarrow \mathbb{R}$ is a linear map with $\mathbb{R}[X]_{2\delta}$ denoting the ring $\mathbb{R}[X]$ restricted to polynomials of degree at most 2δ.[3] In particular for 0/1 problems L vanishes on the truncated ideal generated by $x_i^2 - x_i$. Note that (2) is a relaxation since one can take L to be the evaluation map $f \rightarrow f(x^*)$ for any optimal solution x^*.

Relaxation (2) can be equivalently formulated in terms of *moment matrices* [24]. In the context of this paper, this matrix point of view is more convenient to use and it is described below. In our notation we mainly follow the survey of Laurent [25] (see also [35]).

Variables and Moment Matrix. Throughout this paper, vectors are written as columns. Let N denote the set $\{1, \ldots, n\}$. The collection of all subsets of N is denoted by $\mathcal{P}(N)$. For any integer $t \geq 0$, let $\mathcal{P}_t(N)$ denote the collection of subsets of N having cardinality at most t. Let $y \in \mathbb{R}^{\mathcal{P}(N)}$. For any nonnegative integer $t \leq n$, let $M_t(y)$ denote the matrix with (I, J)-entry $y_{I \cup J}$ for all $I, J \in \mathcal{P}_t(N)$. Matrix $M_t(y)$ is termed in the following as the *t-moment matrix* of y. For a linear function $g(x) = \sum_{i=1}^n g_i \cdot x_i + g_0$, we define $g * y$ as a vector, often called *shift operator*, where the I-th entry is $(g * y)_I = \sum_{i=1}^n g_i y_{I \cup \{i\}} + g_0 y_I$. Let f denote the vector of coefficients of polynomial $f(x)$ (where f_I is the coefficient of monomial $\Pi_{i \in I} x_i$ in $f(x)$).

Definition 1. *The Lasserre relaxation of problem* (1) *at the t-th level, denoted as* $\text{LAS}_t(\mathbb{P})$, *is the following*

---

[3] In [3], $L(p)$ is written $\tilde{\mathbb{E}}[p]$ and called the "pseudo-expectation" of p.

$$\text{LAS}_t(\mathbb{P}): \quad \min\left\{\sum_{I\subseteq N} f_I y_I \mid y \in \mathbb{R}^{\mathcal{P}_{2t+2d}(N)} \text{ and } y \in \mathbb{M}\right\} \tag{3}$$

where $\mathbb{M}$ is the set of vectors $y \in \mathbb{R}^{\mathcal{P}_{2t+2d}(N)}$ that satisfy the following PSD conditions

$$y_\varnothing = 1 \tag{4}$$

$$M_{t+d}(y) \succeq 0 \tag{5}$$

$$M_t(g_\ell * y) \succeq 0 \qquad \ell \in [m] \tag{6}$$

where $d = 0$ if $m = 0$ (no linear constraints) otherwise $d = 1$.

We will use the following known facts (see e.g. [25,35]). Consider any vector $w \in \mathbb{R}^{\mathcal{P}(N)}$ (vector w is intended to be either the vector $y \in \mathbb{R}^{\mathcal{P}(N)}$ of variables or the shifted vector $g * y$ for any $g \in \mathbb{R}^{\mathcal{P}(N)}$). For any $I \in \mathcal{P}(N)$, variables $\{w_I^N : I \subseteq N\}$ are defined as follows:

$$w_I^N := \sum_{H \subseteq N \setminus I} (-1)^{|H|} w_{H \cup I}$$

Note that $w_I = \sum_{I \subseteq J} w_J^N$ (by using inclusion-exclusion principle, see [35]). The latter with $y_\emptyset = 1$ implies that $\sum_{J \subseteq N} y_J^N = 1$, and that the objective function can be rewritten as follows:

$$\sum_{I\subseteq N} f_I y_I = \sum_{I \subseteq N} f(x_I) y_I^N$$

where $f(x_I)$ denotes the value of $f(x)$ when $x_i = 1$ for $i \in I$ and $x_i = 0$ for $i \notin I$.

Congruent transformations are known not to change the sign of the eigenvalues (see e.g. [22]). It follows that in studying the positive-semidefiniteness of matrices we can focus on congruent matrices without loss of generality. Let $D_t(w)$ denote the diagonal matrix in $\mathbb{R}^{\mathcal{P}_t(N) \times \mathcal{P}_t(N)}$ with (I, I)-entry equal to w_I^N for all $I \in \mathcal{P}_t(N)$.

Lemma 1. [25] Matrix $M_n(w)$ is congruent to the diagonal matrix $D_n(w)$.

By Lemma 1, $M_n(y) \succeq 0$ implies that the variables in $\{y_I^N : I \subseteq N\}$ can be interpreted as a probability distribution (see [25,35]), where y_I^N is the probability that the variables with index in I are set to one and the remaining to zero.

Lemma 2. [25] For any polynomial g of degree at most one, $y \in \mathbb{R}^{\mathcal{P}(N)}$ and $z = g * y$ we have $z_I^N = g(x_I) \cdot y_I^N$ where $g(x_I) = \sum_{i \in I} a_i + b$.

Note that, by using Lemma 1 and Lemma 2, it can be easily shown the well known fact that at level n any solution can be written as a convex combination of feasible integral solutions. The latter implies that any integrality gap vanishes at level n.

3   The $(n-1)$-Moment Matrix

In the following we show that $M_{n-1}(w)$ is congruent to the diagonal matrix $D_{n-1}(w)$ perturbed by a rank one matrix, and analyze its eigenvalues. For ease of notation, we will use D to denote $D_{n-1}(w)$ throughout this section.

Lemma 3. *Matrix $M_{n-1}(w)$ is congruent to the matrix $D + w_N^N \cdot vv^\top$, where v is a $|\mathcal{P}_{n-1}(N)|$-dimensional vector with $v_I = (-1)^{n+1-|I|}$ for any $I \in \mathcal{P}_{n-1}(N)$.*

3.1   Positive Semidefiniteness of $M_{n-1}(y)$

In this section we derive the necessary and sufficient conditions for $M_{n-1}(w) \succeq 0$. From Lemma 3 we have that $M_{n-1}(y) \succeq 0 \Leftrightarrow D + w_N^N vv^\top \succeq 0$, where $vv^\top$ is a rank one matrix with entries ± 1.

Lemma 4. *If $w_N^N \neq 0$ then, for any $I \subseteq N$, $\lambda = w_I^N$ is an eigenvalue of the matrix $D + w_N^N vv^\top$ if and only if there is another $J \neq I$ with $w_I^N = w_J^N$, $J \subseteq N$; The remaining eigenvalues are the solutions λ of the following equation*

$$\sum_{N \neq I \subseteq N} \frac{1}{\lambda - w_I^N} = \frac{1}{w_N^N} \tag{7}$$

Proof. Consider the zeroes λ of the characteristic polynomial of $D + w_N^N vv^\top$:

$$\det(\lambda I - (D + w_N^N vv^\top)) = \det(D_\lambda - w_N^N vv^\top) = 0 \tag{8}$$

where $D_\lambda = \lambda I - D$. Applying Cauchy's formula for the determinant of a rank-one pertubation [22, p. 26] we can write this as

$$\det(D_\lambda) - w_N^N v^\top \operatorname{adj}(D_\lambda)v = 0 \tag{9}$$

Consider a solution λ to (9). Exactly one of the following three cases must hold:

1. D_λ is nonsingular, meaning that $\lambda \neq w_I^N$ for all $N \neq I \subseteq N$. Then $\operatorname{adj}(D_\lambda) = (\det D_\lambda)D_\lambda^{-1}$ and the above becomes

   $$\det(D_\lambda)(1 - w_N^N v^\top D_\lambda^{-1} v) = 0$$

   which simplifies to (7).
2. D_λ is singular, and $\lambda = w_I^N$ for exactly one $N \neq I \subseteq N$. Then $\operatorname{adj}(D_\lambda) = \alpha e_I e_I^\top$ for some nonzero α [22, p. 22-23], where $(e_I)_J = 1$ if $I = J$ and $(e_I)_J = 0$ otherwise. Now (9) simplifies to

   $$w_N^N v^\top (\alpha e_I e_I^\top)v = 0$$

   which can only hold if $w_N^N = 0$. Hence such λ cannot be a solution to (8).
3. D_λ is singular and there are more than one $N \neq I \subseteq N$ such that $\lambda = w_I^N$. Then $\operatorname{adj}(D_\lambda) = 0$ [22, p. 22] and λ is a solution to (8). $\qquad\square$

Lemma 5. *Matrix* $D + w_N^N vv^\top$ *is positive-semidefinite if and only if either* $w_I^N \geq 0$ *for all* $I \subseteq N$, *or the following holds*

$$w_K^N < 0, \qquad \text{for exactly one } K \subseteq N, \tag{10}$$

$$w_J^N > 0, \qquad \text{for all } K \neq J \subseteq N, \tag{11}$$

$$\sum_{I \subseteq N} \frac{1}{w_I^N} \leq 0 \tag{12}$$

Proof. If $w_I^N \geq 0$ for all $I \subseteq N$ then $D + w_N^N vv^\top \succeq 0$ since it is the sum of two PSD matrices. Otherwise, there exists $I \subseteq N$ with $w_I^N < 0$ and we distinguish between the following complementary cases.

If there are two different sets $N \neq I, J \subseteq N$ such that $w_I^N = w_J^N < 0$, then by Lemma 4 the matrix $D + w_N^N vv^\top$ has a negative eigenvalue. Therefore we may assume that all the negative entries of D are different from each other. Then by Lemma 4, any potentially negative eigenvalue is given by (7). With this in mind, let $f(\lambda) = \sum_{N \neq I \subseteq N} \frac{1}{\lambda - w_I^N}$ and study the points λ where $f(\lambda)$ intersects the line given by $\frac{1}{w_N^N}$.

There are three cases:

1. For two sets $N \neq I, J \subseteq N$ we have $w_I^N < w_J^N \leq 0$. Then since the function $f(\lambda)$ has vertical asymptotes at the points w_I^N and w_J^N, there must be a point $\lambda < 0$ such that $f(\lambda) = \frac{1}{w_N^N}$ regardless of the value of w_N^N (see Figure 1 (i)).
2. For exactly one $N \neq I \subseteq N$ we have $w_I^N \leq 0$ and $w_N^N < 0$. Then $f(\lambda)$ has one vertical asymptote in $(-\infty, 0]$ and thus the line $\frac{1}{w_N^N}$ crosses the graph of $f(\lambda)$ at least in one $\lambda < 0$ (see Figure 1 (ii)).
3. For exactly one $I \subseteq N$ we have $w_I^N < 0$ and the rest are strictly positive. Then we note that there can be at most one $\lambda < 0$ such that $f(\lambda) = \frac{1}{w_N^N}$. Inspecting the form of the graph shows that there is no intersection in the negative half-plane if and only if $f(0) \geq \frac{1}{w_N^N}$ (see Figure 1 (iii) and (iv) for the case $I = N$). $\qquad\square$

4   Integrality Gaps of Lasserre Hierarchy at Level $n - 1$

In this section we characterize the set of problems $\mathbb{P}$ of the form (1) that can have an integrality gap at level $n - 1$ of the Lasserre relaxation. In particular, we prove that in order to exhibit an integrality gap, a constrained problem can only have constraints each of which rule out only one point of the $\{0, 1\}^n$ hypercube. We fully characterize what this means in the case where the constraints are linear. We also discuss two examples of problems with such constraints, and in particular, we exhibit a simple instance of the MIN KNAPSACK problem that has an unbounded integrality gap. Finally, we show that if $\mathbb{P}$ is an unconstrained problem that has an integrality gap at level $n - 1$, then the objective function of $\mathbb{P}$ must be a polynomial of degree n.

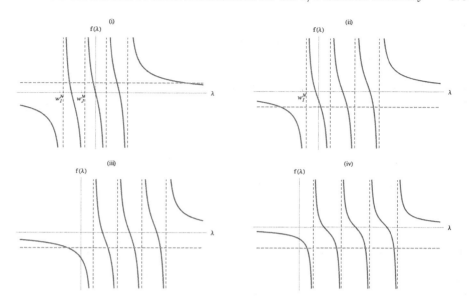

Fig. 1. A conceptual plot of different relevant arrangements of the graph of $f(\lambda)$ and the graph of $\frac{1}{w_N^N}$ (dotted lines)

4.1   Problems with Linear Constraints

In this subsection we focus on 0/1-integer linear programs $\mathbb{P}$ of the form (1). We will assume, w.l.o.g., that if constraint $g(x) \geq 0$ is satisfied by all integral points then it is redundant and no one of these redundant constraints is present.

Theorem 1. *Let $\mathbb{P}$ be a 0/1-integer linear program of the form (1). The Lasserre relaxation $\mathrm{LAS}_{n-1}(\mathbb{P})$ has an integrality gap if and only if there exists a solution $\{y_I^N | I \subseteq N\}$ that satisfies the following conditions:*

$$y_I^N > 0 \qquad \text{for all } I \subseteq N, \tag{13}$$

$$\sum_{I \subseteq N} y_I^N = 1 \tag{14}$$

$$g_\ell(x_{K_\ell})y_{K_\ell}^N < 0 \qquad \text{for exactly one } K_\ell \subseteq N \text{ for each } \ell \in [m], \tag{15}$$

$$g_\ell(x_J)y_J^N > 0 \qquad \text{for all } \ell \in [m], \text{ for all } K_\ell \neq J \subseteq N, \tag{16}$$

$$\sum_{I \subseteq N} \frac{1}{g_\ell(x_I)y_I^N} \leq 0 \qquad \text{for all } \ell \in [m], \tag{17}$$

$$\sum_{I \subseteq N} y_I^N f(x_I) < f(x_{I^*}) \tag{18}$$

where $f(x_{I^*})$ is a minimal integral feasible solution.

Definition 2. *We call $g(x) \geq 0$ a Single Vertex Cutting (SVC) constraint if there exists only one $I \subseteq N$ such that $g(x_I) < 0$ and for every other $I \neq J \subseteq N$ it holds $g(x_J) > 0$.*

Corollary 1. *Let $f(x_{I^*})$ denote the integral optimum of (1). If there is an integrality gap, i.e., $y \in \text{LAS}_{n-1}(\mathbb{P})$ such that $\sum_{I \subseteq N} y_I^N f(x_I) < f(x_{I^*})$, then the constraints in (1) are SVC.*

We are considering only problems with linear constraints over $\{0,1\}^n$, so it is straightforward to characterize the SVC constraints.

Lemma 6. *Let $g(x) = \sum_{i=1}^{n} a_i x_i - b \geq 0$ be a linear SVC constraint. Then $b \neq 0$ and $a_i \neq 0$ for all i, and if P is the set of indices such that $a_i < 0 \Leftrightarrow i \in P$, then $\sum_{i \in P} a_i < b$, but $\sum_{i \in Q} a_i > b$ for all $P \neq Q \subseteq N$.*

4.2   Example Problems with SVC Constraints at Level $n - 1$

As proved in Corollary 1, SVC constraints are in some sense the most difficult constraints to handle for the Lasserre hierarchy. Each such constraint excludes only one point of the $\{0,1\}^n$ hypercube, and thus the feasible set of any integer linear program can be modeled using only these constraints. It follows that if modeled in this way, any integer linear program can potentially have an integrality gap at level $n - 1$ of the Lasserre hierarchy. In this section we give two examples of problems where the Lasserre hierarchy does not converge to the integer polytope even at level $n - 1$.

Unbounded Integrality Gap for the Min Knapsack. One problem where the SVC constraint naturally arises is the KNAPSACK problem. We show that the minimization version of the problem has an unbounded integrality gap at level $n - 1$ of the Lasserre hierarchy. Indeed, consider the following simple instance of the MIN-KNAPSACK:

$$(GapKnap) \min\{\sum_{i=1}^{n} x_i \mid \sum_{i=1}^{n} x_i \geq 1/P, x_i \in \{0,1\} \text{ for } i \in [n]\} \qquad (19)$$

Notice that the optimal *integral* value of $(GapKnap)$ is one. The optimal value of the linear programming relaxation of $(GapKnap)$ is $1/P$, so the integrality gap of the LP is P and can be arbitrarily large.

By using Theorem 1 we prove the following dichotomy-type result. If we allow a "large" P (exponential in the number of variables n), then the Lasserre hierarchy is of no help to limit the unbounded integrality gap of $(GapKnap)$, even at level $(n - 1)$. This analysis is tight since $\text{LAS}_n(GapKnap)$ admits an optimal integral solution with n variables. We also show that the requirement that P is exponential in n is necessary for having a "large" gap at level $(n - 1)$.

Corollary 2. *(Integrality Gap Bounds for* MIN-KNAPSACK*) The integrality gap of $\text{LAS}_{n-1}(GapKnap)$ is k, for any $k \geq 2$ if and only if $P = \Theta(k) \cdot 2^{2n}$.*

Remark 1. We observe that the instance (19) can be easily ruled out by requiring that each coefficient of any variable must be not larger than the constant term in the knapsack constraint. However, even with this pruning step, the integrality gap can be made unbounded up to the last but two levels of the Lasserre hierarchy: add an additional variable x_{n+1} only in the constraint (not in the objective function) and increase the constant term to $1 + 1/P$. Any solution for $\mathrm{LAS}_{n-1}(GapKnap)$ can be easily turned into a feasible solution for the augmented instance by setting the new variables $y_I' = y_{I \setminus \{n+1\}}$ for any $I \in \mathcal{P}_{2t+2}([n+1])$ and observing that any principal submatrix of the new moment matrices has either determinant equal to zero or it is a principal submatrix in the moment matrix of the reduced problem.

Undetected Empty Integer Hull. As discussed at the beginning of this section, any integer linear problem can be modeled using SVC constraints. Formulating the problem in this "pathological" way can potentially hinder the convergence of the Lasserre hierarchy. We demonstrate this by showing an extreme example, where the Lasserre hierarchy cannot detect that the integer hull is empty even at level $n - 1$.

Consider the feasible set given by (exponentially many) inequalities of the form

$$\sum_{i \in P} (1 - x_i) + \sum_{i \in N \setminus P} x_i \geq b \tag{20}$$

for each $P \subseteq N$. Clearly, any integral assignment I such that $x_i = 1$ if $i \in I$ and $x_i = 0$ otherwise, cannot satisfy all of the inequalities when b is positive. However, there exists an assignment of the variables y^N that satisfies the conditions of Theorem 1, and is hence a feasible solution to the Lasserre relaxation of the polytope described above at level $n - 1$, as shown below.

Consider a symmetric solution $y_I^N = \frac{1}{2^n}$ for every $I \subseteq N$ and some constraint of the form (20) corresponding to a given set $P \subseteq N$. Now the variables $z_I^N = g(x_I)y_I^N$ satisfy (15) and (16), and we need to check that it is possible to satisfy (17):

$$\sum_{I \subseteq N} \frac{1}{z_I^N} = \frac{1}{2^n} \sum_{I \subseteq N} \frac{1}{|P \setminus I| + |I \setminus P| - b} \leq 0 \Leftrightarrow \sum_{\varnothing \neq I \subseteq N} \frac{1}{|I| - b} \leq \frac{1}{b}$$

When $0 < b < \frac{1}{2}$, the above is implied by $\sum_{\varnothing \neq I \subseteq N} 2 \leq \frac{1}{b}$, so choosing $b = \frac{1}{2^{n+1}}$ makes (17) satisfied.

4.3   Unconstrained Problems at Level $n - 1$

Let $f : \{0,1\}^n \to \mathbb{R}$ be an objective function of a polynomial minimization problem normalized such that $\min_{x \in \{0,1\}^n} f(x) = 0$ and $\max_{x \in \{0,1\}^n} f(x) = 1$. We start with the conditions that an unconstrained polynomial optimization problem has to satisfy in order do admit a gap at level $n - 1$.

Theorem 2. *Let $\mathbb{P}$ denote an unconstrained polynomial optimization problem of the form* (1). *The Lasserre relaxation* $\text{LAS}_{n-1}(\mathbb{P})$ *has an integrality gap if and only if there exists a solution* $\{y_I^N | I \subseteq N\}$ *that satisfies* (18) *and the following conditions:*

$$\sum_{I \subseteq N} y_I^N = 1 \tag{21}$$

$$y_K^N < 0 \qquad \text{for exactly one } K \subseteq N, \tag{22}$$

$$y_J^N > 0 \qquad \text{for all } K \neq J \subseteq N, \tag{23}$$

$$\sum_{I \subseteq N} \frac{1}{y_I^N} \leq 0 \tag{24}$$

We note that f can always be represented as a multivariate polynomial of degree at most n. The main result of this section is Theorem 3.

Theorem 3. *If f is a function such that f has an integrality gap at level $n-1$, then f is a multivariate polynomial of degree n.*

Proof. We will use some elementary Fourier analysis of boolean functions (see e.g. [31, Ch.1]). To follow an established convention, we switch from studying the function $f : \{0,1\}^n \to \mathbb{R}$ to $h : \{-1,1\}^n \to \mathbb{R}$ via the bijective transform $f(x) = h(1 - 2x)$. Observe that f is of degree t if and only if h is of degree t, and for any $S \subseteq N$ we have $f(x_S) = h(w_S)$, where $w_i = -1$ if $i \in S$ and $w_i = 1$ otherwise.

Assume as before that for some $I_1 \subseteq N$, $h(w_{I_1}) = 1$ and $0 \leq h(w_I) \leq 1$. We assume that $|I_1|$ is even and let $I_2 \subseteq N$ be some fixed set such that $|I_2|$ is odd (the case where $|I_1|$ is odd is symmetric). We assume that h has an integrality gap, so by Lemma 7 (see below) necessarily $\sum_{I \subseteq N} h(w_I) < 2$, which we rewrite in a more convenient form (using $h(w_{I_1}) = 1$)

$$h(w_{I_2}) < 1 - \sum_{I_1 \neq I \neq I_2} h(w_I) \tag{25}$$

Assume now that h has a degree smaller than n, or in other words, its Fourier coefficient $\hat{h}(N)$ is 0:

$$\hat{h}(N) = 2^{-n} \sum_{S \subseteq N} h(w_S)(-1)^{|S|} = 0$$

Removing the normalizing constant and reordering the sum the above implies (using the assumptions on the parity of $|I_1|, |I_2|$)

$$\sum_{\substack{S \neq I_1 \\ |S| \text{ even}}} h(w_S) - \sum_{\substack{S \neq I_2 \\ |S| \text{ odd}}} h(w_S) = -1 + h(w_{I_2}) < - \sum_{I_1 \neq I \neq I_2} h(w_I)$$

by (25). Moving all the h terms to the left hand side yields

$$2 \sum_{\substack{S \neq I_1 \\ |S| \text{ even}}} h(w_S) < 0$$

which contradicts the assumption that $h(w) \geq 0$. □

Lemma 7. *Let $f(x_{I_1}) = 1$ and $0 \leq f(x) \leq 1$ for every $x \in \{0,1\}^n$. If f is such that $\sum_{I \subseteq N} f(x_I) \geq 2$ then there is no gap at level $n-1$.*

We point out that there exists a function of degree n that exhibits an integrality gap at level $n-1$. Consider the function given by

$$f(x) = 1 - \sum_{\varnothing \neq I \subseteq N} (-1)^{|I|} \prod_{i \in I} x_i$$

This function has the value 1 when all the variables are 0, and 0 elsewhere. It is a straightforward application of Theorem 2 to show that $f(x)$ exhibits an integrality gap at level $n-1$. We remark that $f(x)$ can be seen as a constraint satisfaction version of an SVC constraint.

Acknowledgments. Research supported by the Swiss National Science Foundation project 200020_144491/1. We thank Ola Svensson and the anonymous reviewers for helpful comments.

References

1. Arora, S., Barak, B., Steurer, D.: Subexponential algorithms for unique games and related problems. In: FOCS, pp. 563–572 (2010)
2. Arora, S., Rao, S., Vazirani, U.V.: Expander flows, geometric embeddings and graph partitioning. Journal of the ACM **56**(2) (2009)
3. Barak, B., Brandão, F.G.S.L., Harrow, A.W., Kelner, J.A., Steurer, D., Zhou, Y.: Hypercontractivity, sum-of-squares proofs, and their applications. In: STOC, pp. 307–326 (2012)
4. Barak, B., Chan, S.O., Kothari, P.: Sum of squares lower bounds from pairwise independence. In: STOC (2015)
5. Barak, B., Raghavendra, P., Steurer, D.: Rounding semidefinite programming hierarchies via global correlation. In: FOCS, pp. 472–481 (2011)
6. Bateni, M., Charikar, M., Guruswami, V.: Maxmin allocation via degree lower-bounded arborescences. In: STOC, pp. 543–552 (2009)
7. Bhaskara, A., Charikar, M., Vijayaraghavan, A., Guruswami, V., Zhou, Y.: Polynomial integrality gaps for strong sdp relaxations of densest k-subgraph. In: SODA, pp. 388–405 (2012)
8. Cheung, K.K.H.: Computation of the Lasserre ranks of some polytopes. Mathematics of Operations Research **32**(1), 88–94 (2007)
9. Chlamtac, E.: Approximation algorithms using hierarchies of semidefinite programming relaxations. In: FOCS, pp. 691–701 (2007)
10. Chlamtac, E., Singh, G.: Improved approximation guarantees through higher levels of SDP hierarchies. In: Goel, A., Jansen, K., Rolim, J.D.P., Rubinfeld, R. (eds.) APPROX and RANDOM 2008. LNCS, vol. 5171, pp. 49–62. Springer, Heidelberg (2008)
11. Chlamtac, E., Tulsiani, M.: Convex relaxations and integrality gaps. In: Anjos, M.F., Lasserre, J.B. (eds.) Handbook on semidefinite, conic and polynomial optimization. International Series in Operations Research & Management Science, vol. 166, pp. 139–169. Springer, Heidelberg (2012)

12. Cygan, M., Grandoni, F., Mastrolilli, M.: How to sell hyperedges: the hypermatching assignment problem. In: SODA, pp. 342–351 (2013)
13. de la Vega, W.F., Kenyon-Mathieu, C.: Linear programming relaxations of maxcut. In: SODA, pp. 53–61 (2007)
14. Fawzi, H., Saunderson, J., Parrilo, P.: Sparse sum-of-squares certificates on finite abelian groups (2015). CoRR, abs/1503.01207
15. Goemans, M.X., Williamson, D.P.: Improved approximation algorithms for maximum cut and satisfiability problems using semidefinite programming. Journal of the ACM 42(6), 1115–1145 (1995)
16. Gouveia, J., Parrilo, P.A., Thomas, R.R.: Theta bodies for polynomial ideals. SIAM Journal on Optimization 20(4), 2097–2118 (2010)
17. Grigoriev, D.: Complexity of positivstellensatz proofs for the knapsack. Computational Complexity 10(2), 139–154 (2001)
18. Grigoriev, D.: Linear lower bound on degrees of positivstellensatz calculus proofs for the parity. Theoretical Computer Science 259(1–2), 613–622 (2001)
19. Grigoriev, D., Hirsch, E.A., Pasechnik, D.V.: Complexity of semi-algebraic proofs. In: Alt, H., Ferreira, A. (eds.) STACS 2002. LNCS, vol. 2285, p. 419. Springer, Heidelberg (2002)
20. Grigoriev, D., Vorobjov, N.: Complexity of null-and positivstellensatz proofs. Annals of Pure and Applied Logic 113(1–3), 153–160 (2001)
21. Guruswami, V., Sinop, A.K.: Lasserre hierarchy, higher eigenvalues, and approximation schemes for graph partitioning and quadratic integer programming with psd objectives. In: FOCS, pp. 482–491 (2011)
22. Horn, R.A., Johnson, C.R.: Matrix analysis. Cambridge University Press (2013)
23. Karlin, A.R., Mathieu, C., Nguyen, C.T.: Integrality gaps of linear and semi-definite programming relaxations for knapsack. In: Günlük, O., Woeginger, G.J. (eds.) IPCO 2011. LNCS, vol. 6655, pp. 301–314. Springer, Heidelberg (2011)
24. Lasserre, J.B.: Global optimization with polynomials and the problem of moments. SIAM Journal on Optimization 11(3), 796–817 (2001)
25. Laurent, M.: A comparison of the Sherali-Adams, Lovász-Schrijver, and Lasserre relaxations for 0–1 programming. Mathematics of Operations Research 28(3), 470–496 (2003)
26. Laurent, M.: Sums of squares, moment matrices and optimization over polynomials. Emerging Applications of Algebraic Geometry 149, 157–270 (2009)
27. Lee, J.R., Raghavendra, P., Steurer, D.: Lower bounds on the size of semidefinite programming relaxations. In: STOC (to appear, 2015)
28. Lovász, L.: On the shannon capacity of a graph. IEEE Transactions on Information Theory 25, 1–7 (1979)
29. Magen, A., Moharrami, M.: Robust algorithms for on minor-free graphs based on the Sherali-Adams hierarchy. In: APPROX-RANDOM, pp. 258–271 (2009)
30. Nesterov, Y.: Global quadratic optimization via conic relaxation, pp. 363–384. Kluwer Academic Publishers (2000)
31. O'Donnell, R.: Analysis of Boolean Functions. Cambridge University Press (2014)
32. O'Donnell, R., Zhou, Y.: Approximability and proof complexity. In: SODA, pp. 1537–1556 (2013)
33. Parrilo, P.: Structured Semidefinite Programs and Semialgebraic Geometry Methods in Robustness and Optimization. PhD thesis, California Institute of Technology (2000)

34. Raghavendra, P., Tan, N.: Approximating csps with global cardinality constraints using sdp hierarchies. In: SODA, pp. 373–387 (2012)
35. Rothvoß, T.: The lasserre hierarchy in approximation algorithms. Lecture Notes for the MAPSP 2013 - Tutorial, June 2013
36. Schoenebeck, G.: Linear level Lasserre lower bounds for certain k-csps. In: FOCS, pp. 593–602 (2008)
37. Shor, N.: Class of global minimum bounds of polynomial functions. Cybernetics **23**(6), 731–734 (1987)
38. Tulsiani, M.: Csp gaps and reductions in the Lasserre hierarchy. In: STOC, pp. 303–312 (2009)

Replacing Mark Bits with Randomness in Fibonacci Heaps

Jerry Li$^{(\boxtimes)}$ and John Peebles

MIT, Cambridge, MA, USA
{jerryzli,jpeebles}@mit.edu

Abstract. A Fibonacci heap is a deterministic data structure implementing a priority queue with optimal amortized operation costs. An unfortunate aspect of Fibonacci heaps is that they must maintain a "mark bit" which serves only to ensure efficiency of heap operations, not correctness. Karger proposed a simple randomized variant of Fibonacci heaps in which mark bits are replaced by coin flips. This variant still has expected amortized cost $O(1)$ for insert, decrease-key, and merge. Karger conjectured that this data structure has expected amortized cost $O(\log s)$ for delete-min, where s is the number of heap operations.

We give a tight analysis of Karger's randomized Fibonacci heaps, resolving Karger's conjecture. Specifically, we obtain matching upper and lower bounds of $\Theta(\log^2 s / \log \log s)$ for the runtime of delete-min. We also prove a tight lower bound of $\Omega(\sqrt{n})$ on delete-min in terms of the number of heap elements n. The request sequence used to prove this bound also solves an open problem of Fredman on whether cascading cuts are necessary. Finally, we give a simple additional modification to these heaps which yields a tight runtime $O(\log^2 n / \log \log n)$ for delete-min.

1 Introduction

It is natural to explore the space of possible designs for common data structures. Doing so allows one to consider simpler alternative designs and gain more insight into whether particular features of a design are necessary or extraneous.

A natural class of data structures that is amenable to this sort of study is those that store additional information whose sole purpose is to ensure efficiency rather than correctness. The defining characteristic of such *extraneous data* is that the data structure still functions correctly—but perhaps more slowly—if the extraneous data is corrupted.

There are numerous data structures that posses extraneous data. For example, in red-black trees [GS78], the color of a node is extraneous data because even if we adversarially change it, the tree will still answers queries correctly—though perhaps more slowly. The balance factor of nodes in AVL trees and the mark bits of nodes in Fibonacci heaps are also extraneous data [AVL62, FT87].

In this paper, we characterize the extent to which the extraneous "mark bit" data contributes to the performance of Fibonacci heaps. More specifically, we

© Springer-Verlag Berlin Heidelberg 2015
M.M. Halldórsson et al. (Eds.): ICALP 2015, Part I, LNCS 9134, pp. 886–897, 2015.
DOI: 10.1007/978-3-662-47672-7_72

give a tight analysis of what happens to asymptotic performance if one replaces the mark bits with random bits (see Section 2 for details).

This is interesting for three reasons. First, replacing mark bits with random bits simplifies the design of Fibonacci heaps because there is no longer a need to store any mark bits. Second, our results can also be interpreted as an analysis of the performance of Fibonacci heaps under random corruption of mark bits. Third, our results solve an open problem of Fredman on whether cascading cuts are necessary in Fibonacci heaps [Fre05].

1.1 Related Work

The randomized variant of Fibonacci heaps studied in this paper was first proposed by Karger in unpublished work in 2000 [Kar00]. However, Karger's analysis of the performance of these Fibonacci heaps—which we'll call *randomized Fibonacci heaps*—was not tight. Specifically, Karger proved an upper bound of $O(\log^2 s)$ on the expected amortized cost of delete-min where s is the total number of Fibonacci heap operations performed so far. (It is easy to see that the expected amortized cost of all other operations is $O(1)$.) In terms of lower bounds, none better than the trivial sorting lower bound was known.

Following Karger's initial work, the analysis of Karger's randomized Fibonacci heaps has a somewhat amusing history. Hoping to encourage somebody to obtain a tight analysis of delete-min, Karger added this as a recurring bonus problem in MIT's annual graduate algorithms course [Kar13]. As a result, virtually every graduate student to go through MIT's theory group in the past 15 years has at least seen this problem, and many have actively worked on it.

Despite this attention, relatively little progress was made. We initially thought there had been none at all. After posting this paper, however, we were informed of two unpublished results by Price [Pri09] that had never been posted anywhere. These consist of two bounds in terms of s: a "lower bound" weaker than ours and an upper bound that is essentially the same as ours. Price gives an adversary that queries the randomized Fibonacci heap such that it must use $\Omega(\log^2 s / \log \log s)$ expected amortized time per delete-min. However, Price's "lower bound" cheats by allowing the adversary's request sequence to change depending on the random choices made by the randomized Fibonacci heap. To the best of our knowledge, all algorithms that employ Fibonacci heaps don't need to inspect their private state. Thus, in all settings we are aware of, Price's lower bound does not apply. Indeed, such results are not typically described as "lower bounds" in the data structures literature. Moreover, Price does not give any results in terms of n, the number of elements in the randomized Fibonacci heap.

Shortly after this paper was posted, Kaplan, Tarjan, and Zwick posted a paper which analyzes different variants of Fibonacci heaps [KTZ14]. Their work independently solved Fredman's open problem regarding the necessity of cascading cuts using similar techniques to ours.

More broadly, there are several data structures that have been studied which implement priority queues and achieve the same asymptotic performance as

Fibonacci heaps. These include [Pet87], [DGST88], [Høy95], [Tak03], [KT08], [Elm10], [HST11], [Cha13]. Additionally, there are many other works that deal with pairing heaps and their variants; eg., [FSST86], [Pet05], [Elm09]. Pairing heaps offer slightly worse asymptotic performance than Fibonacci heaps but are often faster in practice.

1.2 Terminology

We will differentiate between Fibonacci heaps as defined in [FT87] and Karger's randomized F-heaps by referring to the former as *standard* Fibonacci heaps and the latter as *randomized* Fibonacci heaps. However, when the data structure we are referring to is clear from context, we may simply call it an F-heap.

For variables, s will always refer to the number of operations that have been executed on an F-heap, and n will refer to the number of elements in the F-heap.

1.3 Our Contributions

We fully resolve Karger's question, giving a tight analysis of randomized F-heaps. We give a lower bound of $\Omega(\log^2 s/\log\log s)$ on the worst-case expected amortized runtime of delete-min. We also obtain a matching upper bound of $O(\log^2 s/\log\log s)$. Importantly, our lower bounds employ only non-adaptive request sequences which do not depend on the random outcomes of F-heap operations. Thus, in contrast to Price's work, our results truly are a lower bound on the amortized runtime of Karger's F-heaps.

The above two bounds are in terms of the number of F-heap operations s. In terms of the F-heap size n, we give a lower bound of $\Omega(\sqrt{n})$. (Previous work on pairing F-heaps implies a matching upper bound [FSST86].) The request sequence used to prove this lower bound gives an affirmative answer to the open question posed by Fredman on whether cascading cuts are necessary for performance in F-heaps.

Finally, we give a simple modification that improves the expected amortized performance to $\Theta(\log^2 n/\log\log n)$ by periodically rebuilding the F-heap.

1.4 Roadmap

In Section 2, we review the basic properties of standard F-heaps and define randomized F-heaps. In Section 3, we prove the tight $O(\log^2 s/\log\log s)$ upper bound on the expected amortized cost of delete-min in randomized F-heaps, where s is the number of F-heap operations. In Section 4, we give a tight lower bound of $\Omega(\sqrt{n})$, where n is the F-heap size. This bound serves as a warmup to our more challenging lower bound in the next section. In Section 5, we give a lower bound of $\Omega(\log s/\log\log s)$ on the cost of delete-min. In Section 6, we give a simple modification to Karger's randomized F-heaps which improves the performance of delete-min to $O(\log^2 n/\log\log n)$, replacing the s in the runtime with an n. We also show how to extend our work in Section 4 to yield a matching lower bound. In Section 7, we conclude and give possible directions for future work.

The full version of our paper [LP14] includes additional content which we omitted here for space considerations. In the full version, we explain how we resolve Fredman's open question. We also include proofs and large figures that could not be included in the main paper due to space considerations.

2   Background

A standard Fibonacci heap is a data structure that implements a priority queue and supports the operations insert, merge (or meld), decrease-key, and delete-min. The amortized runtimes of the first three operations is $O(1)$ and the amortized runtime of delete-min is $O(\log n)$ where n is the F-heap size.

We will consider time bounds which depend on the number of operations which have been performed on the F-heap so far, as well as the nukmer of elements in the F-heap. To reason formally about the average time complexity of our data structure, we require a slight extension of the usual definition of amortized runtime.

Definition 1. *We say a data structure has amortized runtime $O(f(n,s))$ if there is some constant $C > 0$ so that for any sequence of k operations on the data structure, the total runtime of the k operations is at most $C \sum_{i=1}^{k} f(n_i, i)$ where n_i is the number of elements in the F-heap at the ith operation.*

We will generally assume that the reader is familiar with the basic design and analysis of F-heaps. Those wishing to review this information may refer to the original paper [FT87] or any typical algorithms textbook.

Recall that each node in an F-heap allocates one bit of data called a mark bit. The only operation that uses the mark bit is the decrease-key operation. Specifically, the decrease-key operation starts by updating the key of the desired node and promoting it into the root list. Then, it starts from the node's former parent and walks up the tree, promoting nodes to the root list until it encounters a node with an unset mark bit. It then sets this node's mark bit and clears the mark bits of all nodes it promoted.

Karger defined *randomized F-heaps* as follows. A randomized F-heap behaves exactly like an F-heap with one exception: how it decides to stop promoting nodes in the decrease key operation. Recall that standard F-heaps look at the mark bit to determine whether to stop walking up the tree. In contrast, randomized F-heaps flip a coin to make this decision. Equivalently, one can think of a randomized F-heap as a simulation of a standard F-heap which intercepts queries to mark bits and returns random bits instead.

We also remark that the manner in which the root list of an F-heap is managed may depend on the specific implementation of the F-heap. However, neither our upper bounds nor our lower bounds depend on any ordering properties of the root list.

3   An $O(\log^2 s/\log\log s)$ Upper Bound

In this section, we upper bound the expected amortized cost of the operations of randomized F-heaps.

Theorem 1. *The expected amortized costs for a randomized F-heap's operations are $O\left(\log s \log n/\log\log s\right) \leq O\left(\log^2 s/\log\log s\right)$ for delete-min and $O(1)$ for everything else.*

We use a simplified version of the potential function introduced in [FT87]: if F is an F-heap, then we let $\Phi(F)$ be the number of root nodes in F. With this amortization, it is easy to see that insert, merge, and decrease-key all run in expected constant time. Thus it suffices to demonstrate that delete-min runs in expected time $O(\log s \log n/\log\log s)$.

Recall the specification for delete-min: we (1) remove the minimum element from the list of roots, (2) add all of its children to the root list, then (3) perform consolidation by rank. If k was the number of roots before the delete-min and r was the maximum rank[1] of any root node in the F-heap before performing step (3) above, then the real work performed is $O(k+r)$. The change in potential is $O(\log n - k)$, so with the correct scaling, the amortized cost of this operation is $O(r + \log n)$. Thus it suffices to show that $r \leq O(\log s \log n/\log\log s)$ in expectation.

We first upper-bound the probability that a node has lost many of its children since the last time it was in the root list. We say a non-root node v in an F-heap is *missing* a child if the child was removed from v and v has not been in the root list since that time.

Lemma 1. *Suppose we have an empty randomized F-heap and we intend to perform s operations on it which will result in an F-heap of size n. Then the probability that every non-root node in the resulting F-heap is missing at most k children is at least $1 - ns2^{-k}$.*

The proof is given in the full paper. As a corollary, we get the following:

Corollary 1. *With probability at least $1 - 1/n$, no node in the F-heap described in the above lemma is missing more than $k = 2\log n + \log s \leq 3\log s$ children.*

For any integer $k \geq 2$, let $f_k(x) = x^k - x^{k-1} - 1$. It is not hard to see that $f_k(x)$ is increasing for $x \geq 1$, has a unique positive root λ_k, and that $\lambda_k > 1$. By by a more involved version of the analysis in the proof of Corollary 1 in [FT87] one can obtain the following result, whose proof we defer to the full paper.

Lemma 2. *Fix k sufficiently large. Suppose a tree in an F-heap with n nodes has the property that no non-root node in the tree is missing more than k children. Then the root has rank $O(\log_{\lambda_k} n)$.*

---

[1] Recall that the rank of a node in an F-heap is the number of children it has.

We also need a technical lemma about the behavior of λ_k, whose proof we defer to the full paper.

Lemma 3. *For k sufficiently large, $1/\log \lambda_k \leq 2k/\log k$.*

Proof (of Theorem 1). Insert, merge, and decrease-key are obviously $O(1)$ so we focus our attention to demonstrating the bound for delete-min. The expected amortized cost of delete-min is at most the maximum rank r of any root node. By Lemma 2, $r \leq O(\log_{\lambda_k} n)$ where k is a bound on how many children are missing from any non-root node in the tree. We can break up $\mathbb{E}[r]$ into two terms and bound them separately. We have,

$$\mathbb{E}[r] = \Pr[k \geq 3 \log s] \cdot \mathbb{E}[r|k \geq 3 \log s] + \Pr[k < 3 \log s] \cdot \mathbb{E}[r|k < 3 \log s].$$

The first term is bounded by $n\Pr[k \geq 3 \log s] \leq 1$ by Corollary 1. The second is bounded by

$$\mathbb{E}[r|k < 3 \log s] \leq \log n / \log \lambda_{3 \log s} = O(\log s \log n / \log \log s)$$

by Lemma 3. Thus, the total expected amortized cost of delete-min is $O(\log s \log n / \log \log s)$.

4   An $\Omega(\sqrt{n})$ Lower Bound

This section is dedicated to the proof of the following lower bound:

Theorem 2. *There exists a request sequence for randomized F-heaps whose expected cost is $\Omega(\sqrt{n})$ per operation on average, where n is the size of the F-heap.*

Our proof also shows that so-called "cascading cuts" are necessary in F-heaps, solving an open problem of Fredman. See the full paper for details.

It is worth clarifying what we mean when we say "$\Omega(\sqrt{n})$ per operation on average" since the F-heap size can change from operation to operation. Formally, this means the sum of the square roots of the F-heap sizes before each operation divided by the number of operations.

Note that the analysis used in Section 2 of [FSST86] proves a matching upper bound. In fact, their upper bound applies for a more general class of pairing-heap-like structures. We also remark that while the expected cost of each operation in the request sequence is $\Omega(\sqrt{n})$ on average per operation, the request sequence has exponential length. We rectify this and obtain a tight bound for the expected amortized cost in terms of s in Section 5.

Notice that the theorem is equivalent to saying that there is a request sequence such that—no matter how one tries to amortize the cost of the operations—there will always be an operation with cost $\Omega(\sqrt{n})$.[2]

---

[2] To see this, let the average per operation cost be c_1 and the maximum amortized cost of an operation be c_2. Then for the total cost c_0 of all operations in a request sequence of length s, we have $c_1 s = c_0 \leq c_2 s$. Thus, $c_1 \leq c_2$.

While it is easy to slightly modify randomized F-heaps to "get around" this lower bound, we include this construction for three reasons. First, it applies to Karger's randomized F-heaps as they were originally formulated. Second, it is a good warm-up to the more complicated construction in Section 5, which is extended in Section 6 to apply even to these modified F-heaps—where s is replaced with n in the statement of the bound. Finally, the request sequence we construct solves Fredman's open question about the necessity of cascading cuts; see the full paper for more details.

The main idea is that by using a very large number of requests, we can force the F-heap into a very bad configuration with high probability. In particular, we exhibit a configuration which we call the *bad state* shown in Figure 1 below.

Formally, the *bad state of rank* $\sqrt{n}$ is an F-heap with trees of rank i for all i from 0 through $\sqrt{n}$ where all trees have height 1, except the rank 0 tree which has height 0. For simplicity, we assume $\sqrt{n}$ is integral. Notice that the total number of F-heap elements is $\Theta(n)$.

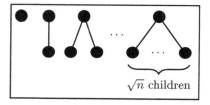

Fig. 1. The bad state of rank $\sqrt{n}$

The bad state has the following two key properties, which we encapsulate in the following two lemmas:

Lemma 4. *There exists a constant length sequence of operations which—when applied to an F-heap in the bad state of rank* $\sqrt{n}$—*returns the F-heap to the bad state and takes* $\Omega(\sqrt{n})$ *time to execute.*

Lemma 5. *There exists a finite length request sequence which, when applied to an empty F-heap, results in an F-heap in the bad state of rank* $\sqrt{n}$ *with probability at least* $1/2$.

Together, these properties imply Theorem 2.

Proof (of Theorem 2 assuming Lemma 4 and Lemma 5). Fix an n. Construct a request sequence as follows. First, use Lemma 5 to construct the first part of the request sequence. With probability at least $1/2$, this result in an F-heap in the bad state of rank $\sqrt{n}$. Moreover, this takes S operations to execute, where S is finite, known, and depends only on n. Then, follow it with S copies of the request sequence guaranteed by Lemma 4. Conditioning on the event that the first part of the request sequence resulted in an F-heap in the bad state of rank $\sqrt{n}$, by Lemma 4, each copy takes $O(\sqrt{n})$ time to execute. Thus we execute $O(S)$ operations on the F-heap, and with probability at least $1/2$, the operations take at least $O(S\sqrt{n})$ time. Therefore the expected average cost of executing this request sequence is $\Omega(\sqrt{n})$ per step on average.

Thus, all that is left is to prove Lemma 4 and Lemma 5 which we do in the following two subsections, respectively.

4.1   Proof of Lemma 4

From the bad state, it is straightforward to force the F-heap to spend $\Omega(\sqrt{n})$ time on a delete-min. In this subsection, we prove this fact.

Proof (of Lemma 4). Consider the following request sequence:

1. Add two elements t_1, t_2 less than every element in the F-heap with $t_1 < t_2$.
2. Delete-min twice.

Applying this procedure to an F-heap in the bad state of rank $\sqrt{n}$ yields the cycle of states shown in a figure in the full paperdeterministically. Notice that the state of the F-heap after applying these operations is unchanged. Moreover, it is clear that the last delete-min operation in the procedure takes $\Theta(\sqrt{n})$ time.

4.2   Proof of Lemma 5

Call a tree of height 1 where the root has c children the *c-star*, so that the bad state consists of one *c*-star, for each $0 \leq c \leq \sqrt{n}$.

We will show how to force the F-heap to construct the bad state by forcing it to construct each *c*-star in the bad state in order from large c to small. Specifically, we will use the following lemma

Lemma 6. *For every c and $\epsilon > 0$, there exists a sequence of operations which (starting from an empty F-heap) results in an F-heap which is a c-star with probability at least $1 - \epsilon$, and which at no point ever constructs a node with rank $> c$.*

We first explain why Lemma 6 implies Lemma 5.

Proof (of Lemma 5 assuming Lemma 6). Our request sequence is obtained by taking the sequences obtained from Lemma 6 for each c from 0 through $\sqrt{n}$ with ϵ sufficiently small, then concatenating the sequences in order from largest c to smallest c. It is easy to see that this sequence results in the desired F-heap.

Proof (of Lemma 6). We proceed by induction on c. For $c = 0$, simply start with an empty F-heap and insert u. This results in the desired F-heap with probability 1. Inductively, suppose the statement is true for $c = k$. Fix $\epsilon > 0$. By induction, there is a request sequence which produces a k-star with probability at least $\sqrt{1-\epsilon}$. Below, we describe a request sequence which constructs a $(k+1)$-star from a k-star with probability at least $\sqrt{1-\epsilon}$. Then by concatenating this request sequence to the one obtained via induction, we produce the desired request sequence. In particular, this request sequence gives rise to the desired F-heap with probability at least $(\sqrt{1-\epsilon})(\sqrt{1-\epsilon}) = 1-\epsilon$ as desired.

Assume the F-heap is a k-star rooted at u. Now insert a node v with $v > u$. This results in the F-heap shown below.

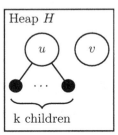

Consider the following procedure which we will apply a large number of times.

1. Add $2^k - 1$ nodes $s_1, \ldots, s_{2^k-1}$ such that $u < v < s_1 < s_2 < \ldots < s_{2^k-1}$.
2. Add a node t smaller than all other nodes in the F-heap and perform a delete-min. (This removes t and consolidates the rest of the nodes.)
3. For all $1 \le i \le 2^k - 1$, decrease the key of s_i to be minimum in the F-heap and delete-min, removing it. The order is arbitrary.

Given an F-heap H as shown in Figure 2, if we apply this procedure over and over again, the state of H after any particular application of the procedure is given by the Markov process shown by the flowchart in Figure 2. A more detailed step-by-step version of the flowchart is given in a figure in the full paper.

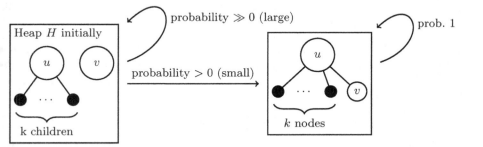

Fig. 2. High-Level description of how one iteration of our procedure works. Each box represents a state of the F-heap and each arrow represents the probability of going from one state to the other after applying steps 1–3 once. After a large number of applications, we will get stuck in the state on the right with high probability.

Notice that H always has a positive probability of gaining a single extra child (and no extra descendants), resulting in the F-heap we are trying to create.

Furthermore, once H enters this state, it will never leave. Notice additionally that in none of these possible transitions do we ever produce a tree with rank greater than $k + 1$. As such, if we apply the procedure a sufficiently large number of times—and provided H had the structure shown in Figure 2—we can construct a sequence of operations that gives the desired resulting H with probability arbitrarily high. By repeating this request sequence sufficiently many times such that this probability is at least $\sqrt{1 - \epsilon}$, we are done.

5   The $\Omega(\log^2 s / \log \log s)$ Lower Bound

This section is devoted to proving the following theorem:

Theorem 3. *There exists a request sequence for randomized F-heaps whose expected cost is $\Omega(\log^2 s / \log \log s)$ per operation on average, where s is the number of F-heap operations.*

Our approach to this bound has the similar structure to Theorem 2: get the F-heap into a "bad" state then have it perform a costly operation repeatedly. However, to prove that bound, we constructed an exponentially long request sequence. The challenge in proving the present bound is that we now need a subexponential length request sequence.

For this bound, the "bad" state we will force the F-heap into is defined as a *generalized bad state* of rank m and is shown in Figure 3.[3] Formally, an F-heap is in a generalized bad state of rank m if it has $m + 1$ root nodes, where the ith root node has rank i, for $0 \leq i \leq m$. Once we get the F-heap into a state of this form, we will use an analog of Lemma 4 to make the F-heap perform costly operations, just as in the proof of Theorem 2.

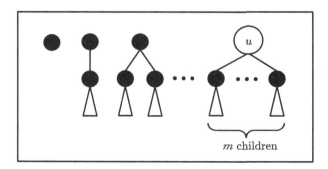

Fig. 3. A generalized bad state

The analogs of Lemma 4 and Lemma 5 we will use are the following:

---

[3] More specifically, we will force the F-heap into a specific known state which is of the form shown in the figure.

Lemma 7. *There exists a constant length sequence of operations which—when applied to an F-heap in a generalized bad state of rank m—returns the F-heap to a generalized bad state of rank m and takes $\Omega(m)$ time to execute.*

Proof. The sequence of operations and proof is exactly the same as in Lemma 4.

Lemma 8. *There exists a request sequence of length $2^{O(\sqrt{m \log m})}$ which, when applied to an empty F-heap, results in an F-heap in a generalized bad state of rank m with probability at least $1/2$.*

Proof (Proof of Theorem 3 assuming Lemma 7 and Lemma 8). Fix an m. Construct a request sequence as follows: use Lemma 8 to construct the first part of the request sequence with length $\ell(m) = 2^{O(\sqrt{m \log m})}$. Follow it with $\ell(m)$ copies of the constant length request sequence given by Lemma 7. This request sequence makes $\Theta(\ell(m))$ requests and takes $\Omega(\ell(m)m)$ time, thus the average time per request is $\Omega(m)$. Letting $s = \ell(m)$ so that $m = \ell^{-1}(s)$, we see that executing s operations takes $\Omega(m) = \Omega(\ell^{-1}(s))$ time per operation on average. Since $\ell^{-1}(s) = \Omega(\log^2 s / \log\log s)$, this completes the proof.

The proof of Lemma 8 is given in the full paper.

6   Going from $\Theta(\log^2 s / \log\log s)$ to $\Theta(\log^2 n / \log\log n)$

In this section, we eliminate the dependence on s in the runtime of randomized F-heaps via a simple change. Specifically, after every operation, we rebuild the F-heap with probability $1/n$. Rebuilding is done as follows: Create a new randomized F-heap, and insert all elements from the old F-heap into the new F-heap. We refer to these self-rebuilding F-heaps as *augmented* randomized F-heaps.

Theorem 4. *The augmented randomized F-heap has worst-case expected amortized runtime $O(\log^2 n / \log\log n)$ for delete-min and $O(1)$ for everything else.*

Theorem 5. *There exists a request sequence for augmented randomized F-heaps whose expected cost is $\Omega(\log^2 n / \log\log n)$ per operation on average, where n is the number of F-heap elements.*

Theorem 4 can be proven by using the same techniques as Theorem 1, so we omit it. Theorem 5 is proven in the full paper.

7   Conclusion and Acknowledgments

This work gave the first tight analysis of randomized F-heaps, resolving a 15 year old question of Karger and a 10 year old open problem of Fredman. We showed that replacing the extraneous mark bit data in F-heaps hurts performance, but only by roughly a log factor. A natural question for further work is whether replacing extraneous data with randomness in other data structures like red-black trees and AVL trees also hurts their performance.

Acknowledgments. We thank David Karger for informing us of this problem and for pointing out that our upper bound was tighter than originally thought.

J.L. was supported by NSF Award # CCF-1217921 and DOE Award # DE-SC0008923 and an Akamai Presidential Fellowship.

J.P.: This material is based upon work supported by the National Science Foundation Graduate Research Fellowship under Grant No. 1122374. An Akamai Presidential Fellowship also supported this work.

References

[AVL62] Adelson-Velskii, G.M., Landis, E.M.: An algorithm for the organization of information. Dokl. Akad. Nauk SSSR **3**, 263–266 (1962)

[Cha13] Chan, T.M.: Quake heaps: a simple alternative to fibonacci heaps. In: Brodnik, A., López-Ortiz, A., Raman, V., Viola, A. (eds.) Ianfest-66. LNCS, vol. 8066, pp. 27–32. Springer, Heidelberg (2013)

[DGST88] Driscoll, J.R., Gabow, H.N., Shrairman, R., Tarjan, R.E.: Relaxed heaps: An alternative to Fibonacci heaps with applications to parallel computation. Commun. ACM **31**(11), 1343–1354 (1988)

[Elm09] Elmasry, A.: Pairing heaps with o(log log n) decrease cost. In: Claire Mathieu, editor, SODA, pp. 471–476. SIAM (2009)

[Elm10] Elmasry, A.: The violation heap: a relaxed Fibonacci-like heap. Discrete Math., Alg. and Appl. **2**(4), 493–504 (2010)

[Fre05] Fredman, M.L.: Binomial, Fibonacci, and pairing heaps. In: Mehta, D.P., Sahni, S. (eds.) Handbook of data structures and applications. Chapman & Hall/CRC, Boca Raton (2005)

[FSST86] Fredman, M.L., Sedgewick, R., Sleator, D.D., Tarjan, R.E.: The pairing heap: A new form of self-adjusting heap. Algorithmica **1**(1), 111–129 (1986)

[FT87] Fredman, M.L., Tarjan, R.E.: Fibonacci heaps and their uses in improved network optimization algorithms. J. ACM **34**(3), 596–615 (1987)

[GS78] Guibas, L.J., Sedgewick, R.: A dichromatic framework for balanced trees. FOCS **1978**, 8–21 (1978)

[Høy95] Høyer, P.: A general technique for implementation of efficient priority queues. In: ISTCS 1995, ISTCS 1995, p. 57-, Washington, DC, IEEE Computer Society (1995)

[HST11] Haeupler, B., Sen, S., Tarjan, R.E.: Rank-pairing heaps. SIAM J. Comput. **40**(6), 1463–1485 (2011)

[Kar00] Karger, D.: untitled manuscript. unpublished (2000)

[Kar13] Karger, D.: personal communication (2013)

[KT08] Kaplan, H., Tarjan, R.E.: Thin heaps, thick heaps. ACM Trans. Algorithms **4**(1), 3:1–3:14 (2008)

[KTZ14] Kaplan, H., Tarjan, R.E., Zwick, U.: Fibonacci heaps revisited (2014). CoRR, abs/1407.5750

[LP14] Li, J., Peebles, J.: Replacing mark bits with randomness in fibonacci heaps (2014). CoRR, abs/1407.2569

[Pet87] Peterson, G.: A balanced tree scheme for meldable heaps with updates. Technical Report GIT-ICS-87-23, Georgia Institute of Technology (1987)

[Pet05] Pettie, S.: Towards a final analysis of pairing heaps. FOCS **2005**, 174–183 (2005)

[Pri09] Price, E.: Randomized Fibonacci heaps. unpublished (2009)

[Tak03] Takaoka, T.: Theory of 2–3 heaps. Discrete Appl. Math. **126**(1), 115–128 (2003)

A PTAS for the Weighted Unit Disk Cover Problem

Jian Li and Yifei Jin[✉]

IIIS, Tsinghua University, Beijing, China
lijian83@mail.tsinghua.edu.cn, jin-yf13@mails.tsinghua.edu.cn

Abstract. We are given a set of weighted unit disks and a set of points in Euclidean plane. The minimum weight unit disk cover (WUDC) problem asks for a subset of disks of minimum total weight that covers all given points. WUDC is one of the geometric set cover problems, which have been studied extensively for the past two decades (for many different geometric range spaces, such as (unit) disks, halfspaces, rectangles, triangles). It is known that the unweighted WUDC problem is NP-hard and admits a polynomial-time approximation scheme (PTAS). For the weighted WUDC problem, several constant approximations have been developed. However, whether the problem admits a PTAS has been an open question. In this paper, we answer this question affirmatively by presenting the first PTAS for WUDC. Our result implies the first PTAS for the minimum weight dominating set problem in unit disk graphs. Combining with existing ideas, our result can also be used to obtain the first PTAS for the maxmimum lifetime coverage problem and an improved constant approximation ratio for the connected dominating set problem in unit disk graphs.

1 Introduction

The set cover (SC) problem is a central problem in theoretical computer science and combinatorial optimziation. In the problem, we are given a ground set U and collection S of subsets of U. Each set $S \in S$ has a non-nagative weight w_S. The goal is to find a subcollection $C \subseteq S$ of minimum total weight such that $\bigcup C$ covers all elements of U. The approximibility of the general SC problem is rather well understood: it is well known that the greedy algorithm is an H_n-approximation ($H_n = \sum_{i=1}^{n} 1/i$) and obtaining a $(1 - \epsilon) \ln n$-approximation for any constant $\epsilon > 0$ is NP-hard [12,19]. In the *geometric set cover problem*, U is a set of points in some Euclidean space $\mathbb{R}^d$, and S consists of geometric objects (e.g., disks, squares, triangles), In such geometric setting, we can hope for better-than-logarithmic approximations due to the special structure of S. Most geometric set cover problems are NP-hard, even for the very simple classes of objects such as unit disks [8,24] (see [6,22] for more examples and exceptions). Approximation algorithms for geometric set cover have been studied extensively for the past two decades, not only because of the importance of the problem per se, but also its rich connections to other important notions and problems, such as VC-dimension [4,9,18], ϵ-net, union complexity [7,32,33], planar separators [20,29], even machine scheduling problems [3].

In this work, we study the geometric set cover problem with one of the simplest class of objects, unit disks. The formal definition of our problem is as follows:

Research supported in part by the National Basic Research Program of China Grant 2015CB358700, 2011CBA00300, 2011CBA00301, the National Natural Science Foundation of China Grant 61202009, 61033001, 61361136003.

© Springer-Verlag Berlin Heidelberg 2015
M.M. Halldórsson et al. (Eds.): ICALP 2015, Part I, LNCS 9134, pp. 898–909, 2015.
DOI: 10.1007/978-3-662-47672-7_73

Definition 1. *Weighted Unit Disk Cover (WUDC): Given a set $\mathcal{D} = \{D_1, \ldots, D_n\}$ of n unit disks and a set $\mathcal{P} = \{P_1, \ldots, P_m\}$ of m points in Euclidean plane $\mathbb{R}^2$. Each disk D_i has a weight $w(D_i)$. Our goal is to choose a subset of disks to cover all points in $\mathcal{P}$, and the total weight of the chosen disks is minimized.*

We note that WUDC is the general version of minimum weight dominating set problem in unit disk graphs (UDG). In fact, several previous results on WUDC were stated in the context of the dominating set problem.

1.1 Previous Results and Our Contribution

We first recall that a polynomial time approximation scheme (PTAS) for a minimization problem is an algorithm $\mathcal{A}$ that takes an input instance, a constant $\epsilon > 0$, returns a solution SOL such that SOL $\leq (1 + \epsilon)$OPT, where OPT is the optimal value, and the running time of $\mathcal{A}$ is polynomially in the size of the input for any fixed constant ϵ.

WUDC is NP-hard, even for the unweighted version (i.e., $w(D_i) = 1$) [8]. For unweighted dominating set in unit disk graphs, Hunt et al. [26] obtained the first PTAS in unit disk graphs. For the more general disk graphs, based on the connection between geometric set cover problem and ϵ-nets, developed in [4,9,18], and the existence of ϵ-net of size $O(1/\epsilon)$ for halfspaces in $\mathbb{R}^3$ [30] (see also [21]), it is possible to achieve a constant factor approximation. As estimated in [29], these constants are at best 20 (A recent result [5] shows that the constant is at most 13). Moreover, there exists a PTAS for unweighted disk cover and minimum dominating set via the local search technique [20,29].

For the general weighted WUDC problem, the story is longer. Ambühl et al. [2] obtained the first approximation for WUDC with a concrete constant 72, without using the ϵ-net machinery. Applying the shifting techique of [23], Huang et al. [25] obtained a $(6 + \epsilon)$-approximation algorithm for WUDC. The approximation factor was later improved to $(5 + \epsilon)$ [10], and to $(4 + \epsilon)$ by several groups [11,16,34]. The current best ratio is 3.63. [1] Besides, the *quasi-uniform sampling method* [7,33] provides another approach to achieve a constant factor approximation for WUDC (even in disk graphs). However, the constant depends on several other constants from rounding LPs and the size of *the union complexity*. Very recently, based on the separator framework of Adamaszek and Wiese [1], Mustafa and Raman [28] obtained a QPTAS (Quasi-polynomial time approximation scheme) for weighted disks in $\mathbb{R}^2$ (in fact, weighted halfspaces in $\mathbb{R}^3$), thus ruling out the APX-hardness of WUDC.

Another closely related work is by Erlebach and van Leeuwen [17], who obtained a PTAS for set cover on weighted unit squares, which is the first PTAS for weighted geometric set cover on any planar objects (except those poly-time solvable cases [6,22]). Although it may seem that their result is quite close to a PTAS for weighted WUDC, as admitted in their paper, their technique is insufficient for handling unit disks and "completely different insight is required".

In light of all the aforementioned results, it seems that we should expect a PTAS for WUDC, but it remains to be an open question (explicitly mentioned as an open problem

---

[1] The algorithm can be found in Du and Wan [14], who attributed the result to a manuscript by Willson et al.

in a number of previous papers, e.g., [2, 14–17, 31]). Our main contribution in this paper is to settle this question affirmatively by presenting the first PTAS for WUDC.

Theorem 1. *There is a polynomial time approximation scheme for the WUDC problem. The running time is $n^{O(1/\epsilon^9)}$.*

Due to the equivalence between WUDC and minimum weight dominating set in unit disk graphs, we immediately have the following corollary.

Corollary 1. *There is a polynomial time approximation scheme for the minimum weight dominating set problem in unit disk graphs.*

We note that the running time $n^{\text{poly}(1/\epsilon)}$ is nearly optimal in light of the negative result by Marx [27], who showed that an EPTAS (i.e., Efficient PTAS, with running time $f(1/\epsilon)\text{poly}(n)$) even for the unweighted dominating set in UDG would contradict the exponential time hypothesis.

Finally, we show that our PTAS for WUDC can be used to obtain improved approximation algorithms for two important problems in wireless sensor networks, the connected dominating set problem and the maximum lifetime coverage problem in UDG.

2 Our Approach - A High Level Overview

By the standard shifting technique[13], it suffices to provide a PTAS for WUDC when all disks are located in a square of constant size (we call it a block, and the constant depends on $1/\epsilon$). This idea is formalized in Huang et al. [25], as follows.

Lemma 1 (Huang ta al. [25]). *Suppose there exists a ρ-approximation for WUDC in a fixed $L \times L$ block, with running time $f(L)$. Then there exists a $(\rho + O(1/L))$-approximation with running time $O(L \cdot n \cdot f(L))$ for WUDC. In particular, setting $L = 1/\epsilon$, there exists a $(\rho+\epsilon)$-approximation for WUDC, with running time $O\left(\frac{1}{\epsilon} \cdot L \cdot f(\frac{1}{\epsilon})\right)$.*

In fact, almost all previous constant factor approximation algorithms for WUDC were obtained by developing constant approximations for a single block of a constant size (which is the main difficulty). The main contribution of the paper is to improve on the previous work [2, 10, 16, 25] for a single block, as in the following lemma.

Lemma 2. *There exists a PTAS for WUDC in a fixed block of size $L \times L$ for $L = 1/\epsilon$. The running time of the PTAS is $n^{O(1/\epsilon^9)}$*

From now on, the approximation error guarantee $\epsilon > 0$ is a fixed constant. Whenever we say a quantity is a constant, the constant may depend on ϵ. We use OPT to represent the optimal solution (and the optimal value) in this block. We use capital letters $A, B, C, \ldots$ to denote points, and small letters $a, b, c, \ldots$ to denote arcs. For two points A and B, we use $|AB|$ to denote the line segment connecting A and B (and its length). We use D_i to denote a disk and D_i to denote its center. For a point A and a real $r > 0$, let $\mathsf{D}(A, r)$ be the disk centered at A with radius r. For a disk D_i, we use $\partial\mathsf{D}_i$ to denote its boundary. We call a segment of $\partial\mathsf{D}_i$ *an arc*.

First, we guess that whether OPT contains more than C disks or not for some constant C. If OPT contains no more than C disks, we enumerate all possible combinations and choose the one which covers all points and has the minimum weight. This takes $O\left(\sum_{i=1}^{C} \binom{n}{i}\right) = O(n^C)$ time, which is polynomial.

The more challenging case is whether OPT contains more than C disks. In this case, we guess (i.e., enumerate all possibilities) the set $\mathcal{G}$ of the C most expensive disks in OPT. There are at most a polynomial number (i.e., $O(n^C)$) possible guesses. Suppose our guess is correct. Then, we delete all disks in $\mathcal{G}$ and all points that are covered by $\mathcal{G}$. Let D_t (with weight w_t) be the cheapest disk in $\mathcal{G}$. We can see that OPT $\geq Cw_t$. Moreover, we can also safely ignore all disks with weight larger than w_t (assuming that our guess is correct). Now, our task is to cover the remaining points with the remaining disks, each having weight at most w_t. We use $\mathcal{D}' = \mathcal{D} \setminus \mathcal{G}$ and $\mathcal{P}' = \mathcal{P} \setminus \mathcal{P}(\mathcal{G})$ to denote the set of the remaining disks and the set of remaining points respectively, where $\mathcal{P}(\mathcal{G})$ denote the set of points covered by some disk in $\mathcal{G}$.

Next, we carefully choose to include in our solution a set $\mathcal{H} \subseteq \mathcal{D}'$ of at most ϵC disks. The purpose of $\mathcal{H}$ is to break the whole instance into many (still a constant) small pieces (substructures), such that each substructure can be solved optimally, via dynamic programming. [2] One difficulty is that the substructures are not independent and may interact with each other (i.e., a disk may appear in more than one substructure). In order to apply the dynamic programming technique to all substructures simultaneously, we have to ensure the orders of the disks in different substructures are consistent with each other. Choosing $\mathcal{H}$ to ensure a globally consistent order of disks is in fact the main technical challenge of the paper.

Suppose we have a set $\mathcal{H}$ which suits our need (i.e., the remaining instance $(\mathcal{D}' \setminus \mathcal{H}, \mathcal{P}' \setminus \mathcal{P}(\mathcal{H}))$ can be solved optimally in polynomial time by dynamic programming). Let $\mathcal{S}$ be the optimal solution of the remaining instance. Our final solution is SOL $= \mathcal{G} \cup \mathcal{H} \cup \mathcal{S}$. First, we can see that $w(\mathcal{S}) \leq w(\mathsf{OPT} - \mathcal{G} - \mathcal{H}) \leq \mathsf{OPT} - w(\mathcal{G})$, since OPT $- \mathcal{G} - \mathcal{H}$ is a feasible solution for the instance $(\mathcal{D}' \setminus \mathcal{H}, \mathcal{P}' \setminus \mathcal{P}(\mathcal{H}))$. Hence, we have that SOL $= w(\mathcal{G}) + w(\mathcal{H}) + w(\mathcal{S}) \leq \mathsf{OPT} + \epsilon C w_t \leq (1 + \epsilon)\mathsf{OPT}$, where the 2nd to last inequality holds because $|\mathcal{H}| \leq \epsilon C$, and the last inequality uses the fact that OPT $\geq w(\mathcal{G}) \geq C w_t$.

Constructing $\mathcal{H}$: Now, we provide a high level sketch for how to construct $\mathcal{H} \subseteq \mathcal{D}'$. First, we partition the block into *small squares* of side length $\mu = O(\epsilon)$ such that any disk centered in a square can cover the whole square and the disks in the same square are close enough. Let the set of small squares be $\Xi = \{\Gamma_{ij}\}_{1 \leq i,j \leq K}$ where $K = L/\mu$. For a small square Γ, let $\mathsf{D}_{s_\Gamma} \in \Gamma$ and $\mathsf{D}_{t_\Gamma} \in \Gamma$ be the furthest pair of disks (i.e., $|D_{s_\Gamma} D_{t_\Gamma}|$ is maximized). We include the pair D_{s_Γ} and D_{t_Γ} in $\mathcal{H}$, for every small square $\Gamma \in \Xi$, and call the pair *the square gadget* for Γ. We only need to focus on covering the remaining points in the *uncovered region* $\mathbb{U}(\mathcal{H})$.

We consider all disks in a small square Γ. The uncovered portion of those disks defined two disjoint connected regions We call such a region, together with all relevant arcs, a *substructure* (formal definition in Section 4). In fact, we can solve the disk covering problem for a single substructure optimally using dynamic programming (which

---

[2] An individual substructure can be solved using a dynamic program similar to [2,22].

is similar to the dynamic program in[2, 22]). It appears that we are almost done, since ("intuitively") all square gadgets have already covered much area of the entire block, and we should be able to use similar dynamic program to handle all such substructures as well. However, the situation is more complicated (than we initially expected) since the arcs are dependent. See Figure 1 for a "not-so-complicated" example. Firstly, there may exist two arcs (*sibling arcs*) which belong to the same disk when the disk is centered in the *core-center area*). The dynamic program has to make decisions for two sibling arcs, which belong to two different substructures (called R-correlated substructures), together. Second, in order to carry out dynamic program, we need a suitable order of all arcs. To ensure such an order exists, we need all substructures interact with each other "nicely".

In particular, besides all square gadgets, we need to add into $\mathcal{H}$ a constant number of extra disks. This is done by a series of "cut" operations. A cut can either break a cycle, or break one substructure into two substructures. To capture how substructures interact, we define an auxiliary graph, call substructure relation graph $\mathfrak{S}$, in which each substructure is a node. The aforementioned R-correlations define a set of blue edges, and geometrically overlapping relation define a set of red edges. Though the cut operations, we can make blue edges form a matching, and red edges also form a matching, and $\mathfrak{S}$ acyclic (we call $\mathfrak{S}$ an acyclic 2-matching). The special structure of $\mathfrak{S}$ allows us to define an ordering of all arcs easily. Together with some other simple properties, we can generalize the dynamic program for one substructure to all substructures.

3 Square Gadgets

We discuss the structure of a square gadget $\mathsf{Gg}(\Gamma)$ associated with the small square Γ. Recall that the square gadget $\mathsf{Gg}(\Gamma) = \mathsf{D}_s \cup \mathsf{D}_t$, where D_s and D_t are the furthest pair of disks in Γ. We can see that for any disk D_i in Γ, there are either one or two arcs of $\partial \mathsf{D}_i$ which are not covered by $\mathsf{Gg}(\Gamma)$. Without loss of generality, assume that $\mathsf{D}_s \mathsf{D}_t$ is horizontal. The line $\mathsf{D}_s \mathsf{D}_t$ divides the whole plane into two half-planes which are denoted by H^+ (the upper half-plane) and H^- (the lower half-plane). $\partial \mathsf{D}_s$ and $\partial \mathsf{D}_t$ intersect at two points P and Q. We need a few definitions which are useful throughout the paper.

1. (Center Area and Core-center Area) Define the *center area* of $\mathsf{Gg}(\Gamma)$ as the intersection of the two disks $\mathsf{D}(\mathsf{D}_s, r_{st})$ and $\mathsf{D}(\mathsf{D}_t, r_{st})$ in the square Γ, where $r_{st} = |\mathsf{D}_s \mathsf{D}_t|$. We use $\mathfrak{C}$ to denote it. Since D_s and D_t are the furthest pair, we can see that every other disk in Γ is centered in the center area $\mathfrak{C}$.

    We define the *core-center area* of $\mathsf{Gg}(\Gamma)$ is the intersection of two unit disks centered at P, Q respectively. Essentially, any unit disk centered in the core-center area has four intersections with the boundary of gadget. Let us denote the area by $\mathfrak{C}_o$.

2. (Active Region) Consider the regions $\left(\bigcup_{\mathsf{D}_i \in \mathfrak{C}_o} \mathsf{D}_i - (\mathsf{D}_s \cup \mathsf{D}_t) \right) \cap H^+$ and $\left(\bigcup_{\mathsf{D}_i \in \mathfrak{C}_o} \mathsf{D}_i - (\mathsf{D}_s \cup \mathsf{D}_t) \right) \cap H^-$. We call each of them an *active region* associated with square Γ. An active region can be covered by disks in the core-center area. We use Ar to denote an active region.

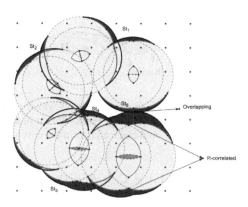

Fig. 1. The general picture of the substructures in a block. The red points are the grid points of squares. Dash green disks are what we have selected in $\mathcal{H}$. There are five substructures in the block.

4 Substructures

Initially, $\mathcal{H}$ includes all square gadgets. In Section 7, we will include in $\mathcal{H}$ a constant number of extra disks. For a set S of disk, we use $\mathbb{R}(S)$ to denote the region covered by disks in S (i.e., $\cup_{D_i \in S} D_i$). Assuming a fixed $\mathcal{H}$, we now describe the basic structure of the uncovered region $\mathbb{R}(\mathcal{D}') - \mathbb{R}(\mathcal{H})$. [3] For ease of notation, we use $\mathbb{U}(\mathcal{H})$ to denote the uncovered region $\mathbb{R}(\mathcal{D}') - \mathbb{R}(\mathcal{H})$. Figure 1 shows an example. Intuitively, the region consists of several "strips" along the boundary of $\mathcal{H}$. Now, we define some notions to describe the structure of those strips.

1. (Baseline) We use $\partial \mathcal{H}$ to denote to be the boundary of $\mathcal{H}$. Consider an arc a whose endpoints P_1, P_2 are on $\partial \mathcal{H}$. We say the arc a cover a point $P \in \partial \mathcal{H}$, if P lies in the segment between P_1 and P_2 along $\partial \mathcal{H}$. We say a point $P \in \partial \mathcal{H}$ *can be covered* if some arc in $\mathcal{D}'$ covers P. A baseline is a consecutive maximal segment of $\partial \mathcal{H}$ that can be covered. We usually use b to denote a baseline.

2. (Substructure) A substructure $\mathsf{St}(\mathsf{b}, \mathcal{A})$ consists of a baseline b and the collection $\mathcal{A}$ of arcs which can cover some point in b. The two endpoints of each arc $a \in \mathcal{A}$ are on b and $\angle(a)$ is less than π. Note that every point of b is covered by some arc in $\mathcal{A}$. Figure 2 illustrates the components of an substructure.

Arc Order: Now we switch our attention to the order of the arcs in a substructure $\mathsf{St}(\mathsf{b}, \mathcal{A})$. Suppose the baseline b starts at point Q_s and ends up at point Q_t. Consider any two points P_1 and P_2 on the baseline b. If P_1 is more close to Q_s than P_2 along the baseline b, we say that P_1 *appears earlier* than P_2 (denoted as $P_1 \prec P_2$). Consider any two arcs a and c in $\mathcal{A}$. The endpoints of arc a are A and B and the endpoints of arc c are C and D. All of points A, B, C, D are on the baseline b. Without loss any generality,

---

[3] Recall that $\mathcal{D}' = \mathcal{D} \setminus \mathcal{G}$ where $\mathcal{G}$ is the C most expensive disks in OPT.

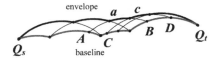

Fig. 2. A substructure. The baseline b consists of the red arcs which are the part of consecutive boundary of $\partial\mathcal{H}$. Q_s, Q_t are the endpoints of b. The black arcs are in the uncovered region. The arc $a \prec c$ since $A \prec C$ and $B \prec D$. The bold black arcs form the envelope.

we assume that $A \prec B$, $C \prec D$ and $A \prec C$. If $B \prec D$, We say arc *a appears earlier* than arc c (denoted as $a \prec c$). Otherwise, we say a and c are incomparable. See Figure 2 for an example. It is easy to see that $\prec$ defines a partial order.

Adjacency: Consider two arcs a (with endpoints $A \prec B$) and c (with endpoints $C \prec D$). If $a \prec c$ and $C \prec B$, we say that a and b are *adjacent* (we can see they must intersect exactly once), and c is the *adjacent successor* of a. Similarly, we can define the adjacent successor of subarc $a[P_1, P_2]$. If c is the adjacent successor of a, meanwhile c intersects with subarc $a[P_1, P_2]$, we say that c is the *adjacent successor* of subarc $a[P_1, P_2]$. Among all adjacent successors of $a[P_1, P_2]$, we call the one whose intersection with $a[P_1, P_2]$ is closest to P_1 the *first adjacent successor* of $a[P_1, P_2]$.

5 Simplifying the Problem

The substructures may overlap in a variety of ways. As we mentioned in Section 2, we need to include in $\mathcal{H}$ more disks in order to make the substructures amenable to the dynamic programming technique. However, this step is somewhat involved and we decide to postpone it to the end of the paper (Section 7). Instead, we present in this section what is the organization of the substructures *after* including more disks in $\mathcal{H}$ and what properties we need for the final dynamic program.

Self-Intersections: In a substructure St, suppose there are two arcs a and c in $\mathcal{A}$ with endpoints A, B and C, D respectively. If $A \prec B \prec C \prec D$ and a and c cover at least one and the same point in $\mathcal{P}$, we say the substructure is *self-intersecting*. So we will eliminate all self-intersections in Section 7. In the rest of the section, we assume all substructures are *non-self-intersecting* and discuss their properties.

Order Consistency: There are two types of relations between substructures which affect how the orientations should be done. One is the overlapping relation and the other is *Remote-Correlation*. See Figure 1 for some examples.

Definition 2 (Remotely correlation). *Consider two substructures* St_u *and* St_l *which are not overlapping. They contain different related active regions of the same gadget. We say that they are* remotely correlated *or* R-correlated.

There are two possible baseline orientations for each substructure (clockwise or anticlockwise around the center of the arc), which gives rise to four possible ways to orient both St_u and St_l. However, there are only two (out of four) of them are consistent (thus we can do dynamic programming on them). More formally, we need the following definition:

As different substructures may interact with each other, we need a dynamic program which can run over all substructures simultaneously. Hence, we need to define a globally consistent ordering of all arcs.

Definition 3 (Global Order Consistency). *We have global order consistency if there is a way to orient the baseline of each substructure, such that the partial orders of the disks for all substructures are consistent in the following sense: It can not happen that $a_i \prec b_i$ in substructure $\mathsf{St}_i(\mathsf{b}_i, \mathcal{A}_i)$ but $a_j \prec b_j$ in $\mathsf{St}_j(\mathsf{b}_j, \mathcal{A}_j)$, where $a_i, a_j \in \partial \mathsf{D}_a$, $b_i, b_j \in \partial \mathsf{D}_b$ and $a_i, b_i \in \mathcal{A}_i$, $a_j, b_j \in \mathcal{A}_j$.*

Substructure Relation Graph $\mathfrak{S}$: we construct an auxiliary graph $\mathfrak{S}$, called the *substructure relation graph*, to capture all R-correlations and Overlapping relations. Each node in $\mathfrak{S}$ represents a substructure. If two substructures are R-correlated, we add a blue edge between the two substructures. If two substructure overlap, we add a red edge.

Definition 4 (Acyclic 2-Matching). *We say the substructure relation graph $\mathfrak{S}$ is an acyclic 2-matching, if $\mathfrak{S}$ is acyclic and is composed by a blue matching and a red matching. In other words, $\mathfrak{S}$ contains only paths, and the red edges and blue edges appear alternately in each path.*

Definition 5 (Point Order Consistency). *Suppose a set $\mathcal{P}_{co}$ of points is covered by both of two overlapping substructures $\mathsf{St}_1(\mathsf{b}_1, \mathcal{A}_1)$ and $\mathsf{St}_2(\mathsf{b}_2, \mathcal{A}_2)$. Consider any two points $P_1, P_2 \in \mathcal{P}_{co}$ and four arcs $a_1, a_2 \in \mathcal{A}_1$, $b_1, b_2 \in \mathcal{A}_2$. Suppose $P_1 \in \mathbb{R}(a_1) \cap \mathbb{R}(b_1)$ and $P_2 \in \mathbb{R}(a_2) \cap \mathbb{R}(b_2)$. But $P_1 \notin \mathbb{R}(a_2) \cup \mathbb{R}(b_2)$ and $P_2 \notin \mathbb{R}(a_1) \cup \mathbb{R}(b_1)$. We say P_1 and P_2 are point-order consistent if $a_1 \prec a_2$ in St_1 and $b_1 \prec b_2$ in St_2. We say the points in $\mathcal{P}_{co}$ satisfy point order consistency if all pair of points in $\mathcal{P}_{co}$ are point-order consistent.*

All introducing all relevant concepts, we can finally state the set of properties we need for the dynamic program.

Lemma 3. *After choosing $\mathcal{H}$, we can ensure the following properties holds:*

P1. *(Active Region Uniqueness) Each substructure contains at most one active region.*
P2. *(Non-self-intersection) Every substructure is non-self-intersecting.*
P3. *(Acyclic 2-Matching) The substructure relation graph $\mathfrak{S}$ is an acyclic 2-matching, i.e., $\mathfrak{S}$ consists of only paths. In each path, red edges and blue edges appear alternately.*
P4. *(Point Order Consistency) Any point is covered by at most two substructures. The points satisfy the point order consistency.*

How to ensure all these properties will be discussed in details in Section 7. Now, everything is in place to describe the dynamic program.

6 Dynamic Programming

Suppose we have already constructed the set $\mathcal{H}$ such that Lemma 3 holds (along with an orientation for each substructure). Without loss any generality, we can assume that the remaining disks can cover all remaining points (otherwise, either the original instance is infeasible or our guess is wrong). In fact, our dynamic program is inspired, and somewhat similar to those in [2, 16, 22].

We can see that we only need to handle each path in $\mathfrak{S}$ separately (since different paths have no interaction at all). Hence, from now on, we simply assume that $\mathfrak{S}$ is a path. Suppose the substructures are $\{St_k(b_k, \mathcal{A}_k)\}_{k \in [m]}$. We use A_k and B_k to denote two endpoints of b_k. Generalizing the previous section, a state for the general DP is , $= \{P_k\}_{k \in [m]}$, where P_k is an intersection point in substructure St_k. Let b_{P_k} and t_{P_k} be the two arcs intersecting at P_k. Suppose $b_{P_k} \prec t_{P_k}$. We call arc b_{P_k} *base-arc* and t_{P_k} *top-arc* for point $\mathbb{q}_k$. Denote the endpoints of b_{P_k} by C_k, C_k' and the endpoints of t_{P_k} by D_k, D_k'. Suppose $b_{P_k}(P, C_k']$ intersects its first successor at P^b (called *base-adjacent point*) and $t_P(P, D_k']$ intersects its first successor at P^t (called *top-adjacent point*). For each $k \in [m]$, we define $St_k^{[P_k]}(b_k[P_k], \mathcal{A}_k[P_k])$ as follows.

- $b_k[P_k]$ is the concatenation of subarc $b_P[P, C_k']$ and the original baseline segment $b_1[C_k', P_t]$. All arcs in $b_1^{[P]}$ have cost zero.
- $\mathcal{A}_k[P_k]$ consists of all arcs $a' \in \mathcal{A}_k$ such that $b_{P_k} \prec a'$ (of course, with the portion covered by $b_k[P]$ subtracted). The cost each such arc is the same as its original cost.

We use $\mathcal{P}(a)$ (or $\mathcal{P}(\mathcal{A})$) to denote the points can be covered by arc a (or arc set $\mathcal{A}$). Let $\mathcal{P}\left[\{P_k\}_{k \in [m]}\right]$ be the point set we need to cover in the subproblem: $\mathcal{P}\left[\{P_k\}_{k \in [m]}\right] = \bigcup_{k \in [m]} \mathcal{P}(\mathcal{A}_k[P_k]) - \bigcup_{k \in [m]} \mathcal{P}(b_{P_k})$. The subproblem $\mathsf{OPT}(\{P_k\}_{k \in [m]})$ is to find, for each substructure St_k, a valid path from P_k to B_k, such that all points in $\mathcal{P}[\{P_k\}_{k \in [m]}]$ can be covered and the total cost is minimized.

The additional challenge for the general case is caused by R-correlations. If two arcs (in two different substructures) belong to the same disk, we say that they are *siblings* of each other. If we processed each substructure independently, some disks would be counted twice. In order to avoid double-counting, we should consider both siblings together, i.e., select them together and pay the disk only once in the DP.

In order to implement the above idea, we need a few more notations. We construct an auxiliary bipartite graph $\mathfrak{B}$. The nodes on one side are all disks in $\mathcal{D}' \setminus \mathcal{H}$, and the node on the other side are substructures. If disk D_i has an arc in the substructure St_j, we add an edge between D_i and St_j. Besides, for each arc of baselines, we add a node to represent it and add an edge between the node and the substructure which contains the arc. Because the weight of any arc of baselines is zero, it shall not induce contradiction that regard them as independent arcs. In fact, there is a 1-1 mapping between the edges in $\mathfrak{B}$ and all arcs.

Fix a state , $= \{P_k\}_{k \in [m]}$. For any arc a in St_k (with intersection point P_k and base-arc b_{P_k}), a has three possible positions: (1) $a \prec b_{P_k}$: we label its corresponding edge with "unprocessed"; (2) $a = b_{P_k}$: we label its corresponding edge with "processing"; (3) Others: we label its corresponding edge with "done". As mentioned before, we need to avoid the situation where one arc becomes the base-arc first (i.e., being added in solution and paid once), and its sibling becomes the base-arc in a later step (hence being paid twice). With the above labeling, we can see that all we need to do is to avoid the states in which one arc is "processing" and its sibling is "unprocessed". If disk D is incident on at least one "processing" edge and not incident on any "unprocessed" edge, we say the D is *ready*. Let $\mathcal{R}$ be the set of ready disks. For each ready disk D, we use $N_p(D)$ to denote the set of neighbors (i.e., substructures) of D connected by "processing" edges. We should consider all substructures in $N_p(D)$ together.

We need in our DP indicator variables to tell us whether a certain transition is feasible: Formally, if $\mathcal{P}[\{P_k\}_{k\in[m]}] = \mathcal{P}[[\mathcal{P}_k][P_i^b]_{\{i\}}]$, let $I_i = 0$. Otherwise, let $I_i = 1$. For ease of notation, for a set $\{e_k\}_{k\in[m]}$ and $S \subseteq [m]$, we write $[e_k][e_i']_S = \{e_k\}_{k\in[m]\setminus S} \cup \{e_i'\}_{i\in S}$. Hence, $[P_k][P_i^b]_{\{i\}} = \{P_k\}_{k\in[m]\setminus i} \cup P_i^b$ and $[P_k][P_i^t]_{N_p(D)} = \{P_k\}_{k\in[m]\setminus N_p(D)} \cup \{P_i^t\}_{i\in N_p(D)}$. Then we have the dynamic program as follows:

$$\mathsf{OPT}\left(\{P_k\}_{k\in[m]}\right) = \min \begin{cases} \min_{i\in[m]}\left\{\mathsf{OPT}\left([P_k][P_i^b]_{\{i\}}\right) + I_i \cdot \infty\right\}, & \text{add no disk} \\ \min_{D\in\mathcal{R}}\left\{\mathsf{OPT}\left([P_k][P_i^t]_{N_p(D)}\right) + w_D\right\}, & \text{add disk D} \end{cases}$$

Note that in the second line, the arc(s) in $N_p(D)$ are base-arcs (w.r.t. state, $(\{P_k\}_{k\in[m]})$.

7 Constructing $\mathcal{H}$

In this section, we describe how to construct the set $\mathcal{H}$ in details. We first include in $\mathcal{H}$ all square gadgets. The boundary of $\mathcal{H}$ consists of several closed curves, as shown in Figure 1. $\mathcal{H}$ and all arcs in the uncovered region $\mathbb{U}(\mathcal{H})$ define a set of substructures.

First, we note that there may exist a closed curve that all points on the curve are covered by some arcs (or informally, we have a cyclic substructure, with the baseline being a cycle). We need to break all such baseline cycles by including a constant number of extra arcs into $\mathcal{H}$. This is easy after we introduce the label-cut operation, and we will spell out all details then. Note that we cannot choose some arbitrary envelope cycle since it may ruin some good properties we want to maintain.

From now on, we assume that all baselines are simple paths. Now, each closed curve contains one or more baselines. So, we have an initial set of well defined substructures. The main purpose of this section is to cut these initial substructures such that Lemma 3 holds.

We will execute a series of operations for constructing $\mathcal{H}$. We first provide below an high level sketch of our algorithm, and outline how the substructures and the substructure relation graph $\mathfrak{S}$ evolve along with the operations.

- First, we deal with active regions. Sometimes, two active region may overlap significantly and become inseparable (formally defined later), they essentially need to be dealt as a single active region. In this case, we merge the two active regions together (we do not need to do anything, but just to pretend that there is only one active region). We can also show that one active region can be merged with at most one other active region. For the rest of cases, two overlapping active region are separable, and we can cut them into at most two non-overlapping active regions, by adding a small number of extra disks in $\mathcal{H}$. After the merging and cutting operations, each substructure contains at most one active region. Hence, the substructures satisfy the property (P1) in Lemma 3. Moreover, we show that if any substructure contains an active region, the substructure is limited in a small region.

- We ensure that each substructure is non-self-intersecting, using a simple greedy algorithm. After this step, (P2) is satisfied.
- In this step, we ensure that substructure relation graph $\mathfrak{G}$ is an acyclic 2-matching (P3). The step has three stages. First, we prove that the set of blue edges forms a matching. Second, we give an algorithm for cutting the substructures which overlap with two or more other substructures. After the cut, each substructure overlaps with no more than one other substructure. So after the first two stages, we can see that $\mathfrak{G}$ is composed of a blue matching and a red matching. At last, we prove that the blue edges and red edges cannot form a cycle, establishing $\mathfrak{G}$ is acyclic.
- The goal of this step is to ensure the point-order consistency (P4). We first show there does not exist a point covered by more than two substructures, when $\mathfrak{G}$ is an acyclic 2-matching. Hence, we only need to handle the case of two overlapping substructures. We show it is enough to break all cycles in a certain planar directed graph. Again, we can add a few more disks to cut all such cycles.
- Lastly, we show that the number of disks added in $\mathcal{H}$ in the above four steps is $O(K^2)$.

References

1. Adamaszek, A., Wiese, A.: Approximation schemes for maximum weight independent set of rectangles. In: FOC, pp. 400–409. IEEE (2013)
2. Ambühl, C., Erlebach, T., Mihalák, M., Nunkesser, M.: Constant-Factor Approximation for Minimum-Weight (Connected) Dominating Sets in Unit Disk Graphs. In: Díaz, J., Jansen, K., Rolim, J.D.P., Zwick, U. (eds.) APPROX 2006 and RANDOM 2006. LNCS, vol. 4110, pp. 3–14. Springer, Heidelberg (2006)
3. Bansal, N., Pruhs, K.: The geometry of scheduling. SICOMP **43**(5), 1684–1698 (2014)
4. Brönnimann, H., Goodrich, M.: Almost optimal set covers in finite vc-dimension. DCG **14**(1), 463–479 (1995)
5. Bus, N., Garg, S., Mustafa, N. H., Ray, S.: Tighter estimates for epsilon-nets for disks. arXiv preprint (2015). arXiv:1501.03246
6. Chan, T.M., Grant, E.: Exact algorithms and apx-hardness results for geometric packing and covering problems. In: CGTA 47, 2, Part A, pp. 112–124 (2014)
7. Chan, T. M., Grant, E., Könemann, J., Sharpe, M.: Weighted capacitated, priority, and geometric set cover via improved quasi-uniform sampling. In: SODA, pp. 1576–1585. SIAM (2012)
8. Clark, B.N., Colbourn, C.J., Johnson, D.S.: Unit disk graphs. Discrete Math. **86**(1–3), 165–177 (1991)
9. Clarkson, K.L., Varadarajan, K.: Improved approximation algorithms for geometric set cover. DCG **37**(1), 43–58 (2007)
10. Dai, D., Yu, C.: A 5 + ϵ-approximation algorithm for minimum weighted dominating set in unit disk graph. TCS **410**(8), 756–765 (2009)
11. Ding, L., Wu, W., Willson, J., Wu, L., Lu, Z., Lee, W.: Constant-approximation for target coverage problem in wireless sensor networks. In: INFOCOM, pp. 1584–1592. IEEE (2012)
12. Dinur, I., Steurer, D.: Analytical approach to parallel repetition. In: STOC, pp. 624–633. ACM (2014)

13. Du, D.-Z., Ko, K., Hu, X.: Design and Analysis of Approximation Algorithms. Springer (2011)
14. Du, D.-Z., Wan, P.-J.: Connected Dominating Set: Theory and Applications, vol. 77. Springer Science & Business Media (2012)
15. Erlebach, T., Grant, T., Kammer, F.: Maximising lifetime for fault-tolerant target coverage in sensor networks. In: SPAA, pp. 187–196. ACM (2011)
16. Erlebach, T., Mihalák, M.: A $(4 + \epsilon)$-Approximation for the Minimum-Weight Dominating Set Problem in Unit Disk Graphs. In: Bampis, E., Jansen, K. (eds.) WAOA 2009. LNCS, vol. 5893, pp. 135–146. Springer, Heidelberg (2010)
17. Erlebach, T., van Leeuwen, E.: Ptas for weighted set cover on unit squares. In: Serna, M., Shaltiel, R., Jansen, K., Rolim, J. (eds.) APPROX. LNCS, vol. 6302, pp. 166–177. Springer, Heidelberg (2010)
18. Even, G., Rawitz, D., Shahar, S.M.: Hitting sets when the vc-dimension is small. Information Processing Letters **95**(2), 358–362 (2005)
19. Feige, U.: A threshold of ln n for approximating set cover. Journal of the ACM (JACM) **45**(4), 634–652 (1998)
20. Gibson, M., Pirwani, I.A.: Algorithms for dominating set in disk graphs: breaking the logn barrier. In: ESA, pp. 243–254. Springer (2010)
21. Har-Peled, S., Kaplan, H., Sharir, M., Smorodinsky, S.: Epsilon-nets for halfspaces revisited (2014). arXiv preprint arXiv:1410.3154
22. Har-Peled, S., Lee, M.: Weighted geometric set cover problems revisited. JoCG **3**(1), 65–85 (2012)
23. Hochbaum, D.S., Maass, W.: Approximation schemes for covering and packing problems in image processing and vlsi. JACM **32**(1), 130–136 (1985)
24. Hochbaum, D.S., Maass, W.: Fast approximation algorithms for a nonconvex covering problem. Journal of Algorithms **8**(3), 305–323 (1987)
25. Huang, Y., Gao, X., Zhang, Z., Wu, W.: A better constant-factor approximation for weighted dominating set in unit disk graph. JCO **18**(2), 179–194 (2009)
26. Hunt, H.B., III, Marathe, M.V., Radhakrishnan, V., Ravi, S.S., Rosenkrantz, D.J., Stearns, R.E.: NC-Approximation Schemes for NP- and PSPACE-Hard Problems for Geometric Graphs (1997)
27. Marx, D.: On the optimality of planar and geometric approximation schemes. In: FOCS, pp. 338–348. IEEE (2007)
28. Mustafa, N.H., Raman, R., Ray, S.: Qptas for geometric set-cover problems via optimal separators (2014). arXiv preprint arXiv:1403.0835
29. Mustafa, N.H., Ray, S.: Ptas for geometric hitting set problems via local search. In: SOCG, pp. 17–22. ACM (2009)
30. Pyrga, E., Ray, S.: New existence proofs ε-nets. In: SOCG, pp. 199–207. ACM (2008)
31. van Leeuwen, E.J.: Optimization and approximation on systems of geometric objects. Phd thesis
32. Varadarajan, K.: Epsilon nets and union complexity. In: SOCG, SCG 2009, pp. 11–16. ACM (2009)
33. Varadarajan, K.: Weighted geometric set cover via quasi-uniform sampling. In: SOCG, pp. 641–648. ACM (2010)
34. Zou, F., Wang, Y., Xu, X.-H., Li, X., Du, H., Wan, P., Wu, W.: New approximations for minimum-weighted dominating sets and minimum-weighted connected dominating sets on unit disk graphs. TCS **412**(3), 198–208 (2011)

Approximating the Expected Values for Combinatorial Optimization Problems over Stochastic Points

Lingxiao Huang[✉] and Jian Li

Institute for Interdisciplinary Information Sciences, Tsinghua University, Beijing, China
huanglingxiao1990@126.com

Abstract. We consider the stochastic geometry model where the location of each node is a random point in a given metric space, or the existence of each node is uncertain. We study the problems of computing the expected lengths of several combinatorial or geometric optimization problems over stochastic points, including closest pair, minimum spanning tree, k-clustering, minimum perfect matching, and minimum cycle cover. We also consider the problem of estimating the probability that the length of closest pair, or the diameter, is at most, or at least, a given threshold. Most of the above problems are known to be #P-hard. We obtain FPRAS (Fully Polynomial Randomized Approximation Scheme) for most of them in both the existential and locational uncertainty models. Our result for stochastic minimum spanning trees in the locational uncertain model improves upon the previously known constant factor approximation algorithm. Our results for other problems are the first known to the best of our knowledge.

1 Introduction

Background: Uncertain or imprecise data are pervasive in applications like sensor monitoring, location based services, data collection and integration [12,14,33]. Consider a temperature monitoring system which collects measures of humidity and wind speed. Since we do not have the perfect sensing instruments, the data obtained are often contaminated with noises[13]. For another example, the locational data collected by the Global-Positioning Systems (GPS) often contains measurement errors [29]. Moreover, many machine learning and prediction algorithms also produce a variety of stochastic models and a large volume of probabilistic data. Thus, managing, analyzing and solving optimization problems over stochastic models and data have recently attracted significant attentions in several research communities (see e.g., [30,33,34]).

In this paper, we study two stochastic geometry models, the locational uncertainty model and the existential uncertainty model, both of which have been studied extensively in recent years (see e.g., [2–4,7,20,21,24–26],some of which will be discussed in the related work section). In fact, a special case of the locational uncertainty model where all points follow the same distribution is a classic topic in stochastic geometry literature (see e.g., [8–10,22,31]). The main interest there has been to derive asymptotics for the

---

Research supported in part by the National Basic Research Program of China Grant 2015CB358700, 2011CBA00300, 2011CBA00301, the National Natural Science Foun- dation of China Grant 61202009, 61033001, 61361136003.

M.M. Halldórsson et al. (Eds.): ICALP 2015, Part I, LNCS 9134, pp. 910–921, 2015.
DOI: 10.1007/978-3-662-47672-7_74

expected values of certain combinatorial problems (e.g., minimum spanning tree). The stochastic geometry model is also of fundamental interest in the area of wireless networks. In many applications, we only have some prior information about the locations of the transmission nodes (e.g., some sensors that will be deployed randomly in a designated area by an aircraft). Such a stochastic wireless network can be captured precisely by this model. See the recent survey [19] and more references therein.

Stochastic Geometry Models: In this paper, we focus on two stochastic geometry models, the locational uncertainty model and existential uncertainty model.

1. (Locational Uncertainty Model) We are given a metric space $\mathcal{P}$. The location of each node $v \in \mathcal{V}$ is a random point in the metric space $\mathcal{P}$ and the probability distribution is given as the input. Formally, we use the term *nodes* to refer to the vertices of the graph, *points* to describe the locations of the nodes in the metric space. We denote the set of nodes as $\mathcal{V} = \{v_1, \ldots, v_n\}$ and the set of points as $\mathcal{P} = \{s_1, \ldots, s_m\}$, where $n = |\mathcal{V}|$ and $m = |\mathcal{P}|$. A realization $\mathbf{r}$ can be represented by an n-dimensional vector $(r_1, \ldots, r_n) \in \mathcal{P}^n$ where point r_i is the location of node v_i for $1 \leq i \leq n$. Let $\mathcal{R}$ denote the set of all possible realizations. We assume that the distributions of the locations of nodes in the metric space $\mathcal{P}$ are independent, thus $\mathbf{r}$ occurs with probability $\Pr[\mathbf{r}] = \prod_{i \in [n]} p_{v_i r_i}$, where p_{vs} represents the probability that the location of node v is point $s \in \mathcal{P}$. The model is also termed as the *locational uncertainty model* in [20].

2. (Existential Uncertainty Model) A closely related model is the *existential uncertainty model* where the location of a node is a fixed point in the given metric space, but the existence of the node is probabilistic. In this model, we use p_i to denote the probability that node v_i exists (if exists, its location is s_i). A realization $\mathbf{r}$ can be represented by a subset $S \subset \mathcal{P}$ and $\Pr[\mathbf{r}] = \prod_{s_i \in S} p_i \prod_{s_i \notin S} (1 - p_i)$.

Problem Formulation: We are interested in following natural problem in the above models: estimating the expected values of certain statistics of combinatorial objects. In this paper, we study several combinatorial or geometry problems in these two models: the closest pair problem, minimum spanning tree, minimum perfect matching (assuming an even number of nodes), k-clustering and minimum cycle cover. We take the minimum spanning tree problem for example. Let MST be the length of the minimum spanning tree (which is a random variable) and MST($\mathbf{r}$) be the length of the minimum spanning tree spanning all points in the realization $\mathbf{r}$. We would like to estimate the following quantity:

$$\mathbb{E}[\mathsf{MST}] = \sum_{\mathbf{r} \in \mathcal{R}} \Pr[\mathbf{r}] \cdot \mathsf{MST}(\mathbf{r}).$$

However, the above formula does not give us an efficient way to estimate the expectation since it involves an exponential number of terms. In fact, computing the exact expected value (for the problems considered in this paper) are either NP-hard or #P-hard. Following many of the theoretical computer science literatures on approximate counting and estimation, our goal is to obtain fully polynomial randomized approximation schemes for computing the expected values.

Table 1. Our results for some problems in different stochastic models

Problems		Existential	Locational
Closest Pair (§2)	$\mathbb{E}[C]$	FPRAS	FPRAS
	$\Pr[C \le 1]$	FPRAS	FPRAS
	$\Pr[C \ge 1]$	Inapprox	Inapprox
Diameter (§2)	$\mathbb{E}[D]$	FPRAS	FPRAS
	$\Pr[D \le 1]$	Inapprox	Inapprox
	$\Pr[D \ge 1]$	FPRAS	FPRAS
Minimum Spanning Tree (§3)	$\mathbb{E}[MST]$	FPRAS[20]	FPRAS
k-Clustering	$\mathbb{E}[kCL]$	FPRAS	Open
Perfect Matching (§4)	$\mathbb{E}[PM]$	N.A.	FPRAS
kth Closest Pair	$\mathbb{E}[kC]$	FPRAS	Open
Cycle Cover	$\mathbb{E}[CC]$	FPRAS	FPRAS
kth Longest m-Nearest Neighbor	$\mathbb{E}[kmNN]$	FPRAS	Open

1.1 Our Contributions

We recall that a *fully polynomial randomized approximation scheme (FPRAS)* for a problem f is a randomized algorithm A that takes an input instance x, a real number $\epsilon > 0$, returns $A(x)$ such that $\Pr[(1-\epsilon)f(x) \le A(x) \le (1+\epsilon)f(x)] \ge \frac{3}{4}$ and its running time is polynomial in both the size of the input n and $1/\epsilon$. Our main contributions can be summarized in Table 1. We need to explain some entries in the table in more details.

1. Closest Pair: We use C to denote the minimum distance of any pair of two nodes. If a realization has less than two nodes, C is zero. Computing $\Pr[C \le 1]$ exactly in the existential model is known to be #P-hard even in an Euclidean plane [21], but no nontrivial algorithmic result is known before. So is computing $\Pr[C \ge 1]$. In fact, it is not hard to show that computing $\Pr[C \ge 1]$ is imapproximable within any factor in a metric space.

   We also consider the problem of computing expected distance $\mathbb{E}[C]$ between the closest pair in the same model. We prove that the problem is #P-hard and give the first known FPRAS in Section 2. Note that an FPRAS for computing $\Pr[C \le 1]$ does not imply an FPRAS for computing $\mathbb{E}[C]$ [1].

2. Diameter: The problem of computing the expected length of the diameter can be reduced to the closest pair problem as follows. Assume that the longest distance between two points in $\mathcal{P}$ is W. We construct the new instance $\mathcal{P}'$ as follows: for any two points $u, v \in \mathcal{P}$, let their distance be $2W - d(u, v)$ in $\mathcal{P}'$. The new instance is still a metric. The sum of the distance of closest pair in $\mathcal{P}$ and the diameter in $\mathcal{P}'$ is exactly $2W$ (if there are at least two realized points). Hence, the answer for the diameter can be easily derived from the answer for closest pair in $\mathcal{P}'$.

3. Minimum Spanning Tree: Computing $\mathbb{E}[MST]$ exactly in both uncertainty models is known to be #P-hard [20]. Kamousi, Chan, and Suri [20] developed an FPRAS

---

[1] To the contrary, an FPRAS for computing $\Pr[C \ge 1]$ or $\Pr[C = 1]$ would imply an FPRAS for computing $\mathbb{E}[C]$ since $\mathbb{E}[C] = \sum_{(s_i, s_j)} \Pr[C = d(s_i, s_j)]d(s_i, s_j) = \int \Pr[C \ge t]dt = \sum_{(s_i, s_j)} \Pr[C \ge d(s_i, s_j)](d(s_i, s_j) - d(s'_i, s'_j))$.

for estimating $\mathbb{E}[\mathsf{MST}]$ in the existential uncertainty model and a constant factor approximation algorithm in the locational uncertainty model.

Estimating $\mathbb{E}[\mathsf{MST}]$ is amendable to several techniques. We obtain an FPRAS for estimating $\mathbb{E}[\mathsf{MST}]$ in the locational uncertainty model using the stoch-core techinque in Section 3. In fact, the idea in [20] can also be extended to give an alternative FPRAS. It is not clear how to extend their idea to other problems.

4. Clustering (k-clustering): In the deterministic k-clustering problem, we want to partition all points into k disjoint subsets such that the spacing of the partition is maximized, where the spacing is defined to be the minimum of any $d(u, v)$ with u, v in different subsets [23]. In fact, the optimal cost of the problem is the length of the $(k-1)$th most expensive edge in the minimum spanning tree [23]. We show how to estimate $\mathbb{E}[\mathsf{kCL}]$ using the HPF (hierarchical partition family) technique.

5. Perfect Matching: We assume that there are even number of nodes to ensure that a perfect matching always exists. Therefore, only the locational uncertainty model is relevant here. We give the first FPRAS for approximating the expected length of minimum perfect matching in Section 4 using a more complicated stoch-core technique.

All of our algorithms run in polynomial time. However, we have not attempted to optimize the exact running time.

Our techniques: Perhaps the simplest and the most commonly used technique for estimating the expectation of a random variable is the Monte Carlo method, that is to use the sample average as the estimate. However, the method is only efficient (i.e., runs in polynomial time) if the variance of the random variable is small (See Lemma 1). To circumvent the difficulty caused by the high variance, a general methodology is to decompose the expectation of the random variable into a convex combination of conditional expectations using the law of total expectation: $\mathbb{E}[X] = \mathbb{E}_Y\big[\mathbb{E}[X \mid Y]\big] = \sum_y \Pr[Y = y]\,\mathbb{E}[X \mid Y = y]$. Hopefully, $\Pr[Y = y]$ can be estimated (or calculated exactly) efficiently, and the random variable X conditioning on each event y has a low variance. However, choosing the events Y to condition on can be tricky.

We develop two new techniques for choosing such events, each being capable of solving a subset of aforementioned problems. In the first technique, we first identify a set $\mathcal{H}$ of points, called the *stoch-core* of the problem, such that (1): with high probability, all nodes realize in $\mathcal{H}$ and (2): conditioning on event (1), the variance is small. Then, we choose Y to be the number of nodes realized to points not in $\mathcal{H}$. We compute the $(1 \pm \epsilon)$-estimates for $Y = 0, 1$ using Monte Carlo by (1) and (2). The problematic part is when Y is large, i.e., many nodes realize to points outside $\mathcal{H}$. Even though the probability of such events is very small, the value of X under such events may be considerably large, thus contributing nontrivially. However, we can show that the contribution of such events is dominated by the first few events and thus can be safely ignored. Choosing appropriate stoch-core is easy for some problems, such as closest pair and minimum spanning tree, while it may require additional idea for other problems such as minimum perfect matching.

Our second technique utilizes a notion called *Hierarchical Partition Family (HPF)*. The HPF has m levels, each representing a clustering of all points. For a combinatorial problem, for which the solution is a set of edges, we define Y to be the highest level such that some edge in the solution is an inter-cluster edge. Informally, conditioning on the information of Y, we can essentially bound the variance of X (hence use the Monte Carlo method). To implement Monte Carlo, we need to be able to take samples efficiently conditioning on Y. We show that such sampling problems can be reduced to, or have connections to, classical approximate counting and sampling problems, such as approximating permanent, counting knapsack.

Due to space constraints, we omit many details, which can be found in the full version of this paper[2].

1.2 Related Work

Several geometric properties of a set of stochastic points have been studied extensively in the literature under the term *stochastic geometry*. For instance, Bearwood et al. [8] shows that if there are n points uniformly and independently distributed in $[0, 1]^2$, the minimal traveling salesman tour visiting them has an expected length $\Omega(\sqrt{n})$. Asymptotic results for minimum spanning trees and minimum matchings on n points uniformly distributed in unit balls are established by Bertsimas and van Ryzin [10]. Similar results can be found in e.g., [9,22,31]. Compared with results in stochastic geometry, we focus on the efficient computation of the statistics, instead of giving explicit mathematical formulas.

Recently, a number of researchers have begun to explore geometric computing under uncertainty and many classical computational geometry problems have been studied in different stochastic/uncertainty models. Agarwal, Cheng, Tao and Yi [4] studied the problem of indexing probabilistic points with continuous distributions for range queries on a line. Agarwal, Efrat, Sankararaman, and Zhang [5] also studied the same problem in the locational uncertainty model under Euclidean metric. The most probable k-nearest neighbor problem and its variants have attracted a lot of attentions in the database community (See e.g., [11]). Several other problems have also been considered recently, such as computing the expected volume of a set of probabilistic rectangles in a Euclidean space [36], convex hulls [2], skylines (Pareto curves) over probabilistic points [1,7], and shape fitting [27].

Kamousi, Chan and Suri [20] initiated the study of estimating the expected length of combinatorial objects in this model. They showed that computing the expected length of the nearest neighbor (NN) graph, the Gabriel graph (GG), the relative neighborhood graph (RNG), and the Delaunay triangulation (DT) can be solved exactly in polynomial time, while computing $\mathbb{E}[\mathsf{MST}]$ is #P-hard and there exists a simple FPRAS for approximating $\mathbb{E}[\mathsf{MST}]$ in the existential model. They also gave a deterministic PTAS for approximating $\mathbb{E}[\mathsf{MST}]$ in an Euclidean plane. In another paper [21], they studied the closest pair and (approximate) nearest neighbor problems (i.e., finding the point with the smallest expected distance from the query point) in the same model.

The *randomly weighted graph* model where the edge weights are independent non-negative variables has also been studied extensively. Frieze [16] and Steele [32] showed that the expected value of the minimum spanning tree on such a graph with identically and independently distributed edges is $\zeta(3)/D$ where $\zeta(3) = \sum_{j=1}^{\infty} 1/j^3$ and D is the derivative of the distribution at 0. Alexopoulos and Jacobson [6] developed algorithms

---
[2] http://arxiv.org/abs/1209.5828

that compute the distribution of MST and the probability that a particular edge belongs to MST when edge lengths follow discrete distributions. However, the running times of their algorithms may be exponential in the worst cases. Recently, Emek, Korman and Shavitt [15] showed that computing the kth moment of a class of properties, including the diameter, radius and minimum spanning tree, admits an FPRAS for each fixed k. Our model differs from their model in that the edge lengths are not independent.

The computational/algorithmic aspects of stochastic geometry have also gained a lot of attention in recent years from the area of wireless networking. In many application scenarios, it is common to assume that the nodes (e.g., sensors) are deployed randomly across a certain area, thereby forming a stochastic network. It is of central importance to study various properties in this network, such as connectivity [17], transmission capacity [18]. We refer interested reader to a recent survey [19] for more references.

1.3 Preliminaries

Before describing our main results, we first consider the straightforward Monte Carlo strategy, which is an important building block in our later developments. Suppose we want to estimate $\mathbb{E}[X]$. In each Monte Carlo iteration, we take a sample (a realization of all nodes), and compute the value of X for the sample. At the end, we output the average over all samples. The number of samples required by this algorithm is suggested by the following standard Chernoff bound.

Lemma 1. (Chernoff Bound) *Let random variables* $X_1, X_2, \ldots, X_N$ *be independent random variables taking on values between 0 and U. Let $X = \frac{1}{N} \sum_{i=1}^{N} X_i$ and μ be the expectation of X, for any $\epsilon > 0$,*

$$\Pr\left[X \in [(1-\epsilon)\mu, (1+\epsilon)\mu]\right] \geq 1 - 2e^{-N\frac{\mu}{U}\epsilon^2/4}.$$

Therefore, for any $\epsilon > 0$, in order to get an $(1 \pm \epsilon)$-approximation with probability $1 - \frac{1}{\text{poly}(n)}$, the number of samples needs to be $O(\frac{U}{\mu\epsilon^2} \log n)$. If $\frac{U}{\mu}$, the ratio between the maximum possible value of X and the expected value $\mathbb{E}[X]$, is bounded by $\text{poly}(m, n, \frac{1}{\epsilon})$, we can use the above Monte Carlo method to estimate $\mathbb{E}[X]$ with a polynomial number of samples. Since we use this condition often, we devote a separate definition to it.

Definition 1. *We call a random variable X poly-bounded if the ratio between the maximum possible value of X and the expected value $\mathbb{E}[X]$ is bounded by $\text{poly}(m, n, \frac{1}{\epsilon})$.*

2 The Closest Pair Problem

2.1 Estimating $\Pr[C \leq 1]$

As a warmup, we first demonstrate how to use the stoch-core technique for the closest pair problem in the existential uncertainty model. Given a set of points $\mathcal{P} = \{s_1, \ldots, s_m\}$ in the metric space, where each point $s_i \in \mathcal{P}$ is present with probability p_i. We use C to denote the distance between the closest pair of vertices in the realized graph. If the realized graph has less than two points, C is zero. The goal is to compute the probability $\Pr[C \leq 1]$.

For a set H of points and a subset $S \subseteq H$, we use $H\langle S \rangle$ to denote the event that among all points in H, all and only points in S are present. For any nonnegative integer i, let $H\langle i \rangle$ denote the event $\bigvee_{S \subseteq H : |S| = i} H\langle S \rangle$, i.e., the event that exactly i points are present in H.

The *stoch-core* of the closest pair problem is simply defined to be $\mathcal{H} = \left\{ s_i \mid p_i \geq \frac{\epsilon}{m^2} \right\}$. Let $\mathcal{F} = \mathcal{P} \setminus \mathcal{H}$. We consider the decomposition

$$\Pr[C \leq 1] = \sum_{i=0}^{|\mathcal{F}|} \Pr[\mathcal{F}\langle i \rangle \wedge C \leq 1] = \sum_{i=0}^{|\mathcal{F}|} \Pr[\mathcal{F}\langle i \rangle] \cdot \Pr[C \leq 1 \mid \mathcal{F}\langle i \rangle].$$

Our algorithm is very simple: estimate the first three terms (i.e., $i = 0, 1, 2$) and use their sum as our final answer.

We can see that $\mathcal{H}$ satisfies the two properties of a stoch-core mentioned in the introduction:

1. The probability that all nodes are realized in $\mathcal{H}$, i.e., $\Pr[\mathcal{F}\langle 0 \rangle]$, is at least $1 - m \cdot \frac{\epsilon}{m^2} = 1 - \frac{\epsilon}{m}$;
2. If there exist two points $s_i, s_j \in \mathcal{H}$ such that $d(s_i, s_j) \leq 1$, we have $\Pr[C \leq 1 \mid \mathcal{F}\langle 0 \rangle] \geq \frac{\epsilon^2}{m^4}$; otherwise, $\Pr[C \leq 1 \mid \mathcal{F}\langle 0 \rangle] = \Pr[\mathcal{H}\langle 0 \rangle \mid \mathcal{F}\langle 0 \rangle] + \Pr[\mathcal{H}\langle 1 \rangle \mid \mathcal{F}\langle 0 \rangle]$. Note that we can compute $\Pr[\mathcal{H}\langle 0 \rangle \mid \mathcal{F}\langle 0 \rangle]$ and $\Pr[\mathcal{H}\langle 1 \rangle \mid \mathcal{F}\langle 0 \rangle]$ in polynomial time. We do not consider this case in the following analysis.

Both properties guarantee that the random variable $I(C \leq 1)$, conditioned on $\mathcal{F}\langle 0 \rangle$, is poly-bounded, hence we can easily get a $(1 \pm \epsilon)$-estimation for $\Pr[\mathcal{F}\langle 0 \rangle \wedge C \leq 1]$ with polynomial many samples with high probability. Similarly, $\Pr[\mathcal{F}\langle i \rangle \wedge C \leq 1]$ can also be estimated with polynomial number of samples for $i = 1, 2$. The algorithm can be found in Algorithm 1.

Algorithm 1. Estimating $\Pr[C \leq 1]$

---

1 Estimate $\Pr[\mathcal{F}\langle 0 \rangle \wedge C \leq 1]$: Take $N_0 = O\big((m/\epsilon)^4 \ln m\big)$ independent samples. Suppose M_0 is the number of samples satisfying $C \leq 1$ and $\mathcal{F}\langle 0 \rangle$. $T_0 \leftarrow \frac{M_0}{N_0}$.

2 Estimate $\Pr[\mathcal{F}\langle 1 \rangle \wedge C \leq 1]$: For each point $s_i \in \mathcal{F}$, take $N_1 = O((m/\epsilon)^4 \ln m)$ independent samples conditioning on the event $\mathcal{F}\langle \{s_i\} \rangle$. Suppose there are M_i samples satisfying $C \leq 1$. $T_1 \leftarrow \sum_{s_i \in \mathcal{F}} p_i M_i / N_1$.

3 Estimate $\Pr[\mathcal{F}\langle 2 \rangle \wedge C \leq 1]$: For each point pair $s_i, s_j \in \mathcal{F}$, take $N_2 = O((m/\epsilon)^4 \ln m)$ independent samples conditioning on the event $\mathcal{F}\langle \{s_i, s_j\} \rangle$. Suppose there are M_{ij} samples satisfying $C \leq 1$. $T_2 \leftarrow \sum_{s_i, s_j \in \mathcal{F}} p_i p_j M_{ij} / N_2$.

4 **Output:** $T_0 + T_1 + T_2$

Lemma 2. *Steps 1,2,3 in Algorithm 1 provide $(1 \pm \epsilon)$-approximations for $\Pr[\mathcal{F}\langle i \rangle \wedge C \leq 1]$ for $i = 0, 1, 2$ respectively, with high probability.*

Theorem 1. *There is an FPRAS for estimating the probability of the distance between the closest pair of nodes is at most 1 in the existential uncertainty model.*

Proof. We only need to show that the contribution from the rest of terms (where more than two points outside stoch-core $\mathcal{H}$ are present) is negligible compared to the third term. Suppose S is the set of all present points such that $C \leq 1$ and there are at least 3 points not in $\mathcal{H}$. Suppose s_i, s_j are the closest pair in S. We associate S with a smaller set $S' \subset S$ by making 1 present point in $(S \cap \mathcal{F}) \setminus \{s_i, s_j\}$ absent (if there are several such S', we choose an arbitrary one). We denote it as $S \sim S'$. We use the notation $S \in F_i$ to denote that the realization S satisfies $(\mathcal{F}\langle i \rangle \wedge C \leq 1)$. Then, we can see that for $i \geq 3$,

$$\Pr[\mathcal{F}\langle i \rangle \wedge C \leq 1] = \sum_{S:S \in F_i} \Pr[S] \leq \sum_{S':S' \in F_{i-1}} \sum_{S:S \sim S'} \Pr[S].$$

For a fixed S', there are at most m different sets S such that $S \sim S'$ and $\Pr[S] \leq \frac{2\epsilon}{m^2}\Pr[S']$ for any such S. Hence, we have that $\sum_{S:S \sim S'} \Pr[S] \leq \frac{2\epsilon}{m}\Pr[S']$. Therefore,

$$\Pr[\mathcal{F}\langle i \rangle \wedge C \leq 1] \leq \frac{2\epsilon}{m} \cdot \sum_{S':S' \in F_{i-1}} \Pr[S'] = \frac{2\epsilon}{m} \cdot \Pr[\mathcal{F}\langle i-1 \rangle \wedge C \leq 1].$$

Hence, overall we have $\sum_{i \geq 3} \Pr[\mathcal{F}\langle i \rangle \wedge C \leq 1] \leq \epsilon \Pr[\mathcal{F}\langle 2 \rangle \wedge C \leq 1]$. This finishes the analysis.

□

2.2 Estimating $\mathbb{E}[C]$

In this section, we consider the problem of estimating $\mathbb{E}[C]$, where C is the distance of the closest pair of present points, in the existential uncertainty model. Now, we introduce our second main technique, the *hierarchical partition family (HPF)* technique, to solve this problem. An HPF is a family Ψ of partitions of $\mathcal{P}$, formally defined as follows.

Definition 2. *(Hierarchical Partition Family (HPF)) Let T be any minimum spanning tree spanning all points of $\mathcal{P}$. Suppose that the edges of T are $e_1, \ldots, e_{m-1}$ with $d(e_1) \geq d(e_2) \geq \ldots \geq d(e_{m-1})$. Let $E_i = \{e_i, e_{i+1}, \ldots, e_{m-1}\}$. The HPF $\Psi(\mathcal{P})$ consists of m partitions $\Gamma_1, \ldots, \Gamma_m$. Γ_1 is the entire point set $\mathcal{P}$. Γ_i consists of i disjoint subsets of $\mathcal{P}$, each corresponding to a connected component of $G_i = G(\mathcal{P}, E_i)$. Γ_m consists of all singleton points in $\mathcal{P}$. It is easy to see that Γ_j is a refinement of Γ_i for $j > i$. Consider two consecutive partitions Γ_i and Γ_{i+1}. Note that G_i contains exactly one more edge (i.e., e_i) than G_{i+1}. Let μ'_{i+1} and μ''_{i+1} be the two components (called the split components) in Γ_{i+1}, each containing an endpoint of e_i. Let $v_i \in \Gamma_i$ be the connected component of G_i that contains e_i. We call v_i the special component in Γ_i. Let $\Gamma'_i = \Gamma_i \setminus v_i$.*

We observe two properties of $\Psi(\mathcal{P})$ that are useful later.

P1. Consider a component $C \in \Gamma_i$. Let s_1, s_2 be two arbitrary points in C. Then $d(s_1, s_2) \leq (m-1)d(e_i)$ (this is because s_1 and s_2 are connected in G_i, and e_i is the longest edge in G_i).

P2. Consider two different components C_1 and C_2 in Γ_i. Let $s_1 \in C_1$ and $s_2 \in C_2$ be two arbitrary points. Then $d(s_1, s_2) \geq d(e_{i-1})$ (this is because the minimum inter-component distance is $d(e_{i-1})$ in G_i).

Let the random variable Y be smallest integer i such that there is at most one present point in each component of Γ_{i+1}. Note that if $Y = i$ then each component of Γ_i contains at most one point, except that the special component v_i contains exactly two present points. The following lemma is a simple consequence of P1 and P2.

Lemma 3. *Conditioning on $Y = i$, it holds that $d(e_i) \leq \mathsf{C} \leq md(e_i)$ (hence, C is poly-bounded).*

Consider the following expansion of $\mathbb{E}[\mathsf{C}]$: $\mathbb{E}[\mathsf{C}] = \sum_{i=1}^{m-1} \Pr[Y = i]\mathbb{E}[\mathsf{C} \mid Y = i]$. For a fixed i, $\Pr[Y = i]$ can be estimated as follows: For a component $C \subset \mathcal{P}$, we use $C\langle j \rangle$ to denote the event that exactly j points in C are present, $C\langle s \rangle$ the event that only s is present in C and $C\langle \leq j \rangle$ the event that no more than j points in C are present. Let μ_i' and μ_i'' be the two split components in Γ_i. Note that

$$\Pr[Y = i] = \Pr[\mu_{i+1}'\langle 1 \rangle] \cdot \Pr[\mu_{i+1}''\langle 1 \rangle] \cdot \prod_{C \in \Gamma_i'} \Pr[C\langle \leq 1 \rangle].$$

The remaining is to show how to estimate $\mathbb{E}[\mathsf{C} \mid Y = i]$. Since C is poly-bounded, it suffices to give an efficient algorithm to take samples conditioning on $Y = i$. This is again not difficult: We take exactly one point $s \in \mu_{i+1}'$ with probability $\Pr[\mu_{i+1}'\langle s \rangle] / \Pr[\mu_{i+1}'\langle 1 \rangle]$. Same for μ_{i+1}''. For each $C \in \Gamma_i'$, take no point from C with probability $\Pr[C\langle 0 \rangle] / \Pr[C\langle \leq 1 \rangle]$; otherwise, take exactly one point $s \in C$ with probability $\Pr[C\langle s \rangle] / \Pr[C\langle \leq 1 \rangle]$. This finishes the description of the FPRAS in the existential uncertainty model.

Theorem 2. *There is an FPRAS for estimating the expected distance between the closest pair of nodes in the existential uncertainty models.*

3 Minimum Spanning Trees

We consider the problem of estimating the expected size of minimum spanning tree in the locational uncertainty model. In this section, we briefly sketch how to solve it using our stoch-core method. Recall that the term nodes refers to the vertices $\mathcal{V}$ of the spanning tree and points describes the locations in $\mathcal{P}$. For ease of exposition, we assume that for each point, there is only one node that may realize at this point.

Recall that we use the notation $v \vDash s$ to denote the event that node v is present at point s. Let $p_{vs} = \Pr[v \vDash s]$. Since node v is realized with certainty, we have $\sum_{s \in \mathcal{P}} p_{vs} = 1$. For each point $s \in \mathcal{P}$, we let $p(s)$ denote the probability that point s is present. For a set H of points, let $p(H) = \sum_{s \in H} p(s)$, i.e., the expected number of points present in H. For a set H of points and a set S of nodes, we use $H\langle S \rangle$ to denote the event that all and only nodes in S are realized to some points in H. If S only contains one node, say v, we use the notation $H\langle v \rangle$ as the shorthand for $H\langle \{v\} \rangle$. Let $H\langle i \rangle$ denote the event

$\bigvee_{S:|S|=i} H\langle S\rangle$, i.e., the event that exactly i nodes are in H. We use $\mathrm{diam}(H)$, called the diameter of H, to denote $\max_{s,t\in H} \mathrm{d}(s,t)$. Let $\mathrm{d}(p, H)$ be the closest distance between point p and any point in H.

Finding stoch-core: We find the stoch-core $\mathcal{H} \leftarrow \mathsf{B}(s, \mathrm{d}(s, t)) = \{s' \in \mathcal{P} \mid \mathrm{d}(s', s) \leq \mathrm{d}(s, t)\}$, where points s and t are the furthest two points among all points r with $p(r) \geq \frac{\epsilon}{16m}$.

Lemma 4. *The stoch-core $\mathcal{H}$ satisfies the following properties:*

Q1. $p(\mathcal{H}) \geq n - \frac{\epsilon}{16} = n - O(\epsilon)$

Q2. $\mathbb{E}[\,\mathsf{MST} \mid \mathcal{H}\langle n\rangle\,] = \Omega\left(\mathrm{diam}(\mathcal{H})\frac{\epsilon^2}{m^2}\right).$

Furthermore, the algorithm runs in linear time.

Estimating $\mathbb{E}[\mathsf{MST}]$: Let $\mathcal{F} = \mathcal{P}\backslash\mathcal{H}$. By the law of total expectation, the expected length of the minimum spanning tree can be expanded as follows: $\mathbb{E}[\mathsf{MST}] = \sum_{i\geq 0} \mathbb{E}[\,\mathsf{MST} \mid \mathcal{F}\langle i\rangle\,] \cdot \Pr[\mathcal{F}\langle i\rangle]$. We only estimate the first two terms $\mathbb{E}[\,\mathsf{MST} \mid \mathcal{F}\langle 0\rangle\,] \cdot \Pr[\mathcal{F}\langle 0\rangle]$ and $\mathbb{E}[\,\mathsf{MST} \mid \mathcal{F}\langle 1\rangle\,] \cdot \Pr[\mathcal{F}\langle 1\rangle]$ and use their sum as our final estimation. Using Properties Q1 and Q2, we can estimate the two terms in polynomial time.

Theorem 3. *There is an FPRAS for estimating the expected length of the minimum spanning tree in the locational uncertainty model.*

4   Minimum Perfect Matchings

In this section, we consider the minimum perfect matching (PM) problem. We use the stoch-core method.

Finding stoch-core: First, we show how to find in poly-time the stoch-core $\mathcal{H}$. See the Pseudo-code in Algorithm 2 for details.

Algorithm 2. Constructing stoch-core $\mathcal{H}$ for Estimating $\mathbb{E}[\mathsf{PM}]$

---

1 Initially, $t \leftarrow 0$ and each point $s \in \mathcal{P}$ is a component $\mathcal{H}_{\{s\}} = \mathsf{B}(s, t)$ by itself.

2 Gradually increase t; If two different components $\mathcal{H}_{S_1}$ and $\mathcal{H}_{S_2}$ intersect (where $\mathcal{H}_S := \cup_{s\in S}\mathsf{B}(s, t)$); Merge them into a new component $\mathcal{H}_{S_1\cup S_2}$.

3 Stop increasing t while the first time the following two conditions are satisfied by components at t:

   Q1. For each node v, there is a unique component $\mathcal{H}_j$ such that
        $p_v(\mathcal{H}_j) \geq 1 - O(\frac{\epsilon}{nm^3})$. We call $\mathcal{H}_j$ the stoch-core of node v, denoted as $\mathcal{H}(v)$.
   Q2. For all j, $|\{v \in \mathcal{V} \mid \mathcal{H}(v) = \mathcal{H}_j\}|$ is even.

4 Output the stopping time T and the components $\mathcal{H}_1, \ldots, \mathcal{H}_k$.

Estimating $\mathbb{E}[\mathsf{PM}]$: We use $\mathcal{H}\langle n\rangle$ to denote the event that for each node v, $v \vDash \mathcal{H}(v)$. We denote the event that there are exactly i nodes which are realized out of their stoch-cores by $\mathcal{F}\langle i\rangle$. Again, we only need to estimate two terms: $\mathbb{E}[\mathsf{PM} \mid \mathcal{F}\langle 0\rangle] \cdot \Pr[\mathcal{F}\langle 0\rangle]$ and $\mathbb{E}[\mathsf{PM} \mid \mathcal{F}\langle 1\rangle] \cdot \Pr[\mathcal{F}\langle 1\rangle]$. Using Properties Q1 and Q2, we can estimate these terms in polynomial time. Our final estimation is simply the sum of the first two terms.

Theorem 4. *Assuming the locational uncertainty model and that the number of nodes is even, there is an FPRAS for estimating the expected length of the minimum perfect matching.*

References

1. Afshani, P., Agarwal, P.K., Arge, L., Larsen, K.G., Phillips, J.M.: (Approximate) Uncertain skylines. In: Proceedings of the 14th International Conference on Database Theory, pp. 186–196. ACM (2011)
2. Agarwal, P.K., Har-Peled, S., Suri, S., Yıldız, H., Zhang, W.: Convex Hulls under Uncertainty. In: Schulz, A.S., Wagner, D. (eds.) ESA 2014. LNCS, vol. 8737, pp. 37–48. Springer, Heidelberg (2014)
3. Agarwal, P.K., Cheng, S.-W., Yi, K.: Range searching on uncertain data. ACM Transactions on Algorithms (TALG) 8(4), 43 (2012)
4. Agarwal, P.K., Cheng, S.W., Tao, Y., Yi, K.: Indexing uncertain data. In: Proceedings of the Twenty-Eighth ACM SIGMOD-SIGACT-SIGART Symposium on Principles of Database Systems, pp. 137–146. ACM (2009)
5. Agarwal, P.K., Efrat, A., Sankararaman, S., Zhang, W.: Nearest-neighbor searching under uncertainty. In: Proceedings of the 31st Symposium on Principles of Database Systems, pp. 225–236. ACM (2012)
6. Alexopoulos, C., Jacobson, J.A.: State space partition algorithms for stochastic systems with applications to minimum spanning trees. Networks 35(2), 118–138 (2000)
7. Atallah, M.J., Qi, Y., Yuan, H.: Asymptotically efficient algorithms for skyline probabilities of uncertain data. ACM Trans. Datab. Syst 32(2), 12 (2011)
8. Beardwood, J., Halton, J.H., Hammersley, J.M.: The shortest path through many points. Proc. Cambridge Philos. Soc. 55, 299–327 (1959)
9. Bern, M.W., Eppstein, D.: Worst-case bounds for suaddictive geometric graphs. In: Symposium on Computational Geometry, pp. 183–188 (1993)
10. Bertsimas, D.J., van Ryzin, G.: An asymptotic determination of the minimum spanning tree and minimum matching constants in geometrical probability. Operations Research Letters 9(4), 223–231 (1990)
11. Cheng, R., Chen, J., Mokbel, M., Chow, C.: Probabilistic verifiers: Evaluating constrained nearest-neighbor queries over uncertain data. In: ICDE (2008)
12. Cheng, R., Chen, J., Xie, X.: Cleaning uncertain data with quality guarantees. Proceedings of the VLDB Endowment 1(1), 722–735 (2008)
13. Deshpande, A., Guestrin, C., Madden, S.R., Hellerstein, J.M., Hong, W.: Model-driven data acquisition in sensor networks. In: Proceedings of the Thirtieth International Conference on Very Large Data Bases, vol. 30, pp. 588–599. VLDB Endowment (2004)
14. Dong, X., Halevy, A.Y., Yu, C.: Data integration with uncertainty. In: Proceedings of the 33rd International Conference on Very Large Data Bases, pp. 687–698. VLDB Endowment (2007)
15. Emek, Y., Korman, A., Shavitt, Y.: Approximating the statistics of various properties in randomly weighted graphs. In: Proceedings of the Twenty-Second Annual ACM-SIAM Symposium on Discrete Algorithms, pp. 1455–1467. SIAM (2011)
16. Frieze, A.M.: On the value of a random minimum spanning tree problem. Discrete Applied Mathematics 10(1), 47–56 (1985)
17. Gupta, P., Kumar, P.R.: Critical power for asymptotic connectivity. In: Proceedings of the 37th IEEE Conference on Decision and Control, vol. 1, pp. 1106–1110. IEEE (1998)
18. Gupta, P., Kumar, P.R.: The capacity of wireless networks. IEEE Transactions on Information Theory 46(2), 388–404 (2000)

19. Haenggi, M., Andrews, J.G., Baccelli, F., Dousse, O., Franceschetti, M.: Stochastic geometry and random graphs for the analysis and design of wireless networks. IEEE Journal on Selected Areas in Communications **27**(7), 1029–1046 (2009)
20. Kamousi, P., Chan, T.M., Suri, S.: Stochastic minimum spanning trees in euclidean spaces. In: Proceedings of the 27th Annual ACM Symposium on Computational Geometry, pp. 65–74. ACM (2011)
21. Kamousi, P., Chan, T.M., Suri, S.: Closest pair and the post office problem for stochastic points. Computational Geometry **47**(2), 214–223 (2014)
22. Karloff, H.J.: How long can a euclidean traveling salesman tour be? In: J. Discrete Math., p. 2(1). SIAM (1989)
23. Kleinberg, J., Eva, T.: Algorithm design. Pearson Education India (2006)
24. Li, J., Deshpande, A.: Ranking continuous probabilistic datasets. Proceedings of the VLDB Endowment **3**(1–2), 638–649 (2010)
25. Li, J., Phillips, J.M., Wang, H.: ϵ-kernel coresets for stochastic points. arXiv preprint arXiv:1411.0194 (2014)
26. Li, J., Wang, H.: Range Queries on Uncertain Data. In: Ahn, H.-K., Shin, C.-S. (eds.) ISAAC 2014. LNCS, vol. 8889, pp. 326–337. Springer, Heidelberg (2014)
27. Löffler, M., Phillips, J.M.: Shape Fitting on Point Sets with Probability Distributions. In: Fiat, A., Sanders, P. (eds.) ESA 2009. LNCS, vol. 5757, pp. 313–324. Springer, Heidelberg (2009)
28. Mainwaring, A., Culler, D., Polastre, J., Szewczyk, R., Anderson, J.: Wireless sensor networks for habitat monitoring. In: Proceedings of the 1st ACM International Workshop on Wireless Sensor Networks and Applications, pp. 88–97. ACM (2002)
29. Pfoser, D., Jensen, C.S.: Capturing the Uncertainty of Moving-Object Representations. In: Güting, R.H., Papadias, D., Lochovsky, F.H. (eds.) SSD 1999. LNCS, vol. 1651, pp. 111–131. Springer, Heidelberg (1999)
30. Shapiro, A., Dentcheva, D., Ruszczyński, A.: Lectures on stochastic programming: modeling and theory, vol. 16. SIAM (2014)
31. Snyder, T.L., Steele, J. M.: A priori bounds on the euclidean traveling salesman. J. Comput., p. 24(3) (1995)
32. Steele, J.M.: On Frieze's $\zeta(3)$ limit for lengths of minimal spanning trees. Discrete Applied Mathematics **18**(1), 99–103 (1987)
33. Suciu, D., Olteanu, D., Ré, C., Koch, C.: Probabilistic databases. Synthesis Lectures on Data Management **3**(2), 1–180 (2011)
34. Swamy, C., Shmoys, D.B.: Approximation algorithms for 2-stage stochastic optimization problems **37**(1), 33–46 (2006)
35. Szewczyk, R., Osterweil, E., Polastre, J., Hamilton, M., Mainwaring, A., Estrin, D.: Habitat monitoring with sensor networks. Communications of the ACM **47**(6), 34–40 (2004)
36. Yıldız, H., Foschini, L., Hershberger, J., Suri, S.: The Union of Probabilistic Boxes: Maintaining the Volume. In: Demetrescu, C., Halldórsson, M.M. (eds.) ESA 2011. LNCS, vol. 6942, pp. 591–602. Springer, Heidelberg (2011)

Deterministic Truncation of Linear Matroids

Daniel Lokshtanov[1], Pranabendu Misra[2]($\boxtimes$),
Fahad Panolan[1,2], and Saket Saurabh[1,2]

[1] University of Bergen, Bergen, Norway
daniello@ii.uib.no
[2] Institute of Mathematical Sciences, Chennai, India
{pranabendu,fahad,saket}@imsc.res.in

Abstract. Let $M = (E, \mathcal{I})$ be a matroid. A *k-truncation* of M is a matroid $M' = (E, \mathcal{I}')$ such that for any $A \subseteq E$, $A \in \mathcal{I}'$ if and only if $|A| \leq k$ and $A \in \mathcal{I}$. Given a linear representation of M we consider the problem of finding a linear representation of the k-truncation of this matroid. This problem can be expressed as the following problem on matrices. Let M be a $n \times m$ matrix over a field $\mathbb{F}$. A *rank k-truncation* of the matrix M is a $k \times m$ matrix M_k (over $\mathbb{F}$ or a related field) such that for every subset $I \subseteq \{1, \ldots, m\}$ of size at most k, the set of columns corresponding to I in M has rank $|I|$ if and only if the corresponding set of columns in M_k has rank $|I|$. A common way to compute a rank k-truncation of a $n \times m$ matrix is to multiply the matrix with a random $k \times n$ matrix (with the entries from a field of an exponential size), yielding a simple randomized algorithm. So a natural question is whether it possible to obtain a rank k-truncation of a matrix, *deterministically.* In this paper we settle this question for matrices over any field in which the field operations can be done efficiently. This includes any finite field and the field of rationals ($\mathbb{Q}$).

Our algorithms are based on the properties of the classical Wronskian determinant, and the folded Wronskian determinant, which was recently introduced by Guruswami and Kopparty [*FOCS, 2013*], and was implicitly present in the work of Forbes and Shpilka [*STOC, 2012*]. These were used in the context of subspace designs, and reducing randomness for polynomial identity testing and other related problems. Our main conceptual contribution in this paper is to show that the Wronskian determinant can also be used to obtain a representation of the truncation of a linear matroid in deterministic polynomial time. Finally, we use our results to derandomize several parameterized algorithms, including an algorithm for computing ℓ-MATROID PARITY, to which several problems like ℓ-MATROID INTERSECTION can be reduced.

D. Lokshtanov is supported by the "BeHard" grant under the recruitment programme of the of Bergen Research Foundation. F. Panolan is supported by the European Research Council under the European Unions Seventh Framework Programme (FP/2007-2013) / ERC Grant Agreement no. 267959. S. Saurabh is supported by "PARAPPROX" ERC starting grant no. 306992.

© Springer-Verlag Berlin Heidelberg 2015
M.M. Halldórsson et al. (Eds.): ICALP 2015, Part I, LNCS 9134, pp. 922–934, 2015.
DOI: 10.1007/978-3-662-47672-7_75

1   Introduction

A *rank k-truncation* of a $n \times m$ matrix M, is a $k \times m$ matrix M_k such that for every subset $I \subseteq \{1, \ldots, m\}$ of size at most k, the set of columns corresponding to I in M_k has rank $|I|$ if and only if the corresponding set of columns in M has rank $|I|$. We can think of finding a rank k-truncation of a matrix as a dimension reduction problem such that linear independence among all sets of columns of size at most k is preserved. This problem is a variant of the more general *dimensionality reduction* problem, which is a basic problem in many areas of computer science such as machine learning, data compression, information processing and others. In dimensionality reduction, we are given a collection of points (vectors) in a high dimensional space, and the objective is to map these points to points in a space of small dimension while preserving some property of the original collection of points. For an example, one could consider the problem of reducing the dimension of the space, while preserving the pairwise distance, for a given collection of points. Using the Johnson-Lindenstrauss Lemma this can be done approximately for any collection of m points, while reducing the dimension of the space to $\mathcal{O}(\log m)$ [4, 22]. In this work, we study dimensionality reduction under the constraint that linear independence of any sub-collection of size up to k of the given set of vectors is preserved. The motivation for this problem comes from *Matroid theory* and its algorithmic applications. For any matroid $M = (E, \mathcal{I})$, a *k-truncation* of M is a matroid $M' = (E, \mathcal{I}')$ such that for any $A \subseteq E$, $A \in I'$ if and only if $|A| \leq k$ and $A \in \mathcal{I}$. Given a linear representation of a matroid $M = (E, \mathcal{I})$ of rank n over a ground set of size m (which has a representation matrix M of dimension $n \times m$), we want to find a linear representation of the k-truncation of the matroid M. In other words, we want to map the set of column vectors of M (which lie in a space of dimension n) to vectors in a space of dimension k such that, any set S of column vectors of M with $|S| \leq k$ are linearly independent if and only if the corresponding set of vectors in the k-dimensional vector space are linearly independent.

A common way to obtain a rank k-truncation of a matrix M, is to left-multiply M by a random matrix of dimension $k \times n$ (with entries from a field of an exponential size). Then using the Schwartz-Zippel Lemma one can show that, the product matrix is a k-truncation of the matrix M with high probability [26]. This raises a natural question of whether there is a deterministic algorithm for computing k-truncation of a matrix. In this paper we settle this question by giving a polynomial time deterministic algorithm to solve this problem. In particular we have the following theorem.

Theorem 1. *Let M be a $n \times m$ matrix over a field $\mathbb{F}$ of rank n. Given a number $k \leq n$, we can compute a matrix M_k over the field $\mathbb{F}(X)$ such that it is a representation of the k-truncation of M, in $\mathcal{O}(mnk)$ field operations over $\mathbb{F}$. Furthermore, given M_k, we can test whether a given set of ℓ columns in M_k are linearly independent in $\mathcal{O}(n^2 k^3)$ field operations over $\mathbb{F}$.*

Observe that, using Theorem 1 we can obtain a deterministic truncation of a matrix over any field where the field operations can be done efficiently.

This includes any finite field ($\mathbb{F}_{p^\ell}$) or field of rationals $\mathbb{Q}$. In particular our result implies that we can find deterministic truncation for important classes of matroids such as graphic matroids, co-graphic matroids, partition matroids and others. We note that for many fields, the k-truncation matrix can be represented over a finite degree extension of $\mathbb{F}$, which is useful in algorithmic applications.

A related notion is the ℓ-*elongation* of a matroid, where $\ell > \mathsf{rank}(M)$. It is defined as the matriod $M' = (E, \mathcal{I}')$ such that $S \subseteq E$ is a basis of M' if and only if, it contains a basis of M and $|S| = \ell$. Note that the rank of the matroid M' is ℓ. We have the following observation and it's corollary.

Observation 2 ([28], page 75). *Let M be a matroid of rank n over a ground set of size m. Let M^*, $T(M, k)$ and $E(M, \ell)$ denote the dual matroid, the k-truncation and the ℓ-elongation of the matroid M, respectively. Then $E(M, \ell) = \{T(M^*, m - \ell)\}^*$, i.e. the ℓ-elongation of M is the dual of the $(m - \ell)$-truncation of the dual of M.*

Corollary 1. *Let M be a linear matroid of rank n, over a ground set of size m, which is representable over a field $\mathbb{F}$. Given a number $\ell \geq n$, we can compute a representation of the ℓ-elongation of M, over the field $\mathbb{F}(X)$ in $\mathcal{O}(mn\ell)$ field operations over $\mathbb{F}$.*

Tools and Techniqes. The main tool used in this work, is the Wronskian determinant and its characterization of the linear independence of a set of polynomials. Given a polynomial $P_j(X)$ and a number ℓ, define $Y_j^\ell = (P_j(X), P_j^{(1)}(X), \ldots, P_j^{(\ell-1)}(X))^T$. Here, $P_j^{(i)}(X)$ is the i-th formal derivative of $P_j(X)$. Formally, the Wronskian matrix of a set of polynomials $P_1(X), \ldots, P_k(X)$ is defined as the $k \times k$ matrix $W(P_1, \ldots, P_k) = [Y_1^k, \ldots, Y_k^k]$. Recall that to get a k-truncation of a linear matroid, we need to map a set of vectors from $\mathbb{F}^n$ to $\mathbb{K}^k$ such that linear independence of any subset of the given vectors of size at most k is preserved. We associate with each vector, a polynomial whose coefficients are the entries of the vector. A known mathematical result states that a set of polynomials $P_1(X), \ldots, P_k(X) \in \mathbb{F}[X]$ are linearly independent over $\mathbb{F}$ if and only if the corresponding Wronskian determinant $\det(W(P_1, \ldots, P_k)) \not\equiv 0$ in $\mathbb{F}[X]$ [2,17,27]. However, this requires that the underlying field be $\mathbb{Q}$ (or $\mathbb{R}$, $\mathbb{C}$), or that it is a finite field whose characteristic is strictly larger than the maximum degree of $P_1(X), \ldots, P_k(X)$.

For fields of small characteristic, we use the notion of α-folded Wronskian, which was introduced by Guruswami and Kopparty [21] in the context of subspace designs, with applications in coding theory. It was also implicitly present in the works of Forbes and Shpilka [12], who used it in reducing randomness for polynomial identity testing and related problems. Let $\mathbb{F}$ be a finite field and α be an element of $\mathbb{F}$. Given a polynomial $P_j(X) \in \mathbb{F}[X]$ and a number ℓ, define $Z_j^\ell = (P_j(X), P_j(\alpha X), \ldots, P_j(\alpha^{\ell-1}X))^T$. Formally, the α-folded Wronskian matrix of a family of polynomials $P_1(X), \ldots, P_k(X)$ is defined as the $k \times k$ matrix $W_\alpha(P_1, \ldots, P_k) = [Z_1^k, \ldots, Z_k^k]$. Let $P_1(X), \ldots, P_k(X)$ be a family of polynomials of degree at most $n - 1$. From, the results of Forbes and Shpilka [12] one can derive that if α is an element of the field $\mathbb{F}$, of order at least n then $P_1(X), \ldots, P_k(X)$ are linearly independent over $\mathbb{F}$ if and only if the α-folded Wronskian determinant $\det(W_\alpha(P_1, \ldots, P_k)) \not\equiv 0$ in $\mathbb{F}[X]$.

Having introduced the tools, we continue to the description of our algorithm. Given a $n \times m$ matrix M over $\mathbb{F}$ and a positive integer k our algorithm for finding a k-truncation of M proceeds as follows. To a column C_i of M we associate a polynomial $P_i(X)$ whose coefficients are the entries of C_i. That is, if $C_i = (c_{1i}, \ldots, c_{ni})^T$ then $P_i(X) = \sum_{j=1}^{n} c_{ji} x^{j-1}$. If the characteristic of the field $\mathbb{F}$ is strictly larger than n or $\mathbb{F} = \mathbb{Q}$ then we return $M_k = [Y_1^k, \ldots, Y_m^k]$ as the required k-truncation of M. In other cases we first compute an $\alpha \in \mathbb{F}$ of order at least n and then return $M_k = [Z_1^k, \ldots, Z_m^k]$. We then use the properties of Wronskian determinant and α-folded Wronskian, to prove the correctness of our algorithm. Observe that when M is a representation of a linear matroid then M_k is a representation of it's k-truncation. Further, each entry of M_k is a polynomial of degree at most $n-1$ in $\mathbb{F}[X]$. Thus, testing whether a set of columns of size at most k is independent, reduces to testing whether a determinant polynomial of degree at most $(n-1)k$ is identically zero or not. This is easily done by evaluating the determinant at $(n-1)k+1$ points in $\mathbb{F}$ and testing if it is zero at all those points.

Our main conceptual contribution in this paper is to show the connection between the Wronskian matrices and the truncation of a linear matroids, which can be used obtain a representation of the truncation in deterministic polynomial time. These matrices are related to the notion of "rank extractors" which have important applications in polynomial identity testing and in the construction of randomness extractors [11,12,14,15]. We believe that these and other related tools could be useful in obtaining other parameterized algorithms, apart from those mentioned in this paper. We note that, one can obtain a different construction of matrix truncation via an earlier result of Gabizon and Raz [15], which was used in construction of randomness extractors.

Applications. Matroid theory has found many algorithmic applications, starting from the characterization of greedy algorithms, to designing fixed parameter tractable (FPT) algorithms and kernelization algorithms. Recently the notion of *representative families* over linear matroids was used in designing fast FPT, as well as kernelization algorithm for several problems [8,10,19,23,24,26,29]. Let us introduce this notion more formally. Let $M = (E, \mathcal{I})$ be a matroid and let $\mathcal{S} = \{S_1, \ldots, S_t\}$ be a family of subsets of E of size p. A subfamily $\widehat{\mathcal{S}} \subseteq \mathcal{S}$ is q-*representative* for $\mathcal{S}$ if for every set $Y \subseteq E$ of size at most q, if there is a set $X \in \mathcal{S}$ disjoint from Y with $X \cup Y \in \mathcal{I}$, then there is a set $\widehat{X} \in \widehat{\mathcal{S}}$ disjoint from Y and $\widehat{X} \cup Y \in \mathcal{I}$. In other words, if a set Y of size at most q can be extended to an independent set of size $|Y| + p$ by adding a subset from $\mathcal{S}$, then it also can be extended to an independent set of size $|Y| + p$ by adding a subset from $\widehat{\mathcal{S}}$ as well. The Two-Families Theorem of Bollobás [1] for extremal set systems and its generalization to subspaces of a vector space of Lovász [25] (see also [13]) imply that every family of sets of size p has a q-representative family with at most

$\binom{p+q}{p}$ sets. Recently, Fomin et. al. [10] gave an efficient randomized algorithm to compute a representative family of size $\binom{p+q}{p}$ in a linear matroid of rank $n > p + q$. This algorithm starts by computing a randomized $(p + q)$-truncation of the given linear matroid and then computes a q-representative family over the truncated matroid deterministically. Therefore one of our motivations to study the k-truncation problem was to find an efficient deterministic computation of a representative family in a linear matroid. Formally, we have

Theorem 3. *Let $M = (E, \mathcal{I})$ be a linear matroid of rank n and let $\mathcal{S}$ be a p-family of independent sets of size t. Let A be a $n \times |E|$ matrix representing M over a field $\mathbb{F}$, and let ω be the exponent of matrix multiplication. Then there are deterministic algorithms computing $\widehat{\mathcal{S}} \subseteq^q_{rep} \mathcal{S}$ as follows.*

1. *A family $\widehat{\mathcal{S}}$ of size $\binom{p+q}{p}$ in $\mathcal{O}\left(\binom{p+q}{p}^2 tp^3n^2 + t\binom{p+q}{q}^\omega np\right) + (n + |E|)^{\mathcal{O}(1)}$, operations over $\mathbb{F}$.*
2. *A family $\widehat{\mathcal{S}}$ of size $np\binom{p+q}{p}$ in $\mathcal{O}\left(\binom{p+q}{p}tp^3n^2 + t\binom{p+q}{q}^{\omega-1}(pn)^{\omega-1}\right) + (n + |E|)^{\mathcal{O}(1)}$ operations over $\mathbb{F}$.*

As a corollary of the above theorem, we obtain a deterministic FPT algorithm for ℓ-MATROID PARITY, derandomizing the main algorithm of Marx [26], to which all other problems are reduced in [26]. In particular this implies a deterministic FPT algorithm for ℓ-MATROID INTERSECTION, certain packing problems and FEEDBACK EDGE SET WITH BUDGET VECTORS. Using our results one can compute, in deterministic polynomial time, the k-truncation of graphic and co-graphic matroids, which has important applications in graph algorithms. Recently, the truncation for co-graphic matroid has been used to obtain deterministic parameterized algorithms, running in time $2^{\mathcal{O}(k)}n^{\mathcal{O}(1)}$ time, for problems where we need to delete k edges that keeps the graph connected and maintain certain parity conditions [20]. These problems include UNDIRECTED EULERIAN EDGE DELETION, DIRECTED EULERIAN EDGE DELETION and UNDIRECTED CONNECTED ODD EDGE DELETION [3,6,7,20].

2   Preliminaries

In this section we give various definitions and notions which we make use of in the paper. We use the following notations: $[n] = \{1, \ldots, n\}$ and $\binom{[n]}{i} = \{X \mid X \subseteq [n], |X| = i\}$.

Fields and Polynomials. In this section we review some definitions and properties of fields. We refer to any graduate text on algebra for more details. The cardinality or the size of a field is called its *order*. For every prime number p and a positive integer ℓ, there exists a finite field of order p^ℓ. Let $\mathbb{F}$ be a finite field and then $\mathbb{F}[X]$ denotes the ring of polynomials in X over $\mathbb{F}$. For the ring

---

Due to space constraints, proofs of some lemmas and some standard definitions have been omitted. These will appear in the full version of the paper.

$\mathbb{F}[X]$, we use $\mathbb{F}(X)$ to denote the *field of fractions* of $\mathbb{F}[X]$. We will use $\mathbb{F}[X]^{<n}$ to denote the set the polynomials in $\mathbb{F}[X]$ of degree $< n$. The *characteristic* of a field, denoted by $\mathrm{char}(\mathbb{F})$, is defined as least positive integer m such that $\sum_{i=1}^{m} 1 = 0$, and is 0 when no such m exists. For a finite field $\mathbb{F}$, $\mathbb{F}^* = \mathbb{F} \setminus \{0\}$ is cyclic group under multiplication. We say that an element $\beta \in \mathbb{F}$ has *order* r, if r is the least integer such that $\beta^r = 1$. All finite fields are obtained as extensions of prime fields, and for any prime p and positive integer ℓ there is exactly one finite field of order p^ℓ up to isomorphism.

Vector and Matrices. A collection of vectors $\{v_1, v_2, \ldots, v_k\}$ are said to be linearly dependent if there exist values $a_1, a_2, \ldots, a_k$, not all zero, from $\mathbb{F}$ such that $\sum_{i=1}^{k} a_i v_i = 0$. Otherwise these vectors are called linearly independent. For a matrix A (or a vector v) by A^T (or v^T) we denoted its *transpose*. The rank of a matrix is the cardinality of the maximum sized collection of columns which are linearly independent. Equivalently, the rank of a matrix is the maximum number k such that there is a $k \times k$ submatrix whose determinant is non-zero. The determinant of a $n \times n$ matrix A is denoted by $\det(A)$. Throughout the paper we use ω to denote the matrix multiplication exponent. The current best known bound on $\omega < 2.373$ [16,30].

Derivatives. Recall the definition of the formal derivative $\frac{d}{dx}$ of a function over $\mathbb{R}$. We denote the k-th formal derivative of a function f by $f^{(k)}$. We can extend this notion to finite fields. Let $\mathbb{F}$ be a finite field and let $\mathbb{F}[X]$ be the ring of polynomials in X over $\mathbb{F}$. Let $P \in \mathbb{F}[X]$ be a polynomial of degree $n-1$, i.e. $P = \sum_{i=0}^{n-1} a_i X^i$ where $a_i \in \mathbb{F}$. Then we define the *formal derivative* of as $P' = \sum_{i=1}^{n-1} i a_i X^{i-1}$. We can extend this definition to the k-th formal derivative of P as $P^{(k)} = (P^{(k-1)})'$. For a polynomial $P(X) \in \mathbb{F}[X]$, the i-th Hasse derivative $D^i(P)$ is defined as the coefficient of Z^i in $P(X+Z)$. Here, $P(X + Z) = \sum_{i=0}^{\infty} D^i(P(X))Z^i$. We note that Hasse derivatives differ from formal derivatives by a multiplicative factor. We refer to [5] and [18] for details.

3   Matroid Truncation

In this section we give the main result of this work. We start by defining the tools required for our algorithm. Let $\mathbb{F}$ be a field. The set of polynomials $P_1(X), P_2(X), \ldots, P_k(X)$ in $\mathbb{F}[X]$ are said to be *linearly independent* over $\mathbb{F}$ if there don't exist $a_1, a_2, \ldots, a_k \in \mathbb{F}$, not all zeros such that $\sum_{i=1}^{k} a_i P_i(X) \equiv 0$. Otherwise they are said to be linearly dependent.

Definition 1. *Let $P(X)$ be a polynomial of degree at most $n-1$ in $\mathbb{F}[X]$. We define the vector v corresponding to the polynomial $P(X)$ as follows: $v[j] = c_j$ where $P(X) = \sum_{j=1}^{n} c_j x^{j-1}$. Similarly given a vector v of length n over $\mathbb{F}$, we define the polynomial $P(X)$ in $\mathbb{F}[X]$ corresponding to the vector v as follows: $P(X) = \sum_{j=1}^{n} v[j] x^{j-1}$.*

Lemma 1. *Let $v_1, \ldots, v_k$ be vectors of length n over $\mathbb{F}$ and let $P_1(X), \ldots, P_k(X)$ be the corresponding polynomials respectively. Then $P_1(X), \ldots, P_k(X)$ are linearly independent over $\mathbb{F}$ if and only if $v_1, \ldots, v_k$ are linearly independent over $\mathbb{F}$.*

Wronskian. Let $\mathbb{F}$ be a field with characteristic at least n. Consider a collection of polynomials $P_1(X), \ldots, P_k(X)$ from $\mathbb{F}[X]$ of degree at most $n - 1$. We define the following matrix, called the *Wronskian*, of $P_1(X), \ldots, P_k(X)$ as follows.

$$W(P_1, \ldots, P_k) = \begin{pmatrix} P_1(X) & P_2(X) & \cdots & P_k(X) \\ P_1^{(1)}(X) & P_2^{(1)}(X) & \cdots & P_k^{(1)}(X) \\ \vdots & \vdots & \ddots & \vdots \\ P_1^{(k-1)}(X) & P_2^{(k-1)}(X) & \cdots & P_k^{(k-1)}(X) \end{pmatrix}_{k \times k}$$

Note that, the determinant of the above matrix actually yields a polynomial. For our purpose we will need the following well known result.

Theorem 4 ([2, 17, 27]). *Let $\mathbb{F}$ be a field and $P_1(X), \ldots, P_k(X)$ be a set of polynomials from $\mathbb{F}[X]^{<n}$ and let $\mathrm{char}(\mathbb{F}) > n$ or $\mathbb{F} = \mathbb{Q}$. Then $P_1(X), \ldots, P_k(X)$ are linearly independent over $\mathbb{F}$ if and only if $\det(W(P_1, \ldots, P_k)) \not\equiv 0$ in $\mathbb{F}[X]$.*

The notion of Wronskian dates back to 1812 [27]. We refer to [2, 17] for some recent variations and proofs. The switch between usual derivatives and Hasse derivatives multiplies the Wronskian determinant by a constant, which is non-zero as long as $n < \mathrm{char}(\mathbb{F})$, and thus this criterion works with both notions. Observe that the Wronskian determinant is a polynomial of degree at most nk in $\mathbb{F}[X]$. Thus to test if such a polynomial is identically zero, we only need to evaluate it at $nk + 1$ arbitrary points of the field $\mathbb{F}$, and check if it is zero at all those points.

Folded Wronskian. The above definition of Wronskian requires us to compute derivatives of degree $(n - 1)$ polynomials, which are well defined only if the underlying field has characteristic greater than $n - 1$. For matrices over fields of small characteristic, we have the notion of *Folded Wronskian*, which is defined as follows. Consider a collection of polynomials $P_1(X), \ldots, P_k(X)$ from $\mathbb{F}[X]$ of degree at most $(n - 1)$. Further, let $\mathbb{F}$ be of order at least $n + 1$, and α be an element of $\mathbb{F}^*$. We define the α-*folded Wronskian*, of $P_1(X), \ldots, P_k(X)$ as follows.

$$W_\alpha(P_1, \ldots, P_k) = \begin{pmatrix} P_1(X) & P_2(X) & \cdots & P_k(X) \\ P_1(\alpha X) & P_2(\alpha X) & \cdots & P_k(\alpha X) \\ \vdots & \vdots & \ddots & \vdots \\ P_1(\alpha^{k-1} X) & P_2(\alpha^{k-1} X) & \cdots & P_k(\alpha^{k-1} X) \end{pmatrix}_{k \times k}$$

As before, the determinant of the above matrix is a polynomial of degree at most nk in $\mathbb{F}[X]$. The following theorem by Forbes and Shpilka [12] shows that the above determinant characterizes the linear independence of the collection of polynomials.

Theorem 5 ([12], Theorem 4.1). [1] *Let $\mathbb{F}$ be a field, α be an element of $\mathbb{F}$ of order $\geq n$ and let $P_1(X), \ldots, P_k(X)$ be a set of polynomials from $\mathbb{F}[X]^{<n}$. Then $P_1(X), \ldots, P_k(X)$ are linearly independent over $\mathbb{F}$ if and only if the α-folded Wronskian determinant $\det(W_\alpha(P_1, \ldots, P_k)) \not\equiv 0$ in $\mathbb{F}[X]$.*

3.1 Deterministic Truncation of Matrices

In this section we look at algorithms for computing k-truncation of matrices. We are given as input a matrix M of over the set of rational numbers $\mathbb{Q}$ or over some finite field $\mathbb{F}$. The following lemma gives us an algorithm to compute the truncation of a matrix using the classical wronskian, over an appropriate field. We shall refer to this as the *classical wronskian method of truncation*.

Lemma 2. *Let M be a $n \times m$ matrix of rank n over a field $\mathbb{F}$, where $\mathbb{F}$ is either $\mathbb{Q}$ or $\operatorname{char}(\mathbb{F}) > n$. Then we can compute a $k \times m$ matrix M_k of rank k over the field $\mathbb{F}(X)$ which is a k-truncation of the matrix M in $\mathcal{O}(mnk)$ operations in $\mathbb{F}$.*

Proof. Let $\mathbb{F}[X]$ be the ring of polynomials in X over $\mathbb{F}$ and let $\mathbb{F}(X)$ be the corresponding field of fractions. Let $C_1, \ldots, C_m$ denote the columns of M. Observe that we have a polynomial $P_i(X)$ corresponding to the column C_i of degree at most $n - 1$, and by Lemma 1 we have that $C_{i_1}, \ldots, C_{i_\ell}$ are linearly independent over $\mathbb{F}$ if and only if $P_{i_1}(X), \ldots, P_{i_\ell}(X)$ are linearly independent over $\mathbb{F}$. Further note that P_i lies in $\mathbb{F}[X]$ and thus also in $\mathbb{F}(X)$. Let D_i be the vector $(P_i(X), P_i^{(1)}(X), \ldots, P_i^{(k-1)}(X))$ of length k with entries from $\mathbb{F}[X]$ (and also in $\mathbb{F}(X)$). Note that the entries of D_i are polynomials of degree at most $n - 1$. Let us define the matrix M_k to be the $(k \times m)$ matrix whose columns are D_i^T, and note that M_k is a matrix with entries from $\mathbb{F}[X]$. We will show that indeed M_k is a k-truncation of the matrix M .

Let $I \subseteq \{1, \ldots, m\}$ such that $|I| = \ell \leq k$. Let $C_{i_1}, \ldots, C_{i_\ell}$ be a linearly independent set of columns of the matrix M over $\mathbb{F}$, where $I = \{i_1, \ldots, i_\ell\}$. We will show that the columns $D_{i_1}^T, \ldots, D_{i_\ell}^T$ are linearly independent in M_k over $\mathbb{F}(X)$. Consider the $k \times \ell$ matrix M_I whose column are the vectors $D_{i_1}^T, \ldots, D_{i_\ell}^T$. We shall show that M_I has rank ℓ by showing that there is a $\ell \times \ell$ submatrix whose determinant is a non-zero polynomial. Let $P_{i_1}(X), \ldots, P_{i_\ell}(X)$ be the polynomials corresponding to the vectors $C_{i_1}, \ldots, C_{i_\ell}$. By Lemma 1 we have that $P_{i_1}(X), \ldots, P_{i_\ell}(X)$ are linearly independent over $\mathbb{F}$. Then by Theorem 4, the $(\ell \times \ell)$ matrix formed by the column vectors $(P_{i_j}(X), P_{i_j}^{(1)}(X), \ldots, P_{i_j}^{(\ell-1)}(X))^T$,

---

[1] We would like to thank the anonymous reviewers, who pointed out the results of Forbes and Shpilka [12], which we were unaware of. In an earlier version of the paper we claimed a new proof of this theorem, which has been removed.

$i_j \in I$, is a non-zero determinant in $\mathbb{F}[X]$. But note that this matrix is a submatrix of M_I. Therefore M_I has rank ℓ in $\mathbb{F}(X)$. Therefore the vectors $D_{i_1}^T, \ldots, D_{i_\ell}^T$ are linearly independent in $\mathbb{F}(X)$. This completes the proof of the forward direction.

Let $I \subseteq \{1, \ldots, m\}$ such that $|I| = \ell \leq k$ and let $D_{i_1}^T, \ldots, D_{i_\ell}^T$ be linearly independent in M_k over $\mathbb{F}(X)$, where $I = \{i_1, \ldots, i_\ell\}$. We will show that the corresponding set of columns $C_{i_1}, \ldots, C_{i_\ell}$ are also linearly independent over $\mathbb{F}$. For a contradiction assume that $C_{i_1}, \ldots, C_{i_\ell}$ are linearly *dependent* over $\mathbb{F}$. Let $P_{i_1}(X), \ldots, P_{i_\ell}(X)$ be the polynomials in $\mathbb{F}[X]$ corresponding to these vectors. Then by Lemma 1 we have that $P_{i_1}(X), \ldots, P_{i_\ell}(X)$ are linearly dependent over $\mathbb{F}$. So there is a tuple $a_{i_1}, \ldots, a_{i_\ell}$ of values of $\mathbb{F}$ such that $\sum_{j=1}^{\ell} a_{i_j} P_{i_j}(X) = 0$. Therefore, for any $d \in \{1, \ldots, \ell-1\}$, we have that $\sum_{j=1}^{\ell} a_{i_j} P_{i_j}^{(d)}(X) = 0$. Now let $D_{i_1}^T, \ldots, D_{i_\ell}^T$ be the column vectors of M_k corresponding to $C_{i_1}, \ldots, C_{i_\ell}$. Note that $\mathbb{F}$ is a subfield of $\mathbb{F}(X)$ and by the above, we have that $\sum_{j=1}^{\ell} a_{i_j} D_{i_j} = 0$. Thus $D_{i_1}^T, \ldots, D_{i_\ell}^T$ are linearly dependent in M_k over $\mathbb{F}(X)$, a contradiction to our assumption.

Thus we have shown that for any $\{i_1, \ldots, i_\ell\} \subseteq \{1, \ldots, m\}$ such that $\ell \leq k$, $C_{i_1}, \ldots, C_{i_\ell}$ are linearly independent over $\mathbb{F}$ if and only if $D_{i_1}, \ldots, D_{i_\ell}$ are linearly independent over $\mathbb{F}(X)$. To estimate the running time, observe that for each C_i we can compute D_i in $\mathcal{O}(kn)$ field operations and thus we can compute M_k in $\mathcal{O}(mnk)$ field operations.

This completes the proof of this lemma. $\square$

Lemma 2 is useful in obtaining k-truncation of matrices which entries are either from the field of large characteristic or from $\mathbb{Q}$. Using Theorem 5, we obtain the following lemma, which allows us to find truncations in fields of small characteristic which have large order. We however require an element of high order of such a field to compute the truncation. In the next lemma we demand a lower bound on the size of the field as we need an element of certain order. We will later see how to remove this requirement from the statement of the next lemma.

Lemma 3. *Let $\mathbb{F}$ be a finite field and α be an element of $\mathbb{F}$ of order at least n. Let M be a $(n \times m)$ matrix of rank n over a field $\mathbb{F}$. Then we can compute a $(k \times m)$ matrix M_k of rank k over the field $\mathbb{F}(X)$ which is a k-truncation of the matrix M in $\mathcal{O}(mnk)$ field operations over $\mathbb{F}$.*

In Lemma 3 we require that α be an element of order at least n. This implies that the order of the field $\mathbb{F}$ must be at least $n+1$. We can ensure these requirements by preprocessing the input before invoking the Lemma 3. Formally, we show the following lemma.

Lemma 4. *Let M be a matrix of dimension $n \times m$ over a finite field $\mathbb{F}$, and of rank n. Let $\mathbb{F} = \mathbb{F}_{p^\ell}$ where $p < n$. Then in polynomial time we can find an extension field $\mathbb{K}$ of order at least $n+1$ and an element α of $\mathbb{K}$ of order at least $n+1$, such that M is a matrix over $\mathbb{K}$ with the same linear independence relationships between its columns as before.*

The next result is useful in finding basis of matrices with entries from $\mathbb{F}[X]$.

Lemma 5. *Let M be a $m \times t$ matrix with entries from $\mathbb{F}[X]^{<n}$ and let $m \leq t$. Let $w : \mathbf{C}(M) \to \mathbb{R}^+$ be a weight function. Then we can compute a minimum weight column basis of M in $\mathcal{O}(m^2 n^2 t + m^\omega n t)$ field operations over $\mathbb{F}$.*

Finally, we combine Lemma 2, Lemma 4, Lemma 3 and Lemma 5 to obtain the proof of our main theorem (namely, Theorem 1).

Observe that, using Theorem 1 we can obtain a deterministic truncation of a matrix over any field where the field operations can be done efficiently. This includes any finite field ($\mathbb{F}_{p^\ell}$) or field of rationals $\mathbb{Q}$. We also remark that, in many instances we can view the truncation as over a finite degree extension of $\mathbb{F}$, which can be computed efficiently. This is useful in algorithmic applications.

4 Application: Computing Representative Families

In this section we give deterministic algorithms to compute representative families of a linear matroid, given its representation matrix.

Definition 1 (q-Representative Family). *Given a matroid $M = (E, \mathcal{I})$ and a family $\mathcal{S}$ of subsets of E, we say that a subfamily $\widehat{\mathcal{S}} \subseteq \mathcal{S}$ is q-representative for $\mathcal{S}$ if the following holds: for every set $Y \subseteq E$ of size at most q, if there is a set $X \in \mathcal{S}$ disjoint from Y with $X \cup Y \in \mathcal{I}$, then there is a set $\widehat{X} \in \widehat{\mathcal{S}}$ disjoint from Y with $\widehat{X} \cup Y \in \mathcal{I}$. If $\widehat{\mathcal{S}} \subseteq \mathcal{S}$ is q-representative for $\mathcal{S}$ we write $\widehat{\mathcal{S}} \subseteq_{rep}^q \mathcal{S}$.*

We say that a family $\mathcal{S} = \{S_1, \ldots, S_t\}$ of sets is a *p-family* if each set in $\mathcal{S}$ is of size p. In [10] the following theorem is proved. See [9, Theorem 4].

Theorem 6 ([9,10]). *Let $M = (E, \mathcal{I})$ be a linear matroid and let $\mathcal{S}$ be a p-family of independent sets of size t. Then there exists $\widehat{\mathcal{S}} \subseteq_{rep}^q \mathcal{S}$ of size $\binom{p+q}{p}$. Furthermore, given a representation A_M of M over a field $\mathbb{F}$, there is a randomized algorithm computing $\widehat{\mathcal{S}} \subseteq_{rep}^q \mathcal{S}$ in $\mathcal{O}\left(\binom{p+q}{p} t p^\omega + t \binom{p+q}{q}^{\omega-1}\right)$ operations over $\mathbb{F}$.*

Fomin et al. [10, Theorem3.1] first gave a deterministic algorithm for computing q-representative of a p-family of independent sets if the rank of the corresponding matroid is $p + q$. To prove Theorem 6 we first compute a randomized k-truncation of $M = (E, \mathcal{I})$ [26], and then compute the representative sets. By using Theorem 1, we may instead obtain a deterministic truncation. Observe that the representation given by Theorem 1 is over $\mathbb{F}(X)$. However, deterministic algorithms to compute basis of matrices over $\mathbb{F}[X]$ are slower compared to the standard algorithms. Therefore, we first show a lemma that allows us to find a set of columns, which contains a basis of the matrix over $\mathbb{F}[X]$, quickly; though the size of the set given by the lemma could be slightly larger than the size of a basis of the matrix. This result together with Theorem 1 imply our Theorem 3. We note that, one can in fact prove Theorem 6 for a "weighted notion of representative family". And as before, by using Theorem 1, we can also obtain a deterministic version of this theorem.

Applications. Marx [26] gave algorithms for several problems based on matroid optimization. The main theorem in his work is Theorem 1.1 [26] on which most applications of [26] are based. This theorem gives a randomized FPT algorithm for the ℓ-MATROID PARITY problem.

ℓ-MATROID PARITY $\hfill$ **Parameter:** k, ℓ
Input: Let $M = (E, \mathcal{I})$ be a linear matroid where the ground set is partitioned into blocks of size ℓ and let A_M be a linear representation M.
Question: is there an independent set that is the union of k blocks?

The proof of the theorem uses an algorithm to find representative sets as a black box. Applying our algorithm (Theorem 3 of this paper) instead gives a deterministic version of Theorem 1.1 of [26].

Proposition 1. *Let $M = (E, \mathcal{I})$ be a linear matroid where the ground set is partitioned into blocks of size ℓ. Given a linear representation A_M of M, it can be determined in $\mathcal{O}(2^{\omega k \ell}\|A_M\|^{\mathcal{O}(1)})$ time whether there is an independent set that is the union of k blocks. ($\|A_M\|$ denotes the length of A_M in the input.)*

We mention an application from [26] which we believe could be useful to obtain single exponential time parameterized and exact algorithms.

ℓ-MATROID INTERSECTION $\hfill$ **Parameter:** k
Input: Let $M_1 = (E, \mathcal{I}_1), \ldots, M_1 = (E, \mathcal{I}_\ell)$ be matroids on the same ground set E given by their representations $A_{M_1}, \ldots, A_{M_\ell}$ over the same field $\mathbb{F}$ and a positive integer k.
Question: Does there exist k element set that is independent in each M_i ($X \in \mathcal{I}_1 \cap \ldots \cap \mathcal{I}_\ell$)?

Proposition 2. *ℓ-MATROID INTERSECTION can be solved in $\mathcal{O}(2^{\omega k \ell}\|A_M\|^{\mathcal{O}(1)})$ time.*

References

1. Bollobás, B.: On generalized graphs. Acta Math. Acad. Sci. Hungar **16**, 447–452 (1965)
2. Bostan, A., Dumas, P.: Wronskians and linear independence. The American Mathematical Monthly **117**(8), 722–727 (2010)
3. Cygan, M., Marx, D., Pilipczuk, M., Pilipczuk, M., Schlotter, I.: Parameterized complexity of eulerian deletion problems. Algorithmica **68**(1), 41–61 (2014)
4. Dasgupta, S., Gupta, A.: An elementary proof of a theorem of Johnson and Lindenstrauss. Random Struct. Algorithms **22**(1), 60–65 (2003). http://dx.doi.org/10.1002/rsa.10073

5. Dvir, Z., Kopparty, S., Saraf, S., Sudan, M.: Extensions to the method of multiplicities, with applications to kakeya sets and mergers. In: FOCS, pp. 181–190. IEEE (2009)
6. Fomin, F.V., Golovach, P.A.: Long circuits and large euler subgraphs. In: Bodlaender, H.L., Italiano, G.F. (eds.) ESA 2013. LNCS, vol. 8125, pp. 493–504. Springer, Heidelberg (2013)
7. Fomin, F.V., Golovach, P.A.: Parameterized complexity of connected even/odd subgraph problems. J. Comput. Syst. Sci. **80**(1), 157–179 (2014)
8. Fomin, F.V., Lokshtanov, D., Panolan, F., Saurabh, S.: Representative sets of product families. In: Schulz, A.S., Wagner, D. (eds.) ESA 2014. LNCS, vol. 8737, pp. 443–454. Springer, Heidelberg (2014)
9. Fomin, F.V., Lokshtanov, D., Saurabh, S.: Efficient computation of representative sets with applications in parameterized and exact algorithms (2013). CoRR abs/1304.4626
10. Fomin, F.V., Lokshtanov, D., Saurabh, S.: Efficient computation of representative sets with applications in parameterized and exact algorithms. In: SODA, pp. 142–151 (2014)
11. Forbes, M.A., Saptharishi, R., Shpilka, A.: Hitting sets for multilinear read-once algebraic branching programs, in any order. In: Shmoys, D.B. (ed.) STOC, pp. 867–875. ACM (2014)
12. Forbes, M.A., Shpilka, A.: On identity testing of tensors, low-rank recovery and compressed sensing. In: STOC, pp. 163–172. ACM (2012)
13. Frankl, P.: An extremal problem for two families of sets. European J. Combin. **3**(2), 125–127 (1982)
14. Gabizon, A.: Deterministic Extraction from Weak Random Sources. Monographs in Theoretical Computer Science. An EATCS Series. Springer (2011)
15. Gabizon, A., Raz, R.: Deterministic extractors for affine sources over large fields. Combinatorica **28**(4), 415–440 (2008)
16. Gall, F.L.: Powers of tensors and fast matrix multiplication. In: Nabeshima, K., Nagasaka, K., Winkler, F., Szántó, Á. (eds.) ISSAC, pp. 296–303. ACM (2014)
17. Garcia, A., Voloch, J.F.: Wronskians and linear independence in fields of prime characteristic. Manuscripta Mathematica **59**(4), 457–469 (1987)
18. Goldschmidt, D.: Algebraic functions and projective curves, vol. 215. Springer (2003)
19. Goyal, P., Misra, N., Panolan, F.: Faster deterministic algorithms for r-dimensional matching using representative sets. In: FSTTCS, pp. 237–248 (2013)
20. Goyal, P., Misra, P., Panolan, F., Philip, G., Saurabh, S.: Finding even subgraphs even faster (2014). CoRR abs/1409.4935
21. Guruswami, V., Kopparty, S.: Explicit subspace designs. In: FOCS, pp. 608–617 (2013)
22. Johnson, W.B., Lindenstrauss, J.: Extensions of lipschitz mappings into a hilbert space. In: Conference in modern analysis and probability, 1982), Contemp. Math., Amer. Math. Soc. vol. 26, pp. 189–206 (1984). http://dx.doi.org/10.1090/conm/026/737400
23. Kratsch, S., Wahlström, M.: Compression via matroids: a randomized polynomial kernel for odd cycle transversal. In: SODA, pp. 94–103. SIAM (2012)
24. Kratsch, S., Wahlström, M.: Representative sets and irrelevant vertices: New tools for kernelization. In: FOCS 2012, pp. 450–459. IEEE (2012)
25. Lovász, L.: Flats in matroids and geometric graphs. In: Combinatorial surveys (Proc. Sixth British Combinatorial Conf., Royal Holloway Coll., Egham), pp. 45–86. Academic Press, London (1977)

26. Marx, D.: A parameterized view on matroid optimization problems. Theor. Comput. Sci. **410**(44), 4471–4479 (2009)
27. Muir, T.: A Treatise on the Theory of Determinants. Dover Publications (1882)
28. Murota, K.: Matrices and matroids for systems analysis, vol. 20. Springer (2000)
29. Shachnai, H., Zehavi, M.: Representative families: a unified tradeoff-based approach. In: Schulz, A.S., Wagner, D. (eds.) ESA 2014. LNCS, vol. 8737, pp. 786–797. Springer, Heidelberg (2014)
30. Williams, V.V.: Multiplying matrices faster than Coppersmith-Winograd. In: STOC 2012, pp. 887–898. ACM (2012)

Linear Time Parameterized Algorithms for Subset Feedback Vertex Set

Daniel Lokshtanov[1], M.S. Ramanujan[1]([✉]), and Saket Saurabh[1,2]

[1] University of Bergen, Bergen, Norway
{daniello,Ramanujan.Sridharan}@ii.uib.no
[2] The Institute of Mathematical Sciences, Chennai, India
saket@imsc.res.in

Abstract. In the SUBSET FEEDBACK VERTEX SET (SUBSET FVS) problem, the input is a graph G on n vertices and m edges, a subset of vertices T, referred to as terminals, and an integer k. The objective is to determine whether there exists a set of at most k vertices intersecting every cycle that contains a terminal. The study of parameterized algorithms for this generalization of the FEEDBACK VERTEX SET problem has received significant attention over the last few years. In fact the parameterized complexity of this problem was open until 2011, when two groups independently showed that the problem is fixed parameter tractable (**FPT**). Using tools from graph minors Kawarabayashi and Kobayashi obtained an algorithm for SUBSET FVS running in time $\mathcal{O}(f(k) \cdot n^2 m)$ [SODA 2012, JCTB 2012]. Independently, Cygan et al. [ICALP 2011, SIDMA 2013] designed an algorithm for SUBSET FVS running in time $2^{\mathcal{O}(k \log k)} \cdot n^{\mathcal{O}(1)}$. More recently, Wahlström obtained the first single exponential time algorithm for SUBSET FVS, running in time $4^k \cdot n^{\mathcal{O}(1)}$ [SODA 2014]. While the $2^{\mathcal{O}(k)}$ dependence on the parameter k is optimal under the Exponential Time Hypothesis (ETH), the dependence of this algorithm as well as those preceding it, on the input size is far from linear.

In this paper we design the first linear time parameterized algorithms for SUBSET FVS. More precisely, we obtain two new algorithms for SUBSET FVS.

– A randomized algorithm for SUBSET FVS running in time $\mathcal{O}(25.6^k k^{\mathcal{O}(1)}(n + m))$.
– A deterministic algorithm for SUBSET FVS running in time $2^{\mathcal{O}(k \log k)}(n + m)$.

In particular, the first algorithm obtains the best possible dependence on both the parameter as well as the input size, up to the constant in the exponent. Both of our algorithms are based on "cut centrality", in the sense that solution vertices are likely to show up in minimum size cuts between vertices sampled from carefully chosen distributions.

1 Introduction

FEEDBACK SET problems constitute one of the most important topics of research in parameterized algorithms [3,4,6–8,10,16,26,27,32,36]. Typically, in these

© Springer-Verlag Berlin Heidelberg 2015
M.M. Halldórsson et al. (Eds.): ICALP 2015, Part I, LNCS 9134, pp. 935–946, 2015.
DOI: 10.1007/978-3-662-47672-7_76

problems, we are given an undirected graph G (or a directed graph) and a positive integer k, and the objective is to "hit" all cycles of the input graph using at most k vertices (or edges or arcs). Recently, there has been a lot of study on the *subset variant* of FEEDBACK SET problems. In these problems, the input also includes a terminal subset $T \subseteq V(G)$ and the goal is to detect the presence of a set, referred to as a *subset feedback vertex set*, that hits all T-*cycles*, that is cycles whose intersection with T is non-empty. In this paper we consider the following problem.

SUBSET FEEDBACK VERTEX SET (SUBSET FVS)
    Instance: A graph G on n vertices and m edges, a vertex subset T,
            and a positive integer k.
    Parameter: k
    Question: Is there a set of k vertices that intersects every T-cycle?

SUBSET FVS generalizes FEEDBACK VERTEX SET as well as the well known MULTIWAY CUT problem. In this paper we explore parameterized algorithms for SUBSET FVS. In parameterized complexity each problem instance has an associated parameter k and a central notion in parameterized complexity is *fixed parameter tractability* (FPT). This means, for a given instance (x, k), solvability in time $\tau(k) \cdot |x|^{\mathcal{O}(1)}$, where τ is an arbitrary function of k.

The study of parameterized algorithms for the SUBSET FVS problem has received significant attention in the last few years. The existence of an FPT algorithm for SUBSET FVS was shown only in 2011, when two groups independently gave FPT algorithms for the problem. Using tools from graph minors Kawarabayashi and Kobayashi obtained an algorithm for SUBSET FVS with running time $\mathcal{O}(f(k) \cdot n^2 m)$ [16] (also see [26]). Independently, Cygan et al. [9,10], combining iterative compression [31] with Gallai's theorem [11] designed an algorithm for SUBSET FVS with running time $2^{\mathcal{O}(k \log k)} \cdot n^{\mathcal{O}(1)}$. Cygan et al. asked whether it is possible to obtain an algorithm for SUBSET FVS running in time $2^{\mathcal{O}(k)} \cdot n^{\mathcal{O}(1)}$. Wahlström [36] resolved this question in the affirmative by giving an algorithm for SUBSET FVS, with running time $4^k \cdot n^{\mathcal{O}(1)}$. It is easy to show that the $2^{\mathcal{O}(k)}$ dependence on the parameter k is optimal under the Exponential Time Hypothesis (ETH) [22]. That is, assuming the ETH, SUBSET FVS does not admit an algorithm with running time $2^{o(k)} \cdot n^{\mathcal{O}(1)}$.

The focus of this paper is the second component of the running time of parameterized algorithms, that is, the running time dependence on the input size n. This direction of research is as old as the existence of parameterized algorithms, with classic results, such as Bodlaender's linear time algorithm for treewidth [1] and the cubic time algorithm of Robertson and Seymour for the disjoint paths problem [35]. A more recent phenomenon is that one strives for linear time parameterized algorithms that do not compromise too much on the dependence of the running time on the parameter k. The gold standard for these results are algorithms with linear dependence on input size as well as provably optimal (under ETH) dependence on the parameter. New results in this direction include parameterized algorithms for problems such as ODD CYCLE TRANSVERSAL [23,33], SUBGRAPH ISOMORPHISM [12], PLANARIZATION [15,25] as well as a single-exponential and linear time parameterized constant factor approximation algorithm for TREEWIDTH [2]. Other recent results include parameterized algorithms with improved dependence on input size for a host of problems [13,17–21].

The running time dependence on the input size for all the previous algorithms for SUBSET FVS is quite far from being linear. Recently, the methods behind the $4^k \cdot n^{\mathcal{O}(1)}$ time algorithm of Wahlström have been applied to give linear time FPT algorithms [24] for several problems, including the edge-deletion variant of UNIQUE LABEL COVER. Interestingly, this approach does not seem to extend to a linear time algorithm for SUBSET FVS. In this paper we design the first linear time parameterized algorithms for SUBSET FVS. The first algorithm is randomized with one-sided error, and obtains linear dependence on n as well as single exponential dependence on k.

Theorem 1. *There is an algorithm that, given an instance (G, T, k) of* SUBSET FVS *runs in time $25.6^k k^{\mathcal{O}(1)}(m + n)$ and either returns a subset feedback vertex set of size at most k or concludes correctly with probability at least $1 - \frac{1}{e}$ that no such set exists, where $m = |E(G)|$ and $n = |V(G)|$ and e is Euler's number.*

The single exponential dependence on k in the running time of Theorem 1 is optimal under the ETH. The second algorithm is deterministic at the cost of a slightly worse dependence on the parameter k.

Theorem 2. *There is an algorithm that given an instance (G, T, k) of* SUBSET FVS *runs in time $\mathcal{O}(2^{\mathcal{O}(k \log k)}(m + n))$ and either returns a subset feedback vertex set of size at most k, or correctly concludes that no such set exists.*

Methodology. Both algorithms begin by applying simple pre-processing rules to ensure that no vertex or edge is irrelevant, and that every vertex is sufficiently connected to the terminals. While the pre-processing rules are quite easy to state, it is surprisingly tricky to apply the rules exhaustively in linear time. We achieve this by using a classic algorithm of Hopcroft and Tarjan [14] to decompose a graph into its 3-connected pieces.

At this point the randomized algorithm of Theorem 1 exploits the following structural insight. Consider a graph G that does have a subset feedback vertex set S of size at most k. $G - S$ has no T-cycles, and a graph without any T-cycles is essentially a forest where some of the terminal-free regions have been replaced by arbitrary graphs. The terminal-free regions may only interact with neighboring regions via single edges. Since a forest has average degree at most 2, at least half the regions interact with at most two other regions in this way. Any such "degree two" region can be separated from the terminals by removing the solution S, as well as the two edges leaving the region in $G - S$. On the other hand the pre-processing rules ensure that every vertex has sufficient flow to the terminals, in particular the rules ensure that each "degree two" region must have at least one neighbor in the solution. From this we infer that the vertices of S appear very frequently in small cuts between vertices in "degree two regions" and terminal vertices. Our algorithm is based on a random process which is likely to produce a vertex v which is in a "degree two region". The algorithm then samples a small set A such that A separates v from the terminal, and with good probability A has a large intersection with the solution S. At this point the algorithm guesses the intersection of the set A with the solution S, removes $A \cap S$ from the graph and starts again. The difficult part of the analysis is to show that whenever the algorithm guesses that a set X is a subset of the solution, the algorithm is correct with probability at least $\frac{1}{2^{O(|X|)}}$.

The deterministic algorithm of Theorem 2 is based on the same ideas as the randomized algorithm, but is quite far from being a "direct derandomization". An attempt at a "direct derandomization" of the algorithm of Theorem 1 could look like this. The randomized algorithm essentially selects a vertex and claims that this vertex is a part of the solution. The analysis basically shows that for any optimal solution S of size at most k, the probability that the randomized algorithm selects a vertex in S is at least $1/25.6$. Suppose that we could compute deterministically for each vertex v, the probability $p(v)$ with which v is selected. We know that $\sum_{v \in S} p(v) \geq \frac{1}{25.6}$. Thus there must be a vertex in $v \in S$ such that $p(v) \geq \frac{1}{25.6k}$. But $\sum_{v \in V(G)} p(v) = 1$, so the number of vertices v such that $p(v) \geq \frac{1}{25.6k}$ is at most $25.6k$. This gives us a candidate set of size $25.6k$ of which a vertex must be in the solution. We can now guess which vertex this is, decrease k by 1, and re-start. The main problem with this approach is that we are unaware of an algorithm to compute $p(v)$ for all vertices v in linear time. The engine behind the algorithm of Theorem 2 is a different random process which also ensures that solution vertices are picked with high probability, but for which the probabilities $p(v)$ are efficiently computable.

2 Preliminaries

The open neighborhood of a vertex v in graph G contains the vertices adjacent to v, and is written as $N_G(v)$. A *separation* of a graph G is a pair (L, R) of subsets of $V(G)$ such that $L \cup R = V(G)$ and there are no edges between $L \setminus R$ and $R \setminus L$ in G. The intersection $L \cap R$ is called the *center* of the separation, and the size $|L \cap R|$ is the *order* of the separation. If P is a path from a vertex in X to a vertex in Y, we say that P is a X-Y path. If X contains a single vertex x, we say that P is a x-Y path. Two paths that do not share any vertices are called vertex disjoint. A cycle which intersects T is called a *T-cycle*. A *subset feedback vertex set* of a graph G and terminal set $T \subseteq V(G)$ is a set $S \subseteq V(G)$ such that $G - S$ does not contain any T-cycles. Contracting an edge uv amounts to removing the vertices u and v from the graph and making a new vertex w which is adjacent to all vertices in $N(u) \cup N(v)$.

Important Separators. We review important separators, as well as some related results. Let G be a graph, let $X, S \subseteq V(G)$ be vertex subsets. We denote by $R_G(X, S)$ the set of vertices of G reachable from X in the graph $G - S$ and we denote by $NR_G(X, S)$ the set of vertices of $G - S$ which are not reachable from

X in the graph $G - S$. We drop the subscript G if it is clear from the context. Let G be a graph and let $X, Y \subset V(G)$ be two disjoint vertex sets. A subset $S \subseteq V(G) \setminus (X \cup Y)$ is called a X-Y *separator* in G if $R_G(X, S) \cap Y = \emptyset$ or in other words there is no path from X to Y in the graph $G - S$. We denote by $\lambda_G(X, Y)$ the size of the smallest X-Y separator in G. If G is clear from context we omit the subscript.

A X-Y separator S_1 is said to **cover** a X-Y separator S if $R(X, S_1) \supset R(X, S)$. Note that the definition of covering is asymmetric; the reachability set is taken from X and not from Y. We say that a X-Y separator S_1 *dominates* another X-Y separator S if S_1 covers S and $|S_1| \leq |S|$. We call a X-Y separator S, *important* if it is minimal and no other X-Y separator dominates it. Important separators were first defined by Marx [29], and have found numerous applications since then [6,28,30,34]. The main reason is that there are not too many important X-Y separators of small size. We now prove a "sampling" version of the algorithm to enumerate important separators given in [5,30].

Lemma 1 ($\star^1$). *There is a randomized algorithm* SampleImp *that given as input a graph G, disjoint and non-adjacent vertex sets X and Y integer k, and rational deletion probability $0 < p < 1$, runs in time $\mathcal{O}(k^{\mathcal{O}(1)}(m+n))$ and outputs a X-Y separator S of size at most k or* fail. *For each important X-Y separator A of size at most k, the probability that $S = A$ is at least $p^{|A|}(1 - p)^{|A|-\lambda(X,Y)}$.*

3   Preprocessing

In this section we describe a linear time preprocessing routine which returns an equivalent instance with certain structural properties.

Lemma 2 ($\star$). *There exists a $\mathcal{O}(k(n + m))$ time algorithm* reduce *that given as input an instance (G, T, k), returns an equivalent instance (G', T', k') with the following properties.*

- $|V(G')| \leq |V(G)|$, $|E(G')| \leq |E(G)|$ *and* $k' \leq k$.
- *Every non-terminal vertex has degree at least 3.*
- *Every terminal vertex has degree at least 2.*
- *Every terminal vertex of degree 2 has only non-terminal neighbors.*
- *Every vertex is in a T-cycle.*
- *For every non-terminal v, either v is adjacent to a terminal or there are at least three internally vertex disjoint paths to T.*
- *Between any pair of vertices there are at most two edges.*
- *Between any pair of non-terminals there is at most one edge.*

We will refer to instances satisfying the conclusions of Lemma 2 as *reduced instances*. Even though reduced instances may contain double edges, these are always incident to a terminal and constitute T-cycles all by themselves. Since every subset feedback vertex set must contain at least one of the two endpoints

---

of such a double edge, our algorithms will quickly get rid of double edges by branching. We start by inspecting the structure of instances that do not contain any T-cycles.

Lemma 3 $(\star)$. *Let G be a graph and $T \subseteq V(G)$ be a set such that G has no T-cycles. Then (a) $G[T]$ is a forest, (b) for every connected component C of $G-T$ and connected component C_T of $G[T]$ there is at most one edge between C and C_T, and (c) contracting all edges with both endpoints non-terminals yields a forest.*

Let G be a graph T be a subset of $V(G)$ such that G does not contain a T-cycle. Next we define the notion of a terminal forest. While the definition might look technical at a first glance, a terminal forest is just the forest obtained from G by contracting all the edges with both endpoints non-terminals, rooting the trees in the forest at arbitrary roots and providing a function χ that maps each vertex v of the forest to the vertex set in G which was contracted into v.

Definition 1. *Let G be a graph T be a subset of $V(G)$ such that G does not contain a T-cycle. A terminal forest of G is a pair (F, χ), where F is a forest of rooted trees and $\chi : V(F) \to T \cup 2^{V(G) \setminus T}$ is a function with the following properties:*

(a) *Each tree T_i in F can be associated to a unique connected component C_i in G; for each $b \in V(T_i)$, $\chi(b) \subseteq C_i$.*
(b) *$\bigcup_{b \in V(F)} \chi(b) = V(G)$ and for any pair of vertices $b, b' \in V(F)$, $\chi(b) \cap \chi(b') = \emptyset$. That is, χ partitions the vertex set $V(G)$.*
(c) *For every $b \in V(F)$, $\chi(b) \neq \emptyset$ and the graph $G[\chi(b)]$ is connected.*
(d) *For every edge $uv \in E(G)$ such that $u, v \notin T$ there exists $b \in V(F)$ such that $uv \in \chi(b)$.*
(e) *For every edge $tv \in E(G)$ such that $t \in T$ and $v \in V(G)$ there exists a $b_t \in V(F)$ and $b_v \in V(F)$ such that $\chi(b_t) = t$, $v \in \chi(b_v)$ and $b_t b_v \in E(F)$.*

From Lemma 3 it follows directly that every graph G that does not have a T-cycle has a terminal forest. For a terminal forest (F, χ) of G we define the function $\chi^- : V(G) \to V(F)$ so that $\chi^-(u)$ is the unique node $b \in V(F)$ so that $u \in \chi(b)$. We extend χ^- to vertex sets of G in the following manner: $\chi^-(S) = \bigcup_{u \in S} \chi^-(u)$.

In Definition 1 we misuse notation - χ can output either vertex sets (non-terminals) or single vertices (terminals). When $\chi(b)$ outputs a terminal t it is sometimes convenient to treat it as the set $\{t\}$ containing the terminal. In the forest F the child-parent and descendant-ancestor relations are well defined. We can extend these relations to vertices in G in the natural way. We will refer to the nodes b of F such that $\chi(b) \in T$ as *terminal nodes*, and to the other nodes as *non-terminal* nodes.

Definition 2. *Let G be a graph and $T \subseteq V(G)$ be such that G has no T-cycles. We say that a terminal $t' \in T$ is an effective descendant of $t \in T \setminus \{t'\}$ if, in the terminal forest (F, χ) of G we have that $\chi^-(t)$ is a descendant of $\chi^-(t')$ and* *there is a path from $\chi^-(t)$ to $\chi^-(t')$ in F with internal vertices disjoint from* $\chi^-(T)$.

To better understand the definition of effective descendants it is helpful to construct the *effective descendant graph*. This is a directed graph with vertex set T. Each vertex in T it has arcs to all of its effective descendants. It is easy to see that the effective descendant graph is obtained from F by contracting, for all terminals t, all edges to t's non-terminal children, and then orienting edges from parents to descendants in F. Furthermore, the effective descendant graph is a forest of rooted trees. Finally, in a graph with no T-cycles, we call a terminal *good* if t has at most one effective descendant. Since any rooted tree has at least as many leaves as vertices with at least two children, we conclude that in a graph with no T-cycles, at least $|T|/2$ terminals are good. We now prove a lemma about the structure of reduced instances, and how a potential subset feedback vertex set interacts with the rest of the graph.

Lemma 4 ($\star$). *Let (G, T, k) be a reduced instance and S be a subset fedback vertex set of G. Let (F, χ) be the terminal forest of $G \setminus S$ then the following holds.*

- *For every leaf b of F, $\chi(b)$ has at least one neighbor in S.*
- *For every non-terminal leaf b in F such that $\chi(b)$ has exactly one neighbor in S, this neighbor is a terminal.*
- *For every non-terminal node $b \in V(F)$ that has degree 2 in F, $\chi(b)$ has at least one neighbor in S.*

4   A Randomized Linear Time Algorithm for SUBSET FVS

Having proved the required structural properties, we now describe the randomized algorithm for SUBSET FVS. The algorithm, called SolveSFVS, is given in Algorithm 1. The algorithm runs in $k^{\mathcal{O}(1)}(n + m)$ time and has one-sided error. Except for applying Lemma 2, whenever the algorithm decreases k by x it also removes x vertices from the graph. Thus, whenever the algorithm outputs success, the input instance is a "yes" instance. The difficult part is to show that if the instance is a "yes" instance then the algorithm returns success with probability at least γ^k, for a constant $\gamma = \frac{1}{25.6}$. The remaining part of the analysis is essentially devoted to the proof of Lemma 5. The algorithm SolveSFVS makes use of some probability constants, $0 < \alpha_t, \alpha_v, \beta, p < 1$. These constants are later set so as to maximize the success probability of the algorithm.

Lemma 5. *If (G, T, k) is a "yes" instance, then Algorithm SolveSFVS outputs success on (G, T, k) with probability at least γ^k for $\gamma = \frac{1}{25.6}$.*

The proof of Lemma 5 is by induction on k. A key ingredient of the proof are three *pushing lemmata* that we show here. Having Lemma 5 at hand, proving Theorem 1 is routine.

Input : An instance (G, T, k) of SUBSET FVS.

Output: success if the algorithm has found a subset feedback vertex set in G of size at most k, or fail.

1   $(G, T, k) \leftarrow$ **reduce**(G, T, k)

2   **if** G *has no T-cycles* **then return** success;

3   **if** $k \leq 0$ **then return** fail;

4   **if** *there exists a double edge uv* **then**

5   pick x from $\{u, v\}$ uniformly at random.

6   **return** SolveSFVS$(G - x, T \setminus \{x\}, k - 1)$

7   **pick** t from T uniformly at random.

8   **with** *probability* $(1 - \alpha_t)$:

9   **return** SolveSFVS$(G - t, T \setminus \{t\}, k - 1)$

10 **if** $|N(t) \cap T| \geq 2$ **then**

11   **with** *probability* $\frac{1}{2}$:

12   pick z from $N(t) \cap T$ uniformly at random.

13   **return** SolveSFVS$(G - z, T \setminus \{z\}, k - 1)$

14   **if** $N(t) \setminus T = \emptyset$ **then return** fail;

15 **pick** v from $N(t) \setminus T$ uniformly at random.

16 **with** *probability* $(1 - \alpha_v)$:

17   **return** SolveSFVS$(G - v, T \setminus \{v\}, k - 1)$

18 **let** $\bar{T} = \{t\}$

19 **insert** each $z \in T \setminus \{t\}$ with probability $1 - \beta$.

20 **let** $G' = G - \{vt \in E(G) \mid t \in \bar{T}\} + (\tau, \{\tau t \mid t \in T\})$

21 $A^* \leftarrow$ SampleImp$(G', v, \bar{T} \cap \{\tau\}, k + 1, p)$

22 **if** $|N(v) \cap \bar{T}| \geq 2$ **or** $A^* \setminus T = \emptyset$ **or** $|A^*| = 1$ **then**

23   **return** SolveSFVS$(G - A^*, T \setminus A^*, k - |A^*|)$

24 **pick** y^* from $A^* \setminus T$ uniformly at random.

25 **return** SolveSFVS$(G - (A^* \setminus \{y^*\}), T \setminus (A^* \setminus \{y^*\}), k - |A| + 1)$

Algorithm 4.1. Algorithm SolveSFVS for SUBSET FVS

We now prove a series of lemmas about how one can modify a solution by "pushing" the solution vertices towards the terminals. In the following two lemma statements, G is a graph, T is a set of terminals and S is a subset feedback vertex set of G. Further, (F, χ) is the terminal forest of $G - S$, b is a terminal node in F and b' is a non-terminal child of b in F. Also, $t = \chi(b)$ and v is the unique neighbor of t in $\chi(b')$ (assuming the existence of b and b', the existence and uniqueness of v follows). Furthermore $A = N(\chi(b')) \cap S$ and $\bar{T} \subseteq T$ is a set of terminals such that $t \in \bar{T}$ and $\bar{T} \cap A = \emptyset$. Finally, G' is the graph obtained from G by adding a new vertex τ, making τ adjacent to all vertices of T, and removing all edges between v and vertices in $\bar{T}$.

Lemma 6. *If b' is a leaf node of F, then A is a v-τ separator in G'. Further, for any v-τ separator A' (in G') that dominates A, $(S \setminus A) \cup A'$ is a subset feedback vertex set of G.*

Proof. To see that A is a v-τ separator in G' observe that $N_{G'}(\chi(b')) = A$ and that $\chi(b') \cap T = \emptyset$. We now move to the second statement, namely that $(S \setminus A) \cup A'$ is a subset feedback vertex set of G.

Suppose not. Then $G - ((S \setminus A) \cup A')$ contains a T-cycle C. Let L' be the set of vertices reachable from v in $G' - A'$, plus A'. Let $R' = (V(G') \setminus L') \cup A'$. (L', R') is a separation in G', since $L' \setminus R'$ are exactly the vertices reachable from v, while $R' \setminus L'$ are the vertices *not* reachable from v in $G' - A'$. A key observation is that since A' dominates A, we have that $\chi(b') \subseteq L' \setminus R'$ and that $A \setminus A' \subseteq L' \setminus R'$.

All edges of G that are not in G' are incident to v. Let $L = L' \cup \{v\}$ and $R = R' \cup \{v\}$, it follows that (L, R) is a separation in G. Since $G - S$ has no T-cycles it follows that C must contain some vertex of $A \setminus A'$. We know that $A \setminus A' \subseteq L' \setminus R'$ and $v \notin A$. Therefore, $A \setminus A' \subseteq L \setminus R$. Furthermore, C is a T-cycle, so C must contain a terminal z. Since A' separates v from τ it follows that $L' \cap T \subseteq A'$. Since v is not a terminal it follows that $L \cap T \subseteq A'$. Since C is disjoint from A' this means that $z \notin L$, so $z \in R \setminus L$. But then C contains a vertex in $L \setminus R$ and a vertex in $R \setminus L$, so $|C \cap L \cap R| \geq 2$. However, the only vertex in $L \cap R$ which is not in A' is v, contradicting that $C \cap A' = \emptyset$.   □

Lemma 7. *If t has a unique effective descendant t', b' is adjacent to $\chi^-(t')$ (in F), v is adjacent to t' and t' is in $\bar{T}$, we conclude the following. A is a v-τ separator in G'. Further, for any v-τ separator A' (in G') that dominates A, $(S \setminus A) \cup A'$ is a subset feedback vertex set of G.*

The proof of Lemma 7 is identical (word by word!) to the proof of Lemma 6, and therefore omitted. For the last lemma statement we need to change the definition of A. Suppose t has a unique effective descendant t', b' is adjacent to $\chi^-(t')$ (in F), v is non-adjacent to t'. Let y be the unique neighbor of t' in $\chi(b')$. We define Q to be the set of vertices reachable from v (including v) in $G - (S \cup \{y, t\})$, and define $A = N(Q) \setminus \{y, t\}$. Just as before, $\bar{T} \subseteq T$ is a set of terminals such that $t \in \bar{T}$ and $\bar{T} \cap A = \emptyset$. Finally, G' is the graph obtained from G by adding a new vertex τ, making τ adjacent to all vertices of T, and removing all edges between v and vertices in $\bar{T}$.

Lemma 8. *$A \cup \{y\}$ is a v-τ separator in G'. Further, for any v-τ separator A' (in G') that dominates $A \cup \{y\}$, if $A' \cap \bar{T} = \emptyset$ then $y \in A'$ and $(S \setminus A) \cup (A' \setminus \{y\})$ is a subset feedback vertex set of G.*

Proof. To see that $A \cup \{y\}$ is a v-τ separator in G' observe that the set of vertices reachable from v in $G' \setminus (A \cup \{y\})$ is exactly Q and that $Q \cap T = \emptyset$. We now show that $y \in A'$. We have that $G'[Q]$ is connected, and contains a neighbor of y, since $G'[\chi(b')]$ is connected. Since A' dominates $A \cup \{y\}$, all vertices in Q are reachable from v in $G - A'$. Since $t' \in \bar{T}$ we have that $t' \notin A'$. If $y \notin A'$ then there is a path from v to τ in $G' - A'$; via Q to y, then to t' by the edge yt' and finally to τ. This contradicts that A' separates v from τ. We conclude that $y \in A'$, and proceed to the last statement, namely that $(S \setminus A) \cup (A' \setminus \{y\})$ is a subset feedback vertex set of G.

Suppose not. Then $G - ((S \setminus A) \cup (A' \setminus \{y\}))$ contains a T-cycle C. Let L' be the set of vertices reachable from v in $G' - A'$, plus A'. Let $R' = (V(G') \setminus L') \cup A'$. (L', R') is a separation in G', since $L' \setminus R'$ are exactly the vertices reachable from v, while $R' \setminus L'$ are the vertices *not* reachable from v in $G' - A'$. A key observation is that since A' dominates $A \cup \{y\}$, we have that $Q \subseteq L' \setminus R'$ and that $A \setminus A' \subseteq L' \setminus R'$.

All edges of G that are not in G' are incident to v. Let $L = L' \cup \{v\}$ and $R = R' \cup \{v\}$, it follows that (L, R) is a separation in G. Since $G - S$ has no T-cycles it follows that C must contain some vertex of $A \setminus (A' \setminus \{y\}) = A \setminus A'$. We know that $A \setminus A' \subseteq L' \setminus R'$ and v is disjoint from A. Therefore, $A \setminus A' \subseteq L \setminus R$. Furthermore, C is a T-cycle, so C must contain a terminal z. Since A' separates v from τ in G' it follows that $L' \cap T \subseteq A'$. Since v is not a terminal it follows that $L \cap T \subseteq A'$. Since C is disjoint from $A' \setminus \{y\}$ and y is not a terminal, this means that $z \notin L$, so $z \in R \setminus L$. Thus C contains a vertex in $L \setminus R$ and a vertex in $R \setminus L$, so $|C \cap L \cap R| \geq 2$. However, the only two vertices in $L \cap R$ which are not in $A' \setminus \{y\}$ are v and y. Thus $C \cap L \cap R = \{v, y\}$.

It follows that C contains path P from v to y with at least one internal vertex, and all of its internal vertices in $R \setminus L$. The internal vertices of P are disjoint from $S \setminus A$ and disjoint from $A \setminus A'$, since $A \setminus A' \subseteq L$. Thus P is disjoint from S. However, all paths between v and y in $G \setminus S$ with at least one internal vertex must intersect $Q \setminus \{v\}$. But $Q \subseteq L$, contradicting that the internal vertices of P are disjoint from L. □

Due to lack of space the deterministic algorithm is omitted in this extended abstract, and may be found in the full version.

References

1. Bodlaender, H.L.: A linear-time algorithm for finding tree-decompositions of small treewidth. SIAM J. Comput. **25**(6), 1305–1317 (1996)
2. Bodlaender, H.L., Drange, P.G., Dregi, M.S., Fomin, F.V., Lokshtanov, D., Pilipczuk, M.: An $O(c^k n)$ 5-approximation algorithm for treewidth. In: FOCS, pp. 499–508 (2013)
3. Cao, Y., Chen, J., Liu, Y.: On feedback vertex set new measure and new structures. In: Kaplan, H. (ed.) SWAT 2010. LNCS, vol. 6139, pp. 93–104. Springer, Heidelberg (2010)
4. Chen, J., Fomin, F.V., Liu, Y., Lu, S., Villanger, Y.: Improved algorithms for feedback vertex set problems. J. Comput. Syst. Sci. **74**(7), 1188–1198 (2008)
5. Chen, J., Liu, Y., Lu, S.: An improved parameterized algorithm for the minimum node multiway cut problem. Algorithmica **55**(1), 1–13 (2009)
6. Chen, J., Liu, Y., Lu, S., O'Sullivan, B., Razgon, I.: A fixed-parameter algorithm for the directed feedback vertex set problem. J. ACM **55**(5) (2008)
7. Chitnis, R., Cygan, M., Hajiaghayi, M., Marx, D.: Directed subset feedback vertex set is fixed-parameter tractable. In: Czumaj, A., Mehlhorn, K., Pitts, A., Wattenhofer, R. (eds.) ICALP 2012, Part I. LNCS, vol. 7391, pp. 230–241. Springer, Heidelberg (2012)

8. Cygan, M., Nederlof, J., Pilipczuk, M., van Rooij, J.M.M., Wojtaszczyk, J.O.: Solving connectivity problems parameterized by treewidth in single exponential time. In: FOCS, pp. 150–159 (2011)
9. Cygan, M., Pilipczuk, M., Pilipczuk, M., Wojtaszczyk, J.O.: Subset feedback vertex set is fixed-parameter tractable. In: Aceto, L., Henzinger, M., Sgall, J. (eds.) ICALP 2011, Part I. LNCS, vol. 6755, pp. 449–461. Springer, Heidelberg (2011)
10. Cygan, M., Wojtaszczyk, J.O.: Subset feedback vertex set is fixed-parameter tractable. SIAM J. Discrete Math. **27**(1), 290–309 (2013)
11. Diestel, R.: Graph Theory, 4th edn. Springer, Heidelberg (2010)
12. Dorn, F.: Planar subgraph isomorphism revisited. In: STACS, pp. 263–274 (2010)
13. Grohe, M., Kawarabayashi, K.-I., Reed, B.A.: A simple algorithm for the graph minor decomposition - logic meets structural graph theory. In: SODA, pp. 414–431 (2013)
14. Hopcroft, J.E., Tarjan, R.E.: Dividing a graph into triconnected components. SIAM J. Comput. **2**(3), 135–158 (1973)
15. Kawarabayashi, K.-I.: Planarity allowing few error vertices in linear time. In: FOCS, pp. 639–648 (2009)
16. Kawarabayashi, K.-I., Kobayashi, Y.: Fixed-parameter tractability for the subset feedback set problem and the s-cycle packing problem. J. Comb. Theory, Ser. B **102**(4), 1020–1034 (2012)
17. Kawarabayashi, K.-I., Kobayashi, Y., Reed, B.A.: The disjoint paths problem in quadratic time. J. Comb. Theory, Ser. B **102**(2), 424–435 (2012)
18. Kawarabayashi, K.-I., Mohar, B.: Graph and map isomorphism and all polyhedral embeddings in linear time. In: STOC, pp. 471–480 (2008)
19. Kawarabayashi, K.-I., Mohar, B., Reed, B.A.: A simpler linear time algorithm for embedding graphs into an arbitrary surface and the genus of graphs of bounded tree-width. In: FOCS, pp. 771–780 (2008)
20. Kawarabayashi, K.-I., Reed, B.A.: A nearly linear time algorithm for the half integral parity disjoint paths packing problem. In: SODA, pp. 1183–1192 (2009)
21. Kawarabayashi, K.-I., Reed, B.A.: An (almost) linear time algorithm for odd cycles transversal. In: SODA, pp. 365–378 (2010)
22. Impagliazzo, R., Paturi, R., Zane, F.: Which problems have strongly exponential complexity? J. Comput. Syst. Sci. **63**(4), 512–530 (2001)
23. Iwata, Y., Oka, K., Yoshida, Y.: Linear-time FPT algorithms via network flow. In: SODA, pp. 1749–1761 (2014)
24. Iwata, Y., Wahlström, M., Yoshida, Y.: Half-integrality, LP-branching and FPT algorithms (2013). CoRR, abs/1310.2841
25. Jansen, B.M.P., Lokshtanov, D., Saurabh, S.: A near-optimal planarization algorithm. In: SODA, pp. 1802–1811 (2014)
26. Kakimura, N., Kawarabayashi, K.-I., Kobayashi, Y.: Erdös-Pósa property and its algorithmic applications: parity constraints, subset feedback set, and subset packing. In: SODA, pp. 1726–1736 (2012)
27. Kociumaka, T., Pilipczuk, M.: Faster deterministic feedback vertex set. Inf. Process. Lett. **114**(10), 556–560 (2014)
28. Lokshtanov, D., Ramanujan, M.S.: Parameterized tractability of multiway cut with parity constraints. In: Czumaj, A., Mehlhorn, K., Pitts, A., Wattenhofer, R. (eds.) ICALP 2012, Part I. LNCS, vol. 7391, pp. 750–761. Springer, Heidelberg (2012)
29. Marx, D.: Parameterized graph separation problems. Theoret. Comput. Sci. **351**(3), 394–406 (2006)
30. Marx, D., Razgon, I.: Fixed-parameter tractability of multicut parameterized by the size of the cutset. SIAM J. Comput. **43**(2), 355–388 (2014)

31. Niedermeier, R.: Invitation to Fixed-Parameter Algorithms. Oxford Lecture Series in Mathematics and its Applications, vol. 31. Oxford University Press, Oxford (2006)
32. Raman, V., Saurabh, S., Subramanian, C.R.: Faster fixed parameter tractable algorithms for finding feedback vertex sets. ACM Transactions on Algorithms **2**(3), 403–415 (2006)
33. Ramanujan, M.S., Saurabh, S.: Linear time parameterized algorithms via skew-symmetric multicuts. In: SODA pp. 1739–1748 (2014)
34. Razgon, I., O'Sullivan, B.: Almost 2-sat is fixed-parameter tractable. J. Comput. Syst. Sci. **75**(8), 435–450 (2009)
35. Robertson, N., Seymour, P.D.: Graph minors. xiii. the disjoint paths problem. J. Comb. Theory, Ser. B **63**(1), 65–110 (1995)
36. Wahlström, M.: Half-integrality, LP-branching and FPT algorithms. In: SODA, pp. 1762–1781 (2014)

An Optimal Algorithm for Minimum-Link Rectilinear Paths in Triangulated Rectilinear Domains

Joseph S.B. Mitchell[1], Valentin Polishchuk[2], Mikko Sysikaski[3], and Haitao Wang[4(✉)]

[1] Stony Brook University, Stony Brook, NY 11794, USA
jsbm@ams.stonybrook.edu
[2] Linköping University, Linköping, Sweden
valentin.polishchuk@liu.se
[3] Google, Zurich, Switzerland
mikko.sysikaski@gmail.com
[4] Utah State University, Logan, UT 84322, USA
haitao.wang@usu.edu

Abstract. We present a new algorithm for finding minimum-link rectilinear paths among h rectilinear obstacles with a total of n vertices in the plane. After the plane is triangulated, for any point s, our algorithm builds an $O(n)$-size data structure in $O(n + h \log h)$ time, such that given any query point t, we can compute a minimum-link rectilinear path from s to t in $O(\log n + k)$ time, where k is the number of edges of the path, and the query time is $O(\log n)$ if we only want to know the value k. The previously best algorithm solves the problem in $O(n \log n)$ time.

1 Introduction

A polygon (or path) is *rectilinear* if all its edges are axis-parallel. Let $\mathcal{P}$ be a set of h disjoint rectilinear obstacles with a total of n vertices in the plane. The plane minus the interior of all obstacles is called the *free space*. The *link distance* of a path is defined to be the number of edges (also called *links*) in the path. A *minimum-link* (or *min-link*) rectilinear path between two points s and t is a rectilinear path from s to t in the free space with the minimum link distance. Our goal is to construct a data structure (called *link distance map*) with respect to a given *source* point s, such that for any query point t, a min-link rectilinear path from s to t can be quickly computed. In the following, we say a link distance map has the *standard query performance* if given any t, the link distance of a min-link s-t path can be computed in $O(\log n)$ time and the actual path can be output in additional time linear in the link distance of the path.

Previous Work. Linear-time algorithms have been given for finding min-link (general polygonal) paths in simple polygons [8,9,20–22]. The link distance map can also be built in linear time [20–22] for simple polygons, with the standard

© Springer-Verlag Berlin Heidelberg 2015
M.M. Halldórsson et al. (Eds.): ICALP 2015, Part I, LNCS 9134, pp. 947–959, 2015.
DOI: 10.1007/978-3-662-47672-7_77

query performance. For polygonal domains, the problem becomes much more difficult. Mitchell, Rote, and Woeginger [16] gave an $O(n^2 \alpha(n) \log^2 n)$ time algorithm for finding min-link paths, where $\alpha(n)$ is the inverse Ackermann function; a link distance map with slightly larger construction time is also given in [16]. As shown in [14], finding min-link paths in polygonal domains is 3SUM-hard.

The rectilinear min-link path problems have also been studied. For simple rectilinear polygons, de Berg [2] presented an algorithm that can build an $O(n)$-size link distance map in $O(n \log n)$ time and $O(n)$ space, with the standard query performance. The construction time was later reduced to $O(n)$ time by Lingas, Maheshwari, and Sack [12], and by Schuierer [19].

For rectilinear polygonal domains, Imai and Asano [10] presented an $O(n \log n)$ time and space algorithm for finding min-link rectilinear paths. Later, Das and Narasimhan [6] described an improved algorithm of $O(n \log n)$ time and $O(n)$ space; Sato, Sakanaka, and Ohtsuki [18] gave a similar algorithm with the same performance. Recently, Mitchell, Polishchuk, and Sysikaski [14] presented a simpler algorithm of $O(n \log n)$ time and $O(n)$ space. Link distance maps of $O(n)$-size can also be built in $O(n \log n)$ time and $O(n)$ space [6,14]. As shown in [6,13], the problem has an $\Omega(n + h \log h)$ time lower bound. Thus, the algorithms in [6,14,18] are optimal only when $h = \Theta(n)$. However, since the value h can be substantially smaller than n, it is desirable to have an algorithm whose running time is bounded by $O(n + f(h))$, where $f(h)$ is a function of h.

Our Results. We consider the rectilinear min-link paths in a rectilinear domain $\mathcal{P}$. After the free space of $\mathcal{P}$ is triangulated, our algorithm builds a link distance map in $O(n + h \log h)$ time and $O(n)$ space, with the standard query performance. The triangulation can be done in $O(n \log n)$ time or $O(n + h \log^{1+\epsilon} h)$ time for any $\epsilon > 0$ [1]. Hence, our result improves the previous $O(n \log n)$ time algorithms [6,14,18], especially when h is substantially smaller than n.

Our Techniques. Our idea is to combine Das and Narasimhan's algorithmic scheme [6] and a *corridor structure* of polygonal domains [4,5,11,17]. The corridor structure partitions the free space of $\mathcal{P}$ into $O(h)$ corridors and $O(h)$ "junction" rectangles that connect all corridors. The algorithm in [6] (which we call the DN algorithm) sweeps the free space, from the source point s, to build the map. The sweep is controlled in a global way so that the time is bounded by $O(n \log n)$. This global sweeping on the entire free space restricts the DN algorithm from being implemented in $O(n + h \log h)$ time because each operation takes $O(\log n)$ time and there are $O(n)$ operations. Using the corridor structure, our algorithm avoids the global sweeping on the entire free space. When the sweep is in junction rectangles, we control the sweep in a global way as in the DN algorithm. However, when the sweep enters a corridor, we process the corridor independently and "locally" without considering the space outside the corridor. Since a corridor is a simple polygon, we are able to design a faster algorithm for processing the sweep in it. When we finish processing a corridor, we arrive at a junction rectangle. Next, we pick an unprocessed junction rectangle that currently has the smallest link distance to s to "resume" the sweep.

This is somewhat similar to Dijkstra's shortest path algorithm. In this way, there are only $O(h)$ operations that need to be performed in logarithmic time each.

We first define notation and review the DN algorithm [6] in Section 2. Our algorithm is presented in Section 3. Due to the space limit, many details are omitted but can be found in the full paper [15].

2   Preliminaries

For simplicity of discussion, let $\mathcal{R}$ be a large rectangle that contains all obstacles of $\mathcal{P}$ and let $\mathcal{F}$ denote the free space of $\mathcal{P}$ in $\mathcal{R}$ (our algorithm can also handle the case where $\mathcal{R}$ is an arbitrary rectilinear polygon). We assume $\mathcal{F}$ has been triangulated. Let s be any point in $\mathcal{F}$. For ease of exposition, we make a general position assumption that no three vertices of $\mathcal{P} \cup \{s\}$ have the same x- or y-coordinate. In the following, "paths" always refer to rectilinear paths in $\mathcal{F}$.

Consider any point $t \in \mathcal{F}$. An s-t path (i.e., a path from s to t) π is called a *horizontal-start-vertical-end* path (or *h-v-path* for short) if the first link of π (i.e., the edge incident to s) is horizontal and the last link of (i.e., the edge incident to t) is vertical. The *h-h-paths*, *v-h-paths*, and *v-v-paths* are defined analogously. To make it consistent, if π is an h-h-path of k links, we also consider it to be an h-v-path of $k + 1$ links (i.e., we enforce an additional edge of zero length at the end of the path), and similarly, it is also considered to be an v-h-path of $k + 1$ links and a v-v-path of $k + 2$ links. A *min-link h-v-path* from s to t is an h-v-path from s to t with the minimum number of links. The *min-link h-h-paths*, *v-h-paths*, and *v-v-paths* are defined similarly. To find a min-link s-t path, we will find the four s-t paths: a min-link h-v-path, a min-link h-h-path, a min-link v-v-path, and a min-link v-h-path, and return the one with minimum link distance.

We will compute four link distance maps of $O(n)$ size each: an *h-h-map*, an *h-v-map*, a *v-h-map*, and a *v-v-map*, defined as follows. The h-h-map is a decomposition of $\mathcal{F}$ into regions such that for any region R, the link distances of the min-link h-h-paths from s to all points in R are the same. The other three maps are defined analogously. Using point location data structures [7], for any query point t, we determine the region containing t in each map and the one with the smallest link distance gives our sought min-link s-t path distance.

The *vertical visibility decomposition* of $\mathcal{F}$, denoted by $VD(\mathcal{F})$, is obtained by extending each vertical edge of the obstacles in $\mathcal{P}$ until it hits either another obstacle or the boundary of $\mathcal{R}$ (e.g., see Fig. 4). We call the above edge extensions the *diagonals*. We consider the point s as a special obstacle and extend a vertical diagonal through s. Since $\mathcal{F}$ has been triangulated, $VD(\mathcal{F})$ can be obtained in $O(n)$ time [1,3]. In $VD(\mathcal{F})$, $\mathcal{F}$ is decomposed into rectangles, also called *cells*. Due to our general position assumption, each vertical side of a cell can contain at most two diagonals. The horizontal visibility decomposition of $\mathcal{F}$, denoted by $HD(\mathcal{F})$, is defined similarly by extending the horizontal edges of $\mathcal{P}$.

Our v-v-map is on $VD(\mathcal{F})$, i.e., for each diagonal d (resp., cell C) of $VD(\mathcal{F})$, the link distances of the min-link v-v-paths from s to all points in d (resp., in C that are not on diagonals) are the same and we denote this distance by $dis_{vv}(d)$

Fig. 1. Illustrating the algorithm: all diagonals are labeled. Note that the three thick dash-dotted diagonals (labeled 5) are swept twice in the 2nd phase.

Fig. 2. Illustrating a cell C where d is on the left side of C.

(resp., $dis_{vv}(C)$). In fact, $dis_{vv}(d) = \min\{dis_{vv}(C_l), dis_{vv}(C_r)\}$, where C_l and C_r are the two cells on the left and right of d, respectively. Our goal is to compute $dis_{vv}(C)$ for each cell C and $dis(d)$ for each diagonal d of $VD(\mathcal{F})$. We also need to maintain some path information to retrieve an actual path for each query.

Similarly, our h-v-map is also on $VD(\mathcal{F})$, but the h-h-map and the v-h-map are both on $HD(\mathcal{F})$. Next we review the DN algorithm [6], but we discuss it in a way that will be helpful for us to introduce our algorithm later in Section 3.

An $O(n \log n)$ Time Algorithm (the DN Algorithm). We compute the v-v-map on $VD(\mathcal{F})$ first. The goal is to compute $dis_{vv}(d)$ for every diagonal d of $VD(\mathcal{F})$. To simplify the notation, we use $dis(\cdot)$ to refer to $dis_{vv}(\cdot)$. Initially, all diagonals have distance value ∞ except $dis(d_s) = 1$, where d_s is the diagonal through s. Note that if a diagonal d is on a side e of a cell, then whenever $dis(d)$ is updated, $dis(e)$ is automatically set to $dis(d)$.

The algorithm has many phases. In the i-th phase for $i \geq 0$, the algorithm determines the set V_i of diagonals d whose distances $dis(d)$ are equal to $2i + 1$, and these diagonals are then "labeled" with distance $2i + 1$ (e.g., see Fig. 1). Initially, $i = 0$, and V_0 consists of the diagonal d_s only. As discussed in [6], if we put light sources on the diagonals in V_{i-1}, then V_i consists of all new diagonals that will get illuminated with light emanating horizontally from the light sources.

Consider a general i-th phase for $i \geq 1$. We assume V_{i-1} has been determined. There are two procedures: *right-sweep* and *left-sweep*. In the right-sweep (resp., left-sweep), we illuminate the diagonals in the rightward (resp., leftward) direction. The right-sweep procedure starts from the *locally-rightmost* diagonals of V_{i-1}, defined as follows. Consider any diagonal d in V_{i-1}. Let C be the cell of $VD(\mathcal{F})$ on the right of d, i.e., d is on the left side of C. Let e_r be the right side of C. If $dis(e_r) \neq 2i - 1$ then d is a *locally-rightmost* diagonal of V_{i-1}. Similarly, the left-sweep starts from the *locally-leftmost* diagonals of V_{i-1}. Both locally-leftmost and locally-rightmost diagonals are referred to as *locally-outmost* diagonals. Below, we first discuss the right-sweep.

For each locally-rightmost diagonal d, we put a rightward "light beam" on d, and let $B(d)$ denote the set of the beam. Initially we insert all locally-rightmost diagonals into a min-heap H_R prioritized by their x-coordinates (i.e., the leftmost

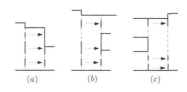

Fig. 3. Illustrating some beam operations in a right-sweep procedure: In (a) and (b), beams are *split*, and some beams are "narrowed" and some beams "terminate" at obstacle edges; in (c), beams are *merged*.

diagonal is at the root). By using H_R, the diagonals involved in the right-sweep will be processed from left to right. If $H_R \neq \emptyset$, we repeatedly do the following.

We obtain the leftmost diagonal d of H_R and remove it from H_R. Let C be the cell on the right of d (e.g., see Fig. 2). We *process* d in the following way. Intuitively we want to propagate the beams of $B(d)$ to other diagonals in C. Let e_l and e_r denote the left and right sides of C, respectively. Note that d is on e_l.

Recall that each cell side has at most two diagonals. If e_l has another diagonal $\hat{d}$ (e.g., see Fig. 2) and $\hat{d}$ has not been labeled (i.e., $dis(\hat{d}) = \infty$), we set $dis(\hat{d}) = 2i + 1$. The beam set $B(\hat{d})$ of $\hat{d}$ is set to $\emptyset$ since no beam from $B(d)$ illuminates $\hat{d}$. Further, although $B(\hat{d}) = \emptyset$, we associate the *leftward direction* with it, because $\hat{d}$ may be a locally-leftmost diagonal of V_i and generate a leftward beam in the next phase. We mark $\hat{d}$ as a locally-leftmost diagonal. If $\hat{d}$ has been labeled, we can show that $dis(\hat{d})$ must be $2i + 1$. In this case, we do nothing on $\hat{d}$.

Next, we consider the diagonals on the right side e_r of C. Depending on the values of $dis(e_r)$, there are several cases. Since we are at the i-th phase, either $dis(e_r) = \infty$ or $dis(e_r) \leq 2i + 1$. If $dis(e_r) < 2i + 1$, then we do nothing on e_r.

If $dis(e_r) = \infty$, we set $dis(e_r) = 2i + 1$. If e_r does not have any diagonals, we are done. Otherwise, for each diagonal d' on e_r, we determine the portions of beams of $B(d)$ that can illuminate d', which are the rightward projections of $B(d)$ on d' (beams of $B(d)$ may be "narrowed" or "split"; see Fig. 3). We use $B(d) \cap d'$ to denote the above portions of $B(d)$. If $B(d) \cap d' = \emptyset$, we mark d' as a locally-rightmost diagonal and set $B(d') = \emptyset$ with the *rightward direction*; otherwise, we set $B(d') = B(d) \cap d'$ and insert d' to H_R.

If $dis(e_r) = 2i + 1$, this case happens because e_r was illuminated by beams from another diagonal $\hat{d}$ on e_l. Hence, each diagonal d' on e_r may already have a non-empty $B(d')$. But d' may receive more beams from $B(d)$. We first determine $B(d) \cap d'$ and then do a "merge" operation by merging $B(d) \cap d'$ with $B(d')$. Finally, we set $B(d')$ to the above merged set of beams (with the rightward direction). If $B(d')$ was empty before the merge and now becomes non-empty, then we insert d' into H_R. If $B(d')$ is still empty after the merge, then we mark d' as a locally-rightmost diagonal. If $B(d')$ was non-empty before the merge, then d' is already in H_R, so we do not need to insert it into H_R again.

The above finishes processing d. The right-sweep is done once H_R becomes empty. We use balanced binary search trees to maintain the beams in $B(d)$ such that "merge" and "split" operations can be performed in logarithmic time each.

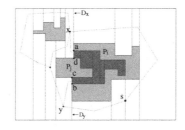

Fig. 4. Illustrating the vertical visibility decomposition (the dashed segments are diagonals) and its dual graph G_{vtd}

Fig. 5. Illustrating the graph G, and the corridor (shaded by slashes) bounded by P_i and P_j

The left-sweep procedure is similar. There is one subtle thing. If the sweep illuminates a diagonal d that has been labeled by the right-sweep, then this is ignored and we proceed as if d were not labeled. As discussed in [6], the reason for this is that the left-sweep may reach more cells than the right-sweep (e.g., see Fig. 1). In this way, each diagonal can be processed at most twice in a phase. But no diagonal can be processed in more than one phase. Also, suppose a diagonal d was marked as a locally-outmost diagonal during the right-sweep; if d is illuminated again in the left-sweep but d is not marked locally-outmost in the left-sweep, then we clear the previous mark on d. After the left-sweep, the remaining locally-outmost diagonals will be used in the next phase.

The above describes the i-th phase of the algorithm. The algorithm is done after all diagonals are labeled. We have only labeled diagonals. We can label cells by an easy adaption of the algorithm. We also need to maintain path information to retrieve a path for each query. The above computes the v-v-map, and the other three maps can be computed similarly. All these details are in [15].

3   Our Improved Algorithm

We first introduce the corridor structure in rectilinear domains, which is similar to that in general polygonal domains [11]. Let G_{vtd} be the dual graph of the vertical visibility decomposition $VD(\mathcal{F})$ (see Fig. 4). Based on G_{vtd}, we obtain a *corridor graph* G as follows (see Fig. 5). First, we remove every degree-one node from G_{vtd} along with its incident edge; repeat this process until no degree-one node exists. Second, remove every degree-two node from G_{vtd} and replace its two incident edges by a single edge; repeat this process until no degree-two node exists. The remaining graph is G. The cells in $VD(\mathcal{F})$ corresponding to the nodes in G are called *junction cells* (see Fig. 5). We consider the diagonal through s as a degenerate junction cell. As in the general polygonal domains [11], the graph G has $O(h)$ nodes and $O(h)$ edges. The removal of all junction cells from $VD(\mathcal{F})$ results in $O(h)$ *corridors*, each of which corresponds to an edge of G.

The boundary of any corridor $\mathcal{C}$ consists of four parts (see Fig. 5): (1) The boundary portion of an obstacle P_i, from a point a to a point b; (2) a diagonal $\overline{bc}$; (3) the boundary portion of an obstacle P_j from c to a point d; (4) a diagonal $\overline{da}$. $\overline{bc}$ and $\overline{ad}$ are called the *doors* of $\mathcal{C}$, which is a simple rectilinear polygon.

We focus on computing the v-v-map on $VD(\mathcal{F})$; other three maps can be computed similarly. Our goal is to label d for each diagonal d, i.e., compute the distance value $dis(d) = dis_{vv}(d)$ (the same algorithm can be used to label cells as well). As before, each diagonal d will maintain a beam set $B(d)$.

We first discuss the main idea. In the DN algorithm, a sweep procedure will enter each corridor though one of its two doors, and the procedure will either sweep the entire corridor and leave the corridor through the other door, or terminate inside the corridor (in which case the sweep "hits" another sweep that entered the corridor through the other door and both sweeps terminate after "collision"). This means that if we can determine the beams and the distance values at the doors of a corridor, then we can process the corridor independently in a more efficient way since the corridor is a simple rectilinear polygon.

In our algorithm, the sweep in the junction cells is still processed and controlled in a global manner in each phase. However, whenever the sweep enters a corridor through one of its doors, the corridor will be processed independently by using our more efficient *corridor-processing* algorithm (i.e., the sweep "jumps" from one junction cell to another through the corridor connecting them). Note that since in the DN algorithm a diagonal may be processed twice in the two sweep procedures in the same phase, here correspondingly an entire corridor may be processed twice in the same phase (this happens *only if* the beams on a door can illuminate the other door directly, and vice versa).

The running time of our algorithm is $O(n + h \log h)$. More specifically, since there are $O(h)$ junction cells, the time spent on processing the diagonals in all junction cells is $O(h \log h)$, and the processing on all corridors takes $O(n+h \log h)$ time because the number of vertices of all corridors is $O(n)$, in addition to another $O(h \log h)$ time spent on maintaining the beams on all diagonals.

3.1   The Algorithm

We only sketch our algorithm; the details are omitted but can be found in [15].

Initially, we set $dis(d)$ to ∞ and $B(d) = \emptyset$ for each diagonal d except that $dis(d_s) = 1$ and $B(d_s) = \{d_s\}$, where d_s is the diagonal through s.

We use a min-heap H to store the diagonals in all junction cells, where the "keys" are the distance values the diagonals currently have (and these values may not be set correctly), with the *smallest* key at the root of H. Since there are $O(h)$ junction cells, the size of H is $O(h)$. Each diagonal d in H is also associated with its beam set $B(d)$ (along with its direction). It is possible that $B(d)$ is empty, in which case d might be a locally-outmost diagonal.

If some diagonals of H have the same keys, we break the ties by the following rules. Consider two diagonals d_1 and d_2 in H with $dis(d_1) = dis(d_2)$. If $B(d_1)$ and $B(d_2)$ are both empty or both non-empty, then we break ties arbitrarily. Otherwise, assume $B(d_1) \neq \emptyset$ but $B(d_2) = \emptyset$. Then, we consider the key of d_1 *smaller* than that of d_2. The reason is as follows. Since $B(d_1) \neq \emptyset$ and $B(d_2) = \emptyset$, the current sweep procedure should be over *before* processing d_2 while the sweep should continue *after* processing d_1, and thus, we should process d_1 before d_2. Therefore, our way of resolving ties in H is crucial and consistent with the DN algorithm. In the following, for any diagonal d, even if d is not in H, we consider $dis(d)$ along with $B(d)$ as the *global-key* of d, and whenever we compare the global-keys of diagonals, we follow the above rules to break ties.

Consider any corridor $\mathcal{C}$ with two doors d and d'. Suppose the beams of $B(d)$ are going inside $\mathcal{C}$, and we want to *process* $\mathcal{C}$ (i.e., compute the v-v-map in $\mathcal{C}$) using the beams of $B(d)$. We say the above way of processing $\mathcal{C}$ is in the *direction* from d to d'. As will be seen later, a corridor may be processed twice: once from d to d' and the other from d' to d. Due to the special geometric structure of the corridor, we have the following observation.

Observation 1. *Suppose d and d' are the two doors of a corridor $\mathcal{C}$, and the direction of the beams of $B(d)$ is towards the inside of $\mathcal{C}$. Then after $\mathcal{C}$ is processed by using $B(d)$, the beam set of d' is not empty.*

Proof. Let $VD(\mathcal{C})$ denote the vertical visibility decomposition of the corridor $\mathcal{C}$. Consider the cell C of $VD(\mathcal{C})$ that contains d'. Without loss of generality, assume d' is on the right side of C. Denote by e_r the right side of C.

First, due to the structure of the corridor, it can be seen that d' is the entire right side of C, i.e., d' is e_r (we omit the detailed proof). Further, note that we obtain the beam set of d' from the rightward beams of the diagonals on the left side of C. Now that d' is the entire right side of C, d' will receive all beams of any diagonal on the left side of C. Hence, the beam set of d' cannot be empty. □

Our algorithm is consistent with the DN algorithm in the sense that after the algorithm finishes, each diagonal in any junction cell is *correctly labeled*, i.e., both its distance value and its beam set are the same as those in the DN algorithm. Let d^* be the diagonal in the root of H. Our algorithm will maintain the following three invariants.

1. The diagonal d^* is correctly labeled. Further, for any other diagonal d in a junction cell, if the global-key of d is no larger than that of d^*, then d has been correctly labeled.
2. For any diagonal d in a junction cell, if $dis(d) \neq \infty$ and the global-key of d is larger than that of d^*, then d is in H.
3. For any corridor $\mathcal{C}$ with two doors d and d', if $\mathcal{C}$ is processed in the direction from d to d', then $\mathcal{C}$ will never be processed from d to d' again in the algorithm (although $\mathcal{C}$ may be processed later in the other direction from d' to d).

Initially $H = \emptyset$. Recall that $dis(d_s) = 1$ and $B(d) = \{d_s\}$. We consider the diagonal d_s through s as a degenerate junction cell. Specifically, we consider d_s as two duplicate diagonals with one generating a rightward beam and the other generating a leftward beam from the entire d_s. We insert these two diagonals into H. As long as H is not empty, we repeatedly do the following.

Let d^* be the diagonal of H with the smallest global-key. We assume $dis(d^*) = 2i+1$ for some integer i. If we were running the DN algorithm, we are currently working on the i-th phase. Let S be the set of all diagonals in H that have the same global-key as d^*. The diagonals of S can be found by continuing the extract-min operations on H in $O(|S| \log |H|)$ time, and after that, diagonals of S are removed from H. There are two cases depending on whether $B(d^*) = \emptyset$.

$B(d^*) \neq \emptyset$. In this case, all diagonals of S have non-empty beam sets. Due to our corridor structure, the sweeps of the i-th phase "paused" at the diagonals in S. To continue the i-th phase, we "resume" the sweeps from these diagonals. Unlike the DN algorithm where we complete the right-sweep before we start the left-sweep, here, before the pause, we may have already done some left-sweep and right-sweep. Hence, the two sweeps may be somehow "interleaved" and our algorithm will need to take care of this situation.

Let S_R (resp., S_L) be the subset of the diagonals of S whose beams are rightward (resp., leftward). Intuitively, the right-sweep (resp., left-sweep) paused at the diagonals in S_R (resp., S_L), and thus, we resume it from the diagonals in S_R (resp., S_L). Below we focus on the right-sweep.

We build another min-heap H_R by inserting the diagonals of S_R, and the "keys" of diagonals in H_R are their x-coordinates such that the leftmost diagonal is at the root. (Similarly, we build a min-heap H_L on S_L for the left-sweep.)

The algorithm essentially performs the i-th phase as the DN algorithm. But since here the right-sweep and left-sweep may be interleaved, some diagonals may have two sets of beams with opposite directions. However, our algorithm makes sure that if a diagonal d has two sets of beams with opposite directions, it will not be in H (i.e., it has been removed from H), but in both H_R and H_L if d has not been processed yet. To differentiate the two sets of beams, we use $B_r(d)$ (resp., $B_l(d)$) to denote the beam set of any diagonal d in H_R (resp., H_L), meaning that the direction of the beams is rightward (resp., leftward).

During the right-sweep, if we find a new diagonal d that has the same global-key as d^*, then d will be inserted to H_R and d will be removed from H if it is already in H. Hence, all diagonals of H_R have the same global-key as d^*.

As long as H_R is not empty, we repeatedly do the following.

We obtain the leftmost diagonal d of H_R and remove it from H_R. The beams of $B_r(d)$ may enter a junction cell or a corridor. If it is the former case, our way of processing d is similar to the DN algorithm, although we need to take care of the situation that the left and right sweeps are interleaved. We skip the details.

Next, we consider the case where beams of $B_r(d)$ enter a corridor $\mathcal{C}$. We process $\mathcal{C}$ using the beams of $B_r(d)$. One may assume we still use the DN algorithm to process $\mathcal{C}$, and later we will replace it by our corridor-processing algorithm. Let δ be the distance value labeled on the other door d' of $\mathcal{C}$ by the above processing and let B' denote the corresponding beam set on d'. Let $dis(d')$ and $B(d')$ be the original distance value and beam set at d' before processing $\mathcal{C}$. By the third algorithm invariant, this is the first time $\mathcal{C}$ is processed in the direction from d to d'. Hence, if $dis(d') \neq \infty$, the value $dis(d')$ must be obtained by the sweep from outside $\mathcal{C}$, i.e., beams in $B(d')$ are towards the inside of $\mathcal{C}$.

Due to the above processing of $\mathcal{C}$, we have obtained another distance value δ and beam set B' for d'. We need to update the label of d' and possibly insert d' to some heap. Depending on the value of $dis(d')$, there are several cases.

1. If $dis(d')$ is ∞, then we set $dis(d') = \delta$ and $B(d') = B'$.
   If $\delta > 2i + 1$, then we insert d' into H.

If $\delta = 2i + 1$, since $dis(d) = 2i + 1$, d' must be illuminated directly by the beams in $B_r(d)$ and the beams of $B(d')$ are still towards right. By Observation 1, $B' \neq \emptyset$. Hence we obtain $B_r(d') = B' \neq \emptyset$ (we set $B_r(d')$ to B' because the beams of B' are rightward). Finally, we insert d' into H_R.

2. If $dis(d') < 2i+1$, then the global-key of d' is smaller than that of d^* because $dis(d^*) = 2i + 1$. By the first algorithm invariant, d' has been correctly labeled. Recall that the direction of $B(d')$ is towards the inside of $\mathcal{C}$. Also by the first algorithm invariant, d is correctly labeled, and the direction of $B_r(d)$ is towards the inside of $\mathcal{C}$. This means that we have computed complete information on the two doors of $\mathcal{C}$ for the min-link v-v-paths from s to the points inside $\mathcal{C}$. Then, we can do a "post-processing" step (to be discussed later) to compute the v-v-map in $\mathcal{C}$ by using the beams of $B_r(d)$ and $B(d')$.

3. If $dis(d') = 2i + 1$, then d' has been labeled in the current phase.

   If $B(d') \neq \emptyset$, then the global-key of d' is the same as that of d^*. By the first algorithm invariant, d' has been correctly labeled. As above, since both d and d' have been correctly labeled, we do a "post-processing" to compute the v-v-map in $\mathcal{C}$ using $B_r(d)$ and $B(d')$.

   If $B(d') = \emptyset$, then the global-key of d' is strictly larger than that of d^*. By the second algorithm invariant, d' is already in H. If $\delta > 2i + 1$, we do nothing. If $\delta = 2i+1$, then as in the above first case, we set $B_r(d') = B' \neq \emptyset$; finally, we insert d' into H_R and remove d' from H.

4. The remaining case is when $dis(d') \neq \infty$ and $dis(d') > 2i + 1$. It can be shown that this case cannot happen.

The above finishes the processing of the diagonal d. The right-sweep procedure is done after the heap H_R becomes empty. Afterwards we do the left-sweep from the diagonals of S_L using the heap H_L in the symmetric way.

$B(d^*) = \emptyset$. In this case, according to our way of comparing global-keys, all diagonals of S have empty beam sets. If we were running the DN algorithm, the diagonals of S would be locally-outmost and we would be about to start the $(i + 1)$-th phase (not the i-th phase). As in the previous case, we run the two sweep procedures starting from the diagonals of S. Let S_R (resp., S_L) be the subset of diagonals of S whose beam directions are rightward (resp., leftward). We build a min-heap H_R (resp., H_L) on the diagonals of S_R (resp., S_L). Below, we only discuss the right-sweep since the left-sweep is similar.

Since we are doing the right-sweep in the $(i+1)$-th phase, each diagonal of S_R will generate a beam from the entire diagonal, and all new diagonals illuminated in the right-sweep will get distance value $2i+3$ instead of $2i+1$. From now on, we associate each diagonal of S_R with the beam, i.e., for each $d \in S_R$, $B_r(d) = \{d\}$. Since each diagonal of S_R originally got an empty beam set (at the end of the i-th phase), by Observation 1, the beam of $B_r(d)$ cannot be towards a junction cell and thus it must be towards the inside of a corridor.

As long as H_R is not empty, we repeatedly do the following.

We obtain the leftmost diagonal d of H_R (which is at the root) and remove it from H_R. Let $\mathcal{C}$ denote the corridor that the beams of $B_r(d)$ enter. We process the corridor $\mathcal{C}$ using the beams of $B_r(d)$, by using our corridor processing algorithm.

Let δ be the distance on the other door d' obtained by the above processing and let B' be the corresponding beam set. Let $dis(d')$ and $B(d')$ be the original distance value and beam set at d'. Again, by the third algorithm invariant, this is the the first time $\mathcal{C}$ is processed in the direction from d to d'; hence, if $dis(d') \neq \infty$, then d' must be labeled by a sweep from outside $\mathcal{C}$ and the beams of $B(d')$ must enter $\mathcal{C}$. Depending on the value of $dis(d')$, we may need to update the label of d' in several cases.

1. If $dis(d') = \infty$, we set $dis(d') = \delta$ and $B(d') = B'$. Note that $\delta \geq 2i + 3$. Hence, the global-key of d' is strictly larger than that of d^*, which has distance value $2i + 1$. We insert d' into H (not H_R).
2. If $dis(d') \leq 2i + 1$, then since $B(d^*) = \emptyset$, the global-key of d' is no larger than that of d^* regardless of whether $B(d')$ is empty or not. By the first algorithm invariant, d' has been correctly labeled. We do a "post-processing" to compute the v-v-map in $\mathcal{C}$ by using the beams of $B_r(d)$ and $B(d')$.
3. The remaining case is when $dis(d') \neq \infty$ and $dis(d') > 2i + 1$. We can show that this case cannot happen.

The above describes the right-sweep procedure. The left-sweep is similar.

This finishes our discussion in the case in which $B(d^*)$ is empty.

The algorithm finishes after all three heaps H, H_L, and H_R become empty.

After that, for each diagonal d in a junction cell, $dis(d)$ and $B(d)$ have been correctly computed. During the algorithm some corridors have been labeled correctly while others are left for post-processing. Specifically, consider any corridor $\mathcal{C}$ and let d_1 and d_2 be its two doors with their beam sets $B(d_1)$ and $B(d_2)$. If $\mathcal{C}$ is not left for a post-processing, then $\mathcal{C}$ has been processed either from d_1 to d_2 or from d_2 to d_1 and the v-v-map in $\mathcal{C}$ has been computed after the processing. Suppose the above processing is from d_1 and d_2. Then, $\mathcal{C}$ is processed using the beams of $B(d_1)$, and $B(d_2)$ is obtained after the processing. Our corridor-processing algorithm on $\mathcal{C}$ runs in $O(m + (h_1 - h_2 + 1)\log h_1)$ time, where m is the number of vertices of $\mathcal{C}$, $h_1 = |B(d_1)|$, and $h_2 = |B(d_2)|$. If $\mathcal{C}$ is left for a post-processing, i.e., to compute the v-v-map in $\mathcal{C}$ by using $B(d_1)$ and $B(d_2)$, our corridor-post-processing algorithm runs in $O(m + h_1 \log h_1 + h_2 \log h_2)$ time. Refer to [15] for the details of these two algorithms. These efforts together lead to an algorithm that can compute the v-v-map on $VD(\mathcal{F})$ in $O(n + h \log h)$ time.

The other three maps can be computed similarly. For computing the h-v-map on $VD(\mathcal{F})$, one difference is on the initial steps. Initially, we let s generate two beams that are two horizontal rays towards right and left, respectively. We set the distance value of d_s to 0, where d_s is the vertical diagonal through s. Then, we consider d_s as two duplicate diagonals associated with the above two beams respectively, and insert the two duplicate diagonals into the heap H. The remaining algorithm is the same as before except that we replace the distance values $2i+1$ and $2i+3$ in the algorithm description with $2i$ and $2i+2$, respectively. The h-h-map and v-h-map on $HD(\mathcal{F})$ can be computed similarly.

Acknowledgments. J. Mitchell is partially supported by grants from Sandia National Labs, the National Science Foundation (CCF-1018388), and the US-Israel Binational Science Foundation (award 2010074). V. Polishchuk is supported by grant 2014-03476 from the Sweden's innovation agency VINNOVA. H. Wang is supported in part by NSF under Grant CCF-1317143.

References

1. Bar-Yehuda, R., Chazelle, B.: Triangulating disjoint Jordan chains. International Journal of Computational Geometry and Applications **4**(4), 475–481 (1994)
2. de Berg, M.: On rectilinear link distance. Computational Geometry: Theory and Applications **1**, 13–34 (1991)
3. Chazelle, B.: Triangulating a simple polygon in linear time. Discrete and Computational Geometry **6**, 485–524 (1991)
4. Chen, D., Inkulu, R., Wang, H.: Two-point L_1 shortest path queries in the plane. In: Proc. of the 30th Annual Symposium on Computational Geometry (SoCG), pp. 406–415 (2014)
5. Chen, Danny Z., Wang, Haitao: A Nearly Optimal Algorithm for Finding L_1 Shortest Paths among Polygonal Obstacles in the Plane. In: Demetrescu, Camil, Halldórsson, Magnús M. (eds.) ESA 2011. LNCS, vol. 6942, pp. 481–492. Springer, Heidelberg (2011)
6. Das, G., Narasimhan, G.: Geometric searching and link distance. In: Proc. of the 2nd Workshop of Algorithms and Data Structures (WADS), pp. 261–272 (1991)
7. Edelsbrunner, H., Guibas, L., Stolfi, J.: Optimal point location in a monotone subdivision. SIAM Journal on Computing **15**(2), 317–340 (1986)
8. Ghosh, S.: Computing the visibility polygon from a convex set and related problems. Journal of Algorithms **12**, 75–95 (1991)
9. Hershberger, J., Snoeyink, J.: Computing minimum length paths of a given homotopy class. Computational Geometry: Theory and Applications **4**, 63–97 (1994)
10. Imai, H., Asano, T.: Efficient algorithms for geometric graph search problems. SIAM Journal on Computing **15**(2), 478–494 (1986)
11. Kapoor, S., Maheshwari, S., Mitchell, J.: An efficient algorithm for Euclidean shortest paths among polygonal obstacles in the plane. Discrete and Computational Geometry **18**(4), 377–383 (1997)
12. Lingas, A., Maheshwari, A., Sack, J.R.: Parallel algorithms for rectilinear link distance problems. Algorithmica **14**, 261–289 (1995)
13. Maheshwari, A., Sack, J.R., Djidjev, H.: Link distance problems. In: Sack, J.-R., Urrutia, J. (eds.) Handbook of Computational Geometry, pp. 519–558. Elsevier, Amsterdam (2000)
14. Mitchell, J., Polishchuk, V., Sysikaski, M.: Minimum-link paths revisited. Computational Geometry: Theory and Applications **47**, 651–667 (2014)
15. Mitchell, J., Polishchuk, V., Sysikaski, M., Wang, H.: An optimal algorithm for minimum-link rectilinear paths in triangulated rectilinear domains (2015). arXiv:1504.06842

16. Mitchell, J., Rote, G., Woeginger, G.: Minimum-link paths among obstacles in the plane. Algorithmica **8**, 431–459 (1992)
17. Mitchell, J., Suri, S.: Separation and approximation of polyhedral objects. Computational Geometry: Theory and Applications **5**, 95–114 (1995)
18. Sato, M., Sakanaka, J., Ohtsuki, T.: A fast line-search method based on a tile plane. In: Proc. of the IEEE International Symposium on Circuits and Systems, pp. 588–597 (1987)
19. Schuierer, S.: An optimal data structure for shortest rectilinear path queries in a simple rectilinear polygon. International Journal of Compututational Geometry and Applications **6**, 205–226 (1996)
20. Suri, S.: A linear time algorithm with minimum link paths inside a simple polygon. Computer Vision, Graphics, and Image Processing **35**(1), 99–110 (1986)
21. Suri, S.: Minimum link paths in polygons and related problems. Ph.D. thesis, Johns Hopkins University, Baltimore, MD (1987)
22. Suri, S.: On some link distance problems in a simple polygon. IEEE Transactions on Robotics and Automation **6**, 108–113 (1990)

Amplification of One-Way Information Complexity via Codes and Noise Sensitivity

Marco Molinaro$^{1(\boxtimes)}$, David P. Woodruff^2, and Grigory Yaroslavtsev^3

^{1} Delft University of Technology, Delft, The Netherlands
m.molinaro@tudelft.nl
^{2} IBM Almaden Research Center, San Jose, USA
dpwoodru@us.ibm.com
^{3} University of Pennsylvania, Philadelphia, PA, USA
grigory@grigory.us

Abstract. We show a new connection between the information complexity of one-way communication problems under product distributions and a relaxed notion of list-decodable codes. As a consequence, we obtain a characterization of the information complexity of one-way problems under product distributions for *any error rate* based on covering numbers. This generalizes the characterization via VC dimension for constant error rates given by Kremer, Nisan, and Ron (CCC, 1999). It also provides an *exponential improvement in the error rate*, yielding tight bounds for a number of problems. In addition, our framework gives a new technique for analyzing the complexity of composition (e.g., XOR and OR) of one-way communication problems, connecting the difficulty of these problems to the *noise sensitivity* of the composing function. Using this connection, we strengthen the lower bounds obtained by Molinaro, Woodruff and Yaroslavtsev (SODA, 2013) for several problems in the distributed and streaming models, obtaining optimal lower bounds for finding the approximate closest pair of a set of points and the approximate largest entry in a matrix product. Finally, to illustrate the utility and simplicity of our framework, we show how it unifies proofs of existing 1-way lower bounds for sparse set disjointness, the indexing problem, the greater than function under product distributions, and the gap-Hamming problem under the uniform distribution.

1 Introduction

We consider the two-party one-way communication complexity model where Alice and Bob want to jointly compute a function $f : \mathcal{X} \times \mathcal{Y} \to \{0,1\}$. More precisely, Alice holds an input $x \in \mathcal{X}$, Bob holds an input $y \in \mathcal{Y}$, and they have access to common random bits; Alice sends a (random) message to Bob, who then tries to output the value $f(x,y)$. The *cost* of a protocol is the maximum (over the inputs and the randomness) number of bits sent by Alice. The goal is to find a randomized protocol of minimum cost that for all inputs computes $f(x,y)$ with probability at least $1 - \alpha$; this minimum cost is denoted by $\mathsf{R}(f)_\alpha^{\rightarrow}$.

© Springer-Verlag Berlin Heidelberg 2015
M.M. Halldórsson et al. (Eds.): ICALP 2015, Part I, LNCS 9134, pp. 960–972, 2015.
DOI: 10.1007/978-3-662-47672-7_78

The one-way communication model has been studied in a number of works, including Yao [23], Papadimitriou and Sipser [19], Ablayev [1], Newman and Szegedy [16], and Kremer et al. [10]. It is particularly relevant to the *data stream* model in which an algorithm sees a stream of elements one at a time, and tries to compute a relation of these elements using as little space (in bits) as possible [15]. One way of lower-bounding the space complexity of data stream algorithms is to set up a one-way communication protocol in which Alice's message consists of the state of the streaming algorithm run on a stream created by Alice. Bob then continues the execution of the streaming algorithm on a stream he creates, and if from the output the players can solve a communication problem f, then the space complexity of the streaming algorithm must be at least the one-way communication complexity of f.

We will consider a distributional version of one-way communication complexity, in which Alice and Bob have inputs $(x, y) \sim \mu \times \nu$, where $\mu \times \nu$ is a *product distribution* on domains $\mathcal{X}$ and $\mathcal{Y}$. That is, Alice's input is drawn from μ, while Bob's input is drawn from ν, and the inputs are independent. We define $\mathsf{R}(f)_\alpha^{\rightarrow,\square}$ to be the maximum, over product distributions $\mu \times \nu$, of $\mathsf{D}(f)_{\mu \times \nu, \alpha}^{\rightarrow}$, where $\mathsf{D}(f)_{\mu \times \nu, \alpha}^{\rightarrow}$ is the minimum cost over deterministic protocols which compute f with error probability at most α when the input is drawn from $\mu \times \nu$. Kremer, Nisan, and Ron [10] show that for constant α and Boolean functions f, $\mathsf{R}(f)_\alpha^{\rightarrow,\square} = \Theta(VC)$, where VC is the VC-dimension of the class $\{f_x : \mathcal{Y} \to \{0,1\} \mid x \in \mathcal{X}\}$ obtained by seeing the rows of the communication matrix of f as functions. Equivalently, VC is the dimension of the largest hypercube which is a submatrix of the communication matrix.

Unfortunately, a characterization for constant α does not suffice for streaming applications. This was the focus of work by Jayram and Woodruff [9], who showed that for a number of streaming problems, such as estimating the empirical entropy and Euclidean norm (and more generally the ℓ_p-norm for $p \leq 2$), the problem requires an extra multiplicative $\log(1/\delta)$ in the space complexity if the algorithm succeeds with probability at least $1 - \delta$. This was shown using one-way communication under a product distribution, and so obtaining the extra $\log(1/\delta)$ factor had to be shown by ways other than resorting to the VC-dimension, since we do not have a general characterization of problems showing how their communication cost scales with the error probability.

Besides single-shot problems, the gap in our understanding of the dependence on the error probability also manifests itself for solving a composition of many copies of a problem simultaneously with constant probability. The authors [13] previously showed that for several streaming problems, the communication cost of solving n copies of a problem simultaneously with probability $2/3$ scales as n times the cost of solving each copy with probability $1 - 1/n$. This composition theorem critically uses that a protocol must obtain a correct output for each of the n instances, and it is unknown if such a statement holds for other composition functions, such as the OR or XOR functions. This has led to $\log n$ factor gaps in the upper and lower bounds for streaming problems such as

- ClosestPair: Alice has n points $p_1, \ldots, p_n$ in $\mathbb{R}^d$, Bob has n points $q_1, \ldots, q_n$ in $\mathbb{R}^d$, and they would like to find a pair p_i, q_j for which $\|p_i - q_j\|_2 \leq (1 + \epsilon) \min_{i', j'} \|p_{i'} - q_{j'}\|_2$, and
- MatrixProduct: Alice has an $n \times d$ matrix A with rows of unit norm, Bob has a $d \times n$ matrix B with columns of unit norm. They want to approximate $\max_{i,j} |AB|_{i,j}$ up to an additive ϵ.

Our Contributions: We introduce the notion of an (α, β)-code and use it to capture the distance between rows of a communication matrix of a function $f : \mathcal{X} \times \mathcal{Y} \to \{0, 1\}$. Informally speaking, this notion says that under Alice's distribution μ, with probability at most β, two independently sampled rows have relative Hamming distance at most α when weighted with respect to Bob's input distribution ν. This notion thus captures the (pairwise) correlation of rows of a communication matrix, with respect to distributions μ and ν. We show that the one-way information cost of protocols under distribution $\mu \times \nu$ with error probability α is $\Omega(\log 1/\beta)$. This result is based on a Fano's inequality for list-decoding that may be of independent interest. This gives a surprisingly generic way of characterizing lower bounds in terms of the error probability.

Characterization Theorem: We use our characterization in terms of codes to obtain a characterization of 1-way communication complexity in terms of *packing numbers*. Here, given a pseudo-metric space $(\mathcal{X}, d)$, the α-packing number is the largest set of points in $\mathcal{X}$ with pairwise distance at least α. We show that $\max_\nu \Omega(\log p_{8\alpha,\nu}) \leq \mathsf{R}(f)_\alpha^{\to, \Box} \leq \max_\nu O(\log p_{\alpha,\nu})$, where $p_{\alpha,\nu}$ is the packing number of the pseudo-metric space $(\{f(x)\}_{x \in \mathcal{X}}, \| \cdot \|_\nu)$, where $\{f(x)\}_{x \in \mathcal{X}}$ is the family of functions corresponding to rows of the communication matrix, and $\| \cdot \|_\nu$ is the weighted relative Hamming distance according to ν. This gives a strengthening of the result of Kremer, Nisan, and Ron [10] since it gives a tight characterization in terms of the error probability α (up to the distinction of α in the upper bound and 8α in the lower bound). We need to resort to packing numbers, since as observed by Jayram and Woodruff [9], there is no characterization possible in terms of the VC-dimension (as used by [10]). However, by relating packing numbers to VC-dimension, we considerably strengthen the result of [10] which states that $(1 - H(\alpha))VC \leq \mathsf{R}(f)_\alpha^{\to, \Box} \leq O(VC \frac{1}{\alpha} \log \frac{1}{\alpha})$, where VC denotes the VC-dimension of f. We obtain the stronger result that $(1 - H(\alpha))VC \leq \mathsf{R}(f)_\alpha^{\to, \Box} \leq O(VC \log(\frac{1}{\alpha}))$. As an example, we use this to show that $\mathsf{R}_\alpha^{\to, \Box}(GT) = \Theta(\log \frac{1}{\alpha})$ where GT is the greater-than function. This is an exponential improvement over the result based on VC-dimension.

Composition Theorem: Next we introduce the notion of noise sensitivity, which captures how a communication problem f whose rows form an (α, β)-code behaves under composition. There is a line of work on understanding how primitive problems behave under composition [3,11,12,17,21]; our work adds to this by characterizing the composition in terms of codes. The noise sensitivity of a composing function g on k inputs with respect to an input distribution μ^k intuitively captures how likely two independent samples of inputs to g from μ^k

result in differing outputs of g. We show that if f is an (α, β)-code with respect to $\mu \times \nu$, then $g \circ f$ is an (α', β')-code with respect to $\mu^k \times \nu^k$ for certain α' and β' related to the noise sensitivity of g, as well as to α and β.

Streaming Applications: As the main application of our composition theorem, we consider the primitive problem f in which Alice holds a string $x \in [k]^m$, Bob has an $\ell \in [k]$ and an index $j \in [m]$, and Bob would like to know if $x_j = \ell$. We show that f is an (α, β)-code for sufficiently good α and β, and we lower bound the noise sensitivity of the OR function. These results imply that solving the OR of k copies of f, denoted $\mathsf{OR}^k \circ f$, with constant probability has one-way communication complexity $\Omega(km \log k)$. For our streaming applications, we further consider an augmented version of this problem, in which Alice has t independent instances of $\mathsf{OR}^k \circ f$, and Bob would like to solve one of these t instances i chosen uniformly at random. Bob is also given Alice's input for the first $i - 1$ instances. For this we show an $\Omega(tkm \log k)$ one-way communication lower bound for constant probability protocols. These results greatly strengthen the results in [13], which could only show this if $\mathsf{OR}^k \circ f$ were replaced with $\mathsf{ALLCOPIES}\ ^k \circ f$, the latter requiring a correct output to all k instances of f rather than just an OR of the k instances. Note that the output of $\mathsf{OR}^k \circ f$ is only a single bit, whereas the output of $\mathsf{ALLCOPIES}\ ^k \circ f$ consists of k bits, making the latter a significantly easier problem. Our result directly improves the streaming application lower bounds in [13], leading to the first tight one-way lower bounds for $\mathsf{ClosestPair}$ and $\mathsf{MatrixProduct}$. The details are in Section 6.

Unified Lower Bounds: To illustrate the power of the framework developed, we recover in a unified way several 1-way lower bounds from the literature, including sparse set disjointness [4,6,20] and indexing [9] under product distributions, and the gap-Hamming problem under the uniform distribution [22].

2   Preliminaries

Information Theory. We use the following notions from information theory (see [5] for more details). Given random variables X, Y and Z on a common probability space, we use $H(X)$ to denote the binary entropy of X and $H(X \mid Y)$ its conditional entropy given Y. The mutual information between X and Y is then defined as $I(X; Y) = H(X) - H(X \mid Y)$, and the conditional mutual information given Z is $I(X; Y \mid Z) = H(X \mid Z) - H(X \mid (Y, Z))$. We will need the *data processing inequality*: for any arbitrary functions $g, h, I(X; Y) \geq I(g(X); h(Y))$.

Distributional and Information Complexity. Consider a function $f : \mathcal{X} \times \mathcal{Y} \to \{0, 1\}$ and a distribution μ over $\mathcal{X} \times \mathcal{Y}$. The one-way *distributional complexity* of f with respect to μ, denoted $\mathsf{D}(f)_{\mu,\alpha}^{\to}$, is the smallest communication cost of a one-way deterministic protocol that outputs $f(x, y)$ on all but an α fraction of inputs weighted according to μ. The one-way *distributional complexity* of f, denoted $\mathsf{D}(f)_{\alpha}^{\to}$, is the supremum of $\mathsf{D}(f)_{\mu,\alpha}^{\to}$ over all distributions μ. The classic Yao's Minimax Theorem [23] shows that randomized and distributional complexity are the same: $R(f)_\alpha^\rightarrow = D(f)_\alpha^\rightarrow$. Motivated by this observation, define the product distribution complexity $R(f)_\alpha^{\rightarrow,\square}$ as the supremum of $D(f)_{\mu \times \nu, \alpha}^\rightarrow$ over all distributions μ for $\mathcal{X}$ and ν for $\mathcal{Y}$.

Now we define information complexity. Again we are given a distribution μ over $\mathcal{X} \times \mathcal{Y}$. Given a *randomized* one-way protocol for computing f, with $A(x, r)$ denoting the message sent by Alice on input x and private randomness r, the *information cost* of this protocol is defined as $I(A(X, R); X \mid Y)$, where the pair (X, Y) is sampled from μ (and R is Alice's randomness, which is independent from X, Y). The *information complexity* with respect to μ, denoted $IC(f)_{\mu,\alpha}^\rightarrow$, is the smallest information cost of a randomized one-way protocol computing $f(X, Y)$ with probability at least $1 - \alpha$ (with respect to $(X, Y) \sim \mu$ and the private randomness of Alce and Bob). Finally the *information complexity* $IC(f)_\alpha^\rightarrow$ is the supremum of $IC(f)_{\mu,\alpha}^\rightarrow$ over all distributions μ. Similarly, the *information complexity over product distributions* $IC(f)_\alpha^{\rightarrow,\square}$ is the supremum of $IC(f)_{\mu \times \nu, \alpha}^\rightarrow$ over all distributions μ on $\mathcal{X}$ and ν on $\mathcal{Y}$. Notice that under a product distribution $(X, Y) \sim \mu \times \nu$ the information cost of a protocol becomes $I(A(X, R); X)$.

We have the following known relationship between information and distributional complexity (which follows from the entropy span bound and non-negativity of entropy): $R(f)_\alpha^{\rightarrow,\square} \geq IC(f)_\alpha^{\rightarrow,\square}$.

Notation. Given a function $f : \mathcal{X} \times \mathcal{Y} \to \{0, 1\}$ and $x \in \mathcal{X}$, we use $f(x) : \mathcal{Y} \to \{0, 1\}$ to denote the function $f(x)(y) = f(x, y)$. We say that $f(x)$ is a *row* of f (i.e., when f is seen as a matrix with rows indexed by $\mathcal{X}$ and columns indexed by $\mathcal{Y}$). Given a distribution ν over a set $\mathcal{Y}$ and a function $v : \mathcal{Y} \to \mathbb{R}$, we define the semi-norm $\|v\|_\nu = \mathbb{E}_{Y \sim \nu}[v(Y)]$. We also use $\|v\|_0$ to denote the number of non-zero entries of v. Finally, given a pseudo-metric space $(\mathcal{X}, d)$ and $x \in \mathcal{X}$, we use $B(x, \alpha)$ to denote the set of points in $\mathcal{X}$ at distance at most α from x.

3   Information Complexity and Relaxed Codes

Definition 3.1. *Consider a pseudo-metric space $(\mathcal{X}, d)$. A subset $\mathcal{C}$ is an (α, β)-code w.r.t. a distribution μ supported on $\mathcal{C}$ if for C, C' chosen independently from μ*

$$\Pr_{C, C'}(d(C, C') \leq \alpha) \leq \beta.$$

The following is the main result of this section which gives a lower bound on the information complexity of communication problems based on (α, β)-codes.

Theorem 3.1. *Consider a communication problem $f : \mathcal{S} \times \mathcal{L} \to \{0, 1\}$. Consider distributions μ (over $\mathcal{S}$) and ν (over $\mathcal{L}$) and suppose that the rows $\{f(s)\}_{s \in \mathcal{S}}$ form an (α, β)-code with respect to μ and the distance $\|.\|_\nu$. Then*

$$IC(f)_{(\mu \times \nu), \frac{\alpha}{8}}^\rightarrow \geq \frac{1}{4} \log \frac{1}{4\beta} - 1.$$

The intuition is that if the rows of the communication problem are quite distinct from each other, a low error protocol allows Bob to recover the identity of the row that Alice's input is indexing, leading to a high information cost.

To make this intuition formal, we start by developing a list-decoding variant of Fano's inequality where a predictor outputs a prediction set, which might be of independent interest; the proof is deferred to the full version of the paper.

Lemma 3.1. *Consider a finite set $\mathcal{X}$ and an arbitrary set $\mathcal{R}$, and let μ and λ be distributions over $\mathcal{X}$ and $\mathcal{R}$ respectively. Also consider a (predictor) function $g : \mathcal{X} \times \mathcal{R} \to 2^{\mathcal{X}}$ such that for some $\beta \in (0,1)$ we have $\Pr_{X \sim \mu, R \sim \lambda}(X \in g(X,R)$ and $\mu(g(X,R)) \leq \beta) \geq p$. Then $\mathrm{I}(X; g(X,R)) \geq p \log \frac{1}{\beta} - 1$.*

The next theorem connects this list-decoding version of Fano's inequality with (α, β)-codes; the mapping M next can be thought as an approximate decoder.

Theorem 3.2. *Consider a finite pseudo-metric space $(\mathcal{X}, d)$. Let $\mathcal{C} \subseteq \mathcal{X}$ be an (α, β)-code with respect to a distribution μ over $\mathcal{C}$. Consider an arbitrary space $\mathcal{R}$ with distribution λ. Consider the random variables $C \sim \mu$, $R \sim \lambda$ and a mapping $M : \mathcal{C} \times \mathcal{R} \to \mathcal{X}$ satisfying $\Pr_{C,R}(d(M(C,R), C) \geq \frac{\alpha}{2}) \leq \frac{1}{4}$. Then*

$$\mathrm{I}(C; M(C,R)) \geq \frac{1}{2} \log \frac{1}{4\beta} - 1.$$

Proof. We employ Lemma 3.1 to the space $\mathcal{C} \times \mathcal{R}$. Construct the predictor $g : \mathcal{C} \times \mathcal{R} \to 2^{\mathcal{C}}$ given by $g(c,r) = B\left(M(c,r), \frac{\alpha}{2}\right)$; notice that $g(c,r)$ only depends on $M(c,r)$. We claim that

$$\Pr_{C \sim \mu, R \sim \lambda}(C \in g(C,R) \text{ and } \mu(g(C,R)) \leq 4\beta) \geq \frac{1}{2}. \tag{1}$$

Let $\mathcal{E}$ denote the event $\{C \in g(C,R)$ and $\mu(g(C,R)) \leq 4\beta\}$, and change the second term to define the event $\mathcal{E}' = \{d(M(C,R), C) \leq \frac{\alpha}{2}$ and $\mu(B(C,\alpha)) \leq 4\beta\}$ (notice that $C \in g(C,R)$ is equivalent to $d(M(C,R), C) \leq \frac{\alpha}{2}$). We claim that $\mathcal{E}'$ implies $\mathcal{E}$: if $\mathcal{E}'$ holds then using its first part and the triangle inequality we get $B(M(C,R), \frac{\alpha}{2}) \subseteq B(C,\alpha)$, so its second part gives $\mu(g(C,R)) = \mu(B(M(C,R), \frac{\alpha}{2})) \leq \mu(B(C,\alpha)) \leq 4\beta$, proving the claim. So to prove inequality (1) it suffices to show $\Pr(\mathcal{E}') \geq \frac{1}{2}$.

Directly from the guarantees of M we have $\Pr(d(M(C,R), C) \leq \frac{\alpha}{2}) \geq \frac{3}{4}$. For $\mu(B(C,\alpha)) \leq 4\beta$, notice that for a random variable $C' \sim \mu$ independent of C we have $\Pr_{C'}(d(c,C') \leq \alpha) = \mu(B(c,\alpha))$ for all $c \in \mathcal{C}$, and since $\mathcal{C}$ is an (α, β)-code, $\beta \geq \Pr_{C,C'}(d(C,C') \leq \alpha) = \mathbb{E}_C[\mu(B(C,\alpha))]$. Then from Markov's inequality we get that $\Pr_C(\mu(B(C,\alpha)) \geq 4\beta) \leq \frac{1}{4}$. Taking a union bound, $\mathcal{E}'$ holds with probability at least $\frac{1}{2}$, thus proving inequality (1).

Then we can apply Lemma 3.1 with $p = \frac{1}{2}$ and 4β to get that $I(C; g(C,R)) \geq \frac{1}{2} \log \frac{1}{4\beta} - 1$. Since $M(C,R)$ determines $g(C,R)$, the data processing inequality implies that $\mathrm{I}(C; M(C,R)) \geq \mathrm{I}(C; g(C,R))$, thus completing the proof. $\qquad \square$

Proof of Theorem 3.1: Consider random variables $(S, L) \sim \mu \times \nu$ and a randomized one-way protocol for $f(S, L)$ with error probability (with respect to S, L and private randomness) at most $\frac{\alpha}{8}$. Let $\mathsf{A}(s, r_A)$ be the message that Alice sends on this protocol over input s and her private randomness r_A, and let $\mathsf{B}(m, \ell, r_B)$ be the output of Bob when he has input ℓ, private randomness r_B and receives message m from Alice. We want to show $\mathrm{I}(S; \mathsf{A}(S, R_A)) \geq \frac{1}{4} \log \frac{1}{4\beta} - 1$.

For that, define $M(f(s), r_A, r_B) : \mathcal{L} \to \{0, 1\}$ by setting $M(f(s), r_A, r_B)(\ell) = \mathsf{B}(\mathsf{A}(s, r_A), \ell, r_B)$ for all $s \in \mathcal{S}$ and $\ell \in \mathcal{L}$. Given the guarantees of the protocol, we have

$$\mathbb{E}_{S \sim \mu, R_A, R_B}[\|M(f(S), R_A, R_B) - f(S)\|_\nu]$$
$$= \Pr_{S \sim \mu, L \sim \nu, R_A, R_B}(M(f(S), R_A, R_B)(L) \neq f(S, L)) \leq \frac{\alpha}{8}.$$

By Markov's inequality, $\Pr_{S \sim \mu, R_A, R_B}\left(\|M(f(S), R_A, R_B) - f(S)\|_\nu \geq \frac{\alpha}{2}\right) \leq \frac{1}{4}$.

Then we can employ Theorem 3.2 with $\mathcal{C}$ set to $\{f(s)\}_{s \in \mathcal{S}}$ to obtain that $\mathrm{I}(f(S); M(f(S), R_A, R_B)) \geq \frac{1}{4} \log \frac{1}{4\beta} - 1$. But the random variable S determines the row $f(S)$ and $(\mathsf{A}(S, R_A), R_B)$ determines the vector $M(f(S), R_A, R_B)$, so by the data processing inequality we get $\mathrm{I}(S; \mathsf{A}(S, R_A), R_B) \geq \frac{1}{4} \log \frac{1}{4\beta} - 1$. Finally, since R_B is independent from S and R_A, we have $\mathrm{I}(S; \mathsf{A}(S, R_A), R_B) = \mathrm{I}(S; \mathsf{A}(S, R_A))$. This concludes the proof of the theorem. □

In the full version of the paper, we show how we can use relaxed codes to recover the lower bounds for k-sparse set disjointness of Dasgupta et al. [6] and for the indexing problem of Jayram and Woodruff [9].

4 Characterization via Packing Numbers

We now show how the lower bounds from the previous section lead to our main characterization theorem of the one-way information complexity under product distributions in terms of packing numbers. Given a pseudo-metric space $(\mathcal{X}, d)$, its α-*packing number* is the size of the largest set of points in $\mathcal{X}$ with pairwise distances at least α; we denote this by $\mathcal{P}(\mathcal{X}, d, \alpha)$. The base of the characterization is a new connection between relaxed codes and packing numbers.

Lemma 4.1. *Consider a pseudo-metric space $(\mathcal{C}, d)$ and an $\alpha \in (0, 1]$. Then $\mathcal{C}$ is an $\left(\alpha, \frac{1}{\mathcal{P}(\mathcal{C}, d, \alpha)}\right)$-code with respect to some distribution μ over $\mathcal{C}$.*

Proof. Let $\mathcal{C}' \subseteq \mathcal{C}$ be a set of size $\mathcal{P}(\mathcal{C}, d, \alpha)$ such that distinct points in $\mathcal{C}'$ have distance at least α. Let μ be the uniform distribution on $\mathcal{C}'$. Then $\Pr_{C, C' \sim \mu}(d(C, C') \leq \alpha) = \Pr_{C, C' \sim \mu}(C = C') = \frac{1}{|\mathcal{C}'|} = \frac{1}{\mathcal{P}(\mathcal{C}, d, \alpha)}$, and hence $\mathcal{C}$ is an $\left(\alpha, \frac{1}{\mathcal{P}(\mathcal{C}, d, \alpha)}\right)$-code with respect to μ. □

Theorem 4.1. *Consider a communication problem $f : \mathcal{S} \times \mathcal{L} \to \{0,1\}$ and let ν be a distribution over $\mathcal{L}$. Let $p_{\alpha,\nu}$ denote the α-packing number of the pseudo-metric space $(\{f(s)\}_{s\in\mathcal{S}}, \|.\|_\nu)$. Then for every $\alpha \in (0,1]$,*

$$\max_\mu IC(f)_{(\mu\times\nu),\frac{\alpha}{8}}^\to \geq \frac{1}{4} \log \frac{p_{\alpha,\nu}}{4} - 1 \tag{2}$$

$$\max_\mu D(f)_{(\mu\times\nu),\alpha}^\to \leq \log p_{\alpha,\nu} + 1, \tag{3}$$

where the $\max_\mu$ range over all distributions over $\mathcal{S}$. In particular, letting p_α^* denote the maximum $p_{\alpha,\nu}$ over all ν, we have for $\alpha \in (0, \frac{1}{8}]$

$$\Omega(\log p_{8\alpha}^*) \leq R(f)_\alpha^{\to,[]} \leq \log p_\alpha^* + 1. \tag{4}$$

Proof. Inequality (2) follows directly from Theorem 3.1 and Lemma 4.1.

For inequality (3), let $\mathcal{S}' \subseteq \mathcal{S}$ be a set of size $p_{\alpha,\nu}$ such that $\|f(s) - f(s')\|_\nu \geq \alpha$ for all distinct $s, s' \in \mathcal{S}'$. The maximality of $\mathcal{S}'$ implies that the balls $\{B(f(s),\alpha)\}_{s\in\mathcal{S}'}$ cover all of $\{f(s)\}_{s\in\mathcal{S}}$. Then Alice and Bob, on inputs s and ℓ respectively, can do the following: Alice uses $\lceil \log p_{\alpha,\nu} \rceil$ bits to send Bob the index of a point $\psi(s)$ in $\mathcal{S}'$ such that $\|f(s) - f(\psi(s))\|_\nu \leq \alpha$; Bob then outputs $f(\psi(s),\ell)$. For any distribution μ, the distributional error of this protocol with respect to $\mu \times \nu$ is at most α: for any $s \in \mathcal{S}$, $\Pr_{L\sim\nu}(f(\psi(s),L) \neq f(s,L)) = \|f(\psi(s)) - f(s)\|_\nu \leq \alpha$. This concludes the proof of inequality (3).

Inequality (4) follows directly by taking a maximum over ν on inequalities (2) and (3) and using the bound $R(f)_\alpha^{\to,[]} \geq IC(f)_\alpha^{\to,[]}$. $\square$

Notice that this characterization implies that Theorem 3.1 is tight up to constants (and up to constants in the error rate) given the right distributions μ and ν.

4.1 Relationship with VC Dimension

We recall the characterization of distributional complexity for *constant error rate* α in terms of VC-Dimension given by [10] and [2]. The *VC-dimension* of a subset $\mathcal{C} \subseteq \{0,1\}^n$ is the largest set of indices $I \subseteq [n]$ such that the projection onto I given by $\{(x_i)_{i\in I} : x \in \mathcal{C}\}$ equals the whole of $\{0,1\}^{|I|}$.

Theorem 4.2 ([2,10]). *Consider a communication problem $f : \mathcal{S} \times \mathcal{L} \to \{0,1\}$ and $\alpha \in (0, \frac{1}{4}]$. Then, if VC denotes the VC-dimension of the rows $\{f(s)\}_{s\in\mathcal{S}}$,*

$$(1 - H(\alpha))VC \leq R(f)_\alpha^{\to,[]} \leq O\left(VC \cdot \frac{1}{\alpha} \log \frac{1}{\alpha}\right). \tag{5}$$

Notice that, *for constant error α*, this characterizes the distributional complexity up to constant factors. Known bounds on the relationship between VC-dimension and packing numbers allow us to directly recover this characterization from Theorem 4.1. First, we need the dual of packing numbers: Given a pseudo-metric space $(\mathcal{X}, d)$, its α-*covering number* is the smallest number of balls $B(x,\alpha)$

of radius α needed to cover $\mathcal{X}$; we denote this by $\mathcal{N}(\mathcal{X}, d, \alpha)$. It is well-known that packing and covering numbers are closely related: for all $\alpha > 0$,

$$\mathcal{N}(\mathcal{X}, d, \alpha) \leq \mathcal{P}(\mathcal{X}, d, \alpha) \leq \mathcal{N}(\mathcal{X}, d, \alpha/2). \tag{6}$$

We have the following relationships between VC-dimension and packing/covering numbers (for completeness we provide a proof of the first one in the appendix).

Lemma 4.2. *Let $\mathcal{C}$ be a subset of $\{0,1\}^n$ and let VC denote its VC-dimension. Then for every $\alpha \in (0, \frac{1}{2}]$,*

$$\max_{\nu} \log \mathcal{N}(\mathcal{C}, \|.\|_{\nu}, \alpha) \geq (1 - H(\alpha))VC,$$

where the maximum is taken over all distributions on $[n]$ and $H(\alpha) = \alpha \log \frac{1}{\alpha} + (1 - \alpha) \log \frac{1}{1-\alpha}$ denotes the binary entropy.

Lemma 4.3 ([7,8]). *Let $\mathcal{C}$ be a subset of $\{0,1\}^n$ and let VC be its VC-dimension. Then for every distribution ν over $[n]$ and $\alpha \in (0, 1]$, we have*

$$\log \mathcal{P}(\mathcal{C}, \|.\|_{\nu}, \alpha) \leq VC \cdot \log \left(\frac{5}{\alpha} \log \frac{10}{\alpha} \right).$$

Using these two lemmas and inequality (6), we get that for $\alpha \in (0, \frac{1}{4}]$

$$(1 - H(\alpha)) \cdot VC \leq \max_{\nu} \log \mathcal{P}(\mathcal{C}, \|.\|_{\nu}, \alpha) \leq VC \cdot \log \left(\frac{5}{\alpha} \log \frac{10}{\alpha} \right).$$

Using these bounds on Theorem 4.1 recovers the VC-dimension characterization from Theorem 4.2; in fact, it gives the improved dependence $O(\log \frac{1}{\epsilon})$ on ϵ.

Corollary 4.1. *Consider a communication problem $f : \mathcal{S} \times \mathcal{L} \to \{0,1\}$ and $\alpha \in (0, \frac{1}{16}]$. Then, letting VC denote the VC-dimension of the rows $\{f(s)\}_{s \in \mathcal{S}}$,*

$$(1 - H(8\alpha)) \cdot \Omega(VC) \leq R(f)_{\alpha}^{\to, []} \leq O \left(VC \cdot \log \frac{1}{\alpha} \right). \tag{7}$$

The greater-than function illustrates the difference between the characterizations in terms of VC-dimension and packing numbers (full version of the paper).

5   Composition of Communication Problems and Noise Sensitivity

In this section we are interested in compositions of communication problems. More precisely, given a communication problem $f : \mathcal{X} \times \mathcal{Y} \to \{0,1\}$ and a composition function $g : \{0,1\}^k \to \{0,1\}$, we use $g \odot f$ to denote the composition $g(f(x_1, y_1), \ldots, f(x_k, y_k))$ (so it is a function mapping $(\mathcal{X} \times \mathcal{Y})^k \to \{0,1\}$). We will used relaxed codes to understand how the composed communication problem $g \odot f$ amplifies the hardness of the base problem f. We will see that the hardness amplification is governed by a generalization of the *noise sensitivity* [18] of g.

Definition 5.1 ((t, γ)-**correlation**). *Given $\gamma \in [0, 1]$, we say that two random variables Z, Z' are γ-correlated if $\Pr(Z = Z') \leq \gamma$. Given $t \in [k]$, we say that two random vectors $(Z_1, \ldots, Z_k)$ and $(Z'_1, \ldots, Z'_k)$ are (t, γ)-correlated if there is a subset $I \subseteq [k]$ of size t such that for all $i \in I$, Z_i and Z'_i are γ-correlated.*

Definition 5.2 ((t, γ)-**Noise sensitivity**). *Consider a function $g : \{0, 1\}^k \to \{0, 1\}$ and fix $t \in [k]$ and $\gamma \in [0, 1]$. Let $\mathfrak{D}$ be a family of distributions over $\{0, 1\}^k$ such that there are (t, γ)-correlated random vectors $\boldsymbol{Z}, \boldsymbol{Z}'$ with distributions in $\mathfrak{D}$. Then the (t, γ)-noise sensitivity of g with respect to $\mathfrak{D}$ is given by*

$$\mathrm{NS}^t_{\gamma, \mathfrak{D}}(g) \triangleq \min_{\boldsymbol{Z}, \boldsymbol{Z}'} \Pr(g(\boldsymbol{Z}) \neq g(\boldsymbol{Z}')),$$

where the minimum is taken over all (t, γ)-correlated random vectors $\boldsymbol{Z}, \boldsymbol{Z}'$ with distributions in $\mathfrak{D}$.

Now we try to give some intuition why noise sensitivity captures how a composition function amplifies the relaxed code of a base function. Consider a communication problem $f : \mathcal{X} \times \mathcal{Y} \to \{0, 1\}$, with a "hard" distribution $\mu \times \nu$, and a composition function $g : \{0, 1\}^k \to \{0, 1\}$. To understand the information complexity of $g \odot f$ under $(\mu \times \nu)^k$, we want to check if it forms an (α, β)-code, which informally means that for "typical" $\boldsymbol{x}, \boldsymbol{x}' \in \mathcal{X}^k$, $\Pr_{\boldsymbol{Y} \sim \nu^k}(g \odot f(\boldsymbol{x}, \boldsymbol{Y}) \neq g \odot f(\boldsymbol{x}', \boldsymbol{Y})) \geq \alpha$. Expanding the left-hand side shows that it is related to the (t, γ)-sensitivity of g, where the noise level γ is given by $\Pr_{Y \sim \nu}(f(x, Y) = f(x', Y))$, again for "typical" $x, x' \in \mathcal{X}$; this noise level is in turn related to how good a relaxed code the rows $\{f(x)\}_{x \in \mathcal{X}}$ are with respect to μ and $\|.\|_\nu$. Formally:

Theorem 5.1. *Consider a communication problem $f : \mathcal{X} \times \mathcal{Y} \to \{0, 1\}$. Let μ and ν be distributions over $\mathcal{X}$ and $\mathcal{Y}$, respectively, such that $\{f(x)\}_{x \in \mathcal{X}}$ forms an (α, β)-code with respect to μ and the distance $\|.\|_\nu$. Let $\mathfrak{D}$ be the set of distributions of the random vectors $(f(x_1, Y_1), \ldots, f(x_k, Y_k))$ with $x_1, \ldots, x_k \in \mathcal{X}$, where $Y_1, \ldots, Y_k$ are independently sampled from ν. Consider a function $g : \{0, 1\}^k \to \{0, 1\}$. Then for $w \in (0, 1 - \beta]$, the rows $\{g \odot f(\boldsymbol{x})\}_{\boldsymbol{x} \in \mathcal{X}^k}$ form an (α_w, β_w)-code with respect to μ^k and the distance $\|.\|_{\nu^k}$, where*

$$\alpha_w = \mathrm{NS}^{k(1-\beta-w)}_{1-\alpha, \mathfrak{D}}(g)$$

$$\beta_w = \left(\frac{e^w}{(1 + w/\beta)^{\beta + w}} \right)^k \leq \left(\frac{e\beta}{w} \right)^{wk}.$$

Proof. It suffices to show that for a $1 - \beta_w$ fraction of the independent random vectors $\boldsymbol{X}, \boldsymbol{X}' \sim \mu^k$, we have $\Pr_{\boldsymbol{Y} \sim \nu^k}(g \odot f(\boldsymbol{X}, \boldsymbol{Y}) \neq g \odot f(\boldsymbol{X}', \boldsymbol{Y})) \geq \mathrm{NS}^{k(1-\beta-w)}_{1-\alpha, \mathfrak{D}}(g)$.

Let $\Omega \subseteq \mathcal{X}^2$ be the set of pairs (x, x') such that $\|f(x) - f(x')\|_\nu > \alpha$, namely $\Pr_{Y \sim \nu}\big(f(x, Y) \neq f(x', Y)\big) > \alpha$. For two vectors $\boldsymbol{x}, \boldsymbol{x}'$ in $\mathcal{X}^k$, let $\#(\boldsymbol{x}, \boldsymbol{x}')$ denote the number of coordinates i such that (x_i, x'_i) belongs to Ω.

Fix any two $\boldsymbol{x}, \boldsymbol{x}'$ in $\mathcal{X}^k$. For $\boldsymbol{Y} = (Y_1, \ldots, Y_k)$ sampled from ν^k, define $Z_i = f(x_i, Y_i)$ and $Z'_i = f(x'_i, Y_i)$. Then by definition of Ω, $\boldsymbol{x}$ and $\boldsymbol{x}'$, we have that the vectors $\boldsymbol{Z} = (Z_1, \ldots, Z_k)$ and $\boldsymbol{Z}' = (Z_1', \ldots, Z_k')$ are $(\#(\boldsymbol{x}, \boldsymbol{x}'), 1 - \alpha)$-correlated with distributions in $\mathfrak{D}$. Then by the definition of (t, γ)-noise sensitivity,

$$\Pr_{\boldsymbol{Y}} \left(g \odot f(\boldsymbol{x}, \boldsymbol{Y}) \neq g \odot f(\boldsymbol{x}', \boldsymbol{Y}) \right) = \Pr_{\boldsymbol{Z}} \left(g(\boldsymbol{Z}) \neq g(\boldsymbol{Z}') \right) \geq \mathrm{NS}_{1-\alpha, \mathfrak{D}}^{\#(\boldsymbol{x}, \boldsymbol{x}')}(g).$$

To show that $\Pr\left(\#(\boldsymbol{X}, \boldsymbol{X}') \geq k(1 - \beta - w)\right)$ is at least $1 - \beta_w$, we observe the following. Since f forms an (α, β)-code, we know that $\Pr\left((\boldsymbol{X}, \boldsymbol{X}') \in \Omega\right) > 1 - \beta$, and thus $\mathbb{E}[\#(\boldsymbol{X}, \boldsymbol{X}')] \geq k(1 - \beta)$. By a multiplicative Chernoff bound (Theorem 4.1 of [14]), we have that the event $k - \#(\boldsymbol{X}, \boldsymbol{X}') > (1 + w/\beta)k\beta$ happens with probability at most $\left(\frac{e^w}{(1+w/\beta)^{\beta+w}}\right)^k = \beta_w$, and hence with probability at least $1 - \beta_w$ we have $\#(\boldsymbol{X}, \boldsymbol{X}') \geq k(1 - \beta - w)$.

To conclude the proof, we show that $\beta_w \leq (e\beta/w)^{wk}$. First, by reducing the denominator we have $\beta_w \leq \left(\frac{e}{1+w/\beta}\right)^{wk}$. But this quantity is at most $\left(\frac{e\beta}{w}\right)^{wk}$, which can be shown using concavity of the map $\beta \mapsto \frac{e}{1+w/\beta}$, and the fact that its derivative at 0 is $\frac{e}{w}$. This concludes the proof. $\qquad\square$

Together with the lower bound of Theorem 3.1 based on relaxed codes, this amplification theorem gives a powerful tool for constructing lower bounds.

5.1 Example: Stronger Direct Sum for XOR

Let $\mathsf{XOR}^k : \{0,1\}^k \rightarrow \{0,1\}$ denote the k-ary XOR function, namely it maps $(z_1, \ldots, z_k) \mapsto \sum_i z_i \mod 2$. The following lower bound on the (t, γ)-Noise sensitivity of XOR^k is proved in the full version of the paper.

Lemma 5.1. *Let* $\mathfrak{prod}$ *denote the set of all product distributions over* $\{0,1\}^k$. *Then* $\mathrm{NS}_{1-\alpha, \mathfrak{prod}}^t(\mathsf{XOR}^k) \geq \frac{1}{2} - \frac{1}{2}(1 - 2\alpha)^t$.

Theorem 5.1 and the above lemma (together with Lemma 4.1 and Theorem 3.1) give a stronger direct sum theorem for XOR (notice the error probability $\frac{1}{k}$ on the right-hand side); details are presented in the full version of the paper.

Corollary 5.1. *For any communication problem* $f : \mathcal{X} \times \mathcal{Y} \rightarrow \{0,1\}$,

$$IC(\mathsf{XOR}^k \odot f)_{\frac{1}{4}}^{\rightarrow, \square} \geq \frac{k}{8} \cdot \left(R(f)_{\frac{1}{k}}^{\rightarrow, \square} - 8 \right).$$

6 Streaming Applications

We have the following tight bounds for streaming [1].

---

[1] Matching upper bounds can be achieved by using n sketches each corresponding to a Johnson-Lindenstrauss transform of dimension $O(1/\epsilon^2 \log n/\delta)$ with arithmetic precision of $O(\log d + \log M)$ bits for closest pair and $O(\log n + \log M)$ for the largest entry in matrix product.

Approximate Closest Pair. This problem is described as follows: Alice has n vectors $\mathbf{v}^1, \mathbf{v}^2, \ldots, \mathbf{v}^n \in [\pm M]^d$, Bob has n vectors $\mathbf{u}^1, \mathbf{u}^2, \ldots, \mathbf{u}^n \in [\pm M]^d$ and a threshold value θ, and his goal is to distinguish (with prob. $1 - \delta$) the cases:

1. For all $i \in [n]$ it holds that $\|\mathbf{u}^i - \mathbf{v}^i\|_p^p \geq (1 + \epsilon)\theta$.
2. There exists i such that $\|\mathbf{u}^i - \mathbf{v}^i\|_p^p \leq (1 - \epsilon)\theta$.

Let $\ell_p(n, d, M, \epsilon, \theta)$ denote this problem.

Theorem 6.1. *Assume n is at least a sufficiently large constant and ϵ is at most a sufficiently small constant. Assume there is a constant $\gamma > 0$ such that $d^{1-\gamma} \geq \frac{1}{\epsilon^2} \log \frac{n}{\delta}$. Then $R_{\delta}^{\rightarrow}(\ell_p(n, d, M, \epsilon, \theta)) \geq \Omega\left(\frac{n}{\epsilon^2} \log \frac{n}{\delta}(\log d + \log M)\right)$ for $p \in \{1, 2\}$.*

Approximating Largest Entry in Matrix Product by Sketching. Given a matrix A, let A_i denote its i-th row and use A^j to denote its j-th column. The goal is to compute a (possibly randomized) $n \times d$ matrix S that has an estimation procedure f_θ satisfying: for every pair of matrices $A, B \in [\pm M]^{n \times n}$, with probability at least $1 - \delta$ over the randomness in the choice of S:

1. $f_\theta(AS, B) = 1$ if $(AB)_{i,j} \geq (1 + \epsilon)\theta$ for some $i, j \in [n]$.
2. $f_\theta(AS, B) = 1$ if $(AB)_{i,j} \leq \theta$ for all $i, j \in [n]$.

Theorem 6.2. *Assume n is a sufficiently large constant and ϵ is at most a sufficiently small constant. Assume there is a constant $\gamma > 0$ such that $n^{1-\gamma} \geq \frac{1}{\epsilon^2} \log \frac{n}{\delta}$. Let S be a (possibly randomized) $n \times d$ matrix that has an estimation procedure f_θ satisfying the properties above. Then the number of bits to specify AS is at least $\Omega(n\frac{1}{\epsilon^2} \log \frac{n}{\delta}(\log n + \log M))$.*

References

1. Ablayev, F.M.: Lower bounds for one-way probabilistic communication complexity and their application to space complexity. Theor. Comput. Sci. **157**(2), 139–159 (1996)
2. Bar-Yossef, Z., Jayram, T.S., Kumar, R., Sivakumar, D.: An information statistics approach to data stream and communication complexity. J. Comput. Syst. Sci. **68**(4), 702–732 (2004)
3. Beals, R., Buhrman, H., Cleve, R., Mosca, M., de Wolf, R.: Quantum lower bounds by polynomials. J. ACM **48**(4), 778–797 (2001)
4. Buhrman, H., García-Soriano, D., Matsliah, A., de Wolf, R.: The non-adaptive query complexity of testing k-parities. Chicago J. Theor. Comput. Sci. (2013)
5. Cover, T.M., Thomas, J.A.: Elements of information theory (2. ed.). Wiley (2006)
6. Dasgupta, Anirban, Kumar, Ravi, Sivakumar, D.: Sparse and lopsided set disjointness via information theory. In: Gupta, Anupam, Jansen, Klaus, Rolim, José, Servedio, Rocco (eds.) APPROX 2012 and RANDOM 2012. LNCS, vol. 7408, pp. 517–528. Springer, Heidelberg (2012)
7. Dudley, R.M.: Central limit theorems for empirical measures. The Annals of Probability **6**(6), 899–929 (1978)

8. Haussler, D.: Decision theoretic generalizations of the PAC model for neural net and other learning applications. Inform. Comput. **100**(1), 78–150 (1992)

9. Jayram, T.S., Woodruff, D.P.: Optimal bounds for johnson-lindenstrauss transforms and streaming problems with sub-constant error. In: SODA (2011)

10. Kremer, I., Nisan, N., Ron, D.: On randomized one-round communication complexity. Computational Complexity, pp. 21–49 (1999)

11. Lee, T., Shraibman, A.: Lower bounds in communication complexity. Foundations and Trends in Theoretical Computer Science **3**(4), 263–399 (2009)

12. Lee, T., Zhang, S.: Composition theorem in communication complexity. In: ICALP (2010)

13. Molinaro, M., Woodruff, D.P., Yaroslavtsev, G.: Beating the direct sum theorem in communication complexity with implications for sketching. In: SODA (2013)

14. Motwani, R., Raghavan, P.: Randomized Algorithms. Cambridge University Press, New York (1995)

15. Muthukrishnan, S.: Data streams: algorithms and applications. Found. Trends Theor. Comput. Sci. **1**(2), 117–236 (2005)

16. Newman, I., Szegedy, M.: Public vs. private coin flips in one round communication games (extended abstract). In: STOC (1996)

17. Nisan, N., Szegedy, M.: On the degree of boolean functions as real polynomials. Computational Complexity **4**, 301–313 (1994)

18. O'Donnell, R.: Analysis of Boolean Functions. Cambridge University Press (2014)

19. Papadimitriou, C.H., Sipser, M.: Communication complexity. J. Comput. Syst. Sci. **28**(2), 260–269 (1984)

20. Saglam, M., Tardos, G.: On the communication complexity of sparse set disjointness and exists-equal problems. In: FOCS (2013)

21. Sherstov, A.: The pattern matrix method. SIAM J. Comput. **40**(6), 1969–2000 (2011)

22. Woodruff, D.P.: The average-case complexity of counting distinct elements. In: ICDT (2009)

23. Yao, A.C.: Lower bounds by probabilistic arguments (extended abstract). In: FOCS (1983)

A $(2 + \epsilon)$-Approximation Algorithm for the Storage Allocation Problem

Tobias Mömke[1] and Andreas Wiese[2][✉]

[1] Saarland University, Saarbrücken, Germany
moemke@cs.uni-saarland.de
[2] Max-Planck-Institut für Informatik, Saarbrücken, Germany
awiese@mpi-inf.mpg.de

Abstract. Packing problems are a fundamental class of problems studied in combinatorial optimization. Three particularly important and well-studied questions in this domain are the Unsplittable Flow on a Path problem (UFP), the Maximum Weight Independent Set of Rectangles problem (MWISR), and the 2-dimensional geometric knapsack problem. In this paper, we study the Storage Allocation Problem (SAP) which is a natural combination of those three questions. Given is a path with edge capacities and a set of tasks that are specified by start and end vertices, demands, and profits. The goal is to select a subset of the tasks that can be drawn as non-overlapping rectangles underneath the capacity profile, the height of a rectangles corresponding to the demand of the respective task. This problem arises naturally in settings where a certain available bandwidth has to be allocated contiguously to selected requests.

While for 2D-knapsack and UFP there are polynomial time $(2 + \epsilon)$-approximation algorithms known [Jansen and Zhang, SODA 2004] [Anagnostopoulos et al., SODA 2014] the best known approximation factor for SAP is $9 + \epsilon$ [Bar-Yehuda, SPAA 2013]. In this paper, we level the understanding of SAP and the other two problems above by presenting a polynomial time $(2 + \epsilon)$-approximation algorithm for SAP. A typically difficult special case of UFP and its variations arises if all input tasks are relatively large compared to the capacity of the smallest edge they are using. For that case, we even obtain a pseudopolynomial time *exact* algorithm for SAP.

1 Introduction

Packing problems belong to the most fundamental problems in combinatorial optimization and approximation algorithms. One very prominent packing problem is the well-known KNAPSACK problem: given is a knapsack with a certain capacity and a set of items I, where each item i is specified by a demand d_i and a profit w_i. The task is to select a subset of the given items $I' \subseteq I$ such that their total demand is bounded by the capacity of the knapsack, the objective being to maximize the obtained profit $w(I') := \sum_{i \in I'} w_i$.

Research partially funded by by the Indo-German Max Planck Center for Computer Science (IMPECS) and Deutsche Forschungsgemeinschaft grant BL511/10-1.

© Springer-Verlag Berlin Heidelberg 2015
M.M. Halldórsson et al. (Eds.): ICALP 2015, Part I, LNCS 9134, pp. 973–984, 2015.
DOI: 10.1007/978-3-662-47672-7_79

There are several natural generalizations of this basic setting. One is to add a second dimension to the problem such that each item i is represented by an axis-parallel rectangle. The problem is then to select a set of items and place their corresponding rectangles non-overlappingly into a rectangular box. This yields the 2-dimensional geometric knapsack problem.

Another natural extension of knapsack is to add a temporal component such that each item i has additionally a start time $s(i)$ and an end time $t(i)$ which specify when it is active, modelling that it stays in the knapsack only during $[s_i, t_i)$. We call the input items *tasks* in this setting. Typically, one models the time horizon by a path where each edge represents a discrete time point and the values $s(i)$ and $t(i)$ represent vertices of the path. Each edge e is equipped with a capacity u_e, modelling the available knapsack capacity at this time (which can differ from edge to edge). For each edge e we denote by T_e the input tasks whose $s(i)$-$t(i)$-path $P(i)$ uses e. For a computed set T' we then require that $d(T_e \cap T') \leq u_e$ for each edge e. This yields the well-studied Unsplittable Flow on a Path problem (UFP).

In this paper, we study the Storage Allocation Problem (SAP) which is a natural combination of UFP and 2-dimensional knapsack: Given the same input as for UFP, the goal is to select a subset T' of the input set T and we want to compute a vertical position $h(i)$ for each task $i \in T'$ such that we can represent the selected tasks by non-overlapping rectangles underneath the capacity profile, the rectangle for each task $i \in T'$ is drawn at height level $h(i)$ and has a width of d_i (see Figure 1(e)). Formally, we require that (i) $h(i) + d_i \leq u_e$ for each task $i \in T'$ and each edge $e \in P(i)$ and (ii) for any two tasks $i, i' \in T'$ if $P(i) \cap P(i') \neq \emptyset$ then $[h(i), h(i) + d_i) \cap [h(i'), h(i') + d_{i'}) = \emptyset$. Observe that any solution satifying conditions (i) and (ii) also satisfies that $d(T' \cap T_e) \leq u_e$, for each edge e. SAP is particularly motivated by settings where tasks need a contiguous portion of an available resource, i.e., a consecutive portion of the computer memory or a frequency bandwidth.

Seen from a different perspective, SAP is an intermediate problem between 2-dimensional knapsack and the Maximum Weight Independent Set of Rectangles problem (MWISR) in which we are also given a set of items in the form of axis-parallel rectangles that we want to select a non-overlapping subset from, but for each rectangle its placement is predetermined. In 2-dimensional knapsack we are allowed to translate the input rectangles in both dimensions, in SAP we can translate them only up and down, and in MWISR they are completely fixed. Also, SAP is related to the Dynamic Storage Allocation problem (DSA) we are given a set of tasks as above and we want to draw their respective rectangles so that the maximum height $\max_{i \in T} h(i) + d_i$ is minimized.

A lot of progress has been made on the packing problems listed above. Specifically, we now have polynomial time $(2 + \epsilon)$-approximation algorithms for 2-dimensional knapsack [28], UFP [4], and DSA [15] and quasi-polynomial time $(1 + \epsilon)$-approximation algorithms for UFP [6], MWISR [1], and 2-dimensional knapsack [2]. The state of the art for SAP is a $(9 + \epsilon)$-approximation in polynomial time [10] which is a best-of-three algorithm. It classifies a task i to be

δ-small if $d_i \leq \delta \cdot b(i)$ where $b(i)$ denotes the *bottleneck capacity* of i which is the minimum capacity of an edge used by i. Similarly, a task i is δ-large if $d_i > \delta \cdot b(i)$. Intuitively, the value δ denote the relative size of the tasks. The mentioned algorithm provides a $(4+\epsilon)$-approximation for δ-small tasks (for some small value δ depending on ϵ), a 3-approximation for $\frac{1}{2}$-large tasks and finally a $(2 + \epsilon)$-approximation for the remaining tasks.

1.1   Our Contribution

In this paper, we level our understanding of SAP and the other packing problems mentioned above in terms of polynomial time approximation algorithms. We present a $(2 + \epsilon)$-approximation algorithm for SAP whose ratio matches the factors of the respective best known polynomial time algorithms for UFP, DSA, and 2D-knapsack. It is a best-of-two algorithm which improves all components of the so far best known $(9 + \epsilon)$-approximation algorithm [10]. First, we show that if tasks are sufficiently small, we can get a $(1 + \epsilon)$-approximation by rounding a suitably defined new LP-relaxation. While such a result is known for UFP [21, Corollary 3.4], it is not clear how to transfer it to SAP, in particular, since the optimal value of the canonical LP for UFP can differ from the best SAP-solution for a given instance by up to a factor 2, even if all input tasks are arbitrarily small. Our key technical contribution here is that we present a way to reduce the overall problem to assigning tasks to rectangular strips underneath the capacity profile. Since tasks are small, at negligible loss we we can ignore the aspect that they are supposed to be drawn as non-overlapping rectangles and we ensure only that the load of each strip is bounded by its capacity. This yields our new LP-formulation for the problem. With a suitable rounding method, we prove that for any $\epsilon > 0$ there exists a $\delta > 0$ such that for SAP-instances with only δ-small tasks we obtain a $(1 + \epsilon)$-approximation algorithm.

Then we study the converse setting where we assume that we are given a constant $\delta > 0$ and a SAP-instance with only δ-large tasks. In the related UFP problem, this is a rather difficult setting and the known PTAS for it [4] is very complex and involved. In this paper we present a very clean and elegant dynamic program for this setting for SAP. In particular, rather than computing an approximation, we solve the problem even *exactly* in pseudopolynomial time. In our DP we guess the tasks in the optimal solution step by step, ordered by their vertical positions in OPT. We prove that by using this order we need to remember only few information from the previous guesses. The pseudopolynomial running time stems from the fact that there are a pseudopolynomial number of possible vertical positions for each task and there are densely packed optimal solutions in which it is not sufficient to allow only fewer values e.g., only powers of $1 + \epsilon$. However, using a result by Knipe [32] about trimming graphs with bounded treewidth, together with an argument by Erlebach et al. [23], we show that there are $(1 + \epsilon)$-approximative solutions in which the task positions come only from a polynomial size subset. This yields a PTAS for δ-large tasks.

We round up our results by showing that any feasible solution for UFP, (i.e., any set of tasks satisfying the edge capacities) can be partitioned into $O(1)$

subsets such that each of them is a feasible solution to SAP. In a sense, this bounds the "price of contiguousness". Moreover, we can also show that if we increase the capacity of each edge by a constant factor, any UFP-solution also yields a SAP solution. This connects well with a result by Gergov [26] which proves an upper bound of 3 for the special case for uniform edge capacities, improving on several earlier results [25, 30, 31].

1.2   Related Work

For the special case of SAP that all edges have the same capacities, a local ratio 7-approximation algorithm is presented by Bar-Noy et al. [8], using an algorithm by Gergov for DSA [26]. This is improved by Bar-Yehuda to a randomized $(2 + \epsilon)$-approximation algorithm and a deterministic $\frac{2e-1}{e-1} + \epsilon \approx 2.582$-approximation for the same special case [9]. In fact, there is a close connection between unsplittable flow and dynamic storage allocation in the case of uniform edge capacities and if all tasks are sufficiently small. A result by Buchsbaum et al. [15] implies that then, if a set of tasks is feasible for UFP then a $(1 - \epsilon)$-fraction of it yields a feasible solution for SAP. However, this connection breaks if edges have different capacities. Chen et al. [22] provide an exact dynamic programming algorithm running in time $O(n(Kn)^K)$ assuming that all demands are integral multiples of $1/K$ and an $(\frac{e}{e-1} + \epsilon)$-approximation if all demands have size $O(1/K)$. As mentioned above, Bar-Yehuda, Beder, and Rawitz [10] present a $(9 + \epsilon)$-approximation algorithm for general SAP with arbitrary edge capacities which is the best known result for this case. For UFP, after a long line of work on the special cases of uniform edge capacities [8, 16, 34], the no-bottleneck-assumption [17, 21] and the general case [4, 6, 14, 20?] the best known results are now are quasi-PTASs due to Bansal et al. [6] and Batra et al. [11] and a polynomial time $(2+\epsilon)$-approximation algorithm by Anagnostopoulos et al. [4]. Recently, Batra et al. [11] presented PTASs for two special cases.

The two-dimensional geometric knapsack problem admits a $(2 + \epsilon)$-approximation algorithm due to Jansen and Zhang [28]. PTASs are known if the size of the knapsack can be increased by a factor $(1 + \epsilon)$ in both dimensions [24] or even only in one of them [27] while the compared optimum has to use the original knapsack. Also, there is a PTAS if the profit of each item equals its area [5]. For MWISR, there are many polynomial time $O(\log n)$-approximations algorithms known [3, 12, 29, 33], and the best known result is a $O(\log n / \log \log n)$-approximation by Chan and Har-Peled [19]. For the unweighted case, there is also a $O(\log \log n)$-approximation by Chalermsook and Chuzhoy [18]. Recently, a quasi-PTAS (for the weighted case) was found [1]. As mentioned above, a result by Buchsbaum et al. [15] for DSA states that if all tasks are sufficiently small then they can be drawn within a height of at most $(1 + \epsilon) \cdot L$ where L denotes the maximum total demand of tasks crossing any edge. Combined with a DP for the other tasks, this yields a $(2+\epsilon)$-approximation algorithm. For bounding the needed height as a function of L the best known bound is from Gergov [26] who shows an upper bound of $3 \cdot L$, improving on previous results [25, 30, 31].

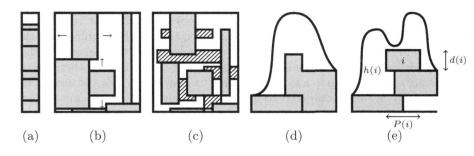

(a)   (b)   (c)   (d)   (e)

Fig. 1. (a) Knapsack problem, (b) two-dimensional knapsack problem, (c) independent set of rectangles, (d) unsplittable flow on a path (UFP), (e) storage allocation problem (SAP)

2   Approximating Small Tasks up to a Factor $1 + \epsilon$

In this section, we prove that for any $\epsilon > 0$ there is a $\delta > 0$ such that for δ-small tasks we can construct a $(1 + \epsilon)$-approximation algorithm. First, we show this result for the special case that the edge capacities are in a constant range. Then, we show how to reduce the general case to this special case.

2.1   Edge Capacities in Constant Range

Let $\epsilon > 0$. Assume that the edge capacities lie in a constant range, i. e., assume that there is a constant U such that $\max_e u_e \leq U \cdot \min_e u_e$. Assume for simplicity that $U \in \mathbb{N}$ and $1/\epsilon \in \mathbb{N}$. We draw a set of strips in the area underneath the capacity profile. A strip is specified by a tuple (k, v_ℓ, v_r) which intuitively represents the rectangle $[v_\ell, v_r] \times \{k \cdot \epsilon \cdot \min_e u_e, (k + 1) \cdot \epsilon \cdot \min_e u_e\}$ where the vertices of the graph are interpreted as integers. Let $\mathcal{S}$ denote the set of all maximally long strips whose respective rectangles fit underneath the capacity profile. Formally, a strip (k, v_ℓ, v_r) is contained in $\mathcal{S}$ if each edge between v_ℓ and v_r has a capacity of at least $(k + 1) \cdot \epsilon \cdot \min_e u_e$ and the edges on the left of v_ℓ and on the right of v_r have a capacity of less than $(k + 1) \cdot \epsilon \cdot \min_e u_e$ or do not exist because v_ℓ/v_r are the left/right-most vertices of the path. For a strip $S = (k, v_\ell, v_r)$ denote by $P(S)$ the set of edges between v_ℓ and v_r.

Instead of aiming directly at selecting a set of tasks T' and finding a non-overlapping drawing of them, we compute a set $T' \subseteq T$ and an assignment $f : T' \to \mathcal{S}$ of them to the strips. We require that for each strip $S \in \mathcal{S}$ that (i) each task $i \in f^{-1}(S)$ fits into S, meaning that $P(i) \subseteq P(S)$, and (ii) the total demand of the tasks in each strip S does not exceed the capacity of S on any edge $e \in P(S)$, i. e., $d(f^{-1}(S) \cap T_e) \leq \epsilon \cdot \min_e u_e$ for each edge $e \in P(S)$.

We call a pair (T', f) satisfying the above a *strip assignment*. It is not true that for any feasible solution (T', h) we can find a strip assignment (T', f) with the same set of tasks T', already because the total capacity of the strips using some edge e might be smaller than u_e. The converse statement is also false, since the second property above is only a relaxation of the requirement that tasks should be drawn as non-overlapping rectangles. Nevertheless, we can show that if tasks are very small compared to the capacity of each strip, then the two notions are equivalent up to a factor $1+\epsilon$. Key to show this is that the unavailable edge capacity is small compared to the total capacity and we assume all tasks to be very small. Also, based on a result by Buchsbaum et al. [15], Bar-Yehuda et al. [9] showed that if the two properties are true for some strip S, then a $(1 - \epsilon)$-fraction of the tasks in $f^{-1}(S)$ can be in fact drawn as non-overlapping rectangles. Using this intuition, we can prove the following lemma.

Lemma 1. *For any $\epsilon > 0$ there is a $\delta_1 > 0$ with the following property. Assume we are given an instance in which every task is δ_1/U-small. Then for any feasible solution (T', h) there is a strip assignment (T'', f') with $w(T'') \geq (1 - O(\epsilon))w(T')$. Conversely, for any strip assignment (T'', f') there is a feasible solution (T', h) with $w(T') \geq (1 - O(\epsilon))w(T'')$.*

Knowing that it is sufficient to compute a good strip assignment, we present now an LP-rounding algorithm for the latter goal. We formulate the problem as an integer program whose LP-relaxation (STRIP-LP) is given below.

$$\max \sum_{i,S} w_i \cdot x_{i,S}$$
$$\text{s.t.} \quad \sum_{i \in T_e} x_{i,S} \cdot d_i \leq \epsilon \cdot \min_e u_e \quad \forall S \in \mathcal{S}, \qquad (2.1)$$
$$\forall e \in P(S)$$
$$\sum_S x_{i,S} \leq 1 \qquad \forall i \in T$$
$$x_{i,S} \geq 0 \qquad \forall i \in T \; \forall S \in \mathcal{S}$$
$$s.t. \, P(i) \subseteq P(S)$$

We compute the optimal feasible solution to the above LP. By losing only a factor $(1 + \epsilon)$, we round it to a strip assignment via randomized rounding with alteration as introduced by Calinescu et al. [16]. Important for this to work is that the demand of each input task is sufficiently small compared to the capacity of each strip. In the rounding, we first sample a preliminary integral solution y such that $\Pr[y_{i,S} = 1] = (1 - \epsilon)x_{i,S}$ and $\Pr[y_{i,S} = 1 \wedge y_{i,S'} = 1] = 0$ for any task i and for any two strips $S, S' \in \mathcal{S}$. Such a distribution can easily be obtained via dependant rounding similar to Bertsimas et al. [13]. Then, intuitively, in an alteration phase for each strip, we consider the tasks in the order of their start vertices and drop a task if it causes a capacity constraint (2.1) to be violated. We can show that the probability that a task is dropped in this alteration phase is bounded by $O(\epsilon)$. In contrast to the method of Calinescu et al. [16] we work with dependent rounding. However, for any pair of tasks the outcomes of the random experiment are still independent and thus the argumentation from [16] still works.

Lemma 2. *For any $\epsilon > 0$ there is a $\delta_2 > 0$ such that given any instance with only δ_2/U-small tasks and a solution x^* to (STRIP-LP), there is a polynomial*

time algorithm computing a strip assignment (T', f) with $w(T') \geq (1-\epsilon) \sum_{i,S} w_i \cdot$
$x^*_{i,S}$.

Together with Lemma 1, we thus obtain a $(1 + O(\epsilon))$-approximation for instances with only δ/U-small tasks where $\delta := \min\{\delta_1, \delta_2\}$. Next, we give a reduction of the general case to this special case.

2.2 Arbitrary Edge Capacities

The key idea is to use some shifting arguments to remove tasks with small total cost from the optimal solution and move some other tasks up into the resulting empty space. As a result, afterwards each task i is drawn at a position $h(i) \in \Omega_\epsilon(b(i))$, i.e., not too far below its bottleneck edge. As a result, we can split the problem into independent subproblems, each having a bounded range of edge capacities.

Lemma 3. *Let $\epsilon > 0$ such that $1/\epsilon$ is an integer. There is a $\delta > 0$ such that for any instance I with only δ-small tasks there is an integer value ℓ with $0 \leq \ell < 1/\epsilon$ and a solution $(\bar{T}, \bar{h})$ to I where*

1. *$w(\bar{T}) \geq (1 - 2\epsilon)\mathrm{OPT}(I)$,*
2. *for each task $i \in \bar{T}$ with $b(i) \geq 2^{\ell+1+r/\epsilon}$ it holds that $\bar{h}(i) \geq 2^{\ell+r/\epsilon}$, and*
3. *for each task $i \in \bar{T}$ with $b(i) < 2^{\ell+1+r/\epsilon}$ it holds that $\bar{h}(i) + d_i \leq 2^{\ell+r/\epsilon}$*

for any $r \in \mathbb{N}_0$.

Proof sketch. Given the optimal solution (T^*, h^*) of the δ-small tasks, we assign the tasks into groups according to the position at which they are drawn. Denote by $T^*_k \subseteq T^*$ all tasks from T^* whose rectangles have non-empty intersection with the horizontal strip $[0, |V|] \times [2^k, 2^{k+1})$.
Formally, we define $T^*_k := \{i \in T^* | [h^*(i), h^*(i) + d_i) \cap [2^k, 2^{k+1}) \neq \emptyset\}$. Note that a task might appear in several of these sets. Let $\bar{k}$ denote the largest index k such that $T^*_k \neq \emptyset$ and consider the sets $T^*_{\bar{k}-1/\epsilon+1}, T^*_{\bar{k}-1/\epsilon+2}, ..., T^*_{\bar{k}}$. If δ is sufficiently small then each task appears in at most two of these groups. We select one set $T^*_{k'} \in \{T^*_{\bar{k}-1/\epsilon}, T^*_{\bar{k}-1/\epsilon+1}, ..., T^*_{\bar{k}}\}$ uniformly at random and remove all its tasks. In expectation we lose at most an 2ϵ-fraction of $w(\bigcup_{k=\bar{k}-1/\epsilon+1}^{\bar{k}} T^*_k)$.
Then, we take all tasks $i \in T^*$ with $b(i) \geq 2^{k'+1}$ and $h(i) + d_i \leq 2^{k'}$ and move them up by $2^{k'}$ units, i.e., we define $\bar{h}^*(i) := h^*(i) + 2^{k'}$ for them. As a result, we obtain the property that for each task $i \in T^* \setminus T^*_{k'}$ with $b(i) \geq 2^{k'+1}$ it holds that $\bar{h}^*(i) \geq 2^{k'}$ and for each task $i' \in T^* \setminus T^*_{k'}$ with $b(i') < 2^{k'+1}$ it holds that $\bar{h}^*(i) + d_i < 2^{k'}$. Iterating the above argument over multiple levels then completes the proof. $\square$

Observe that Lemma 3 decouples the instance into separate subinstances. For each $r \in \mathbb{N}_0$ we have one subinstance consisting of the tasks $T^r := \{i \in T : 2^{\ell+1+r/\epsilon} \leq b(i) < 2^{\ell+1+(r+1)/\epsilon}\}$ for which we are looking for a solution with $2^{\ell+r/\epsilon} \leq \bar{h}(i)$ and $\bar{h}(i) + d_i \leq 2^{\ell+(r+1)/\epsilon}$ for each selected task $i \in T^r$. Thus, we can treat each group T^r independently as an instance where the edge capacities are in a range of $[2^{\ell+r/\epsilon}, 2^{\ell+1+(r+1)/\epsilon} - 2^{\ell+r/\epsilon})$.

We define a constant $\delta' \in O_\epsilon(1)$ such that the algorithm from the previous section gives us a $(1 + \epsilon)$-approximation for the δ'-small tasks in each group T^r. This yields a $(1 + \epsilon)$-approximation algorithm for instances with only $\min\{\delta, \delta'\}$-small tasks where δ is the constant due to Lemma 3.

Theorem 1. *For each $\epsilon > 0$ there is a $\delta > 0$ such that there is a $(1 + \epsilon)$-approximation algorithm for the storage allocation problem if the input consists of δ-small tasks only.*

3   Large Tasks

In this section we present a pseudo-polynomial time exact algorithm for the storage allocation problem for δ-large tasks, for any $\delta > 0$. Subsequently, we show how to turn it into a polynomial time $(1 + \epsilon)$-approximation algorithm. Together with Theorem 1 this yields a polynomial time $(2 + \epsilon)$-approximation algorithm for SAP.

Since all input data are integers we can assume w.l.o.g. that all position values $h(i)$ in the optimal solution are integers: for any optimal solution without this property we can apply "gravity", i.e., decrease the vertical position of all tasks as much as we can. In the resulting solution each arising position of a task is either zero or the sum of the demands of some other tasks and thus an integer.

Our algorithm is a dynamic program. Denote by (OPT, h^*) the optimal solution. In the first step, we guess the task $i_0 \in OPT$ with smallest position $h^*(i_0)$ and all tasks from OPT using its bottleneck edge $e(i_0)$. Denote them by OPT_0. Since all tasks are δ-large, there can be only $1/\delta$ of them. More precisely, we guess these tasks as well as their positions according to h^*. The whole problem splits then into two disjoint subproblems given by the subpath on the left of $e(i_0)$ and the subpath on the right of $e(i_0)$. We recurse on both sides. Consider the left side and let $OPT_L \subseteq OPT$ denote the tasks from OPT whose path is completely contained in the subpath on the left of $e(i_0)$. Note that $OPT_L \cap OPT_0 = \emptyset$. We guess the task $i_1 \in OPT_L$ with smallest position $h^*(i_1)$. Naively, one would like to guess all tasks using $e(i_1)$ and then recurse again. The problem is that $e(i_1)$ might be used by all up to $1/\delta$ tasks in OPT_0 and another $1/\delta$ tasks from OPT_L. Thus, in each recursive step the number of tasks to be remembered would increase by $1/\delta$ while the recursion depth could be even linear. Thus, we could not bound the number of DP-cells by a polynomial. Instead, we first show that the number of tasks $i \in OPT$ (not only OPT_L!) using $e(i_1)$ with $h(i) \geq h^*(i_1)$ is bounded by $1/\delta^2$ as the following lemma implies.

Lemma 4. *Consider any solution (T', h') and a task $i \in T'$. There are at most $1/\delta^2$ tasks $i' \in T'$ such that $e(i) \in P(i')$ and $h(i') \geq h(i)$.*

Proof. For any task i' with $h(i') \geq h(i)$ and $e(i) \in P(i')$, we have that $d_i \leq h(i') \leq b(i') - d_{i'}$. Thus, using that i and i' are δ-large, $d_{i'}/\delta \geq b(i') \geq \delta \cdot b(i) + d_{i'}$ and thus $d_{i'} \geq \delta^2 \cdot b(i)/(1 - \delta) > \delta^2 \cdot b(i)$. □

When recursing on the subpath on the left of $e(i_1)$, we specify the subproblem by its subpath, by the at most $1/\delta^2$ tasks $i \in OPT$ using $e(i_1)$ with $h(i) \geq h^*(i_1)$, and the information that each task in the desired solution to this subproblem has to have a height of at least $h^*(i_1)$.

When continuing with this recursion, each arising subproblem can be characterized by two edges e_L, e_R, by at most $1/\delta^2$ tasks T_L using e_L together with a placement $h(i)$ for each task $i \in T_L$, at most $1/\delta^2$ tasks T_R using e_R together with a placement $h(i)$ for each task $i \in T_R$, and an integer $h_{\min}$. For each such combination we introduce a DP-cell $(e_L, e_R, T_L, T_R, h, h_{\min})$. Formally, it models the following subproblem: assume that we committed to selecting tasks T_L and T_R and assigning heights to them as given by the function h. Now we ask for the maximum profit we can obtain by selecting additional tasks whose paths are contained in the path strictly between e_L and e_R (so excluding e_L and e_R) and assigning heights to them such that $h(i) \geq h_{\min}$ for each selected task i.

For a given DP-cell $C = (e_L, e_R, T_L, T_R, h, h_{\min})$ we denote by (OPT_C, h_C^*) its optimal solution. Observe that $OPT_C \cap T_L = \emptyset = OPT_C \cap T_R$. Let $\hat{i} \in OPT_C$ be the task in OPT_C with minimum height $h_C^*(\hat{i})$. To compute (OPT_C, h_C^*) we guess $\hat{i} \in OPT_C$ and $h_C^*(\hat{i})$. According to Lemma 4 there can be at most $1/\delta^2$ tasks $i \in OPT_C \cup T_L \cup T_R$ using $e(\hat{i})$ such that $h_C^*(i) \geq h_C^*(\hat{i})$. We also guess all those tasks (the tasks from T_L and T_R among them are of course already given) together with their respective heights according to h_C^*. Denote them by $\bar{T}$. We can then show that OPT_C consists of the tasks in $\bar{T}$ together with the tasks in the optimal solutions to the DP-cells $C' := (e_L, e(\hat{i}), T_L, \bar{T}, h', h_C^*(\hat{i}))$ and $C'' := (e(\hat{i}), e_R, \bar{T}, T_R, h'', h_C^*(\hat{i}))$ where the assignments h' and h'' are obtained by inheriting from h the values for the tasks in T_L and T_R, respectively, and taking the guessed values for the tasks in $\bar{T}$. Conversely, we can easily show that we obtain a feasible solution to the original cell C if we combine two arbitrary feasible solutions for C' and C'', respectively, with $\bar{T}$. This proves the correctness of our DP. Since the total number of DP-cells is bounded by $(n \cdot \max_i d_i)^{O(1/\delta^2)}$ we obtain an exact pseudopolynomial time algorithm.

Theorem 2. *Let $\delta > 0$. There is an exact algorithm for instances of SAP with only δ-large tasks whose running time is $(n \cdot \max_i d_i)^{O(1/\delta^2)}$ where n denotes the number of tasks.*

In order to still obtain a PTAS, our strategy is to bound the number of candidate values for the height $h(i)$ of each task i. We will show that there is a set of such candidates of polynomial size such that there is a $(1+\epsilon)$-approximative solution in which each selected task is assigned a height from this set. By allowing only these heights in the above DP computation we then obtain a PTAS.

Our first step is to introduce a polynomial number of heights which we call *anchors lines*. Those are the capacities of the input edges and all powers of $1 + \delta$ between 1 and $\max_e u_e$. Denote by H_0 this set of values. W.l.o.g. from now on we restrictly ourselves to solutions in which for each task the height of its top edge equals the height of an anchor line or its top edge touches the bottom edge of some other task. We call such solutions *top-aligned* solutions. In a given solution, we say that a task is in level 1 if the height of its top edge equals an anchor line. Recursively, a task i is in level $\ell + 1$ if its top edge touches the bottom edge of a task in level ℓ and i is not in level any level $\ell' < \ell$. Our goal is to show that there is a $(1 + \epsilon)$-approximative solution in which each task has a level of at most $c(\epsilon)$ for some constant $c(\epsilon)$ that holds universally for any input instance. Observe that for the heights of the tasks in level ℓ there are only $|H_0| \cdot n^\ell$ possible values that are obtained by recursively defining $H_{k+1} := H_k \cup \{h - d_i | h \in H_k, i \in T\}$ for each k.

To this end, consider an optimal solution (T^*, h^*). We construct the following directed graph $D(T^*, h^*)$. For each task $i \in T^*$ we introduce a vertex in $D(T^*, h^*)$. There is an edge from the vertex for i to the vertex for i' if and only if the following three conditions are satisfied: $P(i) \cap P(i') \neq \emptyset$; $h^*(i) + d_i \leq h^*(i')$; there is no anchor line strictly between $h^*(i) + d_i$ and $h^*(i')$. Using that the tasks are δ-large, we can show that each tasks rectangle is crossed by some anchor line which implies that $D(T^*, h^*)$ is planar. Moreover, the second condition implies that $D(T^*, h^*)$ is acyclic.

Proposition 1. *If the length of the longest chain in $D(T^*, h^*)$ is bounded by some value ℓ, then $h^*(i) \in H_\ell$ for each $i \in T^*$.*

Unfortunately, the length of the longest chain in $D(T^*, h^*)$ might be $\Omega(n)$, e.g., when all tasks are tightly stacked on top of each other. To construct a solution where the latter is bounded, we apply the following theorem to $D(T^*, h^*)$. It can be proven by combining a result from Knipe [32] on trimming weighted graphs with bounded treewidth with the argumentation used in Corollary 2.3 in [23] (see also the discussion in Section 4 in the latter paper).

Theorem 3 ([23, 32]). *Let $\epsilon > 0$. There exists a constant $c(\epsilon) \in \mathbb{N}$ such that for any planar graph $G = (V, E)$ with vertex weights given by a function $w : V \to \mathbb{R}$ there is a set of vertices $V' \subseteq V$ such that $w(V') \geq (1 - \epsilon)w(V)$ and the length of the longest simple path in $G[V']$ is bounded from above by $c(\epsilon)$.*

The above theorem yields a subset of the vertices in $D(T^*, h^*)$ and thus a set of tasks $\bar{T}^* \subseteq T^*$ with $w(\bar{T}^*) \geq (1 - \epsilon)w(T^*)$. Starting from their heights according to the function h^* we construct a top-aligned solution $(T^* \setminus T', h')$ by pushing up each task until the height of its top edge either equals the height of an anchor line or the height of the bottom edge of some other task. Since the length of the longest chain in $D(T^* \setminus T', h')$ is bounded by $c(\epsilon)$ the same upper bound holds for the maximum level of a task. This shows that there is a $(1 - \epsilon)$-approximative solution $(T^* \setminus T', h')$ in which $h'(i) \in H_{c(\epsilon)}$ for each $i \in T^* \setminus T'$. By restricting our DP from the previous section to use only heights in $H_{c(\epsilon)}$ for the tasks we obtain our theorem below.

Theorem 4. *Let $\epsilon > 0$ and $\delta > 0$. There is a polynomial time $(1 + \epsilon)$-approximation algorithm for instances of SAP with only δ-large tasks.*

In order to obtain our overall $(2 + \epsilon)$-approximation algorithm, for any given $\epsilon > 0$ we first choose $\delta > 0$ according to Theorem 1. Then we compute a $(1 + \epsilon)$-approximations for the δ-small and the δ-large tasks using Theorems 1 and 4. Selecting the best of these two solutions yields a $(2 + \epsilon)$-approximation overall.

Theorem 5. *Let $\epsilon > 0$. There is a polynomial time $(2 + \epsilon)$-approximation algorithm for SAP.*

Acknowledgments. We would like to thank Naveen Garg, Amit Kumar, and Jatin Batra for helpful discussions.

References

1. Adamaszek, A., Wiese, A.: Approximation schemes for maximum weight independent set of rectangles. In: 2013 IEEE 54th Annual Symposium on Foundations of Computer Science (FOCS), pp. 400–409. IEEE (2013)
2. Adamaszek, A., Wiese, A.: A quasi-PTAS for the two-dimensional geometric knapsack problem. In: Proceedings of the 26th Annual ACM-SIAM Symposium on Discrete Algorithms, SODA 2015, pp. 1491–1505 (2015)
3. Agarwal, P.K., van Kreveld, M., Suri, S.: Label placement by maximum independent set in rectangles. Computational Geometry **11**, 209–218 (1998)
4. Anagnostopoulos, A., Grandoni, F., Leonardi, S., Wiese, A.: A mazing 2+ϵ approximation for unsplittable flow on a path. In: Proceedings of the 25th Annual ACM-SIAM Symposium on Discrete Algorithms (SODA 2014) (2014)
5. Bansal, N., Caprara, A., Jansen, K., Prädel, L., Sviridenko, M.: A structural lemma in 2-dimensional packing, and its implications on approximability. In: Dong, Y., Du, D.-Z., Ibarra, O. (eds.) ISAAC 2009. LNCS, vol. 5878, pp. 77–86. Springer, Heidelberg (2009)
6. Bansal, N., Chakrabarti, A., Epstein, A., Schieber, B.: A quasi-PTAS for unsplittable flow on line graphs. In: STOC, pp. 721–729. ACM (2006)
7. Bansal, N., Friggstad, Z., Khandekar, R., Salavatipour, R.: A logarithmic approximation for unsplittable flow on line graphs. In: SODA, pp. 702–709 (2009)
8. Bar-Noy, A., Bar-Yehuda, R., Freund, A., Naor, J., Schieber, B.: A unified approach to approximating resource allocation and scheduling. Journal of the ACM (JACM) **48**(5), 1069–1090 (2001)
9. Bar-Yehuda, R., Beder, M., Cohen, Y., Rawitz, D.: Resource allocation in bounded degree trees. Algorithmica **54**(1), 89–106 (2009)
10. Bar-Yehuda, R., Beder, M., Rawitz, D.: A constant factor approximation algorithm for the storage allocation problem. In: Proceedings of the 25th ACM symposium on Parallelism in Algorithms and Architectures, pp. 204–213. ACM (2013)
11. Batra, J., Garg, N., Kumar, A., Mömke, T., Wiese, A.: New approximation schemes for unsplittable flow on a path. In: SODA, pp. 47–58 (2015)
12. Berman, P., DasGupta, B., Muthukrishnan, S., Ramaswami, S.: Improved approximation algorithms for rectangle tiling and packing. In: Proceedings of the Twelfth Annual ACM-SIAM Symposium on Discrete Algorithms, pp. 427–436. Society for Industrial and Applied Mathematics (2001)
13. Bertsimas, D., Teo, C.-P., Vohra, R.: On dependent randomized rounding algorithms. Oper. Res. Lett. **24**(3), 105–114 (1999)
14. Bonsma, P., Schulz, J., Wiese, A.: A constant-factor approximation algorithm for unsplittable flow on paths. SIAM Journal on Computing **43**, 767–799 (2014)
15. Buchsbaum, A.L., Karloff, H., Kenyon, C., Reingold, N., Thorup, M.: Opt versus load in dynamic storage allocation. SIAM Journal on Computing **33**(3), 632–646 (2004)

16. Călinescu, G., Chakrabarti, A., Karloff, H.J., Rabani, Y.: An improved approximation algorithm for resource allocation. ACM Transactions on Algorithms 7, 48:1–48:7 (2011)
17. Chakrabarti, A., Chekuri, C., Gupta, A., Kumar, A.: Approximation algorithms for the unsplittable flow problem. Algorithmica 47, 53–78 (2007)
18. Chalermsook, P., Chuzhoy, J.: Maximum independent set of rectangles. In: Proceedings of the 20th Annual ACM-SIAM Symposium on Discrete Algorithms (SODA 2009), pp. 892–901. SIAM (2009)
19. Chan, T.M., Har-Peled, S.: Approximation algorithms for maximum independent set of pseudo-disks. Discrete & Computational Geometry 48(2), 373–392 (2012)
20. Chekuri, C., Ene, A., Korula, N.: Unsplittable flow in paths and trees and column-restricted packing integer programs. In: APPROX-RANDOM, pp. 42–55 (2009)
21. Chekuri, C., Mydlarz, M., Shepherd, F.: Multicommodity demand flow in a tree and packing integer programs. ACM Transactions on Algorithms 3, (2007)
22. Chen, B., Hassin, R., Tzur, M.: Allocation of bandwidth and storage. IIE Transactions 34(5), 501–507 (2002)
23. Erlebach, T., Hagerup, T., Jansen, K., Minzlaff, M., Wolff, A.: Trimming of graphs, with application to point labeling. Theory of Computing Systems 47(3), 613–636 (2010)
24. Fishkin, A.V., Gerber, O., Jansen, K., Solis-Oba, R.: Packing weighted rectangles into a square. In: Jedrzejowicz, J., Szepietowski, A. (eds.) MFCS 2005. LNCS, vol. 3618, pp. 352–363. Springer, Heidelberg (2005)
25. Gergov, J.: Approximation algorithms for dynamic storage allocation. In: Díaz, J. (ed.) ESA 1996. LNCS, vol. 1136, pp. 52–61. Springer, Heidelberg (1996)
26. Gergov, J.: Algorithms for compile-time memory optimization. In: Proceedings of the Tenth Annual ACM-SIAM Symposium on Discrete Algorithms, pp. 907–908. Society for Industrial and Applied Mathematics (1999)
27. Jansen, K., Solis-Oba, R.: New approximability results for 2-dimensional packing problems. In: Kučera, L., Kučera, A. (eds.) MFCS 2007. LNCS, vol. 4708, pp. 103–114. Springer, Heidelberg (2007)
28. Jansen, K., Zhang, G.: On rectangle packing: maximizing benefits. In: Proceedings of the Fifteenth Annual ACM-SIAM Symposium on Discrete Algorithms, pp. 204–213. Society for Industrial and Applied Mathematics (2004)
29. Khanna, S., Muthukrishnan, S., Paterson, M.: On approximating rectangle tiling and packing. In: Proceedings of the 9th Annual ACM-SIAM Symposium on Discrete Algorithms (SODA 1998), pp. 384–393. SIAM (1998)
30. Kierstead, H.A.: The linearity of first-fit coloring of interval graphs. SIAM Journal on Discrete Mathematics 1(4), 526–530 (1988)
31. Kierstead, H.A.: A polynomial time approximation algorithm for dynamic storage allocation. Discrete Mathematics 88(2), 231–237 (1991)
32. Knipe, D.: Trimming weighted graphs of bounded treewidth. Discrete Applied Mathematics 160(6), 902–912 (2012)
33. Nielsen, F.: Fast stabbing of boxes in high dimensions. Theor. Comp. Sc. 246, 53–72 (2000)
34. Phillips, C.A., Uma, R.N., Wein, J.: Off-line admission control for general scheduling problems. In: Proceedings of the 11th Annual ACM-SIAM Symposium on Discrete Algorithms (SODA 2000), pp. 879–888. ACM (2000)

Shortest Reconfiguration Paths in the Solution Space of Boolean Formulas

Amer E. Mouawad[1], Naomi Nishimura[1], Vinayak Pathak[1(✉)], and Venkatesh Raman[2]

[1] David R. Cheriton School of Computer Science, University of Waterloo, Waterloo, Ontario, Canada
{aabdomou,nishi,vpathak}@uwaterloo.ca
[2] The Institute of Mathematical Sciences, Chennai, India
vraman@imsc.res.in

Abstract. Given a Boolean formula and a satisfying assignment, a flip is an operation that changes the value of a variable in the assignment so that the resulting assignment remains satisfying. We study the problem of computing the shortest sequence of flips (if one exists) that transforms a given satisfying assignment s to another satisfying assignment t of the Boolean formula. Earlier work characterized the complexity of deciding the existence of a sequence of flips between two given satisfying assignments using Schaefer's framework for classification of Boolean formulas. We build on it to provide a trichotomy for the complexity of finding the shortest sequence of flips and show that it is either in P, NP-complete, or PSPACE-complete. Our result adds to the small set of complexity results known for shortest reconfiguration sequence problems by providing an example where the shortest sequence can be found in polynomial time even though the path flips variables that have the same value in both s and t. This is in contrast to all reconfiguration problems studied so far, where polynomial time algorithms for computing the shortest path were known only for cases where the path modified the symmetric difference only. Our proof uses Birkhoff's representation theorem on a set system that we show to be a distributive lattice. The technique is insightful and can perhaps be used for other reconfiguration problems as well.

1 Introduction

1.1 Background and Motivation

Reconfiguration problems are motivated by practical situations where one wants to move from one solution of an optimization problem to another while maintaining feasibility in between [12,14,17]. Each step of the move is dictated by a *reconfiguration step*, which specifies how one solution can be transformed into

A.E. Mouawad and N. Nishimura—Research supported by the Natural Science and Engineering Research Council of Canada.

© Springer-Verlag Berlin Heidelberg 2015
M.M. Halldórsson et al. (Eds.): ICALP 2015, Part I, LNCS 9134, pp. 985–996, 2015.
DOI: 10.1007/978-3-662-47672-7_80

another (for example, in case of a graph problem, by adding or deleting a vertex/edge). Hence reconfiguration problems can be stated concisely in terms of a graph—the *reconfiguration graph*—that has a node for each feasible solution and an undirected edge between two solutions if one can be formed from the other by a single reconfiguration step. Given the motivation mentioned above, reconfiguration problems typically study the complexity of finding a path between two nodes in the reconfiguration graph [4,6,10,12,13]. Most reconfiguration versions of NP-complete decision problems are PSPACE-complete [12] (e.g. maximum independent set) and hence motivated by this, there exists recent work addressing the problems under the framework of parameterized complexity [16,17].

For the problem of satisfiability of Boolean formulas, one defines a reconfiguration step to be a *flip* operation, that changes the value of a variable in a satisfying assignment such that the resulting assignment is also satisfying. Thus in the reconfiguration graph of satisfiability [11], there is a node for each satisfying assignment and an edge whenever the *Hamming distance* between the two assignments, i.e. the number of variables for which the two assignments differ in value, is exactly one. In one of the earliest works on reconfiguration, Gopalan et al. [11] and Schwerdtfeger [19], using Schaefer's [18] framework to classify Boolean formulas, characterized the complexity of determining whether there exists a path between s and t (the st-connectivity problem) in the reconfiguration graph. They define a class of formulas called *tight* and show that st-connectivity is in P for tight formulas and PSPACE-complete otherwise.

1.2   Our Results and Related Work

We study the complexity of computing the *shortest* flip sequence between two satisfying assignments. Since st-connectivity is PSPACE-complete for non-tight formulas, finding the shortest reconfiguration sequence is also PSPACE-complete. We show that the class of tight formulas can be further subdivided into *navigable* formulas, where the shortest reconfiguration sequence can be found in polynomial time, and tight but non-navigable formulas, where it is NP-complete.

Not many results are known for computing the shortest reconfiguration path except for the cases where the algorithm for st-connectivity returns the shortest path itself [11,12,16]. Moreover, the only polynomial-time algorithms known for finding the shortest reconfiguration path have the property that they make no changes to parts of the solution common to s and t. For trees and cactus graphs, the shortest path between maximum independent sets s and t never removes vertices in $s \cap t$ [16]. In the sequence of flips for 2CNF formulas (the only class for which a polynomial-time algorithm for shortest reconfiguration path of satisfiability was previously known), the only variables flipped are those whose values are different in s and t [11].

The problem of computing the shortest reconfiguration sequence of triangulations of a convex polygon is an example where this complexity has been open for more than 40 years [8] while the problem of determining the existence of a reconfiguration sequence is trivially solvable as it is known that one can always transform one triangulation of a polygon to another [15]; here the *flip* operation replaces one diagonal of the given convex polygon with another. There is a lot of work on determining the complexity of finding the shortest reconfiguration sequence for this problem, and it has been settled for some special cases of polygons and point sets [1,5].

Interestingly, one distinction between the triangulations of convex polygons, for which computing the shortest reconfiguration sequence is open, and of simple polygons, where it is NP-complete, is that the former but not the latter has the property that the shortest flip sequence never flips a diagonal shared by s and t. Insights from our results may lead to a better understanding of the role of the symmetric difference in computing shortest reconfiguration paths.

2   Preliminaries and Definitions

We use terminology originally introduced by Schaefer [18] and adapted to reconfiguration by Gopalan et al. [11] and Schwerdtfeger [19].

A *k-ary Boolean logical relation* (or *relation* for short) R is defined as a subset of $\{0,1\}^k$, where $k \geq 1$. Each $i \in \{1, \ldots, k\}$ can be interpreted as a variable of R such that R specifies exactly which assignments of values to the variables are to be considered satisfying.

For any k-ary relation R and positive integer $k' \leq k$, we define a k'-ary *restriction* of R to be any k'-ary relation R' that can be obtained from R by substitution with constants and identification of variables. More precisely, let $X : \{1, \ldots, k\} \rightarrow \{1, \ldots, k'\} \cup \{c_0, c_1\}$ be a mapping from the variables of R to the variables of R' and the constants 0 and 1. Any such X defines a mapping $f_X : \{0,1\}^{k'} \rightarrow \{0,1\}^k$ as follows. For $r \in \{0,1\}^{k'}$, let $f_X(r)$ be the k-bit vector whose i^{th} bit is 0 if $X(i) = c_0$, 1 if $X(i) = c_1$ and equal to the $X(i)^{th}$ bit of r otherwise. We say that a k'-ary relation R' is a restriction of R with respect to $X : \{1, \ldots, k\} \rightarrow \{1, \ldots, k'\} \cup \{c_0, c_1\}$ if $r \in R' \Leftrightarrow f_X(r) \in R$.

A Boolean formula ϕ over a set $\{x_1, \ldots, x_n\}$ of variables defines a relation R_ϕ as follows. For any n-bit vector $v \in \{0,1\}^n$, we interpret v as the assignment to the variables of ϕ where x_i is set to be equal to the ith bit of v. We then say that $v \in R_\phi$ if and only if v is a satisfying assignment.

A *CNF formula* is a Boolean formula of the form $C_1 \wedge \ldots \wedge C_m$, where each C_i, $1 \leq i \leq m$, is a *clause* consisting of a finite disjunction of *literals* (variables or negated variables). A *kCNF formula*, $k \geq 1$, is a CNF formula where each clause has at most k literals. A CNF formula is *Horn* (*dual Horn*) if each clause has at most one positive (negative) literal.

For a finite set of relations $\mathcal{S}$, a *CNF($\mathcal{S}$) formula* over a set of n variables $\{x_1, \ldots, x_n\}$ is a finite collection $\{C_1, \ldots, C_m\}$ of clauses. Each C_i, $1 \leq i \leq m$, is defined by a tuple (R_i, X_i), where R_i is a k_i-ary relation in $\mathcal{S}$ and $X_i : \{1, \ldots, k_i\} \rightarrow \{1, \ldots, n\} \cup \{c_0, c_1\}$ is a function. Each X_i defines a mapping $f_{X_i} : \{0,1\}^n \rightarrow \{0,1\}^{k_i}$ and we say that an assignment v to the variables satisfies ϕ if and only if for all $i \in \{1, \ldots, m\}$, $f_{X_i}(v) \in R_i$. For any variable x_j, we say that x_j *appears in* clause C_i if $X_i(q) = j$ for some $q \in \{1, \ldots, k_i\}$ and for any assignment v to the variables of ϕ, we say that $f_{X_i}(v)$ is the assignment induced by v on R_i.

For example, to represent the class 3CNF in Schaefer's framework, we specify $\mathcal{S}$ as follows. Let $R^0 = \{0,1\}^3 \backslash \{000\}$, $R^1 = \{0,1\}^3 \backslash \{100\}$, $R^2 = \{0,1\}^3 \backslash \{110\}$, $R^3 = \{0,1\}^3 \backslash \{111\}$, and $\mathcal{S} = \{R^0, R^1, R^2, R^3\}$. Since R^i can be used to represent all 3-clauses with exactly i negative literals (regardless of the positions in which they appear in a clause), clearly CNF($\mathcal{S}$) is exactly the class of 3CNF formulas.

Below we define some classes of relations used in the literature and relevant to our work. Note that componentwise bijunctive, OR-free and NAND-free were first defined by Gopalan et al. [11]. Schwerdtfeger [19] later modified them slightly and defined safely component-wise bijunctive, safely OR-free and safely NAND-free. We reuse the names componentwise bijunctive, OR-free and NAND-free for Schwerdtfeger's safely component-wise bijunctive, safely OR-free and safely NAND-free respectively.

Definition 1. *For a k-ary relation R:*

- R *is* bijunctive *if it is the set of satisfying assignments of a 2CNF formula.*
- R *is* Horn *(dual Horn) if it is the set of satisfying assignments of a Horn (dual Horn) formula.*
- R *is* affine *if it is the set of satisfying assignments of a formula $x_{i_1} \oplus \ldots \oplus x_{i_h} \oplus c$, with $i_1, \ldots, i_h \in \{1, \ldots, k\}$ and $c \in \{0,1\}$. Here $\oplus$ denote the exclusive OR operation which evaluates to 1 when exactly one of the values it operates on is 1 and evaluates to 0 otherwise.*
- R *is* componentwise bijunctive *if every connected component of the reconfiguration graph of R and of the reconfiguration graph of every restriction R' of R induces a bijunctive relation.*
- R *is* OR-free *(NAND-free) if there does not exist a restriction R' of R such that $R' = \{01, 10, 11\}$ ($R' = \{01, 10, 00\}$).*

Using his framework, Schaefer showed that SAT($\mathcal{S}$)—the problem of deciding if a CNF($\mathcal{S}$) formula has a satisfying assignment—is in P if every relation in $\mathcal{S}$ is bijunctive, Horn, dual Horn, or affine, and is NP-complete otherwise. The result is remarkable because it divides a large set of problems into two equivalence classes based on their computational complexity, which is the opposite of what one might expect due to Ladner's theorem [2].

Since Schaefer's original paper, a myriad of problems about Boolean formulas have been analyzed, and similar divisions into equivalence classes obtained [7]. Gopalan et al.'s work [11], with corrections presented by Schwerdtfeger [19], shows a dichotomy for the problem of deciding whether a reconfiguration path exists between two satisfying assignments of a CNF($\mathcal{S}$) formula. They call a set $\mathcal{S}$ of relations *tight* if all relations in $\mathcal{S}$ are componentwise bijunctive, or all relations in $\mathcal{S}$ are OR-free, or all relations in $\mathcal{S}$ are NAND-free. They showed that the st-connectivity problem on CNF($\mathcal{S}$) formulas is in P if $\mathcal{S}$ is tight and PSPACE-complete otherwise.

Our trichotomy relies on a new class of formulas that subdivides the tight classes into those for which computing the shortest reconfiguration path can be done in polynomial time and those for which it is NP-complete.

Definition 2. *For a k-ary relation R:*

- *R is* Horn-free *if there does not exist a restriction R′ of R such that R′ =* $\{0,1\}^3 \setminus \{011\}$*, or equivalently, R′ is the set of all satisfying assignments of the clause* $(x \vee \overline{y} \vee \overline{z})$ *for some three variables x, y, and z.*
- *R is* dual-Horn-free *if there does not exist a restriction R′ of R such that R′ =* $\{0,1\}^3 \setminus \{100\}$*, or equivalently, R′ is the set of all satisfying assignments of the clause* $(\overline{x} \vee y \vee z)$ *for some three variables x, y, and z.*

Due to space limitations some proofs have been omitted from the current version of the paper. The corresponding lemmas and theorems have been marked with a star.

Definition 3. *We call a set S of relations* navigable *if one of the following holds:*

(1) All relations in S are OR-free and Horn-free.
(2) All relations in S are NAND-free and dual-Horn-free.
(3) All relations in S are component-wise bijunctive.

It is clear that if S is navigable, then it is also tight. Our main result is the following trichotomy.

Theorem 1. *For a CNF(S) formula ϕ and two satisfying assignments s and t, the problem of computing the shortest reconfiguration path between s and t is in P if S is navigable, NP-complete if S is tight but not navigable, and PSPACE-complete otherwise.*

In the next section, we establish the hardness results; the rest of the paper focuses on our polynomial-time algorithm for navigable formulas. Interestingly, unlike previous classification results, while the NP-completeness result in our case turns out to be relatively easy, the polynomial-time algorithm is quite involved.

3   The Hard Cases

Gopalan et al. [11] showed that if S is not tight, then st-connectivity is PSPACE-complete for CNF(S) formulas. This implies that finding the shortest reconfiguration path is also PSPACE-complete for such classes of formulas.

Theorem 2 (∗). *If S is tight but not navigable, then finding the shortest reconfiguration path on CNF(S) formulas is NP-complete.*

Proof (Sketch). The problem is in NP because the diameter of the reconfiguration graph is polynomial for all tight formulas, as shown by Gopalan et al. [11]. We now prove that it is, in fact, NP-complete.

As S is tight but not navigable, all relations in S are OR-free or all relations in S are NAND-free. Let us first assume that all relations in S are NAND-free. Then, as S is not navigable, there exists a relation which is dual-Horn.

We show a reduction from VERTEX COVER to such a CNF($\mathcal{S}$) formula (we prove the other case by a reduction from INDEPENDENT SET). Given an instance $(G = (V, E), k)$ of VERTEX COVER, we create a variable x_v for each $v \in V$. For each edge $e = (u, v) \in E$, we create two new variables y_e and z_e and the clauses $(y_e \vee \overline{z_e} \vee x_u)$ and $(z_e \vee \overline{y_e} \vee x_v)$. The resulting formula $F(G)$ has $|V| + 2|E|$ variables and $2|E|$ clauses.

It is easy to see that all the relations of $F(G)$ are NAND-free (as we cannot set the values of all but two of their variables to get a NAND relation $R = \{01, 10, 00\}$), however none of them is dual-Horn-free (as each clause is a dual-Horn clause with one negative literal). Hence the formula $F(G)$ is tight but not navigable.

Let s be the satisfying assignment for the formula with all variables set to 0, and let t be the satisfying assignment with all the variables $x_v, v \in V$ set to 0 and the rest set to 1. If G has a vertex cover S of size at most k, then we can form a reconfiguration sequence of length at most $2|E| + 2k$ from s to t by flipping each $x_v, v \in S$ from 0 to 1, flipping the y_e and z_e variables, and then flipping each $x_v, v \in S$ back from 1 to 0. To show that such a reconfiguration sequence exists only if there exists such a vertex cover, we observe that if neither x_u nor x_v has been flipped to 1, neither y_e nor z_e can be flipped to 1 while keeping the formula satisfied at the intermediate steps.   □

4   The Polynomial-time Algorithm for Navigable Formulas

Gopalan et al. gave a polynomial-time algorithm for finding the shortest reconfiguration path in component-wise bijunctive formulas. The path, in this case, flips only variables that have different values in s and t. The NP-completeness proof from the previous section crucially relies on the fact that we need to flip variables with common values; in fact, the hardness lies in deciding precisely which common variables need to be flipped. Thus it is tempting to conjecture that hardness for shortest reconfiguration path is caused by relations where the shortest distance is not always equal to the Hamming distance. This is not the case. The reconfiguration graph for the relation $P_4 = \{000, 001, 101, 111, 110\}$ is a path of length four, where for 000 and 110 the shortest path is of length four but the Hamming distance is two. However, we can find the shortest reconfiguration paths in formulas built out of P_4 in polynomial time, the exact reason for which will become clear in our general description of the algorithm. The intuitive reason is that there are very few candidates for shortest paths; if we restrict our attention to a single clause built out of P_4, then there exists a unique path to follow. It then suffices to determine whether there exist two clauses for which the prescribed paths are in conflict. In general, our proof relies on showing that even if there does not exist a unique path, the set of all possible paths between two satisfying assignments of a navigable formula is not diverse enough to make the problem computationally hard. We show that the set of all possible paths can be characterized using a partial order on the set of flips.

4.1   Notation

Our results make use of two different views of the problem (graph-theoretic and algebraic), and hence two sets of notation.

The graph-theoretic view consists of the reconfiguration graph G_R that has a node for each Boolean string $s \in R$ and an edge whenever the Hamming distance between the two strings is exactly one. We call a path from s to t *monotonically increasing* if the Hamming weights of the vertices on the path increase monotonically as we go from s to t, and define a *monotonically decreasing* path similarly. A path is *canonical* if it consists of a monotonically increasing path followed by a monotonically decreasing path.

The algebraic view consists of a *token system* [9] consisting of a set $\mathscr{S}$ of states and a set τ of tokens. The tokens specify the rules of transition between states. Each token $t \in \tau$ is a function that maps $\mathscr{S}$ to itself. Given a k-ary relation R, we define a token system as follows. The set $\mathscr{S}$ of states consists of all the elements of R and a special state s^* called the *invalid state* that captures all the unsatisfying assignments of the formula. The set τ of tokens is the set $\{x_1^+, \ldots, x_k^+\} \cup \{x_1^-, \ldots, x_k^-\}$, where x_i^+ denotes a flip of variable x_i from 0 to 1, which we call a *positive flip*, and denote the sign of the flip as positive, and x_i^- denotes a flip of variable x_i from 1 to 0, which we call a *negative flip* and denote the sign of the flip as negative.

To complete the description of the token system, we need to specify the function to which each token corresponds. For $x_i^+ \in \tau$ and $s \in \mathscr{S}$, $x_i^+(s^*) = s^*$, $x_i^+(s) = s'$ if the value of variable x_i in s is 0 and the bit string s' obtained on flipping it to 1 lies in R, and $x_i^+(s) = s^*$ if the value of variable x_i in s is 1 or the value of variable x_i in s is 0 and the bit string s' obtained on flipping it to 1 does not lie in R. The function x_i^- is defined analogously. In the rest of this article, we will use the word "flip" instead of "token", and we will use the words "state," "vertex," and "satisfying assignment" interchangeably.

A sequence of flips also defines a function, that is, the composition of all the functions in the sequence. We call a flip sequence *invalid* at a given state s if the sequence applied to s results in invalid state s^*, and *valid* otherwise. Two flip sequences are *equivalent* if they result in the same final state when applied to the same starting state. Finally, we call a flip sequence *canonical* if all positive flips in it occur before all the negative flips. That is, the path from its first state (node) to the last is a canonical path. Note that in any canonical flip sequence, each flip occurs at most once. Given two states $s, t \in \mathscr{S}$, we say that a set $\mathscr{C}$ of flips *transforms* s to t if the elements of $\mathscr{C}$ can be arranged in some order such that the resulting flip sequence transforms s to t. For a given state s and flip set $\mathscr{C}$, we say $\mathscr{C}$ is *valid* if the elements of $\mathscr{C}$ can be arranged in some order such that the resulting flip sequence applied to s results in a valid state.

We describe a flip sequence simply by listing the flips in order. The flip sequence formed by removing flip f from $\mathcal{F}$ is denoted $\mathcal{F} \setminus f$. The flip sequence obtained by reversing $\mathcal{F}$ is $\mathcal{F}^{-1}$, and by performing $\mathcal{F}_1$ followed by $\mathcal{F}_2$ is $\mathcal{F}_1 \cdot \mathcal{F}_2$. We use $\mathscr{C}(\mathcal{F})$ to denote the set of flips that appear in $\mathcal{F}$. A flip sequence (set) consisting of only positive flips will be called a *positive flip sequence (set)*. We use $\mathcal{F}_0$ to denote an empty flip sequence and, by convention, define it to be valid. For a flip sequence $\mathcal{F}$, if $f \in \mathcal{F}$ appears before $f' \in \mathcal{F}$ in the sequence, then we say $f <_{\mathcal{F}} f'$. For a tuple $t = (x_{i_1}, \ldots, x_{i_d})$ of variables and a state s, we use s^t to denote the string of values restricted to $x_{i_1}, \ldots, x_{i_d}$.

4.2   Overview of the Algorithm

Consider, once again, the relation $P_4 = \{000, 001, 101, 111, 110\}$ from Section 4, which we claimed to be navigable. A satisfying assignment to the formula induces, on each clause, a boolean string that consists of the values of the variables appearing in that clause. Similarly, a flip sequence $\mathcal{F}$ induces a flip sequence for each clause C, which is the subsequence of $\mathcal{F}$ that flips a variable that appears in C. Note that $\mathcal{F}$ is valid if and only if the sequence induced on each clause is valid. The relation P_4 satisfies two nice properties:

1. Any valid flip sequence of P_4 is canonical.
2. Let x, y, z be the three variables that represent the three bits of P_4, then there is a total order, namely, $z^+ < x^+ < y^+$, such that any valid positive flip sequence must satisfy this order.

With this observation, formulating an algorithm is easy. Given two satisfying assignments s and t of the formula, find the Boolean string induced on each clause. The shortest flip sequence inside each clause can be computed in constant time. Each clause prescribes a unique order in which the flips of its corresponding flip sequence must be performed. If no two clauses prescribe conflicting orders, then their sequences can be combined into a sequence for the entire formula. If there is a conflict, then we know that no path exists. In general, for navigable formulas, we show, in Lemma 2, that any flip sequence can be transformed into an equivalent canonical flip sequence by rearranging the flips, and, in Lemma 9, that each clause prescribes a partial order instead of a total order. The task of combining the partial orders prescribed by each clause becomes more involved, but can still be done efficiently, as shown in Lemma 11.

We will only consider properties of NAND-free and dual-Horn-free relations, as our algorithm for NAND-free and dual-Horn-free relations can easily be modified to handle OR-free and Horn-free relations (by "reversing" the roles of positive and negative flips).

4.3   The Token System of NAND-free Relations

We begin by proving some useful properties of the token system formed by NAND-free relations.

Lemma 1 (*). *For R a NAND-free relation and $\mathcal{F} = f_1 \ldots f_q$ a valid flip sequence at $s \in R$, if there exists $i \in \{1, \ldots, q-1\}$ such that $f_i = x^-$ is a negative flip and $f_{i+1} = y^+$ is a positive flip, with $x \neq y$, then the sequence $\mathcal{F}' = f_1 \ldots f_{i-1} f_{i+1} f_i \ldots f_q$ is also valid at s and is equivalent to $\mathcal{F}$, i.e., swapping f_i and f_{i+1} results in an equivalent flip sequence.*

Lemma 2 (first proved by Gopalan et al. [11]) shows that any valid flip sequence can be made canonical. Lemma 3 shows that the union of two valid positive flips sets is also a valid flip set.

Lemma 2 (*). *For R a NAND-free relation, if $\mathcal{F}$ is a valid sequence at $s \in R$, then there exists a valid canonical sequence $\mathcal{F}'$ equivalent to $\mathcal{F}$ such that $\mathscr{C}(\mathcal{F}') \subseteq \mathscr{C}(\mathcal{F})$ and, for any two flips $f_1, f_2 \in \mathcal{F}'$ of the same sign, if $f_1 <_{\mathcal{F}'} f_2$ then $f_1 <_{\mathcal{F}} f_2$, i.e., the relative order among flips of the same sign is preserved.*

Lemma 3. *For R a NAND-free relation, if $\mathscr{C}_1$ and $\mathscr{C}_2$ are two positive flip sets that are valid at $s \in R$, then $\mathscr{C}_1 \cup \mathscr{C}_2$ is also a valid flip set at s.*

Proof. Let $u = \mathcal{F}_1(s)$ and $v = \mathcal{F}_2(s)$, where $\mathcal{F}_1$ and $\mathcal{F}_2$ are valid flip sequences such that $\mathscr{C}(F_1) = \mathscr{C}_1$ and $\mathscr{C}(F_2) = \mathscr{C}_2$. Clearly, $\mathcal{F}_1^{-1} \cdot \mathcal{F}_2$ is a valid flip sequence from u to v. Thus, we can apply Lemma 2 to the sequence $\mathcal{F}_1^{-1} \cdot \mathcal{F}_2$ to transform it into the canonical sequence $\mathcal{F}$. Let $\mathcal{F}^+$ denote the prefix of $\mathcal{F}$ that contains all the positive flips. It is clear that $\mathcal{F}_1 \cdot \mathcal{F}^+$ is a valid flip sequence at s and $\mathscr{C}(\mathcal{F}_1 \cdot \mathcal{F}^+) = \mathscr{C}_1 \cup \mathscr{C}_2$. □

Later, we prove a similar lemma for the intersection of two flip sets, but for dual-Horn-free relations. We conclude this subsection with a lemma that shows that if two disjoint flips sets are valid at a state, we can, in some sense, perform the two sets of flips one after the other in either order.

Lemma 4 (*). *For R a NAND-free relation and $\mathcal{F}_1$ and $\mathcal{F}_2$ two positive flip sequences that are valid at $s \in R$, if $\mathscr{C}(\mathcal{F}_1) \cap \mathscr{C}(\mathcal{F}_2) = \emptyset$, then $\mathcal{F}_1$ is valid at $\mathcal{F}_2(s)$ and $\mathcal{F}_2$ is valid at $\mathcal{F}_1(s)$.*

4.4 The Token System of NAND-free and Dual-Horn-free Relations

In this section, we establish stronger properties with the assumption that R is not only NAND-free, but is also dual-Horn-free. We begin by establishing a simple property of relations that are NAND-free and dual-Horn-free.

Lemma 5. *Let R be a NAND-free and dual-Horn-free relation and $s, t_1, t_2 \in R$ be three distinct states such that the flip sequence $\mathcal{F}_1 = x_k^+ x_i^+$ transforms s to t_1, the flip sequence $\mathcal{F}_2 = x_j^+ x_i^+$ transforms s to t_2, and $x_k \neq x_j$. Then the sequence $\mathcal{F}_1' = x_i^+ x_k^+$ also transforms s to t_1 and the sequence $\mathcal{F}_2' = x_i^+ x_j^+$ also transforms s to t_2, i.e., we can swap the flips in both $\mathcal{F}_1$ and $\mathcal{F}_2$.*

Proof. For $u_1 = x_k^+(s)$ and $u_2 = x_j^+(s)$, the sequence $x_j^- x_k^+$ transforms u_2 to u_1. We can reorder the sequence to obtain $x_k^+ x_j^-$, using Lemma 1. For $v = x_k^+(u_2)$, we can use a similar argument to show that x_i^+ is a valid flip at v; we let $w = x_i^+(v)$. The values of variables x_i, x_j, and x_k at states s, u_1, u_2, t_1, t_2, v, and w form exactly the seven satisfying assignments $\{000, 001, 010, 101, 110, 011, 111\}$ of the dual-Horn clause $(\overline{x_i} \vee x_j \vee x_k)$. But since R is dual-Horn-free, there must also exist the state v' for which $x_i = 1, x_j = 0, x_k = 0$. The path $s \to v' \to t_1$ gives the sequence $x_i^+ x_k^+$ and the path $s \to v' \to t_2$ gives the sequence $x_i^+ x_j^+$. ■

The seemingly innocuous lemma above turns out to be very powerful. In the following sequence of lemmas, we build on top of it to eventually prove that the set of all positive valid flip sets starting from an assignment s forms a distributive lattice. The lattice structure then helps us formulate a polynomial-time algorithm for computing the shortest reconfiguration path.

Lemma 6 (*). *Let R be a NAND-free and dual-Horn-free relation and $s, t \in R$ be two satisfying assignments such that $x^+ y^+$ is a valid flip sequence at s and y^+ is a valid flip at t. Furthermore, let $\mathcal{F}$ be a positive flip sequence such that $\mathcal{F}(s) = t$ and $x^+ \notin \mathcal{C}(\mathcal{F})$. Then, the sequence $y^+ x^+$ must also be valid at s.*

Lemma 7 (*). *For R a NAND-free and dual-Horn-free relation, if $\mathcal{F}_1 \cdot x^+ \cdot y^+$ and $\mathcal{F}_2 \cdot y^+$ are both valid positive flip sequences at $s \in R$ such that $x^+ \notin \mathcal{C}(\mathcal{F}_2)$ then $\mathcal{F}_1 \cdot y^+ \cdot x^+$ is also valid at s.*

Next, we show that the set of valid flip sets is closed under intersection.

Lemma 8 (*). *For R a NAND-free and dual-Horn-free relation, if $\mathcal{C}_1$ and $\mathcal{C}_2$ are two positive flip sets that are valid at $s \in R$, then $\mathcal{C}_1 \cap \mathcal{C}_2$ is also a valid flip set at s.*

The above lemma, combined with Lemma 3, shows that the set of valid flip sets starting at s forms a distributive lattice [3]. Using Birkhoff's representation theorem [3] on it directly implies the next lemma. However, for clarity, we also provide an independent proof. Let $\prec$ be a partial order defined on a set $\mathcal{C}$ of flips. We say a set $\mathcal{C}' \subseteq \mathcal{C}$ is *downward closed* if for every $x, y \in \mathcal{C}$, $(y \in \mathcal{C}') \wedge (x \prec y) \implies x \in \mathcal{C}'$. We say that an ordering $\mathcal{F}$ of a subset of elements in $\mathcal{C}$ *obeys* the partial order $\prec$ if (i) $\mathcal{C}(\mathcal{F})$ is downward closed and (ii) for every $x, y \in \mathcal{F}$, $x \prec y \implies x <_{\mathcal{F}} y$.

Lemma 9 (*). *Let R be a NAND-free and dual-Horn-free relation and s be an element of R. Let $\mathscr{P} = \{x^+ \mid x^+ \in \mathcal{C}$ for a positive valid flip set $\mathcal{C}$ at $s\}$. Then there exists a partial order $\prec$ on $\mathscr{P}$ such that any positive flip sequence $\mathcal{F}$ consisting of a subset of $\mathscr{P}$ is a valid flip sequence at s if and only if it obeys $\prec$.*

4.5 The Polynomial Time Algorithm

Let ϕ be a CNF($\mathcal{S}$) formula where every relation in $\mathcal{S}$ is NAND-free and dual-Horn-free, $\{x_1, \ldots, x_n\}$ be the set of variables, and $\{C_1, \ldots, C_m\}$ be the set of clauses in ϕ. We wish to compute the shortest reconfiguration path between s and t in G_ϕ for $s, t \in R_\phi$. Let $\mathscr{P}_s$ and $\mathscr{P}_t$ be the sets of positive flips that occur in any positive flip set valid at s and t, respectively.

In Lemma 10, we generalize Lemma 2 from a single relation to R_ϕ and show that if there exists a valid sequence which transforms s to t, then it can be made canonical. Similarly, Lemma 11 shows that the property of any valid flip sequence for a NAND-free and dual-Horn-free relation being describable by a partial order, as proved in Lemma 9, also applies to CNF($\mathcal{S}$) formulas where every relation in $\mathcal{S}$ is NAND-free and dual-Horn-free.

Lemma 10 (*). *Let ϕ be a CNF($\mathcal{S}$) formula where every relation in $\mathcal{S}$ is NAND-free. For any $s, t \in R_\phi$, if $\mathcal{F}$ is a valid sequence which transforms s to t, then there exists a valid canonical sequence $\mathcal{F}'$ equivalent to $\mathcal{F}$ such that $\mathscr{C}(\mathcal{F}') \subseteq \mathscr{C}(\mathcal{F})$.*

Lemma 11 (*). *Let ϕ be a CNF($\mathcal{S}$) formula where every relation in $\mathcal{S}$ is NAND-free and dual-Horn-free. For any $s, t \in R_\phi$, there exists a partial order $\prec_s$ on $\mathscr{P}_s$ and a partial order $\prec_t$ on $\mathscr{P}_t$ such that any positive flip sequence $\mathcal{F}_s$ consisting of a subset of $\mathscr{P}_s$ is a valid flip sequence at s if and only if it obeys the partial order $\prec_s$ and any positive flip sequence $\mathcal{F}_t$ consisting of a subset of $\mathscr{P}_t$ is a valid flip sequence at t if and only if it obeys the partial order $\prec_t$. Moreover, $\mathscr{P}_s$, $\prec_s$, $\mathscr{P}_t$, and $\prec_t$ can be computed in polynomial time.*

For a set $\mathscr{P}$, a partial order $\prec$ on $\mathscr{P}$, and a subset $A \subseteq \mathscr{P}$, the *smallest lower set* of A is the smallest superset of A that is downward closed. Such a lower set can be constructed in polynomial time by starting with A and including any element f' not in A such that $f' \prec f$ for some $f \in A$. It is clear that any valid flip set that contains A must also contain the smallest lower set of A.

Now the algorithm for finding the shortest reconfiguration path is clear. We start from s and let S be the set of positive flips on the variables that are set to 1 in t and to 0 in s. Then we compute the smallest lower set S' containing S and perform the flips in S' as prescribed by the partial order $\prec_s$ (on $\mathscr{P}_s$) to reach $s' \in R_\phi$. We perform a similar set of flips starting from t to reach $t' \in R_\phi$. If $s' = t'$, we are done. Otherwise, we recursively find the shortest path between s' and t'. The complete algorithm is described in Algorithm 1.

Algorithm 1. SHORTESTPATH(s,t)

Input: A CNF($\mathcal{S}$) formula ϕ where all relations in $\mathcal{S}$ are NAND-free and dual-Horn-free; two satisfying assignments s and t.
Output: Shortest reconfiguration path between s and t.
1: **if** $(s = t)$
2:    **return** $\mathcal{F}_0$ {the empty flip sequence}
3: Let S be the set of positive flips that flip variables assigned 0 in s and 1 in t.
4: Let T be the set of positive flips that flip variables assigned 0 in t and 1 in s.
5: **if** S contains an element not in $\mathscr{P}_s$ or if T contains an element not in $\mathscr{P}_t$
6:    **return** Not connected.
7: Compute the smallest lower set S' of S in $\mathscr{P}_s$ with respect to $\prec_s$.
8: Compute the smallest lower set T' of T in $\mathscr{P}_t$ with respect to $\prec_t$.
9: Let $\mathcal{F}_s$ and $\mathcal{F}_t$ be orderings of S' and T' that obey $\prec_s$ and $\prec_t$, respectively.
10: Let $s' = \mathcal{F}_s(s)$ and $t' = \mathcal{F}_t(t)$.
11: Let $\mathcal{F} = $ SHORTESTPATH(s',t').
12: **return** $\mathcal{F}_s \cdot \mathcal{F} \cdot \mathcal{F}_t^{-1}$.

We are now ready to prove the following theorem.

Theorem 3 (*). *Let S be a navigable set of relations, ϕ be a CNF(S) formula, and s and t be two satisfying assignments of ϕ. We can compute the shortest reconfiguration path between s and t in polynomial time.*

References

1. Aichholzer, O., Mulzer, W., Pilz, A.: Flip distance between triangulations of a simple polygon is NP-complete. In: Bodlaender, H.L., Italiano, G.F. (eds.) ESA 2013. LNCS, vol. 8125, pp. 13–24. Springer, Heidelberg (2013)
2. Arora, S., Barak, B.: Computational Complexity: A Modern Approach, 1st edn. Cambridge University Press, New York, NY, USA (2009)
3. Birkhoff, G.: Rings of sets. Duke Mathematical Journal **3**(3), 443–454 (1937)
4. Bonamy, M., Bousquet, N.: Recoloring bounded treewidth graphs. In: Proceedings of the 7th Latin-American Algorithms, Graphs, and Optimization Symposium (LAGOS) (2013)
5. Bose, P., Lubiw, A., Pathak, V., Verdonschot, S.: Flipping edge-labelled triangulations (2013). CoRR, abs/1310.1166
6. Cereceda, L., van den Heuvel, J., Johnson, M.: Connectedness of the graph of vertex-colourings. Discrete Mathematics **308**(56), 913–919 (2008)
7. Creignou, N., Khanna, S., Sudan, M.: Complexity classifications of boolean constraint satisfaction problems. SIAM (2001)
8. Culik II, K., Wood, D.: A note on some tree similarity measures. Inform. Process. Lett. **15**(1), 39–42 (1982)
9. Eppstein, D., Falmagne, J.-C., Ovchinnikov, S.: Media theory - interdisciplinary applied mathematics. Springer (2008)
10. Fricke, G., Hedetniemi, S.M., Hedetniemi, S.T., Hutson, K.R.: γ-Graphs of Graphs. Discussiones Mathematicae Graph Theory **31**(3), 517–531 (2011)
11. Gopalan, P., Kolaitis, P.G., Maneva, E.N., Papadimitriou, C.H.: The connectivity of boolean satisfiability: computational and structural dichotomies. SIAM Journal on Computing **38**(6), 2330–2355 (2009)
12. Ito, T., Demaine, E.D., Harvey, N.J.A., Papadimitriou, C.H., Sideri, M., Uehara, R., Uno, Y.: On the complexity of reconfiguration problems. Theoretical Computer Science **412**(12–14), 1054–1065 (2011)
13. Ito, T., Kamiński, M., Demaine, E.D.: Reconfiguration of list edge-colorings in a graph. Discrete Applied Mathematics **160**(15), 2199–2207 (2012)
14. Kamiński, M., Medvedev, P., Milanič, M.: Complexity of independent set reconfigurability problems. Theor. Comput. Sci. **439**, 9–15 (2012)
15. Lawson, C.L.: Transforming triangulations. Discrete Mathematics **3**(4), 365–372 (1972)
16. A. E. Mouawad, N. Nishimura, and V. Raman. Vertex cover reconfiguration and beyond, 2014. arXiv:1402.4926
17. Mouawad, A.E., Nishimura, N., Raman, V., Simjour, N., Suzuki, A.: On the parameterized complexity of reconfiguration problems. In: Gutin, G., Szeider, S. (eds.) IPEC 2013. LNCS, vol. 8246, pp. 281–294. Springer, Heidelberg (2013)
18. Schaefer, T.J.: The complexity of satisfiability problems. In: Proceedings of the Tenth Annual ACM Symposium on Theory of Computing, STOC 1978, pp. 216–226. New York, NY, ACM (1978)
19. Schwerdtfeger, K.W.: A computational trichotomy for connectivity of boolean satisfiability (2013). CoRR, abs/1312.4524

Computing the Fréchet Distance
Between Polygons with Holes

Amir Nayyeri[1] and Anastasios Sidiropoulos[2]([✉])

[1] School of Electrical Engineering and Computer Science,
Oregon State University, Corvallis, USA
nayyeria@eecs.oregonstate.edu
[2] Department of Computer Science & Engineering and Department of Mathematics,
The Ohio State University, Columbus, USA
sidiropoulos.1@osu.edu

Abstract. We study the problem of computing the Fréchet distance between subsets of Euclidean space. Even though the problem has been studied extensively for 1-dimensional curves, very little is known for d-dimensional spaces, for any $d \geq 2$. For general polygons in $\mathbb{R}^2$, it has been shown to be NP-hard, and the best known polynomial-time algorithm works only for polygons with at most a single puncture [Buchin *et al.*, 2010]. Generalizing [Buchin *et al.*, 2008] we give a polynomial-time algorithm for the case of arbitrary polygons with a constant number of punctures. Moreover, we show that approximating the Fréchet distance between polyhedral domains in $\mathbb{R}^3$ to within a factor of $n^{1/\log\log n}$ is NP-hard.

1 Introduction

Computing the similarity between two geometric objects is a fundamental problem that arises in several application scenarios, such as computer vision, and graphics (see [8,12,17,24] and the references therein for a more detail account of the various applications). A classical way for estimating such a similarity is the Hausdorff distance between two subsets of a metric space. However, when the objects under consideration are endowed with topological information, it is desirable to use a similarity function that takes this additional structure into account.

One of the most well-studied similarity functions that combines topological and geometric information is the *Fréchet distance*, which we define here for subsets of Euclidean space. Let $X \subset \mathbb{R}^d$ be a *parameter space*, for some $d \geq 1$, and let $h_P : X \to \mathbb{R}^d$, $h_Q : X \to \mathbb{R}^d$ be embeddings[1]. Let $P = h_P(X)$, and

A. Nayyeri—Part of this work was done while the author was a postdoctoral fellow at CMU. Research supported in part by the NSF grants CCF 1065106 and CCF 09-15519.

A. Sidiropoulos—Research supported in part by the NSF grants CCF 1423230 and CAREER 1453472.

[1] In the most general setting, the parameterizations h_P, and h_Q may not be required to be embeddings. However, we restrict our attention here to *simple* polygons with holes, in which case the maps h_P and h_Q are embeddings.

© Springer-Verlag Berlin Heidelberg 2015
M.M. Halldórsson et al. (Eds.): ICALP 2015, Part I, LNCS 9134, pp. 997–1009, 2015.
DOI: 10.1007/978-3-662-47672-7_81

$Q = h_Q(X)$. Then, for a homeomorphism $f : P \to Q$, we define its *Fréchet length* to be $\delta_F(f) = \sup_{x \in P} \|x - f(x)\|_2$, and $\delta_F(P, Q) = \inf_f \delta_F(f)$, where $f : P \to Q$ ranges over all orientation-preserving homeomorphisms[2]. We remark that the Fréchet distance can also be defined for more general ambient spaces, e.g. for surfaces (see, e.g. [19]), but we will restrict our attention to Euclidean space.

1.1   Previous Work on Computing Fréchet Distance

Most of the work on computing the Fréchet distance between two subsets P and Q of some ambient space has been focused on the case where P and Q are one-dimensional curves [1,3,4,7,9,14,15,19]. In contrast, very little is known for computing the Fréchet distance between two-dimensional spaces P and Q. Buchin *et al.* [6] describe the first polynomial-time algorithm to compute the Fréchet distance between two simple polygons in $\mathbb{R}^2$. Buchin *et al.* [5] prove that the problem becomes NP-hard for arbitrary polygons in $\mathbb{R}^2$, and they give a polynomial-time algorithm for polygons with a single puncture. They also prove that computing the Fréchet distance between two terrains in $\mathbb{R}^3$ is NP-hard. It has also been shown by Godau [18] that computing the Fréchet distance between two surfaces is NP-hard, and it is known that this problem is upper semi-computable [2]. Finally, Cook *et al.* [10] gave exact and approximation algorithms for special classes of simply-connected 2-dimensional polygons in $\mathbb{R}^3$.

1.2   Our Contribution

We focus on the problems of computing and approximating the Fréchet distance between two subsets of d-dimensional Euclidean space, and we obtain both upper and lower bounds. Our main contributions are described below.

We present a polynomial-time algorithm for computing the Fréchet distance between two polygons $P, Q \subset \mathbb{R}^2$, with a constant number of punctures. This answers a question of Buchin *et al.* [5], and resolves the main open problem from their paper. Our algorithm uses tools and ideas developed in the context of computing shortest non-crossing walks in the plane [16]. The following summarizes our algorithm for arbitrary polygons in $\mathbb{R}^2$.

Theorem 1 (Exact algorithm for polygonal domains in $\mathbb{R}^2$). *Let P and Q be simple polygons in $\mathbb{R}^2$, with h punctures. There exists a $2^{O(h^2)} n^{O(h)}$ time algorithm for computing $\delta_F(P, Q)$. Moreover, the algorithm outputs a piecewise linear orientation-preserving homeomorphism of minimum Fréchet length.*

The NP-hardness result of [5] leaves open the possibility of an approximation algorithm for the problem. We show that such a result is unlikely in $\mathbb{R}^3$. The formal statement of our result follows.

---

[2] More generally, one can consider homeomorphisms $g : X \to X$, i.e. from the parameter space into itself, and define their Fréchet length to be $\delta_F(g) = \sup_{x \in X} \|h_P(x) - g(h_Q(x))\|_2$. However, we are dealing with *simple* polygonal/polyhedral domains, and therefore the maps h_P, and h_Q are homeomorphisms, which implies that our simpler definition is equivalent.

Theorem 2 (Inapproximability for polyhedral domains in $\mathbb{R}^3$). *Approximating the Fréchet distance between two polyhedral domains in $\mathbb{R}^3$ within a factor of $n^{1/\log\log n}$ is NP-hard.*

We remark that it is not known if there always exists an optimal (or even near-optimal) piecewise linear homeomorphism of polynomial complexity. In fact, it is not known whether any of the problems considered is in NP. The proof of [5] only shows that computing the Fréchet distance between polygons in $\mathbb{R}^2$ is NP-hard, and it is open whether it is in NP.

1.3   Our Techniques

The Exact Algorithm in $\mathbb{R}^2$. Our exact algorithm for computing the Fréchet distance between two polygonal domains starts by picking a small set of diagonals in P and guessing their image in Q. Then, our algorithm cuts P along these diagonals, and Q along their maps, thus reducing the problem to computing the Fréchet distance between two simple polygons. In order to bound the number of possible images of a diagonal, we need to bound the number of possible choices for its endpoints and its homotopy class. To achieve the former, we look into the refined free space diagram [3] and prove that there is a quadratic number of possibilities for each endpoint. For the latter purpose, we exploit ideas from the problem of computing non-crossing walks in a planar arrangement [16]. More specifically, we consider a collection of segments that cut Q into a topological disk. We observe that if the number of crossings of the diagonal maps with any of these segments is sufficiently large, then one of the diagonal maps can be shortcut along a straight line segment without introducing crossings among diagonal maps. Following Buchin *et al.* [6], we observe that such a shortcutting does not increase the Fréchet distance between the diagonals of P and their maps in Q. This shows that the number of homotopy classes is bounded by a function of the number of punctures, and by the above discussion, implies the algorithm.

Inapproximability in $\mathbb{R}^3$. Our inapproximability result is obtained by reducing the Closest Vector Problem under ℓ_∞ norm (CVP$_\infty$) to the problem of computing the Fréchet distance between two polyhedral domains in $\mathbb{R}^3$. Our inapproximability factor follows by a result due to Dinur [13] who showed that CVP$_\infty$ is NP-hard to approximate within a factor of $n^{1/\log\log n}$.

1.4   Organization

The rest of the paper is organized as follows. Section 2 introduces background and notation. Section 3 introduces skeleton maps and explains their use in the computation of Fréchet distance. Section 4 presents the exact algorithm for $\mathbb{R}^2$. The inapproximability result for $\mathbb{R}^3$ is given in the full version of this paper.

2   Background and Notation

Given two points $p, q \in \mathbb{R}^2$, we use $\overline{(p,q)}$ to refer to the line segment with endpoints p and q. We say that a path with endpoints p and q is a (p,q)-*path*. In particular, $\overline{(p,q)}$ is a (p,q)-path. For a simple path $p \in \mathbb{R}^2$ and for points $x, y \in p$, we use $p[x, y]$ to refer to the subpath of p with endpoints x and y. We say that two paths p and q *cross* if for any path p' that is obtained from p by an infinitesimal perturbation, we have $p' \cap q \neq \emptyset$.

Free space diagrams. Let us recall the notion of a free space diagram, introduced by Alt and Godau [3]. Let p, q be two closed curves in $\mathbb{R}^2$, and let $\delta > 0$. The *double free space diagram* F_δ (also denoted as F when δ is clear from the context), is a data structure that is represented by a $[0, 2|p|] \times [0, |q|]$ rectangle, where $|\cdot|$ denotes the curve length. Each point of the double free space diagram corresponds to a pair (x, y) where $x \in p$ and $y \in q$. A pair (x, y) is *feasible* if and only if $\|x - y\|_2 \leq \delta$. The collection of all feasible pairs is called the *feasible subspace* of the free space diagram. An orientation preserving homeomorphism $f : p \to q$ corresponds to a monotonically increasing path ρ in F that has endpoints $(x_0, 0)$ and $(x_0 + |p|, |q|)$, for some $x_0 \in [0, |p|]$. The homeomorphism f has Fréchet length at most δ if it resides within the feasible subspace of F.

Suppose that p and q are piecewise linear curves, and let $p_0, \ldots, p_n$, and $q_1, \ldots, q_m$ be the vertices of p and q, respectively. The vertical lines with x-coordinates corresponding to p_i's and the horizontal lines with y-coordinates corresponding to q_i's partition F into a collection of *cells*. A cell represents all the pairs of points from a specific pair of segments of p and q. The feasible subset of a cell is the intersection of a certain ellipse with the cell, and so it is convex [3].

Consider all points on the vertical segments of F that are also on the boundary of the feasible region and add horizontal lines with their y-coordinates to further refine F. By the convexity of the feasible region inside a cell, it follows that there are at most four such points in each cell and so $O(nm)$ such points overall. Perform the same refinement by adding similar vertical lines as well to obtain a $O(nm) \times O(nm)$ refined grid. Following Alt and Godau we call this diagram the *refined free space diagram* and we denote it by $\mathcal{F}_\delta$ (or just $\mathcal{F}$ when δ is clear from the context). We refer to the cells and segments of this refined diagram as *refined cells* and *refined segments*, respectively. Each vertical refined segment of $\mathcal{F}$ at a vertex of P and each horizontal refined segment of $\mathcal{F}$ at a vertex of Q is either completely feasible or completely infeasible (see Figure 1).

3   Skeletons and Skeleton Maps

Let us now briefly describe the algorithm of [6] for computing the Fréchet distance between two simple polygons in $\mathbb{R}^2$. They first show that a map f_s between the boundaries of a convex polygon and a simple polygon can be extended to a map between the polygons with Fréchet length arbitrarily close to $\delta_F(f_s)$.

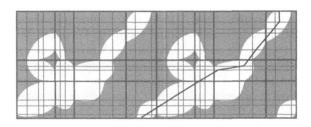

Fig. 1. A path in a refined free space diagram

Lemma 1 (Buchin *et al.* [6]). *Let P be a convex polygon, let Q be a simple polygon in $\mathbb{R}^2$, and let $f_s : \partial P \to \partial Q$ be a homeomorphism. Then, for any $\varepsilon > 0$, there exists a homeomorphism $f : P \to Q$, extending f_s, and such that $\delta_F(f) \leq \delta_F(f_s) + \varepsilon$.*

Given two simple polygons $P, Q \subset \mathbb{R}^2$, Buchin et al. partition P into convex regions. They also obtain a combinatorially equivalent partition of Q into simple regions. Then, they use Lemma 1 to find a collection of maps from each convex region of P to its corresponding simple region of Q. This collection induces the desired homeomorphism between P and Q. Following their idea, we define skeletons and skeleton maps for polygonal domains.

3.1 Skeletons

Let P and Q be polygonal domains with boundaries $\partial P = b_0 \cup \cdots \cup b_h$ and $\partial Q = c_0 \cup \ldots \cup c_h$, respectively, where each b_i, c_j is a closed polygonal curve. Suppose, without loss of generality, that b_0 and c_0 are the outer boundary components.

Let $\Sigma = \{\sigma_1, \sigma_2, \ldots, \sigma_k\}$ be any set of pairwise interior-disjoint straight line segments that partitions P into convex polygons, and let s_i, t_i be the endpoints of σ_i. Σ, in particular, can be the set of diagonals of any triangulation of P. To simplify the exposition, we assume that the endpoints of the diagonals are disjoint. This assumption can be enforced by spreading identical endpoints apart infinitesimally. We refer to the segments σ_i as *diagonals*. We refer to $\mathcal{S}(P) = \partial P \cup \bigcup_{\sigma \in \Sigma} \sigma$ as the *skeleton* of P (see Figure 2).

A continuous map $f_s : \mathcal{S}(P) \to Q$ is called a *skeleton map*. For each $i \in \{1, \ldots, k\}$, let $\gamma_i = f_s(\sigma_i)$. We refer to the paths $\gamma_1, \ldots, \gamma_k$ as the diagonals of Q. We also refer to $\mathcal{S}(Q) = f_s(\mathcal{S}(P))$ (w.r.to f_s) as the skeleton of Q. A skeleton map f_s is called *admissible* if the following conditions hold:

(A1) There exists a permutation $\pi : \{0, \ldots, h\} \to \{0, \ldots, h\}$, with $\pi(0) = 0$, and such that for any $i \in \{0, \ldots, h\}$, the map $f_s|_{b_i}$ is an orientation-preserving homeomorphism between the cycles b_i and $c_{\pi(i)}$.

(A2) For each $i \in \{1, \ldots, k\}$, we have that $f_s|_{\sigma_i}$ is a homeomorphism between σ_i and γ_i. Moreover, the collection of paths $\gamma_1, \ldots, \gamma_k$ is pairwise non-crossing.

(A3) Intuitively, we require that f_s induces a combinatorially equivalent drawing of the planar map corresponding to $\mathcal{S}(P)$. Formally, let $i \in \{1, \ldots, k\}$, and let $v \in \{s_i, t_i\}$ be an endpoint of σ_i. Note that the neighborhood of v in P intersects two segments from some boundary component b_j, and at least one segment from the diagonals of P. Let $\ell_1, \ldots, \ell_t$ be these segments. Then, the circular ordering of $\ell_1, \ldots, \ell_t$ around v is the same as the circular ordering of $f_s(\ell_1), \ldots, f_s(\ell_t)$ around $f_s(v)$ (see Figure 2).

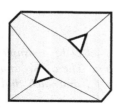 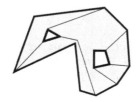 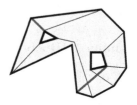

Fig. 2. P and Q, with their diagonals Σ and Γ (left), and the segments that cut Q into a disk drawn in red (right)

The following lemma, which is similar in spirit to Lemma 1, guarantees that a skeleton map of small Fréchet length implies an actual map of small Fréchet length.

Lemma 2. Let $f_s : \mathcal{S}(P) \to \mathcal{S}(Q)$ be an admissible skeleton map, and $\delta_F(f_s)$ be its Fréchet length. Then, for any $\varepsilon > 0$, there exists a homeomorphism $f : P \to Q$, such that $\delta_F(f) \leq \delta_F(f_s) + \varepsilon$.

3.2   Shortcutting Diagonals

Buchin et al. [6] observe that when P and Q are simply-connected polygons, without loss of generality, each diagonal of P can be mapped to a shortest geodesic path within Q. Unfortunately, this is not true in the case of punctured polygons. We now derive a generalization of this property in our setting that takes into account the homotopy class of the diagonals and their images. We first recall the following auxiliary lemma from Buchin et al. [6].

Definition 1 (Shortcutting). Let $\gamma : [0, 1] \to \mathbb{R}^2$ be a path, and let $0 \leq t_1 < t_2 \leq 1$. Let γ' be the path obtained from γ by replacing $\gamma[t_1, t_2]$ with the line segment $\overline{(\gamma(t_1), \gamma(t_2))}$. That is, $\gamma' = \gamma[0, t_1] \circ \overline{(\gamma(t_1), \gamma(t_2))} \circ \gamma[t_2, 1]$, where $\circ$ denotes path composition. Then, we say that γ' is obtained from γ via a shortcutting operation.

Lemma 3 (Buchin et al. [6]). Let $\ell \subset \mathbb{R}^2$ be any line segment, and let γ, γ' be paths, such that γ' is obtained from γ via a shortcutting operation. Then, we have $\delta_F(\ell, \gamma') \leq \delta_F(\ell, \gamma)$.

We will use the following result, implicit in the work of Colin de Verdière and Erickson [11], which in turn follows by a result of Scott and Hass [20].

Lemma 4 (Colin de Verdière and Erickson [11], Scott and Hass [20]). *Let γ_1, γ_2 be simple non-homotopic paths in Q. For any $i \in \{1, 2\}$, let γ_i' be a shortest path in the homotopy class of γ_i. If γ_1 and γ_2 do not cross, then γ_1' and γ_2' do not cross, either.*

Let γ, γ' be two homotopic simple paths. Recall that two homotopic paths have common endpoints. A component of $\mathbb{R}^2 \setminus (\gamma \cup \gamma')$ is called a *bigon* if it is simply connected and its boundary is composed by one subpath of γ and one subpath of γ'. The following lemma is a special case of Lemma 3.1 of Hass and Scott [20].

Lemma 5 (Hass and Scott [20], Lemma 3.1). *Let γ and γ' be distinct homotopic simple paths in a polygonal domain Q. Then, one of the connected components of $\mathbb{R}^2 \setminus (\gamma \cup \gamma')$ is a bigon.*

We use Lemma 5 and Lemma 3 to obtain the following result, which allows us to assume w.l.o.g. that the diagonals of P are mapped into shortest homotopic paths in Q. The high-level technique is similar to the proof of Lemma 4 of Buchin *et al.* [6].

Lemma 6. *Let γ' be a simple path in Q, and let γ be a shortest path in the homotopy class of γ'. Then, there exists a sequence of paths $\gamma^0, \ldots, \gamma^t$, with $\gamma^0 = \gamma'$, $\gamma^t = \gamma$, such that:*

(1) For any $i \in \{0, \ldots, t\}$, γ^i is homotopic to γ'.
(2) For any $i \in \{0, \ldots, t-1\}$, γ^{i+1} is obtained from γ^i via a shortcutting operation.

Corollary 1. *Let $\ell \subset \mathbb{R}^2$ be any line segment, and let γ, γ' be paths, such that γ' is the shortest path in the homotopy class of γ. Then, we have $\delta_F(\ell, \gamma') \leq \delta_F(\ell, \gamma)$.*

Proof. Follows by Lemma 3 and induction on the sequence given by Lemma 6.

Lemma 7. *If there exists an admissible skeleton map $g_s : \mathcal{S}(P) \to Q$, then there exists an admissible skeleton map $g_s' : \mathcal{S}(P) \to Q$ satisfying the following conditions.*

(1) $\delta_F(g_s') \leq \delta_F(g_s)$.
(2) For any $i \in \{1, \ldots, k\}$, $g_s(\sigma_i)$ and $g_s'(\sigma_i)$ are in the same homotopy class. Further, $g_s'(\sigma_i)$ is a shortest path in its homotopy class.

3.3 Bounding the Number of Possible Homotopy Classes of a Diagonal

By the preceding discussion we may restrict our attention to skeleton maps that map every diagonal of P onto some shortest homotopic path in Q. Therefore, in order to determine the image of a diagonal it suffices to guess the endpoints and the homotopy class of its image. In a simply connected polygon, the endpoints of a curve completely determine its homotopy class. Unfortunately, there are infinitely many homotopy classes of curves with the same pair of endpoints in a non-simply connected polygonal domain.

We now derive a bound on the number of possible homotopy classes for images of diagonals of P. Our argument uses tools from the work of Erickson and Nayyeri [16] on computing non-crossing walks in a planar polygonal arrangement (for similar ideas on crossing patterns see Schaefer *et al.* [22,23]).

Let $\{r_1, r_2, \ldots, r_h\}$ be a set of disjoint line segments that cut Q into a topological disk (see Figure 2). The following lemma is implicit in [16].

Lemma 8 (Erickson and Nayyeri [16]). *Let $\{\gamma_1, \ldots, \gamma_k\}$ be a collection of pairwise non-crossing paths in Q with endpoints on ∂Q. Then, there exists a collection of paths $\{\gamma'_1, \ldots, \gamma'_k\}$ satisfying the following conditions.*

(1) For any $i \in \{1, \ldots, k\}$, the path γ'_i is obtained from γ_i via a sequence of zero or more shortcutting operations.

(2) The collection of paths $\{\gamma'_1, \ldots, \gamma'_k\}$ is pairwise non-crossing.

(3) For any $i \in \{1, \ldots, k\}$ and $j \in \{1, \ldots, h\}$, γ'_i crosses r_j at most 2^{2h-2} times.

We next show that we can shortcut the images of the diagonals without violating admissibility.

Lemma 9. *Let $\sigma_1, \ldots, \sigma_k$ be the diagonals of P. Let g_s be an admissible skeleton map. For any $i \in \{1, \ldots, k\}$, let γ'_i be a path in Q that is obtained by performing a sequence of zero or more shortcutting operations on $\gamma_i = g_s(\sigma_i)$. Suppose further that the collection of paths $\{\gamma'_1, \ldots, \gamma'_k\}$ is pairwise non-crossing. Then, there exists an admissible skeleton map $g'_s : \mathcal{S}(P) \to \partial Q \cup (\bigcup_{i=1}^k \gamma'_i)$, with $\delta_F(g'_s) \leq \delta_F(g_s)$, and such that for every $i \in \{1, \ldots, k\}$, $g'_s|_{\sigma_i}$ is a homeomorphism between σ_i and γ'_i.*

Lemma 10. *Let $\sigma_1, \ldots, \sigma_k$ be the diagonals of P. If there exists an admissible skeleton map g_s, then there exists an admissible skeleton map g'_s, with $\delta_F(g'_s) \leq \delta_F(g_s)$, such that for each $i \in \{1, \ldots, k\}$ and $j \in \{1, \ldots, h\}$ the path $g'_s(\sigma_i)$ crosses r_j at most 2^{2h-2} times.*

We next derive an upper bound on the number of possible homotopy classes. The proof of the following lemma uses an argument from Erickson and Nayyeri [16].

Lemma 11. *Let $\sigma_{\iota_1}, \ldots, \sigma_{\iota_h} \in \Sigma$ be a collection of diagonals of P that cut P into a disk. Let g_s be an admissible skeleton map. Then, there exists an efficiently computable set of h-tuples $\mathcal{X} = \{\langle \chi_{i,1}, \ldots, \chi_{i,h} \rangle\}_{i \in I}$, with $|I| = 2^{O(h^2)}$, where each $\chi_{i,j}$ is a homotopy class of paths in Q, satisfying the following. There exists $i \in I$ and an admissible skeleton map g'_s, such that for each $j \in \{1, \ldots, h\}$ the path $g'_s(\sigma_{\iota_j})$ is in the homotopy class $\chi_{i,j}$, and $\delta_F(g'_s) \leq \delta_F(g_s)$.*

3.4   Bounding the Number of Possible Endpoints of a Diagonal

Lemma 7 implies that the image of a diagonal can be computed if its homotopy class and its endpoints are known. Lemma 10 bounds the possibilities for the homotopy class of a diagonal image. The following discretization technique bounds the number of possibilities for each endpoint of a diagonal image.

Lemma 12. *Let $f_s : \mathcal{S}(P) \to Q$ be an admissible skeleton map, and let $\pi : \{0, 1, \ldots, h\} \to \{0, 1, \ldots, h\}$ be the permutation such that for any $i \in \{0, \ldots, h\}$, the map f_s induces a homeomorphism between b_i and $c_{\pi(i)}$. Then, there exists an admissible skeleton map f'_s satisfying the following conditions.*

(1) $\delta_F(f'_s) \leq \delta_F(f_s)$.
(2) For any $i \in \{0, \ldots, h\}$, the map f'_s induces a homeomorphism between b_i and $c_{\pi(i)}$.
(3) For any $i \in \{0, \ldots, h\}$, let $\mathcal{F}^i$ be the refined free space diagram for b_i and $c_{\pi(i)}$, and let ρ'_i be the path in $\mathcal{F}^i$ corresponding to the homeomorphism $f'_s|_{b_i}$. Then for any vertex $x \in b_i$, $\rho'_i(x)$ is an endpoint of a refined vertical segment in $\mathcal{F}^i$.

4   An Exact Algorithm for Polygonal Domains in $\mathbb{R}^2$

In this section we describe an exact polynomial time algorithm to compute the Fréchet distance between two polygonal domains in $\mathbb{R}^2$ with a constant number of boundary components. The proof of the following lemma is essentially identical to the proof for the case of simply connected polygons given by Buchin *et al.* [6], so it is omitted. In light of Lemma 13, for the remainder of this section, we focus on obtaining an algorithm for the decision version of the problem.

Lemma 13. *Let P and Q be polygonal domains in $\mathbb{R}^2$. There is a polynomial size set S of real numbers that contains the value $\delta_F(P, Q)$. Moreover, S can be computed in polynomial time.*

4.1   An Auxiliary Algorithm

We now present an auxiliary algorithm that will be used as a subroutine in our exact algorithm for arbitrary polygons. The input is two simply connected polygons $\widetilde{P}, \widetilde{Q} \subset \mathbb{R}^2$, and a partial homeomorphism between their boundaries. The goal is to compute an orientation preserving homeomorphism between $\widetilde{P}$

and $\widetilde{Q}$ of minimum Fréchet length that extends the given partial homeomorphism of their boundaries.

Let $\mathcal{F}$ the refined free space diagram of $\partial\widetilde{P}$ and $\partial\widetilde{Q}$ (for a fixed $\delta > 0$). Recall that any homeomorphism between $\partial\widetilde{P}$ and $\partial\widetilde{Q}$ corresponds to a path ρ in $\mathcal{F}$. We assume that if there is a homeomorphism of Fréchet length δ then there is one with Fréchet length at most δ whose corresponding path in $\mathcal{F}$ contains the (equivalent) points $(0,0)$ and $(|\partial\widetilde{Q}|, |\partial\widetilde{P}|)$. To enforce this property we pick an arbitrary vertex $v_0 \in \partial\widetilde{P}$ and guess its image under such a homeomorphism, and we shift $\mathcal{F}$ accordingly; Lemma 12 implies that it is enough to consider $O(n^2)$ possibilities for the image of v_0.

Thus, we focus on the first half of $\mathcal{F}$ that is $[0, |\partial\widetilde{P}|] \times [0, |\partial\widetilde{Q}|]$, which we denote by $\mathcal{F}^\ell$. Each point $p \in \partial\widetilde{P}$ (resp. $q \in \partial\widetilde{Q}$) corresponds to exactly one point in $[0, |\partial\widetilde{P}|)$ (resp. $[0, |\partial\widetilde{Q}|)$). In order to simplify the notation, we use p (resp. q) to refer both to a point on $\partial\widetilde{P}$ (resp. $\partial\widetilde{Q}$) and a horizontal (resp. vertical) coordinate in $\mathcal{F}^\ell$. Similarly, we use $\rho(p)$ to refer to a vertical coordinate in $\mathcal{F}^\ell$ as well as the image of p, which is a point on $\partial\widetilde{Q}$. For a pair of points $p, p' \in \partial\widetilde{P}$, we write $p \leq p'$ if $(p, 0)$ is closer than $(p', 0)$ to $(0, 0)$ in $\mathcal{F}^\ell$.

We are now ready to obtain our auxiliary algorithm. We note that Theorem 15 of Buchin *et al.* [6] can be extended to obtain the same result. The precise statement follows.

Lemma 14. *Let $\widetilde{P}, \widetilde{Q} \subset \mathbb{R}^2$ be simply connected polygons, and let $\Sigma = \{\sigma_1, \ldots, \sigma_k\}$ be a set of diagonals of $\widetilde{P}$ that partition it into convex regions. Let $\alpha_1, \ldots, \alpha_t \subset \partial\widetilde{P}$ be pairwise disjoint subpaths of $\partial\widetilde{P}$ that appear in this order in a clockwise traversal of $\partial\widetilde{P}$, and that are internally disjoint from the endpoints of the diagonals in Σ. Similarly, let $\beta_1, \ldots, \beta_t \subset \partial\widetilde{Q}$ be pairwise disjoint subpaths of $\partial\widetilde{Q}$ that appear in this order in a clockwise traversal of $\partial\widetilde{Q}$. For any $i \in \{1, \ldots, t\}$, let $\phi_i : \alpha_i \to \beta_i$ be an orientation-preserving homeomorphism (where every path is considered to be oriented according to a clockwise traversal of $\partial\widetilde{P}$ and $\partial\widetilde{Q}$ respectively). Then, there exists a polynomial-time algorithm which given $\delta > 0$ decides whether there exists an orientation-preserving homeomorphism $f : \widetilde{P} \to \widetilde{Q}$ with $\delta_F(f) \leq \delta$, subject to the constraint that for any $i \in \{1, \ldots, t\}$, we have $f|_{\alpha_i} = \phi_i$. Moreover, if such a homeomorphism exists, the algorithm outputs a homeomorphism f' with $\delta_F(f') \leq \delta + \varepsilon$, for any $\varepsilon > 0$.*

4.2 The Main Algorithm

Proof (Proof of Theorem 1). By Lemma 13 it suffices to obtain an algorithm which given some $\delta \geq 0$ decides whether $\delta_F(P, Q) \leq \delta$. Let $f : P \to Q$ be a homeomorphism with $\delta_F(f) = \delta_F(P, Q)$. Let f_s be the admissible skeleton map obtained by restricting f on the skeleton $\mathcal{S}(P) = \partial P \cup (\bigcup_{\sigma \in \Sigma} \sigma)$. There exists a permutation $\pi : \{0, \ldots, h\} \to \{0, \ldots, h\}$, such that f induces a homeomorphism between b_i and c_i. We guess the permutation π. That is, we run the following procedure for every possible permutation π, and output the best solution found, which results in a multiplicative factor of $O(h!) = 2^{O(h \log h)}$ in the running time.

By Lemma 12 there exists an admissible skeleton map f'_s with $\delta_F(f'_s) \leq \delta_F(f_s)$, and such that for any $i \in \{1, \ldots, h\}$, for every endpoint x of σ_i, with $x \in b_j$ for some $j \in \{1, \ldots, h\}$, we have that $f'_s(\sigma_i)$ is a vertex in the refined free space diagram $\mathcal{F}_j$ that corresponds to the pair of boundary components b_j and c_j. We guess all the endpoints of $f'_s(\sigma_i)$ for all $i \in \{1, \ldots, h\}$. There is a total of at most $n^{O(h)}$ possibilities.

Let Σ' be a subset of segments in Σ that cut P into a topological disc, with $|\Sigma'| = h$. We can compute Σ' by greedily cutting P along diagonals with endpoints on different boundary components (and updating the set of boundary components after each cut). We may assume, after permuting the indices, and without loss of generality, that $\Sigma' = \{\sigma_1, \sigma_2, \ldots, \sigma_h\}$.

By Lemma 11 there exists a collection $\mathcal{X} = \{\langle \chi_{i,1}, \ldots, \chi_{i,h} \rangle\}_{i \in I}$ of efficiently computable h-tuples of homotopy classes of paths in Q, with $|I| = 2^{O(h^2)}$, and an admissible skeleton map f''_s satisfying all the above conditions as f'_s, and such that there exists $i \in I$, such that for every $j \in \{1, \ldots, h\}$, the path $f''_s(\sigma_j)$ is in the homotopy class $\chi_{i,j}$. We compute the set $\mathcal{X}$, and we try all of the $2^{O(h^2)}$ tuples in $\mathcal{X}$, and return the best solution found.

By Lemma 7 there exists a skeleton map f'''_s satisfying all the above conditions as f''_s, and such that for every $i \in \{1, \ldots, h\}$, the path $\gamma'''_i = f'''_s(\sigma_i)$ is shortest in its homotopy class. We compute each path γ'''_i in linear time, using the algorithm of Hershberger and Snoeyink [21]. After computing $\Gamma''' = \{\gamma'''_1, \ldots, \gamma'''_h\}$ we check whether the paths in Γ''' are pairwise non-crossing, and whether cutting Q along Γ''' results in more than one connected component. In either of these cases the algorithm disregards Γ''', and proceeds to the next choice of homotopy classes.

If Γ''' passes the above test, then we compute the homeomorphism $f'''_s|_{\sigma_i}$ between σ_i and γ'''_i, using the algorithm of Alt and Godau [3] for the Fréchet distance between polygonal curves.

Let $\widetilde{P}, \widetilde{Q}$ be the simply connected polygons obtained by cutting P along $\sigma_1, \ldots, \sigma_h$, and along $\gamma'''_1, \ldots, \gamma'''_h$ respectively. The paths $\sigma_1, \ldots, \sigma_h$ correspond to pairwise disjoint paths $\alpha_1, \ldots, \alpha_{2h} \subset \partial \widetilde{P}$. Similarly, the paths $\gamma'''_1, \ldots, \gamma'''_h$ correspond to pairwise disjoint paths $\beta_1, \ldots, \beta_{2h} \subset \partial \widetilde{Q}$. Moreover, the maps $f'''_s|_{\sigma_1}, \ldots, f'''_s|_{\sigma_h}$ induce a collection of homeomoprhisms $\phi_1 : \alpha_1 \to \beta_1, \ldots, \phi_{2h} : \alpha_{2h} \to \beta_{2h}$. By Lemma 14 we can compute in polynomial time a homeomorphism $\widetilde{f} : \widetilde{P} \to \widetilde{Q}$ of minimum Fréchet length. By the above discussion, for the right choice of the permutation π, and the endpoints and homotopy classes of the paths $\gamma'''_1, \ldots, \gamma'''_h$, we have $\delta_F(\widetilde{f}) \leq \delta_F(\widetilde{P}, \widetilde{Q}) \leq \delta_F(P, Q) + \varepsilon$, for any $\varepsilon > 0$. By the construction of $\widetilde{P}$ and $\widetilde{Q}$ the map $\widetilde{f}$ induces a homeomorphism $f : P \to Q$, with $\delta_F(f) \leq \delta_F(\widetilde{f})$, which completes the description of the algorithm. The total running time for all the above steps $2^{O(h \log h)} n^{O(h)} 2^{O(h^2)} n^{O(1)} = 2^{O(h^2)} n^{O(h)}$, concluding the proof.

References

1. Agarwal, P.K., Avraham, R.B, Kaplan, H., Sharir, M.: Computing the discrete Fréchet distance in subquadratic time. In: SODA 2013, pp. 156–167. SIAM (2013)
2. Alt, H., Buchin, M.: Semi-computability of the Fréchet distance between surfaces. In: EWCG 2005, Eindhoven, Netherlands, pp. 45–48
3. Alt, H., Godau, M.: Computing the Fréchet distance between two polygonal curves. Int. J. Comput. Geometry Appl. **5**, 75–91 (1995)
4. Aronov, B., Har-Peled, S., Knauer, C., Wang, Y., Wenk, C.: Fréchet distance for curves, revisited. In: Azar, Y., Erlebach, T. (eds.) ESA 2006. LNCS, vol. 4168, pp. 52–63. Springer, Heidelberg (2006)
5. Buchin, K., Buchin, M., Schulz, A.: Fréchet distance of surfaces: some simple hard cases. In: de Berg, M., Meyer, U. (eds.) ESA 2010, Part II. LNCS, vol. 6347, pp. 63–74. Springer, Heidelberg (2010)
6. Buchin, K., Buchin, M., Wenk, C.: Computing the Fréchet distance between simple polygons. Comp. Geom. Theo. Appl. **41**(1–2), 2–20 (2008)
7. Chambers, E.W., de Verdière, E.C., Erickson, J., Lazard, S., Lazarus, F., Thite, S.: Homotopic Fréchet distance between curves or, walking your dog in the woods in polynomial time. Comput. Geom. Theory Appl. **43**(3), 295–311 (2010)
8. Chazal, F., Lieutier, A., Rossignac, J., Whited, B.: Ball-map: Homeomorphism between compatible surfaces. Int. J. Comput. Geometry Appl. **20**(3), 285–306 (2010)
9. Chen, D., Driemel, A., Guibas, L.J., Nguyen, A., Wenk, C.: Approximate map matching with respect to the Fréchet distance. In: ALENEX 2011, pp. 75–83 (2011)
10. Cook IV, A.F., Driemel, A., Har-Peled, S., Sherette, J., Wenk, C.: Computing the Fréchet distance between folded polygons. In: Dehne, F., Iacono, J., Sack, J.-R. (eds.) WADS 2011. LNCS, vol. 6844, pp. 267–278. Springer, Heidelberg (2011)
11. Éric Colin de Verdière and Jeff Erickson: Tightening non-simple paths and cycles on surfaces. SIAM J. Comput. **39**(8), 3784–3813 (2010)
12. Dey, T.K., Ranjan, P., Wang, Y.: Convergence, stability, and discrete approximation of laplace spectra. In: SODA 2010, pp. 650–663 (2010)
13. Dinur, I.: Approximating svp$_{\text{infinity}}$ to within almost-polynomial factors is np-hard. Theor. Comput. Sci. **285**(1), 55–71 (2002)
14. Driemel, A., Har-Peled, S.: Jaywalking your dog: computing the Fréchet distance with shortcuts. In: SODA 2012, pp. 318–337. SIAM (2012)
15. Driemel, A., Har-Peled, S., Wenk, C.: Approximating the Fréchet distance for realistic curves in near linear time. Discrete & Computational Geometry **48**(1), 94–127 (2012)
16. Erickson, J., Nayyeri, A.: Shortest non-crossing walks in the plane. In: SODA 2011, pp. 297–308. SIAM (2011)
17. Floater, M.S., Hormann, K.: Surface parameterization: a tutorial and survey. In: Advances in Multiresolution for Geometric Modelling, Mathematics and Visualization, pp. 157–186. Springer, Heidelberg (2005)
18. Godau, M.: On the Complexity of Measuring the Similarity Between Geometric Objects in Higher Dimensions. Ph.D thesis, Freie Universität Berlin (1998)
19. Har-Peled, S., Nayyeri, A., Salavatipour, M., Sidiropoulos, A.: How to walk your dog in the mountains with no magic leash. In: SoCG 2012, pp. 121–130. ACM, New York (2012)
20. Hass, J., Scott, P.: Intersections of curves on surfaces. Israel Journal of Mathematics **51**(1–2), 90–120 (1985)

21. Hershberger, J., Snoeyink, J.: Computing minimum length paths of a given homotopy class. Comput. Geom. Theory Appl. **4**(2), 63–97 (1994)
22. Schaefer, M., Sedgwick, E., Štefankovič, D.: Spiraling and folding: The word view. Algorithmica (2009) (in press)
23. Schaefer, M., Štefankovič, D.: Decidability of string graphs. J. Comput. Syst. Sci. **68**(2), 319–334 (2004)
24. van Kaick, O., Zhang, H., Hamarneh, G., Cohen-Or, D.: A survey on shape correspondence. Computer Graphics Forum **30**(6), 1681–1707 (2011)

An Improved Private Mechanism for Small Databases

Aleksandar Nikolov$^{(\boxtimes)}$

Microsoft Research, Redmond, WA 98052, USA
`alenik@microsoft.com`

Abstract. We study the problem of answering a workload of linear queries $\mathcal{Q}$, on a database of size at most $n = o(|\mathcal{Q}|)$ drawn from a universe $\mathcal{U}$ under the constraint of (approximate) differential privacy. Nikolov, Talwar, and Zhang [NTZ13] proposed an efficient mechanism that, for any given $\mathcal{Q}$ and n, answers the queries with average error that is at most a factor polynomial in $\log |\mathcal{Q}|$ and $\log |\mathcal{U}|$ worse than the best possible. Here we improve on this guarantee and give a mechanism whose competitiveness ratio is at most polynomial in $\log n$ and $\log |\mathcal{U}|$, and has no dependence on $|\mathcal{Q}|$. Our mechanism is based on the projection mechanism of [NTZ13], but in place of an ad-hoc noise distribution, we use a distribution which is in a sense optimal for the projection mechanism, and analyze it using convex duality and the restricted invertibility principle.

Keywords: Differential privacy · Convex optimization · Competitive analysis

1 Introduction

The central problem of private data analysis is to characterize to what extent it is possible to compute useful information from statistical data without compromising the privacy of the individuals represented in the dataset. In order to formulate this problem precisely, we need a database model and a definition of what it means to preserve privacy. Following prior work, we model a database as a multiset D of n elements from a universe $\mathcal{U}$, with each database element specifying the data of a single individual. Defining privacy is more subtle. A definition which has received considerable attention in recent years is *differential privacy*, which postulates that a randomized algorithm preserves privacy if its distribution on outputs is almost the same (in an appropriate metric) on any two input databases D and D' that differ in the data of at most a single individual. The formal definition is as follows:

Definition 1 ([DMNS06]). *Two databases D and D' are* neighboring *if the size of their symmetric difference is at most one. A randomized algorithm $\mathcal{M}$ satisfies (ε, δ)-differential privacy if for any two neighboring databases D and D' and any measurable event S in the range of $\mathcal{M}$,*

$$\mathbb{P}[\mathcal{M}(D) \in S] \leq e^{\varepsilon}\mathbb{P}[\mathcal{M}(D') \in S] + \delta.$$

© Springer-Verlag Berlin Heidelberg 2015
M.M. Halldórsson et al. (Eds.): ICALP 2015, Part I, LNCS 9134, pp. 1010–1021, 2015.
DOI: 10.1007/978-3-662-47672-7_82

Differential privacy has a number of desirable properties: it is invariant under post-processing, the privacy loss degrades smoothly under (possibly adaptive) composition, and the privacy guarantees hold in the face of arbitrary side information. We will adopt it as our definition of choice in this paper. We will work in the regime $\delta > 0$, which is often called approximate differential privacy, to distinguish it from pure differential privacy, which is the case $\delta = 0$. Approximate differential privacy provides strong semantic guarantees when δ is $n^{-\omega(1)}$: roughly speaking, it implies that with probability at least $1 - O(n\sqrt{\delta})$, an arbitrarily informed adversary cannot guess from the output of the algorithm if any particular user is represented in the database. See [GKS08] for a precise formulation of this semantic guarantee.

We then turn to the question of understanding the constraints imposed by privacy on the kinds of computation we can perform. We focus on computing answers to a fundamental class of database queries: the *linear queries*, which generalize counting queries. A counting query counts the number of database elements that satisfy a given predicate; a linear query is more general and allows for weighted counts. Formally, a linear query is specified by a function $q: \mathcal{U} \to \mathbb{R}$ ($q: \mathcal{U} \to \{0,1\}$ in the case of counting queries); slightly abusing notation, we define the value of the query as $q(D) \triangleq \sum_{e \in D} q(e)$ (elements of D are counted with multiplicity). We call a set $\mathcal{Q}$ of linear queries a *workload*, and an algorithm that answers a query workload a *mechanism*. Since the work of Dinur and Nissim [DN03], it has been known that answering queries too accurately can lead to very dramatic privacy breaches, and this is true even for counting queries. For example, in [DN03, DMT07] it was shown that answering $\Omega(n)$ random counting queries with error per query $o(\sqrt{n})$ allows an adversary to reconstruct a very accurate representation of a database of size n, which contradicts any reasonable privacy notion. On the other hand, a simple mechanism that adds independent Gaussian noise to each query answer achieves (ε, δ)-differential privacy and answers any set $\mathcal{Q}$ of counting queries with average error $O(\sqrt{|\mathcal{Q}|})$ [DN03, DN04, DMNS06].[1] While this is a useful guarantee for a small number of queries, it quickly loses value when $|\mathcal{Q}|$ is much larger than the database size, and becomes trivial for $\omega(n^2)$ queries. Nevertheless, since the seminal paper of Blum, Ligett and Roth [BLR08], a long line of work [DNR09, DRV10, RR10, HR10, GHRU11, HLM12, GRU12] has shown that even when $|\mathcal{Q}| = \omega(n)$, more sophisticated private mechanisms can achieve error not much larger than $O(\sqrt{n})$. For instance, there exist (ε, δ)-differentially private mechanisms for linear queries that acheive average error $O(\sqrt{n} \log^{1/4} |\mathcal{U}|)$ [GRU12]. There are sets of counting queries for which this bound is tight up to factors polylogarithmic in the size of the database [BUV13].

Specific query workloads allow for error which is much better than the worst-case bounds. Some natural examples are queries counting the number of points in a line interval or d-dimensional axis-aligned box [DNPR10, CSS10, XWG10], or a d-dimensional halfspace [MN12]. It is, therefore, desirable to have mechanisms

---

[1] Here and in the remainder of the introduction we ignore dependence of the error on ε and δ.

whose error bounds adapt *both* to the query workload and to the database size. In particular, if opt$(n, \mathcal{Q})$ is the best possible average error[2] achievable under differential privacy for the workload $\mathcal{Q}$ on databases of size at most n, we would like to have a mechanism with error at most a small factor larger than opt$(n, \mathcal{Q})$ for any n and $\mathcal{Q}$. The first result of this type is due to Nikolov, Talwar, and Zhang [NTZ13], who presented a mechanism running in time polynomial in $|\mathcal{U}|$, $|\mathcal{Q}|$, and n, with error at most polylog$(|\mathcal{Q}|, |\mathcal{U}|) \cdot$ opt$(n, \mathcal{Q})$.

Here we improve the results from [NTZ13]:

Theorem 1 (Informal). *There exists a mechanism that, given a database of size n drawn from a universe $\mathcal{U}$, and a workload $\mathcal{Q}$ of linear queries, runs in time polynomial in $|\mathcal{U}|$, $|\mathcal{Q}|$ and n, and has average error per query at most* polylog$(n, |\mathcal{U}|) \cdot$ opt$(n, \mathcal{Q})$.

Notice that the competitiveness ratio in Theorem 1 is *independent of the number of queries*, which can be significantly larger than both n and $|\mathcal{U}|$. This type of guarantee is easier to prove when $n = \Omega(|\mathcal{Q}|)$, in which case there exist nearly optimal mechanisms that are oblivious of the database size [NTZ13]. Therefore, we focus on the more challenging regime of small databases, i.e. $n = o(|\mathcal{Q}|)$.

It is worth making a couple of remarks about the strength of Theorem 1. First, in many applications the query set $\mathcal{Q}$ is represented compactly and $|\mathcal{U}|$ is exponentially large in the size of a natural representation of the input. In such cases running time polynomial in $|\mathcal{U}|$ may be prohibitive. Nevertheless, our work still gives interesting information theoretic bounds on the optimal error, and, moreover, our mechanism can be a starting point for developing more efficient variants. Moreover, under a plausible complexity theoretic hypothesis, our running time guarantee is the best one can hope for without making further assumptions on $\mathcal{Q}$ [Ull13]. A second remark is that our optimal error guarantees are in terms of *average* error, while many papers in the literature consider worst-case error. Proving a result analogous to Theorem 1 for worst-case error remains an interesting open problem.

Another interesting problem is to remove the dependence on the universe size in the competitiveness ratio. It is plausible that this can be done with the projection mechanism and a well-chosen Gaussian noise distribution, but we would need tighter lower bounds, possibly based on fingerprinting codes as in [BUV13].

Techniques. Following the ideas of [NTZ13], our starting point is a generalization of the well-known Gaussian noise mechanism, which adds appropriately scaled correlated Gaussian noise to the queries. By itself, this mechanism is sufficient to guarantee privacy, but its error is too large when $n = o(|\mathcal{Q}|)$. The main insight of [NTZ13] was to use the knowledge that the database is small to reduce the error via a post-processing step. The post-processing is a form of regression: we find the vector of answers that is closest to the noisy answers while still consistent with the database size bound. (In fact the estimator is slightly more

---

[2] We give a formal definition later.

complicated and related to the hybrid estimator of Zhang [Zha13]). Intuitively, when n is small compared to the number of queries, this regression step cancels a significant fraction of the error.

Our first novel contribution is to analyze the error of this mechanism for arbitrary noise distributions and formulate it as a convex function of the covariance matrix of the noise. Then we write a convex program that captures the problem of finding the covariance matrix for which the performance of the mechanism is optimized on the given query workload and database size bound. We use Gaussian noise with this optimal covariance in place of the recursively constructed ad-hoc noise distribution[3] from [NTZ13]. Finally, we relate the dual of the convex program to a spectral lower bound on $\mathrm{opt}(n, \mathcal{Q})$ via the restricted invertibility principle of Bourgain and Tzafriri [BT87]. We stress that while the restricted invertibility principle was used in [NTZ13] as well, here we need a new argument which works for the optimal covariance matrix we compute and gives a smaller competitiveness ratio.

In addition to the improvement in the competitiveness ratio, we believe our approach here is more direct and more natural and brings a better understanding of the performance of the regression-based mechanism for small databases.

2   Preliminaries

We use capital letters for matrices and lower-case letters for vectors and scalars. We use $\langle \cdot, \cdot \rangle$ for the standard inner product between vectors in $\mathbb{R}^n$. For a matrix $M \in \mathbb{R}^{m \times n}$ and a set $S \subseteq [n]$, we use M_S for the submatrix consisting of the columns of A indexed by elements of S. We use the notation $M \succ 0$ to denote that M is a positive definite matrix, and $M \succeq 0$ to denote that it is positive semidefinite. We use $\sigma_{\min}(M)$ for the smallest singular value of M, i.e. $\sigma_{\min}(M) \triangleq \min_x \|Mx\|_2 / \|x\|_2$. We use $\mathrm{tr}(\cdot)$ for the trace operator, and $\|M\|_2$ for the $\ell_2 \to \ell_2$ operator norm of M, i.e. $\|M\|_2 \triangleq \max_x \|Mx\|_2 / \|x\|_2$.

The distribution of a multivariate Gaussian with mean μ and covariance Σ is denoted $N(\mu, \Sigma)$.

2.1   Histograms, the Query Matrix, and the Sensitivity Polytope

It will be convenient to encode the problem of releasing answers to linear queries using linear-algebraic notation. A common and very useful representation of a database D is the *histogram representation*: the histogram of D is a vector $x \in \mathbb{R}^{\mathcal{U}}$ such that for any $e \in \mathcal{U}$, x_e is equal to the number of copies of e in D. Notice that $\|x\|_1 = n$ and also that if x and x' are respectively the histograms of two neighboring databases D and D', then $\|x - x'\|_1 \leq 1$ (here $\|x\|_1 = \sum_e |x_e|$ is the standard ℓ_1 norm). Linear queries are a linear transformation of x. More concretely, let us define the *query matrix* $A \in \mathbb{R}^{\mathcal{Q} \times \mathcal{U}}$ associated with a set of

---

[3] The distribution in [NTZ13] is independent of the database size bound. This could be a reason why their guarantees scale with $\log |\mathcal{Q}|$ rather than $\log n$.

linear queries $\mathcal{Q}$ by $a_{q,e} = q(e)$. Then it is easy to see that the vector Ax gives the answers to the queries $\mathcal{Q}$ on a database D with histogram x.

Since this does not lead to any loss in generality, for the remainder of this chapter we will assume that databases are given to mechanisms as histograms, and workloads of linear queries are given as query matrices. We will identify the space of size-n databases with histograms in the scaled ℓ_1 ball $nB_1^{\mathcal{U}} \triangleq \{x \in \mathbb{R}^{\mathcal{U}} : \|x\|_1 \leq n\}$, and we will identify neighboring databases with histograms x, x' such that $\|x - x'\|_1 \leq 1$.

The *sensitivity polytope* K_A of a query matrix $A \in \mathbb{R}^{\mathcal{Q} \times \mathcal{U}}$ is the convex hull of the columns of A and the columns of $-A$. Equivalently, $K_A \triangleq AB_1^{\mathcal{U}}$, i.e. the image of the unit ℓ_1 ball in $\mathbb{R}^{\mathcal{U}}$ under multiplication by A. Notice that $nK_A = \{Ax : \|x\|_1 \leq n\}$ is the symmetric convex hull[4] of the possible vectors of query answers to the queries in $\mathcal{Q}$ on databases of size at most n.

2.2   Measures of Error and the Spectral Lower Bound

As our basic notion of error we will consider mean squared error. For a mechanism $\mathcal{M}$ and a subset $X \subseteq \mathbb{R}^{\mathcal{U}}$, let us define the error with respect to the query matrix $A \in \mathbb{R}^{\mathcal{Q} \times \mathcal{U}}$ as

$$\mathrm{err}(\mathcal{M}, X, A) \triangleq \sup_{x \in X} \left(\mathbb{E} \frac{1}{|\mathcal{Q}|} \|Ax - \mathcal{M}(A, x)\|_2^2 \right)^{1/2}.$$

where the expectation is taken over the random coins of $\mathcal{M}$. We also write $\mathrm{err}(\mathcal{M}, nB_1^{\mathcal{U}}, A)$ as $\mathrm{err}(\mathcal{M}, n, A)$. The optimal error achievable by any (ε, δ)-differentially private mechanism for the query matrix A and databases of size up to n is

$$\mathrm{opt}_{\varepsilon,\delta}(n, A) \triangleq \inf_{\mathcal{M}} \mathrm{err}(\mathcal{M}, n, A),$$

where the infimum is taken over all (ε, δ)-differentially private mechanisms $\mathcal{M}$.

Arguing directly about $\mathrm{opt}_{\varepsilon,\delta}(n, A)$ appears difficult. For this reason we use the following spectral lower bound from [NTZ13]. This lower bound was implicit in previous papers, for example [KRSU10].

Theorem 2 ([NTZ13]). *There exists a constant c such that for any query matrix $A \in \mathbb{R}^{\mathcal{Q} \times \mathcal{U}}$, any small enough ε, and any δ small enough with respect to ε, $\mathrm{opt}_{\varepsilon,\delta}(n, A) \geq (c/\varepsilon) \mathrm{SpecLB}(\varepsilon n, A)$, where*

$$\mathrm{SpecLB}(k, A) \triangleq \max_{\substack{S \subseteq \mathcal{U} \\ |S| \leq k}} \sqrt{k/|\mathcal{Q}|} \, \sigma_{\min}(A_S).$$

---

[4] The symmetric convex hull of a set of points $v_1, \ldots, v_N$ is equal to the convex hull of $\pm v_1, \ldots, \pm v_N$.

2.3 Composition and the Gaussian Mechanism

An important basic property of differential privacy is that the privacy guarantees degrade smoothly under composition and are not affected by post-processing.

Lemma 1 ([DMNS06, DKM06]). *Let $\mathcal{M}_1(\cdot)$ satisfy $(\varepsilon_1, \delta_1)$-differential privacy, and $\mathcal{M}_2(x, \cdot)$ satisfy $(\varepsilon_2, \delta_2)$-differential privacy for any fixed x. Then the mechanism $\mathcal{M}_2(\mathcal{M}_1(D), D)$ satisfies $(\varepsilon_1 + \varepsilon_2, \delta_1 + \delta_2)$-differential privacy.*

A basic method to achieve (ε, δ)-differential privacy is the Gaussian mechanism. We use the following generalized variant, introduced in [NTZ13].

Theorem 3 ([DN03, DN04, DMNS06, NTZ13]). *Let $\mathcal{Q}$ be a set of queries with query matrix A, and let $\Sigma \in \mathbb{R}^{\mathcal{Q} \times \mathcal{Q}}$, $\Sigma \succ 0$, be such that $a_e^T \Sigma^{-1} a_e \leq 1$ for all columns a_e of A. Then the mechanism $\mathcal{M}_\Sigma(A, x) = Ax + w$ where $w \sim N(0, c_{\varepsilon,\delta}^2 \Sigma)$ and $c_{\varepsilon,\delta} \triangleq \frac{0.5\sqrt{\varepsilon} + \sqrt{2 \ln(1/\delta)}}{\varepsilon}$ satisfies (ε, δ)-differential privacy.*

3 The Projection Mechanism

A key element in our mechanism is the use of least squares estimation to reduce error on small databases. In this section we introduce and analyze a mechanism based on least squares estimation, similar to the hybrid estimator of [Zha13]. Essentially the same mechanism was used in [NTZ13], but the definition and analysis were tied to a particular noise distribution.

Algorithm 1.. Projection Mechanism $\mathcal{M}_\Sigma^{\mathrm{proj}}$

Input: *(Public)* Query matrix $A \in \mathbb{R}^{\mathcal{Q} \times \mathcal{U}}$; matrix $\Sigma \succ 0$ such that $a_e^T \Sigma^{-1} a_e \leq$ for all columns a_e of A.
Input: *(Private)* Histogram x of a database of size $\|x\|_1 \leq n$.
1: Run the generalized Gaussian mechanism (Theorem 3) to compute $\tilde{y} \triangleq \mathcal{M}_\Sigma(A, x)$;
2: Let Π be the orthogonal projection operator onto the span of the eigenvectors corresponding to the $\lfloor \varepsilon n \rfloor$ largest eigenvalues of Σ
3: Compute $\bar{y} \in n(I - \Pi)K_A$, where K_A is the sensitivity polytope of A, and $\bar{y}$ is

$$\bar{y} = \arg\min\{\|z - (I - \Pi)\tilde{y}\|_2^2 : z \in n(I - \Pi)K_A\}.$$

Output: Vector of answers $\Pi\tilde{y} + \bar{y}$.

As shown in [NTZ13, DNT14], Algorithm 1 can be efficiently implemented using the ellipsoid algorithm or the Frank-Wolfe algorithm.

The next lemma, which we prove in the full version of the paper, gives our analysis of the error of the Projection Mechanism.

Lemma 2. *Assume $\Sigma \succ 0$ is such that $a_e^\intercal \Sigma^{-1} a_e \leq 1$ for all columns a_e of A. Then the Projection Mechanism $\mathcal{M}_\Sigma^{proj}$ in Algorithm 1 is (ε, δ)-differentially private. Moreover, for $\varepsilon = O(1)$,*

$$\text{err}(\mathcal{M}_\Sigma^{proj}, n, A) = O\left(\left(1 + \frac{\sqrt{\log |\mathcal{U}|}}{\sqrt{\log 1/\delta}}\right)^{1/2}\right) \cdot \left(\frac{c_{\varepsilon,\delta}^2}{|\mathcal{Q}|} \sum_{i \leq \varepsilon n} \sigma_i\right)^{1/2},$$

where $\sigma_1 \geq \sigma_2 \geq \ldots \geq \sigma_{|\mathcal{Q}|}$ are the eigenvalues of Σ.

Let us give some intuition for the proof of the lemma. The privacy guarantee is almost immediate from Theorem 3, since the output of $alg_\Sigma^{proj}(A, x)$ is just a post-processing of $\mathcal{M}_\Sigma(A, x)$. To analyze the error of the mechanism, we split the error into two terms: $\mathbb{E}\|\Pi\tilde{y} - \Pi Ax\|_2^2$ and $\mathbb{E}\|\bar{y} - (I - \Pi)Ax\|_2^2$. We prove the following bounds:

$$\mathbb{E}\|\Pi\tilde{y} - \Pi y\|_2^2 = c_{\varepsilon,\delta}^2 \sum_{i=1}^k \sigma_i, \tag{1}$$

$$\mathbb{E}\|\bar{y} - (I - \Pi)y\|_2^2 = O\left(\frac{\sqrt{\log |\mathcal{U}|}}{\sqrt{\log 1/\delta}}\right) c_{\varepsilon,\delta}^2 \sum_{i=1}^k \sigma_i. \tag{2}$$

(1) is a direct calculation; (2) is more challenging and uses an analysis of the least squares estimator which was appeared in the context of differential privacy for the first time in [NTZ13]. The details are similar to the analysis of the small database mechanism in [NTZ13].

4   Optimality of the Projection Mechanism

In this section we show that we can choose a covariance matrix Σ so that $\mathcal{M}_\Sigma^{proj}$ has nearly optimal error:

Theorem 4. *Let ε be a small enough constant and let $\delta = |\mathcal{U}|^{o(1)}$ be small enough with respect to ε. For any query matrix $A \in \mathbb{R}^{\mathcal{Q} \times \mathcal{U}}$, and any database size bound n, there exists a covariance matrix $\Sigma \succ 0$ such that the Projection Mechanism $\mathcal{M}_\Sigma^{proj}$ in Algorithm 1 is (ε, δ)-differentially private and has error*

$$\text{err}(\mathcal{M}, n, A) = O((\log n)(\log 1/\delta)^{1/4}(\log |\mathcal{U}|)^{1/4}) \cdot \frac{1}{\varepsilon} \text{SpecLB}(\varepsilon n, A)$$

$$= O((\log n)(\log 1/\delta)^{1/4}(\log |\mathcal{U}|)^{1/4}) \cdot \text{opt}_{\varepsilon,\delta}(n, A)$$

Moreover, Σ can be computed in time polynomial in $|\mathcal{Q}|$.

Theorem 4 is the formal statement of Theorem 1. (Recall again that Algorithm 1 can be implemented in time polynomial in n, $|\mathcal{Q}|$ and $|\mathcal{U}|$, as shown in [NTZ13, DNT14].)

To prove the theorem, we optimize over the choices of Σ that ensure (ε, δ)-differential privacy, and use convex duality and the restricted invertibility principle to relate the optimal covariance to the spectral lower bound.

4.1   Minimizing the Ky Fan Norm

Recall that for an $m \times m$ matrix $\Sigma \succ 0$ with eigenvalues $\sigma_1 \geq \ldots \geq \ldots \geq \sigma_m$, and a positive integer $k \leq m$, the Ky Fan k-norm is defined as $\|\Sigma\|_{(k)} \triangleq \sigma_1 + \ldots + \sigma_k$. The covariance matrix Σ we use in the projection mechanism will be the one achieving $\min\{\|\Sigma\|_{(k)} : a_e^\mathsf{T} \Sigma^{-1} a_e \leq 1 \ \forall e \in \mathcal{U}\}$, where a_e is the column of the query matrix A associated with the universe element e. This choice is directly motivated by Lemma 2. We can write this optimization problem in the following way.

$$\text{Minimize } \|X^{-1}\|_{(k)} \text{ s.t.} \tag{3}$$

$$X \succ 0 \tag{4}$$

$$\forall e \in \mathcal{U} : a_e^\mathsf{T} X a_e \leq 1. \tag{5}$$

The program above has a geometric meaning. For a positive definite matrix X, the set $E(X) \triangleq \{v \in \mathbb{R}^\mathcal{Q} : v^\mathsf{T} X v\}$ is an ellipsoid centered at the origin. The constraint (5) means that $E(X)$ has to contain all columns of the query matrix A. The objective function (3) is equal to the sum of squared lengths of the k longest major axes of $E(X)$. Therefore, we are looking for the smallest ellipsoid centered at the origin that contains the columns of A, where the "size" of the ellipsoid is the sum of squared lengths of the k longest major axes. We will not use this geometric interpretation in the rest of the paper.

The following lemma is proved in the full version.

Lemma 3. *The objective function (3) and constraints (5) are convex over* $X \succ 0$.

Since the program (3)–(5) is convex, its optimal solution can be approximated in polynomial time within any given degree of accuracy using the ellipsoid algorithm [GLS81].

4.2   The Dual of the Ky Fan Norm Minimization Problem

Our next goal is derive a dual characterization of (3)–(5), which we will then relate to the spectral lower bound SpecLB(k, A). It is useful to work with the dual, because it is a maximization problem, so to prove optimality we just need to show that any feasible solution of the dual gives a lower bound on the optimal error under differential privacy.

To define the dual, we first need to introduce a somewhat complicated function of the singular values of a matrix. The next lemma is needed to argue that this function is well-defined. The lemma was proved in [Nik15].

Lemma 4 ([Nik15]). *Let* $\sigma_1 \geq \ldots \sigma_m \geq 0$ *be non-negative reals, and let* $k \leq m$ *be a positive integer. There exists a unique integer* t, $0 \leq t \leq k - 1$, *such that*

$$\sigma_t > \frac{\sum_{i > t} \sigma_i}{k - t} \geq \sigma_{t+1}, \tag{6}$$

with the convention $\sigma_0 = \infty$.

We now introduce a function which will be used in formulating a dual characterization of (3)–(5).

Definition 2. *Let $\Sigma \succeq 0$ be an $m \times m$ positive semidefinite matrix with singular values $\sigma_1 \geq \ldots \geq \sigma_m$, and let $k \leq m$ be a positive integer. The function $h_k(\Sigma)$ is defined as*

$$h_k(\Sigma) \triangleq \sum_{i=1}^{t} \sigma_i^{1/2} + \sqrt{k-t} \left(\sum_{i>t} \sigma_i \right)^{1/2},$$

where t is the unique integer such that $\sigma_t > \frac{\sum_{i>t} \sigma_i}{k-t} \geq \sigma_{t+1}$.

Lemma 4 guarantees that $h_k(\Sigma)$ is a well-defined real-valued function. The next theorem gives a dual characterization of the optimal value of (3)–(5) in terms of h_k. The theorem is proved in the full version of the paper.

Theorem 5. *Let $A = (a_e)_{e \in \mathcal{U}} \in \mathbb{R}^{\mathcal{Q} \times \mathcal{U}}$ be a rank $|\mathcal{Q}|$ matrix, and let μ be the optimal value of (3)–(5). Then,*

$$\mu^2 = \max h_k(AQA^\mathsf{T})^2 \text{ s.t.} \tag{7}$$
$$Q \succeq 0, \; diagonal, \mathrm{tr}(Q) = 1 \tag{8}$$

4.3 Proof of Theorem 4

Our strategy will be to use the dual formulation in Theorem 5 and the restricted invertibility principle to give a lower bound on $\mathrm{SpecLB}(k, A)$. First we state the restricted invertiblity principle and a consequence of it proved in [NT15].

Theorem 6 ([BT87, SS10]). *Let $\epsilon \in (0, 1)$, let M be an $m \times n$ real matrix, and let W be an $n \times n$ diagonal matrix such that $W \succeq 0$ and $\mathrm{tr}(W) = 1$. For any integer k such that $k \leq \epsilon^2 \mathrm{tr}(MWM^\mathsf{T})/\|MWM^\mathsf{T}\|_2$ there exists a subset $S \subseteq [n]$ of size $|S| = k$ such that $\sigma_{\min}(M_S)^2 \geq (1 - \epsilon)^2 \mathrm{tr}(MWM^\mathsf{T})$.*

For the following lemma, which is a consequence of Theorem 6, we need to recall the definition of the trace (nuclear) norm of a matrix M: $\|M\|_{\mathrm{tr}}$ is equal to the sum of singular values of M.

Lemma 5 ([NT15]). *Let M be an m by n real matrix of rank r, and let $W \succeq 0$ be a diagonal matrix such that $\mathrm{tr}(W) = 1$. Then there exists a submatrix M_S of M, $|S| \leq r$, such that $|S|\sigma_{\min}(M_S)^2 \geq c^2\|MW^{1/2}\|_{\mathrm{tr}}^2/(\log r)^2$, for a universal constant $c > 0$.*

Proof (of Theorem 4). Given a database size n and a query matrix A, we compute the covariance matrix Σ as follows. We compute a matrix X which gives an (approximately) optimal solution to (3)–(5) for $k \triangleq \lfloor \varepsilon n \rfloor$, and we set $\Sigma \triangleq X^{-1}$. Since (3)–(5) is a convex optimization problem, it can be solved in time polynomial in $|\mathcal{Q}|$ to any degree of accuracy using the ellipsoid algorithm [GLS81] (or the algorithm of Overton and Womersley [OW93]). By Lemma 2 and the constraints (5), $\mathcal{M}_{\Sigma}^{\mathrm{proj}}$ is (ε, δ)-differentially private with this choice of Σ.

By Lemma 2,

$$\mathrm{err}(\mathcal{M}_{\Sigma}^{\mathrm{proj}}, n, A) = O\left(\left(1 + \frac{\sqrt{\log |U|}}{\sqrt{\log 1/\delta}}\right)^{1/2}\right) \cdot \frac{c_{\varepsilon,\delta}}{\sqrt{|\mathcal{Q}|}} \|\Sigma\|_{(k)}. \tag{9}$$

By Theorem 5, the optimal solution Q of (7)–(8) satisfies

$$\|\Sigma\|_{(k)} = h_k(AQA^{\mathsf{T}}) = \sum_{i=1}^{t} \lambda_i^{1/2} + \sqrt{k-t} \left(\sum_{i>t} \lambda_i\right)^{1/2},$$

where $\lambda_1 \geq \ldots \geq \lambda_m$ are the eigenvalues of AQA^{T} and t, $0 \leq t < k$, is an integer such that $(k-t)\lambda_t > \sum_{i>t} \lambda_i \geq (k-t)\lambda_{t+1}$. At least one of $\sum_{i=1}^{t} \lambda_i^{1/2}$ and $\sqrt{k-t}\left(\sum_{i>t} \lambda_i\right)^{1/2}$ must be bounded from below by $\frac{1}{2}\|\Sigma\|_{(k)}$. Next we consider these two cases separately.

Assume first that $\sum_{i=1}^{t} \lambda_i^{1/2} \geq \frac{1}{2}\|\Sigma\|_{(k)}$. Let Π be the orthogonal projection operator onto the eigenspace of AQA^{T} corresponding to $\lambda_1, \ldots, \lambda_t$. Then, because $\lambda_1 \geq \ldots \geq \lambda_t$ are the nonzero singular values of $\Pi AQ^{1/2}$, we have $\|\Pi AQ^{1/2}\|_{\mathrm{tr}} = \sum_{i=1}^{t} \lambda_i^{1/2} \geq \frac{1}{2}\|\Sigma\|_{(k)}$. By Lemma 5 applied to the matrices $M = \Pi A$ and $W = Q$, there exists a set $S \subseteq \mathcal{U}$ of size at most $|S| \leq \mathrm{rank}\,\Pi A = t < \varepsilon n$, such that

$$\mathrm{SpecLB}(\varepsilon n, A) \geq \sqrt{\frac{|S|}{|\mathcal{Q}|}} \lambda_{\min}(A_S)$$

$$\geq \sqrt{\frac{|S|}{|\mathcal{Q}|}} \lambda_{\min}(\Pi A_S) \geq \frac{c\|\Pi AQ^{1/2}\|_{\mathrm{tr}}}{(\log \varepsilon n)\sqrt{|\mathcal{Q}|}} \geq \frac{c\|\Sigma\|_{(k)}}{2(\log \varepsilon n)\sqrt{|\mathcal{Q}|}} \tag{10}$$

for an absolute constant c.

For the second case, assume that $\sqrt{k-t}\left(\sum_{i>t} \lambda_i\right)^{1/2} \geq \frac{1}{2}\|\Sigma\|_{(k)}$. Let Π now be an orthogonal projection operator onto the eigenspace of AQA^{T} corresponding to $\lambda_{t+1}, \ldots, \lambda_m$. By the choice of t, we have

$$\frac{\mathrm{tr}(\Pi AQA\Pi)}{\|\Pi AQA\Pi\|_2} = \frac{\sum_{i>t} \lambda_i}{\lambda_{t+1}} \geq k - t.$$

By Theorem 6, applied with $M = \Pi A$, $W = Q$, and $\varepsilon = \frac{1}{2}$, there exists a set $S \subseteq U$ of size $\frac{1}{4}(k-t) < k \leq \varepsilon n$ so that

$$\mathrm{SpecLB}_2(\varepsilon n, A) \geq \sqrt{\frac{|S|}{|\mathcal{Q}|}} \lambda_{\min}(A_S)$$

$$\geq \sqrt{\frac{|S|}{|\mathcal{Q}|}} \lambda_{\min}(\Pi A_S) \geq \frac{\sqrt{k-t}\left(\sum_{i>t} \lambda_i\right)^{1/2}}{4\sqrt{|\mathcal{Q}|}} \geq \frac{\|\Sigma\|_{(k)}}{8\sqrt{|\mathcal{Q}|}}. \tag{11}$$

The theorem follows from (9), the fact that at least one of (10) or (11) holds, and Theorem 2. □

Acknowledgments. The author would like to thank the anonymous reviewers for helpful comments.

References

[BLR08] Blum, A., Ligett, K., Roth, A.: A learning theory approach to non-interactive database privacy. In: Proceedings of the 40th Annual ACM Symposium on Theory of Computing, STOC 2008, pp. 609–618. ACM, New York (2008)

[BT87] Bourgain, J., Tzafriri, L.: Invertibility of large submatrices with applications to the geometry of banach spaces and harmonic analysis. Israel journal of mathematics **57**(2), 137–224 (1987)

[BUV13] Bun, M., Ullman, J., Vadhan, S.: Fingerprinting codes and the price of approximate differential privacy (2013). arXiv preprint arXiv:1311.3158

[CSS10] Hubert Chan, T.-H., Shi, E., Song, D.: Private and continual release of statistics. In: Abramsky, S., Gavoille, C., Kirchner, C., Meyer auf der Heide, F., Spirakis, P.G. (eds.) ICALP 2010. LNCS, vol. 6199, pp. 405–417. Springer, Heidelberg (2010)

[DKM06] Dwork, C., Kenthapadi, K., McSherry, F., Mironov, I., Naor, M.: Our data, ourselves: Privacy via distributed noise generation **4004**, 486–503 (2006)

[DMNS06] Dwork, C., McSherry, F., Nissim, K., Smith, A.: Calibrating noise to sensitivity in private data analysis. In: Halevi, S., Rabin, T. (eds.) TCC 2006. LNCS, vol. 3876, pp. 265–284. Springer, Heidelberg (2006)

[DMT07] Dwork, C., McSherry, F., Talwar, K.: The price of privacy and the limits of lp decoding. In: STOC, pp. 85–94 (2007)

[DN03] Dinur, I., Nissim, K.: Revealing information while preserving privacy, pp. 202–210 (2003)

[DN04] Dwork, C., Nissim, K.: Privacy-preserving datamining on vertically partitioned databases. In: Franklin, M. (ed.) CRYPTO 2004. LNCS, vol. 3152, pp. 528–544. Springer, Heidelberg (2004)

[DNPR10] Dwork, C., Naor, M., Pitassi, T., Rothblum, G.N.: Differential privacy under continual observation. In: Schulman, L.J. (eds.) STOC, pp. 715–724. ACM (2010)

[DNR09] Dwork, C., Naor, M., Reingold, O., Rothblum, G.N., Vadhan, S.: On the complexity of differentially private data release: efficient algorithms and hardness results. In: Proceedings of the 41st Annual ACM Symposium on Theory of computing, pp. 381–390. ACM (2009)

[DNT14] Dwork, C., Nikolov, A., Talwar, K.: Using convex relaxations for efficiently and privately releasing marginals. In: Cheng, S.-W., Devillers, O. (eds.) 30th Annual Symposium on Computational Geometry, SOCG 2014, Kyoto, Japan, June 08–11, 2014, pp. 261. ACM (2014)

[DRV10] Dwork, C., Rothblum, G.N., Vadhan, S.: Boosting and differential privacy. In: Proceedings of the 2010 IEEE 51st Annual Symposium on Foundations of Computer Science, FOCS 2010, pp. 51–60. IEEE Computer Society, Washington (2010)

[GHRU11] Gupta, A., Hardt, M., Roth, A., Ullman, J.: Privately releasing conjunctions and the statistical query barrier. In: STOC, pp. 803–812 (2011)

[GKS08] Ganta, S.R., Kasiviswanathan, S.P., Smith, A.: Composition attacks and auxiliary information in data privacy. In: Li, Y., Liu, B., Sarawagi, S. (eds.) Proceedings of the 14th ACM SIGKDD International Conference on Knowledge Discovery and Data Mining, Las Vegas, Nevada, USA, August 24–27, 2008, pp. 265–273. ACM (2008)

[GLS81] Grötschel, M., Lovász, L., Schrijver, A.: The ellipsoid method and its consequences in combinatorial optimization. Combinatorica 1(2), 169–197 (1981)

[GRU12] Gupta, A., Roth, A., Ullman, J.: Iterative constructions and private data release. In: Cramer, R. (ed.) TCC 2012. LNCS, vol. 7194, pp. 339–356. Springer, Heidelberg (2012)

[HLM12] Hardt, M., Ligett, K., McSherry, F.: A simple and practical algorithm for differentially private data release. In: NIPS (2012, to appear)

[HR10] Hardt, M., Rothblum, G.: A multiplicative weights mechanism for privacy-preserving data analysis. In: Proc. 51st Foundations of Computer Science (FOCS). IEEE (2010)

[KRSU10] Kasiviswanathan, S.P., Rudelson, M., Smith, A., Ullman, J.: The price of privately releasing contingency tables and the spectra of random matrices with correlated rows. In: Proceedings of the 42nd ACM Symposium on Theory of Computing, pp. 775–784. ACM (2010)

[MN12] Muthukrishnan, S., Nikolov, A.: Optimal private halfspace counting via discrepancy. In: Karloff, H.J., Pitassi, T. (eds.) Proceedings of the 44th Symposium on Theory of Computing Conference, STOC 2012, New York, NY, USA, May 19–22, 2012, pp. 1285–1292. ACM (2012)

[Nik15] Nikolov, A.: Randomized rounding for the largest j-simplex problem. In: STOC 2015 (2015, to appear)

[NT15] Nikolov, A., Talwar, K.: Approximating hereditary discrepancy via small width ellipsoids. In: Indyk, P. (ed.) Proceedings of the Twenty-Sixth Annual ACM-SIAM Symposium on Discrete Algorithms, SODA 2015, San Diego, CA, USA, January 4–6, 2015, pp. 324–336. SIAM (2015)

[NTZ13] Nikolov, A., Talwar, K., Zhang, L.: The geometry of differential privacy: the sparse and approximate cases. In: Boneh, D., Roughgarden, T., Feigenbaum, J. (eds.) Symposium on Theory of Computing Conference, STOC 2013, Palo Alto, CA, USA, June 1–4, 2013, pp. 351–360. ACM (2013)

[OW93] Overton, M.L., Womersley, R.S.: Optimality conditions and duality theory for minimizing sums of the largest eigenvalues of symmetric matrices. Math. Programming 62(2, Ser. B), 321–357 (1993)

[RR10] Roth, A., Roughgarden, T.: Interactive privacy via the median mechanism. In: Proceedings of the 42nd ACM Symposium on Theory of Computing, STOC 2010, pp. 765–774. ACM, New York (2010)

[SS10] Spielman, D.A., Srivastava, N.: An elementary proof of the restricted invertibility theorem. Israel Journal of Mathematics, 1–9 (2010)

[Ull13] Ullman, J.: Answering $n^{2+o(1)}$ counting queries with differential privacy is hard. In: STOC (2013)

[XWG10] Xiao, X., Wang, G., Gehrke, J.: Differential privacy via wavelet transforms. In: ICDE, pp. 225–236 (2010)

[Zha13] Zhang, L.: Nearly optimal minimax estimator for high dimensional sparse linear regression. Annals of Statistics (2013, to appear)

Binary Pattern Tile Set Synthesis Is NP-hard

Lila Kari[1], Steffen Kopecki[1], Pierre-Étienne Meunier[2],
Matthew J. Patitz[3][✉], and Shinnosuke Seki[2,4]

[1] Department of Computer Science, University of Western Ontario,
London, ON N6A 1Z8, Canada
{lila,steffen}@csd.uwo.ca
[2] Department of Computer Science, Aalto University,
P.O.Box 15400, 00076 Aalto, Finland
pierre-etienne.meunier@aalto.fi
[3] Department of Computer Science and Computer Engineering,
University of Arkansas, Fayetteville, AR, USA
mpatitz@self-assembly.net
[4] Helsinki Institute for Information Technology (HIIT), Espoo, Finland
s.seki@uec.ac.jp

Abstract. We solve an open problem, stated in 2008, about the feasibility of designing efficient algorithmic self-assembling systems which produce 2-dimensional colored patterns. More precisely, we show that the problem of finding the smallest tile assembly system which will self-assemble an input pattern with 2 colors (i.e., 2-PATS) is **NP**-hard. One crucial lemma makes use of a computer-assisted proof, which is a relatively novel but increasingly utilized paradigm for deriving proofs for complex mathematical problems. This tool is especially powerful for attacking combinatorial problems, as exemplified by the proof for the four color theorem and the recent important advance on the Erdős discrepancy problem using computer programs. In this paper, these techniques will be brought to a new order of magnitude, computational tasks corresponding to one CPU-year. We massively parallelize our program, and provide a full proof of its correctness. Its source code is freely available online.

1 Introduction

The traditional way for mankind to modify the physical world has been via a top-down process of crafting things with tools, in which matter is directly manipulated and shaped by those tools. In this work, we are interested in another crafting paradigm called *self-assembly*, a model of building structures from the

We thank Manuel Bertrand for his infinite patience and helpful assistance with setting up the server and helping debug our network and system problems, and Cécile Barbier, Eric Fede and Kai Poutrain for their assistance with software setup.

S. Kopecki—Supported by the NSERC Discovery Grant R2824A01 and UWO Faculty of Science grant to L.K.

P.-É. Meunier—Supported in part by NSF Grant CCF-1219274.

M.J. Patitz—Supported in part by NSF Grants CCF-1117672 and CCF-1422152.

S. Seki—Supported in part by Academy of Finland, Grant 13266670/T30606.

© Springer-Verlag Berlin Heidelberg 2015
M.M. Halldórsson et al. (Eds.): ICALP 2015, Part I, LNCS 9134, pp. 1022–1034, 2015.
DOI: 10.1007/978-3-662-47672-7_83

bottom up. Via self-assembly, it is possible to design molecular systems so that their components autonomously combine to form structures with nanoscale, even atomic, precision. At this scale, tools are no longer the easiest way to build things, and *programming* the assembly of matter becomes at the same time easier, cheaper, and more powerful.

Using this paradigm, researchers have already built a number of things, such as logic circuits [19, 24], DNA tweezers [32], and molecular robots[16], just to name a few. Such examples demonstrate that self-assembly can be used to manufacture specialized geometrical, mechanical, and computational objects at the nanoscale. Potential future applications of nanoscale self-assembly include the production of new materials with specifically tailored properties (electronic, photonic, etc.) and medical technologies which are capable of diagnosing and even treating diseases in vivo, at the cellular level. Furthermore, studying the processes occurring in self-assembling systems yields precious insights about what is physically, even theoretically, possible in these molecular systems. Questions such as "what is the smallest program capable of performing a given task?" arise naturally in these systems, either from experimental applications, or from more fundamental research on the capabilities of natural systems.

The *abstract Tile Assembly Model* (aTAM) was introduced by Winfree [30] to study the possibilities brought by molecular components built by Seeman [25] using DNA. This model is essentially an asynchronous nondeterministic cellular automaton, and can also be seen as a dynamical variant of Wang tiling [29]. In the aTAM, the basic components are translatable but un-rotatable square *tiles* whose sides are labeled with *glues*, each with an integer *strength*. Growth proceeds from a *seed assembly*, one tile at a time, and at each time step a tile can attach to an existing assembly if the sum of the strengths of the glues on its sides, whose types match the existing assembly, is equal to at least a parameter of the model called the *temperature*.

The problem we study in this paper is the optimization of the design of tile assembly systems in the aTAM which self-assemble to form colored input patterns. DNA tiles can be equipped with proteins [31] and nanoparticles such as gold (Au) [33]. Assemblies of normal tiles as well as tiles thus modified can be considered a *colored pattern*, as a periodic placement of Au nanoparticles on a 2D nanogrid [33] can be considered a 2-colored (i.e., binary) rectangular pattern on which the two colors specify the presence/absence of an Au nanoparticle at the position. Various designs of pattern assemblers have been proposed theoretically and experimentally, see, e.g., [4,6,22,33]. The input for this problem is a rectangular pattern consisting of k colors, and the output is a tile set in the aTAM which self-assembles the pattern. Essentially, each type of tile is assigned a "color", and the goal is to design a system consisting of the minimal number of tile types such that they deterministically self-assemble to form a rectangular assembly in which each tile is assigned the same color as the corresponding location in the pattern. This problem was introduced in [17], and has since then been extensively studied [7,9,11,12,26]. The interest is both theoretical, to determine the computational complexity of designing efficient tile assembly systems, and practical, as the goal of self-assembling patterned substrates onto which a potentially wide variety of molecular components could be attached is a major experimental goal. Known as k-Pats, where k is the number of unique colors in the input pattern, previous work has steadily decreased the value of k for which k-Pats has been shown to be **NP**-hard, from unbounded [7] to 11 [12]. (Additionally, in a variant of k-Pats where the number of tile types of certain colors is restricted, is has been proven to be **NP**-hard for 3 colors [14].) However, the foundational conjecture has been that for $k = 2$, i.e. 2-Pats, the problem is also **NP**-hard. This is our main result, which is thus the terminus of this line of research and a fundamental result in algorithmic self-assembly.

Computer-assisted proofs. In one of its parts (a portion of one direction of the **NP** reduction), our proof of the 2-Pats conjecture requires the solution of a massive combinatorial problem, meaning that one of the lemmas upon which it relies needs a massive exploration of more than $6 \cdot 10^{13}$ cases via a computer program. While this is not a traditional component of mathematical proofs, and may not provide the same level of insight into *why* something is true that a standard proof may, modern hardware and software have now given us the tools to attack combinatorially formidable problems whose proofs, if not augmented by computer programs, would often be impossible or as lacking in their ability to elucidate the reasons for their truth due to explosive case analyses as verification by brute force analysis of a computer program. Indeed, computer science has at the same time introduced combinatorial arguments indicating that most theorems do not have simple proofs, and possible ways to produce *certain facts* anyway, by heavy algorithmic processes. Moreover, the "natural proofs" line of research [1,5,20,23] suggests that understanding "why" complexity classes are separated may be out of reach, and that therefore, the study of these kinds of proofs, and methods to ensure their correctness, are a fundamental direction in computer science today. Asserting the correctness of biological and chemical programs is also an important problem, where "*why*" questions are really not as important as the "*whether*" ones, for instance for therapeutic applications. Computationally intensive proofs are therefore likely to become common in these areas of science.

Historically, Appel and Haken [2,3] were the first to prove a result – the four color theorem – with this kind of method, in 1976. This proof was later simplified in [21]. Since then, important problems in various fields have been solved (fully or partially) with the assistance of computers: the discovery of Mersenne primes [28], the **NP**-hardness of minimum-weight triangulation [18], a special case of Erdős' discrepancy conjecture [15], and the ternary Goldbach conjecture [10], among others. (Over the years, exhaustive exploration and massively parallel programs have also been commonly used in physics, or in combinatorial problems such as solving the Rubik's cube.) However, none of these programs was proven formally, and confidence in the validity of these results thus relies on our trust in the programmers.

Proofs of computer programs. The first rigorous proof of a massive software exploration was for the four colors theorem, recently done in the Coq proof assistant by Gonthier et al. [8]. The order of magnitude of their proof is close to the limits of Coq, and is not comparable with our result, which needs a massively parallel exploration requiring about one CPU-year on very modern, high-end machines (as a sum total over several hundred distributed cores) to complete and verify the correctness of the lemma.

A large parallel cluster was hence employed, which poses a number of new challenges. Indeed, in a sequential program, we often implicitly use the fact that function calls return the output of their computations, which becomes more complicated when using several computers: without using unrealistic hypotheses on the correction of the network and of operating systems, return values could potentially be lost, duplicated or corrupted. Since our program ran for a long time, we cannot make such strong hypotheses, which is why we need to assert the authenticity of messages received by the server by using cryptographic signatures.

Another feature of our proof is the use of a *functional programming language*, OCaml. The conciseness of its code and the proximity of its syntax to mathematical proofs brought us a rigorous proof of the correctness of our program.

The whole framework for carrying out the programmatic part of our proof is reusable for the same kind of tasks in the future.

1.1   Main Result

Our result solves an open problem in the field of DNA self-assembly, the so-called *binary pattern tile set synthesis* (2-PATS) problem [17,26], stated first in 2008. In the general k-PATS for $k \geq 2$, given a placement of k different kinds of nanoparticles, represented in the model as a k-colored rectangular pattern, we are asked to design an optimally small tileset and an L-shaped seed that self-assembles the pattern (see Fig. 1 for an example).

2-PATS has been conjectured to be **NP**-hard since 2008[1]. In [26], Seki proved for the first time the **NP**-hardness of 60-PATS, whose input pattern is allowed to have 60 colors, and the result has since been strengthened to that of 29-PATS [11], and further to 11-PATS [12].

Our main theorem closes this line of research by lowering the number of colors allowed for input patterns to only two. We state the main result of this paper here, although some terms may not formally be defined yet:

Theorem 1. *The* 2-PATS *optimization problem of finding, given a 2 colored rectangular pattern P, the minimal colored tileset (together with an L-shaped seed) that produces a single terminal assembly where the color arrangement is exactly the same as in P, is* **NP**-*hard.*

The main idea of our proof is similar to the strategies adopted by [11,12,26]. We embed the computation of a verifier of solutions for an **NP**-complete problem (in our case, a variant of SAT, which we call M-SAT) in an assembly, which is

---

[1] This problem was claimed to be **NP**-hard in a subsequent paper by the authors of [17] but what they proved was the **NP**-hardness of a different problem (see [27]).

relatively straightforward in Winfree's aTAM. One can indeed engineer a tile assembly system (TAS) in this model, with colored tiles, implementing a verifier of solutions of the variant of SAT, in which a formula F and a variable assignment $\phi \in \{0, 1\}^n$ are encoded in the seed assembly, and a tile of a special color appears after some time if and only if $F(\phi) = 1$. In our actual proof, reported in Sect. 3, we design a set T of 13 tile types and a reduction of a given instance ϕ of M-SAT to a rectangular pattern P_F such that

Property 1. A TAS using tile types in T self-assembles P_F iff F is satisfiable.

Property 2. Any TAS of at most 13 tile types that self-assembles P_F is isomorphic to T.

Therefore, F is solvable if and only if P_F can be self-assembled using at most 13 tile types. In previous works [11,12,26], significant portions of the proofs were dedicated to ensuring their analog of Property 2, and many colors were "wasted" to make the property "manually" checkable (for reference, 33 out of 60 colors just served this purpose for the proof of **NP**-hardness of 60-PATS [26] and 2 out of 11 did that for 11-PATS [12]). Cutting this "waste" causes a combinatorial explosion of cases to test and motivates us to use a computer program to do the verification instead.

Apart from the verification of Property 2 (in Lemma 1), the rest of our proof can be verified as done in traditional mathematical proofs; our proof is in Sect. 3. The verification of Property 2 is done by an algorithm (omitted due to space constraints but described, along with all other proof details, in [13]), which, given a pattern and an integer n, searches for all possible sets of n tile types that self-assemble the pattern. The correctness of the algorithm is proven, and both the (unproven, efficient) C++ code, and the (slower but formally proven) OCaml code implementing the algorithm are freely available online[2]. Both versions were implemented independently and neither is the conversion of the code of the other implementation. The full statistics of the runs are available on demand, and summarized by the Parry user interface: http://pats.lif.univ-mrs.fr.

2 Preliminaries

Let $\mathbb{N}$ be the set of nonnegative integers, and for $n \in \mathbb{N}$, let $[n] = \{0, 1, 2, \ldots, n-1\}$. For $k \geq 1$, a k-*colored pattern* is a partial function from $\mathbb{N}^2$ to the set of (color) indices $[k]$, and a k-*colored rectangular pattern* (of width w and height h) is a pattern whose domain is $[w] \times [h]$.

Let Σ be a glue alphabet. A *(colored) tile type* t is a tuple (g_N, g_W, g_S, g_E, c), where $g_N, g_W, g_S, g_E \in \Sigma$ represent the respective north, west, south, and east glue of t, and $c \in \mathbb{N}$ is a color (index) of t. For instance, the right black tile type in Fig. 1 (Left) is $(1, 1, 0, 0, \text{black})$. We refer to g_N, g_W, g_S, g_E as $t(N), t(W), t(S), t(E)$,

---

[2] http://self-assembly.net/wiki/index.php?title=2PATS-tileset-search (C++ version) and   http://self-assembly.net/wiki/index.php?title=2PATS-search-ocaml (OCaml version)

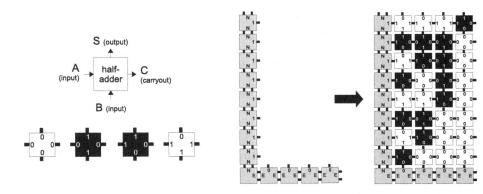

Fig. 1. (Left) Four tile types implement the half-adder with two inputs A, B from the west and south, the output S to the north, and the carryout C to the east. (Right) Copies of the half-adder tiles turn the L-shape seed into the binary counter pattern.

respectively, and by $c(t)$ we denote the color of t. For a set T of tile types, an *assembly* α over T is a partial function from $\mathbb{N}^2$ to T. Its pattern, denoted by $P(\alpha)$, is such that $\text{dom}(P(\alpha)) = \text{dom}(\alpha)$ and $P(\alpha)(x, y) = c(\alpha(x, y))$ for any $(x, y) \in \text{dom}(\alpha)$. Given another assembly β, we say α is a *subassembly* of β if $\text{dom}(\alpha) \subseteq \text{dom}(\beta)$ and, for any $(x, y) \in \text{dom}(\alpha)$, $\beta(x, y) = \alpha(x, y)$.

A *rectilinear tile assembly system* (RTAS) is a pair $\mathcal{T} = (T, \sigma_L)$ of a set T of tile types and an L-shape seed σ_L, which is an assembly over another set of tile types disjoint from T such that $\text{dom}(\sigma)_L = \{(-1, -1)\} \cup ([w] \times \{-1\}) \cup (\{-1\} \times [h])$ for some $w, h \in \mathbb{N}$. The *size* of $\mathcal{T}$ is measured by the number of tile types employed, that is, $|T|$. According to the following general rule that all RTASs obey, it tiles the first quadrant delimited by the seed:

RTAS Tiling Rule: Tile $t \in T$ can attach to an assembly α at position (x, y) if

1. $\alpha(x, y)$ is undefined,
2. both $\alpha(x-1, y)$ and $\alpha(x, y-1)$ are defined,
3. $t(\text{W}) = \alpha(x-1, y)[\text{E}]$ and $t(\text{S}) = \alpha(x, y-1)[\text{N}]$.

The attachment results in a larger assembly β whose domain is $\text{dom}(\alpha) \cup \{(x, y)\}$ such that for any $(x', y') \in \text{dom}(\alpha)$, $\beta(x', y') = \alpha(x, y)$, and $\beta(x, y) = t$. When this attachment takes place in the RTAS $\mathcal{T}$, we write $\alpha \rightarrow_1^{\mathcal{T}} \beta$. Informally speaking, the tile t can attach to the assembly α at (x, y) if on α, both $(x-1, y)$ and $(x, y-1)$ are tiled while (x, y) is not yet, and the west and south glues of t match the east glue of the tile at $(x-1, y)$ and the north glue of the tile at $(x, y-1)$, respectively. This implies that, at the outset, $(0, 0)$ is the sole position where a tile may attach.

Example 1. See Fig. 1 for an RTAS with 4 tile types that self-assembles the binary counter pattern. To its L-shape seed shown there, a black tile of type $(1, 1, 0, 0, \text{black})$ can attach at $(0, 0)$, while no tile of other types can due to glue mismatches. The attachment makes the two positions $(0, 1)$ and $(1,$

0) attachable. Tiling in RTASs thus proceeds from south-west to north-east *rectilinearly* until no attachable position is left.

The set $\mathcal{A}[\mathcal{T}]$ of *producible* assemblies by $\mathcal{T}$ is defined recursively as follows: (1) $\sigma_L \in \mathcal{A}[\mathcal{T}]$, and (2) for $\alpha \in \mathcal{A}[\mathcal{T}]$, if $\alpha \rightarrow_1^{\mathcal{T}} \beta$, then $\beta \in \mathcal{A}[\mathcal{T}]$. A producible assembly $\alpha \in \mathcal{A}[\mathcal{T}]$ is called *terminal* if there is no assembly β such that $\alpha \rightarrow_1^{\mathcal{T}} \beta$. The set of terminal assemblies is denoted by $\mathcal{A}_\square[\mathcal{T}]$. Note that the domain of any producible assembly is a subset of $(\{-1\} \cup [w]) \times (\{-1\} \cup [h])$, starting from the seed σ_L whose domain is $\{(-1, -1)\} \cup ([w] \times \{-1\}) \cup (\{-1\} \times [h])$.

A tile set T is *directed* if for any distinct tile types $t_1, t_2 \in T$, $t_1(\mathtt{W}) \neq t_2(\mathtt{W})$ or $t_1(\mathtt{S}) \neq t_2(\mathtt{S})$ holds. An RTAS $\mathcal{T} = (T, \sigma_L)$ is *directed* if its tile set T is directed (the directedness of RTAS was originally defined in a different but equivalent way). It is clear from the RTAS tiling rule that if $\mathcal{T}$ is directed, then it has exactly one terminal assembly, which we call γ. Let γ' be the subassembly of the terminal assembly such that $\mathrm{dom}(\gamma') \subseteq \mathbb{N}^2$, that is, the tiles on γ' did not originate from the seed σ_L but were tiled by the RTAS. Then we say that $\mathcal{T}$ *uniquely self-assembles the pattern* $P(\gamma')$.

The *pattern self-assembly tile set synthesis* (PATS), proposed by Ma and Lombardi [17], aims at computing the minimum size directed RTAS that uniquely self-assembles a given rectangular pattern. The solution to PATS is required to be directed here, but not originally. However, in [9], it was proved that among all the RTASs that uniquely self-assemble the pattern, the minimum one is directed.

To study the algorithmic complexity of this problem on "real size" particle placement problems, a first restriction that can be placed is on the number of colors allowed for the input patterns, thereby defining the k-PATS problem:

k-COLORED PATS (k-PATS)
GIVEN: a k-colored pattern P
FIND: a smallest directed RTAS that uniquely self-assembles P

The **NP**-hardness of this optimization problem follows from that of its decision variant, which decides, given also an integer m, if such an RTAS is implementable using at most m tile types or not. In the rest of this paper, we use the terminology k-PATS to refer to this decision problem, unless otherwise noted.

3   2-PATS Is NP-hard

We will prove that PATS is **NP**-hard for binary patterns (2-colored patterns). Our proof is a polynomial-time reduction from *monotone satisfiability with few true variables* (M-SAT) to (the decision variant of) 2-PATS. In M-SAT, we consider a number k and a boolean formula F in conjunctive normal form *without negations* and ask whether or not F can be satisfied by only allowing k variables to be true; the **NP**-hardness of M-SAT is proven in [13]. Given an instance of M-SAT we reduce it to a binary pattern $P_{k,F}$ such that a directed RTAS with 13 or less tile types self-assembles $P_{k,F}$ if and only if the answer to the M-SAT instance is yes, i.e., F can be satisfied with exactly k true variables.

We design the pattern $P_{k,F}$ so as to incorporate, as a subpattern, a gadget pattern G shown in Fig. 3. As formally stated in Lemma 1 below, the gadget pattern G has the property that among all the tilesets of size at most 13, exactly one (up to isomorphism) can be employed in a directed RTAS to assemble G, and thus any pattern with G as a subpattern has the same property. Let T be this tileset, shown in Fig. 2. Lemma 1 is verified by an exhaustive search by a computer program whose proof of correctness is omitted due to space constraints (all the other parts of our proof of Theorem 1 are manually checkable).

Lemma 1. *If a directed RTAS whose tileset consists of 13 or less tile types self-assembles the gadget pattern G in Fig. 3, then its tileset is isomorphic to T.*

Due to this property of G, in order to decide the reduced 2-PATS instance $(P_{k,F}, 13)$, it suffices to decide whether a directed RTAS with tileset T self-assembles $P_{k,F}$ or not. This is equivalent to finding an L-shape seed σ_L such that the directed RTAS (T, σ_L) self-assembles $P_{k,F}$. A subtlety of our proof comes from the fact that neither F nor k influence the optimal number of tile types that can assemble $P_{k,F}$ if F is satisfiable.

The tileset T works as an M-SAT verifier, when being used by a directed RTAS. It contains 11 white tile types and 2 black ones.

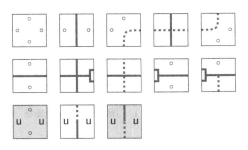

Fig. 2. The tileset T, where the background depicts the color of each tile type and the labels and signals depict the glues (i.e. the glue on a side is equivalent to the label or signal on that side, and the colored signals don't actually appear on the tiles). We refer to the tile types with a gray background as the black tile types. (For better visibility in printouts, the red signals are dotted; blue and green signals can easily be distinguished as blue signals run only horizontally while green signals run only vertically.)

Let us first explain how the RTAS verifies a given M-SAT instance and present its verification visually on its resulting assembly. It does so by "propagating signals" of three kinds (red, green, and blue) via glues from bottom-left to top-right (as the tiles attach in that ordering) and letting them interact with each other. An important fact, that justifies the "signal" vocabulary, is that these signals never fork, i.e. in all the tile types of T, if a signal of type s appears on a west or south glue of a tile $t \in T$, it appears on at most one other side, which is either the east or the north side of t.

We interpret the glues in tile set T as follows. Ten of the white tile types (first and second rows in Fig. 2) simulate three types of signals and their interactions.

Recall that in the RTAS, growth begins from an L-shaped seed and proceeds strictly up and to the right. Therefore, as tiles are added by matching the signals on their bottom and/or left sides, we can think of them as passing the signals to their output (i.e. top and/or right) sides, as indicated by the colored lines showing the signals across each tile. These signals can necessarily, due to the ordering of growth of the assembly and the definitions of the tile types, move only up, right, up and right, or terminate. The signals propagate as follows:

1. blue signals propagate left to right,
2. green signals propagate from bottom to top, and
3. red signals propagate diagonally, bottom left to top right in a wavelike line.

When any two of the signals meet, they simply cross over each other, while the red signal is displaced upwards or rightwards when crossing a blue or green signal, respectively. However when a blue signal crosses a green signal immediately before encountering a red signal, the red signal is destroyed. In order to recognize this configuration, the blue signal is *tagged* when it crosses a green signal; in Fig. 2, the tagging is displayed by the fork in the blue signal. Let us stress that the signals are encoded in the glues of the tiles, and not (at least directly) in their colors.

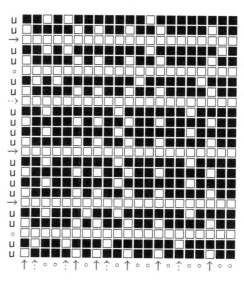

Fig. 3. Binary gadget pattern G, which can only be self-assembled by ≤ 13 tile types by using the tile set T (or one isomorphic to it). To self-assemble G using T one has to use the glues on the L-shaped seed as indicated on the bottom and left. For performance purposes, the bottom row in the pattern was not included in the computerized search; however, because uncovering rows appear in pairs, we add the bottom row here for clarity.

The other three tile types, all with horizontal glues of type u, are used to start rows called "uncovering rows". A major challenge of the reduction is that we cannot force our signals to appear directly in the pattern, because we have only two colors. Instead, we start these "uncovering rows", and make the signals appear in the pattern by their effects on these rows. More specifically, rows with horizontal u glues are always used in pairs:

1. one black tile is above another when no signal is received from below,
2. a white tile is below a black when a green signal is received,
3. a black tile is below a white when a red signal is received.

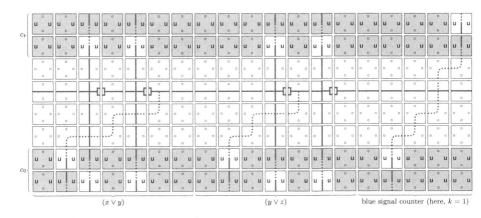

Fig. 4. Subpattern of $P_{k,F}$ for the formula $F = (x \vee y) \wedge (y \vee z)$ with $k = 1$. The position of the blue signal represents the satisfying variable assignment $\phi(y) = 1$. Only the subpattern which encodes F is shown, the gadget pattern and the areas needed to initialize the gadget pattern are omitted here. The different subpatterns shown here are explained in the proof of Theorem 1.

Note that by the definition of the tile set, it's impossible for both signals to be received in the same column. Moreover, blue signals are not "uncovered", since they never reach these rows. Green (resp. red) signals switch to red (resp. green) in the first uncover row, but they switch back to their original state in the second uncover row. This allows the enforcement of the encoding of the three possible values of signals (no signal, green signal, or red signal) with exactly two colors. In our construction, uncovering rows always appear in pairs in order to ensure that the original state of each signal is reestablished after passing through a pair of uncovering rows. In our reduction, we'll use this property to "initialize" a gadget area, above the M-SAT verifier in the pattern, forcing use of tileset T.

An example subassembly which represents the formula $F = (x \vee y) \wedge (y \vee z)$ (without the gadget part) is shown in Fig. 4. A more extensive example of a tile assembly with tileset T can be found in Appendix C of [13] which shows the subpattern of $P_{k,F}$ used for initializing and including the gadget pattern G.

The intuition of the construction of the pattern $P_{k,F}$ and its assembly is that on the vertical arm of the L-shaped seed (i.e., the east border of the upward arm of the seed), variables $x_0, x_1, \ldots, x_{n-1}$ are encoded successively, by the presence of a blue signal if the corresponding variable is set to 1, and a tile with no signal else. Each clause of F is, on the other hand, encoded on the horizontal arm of the L-shape seed as a red signal followed by precisely spaced green signals (intervals between these signals specify which variables are in the clause).

For instance, in Fig. 5, the red signal on the left makes it through (i.e., it is not stopped by a tagged blue signal) and appears in the top uncovering rows, while the one on the right does not. The reason for the red signal being stopped on the right, is that the horizontal spacing between the red and the green signal is

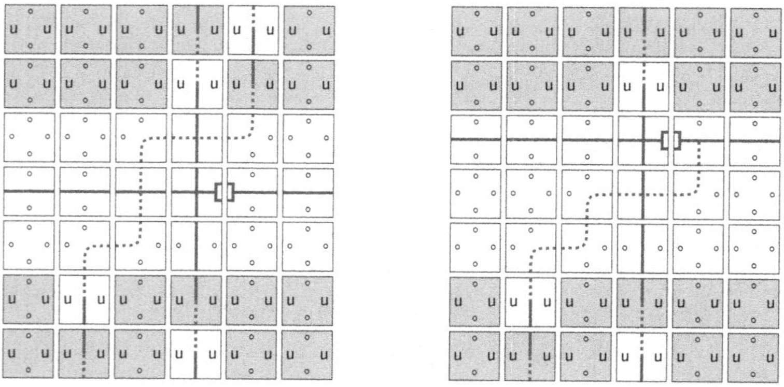

Fig. 5. Example interactions of the signals in the tile set T with uncovering of the configurations: on the left side the red signal can pass through the pattern while the red signal on the right side is destroyed. Note that the position of the blue signal, which is hidden in the horizontal glues, controls whether or not the red signal is destroyed.

"compatible" with the vertical location of blue signal. This compatibility of blue, green, and red signals corresponds to a variable in a clause, represented by the red and green signal, which is set true in the variable assignment, represented by the blue signal. More generally, the absence of red signals on the top uncovering rows c_t means that all the clauses have been satisfied, and the presence of a red signal means that at least one clause could not be satisfied by the assignment. Additionally, note that the positions of blue signals, encoding which variables are set to true in a variable assignment of the M-SAT instance, are not encoded in the pattern, since they travel only through white tiles.

Finally, the part of Fig. 4 which is labeled the "blue signal counter" specifies the number k of true variables in a satisfying variable assignment for F. Note that by the horizontal movement of the red signal from rows c_0 to rows c_t determines the number of blue signals that appear in the white rows in between c_0 and c_t; indeed, the red signal travels one tile to the right in a row without signal, but remains horizontally stationary when passing a row with a blue signal.

References

1. Allender, E., Koucký, M.: Amplifying lower bounds by means of self-reducibility. J. ACM **57**(3), 14:1–14:36 (2010)
2. Appel, K., Haken, W.: Every planar map is four colorable. Part I. discharging. Illinois J. Math. **21**, 429–490 (1977)
3. Appel, K., Haken, W.: Every planar map is four colorable. Part II. reducibility. Illinois J. Math. **21**, 491–567 (1977)
4. Barish, R., Rothemund, P.W.K., Winfree, E.: Two computational primitives for algorithmic self-assembly: Copying and counting. Nano. Lett. **5**(12), 2586–2592 (2005)
5. Chow, T.Y.: Almost-natural proofs. J. Comput. Syst. Sci. **77**(4), 728–737 (2011)

6. Cook, M., Rothemund, P.W.K., Winfree, E.: Self-assembled circuit patterns. In: Chen, J., Reif, J.H. (eds.) DNA 2003. LNCS, vol. 2943, pp. 91–107. Springer, Heidelberg (2004)
7. Czeizler, E., Popa, A.: Synthesizing minimal tile sets for complex patterns in the framework of patterned DNA self-assembly. Theor. Comput. Sci. **499**, 23–37 (2013)
8. Gonthier, G.: Formal proof - the four-color theorem. Not. Am. Math. Soc. **55**(11), 1382–1393 (2008)
9. Göös, M., Lempiäinen, T., Czeizler, E., Orponen, P.: Search methods for tile sets in patterned DNA self-assembly. J. Comput. Syst. Sci. **80**, 297–319 (2014)
10. Helfgott, H.A.: The ternary Goldbach conjecture is true. arXiv:1312.7748 (2013)
11. Johnsen, A.C., Kao, M.-Y., Seki, S.: Computing minimum tile sets to self-assemble color patterns. In: Cai, L., Cheng, S.-W., Lam, T.-W. (eds.) Algorithms and Computation. LNCS, vol. 8283, pp. 699–710. Springer, Heidelberg (2013)
12. Johnsen, A., Kao, M.Y., Seki, S.: A manually-checkable proof for the NP-hardness of 11-colored patterned self-assembly of tile set synthesis. arXiv:1409.1619 (2014)
13. Kari, L., Kopecki, S., Meunier, P.E., Patitz, M.J., Seki, S.: Binary pattern tile set synthesis is NP-hard. arXiv:1404.0967 (2014)
14. Kari, L., Kopecki, S., Seki, S.: 3-color bounded patterned self-assembly. Nat. Comp. (2014) (in Press)
15. Konev, B., Lisitsa, A.: A SAT attack on the Erdös discrepancy conjecture. arXiv: 1402.2184 (2014)
16. Lund, K., Manzo, A.T., Dabby, N., Micholotti, N., Johnson-Buck, A., Nangreave, J., Taylor, S., Pei, R., Stojanovic, M.N., Walter, N.G., Winfree, E., Yan, H.: Molecular robots guided by prescriptive landscapes. Nature **465**, 206–210 (2010)
17. Ma, X., Lombardi, F.: Synthesis of tile sets for DNA self-assembly. IEEE T. Comput. Aid. D. **27**(5), 963–967 (2008)
18. Mulzer, W., Rote, G.: Minimum-weight triangulation is NP-hard. J. ACM 55(2), Article No. 11 (2008)
19. Qian, L., Winfree, E.: Scaling up digital circuit computation with DNA strand displacement cascades. Science **332**(6034), 1196 (2011)
20. Razborov, A.A., Rudich, S.: Natural proofs. In: Proc. STOC 1994, pp. 204–213. ACM, New York (1994)
21. Robertson, N., Sanders, D.P., Seymour, P., Thomas, R.: A new proof of the four-colour theorem. Electron. Res. Announc. AMS. **2**(1), 17–25 (1996)
22. Rothemund, P.W., Papadakis, N., Winfree, E.: Algorithmic self-assembly of DNA Sierpinski triangles. PLoS Biol. **2**(12), 2041–2053 (2004)
23. Rudich, S.: Super-bits, demi-bits, and NP/qpoly-natural proofs. J. Comput. Syst. Sci. **55**, 204–213 (1997)
24. Seelig, G., Soloveichik, D., Zhang, D.Y., Winfree, E.: Enzyme-free nucleic acid logic circuits. Science **314**(5805), 1585–1588 (2006)
25. Seeman, N.C.: Nucleic-acid junctions and lattices. J. Theor. Biol. **99**, 237–247 (1982)
26. Seki, S.: Combinatorial optimization in pattern assembly. In: Mauri, G., Dennunzio, A., Manzoni, L., Porreca, A.E. (eds.) UCNC 2013. LNCS, vol. 7956, pp. 220–231. Springer, Heidelberg (2013)
27. Sterling, A.: https://nanoexplanations.wordpress.com/2011/08/13/dna-self-assembly-of-multicolored-rectangles/
28. Tuckerman, B.: The 24th Mersenne prime. Proc. Nat. Acad. Sci. USA **68**, 2319–2320 (1971)
29. Wang, H.: Proving theorems by pattern recognition - II. AT&T Tech. J. XL(1), 1–41 (1961)

30. Winfree, E.: Algorithmic Self-Assembly of DNA. Ph.D. thesis, California Institute of Technology, June 1998
31. Yan, H., Park, S.H., Finkelson, G., Reif, J.H., LaBean, T.H.: DNA-templated self-assembly of protein arrays and highly conductive nanowires. Science **301**, 1882–1884 (2003)
32. Yurke, B., Turberfield, A.J., Mills, A.P., Simmel, F.C., Neumann, J.L.: A DNA-fuelled molecular machine made of DNA. Nature **406**(6796), 605–608 (2000)
33. Zhang, J., Liu, Y., Ke, Y., Yan, H.: Periodic square-like gold nanoparticle arrays templated by self-assembled 2D DNA nanogrids on a surface. Nano Letters **6**(2), 248–251 (2006)

Near-Optimal Upper Bound on Fourier Dimension of Boolean Functions in Terms of Fourier Sparsity

Swagato Sanyal[(✉)]

School of Technology and Computer Science,
Tata Institute of Fundamental Research, Mumbai, India
swagatos@tcs.tifr.res.in

Abstract. We prove that the Fourier dimension of any Boolean function with Fourier sparsity s is at most $O\left(\sqrt{s}\log s\right)$. This bound is tight up to a factor of $O(\log s)$ as the Fourier dimension and sparsity of the addressing function are quadratically related. We obtain our result by bounding the non-adaptive parity decision tree complexity, which is known to be equivalent to the Fourier dimension. A consequence of our result is that XOR functions have a one way deterministic communication protocol of communication complexity $O(\sqrt{r}\log r)$, where r is the rank of its communication matrix.

1 Introduction

The study of Boolean functions involves studying various properties of Boolean functions and their inter-relationships. Two such properties, which we investigate in this article, are the Fourier dimension and the Fourier sparsity. These two properties were studied by Gopalan *et al.* [2] in the context of property testing. Given a Boolean function $f : \mathbb{F}_2^n \to \{1, -1\}$ with Fourier expansion

$$f(x) = \sum_{\gamma \in \widehat{\mathbb{F}_2^n}} \widehat{f}(\gamma)\chi_\gamma(x),$$

Fourier dimension and Fourier sparsity are defined as follows.

Definition 1 (Fourier dimension and sparsity). *For a Boolean function* $f : \mathbb{F}_2^n \to \{1, -1\}$ *with Fourier expansion*

$$f(x) = \sum_{\gamma \in \widehat{\mathbb{F}_2^n}} \widehat{f}(\gamma)\chi_\gamma(x),$$

the Fourier support *of f, denoted by* $\mathrm{supp}(\widehat{f})$, *is defined as*

$$\mathrm{supp}(\widehat{f}) := \{\gamma \in \widehat{\mathbb{F}_2^n} : \widehat{f}(\gamma) \neq 0\}.$$

© Springer-Verlag Berlin Heidelberg 2015
M.M. Halldórsson et al. (Eds.): ICALP 2015, Part I, LNCS 9134, pp. 1035–1045, 2015.
DOI: 10.1007/978-3-662-47672-7_84

The Fourier sparsity *of f, denoted by* sparsity($\widehat{f}$), *is defined as the size of the Fourier support of f, i.e.,*

$$\text{sparsity}(\widehat{f}) := |\operatorname{supp}(\widehat{f})|,$$

while the Fourier dimension dim($\widehat{f}$) *of f is defined as the dimension of the span of* supp($\widehat{f}$).

The following inequalities easily follow from the definition of Fourier sparsity and dimension.

$$\log_2 \text{sparsity}(\widehat{f}) \leq \dim(\widehat{f}) \leq \text{sparsity}(\widehat{f}). \tag{1}$$

There are functions (e.g., indicator functions of subspaces) for which the first inequality is tight (i.e. holds with equality). For the second inequality, to the best of our knowledge the function having the closest gap between Fourier dimension and sparsity is the *addressing function* $Add_s : \{0,1\}^{\frac{1}{2}\log s + \sqrt{s}} \to \{0,1\}$, defined as

$$Add_s(x, y_1, y_2, \ldots, y_{\sqrt{s}}) := y_x, \quad x \in \{0,1\}^{\frac{1}{2}\log s}, y_i \in \{0,1\}.$$

In other words, at any input (x, y), $Add_s(x, y)$ is the value of the addresee input bit y_x indexed by the addressing variables x. The addressing function[1] has sparsity s and dimension at least $\sqrt{s}$. We prove that that this is the tight upper bound for dim($\widehat{f}$) in terms of sparsity($\widehat{f}$), up to a factor of $O(\log s)$.[2]

Our main result is the following:

Theorem 2. *Let f be a Boolean function with* sparsity($\widehat{f}$) = s. *Then,*

$$\dim(\widehat{f}) = O\left(\sqrt{s}\log s\right).$$

Prior to this work, Gavinsky *et al.* [1] had proved that the sparsity of any Boolean function with full Fourier dimension n is $\Omega(n \log n)$.

Theorem 2 is proved using a lemma of Tsang *et al.* [6] bounding the co-dimension of an affine subspace restricted to which the function reduces to a constant, in terms of Fourier sparsity of the function.

Lemma 3 (Corollary of [6, Lemma 28]). *Let* $f : \mathbb{F}_2^n \to \{1, -1\}$ *be a Boolean function with Fourier sparsity s. Then there is an affine subspace V of* $\mathbb{F}_2^n$ *of co-dimension* $O(\sqrt{s})$ *such that f is constant on V.*

---

[1] To be precise, we should consider the ± 1 valued version of the addressing function described here, where the 0 and 1 in the range are interpreted as $+1$ and -1 respectively.

[2] This is one of the conjectures presented in the open problem session at the Simons workshop on Real Analysis in Testing, Learning and Inapproximability, 2013

Proof idea of Theorem 2: We begin by a simple but crucial observation made by Gopalan *et al.* [2], that the Fourier dimension of a Boolean function is equivalent to its non-adaptive parity decision tree complexity (see Proposition 11). This offers us a potential approach towards upper bounding the Fourier dimension of a Boolean function: exhibiting a shallow non-adaptive parity decision tree of the function.

Towards this end, we first recall the construction of the (adaptive) parity decision tree of Tsang *et al.* [6], which in turn improves on an earlier construction due to Shpilka *et al.* [5, Theorem 1.1]. The broad idea of their construction is as follows: At any point in time, a partial tree is maintained whose leaves are functions which are restrictions of the original function on different affine subspaces. Then a non-constant leaf is picked arbitrarily, and a small set of linear restrictions is obtained by invoking Lemma 3, such that the restricted function at that leaf becomes constant. The next step is observing that if the same function is restricted to all the affine subspaces obtained by setting the same set of parities in all possible ways, the sparsity of each of the corresponding restricted functions is at most half of that of the original function. This is because, in the former restriction, since the function becomes constant, the Fourier coefficients corresponding to non-constant characters must disappear in the restricted space. This can only happen if every non-constant parity gets identified with at least one other parity. This identification leads to halving of the support. Proceeding in this way, a parity decision tree of depth $O(\sqrt{s})$ is obtained.

Note that the choice of parities depends on the leaf (function) chosen, and hence on the outcomes of the preceding queries. Thus the constructed tree is an adaptive one. In this article, we make this tree non-adaptive, at the cost of a logarithmic increase in depth. At each step, we choose an appropriate function (leaf), invoke Lemma 3, and obtain restrictions which make it constant. Then we query the same set of parities at every leaf. Then we argue that this leads to a significant reduction of sparsity. Let $s^{(i)}$ be the Fourier sparsity of the function (leaf) chosen at the i-th step. It can be shown that, in the next step, the size $l^{(i)}$ of the union of the supports of all the leaves falls roughly by $s^{(i)}/4$. From Lemma 3, the number of queries spent in the i-th step is $O(\sqrt{s^{(i)}})$. Using the *Uncertainty Principle* (Theorem 8) one can show that $s^{(i)} \geq \left(l^{(i)}\right)^2/s$. With all these facts it is easy to show that continuing in this fashion, in a small number of steps and making at most $O(\sqrt{s}\log s)$ queries, the size of the union of the Fourier supports of all the leaves becomes so small that we can query all of them, thereby turning all the leaves into constants. The details of the construction of the non-adaptive parity decision tree, and its analysis, is given in Section 3.

Connections to communication complexity and log-rank conjecture: The log-rank conjecture is a long standing and important conjecture in communication complexity. The statement of the conjecture is that the deterministic communication complexity of a Boolean function is asymptotically bounded above by some fixed poly-logarithm of the rank of its communication matrix. The best known upper bound of deterministic communication complexity of a function in terms of the rank is $O(\sqrt{\text{rank}}\log \text{rank})$ due to Lovett [3]. For the special case of XOR functions this result also follows from the work of Tsang et al. [6][3] which improves on the work of Shpilka et al. [5]. A Boolean function $f(x,y)$ on two n bit inputs is an XOR function if there exists a Boolean function F on n bits such that $f(x,y) = F(x \oplus y)$. The rank of the communication matrix of such a function f is known to be equal to the Fourier sparsity s of F. A consequence of Theorem 2 is that XOR functions admit a deterministic one-way protocol of complexity $O(\sqrt{\text{rank}}\log \text{rank}) = O(\sqrt{s}\log s)^4$. We note that both the earlier protocols [3,6] are two-way. The one-way protocol is as follows. Alice and Bob apriori agree on a set of $O(\sqrt{s}\log s)$ monomials S that span the Fourier support of $f(x,y) = F(x \oplus y)$. The existence of S is guaranteed by Theorem 2. Each such monomial M is a product of a monomial M_x in variables in x and a monomial M_y in variables in y. Upon recieving an input x, Alice computes the values of M_x for each $M \in S$, and sends the evaluations to Bob. Bob then, with the help of his input y, can evaluate each monomial $M \in S$. Since every other monomial in $\text{supp}(\widehat{f(x,y)})$ is in the span of S, Bob can compute the values of all those monomials. Finally, Bob evaluates $f(x,y)$ from its Fourier expansion and outputs it.

Some remarks about Lemma 3: Lemma 3 is not believed to be tight. Tsang et al. [6] investigated this question while studying the log-rank conjecture for XOR functions. Tsang et al. [6] suggested a direction towards proving log-rank conjecture for XOR functions. In particular, the authors propose a protocol for such an f based on a parity decision tree of f and show that the communication complexity of the proposed protocol is polylogarithmic in rank of the communication matrix if the following related conjecture is true.

Conjecture 4 ([6, Conjecture 27]). There exists a constant $c > 0$ such that for every Boolean function f with Fourier sparsity s, there exists an affine subspace of co-dimension $O(\log^c s)$ on which f is constant.

Tsang et al. proved the above conjecture for certain classes of functions, which include functions with constant $\mathbb{F}_2$ degree, and prove Lemma 3 for general functions.

We remark that with our proof technique and analysis, any improvement to Lemma 3 (in particular a positive resolution of Conjecture 4), does not yield a better than logarithmic improvement to Theorem 2. If this had not been the case (i.e, our proof actually yielded a super-logarithmic improvement assuming Conjecture 4), then this would have refuted the above conjecture since the addressing function satisfies $\dim = \Theta(\sqrt{\text{sparsity}})$. For further discussion on this topic, the reader is referred to Section 3.

---

[3] For XOR functions, the communication complexity upper bound is in fact $O(\sqrt{\text{rank}})$ which is better than $O(\sqrt{\text{rank}}\log \text{rank})$ by a logarithmic factor.

[4] We thank the anonymous referee for pointing this out.

2   Preliminaries

Let $f : \mathbb{F}_2^n \to \{+1, -1\}$ be a Boolean function. We think of the range $\{+1, -1\}$ as a subset of $\mathbb{R}$. The inputs to f are n variables $x_1, \ldots, x_n$ which take values in $\mathbb{F}_2$. We identify the additive group in $\mathbb{F}_2$ with the group $\{+1, -1\}$ under real number multiplication, and think of the variables as taking $+1$ and -1 values, where 0 and 1 of $\mathbb{F}_2$ get mapped to $+1$ and -1 respectively. We denote this group isomorphism by $(-1)^{(\cdot)}$, i.e., $(-1)^0$ is 1 and $(-1)^1$ is -1. When the x_i's are ± 1, it is well known that every Boolean function $f(x)$ (where x stands for $(x_1, \ldots, x_n)$) can be uniquely written as

$$f(x) = \sum_{S \subseteq [n]} \widehat{f}(S) \prod_{i \in S} x_i.$$

Thus, when the variables are ± 1, f can be written as a multilinear real polynomial. For every $S \subseteq [n]$, the product $\prod_{i \in S} x_i$ is the logical XOR of the bits in S, and $\widehat{f}(S)$ is a real number. These products are exactly the *characters* of $\mathbb{F}_2^n$, which are ± 1 valued versions of the linear forms belonging to the dual vector space $\widehat{\mathbb{F}_2^n}$ of $\mathbb{F}_2^n$. We adopt the following notation in this paper:

$$f(x) = \sum_{\gamma \in \widehat{\mathbb{F}_2^n}} \widehat{f}(\gamma) \chi_\gamma(x).$$

Here, each $\gamma \in \widehat{\mathbb{F}_2^n}$ is a linear function from $\mathbb{F}_2^n$ to $\mathbb{F}_2$, and $\chi_\gamma(\cdot)$ is $(-1)^{\gamma(\cdot)}$.

We recall some standard definitions and facts about the Fourier coefficients.

Definition 5. *Let $f(x) = \sum_{\gamma \in \widehat{\mathbb{F}_2^n}} \widehat{f}(\gamma) \chi_\gamma(x)$ be a Boolean function. The p-th spectral norm $\|\widehat{f}\|_p$ of f is defined as:*

$$\|\widehat{f}\|_p = \left(\sum_{\gamma \in \widehat{\mathbb{F}_2^n}} \left| \widehat{f}(\gamma) \right|^p \right)^{1/p}.$$

Lemma 6 (Parseval's identity). *For a Boolean function f, $\|\widehat{f}\|_2 = 1$.*

The 1st spectral norm of a Boolean function can be bounded in terms of sparsity as follows.

Claim 7. *For a Boolean function f with Fourier sparsity s, $\|\widehat{f}\|_1 \leq \sqrt{s}$.*

Proof.

$$\|\widehat{f}\|_1 \leq \|\widehat{f}\|_2 \cdot \sqrt{s} = \sqrt{s}.$$

The first inequality follows due to Cauchy-Schwarz inequality while the second equality follows from Parseval's identity.   □

For proving our results, we shall use the following version of the Uncertainty Principle. For a proof, the reader is referred to the exercises of chapter 3 of *Analysis of Boolean functions* by O'Donnell [4] where it is given as a hinted exercise.

Theorem 8 (Uncertainty Principle). *Let* $p : \mathbb{R}^n \to \mathbb{R}$ *be a real multilinear non-zero n-variate polynomial with sparsity s (i.e, it has s monomials with non-zero coefficients). Let* U_n *denote the uniform distribution on* $\{+1, -1\}^n$. *Then*

$$\Pr_{x \sim U_n} [p(x) \neq 0] \geq \frac{1}{s}.$$

As stated in the introduction, we need the following theorem due to Tsang *et al.* [6].

Theorem 9 ([6, Lemma28]). *let* $f : \mathbb{F}_2^n \to \{1, -1\}$ *be such that* $\|\widehat{f}\|_1 = A$. *Then there is an affine subspace* V *of* $\mathbb{F}_2^n$ *of co-dimension* $O(A)$ *such that* f *is constant on* V.

Lemma 3 is a simple corollary of this theorem via Claim 7.

For the sake of completion, we provide a proof of an observation made by Gopalan *et al.* [2] connecting the non-adaptive parity decision tree complexity of a function (defined below) and Fourier dimension, that we crucially use in our proofs.

Definition 10 (non-adaptive parity decision tree complexity). *Let* f *be a Boolean function. The* non-adaptive parity decision tree complexity *of* f, *(denoted by* $naDT_\oplus(f)$), *is defined as the minimum integer* t *such that there exist* t *linear forms* $\gamma_1, \ldots, \gamma_t \in \widehat{\mathbb{F}_2^n}$ *such that* f *is a junta of* $\gamma_1, \ldots, \gamma_t$. *In other words, on every input, specifying the outputs of the* γ_i's *specifies the output of* f.

Proposition 11 ([2]). *For a Boolean function* f, $naDT_\oplus(f) = \dim(\widehat{f})$.

Proof. If the outputs of a basis of span of $\text{supp}(\widehat{f})$ are specified, then that clearly specifies the outputs of all characters in $\text{supp}(\widehat{f})$, and hence it specifies the output of the function. Thus $naDT_\oplus(f) \leq \dim(\widehat{f})$.

Now, Let $naDT_\oplus(f) = t$. Let the outputs of $\gamma_1, \ldots, \gamma_t$ specify the output of f. These linear forms are linearly independent as vectors in $\widehat{\mathbb{F}_2^n}$ (else a smaller number of them would decide the output of f). Arbitrarily extend $\gamma_1, \ldots, \gamma_t$ to a basis $\gamma_1, \ldots, \gamma_n$ of $\widehat{\mathbb{F}_2^n}$. For $x = (x_1, \ldots, x_n) \in \mathbb{F}_2^n$, let $L(x) = (\gamma_1(x), \ldots, \gamma_n(x))$. L is easily seen to be an invertible linear transformation from $\mathbb{F}_2^n$ onto itself. Now, $\forall x \in \mathbb{F}_2^n, \forall i = 1, \ldots, n, \gamma_i(x) = (L(x))_i$. Replacing x by $L^{-1}(x)$ we have $\gamma_i(L^{-1}(x)) = x_i$. Now consider the Boolean function $g(x) = f(L^{-1}(x)) = \sum_{\gamma \in \widehat{\mathbb{F}_2^n}} \widehat{f}(\gamma)(-1)^{\gamma(L^{-1}(x))}$. Clearly $\dim(\widehat{g}) = \dim(\widehat{f})$ as L is a full-rank linear transformation. Also, g is completely specified by the outputs of $\gamma_i(L^{-1}(x))$'s for $i = 1, \ldots, t$. Since $\gamma_i(L^{-1}(x)) = x_i$, we have that g is a junta of $x_1, \ldots, x_t$.

Thus all the monomials in $\text{supp}(\widehat{g})$ contain only the variables $x_1, \ldots, x_t$. Thus $\dim(\widehat{f}) = \dim(\widehat{g}) \leq t = \text{naDT}_\oplus(f)$.

The proposition follows by combining the two inequalities. $\qquad\square$

3   Upper Bounding Parity Decision Tree Complexity

In this section, we upper bound the non-adaptive parity decision tree complexity of a Boolean function f with Fourier sparsity s. Consider the following procedure, parametrized by a parameter $\tau \in \mathbb{N}$, that constructs a non-adaptive parity decision tree of f.[5]

> $\text{naDT}_\oplus\text{-PROCEDURE}_\tau(f)$
> Input: Boolean function $f : \mathbb{F}_2^n \to \{+1, -1\}$; Parameter: $\tau \in \mathbb{N}$
> 1. Set $\Gamma \leftarrow \varnothing$, $\mathcal{S} \leftarrow \text{supp}(\widehat{f})$ and $\mathcal{F} \leftarrow \{f\}$.
> 2. While $|\mathcal{S}| > \tau$, do
>   (a) Let g be a function in $\mathcal{F}$ with the largest Fourier sparsity. Let A be a largest affine subspace on which g is constant (breaking ties arbitrarily). Let $\gamma_1, \ldots, \gamma_{n_g}$ be linear functions and $b_1, \ldots, b_{n_g} \in \mathbb{F}_2$ be such that $A = \{x \in \mathbb{F}_2^n : \gamma_1(x) = b_1, \ldots, \gamma_{n_g}(x) = b_{n_g}\}$. Query $\gamma_1, \ldots, \gamma_{n_g}$.
>   (b) Set $\Gamma \leftarrow \Gamma \cup \{\gamma_1, \ldots, \gamma_{n_g}\}$.
>   (c) For each $b = (b_\gamma)_{\gamma \in \Gamma} \in \mathbb{F}_2^{|\Gamma|}$, let V_b be the affine subspace $\{x \in \mathbb{F}_2^n : \forall \gamma \in \Gamma, \gamma(x) = b_\gamma\}$. Set $\mathcal{F} \leftarrow \bigcup_{b \in \mathbb{F}_2^{|\Gamma|}} \{f|_{V_b}\}$.
>   (d) $\mathcal{S} \leftarrow \bigcup_{h \in \mathcal{F}} \text{supp}(\widehat{h})$.
> 3. Query all the parities in $\mathcal{S}$.

Notation: After each iteration of the while loop in the procedure, Γ is the set of parities that have been queried so far, $\mathcal{F}$ is the set of all restrictions of f to the affine subspaces obtained by different assignments to parities in Γ, and $\mathcal{S}$ the union of the Fourier supports of functions in $\mathcal{F}$. Let $\Gamma^{(i)}, \mathcal{F}^{(i)}$ and $\mathcal{S}^{(i)}$ denote $\Gamma, \mathcal{F}$ and $\mathcal{S}$ resepectively at the end of the i-th iteration of the while loop. Let $\Gamma^{(0)} = \varnothing$, $\mathcal{F}^{(0)} = \{f\}$ and $\mathcal{S}^{(0)} = \text{supp}(\widehat{f})$.

For each i, let $b = (b_\gamma)_{\gamma \in \Gamma^{(i)}} \in \mathbb{F}_2^{|\Gamma^{(i)}|}$ and let V_b be the affine subspace defined by linear constraints $\{\gamma(x) = b_\gamma : \gamma \in \Gamma^{(i)}\}$. In V_b, more than one linear functions of the original space may get identified as same.[6] More specifically, δ_1 and δ_2 get identified as same in V_b if and only if $\delta_1 + \delta_2 \in \text{span}\,\Gamma^{(i)}$, i.e. they belong to the same coset of the subspace $\text{span}\,\Gamma^{(i)}$. Thus, $\text{supp}(\widehat{f})$ gets partitioned into equivalence classes, such that for each class, for every $b \in \mathbb{F}_2^{|\Gamma^{(i)}|}$, the linear functions belonging to that class are identified as same in V_b.

---

[5] We will set τ to $\Theta(\sqrt{s})$ to obtain our results.

[6] By 'same' we also include their being negations of each other as the smaller subspace is an affine space and not always a vector space.

Let $l^{(i)}$ denote the number of cosets of the subspace span $\Gamma^{(i)}$ with which supp($\widehat{f}$) has non-empty intersection. For $j = 1, \ldots, l^{(i)}$, let $\beta_j^{(i)}$ be some representative element in supp($\widehat{f}$) of the j-th coset of span $\Gamma^{(i)}$ having non-empty intersection with supp($\widehat{f}$). For each j, let $\beta_j^{(i)} + \alpha_{j,1}^{(i)}, \ldots, \beta_j^{(i)} + \alpha_{j,k_j}^{(i)}$ be the $k_j^{(i)} (\geq 1)$ elements in supp($\widehat{f}$) which are in the same coset of span $\Gamma^{(i)}$ as $\beta_j^{(i)}$.

For each i, j, define the polynomials $P_j^{(i)}(x) := \sum_{l=1}^{k_j} \widehat{f}\left(\beta_j^{(i)} + \alpha_{j,l}^{(i)}\right) \chi_{\alpha_{j,l}^{(i)}}(x)$. Note that the polynomials $P_j^{(i)}$, $j = 1, \ldots, l^{(i)}$, are non-zero.

Given this notation, we can then write the Fourier expansion of f in the following form:

$$f(x) = \sum_{j=1}^{l^{(i)}} P_j^{(i)}(x) \chi_{\beta_j^{(i)}}(x).$$

Notice that each $P_j^{(i)}$ is actually a polynomial in variables in $b = (b_\gamma)_{\gamma \in \Gamma^{(i)}} \in \mathbb{F}_2^{|\Gamma^{(i)}|}$. Thus the value of each $P_j^{(i)}$ is fixed as b is fixed. In other words, each $P_j^{(i)}$ is a constant function in each V_b.

Observation 12. $\forall i, \sum_{j=1}^{l^{(i)}} k_j^{(i)} = s$.

Proposition 13. $|\mathcal{S}^{(i)}| = l^{(i)}$.

Proof. Clearly $|\mathcal{S}^{(i)}| \leq l^{(i)}$. Now, for $j = 1, \ldots, l^{(i)}$, since the polynomial $P_j^{(i)}$ is non-zero, there exists an assignment b to the parities in $\Gamma^{(i)}$ on which $P_j^{(i)}$ evaluates to a non-zero value. Thus the coefficient of $\beta_j^{(i)}$ is non-zero in the restriction of f to the affine subspace obtained by assigning b to the parities in $\Gamma^{(i)}$. Thus for $j = 1, \ldots, l^{(i)}$, $\beta_j^{(i)} \in \mathcal{S}^{(i)}$ which, together with $|\mathcal{S}^{(i)}| \leq l^{(i)}$, implies $|\mathcal{S}^{(i)}| = l^{(i)}$. □

We now argue that after every iteration of the while loop, there exists a function $h \in \mathcal{F}^{(i)}$ which has large Fourier support.

Lemma 14. *After i-th iteration, there exists a $h \in \mathcal{F}^{(i)}$ such that $|\operatorname{supp}(\widehat{h})|$ is at least $(l^{(i)})^2/s$.*

Proof. Consider any function $f|_{V_b} \in \mathcal{F}^{(i)}$. The Fourier decomposition of $f|_{V_b}$ is given by $f|_{V_b} = \sum_{j=1}^{l^{(i)}} P_j^{(i)}(b) \chi_{\beta_j^{(i)}}(x)$. Thus, $|\operatorname{supp}(\widehat{f|_{V_b}})|$ is exactly the number of polynomials $P_j^{(i)}, j = 1, \ldots, l^{(i)}$ such that $P_j^{(i)}(b)$ is non-zero. We analyze this quantity as follows. Pick a $b \in \mathbb{F}_2^{|\Gamma^{(i)}|}$ uniformly at random. For each j,

$j = 1, \ldots, l^{(i)}$, by Theorem 8, $\mathrm{Pr}_b[P_j^{(i)}(b) \neq 0] \geq \frac{1}{k_j^{(i)}}$ (since each $P_j^{(i)}$ is a non-zero polynomial). Thus,

$$
\mathbb{E}_b\left[|\mathrm{supp}(\widehat{f|_{V_b}})|\right] \geq \sum_{j=1}^{l^{(i)}} \frac{1}{k_j^{(i)}} \geq l^{(i)} \cdot \frac{1}{\left(\sum_{j=1}^{l^{(i)}} k_j^{(i)}\right)/l^{(i)}} \qquad \text{[By convexity of } 1/x\text{]}
$$

$$
= \frac{\left(l^{(i)}\right)^2}{s} \qquad\qquad\qquad \text{[By Observation 12].}
$$

Hence, there exists a $h \in \mathcal{F}^{(i)}$ such that $|\mathrm{supp}(\widehat{h})|$ is at least $\left(l^{(i)}\right)^2/s$. $\qquad \square$

Let $g^{(i)}$ be the function chosen in step 2a of the i-th iteration of naDT$_\oplus$-PROCEDURE. Let sparsity$(\widehat{g^{(i)}}) = s^{(i)}$, and $\Delta l^{(i)} = l^{(i-1)} - l^{(i)}$ for $i \geq 1$. The next Lemma proves that, if a function with large Fourier support is picked in step 2a, then that leads to a large reduction in size of $\mathcal{S}$.

Lemma 15. *Assume that* naDT$_\oplus$-PROCEDURE$_\tau(f)$ *is run with* $\tau \geq \sqrt{2s}$. *Assume that it runs for t iterations. Then for $i = 1, \ldots, t$, $\Delta l^{(i)} \geq \frac{s^{(i)}}{4}$.*

Proof. Let $\gamma_1, \ldots, \gamma_{n_{g^{(i)}}}$ be the parities queried in iteration i. Hence there is $b = (b_1, \ldots, b_{n_{g^{(i)}}}) \in (\mathbb{F}_2)^{n_{g^{(i)}}}$ such that $g^{(i)}$ is constant on the affine subspace V_b obtained by setting each γ_j to b_j for $j = 1, \ldots, n_{g^{(i)}}$. Since $g^{(i)}$ is constant on V_b, each non-zero parity in it's Fourier support must disappear in V_b. Thus, for every $b' = (b')_j \in (\mathbb{F}_2)^{n_{g^{(i)}}}$, in the affine space $V_{b'}$ obtained by restricting each γ_j to b'_j, every non-zero parity in $\mathrm{supp}(\widehat{g^{(i)}})$ is matched to some other parity in $\mathrm{supp}(\widehat{g^{(i)}})$. Since $\mathrm{supp}(\widehat{g^{(i)}}) \subseteq \mathcal{S}^{(i-1)}$, it follows that $|\mathcal{S}^{(i)}|$ is at least $\frac{|\mathrm{supp}(\widehat{g^{(i)}})|-1}{2}$ less than $|\mathcal{S}^{(i-1)}|$. By Proposition 13 this implies $\Delta l^{(i)} \geq \frac{s^{(i)}-1}{2}$. Now, $\tau \geq \sqrt{2s}$ implies that for each i, $i = 1, \ldots, t$, $l^{(i-1)} \geq \sqrt{2s}$. From Lemma 14, we have that $s^{(i)} \geq \frac{\left(l^{(i-1)}\right)^2}{s} \geq 2$. Thus $\Delta l^{(i)} \geq \frac{s^{(i)}-1}{2} \geq \frac{s^{(i)}}{4}$. $\qquad \square$

Now we are ready to prove Theorem 2.

Proof (Theorem 2). Run naDT$_\oplus$-PROCEDURE with parameter $\tau = \lceil \sqrt{2s} \rceil$. We first prove that the total number of queries made in the *while* loop of the procedure is $O(\sqrt{s} \log s)$. Assume that the while loop runs for t iterations. Let the number of queries made in step 2a in i-th iteration of the procedure be $\Delta q^{(i)}$. By Lemma 3, $\Delta q^{(i)} = O(\sqrt{s^{(i)}})$. By Lemma 15, $\Delta l^{(i)} \geq \frac{s^{(i)}}{4}$. Hence, $\frac{\Delta q^{(i)}}{\Delta l^{(i)}} = \frac{1}{\Omega(\sqrt{s^{(i)}})}$. From Lemma 14 we have $s^{(i)} \geq \left(l^{(i-1)}\right)^2/s$. Hence $\Delta q^{(i)} = \sqrt{s} \cdot O\left(\Delta l^{(i)}/l^{(i-1)}\right)$. Thus the total number of queries made within the *while* loop of the procedure is

$$\sum_{i=1}^{t} \Delta q^{(i)} = \sqrt{s} \cdot \sum_{i=1}^{t} O\left(\frac{\Delta l^{(i)}}{l^{(i-1)}}\right) = \sqrt{s} \cdot \sum_{i=1}^{t} O\left(\frac{\Delta l^{(i)}}{s - \sum_{j=1}^{i-1} \Delta l^{(j)}}\right)$$

$$\leq \sqrt{s} \cdot \sum_{i=1}^{t} O\left(\frac{1}{s - \sum_{j=1}^{i-1} \Delta l^{(j)}} + \frac{1}{s - \sum_{j=1}^{i-1} \Delta l^{(j)} - 1} + \dots \right.$$

$$\left. \dots + \frac{1}{s - \sum_{j=1}^{i-1} \Delta l^{(j)} - (\Delta l^{(i)} - 1)}\right)$$

$$\leq \sqrt{s} \cdot \sum_{\ell=1}^{s} O\left(\frac{1}{\ell}\right) = O(\sqrt{s} \log s).$$

Finally, the number of queries made in step 3 of the procedure is $O(\sqrt{s})$ as $\tau = O(\sqrt{s})$. From Proposition 11 it follows that $\dim(\widehat{f}) = O(\sqrt{s} \log s)$. ∎

Discussion: As mentioned in the Introduction, a natural approach towards disproving Conjecture 4 is to assume it to be true, and prove that it implies a $o(\sqrt{s})$ upper bound on Fourier dimention. This will refute the conjecture, since, for addressing function (see Section 1), $\dim(\widehat{Add_s}) = \Theta(\sqrt{\text{sparsity}(\widehat{Add_s})})$. However, we cannot disprove the conjecture by an analysis of naDT$_\oplus$-PROCEDURE, assuming the conjecture. To see this let us consider the execution of the procedure on the addressing function. Recall that $Add_s(x, y_1, y_2, \dots, y_{\sqrt{s}}) = y_x$, $x \in \{0,1\}^{\frac{1}{2} \log s}, y_i \in \{0,1\}$. One easily sees that a largest affine subspace V on which the function is constant is the one defined by the constraints $x = x'$, $y_{x'} = b$ where $x' \in \{0,1\}^{\frac{1}{2} \log s}$ and $b \in \{0,1\}$. The function takes the value b everywhere in V. Also, if the addressing bits x and the bit $y_{x'}$ are set to other values than x' and b, the restricted functions in the respective affine subspaces are all constants (if x is set to x') or dictators (on addressee bits). This constitutes the first step of naDT$_\oplus$-PROCEDURE. Since the size of the union of supports of all restricted functions already drops to $\sqrt{s}$, the subsequent steps are querying different dictators on the addresee bits.

The addressing function clearly satisfies Conjecture 4, and all the intermediate functions that are given rise to by naDT$_\oplus$-PROCEDURE are dictators, which also trivially satisfy the conjecture. Thus this rules out the possibility of refuting Conjecture 4 by analysing $\text{naDT}_{\oplus}$-PROCEDURE assuming the conjecture. We notice, however, that if we assume the conjecture, we can improve the upper bound by a factor of $O(\log s)$, to the optimal $O(\sqrt{s})$. The complexity of $\text{naDT}_{\oplus}$-PROCEDURE will then be dominated by the complexity of step 3 which is $O(\sqrt{s})$.

Acknowledgements. The author is grateful to Avishay Tal for noticing a weakness in an earlier analysis of $\text{naDT}_{\oplus}$-PROCEDURE which proved a $O(s^{2/3})$ upper bound, observing that the analysis can be tightened to obtain $O(\sqrt{s}\log s)$ upper bound, and bringing it to the author's notice. The author would like to thank the anonymous referee for pointing out that Theorem 2 implies existence of one-way communication protocols of complexity $O(\sqrt{\text{rank}}\log \text{rank})$ for XOR functions. The author would like to thank Arkadev Chattopadhyay and Prahladh Harsha for many helpful discussions. The author is thankful to Prahladh Harsha for his help in improving the presentation of this article significantly.

References

1. Gavinsky, D., Kirshner, N., de Wolf, R., Samorodnitsky, A.: Private Communication
2. Gopalan, P., O'Donnell, R., Servedio, R.A., Shpilka, A., Wimmer, K.: Testing fourier dimensionality and sparsity. In: Albers, S., Marchetti-Spaccamela, A., Matias, Y., Nikoletseas, S., Thomas, W. (eds.) ICALP 2009, Part I. LNCS, vol. 5555, pp. 500–512. Springer, Heidelberg (2009)
3. Lovett, S.: Communication is bounded by root of rank. In: Symposium on Theory of Computing, STOC 2014, New York, NY, USA, May 31 – June 03, 2014, pp. 842–846 (2014)
4. O'Donnell, R.: Analysis of Boolean Functions. Cambridge University Press (2014). http://www.cambridge.org/de/academic/subjects/computer-science/algorithmics-complexity-computer-algebra-and-computational-g/analysis-boolean-functions
5. Shpilka, A., Tal, A., lee Volk, B..: On the structure of boolean functions with small spectral norm. In: Innovations in Theoretical Computer Science, ITCS 2014, Princeton, NJ, USA, January 12–14, pp. 37–48 (2014). arxiv.org/abs/1304.0371
6. Tsang, H.Y., Wong, C.H., Xie, N., Zhang, S.: Fourier sparsity, spectral norm, and the log-rank conjecture. In: 54th Annual IEEE Symposium on Foundations of Computer Science, FOCS 2013, 26–29 October, 2013, Berkeley, CA, USA, pp. 658–667 (2013)

Condensed Unpredictability

Maciej Skórski[1], Alexander Golovnev[2]([✉]), and Krzysztof Pietrzak[3]

[1] University of Warsaw, Warszawa, Poland
maciej.skorski@gmail.com
[2] New York University, New York, USA
alexgolovnev@gmail.com
[3] IST Austria, Klosterneuburg, Austria
pietrzak@ist.ac.at

Abstract. We consider the task of deriving a key with high HILL entropy (i.e., being computationally indistinguishable from a key with high min-entropy) from an unpredictable source.

Previous to this work, the only known way to transform unpredictability into a key that was ϵ indistinguishable from having min-entropy was via pseudorandomness, for example by Goldreich-Levin (GL) hardcore bits. This approach has the inherent limitation that from a source with k bits of unpredictability entropy one can derive a key of length (and thus HILL entropy) at most $k - 2\log(1/\epsilon)$ bits. In many settings, e.g. when dealing with biometric data, such a $2\log(1/\epsilon)$ bit entropy loss in not an option. Our main technical contribution is a theorem that states that in the high entropy regime, unpredictability implies HILL entropy. Concretely, any variable K with $|K| - d$ bits of unpredictability entropy has the same amount of so called metric entropy (against real-valued, deterministic distinguishers), which is known to imply the same amount of HILL entropy. The loss in circuit size in this argument is exponential in the entropy gap d, and thus this result only applies for small d (i.e., where the size of distinguishers considered is exponential in d).

To overcome the above restriction, we investigate if it's possible to first "condense" unpredictability entropy and make the entropy gap small. We show that any source with k bits of unpredictability can be condensed into a source of length k with $k - 3$ bits of unpredictability entropy. Our condenser simply "abuses" the GL construction and derives a k bit key from a source with k bits of unpredicatibily. The original GL theorem implies nothing when extracting that many bits, but we show that in this regime, GL still behaves like a "condenser" for unpredictability. This result comes with two caveats (1) the loss in circuit size is exponential in k and (2) we require that the source we start with has *no* HILL entropy (equivalently, one can efficiently check if a guess is correct). We leave it as an intriguing open problem to overcome these restrictions or to prove they're inherent.

The full version of the paper is available at http://eprint.iacr.org/2015/384

M. Skórski—Research supported by the WELCOME/2010-4/2 grant.

K. Pietrzak—Research supported by ERC starting grant (259668-PSPC).

M.M. Halldórsson et al. (Eds.): ICALP 2015, Part I, LNCS 9134, pp. 1046–1057, 2015.
DOI: 10.1007/978-3-662-47672-7_85

1   Introduction

Key-derivation considers the following fundamental problem: Given a joint distribution (X, Z) where $X|Z$ (which is short for "X conditioned on Z") is guaranteed to have some kind of entropy, derive a "good" key $K = h(X, S)$ from X by means of some efficient key-derivation function h, possibly using public randomness S.

In practice, one often uses a cryptographic hash function like SHA3 as the key derivation function $h(.)$ [6,17], and then simply assumes that $h(.)$ behaves like a random oracle [2].

In this paper we continue the investigation of key-derivation with provable security guarantees, where we don't make any computational assumption about $h(.)$. This problem is fairly well understood for sources $X|Z$ that have high min-entropy (we'll formally define all the entropy notions used in 2 below), or are computationally indistinguishable from having so (in this case, we say $X|Z$ has high HILL entropy). In the case where $X|Z$ has k bits of min-entropy, we can either use a strong extractor to derive a $k - 2\log \epsilon^{-1}$ key that is ϵ-close to uniform, or a condenser to get a k bit key which is ϵ-close to a variable with $k - \log\log \epsilon^{-1}$ bits of min-entropy. Using extractors/condensers like this also works for HILL entropy, except that now we only get computational guarantees (pseudorandom/high HILL entropy) on the derived key.

Often one has to derive a key from a source $X|Z$ which has no HILL entropy at all. The weakest assumption we can make on $X|Z$ for any kind of key-derivation to be possible, is that X is hard to predict given Z. This has been formalized in [15] by saying that $X|Z$ has k bits of unpredictability entropy, denoted $H_s^{\mathsf{unp}}(X|Z) \geqslant k$, if no circuit of size s can predict X given Z with advantage $\geqslant 2^{-k}$ (to be more general, we allow an additional parameter $\delta \geqslant 0$, and $H_{\delta,s}^{\mathsf{unp}}(X|Z) \geqslant k$ holds if (X, Z) is δ-close to some distribution (Y, Z) with $H_s^{\mathsf{unp}}(Y|Z) \geqslant k$). We will also consider a more restricted notion, where we say that $X|Z$ has k bits of *list*-unpredictability entropy, denoted $H_s^{*\mathsf{unp}}(X|Z) \geqslant k$, if it has k bits of unpredictability entropy relative to an oracle Eq which can be used to verify the correct guess (Eq outputs 1 on input X, and 0 otherwise).[1] We'll discuss this notion in more detail below. For now, let us just mention that for the important special case where it's easy to verify if a guess for X is correct (say, because we condition on $Z = f(X)$ for some one-way function[2] f), the oracle Eq does not help, and thus unpredictability and list-unpredictability coincide. The results proven in this paper imply that from a source $X|Z$ with k bits of list-unpredictability entropy, it's possible to extract a k bit key with $k - 3$ bits of HILL entropy

---

[1] We chose this name as having access to Eq is equivalent to being allowed to output a list of guesses. This is very similar to the well known concept of list-decoding.

[2] To be precise, this only holds for *injective* one-way functions. One can generalise list-unpredictability and let Eq output 1 on some set $\mathcal{X}$, and the adversary wins if she outputs any $X \in \mathcal{X}$. Our results (in particular Theorem 1) also hold for this more general notion, which captures general one-way functions by letting $\mathcal{X} = f^{-1}(f(X))$ be the set of all preimages of $Z = f(X)$.

Proposition 1. *Consider a joint distribution* (X, Z) *over* $\{0,1\}^n \times \{0,1\}^m$ *where*

$$H^{*unp}_{s,\gamma}(X|Z) \geq k \tag{1}$$

Let $S \in \{0,1\}^{n \times k}$ *be uniformly random and* $K = X^T S \in \{0,1\}^k$, *then the unpredictability entropy of* K *is*

$$H^{unp}_{s/2^{2k}poly(m,n),\gamma}(K|Z,S) \geq k - 3 \tag{2}$$

and the HILL entropy of K *is*

$$H^{HILL}_{t,\epsilon+\gamma}(K|Z,S) \geq k - 3 \tag{3}$$

with[3] $t = s \cdot \frac{\epsilon^7}{2^{2k}poly(m,n)}$.

Proposition 1 follows from two results we prove in this paper.

First, in Section 4 we prove Theorem 1 which shows how to "abuse" Goldreich-Levin hardcore bits by generating a k bit key $K = X^T S$ from a source $X|Z$ with k bits of list-unpredictability. The Goldreich-Levin theorem [12] implies nothing about the pseudorandomness of $K|(Z,S)$ when extracting that many bits. Instead, we prove that GL is a good "condenser" for unpredictability entropy: if $X|Z$ has k bits of list-unpredictability entropy, then $K|(Z,S)$ has $k-3$ bits of unpredictability entropy (note that we start with list-unpredictability, but only end up with "normal" unpredictability entropy). This result is used in the first step in Proposition 1, showing that (1) implies (2).

Second, in Section 5 we prove our main result, Theorem 2 which states that any source $X|Z$ which has $|X| - d$ bits of unpredictability entropy, has the same amount of HILL entropy (technically, we show that it implies the same amount of metric entropy against deterministic real-valued distinguishers. This notion implies the same amount of HILL entropy as shown by Barak et al. [1]). The security loss in this argument is exponential in the entropy gap d. Thus, if d is very large, this argument is useless, but if we first condense unpredictability as just explained, we have a gap of only $d = 3$. This result is used in the second step in Proposition 1, showing that (2) implies (3). In the two sections below we discuss two shortcomings of Theorem 1 which we hope can be overcome in future work.[4]

---

[3] We denote with $poly(m, n)$ some fixed polynomial in (n, m), but it can denote different polynomial throughout the paper. In particular, the *poly* here is not the same as in (2) as it hides several extra terms.

[4] After announcing this result at a workshop, we learned that Colin Jia Zheng proved a weaker version of this result. Theorem 4.18 in this PhD thesis, which is available via http://dash.harvard.edu/handle/1/11745716 also states that k bits of unpredictability imply k bits of HILL entropy. Like in our case, the loss in circuit size in his proof is polynomial in ϵ^{-1}, but it's also exponential in n (the length of X), whereas our loss is only exponential in the entropy gap $\Delta = n - k$.

On the Dependency on 2^k in Theorem 1. As outlined above, our first result is Theorem 1, which shows how to condense a source with k bits of list-unpredictability into a k bit key having $k-3$ bits of unpredictability entropy. The loss in circuit size is $2^{2k} poly(m, n)$, and it's not clear if the dependency on 2^k is necessary here, or if one can replace the dependency on 2^k with a dependency on $poly(\epsilon^{-1})$ at the price of an extra ϵ term in the distinguishing advantage. In many settings $\log(\epsilon^{-1})$ is in the order of k, in which case the above difference is not too important. This is for example the case when considering a k bit key for a symmetric primitive like a block-cipher, where one typically assumes the hardness of the cipher to be exponential in the key-length (and thus, if we want ϵ to be in the same order, we have $\log(\epsilon^{-1}) = \Theta(k)$). In other settings, k can be superlinear in $\log(\epsilon^{-1})$, e.g., if the the high entropy string is used to generate an RSA key.

List vs. Normal Unpredictability. Our Theorem 1 shows how to condense a source where $X|Z$ has k bits of *list*-unpredictability entropy into a k bit string with $k-3$ bits unpredictability entropy. It's an open question to which extent it's necessary to assume *list*-unpredictability here, maybe "normal" unpredictability is already sufficient? Note that list-unpredictability is a lower bound for unpredictability as one always can ignore the Eq oracle, i.e., $H_{\epsilon,s}^{unp}(X|Z) \geqslant H_{\epsilon,s}^{*unp}(X|Z)$, and in general, list-unpredictability can be much smaller than unpredictability entropy.[5] Interestingly, we can derive a k bit key with almost k bits of HILL entropy from a source $X|Z$ which k bits unpredictability entropy $H_{\epsilon,s}^{unp}(X|Z) \geqslant k$ in two extreme cases, namely, if either

1. if $X|Z$ has basically no HILL entropy (even against small circuits).
2. or when $X|Z$ has (almost) k bits of (high quality) HILL entropy.

In case 1. we observe that if $H_{\epsilon,t}^{HILL}(X|Z) \approx 0$ for some $t \ll s$, or equivalently, given Z we can efficiently distinguish X from any $X' \neq X$, then the Eq oracle used in the definition of list-unpredictability can be efficiently emulated, which means it's redundant, and thus $X|Z$ has the same amount of list-unpredictability and unpredictability entropy, $H_{s,\epsilon}^{unp}(X|Z) \approx H_{s',\epsilon'}^{*unp}(X|Z)$ for $(\epsilon', s') \approx (\epsilon, s)$. Thus, we can use Theorem 1 to derive a k bit key with $k - O(1)$ bits of HILL entropy in this case. In case 2., we can simply use any condenser for min-entropy to get a key with HILL entropy $k - \log\log \epsilon^{-1}$p As condensing almost all the unpredictability entropy into HILL entropy is possible in the two extreme cases where $X|Z$ has either no or a lot of HILL entropy, it seems conceivable that it's also possible in all the in-between cases (i.e., without making any additional assumptions about $X|Z$ at all).

GL vs. Condensing. Let us stress as this point that, because of the two issues discussed above, our result does not always allow to generate more bits

---

[5] E.g., let X by uniform over $\{0,1\}^n$ and Z arbitrary, but independent of X, then for $s = \exp(n)$ we have $H_s^{unp}(X|Z) = n$ but $H_s^{*unp}(X|Z) = 0$ as we can simply invoke Eq on all $\{0,1\}^n$ until X is found.

with high HILL entropy than just using the Goldreich-Levin theorem. Assuming k bits of unpredictability we get $k - 3$ of HILL, whereas GL will only give $k - 2\log(1/\epsilon)$. But as currently our reduction has a quantitatively larger loss in circuit size than the GL theorem, in order to get HILL entropy of the same quality (i.e., secure against (s, δ) adversaries for some fixed (s, δ)) we must consider the unpredictability entropy of the source $X|Z$ against more powerful adversaries than if we're about to use GL. And in general, the amount of unpredictability (or any other computational) entropy of $X|Z$ can decrease as we consider more powerful adversaries.

2   Entropy Notions

In this section we formally define the different entropy notions considered in this paper. We denote with $\mathcal{D}_s^{rand,\{0,1\}}$ the set of all *probabilistic* circuits of size s with *boolean* output, and $\mathcal{D}_s^{rand,[0,1]}$ denotes the set of all *probabilistic* circuits with *real-valued* output in the range $[0, 1]$. The analogous *deterministic* circuits are denoted $\mathcal{D}_s^{det,\{0,1\}}$ and $\mathcal{D}_s^{det,[0,1]}$. We use $X \sim_{\epsilon,s} Y$ to denote computational indistinguishability of variables X and Y, formally[6]

$$X \sim_{\epsilon,s} Y \iff \forall \mathsf{C} \in \mathcal{D}_s^{rand,\{0,1\}} \; : \; |\Pr[\mathsf{C}(X) = 1] - \Pr[\mathsf{C}(Y) = 1]| \leqslant \epsilon \quad (4)$$

$X \sim_\epsilon Y$ denotes that X and Y have statistical distance ϵ, i.e., $X \sim_{\epsilon,\infty} Y$, and with $X \sim Y$ we denote that they're identically distributed. With U_n we denote the uniform distribution over $\{0, 1\}^n$.

Definition 1. *The **min-entropy** of a random variable X with support $\mathcal{X}$ is*

$$H_\infty(X) = -\log_2 \max_{x \in \mathcal{X}} \Pr[X = x]$$

*For a pair (X, Z) of random variables, the **average min-entropy** of X conditioned on Z is*

$$\widetilde{H}_\infty(X|Z) = -\log_2 \mathop{\mathbb{E}}_{z \leftarrow Z} \max_x \Pr[X = x|Z = z] = -\log_2 \mathop{\mathbb{E}}_{z \leftarrow Z} 2^{-H_\infty(X|Z=z)}$$

HILL entropy is a computational variant of min-entropy, where X (conditioned on Z) has k bits of HILL entropy, if it cannot be distinguished from some Y that (conditioned on Z) has k bits of min-entropy, formally

Definition 2 ([14],[15]). *A random variable X has **HILL entropy** k, denoted by $H_{\epsilon,s}^{\mathsf{HILL}}(X) \geq k$, if there exists a distribution Y satisfying $H_\infty(Y) \geq k$ and $X \sim_{\epsilon,s} Y$.*

*Let (X, Z) be a joint distribution of random variables. Then X has **conditional HILL entropy** k conditioned on Z, denoted by $H_{\epsilon,s}^{\mathsf{HILL}}(X|Z) \geq k$, if there exists a joint distribution (Y, Z) such that $\widetilde{H}_\infty(Y|Z) \geq k$ and $(X, Z) \sim_{\epsilon,s} (Y, Z)$.*

---

[6] Let us mention that the choice of the distinguisher class in (4) irrelevant (up to a small additive difference in circuit size), we can replace $\mathcal{D}_s^{rand,\{0,1\}}$ with any of the three other distinguisher classes.

Barak, Sahaltiel and Wigderson [1] define the notion of metric entropy, which is defined like HILL, but the quantifiers are exchanged. That is, instead of asking for a single distribution (Y, Z) that fools all distinguishers, we only ask that for every distinguisher D, there exists such a distribution. For reasons discussed in Section 2, in the definition below we make the class of distinguishers considered explicit.

Definition 3 ([1],[10]). *Let (X, Z) be a joint distribution of random variables. Then X has* **conditional metric entropy** *k conditioned on Z (against probabilistic boolean distinguishers), denoted by $H_{\epsilon,s}^{\mathsf{Metric},rand,\{0,1\}}(X|Z) \geq k$, if for every $\mathsf{D} \in \mathcal{D}_s^{rand,\{0,1\}}$ there exists a joint distribution (Y, Z) such that $\widetilde{H}_\infty(Y|Z) \geq k$ and*

$$|\Pr[\mathsf{D}(X, Z) = 1] - \Pr[\mathsf{D}(Y, Z) = 1]| \leqslant \epsilon$$

More generally, for class $\in \{rand, det\}, range \in \{[0,1], \{0,1\}\}$, $H_{\epsilon,s}^{\mathsf{Metric},class,range}(X|Z) \geq k$ if for every $\mathsf{D} \in \mathcal{D}_s^{class,range}$ such a (Y, Z) exists.

Like HILL entropy, also unpredictability entropy, which we'll define next, can be seen as a computational variant of min-entropy. Here we don't require indistinguishability as for HILL entropy, but only that the variable is hard to predict.

Definition 4 ([15]). *X has* **unpredictability entropy** *k conditioned on Z, denoted by $H_{\epsilon,s}^{\mathsf{unp}}(X|Z) \geq k$, if (X, Z) is (ϵ, s) indistinguishable from some (Y, Z), where no probabilistic circuit of size s can predict Y given Z with probability better than 2^{-k}, i.e., $H_{s,\epsilon}^{\mathsf{unp}}(X|Z) \geq k$ if and only if*

$$\exists (Y, Z), (X, Z) \sim_{\varepsilon,s} (Y, Z) \ \forall \mathsf{C}, |\mathsf{C}| \leqslant s : \Pr_{(y,z) \leftarrow (Y,Z)}[\mathsf{C}(z) = y] \leqslant 2^{-k} \quad (5)$$

We also define a notion called "list-unpredictability", denoted $H_{\epsilon,s}^{*\mathsf{unp}}(X|Z) \geq k$, which holds if $H_{\epsilon,s}^{\mathsf{unp}}(X|Z) \geq k$ as in (5), but where C additionally gets oracle access to a function $\mathsf{Eq}(.)$ which outputs 1 on input y and 0 otherwise. So, C can efficiently test if some candidate guess for y is correct.[7]

Remark 1 (The ϵ parameter). The ϵ parameter in the definition above is not really necessary, following [16], we added it so we can have a "smooth" notion, which is easier to compare to HILL or smooth min-entropy. If $\epsilon = 0$, we'll simply omit it, then the definition simplifies to

$$H_s^{\mathsf{unp}}(X|Z) \geq k \iff \Pr_{(x,z) \leftarrow (X,Z)}[\mathsf{C}(z) = x] \leqslant 2^{-k}$$

Let us also mention that unpredictability entropy is only interesting if the conditional part Z is not empty as (already for s that is linear in the length of X) we have $H_s^{\mathsf{unp}}(X) = H_\infty(X)$ which can be seen by considering the circuit C (that gets no input as Z is empty) which simply outputs the constant x maximizing $\Pr[X = x]$.

---

[7] We name this notion "list-unpredictability" as we get the same notion when instead of giving C oracle access to $\mathsf{Eq}(.)$, we allow $\mathsf{C}(z)$ to output a list of guesses for y, not just one value, and require that $\Pr_{(y,z) \leftarrow (Y,Z)}[y \in \mathsf{C}(z)] \leqslant 2^{-k}$. This notion is inspired by the well known notion of list-decoding.

Metric vs. HILL. We will use a lemma which states that deterministic real-valued metric entropy implies the same amount of HILL entropy (albeit, with some loss in quality). This lemma has been proven by [1] for the unconditional case, i.e., when Z in the lemma below is empty, it has been observed by [4,10] that the proof also holds in the conditional case as stated below

Lemma 1 ([1,4,10]). *For any joint distribution $(X, Z) \in \{0,1\}^n \times \{0,1\}^m$ and any ϵ, δ, k, s*

$$H_{\epsilon,s}^{\text{Metric},det,[0,1]}(X|Z) \geqslant k \quad \Rightarrow \quad H_{\epsilon+\delta,s\cdot\delta^2/(m+n)}^{\text{HILL}}(X|Z) \geqslant k$$

Note that in Definition 2 of HILL entropy, we only consider security against probabilistic boolean distinguishers (as $\sim_{\epsilon,s}$ was defined this way), whereas in Definiton 3 of metric entropy we make the class of distinguishers explicit. The reason for this is that in the definition of HILL entropy the class of distinguishers considered is irrelevant (except for a small additive degradation in circuit size, cf. [10, Lemma2.1]).[8] Unlike for HILL, for metric entropy the choice of the distinguisher class does matter. In particular, deterministic boolean metric entropy $H_{\epsilon,s}^{\text{Metric},det,\{0,1\}}(X|Y) \geqslant k$ is only known to imply deterministic real-valued metric entropy $H_{\epsilon+\delta,s}^{\text{Metric},det,[0,1]}(X|Y) \geqslant k - \log(\delta^{-1})$, i.e., we must allow for a $\delta > 0$ loss in distinguishing advantage, and this will at the same time result in a loss of $\log(\delta^{-1})$ in the amount of entropy. For this reason, it is crucial that in Theorem 2 we show that unpredictability entropy implies deterministic *real-valued* metric entropy, so we can then apply Lemma 1 to get the same amount of HILL entropy. Dealing with real-valued distinguishers is the main source of technical difficulty in the proof of the Theorem 2, proving the analogous statement for deterministic *boolean* distinguishers is much simpler.

3   Known Results on Provably Secure Key-Derivation

We say that a cryptographic scheme has security α, if no adversary (from some class of adversaries like all polynomial size circuits) can win some security game with advantage $\geqslant \alpha$ if the scheme is instantiated with a uniformly random string.[9] Below we will distinguish between *unpredictability* applications, where the advantage bounds the probability of winning some security game (a typical example are digital signature schemes, where the game captures the existential unforgeability under chosen message attacks), and *indistinguishability* applications, where the advantage bounds the distinguishing advantage from some ideal object (a typical example is the security definition of pseudorandom generators or functions).

---

[8] This easily follows from the fact that in the definition (4) of computational indistinguishability the choice of the distinguisher class is irrelevant.

[9] We'll call this string "key". Though in many settings (in particular when keys are not simply uniform random strings, like in public-key crypto) this string is not used as a key directly, but one rather should think of it as the randomness used to sample the actual keys.

3.1   Key-Derivation from Min-Entropy

Strong Extractors. Let (X, Z) be a source where $\widetilde{H}_\infty(X|Z) \geqslant k$, or equivalently, no adversary can guess X given Z with probability better than 2^{-k} (cf. Def. 1). Consider the case where we want to derive a key $K = h(X, S)$ that is statistically close to uniform given (Z, S). For example, X could be some physical source (like statistics from keystrokes) from which we want to generate almost uniform randomness. Here Z models potential side-information the adversary might have on X. This setting is very well understood, and such a key can be derived using a strong extractor as defined below.

Definition 5 ([18],[5]). *A function* $\mathsf{Ext} : \{0,1\}^n \times \{0,1\}^d \to \{0,1\}^\ell$ *is an average-case* (k, ϵ)-*strong extractor if for every distribution* (X, Z) *over* $\{0,1\}^n \times \{0,1\}^m$ *with* $\widetilde{H}_\infty(X|Z) \geqslant k$ *and* $S \sim U_d$, *the distribution* $(\mathsf{Ext}(X, S), S, Z)$ *has statistical distance* ϵ *to* (U_ℓ, S, Z).

Extractors Ext as above exist with $\ell = k - 2\log(1/\epsilon)$ [14]. Thus, from any (X, Z) where $\widetilde{H}_\infty(X|Z) \geqslant k$ we can extract a key $K = \mathsf{Ext}(X, S)$ of length $k - 2\log(1/\epsilon)$ that is ϵ close to uniform [14]. The entropy gap $2\log(1/\epsilon)$ is optimal by the so called "RT-bound" [19], even if we assume the source is efficiently samplable [7].

If instead of using a uniform ℓ bit key for an α secure scheme, we use a key that is ϵ close to uniform, the scheme will still be at least $\beta = \alpha + \epsilon$ secure. In order to get security β that is of the same order as α, we thus must set $\epsilon \approx \alpha$. When the available amount k of min-entropy is small, for example when dealing with biometric data [3,5], a loss of $2\log(1/\epsilon)$ bits (that's 160 bits for a typical security level $\epsilon = 2^{-80}$) is often unacceptable.

Condensers. The above bound is basically tight for many *indistinguishability* applications like pseudorandom generators or pseudorandom functions.[10] Fortunately, for many applications a close to uniform key is not necessary, and a key $|K|$ with min-entropy $|K| - \Delta$ for some small Δ is basically as good as a uniform one. This is the case for all *unpredictability* applications, which includes OWFs, digital-signatures and MACs.[11] It's not hard to show that if the scheme is α secure with a uniform key it remains at least $\beta = \alpha 2^\Delta$ secure (against the

---

[10] For example, consider a pseudorandom function $\mathsf{F} : \{0,1\}^k \times \{0,1\}^a \to \{0,1\}$ and a key K that is uniform over all keys where $\mathsf{F}(K,0) = 0$, this distribution is $\epsilon \approx 1/2$ close to uniform and has min-entropy $\approx |K| - 1$, but the security breaks completely as one can distinguish $\mathsf{F}(U_k,.)$ from $\mathsf{F}(K,.)$ with advantage $\beta \approx 1/2$ (by quering on input 0, and outputting 1 iff the output is 0).

[11] [8] identify an interesting class of applications called "square-friendly", this class contains all unpredictability applications, and some indistinguishability applications like weak PRFs (which are PRFs that can only be queried on random inputs). This class of applications remains somewhat secure even for a small entropy gap Δ: For $\Delta = 1$ the security is $\beta \approx \sqrt{\alpha}$. This is worse that the $\beta = 2\alpha$ for unpredictability applications, but much better than the complete loss of security $\beta \approx 1/2$ required for some indistinguishability apps like (standard) PRFs.

same class of attackers) if instantiated with any key K that has $|K| - \Delta$ bits of min-entropy.[12] Thus, for unpredictability applications we don't have to extract an almost uniform key, but "condensing" X into a key with $|K| - \Delta$ bits of min-entropy for some small Δ is enough.

[7] show that a $(\log \epsilon + 1)$-wise independent hash function $\mathsf{Cond} : \{0,1\}^n \times \{0,1\}^d \to \{0,1\}^\ell$ is a condenser with the following parameters. For any (X, Z) where $\widetilde{H}_\infty(X|Z) \geqslant \ell$, for a random seed S (used to sample a $(\log \epsilon + 1)$-wise independent hash function), the distribution $(\mathsf{Cond}(X, S), S)$ is ϵ close to a distribution (Y, S) where $\widetilde{H}_\infty(Y|Z) \geqslant \ell - \log \log(1/\epsilon)$. Using such an ℓ bit key (condensed from a source with ℓ bits min-entropy) for an unpredictability application that is α secure (when using a uniform ℓ bit key), we get security $\beta \leqslant \alpha 2^{\log \log(1/\epsilon)} + \epsilon$, which setting $\epsilon = \alpha$ gives $\beta \leqslant \alpha(1 + \log(1/\alpha))$ security, thus, security degrades only by a logarithmic factor.

3.2 Key-Derivation from Computational Entropy

HILL Entropy. As already discussed in the introduction, often we want to derive a key from a distribution (X, Z) where there's no "real" min-entropy at all $\widetilde{H}_\infty(X|Z) = 0$. This is for example the case when Z is the transcript (that can be observed by an adversary) of a key-exchange protocol like Diffie-Hellman, where the agreed value $X = g^{ab}$ is determined by the transcript $Z = (g^a, g^b)$ [11,17]. Another setting where this can be the case is in the context of side-channel attacks, where the leakage Z from a device can completely determine its internal state X. If $X|Z$ has k bits of HILL entropy, i.e., is computationally indistinguishable from having min-entropy k (cf. Def. 2) we can derive keys exactly as described above assuming $X|Z$ had k bits of min-entropy. In particular, if $X|Z$ has $|K| + 2\log(1/\epsilon)$ bits of HILL entropy for some negligible ϵ, we can derive a key K that is pseudorandom, and if $X|Z$ has $|K| + \log \log(1/\epsilon)$ bits of HILL entropy, we can derive a key that is almost as good as a uniform one for any unpredictability application.

Unpredictability Entropy. Clearly, the minimal assumption we must make on a distribution $(X, Z) \in \{0,1\}^n \times \{0,1\}^m$ for any key derivation to be possible at all is that X is hard to compute given Z, that is, $X|Z$ must have some unpredictability entropy as in Definition 4. Goldreich and Levin [12] show how to generate pseudorandom bits from such a source. In particular, the Goldreich-Levin theorem implies that if $X|Z$ has at least $2\log \epsilon^{-1}$ bits of list-unpredictability, then the inner product $R^T X$ of X with a random vector R is ϵ indistinguishable from uniformly random (the loss in circuit size is $poly(n, m)/\epsilon^4$). Using the chain rule

---

[12] Assume some adversary breaks the scheme, say, forges a signature, with advantage β if the key comes from the distribution K. If we sample a uniform key instead, it will have the same distribution as K conditioned on an event that holds with probability $2^{-\Delta}$, and thus this adversary will still break the scheme with probability $\beta/2^\Delta$.

for unpredictability entropy,[13] we can generate an $\ell = k - 2 \log \epsilon^{-1}$ bit long pseudorandom string that is $\ell\epsilon$ indistinguishable (the extra ℓ factor comes from taking the union bound over all bits) from uniform.

Thus, we can turn k bits of list-unpredictability into $k - 2 \log \epsilon^{-1}$ bits of pseudorandom bits (and thus also that much HILL entropy) with quality roughly ϵ. The question whether it's possible to generate significantly more than $k - 2 \log \epsilon^{-1}$ of HILL entropy from a source with k bits of (list-)unpredictability seems to have never been addressed in the literature before. The reason might be that one usually is interested in generating pseudorandom bits (not just HILL entropy), and for this, the $2 \log \epsilon^{-1}$ entropy loss is inherent. The observation that for many applications high HILL entropy is basically as good as pseudorandomness is more recent, and recently gained attention by its usefulness in the context of leakage-resilient cryptography [8, 9].

In this paper we prove that it's in fact possible to turn almost all list-unpredictability into HILL entropy.

4   Condensing Unpredictability

Let $X|Z$ have k bits of list-unpredictability, and assume we start extracting Goldreich-Levin hardcore bits $A_1, A_2, \ldots$ by taking inner products $A_i = R_i^T X$ for random R_i. The first extracted bits $A_1, A_2, \ldots$ will be pseudorandom (given the R_i and Z), but with every extracted bit, the list-unpredictability can also decrease by one bit. As the GL theorem requires at least $2 \log \epsilon^{-1}$ bits of list-unpredictability to extract an ϵ secure pseudorandom bit, we must stop after $k - 2 \log \epsilon^{-1}$ bits. In particular, the more we extract, the worse the pseudorandomness of the extracted string becomes. Unlike the original GL theorem, in our Theorem 1 we only argue about the unpredictability of the extracted string, and unpredictability entropy has the nice property that it can never decrease, i.e., predicting $A_1, \ldots, A_{i+1}$ is always at least as hard as predicting $A_1, \ldots, A_i$. Thus, despite the fact that once i approaches k it becomes easier and easier to predict A_i (given $A_1, \ldots, A_{i-1}, Z$ and the R_i's)[14] this hardness will still add up to $k - O(1)$ bits of unpredictability entropy.

The proof is by contradiction, we assume that $A_1, \ldots, A_k$ can be predicted with advantage 2^{-k+3} (i.e., does not have $k-3$ bits of unpredictability), and then use such a predictor to predict X with advantage $> 2^{-k}$, contradicting the k bit list-unpredictability of $X|Z$. If $A_1, \ldots, A_k$ can be predicted as above, then there must be an index j s.t. A_j can be predicted with good probability conditioned on $A_1, \ldots, A_{j-1}$ being correctly predicted. We then can use the Goldreich-Levin

---

[13] Which states that if $X|Z$ has k bits of list-unpredictability, then for any (A, R) where R is independent of (X, Z), $X|(Z, A, R)$ has $k - |A|$ bits of list-unpredictability entropy. In particular, extracting ℓ inner product bits, decreases the list-unpredictability by at most ℓ.

[14] The only thing we know about the last extracted bit A_k is that it cannot be predicted with advantage $\geqslant 0.75$, more generally, A_{k-j} cannot be predicted with advantage $1/2 + 1/2^{j+2}$.

theorem, which tells us how to find X given such a predictor. Unfortunately, j can be close to k, and to apply the GL theorem, we first need to find the right values for $A_1, \ldots, A_{j-1}$ on which we condition, and also can only use the predictor's guess for A_j if it was correct on the first $j-1$ bits. We have no better strategy for this than trying all possible values, and this is the reason why the loss in circuit size in Theorem 1 depends on 2^k.

In our proof, instead of using the Goldreich-Levin theorem, we will actually use a more fine-grained variant due to Hast which allows to distinguish between errors and erasures, this will give a much better quantitative bound.

Theorem 1 (Condensing Upredictability Entropy). *Consider any distribution (X, Z) over $\{0,1\}^n \times \{0,1\}^m$ where*

$$H_{\epsilon,s}^{*\mathsf{unp}}(X|Z) \geqslant k$$

then for a random $R \leftarrow \{0,1\}^{k \times n}$

$$H_{\epsilon,t}^{\mathsf{unp}}(R.X|Z,R) \geqslant k - \Delta$$

where[15] $t = \frac{s}{2^{2k}\operatorname{poly}(m,n)}$, $\Delta = 3$

5   High Unpredictability Implies Metric Entropy

In this section we state our main results, showing that k bits of unpredictability entropy imply the same amount of HILL entropy, with a loss exponential in the "entropy gap".

Theorem 2 (Unpredictability Entropy Implies HILL Entropy). *For any distribution (X, Z) over $\{0,1\}^n \times \{0,1\}^m$, if $X|Z$ has unpredictability entropy*

$$H_{\gamma,s}^{\mathsf{unp}}(X|Z) \geqslant k \tag{6}$$

then, with $\Delta = n - k$ denoting the entropy gap, $X|Z$ has (real valued, deterministic) metric entropy

$$H_{\epsilon+\gamma,t}^{\mathsf{Metric},det,[0,1]}(X|Z) \geqslant k \quad \text{for} \quad t = \Omega\left(s \cdot \frac{\epsilon^5}{2^{5\Delta}\log^2\left(2^{\Delta}\epsilon^{-1}\right)}\right) \tag{7}$$

By Lemma 1 this further implies that $X|Z$ has, for any $\delta > 0$, HILL entropy

$$H_{\epsilon+\delta+\gamma,\Omega(t\delta^2/(n+m))}^{\mathsf{HILL}}(X|Z) \geqslant k$$

which for $\epsilon = \delta = \gamma$ is $H_{3\epsilon,\Omega(s\cdot\epsilon^7/2^{5\Delta}(n+m)\log^2(2^{\Delta}\epsilon^{-1}))}^{\mathsf{HILL}}(X|Z) \geqslant k$

---

[5] We can set Δ to be any constant > 1 here, but choosing a smaller Δ would imply a smaller t.

References

1. Barak, B., Shaltiel, R., Wigderson, A.: Computational Analogues of Entropy. In: Arora, S., Jansen, K., Rolim, J.D.P., Sahai, A. (eds.) RANDOM 2003 and APPROX 2003. LNCS, vol. 2764, pp. 200–215. Springer, Heidelberg (2003)
2. Bellare, M., Rogaway, P.: Random oracles are practical: A paradigm for designing efficient protocols. In: Ashby, V. (ed.) ACM CCS 1993, pp. 62–73. ACM Press, November 1993
3. Boyen, X., Dodis, Y., Katz, J., Ostrovsky, R., Smith, A.: Secure Remote Authentication Using Biometric Data. In: Cramer, R. (ed.) EUROCRYPT 2005. LNCS, vol. 3494, pp. 147–163. Springer, Heidelberg (2005)
4. Chung, K.-M., Kalai, Y.T., Liu, F.-H., Raz, R.: Memory Delegation. In: Rogaway, P. (ed.) CRYPTO 2011. LNCS, vol. 6841, pp. 151–168. Springer, Heidelberg (2011)
5. Dodis, Y., Ostrovsky, R., Reyzin, L., Smith, A.: Fuzzy Extractors: How to Generate Strong Keys from Biometrics and Other Noisy Data. SIAM Journal on Computing 38(1), 97–139 (2008)
6. Dodis, Y., Gennaro, R., Håstad, J., Krawczyk, H., Rabin, T.: Randomness Extraction and Key Derivation Using the CBC, Cascade and HMAC Modes. In: Franklin, M. (ed.) CRYPTO 2004. LNCS, vol. 3152, pp. 494–510. Springer, Heidelberg (2004)
7. Dodis, Y., Pietrzak, K., Wichs, D.: Key Derivation without Entropy Waste. In: Nguyen, P.Q., Oswald, E. (eds.) EUROCRYPT 2014. LNCS, vol. 8441, pp. 93–110. Springer, Heidelberg (2014)
8. Dodis, Y., Yu, Y.: Overcoming Weak Expectations. In: Sahai, A. (ed.) TCC 2013. LNCS, vol. 7785, pp. 1–22. Springer, Heidelberg (2013)
9. Dziembowski, S., Pietrzak, K.: Leakage-resilient cryptography. In: 49th FOCS, pp. 293–302. IEEE Computer Society Press, October2008
10. Fuller, B., Reyzin, L.: Computational entropy and information leakage. Cryptology ePrint Archive, Report 2012/466 (2012). http://eprint.iacr.org/
11. Gennaro, R., Krawczyk, H., Rabin, T.: Secure Hashed Diffie-Hellman over Non-DDH Groups. In: Cachin, C., Camenisch, J.L. (eds.) EUROCRYPT 2004. LNCS, vol. 3027, pp. 361–381. Springer, Heidelberg (2004)
12. Goldreich, O., Levin, L.A.: A hard-core predicate for all one-way functions. In: 21st ACM STOC. pp. 25–32. ACM Press, May 1989
13. Hast, G.: Nearly one-sided tests and the Goldreich-Levin predicate. In: Biham, E. (ed.) EUROCRYPT 2003. LNCS, vol. 2656, pp. 195–210. Springer, Heidelberg (2003)
14. Håstad, J., Impagliazzo, R., Levin, L.A., Luby, M.: A pseudorandom generator from any one-way function. SIAM Journal on Computing 28(4), 1364–1396 (1999)
15. Hsiao, C.-Y., Lu, C.-J., Reyzin, L.: Conditional Computational Entropy, or Toward Separating Pseudoentropy from Compressibility. In: Naor, M. (ed.) EUROCRYPT 2007. LNCS, vol. 4515, pp. 169–186. Springer, Heidelberg (2007)
16. Hsiao, C.-Y., Lu, C.-J., Reyzin, L.: Conditional Computational Entropy, or Toward Separating Pseudoentropy from Compressibility. In: Naor, M. (ed.) EUROCRYPT 2007. LNCS, vol. 4515, pp. 169–186. Springer, Heidelberg (2007)
17. Krawczyk, H.: Cryptographic Extraction and Key Derivation: The HKDF Scheme. In: Rabin, T. (ed.) CRYPTO 2010. LNCS, vol. 6223, pp. 631–648. Springer, Heidelberg (2010)
18. Nisan, N., Zuckerman, D.: More deterministic simulation in logspace. In: 25th ACM STOC, pp. 235–244. ACM Press, May 1993
19. Radhakrishnan, J., Ta-Shma, A.: Bounds for dispersers, extractors, and depth-two superconcentrators. SIAM J. Discrete Math. 13(1), 2–24 (2000)

Sherali-Adams Relaxations for Valued CSPs

Johan Thapper[1] and Stanislav Živný[2]($\boxtimes$)

[1] Université Paris-Est, Marne-la-Vallée, France
thapper@u-pem.fr
[2] Department of Computer Science, University of Oxford, Oxford, UK
standa.zivny@cs.ox.ac.uk

Abstract. We consider Sherali-Adams linear programming relaxations for solving valued constraint satisfaction problems to optimality. The utility of linear programming relaxations in this context have previously been demonstrated using the lowest possible level of this hierarchy under the name of the basic linear programming relaxation (BLP). It has been shown that valued constraint languages containing only finite-valued weighted relations are tractable if, and only if, the integrality gap of the BLP is 1. In this paper, we demonstrate that almost all of the known tractable languages with arbitrary weighted relations have an integrality gap 1 for the Sherali-Adams relaxation with parameters $(2, 3)$. The result is closely connected to the notion of bounded relational width for the ordinary constraint satisfaction problem and its recent characterisation.

1 Introduction

The constraint satisfaction problem provides a common framework for many theoretical and practical problems in computer science. An instance of the *constraint satisfaction problem* (CSP) consists of a collection of variables that must be assigned labels from a given domain subject to specified constraints. The CSP is NP-complete in general, but tractable fragments can be studied by, following Feder and Vardi [13], restricting the constraint relations allowed in the instances to a fixed, finite set, called the constraint language. The most successful approach to classifying the language-restricted CSP is the so-called algebraic approach [3,5].

An important type of algorithms for CSPs are *consistency methods*. A constraint language is of *bounded relational width* if any CSP instance over this language can be solved by establishing (k, ℓ)-minimality for some fixed integers $1 \leq k \leq \ell$ [1]. The power of consistency methods for constraint languages has recently been fully characterised [3,21] and it has been shown that any constraint language that is of bounded relational width is of relational width at most $(2,3)$[1].

The CSP deals with only feasibility issues: Is there a solution satisfying certain constraints? In this work we are interested in problems that capture both

The authors were supported by London Mathematical Society Grant 41355. Stanislav Živný was supported by a Royal Society University Research Fellowship.

© Springer-Verlag Berlin Heidelberg 2015
M.M. Halldórsson et al. (Eds.): ICALP 2015, Part I, LNCS 9134, pp. 1058–1069, 2015.
DOI: 10.1007/978-3-662-47672-7_86

feasibility and optimisation issues: What is the best solution satisfying certain constraints? Problems of this form can be cast as valued constraint satisfaction problems [16].

An instance of the *valued constraint satisfaction problem* (VCSP) is given by a collection of variables that is assigned labels from a given domain with the goal to *minimise* an objective function given by a sum of weighted relations, each depending on some subset of the variables [8]. The weighted relations can take on finite rational values and positive infinity. The CSP corresponds to the special case of the VCSP when the codomain of all weighted relations is $\{0, \infty\}$.

Like the CSP, the VCSP is NP-hard in general and thus we are interested in the restrictions which give rise to tractable classes of problems. We restrict the *valued constraint language*; that is, all weighted relations in a given instance must belong to a fixed set of weighted relations on the domain. Languages that give rise to classes of problems solvable in polynomial time are called *tractable*, and languages that give rise to classes of problem that are NP-hard are called *intractable*. The computational complexity of Boolean (on a 2-element domain) valued constraint languages [8] and conservative (containing all $\{0, 1\}$-valued unary weighted relations) valued constraint languages [18] have been completely classified with respect to exact solvability.

Every VCSP problem has a natural linear programming (LP) relaxation, proposed independently by a number of authors, e.g. [6], and referred to as the *basic* LP relaxation (BLP) of the VCSP. It is the first level in the Sheralli-Adams hierarchy [24], which provides successively tighter LP relaxations of an integer LP. The BLP has been considered in the context of CSPs for robust approximability [10,20] and constant-factor approximation [9,12]. Higher levels of Sheral-Adams hierarchy have been considered for (in)approximability of CSPs [11,30] but we are not aware of any results related to exact solvability of (valued) CSPs. Semidefinite programming relaxations have also been considered in the context of CSPs for approximability [23] and robust approximability [2].

Consistency methods, and in particular strong 3-consistency has played an important role as a preprocessing step in establishing tractability of valued constraint languages. Cohen et al. proved the tractability of valued constraint languages improved by a symmetric tournament pair (STP) multimorphism via strong 3-consistency preprocessing, and an involved reduction to submodular function minimisation [7]. They also showed that the tractability of any valued constraint language improved by a tournament pair multimorphism via a preprocessing using results on constraint languages invariant under a 2-semilattice polymorphism, which relies on (3, 3)-minimality, and then reducing to the STP case. The only tractable conservative valued constraint languages are those admitting a pair of fractional polymorphisms called STP and MJN [18]; again, the tractability of such languages is proved via a 3-consistency preprocessing reducing to the STP case. It is natural to ask whether this nested use of consistency methods are necessary.

Contributions. In [17,26], the authors showed that the BLP of the VCSP can be used to solve the problem for many valued constraint languages. In [27], it was then shown that for VCSPs with weighted relations taking only finite values, the BLP precisely characterises the tractable (finite-)valued constraint languages; i.e., if BLP fails to solve any instance of some valued constraint language of this type, then this language is NP-hard.

In this paper, we show that a higher-level Sherali-Adams linear programming relaxation [24] suffices to solve most of the previously known tractable valued constraint languages with arbitrary weighted relations, and in particular, all known valued constraint languages that involve some optimisation (and thus do not reduce to constraint languages containing only relations) except for valued constraint languages of generalised weak tournament pair type [29]; such languages are known to be tractable [29] but we do not know whether they are tractable by our linear programming relaxation.

Our main result, Theorem 4, shows that if the support clone of a valued constraint language Γ of finite size contains weak near-unanimity operations of all but finitely many arities, then Γ is tractable via the Sherali-Adams relaxation with parameters $(2, 3)$. This tractability condition is precisely the bounded relational width condition for constraint languages of finite size containing all constants [3,21], and our proof fundamentally relies on the results of Barto and Kozik [3] and Barto [1].

It is folklore that the kth level of Sherali-Adams hierarchy establishes k-consistency for CSPs. We demonstrate that one linear programming relaxation is powerful enough to establish consistency as well as solving an optimisation problem in one go without the need of nested applications of consistency methods. For example, valued constraint languages having a tournament pair multimorphism were previously known to be tractable using ingenious application of various consistency techniques, advanced analysis of constraint networks using modular decompositions, and submodular function minimisation [7]. Here, we show that an even less restrictive condition (having a binary conservative commutative operation in some fractional polymorphism) ensures that the Sherali-Adams relaxation solves all instances to optimum.

Finally, we also give a short proof of the dichotomy theorem for conservative valued constraint languages [18], which previously needed lengthy arguments (although we still rely on Takhanov [25] for a part of the proof).

2   Preliminaries

Valued CSPs. Throughout the paper, let D be a fixed finite set of size at least two. We call D the *domain*, the elements of D *labels* and say that weighted relations take *values*. Let $\overline{\mathbb{Q}} = \mathbb{Q} \cup \{\infty\}$ denote the set of rational numbers with (positive) infinity.

Definition 1. *An m-ary relation over D is any mapping $\phi : D^m \to \{c, \infty\}$ for some $c \in \mathbb{Q}$. We denote by $\mathbf{R}_D$ the set of all relations on D.*

Definition 2. *An m-ary weighted relation over D is any mapping $\phi : D^m \to \overline{\mathbb{Q}}$. We write $ar(\phi) = m$ for the arity of ϕ. We denote by $\mathbf{\Phi}_D$ the set of all weighted relations on D.*

For any m-ary weighted relation $\phi \in \mathbf{\Phi}_D$, we denote by $\mathrm{Feas}(\phi) = \{\mathbf{x} \in D^m \,|\, \phi(\mathbf{x}) < \infty\} \in \mathbf{R}_D$ the underlying m-ary *feasibility relation*, and by $\mathrm{Opt}(\phi) = \{\mathbf{x} \in \mathrm{Feas}(\phi) \,|\, \forall \mathbf{y} \in D^m : \phi(\mathbf{x}) \leq \phi(\mathbf{y})\} \in \mathbf{R}_D$ the m-ary *optimality relation*, which contains the tuples on which ϕ is minimised. A weighted relation $\phi : D^m \to \overline{\mathbb{Q}}$ is called *finite-valued* if $\mathrm{Feas}(\phi) = D^m$.

Definition 3. *Let $V = \{x_1, \ldots, x_n\}$ be a set of variables. A* valued constraint *over V is an expression of the form $\phi(\mathbf{x})$ where $\phi \in \mathbf{\Phi}_D$ and $\mathbf{x} \in V^{ar(\phi)}$. The number m is called the* arity *of the constraint, the weighted relation ϕ is called the* constraint weighted relation, *and the tuple $\mathbf{x}$ the* scope *of the constraint.*

Definition 4. *An instance of the* valued constraint satisfaction problem, *VCSP, is specified by a finite set $V = \{x_1, \ldots, x_n\}$ of variables, a finite set D of labels, and an objective function I expressed as follows: $I(x_1, \ldots, x_n) = \sum_{i=1}^{q} \phi_i(\mathbf{x}_i)$, where each $\phi_i(\mathbf{x}_i)$, $1 \leq i \leq q$, is a valued constraint over V. Each constraint can appear multiple times in I. The goal is to find an* assignment *(or* solution*) of labels to the variables minimising I.*

A solution is called *feasible* (or *satisfying*) if it is of finite value. A VCSP instance I is called *satisfiable* if there is a feasible solution to I. CSPs are a special case of VCSPs with (unweighted) relations with the goal to determine the existence of a feasible solution.

Example 1. In the MIN-UNCUT problem the goal is to find a partition of the vertices of a given graph into two parts so that the number of edges inside the two partitions is minimised. For a graph (V, E) with $V = \{x_1, \ldots, x_n\}$, this NP-hard problem can be expressed as the VCSP instance $I(x_1, \ldots, x_n) = \sum_{(i,j) \in E} \phi_{\mathsf{xor}}(x_i, x_j)$ over the Boolean domain $D = \{0, 1\}$, where $\phi_{\mathsf{xor}} : \{0, 1\}^2 \to \overline{\mathbb{Q}}$ is defined by $\phi_{\mathsf{xor}}(x, y) = 1$ if $x = y$ and $\phi_{\mathsf{xor}}(x, y) = 0$ if $x \neq y$.

Definition 5. *Any set $\Delta \subseteq \mathbf{R}_D$ is called a* constraint language *over D. Any set $\Gamma \subseteq \mathbf{\Phi}_D$ is called a* valued constraint language *over D. We denote by $\mathrm{VCSP}(\Gamma)$ the class of all VCSP instances in which the constraint weighted relations are all contained in Γ. For a constraint language Δ, we denote by $\mathrm{CSP}(\Delta)$ the class $\mathrm{VCSP}(\Delta)$ to emphasise the fact that there is no optimisation involved.*

Definition 6. *A valued constraint language Γ is called* tractable *if $\mathrm{VCSP}(\Gamma')$ can be solved (to optimality) in polynomial time for every finite subset $\Gamma' \subseteq \Gamma$, and Γ is called* intractable *if $\mathrm{VCSP}(\Gamma')$ is NP-hard for some finite $\Gamma' \subseteq \Gamma$.*

Operations and Clones. We recall some basic terminology from universal algebra. Given an m-tuple $\mathbf{x} \in D^m$, we denote its ith entry by $\mathbf{x}[i]$ for $1 \leq i \leq m$. Any mapping $f : D^k \to D$ is called a k-ary *operation*; f is called *conservative* if $f(x_1, \ldots, x_k) \in \{x_1, \ldots, x_k\}$ and *idempotent* if $f(x, \ldots, x) = x$. We will apply a k-ary operation f to k m-tuples $\mathbf{x}_1, \ldots, \mathbf{x}_k \in D^m$ coordinatewise, that is,

$$f(\mathbf{x}_1, \ldots, \mathbf{x}_k) = (f(\mathbf{x}_1[1], \ldots, \mathbf{x}_k[1]), \ldots, f(\mathbf{x}_1[m], \ldots, \mathbf{x}_k[m])). \quad (1)$$

Definition 7. *Let ϕ be an m-ary weighted relation on D. A k-ary operation f on D is a* polymorphism *of ϕ if, for any $\mathbf{x}_1, \ldots, \mathbf{x}_k \in D^m$ with $\mathbf{x}_i \in \text{Feas}(\phi)$ for all $1 \leq i \leq k$, we have that $f(\mathbf{x}_1, \ldots, \mathbf{x}_k) \in \text{Feas}(\phi)$.*

For any valued constraint language Γ over a set D, we denote by $\text{Pol}(\Gamma)$ the set of all operations on D which are polymorphisms of all $\phi \in \Gamma$. We write $\text{Pol}(\phi)$ for $\text{Pol}(\{\phi\})$.

A k-ary *projection* is an operation of the form $\pi_i^{(k)}(x_1, \ldots, x_k) = x_i$ for some $1 \leq i \leq k$. Projections are polymorphisms of all valued constraint languages.

The *composition* of a k-ary operation $f : D^k \to D$ with k ℓ-ary operations $g_i : D^\ell \to D$ for $1 \leq i \leq k$ is the ℓ-ary operation $f[g_1, \ldots, g_k] : D^\ell \to D$ defined by $f[g_1, \ldots, g_k](x_1, \ldots, x_\ell) = f(g_1(x_1, \ldots, x_\ell), \ldots, g_k(x_1, \ldots, x_\ell))$.

We denote by $\mathcal{O}_D$ the set of all finitary operations on D and by $\mathcal{O}_D^{(k)}$ the k-ary operations in $\mathcal{O}_D$. A *clone* of operations, $C \subseteq \mathcal{O}_D$, is a set of operations on D that contains all projections and is closed under composition. It is easy to show that $\text{Pol}(\Gamma)$ is a clone for any valued constraint language Γ.

Definition 8. *A k-ary* fractional operation *ω is a probability distribution over $\mathcal{O}_D^{(k)}$. We define $\text{supp}(\omega) = \{f \in \mathcal{O}_D^{(k)} \mid \omega(f) > 0\}$.*

Definition 9. *Let ϕ be an m-ary weighted relation on D and let ω be a k-ary fractional operation on D. We call ω a* fractional polymorphism *of ϕ (and say that ϕ is* improved *by ω) if $\text{supp}(\omega) \subseteq \text{Pol}(\phi)$ and for any $\mathbf{x}_1, \ldots, \mathbf{x}_k \in D^m$ with $\mathbf{x}_i \in \text{Feas}(\phi)$ for all $1 \leq i \leq k$, we have*

$$\mathbb{E}_{f \sim \omega} [\phi(f(\mathbf{x_1}, \ldots, \mathbf{x_k}))] \leq \text{avg}\{\phi(\mathbf{x_1}), \ldots, \phi(\mathbf{x_k})\}. \tag{2}$$

Definition 10. *For any valued constraint language $\Gamma \subseteq \mathbf{\Phi}_D$, we define $\text{fPol}(\Gamma)$ to be the set of all fractional operations that are fractional polymorphisms of all weighted relations $\phi \in \Gamma$. We write $\text{fPol}(\phi)$ for $\text{fPol}(\{\phi\})$.*

Example 2. A valued constraint language on domain $\{0, 1\}$ is called *submodular* if it has the fractional polymorphism ω defined by $\omega(\min) = \omega(\max) = \frac{1}{2}$, where min and max are the two binary operations that return the smaller and larger of its two arguments respectively with respect to the usual order $0 < 1$.

For a valued constraint language Γ we define $\text{supp}(\Gamma) = \bigcup_{\omega \in \text{fPol}(\Gamma)} \text{supp}(\omega)$.

Lemma 1. *For any valued constraint language Γ, $\text{supp}(\Gamma)$ is a clone.*

We note that Lemma 1 has also been observed in [22] and in [14].

A special case of the following lemma has been observed, in the context of Min-Sol problems [29], by Hannes Uppman.[1]

Lemma 2. *Let Γ be a valued constraint language of finite size on a domain D and let $f \in \text{Pol}(\Gamma)$. Then, $f \in \text{supp}(\Gamma)$ if, and only if, $f \in \text{Pol}(\text{Opt}(I))$ for all instances I of VCSP(Γ).*

---

[1] Private communication.

Cores and Constants. Let $\mathcal{C}_D = \{\{(d)\} \mid d \in D\}$ be the set of constant unary relations on D.

Definition 11. *Let Γ be a valued constraint language with domain D and let $S \subseteq D$. The sub-language $\Gamma[S]$ of Γ induced by S is the valued constraint language defined on domain S and containing the restriction of every weighted relation $\phi \in \Gamma$ onto S.*

Definition 12. *A valued constraint language Γ is a core if all unary operations in $\mathrm{supp}(\Gamma)$ are bijections. A valued constraint language Γ' is a core of Γ if Γ' is a core and $\Gamma' = \Gamma[f(D)]$ for some $f \in \mathrm{supp}(\omega)$ with ω a unary fractional polymorphism of Γ.*

Lemma 3. *Let Γ be a valued constraint language and Γ' a core of Γ. Then, for all instances I of $VCSP(\Gamma)$ and I' of $VCSP(\Gamma')$, where I' is obtained from I by substituting each function in Γ for its restriction in Γ', the optimum of I and I' coincide.*

Lemma 4 ([22]). *Let Γ be a core valued constraint language. The problems $VCSP(\Gamma)$ and $VCSP(\Gamma \cup \mathcal{C}_D)$ are polynomial-time equivalent.*

A special case of Lemma 4 for finite-valued constraint languages was proved by the authors in [27], building on [15], and Lemma 4 can be proved similarly.

3   Sherali-Adams and Valued Relational Width

In this section, we state and prove our main result on the applicability of Sherali-Adams relaxations to VCSPs. First, we define some notions concerning *bounded relational width* which is the basis for our proof.

We write (S, C) for (valued) constraints that involve (unweighted) relations, where S is the scope and C is the constraint relation. For a tuple $\mathbf{x} \in D^S$, we denote by $\pi_{S'}(\mathbf{x})$ its projection onto $S' \subseteq S$. For a constraint (S, C), we define $\pi_{S'}(C) = \{\pi_{S'}(\mathbf{x}) \mid \mathbf{x} \in C\}$.

Let $1 \leq k \leq \ell$ be integers. The following definition is equivalent[2] to the definition of (k, ℓ)-minimality for CSP instances given in [1].

Definition 13. *A CSP-instance $J = (V, D, \{(S_i, C_i)\}_{i=1}^q)$ is said to be (k, ℓ)-minimal if:*

- *For every $S \subseteq V$, $|S| \leq \ell$, there exists $1 \leq i \leq q$ such that $S = S_i$.*
- *For every $i, j \in [q]$ such that $|S_j| \leq k$ and $S_j \subseteq S_i$, $C_j = \pi_{S_j}(C_i)$.*

There is a straightforward polynomial-time algorithm for finding an equivalent (k, ℓ)-minimal instance [1]. This leads to the notion of *relational width*:

---

[2] The two requirements in [1] are: for every $S \subseteq V$ with $|S| \leq \ell$ we have $S \subseteq S_i$ for some $1 \leq i \leq q$; and for every set $W \subseteq V$ with $|W| \leq k$ and every $1 \leq i, j \leq q$ with $W \subseteq S_i$ and $W \subseteq S_j$ we have $\pi_W(C_i) = \pi_W(C_j)$.

Definition 14. *A constraint language Δ has relational width (k, ℓ) if, for every instance $J \in \mathrm{CSP}(\Delta)$, an equivalent (k, ℓ)-minimal instance is non-empty if, and only if, J has a solution.*

A k-ary idempotent operation $f : D^k \to D$ is called a *weak near-unanimity* (WNU) operation if, for all $x, y \in D$, $f(y, x, x, \ldots, x) = f(x, y, x, x, \ldots, x) = f(x, x, \ldots, x, y)$.

Definition 15. *We say that a clone of operations satisfies the* bounded width condition (BWC) *if it contains WNU operations of all but finitely many arities.*

Theorem 1 ([3,21]). *Let Δ be a constraint language of finite size containing all constant unary relations. Then, Δ has bounded relational width if, and only if, $\mathrm{Pol}(\Delta)$ satisfies the BWC.*

Theorem 2 ([1]). *Let Δ be a constraint language. If Δ has bounded relational width, then it has relational width $(2, 3)$.*

Let $I(x_1, \ldots, x_n) = \sum_{i=1}^{q} \phi_i(S_i)$ be an instance of the VCSP, where $S_i \subseteq V = \{x_1, \ldots, x_n\}$ and $\phi_i : D^{|S_i|} \to \overline{\mathbb{Q}}$. First, we make sure that every non-empty $S \subseteq V$ with $|S| \leq \ell$ appears in some term $\phi_i(S)$, possibly by adding constant-0 weighted relations. The Sherali-Adams [24] linear programming relaxation with parameters (k, ℓ) is defined as follows. The variables are $\lambda_i(\mathbf{s})$ for every $i \in [q]$ and tuple $\mathbf{s} \in D^{S_i}$.

$$\min \sum_{i=1}^{q} \sum_{\mathbf{s} \in \mathrm{Feas}(\phi_i)} \lambda_i(\mathbf{s}) \phi_i(\mathbf{s})$$

$$\lambda_j(\mathbf{t}) = \sum_{\mathbf{s} \in D^{S_i}, \pi_{S_j}(\mathbf{s}) = \mathbf{t}} \lambda_i(\mathbf{s}) \qquad \forall i, j \in [q] : S_j \subseteq S_i, |S_j| \leq k, \mathbf{t} \in D^{S_j}$$

$$\sum_{\mathbf{s} \in D^{S_i}} \lambda_i(\mathbf{s}) = 1 \qquad \forall i \in [q]$$

$$\lambda_i(\mathbf{s}) = 0 \qquad \forall i \in [q], \mathbf{s} \notin \mathrm{Feas}(\phi_i)$$

$$\lambda_i(\mathbf{s}) \geq 0 \qquad \forall i \in [q], \mathbf{s} \in D^{S_i}$$

The $\mathrm{SA}(k, \ell)$ optimum is always less than or equal to the VCSP optimum, hence the program is a relaxation. In anticipation of our main theorem, we make the following definition.

Definition 16. *A* valued constraint language *Γ has* valued relational width (k, ℓ) *if, for every instance I of $\mathrm{VCSP}(\Gamma)$, if the $\mathrm{SA}(k, \ell)$-relaxation of I has a feasible solution, then its optimum coincides with the optimum of I.*

For a feasible solution λ of $\mathrm{SA}(k, \ell)$, let $\mathrm{supp}(\lambda_i) = \{\mathbf{s} \in D^{S_i} \mid \lambda_i(\mathbf{s}) > 0\}$.

Lemma 5. *Let I be an instance of $\mathrm{VCSP}(\Gamma)$. Assume that $\mathrm{SA}(k, \ell)$ for I is feasible. Then, there exists an optimal solution λ^* to $\mathrm{SA}(k, \ell)$ such that, for every i, $\mathrm{supp}(\lambda_i^*)$ is closed under every operation in $\mathrm{supp}(\Gamma)$.*

Theorem 3. *Let Γ be a valued constraint language of finite size containing all constant unary relations. If* $\mathrm{supp}(\Gamma)$ *satisfies the BWC, then Γ has valued relational width* $(2,3)$.

Proof. Let I be an instance of VCSP(Γ). The dual of the SA(k,ℓ) relaxation can be written in the following form, with variables z_i for $i \in [q]$ and $y_{j,\mathbf{t},i}$ for $i,j \in [q]$ such that $S_j \subseteq S_i$, $|S_j| \leq k$, and $\mathbf{t} \in D^{S_j}$. The dual variables corresponding to $\lambda_i(\mathbf{s}) = 0$ are eliminated together with the dual inequalities for $i, \mathbf{s} \notin \mathrm{Feas}(\phi_i)$.

$$\max \sum_{i=1}^{q} z_i$$

$$z_i \leq \phi_i(\mathbf{s}) + \sum_{j \in [q], S_j \subseteq S_i} y_{j,\pi_{S_j}(\mathbf{s}),i} - \sum_{j \in [q], S_i \subseteq S_j} y_{i,\mathbf{s},j} \quad \forall i \in [q], |S_i| \leq k, \mathbf{s} \in \mathrm{Feas}(\phi_i)$$

$$z_i \leq \phi_i(\mathbf{s}) + \sum_{\substack{j \in [q], S_j \subseteq S_i \\ |S_j| \leq k}} y_{j,\pi_{S_j}(\mathbf{s}),i} \quad \forall i \in [q], |S_i| > k, \mathbf{s} \in \mathrm{Feas}(\phi_i)$$

It is clear that if I has a feasible solution, then so does the SA(k,ℓ) primal. Assume that the SA($2,3$)-relaxation has a feasible solution. By Lemma 5, there exists an optimal primal solution λ^* such that, for every $i \in [q]$, $\mathrm{supp}(\lambda_i^*)$ is closed under $\mathrm{supp}(\Gamma)$. Let y^*, z^* be an optimal dual solution.

Let $\Delta = \{C_i\}_{i=1}^{q} \cup \{\mathcal{C}_D\}$, where $C_i = \mathrm{supp}(\lambda_i^*)$, and consider the instance $J = (V, D, \{(S_i, C_i)\}_{i=1}^{q})$ of CSP(Δ). We make the following observations:

1. By construction of λ^*, $\mathrm{supp}(\Gamma) \subseteq \mathrm{Pol}(\Delta)$, so Δ contains all constant unary relations and satisfies the BWC. By Theorems 1 and 2, the language Δ has relational width $(2,3)$.
2. The first set of constraints in the primal say that if $i,j \in [q]$, $|S_j| \leq 2$ and $S_j \subseteq S_i$, then $\lambda_j^*(\mathbf{t}) > 0$ (i.e., $\mathbf{t} \in C_j$) iff $\sum_{\mathbf{s} \in D^{S_i}, \pi_{S_j}(\mathbf{s}) = \mathbf{t}} \lambda_i^*(\mathbf{s}) > 0$ (i.e., $\mathbf{t} \in \pi_{S_j}(C_i)$). In other words, J is $(2,3)$-minimal.

These two observations imply that J has a satisfying assignment $\sigma : V \to D$. By complementary slackness, since $\lambda_i^*(\sigma(S_i)) > 0$ for every $i \in [q]$, we must have equality in the corresponding rows in the dual indexed by i and $\sigma(S_i)$. Hence,

$$\sum_{i=1}^{q} z_i^* = \sum_{i=1}^{q} \phi_i(\sigma(S_i)) + \left(\sum_{i=1}^{q} \sum_{\substack{j \in [q], S_j \subseteq S_i \\ |S_j| \leq 2}} y_{j,\pi_{S_j}(\sigma(S_i)),i}^* - \sum_{\substack{i \in [q] \\ |S_i| \leq 2}} \sum_{\substack{j \in [q] \\ S_i \subseteq S_j}} y_{i,\sigma(S_i),j}^*\right)$$

$$(3)$$

By noting that $\pi_{S_j}(\sigma(S_i)) = \sigma(S_j)$, we can rewrite the expression in parenthesis on the right-hand side of (3) as:

$$\sum_{\substack{i,j \in [q], S_j \subseteq S_i \\ |S_j| \leq 2}} y_{j,\sigma(S_j),i}^* - \sum_{\substack{i,j \in [q], S_i \subseteq S_j \\ |S_i| \leq 2}} y_{i,\sigma(S_i),j}^* = 0. \qquad (4)$$

Therefore, $\sum_{i=1}^{q} \sum_{\mathbf{s} \in \mathrm{Feas}(\phi_i)} \lambda_i^*(\mathbf{s})\phi_i(\mathbf{s}) = \sum_{i=1}^{q} z_i^* = \sum_{i=1}^{q} \phi_i(\sigma(S_i))$, where the first equality follows by strong LP-duality, and the second by (3) and (4). Since I was an arbitrary instance of VCSP(Γ), the theorem follows.

4   Generalisations of Known Tractable Languages

In this section, we give some applications of Theorem 3. Firstly, we show that the BWC is preserved by going to a core and the addition of constant unary relations. Hence the BWC guarantees valued relational width $(2,3)$ also for languages not necessarily containing constant unary relations, as required by Theorem 3.

Lemma 6. *Let Γ be a valued constraint language of finite size on domain D and Γ' a core of Γ on domain $D' \subseteq D$. Then, $\mathrm{supp}(\Gamma)$ satisfies the BWC if, and only if, $\mathrm{supp}(\Gamma' \cup \mathcal{C}_{D'})$ satisfies the BWC.*

Theorem 4. *Let Γ be a valued constraint language of finite size. If $\mathrm{supp}(\Gamma)$ satisfies the BWC, then Γ has valued relational width $(2,3)$.*

Secondly, we show that for any VCSP instance over a language of valued relational width $(2,3)$ we can not only compute the value of an optimal solution but we can also find an optimal assignment in polynomial time.

Proposition 1. *Let Γ be a valued constraint language of finite size and I an instance of $VCSP(\Gamma)$. If $\mathrm{supp}(\Gamma)$ satisfies the BWC, then an optimal assignment to I can be found in polynomial time.*

Finally, we show that testing for the BWC is a decidable problem.

Proposition 2. *Testing whether a valued constraint language of finite size satisfies the BWC is decidable.*

Tractable Languages. Here we give some examples of previously studied valued constraint languages and show that they all have valued relational width $(2,3)$.

Example 3. Let ω be a ternary fractional operation defined by $\omega(f) = \omega(g) = \omega(h) = \frac{1}{3}$ for some (not necessarily distinct) majority operations f, g, and h. Cohen et al. proved the tractability of any language improved by ω by a reduction to CSPs with a majority polymorphism [8].

Example 4. Let ω be a ternary fractional operation defined by $\omega(f) = \frac{2}{3}$ and $\omega(g) = \frac{1}{3}$, where $f : \{0,1\}^3 \to \{0,1\}$ is the Boolean majority operation and $g : \{0,1\}^3 \to \{0,1\}$ is the Boolean minority operation. Cohen et al. proved the tractability of any language improved by ω by a simple propagation algorithm [8].

Example 5. Generalising Example 4 from Boolean to arbitrary domains, let ω be a ternary fractional operation such that $\omega(f) = \frac{1}{3}$, $\omega(g) = \frac{1}{3}$, and $\omega(h) = \frac{1}{3}$ for some (not necessarily distinct) conservative majority operations f and g, and a conservative minority operation h; such an ω is called an MJN. Kolmogorov and Živný proved the tractability of any language improved by ω by a 3-consistency algorithm and a reduction, via Example 6, to submodular function minimisation [18].

Corollary 1. *Let Γ be a valued constraint language of finite size such that* supp(Γ) *contains a majority operation. Then, Γ has valued relational width* $(2, 3)$.

Example 6. Let ω be a binary fractional operation defined by $\omega(f) = \omega(g) = \frac{1}{2}$, where f and g are conservative and commutative operations and $f(x, y) \neq g(x, y)$ for every x and y; such an ω is called a *symmetric tournament pair* (STP). Cohen et al. proved the tractability of any language improved by ω by a 3-consistency algorithm and an ingenious reduction to submodular function minimisation [7]. Such languages were shown to be the only tractable languages among conservative finite-valued constraint languages [18].

Corollary 2. *Let Γ be a valued constraint language of finite size such that* supp(Γ) *contains two symmetric tournament operations (that is, binary operations f and g that are both conservative and commutative and $f(x, y) \neq g(x, y)$ for every x and y). Then, Γ has valued relational width* $(2, 3)$.

Example 7. Generalising Example 6, let ω be a binary fractional operation defined by $\omega(f) = \omega(g) = \frac{1}{2}$, where f and g are conservative and commutative operations; such an ω is called a *tournament pair*. Cohen et al. proved the tractability of any language improved by ω by a consistency-reduction relying on Bulatov's result [4], which in turn relies on 3-consistency, to the STP case from Example 6 [7].

Corollary 3. *Let Γ be a valued constraint language of finite size such that* supp(Γ) *contains a tournament operation (that is, a binary conservative and commutative operation). Then, Γ has valued relational width* $(2, 3)$.

Example 8. In this example we denote by $\{\{\ldots\}\}$ a multiset. Let ω be a binary fractional operation on D defined by $\omega(f) = \omega(g) = \frac{1}{2}$ and let μ be a ternary fractional operation on D defined by $\mu(h_1) = \mu(h_2) = \mu(h_3) = \frac{1}{3}$. Moreover, assume that $\{\{f(x, y), g(x, y)\}\} = \{\{x, y\}\}$ for every x and y and $\{\{h_1(x, y, z), h_2(x, y, z), h_3(x, y, z)\}\} = \{\{x, y, z\}\}$ for every x, y, and z. Let Γ be a language on D such that for every two-element subset $\{a, b\} \subseteq D$, either $\omega|_{\{a, b\}}$ is an STP or $\mu|_{\{a, b\}}$ is an MJN. Kolmogorov and Živný proved the tractability of Γ by a 3-consistency algorithm and a reduction, via Example 6, to submodular function minimisation [18]. Such languages were shown to be the only tractable languages among conservative valued constraint languages [18].

Corollary 4. *Let Γ be a valued constraint language of finite size with fractional polymorphisms ω and μ as described in Example 8. Then, Γ has valued relational width* $(2, 3)$.

Dichotomy for Conservative Valued Constraint Languages. A valued constraint language Γ is called *conservative* if Γ contains all unary $\{0, 1\}$-valued weighted relations. Kolmogorov and Živný gave a dichotomy theorem for such languages, showing that they are either NP-hard, or tractable, cf. Example 8. Here we prove this dichotomy using the SA$(2, 3)$-relaxation as the algorithmic tool.

Lemma 7. *Let Γ be a valued constraint language and I be any instance of VCSP(Γ). Then, VCSP($\Gamma \cup \{\text{Opt}(I)\}$) polynomial-time reduces to VCSP(Γ).*

The following theorem was proved by Takhanov [25] with a reduction, essentially amounting to Lemma 7, added in [18].

Theorem 5 ([18,25]). *Let Γ be a conservative valued constraint language. If Pol(Γ) does not contain a majority polymorphism, then Γ is NP-hard.*

Theorem 6. *Let Γ be a conservative valued constraint language. Either Γ is NP-hard, or Γ has valued relational width $(2,3)$.*

Proof. Let F be the set of majority operations in Pol(Γ)\supp(Γ). By Lemma 2, for each $f \in F$, there is an instance I_f of VCSP(Γ) such that $f \notin \text{Pol}(\text{Opt}(I_f))$. Let $\Gamma' = \Gamma \cup \{\text{Opt}(I_f) \mid f \in F\}$. Assume that Pol($\Gamma'$) contains a majority polymorphism f. Then, $f \notin F$, so $f \in \text{supp}(\Gamma)$. From Corollary 1, it follows that Γ has valued relational width $(2,3)$. If Pol(Γ') does not contain a majority polymorphism, then, since Γ is conservative, so is Γ', and hence Γ' is NP-hard by Theorem 5. Therefore, Γ is NP-hard by Lemma 7.

5   Conclusions

We have shown that most previously studied tractable valued constraint languages that are not purely relational fall into the cases covered by Theorem 4. In the full version of this paper, we will prove the converse of Theorem 4, thus giving a precise characterisation of the power of valued relational width $(2,3)$, as well as some computational complexity consequences.

References

1. Barto, L.: The collapse of the bounded width hierarchy. Journal of Logic and Computation (2014)
2. Barto, L., Kozik, M.: Robust Satisfiability of Constraint Satisfaction Problems. In: Proc. STOC 2012, pp. 931–940. ACM (2012)
3. Barto, L., Kozik, M.: Constraint Satisfaction Problems Solvable by Local Consistency Methods. Journal of the ACM 61(1), Article No. 3 (2014)
4. Bulatov, A.: Combinatorial problems raised from 2-semilattices. Journal of Algebra **298**, 321–339 (2006)
5. Bulatov, A., Krokhin, A., Jeavons, P.: Classifying the Complexity of Constraints using Finite Algebras. SIAM Journal on Computing **34**(3), 720–742 (2005)
6. Chekuri, C., Khanna, S., Naor, J., Zosin, L.: A linear programming formulation and approximation algorithms for the metric labeling problem. SIAM J. on Discrete Mathematics **18**(3), 608–625 (2004)
7. Cohen, D.A., Cooper, M.C., Jeavons, P.G.: Generalising submodularity and Horn clauses: Tractable optimization problems defined by tournament pair multimorphisms. Theoretical Computer Science **401**(1–3), 36–51 (2008)
8. Cohen, D.A., Cooper, M.C., Jeavons, P.G., Krokhin, A.A.: The Complexity of Soft Constraint Satisfaction. Artif. Intell. **170**, 983–1016 (2006)

9. Dalmau, V., Krokhin, A., Manokaran, R.: Towards a characterization of constant-factor approximable Min CSPs. In: Proc. SODA 2015 (2015)
10. Dalmau, V., Krokhin, A.A.: Robust Satisfiability for CSPs: Hardness and Algorithmic Results. ACM ToCT 5(4), Article No. 15 (2013)
11. de la Vega, W. F., Kenyon-Mathieu, C.: Linear programming relaxations of maxcut. In: Proc. SODA 2007, pp. 53–61. SIAM (2007)
12. Ene, A., Vondrák, J., Wu, Y.: Local distribution and the symmetry gap: Approximability of multiway partitioning problems. In: SODA 2013, pp. 306–325 (2013)
13. Feder, T., Vardi, M.Y.: The Computational Structure of Monotone Monadic SNP and Constraint Satisfaction: A Study through Datalog and Group Theory. SIAM Journal on Computing 28(1), 57–104 (1998)
14. Fulla, P., Živný, S.: A Galois Connection for Valued Constraint Lan-guages of Infinite Size. In: Proc. ICALP 2015. Springer (2015)
15. Huber, A., Krokhin, A., Powell, R.: Skew bisubmodularity and valued CSPs. SIAM Journal on Computing 43(3), 1064–1084 (2014)
16. Jeavons, P., Krokhin, A., Živný, S.: The complexity of valued constraint satisfaction. Bulletin of the EATCS 113, 21–55 (2014)
17. Kolmogorov, V., Thapper, J., Živný, S.: The power of linear programming for general-valued CSPs. SIAM Journal on Computing 44(1), 1–36 (2015)
18. Kolmogorov, V., Živný, S.: The complexity of conservative valued CSPs. Journal of the ACM 60(2), Article No. 10 (2013)
19. Kozik, M., Krokhin, A., Valeriote, M., Willard, R.: Characterizations of several Maltsev Conditions. Algebra Universalis (2014) (to appear)
20. Kun, G., O'Donnell, R., Tamaki, S., Yoshida, Y., Zhou, Y.: Linear programming, width-1 CSPs, and robust satisfaction. In: ITCS 2012, p. 484–495 (2012)
21. Larose, B., Zádori, L.: Bounded width problems and algebras. Algebra Universalis 56, 439–466 (2007)
22. Kozik, M., Ochremiak, J.: Algebraic Properties of Valued Constraint Satisfaction Problem. In: Proc. ICALP 2015. Springer (2015)
23. Raghavendra, P.: Optimal algorithms and inapproximability results for every CSP? In: Proc. STOC 2008, pp. 245–254. ACM (2008)
24. Sherali, H.D., Adams, W.P.: A hierarchy of relaxations between the continuous and convex hull representations for zero-one programming problems. SIAM Journal of Discrete Mathematics 3(3), 411–430 (1990)
25. Takhanov, R.: A Dichotomy Theorem for the General Minimum Cost Homomorphism Problem. In: Proc. STACS 2010, pp. 657–668 (2010)
26. Thapper, J., Živný, S.: The power of linear programming for valued CSPs. In: Proc. FOCS 2012, pp. 669–678. IEEE (2012)
27. Thapper, J., Živný, S.: The complexity of finite-valued CSPs. In: Proc. STOC 2013, pp. 695–704. ACM (2013) (February 2015). Full version arXiv:1210.2977v3
28. Thapper, J., Zživný, S.: Necessary Conditions on Tractability of Valued Constraint Languages. Technical report, February 2015. arXiv:1502.03482
29. Uppman, H.: The Complexity of Three-Element Min-Sol and Conservative Min-Cost-Hom. In: Fomin, F.V., Freivalds, R., Kwiatkowska, M., Peleg, D. (eds.) ICALP 2013, Part I. LNCS, vol. 7965, pp. 804–815. Springer, Heidelberg (2013)
30. Yoshida, Y., Zhou, Y.: Approximation schemes via Sherali-Adams hierarchy for dense constraint satisfaction problems and assignment problems. In: Proc. ITCS 2014, pp. 423–438. ACM (2014)

Two-sided Online Bipartite Matching and Vertex Cover: Beating the Greedy Algorithm

Yajun Wang[1] and Sam Chiu-wai Wong[2]($\boxtimes$)

[1] Microsoft Research Asia, Beijing, China
[2] University of California, Berkeley, USA
samcwong@berkeley.edu

Abstract. We consider the generalizations of two classical problems, online bipartite matching and ski rental, in the field of online algorithms, and establish a novel connection between them.

In the original setting of online bipartite matching, vertices from only one side of the bipartite graph are online. Motivated by market clearing applications where both buyers and sellers are online, we study the generalization, called two-sided online bipartite matching, in which all vertices can be online. An algorithm for it should maintain a b-matching and try to maximize its size. We show that this problem can be attacked by considering the complementary "dual" problem, two-sided online bipartite vertex cover, which in fact is a generalization of ski rental.

As the greedy algorithm is 1/2-competitive for both problems, the challenge is to beat the ratio of 1/2. In this paper, we present new 0.526-competitive algorithms for both problems under the large budget assumption. A key technical ingredient of our results is a charging-based framework for the design and analysis of water-filling type algorithms. This allows us to systematically establish approximation bounds for various water-filling algorithms.

On the hardness side, we show that no online randomized algorithm achieves a competitive ratio better than 0.570 and 0.625 respectively for these two problems. Our bounds show that the one-sided optimal ratio of $1 - 1/e \approx 0.632$ is indeed unattainable.

1 Introduction

The classical online bipartite matching problem (OBM) studies algorithms which incrementally construct a matching in the presence of online vertex arrival. Informally, while one side of the input bipartite graph is known (offline) initially, vertices on the other side are revealed one after another along with their incident edges. The goal is to maintain a large *monotone* matching from which no edge is ever removed. The optimal competitive ratio for OBM is $1 - 1/e$.

As observed in [3,4], most of the OBM variants studied in the literature share the common feature that vertices of only one side of the bipartite graph arrive

S.C-W. Wong—This research was supported by NSF grants CCF0964033 and CCF1408635, and by Templeton Foundation grant 3966.

© Springer-Verlag Berlin Heidelberg 2015
M.M. Halldórsson et al. (Eds.): ICALP 2015, Part I, LNCS 9134, pp. 1070–1081, 2015.
DOI: 10.1007/978-3-662-47672-7_87

online. While this property indeed holds in many applications, it does not necessarily reflect the reality in general. For example, in the *online market clearing problem* [4], buyers and sellers in a commodity market are represented by the two bipartitions. An edge between a buyer and a seller indicates that the price that the buyer is willing to offer is higher than the seller's valuation. The objective is to maximize the number of trades, or the size of the matching. Here both the buyers and sellers arrive and leave online.

In this paper, this limitation is addressed by allowing all vertices to be online. We introduce the two-sided online bipartite b-matching problem (TOBM) which requires maintaining a large b-matching in an online manner when vertices arrive one at a time. At each step an adversarially chosen vertex v is revealed with its edges incident to the *previously arrived* neighbors. An algorithm must then decide how many copies of each edge should be added to the current b-matching $\mathbf{x}$, while ensuring that the total number of edge copies incident to each vertex u does not exceed its budget b_u. The objective is to maximize the size of $\mathbf{x}$.

As with most other problems in the field, it is clear that the **Greedy** algorithm, which maintains a maximal b-matching, is 1/2-competitive. Thus the central challenge here is to beat **Greedy**. One of our two main results is a 0.526-competitive algorithm for TOBM.

New charging framework. Interestingly, departing from previous approaches to OBM-type problems, we attack the problem by first studying its *dual*. This new approach carries two benefits. Firstly, the structure of the dual suggests a simple class of algorithms analysable under an elegant charging framework, thus circumventing the previous delicate mapping-based argument and the primal-dual method which must simultaneously update both the primal and dual variables. Secondly, the charging-based analysis can be cast back into the primal-dual framework naturally (section 3). In short, our new approach can be viewed as a recipe to systematically engineer primal-dual analyses for OBM-type problems.

Connection to ski rental. As a bonus, the duals of OBM and TOBM, online bipartite vertex cover, are interesting problems in their own right since they generalize the classical ski rental problem. In fact, online bipartite vertex cover can be interpreted as the combinatorial version of ski rental. Thus in a way, our results establish a strong connection between online bipartite matching and ski rental, two of the most well-studied online problems. This connection is somewhat unusual as they do not generalize ski rental but are the dual of its generalization.

Main results. As mentioned above, **Greedy** achieves an approximation factor of 1/2 for just about any online matching and vertex cover type problems. Our main results are 0.526-competitive algorithms for two-sided online bipartite b-matching (under the large budget assumption[1]) and vertex cover. To our knowledge, this is the first successful attempt in breaking the barrier of 1/2 attained by **Greedy**. Our results are also improvements over special cases of the online edge-selection problem [4] and the online edge-arrival problem [24], both of which

---

[1] More precisely, the assumption is needed for only TOBM but not TOBVC.

currently have best competitive ratio $1/2$. Furthermore, our algorithms are also *optimal* with respect to the charging framework described above.

1.1   Preliminaries

As usual, we begin with the definitions of our problems. Let $G = (L \cup R, E)$ be a bipartite graph with *left* vertices L and *right* vertices R. Thus $L \cap R = \emptyset$ and $E \subseteq L \times R$. Let $V = L \cup R$. Each $v \in V$ also has a budget/weight $b_v \geq 0$. We denote by $N(v)$ the neighbors of v. Recall that $\mathbf{x} \in \mathbb{N}_{\geq 0}^E$ is a b-matching if $\sum_{u \in N(v)} x_{uv} \leq b_v \forall v \in V$; x_e is the number of copies of e in the b-matching.

Online Bipartite b-Matching and Vertex Cover (OBM & OBVC) Initially, we are given L and their capacity $b_u, u \in L$. At each step a vertex $v \in R$ is revealed with b_v and its incident edges. An algorithm for OBM maintains a b-matching $\mathbf{x}$, and must irrevocably decide at each step on the values of x_e for each new edge e. The objective is to maximize the size of $\mathbf{x}$.

Similarly, for OBVC we are required to maintain a vertex cover C at all time by only inserting vertices into C, i.e. no vertex removal is allowed. Thus at each step the task is essentially deciding if v or $N(v)$ should be inserted to C. The objective is to minimize the *weight* of C, $b(C) := \sum_{v \in C} b_v$.

Two-sided Online Bipartite b-Matching and Vertex Cover (TOBM & TOBVC) The two-sided version relaxes the constraint that only the right vertices R are online. In this setting, the graph is initially empty and at each step, a new vertex $v \in V$ arrives along with b_v and the edges incident to v and its *already arrived* neighbors. Analogous to OBM & OBVC, an algorithm for TOBM (resp. TOBVC) maintains a b-matching $\mathbf{x}$ (resp. vertex cover) from which no edge (resp. vertex) is ever removed. Furthermore, x_e never changes once initialized. The objectives are exactly the same as before.

Model. We measure the performance of online algorithms via the standard *competitive analysis* and *oblivious adversary model*. Roughly speaking, this means that the adversary chooses the input before the algorithm executes, and a randomized algorithm is c-competitive if the (expected) quality of its solution is within a multiplicative factor of c from optimum.

Fractional vertex cover and b-matching. We will use the LP relaxations for bipartite b-matching and vertex cover extensively. The dual variable y_v is called the *potential* of v.

Primal (b-Matching):	Dual (Vertex Cover):
$\max \sum_{e \in E} x_e$	$\min \sum_{v \in V} b_v y_v$
s.t. $x_v := \sum_{u \in N(v)} x_{uv} \leq b_v, \ \forall v \in V$	s.t. $y_u + y_v \geq 1, \ \forall (u, v) \in E$
$\mathbf{x} \geq 0$	$\mathbf{y} \geq 0$

1.2   Our Contributions and Techniques

Our main results are 0.526-competitive algorithms for TOBM[4] and TOBVC, and an optimal $1 - 1/e$-competitive algorithm for OBVC, beating the baseline $1/2$-competitive **Greedy** and improving over special cases of the online edge-selection problem [4] and the online edge-arrival problem [24]. These results also generalize the classical online bipartite matching and ski rental problems, thus establishing a connection between them. This suggests that there are perhaps more connections of the other work on these two problems in the literature, and understanding them may lead to new insights into some of the open problems.

Tables 1 and 2 summarize our contributions (in **bold**) and the existing state-of-the-art results, for both our problems and their close relatives in the literature. In Table 1, the last three problems are generalizations of the first, whereas in Table 2, the $(n + 1)$-th problem generalizes the n-th.

Table 1. Results for online matching type problems

	online bipartite b-matching	adwords	online bipartite weighted b-matching	two-sided online bipartite b-matching
Comp. ratios	$1 - 1/e \approx 0.632$	$1 - 1/e$ [2]	$1 - 1/e$ [4]	**0.532** [4]
Hardness	$1 - 1/e$	$1 - 1/e$	$1 - 1/e$	**0.625**
References	$[1, 19], [14]^4$	[23]	[7, 11]	**this paper**

Table 2. Results for ski rental type problems

	ski rental	multislope ski rental	online bipartite vertex cover	two-sided online bipartite vertex cover
Comp. ratios	$1 - 1/e$	$1 - 1/e$	$1 - 1/e$	**0.532**
Hardness	$1 - 1/e$	$1 - 1/e$	$1 - 1/e$	**0.570**
References	[18]	[20]	**this paper**[3]	**this paper**

A key technical ingredient of our results is a novel *charging-based framework* for the design and analysis of water-filling type algorithms, first used in [14]. This allows us to systematically establish approximation bounds for different variants of water-filling algorithms.

To give a glimpse into how our charging scheme works, consider the OBVC problem. Upon the arrival of a new $v \in R$, we are to add v or $N(v)$ to the current vertex cover C in order to maintain feasibility. On the other hand, by the definition of vertex cover we know that v or $N(v)$ must be in the optimal solution. Our scheme will charge the increment in the cost of the algorithm's solution to either v or $N(v)$ depending on which of the two are in the optimum.

At a very high level, the elegance of the charging framework relies on the simplicity of deciding between only two choices to process an online vertex. This is also the beauty of starting from the simpler dual problem of vertex cover.

Finally, the optimality of our algorithm under the charging-based framework follows from the solution to a minimax optimization problem, which may be of independent interest.

---

[2] Under the large budget assumption: the ratio converges to the said value as $\min_{v \in V} b_v \to \infty$.

[3] With some work this result can be implied by an earlier paper [5]. See the next section for a discussion.

1.3   Related Work

There are two lines of research related to our work.

Online matching. The online bipartite matching problem was first studied in the seminal paper by Karp et al. [19]. They gave an optimal $1 - 1/e$-competitive algorithm. Subsequent works studied its variants such as b-matching [14], vertex weighted version [1,9], adwords [1,5,8–10,13,23] and online market clearing [4]. Water-filling algorithms have been used for a few variants of the online bipartite matching problem (e.g. [5,14]). Another line of research studies the problem under more relaxed models by assuming certain randomness inherent to the input [12,15,21,22]. Online matching for general graphs have been studied under similar stochastic models [2]. To our knowledge, there is no result on this problem in the more restricted adversarial models other than the naive greedy algorithm, even for just bipartite graphs with vertices from both sides online [4].

Ski rental. The ski rental problem was first studied in [18]. Karlin et al. gave an optimal $\frac{1}{1-1/e}$-competitive algorithm [17]. There are many generalizations of ski rental. Of particular relevance are multislope ski rental [20] and TCP acknowledgment [16], where the competitive ratio $\frac{1}{1-1/e}$ is still achievable. The OBVC problem presented in this paper is also of this nature and, in fact, further generalizes multislope ski rental, as shown in the full version of the paper.

Finally, online vertex cover was studied by Demange et al. [6] in a substantially different model. Their competitive ratios are characterized by $\max_v deg(v)$.

2   One-sided and Two-sided Online Bipartite Vertex Cover

In this section, we study online vertex cover which is the "dual" of online matching. Our results are an *optimal* $1-1/e$-competitive algorithm for Online Bipartite Vertex Cover (OBVC) and a 0.526-competitive algorithm for Two-sided Online Bipartite Vertex Cover (TOBVC). Both problems generalize the well-known ski rental problem, which has an optimal competitive ratio $1 - 1/e$ [17].

We first argue that it suffices to work exclusively with the fractional version. That is, an algorithm can just maintain a *fractional* VC y in a way that y_v never decreases for $v \in V$. This is the consequence of a simple rounding scheme which *randomly* rounds a fractional VC to an integral VC in an online fashion.

Lemma 1. *(Lossless rounding) Any deterministic algorithm for fractional OBVC can be converted to a randomized algorithm for integral OBVC with the same competitive ratio.*

Our algorithm for TOBVC in fact even applies to online *fractional* VC in *general graphs* but this would not imply the same result for integral VC since Lemma 1 holds only for bipartite graphs.

2.1   (One-sided) Online Bipartite Vertex Cover

We present an optimal algorithm for OBVC. We note that with some work, the primal-dual analysis of a water-level algorithm for online fractional matching [5] implies another optimal algorithm for our problem. Our algorithm applies the *water level* paradigm on *vertex cover* instead of *matching*. This difference may appear trivial but it actually has profound consequences as discussed earlier. In particular, because of the structure of vertex cover, our charging-based analysis is considerably simpler and more amenable to generalizations.

For each vertex v, Algorithm 1 maintains a non-decreasing cover *potential* y_v. When an online vertex v arrives, to cover the new edges between v and $N(v)$ are revealed, we must initialize the potential y_v of v and possibly also increase the potentials of its neighbors. Suppose we set $y_v = 1 - y$ for some y. To maintain a feasible vertex cover, any $y_u < y$ for $u \in N(v)$ should be increased to y.

The crux lies in how y is determined. We consider a simple scheme in which y is related to the total potential increment of $N(v)$. More precisely, we require that the total potential increment $b_v(1 - y) + \sum_{u \in N(v):y_u<y} b_u(y - y_u)$ be at most $b_v/(1 - 1/e)$. Such an update rule arises from the need to balance the possibilities whether v or $N(v)$ is in the optimal vertex cover. Intuitively, if v is in the optimum we do not wish to spend more than $b_v/(1 - 1/e)$, which is the "fair" amount of resources v should is entitled to. If, on the other hand, $N(v)$ is in the optimum we should then charge $b_v y_v$ to $N(v)$. The charging analysis below will quantify these statements.

Algorithm 1. Water-filling algorithm for OBVC

Input: L and $b_u, u \in L$
Initialize for each $u \in L$, $y_u = 0$;
for *each online vertex* $v \in R$ **do**
> Maximize $y \leq 1$, s.t., $b_v(1 - y) + \sum_{u \in N(v)} b_u \max\{y - y_u, 0\} \leq b_v/(1 - 1/e)$;
> For each $u \in N(v)$, $y_u \leftarrow \max\{y_u, y\}$;
> $y_v \leftarrow 1 - y$;

end

Analysis. Let C^* be a minimum vertex cover of G. Our strategy is to charge the potential increment to vertices of C^* in such a way that each vertex of $v \in C^*$ is charged at most $b_v/(1 - 1/e)$.

Let v be the current online vertex. Our algorithm sets $y_v = 1 - y$ for some y. Let y_u be the potential of $u \in N(v)$. We consider two cases.

<u>Case 1:</u> $v \in C^*$. We charge the potential increment in $N(v)$ and v to v, which is at most $b_v/(1 - 1/e)$.

<u>Case 2:</u> $v \notin C^*$. Notice that we must have $N(v) \subseteq C^*$. In this case, vertices of $N(v)$ should be responsible for the potential $y_v = 1 - y$. We describe how to charge $b_v(1 - y)$ to $N(v)$.

Let $g(y) = \frac{1}{e-1} + y$. Rewrite $b_v(1 - y) + \sum_{u \in N(v)} b_u \max\{y - y_u, 0\} \leq b_v/(1 - 1/e)$ as

$$\sum_{u \in N(v)} b_u \max\{y - y_u, 0\} \leq b_v \left(\frac{1}{e-1} + y \right) = b_v g(y). \tag{1}$$

Intuitively, if $\sum_{u \in N(v)} b_u(y - y_u) = b_v g(y)$, the most fair scheme should charge $\frac{1-y}{g(y)} b_u(y - y_u)$ to $u \in N(v)$ since the fair "unit charge" is $\frac{b_v(1-y)}{b_v g(y)} = \frac{1-y}{g(y)}$. Since $\frac{1-t}{g(t)}$ is decreasing, $\frac{1-y}{g(y)} b_u(y - y_u)$ can be upper bounded by $b_u \int_{y_u}^{y} \frac{1-t}{g(t)} dt$. This observation motivates the next lemma which constitutes the basis of the major results in this paper.

Lemma 2. *Let* $f : [0,1] \longrightarrow \mathbb{R}_+$ *be continuous s.t.* $\frac{1-t}{f(t)}$ *is decreasing, and* $F(x) = \int_0^x \frac{1-t}{f(t)} dt$. *If* $\sum_{u \in X} b_u(y - y_u) = b_v f(y)$ *for some set* X *and* $y \geq y_u$ *for* $u \in X$, *then*

$$b_v(1 - y) \leq \sum_{u \in X} b_u \left(F(y) - F(y_u) \right).$$

Proof. We have the following

$$\sum_{u \in X} b_u \left(F(y) - F(y_u) \right) = \sum_{u \in X} b_u \int_{y_u}^{y} \frac{1-t}{f(t)} dt \geq \sum_{u \in X} b_u(y - y_u) \frac{1-y}{f(y)} = b_v(1 - y).$$

where the inequality above holds as $\frac{1-t}{f(t)}$ is decreasing. $\square$

Theorem 1. *Algorithm 1 is* $1 - 1/e$-*competitive and hence optimal for* OBVC.

Proof. We charge the potentials used to the vertices of the minimum cover C^*. Let v be an online vertex. The case $v \in C^*$ is trivial as explained before.

Now consider the case $v \notin C^*$. We charge the potential spent on $u \in N(v) \subseteq C^*$ to u itself. The potential spent on v is $b_v y_v = b_v(1-y)$ where y is the potential after processing v. Let $X = \{u \in N(v) \mid y_u < y\}$ be the set of vertices whose potentials increased when processing v. If $y = 1$, we are done as no charging is necessary. If $y < 1$, then we have equality in (1), i.e., $\sum_{u \in X} b_u(y - y_u) = b_v g(y)$ because otherwise y can be bigger without violating the inequality. We charge each $u \in X$ by $G(y) - G(y_u)$, where $G(x) = \int_0^x \frac{1-t}{g(t)} dt$. By Lemma 2, $b_v(1 - y) \leq \sum_{u \in X} b_u(G(y) - G(y_u))$, which is sufficient to account for $b_v(1 - y)$.

In summary, each right vertex $v \in C^*$ is responsible for $b_v/(1-1/e)$ potential. On the other hand, each left vertex $u \in C^*$ is responsible for itself (which contributes at most b_u) as well as the incoming charges from its online neighbors. The sum of these charges can be at most $b_u(G(1) - G(0))$ as the sum $b_u(G(y) - G(y_u))$, taken over the iterations in which y_u increases, telescopes. Therefore the amount of potential charged to a left vertex is bounded by $b_u(1 + G(1) - G(0)) = b_u/(1 - 1/e)$ as $1 + G(1) - G(0) = 1/(1 - 1/e)$.

The total cost of the algorithm is then bounded by $b(C^*)/(1 - 1/e)$, and optimality follows from the fact that OBVC generalizes ski rental (see appendix), which has an optimal ratio of $1 - 1/e$. Finally by Lemma 1, Algorithm 1 can be converted to give a random integral vertex cover with the same performance. $\square$

The reader may wonder how $1 - 1/e$ arises from our charging scheme. By replacing $1/(1 - 1/e)$ by c in our analysis, one can readily establish a ratio of

$$\frac{b(R \cap C^*)}{b(C^*)}c + \frac{b(L \cap C^*)}{b(C^*)}\left(1 + \int_0^1 \frac{1-t}{t+c-1}dt\right) \leq \max\{c, 1 + \int_0^1 \frac{1-t}{t+c-1}dt\}.$$

Setting the two quantities equal will give $c = 1/(1 - /e)$.

2.2   Two-sided Online Bipartite Vertex Cover

In this section we present a 0.526-competitive algorithm for TOBVC, beating the baseline $1/2$-competitive **Greedy** and improving over special cases of the online edge-selection problem [4] and the online edge-arrival problem [24]. Algorithm 2 extends Algorithm 1 in the last section by replacing $g(t) = 1/(e-1) + t$ by any continuous function $f : [0,1] \longrightarrow \mathbb{R}_+$ for which $\frac{1-t}{f(t)}$ is decreasing (see Lemma 2). In other words, in fact we study a class of algorithms parameterized by f.

We must carefully design the function f to obtain a non-trivial competitive ratio. Before getting into the details, we revisit the analysis in the last section to gain some insights. In our charging argument, each vertex in $L \cap C^*$ is responsible for the charges from its neighbors. On the other hand, a vertex in $R \cap C^*$ is only responsible for the potential increment for processing itself. Now if both L and R are online, an online vertex $v \in C^*$ should be responsible for the potential used to process itself when it arrives *and* the charges from future neighbors.

Algorithm 2. Water-filling algorithm for TOBVC

Let T be the set of arrived vertices. Initially $T = \emptyset$;
for *each online vertex v* **do**

    Maximize $y \leq 1$, s.t., $\sum_{u \in N(v) \cap T} b_u \max\{y - y_u, 0\} \leq b_v f(y)$;
    For each $u \in N(v) \cap T$, $y_u \leftarrow \max\{y_u, y\}$;
    $y_v \leftarrow 1 - y$;
    $T \leftarrow T \cup \{v\}$;

end

Informally, the algorithm initializes and updates the potentials y_v in this manner. Upon the arrival of v, if $y < 1$ we spend resources of $b_v f(y)$ on v's neighbors and $1 - y$ on v itself. The previous charging scheme suggests that v is responsible for $b_v f(y)$ if it is in the optimum. However, unlike the one-sided version, there can be future vertices which are neighbors of v. A natural fix is that v be charged by them as well. By Lemma 2, v will take charges at most $b_v \int_{1-y}^1 \frac{1-t}{f(t)}dt$. Setting $z = 1 - y$, the total charges to each $v \in C^*$ are at most

$$b_v \beta(f), \text{ where } \beta(f) := \max_{z \in [0,1]} 1 + f(1-z) + \int_z^1 \frac{1-t}{f(t)}dt.$$

We show how to compute the optimal function $f(\cdot)$ in section 2.2. Now we first formally prove that Algorithm 2 is $1/\beta(f)$-competitive for TOBVC.

Lemma 3. *Let $f : [0,1] \longrightarrow \mathbb{R}_+$ be continuous such that $\frac{1-t}{f(t)}$ is decreasing. We have in Algorithm 2 either $y = 1$ or $\sum_{u \in N(v) \cap T} b_u \max\{y - y_u, 0\} = b_v f(y)$.*

Theorem 2. *Suppose $f : [0,1] \longrightarrow \mathbb{R}_+$ is continuous and $\frac{1-t}{f(t)}$ is decreasing. Let $\beta = \max_{z \in [0,1]} 1 + f(1-z) + \int_z^1 \frac{1-t}{f(t)} dt$ and $F(x) = \int_0^x \frac{1-t}{f(t)} dt$. Then Algorithm 2 is $1/\beta(f)$-competitive for TOBVC.*

Computing the Optimal f. Our challenge now boils down to finding a $f(y)$ to get a small β. In essence, the goal is to solve the following optimization problem

$$\inf_{f \in \mathcal{F}} \max_{z \in [0,1]} 1 + f(1-z) + \int_z^1 \frac{1-t}{f(t)} dt, \tag{2}$$

where $\mathcal{F}$ is the class of positive continuous functions on $[0,1]$ such that $\frac{1-t}{f(t)}$ is decreasing for $f \in \mathcal{F}$. To our knowledge, there is no systematic approach to tackle a minimax optimization problem of this form. A natural way is to express the optimal z in terms of f and use calculus of variation to compute the best f. Unfortunately, in general there is no closed form expression for the optimal z.

To overcome this hurdle, we first disregard the requirement that $\frac{1-t}{f(t)}$ be decreasing. We show that such a relaxation of the optimization problem admits a nice optimality condition: there exists some optimal f such that $1 + f(1-z) + \int_z^1 \frac{1-t}{f(t)} dt$ is constant for *all* z. This property is characterized as follows.

Lemma 4. *Let $r : [0,1] \longrightarrow \mathbb{R}_+$ be a continuous function such that for $\forall p \in [0,1]$, $r(p) + \int_{1-p}^1 \frac{1-x}{r(x)} dx \leq \gamma$ for some $\gamma > 0$. Then there exists a continuous function $f : [0,1] \longrightarrow \mathbb{R}_+$ such that $\forall p \in [0,1]$, $f(p) + \int_{1-p}^1 \frac{1-x}{f(x)} dx \equiv \gamma$.*

It is therefore sufficient to consider functions f that satisfy this optimality condition. Consequently, $f(1-z) = \beta - 1 - \int_z^1 \frac{1-t}{f(t)} dt$ is actually differentiable. Differentiating $1 + f(1-z) + \int_z^1 \frac{1-t}{f(t)} dt$ yields $-f'(1-z) - \frac{1-z}{f(z)} = 0$, i.e.,

$$f(z)f'(1-z) = z - 1.$$

Although this differential equation is atypical as $f(z)$ and $f'(1-z)$ are not taken at the same point, surprisingly it has closed form solutions, as given below.

Lemma 5. *Let r be a non-negative differentiable function on $[0,1]$ and $r(z)r'(1-z) = z - 1$. Then*

$$r(z) = \left(\frac{1+k}{2} - z\right)^{\frac{1+k}{2k}} \left(z + \frac{k-1}{2}\right)^{\frac{k-1}{2k}},$$

where $k \geq 1$. Moreover, $\frac{1-t}{r(t)}$ is decreasing for $t \in [0,1]$.

The final step is just to select the best f from the family of solutions. Since $1 + f(1-z) + \int_z^1 \frac{1-t}{f(t)} dt$ is constant, it suffices to find the smallest $1 + f(0)$, which corresponds to $k \approx 1.997$.

Theorem 3. *Let $f(z) = \left(\frac{1+k}{2} - z\right)^{\frac{1+k}{2k}} \left(z + \frac{k-1}{2}\right)^{\frac{k-1}{2k}}$, where $k \approx 1.997$. Algorithm 2 is then 0.526-competitive for TOBVC.*

Remark: Our algorithm can be viewed as a generalization of **Greedy** because the solution $f(z) = 1 - z$ (with $k = 1$) is equivalent to some variant of it.

3   Two-sided Online Bipartite b-Matching

We give a primal-dual analysis of a variant of the algorithm given in the last section. A by-product of this new analysis is a 0.526-competitive algorithm for TOBM under the large budget assumption, i.e. the competitive ratio tends to 0.526 as $b := \min_{v \in V} b_v \to \infty$. It is known that it suffices to maintain fractional x_{uv} under such an assumption [5,7].

Let $\beta \approx 1.901$ and $f(z)$ be the same as in Theorem 3. Our primal-dual analysis is inspired by the one for online bipartite fractional matching [5]. Algorithm 3 applies to both TOBM and TOBVC. It is very similar to Algorithm 2 when restricted to the dual. To analyze the performance, we claim that the following two invariants hold throughout the execution of the algorithm.

Algorithm 3. Water-filling algorithm for TOBM; β and f as in Theorem 3

Let T be the set of arrived vertices. Initially $T = \emptyset$;
for *each online vertex v* **do**
  Maximize $y \leq 1$, s.t., $\sum_{u \in N(v) \cap T} b_u \max\{y - y_u, 0\} \leq b_v f(y)$;
  Let $X = \{u \in N(v) \cap T \mid y_u < y\}$;
  for *each $u \in X$* **do**
    $y_u \leftarrow y$;
    $x_{uv} \leftarrow \frac{b_u(y - y_u)}{\beta} \left(1 + \frac{1-y}{f(y)}\right)$;
  end
  For each $u \in (N(v) \cap T) \setminus X$, $x_{uv} \leftarrow 0$;
  $y_v \leftarrow 1 - y$, $T \leftarrow T \cup \{v\}$;
end

$$\textbf{Invariant 1: } b_u \cdot \frac{y_u + f(1 - z_u) + \int_{z_u}^{y_u} \frac{1-t}{f(t)} dt}{\beta} \geq x_u$$

Here z_u is the potential of u set upon its arrival, y_u is the current potential of u and $x_u = \sum_{v \in N(u)} x_{uv}$ is the total number of copies of edges incident to u. Since LHS $\leq b_u$ (see last section), the primal is feasible if the invariant holds.

$$\textbf{Invariant 2: } \sum_{u \in T} b_u y_u = \beta \sum_{(u,v) \in E \cap T^2} x_{uv}$$

Invariant 2 guarantees that the primal and dual objective values are always within a factor of β from each other. By weak duality, the algorithm is then simultaneously $1/\beta$-competitive for TOBM and TOBVC.

Lemmas 6 and 7 show that both invariants, which trivially hold initially, are preserved. As the dual is clearly feasible and the primal is feasible since Invariant 1 ensures $x_v \leq b_v$ as discussed earlier. Combining all yields Theorem 4.

Lemma 6 (Invariant 2). *In each iteration of the algorithm, the increase in the dual objective value is exactly β times that of the primal.*

Lemma 7 (Invariant 1). *After processing online vertex v, we have $x_v \leq b_v \frac{y_v + f(1-y_v)}{\beta}$ and $x_u \leq b_u \cdot \frac{y + f(1-z_u) + \int_{z_u}^{y} \frac{1-t}{f(t)} dt}{\beta}$ for $u \in X$.*

Theorem 4. *Our algorithm is* $1/\beta \approx 0.526$-*competitive for* TOBM *and* TOBVC.

As our primal-dual analysis implies the result for TOBVC, it makes sense to question the usefulness of the charging analysis. One advantage is its appealing and intuitive nature which directly explains why the algorithm is competitive. But there are also technical advantages. Recall that we have shown towards the end of section 2.1 that for $c \in [0,1]$, one can obtain for OBVC a ratio of

$$\frac{b(R \cap C^*)}{b(C^*)} c + \frac{b(L \cap C^*)}{b(C^*)}.$$

Such a bound is $1 - 1/e$ in the worst case. Nevertheless in situations where certain information (e.g. stochastic) about $\frac{b(R \cap C^*)}{b(C^*)}$ is available, we can optimize the L.H.S. w.r.t. c and do strictly better than $1 - 1/e$. On the other hand, we can show in a strong sense that *no primal-dual analysis exists for such a result*.

4   Discussion and Open Problems

We presented the first nontrivial algorithm for two-sided online bipartite b-matching and vertex cover. A natural question is whether our competitive ratios 0.526 is optimal. As a first step, the bounds below show that the one-sided optimal ratio of $1 - 1/e \approx 0.632$ is indeed unattainable for the two-sided version.

Theorem 5. *No algorithm has a competitive ratio better than 0.570 for* TOBVC *and 0.625 for* TOBM.

Another interesting problem is to beat the greedy algorithm for TOBM without the large budget assumption. Recently, a connection between the optimal algorithms for integral and fractional OBM was established via the randomized primal-dual method [9]. This is promising as their techniques may be applicable here. Finally, our results are on the *oblivious adversary* model. One may consider weaker models, e.g. stochastic [12,22], random arrival [15,21]. This was done for online matching and should also be possible for online vertex cover.

References

1. Aggarwal, G., Goel, G., Karande, C., Mehta, A.: Online vertex-weighted bipartite matching and single-bid budgeted allocations. In: Proceedings of the Twenty-Second Annual ACM-SIAM Symposium on Discrete Algorithms. SIAM (2011)
2. Bansal, N., Gupta, A., Li, J., Mestre, J., Nagarajan, V., Rudra, A.: When LP is the cure for your matching woes: improved bounds for stochastic matchings. In: de Berg, M., Meyer, U. (eds.) ESA 2010, Part II. LNCS, vol. 6347, pp. 218–229. Springer, Heidelberg (2010)
3. Birnbaum, B., Mathieu, C.: On-line bipartite matching made simple. ACM SIGACT News **39**(1), 80–87 (2008)
4. Blum, A., Sandholm, T., Zinkevich, M.: Online algorithms for market clearing. Journal of the ACM (JACM) **53**(5), 845–879 (2006)
5. Buchbinder, N., Jain, K., Naor, J.S.: Online primal-dual algorithms for maximizing ad-auctions revenue. In: Arge, L., Hoffmann, M., Welzl, E. (eds.) ESA 2007. LNCS, vol. 4698, pp. 253–264. Springer, Heidelberg (2007)

6. Demange, M., Paschos, V.T.: On-line vertex-covering. Theoretical Computer Science **332**(1), 83–108 (2005)
7. Devanur, N.R., Huang, Z., Korula, N., Mirrokni, V.S., Yan, Q.: Whole-page optimization and submodular welfare maximization with online bidders. In: Proceedings of the Fourteenth ACM Conference on Electronic Commerce, pp. 305–322. ACM (2013)
8. Devanur, N.R., Jain, K.: Online matching with concave returns. In: Proceedings of the 44th Symposium on Theory of Computing, pp. 137–144. ACM (2012)
9. Devanur, N.R., Jain, K., Kleinberg, R.D.: Randomized primal-dual analysis of ranking for online bipartite matching. In: SODA 2013: Proceedings of the Thirteenth Annual ACM-SIAM Symposium on Discrete Algorithms (to appear, 2013)
10. Devenur, N.R., Hayes, T.P.: The adwords problem: online keyword matching with budgeted bidders under random permutations. In: EC 2009: Proceedings of the Tenth ACM Conference on Electronic Commerce, pp. 71–78. ACM, New York (2009)
11. Feldman, J., Korula, N., Mirrokni, V., Muthukrishnan, S., Pál, M.: Online ad assignment with free disposal. In: Leonardi, S. (ed.) WINE 2009. LNCS, vol. 5929, pp. 374–385. Springer, Heidelberg (2009)
12. Feldman, J., Mehta, A., Mirrokni, V., Muthukrishnan, S.: Online stochastic matching: Beating 1–1/e. In: 50th Annual IEEE Symposium on Foundations of Computer Science, FOCS 2009, pp. 117–126. IEEE (2009)
13. Goel, G., Mehta, A.: Online budgeted matching in random input models with applications to adwords. In: SODA, vol. 8, pp. 982–991 (2008)
14. Kalyanasundaram, B., Pruhs, K.R.: An optimal deterministic algorithm for online b-matching. Theoretical Computer Science **233**(1), 319–325 (2000)
15. Karande, C., Mehta, A., Tripathi, P.: Online bipartite matching with unknown distributions. In: Proceedings of the 43rd Annual ACM Symposium on Theory of Computing, pp. 587–596. ACM (2011)
16. Karlin, A.R., Kenyon, C., Randall, D.: Dynamic tcp acknowledgement and other stories about e/(e-1). In: Proceedings of the Thirty-Third Annual ACM Symposium on Theory of Computing, pp. 502–509. ACM (2001)
17. Karlin, A.R., Manasse, M.S., McGeoch, L.A., Owicki, S.: Competitive randomized algorithms for nonuniform problems. Algorithmica **11**(6), 542–571 (1994)
18. Karlin, A.R., Manasse, M.S., Rudolph, L., Sleator, D.D.: Competitive snoopy caching. Algorithmica **3**(1), 79–119 (1988)
19. Karp, R.M., Vazirani, U.V., Vazirani, V.V.: An optimal algorithm for on-line bipartite matching. In: Proceedings of the Twenty-Second Annual ACM Symposium on Theory of Computing, pp. 352–358. ACM(1990)
20. Lotker, Z., Patt-Shamir, B., Rawitz, D., Albers, S.: Rent, lease or buy: Randomized algorithms for multislope ski rental. In: 25th International Symposium on Theoretical Aspects of Computer Science (STACS 2008), vol. 1 (2008)
21. Mahdian, M., Yan, Q.: Online bipartite matching with random arrivals: an approach based on strongly factor-revealing lps. In: Proceedings of the 43rd Annual ACM Symposium on Theory of Computing, pp. 597–606. ACM (2011)
22. Manshadi, V.H., Gharan, S.O., Saberi, A.: Online stochastic matching: Online actions based on offline statistics. In: Proceedings of the Twenty-Second Annual ACM-SIAM Symposium on Discrete Algorithms, pp. 1285–1294. SIAM (2011)
23. Mehta, A., Saberi, A., Vazirani, U., Vazirani, V.: Adwords and generalized online matching. Journal of the ACM (JACM) **54**(5), 22 (2007)
24. Mehta, A.: Online matching and ad allocation. Theoretical Computer Science **8**(4) 265–368 (2012)

The Simultaneous Communication
of Disjointness with Applications
to Data Streams

Omri Weinstein[1][(✉)] and David P. Woodruff[2]

[1] Princeton University, Princeton, NJ 08544, USA
oweinste@cs.princeton.edu
[2] IBM Research, Almaden, San Jose, CA, USA
dpwoodru@us.ibm.com

Abstract. We study k-party number-in-hand set disjointness in the simultaneous message-passing model, and show that even if each element $i \in [n]$ is guaranteed to either belong to all k parties or to at most $O(1)$ parties in expectation (and to at most $O(\log n)$ parties with high probability), then $\Omega(n \min(\log 1/\delta, \log k)/k)$ communication is required by any δ-error communication protocol for this problem (assuming $k = \Omega(\log n)$).

We use the strong promise of our lower bound, together with a recent characterization of turnstile streaming algorithms as linear sketches, to obtain new lower bounds for the well-studied problem in data streams of approximating the frequency moments. We obtain a space lower bound of $\Omega(n^{1-2/p}\varepsilon^{-2}\log M \log 1/\delta)$ bits for any algorithm giving a $(1+\varepsilon)$-approximation to the p-th moment $\sum_{i=1}^{n}|x_i|^p$ of an n-dimensional vector $x \in \{\pm M\}^n$ with probability $1-\delta$, for any $\delta \geq 2^{-o(n^{1/p})}$. Our lower bound improves upon a prior $\Omega(n^{1-2/p}\varepsilon^{-2}\log M)$ lower bound which did not capture the dependence on δ, and our bound is optimal whenever $\varepsilon \leq 1/\text{poly}(\log n)$. This is the first example of a lower bound in data streams which uses a characterization in terms of linear sketches to obtain stronger lower bounds than obtainable via the one-way communication model; indeed, our set disjointness lower bound provably cannot hold in the one-way model.

Keywords: Multiparty communication complexity · Information complexity · Frequency moments · Information theory

1   Introduction

Set disjointness is one of the cornerstones of complexity theory. Throughout the years, this communication problem has played a key role in obtaining unconditional lower bounds in many models of computation, including proof complexity,

O. Weinstein and D.P. Woodruff—Research supported by a Simons Fellowship in Theoretical Computer Science and NSF Award CCF-1215990.

M.M. Halldórsson et al. (Eds.): ICALP 2015, Part I, LNCS 9134, pp. 1082–1093, 2015.
DOI: 10.1007/978-3-662-47672-7_88

data streams, data structures and algorithmic game theory (see [CP10] and references therein). Many variants of this problem were studied, starting with the standard two-party model (e.g., [KS92]) and recently in several multiparty communication models (e.g., [She14,BO15]).

Motivated by streaming applications, we study a promise version of number-in-hand multiparty disjointness in the public-coin *simultaneous message passing* model of communication (SMP). In this setting, there are k players each with a bit string $x^i \in \{0,1\}^n$, $i \in [k] = \{1,2,\ldots,k\}$, who are promised that their inputs satisfy one of the following cases:

- (NO instance) for all $j \in [n]$, the number of $i \in [k]$ for which $x^i_j = 1$ is distributed as $\text{Bin}(k,1/k)$, or
- (YES instance) there is a unique $j^* \in [n]$ for which $x^i_{j^*} = 1$ for all $i \in [k]$, and for all $j \neq j^*$, the number of $i \in [n]$ for which $x^i_j = 1$ is distributed as $\text{Bin}(k,1/k)$.

The players simultaneously send a message $M^i(x^i,R)$ to a referee, where R is a public-coin that the players share. The referee then outputs a function $f(M^1(x^1,R),\ldots,M^k(x^k,R),R)$, which should equal 1 if the inputs form a YES instance, and equal 0 otherwise. Notice that if $X \sim \text{Bin}(k,1/k)$, then $\Pr[X > \ell] \leq (e/\ell)^\ell$, and so by a union bound for all coordinates j in a NO instance, the number of $i \in [k]$ for which $X^i_j = 1$ is $O(\log n/\log\log n)$. Thus, for $k = \Omega(\log n/\log\log n)$, and in fact $k = n^{\Omega(1)}$ in our context below, NO and YES instances are distinguishable.

Our First Contribution. We show an $\Omega(n \min(\log 1/\delta, \log k)/k)$ total communication lower bound for any protocol which succeeds with probability at least $1 - \delta$ in solving this promise problem in the public-coin SMP model. This lower bound is optimal up to constant factors whenever $\delta \approx 1/k$ (and $k = poly(n)$), since in this case the entropy of the input of each player is $n \cdot H(1/k) = \Theta(n \cdot \frac{\log k}{k})$; Therefore, if the first $O(1)$ players send their entire inputs to the referee (using standard compression), this yields an $O(n \cdot \frac{\log k}{k})$ total communication protocol with error at most δ for the above promise problem (since in the NO case, the probability that there is an $i \in [n]$ for which t players received a "1" is $k^{-t} < 1/nk$ for large enough constant t by our assumption on the relationship between k and n, hence a union bound finishes the argument).

We then show how this result can be used to obtain strong space lower bounds in the turnstile data stream model. In this model, an integer vector x is initialized to 0^n and undergoes a long sequence of additive updates to its coordinates. The t-th update in the stream has the form $x_i \leftarrow x_i + \delta_t$, where δ_t is an arbitrary (positive or negative) integer. At the end of the stream we are promised that $x \in \{-M, -M+1, \ldots, M\}^n$ for some bound M which is typically assumed to be at least n (and which we assume here).

Approximating the frequency moments $F_p = \sum_{i=1}^n |x_i|^p$ is one of the most fundamental problems in data streams, starting with the seminal work of Alon Matias, and Szegedy [AMS99]. The goal is to output a number $\hat{F}_p \in [(1 - \varepsilon)F_p, (1 + \varepsilon)F_p]$ with probability at least $1 - \delta$ using as little memory in bits as possible. It is known that for $0 < p \leq 2$, $\Theta(\varepsilon^{-1}\log(M)\log 1/\delta)$ bits of space is necessary and sufficient [KNW10, JW13]. Ideas here have been the basis of many other streaming algorithms and lower bounds, with connections to linear algebra [SW11, CDM+13] and information complexity [CKW12, BGPW13].

Perhaps surprisingly, for $p > 2$ a polynomial (in n) amount of space is required [SS02, BYJKS04, CKS03]. The best known upper bound is due to Ganguly and achieves space

$$O(n^{1-2/p}\epsilon^{-2}\log n \cdot \log(M)\log(1/\delta))/\min(\log n, \epsilon^{4/p-2}))).$$

In the case that $\epsilon \leq 1/\mathrm{poly}(\log n)$, this simplifies to $O(n^{1-2/p}\epsilon^{-2}\log M \log(1/\delta))$. On the other hand, if ϵ is a constant, this simplifies to $O(n^{1-2/p}\log n \log M \log(1/\delta))$. The latter complexity is also achieved by algorithms of [AKO10, And]. The lower bound, on the other hand, for any ε, δ is only $\Omega(n^{1-2/p}\varepsilon^{-2}\log M)$ [LW13]. A natural question is whether there are algorithms using less space and achieving a high success probability, that is, if one can do better than just repeating the constant probability data structure and taking a median of $\Theta(\log 1/\delta)$ independent estimates. While there is some work on tightening the bounds in the context of linear sketches over the reals [ANPW13, LW13], these lower bounds do not yield lower bounds in the streaming setting; for more discussion on this, see below.

Our Second Contribution. We prove the following lower bound on the space complexity of the p-th frequency moments problem:

Theorem 1 (Improved Space Lower Bound for Frequency Moments). *For any constant $p > 2$, there exists an absolute constant $\alpha > 1$ such that for any $\varepsilon > n^{-\Omega(1)}$ and $\delta \geq 2^{-o(n^{1/p})}$, any randomized streaming algorithm that obtains a $(1+\varepsilon)$-approximation to F_p in the turnstile streaming model for a vector $x \in [-M, M]^n$ where $M = \Omega(n^{\alpha/p})$, requires $\Omega\left(\varepsilon^{-2} \cdot n^{1-2/p}(\log M)\log 1/\delta\right)$ bits of space.*

Our lower bound is optimal for any $\epsilon \leq 1/\mathrm{poly}(\log n)$. As argued in [LW13], this is an important regime of parameters. Namely, if $\varepsilon = 1\%$, we have that for, e.g., $n = 2^{32}$, $\varepsilon^{-1} \geq \log n$. Our result is a direct strengthening of the $\Omega(n^{1-2/p}\varepsilon^{-2}\log M)$ lower bound of [LW13] which cannot be made sensitive to the error probability δ. Moreover, even for constant ϵ, our lower bound of $\Omega(n^{1-2/p}\log M \log(1/\delta))$ bits improves prior work by a $\log(1/\delta)$ factor. We note that for constant ε, the upper bounds still have space $O(n^{1-2/k}\log n \log M \log(1/\delta))$ bits, so while we obtain an improvement, there is still a gap in this case.

While the ultimate goal in this line of research is to obtain tight space bounds simultaneously for any $\varepsilon, \delta \in (0, 1)$ and $p > 2$, our result is the first to obtain tight bounds simultaneously in ε and δ for a wide range of parameters. Our proof technique is also quite different than previous work, and the first to bypass the limitations of one-way communication complexity. This is necessary since the problem considered in [LW13] has a protocol with information cost $O(n^{1-2/p}\varepsilon^{-2}\log M)$ with 0 error probability, which can be compressed to a protocol with this amount of communication and exponentially small error probability. A description of this protocol can be found in the full version of this paper, including an explanation for why it implies the problem considered in [LW13] does not give stronger lower bounds.

Our Techniques. The key ingredient of our result is proving the aforementioned simultaneous communication lower bound on the promise version of k-party set disjointness. To do so, we use the information complexity paradigm, which allows one to reduce the problem, via a direct sum argument, to the δ-error SMP complexity of a primitive problem – the k-party AND function with the aforementioned promise. We lower bound the information complexity of AND under the NO distribution (an independent bit $\sim Ber(1/k)$), by asking how many independent messages (over her private randomness) the player would need to send in order to convince one that her input is 0 or 1. We use the product structure of Hellinger distance, and relate this quantity to the amount of information a single message of the player reveals via the Maximum Likelihood Estimation principle. To obtain our stronger bound of $\Omega(n \log(1/\delta)/k)$ for any $\delta \geq 2^{-o(n^{1/p})}$, we restrict all players to have the same (randomized) message function. This assumption turns out to be possible in our application, as we observe that linear sketches can in fact be simulated by *symmetric* SMP protocols (see below).

A Reduction to Streaming: To lower bound the space complexity of a streaming algorithm we need a way of relating it to the communication cost of a protocol for this disjointness problem. We use a recent result of Li, Nguyen, and Woodruff [LNW14] showing there is a near-optimal streaming algorithm for any problem in the turnstile model which can be implemented by maintaining $A \cdot x$ in the stream, where A is a matrix with poly(n)-bounded integer entries, and A is sampled from a fixed set of $O(n \log m)$ hardwired matrices. In [LNW14] near-optimal meant up to an $O(\log n)$ multiplicative factor in space, which would not suffice here. However, their proof shows if one maintains $A \cdot x$ mod q, where q is a vector of integers one for each coordinate (which depends on A but not on x), then this is optimal up to a *constant* factor (a formal proof can be found in the full version of this paper). Notice that this need not be optimal for a *specific family of streams*, such as those arising in our communication game, though we use the fact that by results in [LNW14] an algorithm which succeeds with good probability *for any* fixed stream has this form, and therefore we can assume this form in our reduction. This implies a public-coin simultaneous protocol since the players can use the public coin to choose an (A, q) pair, then each communicate $A \cdot x^i$ mod q to the referee, who can combine these (using linearity) to obtain $A \cdot (\sum_{i=1}^{k} x^i)$ mod q. This simulation also implies all players have the same message function, even conditioned on the public coin.

We stress that the use of a public-coin simultaneous communication model is essential for our result, as there is an $O(n/k)$ total communication upper bound with exponentially small error probability in the one-way communication model (for a formal proof of this argument see the full version of this paper).

Given this reduction, one of the player's messages must be $\Omega(n \log(1/\delta)/k^2)$ bits long, which lower bounds the space complexity of the streaming algorithm. By setting $k = \varepsilon n^{1/p}$, and by having the referee add $n^{1/p} e_{j*}$ to the stream, where e_{j*} is the standard unit vector in direction j^*, one can show with probability $1 - \delta$, YES and NO instances differ by a $(1+\varepsilon)$-factor in $F_p(x)$. This is true even given our relaxed definition of disjointness, in which we allow some coordinates to be as large as $\Theta(\log n/ \log \log n)$, provided the average of the k-th powers of these coordinates is $\Theta(1)$. We are not done though, as we seek an extra $\log M$ factor in the lower bound, and for this we superimpose $\Theta(\log M)$ independent copies of this problem at different scales, in a similar fashion to the work of [LW13], and ask the referee to solve a random scaling. There are some technical differences needed to execute this approach in the high $(1-\delta)$ probability regime.

Related Work: We summarize the previous work on the frequency moments problem in Table 1. For a more thorough discussion of related works, in other streaming models, see the full version of this paper. Regarding our communication result, it is noteworthy to mention that Braverman and Oshman [BO15] recently obtained a tight $\Omega(n \log k + k)$ lower bound on the unbounded-round number-in-hand communication complexity of the k-party set disjointness function. Of course, this lower bound applies in particular to simultaneous protocols and is much stronger than the one proven in this paper $(\Omega(n \cdot \log(1/\delta)/k))$. However, this stronger lower bound holds only for distributions which (vastly) violate the promise required for our streaming application, and therefore their lower bound is useless in our context.

Organization. Due to space constraints, this version contains only the proof of our communication lower bound for set disjointness in the SMP model (Section 2). The rest of our results, including the reduction to frequency moments and the proof of Theorem 1, appear in the full version of this paper.

2   Multiparty SMP Complexity of Set-Disjointness

In this section we prove our lower bound on the SMP communication complexity of the k-party Set-Disjointness function. A broader overview of the definitions, tools and properties used below can be found in the Preliminaries section of the full version of this paper.

In what follows, $\mathbf{R}_\delta(f)$ denotes the communication cost of the cheapest randomized SMP protocol which solves the k-party function f with error at most δ over all inputs $(x_1, \ldots, x_k) \in \mathcal{X}^k$, and $\mathrm{IC}_\mu^\delta(f)$ denotes the minimal (external) information cost of an SMP protocol solving f under μ with error at most δ (for a broader overview of the formal definitions and notations used throughout this paper, see the full version of this paper).

We will be interested in a special class of SMP protocols, in which players are restricted to use the same function when sending their messages to the referee. This class will be relevant to our main streaming application (Theorem 1).

Table 1. All results are stated for constant success probability, and can be made to achieve $1 - \delta$ success probability by repeating the data structure independently $O(\log 1/\delta)$ times and taking the median of estimates; this blows up the space by a multiplicative $O(\log 1/\delta)$ factor. Here, $g(p,n) = \min_{c \text{ constant}} g_c(n)$, where $g_1(n) = \log n$, $g_c(n) = \log(g_{c-1}(n))/(1 - 2/p)$. We start the upper bound timeline with [IW05], since that is the first work which achieved an exponent of $1 - 2/p$ for n. For earlier works which achieved worse exponents for n, see [AMS99, CK04, Gan04a, Gan04b]. We note that [AMS99] initiated the problem and obtained an $O(n^{1-1/p}\epsilon^{-2}\log(M))$ bound in the insertion-only model (see also [BO12, BKSV14] for work in the insertion model).

F_p Algorithm	Space Complexity
[IW05]	$O(n^{1-2/p}\epsilon^{-O(1)}\log^{O(1)} n \log(M))$
[BGKS06]	$O(n^{1-2/p}\epsilon^{-2-4/p}\log n \log^2(M))$
[MW10]	$O(n^{1-2/p}\epsilon^{-O(1)}\log^{O(1)} n \log(M))$
[AKO10]	$O(n^{1-2/p}\epsilon^{-2-6/p}\log n \log(M))$
[BO10]	$O(n^{1-2/p}\epsilon^{-2-4/p}\log n \cdot g(p,n)\log(M))$
[And]	$O(n^{1-2/p}\log n \log(M)\epsilon^{-O(1)})$
[Gan11], **Best upper bound**	$O(n^{1-2/p}\epsilon^{-2}\log n \cdot \log(M)/\min(\log n, \epsilon^{4/p-2})))$
[AMS99]	$\Omega(n^{1-5/p})$
[Woo04]	$\Omega(\epsilon^{-2})$
[BYJKS04]	$\Omega(n^{1-2/p-\gamma}\epsilon^{-2/p})$, any constant $\gamma > 0$
[CKS03]	$\Omega(n^{1-2/p}\epsilon^{-2/p})$
[WZ12]	$\Omega(n^{1-2/p}\epsilon^{-4/p}/\log^{O(1)} n)$
[Gan12]	$\Omega(n^{1-2/p}\epsilon^{-2}/\log n)$
[LW13]	$\Omega(n^{1-2/p}\epsilon^{-2}\log(M))$

Definition 1 (Symmetric SMP protocols). *A k-party SMP protocol π is called* symmetric *if for any fixed input* $\mathbf{X} = x$ *and fixing of the public randomness $R = r$,*

$$M_1(x,r) = M_2(x,r) = \ldots = M_k(x,r).$$

For a function f, we denote the randomized public-coin communication complexity of f with respect to symmetric SMP protocols by by $\mathsf{R}_\delta^{\mathsf{SYM}}(f)$. Similarly, we denote by $\mathsf{IC}_\mu^{\mathsf{SYM},\delta}(f)$ the (external) information complexity of f with respect to symmetric SMP protocols (for the formal definition of information cost see the Preliminaries section of the full version).

We use the following distance measures in our arguments.

Definition 2 (Total Variation distance and Hellinger distance). *The Total Variation* distance *between two probability distributions P, Q over the same universe $\mathcal{U}$ is $\Delta(P,Q) := \sup_A |P(A) - Q(A)|$, where A ranges over all measurable events in the probability space.*

The (squared) Hellinger distance *between P and Q is denoted as*

$$h^2(P,Q) = 1 - \sum_{x \in \mathcal{U}} \sqrt{P(x)Q(x)} = \frac{1}{2} \cdot \sum_{x \in \mathcal{U}} \left(\sqrt{P(x)} - \sqrt{Q(x)} \right)^2.$$

By a slight abuse of notation, we sometimes use the above distance measures with random variables instead of their underlying distributions. For example, if A, B are two random variables in the joint probability space $p(a, b)$, then $\Delta(A, B) = \Delta(p(a), p(b))$, and $h(A, B) = h(p(a), p(b))$.

We will prove the following theorem.

Theorem 2 (SMP complexity of multiparty Set-Disjointness). *For any* $\delta \geq n \cdot 2^{-k}$,

$$\mathbf{R}_\delta(\mathsf{Disj}_k^n) \geq \Omega \left(n \cdot \frac{\min\{\log(1/\delta), \log k\}}{k} \right).$$

$$\mathbf{R}_\delta^{\mathsf{SYM}}(\mathsf{Disj}_k^n) \geq \Omega \left(n \cdot \min \left\{ \frac{\log(1/\delta)}{k}, \log k \right\} \right).$$

Recall the k-party Set-Disjointness problem is defined as follows:

Definition 3 (Disj_k^n). *Denote by Disj_k^n the multiparty Set-Disjointness problem in which k players each receive an n-dimensional input vector $\mathbf{X}_j = \{\mathbf{X}_{j,i}\}_{i=1}^n$ (where $\mathbf{X}_{j,i} \in \{0,1\}$). By the end of the protocol, the referee needs to distinguish between the following cases:*

- **(The "NO" case)** $\forall \; i \in [n], \; \sum_j \mathbf{X}_{j,i} < k$, *or*
- **(The "YES" case)** $\exists \; i \in [n]$ *for which* $\sum_j \mathbf{X}_{j,i} = k$.

Denote $\mathsf{AND}_k(x_1, x_2, \ldots, x_k) := \bigwedge_{j=1}^k x_j$. *Note that*

$$\overline{\mathsf{Disj}_k^n}(\mathbf{X}_1, \ldots, \mathbf{X}_k) = \bigvee_{i=1}^n \mathsf{AND}_k (\mathbf{X}_{1,i}, \ldots, \mathbf{X}_{k,i}).$$

We start by defining a "hard" distribution for Disj_k^n which still satisfies the promise (gap) required for our streaming application. Consider the distribution η on n-bit string inputs, defined by the following process.

The Distribution η:

- For each $i \in [n], j \in [k]$ set $\mathbf{X}_{j,i} \sim B(1/k)$, independently at random.
- Pick a uniformly random coordinate $I \in_R [n]$.
- Pick $Z \in_R \{0, 1\}$. If $Z = 1$, set all the values $\mathbf{X}_{j,I}$ to 1, for all $j \in [k]$ (If $Z = 0$, keep all coordinates as before.).
- The referee receives the index I (this feature will only be used in the streaming application (proof of Theorem 1)).

Denote by η_0 the distribution of $\eta \mid "Z = 0"$, and by μ_0 the projection of η_0 on a single coordinate (this is well defined since the distribution over all coordinates is i.i.d). In particular, notice that $\eta_0 = \mu_0^n$ is a product distribution, and for every $i \in [n]$, $\Pr_{\mu_0}[\mathbf{X}_{i,j} = 1$ for all $j \in [k]] = (1/k)^k$. Thus, by a union bound over all n coordinates and our assumption on δ,

$$\Pr_{\mu_0^n}[\mathsf{Disj}_k^n(\mathbf{X}_1, \ldots, \mathbf{X}_k)] \leq n \cdot (1/k)^k \leq n \cdot 2^{-k} \leq \delta. \tag{1}$$

Remark 1. Notice that the "NO" distribution η_0 contains (w.h.p) coordinates $i \in [n]$ for which $\gg 1$ players (in fact, $\Omega(\log n)$ of them) possess the i'th coordinate. This feature is a by-product of the *product* structure of η_0, which will be crucial to our construction and analysis. To best of our knowledge, this is the first paper to show that distributions with such property (where *disjoint* instances in the support have $\omega(1)$ overlapping items in a coordinate, instead of just 1) are still powerful enough to prove lower bounds on the frequency moments problem.

2.1   Direct Sum and the SMP Complexity of AND_k

To prove Theorem 2, we first use a direct sum argument, asserting that under product distributions, solving set disjointness is essentially equivalent to solving n copies of the 1-bit AND_k function. The following direct sum argument is well known (See e.g., [BYJKS04]):

Lemma 1 (Direct sum for Disj_k^n). $\forall \delta \geq n \cdot 2^{-k}$, $\mathsf{IC}_{\eta_0}^{\delta}(\mathsf{Disj}_k^n) \geq n \cdot \mathsf{IC}_{\mu_0}^{2\delta}(\mathsf{AND}_k)$.

We defer the proof of this claim to the full version of this paper. With Claim 1 in hand, it suffices to prove that any (randomized) SMP protocol solving AND_k with error at most δ, must have a large information cost *under μ_0* . This is the content of the next theorem, which is one of our central technical contributions.

Theorem 3. *For every $\delta > 0$,*

$$\mathsf{IC}_{\mu_0}^{\delta}(\mathsf{AND}_k) \geq \Omega\left(\min\left\{\frac{\log 1/\delta}{k}, \frac{\log k}{k}\right\}\right)$$

$$\mathsf{IC}_{\mu_0}^{\mathsf{SYM},\delta}(\mathsf{AND}_k) \geq \Omega\left(\min\left\{\frac{\log 1/\delta}{k}, \log k\right\}\right).$$

Proof. Let π be a (randomized) SMP protocol which solves $\mathsf{AND}_k(\mathbf{X}_1, \ldots, \mathbf{X}_k)$ for all inputs in $\{0,1\}^k$ with success probability at least $1 - \delta$. For the rest of the analysis, we fix the public randomness of the protocol. Indeed, proving the lower bound for every fixing of the tape suffices as the chain rule for mutual information implies $\mathsf{IC}_{\mu_0}(\pi) = \mathbb{E}_R[\mathsf{IC}_{\mu_0}(\pi_R)]$. For each player $j \in [k]$, let M_j denote the transcript of player j's message, and let $M_0^j := M_j | "\mathbf{X}_j = 0"$, $M_1^j := M_j | "\mathbf{X}_j = 1"$ (note that if π is further a symmetric protocol, then M_0^j and M_1^j are the same for every player $j \in [k]$). Since the $\mathbf{X}_j$'s are independent under μ_0, and therefore so are the messages M_j, the chain rule implies that $\mathsf{IC}_{\mu_0}(\pi) = \sum_{j=1}^{k} I(M_j; \mathbf{X}_j)$. We shall argue that $\sum_{j=1}^{k} I(M_j; \mathbf{X}_j) \geq \Omega\left(\frac{\log 1/\delta}{k}, \frac{\log k}{k}\right)$, and if π is further a symmetric protocol, then $\sum_{j=1}^{k} I(M_j; \mathbf{X}_j) \geq \Omega\left(\frac{\log 1/\delta}{k}, \log k\right)$. To this end, let us denote by

$$h^2(M_1^j, M_0^j) := 1 - z_j$$

the (squared) Hellinger distance between player j's message distributions in both cases. There are two cases: if there is a player j for which $z_j = 0$, then $h^2(M_1^j, M_0^j) = 1$, which means that $I(M_j; \mathbf{X}_j) = H(\mathbf{X}_j) = H(1/k) = \Omega(\log(k)/k)$ and thus $\mathsf{IC}_{\mu_0}^{\delta}(\mathsf{AND}_k) \geq \Omega(\log(k)/k)$. Furthermore, if π is symmetric, then $z_1 = z_2 = \ldots = z_j$, which in this case implies by the same reasoning that $I(M_j; \mathbf{X}_j) = \Omega(\log(k)/k)$ for *all* players $j \in [k]$, and thus $\mathsf{IC}_{\mu_0}^{\mathsf{SYM},\delta}(\mathsf{AND}_k) \geq \Omega(\log k)$, as desired.

We may henceforth assume that all z_j's are non-zero, and the rest of the analysis applies for general (not necessarily symmetric) SMP protocols. To this end, let us introduce one final notation: For a *fixed* input $\mathbf{X}_j$, let $M_j^{\oplus t}$ denote (the concatenation of) t *independent* copies of $M_j | \mathbf{X}_j$ (so $M_j^{\oplus t} = (M_0^j)^t$ whenever $\mathbf{X}_j = 0$ and $M_j^{\oplus t} = (M_1^j)^t$ whenever $\mathbf{X}_j = 1$). By the conditional independence of the t copies of M_j (conditioned on $\mathbf{X}_j$) and the product structure of the Hellinger distance (Fact 4 in the full version of this paper), we have that for each $j \in [k]$, the total variation distance between the t-fold message copies in the "YES" and "NO" cases is at least

$$\Delta\left((M_1^j)^t, (M_0^j)^t\right) \geq h^2\left((M_1^j)^t, (M_0^j)^t\right) = 1 - (z_j)^t, \tag{2}$$

where the first inequality follows from the fact that Total Vriation distance upper bounds the (squared) Hellinger distance (see the Preliminaries section of the full version for the formal statement). Set $t_j = O(\log k / \log(1/z_j))$ (note that this is well defined as we assumed $z_j \neq 0$). Thus, for each player $j \in [k]$,

$$\Delta\left((M_1^j)^{t_j}, (M_0^j)^t\right) \geq 1 - \frac{1}{10k}. \tag{3}$$

Equation (3) implies that the error probability of the MLE predictor[1] for predicting $\mathbf{X}_j$ given $M_j^{\oplus t_j}$ is at most $\varepsilon := 1/(10k)$. Therefore, Fano's inequality (Lemma 5 in the full version of the paper) and the data processing inequality together imply that

$$\forall\ j \in [k], \quad I(M_j^{\oplus t_j}; \mathbf{X}_j) \geq H(\mathbf{X}_j) - H(\varepsilon) \geq H\left(\frac{1}{k}\right) - H\left(\frac{1}{10k}\right) \geq \Omega\left(\frac{\log k}{k}\right), \tag{4}$$

---

[1] That is, the predictor which given $M_j^{\oplus t} = m$, outputs $Y := \mathrm{argmax}_{x \in \{0,1\}} \Pr[(M_x^j)^t = m]$.

since $\mathbf{X}_j \sim B(1/k)$ under μ_0, and $H(1/(10k)) \leq \frac{2}{10k}\log(10k) \leq \frac{4}{5}k\log(k)$ (where the first inequality follows since $\forall p \in [0, 1/2]$, it holds that $H(p) \leq p\log(e/p) \leq 2p\log(1/p)$). Now, by the chain rule for mutual information, we have

$$I(M_j^{\oplus t_j}; \mathbf{X}_j) = \sum_{s=1}^{t_j} I((M_j)_s; \mathbf{X}_j | (M_j)_{<s}) \leq \sum_{s=1}^{t_j} I((M_j)_s; \mathbf{X}_j), \tag{5}$$

where the last inequality follows from the fact that $I(A; D|C) = 0 \Rightarrow I(A; B|C) \leq I(A; B|CD)$ (see Fact 6 in the full version of this paper), as the messages $(M_j)_s$ and $(M_j)_{<s}$ are independent conditioned on $\mathbf{X}_i$ (by construction). Notice that $(M_j)_s \sim M_j$ for all $s \in [t]$, as all the messages are equally distributed conditioned on $\mathbf{X}_j$. Combining equations (4) and (5) therefore implies

$$I(M_j; \mathbf{X}_j) \geq \Omega\left(\frac{\log k}{k \cdot t_j}\right) \geq \Omega\left(\frac{\log(1/z_j)}{k}\right), \tag{6}$$

recalling that $t_j = O(\log k/\log(1/z_j))$. Since (6) holds for any player $j \in [k]$, we have

$$\sum_{j=1}^{k} I(M_j; \mathbf{X}_j) \geq \Omega\left(\frac{1}{k} \cdot \sum_{j=1}^{k} \log\left(\frac{1}{z_j}\right)\right). \tag{7}$$

We finish the proof by showing that

$$\sum_{j=1}^{k} \log\left(\frac{1}{z_j}\right) \geq \Omega(\log(1/\delta)). \tag{8}$$

To this end, we first claim that the correctness of π implies that the total variation distance between the transcript distributions of π on the input 0^k and on the input 1^k must be large (notice that below we crucially use the fact that our information complexity definition requires the protocol to be correct on all inputs, so in particular, a δ-error protocol must distinguish with comparable error, between "YES" and "NO" inputs):

Proposition 1. $\Delta(\pi(0^k), \pi(1^k)) \geq 1 - 2\delta$.

Proof. Let $\mathcal{Y}$ be the set of transcripts τ for which $\pi(\tau) = \mathsf{AND}_k(1^k) = 1$. By the correctness assumption, $\Pr[\pi(1^k) \in \mathcal{Y}] \geq 1 - \delta$, and $\Pr[\pi(0^k) \in \mathcal{Y}] \leq \delta$, so the above follows by definition of the total variation distance.

Since μ_0 is a product distribution (the $\mathbf{X}_j$'s are i.i.d), it holds that $\pi(0^k) = \times_{j=1}^{k} M_0^j$, and $\pi(1^k) = \times_{j=1}^{k} M_1^j$. Therefore, recalling that $z_j := 1 - h^2(M_0^j, M_1^j)$ the product structure of the Hellinger distance (Fact 4 in the full version) implies

$$1 - \Pi_{j=1}^{k} z_j = 1 - \Pi_{j=1}^{k}(1 - h^2(M_0^j, M_1^j)) = h^2(\pi(0^k), \pi(1^k)) \geq 1 - 4\sqrt{\delta} \tag{9}$$

where the last transition follows from the combination of Proposition 1 with Corollary 2.4 in full version of the paper (taken with $\alpha = 2\delta$). Rearranging (9), we get $\Pi_{j=1}^{k} z_j \leq 4\sqrt{\delta}$, or equivalently, $\sum_{j=1}^{k} \log\left(\frac{1}{z_j}\right) \geq \frac{1}{2}\log\left(\frac{1}{\delta}\right) - 2 = \Omega\left(\log 1/\delta\right)$, as desired. Combining equations (8) and (7), we conclude that $\mathsf{IC}_{\mu_0}(\pi) \geq \Omega\left(\frac{\log 1/\delta}{k}\right)$, which completes the proof of Theorem 3.

Since communication is always lower bounded by information (Fact 8 in the full version), combining Theorem 3 and Claim 1 directly implies Theorem 2:

Corollary 1. *For any* $\delta \geq n \cdot 2^{-k}$,

$$\mathsf{R}_\delta(\mathsf{Disj}_k^n) \geq \Omega\left(n \cdot \min\left\{\frac{\log 1/\delta}{k}, \frac{\log k}{k}\right\}\right),$$

$$\mathsf{R}_\delta^{\mathsf{SYM}}(\mathsf{Disj}_k^n) \geq \Omega\left(n \cdot \min\left\{\frac{\log 1/\delta}{k}, \log k\right\}\right).$$

References

[AKO10] Andoni, A., Krauthgamer, R., Onak, K.: Streaming algorithms from precision sampling. CoRR, abs/1011.1263 (2010)

[AMS99] Alon, N., Matias, Y., Szegedy, M.: The space complexity of approximating the frequency moments. JCSS **58**(1), 137–147 (1999)

[And] Andoni, A.: High frequency moment via max stability. http://web.mit.edu/andoni/www/papers/fkStable.pdf

[ANPW13] Andoni, A., Nguyên, H.L., Polyanskiy, Y., Wu, Y.: Tight lower bound for linear sketches of moments. In: Fomin, F.V., Freivalds, R., Kwiatkowska, M., Peleg, D. (eds.) ICALP 2013, Part I. LNCS, vol. 7965, pp. 25–32. Springer, Heidelberg (2013)

[BGKS06] Bhuvanagiri, L., Ganguly, S., Kesh, D., Saha, C.: Simpler algorithm for estimating frequency moments of data streams. In: SODA, pp. 708–713 (2006)

[BGPW13] Braverman, M., Garg, A., Pankratov, D., Weinstein, O.: Information lower bounds via self-reducibility. In: Bulatov, A.A., Shur, A.M. (eds.) CSR 2013. LNCS, vol. 7913, pp. 183–194. Springer, Heidelberg (2013)

[BKSV14] Braverman, V., Katzman, J., Seidell, C., Vorsanger, G.: An optimal algorithm for large frequency moments using bits. In: APPROX/RANDOM (2014)

[BO10] Braverman, V., Ostrovsky, R.: Recursive sketching for frequency moments. CoRR, abs/1011.2571 (2010)

[BO12] Braverman, V., Ostrovsky, R.: Approximating large frequency moments with pick-and-drop sampling. CoRR, abs/1212.0202 (2012)

[BO15] Braverman, M., Oshman, R.: The communication complexity of number-in-hand set disjointness with no promise. Electronic Colloquium on Computational Complexity (ECCC) **22**(2) (2015)

[BYJKS04] Bar-Yossef, Z., Jayram, T.S., Kumar, R., Sivakumar, D.: An information statistics approach to data stream and communication complexity. Journal of Computer and System Sciences **68**(4), 702–732 (2004)

[CDM+13] Clarkson, K.L., Drineas, P., Magdon-Ismail, M., Mahoney, M.W., Meng, X., Woodruff, D.P.: The fast cauchy transform and faster robust linear regression. In: SODA (2013)

[CK04] Coppersmith, D., Kumar, R.: An improved data stream algorithm for frequency moments. In: SODA (2004)

[CKS03] Chakrabarti, A., Khot, S., Sun, X.: Near-optimal lower bounds on the multi-party communication complexity of set disjointness. In: CCC, pp. 107–117 (2003)

[CKW12] Chakrabarti, A., Kondapally, R., Wang, Z.: Information complexity versus corruption and applications to orthogonality and gap-hamming. In: Gupta, A., Jansen, K., Rolim, J., Servedio, R. (eds.) APPROX 2012 and RANDOM 2012. LNCS, vol. 7408, pp. 483–494. Springer, Heidelberg (2012)

[CP10] Chattopadhyay, A., Pitassi, T.: The story of set disjointness. SIGACT News **41**(3), 59–85 (2010)

[Gan04a] Ganguly, S.: Estimating frequency moments of data streams using random linear combinations. In: Jansen, K., Khanna, S., Rolim, J.D.P., Ron, D. (eds.) RANDOM 2004 and APPROX 2004. LNCS, vol. 3122, pp. 369–380. Springer, Heidelberg (2004)

[Gan04b] Ganguly, S.: A hybrid algorithm for estimating frequency moments of data streams, Manuscript (2004)

[Gan11] Ganguly, S.: Polynomial estimators for high frequency moments. CoRR, abs/1104.4552 (2011)

[Gan12] Ganguly, S.: A lower bound for estimating high moments of a data stream. CoRR, abs/1201.0253 (2012)

[IW05] Indyk, P., Woodruff, D.: Optimal approximations of the frequency moments of data streams. In: STOC. ACM (2005)

[JW13] Jayram, T.S., Woodruff, D.P.: Optimal bounds for johnson-lindenstrauss transforms and streaming problems with subconstant error. ACM Transactions on Algorithms **9**(3), 26 (2013)

[KNW10] Kane, D.M., Nelson, J., Woodruff, D.P.: On the exact space complexity of sketching and streaming small norms. In: SODA, pp. 1161–1178 (2010)

[KS92] Kalyanasundaram, B., Schnitger, G.: The probabilistic communication complexity of set intersection. SIAM Journal on Discrete Mathematics **5**(4), 545–557 (1992)

[LNW14] Li, Y., Nguyen, H.L., Woodruff, D.P.: Turnstile streaming algorithms might as well be linear sketches. In: STOC, pp. 174–183 (2014)

[LW13] Li, Y., Woodruff, D.P.: A tight lower bound for high frequency moment estimation with small error. In: Raghavendra, P., Raskhodnikova, S., Jansen, K., Rolim, J.D.P. (eds.) RANDOM 2013 and APPROX 2013. LNCS, vol. 8096, pp. 623–638. Springer, Heidelberg (2013)

[MW10] Monemizadeh, M., Woodruff, D.P.: 1-pass relative-error l_p-sampling with applications. In: SODA (2010)

[She14] Sherstov, A.A.: Communication lower bounds using directional derivatives. J. ACM **61**(6), 34 (2014)

[SS02] Saks, M., Sun, X.: Space lower bounds for distance approximation in the data stream model. In: STOC (2002)

[SW11] Sohler, C., Woodruff, D.P.: Subspace embeddings for the l_1-norm with applications. In: STOC, pp. 755–764 (2011)

[Woo04] Woodruff, D.P.: Optimal space lower bounds for all frequency moments In: SODA, pp. 167–175 (2004)

[WZ12] Woodruff, D.P., Zhang, Q.: Tight bounds for distributed functional monitoring. In: STOC, pp. 941–960 (2012)

An Improved Combinatorial Algorithm
for Boolean Matrix Multiplication

Huacheng Yu$^{(\boxtimes)}$

Stanford University, Stanford, USA
yuhch123@gmail.com

Abstract. We present a new combinatorial algorithm for triangle finding and Boolean matrix multiplication that runs in $\hat{O}(n^3/\log^4 n)$ time,
where the $\hat{O}$ notation suppresses poly(loglog) factors. This improves
the previous best combinatorial algorithm by Chan [4] that runs in
$\hat{O}(n^3/\log^3 n)$ time. Our algorithm generalizes the divide-and-conquer
strategy of Chan's algorithm.

Moreover, we propose a general framework for detecting triangles in
graphs and computing Boolean matrix multiplication. Roughly speaking, if we can find the "easy parts" of a given instance efficiently, we can
solve the whole problem faster than n^3.

1 Introduction

Boolean matrix multiplication (BMM) is one of the most fundamental problems
in computer science. It has many applications to triangle finding, transitive closure, context-free grammar parsing, etc [7], [10], [5], [11]. One way to multiply
two Boolean matrices is to treat them as integer matrices, and apply a fast matrix
multiplication algorithm over the integers. Matrix multiplication over fields can
be computed in "truly subcubic time", i.e., computing the product of two $n \times n$
matrices can be done in $O(n^{3-\epsilon})$ additions and multiplication over the field.
For example, the latest generation of such algorithms run in $O(n^{2.373})$ operations [12], [9]. These algorithms are "algebraic", as they rely on the structure of
the field, and in general the ring structure of matrices over the field.

There is a different group of BMM algorithms, often called "combinatorial"
algorithms. They usually reduce the redundancy in computation by exploiting
some combinatorial structure in the Boolean matrices. The "Four Russians"
algorithm by Arlazarov, Dinic, Kronrod, and Faradzhev [1] is the most well-
known combinatorial algorithm for BMM. On the RAM model with word size
$w = \Theta(\log n)$, "Four Russians" algorithm can be implemented in $O(n^3/\log^2 n)$
time. About 40 years later, this result was improved by Bansal and Williams. In
their FOCS'09 [2] paper, they presented an $O(n^3(\log\log n)^2/\log^{9/4} n)$ time combinatorial algorithm for Boolean matrix multiplication, using the weak regularity
lemma for graphs. Recently, Chan presented an $O(n^3 (\log\log n)^3/\log^3 n)$ time
algorithm in his SODA'15 paper [4], improving the running time even further.

H. Yu—Supported in part by NSF CCF-1212372.

© Springer-Verlag Berlin Heidelberg 2015
M.M. Halldórsson et al. (Eds.): ICALP 2015, Part I, LNCS 9134, pp. 1094–1105, 2015.
DOI: 10.1007/978-3-662-47672-7_89

Although these combinatorial algorithms have worse running times than the algebraic ones, they generally have some nice properties. Combinatorial algorithms usually can be generalized in ways that the algebraic ones cannot be. For example, Chan's algorithm partly extends an idea of divide-and-conquer in an algorithm for the offline dominance range reporting problem by Impagliazzo, Lovett, Paturi, and Schneider [8]; the algebraic structure of dominance reporting is completely different from BMM's. Moreover, in practice, these combinatorial algorithms are usually fast and easy to implement, while in contrast, most theoretically fast matrix multiplication algorithms are impractical to implement. Finding a matrix multiplication algorithm that is both "good" in theory and practice is still an important open goal of the area.

In this paper, we generalize the ideas of Impagliazzo et al. [8] and Chan [4], to present a faster combinatorial algorithm for triangle detection: *given an n-node graph, does it contain a triangle?*

Theorem 1. *Given a tripartite graph G on n vertices, we can detect if there is a triangle in G using a combinatorial algorithm in $\hat{O}\left(n^3/\log^4 n\right)$ time on a word RAM with word size $w \geq \Omega(\log n)$.* [1]

Vassilevska Williams and Williams [13] proved that triangle detection and Boolean matrix multiplication are "subcubic equivalent" in the following sense: if there is a $O(n^3/g(n))$ time algorithm for triangle detection on n-node graphs, then we can use it to solve BMM on $n \times n$ matrices in $O(n^3/g(n^{1/3}))$ time. Together with Theorem 1, this gives a fast combinatorial algorithm for Boolean matrix multiplication.

Theorem 2. *There is a combinatorial algorithm to multiply two $n \times n$ Boolean matrices in $\hat{O}\left(n^3/\log^4 n\right)$ time.*

Moreover, we generalize the algorithm, and propose a general framework for solving triangle detection combinatorially.

Definition 1. *The* large subgraph triangle detection problem *with parameters α, β, and γ is: given a tripartite graph $G = (A \cup B \cup C, E)$, output a pair (G', b), where*

- *G' is a subgraph with at least an α-fraction of vertices from A, β-fraction of vertices from B, and γ-fraction of vertices from C, and*
- *$b = 1$ if G' is triangle-free, and $b = 0$ if G' contains a triangle.*

That is, the large subgraph triangle detection problem is to identify a large subgraph G' of G for which we can conclude whether G' is triangle-free or not. This problem is interesting when we can solve it quickly – faster than what is known for standard triangle detection. Additionally, we can show that fast algorithms for large subgraph triangle detection imply fast algorithms for triangle detection in general:

---

[1] We use $\hat{O}(f(n))$ to suppress poly($\log \log f(n)$) factors in the running time.

Theorem 3. *Let n be an integer, $0 < \alpha, \beta, \gamma, c \leq 1$, and G be any tripartite graph on vertex sets A, B, C with at least $\sqrt{n}$ vertices in each part. If there is an algorithm* **L** *for large subgraph triangle detection on every such G that runs in $O(c\alpha\beta\gamma|A||B||C|)$ time, then we can solve triangle detection on n-node graphs in $O(cn^3 + n^{3-\epsilon/2})$ time for any $\epsilon > 0$ such that $\alpha\beta > 10\epsilon(1 + \log\frac{1}{\gamma})$ and $\alpha > 10\epsilon(1 + \log\frac{1}{\beta})$.*

That is, to derive an efficient algorithm for triangle finding, it is sufficient to find and solve an "easy part" of the input. This opens a new direction for attacking this problem.

Related Work. In the work by Bansal and Williams [2], they used the weak regularity lemma of Frieze and Kannan [6] to discover and exploit small substructures in the graph. Generally speaking, a regularity lemma partitions the vertex set of a graph into disjoint sets, so that the edge distribution between any two sets is "close to random." Bansal and Williams enumerate every triple of sets in the partition: if the subgraph induced by the triple is sparse, finding a triangle in this triple is easy. Otherwise, since the induced subgraph is dense and "close to random", the regularity lemma guarantees that it is impossible to check many pairs of vertices *without* finding an edge between them. Integrating the method of Four Russians with the above approach yields an $\hat{O}(n^3/\log^{9/4} n)$ time algorithm for triangle detection.

Chan [4] used a very different approach for triangle detection which we now outline briefly[2]. Consider a tripartite graph on vertex sets (A, B, C) such that $|A| \leq \text{polylog}(|B| + |C|)$ (if this is not the case, partition the set A into polylog$(|B| + |C|)$-size subsets, and solve them independently). If the edge set between A and B is sparse, triangle detection is easy. Otherwise, there is some node $v \in A$ with many neighbors in B. Then the algorithm manually checks every pair of neighbors of v and does two recursive calls. One is on $A \setminus \{v\}$, B and the non-neighbors of v in C, and the other is on $A \setminus \{v\}$, non-neighbors of v in B and neighbors in C. On one hand, this recursive procedure never puts any pairs of neighbors of v in the branch. This guarantees that the algorithm only manually checks every pair in $B \times C$ at most once. On the other hand, the procedure may copy the set of non-neighbors in B when doing a recursive call, which increases the total input size. However, since we have a lower bound on the degree of vertex v to B, the procedure does not copy too many vertices each time. A careful analysis shows that the overhead of the recursion is actually rather tiny.

In Section 2, we show how to extend the idea of divide-and-conquer in Chan's algorithm to get an even faster algorithm for triangle detection (and hence for Boolean matrix multiplication). We give a more intuitive proof of the recursion involving less calculation. In Section 3, we propose a general framework for solving triangle detection, as a starting point for future work in this area.

---

[2] To keep consistency, the following description will be in the language of triangle finding, although his paper originally presented the algorithm in the language of Boolean matrix multiplication.

2   Triangle Detection

Preliminaries and Notations. We shall use the fact that the triangle detection problem in general undirected graphs is time-equivalent to the problem restricted to tripartite graphs, up to a constant factor. (The proof is straightforward.) Henceforth, we will assume the input graph is tripartite, and the tri-partition of its vertices is given to us.

For a graph $G = (V, E)$, a vertex $v \in V$ and a subset of vertices $S \subseteq V$, we denote $d(v, S) = |E \cap (\{v\} \times S)|$, which we call the *degree of v to S*.

Main Results. In what follows, we present an $\hat{O}(n^3 / \log^4 n)$ time combinatorial algorithm for triangle detection.

Suppose we are given a tripartite graph G on vertex sets A, B, C. One (naive) approach to detect if there is a triangle in the graph is: for vertex $v \in A$, and all pairs (u, w) of v's neighbors, check if there is an edge between u and w. The amount of work we do for vertex v is proportional to the number of edges between v and B (the degree of v to B) times the number of edges between v and C (the degree of v to C). This approach is efficient whenever the product of these two degrees is low on average. However, if this is not the case, there must be a vertex $v \in A$ with a large product of degrees. If the enumeration reports no edge between any pair of v's neighbors, we know there has to be a large "non-edge area" between B and C, i.e. between v's neighbors. In the rest of the algorithm, there is no need to look again at any pair of vertices in that area. We implement this idea by recursion: find disjoint subsets of $B \times C$ which together cover all pairs outside the non-edge area, and recurse on them. We show this recursion is actually efficient.

Before stating and proving the efficiency of our main algorithm, we first show that the naive approach proposed above is "Four-Russianizable" as expected. That is, we can apply the Method of Four Russians to speed up the sparse case by a factor of roughly $\log^2 n$.

The following algorithm is a generalization of an algorithm of Bansal and Williams [2].

Lemma 1. *Let $G = (V, E)$ be tripartite on vertex sets A, B, C of sizes k, m, n respectively. If $d(v_i, B)d(v_i, C) \leq \frac{nm}{\Delta^2}$ holds for some Δ and all $v_i \in A$ simultaneously, then we can detect if there is a triangle in G combinatorially in $O(mn\Delta^{6\Delta} + \frac{kmn}{\Delta^4} + k(m + n))$ time, on word RAM with word size $w \geq \Omega(\Delta \log \Delta + \log kmn)$.*

Proof. First, partition the vertices in B(resp. C) into groups $\{B_i\}$(resp. $\{C_i\}$) of sizes Δ^3 arbitrarily, e.g. put $(i\Delta^3 + 1)$-th to $(i + 1)\Delta^3$-th vertex of B(resp. C) in B_i(resp. C_i). Let $\mathcal{S}_B = \{S : |S| \leq \Delta, S \subseteq B_i \text{ for some } B_i\}$, $\mathcal{S}_C = \{S : |S| \leq \Delta, S \subseteq C_i \text{ for some } C_i\}$ be collections of subsets within the same group of B or C with at most Δ vertices. For every $S \in \mathcal{S}_B, S' \in \mathcal{S}_C$, we determine if there is at least one edge between them, and store all the results in a lookup table. This preprocessing takes

$$O\left(\frac{mn}{\Delta^6}\left(\frac{\Delta^3}{\Delta}\right)^2 \cdot \Delta^2 \cdot \Delta^2\right) \leq O(mn\Delta^{6\Delta})$$

time. Note that we can index a subset using $O(\Delta \log \Delta + \log \max\{m, n\}) = O(w)$ bits. This table can be stored in the memory so that one table lookup takes constant time.

With the help of this table, we can check if there is a triangle in G efficiently. We go over all vertices $v_i \in A$, and partition its neighborhood into a minimum number of sets in $\mathcal{S}_B, \mathcal{S}_C$. That is, for every group of vertices, we arbitrarily partition v_i's neighborhood in this group into sets of size exactly Δ and (possibly) one more set of size at most Δ. This generates at most $\frac{m}{\Delta^3} + \frac{d(v_i, B)}{\Delta}$ sets from $\mathcal{S}_B$ and at most $\frac{n}{\Delta^3} + \frac{d(v_i, C)}{\Delta}$ sets from $\mathcal{S}_C$. Using the lookup table, we can detect if there is an edge between any pair of sets from $\mathcal{S}_B$ and $\mathcal{S}_C$ in constant time. Going over all $v_i \in A$ takes

$$O\left(\sum_{i=1}^{k}\left(\frac{m}{\Delta^3} + \frac{d(v_i, B)}{\Delta}\right)\left(\frac{n}{\Delta^3} + \frac{d(v_i, C)}{\Delta}\right) + k(m + n)\right)$$

$$\leq O\left(\frac{kmn}{\Delta^6} + \sum_{i=1}^{k}\frac{m}{\Delta^3} \cdot \frac{n}{\Delta} + \sum_{i=1}^{k}\frac{m}{\Delta} \cdot \frac{n}{\Delta^3} + \sum_{i=1}^{k}\frac{mn}{\Delta^4} + k(m + n)\right)$$

$$\leq O\left(\frac{kmn}{\Delta^4} + k(m + n)\right)$$

time. The total running time is at most $O\left(mn\Delta^{6\Delta} + \frac{kmn}{\Delta^4} + k(m + n)\right)$ as we stated.

$\square$

Using this algorithm for the sparse case as a subroutine, we give a fast combinatorial algorithm for triangle detection.

Theorem 1. *Given a tripartite graph G on n vertices, we can detect if there is a triangle in G using a combinatorial algorithm in $\hat{O}\left(n^3 / \log^4 n\right)$ time on a word RAM with word size $w \geq \Omega(\log n)$.*

Proof. Set parameter $\Delta = \frac{\log n}{100(\log \log n)^2}$. It will remain fixed as we do the recursion, even if the instance size shrinks. The following algorithm detects if there is a triangle in a tripartite graph with vertex sets A, B, C:

Step 0: If $|B| < \Delta^6$ or $|C| < \Delta^6$, solve the instance by exhaustive search and return the answer.

Step 1: If for all vertices $v_i \in A$, $d(v_i, B)d(v_i, C) \leq \frac{|B| \cdot |C|}{\Delta^2}$, we solve the instance by the algorithm in Lemma 1.

Step 2: Otherwise, find a vertex that violates the condition. Without loss of generality, assume v_1 does, $d(v_1, B)d(v_1, C) > \frac{|B| \cdot |C|}{\Delta^2}$.

Step 3: Let B_1(resp. C_1) be v_1's neighborhood in B(resp. C).
If $\frac{|B_1|}{|B|} > \frac{|C_1|}{|C|}$, then recurse on $(A \backslash \{v_1\}, B, C \backslash C_1)$ and $(A \backslash \{v_1\}, B \backslash B_1, C_1)$, else recurse on $(A \backslash \{v_1\}, B \backslash B_1, C)$ and $(A \backslash \{v_1\}, B_1, C \backslash C_1)$.
Return YES if either of the two recursions returned YES.

Step 4: Check all pairs of vertices in $B_1 \times C_1$ for an edge.
Return YES if there is an edge, NO otherwise.

Analysis of the running time.

In each node of the recursion tree, only some of the following four subprocedures are executed:

1. If either $|B|$ or $|C|$ is small, we do exhaustive search, which takes $O(|A||B||C|)$ time.
2. We spend $O(|A|(|B| + |C|))$ time to check whether there is a high degree vertex and generate the inputs for two recursive calls.
3. If A has no high degree nodes, we invoke the algorithm in Lemma 1 which takes

$$O\left(|B||C|\Delta^{6\Delta} + \frac{|A||B||C|}{\Delta^4} + |A|(|B| + |C|)\right)$$

$$= \hat{O}\left(|B||C|n^{o(1)} + |A||B||C|/\log^4 n + |A|(|B| + |C|)\right)$$

time.
4. For every pair of neighbors of v_1 in B and C, we check if they have an edge. This step takes $O(|B_1||C_1|)$ time.

To analyse the total running time, we are going to bound the time we spend on the small-graph case (Subprocedure 1) and the time we spend on the large-graph case (Subprocedure 2,3,4) in the entire execution of the algorithm (the whole recursion tree) separately, and sum up these two cases.

For the case when $|B| \geq \Delta^6$ and $|C| \geq \Delta^6$, Subprocedure 2 is cheap compared to the other steps, taking time $O(|A|(|B| + |C|)) \leq \hat{O}\left(n|B||C|/\log^6 n\right)$. In this case, we mentally charge all the running time to the pairs of vertices in $B \times C$, then sum up over all pairs the cost they need to pay. If A has no high degree vertices, we will run Subprocedure 2 and 3, which takes at most $\hat{O}(|B||C|n^{o(1)} + |A||B||C|/\log^4 n + |A|(|B| + |C|)) \leq \hat{O}\left(n|B||C|/\log^4 n\right)$ time. We charge this running time to all pairs of vertices in $B \times C$ evenly; that is, every pair of vertices gets charged $\hat{O}\left(n/\log^4 n\right)$. If A has a high degree vertex, we will run Subprocedure 2 and 4, which takes at most

$$O\left(|A|(|B| + |C|) + |B_1||C_1|\right)$$

$$\leq \hat{O}\left(\Delta^2|B_1||C_1|n/\log^6 n + |B_1||C_1|\right)$$

$$\leq \hat{O}\left(n|B_1||C_1|/\log^4 n\right)$$

time. We charge this running time to the pairs in $B_1 \times C_1$ evenly, so every pair gets charged $\hat{O}\left(n/\log^4 n\right)$.

Claim. Every pair of vertices is charged at most once, over the entire execution of the algorithm.

The proof of this claim follows from inspection of the algorithm: the above argument only charges pairs of vertices that are not going into the same recursive branch.

There are n^2 pairs at the very beginning. Every pair gets charged at most $\hat{O}\left(n/\log^4 n\right)$. Therefore the running time for the large-graph case is at most $\hat{O}\left(n^3/\log^4 n\right)$.

Next we bound the total running time of Subprocedure 1. This running time is proportional to the number of triples we enumerated in Step 0. Let $T(S)$ be the maximum possible of this number of triples, if we start our recursion from vertex sets A, B, C with $|B||C| \leq S$.

Claim. $T(S) \leq nS$. Moreover, for $S > n\Delta^6$, then

$$T(S) \leq \max_{t>1/\Delta, t'>1/\Delta^2} \{T((1-t')S) + T(t'(1-t)S)\}$$

Proof. $T(S) \leq nS$ is trivial, since we never enumerate any triple more than once. If $S > n\Delta^6$, we have $|B|, |C| > \Delta^6$. That is, we must be starting the recursion from a large-graph case. There is nothing to prove if there is no high degree vertex in A, as we will do no enumeration in Step 0. Otherwise, let $t = \max\left\{\frac{|B_1|}{|B|}, \frac{|C_1|}{|C|}\right\}, t' = \min\left\{\frac{|B_1|}{|B|}, \frac{|C_1|}{|C|}\right\}$, then by the algorithm, we have $tt' > 1/\Delta^2$, in particular, $t > 1/\Delta, t' > 1/\Delta^2$. In the two recursive calls, we have $|B||C|$ values $(1-t')S$ and $t'(1-t)S$ respectively. By the definition of T, we will enumerate at most $T((1-t')S) + T(t'(1-t)S)$ triples. This proves the claim. $\square$

We can upper bound $T(S)$ using this recurrence. Consider the recursion tree $\mathcal{R}$ for $T(S)$. The root has value S. Its left child has value $(1-t')S$ and right child has value $t'(1-t)S$, where t and t' maximize $T((1-t')S) + T(t'(1-t)S)$. We recursively construct the tree for the left child and right child. For a node with value x, we always put $(1-t')x$ in its left child and $t'(1-t)x$ in its right child, for x's optimal parameter t' and t. We expand the tree from nodes with value at least $n^{1.5}(> n\Delta^6)$ recursively. Therefore, we will get a tree with leaf values at most $n^{1.5}$. This tree demonstrates how $T(S)$ is computed according to the recurrence, before we reach $n^{1.5}$. The sum of values of all leaves multiplied by n is an upper bound for $T(S)$, since $T(x) \leq nx$.

We calculate this sum in two cases. For every leaf, there is a unique path from the root to it, in which we follow the left child in some steps, follow the right child in the rest. Consider all leaves such that the unique path from the root to it takes at most $10\Delta\log\Delta(\approx \frac{\log n}{10\log\log n})$ right-child-moves. Since $t' > 1/\Delta^2$, the depth of the tree is at most $\Delta^2\log n$. Therefore, there are at most

$$\binom{\Delta^2\log n}{10\Delta\log\Delta} \cdot 10\Delta\log\Delta \leq 2^{3\Delta\log^2\Delta} \leq n^{0.4}$$

such leaves. By construction, each leaf has value at most $n^{1.5}$, so the sum over all leaves is at most $n^{1.9}$.

For those leaves such that the path from the root takes more than $10\Delta \log \Delta$ right-child-moves, consider a tree $\mathcal{R}'$ with *identical structure* as $\mathcal{R}$, but we set the values of nodes in a different way. We set all ts to 0 but leave all t's unchanged, then calculate the corresponding values. That is, the root still has value S, its left child still has value $(1 - t')S$, and its right child now has value $t'S$. For any node with value x in $\mathcal{R}'$, its left child has value $(1-t')x$ and right child has value $t'x$, for the same ratio t' as the corresponding node in $\mathcal{R}$, although the value x may have changed. Also, leaves now may have values greater than $n^{1.5}$ by the construction of $\mathcal{R}'$. The tree $\mathcal{R}'$ has the following two properties:

1. The sum of values in all leaves is exactly S.
2. For any node such that the path from root to it takes at least k right-child-moves, its value is at least $1/(1 - 1/\Delta)^k$ times the value of the corresponding node in $\mathcal{R}$.

The first property can be proved by induction, since the sum of values of two children is exactly the value of the parent. For the second property, since we require $t > 1/\Delta$ in $\mathcal{R}$, and keep t' unchanged, set t to 0, we will gain a factor of $1/(1 - t) > 1/(1 - 1/\Delta)$ for every right-child-move.

By property 2 above, for all leaves that the path from root takes more than $10\Delta \log \Delta$ right-child-moves, their values in $\mathcal{R}'$ must be at least $1/(1 - 1/\Delta)^{10\Delta \log \Delta} \geq \Delta^{10}$ times the corresponding values in $\mathcal{R}$. However, the sum of these values in $\mathcal{R}'$ is at most S by property 1. Therefore, the sum of values in $\mathcal{R}$ is at most S/Δ^{10}. Summing these two cases up, we prove that $T(S) \leq nS/\Delta^{10} + n^{2.9}$. In particular, $T(n^2) \leq n^3/\Delta^{10} + n^{2.9} = \hat{O}(n^3/\log^{10} n)$.

Finally, we sum up the small-graph and large-graph cases, proving that the algorithm runs in $\hat{O}(n^3/\log^4 n)$ time.

□

Remark. In Lemma 1, we can preprocess for all subsets of size at most $O(\log k/\log\log k)$ instead of $O(\Delta)$. This improves the running time of Theorem 1 from $O\left(\frac{n^3 (\log\log n)^8}{\log^4 n}\right)$ to $O\left(\frac{n^3 (\log\log n)^6}{\log^4 n}\right)$.

Combining the above algorithm with the reduction by Vassilevska Williams and Williams [13], we get an efficient combinatorial algorithm for BMM.

Theorem 4 (Vassilevska Williams and Williams'10). *For any constant c, if we can solve triangle detection on n-node graphs in $O(n^3/\log^c n)$ time, we can also solve Boolean matrix multiplication on $n \times n$ matrices in the same running time.*

Theorem 2. *There is a combinatorial algorithm to multiply two $n \times n$ Boolean matrices in $\hat{O}\left(n^3/\log^4 n\right)$ time.*

3   A General Approach

In this section, we propose a more general approach for triangle finding, which may lead to an even faster combinatorial algorithm.

In the algorithm presented in Section 2, finding a high degree vertex $v \in A$ and its neighborhood B_1, C_1 can be viewed as finding a large easy part of the input. That is, for the subgraph induced by vertices A, B_1, C_1, there is a 2-path for every pair in $B_1 \times C_1$. Thus, we only have to spend $O(|B_1||C_1|)$ time to determine if there is a triangle in it, which is $O(1/|A|)$ time on average for every triple of vertices. After solving this part of the input, we do two recursive calls which together exactly cover the rest of the triples. The high-degree of v guarantees that the easy part we find each time cannot be too small. We will have saved enough time before reaching the case where B or C is close to constant size (in which we basically have no way to beat the exhaustive search). However, if all vertices have low degree, then that instance itself is easy (via Lemma 1). The following theorem generalizes this idea of reducing the triangle detection problem to finding a large subgraph on which triangle detection is easy.

Theorem 3. *Let n be an integer, $0 < \alpha, \beta, \gamma, c \leq 1$, and G be any tripartite graph on vertex sets A, B, C with at least $\sqrt{n}$ vertices in each part. If there is an algorithm $\mathbf{L}$ for large subgraph triangle detection on every such G that runs in $O(c\alpha\beta\gamma|A||B||C|)$ time, then we can solve triangle detection on n-node graphs in $O(cn^3 + n^{3-\epsilon/2})$ time for any $\epsilon > 0$ such that $\alpha\beta > 10\epsilon(1 + \log\frac{1}{\gamma})$ and $\alpha > 10\epsilon(1 + \log\frac{1}{\beta})$.*

Proof. (sketch)

We will prove that the following divide-and-conquer algorithm is efficient:

Step 0. If $|A||B||C| < n^{2.5}$, do exhaustive search on all triples and return the answer.

Step 1. Run $\mathbf{L}$ on $A \cup B \cup C$.
Let G' on $A' \cup B' \cup C'$ be the subgraph $\mathbf{L}$ outputs.
Return YES if G' contains a triangle.

Step 2. Recurse on vertex sets $(A, B, C \setminus C')$, $(A, B \setminus B', C')$, $(A \setminus A', B', C')$.
Return whether any of the three recursive calls returned YES.

For the large-graph case, a similar "charging argument" works here as in the proof of Theorem 1. Note that the input $A \cup B \cup C$ we fed to $\mathbf{L}$ in Step 1 always has at least $\sqrt{n}$ vertices in each part. By our assumption, Step 1 will run in $O(c\alpha\beta\gamma|A||B||C|) \leq O(c|A'||B'||C'|)$ time. In Step 2, the algorithm generates the input for recursive calls, which takes only linear time. We charge the running time of Step 1 and Step 2 to the triples in $A' \times B' \times C'$. On average, each triple is charged $O(c)$ time. Same as the proof of Theorem 1, every triple in $A' \times B' \times C'$

will not go into the same recursive branch together. Therefore, every triple is charged only once in this argument. There are n^3 triples. In total, they are charged $O(cn^3)$ time. This proves that Step 1 and Step 2 take at most $O(cn^3)$ time in the entire algorithm.

For the small-graph case, the time we spent on Step 0 is proportional to the number of triples we enumerated. Let $T(S)$ be the maximum possible value of this number, if we start our recursion with $|A||B||C| = S$. On one hand, we have $T(S) \leq S$, as every triple will be manually checked at most once. Moreover, for $S \geq n^{2.5}$, by the way that the algorithm does recursive calls, we have $T(S) \leq T((1-\gamma)S) + T(\gamma(1-\beta)S) + T(\beta\gamma(1-\alpha)S)$.

We are going to prove $T(S) \leq n^{2.5} \left(\frac{S}{n^{2.5}}\right)^{1-\epsilon}$ by induction on S. For $S \leq n^{2.5}$, the new upper bound automatically holds, since $T(S) \leq S \leq n^{2.5} \left(\frac{S}{n^{2.5}}\right)^{1-\epsilon}$. Otherwise, by the recurrence and induction hypothesis,

$$T(S) \leq n^{2.5} \left(\left(\frac{(1-\gamma)S}{n^{2.5}}\right)^{1-\epsilon} + \left(\frac{\gamma(1-\beta)S}{n^{2.5}}\right)^{1-\epsilon} + \left(\frac{\beta\gamma(1-\alpha)S}{n^{2.5}}\right)^{1-\epsilon} \right)$$

$$\leq n^{2.5} \left(\frac{S}{n^{2.5}}\right)^{1-\epsilon} \left((1-\gamma)^{1-\epsilon} + (\gamma(1-\beta))^{1-\epsilon} + (\beta\gamma(1-\alpha))^{1-\epsilon} \right).$$

The value of $(1-\gamma)^{1-\epsilon} + (\gamma(1-\beta))^{1-\epsilon} + (\beta\gamma(1-\alpha))^{1-\epsilon}$ is always less than 1:

$$(1-\gamma)^{1-\epsilon} + (\gamma(1-\beta))^{1-\epsilon} + (\beta\gamma(1-\alpha))^{1-\epsilon}$$

$$\leq (1 - (1-\epsilon)\gamma) + \gamma^{1-\epsilon}(1 - (1-\epsilon)\beta) + (\beta\gamma)^{1-\epsilon}(1 - (1-\epsilon)\alpha)$$
$$\text{(by } (1-x)^c \leq 1 - cx \text{ when } 0 < c, x < 1)$$

$$= 1 + \gamma\left(-(1-\epsilon) + \gamma^{-\epsilon}\left(1 + \beta\left(-(1-\epsilon) + \beta^{-\epsilon}(1-(1-\epsilon)\alpha)\right)\right)\right)$$

$$\leq 1 + \gamma\left(-(1-\epsilon) + \gamma^{-\epsilon}\left(1 + \beta\left(-(1-\epsilon) + (1 + 2\epsilon\log\frac{1}{\beta})(1 - \frac{2\alpha}{3})\right)\right)\right)$$
$$\text{(by } \epsilon\log\frac{1}{\beta} < 1/10 \text{ and } \epsilon < 1/10)$$

$$\leq 1 + \gamma\left(-(1-\epsilon) + \gamma^{-\epsilon}\left(1 + \beta\left(\epsilon + 2\epsilon\log\frac{1}{\beta} - \frac{2\alpha}{3}\right)\right)\right)$$

$$\leq 1 + \gamma\left(-(1-\epsilon) + (1 + 2\epsilon\log\frac{1}{\gamma})(1 - \frac{\alpha\beta}{3})\right)$$
$$\text{(by } \epsilon\log\frac{1}{\gamma} < 1/10 \text{ and } \epsilon(1 + \log\frac{1}{\beta}) < \alpha/10)$$

$$\leq 1 + \gamma\left(\epsilon + 2\epsilon\log\frac{1}{\gamma} - \frac{\alpha\beta}{3}\right)$$

$$\leq 1 - \frac{\alpha\beta\gamma}{10} < 1.$$
$$\text{(by } \epsilon(1 + \log\frac{1}{\gamma}) < \alpha\beta/10)$$

This proves $T(S) \leq n^{2.5} \left(\frac{S}{n^{2.5}}\right)^{1-\epsilon}$, in particular, $T(n^3) \leq n^{3-\epsilon/2}$. Therefore the total running time of the algorithm is at most $O(cn^3 + n^{3-\epsilon/2})$.

□

Remark. Note that we do not have to restrict ourselves to find an α-fraction of A, β-fraction of B and γ-fraction of C. As long as we can find one part with α-fraction, one with β-fraction and the third with γ-fraction, and adjust the inputs for recursive calls correspondingly, we will be able to solve triangle detection efficiently. In this sense, the algorithm in Section 2 has parameters $c = \frac{1}{\Delta^4}, \alpha = 1, \beta = \frac{1}{\Delta}, \gamma = \frac{1}{\Delta^2}, \epsilon = \Theta(\frac{1}{\Delta \log \Delta})$, while G' is the subgraph induced by (A, B_1, C_1) or the entire graph if all vertices in A have low degree.

4   Conclusion

We have shown how to generalize the idea of divide-and-conquer in Chan's algorithm, and have provided a more intuitive proof of the recursion. The way of analysing the "sublinear" recurrence in Theorem 1, i.e., our $T(S)$ and its analysis, should be able to extend to other problems. We would like to see more applications of this method in proving the efficiency of other combinatorial algorithms that are based on divide-and-conquer.

Also, we would hope to have an $O(n^2)$ time algorithm for triangle detection on tripartite graphs with vertex set sizes n, n, and $\hat{O}(\log^4 n)$. We call this the "lopsided" triangle detection problem, where one side of vertices is very small compared to the others. An argument similar to the reduction by Vassilevska Williams and Williams [13] shows that this would yield an $O(n^2)$ time algorithm for multiplying $n \times n$ and $n \times \hat{O}(\log^4 n)$ Boolean matrices, improving the maximum outer dimension d that $n \times n$ and $n \times d$ Boolean matrices can be multiplied in $O(n^2)$ time, in both the combinatorial and the algebraic world. The current record of d is $\hat{O}(\log^3 n)$ by Chan's algorithm (Chan gave an $O(n^2)$ time algorithm for multiplying $n \times d$ and $d \times n$ matrices, which implies an $O(n^2)$ time triangle finding algorithm), while the record in the algebraic world is merely $O(\log n)$ [3].

Finally, we provide one type of instance for the lopsided triangle detection problem which seems hard to solve in $O(n^2)$ time with our current techniques: a graph G on vertex sets A, B, C with $|A| = |B| = n$ and $|C| = \hat{O}(\log^4 n)$, with roughly $1/\log n$ fraction of edges between A and C, $1/\log n$ fraction of edges between B and C, and constant fraction of edges between A and B. The main difficulty is that the size of C is too small. If we try to do recursion, its size will reach a constant too soon. Once it becomes of constant size, we basically have no way to save anything from exhaustive search.

Acknowledgments. The author would like to thank Ryan Williams for helpful discussions on results and writing of the paper, and the anonymous reviewers for their valuable comments.

References

1. Arlazarov, V.Z., Dinic, E.A., Kronrod, M.A., Faradzhev, I.A.: On economical construction of the transitive closure of a directed graph. Soviet Mathematics Doklady **11**(5), 1209–1210 (1970)
2. Bansal, N., Williams, R.: Regularity lemmas and combinatorial algorithms. In: 50th Annual IEEE Symposium on Foundations of Computer Science, pp. 745–754 (2009)
3. Brockett, R.W., Dobkin, D.: On the number of multiplications required for matrix multiplication. SIAM Journal on Computing **5**, 624–628 (1976)
4. Chan, T.M.: Speeding up the four russians algorithm by about one more logarithmic factor. In: Proceedings of the Twenty-Sixth Annual ACM-SIAM Symposium on Discrete Algorithms, pp. 212–217 (2015)
5. Fischer, M.J., Meyer, A.R.: Boolean matrix multiplication and transitive closure. In: 12th Annual Symposium on Switching and Automata Theory, pp. 129–131 (1971)
6. Frieze, A.M., Kannan, R.: Quick approximation to matrices and applications. Combinatorica **19**(2), 175–220 (1999)
7. Furman, M.E.: Application of a method of fast multiplication of matrices in the problem of finding the transitive closure of a graph. Soviet Mathematics Doklady **11**(5), 1252 (1970)
8. Impagliazzo, R., Lovett, S., Paturi, R., Schneider, S.: 0–1 integer linear programming with a linear number of constraints. CoRR abs/1401.5512 (2014)
9. Le Gall, F.: Powers of tensors and fast matrix multiplication. In: International Symposium on Symbolic and Algebraic Computation, pp. 296–303 (2014)
10. Munro, I.: Efficient determination of the transitive closure of a directed graph. Information Processing Letters **1**(2), 56–58 (1971)
11. Valiant, L.G.: General context-free recognition in less than cubic time. Journal of Computer and System Sciences **10**(2), 308–315 (1975)
12. Vassilevska Williams, V.: Multiplying matrices faster than coppersmith-winograd. In: Proceedings of the 44th Symposium on Theory of Computing Conference, pp 887–898 (2012)
13. Vassilevska Williams, V., Williams, R.: Subcubic equivalences between path matrix and triangle problems. In: 51th Annual IEEE Symposium on Foundations of Computer Science, pp. 645–654 (2010)

Author Index

Abramsky, Samson II-31
Achlioptas, Dimitris II-467
Agrawal, Shweta I-1
Ailon, Nir I-14
Aisenberg, James II-44
Albers, Susanne I-26
Alistarh, Dan II-479
Amanatidis, Georgios I-39
Amarilli, Antoine II-56
Aminof, Benjamin II-375
Anshelevich, Elliot I-52
Aronov, Boris I-65
Avigdor-Elgrabli, Noa I-78
Avin, Chen II-492
Azar, Yossi I-91

Beame, Paul I-103
Behsaz, Babak I-116
Bei, Xiaohui I-129
Beigi, Salman I-143
Beneš, Nikola II-69
Berkholz, Christoph I-155
Bernstein, Aaron I-167
Beyersdorff, Olaf I-180
Bezděk, Peter II-69
Bhangale, Amey I-193
Bhattacharya, Sayan I-206, II-504
Bienvenu, Laurent I-219
Björklund, Andreas I-231, I-243
Bodirsky, Manuel I-256
Bojańczyk, Mikołaj II-427
Bonet, Maria Luisa II-44
Boreale, Michele II-82
Bouajjani, Ahmed II-95
Bourhis, Pierre II-56
Bringmann, Karl II-516
Bun, Mark I-268
Burton, Benjamin A. I-281
Buss, Sam II-44

Canonne, Clément L. I-294
Cao, Yixin I-306
Charron-Bost, Bernadette II-528

Chatterjee, Krishnendu II-108, II-121
Chattopadhyay, Arkadev II-540
Chekuri, Chandra I-318
Chen, Ning I-129, II-552
Chew, Leroy I-180
Ciobanu, Laura II-134
Cohen, Aloni I-331
Cohen, Gil I-343
Cohen, Ilan Reuven I-91
Colcombet, Thomas II-146
Coudron, Matthew I-355
Crǎciun, Adrian II-44
Cseh, Ágnes I-367
Curticapean, Radu I-380
Czyzowicz, Jurek I-393

Dahlgaard, Søren II-564
Dani, Varsha II-575
Datta, Samir II-159
Dell, Holger I-231
Desfontaines, Damien I-219
Diekert, Volker II-134
Disser, Yann I-406
Doron, Dean I-419
Doyen, Laurent II-108
Dubut, Jérémy II-171
Dvořák, Zdeněk I-432

Elder, Murray II-134
Emmi, Michael II-95
Enea, Constantin II-95
Erlebach, Thomas I-444
Etesami, Omid I-143
Etessami, Kousha II-184

Faonio, Antonio I-456
Feldman, Michal II-601
Feldmann, Andreas Emil I-469, II-588
Fijalkow, Nathanaël II-197
Filiot, Emmanuel II-209
Finkel, Olivier II-222
Fomin, Fedor V. I-481, I-494
Fontes, Lila I-506

Frascaria, Dario   I-26
Friedler, Ophir   II-601
Friedrich, Tobias   II-516, II-614
Friggstad, Zachary   I-116
Függer, Matthias   II-528
Fulla, Peter   I-517
Fung, Wai Shing   I-469

Gairing, Martin   II-626
Galanis, Andreas   I-529
Ganguly, Sumit   I-542
Garg, Jugal   I-554
Gąsieniec, Leszek   I-393
Gaspers, Serge   I-567
Gawrychowski, Paweł   I-580, I-593
Gelashvili, Rati   II-479
Georgiadis, Loukas   I-605
Ghaffari, Mohsen   II-638
Gharibian, Sevag   I-617
Giannakopoulos, Yiannis   II-650
Giannopoulou, Archontia C.   I-629
Göbel, Andreas   I-642
Gohari, Amin   I-143
Goldberg, Leslie Ann   I-529, I-642, I-654
Goldreich, Oded   I-666
Goldwasser, Shafi   II-663
Golovnev, Alexander   I-481, I-1046
Goubault, Éric   II-171
Goubault-Larrecq, Jean   II-171
Grohe, Martin   I-155
Große, Ulrike   I-678
Gudmundsson, Joachim   I-678
Gupta, Shalmoli   I-318
Gur, Tom   I-666
Gysel, Rob   I-654

Haase, Christoph   II-234
Hamza, Jad   II-95
Hansen, Thomas Dueholm   I-689
Hemenway, Brett   I-701
Henzinger, Monika   I-206, I-713, I-725
Henzinger, Thomas A.   II-121
Hoefer, Martin   II-504, II-516, II-552
Hoffmann, Michael   I-444
Holmgren, Justin   I-331
Horn, Florian   II-197
Huang, Chien-Chung   I-367, II-504
Huang, Lingxiao   I-910
Husfeldt, Thore   I-231

Ibsen-Jensen, Rasmus   II-121
Im, Sungjin   I-78, I-737
Ishai, Yuval   I-1
Istrate, Gabriel   II-44
Italiano, Giuseppe F.   I-206, I-605

Jagadeesan, Radha   II-31, II-247
Jahanjou, Hamid   I-749
Jain, Rahul   I-506
Jansen, Bart M.P.   I-629
Jerrum, Mark   I-529
Jin, Yifei   I-898
Jurdziński, Marcin   II-260

Kalai, Yael Tauman   II-663
Kamat, Vikram   I-243
Kammer, Frank   I-444
Kaniewski, Jedrzej   I-761
Kannan, Sampath   I-773
Kantor, Erez   II-675
Kaplan, Haim   I-689
Kar, Koushik   I-52
Karbasi, Amin   II-688
Kari, Jarkko   II-273
Kari, Lila   I-1022
Karpinski, Marek   I-785
Kaski, Petteri   I-494
Katz, Matthew J.   I-65
Kavitha, Telikepalli   I-367, II-504
Kawarabayashi, Ken-ichi   II-3
Kawase, Yasushi   I-797
Kayal, Neeraj   I-810
Kerenidis, Iordanis   I-506
Khot, Subhash   I-822
Khurana, Dakshita   I-1
Kiefer, Stefan   II-234
Klein, Felix   II-452
Klimm, Max   I-406
Knauer, Christian   I-678
Knudsen, Mathias Bæk Tejs   II-564
Kobayashi, Yusuke   I-797
Koiran, Pascal   I-810
Kollias, Konstantinos   II-626
Komarath, Balagopal   I-834
Könemann, Jochen   I-469
Kopecki, Steffen   I-1022
Kopparty, Swastik   I-193
Kosowski, Adrian   I-393
Kotsialou, Grammateia   II-626

Koutsoupias, Elias   II-650
Kowalik, Łukasz   I-243
Kozen, Dexter   II-286
Kozik, Marcin   I-846
Kranakis, Evangelos   I-393
Kreutzer, Stephan   II-3
Krinninger, Sebastian   I-713, I-725
Krohmer, Anton   II-614
Kulikov, Alexander S.   I-481
Kulkarni, Raghav   II-159
Künnemann, Marvin   I-859, II-552
Kupec, Martin   I-432
Kuperberg, Denis   II-197, II-299
Kurpisz, Adam   I-872
Kutten, Shay   II-675

Lahav, Ori   II-311
Lapinskas, John   I-654
Laplante, Sophie   I-506
Larsen, Kim G.   II-69
Laura, Luigi   I-605
Laurière, Mathieu   I-506
Lazić, Ranko   II-260
Lee, Troy   I-761
Lengler, Johannes   II-688
Leppänen, Samuli   I-872
Leroux, Jérôme   II-324
Li, Jerry   I-886
Li, Jian   I-898, I-910
Liew, Vincent   I-103
Lin, Chengyu   II-552
Lingas, Andrzej   I-785
Lohrey, Markus   II-337
Loitzenbauer, Veronika   I-713
Lokshtanov, Daniel   I-494, I-629,
    I-922, I-935
Lotker, Zvi   II-492
Lübbecke, Elisabeth   I-406

Mahajan, Meena   I-180
Mamouras, Konstantinos   II-286
Maneth, Sebastian   II-209, II-337
Manthey, Bodo   I-859
Maria, Clément   I-281
Markakis, Evangelos   I-39
Martin, Barnaby   I-256
Mastrolilli, Monaldo   I-872
Mathieu, Claire   I-773
Mazza, Damiano   II-350
Mehta, Ruta   I-554

Meunier, Pierre-Étienne   I-1022
Miao, Peihan   II-552
Michalewski, Henryk   II-362
Mihajlin, Ivan   I-481
Miles, Eric   I-749
Mio, Matteo   II-362
Misra, Pranabendu   I-922
Mitchell, Joseph S.B.   I-947
Molinaro, Marco   I-960
Mömke, Tobias   I-973
Moseley, Benjamin   I-78, I-737
Mottet, Antoine   I-256
Mouawad, Amer E.   I-985
Movahedi, Mahnush   II-575
Mozes, Shay   I-580
Mukherjee, Anish   II-159
Murlak, Filip   II-427
Muscholl, Anca   II-11

Nahum, Yinon   II-492
Nanongkai, Danupon   I-725
Nayyeri, Amir   I-997
Nicholson, Patrick K.   I-593
Nielsen, Jesper Buus   I-456
Nikolov, Aleksandar   I-1010
Nikzad, Afshin   I-39
Nishimura, Naomi   I-985
Nowak, Thomas   II-528

Ochremiak, Joanna   I-846
Otop, Jan   II-121

Panolan, Fahad   I-494, I-922
Park, Sunoo   II-663
Parotsidis, Nikos   I-605
Paskin-Cherniavsky, Anat   I-1
Pathak, Vinayak   I-985
Patitz, Matthew J.   I-1022
Pătrașcu, Mihai   I-103
Pecatte, Timothée   I-810
Peebles, John   I-886
Peleg, David   II-492
Peternek, Fabian   II-337
Petrişan, Daniela   II-286
Pietrzak, Krzysztof   I-1046
Polishchuk, Valentin   I-947
Post, Ian   I-469

Quanrud, Kent   I-318

Rabani, Yuval   I-78
Raman, Venkatesh   I-985
Ramanujan, M.S.   I-935
Raykov, Pavel   II-701
Reynier, Pierre-Alain   II-209
Richerby, David   I-642
Riely, James   II-247
Roland, Jérémie   I-506
Rotbart, Noy   II-564
Rothblum, Ron D.   I-666
Rothenberger, Ralf   II-516
Rubin, Sasha   II-375
Rudra, Atri   II-540

Saberi, Amin   I-39
Sachdeva, Sushant   I-193
Saha, Chandan   I-810
Saia, Jared   II-575
Saket, Rishi   I-822
Salavatipour, Mohammad R.   I-116
Sanyal, Swagato   I-1035
Sarma, Jayalal   I-834
Sauerwald, Thomas   II-516
Saurabh, Saket   I-494, I-629, I-922, I-935
Schewe, Sven   II-388
Schmitz, Sylvain   II-260
Schwentick, Thomas   II-159
Sekar, Shreyas   I-52
Seki, Shinnosuke   I-1022
Senellart, Pierre   II-56
Shen, Alexander   I-219
Shinkar, Igor   I-343
Shukla, Anil   I-180
Sidiropoulos, Anastasios   I-997
Sikora, Jamie   I-617
Silva, Alexandra   II-286
Siminelakis, Paris   II-467
Sivakumar, Rohit   I-116
Skórski, Maciej   I-1046
Skrzypczak, Michał   II-197, II-299
Sledneu, Dzmitry   I-785
Smid, Michiel   I-678
Sorkin, Gregory B.   I-567
Spegni, Francesco   II-375
Spirakis, Paul G.   I-393
Spreer, Jonathan   I-281
Srba, Jiří   II-69
Sreejith, A.V.   II-146
Staton, Sam   II-401
Steger, Angelika   II-688

Stehn, Fabian   I-678
Stein, Cliff   I-167
Stewart, Alistair   II-184
Sunil, K.S.   I-834
Sutre, Grégoire   II-324
Swernofsky, Joseph   II-414
Sysikaski, Mikko   I-947
Szabados, Michal   II-273

Talbot, Jean-Marc   II-209
Tarjan, Robert E.   I-689
Ta-Shma, Amnon   I-419
Terui, Kazushige   II-350
Thaler, Justin   I-268
Thapper, Johan   I-1058
Totzke, Patrick   II-324
Trivedi, Ashutosh   II-388

Uijlen, Sander   II-401
Uznański, Przemysław   I-393

Vafeiadis, Viktor   II-311
Vákár, Matthijs   II-31
Vardi, Moshe Y.   II-108
Varghese, Thomas   II-388
Vazirani, Vijay V.   I-554
Venturi, Daniele   I-456
Vidick, Thomas   I-355
Viola, Emanuele   I-749

Wagner, Lisa   II-504
Wang, Haitao   I-947
Wang, Yajun   I-1070
Wehar, Michael   II-414
Weimann, Oren   I-580
Weinstein, Omri   I-1082
Wiese, Andreas   I-973
Witkowski, Adam   II-427
de Wolf, Ronald   I-761
Wong, Sam Chiu-wai   I-1070
Woodruff, David P.   I-960, I-1082
Wootters, Mary   I-701

Yamaguchi, Yutaro   I-797
Yannakakis, Mihalis   II-184
Yaroslavtsev, Grigory   I-960
Yazdanbod, Sadra   I-554
Young, Maxwell   II-575
Yu, Huacheng   I-1094

Zehavi, Meirav   I-243
Zetzsche, Georg   II-440
Zeume, Thomas   II-159
Zhang, Shengyu   I-129
Zhou, Hang   I-773

Zimmermann, Martin   II-452
Zuleger, Florian   II-375
Zwick, Uri   I-689
Živný, Stanislav   I-517, I-1058

Printed in the United States
by Bookmasters